| Element | Symbol | Atomic Number | Atomic Mass† | Element | Symbol | Atomic Number | Atomic Mass† |
|---------|--------|---------------|--------------|---------|--------|---------------|--------------|
| Actinium | Ac | 89 | (227) | Mendelevium | Md | 101 | (258) |
| Aluminum | Al | 13 | 26.9815386 | Mercury | Hg | 80 | 200.59 |
| Americium | Am | 95 | (243) | Molybdenum | Mo | 42 | 95.94 |
| Antimony | Sb | 51 | 121.760 | Neodymium | Nd | 60 | 144.242 |
| Argon | Ar | 18 | 39.948 | Neon | Ne | 10 | 20.1797 |
| Arsenic | As | 33 | 74.92160 | Neptunium | Np | 93 | (237) |
| Astatine | At | 85 | (210) | Nickel | Ni | 28 | 58.6934 |
| Barium | Ba | 56 | 137.327 | Niobium | Nb | 41 | 92.90638 |
| Berkelium | Bk | 97 | (247) | Nitrogen | N | 7 | 14.0067 |
| Beryllium | Be | 4 | 9.012182 | Nobelium | No | 102 | (259) |
| Bismuth | Bi | 83 | 208.98040 | Osmium | Os | 76 | 190.23 |
| Bohrium | Bh | 107 | (272) | Oxygen | O | 8 | 15.9994 |
| Boron | B | 5 | 10.811 | Palladium | Pd | 46 | 106.42 |
| Bromine | Br | 35 | 79.904 | Phosphorus | P | 15 | 30.973762 |
| Cadmium | Cd | 48 | 112.411 | Platinum | Pt | 78 | 195.084 |
| Calcium | Ca | 20 | 40.078 | Plutonium | Pu | 94 | (244) |
| Californium | Cf | 98 | (251) | Polonium | Po | 84 | (209) |
| Carbon | C | 6 | 12.0107 | Potassium | K | 19 | 39.0983 |
| Cerium | Ce | 58 | 140.116 | Praseodymium | Pr | 59 | 140.90765 |
| Cesium | Cs | 55 | 132.9054519 | Promethium | Pm | 61 | (145) |
| Chlorine | Cl | 17 | 35.453 | Protactinium | Pa | 91 | 231.03588 |
| Chromium | Cr | 24 | 51.9961 | Radium | Ra | 88 | (226) |
| Cobalt | Co | 27 | 58.933195 | Radon | Rn | 86 | (222) |
| Copper | Cu | 29 | 63.546 | Rhenium | Re | 75 | 186.207 |
| Curium | Cm | 96 | (247) | Rhodium | Rh | 45 | 102.90550 |
| Darmstadtium | Ds | 110 | (281) | Roentgenium | Rg | 111 | (280) |
| Dubnium | Db | 105 | (268) | Rubidium | Rb | 37 | 85.4678 |
| Dysprosium | Dy | 66 | 162.500 | Ruthenium | Ru | 44 | 101.07 |
| Einsteinium | Es | 99 | (252) | Rutherfordium | Rf | 104 | (267) |
| Erbium | Er | 68 | 167.259 | Samarium | Sm | 62 | 150.36 |
| Europium | Eu | 63 | 151.964 | Scandium | Sc | 21 | 44.955912 |
| Fermium | Fm | 100 | (257) | Seaborgium | Sg | 106 | (271) |
| Fluorine | F | 9 | 18.9984032 | Selenium | Se | 34 | 78.96 |
| Francium | Fr | 87 | (223) | Silicon | Si | 14 | 28.0855 |
| Gadolinium | Gd | 64 | 157.25 | Silver | Ag | 47 | 107.8682 |
| Gallium | Ga | 31 | 69.723 | Sodium | Na | 11 | 22.98976928 |
| Germanium | Ge | 32 | 72.64 | Strontium | Sr | 38 | 87.62 |
| Gold | Au | 79 | 196.966569 | Sulfur | S | 16 | 32.065 |
| Hafnium | Hf | 72 | 178.49 | Tantalum | Ta | 73 | 180.94788 |
| Hassium | Hs | 108 | (270) | Technetium | Tc | 43 | (98) |
| Helium | He | 2 | 4.002602 | Tellurium | Te | 52 | 127.60 |
| Holmium | Ho | 67 | 164.93032 | Terbium | Tb | 65 | 158.92535 |
| Hydrogen | H | 1 | 1.00794 | Thallium | Tl | 81 | 204.3833 |
| Indium | In | 49 | 114.818 | Thorium | Th | 90 | 232.03806 |
| Iodine | I | 53 | 126.90447 | Thulium | Tm | 69 | 168.93421 |
| Iridium | Ir | 77 | 192.217 | Tin | Sn | 50 | 118.710 |
| Iron | Fe | 26 | 55.845 | Titanium | Ti | 22 | 47.867 |
| Krypton | Kr | 36 | 83.798 | Tungsten | W | 74 | 183.84 |
| Lanthanum | La | 57 | 138.90547 | Uranium | U | 92 | 238.02891 |
| Lawrencium | Lr | 103 | (262) | Vanadium | V | 23 | 50.9415 |
| Lead | Pb | 82 | 207.2 | Xenon | Xe | 54 | 131.293 |
| Lithium | Li | 3 | 6.941 | Ytterbium | Yb | 70 | 173.04 |
| Lutetium | Lu | 71 | 174.967 | Yttrium | Y | 39 | 88.90585 |
| Magnesium | Mg | 12 | 24.3050 | Zinc | Zn | 30 | 65.409 |
| Manganese | Mn | 25 | 54.938045 | Zirconium | Zr | 40 | 91.224 |
| Meitnerium | Mt | 109 | (276) | | | | |

*These atomic masses show as many significant figures as are known for each element. The atomic masses in the periodic table are shown to four significant figures, which is sufficient for solving the problems in this book.

†Approximate values of atomic masses for radioactive elements are given in parentheses.

Second Edition

# Chemistry

## Julia Burdge
*University of Idaho*

CHEMISTRY, SECOND EDITION

Published by McGraw-Hill, a business unit of The McGraw-Hill Companies, Inc., 1221 Avenue of the Americas, New York, NY 10020. Copyright © 2011 by The McGraw-Hill Companies, Inc. All rights reserved. Previous edition © 2009. No part of this publication may be reproduced or distributed in any form or by any means, or stored in a database or retrieval system, without the prior written consent of The McGraw-Hill Companies, Inc., including, but not limited to, in any network or other electronic storage or transmission, or broadcast for distance learning.

Some ancillaries, including electronic and print components, may not be available to customers outside the United States.

This book is printed on acid-free paper.

1 2 3 4 5 6 7 8 9 0 DOW/DOW 1 0 9 8 7 6 5 4 3 2 1 0

ISBN 978–0–07–337564–9
MHID 0–07–337564–0

Publisher: *Ryan Blankenship*
Senior Sponsoring Editor: *Tamara L. Hodge*
Vice-President New Product Launches: *Michael Lange*
Senior Developmental Editor: *Shirley R. Oberbroeckling*
Senior Marketing Manager: *Todd L. Turner*
Senior Project Manager: *Gloria G. Schiesl*
Lead Production Supervisor: *Sandy Ludovissy*
Senior Media Project Manager: *Sandra M. Schnee*
Senior Designer: *David W. Hash*
Cover/Interior Designer: *Greg Nettles/Squarecrow Design*
(USE) Cover Image: *The chemical element bismuth as a synthetic-made crystal. The surface is an iridescent, very thin layer of oxidation. ©Alchemist-hp*
Senior Photo Research Coordinator: *Lori Hancock*
Photo Research: *David Tietz/Editorial Image, LLC*
Supplement Producer: *Mary Jane Lampe*
Compositor: *Precision Graphics*
Typeface: *10/12 Times*
Printer: *R. R. Donnelley*

All credits appearing on page or at the end of the book are considered to be an extension of the copyright page.

**Library of Congress Cataloging-in-Publication Data**

Burdge, Julia R.
  Chemistry / Julia Burdge. -- 2nd ed.
    p. cm.
  Includes index.
  ISBN  978–0–07–337564–9 — ISBN  0–07–337564–0 (hard copy : alk. paper)  1.  Chemistry--
Textbooks.  I. Title.
  QD33.2.B865 2011
  540--dc22
                          2009034024

# ABOUT THE Author

Julia Burdge received her Ph.D. (1994) from the University of Idaho in Moscow, Idaho. Her research and dissertation focused on instrument development for analysis of trace sulfur compounds in air and the statistical evaluation of data near the detection limit.

In 1994 she accepted a position at The University of Akron in Akron, Ohio, as an assistant professor and director of the Introductory Chemistry program. In the year 2000, she was tenured and promoted to associate professor at The University of Akron on the merits of her teaching, service, and research in chemistry education. In addition to directing the general chemistry program and supervising the teaching activities of graduate students, she helped establish a future-faculty development program and served as a mentor for graduate students and postdoctoral associates. Julia has recently relocated back to the Northwest to be near family. She and her children live in Pullman, Washington, home of Washington State University; and she holds an affiliate faculty position in the Chemistry Department at the University of Idaho. She also continues to work with students in Ohio and Florida via an online tutoring program.

Julia and her children are animal lovers and moved three horses, three cats, and a dog with them to the Northwest. They are enjoying the changes of seasons, long horseback rides on the Palouse, and frequent visits with family.

To the people who will always matter the most: Katie, Beau, and Sam.

BRIEF Contents

# Contents

## 1   CHEMISTRY: THE CENTRAL SCIENCE   2

# 2  ATOMS, MOLECULES, AND IONS    34

# 3  STOICHIOMETRY: RATIOS OF COMBINATION    76

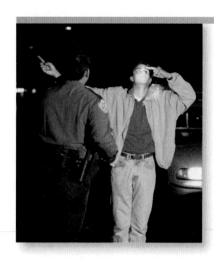

# 4 REACTIONS IN AQUEOUS SOLUTIONS    116

# 5 THERMOCHEMISTRY    170

# 6    Quantum Theory and the Electronic Structure of Atoms    210

# 12 INTERMOLECULAR FORCES AND THE PHYSICAL PROPERTIES OF LIQUIDS AND SOLIDS 492

# 13 PHYSICAL PROPERTIES OF SOLUTIONS 538

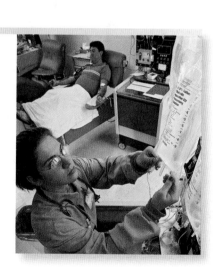

# 14 CHEMICAL KINETICS    578

# 15 CHEMICAL EQUILIBRIUM    628

# 16 Acids and Bases    672

# 17 Acid-Base Equilibria and Solubility Equilibria    726

# 18 ENTROPY, FREE ENERGY, AND EQUILIBRIUM    776

# 19 ELECTROCHEMISTRY    816

# 20 NUCLEAR CHEMISTRY   858

# 21 ENVIRONMENTAL CHEMISTRY   892

# 22 COORDINATION CHEMISTRY    920

# 23 METALLURGY AND THE CHEMISTRY OF METALS    944

# 24 Nonmetallic Elements and Their Compounds   968

# 25 Modern Materials   998

# A Visual Approach Integrating Art and Media

## Sample Problems:

Strong problem-solving skills are essential for student success. I provide a consistent, step-wise approach in every worked example.

---

**Sample Problem 2.7**

Write the empirical formulas for the following molecules: (a) glucose ($C_6H_{12}O_6$), a substance known as blood sugar; (b) adenine ($C_5H_5N_5$), also known as vitamin $B_4$; and (c) nitrous oxide ($N_2O$), a gas that is used as an anesthetic ("laughing gas") and as an aerosol propellant for whipped cream.

**Strategy** To write the empirical formula, the subscripts in the molecular formula must be reduced to the smallest possible whole numbers (without altering the relative numbers of atoms).

**Setup** The molecular formulas in parts (a) and (b) each contain subscripts that are divisible by common numbers. Therefore, we will be able to express the formulas with smaller whole numbers than those in the molecular formulas. In part (c), the molecule has only one O atom, so it is impossible to simplify this formula further.

**Solution** (a) Dividing each of the subscripts in the molecular formula for glucose by 6, we obtain the empirical formula, $CH_2O$. If we had divided the subscripts by 2 or 3, we would have obtained the formulas $C_3H_6O_3$ and $C_2H_4O_2$, respectively. Although the ratio of carbon to hydrogen to oxygen atoms in each of these formulas is correct (1:2:1), neither is the simplest formula because the subscripts are not in the smallest possible whole-number ratio.

(b) Dividing each subscript in the molecular formula of adenine by 5, we get the empirical formula, CHN.

(c) Because the subscripts in the formula for nitrous oxide are already the smallest possible whole numbers, its empirical formula is the same as its molecular formula, $N_2O$.

**Think About It** Make sure that the *ratio* in each empirical formula is the same as that in the corresponding molecular formula and that the subscripts are the smallest possible whole numbers. In part (a), for example, the ratio of C:H:O in the molecular formula is 6:12:6, which is equal to 1:2:1, the ratio expressed in the empirical formula.

**Practice Problem A** Write empirical formulas for the following molecules: (a) caffeine ($C_8H_{10}N_4O_2$), a stimulant found in tea and coffee, (b) butane ($C_4H_{10}$), which is used in cigarette lighters, and (c) glycine ($C_2H_5NO_2$), an amino acid.

Caffeine      Butane      Glycine

**Practice Problem B** For which of the following molecular formulas is the formula shown in parentheses the correct empirical formula? (a) $C_{12}H_{22}O_{11}$ ($C_{12}H_{22}O_{11}$), (b) $C_8H_{12}O_4$ ($C_4H_6O_2$), (c) $H_2O_2$ ($H_2O$)?

---

## Sample Problems:

Strong problem-solving skills are essential for student success. I provide a consistent, step-wise approach in every worked example.

*Strategy:* Read the problem carefully and identify an approach that can be used to answer the question.

*Setup:* Gather the necessary information, including data, constants, and equations.

*Solution:* Following the appropriate strategy, use the information to solve the problem.

*Think About It:* Evaluate the result to make sure that it makes sense. Does it have the right sign, units, and/or magnitude?

*Practice Problem A:* This is a problem similar to the sample problem that students can solve using the same strategy.

*Practice Problem B:* This problem is about the same topic, but is sufficiently different from the Sample Problem that it cannot be solved using the same strategy. These problems encourage students to develop problem-solving strategies on their own—helping them to build the confidence necessary to tackle unfamiliar problems.

---

**Visualizing Chemistry**
**Figure 17.8**

VC 17.5 Which of the following would cause precipitation of the largest amount of AgCl when 0.1 mole is dissolved in the saturated solution shown in Figure 17.8?

a) NaCl
b) CsCl
c) Both would cause precipitation of the same amount.

---

**11.74** Consider the three containers shown, all of which have the same volume and are at the same temperature. (a) Which container has the smallest mole fraction of gas A (red)? (b) Which container has the highest partial pressure of gas B (green)? (c) Which container has the highest total pressure?

(i)      (ii)      (iii)

---

Problems: I provide lots of end-of-chapter problems, grouped by section, graduated in difficulty, and paired so that students have answers to every *other* problem. Many of the end-of-chapter problems are integrated into the online homework, including the conceptual question sets that correspond to the Visualizing Chemistry figures.

---

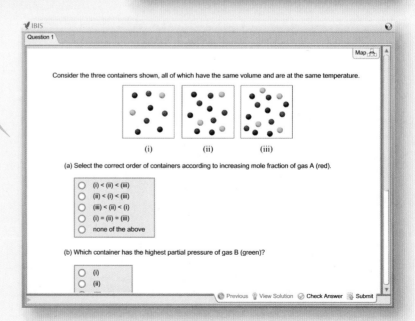

IBIS

Question 1

Map

Consider the three containers shown, all of which have the same volume and are at the same temperature.

(i)      (ii)      (iii)

(a) Select the correct order of containers according to increasing mole fraction of gas A (red).

○ (i) < (ii) < (iii)
○ (ii) < (i) < (iii)
○ (iii) < (ii) < (i)
○ (i) = (ii) = (iii)
○ none of the above

(b) Which container has the highest partial pressure of gas B (green)?

○ (i)
○ (ii)

Previous   View Solution   Check Answer   Submit

## Art:

I have worked closely with artists to develop engaging two-page "Visualizing Chemistry" figures. These pieces are step-wise illustrations of processes and common laboratory procedures, and are designed to make sure students get the right take-away message. Visualizing Chemistry figures have also been made into animations, which I narrate myself. They can be shown in the classroom or viewed by students in the electronic homework. Each figure is accompanied by a series of conceptual end-of-chapter problems, which are also integrated into the online homework.

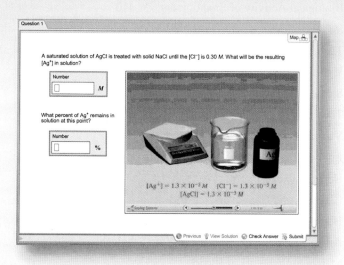

## What's the point?

My book has been developed and written with the *student* in mind. Information is presented clearly in a digestible format with engaging artwork, designed to bring concepts to life; concept art is used consistently to foster better comprehension; and pedagogical features foster the kinds of student activities that lead to competence and confidence in problem-solving .

# LIST OF Applications

# Preface

Welcome to the exciting and dynamic world of Chemistry! My desire to create a new general chemistry textbook grew out of my concern for the interests of students and faculty alike. Having taught general chemistry for many years and having helped new teachers and future faculty develop the skills necessary to teach general chemistry, I believe I have developed a distinct perspective on the common problems and misunderstandings that students encounter while learning the fundamental concepts of chemistry—and that professors encounter while teaching them. I believe that it is possible for a textbook to address many of these issues while conveying the wonder and possibilities that chemistry offers. With this in mind, I have tried to write a text that balances the necessary fundamental concepts with engaging real-life examples and applications, while utilizing a consistent step-by-step problem-solving approach and an innovative art and media program.

## What's New in This Edition?

### New Visualizing Chemistry Figures

I have designed eight new two-page Visualizing Chemistry figures to enhance student understanding of chemical processes and laboratory techniques. The new pieces are:

| | |
|---|---|
| Determination of $\Delta H^\circ_{rxn}$ by Constant-Pressure Calorimetry | Figure 5.9 |
| Determination of Specific Heat by Constant-Pressure Calorimetry | Figure 5.10 |
| Born-Haber Cycle | Figure 8.3 |
| Molar Volume of a Gas | Figure 11.15 |
| Using Equilibrium Tables to Solve Problems | Figure 16.2 |
| Factors That Influence the Entropy of a System | Figure 18.4 |
| Construction of a Galvanic Cell | Figure 19.1 |
| Nuclear Fission and Fusion | Figure 20.7 |

For each Visualizing Chemistry figure, I have also added a series of conceptual end-of-chapter problems. These problems allow students to assess their understanding of the principles in each figure and have been incorporated into the online homework.

### New and Updated Chapter Content

- Chapter 4: New section on serial dilution with Sample and Practice Problems.
- Chapter 7: Incorporation of the concept of diagonal relationships into the main text.
- Chapter 8: New section on percent ionic character with Sample and Practice Problems.
- Chapter 11: New section on the combined gas law with Sample and Practice Problems; and new coverage of deviation from ideal behavior and compressibility factor, Z, including a new figure, inquiry box, and Sample and Practice Problems.
- Chapter 13: Expanded coverage of intermolecular forces, including a new table that includes molecular structures.
- Chapter 16: New section on percent ionization with a new figure (Figure 16.3) showing percent ionization of weak acid as a function of concentration, and new Sample and Practice Problems.
- Chapter 18: New introduction of the concept of entropy, including a new two-page Visualizing Chemistry figure that provides a qualitative illustration of several factors that affect a system's entropy.
- Chapter 25: Introduction of graphene, its properties, and uses.

Having received overwhelmingly positive feedback from students and instructors about the **Section Checkpoint** questions, I have added more—many of them conceptual.

Many new **Sample Problems** and **Practice Problems** have been included in this edition to provide more structure and practice to help students develop strong problem-solving skills. A significant number of these new problems appear in FAQ boxes, designed to help students master important material that is commonly misunderstood. New **FAQ** boxes answer such commonly asked questions as "How can I tell if a proposed reaction mechanism is plausible?" and "Why do some substances dissolve and not others?" Most FAQ boxes contain Sample and Practice Problems, giving students the opportunity to assess and improve their comprehension.

All **Sample Problems** are now followed by two **Practice Problems, A** and **B. Practice Problem B,** while probing comprehension of the same material as **Practice Problem A,** is almost always sufficiently different as to require a different strategy to solve it. This provides students with more varied practice and is intended to foster the development of a robust set of problem-solving skills.

## Modern Content—Solid Science

The world we live in is constantly changing and the science of chemistry continues to expand and evolve to meet the challenges of our modern world. So I have continued developing this text-book to provide a solid grounding in the basic principles of chemistry while setting them within a context of up-to-date information that serves to capture and hold students' attention and prepare them for studies in a variety of fields. I have tried to connect the study of chemistry to the study of other sciences—including physical, biological, environmental, medical, and engineering. My goal is to help students build a solid conceptual understanding and to encourage mastery of chemical conventions—including models, laws and equations, and such universally important principles as nomenclature, stoichiometry, measurement, and scale. While doing so I have integrated coverage of organic chemistry, biochemistry, green chemistry, and other examples to enhance the relevance of fundamental principles.

Toward this end, I have placed my chapter on organic chemistry (10) earlier than you would find it in most texts. It is not an exhaustive chapter, but presents a handful of organic reactions, germane to applications presented in the book, in the context of bonding and molecular structure. One example is the reaction of the hydroxide ion with carbon dioxide to form the hydrogen carbonate ion. Examples such as this are intended to serve both as a functional introduction to organic chemistry and as reinforcement of bonding theories and the importance of hybridization, molecular polarity, and electron density. I believe that this approach will be particularly beneficial to those who go on to take organic chemistry.

Each individual chapter outline serves as an advance organizer for key concepts and is followed by a brief statement of chapter learning objectives—these are two of the many pedagogical devices designed to foster crucial organization and good study habits. Additionally, I have used my own teaching experiences to identify and address common student misconceptions. One way that I have done this is through the use of "student annotations," margin notes written specifically for the student. These notes include "bite-sized" additional information such as common pitfall alerts, analogies to clarify concepts, pertinent reminders, and alternative perspectives.

## Building Problem-Solving Skills

The entirety of the text emphasizes the importance of problem solving as a crucial element in the study of chemistry. Beginning with Chapter 1, a basic guide fosters a consistent approach to solving problems throughout the text. Each **Sample Problem** is divided into four consistently applied steps:

1. **Strategy**   This step lays the basic framework for the problem. We begin by reading the problem thoroughly to determine exactly what is being asked. Next, we determine what skills will be needed, and lay out a plan for solving the problem. Where appropriate, we make an estimate of the sign and/or the ballpark magnitude of the correct result.

2. **Setup**   In this step we gather the necessary information for solving the problem, including information given within the problem itself, equations, constants, and tabulated data.

3. **Solution**   Using the information gathered in the second step, we now calculate the answer to the problem. Attention to units is emphasized in this step, and each answer is reported to the proper number of significant figures

4. **Think About It**   At this stage we consider whether or not the result makes sense. In some cases, the Think About It section shows an alternate route to the same answer. In other cases, it may include information that illustrates the relevance of the problem.

After working through this problem-solving approach in the **Sample Problems,** students are given two Practice Problems to solve. **Practice Problem A** is always very similar to the Sample Problem and can be solved using the same strategy and approach. Although **Practice Problem B** probes comprehension of the same concept as Practice Problem A, it generally is sufficiently different that it cannot be solved using the exact approach used in the Sample Problem. **Practice Problem B** takes problem solving to another level by requiring students to develop a strategy independently. The regular use of the sample problems and practice problems in this text will help students develop a robust and versatile set of problem-solving skills.

Every section of the book that has Sample and Practice Problems ends with a set of **Checkpoint** questions. These are multiple-choice questions designed to help students assess their readiness to move on to the next section. Checkpoints include computational problems and, new to this edition, *conceptual* questions.

# Greater Relevance Through Modern Examples and Applications

I believe that the study of chemistry can be less daunting for students if they know how it applies to interesting, real-life examples. For this reason, I introduce each chapter with a brief and, I hope, *interesting* story that relates the concepts in the chapter to something familiar. Examples include *Chemical Reactions and Chemotherapy (3), Lasers in Medicine (6)* and *Contaminated Infant Formula (12)*. The **Applying What You've Learned** feature at the end of each chapter recalls the subject of the opening story and includes a multi-part exercise requiring students to use several of the skills they have just learned. Each part of the applying-what-you've-learned

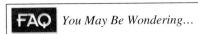

**FAQ** *You May Be Wondering…*

## How Can We Measure Solution Concentrations?

As you may know, white light is actually composed of all the colors of the rainbow. In fact, a rainbow results from the separation of white light by water droplets into the colors or wavelengths that make up the visible spectrum [▶▶ Section 6.1]. Selective absorption of visible light is what makes some solutions appear colored; and, for a solution that is colored, the intensity of color is related to the solution's concentration (see Figure 4.9). This effect gives rise to a type of analysis known as *visible spectrophotometry.* A visible spectrophotometer compares the intensity of light that enters a sample (called the *incident* light) $I_0$, with the intensity of the light that is transmitted through the sample, $I$. *Transmittance (T)* is the ratio of $I$ to $I_0$.

exercise links back to a specific Sample Problem within the chapter. **Bringing Chemistry to Life** segments also work toward this goal, utilizing engaging narrative to further explore applications in the real world, such as *The Stoichiometry of Metabolism (3)* or *Heat Capacity and Hypothermia (5)*. **FAQ boxes** always address a question of interest. These may tackle a topical subject such as *How Important Are Units? (1)* but many also address important fundamental skills such as *How Do I Assign Oxidation Numbers? (4)* or *How Am I Supposed to Remember All These Reactions? (3)*. The end-of-chapter Problem Sets also include a wide range of real-world problems and specific science, medical, and engineering applications. By using so many authentic, modern, and real-world examples, I hope that I have placed the science of chemistry within a human context that will provide for a more engaging learning environment and lead to a fuller understanding of the subject matter and a greater capacity to apply and retain the material.

## Greater Understanding Through Chemical Visualization

This text seeks to enhance student understanding through a variety of both unique and conventional visual techniques. A truly unique element in this text is the inclusion of a distinctive feature entitled **Visualizing Chemistry.** These two-page spreads appear as needed to emphasize fundamental, vitally important principles of chemistry. Setting them apart visually makes them easier to find and revisit as needed throughout the course term. Students asked for more Visualizing Chemistry art and we responded by adding eight new two-page figures. As an example, Chapter 8 includes a **Visualizing Chemistry** figure on the *Born-Haber Cycle* (pp. 304–305) and Chapter 18 on *Factors That Influence the Entropy of a System* (pp. 786–787). Each Visualizing Chemistry figure concludes with a "What's the point?" box that emphasizes the correct take-away message.

There is a series of conceptual end-of-chapter problems for each Visualizing Chemistry figure. These problems allow students to assess their understanding of the principles presented and have been incorporated into the online homework

The use of both **Macro-Micro Art** and **Three-Dimensional Art** builds on the principle of breaking down the complex into simpler, more user-friendly

**Visualizing Chemistry**
**Figure 18.4**

VC 18.1 Consider two gas samples at STP: one consisting of a mole of $F_2$ gas ($S° = 203.34$ J/K · mol) and one consisting of a mole of F gas ($S° = 158.7$ J/K · mol). What factors account for the difference in standard entropies of these two species?

| Volume (i) | Molar mass (ii) | Temperature (iii) | Phase (iv) | Molecular complexity (v) |
|---|---|---|---|---|

a) i, ii, iii, and iv
b) ii and v
c) ii, iv, and v

concepts. Breaking down chemical processes into molecular-level figures makes it easier for students to grasp what is happening on an atomic level. The same theory of breaking down the complex is evident in the treatment of **Hybrid Orbitals.** Through experience, I have learned that this is often a difficult—yet fundamentally crucial—element to grasp. So, for example, I have broken down the process of hybridization through a simple step-by-step visual treatment. **Flow Charts** and a variety of intertextual materials such as **Rewind** and **Fast Forward Buttons** and **Checkpoint** sections are meant to enhance student understanding and comprehension by reinforcing current concepts and connecting new concepts to those covered in other parts of the text.

## Access to Media

In addition to the text itself, students will have access to innovative applications of new educational technologies. Each new Visualizing Chemistry figure has been made into another captivating and pedagogically effective animation for additional reinforcement of subject matter first encountered in the textbook. **MPEG Files** of the Visualizing Chemistry figures will be available for download as **Podcasts,** allowing for convenient viewing to foster increased comprehension.

You will find the electronic homework integrated into the text in numerous places. Each

animation is noted by ⬜, with the Visualizing Chemistry animations noted with ⬜. All Practice Problems B are available in our electronic homework program for practice or assignments. A multitude of end-of-chapter problems are in the electronic homework manager and noted by 🄷.

**IBIS**
**Question 1**

Map

Consider the three containers shown, all of which have the same volume and are at the same temperature.

(i)          (ii)          (iii)

(a) Select the correct order of containers according to increasing mole fraction of gas A (red).

○ (i) < (ii) < (iii)
○ (ii) < (i) < (iii)
○ (iii) < (ii) < (i)
○ (i) = (ii) = (iii)
○ none of the above

(b) Which container has the highest partial pressure of gas B (green)?

○ (i)
○ (ii)

Previous    View Solution    Check Answer    Submit

🄷 11.74 Consider the three containers shown, all of which have the same volume and are at the same temperature. (a) Which container has the smallest mole fraction of gas A (red)? (b) Which container has the highest partial pressure of gas B (green)? (c) Which container has the highest total pressure?

(i)          (ii)          (iii)

For me, this text will always remain a work in progress. I encourage you to contact me with any comments or questions.

Julia Burdge
jburdge@uidaho.edu

# 360° Development Process

A key principle in the development of any general chemistry text is its ability to adapt to teaching specifications in a universal way. The only way to do so is by contacting those universal voices—and learning from their suggestions. Raymond Chang, a veteran author of general chemistry texts, has continued his involvement in the development of this text. Raymond's guidance and expertise proved to be instrumental in the creation of the first edition of this text, and his influential role continued during the development and production of this second edition.

We are confident that our book has the most current content the industry has to offer, thus pushing our desire for accuracy and up-to-date information to the highest standard possible. Extensive and open-minded advice is critical in the production of a superior text.

Here is a brief overview of the initiatives included in the 360° Development Process of *Chemistry*, Second Edition, by Julia Burdge.

**Board of Advisors**   A hand-picked group of trusted teachers, active in the general chemistry course, served as chief advisors and consultants to the author and editorial team during manuscript development. The Board of Advisors reviewed parts of the manuscript, and served as a sounding board for pedagogical, media, and design concerns, and as consultants on organizational changes.

**Symposia**   Every year McGraw-Hill conducts a general chemistry symposium, which is attended by instructors from across the country. These events are an opportunity for editors from McGraw-Hill to gather information about the needs and challenges of instructors teaching these courses. This information helped to improve the book plan for *Chemistry*. These events also offer a forum for the attendees to exchange ideas and experiences with colleagues they might have not otherwise met.

**Focus Group**   In addition to the symposia, we held a focus group specifically for improving the second edition. These selected chemistry professors provided ideas on improvements and suggestions for fine-tuning the content, pedagogy, and art.

**Text Review Panels**   Over 100 teachers and academics from across the country and internationally reviewed the first edition to give feedback on content, design, pedagogy, and organization. This feedback was summarized by the book team and used to guide the direction of the text.

**Accuracy Panel**   A select group of chemistry experts served as the chief advisors for the accuracy and clarity of the text and solutions manual. These individuals reviewed page proofs in the first rounds, and oversaw the writing and accuracy check of the instructor's solutions manuals, test bank, and other ancillary materials.

**Student Focus Groups on Content and Design**   **Two** student class tests and **three** student focus groups provided the editorial team with an understanding of how content and the design of a textbook impact a student's homework and study habits in the general chemistry course.

**Art Development**   Julia Burdge, along with our designer and editors, worked closely with Precision Graphics, an art development company, to create the visual program within this text. Several personal visits to Precision Graphics in Champaign, IL, enabled the author and art team to work together in developing the individual art pieces, art-photo combinations, process boxes, and new animations of chemical processes. Julia's entirely unique **Visualizing Chemistry** figures have been expanded in her second edition based on student and instructor feedback. The end result is a distinctive and innovative visual program that ensures accuracy in relation to textual information, and a style that is uniquely Burdge.

## Board of Advisors

Timothy Brewer   *Eastern Michigan University*
Dana Chatellier   *University of Delaware*
Chris Cheatum   *University of Iowa*
Barbara Cole   *University of Maine*
Gregg Dieckmann   *University of Texas at Dallas*

John Hopkins   *Louisiana State University*
Phillip Klebba   *University of Oklahoma*
Rosemary Loza   *The Ohio State University*
Steven Watkins   *Louisiana State University*
Chris Yerkes   *University of Illinois*

## Burdge Focus Group Participants

Jason Kautz   *University of Nebraska–Lincoln*
Farooq Khan   *University of West Georgia*
Shawn Phillips   *Vanderbilt University*
Michael Ryan   *Marquette University*
Raymond Sadeghi   *University of Texas at San Antonio*

John Sibert   *University of Texas at Dallas*
Sheila Smith   *University of Michigan–Dearborn*
Sherril Soman-Williams   *Grand Valley State University*
Lin Zhu   *Indiana University–Purdue University at Indianapolis*

## Symposia Participants

Chris Bauer   *University of New Hampshire*
Stephen Cabaniss   *University of New Mexico*
Jon Carnahan   *Northern Illinois University*
Chris Cheatum   *University of Iowa*
John DiVincenzo   *Middle Tennessee State University*
William Donovan   *University of Akron*
Mark Freilich   *University of Memphis*
John Hagen   *California Polytechnic State University*
James Hovick   *University of North Carolina at Charlotte*
Wendy Innis-Whitehouse   *University of Texas Pan American*
Michael Jones   *Texas Tech University*
David Laude   *University of Texas at Austin*

Pippa Lock   *McMaster University*
Diana Mason   *University of North Texas*
Maryann McDermott-Jones   *University of Maryland*
Lauren McMills   *Ohio University*
Cortland Pierpont   *University of Colorado*
Jerry Reed-Mundell   *Cleveland State University*
Phil Reid   *University of Washington*
Jimmy Rogers   *University of Texas–Arlington*
Joe Thrasher   *University of Alabama*
Ellen Verdel   *University of South Florida*
Steve Watkins   *Louisiana State University*

## Reviewers

Kaveh Azimi   *Tarrant County College–South*
Anamitro Banerjee   *University of North Dakota*
Rosemary Bartoszek-Loza   *The Ohio State University*
Peter Bell   *Tarleton State University*
Christopher M. Bender   *University of South Carolina Upstate*
Phil Bennett   *Santa Fe Community College*
Mary J. Bojan   *The Pennsylvania State University*
Marcus R. Bond   *Southeast Missouri State University*
Wayne Bosma   *Bradley University*
David A. Boyajian   *Palomar College*
Noland J. Boyd   *Alcorn State University*
Timothy R. Brewer   *Eastern Michigan University*
Bryan E. Breyfogle   *Missouri State University*
Ron Briggs   *Arizona State University*
Stacey A. Buchanan   *Henry Ford Community College*
Erin E. Burke   *Charleston Southern University*
Tara S. Carpenter   *University of Maryland, Baltimore County*
Joe A. Casalnuovo   *California State Polytechnic University, Pomona*
David L. Cedeno   *Illinois State University*
Dana Chatellier   *University of Delaware*
Tabitha R. Chigwada   *West Virginia University*
Henry Choi   *Moorpark College*
Nagash Clarke   *Washtenaw Community College*
William Cleaver   *University of Vermont*
W. Lin Coker, III   *Campbell University*
James A. Collier   *Truckee Meadows Community College*
Carolyn Collins   *College of Southern Nevada*

Christopher Collison   *Rochester Institute of Technology*
Elzbieta Cook   *Louisiana State University*
Steven R. Davis   *University of Mississippi*
Paul A. DiMilla   *Northeastern University*
Donovan A. Dixon   *University of Central Florida*
Paul M. Dixon   *Schoolcraft College*
Mathilda Doerseln Doorley   *Southwest Tennessee Community College*
Michael Doughty   *Southeastern Louisiana University*
Ronald P. Drucker   *City College of San Francisco*
Alhajie A. Dumbuya   *Austin Community College*
Bill Durham   *University of Arkansas*
Amina K. El-Ashmawy   *Collin County Community College*
Huajun Fan   *Prairie View A&M University*
Debra A. Feakes   *Texas State University–San Marcos*
Cheryl Baldwin Frech   *University of Central Oklahoma*
Carlos D. Garcia   *The University of Texas at San Antonio*
Christine Gaudinski   *Aims Community College*
Leanna C. Giancarlo   *University of Mary Washington*
John Goodwin   *Coastal Carolina University*
Bhuvana Gopal   *Ventura County Community College*
Gary M. Gray   *The University of Alabama at Birmingham*
Thomas M. Halasinski   *Middlesex County College*
C. Alton Hassell   *Baylor University*
Michael A. Hauser   *St. Louis Community College–Meramec*
Susan K. Henderson   *Quinnipiac University*
Brad J. Herrick   *Colorado School of Mines*
Paul Higgs   *University of Tennessee at Martin*

Carl B. "Burt" Hollandsworth  *Harding University*
John B. Hopkins  *Louisiana State University*
W. Innis-Whitehouse  *University of Texas–Pan American*
Richard Jarman  *College of DuPage*
Mike Jezercak  *University of Central Oklahoma*
Milton D. Johnston  *Jr.  University of South Florida*
Carole Jubert  *Oregon State University*
Wendy L. Keeney-Kennicutt  *Texas A&M University*
Elizabeth Pearl Kinney  *Madison Area Technical College*
Louis J. Kirschenbaum  *University of Rhode Island*
Andrew Langrehr  *St. Louis Community College–Meramec*
J. Z. Larese  *University of Tennessee, Knoxville*
Terrence A. Lee  *Middle Tennessee State University*
Debbie Leedy  *Glendale Community College*
Brian D. Leskiw  *Youngstown State University*
Margaret Ruth Leslie  *Kent State University*
Lijuan Li  *California State University, Long Beach*
Karen Lou  *Union College*
Rudy Luck  *Michigan Technological University*
Michael Lufaso  *University of North Florida*
Diana Mason  *University of North Texas*
Thomas D. McGrath  *Baylor University*
Craig McLauchlan  *Illinois State University*
Lauren E. H. McMills  *Ohio University*
Madan Mohan  *Austin Community College*
Bruce Moy  *College of Lake County*
Kathy Nabona  *Austin Community College*
Chip Nataro  *Lafayette College*
Edward J. Neth  *University of Connecticut*
Anne-Marie Nickel  *Milwaukee School of Engineering*
Bruce Osterby  *University of Wisconsin–La Crosse*
Mark E. Ott  *Jackson Community College*
Pedro R. Patino  *University of Central Florida*
Shawn T. Phillips  *Vanderbilt University*
Amy M. Pollock  *Michigan State University*
Neil Purdie  *Oklahoma State University*

William Quintana  *New Mexico State University*
Rosemary A. Radar  *Washtenaw Community College*
Ramin Radfar  *Wofford College*
Jimmy Reeves  *University of North Carolina Wilmington*
Michelle Richards-Babb  *West Virginia University*
Gino A. Romeo Jr.  *Yavapai College*
Rebecca J. Rowe  *Colby College*
Steven P. Rowley  *Middlesex County College*
Gillian E. A. Rudd  *Northwestern State University*
Kresimir Rupnik  *Louisiana State University*
Michael A. Russell  *Mt. Hood Community College*
Svein Saebo  *Mississippi State University*
Jamie L. Schneider  *University of Wisconsin–La Crosse*
Mark Schraf  *West Virginia University*
Ingo Schranz  *Henderson State University*
Tom Selegue  *Pima Community College*
J. T. (Dotie) Sipowska  *University of Michigan, Ann Arbor*
Cheryl A. Snyder  *Schoolcraft College*
Jie Song  *University of Michigan–Flint*
Lothar Stahl  *University of North Dakota*
Richard E. Sykora  *University of South Alabama*
Nicholas E. Takach  *The University of Tulsa*
Patricia Metthe Todebush  *Clayton State University*
Joseph L. Toto  *San Diego Mesa College*
Steven Trail  *Elgin Community College*
Sergey Trusov  *Moorpark College*
Rachel Turoscy  *Middlesex County College*
Cyriacus Chris Uzomba  *Austin Community College*
Paul E. Vorndam  *Colorado Community Colleges Online*
Lyle D. Wescott  *The University of Mississippi*
M. Stanley Whittingham  *SUNY at Binghamton*
Troy D. Wood  *University of Buffalo*
Tim Zauche  *University of Wisconsin–Platteville*
Lin Zhu  *IUPUI*
William H. Zoller  *University of Washington*

## Student Class Test Schools

Wayne State University
University of Delaware
Middlesex County College
Louisiana State University

Ozark Tech Community College
Nassau Community College
University of Kansas

## Student Focus Group

University of Illinois
- Stephen CS Barany
- Olivia J. Cornell
- Oliver Hui
- Hanwool Jeong

- Bridget Marry
- Devin Patel
- Tyler Vahldich
- Maura Walsh
- Brett Whitten

# Acknowledgments

I wish to thank the many people who have contributed to the continued development of this text.

Raymond Chang's contributions have again been invaluable. His unfaltering diligence and attention to detail and his ongoing commitment to the quality of this book have had an immeasurable impact on the project.

My Board of Advisors—Tim Brewer, Dana Chatellier, Chris Cheatum, Barbara Cole, Gregg Dieckmann, John Hopkins, Philip Klebba, Rosemary Loza, Steve Watkins and Chris Yerkes—have generously contributed their time, their energy, and their talent to impact the revision of this text. Many have used the book in its first edition and have provided essential feedback for its continued development.

Mike McIntire checked the revised manuscript, solving all of the in-chapter problems and ensuring their accuracy—often going above and beyond the call to contribute to the overall quality of the finished product.

A talented group of people at Precision Graphics worked with me closely on my art program and paging.

My family, as always, continues to be there for me—no matter what.

Finally, I wish to thank my McGraw-Hill family for their continued confidence and support: Publisher Ryan Blankenship, Senior Sponsoring Editor Tami Hodge, Senior Developmental Editor Shirley Oberbroeckling, Senior Project Manager Gloria Schiesl, Senior Designer David Hash, and Senior Marketing Manager Todd Turner.

# ENHANCED SUPPORT FOR Faculty & Students

McGraw-Hill offers various tools and technology products in support of *Chemistry* to both faculty and students alike. Instructors can obtain teaching aides by calling the McGraw-Hill Customer Service Department at 1-800-338-3987, visiting our online catalog at www.mhhe.com, or by contacting their local McGraw-Hill sales representative.

## For the Instructor

### ARIS

The *Assessment, Review, and Instruction System*, also known as ARIS, is an electronic homework and course management system designed for greater flexibility, power, and ease of use than any other system. Whether you are looking for a preplanned course or one you can customize to fit your course needs, ARIS is your solution.

In addition to having access to all student digital learning objects, ARIS enables instructors to:

- Build assignments
- Track student progress
- Be flexible

### McGraw-Hill TEGRITY Campus™

Tegrity Campus is a service that makes class time available all the time by automatically capturing every lecture in a searchable format for students to review when they study and complete assignments. With a simple one-click start-and-stop process, you capture all computer screens and corresponding audio. Students replay any part of any class with easy-to-use browser-based viewing on a PC or Mac.

### Presentation Center

The **Presentation Center** is a complete set of electronic book images and assets for instructors. You can build instructional materials wherever, whenever, and however you want! Accessed from your textbook's ARIS website, the Presentation Center is an online digital library containing photos, artwork, animations, and other media types that can be used to create customized lectures, visually enhanced tests and quizzes, compelling course websites, or attractive printed support materials. All assets are copyrighted by McGraw-Hill Higher Education, but can be used by instructors for classroom purposes. The visual resources in this collection include:

- **Art.** Full-color digital files of all illustrations in the book.
- **Photos.** The photos collection contains digital files of photographs from the text.
- **Tables.** Every table that appears in the text is available electronically.
- **Animations.** Numerous full-color animations illustrating important processes are also provided.
- **PowerPoint® Lecture Outlines.** Ready-made presentations for each chapter of the text.
- **PowerPoint Slides.** All illustrations, photos, and tables are preinserted by chapter into blank PowerPoint slides.

Presentation Center's dynamic search engine allows you to explore by discipline, course, textbook chapter, asset type, or keyword.

### Computerized Test Bank Online

A comprehensive bank of test questions by Debbie Beard (University of Mississippi) is provided within a computerized test bank, enabling you to create paper and online tests or quizzes in this easy-to-use program. Imagine being able to create and access your test or quiz anywhere, at any time.

The test bank contains over 3,000 multiple-choice, true/false, and short answer questions. The questions, which are graded in difficulty, are comparable to the problems in the text.

### Instructors Manual

Written by Nick Flynn from San Angelo State University, the instructor's manual provides literature citations, tips, integration of media, and end-of-chapter problem difficulty levels and categories.

### Student Response System

Wireless technology brings interactivity into the classroom or lecture hall. Instructors and students receive immediate feedback through wireless response pads that are easy to use and engage students. This system can be used by instructors to:

- Take attendance
- Administer quizzes and tests
- Create a lecture with intermittent questions
- Manage lectures and student comprehension through the use of the gradebook
- Integrate interactivity into their PowerPoint presentations

### Content Delivery Flexibility

*Chemistry* by Julia Burdge is available in many formats in addition to the traditional textbook to give instructors and students more choices when deciding on the format of their chemistry text. Choices include:

- **Color Custom by Chapter** For even more flexibility, we offer the Burdge: *Chemistry* text in a full-color, custom version that allows instructors to pick the chapters they want included. Students pay for only what the instructor chooses.
- **eBook** If you or your students are ready for an alternative version of the traditional textbook, McGraw-Hill brings you innovative and inexpensive electronic textbooks. eBooks from McGraw-Hill are smart, interactive, searchable and portable. Included is a powerful suite of built-in tools that allow detailed searching, highlighting, note taking, and student-to-student or instructor-to-student note sharing. In addition, the media-rich eBook for *Chemistry* integrates relevant animations and videos into the textbook content for a true multimedia learning experience.

### Cooperative Chemistry Laboratory Manual

By Melanie Cooper (Clemson University), this innovative guide features open-ended problems designed to simulate experience in a research lab. Working in groups, students investigate one problem over a period of several weeks, so that they might complete three or four projects during the semester, rather than one preprogrammed experiment per class. The emphasis here is on experimental design, analysis, problem solving, and communication.

## For the Students

Students can order supplemental study materials by contacting their campus bookstore, calling 1-800-262-4729, or online at www.shopmcgraw-hill.com.

To help students maximize their learning experience in chemistry, McGraw-Hill offers the following options to students:

### McGraw-Hill ARIS

ARIS (Assessment, Review, and Instruction System) is an electronic study system that offers students a digital portal of knowledge.

Students can readily access a variety of **digital learning objects,** which include:

- Chapter level quizzing
- Animations
- Interactives
- MP3 and MP4 downloads of selected content

## Student Solutions Manual

In this manual by Jon Booze (AccuMedia Publishing Services) and Julia Burdge, the student will find detailed solutions and explanations for the odd-numbered problems in this text.

## Student Study Guide

This study guide, by Lydia Martinez-Rivera (University of Texas at San Antonio), helps students focus their time and energy on important concepts. The study guide offers students a variety of tools:

- Practice questions
- Outline/summary of fundamental concepts—efficient listing of key concepts
- Key terms
- Strategies for difficult concepts

# Chemistry

# 1

# Chemistry
## The Central Science

The "Epidemic Memorial" masks, on display at the Washington State History Museum in Tacoma, Washington, were created by five Native American artists. They represent the effects of smallpox and other diseases on the Native American population.

# How the Scientific Method Helped Defeat Smallpox

To advance understanding of science, researchers use a set of guidelines known as the scientific method. The guidelines involve careful observations, educated reasoning, and the development of hypotheses and theories, which must undergo extensive testing. One of the most compelling examples of the success of the scientific method is the story of smallpox.

Smallpox is one of the diseases classified by the Centers for Disease Control and Prevention (CDC) as a Category A bioterrorism agent. This disease has had an immeasurable impact on human history. During the sixteenth century, European explorers brought smallpox with them to the Americas, devastating native populations and leaving them vulnerable to attack—in effect, shaping the conquest of the New World. In the twentieth century alone the disease killed an estimated half a *billion* people worldwide—leaving many more permanently disfigured, blind, or both.

**Student Annotation:** *Category A* agents are those believed to pose the greatest potential threat to the public and that have a moderate to high potential for large-scale dissemination.

Late in the eighteenth century, an English doctor named Edward Jenner observed that even during outbreaks of smallpox in Europe, milkmaids seldom contracted the disease. He reasoned that when people who had frequent contact with cows contracted *cowpox,* a similar but far less harmful disease, they developed a natural immunity to smallpox. He predicted that intentional exposure to the cowpox virus would produce the same immunity. In 1796 Jenner exposed an 8-year-old boy named James Phipps to the cowpox virus using pus from the cowpox lesions of a milkmaid named Sarah Nelmes. Six weeks later, when Jenner then exposed Phipps to the smallpox virus, the boy did *not* contract the disease. Subsequent experiments using the same technique (later dubbed *vaccination* from the Latin *vacca* meaning "cow") confirmed that immunity to smallpox could be induced.

**Until recently, almost everyone had a smallpox vaccine scar—usually on the upper arm.**

The last naturally occurring case of smallpox occurred in 1977 in Somalia. In 1980 the World Health Organization declared smallpox officially eradicated. This historic triumph over a dreadful disease, one of the greatest medical advances of the twentieth century, began with Jenner's astute observations, inductive reasoning, and careful experimentation—the essential elements of the *scientific method.*

**Student Annotation:** Although naturally occurring smallpox was wiped out worldwide, samples have been kept in research laboratories in the United States and the former Soviet Union, and several countries are now thought to have unauthorized stockpiles of the virus.

At the end of this chapter, you will be able to answer several questions related to the smallpox vaccine [▶▶ Page 25].

## In This Chapter, You Will Learn:

Some of what chemistry is and how it is studied using the scientific method. You will learn about the system of units used by scientists and about expressing and dealing with the numbers that result from scientific measurements.

### Before you begin, you should review

- Basic algebra
- Scientific notation [▶▶ Appendix 1]

## 1.1    The Study of Chemistry

Chemistry often is called the *central science* because knowledge of the principles of chemistry can facilitate understanding of other sciences, including physics, biology, geology, astronomy, oceanography, engineering, and medicine. **Chemistry** is the study of *matter* and the *changes* that matter undergoes. Matter is what makes up our bodies, our belongings, our physical environment, and in fact our universe. **Matter** is anything that has mass and occupies space.

### Chemistry You May Already Know

You may already be familiar with some of the terms used in chemistry. Even if this is your first chemistry course, you may have heard of *molecules* and know them to be tiny pieces of a substance—much too tiny to see. Further, you may know that molecules are made up of *atoms,* even smaller pieces of matter. And even if you don't know what a chemical formula is, you probably know that $H_2O$ is water. You may have used, or at least heard, the term *chemical reaction;* and you are undoubtedly familiar with a variety of chemical reactions, such as those shown in Figure 1.1.

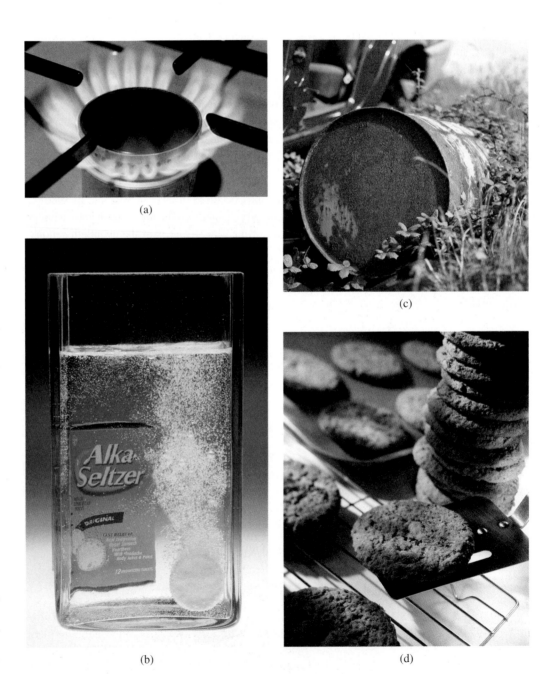

(a)

(c)

**Figure 1.1**    Many familiar processes are chemical reactions: (a) The flame of a gas stove is the combustion of natural gas, which is primarily methane. (b) The bubbles produced when Alka-Seltzer dissolves in water are carbon dioxide, produced by a chemical reaction between two ingredients in the tablets. (c) The formation of rust is a chemical reaction that occurs when iron, water, and oxygen are all present. (d) Many baked goods "rise" as the result of a chemical reaction that produces carbon dioxide.

(b)

(d)

# FAQ   *You May Be Wondering...*

# What Do Molecules Look Like?

Molecules are far too small for us to observe them directly. An effective means of visualizing them is by the use of molecular models. Throughout this book, we will represent matter at the molecular level using *molecular art,* the two-dimensional equivalent of molecular models. In these pictures, atoms are represented as spheres and atoms of particular elements are represented using specific colors. Table 1.1 lists some of the elements that you will encounter most often and the colors used to represent them in this book.

Molecular art can be of *ball-and-stick* models, in which the bonds connecting atoms appear as sticks [Figure 1.2(b)], or of *space-filling* models, in which the atoms appear to overlap one another [Figure 1.2(c)]. Ball-and-stick and space-filling models

illustrate the specific, three-dimensional arrangement of the atoms. The ball-and-stick model does a good job of illustrating the arrangement of atoms, but exaggerates the distances between atoms, relative to their sizes. The space-filling model gives a more accurate picture of these *interatomic* distances but can obscure the details of the three-dimensional arrangement.

**Figure 1.2**   Water represented with a (a) molecular formula, (b) ball-and-stick model, and (c) space-filling model.

| TABLE 1.1 | Colors of Elements Commonly Used in Molecular Art | |
|---|---|---|
| Hydrogen | Sodium | |
| Boron | Phosphorus | |
| Carbon | Sulfur | |
| Nitrogen | Chlorine | |
| Oxygen | Bromine | |
| Fluorine | Iodine | |

The reactions in Figure 1.1 are all things that you can observe at the *macroscopic level.* In other words, these processes and their results are visible to the human eye. In studying chemistry, you will learn to visualize and understand these same processes at the *molecular level.*

Although it can take many different forms, all matter consists of various combinations of atoms of only a relatively small number of simple substances called *elements.* The properties of matter depend on which of these elements it contains and on how the atoms of those elements are arranged.

## The Scientific Method

Experiments are the key to advancing our understanding of chemistry—or any science. Although not all scientists will necessarily take the same approach to experimentation, they all follow a set of guidelines known as the ***scientific method*** to add their results to the larger body of knowledge within a given field. The flowchart in Figure 1.3 illustrates this basic process. The method begins with the gathering of data via observations and experiments. Scientists study these data and try to identify *patterns* or *trends.* When they find a pattern or trend, they may summarize their findings with a ***law,*** a concise verbal or mathematical statement of a reliable relationship between phenomena. Scientists may then formulate a ***hypothesis,*** a tentative explanation for their observations. Further experiments are designed to test the hypothesis. If experiments indicate that the hypothesis is incorrect, the scientists go back to the drawing board, try to come up with

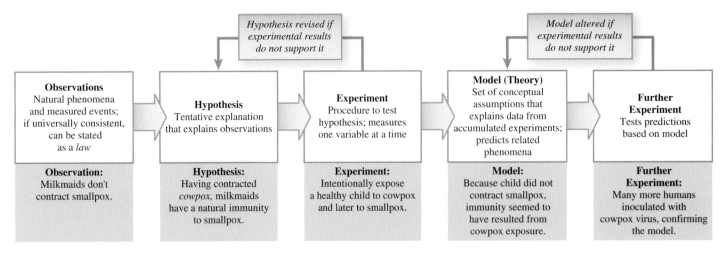

**Figure 1.3**    Flowchart of the scientific method.

a different interpretation of their data, and formulate a new hypothesis. The new hypothesis will then be tested by experiment. When a hypothesis stands the test of extensive experimentation, it may evolve into a theory. A ***theory*** is a unifying principle that explains a body of experimental observations and the laws that are based on them. Theories can also be used to predict related phenomena, so theories are constantly being tested. If a theory is disproved by experiment, then it must be discarded or modified so that it becomes consistent with experimental observations.

## 1.2    Classification of Matter

> **Student Annotation:** Some books refer to *substances* as *pure substances*. These two terms generally mean the same thing although the adjective *pure* is unnecessary in this context because a substance is, by definition, pure.

Chemists classify matter as either a *substance* or a *mixture* of substances. A substance may be further categorized as either an *element* or a *compound*. A ***substance*** is a form of matter that has a definite (constant) composition and distinct properties. Examples are salt (sodium chloride), iron, water, mercury, carbon dioxide, and oxygen. Substances can be either elements (such as iron, mercury, and oxygen) or compounds (such as salt, water, and carbon dioxide). They differ from one another in composition and can be identified by appearance, smell, taste, and other properties.

### States of Matter

> **Student Annotation:** Solids and liquids sometimes are referred to collectively as the *condensed phases*. Liquids and gases sometimes are referred to collectively as *fluids*.

All substances can, in principle, exist as a solid, a liquid, and a gas, the three physical states depicted in Figure 1.4. In a solid, particles are held close together in an orderly fashion with little freedom of motion. As a result, a solid does not conform to the shape of its container. Particles in a liquid are close together but are not held rigidly in position; they are free to move past one another. Thus, a liquid conforms to the shape of the part of the container it fills. In a gas, the particles are separated by distances that are very large compared to the size of the particles. A sample of gas assumes both the shape and the volume of its container.

The three states of matter can be interconverted without changing the chemical composition of the substance. Upon heating, a solid (e.g., ice) will melt to form a liquid (water). Further heating will vaporize the liquid, converting it to a gas (water vapor). Conversely, cooling a gas will cause it to condense into a liquid. When the liquid is cooled further, it will freeze into the solid form. Figure 1.5 shows the three physical states of water.

Multimedia

Matter—three states of matter.

### Elements

An ***element*** is a substance that cannot be separated into simpler substances by chemical means. Iron, mercury, oxygen, and hydrogen are just 4 of the 117 elements that have been identified. Most of the known elements occur naturally on Earth. The others have been produced by scientists via nuclear processes, which are discussed in Chapter 20. As shown in Figure 1.6(a) and (b), an element may consist of atoms or molecules.

For convenience, chemists use symbols of one or two letters to represent the elements. Only the first letter of an element's chemical symbol is capitalized. A list of the elements and their

**Figure 1.4** Molecular-level illustrations of a solid, liquid, and gas.

symbols appears on the inside front cover of this book. The symbols of some elements are derived from their Latin names—for example, Ag from *argentum* (silver), Pb from *plumbum* (lead), and Na from *natrium* (sodium)—while most of them come from their English names—for example, H for hydrogen, Co for cobalt, and Br for bromine.

## Compounds

Most elements can combine with other elements to form compounds. Hydrogen gas, for example, burns in the presence of oxygen gas to form water, which has properties that are distinctly different from those of either hydrogen or oxygen. Thus, water is a **compound,** a substance composed of atoms of two or more elements chemically united in fixed proportions [Figure 1.6(c)]. The elements that make up a compound are called the compound's *constituent elements*. For example, the constituent elements of water are hydrogen and oxygen.

A compound cannot be separated into simpler substances by any physical process. (A physical process [▶▶ Section 1.4] is one that does not change the identity of the matter. Examples of physical processes include boiling, freezing, and filtering.) Instead, the separation of a compound into its constituent elements requires a *chemical reaction*.

## Mixtures

A **mixture** is a combination of two or more substances [Figure 1.6(d)] in which the substances retain their distinct identities. Like pure substances, mixtures can be solids, liquids, or gases. Some familiar examples are mixed nuts, 14-carat gold, apple juice, milk, and air. Mixtures do not have a universal constant composition. Therefore, samples of air collected in different locations will differ in composition because of differences in altitude, pollution, and other factors. Various brands of apple juice may differ in composition because of the use of different varieties of apples, or there may be differences in processing and packaging, and so on.

Mixtures are either *homogeneous* or *heterogeneous*. When we dissolve a teaspoon of sugar in a glass of water, we get a **homogeneous mixture** because the composition of the mixture is

**Figure 1.5** Water as a solid (ice), liquid, and gas. (We can't actually see water vapor, any more than we can see the nitrogen and oxygen that make up most of the air we breathe. When we see steam or clouds, what we are actually seeing is water vapor that has condensed upon encountering cold air.)

**Student Annotation:** A compound may consist of *molecules* or *ions,* which we will discuss in Chapter 2.

(a)

(b)

(c)

(d)

**Figure 1.6**    (a) Isolated atoms of an element. (b) Molecules of an element. (c) Molecules of a compound, consisting of more than one element. (d) A mixture of atoms of an element and molecules of an element and a compound.

(a)                                        (b)

**Figure 1.7**    (a) A heterogeneous mixture contains iron filings and sand. (b) A magnet is used to separate the iron filings from the mixture.

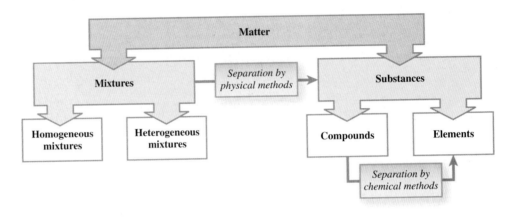

**Figure 1.8**    Flowchart for the classification of matter.

uniform throughout. If we mix sand with iron filings, however, the sand and the iron filings remain distinct and discernible from each other (Figure 1.7). This type of mixture is called a ***heterogeneous mixture*** because the composition is *not* uniform.

Mixtures, whether homogeneous or heterogeneous, can be separated by physical means into pure components without changing the identities of the components. Thus, sugar can be recovered from a water solution by evaporating the solution to dryness. Condensing the vapor will give us back the water component. To separate the sand-iron mixture, we can use a magnet to remove the iron filings from the sand, because sand is not attracted to the magnet [see Figure 1.7(b)]. After separation, the components of the mixture will have the same composition and properties as they did prior to being mixed. The relationships among substances, elements, compounds, and mixtures are summarized in Figure 1.8.

# 1.3    Scientific Measurement

Scientists use a variety of devices to measure the properties of matter. A meterstick is used to measure length; a buret, pipet, graduated cylinder, and volumetric flask are used to measure volume (Figure 1.9); a balance is used to measure mass; and a thermometer is used to measure temperature. Properties that can be measured are called *quantitative* properties because they are expressed using numbers. When we express a measured quantity with a number, though, we must always include the appropriate unit; otherwise, the measurement is meaningless. For example, to say that the depth of a swimming pool is 3 is insufficient to distinguish between one that is 3 *feet* (0.9 meters) and one that is 3 *meters* (9.8 feet) deep. Units are essential to reporting measurements correctly.

The two systems of units with which you are probably most familiar are the *English system* (foot, gallon, pound, etc.) and the *metric system* (meter, liter, kilogram, etc.). Although there has been an increase in the use of metric units in the United States in recent years, English units still are used commonly. For many years scientists recorded measurements in metric units, but in 1960, the General Conference on Weights and Measures, the international authority on units, proposed a revised metric system for universal use by scientists. We will use both metric and revised metric (SI) units in this book.

## SI Base Units

The revised metric system is called the **International System of Units** (abbreviated SI, from the French *Système Internationale d'Unités*). Table 1.2 lists the seven SI base units. All other units of measurement can be derived from these base units. The **SI unit** for *volume,* for instance, is derived by cubing the SI base unit for *length.* The prefixes listed in Table 1.3 are used to denote decimal fractions and multiples of SI units. This enables scientists to tailor the magnitude of a unit to a particular application. For example, the meter (m) is appropriate for describing the dimensions of a classroom, but the kilometer (km), 1000 m, is more appropriate for describing the distance between two cities. Units that you will encounter frequently in the study of chemistry include those for mass, temperature, volume, and density.

**Student Annotation:** According to the U.S. Metric Association (USMA), the United States is "the only significant holdout" with regard to adoption of the metric system. The other countries that continue to use traditional units are Myanmar (formerly Burma) and Liberia.

| TABLE 1.2 | Base SI Units | |
|---|---|---|
| **Base Quantity** | **Name of Unit** | **Symbol** |
| Length | meter | m |
| Mass | kilogram | kg |
| Time | second | s |
| Electric current | ampere | A |
| Temperature | kelvin | K |
| Amount of substance | mole | mol |
| Luminous intensity | candela | cd |

**Student Annotation:** Only one of the seven SI base units, the kilogram, itself contains a prefix.

**Figure 1.9** (a) A volumetric flask is used to prepare a precise volume of a solution for use in the laboratory. (b) A graduated cylinder is used to measure a volume of liquid. It is less precise than the volumetric flask. (c) A volumetric pipette is used to deliver a precise amount of liquid. (d) A burette is used to measure the volume of a liquid that has been added to a container. A reading is taken before and after the liquid is delivered, and the volume delivered is determined by subtracting the first reading from the second.

Volumetric flask
(a)

Graduated cylinder
(b)

Pipette
(c)

Burette
(d)

| TABLE 1.3 | | Prefixes Used with SI Units | |
|---|---|---|---|
| Prefix | Symbol | Meaning | Example |
| Tera- | T | $1 \times 10^{12}$ (1,000,000,000,000) | 1 teragram (Tg) = $1 \times 10^{12}$ g |
| Giga- | G | $1 \times 10^{9}$ (1,000,000,000) | 1 gigawatt (GW) = $1 \times 10^{9}$ W |
| Mega- | M | $1 \times 10^{6}$ (1,000,000) | 1 megahertz (MHz) = $1 \times 10^{6}$ Hz |
| Kilo- | k | $1 \times 10^{3}$ (1,000) | 1 kilometer (km) = $1 \times 10^{3}$ m |
| Deci- | d | $1 \times 10^{-1}$ (0.1) | 1 deciliter (dL) = $1 \times 10^{-1}$ L |
| Centi- | c | $1 \times 10^{-2}$ (0.01) | 1 centimeter (cm) = $1 \times 10^{-2}$ m |
| Milli- | m | $1 \times 10^{-3}$ (0.001) | 1 millimeter (mm) = $1 \times 10^{-3}$ m |
| Micro- | μ | $1 \times 10^{-6}$ (0.000001) | 1 microliter (μL) = $1 \times 10^{-6}$ L |
| Nano- | n | $1 \times 10^{-9}$ (0.000000001) | 1 nanosecond (ns) = $1 \times 10^{-9}$ s |
| Pico- | p | $1 \times 10^{-12}$ (0.000000000001) | 1 picogram (pg) = $1 \times 10^{-12}$ g |

## Mass

Although the terms *mass* and *weight* often are used interchangeably, they do not mean the same thing. Strictly speaking, weight is the force exerted by an object or sample due to gravity. **Mass** is a measure of the amount of matter in an object or sample. Because gravity varies from location to location (gravity on the moon is only about one-sixth that on Earth), the weight of an object varies depending on where it is measured. The mass of an object remains the same regardless of where it is measured. The SI base unit of mass is the kilogram (kg), but in chemistry the smaller gram (g) often is more convenient and is more commonly used:

$$1 \text{ kg} = 1000 \text{ g} = 1 \times 10^{3} \text{ g}$$

## Temperature

There are two temperature scales used in chemistry. Their units are degrees Celsius (°C) and kelvin (K). The Celsius scale was originally defined using the freezing point (0°C) and the boiling point (100°C) of pure water at sea level. As Table 1.2 shows, the SI base unit of temperature is the *kelvin*. Kelvin is known as the *absolute* temperature scale, meaning that the lowest temperature possible is 0 K, a temperature referred to as "absolute zero." No *degree* sign (°) is used to represent a temperature on the Kelvin scale. The theoretical basis of the Kelvin scale has to do with the behavior of gases and is discussed in Chapter 11.

Units of the Celsius and Kelvin scales are equal in magnitude, so *a degree Celsius is equivalent to a kelvin*. Thus, if the temperature of an object increases by 5°C, it also increases by 5 K. Absolute zero on the Kelvin scale is equivalent to –273.15°C on the Celsius scale. We use the following equation to convert a temperature from units of degrees Celsius to kelvin:

**Student Annotation:** Depending on the precision required, the conversion from degrees Celsius to kelvin often is done simply by adding 273, rather than 273.15.

Equation 1.1                                  $$K = °C + 273.15$$

Sample Problem 1.1 illustrates conversions between these two temperature scales.

### Sample Problem 1.1

Normal human body temperature can range over the course of the day from about 36°C in the early morning to about 37°C in the afternoon. Express these two temperatures and the range that they span using the Kelvin scale.

**Strategy** Use Equation 1.1 to convert temperatures from the Celsius scale to the Kelvin scale. Then convert the range of temperatures from degrees Celsius to kelvin, keeping in mind that 1°C is equivalent to 1 K.

**Think About It** Check your math and remember that converting a temperature from degrees Celsius to kelvin is different from converting a *difference* in temperature from degrees Celsius to kelvin.

**Setup** Equation 1.1 is already set up to convert the two temperatures from degrees Celsius to kelvin. No further manipulation of the equation is needed. The range in kelvin will be the same as the range in degrees Celsius.

**Solution** 36°C + 273 = 309 K, 37°C + 273 = 310 K, and the range of 1°C is equal to a range of 1 K.

**Practice Problem A** Express the freezing point of water (0°C), the boiling point of water (100°C), and the range spanned by the two temperatures using the Kelvin scale.

**Practice Problem B** According to the website of the National Aeronautics and Space Administration (NASA), the average temperature of the universe is 2.7 K. Convert this temperature to degrees Celsius.

# Bringing Chemistry to Life

## Fahrenheit Temperature Scale

Outside of scientific circles, the Fahrenheit temperature scale is the one most used in the United States. Before the work of Daniel Gabriel Fahrenheit (German physicist, 1686–1736), there were numerous different, arbitrarily defined temperature scales, none of which gave consistent measurements. Accounts of exactly how Fahrenheit devised his temperature scale vary from source to source. In one account, in 1724, Fahrenheit labeled as 0° the lowest artificially attainable temperature at the time (the temperature of a mixture of ice, water, and ammonium chloride). Using a traditional scale consisting of 12 degrees, he labeled the temperature of a healthy human body as the twelfth degree. On this scale, the freezing point of water occurred at the fourth degree. For better resolution, each degree was further divided into eight smaller degrees. This convention makes the freezing point of water 32° and normal body temperature 96°. Today we consider normal body temperature to be somewhat higher than 96°F.

   The boiling point of water on the Fahrenheit scale is 212°, meaning that there are 180 degrees (212° − 32°) between the freezing and boiling points. This separation is considerably more than the 100 degrees between the freezing point and boiling point of water on the Celsius scale [named after Swedish physicist Ander Celsius (1701–1744)]. Thus, the size of a degree on the Fahrenheit scale is only 100/180 or five-ninths of a degree on the Celsius scale. Equations 1.2 and 1.3 give the relationship between Fahrenheit and Celsius temperatures.

$$\text{temperature in Celsius} = (\text{temperature in Fahrenheit} - 32°\text{F}) \times \frac{5°\text{C}}{9°\text{F}} \qquad \text{Equation 1.2}$$

and

$$\text{temperature in Fahrenheit} = \frac{9°\text{F}}{5°\text{C}} \times (\text{temperature in degrees Celsius}) + 32°\text{F} \qquad \text{Equation 1.3}$$

## Sample Problem 1.2

A body temperature above 39°C constitutes a high fever. Convert this temperature to the Fahrenheit scale.

**Strategy** We are given a temperature in Celsius and are asked to convert it to Fahrenheit.

**Setup** We use Equation 1.3:

$$\text{temperature in Fahrenheit} = \frac{9°\text{F}}{5°\text{C}} \times (\text{temperature in Celsius}) + 32°\text{F}$$

**Solution**

$$\text{temperature in Fahrenheit} = \frac{9°\text{F}}{5°\text{C}} \times 39°\text{C} + 32°\text{F} = 102.2°\text{F}$$

**Think About It** Knowing that "normal" body temperature on the Fahrenheit scale is approximately 99°F (98.6°F is the number most often cited), 102.2°F seems like a reasonable answer.

**Practice Problem A** Convert the temperatures 45.0°C and 90.0°C, and the difference between them, to degrees Fahrenheit.

**Practice Problem B** In Ray Bradbury's 1953 novel *Fahrenheit 451*, 451°F is said to be the temperature at which books, which have been banned in the story, ignite. Convert 451°F to the Celsius scale.

## Derived Units: Volume and Density

There are many quantities, such as volume and density, that require units not included in the base SI units. In these cases, we must combine base units to *derive* appropriate units for the quantity.

The derived SI unit for volume, the meter cubed ($m^3$), is a larger volume than is practical in most laboratory settings. The more commonly used metric unit, the *liter* (L), is derived by cubing the *decimeter* (one-tenth of a meter) and is therefore also referred to as the cubic decimeter ($dm^3$). Another commonly used metric unit of volume is the *milliliter* (mL), which is derived by cubing the centimeter (1/100 of a meter). The milliliter is also referred to as the cubic centimeter ($cm^3$). Figure 1.10 illustrates the relationship between the liter (or $dm^3$) and the milliliter (or $cm^3$).

*Density* is the ratio of mass to volume. Oil floats on water, for example, because, in addition to not mixing with water, oil has a lower density than water. That is, given *equal volumes* of the two liquids, the oil will have a *smaller mass* than the water. Density is calculated using the following equation:

Equation 1.4 $$d = \frac{m}{V}$$

Oil floating on water is a familiar demonstration of density differences.

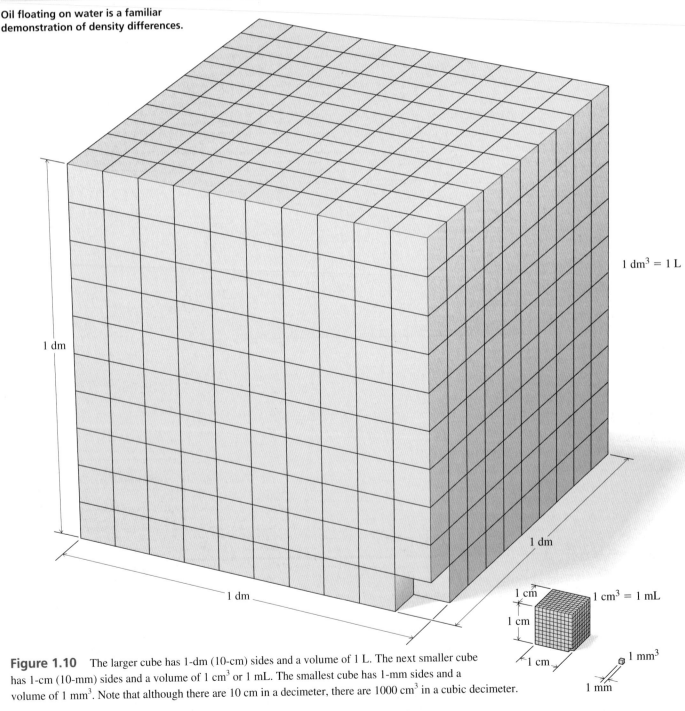

$1\ dm^3 = 1\ L$

1 dm

1 dm

1 dm

1 cm    $1\ cm^3 = 1\ mL$

1 cm

1 cm

$1\ mm^3$

1 mm

**Figure 1.10**   The larger cube has 1-dm (10-cm) sides and a volume of 1 L. The next smaller cube has 1-cm (10-mm) sides and a volume of 1 $cm^3$ or 1 mL. The smallest cube has 1-mm sides and a volume of 1 $mm^3$. Note that although there are 10 cm in a decimeter, there are 1000 $cm^3$ in a cubic decimeter.

where *d, m,* and *V* denote density, mass, and volume, respectively. The SI-derived unit for density is the kilogram per cubic meter ($kg/m^3$). This unit is too large for most common uses, however, so grams per cubic centimeter ($g/cm^3$) and its equivalent, grams per milliliter (g/mL), are used to express the densities of most solids and liquids. Water, for example, has a density of 1.00 $g/cm^3$ at 4°C. Because gas densities generally are very low, we typically express them in units of grams per liter (g/L):

$$1 \text{ g/cm}^3 = 1 \text{ g/mL} = 1000 \text{ kg/m}^3$$

$$1 \text{ g/L} = 0.001 \text{ g/mL}$$

Sample Problem 1.3 illustrates density calculations.

## Sample Problem 1.3

Ice cubes float in a glass of water because solid water is less dense than liquid water. (a) Calculate the density of ice given that, at 0°C, a cube that is 2.0 cm on each side has a mass of 7.36 g, and (b) determine the volume occupied by 23 g of ice at 0°C.

**Strategy** (a) Determine density by dividing mass by volume (Equation 1.4), and (b) use the calculated density to determine the volume occupied by the given mass.

**Setup** (a) We are given the mass of the ice cube, but we must calculate its volume from the dimensions given. The volume of the ice cube is $(2.0 \text{ cm})^3$, or 8.0 $cm^3$. (b) Rearranging Equation 1.4 to solve for volume gives $V = m/d$.

**Solution**

(a) $d = \dfrac{7.36 \text{ g}}{8.0 \text{ cm}^3} = 0.92 \text{ g/cm}^3$   or   0.92 g/mL       (b) $V = \dfrac{23 \text{ g}}{0.92 \text{ g/cm}^3} = 25 \text{ cm}^3$   or   25 mL

**Think About It** For a sample with a density *less* than 1 $g/cm^3$, the number of cubic centimeters should be *greater* than the number of grams. In this case, 25 ($cm^3$) > 23 (g).

**Practice Problem A** Given that 25.0 mL of mercury has a mass of 340.0 g, calculate (a) the density of mercury and (b) the mass of 120.0 mL of mercury.

**Practice Problem B** Calculate (a) the density of a solid substance if a cube measuring 2.33 cm on one side has a mass of 117 g and (b) the mass of a cube of the same substance measuring 7.41 cm on one side.

The box on page 14 illustrates the importance of using units carefully in scientific work.

## Checkpoint 1.3     Scientific Measurement

**1.3.1** The coldest temperature ever recorded on Earth was −128.6°F (recorded at Vostok Station, Antarctica, on July 21, 1983). Express this temperature in degrees Celsius and in kelvins.

a) −89.2°C, −89.2 K

b) −289.1°C, −15.9 K

c) −89.2°C, 183.9 K

d) −173.9°C, 99.3 K

e) −7.0°C, 266.2 K

**1.3.2** What is the density of an object that has a volume of 34.2 $cm^3$ and a mass of 19.6 g?

a) 0.573 $g/cm^3$

b) 1.74 $g/cm^3$

c) 670 $g/cm^3$

d) 53.8 $g/cm^3$

e) 14.6 $g/cm^3$

**1.3.3** A sample of water is heated from room temperature to just below the boiling point. The overall change in temperature is 72°C. Express this temperature change in kelvins.

a) 345 K

b) 72 K

c) 0 K

d) 201 K

e) 273 K

**1.3.4** Given that the density of gold is 19.3 $g/cm^3$, calculate the volume (in $cm^3$) of a gold nugget with a mass of 5.98 g.

a) 3.23 $cm^3$

b) 5.98 $cm^3$

c) 115 $cm^3$

d) 0.310 $cm^3$

e) 13.3 $cm^3$

# FAQ *You May Be Wondering…*

# Why Are Units So Important?

On December 11, 1998, NASA launched the 125-million-dollar Mars Climate Orbiter, which was intended to be the Red Planet's first weather satellite. After a 416-million-mile (mi) journey, the spacecraft was supposed to go into Mars's orbit on September 23, 1999. Instead, it entered Mars's atmosphere about 100 km (62 mi) lower than planned and was destroyed by heat. Mission controllers later determined that the spacecraft was lost because English measurement units were not converted to metric units in the navigation software.

Engineers at Lockheed Martin Corporation, who built the spacecraft, specified its thrust in pounds, which is an English unit of force. Scientists at NASA's Jet Propulsion Laboratory, on the other hand, who were responsible for deployment, had assumed that the thrust data they were given were expressed in *newtons*, a metric unit. To carry out the conversion between pound and new-

ton, we would start with 1 lb = 0.4536 kg and, from Newton's second law of motion,

$$\text{force} = (\text{mass})(\text{acceleration}) = (0.4536 \text{ kg})(9.81 \text{ m/s}^2)$$

$$= 4.45 \text{ kg} \cdot \text{m/s}^2 = 4.45 \text{ N}$$

because 1 newton (N) = 1 kg · m/s². Therefore, instead of converting 1 lb of *force* to 4.45 N, the scientists treated it as a force of 1 N. The considerably smaller engine thrust employed because of the engineers' failure to convert from English to metric units resulted in a lower orbit and the ultimate destruction of the spacecraft.

Commenting on the failure of the Mars mission, one scientist said, "This is going to be the cautionary tale that will be embedded into introduction to the metric system in elementary school, high school, and college science courses till the end of time."

Mars Climate Orbiter during preflight tests.

## 1.4    The Properties of Matter

Substances are identified by their properties as well as by their composition. Properties of a substance may be *quantitative* (measured and expressed with a number) or *qualitative* (not requiring explicit measurement).

### Physical Properties

Color, melting point, boiling point, and physical state are all physical properties. A *physical property* is one that can be observed and measured without changing the *identity* of a substance. For example, we can determine the melting point of ice by heating a block of ice and measuring the temperature at which the ice is converted to water. Liquid water differs from ice in appearance but not in composition; both liquid water and ice are $H_2O$. Melting is a *physical change;* one in which the state of matter changes, but the identity of the matter does not change. We can recover the original ice by cooling the water until it freezes. Therefore, the melting point of a substance is a *physical* property. Similarly, when we say that nitrogen dioxide gas is brown, we are referring to the physical property of color.

## Chemical Properties

The statement "Hydrogen gas burns in oxygen gas to form water" describes a ***chemical property*** of hydrogen, because to observe this property we must carry out a ***chemical change***—burning in oxygen (combustion), in this case. After a chemical change, the original substance (hydrogen gas in this case) will no longer exist. What remains is a different substance (water, in this case). We *cannot* recover the hydrogen gas from the water by means of a physical process, such as boiling or freezing.

Every time we bake cookies, we bring about a chemical change. When heated, the sodium bicarbonate (baking soda) in cookie dough undergoes a chemical change that produces carbon dioxide gas. The gas forms numerous little bubbles in the dough during the baking process, causing the cookies to "rise." Once the cookies are baked, we cannot recover the sodium bicarbonate by cooling the cookies, or by *any* physical process. When we eat the cookies, we cause further chemical changes that occur during digestion and metabolism.

## Extensive and Intensive Properties

All properties of matter are either *extensive* or *intensive*. The measured value of an ***extensive property*** depends on the amount of matter. *Mass* is an extensive property. More matter means more mass. Values of the same extensive property can be added together. For example, two gold nuggets will have a combined mass that is the sum of the masses of each nugget, and the length of two city buses is the sum of their individual lengths. The value of an extensive property depends on the amount of matter.

The value of an ***intensive property*** does *not* depend on the amount of matter. *Density* and *temperature* are intensive properties. Suppose that we have two beakers of water at the same temperature and we combine them to make a single quantity of water in a larger beaker. The density and the temperature of the water in the larger combined quantity will be the same as they were in the two separate beakers. Unlike mass and length, which are additive, temperature, density, and other intensive properties are not additive. Sample Problem 1.4 shows you how to differentiate chemical and physical processes.

---

### Sample Problem 1.4

The diagram in (a) shows a compound made up of atoms of two elements (represented by the green and red spheres) in the liquid state. Which of the diagrams in (b) to (d) represent a physical change, and which diagrams represent a chemical change?

       (a)                    (b)                    (c)                    (d)

**Strategy**  We review the discussion of physical and chemical changes. A physical change does not change the *identity* of a substance, whereas a chemical change *does* change the identity of a substance.

**Setup**  The diagram in (a) shows a substance that consists of molecules of a compound, each of which contains two different atoms, represented by green and red spheres. Diagram (b) contains the same number of red and green spheres, but they are not arranged the same way as in diagram (a).

*(Continued)*

In (b), each molecule is made up of two identical atoms. These are molecules of *elements,* rather than molecules of a compound. Diagram (c) also contains the same numbers of red and green spheres as diagram (a). In (c), however, all the atoms are shown as isolated spheres. These are atoms of elements, rather than molecules of a compound. In diagram (d), the spheres are arranged in molecules, each containing one red and one green sphere. Although the molecules are farther apart in diagram (d), they are the same molecules as shown in diagram (a).

**Solution**  Diagrams (b) and (c) represent chemical changes. Diagram (d) represents a physical change.

> **Think About It**  A chemical change changes the *identity* of matter. A physical change does not.

**Practice Problem A**  Which of the following processes is a physical change? (a) evaporation of water; (b) combination of hydrogen and oxygen gas to produce water; (c) dissolution of sugar in water; (d) separation of sodium chloride (table salt) into its constituent elements, sodium and chlorine; (e) combustion of sugar to produce carbon dioxide and water.

**Practice Problem B**  The diagram in the margin represents the result of a process. Which of the diagrams [(i) to (iii)] could represent the starting material if the process were physical, and which could represent the starting material if the change were chemical?

(i)                    (ii)                    (iii)

---

| **Checkpoint 1.4** | **The Properties of Matter** |
| --- | --- |

**1.4.1**  Which of the following [(a)–(f)] represents a physical change? (Select all that apply.)

**1.4.2**  Which of the following [(a)–(f)] represents a chemical change? (Select all that apply.)

a)

before        after

b)

before        after

c)

before        after

d)

before        after

e)

before        after

f)

before        after

# 1.5 Uncertainty in Measurement

Chemistry makes use of two types of numbers: exact and inexact. *Exact* numbers include numbers with defined values, such as 2.54 in the definition 1 inch (in) = 2.54 cm, 1000 in the definition 1 kg = 1000 g, and 12 in the definition 1 dozen = 12 objects. (The number 1 in each of these definitions is also an exact number.) Exact numbers also include those that are obtained by counting. Numbers measured by any method other than counting are *inexact*.

Measured numbers are inexact because of the measuring devices that are used, the individuals who use them, or both. For example, a ruler that is poorly calibrated will result in measurements that are in error—no matter how carefully it is used. Another ruler may be calibrated properly but have insufficient resolution for the necessary measurement. Finally, whether or not an instrument is properly calibrated or has sufficient resolution, there are unavoidable differences in how different people see and interpret measurements.

## Significant Figures

An inexact number must be reported in such a way as to indicate the uncertainty in its value. This is done using significant figures. **Significant figures** are the *meaningful digits* in a reported number. Consider the measurement of the memory card in Figure 1.11 using the ruler above it. The card's width is between 2 and 3 cm. We may record the width as 2.5 cm, but because there are no gradations between 2 and 3 cm on this ruler, we are *estimating* the second digit. Although we are certain about the 2 in 2.5, we are *not* certain about the 5. The last digit in a measured number is referred to as the *uncertain digit;* and the uncertainty associated with a measured number is generally considered to be ±1 in the place of the last digit. Thus, when we report the width of the memory card to be 2.5 cm, we are implying that its width is 2.5 ± 0.1 cm. Each of the digits in a measured number, including the uncertain digit, is a significant figure. The reported width of the memory card, 2.5 cm, contains two significant figures.

A ruler with millimeter gradations would enable us to be certain about the second digit in this measurement and to estimate a third digit. Now consider the measurement of the memory card using the ruler below it. We may record the width as 2.45 cm. Again, we estimate one digit beyond those we can read. The reported width of 2.45 cm contains three significant figures. Reporting the width as 2.45 cm implies that the width is 2.45 ± 0.01 cm.

The number of significant figures in any number can be determined using the following guidelines:

1. Any digit that is not zero is significant (112.1 has four significant figures).
2. Zeros located between nonzero digits are significant (305 has three significant figures, and 50.08 has four significant figures).
3. Zeros to the left of the first nonzero digit are not significant (0.0023 has two significant figures, and 0.000001 has one significant figure).
4. Zeros to the right of the last nonzero digit are significant if the number contains a decimal point (1.200 has four significant figures).
5. Zeros to the right of the last nonzero digit in a number that does not contain a decimal point may or may not be significant (100 may have one, two, or three significant figures—it is impossible to tell without additional information). To avoid ambiguity in such cases, it is best to express such numbers using scientific notation. If the intended number of significant figures is one, the number is written as $1 \times 10^2$; if the intended number of significant figures is two, the number is written as $1.0 \times 10^2$; and if the intended number of significant figures is three, the number is written as $1.00 \times 10^2$.

Sample Problem 1.5 lets you practice determining the number of significant figures in a number.

**Figure 1.11** The width we report for the memory card depends on which ruler we use to measure it.

**Student Annotation:** It is important not to imply greater certainty in a measured number than is realistic. For example, it would be inappropriate to report the width of the memory card in Figure 1.11 as 2.4500 cm, because this would imply an uncertainty of ±0.0001.

**Student Annotation:** Appendix 1 reviews scientific notation.

## Sample Problem 1.5

Determine the number of significant figures in the following measurements: (a) 443 cm, (b) 15.03 g, (c) 0.0356 kg, (d) $3.000 \times 10^{-7}$ L, (e) 50 mL, (f) 0.9550 m.

**Strategy** All nonzero digits are significant, so the goal will be to determine which of the zeros is significant.

**Setup** Zeros are significant if they appear between nonzero digits or if they appear after a nonzero digit in a number that contains a decimal point. Zeros may or may not be significant if they appear to the right of the last nonzero digit in a number that does not contain a decimal point.

**Solution** (a) 3; (b) 4; (c) 3; (d) 4; (e) 1 or 2, an ambiguous case; (f) 4

**Practice Problem A** Determine the number of significant figures in the following measurements: (a) 1129 m, (b) 0.0003 kg, (c) 1.094 cm, (d) $3.5 \times 10^{12}$ atoms, (e) 150 mL, (f) 9.550 km.

**Practice Problem B** Determine the number of significant figures in each of the following numbers: (a) $3.050 \times 10^{-4}$, (b) 432.00, (c) 8.001, (d) 0.000310, (e) 150, (f) 5.0071.

## Calculations with Measured Numbers

Because we often use one or more measured numbers to calculate a desired result, a second set of guidelines specifies how to handle significant figures in calculations.

1. In addition and subtraction, the answer cannot have more digits to the right of the decimal point than the original number with the smallest number of digits to the right of the decimal point. For example:

$$
\begin{array}{ll}
102.50 & \leftarrow \text{two digits after the decimal point} \\
+\ 0.231 & \leftarrow \text{three digits after the decimal point} \\
\hline
102.731 & \leftarrow \text{round to } 102.73
\end{array}
$$

$$
\begin{array}{ll}
143.29 & \leftarrow \text{two digits after the decimal point} \\
-20.1 & \leftarrow \text{one digit after the decimal point} \\
\hline
123.19 & \leftarrow \text{round to } 123.2
\end{array}
$$

The rounding procedure works as follows. Suppose we want to round 102.13 and 54.86 each to one digit to the right of the decimal point. To begin, we look at the digit(s) that will be dropped. If the leftmost digit to be dropped is less than 5, as in 102.13, we *round down* (to 102.1), meaning that we simply drop the digit(s). If the leftmost digit to be dropped is equal to or greater than 5, as in 54.86, we *round up* (to 54.9), meaning that we add 1 to the preceding digit.

2. In multiplication and division, the number of significant figures in the final product or quotient is determined by the original number that has the smallest number of significant figures. The following examples illustrate this rule:

$$1.4 \times 8.011 = 11.2154 \qquad \leftarrow \text{round to } 11 \text{ (limited by } 1.4 \text{ to } two \text{ significant figures)}$$

$$\frac{11.57}{305.88} = 0.037825290964 \qquad \leftarrow \text{round to } 0.03783 \text{ (limited by } 11.57 \text{ to } four \text{ significant figures)}$$

3. *Exact numbers* can be considered to have an infinite number of significant figures and do not limit the number of significant figures in a calculated result. For example, a penny minted after 1982 has a mass of 2.5 g. If we have three such pennies, the total mass is

$$3 \times 2.5 \text{ g} = 7.5 \text{ g}$$

The answer should *not* be rounded to one significant figure because 3 is an exact number.

4. In calculations with multiple steps, rounding the result of each step can result in "rounding error." Consider the following two-step calculation:

*First step:* $A \times B = C$

*Second step:* $C \times D = E$

Suppose that $A = 3.66$, $B = 8.45$, and $D = 2.11$. The value of $E$ depends on whether we round off $C$ prior to using it in the second step of the calculation.

| Method 1 | Method 2 |
|---|---|
| $C = 3.66 \times 8.45 = 30.9$ | $C = 3.66 \times 8.45 = 30.93$ |
| $E = 30.9 \times 2.11 = 65.2$ | $E = 30.93 \times 2.11 = 65.3$ |

In general, it is best to retain at least one extra digit until the end of a multistep calculation, as shown by method 2, to minimize rounding error.

Sample Problems 1.6 and 1.7 show how significant figures are handled in arithmetic operations.

## Sample Problem 1.6

Perform the following arithmetic operations and report the result to the proper number of significant figures: (a) 317.5 mL + 0.675 mL, (b) 47.80 L − 2.075 L, (c) 13.5 g ÷ 45.18 L, (d) 6.25 cm × 1.175 cm, (e) $5.46 \times 10^2$ g + $4.991 \times 10^3$ g.

**Strategy** Apply the rules for significant figures in calculations, and round each answer to the appropriate number of digits.

**Setup** (a) The answer will contain one digit to the right of the decimal point to match 317.5, which has the fewest digits to the right of the decimal point. (b) The answer will contain two digits to the right of the decimal point to match 47.80. (c) The answer will contain three significant figures to match 13.5, which has the fewest number of significant figures in the calculation. (d) The answer will contain three significant figures to match 6.25. (e) To add numbers expressed in scientific notation, first write both numbers to the same power of 10. That is, $4.991 \times 10^3 = 49.91 \times 10^2$, so the answer will contain two digits to the right of the decimal point (when multiplied by $10^2$) to match both 5.46 and 49.91.

**Solution**

(a)  317.5 mL
$\underline{+\ 0.675 \text{ mL}}$
318.175 mL    ← round to 318.2 mL

(b)  47.80 L
$\underline{-2.075 \text{ L}}$
45.725 L    ← round to 45.73 L

(c) $\dfrac{13.5 \text{ g}}{45.18 \text{ L}} = 0.298804781$ g/L    ← round to 0.299 g/L

(d) 6.25 cm × 1.175 cm = 7.34375 cm$^2$    ← round to 7.34 cm$^2$

(e)    $5.46 \times 10^2$ g
$\underline{+\ 49.91 \times 10^2 \text{ g}}$
$55.37 \times 10^2$ g = $5.537 \times 10^3$ g

**Think About It** It may look as though the rule of addition has been violated in part (e) because the final answer ($5.537 \times 10^3$ g) has three places past the decimal point, not two. However, the rule was applied to get the answer $55.37 \times 10^2$ g, which has *four* significant figures. Changing the answer to correct scientific notation doesn't change the number of significant figures, but in this case it changes the number of places past the decimal point.

**Practice Problem A** Perform the following arithmetic operations, and report the result to the proper number of significant figures: (a) 105.5 L + 10.65 L, (b) 81.058 m − 0.35 m, (c) $3.801 \times 10^{21}$ atoms + $1.228 \times 10^{19}$ atoms, (d) 1.255 dm × 25 dm, (e) 139 g ÷ 275.55 mL.

**Practice Problem B** Perform the following arithmetic operations, and report the result to the proper number of significant figures: (a) 1.0267 cm × 2.508 cm × 12.599 cm, (b) 15.0 kg ÷ 0.036 m$^3$, (c) $1.113 \times 10^{10}$ kg − $1.050 \times 10^9$ kg, (d) 25.75 mL + 15.00 mL, (e) 46 cm$^3$ + 180.5 cm$^3$.

## Sample Problem 1.7

An empty container with a volume of $9.850 \times 10^2$ cm$^3$ is weighed and found to have a mass of 124.6 g. The container is filled with a gas and reweighed. The mass of the container and the gas is 126.5 g. Determine the density of the gas to the appropriate number of significant figures.

**Strategy**  This problem requires two steps: subtraction to determine the mass of the gas, and division to determine its density. Apply the corresponding rule regarding significant figures to each step.

**Setup**  In the subtraction of the container mass from the combined mass of the container and the gas, the result can have only one place past the decimal point: 126.5 g $-$ 124.6 g = 1.9 g. Thus, in the division of the mass of the gas by the volume of the container, the result can have only two significant figures.

**Solution**

$$\begin{array}{r} 126.5 \text{ g} \\ -124.6 \text{ g} \\ \hline \end{array}$$

mass of gas = 1.9 g    ← one place past the decimal point (two significant figures)

$$\text{density} = \frac{1.9 \text{ g}}{9.850 \times 10^2 \text{ cm}^3} = 0.00193 \text{ g/cm}^3 \leftarrow \text{round to } 0.0019 \text{ g/cm}^3$$

The density of the gas is $1.9 \times 10^{-3}$ g/cm$^3$.

**Think About It**  In this case, although each of the three numbers we started with has *four* significant figures, the solution has only *two* significant figures.

**Practice Problem A**  An empty container with a volume of 150.0 cm$^3$ is weighed and found to have a mass of 72.5 g. The container is filled with a liquid and reweighed. The mass of the container and the liquid is 194.3 g. Determine the density of the liquid to the appropriate number of significant figures.

**Practice Problem B**  Another empty container with an unknown volume is weighed and found to have a mass of 81.2 g. The container is then filled with a liquid with a density of 1.015 g/cm$^3$ and reweighed. The mass of the container and the liquid is 177.9 g. Determine the volume of the container to the appropriate number of significant figures.

(a)

(b)

(c)

**Figure 1.12**  The distribution of papers shows the difference between accuracy and precision.
(a) Good accuracy and good precision.
(b) Poor accuracy but good precision.
(c) Poor accuracy and poor precision.

## Accuracy and Precision

Accuracy and precision are two ways to gauge the quality of a set of measured numbers. Although the difference between the two terms may be subtle, it is important. ***Accuracy*** tells us how close a measurement is to the *true* value. ***Precision*** tells us how closely multiple measurements of the same thing are to one another (Figure 1.12).

Suppose that three students are asked to determine the mass of an aspirin tablet. Each student weighs the aspirin tablet three times. The results (in grams) are

|               | **Student A** | **Student B** | **Student C** |
|---------------|---------------|---------------|---------------|
|               | 0.335         | 0.357         | 0.369         |
|               | 0.331         | 0.375         | 0.373         |
|               | 0.333         | 0.338         | 0.371         |
| Average value | 0.333         | 0.357         | 0.371         |

The true mass of the tablet is 0.370 g. Student A's results are more precise than those of student B, but neither set of results is very accurate. Student C's results are both precise (very small deviation of individual masses from the average mass) and accurate (average value very close to the true value). Figure 1.13 shows all three students' results in relation to the true mass of the tablet. Highly accurate measurements are usually precise, as well, although highly precise measurements do not necessarily guarantee accurate results. For example, an improperly calibrated meterstick or a faulty balance may give precise readings that are significantly different from the correct value.

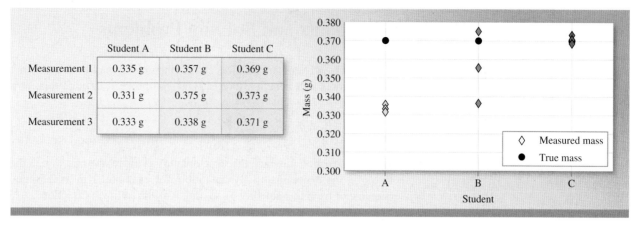

|  | Student A | Student B | Student C |
|---|---|---|---|
| Measurement 1 | 0.335 g | 0.357 g | 0.369 g |
| Measurement 2 | 0.331 g | 0.375 g | 0.373 g |
| Measurement 3 | 0.333 g | 0.338 g | 0.371 g |

**Figure 1.13**   Graphing the students' data illustrates the difference between precision and accuracy. Student A's results are precise (values are close to one another) but not accurate because the average value is far from the true value. Student B's results are neither precise nor accurate. Student C's results are both precise and accurate.

## Checkpoint 1.5    Uncertainty in Measurement

**1.5.1**   What volume of water does the graduated cylinder contain (to the proper number of significant figures)?

a) 32.2 mL

b) 30.25 mL

c) 32.5 mL

d) 32.50 mL

e) 32.500 mL

**1.5.2**   Which of the following is the sum of the following numbers to the correct number of significant figures?

$$3.115 + 0.2281 + 712.5 + 45 =$$

a) 760.8431

b) 760.843

c) 760.84

d) 760.8

e) 761

**1.5.3**   The true dependence of $y$ on $x$ is represented by the line. Three students measured $y$ as a function of $x$ and plotted their data on the graph. Which set of data has the best accuracy and which has the best precision, respectively?

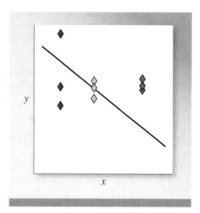

a) red, green

b) green, green

c) green, purple

d) purple, purple

e) purple, green

**1.5.4**   What is the result of the following calculation to the correct number of significant figures?

$$(6.266 - 6.261) \div 522.0 =$$

a) $9.5785 \times 10^{-6}$

b) $9.579 \times 10^{-6}$

c) $9.58 \times 10^{-6}$

d) $9.6 \times 10^{-6}$

e) $1 \times 10^{-5}$

## 1.6   Using Units and Solving Problems

Solving problems correctly in chemistry requires careful manipulation of both numbers and units. Paying attention to the units will benefit you greatly as you proceed through this, or any other, science course.

### Conversion Factors

A ***conversion factor*** is a fraction in which the same quantity is expressed one way in the numerator and another way in the denominator. By definition, for example, 1 in = 2.54 cm. We can derive a conversion factor from this equality by writing it as the following fraction:

$$\frac{1 \text{ in}}{2.54 \text{ cm}}$$

Because the numerator and denominator express the same length, this fraction is equal to 1; as a result, we can equally well write the conversion factor as

$$\frac{2.54 \text{ cm}}{1 \text{ in}}$$

Because both forms of this conversion factor are equal to 1, we can multiply a quantity by either form without changing the value of that quantity. This is useful for changing the units in which a given quantity is expressed—something you will do often throughout this text. For instance, if we need to convert a length from inches to centimeters, we multiply the length in inches by the appropriate conversion factor.

$$12.00 \text{ in} \times \frac{2.54 \text{ cm}}{1 \text{ in}} = 30.48 \text{ cm}$$

We chose the form of the conversion factor that cancels the unit inches and produces the desired unit, centimeters. The result contains four significant figures because exact numbers, such as those obtained from definitions, do not limit the number of significant figures in the result of a calculation. Thus, the number of significant figures in the answer to this calculation is based on the number 12.00, not the number 2.54.

**Student Annotation:** Because 1 in is defined as exactly 2.54 cm, neither the number 1 nor the number 2.54 limits the number of significant figures in a calculation where this conversion factor is used.

### Dimensional Analysis—Tracking Units

The use of conversion factors in problem solving is called ***dimensional analysis*** or the *factor-label method*. Many problems require the use of more than one conversion factor. The conversion of 12.00 into meters, for example, takes two steps: one to convert inches to centimeters, which we have already demonstrated, and one to convert centimeters to meters. The additional conversion factor required is derived from the equality

$$1 \text{ m} = 100 \text{ cm}$$

and is expressed as either

$$\frac{100 \text{ cm}}{1 \text{ m}} \quad \text{or} \quad \frac{1 \text{ m}}{100 \text{ cm}}$$

We must choose the conversion factor that will introduce the unit meter and cancel the unit centimeter (i.e., the one on the right). We can set up a problem of this type as the following series of unit conversions so that it is unnecessary to calculate an intermediate answer at each step:

**Student Annotation:** If we had accidentally used the reciprocal of the conversion from centimeters to meters, the result would have been 3048 cm²/m, which would make no sense—both because the units are nonsensical and because the numerical result is not reasonable. You know that 12 inches is a foot and that a foot is not equal to *thousands* of meters!

$$12.00 \text{ in} \times \frac{2.54 \text{ cm}}{1 \text{ in}} \times \frac{1 \text{ m}}{100 \text{ cm}} = 0.3048 \text{ m}$$

Careful tracking of units and their cancellation can be a valuable tool in checking your work. If we had accidentally used the *reciprocal* of one of the conversion factors, the resulting units would have been something other than meters. Unexpected or nonsensical units can reveal an error in your problem-solving strategy.

Sample Problem 1.8 shows how to derive conversion factors and use them to do unit conversions.

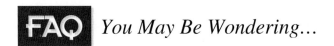 *You May Be Wondering...*

# How Can I Enhance My Chances of Success in Chemistry Class?

Success in a chemistry class depends largely on problem-solving ability. The sample problems throughout the text are designed to help you develop problem-solving skills. Each is divided into four steps: Strategy, Setup, Solution, and Think About It.

**Strategy:** Read the problem carefully and determine what is being asked and what information is provided. The Strategy step is where you should think about what skills are required and lay out a plan for solving the problem. Give some thought to what you expect the result to be. If you are asked to determine the number of atoms in a sample of matter, for example, you should expect the answer to be a whole number. Determine what, if any, units should be associated with the result. When possible, make a ballpark estimate of the magnitude of the correct result, and make a note of your estimate.

**Setup:** Next, gather the information necessary to solve the problem. Some of the information will have been given in the problem itself. Other information, such as equations, constants, and tabulated data (including atomic masses) should also be brought together in this step. Write down and label clearly all of the information you will use to solve the problem. Be sure to write appropriate units with each piece of information.

**Solution:** Using the necessary equations, constants, and other information, calculate the answer to the problem. Pay particular attention to the units associated with each number, tracking and canceling units carefully throughout the calculation. In the event that multiple calculations are required, label any intermediate results, but don't round to

the necessary number of significant figures until the final calculation. Always carry at least one extra significant figure in intermediate calculations. Make sure that the final answer has the correct number of significant figures.

**Think About It:** Consider your calculated result and ask yourself whether or not it makes sense. Compare the units and the magnitude of your result with your ballpark estimate from the Strategy step. If your result does not have the appropriate units, or if its magnitude or sign is not reasonable, check your solution for possible errors. A very important part of problem solving is being able to judge whether the answer is reasonable. It is relatively easy to spot a wrong sign or incorrect units, but you should also develop a sense of magnitude and be able to tell when an answer is either way too big or way too small. For example, if a problem asks how many molecules are in a sample and you calculate a number that is less than 1, you should know that it cannot be correct.

Finally, each sample problem is followed by at least one practice problem. This typically is a very similar problem that can be solved using the same strategy. Most sample problems also have a *second* practice problem that tests the same skills, but requires an approach slightly different from the one used to solve the preceding sample and practice problems.

Regular use of the sample problems and practice problems in this text can help you develop an effective set of problem-solving skills. They can also help you assess whether you are ready to move on to the next new concepts. If you struggle with the practice problems, then you probably need to review the corresponding sample problem and the concepts that led up to it.

---

## Sample Problem 1.8

The Food and Drug Administration (FDA) recommends that dietary sodium intake be no more than 2400 mg per day. What is this mass in pounds (lb), if 1 lb = 453.6 g?

**Strategy** This problem requires a two-step dimensional analysis, because we must convert milligrams to grams and then grams to pounds.

Assume the number 2400 has four significant figures.

**Setup** The necessary conversion factors are derived from the equalities 1 g = 1000 mg and 1 lb = 453.6 g.

$$\frac{1\ g}{1000\ mg} \quad \text{or} \quad \frac{1000\ mg}{1\ g} \quad \text{and} \quad \frac{1\ lb}{453.6\ g} \quad \text{or} \quad \frac{453.6\ g}{1\ lb}$$

From each pair of conversion factors, we select the one that will result in the proper unit cancellation.

**Solution**

$$2400\ \text{mg} \times \frac{1\ g}{1000\ \text{mg}} \times \frac{1\ lb}{453.6\ g} = 0.005291\ lb$$

**Student Annotation:** Because pounds are much larger than milligrams, a given mass will be a much smaller number of pounds than of milligrams.

**Think About It** Make sure that the magnitude of the result is reasonable and that the units have canceled properly. If we had mistakenly multiplied by 1000 and 453.6 instead of dividing by them, the result (2400 mg × 1000 mg/g × 453.6 g/lb = 1.089 × 10⁹ mg²/lb) would be unreasonably large—and the units would not have canceled properly.

*(Continued)*

**Practice Problem A** The American Heart Association recommends that healthy adults limit dietary cholesterol to no more than 300 mg per day. Convert this mass of cholesterol to ounces (1 oz = 28.3459 g). Assume 300 mg has just one significant figure.

**✋Practice Problem B** An object has a mass of 24.98 g. What is its mass in ounces?

---

Sample Problem 1.9 shows how to handle problems in which conversion factors are squared or cubed in dimensional analysis.

### Sample Problem 1.9

An average adult has 5.2 L of blood. What is the volume of blood in cubic meters?

**Strategy** There are several ways to solve a problem such as this. One way is to convert liters to cubic centimeters and then cubic centimeters to cubic meters.

**Setup** $1\ L = 1000\ cm^3$ and $1\ cm = 1 \times 10^{-2}\ m$. When a unit is raised to a power, the corresponding conversion factor must also be raised to that power in order for the units to cancel appropriately.

**Solution**

$$5.2\ L \times \frac{1000\ cm^3}{1\ L} \times \left(\frac{1 \times 10^{-2}\ m}{1\ cm}\right)^3 = 5.2 \times 10^{-3}\ m^3$$

**Think About It** Based on the preceding conversion factors, $1\ L = 1 \times 10^{-3}\ m^3$. Therefore, 5 L of blood would be equal to $5 \times 10^{-3}\ m^3$, which is close to the calculated answer.

**Practice Problem A** The density of silver is 10.5 g/cm³. What is its density in kg/m³?

**✋Practice Problem B** The density of mercury is 13.6 g/cm³. What is its density in mg/mm³?

---

### Checkpoint 1.6    Using Units and Solving Problems

**1.6.1** The density of lithium metal is 535 kg/m³. What is this density in g/cm³?

  a) 0.000535 g/cm³

  b) 0.535 g/cm³

  c) 0.0535 g/cm³

  d) 0.54 g/cm³

  e) 53.5 g/cm³

**1.6.2** Convert 43.1 cm³ to liters.

  a) 43.1 L

  b) 43,100 L

  c) 0.0431 L

  d) 4310 L

  e) 0.043 L

**1.6.3** What is the volume of a 5.75-g object that has a density of 3.97 g/cm³?

  a) 1.45 cm³

  b) 0.690 cm³

  c) 22.8 cm³

  d) 0.0438 cm³

  e) 5.75 cm³

**1.6.4** How many cubic centimeters are there in a cubic meter?

  a) 10

  b) 100

  c) 1000

  d) $1 \times 10^4$

  e) $1 \times 10^6$

# Applying What You've Learned

Although naturally occurring smallpox was eradicated by a superbly coordinated effort including the World Health Organization and health-care providers worldwide, the classification of smallpox as a Category A bioterrorism agent has renewed interest in its treatment and prevention. Moreover, although vaccination against the disease is considered relatively safe for most individuals, it is not entirely without risk.

The CDC estimates that 14 to 52 out of every million people who are vaccinated for smallpox will suffer serious, potentially life-threatening reactions to the vaccine. In these cases, immediate medical attention is required. The first course of treatment is with *vaccinia immune globulin* (VIG). If a patient does not respond to treatment with VIG, a second option is cidofovir, a drug that currently is approved by the Food and Drug Administration (FDA) to treat specific viral infections of the eye in individuals with compromised immune systems.

Cidofovir, marketed under the name Vistide, is distributed in vials containing 375 mg of the drug dissolved in 5 mL of water. The manufacturer specifies that the drug should be kept at room temperature (68°F–77°F). The vial contents are first diluted with saline and then administered intravenously with a recommended dosage of 5 mg cidofovir per kilogram of body weight.

**Student Annotation:** Smallpox vaccine is made from the *vaccinia* virus.

**Student Annotation:** Both drugs are available for use in the treatment of a serious reaction to smallpox vaccine only through the FDA's Investigational New Drug (IND) protocol.

## Problems:

a)  Convert cidofovir's recommended storage-temperature range to the Celsius scale. [◄◄ Sample Problem 1.2]

b)  If the fluid in a single vial of cidofovir has a volume of 5.00 mL and a mass of 5.89 g, what is the density of the fluid? [◄◄ Sample Problem 1.3] (Report the density to the appropriate number of significant figures. [◄◄ Sample Problem 1.6])

c)  What mass of cidofovir should be administered to a 177-lb man (1 lb = 0.4536 kg)? [◄◄ Sample Problem 1.8]

d)  Convert the density in part (b) to g/L and to $kg/m^3$. [◄◄ Sample Problem 1.9]

Cidofovir

# CHAPTER SUMMARY

## Section 1.1

- *Chemistry* is the study of *matter* and the changes matter undergoes.

- Chemists go about research using a set of guidelines and practices known as the *scientific method*, in which observations give rise to *laws*, data give rise to *hypotheses*, hypotheses are tested with experiments, and successful hypotheses give rise to *theories*, which are further tested by experiment.

## Section 1.2

- All matter exists either as a *substance* or as a mixture of substances. Substances may be *elements* (containing only one kind of atom) or *compounds* (containing two or more kinds of atoms). A *mixture* may be *homogeneous* (a solution) or *heterogeneous*. Mixtures may be separated using physical processes. Compounds can be separated into their constituent elements using chemical processes. Elements cannot be separated into simpler substances.

## Section 1.3

- Scientists use a system of units referred to as the *International System of Units* or *SI units*.

- There are seven *base* SI units including the kilogram (for *mass*) and the *kelvin* (for temperature). SI units for such quantities as volume and *density* are derived from the base units.

## Section 1.4

- Substances are identified by their *quantitative* (involving numbers) and *qualitative* (not involving numbers) properties.

- *Physical properties* are those that can be determined without the matter in question undergoing a chemical change. A *physical change* is one in which the identity of the matter involved does not change.

- *Chemical properties* are determined only as the result of a *chemical change*, in which the original substance is converted to a different substance. Physical and chemical properties may be *extensive* (dependent on the amount of matter) or *intensive* (independent of the amount of matter).

## Section 1.5

- Measured numbers are *inexact*. Numbers obtained by counting or that are part of a definition are *exact* numbers.

- *Significant figures* are used to specify the uncertainty in a measured number or in a number calculated using measured numbers. Significant figures must be carried through calculations such that the implied uncertainty in the final answer is reasonable.

- *Accuracy* refers to how close measured numbers are to a *true* value. *Precision* refers to how close measured numbers are to *one another*.

## Section 1.6

- A *conversion factor* is a fraction in which the numerator and denominator are the same quantity expressed in different units. Multiplying by a conversion factor is *unit conversion*.

- *Dimensional analysis* is a series of unit conversions used in the solution of a multistep problem.

# KEY WORDS

| | | | |
|---|---|---|---|
| Accuracy, 20 | Element, 6 | Law, 5 | Quantitative property, 14 |
| Chemical change, 15 | Extensive property, 15 | Mass, 10 | Scientific method, 5 |
| Chemical property, 15 | Heterogeneous mixture, 8 | Matter, 4 | Significant figures, 17 |
| Chemistry, 4 | Homogeneous mixture, 7 | Mixture, 7 | SI unit, 9 |
| Compound, 7 | Hypothesis, 5 | Physical change, 15 | Substance, 6 |
| Conversion factor, 22 | Intensive property, 15 | Physical property, 14 | Theory, 6 |
| Density, 12 | International System of Units, 9 | Precision, 20 | |
| Dimensional analysis, 22 | Kelvin, 10 | Qualitative property, 14 | |

# KEY EQUATIONS

1.1 $\quad K = °C + 273.15$

1.2 $\quad \text{temperature in degrees Celsius} = (\text{temperature in degrees Fahrenheit} - 32°F) \times \dfrac{5°C}{9°F}$

1.3 $\quad \text{temperature in degrees Fahrenheit} = \dfrac{9°F}{5°C} \times (\text{temperature in degrees Celsius}) + 32°F$

1.4 $\quad d = \dfrac{m}{V}$

# QUESTIONS AND PROBLEMS

## Section 1.1: The Study of Chemistry

### Review Questions

1.1 Define the terms *chemistry* and *matter*.

1.2 Explain what is meant by the scientific method.

1.3 What is the difference between a hypothesis and a theory?

### Problems

1.4 Classify each of the following statements as a hypothesis, law, or theory. (a) Beethoven's contribution to music would have been much greater if he had married. (b) An autumn leaf gravitates toward the ground because there is an attractive force between the leaf and Earth. (c) All matter is composed of very small particles called atoms.

**1.5** Classify each of the following statements as a hypothesis, law, or theory. (a) The force acting on an object is equal to its mass times its acceleration. (b) The universe as we know it started with a big bang. (c) There are many civilizations more advanced than ours on other planets.

1.6 Identify the elements present in the following molecules (see Table 1.1).

(a)          (b)          (c)          (d)

**1.7** Identify the elements present in the following molecules (see Table 1.1).

(a)          (b)          (c)          (d)

## Section 1.2: Classification of Matter

### Review Questions

1.8 Give an example for each of the following terms: (a) matter, (b) substance, (c) mixture.

1.9 Give an example of a homogeneous mixture and an example of a heterogeneous mixture.

1.10 Give an example of an element and a compound. How do elements and compounds differ?

1.11 What is the number of known elements?

### Problems

1.12 Give the names of the elements represented by the chemical symbols Li, F, P, Cu, As, Zn, Cl, Pt, Mg, U, Al, Si, Ne (see the table inside the front cover).

**1.13** Give the chemical symbols for the following elements: (a) potassium, (b) tin, (c) chromium, (d) boron, (e) barium, (f) plutonium, (g) sulfur, (h) argon, (i) mercury (see the table inside the front cover).

1.14 Classify each of the following substances as an element or a compound: (a) hydrogen, (b) water, (c) gold, (d) sugar.

1.15 Classify each of the following as an element, a compound, a homogeneous mixture, or a heterogeneous mixture: (a) seawater, (b) helium gas, (c) sodium chloride (salt), (d) a bottle of soft drink, (e) a milkshake, (f) air in a bottle, (g) concrete.

1.16 Identify each of the diagrams shown here as a solid, liquid, gas, or mixture of two substances.

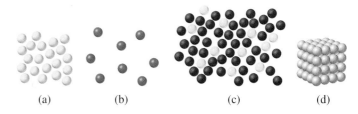

(a)          (b)          (c)          (d)

**1.17** Identify each of the diagrams shown here as an element or a compound.

(a)          (b)          (c)          (d)

## Section 1.3: Scientific Measurement

### Review Questions

1.18 Name the SI base units that are important in chemistry, and give the SI units for expressing the following: (a) length, (b) volume, (c) mass, (d) time, (e) temperature.

1.19 Write the numbers represented by the following prefixes: (a) mega-, (b) kilo-, (c) deci-, (d) centi-, (e) milli-, (f) micro-, (g) nano-, (h) pico-.

1.20 What units do chemists normally use for the density of liquids and solids? For the density of gas? Explain the differences.

1.21 What is the difference between mass and weight? If a person weighs 168 lb on Earth, about how much would the person weigh on the moon?

1.22 Describe the three temperature scales used in the laboratory and in everyday life: the Fahrenheit, Celsius, and Kelvin scales.

### Problems

**1.23** Bromine is a reddish-brown liquid. Calculate its density (in g/mL) if 586 g of the substance occupies 188 mL.

1.24 The density of ethanol, a colorless liquid that is commonly known as grain alcohol, is 0.798 g/mL. Calculate the mass of 17.4 mL of the liquid.

1.25 Convert the following temperatures to degrees Celsius or Fahrenheit: (a) 95°F, the temperature on a hot summer day; (b) 12°F, the temperature on a cold winter day; (c) a 102°F fever; (d) a furnace operating at 1852°F; (e) −273.15°C (theoretically the lowest attainable temperature).

1.26    (a) Normally the human body can endure a temperature of 105°F for only short periods of time without permanent damage to the brain and other vital organs. What is this temperature in degrees Celsius? (b) Ethylene glycol is a liquid organic compound that is used as an antifreeze in car radiators. It freezes at −11.5°C. Calculate its freezing temperature in degrees Fahrenheit. (c) The temperature on the surface of the sun is about 6300°C. What is this temperature in degrees Fahrenheit?

1.27    The density of water at 40°C is 0.992 g/mL. What is the volume of 2.50 g of water at this temperature?

1.28    The density of platinum (Pt) is 21.5 $g/cm^3$ at 25°C. What is the volume of 87.6 g of Pt at this temperature?

1.29    Convert the following temperatures to kelvin: (a) 115.21°C, the melting point of sulfur; (b) 37°C, the normal body temperature; (c) 357°C, the boiling point of mercury.

1.30    Convert the following temperatures to degrees Celsius: (a) 77 K, the boiling point of liquid nitrogen, (b) 4.22 K, the boiling point of liquid helium, (c) 600.61 K, the melting point of lead.

## Section 1.4: The Properties of Matter

### Review Questions

1.31    What is the difference between qualitative data and quantitative data?

1.32    Using examples, explain the difference between a physical property and a chemical property.

1.33    How does an intensive property differ from an extensive property?

1.34    Determine which of the following properties are intensive and which are extensive: (a) length, (b) volume, (c) temperature, (d) mass.

### Problems

1.35    Classify the following as qualitative or quantitative statements, giving your reasons. (a) The sun is approximately 93 million mi from Earth. (b) Leonardo da Vinci was a better painter than Michelangelo. (c) Ice is less dense than water. (d) Butter tastes better than margarine. (e) A stitch in time saves nine.

1.36    Determine whether the following statements describe chemical or physical properties: (a) Oxygen gas supports combustion. (b) Fertilizers help to increase agricultural production. (c) Water boils below 100°C on top of a mountain. (d) Lead is denser than aluminum. (e) Uranium is a radioactive element.

1.37    Determine whether each of the following describes a physical change or a chemical change: (a) The helium gas inside a balloon tends to leak out after a few hours. (b) A flashlight beam slowly gets dimmer and finally goes out. (c) Frozen orange juice is reconstituted by adding water to it. (d) The growth of plants depends on the sun's energy in a process called photosynthesis. (e) A spoonful of salt dissolves in a bowl of soup.

1.38    A student pours 44.3 g of water at 10°C into a beaker containing 115.2 g of water at 10°C. What are the final mass, temperature, and density of the combined water? The density of water at 10°C is 1.00 g/mL.

1.39    A 37.2-g sample of lead (Pb) pellets at 20°C is mixed with a 62.7-g sample of lead pellets at the same temperature. What are the final mass, temperature, and density of the combined sample? The density of Pb at 20°C is 11.35 $g/cm^3$.

## Section 1.5: Uncertainty in Measurement

### Review Questions

1.40    Comment on whether each of the following statements represents an exact number: (a) 50,247 tickets were sold at a sporting event, (b) 509.2 mL of water was used to make a birthday cake, (c) 3 dozen eggs were used to make a breakfast, (d) 0.41 g of oxygen was inhaled in each breath, (e) Earth orbits the sun every 365.2564 days.

1.41    What is the advantage of using scientific notation over decimal notation?

1.42    Define *significant figure*. Discuss the importance of using the proper number of significant figures in measurements and calculations.

1.43    Distinguish between the terms *accuracy* and *precision*. In general, explain why a precise measurement does not always guarantee an accurate result.

### Problems

1.44    Express the following numbers in scientific notation: (a) 0.000000027, (b) 356, (c) 47,764, (d) 0.096.

1.45    Express the following numbers as decimals: (a) $1.52 \times 10^{-2}$, (b) $7.78 \times 10^{-8}$, (c) $1 \times 10^{-6}$, (d) $1.6001 \times 10^3$.

1.46    Express the answers to the following calculations in scientific notation:

(a) $145.75 + (2.3 \times 10^{-1})$
(b) $79,500 \div (2.5 \times 10^2)$
(c) $(7.0 \times 10^{-3}) - (8.0 \times 10^{-4})$
(d) $(1.0 \times 10^4) \times (9.9 \times 10^6)$

1.47    Express the answers to the following calculations in scientific notation:

(a) $0.0095 + (8.5 \times 10^{-3})$
(b) $653 \div (5.75 \times 10^{-8})$
(c) $850,000 - (9.0 \times 10^5)$
(d) $(3.6 \times 10^{-4}) \times (3.6 \times 10^6)$

1.48    Determine the number of significant figures in each of the following measurements: (a) 4867 mi, (b) 56 mL, (c) 60,104 tons, (d) 2900 g, (e) 40.2 $g/cm^3$, (f) 0.0000003 cm, (g) 0.7 min, (h) $4.6 \times 10^{19}$ atoms.

1.49    Determine the number of significant figures in each of the following measurements: (a) 0.006 L, (b) 0.0605 dm, (c) 60.5 mg, (d) 605.5 $cm^2$, (e) $9.60 \times 10^3$ g, (f) 6 kg, (g) 60 m.

1.50    Carry out the following operations as if they were calculations of experimental results, and express each answer in the correct units with the correct number of significant figures:

(a) 5.6792 m + 0.6 m + 4.33 m
(b) 3.70 g − 2.9133 g
(c) 4.51 cm × 3.6666 cm

1.51    Carry out the following operations as if they were calculations of experimental results, and express each answer in the correct units with the correct number of significant figures:

(a) 7.310 km ÷ 5.70 km
(b) $(3.26 \times 10^{-3} \text{ mg}) - (7.88 \times 10^{-5} \text{ mg})$
(c) $(4.02 \times 10^6 \text{ dm}) + (7.74 \times 10^7 \text{ dm})$

1.52    Three students (A, B, and C) are asked to determine the volume of a sample of ethanol. Each student measures the volume three times with a graduated cylinder. The results in milliliters are: A (87.1, 88.2, 87.6); B (86.9, 87.1, 87.2); C (87.6, 87.8, 87.9). The true volume is 87.0 mL. Comment on the precision and the accuracy of each student's results.

1.53    Three apprentice tailors (X, Y, and Z) are assigned the task of measuring the seam of a pair of trousers. Each one makes three measurements. The results in inches are X (31.5, 31.6, 31.4); Y (32.8, 32.3, 32,7); Z (31.9, 32.2, 32.1). The true length is 32.0 in. Comment on the precision and the accuracy of each tailor's measurements.

## Section 1.6: Using Units and Solving Problems

### Problems

1.54    Carry out the following conversions: (a) 22.6 m to decimeters, (b) 25.4 mg to kilograms, (c) 556 mL to liters, (d) 10.6 kg/m$^3$ to g/cm$^3$.

1.55    Carry out the following conversions: (a) 242 lb to milligrams, (b) 68.3 cm$^3$ to cubic meters, (c) 7.2 m$^3$ to liters, (d) 28.3 µg to pounds.

1.56    The average speed of helium at 25°C is 1255 m/s. Convert this speed to miles per hour (mph).

1.57    How many seconds are there in a solar year (365.24 days)?

1.58    How many minutes does it take light from the sun to reach Earth? (The distance from the sun to Earth is 93 million mi; the speed of light is 3.00 × 10$^8$ m/s.)

1.59    A slow jogger runs a mile in 13 min. Calculate the speed in (a) in/s, (b) m/min, (c) km/h (1 mi = 1609 m; 1 in = 2.54 cm).

1.60    A 6.0-ft person weighs 168 lb. Express this person's height in meters and weight in kilograms (1 lb = 453.6 g; 1 m = 3.28 ft).

1.61    The current speed limit in some states in the United States is 55 mph. What is the speed limit in kilometers per hour (1 mi = 1609 m)?

1.62    For a fighter jet to take off from the deck of an aircraft carrier, it must reach a speed of 62 m/s. Calculate the speed in miles per hour.

1.63    The "normal" lead content in human blood is about 0.40 part per million (that is, 0.40 g of lead per million grams of blood). A value of 0.80 part per million (ppm) is considered to be dangerous. How many grams of lead are contained in 6.0 × 10$^3$ g of blood (the amount in an average adult) if the lead content is 0.62 ppm?

1.64    Carry out the following conversions: (a) 1.42 light-years to miles (a light-year is an astronomical measure of distance—the distance traveled by light in a year, or 365 days; the speed of light is 3.00 × 10$^8$ m/s), (b) 32.4 yd to centimeters, (c) 3.0 × 10$^{10}$ cm/s to ft/s.

1.65    Carry out the following conversions: (a) 185 nm to meters, (b) 4.5 billion years (roughly the age of Earth) to seconds (assume 365 days in a year), (c) 71.2 cm$^3$ to cubic meters, (d) 88.6 m$^3$ to liters.

1.66    Aluminum is a lightweight metal (density = 2.70 g/cm$^3$) used in aircraft construction, high-voltage transmission lines, beverage cans, and foils. What is its density in kg/m$^3$?

1.67    The density of ammonia gas under certain conditions is 0.625 g/L. Calculate its density in g/cm$^3$.

1.68    (a) Carbon monoxide (CO) is a poisonous gas because it binds very strongly to the oxygen carrier hemoglobin in blood. A concentration of 8.00 × 10$^2$ ppm by volume of carbon monoxide is considered lethal to humans. Calculate the volume in liters occupied by carbon monoxide in a room that measures 17.6 m long, 8.80 m wide, and 2.64 m high at this concentration. (b) Prolonged exposure to mercury (Hg) vapor can cause neurological disorder and respiratory problems. For safe air quality control, the concentration of mercury vapor must be under 0.050 mg/m$^3$. Convert this number to g/L. (c) The general test for type II diabetes is that the blood sugar (glucose) level should be below 120 mg per deciliter (mg/dL). Convert this number to micrograms per milliliter (µg/mL).

1.69    The average time it takes for a molecule to diffuse a distance of x cm is given by

$$t = \frac{x^2}{2D}$$

where t is the time in seconds and D is the diffusion coefficient. Given that the diffusion coefficient of glucose is 5.7 × 10$^{-7}$ cm$^2$/s, calculate the time it would take for a glucose molecule to diffuse 10 µm, which is roughly the size of a cell.

1.70    A human brain weighs about 1 kg and contains about 10$^{11}$ cells. Assuming that each cell is completely filled with water (density = 1 g/mL), calculate the length of one side of such a cell if it were a cube. If the cells are spread out into a thin layer that is a single cell thick, what is the surface area in square meters?

## Additional Problems

1.71    Using the appropriate number of significant figures, report the length of the blue rectangle (a) using the ruler shown above the rectangle and (b) using the ruler shown below the rectangle.

1.72    A piece of metal with a mass of 13.2 g was dropped into a graduated cylinder containing 17.00 mL of water. The graduated cylinder after the addition of the metal is shown. Determine the density of the metal to the appropriate number of significant figures.

1.73    Which of the following statements describe physical properties and which describe chemical properties? (a) Iron has a tendency to rust. (b) Rainwater in industrialized regions tends to be acidic. (c) Hemoglobin molecules have a red color. (d) When a glass of water is left out in the sun, the water gradually disappears. (e) Carbon dioxide in air is converted to more complex molecules by plants during photosynthesis.

**1.74** In determining the density of a rectangular metal bar, a student made the following measurements: length, 8.53 cm; width, 2.4 cm; height, 1.0 cm; mass, 52.7064 g. Calculate the density of the metal to the correct number of significant figures.

**1.75** Calculate the mass of each of the following: (a) a sphere of gold with a radius of 10.0 cm (volume of a sphere with a radius $r$ is $V = \frac{4}{3}\pi r^3$; density of gold = 19.3 g/cm$^3$), (b) a cube of platinum of edge length 0.040 mm (density = 21.4 g/cm$^3$), (c) 50.0 mL of ethanol (density = 0.798 g/mL).

**1.76** A cylindrical glass tube 12.7 cm in length is filled with mercury (density = 13.6 g/mL). The mass of mercury needed to fill the tube is 105.5 g. Calculate the inner diameter of the tube (volume of a cylinder of radius $r$ and length $h$ is $V = \pi r^2 h$).

**1.77** The following procedure was used to determine the volume of a flask. The flask was weighed dry and then filled with water. If the masses of the empty flask and filled flask were 56.12 g and 87.39 g, respectively, and the density of water is 0.9976 g/cm$^3$, calculate the volume of the flask in cubic centimeters.

**1.78** The speed of sound in air at room temperature is about 343 m/s. Calculate this speed in miles per hour (1 mi = 1609 m).

**1.79** A piece of silver (Ag) metal weighing 194.3 g is placed in a graduated cylinder containing 242.0 mL of water. The volume of water now reads 260.5 mL. From these data calculate the density of silver.

**1.80** The experiment described in Problem 1.79 is a crude but convenient way to determine the density of some solids. Describe a similar experiment that would enable you to measure the density of ice. Specifically, what would be the requirements for the liquid used in your experiment?

**1.81** A lead sphere has a mass of $1.20 \times 10^4$ g, and its volume is $1.05 \times 10^3$ cm$^3$. Calculate the density of lead.

**1.82** Lithium is the least dense metal known (density = 0.53 g/cm$^3$). What is the volume occupied by $1.20 \times 10^3$ g of lithium?

**1.83** The medicinal thermometer commonly used in homes can be read to $\pm0.1°$F, whereas those in the doctor's office may be accurate to $\pm0.1°$C. Percent error is often expressed as the absolute value of the difference between the true value and the experimental value, divided by the true value:

$$\text{percent error} = \frac{|\text{true value} - \text{experimental value}|}{\text{true value}} \times 100\%$$

The vertical lines indicate absolute value. In degrees Celsius, express the percent error expected from each of these thermometers in measuring a person's body temperature of 38.9°C.

**1.84** Vanillin (used to flavor vanilla ice cream and other foods) is the substance whose aroma the human nose detects in the smallest amount. The threshold limit is $2.0 \times 10^{-11}$ g per liter of air. If the current price of 50 g of vanillin is $112, determine the cost to supply enough vanillin so that the aroma could be detected in a large aircraft hangar with a volume of $5.0 \times 10^7$ ft$^3$.

**1.85** At what temperature does the numerical reading on a Celsius thermometer equal that on a Fahrenheit thermometer?

**1.86** Suppose that a new temperature scale has been devised on which the melting point of ethanol (−117.3°C) and the boiling point of ethanol (78.3°C) are taken as 0°S and 100°S, respectively, where

S is the symbol for the new temperature scale. Derive an equation relating a reading on this scale to a reading on the Celsius scale. What would this thermometer read at 25°C?

**1.87** A resting adult requires about 240 mL of pure oxygen per minute and breathes about 12 times every minute. If inhaled air contains 20 percent oxygen by volume and exhaled air 16 percent, what is the volume of air per breath? (Assume that the volume of inhaled air is equal to that of exhaled air.)

**1.88** (a) Referring to Problem 1.87, calculate the total volume (in liters) of air an adult breathes in a day. (b) In a city with heavy traffic, the air contains $2.1 \times 10^{-6}$ L of carbon monoxide (a poisonous gas) per liter. Calculate the average daily intake of carbon monoxide in liters by a person.

**1.89** The total volume of seawater is $1.5 \times 10^{21}$ L. Assume that seawater contains 3.1 percent sodium chloride by mass and that its density is 1.03 g/mL. Calculate the total mass of sodium chloride in kilograms and in tons (1 ton = 2000 lb; 1 lb = 453.6 g).

**1.90** Magnesium (Mg) is a valuable metal used in alloys, in batteries, and in the manufacture of chemicals. It is obtained mostly from seawater, which contains about 1.3 g of Mg for every kilogram of seawater. Referring to Problem 1.89, calculate the volume of seawater (in liters) needed to extract $8.0 \times 10^4$ tons of Mg, which is roughly the annual production in the United States.

**1.91** A student is given a crucible and asked to prove whether it is made of pure platinum. She first weighs the crucible in air and then weighs it suspended in water (density = 0.9986 g/mL). The readings are 860.2 g and 820.2 g, respectively. Based on these measurements and given that the density of platinum is 21.45 g/cm$^3$, what should her conclusion be? (*Hint:* An object suspended in a fluid is buoyed up by the mass of the fluid displaced by the object. Neglect the buoyancy of air.)

**1.92** The surface area and average depth of the Pacific Ocean are $1.8 \times 10^8$ km$^2$ and $3.9 \times 10^3$ m, respectively. Calculate the volume of water in the ocean in liters.

**1.93** The unit "troy ounce" is often used for precious metals such as gold (Au) and platinum (Pt) (1 troy ounce = 31.103 g). (a) A gold coin weighs 2.41 troy ounces. Calculate its mass in grams. (b) Is a troy ounce heavier or lighter than an ounce (1 lb = 16 oz; 1 lb = 453.6 g)?

**1.94** Osmium (Os) is the densest element known (density = 22.57 g/cm$^3$). Calculate the mass in pounds and in kilograms of an Os sphere 15 cm in diameter (about the size of a grapefruit) (volume of a sphere of radius $r$ is $\frac{4}{3}\pi r^3$).

**1.95** Calculate the percent error for the following measurements: (a) The density of alcohol (ethanol) is found to be 0.802 g/mL (true value = 0.798 g/mL). (b) The mass of gold in an earring is analyzed to be 0.837 g (true value = 0.864 g).

**1.96** The natural abundances of elements in the human body, expressed as percent by mass, are oxygen (O), 65 percent; carbon (C), 18 percent; hydrogen (H), 10 percent; nitrogen (N), 3 percent; calcium (Ca), 1.6 percent; phosphorus (P), 1.2 percent; all other elements, 1.2 percent. Calculate the mass in grams of each element in the body of a 62-kg person.

**1.97** The men's world record for running a mile outdoors (as of 1999) is 3 min 43.13 s. At this rate, how long would it take to run a 1500-m race (1 mi = 1609 m)?

**1.98** Venus, the second closest planet to the sun, has a surface temperature of $7.3 \times 10^2$ K. Convert this temperature to degrees Celsius and degrees Fahrenheit.

**1.99** Chalcopyrite, the principal ore of copper (Cu), contains 34.63 percent Cu by mass. How many grams of Cu can be obtained from $5.11 \times 10^3$ kg of the ore?

**1.100** It has been estimated that $8.0 \times 10^4$ tons of gold (Au) have been mined. Assume gold costs $900 per Troy ounce. What is the total worth of this quantity of gold? (1 Troy ounce = 31.103 g)

**1.101** A 1.0-mL volume of seawater contains about $4.0 \times 10^{-12}$ g of gold. The total volume of ocean water is $1.5 \times 10^{21}$ L. Calculate the total amount of gold (in grams) that is present in seawater and the worth of the gold in dollars (see Problem 1.100). With so much gold out there, why hasn't someone become rich by mining gold from the ocean?

**1.102** Measurements show that 1.0 g of iron (Fe) contains $1.1 \times 10^{22}$ Fe atoms. How many Fe atoms are in 4.9 g of Fe, which is the total amount of iron in the body of an average adult?

**1.103** The thin outer layer of Earth, called the crust, contains only 0.50 percent of Earth's total mass and yet is the source of almost all the elements (the atmosphere provides elements such as oxygen, nitrogen, and a few other gases). Silicon (Si) is the second most abundant element in Earth's crust (27.2 percent by mass). Calculate the mass of silicon in kilograms in Earth's crust (mass of Earth = $5.9 \times 10^{21}$ tons; 1 ton = 2000 lb; 1 lb = 453.6 g).

**1.104** The radius of a copper (Cu) atom is roughly $1.3 \times 10^{-10}$ m. How many times can you divide evenly a 10-cm-long piece of copper wire until it is reduced to two separate copper atoms? (Assume there are appropriate tools for this procedure and that copper atoms are lined up in a straight line, in contact with each other. Round off your answer to an integer.)

**1.105** One gallon of gasoline in an automobile's engine produces on the average 9.5 kg of carbon dioxide, which is a greenhouse gas; that is, it promotes the warming of Earth's atmosphere. Calculate the annual production of carbon dioxide in kilograms if there are 40 million cars in the United States and each car covers a distance of 5000 mi at a consumption rate of 20 miles per gallon.

**1.106** A sheet of aluminum (Al) foil has a total area of 1.000 ft$^2$ and a mass of 3.636 g. What is the thickness of the foil in millimeters (density of Al = 2.699 g/cm$^3$)?

**1.107** Comment on whether each of the following is a homogeneous mixture or a heterogeneous mixture: (a) air in a closed bottle, (b) air over New York City.

**1.108** Chlorine is used to disinfect swimming pools. The accepted concentration for this purpose is 1 ppm chlorine, or 1 g of chlorine per million grams of water. Calculate the volume of a chlorine solution (in milliliters) a homeowner should add to her swimming pool if the solution contains 6.0 percent chlorine by mass and there are $2.0 \times 10^4$ gallons (gal) of water in the pool (1 gal = 3.79 L; density of liquids = 1.0 g/mL).

**1.109** The world's total petroleum reserve is estimated at $2.0 \times 10^{22}$ joules [a joule (J) is the unit of energy where 1 J = 1 kg $\cdot$ m$^2$/s$^2$]. At the present rate of consumption, $1.8 \times 10^{20}$ joules per year (J/yr), how long would it take to exhaust the supply?

**1.110** In water conservation, chemists spread a thin film of a certain inert material over the surface of water to cut down on the rate of evaporation of water in reservoirs. This technique was pioneered by Benjamin Franklin three centuries ago. Franklin found that 0.10 mL of oil could spread over the surface of water about 40 m$^2$ in area. Assuming that the oil forms a *monolayer,* that is, a layer that is only one molecule thick, estimate the length of each oil molecule in nanometers (1 nm = $1 \times 10^{-9}$ m).

**1.111** Fluoridation is the process of adding fluorine compounds to drinking water to help fight tooth decay. A concentration of 1 ppm of fluorine is sufficient for the purpose (1 ppm means one part per million, or 1 g of fluorine per 1 million g of water). The compound normally chosen for fluoridation is sodium fluoride, which is also added to some toothpastes. Calculate the quantity of sodium fluoride in kilograms needed per year for a city of 50,000 people if the daily consumption of water per person is 150 gal. What percent of the sodium fluoride is "wasted" if each person uses only 6.0 L of water a day for drinking and cooking (sodium fluoride is 45.0 percent fluorine by mass; 1 gal = 3.79 L; 1 year = 365 days; 1 ton = 2000 lb; 1 lb = 453.6 g; density of water = 1.0 g/mL)?

**1.112** A gas company in Massachusetts charges $1.30 for 15.0 ft$^3$ of natural gas. (a) Convert this rate to dollars per liter of gas. (b) If it takes 0.304 ft$^3$ of gas to boil a liter of water, starting at room temperature (25°C), how much would it cost to boil a 2.1-L kettle of water?

**1.113** Pheromones are compounds secreted by females of many insect species to attract mates. Typically, $1.0 \times 10^{-8}$ g of a pheromone is sufficient to reach all targeted males within a radius of 0.50 mi. Calculate the density of the pheromone (in grams per liter) in a cylindrical air space having a radius of 0.50 mi and a height of 40 ft (volume of a cylinder of radius $r$ and height $h$ is $\pi r^2 h$).

**1.114** A bank teller is asked to assemble $1 sets of coins for his clients. Each set is made up of three quarters, one nickel, and two dimes. The masses of the coins are quarter, 5.645 g; nickel, 4.967 g; and dime, 2.316 g. What is the maximum number of sets that can be assembled from 33.871 kg of quarters, 10.432 kg of nickels, and 7.990 kg of dimes? What is the total mass (in grams) of the assembled sets of coins?

**1.115** A graduated cylinder is filled to the 40.00-mL mark with a mineral oil. The masses of the cylinder before and after the addition of the mineral oil are 124.966 g and 159.446 g, respectively. In a separate experiment, a metal ball bearing of mass 18.713 g is placed in the cylinder and the cylinder is again filled to the 40.00-mL mark with the mineral oil. The combined mass of the ball bearing and mineral oil is 50.952 g. Calculate the density and radius of the ball bearing (volume of a sphere of radius $r$ is $\frac{4}{3}\pi r^3$).

**1.116** Bronze is an alloy made of copper (Cu) and tin (Sn). Calculate the mass of a bronze cylinder of radius 6.44 cm and length 44.37 cm. The composition of the bronze is 79.42 percent Cu and 20.58 percent Sn and the densities of Cu and Sn are 8.94 g/cm$^3$ and 7.31 g/cm$^3$, respectively. What assumption should you make in this calculation?

**1.117** A chemist in the nineteenth century prepared an unknown substance. In general, do you think it would be more difficult to prove that it is an element or a compound? Explain.

**1.118** A chemist mixes two liquids A and B to form a homogeneous mixture. The densities of the liquids are 2.0514 g/mL for A and 2.6678 g/mL for B. When she drops a small object into the mixture, she finds that the object becomes suspended in the liquid; that is, it neither sinks nor floats. If the mixture is made of 41.37 percent A and 58.63 percent B by volume, what is the density of the object? Can this procedure be used in general to determine the densities of solids? What assumptions must be made in applying this method?

**1.119** You are given a liquid. Briefly describe the steps you would take to show whether it is a pure substance or a homogeneous mixture.

**1.120** TUMS is a popular remedy for acid indigestion. A typical TUMS tablet contains calcium carbonate plus some inert substances. When ingested, it reacts with the gastric juice (hydrochloric acid) in the stomach to give off carbon dioxide gas. When a 1.328-g tablet reacted with 40.00 mL of hydrochloric acid (density = 1.140 g/mL), carbon dioxide gas was given off and the resulting solution weighed 46.699 g. Calculate the number of liters of carbon dioxide gas released if its density is 1.81 g/L.

**1.121** A 250-mL glass bottle was filled with 242 mL of water at 20°C and tightly capped. It was then left outdoors overnight, where the average temperature was –5°C. Predict what would happen. The density of water at 20°C is 0.998 g/cm³ and that of ice at –5°C is 0.916 g/cm³.

# Pre-Professional Practice Exam Problems: Verbal Reasoning

English writer and essayist Lady Mary Wortley Montagu (1689–1762) traveled extensively and was fascinated by the customs in other countries. While in Turkey, she observed the practice of "engrafting" wherein people were inoculated against smallpox by intentional exposure to a mild form of the disease. She was so convinced of the efficacy and the safety of engrafting, that she had both of her children inoculated. She herself had survived smallpox as a child. Lady Montagu campaigned for the practice when she returned to England, and despite opposition from doctors and religious leaders, inoculation came into common use. It remained the primary defense against the scourge of smallpox for decades—until Jenner developed the practice of vaccination.

1.  The main point of the passage is that

    a) Lady Montagu survived smallpox as a child.

    b) Lady Montagu brought the practice of engrafting from Turkey to England.

    c) Doctors in eighteenth-century England were opposed to the practice of engrafting.

    d) Jenner developed the practice of vaccination.

2.  Based on the passage, Lady Montagu was most likely

    a) a doctor.

    b) Turkish.

    c) severely scarred by smallpox.

    d) a member of a prominent British family.

3.  The author refers to Lady Montagu having survived smallpox to

    a) explain why Lady Montagu was fascinated by the practice of engrafting.

    b) compare Lady Montagu to the doctors and religious leaders in England.

    c) explain why Lady Montagu herself did not undergo the engrafting procedure.

    d) emphasize Lady Montagu's fascination with other cultures.

4.  Based on the passage, the author most likely thinks that Lady Montagu was

    a) educated and influential.

    b) inconsequential in the prevention of smallpox in England.

    c) trained in science and medicine.

    d) married to the British ambassador to Turkey.

# ANSWERS TO IN-CHAPTER MATERIALS

## Practice Problems

**1.1A** 273 K and 373 K, range = 100 K. **1.1B** –270.5°C. **1.2A** 113°F, 194°F; difference = 81°F. **1.2B** 233°C. **1.3A** (a) 14 g/mL, (b) $1.6 \times 10^3$ g. **1.3B** (a) 9.25 g/cm$^3$, (b) $3.76 \times 10^3$ g. **1.4A** (a) and (c). **1.4B** iii, i. **1.5A** (a) 4, (b) 1, (c) 4, (d) 2, (e) 2 or 3, (f) 4. **1.5B** (a) 4, (b) 5, (c) 4, (d) 3, (e) ambiguous, (f) 5. **1.6A** (a) 116.2 L, (b) 80.71 m, (c) $3.813 \times 10^{21}$ atoms, (d) 31 dm$^2$, (e) 0.504 g/mL. **1.6B** (a) 32.44 cm$^3$, (b) $4.2 \times 10^2$ kg/m$^3$, (c) $1.008 \times 10^{10}$ kg, (d) 40.75 mL, (e) 227 cm$^3$. **1.7A** 0.8120 g/cm$^3$. **1.7B** 95.3 cm$^3$. **1.8A** 0.01 oz. **1.8B** 0.8813 oz. **1.9A** $1.05 \times 10^4$ kg/m$^3$. **1.9B** 13.6 mg/mm$^3$.

## Checkpoints

**1.3.1** c. **1.3.2** a. **1.3.3** b. **1.3.4** d. **1.4.1** b, c, e. **1.4.2** a, d, f. **1.5.1** c. **1.5.2** e. **1.5.3** c. **1.5.4** e. **1.6.1** b. **1.6.2** c. **1.6.3** a. **1.6.4** e.

## Applying What You've Learned

**a)** The recommended storage-temperature range for cidofovir is 20°C–25°C. **b)** The density of the fluid in a vial is 1.18 g/mL. (The density should be reported to three significant figures.) **c)** The recommended dosage of cidofovir for a 177-lb man is $4 \times 10^2$ mg or 0.4 g. **d)** $1.18 \times 10^3$ g/L, $1.18 \times 10^3$ kg/m$^3$.

# 2

# Atoms, Molecules, and Ions

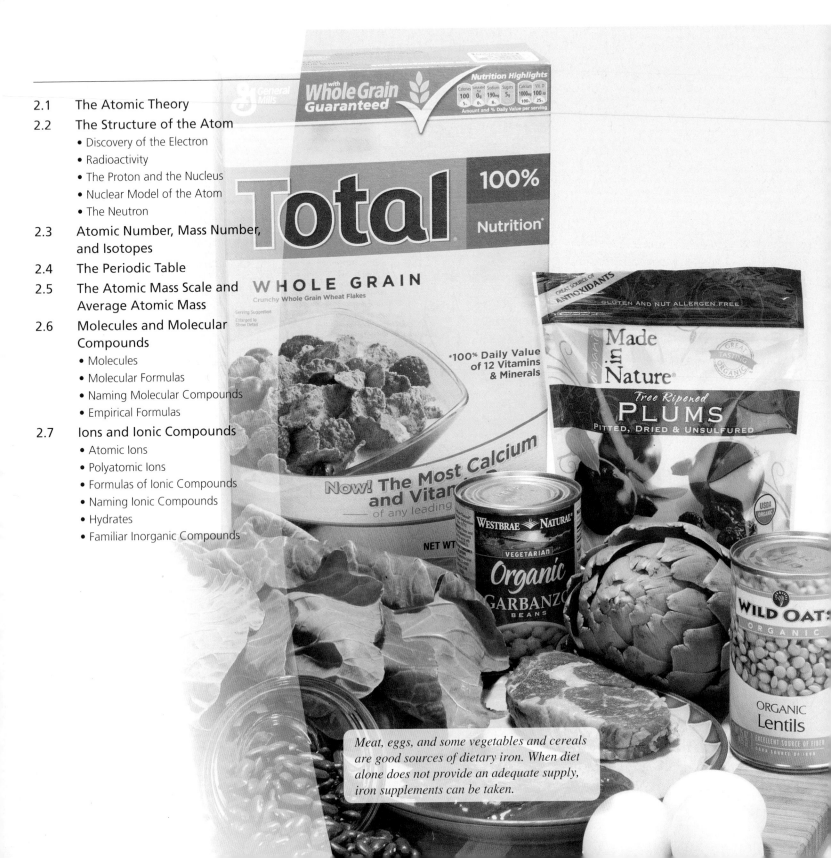

Meat, eggs, and some vegetables and cereals are good sources of dietary iron. When diet alone does not provide an adequate supply, iron supplements can be taken.

# How Certain Atoms, Molecules, and Ions Can Affect Human Health

Atoms, molecules, and ions make up the substances we encounter every day. Some of these substances are important components of a balanced diet. An estimated 25 percent of the world's population suffers from iron deficiency, the most common nutritional deficiency in the world. Iron is necessary for the production of hemoglobin, the component in red blood cells responsible for the transport of oxygen. An inadequate supply of iron and the resulting shortage of hemoglobin can cause iron deficiency anemia (IDA). Some of the symptoms of IDA are fatigue, weakness, pale color, poor appetite, headache, and light-headedness.

Although IDA can be caused by loss of blood or by poor absorption of iron, the most common cause is insufficient iron in the diet. Dietary iron comes from such sources as meat, eggs, leafy green vegetables, dried beans, and dried fruits. Some breakfast cereals, such as Total, are fortified with iron in the form of iron metal, also known as *elemental* or *reduced* iron. The absorption of dietary iron can be enhanced by the intake of vitamin C (ascorbic acid). When the diet fails to provide enough iron, a nutritional supplement may be necessary to prevent a deficiency. Many supplements provide iron in the form of a compound called *ferrous sulfate.*

Elemental iron, ascorbic acid, and the iron in ferrous sulfate are examples of some of the *atoms, molecules,* and *ions* that are essential for human health.

**Student Annotation:** Iron absorption can be diminished by certain disorders, such as Crohn's disease, and by some medications.

At the end of this chapter you will be able to solve a series of problems involving iron, iron sulfate, and ascorbic acid [▶▶ Page 67].

## In This Chapter, You Will Learn:

What atoms are made of and how they are arranged in molecules and ions that make up the substances that we encounter every day. You will also learn how to associate the name of a substance with its chemical formula.

### Before you begin, you should review

- Significant figures [◀◀ Section 1.5]
- Dimensional analysis [◀◀ Section 1.6]

**2.1**  **The Atomic Theory**

In the fifth century B.C. the Greek philosopher Democritus proposed that all matter consists of very small, indivisible particles, which he named *atomos* (meaning uncuttable or indivisible). Although Democritus's idea was not accepted by many of his contemporaries (notably Plato and Aristotle), somehow it endured. Experimental evidence from early scientific investigations provided support for the notion of "atomism" and gradually gave rise to the modern definitions of elements and compounds. In 1808, an English scientist and schoolteacher, John Dalton[1] (Figure 2.1), formulated a precise definition of the indivisible building blocks of matter that we call atoms.

Dalton's work marked the beginning of the modern era of chemistry. The hypotheses about the nature of matter on which Dalton's atomic theory is based can be summarized as follows:

1. Elements are composed of extremely small particles called atoms. All atoms of a given element are identical, having the same size, mass, and chemical properties. The atoms of one element are different from the atoms of all other elements.

2. Compounds are composed of atoms of more than one element. In any given compound, the same types of atoms are always present in the same relative numbers.

3. A chemical reaction rearranges atoms in chemical compounds; it does not create or destroy them.

Figure 2.2 is a schematic representation of these hypotheses.

Dalton's concept of an atom was far more detailed and specific than that of Democritus. The first hypothesis states that atoms of one element are different from atoms of all other elements. Dalton made no attempt to describe the structure or composition of atoms—he had no idea what an atom was really like. He did realize, though, that the different properties shown by elements such as hydrogen and oxygen could be explained by assuming that hydrogen atoms were not the same as oxygen atoms.

The second hypothesis suggests that, to form a certain compound, we not only need atoms of the right *kinds* of elements, but specific *numbers* of these atoms as well. This idea is an extension of a law published in 1799 by Joseph Proust, a French chemist. According to Proust's **law of definite proportions,** different samples of a given compound always contain the same elements in the same mass *ratio.* Thus, if we were to analyze samples of carbon dioxide gas obtained from different sources, such as the exhaust from a car in Mexico City or the air above a pine forest in northern

**Figure 2.1**    (a) John Dalton. (b) In addition to his other work, Dalton devised a system of symbols, which he used to represent the elements. (c) Dalton's only recreation is said to have been a nightly pint of ale and lawn bowling every Thursday afternoon. Note how closely the game balls resemble modern-day molecular models.

(a)

(b)

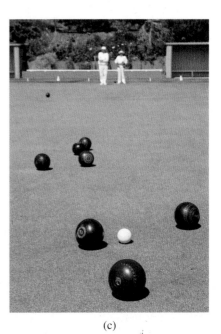

(c)

---

1. John Dalton (1766–1844). English chemist, mathematician, and philosopher. In addition to the atomic theory, Dalton formulated several gas laws and gave the first detailed description of the type of color blindness, now called "Daltonism," from which he suffered. He also assigned relative weights to the elements, many of which differ considerably from those that we use today. He was described by his friends as awkward and without social grace.

(a)                              (b)                              (c)

**Figure 2.2** This represents a chemical reaction between the elements oxygen and carbon. (a) Oxygen does not exist as isolated atoms under ordinary conditions, but rather as *molecules,* each of which consists of *two* oxygen atoms. Note that the oxygen atoms (red spheres) appear all to be identical to one another (hypothesis 1). (b) Likewise, the carbon atoms (black spheres) all appear to be identical to one another. Carbon also exists in the form of molecules that are more varied and complex than those of oxygen. The carbon has been represented as isolated atoms to simplify the figure. (c) The compound $CO_2$ forms when each carbon atom combines with two oxygen atoms (hypothesis 2). Finally, the reaction results in the rearrangement of the atoms, but all the atoms present before the reaction (left of the arrow) are also present after the reaction (right of the arrow) (hypothesis 3).

Maine, each sample would contain the same ratio by mass of oxygen to carbon. Consider the following results of the analysis of three samples of carbon dioxide, each from a different source:

| Sample | Mass of O (g) | Mass of C (g) | Ratio (g O : g C) |
|---|---|---|---|
| 123  g carbon dioxide | 89.4 | 33.6 | 2.66:1 |
| 50.5 g carbon dioxide | 36.7 | 13.8 | 2.66:1 |
| 88.6 g carbon dioxide | 64.4 | 24.2 | 2.66:1 |

In any sample of pure carbon dioxide, there are 2.66 g of oxygen for every gram of carbon present. This constant mass ratio can be explained by assuming that the elements exist in tiny particles of fixed mass (atoms), and that compounds are formed by the combination of fixed numbers of each type of particle.

Dalton's second hypothesis also supports the *law of multiple proportions.* According to this law, if two elements can combine to form more than one compound with each other, the masses of one element that combine with a fixed mass of the other element are in ratios of small whole numbers. That is, different compounds made up of the same elements differ in the number of atoms of each kind that combine. For example, carbon combines with oxygen to form carbon dioxide and carbon monoxide. In any sample of pure carbon monoxide, there are 1.33 g of oxygen for every gram of carbon.

| Sample | Mass of O (g) | Mass of C (g) | Ratio (g O : g C) |
|---|---|---|---|
| 16.3 g carbon monoxide | 9.31 | 6.99 | 1.33:1 |
| 25.9 g carbon monoxide | 14.8 | 11.1 | 1.33:1 |
| 88.4 g carbon monoxide | 50.5 | 37.9 | 1.33:1 |

Thus, the ratio of oxygen to carbon in carbon *di*oxide is 2.66; and the ratio of oxygen to carbon in carbon *mon*oxide is 1.33. According to the law of multiple proportions, the ratio of two such ratios can be expressed as small whole numbers.

$$\frac{\text{ratio of O to C in carbon dioxide}}{\text{ratio of O to C in carbon monoxide}} = \frac{2.66}{1.33} = 2:1$$

For samples containing equal masses of carbon, the ratio of oxygen in carbon dioxide to oxygen in carbon monoxide is 2:1. Modern measurement techniques indicate that one atom of carbon combines with two atoms of oxygen in carbon dioxide and with one atom of oxygen in carbon monoxide. This result is consistent with the law of multiple proportions (Figure 2.3).

Dalton's third hypothesis is another way of stating the *law of conservation of mass,*[2] which is that matter can be neither created nor destroyed. Because matter is made up of atoms that are unchanged in a chemical reaction, it follows that mass must be conserved as well.

Multimedia
Law of conservation of mass.

2. According to Albert Einstein, mass and energy are alternate aspects of a single entity called mass-energy. Chemical reactions usually involve a gain or loss of heat and other forms of energy. Thus, when energy is lost in a reaction, for example, mass is also lost. Except for nuclear reactions (see Chapter 20), however, changes of mass in chemical reactions are far too small to detect. Therefore, for all practical purposes mass is conserved.

**Figure 2.3**     An illustration of the law of multiple proportions.

Carbon dioxide

$$\frac{O}{C} = \frac{2}{1}$$

$$= \frac{2}{1}$$

Carbon monoxide

$$\frac{O}{C} = \frac{1}{1}$$

$$\frac{\text{O to C ratio in CO}_2}{\text{O to C ratio in CO}}$$

Dalton's brilliant insight into the nature of matter was the main stimulus for the rapid progress of chemistry during the nineteenth century.

Sample Problem 2.1 shows how some common compounds obey the law of multiple proportions.

### Sample Problem 2.1

(a) Both water ($H_2O$) and hydrogen peroxide ($H_2O_2$) are composed of hydrogen and oxygen. When water is decomposed into its constituent elements, it produces 0.125 g hydrogen for every gram of oxygen. When hydrogen peroxide is decomposed, it produces 0.063 g hydrogen for every gram of oxygen. Determine the whole number ratio of g H : 1.00 g O in water to g H : 1.00 g O in hydrogen peroxide to show how these data illustrate the law of multiple proportions. (b) Sulfur and oxygen can combine to form several compounds including sulfur dioxide ($SO_2$) and sulfur trioxide ($SO_3$). Sulfur dioxide contains 0.9978 g oxygen for every gram of sulfur. Sulfur trioxide contains 1.497 g oxygen for every gram of sulfur. Determine the whole number ratio of g O : 1.00 g S in sulfur dioxide to g O : 1.00 g S in sulfur trioxide.

**Strategy**  For two compounds, each consisting of just two different elements, we are given the mass ratio of one element to the other. In each case, to show how the information given illustrates the law of multiple proportions, we divide the larger ratio by the smaller ratio.

**Setup**  (a) The mass ratio of hydrogen to oxygen is higher in water than it is in hydrogen peroxide. Therefore, we divide the number of grams of hydrogen per gram of oxygen given for *water* by that given for *hydrogen peroxide*. (b) The mass ratio of oxygen to sulfur is higher in sulfur trioxide than it is in sulfur dioxide. Therefore, we divide the number of grams of oxygen per gram of sulfur given for *sulfur trioxide* by that given for *sulfur dioxide*.

**Think About It**  When the result of such calculations is not a whole number, we must decide whether the result is close enough to *round* to a whole number as in part (a), or whether to multiply through to get whole numbers as in part (b). Only numbers that are *very* close to whole can be rounded. For example, numbers ending in approximately .25, .33, or .5 should be multiplied by 4, 3, or 2, respectively, to give a whole number.

**Solution**

(a) $\dfrac{\text{g H : 1.00 g O in water}}{\text{g H : 1.00 g O in hydrogen peroxide}} = \dfrac{0.125}{0.063} = 1.98{:}1 \approx 2{:}1$

(b) $\dfrac{\text{g O : 1.00 g S in sulfur trioxide}}{\text{g O : 1.00 g S in sulfur dioxide}} = \dfrac{1.497}{0.9978} = 1.50{:}1$. Multiplying through by 2 gives 3:2.

**Practice Problem A**  In each case, calculate the appropriate ratio to show that the information given is consistent with the law of multiple proportions. (a) Both ammonia ($NH_3$) and hydrazine ($N_2H_4$) are composed of nitrogen and hydrogen. Ammonia contains 0.2158 g hydrogen for every gram of nitrogen. Hydrazine contains 0.1439 g hydrogen for every gram of nitrogen. (b) Two of the compounds that consist of nitrogen and oxygen are nitric oxide, also known as nitrogen monoxide (NO) and nitrous oxide ($N_2O$), which is also known as dinitrogen monoxide. Nitric oxide contains 1.142 g oxygen for every gram of nitrogen. Nitrous oxide contains 0.571 g oxygen for every gram of nitrogen.

**Practice Problem B**  (a) Two of the simplest compounds containing just carbon and hydrogen are methane and ethane. Given that methane contains 0.3357 g hydrogen for every 1.00 g carbon and that the ratio of g hydrogen/1.00 g carbon in methane to g hydrogen/1.00 g carbon in ethane is 4:3, determine the number of grams of hydrogen per gram of carbon in ethane. (b) Xenon (Xe) and fluorine (F) can combine to form several different compounds, including $XeF_2$, which contains 0.2894 g fluorine for every gram of xenon. Use the law of multiple proportions to determine $n$, which represents the number of F atoms in another compound, $XeF_n$, given that it contains 0.8682 g of F for every gram of Xe.

**Checkpoint 2.1** | **The Atomic Theory**

**2.1.1**

For the two compounds pictured, evaluate the following ratio:

$$\frac{\text{g yellow} : 1.00 \text{ g blue (right)}}{\text{g yellow} : 1.00 \text{ g blue (left)}}$$

a) 2:1

b) 3:2

c) 3:4

d) 5:2

e) 5:4

**2.1.2** For which of the following pairs of compounds is the ratio

$$\frac{\text{g yellow} : 1.00 \text{ g blue (right)}}{\text{g yellow} : 1.00 \text{ g blue (left)}}$$

equal to 4:1?

a)

b)

c)

d)

e)

# 2.2 | The Structure of the Atom

On the basis of Dalton's atomic theory, we can define an *atom* as the basic unit of an element that can enter into chemical combination. Dalton imagined an atom that was both extremely small and indivisible. However, a series of investigations that began in the 1850s and extended into the twentieth century clearly demonstrated that atoms actually possess internal structure; that is, they are made up of even smaller particles, which are called *subatomic particles.* This research led to the discovery of electrons, protons, and neutrons.

## Discovery of the Electron

Many scientists in the 1890s studied *radiation,* the emission and transmission of energy through space in the form of waves. Information gained from this research contributed greatly to our understanding of atomic structure. One device used to investigate this phenomenon was a cathode ray tube, the forerunner of the tubes used in older televisions and computer monitors (Figure 2.4).

Multimedia

Cathode ray tube experiment.

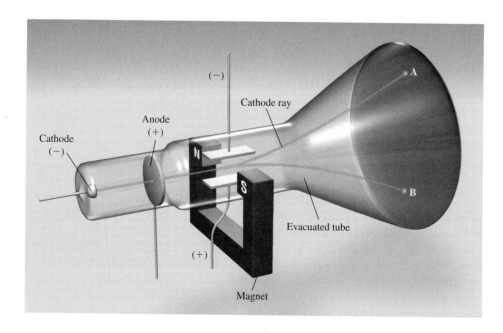

**Figure 2.4** A cathode ray tube with an electric field perpendicular to the direction of the cathode rays and an external magnetic field. The symbols N and S denote the north and south poles of the magnet. The cathode rays will strike the end of the tube at point A in the presence of a magnetic field and at point B in the presence of an electric field. (In the absence of any external field—or when the effects of the electric field and magnetic field cancel each other—the cathode rays will not be deflected but will travel in a straight line and strike the middle of the circular screen.)

(a)                              (b)                              (c)

**Figure 2.5** (a) A cathode ray produced in a discharge tube. The ray itself is invisible, but the fluorescence of a zinc sulfide coating on the glass causes it to appear green. (b) The cathode ray bends toward one pole of a magnet and (c) bends away from the opposite pole.

A cathode ray tube consists of two metal plates sealed inside a glass tube from which most of the air has been evacuated. When the metal plates are connected to a high-voltage source, the negatively charged plate, called the *cathode*, emits an invisible ray. The cathode ray is drawn to the positively charged plate, called the *anode*, where it passes through a hole and continues traveling to the other end of the tube. When the ray strikes the specially phosphor-coated surface, it produces a bright light.

Because consistent results are observed regardless of the composition of the cathode, cathode rays were presumed to be a component of all matter. Furthermore, because the path of the cathode rays could be deflected by magnetic and electric fields, as shown in Figure 2.4, they must be streams of charged *particles*. According to electromagnetic theory, a moving charged body behaves like a magnet and can interact with electric and magnetic fields through which it passes. Because the cathode ray is attracted by the plate bearing positive charges and repelled by the plate bearing negative charges, it must consist of negatively charged particles. We know these negatively charged particles as **electrons**. Figure 2.5 shows the deflection of a stream of electrons away from the south pole of a bar magnet. This effect on the path of electrons is what causes a television picture to become distorted temporarily when a magnet is brought close to the screen.

An English physicist, J. J. Thomson, used a cathode ray tube and his knowledge of electromagnetic theory to determine the ratio of electric charge to the mass of an individual electron. The number he calculated was $1.76 \times 10^8$ C/g, where C stands for *coulomb*, which is the derived SI unit of electric charge. (In SI base units, $1 \text{ C} = 1 \text{ A} \cdot \text{s}$.) Thereafter, in a series of experiments carried out between 1908 and 1917, American R. A. Millikan succeeded in measuring the charge of the electron with great precision. In his experiment, Millikan examined the motion of single tiny drops of oil that picked up static charge from particles in the air. He suspended the charged drops in air by applying an electric field and followed their motions through a microscope (Figure 2.6). His work proved that the charge on each electron was exactly the same: $-1.6022 \times 10^{-19}$ C. Using Thomson's charge-to-mass ratio and the charge from his own experiments, Millikan calculated the mass of an electron as follows:

**Student Annotation:** Recall that A and s are the SI base units *ampere* and *second,* respectively [◄◄ Section 1.3, Table 1.2].

**Student Annotation:** Millikan actually determined the charge on droplets with multiple charges attached. The charges he determined were always *multiples* of $-1.6022 \times 10^{-19}$ C.

$$\text{mass of an electron} = \frac{\text{charge}}{\text{charge/mass}} = \frac{-1.6022 \times 10^{-19} \text{ C}}{-1.76 \times 10^8 \text{ C/g}} = 9.10 \times 10^{-28} \text{ g}$$

This is an *extremely* small mass.

## Radioactivity

In 1895, the German physicist Wilhelm Röntgen noticed that cathode rays caused glass and metals to emit yet another type of ray. This highly energetic radiation penetrated matter, darkened covered photographic plates, and caused a variety of substances to *fluoresce* (give off light). These rays were *not* deflected by a magnet, however, so unlike cathode rays, they could not contain charged particles. Röntgen called them X rays because of their mysterious nature.

Not long after Röntgen's discovery, Antoine Becquerel, a professor of physics in Paris, began to study the fluorescent properties of substances. Purely by accident, he found that exposing thickly wrapped photographic plates (so no light could get in) to a uranium compound caused them to darken, even without the stimulation of cathode rays. Like X rays, the rays from the uranium compound were highly energetic and could not be deflected by a magnet, but they differed from

**Figure 2.6**   Schematic diagram of Millikan's oil-drop experiment.

Multimedia

Millikan oil-drop experiment.

X rays because they arose spontaneously. One of Becquerel's students, Marie Curie, suggested the name *radioactivity* to describe this spontaneous emission of particles and/or radiation. Today, we use the term *radioactive* to describe any element that spontaneously emits radiation.

Three types of rays are produced by the breakdown, or *decay,* of radioactive substances such as uranium. Two of the three are deflected by oppositely charged metal plates (Figure 2.7). *Alpha (α) rays* consist of positively charged particles, called *α particles,* that are deflected *away* from the positively charged plate. *Beta (β) rays,* or *β particles,* are *electrons,* so they are deflected away from the *negatively* charged plate. The third type of radioactive radiation consists of high-energy *gamma (γ) rays.* Like X rays, γ rays have no charge and are unaffected by external electric or magnetic fields.

## The Proton and the Nucleus

By the early 1900s, scientists knew that atoms contained electrons but were electrically neutral overall. To be neutral, an atom must contain equal amounts of positive and negative charge. Thomson proposed, therefore, that an atom could be thought of as a sphere of positively charged matter in which negatively charged electrons were embedded uniformly, like the chocolate chips in a scoop of mint chocolate chip ice cream. This so-called plum-pudding model was the accepted theory for a number of years.

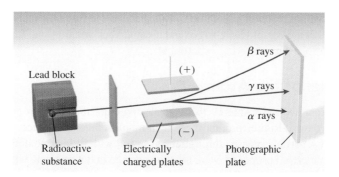

**Figure 2.7**   Three types of rays emitted by radioactive elements. So-called β rays actually consist of negatively charged particles (electrons) and are therefore attracted by the positively charged plate. The opposite holds true for so-called α rays—they are actually positively charged particles and are drawn to the negatively charged plate. Because γ rays are not particles and have no charge, their path is unaffected by an external electric field.

**Figure 2.8**  Rutherford's experimental design for measuring the scattering of $\alpha$ particles by a piece of gold foil. The plum-pudding model predicted that the $\alpha$ particles would all pass through the gold foil undeflected. The actual result: Most of the $\alpha$ particles do pass through the gold foil with little or no deflection, but a few are deflected at large angles. Occasionally an $\alpha$ particle bounces off the foil back toward the source. The nuclear model explains the results of Rutherford's experiments.

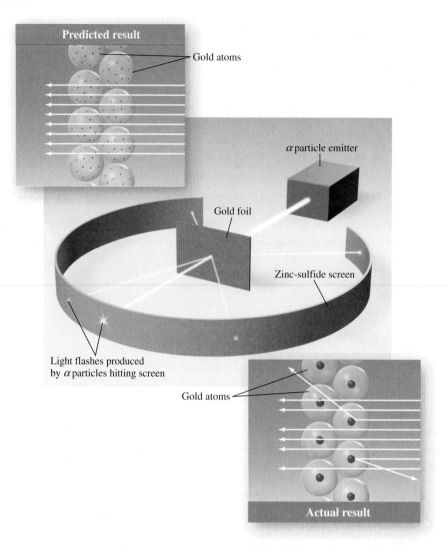

In 1910 the New Zealand physicist Ernest Rutherford, who had studied with Thomson at Cambridge University, decided to use $\alpha$ particles to probe the structure of atoms. Together with his associate Hans Geiger and an undergraduate named Ernest Marsden, Rutherford carried out a series of experiments using very thin foils of gold and other metals as targets for $\alpha$ particles from a radioactive source. They observed that the majority of particles penetrated the foil either completely undeflected or with only a small angle of deflection. Every now and then, however, an $\alpha$ particle was scattered (or deflected) at a large angle. In some instances, the $\alpha$ particle actually bounced back in the direction from which it had come! This was an extraordinarily surprising finding. In Thomson's model the positive charge of the atom was so diffuse that the relatively massive, positively charged particles should all have passed through the foil with little or no deflection. To quote Rutherford's initial reaction when told of this discovery: "It was as incredible as if you had fired a 15-inch shell at a piece of tissue paper and it came back and hit you." Figure 2.8 illustrates the results of Rutherford's $\alpha$-scattering experiment.

## Nuclear Model of the Atom

Rutherford later explained the results of the $\alpha$-scattering experiment by proposing a new model for the atom. According to Rutherford, most of the atom must be empty space. This would explain why the majority of $\alpha$ particles passed through the gold foil with little or no deflection. The atom's positive charges, Rutherford proposed, were all concentrated in the ***nucleus,*** which is an extremely dense central *core* within the atom. Whenever an $\alpha$ particle came close to a nucleus in the scattering experiment, it experienced a large repulsive force and therefore a large deflection. Moreover, an $\alpha$ particle traveling directly toward a nucleus would be completely repelled and its direction would be reversed.

The positively charged particles in the nucleus are called **protons.** In separate experiments, it was found that each proton carried the same *quantity* of charge as an electron (just opposite in sign) but had a mass of $1.67262 \times 10^{-24}$ g. Although this is an extremely small value, it is nearly 2000 times the mass of an electron.

Based on these data, the atom was believed to consist of a nucleus that accounted for most of the mass of the atom, but which occupied only a tiny fraction of its volume. We express atomic (and molecular) dimensions using the SI unit *picometer (pm)*, where

$$1 \text{ pm} = 1 \times 10^{-12} \text{ m}$$

A typical atomic radius is about 100 pm, whereas the radius of an atomic nucleus is only about $5 \times 10^{-3}$ pm. You can appreciate the relative sizes of an atom and its nucleus by imagining that if an atom were the size of the New Orleans Superdome, the volume of its nucleus would be comparable to that of a marble. While the protons are confined to the nucleus of the atom, the electrons are distributed *around* the nucleus at relatively large distances from it.

The concept of atomic radius is useful experimentally, but you should not get the impression that atoms have well-defined boundaries or surfaces. We will learn in Chapter 6 that the outer regions of atoms are relatively "fuzzy," not sharply defined.

## The Neutron

Rutherford's model of atomic structure left one major problem unsolved. It was known that hydrogen, the simplest atom, contained only one proton and that the helium atom contained two protons. Therefore, the ratio of the mass of a helium atom to that of a hydrogen atom should be 2:1. (Because electrons are much lighter than protons, their contribution to atomic mass can be ignored.) In reality, however, the ratio is 4:1. Rutherford and others postulated that there must be another type of subatomic particle in the atomic nucleus, the proof of which was provided by James Chadwick, an English physicist, in 1932. When Chadwick bombarded a thin sheet of beryllium with $\alpha$ particles, a very high energy radiation was emitted by the metal that was not deflected by either electric or magnetic fields. Although similar to $\gamma$ rays, later experiments showed that the rays actually consisted of a third type of subatomic particle, which Chadwick named **neutrons,** because they were electrically neutral particles having a mass slightly greater than that of protons. The mystery of the mass ratio could now be explained. A typical helium nucleus consists of two protons and two neutrons, whereas a typical hydrogen nucleus contains only a proton; the mass ratio, therefore, is 4:1.

Figure 2.9 shows the location of the elementary particles (protons, neutrons, and electrons) in an atom. There are other subatomic particles, but the electron, the proton, and the neutron are the three fundamental components of the atom that are important in chemistry. Table 2.1 lists the masses and charges of these three elementary particles.

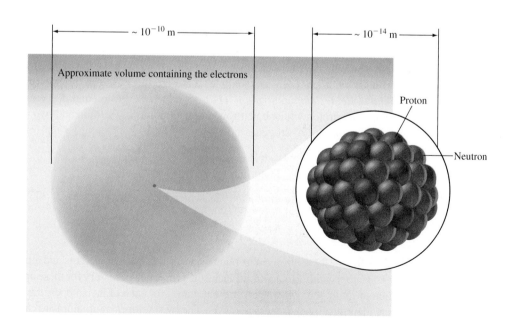

**Figure 2.9**   The protons and neutrons in an atom are contained in the tiny volume of the nucleus. Electrons are distributed within the sphere surrounding the nucleus.

| TABLE 2.1 | Masses and Charges of Subatomic Particles | | |
|---|---|---|---|
| **Particle** | **Mass (g)** | **Charge (C)** | **Charge Unit** |
| Electron* | $9.10938 \times 10^{-28}$ | $-1.6022 \times 10^{-19}$ | $-1$ |
| Proton | $1.67262 \times 10^{-24}$ | $+1.6022 \times 10^{-19}$ | $+1$ |
| Neutron | $1.67493 \times 10^{-24}$ | $0$ | $0$ |

*More refined measurements have resulted in a small change to Millikan's original value.

## 2.3   Atomic Number, Mass Number, and Isotopes

All atoms can be identified by the number of protons and neutrons they contain. The *atomic number (Z)* is the number of protons in the nucleus of each atom of an element. It also indicates the number of *electrons* in the atom—because atoms are *neutral* and contain the same number of protons and electrons. The chemical identity of an atom can be determined solely from its atomic number. For example, the atomic number of nitrogen is 7. Thus, each nitrogen atom has seven protons and seven electrons. Or, viewed another way, every atom in the universe that contains seven protons is a nitrogen atom.

The *mass number (A)* is the total number of neutrons *and* protons present in the nucleus of an atom of an element. Except for the most common form of hydrogen, which has one proton and no neutrons, all atomic nuclei contain both protons and neutrons. In general the mass number is given by

> **Student Annotation:** Collectively, protons and neutrons are called *nucleons.* A nucleon is a particle within the nucleus.

mass number ($A$) = number of protons ($Z$) + number of neutrons

The number of neutrons in an atom equals the difference between the mass number and the atomic number, or ($A - Z$). For example, the mass number of fluorine is 19 and the atomic number is 9 (indicating 9 protons in the nucleus). Thus, the number of neutrons in an atom of fluorine is $19 - 9 = 10$. The atomic number, number of neutrons, and mass number all must be positive integers (whole numbers).

The accepted way to denote the atomic number and mass number of an atom of an element (X) is as follows:

Mass number
(number of protons + neutrons)

$^{A}_{Z}\text{X}$ ← Element symbol

Atomic number
(number of protons)

> **Student Annotation:** Because these symbols designate isotopes by specifying numbers of nucleons, they are sometimes referred to as *nuclear* symbols.

Contrary to the first hypothesis of Dalton's atomic theory, atoms of a given element do not all have the same mass. Most elements have two or more *isotopes,* atoms that have the same atomic number ($Z$) but different mass numbers ($A$). For example, there are three isotopes of hydrogen, called *hydrogen* (or *protium*), *deuterium,* and *tritium.* Hydrogen has one proton and no neutrons in its nucleus, deuterium has one proton and one neutron, and tritium has one proton and two neutrons. Thus, to represent the isotopes of hydrogen, we write

$^{1}_{1}\text{H}$          $^{2}_{1}\text{H}$          $^{3}_{1}\text{H}$
hydrogen     deuterium     tritium

Similarly, the two common isotopes of uranium ($Z = 92$), which have mass numbers of 235 and 238, respectively, can be represented as follows:

$^{235}_{92}\text{U}$          $^{238}_{92}\text{U}$

The first isotope, with $235 - 92 = 143$ neutrons in its nucleus, is used in nuclear reactors and atomic bombs, whereas the second isotope, with 146 neutrons, lacks the properties necessary for these applications. With the exception of hydrogen, which has different names for each of its isotopes, the isotopes of other elements are identified by their mass numbers. The two isotopes of uranium are called uranium-235 (pronounced "uranium two thirty-five") and uranium-238 (pronounced "uranium two thirty-eight"). Because the subscripted atomic number can be determined from the elemental symbol, it may be omitted from these representations without the loss of any information. The symbols $^3$H and $^{235}$U are sufficient to specify the isotopes tritium and uranium-235, respectively.

The chemical properties of an element are determined primarily by the protons and electrons in its atoms; neutrons do not take part in chemical changes under normal conditions. Therefore, isotopes of the same element exhibit similar chemical properties, forming the same types of compounds and displaying similar reactivities.

Sample Problem 2.2 shows how to calculate the number of protons, neutrons, and electrons using atomic numbers and mass numbers.

---

### Sample Problem 2.2

Determine the numbers of protons, neutrons, and electrons in each of the following species: (a) $^{35}_{17}$Cl, (b) $^{37}_{17}$Cl, (c) $^{41}$K, and (d) carbon-14.

**Strategy** Recall that the superscript denotes the mass number ($A$), and the subscript denotes the atomic number ($Z$). In the case where no subscript is shown, as in parts (c) and (d), the atomic number can be deduced from the elemental symbol or name. For the purpose of determining the number of electrons, remember that atoms are neutral, so the number of electrons equals the number of protons.

**Setup** Number of protons = $Z$, number of neutrons = $A - Z$, and number of electrons = number of protons. Recall that the 14 in carbon-14 is the mass number.

**Solution**
(a) The atomic number is 17, so there are 17 protons. The mass number is 35, so the number of neutrons is $35 - 17 = 18$. The number of electrons equals the number of protons, so there are 17 electrons.
(b) Again, the atomic number is 17, so there are 17 protons. The mass number is 37, so the number of neutrons is $37 - 17 = 20$. The number of electrons equals the number of protons, so there are 17 electrons, too.
(c) The atomic number of K (potassium) is 19, so there are 19 protons. The mass number is 41, so there are $41 - 19 = 22$ neutrons. There are 19 electrons.
(d) Carbon-14 can also be represented as $^{14}$C. The atomic number of carbon is 6, so there are 6 protons and 6 electrons. There are $14 - 6 = 8$ neutrons.

> **Think About It** Verify that the number of protons and the number of neutrons for each example sum to the mass number that is given. In part (a), for example, there are 17 protons and 18 neutrons, which sum to give a mass number of 35, the value given in the problem. In part (b), 17 protons + 20 neutrons = 37. In part (c), 19 protons + 22 neutrons = 41. In part (d), 6 protons + 8 neutrons = 14.

**Practice Problem A** How many protons, neutrons, and electrons are there in an atom of (a) $^{10}_{5}$B, (b) $^{36}$Ar, (c) $^{85}_{38}$Sr, and (d) carbon-11?

**Practice Problem B** Give the correct symbols to identify an atom that contains (a) 4 protons, 4 electrons, and 5 neutrons; (b) 23 protons, 23 electrons, and 28 neutrons; (c) 54 protons, 54 electrons, and 70 neutrons; and (d) 31 protons, 31 electrons, and 38 neutrons.

---

### Checkpoint 2.3    Atomic Number, Mass Number, and Isotopes

**2.3.1** How many neutrons are there in an atom of $^{60}$Ni?

a) 60
b) 30
c) 28
d) 32
e) 29

**2.3.2** What is the *mass number* of an oxygen atom with nine neutrons in its nucleus?

a) 8
b) 9
c) 17
d) 16
e) 18

---

## 2.4  The Periodic Table

More than half of the elements known today were discovered between 1800 and 1900. During this period, chemists noted that the physical and chemical properties of certain groups of elements were similar to one another. These similarities, together with the need to organize the large volume of available information about the structure and properties of elemental substances, led to the

| 1A 1 | | | | | | | | | | | | | | | | | | 8A 18 |
|---|---|---|---|---|---|---|---|---|---|---|---|---|---|---|---|---|---|---|
| 1 **H** Hydrogen | 2A 2 | | | | | | | | | | | | 3A 13 | 4A 14 | 5A 15 | 6A 16 | 7A 17 | 2 **He** Helium |
| 3 **Li** Lithium | 4 **Be** Beryllium | | | | | | | | | | | | 5 **B** Boron | 6 **C** Carbon | 7 **N** Nitrogen | 8 **O** Oxygen | 9 **F** Fluorine | 10 **Ne** Neon |
| 11 **Na** Sodium | 12 **Mg** Magnesium | 3B 3 | 4B 4 | 5B 5 | 6B 6 | 7B 7 | 8B 8 | 9 | 10 | 1B 11 | 2B 12 | | 13 **Al** Aluminum | 14 **Si** Silicon | 15 **P** Phosphorus | 16 **S** Sulfur | 17 **Cl** Chlorine | 18 **Ar** Argon |
| 19 **K** Potassium | 20 **Ca** Calcium | 21 **Sc** Scandium | 22 **Ti** Titanium | 23 **V** Vanadium | 24 **Cr** Chromium | 25 **Mn** Manganese | 26 **Fe** Iron | 27 **Co** Cobalt | 28 **Ni** Nickel | 29 **Cu** Copper | 30 **Zn** Zinc | | 31 **Ga** Gallium | 32 **Ge** Germanium | 33 **As** Arsenic | 34 **Se** Selenium | 35 **Br** Bromine | 36 **Kr** Krypton |
| 37 **Rb** Rubidium | 38 **Sr** Strontium | 39 **Y** Yttrium | 40 **Zr** Zirconium | 41 **Nb** Niobium | 42 **Mo** Molybdenum | 43 **Tc** Technetium | 44 **Ru** Ruthenium | 45 **Rh** Rhodium | 46 **Pd** Palladium | 47 **Ag** Silver | 48 **Cd** Cadmium | | 49 **In** Indium | 50 **Sn** Tin | 51 **Sb** Antimony | 52 **Te** Tellurium | 53 **I** Iodine | 54 **Xe** Xenon |
| 55 **Cs** Cesium | 56 **Ba** Barium | 71 **Lu** Lutetium | 72 **Hf** Hafnium | 73 **Ta** Tantalum | 74 **W** Tungsten | 75 **Re** Rhenium | 76 **Os** Osmium | 77 **Ir** Iridium | 78 **Pt** Platinum | 79 **Au** Gold | 80 **Hg** Mercury | | 81 **Tl** Thallium | 82 **Pb** Lead | 83 **Bi** Bismuth | 84 **Po** Polonium | 85 **At** Astatine | 86 **Rn** Radon |
| 87 **Fr** Francium | 88 **Ra** Radium | 103 **Lr** Lawrencium | 104 **Rf** Rutherfordium | 105 **Db** Dubnium | 106 **Sg** Seaborgium | 107 **Bh** Bohrium | 108 **Hs** Hassium | 109 **Mt** Meitnerium | 110 **Ds** Darmstadtium | 111 **Rg** Roentgenium | 112 – – | | 113 – – | 114 – – | 115 – – | 116 – – | 117 – – | 118 – – |

| 57 **La** Lanthanum | 58 **Ce** Cerium | 59 **Pr** Praseodymium | 60 **Nd** Neodymium | 61 **Pm** Promethium | 62 **Sm** Samarium | 63 **Eu** Europium | 64 **Gd** Gadolinium | 65 **Tb** Terbium | 66 **Dy** Dysprosium | 67 **Ho** Holmium | 68 **Er** Erbium | 69 **Tm** Thulium | 70 **Yb** Ytterbium |
|---|---|---|---|---|---|---|---|---|---|---|---|---|---|
| 89 **Ac** Actinium | 90 **Th** Thorium | 91 **Pa** Protactinium | 92 **U** Uranium | 93 **Np** Neptunium | 94 **Pu** Plutonium | 95 **Am** Americium | 96 **Cm** Curium | 97 **Bk** Berkelium | 98 **Cf** Californium | 99 **Es** Einsteinium | 100 **Fm** Fermium | 101 **Md** Mendelevium | 102 **No** Nobelium |

**Figure 2.10**    The modern periodic table. The elements are arranged according to atomic number, which is shown above each element's symbol. With the exception of hydrogen (H), nonmetals appear at the far right of the table. The two rows of metals beneath the main body of the table are set apart to keep the table from being too wide. Actually, lanthanum (57) should follow barium (56), and actinium (89) should follow radium (88). The 1–18 group designation has been recommended by the International Union of Pure and Applied Chemistry (IUPAC) but is not yet in wide use. In this text, we generally use the standard U.S. notation for group numbers (1A–8A and 1B–8B). No names have yet been assigned to elements 112–116 and 118. As of this writing, the synthesis of element 117 has not yet been reported.

development of the *periodic table,* a chart in which elements having similar chemical and physical properties are grouped together. Figure 2.10 shows the modern periodic table in which the elements are arranged by atomic number (shown above the element symbol) in horizontal rows called *periods* and in vertical columns called *groups* or *families.* Elements in the same *group* tend to have similar physical and chemical properties.

The elements can be categorized as metals, nonmetals, or metalloids. A *metal* is a good conductor of heat and electricity, whereas a *nonmetal* is usually a poor conductor of heat and electricity. A *metalloid* has properties that are intermediate between those of metals and nonmetals. Figure 2.10 shows that the majority of known elements are metals; only 17 elements are nonmetals, and fewer than 10 elements are metalloids. Although most sources, including this text, designate the elements B, Si, Ge, As, Sb, and Te as metalloids, sources vary for the elements Po and At. In this text, we classify both Po and At as metalloids. From left to right across any period, the physical and chemical properties of the elements change gradually from metallic to nonmetallic.

Elements are often referred to collectively by their periodic table group number (Group 1A, Group 2A, and so on). For convenience, however, some element groups have been given special names. The Group 1A elements, with the exception of H (i.e., Li, Na, K, Rb, Cs, and Fr), are called *alkali metals,* and the

Alkali metals

Alkaline earth metals

Transition metals

Chalcogens

Halogens

Noble gases

Group 2A elements (Be, Mg, Ca, Sr, Ba, and Ra) are called **alkaline earth metals.** Elements in Group 6A (O, S, Se, Te, and Po) are sometimes referred to as the **chalcogens.** Elements in Group 7A (F, Cl, Br, I, and At) are known as **halogens,** and elements in Group 8A (He, Ne, Ar, Kr, Xe, and Rn) are called **noble gases,** or rare gases. The elements in Group 1B and Groups 3B–8B collectively are called the **transition elements** or **transition metals.**

The periodic table is a handy tool that correlates the properties of the elements in a systematic way and helps us to predict chemical behavior. At the turn of the twentieth century, the periodic table was deemed "the most predictive tool in all of science." We will take a more detailed look at this keystone of chemistry in Chapter 7.

The following Bringing Chemistry to Life box describes the distribution of the elements in Earth's crust.

## Bringing Chemistry to Life

### Distribution of Elements on Earth

Earth's crust extends from the surface to a depth of about 40 km (about 25 mi). Because of technical difficulties, scientists have been unable to study the inner portions of Earth as easily and as thoroughly as the crust. Nevertheless, it is believed that there is a solid core consisting mostly of iron at the center of Earth. Surrounding the core is a layer called the *mantle,* which consists of hot fluid containing iron, carbon, silicon, and sulfur.

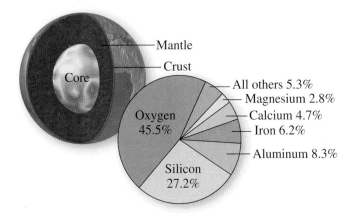

Of the 83 elements that are found in nature, 12 make up 99.7 percent of Earth's crust by mass. They are, in decreasing order of natural abundance, oxygen (O), silicon (Si), aluminum (Al), iron (Fe), calcium (Ca), magnesium (Mg), sodium (Na), potassium (K), titanium (Ti), hydrogen (H), phosphorus (P), and manganese (Mn). When discussing the natural abundance of the elements, keep in mind that the elements are unevenly distributed throughout Earth's crust, and most elements are combined chemically with other elements. Some of the elements that can be found in nature in pure form include gold, copper, and sulfur.

---

### Checkpoint 2.4    The Periodic Table

**2.4.1** Which of the following series of elemental symbols lists a nonmetal, a metal, and a metalloid?

a) Ca, Cu, Si

b) K, Mg, B

c) Br, Ba, Ge

d) O, Na, S

e) Ag, Cr, As

**2.4.2** Which of the following elements would you expect to have properties most similar to those of chlorine (Cl)?

a) Cu

b) F

c) Na

d) Cr

e) S

**FAQ**  *You May Be Wondering...*

# How Are Atomic Masses Measured?

The most direct and most accurate method for determining atomic and molecular masses is mass spectrometry. In a *mass spectrometer,* such as that depicted in Figure 2.11, a gaseous sample is bombarded by a stream of high-energy electrons. Collisions between the electrons and the gaseous atoms (or molecules) produce positively charged species, called *ions,* by dislodging an electron from the atoms or molecules. These positive ions (of mass *m* and charge *e*) are accelerated as they pass through two oppositely charged plates. The emerging ions are deflected into a circular path by a magnet. The radius of the path depends on the charge-to-mass ratio (i.e., *e/m*). Ions with a small *e/m* ratio trace a wider arc than those having a larger *e/m* ratio, so ions with equal charges but different masses are separated from one another. The mass of each ion (and hence its parent atom or molecule) is determined from the magnitude of its deflection. Eventually the ions arrive at the detector, which registers a current for each type of ion. The amount of current generated is directly proportional to the number of ions, so it enables us to determine the relative abundance of isotopes.

**Figure 2.11**    Schematic diagram of one type of mass spectrometer.

## 2.5    The Atomic Mass Scale and Average Atomic Mass

In some experimental work it is important to know the masses of individual atoms, which depend on the number of protons, neutrons, and electrons they contain. However, even the smallest speck of dust that our unaided eyes can perceive consists of as many as $1 \times 10^{16}$ atoms! Although we cannot weigh a single atom, it is possible to determine experimentally the mass of one atom *relative* to another. The first step is to assign a value to the mass of one atom of a given element so that it can be used as a standard.

The first mass spectrometer, developed in the 1920s by the English physicist F. W. Aston, was crude by today's standards. Nevertheless, it provided indisputable evidence of the existence of isotopes, such as neon-20 (natural abundance 90.48 percent) and neon-22 (natural abundance 9.25 percent). When more sophisticated and sensitive mass spectrometers became available, scientists were surprised to discover that neon has a *third* stable isotope (neon-21) with natural abundance 0.27 percent (Figure 2.12). This example illustrates how very important experimental accuracy is to a quantitative science like chemistry. Early experiments failed to detect neon-21 because its natural abundance was so small. Only 27 in 10,000 Ne atoms are neon-21.

Sample Problem 2.3 shows how to calculate the average atomic mass of oxygen.

**Figure 2.12** Mass spectrum of neon.

## Sample Problem 2.3

Oxygen is the most abundant element in both Earth's crust and the human body. The atomic masses of its three stable isotopes, $^{16}_{8}O$ (99.757 percent), $^{17}_{8}O$ (0.038 percent), and $^{18}_{8}O$ (0.205 percent), are 15.9949, 16.9991, and 17.9992 amu, respectively. Calculate the average atomic mass of oxygen using the relative abundances given in parentheses. Report the result to four significant figures.

**Strategy** Each isotope contributes to the average atomic mass based on its relative abundance. Multiplying the mass of each isotope by its fractional abundance (percent value divided by 100) will give its contribution to the average atomic mass.

**Setup** Each percent abundance must be converted to a fractional abundance: 99.757 percent to 99.757/100 or 0.99757, 0.038 percent to 0.038/100 or 0.00038, and 0.205 percent to 0.205/100 or 0.00205. Once we find the contribution to the average atomic mass for each isotope, we can then add the contributions together to obtain the average atomic mass.

**Solution**

(0.99757)(15.9949 amu) + (0.00038)(16.9991 amu) + (0.00205)(17.9992) = 15.999 amu

> **Think About It** The average atomic mass should be closest to the atomic mass of the most abundant isotope (oxygen-16, in this case) and should, to the appropriate number of significant figures, be the same number that appears in the periodic table on the inside front cover of this book (16.00 amu, in this case)!

**Practice Problem A** The atomic masses of the two stable isotopes of copper, $^{63}_{29}Cu$ (69.17 percent) and $^{65}_{29}Cu$ (30.83 percent), are 62.929599 and 64.927793 amu, respectively. Calculate the average atomic mass of copper.

**Practice Problem B** The average atomic mass of nitrogen is 14.0067. The atomic masses of the two stable isotopes of nitrogen, $^{14}N$ and $^{15}N$, are 14.003074002 and 15.00010897 amu, respectively. Use this information to determine the percent abundance of each nitrogen isotope.

According to international agreement, **atomic mass** is the mass of an atom in atomic mass units. One **atomic mass unit (amu)** is defined as a mass exactly equal to one-twelfth the mass of one carbon-12 atom. Carbon-12 is the carbon isotope that has six protons and six neutrons. Setting the atomic mass of carbon-12 at 12 amu provides the standard for measuring the atomic mass of the other elements. For example, experiments have shown that a hydrogen atom ($^{1}H$) is only 8.3985 percent as massive as the carbon-12 atom. Thus, if the mass of one carbon-12 atom is exactly 12 amu, the atomic mass of hydrogen must be 0.083985 × 12 amu, or 1.0078 amu. (Because the mass of the carbon atom is an exact number [◄◄ Section 1.5], it does not limit the number of significant figures in the calculated result.) Similar calculations show that the atomic mass of fluorine-19 is 18.9984 amu and that of oxygen-16 is 15.9949 amu. Thus, although we cannot measure the mass of a single oxygen-16 atom, we know that it is approximately 16 times as massive as a hydrogen-1 atom.

**Student Annotation:** The atomic masses in the periodic table are *average atomic masses*. The term *atomic weight* is sometimes used to mean average atomic mass.

When you look up the atomic mass of carbon in a table such as the one on the inside front cover of this book, you will find that its value is 12.01 amu, not 12.00 amu. The difference arises because most naturally occurring elements (including carbon) have more than one isotope. This means that when we measure the atomic mass of an element, we must generally settle for the average mass of the naturally occurring mixture of isotopes. For example, the natural abundances of carbon-12 and carbon-13 are 98.93 percent and 1.07 percent, respectively. The atomic mass of carbon-13 has been determined to be 13.003355 amu. Thus, the average atomic mass of natural carbon can be calculated as follows:

$$(0.9893)(12.00000 \text{ amu}) + (0.0107)(13.003355 \text{ amu}) = 12.01 \text{ amu}$$

In calculations involving percentages, we need to convert each percent abundance to a fractional abundance. For example, 98.93 percent becomes 98.93/100, or 0.9893. Because there are many more carbon-12 atoms than carbon-13 atoms in naturally occurring carbon, the average atomic mass is much closer to the mass of carbon-12 than to that of carbon-13.

When we say that the atomic mass of carbon is 12.01 amu, we are referring to the average value. If we could examine an individual atom of naturally occurring carbon, we would find either an atom of atomic mass of exactly 12 amu or one of 13.003355 amu, but never one of 12.01 amu.

The atomic masses of many isotopes have been accurately determined to five or six significant figures. For most purposes, though, we will use average atomic masses, which are generally given to four significant figures (see the table of atomic masses on the inside front cover). For simplicity, we will omit the word *average* when we discuss the atomic masses of the elements.

---

**Checkpoint 2.5**    **The Atomic Mass Scale and Average Atomic Mass**

**2.5.1** Boron has two naturally occurring isotopes, $^{10}$B and $^{11}$B, which have masses 10.0129 and 11.0093 amu, respectively. Given the average atomic mass of boron (10.81 amu), determine the percent abundance of each isotope.

a) 50% $^{10}$B, 50% $^{11}$B

b) 20% $^{10}$B, 80% $^{11}$B

c) 98% $^{10}$B, 2% $^{11}$B

d) 93% $^{10}$B, 7% $^{11}$B

e) 22% $^{10}$B, 78% $^{11}$B

**2.5.2** The two naturally occurring isotopes of antimony, $^{121}$Sb (57.21 percent) and $^{123}$Sb (42.79 percent), have masses of 120.904 and 122.904 amu, respectively. What is the average atomic mass of Sb?

a) 121.90 amu

b) 122.05 amu

c) 121.76 amu

d) 121.34 amu

e) 122.18 amu

---

## 2.6    Molecules and Molecular Compounds

Out of all the elements, only the six noble gases in Group 8A of the periodic table (He, Ne, Ar, Kr, Xe, and Rn) exist in the form of isolated atoms under ordinary conditions. For this reason, they are called *monatomic* (meaning a single atom) gases. Most matter is composed of *molecules*, which we discuss in this section, or *ions*, which we will discuss in Section 2.7.

### Molecules

A *molecule* is a combination of at least two atoms in a specific arrangement held together by electrostatic forces known as *covalent chemical bonds* [▶▶ Section 8.3]. A molecule may contain atoms of the same element, or it may contain atoms of two or more elements joined in a fixed ratio, in accordance with the law of definite proportions [◀◀ Section 2.1]. Thus, a molecule can be an *element* or it can be a *compound,* which, by definition, is made up of two or more elements [◀◀ Section 1.2]. Hydrogen gas, for example, is an element, but it consists of molecules, each of which is made up of two H atoms. Water, on the other hand, is a compound that consists of molecules, each of which contains two H atoms and one O atom.

The hydrogen molecule, symbolized as $H_2$, is called a ***diatomic molecule*** because it contains *two* atoms. Other elements that normally exist as diatomic molecules are nitrogen ($N_2$), oxygen ($O_2$), and the Group 7A elements—fluorine ($F_2$), chlorine ($Cl_2$), bromine ($Br_2$), and iodine ($I_2$). These are known as ***homonuclear*** diatomic molecules because both atoms in each molecule are of the same element. A diatomic molecule can also contain atoms of different elements. Examples of these ***heteronuclear*** diatomic molecules include hydrogen chloride (HCl) and carbon monoxide (CO).

Most molecules contain more than two atoms. They can all be atoms of the same element, as in ozone ($O_3$) and white phosphorus ($P_4$), or they can be combinations of two or more different elements, as in water ($H_2O$) and methane ($CH_4$). Molecules containing more than two atoms are called ***polyatomic molecules.***

Homonuclear diatomic     Heteronuclear diatomic     Polyatomic

## Molecular Formulas

A ***chemical formula*** denotes the composition of the substance. A ***molecular formula*** shows the exact number of atoms of each element in a molecule. In our discussion of molecules, each example was given with its molecular formula in parentheses. Thus, $H_2$ is the molecular formula for hydrogen, $O_2$ is that for oxygen, $O_3$ is that for ozone, and $H_2O$ is that of water. The subscript numeral indicates the number of atoms of an element present in the molecule. There is no subscript for O in $H_2O$ because there is only one atom of oxygen in a molecule of water. The number "one" is never used as a subscript in a chemical formula. Oxygen ($O_2$) and ozone ($O_3$) are allotropes of oxygen. An ***allotrope*** is one of two or more distinct forms of an element. Two of the allotropic forms of the element carbon—diamond and graphite—have dramatically different properties (and *prices*).

We can also represent molecules with *structural formulas*. The ***structural formula*** shows not only the elemental composition, but also the general arrangement of atoms within the molecule. In the case of water, each of the hydrogen atoms is connected to the oxygen atom. Figure 2.13 shows the molecular formula, structural formula, and molecular models (both ball-and-stick and space-filling variety) for water. Note that the chemical bond between two atoms can be represented with either a pair of dots or a line.

In Chapters 8 and 9, you will learn how to use the molecular formula to deduce the structural formula and the three-dimensional arrangement of a molecule. We will use all of these methods for representing molecules throughout the book, so you should be familiar with each method and with the information it provides.

Sample Problem 2.4 shows how to write a molecular formula from the corresponding molecular model.

$H_2O$

H:O:H

H—O—H

**Figure 2.13** Several ways to represent the water molecule.

---

### Sample Problem 2.4

Write the molecular formula of ethanol based on its ball-and-stick model, shown here.

Ethanol

**Strategy** Refer to the labels on the atoms (or see Table 1.1).

**Setup** There are *two* carbon atoms, *six* hydrogen atoms, and *one* oxygen atom, so the subscript on C will be 2 and the subscript on H will be 6, and there will be no subscript on O.

**Solution** $C_2H_6O$

---

**Think About It** Often the molecular formula for an organic compound such as ethanol is written so that the formula more closely resembles the actual arrangement of atoms in the molecule. Thus, the molecular formula for ethanol is commonly written as $C_2H_5OH$.

*(Continued)*

**Practice Problem A** Chloroform was used as an anesthetic for childbirth and surgery during the nineteenth century. Write the molecular formula for chloroform based on the molecular model shown below.

**⌐Practice Problem B** Write the molecular formula for acetone based on the molecular model shown below.

**Student Annotation:** Note that acetone contains a *double* bond. Multiple bonds between atoms will be discussed in Chapters 8 and 9.

Acetone                Chloroform

## Naming Molecular Compounds

When chemistry was a young science, it was possible for practitioners to memorize the names of the relatively small number of known compounds. Often a compound's name was derived from its physical appearance, properties, origin, or application—for example, milk of magnesia, laughing gas, formic acid (*formica* is the Latin word for ant, and *formic acid* is the compound responsible for the sting of an ant bite), and baking soda.

Today there are many millions of known compounds, and many more being made every year, so it would be impossible to memorize all their names. Fortunately, it is unnecessary, because over the years chemists have devised a convenient system for naming chemical substances. The rules are the same worldwide, facilitating communication among scientists and providing a useful way of labeling an overwhelming variety of substances. Mastering these rules now will benefit you tremendously as you progress through your chemistry course. You must be able to name a compound, given its chemical formula, and you must be able to write the chemical formula of a compound, given its name.

**Student Annotation:** The term *binary* can also refer to *ionic* compounds, which we discuss in Section 2.7.

We begin our discussion of chemical *nomenclature,* the naming of chemical compounds, with ***binary*** molecular compounds, substances that consist of just two different elements. Most are composed of two nonmetals (see Figure 2.10). To name such a compound, we first name the element that appears first in the formula. For HCl that would be hydrogen. We then name the second element, changing the ending of its name to *–ide*. For HCl, the second element is chlorine, so we would change chlorine to chloride. Thus, the systematic name of HCl is *hydrogen chloride*. Similarly, HI is hydrogen iodide (iod*ine* → iod*ide*) and SiC is silicon carbide (carb*on* → carb*ide*).

It is quite common for one pair of elements to form several different binary molecular compounds. In these cases, confusion in naming the compounds is avoided by the use of Greek prefixes to denote the number of atoms of each element present. Some of the Greek prefixes are listed in Table 2.2, and several compounds named using prefixes are listed in Table 2.3.

The prefix *mono–* is generally omitted for the first element. $SO_2$, for example, is named *sulfur dioxide,* not *monosulfur dioxide.* Thus, the absence of a prefix for the first element usually means there is only one atom of that element present in the molecule. In addition, for ease of pronunciation, we usually eliminate the last letter of a prefix that ends in "o" or "a" when naming an oxide. Thus, $N_2O_5$ is *dinitrogen pentoxide,* rather than *dinitrogen pentaoxide.*

Sample Problem 2.5 gives you some practice naming binary molecular compounds from their formulas.

| **TABLE 2.2** | Greek Prefixes | | |
|---|---|---|---|
| **Prefix** | **Meaning** | **Prefix** | **Meaning** |
| Mono– | 1 | Hexa– | 6 |
| Di– | 2 | Hepta– | 7 |
| Tri– | 3 | Octa– | 8 |
| Tetra– | 4 | Nona– | 9 |
| Penta– | 5 | Deca– | 10 |

| **TABLE 2.3** | Some Compounds Named Using Greek Prefixes | | |
|---|---|---|---|
| **Compound** | **Name** | **Compound** | **Name** |
| CO | Carbon monoxide | $SO_3$ | Sulfur trioxide |
| $CO_2$ | Carbon dioxide | $NO_2$ | Nitrogen dioxide |
| $SO_2$ | Sulfur dioxide | $N_2O_5$ | Dinitrogen pentoxide |

## Sample Problem 2.5

Name the following binary molecular compounds: (a) $NF_3$ and (b) $N_2O_4$.

**Strategy** Each compound will be named using the systematic nomenclature including, where necessary, appropriate Greek prefixes.

**Setup** With binary compounds, we start with the name of the element that appears *first* in the formula, and we change the ending of the *second* element's name to *–ide*. We use prefixes, where appropriate, to indicate the number of atoms of each element. In part (a) the molecule contains one nitrogen atom and three fluorine atoms. We will omit the prefix *mono–* for nitrogen because it is the first element listed in the formula, and we will use the prefix *tri–* to denote the number of fluorine atoms. In part (b) the molecule contains two nitrogen atoms and four oxygen atoms, so we will use the prefixes *di–* and *tetra–* in naming the compound. Recall that in naming an oxide, the last letter of a prefix that ends in "a" or "o" is omitted.

**Solution** (a) nitrogen trifluoride and (b) dinitrogen tetroxide

**Think About It** Make sure that the prefixes match the subscripts in the molecular formulas and that the word *oxide* is not preceded immediately by an "a" or an "o."

**Practice Problem A** Name the following binary molecular compounds: (a) $Cl_2O$ and (b) $SiCl_4$.

**Practice Problem B** Name the following binary molecular compounds: (a) $ClO_2$ and (b) $CBr_4$.

Writing the formula for a molecular compound, given its systematic name, is usually straightforward. For instance, the name *boron trichloride* indicates the presence of one boron atom (no prefix) and three chlorine atoms (tri–), so the corresponding molecular formula is $BCl_3$. Note once again that the order of the elements is the same in both the name and the formula.

Sample Problem 2.6 gives you some practice determining the formulas of binary molecular compounds from their names.

## Sample Problem 2.6

Write the chemical formulas for the following binary molecular compounds: (a) sulfur tetrafluoride and (b) tetraphosphorus decasulfide.

**Strategy** The formula for each compound will be deduced using the systematic nomenclature guidelines.

**Setup** In part (a) there is no prefix for sulfur, so there is only one sulfur atom in a molecule of the compound. Therefore, we will use no prefix for the S in the formula. The prefix *tetra–* means that there are four fluorine atoms. In part (b) the prefixes *tetra–* and *deca–* denote four and ten, respectively.

**Solution** (a) $SF_4$ and (b) $P_4S_{10}$

**Think About It** Double-check that the subscripts in the formulas match the prefixes in the compound names: (a) 4 = tetra and (b) 4 = tetra and 10 = deca.

**Practice Problem A** Give the molecular formula for each of the following compounds: (a) carbon disulfide and (b) dinitrogen trioxide.

**Practice Problem B** Give the molecular formula for each of the following compounds: (a) sulfur hexafluoride and (b) disulfur decafluoride.

The names of molecular compounds containing hydrogen do not usually conform to the systematic nomenclature guidelines. Traditionally, many of these compounds are called either by their common, nonsystematic names or by names that do not indicate explicitly the number of H atoms present:

| | | | |
|---|---|---|---|
| $B_2H_6$ | Diborane | $PH_3$ | Phosphine |
| $SiH_4$ | Silane | $H_2O$ | Water |
| $NH_3$ | Ammonia | $H_2S$ | Hydrogen sulfide |

**Student Annotation:** Binary compounds containing carbon and hydrogen are *organic* compounds and do not follow the same naming conventions as other molecular compounds. Organic compounds and their nomenclature are discussed in detail in Chapter 10.

| TABLE 2.4 | Some Simple Acids | |
|---|---|---|
| **Formula** | **Binary Compound Name** | **Acid Name** |
| HF | Hydrogen fluoride | Hydrofluoric acid |
| HCl | Hydrogen chloride | Hydrochloric acid |
| HBr | Hydrogen bromide | Hydrobromic acid |
| HI | Hydrogen iodide | Hydroiodic acid |
| HCN* | Hydrogen cyanide | Hydrocyanic acid |

\* Although HCN is not a *binary* compound, it is included in this table because it is similar chemically to HF, HCl, HBr, and HI.

**Student Annotation:** In Chapter 16 we will explore acids and bases in greater detail; and we will see that there are other ways to define the terms *acid* and *base*.

**Student Annotation:** An ion is a *charged* species.

**Student Annotation:** Some carbon-containing species such as $CN^-$ and $CO_3^{2-}$ are considered inorganic.

Even the order in which the elements are written in these hydrogen-containing compounds is irregular. In water and hydrogen sulfide, H is written first, whereas it is written last in the other compounds.

*Acids* make up another important class of molecular compounds. One definition of an ***acid*** is a substance that produces hydrogen ions ($H^+$) when dissolved in water. Several binary molecular compounds produce hydrogen ions when dissolved in water and are, therefore, acids. In these cases, two different names can be assigned to the same chemical formula. For example, HCl, *hydrogen chloride,* is a gaseous compound. When it is dissolved in water, however, we call it *hydrochloric acid.* The rules for naming simple acids of this type are as follows: remove the *–gen* ending from hydrogen (leaving *hydro–*), change the *–ide* ending on the second element to *–ic,* combine the two words, and add the word *acid.*

$$\text{hydro}gen \text{ chlor}ide + \text{ –}ic \text{ acid} \longrightarrow \text{hydrochloric acid}$$

Likewise, hydrogen fluoride (HF) becomes *hydrofluoric acid.* Table 2.4 lists these and other examples.

For a compound to produce hydrogen ions upon dissolving, it must contain at least one ***ionizable hydrogen atom.*** An ionizable hydrogen atom is one that separates from the molecule upon dissolving and becomes a hydrogen ion ($H^+$).

So far our discussion of nomenclature has focused on ***inorganic compounds,*** which are generally defined as compounds that do not contain carbon. Another important class of molecular substances is *organic* compounds, which have their own system of nomenclature. ***Organic compounds*** contain carbon and hydrogen, sometimes in combination with other elements such as oxygen, nitrogen, sulfur, and the halogens. The simplest organic compounds are those that contain only carbon and hydrogen and are known as ***hydrocarbons.*** Among hydrocarbons, the simplest examples are compounds known as ***alkanes.*** The name of an alkane depends on the number of carbon atoms in the molecule. Table 2.5 gives the molecular formulas, systematic names of some of the simplest alkanes, and ball-and-stick models.

Many organic compounds are derivatives of alkanes in which one of the H atoms has been replaced by a group of atoms known as a ***functional group.*** The functional group determines many of the chemical properties of a compound because it typically is where a chemical reaction *occurs.* Table 2.6 lists the names and provides space-filling models of several important functional groups.

Ethanol, for example, the alcohol in alcoholic beverages, is ethane ($C_2H_6$) with one of the hydrogen atoms replaced by an alcohol ($-OH$) group. Its name is derived from that of *ethane,* indicating that it contains *two* carbon atoms.

The molecular formula of ethanol can also be written $C_2H_6O$, but $C_2H_5OH$ conveys more information about the structure of the molecule. Organic compounds and several functional groups are discussed in greater detail in Chapter 10.

Ethanol

## Empirical Formulas

In addition to the ways we have learned so far, molecular substances can also be represented using ***empirical formulas.*** The word *empirical* means "from experience" or, in the context of chemical formulas, "from experiment." The empirical formula tells what elements are present in a molecule and in what whole-number ratio they are combined. For example, the molecular formula of hydrogen peroxide is $H_2O_2$, but its empirical formula is simply HO. Hydrazine, which has been used as a rocket fuel, has the molecular formula $N_2H_4$, so its empirical formula is $NH_2$. Although the ratio of nitrogen to hydrogen is 1:2 in both the molecular formula ($N_2H_4$) and the empirical formula

| TABLE 2.5 | Formulas, Names, and Models of Some Simple Alkanes | |
|---|---|---|
| **Formula** | **Name** | **Molecular Model** |
| $CH_4$ | Methane | |
| $C_2H_6$ | Ethane | |
| $C_3H_8$ | Propane | |
| $C_4H_{10}$ | Butane | |
| $C_5H_{12}$ | Pentane | |
| $C_6H_{14}$ | Hexane | |
| $C_7H_{16}$ | Heptane | |
| $C_8H_{18}$ | Octane | |
| $C_9H_{20}$ | Nonane | |
| $C_{10}H_{22}$ | Decane | |

($NH_2$), only the molecular formula tells us the actual number of N atoms (two) and H atoms (four) present in a hydrazine molecule.

In many cases, the empirical and molecular formulas are identical. In the case of water, for example, there is no combination of smaller whole numbers that can convey the ratio of two H atoms for every one O atom, so the empirical formula is the same as the molecular formula: $H_2O$. Table 2.7 lists the molecular and empirical formulas for several compounds.

Empirical formulas are the *simplest* chemical formulas; they are written by reducing the subscripts in molecular formulas to the smallest possible whole numbers (without altering the relative numbers of atoms). Molecular formulas are the *true* formulas of molecules. As we will see in Chapter 3, when chemists analyze an unknown compound, the first step is usually the determination of the compound's empirical formula.

Sample Problem 2.7 (p. 57) lets you practice determining empirical formulas from molecular formulas.

**TABLE 2.6**    Organic Functional Groups

| Name | Functional Group | Molecular Model |
|------|------------------|-----------------|
| Alcohol | $-OH$ | |
| Aldehyde | $-CHO$ | |
| Carboxylic acid | $-COOH$ | |
| Amine | $-NH_2$ | |

**TABLE 2.7**    Molecular and Empirical Formulas

| Compound | Molecular Formula | Model | Empirical Formula | Molecular Model |
|----------|-------------------|-------|-------------------|-----------------|
| Water | $H_2O$ | | $H_2O$ | |
| Hydrogen peroxide | $H_2O_2$ | | $HO$ | |
| Ethane | $C_2H_6$ | | $CH_3$ | |
| Propane | $C_3H_8$ | | $C_3H_8$ | |
| Acetylene | $C_2H_2$ | | $CH$ | |
| Benzene | $C_6H_6$ | | $CH$ | |

## Sample Problem 2.7

Write the empirical formulas for the following molecules: (a) glucose ($C_6H_{12}O_6$), a substance known as blood sugar; (b) adenine ($C_5H_5N_5$), also known as vitamin $B_4$; and (c) nitrous oxide ($N_2O$), a gas that is used as an anesthetic ("laughing gas") and as an aerosol propellant for whipped cream.

**Strategy** To write the empirical formula, the subscripts in the molecular formula must be reduced to the smallest possible whole numbers (without altering the relative numbers of atoms).

**Setup** The molecular formulas in parts (a) and (b) each contain subscripts that are divisible by common numbers. Therefore, we will be able to express the formulas with smaller whole numbers than those in the molecular formulas. In part (c), the molecule has only one O atom, so it is impossible to simplify this formula further.

**Solution** (a) Dividing each of the subscripts in the molecular formula for glucose by 6, we obtain the empirical formula, $CH_2O$. If we had divided the subscripts by 2 or 3, we would have obtained the formulas $C_3H_6O_3$ and $C_2H_4O_2$, respectively. Although the ratio of carbon to hydrogen to oxygen atoms in each of these formulas is correct (1:2:1), neither is the simplest formula because the subscripts are not in the smallest possible whole-number ratio.

(b) Dividing each subscript in the molecular formula of adenine by 5, we get the empirical formula, CHN.

(c) Because the subscripts in the formula for nitrous oxide are already the smallest possible whole numbers, its empirical formula is the same as its molecular formula, $N_2O$.

> **Think About It** Make sure that the *ratio* in each empirical formula is the same as that in the corresponding molecular formula and that the subscripts are the smallest possible whole numbers. In part (a), for example, the ratio of C:H:O in the molecular formula is 6:12:6, which is equal to 1:2:1, the ratio expressed in the empirical formula.

**Practice Problem A** Write empirical formulas for the following molecules: (a) caffeine ($C_8H_{10}N_4O_2$), a stimulant found in tea and coffee, (b) butane ($C_4H_{10}$), which is used in cigarette lighters, and (c) glycine ($C_2H_5NO_2$), an amino acid.

|  Caffeine  |  Butane  |  Glycine  |

**Practice Problem B** For which of the following molecular formulas is the formula shown in parentheses the correct empirical formula? (a) $C_{12}H_{22}O_{11}$ ($C_{12}H_{22}O_{11}$), (b) $C_8H_{12}O_4$ ($C_4H_6O_2$), (c) $H_2O_2$ ($H_2O$)?

## Checkpoint 2.6    Molecules and Molecular Compounds

**2.6.1** What is the correct systematic name of $PCl_5$?

a) Phosphorus chloride

b) Phosphorus pentachloride

c) Monophosphorus chloride

d) Pentachlorophosphorus

e) Pentaphosphorus chloride

**2.6.2** What is the name of the compound shown?

a) Methane

b) Carbon tetrahydrogen monoxide

c) Methanol

d) Methane monoxide

e) Tetrahydrogen carbon monoxide

**2.6.3** What is the correct formula for the compound carbon tetrachloride?

a) $C_4Cl_4$

b) $C_4Cl$

c) $CCl_4$

d) $CCl_2$

e) $CCl$

**2.6.4** What is the empirical formula of the compound shown?

a) $C_6H_6$

b) $C_4H_4$

c) $C_2H_2$

d) $CH_2$

e) $CH$

<h2>2.7    Ions and Ionic Compounds</h2>

The number of positively charged protons in the nucleus of an atom remains the same during ordinary chemical reactions, but negatively charged electrons may be lost or gained—resulting in the formation of *ions*. An **ion** is an atom (or a *group* of atoms) that has a net positive or negative charge. The ions that make up an ionic compound are held together by strong electrostatic forces known as *ionic bonds* [▶ Section 8.2].

### Atomic Ions

An **atomic ion** or **monatomic ion** is one that consists of just *one* atom with a positive or negative charge. The *loss* of one or more electrons from an atom yields a **cation,** an ion with a net *positive* charge. For example, a sodium atom (Na) can readily lose an electron to become a sodium cation, which is represented by $Na^+$:

| Na Atom | $Na^+$ Ion |
|---|---|
| 11 protons | 11 protons |
| 11 electrons | 10 electrons |

An **anion** is an ion whose net charge is *negative* due to an *increase* in the number of electrons. A chlorine atom (Cl), for instance, can gain an electron to become a chloride ion ($Cl^-$):

| Cl Atom | $Cl^-$ Ion |
|---|---|
| 17 protons | 17 protons |
| 17 electrons | 18 electrons |

Sodium chloride (NaCl), ordinary table salt, is called an **ionic compound** because it consists of *cations* ($Na^+$) and *anions* ($Cl^-$).

An atom can lose or gain more than one electron. Examples include $Mg^{2+}$, $Fe^{3+}$, $S^{2-}$, and $N^{3-}$. Figure 2.14 shows the charges of many more monatomic ions from across the periodic table. With very few exceptions, metals tend to form cations and nonmetals form anions. The charges on monatomic ions of elements in Groups 1A through 7A of the periodic table are fairly predictable. The cations that form from elements of groups 1A, 2A, and 3A have charges equal to their respective group numbers. Most of the anions that form from elements of groups 4A through 7A have charges equal to the corresponding group number minus 8. For example, the monatomic anion formed by oxygen (group 6A) has a charge of $6 - 8 = -2$. You should be able to determine the charges on ions of elements in groups 1A through 7A (for any element that forms only one common ion) using only a periodic table.

A monatomic cation is named simply by adding the word *ion* to the name of the element. Thus, the ion of potassium ($K^+$) is known as potassium ion. Similarly, the cations formed by the elements magnesium and aluminum ($Mg^{2+}$ and $Al^{3+}$) are called magnesium ion and aluminum ion, respectively. It is not necessary for the name to specify the charge on these ions because their charges are equal to their group numbers.

**Student Annotation:** Note that a multiple charge is denoted with the number followed by the sign; thus, 2+ not +2.

**Figure 2.14** Common monatomic ions arranged by their positions in the periodic table. Note that mercury(I), $Hg_2^{2+}$, is actually a *poly*atomic ion.

Certain metals, especially the *transition metals,* can form cations of more than one possible charge. Iron, for example, can form $Fe^{2+}$ and $Fe^{3+}$. An older nomenclature system that is still in limited use assigns the ending *–ous* to the cation with the *smaller* positive charge and the ending *–ic* to the cation with the *greater* positive charge:

$$Fe^{2+}: \text{ferrous ion}$$
$$Fe^{3+}: \text{ferric ion}$$

This method of naming ions has some distinct limitations. First, the *–ous* and *–ic* suffixes indicate the *relative* charges of the two cations involved, not the *actual* charges. Thus, $Fe^{3+}$ is the ferric ion, but $Cu^{2+}$ is the cupric ion. In addition, the *–ous* and *–ic* endings make it possible to name only two cations with different charges. Some metals, such as manganese (Mn), can form cations with three or more different charges.

Therefore, it has become increasingly common to designate different cations with Roman numerals, using the Stock[3] system. In this system, the Roman numeral I indicates a positive charge of one, II means a positive charge of two, and so on, as shown for manganese:

$$Mn^{2+}: \text{manganese(II) ion}$$
$$Mn^{3+}: \text{manganese(III) ion}$$
$$Mn^{4+}: \text{manganese(IV) ion}$$

These names are pronounced "manganese-two ion," "manganese-three ion," and "manganese-four ion," respectively. Using the Stock system, the ferrous and ferric ions are iron(II) and iron(III), respectively. To avoid confusion, and in keeping with modern practice, we will use the Stock system to name compounds in this textbook.

A monatomic anion is named by changing the ending of the element's name to *–ide*, and adding the word *ion*. Thus, the anion of chlorine ($Cl^-$), is called *chloride ion*. The anions of carbon, nitrogen, and oxygen ($C^{4-}$, $N^{3-}$, and $O^{2-}$) are called *carbide, nitride,* and *oxide,* respectively. Because there is only one possible charge for an ion formed from a nonmetal, it is unnecessary for the ion's name to specify its charge. Table 2.8 lists alphabetically a number of common monatomic ions.

## Polyatomic Ions

Ions that consist of a combination of two or more atoms are called **polyatomic ions.** Because these ions are commonly encountered in general chemistry, you must know the names, formulas, and charges of the polyatomic ions listed in Table 2.9. Although most of the common polyatomic ions are anions, a few are cations.

---

3. Alfred E. Stock (1876–1946). German chemist. Stock did most of his research in the synthesis and characterization of boron, beryllium, and silicon compounds. He was a pioneer in the study of mercury poisoning.

| TABLE 2.8 | Names and Formulas of Some Common Monatomic Ions |
| --- | --- |

| Name | Formula |
| --- | --- |
| **Cations** | |
| aluminum | $Al^{3+}$ |
| barium | $Ba^{2+}$ |
| cadmium | $Cd^{2+}$ |
| calcium | $Ca^{2+}$ |
| cesium | $Cs^+$ |
| chromium(III) | $Cr^{3+}$ |
| cobalt(II) | $Co^{2+}$ |
| copper(I) | $Cu^+$ |
| copper(II) | $Cu^{2+}$ |
| hydrogen | $H^+$ |
| iron(II) | $Fe^{2+}$ |
| iron(III) | $Fe^{3+}$ |
| lead(II) | $Pb^{2+}$ |
| lithium | $Li^+$ |
| magnesium | $Mg^{2+}$ |
| manganese(II) | $Mn^{2+}$ |
| mercury(II) | $Hg^{2+}$ |
| potassium | $K^+$ |
| silver | $Ag^+$ |
| sodium | $Na^+$ |
| strontium | $Sr^{2+}$ |
| tin(II) | $Sn^{2+}$ |
| zinc | $Zn^{2+}$ |
| **Anions** | |
| bromide | $Br^-$ |
| chloride | $Cl^-$ |
| fluoride | $F^-$ |
| hydride | $H^-$ |
| iodide | $I^-$ |
| nitride | $N^{3-}$ |
| oxide | $O^{2-}$ |
| sulfide | $S^{2-}$ |

**Student Annotation:** Like the atoms in a molecule, the atoms that make up a polyatomic ion are held together by covalent chemical bonds [▶▶ Section 8.3].

| TABLE 2.9 | Common Polyatomic Ions |
|---|---|
| **Name** | **Formula/Charge** |
| **Cations** | |
| acetate | $C_2H_3O_2^-$ |
| ammonium | $NH_4^+$ |
| hydronium | $H_3O^+$ |
| mercury(I) | $Hg_2^{2+}$ |
| **Anions** | |
| azide | $N_3^-$ |
| carbonate | $CO_3^{2-}$ |
| chlorate | $ClO_3^-$ |
| chlorite | $ClO_2^-$ |
| chromate | $CrO_4^{2-}$ |
| cyanide | $CN^-$ |
| dichromate | $Cr_2O_7^{2-}$ |
| dihydrogen phosphate | $H_2PO_4^-$ |
| hydrogen carbonate or bicarbonate | $HCO_3^-$ |
| hydrogen phosphate | $HPO_4^{2-}$ |
| hydrogen sulfate or bisulfate | $HSO_4^-$ |
| hydroxide | $OH^-$ |
| hypochlorite | $ClO^-$ |
| nitrate | $NO_3^-$ |
| nitrite | $NO_2^-$ |
| oxalate | $C_2O_4^{2-}$ |
| perchlorate | $ClO_4^-$ |
| permanganate | $MnO_4^-$ |
| peroxide | $O_2^{2-}$ |
| phosphate | $PO_4^{3-}$ |
| phosphite | $PO_3^{3-}$ |
| sulfate | $SO_4^{2-}$ |
| sulfite | $SO_3^{2-}$ |
| thiocyanate | $SCN^-$ |

**Student Annotation:** Some oxoanions occur in series of ions that contain the same central atom and have the same charge, but contain different numbers of oxygen atoms. The FAQ box on page 62 gives more information about how ions in these series are named.

| | |
|---|---|
| hypochlorite | $ClO^-$ |
| chlorite | $ClO_2^-$ |
| chlorate | $ClO_3^-$ |
| perchlorate | $ClO_4^-$ |
| nitrate | $NO_3^-$ |
| nitrite | $NO_2^-$ |
| phosphate | $PO_4^{3-}$ |
| phosphite | $PO_3^{3-}$ |
| sulfate | $SO_4^{2-}$ |
| sulfite | $SO_3^{2-}$ |

## Formulas of Ionic Compounds

The formulas of ionic compounds are generally empirical formulas because ionic compounds consist of vast arrays of interspersed cations and anions, not discrete molecular units. For example, solid sodium chloride (NaCl) consists of equal numbers of $Na^+$ and $Cl^-$ ions arranged in a three-dimensional network of alternating cations and anions (Figure 2.15). There is a 1:1 ratio of cations to anions, so the compound is electrically neutral. As a result, the empirical formula of sodium chloride is NaCl. As you can see in Figure 2.15, no $Na^+$ ion in NaCl is associated with just one particular $Cl^-$ ion. In fact, each $Na^+$ ion is surrounded by six $Cl^-$ ions and vice versa. In other ionic compounds, the actual structure may be different, but the arrangement of cations and anions is such that the compounds are all electrically neutral. The charges on the cation and anion are not shown in the formula for an ionic compound.

For ionic compounds to be electrically neutral, the sum of the charges on the cation and anion in each formula unit must be zero. If the charges on the cations and anions are numerically

**Student Annotation:** A three-dimensional array of cations and anions is called a *lattice*.

Multimedia

Electron transfer between sodium and chlorine.

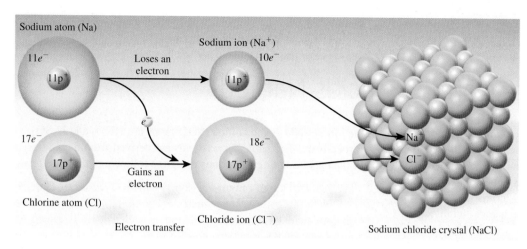

Electron transfer    Chloride ion (Cl⁻)    Sodium chloride crystal (NaCl)

**Figure 2.15** An electron is transferred from the sodium atom to the chlorine atom, giving a sodium ion and a chloride ion. The oppositely charged ions are attracted to each other and form a solid lattice.

different, you can apply the following guideline to make the formula electrically neutral (and thus obtain the empirical formula): Write a subscript for the cation that is numerically equal to the charge on the anion and a subscript for the anion that is numerically equal to the charge on the cation. If the charges are numerically equal, then no subscripts are necessary.

Let us consider some examples.

**Potassium Bromide**  The potassium ion ($K^+$) and the bromide ion ($Br^-$) combine to form the ionic compound *potassium bromide*. The sum of the charges is $1 + (-1) = 0$, so no subscripts are necessary. The formula is KBr.

**Zinc Iodide**  The zinc ion ($Zn^{2+}$) and the iodide ion ($I^-$) combine to form *zinc iodide*. The sum of the charges of one $Zn^{2+}$ ion and one $I^-$ ion is $+2 + (-1) = +1$. To make the charges add up to zero, we multiply the $-1$ charge of the anion by 2 and add the subscript "2" to the symbol for iodine. Thus, the formula for zinc iodide is $ZnI_2$.

**Ammonium Chloride**  The cation is $NH_4^+$ and the anion is $Cl^-$. The sum of the charges is $1 + (-1) = 0$, so the ions combine in a 1:1 ratio and the resulting formula is $NH_4Cl$.

**Aluminum Oxide**  The cation is $Al^{3+}$ and the anion is $O^{2-}$. The following diagram can be used to determine the subscripts for this compound:

$$Al^{3+} \qquad O^{2-}$$
$$Al_2O_3$$

The sum of the charges for aluminum oxide is $2(+3) + 3(-2) = 0$. Thus, the formula is $Al_2O_3$.

**Calcium Phosphate**  The cation is $Ca^{2+}$ and the anion is $PO_4^{3-}$. The following diagram can be used to determine the subscripts:

$$Ca^{2+} \qquad PO_4^{3-}$$
$$Ca_3(PO_4)_2$$

The sum of the charges is $3(+2) + 2(-3) = 0$. Thus, the formula for calcium phosphate is $Ca_3(PO_4)_2$. When we add a subscript to a polyatomic ion, we must first put parentheses around the ion's formula to indicate that the subscript applies to *all* the atoms in the polyatomic ion.

## Naming Ionic Compounds

An ionic compound is named using the name of the cation followed by the name of the anion, eliminating the word *ion* from each. Several examples were given earlier in the Formulas of Ionic Compounds section. Other examples are sodium cyanide (NaCN), potassium permanganate

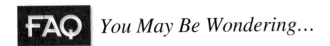

*You May Be Wondering...*

# How Are Oxoanions and Oxoacids Named?

**Oxoanions** are polyatomic anions that contain one or more oxygen atoms and one atom (the "central atom") of another element. Examples include the chlorate ($ClO_3^-$), nitrate ($NO_3^-$), and sulfate ($SO_4^{2-}$) ions. Often, two or more oxoanions have the same central atom but different numbers of O atoms (e.g., $NO_3^-$ and $NO_2^-$). Starting with the oxoanions whose names end in *–ate*, we can name these ions as follows:

1. The ion with one *more* O atom than the *–ate* ion is called the *per . . . ate* ion. Thus, $ClO_3^-$ is the chlorate ion, so $ClO_4^-$ is the *perchlorate ion.*
2. The ion with one *less* O atom than the *–ate* anion is called the *–ite* ion. Thus, $ClO_2^-$ is the *chlorite ion.*
3. The ion with *two* fewer O atoms than the *–ate* ion is called the *hypo . . . ite* ion. Thus, $ClO^-$ is the *hypochlorite ion.*

At a minimum, you must commit to memory the formulas and charges of the oxoanions whose names end in *–ate* so that you can apply these guidelines when necessary.

In addition to the simple acids discussed in Section 2.6, there is another important class of acids known as **oxoacids,** which ionize to produce hydrogen ions and the corresponding oxoanions. The formula of an oxoacid can be determined by adding enough $H^+$ ions to the corresponding oxoanion to yield a formula with no net charge. For example, the formulas of oxoacids based on

the nitrate ($NO_3^-$) and sulfate ($SO_4^{2-}$) ions are $HNO_3$ and $H_2SO_4$, respectively. The names of oxoacids are derived from the names of the corresponding oxoanions using the following guidelines:

1. An acid based on an *–ate* ion is called . . . *ic* acid. Thus, $HClO_3$ is called *chloric acid.*
2. An acid based on an *–ite* ion is called . . . *ous* acid. Thus, $HClO_2$ is called *chlorous acid.*
3. Prefixes in oxoanion names are retained in the names of the corresponding oxoacids. Thus, $HClO_4$ and $HClO$ are called *perchloric acid* and *hypochlorous acid,* respectively.

Many oxoacids, such as $H_2SO_4$ and $H_3PO_4$, are **polyprotic**—meaning that they have more than one ionizable hydrogen atom. In these cases, the names of anions in which one or more (but not all) of the hydrogen ions have been removed must indicate the number of H ions that remain, as shown for the anions derived from phosphoric acid:

| | |
|---|---|
| $H_3PO_4$ | phosphoric acid |
| $H_2PO_4^-$ | dihydrogen phosphate ion |
| $HPO_4^{2-}$ | hydrogen phosphate ion |
| $PO_4^{3-}$ | phosphate ion |

Sample Problems 2.8 and 2.9 test your ability to name and identify oxoanions and oxoacids.

($KMnO_4$), and ammonium sulfate [$(NH_4)_2SO_4$]. Unlike the naming of molecular compounds, no Greek prefixes are used. For example, $Li_2CO_3$ is lithium carbonate, not dilithium carbonate, even though there are two lithium ions for every carbonate ion. Prefixes are unnecessary because the ions have known charges. Lithium ion always has a charge of +1, and carbonate ion always has a charge of −2. The only ratio in which they can combine to form a neutral compound is two $Li^+$ ions for every one $CO_3^{2-}$ ion. Therefore, the name *lithium carbonate* is sufficient to convey the compound's empirical formula.

In cases where a metal cation may have more than one possible charge, recall that the charge is indicated in the name of the ion with a Roman numeral in parentheses. Thus, the compounds $FeCl_2$ and $FeCl_3$ are named *iron(II) chloride* and *iron(III) chloride*, respectively. (These are pronounced "iron-two chloride" and "iron-three chloride.") Figure 2.16 summarizes the steps for naming molecular and ionic compounds.

Sample Problems 2.10 and 2.11 illustrate how to name ionic compounds and write formulas for ionic compounds based on the information given in Figure 2.16 and Tables 2.8 and 2.9.

**Student Annotation:** In general, we determine whether a compound is molecular or ionic by asking these questions: Does the compound consist of only nonmetals? Then it is probably molecular. Does the compound contain either a metal cation or the $NH_4^+$ ion? Then it is probably ionic. (Note that it is possible for an ionic compound to contain only nonmetals if it contains the $NH_4^+$ ion.)

## Sample Problem 2.8

Name the following species: (a) $BrO_4^-$, (b) $HCO_3^-$, and (c) $H_2CO_3$.

**Strategy** Each species is either an oxoanion or an oxoacid. Identify the "reference oxoanion" (the one with the *–ate* ending) for each, and apply the rules to determine appropriate names.

**Setup** (a) Chlorine, bromine, and iodine (members of group 7A) all form analogous series of oxoanions with one to four oxygen atoms. Thus, the reference oxoanion is bromate ($BrO_3^-$), which is analogous to chlorate ($ClO_3^-$). In parts (b) and (c), $HCO_3^-$ and $H_2CO_3$ have one and two more hydrogens, respectively, than the carbonate ion ($CO_3^{2-}$).

**Solution** (a) $BrO_4^-$ has one more O atom than the bromate ion ($BrO_3^-$), so $BrO_4^-$ is the *perbromate* ion. (b) $CO_3^{2-}$ is the carbonate ion. Because $HCO_3^-$ has one ionizable hydrogen atom, it is called the *hydrogen carbonate ion.* (c) With two ionizable hydrogen atoms and no charge on the compound, $H_2CO_3$ is *carbonic acid.*

**Practice Problem A** Name the following species: (a) HBrO, (b) $HSO_4^-$, and (c) $H_2C_2O_4$.

**Practice Problem B** Name the following species: (a) $HIO_3$, (b) $HCrO_4^-$, and (c) $HC_2O_4^-$.

**Think About It** Remembering all these names and formulas is greatly facilitated by memorizing the common ions that end in *–ate*.

| | |
|---|---|
| chlorate | $ClO_3^-$ |
| iodate | $IO_3^-$ |
| bromate | $BrO_3^-$ |
| sulfate | $SO_4^{2-}$ |
| phosphate | $PO_4^{3-}$ |
| nitrate | $NO_3^-$ |
| carbonate | $CO_3^{2-}$ |
| oxalate | $C_2O_4^{2-}$ |
| chromate | $CrO_4^{2-}$ |
| permanganate | $MnO_4^-$ |

## Sample Problem 2.9

Determine the formula of sulfurous acid.

**Strategy** The *–ous* ending in the name of an acid indicates that the acid is derived from an oxoanion ending in *–ite.* Determine the formula and charge of the oxoanion, and add enough hydrogens to make a neutral formula.

**Setup** The sulfite ion is $SO_3^{2-}$.

**Solution** The formula of sulfurous acid is $H_2SO_3$.

**Practice Problem A** Determine the formula of perbromic acid. (Refer to the information in Sample Problem 2.8.)

**Practice Problem B** Determine the formula of chromic acid.

## Sample Problem 2.10

Name the following ionic compounds: (a) MgO, (b) $Al(OH)_3$, and (c) $Fe_2(SO_4)_3$.

**Strategy** Begin by identifying the cation and the anion in each compound, and then combine the names for each, eliminating the word *ion.*

**Setup** MgO contains $Mg^{2+}$ and $O^{2-}$, the magnesium ion and the oxide ion; $Al(OH)_3$ contains $Al^{3+}$ and $OH^-$, the aluminum ion and the hydroxide ion; and $Fe_2(SO_4)_3$ contains $Fe^{3+}$ and $SO_4^{2-}$, the iron(III) ion and the sulfate ion. We know that the iron in $Fe_2(SO_4)_3$ is iron(III), $Fe^{3+}$, because it is combined with the sulfate ion in a 2:3 ratio.

**Solution** (a) Combining the cation and anion names, and eliminating the word *ion* from each of the individual ions' names, we get *magnesium oxide* as the name of MgO; (b) $Al(OH)_3$ is *aluminum hydroxide;* and (c) $Fe_2(SO_4)_3$ is *iron(III) sulfate.*

**Think About It** Be careful not to confuse the subscript in a formula with the charge on the metal ion. In part (c), for example, the subscript on Fe is 2, but this is an iron(III) compound.

**Practice Problem A** Name the following ionic compounds: (a) $Na_2SO_4$, (b) $Cu(NO_3)_2$, (c) $Fe_2(CO_3)_3$.

**Practice Problem B** Name the following ionic compounds: (a) $K_2Cr_2O_7$, (b) $Li_2C_2O_4$, (c) $CuNO_3$.

**Figure 2.16**   Steps for naming molecular and ionic compounds.

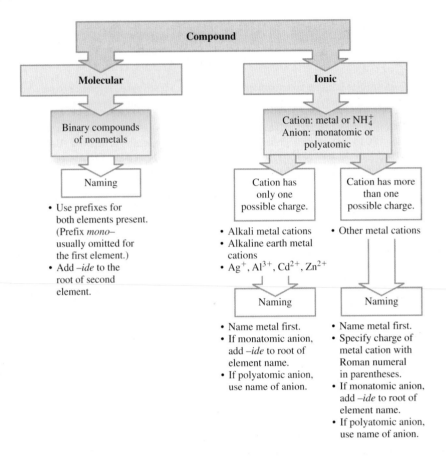

### Sample Problem 2.11

Deduce the formulas of the following ionic compounds: (a) mercury(I) chloride, (b) lead(II) chromate, and (c) potassium hydrogen phosphate.

**Strategy**  Identify the ions in each compound, and determine their ratios of combination using the charges on the cation and anion in each.

**Setup**  (a) Mercury(I) chloride is a combination of $Hg_2^{2+}$ and $Cl^-$. [Mercury(I) is one of the few cations listed in Table 2.9.] In order to produce a neutral compound, these two ions must combine in a 1:2 ratio. (b) Lead(II) chromate is a combination of $Pb^{2+}$ and $CrO_4^{2-}$. These ions combine in a 1:1 ratio. (c) Potassium hydrogen phosphate is a combination of $K^+$ and $HPO_4^{2-}$. These ions combine in a 2:1 ratio.

**Solution**  The formulas are (a) $Hg_2Cl_2$, (b) $PbCrO_4$, and (c) $K_2HPO_4$.

**Think About It** Make sure that the charges sum to zero in each compound formula. In part (a), for example, $Hg_2^{2+} + 2Cl^-$ = $(2+) + 2(-1) = 0$; in part (b), $(+2) + (-2) = 0$; and in part (c), $2(+1) + (-2) = 0$.

**Practice Problem A** Deduce the formulas of the following ionic compounds: (a) lead(II) chloride, (b) magnesium carbonate, and (c) ammonium phosphate.

**Practice Problem B** Deduce the formulas of the following ionic compounds: (a) iron(III) sulfide, (b) mercury(II) nitrate, and (c) potassium sulfite.

## Hydrates

*Hydrates* are compounds that have a specific number of water molecules within their solid structure. In its normal state, for example, each unit of copper(II) sulfate has five water molecules associated with it. The systematic name for this compound is copper(II) sulfate pentahydrate, and its formula is written as $CuSO_4 \cdot 5H_2O$. The water molecules can be driven off by heating. When this

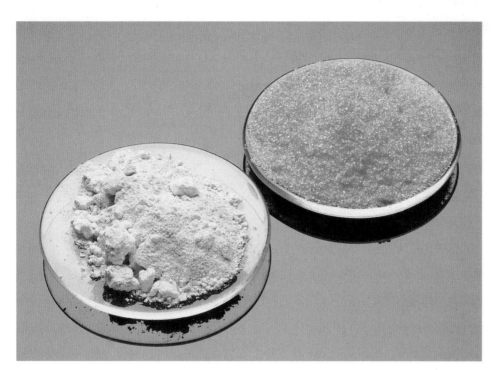

occurs, the resulting compound is $CuSO_4$, which is sometimes called anhydrous copper(II) sulfate; *anhydrous* means that the compound no longer has water molecules associated with it. Hydrates and the corresponding anhydrous compounds often have distinctly different physical and chemical properties (Figure 2.17).

Some other hydrates are

| | |
|---|---|
| $BaCl_2 \cdot 2H_2O$ | barium chloride dihydrate |
| $LiCl \cdot H_2O$ | lithium chloride monohydrate |
| $MgSO_4 \cdot 7H_2O$ | magnesium sulfate heptahydrate |
| $Sr(NO_3)_2 \cdot 4H_2O$ | strontium nitrate tetrahydrate |

## Familiar Inorganic Compounds

Some compounds are better known by their common names than by their systematic chemical names. Familiar examples are listed in Table 2.10.

| TABLE 2.10 | Common and Systematic Names of Some Familiar Inorganic Compounds | |
|---|---|---|
| **Formula** | **Common Name** | **Systematic Name** |
| $H_2O$ | Water | Dihydrogen monoxide |
| $NH_3$ | Ammonia | Trihydrogen nitride |
| $CO_2$ | Dry ice | Solid carbon dioxide |
| $NaCl$ | Salt | Sodium chloride |
| $N_2O$ | Nitrous oxide, laughing gas | Dinitrogen monoxide |
| $CaCO_3$ | Marble, chalk, limestone | Calcium carbonate |
| $NaHCO_3$ | Baking soda | Sodium hydrogen carbonate |
| $MgSO_4 \cdot 7H_2O$ | Epsom salt | Magnesium sulfate heptahydrate |
| $Mg(OH)_2$ | Milk of magnesia | Magnesium hydroxide |

**Checkpoint 2.7    Ions and Ionic Compounds**

**2.7.1**  What is the correct name of the compound $PbSO_4$?

a) Lead sulfate

b) Lead(I) sulfate

c) Lead(II) sulfate

d) Monolead sulfate

e) Lead monosulfate

**2.7.2**  What is the correct formula for the compound iron(III) carbonate?

a) $FeCO_3$

b) $Fe_3CO_3$

c) $Fe_2CO_3$

d) $Fe_2(CO_3)_3$

e) $Fe_3(CO_3)_2$

**2.7.3**  Which of the following is the correct formula for nitrous acid?

a) HNO

b) $HN_2O$

c) $N_2O$

d) $HNO_2$

e) $HNO_3$

**2.7.4**  What is the formula of nickel(II) nitrate hexahydrate?

a) $NiNO_3 \cdot 6H_2O$

b) $Ni_2NO_3 \cdot 6H_2O$

c) $Ni(NO_3)_2 \cdot 6H_2O$

d) $NiNO_3 \cdot 12H_2O$

e) $Ni(NO_3)_2 \cdot 12H_2O$

**2.7.5**  What is the correct formula for sodium nitride?

a) NaN

b) $NaN_3$

c) $Na_3N$

d) $NaNO_3$

e) $NaNO_2$

**2.7.6**  What is the correct name of the compound $Hg_2CrO_4$?

a) Mercury(I) chromate

b) Mercury(II) chromate

c) Mercury dichromate

d) Dimercury chromate

e) Monomercury chromate

# Applying What You've Learned

Although iron is an essential element, it is also a potentially toxic substance. Hemochromatosis is one of the most common hereditary disorders, causing "iron overload" or the storage of excess iron in the tissues and organs. Individuals with hemochromatosis often must undergo periodic phlebotomy (removal of blood) in order to remove excess stored iron, which would otherwise cause irreversible damage to internal organs including the liver and kidneys. Those who have a tendency to store too much iron are advised to avoid combining iron-rich foods with substances that enhance iron absorption, such as ascorbic acid (vitamin C).

Ascorbic acid

Because of iron's toxicity, iron supplements are potentially dangerous, especially to children. In fact, iron poisoning is the most common toxicological emergency in young children—due in part to the resemblance many iron supplements bear to candy. Most vitamins that contain iron are sold with childproof caps to help prevent accidental overdose. The Food and Drug Administration (FDA) requires supplements containing more than 30 mg of iron per dose to be sold in single-dose blister packs to make it more difficult for a child to consume a dangerous amount.

## Problems:

a) Iron has four naturally occurring isotopes: $^{54}$Fe (53.9396 amu), $^{56}$Fe (55.9349 amu), $^{57}$Fe (56.9354 amu), and $^{58}$Fe (57.9333 amu). For each isotope, determine the number of neutrons in the nucleus. [◀◀ Sample Problem 2.2]

b) Calculate the average atomic mass of iron given that the natural abundances of the four isotopes are 5.845, 91.754, 2.119, and 0.282 percent, respectively. [◀◀ Sample Problem 2.3]

c) Write the molecular formula for ascorbic acid (see the ball-and-stick model). [◀◀ Sample Problem 2.4]

d) Determine the empirical formula of ascorbic acid. [◀◀ Sample Problem 2.7]

e) Write the formula for ferrous sulfate [iron(II) sulfate]. [◀◀ Sample Problem 2.11]

# CHAPTER SUMMARY

## Section 2.1

- Dalton's atomic theory states that all matter is made up of tiny indivisible, immutable particles called *atoms*. Compounds form, moreover, when atoms of different elements combine in fixed ratios. According to the *law of definite proportions,* any sample of a given compound will always contain the same elements in the same mass ratio.

- The *law of multiple proportions* states that if two elements can form more than one compound with one another, the mass ratio of one will be related to the mass ratio of the other by a small whole number.

- The *law of conservation of mass* states that matter can be neither created nor destroyed.

## Section 2.2

- On the basis of Dalton's atomic theory, the *atom* is the basic unit of an element. Studies with *radiation* indicated that atoms contained subatomic particles, one of which was the *electron.*

- Experiments with *radioactivity* have shown that some atoms give off different types of radiation, called *alpha (α) rays, beta (β) rays,* and *gamma (γ) rays.* Alpha rays are composed of *α particles,* which are actually helium nuclei. Beta rays are composed of *β particles,* which are actually electrons. Gamma rays are high-energy radiation.

- Most of the mass of an atom resides in a tiny, dense region known as the *nucleus.* The nucleus contains positively charged particles called *protons* and electrically neutral particles called *neutrons.* The charge on a proton is equal in magnitude but opposite in sign to the charge on an electron. The electrons occupy the relatively large volume around the nucleus. A neutron has a slightly greater mass than a proton, but each is almost 2000 times as massive as an electron.

## Section 2.3

- The *atomic number (Z)* is the number of protons in the nucleus of an atom. The atomic number determines the identity of the atom. The *mass number (A)* is the sum of the protons and neutrons in the nucleus.

- Protons and neutrons are referred to collectively as *nucleons.*

- Atoms with the same atomic number but different mass numbers are called *isotopes.*

## Section 2.4

- The *periodic table* arranges the elements in rows (*periods*) and columns (*groups* or *families*). Elements in the same group exhibit similar properties.

- All elements fall into one of three categories: *metal, nonmetal,* or *metalloid.*

- Some of the groups have special names including *alkali metals* (Group 1A, except hydrogen), *alkaline earth metals* (Group 2A), *chalcogens* (Group 6A), *halogens* (Group 7A), *noble gases* (Group 8A), and *transition elements* or *transition metals* (Group 1B and Groups 3B–8B).

## Section 2.5

- *Atomic mass* is the mass of an atom in atomic mass units. One *atomic mass unit (amu),* is exactly one-twelfth the mass of a carbon-12 atom.

- The periodic table contains the *average* atomic mass (sometimes called the *atomic weight*) of each element.

## Section 2.6

- A *molecule* is an electrically neutral group of two or more atoms. Molecules consisting of just two atoms are called *diatomic.* Diatomic molecules may be *homonuclear* (just one kind of atom) or *heteronuclear* (two kinds of atoms). In general, molecules containing more than two atoms are called *polyatomic.*

- A *chemical formula* denotes the composition of a substance. A *molecular formula* specifies the exact numbers of atoms in a molecule of a compound. A *structural formula* shows the arrangement of atoms in a substance.

- An *allotrope* is one of two or more different forms of an element.

- Molecular compounds are named according to a set of rules, including the use of Greek prefixes to specify the number of each kind of atom in the molecule.

- *Binary* compounds are those that consist of *two* elements. An *acid* is a substance that generates hydrogen ions when it dissolves in water. An *ionizable hydrogen atom* is one that can be removed in water to become a hydrogen ion ($H^+$).

- *Inorganic compounds* are generally those that do not contain carbon. *Organic compounds* contain carbon and hydrogen, sometimes in combination with other elements. *Hydrocarbons* contain only carbon and hydrogen. The simplest hydrocarbons are the *alkanes.* A *functional group* is a group of atoms that determines the chemical properties of an organic compound.

- *Empirical formulas* express, in the smallest possible whole numbers, the ratio of the combination of atoms of the elements in a compound. The empirical and molecular formulas of a compound may or may not be identical.

## Section 2.7

- An *ion* is an atom or group of atoms with a net charge. An *atomic ion* or a *monatomic ion* consists of just one atom.

- An ion with a net positive charge is a *cation.* An ion with a net negative charge is an *anion.* An *ionic compound* is one that consists of cations and anions in an electrically neutral combination. A three-dimensional array of alternating cations and anions is called a *lattice.*

- Ionic compounds are named using rules similar to those for molecular compounds. In general, prefixes are *not* used to denote the number of ions in the names of ionic compounds.

- *Polyatomic ions* are those that contain more than one atom chemically bonded together. *Oxoanions* are polyatomic ions that contain one or more oxygen atoms.

- *Oxoacids* are acids based on oxoanions. Acids with more than one ionizable hydrogen atom are called *polyprotic.*

- *Hydrates* are compounds whose formulas include a specific number of water molecules.

# KEY WORDS

| | | | |
|---|---|---|---|
| Acid, 54 | Cation, 58 | Ionic compound, 58 | Nucleus, 42 |
| Alkali metal, 46 | Chalcogens, 47 | Ionizable hydrogen atom, 54 | Organic compounds, 54 |
| Alkaline earth metal, 47 | Chemical formula, 51 | Isotope, 44 | Oxoacid, 62 |
| Alkane, 54 | Diatomic molecule, 51 | Lattice, 60 | Oxoanion, 62 |
| Allotrope, 51 | Electron, 39 | Law of conservation of mass, 37 | Period, 46 |
| $\alpha$ particle, 41 | Empirical formula, 54 | Law of definite proportions, 36 | Periodic table, 46 |
| Alpha ($\alpha$) ray, 41 | Family, 46 | Law of multiple proportions, 37 | Polyatomic ion, 59 |
| Anion, 58 | Functional group, 54 | Mass number ($A$), 44 | Polyatomic molecule, 51 |
| Atom, 39 | Gamma ($\gamma$) rays, 41 | Metal, 46 | Polyprotic acid, 62 |
| Atomic ion, 58 | Group, 46 | Metalloid, 46 | Proton, 43 |
| Atomic mass, 49 | Halogens, 47 | Molecular formula, 51 | Radiation, 39 |
| Atomic mass unit (amu), 49 | Heteronuclear, 51 | Molecule, 50 | Radioactivity, 41 |
| Atomic number ($Z$), 44 | Homonuclear, 51 | Monatomic ion, 58 | Structural formula, 51 |
| Atomic weight, 50 | Hydrate, 64 | Neutron, 43 | Transition elements, 47 |
| $\beta$ particle, 41 | Hydrocarbon, 54 | Noble gases, 47 | Transition metals, 47 |
| Beta ($\beta$) ray, 41 | Inorganic compounds, 54 | Nonmetal, 46 | |
| Binary, 52 | Ion, 58 | Nucleons, 44 | |

# QUESTIONS AND PROBLEMS

## Section 2.1: The Atomic Theory

### Review Questions

**2.1**   What are the hypotheses on which Dalton's atomic theory is based?

**2.2**   State the laws of definite proportions and multiple proportions. Illustrate each with an example.

**2.3**   The elements nitrogen and oxygen can form a variety of different compounds. Two such compounds, NO and $N_2O_4$, were decomposed into their constituent elements. One produced 0.8756 g N for every gram of O; the other produced 0.4378 g N for every gram of O. Show that these results are consistent with the law of multiple proportions.

**2.4**   Two different compounds, each containing only phosphorus and chlorine, were decomposed into their constituent elements. One produced 0.2912 g P for every gram of Cl; the other produced 0.1747 g P for every gram of Cl. Show that these results are consistent with the law of multiple proportions.

**2.5**   Sulfur reacts with fluorine to produce three different compounds. The mass ratio of fluorine to sulfur for each compound is given in the following table:

| Compound | mass F : mass S |
|---|---|
| $S_2F_{10}$ | 2.962 |
| $SF_4$ | 2.370 |
| $SF_6$ | 3.555 |

Show that these data are consistent with the law of multiple proportions.

**2.6**   Both FeO and $Fe_2O_3$ contain only iron and oxygen. The mass ratio of oxygen to iron for each compound is given in the following table:

| Compound | mass O : mass Fe |
|---|---|
| FeO | 0.2865 |
| $Fe_2O_3$ | 0.4297 |

Show that these data are consistent with the law of multiple proportions.

### Problems

**2.7**   For the two compounds pictured, evaluate the following ratio:

$$\frac{\text{g blue} : 1.00\text{g red (right)}}{\text{g blue} : 1.00 \text{ g red (left)}}$$

**2.8**   For the two compounds pictured, evaluate the following ratio:

$$\frac{\text{g green} : 1.00 \text{ g yellow (right)}}{\text{g green} : 1.00 \text{ g yellow (left)}}$$

## Section 2.2: The Structure of the Atom

### Review Questions

**2.9**   Define the following terms: (a) $\alpha$ particle, (b) $\beta$ particle, (c) $\gamma$ ray, (d) X ray.

**2.10**   Name the types of radiation known to be emitted by radioactive elements.

**2.11**   Compare the properties of the following: $\alpha$ particles, cathode rays, protons, neutrons, and electrons.

**2.12**   Describe the contributions of the following scientists to our knowledge of atomic structure: J. J. Thomson, R. A. Millikan, Ernest Rutherford, and James Chadwick.

**2.13**   Describe the experimental basis for believing that the nucleus occupies a very small fraction of the volume of the atom.

## Problems

**2.14**   The diameter of a neutral helium atom is about $1 \times 10^2$ pm. Suppose that we could line up helium atoms side by side in contact with one another. Approximately how many atoms would it take to make the distance 1 cm from end to end?

**2.15**   Roughly speaking, the radius of an atom is about 10,000 times greater than that of its nucleus. If an atom were magnified so that the radius of its nucleus became 2.0 cm, about the size of a marble, what would be the radius of the atom in miles (1 mi = 1609 m)?

## Section 2.3: Atomic Number, Mass Number, and Isotopes

### Review Questions

**2.16**   Use the helium-4 isotope to define atomic number and mass number. Why does knowledge of the atomic number enable us to deduce the number of electrons present in an atom?

**2.17**   Why do all atoms of an element have the same atomic number, although they may have different mass numbers?

**2.18**   What do we call atoms of the same elements with different mass numbers?

**2.19**   Explain the meaning of each term in the symbol $^{A}_{Z}X$.

### Problems

**2.20**   What is the mass number of an iron atom that has 28 neutrons?

**2.21**   Calculate the number of neutrons of $^{239}$Pu.

**2.22**   For each of the following species, determine the number of protons and the number of neutrons in the nucleus: $^{3}_{2}$He, $^{4}_{2}$He, $^{24}_{12}$Mg, $^{25}_{12}$Mg, $^{48}_{22}$Ti, $^{79}_{35}$Br, $^{195}_{78}$Pt.

**2.23**   Indicate the number of protons, neutrons, and electrons in each of the following species: $^{15}_{7}$N, $^{33}_{16}$S, $^{63}_{29}$Cu, $^{84}_{38}$Sr, $^{130}_{56}$Ba, $^{186}_{74}$W, $^{202}_{80}$Hg.

**2.24**   Write the appropriate symbol for each of the following isotopes: (a) $Z = 11, A = 23$; (b) $Z = 28, A = 64$; (c) $Z = 50, A = 115$; (d) $Z = 20, A = 42$.

**2.25**   Write the appropriate symbol for each of the following isotopes: (a) $Z = 74, A = 186$; (b) $Z = 80, A = 201$; (c) $Z = 34, A = 76$; (d) $Z = 94, A = 239$.

**2.26**   Determine the mass number of (a) a boron atom with 5 neutrons, (b) a magnesium atom with 14 neutrons, (c) a bromine atom with 46 neutrons, and (d) a mercury atom with 116 neutrons.

**2.27**   Determine the mass number of (a) a fluorine atom with 10 neutrons, (b) a sulfur atom with 18 neutrons, (c) an arsenic atom with 42 neutrons, and (d) a platinum atom with 114 neutrons.

**2.28**   The following radioactive isotopes are used in medicine for imaging organs, studying blood circulation, treating cancer, and so on. Give the number of neutrons present in each isotope: $^{198}$Au, $^{47}$Ca, $^{60}$Co, $^{18}$F, $^{125}$I, $^{131}$I, $^{42}$K, $^{43}$K, $^{24}$Na, $^{32}$P, $^{85}$Sr, $^{99}$Tc.

## Section 2.4: The Periodic Table

### Review Questions

**2.29**   What is the periodic table, and what is its significance in the study of chemistry?

**2.30**   State two differences between a metal and a nonmetal.

**2.31**   Write the names and symbols for four elements in each of the following categories: (a) nonmetal, (b) metal, (c) metalloid.

**2.32**   Give two examples of each of the following: (a) alkali metals, (b) alkaline earth metals, (c) halogens, (d) noble gases, (e) chalcogens, (f) transition metals.

**2.33**   The explosion of an atomic bomb in the atmosphere releases many radioactive isotopes into the environment. One of the isotopes is $^{90}$Sr. Via a relatively short food chain, it can enter the human body. Considering the position of strontium in the periodic table, explain why it is particularly harmful to humans.

### Problems

**2.34**   Elements whose names end with –*ium* are usually metals; sodium is one example. Identify a nonmetal whose name also ends with –*ium*.

**2.35**   Describe the changes in properties (from metals to nonmetals or from nonmetals to metals) as we move (a) down a periodic group and (b) across the periodic table from left to right.

**2.36**   Consult a handbook of chemical and physical data (ask your instructor where you can locate a copy of the handbook) to find (a) two metals less dense than water, (b) two metals more dense than mercury, (c) the densest known solid metallic element, and (d) the densest known solid nonmetallic element.

**2.37**   Group the following elements in pairs that you would expect to show similar chemical properties: K, F, P, Na, Cl, and N.

**2.38**   Group the following elements in pairs that you would expect to show similar chemical properties: I, Ba, O, Br, S, and Ca.

**2.39**   Write the symbol for each of the following biologically important elements in the given periodic table: iron (present in hemoglobin for transporting oxygen), iodine (present in the thyroid gland), sodium (present in intracellular and extracellular fluids), phosphorus (present in bones and teeth), sulfur (present in proteins), and magnesium (present in chlorophyll molecules).

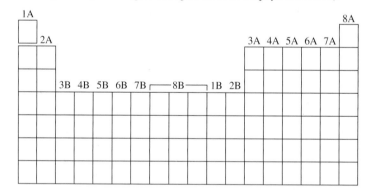

## Section 2.5: The Atomic Mass Scale and Average Atomic Mass

### Review Questions

**2.40**   What is an atomic mass unit? Why is it necessary to introduce such a unit?

**2.41**   What is the mass (in amu) of a carbon-12 atom? Why is the atomic mass of carbon listed as 12.01 amu in the table on the inside front cover of this book?

**2.42**   Explain clearly what is meant by the statement "The atomic mass of gold is 197.0 amu."

**2.43**   What information would you need to calculate the average atomic mass of an element?

## Problems

**2.44** The atomic masses of $^{35}_{17}Cl$ (75.53 percent) and $^{37}_{17}Cl$ (24.47 percent) are 34.968 and 36.956 amu, respectively. Calculate the average atomic mass of chlorine. The percentages in parentheses denote the relative abundances.

**2.45** The atomic masses of $^{204}Pb$ (1.4 percent), $^{206}Pb$ (24.1 percent), $^{207}Pb$ (22.1 percent), and $^{208}Pb$ (52.4 percent) are 203.973020, 205.974440, 206.975872, and 207.976627 amu, respectively. Calculate the average atomic mass of lead. The percentages in parentheses denote the relative abundances.

**2.46** The atomic masses of $^{203}Tl$ and $^{205}Tl$ are 202.972320 and 204.974401 amu, respectively. Calculate the natural abundances of these two isotopes. The average atomic mass of thallium is 204.4 amu.

**2.47** The atomic masses of $^6Li$ and $^7Li$ are 6.0151 amu and 7.0160 amu, respectively. Calculate the natural abundances of these two isotopes. The average atomic mass of Li is 6.941 amu.

**2.48** What is the mass in grams of 13.2 amu?

**2.49** How many atomic mass units are there in 8.4 g?

## Section 2.6: Molecules and Molecular Compounds

### Review Questions

**2.50** What is the difference between an atom and a molecule?

**2.51** What are allotropes? Give an example. How are allotropes different from isotopes?

**2.52** Describe the two commonly used molecular models.

**2.53** What does a chemical formula represent? Determine the ratio of the atoms in the following molecular formulas: (a) NO, (b) $NCl_3$, (c) $N_2O_4$, (d) $P_4O_6$.

**2.54** Define molecular formula and empirical formula. What are the similarities and differences between the empirical formula and molecular formula of a compound?

**2.55** Give an example of a case in which two molecules have different molecular formulas but the same empirical formula.

**2.56** What is the difference between inorganic compounds and organic compounds?

**2.57** Give one example each for a binary compound and a ternary compound. (A ternary compound is one that contains three different elements.)

**2.58** Explain why the formula HCl can represent two different chemical systems.

### Problems

**2.59** For each of the following diagrams, determine whether it represents diatomic molecules, polyatomic molecules, molecules that are not compounds, molecules that are compounds, or an elemental form of the substance.

(a)                    (b)                    (c)

**2.60** For each of the following diagrams, determine whether it represents diatomic molecules, polyatomic molecules, molecules that are not compounds, molecules that are compounds, or an elemental form of the substance.

(a)                    (b)                    (c)

**2.61** Identify the following as elements or compounds: $NH_3$, $N_2$, $S_8$, NO, CO, $CO_2$, $H_2$, $SO_2$.

**2.62** Give two examples of each of the following: (a) a diatomic molecule containing atoms of the same element, (b) a diatomic molecule containing atoms of different elements, (c) a polyatomic molecule containing atoms of the same element, (d) a polyatomic molecule containing atoms of different elements.

**2.63** Write the empirical formulas of the following compounds: (a) $C_2N_2$, (b) $C_6H_6$, (c) $C_9H_{20}$, (d) $P_4O_{10}$, (e) $B_2H_6$.

**2.64** Write the empirical formulas of the following compounds: (a) $Al_2Br_6$, (b) $Na_2S_2O_4$, (c) $N_2O_5$, (d) $K_2Cr_2O_7$, (e) $H_2C_2O_4$.

**2.65** Write the molecular formula of alanine, an amino acid used in protein synthesis. The color codes are black (carbon), blue (nitrogen), red (oxygen), and white (hydrogen).

**2.66** Write the molecular formula of ethanol. The color codes are black (carbon), red (oxygen), and white (hydrogen).

**2.67** Name the following binary molecular compounds: (a) $NCl_3$, (b) $IF_7$, (c) $P_4O_6$, (d) $S_2Cl_2$.

**2.68** Write chemical formulas for the following molecular compounds: (a) phosphorus tribromide, (b) dinitrogen tetrafluoride, (c) xenon tetroxide, (d) selenium trioxide.

**2.69** Write the molecular formulas and names of the following compounds.

(a)                    (b)                    (c)

**2.70** Write the molecular formulas and names of the following compounds.

(a)  (b)  (c)

## Section 2.7: Ions and Ionic Compounds

### Review Questions

**2.71** Give an example of each of the following: (a) a monatomic cation, (b) a monatomic anion, (c) a polyatomic cation, (d) a polyatomic anion.

**2.72** What is an ionic compound? How is electrical neutrality maintained in an ionic compound?

**2.73** Explain why the chemical formulas of ionic compounds are usually the same as their empirical formulas.

**2.74** What is the Stock system? What are its advantages over the older system of naming cations?

### Problems

**2.75** Give the number of protons and electrons in each of the following common ions: $Na^+$, $Ca^{2+}$, $Al^{3+}$, $Fe^{2+}$, $I^-$, $F^-$, $S^{2-}$, $O^{2-}$, $N^{3-}$.

**2.76** Give the number of protons and electrons in each of the following common ions: $K^+$, $Mg^{2+}$, $Fe^{3+}$, $Br^-$, $Mn^{2+}$, $C^{4-}$, $Cu^{2+}$.

**2.77** Write the formulas for the following ionic compounds: (a) sodium oxide, (b) iron sulfide (containing the $Fe^{2+}$ ion), (c) cobalt sulfate (containing the $Co^{3+}$ and $SO_4^{2-}$ ions), (d) barium fluoride.

**2.78** Write the formulas for the following ionic compounds: (a) copper bromide (containing the $Cu^+$ ion), (b) manganese oxide (containing the $Mn^{3+}$ ion), (c) mercury iodide (containing the $Hg_2^{2+}$ ion), (d) magnesium phosphate (containing the $PO_4^{3-}$ ion).

**2.79** Which of the following compounds are likely to be ionic? Which are likely to be molecular? $SiCl_4$, $LiF$, $BaCl_2$, $B_2H_6$, $KCl$, $C_2H_4$

**2.80** Which of the following compounds are likely to be ionic? Which are likely to be molecular? $CH_4$, $NaBr$, $BaF_2$, $CCl_4$, $ICl$, $CsCl$, $NF_3$

**2.81** Name the following compounds: (a) $KH_2PO_4$, (b) $K_2HPO_4$, (c) $HBr$ (gas), (d) $HBr$ (in water), (e) $Li_2CO_3$, (f) $K_2Cr_2O_7$, (g) $NH_4NO_2$, (h) $HIO_3$, (i) $PF_5$, (j) $P_4O_6$, (k) $CdI_2$, (l) $SrSO_4$, (m) $Al(OH)_3$.

**2.82** Name the following compounds: (a) $KClO$, (b) $Ag_2CO_3$, (c) $HNO_2$, (d) $KMnO_4$, (e) $CsClO_3$, (f) $KNH_4SO_4$, (g) $FeO$, (h) $Fe_2O_3$, (i) $TiCl_4$, (j) $NaH$, (k) $Li_3N$, (l) $Na_2O$, (m) $Na_2O_2$.

**2.83** Write the formulas for the following compounds: (a) rubidium nitrite, (b) potassium sulfide, (c) sodium hydrogen sulfide, (d) magnesium phosphate, (e) calcium hydrogen phosphate, (f) lead(II) carbonate, (g) tin(II) fluoride, (h) ammonium sulfate, (i) silver perchlorate, (j) boron trichloride.

**2.84** Write the formulas for the following compounds: (a) copper(I) cyanide, (b) strontium chlorite, (c) perbromic acid, (d) hydroiodic acid, (e) disodium ammonium phosphate, (f) potassium dihydrogen

phosphate, (g) iodine heptafluoride, (h) tetraphosphorus decasulfide, (i) mercury(II) oxide, (j) mercury(I) iodide, (k) selenium hexafluoride.

**2.85** In the diagrams shown here, match each of the drawings with the following ionic compounds: $Al_2O_3$, $LiH$, $Na_2S$, $Mg(NO_3)_2$. (Green spheres represent cations and red spheres represent anions.)

(a)  (b)  (c)  (d)

**2.86** Given the formulas for the ionic compounds, draw the correct ratio of cations to anions as shown in Problem 2.85: (a) $BaSO_4$, (b) $CaF_2$, (c) $Mg_3N_2$, (d) $K_2O$.

## Additional Problems

**2.87** Define the following terms: acids, bases, oxoacids, oxoanions, and hydrates.

**2.88** A sample of a uranium compound is found to be losing mass gradually. Explain what is happening to the sample.

**2.89** In which one of the following pairs do the two species resemble each other most closely in chemical properties: (a) $_1^1H$ and $_1^1H^+$, (b) $_7^{14}N$ and $_7^{14}N^{3-}$, (c) $_6^{12}C$ and $_6^{13}C$? Explain.

**2.90** One isotope of a metallic element has mass number 65 and 35 neutrons in the nucleus. The cation derived from the isotope has 28 electrons. Write the symbol for this cation.

**2.91** One isotope of a nonmetallic element has mass number 127 and 74 neutrons in the nucleus. The anion derived from the isotope has 54 electrons. Write the symbol for this anion.

**2.92** The following table gives numbers of electrons, protons, and neutrons in atoms or ions of a number of elements. Answer the following: (a) Which of the species are neutral? (b) Which are negatively charged? (c) Which are positively charged? (d) What are the conventional symbols for all the species?

| | **Atom or Ion of Element** | | | | | | |
| | **A** | **B** | **C** | **D** | **E** | **F** | **G** |
|---|---|---|---|---|---|---|---|
| Number of electrons | 5 | 10 | 18 | 28 | 36 | 5 | 9 |
| Number of protons | 5 | 7 | 19 | 30 | 35 | 5 | 9 |
| Number of neutrons | 5 | 7 | 20 | 36 | 46 | 6 | 10 |

**2.93** What is wrong with or ambiguous about the phrase "four molecules of NaCl"?

**2.94** The following phosphorus sulfides are known: $P_4S_3$, $P_4S_7$, and $P_4S_{10}$. Do these compounds obey the law of multiple proportions?

**2.95** Which of the following are elements, which are molecules but not compounds, which are compounds but not molecules, and which are both compounds and molecules? (a) $SO_2$, (b) $S_8$, (c) Cs, (d) $N_2O_5$, (e) O, (f) $O_2$, (g) $O_3$, (h) $CH_4$, (i) KBr, (j) S, (k) $P_4$, (l) LiF.

**2.96** What is wrong with the name (given in parentheses or brackets) for each of the following compounds: (a) $BaCl_2$ (barium dichloride), (b) $Fe_2O_3$ [iron(II) oxide], (c) $CsNO_2$ (cesium nitrate), (d) $Mg(HCO_3)_2$ [magnesium(II) bicarbonate]?

**2.97** Discuss the significance of assigning an atomic mass of exactly 12 amu to the carbon-12 isotope.

**2.98** Determine what is wrong with the chemical formula and write the correct chemical formula for each of the following compounds: (a) $(NH_3)_2CO_3$ (ammonium carbonate), (b) CaOH (calcium hydroxide), (c) $CdSO_3$ (cadmium sulfide), (d) $ZnCrO_4$ (zinc dichromate).

**2.99** Fill in the blanks in the table:

| Symbol | | $^{54}_{26}Fe^{2+}$ | | | |
|---|---|---|---|---|---|
| Protons | 5 | | | 79 | 86 |
| Neutrons | 6 | | 16 | 117 | 136 |
| Electrons | 5 | | 18 | 79 | |
| Net charge | | | | $-3$ | 0 |

**2.100** (a) Which elements are most likely to form ionic compounds? (b) Which metallic elements are most likely to form cations with different charges?

**2.101** Write the formula of the common ion derived from each of the following: (a) Li, (b) S, (c) I, (d) N, (e) Al, (f) Cs, (g) Mg.

**2.102** Which of the following symbols provides more information about the atom: $^{23}$Na or $_{11}$Na? Explain.

**2.103** Write the chemical formulas and names of the binary acids and oxoacids that contain Group 7A elements. Do the same for elements in Groups 3A, 4A, 5A, and 6A.

**2.104** Determine the molecular and empirical formulas of the compounds shown here. (Black spheres are carbon, and white spheres are hydrogen.)

(a)          (b)          (c)          (d)

**2.105** For the noble gases (the Group 8A elements) $^{4}_{2}$He, $^{20}_{10}$Ne, $^{40}_{18}$Ar, $^{84}_{36}$Kr, and $^{132}_{54}$Xe, (a) determine the number of protons and neutrons in the nucleus of each atom, and (b) determine the ratio of neutrons to protons in the nucleus of each atom. Describe any general trend you discover in the way this ratio changes with increasing atomic number.

**2.106** List the elements that exist as gases at room temperature. (*Hint:* Most of these elements can be found in Groups 5A, 6A, 7A, and 8A.)

**2.107** The Group 1B metals, Cu, Ag, and Au, are called coinage metals. What chemical properties make them especially suitable for making coins and jewelry?

**2.108** The elements in Group 8A of the periodic table are called noble gases. Can you suggest what "noble" means in this context?

**2.109** The formula for calcium oxide is CaO. What are the formulas for magnesium oxide and strontium oxide?

**2.110** A common mineral of barium is barytes, or barium sulfate ($BaSO_4$). Because elements in the same periodic group have similar chemical properties, we might expect to find some radium sulfate ($RaSO_4$) mixed with barytes since radium is the last member of Group 2A. However, the only source of radium compounds in nature is in uranium minerals. Why?

**2.111** List five elements each that are (a) named after places, (b) named after people, (c) named after a color. (Consult www.Google.com, www.Wikipedia.com, or www.Webelements.com.)

**2.112** Name the only country that is named after an element. (*Hint:* This country is in South America.)

**2.113** Fluorine reacts with hydrogen (H) and deuterium (D) to form hydrogen fluoride (HF) and deuterium fluoride (DF), where deuterium ($^{2}_{1}$H) is an isotope of hydrogen. Would a given amount of fluorine react with different masses of the two hydrogen isotopes? Does this violate the law of definite proportion? Explain.

**2.114** Predict the formula and name of a binary compound formed from the following elements: (a) Na and H, (b) B and O, (c) Na and S, (d) Al and F, (e) F and O, (f) Sr and Cl.

**2.115** Identify each of the following elements: (a) a halogen whose anion contains 36 electrons, (b) a radioactive noble gas with 86 protons, (c) a Group 6A element whose anion contains 36 electrons, (d) an alkali metal cation that contains 36 electrons, (e) a Group 4A cation that contains 80 electrons.

**2.116** Show the locations of (a) alkali metals, (b) alkaline earth metals, (c) the halogens, and (d) the noble gases in the given outline of a periodic table. Also draw dividing lines between metals and metalloids and between metalloids and nonmetals.

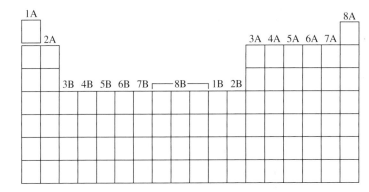

**2.117** Fill in the blanks in the table.

| Cation | Anion | Formula | Name |
|---|---|---|---|
| | | | Magnesium bicarbonate |
| | | $SrCl_2$ | |
| $Fe^{3+}$ | $NO_2^-$ | | |
| | | | Manganese(II) chlorate |
| | | $SnBr_4$ | |
| $Co^{2+}$ | $PO_4^{3-}$ | | |
| $Hg_2^{2+}$ | $I^-$ | | |
| | | $Cu_2CO_3$ | |
| | | | Lithium nitride |
| $Al^{3+}$ | $S^{2-}$ | | |

**2.118** Some compounds are better known by their common names than by their systematic chemical names. Give the chemical formulas of the following substances: (a) dry ice, (b) salt, (c) laughing gas, (d) marble (chalk, limestone), (e) baking soda, (f) ammonia, (g) water, (h) milk of magnesia, (i) epsom salt.

**2.119** In the footnote on page 37 it was pointed out that mass and energy are alternate aspects of a single entity called mass-energy. The relationship between these two physical quantities is Einstein's equation, $E = mc^2$, where $E$ is energy, $m$ is mass, and $c$ is the speed of light. In a combustion experiment, it was found that 12.096 g of hydrogen molecules combined with 96.000 g of oxygen molecules to form water and released $1.715 \times 10^3$ kJ of heat. Use Einstein's equation to calculate the corresponding mass change in this process, and comment on whether or not the law of conservation of mass holds for ordinary chemical processes.

2.120 (a) Describe Rutherford's experiment and how the results revealed the nuclear structure of the atom. (b) Consider the $^{23}$Na atom. Given that the radius and mass of the nucleus are $3.04 \times 10^{-15}$ m and $3.82 \times 10^{-23}$ g, respectively, calculate the density of the nucleus in g/cm$^3$. The radius of a $^{23}$Na atom is 186 pm. Calculate the density of the space occupied by the electrons outside the nucleus in the sodium atom. Do your results support Rutherford's model of an atom? [The volume of a sphere of radius $r$ is $\frac{4}{3}\pi r^3$.]

**2.121** Draw all possible structural formulas of the following hydrocarbons: $CH_4$, $C_2H_6$, $C_3H_8$, $C_4H_{10}$, $C_5H_{12}$.

2.122 Draw two different structural formulas based on the molecular formula $C_2H_6O$. Is the fact that you can have more than one compound with the same molecular formula consistent with Dalton's atomic theory?

**2.123** Ethane and acetylene are two gaseous hydrocarbons. Chemical analyses show that in one sample of ethane, 2.65 g of carbon are combined with 0.665 g of hydrogen, and in one sample of acetylene, 4.56 g of carbon are combined with 0.383 g of hydrogen. (a) Are these results consistent with the law of multiple proportions? (b) Write reasonable molecular formulas for these compounds.

2.124 A cube made of platinum (Pt) has an edge length of 1.0 cm. (a) Calculate the number of Pt atoms in the cube. (b) Atoms are spherical in shape. Therefore, the Pt atoms in the cube cannot fill all the available space. If only 74 percent of the space inside the cube is taken up by Pt atoms, calculate the radius in picometers of a Pt atom. The density Pt is 21.45 g/cm$^3$, and the mass of a single Pt atom is $3.240 \times 10^{-22}$ g. [The volume of a sphere of radius $r$ is $\frac{4}{3}\pi r^3$.]

**2.125** A monatomic ion has a charge of +2. The nucleus of the parent atom has a mass number of 55. If the number of neutrons in the nucleus is 1.2 times that of the number of protons, what is the name and symbol of the element?

2.126 In the following $2 \times 2$ crossword, each letter must be correct in four ways: horizontally, vertically, diagonally, and by itself. When the puzzle is complete, the four spaces will contain the overlapping symbols of 10 elements. Use capital letters for each square. There is only one correct solution.

| 1 | 2 |
|---|---|
| 3 | 4 |

*Horizontal*
1–2: Two-letter symbol for a metal used in ancient times
3–4: Two-letter symbol for a metal that burns in air and is found in Group 5A

*Vertical*
1–3: Two-letter symbol for a metalloid
2–4: Two-letter symbol for a metal used in U.S. coins

*Single Square*
1: A colorful nonmetal
2: A colorless gaseous nonmetal
3: An element that makes fireworks green
4: An element that has medicinal uses

*Diagonal*
1–4: Two-letter symbol for an element used in electronics
2–3: Two-letter symbol for a metal used with Zr to make wires for superconducting magnets

**2.127** Name the given acids.

2.128 (a) Assuming an atomic nucleus is spherical in shape, show that its radius $r$ is proportional to the cube root of the mass number (A). (b) In general, the radius of a nucleus is given by $r = r_0 A^{1/3}$, where $r_0$ is a proportionality constant given by $1.2 \times 10^{-15}$ m. Calculate the volume of the Li nucleus. (c) Given that the radius of a $^{7}_{3}$Li atom is 152 pm, calculate what fraction of the atom's volume is occupied by its nucleus. Does your result support Rutherford's model of the atom?

# PRE-PROFESSIONAL PRACTICE EXAM PROBLEMS:
# PHYSICAL AND BIOLOGICAL SCIENCES

Carbon-14, a radioactive isotope of carbon, is used to determine the ages of fossils in a technique called *carbon dating*. Carbon-14 is produced in the upper atmosphere when nitrogen-14 atoms are bombarded by neutrons from cosmic rays. $^{14}C$ undergoes a process called $\beta$ emission in which a neutron in the nucleus decays to form a proton and an electron. The electron, or $\beta$ particle, is ejected from the nucleus. Because the production and decay of $^{14}C$ occur simultaneously, the total amount of $^{14}C$ in the atmosphere is constant. Plants absorb $^{14}C$ in the form of $CO_2$ and animals consume plants and other animals. Thus, all living things contain a constant ratio of $^{12}C$ to $^{14}C$. When a living thing dies, the $^{14}C$ it contains continues to decay but because replenishment ceases, the ratio of $^{12}C$ to $^{14}C$ changes over time. Scientists use the $^{12}C$ to $^{14}C$ ratio to determine the age of material that was once living.

1. If atmospheric conditions were to change such that $^{14}C$ were produced at twice the current rate,

   a) the world's supply of $^{14}N$ would be consumed completely.

   b) the $^{12}C$ to $^{14}C$ ratio in living things would increase.

   c) the $^{12}C$ to $^{14}C$ ratio in living things would decrease.

   d) the $^{12}C$ to $^{14}C$ ratio in living things would not change.

2. When a $^{14}N$ nucleus is bombarded by a neutron to produce a $^{14}C$ nucleus, what else is produced?

   a) Nothing
   b) Another neutron
   c) An electron
   d) A proton

3. Based on the description of $\beta$ emission in the passage, what nucleus results from the decay of a $^{14}C$ nucleus by $\beta$ emission?

   a) $^{14}N$

   b) $^{13}N$

   c) $^{12}C$

   d) $^{13}C$

4. The accuracy of carbon dating depends on the assumption that

   a) $^{14}C$ is the only radioactive species in the material being tested.

   b) the rate of decay of $^{14}C$ is constant.

   c) $^{12}C$ and $^{14}C$ undergo radioactive decay at the same rate.

   d) each $^{14}C$ nucleus decays to give a $^{12}C$ nucleus.

# ANSWERS TO IN-CHAPTER MATERIALS

## Practice Problems

**2.1A** (a) 3:2, (b) 2:1. **2.1B** (a) 0.2518 g, (b) $n = 6$ ($XeF_6$). **2.2A** (a) p = 5, n = 5, $e$ = 5. (b) p = 18, n = 18, $e$ = 18. (c) p = 38, n = 47, $e$ = 38. (d) p = 6, n = 5, $e$ = 6. **2.2B** (a) $^{9}_{4}Be$, (b) $^{51}_{23}V$, (c) $^{124}_{54}Xe$, (d) $^{69}_{31}Ga$.
**2.3A** 63.55 amu. **2.3B** 99.64% $^{14}N$, 0.36% $^{15}N$. **2.4A** $CHCl_3$. **2.4B** $C_3H_6O$.
**2.5A** (a) dichlorine monoxide, (b) silicon tetrachloride. **2.5B** (a) chlorine dioxide, (b) carbon tetrabromide. **2.6A** (a) $CS_2$, (b) $N_2O_3$. **2.6B** (a) $SF_6$, (b) $S_2F_{10}$. **2.7A** (a) $C_4H_5N_2O$, (b) $C_2H_5$, (c) $C_2H_5NO_2$. **2.7B** (a).
**2.8A** (a) hypobromous acid, (b) hydrogen sulfate ion, (c) oxalic acid.
**2.8B** (a) iodic acid, (b) hydrogen chromate ion, (c) hydrogen oxalate ion. **2.9A** $HBrO_4$. **2.9B** $H_2CrO_4$. **2.10A** (a) sodium sulfate, (b) copper(II) nitrate, (c) iron(III) carbonate. **2.10B** (a) potassium dichromate, (b) lithium oxalate, (c) copper(I) nitrate. **2.11A** (a) $PbCl_2$, (b) $MgCO_3$, (c) $(NH_4)_3PO_4$.
**2.11B** (a) $Fe_2S_3$, (b) $Hg(NO_3)_2$, (c) $K_2SO_3$.

## Checkpoints

**2.1.1** e. **2.1.2** b. **2.3.1** d. **2.3.2** c. **2.4.1** c. **2.4.2** b. **2.5.1** b. **2.5.2** c. **2.6.1** b.
**2.6.2** c. **2.6.3** c. **2.6.4** e. **2.7.1** c. **2.7.2** d. **2.7.3** d. **2.7.4** c. **2.7.5** c. **2.7.6** a.

## Applying What You've Learned

**a)** There are $54 - 26 = 28$ neutrons in the $^{54}Fe$ nucleus, 30 neutrons in the $^{56}Fe$ nucleus, 31 neutrons in the $^{57}Fe$ nucleus, and 32 neutrons in the $^{58}Fe$ nucleus. **b)** The average atomic mass of iron is 55.845 amu. **c)** The molecular formula for ascorbic acid is $C_6H_8O_6$. **d)** The empirical formula for ascorbic acid is $C_3H_4O_3$. **e)** The formula for ferrous sulfate is $FeSO_4$.

# 3

# Stoichiometry
## Ratios of Combination

Lance Armstrong, seven-time winner of the prestigious Tour de France race, was diagnosed with advanced testicular cancer in 1996. His successful course of treatment included the drug Platinol or cisplatin.

# Stoichiometry's Importance in the Manufacture of Drugs

Stoichiometry refers to the quantitative relationships between the substances that are consumed and produced by chemical reactions. These quantitative relationships are important in the development of large-scale production of such things as chemotherapeutic drugs for the treatment of cancer.

One of cancer chemotherapy's greatest success stories began with an accidental discovery. In 1964, Barnett Rosenberg and his research group at Michigan State University were studying the effect of an electric field on the growth of bacteria. Using platinum electrodes, they passed an electric current through a bacterial culture. To their surprise, the cells in the culture stopped dividing. The researchers determined that cisplatin, $Pt(NH_3)_2Cl_2$, a compound containing platinum from the electrodes, was responsible. Furthermore, they reasoned that because cancer is the result of the uncontrolled division of abnormal cells, the compound might be useful as an anticancer drug.

Platinol, the name under which cisplatin is marketed, was approved by the FDA in 1978 for the treatment of metastatic testicular and ovarian cancers. Today it is one of the most widely prescribed cancer drugs—being used also for cancers of the bladder, lung, and stomach—and is probably best known for the role it played in seven-time Tour de France winner Lance Armstrong's battle with testicular cancer. Cisplatin works by attaching itself to the DNA of cancer cells and preventing their replication. The damaged cells are then destroyed by the body's immune system. Unfortunately, cisplatin can cause serious side effects, including severe kidney damage. Ongoing research efforts are directed toward finding related compounds that are less toxic.

In 1964, cisplatin was produced accidentally when platinum electrodes reacted with ammonia molecules and chloride ions that were present in a bacterial culture. Today, manufacturers use the principles of stoichiometry to produce cisplatin in the most efficient, economical way possible.

At the end of this chapter, you will be able to solve several problems related to the drug cisplatin. [▶▶ page 105]

## In This Chapter, You Will Learn:

How we use chemical equations to represent chemical *reactions*. You will also learn how balanced chemical equations are used to solve a variety of problems.

### Before you begin, you should review

• Average atomic mass [◀◀ Section 2.5]
• Molecular formulas [◀◀ Section 2.6]

# 3.1 Molecular and Formula Masses

Using atomic masses from the periodic table and a molecular formula, we can determine the **molecular mass,** which is the mass in atomic mass units (amu) of an individual molecule. The molecular mass is simply the sum of the atomic masses of the atoms that make up the molecule. We multiply the atomic mass of each element by the number of atoms of that element in the molecule and then sum the masses for each element present. For example,

$$\text{molecular mass of } H_2O = 2(\text{atomic mass of H}) + \text{atomic mass of O}$$

$$= 2(1.008 \text{ amu}) + 16.00 \text{ amu} = 18.02 \text{ amu}$$

Because the atomic masses on the periodic table are average atomic masses, the result of such a determination is an average molecular mass, sometimes referred to as the **molecular weight.** As with the term *atomic* mass, we will use the term *molecular* mass in this text.

Although an ionic compound does not have a molecular mass, we can use its empirical formula to determine its **formula mass** (the mass of a "formula unit"), sometimes called the **formula weight.** Sample Problem 3.1 illustrates how to determine molecular mass and formula mass.

Caffeine

---

**Sample Problem 3.1**

Calculate the molecular mass or the formula mass, as appropriate, for each of the following compounds: (a) propane, ($C_3H_8$), (b) lithium hydroxide, (LiOH), and (c) barium acetate, [$Ba(C_2H_3O_2)_2$].

**Strategy** Determine the molecular mass (for each molecular compound) or formula mass (for each ionic compound) by summing all the atomic masses.

**Setup** Using the formula for each compound, determine the number of atoms of each element present. A molecule of propane contains three C atoms and eight H atoms. The compounds in parts (b) and (c) are ionic and will therefore have formula masses rather than molecular masses. A formula unit of lithium hydroxide contains one Li atom, one O atom, and one H atom. A formula unit of barium acetate contains one Ba atom, four C atoms, six H atoms, and four O atoms. (Remember that the subscript after the parentheses means that there are two acetate ions, each of which contains two C atoms, three H atoms, and two O atoms.)

**Solution** For each compound, multiply the number of atoms by the atomic mass of each element and then sum the calculated values.

(a) The molecular mass of propane is 3(12.01 amu) + 8(1.008 amu) = 44.09 amu.
(b) The formula mass of lithium hydroxide is 6.941 amu + 16.00 amu + 1.008 amu = 23.95 amu.
(c) The formula mass of barium acetate is 137.3 amu + 4(12.01 amu) + 6(1.008 amu) + 4(16.00 amu) = 255.4 amu.

**Think About It** Double-check that you have counted the number of atoms correctly for each compound and that you have used the proper atomic masses from the periodic table.

**Practice Problem A** Calculate the molecular or formula mass of each of the following compounds: (a) magnesium chloride ($MgCl_2$), (b) sulfuric acid ($H_2SO_4$), and (c) ibuprofen ($C_{13}H_{18}O_2$).

**Practice Problem B** Calculate the molecular or formula mass of each of the following compounds: (a) calcium carbonate ($CaCO_3$), (b) nitrous acid ($HNO_2$), and (c) caffeine ($C_8H_{10}N_4O_2$).

---

**Checkpoint 3.1**   **Molecular and Formula Masses**

**3.1.1** What is the molecular mass of citric acid ($H_3C_6H_5O_7$)?

a) 192.12 amu

b) 189.10 amu

c) 132.07 amu

d) 29.02 amu

e) 89.07 amu

**3.1.2** What is the formula mass of calcium citrate [$Ca_3(C_6H_5O_7)_2$]?

a) 309.34 amu

b) 69.10 amu

c) 229.18 amu

d) 498.44 amu

e) 418.28 amu

Citric acid

# 3.2 | Percent Composition of Compounds

The formula of a compound indicates the number of atoms of each element in a unit of the compound. From a molecular or empirical formula, we can calculate what percent of the total mass is contributed by each element in a compound. A list of the percent by mass of each element in a compound is known as the compound's **percent composition by mass.** One way that the purity of a compound can be verified is by comparing its percent composition by mass, determined experimentally, with its calculated percent composition. Percent composition is calculated by dividing the mass of each element in a unit of the compound by the molecular or formula mass of the compound and then multiplying by 100 percent. Mathematically, the percent by mass of an element in a compound is expressed as

$$\text{percent by mass of an element} = \frac{n \times \text{atomic mass of element}}{\text{molecular or formula mass of compound}} \times 100\% \quad \text{Equation 3.1}$$

where $n$ is the number of atoms of the element in a molecule or formula unit of the compound. For example, in a molecule of hydrogen peroxide ($H_2O_2$), there are two H atoms and two O atoms. The atomic masses of H and O are 1.008 and 16.00 amu, respectively, so the molecular mass of $H_2O_2$ is 34.02 amu. Therefore, the percent composition of $H_2O_2$ is calculated as follows:

$$\%H = \frac{2 \times 1.008 \text{ amu H}}{34.02 \text{ amu } H_2O_2} \times 100\% = 5.926\%$$

$$\%O = \frac{2 \times 16.00 \text{ amu O}}{34.02 \text{ amu } H_2O_2} \times 100\% = 94.06\%$$

The sum of percentages is 5.296% + 94.06% = 99.99%. The small discrepancy from 100 percent is due to rounding of the atomic masses of the elements. We could equally well have used the empirical formula of hydrogen peroxide (HO) for the calculation. In this case, we would have used the *empirical formula mass,* 17.01 amu, in place of the molecular mass.

$$\%H = \frac{1.008 \text{ amu H}}{17.01 \text{ amu}} \times 100\% = 5.926\%$$

$$\%O = \frac{16.00 \text{ amu O}}{17.01 \text{ amu}} \times 100\% = 94.06\%$$

Because both the molecular formula and the empirical formula tell us the composition of the compound, they both give the same percent composition by mass. Sample Problem 3.2 shows how to calculate percent composition by mass.

## Sample Problem 3.2

Lithium carbonate ($Li_2CO_3$) was the first "mood-stabilizing" drug approved by the FDA for the treatment of mania and manic-depressive illness, also known as bipolar disorder. Calculate the percent composition by mass of lithium carbonate.

**Strategy** Use Equation 3.1 to determine the percent by mass contributed by each element in the compound.

**Setup** Lithium carbonate is an ionic compound that contains Li, C, and O. In a formula unit, there are two Li atoms, one C atom, and three O atoms with atomic masses 6.941, 12.01, and 16.00 amu, respectively. The formula mass of $Li_2CO_3$ is 2(6.941 amu) + 12.01 amu + 3(16.00 amu) = 73.89 amu.

**Solution**
For each element, multiply the number of atoms by the atomic mass, divide by the formula mass, and multiply by 100 percent.

$$\%Li = \frac{2 \times 6.941 \text{ amu Li}}{73.89 \text{ amu } Li_2CO_3} \times 100\% = 18.79\%$$

$$\%C = \frac{12.01 \text{ amu C}}{73.89 \text{ amu } Li_2CO_3} \times 100\% = 16.25\%$$

$$\%O = \frac{3 \times 16.00 \text{ amu O}}{73.89 \text{ amu } Li_2CO_3} \times 100\% = 64.96\%$$

**Think About It** Make sure that the percent composition results for a compound sum to approximately 100. (In this case, the results sum to exactly 100 percent—18.79% + 16.25% + 64.96% = 100.00%—but remember that because of rounding, the percentages may sum to very slightly more or very slightly less.)

*(Continued)*

**Practice Problem A** Determine the percent composition by mass of acetaminophen ($C_8H_9NO_2$), the active ingredient in over-the-counter pain relievers such as Tylenol (Figure 3.1).

**Practice Problem B** Determine the percent composition by mass of the artificial sweetener aspartame ($C_{14}H_{18}N_2O_5$).

Acetaminophen

**Figure 3.1**   Acetaminophen is the active ingredient in Tylenol, a nonaspirin pain reliever.

Aspirin

## Checkpoint 3.2    Percent Composition of Compounds

**3.2.1**  What is the percent composition by mass of aspirin ($C_9H_8O_4$)?

a) 44.26% C, 3.28% H, 52.46% O

b) 60.00% C, 4.47% H, 35.53% O

c) 41.39% C, 3.47% H, 55.14% O

d) 42.86% C, 6.35% H, 50.79% O

e) 42.86%C, 38.09% H, 19.05% O

**3.2.2**  What is the percent composition by mass of sodium bicarbonate ($NaHCO_3$)?

a) 20.89% Na, 2.75% H, 32.74% C, 43.62% O

b) 44.20% Na, 1.94% H, 23.09% C, 30.76% O

c) 21.28% Na, 0.93% H, 33.35% C, 44.43% O

d) 44.20% Na, 1.94% H, 23.09% C, 30.76% O

e) 27.37% Na, 1.20% H, 14.30% C, 57.14% O

# 3.3   Chemical Equations

A chemical reaction, as described in the third hypothesis of Dalton's atomic theory [◀◀ Section 2.1], is the rearrangement of atoms in a sample of matter. Examples include the rusting of iron and the explosive combination of hydrogen and oxygen gases to produce water. A ***chemical equation*** uses chemical symbols to denote what occurs in a chemical reaction. We have seen how chemists represent elements and compounds using chemical symbols. Now we will look at how chemists represent chemical reactions using chemical equations.

## Interpreting and Writing Chemical Equations

A chemical equation represents a *chemical statement*. When you encounter a chemical equation, you may find it useful to read it as though it were a sentence.

Read

$$NH_3 + HCl \longrightarrow NH_4Cl$$

as "Ammonia and hydrogen chloride react to produce ammonium chloride."

Read

$$CaCO_3 \longrightarrow CaO + CO_2$$

as "Calcium carbonate reacts to produce calcium oxide and carbon dioxide." Thus, the plus signs can be interpreted simply as the word *and*, and the arrows can be interpreted as the phrase "react(s) to produce."

In addition to interpreting chemical equations, you must also be able to write chemical equations to represent reactions. For example, the equation for the process by which sulfur and oxygen react to produce sulfur dioxide is written as

$$S + O_2 \longrightarrow SO_2$$

Likewise, we write the equation for the reaction of sulfur trioxide and water to produce sulfuric acid as

$$SO_3 + H_2O \longrightarrow H_2SO_4$$

Each chemical species that appears to the left of the arrow is called a ***reactant.*** Reactants are those substances that are *consumed* in the course of a chemical reaction. Each species that appears to the right of the arrow is called a ***product.*** Products are the substances that *form* during the course of a chemical reaction.

Chemists usually indicate the physical states of reactants and products with italicized letters in parentheses following each species in the equation. Gases, liquids, and solids are labeled with $(g)$, $(l)$, and $(s)$, respectively. Chemical species that are dissolved in water are said to be ***aqueous*** and are labeled $(aq)$. The equation examples given previously can be written as follows:

$$NH_3(g) + HCl(g) \longrightarrow NH_4Cl(s) \qquad S(s) + O_2(g) \longrightarrow SO_2(g)$$

$$CaCO_3(s) \longrightarrow CaO(s) + CO_2(g) \qquad SO_3(g) + H_2O(l) \longrightarrow H_2SO_4(aq)$$

In some cases, the physical state must be expressed more specifically. Carbon, for example, exists in two different solid forms: diamond and graphite. Rather than simply write $C(s)$, we must specify the form of solid carbon by writing C(diamond) or C(graphite).

$$C(graphite) + O_2(g) \longrightarrow CO_2(g)$$

$$C(diamond) + O_2(g) \longrightarrow CO_2(g)$$

Chemical equations are also used to represent *physical* processes. Sucrose ($C_{12}H_{22}O_{11}$) dissolving in water, for example, is a physical process [◀◀ Section 1.2] that can be represented with the following chemical equation:

$$C_{12}H_{22}O_{11}(s) \xrightarrow{H_2O} C_{12}H_{22}O_{11}(aq)$$

The $H_2O$ over the arrow in the equation denotes the process of dissolving a substance in water. Although formulas or symbols often are omitted for simplicity, they can be written over an arrow in a chemical equation to indicate the conditions under which the reaction takes place. For example, in the chemical equation

$$2KClO_3(s) \xrightarrow{\Delta} 2KCl(s) + 3O_2(g)$$

the symbol $\Delta$ indicates that the addition of heat is necessary to make $KClO_3$ react to form $KCl$ and $O_2$.

## Balancing Chemical Equations

Based on what you have learned so far, the chemical equation for the explosive reaction of hydrogen gas with oxygen gas to form liquid water would be

$$H_2(g) \quad + \quad O_2(g) \quad \longrightarrow \quad H_2O(l)$$

This equation as written violates the law of conservation of mass, however, because four atoms (two H and two O) react to produce only three atoms (two H and one O).

**Student Annotation:** Some problems in later chapters can be solved only if the states of reactants and products are specified. It is a good idea to get in the habit now of including the physical states of reactants and products in the chemical equations that you write.

**Student Annotation:** Students sometimes fear that they will be asked to determine the products of an unfamiliar chemical reaction—and will be unable to do so. In this chapter and in Chapter 4, you will learn how to deduce the products of several different types of reactions.

The equation must be *balanced* so that the same number of each kind of atom appears on both sides of the reaction arrow. Balancing is achieved by writing appropriate ***stoichiometric coefficients*** (often referred to simply as *coefficients*) to the left of the chemical formulas. In this case, we write a coefficient of 2 to the left of both the $H_2(g)$ and the $H_2O(l)$:

$$2H_2(g) \quad + \quad O_2(g) \quad \longrightarrow \quad 2H_2O(l)$$

There are now four H atoms and two O atoms on each side of the arrow. When balancing a chemical equation, we can change only the coefficients that precede the chemical formulas, *not* the subscripts within the chemical formulas. Changing the subscripts would change the formulas for the species involved in the reaction. For example, changing the product from $H_2O$ to $H_2O_2$ would result in equal numbers of each kind of atom on both sides of the equation, but the equation we set out to balance represented the combination of hydrogen gas and oxygen gas to form water, not hydrogen peroxide. Additionally, we cannot add reactants or products to the chemical equation for the purpose of balancing it. To do so would result in an equation that represents the wrong reaction. The chemical equation must be made quantitatively correct without changing its qualitative chemical statement.

Balancing a chemical equation requires something of a trial-and-error approach. You may find that you change the coefficient for a particular reactant or product, only to have to change it again later in the process. In general, it will facilitate the balancing process if you do the following:

**Student Annotation:** When we study electrochemistry in detail, we will learn a method for balancing certain equations that *does* allow the addition of $H_2O$, $H^+$, and $OH^-$ [▸▸ Section 19.1].

1.  Change the coefficients of compounds (e.g., $CO_2$) before changing the coefficients of elements (e.g., $O_2$).
2.  Treat polyatomic ions that appear on both sides of the equation (e.g., $CO_3^{2-}$) as units, rather than counting their constituent atoms individually.
3.  Count atoms and/or polyatomic ions carefully, and track their numbers each time you change a coefficient.

**Student Annotation:** *Combustion* refers to burning in the presence of oxygen. Combustion of a hydrocarbon such as butane produces carbon dioxide and water.

To balance the chemical equation for the combustion of butane, we first take an inventory of the numbers of each type of atom on each side of the arrow.

$$C_4H_{10}(g) + O_2(g) \longrightarrow CO_2(g) + H_2O(l)$$

$$4-C-1$$

$$10-H-2$$

$$2-O-3$$

Initially, there are four C atoms on the left and one on the right; ten H atoms on the left and two on the right; and two O atoms on the left with three on the right. As a first step, we will place a coefficient of 4 in front of $CO_2(g)$ on the product side.

$$C_4H_{10}(g) + O_2(g) \longrightarrow \mathbf{4}CO_2(g) + H_2O(l)$$

$$4-C-4$$

$$10-H-2$$

$$2-O-9$$

This changes the tally of atoms as shown. Thus, the equation is balanced for carbon, but not for hydrogen or oxygen. Next, we place a coefficient of 5 in front of $H_2O(l)$ on the product side and tally the atoms on both sides again.

$$C_4H_{10}(g) + O_2(g) \longrightarrow \mathbf{4}CO_2(g) + \mathbf{5}H_2O(l)$$

$$4-C-4$$

$$10-H-10$$

$$2-O-13$$

Now the equation is balanced for carbon and hydrogen. Only oxygen remains to be balanced. There are 13 O atoms on the product side of the equation (eight in $CO_2$ molecules and another five in $H_2O$ molecules), so we need 13 O atoms on the reactant side. Because each oxygen molecule contains two O atoms, we will have to place a coefficient of $\frac{13}{2}$ in front of $O_2(g)$:

$$C_4H_{10}(g) + \tfrac{13}{2}O_2(g) \longrightarrow 4CO_2(g) + 5H_2O(l)$$

$$4-C-4$$

$$10-H-10$$

$$13-O-13$$

With equal numbers of each kind of atom on both sides of the equation, this equation is now balanced. For now, however, you should practice balancing equations with the smallest possible *whole* number coefficients. Multiplying each coefficient by 2 gives all whole numbers and a final balanced equation:

$$2C_4H_{10}(g) + 13O_2(g) \longrightarrow 8CO_2(g) + 10H_2O(l)$$

$$8-C-8$$

$$20-H-20$$

$$26-O-26$$

**Student Annotation:** A balanced equation is, in a sense, a mathematical equality. We can multiply or divide through by any number, and the equality will still be valid.

Sample Problem 3.3 lets you practice writing and balancing a chemical equation.

## Sample Problem 3.3

Write and balance the chemical equation for the aqueous reaction of barium hydroxide and perchloric acid to produce aqueous barium perchlorate and water.

**Strategy** Determine the formulas and physical states of all reactants and products, and use them to write a chemical equation that makes the correct chemical statement. Finally, adjust coefficients in the resulting chemical equation to ensure that there are identical numbers of each type of atom on both sides of the reaction arrow.

**Setup** The reactants are $Ba(OH)_2$ and $HClO_4$, and the products are $Ba(ClO_4)_2$ and $H_2O$ [◀◀ Sections 2.6 and 2.7]. Because the reaction is aqueous, all species except $H_2O$ will be labeled $(aq)$ in the equation. Being a liquid, $H_2O$ will be labeled $(l)$.

**Solution** The chemical statement "barium hydroxide and perchloric acid react to produce barium perchlorate and water" can be represented with the following unbalanced equation:

$$Ba(OH)_2(aq) + HClO_4(aq) \longrightarrow Ba(ClO_4)_2(aq) + H_2O(l)$$

Perchlorate ions $(ClO_4^-)$ appear on both sides of the equation, so count them as units, rather than count the individual atoms they contain. Thus, the tally of atoms and polyatomic ions is

$$1-Ba-1$$
$$2-O-1 \quad \text{(not including O atoms in } ClO_4^- \text{ ions)}$$
$$3-H-2$$
$$1-ClO_4^--2$$

The barium atoms are already balanced, and placing a coefficient of 2 in front of $HClO_4(aq)$ balances the number of perchlorate ions.

$$Ba(OH)_2(aq) + 2HClO_4(aq) \longrightarrow Ba(ClO_4)_2(aq) + H_2O(l)$$

$$1-Ba-1$$
$$2-O-1 \quad \text{(not including O atoms in } ClO_4^- \text{ ions)}$$
$$4-H-2$$
$$2-ClO_4^--2$$

*(Continued)*

**Think About It** Check to be sure the equation is balanced by counting all the atoms individually.

1−Ba−1

10− O −10

4− H −4

2− Cl−2

Placing a coefficient of 2 in front of $H_2O(l)$ balances both the O and H atoms, giving us the final balanced equation:

$$Ba(OH)_2(aq) + 2HClO_4(aq) \longrightarrow Ba(ClO_4)_2(aq) + 2H_2O(l)$$

1−Ba−1

2−O−2   (not including O atoms in $ClO_4^-$ ions)

4−H−4

2−$ClO_4^-$−2

**Practice Problem A** Write and balance the chemical equation that represents the combustion of propane (i.e., the reaction of propane gas with oxygen gas to produce carbon dioxide gas and liquid water).

**Practice Problem B** Write and balance the chemical equation that represents the reaction of sulfuric acid with sodium hydroxide to form water and sodium sulfate.

## Bringing Chemistry to Life

### The Stoichiometry of Metabolism

The carbohydrates and fats we eat are broken down into small molecules in the digestive system. Carbohydrates are broken down into simple sugars such as glucose ($C_6H_{12}O_6$), and fats are broken down into fatty acids (carboxylic acids that contain hydrocarbon chains) and glycerol ($C_3H_8O_3$). The small molecules produced in the digestion process are subsequently consumed by a series of complex biochemical reactions. Although the metabolism of simple sugars and fatty acids involves relatively complex processes, the results are essentially the same as that of combustion—that is, simple sugars and fatty acids react with oxygen to produce carbon dioxide, water, and energy. The balanced chemical equation for the metabolism of glucose is

Glycerol

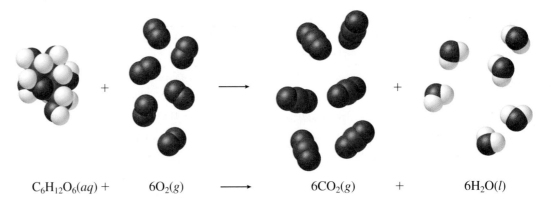

$$C_6H_{12}O_6(aq) + \quad 6O_2(g) \longrightarrow \quad 6CO_2(g) \quad + \quad 6H_2O(l)$$

### Sample Problem 3.4

Butyric acid (also known as butanoic acid, $C_4H_8O_2$) is one of many compounds found in milk fat. First isolated from rancid butter in 1869, butyric acid has received a great deal of attention in recent years as a potential anticancer agent. Assuming that the only products are $CO_2$ and $H_2O$, write and balance the equation for the metabolism of butyric acid.

**Strategy** Write an unbalanced equation to represent the combination of reactants and formation of products as stated in the problem, and then balance the equation.

**Setup** Metabolism in this context refers to the combination of $C_4H_8O_2$ with $O_2$ to produce $CO_2$ and $H_2O$.

Butyric acid

**Solution**

$$C_4H_8O_2(aq) + O_2(g) \longrightarrow CO_2(g) + H_2O(l)$$

Balance the number of C atoms by changing the coefficient for $CO_2$ from 1 to 4.

$$C_4H_8O_2(aq) + O_2(g) \longrightarrow 4CO_2(g) + H_2O(l)$$

Balance the number of H atoms by changing the coefficient for $H_2O$ from 1 to 4.

$$C_4H_8O_2(aq) + O_2(g) \longrightarrow 4CO_2(g) + 4H_2O(l)$$

Finally, balance the number of O atoms by changing the coefficient for $O_2$ from 1 to 5.

$$C_4H_8O_2(aq) + 5O_2(g) \longrightarrow 4CO_2(g) + 4H_2O(l)$$

**Think About It** Count the number of each type of atom on each side of the reaction arrow to verify that the equation is properly balanced. There are 4 C, 8 H, and 12 O in the reactants and in the products, so the equation is balanced.

**Practice Problem A** Another compound found in milk fat that appears to have anticancer and anti-obesity properties is conjugated linoleic acid (CLA; $C_{18}H_{32}O_2$). Assuming again that the only products are $CO_2$ and $H_2O$, write and balance the equation for the metabolism of CLA.

Conjugated linoleic acid

**Practice Problem B** Write and balance the equation for the combination of ammonia gas with solid copper(II) oxide to produce copper metal, nitrogen gas, and liquid water.

## Checkpoint 3.3    Chemical Equations

**3.3.1** What are the stoichiometric coefficients in the following equation when it is balanced?

$$CH_4(g) + H_2O(g) \longrightarrow H_2(g) + CO_2(g)$$

a) 1, 2, 2, 2

b) 2, 1, 1, 2

c) 1, 2, 2, 1

d) 2, 2, 2, 1

e) 1, 2, 4, 1

**3.3.2** Which chemical equation represents the reaction shown (blue spheres are nitrogen, and white spheres are hydrogen)?

a) $6N(g) + 18H(g) \longrightarrow 6NH_3(g)$

b) $3N_2(g) + 3H_2(g) \longrightarrow 2NH_3(g)$

c) $2N_2(g) + 3H_2(g) \longrightarrow 2NH_3(g)$

d) $N_2(g) + 3H_2(g) \longrightarrow 2NH_3(g)$

e) $3N_2(g) + 6H_2(g) \longrightarrow 6NH_3(g)$

| **Checkpoint 3.3** | **Chemical Equations** *(continued)* |

**3.3.3** Which is the correctly balanced form of the given equation?

$$S(s) + O_3(g) \longrightarrow SO_2(g)$$

a) $S(s) + O_3(g) \longrightarrow SO_3(g)$

b) $3S(s) + 6O_3(g) \longrightarrow 3SO_2(g)$

c) $3S(s) + O_3(g) \longrightarrow 3SO_2(g)$

d) $3S(s) + 2O_3(g) \longrightarrow SO_2(g)$

e) $3S(s) + 2O_3(g) \longrightarrow 3SO_2(g)$

**3.3.4** Carbon monoxide reacts with oxygen to produce carbon dioxide according to the following balanced equation:

$$2CO(g) + O_2(g) \longrightarrow 2CO_2(g)$$

A reaction vessel containing the reactants is pictured.

Which of the following represents the contents of the reaction vessel when the reaction is complete?

a)

b)

c)

d)

## 3.4 The Mole and Molar Masses

A balanced chemical equation tells us not only what species are consumed and produced in a chemical reaction, but also in what relative amounts. In the combination of hydrogen and oxygen to produce water, the balanced chemical equation tells us that two $H_2$ molecules react with one $O_2$ molecule to produce two $H_2O$ molecules. When carrying out a reaction such as this, however, chemists do not work with individual molecules. Rather, they work with macroscopic quantities that contain enormous numbers of molecules. Regardless of the number of molecules involved, though, the molecules combine in the ratio specified in the balanced equation. Twenty molecules of hydrogen would combine with ten molecules of oxygen, ten million molecules of hydrogen would combine with five million molecules of oxygen, and so on. Clearly it would not be convenient to express the quantities used in the laboratory in terms of the numbers of molecules involved! Instead, chemists use a unit of measurement called the *mole*.

### The Mole

If you're getting doughnuts for yourself, you probably buy them individually, but if you're getting them for your entire dorm floor, you had better buy them by the dozen. One dozen doughnuts contains exactly 12 doughnuts. In fact, a dozen of anything contains exactly 12 of that thing. Pencils typically come 12 to a box, and 12 such boxes may be shipped to the campus bookstore in a bigger box. The bigger box contains one gross of pencils (144). Whether a dozen or a gross, each is a convenient quantity that contains a reasonable, specific, exact number of items.

Chemists, too, have adopted such a number to make it easy to express the number of molecules (or atoms or ions) in a typical macroscopic sample of matter. Atoms and molecules are so much smaller than doughnuts or pencils, though, that the number used by chemists is significantly bigger than a dozen or a gross. The quantity used by chemists is the *mole* (mol), which is defined as the amount of a substance that contains as many elementary entities (atoms, molecules, formula units, etc.) as there are atoms in exactly 0.012 kg (12 g) of carbon-12. The number of atoms in exactly 12 g of carbon-12, which is determined experimentally, is known as ***Avogadro's number*** ($N_A$), in honor of the Italian scientist Amedeo Avogadro.[1] The currently accepted value

**Student Annotation:**
1 dozen = 12
1 gross = 144
1 mole = $6.022 \times 10^{23}$

**Student Annotation:** Avogadro's number is almost unimaginably big. If $6.022 \times 10^{23}$ pennies were distributed equally among every inhabitant of the United States, each man, woman, and child would have over 20 trillion dollars! Furthermore, each person's share of pennies, neatly stacked, would occupy approximately the same volume as 760 Empire State Buildings.

---

1. Lorenzo Romano Amedeo Carlo Avogadro di Quaregua e di Cerret (1776–1856). Italian mathematical physicist. He practiced law for many years before he became interested in science. His most famous work, now known as Avogadro's law (see Chapter 11), was largely ignored during his lifetime, although in the late nineteenth century, it became the basis for determining atomic masses.

of Avogadro's number is $6.0221418 \times 10^{23}$, although we usually round it to $6.022 \times 10^{23}$. Thus, a dozen doughnuts contains 12 doughnuts; a gross of pencils contains 144 pencils; and a mole of $O_2$ gas, an amount that at room temperature and ordinary pressure would fill slightly more than half of a 10-gallon fish tank, contains $6.022 \times 10^{23}$ $O_2$ molecules. Figure 3.2 shows samples containing one mole each of several common substances.

Consider again the formation of water from hydrogen and oxygen gases. No matter how large the number of molecules involved, the ratio is always *two* $H_2$ combining with *one* $O_2$ to form *two* $H_2O$. Therefore, just as two molecules of hydrogen combine with one molecule of oxygen to form two molecules of water, two moles of hydrogen molecules ($2 \times N_A$ or $12.044 \times 10^{23}$ $H_2$ molecules) combine with one mole of oxygen molecules ($6.022 \times 10^{23}$ $O_2$ molecules) to form two moles of water molecules ($12.044 \times 10^{23}$ $H_2O$ molecules). Thus, we can now interpret the chemical equation for this reaction in terms of moles. Figure 3.3 depicts this reaction on both the molecular and macroscopic scales.

**Figure 3.2**  One mole each of some familiar substances: (*left*) sulfur, (*middle*) iron, (*right*) mercury, and (*in balloon*) helium.

$$2H_2 \qquad + \qquad O_2 \qquad \longrightarrow \qquad 2H_2O$$

(a)

(b)

(c)

**Figure 3.3**  The reaction of $H_2$ and $O_2$ to form $H_2O$ shown as (a) a balanced chemical equation, (b) molecular models, and (c) an experiment (at 0°C and atmospheric pressure) in which 44.8 L of $H_2$ are combined with 22.4 L of $O_2$ to produce 36.0 mL of liquid $H_2O$.

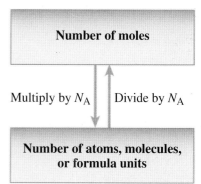

| Number of moles |
|:---:|

Multiply by $N_A$      Divide by $N_A$

| Number of atoms, molecules, or formula units |
|:---:|

$N_A$ = Avogadro's number

Note also that the subscripts in a chemical formula denote the ratio of combination of atoms in a substance. Just as a water molecule contains two H atoms and one O atom, a mole of water molecules contains two moles of H atoms and one mole of O atoms. The ratio of combination stays the same, regardless of the size of the sample.

Sample Problem 3.5 lets you practice converting between moles and atoms.

## Sample Problem 3.5

Calcium is the most abundant metal in the human body. A typical human body contains roughly 30 moles of calcium. Determine (a) the number of Ca atoms in 30.00 moles of calcium and (b) the number of moles of calcium in a sample containing $1.00 \times 10^{20}$ Ca atoms.

**Strategy** Use Avogadro's number to convert from moles to atoms and from atoms to moles.

**Setup** When the number of moles is known, we multiply by Avogadro's number to convert to atoms. When the number of atoms is known, we divide by Avogadro's number to convert to moles.

**Solution**

(a) $30.00 \; \text{mol Ca} \times \dfrac{6.022 \times 10^{23} \; \text{Ca atoms}}{1 \; \text{mol Ca}} = 1.807 \times 10^{25} \; \text{Ca atoms}$

(b) $1.00 \times 10^{20} \; \text{Ca atoms} \times \dfrac{1 \; \text{mol Ca}}{6.022 \times 10^{23} \; \text{Ca atoms}} = 1.66 \times 10^{-4} \; \text{mol Ca}$

**Think About It** Make sure that units cancel properly in each solution and that the result makes sense. In part (a), for example, the number of moles (30) is greater than one, so the number of atoms is greater than Avogadro's number. In part (b), the number of atoms ($1 \times 10^{20}$) is less than Avogadro's number, so there is less than a mole of substance.

**Practice Problem A** Potassium is the second most abundant metal in the human body. Calculate (a) the number of atoms in 7.31 moles of potassium and (b) the number of moles of potassium that contains $8.91 \times 10^{25}$ atoms.

**Practice Problem B** Calculate (a) the number of atoms in $1.05 \times 10^{-6}$ mole of helium and (b) the number of moles of helium that contains $2.33 \times 10^{21}$ atoms.

## Determining Molar Mass

Although chemists often wish to combine substances in specific mole ratios, there is no direct way to measure the number of moles in a sample of matter. Instead, chemists determine how many moles there are of a substance by measuring its mass (usually in grams). The molar mass of the substance is then used to convert from grams to moles.

The ***molar mass*** ($\mathcal{M}$) of a substance is the mass in grams of 1 mole of the substance. By definition, the mass of a mole of carbon-12 is exactly 12 g. Note that the molar mass of carbon is numerically equal to its atomic mass. Likewise, the atomic mass of calcium is 40.08 amu and its molar mass is 40.08 g, the atomic mass of sodium is 22.99 amu and its molar mass is 22.99 g, and so on. In general, an element's molar mass in grams is numerically equal to its atomic mass in atomic mass units. The molar mass (in grams) of any compound is numerically equal to its molecular or formula mass (in amu). The molar mass of water, for example, is 18.02 g, and the molar mass of sodium chloride (NaCl) is 58.44 g.

When it comes to expressing the molar mass of elements such as oxygen and hydrogen, we have to be careful to specify what form of the element we mean. For instance, the element oxygen exists predominantly as diatomic molecules ($O_2$). Thus, if we say 1 mole of oxygen and by *oxygen* we mean $O_2$, the molecular mass is 32.00 amu and the molar mass is 32.00 g. If on the other hand we mean a mole of atomic oxygen (O), then the molar mass is only 16.00 g, which is numerically equal to the atomic mass of O (16.00 amu). You should be able to tell from the context which form of an element is intended, as the following examples illustrate:

**Student Annotation:** Because the number of moles specifies the number of particles (atoms, molecules, or ions), using molar mass as a conversion factor, in effect, allows us to count the particles in a sample of matter by weighing the sample.

**Student Annotation:** Recall from Section 2.5 that

$$1 \; \text{amu} = 1.661 \times 10^{-24} \; \text{g}$$

This is the reciprocal of Avogadro's number. Expressed another way:

$$1 \; \text{g} = 6.022 \times 10^{23} \; \text{amu}$$

In effect, there is 1 mole of atomic mass units in a gram.

| Context | *Oxygen* Means | Molar Mass |
|---|:---:|:---:|
| How many moles of oxygen react with 2 moles of hydrogen to produce water? | $O_2$ | 32.00 g |
| How many moles of oxygen are there in 1 mole of water? | O | 16.00 g |
| Air is approximately 21% oxygen. | $O_2$ | 32.00 g |
| Many organic compounds contain oxygen. | O | 16.00 g |

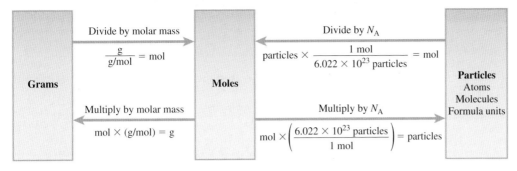

**Figure 3.4** Flowchart for conversions among mass, moles, and number of particles.

Although the term molar mass specifies the mass of 1 mole of a substance, making the appropriate units simply grams (g), we usually express molar masses in units of grams per mole (g/mol) to facilitate calculations involving moles.

## Interconverting Mass, Moles, and Numbers of Particles

Molar mass is the conversion factor that we use to convert from mass ($m$) to moles ($n$), and vice versa. We use Avogadro's number to convert from number of moles to number of particles ($N$), and vice versa. *Particles* in this context may refer to atoms, molecules, ions, or formula units. Figure 3.4 summarizes the operations involved in these conversions.

Sample Problems 3.6 and 3.7 illustrate how the conversions are done.

### Sample Problem 3.6

Determine (a) the number of moles of C in 10.00 g of naturally occurring carbon and (b) the mass of 0.905 mole of sodium chloride.

**Strategy** Use molar mass to convert from mass to moles and to convert from moles to mass.

**Setup** The molar mass of carbon is 12.01 g/mol. The molar mass of a compound is numerically equal to its formula mass. The molar mass of sodium chloride (NaCl) is 58.44 g/mol.

**Solution**

(a) $10.00 \text{ g C} \times \dfrac{1 \text{ mol C}}{12.01 \text{ g C}} = 0.8326 \text{ mol C}$

(b) $0.905 \text{ mol NaCl} \times \dfrac{58.44 \text{ g NaCl}}{1 \text{ mol NaCl}} = 52.9 \text{ g NaCl}$

> **Think About It** Always double-check unit cancellations in problems such as these—errors are common when molar mass is used as a conversion factor. Also make sure that the results make sense. In both cases, a mass smaller than the molar mass corresponds to less than a mole of substance.

**Practice Problem A** (a) Determine the mass in grams of 2.75 moles of glucose ($C_6H_{12}O_6$). (b) Determine the number of moles in 59.8 g of sodium nitrate (NaNO₃).

**Practice Problem B** (a) Determine the mass of sodium metal that contains the same number of moles as 87.2 g of copper metal. (b) Determine the number of moles of helium that has the same mass as 4.505 moles of neon.

### Sample Problem 3.7

(a) Determine the number of water molecules and the numbers of H and O atoms in 3.26 g of water.
(b) Determine the mass of $7.92 \times 10^{19}$ carbon dioxide molecules.

**Strategy** Use molar mass and Avogadro's number to convert from mass to molecules, and vice versa. Use the molecular formula of water to determine the numbers of H and O atoms.

*(Continued)*

**Setup** (a) Starting with mass (3.26 g of water), we use molar mass (18.02 g/mol) to convert to moles of water. From moles, we use Avogadro's number to convert to number of water molecules. In part (b), we reverse the process in part (a) to go from number of molecules to mass of carbon dioxide.

**Solution**

(a) $3.26 \text{ g H}_2\text{O} \times \dfrac{1 \text{ mol H}_2\text{O}}{18.02 \text{ g H}_2\text{O}} \times \dfrac{6.022 \times 10^{23} \text{ H}_2\text{O molecules}}{1 \text{ mol H}_2\text{O}} = 1.09 \times 10^{23} \text{ H}_2\text{O molecules}$

Using the molecular formula, we can determine the number of H and O atoms in 3.26 g of $H_2O$ as follows:

$$1.09 \times 10^{23} \text{ H}_2\text{O molecules} \times \dfrac{2 \text{ H atoms}}{1 \text{ H}_2\text{O molecule}} = 2.18 \times 10^{23} \text{ H atoms}$$

$$1.09 \times 10^{23} \text{ H}_2\text{O molecules} \times \dfrac{1 \text{ O atom}}{1 \text{ H}_2\text{O molecule}} = 1.09 \times 10^{23} \text{ O atoms}$$

**Think About It** Again, check the cancellation of units carefully and make sure that the magnitudes of your results are reasonable.

(b) $7.92 \times 10^{19} \text{ CO}_2 \text{ molecules} \times \dfrac{1 \text{ mol CO}_2}{6.022 \times 10^{23} \text{ CO}_2 \text{ molecules}} \times \dfrac{44.01 \text{ g CO}_2}{1 \text{ mol CO}_2} = 5.79 \times 10^{-3} \text{ g CO}_2$

**Practice Problem A** (a) Calculate the number of oxygen molecules and the number of oxygen atoms in 35.5 g of $O_2$. (b) Calculate the mass of $9.95 \times 10^{14}$ $SO_3$ molecules.

**Practice Problem B** (a) Determine the number of oxygen atoms and hydrogen atoms in 1.00 kg of water. (b) Calculate the mass of calcium carbonate that contains 1.00 mol of oxygen atoms.

## Empirical Formula from Percent Composition

In Section 3.2, we learned how to use the chemical formula (either molecular or empirical) to determine the percent composition by mass. With the concepts of the mole and molar mass, we can now use the experimentally determined percent composition to determine the empirical formula of a compound. Sample Problem 3.8 shows how to do this.

### Sample Problem 3.8

Determine the empirical formula of a compound that is 30.45 percent nitrogen and 69.55 percent oxygen by mass.

**Strategy** Assume a 100-g sample so that the mass percentages of nitrogen and oxygen given in the problem statement correspond to the masses of N and O in the compound. Then, using the appropriate molar masses, convert the grams of each element to moles. Use the resulting numbers as subscripts in the empirical formula, reducing them to the lowest possible whole numbers for the final answer.

**Setup** The empirical formula of a compound consisting of N and O is $N_xO_y$. The molar masses of N and O are 14.01 and 16.00 g/mol, respectively. One hundred grams of a compound that is 30.45 percent nitrogen and 69.55 percent oxygen by mass contains 30.45 g N and 69.55 g O.

**Solution**

$$30.45 \text{ g N} \times \dfrac{1 \text{ mol N}}{14.01 \text{ g N}} = 2.173 \text{ mol N}$$

$$69.55 \text{ g O} \times \dfrac{1 \text{ mol O}}{16.00 \text{ g O}} = 4.347 \text{ mol O}$$

**Think About It** Use the method described in Sample Problem 3.2 to calculate the percent composition of the empirical formula $NO_2$ and verify that it is the same as that given in this problem.

This gives a formula of $N_{2.173}O_{4.347}$. Dividing both subscripts by the smaller of the two to get the smallest possible whole numbers (2.173/2.173 = 1, 4.347/2.173 ≈ 2) gives an empirical formula of $NO_2$. This may or may not be the molecular formula of the compound because both $NO_2$ and $N_2O_4$ have this empirical formula. Without knowing the molar mass, we cannot be sure which one it is.

**Practice Problem A** Determine the empirical formula of a compound that is 52.15 percent C, 13.13 percent H, and 34.73 percent O by mass.

**Practice Problem B** Determine the empirical formula of a compound that is 85.63 percent C and 14.37 percent H by mass.

---

| Checkpoint 3.4 | The Mole and Molar Masses |
|---|---|

**3.4.1** How many molecules are in 30.1 g of sulfur dioxide ($SO_2$)?

a) $1.81 \times 10^{25}$

b) $2.83 \times 10^{23}$

c) $6.02 \times 10^{23}$

d) $1.02 \times 10^{24}$

e) $5.00 \times 10^{-23}$

**3.4.2** How many moles of hydrogen are there in 6.50 g of ammonia ($NH_3$)?

a) 0.382 mol

b) 1.39 mol

c) 0.215 mol

d) 1.14 mol

e) 2.66 mol

**3.4.3** Determine the empirical formula of a compound that has the following composition: 92.3 percent C and 7.7 percent H.

a) CH

b) $C_2H_3$

c) $C_4H_6$

d) $C_6H_7$

e) $C_4H_3$

**3.4.4** Determine the empirical formula of a compound that has the following composition: 48.6 percent C, 8.2 percent H, and 43.2 percent O.

a) $C_3H_8O$

b) $C_3H_6O$

c) $C_2H_5O_2$

d) $C_2H_6O$

e) $C_3H_6O_2$

---

# 3.5 Combustion Analysis

As we saw in Section 3.4, knowing the mass of each element contained in a sample of a substance enables us to determine the empirical formula of the substance. One common, practical use of this ability is the experimental determination of empirical formula by **combustion analysis.**

Combustion analysis of organic compounds (containing carbon, hydrogen, and sometimes oxygen) is carried out using an apparatus like the one shown in Figure 3.5. A sample of known mass is placed in the furnace and heated in the presence of oxygen. The carbon dioxide and water produced from carbon and hydrogen, respectively, in the combustion reaction are collected in "traps," which are weighed before and after the combustion. The difference in mass of each trap before and after the reaction is the mass of the collected product. Knowing the mass of each product, we can determine the percent composition of the compound. And, from percent composition, we can determine the empirical formula.

## Determination of Empirical Formula

When a compound such as glucose is burned in a combustion analysis apparatus, carbon dioxide ($CO_2$) and water ($H_2O$) are produced. Because only oxygen gas is added to the reaction, the carbon and hydrogen present in the products must have come from the glucose. The oxygen in the products may have come from the glucose, but it may also have come from the added oxygen. Suppose that in one such experiment the combustion of 18.8 g of glucose produced 27.6 g of $CO_2$ and 11.3 g of $H_2O$. We can calculate the mass of carbon and hydrogen in the original 18.8-g sample of glucose as follows:

$$\text{mass of C} = 27.6 \text{ g } CO_2 \times \frac{1 \text{ mol } CO_2}{44.01 \text{ g } CO_2} \times \frac{1 \text{ mol C}}{1 \text{ mol } CO_2} \times \frac{12.01 \text{ g C}}{1 \text{ mol C}} = 7.53 \text{ g C}$$

$$\text{mass of H} = 11.3 \text{ g } H_2O \times \frac{1 \text{ mol } H_2O}{18.02 \text{ g } H_2O} \times \frac{2 \text{ mol H}}{1 \text{ mol } H_2O} \times \frac{1.008 \text{ g H}}{1 \text{ mol H}} = 1.26 \text{ g H}$$

**Figure 3.5**   Schematic of a combustion analysis apparatus. $CO_2$ and $H_2O$ produced in combustion are trapped and weighed. The amounts of these products are used to determine how much carbon and hydrogen the combusted sample contained. (CuO is used to ensure complete combustion of all carbon to $CO_2$.)

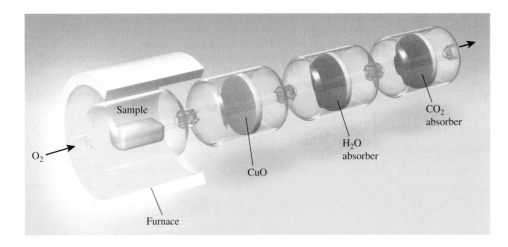

Thus, 18.8 g of glucose contains 7.53 g of carbon and 1.26 g of hydrogen. The remaining mass [18.8 g − (7.53 g + 1.26 g) = 10.0 g] is oxygen.

The number of moles of each element present in 18.8 g of glucose is

$$\text{moles of C} = 7.53 \text{ g C} \times \frac{1 \text{ mol C}}{12.01 \text{ g C}} = 0.627 \text{ mol C}$$

$$\text{moles of H} = 1.26 \text{ g H} \times \frac{1 \text{ mol H}}{1.008 \text{ g H}} = 1.25 \text{ mol H}$$

$$\text{moles of O} = 10.0 \text{ g O} \times \frac{1 \text{ mol O}}{16.00 \text{ g O}} = 0.626 \text{ mol O}$$

**Student Annotation:** Determination of an empirical formula from combustion data can be especially sensitive to rounding error. When solving problems such as these, don't round until the very end.

The empirical formula of glucose can therefore be written $C_{0.627}H_{1.25}O_{0.626}$. Because the numbers in an empirical formula must be integers, we divide each of the subscripts by the smallest subscript, 0.626 (0.627/0.626 ≈ 1, 1.25/0.626 ≈ 2, and 0.626/0.626 = 1), and obtain $CH_2O$ for the empirical formula.

## Determination of Molecular Formula

The empirical formula gives only the ratio of combination of the atoms in a molecule, so there may be numerous compounds with the same empirical formula. If we know the approximate molar mass of the compound, though, we can determine the molecular formula from the empirical formula. For instance, the molar mass of glucose is about 180 g. The *empirical-formula mass* of $CH_2O$ is about 30 g [12.01 g + 2(1.008 g) + 16.00 g]. To determine the molecular formula, we first divide the molar mass by the empirical-formula mass: 180 g/30 g = 6. This tells us that there are six empirical-formula units per molecule in glucose. Multiplying each subscript by 6 (recall that when none is shown, the subscript is understood to be a 1) gives the molecular formula, $C_6H_{12}O_6$.

Sample Problem 3.9 shows how to determine the molecular formula of a compound from its combustion data and molar mass.

Glucose

Benzene

### Sample Problem 3.9

Combustion of a 5.50-g sample of benzene produces 18.59 g $CO_2$ and 3.81 g $H_2O$. Determine the empirical formula and the molecular formula of benzene, given that its molar mass is approximately 78 g/mol.

**Strategy**  From the product masses, determine the mass of C and the mass of H in the 5.50-g sample of benzene. Sum the masses of C and H; the difference between this sum and the original sample mass is the mass of O in the sample (if O is in fact present in benzene). Convert the mass of each element to moles, and use the results as subscripts in a chemical formula. Convert the subscripts to whole numbers by dividing each by the smallest subscript. This gives the empirical formula. To calculate the molecular formula, first divide the molar mass given in the problem statement by the empirical-formula mass. Then, multiply the subscripts in the empirical formula by the resulting number to obtain the subscripts in the molecular formula.

**Setup** The necessary molar masses are $CO_2$, 44.01 g/mol; $H_2O$, 18.02 g/mol; C, 12.01 g/mol; H, 1.008 g/mol; and O, 16.00 g/mol.

**Solution** We calculate the mass of carbon and the mass of hydrogen in the products (and therefore in the original 5.50-g sample) as follows:

$$\text{mass of C} = 18.59 \text{ g } CO_2 \times \frac{1 \text{ mol } CO_2}{44.01 \text{ g } CO_2} \times \frac{1 \text{ mol C}}{1 \text{ mol } CO_2} \times \frac{12.01 \text{ g C}}{1 \text{ mol C}} = 5.073 \text{ g C}$$

$$\text{mass of H} = 3.81 \text{ g } H_2O \times \frac{1 \text{ mol } H_2O}{18.02 \text{ g } H_2O} \times \frac{2 \text{ mol H}}{1 \text{ mol } H_2O} \times \frac{1.008 \text{ g H}}{1 \text{ mol H}} = 0.426 \text{ g H}$$

The total mass of products is 5.073 g + 0.426 g = 5.499 g. Because the combined masses of C and H account for the entire mass of the original sample (5.499 g ≈ 5.50 g), this compound must not contain O.
Converting mass to moles for each element present in the compound,

$$\text{moles of C} = 5.073 \text{ g C} \times \frac{1 \text{ mol C}}{12.01 \text{ g C}} = 0.4224 \text{ mol C}$$

$$\text{moles of H} = 0.426 \text{ g H} \times \frac{1 \text{ mol H}}{1.008 \text{ g H}} = 0.423 \text{ mol H}$$

gives the formula $C_{0.4224}H_{0.423}$. Converting the subscripts to whole numbers (0.4224/0.4224 = 1; 0.423/0.4224 ≈ 1) gives the empirical formula, CH.
Finally, dividing the approximate molar mass (78 g/mol) by the empirical-formula mass (12.01 g/mol + 1.008 g/mol = 13.02 g/mol) gives 78/13.02 ≈ 6. Then, multiplying both subscripts in the empirical formula by 6 gives the molecular formula, $C_6H_6$.

> **Think About It** Use the molecular formula to determine the molar mass and make sure that the result agrees with the molar mass given in the problem. For $C_6H_6$ the molar mass is 6(12.01 g/mol) + 6(1.008 g/mol) = 78.11 g/mol, which agrees with the 78 g/mol given in the problem statement.

**Practice Problem A** The combustion of a 28.1-g sample of ascorbic acid (vitamin C) produces 42.1 g $CO_2$ and 11.5 g $H_2O$. Determine the empirical and molecular formulas of ascorbic acid. The molar mass of ascorbic acid is approximately 176 g/mol.

**Practice Problem B** Determine the mass of $CO_2$ and the mass of $H_2O$ produced by the combustion of 1.05 g of a compound with the empirical formula $CH_4O$.

---

## Checkpoint 3.5 Combustion Analysis

**3.5.1** What is the empirical formula of a compound containing C, H, and O if combustion of 1.23 g of the compound yields 1.8 g $CO_2$ and 0.74 g $H_2O$?

a) $CH_3O$

b) $C_2H_3O$

c) CHO

d) $C_2H_3O_2$

e) $CH_2O$

**3.5.2** What are the empirical and molecular formulas of a hydrocarbon if combustion of 2.10 g of the compound yields 6.59 g $CO_2$ and 2.70 g $H_2O$ and its molar mass is about 70 g/mol?

a) CH, $C_6H_6$

b) CH, $C_5H_5$

c) $CH_2$, $C_6H_{12}$

d) $CH_2$, $C_5H_{10}$

e) $CH_2$, $C_3H_6$

**3.5.3** Determine the masses of $CO_2$ and $H_2O$ produced by the combustion of 0.986 g of a compound with empirical formula $C_3H_6O_2$.

a) 1.76 g $CO_2$, 0.719 g $H_2O$

b) 0.480 g $CO_2$, 0.081 g $H_2O$

c) 1.76 g $CO_2$, 1.44 g $H_2O$

d) 0.329 g $CO_2$, 0.657 g $H_2O$

e) 0.657 g $CO_2$, 0.329 g $H_2O$

**3.5.4** How is it possible for the combined masses of $CO_2$ and $H_2O$ produced in combustion to be greater than the mass of the compound that is burned?

a) There is experimental error.

b) Different balances are used to determine mass before and after combustion.

c) Combustion violates the law of conservation of mass.

d) Oxygen consumed in the combustion contributes to the mass of products.

e) $CO_2$ and $H_2O$ have greater molar masses than the compound that is burned.

**Multimedia**

Matter—stoichiometry.

## 3.6   Calculations with Balanced Chemical Equations

Often we would like to predict how much of a particular product will form from a given amount of a reactant. Other times, we perform an experiment, measure the amount of product formed, and use this information to deduce the quantity or composition of a reactant. Balanced chemical equations can be powerful tools for this type of problem solving.

### Moles of Reactants and Products

Based on the equation for the reaction of carbon monoxide with oxygen to produce carbon dioxide,

$$2CO(g) \;+\; O_2(g) \longrightarrow 2CO_2(g)$$

2 moles of CO combine with 1 mole of $O_2$ to produce 2 moles of $CO_2$. In stoichiometric calculations, we say that 2 moles of CO are *equivalent* to 2 moles of $CO_2$, which can be represented as

$$2 \text{ mol CO} \simeq 2 \text{ mol CO}_2$$

where the symbol $\simeq$ means "is stoichiometrically equivalent to" or simply "is equivalent to." The ratio of moles of CO consumed to moles of $CO_2$ produced is 2:2 or 1:1. Regardless of the number of moles of CO consumed in the reaction, the same number of moles of $CO_2$ will be produced. We can use this constant ratio as a conversion factor that can be written as

$$\frac{2 \text{ mol CO}}{2 \text{ mol CO}_2} \quad \text{or} \quad \frac{1 \text{ mol CO}}{1 \text{ mol CO}_2}$$

The ratio can also be written as the reciprocal,

$$\frac{2 \text{ mol CO}_2}{2 \text{ mol CO}} \quad \text{or} \quad \frac{1 \text{ mol CO}_2}{1 \text{ mol CO}}$$

These conversion factors enable us to determine how many moles of $CO_2$ will be produced upon reaction of a given amount of CO, or how much CO is necessary to produce a specific amount of $CO_2$. Consider the complete reaction of 3.82 moles of CO to form $CO_2$. To calculate the number of moles of $CO_2$ produced, we use the conversion factor with moles of $CO_2$ in the numerator and moles of CO in the denominator.

$$\text{moles } CO_2 \text{ produced} = 3.82 \text{ mol CO} \times \frac{1 \text{ mol CO}_2}{1 \text{ mol CO}} = 3.82 \text{ mol CO}_2$$

Similarly, we can use other ratios represented in the balanced equation as conversion factors. For example, we have 1 mol $O_2 \simeq 2$ mol $CO_2$ and 2 mol CO $\simeq 1$ mol $O_2$. The corresponding conversion factors allow us to calculate the amount of $CO_2$ produced upon reaction of a given amount of $O_2$, and the amount of one reactant necessary to react completely with a given amount of the other. Using the preceding example, we can determine the **stoichiometric amount** of $O_2$ (how many moles of $O_2$ are needed to react with 3.82 moles of CO).

$$\text{moles } O_2 \text{ needed} = 3.82 \text{ mol CO} \times \frac{1 \text{ mol O}_2}{2 \text{ mol CO}} = 1.91 \text{ mol O}_2$$

Sample Problem 3.10 illustrates how to determine reactant and product amounts using a balanced chemical equation.

| Sample Problem 3.10 |
| --- |

Urea [$(NH_2)_2CO$] is a by-product of protein metabolism. This waste product is formed in the liver and then filtered from the blood and excreted in the urine by the kidneys. Urea can be synthesized in the laboratory by the combination of ammonia and carbon dioxide according to the equation

$$2NH_3(g) + CO_2(g) \longrightarrow (NH_2)_2CO(aq) + H_2O(l)$$

(a) Calculate the amount of urea that will be produced by the complete reaction of 5.25 moles of ammonia. (b) Determine the stoichiometric amount of carbon dioxide required to react with 5.25 moles of ammonia.

**Strategy** Use the balanced chemical equation to determine the correct stoichiometric conversion factors, and then multiply by the number of moles of ammonia given.

**Setup** According to the balanced chemical equation, the conversion factor for ammonia and urea is either

$$\frac{2 \text{ mol } NH_3}{1 \text{ mol } (NH_2)_2CO} \quad \text{or} \quad \frac{1 \text{ mol } (NH_2)_2CO}{2 \text{ mol } NH_3}$$

To multiply by moles of $NH_3$ and have the units cancel properly, we use the conversion factor with moles of $NH_3$ in the denominator.

Similarly, the conversion factor for ammonia and carbon dioxide can be written as

$$\frac{2 \text{ mol } NH_3}{1 \text{ mol } CO_2} \quad \text{or} \quad \frac{1 \text{ mol } CO_2}{2 \text{ mol } NH_3}$$

Again, we select the conversion factor with ammonia in the denominator so that moles of $NH_3$ will cancel in the calculation.

**Solution**

(a) moles $(NH_2)_2CO$ produced = $5.25 \text{ mol } NH_3 \times \dfrac{1 \text{ mol } (NH_2)_2CO}{2 \text{ mol } NH_3}$ = 2.63 mol $(NH_2)_2CO$

(b) moles $CO_2$ required = $5.25 \text{ mol } NH_3 \times \dfrac{1 \text{ mol } CO_2}{2 \text{ mol } NH_3}$ = 2.63 mol $CO_2$

**Think About It** As always, check to be sure that units cancel properly in the calculation. Also, the balanced equation indicates that there will be *fewer* moles of urea produced than ammonia consumed. Therefore, your calculated number of moles of urea (2.63) should be *smaller* than the number of moles given in the problem (5.25). Similarly, the stoichiometric coefficients in the balanced equation are the same for carbon dioxide and urea, so your answers to this problem should also be the same for both species.

**Practice Problem A** Nitrogen and hydrogen react to form ammonia according to the following balanced equation: $N_2(g) + 3H_2(g) \longrightarrow 2NH_3(g)$. Calculate the number of moles of hydrogen required to react with 0.0880 mole of nitrogen, and the number of moles of ammonia that will form.

**Practice Problem B** Tetraphosphorus decoxide ($P_4O_{10}$) reacts with water to produce phosphoric acid. Write and balance the equation for this reaction, and determine the number of moles of each reactant required to produce 5.80 moles of phosphoric acid.

## Mass of Reactants and Products

Balanced chemical equations give us the relative amounts of reactants and products in terms of moles. However, because we measure reactants and products in the laboratory by weighing them, most often such calculations start with mass rather than the number of moles. Sample Problem 3.11 illustrates how to determine amounts of reactants and products in terms of grams.

## Sample Problem 3.11

Nitrous oxide ($N_2O$), also known as "laughing gas," is commonly used as an anesthetic in dentistry. It is manufactured by heating ammonium nitrate. The balanced equation is

$$NH_4NO_3(s) \xrightarrow{\Delta} N_2O(g) + 2H_2O(g)$$

(a) Calculate the mass of ammonium nitrate that must be heated in order to produce 10.0 g of nitrous oxide. (b) Determine the corresponding mass of water produced in the reaction.

**Strategy**  For part (a), use the molar mass of nitrous oxide to convert the given mass of nitrous oxide to moles, use the appropriate stoichiometric conversion factor to convert to moles of ammonium nitrate, and then use the molar mass of ammonium nitrate to convert to grams of ammonium nitrate. For part (b), use the molar mass of nitrous oxide to convert the given mass of nitrous oxide to moles, use the stoichiometric conversion factor to convert from moles of nitrous oxide to moles of water, and then use the molar mass of water to convert to grams of water.

**Setup**  The molar masses are as follows: 80.05 g/mol for $NH_4NO_3$, 44.02 g/mol for $N_2O$, and 18.02 g/mol for $H_2O$. The conversion factors from nitrous oxide to ammonium nitrate and from nitrous oxide to water are, respectively:

$$\frac{1 \text{ mol } NH_4NO_3}{1 \text{ mol } N_2O} \quad \text{and} \quad \frac{2 \text{ mol } H_2O}{1 \text{ mol } N_2O}$$

**Solution**

(a)
$$10.0 \text{ g } N_2O \times \frac{1 \text{ mol } N_2O}{44.02 \text{ g } N_2O} = 0.227 \text{ mol } N_2O$$

$$0.227 \text{ mol } N_2O \times \frac{1 \text{ mol } NH_4NO_3}{1 \text{ mol } N_2O} = 0.227 \text{ mol } NH_4NO_3$$

$$0.227 \text{ mol } NH_4NO_3 \times \frac{80.05 \text{ g } NH_4NO_3}{1 \text{ mol } NH_4NO_3} = 18.2 \text{ g } NH_4NO_3$$

Thus, 18.2 g of ammonium nitrate must be heated in order to produce 10.0 g of nitrous oxide.

(b) Starting with the number of moles of nitrous oxide determined in the first step of part (a),

$$0.227 \text{ mol } N_2O \times \frac{2 \text{ mol } H_2O}{1 \text{ mol } N_2O} = 0.454 \text{ mol } H_2O$$

$$0.454 \text{ mol } H_2O \times \frac{18.02 \text{ g } H_2O}{1 \text{ mol } H_2O} = 8.18 \text{ g } H_2O$$

Therefore, 8.18 g of water will also be produced in the reaction.

**Think About It**  Use the law of conservation of mass to check your answers. Make sure that the combined mass of both products is equal to the mass of reactant you determined in part (a). In this case (rounded to the appropriate number of significant figures), 10.0 g + 8.18 g = 18.2 g. Remember that small differences may arise as the result of rounding.

**Practice Problem A**  Calculate the mass of water produced by the metabolism of 56.8 g of glucose. (See the box on page 84 for the necessary equation.)

**Practice Problem B**  What mass of glucose must be metabolized in order to produce 175 g of water?

## Checkpoint 3.6   Calculations with Balanced Chemical Equations

**3.6.1**  How many moles of LiOH will be produced if 0.550 mol Li reacts according to the following equation?

$$2Li(s) + 2H_2O(l) \longrightarrow 2LiOH(aq) + H_2(g)$$

a) 0.550 mol

b) 1.10 mol

c) 0.275 mol

d) 2.20 mol

e) 2.00 mol

**3.6.2**  Determine the stoichiometric amount (in grams) of $O_2$ necessary to react with 5.71 g Al according to the following equation:

$$4Al(s) + 3O_2(g) \longrightarrow 2Al_2O_3(s)$$

a) 5.08 g

b) 9.03 g

c) 2.54 g

d) 4.28 g

e) 7.61 g

# 3.7 Limiting Reactants

When a chemist carries out a reaction, the reactants usually are not present in stoichiometric amounts. Because the goal of a reaction is usually to produce the maximum quantity of a useful compound from the starting materials, an excess of one reactant is commonly supplied to ensure that the more expensive or more important reactant is converted completely to the desired product. Consequently, some of the reactant supplied in excess will be left over at the end of the reaction. The reactant used up first in a reaction is called the **limiting reactant,** because the amount of this reactant *limits* the amount of product that can form. When all the limiting reactant has been consumed, no more product can be formed. **Excess reactants** are those present in quantities *greater* than necessary to react with the quantity of the limiting reactant.

The concept of a limiting reactant applies to everyday tasks, too, such as making ham sandwiches. Suppose you want to make the maximum number of ham sandwiches possible, each of which will consist of two slices of bread and one slice of ham. If you have eight slices of bread and six slices of ham, how many sandwiches can you make? The answer is four, because after making four sandwiches you will be out of bread. You will have two slices of ham left over, but without additional bread you will be unable to make any more sandwiches. In this case, bread is the limiting reactant and ham is the excess reactant.

**Student Annotation:** Limiting reactants and excess reactants are also referred to as *limiting reagents* and *excess reagents.*

Multimedia

Limiting reagent in reaction of NO and $O_2$.

## Determining the Limiting Reactant

In problems involving limiting reactants, the first step is to determine which is the limiting reactant. After the limiting reactant has been identified, the rest of the problem can be solved using the approach outlined in Section 3.6. Consider the formation of methanol ($CH_3OH$) from carbon monoxide and hydrogen:

$$CO(g) + 2H_2(g) \longrightarrow CH_3OH(l)$$

Suppose that initially we have 5 moles of CO and 8 moles of $H_2$, the ratio shown in Figure 3.6(a).

We can use the stoichiometric conversion factors to determine how many moles of $H_2$ are necessary for all the CO to react. From the balanced equation, we have 1 mol CO ≏ 2 mol $H_2$. Therefore, the amount of $H_2$ necessary to react with 5 mol CO is

$$\text{moles of } H_2 = 5 \text{ mol CO} \times \frac{2 \text{ mol } H_2}{1 \text{ mol CO}} = 10 \text{ mol } H_2$$

Because there are only 8 moles of $H_2$ available, there is insufficient $H_2$ to react with all the CO. Therefore, $H_2$ is the limiting reactant and CO is the excess reactant. $H_2$ will be used up first, and when it is gone, the formation of methanol will cease and there will be some CO left over, as shown in Figure 3.6(b). To determine how much CO will be left over when the reaction is complete, we must first calculate the amount of CO that will react with all 8 moles of $H_2$:

$$\text{moles of } CO = 8 \text{ mol } H_2 \times \frac{1 \text{ mol CO}}{2 \text{ mol } H_2} = 4 \text{ mol CO}$$

Thus, there will be 4 moles of CO consumed and 1 mole (5 mol − 4 mol) left over. Sample Problem 3.12 illustrates how to combine the concept of a limiting reactant with the conversion between mass and moles. Figure 3.7 (pp. 98–99) illustrates the steps for this type of calculation.

Media Player/MPEG Animation: Figure 3.7, Limiting Reactant Problems, pp. 98–99.

(a)                                         (b)

**Figure 3.6** The reaction of (a) $H_2$ and CO to form (b) $CH_3OH$. Each molecule represents 1 mole of substance. In this case, $H_2$ is the limiting reactant and there is 1 mole of CO remaining when the reaction is complete.

Figure 3.7
# Limiting Reactant Problems

**START**

Determine what mass of $NH_3$ forms when 84.06 g $N_2$ and 22.18 g $H_2$ react according to the equation:

$$N_2 \ + \ 3H_2 \ \longrightarrow \ 2NH_3$$

Convert to moles.

$$\frac{84.06 \text{ g } N_2}{28.02 \text{ g/mol}} = 3.000 \text{ mol } N_2$$

$$\frac{22.18 \text{ g } H_2}{2.016 \text{ g/mol}} = 11.00 \text{ mol } H_2$$

Determine moles $NH_3$.

Total mass before reaction:

$$84.06 \text{ g } N_2 \ + \ 22.18 \text{ g } H_2 \ = \ 106.24 \text{ g}$$

Compare the total mass *after* the reaction with the total mass *before* the reaction. The small difference between the masses before and after is due to rounding.

$$\sum \text{ after reaction } = \ 102.2 \text{ g } NH_3 \ + \ 4.03 \text{ g } H_2 \ = \ 106.2 \text{ g}$$

Add the mass of the product and the mass of leftover excess reactant to get the total mass after reaction.

Use coefficients as conversion factors.

**Method 1**

$$3.000 \text{ mol N}_2 \times \frac{2 \text{ mol NH}_3}{1 \text{ mol N}_2} = 6.000 \text{ mol NH}_3$$

$$11.00 \text{ mol H}_2 \times \frac{2 \text{ mol NH}_3}{3 \text{ mol H}_2} = 7.333 \text{ mol NH}_3$$

or

**Method 2**

Rewrite the balanced equation using actual amounts. According to the balanced equation, 3.667 mol $N_2$ are required to react with 11.00 mol $H_2$.

$$3.000 \text{ N}_2 + 9.000 \text{ H}_2 \longrightarrow 6.000 \text{ NH}_3$$

$$3.667 \text{ N}_2 + 11.00 \text{ H}_2 \longrightarrow 7.333 \text{ NH}_3$$

Either way, the *smaller* amount of product is correct.

6.000 mol $NH_3$

Convert to grams.

$$6.000 \text{ mol NH}_3 \times \frac{17.03 \text{ g NH}_3}{1 \text{ mol NH}_3} = 102.2 \text{ g NH}_3$$

CHECK

$N_2$ was the limiting reactant. Calculate how much $H_2$ is left over.

$$2.00 \text{ mol H}_2 \times \frac{2.016 \text{ g H}_2}{1 \text{ mol H}_2} = 4.03 \text{ g H}_2$$

Convert to grams.

11.00 mol initially
− 9.00 mol consumed
2.00 mol $H_2$ remaining

## What's the point?

There is more than one correct method for solving many types of problems. This limiting reactant problem shows two different routes to the correct answer, and shows how the result can be compared to the information given in the problem to determine whether or not it is reasonable and correct.

The reaction of sodium bicarbonate and citric acid produces Alka-Seltzer's effervescence.

## Sample Problem 3.12

Alka-Seltzer tablets contain aspirin, sodium bicarbonate, and citric acid. When they come into contact with water, the sodium bicarbonate ($NaHCO_3$) and citric acid ($H_3C_6H_5O_7$) react to form carbon dioxide gas, among other products.

$$3NaHCO_3(aq) + H_3C_6H_5O_7(aq) \longrightarrow 3CO_2(g) + 3H_2O(l) + Na_3C_6H_5O_7(aq)$$

The formation of $CO_2$ causes the trademark fizzing when the tablets are dropped into a glass of water. An Alka-Seltzer tablet contains 1.700 g of sodium bicarbonate and 1.000 g of citric acid. Determine, for a single tablet dissolved in water, (a) which ingredient is the limiting reactant, (b) what mass of the excess reactant is left over when the reaction is complete, and (c) what mass of $CO_2$ forms.

**Strategy**  Convert each of the reactant masses to moles. Use the balanced equation to write the necessary stoichiometric conversion factor and determine which reactant is limiting. Again, using the balanced equation, write the stoichiometric conversion factors to determine the number of moles of excess reactant remaining and the number of moles of $CO_2$ produced. Finally, use the appropriate molar masses to convert moles of excess reactant and moles of $CO_2$ to grams.

**Setup**  The required molar masses are 84.01 g/mol for $NaHCO_3$, 192.12 g/mol for $H_3C_6H_5O_7$, and 44.01 g/mol for $CO_2$. From the balanced equation we have 3 mol $NaHCO_3$ ≏ 1 mol $H_3C_6H_5O_7$, 3 mol $NaHCO_3$ ≏ 3 mol $CO_2$, and 1 mol $H_3C_6H_5O_7$ ≏ 3 mol $CO_2$. The necessary stoichiometric conversion factors are therefore:

$$\frac{3 \text{ mol } NaHCO_3}{1 \text{ mol } H_3C_6H_5O_7} \qquad \frac{1 \text{ mol } H_3C_6H_5O_7}{3 \text{ mol } NaHCO_3} \qquad \frac{3 \text{ mol } CO_2}{3 \text{ mol } NaHCO_3} \qquad \frac{3 \text{ mol } CO_2}{1 \text{ mol } H_3C_6H_5O_7}$$

**Solution**

$$1.700 \text{ g } NaHCO_3 \times \frac{1 \text{ mol } NaHCO_3}{84.01 \text{ g } NaHCO_3} = 0.02024 \text{ mol } NaHCO_3$$

$$1.000 \text{ g } H_3C_6H_5O_7 \times \frac{1 \text{ mol } H_3C_6H_5O_7}{192.12 \text{ g } H_3C_6H_5O_7} = 0.005205 \text{ mol } H_3C_6H_5O_7$$

(a) To determine which reactant is limiting, calculate the amount of citric acid necessary to react completely with 0.02024 mol sodium bicarbonate.

$$0.02024 \text{ mol } NaHCO_3 \times \frac{1 \text{ mol } H_3C_6H_5O_7}{3 \text{ mol } NaHCO_3} = 0.006745 \text{ mol } H_3C_6H_5O_7$$

The amount of $H_3C_6H_5O_7$ required to react with 0.02024 mol of $NaHCO_3$ is more than a tablet contains. Therefore, citric acid is the limiting reactant and sodium bicarbonate is the excess reactant.

(b) To determine the mass of excess reactant ($NaHCO_3$) left over, first calculate the amount of $NaHCO_3$ that will react:

$$0.005205 \text{ mol } H_3C_6H_5O_7 \times \frac{3 \text{ mol } NaHCO_3}{1 \text{ mol } H_3C_6H_5O_7} = 0.01562 \text{ mol } NaHCO_3$$

Thus, 0.01562 mole of $NaHCO_3$ will be consumed, leaving 0.00462 mole unreacted. Convert the unreacted amount to grams as follows:

$$0.00462 \text{ mol } NaHCO_3 \times \frac{84.01 \text{ g } NaHCO_3}{1 \text{ mol } NaHCO_3} = 0.388 \text{ g } NaHCO_3$$

(c) To determine the mass of $CO_2$ produced, first calculate the number of moles of $CO_2$ produced from the number of moles of limiting reactant ($H_3C_6H_5O_7$) consumed:

$$0.005205 \text{ mol } H_3C_6H_5O_7 \times \frac{3 \text{ mol } CO_2}{1 \text{ mol } H_3C_6H_5O_7} = 0.01562 \text{ mol } CO_2$$

Convert this amount to grams as follows:

$$0.01562 \text{ mol } CO_2 \times \frac{44.01 \text{ g } CO_2}{1 \text{ mol } CO_2} = 0.6874 \text{ g } CO_2$$

To summarize the results: (a) citric acid is the limiting reactant, (b) 0.388 g sodium bicarbonate remains unreacted, and (c) 0.6874 g carbon dioxide is produced.

**Think About It**  In a problem such as this, it is a good idea to check your work by calculating the amounts of the other products in the reaction. According to the law of conservation of mass, the combined starting mass of the two reactants (1.700 g + 1.000 g = 2.700 g) should equal the sum of the masses of products and leftover excess reactant. In this case, the masses of $H_2O$ and $Na_3C_6H_5O_7$ produced are 0.2815 g and 1.343 g, respectively. The mass of $CO_2$ produced is 0.6874 g [from part (c)] and the amount of excess $NaHCO_3$ is 0.388 g [from part (b)]. The total, 0.2815 g + 1.343 g + 0.6874 g + 0.388 g, is 2.700 g, identical to the total mass of the reactants.

**Practice Problem A** Ammonia is produced by the reaction of nitrogen and hydrogen according to the equation, $N_2(g) + 3H_2(g) \longrightarrow 2NH_3(g)$. Calculate the mass of ammonia produced when 35.0 g of nitrogen react with 12.5 g of hydrogen. Which is the excess reactant and how much of it will be left over when the reaction is complete?

**Practice Problem B** Potassium hydroxide and phosphoric acid react to form potassium phosphate and water according to the equation: $3KOH(aq) + H_3PO_4(aq) \longrightarrow K_3PO_4(aq) + 3H_2O(l)$. Determine the starting mass of each reactant if 55.7 g $K_3PO_4$ is produced and 89.8 g $H_3PO_4$ remains unreacted.

## Reaction Yield

When you use stoichiometry to calculate the amount of product formed in a reaction, you are calculating the ***theoretical yield*** of the reaction. The theoretical yield is the amount of product that forms when *all* the limiting reactant reacts to form the desired product. It is the *maximum* obtainable yield, predicted by the balanced equation. In practice, the ***actual yield***—the amount of product actually obtained from a reaction—is almost always less than the theoretical yield. There are many reasons for the difference between the actual and theoretical yields. For instance, some of the reactants may not react to form the desired product. They may react to form different products, in something known as *side reactions,* or they may simply remain unreacted. In addition, it may be difficult to isolate and recover all the product at the end of the reaction. Chemists often determine the efficiency of a chemical reaction by calculating its ***percent yield,*** which tells *what percentage the actual yield is of the theoretical yield*. It is calculated as follows:

$$\% \text{ yield} = \frac{\text{actual yield}}{\text{theoretical yield}} \times 100\% \qquad \text{Equation 3.2}$$

Percent yields may range from a tiny fraction to 100 percent. (They cannot exceed 100 percent.) Chemists try to maximize percent yield in a variety of ways. Factors that can affect percent yield, including temperature and pressure, are discussed in Chapter 15. Sample Problem 3.13 shows how to calculate the percent yield of a pharmaceutical manufacturing process.

### Sample Problem 3.13

Aspirin, acetylsalicylic acid ($C_9H_8O_4$), is the most commonly used pain reliever in the world. It is produced by the reaction of salicylic acid ($C_7H_6O_3$) and acetic anhydride ($C_4H_6O_3$) according to the following equation:

| $C_7H_6O_3$ | + | $C_4H_6O_3$ | $\longrightarrow$ | $C_9H_8O_4$ | + | $HC_2H_3O_2$ |
| salicylic acid | | acetic anhydride | | acetylsalicylic acid | | acetic acid |

In a certain aspirin synthesis, 104.8 g of salicylic acid and 110.9 g of acetic anhydride are combined. Calculate the percent yield of the reaction if 105.6 g of aspirin are produced.

**Strategy** Convert reactant grams to moles, and determine which is the limiting reactant. Use the balanced equation to determine the number of moles of aspirin that can be produced, and convert this number of moles to grams for the theoretical yield. Use the actual yield (given in the problem) and the calculated theoretical yield to calculate the percent yield.

**Setup** The necessary molar masses are 138.12 g/mol for salicylic acid, 102.09 g/mol for acetic anhydride, and 180.15 g/mol for aspirin.

*(Continued)*

**Solution**

$$104.8 \text{ g } C_7H_6O_3 \times \frac{1 \text{ mol } C_7H_6O_3}{138.12 \text{ g } C_7H_6O_3} = 0.7588 \text{ mol } C_7H_6O_3$$

$$110.9 \text{ g } C_4H_6O_3 \times \frac{1 \text{ mol } C_4H_6O_3}{102.09 \text{ g } C_4H_6O_3} = 1.086 \text{ mol } C_4H_6O_3$$

Because the two reactants combine in a 1:1 mole ratio, the reactant present in the smallest number of moles (in this case, salicylic acid) is the limiting reactant. According to the balanced equation, one mole of aspirin is produced for every mole of salicylic acid consumed.

$$1 \text{ mol salicylic acid } (C_7H_6O_3) \simeq 1 \text{ mol aspirin } (C_9H_8O_4)$$

Therefore, the theoretical yield of aspirin is 0.7588 mol. We convert this to grams using the molar mass of aspirin:

$$0.7588 \text{ mol } C_9H_8O_4 \times \frac{180.15 \text{ g } C_9H_8O_4}{1 \text{ mol } C_9H_8O_4} = 136.7 \text{ g } C_9H_8O_4$$

Thus, the theoretical yield is 136.7 g. If the actual yield is 105.6 g, the percent yield is

$$\% \text{ yield} = \frac{105.6 \text{ g}}{136.7 \text{ g}} \times 100\% = 77.25\% \text{ yield}$$

**Think About It** Make sure you have used the proper molar masses and remember that percent yield can never exceed 100 percent.

**Practice Problem A** Diethyl ether is produced from ethanol according to the following equation:

$$2CH_3CH_2OH(l) \longrightarrow CH_3CH_2OCH_2CH_3(l) + H_2O(l)$$

Calculate the percent yield if 68.6 g of ethanol reacts to produce 16.1 g of ether.

**Practice Problem B** What mass of ether will be produced if 221 g of ethanol reacts with a 68.9 percent yield?

## Checkpoint 3.7    Limiting Reactants

**3.7.1** What mass of $CaSO_4$ is produced according to the given equation when 5.00 g of each reactant are combined?

$$CaF_2(s) + H_2SO_4(aq) \longrightarrow CaSO_4(s) + 2HF(g)$$

a) 10.0 g

b) 11.6 g

c) 6.94 g

d) 8.72 g

e) 5.02 g

**3.7.2** What is the percent yield for a process in which 10.4 g $CH_3OH$ reacts and 10.1 g $CO_2$ forms according to the following equation?

$$2CH_3OH(l) + 3O_2(g) \longrightarrow 2CO_2(g) + 4H_2O(l)$$

a) 97.1%

b) 70.7%

c) 52.1%

d) 103%

e) 37.9%

**3.7.3** How many moles of $NH_3$ can be produced by the combination of 3.0 mol $N_2$ and 1.5 mol $H_2$?

a) 2.0 mol

b) 1.5 mol

c) 0.50 mol

d) 6.0 mol

e) 1.0 mol

**3.7.4** What mass of water is produced by the reaction of 50.0 g $CH_3OH$ with an excess of $O_2$ when the yield is 53.2%?

$$2CH_3OH(g) + 3O_2(g) \longrightarrow 2CO_2(g) + 4H_2O(l)$$

a) 28.1 g

b) 56.2 g

c) 29.9 g

d) 15.0 g

e) 26.6 g

**3.7.5** Reactants A (red) and B (blue) combine to form a single product C (purple) according to the equation $2A + B \longrightarrow C$. What is the limiting reactant in the reaction vessel shown?

a) A

b) B

c) C

d) None. Reactants are present in stoichiometric amounts.

**3.7.6** Which of the following represents the contents of the reaction vessel in Checkpoint 3.7.5 after the reaction is complete?

a)

b)

c)

d)

e)

# How Am I Supposed to Remember All These Reactions?

As you continue to study chemistry, you will encounter a wide variety of chemical reactions. The sheer number of different reactions can seem daunting at times, but most of them fall into a relatively small number of categories. Becoming familiar with several reaction types and learning to recognize patterns of reactivity will help you make sense out of the reactions in this book. Three of the most commonly encountered reaction types are *combination, decomposition,* and *combustion.*

*Combination.* A reaction in which two or more reactants combine to form a single product is known as a **combination reaction.** Examples include the reaction of ammonia and hydrogen chloride to form ammonium chloride,

$$NH_3(g) + HCl(g) \longrightarrow NH_4Cl(s)$$

and the reaction of nitrogen and hydrogen gases to form ammonia,

$$N_2(g) + 3H_2(g) \longrightarrow 2NH_3(g)$$

*Decomposition.* A reaction in which two or more products form from a single reactant is known as a **decomposition reaction.** A decomposition reaction is essentially the opposite of a combination reaction. Examples of this type of reaction include the decomposition of calcium carbonate to produce calcium oxide and carbon dioxide gas,

$$CaCO_3(s) \xrightarrow{\Delta} CaO(s) + CO_2(g)$$

and the decomposition of hydrogen peroxide to produce water and oxygen gas,

$$2H_2O_2(aq) \longrightarrow 2H_2O(l) + O_2(g)$$

*Combustion.* As you learned in Section 3.3, a **combustion reaction** is one in which a substance burns in the presence of oxygen. Combustion of a compound that contains C and H (or C, H, and O) produces carbon dioxide gas and water. By convention, we will consider the water produced in a combustion reaction to be *liquid* water. Examples of this type of combustion are the combustion of formaldehyde,

$$CH_2O(l) + O_2(g) \longrightarrow CO_2(g) + H_2O(l)$$

and the combustion of methane,

$$CH_4(g) + 2O_2(g) \longrightarrow CO_2(g) + 2H_2O(l)$$

Although these combustion reactions are shown here as balanced equations, oxygen is generally supplied in excess in such processes to ensure complete combustion.

---

## Sample Problem 3.14

Determine whether each of the following equations represents a combination reaction, a decomposition reaction, or a combustion reaction: (a) $H_2(g) + Br_2(g) \longrightarrow 2HBr(g)$, (b) $2HCO_2H(l) + O_2(g) \longrightarrow 2CO_2(g) + 2H_2O(l)$, (c) $2KClO_3(s) \longrightarrow 2KCl(s) + 3O_2(g)$.

**Strategy** Look at the reactants and products in each balanced equation to see if two or more reactants combine into one product (a combination reaction), if one reactant splits into two or more products (a decomposition reaction), or if the main products formed are carbon dioxide gas and water (a combustion reaction).

**Setup** The equation in part (a) depicts *two reactants* and *one product.* The equation in part (b) represents a combination with $O_2$ of a compound containing C, H, and O to produce $CO_2$ and $H_2O$. The equation in part (c) represents *two products* being formed from a *single reactant.*

**Solution** These equations represent (a) a combination reaction, (b) a combustion reaction, and (c) a decomposition reaction.

> **Think About It** Make sure that a reaction identified as a combination has only one product [as in part (a)], a reaction identified as a decomposition has only one reactant [as in part (b)], and a reaction identified as a combustion produces only $CO_2$ and $H_2O$ [as in part (c)].

**Practice Problem A** Identify each of the following as a combination, decomposition, or combustion reaction: (a) $C_2H_4O_2(l) + 2O_2(g) \longrightarrow 2CO_2(g) + 2H_2O(l)$, (b) $2Na(s) + Cl_2(g) \longrightarrow 2NaCl(s)$, (c) $2NaH(s) \longrightarrow 2Na(s) + H_2(g)$.

**Practice Problem B** Using the chemical species $A_2$, B, and AB, write a balanced equation for a combination reaction.

# Applying What You've Learned

Cisplatin [$Pt(NH_3)_2Cl_2$] is sometimes called "the penicillin of cancer drugs" because it is effective in the treatment of a wide variety of cancers. It is prepared by the reaction of ammonium tetrachloroplatinate(II) with ammonia:

**Student Annotation:** Note that the equation is not balanced.

$(NH_4)_2PtCl_4(aq)$          $+$          $NH_3(aq)$          $\longrightarrow$          $Pt(NH_3)_2Cl_2(s)$          $+$          $NH_4Cl(aq)$

This is a very expensive process because $(NH_4)_2PtCl_4$ contains platinum, a precious metal that historically has sold for roughly twice the price of gold. Using a large excess of ammonia helps manufacturers maximize conversion of the high-priced reactant to the desired product.

## Problems:

a)  Determine the molecular mass and percent composition by mass for ammonium tetrachloroplatinate(II) and for cisplatin. [◄◄ Sample Problems 3.1 and 3.2]

b)  Balance the equation for the production of cisplatin from ammonium tetrachloroplatinate(II) and ammonia. [◄◄ Sample Problem 3.3]

c)  Determine the number of each type of atom in 50.00 g of cisplatin. [◄◄ Sample Problem 3.7]

d)  In a particular process, 172.5 g $(NH_4)_2PtCl_4$ is combined with an excess of $NH_3$. Assuming all the limiting reactant is converted to product, how many grams of $Pt(NH_3)_2Cl_2$ will be produced? [◄◄ Sample Problem 3.11]

e)  If the actual amount of $Pt(NH_3)_2Cl_2$ produced in part (d) is 129.6 g, what is the percent yield? [◄◄ Sample Problem 3.13]

# CHAPTER SUMMARY

### Section 3.1

- *Molecular mass* is calculated by summing the masses of all atoms in a molecule. *Molecular weight* is another term for molecular mass.

- For ionic compounds, we use the analogous terms *formula mass* and *formula weight.*

- Molecular masses, molecular weights, formula masses, and formula weights are expressed in atomic mass units (amu).

### Section 3.2

- Molecular or formula mass can be used to determine *percent composition by mass* of a compound.

### Section 3.3

- A *chemical equation* is a written representation of a chemical reaction or a physical process. Chemical species on the left side of the equation are called *reactants,* while those on the right side of the equation are called *products.*

- The physical state of each reactant and product is specified in parentheses as $(s)$, $(l)$, $(g)$, or $(aq)$ for *solid, liquid, gas,* and *aqueous* (dissolved in water), respectively.

- Chemical equations are balanced only by changing the *stoichiometric coefficients* of the reactants and/or products, and never by changing the formulas of the reactants and/or products (i.e., by changing their subscripted numbers).

### Section 3.4

- A *mole* is the amount of a substance that contains $6.022 \times 10^{23}$ [*Avogadro's number* $(N_A)$] of elementary particles (atoms, molecules, ions, formula units, etc.).

- *Molar mass* $(\mathcal{M})$ is the mass of one mole of a substance, usually expressed in grams. The molar mass of a substance in grams is numerically equal to the *atomic, molecular,* or *formula* mass of the substance in amu.

- Molar mass and Avogadro's number can be used to interconvert among *mass, moles,* and *number of particles* (atoms, molecules, ions, formula units, etc.).

### Section 3.5

- *Combustion analysis* is used to determine the empirical formula of a compound. The *empirical formula* can be used to calculate percent composition.

- The empirical formula and molar mass can be used to determine the molecular formula.

### Section 3.6

- A balanced chemical equation can be used to determine how much product will form from given amounts of reactants, how much of one reactant is necessary to react with a given amount of another, or how much reactant is required to produce a specified amount of product. Reactants that are combined in exactly the ratio specified by the balanced equation are said to be "combined in *stoichiometric amounts.*"

### Section 3.7

- The *limiting reactant* is the reactant that is consumed completely in a chemical reaction. An *excess reactant* is the reactant that is not consumed completely. The maximum amount of product that can form depends on the amount of limiting reactant.

- The *theoretical yield* of a reaction is the amount of product that will form if all the limiting reactant is consumed by the desired reaction.

- The *actual yield* is the amount of product actually recovered.

- *Percent yield* [(actual/theoretical) $\times$ 100%] is a measure of the efficiency of a chemical reaction.

- *Combustion* (in which a substance burns in the presence of oxygen), *combination* (in which two or more reactants combine to form a single product), and *decomposition* (in which a reactant splits apart to form two or more products) are three types of reactions that are commonly encountered.

# KEY WORDS

Actual yield, 101

Aqueous, 81

Avogadro's number, 86

Chemical equation, 80

Combination reaction, 104

Combustion analysis, 91

Combustion reaction, 104

Decomposition reaction, 104

Excess reactant, 97

Formula mass, 78

Formula weight, 78

Limiting reactant, 97

Molar mass, 88

Mole, 87

Molecular mass, 78

Molecular weight, 78

Percent composition by mass, 79

Percent yield, 101

Product, 81

Reactant, 81

Stoichiometric amount, 94

Stoichiometric coefficients, 82

Theoretical yield, 101

# KEY EQUATIONS

3.1    $\text{percent by mass of an element} = \dfrac{n \times \text{atomic mass of element}}{\text{molecular or formula mass of compound}} \times 100\%$

3.2    $\% \text{ yield} = \dfrac{\text{actual yield}}{\text{theoretical yield}} \times 100\%$

# QUESTIONS AND PROBLEMS

## Section 3.1: Molecular and Formula Masses

### Review Questions

3.1 What is meant by the term *molecular mass,* and why is the molecular mass that we calculate generally an average molecular mass?

3.2 Explain the difference between the terms *molecular mass* and *formula mass.* To what type of compound does each term refer?

### Problems

**3.3** Calculate the molecular mass (in amu) of each of the following substances: (a) $CH_3Cl$, (b) $N_2O_4$, (c) $SO_2$, (d) $C_6H_{12}$, (e) $H_2O_2$, (f) $C_{12}H_{22}O_{11}$, (g) $NH_3$.

3.4 Calculate the molecular mass (in amu) of each of the following substances: (a) $C_6H_6O$, (b) $H_2SO_4$, (c) $C_6H_6$, (d) $C_6H_{12}O_6$, (e) $BCl_3$, (f) $N_2O_5$, (g) $H_3PO_4$.

3.5 Calculate the molecular mass or formula mass (in amu) of each of the following substances: (a) $CH_4$, (b) $NO_2$, (c) $SO_3$, (d) $C_6H_6$, (e) NaI, (f) $K_2SO_4$, (g) $Ca_3(PO_4)_2$.

3.6 Calculate the molecular mass or formula mass (in amu) of each of the following substances: (a) $Li_2CO_3$, (b) $C_2H_6$, (c) $NF_2$, (d) $Al_2O_3$, (e) $Fe(NO_3)_3$, (f) $PCl_5$, (g) $Mg_3N_2$.

## Section 3.2: Percent Composition of Compounds

### Review Questions

3.7 Use ammonia ($NH_3$) to explain what is meant by the percent composition by mass of a compound.

3.8 Describe how the knowledge of the percent composition by mass of an unknown compound can help us identify the compound.

### Problems

**3.9** Tin (Sn) exists in Earth's crust as $SnO_2$. Calculate the percent composition by mass of Sn and O in $SnO_2$.

3.10 For many years chloroform ($CHCl_3$) was used as an inhalation anesthetic in spite of the fact that it is also a toxic substance that may cause severe liver, kidney, and heart damage. Calculate the percent composition by mass of this compound.

**3.11** All the substances listed here are fertilizers that contribute nitrogen to the soil. Which of these is the richest source of nitrogen on a mass percentage basis?

(a) Urea [$(NH_2)_2CO$]
(b) Ammonium nitrate ($NH_4NO_3$)
(c) Guanidine [$HNC(NH_2)_2$]
(d) Ammonia ($NH_3$)

3.12 The anticaking agent added to Morton salt is calcium silicate ($CaSiO_3$). This compound can absorb up to 2.5 times its mass of water and still remain a free-flowing powder. Calculate the percent composition of $CaSiO_3$.

**3.13** Tooth enamel is $Ca_5(PO_4)_3(OH)$. Calculate the percent composition of the elements present.

## Section 3.3: Chemical Equations

### Review Questions

3.14 Use the formation of water from hydrogen and oxygen to explain the following terms: *chemical reaction, reactant,* and *product.*

3.15 What is the difference between a chemical reaction and a chemical equation?

3.16 Why must a chemical equation be balanced? What law is obeyed by a balanced chemical equation?

3.17 Write the symbols used to represent gas, liquid, solid, and the aqueous phase in chemical equations.

### Problems

3.18 Write an unbalanced equation to represent each of the following reactions: (a) nitrogen and oxygen react to form nitrogen dioxide, (b) dinitrogen pentoxide reacts to form dinitrogen tetroxide and oxygen, (c) ozone reacts to form oxygen, (d) chlorine and sodium iodide react to form iodine and sodium chloride, and (e) magnesium and oxygen react to form magnesium oxide.

**3.19** Write an unbalanced equation to represent each of the following reactions: (a) potassium hydroxide and phosphoric acid react to form potassium phosphate and water; (b) zinc and silver chloride react to form zinc chloride and silver; (c) sodium hydrogen carbonate reacts to form sodium carbonate, water, and carbon dioxide; (d) ammonium nitrite reacts to form nitrogen and water; and (e) carbon dioxide and potassium hydroxide react to form potassium carbonate and water.

3.20 For each of the following unbalanced chemical equations, write the corresponding chemical statement.

(a) $S_8 + O_2 \longrightarrow SO_2$
(b) $CH_4 + O_2 \longrightarrow CO_2 + H_2O$
(c) $N_2 + H_2 \longrightarrow NH_3$
(d) $P_4O_{10} + H_2O \longrightarrow H_3PO_4$
(e) $S + HNO_3 \longrightarrow H_2SO_4 + NO_2 + H_2O$

**3.21** For each of the following unbalanced chemical equations, write the corresponding chemical statement.

(a) $K + H_2O \longrightarrow KOH + H_2$
(b) $Ba(OH)_2 + HCl \longrightarrow BaCl_2 + H_2O$
(c) $Cu + HNO_3 \longrightarrow Cu(NO_3)_2 + NO + H_2O$
(d) $Al + H_2SO_4 \longrightarrow Al_2(SO_4)_3 + H_2$
(e) $HI \longrightarrow H_2 + I_2$

3.22 Balance the following equations using the method outlined in Section 3.3.

(a) $C + O_2 \longrightarrow CO$
(b) $CO + O_2 \longrightarrow CO_2$
(c) $H_2 + Br_2 \longrightarrow HBr$
(d) $K + H_2O \longrightarrow KOH + H_2$
(e) $Mg + O_2 \longrightarrow MgO$
(f) $O_3 \longrightarrow O_2$
(g) $H_2O_2 \longrightarrow H_2O + O_2$
(h) $N_2 + H_2 \longrightarrow NH_3$
(i) $Zn + AgCl \longrightarrow ZnCl_2 + Ag$
(j) $S_8 + O_2 \longrightarrow SO_2$
(k) $NaOH + H_2SO_4 \longrightarrow Na_2SO_4 + H_2O$
(l) $Cl_2 + NaI \longrightarrow NaCl + I_2$
(m) $KOH + H_3PO_4 \longrightarrow K_3PO_4 + H_2O$
(n) $CH_4 + Br_2 \longrightarrow CBr_4 + HBr$

**3.23**    Balance the following equations using the method outlined in Section 3.3.

(a) $N_2O_5 \longrightarrow N_2O_4 + O_2$
(b) $KNO_3 \longrightarrow KNO_2 + O_2$
(c) $NH_4NO_3 \longrightarrow N_2O + H_2O$
(d) $NH_4NO_2 \longrightarrow N_2 + H_2O$
(e) $NaHCO_3 \longrightarrow Na_2CO_3 + H_2O + CO_2$
(f) $P_4O_{10} + H_2O \longrightarrow H_3PO_4$
(g) $HCl + CaCO_3 \longrightarrow CaCl_2 + H_2O + CO_2$
(h) $Al + H_2SO_4 \longrightarrow Al_2(SO_4)_3 + H_2$.
(i) $CO_2 + KOH \longrightarrow K_2CO_3 + H_2O$
(j) $CH_4 + O_2 \longrightarrow CO_2 + H_2O$
(k) $Be_2C + H_2O \longrightarrow Be(OH)_2 + CH_4$
(l) $Cu + HNO_3 \longrightarrow Cu(NO_3)_2 + NO + H_2O$
(m) $S + HNO_3 \longrightarrow H_2SO_4 + NO_2 + H_2O$
(n) $NH_3 + CuO \longrightarrow Cu + N_2 + H_2O$

**3.24**    Which of the following equations best represents the reaction shown in the diagram?

(a) $8A + 4B \longrightarrow C + D$
(b) $4A + 8B \longrightarrow 4C + 4D$
(c) $2A + B \longrightarrow C + D$
(d) $4A + 2B \longrightarrow 4C + 4D$
(e) $2A + 4B \longrightarrow C + D$

**3.25**    Which of the following equations best represents the reaction shown in the diagram?

(a) $A + B \longrightarrow C + D$
(b) $6A + 4B \longrightarrow C + D$
(c) $A + 2B \longrightarrow 2C + D$
(d) $3A + 2B \longrightarrow 2C + D$
(e) $3A + 2B \longrightarrow 4C + 2D$

## Section 3.4: The Mole and Molar Masses

### Review Questions

**3.26**    Define the term *mole*. What is the unit for mole in calculations? What does the mole have in common with the pair, the dozen, and the gross? What does Avogadro's number represent?

**3.27**    What is the molar mass of an atom? What are the commonly used units for molar mass?

**3.28**    What does the word *empirical* in empirical formula mean?

**3.29**    If we know the empirical formula of a compound, what additional information do we need to determine its molecular formula?

### Problems

**3.30**    Earth's population is about 6.5 billion. Suppose that every person on Earth participates in a process of counting identical particles at the rate of two particles per second. How many years would it

take to count $6.0 \times 10^{23}$ particles? Assume that there are 365 days in a year.

**3.31**    The thickness of a piece of paper is 0.0036 in. Suppose a certain book has an Avogadro's number of pages; calculate the thickness of the book in light-years. (*Hint:* See Problem 1.64 for the definition of light-year.)

**3.32**    How many atoms are there in 5.10 moles of sulfur (S)?

**3.33**    How many moles of cobalt (Co) atoms are there in $6.00 \times 10^9$ (6 billion) Co atoms?

**3.34**    How many moles of calcium (Ca) atoms are in 77.4 g of Ca?

**3.35**    How many grams of gold (Au) are there in 15.3 moles of Au?

**3.36**    What is the mass in grams of a single atom of each of the following elements: (a) Hg, (b) Ne?

**3.37**    What is the mass in grams of a single atom of each of the following elements: (a) As, (b) Ni?

**3.38**    What is the mass in grams of $1.00 \times 10^{12}$ lead (Pb) atoms?

**3.39**    How many atoms are present in 3.14 g of copper (Cu)?

**3.40**    Which of the following has more atoms: 1.10 g of hydrogen atoms or 14.7 g of chromium atoms?

**3.41**    Which of the following has a greater mass: two atoms of lead or $5.1 \times 10^{-23}$ mole of helium?

**3.42**    Calculate the molar mass of the following substances: (a) $Li_2CO_3$, (b) $CS_2$, (c) $CHCl_3$ (chloroform), (d) $C_6H_8O_6$ (ascorbic acid, or vitamin C), (e) $KNO_3$, (f) $Mg_3N_2$.

**3.43**    Calculate the molar mass of a compound if 0.372 mole of it has a mass of 152 g.

**3.44**    How many molecules of ethane ($C_2H_6$) are present in 0.334 g of $C_2H_6$?

**3.45**    Calculate the number of C, H, and O atoms in 1.50 g of glucose ($C_6H_{12}O_6$), a sugar.

**3.46**    Urea [$(NH_2)_2CO$] is used for fertilizer and many other things. Calculate the number of N, C, O, and H atoms in $1.68 \times 10^4$ g of urea.

**3.47**    Pheromones are a special type of compound secreted by the females of many insect species to attract the males for mating. One pheromone has the molecular formula $C_{19}H_{38}O$. Normally, the amount of this pheromone secreted by a female insect is about $1.0 \times 10^{-12}$ g. How many molecules are there in this quantity?

**3.48**    The density of water is 1.00 g/mL at 4°C. How many water molecules are present in 2.56 mL of water at this temperature?

**3.49**    Cinnamic alcohol is used mainly in perfumery, particularly in soaps and cosmetics. Its molecular formula is $C_9H_{10}O$. (a) Calculate the percent composition by mass of C, H, and O in cinnamic alcohol. (b) How many molecules of cinnamic alcohol are contained in a sample of mass 0.469 g?

**3.50**    Allicin is the compound responsible for the characteristic smell of garlic. An analysis of the compound gives the following percent composition by mass: C: 44.4 percent; H: 6.21 percent; S: 39.5 percent; O: 9.86 percent. Calculate its empirical formula. What is its molecular formula given that its molar mass is about 162 g?

**3.51** Peroxyacylnitrate (PAN) is one of the components of smog. It is a compound of C, H, N, and O. Determine the percent composition of oxygen and the empirical formula from the following percent composition by mass: 19.8 percent C, 2.50 percent H, 11.6 percent N. What is its molecular formula given that its molar mass is about 120 g?

**3.52** The formula for rust can be represented by $Fe_2O_3$. How many moles of Fe are present in 24.6 g of the compound?

**3.53** How many grams of sulfur (S) are needed to react completely with 246 g of mercury (Hg) to form HgS?

**3.54** Calculate the mass in grams of iodine ($I_2$) that will react completely with 20.4 g of aluminum (Al) to form aluminum iodide ($AlI_3$).

**3.55** Tin(II) fluoride ($SnF_2$) is often added to toothpaste as an ingredient to prevent tooth decay. What is the mass of F in grams in 24.6 g of the compound?

**3.56** Determine the empirical formulas of the compounds with the following compositions: (a) 2.1 percent H, 65.3 percent O, 32.6 percent S, (b) 20.2 percent Al, 79.8 percent Cl.

**3.57** Determine the empirical formulas of the compounds with the following compositions: (a) 40.1 percent C, 6.6 percent H, 53.3 percent O, (b) 18.4 percent C, 21.5 percent N, 60.1 percent K.

**3.58** The empirical formula of a compound is CH. If the molar mass of this compound is about 78 g, what is its molecular formula?

**3.59** The molar mass of caffeine is 194.19 g. Is the molecular formula of caffeine $C_4H_5N_2O$ or $C_8H_{10}N_4O_2$?

**3.60** Monosodium glutamate (MSG), a food-flavor enhancer, has been blamed for "Chinese restaurant syndrome," the symptoms of which are headaches and chest pains. MSG has the following composition by mass: 35.51 percent C, 4.77 percent H, 37.85 percent O, 8.29 percent N, and 13.60 percent Na. What is its molecular formula if its molar mass is about 169 g?

**3.61** Toxicologists use the term $LD_{50}$ to describe the number of grams of a substance per kilogram of body weight that is a lethal dose for 50 percent of test animals. Calculate the number of arsenic(VI) oxide molecules corresponding to an $LD_{50}$ value of 0.015 for a 184-lb man, assuming that the test animals and humans have the same $LD_{50}$.

**3.62** Chemical analysis shows that the oxygen-carrying protein hemoglobin is 0.34 percent Fe by mass. What is the minimum possible molar mass of hemoglobin? The actual molar mass of hemoglobin is about 65,000 g. How would you account for the discrepancy between your minimum value and the experimental value?

## Section 3.5: Combustion Analysis

### Review Questions

**3.63** In combustion analysis, is the combined mass of the products ($CO_2$ and $H_2O$) less than, equal to, or greater than the combined mass of the compound that is combusted and the $O_2$ that reacts with it? Explain.

**3.64** Explain why, in combustion analysis, we cannot determine the amount of oxygen in the sample directly from the amount of oxygen in the products $H_2O$ and $CO_2$.

### Problems

**3.65** Menthol is a flavoring agent extracted from peppermint oil. It contains C, H, and O. In one combustion analysis, 10.00 mg of the substance yields 11.53 mg $H_2O$ and 28.16 mg $CO_2$. What is the empirical formula of menthol?

**3.66** Determine the empirical formula of an organic compound containing only C and H given that combustion of a 1.50-g sample of the compound produces 4.71 g $CO_2$ and 1.93 g $H_2O$.

**3.67** Lactic acid, which consists of C, H, and O, has long been thought to be responsible for muscle soreness following strenuous exercise. Determine the empirical formula of lactic acid given that combustion of a 10.0-g sample produces 14.7 g $CO_2$ and 6.00 g $H_2O$.

**3.68** Perchloroethylene, also known as "perc," is the solvent used in most dry cleaning. It consists of C and Cl. Determine the empirical formula of perchloroethylene given that a 2.58-g sample yields 1.37 g $CO_2$ in a combustion analysis experiment.

**3.69** Paradichlorobenzene, a common ingredient in solid air fresheners, contains only C, H, and Cl and has a molar mass of about 147 g. Given that combustion of 1.68 g of this compound produces 3.02 g $CO_2$ and 0.412 g $H_2O$, determine its empirical and molecular formulas.

**3.70** Ascorbic acid (vitamin C) contains C, H, and O. In one combustion analysis, 5.24 g of ascorbic acid yields 7.86 g $CO_2$ and 2.14 g $H_2O$. Calculate the empirical formula and molecular formula of ascorbic acid given that its molar mass is about 176 g.

**3.71** The amino acid cysteine plays an important role in the three-dimensional structure of proteins by forming "disulfide bridges." The percent composition of cysteine is 29.74 percent C, 5.82 percent H, 26.41 percent O, 11.56 percent N, and 26.47 percent S. What is the molecular formula if its molar mass is approximately 121 g?

## Section 3.6: Calculations with Balanced Chemical Equations

### Review Questions

**3.72** On what law is stoichiometry based? Why is it essential to use balanced equations in solving stoichiometric problems?

**3.73** Describe the steps involved in balancing a chemical equation.

### Problems

**3.74** Consider the combustion of carbon monoxide (CO) in oxygen gas:

$$2CO(g) + O_2(g) \longrightarrow 2CO_2(g)$$

Starting with 3.60 moles of CO, calculate the number of moles of $CO_2$ produced if there is enough oxygen gas to react with all the CO.

**3.75** Silicon tetrachloride ($SiCl_4$) can be prepared by heating Si in chlorine gas:

$$Si(s) + 2Cl_2(g) \longrightarrow SiCl_4(l)$$

In one reaction, 0.507 mole of $SiCl_4$ is produced. How many moles of molecular chlorine were used in the reaction?

**3.76** Ammonia is a principal nitrogen fertilizer. It is prepared by the reaction between hydrogen and nitrogen:

$$3H_2(g) + N_2(g) \longrightarrow 2NH_3(g)$$

In a particular reaction, 6.0 moles of $NH_3$ were produced. How many moles of $H_2$ and how many moles of $N_2$ were consumed to produce this amount of $NH_3$?

**3.77** Consider the combustion of butane ($C_4H_{10}$):

$$2C_4H_{10}(g) + 13O_2(g) \longrightarrow 8CO_2(g) + 10H_2O(l)$$

In a particular reaction, 5.0 moles of $C_4H_{10}$ react with an excess of $O_2$. Calculate the number of moles of $CO_2$ formed.

**3.78** The annual production of sulfur dioxide from burning coal and fossil fuels, auto exhaust, and other sources is about 26 million tons. The equation for the reaction is

$$S(s) + O_2(g) \longrightarrow SO_2(g)$$

How much sulfur (in tons), present in the original materials, would result in that quantity of $SO_2$?

**3.79** When baking soda (sodium bicarbonate or sodium hydrogen carbonate, $NaHCO_3$) is heated, it releases carbon dioxide gas, which is responsible for the rising of cookies, doughnuts, and bread. (a) Write a balanced equation for the decomposition of the compound (one of the products is $Na_2CO_3$). (b) Calculate the mass of $NaHCO_3$ required to produce 20.5 g of $CO_2$.

**3.80** When potassium cyanide (KCN) reacts with acids, a deadly poisonous gas, hydrogen cyanide (HCN), is given off. Here is the equation:

$$KCN(aq) + HCl(aq) \longrightarrow KCl(aq) + HCN(g)$$

If a sample of 0.140 g of KCN is treated with an excess of HCl, calculate the amount of HCN formed, in grams.

**3.81** Fermentation is a complex chemical process of winemaking in which glucose is converted into ethanol and carbon dioxide:

$$\underset{\text{glucose}}{C_6H_{12}O_6} \longrightarrow \underset{\text{ethanol}}{2C_2H_5OH} + 2CO_2$$

Starting with 500.4 g of glucose, what is the maximum amount of ethanol in grams and in liters that can be obtained by this process (density of ethanol = 0.789 g/mL)?

**3.82** Each copper(II) sulfate unit is associated with five water molecules in crystalline copper(II) sulfate pentahydrate ($CuSO_4 \cdot 5H_2O$). When this compound is heated in air above 100°C, it loses the water molecules and also its blue color:

$$CuSO_4 \cdot 5H_2O \longrightarrow CuSO_4 + 5H_2O$$

If 9.60 g of $CuSO_4$ is left after heating 15.01 g of the blue compound, calculate the number of moles of $H_2O$ originally present in the compound.

**3.83** For many years the extraction of gold—that is, the separation of gold from other materials—involved the use of potassium cyanide:

$$4Au + 8KCN + O_2 + 2H_2O \longrightarrow 4KAu(CN)_2 + 4KOH$$

What is the minimum amount of KCN in moles needed to extract 29.0 g (about an ounce) of gold?

**3.84** Limestone ($CaCO_3$) is decomposed by heating to quicklime (CaO) and carbon dioxide. Calculate how many grams of quicklime can be produced from 1.0 kg of limestone.

**3.85** Nitrous oxide ($N_2O$) is also called "laughing gas." It can be prepared by the thermal decomposition of ammonium nitrate ($NH_4NO_3$). The other product is $H_2O$. (a) Write a balanced

equation for this reaction. (b) How many grams of $N_2O$ are formed if 0.46 mole of $NH_4NO_3$ is used in the reaction?

**3.86** The fertilizer ammonium sulfate [$(NH_4)_2SO_4$] is prepared by the reaction between ammonia ($NH_3$) and sulfuric acid:

$$2NH_3(g) + H_2SO_4(aq) \longrightarrow (NH_4)_2SO_4(aq)$$

How many kilograms of $NH_3$ are needed to produce $1.00 \times 10^5$ kg of $(NH_4)_2SO_4$?

**3.87** A common laboratory preparation of oxygen gas is the thermal decomposition of potassium chlorate ($KClO_3$). Assuming complete decomposition, calculate the number of grams of $O_2$ gas that can be obtained from 46.0 g of $KClO_3$. (The products are KCl and $O_2$.)

## Section 3.7: Limiting Reactants

### Review Questions

**3.88** Define *limiting reactant* and *excess reactant*. What is the significance of the limiting reactant in predicting the amount of the product obtained in a reaction? Can there be a limiting reactant if only one reactant is present?

**3.89** Give an everyday example that illustrates the limiting reactant concept.

**3.90** Why is the theoretical yield of a reaction determined only by the amount of the limiting reactant?

**3.91** Why is the actual yield of a reaction almost always smaller than the theoretical yield?

### Problems

**3.92** Consider the reaction

$$2A + B \longrightarrow C$$

(a) In the diagram here that represents the reaction, which reactant, A or B, is the limiting reactant? (b) Assuming a complete reaction, draw a molecular-model representation of the amounts of reactants and products left after the reaction. The atomic arrangement in C is ABA.

**3.93** Consider the reaction

$$N_2 + 3H_2 \longrightarrow 2NH_3$$

Assuming each model represents one mole of the substance, show the number of moles of the product and the excess reactant left after the complete reaction.

**3.94** Nitric oxide (NO) reacts with oxygen gas to form nitrogen dioxide ($NO_2$), a dark-brown gas:

$$2NO(g) + O_2(g) \longrightarrow 2NO_2(g)$$

In one experiment, 0.886 mole of NO is mixed with 0.503 mole of $O_2$. Determine which of the two reactants is the limiting reactant. Calculate also the number of moles of $NO_2$ produced.

**3.95** Consider the reaction pictured, where each red sphere represents an oxygen atom and each blue sphere represents a nitrogen atom. Write the balanced equation and identify the limiting reactant.

3.96 Products of the combustion analysis of a hydrocarbon are represented as shown. Determine the empirical formula of the hydrocarbon.

**3.97** Consider the reaction

$$MnO_2 + 4HCl \longrightarrow MnCl_2 + Cl_2 + 2H_2O$$

If 0.86 mole of $MnO_2$ and 48.2 g of HCl react, which reactant will be used up first? How many grams of $Cl_2$ will be produced?

3.98 Hydrogen fluoride is used in the manufacture of Freons (which destroy ozone in the stratosphere) and in the production of aluminum metal. It is prepared by the reaction

$$CaF_2 + H_2SO_4 \longrightarrow CaSO_4 + 2HF$$

In one process, 6.00 kg of $CaF_2$ is treated with an excess of $H_2SO_4$ and yields 2.86 kg of HF. Calculate the percent yield of HF.

**3.99** Nitroglycerin ($C_3H_5N_3O_9$) is a powerful explosive. Its decomposition may be represented by

$$4C_3H_5N_3O_9 \longrightarrow 6N_2 + 12CO_2 + 10H_2O + O_2$$

This reaction generates a large amount of heat and gaseous products. It is the sudden formation of these gases, together with their rapid expansion, that produces the explosion. (a) What is the maximum amount of $O_2$ in grams that can be obtained from $2.00 \times 10^2$ g of nitroglycerin? (b) Calculate the percent yield in this reaction if the amount of $O_2$ generated is found to be 6.55 g.

3.100 Titanium(IV) oxide ($TiO_2$) is a white substance produced by the action of sulfuric acid on the mineral ilmenite ($FeTiO_3$):

$$FeTiO_3 + H_2SO_4 \longrightarrow TiO_2 + FeSO_4 + H_2O$$

Its opaque and nontoxic properties make it suitable as a pigment in plastics and paints. In one process, $8.00 \times 10^3$ kg of $FeTiO_3$ yielded $3.67 \times 10^3$ kg of $TiO_2$. What is the percent yield of the reaction?

**3.101** Ethylene ($C_2H_4$), an important industrial organic chemical, can be prepared by heating hexane ($C_6H_{14}$) at 800°C:

$$C_6H_{14} \xrightarrow{\Delta} C_2H_4 + \text{other products}$$

If the yield of ethylene production is 42.5 percent, what mass of hexane must be used to produce 481 g of ethylene?

3.102 When heated, lithium reacts with nitrogen to form lithium nitride:

$$6Li(s) + N_2(g) \longrightarrow 2Li_3N(s)$$

What is the theoretical yield of $Li_3N$ in grams when 12.3 g of Li is heated with 33.6 g of $N_2$? If the actual yield of $Li_3N$ is 5.89 g, what is the percent yield of the reaction?

**3.103** Disulfide dichloride ($S_2Cl_2$) is used in the vulcanization of rubber, a process that prevents the slippage of rubber molecules past one another when stretched. It is prepared by heating sulfur in an atmosphere of chlorine:

$$S_8(l) + 4Cl_2(g) \longrightarrow 4S_2Cl_2(l)$$

What is the theoretical yield of $S_2Cl_2$ in grams when 4.06 g of $S_8$ is heated with 6.24 g of $Cl_2$? If the actual yield of $S_2Cl_2$ is 6.55 g, what is the percent yield?

3.104 Determine whether each of the following equations represents a combination reaction, a decomposition reaction, or a combustion reaction: (a) $2NaHCO_3 \longrightarrow Na_2CO_3 + CO_2 + H_2O$, (b) $NH_3 + HCl \longrightarrow NH_4Cl$, (c) $2CH_3OH + 3O_2 \longrightarrow 2CO_2 + 4H_2O$.

**3.105** Determine whether each of the following equations represents a combination reaction, a decomposition reaction, or a combustion reaction: (a) $C_3H_8 + 5O_2 \longrightarrow 3CO_2 + 4H_2O$, (b) $2NF_2 \longrightarrow N_2F_4$, (c) $CuSO_4 \cdot 5H_2O \longrightarrow CuSO_4 + 5H_2O$.

### Additional Problems

3.106 The diagram represents the products ($CO_2$ and $H_2O$) formed after the combustion of a hydrocarbon (a compound containing only C and H atoms). Write an equation for the reaction. (*Hint:* The molar mass of the hydrocarbon is about 30 g.)

3.107 Consider the reaction of hydrogen gas with oxygen gas:

$$2H_2(g) + O_2(g) \longrightarrow 2H_2O(g)$$

Assuming a complete reaction, which of the diagrams (a–d) shown here represents the amounts of reactants and products left after the reaction?

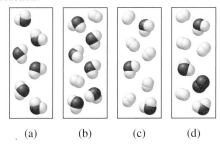

(a)    (b)    (c)    (d)

3.108 Cysteine, shown here, is one of the 20 amino acids found in proteins in humans. Write the molecular formula of cysteine, and calculate its molar mass.

**3.109**  Isoflurane, shown here, is a common inhalation anesthetic. Write its molecular formula, and calculate its molar mass.

**3.110**  Industrially, nitric acid is produced by the Ostwald process represented by the following equations:

$$4NH_3(g) + 5O_2(g) \longrightarrow 4NO(g) + 6H_2O(l)$$
$$2NO(g) + O_2(g) \longrightarrow 2NO_2(g)$$
$$2NO_2(g) + H_2O(l) \longrightarrow HNO_3(aq) + HNO_2(aq)$$

What mass of $NH_3$ (in grams) must be used to produce 1.00 ton of $HNO_3$ by the Ostwald process, assuming an 80 percent yield in each step (1 ton = 2000 lb; 1 lb = 453.6 g)?

**3.111**  A sample of a compound of Cl and O reacts with an excess of $H_2$ to give 0.233 g of HCl and 0.403 g of $H_2O$. Determine the empirical formula of the compound.

**3.112**  The atomic mass of element X is 33.42 amu. A 27.22-g sample of X combines with 84.10 g of another element Y to form a compound XY. Calculate the atomic mass of Y.

**3.113**  How many moles of O are needed to combine with 0.212 mole of C to form (a) CO and (b) $CO_2$?

**3.114**  The aluminum sulfate hydrate $[Al_2(SO_4)_3 \cdot xH_2O]$ contains 8.10 percent Al by mass. Calculate $x$, that is, the number of water molecules associated with each $Al_2(SO_4)_3$ unit.

**3.115**  The explosive nitroglycerin $(C_3H_5N_3O_9)$ has also been used as a drug to treat heart patients to relieve pain (angina pectoris). We now know that nitroglycerin produces nitric oxide (NO), which causes muscles to relax and allows the arteries to dilate. If each nitroglycerin molecule releases one NO per atom of N, calculate the mass percent of NO in nitroglycerin.

**3.116**  The compound 2,3-dimercaptopropanol $(HSCH_2CHSHCH_2OH)$, commonly known as British Anti-Lewisite (BAL), was developed during World War I as an antidote to arsenic-containing poison gas. (a) If each BAL molecule binds one arsenic (As) atom, how many As atoms can be removed by 1.0 g of BAL? (b) BAL can also be used to remove poisonous heavy metals like mercury (Hg) and lead (Pb). If each BAL binds one Hg atom, calculate the mass percent of Hg in a BAL-Hg complex. (An H atom is removed when a BAL molecule binds an Hg atom.)

**3.117**  Mustard gas $(C_4H_8Cl_2S)$ is a poisonous gas that was used in World War I and banned afterward. It causes general destruction of body tissues, resulting in the formation of large water blisters. There is no effective antidote. Calculate the percent composition by mass of the elements in mustard gas.

**3.118**  The carat is the unit of mass used by jewelers. One carat is exactly 200 mg. How many carbon atoms are present in a 2-carat diamond?

**3.119**  An iron bar weighed 664 g. After the bar had been standing in moist air for a month, exactly one-eighth of the iron turned to rust $(Fe_2O_3)$. Calculate the final mass of the iron bar and rust.

**3.120**  A certain metal oxide has the formula MO where M denotes the metal. A 39.46-g sample of the compound is strongly heated in an atmosphere of hydrogen to remove oxygen as water molecules. At the end, 31.70 g of the metal is left over. If O has an atomic mass of 16.00 amu, calculate the atomic mass of M and identify the element.

**3.121**  An impure sample of zinc (Zn) is treated with an excess of sulfuric acid $(H_2SO_4)$ to form zinc sulfate $(ZnSO_4)$ and molecular hydrogen $(H_2)$. (a) Write a balanced equation for the reaction. (b) If 0.0764 g of $H_2$ is obtained from 3.86 g of the sample, calculate the percent purity of the sample. (c) What assumptions must you make in part (b)?

**3.122**  One of the reactions that occurs in a blast furnace, where iron ore is converted to cast iron, is

$$Fe_2O_3 + 3CO \longrightarrow 2Fe + 3CO_2$$

Suppose that $1.64 \times 10^3$ kg of Fe is obtained from a $2.62 \times 10^3$-kg sample of $Fe_2O_3$. Assuming that the reaction goes to completion, what is the percent purity of $Fe_2O_3$ in the original sample?

**3.123**  Carbon dioxide $(CO_2)$ is the gas that is mainly responsible for global warming (the greenhouse effect). The burning of fossil fuels is a major cause of the increased concentration of $CO_2$ in the atmosphere. Carbon dioxide is also the end product of metabolism (see Sample Problem 3.4). Using glucose as an example of food, calculate the annual human production of $CO_2$ in grams, assuming that each person consumes $5.0 \times 10^2$ g of glucose per day, that the world's population is 6.5 billion, and that there are 365 days in a year.

**3.124**  Carbohydrates are compounds containing carbon, hydrogen, and oxygen in which the hydrogen to oxygen ratio is 2:1. A certain carbohydrate contains 40.0 percent carbon by mass. Calculate the empirical and molecular formulas of the compound if the approximate molar mass is 178 g.

**3.125**  Which of the following has the greater mass: 0.72 g of $O_2$ or 0.0011 mole of chlorophyll $(C_{55}H_{72}MgN_4O_5)$?

**3.126**  Analysis of a metal chloride $XCl_3$ shows that it contains 67.2 percent Cl by mass. Calculate the molar mass of X, and identify the element.

**3.127**  Hemoglobin $(C_{2952}H_{4664}N_{812}O_{832}S_8Fe_4)$ is the oxygen carrier in blood. (a) Calculate its molar mass. (b) An average adult has about 5.0 L of blood. Every milliliter of blood has approximately $5.0 \times 10^9$ erythrocytes, or red blood cells, and every red blood cell has about $2.8 \times 10^8$ hemoglobin molecules. Calculate the mass of hemoglobin molecules in grams in an average adult.

**3.128**  Myoglobin stores oxygen for metabolic processes in muscle. Chemical analysis shows that it contains 0.34 percent Fe by mass. What is the molar mass of myoglobin? (There is one Fe atom per molecule.)

**3.129**  Calculate the number of cations and anions in each of the following compounds: (a) 8.38 g of KBr, (b) 5.40 g of $Na_2SO_4$, (c) 7.45 g of $Ca_3(PO_4)_2$.

**3.130**  A mixture of NaBr and $Na_2SO_4$ contains 29.96 percent Na by mass. Calculate the percent by mass of each compound in the mixture.

**3.131**  Aspirin or acetylsalicylic acid is synthesized by combining salicylic acid with acetic anhydride:

$$C_7H_6O_3 + C_4H_6O_3 \longrightarrow C_9H_8O_4 + HC_2H_3O_2$$

salicylic acid   acetic anhydride      aspirin      acetic acid

(a) How much salicylic acid is required to produce 0.400 g of aspirin (about the content in a tablet), assuming acetic anhydride is present in excess? (b) Calculate the amount of salicylic acid needed if only 74.9 percent of salicylic is converted to aspirin. (c) In one experiment, 9.26 g of salicylic acid reacts with 8.54 g of acetic anhydride. Calculate the theoretical yield of aspirin and the percent yield if only 10.9 g of aspirin is produced.

**3.132** Calculate the percent composition by mass of all the elements in calcium phosphate $[Ca_3(PO_4)_2]$, a major component of bone.

**3.133** Lysine, an essential amino acid in the human body, contains C, H, O, and N. In one experiment, the complete combustion of 2.175 g of lysine gave 3.94 g $CO_2$ and 1.89 g $H_2O$. In a separate experiment, 1.873 g of lysine gave 0.436 g $NH_3$. (a) Calculate the empirical formula of lysine. (b) The approximate molar mass of lysine is 150 g. What is the molecular formula of the compound?

**3.134** Does 1 g of hydrogen molecules contain as many H atoms as 1 g of hydrogen atoms?

**3.135** Avogadro's number has sometimes been described as a conversion factor between amu and grams. Use the fluorine atom (19.00 amu) as an example to show the relationship between the atomic mass unit and the gram.

**3.136** The natural abundances of the two stable isotopes of hydrogen (hydrogen and deuterium) are 99.99 percent $^1_1H$ and 0.01 percent $^2_1H$. Assume that water exists as either $H_2O$ or $D_2O$. Calculate the number of $D_2O$ molecules in exactly 400 mL of water (density 1.00 g/mL).

**3.137** In the formation of carbon monoxide, CO, it is found that 2.445 g of carbon combine with 3.257 g of oxygen. What is the atomic mass of oxygen if the atomic mass of carbon is 12.01 amu?

**3.138** What mole ratio of molecular chlorine ($Cl_2$) to molecular oxygen ($O_2$) would result from the breakup of the compound $Cl_2O_7$ into its constituent elements?

**3.139** Which of the following substances contains the greatest mass of chlorine: (a) 5.0 g $Cl_2$, (b) 60.0 g $NaClO_3$, (c) 0.10 mol KCl, (d) 30.0 g $MgCl_2$, (e) 0.50 mol $Cl_2$?

**3.140** A compound made up of C, H, and Cl contains 55.0 percent Cl by mass. If 9.00 g of the compound contain $4.19 \times 10^{23}$ H atoms, what is the empirical formula of the compound?

**3.141** Platinum forms two different compounds with chlorine. One contains 26.7 percent Cl by mass, and the other contains 42.1 percent Cl by mass. Determine the empirical formulas of the two compounds.

**3.142** Heating 2.40 g of the oxide of metal X (molar mass of X = 55.9 g/mol) in carbon monoxide (CO) yields the pure metal and carbon dioxide. The mass of the metal product is 1.68 g. From the data given, show that the simplest formula of the oxide is $X_2O_3$ and write a balanced equation for the reaction.

**3.143** A compound X contains 63.3 percent manganese (Mn) and 36.7 percent O by mass. When X is heated, oxygen gas is evolved and a new compound Y containing 72.0 percent Mn and 28.0 percent O is formed. (a) Determine the empirical formulas of X and Y. (b) Write a balanced equation for the conversion of X to Y.

**3.144** The formula of a hydrate of barium chloride is $BaCl_2 \cdot xH_2O$. If 1.936 g of the compound gives 1.864 g of anhydrous $BaSO_4$ upon treatment with sulfuric acid, calculate the value of x.

**3.145** It is estimated that the day Mt. St. Helens erupted (May 18, 1980), about $4.0 \times 10^5$ tons of $SO_2$ were released into the atmosphere. If all the $SO_2$ were eventually converted to sulfuric acid, how many tons of $H_2SO_4$ were produced?

**3.146** A mixture of $CuSO_4 \cdot 5H_2O$ and $MgSO_4 \cdot 7H_2O$ is heated until all the water is lost. If 5.020 g of the mixture gives 2.988 g of the anhydrous salts, what is the percent by mass of $CuSO_4 \cdot 5H_2O$ in the mixture?

**3.147** When 0.273 g of Mg is heated strongly in a nitrogen ($N_2$) atmosphere, a chemical reaction occurs. The product of the reaction weighs 0.378 g. Calculate the empirical formula of the compound containing Mg and N. Name the compound.

**3.148** A mixture of methane ($CH_4$) and ethane ($C_2H_6$) of mass 13.43 g is completely burned in oxygen. If the total mass of $CO_2$ and $H_2O$ produced is 64.84 g, calculate the fraction of $CH_4$ in the mixture.

**3.149** Leaded gasoline contains an additive to prevent engine "knocking." On analysis, the additive compound is found to contain carbon, hydrogen, and lead (Pb) (hence, "leaded gasoline"). When 51.36 g of this compound is burned in an apparatus such as that shown in Figure 3.5, 55.90 g of $CO_2$ and 28.61 g of $H_2O$ are produced. Determine the empirical formula of the gasoline additive.

**3.150** Because of its detrimental effect on the environment, the lead compound described in Problem 3.149 has been replaced in recent years by methyl *tert*-butyl ether (a compound of C, H, and O) to enhance the performance of gasoline. (As of 1999, this compound is also being phased out because of its contamination of drinking water.) When 12.1 g of the compound is burned in an apparatus like the one shown in Figure 3.5, 30.2 g of $CO_2$ and 14.8 g of $H_2O$ are formed. What is the empirical formula of the compound?

**3.151** Suppose you are given a cube made of magnesium (Mg) metal of edge length 1.0 cm. (a) Calculate the number of Mg atoms in the cube. (b) Atoms are spherical in shape. Therefore, the Mg atoms in the cube cannot fill all the available space. If only 74 percent of the space inside the cube is taken up by Mg atoms, calculate the radius in picometers of an Mg atom. (The density of Mg is 1.74 g/cm$^3$, and the volume of a sphere of radius r is $\frac{4}{3}\pi r^3$.)

**3.152** A certain sample of coal contains 1.6 percent sulfur by mass. When the coal is burned, the sulfur is converted to sulfur dioxide. To prevent air pollution, this sulfur dioxide is treated with calcium oxide (CaO) to form calcium sulfite ($CaSO_3$). Calculate the daily mass (in kilograms) of CaO needed by a power plant that uses $6.60 \times 10^6$ kg of coal per day.

**3.153** Air is a mixture of many gases. However, in calculating its molar mass we need consider only the three major components: nitrogen, oxygen, and argon. Given that one mole of air at sea level is made up of 78.08 percent nitrogen, 20.95 percent oxygen, and 0.97 percent argon, what is the molar mass of air?

**3.154** A die has an edge length of 1.5 cm. (a) What is the volume of one mole of such dice? (b) Assuming that the mole of dice could be packed in such a way that they were in contact with one another, forming stacking layers covering the entire surface of Earth, calculate the height in meters the layers would extend outward. [The radius (r) of Earth is 6371 km, and the area of a sphere is $4\pi r^2$.]

**3.155** The following is a crude but effective method for estimating the *order of magnitude* of Avogadro's number using stearic acid $(C_{18}H_{36}O_2)$. When stearic acid is added to water, its molecules collect at the surface and form a monolayer; that is, the layer is only one molecule thick. The cross-sectional area of each stearic acid molecule has been measured to be $0.21 \text{ nm}^2$. In one experiment it is found that $1.4 \times 10^{-4}$ g of stearic acid is needed to form a monolayer over water in a dish of diameter 20 cm. Based on these measurements, what is Avogadro's number? (The area of a circle of radius $r$ is $\pi r^2$.)

3.156 Octane $(C_8H_{18})$ is a component of gasoline. Complete combustion of octane yields $H_2O$ and $CO_2$. Incomplete combustion produces $H_2O$ and CO, which not only reduces the efficiency of the engine using the fuel but is also toxic. In a certain test run, 1.000 gallon (gal) of octane is burned in an engine. The total mass of CO, $CO_2$, and $H_2O$ produced is 11.53 kg. Calculate the efficiency of the process; that is, calculate the fraction of octane converted to $CO_2$. The density of octane is 2.650 kg/gal.

**3.157** Industrially, hydrogen gas can be prepared by combining propane gas $(C_3H_8)$ with steam at about 400°C. The products are carbon monoxide (CO) and hydrogen gas $(H_2)$. (a) Write a balanced equation for the reaction. (b) How many kilograms of $H_2$ can be obtained from $2.84 \times 10^3$ kg of propane?

3.158 A reaction having a 90 percent yield may be considered a successful experiment. However, in the synthesis of complex molecules such as chlorophyll and many anticancer drugs, a chemist often has to carry out multiple-step syntheses. What is the overall percent yield for such a synthesis, assuming it is a 30-step reaction with a 90 percent yield at each step?

**3.159** Potash is any potassium mineral that is used for its potassium content. Most of the potash produced in the United States goes into fertilizer. The major sources of potash are potassium chloride (KCl) and potassium sulfate $(K_2SO_4)$. Potash production is often reported as the potassium oxide $(K_2O)$ equivalent or the amount of $K_2O$ that could be made from a given mineral. (a) If KCl costs $0.55 per kg, for what price (dollar per kg) must $K_2SO_4$ be sold to supply the same amount of potassium on a per dollar basis? (b) What mass (in kg) of $K_2O$ contains the same number of moles of K atoms as 1.00 kg of KCl?

3.160 A 21.496-g sample of magnesium is burned in air to form magnesium oxide and magnesium nitride. When the products are treated with water, 2.813 g of gaseous ammonia is generated. Calculate the amounts of magnesium nitride and magnesium oxide formed.

3.161 A certain metal M forms a bromide containing 53.79 percent Br by mass. What is the chemical formula of the compound?

3.162 A sample of iron weighing 15.0 g was heated with potassium chlorate $(KClO_3)$ in an evacuated container. The oxygen generated from the decomposition of $KClO_3$ converted some of the Fe to $Fe_2O_3$. If the combined mass of Fe and $Fe_2O_3$ was 17.9 g, calculate the mass of $Fe_2O_3$ formed and the mass of $KClO_3$ decomposed.

**3.163** A sample containing NaCl, $Na_2SO_4$, and $NaNO_3$ gives the following elemental analysis: 32.08 percent Na, 36.01 percent O, 19.51 percent Cl. Calculate the mass percent of each compound in the sample.

3.164 A sample of 10.0 g of sodium reacts with oxygen to form 13.83 g of sodium oxide $(Na_2O)$ and sodium peroxide $(Na_2O_2)$. Calculate the percent composition of the product mixture.

**3.165** The depletion of ozone $(O_3)$ in the stratosphere has been a matter of great concern among scientists in recent years. It is believed that ozone can react with nitric oxide (NO) that is discharged from high-altitude jet planes. The reaction is

$$O_3 + NO \longrightarrow O_2 + NO_2$$

If 0.740 g of $O_3$ reacts with 0.670 g of NO, how many grams of $NO_2$ will be produced? Which compound is the limiting reactant? Calculate the number of moles of the excess reactant remaining at the end of the reaction.

3.166 Propane $(C_3H_8)$ is a minor component of natural gas and is used in domestic cooking and heating. (a) Balance the following equation representing the combustion of propane in air:

$$C_3H_8 + O_2 \longrightarrow CO_2 + H_2O$$

(b) How many grams of carbon dioxide can be produced by burning 3.65 moles of propane? Assume that oxygen is the excess reactant in this reaction.

**3.167** What is wrong or ambiguous with each of the following statements? (a) $NH_4NO_2$ is the limiting reactant in the reaction

$$NH_4NO_2(s) \longrightarrow N_2(g) + 2H_2O(l)$$

(b) The limiting reactants for the reaction shown here are $NH_3$ and NaCl.

$$NH_3(aq) + NaCl(aq) + H_2CO_3(aq) \longrightarrow NaHCO_3(aq) + NH_4Cl(aq)$$

# PRE-PROFESSIONAL PRACTICE EXAM PROBLEMS: PHYSICAL AND BIOLOGICAL SCIENCES

The first step in producing phosphorus fertilizer is the treatment of fluorapatite, a phosphate rock, with sulfuric acid to yield calcium dihydrogen phosphate, calcium sulfate, and hydrogen fluoride gas. In one experiment, a chemist combines 1.00 kg of each reactant.

1. Select the correct balanced equation to represent the reaction.

   a) $2CaPO_4F(s) + 2H_2SO_4(aq)$
   $$\longrightarrow CaH_2PO_4(aq) + CaSO_4(aq) + HF(g)$$
   b) $2CaPO_4F(s) + H_2SO_4(aq)$
   $$\longrightarrow CaH_2PO_4(aq) + CaSO_4(aq) + HF(g)$$
   c) $Ca_5(PO_4)_3F(s) + 3H_2SO_4(aq)$
   $$\longrightarrow 3Ca(H_2PO_4)_2(aq) + 2CaSO_4(aq) + HF(g)$$
   d) $2Ca_5(PO_4)_3F(s) + 7H_2SO_4(aq)$
   $$\longrightarrow 3Ca(H_2PO_4)_2(aq) + 7CaSO_4(aq) + 2HF(g)$$

2. Assuming that all the limiting reactant is converted to products, what mass of calcium dihydrogen phosphate is produced?

   a) 464 g

   b) 696 g

   c) 92.8 g

   d) 199 g

3. What mass of the excess reactant remains unreacted?

   a) 319 g

   b) 681 g

   c) 406 g

   d) 490 g

4. What is the percent yield of the reaction if 197 g of calcium dihydrogen phosphate is produced?

   a) 99.0 percent

   b) 42.5 percent

   c) 28.3 percent

   d) 212 percent

# ANSWERS TO IN-CHAPTER MATERIALS

## Answers to Practice Problems

**3.1A** (a) 95.21 amu, (b) 98.09 amu, (c) 206.27 amu. **3.1B** (a) 100.09 amu, (b) 47.02 amu, (c) 194.20 amu. **3.2A** 63.56% C, 6.002% H, 9.268% N, 21.17% O. **3.2B** 57.13% C, 6.165% H, 9.521% N, 27.18% O. **3.3A** $C_3H_8(g) + 5O_2(g) \longrightarrow 3CO_2(g) + 4H_2O(l)$. **3.3B** $H_2SO_4(aq) + 2NaOH(aq) \longrightarrow Na_2SO_4(aq) + 2H_2O(l)$. **3.4A** $C_{18}H_{32}O_2(aq) + 25O_2(g) \longrightarrow 18CO_2(g) + 16H_2O(l)$. **3.4B** $2NH_3(g) + 3CuO(s) \longrightarrow 3Cu(s) + N_2(g) + 3H_2O(l)$. **3.5A** (a) $4.40 \times 10^{24}$ atoms K, (b) $1.48 \times 10^2$ mol K. **3.5B** (a) $6.32 \times 10^{17}$ atoms He, (b) $3.87 \times 10^{-3}$ mol He. **3.6A** (a) 495 g, (b) 0.704 mol. **3.6B** (a) 31.5 g, (b) 22.71 mol. **3.7A** (a) $6.68 \times 10^{23}$ $O_2$ molecules, $1.34 \times 10^{24}$ O atoms, (b) $1.32 \times 10^{-7}$ g. **3.7B** (a) $3.34 \times 10^{25}$ O atoms, $6.68 \times 10^{25}$ H atoms, (b) 33.4 g. **3.8A** $C_3H_8O$. **3.8B** $CH_2$. **3.9A** $C_3H_4O_3$ and $C_6H_8O_6$. **3.9B** 1.44 g $CO_2$, 1.18 g $H_2O$. **3.10A** 0.264 mol $H_2$ and 0.176 mol $NH_3$. **3.10B** $P_4O_{10} + 6H_2O \longrightarrow 4H_3PO_4$; 1.45 mol $P_4O_{10}$, 8.70 mol $H_2O$. **3.11A** 34.1 g. **3.11B** 292 g. **3.12A** 42.6 g ammonia; nitrogen is limiting reactant and 4.95 g is hydrogen left over. **3.12B** 44.2 g KOH, 115.5 g $H_3PO_4$. **3.13A** 29.2%. **3.13B** 122 g. **3.14A** (a) combustion, (b) combination, (c) decomposition. **3.14B** $A_2 + 2B \longrightarrow 2AB$.

## Answers to Checkpoints

**3.1.1** a. **3.1.2** d. **3.2.1** b. **3.2.2** e. **3.3.1** e. **3.3.2** d. **3.3.3** e. **3.3.4** d. **3.4.1** b. **3.4.2** d. **3.4.3** a. **3.4.4** e. **3.5.1** e. **3.5.2** d. **3.5.3** a. **3.5.4** d. **3.6.1** a. **3.6.2** a. **3.7.1** c. **3.7.2** b. **3.7.3** e. **3.7.4** c. **3.7.5** a. **3.7.6** c.

## Answers to Applying What You've Learned

**a)** The molecular masses of ammonium tetrachloroplatinate(II) and cisplatin are 373.0 and 300.1 amu, respectively. The percent composition by mass of ammonium tetrachloroplatinate(II) is 7.512% N, 2.162% H, 52.31% Pt, and 38.02% Cl; and the percent composition by mass of cisplatin is 9.337% N, 2.015% H, 65.01% Pt, and 23.63% Cl. **b)** The balanced equation is $(NH_4)_2PtCl_4(aq) + 2NH_3(aq) \longrightarrow Pt(NH_3)_2Cl_2(s) + 2NH_4Cl(aq)$. **c)** 50.00 g of cisplatin contains $1.003 \times 10^{23}$ Pt atoms, $2.007 \times 10^{23}$ N atoms, $6.020 \times 10^{23}$ H atoms, and $2.007 \times 10^{23}$ Cl atoms. **d)** If all the limiting reactant is converted to product, 138.8 g of cisplatin will be produced. **e)** An actual yield of 129.6 g constitutes a 93.37% yield.

# 4

# Reactions in Aqueous Solutions

A traditional sobriety test for a driver suspected of being intoxicated may have included instructing the driver to walk a straight line or touch his or her own nose. Today it is common for the more quantitative method of the Breathalyzer test to be used.

# How An Aqueous Reaction Is Used to Prosecute Drunk Drivers

Chemical reactions can be used to verify the presence of a substance or to measure the amount of a substance present. This type of analysis can be used to determine the amount of alcohol in a person's blood.

Every year in the United States tens of thousands of people are killed and half a million more are injured as a result of drunk driving. In recent years, most states have lowered the legal limit of blood alcohol concentration (BAC) from 0.10 to 0.08 percent. Despite stiffer penalties for drunk-driving offenses and high-profile campaigns to educate the public about the dangers of driving while intoxicated, law enforcement agencies still must devote a great deal of work to removing drunk drivers from America's roads.

The police often use a device called a *Breathalyzer* to test drivers suspected of being drunk. In one type of device the breath of a driver suspected of driving under the influence of alcohol is bubbled through an orange solution containing potassium dichromate ($K_2Cr_2O_7$) and sulfuric acid ($H_2SO_4$). The alcohol in the driver's breath reacts with the dichromate ion to produce acetic acid ($HC_2H_3O_2$), which is colorless, and green chromium(III) sulfate [$Cr_2(SO_4)_3$]. The degree of color change from orange to green indicates the alcohol concentration in the breath sample, which is used to estimate the BAC.

The basis for the Breathalyzer test is a relatively simple chemical reaction called an oxidation-reduction reaction. This is one of several important types of reactions that can occur in *aqueous solution*.

At the end of this chapter, you will be able to solve several problems related to the aqueous reaction used in the Breathalyzer test. [▶▶ Page 159].

## In This Chapter, You Will Learn:

About some of the properties of aqueous solutions and about several different types of reactions that can occur between dissolved substances. You will also learn how to express the concentration of a solution and how concentration can be useful in solving quantitative problems.

### Before you begin, you should review

- Identifying compounds as either *molecular* or *ionic*
  [◀◀ Sections 2.6 and 2.7]

- Names, formulas, and charges of the common polyatomic ions
  [◀◀ Table 2.9]

**Multimedia**

Solutions—strong, weak, and nonelectrolytes.

# 4.1 General Properties of Aqueous Solutions

A *solution* is a homogeneous mixture [◄◄ Section 1.2] of two or more substances. Solutions may be gaseous (such as air), solid (such as brass), or liquid (such as saltwater). Usually, the substance present in the largest amount is referred to as the *solvent* and any substance present in a smaller amount is called the *solute.* For example, if we dissolve a teaspoon of sugar in a glass of water, water is the solvent and sugar is the solute. In this chapter we will focus on the properties of aqueous solutions—those in which water is the solvent. Throughout the remainder of this chapter, unless otherwise noted, *solution* will refer specifically to an aqueous solution.

## Electrolytes and Nonelectrolytes

> **Student Annotation:** A substance that *dissolves* in a particular solvent is said to be "soluble" in that solvent. In this chapter, we will use the word *soluble* to mean "water-soluble."

You have probably heard of electrolytes in the context of sports drinks such as Gatorade. Electrolytes in body fluids are necessary for the transmission of electrical impulses, which are critical to physiological processes such as nerve impulses and muscle contractions. In general, an *electrolyte* is a substance that dissolves in water to yield a solution that conducts electricity. By contrast, a *nonelectrolyte* is a substance that dissolves in water to yield a solution that does *not* conduct electricity. Every water-soluble substance fits into one of these two categories.

The difference between an aqueous solution that conducts electricity and one that does not is the presence or absence of *ions.* As an illustration, consider solutions of sugar and salt. The physical processes of sugar (sucrose, $C_{12}H_{22}O_{11}$) dissolving in water and salt (sodium chloride, NaCl) dissolving in water can be represented with the following chemical equations:

$$C_{12}H_{22}O_{11}(s) \xrightarrow{H_2O} C_{12}H_{22}O_{11}(aq) \quad \text{and} \quad NaCl(s) \xrightarrow{H_2O} Na^+(aq) + Cl^-(aq)$$

Note that while the sucrose molecules remain intact upon dissolving, becoming aqueous sucrose molecules, the sodium chloride dissociates, producing aqueous sodium ions and aqueous chloride ions. *Dissociation* is the process by which an ionic compound, upon dissolution, breaks apart into its constituent ions. It is the presence of ions that allows the solution of sodium chloride to conduct electricity. Thus, sodium chloride is an *electrolyte* and sucrose is a *nonelectrolyte.*

Like sucrose, which is a molecular compound [◄◄ Section 2.6], many water-soluble molecular compounds are nonelectrolytes. Some molecular compounds are electrolytes, however, because they ionize on dissolution. *Ionization* is the process by which a molecular compound forms ions when it dissolves. Recall from Chapter 2 that acids are compounds that dissolve in water to produce hydrogen ions ($H^+$) [◄◄ Section 2.6]. HCl, for example, ionizes to produce $H^+$ ions and $Cl^-$ ions.

$$HCl(g) \xrightarrow{H_2O} H^+(aq) + Cl^-(aq)$$

> **Student Annotation:** Bases may be *molecular,* like ammonia ($NH_3$), or *ionic,* like sodium hydroxide (NaOH).

Acids constitute one of two important classes of molecular compounds that are electrolytes. Molecular bases constitute the other one. A *base* is a compound that dissolves in water to produce hydroxide ions ($OH^-$). Ammonia ($NH_3$), for example, ionizes in water to produce ammonium ($NH_4^+$) and hydroxide ($OH^-$) ions.

$$NH_3(g) + H_2O(l) \rightleftharpoons NH_4^+(aq) + OH^-(aq)$$

## Strong Electrolytes and Weak Electrolytes

In a solution of sodium chloride, *all* the dissolved compound exists in the form of ions. Thus, NaCl, which is an ionic compound [◄◄ Section 2.7], is said to have *dissociated completely.* An electrolyte that dissociates completely is known as a *strong electrolyte.* All water-soluble ionic compounds dissociate completely upon dissolving, so all water-soluble ionic compounds are strong electrolytes.

The list of molecular compounds that are strong electrolytes is fairly short. It comprises the seven strong acids, which are listed in Table 4.1. A strong acid ionizes completely, resulting in a solution that contains hydrogen ions and the corresponding anions but essentially no acid molecules.

Most of the molecular compounds that are electrolytes are weak electrolytes. A *weak electrolyte* is a compound that produces ions upon dissolving but exists in solution *predominantly* as molecules that are *not* ionized. Most acids (except those listed in Table 4.1) are weak electrolytes. Acetic acid ($HC_2H_3O_2$) is not one of the strong acids listed in Table 4.1, so it is a weak acid. Its ionization in water is represented by the following chemical equation:

$$HC_2H_3O_2(l) \rightleftharpoons H^+(aq) + C_2H_3O_2^-(aq)$$

| TABLE 4.1 | The Strong Acids |
|---|---|

| Acid | Ionization Equation |
|---|---|
| Hydrochloric acid | $HCl(aq) \longrightarrow H^+(aq) + Cl^-(aq)$ |
| Hydrobromic acid | $HBr(aq) \longrightarrow H^+(aq) + Br^-(aq)$ |
| Hydroiodic acid | $HI(aq) \longrightarrow H^+(aq) + I^-(aq)$ |
| Nitric acid | $HNO_3(aq) \longrightarrow H^+(aq) + NO_3^-(aq)$ |
| Chloric acid | $HClO_3(aq) \longrightarrow H^+(aq) + ClO_3^-(aq)$ |
| Perchloric acid | $HClO_4(aq) \longrightarrow H^+(aq) + ClO_4^-(aq)$ |
| Sulfuric acid* | $H_2SO_4(aq) \longrightarrow H^+(aq) + HSO_4^-(aq)$ |
| | $HSO_4^-(aq) \rightleftharpoons H^+(aq) + SO_4^{2-}(aq)$ |

*Note that although each sulfuric acid molecule has two ionizable hydrogen atoms, it only undergoes the first ionization completely, effectively producing one $H^+$ ion and one $HSO_4^-$ ion per $H_2SO_4$ molecule. The second ionization happens only to a very small extent.

Note the use of the double arrow, $\rightleftharpoons$, in this equation and in two earlier equations, including one in Table 4.1. This denotes a reaction that occurs in both directions and does not result in *all* the reactant(s) (e.g., acetic acid) being converted permanently to product(s) (e.g., hydrogen ions and acetate ions). Instead, forward and reverse reactions both occur, and a state of *dynamic chemical equilibrium* is established.

Although acetic acid molecules ionize, the resulting ions have a strong tendency to recombine to form acetic acid molecules again. Eventually, the ions produced by the ionization will be recombining at the same rate at which they are produced, and there will be no further change in the numbers of acetic acid molecules, hydrogen ions, or acetate ions. Because there is a stronger tendency for the ions to recombine than for the molecules to *ionize*, at any given point in time, most of the dissolved acetic acid exists as molecules that are not ionized (reactant). Only a very small percentage exists in the form of hydrogen ions and acetate ions (products).

The ionization of a weak base, while similar in many ways to the ionization of a weak acid, requires some additional explanation. Ammonia ($NH_3$) is a common weak base. The ionization of ammonia in water is represented by the equation

$$NH_3(g) + H_2O(l) \rightleftharpoons NH_4^+(aq) + OH^-(aq)$$

Note that the ammonia molecule does not ionize by breaking apart into ions. Rather, it does so by ionizing a *water* molecule. The $H^+$ ion from a water molecule attaches to an ammonia molecule, producing an ammonium ion ($NH_4^+$) and leaving what remains of the water molecule, the $OH^-$ ion, in solution.

$$NH_3(g) \quad + \quad H_2O(l) \quad \rightleftharpoons \quad NH_4^+(aq) \quad + \quad OH^-(aq)$$

As with the ionization of a weak acid, the reverse process predominates and at any given point in time, there will be far more $NH_3$ molecules present than there will be $NH_4^+$ and $OH^-$ ions.

We can distinguish between electrolytes and nonelectrolytes experimentally using an apparatus like the one pictured in Figure 4.1. A lightbulb is connected to a battery using a circuit that includes the contents of the beaker. For the bulb to light, electric current must flow from one electrode to the other. Pure water is a very poor conductor of electricity because $H_2O$ ionizes to only a minute extent. There are virtually no ions in pure water to conduct the current, so $H_2O$ is considered a nonelectrolyte. If we add a small amount of salt (sodium chloride), however, the lightbulb will begin to glow as soon as the salt dissolves in the water. Sodium chloride dissociates completely in water to give $Na^+$ and $Cl^-$ ions. Because the NaCl solution conducts electricity, we say that NaCl is an electrolyte.

If the solution contains a nonelectrolyte, as it does in Figure 4.1(a), the bulb will not light. If the solution contains an electrolyte, as it does in Figure 4.1(b) and (c), the bulb will light. The

**Student Annotation:** In a state of *dynamic chemical equilibrium,* or simply *equilibrium,* both forward and reverse reactions continue to occur. However, because they are occurring at the same *rate,* no net change is observed over time in the amounts of reactants or products. Chemical equilibrium is the subject of Chapters 15 to 17.

| To (+) | To (−) | To (+) | To (−) |
| electrode | electrode | electrode | electrode |
| (a) | (b) | (c) | |

**Figure 4.1**    An apparatus for distinguishing between electrolytes and nonelectrolytes and between weak electrolytes and strong electrolytes. A solution's ability to conduct electricity depends on the number of ions it contains. (a) Pure water contains almost no ions and does not conduct electricity; therefore the lightbulb is not lit. (b) A weak electrolyte solution such as HF(*aq*) contains a small number of ions, and the lightbulb is dimly lit. (c) A strong electrolyte solution such as NaCl(*aq*) contains a large number of ions, and the lightbulb is brightly lit. The molar amounts of dissolved substances in the beakers in (b) and (c) are equal.

cations in solution are attracted to the negative electrode, and the anions are attracted to the positive electrode. This movement sets up an electric current that is equivalent to the flow of electrons along a metal wire. How brightly the bulb burns depends upon the number of ions in solution. In Figure 4.1(b), the solution contains a *weak* electrolyte and therefore a relatively small number of ions, so the bulb lights only weakly. The solution in Figure 4.1(c) contains a *strong* electrolyte, which produces a relatively large number of ions, so the bulb lights brightly.

## Bringing Chemistry to Life

### The Invention of Gatorade

In 1965, University of Florida (UF) assistant coach Dwayne Douglas was concerned about the health of Gators football players. He noted that during practices and games in hot weather the players (1) lost a great deal of weight, (2) seldom needed to urinate, and (3) had limited stamina, especially during the second half of a practice or game. He consulted Dr. Robert Cade, researcher and kidney-disease specialist at UF's medical college, who embarked on a project to identify the cause of the athletes' lack of endurance. It was found that after a period of intense activity accompanied by profuse sweating, the players had low blood sugar, low blood volume, and an imbalance of electrolytes—all of which contributed to heat exhaustion. Cade and his research fellows theorized that the depletion of sugar, water, and electrolytes might be remedied by having the athletes drink a solution containing just the right amounts of each. Using this theory, they developed a beverage containing water, sugar, and sodium and potassium salts similar to those present in sweat. By all accounts, the beverage tasted so bad that no one would drink it. Mary Cade, Robert Cade's wife, suggested adding lemon juice to make the concoction more palatable—and the drink that would become Gatorade was born. In their 1966 season the Gators earned a reputation as the "second-half" team, often coming from behind in the third or fourth quarter. Gators coach Ray Graves attributed his team's newfound late-in-the-game strength to the newly developed sideline beverage that replenished blood sugar, blood volume, and electrolyte balance. Sports drinks are now a multibillion dollar industry, and there are several popular brands, although Gatorade still maintains a large share of the market.

## Sample Problem 4.1

Sports drinks typically contain sucrose ($C_{12}H_{22}O_{11}$), fructose ($C_6H_{12}O_6$), sodium citrate ($Na_3C_6H_5O_7$), potassium citrate ($K_3C_6H_5O_7$), and ascorbic acid ($H_2C_6H_6O_6$), among other ingredients. Classify each of these ingredients as a nonelectrolyte, a weak electrolyte, or a strong electrolyte.

**Strategy** Identify each compound as ionic or molecular; identify each molecular compound as acid, base, or neither; and identify each acid as strong or weak.

**Setup** Sucrose and fructose contain no cations and are therefore molecular compounds—neither is an acid or a base. Sodium citrate and potassium citrate contain metal cations and are therefore ionic compounds. Ascorbic acid is an acid that does not appear on the list of strong acids in Table 4.1, so ascorbic acid is a weak acid.

**Solution** Sucrose and fructose are nonelectrolytes. Sodium citrate and potassium citrate are strong electrolytes. Ascorbic acid is a weak electrolyte.

> **Think About It** Remember that any soluble ionic compound is a strong electrolyte, whereas most molecular compounds are non-electrolytes or weak electrolytes. The only molecular compounds that are strong electrolytes are the strong acids listed in Table 4.1.

**Practice Problem A** A so-called enhanced water contains citric acid ($H_3C_6H_5O_7$), magnesium lactate [$Mg(C_3H_5O_3)_2$], calcium lactate [$Ca(C_3H_5O_3)_2$], and potassium phosphate ($K_3PO_4$). Classify each of these compounds as a nonelectrolyte, a weak electrolyte, or a strong electrolyte.

**Practice Problem B** These diagrams show aqueous solutions of three different compounds: (a) $AB_2$, (b) $AC_2$, and (c) $AD_2$. Arrange the compounds in order of increasing electrolyte strength.

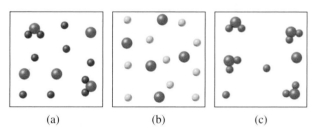

(a)                    (b)                    (c)

## Checkpoint 4.1   General Properties of Aqueous Solutions

**4.1.1** Soluble ionic compounds are _____.

a) always nonelectrolytes

b) always weak electrolytes

c) always strong electrolytes

d) never strong electrolytes

e) sometimes nonelectrolytes

**4.1.2** Soluble molecular compounds are _____.

a) always nonelectrolytes

b) always weak electrolytes

c) always strong electrolytes

d) never strong electrolytes

e) sometimes strong electrolytes

**4.1.3** Which of the following compounds is a weak electrolyte?

a) LiCl

b) $(C_2H_5)_2NH$

c) $KNO_3$

d) NaI

e) $HNO_3$

**4.1.4** Which of the following compounds is a strong electrolyte?

a) HF

b) $H_2CO_3$

c) NaF

d) $NH_3$

e) $H_2O$

*You May Be Wondering...*

# How Can I Tell if a Compound Is an Electrolyte?

While the experimental method described in Figure 4.1 can be useful, often you will have to characterize a compound as a non-electrolyte, a weak electrolyte, or a strong electrolyte just by looking at its formula. A good first step is to determine whether the compound is *ionic* or *molecular.*

An ionic compound contains a *cation* (which is either a metal ion or the ammonium ion) and an *anion* (which may be atomic or polyatomic). A binary compound that contains a metal and a nonmetal is almost always ionic. This is a good time to review the polyatomic anions in Table 2.8 [◄◄ Section 2.7]. You will need to be able to recognize them in the formulas of compounds. Any ionic compound that dissolves in water is a strong electrolyte.

If a compound does not contain a metal cation or the ammonium cation, it is molecular. In this case, you will need to determine whether or not the compound is an acid. Acids generally can be recognized by the way their formulas are written, with the ionizable hydrogens written first. $HC_2H_3O_2$, $H_2CO_3$, and $H_3PO_4$ are acetic acid, carbonic acid, and phosphoric acid, respectively. Formulas of carboxylic acids, such as acetic acid, often are written with their ionizable hydrogen atoms *last* to keep the functional group together in the formula. Thus, either $HC_2H_3O_2$ or $CH_3COOH$ is correct for acetic acid. To make it easier to identify compounds as acids, in this chapter we will write all acid formulas with the ionizable H atom(s) first. If a compound is an acid, it is an electrolyte. If it is one of the acids listed in Table 4.1, it is a strong acid and therefore a strong electrolyte. Any acid not listed in Table 4.1 is a weak acid and therefore a weak electrolyte.

If a molecular compound is not an acid, you must then consider whether or not it is a weak base. Many weak bases are relat-

ed to ammonia in that they consist of a nitrogen atom bonded to hydrogen and/or carbon atoms. Examples include methylamine ($CH_3NH_2$), pyridine ($C_5H_5N$), and hydroxylamine ($NH_2OH$). Weak bases are weak electrolytes.

If a molecular compound is neither an acid nor a weak base, it is a nonelectrolyte. The flowchart in Figure 4.2 can be useful for classification of water-soluble compounds.

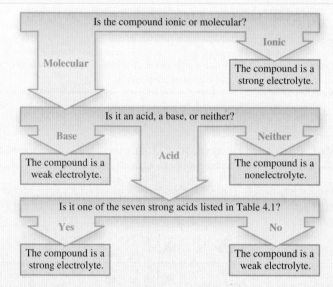

**Figure 4.2**   Flowchart for determining if a compound is a strong electrolyte, a weak electrolyte, or a nonelectrolyte.

---

## 4.2   Precipitation Reactions

When an aqueous solution of lead(II) nitrate [$Pb(NO_3)_2$] is added to an aqueous solution of sodium iodide (NaI), a yellow insoluble solid—lead(II) iodide ($PbI_2$)—forms. Sodium nitrate ($NaNO_3$), the other reaction product, remains in solution. Figure 4.3 (page 124) shows this reaction in progress. An insoluble solid product that separates from a solution is called a ***precipitate,*** and a chemical reaction in which a precipitate forms is called a ***precipitation reaction.***

Precipitation reactions usually involve ionic compounds, but a precipitate does not form every time two solutions of electrolytes are combined. Instead, whether or not a precipitate forms when two solutions are mixed depends on the solubility of the products.

### Solubility Guidelines for Ionic Compounds in Water

When an ionic substance such as sodium chloride dissolves in water, the water molecules remove individual ions from the three-dimensional solid structure and surround them. This process, called

Multimedia
Precipitation of $BaSO_4$.

## Sample Problem 4.2

Classify each of the following compounds as a nonelectrolyte, a weak electrolyte, or a strong electrolyte: (a) methanol ($CH_3OH$), (b) sodium hydroxide (NaOH), (c) ethylamine ($C_2H_5NH_2$), and (d) hydrofluoric acid (HF).

**Strategy** Classify each compound as ionic or molecular. Soluble ionic compounds are strong electrolytes. Classify each molecular compound as an acid, base, or neither. Molecular compounds that are neither acids nor bases are nonelectrolytes. Molecular compounds that are bases are weak electrolytes. Finally, classify acids as either strong or weak. Strong acids are strong electrolytes, and weak acids are weak electrolytes.

**Setup** (a) Methanol contains neither a metal cation nor the ammonium ion. It is therefore molecular. Its formula does not begin with H, so it is probably not an acid, and it does not contain a nitrogen atom, so it is not a weak base. Molecular compounds that are neither acids nor bases are *nonelectrolytes.*

(b) Sodium hydroxide contains a metal cation ($Na^+$) and is therefore ionic. It is also one of the strong bases.

(c) Ethylamine contains no cations and is therefore molecular. It is also a nitrogen-containing base, similar to ammonia.

(d) Hydrofluoric acid is, as its name suggests, an acid. However, it is not on the list of strong acids in Table 4.1 and is, therefore, a weak acid.

**Solution** (a) Nonelectrolyte

(b) Strong electrolyte

(c) Weak electrolyte

(d) Weak electrolyte

**Think About It** Make sure that you have correctly identified compounds that are ionic and compounds that are molecular. Remember that strong acids are strong electrolytes, weak acids and weak bases are weak electrolytes, and strong bases are strong electrolytes (by virtue of their being soluble ionic compounds). Molecular compounds, with the exceptions of acids and weak bases, are nonelectrolytes.

**Practice Problem A** Identify the following compounds as nonelectrolytes, weak electrolytes, or strong electrolytes: ethanol ($C_2H_5OH$), nitrous acid ($HNO_2$), and sodium hydrogen carbonate ($NaHCO_3$, also known as *bicarbonate*).

**Practice Problem B** Identify the following compounds as nonelectrolytes, weak electrolytes, or strong electrolytes: phosphorous acid ($H_3PO_3$), hydrogen peroxide ($H_2O_2$), and ammonium sulfate [$(NH_4)_2SO_4$].

*hydration,* is shown in Figure 4.4 (page 124). Water is an excellent solvent for ionic compounds because $H_2O$ is a *polar* molecule; that is, its electrons are distributed such that there is a partial negative charge on the oxygen atom, denoted by the δ− symbol, and partial positive charges, denoted by the δ+ symbol, on each of the hydrogen atoms. The oxygen atoms in the surrounding water molecules are attracted to the cations, while the hydrogen atoms are attracted to the anions. These attractions explain the orientation of water molecules around each of the ions in solution. The surrounding water molecules prevent the cations and anions from recombining.

*Solubility* is defined as the maximum amount of solute that will dissolve in a given quantity of solvent at a specific temperature. Not all ionic compounds dissolve in water. Whether or not an ionic compound is water soluble depends on the relative magnitudes of the water molecules' attraction to the ions, and the ions' attraction for each other. We will learn more about the magnitudes of attractive forces in ionic compounds in Chapter 8, but for now it is useful to learn some guidelines that enable us to predict the solubility of ionic compounds. Table 4.2 lists groups of compounds that are *soluble* and shows the *insoluble* exceptions. Table 4.3 lists groups of compounds that are *insoluble* and shows the *soluble* exceptions.

Sample Problem 4.3 gives you some practice applying the solubility guidelines.

**Student Annotation:** The partial charges on the oxygen atom and the hydrogen atoms sum to zero. Water molecules, although polar, have no *net* charge. You will learn more about partial charges and molecular polarity in Chapters 8 and 9.

**Student Annotation:** If the water molecules' attraction for the ions exceeds the ions' attraction to one another, then the ionic compound will dissolve. If the ions' attraction to each other exceeds the water molecules' attraction to the ions, then the compound won't dissolve.

Multimedia

Chemical reactions—predicting precipitation reactions (interactive).

The addition of
a colorless
NaI(*aq*) solution...

Na$^+$

I$^-$

Pb$^{2+}$

NO$_3^-$

to a colorless Pb(NO$_3$)$_2$(*aq*) solution...

produces PbI$_2$(*s*), a yellow precipitate...

which settles out of solution.
The remaining solution contains
Na$^+$ and NO$_3^-$ ions.

**Figure 4.3**     A colorless aqueous solution of NaI is added to a colorless aqueous solution of Pb(NO$_3$)$_2$. A yellow precipitate, PbI$_2$, forms. Na$^+$ and NO$_3^-$ ions remain in solution.

**Figure 4.4**     Hydration of anions and cations of a soluble ionic compound. Water molecules surround each anion with their partial positive charges (H atoms) oriented toward the negatively charged anion; and they surround each cation with their partial negative charges (O atoms) oriented toward the positively charged cation.

| **TABLE 4.2** | **Solubility Guidelines: Soluble Compounds** |
|---|---|

| **Water-Soluble Compounds** | **Insoluble Exceptions** |
|---|---|
| Compounds containing an alkali metal cation ($Li^+$, $Na^+$, $K^+$, $Rb^+$, $Cs^+$) or the ammonium ion ($NH_4^+$) | |
| Compounds containing the nitrate ion ($NO_3^-$), acetate ion ($C_2H_3O_2^-$), or chlorate ion ($ClO_3^-$) | |
| Compounds containing the chloride ion ($Cl^-$), bromide ion ($Br^-$), or iodide ion ($I^-$) | Compounds containing $Ag^+$, $Hg_2^{2+}$, or $Pb^{2+}$ |
| Compounds containing the sulfate ion ($SO_4^{2-}$) | Compounds containing $Ag^+$, $Hg_2^{2+}$, $Pb^{2+}$, $Ca^{2+}$, $Sr^{2+}$, or $Ba^{2+}$ |

**Student Annotation:** Some books list fewer exceptions to these solubility rules. In fact, ionic compounds listed here as "insoluble" are actually *very slightly* soluble. It is *how* soluble a compound must be to be called "soluble" that may vary from book to book.

**Student Annotation:** Ionic compounds often are classified according to the anions they contain. Compounds that contain the chloride ion are called *chlorides*, compounds containing the nitrate ion are called *nitrates*, and so on.

| **TABLE 4.3** | **Solubility Guidelines: Insoluble Compounds** |
|---|---|

| **Water-Insoluble Compounds** | **Soluble Exceptions** |
|---|---|
| Compounds containing the carbonate ion ($CO_3^{2-}$), phosphate ion ($PO_4^{3-}$), chromate ion ($CrO_4^{2-}$), or sulfide ion ($S^{2-}$) | Compounds containing $Li^+$, $Na^+$, $K^+$, $Rb^+$, $Cs^+$, or $NH_4^+$ |
| Compounds containing the hydroxide ion ($OH^-$) | Compounds containing $Li^+$, $Na^+$, $K^+$, $Rb^+$, $Cs^+$, or $Ba^{2+}$ |

**Student Annotation:** Note the reoccurrence of the same three groups of ions in the exceptions columns in Tables 4.2 and 4.3: Group 1A or the ammonium cation; $Ag^+$, $Hg_2^{2+}$, or $Pb^{2+}$; and the heavier Group 2A cations.

## Sample Problem 4.3

Classify each of the following compounds as soluble or insoluble in water: (a) $AgNO_3$, (b) $CaSO_4$, (c) $K_2CO_3$.

**Strategy** Use the guidelines in Tables 4.2 and 4.3 to determine whether or not each compound is expected to be water soluble.

**Setup** (a) $AgNO_3$ contains the nitrate ion ($NO_3^-$). According to Table 4.2, *all* compounds containing the nitrate ion are soluble.

(b) $CaSO_4$ contains the sulfate ion ($SO_4^{2-}$). According to Table 4.2, compounds containing the sulfate ion are soluble unless the cation is $Ag^+$, $Hg_2^{2+}$, $Pb^{2+}$, $Ca^{2+}$, $Sr^{2+}$, or $Ba^{2+}$. Thus, the $Ca^{2+}$ ion is one of the insoluble exceptions.

(c) $K_2CO_3$ contains an alkali metal cation ($K^+$) for which, according to Table 4.2, there are no insoluble exceptions. Alternatively, Table 4.3 shows that most compounds containing the carbonate ion ($CO_3^{2-}$) are insoluble—but compounds containing a Group 1A cation such as $K^+$ are soluble exceptions.

**Solution** (a) Soluble, (b) insoluble, (c) soluble.

> **Think About It** Check the ions in each compound against the information in Tables 4.2 and 4.3 to confirm that you have drawn the right conclusions.

**Practice Problem A** Classify each of the following compounds as soluble or insoluble in water: (a) $PbCl_2$, (b) $(NH_4)_3PO_4$, (c) $Fe(OH)_3$.

**Practice Problem B** Classify each of the following compounds as soluble or insoluble in water: (a) $MgBr_2$, (b) $Ca_3(PO_4)_2$, (c) $KClO_3$.

## Molecular Equations

The reaction shown in Figure 4.3 can be represented with the chemical equation

$$Pb(NO_3)_2(aq) + 2NaI(aq) \longrightarrow 2NaNO_3(aq) + PbI_2(s)$$

Based on this chemical equation, the metal cations seem to exchange anions. That is, the $Pb^{2+}$ ion, originally paired with $NO_3^-$ ions, ends up paired with $I^-$ ions; similarly, each $Na^+$ ion, originally paired with an $I^-$ ion, ends up paired with an $NO_3^-$ ion. This equation, as written, is called a ***molecular equation,*** which is a chemical equation written with all compounds represented

**Student Annotation:** Reactions in which compounds exchange ions are sometimes called *metathesis* or *double replacement* reactions.

by their chemical formulas, making it look as though they exist in solution as molecules or formula units.

You now know enough chemistry to predict the products of this type of chemical reaction! Simply write the formulas for the reactants, and then write formulas for the compounds that would form if the cations in the reactants were to trade anions. For example, if you want to write the equation for the reaction that occurs when solutions of sodium sulfate and barium hydroxide are combined, you would first write the formulas of the reactants [◄◄ Section 2.7]:

$$Na_2SO_4(aq) + Ba(OH)_2(aq) \longrightarrow$$

Then you would write the formula for one product by combining the cation from the first reactant ($Na^+$) with the anion from the second ($OH^-$); and write the formula for the other product by combining the cation from the second reactant ($Ba^{2+}$) with the anion from the first ($SO_4^{2-}$). Thus, the equation is

$$Na_2SO_4(aq) + Ba(OH)_2(aq) \longrightarrow 2NaOH + BaSO_4$$

Although we have balanced the equation [◄◄ Section 3.3], we have not yet put phases in parentheses for the products.

The final step in predicting the outcome of such a reaction is to determine which of the products, if any, will precipitate from solution. We do this using the solubility guidelines for ionic compounds (Tables 4.2 and 4.3). The first product (NaOH) contains a Group 1A cation ($Na^+$) and will therefore be soluble. We indicate its phase as ($aq$). The second product ($BaSO_4$) contains the sulfate ion ($SO_4^{2-}$). Sulfate compounds are soluble unless the cation is $Ag^+$, $Hg_2^{2+}$, $Pb^{2+}$, $Ca^{2+}$, $Sr^{2+}$, or $Ba^{2+}$. $BaSO_4$ is therefore insoluble and will precipitate. We indicate its phase as ($s$):

$$Na_2SO_4(aq) + Ba(OH)_2(aq) \longrightarrow 2NaOH(aq) + BaSO_4(s)$$

## Ionic Equations

Although molecular equations are useful, especially from the standpoint of knowing which solutions to combine in the laboratory, they are in a sense unrealistic. Soluble ionic compounds are *strong electrolytes* [◄◄ Section 4.1]. As such, they exist in solution as hydrated *ions*, rather than as formula units. Thus, it would be more realistic to represent the aqueous species in the reaction of $Na_2SO_4(aq)$ with $Ba(OH)_2(aq)$ as follows:

$$Na_2SO_4(aq) \longrightarrow 2Na^+(aq) + SO_4^{2-}(aq)$$

$$Ba(OH)_2(aq) \longrightarrow Ba^{2+}(aq) + 2OH^-(aq)$$

$$NaOH(aq) \longrightarrow Na^+(aq) + OH^-(aq)$$

If we were to rewrite the equation, representing the dissolved compounds as hydrated ions, it would be

$$2Na^+(aq) + SO_4^{2-}(aq) + Ba^{2+}(aq) + 2OH^-(aq) \longrightarrow 2Na^+(aq) + 2OH^-(aq) + BaSO_4(s)$$

This version of the equation is called an ***ionic equation,*** a chemical equation in which any compound that exists completely or predominantly as ions in solution is represented as those ions. Species that are insoluble or that exist in solution completely or predominantly as molecules are represented with their chemical formulas, as they were in the molecular equation.

## Net Ionic Equations

$Na^+(aq)$ and $OH^-(aq)$ both appear as reactants and products in the ionic equation for the reaction of $Na_2SO_4(aq)$ with $Ba(OH)_2(aq)$. Ions that appear on both sides of the equation arrow are called ***spectator ions*** because they do not participate in the reaction. Spectator ions cancel one another, just as identical terms on both sides of an algebraic equation cancel one another, so we need not show spectator ions in chemical equations.

$$\cancel{2Na^+(aq)} + SO_4^{2-}(aq) + Ba^{2+}(aq) + \cancel{2OH^-(aq)} \longrightarrow \cancel{2Na^+(aq)} + \cancel{2OH^-(aq)} + BaSO_4(s)$$

**Student Annotation:** Although the reactants may be written in either order in the net ionic equation, it is common for the cation to be shown first and the anion second.

Eliminating the spectator ions yields the following equation:

$$Ba^{2+}(aq) + SO_4^{2-}(aq) \longrightarrow BaSO_4(s)$$

This version of the equation is called a *net ionic equation,* which is a chemical equation that includes only the species that are actually involved in the reaction. The net ionic equation, in effect, tells us what actually happens when we combine solutions of sodium sulfate and barium hydroxide.

The steps necessary to determine the molecular, ionic, and net ionic equations for a precipitation reaction are as follows:

1. Write and balance the molecular equation, predicting the products by assuming that the cations trade anions.
2. Write the ionic equation by separating strong electrolytes into their constituent ions.
3. Write the net ionic equation by identifying and canceling spectator ions on both sides of the equation.

If both products of a reaction are strong electrolytes, all the ions in solution are spectator ions. In this case, there is no net ionic equation and no reaction takes place.

Sample Problem 4.4 illustrates the stepwise determination of molecular, ionic, and net ionic equations.

## Sample Problem 4.4

Write the molecular, ionic, and net ionic equations for the reaction that occurs when aqueous solutions of lead acetate [$Pb(C_2H_3O_2)_2$], and calcium chloride ($CaCl_2$), are combined.

**Strategy** Predict the products by exchanging ions and balance the equation. Determine which product will precipitate based on the solubility guidelines in Tables 4.2 and 4.3. Rewrite the equation showing strong electrolytes as ions. Identify and cancel spectator ions.

**Setup** The products of the reaction are $PbCl_2$ and $Ca(C_2H_3O_2)_2$. $PbCl_2$ is insoluble, because $Pb^{2+}$ is one of the insoluble exceptions for chlorides, which are generally soluble. $Ca(C_2H_3O_2)_2$ is soluble because all acetates are soluble.

**Solution** Molecular equation:

$$Pb(C_2H_3O_2)_2(aq) + CaCl_2(aq) \longrightarrow PbCl_2(s) + Ca(C_2H_3O_2)_2(aq)$$

Ionic equation:

$$Pb^{2+}(aq) + 2C_2H_3O_2^-(aq) + Ca^{2+}(aq) + 2Cl^-(aq) \longrightarrow PbCl_2(s) + Ca^{2+}(aq) + 2C_2H_3O_2^-(aq)$$

Net ionic equation:

$$Pb^{2+}(aq) + 2Cl^-(aq) \longrightarrow PbCl_2(s)$$

> **Think About It** Remember that the charges on ions in a compound must sum to zero. Make sure that you have written correct formulas for the products and that each of the equations you have written is balanced. If you find that you are having trouble balancing an equation, check to make sure you have correct formulas for the products.

**Practice Problem A** Write the molecular, ionic, and net ionic equations for the combination of $Sr(NO_3)_2(aq)$ and $Li_2SO_4(aq)$.

**Practice Problem B** Write the molecular, ionic, and net ionic equations for the combination of $KNO_3(aq)$ and $BaCl_2(aq)$.

## Checkpoint 4.2    Precipitation Reactions

**4.2.1** Which of the following are water soluble? (Choose all that apply.)

a) $Na_2S$

b) $Ba(C_2H_3O_2)_2$

c) $CaCO_3$

d) $CuBr_2$

e) $Hg_2Cl_2$

**4.2.2** Which of the following are water insoluble? (Choose all that apply.)

a) $Ag_2CrO_4$

b) $Li_2CO_3$

c) $Ca_3(PO_4)_2$

d) $BaSO_4$

e) $ZnCl_2$

*(Continued)*

## Checkpoint 4.2    Precipitation Reactions (*continued*)

**4.2.3** What are the spectator ions in the ionic equation for the combination of $Li_2CO_3(aq)$ and $Ba(OH)_2(aq)$?

a) $CO_3^{2-}$ and $OH^-$

b) $Li^+$ and $OH^-$

c) $Li^+$ and $Ba^{2+}$

d) $Ba^{2+}$ and $OH^-$

e) $Ba^{2+}$ and $CO_3^{2-}$

**4.2.4** Select the correct net ionic equation for the combination of $Fe(NO_3)_2(aq)$ and $Na_2CO_3(aq)$.

a) $Na^+(aq) + CO_3^{2-}(aq) \longrightarrow NaCO_3(s)$

b) $Fe^{2+}(aq) + CO_3^{2-}(aq) \longrightarrow FeCO_3(s)$

c) $2Na^+(aq) + CO_3^{2-}(aq) \longrightarrow Na_2CO_3(s)$

d) $Fe^{2+}(aq) + 2NO_3^-(aq) \longrightarrow Fe(NO_3)_2(s)$

e) $Na^+(aq) + NO_3^-(aq) \longrightarrow NaNO_3(s)$

**4.2.5** Which reaction is represented by the net ionic equation for the combination of aqueous solutions of $LiOH$ and $Cu(NO_3)_2$?

(e)    There is no net ionic equation.  No reaction occurs.

**4.2.6** Which reaction is represented by the net ionic equation for the combination of aqueous solutions of $LiNO_3$ and $NaC_2H_3O_2$?

(e)    There is no net ionic equation.  No reaction occurs.

## 4.3    Acid-Base Reactions

Another type of reaction occurs when two solutions, one containing an acid and one containing a base, are combined. We frequently encounter acids and bases in everyday life (Figure 4.5). Ascorbic acid, for instance, is also known as vitamin C; acetic acid is the component responsible for the sour taste and characteristic smell of vinegar; and hydrochloric acid is the acid in muriatic acid and is also the principal ingredient in gastric juice (stomach acid). Ammonia, found in many cleaning products, and sodium hydroxide, found in drain cleaner, are common bases. Acid-base chemistry is extremely important to biological processes. Let's look again at the properties of acids and bases, and then look at acid-base reactions.

### Strong Acids and Bases

Media Player/
MPEG Content

Acids and bases—dissociation
of strong and weak acids.

As we saw in Section 4.1, the seven strong acids—those that ionize completely in solution—are listed in Table 4.1. All other acids are weak acids. The strong bases are the hydroxides of Group 1A and heavy Group 2A metals. These are soluble ionic compounds, which dissociate completely

**Figure 4.5**  Some common acids and bases. From left to right: Sodium hydroxide (NaOH), ascorbic acid ($C_6H_8O_6$ or, with its ionizable hydrogens written first, $H_2C_6H_6O_6$), hydrochloric acid (HCl), acetic acid ($HC_2H_3O_2$), and ammonia ($NH_3$). HCl and NaOH are both strong electrolytes and exist in solution entirely as ions. Water molecules are not shown.

| TABLE 4.4 | Strong Acids and Strong Bases | |
|---|---|---|
| | **Strong Acids** | **Strong Bases** |
| | HCl | LiOH |
| | HBr | NaOH |
| | HI | KOH |
| | $HNO_3$ | RbOH |
| | $HClO_3$ | CsOH |
| | $HClO_4$ | $Ca(OH)_2$ |
| | $H_2SO_4$ | $Sr(OH)_2$ |
| | | $Ba(OH)_2$ |

**Student Annotation:** Although three of the Group 2A hydroxides [$Ca(OH)_2$, $Sr(OH)_2$, and $Ba(OH)_2$] are typically classified as strong bases, only $Ba(OH)_2$ is sufficiently soluble to be used commonly in the laboratory. For any ionic compound, what does dissolve—even if it is only a tiny amount—dissociates completely.

and exist entirely as ions in solution. Thus, both strong acids and strong bases are strong electrolytes. Table 4.4 lists the strong acids and strong bases. It is important that you know these compounds.

## Brønsted Acids and Bases

In Section 2.6 we defined an acid as a substance that ionizes in water to produce $H^+$ ions, and a base as a substance that ionizes (or dissociates, in the case of an ionic base) in water to produce $OH^-$ ions. These definitions are attributed to the Swedish chemist Svante Arrhenius.[1] Although

---

1. Svante August Arrhenius (1859–1927). Swedish chemist. Arrhenius made important contributions to the study of chemical kinetics and electrolyte solutions. (He also speculated that life had come to Earth from other planets.) Arrhenius was awarded the Nobel Prize in Chemistry in 1903.

the ***Arrhenius acid*** and ***Arrhenius base*** definitions are useful, they are restricted to the behavior of compounds in aqueous solution. More inclusive definitions were proposed by the Danish chemist Johannes Brønsted[2] in 1932. A ***Brønsted acid*** is a proton *donor,* and a ***Brønsted base*** is a proton *acceptor.* In this context, the word *proton* refers to a hydrogen atom that has lost its electron—also known as a *hydrogen ion* ($H^+$). Consider the ionization of the weak base ammonia ($NH_3$).

**Student Annotation:** The H atom consists of a proton and an electron. When the electron is lost, all that remains is the proton—hence the use of the term *proton* in this context.

$$NH_3(aq) + H_2O(l) \rightleftharpoons NH_4^+(aq) + OH^-(aq)$$

This equation shows that $NH_3$ is a base in the Arrhenius sense; that is, it produces $OH^-$ in solution. It is also a base in the Brønsted sense because it accepts a proton ($H^+$) from the water molecule to become the ammonium ion ($NH_4^+$).

$$NH_3(aq) \quad + \quad H_2O(l) \quad \rightleftharpoons \quad NH_4^+(aq) \quad + \quad OH^-(aq)$$

Now consider the ionization of hydrofluoric acid (HF), a weak acid.

$$HF(aq) \rightleftharpoons H^+(aq) + F^-(aq)$$

HF is an acid in the Arrhenius sense because it produces $H^+$ in solution. It is also an acid in the Brønsted sense because it donates a proton to the aqueous solution. An aqueous proton ($H^+$), how-ever, does not exist as an isolated species in solution. Rather, it is hydrated just as other aqueous ions are [◀◀ Section 4.2]. The proton, being positively charged, is strongly attracted to the partial negative charge on the oxygen atom in a water molecule. Thus, it is convenient and more *realistic* for us to represent the ionization of HF with the equation

**Student Annotation:** In reality, aqueous protons are surrounded by water molecules, just as other ions are. We use $H_3O^+$ in chemical equations, showing just one of the water molecules involved, to emphasize that the proton is hydrated in solution.

$$HF(aq) + H_2O(l) \rightleftharpoons H_3O^+(aq) + F^-(aq)$$

where we include the water molecule to which the proton becomes attached, both before and after the ionization. (We could show $H_3O^+$ as $H_2O \cdot H^+$ to emphasize that it is a water molecule attached to a proton.) With the equation written this way, we show that HF donates a proton to $H_2O$, thus converting the $H_2O$ molecule to the ***hydronium ion*** ($H_3O^+$).

**Student Annotation:** The terms and symbols *hydrogen ion, proton, hydronium ion,* $H^+$, and $H_3O^+$ all refer to the same aqueous species and are used interchangeably.

$$HF(aq) \quad + \quad H_2O(l) \quad \rightleftharpoons \quad H_3O^+(aq) \quad + \quad F^-(aq)$$

Both $H^+$ and $H_3O^+$ will be used in chemical equations throughout the text. You should be aware that they refer to the same aqueous species.

The Brønsted definitions of acids and bases are not restricted to species in aqueous solution. In fact, Brønsted acid-base reactions sometimes take place in the gas phase. For example, in the reaction between HCl and $NH_3$ gases, HCl acts as the Brønsted acid, donating its proton to $NH_3$, which, by accepting the proton, acts as a Brønsted base. The products of this proton transfer are the chloride ion ($Cl^-$) and the ammonium ion ($NH_4^+$), which subsequently combine to form the ionic solid ammonium chloride.

$$HCl(g) + NH_3(g) \longrightarrow NH_4Cl(s)$$

Most of the strong acids are ***monoprotic acids,*** meaning that each acid molecule has one pro-ton to donate. One of the strong acids, $H_2SO_4$, is a ***diprotic acid,*** meaning that each acid molecule has two protons that it can donate. Other diprotic acids include oxalic acid ($H_2C_2O_4$) and carbonic acid ($H_2CO_3$). There are also ***triprotic acids,*** those with three protons, although they are relatively less common than mono- or diprotic acids. Examples include phosphoric acid ($H_3PO_4$) and citric acid ($H_3C_6H_5O_7$). In general, acids with more than one proton are called ***polyprotic acids.***

Of the polyprotic acids, only sulfuric acid is a strong acid. Recall from Table 4.1, though, that $H_2SO_4$ is strong only in its first ionization in water. Although $H_2SO_4$ ionizes completely to

---

2. Johannes Nicolaus Brønsted (1879–1947). Danish chemist. In addition to his theory of acids and bases, Brønsted worked on thermodynamics and the separation of mercury into its isotopes. In some books, Brønsted acids and bases are called Brønsted-Lowry acids and bases. Thomas Martin Lowry (1874–1936). English chemist. Brønsted and Lowry developed essentially the same acid-base theory independently in 1923.

yield $H^+$ and $HSO_4^-$, the subsequent ionization of the hydrogen sulfate ion ($HSO_4^-$) happens only to a very small extent. Note the single and double arrows in the following two equations.

$$H_2SO_4(aq) \longrightarrow H^+(aq) + HSO_4^-(aq)$$

$$HSO_4^-(aq) \rightleftharpoons H^+(aq) + SO_4^{2-}(aq)$$

For all other polyprotic acids, each ionization is incomplete and is represented by an equation with a double arrow. The first, second, and third ionizations of phosphoric acid are represented as

$$H_3PO_4(aq) \rightleftharpoons H^+(aq) + H_2PO_4^-(aq)$$

$$H_2PO_4^-(aq) \rightleftharpoons H^+(aq) + HPO_4^{2-}(aq)$$

$$HPO_4^{2-}(aq) \rightleftharpoons H^+(aq) + PO_4^{3-}(aq)$$

Some of each of the species shown is present in a solution of phosphoric acid. Because each successive ionization happens to a smaller and smaller extent, the relative concentrations of species in solution are as follows:

$$[H_3PO_4] > [H^+] \approx [H_2PO_4^-] > [HPO_4^{2-}] > [PO_4^{3-}]$$

> **Student Annotation:** These relative concentrations are true only in an aqueous solution of phosphoric acid that contains no other dissolved compounds. We will look in detail at how to determine concentrations in aqueous solutions of polyprotic acids in Chapter 16.

Just as some acids produce more than one $H^+$ ion, some strong bases produce more than one $OH^-$ ion. Barium hydroxide, for example, dissociates to produce 2 moles of hydroxide ion for every mole of $Ba(OH)_2$ dissolved.

$$Ba(OH)_2(s) \xrightarrow{H_2O} Ba^{2+}(aq) + 2OH^-(aq)$$

Compounds such as this are referred to as *dibasic* bases, indicating that they produce 2 moles of hydroxide per mole of compound. Those that produce only 1 mole of hydroxide per mole of compound, such as NaOH, are called *monobasic* bases.

## Acid-Base Neutralization

A *neutralization reaction* is a reaction between an acid and a base. In general, an aqueous acid-base reaction produces water and a *salt,* which is an ionic compound made up of the cation from a base and the anion from an acid. The substance we know as table salt, NaCl, is a familiar example. It is a product of the following acid-base reaction:

$$HCl(aq) + NaOH(aq) \longrightarrow H_2O(l) + NaCl(aq)$$

However, because the acid, base, and salt are all strong electrolytes, they exist entirely as ions in solution. The ionic equation is

$$H^+(aq) + Cl^-(aq) + Na^+(aq) + OH^-(aq) \longrightarrow H_2O(l) + Na^+(aq) + Cl^-(aq)$$

The net ionic equation is

$$H^+(aq) + OH^-(aq) \longrightarrow H_2O(l)$$

Both $Na^+$ and $Cl^-$ are spectator ions. If we were to carry out the preceding reaction using stoichiometric amounts [◄◄ Section 3.6] of HCl and NaOH, the result would be neutral saltwater with no leftover acid or base.

The following are also examples of acid-base neutralization reactions, represented by molecular equations:

$$HNO_3(aq) + KOH(aq) \longrightarrow H_2O(l) + KNO_3(aq)$$

$$H_2SO_4(aq) + 2NaOH(aq) \longrightarrow 2H_2O(l) + Na_2SO_4(aq)$$

$$2HC_2H_3O_2(aq) + Ba(OH)_2(aq) \longrightarrow 2H_2O(l) + Ba(C_2H_3O_2)_2(aq)$$

$$HCl(aq) + NH_3(aq) \longrightarrow NH_4Cl(aq)$$

The last equation looks different because it does not show water as a product. Recall, however, that $NH_3(aq)$ ionizes to give $NH_4^+(aq)$ and $OH^-(aq)$. If we include these two species as reactants in place of $NH_3(aq)$, the equation becomes

$$HCl(aq) + NH_4^+(aq) + OH^-(aq) \longrightarrow H_2O(l) + NH_4Cl(aq)$$

Sample Problem 4.5 involves an acid-base neutralization reaction.

Multimedia

Acids and bases—neutralization reaction of NaOH and HCl.

> **Student Annotation:** A compound in which the anion is oxide ($O^{2-}$) or hydroxide ($OH^-$) is not considered a salt.

> **Student Annotation:** Acid-base neutralization reactions, like precipitation reactions [◄◄ Section 4.2], are *metathesis* reactions, where two species exchange ions.

## Sample Problem 4.5

Milk of magnesia, an over-the-counter laxative, is a mixture of magnesium hydroxide [$Mg(OH)_2$] and water. Because $Mg(OH)_2$ is insoluble in water (see Table 4.3), milk of magnesia is a *suspension* rather than a solution. The undissolved solid is responsible for the milky appearance of the product. When acid such as HCl is added to milk of magnesia, the suspended $Mg(OH)_2$ dissolves, and the result is a clear, colorless solution. Write and balance the molecular equation, and then give the ionic and net ionic equations for this reaction.

(a) **Milk of magnesia**          (b) **Addition of HCl**          (c) **Resulting clear solution**

**Strategy**  Determine the products of the reaction; then write and balance the equation. Remember that one of the reactants, $Mg(OH)_2$, is a solid. Identify any strong electrolytes and rewrite the equation showing strong electrolytes as ions. Identify and cancel the spectator ions.

**Setup**  Because this is an acid-base neutralization reaction, one of the products is water. The other product is a salt comprising the cation from the base, $Mg^{2+}$, and the anion from the acid, $Cl^-$. For the formula to be neutral, these ions combine in a 1:2 ratio, giving $MgCl_2$ as the formula of the salt.

**Solution**

$$Mg(OH)_2(s) + 2HCl(aq) \longrightarrow 2H_2O(l) + MgCl_2(aq)$$

Of the species in the molecular equation, only HCl and $MgCl_2$ are strong electrolytes. Therefore, the ionic equation is

$$Mg(OH)_2(s) + 2H^+(aq) + 2Cl^-(aq) \longrightarrow 2H_2O(l) + Mg^{2+}(aq) + 2Cl^-(aq)$$

$Cl^-$ is the only spectator ion. The net ionic equation is

$$Mg(OH)_2(s) + 2H^+(aq) \longrightarrow 2H_2O(l) + Mg^{2+}(aq)$$

**Think About It** Make sure your equation is balanced and that you only show strong electrolytes as ions. $Mg(OH)_2$ is *not* shown as aqueous ions because it is insoluble.

**Practice Problem A**  Write and balance the molecular equation and then give the ionic and net ionic equations for the neutralization reaction between $Ba(OH)_2(aq)$ and $HF(aq)$.

**Practice Problem B**  Write and balance the molecular equation and then give the ionic and net ionic equations for the neutralization reaction between $NH_3(aq)$ and $H_2SO_4(aq)$.

## Checkpoint 4.3     Acid-Base Reactions

**4.3.1**  Identify the Brønsted acid in the following equation:

$$H_2SO_4(aq) + 2NH_3(aq) \longrightarrow (NH_4)_2SO_4(aq)$$

a) $H_2SO_4(aq)$

b) $NH_3(aq)$

c) $H_2O(l)$

d) $(NH_4)_2SO_4$

e) This equation does not contain a Brønsted acid.

**4.3.2**  Identify the Brønsted base in the following equation:

$$HCl(aq) + NO_2^-(aq) \longrightarrow HNO_2(aq) + Cl^-(aq)$$

a) $HCl(aq)$

b) $NO_2^-(aq)$

c) $HNO_2(aq)$

d) $Cl^-(aq)$

e) $H_2O(l)$

**4.3.3** Which of the following is the correct net ionic equation for the reaction of $H_2SO_4$ and KOH?

a) $H_2(aq) + 2OH^-(aq) \longrightarrow 2H_2O(l)$

b) $2H^+(aq) + 2OH^-(aq) \longrightarrow 2H_2O(l)$

c) $2H^+(aq) + OH^-(aq) \longrightarrow H_2O(l)$

d) $H_2SO_4(aq) + 2OH^-(aq) \longrightarrow 2H_2O(l) + SO_4^{2-}(aq)$

e) $H^+(aq) + HSO_4^-(aq) + 2OH^-(aq) \longrightarrow 2H_2O(l) + SO_4^{2-}(aq)$

**4.3.4** Which of the following is the correct net ionic equation for the reaction of HF and LiOH?

a) $HF(aq) + LiOH(aq) \longrightarrow H_2O(aq) + LiF(aq)$

b) $H^+(aq) + F^-(aq) + Li^+(aq) + OH^-(aq) \longrightarrow$ $H_2O(aq) + Li^+(aq) + F^-(aq)$

c) $H^+(aq) + OH^-(aq) \longrightarrow H_2O(l)$

d) $HF(aq) + OH^-(aq) \longrightarrow H_2O(l) + F^-(aq)$

e) $H^+(aq) + Li^+(aq) + OH^-(aq) \longrightarrow H_2O(l) + Li^+(aq)$

# 4.4 Oxidation-Reduction Reactions

In Sections 4.2 and 4.3 we encountered two types of chemical reactions that can occur when two electrolyte solutions are combined: *precipitation,* in which ionic compounds exchange ions, and *acid-base neutralization,* in which a proton is transferred from an acid to a base. In this section, we will learn about **oxidation-reduction reactions,** commonly called *redox* reactions. A **redox reaction** is a chemical reaction in which *electrons* are transferred from one reactant to another. For example, if we place a piece of zinc metal into a solution that contains copper ions, the following reaction will occur:

$$Zn(s) + Cu^{2+}(aq) \longrightarrow Zn^{2+}(aq) + Cu(s)$$

This reaction is shown in Figure 4.6. In this process, zinc atoms are *oxidized* (they lose electrons) and copper ions are *reduced* (they gain electrons). Each zinc atom loses two electrons to become a zinc ion,

$$Zn(s) \longrightarrow Zn^{2+}(aq) + 2e^-$$

and each copper ion gains two electrons to become a copper atom.

$$Cu^{2+}(aq) + 2e^- \longrightarrow Cu(s)$$

**Figure 4.6** Oxidation of zinc in a solution of copper(II) sulfate.

(a)
Zinc atoms enter the solution as zinc ions ($Zn^{2+}$). Copper ions are reduced to copper atoms on the surface of the metal.

(b)
Cu atoms have replaced Zn atoms in the solid, and $Zn^{2+}$ ions have replaced $Cu^{2+}$ ions in solution.

These two equations show electrons as a product in the zinc reaction and as a reactant in the copper reaction. Each of these two equations represents a ***half-reaction,*** the oxidation or the reduction reaction in a redox reaction. The sum of the two half-reaction equations is the overall equation for the redox reaction:

$$Zn(s) \longrightarrow Zn^{2+}(aq) + 2e^-$$
$$+ \; Cu^{2+}(aq) + 2e^- \longrightarrow Cu(s)$$
$$\overline{Zn(s) + Cu^{2+}(aq) + 2e^- \longrightarrow Zn^{2+}(aq) + Cu(s) + 2e^-}$$

Although these two processes can be represented by separate equations, they cannot occur separately. For one species to gain electrons, another must lose them, and vice versa.

***Oxidation*** is the *loss* of electrons. The opposite process, the *gain* of electrons, is called ***reduction.*** In the reaction of Zn with $Cu^{2+}$, Zn is called the ***reducing agent*** because it donates electrons, causing $Cu^{2+}$ to be reduced. $Cu^{2+}$ is called the ***oxidizing agent,*** on the other hand, because it accepts electrons, causing Zn to be oxidized.

Another example of a redox reaction is the formation of calcium oxide (CaO) from its constituent elements.

$$2Ca(s) + O_2(g) \longrightarrow 2CaO(s)$$

In this reaction, each calcium atom loses two electrons (is oxidized) and each oxygen atom gains two electrons (is reduced). The corresponding half-reactions are

$$2Ca \longrightarrow 2Ca^{2+} + 4e^-$$
$$O_2 + 4e^- \longrightarrow 2O^{2-}$$

The resulting $Ca^{2+}$ and $O^{2-}$ ions combine to form CaO.

Redox reactions take place because atoms of different elements have different tendencies to gain electrons. Oxygen, for instance, has a much greater tendency to gain electrons than does calcium. Calcium, being a metal, has a significant tendency to lose electrons. Compounds that form between elements with significantly different tendencies to gain electrons generally are ionic. By knowing the charges on the monatomic ions in such compounds, we can keep track of the electrons that have been lost and gained.

## Oxidation Numbers

When elements of similar abilities to gain electrons combine, they tend to form molecular compounds, as in the formation of HF and $NH_3$ from their respective elements:

$$H_2(g) + F_2(g) \longrightarrow 2HF(g)$$
$$N_2(g) + 3H_2(g) \longrightarrow 2NH_3(g)$$

In the formation of hydrogen fluoride (HF), therefore, fluorine does not gain an electron per se—and hydrogen does not lose one. Experimental evidence shows, however, that there is a partial transfer of electrons from H to F. Oxidation numbers provide us with a way to "balance the books" with regard to electrons in a chemical equation. The ***oxidation number,*** also called the ***oxidation state,*** is the charge an atom would have *if* electrons were transferred completely. For example, we can rewrite the preceding equations for the formation of HF and $NH_3$ as follows:

$$H_2(g) \quad + \quad F_2(g) \quad \longrightarrow \quad 2HF(g)$$

| $\boxed{0}$ over $\boxed{0}$ | $\boxed{0}$ over $\boxed{0}$ | $\boxed{+1}\boxed{-1}$ over $\boxed{+1}\boxed{-1}$ |

$$N_2(g) \quad + \quad 3H_2(g) \quad \longrightarrow \quad 2NH_3(g)$$

| $\boxed{0}$ over $\boxed{0}$ | $\boxed{0}$ over $\boxed{0}$ | $\boxed{-3}\boxed{+1}$ over $\boxed{-3}\boxed{+3}$ |

The numbers in circles below each element are the oxidation numbers, and the numbers in the boxes are each atom's contribution to overall charge. In both of the reactions shown, the reactants are all homonuclear diatomic molecules. Thus, we would expect *no* transfer of electrons from one atom to the other and the oxidation number of each is zero. For the product molecules, however,

**Student Annotation:** The term *oxidation* originally was used by chemists to mean "reaction with oxygen." Like definitions of acids and bases, though, it has been redefined to include any reaction in which electrons are lost.

Multimedia
Chemical reactions—formation of $Ag_2S$ by oxidation reduction.

Multimedia
Chemical reactions—vanadium.

**Student Annotation:** The number in a circle under each element's symbol contains the oxidation state of the element. The number in a square contains the total contribution to charge by that element. For more on the use of this convention, see the "You May Be Wondering" box on page 137.

for the sake of determining oxidation numbers, we assume that complete electron transfer has taken place and that each atom has either gained or lost one or more electrons. The oxidation numbers reflect the number of electrons assumed to have been transferred.

Oxidation numbers enable us to identify elements that are oxidized and reduced at a glance. The elements that show an *increase* in oxidation number—hydrogen in the preceding examples—are oxidized, whereas the elements that show a *decrease* in oxidation number—fluorine and nitrogen—are reduced.

In a sense, each atom's oxidation number makes a contribution to the overall charge on the species. Note that the oxidation numbers in both HF [$(+1) + (-1) = 0$] and $NH_3$ [$(-3) + 3(+1) = 0$] sum to zero. Because compounds are electrically neutral, the oxidation numbers in any compound will sum to zero. For a polyatomic ion, oxidation numbers must sum to the charge on the ion. (The oxidation number of a monatomic ion is equal to its charge.)

## Oxidation of Metals in Aqueous Solutions

Recall from the beginning of this section that zinc metal reacts with aqueous copper ions to form aqueous zinc ions and copper metal. One way that this reaction might be carried out is for zinc metal to be immersed in a solution of copper(II) sulfate ($CuSO_4$), as depicted in Figure 4.6. The molecular equation for this reaction is

$$Zn(s) \quad + \quad CuCl_2(aq) \quad \longrightarrow \quad ZnCl_2(aq) \quad + \quad Cu(s)$$

This is an example of a **displacement reaction.** Zinc *displaces,* or *replaces,* copper in the dissolved salt by being oxidized from Zn to $Zn^{2+}$. Copper is displaced from the salt (and removed from solution) by being reduced from $Cu^{2+}$ to Cu. Chloride ($Cl^-$), which is neither oxidized nor reduced, is a spectator ion in this reaction.

What would happen, then, if we placed copper metal into a solution containing zinc chloride ($ZnCl_2$)? Would Cu(s) be oxidized to $Cu^{2+}(aq)$ by $Zn^{2+}(aq)$ the way Zn(s) is oxidized to $Zn^{2+}(aq)$ by $Cu^{2+}(aq)$? The answer is no. In fact, no reaction would occur if we were to immerse copper metal into an aqueous solution of $ZnCl_2$.

$$Cu(s) + ZnCl_2(aq) \longrightarrow \text{no reaction}$$

No reaction occurs between Cu(s) and $Zn^{2+}(aq)$, whereas a reaction does occur between Zn(s) and $Cu^{2+}(aq)$ because zinc is more easily oxidized than copper.

The **activity series** (Table 4.5) is a list of metals (and hydrogen) arranged from top to bottom in order of decreasing ease of oxidation. The second column shows the oxidation half-reaction corresponding to each element in the first column. Note the positions of zinc and copper in the table. Zinc appears higher in the table and is therefore oxidized more easily. In fact, an element in the series will be oxidized by the ions of any element that appears below it. According to Table 4.5, therefore, zinc metal will be oxidized by a solution containing any of the following ions: $Cr^{3+}$, $Fe^{2+}$, $Cd^{2+}$, $Co^{2+}$, $Ni^{2+}$, $Sn^{2+}$, $H^+$, $Cu^{2+}$, $Ag^+$, $Hg^{2+}$, $Pt^{2+}$, or $Au^{3+}$. On the other hand, zinc will not be oxidized by a solution containing $Mn^{2+}$, $Al^{3+}$, $Mg^{2+}$, $Na^+$, $Ca^{2+}$, $Ba^{2+}$, $K^+$, or $Li^+$ ions.

Metals listed at the top of the activity series are called the *active metals*. These include the alkali and alkaline earth metals. These metals are so reactive that they are not found in nature in their elemental forms. Metals at the bottom of the series, such as copper, silver, platinum, and gold, are called the *noble metals* because they have very little tendency to react. These are the metals most often used for jewelry and coins.

## Balancing Simple Redox Equations

To learn how to balance redox equations, let's revisit the practice of balancing equations. In Chapter 3, you learned to balance equations by counting the number of each kind of atom on each side of the equation arrow. For the purpose of balancing redox equations, it is also necessary to count electrons. For example, consider the net ionic equation for the reaction of chromium metal with nickel ion:

$$Cr(s) + Ni^{2+}(aq) \longrightarrow Cr^{3+}(aq) + Ni(s)$$

Although this equation has equal numbers of each type of atom on both sides, it is not balanced because there is a charge of $+2$ on the reactant side and a charge of $+3$ on the product side. To balance it, we can separate it into its half-reactions.

Some metals such as copper are so unreactive that they are found in nature in the uncombined state.

**Student Annotation:** When a metal is *oxidized* by an aqueous solution, it becomes an aqueous ion.

**TABLE 4.5**    Activity Series

| Element | Oxidation Half-Reaction |
|---|---|
| Lithium | $Li \longrightarrow Li^+ + e^-$ |
| Potassium | $K \longrightarrow K^+ + e^-$ |
| Barium | $Ba \longrightarrow Ba^{2+} + 2e^-$ |
| Calcium | $Ca \longrightarrow Ca^{2+} + 2e^-$ |
| Sodium | $Na \longrightarrow Na^+ + e^-$ |
| Magnesium | $Mg \longrightarrow Mg^{2+} + 2e^-$ |
| Aluminum | $Al \longrightarrow Al^{3+} + 3e^-$ |
| Manganese | $Mn \longrightarrow Mn^{2+} + 2e^-$ |
| Zinc | $Zn \longrightarrow Zn^{2+} + 2e^-$ |
| Chromium | $Cr \longrightarrow Cr^{3+} + 3e^-$ |
| Iron | $Fe \longrightarrow Fe^{2+} + 2e^-$ |
| Cadmium | $Cd \longrightarrow Cd^{2+} + 2e^-$ |
| Cobalt | $Co \longrightarrow Co^{2+} + 2e^-$ |
| Nickel | $Ni \longrightarrow Ni^{2+} + 2e^-$ |
| Tin | $Sn \longrightarrow Sn^{2+} + 2e^-$ |
| Lead | $Pb \longrightarrow Pb^{2+} + 2e^-$ |
| Hydrogen | $H_2 \longrightarrow 2H^+ + 2e^-$ |
| Copper | $Cu \longrightarrow Cu^{2+} + 2e^-$ |
| Silver | $Ag \longrightarrow Ag^+ + e^-$ |
| Mercury | $Hg \longrightarrow Hg^{2+} + 2e^-$ |
| Platinum | $Pt \longrightarrow Pt^{2+} + 2e^-$ |
| Gold | $Au \longrightarrow Au^{3+} + 3e^-$ |

*Increasing ease of oxidation* ↑

**Multimedia**

Sodium, lithium, and potassium reaction series.

$$Cr(s) \longrightarrow Cr^{3+}(aq) + 3e^-$$

$$Ni^{2+}(aq) + 2e^- \longrightarrow Ni(s)$$

When we add half-reactions to get the overall reaction, the electrons must cancel. Because any electrons lost by one species must be gained by the other, electrons may *not* appear in an overall chemical equation. Therefore, prior to adding these two half-reactions, we must multiply the chromium half-reaction by 2

$$2[Cr(s) \longrightarrow Cr^{3+}(aq) + 3e^-]$$

and the nickel half-reaction by 3.

$$3[Ni^{2+}(aq) + 2e^- \longrightarrow Ni(s)]$$

Then when we add the half-reactions, the electrons cancel and we get the balanced overall equation.

$$2Cr(s) \longrightarrow 2Cr^{3+}(aq) + \cancel{6e^-}$$
$$+ \ 3Ni^{2+}(aq) + \cancel{6e^-} \longrightarrow 3Ni(s)$$
$$\overline{2Cr(s) + 3Ni^{2+}(aq) \longrightarrow 2Cr^{3+}(aq) + 3Ni(s)}$$

This is known as the **half-reaction method** of balancing redox equations. We will use this method extensively when we examine more complex redox reactions in Chapter 19.

The activity series enables us to predict whether or not a metal will be oxidized by a solution containing a particular salt or by an acid. Sample Problems 4.7 and 4.8 on page 139 give you more practice making such predictions and balancing redox equations.

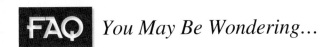

## How Do I Assign Oxidation Numbers?

The following guidelines will help you assign oxidation numbers. There are essentially two rules:

1. The oxidation number of any element, in its elemental form, is zero.
2. The oxidation numbers in any chemical species must sum to the overall charge on the species. That is, oxidation numbers must sum to zero for any molecule and must sum to the charge on any polyatomic ion. The oxidation number of a monatomic ion is equal to the charge on the ion.

In addition to these two rules, it is necessary to know the elements that always, or nearly always, have the same oxidation number. Table 4.6 lists elements whose oxidation numbers are "reliable," in order of decreasing reliability.

 To determine oxidation numbers in a compound or a polyatomic ion, you must use a stepwise, systematic approach. Draw a circle under each element's symbol in the chemical formula. Then draw a square under each circle. In the circle, write the oxidation number of the element; in the square, write the total contribution to charge by that element. Start with the oxidation numbers you know, and use them to figure out the ones you don't know. Here is an example:

$KMnO_4$

Oxidation number
Total contribution to charge

Fill in the oxidation number first for the element that appears highest on the list in Table 4.6. Potassium (K) is a Group 1A metal. In its compounds, it always has the oxidation number $+1$. We write $+1$ in the circle beneath the K. Because there is only one K atom in this formula, the total contribution to charge is also $+1$, so we also write $+1$ in the square beneath the K.

$KMnO_4$

Oxidation number   $+1$
Total contribution to charge   $+1$

Next on the list is oxygen (O). In compounds, O usually has the oxidation number $-2$, so we assign it $-2$. Because there are four O atoms in the formula, the total contribution to charge by O atoms is $4(-2) = -8$.

$KMnO_4$

Oxidation number   $+1$   $-2$
Total contribution to charge   $+1$   $-8$

The numbers in the squares, all the contributions to overall charge, must sum to zero. This requires putting $+7$ in the box beneath the Mn atom. Because there is just one Mn atom in this formula, the contribution to charge is the same as the oxidation number. Thus, $(+1) + (+7) + (-8) = 0$.

$KMnO_4$

Oxidation number   $+1$ $+7$ $-2$
Total contribution to charge   $+1$ $+7$ $-8$

Sample Problem 4.6 lets you determine oxidation numbers in three more compounds and a polyatomic ion.

| TABLE 4.6 | Elements with Reliable Oxidation Numbers in Compounds or Polyatomic Ions | |
|---|---|---|
| **Element** | **Oxidation Number** | **Exceptions** |
| Fluorine | $-1$ | |
| Group 1A or 2A metal | $+1$ or $+2$, respectively | |
| Hydrogen | $+1$ | Any combination with a Group 1A or 2A metal to form a metal hydride. Examples: LiH and $CaH_2$—the oxidation number of H is $-1$ in both examples. |
| Oxygen | $-2$ | Any combination with something higher on the list that necessitates its having a different oxidation number (see rule 2 for assigning oxidation numbers). Examples: $H_2O_2$ and $KO_2$—the oxidation number of O for $H_2O_2$ is $-1$ and for $KO_2$ is $-\frac{1}{2}$. |
| Group 7A (other than fluorine) | $-1$ | Any combination with something higher on the list that necessitates its having a different oxidation number (see rule 2 for assigning oxidation numbers). Examples: ClF, $BrO_4^-$, and $IO_3^-$—the oxidation numbers of Cl, Br, and I are $+1$, $+7$, and $+5$, respectively. Remember that these exceptions do not apply to fluorine, which *always* has an oxidation state of $-1$ when it is part of a compound. |

*You May Be Wondering...(continued)*

**Sample Problem 4.6**

Determine the oxidation number of each atom in the following compounds and ion: (a) $SO_2$, (b) NaH, (c) $CO_3^{2-}$, (d) $N_2O_5$.

**Strategy** For each compound, assign an oxidation number first to the element that appears higher in Table 4.6 (page 137). Then use rule 2 to determine the oxidation number of the other element.

**Setup** (a) O appears in Table 4.6 but S does not, so we assign oxidation number $-2$ to O. Because there are two O atoms in the molecule, the total contribution to charge by O is $2(-2) = -4$. The lone S atom must therefore contribute $+4$ to the overall charge.

(b) Both Na and H appear in Table 4.6, but Na appears higher in the table, so we assign the oxidation number $+1$ to Na. This means that H must contribute $-1$ to the overall charge.

(c) We assign the oxidation number $-2$ to O. Because there are three O atoms in the carbonate ion, the total contribution to charge by O is $-6$. To have the contributions to charge sum to the charge on the ion ($-2$), the C atom must contribute $+4$.

(d) We assign the oxidation number $-2$ to O. Because there are five O atoms in the $N_2O_5$ molecule, the total contribution to charge by O is $-10$. To have the contributions to charge sum to zero, the contribution by N must be $+10$, and because there are two N atoms, each one must contribute $+5$. Therefore, the oxidation number of N is $+5$.

**Solution**
(a) In $SO_2$, the oxidation numbers of S and O are $+4$ and $-2$, respectively.

$$SO_2$$

$$\boxed{+4}\boxed{-2}$$
$$\boxed{+4}\boxed{-4}$$

(b) In NaH, the oxidation numbers of Na and H are $+1$ and $-1$, respectively.

$$NaH$$

$$\boxed{+1}\boxed{-1}$$
$$\boxed{+1}\boxed{-1}$$

(c) In $CO_3^{2-}$, the oxidation numbers of C and O are $+4$ and $-2$, respectively.

$$CO_3^{2-}$$

$$\boxed{+4}\boxed{-2}$$
$$\boxed{+4}\boxed{-6}$$

(d) In $N_2O_5$, the oxidation numbers of N and O are $+5$ and $-2$, respectively.

$$N_2O_5$$

$$\boxed{+5}\boxed{-2}$$
$$\boxed{+10}\boxed{-10}$$

**Think About It** Use the circle and square system to verify that the oxidation numbers you have assigned do indeed sum to the overall charge on each species.

**Practice Problem A** Assign oxidation numbers to each atom in the following compounds and ion: $H_2O_2$, $ClO_3^-$, $H_2SO_4$.

**Practice Problem B** Assign oxidation numbers to each atom in the following species: $O_2^{2-}$, $ClO^-$, $MnO_2$.

## Sample Problem 4.7

Predict which of the following reactions will occur, and for those that will occur, write the net ionic equation and indicate which element is oxidized and which is reduced: (a) $Fe(s) + PtCl_2(aq) \longrightarrow$ ? (b) $Cr(s) + AuCl_3(aq) \longrightarrow$ ? (c) $Pb(s) + Zn(NO_3)_2(aq) \longrightarrow$ ?

**Strategy** Recognize that the salt in each equation (the compound on the reactant side) is a strong electrolyte. What is important is the identity of the metal cation *in* the salt. For each equation, compare the positions in Table 4.5 (page 136) of the solid metal and the metal cation from the salt to determine whether or not the solid metal will be oxidized. If the cation appears lower in the table, the solid metal will be oxidized (i.e., the reaction will occur). If the cation appears higher in the table, the solid metal will not be oxidized (i.e., no reaction will occur).

**Setup** (a) The cation in $PtCl_2$ is $Pt^{2+}$. Platinum appears lower in Table 4.5 than iron, so $Pt^{2+}(aq)$ will oxidize $Fe(s)$.

(b) The cation in $AuCl_3$ is $Au^{3+}$. Gold appears lower in Table 4.5 than chromium, so $Au^{3+}(aq)$ will oxidize $Cr(s)$.

(c) The cation in $Zn(NO_3)_2$ is $Zn^{2+}$. Zinc appears higher in Table 4.5 than lead, so $Zn^{2+}(aq)$ will not oxidize $Pb(s)$.

**Solution** (a) $Fe(s) + Pt^{2+}(aq) \longrightarrow Fe^{2+}(aq) + Pt(s)$; iron is oxidized (0 to +2) and platinum is reduced (+2 to 0).

(b) $Cr(s) + Au^{3+}(aq) \longrightarrow Cr^{3+}(aq) + Au(s)$; chromium is oxidized (0 to +3) and gold is reduced (+3 to 0).

(c) No reaction.

> **Think About It** Check your conclusions by working each problem backward. For part (b), for example, write the net ionic equation in reverse, using the products as the reactants: $Au(s) + Cr^{3+}(aq) \longrightarrow$ ? Now compare the positions of gold and chromium in Table 4.5 again. Chromium is higher, so chromium(III) ions cannot oxidize gold. This confirms your conclusion that the forward reaction (the oxidation of chromium by gold ions) will occur.

**Practice Problem A** Predict which of the following reactions will occur, and for those that will occur, write the net ionic equation and indicate which element is oxidized and which is reduced: (a) $Co(s) + BaI_2(aq) \longrightarrow$ ? (b) $Sn(s) + CuBr_2(aq) \longrightarrow$ ? (c) $Ag(s) + NaCl(aq) \longrightarrow$ ?

**Practice Problem B** Predict which of the following reactions will occur, and for those that will occur, write the net ionic equation and indicate which element is oxidized and which is reduced: (a) $Ni(s) + Cu(NO_3)_2(aq) \longrightarrow$ ? (b) $Ag(s) + KCl(aq) \longrightarrow$ ? (c) $Al(s) + AuCl_3(aq) \longrightarrow$ ?

## Sample Problem 4.8

Predict which of the following reactions will occur, and for those that will occur, balance the equation and indicate which element is oxidized and which is reduced: (a) $Al(s) + CaCl_2(aq) \longrightarrow$ ? (b) $Cr(s) + Pb(C_2H_3O_2)_2(aq) \longrightarrow$ ? (c) $Sn(s) + HI(aq) \longrightarrow$ ?

**Strategy** As in Sample Problem 4.7, identify the cation in the aqueous species and for each equation, compare the positions in Table 4.5 of the solid metal and the cation to determine whether or not the solid metal will be oxidized. If the cation appears lower in the table, the reaction will occur.

**Setup** (a) The cation in $CaCl_2$ is $Ca^{2+}$. Calcium appears higher in Table 4.5 than aluminum, so $Ca^{2+}(aq)$ will not oxidize $Al(s)$. (b) The cation in $Pb(C_2H_3O_2)_2$ is $Pb^{2+}$. Lead appears lower in Table 4.5 than chromium, so $Pb^{2+}(aq)$ will oxidize $Cr(s)$. (c) The cation in HI is $H^+$. Hydrogen appears lower in Table 4.5 than tin, so $H^+(aq)$ will oxidize $Sn(s)$.

**Solution** (a) No reaction.

(b) The two half-reactions are represented by the following:

Oxidation: $Cr(s) \longrightarrow Cr^{3+}(aq) + 3e^-$

Reduction: $Pb^{2+}(aq) + 2e^- \longrightarrow Pb(s)$

To balance the charges, we must multiply the oxidation half-reaction by 2 and the reduction half-reaction by 3:

$$2 \times [Cr(s) \longrightarrow Cr^{3+}(aq) + 3e^-] = 2Cr(s) \longrightarrow 2Cr^{3+}(aq) + 6e^-$$

$$3 \times [Pb^{2+}(aq) + 2e^- \longrightarrow Pb(s)] = 3Pb^{2+}(aq) + 6e^- \longrightarrow 3Pb(s)$$

*(Continued)*

We can then add the two half-reactions, canceling the electrons on both sides to get

$$2Cr(s) + 3Pb^{2+}(aq) \longrightarrow 2Cr^{3+}(aq) + 3Pb(s)$$

The overall, balanced molecular equation is

$$2Cr(s) + 3Pb(C_2H_3O_2)_2(aq) \longrightarrow 2Cr(C_2H_3O_2)_3(aq) + 3Pb(s)$$

Chromium is oxidized (0 to +3) and lead is reduced (+2 to 0).

(c) The two half-reactions are as follows:

Oxidation: $Sn(s) \longrightarrow Sn^{2+}(aq) + 2e^-$

Reduction: $2H^+(aq) + 2e^- \longrightarrow H_2(g)$

Adding the two half-reactions and canceling the electrons on both sides yields

$$Sn(s) + 2H^+(aq) \longrightarrow Sn^{2+}(aq) + H_2(g)$$

The overall, balanced molecular equation is

$$Sn(s) + 2HI(aq) \longrightarrow SnI_2(aq) + H_2(g)$$

Tin is oxidized (0 to +2) and hydrogen is reduced (+1 to 0).

**Student Annotation:** Reactions in which hydrogen ion is reduced to hydrogen gas are known as **hydrogen displacement** reactions.

> **Think About It** Check your conclusions by working each problem backward. Write each equation in reverse and compare the positions of the elements in the activity series.

**Practice Problem A** Predict which of the following reactions will occur, and for those that will occur, give the overall, balanced molecular equation and indicate which element is oxidized and which is reduced. (a) $Mg(s) + Cr(C_2H_3O_2)_3(aq) \longrightarrow$ ? (b) $Cu(s) + HBr(aq) \longrightarrow$ ? (c) $Cd(s) + AgNO_3(aq) \longrightarrow$ ?

**Practice Problem B** Predict which of the following reactions will occur, and for those that will occur, indicate which element is oxidized and which is reduced. (a) $Pt(s) + Cu(NO_3)_2(aq) \longrightarrow$ ? (b) $Ag(s) + AuCl_3(aq) \longrightarrow$ ? (c) $Sn(s) + HNO_3(aq) \longrightarrow$ ?

## Other Types of Redox Reactions

Several of the reaction types that you have already encountered are also redox reactions.

### Combination Reactions

Combination reactions such as the formation of ammonia from its constituent elements can involve oxidation and reduction.

$$N_2(g) \quad + \quad 3H_2(g) \quad \longrightarrow \quad 2NH_3(g)$$

In this reaction, nitrogen is reduced from 0 to −3, while hydrogen is oxidized from 0 to +1. Other examples of combination reactions include those shown in Figure 4.7.

### Decomposition

Decomposition can also be a redox reaction, as illustrated by the following examples:

$$2NaH(s) \longrightarrow 2Na(s) + H_2(g)$$

$$2K\,Cl\,O_3(s) \longrightarrow 2KCl(s) + 3O_2(g)$$

$$2H_2O_2(aq) \longrightarrow 2H_2O(l) + O_2(g)$$

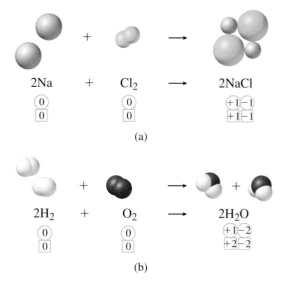

$$2Na + Cl_2 \longrightarrow 2NaCl$$

(a)

$$2H_2 + O_2 \longrightarrow 2H_2O$$

(b)

**Figure 4.7** (a) Reaction between sodium and chlorine to form sodium chloride, and (b) reaction between hydrogen and oxygen to form water. For each element, the oxidation number appears in the circle and the total contribution to charge appears in the square below it.

The decomposition of hydrogen peroxide, shown in the preceding equation, is an example of a **disproportionation reaction,** in which one element undergoes both oxidation and reduction. In the case of $H_2O_2$, the oxidation number of O is initially $-1$. In the products of the decomposition, O has an oxidation number of $-2$ in $H_2O$ and of 0 in $O_2$.

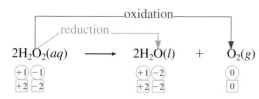

$$2H_2O_2(aq) \longrightarrow 2H_2O(l) + O_2(g)$$

Finally, **combustion** [◀◀ Section 3.3] is a redox process.

$$CH_4(g) + 2O_2(g) \longrightarrow CO_2(g) + 2H_2O(l)$$

Figure 4.8 shows the known oxidation numbers of elements in compounds—arranged according to their positions in the periodic table.

---

## Checkpoint 4.4    Oxidation-Reduction Reactions

**4.4.1** Determine the oxidation number of sulfur in each of the following species: $H_2S$, $HSO_3^-$, $SCl_2$, and $S_8$.

a) $+2, +6, -2, +\frac{1}{4}$

b) $-2, +3, +2, 0$

c) $-2, +5, +2, -\frac{1}{4}$

d) $-1, +4, +2, 0$

e) $-2, +4, +2, 0$

**4.4.2** What species is the reducing agent in the following equation?

$$Mg(s) + 2HCl(aq) \longrightarrow MgCl_2(aq) + H_2(g)$$

a) $Mg(s)$

b) $H^+(aq)$

c) $Cl^-(aq)$

d) $Mg^{2+}(aq)$

e) $H_2(g)$

**4.4.3** Which of the following equations represents a redox reaction? (Choose all that apply.)

a) $2Mg(s) + O_2(g) \longrightarrow 2MgO(s)$

b) $Cu(s) + PtCl_2(aq) \longrightarrow CuCl_2(aq) + Pt(s)$

c) $NH_4Cl(aq) + AgNO_3(aq) \longrightarrow NH_4NO_3(aq) + AgCl(s)$

d) $2NaN_3(s) \longrightarrow 2Na(s) + 3N_2(g)$

e) $CaCO_3(s) \longrightarrow CaO(s) + CO_2(g)$

**4.4.4** According to the activity series, which of the following redox reactions will occur? (Choose all that apply.)

a) $Fe(s) + NiBr_2(aq) \longrightarrow FeBr_2(aq) + Ni(s)$

b) $Sn(s) + Pb(NO_3)_2(aq) \longrightarrow Sn(NO_3)_2(aq) + Pb(s)$

c) $Mg(s) + BaI_2(aq) \longrightarrow MgI_2(aq) + Ba(s)$

d) $Pb(s) + PtCl_2(aq) \longrightarrow PbCl_2(aq) + Pt(s)$

e) $Zn(s) + CaBr_2(aq) \longrightarrow ZnBr_2(aq) + Ca(s)$

**Figure 4.8** Periodic table showing oxidation numbers for each element. The most common oxidation numbers are shown in red.

## 4.5 Concentration of Solutions

One of the factors that can influence reactions in aqueous solution is concentration. The **concentration** of a solution is the amount of solute dissolved in a given quantity of solvent or solution. Consider the two solutions of iodine pictured in Figure 4.9. The solution on the left is more concentrated than the one on the right—that is, it contains a higher ratio of solute to solvent. By contrast, the solution on the right is more dilute. The color is more intense in the more concentrated solution. Often the concentrations of reactants determine how fast a chemical reaction occurs. For example, the reaction of magnesium metal and acid [◀◀ Section 4.4] happens faster if the concentration of acid is greater. As we will see in Chapter 13, there are several different ways to express the concentration of a solution. In this chapter, we introduce only molarity, which is one of the most commonly used units of concentration.

### Molarity

**Molarity,** or **molar concentration,** symbolized $M$, is defined as the number of moles of solute per liter of solution. Thus, 1 L of a 1.5 molar solution of glucose ($C_6H_{12}O_6$), written as 1.5 $M$ $C_6H_{12}O_6$, contains 1.5 moles of dissolved glucose. Half a liter of the same solution would contain 0.75 mole

**Student Annotation:** The qualitative terms *concentrated* and *dilute* are relative terms, like *expensive* and *cheap*.

Multimedia

Solutions—preparation of solutions.

**Student Annotation:** Molarity can equally well be defined as millimoles per milliliter (mmol/mL), which can simplify some calculations.

**Figure 4.9** Two solutions of iodine in benzene. The solution on the left is more concentrated. The solution on the right is more dilute.

Concentrated solution:
More solute particles per unit volume

Dilute solution:
Fewer solute particles per unit volume

of dissolved glucose, a milliliter of the solution would contain $1.5 \times 10^{-3}$ mole of dissolved glucose, and so on.

$$\text{molarity} = \frac{\text{moles solute}}{\text{liters solution}} \qquad \text{Equation 4.1}$$

To calculate the molarity of a solution, we divide the number of moles of solute by the volume of the solution in liters.

Equation 4.1 can be rearranged in three ways to solve for any of the three variables: molarity ($M$), moles of solute (mol), or volume of solution in liters (L).

$$(1)\ M = \frac{\text{mol}}{\text{L}} \qquad (2)\ \text{L} = \frac{\text{mol}}{M} \qquad (3)\ \text{mol} = M \times \text{L}$$

**Student Annotation:** Students sometimes have difficulty seeing how units cancel in these equations. It may help to write $M$ as mol/L until you become completely comfortable with these equations.

Sample Problem 4.9 illustrates how to use these equations to solve for molarity, volume of solution, and moles of solute.

## Sample Problem 4.9

For an aqueous solution of glucose ($C_6H_{12}O_6$), determine (a) the molarity of 2.00 L of a solution that contains 50.0 g of glucose, (b) the volume of this solution that would contain 0.250 mole of glucose, and (c) the number of moles of glucose in 0.500 L of this solution.

**Strategy** Convert the mass of glucose given to moles, and use the equations for interconversions of $M$, liters, and moles to calculate the answers.

**Student Annotation:** A common way to state the concentration of this solution is to say, "This solution is 0.139 $M$ in glucose."

**Setup** The molar mass of glucose is 180.2 g.

$$\text{moles of glucose} = \frac{50.0\ \text{g}}{180.2\ \text{g/mol}} = 0.277\ \text{mol}$$

**Solution** (a) molarity $= \dfrac{0.277\ \text{mol}\ C_6H_{12}O_6}{2.00\ \text{L solution}} = 0.139\ M$

(b) volume $= \dfrac{0.250\ \text{mol}\ C_6H_{12}O_6}{0.139\ M} = 1.80\ \text{L}$

(c) moles of $C_6H_{12}O_6$ in 0.500 L $= 0.500\ \text{L} \times 0.139\ M = 0.0695\ \text{mol}$

**Think About It** Check to see that the magnitudes of your answers are logical. For example, the mass given in the problem corresponds to 0.277 mole of solute. If you are asked, as in part (b), for the volume that contains a number of moles smaller than 0.277, make sure your answer is smaller than the original volume.

*(Continued)*

**Practice Problem A** For an aqueous solution of sucrose ($C_{12}H_{22}O_{11}$), determine (a) the molarity of 5.00 L of a solution that contains 235 g of sucrose, (b) the volume of this solution that would contain 1.26 mole of sucrose, and (c) the number of moles of sucrose in 1.89 L of this solution.

**Practice Problem B** For an aqueous solution of sodium chloride (NaCl), determine (a) the molarity of 3.75 L of a solution that contains 155 g of sodium chloride, (b) the volume of this solution that would contain 4.58 moles of sodium chloride, and (c) the number of moles of sodium chloride in 22.75 L of this solution.

Media Player/MPEG Animation: Figure 4.10, Preparing a Solution from a Solid, pp. 146–147.

**Student Annotation:** It is important to remember that molarity is defined in terms of the volume of solution, not the volume of solvent. In many cases, these two are not the same.

The procedure for preparing a solution of known molarity is shown in Figure 4.10 (pp. 146–147). First, the solute is weighed accurately and transferred, often with a funnel, to a volumetric flask of the desired volume. Next, water is added to the flask, which is then swirled to dissolve the solid. After all the solid has dissolved, more water is added slowly to bring the level of solution exactly to the volume mark. Finally, the flask is capped and inverted to ensure thorough mixing. Knowing the volume of the solution in the flask and the quantity of compound dissolved, we can determine the molarity of the solution using Equation 4.1. Note that this procedure does not require that we know the exact amount of water added. Because of the way molarity is defined, it is important only that we know the final volume of the *solution*.

## Dilution

Multimedia
Solutions—dilution.

Concentrated "stock" solutions of commonly used substances typically are kept in the laboratory stockroom. Often we need to dilute these stock solutions before using them. ***Dilution*** is the process of preparing a less concentrated solution from a more concentrated one. Suppose that we want to prepare 1.00 L of a 0.400 *M* $KMnO_4$ solution from a solution of 1.00 *M* $KMnO_4$. For this purpose we need 0.400 mole of $KMnO_4$. Because there is 1.00 mole of $KMnO_4$ in 1.00 L of a 1.00 *M* $KMnO_4$ solution, there is 0.400 mole of $KMnO_4$ in 0.400 L of the same solution:

$$\frac{1.00 \text{ mol } KMnO_4}{1.00 \text{ L of solution}} = \frac{0.400 \text{ mol } KMnO_4}{0.400 \text{ L of solution}}$$

Therefore, we must withdraw 400 mL from the 1.00 *M* $KMnO_4$ solution and dilute it to 1.00 L by adding water (in a 1.00-L volumetric flask). This method gives us 1.00 L of the desired 0.400 *M* $KMnO_4$.

In carrying out a dilution process, it is useful to remember that adding more solvent to a given amount of the stock solution changes (decreases) the concentration of the solution without changing the number of moles of solute present in the solution (Figure 4.11, p. 148).

**Equation 4.2**    moles of solute before dilution = moles of solute after dilution

Using arrangement (3) of Equation 4.1, we can calculate the number of moles of solute:

$$\text{moles of solute} = \frac{\text{moles of solute}}{\text{liters of solution}} \times \text{liters of solution}$$

Because the number of moles of solute before the dilution is the same as that after dilution, we can write

**Equation 4.3**    $$M_c \times L_c = M_d \times L_d$$

where the subscripts c and d stand for *concentrated* and *dilute*, respectively. Thus, by knowing the molarity of the concentrated stock solution ($M_c$) and the desired final molarity ($M_d$) and volume ($L_d$) of the dilute solution, we can calculate the volume of stock solution required for the dilution ($L_c$).

Because most volumes measured in the laboratory are in milliliters rather than liters, it is worth pointing out that Equation 4.3 can also be written with volumes of the concentrated and dilute solutions in milliliters.

$$M_c \times mL_c = M_d \times mL_d \qquad \text{Equation 4.4}$$

**Student Annotation:** Students sometimes resist using the unit *millimole*. However, using the $M_c \times mL_c = M_d \times mL_d$ form in Equation 4.4 often reduces the number of steps in a problem, thereby reducing the number of opportunities to make calculation errors.

In this form of the equation, the product of each side is in millimoles (mmol) rather than moles. We apply Equation 4.4 in Sample Problem 4.10.

---

## Sample Problem 4.10

What volume of 12.0 $M$ HCl, a common laboratory stock solution, must be used to prepare 250.0 mL of 0.125 $M$ HCl?

**Strategy** Use Equation 4.4 to determine the volume of 12.0 $M$ HCl required for the dilution.

**Setup** $M_c = 12.0\ M$, $M_d = 0.125\ M$, $mL_d = 250.0$ mL.

**Solution**

$$12.0\ M \times mL_c = 0.125\ M \times 250.0\ \text{mL}$$

$$mL_c = \frac{0.125\ M \times 250.0\ \text{mL}}{12.0\ M} = 2.60\ \text{mL}$$

**Student Annotation:** It is very important to note that, for safety, when diluting a concentrated acid, the acid must be added to the water, and *not* the other way around.

**Think About It** Plug the answer into Equation 4.4, and make sure that the product of concentration and volume is the same on both sides of the equation.

**Practice Problem A** What volume of 6.0 $M$ $H_2SO_4$ is needed to prepare 500.0 mL of a solution that is 0.25 $M$ in $H_2SO_4$?

**Practice Problem B** What volume of 0.20 $M$ $H_2SO_4$ can be prepared by diluting 125 mL of 6.0 $M$ $H_2SO_4$?

---

## Serial Dilution

A series of dilutions may be used in the laboratory to prepare a number of increasingly dilute solutions from a stock solution. The method involves preparing a solution as described on page 146 and diluting a portion of the prepared solution to make a *more* dilute solution. For example, we could use the 0.400 $M$ $KMnO_4$ solution described on page 144 to prepare a series of five increasingly dilute solutions, with the concentration decreasing by a factor of 10 at each stage. Using a volumetric pipette, we withdraw 10.00 mL of the 0.400 $M$ solution and deliver it into a 100.00-mL volumetric flask as shown in Figure 4.12(a). We then dilute to the volumetric mark and cap and invert the flask to ensure complete mixing. The concentration of the newly prepared solution is determined using Equation 4.4, where $M_c$ is 0.400 $M$, and $mL_c$ and $mL_d$ are 10.00 mL and 100.00 mL, respectively.

$$0.400\ M \times 10.00\ \text{mL} = M_d \times 100.00\ \text{mL}$$

$$M_d = 0.0400\ M \quad \text{or} \quad 4.00 \times 10^{-2}\ M$$

Repeating this process four more times, each time using the most recently prepared solution as the "concentrated" solution and diluting 10.00 mL to 100.00 mL, we get five $KMnO_4$ solutions with concentrations $4.00 \times 10^{-2}\ M$, $4.00 \times 10^{-3}\ M$, $4.00 \times 10^{-4}\ M$, $4.00 \times 10^{-5}\ M$, and $4.00 \times 10^{-6}\ M$ [Figure 4.12(b)]. This type of serial dilution is commonly used to prepare "standard" solutions with precisely known concentrations, for quantitative analysis.

Figure 4.10

# Preparing a Solution from a Solid

Weigh out the solid $KMnO_4$.
(The tare function on a digital
balance automatically subtracts
the mass of the weighing paper.)

The mass likely will not be
exactly the calculated number.

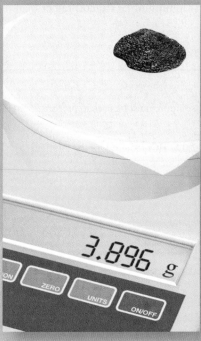

Transfer the weighed $KMnO_4$
to the volumetric flask.

Calculate the mass of $KMnO_4$ necessary
for the target concentration of 0.1 $M$.

$$\frac{0.1 \text{ mol}}{\text{L}} \times 0.2500 \text{ L} = 0.02500 \text{ mol}$$

$$0.02500 \text{ mol} \times \frac{158.04 \text{ g}}{\text{mol}} = 3.951 \text{ g } KMnO_4$$

Add water sufficient to dissolve the $KMnO_4$.

Swirl the flask to dissolve the solid.

Add more water.

Fill exactly to the calibration mark using a wash bottle or eye dropper.

After capping and inverting the flask to ensure complete mixing, we calculate the actual concentration of the prepared solution.

$$3.896 \text{ g KMnO}_4 \times \frac{1 \text{ mol}}{158.04 \text{ g}} = 0.024652 \text{ mol}$$

$$\frac{0.024652 \text{ mol}}{0.2500 \text{ L}} = 0.09861 \ M$$

## What's the point?

The goal is to prepare a solution of precisely known concentration, with that concentration being very close to the target concentration of 0.1 $M$. Note that because 0.1 is a *specified* number, it does not limit the number of significant figures in our calculations.

**Figure 4.11**   Dilution changes the concentration of a solution; it does not change the number of moles of solute in the solution.

Add solvent →

Before dilution:
More solute particles per unit volume

After dilution:
Fewer solute particles per unit volume

(a)

**Figure 4.12**   Serial dilution. (a) A solution of precisely known concentration is prepared in a volumetric flask. A precise volume of the solution is transferred to a second volumetric flask and subsequently diluted. (b) A precise volume of the second solution is transferred to a third volumetric flask and diluted. The process is repeated several times, each time producing more dilute solution. In this example, the concentration is reduced by a factor of 10 at each stage.

(b)

Sample Problem 4.11 illustrates the method of serial dilution to prepare a series of standard HCl solutions.

---

**Sample Problem 4.11**

Starting with a 2.00 $M$ stock solution of hydrochloric acid, four standard solutions (1 to 4) are prepared by sequentially diluting 10.00 mL of each solution to 250.00 mL. Determine (a) the concentrations of all four standard solutions and (b) the number of moles of HCl in each solution.

**Strategy** In part (a), because the volumes are all given in milliliters, we will use Equation 4.4, rearranged to solve for $M_d$, to determine the molar concentration of each standard solution. In part (b), Equation 4.1, rearranged to solve for moles, can be used to calculate the number of moles in each. We must remember to convert each solution's volume to liters so that units will cancel properly.

**Setup** (a) $M_d = \dfrac{M_c \times mL_c}{mL_d}$; (b) mol = $M \times$ L, 250.00 mL = $2.500 \times 10^{-1}$ L.

**Solution** (a) $M_{d1} = \dfrac{2.00\ M \times 10.00\ mL}{250.00\ mL} = 8.00 \times 10^{-2}\ M$

$M_{d2} = \dfrac{8.00 \times 10^{-2}\ M \times 10.00\ mL}{250.00\ mL} = 3.20 \times 10^{-3}\ M$

$M_{d3} = \dfrac{3.20 \times 10^{-3}\ M \times 10.00\ mL}{250.00\ mL} = 1.28 \times 10^{-4}\ M$

$M_{d4} = \dfrac{1.28 \times 10^{-4}\ M \times 10.00\ mL}{250.00\ mL} = 5.12 \times 10^{-6}\ M$

(b) $mol_1 = 8.00 \times 10^{-2}\ M \times 2.500 \times 10^{-1}\ L = 2.00 \times 10^{-2}$ mol
$mol_2 = 3.20 \times 10^{-3}\ M \times 2.500 \times 10^{-1}\ L = 8.00 \times 10^{-4}$ mol
$mol_3 = 1.28 \times 10^{-4}\ M \times 2.500 \times 10^{-1}\ L = 3.20 \times 10^{-5}$ mol
$mol_4 = 5.12 \times 10^{-6}\ M \times 2.500 \times 10^{-1}\ L = 1.28 \times 10^{-6}$ mol

**Think About It** Serial dilution is one of the fundamental practices of homeopathy. Some remedies undergo so many serial dilutions that very few (if any) molecules of the original substance still exist in the final preparation.

---

**Practice Problem A** Starting with a 6.552 $M$ stock solution of $HNO_3$, five standard solutions are prepared via serial dilution. At each stage, 25.00 mL of solution is diluted to 100.00 mL. Determine (a) the concentration of and (b) the number of moles of $HNO_3$ in each standard solution.

**Practice Problem B** Five standard solutions of HBr are prepared by serial dilution in which, at each stage, 10.00 mL is diluted to 150.00 mL. Given that the concentration of the most dilute solution is $3.22 \times 10^{-6}\ M$, determine the concentration of the original HBr stock solution.

## Solution Stoichiometry

Soluble ionic compounds such as $KMnO_4$ are strong electrolytes, so they undergo complete dissociation upon dissolution and exist in solution entirely as ions. $KMnO_4$ dissociates, for example, to give 1 mole of potassium ion and 1 mole of permanganate ion for every mole of potassium permanganate. Thus, a 0.400 $M$ solution of $KMnO_4$ will be 0.400 $M$ in $K^+$ and 0.400 $M$ in $MnO_4^-$.

In the case of a soluble ionic compound with other than a 1:1 combination of constituent ions, we must use the subscripts in the chemical formula to determine the concentration of each ion in solution. Sodium sulfate ($Na_2SO_4$) dissociates, for example, to give twice as many sodium ions as sulfate ions.

$$Na_2SO_4(s) \xrightarrow{H_2O} 2Na^+(aq) + SO_4^{2-}(aq)$$

Therefore, a solution that is 0.35 $M$ in $Na_2SO_4$ is actually 0.70 $M$ in $Na^+$ and 0.35 $M$ in $SO_4^{2-}$. Frequently, molar concentrations of dissolved species are expressed using square brackets. Thus, the concentrations of species in a 0.35 $M$ solution of $Na_2SO_4$ can be expressed as follows: $[Na^+] = 0.70\ M$ and $[SO_4^{2-}] = 0.35\ M$.

Sample Problem 4.13 (page 152) lets you practice relating concentrations of compounds and concentrations of individual ions using solution stoichiometry.

**Student Annotation:** If we only need to express the concentration of the compound, rather than the concentrations of the individual ions, we could express the concentration of this solution as $[Na_2SO_4] = 0.35\ M$.

**Student Annotation:** Square brackets around a chemical species can be read as "the concentration of" that species. For example, $[Na^+]$ is read as "the concentration of sodium ion."

**FAQ** *You May Be Wondering…*

# How Can We Measure Solution Concentrations?

As you may know, white light is actually composed of all the colors of the rainbow. In fact, a rainbow results from the separation of white light by water droplets into the colors or wavelengths that make up the visible spectrum [▶▶ Section 6.1]. Selective absorption of visible light is what makes some solutions appear colored; and, for a solution that is colored, the intensity of color is related to the solution's concentration (see Figure 4.9). This effect gives rise to a type of analysis known as *visible spectrophotometry.* A visible spectrophotometer compares the intensity of light that enters a sample (called the *incident* light) $I_0$, with the intensity of the light that is transmitted through the sample, $I$. *Transmittance* (T) is the ratio of $I$ to $I_0$.

Equation 4.5         $T = \dfrac{I}{I_0}$

*Absorbance* (A) measures how much light is absorbed by the solution and is defined as the negative logarithm of transmittance.

Equation 4.6     $A = -\log T = -\log \dfrac{I}{I_0}$

**Student Annotation:** Both transmittance and absorbance are *unitless* quantities.

Plotting absorbance as a function of wavelength gives an *absorption spectrum.* The absorption spectrum, that is, the characteristic absorption over a range of wavelengths, can serve as a sort of fingerprint for the identification of a compound in solution.

The quantitative relationship between absorbance and a solution's concentration is called the *Beer-Lambert law* and is expressed as

Equation 4.7             $A = \varepsilon bc$

where    $\varepsilon$ = proportionality constant called the *molar absorptivity*
    $b$ = path length of solution (in cm) through which light travels
    $c$ = molar concentration of solution

The molar absorptivity is specific to a chemical species and is a measure of how strongly the species absorbs light at a particular wavelength. Figure 4.13 shows how absorbance depends on path length and concentration. Quantitative analysis using visible spectrophotometry generally requires selection of the appropriate wavelength for analysis (usually the wavelength at which absorbance is highest), determination of absorbance for a series of solutions of known concentration (the *standards*), construction of a calibration curve (Figure 4.14), and calculation of an unknown concentration using the calibration curve.

Sample Problem 4.12 shows how visible spectrophotometry is used to determine the concentration of a colored solution.

**Figure 4.13**    (a) A colored solution absorbs some of the incident visible light, diminishing the light's intensity from $I_0$ to $I$. (b) The intensity is reduced more when the light travels through a longer path length of the same solution or (c) when the light travels through the same path length of a more concentrated solution.

**Figure 4.14**    A calibration curve with absorbance (A) on the y axis and molar concentration on the x axis. Linear regression is done using a spreadsheet or graphing calculator to generate the line that best fits all the calibration data. An unknown concentration can be determined by drawing a dashed line from the point on the calibration line corresponding to the measured absorbance to the x axis, as shown. In this case, a measured absorbance of 0.83 corresponds to a concentration of 1.1 M.

## Sample Problem 4.12

Absorbance values for five standard $KMnO_4$ solutions were determined at a wavelength of 528 nm. The data are tabulated here.

| $[KMnO_4]$ $(M)$ | $A$ |
|---|---|
| 0.0500 | 0.172 |
| 0.100 | 0.310 |
| 0.150 | 0.502 |
| 0.200 | 0.680 |
| 0.250 | 0.822 |

The absorbance of a $KMnO_4$ solution with unknown concentration, analyzed under the same conditions, was 0.550. Determine the concentration of the unknown solution.

**Strategy** Using either a graphing calculator or an Excel spreadsheet, we graph the tabulated data to verify that the relationship between absorbance and concentration is linear and that the line goes through the origin. We then use the line and the measured absorbance to determine the concentration of the unknown solution.

**Student Annotation:** Equation 4.7 takes the form of the linear equation $y = mx + b$, where $y$ is the absorbance, $m$ (the slope) is the product of molar absorptivity and path length, $x$ is the molar concentration, and $b$ (the $y$-intercept) is zero.

**Setup** Graphing the tabulated data with Excel gives the following plot:

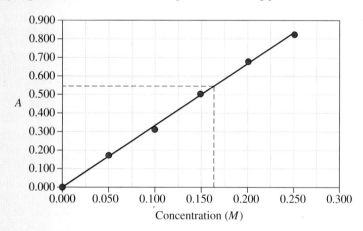

**Solution** Solving for the unknown concentration gives 0.165 $M$.

**Think About It** Absorbance expresses the magnitude of light absorption by a sample. It would be significantly harder to determine unknown concentrations if we were to plot *transmittance* as a function of concentration because the relationship between concentration and transmittance is not linear.

**Practice Problem A** Absorbance values for five standard solutions of a colored solute were determined at 410 nm with a 1.00-cm path length, giving the following table of data:

| Solute concentration $(M)$ | $A$ |
|---|---|
| 0.250 | 0.165 |
| 0.500 | 0.317 |
| 0.750 | 0.510 |
| 1.000 | 0.650 |
| 1.250 | 0.837 |

The absorbance of a solution of unknown concentration containing the same solute was 0.400. What is the concentration of the unknown solution?

**Practice Problem B** Referring to Practice Problem A, (a) determine the absorbance values you would expect for solutions with the following concentrations: 0.4 $M$, 0.6 $M$, 0.8 $M$, 1.1 $M$. (b) Using the data in the table in Practice Problem A, calculate the average molar absorptivity of the compound and determine the units of molar absorptivity.

## Sample Problem 4.13

Using square-bracket notation, express the concentration of (a) chloride ion in a solution that is 1.02 $M$ in $AlCl_3$, (b) nitrate ion in a solution that is 0.451 $M$ in $Ca(NO_3)_2$, and (c) $Na_2CO_3$ in a solution in which $[Na^+] = 0.124\ M$.

**Strategy**   Use the concentration given in each case and the stoichiometry indicated in the corresponding chemical formula to determine the concentration of the specified ion or compound.

**Setup**   (a) There are 3 moles of $Cl^-$ ion for every 1 mole of $AlCl_3$,

$$AlCl_3(s) \xrightarrow{H_2O} Al^{3+}(aq) + 3Cl^-(aq)$$

so the concentration of $Cl^-$ will be three times the concentration of $AlCl_3$.

(b) There are 2 moles of nitrate ion for every 1 mole of $Ca(NO_3)_2$,

$$Ca(NO_3)_2(s) \xrightarrow{H_2O} Ca^{2+}(aq) + 2NO_3^-(aq)$$

so $[NO_3^-]$ will be twice $[Ca(NO_3)_2]$.

(c) There is 1 mole of $Na_2CO_3$ for every 2 moles of sodium ion,

$$Na_2CO_3(s) \xrightarrow{H_2O} 2Na^+(aq) + CO_3^{2-}(aq)$$

so $[Na_2CO_3]$ will be half of $[Na^+]$. (Assume that $Na_2CO_3$ is the only source of $Na^+$ ions in this solution.)

**Solution**

(a) $[Cl^-] = [AlCl_3] \times \dfrac{3\ \text{mol } Cl^-}{1\ \text{mol } AlCl_3}$

$\qquad = \dfrac{1.02\ \text{mol } AlCl_3}{L} \times \dfrac{3\ \text{mol } Cl^-}{1\ \text{mol } AlCl_3}$

$\qquad = \dfrac{3.06\ \text{mol } Cl^-}{L}$

$\qquad = 3.06\ M$

(b) $[NO_3^-] = [Ca(NO_3)_2] \times \dfrac{2\ \text{mol } NO_3^-}{1\ \text{mol } Ca(NO_3)_2}$

$\qquad = \dfrac{0.451\ \text{mol } Ca(NO_3)_2}{L} \times \dfrac{2\ \text{mol } NO_3^-}{1\ \text{mol } Ca(NO_3)_2}$

$\qquad = \dfrac{0.902\ \text{mol } NO_3^-}{L}$

$\qquad = 0.902\ M$

(c) $[Na_2CO_3] = [Na^+] \times \dfrac{1\ \text{mol } Na_2CO_3}{2\ \text{mol } Na^+}$

$\qquad = \dfrac{0.124\ \text{mol } Na^+}{L} \times \dfrac{1\ \text{mol } Na_2CO_3}{2\ \text{mol } Na^+}$

$\qquad = \dfrac{0.0620\ \text{mol } Na_2CO_3}{L}$

$\qquad = 0.0620\ M$

**Think About It**   Make sure that units cancel properly. Remember that the concentration of an ion can never be less than the concentration of its dissolved parent compound. It will always be the concentration of the parent compound times its stoichiometric subscript in the chemical formula.

**Practice Problem A**   Using the square-bracket notation, express the concentrations of ions in a solution that is 0.750 $M$ in aluminum sulfate $[Al_2(SO_4)_3]$.

**Practice Problem B**   Using the square-bracket notation, express the concentration of chloride ions in a solution that is 0.250 $M$ in sodium chloride (NaCl) and 0.25 $M$ in magnesium chloride ($MgCl_2$).

## Checkpoint 4.5     Concentration of Solutions

**4.5.1** Calculate the molar concentration of a solution prepared by dissolving 58.5 g NaOH in enough water to yield 1.25 L of solution.

a) 1.46 *M*

b) 46.8 *M*

c) $2.14 \times 10^{-2}$ *M*

d) 1.17 *M*

e) 0.855 *M*

**4.5.2** What mass of glucose ($C_6H_{12}O_6$) in grams must be used in order to prepare 500 mL of a solution that is 2.50 *M* in glucose?

a) 225 g

b) 125 g

c) 200 g

d) 1.25 g

e) 625 g

**4.5.3** What volume in mL of a 1.20 *M* HCl solution must be diluted in order to prepare 1.00 L of 0.0150 *M* HCl?

a) 15.0 mL

b) 12.5 mL

c) 12.0 mL

d) 85.0 mL

e) 115 mL

**4.5.4** A solution that is 0.18 *M* in $Na_2CO_3$ is also _____. (Choose all that apply.)

a) 0.18 *M* in $CO_3^{2-}$

b) 0.18 *M* in $Na^+$

c) 0.09 *M* in $Na^+$

d) 0.09 *M* in $CO_3^{2-}$

e) 0.36 *M* in $Na^+$

**4.5.5** Which best represents the before-and-after molecular-level view of the dilution of a concentrated stock solution?

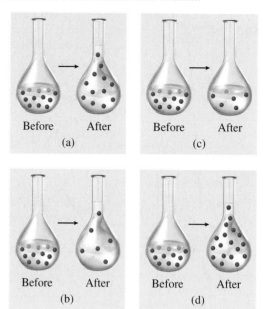

**4.5.6** Which best represents an aqueous solution of sodium sulfate?

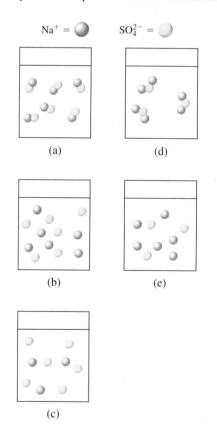

Aqueous Reactions and Chemical Analysis

**Student Annotation:** Experiments that measure the amount of a substance present are called *quantitative analysis.*

Certain aqueous reactions are useful for determining how much of a particular substance is present in a sample. For example, if we want to know the concentration of lead in a sample of water, or if we need to know the concentration of an acid, knowledge of precipitation reactions, acid-base reactions, and solution stoichiometry will be useful. Two common types of such quantitative analyses are *gravimetric analysis* and *acid-base titration.*

## Gravimetric Analysis

*Gravimetric analysis* is an analytical technique based on the measurement of mass. One type of gravimetric analysis experiment involves the formation and isolation of a precipitate, such as $AgCl(s)$:

**Student Annotation:** According to the information in Table 4.2, AgCl is an insoluble exception to the chlorides, which typically are soluble.

$$AgNO_3(aq) + NaCl(aq) \longrightarrow NaNO_3(aq) + AgCl(s)$$

This reaction is often used in gravimetric analysis because the reactants can be obtained in pure form. The net ionic equation is

$$Ag^+(aq) + Cl^-(aq) \longrightarrow AgCl(s)$$

Suppose, for example, that we wanted to test the purity of a sample of NaCl by determining the percent by mass of Cl. First, we would accurately weigh out some NaCl and dissolve it in water. To this mixture, we would add enough $AgNO_3$ solution to cause the precipitation of all the $Cl^-$ ions present in solution as AgCl. (In this procedure NaCl is the limiting reagent and $AgNO_3$ is the excess reagent.) We would then separate, dry, and weigh the AgCl precipitate. From the measured mass of AgCl, we would be able to calculate the mass of Cl using the percent by mass of Cl in AgCl. Because all the Cl in the precipitate came from the dissolved NaCl, the amount of Cl that we calculate is the amount that was present in the original NaCl sample. We could then calculate the percent by mass of Cl in the NaCl and compare it to the known composition of NaCl to determine its purity.

Gravimetric analysis is a highly accurate technique, because the mass of a sample can be measured accurately. However, this procedure is applicable only to reactions that go to completion or have nearly 100 percent yield. In addition, if AgCl were soluble to any significant degree, it would not be possible to remove all the $Cl^-$ ions from the original solution, and the subsequent calculation would be in error. Sample Problem 4.14 shows the calculations involved in a gravimetric experiment.

---

### Sample Problem 4.14

A 0.8633-g sample of an ionic compound containing chloride ions and an unknown metal cation is dissolved in water and treated with an excess of $AgNO_3$. If 1.5615 g of AgCl precipitate forms, what is the percent by mass of Cl in the original compound?

**Strategy** Using the mass of AgCl precipitate and the percent composition of AgCl, determine what mass of chloride the precipitate contains. The chloride in the precipitate was originally in the unknown compound. Using the mass of chloride and the mass of the original sample, determine the percent Cl in the compound.

**Setup** To determine the percent Cl in AgCl, divide the molar mass of Cl by the molar mass of AgCl:

$$\frac{35.45 \text{ g}}{35.45 \text{ g} + 107.9 \text{ g}} \times 100\% = 24.73\%$$

The mass of Cl in the precipitate is $0.2473 \times 1.5615$ g = 0.3862 g.

**Solution** The percent Cl in the unknown compound is the mass of Cl in the precipitate divided by the mass of the original sample:

$$\frac{0.3862 \text{ g}}{0.8633 \text{ g}} \times 100\% = 44.73\% \text{ Cl}$$

**Think About It** Pay close attention to which numbers correspond to which quantities. It is easy in this type of problem to lose track of which mass is the precipitate and which is the original sample. Dividing by the wrong mass at the end will result in an incorrect answer.

**Practice Problem A** A 0.5620-g sample of an ionic compound containing the bromide ion ($Br^-$) is dissolved in water and treated with an excess of $AgNO_3$. If the mass of the AgBr precipitate that forms is 0.8868 g, what is the percent by mass of Br in the original compound?

**Practice Problem B** A sample that is 63.9 percent chloride by mass is dissolved in water and treated with an excess of $AgNO_3$. If the mass of the AgCl precipitate that forms is 1.085 g, what was the mass of the original sample?

Gravimetric analysis is a quantitative method, not a qualitative one, so it does not establish the identity of the unknown substance. Thus, the results in Sample Problem 4.14 do *not* identify the cation. However, knowing the percent by mass of Cl greatly helps us narrow the possibilities. Because no two compounds containing the same anion (or cation) have the same percent composition by mass, comparison of the percent by mass obtained from gravimetric analysis with that calculated from a series of known compounds could reveal the identity of the unknown compounds.

## Acid-Base Titrations

Quantitative studies of acid-base neutralization reactions are most conveniently carried out using a technique known as titration. In *titration*, a solution of accurately known concentration, called a *standard solution*, is added gradually to another solution of unknown concentration, until the chemical reaction between the two solutions is complete, as shown in Figure 4.15. If we know the volumes of the standard and unknown solutions used in the titration, along with the concentration of the standard solution, we can calculate the concentration of the unknown solution.

A solution of the strong base sodium hydroxide can be used as the standard solution in a titration, but it must first be *standardized,* because sodium hydroxide in solution reacts with carbon dioxide in the air, making its concentration unstable over time. We can standardize the sodium hydroxide solution by titrating it against an acid solution of accurately known concentration. The acid often chosen for this task is a monoprotic acid called potassium hydrogen phthalate (KHP), for which the molecular formula is $KHC_8H_4O_4$. KHP is a white, soluble solid that is commercially available in highly pure form. The reaction between KHP and sodium hydroxide is

$$KHC_8H_4O_4(aq) + NaOH(aq) \longrightarrow KNaC_8H_4O_4(aq) + H_2O(l)$$

$HC_8H_4O_4^-$

and the net ionic equation is

$$HC_8H_4O_4^-(aq) + OH^-(aq) \longrightarrow C_8H_4O_4^{2-}(aq) + H_2O(l)$$

To standardize a solution of NaOH with KHP, a known amount of KHP is transferred to an Erlenmeyer flask and some distilled water is added to make up a solution. Next, NaOH solution is carefully added to the KHP solution from a burette until all the acid has reacted with the base. This point in the titration, where the acid has been completely neutralized, is called the *equivalence point.* It is usually signaled by the *endpoint,* where an indicator causes a sharp change in the color of the solution. In acid-base titrations, *indicators* are substances that have distinctly different colors in acidic and basic media. One commonly used indicator is phenolphthalein, which is colorless in acidic and neutral solutions but reddish pink in basic solutions. At the equivalence point, all the KHP present has been neutralized by the added NaOH and the solution is still colorless. However, if we add just one more drop of NaOH solution from the buret, the solution will be basic and will immediately turn pink. Sample Problem 4.15 illustrates just such a titration.

Multimedia

Chemical reactions—titrations.

**Student Annotation:** *Standardization* in this context is the meticulous determination of concentration.

**Student Annotation:** Note that KHP is a *monoprotic* acid, so it reacts in a 1:1 ratio with hydroxide ion.

**Student Annotation:** The endpoint in a titration is used to approximate the equivalence point. A careful choice of indicators, which we will discuss in Chapter 16, helps make this approximation reasonable. Phenolphthalein, although very common, is not appropriate for every acid-base titration.

**Figure 4.15**    Apparatus for titration.

## Sample Problem 4.15

In a titration experiment, a student finds that 25.49 mL of an NaOH solution is needed to neutralize 0.7137 g of KHP. What is the concentration (in $M$) of the NaOH solution?

**Strategy**    Using the mass given and the molar mass of KHP, determine the number of moles of KHP. Recognize that the number of moles of NaOH in the volume given is equal to the number of moles of KHP. Divide moles of NaOH by volume (in liters) to get molarity.

**Setup**    The molar mass of KHP ($KHC_8H_4O_4$) = [39.1 g + 5(1.008 g) + 8(12.01 g) + 4(16.00 g)] = 204.2 g/mol.

**Solution**

$$\text{moles of KHP} = \frac{0.7137 \text{ g}}{204.2 \text{ g/mol}} = 0.003495 \text{ mol}$$

Because moles of KHP = moles of NaOH, then moles of NaOH = 0.003495 mol.

$$\text{molarity of NaOH} = \frac{0.003495 \text{ mol}}{0.02549 \text{ L}} = 0.1371 \text{ } M$$

**Think About It**    Remember that molarity can also be defined as mmol/mL. Try solving the problem again using millimoles and make sure you get the same answer.

$$0.003495 \text{ mol} = 3.495 \times 10^{-3} \text{ mol}$$
$$= 3.495 \text{ mmol}$$

and

$$\frac{3.495 \text{ mmol}}{25.49 \text{ mL}} = 0.1371 \text{ } M$$

**Practice Problem A**    How many grams of KHP are needed to neutralize 22.36 mL of a 0.1205 $M$ NaOH solution?

**Practice Problem B**    What volume (in mL) of a 0.2550 $M$ NaOH solution can be neutralized by 10.75 g of KHP?

The reaction between NaOH and KHP is a relatively simple acid-base neutralization. Suppose, though, that instead of KHP, we wanted to use a diprotic acid such as $H_2SO_4$ for the titration. The reaction is represented by

$$2NaOH(aq) + H_2SO_4(aq) \longrightarrow Na_2SO_4(aq) + 2H_2O(l)$$

Because 2 mol NaOH $\backsimeq$ 1 mol $H_2SO_4$, we need twice as much NaOH to react completely with an $H_2SO_4$ solution of the *same* molar concentration and volume as a monoprotic acid such as HCl. On the other hand, we would need twice the amount of HCl to neutralize a $Ba(OH)_2$ solution compared to an NaOH solution having the same concentration and volume because 1 mole of $Ba(OH)_2$ yields 2 moles of $OH^-$ ions:

$$2HCl(aq) + Ba(OH)_2(aq) \longrightarrow BaCl_2(aq) + 2H_2O(l)$$

In any acid-base titration, regardless of what acid and base are reacting, the total number of moles of $H^+$ ions that have reacted at the equivalence point must be equal to the total number of moles of $OH^-$ ions that have reacted. Sample Problem 4.16 explores the titration of an NaOH solution with a diprotic acid.

## Sample Problem 4.16

What volume (in mL) of a 0.203 $M$ NaOH solution is needed to neutralize 25.0 mL of a 0.188 $M$ $H_2SO_4$ solution?

**Strategy** First, write and balance the chemical equation that corresponds to the neutralization reaction:

$$2NaOH(aq) + H_2SO_4(aq) \longrightarrow 2H_2O(l) + Na_2SO_4(aq)$$

The base and the diprotic acid combine in a 2:1 ratio: $2NaOH \simeq H_2SO_4$. Use the molarity and the volume given to determine the number of millimoles of $H_2SO_4$. Use the number of millimoles of $H_2SO_4$ to determine the number of millimoles of NaOH. Using millimoles of NaOH and the concentration given, determine the volume of NaOH that will contain the correct number of millimoles.

**Student Annotation:**
Remember: molarity × mL = millimoles. This saves steps in titration problems.

**Setup** The necessary conversion factors are:

From the balanced equation: $\dfrac{2\,\text{mmol NaOH}}{1\,\text{mmol }H_2SO_4}$

From the molarity of the NaOH given: $\dfrac{1\,\text{mL NaOH}}{0.203\,\text{mmol NaOH}}$

**Solution**

$$\text{millimoles of }H_2SO_4 = 0.188\,M \times 25.0\,\text{mL} = 4.70\,\text{mmol}$$

$$\text{millimoles of NaOH required} = 4.70\,\cancel{\text{mmol }H_2SO_4} \times \frac{2\,\text{mmol NaOH}}{1\,\cancel{\text{mmol }H_2SO_4}} = 9.40\,\text{mmol NaOH}$$

$$\text{volume of 0.203 }M\text{ NaOH} = 9.40\,\cancel{\text{mmol NaOH}} \times \frac{1\,\text{mL NaOH}}{0.203\,\cancel{\text{mmol NaOH}}} = 46.3\,\text{mL}$$

**Think About It** Notice that the two concentrations 0.203 $M$ and 0.188 $M$ are similar. Both round to the same value (~0.20 $M$) to two significant figures. Therefore, the titration of a diprotic acid with a monobasic base of roughly equal concentration should require roughly twice as much base as the beginning volume of acid: $2 \times 25.0\,\text{mL} \approx 46.3\,\text{mL}$.

**Practice Problem A** How many milliliters of a 1.42 $M$ $H_2SO_4$ solution are needed to neutralize 95.5 mL of a 0.336 $M$ KOH solution?

**⏲Practice Problem B** How many milliliters of a 0.211 $M$ HCl solution are needed to neutralize 275 mL of a 0.0350 $M$ $Ba(OH)_2$ solution?

Sample Problem 4.17 shows how titration with a standard base can be used to determine the molar mass of an unknown acid.

## Sample Problem 4.17

A 0.1216-g sample of a monoprotic acid is dissolved in 25 mL water, and the resulting solution is titrated with 0.1104 $M$ NaOH solution. A 12.5-mL volume of the base is required to neutralize the acid. Calculate the molar mass of the acid.

**Strategy** Using the concentration and volume of the base, we can determine the number of moles of base required to neutralize the acid. We then determine the number of moles of acid and divide the mass of the acid by the number of moles to get molar mass.

**Setup** Because the acid is monoprotic, it will react in a 1:1 ratio with the base; therefore, the number of moles of acid will be equal to the number of moles of base. The volume of base in liters is 0.0125 L.

*(Continued)*

### Solution

$$\text{moles of base} = 0.0125 \text{ L} \times 0.1104 \text{ mol/L} = 0.00138$$

Because moles of base = moles of acid, the moles of acid = 0.00138 mol. Therefore,

$$\text{molar mass of the acid} = \frac{0.1216 \text{ g}}{0.00138 \text{ mol}} = 88.1 \text{ g/mol}$$

> **Think About It** For this technique to work, we must know whether the acid is monoprotic, diprotic, or polyprotic. A diprotic acid, for example, would combine in a 1:2 ratio with the base, and the result would have been a molar mass twice as large.

**Practice Problem A** What is the molar mass of a monoprotic acid if 28.1 mL of 0.0788 *M* NaOH is required to neutralize a 0.205-g sample?

**Practice Problem B** What is the molar mass of a diprotic acid if 30.5 mL of 0.1112 *M* NaOH is required to neutralize a 0.1365-g sample?

---

## Checkpoint 4.6    Aqueous Reactions and Chemical Analysis

**4.6.1** What mass of AgCl will be recovered if a solution containing 5.00 g of NaCl is treated with enough $AgNO_3$ to precipitate all the chloride ion?

a) 12.3 g

b) 5.00 g

c) 3.03 g

d) 9.23 g

e) 10.0 g

**4.6.2** A 10.0-g sample of an unknown ionic compound is dissolved, and the solution is treated with enough $AgNO_3$ to precipitate all the chloride ion. If 30.1 g of AgCl is recovered, which of the following compounds could be the unknown?

a) NaCl

b) $NaNO_3$

c) $BaCl_2$

d) $MgCl_2$

e) KCl

**4.6.3** Which of the following best represents the contents of a beaker in which equal volumes of 0.10 *M* $BaCl_2$ and 0.10 *M* $AgNO_3$ were combined?

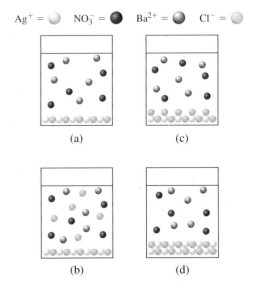

**4.6.4** If 25.0 mL of an $H_2SO_4$ solution requires 39.9 mL of 0.228 *M* NaOH to neutralize, what is the concentration of the $H_2SO_4$ solution?

a) 0.728 *M*

b) 0.364 *M*

c) 0.182 *M*

d) 0.228 *M*

e) 0.910 *M*

**4.6.5** What volume of 0.144 *M* $H_2SO_4$ is required to neutralize 25.0 mL of 0.0415 *M* $Ba(OH)_2$?

a) 7.20 mL

b) 3.60 mL

c) 14.4 mL

d) 50.0 mL

e) 12.5 mL

**4.6.6** Which of the following best represents the contents of a beaker in which equal volumes of 0.10 *M* NaCl and 0.10 *M* $Pb(NO_3)_2$ were combined?

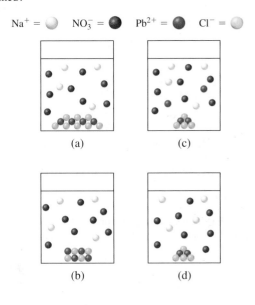

## Applying What You've Learned

The balanced overall equation for the Breathalyzer reaction is

$$3CH_3CH_2OH(g) + 2K_2Cr_2O_7(aq) + 8H_2SO_4(aq) \longrightarrow$$
$$3HC_2H_3O_2(aq) + 2Cr_2(SO_4)_3(aq) + 2K_2SO_4(aq) + 11H_2O(l)$$

**Problems:**

a)  Classify each of the species in the Breathalyzer reaction as a strong electrolyte, weak electrolyte, or nonelectrolyte. [◄◄ Sample Problem 4.2]

b)  Write the ionic and net ionic equations for the Breathalyzer reaction. [◄◄ Sample Problem 4.4]

c)  Determine the oxidation number of each element in the overall equation. [◄◄ Sample Problem 4.6]

d)  One manufacturer of Breathalyzers specifies a potassium dichromate concentration of 0.025 percent weight per volume (0.025 g $K_2Cr_2O_7$ per 100 mL of solution). Express this concentration in terms of molarity. [◄◄ Sample Problem 4.9]

e)  What volume of 0.014 $M$ stock solution of $K_2Cr_2O_7$ would have to be diluted to 250 mL to make a solution of the specified concentration? [◄◄ Sample Problem 4.10]

f)  Using square-bracket notation, express the molarity of each ion in a $K_2Cr_2O_7$ solution of the specified concentration. [◄◄ Sample Problem 4.13]

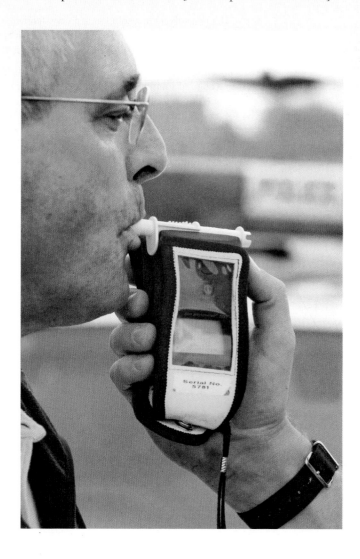

# CHAPTER SUMMARY

## Section 4.1

- A *solution* is a homogeneous mixture consisting of a *solvent* and one or more dissolved species called *solutes.*

- An *electrolyte* is a compound that dissolves in water to give an electrically conducting solution. *Nonelectrolytes* dissolve to give nonconducting solutions. Acids and *bases* are electrolytes.

- Electrolytes may be ionic or molecular. Ionic electrolytes undergo *dissociation* in solution; molecular electrolytes undergo *ionization.* *Strong electrolytes* dissociate (or ionize) completely. *Weak electrolytes* ionize only partially.

## Section 4.2

- A *precipitation reaction* results in the formation of an insoluble product called a *precipitate.* From general guidelines about solubilities of ionic compounds, we can predict whether a precipitate will form in a reaction.

- *Hydration* is the process in which water molecules surround solute particles.

- *Solubility* is the amount of solute that will dissolve in a specified amount of a given solvent at a specified temperature.

- A *molecular equation* represents a reaction as though none of the reactants or products has dissociated or ionized.

- An *ionic equation* represents the strong electrolytes in a reaction as ions.

- A *spectator ion* is one that is *not* involved in the reaction. Spectator ions appear on both sides of the ionic equation. A *net ionic equation* is an ionic equation from which spectator ions have been eliminated.

## Section 4.3

- The hydrogen ion in solution is more realistically represented as the *hydronium ion* ($H_3O^+$). The terms *hydrogen ion, hydronium ion,* and *proton* are used interchangeably in the context of acid-base reactions.

- *Arrhenius acids* ionize in water to give $H^+$ ions, whereas *Arrhenius bases* ionize (or *dissociate*) in water to give $OH^-$ ions. *Brønsted acids* donate protons ($H^+$ ions), whereas *Brønsted bases* accept protons.

- Brønsted acids may be *monoprotic, diprotic,* or *triprotic,* depending on the number of ionizable hydrogen atoms they have. In general, an acid with more than one ionizable hydrogen atom is called *polyprotic.*

- The reaction of an acid and a base is a *neutralization reaction.* The products of a neutralization reaction are water and a *salt.*

## Section 4.4

- *Oxidation-reduction,* or *redox, reactions* are those in which *electrons* are exchanged. Oxidation and reduction always occur *simultaneously.* You cannot have one without the other.

- *Oxidation* is the loss of electrons; *reduction* is the gain of electrons. In a redox reaction, the *oxidizing agent* is the reactant that gets reduced and the *reducing agent* is the reactant that gets oxidized.

- *Oxidation numbers* or *oxidation states* help us keep track of charge distribution and are assigned to all atoms in a compound or ion according to specific rules.

- Many redox reactions can be further classified as *combination, decomposition, displacement, hydrogen displacement, combustion,* or *disproportionation* reactions. The *activity series* can be used to determine whether or not a displacement reaction will occur.

- A *half-reaction* is a chemical equation representing only the oxidation or only the reduction of an oxidation-reduction reaction. Redox equations, which must be balanced for both mass and charge, can be balanced using the *half-reaction method.*

## Section 4.5

- The *concentration* of a solution is the amount of solute dissolved in a given amount of solution. *Molarity* (*M*) or *molar concentration* expresses concentration as the number of moles of solute in 1 L of solution.

- Adding a solvent to a solution, a process known as *dilution,* decreases the concentration (molarity) of the solution without changing the total number of moles of solute present in the solution.

- *Visible spectrophotometry* can be used to measure the concentration of a colored solution. *Transmittance* (*T*) is the ratio of transmitted light (*I*) to incident light ($I_0$). *Absorbance* (*A*) is the negative log of transmittance. A plot of absorbance against wavelength is an *absorption spectrum.* The *Beer-Lambert law* relates the absorbance of a solution to its concentration and the path length through which the light passes. The *molar absorptivity* is the proportionality constant in the Beer-Lambert law.

## Section 4.6

- *Gravimetric analysis* often involves a precipitation reaction.

- Acid-base *titration* involves an acid-base reaction. Typically, a solution of known concentration (a *standard solution*) is added gradually to a solution of unknown concentration with the goal of determining the unknown concentration.

- The point at which the reaction in the titration is complete is called the *equivalence point.* An *indicator* is a substance that changes color at or near the equivalence point of a titration. The point at which the indicator changes color is called the *endpoint* of the titration.

# KEY WORDS

Absorbance, 150

Absorption spectrum, 150

Activity series, 135

Arrhenius acid, 130

Arrhenius base, 130

Base, 118

Beer-Lambert law, 150

Brønsted acid, 130

Brønsted base, 130

Combustion, 141

Concentration, 142

Dilution, 144

Diprotic acid, 130

Displacement reaction, 135

Disproportionation reaction, 141

Dissociation, 118

Electrolyte, 118

Endpoint, 155

Equivalence point, 155

Gravimetric analysis, 154

# KEY EQUATIONS

4.1    $\text{molarity} = \dfrac{\text{moles solute}}{\text{liters solution}}$

4.2    moles of solute before dilution = moles of solute after dilution

4.3    $M_c \times L_c = M_d \times L_d$

4.4    $M_c \times mL_c = M_d \times mL_d$

4.5    $T = \dfrac{I}{I_0}$

4.6    $A = -\log T = -\log \dfrac{I}{I_0}$

4.7    $A = \varepsilon bc$

# QUESTIONS AND PROBLEMS

## Section 4.1: General Properties of Aqueous Solutions

### Review Questions

4.1    Define *solute, solvent,* and *solution* by describing the process of dissolving a solid in a liquid.

4.2    What is the difference between a nonelectrolyte and an electrolyte? Between a weak electrolyte and a strong electrolyte?

4.3    What is the difference between the symbols $\longrightarrow$ and $\rightleftharpoons$ in chemical equations?

4.4    Water is an extremely weak electrolyte and therefore cannot conduct electricity. Why are we often cautioned not to operate electrical appliances when our hands are wet?

4.5    Lithium fluoride (LiF) is a strong electrolyte. What species are present in LiF(*aq*)?

### Problems

4.6    The aqueous solutions of three compounds are shown in the diagrams. Identify each compound as a nonelectrolyte, a weak electrolyte, or a strong electrolyte.

(a)

(b)

(c)

4.7    Which of the following diagrams best represents the hydration of NaCl when dissolved in water? The $Cl^-$ ion is larger in size than the $Na^+$ ion.

(a)

(b)

(c)

4.8    Identify each of the following substances as a strong electrolyte, weak electrolyte, or nonelectrolyte: (a) $H_2O$, (b) KCl, (c) $HNO_3$, (d) $HC_2H_3O_2$, (e) $C_{12}H_{22}O_{11}$.

4.9    Identify each of the following substances as a strong electrolyte, weak electrolyte, or nonelectrolyte: (a) $Ba(NO_3)_2$, (b) Ne, (c) $NH_3$, (d) NaOH.

4.10    The passage of electricity through an electrolyte solution is caused by the movement of (a) electrons only, (b) cations only, (c) anions only, (d) both cations and anions.

4.11    Predict and explain which of the following systems are electrically conducting: (a) solid NaCl, (b) molten NaCl, (c) an aqueous solution of NaCl.

4.12    You are given a water-soluble compound X. Describe how you would determine whether it is an electrolyte or a nonelectrolyte. If it is an electrolyte, how would you determine whether it is strong or weak?

**4.13**    Explain why a solution of HCl in benzene does not conduct electricity but in water it does.

## Section 4.2: Precipitation Reactions

### Review Questions

4.14    Describe hydration. What properties of water enable its molecules to interact with ions in solution?

4.15    What is the difference between an ionic equation and a molecular equation?

4.16    What is the advantage of writing net ionic equations?

### Problems

⏴4.17    Two aqueous solutions of $AgNO_3$ and $NaCl$ are mixed. Which of the following diagrams best represents the mixture?

(a)            (b)            (c)            (d)

⏴4.18    Two aqueous solutions of $KOH$ and $MgCl_2$ are mixed. Which of the following diagrams best represents the mixture?

(a)            (b)            (c)            (d)

⏴4.19    Characterize the following compounds as soluble or insoluble in water: (a) $Ca_3(PO_4)_2$, (b) $Mn(OH)_2$, (c) $AgClO_3$, (d) $K_2S$.

⏴4.20    Characterize the following compounds as soluble or insoluble in water: (a) $CaCO_3$, (b) $ZnSO_4$, (c) $Hg(NO_3)_2$, (d) $HgSO_4$, (e) $NH_4ClO_4$.

⏴4.21    Write ionic and net ionic equations for the following reactions:

(a) $AgNO_3(aq) + Na_2SO_4(aq) \longrightarrow$
(b) $BaCl_2(aq) + ZnSO_4(aq) \longrightarrow$
(c) $(NH_4)_2CO_3(aq) + CaCl_2(aq) \longrightarrow$

⏴4.22    Write ionic and net ionic equations for the following reactions:

(a) $Na_2S(aq) + ZnCl_2(aq) \longrightarrow$
(b) $K_3PO_4(aq) + 3Sr(NO_3)_2(aq) \longrightarrow$
(c) $Mg(NO_3)_2(aq) + 2NaOH(aq) \longrightarrow$

⏴4.23    Which of the following processes will likely result in a precipitation reaction? (a) Mixing an $NaNO_3$ solution with a $CuSO_4$ solution. (b) Mixing a $BaCl_2$ solution with a $K_2SO_4$ solution. Write a net ionic equation for the precipitation reaction.

## Section 4.3: Acid-Base Reactions

### Review Questions

4.24    List the general properties of acids and bases.

4.25    Give Arrhenius's and Brønsted's definitions of an acid and a base. Why are Brønsted's definitions more useful in describing acid-base properties?

4.26    Give an example of a monoprotic acid, a diprotic acid, and a triprotic acid.

4.27    What are the characteristics of an acid-base neutralization reaction?

4.28    What factors qualify a compound as a salt? Specify which of the following compounds are salts: $CH_4$, $NaF$, $NaOH$, $CaO$, $BaSO_4$, $HNO_3$, $NH_3$, $KBr$.

4.29    Identify the following as a weak or strong acid or base: (a) $NH_3$, (b) $H_3PO_4$, (c) $LiOH$, (d) $HCOOH$ (formic acid), (e) $H_2SO_4$, (f) $HF$, (g) $Ba(OH)_2$.

### Problems

⏴4.30    Identify each of the following species as a Brønsted acid, base, or both: (a) HI, (b) $C_2H_3O_2^-$, (c) $H_2PO_4^-$, (d) $HSO_4^-$.

**4.31**    Identify each of the following species as a Brønsted acid, base, or both: (a) $PO_4^{3-}$, (b) $ClO_2^-$, (c) $NH_4^+$, (d) $HCO_3^-$.

⏴4.32    Balance the following equations and write the corresponding ionic and net ionic equations (if appropriate):

(a) $HBr(aq) + NH_3(aq) \longrightarrow$
(b) $Ba(OH)_2(aq) + H_3PO_4(aq) \longrightarrow$
(c) $HClO_4(aq) + Mg(OH)_2(s) \longrightarrow$

⏴4.33    Balance the following equations and write the corresponding ionic and net ionic equations (if appropriate):

(a) $HC_2H_3O_2(aq) + KOH(aq) \longrightarrow$
(b) $H_2CO_3(aq) + NaOH(aq) \longrightarrow$
(c) $HNO_3(aq) + Ba(OH)_2(aq) \longrightarrow$

## Section 4.4: Oxidation-Reduction Reactions

### Review Questions

4.34    Give an example of a combination redox reaction, a decomposition redox reaction, and a displacement redox reaction.

4.35    All combustion reactions are redox reactions. True or false? Explain.

4.36    What is an oxidation number? How is it used to identify redox reactions? Explain why, except for ionic compounds, the oxidation number does not have any physical significance.

4.37    (a) Without referring to Figure 4.8, give the oxidation numbers of the alkali and alkaline earth metals in their compounds. (b) Give the highest oxidation numbers that the Groups 3A–7A elements can have.

4.38    How is the activity series organized? How is it used in the study of redox reactions?

4.39    Use the following reaction to define the terms *redox reaction, half-reaction, oxidizing agent,* and *reducing agent:*
$4Na(s) + O_2(g) \longrightarrow 2Na_2O(s)$.

4.40    Is it possible to have a reaction in which oxidation occurs and reduction does not? Explain.

## Problems

**4.41** For the complete redox reactions given here, break down each reaction into its half-reactions, identify the oxidizing agent, and identify the reducing agent.

(a) $2Sr + O_2 \longrightarrow 2SrO$
(b) $2Li + H_2 \longrightarrow 2LiH$
(c) $2Cs + Br_2 \longrightarrow 2CsBr$
(d) $3Mg + N_2 \longrightarrow Mg_3N_2$

**4.42** For the complete redox reactions given here, write the half-reactions and identify the oxidizing and reducing agents:

(a) $4Fe + 3O_2 \longrightarrow 2Fe_2O_3$
(b) $Cl_2 + 2NaBr \longrightarrow 2NaCl + Br_2$
(c) $Si + 2F_2 \longrightarrow SiF_4$
(d) $H_2 + Cl_2 \longrightarrow 2HCl$

**4.43** Arrange the following species in order of increasing oxidation number of the sulfur atom: (a) $H_2S$, (b) $S_8$, (c) $H_2SO_4$, (d) $S^{2-}$, (e) $HS^-$, (f) $SO_2$, (g) $SO_3$.

**4.44** Phosphorus forms many oxoacids. Indicate the oxidation number of phosphorus in each of the following acids: (a) $HPO_3$, (b) $H_3PO_2$, (c) $H_3PO_3$, (d) $H_3PO_4$, (e) $H_4P_2O_7$, (f) $H_5P_3O_{10}$.

**4.45** Give the oxidation numbers for the underlined atoms in the following molecules and ions: (a) $Cl\underline{F}$, (b) $\underline{I}F_7$, (c) $\underline{C}H_4$, (d) $\underline{C}_2H_2$, (e) $\underline{C}_2H_4$, (f) $K_2\underline{Cr}O_4$, (g) $K_2\underline{Cr}_2O_4$, (h) $K\underline{Mn}O_4$, (i) $NaH\underline{C}O_3$, (j) $\underline{Li}_2$, (k) $Na\underline{I}O_3$, (l) $K\underline{O}_2$, (m) $\underline{P}F_6^-$, (n) $K\underline{Au}Cl_4$.

**4.46** Give the oxidation number for the following species: $H_2$, $Se_8$, $P_4$, O, U, $As_4$, $B_{12}$.

**4.47** Give the oxidation numbers for the underlined atoms in the following molecules and ions: (a) $\underline{Cs}_2O$, (b) $Ca\underline{I}_2$, (c) $\underline{Al}_2O_3$, (d) $H_3\underline{As}O_3$, (e) $\underline{Ti}O_2$, (f) $\underline{Mo}O_4^{2-}$, (g) $\underline{Pt}Cl_4^{2-}$, (h) $\underline{Pt}Cl_6^{2-}$, (i) $\underline{Sn}F_2$, (j) $\underline{Cl}F_3$, (k) $\underline{Sb}F_6^-$.

**4.48** Give the oxidation numbers for the underlined atoms in the following molecules and ions: (a) $Mg_3\underline{N}_2$, (b) $Cs\underline{O}_2$, (c) $Ca\underline{C}_2$, (d) $\underline{C}O_3^{2-}$, (e) $\underline{C}_2O_4^{2-}$, (f) $Zn\underline{O}_2^{2-}$, (g) $Na\underline{B}H_4$, (h) $\underline{W}O_4^{2-}$.

**4.49** Nitric acid is a strong oxidizing agent. State which of the following species is *least* likely to be produced when nitric acid reacts with a strong reducing agent such as zinc metal, and explain why: $N_2O$, $NO$, $NO_2$, $N_2O_4$, $N_2O_5$, $NH_4^+$.

**4.50** Determine which of the following metals can react with acid: (a) Au, (b) Ni, (c) Zn, (d) Ag, (e) Pt.

**4.51** One of the following oxides does not react with molecular oxygen: $NO$, $N_2O$, $SO_2$, $SO_3$, $P_4O_6$. Based on oxidation numbers, which one is it? Explain.

**4.52** Predict the outcome of the reactions represented by the following equations by using the activity series, and balance the equations.

(a) $Cu(s) + HCl(aq) \longrightarrow$
(b) $Au(s) + NaBr(aq) \longrightarrow$
(c) $Mg(s) + CuSO_4(aq) \longrightarrow$
(d) $Zn(s) + KBr(aq) \longrightarrow$

**4.53** Classify the following redox reactions as combination, decomposition, or displacement:

(a) $2H_2O_2 \longrightarrow 2H_2O + O_2$
(b) $Mg + 2AgNO_3 \longrightarrow Mg(NO_3)_2 + 2Ag$
(c) $NH_4NO_2 \longrightarrow N_2 + 2H_2O$
(d) $H_2 + Br_2 \longrightarrow 2HBr$

**4.54** Classify the following redox reactions as combination, decomposition, or displacement:

(a) $P_4 + 10Cl_2 \longrightarrow 4PCl_5$
(b) $2NO \longrightarrow N_2 + O_2$
(c) $Cl_2 + 2KI \longrightarrow 2KCl + I_2$
(d) $3HNO_2 \longrightarrow HNO_3 + H_2O + 2NO$

## Section 4.5: Concentration of Solutions

### Visualizing Chemistry
**Figure 4.10 Preparing a Solution from a Solid**

**VC 4.1** Which of the following would result in the actual concentration of the prepared solution being higher than the final, calculated value?

a) Loss of some of the solid during transfer to the volumetric flask.
b) Neglecting to add the last bit of water with the wash bottle to fill to the volumetric mark.
c) Neglecting to tare the balance with the weigh paper on the pan.

**VC 4.2** Why can't we prepare the solution by first filling the volumetric flask to the mark and then adding the solid?

a) The solid would not all dissolve.
b) The solid would not all fit into the flask.
c) The final volume would not be correct.

**VC 4.3** What causes the concentration of the prepared solution not to be exactly 0.1 *M*?

a) Rounding error in the calculations.
b) The volume of the flask is not exactly 250 mL.
c) The amount of solid weighed out is not exactly the calculated mass.

**VC 4.4** The volumetric flask used to prepare a solution from a solid is shown before and after the last of the water has been added.

(i)          (ii)

Which of the following statements is true?

a) The concentration of solute is greater in i than in ii.
b) The concentration of solute is smaller in i than in ii.
c) The concentration of solute in i is equal to the concentration of solute in ii.

## Review Questions

**4.55** Write the equation for calculating molarity. Why is molarity a convenient concentration unit in chemistry?

**4.56** Describe the steps involved in preparing a solution of known molar concentration using a volumetric flask.

**4.57**    Describe the basic steps involved in diluting a solution of known concentration.

**4.58**    Write the equation that enables us to calculate the concentration of a diluted solution. Give units for all the terms.

*Problems*

**4.59**    Calculate the mass of KI in grams required to prepare $5.00 \times 10^2$ mL of a 2.80 *M* solution.

**4.60**    Describe how you would prepare 250 mL of a 0.707 *M* $NaNO_3$ solution.

**4.61**    How many moles of $MgCl_2$ are present in 60.0 mL of a 0.100 *M* $MgCl_2$ solution?

**4.62**    How many grams of KOH are present in 35.0 mL of a 5.50 *M* KOH solution?

**4.63**    Calculate the molarity of each of the following solutions:
(a) 29.0 g of ethanol ($C_2H_5OH$) in 545 mL of solution,
(b) 15.4 g of sucrose ($C_{12}H_{22}O_{11}$) in 74.0 mL of solution,
(c) 9.00 g of sodium chloride (NaCl) in 86.4 mL of solution.

**4.64**    Calculate the molarity of each of the following solutions:
(a) 6.57 g of methanol ($CH_3OH$) in $1.50 \times 10^2$ mL of solution,
(b) 10.4 g of calcium chloride ($CaCl_2$) in $2.20 \times 10^2$ mL of solution, (c) 7.82 g of naphthalene ($C_{10}H_8$) in 85.2 mL of benzene solution.

**4.65**    Calculate the volume in milliliters of a solution required to provide the following: (a) 2.14 g of sodium chloride from a 0.270 *M* solution, (b) 4.30 g of ethanol from a 1.50 *M* solution, (c) 0.85 g of acetic acid ($HC_2H_3O_2$) from a 0.30 *M* solution.

**4.66**    Determine how many grams of each of the following solutes would be needed to make $2.50 \times 10^2$ mL of a 0.100 *M* solution:
(a) cesium iodide (CsI), (b) sulfuric acid ($H_2SO_4$), (c) sodium carbonate ($Na_2CO_3$), (d) potassium dichromate ($K_2Cr_2O_7$),
(e) potassium permanganate ($KMnO_4$).

**4.67**    Describe how to prepare 1.00 L of a 0.646 *M* HCl solution, starting with a 2.00 *M* HCl solution.

**4.68**    Water is added to 25.0 mL of a 0.866 *M* $KNO_3$ solution until the volume of the solution is exactly 500 mL. What is the concentration of the final solution?

**4.69**    How would you prepare 60.0 mL of 0.200 *M* $HNO_3$ from a stock solution of 4.00 *M* $HNO_3$?

**4.70**    You have 505 mL of a 0.125 *M* HCl solution and you want to dilute it to exactly 0.100 *M*. How much water should you add?

**4.71**    (a) Determine the chloride ion concentration in each of the following solutions: 0.150 *M* $BaCl_2$, 0.566 *M* NaCl, 1.202 *M* $AlCl_3$. (b) What is the concentration of a $Sr(NO_3)_2$ solution that is 2.55 *M* in nitrate ion?

**4.72**    (a) What is the $Na^+$ concentration in each of the following solutions: 3.25 *M* sodium sulfate, 1.78 *M* sodium carbonate, 0.585 *M* sodium bicarbonate? (b) What is the concentration of a lithium carbonate solution that is 0.595 *M* in $Li^+$?

**4.73**    Determine the resulting nitrate ion concentration when 95.0 mL of 0.992 *M* potassium nitrate and 155.5 mL of 1.570 *M* calcium nitrate are combined.

**4.74**    What volume of 0.112 *M* ammonium sulfate contains 5.75 g of ammonium ion?

## Section 4.6: Aqueous Reactions and Chemical Analysis

*Review Questions*

**4.75**    Describe the basic steps involved in gravimetric analysis. How does this procedure help us determine the identity of a compound or the purity of a compound if its formula is known?

**4.76**    Distilled water must be used in the gravimetric analysis of chlorides. Why?

**4.77**    Describe the basic steps involved in an acid-base titration. Why is this technique of great practical value?

**4.78**    How does an acid-base indicator work?

**4.79**    A student carried out two titrations using an NaOH solution of unknown concentration in the burette. In one titration she weighed out 0.2458 g of KHP (see page 155) and transferred it to an Erlenmeyer flask. She then added 20.00 mL of distilled water to dissolve the acid. In the other titration she weighed out 0.2507 g of KHP but added 40.00 mL of distilled water to dissolve the acid. Assuming no experimental error, would she obtain the same result for the concentration of the NaOH solution?

**4.80**    Would the volume of a 0.10 *M* NaOH solution needed to titrate 25.0 mL of a 0.10 *M* $HNO_2$ (a weak acid) solution be different from that needed to titrate 25.0 mL of a 0.10 *M* HCl (a strong acid) solution?

*Problems*

**4.81**    If 30.0 mL of 0.150 *M* $CaCl_2$ is added to 15.0 mL of 0.100 *M* $AgNO_3$, what is the mass in grams of AgCl precipitate?

**4.82**    A sample of 0.6760 g of an unknown compound containing barium ions ($Ba^{2+}$) is dissolved in water and treated with an excess of $Na_2SO_4$. If the mass of the $BaSO_4$ precipitate formed is 0.4105 g, what is the percent by mass of Ba in the original unknown compound?

**4.83**    How many grams of NaCl are required to precipitate most of the Ag ions from $2.50 \times 10^2$ mL of a 0.0113 *M* $AgNO_3$ solution? Write the net ionic equation for the reaction.

**4.84**    Diagram (a) shows a solution of a base and an acid before the neutralization reaction. Each of the after-reaction diagrams, (b)–(d), shows the products of reaction with one of the following acids: HCl, $H_2SO_4$, $H_3PO_4$. Determine which diagram corresponds to which acid. Blue spheres = $OH^-$ ions, red spheres = acid molecules, green spheres = anions of the acids. Assume all the acid-base neutralization reactions go to completion.

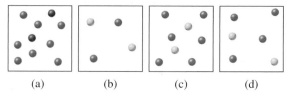

(a)          (b)          (c)          (d)

**4.85**    Diagram (a) shows a mixture of HCl and a base before the neutralization reaction. Of diagrams (b)–(d), which represents the products of reaction when the base is sodium hydroxide and which represents the products when the base is barium hydroxide?

Blue spheres = base, red spheres = $H^+$, grey spheres = cations of the bases. Assume all the acid-base neutralization reactions go to completion.

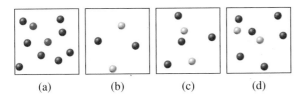

(a)     (b)     (c)     (d)

**4.86** Calculate the concentration (in molarity) of an NaOH solution if 25.0 mL of the solution is needed to neutralize 17.4 mL of a 0.312 $M$ HCl solution.

**4.87** Calculate the volume in milliliters of a 1.420 $M$ NaOH solution required to titrate the following solutions:

(a) 25.00 mL of a 2.430 $M$ HCl solution
(b) 25.00 mL of a 4.500 $M$ $H_2SO_4$ solution
(c) 25.00 mL of a 1.500 $M$ $H_3PO_4$ solution

**4.88** What volume of a 0.500 $M$ HCl solution is needed to neutralize each of the following:

(a) 10.0 mL of a 0.300 $M$ NaOH solution
(b) 10.0 mL of a 0.200 $M$ $Ba(OH)_2$ solution

## Additional Problems

**4.89** Classify the following reactions according to the types discussed in the chapter:

(a) $Cl_2 + 2OH^- \longrightarrow Cl^- + ClO^- + H_2O$
(b) $Ca^{2+} + CO_3^{2-} \longrightarrow CaCO_3$
(c) $NH_3 + H^+ \longrightarrow NH_4^+$
(d) $2CCl_4 + CrO_4^{2-} \longrightarrow 2COCl_2 + CrO_2Cl_2 + 2Cl^-$
(e) $Ca + F_2 \longrightarrow CaF_2$
(f) $2Li + H_2 \longrightarrow 2LiH$
(g) $Ba(NO_3)_2 + Na_2SO_4 \longrightarrow 2NaNO_3 + BaSO_4$
(h) $CuO + H_2 \longrightarrow Cu + H_2O$
(i) $Zn + 2HCl \longrightarrow ZnCl_2 + H_2$
(j) $2FeCl_2 + Cl_2 \longrightarrow 2FeCl_3$

**4.90** Oxygen ($O_2$) and carbon dioxide ($CO_2$) are colorless and odorless gases. Suggest two chemical tests that would allow you to distinguish between these two gases.

**4.91** Which of the following aqueous solutions would you expect to be the best conductor of electricity at 25°C? Explain your answer.

(a) 0.20 $M$ NaCl
(b) 0.60 $M$ $HC_2H_3O_2$
(c) 0.25 $M$ HCl
(d) 0.20 $M$ $Mg(NO_3)_2$

**4.92** A $5.00 \times 10^2$ mL sample of 2.00 $M$ HCl solution is treated with 4.47 g of magnesium. Calculate the concentration of the acid solution after all the metal has reacted. Assume that the volume remains unchanged.

**4.93** Calculate the volume of a 0.156 $M$ $CuSO_4$ solution that would react with 7.89 g of zinc.

**4.94** Sodium carbonate ($Na_2CO_3$) is available in very pure form and can be used to standardize acid solutions. What is the molarity of an HCl solution if 28.3 mL of the solution is required to react with 0.256 g of $Na_2CO_3$?

**4.95** Identify each of the following compounds as a nonelectrolyte, a weak electrolyte, or a strong electrolyte: (a) ethanolamine ($C_2H_5ONH_2$), (b) potassium fluoride (KF), (c) ammonium nitrate ($NH_4NO_3$), (d) isopropanol ($C_3H_7OH$).

**4.96** Identify each of the following compounds as a nonelectrolyte, a weak electrolyte, or a strong electrolyte: (a) lactose ($C_{12}H_{22}O_{11}$), (b) lactic acid ($HC_3H_5O_3$), (c) dimethylamine [$(CH_3)_2NH$], (d) barium hydroxide [$Ba(OH)_2$].

**4.97** Determine the predominant species (there may be more than one) in an aqueous solution for each of the compounds in Problem 4.95.

**4.98** Determine the predominant species (there may be more than one) in an aqueous solution for each of the compounds in Problem 4.96.

**4.99** A 3.664-g sample of a monoprotic acid was dissolved in water. It took 20.27 mL of a 0.1578 $M$ NaOH solution to neutralize the acid. Calculate the molar mass of the acid.

**4.100** Acetic acid ($HC_2H_3O_2$) is an important ingredient of vinegar. A sample of 50.0 mL of a commercial vinegar is titrated against a 1.00 $M$ NaOH solution. What is the concentration (in $M$) of acetic acid present in the vinegar if 5.75 mL of the base is needed for the titration?

**4.101** A 15.00-mL solution of potassium nitrate ($KNO_3$) was diluted to 125.0 mL, and 25.00 mL of this solution was then diluted to 1.000 $\times 10^3$ mL. The concentration of the final solution is 0.00383 $M$. Calculate the concentration of the original solution.

**4.102** When 2.50 g of a zinc strip was placed in an $AgNO_3$ solution, silver metal formed on the surface of the strip. After some time had passed, the strip was removed from the solution, dried, and weighed. If the mass of the strip was 3.37 g, calculate the mass of Ag and Zn metals present.

**4.103** Calculate the mass of the precipitate formed when 2.27 L of 0.0820 $M$ $Ba(OH)_2$ is mixed with 3.06 L of 0.0664 $M$ $Na_2SO_4$.

**4.104** Calculate the concentration of the acid (or base) remaining in solution when 10.7 mL of 0.211 $M$ $HNO_3$ is added to 16.3 mL of 0.258 $M$ NaOH.

**4.105** (a) Describe a preparation for magnesium hydroxide [$Mg(OH)_2$] and predict its solubility. (b) Milk of magnesia contains mostly $Mg(OH)_2$ and is effective in treating acid (mostly hydrochloric acid) indigestion. Calculate the volume of a 0.035 $M$ HCl solution (a typical acid concentration in an upset stomach) needed to react with two spoonfuls (approximately 10 mL) of milk of magnesia [at 0.080 g $Mg(OH)_2$/mL].

**4.106** A quantitative definition of solubility is the number of grams of a solute that will dissolve in a given volume of water at a particular temperature. Describe an experiment that would enable you to determine the solubility of a soluble compound.

**4.107** A 60.0-mL 0.513 $M$ glucose ($C_6H_{12}O_6$) solution is mixed with 120.0 mL of a 2.33 $M$ glucose solution. What is the concentration of the final solution? Assume the volumes are additive.

**4.108** An ionic compound X is only slightly soluble in water. What test would you employ to show that the compound does indeed dissolve in water to a certain extent?

**4.109** You are given a colorless liquid. Describe three chemical tests you would perform on the liquid to show that it is water.

4.110   Using the apparatus shown in Figure 4.1, a student found that a sulfuric acid solution caused the lightbulb to glow brightly. However, after the addition of a certain amount of a barium hydroxide [Ba(OH)$_2$] solution, the light began to dim even though Ba(OH)$_2$ is also a strong electrolyte. Explain.

**4.111**   Which of the diagrams shown corresponds to the reaction between AgOH(s) and HNO$_3$(aq)? Write a balanced equation for the reaction. (For simplicity, water molecules are not shown.)

(a)              (b)              (c)

● Ag$^+$
● NO$_3^-$

4.112   Which of the diagrams shown corresponds to the reaction between Ba(OH)$_2$(aq) and H$_2$SO$_4$(aq)? Write a balanced equation for the reaction. (For simplicity, water molecules are not shown.)

(a)              (b)              (c)

● Ba$^{2+}$
● SO$_4^{2-}$

**4.113**   You are given a soluble compound of an unknown molecular formula. (a) Describe three tests that would show that the compound is an acid. (b) Once you have established that the compound is an acid, describe how you would determine its molar mass using an NaOH solution of known concentration. (Assume the acid is monoprotic.) (c) How would you find out whether the acid is weak or strong? You are provided with a sample of NaCl and an apparatus like that shown in Figure 4.1 for comparison.

4.114   You are given two colorless solutions, one containing NaCl and the other sucrose (C$_{12}$H$_{22}$O$_{11}$). Suggest a chemical and a physical test that would allow you to distinguish between these two solutions.

4.115   The concentration of lead ions (Pb$^{2+}$) in a sample of polluted water that also contains nitrate ions (NO$_3^-$) is determined by adding solid sodium sulfate (Na$_2$SO$_4$) to exactly 500 mL of the water. (a) Write the molecular and net ionic equations for the reaction. (b) Calculate the molar concentration of Pb$^{2+}$ if 0.00450 g of Na$_2$SO$_4$ was needed for the complete precipitation of Pb$^{2+}$ ions as PbSO$_4$.

4.116   Hydrochloric acid is not an oxidizing agent in the sense that sulfuric acid and nitric acid are. Explain why the chloride ion is not a strong oxidizing agent like SO$_4^{2-}$ and NO$_3^-$.

**4.117**   Explain how you would prepare potassium iodide (KI) by means of (a) an acid-base reaction and (b) a reaction between an acid and a carbonate compound.

4.118   Sodium reacts with water to yield hydrogen gas. Why is this reaction not used in the laboratory preparation of hydrogen?

**4.119**   Describe how you would prepare the following compounds: (a) Mg(OH)$_2$, (b) AgI, (c) Ba$_3$(PO$_4$)$_2$.

4.120   Someone spilled concentrated sulfuric acid on the floor of a chemistry laboratory. To neutralize the acid, would it be preferable to pour concentrated sodium hydroxide solution or spray solid sodium bicarbonate over the acid? Explain your choice and the chemical basis for the action.

**4.121**   Describe in each case how you would separate the cations or anions in the following aqueous solutions: (a) NaNO$_3$ and Ba(NO$_3$)$_2$, (b) Mg(NO$_3$)$_2$ and KNO$_3$, (c) KBr and KNO$_3$, (d) K$_3$PO$_4$ and KNO$_3$, (e) Na$_2$CO$_3$ and NaNO$_3$.

4.122   The following are common household compounds: salt (NaCl), sugar (sucrose), vinegar (contains acetic acid), baking soda (NaHCO$_3$), washing soda (Na$_2$CO$_3$ · 10H$_2$O), boric acid (H$_3$BO$_3$, used in eyewash), Epsom salts (MgSO$_4$ · 7H$_2$O), sodium hydroxide (used in drain openers), ammonia, milk of magnesia [Mg(OH)$_2$], and calcium carbonate. Based on what you have learned in this chapter, describe tests that would allow you to identify each of these compounds.

**4.123**   Sulfites (compounds containing the SO$_3^{2-}$ ions) are used as preservatives in dried fruits and vegetables and in wine making. In an experiment to test the presence of sulfite in fruit, a student first soaked several dried apricots in water overnight and then filtered the solution to remove all solid particles. She then treated the solution with hydrogen peroxide (H$_2$O$_2$) to oxidize the sulfite ions to sulfate ions. Finally, the sulfate ions were precipitated by treating the solution with a few drops of a barium chloride (BaCl$_2$) solution. Write a balanced equation for each of the preceding steps.

4.124   A 0.8870-g sample of a mixture of NaCl and KCl is dissolved in water, and the solution is then treated with an excess of AgNO$_3$ to yield 1.913 g of AgCl. Calculate the percent by mass of each compound in the mixture.

4.125   Chlorine forms a number of oxides with the following oxidation numbers: +1, +3, +4, +6, and +7. Write a formula for each of these compounds.

4.126   A useful application of oxalic acid is the removal of rust (Fe$_2$O$_3$) from, say, bathtub rings according to the reaction

$$Fe_2O_3(s) + 6H_2C_2O_4(aq) \longrightarrow 2Fe(C_2O_4)_3^{3-}(aq) + 3H_2O + 6H^+(aq)$$

Calculate the number of grams of rust that can be removed by $5.00 \times 10^2$ mL of a 0.100 M solution of oxalic acid.

**4.127**   Acetylsalicylic acid (HC$_9$H$_7$O$_4$) is a monoprotic acid commonly known as "aspirin." A typical aspirin tablet, however, contains only a small amount of the acid. In an experiment to determine its composition, an aspirin tablet was crushed and dissolved in water. It took 12.25 mL of 0.1466 M NaOH to neutralize the solution. Calculate the number of grains of aspirin in the tablet (one grain = 0.0648 g).

4.128   A 0.9157-g mixture of CaBr$_2$ and NaBr is dissolved in water, and AgNO$_3$ is added to the solution to form AgBr precipitate. If the mass of the precipitate is 1.6930 g, what is the percent by mass of NaBr in the original mixture?

**4.129** Hydrogen halides (HF, HCl, HBr, HI) are highly reactive compounds that have many industrial and laboratory uses. (a) In the laboratory, HF and HCl can be generated by combining $CaF_2$ and NaCl with concentrated sulfuric acid. Write appropriate equations for the reactions. (*Hint:* These are not redox reactions.) (b) Why is it that HBr and HI cannot be prepared similarly, that is, by combining NaBr and NaI with concentrated sulfuric acid? (*Hint:* $H_2SO_4$ is a stronger oxidizing agent than both $Br_2$ and $I_2$.) (c) HBr can be prepared by reacting phosphorus tribromide ($PBr_3$) with water. Write an equation for this reaction.

**4.130** A 325-mL sample of solution contains 25.3 g of $CaCl_2$. (a) Calculate the molar concentration of $Cl^-$ in this solution. (b) How many grams of $Cl^-$ are in 0.100 L of this solution?

**4.131** Phosphoric acid ($H_3PO_4$) is an important industrial chemical used in fertilizers, detergents, and the food industry. It is produced by two different methods. In the *electric furnace method* elemental phosphorus ($P_4$) is burned in air to form $P_4O_{10}$, which is then combined with water to give $H_3PO_4$. In the *wet process* the mineral phosphate rock [$Ca_5(PO_4)_3F$] is combined with sulfuric acid to give $H_3PO_4$ (and HF and $CaSO_4$). Write equations for these processes, and classify each step as precipitation, acid-base, or redox reaction.

**4.132** Ammonium nitrate ($NH_4NO_3$) is one of the most important nitrogen-containing fertilizers. Its purity can be analyzed by titrating a solution of $NH_4NO_3$ with a standard NaOH solution. In one experiment a 0.2041-g sample of industrially prepared $NH_4NO_3$ required 24.42 mL of 0.1023 *M* NaOH for neutralization. (a) Write a net ionic equation for the reaction. (b) What is the percent purity of the sample?

**4.133** Potassium superoxide ($KO_2$) is used in some self-containing breathing equipment by firefighters. It reacts with carbon dioxide in respired (exhaled) air to form potassium carbonate and oxygen gas. (a) Write an equation for the reaction. (b) What is the oxidation number of oxygen in the $O_2^-$ ion? (c) How many liters of respired air can react with 7.00 g of $KO_2$ if each liter of respired air contains 0.063 g of $CO_2$?

**4.134** Barium sulfate ($BaSO_4$) has important medical uses. The dense salt absorbs X rays and acts as an opaque barrier. Thus, X-ray examination of a patient who has swallowed an aqueous suspension of $BaSO_4$ particles allows the radiologist to diagnose an ailment of the patient's digestive tract. Given the following starting compounds, describe how you would prepare $BaSO_4$ by neutralization and by precipitation: $Ba(OH)_2$, $BaCl_2$, $BaCO_3$, $H_2SO_4$, and $K_2SO_4$.

**4.135** Is the following reaction a redox reaction? Explain.

$$3O_2(g) \longrightarrow 2O_3(g)$$

**4.136** What is the oxidation number of O in HFO?

**4.137** Draw molecular models to represent the following acid-base reactions:

(a) $OH^- + H_3O^+ \longrightarrow 2H_2O$
(b) $NH_4^+ + NH_2^- \longrightarrow 2NH_3$

Identify the Brønsted acid and base in each case.

**4.138** On standing, a concentrated nitric acid gradually turns yellow. Explain. (*Hint:* Nitric acid slowly decomposes. Nitrogen dioxide is a colored gas.)

**4.139** When preparing a solution of known concentration, explain why one must first dissolve the solid completely before adding enough solvent to fill the volumetric flask to the mark.

**4.140** Can the following decomposition reaction be characterized as an acid-base reaction? Explain.

$$NH_4Cl(s) \longrightarrow NH_3(g) + HCl(g)$$

**4.141** Give a chemical explanation for each of the following: (a) When calcium metal is added to a sulfuric acid solution, hydrogen gas is generated. After a few minutes, the reaction slows down and eventually stops even though none of the reactants is used up. Explain. (b) In the activity series, aluminum is above hydrogen, yet the metal appears to be unreactive toward hydrochloric acid. Why? (*Hint:* Al forms an oxide, $Al_2O_3$, on the surface.) (c) Sodium and potassium lie above copper in the activity series. Explain why $Cu^{2+}$ ions in a $CuSO_4$ solution are not converted to metallic copper upon the addition of these metals. (d) A metal M reacts slowly with steam. There is no visible change when it is placed in a pale green iron(II) sulfate solution. Where should we place M in the activity series? (e) Before aluminum metal was obtained by electrolysis, it was produced by reducing its chloride ($AlCl_3$) with an active metal. What metals would you use to produce aluminum in that way?

**4.142** The recommended procedure for preparing a very dilute solution is not to weigh out a very small mass or measure a very small volume of a stock solution. Instead, it is done by a series of dilutions. A sample of 0.8214 g of $KMnO_4$ was dissolved in water and made up to the volume in a 500-mL volumetric flask. A 2.000-mL sample of this solution was transferred to a 1000-mL volumetric flask and diluted to the mark with water. Next, 10.00 mL of the diluted solution was transferred to a 250-mL flask and diluted to the mark with water. (a) Calculate the concentration (in molarity) of the final solution. (b) Calculate the mass of $KMnO_4$ needed to directly prepare the final solution.

**4.143** The following "cycle of copper" experiment is performed in some general chemistry laboratories. The series of reactions starts with copper and ends with metallic copper. The steps are as follows: (1) A piece of copper wire of known mass is allowed to react with concentrated nitric acid [the products are copper(II) nitrate, nitrogen dioxide, and water]. (2) The copper(II) nitrate is treated with a sodium hydroxide solution to form copper(II) hydroxide precipitate. (3) On heating, copper(II) hydroxide decomposes to yield copper(II) oxide. (4) The copper(II) oxide is combined with concentrated sulfuric acid to yield copper(II) sulfate. (5) Copper(II) sulfate is treated with an excess of zinc metal to form metallic copper. (6) The remaining zinc metal is removed by treatment with hydrochloric acid, and metallic copper is filtered, dried, and weighed. (a) Write a balanced equation for each step and classify the reactions. (b) Assuming that a student started with 65.6 g of copper, calculate the theoretical yield at each step. (c) Considering the nature of the steps, comment on why it is possible to recover most of the copper used at the start.

**4.144** Use the periodic table framework given here to show the names and positions of two metals that can (a) displace hydrogen from cold water and (b) displace hydrogen from acid. (c) Also show two metals that do not react with either water or acid.

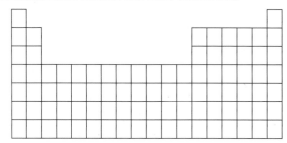

**4.145** A 22.02-mL solution containing 1.615 g $Mg(NO_3)_2$ is mixed with a 28.64-mL solution containing 1.073 g NaOH. Calculate the concentrations of the ions remaining in solution after the reaction is complete. Assume volumes are additive.

4.146 Because the acid-base and precipitation reactions discussed in this chapter all involve ionic species, their progress can be monitored by measuring the electrical conductance of the solution. Match each of the following reactions with one of the diagrams shown here. The electrical conductance is shown in arbitrary units.

(1) A 1.0 *M* KOH solution is added to 1.0 L of 1.0 *M* $HC_2H_3O_2$.
(2) A 1.0 *M* NaOH solution is added to 1.0 L of 1.0 *M* HCl.
(3) A 1.0 *M* $BaCl_2$ solution is added to 1.0 L of 1.0 *M* $K_2SO_4$.
(4) A 1.0 *M* NaCl solution is added to 1.0 L of 1.0 *M* $AgNO_3$.
(5) A 1.0 *M* $HC_2H_3O_2$ solution is added to 1.0 L of 1.0 *M* $NH_3$.

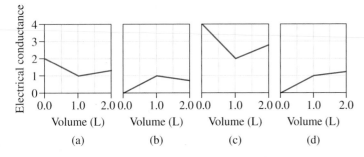

(a)        (b)        (c)        (d)

**4.147** A volume of 35.2 mL of a 1.66 *M* $KMnO_4$ solution is mixed with 16.7 mL of a 0.892 *M* $KMnO_4$ solution. Calculate the concentration of the final solution.

4.148 A volume of 46.2 mL of a 0.568 *M* calcium nitrate $[Ca(NO_3)_2]$ solution is mixed with 80.5 mL of a 1.396 *M* calcium nitrate solution. Calculate the concentration of the final solution.

**4.149** The general test for Type 2 diabetes is that the blood sugar (glucose, $C_6H_{12}O_6$) level should be below 120 mg per deciliter. Convert this concentration to molarity.

4.150 The current maximum level of fluoride that the EPA allows in U.S. drinking water is 4 mg/L. Convert this concentration to molarity.

**4.151** The concentration of $Cu^{2+}$ ions in the water (which also contains sulfate ions) discharged from a certain industrial plant is determined by adding excess sodium sulfide ($Na_2S$) solution to 0.800 L of the water. The molecular equation is

$$Na_2S(aq) + CuSO_4(aq) \longrightarrow Na_2SO_4(aq) + CuS(s)$$

Write the net ionic equation and calculate the molar concentration of $Cu^{2+}$ in the water sample if 0.0177 g of solid CuS is formed.

**4.152** Chemical tests of four metals A, B, C, and D show the following results.

(a) Only B and C react with 0.5 *M* HCl to give $H_2$ gas.
(b) When B is added to a solution containing the ions of the other metals, metallic A, C, and D are formed.
(c) A reacts with 6 *M* $HNO_3$ but D does not.

Arrange the metals in the increasing order as reducing agents. Suggest four metals that fit these descriptions.

# PRE-PROFESSIONAL PRACTICE EXAM PROBLEMS: PHYSICAL AND BIOLOGICAL SCIENCES

One of the atmospheric pollutants that contributes to acid rain is nitrogen dioxide ($NO_2$). A major source of $NO_2$ is automobile exhaust. When nitrogen in the air is burned in an internal combustion engine, it is converted to nitric oxide (NO):

$$N_2(g) + O_2(g) \longrightarrow 2NO(g) \qquad \text{[Equation I]}$$

The NO is then oxidized in the atmosphere to form $NO_2$:

$$2NO(g) + O_2(g) \longrightarrow 2NO_2(g) \qquad \text{[Equation II]}$$

$NO_2$ is a brown gas that is largely responsible for the orange-brown haze visible over some densely populated cities. It exists in equilibrium with its colorless dimeric form:

$$2NO_2(g) \rightleftharpoons N_2O_4(g) \qquad \text{[Equation III]}$$

Atmospheric $NO_2$ reacts with water to form nitric and nitrous acids:

$$2NO_2(g) + H_2O(l) \longrightarrow HNO_3(aq) + HNO_2(aq) \quad \text{[Equation IV]}$$

1. What is the reducing agent in Equation I?

a) $N_2$
b) $O_2$
c) NO
d) There is no reducing agent in Equation I.

2. According to Equation IV, if 5 g of $NO_2$ reacts with 1355 mL of water, what will be the resulting concentration of $HNO_3$ and the resulting concentration of $HNO_2$, respectively? (Assume that the volume of the resulting solution is equal to the volume of water.)

a) 0.04 *M* and 0.01 *M*
b) 0.04 *M* and 0.04 *M*
c) 0.08 *M* and 0.02 *M*
d) 0.08 *M* and 0.08 *M*

3. What type of reaction does Equation IV represent?

a) Disproportionation
b) Decomposition
c) Combination
d) Combustion

4. What is the correct sequence of oxidation numbers for N in $N_2$, $NO_2$, $N_2O_4$, $HNO_3$, and $HNO_2$?

a) +1, +4, +2, +6, +4
b) 0, −2, +4, +5, +5
c) 0, +4, +4, +5, +3
d) +1, +2, +4, +6, +4

# ANSWERS TO IN-CHAPTER MATERIALS

## Answers to Practice Problems

**4.1A** weak electrolyte, strong electrolyte, strong electrolyte, and strong electrolyte. **4.1B** $AD_2 < AB_2 < AC_2$. **4.2A** nonelectrolyte, weak electrolyte, and strong electrolyte. **4.2B** weak electrolyte, nonelectrolyte, and strong electrolyte. **4.3A** insoluble, soluble, and insoluble. **4.3B** soluble, insoluble, and soluble. **4.4A** $Sr(NO_3)_2(aq) + Li_2SO_4(aq) \longrightarrow$ $SrSO_4(s) + 2LiNO_3(aq)$, $Sr^{2+}(aq) + 2NO_3^-(aq) + 2Li^+(aq) + SO_4^{2-}(aq)$ $\longrightarrow SrSO_4(s) + 2Li^+(aq) + 2NO_3^-(aq)$, $Sr^{2+}(aq) + SO_4^{2-}(aq)$ $\longrightarrow SrSO_4(s)$. **4.4B** $2KNO_3(aq) + BaCl_2(aq) \longrightarrow 2KCl(aq) +$ $Ba(NO_3)_2(aq)$, $2K^+(aq) + 2NO_3^-(aq) + Ba^{2+}(aq) + 2Cl^-(aq) \longrightarrow$ $2K^+(aq) + 2Cl^-(aq) + Ba^{2+}(aq) + 2NO_3^-(aq)$, no net ionic equation (no reaction). **4.5A** $Ba(OH)_2(aq) + 2HF(aq) \longrightarrow BaF_2(s) + 2H_2O(l)$, $Ba^{2+}(aq) + 2OH^-(aq) + 2HF(aq) \longrightarrow BaF_2(s) + 2H_2O(l)$, $Ba^{2+}(aq)$ $+ 2OH^-(aq) + 2HF(aq) \longrightarrow BaF_2(s) + 2H_2O(l)$. **4.5B** $2NH_3(aq) +$ $H_2SO_4(aq) \longrightarrow (NH_4)_2SO_4(aq)$, $2NH_3(aq) + H^+(aq) + HSO_4^-(aq)$ $\longrightarrow 2NH_4^+(aq) + SO_4^{2-}(aq)$, $2NH_3(aq) + H^+(aq) + HSO_4^-(aq) \longrightarrow$ $2NH_4^+(aq) + SO_4^{2-}(aq)$. **4.6A** $H = +1, O = -1, Cl = +5, O = -2$, $H = +1, S = +6, O = -2$. **4.6B** $O = -1, Cl = +1, O = -2, Mn = +4$, $O = -2$. **4.7A** (a) no reaction, (b) $Sn(s) + Cu^{2+}(aq) \longrightarrow Sn^{2+}(aq)$ $+ Cu(s)$, (c) no reaction. **4.7B** (a) $Ni(s) + Cu^{2+}(aq) \longrightarrow Ni^{2+}(aq) +$ $Cu(s)$, (b) no reaction, (c) $Al(s) + Au^{3+}(aq) \longrightarrow Al^{3+}(aq) + Au(s)$. **4.8A** (a) $3Mg(s) + 2Cr(C_2H_3O_2)_3(aq) \longrightarrow 3Mg(C_2H_3O_2)_2(aq) +$ $2Cr(s)$, Mg is oxidized and Cr is reduced; (b) no reaction; (c) $Cd(s) +$ $2AgNO_3(aq) \longrightarrow 2Ag(s) + Cd(NO_3)_2(aq)$, Cd is oxidized and Ag is reduced. **4.8B** (a) no reaction, (b) Au is reduced and Ag is oxidized,

(c) H is reduced and Sn is oxidized. **4.9A** (a) 0.137 $M$, (b) 9.18 L, (c) 0.259 mol. **4.9B** (a) 0.707 $M$, (b) 6.48 L, (c) 16.1 mol. **4.10A** 21 mL. **4.10B** 3.8 mL. **4.11A** 1.638 $M$, 0.4095 $M$, 0.1024 $M$, $2.559 \times 10^{-2}$ $M$, $6.398 \times 10^{-3}$ $M$. (b) 0.1638 mol, $4.095 \times 10^{-2}$ mol, $1.024 \times 10^{-2}$ mol, $2.559 \times 10^{-3}$ mol, $6.398 \times 10^{-4}$ mol. **4.11B** 2.45 $M$. **4.12A** 0.60 $M$.

**4.12B** (a) 0.26, 0.40, 0.53, 0.73. (b) 0.656 L/mol · cm. **4.13A** $[Al^{3+}] =$ 1.50 $M$, $[SO_4^{2-}] = 2.25$ $M$. **4.13B** $[Cl^-] = 0.250$ $M$ and 0.50 $M$. **4.14A** 67.13%. **4.14B** 0.420 g. **4.15A** 0.5502 g KHP. **4.15B** 206.4 mL. **4.16A** 11.3 mL. **4.16B** 91.2 mL. **4.17A** 92.6 g/mol. **4.17B** 80.5 g/mol.

## Answers to Checkpoints

**4.1.1** c. **4.1.2** e. **4.1.3** b. **4.1.4** c. **4.2.1** a, b, d. **4.2.2** a, c, d. **4.2.3** b. **4.2.4** b. **4.2.5** d. **4.2.6** e. **4.3.1** a. **4.3.2** b. **4.3.3** e. **4.3.4** d. **4.4.1** e. **4.4.2** a. **4.4.3** a, b, d. **4.4.4** a, d. **4.5.1** d. **4.5.2** a. **4.5.3** b. **4.5.4** a, e. **4.5.5** a. **4.5.6** e. **4.6.1** a. **4.6.2** d. **4.6.3** b. **4.6.4** c. **4.6.5** a. **4.6.6** c.

## Answers to Applying What You've Learned

**a)** $CH_3CH_2OH$ is a nonelectrolyte, $K_2Cr_2O_7$ is a strong electrolyte, and $H_2SO_4$ is a strong electrolyte. $HC_2H_3O_2$ is a weak electrolyte, $Cr_2(SO_4)_3$ is a strong electrolyte, $K_2SO_4$ is a strong electrolyte, and $H_2O$ is a nonelectrolyte.
**b)** The ionic equation is
$$3CH_3CH_2OH(g) + 4K^+(aq) + 2Cr_2O_7^{2-}(aq) + 8H^+(aq) + 8HSO_4^-(aq) \longrightarrow$$
$$3HC_2H_3O_2(aq) + 4Cr^{3+}(aq) + 6SO_4^{2-}(aq) + 4K^+(aq) + 2SO_4^{2-}(aq) + 11H_2O(l)$$
The net ionic equation is
$$3CH_3CH_2OH(g) + 2Cr_2O_7^{2-}(aq) + 8H^+(aq) + 8HSO_4^-(aq) \longrightarrow 3HC_2H_3O_2(aq) + 4Cr^{3+}(aq) + 8SO_4^{2-}(aq) + 11H_2O(l)$$
**c)** The oxidation state of each element in the overall equation is as follows:

| $CH_3CH_2OH$ | + | $K_2Cr_2O_7$ | + | $H_2SO_4$ | $\longrightarrow$ | $HC_2H_3O_2$ | + | $Cr_2(SO_4)_3$ | + | $K_2SO_4$ | + | $H_2O$ |

| $-2$ $+1$ $-2$ $+1$ $-2$ $+1$ | $+1$ $+6$ $-2$ | $+1$ $+6$ $-2$ | $+1$ $0$ $+1$ $-2$ | $+3$ $+6$ $-2$ | $+1$ $+6$ $-2$ | $+1$ $-2$ |
| $-2$ $+3$ $-2$ $+2$ $-2$ $+1$ | $+2$ $+12$ $-14$ | $+2$ $+6$ $-8$ | $+1$ $0$ $+3$ $-4$ | $+6$ $+18$ $-24$ | $+2$ $+6$ $-8$ | $+2$ $-2$ |

**d)** The molar concentration of a 0.025 percent weight by volume solution of $K_2Cr_2O_7$ is $8.5 \times 10^{-4}$ $M$.
**e)** 15 mL of a 0.014 $M$ stock solution of $K_2Cr_2O_7$ would be required to prepare 250 mL of $8.5 \times 10^{-4}$ $M$ $K_2Cr_2O_7$.
**f)** $[K^+] = 1.7 \times 10^{-3}$ $M$; $[Cr_2O_7^{2-}] = 8.5 \times 10^{-4}$ $M$.

# 5

# Thermochemistry

*Doughnuts, although typically high in fat and Calories, are very popular. The energy content of food such as doughnuts is determined by burning a sample of the food in a calorimeter.*

# How Thermochemistry Helped Uncover Deceptive Labeling of "Diet" Doughnuts

The Food and Drug Administration (FDA) and the U.S. Department of Agriculture (USDA) require "Nutrition Facts" labels on nearly all prepared foods. These labels indicate such things as serving size, grams of fat per serving, and Calories per serving. Health-conscious consumers typically are avid readers of food labels and use them to make informed dietary choices. Furthermore, because dieting to lose weight is something of a national obsession, many Americans are willing to pay premium prices for packaged foods that both taste good *and* are low in fat and Calories. For this reason, it is important that the labels on packaged foods contain accurate information.

In 1997, spurred by numerous complaints from consumers and health-care professionals, the FDA launched an investigation into the practices of a weight-loss product company with packaging facilities in Kentucky and Illinois. The company sold "low-fat, carob-coated" doughnuts that, according to package labeling, each contained just 3 g of fat and 135 Calories. According to an analysis by the FDA, the doughnuts were actually *chocolate*-coated and each contained 18 g of fat and 530 Calories! As the FDA investigation revealed, the company had simply purchased ordinary doughnuts from a Chicago bakery and repackaged them as "diet" doughnuts, marking up the price by as much as 200 percent. The investigation led to a criminal indictment.

Experimental determination of the caloric content of food is done using the techniques of *thermochemistry*. A weighed sample of the food is burned in a bomb calorimeter, and the heat generated by the combustion, which is equal to the energy content of the food, is determined by measuring the resulting temperature increase in the calorimeter.

**Nutrition Facts**
Serving Size 3/4 Cup (27g)

| Amount Per Serving | Cereal | With 1/2 Cup Skim Milk |
|---|---|---|
| **Calories** | 90 | 130 |
| Calories from Fat | 10 | 10 |
| | **% Daily Value** | |
| **Total Fat** 1g* | **2%** | **2%** |
| Saturated Fat 0g | **0%** | **0%** |
| Trans Fat 0g | **0%** | **0%** |
| **Cholesterol** 0mg | **0%** | **0%** |
| **Sodium** 190mg | **8%** | **11%** |
| **Potassium** 85mg | **2%** | **8%** |
| **Total Carbohydrate** 23g | **8%** | **10%** |
| Dietary Fiber 5g | **20%** | **20%** |
| Sugars 5g | | |
| **Protein** 2g | | |
| Vitamin A | 0% | 4% |
| Vitamin C | 10% | 15% |
| Calcium | 0% | 15% |
| Iron | 2% | 2% |

**Student Annotation:** Carob is a chocolate substitute made from the seed pods of the Mediterranean Carob tree.

**Student Annotation:** Cell respiration is the complex, multistep process by which living things convert the food they eat into energy, carbon dioxide, and water. Combustion occurs in a single step but is the same overall process, with the same products, as cell respiration.

At the end of this chapter, you will be able to solve a series of problems related to the fat and calorie content of doughnuts [▶▶ page 200].

## In This Chapter, You Will Learn:

About the energy changes associated with chemical reactions and physical processes. You will also learn how calorimetry is used to measure the energy released or absorbed during a process.

### Before you begin, you should review

• Tracking units [◀◀ Section 1.6]

• Balancing chemical equations [◀◀ Section 3.3]

## 5.1    Energy and Energy Changes

You have already learned that matter can undergo physical changes and chemical changes [◄◄ Section 1.4]. The melting of ice, for example, is a physical change that can be represented by the following equation:

$$H_2O(s) \longrightarrow H_2O(l)$$

The formation of water from its constituent elements, represented by the following equation, is an example of a chemical change:

$$2H_2(g) + O_2(g) \longrightarrow 2H_2O(l)$$

In each case, there is energy involved in the change. Energy (in the form of heat) must be *supplied* to melt ice, whereas energy (in the form of heat and light) is *produced* by the explosive combination of hydrogen and oxygen gases. In fact, every change that matter undergoes is accompanied by either the absorption or the release of energy. In this chapter, we will focus on the energy changes associated with physical and chemical processes.

### Forms of Energy

*Energy* is usually defined as the capacity to do work or transfer heat. All forms of energy are either kinetic or potential. *Kinetic energy* is the energy that results from *motion*. It is calculated with the equation

Equation 5.1                             $E_k = \frac{1}{2}mu^2$

**Student Annotation:** Some textbooks use the letter *v* to denote velocity.

where $m$ is the mass of the object and $u$ is its velocity. One form of kinetic energy of particular interest to chemists is *thermal energy*, which is the energy associated with the random motion of atoms and molecules. We can monitor changes in thermal energy by measuring temperature changes.

*Potential energy* is the energy possessed by an object by virtue of its position. The two forms of potential energy of greatest interest to chemists are chemical energy and electrostatic energy. *Chemical energy* is energy stored within the structural units (molecules or polyatomic ions) of chemical substances. The amount of chemical energy in a sample of matter depends on the types and arrangements of atoms in the structural units that make up the sample.

*Electrostatic energy* is potential energy that results from the interaction of charged particles. Oppositely charged particles *attract* each other, whereas particles of like charges *repel* each other. The magnitude of the resulting electrostatic potential energy is proportional to the product of the two charges ($Q_1$ and $Q_2$) divided by the distance ($d$) between them.

Equation 5.2                             $E_{el} \propto \dfrac{Q_1Q_2}{d}$

If the charges $Q_1$ and $Q_2$ are opposite (i.e., one positive and one negative), the result is a negative value for $E_{el}$, which indicates *attraction*. Like charges (i.e., both positive or both negative) result in a positive value for $E_{el}$, indicating *repulsion*.

Kinetic and potential energy are interconvertible—that is, one form can be converted to the other. For example, dropping an object and allowing it to fall converts potential energy to kinetic energy. Likewise, a chemical reaction that gives off heat converts chemical energy (potential) to thermal energy (kinetic). Although energy can assume many different forms that are interconvertible, the total amount of energy in the universe is constant. When energy of one form disappears, the same amount of energy must appear in another form or forms. This principle is known as the *law of conservation of energy.*

### Energy Changes in Chemical Reactions

To analyze energy changes associated with chemical reactions, we must first define the *system,* the specific part of the universe that is of interest to us. For chemists, systems usually include the substances involved in chemical and physical changes. In an acid-base neutralization experiment, for example, the system may be the contents of a beaker in which HCl and NaOH react with each other. The rest of the universe outside the system, including the beaker itself, constitutes the *surroundings.*

Many chemical reactions are carried out for the purpose of exploiting the associated energy change, rather than for the purpose of obtaining the products of the reactions. For example, combustion reactions involving fossil fuels are carried out for the thermal energy they produce, not for their products, which are carbon dioxide and water.

It is important to distinguish between thermal energy and heat. **Heat** is the transfer of thermal energy between two bodies that are at different temperatures. Although the term *heat* by itself implies the transfer of energy, we customarily talk of "heat flow," meaning "heat absorbed" or "heat released," when describing the energy changes that occur during a process. **Thermochemistry** is the study of heat (the transfer of thermal energy) in chemical reactions.

The combustion of hydrogen gas in oxygen is one of many chemical reactions that release considerable quantities of energy (Figure 5.1):

$$2H_2(g) + O_2(g) \longrightarrow 2H_2O(l) + \text{energy}$$

In this case, we label the mixture of reactants and product (hydrogen, oxygen, and water molecules) the *system*. Because energy cannot be created or destroyed, any energy released by the system must be gained by the surroundings. Thus, the heat generated by the combustion process is transferred from the system to its surroundings. This reaction is an example of an **exothermic process,** which is any process that gives off heat—that is, transfers thermal energy *from* the system *to* the surroundings. Figure 5.2(a) shows the energy change for the combustion of hydrogen gas.

Next, consider the decomposition of mercury(II) oxide (HgO) at high temperatures:

$$\text{energy} + 2HgO(s) \longrightarrow 2Hg(l) + O_2(g)$$

This reaction is an **endothermic process** because heat has to be supplied to the system (i.e., to HgO) by the surroundings [Figure 5.2(b)] for the reaction to occur. Thus, thermal energy is transferred *from* the surroundings *to* the system in an endothermic process.

According to Figure 5.2, the energy of the products of an exothermic reaction is lower than the energy of the reactants. The difference in energy between the reactants $H_2$ and $O_2$ and the product $H_2O$ is the heat released by the system to the surroundings. In an endothermic reaction, on the other hand, the energy of the products is higher than the energy of the reactants. Here, the difference between the energy of the reactant HgO and the products Hg and $O_2$ is the heat absorbed by the system from the surroundings.

## Units of Energy

The SI unit of energy is the *joule* (J), named for the physicist James Joule.[1] The joule is a fairly small quantity of energy. It is the amount of kinetic energy possessed by a 2-kg mass moving at a speed of 1 m/s.

$$E_k = \tfrac{1}{2}mu^2 = \tfrac{1}{2}(2 \text{ kg})(1 \text{ m/s})^2 = 1 \text{ kg} \cdot \text{m}^2/\text{s}^2 = 1 \text{ J}$$

**Figure 5.1** The Hindenburg, a German airship filled with hydrogen gas, was destroyed in a horrific fire as it landed at Lakehurst, New Jersey, in 1937.

Multimedia

Thermochemistry—heat flow in endothermic and exothermic reactions.

**Student Annotation:** In these examples, all the energy released or absorbed by the system is in the form of heat. In Section 5.2, we will consider examples in which some of the energy is in the form of work.

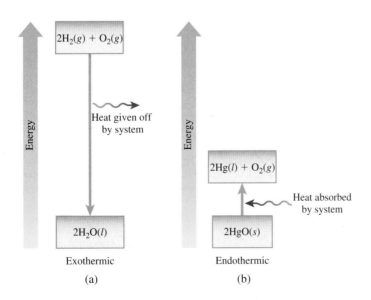

**Figure 5.2** (a) An exothermic reaction. Heat is given off by the system. (b) An endothermic reaction. Heat is absorbed by the system.

1. James Prescott Joule (1818–1889). English physicist. As a young man, Joule was tutored by John Dalton. He is most famous for determining the mechanical equivalent of heat, the conversion between mechanical energy and thermal energy.

The joule can also be defined as the amount of energy exerted when a force of 1 newton (N) is applied over a distance of 1 meter.

$$1\,J = 1\,N \cdot m$$

where

$$1\,N = 1\,kg \cdot m/s^2$$

Because the magnitude of a joule is so small, we very often express the energy changes in chemical reactions using the unit kilojoule (kJ).

$$1\,kJ = 1000\,J$$

Sample Problem 5.1 shows how to calculate kinetic and potential energies.

## Sample Problem 5.1

(a) Calculate the kinetic energy of a helium atom moving at a speed of 125 m/s. (b) How much greater is the magnitude of electrostatic attraction between an electron and a nucleus containing three protons versus that between an electron and a nucleus containing one proton? (Assume that the distance between the nucleus and the electron is the same in each case.)

**Strategy** (a) Use Equation 5.1 ($E_k = \frac{1}{2}mu^2$) to calculate the kinetic energy of an atom. We will need to know the mass of the atom in kilograms.

(b) Use Equation 5.2 ($E_{el} \propto Q_1Q_2/d$) to compare the electrostatic potential energy between the two charged particles in each case.

**Setup**
(a) The mass of a helium atom is 4.003 amu. Its mass in kilograms is

$$4.003\ \cancel{amu} \times \frac{1.661 \times 10^{-24}\ \cancel{g}}{1\ \cancel{amu}} \times \frac{1\ kg}{1 \times 10^3\ \cancel{g}} = 6.649 \times 10^{-27}\ kg$$

(b) The charge on a nucleus with three protons is +3; the charge on a nucleus with one proton is +1. In each case, the electron's charge is −1. Although we are not given the distance between the opposite charges in either case, we are told that the distances in both cases are equal. We can write Equation 5.2 for each case and divide one by the other to determine the relative magnitudes of the results.

**Solution**
(a) $E_k = \frac{1}{2}mu^2$

$$= \frac{1}{2}(6.649 \times 10^{-27}\ kg)(125\ m/s)^2$$

$$= 5.19 \times 10^{-23}\ kg \cdot m^2/s^2 = 5.19 \times 10^{-23}\ J$$

**Student Annotation:** Remember that the base units of the joule are kg · m²/s².

(b) $E_{el} \propto \dfrac{Q_1Q_2}{d}$

$$\frac{E_{el}\ \text{where}\ Z = +3}{E_{el}\ \text{where}\ Z = -1} = \frac{E_{el} \propto \dfrac{(+3)(-1)}{d}}{E_{el} \propto \dfrac{(+1)(-1)}{d}} = 3$$

The electrostatic potential energy between charges of +3 and −1 is three times that between charges of +1 and −1.

**Think About It** We expect the energy of an atom, even a fast-moving one, to be extremely small. And we expect the attraction between charges of larger magnitude to be greater than that between charges of smaller magnitude.

**Practice Problem A** (a) Calculate the energy in joules of a 5.25-g object moving at a speed of 655 m/s, and (b) determine how much greater the electrostatic energy is between charges of +2 and −2 than it is between charges of +1 and −1 (assume that the distance between the charges is the same in each case).

**Practice Problem B** (a) Calculate the velocity (in m/s) of a 0.340-g object that has $E_k = 23.5$ J, and (b) determine which of the following pairs of charged particles has the greater electrostatic energy between them: charges of +1 and −2 separated by a distance of $d$ or charges of +2 and −2 separated by a distance of $2d$.

Another unit used to express energy is the calorie (cal). Although the calorie is not an SI unit, its use is still quite common. The calorie is defined in terms of the joule:

$$1 \text{ cal} \equiv 4.184 \text{ J}$$

Because this is a definition, the number 4.184 is an exact number, which does not limit the number of significant figures in a calculation [◄◄ Section 1.5]. You may be familiar with the term *calorie* from nutrition labels. In fact, the "calories" listed on food packaging are really *kilocalories*. Often the distinction is made by capitalizing the "C" in "calorie" when it refers to the energy content of food:

$$1 \text{ Cal} \equiv 1000 \text{ cal}$$

and

$$1 \text{ Cal} \equiv 4184 \text{ J}$$

**Student Annotation:** The triple equal sign is used to denote a definition. Recall that numbers with defined values are exact numbers [◄◄ Section 1.5], which do not limit the number of significant figures in a calculation.

## Checkpoint 5.1    Energy and Energy Changes

**5.1.1**  Calculate the kinetic energy of a 5.0-kg mass moving at 26 m/s.

a) $1.7 \times 10^3$ J

b) $3.4 \times 10^3$ J

c) 130 J

d) 65 J

e) $13 \times 10^3$ J

**5.1.2**  How much greater is the electrostatic potential energy between particles with charges $+3$ and $-3$ than between particles with charges $+1$ and $-1$? (Assume the same distance between particles.)

a) 3 times

b) 9 times

c) 6 times

d) 4 times

e) 30 times

**5.1.3**  Calculate the number of calories in 723.01 J.

a) 172.80 cal

b) 172.8 cal

c) 3025.1 cal

d) 3025 cal

e) 0.173 cal

**5.1.4**  The label on packaged food indicates that it contains 215 Cal per serving. Convert this amount of energy to joules.

a) 51.4 J

b) $5.14 \times 10^4$ J

c) $5.14 \times 10^{-2}$ J

d) $9.00 \times 10^2$ J

e) $9.00 \times 10^5$ J

**5.1.5**  Arrange the following pairs of charged particles in order of increasing electrostatic potential energy.

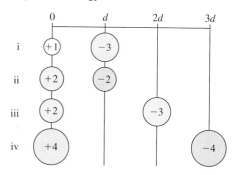

a) i $<$ ii $<$ iii $<$ iv

b) iv $<$ iii $<$ ii $<$ i

c) i $=$ iii $<$ ii $<$ iv

d) ii $<$ i $=$ iii $<$ iv

e) iv $<$ i $=$ ii $<$ iii

**5.1.6**  Which of the following changes would double the magnitude of the electrostatic attraction between two oppositely charged particles? (Select all that apply.)

a) Doubling both charges

b) Doubling one of the charges

c) Doubling the distance between the charges

d) Reducing the distance between the charges by half

e) Doubling both charges and doubling the distance between the charges

# 5.2    Introduction to Thermodynamics

Thermochemistry is part of a broader subject called ***thermodynamics,*** which is the scientific study of the interconversion of heat and other kinds of energy. The laws of thermodynamics provide useful guidelines for understanding the energetics and directions of processes. In this section we will introduce the first law of thermodynamics, which is particularly relevant to the study of thermochemistry. We will continue our discussion of thermodynamics in Chapter 18.

We have defined a system as the part of the universe we are studying. There are three types of systems. An ***open system*** can exchange mass and energy with its surroundings. For example, an open system may consist of a quantity of water in an open container, as shown in Figure 5.3(a). If we close the flask, as in Figure 5.3(b), so that no water vapor can escape from or condense into the container, we create a ***closed system,*** which allows the transfer of *energy* but not *mass*. By placing the water in an insulated container, as shown in Figure 5.3(c), we can construct an ***isolated system,*** which does not exchange either mass or energy with its surroundings.

**Student Annotation:** The energy exchanged between open systems or closed systems and their surroundings is usually in the form of heat.

## States and State Functions

In thermodynamics, we study changes in the ***state of a system,*** which is defined by the values of all relevant macroscopic properties, such as composition, energy, temperature, pressure, and volume. Energy, pressure, volume, and temperature are said to be ***state functions***—properties that are determined by the state of the system, regardless of how that condition was achieved. In other words, when the state of a system changes, the magnitude of change in any state function depends only on the initial and final states of the system and not on how the change is accomplished.

Consider, for example, your position in a six-story building. Your elevation depends upon which floor you are on. If you change your elevation by taking the stairs from the ground floor up to the fourth floor, the change in your elevation depends only upon your initial state (the ground floor—the floor you started on) and your final state (the fourth floor—the floor you went to). It does not depend on whether you went directly to the fourth floor or up to the sixth and then down to the fourth floor. Your overall change in elevation is the same either way because it depends only on your initial and final elevations. Thus, change in elevation is a state function.

The amount of effort it takes to get from the ground floor to the fourth floor, on the other hand, depends on how you get there. More effort has to be exerted to go from the ground floor to the sixth floor and back down to the fourth floor than to go from the ground floor to the fourth floor directly. The effort required for this change in elevation is *not* a state function. Furthermore, if you subsequently return to the ground floor, your overall change in elevation will be zero, because your initial and final states are the same, but the amount of effort you exerted going from the ground floor to the fourth floor and back to the ground floor is *not* zero. Even though your initial and final states are the same, you do not get back the effort that went into climbing up and down the stairs.

Energy is a state function, too. Using potential energy as an example, your net increase in gravitational potential energy is always the same, regardless of how you get from the ground floor to the fourth floor of a building (Figure 5.4).

**Figure 5.3**    (a) An open system allows exchange of both energy and matter with the surroundings. (b) A closed system allows exchange of energy but not matter. (c) An isolated system does not allow exchange of energy or matter. (This flask is enclosed by an insulating vacuum jacket.)

**Figure 5.4** The change in elevation that occurs when a person goes from the ground floor to the fourth floor in a building does not depend on the path taken.

## The First Law of Thermodynamics

The *first law of thermodynamics,* which is based on the law of conservation of energy, states that energy can be converted from one form to another but cannot be created or destroyed. It would be impossible to demonstrate this by measuring the total amount of energy in the universe; in fact, just determining the total energy content of a small sample of matter would be extremely difficult. Fortunately, because energy is a state function, we can demonstrate the first law by measuring the change in the energy of a system between its initial state and its final state in a process. The change in internal energy, $\Delta U$, is given by

$$\Delta U = U_f - U_i$$

where $U_i$ and $U_f$ are the internal energies of the system in the initial and final states, respectively.

The internal energy of a system has two components: kinetic energy and potential energy. The kinetic energy component consists of various types of molecular motion and the movement of electrons within molecules. Potential energy is determined by the attractive interactions between electrons and nuclei and by repulsive interactions between electrons and between nuclei in individual molecules, as well as by interactions between molecules. It is impossible to measure all these contributions accurately, so we cannot calculate the total energy of a system with any certainty. Changes in energy, on the other hand, can be determined experimentally.

Consider the reaction between 1 mole of sulfur and 1 mole of oxygen gas to produce 1 mole of sulfur dioxide:

$$S(s) + O_2(g) \longrightarrow SO_2(g)$$

In this case our system is composed of the reactant molecules and the product molecules. We do not know the internal energy content of either the reactants or the product, but we can accurately measure the *change* in energy content, $\Delta U$, given by

$$\Delta U = U(\text{products}) - U(\text{reactants})$$

$$= \text{energy content of 1 mol } SO_2(g) - \text{energy content of 1 mol } S(s) \text{ and 1 mol } O_2(g)$$

This reaction gives off heat. Therefore, the energy of the product is less than that of the reactants, and $\Delta U$ is negative.

The release of heat that accompanies this reaction indicates that some of the chemical energy contained in the system has been converted to thermal energy. Furthermore, the thermal energy released by the system is absorbed by the surroundings. The transfer of energy from the system to the surroundings does not change the total energy of the universe. That is, the sum of the energy changes is zero:

$$\Delta U_{sys} + \Delta U_{surr} = 0$$

where the subscripts "sys" and "surr" denote system and surroundings, respectively. Thus, if a system undergoes an energy change $\Delta U_{sys}$, the rest of the universe, or the surroundings, must undergo a change in energy that is equal in magnitude but opposite in sign:

$$\Delta U_{sys} = -\Delta U_{surr}$$

**Student Annotation:** The symbol $\Delta$ is commonly used to mean *final* minus *initial*.

**Student Annotation:** Elemental sulfur exists as $S_8$ molecules, but we typically represent it simply as S to simplify chemical equations.

Energy released in one place must be gained somewhere else. Furthermore, because energy can be changed from one form to another, the energy lost by one system can be gained by another system in a different form. For example, the energy released by burning coal in a power plant may ultimately turn up in our homes as electric energy, heat, light, and so on.

## Work and Heat

**Student Annotation:** The units for heat and work are the same as those for energy: joules, kilojoules, or calories.

Recall from Section 5.1 that energy is defined as the capacity to do work or transfer heat. When a system releases or absorbs heat, its internal energy changes. Likewise, when a system does work on its surroundings, or when the surroundings do work on the system, the system's internal energy also changes. The overall change in the system's internal energy is given by

Equation 5.3    $$\Delta U = q + w$$

where $q$ is heat (released or absorbed by the system) and $w$ is work (done *on* the system or done *by* the system). Note that it is possible for the heat and work components to cancel each other out and for there to be no change in the system's internal energy.

**Student Annotation:** Interestingly, although neither $q$ nor $w$ is a state function (each depends on the path between the initial and final states of the system), their sum, $\Delta U$, *is always* the same because $U$ is a state function.

In chemistry, we are normally interested in the energy changes associated with the system rather than the surroundings. Therefore, unless otherwise indicated, $\Delta U$ will refer specifically to $\Delta U_{sys}$. The sign conventions for $q$ and $w$ are as follows: $q$ is positive for an endothermic process and negative for an exothermic process, and $w$ is positive for work done on the system by the surroundings and negative for work done by the system on the surroundings. Table 5.1 summarizes the sign conventions for $q$ and $w$.

The drawings in Figure 5.5 illustrate the logic behind the sign conventions for $q$ and $w$. If a system releases heat to the surroundings or does work on the surroundings [Figure 5.5(a)], we would expect its internal energy to decrease because they are energy-depleting processes. For this reason, both $q$ and $w$ are negative. Conversely, if heat is added to the system or if work is done on the system [Figure 5.5(b)], then the internal energy of the system increases. In this case, both $q$ and $w$ are positive.

Sample Problem 5.2 shows how to determine the overall change in the internal energy of a system.

| TABLE 5.1 | Sign Conventions for Heat ($q$) and Work ($w$) | |
|---|---|---|
| **Process** | | **Sign** |
| Heat absorbed by the system (endothermic process) | | $q$ is positive |
| Heat released by the system (exothermic process) | | $q$ is negative |
| Work done on the system by the surroundings (for example, a volume decrease) | | $w$ is positive |
| Work done by the system on the surroundings (for example, a volume increase) | | $w$ is negative |

**Figure 5.5** (a) When heat is released by the system (to the surroundings), $q$ is negative. When work is done by the system (on the surroundings), $w$ is negative. (b) When heat is absorbed by the system (from the surroundings), $q$ is positive. When work is done on the system (by the surroundings), $w$ is positive.

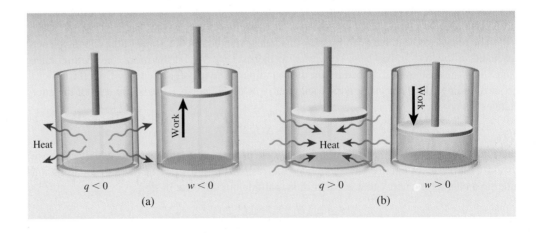

## Sample Problem 5.2

Calculate the overall change in internal energy, $\Delta U$, (in joules) for a system that absorbs 188 J of heat and does 141 J of work on its surroundings.

**Strategy** Combine the two contributions to internal energy using Equation 5.3 and the sign conventions for $q$ and $w$.

**Setup** The system absorbs heat, so $q$ is *positive*. The system does work on the surroundings, so $w$ is negative.

**Solution**

$$\Delta U = q + w = 188 \text{ J} + (-141 \text{ J}) = 47 \text{ J}$$

**Think About It** Consult Table 5.1 to make sure you have used the proper sign conventions for $q$ and $w$.

**Practice Problem A** Calculate the change in total internal energy for a system that releases $1.34 \times 10^4$ kJ of heat and does $2.98 \times 10^4$ kJ of work on the surroundings.

**Practice Problem B** Calculate the magnitude of $q$ for a system that does $7.05 \times 10^5$ kJ of work on its surroundings and for which the change in total internal energy is $-9.55 \times 10^3$ kJ. Indicate whether heat is absorbed or released by the system.

## Checkpoint 5.2    Introduction to Thermodynamics

**5.2.1** Calculate the overall change in internal energy for a system that releases 43 J in heat in a process in which no work is done.

a) 43 J

b) $-2.3 \times 10^{-2}$ J

c) 0 J

d) $2.3 \times 10^{-2}$ J

e) $-43$ J

**5.2.2** Calculate $w$, and determine whether work is done by the system or on the system when 928 kJ of heat is released and $\Delta U = -1.47 \times 10^3$ kJ.

a) $w = -1.36 \times 10^6$ kJ, done by the system

b) $w = 1.36 \times 10^6$ kJ, done on the system

c) $w = -5.4 \times 10^2$ kJ, done by the system

d) $w = 2.4 \times 10^3$ kJ, done on the system

e) $w = -2.4 \times 10^3$ kJ, done by the system

**5.2.3** Which of the following is a closed system? (Select all that apply.)

a) Pot of boiling water on a stove

b) Pot of cold water on a countertop

c) Helium-filled Mylar balloon

d) Styrofoam cup full of hot coffee (without lid)

e) Glass of iced tea

**5.2.4** Which of the following could be described as a state function? (Select all that apply.)

a) Dieter's weight

b) Calories burned by a person walking to the health club

c) Distance traveled by a person walking to the health club

d) Distance between two points on a map

e) Energy required to heat a given amount of water from 25°C to the boiling point.

## 5.3  Enthalpy

To calculate $\Delta U$, we must know the values and signs of both $q$ and $w$. As we will see in Section 5.4, we determine $q$ by measuring temperature changes. To determine $w$, we need to know whether the reaction occurs under constant-volume or constant-pressure conditions.

### Reactions Carried Out at Constant Volume or at Constant Pressure

Imagine carrying out the decomposition of sodium azide ($NaN_3$) in two different experiments. In the first experiment, the reactant is placed in a metal cylinder with a fixed volume. When detonated, the $NaN_3$ reacts, generating a large quantity of $N_2$ gas inside the closed, fixed-volume container.

**Student Annotation:** The explosive decomposition of $NaN_3$ is the reaction that inflates air bags in cars.

$$2NaN_3(s) \longrightarrow 2Na(s) + 3N_2(g)$$

The effect of this reaction will be an increase in the pressure inside the container, similar to what happens if you shake a bottle of soda vigorously prior to opening it.

Now imagine carrying out the same reaction in a metal cylinder with a movable piston. As this explosive decomposition proceeds, the piston in the metal cylinder will move. The gas produced in the reaction pushes the cylinder upward, thereby increasing the volume of the container and preventing any increase in pressure. This is a simple example of mechanical work done by a chemical reaction. Specifically, this type of work is known as *pressure-volume*, or *PV*, work. The amount of work done by such a process is given by

Equation 5.4 $$w = -P\Delta V$$

where $P$ is the external, opposing pressure and $\Delta V$ is the change in the volume of the container as the result of the piston being pushed upward. In keeping with the sign conventions in Table 5.1, an increase in volume results in a negative value for $w$, whereas a decrease in volume results in a positive value for $w$. Figure 5.6 illustrates this reaction (a) being carried out at a constant volume, and (b) at a constant pressure.

When a chemical reaction is carried out at constant volume, then no *PV* work can be done because $\Delta V = 0$ in Equation 5.4. From Equation 5.3 it follows that

Equation 5.5 $$\Delta U = q - P\Delta V$$

and, because $P\Delta V = 0$ at constant volume,

Equation 5.6 $$q_V = \Delta U$$

We add the subscript "$V$" to indicate that this is a constant-volume process. This equality may seem strange at first. We said earlier that $q$ is *not* a state function. However, for a process carried out under constant-volume conditions, $q$ can have only one specific value, which is equal to $\Delta U$. In other words, while $q$ is *not* a state function, $q_V$ *is* one.

NaN$_3$                    Increased
                          pressure

(a)

Work        $\Delta V$

Increased
volume

(b)

**Figure 5.6**    (a) The explosive decomposition of NaN$_3$ at constant volume results in an increase in pressure inside the vessel. (b) The decomposition at constant pressure, in a vessel with a movable piston, results in an increase in volume. The resulting change in volume, $\Delta V$, can be used to calculate the work done by the system.

Constant-volume conditions are often inconvenient and sometimes impossible to achieve. Most reactions occur in open containers, under conditions of constant pressure (usually at whatever the atmospheric pressure happens to be where the experiments are conducted). In general, for a constant-pressure process, we write

$$\Delta U = q + w$$
$$= q_P - P\Delta V$$

or

$$q_P = \Delta U + P\Delta V \qquad \text{Equation 5.7}$$

where the subscript "$P$" denotes constant pressure.

## Enthalpy and Enthalpy Changes

There is a thermodynamic function of a system called **enthalpy** ($H$), which is defined as

$$H = U + PV \qquad \text{Equation 5.8}$$

where $U$ is the internal energy of the system and $P$ and $V$ are the pressure and volume of the system, respectively. Because $U$ and $PV$ have energy units, enthalpy also has energy units. Furthermore, $U$, $P$, and $V$ are all state functions—that is, the changes in $(U + PV)$ depend only on the initial and final states. It follows, therefore, that the change in $H$, or $\Delta H$, also depends only on the initial and final states. Thus, $H$ is a state function.

For any process, the *change* in enthalpy is given by

$$\Delta H = \Delta U + \Delta(PV) \qquad \text{Equation 5.9}$$

If the pressure is held constant, then

$$\Delta H = \Delta U + P\Delta V \qquad \text{Equation 5.10}$$

If we solve Equation 5.10 for $\Delta U$,

$$\Delta U = \Delta H - P\Delta V$$

Then, substituting the result for $\Delta U$ into Equation 5.7, we obtain

$$q_P = (\Delta H - P\Delta V) + P\Delta V$$

The $P\Delta V$ terms cancel, and for a constant-pressure process, the heat exchanged between the system and the surroundings is equal to the enthalpy change:

$$q_P = \Delta H \qquad \text{Equation 5.11}$$

Again, $q$ is *not* a state function, but $q_P$ *is;* that is, the heat change at constant pressure can have only one specific value and is equal to $\Delta H$.

We now have two quantities—$\Delta U$ and $\Delta H$—that can be associated with a reaction. If the reaction occurs under constant-volume conditions, then the heat change, $q_V$, is equal to $\Delta U$. If the reaction is carried out at constant pressure, on the other hand, the heat change, $q_P$, is equal to $\Delta H$.

Because most laboratory reactions are constant-pressure processes, the heat exchanged between the system and surroundings is equal to the change in enthalpy for the process. For any reaction, we define the change in enthalpy, called the **enthalpy of reaction** ($\Delta H$), as the difference between the enthalpies of the products and the enthalpies of the reactants:

$$\Delta H = H(\text{products}) - H(\text{reactants}) \qquad \text{Equation 5.12}$$

The enthalpy of reaction can be positive or negative, depending on the process. For an endothermic process (where heat is absorbed by the system from the surroundings), $\Delta H$ is positive (i.e., $\Delta H > 0$). For an exothermic process (where heat is released by the system to the surroundings), $\Delta H$ is negative (i.e., $\Delta H < 0$).

**Student Annotation:** The SI unit of pressure is the *pascal* (Pa), which in SI base units is $1 \text{ kg}/(\text{m} \cdot \text{s}^2)$. Volume in SI base units is cubic meters ($\text{m}^3$). Therefore, multiplying units of *pressure* by units of *volume* gives ($1 \text{ kg}/\text{m} \cdot \text{s}^2$) $\times \text{m}^3 = 1 \text{ kg} \cdot \text{m}^2/\text{s}^2$, which is the definition of the *joule* (J). Thus, $PV$ has units of *energy*.

**Student Annotation:** The enthalpy of reaction is often symbolized by $\Delta H_{rxn}$. The subscript can be changed to denote a specific type of reaction or physical process: $\Delta H_{vap}$ can be used for the enthalpy of *vaporization*, for example.

We will now apply the idea of enthalpy changes to two common processes, the first involving a physical change and the second involving a chemical change.

## Thermochemical Equations

Under ordinary atmospheric conditions at sea level, ice melts to form liquid water when exposed to temperatures above 0°C. Measurements show that for every mole of ice converted to liquid water under these conditions, 6.01 kJ of heat energy is absorbed by the system (the ice). Because the pressure is constant, the heat change is equal to the enthalpy change, $\Delta H$. This is an *endothermic* process ($\Delta H > 0$), because heat is absorbed by the ice from its surroundings (Figure 5.7a). The equation for this physical change is

**Student Annotation:** Although, strictly speaking, it is unnecessary to include the sign of a positive number, we will include the sign of all positive $\Delta H$ values to emphasize the thermochemical sign convention.

$$H_2O(s) \longrightarrow H_2O(l) \qquad \Delta H = +6.01 \text{ kJ/mol}$$

The "per mole" in the unit for $\Delta H$ means that this is the enthalpy change *per mole of the reaction (or process) as it is written*—that is, when 1 mole of ice is converted to 1 mole of liquid water.

Now consider the combustion of methane ($CH_4$), the principal component of natural gas:

$$CH_4(g) + 2O_2(g) \longrightarrow CO_2(g) + 2H_2O(l) \qquad \Delta H = -890.4 \text{ kJ/mol}$$

From experience we know that burning natural gas releases heat to the surroundings, so it is an exothermic process. Under constant-pressure conditions, this heat change is equal to the enthalpy change and $\Delta H$ must have a negative sign [Figure 5.7(b)]. Again, the "per mole" in the units for $\Delta H$ means that when 1 mole of $CH_4$ reacts with 2 moles of $O_2$ to yield 1 mole of $CO_2$ and 2 moles of liquid $H_2O$, 890.4 kJ of heat is released to the surroundings.

**Student Annotation:** When you specify that a particular amount of heat is released, it is not necessary to include a negative sign.

It is important to keep in mind that the $\Delta H$ value in kJ/mol does not mean *per mole* of a particular reactant or product. It refers to all the species in a reaction in the molar amounts specified by the coefficients in the balanced equation. Thus, for the combustion of methane, the $\Delta H$ value of $-890.4$ kJ/mol can be expressed in any of the following ways:

$$\frac{-890.4 \text{ kJ}}{1 \text{ mol } CH_4} \qquad \frac{-890.4 \text{ kJ}}{2 \text{ mol } O_2} \qquad \frac{-890.4 \text{ kJ}}{1 \text{ mol } CO_2} \qquad \frac{-890.4 \text{ kJ}}{2 \text{ mol } H_2O}$$

Although the importance of expressing $\Delta H$ in units of kJ/mol (rather than just kilojoules) will become apparent when we study thermodynamics in greater detail [▶▶ Chapter 18], you should learn and become comfortable with this convention now.

The equations for the melting of ice and the combustion of methane are examples of ***thermochemical equations***, which are chemical equations that show the enthalpy changes as well as the mass relationships. It is essential to specify a balanced chemical equation when quoting the enthalpy change of a reaction. The following guidelines are helpful in interpreting, writing, and manipulating thermochemical equations:

1. When writing thermochemical equations, we must always specify the physical states of all reactants and products, because they help determine the actual enthalpy changes. In the

**Figure 5.7** (a) Melting 1 mole of ice at 0°C, an endothermic process, results in an enthalpy increase of 6.01 kJ ($\Delta H = +6.01$ kJ/mol). (b) The burning of 1 mole of methane in oxygen gas, an exothermic process, results in an enthalpy decrease in the system of 890.4 kJ ($\Delta H = -890.4$ kJ/mol). The enthalpy diagrams of these two processes are not shown to the same scale.

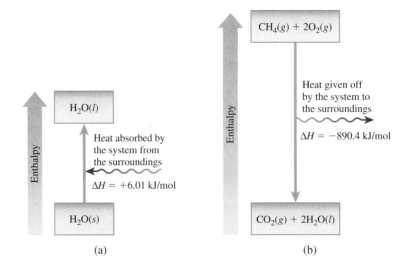

equation for the combustion of methane, for example, changing the liquid water product to water vapor changes the value of $\Delta H$:

$$CH_4(g) + 2O_2(g) \longrightarrow CO_2(g) + 2H_2O(g) \qquad \Delta H = -802.4 \text{ kJ/mol}$$

The enthalpy change is $-802.4$ kJ rather than $-890.4$ kJ because 88.0 kJ is needed to convert 2 moles of liquid water to 2 moles of water vapor; that is,

$$2H_2O(l) \longrightarrow 2H_2O(g) \qquad \Delta H = +88.0 \text{ kJ/mol}$$

**Student Annotation:** Remember that the "per mole" in this context refers to per mole of reaction—not, in this case, per mole of water.

2. If we multiply both sides of a thermochemical equation by a factor $n$, then $\Delta H$ must also change by the same factor. Thus, for the melting of ice, if $n = 2$, we have

$$2H_2O(s) \longrightarrow 2H_2O(l) \qquad \Delta H = 2(6.01 \text{ kJ/mol}) = +12.02 \text{ kJ/mol}$$

3. When we reverse a chemical equation, we change the roles of reactants and products. Consequently, the magnitude of $\Delta H$ for the equation remains the same, but its sign changes. For example, if a reaction consumes thermal energy from its surroundings (i.e., if it is endothermic), then the reverse reaction must release thermal energy back to its surroundings (i.e., it must be exothermic) and the enthalpy change expression must also change its sign. Thus, reversing the melting of ice and the combustion of methane, the thermochemical equations become

$$H_2O(l) \longrightarrow H_2O(s) \qquad \Delta H = -6.01 \text{ kJ/mol}$$

$$CO_2(g) + 2H_2O(l) \longrightarrow CH_4(g) + 2O_2(g) \qquad \Delta H = +890.4 \text{ kJ/mol}$$

What was an endothermic process becomes an exothermic process when reversed, and vice versa.

Sample Problem 5.3 illustrates the use of a thermochemical equation to relate the mass of a product to the energy consumed in the reaction.

## Sample Problem 5.3

Given the thermochemical equation for photosynthesis,

$$6H_2O(l) + 6CO_2(g) \longrightarrow C_6H_{12}O_6(s) + 6O_2(g) \qquad \Delta H = +2803 \text{ kJ/mol}$$

calculate the solar energy required to produce 75.0 g of $C_6H_{12}O_6$.

**Strategy** The thermochemical equation shows that for every mole of $C_6H_{12}O_6$ produced, 2803 kJ is absorbed. We need to find out how much energy is absorbed for the production of 75.0 g of $C_6H_{12}O_6$. We must first find out how many moles there are in 75.0 g of $C_6H_{12}O_6$.

**Setup** The molar mass of $C_6H_{12}O_6$ is 180.2 g/mol, so 75.0 g of $C_6H_{12}O_6$ is

$$75.0 \text{ g} \times \frac{1 \text{ mol } C_6H_{12}O_6}{180.2 \text{ g}} = 0.416 \text{ mol}$$

We will multiply the thermochemical equation, including the enthalpy change, by 0.416, in order to write the equation in terms of the appropriate amount of $C_6H_{12}O_6$.

**Solution**

$$(0.416 \text{ mol})[6H_2O(l) + 6CO_2(g) \longrightarrow C_6H_{12}O_6(s) + 6O_2(g)]$$

and $(0.416 \text{ mol})(\Delta H) = (0.416 \text{ mol})(2803 \text{ kJ/mol})$ gives

$$2.50H_2O(l) + 2.50CO_2(g) \longrightarrow 0.416C_6H_{12}O_6(s) + 2.50O_2(g) \qquad \Delta H = +1.17 \times 10^3 \text{ kJ}$$

Therefore, $1.17 \times 10^3$ kJ of energy in the form of sunlight is consumed in the production of 75.0 g of $C_6H_{12}O_6$. Note that the "per mole" units in $\Delta H$ are canceled when we multiply the thermochemical equation by the number of moles of $C_6H_{12}O_6$.

**Think About It** The specified amount of $C_6H_{12}O_6$ is less than half a mole. Therefore, we should expect the associated enthalpy change to be less than half that specified in the thermochemical equation for the production of 1 mole of $C_6H_{12}O_6$.

**Practice Problem A** Calculate the solar energy required to produce 5255 g of $C_6H_{12}O_6$.

**Practice Problem B** Calculate the mass (in grams) of $O_2$ that is produced by photosynthesis when $2.49 \times 10^4$ kJ of solar energy is consumed.

**5.3.1** Given the thermochemical equation: $H_2(g)$ + $Br_2(l) \longrightarrow 2HBr(g)$, $\Delta H = -72.4$ kJ/mol, calculate the amount of heat released when a kilogram of $Br_2(l)$ is consumed in this reaction.

    a) $7.24 \times 10^4$ kJ

    b) 453 kJ

    c) 906 kJ

    d) 227 kJ

    e) 724 kJ

**5.3.2** Given the thermochemical equation: $2Cu_2O(s) \longrightarrow 4Cu(s) + O_2(g)$, $\Delta H = +333.8$ kJ/mol, calculate the mass of copper produced when $1.47 \times 10^4$ kJ is consumed in this reaction.

    a) 11.2 kg

    b) 176 kg

    c) 44.0 kg

    d) 334 kg

    e) 782 kg

# 5.4  Calorimetry

In the study of thermochemistry, heat changes that accompany physical and chemical processes are measured with a *calorimeter,* a closed container designed specifically for this purpose. We begin our discussion of *calorimetry,* the measurement of heat changes, by defining two important terms: *specific heat* and *heat capacity.*

## Specific Heat and Heat Capacity

The ***specific heat*** (*s*) of a substance is the amount of heat required to raise the temperature of 1 g of the substance by 1°C. The ***heat capacity*** (*C*) is the amount of heat required to raise the temperature of an *object* by 1°C. We can use the specific heat of a substance to determine the heat capacity of a specified quantity of that substance. For example, we can use the specific heat of water, 4.184 J/(g · °C), to determine the heat capacity of a kilogram of water:

> **Student Annotation:** Although heat capacity is typically given for an object rather than for a substance—the "object" may be a given quantity of a particular substance.

$$\text{heat capacity of 1 kg of water} = \frac{4.184\ \text{J}}{1\ \text{g} \cdot {}^\circ\text{C}} \times 1000\ \text{g} = 4184 \quad \text{or} \quad 4.184 \times 10^3\ \text{J/}^\circ\text{C}$$

Note that specific heat has the units J/(g · °C) and heat capacity has the units J/°C. Table 5.2 shows the specific heat values of some common substances. If we know the specific heat and the amount of a substance, then the change in the sample's temperature ($\Delta T$) will tell us the amount of heat ($q$) that has been absorbed or released in a particular process. One equation for calculating the heat associated with a temperature change is given by

| Equation 5.13 | $q = ms\Delta T$ |
|---|---|

where $m$ is the mass of the substance undergoing the temperature change, $s$ is the specific heat, and $\Delta T$ is the temperature change: $\Delta T = T_{\text{final}} - T_{\text{initial}}$. Another equation for calculating the heat associated with a temperature change is given by

| TABLE 5.2 | Specific Heat Values of Some Common Substances |
|---|---|
| **Substance** | **Specific Heat (J/g · °C)** |
| Al(*s*) | 0.900 |
| Au(*s*) | 0.129 |
| C (graphite) | 0.720 |
| C (diamond) | 0.502 |
| Cu(*s*) | 0.385 |
| Fe(*s*) | 0.444 |
| Hg(*l*) | 0.139 |
| $H_2O(l)$ | 4.184 |
| $C_2H_5OH(l)$ (ethanol) | 2.46 |

Media Player/MPEG Animation: Figure 5.9, Determination of $\Delta H^\circ_{\text{rxn}}$ by Constant-Pressure Calorimetry, pp. 186–187.

$$q = C\Delta T \qquad \text{Equation 5.14}$$

where $C$ is the heat capacity and $\Delta T$ is the temperature change. The sign convention for $q$ is the same as that for an enthalpy change: $q$ is positive for endothermic processes and negative for exothermic processes.

Sample Problem 5.4 shows how to use the specific heat of a substance to calculate the amount of heat needed to raise the temperature of the substance by a particular amount.

---

### Sample Problem 5.4

Calculate the amount of heat (in kJ) required to heat 255 g of water from 25.2°C to 90.5°C.

**Strategy** Use Equation 5.13 ($q = ms\Delta T$) to calculate $q$.

**Setup** $m = 255$ g, $s = 4.184$ J/g · °C, and $\Delta T = 90.5°C - 25.2°C = 65.3°C$.

**Solution**

$$q = \frac{4.184\ \text{J}}{\text{g} \cdot °\text{C}} \times 255\ \text{g} \times 65.3°\text{C} = 6.97 \times 10^4\ \text{J} \quad \text{or} \quad 69.7\ \text{kJ}$$

**Practice Problem A** Calculate the amount of heat (in kJ) required to heat 1.01 kg of water from 0.05°C to 35.81°C.

**Practice Problem B** What will be the final temperature of a 514-g sample of water, initially at 10.0°C, after 90.8 kJ have been added to it?

## Constant-Pressure Calorimetry

A crude constant-pressure calorimeter can be constructed from two Styrofoam coffee cups, as shown in Figure 5.8. This device, called a coffee-cup calorimeter, can be used to measure the heat exchanged between the system and surroundings for a variety of reactions, such as acid-base neutralization, heat of solution, and heat of dilution. Because the pressure is constant, the heat change for the process ($q$) is equal to the enthalpy change ($\Delta H$). In such experiments, we consider the reactants and products to be the system, and the water in the calorimeter to be the surroundings. We neglect the small heat capacity of the Styrofoam cups in our calculations. In the case of an exothermic reaction, the heat released by the system is absorbed by the water (surroundings), thereby increasing its temperature. Knowing the mass of the water in the calorimeter, the specific heat of water, and the change in temperature, we can calculate $q_P$ of the system using the equation

$$q_{\text{sys}} = -ms\Delta T \qquad \text{Equation 5.15}$$

Note that the minus sign makes $q_{\text{sys}}$ a negative number if $\Delta T$ is a positive number (i.e., if the temperature goes up). This is in keeping with the sign conventions listed in Table 5.1. A negative $\Delta H$ or a negative $q$ indicates an exothermic process, whereas a positive $\Delta H$ or a positive $q$ indicates an endothermic process. Table 5.3 lists some of the reactions that can be studied with a constant-pressure calorimeter. Figure 5.9 shows how constant-pressure calorimetry can be used to determine $\Delta H$ for a reaction.

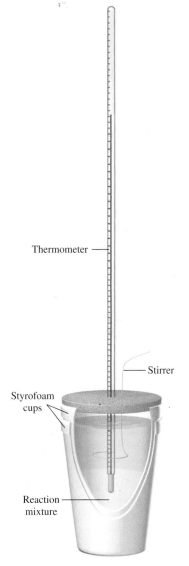

**Figure 5.8** A constant-pressure coffee-cup calorimeter made of two Styrofoam cups. The nested cups help to insulate the reaction mixture from the surroundings. Two solutions of known volume containing the reactants at the same temperature are carefully mixed in the calorimeter. The heat produced or absorbed by the reaction can be determined by measuring the temperature change.

| TABLE 5.3 | Heats of Some Typical Reactions and Physical Processes Measured at Constant Pressure | |
|---|---|---|
| **Type of Reaction** | **Example** | **$\Delta H$ (kJ/mol)** |
| Heat of neutralization | $HCl(aq) + NaOH(aq) \longrightarrow H_2O(l) + NaCl(aq)$ | −56.2 |
| Heat of ionization | $H_2O(l) \longrightarrow H^+(aq) + OH^-(aq)$ | +56.2 |
| Heat of fusion | $H_2O(s) \longrightarrow H_2O(l)$ | +6.01 |
| Heat of vaporization | $H_2O(l) \longrightarrow H_2O(g)$ | +44.0* |

*Measured at 25°C. At 100°C, the value is +40.79 kJ.

Figure 5.9

# Determination of $\Delta H^\circ_{rxn}$ by Constant-Pressure Calorimetry

One at a time, we pour the solutions into the calorimeter.

START

We start with 50.0 mL each of 1.00 $M$ HCl and 1.00 $M$ NaOH. Both solutions are at room temperature, which in this example is 25.0°C. The net ionic equation that represents the reaction is

$$H^+(aq) + OH^-(aq) \longrightarrow H_2O(l)$$

We have 0.0500 L × 1.00 $M$ = 0.0500 mol of each reactant.

When both solutions have been added, we cap the calorimeter to prevent loss of energy to the environment and use the stirrer to ensure that the solutions are mixed thoroughly.

As the reaction proceeds, the temperature of the water increases as it absorbs the energy given off by the reaction. We record the maximum water temperature as 31.7°C.

Assuming the density and specific heat of the solution to be the same as those of water (1 g/mL, 4.184 J/g · °C), we calculate $q_{soln}$ as follows:

$q_{soln}$ = specific heat of water × mass of water × temperature change

$$= \frac{4.184 \text{ J}}{1 \text{ g} \cdot °C} \times 100.0 \text{ g} \times (31.7 - 25.0)°C = 2803 \text{ J}$$

Assuming that the heat capacity of the calorimeter is negligible, we know that $q_{soln} = -q_{rxn}$, and we can write

$q_{rxn} = -2803 \text{ J}$

This is the heat of reaction when 0.0500 mol $H^+$ reacts with 0.0500 mol $OH^-$. To determine $\Delta H_{rxn}$, we divide $q_{rxn}$ by the number of moles. (These reactants are present in stoichiometric amounts. If there were a *limiting* reactant, we would divide by the number of moles of limiting reactant.)

$$\Delta H_{rxn} = \frac{-2803 \text{ J}}{0.0500 \text{ mol}} = -5.61 \times 10^4 \text{ J/mol or } -56.1 \text{ kJ/mol}$$

This result is very close to the number we get using Equation 5.12 and the data in Appendix 2.

## What's the point?

Constant-pressure calorimetry can be used to determine $\Delta H_{rxn}$—the heat of reaction for the reactant quantities specified by the balanced equation. However, when we carry out calorimetry experiments in the laboratory, we typically use much smaller quantities of reactants than those represented in a chemical equation. By measuring the temperature change of the surroundings (a known quantity of water in which the reactants are dissolved), we can determine $q_{rxn}$ for the reactant quantities in the experiment. We can then divide $q_{rxn}$ by the number of moles of reactant to determine $\Delta H_{rxn}$.

Constant-pressure calorimetry can also be used to determine the heat capacity of an object or the specific heat of a substance. Suppose, for example, that we have a lead pellet with a mass of 26.47 g originally at 89.98°C. We drop the pellet into a constant-pressure calorimeter containing 100.0 g of water at 22.50°C. The temperature of the water increases to 23.17°C. In this case, we consider the pellet to be the system and the water to be the surroundings. Because it is the temperature of the surroundings that we measure and because $q_{sys} = -q_{surr}$, we can rewrite Equation 5.15 as

**Student Annotation:** The temperature of the water stops changing when it and the temperature of the lead pellet are equal. Therefore, the final temperature of the pellet is also 23.17°C.

$$q_{surr} = ms\Delta T$$

Thus, $q_{water}$ of the water is

$$q_{water} = \frac{4.184 \text{ J}}{\text{g} \cdot °\text{C}} \times 100.0 \text{ g} \times (23.17°\text{C} - 22.50°\text{C}) = 280 \text{ J}$$

and $q_{pellet}$ is $-280$ J. The negative sign indicates that heat is *released* by the pellet. Dividing $q_{pellet}$ by the temperature change ($\Delta T$) gives us the heat capacity of the pellet ($C_{pellet}$).

$$C_{pellet} = \frac{-280 \text{ J}}{23.17°\text{C} - 89.95°\text{C}} = 4.19 \text{ J/°C}$$

Furthermore, because we know the *mass* of the pellet, we can determine the specific heat of lead ($s_{lead}$):

$$s_{Pb} = \frac{C_{pellet}}{m_{pellet}} = \frac{4.19 \text{ J/°C}}{26.47 \text{ g}} = 0.158 \text{ J/g} \cdot °\text{C} \quad \text{or} \quad 0.16 \text{ J/g} \cdot °\text{C}$$

Media Player/MPEG Animation: Figure 5.10, Determination of Specific Heat by Constant-Pressure Calorimetry, pp. 190–191.

Sample Problem 5.5 shows how to use constant-pressure calorimetry to calculate the heat capacity ($C$) of an object. Figure 5.10 (pp. 190–191) illustrates the process of determining the specific heat of a metal using constant-pressure calorimetry.

### Sample Problem 5.5

A metal pellet with a mass of 100.0 g, originally at 88.4°C, is dropped into 125 g of water originally at 25.1°C. The final temperature of both the pellet and the water is 31.3°C. Calculate the heat capacity $C$ (in J/°C) of the pellet.

**Strategy**  Use Equation 5.13 ($q = ms\Delta T$) to determine the heat absorbed by the water; then use Equation 5.14 ($q = C\Delta T$) to determine the heat capacity of the metal pellet.

**Setup**  $m_{water} = 125$ g, $s_{water} = 4.184$ J/g · °C, and $\Delta T_{water} = 31.3°\text{C} - 25.1°\text{C} = 6.2°\text{C}$. The heat absorbed by the water must be released by the pellet: $q_{water} = -q_{pellet}$ · $m_{pellet} = 100.0$ g and $\Delta T_{pellet} = 31.3°\text{C} - 88.4°\text{C} = -57.1°\text{C}$.

**Solution**  From Equation 5.13, we have

$$q_{water} = \frac{4.184 \text{ J}}{\text{g} \cdot °\text{C}} \times 125 \text{ g} \times 6.2°\text{C} = 3242.6 \text{ J}$$

Thus,

$$q_{pellet} = -3242.6 \text{ J}$$

From Equation 5.14, we have

$$-3242.6 \text{ J} = C_{pellet} \times (-57.1°\text{C})$$

Thus,

$$C_{pellet} = 57 \text{ J/°C}$$

**Think About It**  The units cancel properly to give appropriate units for heat capacity. Moreover, $\Delta T_{pellet}$ is a negative number because the temperature of the pellet decreases.

**Practice Problem A**  What would the final temperature be if the pellet from Sample Problem 5.5, initially at 95°C, were dropped into a 218-g sample of water, initially at 23.8°C?

**Practice Problem B**  What mass of water could be warmed from 23.8°C to 46.3°C by the pellet in Sample Problem 5.5 initially at 116°C?

## Bringing Chemistry to Life

### Heat Capacity and Hypothermia

Like a warm metal pellet, the human body loses heat when it is immersed in cold water. Because we are warm-blooded animals, our body temperature is maintained at around 37°C. The human body is about 70 percent water by mass and water has a very high specific heat, so fluctuations in body temperature normally are very small. An air temperature of 25°C (often described as "room temperature") feels warm to us because air has a small specific heat (about 1 J/g · °C) and a low density. Consequently, very little heat is lost from the body to the surrounding air. The situation is drastically different if the body is immersed in water at 25°C. The heat lost by the human body when immersed in water can be thousands of times greater than that lost to air of the same temperature.

*Hypothermia* occurs when the body's mechanisms for producing and conserving heat are exceeded by loss of heat to the surroundings. Although hypothermia is dangerous and potentially deadly, there are certain circumstances under which it may actually be beneficial. A colder body temperature slows down all the normal biochemical processes, reducing the brain's need for oxygen, and prolonging the time period during which resuscitation efforts can be effective. Occasionally we hear about a seemingly miraculous recovery of a near-drowning victim who was submerged for a long period of time. These victims are usually small children who were submerged in icy water. The small size and, therefore, small heat capacity of a child allows for rapid cooling and may afford some protection from hypoxia—the lack of oxygen that causes death in drowning victims.

**Student Annotation:** Hypothermia routinely is induced in patients undergoing open heart surgery, drastically reducing the body's need for oxygen. Under these conditions, the heart can be stopped for the duration of the surgery.

The lowest body temperature ever recorded for a human (who survived) was 13.7°C. Anna Bagenholm, 29, spent over an hour submerged after falling headfirst through the ice on a frozen river. Although she was clinically dead when she was pulled from the river, she has made a full recovery. There's a saying among doctors who treat hypothermia patients: "No one is dead until they're *warm* and dead."

Figure 5.10

# Determination of Specific Heat by Constant-Pressure Calorimetry

We add metal shot (125.0 g at 100.0°C) to the water, and we cap the calorimeter to prevent loss of energy to the environment.

We place 100.0 mL (100.0 g) of water in the calorimeter. The temperature of this water is 25.0°C.

We place 125.0 g of metal shot in a test tube and immerse it in boiling water long enough to heat all of the metal to the boiling point of water (100.0°C).

As energy is transferred from the metal shot to the water, the temperature of the water increases and the temperature of the metal shot decreases. We use the stirrer to ensure thorough mixing. The thermometer measures the temperature of the water.

When the temperature of the metal shot and the water are equal, the temperature of the water has reached a maximum value. We record this temperature as 34.1°C.

We know that $q_{\text{water}} = -q_{\text{metal}}$.
Substituting in the information given we write:

$q_{\text{water}}$ = specific heat of water × mass of water × temperature change

$$= \frac{4.184 \text{ J}}{1 \text{ g} \cdot {}^\circ\text{C}} \times 100.0 \text{ g} \times (34.1 - 25.0){}^\circ\text{C} = 3807 \text{ J}$$

$q_{\text{metal}}$ = specific heat of metal × mass of metal × temperature change

$$= x \times 125.0 \text{ g} \times (34.1 - 100.0){}^\circ\text{C} = -8238 \, x \text{ g} \cdot {}^\circ\text{C}$$

and

$$3807 \text{ J} = -(-8238 \, x) \text{ g} \cdot {}^\circ\text{C}$$

$$x = \frac{3807 \text{ J}}{8238 \text{ g} \cdot {}^\circ\text{C}} = 0.46 \text{ J/g} \cdot {}^\circ\text{C}$$

The specific heat of the metal is therefore 0.46 J/g · °C.

## What's the point?

We can determine the specific heat of a metal by combining a known mass of the metal at a known temperature with a known mass of water at a known temperature. Assuming the calorimeter has a negligible heat capacity, the amount of energy lost by the hotter metal is equal to the amount of energy gained by the cooler water.

# What If the Heat Capacity of the Calorimeter Isn't Negligible?

Although we usually assume that no heat is absorbed by the Styrofoam cups we use in the laboratory for constant-pressure calorimetry, in reality, the calorimeter generally does absorb a small portion of the heat produced by a chemical reaction. We can determine the heat capacity of a coffee-cup calorimeter by combining reactant solutions with precisely known concentrations and masses. Once we have determined the heat capacity, we can correct for the heat absorbed by the calorimeter when we carry out other experiments.

Consider an experiment in which we combine 50.0 mL of 0.250 $M$ HCl($aq$) with 50.0 mL of 0.250 $M$ NaOH. The calorimeter and both solutions are initially at 23.50°C, and the density and specific heat of the combined solution are the same as that of water (1.000 g/mL and 4.184 J/g · °C, respectively). We can determine the amount of heat the reaction will generate using moles of reactants and the heat of neutralization value from Table 5.3. We have 0.0500 L × 0.250 $M$ = 0.0125 mole of each reactant. According to Table 5.3, the heat of neutralization is −56.2 kJ/mol. Thus, we expect the enthalpy change of the system to be 0.0125 mol × (−56.2 kJ/mol) = −0.703 kJ. Converting this to joules and using Equation 5.15, we calculate the temperature change we expect from the combination of these reactants.

$$q_{sys} = -703 \text{ J} = -100.0 \text{ g} \times 4.184 \text{ J/g} \cdot °C \times \Delta T$$

$$\Delta T = \frac{-703 \text{ J}}{(-100.0 \text{ g})(4.184 \text{ J/g} \cdot °C)} = 1.68°C$$

Thus, we expect the temperature to increase by 1.68°C. However, the measured final temperature is 25.09°C, an increase of only 1.59°C. The water temperature increased by less than expected because the calorimeter absorbed part of the heat produced by the neutralization reaction. We can determine how much heat the calorimeter absorbed by using Equation 5.15 again. This time, we use the measured temperature change to calculate the amount of heat absorbed by the water.

$$q_{surr} = 100.0 \text{ g} \times 4.184 \text{ J/g} \cdot °C \times 1.59°C = 665 \text{ J}$$

We know that the reaction produced 703 J but the water absorbed only 665 J. The remaining energy, 703 − 665 = 38 J, is $q_{calorimeter}$, the heat absorbed by the calorimeter itself. We use Equation 5.14 to calculate the heat capacity of the calorimeter.

$$38 \text{ J} = C_{calorimeter} \times 1.59°C$$

$$C_{calorimeter} = 23.9 \text{ J/°C}$$

Sample Problem 5.6 shows how to take into account the heat capacity of a calorimeter.

Student Annotation: The calorimeter had the same initial temperature and the same final temperature as the solutions.

## Sample Problem 5.6

A 12.1-g piece of aluminum at 81.7°C is added to a sample of water at 23.4°C in a constant-pressure calorimeter. If the final temperature of the water is 24.9°C, and the heat capacity of the calorimeter is 19.8 J/°C, calculate the mass of the water in the calorimeter.

**Strategy** The first law of thermodynamics dictates that the energy given off by a system must be absorbed by the surroundings. In this case, the surroundings consist of the water and the calorimeter, both of which absorb the heat given off by the warm aluminum, which we will call the system. We should first calculate the amount of heat given off by the aluminum ($q_{Al}$) using Equation 5.13. Because $q_{surr} = -q_{sys}$, we know that the sum of $q_{water}$ and $q_{calorimeter}$ must equal $-q_{Al}$.

**Setup** The specific heat of aluminum, from Table 5.2, is 0.900 J/g · °C. $\Delta T_{Al}$ is 24.9°C − 81.7°C = −56.8°C. $\Delta T_{water}$ and $\Delta T_{calorimeter}$ are both 24.9°C − 23.4°C = 1.5°C. The specific heat of water is 4.184 J/g · °C.

**Solution**
$$q_{Al} = 12.1 \text{ g} \times (-56.8°C) \times 0.900 \text{ J/g} \cdot °C = -618.6 \text{ J}$$

$$q_{water} + q_{calorimeter} = 618.6 \text{ J}$$

Let the mass of water be $x$ g.

$$x \text{ g} \times \frac{4.184 \text{ J}}{\text{g} \cdot °C} \times 1.5°C + \frac{19.8 \text{ J}}{°C} \times 1.5°C = 618.6 \text{ J}$$

$$6.28 \text{ J } x + 29.7 \text{ J} = 618.6 \text{ J}$$

$$6.28 \text{ J } x = 618.6 \text{ J} - 29.7 \text{ J} = 588.9 \text{ J}$$

$$x = 93.8$$

The calorimeter contains 93.8 g of water.

**Think About It** If we had neglected the heat capacity of the calorimeter, we would have concluded that there was a slightly larger mass of water in the calorimeter. In that case, the solution would have been

$$x \text{ g} \times \frac{4.184 \text{ J}}{\text{g} \cdot °C} \times 1.5°C = 618.6 \text{ J}$$

and $x = 98.5$

**Practice Problem A** A 50.75-g sample of water at 75.6°C is added to a sample of water at 24.1°C in a constant-pressure calorimeter. If the final temperature of the combined water is 39.4°C and the heat capacity of the calorimeter is 26.3 J/°C, calculate the mass of the water originally in the calorimeter.

**Practice Problem B** A 25.95-g sample of methanol at 35.6°C is added to a 38.65-g sample of ethanol at 24.7°C in a constant-pressure calorimeter. If the final temperature of the combined liquids is 28.5°C and the heat capacity of the calorimeter is 19.3 J/°C, determine the specific heat of methanol.

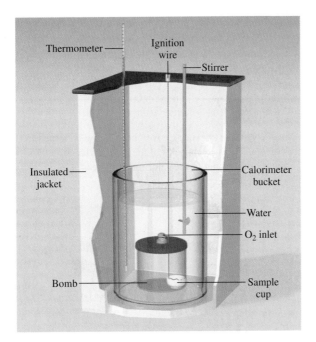

**Figure 5.11**  A constant-volume
bomb calorimeter. The calorimeter is
filled with oxygen gas at high pressure
before it is placed in the bucket. The
sample is ignited electrically, and
the heat produced by the reaction
is determined by measuring the
temperature increase in the known
amount of water surrounding the bomb.

## Constant-Volume Calorimetry

The heat of combustion is usually measured using constant-volume calorimetry. Typically, a known mass of the compound to be analyzed is placed in a steel container called a *constant-volume bomb*, or simply a *bomb*, which is pressurized with oxygen. The closed bomb is then immersed in a known amount of water in an insulated container, as shown in Figure 5.11. (Together, the steel bomb and the water in which it is submerged constitute the *calorimeter*.) The sample is ignited electrically, and the heat released by the combustion of the sample is absorbed by the bomb and the water and can be determined by measuring the increase in temperature of the water. The special design of this type of calorimeter allows us to assume that no heat (or mass) is lost to the surroundings during the time it takes to carry out the reaction and measure the temperature change. Therefore, we can call the bomb and the water in which it is submerged an *isolated* system. Because no heat enters or leaves the system during the process, the heat change of the system overall ($q_{system}$) is zero and we can write

$$q_{cal} = -q_{rxn}$$

where $q_{cal}$ and $q_{rxn}$ are the heat changes for the calorimeter and the reaction, respectively. Thus,

$$q_{rxn} = -q_{cal}$$

To calculate $q_{cal}$, we need to know the heat capacity of the calorimeter ($C_{cal}$) and the change in temperature, that is,

$$q_{cal} = C_{cal}\Delta T \qquad \text{Equation 5.16}$$

And, because $q_{rxn} = -q_{cal}$,

$$q_{rxn} = -C_{cal}\Delta T \qquad \text{Equation 5.17}$$

The heat capacity of the calorimeter ($C_{cal}$) is determined by burning a substance with an accurately known heat of combustion. For example, it is known that the combustion of a 1.000-g sample of benzoic acid ($C_6H_5COOH$) releases 26.38 kJ of heat. If the measured temperature increase is 4.673°C, then the heat capacity of the calorimeter is given by

$$C_{cal} = \frac{q_{cal}}{\Delta T} = \frac{26.38 \text{ kJ}}{4.673°C} = 5.645 \text{ kJ/}°C$$

Once $C_{cal}$ has been determined, the calorimeter can be used to measure the heat of combustion of other substances. Because a reaction in a bomb calorimeter occurs under constant-volume rather than constant-pressure conditions, the measured heat change corresponds to the *internal energy* change ($\Delta U$) rather than to the *enthalpy* change ($\Delta H$) (see Equations 5.6 and 5.11). It is possible to

Benzoic acid

correct the measured heat changes so that they correspond to $\Delta H$ values, but the corrections usually are quite small, so we will not concern ourselves with the details here.

Sample Problem 5.7 shows how to use constant-volume calorimetry to determine the energy content per gram of a substance.

## Sample Problem 5.7

A Famous Amos bite-sized chocolate chip cookie weighing 7.25 g is burned in a bomb calorimeter to determine its energy content. The heat capacity of the calorimeter is 39.97 kJ/°C. During the combustion, the temperature of the water in the calorimeter increases by 3.90°C. Calculate the energy content (in kJ/g) of the cookie.

**Nutrition facts label on a box of Famous Amos chocolate chip cookies.**

**Strategy**  Use Equation 5.17 ($q_{rxn} = -C_{cal}\Delta T$) to calculate the heat released by the combustion of the cookie. Divide the heat released by the mass of the cookie to determine its energy content per gram.

**Setup**  $C_{cal}$ = 39.97 kJ/°C and $\Delta T$ = 3.90°C.

**Solution**  From Equation 5.17 we have

$$q_{rxn} = -C_{cal}\Delta T = -(39.97 \text{ kJ/°C})(3.90°C) = -1.559 \times 10^2 \text{ kJ}$$

Because energy content is a positive quantity, we write

$$\text{energy content per gram} = \frac{1.559 \times 10^2 \text{ kJ}}{7.25 \text{ g}} = 21.5 \text{ kJ/g}$$

**Think About It**  According to the label on the cookie package, a serving size is four cookies, or 29 g, and each serving contains 150 Cal. Convert the energy per gram to Calories per serving to verify the result.

$$\frac{21.5 \text{ kJ}}{g} \times \frac{1 \text{ Cal}}{4.184 \text{ kJ}} \times \frac{29 \text{ g}}{\text{serving}} =$$
$$1.5 \times 10^2 \text{ Cal/serving}$$

**Student Annotation:** The negative sign in the result indicates that heat is released by the combustion.

**Practice Problem A**  A serving of Grape-Nuts cereal (5.80 g) is burned in a bomb calorimeter with a heat capacity of 43.7 kJ/°C. During the combustion, the temperature of the water in the calorimeter increased by 1.92°C. Calculate the energy content (in kJ/g) of Grape-Nuts.

**Practice Problem B**  The energy content of raisin bread is 13.1 kJ/g. Calculate the temperature increase when a slice of raisin bread (32.0 g) is burned in the calorimeter in Sample Problem 5.7.

---

| **Checkpoint 5.4** | **Calorimetry** |

**5.4.1** A 1.000-g sample of benzoic acid is burned in a calorimeter to determine its heat capacity, $C_{cal}$. The reaction gives off 26.42 kJ of heat and the temperature of the water in the calorimeter increases from 23.40°C to 27.20°C. What is the heat capacity of the calorimeter?

  a) 3.80 kJ/°C

  b) 6.95 kJ/°C

  c) 100 kJ/°C

  d) 0.144 kJ/°C

  e) 7.81 kJ/°C

**5.4.2** One-gram samples of Al, Fe, and Au are each heated from 40°C to 75°C. Arrange the metals in order of increasing amount of heat absorbed in the process.

  a) Al < Fe < Au

  b) Al < Au < Fe

  c) Fe < Au < Al

  d) Au < Fe < Al

  e) All the metals absorbed the same amount of heat.

**5.4.3** A reaction, carried out in a bomb calorimeter with $C_{cal} = 5.01$ kJ/°C, gives off 318 kJ of heat. The initial temperature of the water is 24.8°C. What is the final temperature of the water in the calorimeter?

  a) 88.3°C

  b) 63.5°C

  c) 29.8°C

  d) 162°C

  e) 76.7°C

**5.4.4** Quantities of 50.0 mL of 1.00 $M$ HCl and 50.0 mL of 1.00 $M$ NaOH are combined in a constant-pressure calorimeter. Both solutions are initially at 24.4°C. Calculate the final temperature of the combined solutions. (Use the data from Table 5.3. Assume that the mass of the combined solutions is 100.0 g and that the solution's specific heat is the same as that for water, 4.184 J/g · °C.) The heat capacity of the calorimeter is negligibly small.

  a) 31.1°C

  b) 29.0°C

  c) 44.2°C

  d) 91.8°C

  e) 35.7°C

## 5.5   Hess's Law

Because enthalpy is a state function, the change in enthalpy that occurs when reactants are converted to products in a reaction is the same whether the reaction takes place in one step or in a series of steps. This observation is called **Hess's law.**[2] An analogy for Hess's law can be made to the floors in a building. Suppose, for example, that you take the elevator from the first floor to the sixth floor of the building. The net gain in your gravitational potential energy (which is analogous to the enthalpy change for the overall process) is the same whether you go directly there or stop at each floor on your way up (breaking the trip into a series of steps).

    Recall from Section 5.3 that the enthalpy change for the combustion of a mole of methane depends on whether the product water is liquid or gas. More heat is given off by the reaction that produces liquid water. We can use this example to illustrate Hess's law by envisioning the first of these reactions happening in two steps. In step 1, methane and oxygen are converted to carbon dioxide and liquid water, releasing heat.

$$CH_4(g) + 2O_2(g) \longrightarrow CO_2(g) + 2H_2O(l) \qquad \Delta H = -890.4 \text{ kJ/mol}$$

In step 2, the liquid water is vaporized, which requires an input of heat.

$$2H_2O(l) \longrightarrow 2H_2O(g) \qquad \Delta H = +88.0 \text{ kJ/mol}$$

We can add balanced chemical equations just as we can add algebraic equalities, canceling identical items on opposite sides of the equation arrow:

$$
\begin{aligned}
CH_4(g) + 2O_2(g) &\longrightarrow CO_2(g) + \cancel{2H_2O(l)} & \Delta H &= -890.4 \text{ kJ/mol} \\
+ \qquad \cancel{2H_2O(l)} &\longrightarrow 2H_2O(g) & \Delta H &= +88.0 \text{ kJ/mol} \\
\hline
CH_4(g) + 2O_2(g) &\longrightarrow CO_2(g) + 2H_2O(g) & \Delta H &= -802.4 \text{ kJ/mol}
\end{aligned}
$$

When we add thermochemical equations, we add the $\Delta H$ values as well. This gives us the overall enthalpy change for the net reaction. Using this method, we can deduce the enthalpy changes for many reactions, some of which may not be possible to carry out directly. In general, we apply Hess's law by arranging a series of chemical equations (corresponding to a series of steps) in

---

2. Germain Henri Hess (1802–1850). Swiss chemist. Hess was born in Switzerland but spent most of his life in Russia. For formulating Hess's law, he is called the father of thermochemistry.

such a way that they sum to the desired overall equation. Often, in applying Hess's law, we must manipulate the equations involved, multiplying by appropriate coefficients, reversing equations, or both. It is important to follow the guidelines [◄◄ Section 5.3] for the manipulation of thermochemical equations and to make the corresponding change to the enthalpy change of each step.

Sample Problem 5.8 illustrates the use of this method for determining $\Delta H$.

---

## Sample Problem 5.8

Given the following thermochemical equations,

$$NO(g) + O_3(g) \longrightarrow NO_2(g) + O_2(g) \qquad \Delta H = -198.9 \text{ kJ/mol}$$
$$O_3(g) \longrightarrow \tfrac{3}{2}O_2(g) \qquad \Delta H = -142.3 \text{ kJ/mol}$$
$$O_2(g) \longrightarrow 2O(g) \qquad \Delta H = +495 \text{ kJ/mol}$$

determine the enthalpy change for the reaction

$$NO(g) + O(g) \longrightarrow NO_2(g)$$

**Strategy** Arrange the given thermochemical equations so that they sum to the desired equation. Make the corresponding changes to the enthalpy changes, and add them to get the desired enthalpy change.

**Setup** The first equation has NO as a reactant with the correct coefficient, so we use it as is.

$$NO(g) + O_3(g) \longrightarrow NO_2(g) + O_2(g) \qquad \Delta H = -198.9 \text{ kJ/mol}$$

The second equation must be reversed so that the $O_3$ introduced by the first equation will cancel ($O_3$ is not part of the overall chemical equation). We also must change the sign on the corresponding $\Delta H$ value.

$$\tfrac{3}{2}O_2(g) \longrightarrow O_3(g) \qquad \Delta H = +142.3 \text{ kJ/mol}$$

These two steps sum to give the following:

$$NO(g) + \cancel{O_3(g)} \longrightarrow NO_2(g) + \cancel{O_2(g)} \qquad \Delta H = -198.9 \text{ kJ/mol}$$
$$+ \tfrac{1}{2}O_2(g)\, \cancel{\tfrac{3}{2}O_2(g)} \longrightarrow \cancel{O_3(g)} \qquad \Delta H = +142.3 \text{ kJ/mol}$$
$$\overline{NO(g) + \tfrac{1}{2}O_2(g) \longrightarrow NO_2(g) \qquad \Delta H = -56.6 \text{ kJ/mol}}$$

We then replace the $\tfrac{1}{2}O_2$ on the left with O by incorporating the last equation. To do so, we divide the third equation by 2 and reverse its direction. As a result, we must also divide its $\Delta H$ value by 2 and change its sign.

$$O(g) \longrightarrow \tfrac{1}{2}O_2\,(g) \qquad \Delta H = -247.5 \text{ kJ/mol}$$

Finally, we sum all the steps and add their enthalpy changes.

**Solution**

$$NO(g) + \cancel{O_3(g)} \longrightarrow NO_2(g) + \cancel{O_2(g)} \qquad \Delta H = -198.9 \text{ kJ/mol}$$
$$\cancel{\tfrac{3}{2}O_2(g)} \longrightarrow \cancel{O_3(g)} \qquad \Delta H = +142.3 \text{ kJ/mol}$$
$$+ \qquad O(g) \longrightarrow \cancel{\tfrac{1}{2}O_2(g)} \qquad \Delta H = -247.5 \text{ kJ/mol}$$
$$\overline{NO(g) + O(g) \longrightarrow NO_2(g) \qquad \Delta H = -304 \text{ kJ/mol}}$$

| **Think About It** Double-check the cancellation of identical items. |
|---|

**Practice Problem A** Use the thermochemical equations provided in Sample Problem 5.8 to determine the enthalpy change for the reaction $2NO(g) + 4O(g) \longrightarrow 2NO_2(g) + O_2(g)$.

⬆ **Practice Problem B** Use the thermochemical equations provided in Sample Problem 5.8 to determine the enthalpy change for the reaction $2NO_2(g) \longrightarrow 2NO(g) + O_2(g)$.

---

## Checkpoint 5.5     Hess's Law

**5.5.1** Given the following information:

$$2H_2(g) + O_2(g) \longrightarrow 2H_2O(g) \qquad \Delta H = -483.6 \text{ kJ/mol}$$
$$3O_2(g) \longrightarrow 2O_3(g) \qquad \Delta H = +284.6 \text{ kJ/mol}$$

what is $\Delta H$ for $3H_2(g) + O_3(g) \longrightarrow 3H_2O(g)$?

a) $-199$ kJ/mol

b) $-1010$ kJ/mol

c) $-867.7$ kJ/mol

d) $+768.2$ kJ/mol

e) $-440.8$ kJ/mol

**5.5.2** Given the following information:

$$P_4(s) + 3O_2(g) \longrightarrow P_4O_6(s) \qquad \Delta H = -1640.1 \text{ kJ/mol}$$
$$P_4(s) + 5O_2(g) \longrightarrow P_4O_{10}(s) \qquad \Delta H = -2940.1 \text{ kJ/mol}$$

what is the value of $\Delta H_{rxn}$ for $P_4O_6(s) + 2O_2(g) \longrightarrow P_4O_{10}(s)$?

a) $-1300.0$ kJ/mol

b) $+4580.2$ kJ/mol

c) $-4580.2$ kJ/mol

d) $+982.6$ kJ/mol

e) $-982.6$ kJ/mol

So far we have learned that we can determine the enthalpy change that accompanies a reaction by measuring the heat absorbed or released (at constant pressure). According to Equation 5.11, $\Delta H$ can also be calculated if we know the enthalpies of all reactants and products. However, there is no way to measure the *absolute* value of the enthalpy of a substance. Only values *relative* to an arbitrary reference can be determined. This problem is similar to the one geographers face in expressing the elevations of specific mountains or valleys. Rather than trying to devise some type of "absolute" elevation scale (perhaps based on the distance from the center of Earth), by common agreement all geographical heights and depths are expressed relative to sea level, an arbitrary reference with a defined elevation of "zero" meters or feet. Similarly, chemists have agreed on an arbitrary reference point for enthalpy.

The "sea level" reference point for all enthalpy expressions is called the **standard enthalpy of formation** ($\Delta H_f^\circ$), which is defined as the heat change that results when 1 mole of a compound is formed from its constituent elements in their standard states. The superscripted degree sign denotes standard-state conditions, and the subscripted f stands for *formation*. The phrase "in their standard states" refers to the most stable form of an element under standard conditions, meaning at ordinary atmospheric pressure. The element oxygen, for example, can exist as atomic oxygen (O), diatomic oxygen ($O_2$), or ozone ($O_3$). By far the most stable form at ordinary atmospheric pressure, though, is diatomic oxygen. Thus, the standard state of oxygen is $O_2(g)$. Although the standard state does not specify a temperature, we will always use $\Delta H_f^\circ$ values measured at 25°C.

Appendix 2 lists the standard enthalpies of formation for a number of elements and compounds. By convention, the standard enthalpy of formation of any element in its most stable form is zero. Again, using the element oxygen as an example, we can write $\Delta H_f^\circ(O_2) = 0$, but $\Delta H_f^\circ(O_3) \neq 0$ and $\Delta H_f^\circ(O) \neq 0$. Similarly, graphite is a more stable allotropic form of carbon than diamond under standard conditions and 25°C, so we have $\Delta H_f^\circ(\text{graphite}) = 0$ and $\Delta H_f^\circ(\text{diamond}) \neq 0$.

The importance of the standard enthalpies of formation is that once we know their values, we can readily calculate the **standard enthalpy of reaction** ($\Delta H_{rxn}^\circ$), defined as the enthalpy of a reaction carried out under standard conditions. For example, consider the hypothetical reaction

$$a\text{A} + b\text{B} \longrightarrow c\text{C} + d\text{D}$$

where $a$, $b$, $c$, and $d$ are stoichiometric coefficients. For this reaction $\Delta H_{rxn}^\circ$ is given by

$$\Delta H_{rxn}^\circ = [c\Delta H_f^\circ(\text{C}) + d\Delta H_f^\circ(\text{D})] - [a\Delta H_f^\circ(\text{A}) + b\Delta H_f^\circ(\text{B})] \qquad \text{Equation 5.18}$$

We can generalize Equation 5.18 as

$$\Delta H_{rxn}^\circ = \Sigma n\Delta H_f^\circ(\text{products}) - \Sigma m\Delta H_f^\circ(\text{reactants}) \qquad \text{Equation 5.19}$$

where $m$ and $n$ are the stoichiometric coefficients for the reactants and products, respectively, and $\Sigma$ (sigma) means "the sum of." In these calculations the stoichiometric coefficients are treated as numbers without units. Thus, the result has units of kJ/mol, where again, "per mole" means per mole of reaction as written. To use Equation 5.19 to calculate $\Delta H_{rxn}^\circ$, we must know the $\Delta H_f^\circ$ values of the compounds that take part in the reaction. These values, tabulated in Appendix 2, are determined either by the direct method or the indirect method.

The *direct* method of measuring $\Delta H_f^\circ$ works for compounds that can be synthesized from their elements easily and safely. Suppose we want to know the enthalpy of formation of carbon dioxide. We must measure the enthalpy of the reaction when carbon (graphite) and molecular oxygen in their standard states are converted to carbon dioxide in its standard state:

$$\text{C(graphite)} + O_2(g) \longrightarrow CO_2(g) \qquad \Delta H_{rxn}^\circ = -393.5 \text{ kJ/mol}$$

We know from experience that this combustion goes to completion. Thus, from Equation 5.19 we can write

$$\Delta H_{rxn}^\circ = \Delta H_f^\circ(CO_2) - [\Delta H_f^\circ(\text{graphite}) + \Delta H_f^\circ(O_2)] = -393.5 \text{ kJ/mol}$$

Because graphite and $O_2$ are the most stable allotropic forms of their respective elements, $\Delta H_f^\circ(\text{graphite})$ and $\Delta H_f^\circ(O_2)$ are both zero. Therefore,

$$\Delta H_{rxn}^\circ = \Delta H_f^\circ(CO_2) = -393.5 \text{ kJ/mol}$$

or

$$\Delta H_f^\circ(CO_2) = -393.5 \text{ kJ/mol}$$

Arbitrarily assigning a value of zero to $\Delta H_f^\circ$ for each element in its standard state does *not* affect the outcome of these calculations. Remember, in thermochemistry we are interested only in enthalpy changes because they can be determined experimentally, whereas the absolute enthalpy values cannot. The choice of a zero "reference level" for enthalpy is intended to simplify the calculations. Referring again to the terrestrial altitude analogy, we find that Mt. Everest (the highest peak in the world) is 8708 ft higher than Mt. Denali (the highest peak in North America). This difference in altitude would be the same whether we had chosen sea level or the center of Earth as our reference elevation.

Other compounds that can be studied by the direct method are $SF_6$, $P_4O_{10}$, and $CS_2$. The equations representing their syntheses are

$$S(rhombic) + 3F_2(g) \longrightarrow SF_6(g)$$

$$P_4(white) + 5O_2(g) \longrightarrow P_4O_{10}(s)$$

$$C(graphite) + 2S(rhombic) \longrightarrow CS_2(l)$$

S(rhombic) and P(white) are the most stable allotropes of sulfur and phosphorus, respectively, at 1 atm and 25°C, so their $\Delta H_f^\circ$ values are zero.

Sample Problem 5.9 shows how $\Delta H_f^\circ$ values can be used to determine $\Delta H_{rxn}^\circ$.

## Sample Problem 5.9

Using data from Appendix 2, calculate $\Delta H_{rxn}^\circ$ for $Ag^+(aq) + Cl^-(aq) \longrightarrow AgCl(s)$.

**Strategy** Use Equation 5.19 [$\Delta H_{rxn}^\circ = \Sigma n \Delta H_f^\circ(\text{products}) - \Sigma m \Delta H_f^\circ(\text{reactants})$] and $\Delta H_f^\circ$ values from Appendix 2 to calculate $\Delta H_{rxn}^\circ$.

**Setup** The $\Delta H_f^\circ$ values for $Ag^+(aq)$, $Cl^-(aq)$, and $AgCl(s)$ are $+105.9$, $-167.2$, and $-127.0$ kJ/mol, respectively.

**Solution** Using Equation 5.19,

$$\Delta H_{rxn}^\circ = \Delta H_f^\circ(AgCl) - [\Delta H_f^\circ(Ag^+) + \Delta H_f^\circ(Cl^-)]$$

$$= -127.0 \text{ kJ/mol} - [(+105.9 \text{ kJ/mol}) + (-167.2 \text{ kJ/mol})]$$

$$= -127.0 \text{ kJ/mol} - (-61.3 \text{ kJ/mol}) = -65.7 \text{ kJ/mol}$$

**Think About It** Watch out for misplaced or missing minus signs. This is an easy place to lose track of them.

**Practice Problem A** Using data from Appendix 2, calculate $\Delta H_{rxn}^\circ$ for $CaCO_3(s) \longrightarrow CaO(s) + CO_2(g)$.

**Practice Problem B** Using data from Appendix 2, calculate $\Delta H_{rxn}^\circ$ for $2SO(g) + \frac{2}{3}O_3(g) \longrightarrow 2SO_2(g)$.

Many compounds cannot be synthesized from their elements directly. In some cases, the reaction proceeds too slowly, or side reactions produce substances other than the desired compound. In these cases $\Delta H_f^\circ$ can be determined by an *indirect* approach, using Hess's law. If we know a series of reactions for which $\Delta H_{rxn}^\circ$ can be measured, and we can arrange them in such a way as to have them sum to the equation corresponding to the formation of the compound of interest, we can calculate $\Delta H_f^\circ$ for the compound.

Sample Problem 5.10 shows how to use Hess's law to calculate the $\Delta H_f^\circ$ value by the indirect method for a compound that cannot be produced easily from its constituent elements.

## Sample Problem 5.10

Given the following information, calculate the standard enthalpy of formation of acetylene ($C_2H_2$) from its constituent elements:

| | | |
|---|---|---|
| $C(graphite) + O_2(g) \longrightarrow CO_2(g)$ | $\Delta H_{rxn}^\circ = -393.5$ kJ/mol | (1) |
| $H_2(g) + \frac{1}{2}O_2(g) \longrightarrow H_2O(l)$ | $\Delta H_{rxn}^\circ = -285.8$ kJ/mol | (2) |
| $2C_2H_2(g) + 5O_2(g) \longrightarrow 4CO_2(g) + 2H_2O(l)$ | $\Delta H_{rxn}^\circ = -2598.8$ kJ/mol | (3) |

**Strategy** Arrange the equations that are provided so that they will sum to the desired equation. This may require reversing or multiplying one or more of the equations. For any such change, the corresponding change must also be made to the $\Delta H^{\circ}_{rxn}$ value.

**Setup** The equation corresponding to the standard enthalpy of formation of acetylene is

$$2C(graphite) + H_2(g) \longrightarrow C_2H_2(g)$$

We multiply Equation (1) and its $\Delta H^{\circ}_{rxn}$ value by 2:

$$2C(graphite) + 2O_2(g) \longrightarrow 2CO_2(g) \qquad \Delta H^{\circ}_{rxn} = -787.0 \text{ kJ/mol}$$

We include Equation (2) and its $\Delta H^{\circ}_{rxn}$ value as is:

$$H_2(g) + \tfrac{1}{2}O_2(g) \longrightarrow H_2O(l) \qquad \Delta H^{\circ}_{rxn} = -285.8 \text{ kJ/mol}$$

We reverse Equation (3) and divide it by 2 (i.e., multiply through by $\tfrac{1}{2}$):

$$2CO_2(g) + H_2O(l) \longrightarrow C_2H_2(g) + \tfrac{5}{2}O_2(g) \qquad \Delta H^{\circ}_{rxn} = +1299.4 \text{ kJ/mol}$$

**Student Annotation:** The original $\Delta H^{\circ}_{rxn}$ value of Equation (3) has its sign reversed and it is divided by 2.

**Solution** Summing the resulting equations and the corresponding $\Delta H^{\circ}_{rxn}$ values:

| | |
|---|---|
| $2C(graphite) + 2O_2(g) \longrightarrow 2CO_2(g)$ | $\Delta H^{\circ}_{rxn} = -787.0 \text{ kJ/mol}$ |
| $H_2(g) + \tfrac{1}{2}O_2(g) \longrightarrow H_2O(l)$ | $\Delta H^{\circ}_{rxn} = -285.8 \text{ kJ/mol}$ |
| $+ \quad 2CO_2(g) + H_2O(l) \longrightarrow C_2H_2(g) + \tfrac{5}{2}O_2(g)$ | $\Delta H^{\circ}_{rxn} = +1299.4 \text{ kJ/mol}$ |
| $2C(graphite) + H_2(g) \longrightarrow C_2H_2(g)$ | $\Delta H^{\circ}_{f} = +226.6 \text{ kJ/mol}$ |

**Think About It** Watch out for misplaced or missing minus signs. This is an easy place to lose track of them.

**Practice Problem A** Use the following data to calculate $\Delta H^{\circ}_{f}$ for $CS_2(l)$:

| | |
|---|---|
| $C(graphite) + O_2(g) \longrightarrow CO_2(g)$ | $\Delta H^{\circ}_{rxn} = -393.5 \text{ kJ/mol}$ |
| $S(rhombic) + O_2(g) \longrightarrow SO_2(g)$ | $\Delta H^{\circ}_{rxn} = -296.4 \text{ kJ/mol}$ |
| $CS_2(l) + 3O_2(g) \longrightarrow CO_2(g) + 2SO_2(g)$ | $\Delta H^{\circ}_{rxn} = -1073.6 \text{ kJ/mol}$ |

**Practice Problem B** $\Delta H^{\circ}_{f}$ of hydrogen chloride [HCl(g)] is $-92.3$ kJ/mol. Given the following data, determine the identity of the two missing products and calculate $\Delta H^{\circ}_{rxn}$ for Equation 3. [*Hint:* Start by writing the chemical equation that corresponds to $\Delta H^{\circ}_{f}$ for HCl(g).]

| | | |
|---|---|---|
| $N_2(g) + 4H_2(g) + Cl_2(g) \longrightarrow 2NH_4Cl(s)$ | $\Delta H^{\circ}_{rxn} = -630.78 \text{ kJ/mol}$ | (1) |
| $N_2(g) + 3H_2(g) \longrightarrow 2NH_3(g)$ | $\Delta H^{\circ}_{rxn} = -92.6 \text{ kJ/mol}$ | (2) |
| $NH_4Cl(s) \longrightarrow$ | | (3) |

---

## Checkpoint 5.6  Standard Enthalpies of Formation

**5.6.1** Using data from Appendix 2, calculate $\Delta H^{\circ}_{rxn}$ for $H_2(g) + F_2(g) \longrightarrow 2HF(g)$.

a) $-271.6$ kJ/mol

b) $-543.2$ kJ/mol

c) $+271.6$ kJ/mol

d) 0 kJ/mol

e) $-135.8$ kJ/mol

**5.6.2** Using data from Appendix 2, calculate $\Delta H^{\circ}_{rxn}$ for $2NO_2(g) \longrightarrow N_2O_4(g)$.

a) $-24.19$ kJ/mol

b) $-33.85$ kJ/mol

c) $+9.66$ kJ/mol

d) $+67.7$ kJ/mol

e) $-58.04$ kJ/mol

**5.6.3** Which of the following $\Delta H^{\circ}_{rxn}$ values is a $\Delta H^{\circ}_{f}$ value? (Select all that apply.)

a) $H_2(g) + Br_2(l) \longrightarrow 2HBr(g)$ $\qquad \Delta H^{\circ}_{rxn} = -72.4$ kJ/mol

b) $4Al(s) + 3O_2(g) \longrightarrow 2Al_2O_3(s)$ $\qquad \Delta H^{\circ}_{rxn} = -3339.6$ kJ/mol

c) $Ag(s) + \tfrac{1}{2}Cl_2(g) \longrightarrow AgCl(s)$ $\qquad \Delta H^{\circ}_{rxn} = -127.0$ kJ/mol

d) $Cu^{2+}(aq) + SO_4^{2-}(aq) \longrightarrow CuSO_4(s)$ $\qquad \Delta H^{\circ}_{rxn} = +73.25$ kJ/mol

e) $\tfrac{1}{2}H_2(g) + \tfrac{1}{2}N_2(g) + \tfrac{3}{2}O_2(g) \longrightarrow HNO_3(l)$ $\qquad \Delta H^{\circ}_{rxn} = -173.2$ kJ/mol

**5.6.4** Using the following data, calculate $\Delta H^{\circ}_{f}$ for CO(g):

| | |
|---|---|
| $C(graphite) + O_2(g) \longrightarrow CO_2(g)$ | $\Delta H^{\circ}_{rxn} = -393.5 \text{ kJ/mol}$ |
| $CO(g) + \tfrac{1}{2}O_2(g) \longrightarrow CO_2(g)$ | $\Delta H^{\circ}_{rxn} = -283.0 \text{ kJ/mol}$ |

a) $-393.5$ kJ/mol

b) $-676.5$ kJ/mol

c) $+676.5$ kJ/mol

d) $+110.5$ kJ/mol

e) $-110.5$ kJ/mol

# Applying What You've Learned

One of the most popular approaches to dieting in recent years has been to reduce dietary fat. One reason many people want to avoid eating fat is its high Calorie content. Compared to carbohydrates and proteins, each of which contains an average of 4 Calories per gram (17 kJ/g), fat contains 9 Calories per gram (38 kJ/g). Tristearin, a typical fat, is metabolized (or combusted) according to the following equation:

$$C_{57}H_{110}O_6(s) + 81.5O_2(g) \longrightarrow 57CO_2(g) + 55H_2O(l) \qquad \Delta H° = -37{,}760 \text{ kJ/mol}$$

Although the food industry has succeeded in producing low-fat versions of nearly everything we eat, it has thus far failed to produce a palatable low-fat doughnut. The flavor, texture, and what the industry calls "mouth feel" of a doughnut depends largely on the process of deep-fat frying. Fortunately for people in the doughnut business, though, high fat content has not diminished the popularity of doughnuts.

**Many popular foods are available in low-fat varieties.**

## Problems:

According to information obtained from www.krispykreme.com, a Krispy Kreme original glazed doughnut weighs 52 g and contains 200 Cal and 12 g of fat.

a) Assuming that the fat in the doughnut is metabolized according to the given equation for tristearin, calculate the number of Calories in the reported 12 g of fat in each doughnut. [◄◄ Sample Problem 5.3]

b) If all the energy contained in a Krispy Kreme doughnut (not just in the fat) were transferred to 6.00 kg of water originally at 25.5°C, what would be the final temperature of the water? [◄◄ Sample Problem 5.4]

c) When a Krispy Kreme apple fritter weighing 101 g is burned in a bomb calorimeter with $C_{cal}$ = 95.3 kJ/°C, the measured temperature increase is 16.7°C. Calculate the number of Calories in a Krispy Kreme apple fritter. [◄◄ Sample Problem 5.7]

d) What would the $\Delta H°$ value be for the metabolism of 1 mole of the fat tristearin if the water produced by the reaction were gaseous instead of liquid? [◄◄ Sample Problem 5.8] [*Hint:* Use data from Appendix 2 to determine the $\Delta H°$ value for the reaction $H_2O(l) \longrightarrow H_2O(g)$. [◄◄ Sample Problem 5.9]]

**Nutrition Facts**

Serving Size 1 donut (about 52g)
Servings Per Container 12

| Amount Per Serving | | |
|---|---|---|
| **Calories** 200 | **Calories From Fat** 100 | |
| | | **%Daily Value*** |
| **Total Fat** 12g | | 18% |
| Saturated Fat 3g | | 15% |
| *Trans* Fat 4g | | |
| **Cholesterol** 5mg | | 1% |
| **Sodium** 95mg | | 4% |
| **Total Carbohydrate** 22g | | 7% |
| Dietary Fiber <1g | | 1% |
| Sugars 10g | | |
| **Protein** 2g | | |
| Vitamin A 0% | • | Vitamin C 2% |
| Calcium 6% | • | Iron 4% |

*Percent of Daily Values (DV) are based on a 2,000 calorie diet.

**Nutrition facts label for Krispy Kreme original glazed doughnuts.**

# Chapter Summary

## Section 5.1

- *Energy* is the capacity to do work or transfer heat. Energy may be *kinetic energy* (the energy associated with *motion*) or *potential energy* (energy possessed by virtue of *position*). *Thermal energy* is a form of kinetic energy. *Chemical energy* and *electrostatic energy* are forms of potential energy.

- The *law of conservation of energy* states that energy can neither be created nor destroyed. The SI unit of energy is the *joule* (J).

- The *system* is the particular part of the universe that we are interested in studying—such as the reactants and products in a chemical reaction. The term *surroundings* refers to the rest of the universe. System + surroundings = universe.

- *Heat* refers to the flow of thermal energy between two bodies at different temperatures. *Thermochemistry* is the study of the heat associated with chemical reactions and physical processes.

- In an *exothermic process,* heat is released to the surroundings, so the energy of the system decreases. In an *endothermic process,* heat is absorbed from the surroundings, so the energy of the system increases.

## Section 5.2

- *Thermodynamics* is the study of the conversions among different types of energy. Thermochemistry is a branch of thermodynamics.

- An *open system* is one that can exchange both matter and energy with its surroundings. A *closed system* is one that can exchange energy but not matter with its surroundings. An *isolated system* is one that cannot exchange either energy or matter with its surroundings.

- The *state of a system* is defined by the values of all relevant macroscopic properties, such as temperature, volume, and pressure. A *state function* is one whose value depends only on the state of the system and not on how that state was achieved. State functions include energy, pressure, volume, and temperature.

- The *first law of thermodynamics* states that energy cannot be created or destroyed, but it can be changed from one form to another. The first law of thermodynamics is based on the law of conservation of energy.

## Section 5.3

- *Enthalpy* is the heat exchanged between the system and surroundings at constant pressure. It is a state function. *Enthalpy of reaction* ($\Delta H_{rxn}$) is the heat exchanged at constant pressure for a specific reaction.

- A *thermochemical equation* is a balanced chemical equation for which the enthalpy change ($\Delta H_{rxn}$) is given.

## Section 5.4

- *Calorimetry* is the science of measuring temperature changes to determine heats associated with chemical reactions. Calorimetry may be carried out at constant pressure (in a coffee-cup calorimeter) or at constant volume (in a bomb calorimeter).

- The *specific heat* (s) of a substance is the amount of heat required to increase the temperature of 1 g of the substance by 1°C. The *heat capacity* of an object is the amount of heat required to increase the temperature of the object by 1°C.

## Section 5.5

- *Hess's law* states that the enthalpy change for a reaction that occurs in a series of steps is equal to the sum of the enthalpy changes of the individual steps. Hess's law is valid because enthalpy is a state function.

## Section 5.6

- The *standard enthalpy of formation* ($\Delta H_f^\circ$) is the enthalpy change associated with the formation of 1 mole of a substance from its constituent elements, each in their standard states. The *standard enthalpy of reaction* ($\Delta H_{rxn}^\circ$) can be calculated for any reaction using tabulated standard enthalpies of formation ($\Delta H_f^\circ$) of the products and reactants.

# Key Words

| | | | |
|---|---|---|---|
| Calorimetry, 184 | Exothermic process, 173 | Law of conservation of energy, 172 | State of a system, 176 |
| Chemical energy, 172 | First law of thermodynamics, 177 | Open system, 176 | Surroundings, 172 |
| Closed system, 176 | Heat, 173 | Potential energy, 172 | System, 172 |
| Electrostatic energy, 172 | Heat capacity, 184 | Specific heat, 184 | Thermal energy, 172 |
| Endothermic process, 173 | Hess's law, 195 | Standard enthalpy of formation, 197 | Thermochemical equation, 182 |
| Energy, 172 | Isolated system, 176 | | Thermochemistry, 173 |
| Enthalpy, 181 | Joule, 173 | Standard enthalpy of reaction, 197 | Thermodynamics, 176 |
| Enthalpy of reaction, 181 | Kinetic energy, 172 | State function, 176 | |

# KEY EQUATIONS

5.1    $E_k = \frac{1}{2}mu^2$

5.2    $E_{el} \propto \dfrac{Q_1 Q_2}{d}$

5.3    $\Delta U = q + w$

5.4    $w = -P\Delta V$

5.5    $\Delta U = q - P\Delta V$

5.6    $q_V = \Delta U$

5.7    $q_P = \Delta U + P\Delta V$

5.8    $H = U + PV$

5.9    $\Delta H = \Delta U + \Delta(PV)$

5.10    $\Delta H = \Delta U + P\Delta V$

5.11    $q_P = \Delta H$

5.12    $\Delta H = H(\text{products}) - H(\text{reactants})$

5.13    $q = ms\Delta T$

5.14    $q = C\Delta T$

5.15    $q_{sys} = -ms\Delta T$

5.16    $q_{cal} = C_{cal}\Delta T$

5.17    $q_{rxn} = -C_{cal}\Delta T$

5.18    $\Delta H^{\circ}_{rxn} = [c\Delta H^{\circ}_f(C) + d\Delta H^{\circ}_f(D)] - [a\Delta H^{\circ}_f(A) + b\Delta H^{\circ}_f(B)]$

5.19    $\Delta H^{\circ}_{rxn} = \Sigma n\Delta H^{\circ}_f(\text{products}) - \Sigma m\Delta H^{\circ}_f(\text{reactants})$

# QUESTIONS AND PROBLEMS

## Section 5.1: Energy and Energy Changes

### Review Questions

5.1    Define these terms: *system, surroundings, thermal energy, chemical energy, potential energy, kinetic energy, law of conservation of energy.*

5.2    What is heat? How does heat differ from thermal energy? Under what condition is heat transferred from one system to another?

5.3    What are the units for energy commonly employed in chemistry?

5.4    A truck initially traveling at 60 km/h is brought to a complete stop at a traffic light. Does this change violate the law of conservation of energy? Explain.

5.5    These are various forms of energy: chemical, heat, light, mechanical, and electrical. Suggest several ways of converting one form of energy to another.

5.6    Describe the interconversions of forms of energy occurring in these processes: (a) You throw a softball up into the air and catch it. (b) You switch on a flashlight. (c) You ride the ski lift to the top of the hill and then ski down. (d) You strike a match and let it burn completely.

5.7    Define these terms: *thermochemistry, exothermic process, endothermic process.*

5.8    Stoichiometry is based on the law of conservation of mass. On what law is thermochemistry based?

5.9    Describe two exothermic processes and two endothermic processes.

5.10    Decomposition reactions are usually endothermic, whereas combination reactions are usually exothermic. Give a qualitative explanation for these trends.

## Section 5.2: Introduction to Thermodynamics

### Review Questions

5.11    On what law is the first law of thermodynamics based? Explain the sign conventions in the equation

$$\Delta U = q + w$$

5.12    Explain what is meant by a state function. Give two examples of quantities that are state functions and two that are not state functions.

### Problems

5.13    The work done to compress a gas is 74 J. As a result, 26 J of heat is given off to the surroundings. Calculate the change in energy of the gas.

5.14    In a gas expansion, 87 J of heat is absorbed from the surroundings and the energy of the system decreases by 128 J. Calculate the work done.

5.15    Calculate *w*, and determine whether work is done *by* the system or *on* the system when 67 J of heat is released and $\Delta U = 510$ J.

5.16    Calculate *q*, and determine whether heat is absorbed or released when a system does work on the surroundings equal to 64 J and $\Delta U = -213$ J.

## Section 5.3: Enthalpy

### Review Questions

5.17    Consider these changes.

(a) $Hg(l) \longrightarrow Hg(g)$
(b) $3O_2(g) \longrightarrow 2O_3(g)$
(c) $CuSO_4 \cdot 5H_2O(s) \longrightarrow CuSO_4(s) + 5H_2O(g)$
(d) $H_2(g) + F_2(g) \longrightarrow 2HF(g)$

At constant pressure, in which of the reactions is work done by the system on the surroundings? By the surroundings on the system? In which of them is no work done?

5.18    Define these terms: *enthalpy* and *enthalpy of reaction.* Under what condition is the heat of a reaction equal to the enthalpy change of the same reaction?

5.19    In writing thermochemical equations, why is it important to indicate the physical state (that is, gaseous, liquid, solid, or aqueous) of each substance?

5.20    Explain the meaning of this thermochemical equation:

$$4NH_3(g) + 5O_2(g) \longrightarrow 4NO(g) + 6H_2O(g)$$
$$\Delta H = -904 \text{ kJ/mol}$$

5.21   Consider this reaction:

$$2CH_3OH(l) + 3O_2(g) \longrightarrow 4H_2O(l) + 2CO_2(g)$$
$$\Delta H = -1452.8 \text{ kJ/mol}$$

What is the value of $\Delta H$ if (a) the equation is multiplied throughout by 2; (b) the direction of the reaction is reversed so that the products become the reactants, and vice versa; (c) water vapor instead of liquid water is formed as the product?

## Problems

 5.22   A sample of nitrogen gas expands in volume from 1.6 to 5.4 L at constant temperature. Calculate the work done in joules if the gas expands (a) against a vacuum, (b) against a constant pressure of 0.80 atm, and (c) against a constant pressure of 3.7 atm. (See Equation 5.4. 1 L · atm = 101.3 J.)

5.23   A gas expands in volume from 26.7 to 89.3 mL at constant temperature. Calculate the work done (in joules) if the gas expands (a) against a vacuum, (b) against a constant pressure of 1.5 atm, and (c) against a constant pressure of 2.8 atm. (1 L · atm = 101.3 J.)

5.24   A gas expands and does $PV$ work on the surroundings equal to 325 J. At the same time, it absorbs 127 J of heat from the surroundings. Calculate the change in energy of the gas.

5.25   The first step in the industrial recovery of zinc from the zinc sulfide ore is roasting, that is, the conversion of ZnS to ZnO by heating:

$$2ZnS(s) + 3O_2(g) \longrightarrow 2ZnO(s) + 2SO_2(g)$$
$$\Delta H = -879 \text{ kJ/mol}$$

Calculate the heat evolved (in kJ) per gram of ZnS roasted.

5.26   Determine the amount of heat (in kJ) given off when $1.26 \times 10^4$ g of $NO_2$ are produced according to the equation

$$2NO(g) + O_2(g) \longrightarrow 2NO_2(g) \quad \Delta H = -114.6 \text{ kJ/mol}$$

5.27   Consider the reaction

$$2H_2O(g) \longrightarrow 2H_2(g) + O_2(g) \quad \Delta H = +483.6 \text{ kJ/mol}$$

at a certain temperature. If the increase in volume is 32.7 L against an external pressure of 1.00 atm, calculate $\Delta U$ for this reaction. (1 L · atm = 101.3 J.)

5.28   Consider the reaction

$$H_2(g) + Cl_2(g) \longrightarrow 2HCl(g) \quad \Delta H = -184.6 \text{ kJ/mol}$$

If 3 moles of $H_2$ react with 3 moles of $Cl_2$ to form HCl, calculate the work done (in joules) against a pressure of 1.0 atm. What is $\Delta U$ for this reaction? Assume the reaction goes to completion and that $\Delta V = 0$. (1 L · atm = 101.3 J.)

5.29   For most biological processes, the changes in internal energy are approximately equal to the changes in enthalpy. Explain.

## Section 5.4: Calorimetry

### Visualizing Chemistry
### Figure 5.9 and Figure 5.10

VC 5.1   Referring to Figure 5.9, which of the following would result in the calculated value of $\Delta H_{rxn}$ being too high?

a) Spilling some of one of the reactant solutions before adding it to the calorimeter.
b) Reading the final temperature before it reached its maximum value.
c) Misreading the thermometer at the beginning of the experiment and recording too low an initial temperature.

VC 5.2   How would the $\Delta H_{rxn}$ calculated in Figure 5.9 be affected if the concentration of one of the reactant solutions were twice as high as it was supposed to be?

a) The calculated $\Delta H_{rxn}$ would not be affected.
b) The calculated $\Delta H_{rxn}$ would be too low.
c) The calculated $\Delta H_{rxn}$ would be too high.

VC 5.3   For an exothermic reaction like the one depicted in Figure 5.9, if the heat capacity of the calorimeter is not negligibly small, the heat absorbed by the water will be _____ the heat given off by the reaction.

a) greater than
b) less than
c) equal to

VC 5.4   Referring to Figure 5.9, how would the results of the experiment have been different if the reaction had been endothermic?

a) The results would have been the same.
b) There would have been a smaller temperature increase.
c) There would have been a temperature decrease.

VC 5.5   What would happen to the specific heat calculated in Figure 5.10 if some of the warm metal shot were lost during the transfer to the calorimeter?

a) It would not affect the calculated value of specific heat.
b) It would cause the calculated value of specific heat to be too high.
c) It would cause the calculated value of specific heat to be too low.

VC 5.6   What would happen to the specific heat calculated in Figure 5.10 if the test tube containing the metal shot were left in the boiling water for longer than the recommended time?

a) It would not affect the calculated value of specific heat.
b) It would cause the calculated value of specific heat to be too high.
c) It would cause the calculated value of specific heat to be too low.

VC 5.7   What would happen to the specific heat calculated in Figure 5.10 if some of the water were spilled prior to being added to the calorimeter?

a) It would not affect the calculated value of specific heat.
b) It would cause the calculated value of specific heat to be too high.
c) It would cause the calculated value of specific heat to be too low.

VC 5.8   Referring to the process depicted in Figure 5.10, which of the following must be known precisely for the calculated specific heat to be accurate.

a) The mass of the boiling water.
b) The temperature of the metal shot before it is immersed in the boiling water.
c) The mass of the water that is added to the calorimeter.

## Review Questions

5.30   What is the difference between specific heat and heat capacity? What are the units for these two quantities? Which is the intensive property and which is the extensive property?

5.31   Consider two metals A and B, each having a mass of 100 g and an initial temperature of 20°C. The specific heat of A is larger than that of B. Under the same heating conditions, which metal would take longer to reach a temperature of 21°C?

**5.32**   Define *calorimetry* and describe two commonly used calorimeters. In a calorimetric measurement, why is it important that we know the heat capacity of the calorimeter? How is this value determined?

**5.33**   Consider the following data:

| Metal | Al | Cu |
|---|---|---|
| Mass (g) | 10 | 30 |
| Specific heat (J/g · °C) | 0.900 | 0.385 |
| Temperature (°C) | 40 | 60 |

When these two metals are placed in contact, which of the following will take place?

(a) Heat will flow from Al to Cu because Al has a larger specific heat.
(b) Heat will flow from Cu to Al because Cu has a larger mass.
(c) Heat will flow from Cu to Al because Cu has a larger heat capacity.
(d) Heat will flow from Cu to Al because Cu is at a higher temperature.
(e) No heat will flow in either direction.

*Problems*

**5.34**   A piece of silver with a mass of 362 g has a heat capacity of 85.7 J/°C. What is the specific heat of silver?

**5.35**   A 6.22-kg piece of copper metal is heated from 20.5°C to 324.3°C. Calculate the heat absorbed (in kJ) by the metal.

**5.36**   Calculate the amount of heat liberated (in kJ) from 366 g of mercury when it cools from 77.0°C to 12.0°C.

**5.37**   A sheet of gold weighing 10.0 g and at a temperature of 18.0°C is placed flat on a sheet of iron weighing 20.0 g and at a temperature of 55.6°C. What is the final temperature of the combined metals? Assume that no heat is lost to the surroundings. (*Hint:* The heat gained by the gold must be equal to the heat lost by the iron. The specific heats of the metals are given in Table 5.2.)

**5.38**   A 0.1375-g sample of solid magnesium is burned in a constant-volume bomb calorimeter that has a heat capacity of 3024 J/°C. The temperature increases by 1.126°C. Calculate the heat given off by the burning Mg, in kJ/g and in kJ/mol.

**5.39**   A quantity of $2.00 \times 10^2$ mL of 0.862 $M$ HCl is mixed with $2.00 \times 10^2$ mL of 0.431 $M$ Ba(OH)$_2$ in a constant-pressure calorimeter of negligible heat capacity. The initial temperature of the HCl and Ba(OH)$_2$ solutions is the same at 20.48°C. For the process

$$H^+(aq) + OH^-(aq) \longrightarrow H_2O(l)$$

the heat of neutralization is −56.2 kJ/mol. What is the final temperature of the mixed solution? Assume the specific heat of the solution is the same as that for pure water.

**5.40**   The fuel value of hamburger is approximately 3.6 kcal/g. If a man eats 1 pound of hamburger for lunch and if none of the energy is stored in his body, estimate the amount of water that would have to be lost in perspiration to keep his body temperature constant. The heat of vaporization of water may be taken as 2.41 kJ/g (1 lb = 453.6 g).

**5.41**   Metabolic activity in the human body releases approximately $1.0 \times 10^4$ kJ of heat per day. Assuming the body contains 50 kg of water, how fast would the body temperature rise if it were an isolated system? How much water must the body eliminate as perspiration to maintain the normal body temperature (98.6°F)? The heat of vaporization of water may be taken as 2.41 kJ/g. Comment on your results.

**5.42**   A hemoglobin molecule (molar mass = 65,000 g) can bind up to four oxygen molecules. In a certain experiment a 0.085-L solution containing 6.0 g of deoxyhemoglobin (hemoglobin without oxygen molecules bound to it) was combined with an excess of oxygen in a constant-pressure calorimeter of negligible heat capacity. Calculate the enthalpy of reaction per mole of oxygen bound if the temperature rose by 0.044°C. Assume the solution is dilute so that the specific heat of the solution is equal to that of water.

## Section 5.5: Hess's Law

*Review Questions*

**5.43**   State Hess's law. Explain, with one example, the usefulness of Hess's law in thermochemistry.

**5.44**   Describe how chemists use Hess's law to determine the $\Delta H_f^\circ$ of a compound by measuring its heat (enthalpy) of combustion.

*Problems*

**5.45**   From these data,

$$S(\text{rhombic}) + O_2(g) \longrightarrow SO_2(g) \qquad \Delta H_{rxn}^\circ = -296.06 \text{ kJ/mol}$$
$$S(\text{monoclinic}) + O_2(g) \longrightarrow SO_2(g) \qquad \Delta H_{rxn}^\circ = -296.36 \text{ kJ/mol}$$

calculate the enthalpy change for the transformation

$$S(\text{rhombic}) \longrightarrow S(\text{monoclinic})$$

(Monoclinic and rhombic are different allotropic forms of elemental sulfur.)

**5.46**   From the following data,

$$C(\text{graphite}) + O_2(g) \longrightarrow CO_2(g) \qquad \Delta H_{rxn}^\circ = -393.5 \text{ kJ/mol}$$
$$H_2(g) + \tfrac{1}{2}O_2(g) \longrightarrow H_2O(l) \qquad \Delta H_{rxn}^\circ = -285.8 \text{ kJ/mol}$$
$$2C_2H_6(g) + 7O_2(g) \longrightarrow 4CO_2(g) + 6H_2O(l)$$
$$\Delta H_{rxn}^\circ = -3119.6 \text{ kJ/mol}$$

calculate the enthalpy change for the reaction

$$2C(\text{graphite}) + 3H_2(g) \longrightarrow C_2H_6(g)$$

**5.47**   From the following heats of combustion,

$$CH_3OH(l) + \tfrac{3}{2}O_2(g) \longrightarrow CO_2(g) + 2H_2O(l)$$
$$\Delta H_{rxn}^\circ = -726.4 \text{ kJ/mol}$$
$$C(\text{graphite}) + O_2(g) \longrightarrow CO_2(g) \qquad \Delta H_{rxn}^\circ = -393.5 \text{ kJ/mol}$$
$$H_2(g) + \tfrac{1}{2}O_2(g) \longrightarrow H_2O(l) \qquad \Delta H_{rxn}^\circ = -285.8 \text{ kJ/mol}$$

calculate the enthalpy of formation of methanol (CH$_3$OH) from its elements:

$$C(\text{graphite}) + 2H_2(g) + \tfrac{1}{2}O_2(g) \longrightarrow CH_3OH(l)$$

**5.48**   Calculate the standard enthalpy change for the reaction

$$2Al(s) + Fe_2O_3(s) \longrightarrow 2Fe(s) + Al_2O_3(s)$$

given that

$$2Al(s) + \tfrac{3}{2}O_2(g) \longrightarrow Al_2O_3(s) \quad \Delta H_{rxn}^\circ = -1669.8 \text{ kJ/mol}$$
$$2Fe(s) + \tfrac{3}{2}O_2(g) \longrightarrow Fe_2O_3(s) \quad \Delta H_{rxn}^\circ = -822.2 \text{ kJ/mol}$$

## Section 5.6: Standard Enthalpies of Formation

*Review Questions*

**5.49**   What is meant by the standard-state condition?

**5.50**   How are the standard enthalpies of an element and of a compound determined?

**5.51**   What is meant by the standard enthalpy of a reaction?

**5.52**   Write the equation for calculating the enthalpy of a reaction. Define all the terms.

*Problems*

**5.53** Which of the following standard enthalpy of formation values is not zero at 25°C: Na(monoclinic), Ne(g), CH₄(g), S₈(monoclinic), Hg(l), H(g)?

**5.54** The $\Delta H_f^\circ$ values of the two allotropes of oxygen, $O_2$ and $O_3$, are 0 and 142.2 kJ/mol, respectively, at 25°C. Which is the more stable form at this temperature?

**5.55** Which is the more negative quantity at 25°C: $\Delta H_f^\circ$ for $H_2O(l)$ or $\Delta H_f^\circ$ for $H_2O(g)$?

**5.56** Predict the value of $\Delta H_f^\circ$ (greater than, less than, or equal to zero) for these elements at 25°C: (a) $Br_2(g)$, $Br_2(l)$; (b) $I_2(g)$, $I_2(s)$.

**5.57** In general, compounds with negative $\Delta H_f^\circ$ values are more stable than those with positive $\Delta H_f^\circ$ values. $H_2O_2(l)$ has a negative $\Delta H_f^\circ$ (see Appendix 2). Why, then, does $H_2O_2(l)$ have a tendency to decompose to $H_2O(l)$ and $O_2(g)$?

**5.58** Suggest ways (with appropriate equations) that would allow you to measure the $\Delta H_f^\circ$ values of $Ag_2O(s)$ and $CaCl_2(s)$ from their elements. No calculations are necessary.

**5.59** Calculate the heat of decomposition for this process at constant pressure and 25°C:

$$CaCO_3(s) \longrightarrow CaO(s) + CO_2(g)$$

(Look up the standard enthalpy of formation of the reactant and products in Appendix 2.)

**5.60** The standard enthalpies of formation of ions in aqueous solutions are obtained by arbitrarily assigning a value of zero to $H^+$ ions; that is, $\Delta H_f^\circ[H^+(aq)] = 0$. (a) For the following reaction

$$HCl(g) \xrightarrow{H_2O} H^+(aq) + Cl^-(aq) \qquad \Delta H^\circ = -74.9 \text{ kJ/mol}$$

calculate $\Delta H_f^\circ$ for the $Cl^-$ ions. (b) Given that $\Delta H_f^\circ$ for $OH^-$ ions is $-229.6$ kJ/mol, calculate the enthalpy of neutralization when 1 mole of a strong monoprotic acid (such as HCl) is titrated by 1 mole of a strong base (such as KOH) at 25°C.

**5.61** Calculate the heats of combustion for the following reactions from the standard enthalpies of formation listed in Appendix 2:

(a) $2H_2(g) + O_2(g) \longrightarrow 2H_2O(l)$
(b) $2C_2H_2(g) + 5O_2(g) \longrightarrow 4CO_2(g) + 2H_2O(l)$

**5.62** Calculate the heats of combustion for the following reactions from the standard enthalpies of formation listed in Appendix 2:

(a) $C_2H_4(g) + 3O_2(g) \longrightarrow 2CO_2(g) + 2H_2O(l)$
(b) $2H_2S(g) + 3O_2(g) \longrightarrow 2H_2O(l) + 2SO_2(g)$

**5.63** Methanol, ethanol, and *n*-propanol are three common alcohols. When 1.00 g of each of these alcohols is burned in air, heat is liberated as follows: (a) methanol ($CH_3OH$), $-22.6$ kJ; (b) ethanol ($C_2H_5OH$), $-29.7$ kJ; (c) *n*-propanol ($C_3H_7OH$), $-33.4$ kJ. Calculate the heats of combustion of these alcohols in kJ/mol.

**5.64** The standard enthalpy change for the following reaction is 436.4 kJ/mol:

$$H_2(g) \longrightarrow H(g) + H(g)$$

Calculate the standard enthalpy of formation of atomic hydrogen (H).

**5.65** From the standard enthalpies of formation, calculate $\Delta H_{rxn}^\circ$ for the reaction

$$C_6H_{12}(l) + 9O_2(g) \longrightarrow 6CO_2(g) + 6H_2O(l)$$

For $C_6H_{12}(l)$, $\Delta H_f^\circ = -151.9$ kJ/mol.

**5.66** Pentaborane-9 ($B_5H_9$) is a colorless, highly reactive liquid that will burst into flames when exposed to oxygen. The reaction is

$$2B_5H_9(l) + 12O_2(g) \longrightarrow 5B_2O_3(s) + 9H_2O(l)$$

Calculate the kilojoules of heat released per gram of the compound reacted with oxygen. The standard enthalpy of formation of $B_5H_9$ is 73.2 kJ/mol.

**5.67** Determine the amount of heat (in kJ) given off when $1.26 \times 10^4$ g of ammonia is produced according to the equation

$$N_2(g) + 3H_2(g) \longrightarrow 2NH_3(g) \qquad \Delta H_{rxn}^\circ = -92.6 \text{ kJ/mol}$$

Assume that the reaction takes place under standard-state conditions at 25°C.

**5.68** Using the data in Appendix 2, calculate the enthalpy change for the gaseous reaction shown here. (*Hint:* First determine the limiting reactant.)

## Additional Problems

**5.69** The convention of arbitrarily assigning a zero enthalpy value for the most stable form of each element in the standard state at 25°C is a convenient way of dealing with enthalpies of reactions. Explain why this convention cannot be applied to nuclear reactions.

**5.70** Consider the following two reactions:

$$A \longrightarrow 2B \qquad \Delta H_{rxn}^\circ = H_1$$
$$A \longrightarrow C \qquad \Delta H_{rxn}^\circ = H_2$$

Determine the enthalpy change for the process

$$2B \longrightarrow C$$

**5.71** The standard enthalpy change $\Delta H^\circ$ for the thermal decomposition of silver nitrate according to the following equation is $+78.67$ kJ:

$$AgNO_3(s) \longrightarrow AgNO_2(s) + \tfrac{1}{2}O_2(g)$$

The standard enthalpy of formation of $AgNO_3(s)$ is $-123.02$ kJ/mol. Calculate the standard enthalpy of formation of $AgNO_2(s)$.

**5.72** Hydrazine ($N_2H_4$) decomposes according to the following reaction:

$$3N_2H_4(l) \longrightarrow 4NH_3(g) + N_2(g)$$

(a) Given that the standard enthalpy of formation of hydrazine is 50.42 kJ/mol, calculate $\Delta H^\circ$ for its decomposition. (b) Both hydrazine and ammonia burn in oxygen to produce $H_2O(l)$ and $N_2(g)$. Write balanced equations for each of these processes, and calculate $\Delta H^\circ$ for each of them. On a mass basis (per kg), which would be the better fuel: hydrazine or ammonia?

**5.73** Consider the reaction

$$N_2(g) + 3H_2(g) \longrightarrow 2NH_3(g) \qquad \Delta H = -92.6 \text{ kJ/mol}$$

When 2 moles of $N_2$ react with 6 moles of $H_2$ to form 4 moles of $NH_3$ at 1 atm and a certain temperature, there is a decrease in volume equal to 98 L. Calculate $\Delta U$ for this reaction. (The conversion factor is 1 L · atm = 101.3 J.)

**5.74** Calculate the heat released when 2.00 L of $Cl_2(g)$ with a density of 1.88 g/L reacts with an excess of sodium metal at 25°C and 1 atm to form sodium chloride.

**5.75** Photosynthesis produces glucose ($C_6H_{12}O_6$) and oxygen from carbon dioxide and water:

$$6CO_2(g) + 6H_2O(l) \longrightarrow C_6H_{12}O_6(s) + 6O_2(g)$$

(a) How would you determine experimentally the $\Delta H^{\circ}_{rxn}$ value for this reaction? (b) Solar radiation produces about $7.0 \times 10^{14}$ kg of glucose a year on Earth. What is the corresponding $\Delta H^{\circ}$ change?

**5.76** Consider the reaction

$$2Na(s) + 2H_2O(l) \longrightarrow 2NaOH(aq) + H_2(g)$$

When 2 moles of Na react with water at 25°C and 1 atm, the volume of $H_2$ formed is 24.5 L. Calculate the work done in joules when 0.34 g of Na reacts with water under the same conditions. (The conversion factor is 1 L · atm = 101.3 J.)

**5.77** You are given the following data:

$$H_2(g) \longrightarrow 2H(g) \qquad \Delta H^{\circ} = 436.4 \text{ kJ/mol}$$
$$Br_2(g) \longrightarrow 2Br(g) \qquad \Delta H^{\circ} = 192.5 \text{ kJ/mol}$$
$$H_2(g) + Br_2(g) \longrightarrow 2HBr(g) \qquad \Delta H^{\circ} = -72.4 \text{ kJ/mol}$$

Calculate $\Delta H^{\circ}$ for the reaction

$$H(g) + Br(g) \longrightarrow HBr(g)$$

**5.78** Methanol ($CH_3OH$) is an organic solvent and is also used as a fuel in some automobile engines. From the following data, calculate the standard enthalpy of formation of methanol:

$$2CH_3OH(l) + 3O_2(g) \longrightarrow 2CO_2(g) + 4H_2O(l)$$
$$\Delta H^{\circ}_{rxn} = -1452.8 \text{ kJ/mol}$$

**5.79** A 44.0-g sample of an unknown metal at 99.0°C was placed in a constant-pressure calorimeter containing 80.0 g of water at 24.0°C. The final temperature of the system was found to be 28.4°C. Calculate the specific heat of the metal. (The heat capacity of the calorimeter is 12.4 J/°C.)

**5.80** A student mixes 88.6 g of water at 74.3°C with 57.9 g of water at 24.8°C in an insulated flask. What is the final temperature of the combined water?

**5.81** Producer gas (carbon monoxide) is prepared by passing air over red-hot coke:

$$C(s) + \tfrac{1}{2}O_2(g) \longrightarrow CO(g)$$

Water gas (a mixture of carbon monoxide and hydrogen) is prepared by passing steam over red-hot coke:

$$C(s) + H_2O(g) \longrightarrow CO(g) + H_2(g)$$

For many years, both producer gas and water gas were used as fuels in industry and for domestic cooking. The large-scale preparation of these gases was carried out alternately, that is, first producer gas, then water gas, and so on. Using thermochemical reasoning, explain why this procedure was chosen.

**5.82** Compare the heat produced by the complete combustion of 1 mole of methane ($CH_4$) with a mole of water gas (0.50 mole $H_2$ and 0.50 mole CO) under the same conditions. On the basis of your answer, would you prefer methane over water gas as a fuel? Can you suggest two other reasons why methane is preferable to water gas as a fuel?

**5.83** The so-called hydrogen economy is based on hydrogen produced from water using solar energy. The gas is then burned as a fuel:

$$2H_2(g) + O_2(g) \longrightarrow 2H_2O(l)$$

A primary advantage of hydrogen as a fuel is that it is nonpolluting. A major disadvantage is that it is a gas and therefore is harder to store than liquids or solids. Calculate the number of moles of $H_2$ required to produce an amount of energy equivalent

to that produced by the combustion of a gallon of octane ($C_8H_{18}$). The density of octane is 2.66 kg/gal, and its standard enthalpy of formation is −249.9 kJ/mol.

**5.84** Ethanol ($C_2H_5OH$) and gasoline (assumed to be all octane, $C_8H_{18}$) are both used as automobile fuel. If gasoline is selling for $2.20/gal, what would the price of ethanol have to be in order to provide the same amount of heat per dollar? The density and $\Delta H^{\circ}_f$ of octane are 0.7025 g/mL and −249.9 kJ/mol, respectively, and of ethanol are 0.7894 g/mL and −277.0 kJ/mol, respectively (1 gal = 3.785 L).

**5.85** The combustion of how many moles of ethane ($C_2H_6$) would be required to heat 855 g of water from 25.0°C to 98.0°C?

**5.86** If energy is conserved, how can there be an energy crisis?

**5.87** The heat of vaporization of a liquid ($\Delta H_{vap}$) is the energy required to vaporize 1.00 g of the liquid at its boiling point. In one experiment, 60.0 g of liquid nitrogen (boiling point = −196°C) is poured into a Styrofoam cup containing $2.00 \times 10^2$ g of water at 55.3°C. Calculate the molar heat of vaporization of liquid nitrogen if the final temperature of the water is 41.0°C.

**5.88** Explain the cooling effect experienced when ethanol is rubbed on your skin, given that

$$C_2H_5OH(l) \longrightarrow C_2H_5OH(g) \qquad \Delta H^{\circ} = 42.2 \text{ kJ/mol}$$

**5.89** For which of the following reactions does $\Delta H^{\circ}_{rxn} = \Delta H^{\circ}_f$?

(a) $H_2(g) + S(\text{rhombic}) \longrightarrow H_2S(g)$
(b) $C(\text{diamond}) + O_2(g) \longrightarrow CO_2(g)$
(c) $H_2(g) + CuO(s) \longrightarrow H_2O(l) + Cu(s)$
(d) $O(g) + O_2(g) \longrightarrow O_3(g)$

**5.90** Calculate the work done (in joules) when 1.0 mole of water is frozen at 0°C and 1.0 atm. The volumes of 1 mole of water and ice at 0°C are 0.0180 and 0.0196 L, respectively. (The conversion factor is 1 L · atm = 101.3 J.)

**5.91** A certain gas initially at 0.050 L undergoes expansion until its volume is 0.50 L. Calculate the work done (in joules) by the gas if it expands (a) against a vacuum and (b) against a constant pressure of 0.20 atm. (The conversion factor is 1 L · atm = 101.3 J.)

**5.92** Calculate the standard enthalpy of formation for diamond, given that

$$C(\text{graphite}) + O_2(g) \longrightarrow CO_2(g) \qquad \Delta H^{\circ} = -393.5 \text{ kJ/mol}$$
$$C(\text{diamond}) + O_2(g) \longrightarrow CO_2(g) \qquad \Delta H^{\circ} = -395.4 \text{ kJ/mol}$$

**5.93** (a) For most efficient use, refrigerator freezer compartments should be fully packed with food. What is the thermochemical basis for this recommendation? (b) Starting at the same temperature, tea and coffee remain hot longer in a thermal flask than chicken noodle soup. Explain.

**5.94** Calculate the standard enthalpy change for the fermentation process, in which glucose ($C_6H_{12}O_6$) is converted to ethanol ($C_2H_5OH$) and carbon dioxide.

**5.95** Portable hot packs are available for skiers and people engaged in other outdoor activities in a cold climate. The air-permeable paper packet contains a mixture of powdered iron, sodium chloride, and other components, all moistened by a little water. The exothermic reaction that produces the heat is a very common one—the rusting of iron:

$$4Fe(s) + 3O_2(g) \longrightarrow 2Fe_2O_3(s)$$

When the outside plastic envelope is removed, $O_2$ molecules penetrate the paper, causing the reaction to begin. A typical packet contains 250 g of iron to warm your hands or feet for up to 4 hours. How much heat (in kJ) is produced by this reaction? (*Hint:* See Appendix 2 for $\Delta H_f^\circ$ values.)

5.96    A man ate 0.50 pound of cheese (an energy intake of $4 \times 10^3$ kJ). Suppose that none of the energy was stored in his body. What mass (in grams) of water would he need to perspire in order to maintain his original temperature? (It takes 44.0 kJ to vaporize 1 mole of water.)

5.97    The total volume of the Pacific Ocean is estimated to be $7.2 \times 10^8$ km$^3$. A medium-sized atomic bomb produces $1.0 \times 10^{15}$ J of energy upon explosion. Calculate the number of atomic bombs needed to release enough energy to raise the temperature of the water in the Pacific Ocean by 1°C.

5.98    A woman expends 95 kJ of energy walking a kilometer. The energy is supplied by the metabolic breakdown of food, which has an efficiency of 35 percent. How much energy does she save by walking the kilometer instead of driving a car that gets 8.2 km per liter of gasoline (approximately 20 mi/gal)? The density of gasoline is 0.71 g/mL, and its enthalpy of combustion is −49 kJ/g.

5.99    The carbon dioxide exhaled by sailors in a submarine is often removed by reaction with an aqueous lithium hydroxide solution. (a) Write a balanced equation for this process. (*Hint:* The products are water and a soluble salt.) (b) If every sailor consumes $1.2 \times 10^4$ kJ of energy every day and assuming that this energy is totally supplied by the metabolism of glucose ($C_6H_{12}O_6$), calculate the amounts of $CO_2$ produced and LiOH required to purify the air.

5.100   The enthalpy of combustion of benzoic acid ($C_6H_5COOH$) is commonly used as the standard for calibrating constant-volume bomb calorimeters; its value has been accurately determined to be −3226.7 kJ/mol. When 1.9862 g of benzoic acid are burned in a calorimeter, the temperature rises from 21.84°C to 25.67°C. What is the heat capacity of the bomb? (Assume that the quantity of water surrounding the bomb is exactly 2000 g.)

5.101   Calcium oxide (CaO) is used to remove sulfur dioxide generated by coal-burning power stations:

$$2CaO(s) + 2SO_2(g) + O_2(g) \longrightarrow 2CaSO_4(s)$$

Calculate the enthalpy change if $6.6 \times 10^5$ g of $SO_2$ is removed by this process.

5.102   Glauber's salt, sodium sulfate decahydrate ($Na_2SO_4 \cdot 10H_2O$), undergoes a phase transition (that is, melting or freezing) at a convenient temperature of about 32°C:

$$Na_2SO_4 \cdot 10H_2O(s) \longrightarrow Na_2SO_4 \cdot 10H_2O(l)$$
$$\Delta H^\circ = 74.4 \text{ kJ/mol}$$

As a result, this compound is used to regulate the temperature in homes. It is placed in plastic bags in the ceiling of a room. During the day, the endothermic melting process absorbs heat from the surroundings, cooling the room. At night, it gives off heat as it freezes. Calculate the mass of Glauber's salt in kilograms needed to lower the temperature of air in a room by 8.2°C. The mass of air in the room is 605.4 kg; the specific heat of air is 1.2 J/g · °C.

5.103   An excess of zinc metal is added to 50.0 mL of a 0.100 *M* AgNO$_3$ solution in a constant-pressure calorimeter like the one pictured in Figure 5.8. As a result of the reaction

$$Zn(s) + 2Ag^+(aq) \longrightarrow Zn^{2+}(aq) + 2Ag(s)$$

the temperature rises from 19.25°C to 22.17°C. If the heat capacity of the calorimeter is 98.6 J/°C, calculate the enthalpy change for the given reaction on a molar basis. Assume that the density and specific heat of the solution are the same as those for water, and ignore the specific heats of the metals.

5.104   (a) A person drinks four glasses of cold water (3.0°C) every day. The volume of each glass is $2.5 \times 10^2$ mL. How much heat (in kJ) does the body have to supply to raise the temperature of the water to 37°C, the body temperature? (b) How much heat would your body lose if you were to ingest $8.0 \times 10^2$ g of snow at 0°C to quench your thirst? (The amount of heat necessary to melt snow is 6.01 kJ/mol.)

5.105   A driver's manual states that the stopping distance quadruples as the speed doubles; that is, if it takes 30 ft to stop a car moving at 25 mph, then it would take 120 ft to stop a car moving at 50 mph. Justify this statement by using mechanics and the first law of thermodynamics. [Assume that when a car is stopped, its kinetic energy ($\frac{1}{2}mu^2$) is totally converted to heat.]

5.106   At 25°C the standard enthalpy of formation of HF($aq$) is −320.1 kJ/mol; of OH$^-$($aq$), it is −229.6 kJ/mol; of F$^-$($aq$), it is −329.1 kJ/mol; and of $H_2O(l)$, it is −285.8 kJ/mol.

(a) Calculate the standard enthalpy of neutralization of HF($aq$):

$$HF(aq) + OH^-(aq) \longrightarrow F^-(aq) + H_2O(l)$$

(b) Using the value of −56.2 kJ as the standard enthalpy change for the reaction

$$H^+(aq) + OH^-(aq) \longrightarrow H_2O(l)$$

calculate the standard enthalpy change for the reaction

$$HF(aq) \longrightarrow H^+(aq) + F^-(aq)$$

5.107   Why are cold, damp air and hot, humid air more uncomfortable than dry air at the same temperatures? [The specific heats of water vapor and air are approximately 1.9 J/(g · °C) and 1.0 J/(g · °C), respectively.]

5.108   From the enthalpy of formation for $CO_2$ and the following information, calculate the standard enthalpy of formation for carbon monoxide (CO).

$$CO(g) + \frac{1}{2}O_2(g) \longrightarrow CO_2(g) \qquad \Delta H^\circ = -283.0 \text{ kJ/mol}$$

Why can't we obtain the standard enthalpy of formation directly by measuring the enthalpy of the following reaction?

$$C(graphite) + \frac{1}{2}O_2(g) \longrightarrow CO(g)$$

5.109   A 46-kg person drinks 500 g of milk, which has a "caloric" value of approximately 3.0 kJ/g. If only 17 percent of the energy in milk is converted to mechanical work, how high (in meters) can the person climb based on this energy intake? [*Hint:* The work done in ascending is given by *mgh*, where *m* is the mass (in kg), *g* is the gravitational acceleration (9.8 m/s$^2$), and *h* is the height (in meters).]

5.110   The height of Niagara Falls on the American side is 51 m. (a) Calculate the potential energy of 1.0 g of water at the top of the falls relative to the ground level. (b) What is the speed of the falling water if all the potential energy is converted to kinetic energy? (c) What would be the increase in temperature of the water if all the kinetic energy were converted to heat? (See Problem 5.109 for information.)

5.111   In the nineteenth century two scientists named Dulong and Petit noticed that for a solid element, the product of its molar mass and its specific heat is approximately 25 J/°C. This observation, now called Dulong and Petit's law, was used to estimate the specific heat of metals. Verify the law for the metals listed in Table 5.2. The law does not apply to one of the metals. Which one is it? Why?

5.112   Determine the standard enthalpy of formation of ethanol ($C_2H_5OH$) from its standard enthalpy of combustion (−1367.4 kJ/mol).

**5.113** Acetylene ($C_2H_2$) and benzene ($C_6H_6$) have the same empirical formula. In fact, benzene can be made from acetylene as follows:

$$3C_2H_2(g) \longrightarrow C_6H_6(l)$$

The enthalpies of combustion for $C_2H_2$ and $C_6H_6$ are $-1299.4$ and $-3267.4$ kJ/mol, respectively. Calculate the standard enthalpies of formation of $C_2H_2$ and $C_6H_6$ and hence the enthalpy change for the formation of $C_6H_6$ from $C_2H_2$.

**5.114** Ice at 0°C is placed in a Styrofoam cup containing 361 g of a soft drink at 23°C. The specific heat of the drink is about the same as that of water. Some ice remains after the ice and soft drink reach an equilibrium temperature of 0°C. Determine the mass of ice that has melted. Ignore the heat capacity of the cup. (*Hint:* It takes 334 J to melt 1 g of ice at 0°C.)

**5.115** A gas company in Massachusetts charges 27 cents for a mole of natural gas ($CH_4$). Calculate the cost of heating 200 mL of water (enough to make a cup of coffee or tea) from 20°C to 100°C. Assume that only 50 percent of the heat generated by the combustion is used to heat the water; the rest of the heat is lost to the surroundings.

**5.116** How much metabolic energy must a 5.2-g hummingbird expend to fly to a height of 12 m? (See the hint in Problem 5.109.)

**5.117** Acetylene ($C_2H_2$) can be made by combining calcium carbide ($CaC_2$) with water. (a) Write an equation for the reaction. (b) What is the maximum amount of heat (in joules) that can be obtained from the combustion of acetylene, starting with 74.6 g of $CaC_2$?

**5.118** The average temperature in deserts is high during the day but quite cool at night, whereas that in regions along the coastline is more moderate. Explain.

**5.119** Both glucose and fructose are simple sugars with the same molecular formula of $C_6H_{12}O_6$. Sucrose ($C_{12}H_{22}O_{11}$), or table sugar, consists of a glucose molecule bonded to a fructose molecule (a water molecule is eliminated in the formation of sucrose). (a) Calculate the energy released when a 2.0-g glucose tablet is burned in air. (b) To what height can a 65-kg person climb after ingesting such a tablet, assuming only 30 percent of the energy released is available for work. (See the hint for Problem 5.109.) Repeat the calculations for a 2.0-g sucrose tablet.

**5.120** About $6.0 \times 10^{13}$ kg of $CO_2$ is fixed (converted to more complex organic molecules) by photosynthesis every year. (a) Assuming all the $CO_2$ ends up as glucose ($C_6H_{12}O_6$), calculate the energy (in kJ) stored by photosynthesis per year. (b) A typical coal-burning electric power station generates about $2.0 \times 10^6$ W per year. How many such stations are needed to generate the same amount of energy as that captured by photosynthesis (1 W = 1 J/s)?

**5.121** When 1.034 g of naphthalene ($C_{10}H_8$) is burned in a constant-volume bomb calorimeter at 298 K, 41.56 kJ of heat is evolved. Calculate $\Delta U$ and $w$ for the reaction on a molar basis.

**5.122** From a thermochemical point of view, explain why a carbon dioxide fire extinguisher or water should not be used on a magnesium fire.

**5.123** Consider the reaction

$$2H_2(g) + O_2(g) \longrightarrow 2H_2O(l)$$

Under atmospheric conditions (1.00 atm) it was found that the formation of water resulted in a decrease in volume equal to 73.4 L. Calculate $\Delta U$ for the process. $\Delta H = -571.6$ kJ/mol. (The conversion factor is 1 L · atm = 101.3 J.)

**5.124** *Lime* is a term that includes calcium oxide (CaO, also called quicklime) and calcium hydroxide [$Ca(OH)_2$, also called slaked lime]. It is used in the steel industry to remove acidic impurities, in air-pollution control to remove acidic oxides such as $SO_2$, and in water treatment. Quicklime is made industrially by heating limestone ($CaCO_3$) above 2000°C:

$$CaCO_3(s) \longrightarrow CaO(s) + CO_2(g) \qquad \Delta H° = 177.8 \text{ kJ/mol}$$

Slaked lime is produced by treating quicklime with water:

$$CaO(s) + H_2O(l) \longrightarrow Ca(OH)_2(s) \qquad \Delta H° = -65.2 \text{ kJ/mol}$$

The exothermic reaction of quicklime with water and the rather small specific heats of both quicklime [0.946 J/(g · °C)] and slaked lime [1.20 J/(g · °C)] make it hazardous to store and transport lime in vessels made of wood. Wooden sailing ships carrying lime would occasionally catch fire when water leaked into the hold. (a) If a 500.0-g sample of water reacts with an equimolar amount of CaO (both at an initial temperature of 25°C), what is the final temperature of the product, $Ca(OH)_2$? Assume that the product absorbs all the heat released in the reaction. (b) Given that the standard enthalpies of formation of CaO and $H_2O$ are $-635.6$ and $-285.8$ kJ/mol, respectively, calculate the standard enthalpy of formation of $Ca(OH)_2$.

**5.125** A 4.117-g impure sample of glucose ($C_6H_{12}O_6$) was burned in a constant-volume calorimeter having a heat capacity of 19.65 kJ/°C. If the rise in temperature is 3.134°C, calculate the percent by mass of the glucose in the sample. Assume that the impurities are unaffected by the combustion process and that $\Delta U = \Delta H$. See Appendix 2 for thermodynamic data.

**5.126** The combustion of 0.4196 g of a hydrocarbon releases 17.55 kJ of heat. The masses of the products are $CO_2 = 1.419$ g and $H_2O = 0.290$ g. (a) What is the empirical formula of the compound? (b) If the approximate molar mass of the compound is 76 g/mol, calculate its standard enthalpy of formation.

**5.127** In a constant-pressure calorimetry experiment, a reaction gives off 21.8 kJ of heat. The calorimeter contains 150 g of water, initially at 23.4°C. What is the final temperature of the water? The heat capacity of the calorimeter is negligibly small.

**5.128** At 850°C, $CaCO_3$ undergoes substantial decomposition to yield CaO and $CO_2$. Assuming that the $\Delta H_f°$ values of the reactant and products are the same at 850°C as they are at 25°C, calculate the enthalpy change (in kJ) if 66.8 g of $CO_2$ is produced in one reaction.

**5.129** For reactions in condensed phases (liquids and solids), the difference between $\Delta H$ and $\Delta U$ is usually quite small. This statement holds for reactions carried out under atmospheric conditions. For certain geochemical processes, however, the external pressure may be so great that $\Delta H$ and $\Delta U$ can differ by a significant amount. A well-known example is the slow conversion of graphite to diamond under Earth's surface. Calculate $\Delta H - \Delta U$ for the conversion of 1 mole of graphite to 1 mole of diamond at a pressure of 50,000 atm. The densities of graphite and diamond are 2.25 g/cm³ and 3.52 g/cm³, respectively.

**5.130** Which of the constant-pressure processes given here has the smallest difference between $\Delta H$ and $\Delta U$: (a) water $\longrightarrow$ water vapor, (b) water $\longrightarrow$ ice, (c) ice $\longrightarrow$ water vapor? Explain.

**5.131** Construct a table with the headings $q$, $w$, $\Delta U$, and $\Delta H$. For each of the following processes, deduce whether each of the quantities listed is positive (+), negative (–), or zero (0): (a) freezing of benzene, (b) reaction of sodium with water, (c) boiling of liquid ammonia, (d) melting of ice, (e) expansion of a gas at constant temperature.

5.132 Give an example for each of the following situations: (a) adding heat to a system raises its temperature, (b) adding heat to a system does not change its temperature, and (c) a system's temperature changes despite no heat being added to it or removed from it.

5.133 Metabolic activity in the human body releases approximately $1.0 \times 10^4$ kJ of heat per day. Assume that a 55-kg body has the same specific heat as water; how much would the body temperature rise if it were an isolated system? How much water must the body eliminate as perspiration to maintain the normal body temperature (98.6°F)? Comment on your results. (The heat of vaporization of water is 2.41 kJ/g.)

5.134 A 3.52-g sample of ammonium nitrate ($NH_4NO_3$) was added to 80.0 mL of water in a constant-pressure calorimeter of negligible heat capacity. As a result, the temperature of the solution decreased from 21.6°C to 18.1°C. Calculate the heat of solution ($\Delta H_{soln}$) in kJ/mol:

$$NH_4NO_3(s) \longrightarrow NH_4^+(aq) + NO_3^-(aq)$$

Assume the specific heat of the solution is the same as that of water.

5.135 A quantity of 50.0 mL of 0.200 $M$ Ba(OH)$_2$ is mixed with 50.0 mL of 0.400 $M$ HNO$_3$ in a constant-pressure calorimeter having a heat capacity of 496 J/°C. The initial temperature of both solutions is the same at 22.4°C. What is the final temperature of the mixed solution? Assume that the specific heat of the solutions is the same as that of water and the molar heat of neutralization is −56.2 kJ/mol.

5.136 A quantity of 85.0 mL of 0.600 $M$ HCl is mixed with 85.0 mL of 0.600 $M$ KOH in a constant-pressure calorimeter. The initial temperature of both solutions is the same at 17.35°C, and the final temperature of the mixed solution is 19.02°C. What is the heat capacity of the calorimeter? Assume that the specific heat of the solutions is the same as that of water and the molar heat of neutralization is −56.2 kJ/mol.

# PRE-PROFESSIONAL PRACTICE EXAM PROBLEMS: PHYSICAL AND BIOLOGICAL SCIENCES

A bomb calorimeter was calibrated by burning 1.013 g of benzoic acid ($C_7H_6O_2$) ($\Delta U_{comb} = 3.221 \times 10^3$ kJ/mol). The temperature change in the calorimeter during the calibration combustion was 5.19°C. A nutritional chemist then used the calibrated calorimeter to determine the energy content of food. The chemist carefully dried a sample of food and placed 0.8996 g of the sample in the calorimeter with sufficient oxygen for the combustion to go to completion. Combustion of the food sample caused the temperature of the calorimeter to increase by 4.42°C.

1. Approximately how many moles of $O_2$ gas were consumed in the calibration combustion?

   a) 0.008
   b) 0.1
   c) 0.2
   d) 0.06

2. What is the heat capacity ($C_V$) of the calorimeter?

   a) 5.15 kJ/°C
   b) 5.08 kJ/°C
   c) 5.12 kJ/°C
   d) 4.97 kJ/°C

3. What is the energy content of the food?

   a) 22.8 kJ/g
   b) 4.97 kJ/g
   c) 25.3 kJ/g
   d) 0.201 kJ/g

4. What would be the effect on the result if the food sample were not completely dried prior to being placed in the calorimeter?

   a) The combustion of the sample would be incomplete.
   b) The calculated energy content per gram would be too low.
   c) The calculated energy content per gram would be too high.
   d) There would be no effect on the result.

# ANSWERS TO IN-CHAPTER MATERIALS

## Answers to Practice Problems

**5.1A** (a) $1.13 \times 10^3$ J, (b) 4. **5.1B** (a) 372 m/s, (b) neither. **5.2A** $-4.32 \times 10^4$ kJ. **5.2B** $6.95 \times 10^5$ kJ, heat is absorbed. **5.3A** $8.174 \times 10^4$ kJ. **5.3B** $1.71 \times 10^3$ g. **5.4A** 151 kJ. **5.4B** 52.2°C. **5.5A** 28°C. **5.5B** 42 g. **5.6A** 114 g. **5.6B** 2.36 J/g · °C. **5.7A** 14.5 kJ/g. **5.7B** 10.5°C rise. **5.8A** −1103 kJ/mol. **5.8B** 113.2 kJ/mol. **5.9A** 177.8 kJ/mol. **5.9B** −697.6 kJ/mol. **5.10A** 87.3 kJ/mol. **5.10B** $NH_3(g) + HCl(g)$, 176.8 kJ/mol.

## Answers to Checkpoints

**5.1.1** a. **5.1.2** b. **5.1.3** a. **5.1.4** e. **5.1.5** c. **5.1.6** b, d, e. **5.2.1** e. **5.2.2** c. **5.2.3** c. **5.2.4** a, d, e. **5.3.1** b. **5.3.2** a. **5.4.1** b. **5.4.2** d. **5.4.3** a. **5.4.4** a. **5.5.1** c. **5.5.2** a. **5.6.1** b. **5.6.2** e. **5.6.3** c, e. **5.6.4** e.

## Answers to Applying What You've Learned

**a)** $1.2 \times 10^2$ Cal. **b)** 59°C. **c)** 380 Cal. **d)** −35,340 kJ/mol (using $\Delta H_f^\circ$ values from Appendix 2, $H_2O(l) \longrightarrow H_2O(g)$, $\Delta H^\circ = +44.0$ kJ/mol).

# 6

# Quantum Theory and the Electronic Structure of Atoms

*Lasers are used in a variety of surgical procedures, including cosmetic procedures. LASIK surgery, shown here, is done to improve eyesight.*

# How Our Understanding of Electronic Structure Has Advanced Medicine

Over the past two decades, the use of lasers has revolutionized many medical procedures, including the excision of malignant tumors, the treatment of the symptoms associated with an enlarged prostate, and a wide variety of cosmetic procedures. The benefits of laser surgery typically include smaller incisions, less bleeding, less pain, and shorter recovery times than with traditional surgical methods. Among the most popular laser procedures is LASIK surgery to improve eyesight. In LASIK, a nearly circular incision is made in the cornea, creating a hinged flap on the surface of the eyeball. The flap is then lifted, and a laser is used to reshape the cornea by selective ablation or vaporization of the underlying corneal tissue. The flap is then replaced, conforming to the reshaped cornea. Many patients report an immediate improvement in vision, and typical recovery times are very short.

**Student Annotation:** LASIK is an acronym for LASer-assisted In situ Keratomileusis.

Corneal flap

Hinge

The light emitted by a laser is the result of electronic transitions and is powerful enough to vaporize biological tissue. Although the nuclear model that Rutherford proposed based on his gold-foil experiment specified the location of the protons and the neutrons in an atom, it failed to describe the location or behavior of the *electrons*. Early in the twentieth century, the application of a radical new theory in physics called *quantum theory*, and the ingenious interpretation of experimental evidence by Max Planck, Albert Einstein, and others, led to our current understanding of the *electronic structure of atoms*. This understanding of electronic structure is what makes such things as lasers possible.

Laser pulses

Cornea flattened

Flap replaced

Central cornea flattened

At the end of this chapter, you will be able to answer a variety of questions about the nature of light and the electronic structure of the atom. [▶▶ page 247]

## In This Chapter, You Will Learn:

About some of the properties of electromagnetic radiation or *light* and how these properties have been used to study and elucidate the electronic structure of atoms. You will also learn how to determine the arrangement of electrons in a particular atom.

### Before you begin, you should review

- Tracking units [◀◀ Section 1.6]
- The nuclear model of the atom [◀◀ Section 2.2]

# 6.1 The Nature of Light

When we say "light," we generally mean *visible* light, which is the light we can detect with our eyes. Visible light, however, is only a small part of the continuum of radiation that comprises the ***electromagnetic spectrum.*** In addition to visible light, the electromagnetic spectrum includes radio waves, microwave radiation, infrared and ultraviolet radiation, X rays, and gamma rays, as shown in Figure 6.1. Some of these terms may be familiar to you. For instance, the danger of exposure to ultraviolet radiation is why you need to use sunscreen. You may have used microwave radiation from a microwave oven to reheat food or to pop popcorn; you may have had X rays during a routine dental checkup or after breaking a bone; and you may recall from Chapter 2 that gamma rays are emitted from some radioactive materials. Although these phenomena may seem very different from each other and from visible light, they all are the transmission of energy in the form of *waves*.

## Properties of Waves

The fundamental properties of waves are illustrated in Figure 6.2. Waves are characterized by their wavelength, frequency, and amplitude. ***Wavelength*** $\lambda$ (lambda) is the distance between identical points on successive waves (e.g., successive peaks or successive troughs). The ***frequency*** $\nu$ (nu) is the *number* of waves that pass through a particular point in 1 second. ***Amplitude*** is the vertical distance from the midline of a wave to the top of the peak or the bottom of the trough.

> **Student Annotation:** The speed of light is *defined* to be 2.99792458 × 10⁸ m/s. It is, therefore, an *exact* number and usually does not limit the number of significant figures in a calculated result. In most calculations, however, the speed of light is rounded to three significant figures: $c = 3.00 \times 10^8$ m/s.

The speed of a wave depends on the type of wave and the nature of the medium through which the wave is traveling (e.g., air, water, or a vacuum). The speed of light through a vacuum, $c$, is $2.99792458 \times 10^8$ m/s. The speed, wavelength, and frequency of a wave are related by the equation

Equation 6.1 $\qquad\qquad\qquad c = \lambda \nu$

> **Student Annotation:** Frequency is expressed as cycles per second, or simply *reciprocal seconds* (s⁻¹), which is also known as *hertz* (Hz).

where $\lambda$ and $\nu$ are expressed in meters (m) and reciprocal seconds (s$^{-1}$), respectively. While wavelength in meters is convenient for this equation, the units customarily used to express the wavelength of electromagnetic radiation depend on the type of radiation and the magnitude of the corresponding wavelength. The wavelength of visible light, for instance, is on the order of nanometers (nm, or $10^{-9}$ m), and that of microwave radiation is on the order of centimeters (cm, or $10^{-2}$ m).

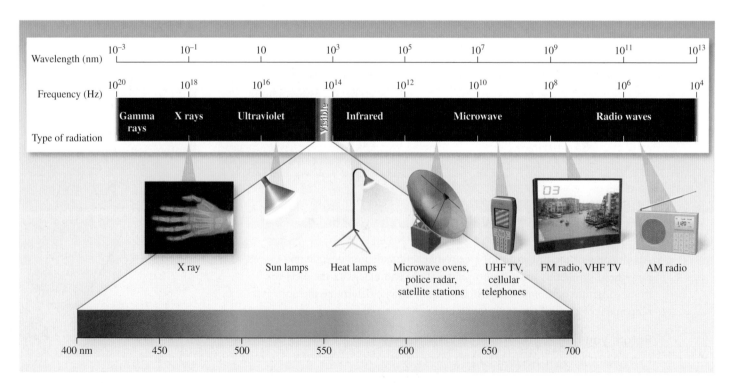

**Figure 6.1**    Electromagnetic spectrum. Each type of radiation is spread over a specific range of wavelengths (and frequencies). Visible light ranges from 400 nm (violet) to 700 nm (red).

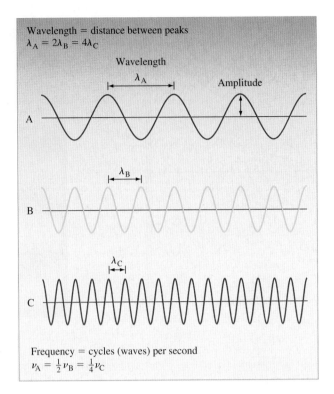

Wavelength = distance between peaks
$\lambda_A = 2\lambda_B = 4\lambda_C$

Wavelength

$\lambda_A$

Amplitude

A

$\lambda_B$

B

$\lambda_C$

C

Frequency = cycles (waves) per second
$\nu_A = \frac{1}{2}\nu_B = \frac{1}{4}\nu_C$

**Figure 6.2** Characteristics of waves: wavelength, amplitude, and frequency.

## The Electromagnetic Spectrum

In 1873 James Clerk Maxwell proposed that visible light consisted of electromagnetic waves. According to Maxwell's theory, an **electromagnetic wave** has an electric field component and a magnetic field component. These two components have the same wavelength and frequency, and hence the same speed, but they travel in mutually perpendicular planes (Figure 6.3). The significance of Maxwell's theory is that it provides a mathematical description of the general behavior of light. In particular, his model accurately describes how energy in the form of radiation can be propagated through space as oscillating electric and magnetic fields.

## The Double-Slit Experiment

A simple yet convincing demonstration of the wave nature of light is the phenomenon of *interference.* When a light source passes through a narrow opening, called a *slit,* a bright line is generated in the path of the light through the slit. When the same light source passes through two closely spaced slits, however, as shown in Figure 6.4, the result is not two bright lines, one in the path of each slit, but rather a series of light and dark lines known as an *interference pattern.* When the light sources recombine after passing through the slits, they do so *constructively* where the two waves are *in phase* (giving rise to the light lines) and *destructively* where the waves are *out of phase* (giving rise to the dark lines). Constructive interference and destructive interference are properties of waves.

The various types of electromagnetic radiation in Figure 6.1 differ from one another in wavelength and frequency. Radio waves, which have long wavelengths and low frequencies, are

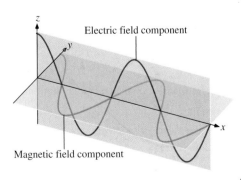

$z$

Electric field component

$y$

$x$

Magnetic field component

**Figure 6.3** Electric field and magnetic field components of an electromagnetic wave. These two components have the same wavelength, frequency, and amplitude, but they vibrate in two mutually perpendicular planes.

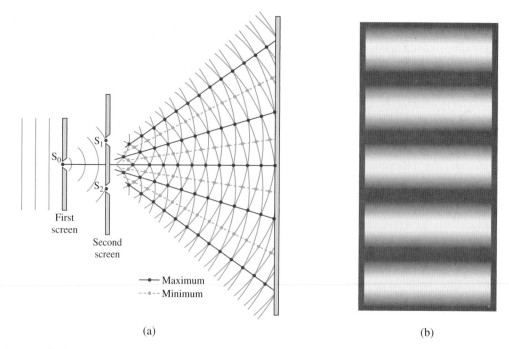

**Figure 6.4**   Double-slit experiment. (a) Red lines correspond to the maximum intensity resulting from constructive interference. Dashed blue lines correspond to the minimum intensity resulting from destructive interference. (b) Interference pattern with alternating bright and dark lines.

emitted by large antennas, such as those used by broadcasting stations. The shorter, visible light waves are produced by the motions of electrons within atoms and molecules. The shortest waves, which also have the highest frequency, are $\gamma$ (gamma) rays, which result from nuclear processes [◄◄ Section 2.2]. As we will see shortly, the higher the frequency, the more energetic the radiation. Thus, ultraviolet radiation, X rays, and $\gamma$ rays are high-energy radiation, whereas infrared radiation, microwave radiation, and radio waves are low-energy radiation.

Sample Problem 6.1 illustrates the conversion between wavelength and frequency.

---

## Sample Problem 6.1

One type of laser used in the treatment of vascular skin lesions is a neodymium-doped yttrium aluminum garnet or Nd:YAG laser. The wavelength commonly used in these treatments is 532 nm. What is the frequency of this radiation?

**Strategy**  Wavelength and frequency are related by Equation 6.1 ($c = \lambda\nu$), so we must rearrange Equation 6.1 to solve for frequency. Because we are given the wavelength of the electromagnetic radiation in nanometers, we must convert this wavelength to meters and use $c = 3.00 \times 10^8$ m/s.

**Setup**  Solving for frequency gives $\nu = c/\lambda$. Next we convert the wavelength to meters:

$$\lambda \text{ (in meters)} = 532 \ \cancel{\text{nm}} \times \frac{1 \times 10^{-9} \text{ m}}{1 \ \cancel{\text{nm}}} = 5.32 \times 10^{-7} \text{ m}$$

**Think About It** Make sure your units cancel properly. A common error in this type of problem is neglecting to convert wavelength to meters.

**Solution**

$$\nu = \frac{3.00 \times 10^8 \text{ m/s}}{5.32 \times 10^{-7} \text{ m}} = 5.64 \times 10^{14} \text{ s}^{-1}$$

---

**Practice Problem A**  What is the wavelength (in meters) of an electromagnetic wave whose frequency is $1.61 \times 10^{12}$ s$^{-1}$?

**Practice Problem B**  What is the frequency (in reciprocal seconds) of electromagnetic radiation with a wavelength of 1.03 cm?

---

| **Checkpoint 6.1** | **The Nature of Light** |

**6.1.1** Calculate the wavelength of light with frequency $3.45 \times 10^{14} \text{ s}^{-1}$.

   a) $1.15 \times 10^{-6}$ nm

   b) $1.04 \times 10^{23}$ nm

   c) $8.70 \times 10^{2}$ nm

   d) 115 nm

   e) $9.66 \times 10^{-24}$ nm

**6.1.2** Calculate the frequency of light with wavelength 126 nm.

   a) $2.38 \times 10^{15} \text{ s}^{-1}$

   b) $4.20 \times 10^{-16} \text{ s}^{-1}$

   c) $37.8 \text{ s}^{-1}$

   d) $2.65 \times 10^{-2} \text{ s}^{-1}$

   e) $3.51 \times 10^{19} \text{ s}^{-1}$

**6.1.3** When traveling through a translucent medium, such as glass, light travels more slowly than it travels through a vacuum. Red light with a wavelength of 684 nm travels through Pyrex glass with a frequency of $2.92 \times 10^{14} \text{ s}^{-1}$. Calculate the speed of this light.

   a) $3.00 \times 10^{8}$ m/s

   b) $2.00 \times 10^{8}$ m/s

   c) $2.92 \times 10^{6}$ m/s

   d) $4.23 \times 10^{7}$ m/s

   e) $2.23 \times 10^{8}$ m/s

**6.1.4** Of the waves pictured, _____ has the greatest frequency, _____ has the greatest wavelength, and _____ has the greatest amplitude.

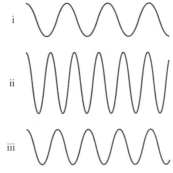

   a) i, ii, iii

   b) i, iii, ii

   c) ii, i, ii

   d) ii, i, iii

   e) ii, iii, ii

---

| **6.2** | ## Quantum Theory |

Early attempts by nineteenth-century physicists to figure out the structure of the atom met with only limited success. This was largely because they were attempting to understand the behavior of subatomic particles using the laws of classical physics that govern the behavior of macroscopic objects. It took a long time to realize—and an even longer time to accept—that the properties of atoms are *not* governed by the same physical laws as larger objects.

### Quantization of Energy

When a solid is heated, it emits electromagnetic radiation, known as ***blackbody radiation,*** over a wide range of wavelengths. The red glow of the element of an electric stove and the bright white light of a tungsten lightbulb are examples of blackbody radiation. Measurements taken in the latter part of the nineteenth century showed that the amount of energy given off by an object at a certain temperature depends on the wavelength of the emitted radiation. Attempts to account for this dependence in terms of established wave theory and thermodynamic laws were only partially successful. One theory was able to explain short-wavelength dependence but failed to account for the longer wavelengths. Another theory accounted for the longer wavelengths but failed for short wavelengths. With no *one* theory that could explain both observations, it seemed that something fundamental was missing from the laws of classical physics.

In 1900, Max Planck[1] provided the solution and launched a new era in physics with an idea that departed drastically from accepted concepts. Classical physics assumed that radiant energy was continuous; that is, it could be emitted or absorbed in any amount. Based on data from blackbody radiation experiments, Planck proposed that radiant energy could only be emitted or absorbed in discrete quantities, like small packages or bundles. Planck gave the name ***quantum*** to the smallest quantity of energy that can be emitted (or absorbed) in the form of electromagnetic radiation. The energy $E$ of a single quantum of energy is given by

---

1. Max Karl Ernst Ludwig Planck (1858–1947). German physicist. Planck received the Nobel Prize in Physics in 1918 for his quantum theory. He also made significant contributions in thermodynamics and other areas of physics.

Equation 6.2                                     $E = h\nu$

where $h$ is called *Planck's constant* and $\nu$ is the *frequency* of the radiation. The value of Planck's constant is $6.63 \times 10^{-34}$ J · s.

According to quantum theory, energy is always emitted in whole-number multiples of $h\nu$. At the time Planck presented his theory, he could not explain why energies should be fixed or quantized in this manner. Starting with this hypothesis, however, he had no difficulty correlating the experimental data for the emission by solids over the entire range of wavelengths; the experimental data supported his new *quantum theory*.

The idea that energy is *quantized* rather than *continuous* may seem strange, but the concept of quantization has many everyday analogies. For example, vending machines dispense cans or bottles of soft drinks only in whole numbers (you can't buy part of a can or bottle from a machine). Each can or bottle is a quantum of its soft drink. Even processes in living systems involve quantized phenomena. The eggs laid by hens are quanta (hens lay only whole eggs). Similarly, when a dog or cat gives birth to a litter, the number of offspring is always an integer. Each puppy or kitten is a quantum of that animal. Planck's quantum theory revolutionized physics. Indeed, the flurry of research that ensued altered our concept of nature forever.

## Bringing Chemistry to Life

### Laser Pointers

The laser pointers that have become so common typically emit radiation in the red region of the visible spectrum with output wavelengths ranging from 630 to 680 nm. Low prices and availability have made the devices popular not only with instructors and other speakers, but with teenagers and even children—raising some significant safety concerns. Although the human blink reflex generally is sufficient to protect against serious or permanent injury to the eye by one of these devices, intentional prolonged exposure of the eye to the beam from a laser pointer can be dangerous. Of particular concern are the new green laser pointers that emit a wavelength of 532 nm. The lasers in these devices also produce radiation in the infrared region of the electromagnetic spectrum (1064 nm), but they are equipped with filters to prevent the emission of infrared radiation. However, some of the inexpensive imported lasers do not bear adequate safety labeling and the filters are easily removed—potentially resulting in the emission of dangerous radiation.

### Sample Problem 6.2

How much more energy per photon is there in green light of wavelength 532 nm than in red light of wavelength 635 nm?

**Strategy** Convert each wavelength to frequency using Equation 6.1, and then calculate the energy per photon using Equation 6.2.

**Setup** Convert the wavelengths to meters:

$$532 \text{ nm} \times \frac{1 \times 10^{-9} \text{ m}}{1 \text{ nm}} = 5.32 \times 10^{-7} \text{ m}$$

$$634 \text{ nm} \times \frac{1 \times 10^{-9} \text{ m}}{1 \text{ nm}} = 6.35 \times 10^{-7} \text{ m}$$

Planck's constant, $h$, is $6.63 \times 10^{-34}$ J · s.

**Solution** For 532 nm,

$$\nu = \frac{c}{\lambda} = \frac{3.00 \times 10^8 \text{ m/s}}{5.32 \times 10^{-7} \text{ m}} = 5.64 \times 10^{14} \text{ s}^{-1}$$

and

$$E = h\nu = (6.63 \times 10^{-34} \text{ J} \cdot \text{s})(5.64 \times 10^{14} \text{ s}^{-1}) = 3.74 \times 10^{-19} \text{ J}$$

The energy of a single photon with wavelength 532 nm is $3.74 \times 10^{-19}$ J.

Following the same procedure, the energy of a photon of wavelength 635 nm is $3.13 \times 10^{-19}$ J. The difference between them is $(3.74 \times 10^{-19} \text{ J}) - (3.13 \times 10^{-19} \text{ J}) = 6.1 \times 10^{-20}$ J. Therefore, a photon of green light ($\lambda$ = 532 nm) has $6.1 \times 10^{-20}$ J *more* energy than a photon of red light ($\lambda$ = 635 nm).

**Practice Problem A** Calculate the difference in energy (in joules) between a photon with $\lambda$ = 680 nm (red) and a photon with $\lambda$ = 442 nm (blue).

**Practice Problem B** In what region of the electromagnetic spectrum is a photon found that possesses twice as much energy as one in the blue region ($\lambda$ = 442 nm) of the visible spectrum?

**Think About It** As wavelength decreases, frequency increases. Energy, being directly proportional to frequency, also increases.

## Photons and the Photoelectric Effect

In 1905, only 5 years after Planck presented his quantum theory, Albert Einstein[2] used the theory to explain another mysterious physical phenomenon, the ***photoelectric effect,*** in which electrons are ejected from the surface of a metal exposed to light of at least a certain minimum frequency, called the *threshold frequency* (Figure 6.5). The number of electrons ejected was proportional to the intensity (or brightness) of the light, but the energies of the ejected electrons were not. Below the threshold frequency no electrons were ejected no matter how intense the light.

The photoelectric effect could not be explained by the wave theory of light, which associated the energy of light with its intensity. Einstein, however, made an extraordinary assumption. He suggested that a beam of light is really a stream of particles. These *particles of light* are now called ***photons.*** Using Planck's quantum theory of radiation as a starting point, Einstein deduced that each *photon* must possess energy $E$ given by the equation

$$E_{\text{photon}} = h\nu$$

where $h$ is Planck's constant and $\nu$ is the frequency of the light. Electrons are held in a metal by attractive forces, and so removing them from the metal requires light of a sufficiently high frequency (which corresponds to a sufficiently high energy) to break them free. Shining a beam of light onto a metal surface can be thought of as shooting a beam of particles—*photons*—at the metal atoms. If the frequency of the photons is such that $h\nu$ exactly equals the energy that binds the electrons in the metal, then the light will have just enough energy to knock the electrons loose. If we use light of a higher frequency, then not only will the electrons be knocked loose, but they will also acquire some kinetic energy. This situation is summarized by the equation

Equation 6.3 $$h\nu = \text{KE} + W$$

where KE is the kinetic energy of the ejected electron and $W$ is the binding energy of the electron in the metal. [Binding energies are typically given in units of electron volts (eV), where 1 eV = $1.602 \times 10^{-19}$ J.] Rewriting Equation 6.3 as

$$\text{KE} = h\nu - W$$

shows that the more energetic the photon (i.e., the higher its frequency), the greater the kinetic energy of the ejected electron. If the frequency of light is below the threshold frequency, the photon will simply bounce off the surface and no electrons will be ejected. If the frequency is equal to the threshold frequency, it will dislodge the most loosely held electron. Above the threshold frequency, it will not only dislodge the electron, but also impart some kinetic energy to the ejected electron.

Now consider two beams of light having the same frequency (greater than the threshold frequency) but different intensities. The more intense beam of light consists of a larger number of photons, so it ejects more electrons from the metal's surface than the weaker beam of light. Thus, the more intense the light, the greater the number of electrons emitted by the target metal; the higher the frequency of the light, the greater the kinetic energy of the ejected electrons.

**Figure 6.5** Apparatus for studying the photoelectric effect. Light of a certain frequency falls on a clean metal surface. Ejected electrons are attracted toward the positive electrode. The flow of electrons is registered by a detecting meter.

---

2. Albert Einstein (1879–1955). German-born American physicist. Regarded by many as one of the two greatest physicists the world has known (the other is Isaac Newton). The three papers (on special relativity, Brownian motion, and the photoelectric effect) that he published in 1905 while employed as a technical assistant in the Swiss patent office in Berne have profoundly influenced the development of physics. He received the Nobel Prize in Physics in 1921 for his explanation of the photoelectric effect.

## *You May Be Wondering…*

# Where Have I Encountered the Photoelectric Effect?

Chances are good that you encounter the photoelectric effect regularly. Some everyday applications include the type of device that prevents a garage door from closing when something is in the door's path and motion-detection systems used in museums and other high-security environments. These sorts of devices work simply by responding to an interruption in a beam of light. In each case, a beam of light normally shines on a *photocathode,* a surface that emits photoelectrons, and the photoelectrons, accelerated toward an anode by high voltage, constitute a current. When the beam of light is interrupted, the flow of photoelectrons stops and the current is cut off. In the case of the garage-door safety device, when the current stops, the movement of the door stops. In the case of the motion-detection systems, when the current stops, an alarm may sound or a light may turn on.

One of the more exotic uses of the photoelectric effect is in night-vision goggles. Although you may never have looked through night-vision goggles, you have probably seen night-vision images on the news or in a suspense-filled movie such as *Silence of the Lambs.* Typically, the night-vision images we see are from what are known as third-generation night-vision devices. These devices use a photocathode material (gallium arsenide) that emits photoelectrons when struck by photons in the *infrared* region of the electromagnetic spectrum. (First- and second-generation devices used different photocathode materials and relied more on the amplification of low-level visible light.) The photoelectrons emitted by the photocathode enter a microchannel plate (MCP), an array of tiny parallel tubes, where each strikes the internal surface of a tube causing many more electrons to be ejected—a process called *secondary emission.* This effectively amplifies the current generated by each photoelectron. The amplified current is then accelerated by high voltage toward a phosphorus screen, where incident electrons cause the emission of visible light, generating the familiar green glow of night vision.

(a)

(b)

(a) Night-vision goggles schematic. 1. Photocathode (gallium arsenide). 2. Microchannel plate (MCP). 3. High-voltage source. 4. Phosphorus screen. (b) The green glow of night vision.

Sample Problem 6.3 shows how to determine the energy of a single photon of light of a given wavelength and how to determine the maximum kinetic energy of an electron ejected via the photoelectric effect.

Einstein's theory of light posed a dilemma for scientists. On the one hand, it explains the photoelectric effect. On the other hand, the particle theory of light is inconsistent with the known wavelike properties of light. The only way to resolve the dilemma is to accept the idea that light possesses properties characteristic of both particles *and* waves. Depending on the experiment, light behaves either as a wave or as a stream of particles. This concept was totally alien to the way physicists had thought about matter and radiation, and it took a long time for them to accept it. We will see in Section 6.4 that possessing properties of both particles and waves is not unique to light but ultimately is characteristic of all matter, including electrons.

**Sample Problem 6.3**

Calculate the energy (in joules) of (a) a photon with a wavelength of $5.00 \times 10^4$ nm (infrared region) and (b) a photon with a wavelength of 52 nm (ultraviolet region). (c) Calculate the kinetic energy of an electron ejected by the photon in part (b) from a metal with a binding energy of 3.7 eV.

**Strategy** In parts (a) and (b), we are given the wavelength of light. Use Equation 6.1 to convert wavelength to frequency; then use Equation 6.2 to determine the energy of the photon for each wavelength. In part (c), we are asked to determine the kinetic energy of an ejected electron. For this we use Equation 6.3. The binding energy, given in electron volts, must be converted to joules for units to cancel.

**Setup** The wavelengths must be converted from nanometers to meters:

(a) $5.00 \times 10^4 \text{ nm} \times \dfrac{1 \times 10^{-9} \text{ m}}{1 \text{ nm}} = 5.00 \times 10^{-5}$ m

(b) $52 \text{ nm} \times \dfrac{1 \times 10^{-9} \text{ m}}{1 \text{ nm}} = 5.2 \times 10^{-8}$ m

Planck's constant, $h$, is $6.63 \times 10^{-34}$ J $\cdot$ s.

(c) $W = 3.7 \text{ eV} \times \dfrac{1.602 \times 10^{-19} \text{ J}}{1 \text{ eV}} = 5.9 \times 10^{-19}$ J

**Solution** (a) $v = \dfrac{c}{\lambda} = \dfrac{3.00 \times 10^8 \text{ m/s}}{5.00 \times 10^{-5} \text{ m}} = 6.00 \times 10^{12} \text{ s}^{-1}$

and

$E = hv = (6.63 \times 10^{-34} \text{ J} \cdot \text{s})(6.00 \times 10^{12} \text{ s}^{-1}) = 3.98 \times 10^{-21}$ J

This is the energy of a single photon with wavelength $5.00 \times 10^4$ nm.

(b) Following the same procedure as in part (a), the energy of a photon of wavelength 52 nm is $3.8 \times 10^{-18}$ J.

(c) KE $= hv - W = 3.8 \times 10^{-18} \text{ J} - 5.9 \times 10^{-19} \text{ J} = 3.2 \times 10^{-18}$ J.

**Think About It** Remember that frequency and wavelength are inversely proportional (Equation 6.1). Thus, as wavelength *decreases*, frequency and energy *increase*. Note that the binding energy becomes less significant as the energy of the incident photon increases.

**Practice Problem A** Calculate the energy (in joules) of (a) a photon with wavelength $2.11 \times 10^2$ nm, and (b) a photon with frequency $1.78 \times 10^8 \text{ s}^{-1}$. (c) Calculate the kinetic energy of an electron ejected by the photon in part (a) from a metal with a binding energy of 4.66 eV.

**Practice Problem B** (a) Calculate the wavelength (in nm) of light with energy $1.89 \times 10^{-20}$ J per photon. (b) For light of wavelength 410 nm, calculate the number of photons per joule. (c) Determine the binding energy (in eV) of a metal if the kinetic energy possessed by an ejected electron [using one of the photons in part (b)] is $2.93 \times 10^{-19}$ J.

**Checkpoint 6.2** **Quantum Theory**

**6.2.1** Calculate the energy per photon of light with wavelength 650 nm.

a) $1.29 \times 10^{-31}$ J

b) $4.31 \times 10^{-40}$ J

c) $1.02 \times 10^{-27}$ J

d) $1.44 \times 10^{-48}$ J

e) $3.06 \times 10^{-19}$ J

**6.2.2** Calculate the wavelength of light that has energy $1.32 \times 10^{-23}$ J/photon.

a) $5.02 \times 10^{-9}$ cm

b) $6.64 \times 10^3$ cm

c) $2.92 \times 10^{-63}$ cm

d) 1.51 cm

e) 66.4 cm

**Checkpoint 6.2    Quantum Theory (*continued*)**

6.2.3  Which of the following is characteristic of high-energy radiation?

a) High wavelength

b) High velocity

c) High frequency

d) High amplitude

e) All of the above

6.2.4  A clean metal surface is irradiated with light of three different wavelengths: $\lambda_1$, $\lambda_2$, and $\lambda_3$. The kinetic energies of the ejected electrons are as follows: $\lambda_1$: $2.9 \times 10^{-20}$ J; $\lambda_2$: approximately zero; $\lambda_3$: $4.2 \times 10^{-19}$ J. Arrange the light in order of increasing wavelength.

a) $\lambda_1 < \lambda_2 < \lambda_3$

b) $\lambda_2 < \lambda_1 < \lambda_3$

c) $\lambda_3 < \lambda_2 < \lambda_1$

d) $\lambda_3 < \lambda_1 < \lambda_2$

e) $\lambda_2 < \lambda_3 < \lambda_1$

# 6.3   Bohr's Theory of the Hydrogen Atom

In addition to explaining the photoelectric effect, Planck's quantum theory and Einstein's ideas made it possible for scientists to unravel another nineteenth-century mystery in physics: atomic line spectra.

In the seventeenth century, Newton had shown that sunlight is composed of various color components that can be recombined to produce white light. Since that time, chemists and physicists have studied the characteristics of such *emission spectra.* The emission spectrum of a substance can be seen by energizing a sample of material with either thermal energy or some other form of energy (such as a high-voltage electrical discharge if the substance is a gas). A "red-hot" or "white-hot" iron bar freshly removed from a fire produces a characteristic glow. The glow is the visible portion of its emission spectrum. The heat given off by the same iron bar is another portion of its emission spectrum—the infrared region. A feature common to the emission spectrum of the sun and that of a heated solid is that both are continuous; that is, all wavelengths of visible light are present in each spectrum (Figure 6.6).

**Student Annotation:** If you have ever seen a rainbow, you are familiar with this phenomenon. The rainbow is the visible portion of the sun's emission spectrum.

## Atomic Line Spectra

Unlike those of the sun or a white-hot iron bar, the emission spectra of atoms in the gas phase do not show a continuous spread of wavelengths from red to violet; rather, the atoms produce bright lines in distinct parts of the visible spectrum. These *line spectra* are the emission of light only at

(a)

(b)

**Figure 6.6**   The visible white light emitted by (a) the sun and (b) a white-hot iron bar. In each case, the white light is the combination of all visible wavelengths (see Figure 6.1).

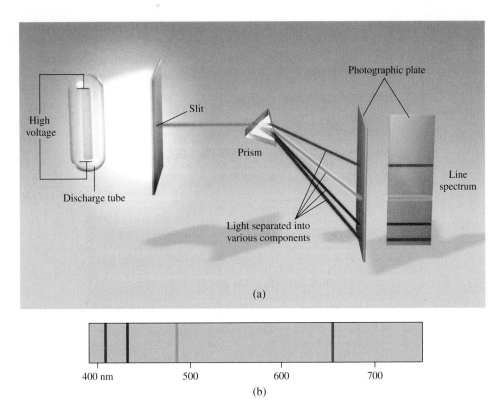

(a)

(b)

**Figure 6.7** (a) Experimental arrangement for studying the emission spectra of atoms and molecules. The gas being studied is in a discharge tube containing two electrodes. As electrons flow from the negative electrode to the positive electrode, they collide with the gas particles. The collisions lead to the emission of light by the atoms (or molecules). The emitted light is separated into its components by a prism. (b) Line emission spectrum of hydrogen.

*specific wavelengths.* Figure 6.7 is a schematic diagram of a discharge tube that is used to study emission spectra.

Every element has a unique emission spectrum, so the characteristic lines in atomic spectra can be used in chemical analysis to identify elements, much as fingerprints are used to identify people. When the lines of the emission spectrum of a known element exactly match the lines of the emission spectrum of an unknown sample, the identity of the element in the sample is established. Although the procedure of identifying elements by their line spectra had been used for many years in chemical analysis, the origin of the spectral lines was not understood until early in the twentieth century. Figure 6.8 shows the emission spectra of several elements.

In 1885, Johann Balmer[3] developed a remarkably simple equation that could be used to calculate the wavelengths of the four visible lines in the emission spectrum of hydrogen. Johannes Rydberg[4] developed Balmer's equation further, yielding an equation that could calculate not only the *visible* wavelengths, but those of *all* hydrogen's spectral lines:

$$\frac{1}{\lambda} = R_\infty \left( \frac{1}{n_1^2} - \frac{1}{n_2^2} \right) \qquad \text{Equation 6.4}$$

In Equation 6.4, now known as the *Rydberg equation,* $\lambda$ is the wavelength of a line in the spectrum; $R_\infty$ is the Rydberg constant ($1.09737316 \times 10^7 \, \text{m}^{-1}$); and $n_1$ and $n_2$ are positive integers, where $n_2 > n_1$.

**Student Annotation:** The Rydberg equation is a mathematical relationship that was derived from experimental data. Although it predates quantum theory by decades, it agrees remarkably well with it for one-electron systems such as the hydrogen atom.

## The Line Spectrum of Hydrogen

In 1913, not long after Planck's and Einstein's discoveries, a theoretical explanation of the emission spectrum of the hydrogen atom was presented by the Danish physicist Niels Bohr.[5] Bohr's treatment is very complex and is no longer considered to be correct in all its details. We will concentrate only on his important assumptions and final results, which account for the observed spectral lines and provide an important step toward the understanding of quantum theory.

3. Johann Jakob Balmer (1825–1898). Swiss mathematician. From 1859 until his death in 1898 Balmer taught math at a secondary school for girls in Basel, Switzerland. Although physicists did not understand why his equation worked until long after his death, the visible series of lines in the spectrum of hydrogen is named for him.

4. Johannes Robert Rydberg (1854–1919). Swedish mathematician and physicist. Rydberg analyzed many atomic spectra in an effort to understand the periodic properties of elements. Although he was nominated twice for the Nobel Prize in Physics, he never received it.

5. Niels Henrik David Bohr (1885–1962). Danish physicist. One of the founders of modern physics, he received the Nobel Prize in Physics in 1922 for his theory explaining the line spectrum of hydrogen.

**Figure 6.8** Emission spectra of several elements.

Bright-line Spectra

Lithium (Li)

Sodium (Na)

Potassium (K)

Calcium (Ca)

Strontium (Sr)

Barium (Ba)

Hydrogen (H)

Helium (He)

Neon (Ne)

Argon (Ar)

Alkali metals

Alkaline earth metals

Gases

Media Player/MPEG Animation: Figure 6.10, Emission Spectrum of Hydrogen, pp. 224–225.

When Bohr first approached this problem, physicists already knew that the atom contains electrons and protons. They thought of an atom as an entity in which electrons whirled around the nucleus in circular orbits at high velocities. This was an appealing description because it resembled the familiar model of planetary motion around the sun. However, according to the laws of classical physics, an electron moving in an orbit of a hydrogen atom would experience an acceleration toward the nucleus by radiating away energy in the form of electromagnetic waves. Thus, such an electron would quickly spiral into the nucleus and annihilate itself with the proton. To explain why this does not happen, Bohr postulated that the electron is allowed to occupy only certain orbits of specific energies. In other words, the energies of the electron are quantized. An electron in any of the allowed orbits will not radiate energy and therefore will not spiral into the nucleus.

Bohr attributed the emission of radiation by an energized hydrogen atom to the electron dropping from a higher-energy orbit to a lower one and giving up a quantum of energy (a photon) in the form of light (Figure 6.10, pp. 224–225). Using arguments based on electrostatic interaction and Newton's laws of motion, Bohr showed that the energies that the electron in the hydrogen atom can possess are given by

Equation 6.5
$$E_n = -2.18 \times 10^{-18} \text{ J} \left(\frac{1}{n^2}\right)$$

where $n$ is an integer with values $n = 1, 2, 3$, and so on. The negative sign in Equation 6.5 is an arbitrary convention, signifying that the energy of the electron in the atom is *lower* than the energy of a *free electron*, which is an electron that is infinitely far from the nucleus. The energy of a free electron is arbitrarily assigned a value of zero. Mathematically, this corresponds to setting $n$ equal to infinity in Equation 6.5:

$$E_\infty = -2.18 \times 10^{-18} \text{ J} \left(\frac{1}{\infty^2}\right) = 0$$

As the electron gets closer to the nucleus (as $n$ decreases), $E_n$ becomes larger in absolute value, but also more negative. The most negative value, then, is reached when $n = 1$, which corresponds to the most stable energy state. We call this the **ground state,** the *lowest* energy state of an atom. The stability of the electron diminishes as $n$ increases. Each energy state in which $n > 1$ is called an **excited state.** Each excited state is higher in energy than the ground state. In the hydrogen atom, an electron for which $n$ is greater than 1 is said to be in an excited state.

The radius of each circular orbit in Bohr's model depends on $n^2$. Thus, as $n$ increases from 1 to 2 to 3, the orbit radius increases very rapidly. The higher the excited state, the farther away the electron is from the nucleus (and the less tightly held it is by the nucleus).

Bohr's theory enables us to explain the line spectrum of the hydrogen atom. Radiant energy absorbed by the atom causes the electron to move from the ground state ($n = 1$) to an excited state ($n > 1$). Conversely, radiant energy (in the form of a photon) is *emitted* when the electron moves from a higher-energy excited state to a lower-energy excited state or the ground state.

The quantized movement of the electron from one energy state to another is analogous to the movement of a tennis ball either up or down a set of stairs (Figure 6.9). The ball can be on any of several steps but never between steps. The journey from a lower step to a higher one is an energy-requiring process, whereas movement from a higher step to a lower step is an energy-releasing process. The quantity of energy involved in either type of change is determined by the distance between the beginning and ending steps. Similarly, the amount of energy needed to move an electron in the Bohr atom depends on the difference in energy levels between the initial and final states.

**Figure 6.9** Mechanical analogy for the emission processes. The ball can rest on any step but not between steps.

To apply Equation 6.5 to the emission process in a hydrogen atom, let us suppose that the electron is initially in an excited state characterized by $n_i$. During emission, the electron drops to a lower-energy state characterized by $n_f$ (the subscripts i and f denote the *initial* and *final* states, respectively). This lower-energy state may be the ground state, but it can be any state lower than the initial excited state. The difference between the energies of the initial and final states is

$$\Delta E = E_f - E_i$$

From Equation 6.5,

$$E_f = -2.18 \times 10^{-18} \text{ J} \left(\frac{1}{n_f^2}\right)$$

and

$$E_i = -2.18 \times 10^{-18} \text{ J} \left(\frac{1}{n_i^2}\right)$$

Therefore,

$$\Delta E = \left(\frac{-2.18 \times 10^{-18} \text{ J}}{n_f^2}\right) - \left(\frac{-2.18 \times 10^{-18} \text{ J}}{n_i^2}\right)$$

$$= -2.18 \times 10^{-18} \text{ J} \left(\frac{1}{n_f^2} - \frac{1}{n_i^2}\right)$$

**Student Annotation:** When $n_i > n_f$, $\Delta E$ is *negative*, indicating energy is *emitted*. When $n_f > n_i$, $\Delta E$ is *positive*, indicating energy is *absorbed*.

Because this transition results in the emission of a photon of frequency $\nu$ and energy $h\nu$, we can write

$$\Delta E = h\nu = -2.18 \times 10^{-18} \text{ J} \left(\frac{1}{n_f^2} - \frac{1}{n_i^2}\right) \qquad \text{Equation 6.6}$$

A photon is emitted when $n_i > n_f$. Consequently, the term in parentheses is *positive*, making $\Delta E$ *negative* (energy is lost to the surroundings). A photon is absorbed when $n_f > n_i$, making the term in parentheses *negative*, so $\Delta E$ is *positive*. Each spectral line in the emission spectrum of hydrogen corresponds to a particular transition in a hydrogen atom. When we study a large number of hydrogen atoms, we observe all possible transitions and hence the corresponding spectral lines. The brightness of a spectral line depends on how many photons of the same wavelength are emitted.

To calculate the wavelength of an emission line, we substitute $c/\lambda$ for $\nu$ and then divide both sides of Equation 6.6 by $hc$. In addition, because wavelength (and frequency) can only have positive values, we take the absolute value of the right side of the equation. (In this case, we do so simply by eliminating the negative sign.)

**Student Annotation:** Because $2.18 \times 10^{-18}$ J/$hc$ = $1.096 \times 10^7$ m$^{-1}$, which to three significant figures is equal to $R_\infty$, this equation is essentially the same as the Rydberg equation (Equation 6.4).

$$\frac{1}{\lambda} = \frac{2.18 \times 10^{-18} \text{ J}}{hc} \left(\frac{1}{n_f^2} - \frac{1}{n_i^2}\right) \qquad \text{Equation 6.7}$$

The emission spectrum of hydrogen includes a wide range of wavelengths from the infrared to the ultraviolet. Table 6.1 (p. 226) lists the series of transitions in the hydrogen spectrum, each with a different value of $n_f$. The series are named after their discoverers (Lyman, Balmer, Paschen, and Brackett). The Balmer series was the first to be studied because some of its lines occur in the visible region.

Figure 6.10

# Emission Spectrum of Hydrogen

Hydrogen atoms in the ground state are excited by the addition of energy. Electrons that have been promoted to higher energy levels ($n > 1$) return to lower energy levels and emit the excess energy as electromagnetic radiation.

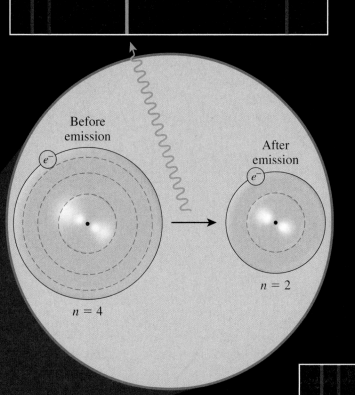

The transitions shown are as follows:

- $n = 3$ to $n = 2$
- $n = 4$ to $n = 2$
- $n = 5$ to $n = 2$
- $n = 6$ to $n = 2$

Electrons in H atoms may be promoted to excited states other than 3, 4, 5, or 6; and electrons in excited states may return to a state other than $n = 2$. However, the transitions shown are the ones that give rise to the visible lines in the hydrogen emission spectrum.

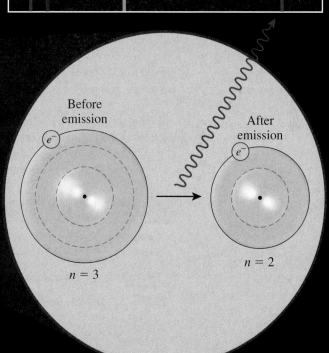

## What's the point?

Each line in the visible emission spectrum of hydrogen is the result of an electronic transition from a higher excited state ($n = 3, 4, 5,$ or 6) to a lower excited state ($n = 2$). The energy gap between the initial and final states determines the wavelength of the light emitted.

| TABLE 6.1 | Emission Series in the Hydrogen Spectrum | | |
|-----------|-----------------|-------------|-----------------|
| **Series** | $n_f$ | $n_i$ | **Spectrum Region** |
| Lyman | 1 | 2, 3, 4, . . . | Ultraviolet |
| Balmer | 2 | 3, 4, 5, . . . | Visible and ultraviolet |
| Paschen | 3 | 4, 5, 6, . . . | Infrared |
| Brackett | 4 | 5, 6, 7, . . . | Infrared |

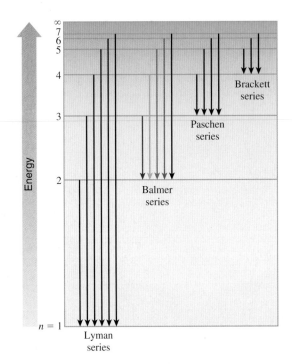

**Figure 6.11**    Energy levels in the hydrogen atom and the various emission series. Each series terminates at a different value of $n$.

Figure 6.11 shows transitions associated with spectral lines in each of the emission series. Each horizontal line represents one of the allowed energy levels for the electron in a hydrogen atom. The energy levels are labeled with their $n$ values.

Sample Problem 6.4 illustrates the use of Equation 6.7.

## Sample Problem 6.4

Calculate the wavelength (in nm) of the photon emitted when an electron transitions from the $n = 4$ state to the $n = 2$ state in a hydrogen atom.

**Strategy** Use Equation 6.7 to calculate $\lambda$.

**Setup** According to the problem, the transition is from $n = 4$ to $n = 2$, so $n_i = 4$ and $n_f = 2$. The required constants are $h = 6.63 \times 10^{-34}$ J · s and $c = 3.00 \times 10^8$ m/s.

**Solution**

$$\frac{1}{\lambda} = \frac{2.18 \times 10^{-18} \text{ J}}{hc} \left( \frac{1}{n_f^2} - \frac{1}{n_i^2} \right)$$

$$= \frac{2.18 \times 10^{-18} \text{ J}}{(6.63 \times 10^{-34} \text{ J} \cdot \text{s})(3.00 \times 10^8 \text{ m/s})} \left( \frac{1}{2^2} - \frac{1}{4^2} \right)$$

$$= 2.055 \times 10^6 \text{ m}^{-1}$$

**Student Annotation:** Remember to keep at least one extra digit in intermediate answers to avoid rounding error in the final result [◄◄ Section 1.5].

Therefore,

$$\lambda = 4.87 \times 10^{-7} \text{ m} = 487 \text{ nm}$$

**Practice Problem A** What is the wavelength (in nm) of a photon emitted during a transition from the $n = 3$ state to the $n = 1$ state in the H atom?

**Practice Problem B** What is the value of $n_i$ for an electron that emits a photon of wavelength 93.14 nm when it returns to the ground state in the H atom?

**Think About It** Look again at the line spectrum of hydrogen in Figure 6.7 and make sure that your result matches one of them. Note that for an emission, $n_i$ is always greater than $n_f$, and Equation 6.7 gives a positive result.

## Bringing Chemistry to Life

### Lasers

The word *laser* is an acronym for *l*ight *a*mplification by *s*timulated *e*mission of *r*adiation. It is a special type of emission that may involve electronic transitions in *atoms* or *molecules*. The first laser, developed in 1960, was a ruby laser. Ruby is a deep-red mineral consisting of corundum ($Al_2O_3$), in which some of the $Al^{3+}$ ions have been replaced by $Cr^{3+}$ ions. A cylindrical ruby crystal is positioned between two perfectly parallel mirrors, one of which is only partially reflective. A light source called a flashlamp is used to excite the chromium atoms to a higher energy level. The excited atoms are unstable, so at a given instant some of them return to the ground state by emitting a photon in the red region of the spectrum ($\lambda = 694.3$ nm). Spontaneous emission of photons occurs in all directions, but photons emitted directly at either mirror will be reflected back. As the reflected photons pass back through the ruby crystal, they stimulate the emission of more photons in the same direction. These photons, in turn, are reflected back through the crystal, stimulating still more emissions, and so on. Because the light waves are *in phase*—that is, their maxima coincide and their minima coincide—the photons enhance one another, increasing their power with each passage between the mirrors. When the light reaches a certain intensity, it emerges from the partially reflective mirror as a laser beam. Laser light is characterized by three properties: It is *intense,* it has a precisely known wavelength and therefore *energy,* and it is *coherent.*

**Student Annotation:** *Coherent* means that the light waves are all in phase.

The applications of lasers are quite numerous. In addition to their use in surgery, their high intensity and ease of focus make them suitable for drilling holes in metals, welding, and carrying out nuclear fusion. The fact that they are highly directional and have precisely known wavelengths makes them very useful for telecommunications. Lasers are also used in isotope separation, in chemical analysis, in holography (three-dimensional photography), in compact disc players, and in supermarket scanners.

| **Checkpoint 6.3** | **Bohr's Theory of the Hydrogen Atom** |

**6.3.1** Calculate the energy of an electron in the $n = 3$ state in a hydrogen atom.

a) $2.42 \times 10^{-19}$ J

b) $-2.42 \times 10^{-19}$ J

c) $7.27 \times 10^{-19}$ J

d) $-7.27 \times 10^{-19}$ J

e) $-6.54 \times 10^{-18}$ J

**6.3.2** Calculate $\Delta E$ of an electron that goes from $n = 1$ to $n = 5$.

a) $8.72 \times 10^{-20}$ J

b) $-8.72 \times 10^{-20}$ J

c) $5.45 \times 10^{-17}$ J

d) $2.09 \times 10^{-18}$ J

e) $-2.09 \times 10^{-18}$ J

**6.3.3** What is the wavelength of light emitted when an electron in a hydrogen atom goes from $n = 5$ to $n = 3$?

a) $4.87 \times 10^{-7}$ m

b) $6.84 \times 10^{-7}$ m

c) $1.28 \times 10^{-6}$ m

d) $3.65 \times 10^{-7}$ m

e) $1.02 \times 10^{-7}$ m

**6.3.4** Which wavelength corresponds to the transition of an electron in a hydrogen atom from $n = 2$ to $n = 1$?

a) 182 nm

b) 91.2 nm

c) 724 nm

d) 812 nm

e) 122 nm

# 6.4    Wave Properties of Matter

Bohr's theory was both fascinating and puzzling. It fit the experimental data for hydrogen, but physicists did not understand the underlying principle. Why, for example, was an electron restricted to orbiting the nucleus at certain fixed distances? For a decade no one, not even Bohr himself, could offer a logical explanation. In 1924 Louis de Broglie[6] provided a solution to this puzzle. De Broglie reasoned that if energy (light) can, under certain circumstances, behave like a stream of particles (photons), then perhaps particles such as electrons can, under certain circumstances, exhibit wave-like properties.

## The de Broglie Hypothesis

In developing his revolutionary theory, de Broglie incorporated his observations of macroscopic phenomena that exhibited quantized behavior. For example, a guitar string has certain discrete frequencies of vibration, like those shown in Figure 6.12(a). The waves generated by plucking a guitar string are *standing* or *stationary waves* because they do not travel along the string. Some points on the string, called **nodes,** do not move at all; that is, the amplitude of the wave at these points is *zero.* There is a node at each end, and there may be one or more nodes between the ends. The greater the frequency of vibration, the shorter the wavelength of the standing wave and the greater the number of nodes. According to de Broglie, an electron in an atom behaves like a *standing wave.* However, as Figure 6.12 shows, only certain wavelengths are possible or *allowed.*

De Broglie argued that if an electron does behave like a standing wave in the hydrogen atom, the wavelength must fit the circumference of the orbit exactly; that is, the circumference of the orbit must be an integral multiple of the wavelength, as shown in Figure 6.12(b). Otherwise, the wave would partially cancel itself by destructive interference on each successive orbit, quickly reducing its amplitude to zero.

The relationship between the circumference of an allowed orbit $(2\pi r)$ and the wavelength $(\lambda)$ of the electron is given by

Equation 6.8    $2\pi r = n\lambda$

---

6. Louis Victor Pierre Raymond Duc de Broglie (1892–1977). French physicist. A member of an old and noble family in France, he held the title of a prince. In his doctoral dissertation, he proposed that matter and radiation have the properties of both wave and particle. For this work, de Broglie was awarded the Nobel Prize in Physics in 1929.

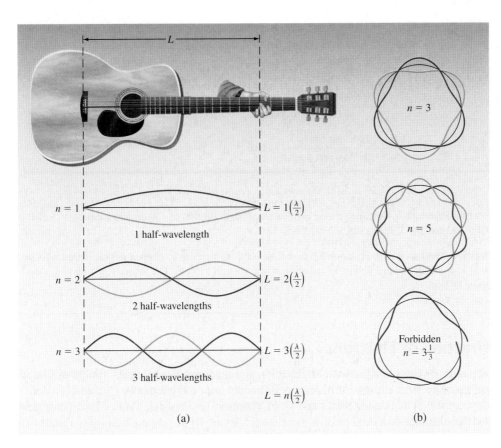

**Figure 6.12**   Standing waves of a vibrating guitar string. The length of the string must be equal to a whole number times one-half the wavelength ($\lambda/2$). (b) In a circular orbit, only whole number multiples of wavelengths are allowed. Any fractional number of wavelengths would result in cancellation of the wave due to destructive interference.

where $r$ is the radius of the orbit, $\lambda$ is the wavelength of the electron wave, and $n$ is a positive integer (1, 2, 3, . . .). Because $n$ is an integer, $r$ can have only certain values (integral multiples of $\lambda$) as $n$ increases from 1 to 2 to 3 and so on. And, because the energy of the electron depends on the size of the orbit (or the value of $r$), the energy can have only certain values, too. Thus, the energy of the electron in a hydrogen atom, if it behaves like a standing wave, must be quantized.

De Broglie's reasoning led to the conclusion that waves can behave like particles and particles can exhibit wavelike properties. De Broglie deduced that the particle and wave properties are related by the following expression:

$$\lambda = \frac{h}{mu}$$

Equation 6.9

where $\lambda$, $m$, and $u$ are the wavelength associated with a moving particle, its mass, and its velocity, respectively. Equation 6.9 implies that a particle in motion can be treated as a wave, and a wave can exhibit the properties of a particle. To help you remember this important point, notice that the left side of Equation 6.9 involves the wavelike property of wavelength, whereas the right side involves mass, a property of particles. A wavelength calculated using Equation 6.9 is usually referred to specifically as a ***de Broglie wavelength.*** Likewise, we will refer to a *mass* calculated using Equation 6.9 as a *de Broglie mass.*

Sample Problem 6.5 illustrates how de Broglie's theory and Equation 6.9 can be applied.

**Student Annotation:** Mass ($m$) must be expressed in kilograms for units to cancel properly in Equation 6.9.

## Sample Problem 6.5

Calculate the de Broglie wavelength of the "particle" in the following two cases: (a) a 25-g bullet traveling at 612 m/s, and (b) an electron ($m = 9.109 \times 10^{-31}$ kg) moving at 63.0 m/s.

**Strategy**  Use Equation 6.9 to calculate the de Broglie wavelengths. Remember that the mass in Equation 6.9 must be expressed in kilograms for the units to cancel properly.

**Setup**  Planck's constant, $h$, is $6.63 \times 10^{-34}$ J $\cdot$ s or, for the purpose of making the unit cancellation obvious, $6.63 \times 10^{-34}$ kg $\cdot$ m²/s.

**Student Annotation:** 1 J = 1 kg $\cdot$ m²/s².

*(Continued)*

**Think About It** While you are new at solving these problems, always write out the units of Planck's constant ($J \cdot s$) as $kg \cdot m^2/s$. This will enable you to check your unit cancellations and detect common errors such as expressing mass in grams rather than kilograms. Note that the calculated wavelength of a macroscopic object, even one as small as a bullet, is extremely small. An object must be at least as small as a subatomic particle for its wavelength to be large enough for us to observe.

**Student Annotation:** Momentum has units of $kg \cdot m/s$ or $N \cdot s$, where N is the *newton*, the SI unit of *force*. The newton is a derived SI unit: $1 N = 1 kg \cdot m/s^2$.

**Solution**

(a) $25 \text{ g} \times \dfrac{1 \text{ kg}}{1000 \text{ g}} = 0.025 \text{ kg}$

$$\lambda = \frac{h}{mu} = \frac{6.63 \times 10^{-34} \text{ kg} \cdot \text{m}^2/\text{s}}{(0.025 \text{ kg})(612 \text{ m/s})} = 4.3 \times 10^{-35} \text{ m}$$

(b) $\lambda = \dfrac{h}{mu} = \dfrac{6.63 \times 10^{-34} \text{ kg} \cdot \text{m}^2/\text{s}}{(9.109 \times 10^{-31} \text{ kg})(63.0 \text{ m/s})} = 1.16 \times 10^{-5} \text{ m}$

**Practice Problem A** Calculate the de Broglie wavelength (in nm) of a hydrogen atom ($m = 1.674 \times 10^{-27}$ kg) moving at 1500 cm/s.

**Practice Problem B** Use Equation 6.9 to calculate the *momentum, p* (defined as mass times velocity, $m \times u$) associated with a photon of radiation of wavelength 810 nm. The velocity of a photon is the speed of light, *c*.

## Diffraction of Electrons

Shortly after de Broglie introduced his equation and predicted that electrons should exhibit wave properties, successful electron diffraction experiments were carried out by Clinton Davisson[7] and Lester Germer[8] in the United States and G. P. Thomson[9] in England. These experiments demonstrated that electrons do indeed possess wavelike properties. By directing a beam of electrons (which are most definitely particles) through a thin piece of gold foil, Thomson obtained a set of concentric rings on a screen, similar to the diffraction pattern observed when X rays (which are most definitely waves) were used. Figure 6.13 shows X-ray and electron diffraction patterns for aluminum.

(a)

(b)

**Figure 6.13** (a) X-ray diffraction pattern of aluminum foil. (b) Electron diffraction of aluminum foil. The similarity of these two patterns shows that electrons can behave like X rays and display wave properties.

| Checkpoint 6.4 | **Wave Properties of Matter** |

6.4.1 Calculate the de Broglie wavelength associated with a helium-4 atom (4.00 amu) moving at $3.0 \times 10^6$ m/s.

a) $2.0 \times 10^{-20}$ m

b) $3.3 \times 10^{-11}$ m

c) $3.3 \times 10^{-14}$ m

d) $1.8 \times 10^{-19}$ m

e) $6.6 \times 10^{-27}$ m

6.4.2 What is the momentum of a photon of light with $\lambda = 122$ nm? (The velocity of a photon is the speed of light, *c*.)

a) $5.43 \times 10^{-27}$ kg $\cdot$ m/s

b) $6.63 \times 10^{-34}$ kg $\cdot$ m/s

c) $6.00 \times 10^{-17}$ kg $\cdot$ m/s

d) $8.64 \times 10^{-11}$ kg $\cdot$ m/s

e) $2.71 \times 10^{-27}$ kg $\cdot$ m/s

## 6.5 Quantum Mechanics

The discovery that waves could have matterlike properties and that matter could have wavelike properties was revolutionary. Although scientists had long believed that energy and matter were distinct entities, the distinction between them, at least at the atomic level, was no longer clear. Bohr's theory was tremendously successful in explaining the line spectrum of hydrogen, but it failed to explain the spectra of atoms with more than one electron. The electron appeared to behave as a particle in some circumstances and as a wave in others. Neither description could completely explain the behavior of electrons in atoms. This left scientists frustrated in their quest to understand exactly where the electrons in an atom are.

7. Clinton Joseph Davisson (1881–1958). American physicist. He and G. P. Thomson shared the Nobel Prize in Physics in 1937 for demonstrating the wave properties of electrons.

8. Lester Halbert Germer (1896–1972). American physicist. Discoverer (with Davisson) of the wave properties of electrons.

9. George Paget Thomson (1892–1975). English physicist. Son of J. J. Thomson, he received the Nobel Prize in Physics in 1937, along with Clinton Davisson, for demonstrating the wave properties of electrons.

## The Uncertainty Principle

To describe the problem of trying to locate a subatomic particle that behaves like a wave, Werner Heisenberg[10] formulated what is now known as the **Heisenberg uncertainty principle:** It is impossible to know simultaneously both the *momentum p* and the *position x* of a particle with certainty. Stated mathematically,

$$\Delta x \cdot \Delta p \geq \frac{h}{4\pi}$$
Equation 6.10

For a particle of mass *m*,

$$\Delta x \cdot m\Delta u \geq \frac{h}{4\pi}$$
Equation 6.11

where $\Delta x$ and $\Delta u$ are the uncertainties in measuring the position and velocity of the particle, respectively. The $\geq$ signs have the following meaning. If the measured uncertainties of position and velocity are large (say, in a crude experiment), their product can be substantially greater than $h/4\pi$ (hence the $>$ sign). The significance of Equation 6.11 is that even in the most favorable conditions for measuring position and velocity, the product of the uncertainties can never be less than $h/4\pi$ (hence the $=$ sign). Thus, making measurement of the velocity of a particle *more* precise (i.e., making $\Delta u$ a *small* quantity) means that the position must become correspondingly *less* precise (i.e., $\Delta x$ will become *larger*). Similarly, if the position of the particle is known more precisely, its velocity measurement must become less precise.

If the Heisenberg uncertainty principle is applied to the hydrogen atom, we find that the electron cannot orbit the nucleus in a well-defined path, as Bohr thought. If it did, we could determine precisely both the position of the electron (from the radius of the orbit) and its speed (from its kinetic energy) at the same time. This would violate the uncertainty principle.

Sample Problem 6.6 shows how the Heisenberg uncertainty principle can be applied.

### Sample Problem 6.6

An electron in a hydrogen atom is known to have a velocity of $5 \times 10^6$ m/s $\pm$ 1 percent. Using the uncertainty principle, calculate the minimum uncertainty in the position of the electron and, given that the diameter of the hydrogen atom is less than 1 angstrom (Å), comment on the magnitude of this uncertainty compared to the size of the atom.

**Strategy** The uncertainty in the velocity, 1 percent of $5 \times 10^6$ m/s, is $\Delta u$. Using Equation 6.11, calculate $\Delta x$ and compare it with the diameter of the hydrogen atom. Recall that 1 Å is equal to $1 \times 1^{-10}$ m. [◀◀ Section 2.2]

**Setup** The mass of an electron (from Table 2.1, rounded to three significant figures and converted to kilograms) is $9.11 \times 10^{-31}$ kg. Planck's constant, *h*, is $6.63 \times 10^{-34}$ kg · m$^2$/s.

**Solution**
$$\Delta u = 0.01 \times 5 \times 10^6 \text{ m/s} = 5 \times 10^4 \text{ m/s}$$

$$\Delta x = \frac{h}{4\pi \cdot m\Delta u}$$

Therefore,

$$\Delta x = \frac{6.63 \times 10^{-34} \text{ kg} \cdot \text{m}^2\text{/s}}{4\pi(9.11 \times 10^{-31} \text{ kg})(5 \times 10^4 \text{ m/s})} \geq 1 \times 10^{-9} \text{ m}$$

The *minimum* uncertainty in the position *x* is $1 \times 10^{-9}$ m = 10 Å. The uncertainty in the electron's position is 10 times larger than the atom!

**Think About It** A common error is expressing the mass of the particle in grams instead of kilograms, but you should discover this inconsistency if you check your unit cancellation carefully. Remember that if one uncertainty is small, the other must be large. The uncertainty principle applies in a practical way only to submicroscopic particles. In the case of a macroscopic object, where the mass is much larger than that of an electron, small uncertainties, relative to the size of the object, are possible for both position and velocity.

**Practice Problem A** Calculate the minimum uncertainty in the position of the 25-g bullet from Sample Problem 6.5 if the uncertainty in its velocity is (a) $\pm$1 percent and (b) $\pm$0.01 percent.

**Practice Problem B** (a) Calculate the minimum uncertainty in the momentum of an object for which the uncertainty in position is 3 Å. To what minimum uncertainty in velocity does this correspond if the particle is (b) a neutron (mass = 1.0087 amu) and (c) an electron (mass = $5.486 \times 10^{-4}$ amu)?

---

10. Werner Karl Heisenberg (1901–1976). German physicist. One of the founders of modern quantum theory. Heisenberg received the Nobel Prize in Physics in 1932.

## The Schrödinger Equation

Bohr made a significant contribution to our understanding of atoms, and his suggestion that the energy of an electron in an atom is quantized remains unchallenged, but his theory did not provide a complete description of the behavior of electrons in atoms. In 1926 the Austrian physicist Erwin Schrödinger,[11] using a complex mathematical technique, formulated an equation that describes the behavior and energies of submicroscopic particles in general, an equation analogous to Newton's laws of motion for macroscopic objects. The *Schrödinger equation* requires advanced calculus to solve, and we will not discuss it here. The equation, however, incorporates both particle behavior, in terms of mass *m,* and wave behavior, in terms of a *wave function* $\psi$ (psi), which depends on the location in space of the system (such as an electron in an atom).

The wave function itself has no direct physical meaning. However, the probability of finding the electron in a certain region in space is proportional to the square of the wave function, $\psi^2$. The idea of relating $\psi^2$ to probability stemmed from a wave theory analogy. According to wave theory, the intensity of light is proportional to the square of the amplitude of the wave, or $\psi^2$. The most likely place to find a photon is where the intensity is greatest—that is, where the value of $\psi^2$ is greatest. A similar argument associates $\psi^2$ with the likelihood of finding an electron in regions surrounding the nucleus.

Schrödinger's equation launched an entirely new field, called *quantum mechanics* (or *wave mechanics*), and began a new era in physics and chemistry. We now refer to the developments in quantum theory from 1913—when Bohr presented his model of the hydrogen atom—to 1926 as "old quantum theory."

## The Quantum Mechanical Description of the Hydrogen Atom

The Schrödinger equation specifies the possible energy states the electron can occupy in a hydrogen atom and identifies the corresponding wave functions ($\psi$). These energy states and wave functions are characterized by a set of *quantum numbers* (to be discussed shortly), with which we can construct a comprehensive model of the hydrogen atom.

Although quantum mechanics does not allow us to specify the exact location of an electron in an atom, it does define the region where the electron is most likely to be at a given time. The concept of *electron density* gives the probability that an electron will be found in a particular region of an atom. The square of the wave function, $\psi^2$, defines the distribution of electron density in three-dimensional space around the nucleus. Regions of high electron density represent a high probability of locating the electron (Figure 6.14).

To distinguish the quantum mechanical description of an atom from Bohr's model, we speak of an atomic *orbital,* rather than an orbit. An *atomic orbital* can be thought of as the wave function of an electron in an atom. When we say that an electron is in a certain orbital, we mean that the distribution of the electron density or the probability of locating the electron in space is described by the square of the wave function associated with that orbital. An atomic orbital, therefore, has a characteristic energy, as well as a characteristic distribution of electron density.

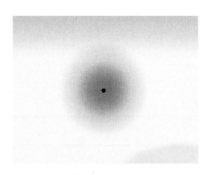

**Figure 6.14**    Representation of the electron density distribution surrounding the nucleus in the hydrogen atom. It shows a higher probability of finding the electron closer to the nucleus.

---

### Checkpoint 6.5    Quantum Mechanics

**6.5.1**  What is the minimum uncertainty in the position of an electron moving at a speed of $4 \times 10^6$ m/s $\pm$ 1 percent? (The mass of an electron is $9.11 \times 10^{-31}$ kg.)

a) $2 \times 10^{-8}$ m

b) $1 \times 10^{-9}$ m

c) $6 \times 10^{-9}$ m

d) $7 \times 10^{-8}$ m

e) $1 \times 10^{-12}$ m

**6.5.2**  What is the minimum uncertainty in the position of a proton moving at a speed of $4 \times 10^6$ m/s $\pm$ 1 percent? (The mass of a proton is $1.67 \times 10^{-27}$ kg.)

a) $1 \times 10^{-13}$ m

b) $8 \times 10^{-10}$ m

c) $4 \times 10^{-11}$ m

d) $3 \times 10^{-12}$ m

e) $8 \times 10^{-13}$ m

---

11. Erwin Schrödinger (1887–1961). Austrian physicist. Schrödinger formulated wave mechanics, which laid the foundation for modern quantum theory. He received the Nobel Prize in Physics in 1933.

## 6.6 Quantum Numbers

In Bohr's model of the hydrogen atom, only one number, $n$, was necessary to describe the location of the electron. In quantum mechanics, three *quantum numbers* are required to describe the *distribution of electron density* in an atom. These numbers are derived from the mathematical solution of Schrödinger's equation for the hydrogen atom. They are called the *principal* quantum number, the *angular momentum* quantum number, and the *magnetic* quantum number. Each atomic orbital in an atom is characterized by a unique set of these three quantum numbers.

**Student Annotation:** The three quantum numbers $n$, $\ell$, and $m_\ell$ specify the *size*, *shape*, and *orientation* of an orbital, respectively.

### Principal Quantum Number (*n*)

The *principal quantum number* (*n*) designates the *size* of the orbital. The larger $n$ is, the greater the average distance of an electron in the orbital from the nucleus and therefore the larger the orbital. The principal quantum number can have integral values of 1, 2, 3, and so forth, and it corresponds to the quantum number in Bohr's model of the hydrogen atom. Recall from Equation 6.5 that in a hydrogen atom, the value of $n$ alone determines the energy of an orbital. (As we will see shortly, this is *not* the case for an atom that contains more than one electron.)

### Angular Momentum Quantum Number ($\ell$)

The *angular momentum quantum number* ($\ell$) describes the *shape* of the atomic orbital (see Section 6.7). The values of $\ell$ are integers that depend on the value of the principal quantum number, $n$. For a given value of $n$, the possible values of $\ell$ range from 0 to $n - 1$. If $n = 1$, there is only one possible value of $\ell$; that is, 0 ($n - 1$ where $n = 1$). If $n = 2$, there are two values of $\ell$: 0 and 1. If $n = 3$, there are three values of $\ell$: 0, 1, and 2. The value of $\ell$ is designated by the letters $s$, $p$, $d$, and $f$ as follows:[12]

| $\ell$ | 0 | 1 | 2 | 3 |
|---|---|---|---|---|
| Orbital designation | $s$ | $p$ | $d$ | $f$ |

Thus, if $\ell = 0$, we have an $s$ orbital; if $\ell = 1$, we have a $p$ orbital; and so on.

A collection of orbitals with the same value of $n$ is frequently called a *shell*. One or more orbitals with the same $n$ and $\ell$ values are referred to as a *subshell*. For example, the shell designated by $n = 2$ is composed of two subshells: $\ell = 0$ and $\ell = 1$ (the allowed values of $\ell$ for $n = 2$). These subshells are called the $2s$ and $2p$ subshells where 2 denotes the value of $n$, and $s$ and $p$ denote the values of $\ell$.

### Magnetic Quantum Number ($m_\ell$)

The *magnetic quantum number* ($m_\ell$) describes the *orientation* of the orbital in space (see Section 6.7). Within a subshell, the value of $m_\ell$ depends on the value of $\ell$. For a certain value of $\ell$, there are $(2\ell + 1)$ integral values of $m_\ell$ as follows:

$$-\ell, \ldots 0, \ldots, +\ell$$

If $\ell = 0$, there is only one possible value of $m_\ell$: 0. If $\ell = 1$, then there are *three* values of $m_\ell$: $-1$, 0, and $+1$. If $\ell = 2$, there are *five* values of $m_\ell$, namely, $-2$, $-1$, 0, $+1$, and $+2$, and so on. The number of $m_\ell$ values indicates the number of *orbitals* in a subshell with a particular $\ell$ value; that is, each $m_\ell$ value refers to a different orbital.

Table 6.2 summarizes the allowed values of the three quantum numbers, $n$, $\ell$, and $m_\ell$, and Figure 6.15 illustrates schematically how the allowed values of quantum numbers give rise to the number of subshells and orbitals in each shell of an atom.

Sample Problem 6.7 gives you some practice with the allowed values of quantum numbers.

**Student Annotation:** The number of subshells in a shell is equal to $n$. The number of orbitals in a shell is equal to $n^2$. The number of orbitals in a subshell is equal to $2\ell + 1$.

---

12. The unusual sequence of letters ($s$, $p$, $d$, and $f$) has an historical origin. Physicists who studied atomic emission spectra tried to correlate their observations of spectral lines with the energy states involved in the transitions. They described the emission lines as *s*harp, *p*rincipal, *d*iffuse, and *f*undamental.

| TABLE 6.2 | Allowed Values of the Quantum Numbers $n$, $\ell$, and $m_\ell$ | | |
|---|---|---|---|
| **When $n$ is** | **$\ell$ can be** | **When $\ell$ is** | **$m_\ell$ can be** |
| 1 | only 0 | 0 | only 0 |
| 2 | 0 or 1 | 0 | only 0 |
|   |   | 1 | $-1$, 0, or $+1$ |
| 3 | 0, 1, or 2 | 0 | only 0 |
|   |   | 1 | $-1$, 0, or $+1$ |
|   |   | 2 | $-2$, $-1$, 0, $+1$, or $+2$ |
| 4 | 0, 1, 2, or 3 | 0 | only 0 |
|   |   | 1 | $-1$, 0, or $+1$ |
|   |   | 2 | $-2$, $-1$, 0, $+1$, or $+2$ |
|   |   | 3 | $-3$, $-2$, $-1$, 0, $+1$, $+2$, or $+3$ |

**Figure 6.15**    Illustration of how quantum numbers designate shells, subshells, and orbitals.

---

### Sample Problem 6.7

What are the possible values for the magnetic quantum number ($m_\ell$) when the principal quantum number ($n$) is 3 and the angular momentum quantum number ($\ell$) is 1?

**Strategy** Use the rules governing the allowed values of $m_\ell$. Recall that the possible values of $m_\ell$ depend on the value of $\ell$, not on the value of $n$.

**Setup** The possible values of $m_\ell$ are $-\ell, \ldots, 0, \ldots, +\ell$.

**Solution** The possible values of $m_\ell$ are $-1$, 0, and $+1$.

**Think About It** Consult Table 6.2 to make sure your answer is correct. Table 6.2 confirms that it is the value of $\ell$, not the value of $n$, that determines the possible values of $m_\ell$.

**Practice Problem A** (a) What are the possible values for $m_\ell$ when the principal quantum number ($n$) is 2 and the angular momentum quantum number ($\ell$) is 0? (b) What are the possible values for $m_\ell$ when the principal quantum number ($n$) is 3 and the angular momentum quantum number ($\ell$) is 2?

**Practice Problem B** (a) What is the lowest possible value of the principal quantum number ($n$) when the angular momentum quantum number ($\ell$) is 1? (b) What are the possible values of the angular momentum quantum number ($\ell$) when the magnetic quantum number ($m_\ell$) is 0?

---

### Electron Spin Quantum Number ($m_s$)

Whereas three quantum numbers are sufficient to describe an atomic orbital, an additional quantum number becomes necessary to describe an electron that occupies the orbital.

Experiments on the emission spectra of hydrogen and sodium atoms indicated that each line in the emission spectra could be split into two lines by the application of an external magnetic field. The only way physicists could explain these results was to assume that electrons act like tiny magnets. If electrons are thought of as spinning on their own axes, as Earth does, their magnetic properties can be accounted for. According to electromagnetic theory, a spinning charge generates a magnetic field, and it is this motion that causes an electron to behave like a magnet. Figure 6.16 shows the two possible spinning motions of an electron, one clockwise and the other counterclockwise. To specify the electron's spin, we use the ***electron spin quantum number ($m_s$).*** Because there are two possible directions of spin, opposite each other, $m_s$ has *two* possible values: $+\frac{1}{2}$ and $-\frac{1}{2}$.

Conclusive proof of electron spin was established by Otto Stern[13] and Walther Gerlach[14] in 1924. Figure 6.17 shows the basic experimental arrangement. A beam of gaseous atoms generated in a hot furnace passes through a nonuniform magnetic field. The interaction between an electron and the magnetic field causes the atom to be deflected from its straight-line path. Because the direction of spin is random, the electrons in *half* of the atoms will be spinning in one direction. Those atoms will be deflected in one way. The electrons in the other half of the atoms will be spinning in the *opposite* direction. Those atoms will be deflected in the other direction. Thus, two spots of equal intensity are observed on the detecting screen.

To summarize, we can designate an *orbital* in an atom with a set of *three* quantum numbers. These three quantum numbers indicate the size ($n$), shape ($\ell$), and orientation ($m_\ell$) of the orbital. A fourth quantum number ($m_s$) is necessary to designate the spin of an electron in the orbital.

### Checkpoint 6.6    Quantum Numbers

**6.6.1** Which of the following is a legitimate set of three quantum numbers: $n$, $\ell$, and $m_\ell$? (Select all that apply.)

a) 1, 0, 0

b) 2, 0, 0

c) 1, 0, +1

d) 2, 1, +1

e) 2, 2, −1

**6.6.2** How many orbitals are there in a subshell designated by the quantum numbers $n = 3$, $\ell = 2$?

a) 2

b) 3

c) 5

d) 7

e) 10

**6.6.3** How many subshells are there in the shell designated by $n = 3$?

a) 1

b) 2

c) 3

d) 6

e) 9

**6.6.4** What is the total number of orbitals in the shell designated by $n = 3$?

a) 1

b) 2

c) 3

d) 6

e) 9

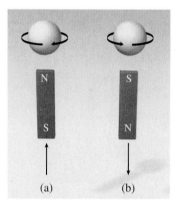

(a)                    (b)

**Figure 6.16** (a) Clockwise and (b) counterclockwise spins of an electron. The magnetic fields generated by these two spinning motions are analogous to those from the two magnets. The upward and downward arrows are used to denote the direction of spin.

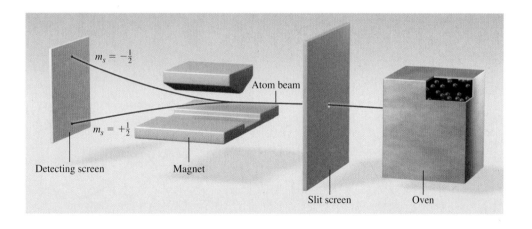

**Figure 6.17** Experimental arrangement for demonstrating electron spin. A beam of atoms is directed through a magnetic field. When a hydrogen atom, with a single electron, passes through the field, it is deflected in one direction or the other, depending on the direction of the electron's spin. In a stream consisting of many atoms, there will be equal distributions of the two kinds of spins, so two spots of equal intensity are detected on the screen.

---

13. Otto Stern (1888–1969). German physicist. He made important contributions to the study of the magnetic properties of atoms and the kinetic theory of gases. Stern was awarded the Nobel Prize in Physics in 1943.

14. Walther Gerlach (1889–1979). German physicist. Gerlach's main area of research was in quantum theory.

## 6.7  Atomic Orbitals

Strictly speaking, an atomic orbital does not have a well-defined shape because the wave function characterizing the orbital extends from the nucleus to infinity. In that sense, it is difficult to say what an orbital looks like. On the other hand, it is certainly useful to think of orbitals as having specific shapes. Being able to visualize atomic orbitals is essential to understanding the formation of chemical bonds and molecular geometry, which are discussed in Chapters 8 and 9. In this section, we will look at each type of orbital separately.

### *s* Orbitals

For any value of the principal quantum number ($n$), the value 0 is possible for the angular momentum quantum number ($\ell$), corresponding to an *s* subshell. Furthermore, when $\ell = 0$, the magnetic quantum number ($m_\ell$) has only one possible value, 0, corresponding to an *s* orbital. Therefore, there is an *s* subshell in every shell, and each *s* subshell contains just one orbital, an **s orbital.**

**Student Annotation:** The radial probability distribution can be thought of as a map of "where an electron spends most of its time."

Figure 6.18 illustrates three ways to represent the distribution of electrons: the probability density, the spherical distribution of electron density, and the radial probability distribution (the

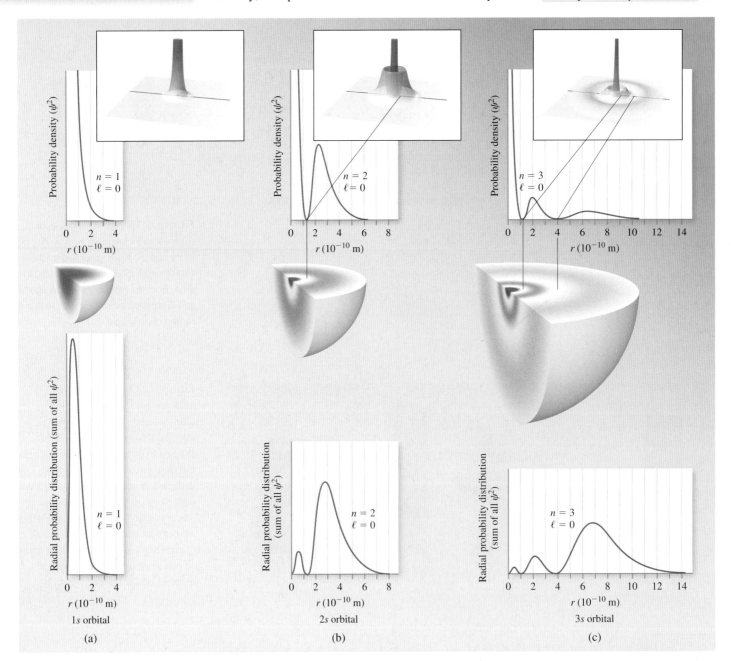

**Figure 6.18**    From top to bottom, the *probability density* and corresponding relief map, the distribution of electron density represented spherically with shading corresponding to the relief map above, and the *radial probability distribution* for (a) the 1*s*, (b) the 2*s*, and (c) the 3*s* orbitals of hydrogen.

probability of finding the electron as a function of distance from the nucleus) for the 1s, 2s, and 3s orbitals of hydrogen. The boundary surface (the outermost surface of the spherical representation) is a common way to represent atomic orbitals, incorporating the volume in which there is about a 90 percent probability of finding the electron at any given time.

All s orbitals are spherical in shape but differ in size, which increases as the principal quantum number increases. The radial probability distribution for the 1s orbital exhibits a maximum at 52.9 pm (0.529 Å) from the nucleus. The radial probability distribution plots for the 2s and 3s orbitals exhibit two and three maxima, respectively, with the greatest probability occurring at a greater distance from the nucleus as n increases. Between the two maxima for the 2s orbital there is a point on the plot where the probability drops to zero. This corresponds to a *node* in the electron density, where the standing wave has zero amplitude. There are two such nodes in the radial probability distribution plot of the 3s orbital.

Although the boundary surface diagram of an s orbital does not show the number of nodes, the most important features of atomic orbitals, for our purposes, are their overall shapes and *relative* sizes. These features are adequately represented by boundary surface diagrams.

**Student Annotation:** Interestingly, this distance is equal to the radius of the $n = 1$ orbit in the Bohr model of the hydrogen atom and is defined as the *Bohr radius.*

## p Orbitals

When the principal quantum number ($n$) is 2 or greater, the value 1 is possible for the angular momentum quantum number ($\ell$), corresponding to a $p$ subshell. And, when $\ell = 1$, the magnetic quantum number ($m_\ell$) has three possible values: $-1, 0,$ and $+1$, each corresponding to a different **p orbital.** Therefore, there is a $p$ subshell in every shell for which $n \geq 2$, and each $p$ subshell contains three $p$ orbitals. These three $p$ orbitals are labeled $2p_x$, $2p_y$, and $2p_z$ (Figure 6.19), with the subscripted letters indicating the axis along which each orbital is oriented. These three $p$ orbitals are identical in size, shape, and energy; they differ from one another only in orientation. Note, however, that there is no simple relation between the values of $m_\ell$ and the $x$, $y$, and $z$ directions. For our purpose, you need only remember that because there are three possible values of $m_\ell$, there are three $p$ orbitals with different orientations.

The boundary surface diagrams of $p$ orbitals in Figure 6.19 show that each $p$ orbital can be thought of as two lobes on opposite sides of the nucleus. Like $s$ orbitals, $p$ orbitals increase in size from $2p$ to $3p$ to $4p$ orbital and so on.

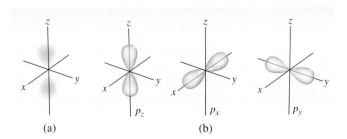

(a)                                    (b)

**Figure 6.19**   (a) Electron distribution in a $p$ orbital. (b) Boundary surfaces for the $p_x$, $p_y$, and $p_z$ orbitals.

## d Orbitals and Other Higher-Energy Orbitals

When the principal quantum number ($n$) is 3 or greater, the value 2 is possible for the angular momentum quantum number ($\ell$), corresponding to a $d$ subshell. When $\ell = 2$, the magnetic quantum number ($m_\ell$) has *five* possible values, $-2, -1, 0, +1,$ and $+2$, each corresponding to a different **d orbital.** Again there is no direct correspondence between a given orientation and a particular $m_\ell$ value. All the $3d$ orbitals in an atom are identical in energy and are labeled with subscripts denoting their orientation with respect to the $x$, $y$, and $z$ axes and to the planes defined by them. The $d$ orbitals that have higher principal quantum numbers ($4d$, $5d$, etc.) have shapes similar to those shown for the $3d$ orbitals in Figure 6.20.

**Figure 6.20**   Boundary surfaces for the $d$ orbitals.

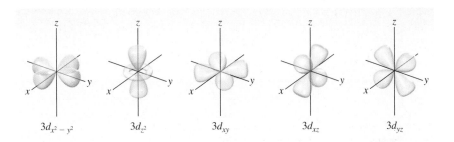

$3d_{x^2-y^2}$          $3d_{z^2}$          $3d_{xy}$          $3d_{xz}$          $3d_{yz}$

The *f orbitals* are important when accounting for the behavior of elements with atomic numbers greater than 57, but their shapes are difficult to represent. In general chemistry we will not concern ourselves with the shapes of orbitals having $\ell$ values greater than 2.

Sample Problem 6.8 shows how to label orbitals with quantum numbers.

---

## Sample Problem 6.8

List the values of $n$, $\ell$, and $m_\ell$ for each of the orbitals in a 4d subshell.

**Strategy** Consider the significance of the number and the letter in the 4d designation and determine the values of $n$ and $\ell$. There are multiple possible values for $m_\ell$, which will have to be deduced from the value of $\ell$.

**Setup** The integer at the beginning of an orbital designation is the principal quantum number ($n$). The letter in an orbital designation gives the value of the angular momentum quantum number ($\ell$). The magnetic quantum number ($m_s$) can have integral values of $-\ell, \ldots, 0, \ldots, +\ell$.

**Think About It** Consult Figure 6.15 to verify your answers.

**Solution** The values of $n$ and $\ell$ are 4 and 2, respectively, so the possible values of $m_\ell$ are $-2$, $-1$, 0, $+1$, and $+2$.

**Practice Problem A** Give the values of $n$, $\ell$, and $m_\ell$ for the orbitals in a 3d subshell.

**Practice Problem B** Using quantum numbers, explain why there is no 2d subshell.

---

## Energies of Orbitals

The energies of orbitals in the hydrogen atom, or any one-electron ion, depend only on the value of the principal quantum number ($n$), and energy increases as $n$ increases. For this reason, orbitals in the same shell have the same energy regardless of their subshell (Figure 6.21).

$$1s < 2s = 2p < 3s = 3p = 3d < 4s = 4p = 4d = 4f$$

**Student Annotation:** Recall that Bohr, whose model only *had* one quantum number, was able to calculate the energies of electrons in the hydrogen atom accurately.

Thus, all four orbitals (one 2s and three 2p) in the second shell have the same energy; all nine orbitals (one 3s, three 3p, and five 3d) in the third shell have the same energy; and all sixteen orbitals (one 4s, three 4p, five 4d, and seven 4f) in the fourth shell have the same energy. The energy picture is more complex for many-electron atoms than it is for hydrogen, as is discussed in Section 6.8.

**Figure 6.21** Orbital energy levels in the hydrogen atom. Each box represents one orbital. Orbitals with the same principal quantum number ($n$) all have the same energy.

## Checkpoint 6.7 — Atomic Orbitals

**6.7.1** How many orbitals are there in the 5*f* subshell?

a) 5

b) 7

c) 14

d) 16

e) 28

**6.7.2** The energy of an orbital in the hydrogen atom depends on _____.

a) *n*, ℓ, and $m_\ell$

b) *n* and ℓ

c) *n* only

d) ℓ only

e) $m_\ell$ only

**6.7.3** In a hydrogen atom, which orbitals are higher in energy than a 3*s* orbital? (Select all that apply.)

a) 3*p*

b) 4*s*

c) 2*p*

d) 3*d*

e) 4*p*

**6.7.4** Which of the following sets of quantum numbers, *n*, ℓ, and $m_\ell$, corresponds to a 3*p* orbital?

a) 3, 0, 0

b) 3, 1, 0

c) 3, 2, −1

d) 1, 1, −2

e) 1, 3, 1

## 6.8 Electron Configuration

The hydrogen atom is a particularly simple system because it contains only one electron. The electron may reside in the 1*s* orbital (the *ground state*), or it may be found in some higher-energy orbital (an *excited state*). With many-electron systems, we need to know the ground-state ***electron configuration***—that is, how the electrons are distributed in the various atomic orbitals. To do this, we need to know the relative energies of atomic orbitals in a many-electron system, which differ from those in a one-electron system such as hydrogen.

### Energies of Atomic Orbitals in Many-Electron Systems

Consider the two emission spectra shown in Figure 6.22. The spectrum of helium contains more lines than that of hydrogen. This indicates that there are more possible transitions, corresponding to emission in the visible range, in a helium atom than in a hydrogen atom. This is due to the *splitting* of energy levels caused by electrostatic interactions between helium's two electrons.

    Figure 6.23 shows the general order of orbital energies in a many-electron atom. In contrast to the hydrogen atom, in which the energy of an orbital depends only on the value of *n* (Figure 6.21), the energy of an orbital in a many-electron system depends on both the value of *n* and the value of ℓ. For example, 3*p* orbitals all have the same energy, but they are higher in energy than the 3*s* orbital and lower in energy than the 3*d* orbitals. In a many-electron atom, for a given value of *n*, the energy of an orbital increases with increasing value of ℓ. One important consequence of the splitting of energy levels is the relative energies of *d* orbitals in one shell and the *s* orbital in the next higher shell. As Figure 6.23 shows, the 4*s* orbital is lower in energy than the 3*d* orbitals. Likewise, the 5*s* orbital is lower in energy than the 4*d* orbital, and so on. This fact becomes important when we determine how the electrons in an atom populate the atomic orbitals.

**Student Annotation:** "Splitting" of energy levels refers to the splitting of a shell into subshells of different energies, as shown in Figure 6.23.

Hydrogen

| | | | | |
|---|---|---|---|---|
| 400 nm | 500 | 600 | 700 |

Helium

| | | | | |
|---|---|---|---|---|
| 400 nm | 500 | 600 | 700 |

**Figure 6.22** Comparison of the emission spectra of H and He.

**Figure 6.23** Orbital energy levels in many-electron atoms. For a given value of $n$, orbital energy increases with the value of $\ell$.

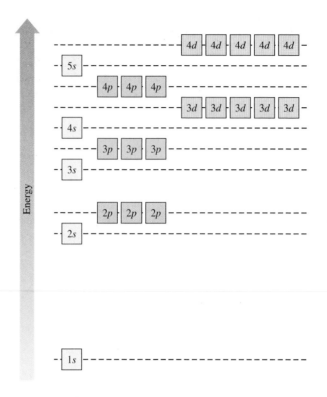

## The Pauli Exclusion Principle

According to the **Pauli**[15] **exclusion principle,** no two electrons in the same atom can have the same four quantum numbers. If two electrons in an atom have the same $n$, $\ell$, and $m_\ell$ values (meaning that they occupy the same *orbital*), then they must have different values of $m_s$; that is, one must have $m_s = +\frac{1}{2}$ and the other must have $m_s = -\frac{1}{2}$. Because there are only two possible values for $m_s$, and no two electrons in the same orbital may have the same value for $m_s$, a maximum of *two* electrons may occupy an atomic orbital, and these two electrons must have opposite spins.

**Student Annotation:** Two electrons in the same orbital with opposite spins are said to have *paired spins.*

We can indicate the arrangement of electrons in atomic orbitals with labels that identify each orbital (or subshell) and the number of electrons in it. Thus, we could describe a hydrogen atom in the ground state using $1s^1$.

**Student Annotation:** $1s^1$ is read as "one s one."

denotes the principal quantum number $n$

denotes the number of electrons in the orbital or subshell

denotes the angular momentum quantum number $\ell$

We can also represent the arrangement of electrons in an atom using *orbital diagrams,* in which each orbital is represented by a labeled box. The orbital diagram for a hydrogen atom in the ground state is

H   [↑]
    $1s^1$

The upward arrow denotes one of the two possible spins (one of the two possible $m_s$ values) of the electron in the hydrogen atom (the other possible spin is indicated with a downward arrow). Under certain circumstances, as we will see shortly, it is useful to indicate the explicit locations of electrons.

**Student Annotation:** The *ground state* for a many-electron atom is the one in which all the electrons occupy orbitals of the lowest possible energy.

The orbital diagram for a helium atom in the ground state is

He   [↑↓]
     $1s^2$

15. Wolfgang Pauli (1900–1958). Austrian physicist. One of the founders of quantum mechanics, Pauli was awarded the Nobel Prize in Physics in 1945.

The label $1s^2$ indicates there are *two* electrons in the $1s$ orbital. Note also that the arrows in the box point in opposite directions, representing opposite electron spins. Generally when an orbital diagram includes an orbital with a single electron, we represent it with an upward arrow—although we could represent it equally well with a downward arrow. The choice is arbitrary and has no effect on the energy of the electron.

## The Aufbau Principle

We can continue the process of writing electron configurations for elements based on the order of orbital energies and the Pauli exclusion principle. This process is based on the *Aufbau principle,* which makes it possible to "build" the periodic table of the elements and determine their electron configurations by steps. Each step involves adding one proton to the nucleus and one electron to the appropriate atomic orbital. Through this process we gain a detailed knowledge of the electron configurations of the elements. As we will see in later chapters, knowledge of electron configurations helps us understand and predict the properties of the elements. It also explains why the elements fit into the periodic table the way they do.

After helium, the next element in the periodic table is lithium, which has three electrons. Because of the restrictions imposed by the Pauli exclusion principle, an orbital can accommodate no more than two electrons. Thus, the third electron cannot reside in the $1s$ orbital. Instead, it must reside in the next available orbital with the lowest possible energy. According to Figure 6.23, this is the $2s$ orbital. Therefore, the electron configuration of lithium is $1s^2 2s^1$, and the orbital diagram is

$$\text{Li} \quad \boxed{\uparrow\downarrow} \quad \boxed{\uparrow}$$
$$\qquad\quad 1s^2 \qquad 2s^1$$

Similarly, we can write the electron configuration of beryllium as $1s^2 2s^2$ and represent it with the orbital diagram

$$\text{Be} \quad \boxed{\uparrow\downarrow} \quad \boxed{\uparrow\downarrow}$$
$$\qquad\quad 1s^2 \qquad 2s^2$$

With both the $1s$ and the $2s$ orbitals filled to capacity, the next electron, which is needed for the electron configuration of boron, must reside in the $2p$ subshell. Because all three $2p$ orbitals are of equal energy, or *degenerate,* the electron can occupy any one of them. By convention, we usually show the first electron to occupy the $p$ subshell in the first empty box in the orbital diagram.

$$\text{B} \quad \boxed{\uparrow\downarrow} \quad \boxed{\uparrow\downarrow} \quad \boxed{\uparrow\,|\,\,|\,\,}$$
$$\qquad 1s^2 \qquad 2s^2 \qquad\quad 2p^1$$

## Hund's Rule

Will the sixth electron, which is needed to represent the electron configuration of carbon, reside in the $2p$ orbital that is already half occupied, or will it reside in one of the other, empty $2p$ orbitals? According to *Hund's*[16] *rule,* the most stable arrangement of electrons in orbitals of equal energy is the one in which the number of electrons with the same spin is maximized. As we have seen, no two electrons in any orbital may have the same spin, so maximizing the number of electrons with the same spin requires putting the electrons in separate orbitals. Accordingly, in any subshell, an electron will occupy an empty orbital rather than one that already contains an electron.

The electron configuration of carbon is, therefore, $1s^2 2s^2 2p^2$, and its orbital diagram is

$$\text{C} \quad \boxed{\uparrow\downarrow} \quad \boxed{\uparrow\downarrow} \quad \boxed{\uparrow\,|\,\uparrow\,|\,\,}$$
$$\qquad 1s^2 \qquad 2s^2 \qquad\quad 2p^2$$

Similarly, the electron configuration of nitrogen is $1s^2 2s^2 2p^3$, and its orbital diagram is

$$\text{N} \quad \boxed{\uparrow\downarrow} \quad \boxed{\uparrow\downarrow} \quad \boxed{\uparrow\,|\,\uparrow\,|\,\uparrow}$$
$$\qquad 1s^2 \qquad 2s^2 \qquad\quad 2p^3$$

---

16. Frederick Hund (1896–1997). German physicist. Hund's work was mainly in quantum mechanics. He also helped to develop the molecular orbital theory of chemical bonding.

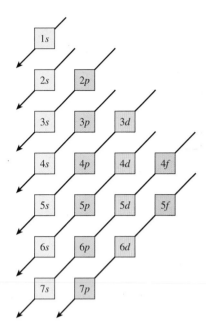

**Figure 6.24** A simple way to remember the order in which orbitals fill with electrons.

Multimedia

Atomic Structure—electron configurations.

Once all the 2p orbitals are singly occupied, additional electrons will have to pair with those already in the orbitals. Thus, the electron configurations and orbital diagrams for O, F, and Ne are as follows:

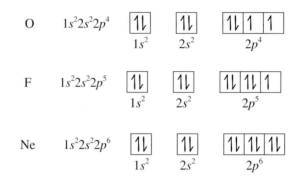

## General Rules for Writing Electron Configurations

Based on the preceding examples we can formulate the following general rules for determining the electron configuration of an element in the ground state:

1. Electrons will reside in the available orbitals of the lowest possible energy.
2. Each orbital can accommodate a maximum of two electrons.
3. Electrons will not pair in degenerate orbitals if an empty orbital is available.
4. Orbitals will fill in the order indicated in Figure 6.23. Figure 6.24 provides a simple way for you to remember the proper order.

Sample Problem 6.9 illustrates the procedure for determining the ground-state electron configuration of an atom.

<hr>

### Sample Problem 6.9

Write the electron configuration and give the orbital diagram of a calcium (Ca) atom ($Z = 20$).

**Strategy** Use the general rules given and the Aufbau principle to "build" the electron configuration of a calcium atom and represent it with an orbital diagram.

**Setup** Because $Z = 20$, we know that a Ca atom has 20 electrons. They will fill orbitals in the order designated in Figure 6.23, obeying the Pauli exclusion principle and Hund's rule. Orbitals will fill in the following order: $1s$, $2s$, $2p$, $3s$, $3p$, $4s$. Each $s$ subshell can contain a maximum of two electrons, whereas each $p$ subshell can contain a maximum of six electrons.

**Solution**

Ca     $1s^2 2s^2 2p^6 3s^2 3p^6 4s^2$     $\boxed{\uparrow\downarrow}$  $\boxed{\uparrow\downarrow}$  $\boxed{\uparrow\downarrow}\boxed{\uparrow\downarrow}\boxed{\uparrow\downarrow}$  $\boxed{\uparrow\downarrow}$  $\boxed{\uparrow\downarrow}\boxed{\uparrow\downarrow}\boxed{\uparrow\downarrow}$  $\boxed{\uparrow\downarrow}$

$\qquad\qquad\qquad\qquad\qquad\qquad\quad 1s^2 \qquad 2s^2 \qquad 2p^6 \qquad\quad 3s^2 \qquad 3p^6 \qquad\quad 4s^2$

**Think About It** Look at Figure 6.23 again to make sure you have filled the orbitals in the right order and that the sum of electrons is 20. Remember that the $4s$ orbital fills before the $3d$ orbitals.

**Practice Problem A** Write the electron configuration and give the orbital diagram of a rubidium (Rb) atom ($Z = 37$).

**Practice Problem B** Write the electron configuration and give the orbital diagram of a bromine (Br) atom ($Z = 35$).

**Checkpoint 6.8** **Electron Configuration**

**6.8.1** Which of the following electron configurations correctly represents the Ti atom?

a) $1s^2 \, 2s^2 \, 2p^6 \, 3s^2 \, 3p^6 \, 3d^4$

b) $1s^2 \, 2s^2 \, 2p^6 \, 3s^2 \, 3p^6 \, 4s^2 \, 3d^2$

c) $1s^2 \, 2s^2 \, 2p^6 \, 3s^2 \, 3p^6 \, 4s^2 \, 3d^{10}$

d) $1s^2 \, 2s^2 \, 2p^6 \, 3s^2 \, 3p^6 \, 3d^{10}$

e) $1s^2 \, 2s^2 \, 2p^6 \, 3s^2 \, 3p^6 \, 4s^4$

**6.8.2** What element is represented by the following electron configuration? $1s^2 \, 2s^2 \, 2p^6 \, 3s^2 \, 3p^6 \, 4s^2 \, 3d^{10} \, 4p^4$

a) Br

b) As

c) S

d) Se

e) Te

**6.8.3** Which orbital diagram is correct for the ground state S atom?

# 6.9 Electron Configurations and the Periodic Table

The electron configurations of all elements except hydrogen and helium can be represented using a **noble gas core,** which shows in brackets the electron configuration of the noble gas element that most recently precedes the element in question, followed by the electron configuration in the outermost occupied subshells. Figure 6.25 gives the outermost ground-state electron configurations of elements from H ($Z = 1$) through Rg ($Z = 111$). Notice the similar pattern of electron configurations in the elements lithium ($Z = 3$) through neon ($Z = 10$) and those of sodium ($Z = 11$) through argon ($Z = 18$). Both Li and Na, for example, have the configuration $ns^1$ in their outermost occupied subshells. For Li, $n = 2$; for Na, $n = 3$. Both F and Cl have electron configuration $ns^2np^5$, where $n = 2$ for F and $n = 3$ for Cl, and so on.

As mentioned in Section 6.8, the $4s$ subshell is filled before the $3d$ subshell in a many-electron atom (see Figure 6.23). Thus, the electron configuration of potassium ($Z = 19$) is $1s^2 2s^2 2p^6 3s^2 3p^6 4s^1$. Because $1s^2 2s^2 2p^6 3s^2 3p^6$ is the electron configuration of argon, we can simplify the electron configuration of potassium by writing $[Ar]4s^1$, where $[Ar]$ denotes the "argon core."

$$K \qquad \underbrace{1s^2 2s^2 2p^6 3s^2 3p^6}_{[Ar]} 4s^1 \qquad \longrightarrow \qquad [Ar]\, 4s^1$$

The placement of the outermost electron in the $4s$ orbital (rather than in the $3d$ orbital) of potassium is strongly supported by experimental evidence. The physical and chemical properties of potassium are very similar to those of lithium and sodium, the first two alkali metals. In both lithium and sodium, the outermost electron is in an $s$ orbital (there is no doubt that their outermost electrons occupy $s$ orbitals because there is no $1d$ or $2d$ subshell). Based on its similarities to the other alkali metals, we expect potassium to have an analogous electron configuration; that is, we expect the last electron in potassium to occupy the $4s$ rather than the $3d$ orbital.

The elements from Group 3B through Group 1B are *transition metals* [◄◄ Section 2.4]. Transition metals either have incompletely filled $d$ subshells or readily give rise to cations that have incompletely filled $d$ subshells. In the first transition metal series, from scandium ($Z = 21$) through copper ($Z = 29$), additional electrons are placed in the $3d$ orbitals according to Hund's rule. However, there are two anomalies. The electron configuration of chromium ($Z = 24$) is $[Ar]4s^1 3d^5$ and not $[Ar]4s^2 3d^4$, as we might expect. A similar break in the pattern is observed for

**Student Annotation:** Although zinc and the other elements in Group 2B sometimes are included under the heading "transition metals," they neither have nor readily acquire partially filled $d$ subshells. Strictly speaking, they are *not* transition metals.

**Figure 6.25**    Outermost ground-state electron configurations for the known elements.

**Student Annotation:** Electron configurations such as these may also be written with the *d* subshell first. For example, [Ar]$4s^13d^{10}$ can also be written as [Ar]$3d^{10}4s^1$. Either way is acceptable.

copper, whose electron configuration is [Ar]$4s^13d^{10}$ rather than [Ar]$4s^23d^9$. The reason for these anomalies is that a slightly greater stability is associated with the half-filled ($3d^5$) and completely filled ($3d^{10}$) subshells.

For elements Zn ($Z = 30$) through Kr ($Z = 36$), the $3d$, $4s$, and $4p$ subshells fill in a straightforward manner. With rubidium ($Z = 37$), electrons begin to enter the $n = 5$ energy level.

Some of the electron configurations in the second transition metal series [yttrium ($Z = 39$) through silver ($Z = 47$)] are also irregular, but the details of many of these irregularities are beyond the scope of this text and we will not be concerned with them.

The sixth period of the periodic table begins with cesium ($Z = 55$) and barium ($Z = 56$), whose electron configurations are [Xe]$6s^1$ and [Xe]$6s^2$, respectively. Following barium, there is a gap in the periodic table where the ***lanthanide (rare earth) series*** belongs. The lanthanides are a series of 14 elements that have incompletely filled $4f$ subshells or that readily give rise to *cations* that have incompletely filled 4$f$ subshells. The lanthanides (and the actinides, to be discussed next) are shown at the bottom of the periodic table to keep the table from being too wide.

Lanthanides 6: La Ce Pr Nd Pm Sm Eu Gd Tb Dy Ho Er Tm Yb

Actinides 7: Ac Th Pa U Np Pu Am Cm Bk Cf Es Fm Md No

In theory, the lanthanides arise from the filling of the seven degenerate 4$f$ orbitals. In reality, however, the energies of the 5$d$ and 4$f$ orbitals are very close and the electron configurations of these elements sometimes involve 5$d$ electrons. For example, in lanthanum itself ($Z = 57$) the 4$f$ orbital is slightly higher in energy than the 5$d$ orbital. Thus, lanthanum's electron configuration is $[Xe]6s^2 5d^1$ rather than $[Xe]6s^2 4f^1$.

After the 4$f$ subshell is completely filled, the next electron enters the 5$d$ subshell of lutetium ($Z = 71$). This series of elements, including lutetium and hafnium ($Z = 72$) and extending through mercury ($Z = 80$), is characterized by the filling of the 5$d$ subshell. The 6$p$ subshells are filled next, which takes us to radon ($Z = 86$).

The last row of elements begins with francium ($Z = 87$; electron configuration $[Rn]7s^1$) and radium ($Z = 88$; electron configuration $[Rn]7s^2$), and then continues with the **actinide series,** which starts at actinium ($Z = 89$) and ends with nobelium ($Z = 102$). The actinide series has partially filled 5$f$ and/or 6$d$ subshells. The elements lawrencium ($Z = 103$) through darmstadtium ($Z = 110$) have a filled 5$f$ subshell and are characterized by the filling of the 6$d$ subshell.

With few exceptions, you should be able to write the electron configuration of any element, using Figure 6.23 (or Figure 6.24) as a guide. Elements that require particular care are the transition metals, the lanthanides, and the actinides. You may notice from looking at the electron configurations of gadolinium ($Z = 64$) and curium ($Z = 96$) that half-filled $f$ subshells also appear to exhibit slightly enhanced stability. As we noted earlier, at larger values of the principal quantum number $n$, the order of subshell filling may be irregular due to the closeness of the energy levels.

Figure 6.26 groups the elements according to the type of subshell in which the outermost electrons are placed. Elements whose outermost electrons are in an $s$ subshell are referred to as $s$-block elements, those whose outermost electrons are in a $p$ subshell are referred to as $p$-block elements, and so on.

Sample Problem 6.10 shows how to write electron configurations.

**Student Annotation:** When $n = 4$, $\ell$ can equal 3, corresponding to an $f$ subshell. There are seven possible values for $m_\ell$ when $\ell = 3$: $-3, -2, -1, 0, +1, +2,$ and $+3$. Therefore, there are seven $f$ orbitals.

**Student Annotation:** Most of these elements are not found in nature but have been synthesized in nuclear reactions, which are the subject of Chapter 20.

| 1s | | | | |
| 2s | | | | 2p |
| 3s | | | | 3p |
| 4s | | | 3d | 4p |
| 5s | | | 4d | 5p |
| 6s | | 4f | 5d | 6p |
| 7s | | 5f | 6d | 7p |

**Figure 6.26**    Classification of groups of elements in the periodic table according to the type of subshell being filled with electrons.

## Sample Problem 6.10

Without referring to Figure 6.25, write the electron configuration for an arsenic atom ($Z = 33$) in the ground state.

**Strategy**  Use Figure 6.23 or Figure 6.24 to determine the order in which the subshells will fill, and then assign electrons to the appropriate subshells.

**Setup**  The noble gas core for As is [Ar], where $Z = 18$ for Ar. The order of filling beyond the noble gas core is $4s$, $3d$, and $4p$. Fifteen electrons must go into these subshells because there are $33 - 18 = 15$ electrons in As beyond its noble gas core.

**Solution**  $[Ar]4s^2 3d^{10} 4p^3$

> **Think About It**  Arsenic is a *p*-block element; therefore, we should expect its outermost electrons to reside in a *p* subshell.

**Practice Problem A**  Without referring to Figure 6.25, write the electron configuration for a radium atom ($Z = 88$) in the ground state.

**Practice Problem B**  Without referring to Figure 6.25, determine the identity of the element with the following electron configuration:

$$[Xe]6s^2 4f^{14} 5d^{10} 6p^5$$

## Checkpoint 6.9    Electron Configurations and the Periodic Table

**6.9.1**  Which of the following electron configurations correctly represents the Ag atom?

a) $[Kr]5s^2 4d^9$

b) $[Kr]5s^2 4d^{10}$

c) $[Kr]5s^1 4d^{10}$

d) $[Xe]5s^2 4d^9$

e) $[Xe]5s^1 4d^{10}$

**6.9.2**  What element is represented by the following electron configuration: $[Kr]5s^2 4d^{10} 5p^5$?

a) Tc

b) Br

c) I

d) Xe

e) Te

**6.9.3**  Which of the following is a *d*-block element? (Select all that apply.)

a) Sb

b) Au

c) Ca

d) Zn

e) U

**6.9.4**  Which of the following is a *p*-block element? (Select all that apply.)

a) Pb

b) C

c) Sr

d) Xe

e) Na

# Applying What You've Learned

Two of the most commonly used lasers in medicine are the $CO_2$ laser, which emits a beam at 10,600 nm, and the Ar ion laser, which emits beams at 488 nm and 514 nm. The $CO_2$ laser is used to remove benign skin lesions such as warts and moles; to remove tumors from especially sensitive areas such as the brain and spinal cord; to resurface scars, skin irregularities, and wrinkles for cosmetic purposes; and as a "laser scalpel" in situations where there may be a risk of excessive bleeding. Ar laser light is strongly absorbed by melanin, the pigment found in human skin, making it useful in the removal of dark-colored skin anomalies. Ar lasers are also used in some surgeries on the retina and the inner ear.

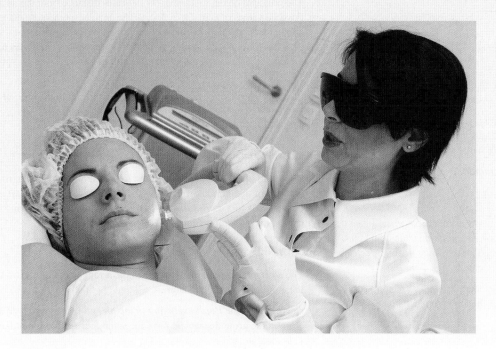

## Problems:

a)  Calculate the frequency and the energy per photon of the light emitted by a $CO_2$ laser. [◀◀ Sample Problems 6.1 and 6.2]

b)  Calculate the difference in energy per photon in the two wavelengths emitted by the Ar laser. Which of the wavelengths is more energetic?
    [◀◀ Sample Problem 6.3]

c)  Calculate the de Broglie wavelength of an argon atom moving at 1000 m/s.
    [◀◀ Sample Problem 6.5]

d)  Give the values of $n$, $\ell$, and $m_\ell$ for the $3p$ orbitals in an Ar atom.
    [◀◀ Sample Problem 6.8]

e)  Write the electron configuration of Ar. [◀◀ Sample Problems 6.9 and 6.10]

# CHAPTER SUMMARY

## Section 6.1

- What we commonly refer to as "light" is actually the visible portion of the *electromagnetic spectrum.* All light has certain common characteristics including wavelength, frequency, and amplitude.

- *Wavelength* ($\lambda$) is the distance between two crests or two troughs of a wave. *Frequency* ($\nu$) is the number of waves that pass a point per unit time. *Amplitude* is the distance between the midpoint and crest or trough of a wave.

- *Electromagnetic waves* have both electric and magnetic components that are both mutually perpendicular and in phase.

## Section 6.2

- *Blackbody radiation* is the electromagnetic radiation given off by a solid when it is heated.

- Max Planck proposed that energy, like matter, was composed of tiny, indivisible "packages" called *quanta. Quanta* is the plural of *quantum.*

- Albert Einstein used Planck's revolutionary quantum theory to explain the *photoelectric effect,* in which electrons are emitted when light of a certain minimum frequency shines on a metal surface.

- A *quantum* of light is referred to as a *photon.*

## Section 6.3

- An *emission spectrum* is the light given off by an object when it is excited thermally. Emission spectra may be *continuous,* including all the wavelengths within a particular range, or they may be *line spectra,* consisting only of certain discrete wavelengths.

- The *ground state* is the lowest possible energy state for an atom. An *excited state* is any energy level higher than the ground state.

## Section 6.4

- A *node* is a point at which a standing wave has zero amplitude.

- Having observed that light could exhibit particle-like behavior, de Broglie proposed that matter might also exhibit wavelike behavior. The *de Broglie wavelength* is the wavelength associated with a particle of very small mass. Soon after de Broglie's proposal, experiments showed that electrons could exhibit diffraction—a property of waves.

## Section 6.5

- According to the *Heisenberg uncertainty principle,* the product of the uncertainty of the *location* and the uncertainty of the *momentum* of a very small particle must have a certain minimum value. It is thus impossible to know simultaneously both the location and momentum of an electron.

- The *electron density* gives the probability of finding an electron in a particular region in an atom. An *atomic orbital* is the region of three-dimensional space, defined by $\psi^2$ (the square of the wave function, $\psi$), where the probability of finding an electron is high. An atomic orbital can accommodate a maximum of *two* electrons.

## Section 6.6

- An atomic orbital is defined by three *quantum numbers:* the *principal quantum number (n),* the *angular momentum quantum number ($\ell$),* and the *magnetic quantum number ($m_\ell$).*

- The principal quantum number ($n$) indicates distance from the nucleus. Possible values of $n$ are $(1, 2, 3, \ldots)$ The angular momentum quantum number ($\ell$) indicates the shape of the orbital. Possible values of $\ell$ are $(0, 1, \ldots, n-1)$. The magnetic quantum number ($m_\ell$) indicates the orbital's orientation in space. Possible values of $m_\ell$ are $(-\ell, \ldots, 0, \ldots, +\ell)$.

- Two electrons that occupy the same atomic orbital in the ground state must have different *electron spin quantum numbers ($m_s$),* either $+\frac{1}{2}$ or $-\frac{1}{2}$.

## Section 6.7

- The value of the angular momentum quantum number ($\ell$) determines the type of the atomic orbital: $\ell = 0$ corresponds to an *s orbital,* $\ell = 1$ corresponds to a *p orbital,* $\ell = 2$ corresponds to a *d orbital,* and $\ell = 3$ corresponds to an *f orbital.*

## Section 6.8

- The *electron configuration* specifies the arrangement of electrons in the atomic orbitals of an atom.

- According to the *Pauli exclusion principle,* no two electrons in an atom in the ground state can have the same four quantum numbers, $n$, $\ell$, $m_\ell$, and $m_s$.

- The *Aufbau principle* describes the theoretical, sequential building up of the elements in the periodic table by the stepwise addition of protons and electrons.

- Atomic orbitals that have the same energy are called *degenerate.* According to *Hund's rule,* degenerate orbitals must all contain one electron before any can contain two electrons.

## Section 6.9

- The *noble gas core* makes it possible to abbreviate the writing of electron configurations.

- The *lanthanide (rare earth) series* and *actinide series* appear at the bottom of the periodic table. They represent the filling of f orbitals.

# KEY WORDS

Actinide series, 245

Amplitude, 212

Angular momentum quantum number ($\ell$), 233

Atomic orbital, 232

Aufbau principle, 241

Blackbody radiation, 215

d orbital, 237

de Broglie wavelength, 229

Degenerate, 241

Electromagnetic spectrum, 212

Electromagnetic wave, 213

Electron configuration, 239

Electron density, 232

Electron spin quantum number ($m_s$), 235

Emission spectra, 220

Excited state, 222

f orbital, 238

Frequency ($\nu$), 212

Ground state, 222

Heisenberg uncertainty principle, 231

Hund's rule, 241

# KEY EQUATIONS

6.1    $c = \lambda \nu$

6.2    $E = h\nu$

6.3    $h\nu = \text{KE} + W$

6.4    $\dfrac{1}{\lambda} = R_\infty \left( \dfrac{1}{n_1^2} - \dfrac{1}{n_2^2} \right)$

6.5    $E_n = -2.18 \times 10^{-18} \, \text{J} \left( \dfrac{1}{n^2} \right)$

6.6    $\Delta E = h\nu = -2.18 \times 10^{-18} \, \text{J} \left( \dfrac{1}{n_f^2} - \dfrac{1}{n_i^2} \right)$

6.7    $\dfrac{1}{\lambda} = \dfrac{-2.18 \times 10^{-18} \, \text{J}}{hc} \left( \dfrac{1}{n_f^2} - \dfrac{1}{n_i^2} \right)$

6.8    $2\pi r = n\lambda$

6.9    $\lambda = \dfrac{h}{mu}$

6.10    $\Delta x \cdot \Delta p \geq \dfrac{h}{4\pi}$

6.11    $\Delta x \cdot m\Delta u \geq \dfrac{h}{4\pi}$

# QUESTIONS AND PROBLEMS

## Section 6.1: The Nature of Light

### Review Questions

6.1    What is a wave? Using a diagram, define the following terms associated with waves: wavelength, frequency, amplitude.

6.2    What are the units for wavelength and frequency of electromagnetic waves? What is the speed of light in meters per second and miles per hour?

6.3    List the types of electromagnetic radiation, starting with the radiation having the longest wavelength and ending with the radiation having the shortest wavelength.

6.4    Give the high and low wavelength values that define the visible region of the electromagnetic spectrum.

### Problems

6.5    (a) What is the wavelength (in nm) of light having a frequency of $8.6 \times 10^{13}$ Hz? (b) What is the frequency (in Hz) of light having a wavelength of 566 nm?

6.6    (a) What is the frequency of light having a wavelength of 456 nm? (b) What is the wavelength (in nm) of radiation having a frequency of $2.45 \times 10^9$ Hz? (This is the type of radiation used in microwave ovens.)

6.7    The average distance between Mars and Earth is about $1.3 \times 10^8$ miles. How long would it take video images transmitted from the Mars *Spirit* rover on Mars' surface to reach Earth (1 mile = 1.61 km)?

6.8    How many minutes would it take a radio wave to travel from the planet Venus to Earth? (The average distance from Venus to Earth = 28 million miles.)

6.9    The SI unit of time is the second, which is defined as 9,192,631,770 cycles of radiation associated with a certain emission process in the cesium atom. Calculate the wavelength of this radiation (to three significant figures). In which region of the electromagnetic spectrum is this wavelength found?

6.10    The SI unit of length is the meter, which is defined as the length equal to 1,650,763.73 wavelengths of the light emitted by a particular energy transition in krypton atoms. Calculate the frequency of the light to three significant figures.

## Section 6.2: Quantum Theory

### Review Questions

6.11    Briefly explain Planck's quantum theory and explain what a quantum is. What are the units for Planck's constant?

6.12    Give two everyday examples that illustrate the concept of quantization.

6.13    Explain what is meant by the photoelectric effect.

6.14    What are photons? What role did Einstein's explanation of the photoelectric effect play in the development of the particle-wave interpretation of the nature of electromagnetic radiation?

*Problems*

6.15    A photon has a wavelength of 705 nm. Calculate the energy of the photon in joules.

6.16    The blue color of the sky results from the scattering of sunlight by molecules in the air. The blue light has a frequency of about $7.5 \times 10^{14}$ Hz. (a) Calculate the wavelength (in nm) associated with this radiation, and (b) calculate the energy (in joules) of a single photon associated with this frequency.

6.17    A photon has a frequency of $6.5 \times 10^9$ Hz. (a) Convert this frequency into wavelength (nm). Does this frequency fall in the visible region? (b) Calculate the energy (in joules) of this photon. (c) Calculate the energy (in joules) of 1 mole of photons all with this frequency.

6.18    What is the wavelength (in nm) of radiation that has an energy content of $2.13 \times 10^3$ kJ/mol? In which region of the electromagnetic spectrum is this radiation found?

6.19    When copper is bombarded with high-energy electrons, X rays are emitted. Calculate the energy (in joules) associated with the photons if the wavelength of the X rays is 0.154 nm.

6.20    A particular form of electromagnetic radiation has a frequency of $9.87 \times 10^{15}$ Hz. (a) What is its wavelength in nanometers? In meters? (b) To what region of the electromagnetic spectrum would you assign it? (c) What is the energy (in joules) of one quantum of this radiation?

6.21    Photosynthesis makes use of visible light to bring about chemical changes. Explain why heat energy in the form of infrared radiation is ineffective for photosynthesis.

6.22    The retina of a human eye can detect light when radiant energy incident on it is at least $4.0 \times 10^{-17}$ J. For light of 585-nm wavelength, how many photons does this energy correspond to?

6.23    The radioactive $^{60}$Co isotope is used in nuclear medicine to treat certain types of cancer. Calculate the wavelength and frequency of an emitted gamma particle having the energy of $1.29 \times 10^{11}$ J/mol.

## Section 6.3: Bohr's Theory of the Hydrogen Atom

**Visualizing Chemistry**

**Figure 6.10**

VC 6.1    Which of the following best explains why we see only four lines in the emission spectrum of hydrogen?

   a) Hydrogen has only four different electronic transitions.
   b) Only four of hydrogen's electronic transitions correspond to visible wavelengths.
   c) The other lines in hydrogen's emission spectrum can't be seen easily against the black background.

VC 6.2    One way to see the emission spectrum of hydrogen is to view the hydrogen in an electric discharge tube through a *spectroscope*, a device that separates the wavelengths. Why can we not view the emission spectrum simply by pointing the spectroscope at a sample of hydrogen confined in a glass tube or flask?

   a) Without the electrons being in excited states, there would be no emission of light.
   b) The glass would make it impossible to see the emission spectrum.
   c) Hydrogen alone does not exhibit an emission spectrum—it must be combined with oxygen.

VC 6.3    How many lines would we see in the emission spectrum of hydrogen if the downward transitions from excited states all ended at $n = 1$ and no transitions ended at $n = 2$?

   a) We would still see four lines.
   b) We would see five lines.
   c) We would not see any lines.

VC 6.4    For a hydrogen atom in which the electron has been excited to $n = 4$, how many different transitions can occur as the electron eventually returns to the ground state?

   a) 1.
   b) 2.
   c) 3.

*Review Questions*

6.24    What are emission spectra? How do line spectra differ from continuous spectra?

6.25    What is an energy level? Explain the difference between ground state and excited state.

6.26    Briefly describe Bohr's theory of the hydrogen atom and how it explains the appearance of an emission spectrum. How does Bohr's theory differ from concepts of classical physics?

6.27    Explain the meaning of the negative sign in Equation 6.5.

*Problems*

6.28    Explain why elements produce their own characteristic colors when they emit photons.

6.29    Some copper compounds emit green light when they are heated in a flame. How would you determine whether the light is of one wavelength or a mixture of two or more wavelengths?

6.30    Is it possible for a fluorescent material to emit radiation in the ultraviolet region after absorbing visible light? Explain your answer.

6.31    Explain how astronomers are able to tell which elements are present in distant stars by analyzing the electromagnetic radiation emitted by the stars.

6.32    Consider the following energy levels of a hypothetical atom:

   $E_4$: $-1.0 \times 10^{-19}$ J

   $E_3$: $-5.0 \times 10^{-19}$ J

   $E_2$: $-10 \times 10^{-19}$ J

   $E_1$: $-15 \times 10^{-19}$ J

   (a) What is the wavelength of the photon needed to excite an electron from $E_1$ to $E_4$? (b) What is the energy (in joules) a photon must have to excite an electron from $E_2$ to $E_3$? (c) When an electron drops from the $E_3$ level to the $E_1$ level, the atom is said to undergo emission. Calculate the wavelength of the photon emitted in this process.

**6.33**    The first line of the Balmer series occurs at a wavelength of 656.3 nm. What is the energy difference between the two energy levels involved in the emission that results in this spectral line?

**6.34**    Calculate the wavelength (in nm) of a photon emitted by a hydrogen atom when its electron drops from the $n = 7$ state to the $n = 2$ state.

**6.35**    Calculate the frequency (Hz) and wavelength (nm) of the emitted photon when an electron drops from the $n = 4$ to the $n = 3$ level in a hydrogen atom.

**6.36**    Careful spectral analysis shows that the familiar yellow light of sodium lamps (such as street lamps) is made up of photons of two wavelengths, 589.0 nm and 589.6 nm. What is the difference in energy (in joules) between photons with these wavelengths?

**6.37**    An electron in the hydrogen atom makes a transition from an energy state of principal quantum number $n_i$ to the $n = 1$ state. If the photon emitted has a wavelength of 94.9 nm, what is the value of $n_i$?

## Section 6.4: Wave Properties of Matter

### Review Questions

**6.38**    How does de Broglie's hypothesis account for the fact that the energies of the electron in a hydrogen atom are quantized?

**6.39**    Why is Equation 6.9 meaningful only for submicroscopic particles, such as electrons and atoms, and not for macroscopic objects?

**6.40**    Does a baseball in flight possess wave properties? If so, why can we not determine its wave properties?

### Problems

**6.41**    Thermal neutrons are neutrons that move at speeds comparable to those of air molecules at room temperature. These neutrons are most effective in initiating a nuclear chain reaction among $^{235}U$ isotopes. Calculate the wavelength (in nm) associated with a beam of neutrons moving at $7.00 \times 10^2$ m/s (mass of a neutron = $1.675 \times 10^{-27}$ kg).

**6.42**    Protons can be accelerated to speeds near that of light in particle accelerators. Estimate the wavelength (in nm) of such a proton moving at $2.90 \times 10^8$ m/s (mass of a proton = $1.673 \times 10^{-27}$ kg).

**6.43**    What is the de Broglie wavelength (in cm) of a 12.4-g hummingbird flying at $1.20 \times 10^2$ mph (1 mile = 1.61 km)?

**6.44**    What is the de Broglie wavelength (in nm) associated with a 2.5-g Ping-Pong ball traveling at 15 mph?

## Section 6.5: Quantum Mechanics

### Review Questions

**6.45**    What are the inadequacies of Bohr's theory?

**6.46**    What is the Heisenberg uncertainty principle? What is the Schrödinger equation?

**6.47**    What is the physical significance of the wave function?

**6.48**    How is the concept of electron density used to describe the position of an electron in the quantum mechanical treatment of an atom?

**6.49**    What is an atomic orbital? How does an atomic orbital differ from an orbit?

### Problems

**6.50**    Alveoli are tiny sacs of air in the lungs. Their average diameter is $5.0 \times 10^{-5}$ m. Calculate the uncertainty in the velocity of an oxygen molecule ($5.3 \times 10^{-26}$ kg) trapped within a sac. (*Hint:* The maximum uncertainty in the position of the molecule is given by the diameter of the sac.)

**6.51**    Suppose that photons of blue light (430 nm) are used to locate the position of a 2.80-g Ping-Pong ball in flight and that the uncertainty in the position is equal to one wavelength. What is the minimum uncertainty in the speed of the Ping-Pong ball? Comment on the magnitude of your result.

## Section 6.6: Quantum Numbers

### Review Questions

**6.52**    Describe the four quantum numbers used to characterize an electron in an atom.

**6.53**    Which quantum number defines a shell? Which quantum numbers define a subshell?

**6.54**    Which of the four quantum numbers ($n$, $\ell$, $m_\ell$, $m_s$) determine (a) the energy of an electron in a hydrogen atom and in a many-electron atom, (b) the size of an orbital, (c) the shape of an orbital, (d) the orientation of an orbital in space?

### Problems

**6.55**    An electron in a certain atom is in the $n = 2$ quantum level. List the possible values of $\ell$ and $m_\ell$ that it can have.

**6.56**    An electron in an atom is in the $n = 3$ quantum level. List the possible values of $\ell$ and $m_\ell$ that it can have.

**6.57**    List all the possible subshells and orbitals associated with the principal quantum number $n$, if $n = 4$.

**6.58**    List all the possible subshells and orbitals associated with the principal quantum number $n$, if $n = 5$.

## Section 6.7: Atomic Orbitals

### Review Questions

**6.59**    Describe the shapes of $s$, $p$, and $d$ orbitals. How are these orbitals related to the quantum numbers $n$, $\ell$, and $m_\ell$?

**6.60**    List the hydrogen orbitals in increasing order of energy.

**6.61**    Describe the characteristics of an $s$ orbital, $p$ orbital, and $d$ orbital. Which of the following orbitals do not exist: $1p$, $2s$, $2d$, $3p$, $3d$, $3f$, $4g$?

**6.62**    Why is a boundary surface diagram useful in representing an atomic orbital?

### Problems

**6.63**    Give the values of the quantum numbers associated with the following orbitals: (a) $2p$, (b) $3s$, (c) $5d$.

**6.64**    Give the values of the four quantum numbers of an electron in the following orbitals: (a) $3s$, (b) $4p$, (c) $3d$.

**6.65**    Discuss the similarities and differences between a $1s$ and a $2s$ orbital.

**6.66**    What is the difference between a $2p_x$ and a $2p_y$ orbital?

**6.67**    Why do the $3s$, $3p$, and $3d$ orbitals have the same energy in a hydrogen atom but different energies in a many-electron atom?

6.68    Make a chart of all allowable orbitals in the first four principal energy levels of the hydrogen atom. Designate each by type (for example, s, p), and indicate how many orbitals of each type there are.

6.69    For each of the following pairs of hydrogen orbitals, indicate which is higher in energy: (a) 1s, 2s; (b) 2p, 3p; (c) $3d_{xy}$, $3d_{yz}$; (d) 3s, 3d; (e) 4f, 5s.

6.70    Which orbital in each of the following pairs is lower in energy in a many-electron atom: (a) 2s, 2p; (b) 3p, 3d; (c) 3s, 4s; (d) 4d, 5f?

## Section 6.8: Electron Configuration

### Review Questions

6.71    What is electron configuration? Describe the roles that the Pauli exclusion principle and Hund's rule play in writing the electron configuration of elements.

6.72    Explain the meaning of the symbol $4d^6$.

6.73    State the Aufbau principle, and explain the role it plays in classifying the elements in the periodic table.

### Problems

6.74    Calculate the total number of electrons that can occupy (a) one s orbital, (b) three p orbitals, (c) five d orbitals, (d) seven f orbitals.

6.75    What is the total number of electrons that can be held in all orbitals having the same principal quantum number n?

6.76    Determine the maximum number of electrons that can be found in each of the following subshells: 3s, 3d, 4p, 4f, 5f.

6.77    Indicate the total number of (a) p electrons in N (Z = 7), (b) s electrons in Si (Z = 14), and (c) 3d electrons in S (Z = 16).

6.78    Indicate which of the following sets of quantum numbers in an atom are unacceptable and explain why: (a) $(1, 1, +\frac{1}{2}, -\frac{1}{2})$, (b) $(3, 0, -1, +\frac{1}{2})$, (c) $(2, 0, +1, +\frac{1}{2})$, (d) $(4, 3, -2, +\frac{1}{2})$, (e) (3, 2, +1, 1).

6.79    The ground-state electron configurations listed here are incorrect. Explain what mistakes have been made in each and write the correct electron configurations.
Al: $1s^2 2s^2 2p^4 3s^2 3p^3$
B: $1s^2 2s^2 2p^5$
F: $1s^2 2s^2 2p^6$

6.80    Indicate the number of unpaired electrons present in each of the following atoms: B, Ne, P, Sc, Mn, Se, Kr, Fe, Cd, I, Pb.

6.81    The electron configuration of a neutral atom is $1s^2 2s^2 2p^6 3s^2$. Write a complete set of quantum numbers for each of the electrons. Name the element.

6.82    Which of the following species has the greatest number of unpaired electrons: $S^+$, S, or $S^-$? Explain how you arrive at your answer.

## Section 6.9: Electron Configurations and the Periodic Table

### Review Questions

6.83    Describe the characteristics of transition metals.

6.84    What is the noble gas core? How does it simplify the writing of electron configurations?

6.85    What are the group and period of the element osmium?

6.86    Define the following terms and give an example of each: lanthanides, actinides.

6.87    Explain why the ground-state electron configurations of Cr and Cu are different from what we might expect.

6.88    Write the electron configuration of a xenon core.

6.89    Comment on the correctness of the following statement: The probability of finding two electrons with the same four quantum numbers in an atom is zero.

### Problems

6.90    Use the Aufbau principle to obtain the ground-state electron configuration of selenium.

6.91    Use the Aufbau principle to obtain the ground-state electron configuration of technetium.

6.92    Write the ground-state electron configurations for the following elements: B, V, C, As, I, Au.

6.93    Write the ground-state electron configurations for the following elements: Ge, Fe, Zn, Ni, W, Tl.

## Additional Problems

6.94    When a compound containing cesium ion is heated in a Bunsen burner flame, photons with an energy of $4.30 \times 10^{-19}$ J are emitted. What color is the cesium flame?

6.95    Discuss the current view of the correctness of the following statements. (a) The electron in the hydrogen atom is in an orbit that never brings it closer than 100 pm to the nucleus. (b) Atomic absorption spectra result from transitions of electrons from lower to higher energy levels. (c) A many-electron atom behaves somewhat like a solar system that has a number of planets.

6.96    Distinguish carefully between the following terms: (a) wavelength and frequency, (b) wave properties and particle properties, (c) quantization of energy and continuous variation in energy.

6.97    What is the maximum number of electrons in an atom that can have the following quantum numbers? Specify the orbitals in which the electrons would be found. (a) $n = 2$, $m_s = +\frac{1}{2}$; (b) $n = 4$, $m_\ell = +1$; (c) $n = 3$, $\ell = 2$; (d) $n = 2$, $\ell = 0$, $m_s = -\frac{1}{2}$; (e) $n = 4$, $\ell = 3$, $m_\ell = -2$.

6.98    Identify the following individuals and their contributions to the development of quantum theory: Bohr, de Broglie, Einstein, Planck, Heisenberg, Schrödinger.

6.99    In a photoelectric experiment a student uses a light source whose frequency is greater than that needed to eject electrons from a certain metal. However, after continuously shining the light on the same area of the metal for a long period of time the student notices that the maximum kinetic energy of ejected electrons begins to decrease, even though the frequency of the light is held constant. How would you account for this behavior?

6.100    A baseball pitcher's fastballs have been clocked at about 100 mph. (a) Calculate the wavelength of a 0.141-kg baseball (in nm) at this speed. (b) What is the wavelength of a hydrogen atom at the same speed (1 mile = 1609 m)?

6.101    A ruby laser produces radiation of wavelength 633 nm in pulses whose duration is $1.00 \times 10^{-9}$ s. (a) If the laser produces 0.376 J of energy per pulse, how many photons are produced in each pulse? (b) Calculate the power (in watts) delivered by the laser per pulse (1 W = 1 J/s).

6.102    A 368-g sample of water absorbs infrared radiation at $1.06 \times 10^4$ nm from a carbon dioxide laser. Suppose all the absorbed radiation is converted to heat. Calculate the number of photons at this wavelength required to raise the temperature of the water by 5.00°C.

6.103 Photodissociation of water

$$H_2O(l) + h\nu \longrightarrow H_2(g) + \tfrac{1}{2}O_2(g)$$

has been suggested as a source of hydrogen. The $\Delta H^\circ_{rxn}$ for the reaction, calculated from thermochemical data, is 285.8 kJ per mole of water decomposed. Calculate the maximum wavelength (in nm) that would provide the necessary energy. In principle, is it feasible to use sunlight as a source of energy for this process?

6.104 Spectral lines of the Lyman and Balmer series do not overlap. Verify this statement by calculating the longest wavelength associated with the Lyman series and the shortest wavelength associated with the Balmer series (in nm).

6.105 Only a fraction of the electric energy supplied to a tungsten lightbulb is converted to visible light. The rest of the energy shows up as infrared radiation (that is, heat). A 75-W lightbulb converts 15.0 percent of the energy supplied to it into visible light (assume the wavelength to be 550 nm). How many photons are emitted by the lightbulb per second (1 W = 1 J/s)?

6.106 Certain sunglasses have small crystals of silver chloride (AgCl) incorporated in the lenses. When the lenses are exposed to light of the appropriate wavelength, the following reaction occurs:

$$AgCl \longrightarrow Ag + Cl$$

The Ag atoms formed produce a uniform gray color that reduces the glare. If $\Delta H$ for the preceding reaction is 248 kJ/mol, calculate the maximum wavelength of light that can induce this process.

6.107 The He ion contains only one electron and is therefore a hydrogen-like ion. Calculate the wavelengths, in increasing order, of the first four transitions in the Balmer series of the $He^+$ ion. Compare these wavelengths with the same transitions in an H atom. Comment on the differences. (The Rydberg constant for He is $4.39 \times 10^7$ m$^{-1}$.)

6.108 The retina of a human eye can detect light when radiant energy incident on it is at least $4.0 \times 10^{-17}$ J. For light of 575-nm wavelength, how many photons does this correspond to?

6.109 An electron in an excited state in a hydrogen atom can return to the ground state in two different ways: (a) via a direct transition in which a photon of wavelength $\lambda_1$ is emitted and (b) via an intermediate excited state reached by the emission of a photon of wavelength $\lambda_2$. This intermediate excited state then decays to the ground state by emitting another photon of wavelength $\lambda_3$. Derive an equation that relates $\lambda_1$ to $\lambda_2$ and $\lambda_3$.

6.110 A photoelectric experiment was performed by separately shining a laser at 450 nm (blue light) and a laser at 560 nm (yellow light) on a clean metal surface and measuring the number and kinetic energy of the ejected electrons. Which light would generate more electrons? Which light would eject electrons with greater kinetic energy? Assume that the same amount of energy is delivered to the metal surface by each laser and that the frequencies of the laser lights exceed the threshold frequency.

6.111 The electron configurations described in this chapter all refer to gaseous atoms in their ground states. An atom may absorb a quantum of energy and promote one of its electrons to a higher-energy orbital. When this happens, we say that the atom is in an excited state. The electron configurations of some excited atoms are given. Identify these atoms and write their ground-state configurations:

(a) $1s^1 2s^1$
(b) $1s^2 2s^2 2p^2 3d^1$
(c) $1s^2 2s^2 2p^6 4s^1$
(d) $[Ar]4s^1 3d^{10} 4p^4$
(e) $[Ne]3s^2 3p^4 3d^1$

6.112 Draw the shapes (boundary surfaces) of the following orbitals: (a) $2p_y$, (b) $3d_{z^2}$, (c) $3d_{x^2-y^2}$. (Show coordinate axes in your sketches.)

6.113 Draw orbital diagrams for atoms with the following electron configurations:

(a) $1s^2 2s^2 2p^5$
(b) $1s^2 2s^2 2p^6 3s^2 3p^3$
(c) $1s^2 2s^2 2p^6 3s^2 3p^6 4s^2 3d^7$

6.114 If Rutherford and his coworkers had used electrons instead of alpha particles to probe the structure of the nucleus (see Chapter 2), how would the result have changed?

6.115 Scientists have found interstellar hydrogen atoms with quantum number $n$ in the hundreds. Calculate the wavelength of light emitted when a hydrogen atom undergoes a transition from $n = 236$ to $n = 235$. In what region of the electromagnetic spectrum does this wavelength fall?

6.116 Ionization energy is the minimum energy required to remove an electron from an atom. It is usually expressed in units of kJ/mol, that is, the energy in kilojoules required to remove one mole of electrons from one mole of atoms. (a) Calculate the ionization energy for the hydrogen atom. (b) Repeat the calculation, assuming in this second case that the electrons are removed from the $n = 2$ state, instead of from the ground state.

6.117 An electron in a hydrogen atom is excited from the ground state to the $n = 4$ state. Comment on the correctness of the following statements (true or false).

(a) $n = 4$ is the first excited state.
(b) It takes more energy to ionize (remove) the electron from $n = 4$ than from the ground state.
(c) The electron is farther from the nucleus (on average) in $n = 4$ than in the ground state.
(d) The wavelength of light emitted when the electron drops from $n = 4$ to $n = 1$ is longer than that from $n = 4$ to $n = 2$.
(e) The wavelength the atom absorbs in going from $n = 1$ to $n = 4$ is the same as that emitted as it goes from $n = 4$ to $n = 1$.

6.118 The ionization energy of a certain element is 412 kJ/mol (see Problem 6.116). However, when the atoms of this element are in the first excited state, the ionization energy is only 126 kJ/mol. Based on this information, calculate the wavelength of light emitted in a transition from the first excited state to the ground state.

6.119 How many photons at 586 nm must be absorbed to melt $5.0 \times 10^2$ g of ice? On average, how many $H_2O$ molecules does one photon convert from ice to water? (*Hint:* It takes 334 J to melt 1 g of ice at 0°C.)

6.120 Portions of orbital diagrams representing the ground-state electron configurations of certain elements are shown here. Which of them violate the Pauli exclusion principle? Which violate Hund's rule?

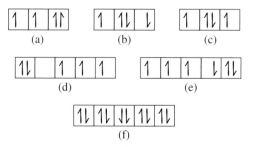

6.121 The UV light that is responsible for tanning the skin falls in the 320- to 400-nm region. Calculate the total energy (in joules) absorbed by a person exposed to this radiation for 2.5 h, given that there are $2.0 \times 10^{16}$ photons hitting Earth's surface per square centimeter per second over a 80-nm (320 to 400 nm) range and that the exposed body area is 0.45 m². Assume that only half of the radiation is absorbed and the other half is reflected by the body. (*Hint:* Use an average wavelength of 360 nm in calculating the energy of a photon.)

6.122    The sun is surrounded by a white circle of gaseous material called the corona, which becomes visible during a total eclipse of the sun. The temperature of the corona is in the millions of degrees Celsius, which is high enough to break up molecules and remove some or all of the electrons from atoms. One way astronomers have been able to estimate the temperature of the corona is by studying the emission lines of ions of certain elements. For example, the emission spectrum of $Fe^{14+}$ ions has been recorded and analyzed. Knowing that it takes $3.5 \times 10^4$ kJ/mol to convert $Fe^{13+}$ to $Fe^{14+}$, estimate the temperature of the sun's corona. (*Hint:* The average kinetic energy of one mole of a gas is $\frac{3}{2}RT$.)

**6.123**    When an electron makes a transition between energy levels of a hydrogen atom, there are no restrictions on the initial and final values of the principal quantum number $n$. However, there is a quantum mechanical rule that restricts the initial and final values of the orbital angular momentum $\ell$. This is the *selection rule*, which states that $\Delta\ell = \pm 1$; that is, in a transition, the value of $\ell$ can only increase or decrease by 1. According to this rule, which of the following transitions are allowed: (a) $1s \longrightarrow 2s$, (b) $2p \longrightarrow 1s$, (c) $1s \longrightarrow 3d$, (d) $3d \longrightarrow 4f$, (e) $4d \longrightarrow 3s$? In view of this selection rule, explain why it is possible to observe the various emission series shown in Figure 6.11.

6.124    In 1996 physicists created an anti-atom of hydrogen. In such an atom, which is the antimatter equivalent of an ordinary atom, the electric charges of all the component particles are reversed. Thus the nucleus of an anti-atom is made of an antiproton, which has the same mass as a proton but bears a negative charge, while the electron is replaced by an anti-electron (also called a positron) with the same mass as an electron, but bearing a positive charge. Would you expect the energy levels, emission spectra, and atomic orbitals of an antihydrogen atom to be different from those of a hydrogen atom? What would happen if an anti-atom of hydrogen collided with a hydrogen atom?

**6.125**    In an electron microscope, electrons are accelerated by passing them through a voltage difference. The kinetic energy thus acquired by the electrons is equal to the voltage times the charge on the electron. Thus a voltage difference of 1 volt imparts a kinetic energy of $1.602 \times 10^{-19}$ volt-coulomb or $1.602 \times 10^{-19}$ J. Calculate the wavelength associated with electrons accelerated by $5.00 \times 10^3$ volts.

**6.126**    A microwave oven operating at $1.22 \times 10^8$ nm is used to heat 150 mL of water (roughly the volume of a tea cup) from 20°C to 100°C. Calculate the number of photons needed if 92.0 percent of microwave energy is converted to the thermal energy of water.

**6.127**    Calculate the wavelength and frequency of an emitted gamma particle having the energy of $3.14 \times 10^{11}$ J/mol.

**6.128**    (a) An electron in the ground state of the hydrogen atom moves at an average speed of $5 \times 10^6$ m/s. If the speed is known to an uncertainty of 20 percent, what is the minimum uncertainty in its position? Given that the radius of the hydrogen atom in the ground state is $5.29 \times 10^{-11}$ m, comment on your result. The mass of an electron is $9.1094 \times 10^{-31}$ kg. (b) A 0.15-kg baseball thrown at 100 mph has a momentum of 6.7 kg · m/s. If the uncertainty in measuring the momentum is $1.0 \times 10^{-7}$ of the momentum, calculate the uncertainty in the baseball's position.

**6.129**    Calculate the energies needed to remove an electron from the $n = 1$ state and the $n = 5$ state in the $Li^{2+}$ ion. What is the wavelength (in nm) of the emitted photon in a transition from $n = 5$ to $n = 1$?

Solving Equation 6.4 for energy gives $\Delta E = R_\infty hc \left( \dfrac{1}{n_1^2} - \dfrac{1}{n_2^2} \right)$,

where the Rydberg constant $R_\infty$ for hydrogen-like atoms is $1.097 \times 10^7$ m$^{-1} \cdot Z^2$, and $Z$ is the atomic number.

6.130    According to Einstein's special theory of relativity, the mass of a moving particle, $m_{\text{moving}}$, is related to its mass at $m_{\text{rest}}$, by the following equation

$$m_{\text{moving}} = \frac{m_{\text{rest}}}{\sqrt{1 - (u/c)^2}}$$

where $u$ and $c$ are the speeds of the particle and light, respectively. (a) In particle accelerators, protons, electrons, and other charged particles are often accelerated to speeds close to the speed of light. Calculate the wavelength (in nm) of a proton moving at 50.0 percent the speed of light. The mass of a proton is $1.67 \times 10^{-27}$ kg. (b) Calculate the mass of a $6.0 \times 10^{-2}$ kg tennis ball moving at 63 m/s. Comment on your results.

**6.131**    The mathematical equation for studying the photoelectric effect is

$$h\nu = W + \tfrac{1}{2} m_e u^2$$

where $\nu$ is the frequency of light shining on the metal; $W$ is the energy needed to remove an electron from the metal; and $m_e$ and $u$ are the mass and speed of the ejected electron, respectively. In an experiment, a student found that a maximum wavelength of 351 nm is needed to just dislodge electrons from a zinc metal surface. Calculate the velocity (in m/s) of an ejected electron when the student employed light with a wavelength of 313 nm.

6.132    In the beginning of the twentieth century, some scientists thought that a nucleus may contain both electrons and protons. Use the Heisenberg uncertainty principle to show that an electron cannot be confined within a nucleus. Repeat the calculation for a proton. Comment on your results. Assume the radius of a nucleus to be $1.0 \times 10^{-15}$ m. The masses of an electron and a proton are $9.109 \times 10^{-31}$ kg and $1.673 \times 10^{-27}$ kg, respectively. (*Hint:* Treat the radius of the nucleus as the uncertainty in position.)

6.133    Blackbody radiation is the term used to describe the dependence of the radiation energy emitted by an object on wavelength at a certain temperature. Planck proposed the quantum theory to account for the dependence. Shown in the figure is a plot of the radiation energy emitted by our sun versus wavelength. This curve is characteristic of objects at about 6000 K, which is the temperature at the surface of the sun. At a higher temperature, the curve has a similar shape but the maximum will shift to a shorter wavelength. (a) What does this curve reveal about two

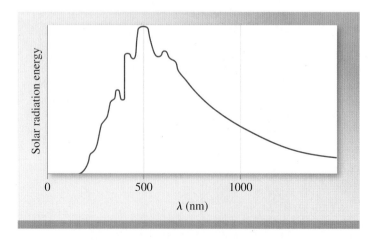

consequences of great biological significance on Earth? (b) How are astronomers able to determine the temperature at the surface of stars in general?

6.134 All molecules undergo vibrational motions. Quantum mechanical treatment shows that the vibrational energy $E_{vib}$ of a diatomic molecule such as HCl is given by

$$E_{vib} = \left(n + \frac{1}{2}\right)h\nu$$

where $n$ is a quantum number ($n = 0, 1, 2, 3, \ldots$) and $\nu$ is the fundamental frequency of vibration. (a) Sketch the first three vibrational energy levels for HCl. (b) Calculate the energy required to excite an HCl molecule from the ground level to the first excited level. The fundamental frequency of vibration for

HCl is $8.66 \times 10^{13}$ s$^{-1}$. (c) The fact that the lowest vibrational energy in the ground level is not zero but equal to $\frac{1}{2}h\nu$ means that molecules will vibrate at all temperatures, including absolute zero. Use the Heisenberg uncertainty principle to justify this prediction. (*Hint:* Consider a molecule that is not vibrating and start by predicting the uncertainty in its momentum.)

6.135 The wave function for the 2s orbital in the hydrogen atom is

$$\Psi_{2s} = \frac{1}{\sqrt{2a_0^3}}\left(1 - \frac{\rho}{2}\right)e^{-\rho/2}$$

where $a_0$ is the value of the radius of the first Bohr orbit, equal to 0.529 nm; $\rho$ is $Z(r/a_0)$; and $r$ is the distance from the nucleus in meters. Calculate the distance from the nucleus (in nm) of the node of the 2s wave function.

# PRE-PROFESSIONAL PRACTICE EXAM PROBLEMS: PHYSICAL AND BIOLOGICAL SCIENCES

According to Wien's law, the wavelength of maximum intensity in blackbody radiation, $\lambda_{max}$, is inversely proportional to the temperature of the radiating body. Mathematically,

$$\lambda_{max} = \frac{b}{T}$$

where $b$ is Wien's displacement constant ($2.898 \times 10^6$ nm · K) and $T$ is the temperature of the radiating body in kelvins. The sun, composed primarily of hydrogen, emits a continuous spectrum from the region known as the photosphere, with the most intense emission occurring at approximately 500 nm.

1.  What is the approximate surface temperature of the sun?

    a) 60,000 K
    b) 6000 K
    c) 1500 K
    d) 500 K

2.  What is the frequency of the sun's $\lambda_{max}$?

    a) $6.0 \times 10^{14}$ s$^{-1}$
    b) $5.0 \times 10^{14}$ s$^{-1}$
    c) $2.5 \times 10^2$ s$^{-1}$
    d) $5.9 \times 10^{-14}$ s$^{-1}$

3.  What is the energy of a photon with wavelength $\lambda_{max}$ for a blackbody at 5200 K?

    a) $5.4 \times 10^{-14}$ J
    b) $5.4 \times 10^{14}$ J
    c) $3.6 \times 10^{-19}$ J
    d) $5.6 \times 10^{-7}$ J

4.  The visible region of the electromagnetic spectrum includes wavelengths from about 400 nm to about 700 nm. What is the minimum approximate temperature that would be required for the maximum intensity wavelength emitted by a blackbody to occur in the ultraviolet region of the spectrum?

    a) 71,000 K
    b) 9100 K
    c) 7300 K
    d) 64,000 K

# ANSWERS TO IN-CHAPTER MATERIALS

## Answers to Practice Problems

**6.1A** $1.86 \times 10^{-4}$ m. **6.1B** $2.91 \times 10^{10}$ s$^{-1}$. **6.2A** $1.58 \times 10^{-19}$ J.
**6.2B** UV. **6.3A** (a) $9.43 \times 10^{-19}$ J, (b) $1.18 \times 10^{-25}$ J, (c) $1.96 \times 10^{-19}$ J.
**6.3B** (a) $1.05 \times 10^4$ nm, (b) $2.1 \times 10^{18}$, (c) 1.2 eV. **6.4A** 103 nm.
**6.4B** 7. **6.5A** 26 nm. **6.5B** $8.2 \times 10^{-28}$ kg · m/s. **6.6A** (a) $3 \times 10^{-34}$ m,
(b) $3 \times 10^{-32}$ m. **6.6B** (a) $\pm 2 \times 10^{-25}$ kg · m/s, (b) $\pm 1 \times 10^2$ m/s,
(c) $\pm 2 \times 10^5$ m/s. **6.7A** Only 0, (b) $-2, -1, 0, +1, +2$. **6.7B** (a) 2,
(b) 0, 1, 2, 3. **6.8A** $n = 3$, $\ell = 2$, $m_\ell = -2, -1, 0, +1, +2$.
**6.8B** For a $d$ orbital, $\ell = 2$, but when $n = 2$, $\ell$ cannot be 2.
**6.9A** $1s^2 2s^2 2p^6 3s^2 3p^6 4s^2 3d^{10} 4p^6 5s^1$. **6.9B** $1s^2 2s^2 2p^6 3s^2 3p^6 4s^2 3d^{10} 4p^5$.
**6.10A** [Rn]$7s^2$. **6.10B** At.

## Answers to Checkpoints

**6.1.1** c. **6.1.2** a. **6.1.3** b. **6.1.4** c. **6.2.1** e. **6.2.2** d. **6.2.3** c. **6.2.4** d. **6.3.1** b.
**6.3.2** d. **6.3.3** c. **6.3.4** e. **6.4.1** c. **6.4.2** a. **6.5.1** b. **6.5.2** e. **6.6.1** a, b, d.
**6.6.2** c. **6.6.3** c. **6.6.4** e. **6.7.1** b. **6.7.2** c. **6.7.3** b, e. **6.7.4** b. **6.8.1** b.
**6.8.2** d. **6.8.3** b. **6.9.1** c. **6.9.2** c. **6.9.3** b, d. **6.9.4** a, b, d.

## Answers to Applying What You've Learned

**a)** For light with $\lambda = 10,600$ nm, $\nu = 2.83 \times 10^{13}$ s$^{-1}$ and $E = 1.88 \times 10^{-20}$ J/photon. **b)** The difference in energy between the two argon wavelengths is $2.06 \times 10^{-20}$ J/photon. The light with $\lambda = 488$ nm is the more energetic of the two. **c)** de Broglie wavelength $= 1 \times 10^{-11}$ m. **d)** For the $3p$ orbitals in Ar (or in any atom with $3p$ orbitals), $n = 3$, $\ell = 1$, and $m_\ell = -1, 0$, and $+1$. **e)** The electron configuration of Ar is $1s^2 2s^2 2p^6 3s^2 3p^6$ or [Ne]$3s^2 3p^6$.

# 7

# Electron Configuration and the Periodic Table

*Table salt, one of the most common seasonings, contains a Group 1A metal.*

# What an Element's Position in the Periodic Table Can Tell Us

Elements in the same group of the periodic table tend to exhibit similar physical and chemical properties. The alkali metals, for example, Li, Na, and K, all form chloride salts that have similar properties. Because of their similarities to sodium chloride (NaCl), lithium chloride (LiCl) and potassium chloride (KCl) have both been used as salt substitutes for people on low-sodium diets. Despite their similarities, only one of these salt substitutes proved to be safe. In the 1940s, a group of cardiac patients on low-sodium diets were given lithium chloride to use in place of sodium chloride. Lithium chloride looks just like sodium chloride and has a very similar taste. The resulting unregulated intake of lithium had disastrous results. Many of the patients developed symptoms of severe lithium toxicity and several of them died. Today, salt substitutes such as Morton Salt Substitute and Nu-Salt contain *potassium* chloride, which also looks and tastes like sodium chloride. The similarities in appearance and flavor of these compounds are largely due to the chemical similarity of the Group 1A elements they contain.

Although some of their biochemical properties vary, the physical and chemical properties of the alkali metals are quite similar. This is why they were placed in the same group in the periodic table in the first place. It turns out they belong in the same group because they have the same valence electron configuration.

At the end of this chapter, you will be able to answer a series of questions about the alkali metals and the ions that they form [▶▶ Page 288].

## In This Chapter, You Will Learn:

Some of the chemical and physical properties of the elements and how these properties are related to each element's position in the periodic table.

### Before you begin, you should review

- Atomic ions [◀◀ Section 2.7]
- Electron configurations and the periodic table [◀◀ Section 6.9]

## 7.1   Development of the Periodic Table

In the nineteenth century, chemists had only a vague idea of atoms and molecules and did not yet know about electrons and protons. Nevertheless, they devised the periodic table using their knowledge of atomic masses. Accurate measurements of the atomic masses of many elements had already been made. Arranging elements according to their atomic masses in a periodic table seemed logical to those chemists, who believed that chemical behavior should somehow be related to atomic mass.

In 1864 the English chemist John Newlands[1] noticed that when the elements were arranged in order of atomic mass, every eighth element had similar properties. Newlands referred to this peculiar relationship as the *law of octaves*. However, this "law" turned out to be inadequate for elements beyond calcium, and Newlands's work was not accepted by the scientific community.

In 1869 the Russian chemist Dmitri Mendeleev[2] and the German chemist Lothar Meyer[3] independently proposed a much more extensive tabulation of the elements based on the regular, periodic recurrence of properties—a phenomenon known as *periodicity*.

Mendeleev's classification system was a great improvement over Newlands's for two reasons. First, it grouped the elements together more accurately, according to their properties. Second, and equally important, it made it possible to predict the properties of several elements that had not yet been discovered. For example, Mendeleev proposed the existence of an unknown element that he called eka-aluminum and predicted a number of its properties. (*Eka* is a Sanskrit word meaning "first"; thus, eka-aluminum would be the first element under aluminum in the same group.) When gallium was discovered 4 years later, its properties matched the predicted properties of eka-aluminum remarkably well:

|  | **Eka-Aluminum (Ea)** | **Gallium (Ga)** |
|---|---|---|
| Atomic mass | 68 amu | 69.9 amu |
| Melting point | Low | 30.15°C |
| Density | 5.9 g/cm$^3$ | 5.94 g/cm$^3$ |
| Formula of oxide | $Ea_2O_3$ | $Ga_2O_3$ |

Mendeleev's periodic table included 66 known elements. By 1900, some 30 more had been added to the list, filling in some of the empty spaces. Figure 7.1 gives the time period during which each element was discovered.

Although this periodic table was remarkably successful, the early versions had some inconsistencies that were impossible to overlook. For example, the atomic mass of argon (39.95 amu) is greater than that of potassium (39.10 amu), but argon comes before potassium in the periodic table. If elements were arranged solely according to increasing atomic mass, argon would appear in the position occupied by potassium in our modern periodic table (see the inside front cover). No chemist would place argon, a gas with no tendency to react, in the same group as lithium and sodium, two highly reactive metals! This and other discrepancies suggested that some fundamental property other than atomic mass must be the basis of periodicity. The fundamental property turned out to be the number of protons in an atom's nucleus, something that could not have been known by Mendeleev and his contemporaries.

In 1913 a young English physicist, Henry Moseley,[4] discovered a correlation between what he called *atomic number* and the frequency of X rays generated by bombarding an element with high-energy electrons. Moseley noticed that, in general, the frequencies of X rays emitted from the elements increased with increasing atomic mass. Among the few exceptions he found were argon and potassium. Although argon has a greater atomic mass than potassium, the X-ray emission from potassium indicated that it has the greater atomic number. Ordering the periodic table using atomic

---

1. John Alexander Reina Newlands (1838–1898). English chemist. Newlands's work was a step in the right direction in the classification of the elements. Unfortunately, because of its shortcomings, he was subjected to much criticism, and even ridicule. At one meeting he was asked if he had ever examined the elements according to the order of their initial letters! Nevertheless, in 1887 Newlands was honored by the Royal Society of London for his contribution.

2. Dmitri Ivanovich Mendeleev (1836–1907). Russian chemist. His work on the periodic classification of elements is regarded by many as the most significant achievement in chemistry in the nineteenth century.

3. Julius Lothar Meyer (1830–1895). German chemist. In addition to his contribution to the periodic table, Meyer also discovered the chemical affinity of hemoglobin for oxygen.

4. Henry Gwyn-Jeffreys Moseley (1887–1915). English physicist. Moseley discovered the relationship between X-ray spectra and atomic number. A lieutenant in the Royal Engineers, he was killed in action at the age of 28 during the British campaign in Gallipoli, Turkey.

□ Ancient times □ 1735–1843 □ 1894–1918

□ Middle Ages–1700 □ 1843–1886 □ 1923–1961 □ 1965–

**Figure 7.1** Periodic table of elements classified by dates of discovery.

| H | | | | | | | | | | | | | | | | | He |
|---|---|---|---|---|---|---|---|---|---|---|---|---|---|---|---|---|---|
| Li | Be | | | | | | | | | | | B | C | N | O | F | Ne |
| Na | Mg | | | | | | | | | | | Al | Si | P | S | Cl | Ar |
| K | Ca | Sc | Ti | V | Cr | Mn | Fe | Co | Ni | Cu | Zn | Ga | Ge | As | Se | Br | Kr |
| Rb | Sr | Y | Zr | Nb | Mo | Tc | Ru | Rh | Pd | Ag | Cd | In | Sn | Sb | Te | I | Xe |
| Cs | Ba | Lu | Hf | Ta | W | Re | Os | Ir | Pt | Au | Hg | Tl | Pb | Bi | Po | At | Rn |
| Fr | Ra | Lr | Rf | Db | Sg | Bh | Hs | Mt | Ds | Rg | | | | | | | |

| La | Ce | Pr | Nd | Pm | Sm | Eu | Gd | Tb | Dy | Ho | Er | Tm | Yb |
|---|---|---|---|---|---|---|---|---|---|---|---|---|---|
| Ac | Th | Pa | U | Np | Pu | Am | Cm | Bk | Cf | Es | Fm | Md | No |

number enabled scientists to make sense out of the discrepancies that had puzzled them earlier. Moseley concluded that the atomic number was equal to the number of protons in the nucleus and to the number of electrons in an atom.

Entries in modern periodic tables usually include an element's atomic number along with its symbol. Electron configurations of elements help to explain the periodic recurrence of physical and chemical properties. The importance and usefulness of the periodic table lie in the fact that we can use our understanding of the general properties and trends within a group or a period to predict with considerable accuracy the properties of any element, even though that element may be unfamiliar to us.

Sample Problem 7.1 shows how the periodic table can be used to predict similarities and differences in the properties of elements.

## Sample Problem 7.1

What elements would you expect to exhibit properties most similar to those of chlorine?

**Strategy** Because elements in the same group tend to have similar properties, you should identify elements in the same group as chlorine.

**Setup** Chlorine is a member of Group 7A.

**Solution** Fluorine, bromine, and iodine, the other nonmetals in Group 7A, should have properties most similar to those of chlorine.

**Think About It** Astatine (At) is also in Group 7A. Astatine, though, is classified as a metalloid, and we have to be careful comparing nonmetals to *metalloids* (or to *metals*). As a metalloid, the properties of astatine should be less similar to those of chlorine than the other members of Group 7A.

**Practice Problem A** What element(s) would you expect to exhibit properties most similar to those of silicon (Si)?

**Practice Problem B** Arrange the following Group 5A elements in order of increasing similarity of properties to N: As, Bi, and P.

**Student Annotation:** Actually, astatine is radioactive and very little is known about its properties.

## Bringing Chemistry to Life

### The Chemical Elements of Life

Of the 117 elements currently known, relatively few are essential to living systems. Actually, only six nonmetallic elements form the building blocks of cells: C, H, O, N, P, and S. They are the constituent elements of proteins, nucleic acids, and carbohydrates. Although the natural abundance of carbon is quite low (about 0.1 percent by mass of Earth's crust), it is present in nearly all biological molecules. Carbon is perhaps the most versatile element because it has the ability to form various types of chemical bonds. Carbon atoms can form bonds to each other, linking up to form an enormous variety of chains and ring structures.

The metals play several different roles in living systems. As cations ($Na^+$, $K^+$, $Ca^{2+}$, and $Mg^{2+}$), they serve to maintain the balance between intracellular and extracellular fluids, nerve transmissions, and other activities. They are also needed for protein functions. For example, the $Fe^{2+}$ ion binds oxygen in hemoglobin molecules and $Cu^{2+}$, $Zn^{2+}$, and $Mg^{2+}$ ions are essential for enzyme activity. In addition, calcium in the form of $Ca_5(PO_4)_3(OH)$ and $Ca_3(PO_4)_2$ is an essential component of teeth and bones.

The periodic table shown highlights the essential elements in the human body. Of special interest are the *trace elements,* such as iron (Fe), copper (Cu), zinc (Zn), iodine (I), cobalt (Co), selenium (Se), and fluorine (F), which together make up about 0.1 percent of the body's mass. Although the trace elements are present in very small amounts, they are crucial for our health. In many cases, however, their exact biological role is still not fully understood.

These elements are necessary for biological functions such as growth, the transport of oxygen for metabolism, and defense against disease. There is a delicate balance in the amounts of these elements in our bodies. Too much or too little over an extended period of time can lead to serious illness, retardation, or even death.

## Checkpoint 7.1    Development of the Periodic Table

**7.1.1** Which of the following elements would you expect to have chemical properties most similar to those of S?

a) P

b) Cl

c) Se

d) Na

e) Sr

**7.1.2** Which of the following elements would you expect to have properties similar to those of Ba?

a) Sr

b) Rb

c) Na

d) K

e) B

**7.1.3** Element 117 is not known to exist. However, based on its position in the periodic table, which of the following properties would you expect it to exhibit?

(i) gaseous at room temperature
(ii) unreactive
(iii) properties similar to At
(iv) properties similar to Ra

a) i and iv

b) i, ii, and iv

c) iii only

d) iv only

e) i and iii

## 7.2 The Modern Periodic Table

Figure 7.2 shows the modern periodic table together with the outermost ground-state electron configurations of the elements. (The electron configurations of the elements are also given in Figure 6.25.) Starting with hydrogen, the electronic subshells are filled in the order shown in Figure 6.23 [◄◄ Section 6.8].

### Classification of Elements

Based on the type of subshell containing the outermost electrons, the elements can be divided into categories—the main group elements, the noble gases, the transition elements (or transition metals), the lanthanides, and the actinides. The *main group elements* (also called the *representative elements*) are the elements in Groups 1A through 7A. With the exception of helium, each of the *noble gases* (the Group 8A elements) has a completely filled $p$ subshell. The electron configurations are $1s^2$ for helium and $ns^2np^6$ for the other noble gases, where $n$ is the principal quantum number for the outermost shell.

The transition metals are the elements in Groups 1B and 3B through 8B. Transition metals either have incompletely filled $d$ subshells or readily produce cations with incompletely filled $d$ subshells. According to this definition, the elements of Group 2B are *not* transition metals. They are $d$-block elements, though, so they generally are included in the discussion of transition metals.

**Student Annotation:** In this context, *outermost* electrons refers to those that are placed in orbitals *last* using the Aufbau principle [◄◄ Section 6.8].

**Student Annotation:** The metals of Group 2B typically form +2 ions, although they can also form +1 ions. In either case, the electron configuration includes a completed $d$ subshell [►► Section 7.6]. Thus, strictly speaking, Zn, Cd, and Hg are *not* transition metals.

**Figure 7.2** Valence electron configurations of the elements. For simplicity, the filled $f$-subshells are not shown in elements 72 through 86, 104 through 116, and 118.

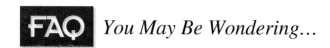

## *You May Be Wondering…*

# Why Are There Two Different Sets of Numbers at the Top of the Periodic Table?

The numbering of the transition metal groups 3B through 7B indicates the similarity between the outer electron configurations of these elements and those of the corresponding main group elements. For example, scandium (Sc; Group 3B) and gallium (Ga; Group 3A) each have three outer electrons. However, because their outer electrons reside in different types of atomic orbitals (*s* and *d* orbitals in the case of Sc; *s* and *p* orbitals in the case of Ga), they belong in different groups. The elements of Groups 1B and 2B have filled *d* subshells [◄◄ Section 6.9]. Unlike the elements of Group 2B, elements of Group 1B form cations with incompletely filled *d* subshells. Their group numbers correspond to the one and two electrons, respectively, that they have in *s* orbitals—just like the main group elements in Groups 1A and 2A. The elements iron (Fe), cobalt (Co), and nickel (Ni), and the elements that appear beneath them in the periodic table, cannot be classified in this way and are all placed in Group 8B.

The designation of A and B groups is not universal. In Europe, the practice is to use B for main group elements and A for transition or *d*-block elements, which is just the opposite of the American convention. The International Union of Pure and Applied Chemistry (IUPAC) has recommended eliminating ambiguity by numbering the columns sequentially with Arabic numerals 1 through 18 (see Figure 7.2). The proposal has not been accepted universally, and many modern periodic tables retain the traditional group designations. Periodic tables in this text display both the IUPAC-recommended Arabic numerals and the traditional American numbering system. Discussions in the text will refer to the traditional American group numbers.

**Figure 7.3**  Periodic table with color-coding of main group elements, noble gases, transition metals, group 2B metals, lanthanides, and actinides.

The lanthanides and actinides are sometimes called *f*-block transition elements because they have incompletely filled *f* subshells. Figure 7.3 distinguishes the groups of elements discussed here.

There is a distinct pattern to the electron configurations of the elements in a particular group. See, for example, the electron configurations of Groups 1A and 2A in Table 7.1. Each member of Group 1A has a noble gas core plus one additional electron, giving each alkali metal the general electron configuration of [noble gas]$ns^1$. Similarly, the Group 2A alkaline earth metals have a noble gas core and an outer electron configuration of $ns^2$.

The outermost electrons of an atom are called **valence electrons,** which are the ones involved in the formation of chemical bonds between atoms. The similarity of the valence electron configurations (i.e., they have the same number and type of valence electrons) is what makes the elements in the same group resemble one another chemically. This observation holds true for the other main group elements as well. For instance, the halogens (Group 7A) all have outer electron configurations of $ns^2np^5$, and they have similar properties.

| TABLE 7.1 | Electron Configurations of Group 1A and Group 2A Elements |
|---|---|

| | Group 1A | | | Group 2A | |
|---|---|---|---|---|---|
| Li | $[He]2s^1$ | | Be | $[He]2s^2$ | |
| Na | $[Ne]3s^1$ | | Mg | $[Ne]3s^2$ | |
| K | $[Ar]4s^1$ | | Ca | $[Ar]4s^2$ | |
| Rb | $[Kr]5s^1$ | | Sr | $[Kr]5s^2$ | |
| Cs | $[Xe]6s^1$ | | Ba | $[Xe]6s^2$ | |
| Fr | $[Rn]7s^1$ | | Ra | $[Rn]7s^2$ | |

**Student Annotation:** Although hydrogen's electron configuration is $1s^1$ [◄◄ Section 6.9, Figure 6.25], it is a nonmetal and is not really a member of Group 1A.

In predicting properties for Groups 3A through 7A, we must take into account that each of these groups contains elements on both sides of the line that divides metals and nonmetals. For example, the elements in Group 4A all have the same outer electron configuration, $ns^2np^2$, but there is considerable variation in chemical properties among these elements because carbon is a nonmetal, silicon and germanium are metalloids, and tin and lead are metals.

As a group, the noble gases behave very similarly. The noble gases are chemically inert because they all have completely filled outer $ns$ and $np$ subshells, a condition that imparts unusual stability. With the exception of krypton and xenon, they do not undergo chemical reactions.

Although the outer electron configuration of the transition metals is not always the same within a group and there is often no regular pattern in the way the electron configuration changes from one metal to the next in the same period, all transition metals share many characteristics (multiple oxidation states, richly colored compounds, magnetic properties, and so on) that set them apart from other elements. These properties are similar because all these metals have incompletely filled $d$ subshells. Likewise, the lanthanide and actinide elements resemble one another because they have incompletely filled $f$ subshells.

Sample Problem 7.2 shows how to determine the electron configuration from the number of electrons in an atom.

## Sample Problem 7.2

Without using a periodic table, give the ground-state electron configuration and block designation ($s$-, $p$-, $d$-, or $f$-block) of an atom with (a) 17 electrons, (b) 37 electrons, and (c) 22 electrons. Classify each atom as a main group element or transition metal.

**Strategy** Use the Aufbau principle discussed in Section 6.8. Start writing each electron configuration with principal quantum number $n = 1$, and then continue to assign electrons to orbitals in the order presented in Figure 6.23 until all the electrons have been accounted for.

**Setup** According to Figure 6.23, orbitals fill in the following order: $1s$, $2s$, $2p$, $3s$, $3p$, $4s$, $3d$, $4p$, $5s$, $4d$, $5p$, $6s$, and so on. Recall that an $s$ subshell contains one orbital, a $p$ subshell contains three orbitals, and a $d$ subshell contains five orbitals. Remember, too, that each orbital can accommodate a maximum of two electrons. The block designation of an element corresponds to the type of subshell occupied by the last electrons added to the configuration according to the Aufbau principle.

**Solution**

(a) $1s^2 2s^2 2p^6 3s^2 3p^5$, $p$-block, main group

(b) $1s^2 2s^2 2p^6 3s^2 3p^6 4s^2 3d^{10} 4p^6 5s^1$, $s$-block, main group

(c) $1s^2 2s^2 2p^6 3s^2 3p^6 4s^2 3d^2$, $d$-block, transition metal

**Think About It** Consult Figure 6.25 on page 244 to confirm your answers.

**Practice Problem A** Without using a periodic table, give the ground-state electron configuration and block designation ($s$-, $p$-, $d$-, or $f$-block) of an atom with (a) 15 electrons, (b) 20 electrons, and (c) 35 electrons.

**Practice Problem B** Identify the elements represented by (a) $1s^2 2s^2 2p^6 3s^2 3p^1$, (b) $1s^2 2s^2 2p^6 3s^2 3p^6 4s^2 3d^{10}$, and (c) $1s^2 2s^2 2p^6 3s^2 3p^6 4s^2 3d^{10} 4p^6 5s^2$.

## Representing Free Elements in Chemical Equations

Having classified the elements according to their ground-state electron configurations, we can now learn how chemists represent elements in chemical equations.

**Metals**   Because metals do not exist in discrete molecular units but rather in complex, three-dimensional networks of atoms, we always use their empirical formulas in chemical equations. The empirical formulas are the same as the symbols that represent the elements. For example, the empirical formula for iron is Fe, the same as the symbol for the element.

**Nonmetals**   There is no single rule regarding the representation of nonmetals in chemical equations. Carbon, for example, exists in several allotropic forms. Regardless of the allotrope, we use its empirical formula C to represent elemental carbon in chemical equations. Often the symbol C will be followed by the specific allotrope in parentheses as in the equation representing the conversion of graphite to diamond, two of carbon's allotropic forms:

**Student Annotation:** Recall that allotropes are different forms of the same element [◀◀ Section 2.6].

$$C(\text{graphite}) \longrightarrow C(\text{diamond})$$

For nonmetals that exist as polyatomic molecules, we generally use the molecular formula in equations: $H_2$, $N_2$, $O_2$, $F_2$, $Cl_2$, $Br_2$, $I_2$, and $P_4$, for example. In the case of sulfur, however, we usually use the empirical formula S rather than the molecular formula $S_8$. Thus, instead of writing the equation for the combustion of sulfur as

$$S_8(s) + 8O_2(g) \longrightarrow 8SO_2(g)$$

we usually write

$$S(s) + O_2(g) \longrightarrow SO_2(g)$$

although, technically, both ways are correct.

**Noble Gases**   All the noble gases exist as isolated atoms, so we use their symbols: He, Ne, Ar, Kr, Xe, and Rn.

**Metalloids**   The metalloids, like the metals, all have complex three-dimensional networks, so we also represent them with their empirical formulas—that is, their symbols: B, Si, Ge, and so on.

---

**Checkpoint 7.2**    **The Modern Periodic Table**

**7.2.1**   Which electron configuration is correct for a germanium (Ge) atom in the ground state?

a) $1s^2 2s^2 2p^6 3s^2 3p^6 4s^2 4p^2$

b) $1s^2 2s^2 2p^6 3s^2 3p^6 4s^2 3d^{10} 4p^2$

c) $1s^2 2s^2 2p^6 3s^2 3p^2$

d) $1s^2 2s^2 2p^6 3s^2 3p^6 4s^2 4p^2 4d^{10}$

e) $1s^2 2s^2 2p^6 3s^2 3p^2 3d^{10}$

**7.2.2**   Which of the following equations correctly represent the chemical reaction in which graphite combines with sulfur to form carbon disulfide gas [$CS_2(g)$]? (Select all that apply.)

a) $C(\text{graphite}) + 2S(s) \longrightarrow CS_2(g)$

b) $C(\text{graphite}) + S_2(s) \longrightarrow CS_2(g)$

c) $C(\text{graphite}) + S_8(s) \longrightarrow CS_8(g)$

d) $C(\text{graphite}) + \frac{1}{4}S_8(s) \longrightarrow CS_2(g)$

e) $4C(\text{graphite}) + S_8(s) \longrightarrow 4CS_2(g)$

---

## 7.3    Effective Nuclear Charge

As we have seen, the electron configurations of the elements show a periodic variation with increasing atomic number. In this and the next few sections, we will examine how electron configuration explains the periodic variation of physical and chemical properties of the elements. We begin by introducing the concept of *effective nuclear charge.*

Nuclear charge (Z) is simply the number of protons in the nucleus of an atom. ***Effective nuclear charge ($Z_{eff}$)*** is the actual magnitude of positive charge that is "experienced" by an electron in the atom. The only atom in which the nuclear charge and effective nuclear charge are the same is hydrogen, which has only one electron. In all other atoms, the electrons are simultaneously attracted to the nucleus and repelled by one another. This results in a phenomenon known as ***shielding.*** An electron in a many-electron atom is partially shielded from the positive charge of the nucleus by the other electrons in the atom.

**Student Annotation:** *Shielding* is also known as *screening.*

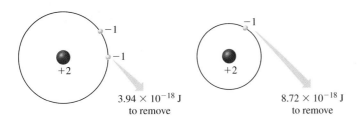

**Figure 7.4**   Removal of the first electron in He requires less energy than removal of the second electron because of shielding.

One way to illustrate how electrons in an atom shield one another is to consider the amounts of energy required to remove the two electrons from a helium atom, shown in Figure 7.4. Experiments show that it takes $3.94 \times 10^{-18}$ J to remove the first electron but $8.72 \times 10^{-18}$ J to remove the second one. There is no shielding once the first electron is removed, so the second electron feels the full effect of the $+2$ nuclear charge and is more difficult to remove.

Although all the electrons in an atom shield one another to some extent, those that are most effective at shielding are the *core* electrons. As a result, the value of $Z_{eff}$ increases steadily from left to right across a period of the periodic table because the number of core electrons remains the same (only the number of protons, $Z$, and the number of *valence* electrons increases).

> **Student Annotation:** Core electrons are those in the completed inner shells.

As we move to the right across period 2, the nuclear charge increases by 1 with each new element, but the *effective* nuclear charge increases only by an average of 0.64. (If the valence electrons did *not* shield one another, the effective nuclear charge would also increase by 1 each time a proton was added to the nucleus.)

|  | **Li** | **Be** | **B** | **C** | **N** | **O** | **F** |
|---|---|---|---|---|---|---|---|
| $Z$ | 3 | 4 | 5 | 6 | 7 | 8 | 9 |
| $Z_{eff}$ (felt by valence electrons) | 1.28 | 1.91 | 2.42 | 3.14 | 3.83 | 4.45 | 5.10 |

In general, the effective nuclear charge is given by

$$Z_{eff} = Z - \sigma \qquad \text{Equation 7.1}$$

where $\sigma$ is the shielding constant. The shielding constant is greater than zero but smaller than $Z$.

The change in $Z_{eff}$ as we move from the top of a group to the bottom is generally less significant than the change as we move across a period. Although each step down a group represents a large increase in the nuclear charge, there is also an additional shell of core electrons to shield the valence electrons from the nucleus. Consequently, the *effective* nuclear charge changes less than the nuclear charge as we move down a column of the periodic table.

## 7.4   Periodic Trends in Properties of Elements

Several physical and chemical properties of the elements depend on effective nuclear charge. To understand the trends in these properties, it is helpful to visualize the electrons of an atom in *shells.* Recall that the value of the principal quantum number ($n$) increases as the distance from the nucleus increases [◄◄ Section 6.7]. If we take this statement literally, and picture all the electrons in a shell at the same distance from the nucleus, the result is a sphere of uniformly distributed negative charge, with its distance from the nucleus depending on the value of $n$. With this as a starting point, we will examine the periodic trends in atomic radius, ionization energy, and electron affinity.

(a)

(b)

**Figure 7.5**   (a) Atomic radius in metals is defined as half the distance between adjacent metal atoms. (b) Atomic radius in nonmetals is defined as half the distance between bonded identical atoms in a molecule.

### Atomic Radius

Intuitively, we think of the *atomic radius* as the distance between the nucleus of an atom and its valence shell (i.e., the outermost shell that is occupied by one or more electrons), because we usually envision atoms as spheres with discrete boundaries. According to the quantum mechanical model of the atom, though, there is no specific distance from the nucleus beyond which an electron may not be found [◄◄ Section 6.7]. Therefore, the atomic radius requires a specific definition.

There are two ways in which the atomic radius is commonly defined. One is the **metallic radius,** which is half the distance between the nuclei of two adjacent, identical metal atoms [Figure 7.5(a)]. The other is the **covalent radius,** which is half the distance between adjacent, identical nuclei in a molecule [Figure 7.5(b)].

**Multimedia**
Periodic Table—atomic radius.

Figure 7.6 shows the atomic radii of the main group elements according to their positions in the periodic table. There are two distinct trends. The atomic radius *decreases* as we move from left to right across a period and *increases* from top to bottom as we move down within a group. The increase down a group is fairly easily explained. As we step down a column, the outermost occupied shell has an ever-increasing value of *n,* so it lies farther from the nucleus, making the radius bigger.

Now let's try to understand the decrease in radius from left to right across a period. Although this trend may at first seem counterintuitive, given that the number of valence electrons is increasing with each new element, consider the shell model in which all the electrons in a shell form a uniform sphere of negative charge around the nucleus at a distance specified by the value of *n.* As we move from left to right across a period, the effective nuclear charge increases and each step to the right adds another electron to the valence shell. Coulomb's law dictates that there will be a more powerful attraction between the nucleus and the valence shell when the magnitudes of both charges increase. The result is that as we step across a period the valence shell is drawn closer to the nucleus, making the atomic radius smaller. Figure 7.7 shows how the effective nuclear charge, charge on the valence shell, and atomic radius vary across period 2. We can picture the valence shells in all the atoms as

**Student Annotation:** Although the overall trend in atomic size for transition elements is also to decrease from left to right and increase from top to bottom, the observed radii do not vary in as regular a way as do the main group elements.

**Figure 7.6** Atomic radii of the elements (in picometers).

Increasing atomic radius

Increasing atomic radius

| | 1A | 2A | 3A | 4A | 5A | 6A | 7A | 8A |
|---|---|---|---|---|---|---|---|---|
| | H 37 | | | | | | | He 31 |
| | Li 152 | Be 112 | B 85 | C 77 | N 75 | O 73 | F 72 | Ne 70 |
| | Na 186 | Mg 160 | Al 143 | Si 118 | P 110 | S 103 | Cl 99 | Ar 98 |
| | K 227 | Ca 197 | Ga 135 | Ge 123 | As 120 | Se 117 | Br 114 | Kr 112 |
| | Rb 248 | Sr 215 | In 166 | Sn 140 | Sb 141 | Te 143 | I 133 | Xe 131 |
| | Cs 265 | Ba 222 | Ti 171 | Pb 175 | Bi 155 | Po 164 | At 142 | Rn 140 |

**Figure 7.7** Atomic radius decreases from left to right across a period because of the increased electrostatic attraction between the effective nuclear charge and the charge on the valence shell. The white circle shows the atomic size in each case. The comparison of attractive forces between the nuclei and valence shells is done using Coulomb's law [▶▶ Page 272].

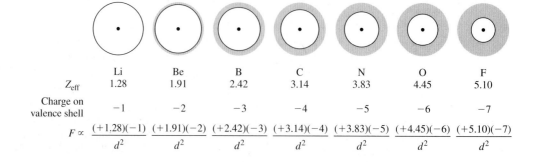

| | Li | Be | B | C | N | O | F |
|---|---|---|---|---|---|---|---|
| $Z_{eff}$ | 1.28 | 1.91 | 2.42 | 3.14 | 3.83 | 4.45 | 5.10 |
| Charge on valence shell | $-1$ | $-2$ | $-3$ | $-4$ | $-5$ | $-6$ | $-7$ |
| $F \propto$ | $\dfrac{(+1.28)(-1)}{d^2}$ | $\dfrac{(+1.91)(-2)}{d^2}$ | $\dfrac{(+2.42)(-3)}{d^2}$ | $\dfrac{(+3.14)(-4)}{d^2}$ | $\dfrac{(+3.83)(-5)}{d^2}$ | $\dfrac{(+4.45)(-6)}{d^2}$ | $\dfrac{(+5.10)(-7)}{d^2}$ |

being initially at the same distance (determined by *n*) from the nuclei, but being pulled closer by a larger attractive force resulting from increases in both $Z_{eff}$ and the number of valence electrons.

Sample Problem 7.3 shows how to use these trends to compare the atomic radii of different elements.

---

### Sample Problem 7.3

Referring only to a periodic table, arrange the elements P, S, and O in order of increasing atomic radius.

**Strategy** Use the left-to-right (decreasing) and top-to-bottom (increasing) trends to compare the atomic radii of two of the three elements at a time.

**Setup** Sulfur is to the right of phosphorus in the third row, so sulfur should be smaller than phosphorus. Oxygen is above sulfur in Group 6A, so oxygen should be smaller than sulfur.

**Solution:** O < S < P.

**Practice Problem A** Referring only to a periodic table, arrange the elements Ge, Se, and F in order of increasing atomic radius.

**Practice Problem B** For which of the following pairs of elements can the atomic radii *not* be compared using the periodic table alone: P and Se, Se and Cl, or P and O?

> **Think About It** Consult Figure 7.6 to confirm the order. Note that there are circumstances under which the trends alone will be insufficient to compare the radii of two elements. Using only a periodic table, for example, it would not be possible to determine that chlorine ($r = 99$ pm) has a larger radius than oxygen ($r = 73$ pm).

---

## Ionization Energy

*Ionization energy (IE)* is the minimum energy required to remove an electron from an atom in the gas phase. Typically, we express ionization energy in kJ/mol, the number of kilojoules required to remove a mole of electrons from a mole of gaseous atoms. Sodium, for example, has an ionization energy of 495.8 kJ/mol, meaning that the energy input required to drive the process

$$Na(g) \longrightarrow Na^+(g) + e^-$$

is 495.8 kJ/mol. Specifically, this is the *first* ionization energy of sodium, $IE_1(Na)$, which corresponds to the removal of the most loosely held electron. Figure 7.8(a) shows the first ionization energies of the main group elements according to their positions in the periodic table. Figure 7.8(b) shows a graph of $IE_1$ as a function of atomic number.

> **Student Annotation:** As with atomic radius, ionization energy changes in a similar but somewhat less regular way among the transition elements.

**Figure 7.8**   (a) First ionization energies (in kJ/mol) of the main group elements. (b) First ionization energy as a function of atomic number.

In general, as effective nuclear charge increases, ionization energy also increases. Thus, $IE_1$ increases from left to right across a period. Despite this trend, the graph in Figure 7.8(b) shows that $IE_1$ for a Group 3A element is smaller than that for the corresponding Group 2A element. Likewise, $IE_1$ for a Group 6A element is smaller than that for the corresponding Group 5A element. Both of these *interruptions* of the upward trend in $IE_1$ can be explained by using electron configuration.

Recall that the energy of an electron in a many-electron system depends not only on the principal quantum number ($n$), but also on the angular momentum quantum number ($\ell$) [◀◀ Section 6.8, Figure 6.23]. Within a given shell, electrons with the higher value of $\ell$ have a higher energy (are less tightly held by the nucleus) and are therefore *easier* to remove. Figure 7.9(a) shows the relative energies of an $s$ subshell ($\ell = 0$) and a $p$ subshell ($\ell = 1$). Ionization of an element in Group 2A requires the removal of an electron from an $s$ orbital, whereas ionization of an element in Group 3A requires the removal of an electron from a $p$ orbital; therefore, the element in Group 3A has a lower ionization energy than the element in Group 2A.

As for the decrease in ionization energy in elements of Group 6A compared to those in Group 5A, both ionizations involve the removal of a $p$ electron, but the ionization of an atom in Group 6A involves the removal of a *paired* electron. The repulsive force between two electrons in the same orbital makes it easier to remove one of them, making the ionization energy for the Group 6A element actually lower than that for the Group 5A element. [See Figure 7.9(b).]

The first ionization energy $IE_1$ decreases as we move from top to bottom within a group due to the increasing atomic radius. Although the effective nuclear charge does not change significantly as we step down a group, the atomic radius increases because the value of $n$ for the valence shell increases. According to Coulomb's law, the attractive force between a valence electron and the effective nuclear charge gets *weaker* as the distance between them increases. This makes it easier to remove an electron, and so $IE_1$ decreases.

It is possible to remove additional electrons in subsequent ionizations, giving $IE_2$, $IE_3$, and so on. The second and third ionizations of sodium, for example, can be represented, respectively, as

$$Na^+(g) \longrightarrow Na^{2+}(g) + e^- \quad \text{and} \quad Na^{2+}(g) \longrightarrow Na^{3+}(g) + e^-$$

However, the removal of successive electrons requires ever-increasing amounts of energy because it is harder to remove an electron from a cation than from an atom (and it gets even harder as the charge on the cation increases). Table 7.2 lists the ionization energies of the elements in period

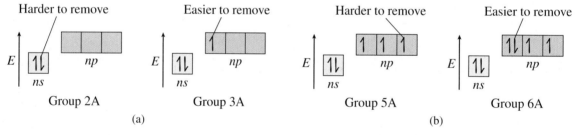

**Figure 7.9**    (a) It is harder to remove an electron from an $s$ orbital than it is to remove an electron from a $p$ orbital with the same principal quantum number. (b) Within a $p$ subshell, it is easier to remove an electron from a doubly occupied orbital than from a singly occupied orbital.

| TABLE 7.2 | | Ionization Energies (in kJ/mol) for Elements 3 through 11* | | | | | | | | | |
|---|---|---|---|---|---|---|---|---|---|---|---|
| | Z | $IE_1$ | $IE_2$ | $IE_3$ | $IE_4$ | $IE_5$ | $IE_6$ | $IE_7$ | $IE_8$ | $IE_9$ | $IE_{10}$ |
| Li | 3 | 520 | 7,298 | 11,815 | | | | | | | |
| Be | 4 | 899 | 1,757 | 14,848 | 21,007 | | | | | | |
| B | 5 | 800 | 2,427 | 3,660 | 25,026 | 32,827 | | | | | |
| C | 6 | 1,086 | 2,353 | 4,621 | 6,223 | 37,831 | 47,277 | | | | |
| N | 7 | 1,402 | 2,856 | 4,578 | 7,475 | 9,445 | 53,267 | 64,360 | | | |
| O | 8 | 1,314 | 3,388 | 5,301 | 7,469 | 10,990 | 13,327 | 71,330 | 84,078 | | |
| F | 9 | 1,681 | 3,374 | 6,050 | 8,408 | 11,023 | 15,164 | 17,868 | 92,038 | 106,434 | |
| Ne | 10 | 2,080 | 3,952 | 6,122 | 9,371 | 12,177 | 15,238 | 19,999 | 23,069 | 115,380 | 131,432 |
| Na | 11 | 496 | 4,562 | 6,910 | 9,543 | 13,354 | 16,613 | 20,117 | 25,496 | 28,932 | 141,362 |

*Cells shaded with blue represent the removal of core electrons.

2 and of sodium. These data show that it takes much more energy to remove core electrons than to remove valence electrons. There are two reasons for this. First, core electrons are closer to the nucleus, and second, core electrons experience a greater effective nuclear charge because there are fewer filled shells shielding them from the nucleus. Both of these factors contribute to a greater attractive force between the electrons and the nucleus, which must be overcome to remove the electrons.

Sample Problem 7.4 shows how to use these trends to compare first ionization energies, and subsequent ionization energies, of specific atoms.

---

**Sample Problem 7.4**

Would you expect Na or Mg to have the greater first ionization energy ($IE_1$)? Which should have the greater second ionization energy ($IE_2$)?

**Strategy** Consider effective nuclear charge and electron configuration to compare the ionization energies. Effective nuclear charge increases from left to right in a period (thus increasing $IE$), and it is more difficult to remove a paired core electron than an unpaired valence electron.

**Setup** Na is in Group 1A, and Mg is beside it in Group 2A. Na has one valence electron, and Mg has two valence electrons.

**Solution** $IE_1(Mg) > IE_1(Na)$ because Mg is to the right of Na in the periodic table (i.e., Mg has the greater effective nuclear charge, so it is more difficult to remove its electron). $IE_2(Na) > IE_2(Mg)$ because the second ionization of Mg removes a valence electron, whereas the second ionization of Na removes a core electron.

> **Think About It** The first ionization energies of Na and Mg are 496 and 738 kJ/mol, respectively. The second ionization energies of Na and Mg are 4562 and 1451 kJ/mol, respectively.

**Practice Problem A** Which element, Mg or Al, will have the higher first ionization energy and which will have the higher third ionization energy?

**Practice Problem B** Explain why Rb has a lower $IE_1$ than Sr, but Sr has a lower $IE_2$ than Rb.

---

## Electron Affinity

*Electron affinity (EA)* is the energy released (the negative of the enthalpy change $\Delta H$) when an atom in the gas phase accepts an electron. Consider the process in which a gaseous chlorine atom accepts an electron:

$$Cl(g) + e^- \longrightarrow Cl^-(g) \qquad \Delta H = -349.0 \text{ kJ/mol}$$

A negative value of $\Delta H$ indicates an exothermic process [◄◄ Section 5.3], so 349.0 kJ/mol of energy is released (the definition of electron affinity) when a mole of gaseous chlorine atoms accepts a mole of electrons. A positive electron affinity indicates a process that is energetically favorable. In general, the larger and more positive the $EA$ value, the more favorable the process and the more apt it is to occur. Figure 7.10 shows electron affinities for the main group elements.

Like ionization energy, electron affinity increases from left to right across a period. This trend in $EA$ is due to the increase in effective nuclear charge from left to right (i.e., it becomes progressively easier to add a negatively charged electron as the positive charge of the element's nucleus increases). There are also periodic interruptions of the upward trend of $EA$ from left to right, similar to those observed for $IE_1$, although they do *not* occur for the same elements. For example, the $EA$ of a Group 2A element is lower than that for the corresponding Group 1A element, and the $EA$ of a Group 5A element is lower than that for the corresponding Group 4A element. These exceptions to the trend are due to the electron configurations of the elements involved.

It is harder to add an electron to a Group 2A element ($ns^2$) than to the Group 1A element ($ns^1$) in the same period because the electron added to the Group 2A element is placed in an orbital of higher energy (a $p$ orbital versus an $s$ orbital). Likewise, it is harder to add an electron to a Group 5A element ($ns^2np^3$) than to the corresponding Group 4A element ($ns^2np^2$) because the electron added to the Group 5A element must be placed in an orbital that already contains an electron. Figure 7.11 illustrates these points.

Just as more than one electron can be removed from an atom, more than one electron can also be added to an atom. While many first electron affinities are positive, subsequent electron

> **Student Annotation:** Some books define electron affinity as $\Delta H$, rather than the negative of $\Delta H$ for the process of adding an electron. This simply changes the sign of $EA$ relative to what we show here.

> **Student Annotation:** Although $IE$ and $EA$ both increase from left to right across a period, an increase in $IE$ means that it is *less* likely that an electron will be *removed* from an atom. An increase in $EA$, on the other hand, means that it is *more* likely that an electron will be *accepted* by an atom.

> **Student Annotation:** There is a much less significant and less regular variation in electron affinities from top to bottom within a group. See Figure 7.11(a).

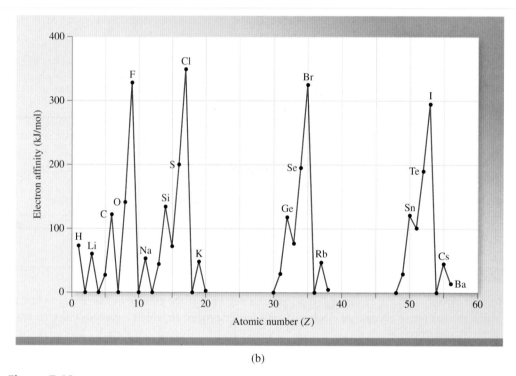

**Figure 7.10**    (a) Electron affinities (kJ/mol) of the main group elements. (b) Electron affinity as a function of atomic number.

affinities are always negative. Considerable energy is required to overcome the repulsive forces between the electron and the negatively charged ion. The addition of two electrons to a gaseous oxygen atom can be represented as:

| Process | $\Delta H$ (kJ/mol) | Electron Affinity |
|---|---|---|
| $O(g) + e^- \longrightarrow O^-(g)$ | $-141$ | $EA_1 = 141$ kJ/mol |
| $O^-(g) + e^- \longrightarrow O^{2-}(g)$ | $744$ | $EA_2 = -744$ kJ/mol |

The term *second electron affinity* may seem like something of a misnomer, because an anion in the gas phase has no real "affinity" for an electron. As will be discussed in Chapter 8, a significantly *endothermic* process such as the addition of an electron to a gaseous $O^-$ ion happens only in concert with one or more *exothermic* processes that more than compensate for the required energy input.

Sample Problem 7.5 lets you practice using the periodic table to compare the electron affinities of elements.

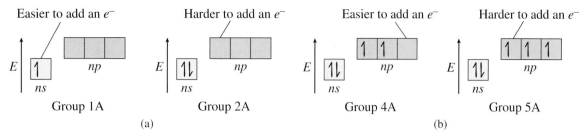

**Figure 7.11** (a) It is easier to add an electron to an *s* orbital than to add one to a *p* orbital with the same principal quantum number. (b) Within a *p* subshell, it is easier to add an electron to an empty orbital than to add one to an orbital that already contains an electron.

## Sample Problem 7.5

For each pair of elements, indicate which one you would expect to have the greater first electron affinity, $EA_1$: (a) Al or Si, (b) Si or P.

**Strategy** Consider the effective nuclear charge and electron configuration to compare the electron affinities. The effective nuclear charge increases from left to right in a period (thus generally increasing *EA*), and it is more difficult to add an electron to a partially occupied orbital than to an empty one. Writing out orbital diagrams for the valence electrons is helpful for this type of problem.

**Setup** (a) Al is in Group 3A and Si is beside it in Group 4A. Al has three valence electrons ([Ne]$3s^2 3p^1$), and Si has four valence electrons ([Ne]$3s^2 3p^2$).

(b) P is in Group 5A (to the right of Si), so it has five valence electrons ([Ne]$3s^2 3p^3$).

**Solution** (a) $EA_1(\text{Si}) > EA_1(\text{Al})$ because Si is to the right of Al and therefore has a greater effective nuclear charge.

(b) $EA_1(\text{Si}) > EA_1(\text{P})$ because although P is to the right of Si in the third period of the periodic table (giving P the larger $Z_{\text{eff}}$), adding an electron to a P atom requires placing it in a $3p$ orbital that is partially occupied. The energy cost of *pairing* electrons outweighs the energy advantage of adding an electron to an atom with a larger effective nuclear charge.

$3s^2 \qquad 3p^1$
Valence orbital diagram for Al

$3s^2 \qquad 3p^2$
Valence orbital diagram for Si

$3s^2 \qquad 3p^3$
Valence orbital diagram for P

**Think About It** The first electron affinities of Al, Si, and P are 42.5, 134, and 72.0 kJ/mol, respectively.

**Practice Problem A** Would you expect Mg or Al to have the higher $EA_1$?

**Practice Problem B** Why is the $EA_1$ for Ge greater than the $EA_1$ for As?

## Metallic Character

Metals tend to

- Be shiny, lustrous, and malleable
- Be good conductors of both heat and electricity
- Have low ionization energies (so they commonly form *cations*)
- Form ionic compounds with chlorine (metal chlorides)
- Form basic, ionic compounds with oxygen (metal oxides)

Nonmetals, on the other hand, tend to

- Vary in color and lack the shiny appearance associated with metals
- Be brittle, rather than malleable
- Be poor conductors of both heat and electricity
- Form acidic, molecular compounds with oxygen
- Have high electron affinities (so they commonly form *anions*)

**Student Annotation:** *Malleability* is the property that allows metals to be pounded into thin sheets. *Ductility,* the capacity to be drawn out into wires, is another characteristic of metals.

Metallic character increases from top to bottom in a group and decreases from left to right within a period.

*Metalloids* are elements with properties intermediate between those of metals and nonmetals. Because the definition of metallic character depends on a combination of properties, there may be some variation in the elements identified as metalloids in different sources. Astatine (At), for example, is listed as a metalloid in some sources and a nonmetal in others.

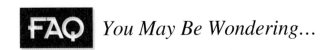

### What Causes the Periodic Trends in Properties?

Many of the periodic trends in properties of the elements can be explained using **Coulomb's law,** which states that the force ($F$) between two charged objects ($Q_1$ and $Q_2$) is directly proportional to the product of the two charges and *inversely* proportional to the distance ($d$) between the objects squared. Recall that the *energy* between two oppositely charged particles ($E_{el}$) is inversely proportional to $d$ [◄◄ Section 5.1, Equation 5.2]. The SI unit of force is the newton (1 N = 1 kg · m/s$^2$), and the SI unit of energy is the joule (1 J = 1 kg · m$^2$/s$^2$).

$$F \propto \frac{Q_1 \times Q_2}{d^2}$$

When the charges have opposite signs, $F$ is negative—indicating an *attractive* force between the objects. When the charges have the same sign, $F$ is positive—indicating a *repulsive* force. Table 7.3 shows how the magnitude of the attractive force between two oppositely charged objects at a fixed distance from each other varies with changes in the magnitudes of the charges.

Sample Problem 7.6 illustrates how Coulomb's law can be used to compare the magnitudes of attractive forces between charged objects.

| **TABLE 7.3** | Attractive Force between Oppositely Charged Objects at a Fixed Distance ($d = 1$) from Each Other | | |
|---|---|---|---|
| **$Q_1$** | **$Q_2$** | | **Attractive force is proportional to** |
| +1 | −1 | | 1 |
| +2 | −2 | | 4 |
| +3 | −3 | | 9 |

The product of a positive number and a negative number is a negative number. When we are simply comparing the magnitudes of attractive forces, however, it is unnecessary to include the sign.

---

### Sample Problem 7.6

For carbon and nitrogen, use the effective nuclear charges given in Figure 7.7 and the atomic radii given in Figure 7.6 to compare the attractive force between the nucleus in each atom and the valence electron that would be removed by the first ionization.

**Strategy** Use Coulomb's law to calculate a number to which the attractive force will be proportional in each case.

**Setup** From Figure 7.7, the effective nuclear charges of C and N are 3.14 and 3.83, respectively, and the radii of C and N are 77 pm and 75 pm, respectively. The first ionization energies are 1086 kJ/mol (C) and 1402 kJ/mol (N). The charge on the valence electron in each case is −1.

**Solution** For C: $F \propto \dfrac{3.14 \times (-1)}{(77 \text{ pm})^2} = -5.3 \times 10^{-4}$

For N: $F \propto \dfrac{3.83 \times (-1)}{(75 \text{ pm})^2} = -6.8 \times 10^{-4}$

Note that in this type of comparison, it doesn't matter what units we use for the distance between the charges. We are not trying to calculate a particular attractive force, only to compare the magnitudes of these two attractive forces.

> **Think About It** Remember that the negative sign simply indicates that the force is attractive rather than repulsive. The calculated number for nitrogen is about 28 percent larger than that for carbon.

**Practice Problem A** Between which two charges is the attractive force larger: +3.26 and −1.15 separated by a distance of 1.5 pm, or +2.84 and −3.63 separated by a distance of 2.5 pm?

**Practice Problem B** What must the distance be between charges of +2.25 and −1.86 for the attractive force between them to be the same as that between charges of +4.06 and −2.11 separated by a distance of 2.16 pm?

| Checkpoint 7.4 | **Periodic Trends in Properties of Atoms** |

**7.4.1** Arrange the elements Ca, Sr, and Ba in order of increasing $IE_1$.

a) Ca < Sr < Ba

b) Ba < Sr < Ca

c) Ba < Ca < Sr

d) Sr < Ba < Ca

e) Sr < Ca < Ba

**7.4.2** Arrange the elements Li, Be, and B in order of increasing $IE_2$.

a) Li < Be < B

b) Li < B < Be

c) Be < B < Li

d) Be < Li < B

e) B < Be < Li

**7.4.3** For each of the following pairs of elements, indicate which will have the greater $EA_1$: Rb or Sr, C or N, O or F:

a) Rb, C, O

b) Sr, N, F

c) Sr, C, F

d) Sr, N, O

e) Rb, C, F

**7.4.4** Which element, K or Ca, will have the greater $IE_1$, which will have the greater $IE_2$, and which will have the greater $EA_1$?

a) Ca, K, K

b) K, K, Ca

c) K, Ca, K

d) Ca, Ca, K

e) Ca, Ca, Ca

## 7.5 Electron Configuration of Ions

Because many ionic compounds are made up of monatomic anions and cations, it is helpful to know how to write the electron configurations of these ionic species. Just as for atoms, we use the Pauli exclusion principle and Hund's rule to write the ground-state electron configurations of cations and anions.

Recall from Chapter 2 that we can use the periodic table to predict the charges on many of the ions formed by main group elements. Elements in Groups 1A and 2A, for example, form ions with charges of $+1$ and $+2$, respectively. Elements in Groups 6A and 7A form ions with charges of $-2$ and $-1$, respectively. Knowing something about electron configurations enables us to explain these charges.

### Ions of Main Group Elements

In Section 7.4, we learned about the tendencies of atoms to lose or gain electrons. In every period of the periodic table, the element with the highest $IE_1$ is the Group 8A element, the noble gas. [See Figure 7.8(b).] Also, Group 8A is the only group in which none of the members has *any* tendency to accept an electron; that is, they all have negative $EA$ values. [See Figure 7.10(b).] High ionization energies and low electron affinities make the noble gases almost completely unreactive. Ultimately, the cause of this lack of reactivity is electron configuration. The $1s^2$ configuration of He and the $ns^2np^6$ ($n \geq 2$) valence electron configurations of the other noble gases are extraordinarily stable. Other main group elements tend to either lose or gain the number of electrons needed to achieve the same number of electrons as the nearest noble gas. Species with identical electron configurations are called ***isoelectronic.***

To write the electron configuration of an ion formed by a main group element, we first write the configuration for the atom and either add or remove the appropriate number of electrons. Electron configurations for the sodium and chloride ions are

$$\text{Na: } 1s^2 2s^2 2p^6 3s^1 \longrightarrow \text{Na}^+\text{: } 1s^2 2s^2 2p^6 \quad \text{(10 electrons total, isoelectronic with Ne)}$$

$$\text{Cl: } 1s^2 2s^2 2p^6 3s^2 3p^5 \longrightarrow \text{Cl}^-\text{: } 1s^2 2s^2 2p^6 3s^2 3p^6 \quad \text{(18 electrons total, isoelectronic with Ar)}$$

We can also write electron configurations for ions using the noble gas core.

$$\text{Na: } [\text{Ne}]3s^1 \longrightarrow \text{Na}^+\text{: } [\text{Ne}]$$

$$\text{Cl: } [\text{Ne}]3s^2 3p^5 \longrightarrow \text{Cl}^-\text{: } [\text{Ne}]3s^2 3p^6 \quad \text{or} \quad [\text{Ar}]$$

Sample Problem 7.7 gives you some practice writing electron configurations for the ions of main group elements.

**Student Annotation:** It is a common error to mistake species with the same valence electron configuration for isoelectronic species. For example, $F^-$ and Ne are isoelectronic. $F^-$ and $Cl^-$ are not.

## Sample Problem 7.7

Write electron configurations for the following ions of main group elements: (a) $N^{3-}$, (b) $Ba^{2+}$, and (c) $Be^{2+}$.

**Strategy** First write electron configurations for the atoms. Then add electrons (for anions) or remove electrons (for cations) to account for the charge.

**Setup** (a) $N^{3-}$ forms when N ($1s^2 2s^2 2p^3$ or $[He]2s^2 2p^3$), a main group nonmetal, gains three electrons.

(b) $Ba^{2+}$ forms when Ba ($1s^2 2s^2 2p^6 3s^2 3p^6 4s^2 3d^{10} 4p^6 5s^2 4d^{10} 5p^6 6s^2$ or $[Xe]6s^2$) loses two electrons.

(c) $Be^{2+}$ forms when Be ($1s^2 2s^2$ or $[He]2s^2$) loses two electrons.

**Solution** (a) $[He]2s^2 2p^6$ or [Ne]

(b) $[Kr]5s^2 4d^{10} 5p^6$ or [Xe]

(c) $1s^2$ or [He]

**Think About It** Be sure to add electrons to form an anion, and remove electrons to form a cation.

**Practice Problem A** Write electron configurations for (a) $O^{2-}$, (b) $Ca^{2+}$, and (c) $Se^{2-}$.

**Practice Problem B** List all the species (atoms and/or ions) that are likely to have the following electron configuration: $1s^2 2s^2 2p^6$.

## Ions of *d*-Block Elements

Recall from Section 6.8 that the 4*s* orbital fills before the 3*d* orbitals for the elements in the first row of the *d*-block (Sc to Zn) [◄◄ Section 6.8]. Following the pattern for writing electron configurations for main group ions, then, we might expect the two electrons lost in the formation of the $Fe^{2+}$ ion to come from the 3*d* subshell. It turns out, though, that an atom always loses electrons first from the shell with the *highest* value of *n*. In the case of Fe, that would be the 4*s* subshell.

**Student Annotation:** This explains, in part, why many of the transition metals can form ions with a +2 charge.

$$Fe: [Ar]4s^2 3d^6 \longrightarrow Fe^{2+}: [Ar]3d^6$$

Iron can also form the $Fe^{3+}$ ion, in which case the third electron is removed from the 3*d* subshell.

$$Fe: [Ar]4s^2 3d^6 \longrightarrow Fe^{3+}: [Ar]3d^5$$

In general, when a *d*-block element becomes an ion, it loses electrons first from the *ns* subshell and then from the $(n-1)d$ subshell.

Sample Problem 7.8 gives you some practice writing electron configurations for the ions of *d*-block elements.

## Sample Problem 7.8

Write electron configurations for the following ions of *d*-block elements: (a) $Zn^{2+}$, (b) $Mn^{2+}$, and (c) $Cr^{3+}$.

**Strategy** First write electron configurations for the atoms. Then add electrons (for anions) or remove electrons (for cations) to account for the charge. The electrons removed from a *d*-block element must come first from the outermost *s* subshell, not the partially filled *d* subshell.

**Student Annotation:** Remember that the electron configuration of Cr is anomalous in that it has only one 4*s* electron, making its *d* subshell half filled [◄◄ Section 6.9].

**Setup** (a) $Zn^{2+}$ forms when Zn ($1s^2 2s^2 2p^6 3s^2 3p^6 4s^2 3d^{10}$ or $[Ar]4s^2 3d^{10}$) loses two electrons.

(b) $Mn^{2+}$ forms when Mn ($1s^2 2s^2 2p^6 3s^2 3p^6 4s^2 3d^5$ or $[Ar]4s^2 3d^5$) loses two electrons.

(c) $Cr^{3+}$ forms when Cr ($1s^2 2s^2 2p^6 3s^2 3p^6 4s^1 3d^5$ or $[Ar]4s^1 3d^5$) loses three electrons—one from the 4*s* subshell and two from the 3*d* subshell.

**Solution** (a) $[Ar]3d^{10}$

(b) $[Ar]3d^5$

(c) $[Ar]3d^3$

**Think About It** Be sure to add electrons to form an anion and remove electrons to form a cation. Also, double-check to make sure that electrons removed from a *d*-block element come first from the *ns* subshell and then, if necessary, from the $(n-1)d$ subshell.

**Practice Problem A** Write electron configurations for (a) $Co^{3+}$, (b) $Cu^{2+}$, and (c) $Ag^+$.

**Practice Problem B** What common *d*-block ion (see Figure 2.14) is isoelectronic with $Zn^{2+}$?

## Checkpoint 7.5 Electron Configuration of Ions

**7.5.1** Which of the following ions are isoelectronic with a noble gas? (Select all that apply.)

a) $Mn^{2+}$

b) $Ca^{2+}$

c) $Br^-$

d) $O^{2+}$

e) $F^-$

**7.5.2** Which of the following pairs are isoelectronic with each other? (Select all that apply.)

a) $Ca^{2+}$ and $Sr^{2+}$

b) $O^{2-}$ and $Mg^{2+}$

c) $I^-$ and $Kr$

d) $S^{2-}$ and $Cl^-$

e) $He$ and $H^+$

**7.5.3** Select the correct ground-state electron configuration for $Ti^{2+}$.

a) $[Ar]4s^23d^2$

b) $[Ar]4s^23d^4$

c) $[Ar]4s^2$

d) $[Ar]3d^2$

e) $[Ar]4s^13d^1$

**7.5.4** Select the correct ground-state electron configuration for $S^{2-}$.

a) $[Ne]3p^4$

b) $[Ne]3s^23p^6$

c) $[Ne]3s^23p^2$

d) $[Ne]3p^6$

e) $[Ne]$

**7.5.5** Which periodic table's highlighted portion includes elements that can form an isoelectronic series? (Select all that apply.)

a)   b)   c)

d)   e)   f)

---

## 7.6 Ionic Radius

When an atom gains or loses one or more electrons to become an ion, its radius changes. The *ionic radius,* the radius of a cation or an anion, affects the physical and chemical properties of an ionic compound. The three-dimensional structure of an ionic compound, for example, depends on the relative sizes of its cations and anions.

### Comparing Ionic Radius with Atomic Radius

When an atom loses an electron and becomes a cation, its radius decreases due in part to a reduction in electron-electron repulsions (and consequently a reduction in shielding) in the valence shell. A significant decrease in radius occurs when *all* of an atom's valence electrons are removed. This is the case with ions of most main group elements, which are isoelectronic with the noble gases preceding them. Consider Na, which loses its $3s$ electron to become $Na^+$:

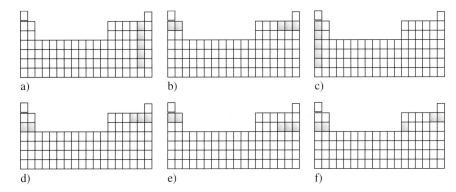

Na: $1s^22s^22p^63s^1$ $\longrightarrow$ Na$^+$: $1s^22s^22p^6$

Multimedia

Periodic Table—atomic and ionic radii.

**Valence orbitals in Na and in Na⁺**

**Figure 7.12** A comparison of atomic and ionic radii (in picometers) for main group elements and their common ions (those that are isoelectronic with noble gases).

The valence electron of Na has a principal quantum number of $n = 3$. When it has been removed, the resulting $Na^+$ ion no longer has any electrons in the $n = 3$ shell. The outermost electrons of the $Na^+$ ion have a principal quantum number of $n = 2$. Because the value of $n$ determines the distance from the nucleus, this corresponds to a smaller radius.

When an atom gains one or more electrons and becomes an anion, its radius increases due to increased electron-electron repulsions. Adding an electron causes the rest of the electrons in the valence shell to spread out and take up more space to maximize the distance between them.

Figure 7.12 shows the ionic radii for those ions of main group elements that are isoelectronic with noble gases and compares them to the radii of the parent atoms. Note that the ionic radius, like the atomic radius, increases from top to bottom in a group.

## Isoelectronic Series

An *isoelectronic series* is a series of two or more species that have identical electron configurations, but different nuclear charges. For example, $O^{2-}$, $F^-$, and Ne constitute an isoelectronic series. Although these three species have identical electron configurations, they have different radii. In an isoelectronic series, the species with the smallest nuclear charge (i.e., the smallest atomic number $Z$) will have the largest radius. The species with the largest nuclear charge (i.e., the largest $Z$) will have the smallest radius.

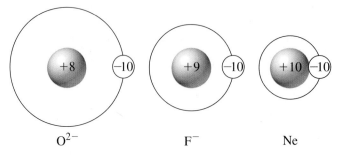

$$O^{2-} \qquad F^- \qquad Ne$$

Sample Problem 7.9 shows how to identify members of an isoelectronic series and how to arrange them according to radius.

### Sample Problem 7.9

Identify the isoelectronic series in the following group of species, and arrange them in order of increasing radius: $K^+$, Ne, Ar, Kr, $P^{3-}$, $S^{2-}$, and $Cl^-$.

**Strategy** Isoelectronic series are species with identical electron configurations but different nuclear charges. Determine the number of electrons in each species. The radii of isoelectronic series members decreases with increasing nuclear charge.

**Setup** The number of electrons in each species is as follows: 18 ($K^+$), 10 (Ne), 18 (Ar), 36 (Kr), 18 ($P^{3-}$), 18 ($S^{2-}$), and 18 ($Cl^-$). The nuclear charges of the species with 18 electrons are $+19$ ($K^+$), $+18$ (Ar), $+15$ ($P^{3-}$), $+16$ ($S^{2-}$), and $+17$ ($Cl^-$).

**Solution** The isoelectronic series includes $K^+$, Ar, $P^{3-}$, $S^{2-}$, and $Cl^-$. In order of increasing radius: $K^+ < Ar < Cl^- < S^{2-} < P^{3-}$.

> **Think About It** Consult Figure 7.12 to check your result. With identical electron configurations, the attractive force between the valence electrons and the nucleus will be strongest for the largest nuclear charge. Thus, the larger the nuclear charge, the closer in the valence electrons will be pulled and the smaller the radius will be.

**Practice Problem A** Arrange the following isoelectronic series in order of increasing radius: $Se^{2-}$, $Br^-$, Kr, and $Rb^+$.

**Practice Problem B** List all the common ions that are isoelectronic with Ne.

### Checkpoint 7.6    Ionic Radius

**7.6.1** Which of the following species are isoelectronic with Kr? (Select all that apply.)

a) He

b) Ne

c) Ar

d) $Br^-$

e) $Rb^+$

**7.6.2** Which of the following are arranged correctly in order of increasing radius? (Select all that apply.)

a) $F^- < Cl^- < Br^- < I^-$

b) $O^{2-} < F^- < Ne < Na^+$

c) $Ca^{2+} < K^+ < Ar < Cl^-$

d) $Rb^+ < K^+ < Na^+ < Li^+$

e) $Sr^{2+} < Ca^{2+} < Mg^{2+} < Be^{2+}$

**7.6.3** Which of the following is the most realistic representation of an atom from Group 6A becoming an ion?

**7.6.4** Which of the following is the most realistic representation of an atom from Group 1A becoming an ion?

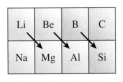
## 7.7    Periodic Trends in Chemical Properties of the Main Group Elements

Ionization energy and electron affinity enable us to understand the types of reactions that elements undergo and the types of compounds they form. These two parameters actually measure similar things. Ionization energy is a measure of how powerfully an atom attracts its own electrons, while electron affinity is a measure of how powerfully an atom can attract electrons from another source. As a very simple example of how this helps us understand a chemical reaction, consider the combination of a sodium atom and a chlorine atom, shown in Figure 7.13.

Sodium, with its low ionization energy, has a relatively weak attraction for its one valence electron. Chlorine, with its energetically favorable electron affinity, has the ability to attract electrons from another source. In this case, the electron that is loosely held by the Na atom, and powerfully attracted by the Cl atom, is transferred from Na to Cl, thus producing a sodium ion ($Na^+$) and a chloride ion ($Cl^-$). According to Coulomb's law, oppositely charged objects attract each other. The positively charged sodium ion and the negatively charged chloride ion are drawn together by electrostatic attraction, and the result is the formation of the solid ionic compound sodium chloride (NaCl).

### General Trends in Chemical Properties

Before we examine the elements in individual groups, let's identify some overall trends. We have said that elements in the same group resemble one another in chemical behavior because they have similar valence electron configurations. This statement, although correct in the general sense, must be applied with caution. Chemists have long known that the properties of the first member of each group (Li, Be, B, C, N, O, and F) are different from those of the rest of the members of the same group. Lithium, for example, exhibits many, but not all, of the properties characteristic of the Group 1A (alkali) metals. For example, unlike the other Group 1A elements, Li reacts with the $O_2$ and $N_2$ in air to form a simple oxide ($Li_2O$) and nitride ($Li_3N$), respectively. Similarly, beryllium is a somewhat atypical member of Group 2A (alkaline earth metals) in that it forms covalent compounds, and so on. The differences can be attributed to the unusually small size of the first element in each group (see Figure 7.6).

Another trend in the chemical behavior of main group elements is the diagonal relationship. *Diagonal relationships* refer to similarities between pairs of elements in different groups and periods of the periodic table. Specifically, the first three members of the second period (Li, Be, and B) exhibit many similarities to the elements located diagonally below them in the periodic table (Mg, Al, and Si). The reason for this phenomenon is the similarity of charge densities of their cations. (Charge density is the charge on an ion divided by its volume.) Cations with comparable charge densities react similarly with anions and therefore form the same types of compounds. Thus, the chemistry of lithium resembles that of magnesium in some ways; the same holds for beryllium and aluminum and for boron and silicon. Each of these pairs is said to exhibit a diagonal relationship.

A comparison of the properties of elements in the same group is most valid if the elements in question have a similar metallic (or nonmetallic) character. The elements in Groups 1A and 2A, for example, are all metals, whereas those in Groups 7A and 8A are all nonmetals. We have to be more careful when comparing the elements of Groups 3A through 6A, though, because a single group may contain metals, metalloids, and nonmetals. In these groups, we should expect a greater variation in chemical properties even though all group members have similar valence electron configurations.

| Li | Be | B | C |
|----|----|---|---|
| Na | Mg | Al | Si |

Diagonal relationships

**Figure 7.13** Formation of NaCl from its constituent elements. Note that although the charges are not all shown, the solid consists of a three-dimensional array of alternating oppositely-charged ions.

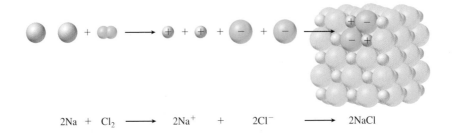

$$2Na + Cl_2 \longrightarrow 2Na^+ + 2Cl^- \longrightarrow 2NaCl$$

### Hydrogen ($1s^1$)

There is no completely suitable position for hydrogen in the periodic table (it really belongs in a group by itself). Traditionally hydrogen is shown at the top of Group 1A, because, like the alkali metals, it has a single *s* valence electron and forms a cation with a charge of $+1$ ($H^+$), which is hydrated in solution. On the other hand, hydrogen also forms the *hydride* ion ($H^-$) in ionic compounds such as NaH and $CaH_2$. In this respect, hydrogen resembles the members of Group 7A (halogens), all of which form $-1$ anions ($F^-$, $Cl^-$, $Br^-$, and $I^-$) in ionic compounds. Ionic hydrides react with water to produce hydrogen gas and the corresponding metal hydroxides:

$$2NaH(s) + 2H_2O(l) \longrightarrow 2NaOH(aq) + 2H_2(g)$$

$$CaH_2(s) + 2H_2O(l) \longrightarrow Ca(OH)_2(aq) + 2H_2(g)$$

The most important compound of hydrogen is water, which forms when hydrogen burns in air:

$$2H_2(g) + O_2(g) \longrightarrow 2H_2O(l)$$

## Properties of the Active Metals

### Group 1A Elements ($ns^1$, $n \geq 2$)

Figure 7.14 shows samples of the Group 1A elements. These elements all have low ionization energies, making it easy for them to become $M^+$ ions. In fact, these metals are so reactive that they are never found in nature in the pure elemental state. They react with water to produce hydrogen gas and the corresponding metal hydroxide:

$$2M(s) + 2H_2O(l) \longrightarrow 2MOH(aq) + H_2(g)$$

Lithium                                Sodium                              Potassium

Rubidium                               Cesium                     **Figure 7.14**   Group 1A elements.

where M denotes an alkali metal. When exposed to air, they gradually lose their shiny appearance as they react with oxygen to form metal oxides. Lithium forms lithium oxide (containing the oxide ion, $O^{2-}$):

$$4Li(s) + O_2(g) \longrightarrow 2Li_2O(s)$$

The other alkali metals all form oxides or *peroxides* (containing the peroxide ion, $O_2^{2-}$):

$$2Na(s) + O_2(g) \longrightarrow Na_2O_2(s)$$

Potassium, rubidium, and cesium also form *superoxides* (containing the superoxide ion, $O_2^-$):

$$K(s) + O_2(g) \longrightarrow KO_2(s)$$

The type of oxide that forms when an alkali metal reacts with oxygen has to do with the stability of the various oxides. Because these oxides are all ionic compounds, their stability depends on how strongly the cations and anions attract one another. Lithium tends to form predominantly the oxide because lithium oxide is more stable than lithium peroxide.

### Group 2A Elements ($ns^2$, $n \geq 2$)

Figure 7.15 shows samples of the Group 2A elements. As a group, the alkaline earth metals are somewhat less reactive than the alkali metals. Both the first and the second ionization energies decrease (and metallic character increases) from beryllium to barium. Group 2A elements tend to form $M^{2+}$ ions, where M denotes an alkaline earth metal atom.

The reactions of alkaline earth metals with water vary considerably. Beryllium does not react with water; magnesium reacts slowly with steam; and calcium, strontium, and barium react vigorously with cold water.

**Student Annotation:** Because they have less metallic character than the other Group 2A elements, beryllium and magnesium form some molecular compounds such as $BeH_2$ and $MgH_2$.

**Figure 7.15**   Group 2A elements.

Beryllium

Magnesium

Calcium

Strontium

Barium

$$Ca(s) + 2H_2O(l) \longrightarrow Ca(OH)_2(s) + H_2(g)$$

$$Sr(s) + 2H_2O(l) \longrightarrow Sr(OH)_2(s) + H_2(g)$$

$$Ba(s) + 2H_2O(l) \longrightarrow Ba(OH)_2(aq) + H_2(g)$$

The reactivity of the alkaline earth metals toward oxygen also increases from Be to Ba. Beryllium and magnesium form oxides (BeO and MgO) only at elevated temperatures, whereas CaO, SrO, and BaO form at room temperature.

Magnesium reacts with aqueous acid to produce hydrogen gas:

$$Mg(s) + 2H^+(aq) \longrightarrow Mg^{2+}(aq) + H_2(g)$$

Calcium, strontium, and barium also react with aqueous acid solutions to produce hydrogen gas. However, because the metals also react with water, the two different reactions (with $H^+$ and with $H_2O$) occur simultaneously.

## Properties of Other Main Group Elements

### Group 3A Elements ($ns^2np^1$, $n \geq 2$)

Figure 7.16 shows samples of the first four Group 3A elements. Boron, the first member of the group, is a metalloid; the others (Al, Ga, In, and Tl) are metals. Boron does not form binary ionic compounds and is unreactive toward both oxygen and water. Aluminum, the next element in the group, readily forms aluminum oxide when exposed to air:

$$4Al(s) + 3O_2(g) \longrightarrow 2Al_2O_3(s)$$

The aluminum oxide forms a protective coating, preventing the underlying metal from reacting further. This fact makes it possible to use aluminum for structural materials, such as aluminum siding and the shells of airplanes. Without the protective coating, layer after layer of Al atoms would become oxidized, and the structure would eventually crumble.

**Figure 7.16** Group 3A elements.

Boron

Aluminum

Gallium

Indium

Aluminum forms the $Al^{3+}$ ion. It reacts with hydrochloric acid according to the equation:

$$2Al(s) + 6H^+(aq) \longrightarrow 2Al^{3+}(aq) + 3H_2(g)$$

The other Group 3A metals (Ga, In, and Tl) can form both $M^+$ and $M^{3+}$ ions. As we move down the group, the $M^+$ ion becomes the more stable of the two.

The metallic elements in Group 3A also form many molecular compounds. For example, aluminum reacts with hydrogen to form $AlH_3$, which has properties similar to those of $BeH_2$. The progression of properties across the second row of the periodic table illustrates the gradual shift from metallic to nonmetallic character in the main group elements.

### Group 4A Elements ($ns^2np^2$, $n \geq 2$)

Figure 7.17 shows samples of the Group 4A elements. Carbon, the first member of the group, is a nonmetal, whereas silicon and germanium, the next two members, are metalloids. Tin and lead, the last two members of the group, are metals. They do not react with water, but they do react with aqueous acid to produce hydrogen gas:

$$Sn(s) + 2H^+(aq) \longrightarrow Sn^{2+}(aq) + H_2(g)$$
$$Pb(s) + 2H^+(aq) \longrightarrow Pb^{2+}(aq) + H_2(g)$$

The Group 4A elements form compounds in both the +2 and +4 oxidation states. For carbon and silicon, the +4 oxidation state is the more stable one. For example, $CO_2$ is more stable than CO, and $SiO_2$ is a stable compound, but SiO does not exist under ordinary conditions. As we move down the group, however, the relative stability of the two oxidation states is reversed. In tin compounds the +4 oxidation state is only slightly more stable than the +2 oxidation state. In lead compounds the +2 oxidation state is the more stable one. The outer electron configuration of lead is $6s^26p^2$, and lead tends to lose only the $6p$ electrons to form $Pb^{2+}$ rather than both the $6p$ and $6s$ electrons to form $Pb^{4+}$.

**Student Annotation:** Carbon, being a non-metal, achieves its +4 oxidation state without actually losing four electrons, which would be very "expensive" in terms of the energy required. For the same reason, compounds containing metals in very high oxidation states also tend to be molecular rather than ionic.

### Group 5A Elements ($ns^2np^3$, $n \geq 2$)

Figure 7.18 shows samples of the Group 5A elements. Nitrogen and phosphorus are nonmetals, arsenic and antimony are metalloids, and bismuth is a metal. Because Group 5A contains elements in all three categories, we expect greater variation in their chemical properties.

|  |  |  |
|---|---|---|
| Carbon (graphite) | Carbon (diamond) | Silicon |
| Germanium | Tin | Lead |

**Figure 7.17**    Group 4A elements.

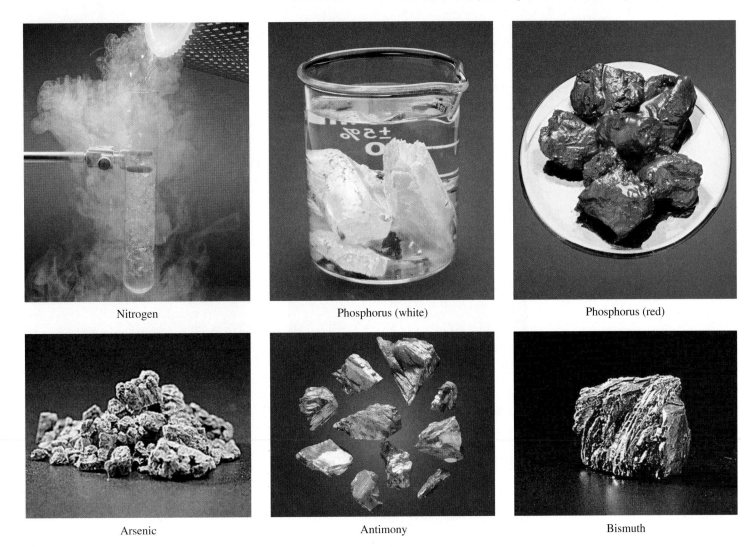

| | | |
|---|---|---|
| Nitrogen | Phosphorus (white) | Phosphorus (red) |
| Arsenic | Antimony | Bismuth |

**Figure 7.18** Group 5A elements.

Elemental nitrogen is a diatomic gas ($N_2$). It forms a variety of oxides (NO, $N_2O$, $NO_2$, $N_2O_4$, and $N_2O_5$), all of which are gases except for $N_2O_5$, which is a solid at room temperature. Nitrogen has a tendency to accept three electrons to form the nitride ion ($N^{3-}$). Most metal nitrides, such as $Li_3N$ and $Mg_3N_2$, are ionic compounds. Phosphorus exists as individual $P_4$ molecules (white phosphorus) or chains of $P_4$ molecules (red phosphorus). It forms two solid oxides with the formulas $P_4O_6$ and $P_4O_{10}$. The industrially important oxoacids nitric acid and phosphoric acid form when $N_2O_5$ and $P_4O_{10}$, respectively, react with water:

$$N_2O_5(s) + H_2O(l) \longrightarrow 2HNO_3(aq)$$
$$P_4O_{10}(s) + 6H_2O(l) \longrightarrow 4H_3PO_4(aq)$$

Arsenic, antimony, and bismuth have extensive three-dimensional structures. Bismuth is far less reactive than metals in the preceding groups.

### Group 6A Elements ($ns^2np^4$, $n \geq 2$)

Figure 7.19 shows samples of the first three Group 6A elements. The first three members of the group (oxygen, sulfur, and selenium) are nonmetals, whereas the last two (tellurium and polonium) are metalloids. Oxygen is a colorless, odorless, diatomic gas; elemental sulfur and selenium exist as the molecules $S_8$ and $Se_8$, respectively; and tellurium and polonium have more extensive three-dimensional structures. (Polonium is a radioactive element that is difficult to study in the laboratory.) Oxygen has a tendency to accept two electrons to form the oxide ion ($O^{2-}$) in many compounds. Sulfur, selenium, and tellurium also form ions by accepting two electrons: $S^{2-}$, $Se^{2-}$, and $Te^{2-}$. The elements in Group 6A (especially oxygen) form a large number of molecular compounds with nonmetals. The important compounds of sulfur are $SO_2$, $SO_3$, and $H_2S$. Sulfuric acid, an oxoacid, forms when sulfur trioxide reacts with water:

$$SO_3(g) + H_2O(l) \longrightarrow H_2SO_4(aq)$$

**Student Annotation:** The bismuth shown on the cover appears colorful because of a thin surface layer of bismuth oxide ($B_2O_3$).

**Figure 7.19**   Group 6A elements.

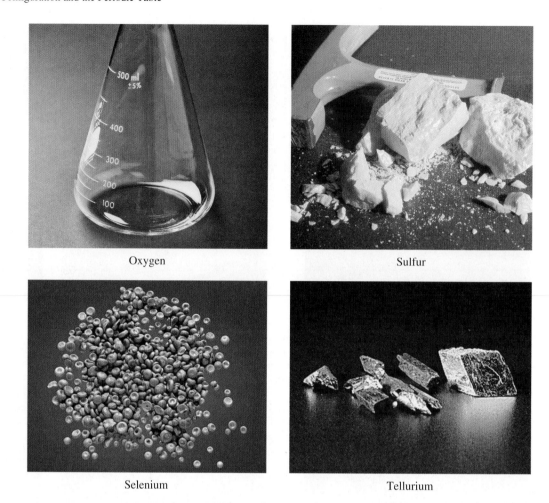

Oxygen

Sulfur

Selenium

Tellurium

### Group 7A Elements ($ns^2np^5$, $n \geq 2$)

Figure 7.20 shows samples of the first four Group 7A elements. All the halogens are nonmetals with the general formula $X_2$, where X denotes a halogen element. Like the Group 1A metals, the Group 7A nonmetals are too reactive to be found in nature in the elemental form. (Astatine, the last member of Group 7A, is radioactive. Very little is known about its properties.)

The halogens have high ionization energies and large, energetically favorable electron affinities. Anions derived from the halogens ($F^-$, $Cl^-$, $Br^-$, and $I^-$) are called *halides*. The vast majority of alkali metal halides are ionic compounds. The halogens also form many molecular compounds among themselves, such as ICl and $BrF_3$, and with nonmetals in other groups, such as $NF_3$, $PCl_5$, and $SF_6$. The halogens react with hydrogen to form hydrogen halides:

$$H_2(g) + X_2(g) \longrightarrow 2HX(g)$$

This reaction is explosive when it involves fluorine, but it becomes less and less violent as we substitute chlorine, bromine, and iodine. The hydrogen halides dissolve in water to form hydrohalic acids. Hydrofluoric acid (HF) is a weak acid (meaning it is a weak electrolyte), but the other hydrohalic acids (HCl, HBr, and HI) are all strong acids (strong electrolytes).

Fluorine

Chlorine

Bromine

Iodine

**Figure 7.20**   Group 7A elements.

## Group 8A Elements ($ns^2np^6$, $n \geq 2$)

Figure 7.21 shows samples of the Group 8A elements. All the noble gases exist as monatomic species. With the exception of helium, which has the electron configuration $1s^2$, their atoms have completely filled outer $ns$ and $np$ subshells. Their electron configurations give the noble gases their great stability. The Group 8A ionization energies are among the highest of all the elements (see Figure 7.8). Their electron affinities are all less than zero (Figure 7.10), so they have no tendency to accept extra electrons.

For years the noble gases were called *inert gases* because they were not known to react with anything. Beginning in 1963, however, compounds were prepared from xenon and krypton by exposing them to very strong oxidizing agents such as fluorine and oxygen. Some of the compounds that have been prepared are $XeF_4$, $XeO_3$, $XeOF_4$, and $KrF_2$. Although the chemistry of the noble gases is interesting, their compounds are not involved in any natural biological processes and they currently have no major commercial applications.

**Student Annotation:** Note that the common ions formed by the other main group elements are those that make them isoelectronic with a noble gas. In Group 1A elements, for example, each atom *loses* one electron to become isoelectronic with the noble gas that immediately precedes it; in Group 7A elements, each atom *gains* one electron to become isoelectronic with the noble gas that immediately follows it; and so on.

## Comparison of Group 1A and Group 1B Elements

Although the outer electron configurations of Groups 1A and 1B are similar (members of both groups have a single valence electron in an *s* orbital), their chemical properties are very different.

The first ionization energies of Cu, Ag, and Au are 745, 731, and 890 kJ/mol, respectively. Because these values are considerably larger than those of the alkali metals, the Group 1B elements are much less reactive. The higher ionization energies of the Group 1B elements result from incomplete shielding of the nucleus by the inner *d* electrons (compared with the more effective

**Figure 7.21** Discharge tubes containing Group 8A elements.

Helium

Neon

Argon

Krypton

Xenon

shielding by the completely filled noble gas cores). Consequently, the outer *s* electrons of the Group 1B elements are more strongly attracted by the nucleus. In fact, copper, silver, and gold are so unreactive that they are usually found in the uncombined state in nature. The inertness, rarity, and attractive appearance of these metals make them valuable in the manufacture of coins and jewelry. For this reason, these metals are also known as "coinage metals." The differences in the chemistry of the elements in Group 2A from that of the elements in Group 2B can be explained in a similar way.

**X-ray image of a human skeleton**

## Bringing Chemistry to Life

### Radioactive Bone

One of the consequences of two chemical species having similar properties is that human physiology sometimes mistakes one species for another. Healthy bones require constant replenishment of calcium. Our dietary calcium comes primarily from dairy products but can also be found in some vegetables—especially dark, leafy greens such as spinach and kale. Strontium-90, a radioactive isotope, is found in the fallout of atomic bomb explosions and is a component in the waste generated by nuclear power facilities. Strontium-90 released into the atmosphere will eventually settle on land and water, and it can reach our bodies through ingestion—especially of vegetation—and through the inhalation of airborne particles. Because calcium and strontium are chemically similar, $Sr^{2+}$ ions are mistaken by the body for $Ca^{2+}$ ions and are incorporated into the bones. Constant exposure to the radiation emitted by strontium-90 affects not only the bone and surrounding soft tissue, but also the bone marrow, damaging and destroying stem cells vital to the immune system. Such long-term exposure leads to an increased risk of leukemia and other cancers.

| 1A 1 | | | | | | | | | | | | 8A 18 |
|---|---|---|---|---|---|---|---|---|---|---|---|---|
| 1 | 2A 2 | | | | | | | | 3A 13 4A 14 5A 15 6A 16 7A 17 | | | 1 |
| 2 | | | | | | | | | | | | 2 |
| 3 | | 3B 3 4B 4 5B 5 6B 6 7B 7 | 8B 8 9 10 | 1B 11 2B 12 | | | | | | | | 3 |
| 4 | **Ca** | | | | | | | | | | | 4 |
| 5 | **Sr** | | | | | | | | | | | 5 |
| 6 | | | | | | | | | | | | 6 |
| 7 | | | | | | | | | | | | 7 |

**Ca and Sr, being in the same group, have similar chemical properties.**

## Variation in Properties of Oxides Within a Period

One way to compare the properties of the main group elements across a period is to examine the properties of a series of similar compounds. Because oxygen combines with almost all elements, we will compare the properties of oxides of the third-period elements to see how metals differ from metalloids and nonmetals. Some elements in the third period (P, S, and Cl) form several types of oxides, but for simplicity we will consider only those oxides in which the elements have the highest oxidation number. Table 7.4 lists a few general characteristics of these oxides and some specific physical properties of the oxides of third-period elements.

The tendency of oxygen to form the oxide ion is greatly favored when oxygen combines with metals that have low ionization energies, such as those in Groups 1A and 2A and aluminum. Thus, $Na_2O$, $MgO$, and $Al_2O_3$ are ionic compounds, as evidenced by their high melting points and boiling points. They have extensive three-dimensional structures in which each cation is surrounded by a specific number of anions, and vice versa. As the ionization energies of the elements increase from

| TABLE 7.4 | Some Properties of Oxides of the Third-Period Elements | | | | | | |
|---|---|---|---|---|---|---|---|
| | $Na_2O$ | MgO | $Al_2O_3$ | $SiO_2$ | $P_4O_{10}$ | $SO_3$ | $Cl_2O_7$ |
| Type of compound | ← | Ionic | → | ← | Molecular | | → |
| Structure | ← | Extensive three-dimensional | → | ← Discrete molecular units → | | | |
| Melting point (°C) | 1275 | 2800 | 2045 | 1610 | 580 | 16.8 | −91.5 |
| Boiling point (°C) | ? | 3600 | 2980 | 2230 | ? | 44.8 | 82 |
| Acid-base nature | Basic | Basic | Amphoteric | ← | | Acidic | → |

left to right, so does the molecular nature of the oxides that form. Silicon is a metalloid; its oxide ($SiO_2$) also has a huge three-dimensional network, although it is not an ionic compound. The oxides of phosphorus, sulfur, and chlorine are molecular compounds composed of small discrete units. The weak attractions among these molecules result in relatively low melting points and boiling points.

Most oxides can be classified as acidic or basic depending on whether they *produce* acids or bases when dissolved in water (or whether they *react* as acids or bases). Some oxides are ***amphoteric***, which means that they display both acidic and basic properties. The first two oxides of the third period, $Na_2O$ and MgO, are basic oxides. For example, $Na_2O$ reacts with water to form the base sodium hydroxide:

$$Na_2O(s) + H_2O(l) \longrightarrow 2NaOH(aq)$$

Magnesium oxide is quite insoluble; it does not react with water to any appreciable extent. However, it does react with acids in a manner that resembles an acid-base reaction:

$$MgO(s) + 2HCl(aq) \longrightarrow MgCl_2(aq) + H_2O(l)$$

The products of this reaction are a salt ($MgCl_2$) and water, the same kind of products that are obtained in an acid-base neutralization.

Aluminum oxide is even less soluble than magnesium oxide. It, too, does not react with water, but it exhibits the properties of a base by reacting with acids:

$$Al_2O_3(s) + 6HCl(aq) \longrightarrow 2AlCl_3(aq) + 3H_2O(l)$$

It also exhibits acidic properties by reacting with bases:

$$Al_2O_3(s) + 2NaOH(aq) + 3H_2O(l) \longrightarrow 2NaAl(OH)_4(aq)$$

Thus, $Al_2O_3$ is classified as an amphoteric oxide because it has properties of both acids and bases. Other amphoteric oxides are ZnO, BeO, and $Bi_2O_3$.

Silicon dioxide is insoluble and does not react with water. It has acidic properties, however, because it reacts with a very concentrated aqueous base:

$$SiO_2(s) + 2OH^-(aq) \longrightarrow SiO_3^{2-}(aq) + H_2O(l)$$

For this reason, concentrated aqueous, strong bases such as sodium hydroxide (NaOH) should *not* be stored in Pyrex glassware, which is made of $SiO_2$.

The remaining third-period oxides ($P_4O_{10}$, $SO_3$, and $Cl_2O_7$) are acidic. They react with water to form phosphoric acid, sulfuric acid, and perchloric acid, respectively:

$$P_4O_{10}(s) + 6H_2O(aq) \longrightarrow 4H_3PO_4(aq)$$

$$SO_3(g) + H_2O(l) \longrightarrow 2H_2SO_4(aq)$$

$$Cl_2O_7(l) + H_2O(l) \longrightarrow 2HClO_4(aq)$$

**Student Annotation:** Certain oxides such as CO and NO are neutral; that is, they do not react with water to produce acidic or basic solutions. In general, oxides of nonmetals are either acidic or neutral.

This brief examination of oxides of the third-period elements shows that as the metallic character of the elements decreases from left to right across the period, their oxides change from basic to amphoteric to acidic. Metal oxides are usually basic, and most oxides of nonmetals are acidic. The intermediate properties of the oxides (as demonstrated by the amphoteric oxides) are exhibited by elements whose positions are intermediate within the period. Because the metallic character of the elements increases from top to bottom within a group of main group elements, the oxides of elements with higher atomic numbers are more basic than the lighter elements.

# Applying What You've Learned

In 1949, the Australian psychiatrist John Cade published the results of his studies showing that lithium was useful in the treatment of "manic episodes," one of the phases of what is known today as bipolar disorder. Although the research had shown real promise, its publication coincided with news of the lithium poisoning and resulting deaths of a group of cardiac patients that had used lithium chloride as a dietary salt substitute. Reports of this disaster prompted drug manufacturers to withdraw all lithium salts from the market, and for a time, any medical use of lithium was viewed as too dangerous even to consider. Additional research in Europe and the United States resulted in the gradual acceptance of lithium as a potentially valuable psychiatric therapy. The FDA approved lithium carbonate ($Li_2CO_3$) in 1970 for the treatment of manic illness, and in 1974 for the treatment of bipolar disorder.

**Problems:**

a)   Without referring to a periodic table, write the electron configuration of lithium ($Z = 3$). [◄◄ Sample Problem 7.2]

b)   Referring only to a periodic table, arrange Li and the other alkali metals (not including Fr) in order of increasing atomic radius. [◄◄ Sample Problem 7.3]

c)   Again referring only to the periodic table, arrange the members of Group 1A (not including Fr) in order of increasing ionization energy ($IE_1$). [◄◄ Sample Problem 7.4]

d)   Write the electron configuration for each of the alkali metal cations. [◄◄ Sample Problem 7.7]

e)   For each alkali metal cation in part (d), identify an isoelectronic series consisting of a noble gas and, where appropriate, one or more common ions (see Figure 2.14). [◄◄ Sample Problem 7.8] List each series in order of increasing radius. [◄◄ Sample Problem 7.9]

# Chapter Summary

## Section 7.1

- The modern periodic table was devised independently by Dmitri Mendeleev and Lothar Meyer in the nineteenth century. The elements that were known at the time were grouped based on their physical and chemical properties. Using his arrangements of the elements, Mendeleev successfully predicted the existence of elements that had not yet been discovered.

- Early in the twentieth century, Henry Moseley refined the periodic table with the concept of the *atomic number,* thus resolving a few inconsistencies in the tables proposed by Mendeleev and Meyer.

- Elements in the same group of the periodic table tend to have similar physical and chemical properties.

## Section 7.2

- The periodic table can be divided into the **main group elements** (also known as the *representative elements*) and the *transition metals.* It is further divided into smaller groups or *columns* of elements that all have the same configuration of **valence electrons.**

- The 18 columns of the periodic table are labeled 1A through 8A (*s*- and *p*-block elements) and 1B through 8B (*d*-block elements), or by the numbers 1 through 18.

## Section 7.3

- **Effective nuclear charge ($Z_{eff}$)** is the nuclear charge that is "felt" by the valence electrons. It is usually lower than the nuclear charge due to **shielding** by the core electrons.

- According to **Coulomb's law,** the attractive force ($F$) between two oppositely charged particles ($Q_1$ and $Q_2$) is directly proportional to the product of the charges and inversely proportional to the distance ($d$) between the objects squared: ($F \propto Q_1 \cdot Q_2/d^2$).

## Section 7.4

- **Atomic radius** is the distance between an atom's nucleus and its valence shell. The atomic radius of a metal atom is defined as the **metallic radius,** which is one-half the distance between adjacent, identical nuclei in a metal solid. The atomic radius of a nonmetal is defined as the **covalent radius,** which is one-half the distance between adjacent, identical nuclei in a molecule. In general, atomic radii *decrease* from left to right across a *period* of the periodic table and *increase* from top to bottom down a *group.*

- **Ionization energy (IE)** is the energy required to remove an electron from a gaseous atom. The first ionization energy ($IE_1$) is smaller than subsequent ionization energies [e.g., second ($IE_2$), third ($IE_3$), and so on]. The first ionization of any atom removes a valence electron. Ionization energies increase dramatically when core electrons are being removed.

- First ionization energies ($IE_1$ values) tend to increase across a period and decrease down a group. Exceptions to this trend can be explained based upon the electron configuration of the element.

- **Electron affinity (EA)** is the energy released when an atom in the gas phase accepts an electron. *EA* is equal to $-\Delta H$ for the process $A(g) + e^- \longrightarrow A^-(g)$.

- Electron affinities tend to increase across a period. As with first ionization energies, exceptions to the trend can be explained based on the electron configuration of the element.

- Metals tend to be shiny, lustrous, malleable, ductile, and conducting (for both heat and electricity). Metals typically lose electrons to form cations, and they tend to form ionic compounds (including basic oxides).

- Nonmetals tend to be brittle and not good conductors (for either heat or electricity). They can gain electrons to form anions but they commonly form molecular compounds (including acidic oxides).

- In general, metallic character decreases across a period and increases down a group of the periodic table. **Metalloids** are elements with properties intermediate between metals and nonmetals.

## Section 7.5

- Ions of main group elements are **isoelectronic** with noble gases. When a *d*-block element loses one or more electrons, it loses them first from the shell with the highest principal quantum number (e.g., electrons in the 4*s* subshell are lost before electrons in the 3*d* subshell).

## Section 7.6

- **Ionic radius** is the distance between the nucleus and valence shell of a cation or an anion. A cation is smaller than its parent atom. An anion is larger than its parent atom.

- An **isoelectronic series** consists of one or more ions and a noble gas, all of which have identical electron configurations. Within an isoelectronic series, the greater the nuclear charge, the smaller the radius.

## Section 7.7

- A **diagonal relationship** describes similarities in the chemical properties of elements that are in different groups, but that are positioned diagonally from each other in the periodic table.

- Although members of a group in the periodic table exhibit similar chemical and physical properties, the first member of each group tends to be significantly different from the other members. Hydrogen is essentially a group unto itself.

- The alkali metals (Group 1A) tend to be highly reactive toward oxygen, water, and acid. Group 2A metals are less reactive than Group 1A metals, but the heavier members all react with water to produce metal hydroxides and hydrogen gas. Groups that contain both metals and nonmetals (e.g., Groups 4A, 5A, and 6A) tend to show greater variability in their physical and chemical properties.

- **Amphoteric** oxides, such as $Al_2O_3$, are those that exhibit both acidic and basic behavior.

# Key Words

# KEY EQUATION

$$7.1 \qquad Z_{eff} = Z - \sigma$$

# QUESTIONS AND PROBLEMS

## Section 7.1: Development of the Periodic Table

### Review Questions

7.1    Briefly describe the significance of Mendeleev's periodic table.

7.2    What is Moseley's contribution to the modern periodic table?

7.3    Describe the general layout of a modern periodic table.

7.4    What is the most important relationship among elements in the same group in the periodic table?

## Section 7.2: The Modern Periodic Table

### Review Questions

7.5    Classify each of the following elements as a metal, a nonmetal, or a metalloid: As, Xe, Fe, Li, B, Cl, Ba, P, I, Si.

7.6    Compare the physical and chemical properties of metals and nonmetals.

7.7    Draw a rough sketch of a periodic table (no details are required). Indicate regions where metals, nonmetals, and metalloids are located.

7.8    What is a main group element? Give names and symbols of four main group elements.

7.9    Without referring to a periodic table, write the name and give the symbol for one element in each of the following groups: 1A, 2A, 3A, 4A, 5A, 6A, 7A, 8A, transition metals.

7.10    Indicate whether the following elements exist as atomic species, molecular species, or extensive three-dimensional structures in their most stable states at room temperature, and write the molecular or empirical formula for each one: phosphorus, iodine, magnesium, neon, carbon, sulfur, cesium, and oxygen.

7.11    You are given a sample of a dark, shiny solid and asked to determine whether it is the nonmetal *iodine* or a *metallic* element. What test could you do that would enable you to answer the question without destroying the sample?

7.12    What are valence electrons? For main group elements, the number of valence electrons of an element is equal to its group number. Show that this is true for the following elements: Al, Sr, K, Br, P, S, C.

7.13    Write the outer electron configurations for the (a) alkali metals, (b) alkaline earth metals, (c) halogens, (d) noble gases.

7.14    Use the first-row transition metals (Sc to Cu) as an example to illustrate the characteristics of the electron configurations of transition metals.

7.15    Arsenic is not an essential element for the human body. Based on its position in the periodic table, suggest a reason for its toxicity.

### Problems

7.16    In the periodic table, the element hydrogen is sometimes grouped with the alkali metals and sometimes with the halogens. Explain why hydrogen can resemble the Group 1A and the Group 7A elements.

**7.17**    A neutral atom of a certain element has 16 electrons. Consulting only the periodic table, identify the element and write its ground-state electron configuration.

7.18    Group the following electron configurations in pairs that would represent elements with similar chemical properties:

(a) $1s^2 2s^2 2p^6 3s^2$
(b) $1s^2 2s^2 2p^3$
(c) $1s^2 2s^2 2p^6 3s^2 3p^6 4s^2 3d^{10} 4p^6$
(d) $1s^2 2s^2$
(e) $1s^2 2s^2 2p^6$
(f) $1s^2 2s^2 2p^6 3s^2 3p^3$

**7.19**    Group the following electron configurations in pairs that would represent elements with similar chemical properties:

(a) $1s^2 2s^2 2p^5$
(b) $1s^2 2s^1$
(c) $1s^2 2s^2 2p^6$
(d) $1s^2 2s^2 2p^6 3s^2 3p^5$
(e) $1s^2 2s^2 2p^6 3s^2 3p^6 4s^1$
(f) $1s^2 2s^2 2p^6 3s^2 3p^6 4s^2 3d^{10} 4p^6$

7.20    Without referring to a periodic table, write the electron configuration of elements with the following atomic numbers: (a) 9, (b) 20, (c) 26, (d) 33.

**7.21**    Specify the group of the periodic table in which each of the following elements is found: (a) $[Ne]3s^1$, (b) $[Ne]3s^2 3p^3$, (c) $[Ne]3s^2 3p^6$, (d) $[Ar]4s^2 3d^8$.

## Section 7.3: Effective Nuclear Charge

### Review Questions

7.22    Explain the term *effective nuclear charge*.

7.23    Explain why the atomic radius of Be is smaller than that of Li.

### Problems

7.24    The electron configuration of B is $1s^2 2s^2 2p^1$. (a) If each core electron (that is, the $1s$ electrons) were totally effective in shielding the valence electrons (that is, the $2s$ and $2p$ electrons) from the nucleus and the valence electrons did not shield one another, what would be the shielding constant ($\sigma$) and the effective nuclear charge ($Z_{eff}$) for the $2s$ and $2p$ electrons? (b) In reality, the shielding constants for the $2s$ and $2p$ electrons in B are slightly different. They are 2.42 and 2.58, respectively. Calculate $Z_{eff}$ for these electrons, and explain the differences from the values you determined in part (a).

**7.25** The electron configuration of C is $1s^2 2s^2 2p^2$. (a) If each core electron (that is, the $1s$ electrons) were totally effective in shielding the valence electrons (that is, the $2s$ and $2p$ electrons) from the nucleus and the valence electrons did not shield one another, what would be the shielding constant ($\sigma$) and the effective nuclear charge ($Z_{eff}$) for the $2s$ and $2p$ electrons? (b) In reality, the shielding constants for the $2s$ and $2p$ electrons in C are slightly different. They are 2.78 and 2.86, respectively. Calculate $Z_{eff}$ for these electrons, and explain the differences from the values you determined in part (a).

## Section 7.4: Periodic Trends in Properties of Elements

### Review Questions

**7.26** Define *atomic radius*. Does the size of an atom have a precise meaning?

**7.27** How does atomic radius change (a) from left to right across a period and (b) from top to bottom in a group?

**7.28** Define *ionization energy*. Explain why ionization energy measurements are usually made when atoms are in the gaseous state. Why is the second ionization energy always greater than the first ionization energy for any element?

**7.29** Sketch the outline of the periodic table, and show group and period trends in the first ionization energy of the elements. What types of elements have the highest ionization energies and what types have the lowest ionization energies?

**7.30** (a) Define *electron affinity*. (b) Explain why electron affinity measurements are made with gaseous atoms. (c) Ionization energy is always a positive quantity, whereas electron affinity may be either positive or negative. Explain.

**7.31** Explain the trends in electron affinity from aluminum to chlorine (see Figure 7.10).

### Problems

**7.32** On the basis of their positions in the periodic table, select the atom with the larger atomic radius in each of the following pairs: (a) Na, Si; (b) Ba, Be; (c) N, F; (d) Br, Cl; (e) Ne, Kr.

**7.33** Arrange the following atoms in order of increasing atomic radius: Na, Al, P, Cl, Mg.

**7.34** Which is the largest atom in the third period of the periodic table?

**7.35** Which is the smallest atom in Group 7A?

**7.36** Based on size, identify the spheres shown as Na, Mg, O, and S.

**7.37** Based on size, identify the spheres shown as K, Ca, S, and Se.

**7.38** Why is the radius of the lithium atom considerably larger than the radius of the hydrogen atom?

**7.39** Use the second period of the periodic table as an example to show that the size of atoms decreases as we move from left to right. Explain the trend.

**7.40** Arrange the following in order of increasing first ionization energy: Na, Cl, Al, S, and Cs.

**7.41** Arrange the following in order of increasing first ionization energy: F, K, P, Ca, and Ne.

**7.42** Use the third period of the periodic table as an example to illustrate the change in first ionization energies of the elements as we move from left to right. Explain the trend.

**7.43** In general, the first ionization energy increases from left to right across a given period. Aluminum, however, has a lower first ionization energy than magnesium. Explain.

**7.44** The first and second ionization energies of K are 419 and 3052 kJ/mol, and those of Ca are 590 and 1145 kJ/mol, respectively. Compare their values and comment on the differences.

**7.45** Two atoms have the electron configurations $1s^2 2s^2 2p^6$ and $1s^2 2s^2 2p^6 3s^1$. The first ionization energy of one is 2080 kJ/mol, and that of the other is 496 kJ/mol. Match each ionization energy with one of the given electron configurations. Justify your choice.

**7.46** A hydrogen-like ion is an ion containing only one electron. The energies of the electron in a hydrogen-like ion are given by

$$E_n = -(2.18 \times 10^{-18} \text{ J})Z^2\left(\frac{1}{n^2}\right)$$

where $n$ is the principal quantum number and $Z$ is the atomic number of the element. Calculate the ionization energy (in kJ/mol) of the He$^+$ ion.

**7.47** Plasma is a state of matter consisting of positive gaseous ions and electrons. In the plasma state, a mercury atom could be stripped of its 80 electrons and therefore would exist as Hg$^{80+}$. Use the equation in Problem 7.46 to calculate the energy required for the last ionization step, that is,

$$\text{Hg}^{79+}(g) \longrightarrow \text{Hg}^{80+}(g) + e^-$$

**7.48** Arrange the elements in each of the following groups in order of increasing electron affinity: (a) Li, Na, K; (b) F, Cl, Br, I.

**7.49** Specify which of the following elements you would expect to have the greatest electron affinity: He, K, Co, S, Cl.

**7.50** Considering their electron affinities, do you think it is possible for the alkali metals to form an anion like M$^-$, where M represents an alkali metal?

**7.51** Explain why alkali metals have a greater affinity for electrons than alkaline earth metals.

## Section 7.5: Electron Configuration of Ions

### Review Questions

**7.52** How does the electron configuration of ions derived from main group elements give them stability?

**7.53** What do we mean when we say that two ions or an atom and an ion are *isoelectronic*?

**7.54** Is it possible for the atoms of one element to be isoelectronic with the atoms of another element? Explain.

**7.55** Give three examples of first-row transition metal (Se to Cu) ions that are isoelectronic with argon.

### Problems

**7.56** A M$^{2+}$ ion derived from a metal in the first transition metal series has four electrons in the $3d$ subshell. What element might M be?

**7.57** A metal ion with a net $+3$ charge has five electrons in the $3d$ subshell. Identify the metal.

**7.58** Write the ground-state electron configurations of the following ions: (a) $Li^+$, (b) $H^-$, (c) $N^{3-}$, (d) $F^-$, (e) $S^{2-}$, (f) $Al^{3+}$, (g) $Se^{2-}$, (h) $Br^-$, (i) $Rb^+$, (j) $Sr^{2+}$, (k) $Sn^{2+}$, (l) $Te^{2-}$, (m) $Ba^{2+}$, (n) $Pb^{2+}$, (o) $In^{3+}$, (p) $Tl^+$, (q) $Tl^{3+}$.

**7.59** Write the ground-state electron configurations of the following ions, which play important roles in biochemical processes in our bodies: (a) $Na^+$, (b) $Mg^{2+}$, (c) $Cl^-$, (d) $K^+$, (e) $Ca^{2+}$, (f) $Fe^{2+}$, (g) $Cu^{2+}$, (h) $Zn^{2+}$.

**7.60** Write the ground-state electron configurations of the following transition metal ions: (a) $Sc^{3+}$, (b) $Ti^{4+}$, (c) $V^{5+}$, (d) $Cr^{3+}$, (e) $Mn^{2+}$, (f) $Fe^{2+}$, (g) $Fe^{3+}$, (h) $Co^{2+}$, (i) $Ni^{2+}$, (j) $Cu^+$, (k) $Cu^{2+}$, (l) $Ag^+$, (m) $Au^+$, (n) $Au^{3+}$, (o) $Pt^{2+}$.

**7.61** Name the ions with three charges that have the following electron configurations: (a) $[Ar]3d^3$, (b) $[Ar]$, (c) $[Kr]4d^6$, (d) $[Xe]4f^{14}5d^6$.

**7.62** Which of the following species are isoelectronic with each other: $C$, $Cl^-$, $Mn^{2+}$, $B^-$, $Ar$, $Zn$, $Fe^{3+}$, $Ge^{2+}$?

**7.63** Group the species that are isoelectronic: $Be^{2+}$, $F^-$, $Fe^{2+}$, $N^{3-}$, $He$, $S^{2-}$, $Co^{3+}$, $Ar$.

**7.64** Thallium (Tl) is a neurotoxin and exists mostly in the Tl(I) oxidation state in its compounds. Aluminum (Al), which causes anemia and dementia, is only stable in the Al(III) form. The first, second, and third ionization energies of Tl are 589, 1971, and 2878 kJ/mol, respectively. The first, second, and third ionization energies of Al are 577.5, 1817, and 2745 kJ/mol, respectively. Plot the ionization energies of Al and Tl versus atomic number and explain the trends.

### Section 7.6: Ionic Radius

#### Review Questions

**7.65** Define *ionic radius*. How does the size of an atom change when it is converted to (a) an anion and (b) a cation?

**7.66** Explain why, for isoelectronic ions, the anions are larger than the cations.

#### Problems

**7.67** Indicate which one of the two species in each of the following pairs is smaller: (a) $Cl$ or $Cl^-$, (b) $Na$ or $Na^+$, (c) $O^{2-}$ or $S^{2-}$, (d) $Mg^{2+}$ or $Al^{3+}$, (e) $Au^+$ or $Au^{3+}$.

**7.68** List the following ions in order of increasing ionic radius: $N^{3-}$, $Na^+$, $F^-$, $Mg^{2+}$, $O^{2-}$.

**7.69** Explain which of the following cations is larger, and why: $Cu^+$ or $Cu^{2+}$.

**7.70** Explain which of the following anions is larger, and why: $Se^{2-}$ or $Te^{2-}$.

**7.71** Both $Mg^{2+}$ and $Ca^{2+}$ are important biological ions. One of their functions is to bind to the phosphate group of ATP molecules or amino acids of proteins. For Group 2A metals in general, the tendency for binding to the anions increases in the order $Ba^{2+} < Sr^{2+} < Ca^{2+} < Mg^{2+}$. Explain this trend.

### Section 7.7: Periodic Trends in Chemical Properties of the Main Group Elements

#### Review Questions

**7.72** Why do members of a group exhibit similar chemical properties?

**7.73** Which elements are more likely to form acidic oxides? Basic oxides? Amphoteric oxides?

#### Problems

**7.74** Give the physical states (gas, liquid, or solid) of the main group elements in the fourth period (K, Ca, Ga, Ge, As, Se, Br) at room temperature.

**7.75** The boiling points of neon and krypton are $-246.1°C$ and $-153.2°C$, respectively. Using these data, estimate the boiling point of argon.

**7.76** Use the alkali metals and alkaline earth metals as examples to show how we can predict the chemical properties of elements simply from their electron configurations.

**7.77** Based on your knowledge of the chemistry of the alkali metals, predict some of the chemical properties of francium, the last member of the group.

**7.78** As a group, the noble gases are very stable chemically (only Kr and Xe are known to form compounds). Why?

**7.79** Why are Group 1B elements more stable than Group 1A elements even though they seem to have the same outer electron configuration, $ns^1$, where $n$ is the principal quantum number of the outermost shell?

**7.80** How do the chemical properties of oxides change from left to right across a period? How do they change from top to bottom within a particular group?

**7.81** Write balanced equations for the reactions between each of the following oxides and water: (a) $Li_2O$, (b) $CaO$, (c) $SO_3$.

**7.82** Write formulas for and name the binary hydrogen compounds of the second-period elements (Li to F). Describe how the physical and chemical properties of these compounds change from left to right across the period.

**7.83** Which oxide is more basic, MgO or BaO? Why?

### Additional Problems

**7.84** State whether each of the following properties of the main group elements generally increases or decreases (a) from left to right across a period and (b) from top to bottom within a group: metallic character, atomic size, ionization energy, acidity of oxides.

**7.85** Referring to the periodic table, name (a) the halogen in the fourth period, (b) an element similar to phosphorus in chemical properties, (c) the most reactive metal in the fifth period, (d) an element that has an atomic number smaller than 20 and is similar to strontium.

**7.86** Write equations representing the following processes:

(a) The electron affinity of $S^-$
(b) The third ionization energy of titanium
(c) The electron affinity of $Mg^{2+}$
(d) The ionization energy of $O^{2-}$

**7.87** Arrange the following isoelectronic species in order of increasing ionization energy: $O^{2-}$, $F^-$, $Na^+$, $Mg^{2+}$.

**7.88** Write the empirical (or molecular) formulas of compounds that the elements in the third period (sodium to chlorine) should form with (a) molecular oxygen and (b) molecular chlorine. In each case indicate whether you would expect the compound to be ionic or molecular in character.

**7.89** Element M is a shiny and highly reactive metal (melting point 63°C), and element X is a highly reactive nonmetal (melting point $-7.2°C$). They react to form a compound with the empirical formula MX, a colorless, brittle white solid that melts at 734°C. When dissolved in water or when in the molten state, the substance conducts electricity. When chlorine gas is bubbled through an aqueous solution containing MX, a reddish-brown liquid appears and $Cl^-$ ions are formed. From these observations, identify M and X. (You may need to consult a handbook of chemistry for the melting-point values.)

**7.90** Match each of the elements on the right with its description on the left:

(a) A dark-red liquid                                  Calcium (Ca)
(b) A colorless gas that burns in oxygen gas           Gold (Au)
(c) A metal that reacts violently with water           Hydrogen ($H_2$)
(d) A shiny metal that is used in jewelry              Argon (Ar)
(e) An inert gas                                       Bromine ($Br_2$)

**7.91** Arrange the following species in isoelectronic pairs: $O^+$, Ar, $S^{2-}$, Ne, Zn, $Cs^+$, $N^{3-}$, $As^{3+}$, N, Xe.

**7.92** In which of the following are the species written in decreasing order by size of radius: (a) Be, Mg, Ba, (b) $N^{3-}$, $O^{2-}$, $F^-$, (c) $Tl^{3+}$, $Tl^{2+}$, $Tl^+$?

**7.93** Which of the following properties show a clear periodic variation: (a) first ionization energy, (b) molar mass of the elements, (c) number of isotopes of an element, (d) atomic radius?

**7.94** When carbon dioxide is bubbled through a clear calcium hydroxide solution, the solution appears milky. Write an equation for the reaction, and explain how this reaction illustrates that $CO_2$ is an acidic oxide.

**7.95** You are given four substances: a fuming red liquid, a dark metallic-looking solid, a pale-yellow gas, and a yellow-green gas that attacks glass. You are told that these substances are the first four members of Group 7A, the halogens. Name each one.

**7.96** For each pair of elements listed, give three properties that show their chemical similarity: (a) sodium and potassium and (b) chlorine and bromine.

**7.97** Name the element that forms compounds, under appropriate conditions, with every other element in the periodic table except He, Ne, and Ar.

**7.98** Explain why the first electron affinity of sulfur is 200 kJ/mol but the second electron affinity is −649 kJ/mol.

**7.99** The $H^-$ ion and the He atom have two $1s$ electrons each. Which of the two species is larger? Explain.

**7.100** Predict the products of the following oxides with water: $Na_2O$, BaO, $CO_2$, $N_2O_5$, $P_4O_{10}$, $SO_3$. Write an equation for each of the reactions. Specify whether the oxides are acidic, basic, or amphoteric.

**7.101** Write the formulas and names of the oxides of the second-period elements (Li to N). Identify the oxides as acidic, basic, or amphoteric. Use the highest oxidation state of each element.

**7.102** State whether each of the following elements is a gas, liquid, or solid under atmospheric conditions. Also state whether it exists in the elemental form as atoms, molecules, or a three-dimensional network: Mg, Cl, Si, Kr, O, I, Hg, Br.

**7.103** What factors account for the unique nature of hydrogen?

**7.104** The air in a manned spacecraft or submarine needs to be purified of exhaled carbon dioxide. Write equations for the reactions between carbon dioxide and (a) lithium oxide ($Li_2O$), (b) sodium peroxide ($Na_2O_2$), and (c) potassium superoxide ($KO_2$).

**7.105** The formula for calculating the energies of an electron in a hydrogen-like ion is given in Problem 7.46. This equation can be applied only to one-electron atoms or ions. One way to modify it for more complex species is to replace Z with $Z - \sigma$ or $Z_{eff}$. Calculate the value of $\sigma$ if the first ionization energy of helium is $3.94 \times 10^{-18}$ J per atom. (Disregard the minus sign in the given equation in your calculation.)

**7.106** Why do noble gases have negative electron affinity values?

**7.107** The atomic radius of K is 227 pm and that of $K^+$ is 138 pm. Calculate the percent decrease in volume that occurs when K($g$) is converted to $K^+(g)$. (The volume of a sphere is $\frac{4}{3}\pi r^3$, where $r$ is the radius of the sphere.)

**7.108** The atomic radius of F is 72 pm and that of $F^-$ is 133 pm. Calculate the percent increase in volume that occurs when F($g$) is converted to $F^-(g)$. (See Problem 7.107 for the volume of a sphere.)

**7.109** A technique called photoelectron spectroscopy is used to measure the ionization energy of atoms. A gaseous sample is irradiated with UV light, and electrons are ejected from the valence shell. The kinetic energies of the ejected electrons are measured. Because the energy of the UV photon and the kinetic energy of the ejected electron are known, we can write

$$h\nu = IE + \tfrac{1}{2}mu^2$$

where $\nu$ is the frequency of the UV light, and $m$ and $u$ are the mass and velocity of the electron, respectively. In one experiment the kinetic energy of the ejected electron from potassium is found to be $5.34 \times 10^{-19}$ J using a UV source of wavelength 162 nm. Calculate the ionization energy of potassium. How can you be sure that this ionization energy corresponds to the electron in the valence shell (that is, the most loosely held electron)?

**7.110** The energy needed for the following process is $1.96 \times 10^4$ kJ/mol:

$$Li(g) \longrightarrow Li^{3+}(g) + 3e^-$$

If the first ionization energy of lithium is 520 kJ/mol, calculate the second ionization energy of lithium, that is, the energy required for the process

$$Li^+(g) \longrightarrow Li^{2+}(g) + e^-$$

(*Hint:* You need the equation in Problem 7.46.)

**7.111** A student is given samples of three elements, X, Y, and Z, which could be an alkali metal, a member of Group 4A, or a member of Group 5A. She makes the following observations: Element X has a metallic luster and conducts electricity. It reacts slowly with hydrochloric acid to produce hydrogen gas. Element Y is a light yellow solid that does not conduct electricity. Element Z has a metallic luster and conducts electricity. When exposed to air, it slowly forms a white powder. A solution of the white powder in water is basic. What can you conclude about the elements from these observations?

**7.112** What is the electron affinity of the $Na^+$ ion?

**7.113** The ionization energies of sodium (in kJ/mol), starting with the first and ending with the eleventh, are 496, 4562, 6910, 9543, 13,354, 16,613, 20,117, 25,496, 28,932, 141,362, 159,075. Plot the log of ionization energy ($y$ axis) versus the number of ionization ($x$ axis); for example, log 496 is plotted versus 1 (labeled $IE_1$, the first ionization energy), log 4562 is plotted versus 2 (labeled $IE_2$, the second ionization energy), and so on. (a) Label $IE_1$ through $IE_{11}$ with the electrons in orbitals such as $1s$, $2s$, $2p$, and $3s$. (b) What can you deduce about electron shells from the breaks in the curve?

**7.114** Experimentally, the electron affinity of an element can be determined by using a laser light to ionize the anion of the element in the gas phase:

$$X^-(g) + h\nu \longrightarrow X(g) + e^-$$

Referring to Figure 7.10, calculate the photon wavelength (in nm) corresponding to the electron affinity for chlorine. In what region of the electromagnetic spectrum does this wavelength fall?

**7.115** Explain, in terms of their electron configurations, why $Fe^{2+}$ is more easily oxidized to $Fe^{3+}$ than $Mn^{2+}$ is to $Mn^{3+}$.

**7.116** Write the formulas and names of the hydrides of the following second-period elements: Li, C, N, O, F. Predict their reactions with water.

**7.117** Based on knowledge of the electronic configuration of titanium, state which of the following compounds of titanium is unlikely to exist: $K_3TiF_6$, $K_2Ti_2O_5$, $TiCl_3$, $K_2TiO_4$, $K_2TiF_6$.

**7.118** In halogen displacement reactions a halogen element can be generated by oxidizing its anions with a halogen element that lies above it in the periodic table. This means that there is no way to prepare elemental fluorine, because it is the first member of Group 7A. Indeed, for years the only way to prepare elemental fluorine was to oxidize $F^-$ ions by electrolytic means. Then, in 1986, a chemist reported that by combining potassium hexafluoromanganate(IV) ($K_2MnF_6$) with antimony pentafluoride ($SbF_5$) at 150°C, he had generated elemental fluorine. Balance the following equation representing the reaction:

$$K_2MnF_6 + SbF_5 \longrightarrow KSbF_6 + MnF_3 + F_2$$

**7.119** Write a balanced equation for the preparation of (a) molecular oxygen, (b) ammonia, (c) carbon dioxide, (d) molecular hydrogen, (e) calcium oxide. Indicate the physical state of the reactants and products in each equation.

**7.120** Write chemical formulas for oxides of nitrogen with the following oxidation numbers: +1, +2, +3, +4, +5. (*Hint:* There are two oxides of nitrogen with a +4 oxidation number.)

**7.121** Most transition metal ions are colored. For example, a solution of $CuSO_4$ is blue. How would you show that the blue color is due to the hydrated $Cu^{2+}$ ions and not the $SO_4^{2-}$ ions?

**7.122** In general, atomic radius and ionization energy have opposite periodic trends. Why?

**7.123** Explain why the electron affinity of nitrogen is approximately zero, while the elements on either side, carbon and oxygen, have substantial positive electron affinities.

**7.124** Consider the halogens chlorine, bromine, and iodine. The melting point and boiling point of chlorine are −101.5°C and −34.0°C and those of iodine are 113.7°C and 184.3°C, respectively. Thus chlorine is a gas and iodine is a solid under room conditions. Estimate the melting point and boiling point of bromine. Compare your values with those from the webelements.com website.

**7.125** Although it is possible to determine the second, third, and higher ionization energies of an element, the same cannot usually be done with the electron affinities of an element. Explain.

**7.126** Little is known of the chemistry of astatine, the last member of Group 7A. Describe the physical characteristics that you would expect this halogen to have. Predict the products of the reaction between sodium astatide (NaAt) and sulfuric acid. (*Hint:* Sulfuric acid is an oxidizing agent.)

**7.127** As discussed in the chapter, the atomic mass of argon is greater than that of potassium. This observation created a problem in the early development of the periodic table because it meant that argon should be placed after potassium. (a) How was this difficulty resolved? (b) From the following data, calculate the average atomic masses of argon and potassium: Ar-36 (35.9675 amu, 0.337 percent), Ar-38 (37.9627 amu, 0.063 percent), Ar-40 (39.9624 amu, 99.60 percent), K-39 (38.9637 amu, 93.258 percent), K-40 (39.9640 amu, 0.0117 percent), K-41 (40.9618 amu, 6.730 percent).

**7.128** Calculate the maximum wavelength of light (in nm) required to ionize a single sodium atom.

**7.129** Predict the atomic number and ground-state electron configuration of the next member of the alkali metals after francium.

**7.130** Why do elements that have high ionization energies also have more positive electron affinities? Which group of elements would be an exception to this generalization?

**7.131** The first four ionization energies of an element are approximately 738, 1450, $7.7 \times 10^3$, and $1.1 \times 10^4$ kJ/mol. To which periodic group does this element belong? Explain your answer.

**7.132** Some chemists think that helium should properly be called "helon." Why? What does the ending in helium (-ium) suggest?

**7.133** (a) The formula of the simplest hydrocarbon is $CH_4$ (methane). Predict the formulas of the simplest compounds formed between hydrogen and the following elements: silicon, germanium, tin, and lead. (b) Sodium hydride (NaH) is an ionic compound. Would you expect rubidium hydride (RbH) to be more or less ionic than NaH? (c) Predict the reaction between radium (Ra) and water. (d) When exposed to air, aluminum forms a tenacious oxide ($Al_2O_3$) coating that protects the metal from corrosion. Which metal in Group 2A would you expect to exhibit similar properties? (See Page 278.)

**7.134** Match each of the elements on the right with its description on the left:

(a) A pale yellow gas that reacts with water.      Nitrogen ($N_2$)
(b) A soft metal that reacts with water to          Boron (B)
    produce hydrogen.                                Fluorine ($F_2$)
(c) A metalloid that is hard and has a high         Aluminum (Al)
    melting point.                                   Sodium (Na)
(d) A colorless, odorless gas.
(e) A metal that is more reactive than iron,
    but does not corrode in air.

**7.135** Write at least two paragraphs describing the importance of the periodic table. Pay particular attention to the significance of the position of an element in the table and how the position relates to the chemical and physical properties of the element.

**7.136** On one graph, plot the effective nuclear charge (shown in parentheses) and atomic radius (see Figure 7.6) versus atomic number for the second-period elements: Li(1.28), Be(1.91), B(2.42), C(3.14), N(3.83), O(4.45), F(5.10), Ne(5.76). Comment on the trends.

**7.137** One allotropic form of an element X is a colorless crystalline solid. The reaction of X with an excess amount of oxygen produces a colorless gas. This gas dissolves in water to yield an acidic solution. Choose one of the following elements that matches X: (a) sulfur, (b) phosphorus, (c) carbon, (d) boron, (e) silicon.

**7.138** The ionization energy of a certain element is 412 kJ/mol. When the atoms of this element are in the first excited state, however, the ionization energy is only 126 kJ/mol. Based on this information, calculate the wavelength of light emitted in a transition from the first excited state to the ground state.

**7.139** One way to estimate the effective charge ($Z_{eff}$) of a many-electron atom is to use the equation $IE_1 = (1312 \text{ kJ/mol})(Z_{eff}^2/n^2)$, where $IE_1$ is the first ionization energy and $n$ is the principal quantum number of the shell in which the electron resides. Use this equation to calculate the effective charges of Li, Na, and K. Also calculate $Z_{eff}/n$ for each metal. Comment on your results.

**7.140** Use your knowledge of thermochemistry to calculate the $\Delta H$ for the following processes: (a) $Cl^-(g) \longrightarrow Cl^+(g) + 2e^-$ and (b) $K^+(g) + 2e^- \longrightarrow K^-(g)$.

**7.141** To prevent the formation of oxides, peroxides, and superoxides, alkali metals are sometimes stored in an inert atmosphere. Which of the following gases should not be used for lithium: Ne, Ar, $N_2$, Kr? Explain. (*Hint:* As mentioned in the chapter, Li and Mg exhibit a diagonal relationship. Compare the common compounds of these two elements.)

# PRE-PROFESSIONAL PRACTICE EXAM PROBLEMS: PHYSICAL AND BIOLOGICAL SCIENCES

These questions are not based on a descriptive passage.

1. A halogen has valence electrons in which orbitals?

   a) $s$
   b) $s$ and $p$
   c) $p$
   d) $s$, $p$, and $d$

2. How many subshells does a shell with principal quantum number $n$ contain?

   a) $n$
   b) $n^2$
   c) $n - 1$
   d) $2n - 1$

3. In a shell that contains an $f$ subshell, what is the ratio of $f$ orbitals to $s$ orbitals?

   a) 14:1
   b) 7:1
   c) 7:3
   d) 7:5

4. What is the maximum number of electrons that can be in the $n = 3$ shell?

   a) 2
   b) 6
   c) 8
   d) 18

# ANSWERS TO IN-CHAPTER MATERIALS

**Answers to Practice Problems**

**7.1A** Ge. **7.1B** Bi < As < P. **7.2A** (a) $1s^2 2s^2 2p^6 3s^2 3p^3$, $p$-block, (b) $1s^2 2s^2 2p^6 3s^2 3p^6 4s^2$, $s$-block, (c) $1s^2 2s^2 2p^6 3s^2 3p^6 4s^2 3d^{10} 4p^5$, $p$-block. **7.2B** (a) Al, (b) Zn, (c) Sr. **7.3A** F < Se < Ge. **7.3B** P and Se. **7.4A** Mg, Mg. **7.4B** Rb has a smaller $Z_{eff}$, $IE_2$ for Rb corresponds to the removal of a core electron. **7.5A** Al. **7.5B** Adding an electron to As involves pairing. **7.6A** The attractive force is slightly larger between $+3.26$ and $-1.15$ separated by 1.5 pm. **7.6B** 1.51 pm. **7.7A** (a) [Ne], (b) [Ar], (c) [Kr]. **7.7B** $N^{3-}$, $O^{2-}$, $F^-$, Ne, $Na^+$, $Mg^{2+}$, $Al^{3+}$. **7.8A** (a) [Ar] $3d^6$, (b) [Ar] $3d^9$, (c) [Kr] $4d^{10}$. **7.8B** $Cu^+$. **7.9A** $Rb^+ < Kr < Br^- < Se^{2-}$. **7.9B** $F^-$, $O^{2-}$, $N^{3-}$, $Na^+$, $Mg^{2+}$, $Al^{3+}$.

**Answers to Checkpoints**

**7.1.1** c. **7.1.2** a. **7.1.3** c. **7.2.1** b. **7.2.2** a, d, e. **7.4.1** b. **7.4.2** c. **7.4.3** e. **7.4.4** a. **7.5.1** b, c, e. **7.5.2** b, d. **7.5.3** d. **7.5.4** b. **7.5.5** d, f. **7.6.1** d, e. **7.6.2** a, c. **7.6.3** c. **7.6.4** d.

**Answers to Applying What You've Learned**

a) $1s^2 2s^1$. b) In order of increasing atomic radius: Li < Na < K < Rb < Cs. c) In order of increasing ionization energy ($IE_1$): Cs < Rb < K < Na < Li. d) $Li^+$: $1s^2$ or [He]; $Na^+$: $1s^2 2s^2 2p^6$ or [Ne]; $K^+$: $1s^2 2s^2 2p^6 3s^2 3p^6$ or [Ar]; $Rb^+$: $1s^2 2s^2 2p^6 3s^2 3p^6 4s^2 3d^{10} 4p^6$ or [Kr]; $Cs^+$: $1s^2 2s^2 2p^6 3s^2 3p^6 4s^2 3d^{10} 4p^6 5s^2 4d^{10} 5p^6$ or [Xe]. e) Isoelectronic with $Li^+$: He. Isoelectronic with $Na^+$: $Al^{3+}$, $Mg^{2+}$, Ne, $F^-$, $O^{2-}$, and $N^{3-}$. Isoelectronic with $K^+$: $Ca^{2+}$, Ar, $Cl^-$, $S^{2-}$, and $P^{3-}$. Isoelectronic with $Rb^+$: $Sr^{2+}$, Kr, $Br^-$, and $Se^{2-}$. Isoelectronic with $Cs^+$: $Ba^{2+}$, Xe, $I^-$, and $Te^{2-}$.

# 8

# Chemical Bonding I
## Basic Concepts

Dynamite, a stabilized form of the explosive nitroglycerin, is used to blast through solid rock.

# What Chemical Bonds Have to Do with Explosives

When the Swedish chemist Alfred Nobel died in 1896, his will specified that the bulk of his considerable fortune was to be used to establish the prizes that bear his name. The prizes, given annually in five categories (Chemistry, Physics, Physiology or Medicine, Literature, and Peace), are intended to recognize significant contributions to the betterment of humankind. In life, Nobel had been a prolific scientist and entrepreneur. He held more than 300 patents, including the one for dynamite, a stabilized form of the explosive nitroglycerin. His extensive work on the development and manufacture of explosives earned him the title "merchant of death" and was the cause of personal tragedy when his younger brother was killed in an explosion at one of the family's factories.

Ironically, toward the end of his life, Nobel developed heart-disease related chest pain (angina pectoris) and was directed by his physician to take nitroglycerin orally, which Nobel refused to do. Glyceryl trinitrate, the name used by the medical community—perhaps to avoid the impression that doctors are prescribing explosives to patients—still is widely used to treat the symptoms of heart disease and other medical conditions. One of the more recently developed and intriguing uses of nitroglycerin is its placement in the tip of a condom to stimulate an erection.

Both the explosive nature of nitroglycerin and its effectiveness in treating such conditions as heart disease and erectile dysfunction can be illuminated by an understanding of the basic concepts of *chemical bonding*.

**Student Annotation:** A sixth prize, the *Nobel Memorial Prize in Economics,* is awarded along with the others but was not specified in Nobel's will, and it is not really a Nobel Prize.

Nitroglycerin

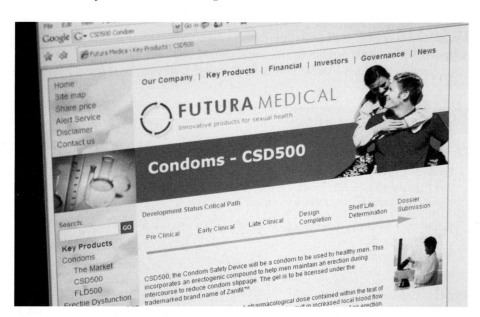

At the end of this chapter, you will be able to answer a series of questions about nitric oxide and nitroglycerin [▶▶ Page 329].

## In This Chapter, You Will Learn:

About two different types of chemical bonds, called *ionic* and *covalent*. You will also learn how to represent atoms and atomic ions using Lewis dot symbols and how to represent molecules and polyatomic ions using Lewis *structures*.

### Before you begin, you should review

- Atomic radius [◀◀ Section 7.4]
- Ions of main group elements [◀◀ Section 7.5]

## 8.1    Lewis Dot Symbols

The development of the periodic table and the concept of electron configuration gave chemists a way to explain the formation of compounds. The explanation, formulated by Gilbert Lewis,[1] is that atoms combine to achieve a more stable electron configuration. Maximum stability results when an atom is *isoelectronic* with a noble gas.

**Student Annotation:** Remember that for two species to be isoelectronic they must have exactly the same electron configuration [◄◄ Section 7.5].

When atoms interact to form compounds, it is their *valence electrons* that actually interact. Therefore, it is helpful to have a method for depicting the valence electrons of the atoms involved. This can be done using Lewis dot symbols. A ***Lewis dot symbol*** consists of the element's symbol surrounded by dots, where each dot represents a valence electron. For the main group elements, the number of dots in the Lewis dot symbol is the same as the group number, as shown in Figure 8.1. (Because they have incompletely filled inner shells, transition metals typically are not represented with Lewis dot symbols.)

Figure 8.1 shows that dots are placed above and below as well as to the left and right of the symbol. The exact order in which the dots are placed around the element symbol is not important, but the number of dots is. Thus, any of the following would be correct for the Lewis dot symbol for boron:

$$\dot{\text{B}}\cdot \qquad \dot{\text{B}}\cdot \qquad \cdot\dot{\text{B}}\cdot \qquad \cdot\dot{\text{B}}$$

When writing Lewis dot symbols, though, we do not "pair" dots until absolutely necessary. Thus, we would *not* represent boron with a pair of dots on one side and a single dot on another.

For main group metals such as Na or Mg, the number of dots in the Lewis dot symbol is the number of electrons that are lost when the atom forms a cation that is isoelectronic with the preceding noble gas. For nonmetals of the second period (B through F), the number of unpaired dots is the number of bonds the atom can form.

**Student Annotation:** As we will see shortly [►► Section 8.8], a larger nonmetal atom (one in the third period or beyond) can actually form as many bonds as the total number of dots in its Lewis dot symbol.

In addition to atoms, we can also represent atomic ions with Lewis dot symbols. To do so, we simply add (for anions) or subtract (for cations) the appropriate number of dots from the Lewis dot symbol of the atom and include the ion's charge.

Sample Problem 8.1 shows how to use Lewis dot symbols to represent atomic ions.

---

### Sample Problem 8.1

Write Lewis dot symbols for (a) fluoride ion ($F^-$), (b) potassium ion ($K^+$), and (c) sulfide ion ($S^{2-}$).

**Strategy** Starting with the Lewis dot symbol for each element, add dots (for anions) or remove dots (for cations) as needed to achieve the correct charge on each ion. Don't forget to include the appropriate charge on the Lewis dot symbol.

**Setup** The Lewis dot symbols for F, K, and S are $:\ddot{\text{F}}\cdot$, $\text{K}\cdot$, and $\cdot\ddot{\text{S}}\cdot$, respectively.

**Solution** (a) $\left[:\ddot{\text{F}}:\right]^-$

(b) $K^+$

(c) $\left[:\ddot{\text{S}}:\right]^{2-}$

**Think About It** For ions that are isoelectronic with noble gases, cations should have no dots remaining around the element symbol, whereas anions should have eight dots around the element symbol. Note, too, that we put square brackets around the Lewis dot symbol for an anion and place the negative charge outside the brackets.

**Practice Problem A** Write Lewis dot symbols for (a) $Ca^{2+}$, (b) $N^{3-}$, and (c) $I^-$.

**Practice Problem B** Indicate the charge on each of the ions represented by the following Lewis dot symbols: (a) $\left[:\ddot{\text{O}}:\right]^?$, (b) $H^?$, and (c) $\left[:\ddot{\text{P}}:\right]^?$.

---

1. Gilbert Newton Lewis (1875–1946). American chemist. Lewis made many important contributions in the areas of chemical bonding, thermodynamics, acids and bases, and spectroscopy. Despite the significance of Lewis's work, he was never awarded a Nobel Prize.

| 1A 1 | | | | | | | | | | | | | | | | | 8A 18 |
|---|---|---|---|---|---|---|---|---|---|---|---|---|---|---|---|---|---|
| ·H | 2A 2 | | | | | | | | | | | 3A 13 | 4A 14 | 5A 15 | 6A 16 | 7A 17 | He: |
| ·Li | ·Be· | | | | | | | | | | | ·Ḃ· | ·Ċ· | ·N̈· | ·Ö· | :F̈· | :N̈e: |
| ·Na | ·Mg· | 3B 3 | 4B 4 | 5B 5 | 6B 6 | 7B 7 | | 8B 8 | 9 | 10 | 1B 11 | 2B 12 | ·Äl· | ·Ṡi· | ·P̈· | ·S̈· | :C̈l· | :Är: |
| ·K | ·Ca· | | | | | | | | | | | | ·Ga· | ·Ġe· | ·Äs· | ·S̈e· | :B̈r· | :K̈r: |
| ·Rb | ·Sr· | | | | | | | | | | | | ·İn· | ·Ṡn· | ·S̈b· | ·T̈e· | ·Ï· | :Ẍe: |
| ·Cs | ·Ba· | | | | | | | | | | | | ·T̈l· | ·Pb· | ·B̈i· | ·Po· | :Ät· | :R̈n: |
| ·Fr | ·Ra· | | | | | | | | | | | | | | | | |

**Figure 8.1** Lewis dot symbols of the main group elements.

---

**Checkpoint 8.1    Lewis Dot Symbols**

**8.1.1** Using only a periodic table, determine the correct Lewis dot symbol for a silicon (Si) atom.

a) :Si:

b) ·Ṡi·

c) :Ṡi·

d) ·Si·

e) :Si

**8.1.2** Using only a periodic table, determine the correct Lewis dot symbol for the bromide ion ($Br^-$).

a) $\left[ :\ddot{B}r· \right]^-$

b) $\left[ :\ddot{B}r \right]^-$

c) $\left[ ·\ddot{B}r· \right]^-$

d) $\left[ Br· \right]^-$

e) $\left[ :\ddot{B}r: \right]^-$

**8.1.3** To which group does element X belong if its Lewis symbol is ·Ẍ·?

a) 1A

b) 2A

c) 3A

d) 4A

e) 5A

f) 6A

g) 7A

h) It is not possible to tell.

**8.1.4** To which group does element Y belong if the Lewis symbol for its anion is $\left[ :\ddot{Y}: \right]^{m-}$, where $m$ represents the charge?

a) 4A

b) 5A

c) 6A

d) 7A

e) It is not possible to tell.

---

# 8.2    Ionic Bonding

Recall from Chapter 7 that atoms of elements with low ionization energies tend to form cations, while those with high positive electron affinities tend to form anions. *Ionic bonding* refers to the electrostatic attraction that holds these oppositely charged ions together in an ionic compound, such as $K^+$ and $I^-$ in potassium iodide (KI), the "iodizing" ingredient in iodized salt. The electron configuration of potassium is $[Ar]4s^1$, and that of iodine is $[Kr]5s^2 4d^{10} 5p^5$. When potassium and iodine atoms come into contact with each other, the valence electron of potassium is transferred to the iodine atom. We can imagine these processes taking place separately and represent each process using Lewis dot symbols.

**Student Annotation:** Iodized salt is used to prevent the devastating effects of iodine deficiency disorders, which include stillbirths, birth defects, and mental retardation.

$$·K \longrightarrow K^+ + e^-$$

$$:\ddot{I}· + e^- \longrightarrow \left[ :\ddot{I}: \right]^-$$

The sum of these two equations is

$$K\cdot + \ :\!\overset{\displaystyle ..}{\underset{\displaystyle ..}{I}}\!\cdot \ \longrightarrow \ K^+ + \left[\overset{\displaystyle ..}{\underset{\displaystyle ..}{:\!I\!:}}\right]^-$$

The electrostatic attraction between the resulting cation and anion draws them together to form the electrically neutral compound KI.

The net energy change associated with the formation of $K^+$ and $I^-$ ions is *endothermic;* that is, energy must be supplied for the overall transfer of an electron from K to I to take place. The ionization of potassium requires the input of 419 kJ/mol, and the electron affinity of iodine is only 295 kJ/mol [◄◄ Section 7.4]. If a mole of KI were to form as we have described it, an input of [419 + (−295)] = 124 kJ of energy would be required. Ionic compounds do form, though, and often in vigorous reactions. In fact, the formation of ionic bonds is highly *exothermic* and more than compensates for the energy input required to transfer electrons from metal atoms to nonmetal atoms. We can quantify the energy change associated with the formation of ionic bonds with *lattice energy.*

## Lattice Energy

*Lattice energy* is the amount of energy required to convert a mole of ionic solid to its constituent ions in the gas phase. For example, the lattice energy of potassium iodide is 632 kJ/mol. Thus, it takes 632 kJ of energy to convert 1 mole of KI(*s*) to 1 mole each of $K^+(g)$ and $I^-(g)$.

$$KI(s) \longrightarrow K^+(g) + I^-(g) \qquad \Delta H = 632 \text{ kJ/mol}$$

The magnitude of lattice energy is a measure of an ionic compound's stability. The greater the lattice energy, the more stable the compound. Table 8.1 lists the lattice energies for some ionic compounds.

Lattice energy depends on the magnitudes of the charges and on the distance between them. For example, LiI, NaI, and KI all have the same anion ($I^-$) and all have cations with the same charge (+1). The trend in their lattice energies (LiI > NaI > KI) can be explained on the basis of ionic radius. The radii of alkali metal ions increase as we move down a group in the periodic table ($r_{Li^+} < r_{Na^+} < r_{K^+}$) [◄◄ Section 7.6]. Knowing the radius of each ion, we can use Coulomb's law to compare the attractive forces between the ions in these three compounds:

| | | |
|---|---|---|
| $Li^+$ | $I^-$ | |
| 0.76 Å | 2.20 Å | |

$$F \propto \frac{(+1)\times(-1)}{(0.76+2.20)^2} \propto -0.11$$

Largest lattice energy
(732 kJ/mol)

| | | |
|---|---|---|
| $Na^+$ | $I^-$ | |
| 1.02 Å | 2.20 Å | |

$$F \propto \frac{(+1)\times(-1)}{(1.02+2.20)^2} \propto -0.10$$

Intermediate lattice energy
(686 kJ/mol)

| | | |
|---|---|---|
| $K^+$ | $I^-$ | |
| 1.38 Å | 2.20 Å | |

$$F \propto \frac{(+1)\times(-1)}{(1.38+2.20)^2} \propto -0.08$$

Smallest lattice energy
(632 kJ/mol)

LiI, with the smallest distance between ions, has the strongest attractive forces. It should therefore have the largest lattice energy. KI, with the largest distance between ions, has the weakest attractive forces, and should have the smallest lattice energy. NaI has an intermediate distance between ions and should have an intermediate lattice energy. Thus, Coulomb's law correctly predicts the relative magnitudes of the lattice energies of LiI, NaI, and KI.

Now consider the compounds LiF and MgO. With the distances between ions roughly equal (0.76 + 1.33 = 2.09  for LiF versus 0.72  + 1.40 = 2.12 Å for MgO) and the magnitude of each

| Compound | Lattice Energy (kJ/mol) | Melting Point (°C) |
|---|---|---|
| LiF | 1017 | 845 |
| LiCl | 860 | 610 |
| LiBr | 787 | 550 |
| LiI | 732 | 450 |
| NaCl | 787 | 801 |
| NaBr | 736 | 750 |
| NaI | 686 | 662 |
| KCl | 699 | 772 |
| KBr | 689 | 735 |
| KI | 632 | 680 |
| $MgCl_2$ | 2527 | 714 |
| $Na_2O$ | 2570 | Sub* |
| MgO | 3890 | 2800 |

**TABLE 8.1** Lattice Energies of Selected Ionic Compounds

*$Na_2O$ sublimes at 1275°C.

charge increased by a factor of 2 in MgO as compared to LiF, the lattice energy of MgO is roughly four times as large as that of LiF:

$$F \propto \frac{(+1) \times (-1)}{(0.76 + 1.33)^2} \propto -0.23$$

Smaller lattice energy
(1017 kJ/mol)

$$F \propto \frac{(+2) \times (-2)}{(0.72 + 1.40)^2} \propto -0.89$$

Larger lattice energy
(3890 kJ/mol)

Sample Problem 8.2 shows how to use ionic radii and Coulomb's law to compare lattice energies for ionic compounds.

## Sample Problem 8.2

Arrange MgO, CaO, and SrO in order of increasing lattice energy.

**Strategy** Consider the charges on the ions and the distances between them. Apply Coulomb's law to determine the relative lattice energies.

**Setup** MgO, CaO, and SrO all contain the same anion ($O^{2-}$), and all contain cations with the same charge (+2). In this case, then, the distance between ions will determine the relative lattice energies. Recall that lattice energy increases as the distance between ions decreases (because the force between oppositely charged particles increases as the distance between them decreases). Because all three compounds contain the same anion, we need only consider the radii of the cations when determining the distance between ions. From Figure 7.6, the ionic radii are 0.72 Å ($Mg^{2+}$), 1.00 Å ($Ca^{2+}$), and 1.18 Å ($Sr^{2+}$).

*(Continued)*

**Think About It** Mg, Ca, and Sr are all Group 2A metals, so we could have predicted this result without knowing their radii. Recall that ionic radii increase as we move down a column in the periodic table, and charges that are farther apart are more easily separated (meaning the lattice energy will be smaller). The lattice energies of SrO, CaO, and MgO are 3217, 3414, and 3890 kJ/mol, respectively.

**Solution** MgO has the smallest distance between ions, whereas SrO has the largest distance between ions. Therefore, in order of increasing lattice energy: SrO < CaO < MgO.

**Practice Problem A** Determine which compound has the larger lattice energy: $MgCl_2$ or $SrCl_2$.

**Practice Problem B** Arrange the compounds NaF, MgO, and AlN in order of increasing lattice energy.

Although lattice energy is a useful measure of an ionic compound's stability, it is *not* a quantity that we can measure directly. Instead, we use various thermodynamic quantities that can be measured, and calculate lattice energy using Hess's law [◀◀ Section 5.5].

## The Born-Haber Cycle

We have described the formation of an ionic compound as though it happens when gaseous ions coalesce into a solid. In fact, the reactions that produce ionic solids generally do not occur this way. Figure 8.2 illustrates the formation of sodium chloride (NaCl) from its constituent elements.

We can imagine the reaction of Na($s$) and $Cl_2$($g$) to form NaCl($s$) as taking place in a series of steps for which most of the energy changes can be measured directly (See Table 8.2). Using these energy changes and the enthalpy of formation for NaCl, we can calculate the lattice energy using Hess's law. This method of determining the lattice energy is known as the **_Born-Haber cycle._**

The net reaction resulting from the series of steps in Table 8.2 is

$$Na(s) + \tfrac{1}{2}Cl_2(g) \longrightarrow Na^+(g) + Cl^-(g)$$

The final step in the formation of NaCl($g$) would be the coalescence of $Na^+$($g$) + $Cl^-$($g$). This is the step for which we cannot measure the energy change directly. However, we *can* measure the standard heat of formation of NaCl($s$). (It is tabulated in Appendix 2 as −410.9 kJ/mol.) Although the formation of NaCl($s$) from its constituent elements is not actually a step in our imaginary process, knowing its value enables us to calculate the lattice energy of NaCl. Figure 8.3 illustrates how this is done using all of these thermodynamic data and Hess's law.

Sample Problem 8.3 shows how to use the Born-Haber cycle to calculate the lattice energy.

Media Player/MPEG Animation: Figure 8.3, Born-Haber Cycle, pp. 304–305.

---

## Sample Problem 8.3

Using data from Figures 7.8 and 7.10 and Appendix 2, calculate the lattice energy of cesium chloride (CsCl).

**Strategy** Using Figure 8.3 as a guide, combine the pertinent thermodynamic data and use Hess's law to calculate the lattice energy.

**Setup** From Figure 7.8, $IE_1$(Cs) = 376 kJ/mol. From Figure 7.10, $EA_1$(Cl) = 349.0 kJ/mol. From Appendix 2, $\Delta H_f^\circ$[Cs($g$)] = 76.50 kJ/mol, $\Delta H_f^\circ$[Cl($g$)] = 121.7 kJ/mol, and $\Delta H_f^\circ$[CsCl($s$)] = −422.8 kJ/mol. Because we are interested in magnitudes only, we can use the absolute values of the thermodynamic data. And, because only the standard heat of formation of CsCl($s$) is a negative number, it is the only one for which the sign changes.

**Solution**

$$\{\Delta H_f^\circ[Cs(g)] + \Delta H_f^\circ[Cl(g)] + IE_1(Cs) + |\Delta H_f^\circ[CsCl(s)]|\} - EA_1(Cl) = \text{lattice energy}$$

$$= (76.50 \text{ kJ/mol} + 121.7 \text{ kJ/mol} + 376 \text{ kJ/mol} + 422.8 \text{ kJ/mol}) - 349.0 \text{ kJ/mol}$$

$$= 648 \text{ kJ/mol}$$

**Think About It** Compare this value to that for NaCl in Figure 8.3 (787 kJ/mol). Both compounds contain the same anion ($Cl^-$) and both have cations with the same charge (+1), so the relative sizes of the cations will determine the relative strengths of their lattice energies. Because $Cs^+$ is larger than $Na^+$, the lattice energy of CsCl is smaller than that of NaCl.

**Practice Problem A** Using data from Figures 7.8 and 7.10 and Appendix 2, calculate the lattice energy of rubidium iodide (RbI).

**Practice Problem B** The lattice energy of MgO is 3890 kJ/mol, and the second ionization energy ($IE_2$) of Mg is 1450.6 kJ/mol. Using these data, as well as data from Figures 7.8 and 7.10 and Appendix 2, determine the second electron affinity for oxygen, $EA_2$(O).

$$Cl_2 + 2Na \longrightarrow 2NaCl$$

**Figure 8.2** Sodium metal and chlorine gas combine to produce sodium chloride in a highly exothermic reaction.

| TABLE 8.2 | Hypothetical Steps in the Formation of $Na^+(g)$ and $Cl^-(g)$ from $Na(s)$ and $Cl_2(g)$ |
| --- | --- |
| **Chemical Equation** | **Energy Change (kJ/mol)** |
| $Na(s) \longrightarrow Na(g)$ | 107.7* |
| $\frac{1}{2}Cl_2(g) \longrightarrow Cl(g)$ | 121.7† |
| $Na(g) \longrightarrow Na^+(g) + e^-$ | 495.9‡ |
| $Cl(g) + e^- \longrightarrow Cl^-(g)$ | −349§ |

*Standard heat of formation ($\Delta H_f^\circ$) of $Na(g)$ from Appendix 2.
†Standard heat of formation ($\Delta H_f^\circ$) of $Cl(g)$ from Appendix 2.
‡First ionization energy ($IE_1$) of Na from Figure 7.8.
§$\Delta H^\circ$ for this process is negative. (Recall that by definition, *EA* is the amount of energy *released* [◀◀ Section 7.4]. This $\Delta H$ is equal to $-EA$.)

## Checkpoint 8.2    Ionic Bonding

**8.2.1** Will the lattice energy of KF be larger or smaller than that of LiF, larger or smaller than that of KCl, and larger or smaller than that of KI?

a) larger, larger, and smaller

b) smaller, larger, and smaller

c) smaller, larger, and larger

d) smaller, smaller, and smaller

e) larger, smaller, and larger

**8.2.2** Using the following data, calculate the lattice energy of KF:
$\Delta H_f^\circ[K(g)] = 89.99$ kJ/mol, $\Delta H_f^\circ[F(g)] = 80.0$ kJ/mol,
$IE_1(K) = 418.8$ kJ/mol, $\Delta H_f^\circ[KF(s)] = -547$ kJ/mol,
and $EA_1(F) = 328$ kJ/mol.

a) 808 kJ/mol

b) −286 kJ/mol

c) 261 kJ/mol

d) 1355 kJ/mol

e) −261 kJ/mol

**8.2.3** Lattice energies are graphed for three series of compounds in which the ion charges are $+2, -2$; $+2, -1$; and $+1, -1$. The ions in each series of compounds are separated by different distances. Identify the series.

a) red $+2, -2$; blue $+1, -1$; green $+2, -1$

b) red $+2, -2$; blue $+2, -1$; green $+1, -1$

c) red $+2, -1$; blue $+2, -2$; green $+1, -1$

d) red $+1, -1$; blue $+2, -1$; green $+2, -2$

e) red $+2, -1$; blue $+1, -1$; green $+2, -2$

**Figure 8.3**

# Born-Haber Cycle

$IE_1$(Na)

**4**

$\Delta H_f^\circ$ [Cl($g$)]

**3**

$\Delta H_f^\circ$ [Na($g$)]

**2**

$\Delta H_f^\circ$ [NaCl($s$)]

**1**

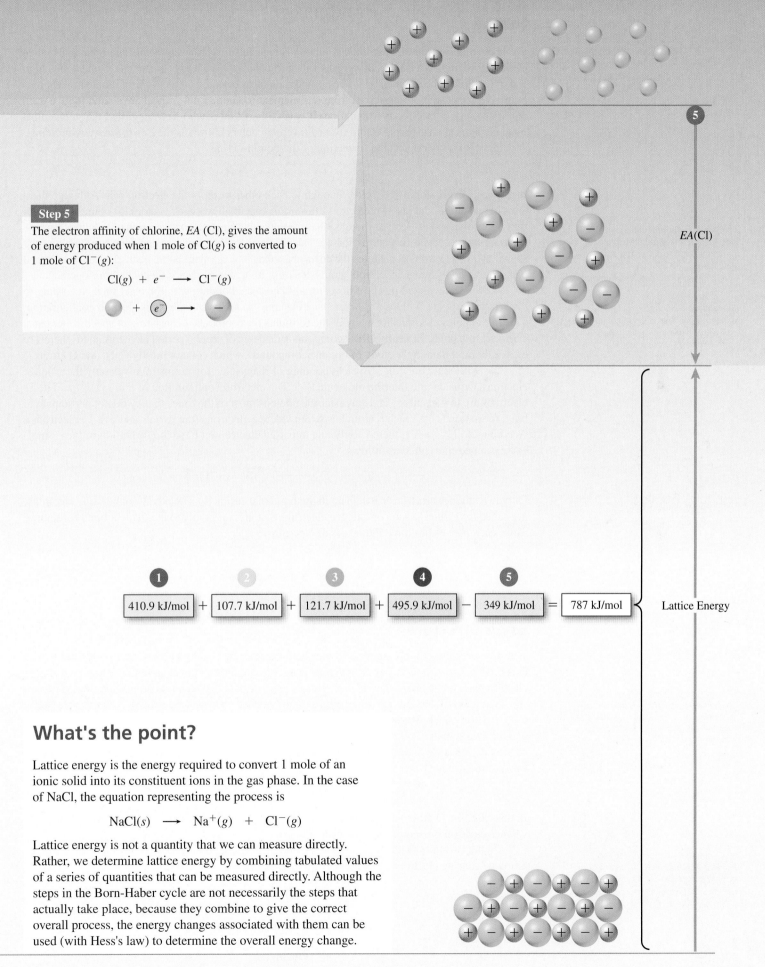

5

**Step 5**

The electron affinity of chlorine, *EA* (Cl), gives the amount of energy produced when 1 mole of Cl(*g*) is converted to 1 mole of Cl$^-$(*g*):

$$Cl(g) + e^- \longrightarrow Cl^-(g)$$

*EA*(Cl)

| 1 | | 2 | | 3 | | 4 | | 5 | | |
|---|---|---|---|---|---|---|---|---|---|---|
| 410.9 kJ/mol | + | 107.7 kJ/mol | + | 121.7 kJ/mol | + | 495.9 kJ/mol | − | 349 kJ/mol | = | 787 kJ/mol |

Lattice Energy

## What's the point?

Lattice energy is the energy required to convert 1 mole of an ionic solid into its constituent ions in the gas phase. In the case of NaCl, the equation representing the process is

$$NaCl(s) \longrightarrow Na^+(g) + Cl^-(g)$$

Lattice energy is not a quantity that we can measure directly. Rather, we determine lattice energy by combining tabulated values of a series of quantities that can be measured directly. Although the steps in the Born-Haber cycle are not necessarily the steps that actually take place, because they combine to give the correct overall process, the energy changes associated with them can be used (with Hess's law) to determine the overall energy change.

## 8.3 | Covalent Bonding

We learned in Section 8.2 that ionic compounds tend to form between metals and nonmetals when electrons are transferred from an element with a low ionization energy (the metal) to one with a high electron affinity (the nonmetal). When compounds form between elements with more similar properties, electrons are not transferred from one element to another but instead are shared to give each atom a noble gas electron configuration. It was Gilbert Lewis who first suggested that a chemical bond involves atoms sharing electrons, and this approach is known as the ***Lewis theory of bonding.***

Lewis theory depicts the formation of the bond in $H_2$ as

$$H\cdot + \cdot H \longrightarrow H\text{:}H$$

In essence, two H atoms move close enough to each other to *share* the electron pair. Although there are still two atoms and just two electrons, this arrangement allows each H atom to "count" both electrons as its own and to "feel" as though it has the noble gas electron configuration of helium. This type of arrangement, where two atoms share a pair of electrons, is known as ***covalent bonding,*** and the shared pair of electrons constitutes the ***covalent bond***. In a covalent bond, each electron in a shared pair is attracted to the nuclei of both atoms. It is this attraction that holds the two atoms together.

Lewis summarized much of his theory of chemical bonding with the octet rule. According to the ***octet rule,*** atoms will lose, gain, or share electrons to achieve a noble gas electron configuration. This rule enables us to predict many of the formulas for compounds consisting of specific elements. The octet rule holds for nearly all the compounds made up of second period elements and is therefore especially important in the study of organic compounds, which contain mostly C, N, and O atoms.

As with ionic bonding, covalent bonding of many-electron atoms involves only the valence electrons. Consider the fluorine molecule ($F_2$). The electron configuration of F is $1s^2 2s^2 2p^5$. The $1s$ electrons are low in energy and stay near the nucleus most of the time, so they do not participate in bond formation. Thus, each F atom has seven valence electrons (the two $2s$ and five $2p$ electrons). According to Figure 8.1, there is only one unpaired electron on F, so the formation of the $F_2$ molecule can be represented as follows:

$$:\!\ddot{F}\!\cdot \; + \; \cdot\!\ddot{F}\!: \; \longrightarrow \; :\!\ddot{F}\!:\!\ddot{F}\!: \; \text{ or } \; :\!\ddot{F}\!-\!\ddot{F}\!:$$

Only two valence electrons participate in the bond that forms $F_2$. The other, nonbonding electrons, are called ***lone pairs***—pairs of valence electrons that are not involved in covalent bond formation. Thus, each F in $F_2$ has three lone pairs of electrons.

$$\longrightarrow :\!\ddot{F}\!-\!\ddot{F}\!: \longleftarrow\text{lone pair}$$

### Lewis Structures

The structures used to represent molecules held together by covalent bonds, such as $H_2$ and $F_2$, are called *Lewis structures*. A ***Lewis structure*** is a representation of covalent bonding in which shared electron pairs are shown either as dashes or as pairs of dots between two atoms, and lone pairs are shown as pairs of dots on individual atoms. Only valence electrons are shown in a Lewis structure.

To draw the Lewis structure of the water molecule, recall that the Lewis dot symbol for oxygen has two unpaired dots (Figure 8.1), meaning that it has two unpaired electrons and can form two bonds. Because hydrogen has only one electron, it can form only one covalent bond. Thus, the Lewis structure for water is

$$H\text{:}\ddot{O}\text{:}H \quad \text{ or } \quad H-\ddot{O}-H$$

In this case, the O atom has two lone pairs. The hydrogen atom has no lone pairs because its only valence electron is used to form a covalent bond.

In the $F_2$ and $H_2O$ molecules, the F, H, and O atoms each achieve a stable noble gas configuration by sharing electrons, thus illustrating the octet rule:

F with 8 $e^-$ —(:$\ddot{F}$:$\ddot{F}$:)— F with 8 $e^-$

H with 2 $e^-$ —(H:$\ddot{O}$:H)— H with 2 $e^-$

O with 8 $e^-$

The octet rule works best for elements in the second period of the periodic table. These elements have only 2*s* and 2*p* valence subshells, which can hold a total of eight electrons. When an atom of one of these elements forms a covalent compound, it can attain the noble gas electron configuration [Ne] by sharing electrons with other atoms in the same compound. In Section 8.8, we will discuss some important exceptions to the octet rule.

## Multiple Bonds

Atoms can form several different types of covalent bonds, such as single bonds and multiple bonds. In a ***single bond,*** two atoms are held together by one electron pair. ***Multiple bonds*** form, on the other hand, when two atoms share two or more pairs of electrons. A multiple bond in which the atoms share two pairs of electrons is called a ***double bond.*** Double bonds are found in molecules such as carbon dioxide ($CO_2$) and ethylene ($C_2H_4$):

Each O has 8 $e^-$ — (O::C::O)   or   $\ddot{O}=C=\ddot{O}$

C has 8 $e^-$

Each H has 2 $e^-$ — H:C::C:H   or   figure

Each C has 8 $e^-$

A ***triple bond*** arises when two atoms share three pairs of electrons, as in molecules such as nitrogen ($N_2$) and acetylene ($C_2H_2$):

Each N has 8 $e^-$

(:N:::N:)   or   :N≡N:

Each H has 2 $e^-$   —(H:C:::C:H)   or   H—C≡C—H

Each C has 8 $e^-$

In ethylene and acetylene, all the valence electrons are used in bonding; there are no lone pairs on the carbon atoms. In fact, with the important exception of carbon monoxide (CO), most stable molecules containing carbon do not have lone pairs on the carbon atoms.

Multiple bonds are shorter than single bonds. *Bond length* is defined as the distance between the nuclei of two covalently bonded atoms in a molecule (Figure 8.4). Table 8.3 lists some experimentally determined bond lengths. For a given pair of atoms, such as carbon and nitrogen, triple bonds are shorter than double bonds, and double bonds are shorter than single bonds. The shorter multiple bonds are also stronger than single bonds, as we will see in Section 8.9.

## Comparison of Ionic and Covalent Compounds

Ionic and covalent compounds differ markedly in their general physical properties because of differences in the nature of their bonds. There are two types of attractive forces in covalent compounds, the *intramolecular* bonding force that holds the atoms together in a molecule, and the *intermolecular* forces between molecules. Bond enthalpy, discussed in Section 8.9, can be used to quantify the intramolecular bonding force. Intermolecular forces are usually quite weak compared to the forces holding atoms together within a molecule, so molecules of a covalent compound are not held together tightly. As a result, covalent compounds are usually gases, liquids, or low-melting solids.

On the other hand, the electrostatic forces holding ions together in an ionic compound are usually very strong, so ionic compounds are solids at room temperature and have high melting points. Many ionic compounds are soluble in water, and the resulting aqueous solutions conduct electricity because the compounds are strong electrolytes. Most covalent compounds are insoluble in water, or if they do dissolve, their aqueous solutions generally do not conduct electricity, because the compounds are nonelectrolytes. Molten ionic compounds conduct electricity because they contain mobile cations and anions; liquid or molten covalent compounds do *not* conduct electricity because no ions are present. Table 8.4 compares some of the properties of a typical ionic compound, sodium chloride (NaCl), with those of a covalent compound, carbon tetrachloride ($CCl_4$).

$N_2$
Bond length 1.10 Å

CO
Bond length 1.13 Å

**Figure 8.4**  Bond length is the distance between the nuclei of two bonded atoms.

| TABLE 8.3 | Average Bond Lengths of Some Common Single, Double, and Triple Bonds |
|---|---|
| **Bond Type** | **Bond Length (pm)** |
| C—H | 107 |
| O—H | 96 |
| C—O | 143 |
| C=O | 121 |
| C≡O | 113 |
| C—C | 154 |
| C=C | 133 |
| C≡C | 120 |
| C—N | 143 |
| C=N | 138 |
| C≡N | 116 |
| N—N | 147 |
| N=N | 124 |
| N≡N | 110 |
| N—O | 136 |
| N=O | 122 |
| O—O | 148 |
| O=O | 121 |

| TABLE 8.4 | Comparison of Some Properties of an Ionic Compound (NaCl) and a Covalent Compound ($CCl_4$) | |
|---|---|---|
| **Property** | **NaCl** | **$CCl_4$** |
| Appearance | White solid | Colorless liquid |
| Melting point (°C) | 801 | −23 |
| Molar heat of fusion* (kJ/mol) | 30.2 | 2.5 |
| Boiling point (°C) | 1413 | 76.5 |
| Molar heat of vaporization* (kJ/mol) | 600 | 30 |
| Density (g/cm³) | 2.17 | 1.59 |
| Solubility in water | High | Very low |
| Electrical conductivity | | |
|    Solid | Poor | Poor |
|    Liquid | Good | Poor |
|    Aqueous | Good | Poor |

*The *molar heat of fusion* and *molar heat of vaporization* are the amounts of heat needed to melt 1 mole of the solid and to vaporize 1 mole of the liquid, respectively.

# 8.4 Electronegativity and Polarity

So far, we have described chemical bonds as either *ionic,* when they occur between a metal and a nonmetal, or *covalent,* when they occur between nonmetals. In fact, ionic and covalent bonds are simply the extremes in a spectrum of bonding. Bonds that fall between these two extremes are *polar,* meaning that electrons are shared but are not shared equally. Such bonds are referred to as *polar covalent bonds.* The following shows a comparison of the different types of bonds, where M and X represent two different elements:

| M:X | $M^{\delta+}X^{\delta-}$ | $M^+X^-$ |
|---|---|---|
| **Pure covalent bond** | **Polar covalent bond** | **Ionic bond** |
| Neutral atoms held together by *equally* shared electrons | Partially charged atoms held together by *unequally* shared electrons | Oppositely charged ions held together by electrostatic attraction |

    To illustrate the spectrum of bonding, let's consider three substances: $H_2$, HF, and NaF. In the $H_2$ molecule, where the two bonding atoms are identical, the electrons are shared equally. That is, the electrons in the covalent bond spend roughly the same amount of time in the vicinity of each H atom. In the HF molecule, on the other hand, where the two bonding atoms are different, the electrons are *not* shared equally. They spend more time in the vicinity of the F atom than in the vicinity of the H atom. (The δ symbol is used to denote partial charges on the atoms.) In NaF, the electrons are not shared at all but rather are transferred from sodium to fluorine.

    One way to visualize the distribution of electrons in species such as $H_2$, HF, and NaF is to use electrostatic potential models (Figure 8.5). These models show regions where electrons spend a lot of time in red, and regions where electrons spend very little time in blue. (Regions where electrons spend a moderate amount of time appear in green.)

**Student Annotation:** In fact, the transfer of electrons in NaF is *nearly* complete. Even in an ionic bond, the electrons in question spend a small amount of time near the cation.

**Figure 8.5** Electron density maps show the distribution of charge in a covalent species ($H_2$), a polar covalent species (HF), and an ionic species (NaF). The most electron-rich regions are red; the most electron-poor regions are blue.

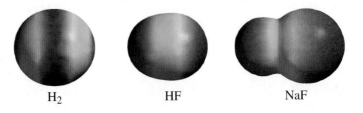

$H_2$        HF        NaF

# Electronegativity

*Electronegativity* is the ability of an atom in a compound to draw electrons to itself. It determines where electrons in a compound spend most of their time. Elements with high electronegativity have a greater tendency to attract electrons than do elements with low electronegativity. Electronegativity is related to electron affinity and ionization energy. An atom such as fluorine, which has a high electron affinity (tends to accept electrons) and a high ionization energy (does not lose electrons easily), has a high electronegativity. Sodium, on the other hand, has a low electron affinity, a low ionization energy, and therefore a low electronegativity.

Electronegativity is a relative concept, meaning that an element's electronegativity can be measured only in relation to the electronegativity of other elements. Linus Pauling[2] devised a method for calculating the relative electronegativities of most elements. These values are shown in Figure 8.6. In general, electronegativity increases from left to right across a period in the periodic table, as the metallic character of the elements decreases. Within each group, electronegativity decreases with increasing atomic number and increasing metallic character. The transition metals do not follow these trends. The most electronegative elements (the halogens, oxygen, nitrogen, and sulfur) are found in the upper right-hand corner of the periodic table, and the least electronegative elements (the alkali and alkaline earth metals) are clustered near the lower left-hand corner. These trends are readily apparent in the graph in Figure 8.7.

Multimedia

Periodic Table—electronegativity.

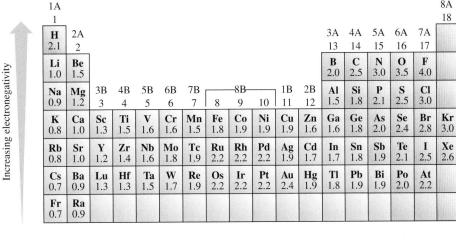

**Figure 8.6** Electronegativities of common elements.

**Figure 8.7** Variation of electronegativity with atomic number.

---

2. Linus Carl Pauling (1901–1994). American chemist. Regarded by many as the most influential chemist of the twentieth century. Pauling received the Nobel Prize in Chemistry in 1954 for his work on protein structure, and the Nobel Peace Prize in 1962 for his tireless campaign against the testing and proliferation of nuclear arms. He is the only person ever to have received two unshared Nobel Prizes.

Electronegativity and electron affinity are related but distinct concepts. Both indicate the tendency of an atom to attract electrons. Electron affinity, however, refers to an isolated atom's ability to attract an additional electron in the gas phase, whereas electronegativity refers to the ability of an atom in a chemical bond (with another atom) to attract the shared electrons. Electron affinity, moreover, is an experimentally measurable quantity, whereas electronegativity is an estimated number that cannot be measured directly.

Atoms of elements with widely different electronegativities tend to form ionic compounds with each other because the atom of the less electronegative element gives up its electrons to the atom of the more electronegative element. Atoms of elements with comparable electronegativities tend to form polar covalent bonds, or simply polar bonds, with each other because the shift in electron density from one atom to the other is usually small. Only atoms of the same element, which have the same electronegativity, can be joined by a pure covalent bond.

There is no sharp distinction between *nonpolar covalent* and *polar covalent* or between *polar covalent* and *ionic,* but the following guidelines can help distinguish among them:

- A bond between atoms whose electronegativities differ by less than 0.5 is generally considered purely covalent or **nonpolar.**
- A bond between atoms whose electronegativities differ by the range of 0.5 to 2.0 is generally considered *polar covalent.*
- A bond between atoms whose electronegativities differ by 2.0 or more is generally considered *ionic.*

Sometimes chemists describe bonds using the term *percent ionic character* [▶▶ Page 312]. A purely ionic bond would have 100 percent ionic character (although no such bond is known). A purely covalent, nonpolar bond has 0 percent ionic character.

Sample Problem 8.4 shows how to use electronegativities to identify a chemical bond as nonpolar, polar, or ionic.

**Multimedia**

Chemical Bonding—ionic covalent and polar covalent bonds.

---

## Sample Problem 8.4

Classify the following bonds as nonpolar, polar, or ionic: (a) the bond in ClF, (b) the bond in CsBr, and (c) the carbon-carbon double bond in $C_2H_4$.

**Strategy**  Using the information in Figure 8.6, determine which bonds have identical, similar, and widely different electronegativities.

**Setup**  Electronegativity values from Figure 8.6 are: Cl (3.0), F (4.0), Cs (0.7), Br (2.8), C (2.5).

**Solution**  (a) The difference between the electronegativities of F and Cl is $4.0 - 3.0 = 1.0$, making the bond in ClF polar.

(b) In CsBr, the difference is $2.8 - 0.7 = 2.1$, making the bond ionic.

(c) In $C_2H_4$, the two atoms are identical. (Not only are they the same element, but each C atom is bonded to two H atoms.) The carbon-carbon double bond in $C_2H_4$ is nonpolar.

> **Think About It**  By convention, the difference in electronegativity is always calculated by subtracting the smaller number from the larger one, so the result is always positive.

**Practice Problem A**  Classify the following bonds as nonpolar, polar, or ionic: (a) the bonds in $H_2S$, (b) the H—O bonds in $H_2O_2$, and (c) the O—O bond in $H_2O_2$.

**Practice Problem B**  Electrostatic potential maps are shown for HCl and LiH. Determine which diagram is which. (The H atom is shown on the left in both.)

| TABLE 8.5 | Bond Lengths and Dipole Moments of the Hydrogen Halides | |
|---|---|---|
| **Molecule** | **Bond Length (Å)** | **Dipole Moment (D)** |
| HF | 0.92 | 1.82 |
| HCl | 1.27 | 1.08 |
| HBr | 1.41 | 0.82 |
| HI | 1.61 | 0.44 |

## Dipole Moment, Partial Charges, and Percent Ionic Character

The shift of electron density in a polar bond is symbolized by placing a crossed arrow (a dipole arrow) above the Lewis structure to indicate the direction of the shift. For example,

$$\overset{\overset{+\longrightarrow}{}}{H-\ddot{\underset{\cdot\cdot}{F}}:}$$

The consequent charge separation can be represented as

$$\overset{\delta+}{H}-\overset{\delta-}{\underset{\cdot\cdot}{\ddot{F}}}:$$

A quantitative measure of the polarity of a bond is its ***dipole moment (μ),*** which is calculated as the product of the charge (Q) and the distance (r) between the charges:

$$\mu = Q \times r \qquad \text{Equation 8.1}$$

For a diatomic molecule containing a polar bond to be electrically neutral, the partial positive and partial negative charges must have the same magnitude. Therefore, the Q term in Equation 8.1 refers to the magnitude of the partial charges and the calculated value of $\mu$ is always positive. Dipole moments are usually expressed in debye units (D), named for Peter Debye.[3] In terms of more familiar SI units,

$$1\,D = 3.336 \times 10^{-30}\,C \cdot m$$

where C is coulombs and m is meters. Table 8.5 lists several polar diatomic molecules, their bond lengths, and their experimentally measured dipole moments.

Sample Problem 8.5 shows how to use bond lengths and dipole moments to determine the magnitude of the partial charges in a polar diatomic molecule.

**Student Annotation:** The distance, $r$, between partial charges in a polar diatomic molecule is the bond length expressed in meters. Bond lengths are usually given in angstroms (Å) or picometers (pm), so it is generally necessary to convert to meters.

**Student Annotation:** We usually express the charge on an electron as −1. This refers to *units of electronic charge.* However, remember that the charge on an electron can also be expressed in *coulombs* [◄◄ Section 2.2]. The conversion factor between the two is necessary to calculate dipole moments: $1e^- = 1.6022 \times 10^{-19}\,C.$

### Sample Problem 8.5

Burns caused by hydrofluoric acid [HF($aq$)] are unlike any other acid burns and present unique medical complications. HF solutions typically penetrate the skin and damage internal tissues, including bone, often with minimal surface damage. Less concentrated solutions actually can cause greater injury than more concentrated ones by penetrating more deeply before causing injury, thus delaying the onset of symptoms and preventing timely treatment. Determine the magnitude of the partial positive and partial negative charges in the HF molecule.

**Strategy** Rearrange Equation 8.1 to solve for Q. Convert the resulting charge in coulombs to charge in units of electronic charge.

**Setup** According to Table 8.5, $\mu = 1.82$ D and $r = 0.92$ Å for HF. The dipole moment must be converted from debye to C · m and the distance between the ions must be converted to meters.

$$\mu = 1.82\,D \times \frac{3.336 \times 10^{-30}\,C \cdot m}{1\,D} = 6.07 \times 10^{-30}\,C \cdot m$$

$$r = 0.92\,\text{Å} \times \frac{1 \times 10^{-10}\,m}{1\,\text{Å}} = 9.2 \times 10^{-11}\,m$$

*(Continued)*

**Student Annotation:** Hydrofluoric acid has several important industrial applications, including the etching of glass and the manufacture of electronic components.

---

3. Peter Joseph William Debye (1884–1966). American chemist and physicist of Dutch origin. Debye made many significant contributions to the study of molecular structure, polymer chemistry, X-ray analysis, and electrolyte solutions. He was awarded the Nobel Prize in Chemistry in 1936.

**Solution** In coulombs:

$$Q = \frac{\mu}{r} = \frac{6.07 \times 10^{-30}\, \text{C} \cdot \text{m}}{9.2 \times 10^{-11}\, \text{m}} = 6.598 \times 10^{-20}\, \text{C}$$

In units of electronic charge:

$$6.598 \times 10^{-20}\, \text{C} \times \frac{1e^-}{1.6022 \times 10^{-19}\, \text{C}} = 0.41e^-$$

Therefore, the partial charges in HF are $+0.41$ and $-0.41$ on H and F, respectively.

$$\overset{+0.41}{\text{H}}\!-\!\overset{-0.41}{\ddot{\underset{\displaystyle ..}{\text{F}}}\!:}$$

**Practice Problem A** Using data from Table 8.5, determine the magnitude of the partial charges in HBr.

**Practice Problem B** Given that the partial charges on C and O in carbon monoxide are $+0.020$ and $-0.020$, respectively, calculate the dipole moment of CO. (The distance between the partial charges, $r$, is 113 pm.)

Although the designations "covalent," "polar covalent," and "ionic" can be useful, sometimes chemists wish to describe and compare chemical bonds with more precision. For this purpose, we can use Equation 8.1 to calculate the dipole moment we would expect if the charges on the atoms were discrete instead of partial; that is, if an electron had actually been transferred from one atom to the other. Comparing this calculated dipole moment with the measured value gives us a quantitative way to describe the nature of a bond using the term **percent ionic character,** which is defined as the ratio of observed $\mu$ to calculated $\mu$, multiplied by 100.

**Equation 8.2**    $\text{percent ionic character} = \dfrac{\mu\ (\text{observed})}{\mu\ (\text{calculated assuming discrete charges})} \times 100\%$

Figure 8.8 illustrates the relationship between percent ionic character and the electronegativity difference in a heteronuclear diatomic molecule.

Sample Problem 8.6 shows how to calculate percent ionic character using Equation 8.2.

## Sample Problem 8.6

Using data from Table 8.5, calculate the percent ionic character of the bond in HI.

**Strategy** Use Equation 8.1 to calculate the dipole moment in HI assuming that the charges on H and I are $+1$ and $-1$, respectively; and use Equation 8.2 to calculate percent ionic character. The magnitude of the charges must be expressed as coulombs ($1\, e^- = 1.6022 \times 10^{-19}\, \text{C}$); the bond length ($r$) must be expressed as meters ($1\, \text{Å} = 1 \times 10^{-10}\, \text{m}$); and the calculated dipole moment should be expressed as debyes ($1\, \text{D} = 3.336 \times 10^{-30}\, \text{C} \cdot \text{m}$).

**Setup** From Table 8.5, the bond length in HI is $1.61\text{Å}$ ($1.61 \times 10^{-10}\, \text{m}$) and the measured dipole moment of HI is 0.44 D.

**Solution** The dipole moment we would expect if the magnitude of charges were $1.6022 \times 10^{-19}\, \text{C}$ is

$$\mu = Q \times r = (1.6022 \times 10^{-19}\, \text{C}) \times (1.61 \times 10^{-10}\, \text{m}) = 2.58 \times 10^{-29}\, \text{C} \cdot \text{m}$$

Converting to debyes gives

$$2.58 \times 10^{-29}\, \cancel{\text{C} \cdot \text{m}} \times \frac{1\, \text{D}}{3.336 \times 10^{-30}\, \cancel{\text{C} \cdot \text{m}}} = 7.73\, \text{D}$$

The percent ionic character of the H—I bond is

$$\frac{0.44\, \text{D}}{7.73\, \text{D}} \times 100\% = 5.7\%$$

**Practice Problem A** Using data from Table 8.5, calculate the percent ionic character of the bond in HF.

**Practice Problem B** Using information from Figure 7.12, and given that the NaI bond has 59.7 percent ionic character, determine the measured dipole moment of NaI.

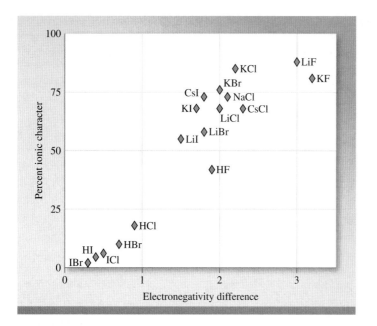

**Figure 8.8** Relationship between percent ionic character and electronegativity difference.

### Checkpoint 8.4 — Electronegativity and Polarity

**8.4.1** In which of the following molecules are the bonds *most* polar?

a) $H_2Se$

b) $H_2O$

c) $CO_2$

d) $BCl_3$

e) $PCl_5$

**8.4.2** Using data from Table 8.5, calculate the magnitude of the partial charges in HI.

a) 0.39

b) 1.8

c) 0.057

d) 0.60

e) 0.15

**8.4.3** Arrange molecules A through E in order of increasing percent ionic character.

a) $A < B < C < D < E$

b) $A = B < C < D < E$

c) $A = B < C < D = E$

d) $A < B < C < D = E$

e) $A < B = C = D < E$

**8.4.4** Using data from Table 8.5, calculate the percent ionic character of HCl.

a) 85.0 percent

b) 6.10 percent

c) 17.7 percent

d) 2.03 percent

e) 20.3 percent

## 8.5 Drawing Lewis Structures

Although the octet rule and Lewis structures alone do not present a complete picture of covalent bonding, they do help us account for some of the properties of molecules. In addition, Lewis structures provide a starting point for the bonding theories that we will examine in Chapter 9. It is crucial, therefore, that you learn a system for drawing correct Lewis structures for molecules and polyatomic ions. The basic steps are as follows:

1. From the molecular formula, draw the skeletal structure of the compound, using chemical symbols and placing bonded atoms next to one another. For simple compounds, this step is fairly easy. Often there will be a unique central atom surrounded by a group of other identical atoms. In general, the *least* electronegative atom will be the central atom. Draw a single covalent bond (dash) between the central atom and each of the surrounding atoms. (For more complex compounds whose structures might not be obvious, you may need to be given information in addition to the molecular formula.)

**Student Annotation:** H cannot be a central atom because it only forms *one* covalent bond.

2. Count the total number of valence electrons present. Remember that an element's *group number* (1A–8A) gives the number of valence electrons it contributes to the total number. For polyatomic ions, add electrons to the total number to account for negative charges; subtract electrons from the total number to account for positive charges.

3. For each bond (dash) in the skeletal structure, subtract two electrons from the total valence electrons (determined in step 2) to determine the number of remaining electrons.

4. Use the remaining electrons to complete the octets of the terminal atoms (those bonded to the central atom) by placing pairs of electrons on each atom. (Remember that an H atom only requires two electrons to complete its valence shell.) If there is more than one type of terminal atom, complete the octets of the most electronegative atoms first.

5. If any electrons remain after step 4, place them in pairs on the central atom.

6. If the central atom has fewer than eight electrons after completing steps 1 to 5, move one or more pairs from the terminal atoms to form multiple bonds between the central atom and the terminal atoms.

**Student Annotation:**
To summarize the process of drawing Lewis structures:

1. Draw the skeletal structure.
2. Count the valence electrons.
3. Subtract two electrons for each bond.
4. Distribute the remaining electrons.
5. Complete the octets of all atoms.
6. Use multiple bonds if necessary.

## Steps for Drawing Lewis Structures

| Step | CH$_4$ | CCl$_4$ | H$_2$O | O$_2$ | CN$^-$ |
|------|--------|---------|--------|-------|--------|
| 1 | H—C—H (with H above and below) | Cl—C—Cl (with Cl above and below) | H—O—H | O—O | C—N |
| 2 | 8 | 32 | 8 | 12 | 10 |
| 3 | $8 - 8 = 0$ | $32 - 8 = 24$ | $8 - 4 = 4$ | $12 - 2 = 10$ | $10 - 2 = 8$ |
| 4 | H—C—H (with H above and below) | :Cl̈—C—Cl̈: (with :Cl̈: above and below) | H—O—H | :Ö—Ö: | :C—N̈: |
| 5 | — | — | H—Ö—H | — | — |
| 6 | — | — | — | :O=O: | [:C≡N:]$^-$ |

**Student Annotation:** Like Lewis dot symbols for atomic anions, Lewis structures for polyatomic anions are enclosed by square brackets.

Sample Problem 8.7 shows how to draw a Lewis structure.

### Sample Problem 8.7

Draw the Lewis structure for carbon disulfide (CS$_2$).

**Strategy** Use the procedure described in steps 1 through 6 above for drawing Lewis structures.

**Setup**

Step 1:    C and S have identical electronegativities. We will draw the skeletal structure with the unique atom, C, at the center.

$$S—C—S$$

Step 2: The total number of valence electrons is 16, six from each S atom and four from the C atom [2(6) + 4 = 16].

Step 3: Subtract four electrons to account for the bonds in the skeletal structure, leaving us 12 electrons to distribute.

Step 4: Distribute the 12 remaining electrons as three lone pairs on each S atom.

$$:\ddot{S}-C-\ddot{S}:$$

Step 5: There are no electrons remaining after step 4, so step 5 does not apply.

Step 6: To complete carbon's octet, use one lone pair from each S atom to make a double bond to the C atom.

**Solution**

$$:\ddot{S}=C=\ddot{S}:$$

> **Think About It** Counting the total number of valence electrons should be relatively simple to do, but it is often done hastily and is therefore a potential source of error in this type of problem. Remember that the number of valence electrons for each element is equal to the group number of that element.

**Practice Problem A** Draw the Lewis structure for $NF_3$.

**Practice Problem B** Draw the Lewis structure for $ClO_3^-$.

---

### Checkpoint 8.5    Drawing Lewis Structures

**8.5.1** Identify the correct Lewis structure for formic acid (HCOOH).

a) $H-\ddot{C}-\ddot{O}-\ddot{O}-H$

b) $H-\ddot{O}=C=\ddot{O}-H$

c) $H-\ddot{O}-\ddot{C}-\ddot{O}-H$

d) $H-\overset{\overset{\displaystyle :O:}{||}}{C}-\ddot{O}-H$

e) $:\ddot{O}-\overset{\overset{\displaystyle :\ddot{O}:}{|}}{\underset{\underset{\displaystyle H}{|}}{C}}-H$

**8.5.2** Identify the correct Lewis structure for hydrogen peroxide ($H_2O_2$).

a) $H-\ddot{O}=\ddot{O}-H$

b) $H-O\equiv O-H$

c) $H-\ddot{O}-\ddot{O}-H$

d) $H=O=O=H$

e) $H-H-\ddot{O}-\ddot{O}:$

---

## 8.6    Lewis Structures and Formal Charge

So far you have learned two different methods of electron "bookkeeping." In Chapter 4, you learned about oxidation numbers [◄◄ Section 4.4], and in Section 8.4, you learned how to calculate partial charges. There is one additional commonly used method of electron bookkeeping—namely, *formal charge,* which can be used to determine the most plausible Lewis structures when more than one possibility exists for a compound. Formal charge is determined by comparing the number of electrons associated with an atom in a Lewis structure with the number of electrons that would be associated with the isolated atom. In an isolated atom, the number of electrons associated with the atom is simply the number of valence electrons. (As usual, we need not be concerned with the core electrons.)

Multimedia

Chemical Bonding—formal charge calculations.

To determine the number of electrons associated with an atom in a Lewis structure, keep in mind the following:

- All the atom's nonbonding electrons are associated with the atom.
- Half of the atom's bonding electrons are associated with the atom.

$$\text{formal charge} = \text{valence electrons} - \text{associated electrons} \qquad \text{Equation 8.3}$$

An atom's formal charge is calculated as follows: We can illustrate the concept of formal charge using the ozone molecule ($O_3$). Use the step-by-step method for drawing Lewis structures to draw the Lewis structure for ozone, and then determine the formal charge on each O atom by subtracting the number of associated electrons from the number of valence electrons.

$$2 \text{ unshared} + \frac{6 \text{ shared}}{2} = 5\,e^-$$

$$4 \text{ unshared} + \frac{4 \text{ shared}}{2} = 6\,e^- \qquad 6 \text{ unshared} + \frac{2 \text{ shared}}{2} = 7\,e^-$$

| | | | |
|---|---|---|---|
| Valence $e^-$ | 6 | 6 | 6 |
| $e^-$ associated with atom | 6 | 5 | 7 |
| Difference (formal charge) | 0 | +1 | -1 |

$$:\!\overset{\cdot\cdot}{O}\!=\!\overset{\cdot\cdot}{O}\!-\!\overset{\cdot\cdot}{O}\!:$$

As with oxidation numbers, the sum of the formal charges must equal the overall charge on the species [◀◀ Section 4.4]. Because $O_3$ is a molecule, its formal charges must sum to zero. For ions, the formal charges must sum to the overall charge on the ion.

Formal charges do *not* represent actual charges on atoms in a molecule. In the $O_3$ molecule, for example, there is no evidence that the central atom bears a net +1 charge or that one of the terminal atoms bears a −1 charge. Assigning formal charges to the atoms in the Lewis structure merely helps us keep track of the electrons involved in bonding in the molecule.

Sample Problem 8.8 lets you practice determining formal charges.

---

### Sample Problem 8.8

The widespread use of fertilizers has resulted in the contamination of some groundwater with nitrates, which are potentially harmful. Nitrate toxicity is due primarily to its conversion in the body to nitrite ($NO_2^-$), which interferes with the ability of hemoglobin to transport oxygen. Determine the formal charges on each atom in the nitrate ion ($NO_3^-$).

**Strategy** Use steps 1 through 6 on page 314 for drawing Lewis stuctures to draw the Lewis structure of $NO_3^-$. For each atom, subtract the associated electrons from the valence electrons.

**Setup**

$$\left[\begin{array}{c} :\!\overset{\cdot\cdot}{O}\!: \\ | \\ :\!\overset{\cdot\cdot}{O}\!-\!N\!=\!\overset{\cdot\cdot}{O}\!: \end{array}\right]^-$$

The N atom has five valence electrons and four associated electrons (one from each single bond and two from the double bond). Each singly bonded O atom has six valence electrons and seven associated electrons (six in three lone pairs and one from the single bond). The doubly bonded O atom has six valence electrons and six associated electrons (four in two lone pairs and two from the double bond).

**Solution** The formal charges are as follows: +1 (N atom), −1 (singly bonded O atoms), and 0 (doubly bonded O atom).

**Think About It** The sum of formal charges $(+1) + (-1) + (-1) + (0) = -1$ is equal to the overall charge on the nitrate ion.

---

**Practice Problem A** Determine the formal charges on each atom in the carbonate ion ($CO_3^{2-}$).

**Practice Problem B** Determine the formal charges and use them to determine the overall charge, if any, on the species represented by the following Lewis structure:

$$\left[\begin{array}{c} :\!\overset{\cdot\cdot}{O}\!: \\ | \\ :\!\overset{\cdot\cdot}{O}\!-\!S\!-\!\overset{\cdot\cdot}{O}\!: \end{array}\right]^?$$

---

Sometimes, there is more than one possible skeletal arrangement of atoms for the Lewis structure for a given species. In such cases, we often can select the best skeletal arrangement by using formal charges and the following guidelines:

- For molecules, a Lewis structure in which all formal charges are zero is preferred to one in which there are nonzero formal charges.

- Lewis structures with small formal charges (0 and ±1) are preferred to those with large formal charges (±2, ±3, and so on).
- The best skeletal arrangement of atoms will give rise to Lewis structures in which the formal charges are consistent with electronegativities. For example, the more electronegative atoms should have the more negative formal charges.

Sample Problem 8.9 shows how formal charge can be used to determine the best skeletal arrangement of atoms for the Lewis structure of a molecule or polyatomic ion.

---

**Sample Problem 8.9**

Formaldehyde ($CH_2O$), which can be used to preserve biological specimens, is commonly sold as a 37% aqueous solution. Use formal charges to determine which skeletal arrangement of atoms shown here is the best choice for the Lewis structure of $CH_2O$.

$$H-C-O-H \qquad \begin{matrix} O \\ \| \\ H-C-H \end{matrix}$$

**Strategy** Complete the Lewis structures for each of the $CH_2O$ skeletons shown and determine the formal charges on the atoms in each one.

**Setup** The completed Lewis structures for the skeletons shown are:

$$H-\ddot{C}-\ddot{O}-H \qquad \begin{matrix} \cdot\ddot{O}\cdot \\ \| \\ H-C-H \end{matrix}$$

In the structure on the left, the formal charges are as follows:

Both H atoms: 1 valence $e^-$ − 1 associated $e^-$ (from single bond) = 0
C atom: 4 valence $e^-$ − 5 associated $e^-$ (two in the lone pair, one from the single bond, and two from the double bond) = −1
O atom: 6 valence $e^-$ − 5 associated $e^-$ (two from the lone pair, one from the single bond, and two from the double bond) = +1

$$H-\ddot{C}=\ddot{O}-H$$
Formal charges     0    −1   +1   0

In the structure on the right, the formal charges are as follows:

Both H atoms: 1 valence $e^-$ − 1 associated $e^-$ (from single bond) = 0
C atom: 4 valence $e^-$ − 4 associated $e^-$ (one from each single bond, and two from the double bond) = 0
O atom: 6 valence $e^-$ − 6 associated $e^-$ (four from the two lone pairs and two from the double bond) = 0

$$\begin{matrix} \cdot\ddot{O}\cdot \\ \| \\ H-C-H \end{matrix}$$
Formal charges all zero

**Solution** Of the two possible arrangements, the structure on the left has an O atom with a positive formal charge, which is inconsistent with oxygen's high electronegativity. Therefore, the structure on the right, in which both H atoms are attached directly to the C atom and all atoms have a formal charge of zero, is the better choice for the Lewis structure of $CH_2O$.

> **Think About It** For a molecule, formal charges of zero are preferred. When there are nonzero formal charges, they should be consistent with the electronegativities of the atoms in the molecule. A positive formal charge on oxygen, for example, is inconsistent with oxygen's high electronegativity.

**Practice Problem A** Two possible arrangements are shown for the Lewis structure of a carboxyl group, −COOH. Use formal charges to determine which of the two arrangements is better.

$$\begin{matrix} \cdot\ddot{O}\cdot \\ \| \\ -C-\ddot{O}-H \end{matrix} \qquad -\ddot{C}-\ddot{O}=\ddot{O}-H$$

**Practice Problem B** Use Lewis structures and formal charges to determine the best skeletal arrangement of atoms in $NCl_2^-$.

| Checkpoint 8.6 | Lewis Structures and Formal Charge |

**8.6.1** Determine the formal charges on H, C, and N, respectively, in HCN.

a) 0, +1, and −1

b) −1, +1, and 0

c) 0, −1, and +1

d) 0, +1, and +1

e) 0, 0, and 0

**8.6.2** Which of the Lewis structures shown is most likely preferred for $NCO^-$?

a) $\left[ :N=C=\ddot{O}: \right]^-$

b) $\left[ :\ddot{N}-C\equiv O: \right]^-$

c) $\left[ :N\equiv C-\ddot{O}: \right]^-$

d) $\left[ :\ddot{N}-C=\ddot{O}: \right]^-$

e) $\left[ :N=C-\ddot{O}: \right]^-$

## 8.7   Resonance

Our drawing of the Lewis structure for ozone ($O_3$) satisfied the octet rule for the central O atom because we placed a double bond between it and one of the two terminal O atoms. In fact, we can put the double bond at either end of the molecule, as shown by the following two equivalent Lewis structures:

$$\ddot{O}=\ddot{O}-\ddot{O}: \quad \longleftrightarrow \quad :\ddot{O}-\ddot{O}=\ddot{O}$$

A single bond between O atoms should be longer than a double bond between O atoms, but experimental evidence indicates that both of the bonds in $O_3$ are equal in length (128 pm). Because neither one of these two Lewis structures accounts for the known bond lengths in $O_3$, we use both Lewis structures to represent the ozone molecule.

Each of the Lewis structures is called a resonance structure. A **resonance structure** is one of two or more Lewis structures for a single molecule that cannot be represented accurately by only one Lewis structure. The double-headed arrow indicates that the structures shown are resonance structures. Like the medieval European traveler to Africa who described a rhinoceros as a cross between a griffin and a unicorn (two familiar but imaginary animals), we describe ozone, a real molecule, in terms of two familiar but nonexistent structures.

A common misconception about resonance is that a molecule such as ozone somehow shifts quickly back and forth from one resonance structure to the other. Neither resonance structure, though, adequately represents the actual molecule, which has its own unique, stable structure. "Resonance" is a human invention, designed to address the limitations of a simple bonding model. To extend the animal analogy, a rhinoceros is a distinct, real creature, not some oscillation between the mythical griffin and unicorn!

The carbonate ion provides another example of resonance:

$$\left[ \begin{matrix} :\ddot{O}: \\ | \\ :\ddot{O}=C-\ddot{O}: \end{matrix} \right]^{2-} \longleftrightarrow \left[ \begin{matrix} \cdot\ddot{O} \\ || \\ :\ddot{O}-C-\ddot{O}: \end{matrix} \right]^{2-} \longleftrightarrow \left[ \begin{matrix} :\ddot{O}: \\ | \\ :\ddot{O}-C=\ddot{O} \end{matrix} \right]^{2-}$$

According to experimental evidence, all three carbon-oxygen bonds in $CO_3^{2-}$ are equivalent. Therefore, the properties of the carbonate ion are best explained by considering its resonance structures together.

The concept of resonance applies equally well to organic systems. A good example is the benzene molecule ($C_6H_6$):

If one of these resonance structures corresponded to the actual structure of benzene, there would be two different bond lengths between adjacent C atoms, one with the properties of a single bond and the other with the properties of a double bond. In fact, the distance between all adjacent C atoms in benzene is 140 pm, which is shorter than a C—C bond (154 pm) and longer than a C=C bond (133 pm).

A simpler way of drawing the structure of the benzene molecule and other compounds containing the benzene ring is to show only the skeleton and not the carbon and hydrogen atoms. By this convention, the resonance structures are represented by

Note that the C atoms at the corners of the hexagon and the H atoms are not shown, although they are understood to be there. Only the bonds between the C atoms are shown.

Resonance structures differ only in the positions of their *electrons*—not in the positions of their atoms. Thus, $:N=N=\ddot{O}:$ and $:N\equiv N-\ddot{O}:$ are resonance structures of each other, whereas $:N=N=\ddot{O}:$ and $:N=O=N:$ are not.

Sample Problem 8.10 shows how to draw resonance structures.

**Student Annotation:** The representation of organic compounds is discussed in more detail in Chapter 10.

## Sample Problem 8.10

High oil and gasoline prices have renewed interest in alternative methods of producing energy, including the "clean" burning of coal. Part of what makes "dirty" coal *dirty* is its high sulfur content. Burning dirty coal produces sulfur dioxide ($SO_2$), among other pollutants. Sulfur dioxide is oxidized in the atmosphere to form sulfur trioxide ($SO_3$), which subsequently combines with water to produce sulfuric acid—a major component of acid rain. Draw all possible resonance structures of sulfur trioxide.

**Strategy** Draw two or more Lewis structures for $SO_3$ in which the atoms are arranged the same way but the electrons are arranged differently.

**Setup** Following the steps for drawing Lewis structures, we determine that a correct Lewis structure for $SO_3$ contains two sulfur-oxygen single bonds and one sulfur-oxygen double bond.

$$:\ddot{O}:$$
$$|$$
$$:O=S-\ddot{O}:$$

But the double bond can be put in any one of three positions in the molecule.

**Solution**
$$\begin{array}{ccccc}
:\ddot{O}: & & \ddot{O} & & :\ddot{O}: \\
| & & || & & | \\
:O=S-\ddot{O}: & \longleftrightarrow & :\ddot{O}-S-\ddot{O}: & \longleftrightarrow & :\ddot{O}-S=\ddot{O}:
\end{array}$$

**Think About It** Always make sure that resonance structures differ only in the positions of the electrons, *not* in the positions of the atoms.

**Practice Problem A** Draw all possible resonance structures for the nitrate ion ($NO_3^-$).

**Practice Problem B** Draw three resonance structures for the thiocyanate ion ($NCS^-$), and determine the formal charges in each resonance structure. Based on formal charges, list the resonance structures in order of increasing relative importance.

## Checkpoint 8.7     Resonance

8.7.1 Indicate which of the following are resonance structures of $:\ddot{C}l—Be—\ddot{C}l:$ (select all that apply).

   a) $:Cl=Be=\ddot{C}l:$

   b) $:Cl\equiv Be—\ddot{C}l:$

   c) $:Be\equiv Cl—\ddot{C}l:$

   d) $:\ddot{C}l—Be\equiv Cl:$

   e) $:Be=Cl=\ddot{C}l:$

8.7.2 How many resonance structures can be drawn for the nitrite ion ($NO_2^-$)? (N and O must obey the octet rule.)

   a) 1

   b) 2

   c) 3

   d) 4

   e) 5

## 8.8    Exceptions to the Octet Rule

The octet rule almost always holds for second-period elements. Exceptions to the octet rule fall into three categories:

1. The central atom has fewer than eight electrons due to a shortage of electrons.
2. The central atom has fewer than eight electrons due to an odd number of electrons.
3. The central atom has more than eight electrons.

### Incomplete Octets

In some compounds the number of electrons surrounding the central atom in a stable molecule is fewer than eight. Beryllium, for example, which is the Group 2A element in the second period, has the electron configuration $[He]2s^2$. Thus, it has two valence electrons in the $2s$ orbital. In the gas phase, beryllium hydride ($BeH_2$) exists as discrete molecules. The Lewis structure of $BeH_2$ is

$$H-Be-H$$

Only four electrons surround the Be atom, so there is no way to satisfy the octet rule for beryllium in this molecule.

Elements in Group 3A also tend to form compounds in which they are surrounded by fewer than eight electrons. Boron, for example, has the electron configuration $[He]2s^2 2p^1$, so it has only three valence electrons. Boron reacts with the halogens to form a class of compounds having the general formula $BX_3$, where X is a halogen atom. Thus, there are only six electrons around the boron atom in boron trifluoride:

$$\ddot{\underset{\displaystyle :\ddot{F}:}{:\ddot{F}-B-\ddot{F}:}}$$

We actually *can* satisfy the octet rule for boron in $BF_3$ by using a lone pair on one of the F atoms to form a double bond between the F atom and boron. This gives rise to three additional resonance structures:

$$:\ddot{F}=B-\ddot{F}: \quad \longleftrightarrow \quad :\ddot{F}-B-\ddot{F}: \quad \longleftrightarrow \quad :\ddot{F}-B=\ddot{F}:$$

Although these resonance structures result in boron carrying a negative formal charge while fluorine carries a positive formal charge, a situation that is inconsistent with the electronegativities of the atoms involved, the experimentally determined bond length in $BF_3$ (130.9 pm) is shorter than a single bond (137.3 pm). The shorter bond length would appear to support the idea behind the three resonance structures.

On the other hand, boron trifluoride combines with ammonia in a reaction that is better represented using the Lewis structure in which boron has only six valence electrons around it:

It seems, then, that the properties of $BF_3$ are best explained by all four resonance structures.

The B—N bond in $F_3B-NH_3$ is different from the covalent bonds discussed so far in the sense that both electrons are contributed by the N atom. This type of bond is called a ***coordinate covalent bond*** (also referred to as a ***dative bond***), which is defined as a covalent bond in which one of the atoms donates both electrons. Although the properties of a coordinate covalent bond do not differ from those of a normal covalent bond (i.e., the electrons are shared in both cases), the distinction is useful for keeping track of valence electrons and assigning formal charges.

### Odd Numbers of Electrons

Some molecules, such as nitrogen dioxide ($NO_2$), contain an odd number of electrons.

$$:\ddot{O}=\dot{N}-\ddot{O}:$$

Because we need an even number of electrons for every atom in a molecule to have a complete octet, the octet rule cannot be obeyed for all the atoms in these molecules. Molecules with an odd number of electrons are sometimes referred to as *free radicals* (or just *radicals*). Many radicals are highly reactive, because there is a tendency for the unpaired electron to form a covalent bond with an unpaired electron on another molecule. When two nitrogen dioxide molecules collide, for example, they form dinitrogen tetroxide, a molecule in which the octet rule *is* satisfied for both the N and O atoms.

## Bringing Chemistry to Life

### The Power of Radicals

Beginning about a week after the September 11, 2001 attacks, letters containing anthrax bacteria were mailed to several news media offices and to two U.S. senators. Of the 22 people who subsequently contracted anthrax, five died. Anthrax is a spore-forming bacterium (*Bacillus anthracis*) and, like smallpox, is classified by the CDC as a *Category A bioterrorism agent*. Spore-forming bacteria are notoriously difficult to kill, making the cleanup of the buildings contaminated by anthrax costly and time-consuming. The American Media Inc. (AMI) building in Boca Raton, Florida, was not deemed safe to enter until July of 2004, after it had been treated with chlorine dioxide ($ClO_2$), the only structural fumigant approved by the Environmental Protection Agency (EPA) for anthrax decontamination. The effectiveness of $ClO_2$ in killing anthrax and other hardy biological agents stems in part from its being a *radical*, meaning that it contains an odd number of electrons.

Sample Problem 8.11 lets you practice drawing Lewis structures for species with odd numbers of electrons.

The American Media Inc. building in Boca Raton, Florida.

Severe flooding in New Orleans after Hurricane Katrina in 2005.

### Sample Problem 8.11

Draw the Lewis structure of chlorine dioxide ($ClO_2$).

**Strategy** The skeletal structure is

$$O-Cl-O$$

This puts the unique atom, Cl, in the center and puts the more electronegative O atoms in terminal positions.

**Setup** There are a total of 19 valence electrons (seven from the Cl and six from each of the two O atoms). We subtract four electrons to account for the two bonds in the skeleton, leaving us with 15 electrons to distribute as follows: three lone pairs on each O atom, one lone pair on the Cl atom, and the last remaining electron also on the Cl atom.

**Solution** $:\ddot{O}-\ddot{C}l-\ddot{O}:$

**Think About It** $ClO_2$ is used primarily to bleach wood pulp in the manufacture of paper, but it is also used to bleach flour, disinfect drinking water, and deodorize certain industrial facilities. Recently, it has been used to eradicate the toxic mold in homes in New Orleans that were damaged by the devastating floodwaters of Hurricane Katrina in 2005.

**Practice Problem A** Draw the Lewis structure for the OH species.

**Practice Problem B** Draw the Lewis structure for the $NS_2$ molecule.

**Student Annotation:** The OH species is a *radical*, not to be confused with the hydroxide *ion* ($OH^-$).

## Expanded Octets

Atoms of the second-period elements cannot have more than eight valence electrons around them, but atoms of elements in and beyond the third period of the periodic table can. In addition to the $3s$ and $3p$ orbitals, elements in the third period also have $3d$ orbitals that can be used in bonding. These orbitals enable an atom to form an *expanded octet*. One compound in which there is an expanded octet is sulfur hexafluoride, a very stable compound. The electron configuration of sulfur is $[Ne]3s^23p^4$. In $SF_6$, each of sulfur's 6 valence electrons forms a covalent bond with a fluorine atom, so there are 12 electrons around the central sulfur atom:

$$\ddot{\underset{\ddot{\,}}{F}}\underset{\overset{\displaystyle |}{\underset{\displaystyle |}{\overset{\displaystyle \ddot{F}}{\underset{\displaystyle \ddot{F}}{S}}}}}{\overset{\displaystyle \ddot{F}}{\underset{\displaystyle}{}}}\ddot{F}$$

In Chapter 9 we will see that these 12 electrons, or six bonding pairs, are accommodated in six orbitals that originate from the one $3s$, the three $3p$, and two of the five $3d$ orbitals. Sulfur also forms many compounds in which it does obey the octet rule. In sulfur dichloride, for instance, S is surrounded by only eight electrons:

$$:\ddot{C}l-\ddot{S}-\ddot{C}l:$$

Sample Problems 8.12 and 8.13 involve compounds that do not obey the octet rule.

---

## Sample Problem 8.12

Draw the Lewis structure of boron triiodide ($BI_3$).

**Strategy** Follow the step-by-step procedure for drawing Lewis structures. The skeletal structure is

$$\underset{\displaystyle I-B-I}{\overset{\displaystyle I}{\overset{\displaystyle |}{}}}$$

**Setup** There are a total of 24 valence electrons in $BI_3$ (three from the B and seven from each of the three I atoms). We subtract six electrons to account for the three bonds in the skeleton, leaving 18 electrons to distribute as three lone pairs on each I atom.

**Solution**

$$\underset{\displaystyle :\ddot{I}-B-\ddot{I}:}{\overset{\displaystyle :\ddot{I}:}{\overset{\displaystyle |}{}}}$$

**Think About It** Boron is one of the elements that does not always follow the octet rule. Like $BF_3$, however, $BI_3$ can be drawn with a double bond to satisfy the octet of boron. This gives rise to a total of four resonance structures:

$$\underset{:\ddot{I}-B-\ddot{I}:}{\overset{:\ddot{I}:}{|}} \longleftrightarrow \underset{\ddot{I}=B-\ddot{I}:}{\overset{:\ddot{I}:}{|}} \longleftrightarrow \underset{:\ddot{I}-B-\ddot{I}:}{\overset{\cdot\ddot{I}\cdot}{||}} \longleftrightarrow \underset{:\ddot{I}-B=\ddot{I}\cdot}{\overset{:\ddot{I}:}{|}}$$

**Practice Problem A** Draw the Lewis structure of beryllium fluoride ($BeF_2$).

**Practice Problem B** Draw the Lewis structure of boron trichloride ($BCl_3$).

---

**Student Annotation:** $BeF_2$ is also known as beryllium *di*fluoride, but because this is the only compound that beryllium forms with fluorine, the name *beryllium fluoride* is unambiguous.

## Sample Problem 8.13

Draw the Lewis structure of arsenic pentafluoride ($AsF_5$).

**Strategy** Follow the steps for drawing Lewis structures. The skeletal structure already has more than an octet around the As atom.

$$\underset{\displaystyle F}{\overset{\displaystyle F}{\underset{\displaystyle F}{\overset{\displaystyle |}{F-\underset{\displaystyle |}{As}-F}}}}$$

**Setup** There are 40 total valence electrons [five from As (Group 5A) and seven from each of the five F atoms (Group 7A)]. We subtract 10 electrons to account for the five bonds in the skeleton, leaving 30 to be distributed. Next, place three lone pairs on each F atom, thereby completing all their octets and using up all the electrons.

**Solution**

$$\underset{\displaystyle :\ddot{F}:}{\overset{\displaystyle :\ddot{F}:}{\underset{\displaystyle :\ddot{F}:}{\overset{\displaystyle |}{:\ddot{F}-\underset{\displaystyle |}{As}-\ddot{F}:}}}}$$

**Think About It** Always make sure that the number of electrons represented in your final Lewis structure matches the total number of valence electrons you are supposed to have.

**Practice Problem A** Draw the Lewis structure of phosphorus pentachloride ($PCl_5$).

**Practice Problem B** Draw the Lewis structure of antimony pentafluoride ($SbF_5$).

When drawing Lewis structures of compounds containing a central atom from the third period and beyond, the octet rule may be satisfied for all the atoms before all the valence electrons have been used up. When this happens, the extra electrons should be placed as lone pairs on the central atom. Sample Problem 8.14 illustrates this approach.

## Sample Problem 8.14

Draw the Lewis structure of xenon tetrafluoride ($XeF_4$).

**Strategy** Follow the steps for drawing Lewis structures. The skeletal structure is

$$
\begin{array}{c}
\text{F} \\
| \\
\text{F}-\text{Xe}-\text{F} \\
| \\
\text{F}
\end{array}
$$

**Setup** There are 36 total valence electrons (eight from Xe and seven from each of the four F atoms). We subtract eight electrons to account for the bonds in the skeleton, leaving 28 to distribute. We first complete the octets of all four F atoms. When this is done, four electrons remain, so we place two lone pairs on the Xe atom.

**Solution**

$$
\begin{array}{c}
:\!\ddot{\text{F}}\!: \\
| \\
:\!\ddot{\text{F}}\!-\!\ddot{\text{X}}\text{e}\!-\!\ddot{\text{F}}\!: \\
| \\
:\!\ddot{\text{F}}\!:
\end{array}
$$

**Think About It** Atoms beyond the second period can accommodate more than an octet of electrons, whether those electrons are used in bonds or reside on the atom as lone pairs.

**Practice Problem A** Draw the Lewis structure of the iodine tetrachloride ion ($ICl_4^-$).

**Practice Problem B** Draw the Lewis structure of krypton difluoride ($KrF_2$).

## Checkpoint 8.8    Exceptions to the Octet Rule

**8.8.1** In which of the following species does the central atom *not* obey the octet rule?

a) $ClO_2^-$

b) $CO_2$

c) $BrO_3^-$

d) HCN

e) $ICl_4^-$

**8.8.2** Which elements cannot have more than an octet of electrons? (Select all that apply.)

a) N

b) C

c) S

d) Br

e) O

**8.8.3** In which species does the central atom *obey* the octet rule? (Select all that apply.)

a) $I_3^-$

b) $BH_3$

c) $AsF_6^-$

d) $NO_2$

e) $ClO_2^-$

**8.8.4** How many lone pairs are there on the central atom in the Lewis structure of $ICl_2^-$?

a) 0

b) 1

c) 2

d) 3

e) 4

*You May Be Wondering...*

# Which Is More Important: Formal Charge or the Octet Rule?

We have learned that although central atoms from the third period (and beyond) often obey the octet rule, it is not necessary for them to do so. Rather, they can have what we call "expanded octets," meaning they are surrounded by more than eight electrons.

The sulfate ion ($SO_4^{2-}$), for example, can be represented by a resonance structure that obeys the octet rule (structure I), or by a structure that does not (structure II):

|  |  |
|---|---|
| Structure I | Structure II |

In structure I, although the octet rule is obeyed for the central atom, there are nonzero formal charges (shown in blue) on all the atoms. In structure II, by relocating a lone pair from each of two oxygen atoms and creating two double bonds, we change three of the formal charges to zero.

Which of these structures is better (or "more important") has been the subject of some debate among educators over the past two decades. Although some chemists may have a strong preference for one or the other, it is important for you to understand that *both* are valid Lewis structures and you should be able to draw both types of structures.

Sample Problem 8.15 gives you some practice drawing resonance structures for species that can obey the octet rule but also can be drawn with expanded octets to minimize formal charges.

## Sample Problem 8.15

Draw two resonance structures for sulfurous acid ($H_2SO_3$): one that obeys the octet rule for the central atom, and one that minimizes the formal charges. Determine the formal charge on each atom in both structures.

**Strategy** Begin by drawing the skeletal structure and counting the total number of valence electrons. Use the steps outlined in Section 8.5 to draw the first structure and reposition one or more lone pairs to adjust the formal charges for the second structure.

**Setup** The skeletal structure is

$$\text{H—O—}\overset{\displaystyle O}{\underset{\displaystyle |}{S}}\text{—O—H}$$

Note that each hydrogen in an oxoacid is attached to an oxygen atom and *not* directly to the central atom. The total number of valence electrons is 26 (six from S, six from each O, and one from each H).

**Solution** Following the procedure outlined on page 314, we get the first structure:

From the top O atom, which has three lone pairs, we reposition one lone pair to create a double bond between O and S to get the second structure:

Incorporating the double bond results in every atom having a formal charge of zero.

**Think About It** In some species, such as the sulfate ion, it is possible to incorporate too many double bonds. Structures with *three* and *four* double bonds to sulfur would give formal charges on S and O that are inconsistent with the electronegativities of these elements. In general, if you are trying to minimize formal charges by expanding the central atom's octet, only add enough double bonds to make the formal charge on the central atom zero.

**Practice Problem A** Draw three resonance structures for the hydrogen sulfite ion ($HSO_3^-$)—one that obeys the octet rule for the central atom, and two that expand the octet of the central atom. Calculate the formal charges on all atoms in each structure and determine which, if any, of the resonance structures has formal charges that are inconsistent with the elements' electronegativities.

**Practice Problem B** Draw two resonance structures for each species—one that obeys the octet rule, and one in which the formal charge on the central atom is zero: $PO_4^{3-}$, $HClO_3$, $SO_3$, $SO_2$.

| 8.9 | # Bond Enthalpy

One measure of the stability of a molecule is its **bond enthalpy**, which is the enthalpy change associated with breaking a particular bond in 1 mole of gaseous molecules. (Bond enthalpies in solids and liquids are affected by neighboring molecules.) The experimentally determined bond enthalpy of the diatomic hydrogen molecule, for example, is

$$H_2(g) \longrightarrow H(g) + H(g) \qquad \Delta H° = 436.4 \text{ kJ/mol}$$

According to this equation, breaking the covalent bonds in 1 mole of gaseous $H_2$ molecules requires 436.4 kJ of energy. For the less stable chlorine molecule,

$$Cl_2(g) \longrightarrow Cl(g) + Cl(g) \qquad \Delta H° = 242.7 \text{ kJ/mol}$$

Bond enthalpies can also be directly measured for heteronuclear diatomic molecules, such as HCl,

$$HCl(g) \longrightarrow H(g) + Cl(g) \qquad \Delta H° = 431.9 \text{ kJ/mol}$$

as well as for molecules containing multiple bonds:

$$O_2(g) \longrightarrow O(g) + O(g) \qquad \Delta H° = 498.7 \text{ kJ/mol}$$

$$N_2(g) \longrightarrow N(g) + N(g) \qquad \Delta H° = 941.4 \text{ kJ/mol}$$

Measuring the strength of covalent bonds in polyatomic molecules is more complicated. For example, measurements show that the energy needed to break the first O−H bond in $H_2O$ is different from that needed to break the second O−H bond:

$$H_2O(g) \longrightarrow H(g) + OH(g) \qquad \Delta H° = 502 \text{ kJ/mol}$$

$$OH(g) \longrightarrow H(g) + O(g) \qquad \Delta H° = 427 \text{ kJ/mol}$$

In each case, an O−H bond is broken, but the first step requires the input of more energy than the second. The difference between the two $\Delta H°$ values suggests that the second O−H bond itself undergoes change, because of the changes in its chemical environment.

We can now understand why the bond enthalpy of the same O−H bond in two different molecules, such as methanol ($CH_3OH$) and water ($H_2O$), will not be the same: their environments are different. For polyatomic molecules, therefore, we speak of the *average* bond enthalpy of a particular bond. For example, we can measure the enthalpy of the O−H bond in 10 different polyatomic molecules and obtain the average O−H bond enthalpy by dividing the sum of the bond enthalpies by 10. Table 8.6 lists the average bond enthalpies of a number of diatomic and polyatomic molecules. As we noted earlier, triple bonds are stronger than double bonds, and double bonds are stronger than single bonds.

A comparison of the thermochemical changes that take place during a number of reactions reveals a strikingly wide variation in the enthalpies of different reactions. For example, the combustion of hydrogen gas in oxygen gas is fairly *exothermic*:

$$H_2(g) + \tfrac{1}{2}O_2(g) \longrightarrow H_2O(l) \qquad \Delta H° = -285.8 \text{ kJ/mol}$$

The formation of glucose from carbon dioxide and water, on the other hand, best achieved by photosynthesis, is highly *endothermic*:

$$6CO_2(g) + 6H_2O(l) \longrightarrow C_6H_{12}O_6(s) + 6O_2(g) \qquad \Delta H° = 2801 \text{ kJ/mol}$$

We can account for such variations by looking at the stability of individual reactant and product molecules. After all, most chemical reactions involve the making and breaking of bonds. Therefore, knowing the bond enthalpies and hence the stability of molecules reveals something about the thermochemical nature of the reactions that molecules undergo.

In many cases, it is possible to predict the approximate enthalpy of a reaction by using the average bond enthalpies. Because energy is always required to break chemical bonds and chemical bond formation is always accompanied by a release of energy, we can estimate the enthalpy of a reaction by counting the total number of bonds broken and formed in the reaction and recording all the corresponding enthalpy changes. The enthalpy of reaction in the gas phase is given by

$$\Delta H° = \Sigma BE(\text{reactants}) - \Sigma BE(\text{products}) \qquad \text{Equation 8.4}$$

$$= \text{total energy } input \text{ (to } break \text{ bonds)} - \text{total energy } released \text{ (by bond } formation)$$

| TABLE 8.6 | Bond Enthalpies | | |
|---|---|---|---|
| **Bond** | **Bond Enthalpy (kJ/mol)** | **Bond** | **Bond Enthalpy (kJ/mol)** |
| H—H* | 436.4 | N—N | 193 |
| H—N | 393 | N=N | 418 |
| H—O | 460 | N≡N | 941.4 |
| H—S | 368 | N—O | 176 |
| H—P | 326 | N=O | 607 |
| H—F | 568.2 | N—F | 272 |
| H—Cl | 431.9 | N—Cl | 200 |
| H—Br | 366.1 | N—Br | 243 |
| H—I | 298.3 | N—I | 159 |
| C—H | 414 | O—O | 142 |
| C—C | 347 | O=O | 498.7 |
| C=C | 620 | O—P | 502 |
| C≡C | 812 | O=S | 469 |
| C—N | 276 | O—F | 190 |
| C=N | 615 | O—Cl | 203 |
| C≡N | 891 | O—Br | 234 |
| C—O | 351 | O—I | 234 |
| C=O† | 745 | P—P | 197 |
| C≡O | 1070 | P=P | 489 |
| C—P | 263 | S—S | 268 |
| C—S | 255 | S=S | 352 |
| C=S | 477 | F—F | 156.9 |
| C—F | 453 | Cl—Cl | 242.7 |
| C—Cl | 339 | Cl—F | 193 |
| C—Br | 276 | Br—Br | 192.5 |
| C—I | 216 | I—I | 151.0 |

**Student Annotation:** Bond enthalpies for diatomic molecules have more significant figures than those for polyatomic molecules. Those for polyatomic molecules are average values based on the bonds in more than one compound.

*Bond enthalpies shown in red are for diatomic molecules.

†The C=O bond enthalpy in $CO_2$ is 799 kJ/mol.

**Student Annotation:** To many students, Equation 8.4 appears to be backward. Ordinarily you calculate a change as *final minus initial*. Here we are determining the difference between the amount of heat we have to add to break reactant bonds and the amount of heat released when product bonds form. The sign of the final answer tells us if the process is endothermic (+) or exothermic (−) overall.

where BE stands for average bond enthalpy and $\Sigma$ is the summation sign. As written, Equation 8.4 takes care of the sign convention for $\Delta H°$. Thus, if the total energy input needed to break bonds in the reactants is less than the total energy released when bonds are formed in the products, then $\Delta H°$ is negative and the reaction is exothermic [Figure 8.9(a)]. On the other hand, if less energy is released (bond making) than absorbed (bond breaking), $\Delta H°$ is positive and the reaction is endothermic [Figure 8.9(b)].

If all the reactants and products are diatomic molecules, then the equation for the enthalpy of reaction will yield accurate results because the bond enthalpies of diatomic molecules are accurately known. If some or all of the reactants and products are polyatomic molecules, the equation will yield only approximate results because the bond enthalpies used will be averages.

Sample Problem 8.16 shows how to estimate enthalpies of reaction using bond enthalpies.

## Sample Problem 8.16

Use bond enthalpies from Table 8.6 to estimate the enthalpy of reaction for the combustion of methane:

$$CH_4(g) + 2O_2(g) \longrightarrow CO_2(g) + 2H_2O(l)$$

**Strategy** Draw Lewis structures to determine what bonds are to be broken and what bonds are to be formed.

**Setup**

Bonds to break: 4 C—H and 2 O=O.
Bonds to form: 2 C=O and 4 H—O.
Bond enthalpies from Table 8.6: 414 kJ/mol (C—H), 498.7 kJ/mol (O=O), 799 kJ/mol (C=O in $CO_2$), and 460 kJ/mol (H—O).

**Solution**

$$[4(414 \text{ kJ/mol}) + 2(498.7 \text{ kJ/mol})] - [2(799 \text{ kJ/mol}) + 4(460 \text{ kJ/mol})] = -785 \text{ kJ/mol}.$$

**Practice Problem A** Use bond enthalpies from Table 8.6 to estimate the enthalpy of reaction for the combination of carbon monoxide and oxygen to produce carbon dioxide:

$$2CO(g) + O_2(g) \longrightarrow 2CO_2(g)$$

**Practice Problem B** Using the following chemical equation, data from Table 8.6, and data from Appendix 2, determine the P—Cl bond enthalpy:

$$PH_3(g) + 3HCl(g) \longrightarrow PCl_3(g) + 3H_2(g)$$

**Student Annotation:** Don't skip the step of drawing Lewis structures. This is the only way to know for certain what types and numbers of bonds must be broken and formed.

**Think About It** Use Equation 5.19 [◀◀ Section 5.6] and data from Appendix 2 to calculate this enthalpy of reaction again; then compare your results using the two approaches. The difference in this case is due to two things: Most tabulated bond enthalpies are averages and, by convention, we show the product of combustion as liquid water—but average bond enthalpies apply to species in the gas phase, where there is little or no influence exerted by neighboring molecules.

**Student Annotation:** Remember that heats of reaction are expressed in kJ/mol, where the "per mole" refers to *per mole of reaction as written* [◀◀ Section 5.3].

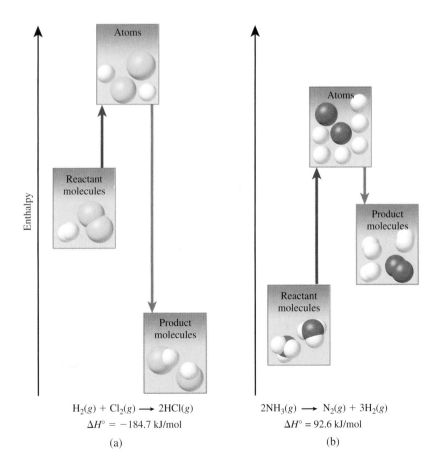

**Figure 8.9** Enthalpy changes in (a) an exothermic reaction and (b) an endothermic reaction. The $\Delta H°$ values are calculated using Equation 5.19 and tabulated $\Delta H_f°$ values from Appendix 2.

$$H_2(g) + Cl_2(g) \longrightarrow 2HCl(g)$$
$$\Delta H° = -184.7 \text{ kJ/mol}$$
(a)

$$2NH_3(g) \longrightarrow N_2(g) + 3H_2(g)$$
$$\Delta H° = 92.6 \text{ kJ/mol}$$
(b)

## Checkpoint 8.9    Bond Enthalpy

**8.9.1** Use data from Table 8.6 to estimate $\Delta H_{rxn}$ for the decomposition of hydrogen peroxide to form water and oxygen.

$$H_2O_2(l) \longrightarrow H_2O(l) + \tfrac{1}{2}O_2(g)$$

a) $-71$ kJ/mol

b) $+357$ kJ/mol

c) $-357$ kJ/mol

d) $-107$ kJ/mol

e) $+71$ kJ/mol

**8.9.2** Use data from Table 8.6 to estimate $\Delta H_{rxn}$ for the reaction of fluorine and chlorine to produce ClF.

$$F_2(g) + Cl_2(g) \longrightarrow 2ClF(g)$$

a) $-77.5$ kJ/mol

b) $-206.6$ kJ/mol

c) $206.6$ kJ/mol

d) $-13.6$ kJ/mol

e) $13.6$ kJ/mol

**8.9.3** Use bond enthalpies to determine $\Delta H_{rxn}$ for the reaction shown.

a) $-1139$ kJ/mol

b) $+1139$ kJ/mol

c) $-346$ kJ/mol

d) $+114$ kJ/mol

e) $-114$ kJ/mol

**8.9.4** Use bond enthalpies to determine $\Delta H_{rxn}$ for the reaction shown.

a) $-1028$ kJ/mol

b) $-200$ kJ/mol

c) $-392$ kJ/mol

d) $+392$ kJ/mol

e) $+200$ kJ/mol

# Applying What You've Learned

Researchers in the early 1990s made the sensational announcement that nitric oxide (NO), which had long been thought of only as a component of air pollution, turns out to play an important role in human physiology. They found that NO serves as a *signal* molecule, being produced *in vivo* and regulating a wide variety of cell functions in the body including in the cardiovascular, nervous, and immune systems.

The discovery of the biological role of nitric oxide has shed light on how nitroglycerin works as a drug. For many years, nitroglycerin tablets have been prescribed for cardiac patients to relieve the pain caused by brief interruptions in the flow of blood to the heart, although how it worked was not understood. We now know that nitroglycerin produces nitric oxide in the body, which causes muscles to relax and allows the arteries to dilate.

Research continues to uncover the role nitric oxide plays in biological processes, and medicine continues to find new uses for this molecule.

**Problems:**

a) Without consulting Figure 8.1, give the Lewis dot symbols for N and O. [◄◄ Sample Problem 8.1]

b) Classify the bond in NO as nonpolar, polar, or ionic. [◄◄ Sample Problem 8.4]

c) Given the experimentally determined dipole moment (0.16 D) and the bond length (1.15 Å), determine the magnitude of the partial charges in the NO molecule. [◄◄ Sample Problem 8.5]

d) Draw the Lewis structures for NO and for nitroglycerin ($C_3H_5N_3O_9$). [◄◄ Sample Problem 8.7]

e) Determine the formal charges on each atom in NO and in nitroglycerin. [◄◄ Sample Problem 8.8]

f) Nitroglycerin decomposes explosively to give carbon dioxide, water, nitrogen, and oxygen. Given the balanced equation for this reaction,

$$C_3H_5(NO_3)_3 \longrightarrow 3CO_2 + 2.5H_2O + 1.5N_2 + 0.25O_2$$

use bond enthalpies to estimate $\Delta H°$ for the reaction. [◄◄ Sample Problem 8.16]

# CHAPTER SUMMARY

## Section 8.1

* A *Lewis dot symbol* depicts an atom or an atomic ion of a main group element with dots (representing the valence electrons) arranged around the element's symbol. Main group atoms lose or gain one or more electrons to become isoelectronic with noble gases.

## Section 8.2

* The electrostatic attraction that holds ions together in an ionic compound is referred to as *ionic bonding.*
* *Lattice energy* is the amount of energy required to convert a mole of ionic solid to its constituent ions in the gas phase. Lattice energy cannot be measured directly, but is determined using the *Born-Haber cycle* and thermodynamic quantities that can be measured directly.

## Section 8.3

* According to the *Lewis theory of bonding, covalent bonding* results when atoms *share* valence electrons. The atoms in molecules and those in polyatomic ions are held together by *covalent bonds.*
* According to the *octet rule,* atoms will lose, gain, or share electrons to achieve a noble gas configuration. Pairs of valence electrons that are *not* involved in the covalent bonding in a molecule or polyatomic ion (i.e., valence electrons that are not shared) are called *lone pairs.*
* *Lewis structures* are drawn to represent molecules and polyatomic ions, showing the arrangement of atoms and the positions of all valence electrons. Lewis structures represent the shared pairs of valence electrons either as two dots, ··, or as a single dash, —. Any *unshared* electrons are represented as dots.
* One shared pair of electrons between atoms constitutes a *single bond. Multiple bonds* form between atoms that share more than one pair of electrons. Two shared pairs constitute a *double bond,* and three shared pairs constitute a *triple bond.*

## Section 8.4

* Bonds in which electrons are not shared equally are *polar* and are referred to as *polar covalent bonds.*
* *Electronegativity* is an atom's ability to draw shared electrons toward itself. Bonds between elements of widely different electronegativities ($\Delta \geq 2.0$) are *ionic.* Covalent bonds between atoms with significantly different electronegativities ($0.5 \leq \Delta < 2.0$) are *polar.* Bonds between atoms with very similar electronegativities ($\Delta < 0.5$) are *nonpolar.*
* *Percent ionic character* quantifies the polarity of a bond, and is determined by comparing the measured dipole moment to the one predicted by assuming that the bonded atoms have discrete charges.
* The *dipole moment ($\mu$)* is a quantitative measure of the polarity of a bond.

## Section 8.5

* Lewis structures of molecules or polyatomic ions can be drawn using the following step-by-step procedure:
    1. Use the molecular formula to draw the skeletal structure.
    2. Count the total number of valence electrons, adding electrons to account for a negative charge and subtracting electrons to account for a positive charge.
    3. Subtract two electrons for each bond in the skeletal structure.
    4. Distribute the remaining valence electrons to complete octets, completing the octets of the more electronegative atoms first.
    5. Place any remaining electrons on the central atom.
    6. Include double or triple bonds, if necessary, to complete the octets of all atoms.

## Section 8.6

* *Formal charge* is a way of keeping track of the valence electrons in a species. Formal charges should be consistent with electronegativities and can be used to determine the best arrangement of atoms and electrons for a Lewis structure.

## Section 8.7

* *Resonance structures* are two or more equally correct Lewis structures that differ in the positions of the electrons but *not* in the positions of the atoms. Different resonance structures of a compound can be separated by a resonance arrow, ⟷.

## Section 8.8

* In an ordinary covalent bond, each atom contributes one electron to the shared pair of electrons. In cases where just one of the atoms contributes *both* of the electrons, the bond is called a *coordinate covalent bond* or a *dative bond.*
* A species that contains an odd number of electrons is called a *free radical.*

## Section 8.9

* *Bond enthalpy* is the energy required to break 1 mole of a particular type of bond. Bond enthalpies are a measure of the stability of covalent bonds and can be used to estimate the enthalpy change for a reaction.

# KEY WORDS

Bond enthalpy,  325
Born-Haber cycle,  302
Coordinate covalent bond,  320
Covalent bond,  306
Covalent bonding,  306
Dative bond,  320
Dipole moment ($\mu$),  311

Double bond,  307
Electronegativity,  309
Formal charge,  315
Free radical,  321
Ionic bonding,  299
Lattice energy,  300
Lewis dot symbol,  298

Lewis structure,  306
Lewis theory of bonding,  306
Lone pair,  306
Multiple bond,  307
Nonpolar,  310
Octet rule,  306
Percent ionic character,  312

Polar,  308
Polar covalent bond,  308
Resonance structure,  318
Single bond,  307
Triple bond,  307

# KEY EQUATIONS

8.1    $\mu = Q \times r$

8.2    percent ionic character $= \dfrac{\mu \text{ (observed)}}{\mu \text{ (calculated assuming discrete charges)}} = 100\%$

8.3    formal charge = valence electrons − associated electrons

8.4    $\Delta H° = \Sigma BE(\text{reactants}) - \Sigma BE(\text{products})$

# QUESTIONS AND PROBLEMS

## Section 8.1: Lewis Dot Symbols

### Review Questions

8.1    What is a Lewis dot symbol? What elements do we generally represent with Lewis symbols?

8.2    Use the second member of each group from Group 1A to Group 7A to show that the number of valence electrons on an atom of the element is the same as its group number.

8.3    Without referring to Figure 8.1, write Lewis dot symbols for atoms of the following elements: (a) Be, (b) K, (c) Ca, (d) Ga, (e) O, (f) Br, (g) N, (h) I, (i) As, (j) F.

8.4    Write Lewis dot symbols for the following ions: (a) $Li^+$, (b) $Cl^-$, (c) $S^{2-}$, (d) $Sr^{2+}$, (e) $N^{3-}$.

8.5    Write Lewis dot symbols for the following atoms and ions: (a) I, (b) $I^-$, (c) S, (d) $S^{2-}$, (e) P, (f) $P^{3-}$, (g) Na, (h) $Na^+$, (i) Mg, (j) $Mg^{2+}$, (k) Al, (l) $As^{3+}$, (m) Pb, (n) $Pb^{2+}$.

## Section 8.2: Ionic Bonding

### Visualizing Chemistry
### Figure 8.3

VC 8.1    What additional information would you need to calculate the lattice energy for a compound if the charges on the cation and anion were +2 and −1, respectively, rather than +1 and −1?

a) No additional information is needed.
b) $IE_2$ of the cation.
c) $IE_2$ of the cation and $EA_2$ of the anion.

VC 8.2    What additional information would you need to calculate the lattice energy for a compound if the charges on the cation and anion were +2 and −2, respectively, rather than +1 and −1?

a) No additional information is needed.
b) $IE_2$ of the cation.
c) $IE_2$ of the cation and $EA_2$ of the anion.

VC 8.3    How would the magnitude of the lattice energy calculated using the Born-Haber cycle change if the charges on the cation and anion were +2 and −2, respectively, rather than +1 and −1?

a) Lattice energy would increase.
b) Lattice energy would decrease.
c) Whether lattice energy would increase or decrease depends on the relative magnitudes of $IE_2$ of the cation and $EA_2$ of the anion.

VC 8.4    What law enables us to use the Born-Haber cycle to calculate lattice energy?

a) Coulomb's law.
b) Hess's law.
c) Law of multiple proportions.

### Review Questions

8.6    Explain what *ionic bonding* is.

8.7    Explain how ionization energy and electron affinity determine whether atoms of elements will combine to form ionic compounds.

8.8    Name five metals and five nonmetals that are very likely to form ionic compounds. Write formulas for compounds that might result from the combination of these metals and nonmetals. Name these compounds.

8.9    Name one ionic compound that contains only nonmetallic elements.

8.10    Name one ionic compound that contains a polyatomic cation and a polyatomic anion (see Table 2.9).

8.11    Explain why ions with charges greater than ±3 are seldom found in ionic compounds.

8.12    The term *molar mass* was introduced in Chapter 3. Molar mass is numerically equivalent to molecular mass, although the units are different, for a covalent compound. What is the advantage of using the term molar mass when we discuss ionic compounds?

8.13    In which of the following states would NaCl be electrically conducting: (a) solid, (b) molten (that is, melted), (c) dissolved in water? Explain.

8.14    Beryllium forms a compound with chlorine that has the empirical formula $BeCl_2$. How would you determine whether it is an ionic compound? (The compound is not soluble in water.)

8.15    What is *lattice energy*, and what does it indicate about the stability of an ionic compound?

8.16    Explain how the lattice energy of an ionic compound such as KCl can be determined using the Born-Haber cycle. On what law is this procedure based?

8.17    Specify which compound in each of the following pairs of ionic compounds should have the higher lattice energy: (a) KCl or MgO, (b) LiF or LiBr, (c) $Mg_3N_2$ or NaCl. Explain your choice.

8.18    Specify which compound in each of the following pairs of ionic compounds should have the higher lattice energy: (a) AlN or CaO, (b) NaF or CsF, (c) $MgCl_2$ or $MgF_2$. Explain your choice.

### Problems

8.19    An ionic bond is formed between a cation $A^+$ and an anion $B^-$. Based on Coulomb's law

$$E \propto \frac{Q_1 \times Q_2}{d}$$

how would the energy of the ionic bond be affected by the following changes: (a) doubling the radius of $A^+$, (b) tripling the charge on $A^+$, (c) doubling the charges on $A^+$ and $B^-$, (d) decreasing the radii of $A^+$ and $B^-$ to half their original values?

**8.20** Give the empirical formulas and names of the compounds formed from the following pairs of ions: (a) $Rb^+$ and $I^-$, (b) $Cs^+$ and $SO_4^{2-}$, (c) $Sr^{2+}$ and $N^{3-}$, (d) $Al^{3+}$ and $S^{2-}$.

**8.21** Use Lewis dot symbols to show the transfer of electrons between the following atoms to form cations and anions: (a) Na and F, (b) K and S, (c) Ba and O, (d) Al and N.

**8.22** Write the Lewis dot symbols of the reactants and products in the following reactions. (First balance the equations.)

(a) $Sr + Se \longrightarrow SrSe$       (c) $Li + N_2 \longrightarrow Li_3N$
(b) $Ca + H_2 \longrightarrow CaH_2$       (d) $Al + S \longrightarrow Al_2S_3$

**8.23** Use the Born-Haber cycle outlined in Section 8.2 for NaCl to calculate the lattice energy of LiCl. Use data from Figures 7.8 and 7.10 and Appendix 2.

**8.24** Calculate the lattice energy of $CaCl_2$. Use data from Figures 7.8 and 7.10 and Appendix 2. (The second ionization energy of Ca, $IE_2$, is 1145 kJ/mol.)

## Section 8.3: Covalent Bonding

### Review Questions

**8.25** Describe Lewis's contribution to our understanding of the covalent bond.

**8.26** Use an example to illustrate each of the following terms: *lone pair, Lewis structure, octet rule, bond length.*

**8.27** What is the difference between a Lewis *symbol* and a Lewis *structure?*

**8.28** How many lone pairs are on the underlined atoms in these compounds: H<u>Br</u>, H<u>2</u>S, <u>C</u>H<u>4</u>?

**8.29** Compare single, double, and triple bonds in a molecule, and give an example of each. For the same bonding atoms, how does the bond length change from single bond to triple bond?

**8.30** Compare the properties of ionic compounds and covalent compounds.

**8.31** Summarize the essential features of the Lewis octet rule. The octet rule applies mainly to the second period elements. Explain.

### Problems

**8.32** For each of the following pairs of elements, state whether the binary compound they form is likely to be ionic or covalent. Write the empirical formula and name of the compound: (a) I and Cl, (b) Mg and F.

**8.33** For each of the following pairs of elements, state whether the binary compound they form is likely to be ionic or covalent. Write the empirical formula and name of the compound: (a) B and F, (b) K and Br.

## Section 8.4: Electronegativity and Polarity

### Review Questions

**8.34** Define *electronegativity,* and explain the difference between electronegativity and electron affinity. Describe in general how the electronegativities of the elements change according to their position in the periodic table.

**8.35** What is a *polar covalent bond?* Name two compounds that contain one or more polar covalent bonds.

### Problems

**8.36** List the following bonds in order of increasing ionic character: the lithium-to-fluorine bond in LiF, the potassium-to-oxygen bond in $K_2O$, the nitrogen-to-nitrogen bond in $N_2$, the sulfur-to-oxygen bond in $SO_2$, the chlorine-to-fluorine bond in $ClF_3$.

**8.37** Arrange the following bonds in order of increasing ionic character: carbon to hydrogen, fluorine to hydrogen, bromine to hydrogen, sodium to chlorine, potassium to fluorine, lithium to chlorine.

**8.38** Four atoms are arbitrarily labeled D, E, F, and G. Their electronegativities are as follows: D = 3.8, E = 3.3, F = 2.8, and G = 1.3. If the atoms of these elements form the molecules DE, DG, EG, and DF, how would you arrange these molecules in order of increasing covalent bond character?

**8.39** List the following bonds in order of increasing ionic character: cesium to fluorine, chlorine to chlorine, bromine to chlorine, silicon to carbon.

**8.40** Classify the following bonds as covalent, polar covalent, or ionic, and explain: (a) the CC bond in $H_3CCH_3$, (b) the KI bond in KI, (c) the NB bond in $H_3NBCl_3$, (d) the CF bond in $CF_4$.

**8.41** Classify the following bonds as covalent, polar covalent, or ionic, and explain: (a) the SiSi bond in $Cl_3SiSiCl_3$, (b) the SiCl bond in $Cl_3SiSiCl_3$, (c) the CaF bond in $CaF_2$, (d) the NH bond in $NH_3$.

## Section 8.5: Drawing Lewis Structures

### Problems

**8.42** Draw Lewis structures for the following molecules and ions: (a) $NCl_3$, (b) OCS, (c) $H_2O_2$, (d) $CH_3COO^-$, (e) $CN^-$, (f) $CH_3CH_2NH_3^+$.

**8.43** Draw Lewis structures for the following molecules and ions: (a) $OF_2$, (b) $N_2F_2$, (c) $Si_2H_6$, (d) $OH^-$, (e) $CH_2ClCOO^-$, (f) $CH_3NH_3^+$.

**8.44** Draw Lewis structures for the following molecules: (a) ICl, (b) $PH_3$, (c) $P_4$ (each P is bonded to three other P atoms), (d) $H_2S$, (e) $N_2H_4$, (f) $HClO_3$.

**8.45** Draw Lewis structures for the following molecules: (a) $COBr_2$ (C is bonded to O and Br atoms), (b) $H_2Se$, (c) $NH_2OH$, (d) $CH_3NH_2$, (e) $CH_3CH_2Br$, (f) $NCl_3$.

## Section 8.6: Lewis Structures and Formal Charge

### Review Questions

**8.46** Explain the concept of *formal charge.* Do formal charges represent an actual separation of charges?

### Problems

**8.47** Draw Lewis structures for the following ions: (a) $NO_2^+$, (b) $SCN^-$, (c) $S_2^{2-}$, (d) $ClF_2^+$. Show formal charges.

**8.48** Draw Lewis structures for the following ions: (a) $O_2^{2-}$, (b) $C_2^{2-}$, (c) $NO^+$, (d) $NH_4^+$. Show formal charges.

**8.49** The skeletal structure of acetic acid shown here is correct, but some of the bonds are wrong. (a) Identify the incorrect bonds and explain what is wrong with them. (b) Write the correct Lewis structure for acetic acid.

$$H=\overset{\overset{\displaystyle H}{|}}{\underset{\underset{\displaystyle H}{|}}{C}}-\overset{\overset{\displaystyle :\ddot{O}:}{|}}{C}-\ddot{O}-H$$

8.50 The following Lewis structures are incorrect. Explain what is wrong with each one, and give a correct Lewis structure for the molecule. (Relative positions of atoms are shown correctly.)

(a) $H-\ddot{C}=\ddot{N}$  (d) $\ddot{:}F\diagdown\underset{:F:}{B}\diagup F\ddot{:}$  (g) $\ddot{:}F\diagdown\underset{:F:}{N}\diagup F\ddot{:}$

(b) $H=C=C=H$  (e) $H-\ddot{O}=\ddot{F}:$

(c) $\ddot{O}-Sn-\ddot{O}$  (f) $\underset{\cdot\ddot{O}\cdot}{\overset{H}{\underset{|}{\ddot{C}}}}-\ddot{F}:$

## Section 8.7: Resonance

### Review Questions

8.51 What is a *resonance structure*? Is it possible to isolate one resonance structure of a compound for analysis? Explain.

8.52 What are the rules for writing resonance structures?

### Problems

**8.53** Draw Lewis structures for the following species, including all resonance forms, and show formal charges: (a) $HCO_2^-$, (b) $CH_2NO_2^-$. The relative positions of the atoms are as follows:

$$\begin{array}{ccccc} & O & & H & O \\ H & C & & & C & N \\ & O & & H & O \end{array}$$

8.54 Draw three resonance structures for the chlorate ion ($ClO_3^-$). Show formal charges.

8.55 Draw three resonance structures for hydrazoic acid ($HN_3$). The atomic arrangement is HNNN. Show formal charges.

8.56 Draw two resonance structures for diazomethane ($CH_2N_2$). Show formal charges. The skeletal structure of the molecule is

$$\begin{array}{c} H \\ \phantom{xx}C\phantom{x}N\phantom{x}N \\ H \end{array}$$

**8.57** Draw three reasonable resonance structures for the $OCN^-$ ion. Show formal charges and rank the importance of the structures.

8.58 Draw three resonance structures for the molecule $N_2O$ in which the atoms are arranged in the order NNO. Indicate formal charges and arrange the resonance structures in order of increasing relative importance.

**8.59** Draw a resonance structure of the adenine molecule shown here, which is part of the DNA structure. Show all the lone pairs and label the formal charges.

## Section 8.8: Exceptions to the Octet Rule

### Review Questions

8.60 Why does the octet rule not hold for many compounds containing elements in the third period of the periodic table and beyond?

8.61 Give three examples of compounds that do not satisfy the octet rule. Write a Lewis structure for each.

8.62 Because fluorine has seven valence electrons ($2s^2 2p^5$), seven covalent bonds in principle could form around the atom. Such a compound might be $FH_7$ or $FCl_7$. These compounds have never been prepared. Why?

8.63 What is a *coordinate covalent bond?* Is it different from an ordinary covalent bond?

8.64 Because the central atom in each case is from the same group of the periodic table, the Lewis structures we draw for $SO_2$ and $O_3$ are essentially the same. Explain why we can draw a resonance structure for $SO_2$ in which the formal charge on the central atom is zero, but we cannot do this for $O_3$.

8.65 What is the advantage of drawing the Lewis structures of oxoanions and oxoacids using expanded octets?

### Problems

8.66 The $AlI_3$ molecule has an incomplete octet around Al. Draw three resonance structures of the molecule in which the octet rule is satisfied for both the Al and the I atoms. Show formal charges.

**8.67** In the vapor phase, beryllium chloride consists of discrete $BeCl_2$ molecules. Is the octet rule satisfied for Be in this compound? If not, can you form an octet around Be by drawing another resonance structure? How plausible is this structure?

8.68 Of the noble gases, only Kr, Xe, and Rn are known to form a few compounds with O and/or F. Write Lewis structures for the following molecules: (a) $XeF_2$, (b) $XeF_4$, (c) $XeF_6$, (d) $XeOF_4$, (e) $XeO_2F_2$. In each case Xe is the central atom.

**8.69** Write a Lewis structure for $SbCl_5$. Does this molecule obey the octet rule?

8.70 Write Lewis structures for $SeF_4$ and $SeF_6$. Is the octet rule satisfied for Se?

8.71 Write Lewis structures for the reaction
$$AlCl_3 + Cl^- \longrightarrow AlCl_4^-$$
What kind of bond joins Al and Cl in the product?

8.72 Draw two resonance structures for the bromate ion ($BrO_3^-$), one that obeys the octet rule and one in which the formal charge on the central atom is zero.

**8.73** Draw two resonance structures for the sulfite ion ($SO_3^{2-}$), one that obeys the octet rule and one in which the formal charge on the central atom is zero.

## Section 8.9: Bond Enthalpy

### Review Questions

8.74 What is *bond enthalpy?* Bond enthalpies of polyatomic molecules are average values, whereas those of diatomic molecules can be accurately determined. Why?

8.75 Explain why the bond enthalpy of a molecule is usually defined in terms of a gas-phase reaction. Why are bond-breaking processes always endothermic and bond-forming processes always exothermic?

## Problems

8.76 From the following data, calculate the average bond enthalpy for the N—H bond:

$$NH_3(g) \longrightarrow NH_2(g) + H(g) \qquad \Delta H° = 435 \text{ kJ/mol}$$
$$NH_2(g) \longrightarrow NH(g) + H(g) \qquad \Delta H° = 381 \text{ kJ/mol}$$
$$NH(g) \longrightarrow N(g) + H(g) \qquad \Delta H° = 360 \text{ kJ/mol}$$

8.77 For the reaction

$$O(g) + O_2(g) \longrightarrow O_3(g) \qquad \Delta H° = -107.2 \text{ kJ/mol}$$

Calculate the average bond enthalpy in $O_3$.

8.78 The bond enthalpy of $F_2(g)$ is 156.9 kJ/mol. Calculate $\Delta H_f°$ for $F(g)$.

8.79 For the reaction

$$2C_2H_6(g) + 7O_2(g) \longrightarrow 4CO_2(g) + 6H_2O(g)$$

(a) Predict the enthalpy of reaction from the average bond enthalpies in Table 8.6. (b) Calculate the enthalpy of reaction from the standard enthalpies of formation (see Appendix 2) of the reactant and product molecules, and compare the result with your answer for part (a).

8.80 Use average bond enthalpies from Table 8.6 to estimate $\Delta H_{rxn}$ for the following reaction.

8.81 Use average bond enthalpies from Table 8.6 to estimate $\Delta H_{rxn}$ for the following reaction.

## Additional Problems

8.82 Classify the following substances as ionic compounds or covalent compounds containing discrete molecules: $CH_4$, KF, CO, $SiCl_4$, $BaCl_2$.

8.83 Which of the following are ionic compounds and which are covalent compounds: RbCl, $PF_5$, $BrF_3$, $KO_2$, $CI_4$?

8.84 Match each of the following energy changes with one of the processes given: ionization energy, electron affinity, bond enthalpy, and standard enthalpy of formation.

(a) $F(g) + e^- \longrightarrow F^-(g)$     (c) $Na(g) \longrightarrow Na^+(g) + e^-$
(b) $F_2(g) \longrightarrow 2F(g)$     (d) $Na(s) + \frac{1}{2}F_2(g) \longrightarrow NaF(s)$

8.85 The formulas for the fluorides of the third-period elements are NaF, $MgF_2$, $AlF_3$, $SiF_4$, $PF_5$, $SF_6$, and $ClF_3$. Classify these compounds as covalent or ionic.

8.86 Use ionization energy (see Figure 7.8) and electron affinity (see Figure 7.10) values to calculate the energy change (in kJ/mol) for the following reactions:

(a) $Li(g) + I(g) \longrightarrow Li^+(g) + I^-(g)$
(b) $Na(g) + F(g) \longrightarrow Na^+(g) + F^-(g)$
(c) $K(g) + Cl(g) \longrightarrow K^+(g) + Cl^-(g)$

8.87 Describe some characteristics of an ionic compound such as KF that would distinguish it from a covalent compound such as benzene ($C_6H_6$).

8.88 Write Lewis structures for $BrF_3$, $ClF_5$, and $IF_7$. Identify those in which the octet rule is not obeyed.

8.89 Write three reasonable resonance structures for the azide ion $N_3^-$ in which the atoms are arranged as NNN. Show formal charges.

8.90 The amide group plays an important role in determining the structure of proteins:

$$-\overset{..}{\underset{|}{N}}-\overset{\overset{\displaystyle ..\!\!:\ddot{O}}{\|}}{C}-$$
$$\phantom{-----}H$$

Draw another resonance structure for this group. Show formal charges.

8.91 Give an example of an ion or molecule containing Al that (a) obeys the octet rule, (b) has an expanded octet, and (c) has an incomplete octet.

8.92 Draw four reasonable resonance structures for the $PO_3F^{2-}$ ion. The central P atom is bonded to the three O atoms and to the F atom. Show formal charges.

8.93 Attempts to prepare the compounds $CF_2$, $LiO_2$, $CsCl_2$, and $PI_5$ as stable species under atmospheric conditions have failed. Suggest possible reasons for the failure.

8.94 Draw reasonable resonance structures for the following ions: (a) $HSO_4^-$, (b) $PO_4^{3-}$, (c) $HPO_3^{2-}$, (d) $IO_3^-$.

8.95 Are the following statements true or false? (a) Formal charges represent an actual separation of charges. (b) $\Delta H_{rxn}°$ can be estimated from the bond enthalpies of reactants and products. (c) All second-period elements obey the octet rule in their compounds. (d) The resonance structures of a molecule can be separated from one another in the laboratory.

8.96 A rule for drawing plausible Lewis structures is that the central atom is generally less electronegative than the surrounding atoms. Explain why this is so.

8.97 Using the following information and the fact that the average C—H bond enthalpy is 414 kJ/mol, estimate the standard enthalpy of formation of methane ($CH_4$).

$$C(s) \longrightarrow C(g) \qquad \Delta H_{rxn}° = 716 \text{ kJ/mol}$$
$$2H_2(g) \longrightarrow 4H(g) \qquad \Delta H_{rxn}° = 872.8 \text{ kJ/mol}$$

8.98 Based on changes in enthalpy, which of the following reactions will occur more readily?

(a) $Cl(g) + CH_4(g) \longrightarrow CH_3Cl(g) + H(g)$
(b) $Cl(g) + CH_4(g) \longrightarrow CH_3(g) + HCl(g)$

8.99 Which of the following molecules has the shortest nitrogen-to-nitrogen bond: $N_2H_4$, $N_2O$, $N_2$, $N_2O_4$? Explain.

8.100 Most organic acids can be represented as RCOOH, where COOH is the carboxyl group and R is the rest of the molecule. [For example, R is $CH_3$ in acetic acid ($CH_3COOH$).] (a) Draw a Lewis structure for the carboxyl group. (b) Upon ionization, the carboxyl group is converted to the carboxylate group (COO$^-$). Draw resonance structures for the carboxylate group.

**8.101** Which of the following species are isoelectronic: $NH_4^+$, $C_6H_6$, $CO$, $CH_4$, $N_2$, $B_3N_3H_6$?

8.102 The following is a simplified (skeletal) structure of the amino acid histidine. Draw a complete Lewis structure of the molecule.

**8.103** The following is a simplified (skeletal) structure of the amino acid tryptophan. Draw a complete Lewis structure of the molecule.

8.104 The following species have been detected in interstellar space: (a) CH, (b) OH, (c) $C_2$, (d) HNC, (e) HCO. Draw Lewis structures for these species.

**8.105** The amide ion ($NH_2^-$) is a Brønsted base. Use Lewis structures to represent the reaction between the amide ion and water.

8.106 Draw Lewis structures for the following organic molecules: (a) tetrafluoroethylene ($C_2F_4$), (b) propane ($C_3H_8$), (c) butadiene ($CH_2CHCHCH_2$), (d) propyne ($CH_3CCH$), (e) benzoic acid ($C_6H_5COOH$). (To draw $C_6H_5COOH$, replace an H atom in benzene with a COOH group.)

**8.107** The triiodide ion ($I_3^-$) in which the I atoms are arranged in a straight line is stable, but the corresponding $F_3^-$ ion does not exist. Explain.

8.108 Compare the bond enthalpy of $F_2$ with the overall energy change for the following process:

$$F_2(g) \longrightarrow F^+(g) + F^-(g)$$

Which is the preferred dissociation for $F_2$, energetically speaking?

8.109 Methyl isocyanate ($CH_3NCO$) is used to make certain pesticides. In December 1984, water leaked into a tank containing this substance at a chemical plant, producing a toxic cloud that killed thousands of people in Bhopal, India. Draw Lewis structures for $CH_3NCO$, showing formal charges.

8.110 The chlorine nitrate ($ClONO_2$) molecule is believed to be involved in the destruction of ozone in the Antarctic stratosphere. Draw a plausible Lewis structure for this molecule.

**8.111** Several resonance structures for the molecule $CO_2$ are shown here. Explain why some of them are likely to be of little importance in describing the bonding in this molecule.

(a) $:\!\ddot{O}\!=\!C\!=\!\ddot{O}\!:$    (c) $:\!O\!\equiv\!\ddot{C}\!-\!\ddot{O}\!:$

(b) $:\!O\!\equiv\!C\!-\!\ddot{O}\!:$    (d) $:\!\ddot{O}\!-\!C\!-\!\ddot{O}\!:$

8.112 For each of the following organic molecules draw a Lewis structure in which the carbon atoms are bonded to each other by single bonds: (a) $C_2H_6$, (b) $C_4H_{10}$, (c) $C_5H_{12}$. For parts (b) and (c), show only structures in which each C atom is bonded to no more than two other C atoms.

8.113 Draw Lewis structures for the following chlorofluorocarbons (CFCs), which are partly responsible for the depletion of ozone in the stratosphere: (a) $CFCl_3$, (b) $CF_2Cl_2$, (c) $CHF_2Cl$, (d) $CF_3CHF_2$.

8.114 Draw Lewis structures for the following organic molecules: $C_2H_3F$, $C_3H_6$, $C_4H_8$. In each there is one $C=C$ bond, and the rest of the carbon atoms are joined by $C-C$ bonds.

8.115 Calculate $\Delta H°$ for the reaction

$$H_2(g) + I_2(g) \longrightarrow 2HI(g)$$

using (a) Equation 8.4 and (b) Equation 5.19, given that $\Delta H_f°$ for $I_2(g)$ is 61.0 kJ/mol.

8.116 Draw Lewis structures for the following organic molecules: (a) methanol ($CH_3OH$); (b) ethanol ($CH_3CH_2OH$); (c) tetraethyl lead [$Pb(CH_2CH_3)_4$], which is used in "leaded gasoline"; (d) methylamine ($CH_3NH_2$), which is used in tanning; (e) mustard gas ($ClCH_2CH_2SCH_2CH_2Cl$), a poisonous gas used in World War I; (f) urea [$(NH_2)_2CO$], a fertilizer; and (g) glycine ($NH_2CH_2COOH$), an amino acid.

8.117 Write Lewis structures for the following four isoelectronic species: (a) CO, (b) $NO^+$, (c) $CN^-$, (d) $N_2$. Show formal charges.

8.118 Oxygen forms three types of ionic compounds in which the anions are oxide ($O^{2-}$), peroxide ($O_2^{2-}$), and superoxide ($O_2^-$). Draw Lewis structures of these ions.

8.119 Comment on the correctness of the statement, "All compounds containing a noble gas atom violate the octet rule."

8.120 Write three resonance structures for (a) the cyanate ion ($NCO^-$) and (b) the isocyanate ion ($CNO^-$). In each case, rank the resonance structures in order of increasing importance.

8.121 (a) From the following data calculate the bond enthalpy of the $F_2^-$ ion.

| | |
|---|---|
| $F_2(g) \longrightarrow 2F(g)$ | $\Delta H°_{rxn} = 156.9$ kJ/mol |
| $F^-(g) \longrightarrow F(g) + e^-$ | $\Delta H°_{rxn} = 333$ kJ/mol |
| $F_2^-(g) \longrightarrow F_2(g) + e^-$ | $\Delta H°_{rxn} = 290$ kJ/mol |

(b) Explain the difference between the bond enthalpies of $F_2$ and $F_2^-$.

8.122 The resonance concept is sometimes described by analogy to a mule, which is a cross between a horse and a donkey. Compare this analogy with the one used in this chapter, that is, the description of a rhinoceros as a cross between a griffin and a unicorn. Which description is more appropriate? Why?

8.123 The $N-O$ bond distance in nitric oxide is 115 pm, which is intermediate between a triple bond (106 pm) and a double bond (120 pm). (a) Draw two resonance structures for NO, and comment on their relative importance. (b) Is it possible to draw a resonance structure having a triple bond between the atoms?

8.124 Although nitrogen dioxide ($NO_2$) is a stable compound, there is a tendency for two such molecules to combine to form dinitrogen tetroxide ($N_2O_4$). Why? Draw four resonance structures of $N_2O_4$, showing formal charges.

8.125 In the gas phase, aluminum chloride exists as a dimer (a unit of two) with the formula $Al_2Cl_6$. Its skeletal structure is given by

Complete the Lewis structure and indicate the coordinate covalent bonds in the molecule. Does this dimer possess a dipole moment? Explain.

8.126  Draw a Lewis structure for nitrogen pentoxide ($N_2O_5$) in which each N is bonded to three O atoms.

8.127  The hydroxyl radical (OH) plays an important role in atmospheric chemistry. It is highly reactive and has a tendency to combine with an H atom from other compounds, causing them to break up. Thus OH is sometimes called a "detergent" radical because it helps to clean up the atmosphere. (a) Draw the Lewis structure for the radical. (b) Refer to Table 8.6 and explain why the radical has a high affinity for H atoms. (c) Estimate the enthalpy change for the following reaction:

$$OH(g) + CH_4(g) \longrightarrow CH_3(g) + H_2O(g)$$

(d) The radical is generated when sunlight hits water vapor. Calculate the maximum wavelength (in nm) required to break an O—H bond in $H_2O$.

8.128  Vinyl chloride ($C_2H_3Cl$) differs from ethylene ($C_2H_4$) in that one of the H atoms is replaced with a Cl atom. Vinyl chloride is used to prepare poly(vinyl chloride), which is an important polymer used in pipes. (a) Draw the Lewis structure of vinyl chloride. (b) The repeating unit in poly(vinyl chloride) is $-CH_2-CHCl-$. Draw a portion of the molecule showing three such repeating units. (c) Calculate the enthalpy change when $1.0 \times 10^3$ kg of vinyl chloride forms poly(vinyl chloride).

8.129  Experiments show that it takes 1656 kJ/mol to break all the bonds in methane ($CH_4$) and 4006 kJ/mol to break all the bonds in propane ($C_3H_8$). Based on these data, calculate the average bond enthalpy of the C—C bond.

8.130  In 1998 scientists using a special type of electron microscope were able to measure the force needed to break a *single* chemical bond. If $2.0 \times 10^{-9}$ N was needed to break a C—Si bond, estimate the bond enthalpy in kJ/mol. Assume that the bond has to be stretched by a distance of 2 Å ($2 \times 10^{-10}$ m) before it is broken.

8.131  The American chemist Robert S. Mulliken suggested a different definition for the electronegativity (EN) of an element, given by

$$EN = \frac{IE_1 + EA}{2}$$

where $IE_1$ is the first ionization energy and $EA$ is the electron affinity of the element. Calculate the electronegativities of O, F, and Cl using the preceding equation. Compare the electronegativities of these elements on the Mulliken and Pauling scales. (To convert to the Pauling scale, divide each EN value by 230 kJ/mol.)

8.132  Among the common inhaled anesthetics are:

Halothane ($CF_3CHClBr$)          Isoflurane ($CF_3CHClOCHF_2$)
Enflurane ($CHFClCF_2OCHF_2$)   Methoxyflurane ($CHCl_2CF_2OCH_3$)

Draw Lewis structures of these molecules.

8.133  A student in your class claims that magnesium oxide actually consists of $Mg^+$ and $O^-$ ions, not $Mg^{2+}$ and $O^{2-}$ ions. Suggest some experiments one could do to show that your classmate is wrong.

8.134  From the lattice energy of KCl in Table 8.6, and the ionization energy of K and electron affinity on Cl in Figures 7.8 and 7.10, respectively, calculate the $\Delta H°$ for the reaction

$$K(g) + Cl(g) \longrightarrow KCl(s)$$

8.135  The species $H_3^+$ is the simplest polyatomic ion. The geometry of the ion is that of an equilateral triangle. (a) Draw three resonance structures to represent the ion. (b) Given the following information

$$2H + H^+ \longrightarrow H_3^+ \qquad \Delta H° = -849 \text{ kJ/mol}$$
$$H_2 \longrightarrow 2H \qquad \Delta H° = 436.4 \text{ kJ/mol}$$

calculate $\Delta H°$ for the reaction

$$H^+ + H_2 \longrightarrow H_3^+$$

8.136  The bond enthalpy of the C—N bond in the amide group of proteins (see Problem 8.90) can be treated as an average of C—N and C=N bonds. Calculate the maximum wavelength of light needed to break the bond.

8.137  In 1999 an unusual cation containing only nitrogen ($N_5^+$) was prepared. Draw three resonance structures of the ion, showing formal charges. (*Hint:* The N atoms are joined in a linear fashion.)

8.138  Write the formulas of the binary hydrides for the second-period elements from Li to F. Identify the bonding in each as covalent, polar covalent, or ionic.

8.139  Using Table 8.6, compare the following bond enthalpies: C—C in $C_2H_6$, N—N in $N_2H_4$, and O—O in $H_2O_2$. What effect do lone pairs on adjacent atoms appear to have on bond enthalpy?

8.140  Do a Web search of the following ionic compounds and give brief descriptions of their medical uses: $AgNO_3$, $BaSO_4$, $CaSO_4$, KI, $Li_2CO_3$, $Mg(OH)_2$, $NaHCO_3$, NaF, $TiO_2$, ZnO.

# PRE-PROFESSIONAL PRACTICE EXAM PROBLEMS: PHYSICAL AND BIOLOGICAL SCIENCES

Nitrous oxide ($N_2O$) is an anesthetic commonly used for dental procedures. Because of the euphoria caused by inhaling it, $N_2O$ is commonly known as "laughing gas." It is licensed for use as a food additive and as an aerosol propellant. It is used to displace air from potato chip bags to extend shelf life and as the propellant in whipped cream canisters. In recent years $N_2O$ has become popular as a recreational drug, due in part to its ready availability to consumers. Although $N_2O$ is legal, it is regulated by the FDA; its sale and distribution for the purpose of human consumption are not permitted.

| Bond | Bond Enthalpy (kJ/mol) |
|------|------------------------|
| N—O | 176 |
| N=O | 607 |
| N—N | 193 |
| N=N | 418 |
| N≡N | 941.4 |
| O=O | 498.7 |

1. Which of the following Lewis structures are possible for $N_2O$?

   :N≡N—Ö:    :N̈=N=Ö:    :N̈=O=N̈:    :N̈—O—N̈:

      I            II           III         IV

   a) I only
   b) I and II
   c) I, II, and III
   d) I, II, III, and IV

2. Use formal charges to choose the best of the resonance structures shown.

   a) I
   b) II
   c) III
   d) IV

3. Using the best resonance structure and the average bond enthalpies given, determine the $\Delta H_f^\circ$ for $N_2O$.

   a) 73 kJ/mol
   b) −73 kJ/mol
   c) 166 kJ/mol
   d) −166 kJ/mol

4. Why does the calculated $\Delta H_f^\circ$ value differ from the tabulated value of 81.56 kJ/mol?

   a) The tabulated value is wrong.
   b) None of the resonance structures depicts the bonds realistically.
   c) To a reasonable number of significant figures, the calculated and tabulated values are the same.
   d) Using bond enthalpies gives only an estimate of $\Delta H_{rxn}^\circ$.

# ANSWERS TO IN-CHAPTER MATERIALS

## Practice Problems

**8.1A** (a) $Ca^{2+}$, (b) $\left[:\ddot{N}:\right]^{3-}$, (c) $\left[:\ddot{I}:\right]^{-}$. **8.1B** (a) 2−, (b) +, (c) 3−.
**8.2A** $MgCl_2$. **8.2B** NaF < MgO < AlN. **8.3A** 629 kJ/mol.
**8.3B** −841 kJ/mol. **8.4A** (a) nonpolar, (b) polar, (c) nonpolar.
**8.4B** Left, LiCl. Right, HCl. **8.5A** 0.12. **8.5B** 0.11 D. **8.6A** 41 percent.

**8.6B** 9.23 D. **8.7A** :F̈—N̈—F̈: **8.7B** $\left[:\ddot{O}—Cl—\ddot{O}:\right]^{-}$

**8.8A** C atom = 0, double-bonded O atom = 0, single-bonded O atoms = −1. **8.8B** S atom = +1, O atoms = −1,

overall charge = −2. **8.9A** —C̈—Ö—H **8.9B** Cl—N—Cl

**8.10A** $\left[\ddot{O}=N—\ddot{O}:\right]^{-} \rightleftharpoons \left[\ddot{O}—N—\ddot{O}:\right]^{-} \rightleftharpoons \left[:\ddot{O}—N=\ddot{O}:\right]^{-}$

**8.10B** $\left[:\ddot{N}—C≡S:\right]^{-} \rightleftharpoons \left[:N=C=\ddot{S}:\right]^{-} \rightleftharpoons \left[:N≡C—\ddot{S}:\right]^{-}$

**8.11A** ·Ö—H **8.11B** :S̈=N—S̈: **8.12A** :F̈—Be—F̈: **8.12B** :C̈l—B—C̈l:

**8.13A** Cl—P—Cl with Cl below **8.13B** F—Sb—F with F **8.14A** $\left[:C̈l—\ddot{I}—C̈l:\right]^{-}$

**8.14B** :F̈—K̈r—F̈:

**8.15A**

$\left[H—\ddot{O}—\overset{-1}{\underset{+1}{S}}—\ddot{O}:^{-1}\right]^{-} \left[H—\ddot{O}—S—\ddot{O}:^{-1}\right]^{-} \left[H—\ddot{O}—S=\ddot{O}:\right]^{-}$

The structure with two double bonds is inconsistent with the electronegativities of S and O. The negative formal charge should be on O, not on S.

**8.15B**

$\left[:\ddot{O}—P—\ddot{O}:\right]^{3-} \left[:\ddot{O}—P—\ddot{O}:\right]^{3-}$    H—Ö—Cl—Ö:    H—Ö—Cl=Ö:

:Ö—S—Ö:    :O=S=O:      :O=S—Ö:    :O=S=O:

**8.16A** −557 kJ/mol. **8.16B** 328 kJ/mol.

## Answers to Checkpoints

**8.1.1** b. **8.1.2** e. **8.1.3** e. **8.1.4** e. **8.2.1** c. **8.2.2** a. **8.2.3** b. **8.4.1** b. **8.4.2** c.
**8.4.3** d. **8.4.4** c. **8.5.1** d. **8.5.2** c. **8.6.1** e. **8.6.2** c. **8.7.1** a, b, d. **8.7.2** b.
**8.8.1** e. **8.8.2** a, b, e. **8.8.3** e. **8.8.4** d. **8.9.1** a. **8.9.2** e. **8.9.3** e. **8.9.4** b.

## Answers to Applying What You've Learned

**a)** ·N̈· ·Ö· **b)** The difference in electronegativity is (3.5 − 3.0) = 0.5, making this a polar covalent bond. **c)** The partial charges on N and O in NO are +0.029 and −0.029, respectively. **d)** N=Ö: and

**e)** Formal charges are indicated on each Lewis structure in part (d). **f)** $\Delta H^\circ$ for the explosive decomposition of nitroglycerin is $-1.94 \times 10^3$ kJ/mol.

# 9

# Chemical Bonding II
## Molecular Geometry and Bonding Theories

Molecules responsible for the aroma of a substance, such as freshly-baked pie, have distinctive shapes. Their shapes enable molecules to stimulate our sense of smell.

# How Molecular Shape Affects Our Perception of the World

Research has revealed that the geometry or *shape* of a molecule determines some of its important properties. The 2004 Nobel Prize in Physiology or Medicine was awarded to Richard Axel and Linda Buck for their discoveries of odorant receptors and the organization of the olfactory system. Their work shed new light on how our sense of smell works. Although it is somewhat more complicated than can be explained in detail here, the odor of a molecule depends in large part on which of the olfactory receptors it stimulates. *Olfactory receptors* are specialized proteins, located on hairlike cilia in the back of the nose. Stimulation of a receptor triggers an electrical signal that travels along nerve fibers to the brain, where the scent is identified. Which olfactory receptors a molecule stimulates depends on its shape.

We are thought to have receptors for at least seven different, fundamental scents: floral, pepperminty, musky, pungent, camphorlike, ethereal, and putrid. Different parts of a larger molecule may have distinctly different shapes, enabling it to stimulate more than one type of olfactory receptor. This results in the perception of a combination of odors. For example, parts of the benzaldehyde molecule are shaped such that they stimulate the camphorlike, floral, and pepperminty receptors. This is a combination that we perceive as the smell of almonds. Such combinations enable us to recognize thousands of different smells.

This discussion illustrates the importance of *molecular geometry* to the most enigmatic of our senses—smell. Many biochemical processes are specific in that they depend on the shapes of the molecules involved. *Chemical bonding theories* help us to predict and/or explain these shapes.

Benzaldehyde

At the end of this chapter, you will be able to answer several questions about the geometry and structure of a molecule with a distinctive, musky odor [▶▶ Page 378].

## In This Chapter, You Will Learn:

How to determine the three-dimensional shape of a molecule and how the interactions of atomic orbitals give rise to chemical bonds.

### Before you begin, you should review

- Shapes of atomic orbitals [◀◀ Section 6.7]
- Electron configurations of atoms [◀◀ Section 6.8]
- Drawing Lewis structures [◀◀ Section 8.5]

# 9.1   Molecular Geometry

Many familiar chemical and biochemical processes depend heavily on the three-dimensional shapes of the molecules and/or ions involved. Our sense of smell is one example; the effectiveness of a particular drug is another. Although the shape of a molecule or polyatomic ion must be determined experimentally, we can predict their shapes reasonably well using Lewis structures [◄◄ Section 8.5] and the *valence-shell electron-pair repulsion (VSEPR)* model. In this section, we will focus primarily on determining the shapes of molecules of the general type $AB_x$, where A is a central atom surrounded by $x$ B atoms and $x$ can have integer values of 2 to 6. For example, $NH_3$ is an $AB_3$ molecule in which A is nitrogen, B is hydrogen, and $x = 3$. Its Lewis structure is shown in the margin.

Table 9.1 lists examples of each type of $AB_x$ molecule and polyatomic ion that we will consider. Throughout this chapter, we will discuss concepts that apply both to molecules and to polyatomic ions, but we will usually refer to them collectively as "molecules."

Having the molecular formula alone is insufficient to predict the shape of a molecule. For instance, $AB_2$ molecules may be linear or bent:

Linear                    Bent

Moreover, $AB_3$ molecules may be planar, pyramidal, or T-shaped:

Trigonal planar     Trigonal pyramidal     T-shaped

To determine shape, we must start with a correct Lewis structure and apply the VSEPR model.

## The VSEPR Model

The basis of the VSEPR model is that electron pairs in the valence shell of an atom *repel* one another. As we learned in Chapter 8, there are two types of electron pairs: bonding pairs and nonbonding pairs (also known as lone pairs). Furthermore, bonding pairs may be found in single bonds or in multiple bonds. For clarity, we will refer to electron *domains* instead of electron pairs when we use the VSEPR model. An *electron domain* in this context is a lone pair or a bond, regardless of whether the bond is single, double, or triple. Consider the following examples:

|  | $CO_2$ | $O_3$ | $NH_3$ | $PCl_5$ | $XeF_4$ |
|---|---|---|---|---|---|
|  | 2 double bonds | 1 single bond 1 double bond + 1 lone pair | 3 single bonds + 1 lone pair | 5 single bonds | 4 single bonds + 2 lone pairs |
| Total number of electron domains on central atom | 2 electron domains | 3 electron domains | 4 electron domains | 5 electron domains | 6 electron domains |

H—N̈—H
|
H

Lewis structure of $NH_3$

**Multimedia**

Chemical bonding—valence-shell electron-pair repulsion theory.

| TABLE 9.1 | Examples of $AB_x$ Molecules and Polyatomic Ions |
|---|---|
| $AB_2$ | $BeCl_2$, $SO_2$, $H_2O$, $NO_2^-$ |
| $AB_3$ | $BF_3$, $NH_3$, $ClF_3$, $SO_3^{2-}$ |
| $AB_4$ | $CCl_4$, $NH_4^+$, $SF_4$, $XeF_4$, $ClO_4^-$ |
| $AB_5$ | $PCl_5$, $IF_5$, $SbF_5$, $BrF_5$ |
| $AB_6$ | $SF_6$, $UF_6$, $TiCl_6^{3-}$ |

**Student Annotation:** Note that you cannot tell the shape of a molecule or ion simply from its formula—you must apply VSEPR theory.

Note the number of electron domains on the central atom in each molecule. The VSEPR model predicts that because these electron domains repel one another, they will arrange themselves to be as far apart as possible, thus minimizing the repulsive interactions between them.

We can visualize the arrangement of electron domains using balloons, as shown in Figure 9.1. Like the B atoms in our $AB_x$ molecules, the balloons are all connected to a central, fixed point, which represents the central atom (A). When they are as far apart as possible, they adopt the five geometries shown in the figure. When there are only two balloons, they orient themselves to point in opposite directions [Figure 9.1(a)]. With three balloons the arrangement is a trigonal plane [Figure 9.1(b)]. With four balloons the arrangement adopted is a tetrahedron [Figure 9.1(c)]. With five balloons, three of them adopt positions in a trigonal plane whereas the other two point opposite to each other, forming an axis that is perpendicular to the trigonal plane [Figure 9.1(d)]. This geometry is called a trigonal bipyramid. Finally, with six balloons the arrangement is an octahedron, which is essentially a square bipyramid [Figure 9.1(e)]. Each of the $AB_x$ molecules we consider will have one of these five electron-domain geometries: linear, trigonal planar, tetrahedral, trigonal bipyramidal, or octahedral.

**Student Annotation:** It is essential that you are able to draw the correct Lewis structure and that you count the electron domains on the central atom carefully. This will be your "blueprint" for determining the shape of a molecule or polyatomic ion.

(a)

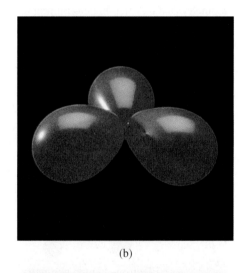

(b)

**Figure 9.1**  The arrangements adopted by (a) two, (b) three, (c) four, (d) five, and (e) six balloons.

(c)

(d)

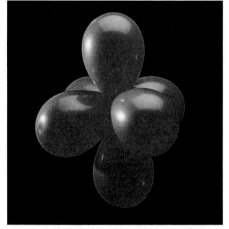

(e)

## Electron-Domain Geometry and Molecular Geometry

It is important to distinguish between the ***electron-domain geometry,*** which is the arrangement of electron domains (bonds and lone pairs) around the central atom, and the ***molecular geometry,*** which is the arrangement of bonded *atoms.* Figure 9.2 illustrates the molecular geometries of $AB_x$ molecules in which all the electron domains are bonds—that is, there are no lone pairs on any of the central atoms. In these cases, the molecular geometry is the same as the electron-domain geometry.

In an $AB_x$ molecule, a ***bond angle*** is the angle between two adjacent A—B bonds. In an $AB_2$ molecule there are only two bonds and therefore only one bond angle, and, provided that there are no lone pairs on the central atom, the bond angle is 180°. $AB_3$ and $AB_4$ molecules have three and four bonds, respectively. However, in each case there is only one bond angle possible between any two A—B bonds. In an $AB_3$ molecule, the bond angle is 120°, and in an $AB_4$ molecule, the bond angle is 109.5°—again, provided that there are no lone pairs on the central atoms. Similarly, in an $AB_6$ molecule the bond angles between adjacent bonds are all 90°. (The angle between any two A—B bonds that point in opposite directions is 180°.)

$AB_5$ molecules contain two different bond angles between adjacent bonds. The reason for this is that, unlike those in the other $AB_x$ molecules, the positions occupied by bonds in a trigonal bipyramid are not all equivalent. The three bonds that are arranged in a trigonal plane are referred to as ***equatorial.*** The bond angle between any two of the three equatorial bonds is 120°. The two bonds that form an axis perpendicular to the trigonal plane are referred to as ***axial.*** The bond angle between either of the axial bonds and any one of the equatorial bonds is 90°. (As in the case of the $AB_6$ molecule, the angle between any two A—B bonds that point in opposite directions is 180°.) Figure 9.2 illustrates all these bond angles. The angles shown in the figure are the bond angles that are observed when all the electron domains on the central atom are identical. As we will see later in this section, the bond angles in many molecules will differ slightly from these *ideal* values.

When the central atom in an $AB_x$ molecule bears one or more lone pairs, the electron-domain geometry and the molecular geometry are no longer the same. However, we still use the electron-domain geometry as a first step in determining the molecular geometry. The first step in determining the molecular geometry of $O_3$ (or any species), for example, is to draw its Lewis structure. Two different resonance structures can be drawn for $O_3$:

$$\ddot{\text{O}}=\ddot{\text{O}}-\ddot{\text{O}}: \longleftrightarrow :\ddot{\text{O}}-\ddot{\text{O}}=\ddot{\text{O}}:$$

Either one can be used to determine its geometry.

The next step is to count the electron domains on the central atom. In this case, there are three: one single bond, one double bond, and one lone pair. Using the VSEPR model, we first determine the electron-domain geometry. According to the information in Figure 9.2, three electron domains on the central atom will be arranged in a trigonal plane. Molecular geometry, however, is dictated by the arrangement of *atoms.* If we consider only the positions of the three atoms in this molecule, the molecular geometry (the molecule's shape) is *bent.*

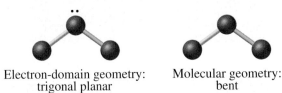

Electron-domain geometry:      Molecular geometry:
      trigonal planar                            bent

In addition to the five basic geometries depicted in Figure 9.2, you must be familiar with how molecular geometry can differ from electron-domain geometry. Table 9.2 shows the common

Media Player/
MPEG Content

Chemical bonding—VSEPR
and molecular geometry
(interactive).

**Student Annotation:** It would be impossible to overstate the importance of being able to draw Lewis structures correctly, especially for students who will go on to study organic chemistry.

**Figure 9.2**   Five geometries of $AB_x$ molecules in which all the electron domains are bonds.

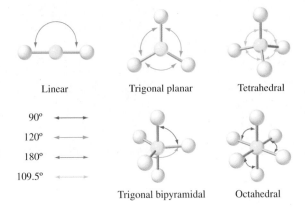

Linear          Trigonal planar          Tetrahedral

90°
120°
180°
109.5°

Trigonal bipyramidal          Octahedral

| TABLE 9.2 | | Electron-Domain and Molecular Geometries of Molecules with Lone Pairs on the Central Atom | | | | |
|---|---|---|---|---|---|---|
| Total Number of Electron Domains | Type of Molecule | Electron-Domain Geometry | Number of Lone Pairs | Placement of Lone Pairs | Molecular Geometry | Example |
| 3 | $AB_2$ | Trigonal planar | 1 | | Bent | $SO_2$ |
| 4 | $AB_3$ | Tetrahedral | 1 | | Trigonal pyramidal | $NH_3$ |
| 4 | $AB_2$ | Tetrahedral | 2 | | Bent | $H_2O$ |
| 5 | $AB_4$ | Trigonal bipyramidal | 1 | | Seesaw-shaped | $SF_4$ |
| 5 | $AB_3$ | Trigonal bipyramidal | 2 | | T-shaped | $ClF_3$ |
| 5 | $AB_2$ | Trigonal bipyramidal | 3 | | Linear | $IF_2^-$ |
| 6 | $AB_5$ | Octahedral | 1 | | Square pyramidal | $BrF_5$ |
| 6 | $AB_4$ | Octahedral | 2 | | Square planar | $XeF_4$ |

molecular geometries where there are one or more lone pairs on the central atom. Note the positions occupied by the lone pairs in the trigonal bipyramidal electron-domain geometry. When there are lone pairs on the central atom in a trigonal bipyramid, the lone pairs preferentially occupy equatorial positions because repulsion is greater when the angle between electron domains is 90° or less. Placing a lone pair in an *axial* position would put it at 90° to three other electron domains. Placing it in an *equatorial* position puts it at 90° to only two other domains, thus minimizing the number of strong repulsive interactions.

All positions are equivalent in the octahedral geometry, so one lone pair on the central atom can occupy any of the positions. If there is a second lone pair in this geometry, though, it must occupy the position opposite the first. This arrangement minimizes the repulsive forces between the two lone pairs (they are 180° apart instead of 90° apart).

In summary, the steps to determine the electron-domain and molecular geometries are as follows:

1. Draw the Lewis structure of the molecule or polyatomic ion.
2. Count the number of electron domains on the central atom.
3. Determine the electron-domain geometry by applying the VSEPR model.
4. Determine the molecular geometry by considering the positions of the atoms only.

Sample Problem 9.1 shows how to determine the shape of a molecule or polyatomic ion.

## Sample Problem 9.1

Determine the shapes of (a) $SO_3$ and (b) $ICl_4^-$.

**Strategy** Use Lewis structures and the VSEPR model to determine first the electron-domain geometry and then the molecular geometry (shape).

**Setup** (a) The Lewis structure of $SO_3$ is:

$$:\overset{..}{\underset{|}{O}}:$$
$$:\overset{..}{O}=\overset{|}{S}-\overset{..}{\underset{..}{O}}:$$

There are three electron domains on the central atom: one double bond and two single bonds.

(b) The Lewis structure of $ICl_4^-$ is:

$$\left[ \begin{array}{c} :\overset{..}{\underset{|}{Cl}}: \\ :\overset{..}{\underset{..}{Cl}}-\overset{..}{\underset{|}{I}}-\overset{..}{\underset{..}{Cl}}: \\ :\overset{..}{\underset{..}{Cl}}: \end{array} \right]^-$$

There are six electron domains on the central atom in $ICl_4^-$: four single bonds and two lone pairs.

**Solution**
(a) According to the VSEPR model, three electron domains will be arranged in a trigonal plane. Since there are no lone pairs on the central atom in $SO_3$, the molecular geometry is the same as the electron-domain geometry. Therefore, the shape of $SO_3$ is trigonal planar.

Electron-domain geometry: trigonal planar ⟶ Molecular geometry: trigonal planar

(b) Six electron domains will be arranged in an octahedron. Two lone pairs on an octahedron will be located on opposite sides of the central atom, making the shape of $ICl_4^-$ square planar.

Electron-domain geometry: octahedral    ⟶    Molecular geometry: square planar

**Practice Problem A** Determine the shapes of (a) $CO_2$ and (b) $SCl_2$.

**Practice Problem B** (a) From what group must the terminal atoms come in an $AB_x$ molecule where the central atom is from Group 6A, for the electron-domain geometry and the molecular geometry both to be trigonal planar? (b) From what group must the terminal atoms come in an $AB_x$ molecule where the central atom is from Group 7A, for the electron-domain geometry to be octahedral and the molecular geometry to be square pyramidal?

## Deviation from Ideal Bond Angles

Some electron domains are better than others at repelling neighboring domains. As a result, the bond angles may be slightly different from those shown in Figure 9.2. For example, the electron-domain geometry of ammonia ($NH_3$) is tetrahedral, so we might predict the H—N—H bond angles to be 109.5°. In fact, the bond angles are about 107°, slightly smaller than predicted. The lone pair on the nitrogen atom repels the N—H bonds more strongly than the bonds repel one another. It therefore "squeezes" them closer together than the ideal tetrahedral angle of 109.5°.

In effect, a lone pair takes up more *space* than the bonding pairs. This can be understood by considering the attractive forces involved in determining the location of the electron pairs. A lone pair on a central atom is attracted only to the nucleus of that atom. A bonding pair of electrons, on the other hand, is simultaneously attracted by the nuclei of both of the bonding atoms. As a result, the lone pair has more freedom to spread out and greater capacity to repel other electron domains. Also, because they contain more electron density, multiple bonds repel more strongly than single bonds. Consider the bond angles in each of the following examples:

| $NH_3$ | $CH_2O$ | $SF_4$ | $BrF_5$ |

## Geometry of Molecules with More Than One Central Atom

Thus far we have considered the geometries of molecules having only one central atom. We can determine the overall geometry of more complex molecules by treating them as though they have multiple central atoms. Methanol ($CH_3OH$), for example, has a central C atom and a central O atom, as shown in the following Lewis structure:

$$H-\overset{\overset{\displaystyle H}{|}}{\underset{\underset{\displaystyle H}{|}}{C}}-\overset{..}{\underset{..}{O}}-H$$

Both the C and the O atoms are surrounded by four electron domains. In the case of C, they are three C—H bonds and one C—O bond. In the case of O, they are one O—C bond, one O—H bond, and two lone pairs. In each case, the electron-domain geometry is tetrahedral. However, the molecular geometry of the C part of the molecule is *tetrahedral*, whereas the molecular geometry of the O part of the molecule is *bent*. Note that although the Lewis structure makes it appear as though there is a 180° angle between the O—C and O—H bonds, the angle is actually approximately 109.5°, the angle in a tetrahedral arrangement of electron domains.

Sample Problem 9.2 shows how to determine when bond angles differ from ideal values.

**Student Annotation:** When we specify the geometry of a particular portion of a molecule, we refer to it as the geometry "about" a particular atom. In methanol, for example, we say that the geometry is *tetrahedral about the C atom* and *bent about the O atom*.

## Sample Problem 9.2

Acetic acid, the substance that gives vinegar its characteristic smell and sour taste, is sometimes used in combination with corticosteroids to treat certain types of ear infections. Its Lewis structure is

$$H-\overset{\overset{\displaystyle H}{|}}{\underset{\underset{\displaystyle H}{|}}{C}}-\overset{\overset{\displaystyle \overset{..}{O}}{||}}{C}-\overset{..}{\underset{..}{O}}-H$$

*(Continued)*

Determine the molecular geometry about each of the central atoms, and determine the approximate value of each of the bond angles in the molecule. Which if any of the bond angles would you expect to be smaller than the ideal values?

**Strategy**  Identify the central atoms and count the number of electron domains around each of them. Use the VSEPR model to determine each electron-domain geometry, and the information in Table 9.2 to determine the molecular geometry about each central atom.

**Setup**  The leftmost C atom is surrounded by four electron domains: one C—C bond and three C—H bonds. The middle C atom is surrounded by three electron domains: one C—C bond, one C—O bond, and one C=O (double) bond. The O atom is surrounded by four electron domains: one O—C bond, one O—H bond, and two lone pairs.

**Solution**  The electron-domain geometry of the leftmost C is tetrahedral. Because all four electron domains are bonds, the molecular geometry of this part of the molecule is also tetrahedral. The electron-domain geometry of the middle C is trigonal planar. Again, because all the domains are bonds, the molecular geometry is also trigonal planar. The electron-domain geometry of the O atom is tetrahedral. Because two of the domains are lone pairs, the molecular geometry about the O atom is bent.

Bond angles are determined using electron-domain geometry. Therefore, the approximate bond angles about the leftmost C are 109.5°, those about the middle C are 120°, and those about the O are 109.5°. The angle between the two single bonds on the middle carbon will be *less* than 120° because the double bond repels the single bonds more strongly than they repel each other. Likewise, the bond angle between the two bonds on the O will be less than 109.5° because the lone pairs on O repel the single bonds more strongly than they repel each other and push the two bonding pairs closer together. The angles are labeled as follows:

**Think About It**  Compare these answers with the information in Figure 9.2 and Table 9.2.

**Practice Problem A**  Ethanolamine ($HOCH_2CH_2NH_2$) has a smell similar to ammonia and is commonly found in biological tissues. Its Lewis structure is

Determine the molecular geometry about each central atom and label all the bond angles. Cite any expected deviations from ideal bond angles.

**Practice Problem B**  The bond angle in $NH_3$ is significantly smaller than the ideal bond angle of 109.5° because of the lone pair on the central atom. Explain why the bond angle in $SO_2$ is very close to 120° despite there being a lone pair on the central atom.

---

### Checkpoint 9.1     Molecular Geometry

**9.1.1**  What are the electron-domain geometry and molecular geometry of $CO_3^{2-}$?

a) tetrahedral, trigonal planar

b) tetrahedral, trigonal pyramidal

c) trigonal pyramidal, trigonal pyramidal

d) trigonal planar, trigonal planar

e) tetrahedral, tetrahedral

**9.1.2**  What are the electron-domain geometry and molecular geometry of $ClO_3^-$?

a) tetrahedral, trigonal planar

b) tetrahedral, trigonal pyramidal

c) trigonal pyramidal, trigonal pyramidal

d) trigonal planar, trigonal planar

e) tetrahedral, tetrahedral

**9.1.3** What is the approximate value of the bond angle indicated?

$$H-C\!=\!C-H$$
(with H H above the carbons)

a) $< 90°$
b) $< 109.5°$
c) $> 109.5°$
d) $> 120°$
e) $< 120°$

**9.1.4** What is the approximate value of the bond angle indicated?

$$H-C-\ddot{O}-H$$
(with H above and H below the carbon)

a) $< 180°$
b) $> 180°$
c) $< 109.5°$
d) $> 109.5°$
e) $< 90°$

**9.1.5** Which of the following shows a deviation from ideal bond angles that is not possible for an $AB_x$ molecule?

T-shaped (i) — $<90°$
Seesaw-shaped (ii) — $>120°$
Trigonal pyramidal (iii) — $<109.5°$
Square pyramidal (iv) — $>90°$

a) ii only
b) i and iii
c) i, ii, and iv
d) ii and iv
e) All these deviations from ideal bond angles are possible.

# 9.2 Molecular Geometry and Polarity

Molecular geometry is tremendously important in understanding the physical and chemical behavior of a substance. *Molecular polarity,* for example, is one of the most important consequences of molecular geometry, because molecular polarity influences physical, chemical, and biological properties. Recall from Section 8.4 that a bond between two atoms of different electronegativities is polar and that a diatomic molecule containing a polar bond is a *polar molecule.* Whether a molecule made up of three or more atoms is polar depends not only on the polarity of the individual bonds, but also on its molecular geometry.

Each of the $CO_2$ and $H_2O$ molecules contains two identical atoms bonded to a central atom and two polar bonds. However, only one of these molecules is polar. To understand why, think of each individual bond dipole as a vector. The overall dipole moment of the molecule is determined by vector addition of the individual bond dipoles.

In the case of $CO_2$, we have two identical vectors pointing in opposite directions. When the vectors are placed on a Cartesian coordinate system, they have no *y* component and their *x* components are equal in magnitude but opposite in sign. The sum of these two vectors is zero in both the *x* and *y* directions. Thus, although the *bonds* in $CO_2$ are polar, the *molecule* is nonpolar.

**Multimedia**
Chemical bonding—molecular geometry and polarity (interactive).

**Student Annotation:** Recall that we can represent an individual bond dipole using a crossed arrow that points toward the more electronegative atom [◄◄ Section 8.4].

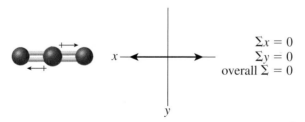

$\Sigma x = 0$
$\Sigma y = 0$
overall $\Sigma = 0$

The vectors representing the bond dipoles in water, although equal in magnitude and opposite in the *x* direction, are not opposite in the *y* direction. Therefore, although their *x* components sum to zero, their *y* components do not. This means that there is a net resultant dipole and $H_2O$ is *polar.*

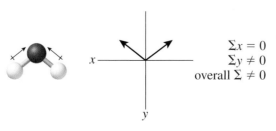

$\Sigma x = 0$
$\Sigma y \ne 0$
overall $\Sigma \ne 0$

## *You May Be Wondering…*

# Can More Complex Molecules Contain Polar Bonds and Still Be Nonpolar?

In $AB_x$ molecules where $x \geq 3$, it may be less obvious whether the individual bond dipoles cancel one another. Consider the molecule $BF_3$, for example, which has a trigonal planar geometry:

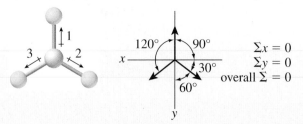

We will simplify the math in this analysis by assigning the vectors representing the three identical B—F bonds an arbitrary magnitude of 1.00. The $x, y$ coordinates for the end of arrow 1 are (0, 1.00). Determining the coordinates for the ends of arrows 2 and 3 requires the use of trigonometric functions. You may have learned the mnemonic SOH CAH TOA, where the letters stand for

Sin = Opposite over Hypotenuse

Cos = Adjacent over Hypotenuse

Tan = Opposite over Adjacent

The $x$ coordinate for the end of arrow 2 corresponds to the length of the line *opposite* the 60° angle. The hypotenuse of the triangle has a length of 1.00 (the arbitrarily assigned value). Therefore, using SOH,

$$\sin 60° = 0.866 = \frac{\text{opposite}}{\text{hypotenuse}} = \frac{\text{opposite}}{1}$$

so the $x$ coordinate for the end of arrow 2 is 0.866.

The magnitude of the $y$ coordinate corresponds to the length of the line *adjacent* to the 60° angle. Using TOA,

$$\tan 60° = 1.73 = \frac{\text{opposite}}{\text{adjacent}} = \frac{0.866}{\text{adjacent}}$$

$$\text{adjacent} = \frac{0.866}{1.73} = 0.500$$

so the $y$ coordinate for the end of arrow 2 is $-0.500$. (The trigonometric formula gives us the length of the side. We know from the diagram that the sign of this $y$ component is negative.)

Arrow 3 is similar to arrow 2. Its $x$ component is equal in magnitude but opposite in sign, and its $y$ component is the same magnitude and sign as that for arrow 2. Therefore, the $x$ and $y$ coordinates for all three vectors are

|         | $x$      | $y$      |
|---------|----------|----------|
| Arrow 1 | 0        | 1        |
| Arrow 2 | 0.866    | $-0.500$ |
| Arrow 3 | $-0.866$ | $-0.500$ |
| Sum =   | 0        | 0        |

Because the individual bond dipoles (represented here as the vectors) sum to zero, the molecule is nonpolar overall.

Although it is somewhat more complicated, a similar analysis can be done to show that all $x$, $y$, and $z$ coordinates sum to zero when there are four identical polar bonds arranged in a tetrahedron about a central atom. In fact, any time there are identical bonds symmetrically distributed around a central atom, with no lone pairs on the central atom, the molecule will be nonpolar overall, even if the bonds themselves are polar.

In cases where the bonds are distributed symmetrically around the central atom, the nature of the atoms surrounding the central atom determines whether the molecule is polar overall. For example, $CCl_4$ and $CHCl_3$ have the same molecular geometry (tetrahedral), but $CCl_4$ is nonpolar because the bond dipoles cancel one another. In $CHCl_3$, however, the bonds are not all

Dipole moments can be used to distinguish between molecules that have the same chemical formula but different arrangements of atoms. Such compounds are called ***structural isomers.*** For example, there are two structural isomers of dichloroethylene ($C_2H_2Cl_2$). Because the individual bond dipoles sum to zero in *trans*-dichloroethylene, the *trans* isomer is nonpolar:

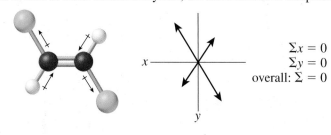

identical, and therefore the bond dipoles do not sum to zero. The $CHCl_3$ molecule is polar.

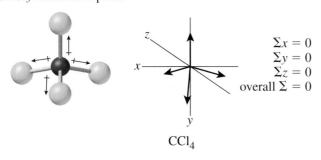

$\Sigma x = 0$
$\Sigma y = 0$
$\Sigma z = 0$
overall $\Sigma = 0$

$CCl_4$

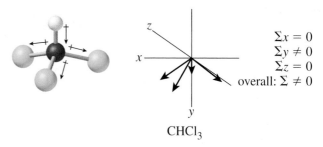

$\Sigma x = 0$
$\Sigma y \neq 0$
$\Sigma z = 0$
overall: $\Sigma \neq 0$

$CHCl_3$

Sample Problem 9.3 shows you how to determine whether a molecule is polar.

---

### Sample Problem 9.3

Determine whether (a) $PCl_5$ and (b) $H_2CO$ (C double bonded to O) are polar.

**Strategy**  For each molecule, draw the Lewis structure, use the VSEPR model to determine its molecular geometry, and then determine whether the individual bond dipoles cancel.

**Setup**  (a) The Lewis structure of $PCl_5$ is

$$\ddot{C}l \underset{\ddot{C}l}{\overset{:\ddot{C}l:}{\underset{:\ddot{C}l:}{P}}} :\ddot{C}l$$

With five identical electron domains around the central atom, the electron-domain and molecular geometries are trigonal bipyramidal. The equatorial bond dipoles will cancel one another, just as in the case of $BF_3$, and the axial bond dipoles will also cancel each other.

(b) The Lewis structure of $H_2CO$ is

$$H \overset{\cdot \ddot{O} \cdot}{\underset{C}{\|}} H$$

The bond dipoles, although symmetrically distributed around the C atom, are not identical and therefore will not sum to zero.

**Solution**  (a) $PCl_5$ is nonpolar.

(b) $H_2CO$ is polar.

> **Think About It**  Make sure that your Lewis structures are correct and that you count electron domains on the central atom carefully. This will give you the correct electron-domain and molecular geometries. Molecular polarity depends both on individual bond dipoles and molecular geometry.

**Practice Problem A**  Determine whether (a) $CH_2Cl_2$ and (b) $XeF_4$ are polar.

**Practice Problem B**  For which molecular geometries (linear, bent, trigonal planar, trigonal pyramidal, tetrahedral, square planar, T-shaped, seesaw-shaped, trigonal bipyramidal, square pyramidal, octahedral) can an $AB_x$ molecule be nonpolar if there are (a) two different types of terminal atoms and (b) three different types of terminal atoms?

---

The bond dipoles in the *cis* isomer do not cancel one another, so *cis*-dichloroethylene is polar:

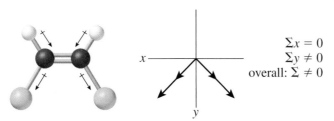

$\Sigma x = 0$
$\Sigma y \neq 0$
overall: $\Sigma \neq 0$

Because of the difference in polarity, these two isomers can be distinguished experimentally by measuring the dipole moment.

*You May Be Wondering…*

# How Are Electrons in Orbitals Represented?

In Chapter 6 we learned that although an electron is a particle with a known mass, it exhibits wavelike properties. The quantum mechanical model of the atom, which gives rise to the familiar shapes of $s$ and $p$ atomic orbitals, treats electrons in atoms as waves, rather than particles. Therefore, rather than use arrows to denote the locations and spins of electrons, we will adopt a convention whereby a singly occupied orbital will appear as a light color and a doubly occupied

orbital will appear as a darker version of the same color. In the representations of orbitals that follow, atomic $s$ orbitals will be represented as yellow, and atomic $p$ orbitals will be represented as light blue if singly occupied and darker blue if doubly occupied. (Empty $p$ orbitals will appear white.) When two electrons occupy the same atomic orbital, their spins are paired. Paired electrons occupy the same orbital and have opposite spins [◄◄ Section 6.6].

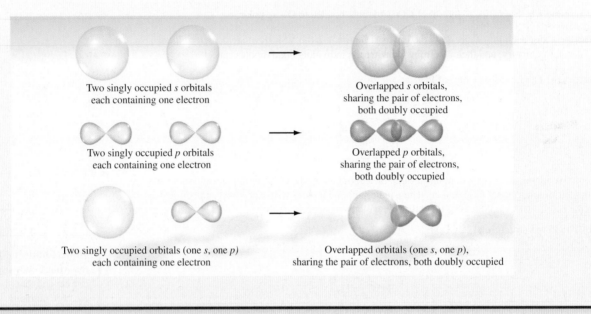

Two singly occupied $s$ orbitals each containing one electron

Overlapped $s$ orbitals, sharing the pair of electrons, both doubly occupied

Two singly occupied $p$ orbitals each containing one electron

Overlapped $p$ orbitals, sharing the pair of electrons, both doubly occupied

Two singly occupied orbitals (one $s$, one $p$) each containing one electron

Overlapped orbitals (one $s$, one $p$), sharing the pair of electrons, both doubly occupied

**Checkpoint 9.2**    **Molecular Geometry and Polarity**

**9.2.1**  Identify the polar molecules in the following group: HBr, $CH_4$, $CS_2$.

    a) HBr only

    b) HBr and $CS_2$

    c) HBr, $CH_4$, and $CS_2$

    d) $CH_4$ and $CS_2$

    e) $CH_4$ only

**9.2.2**  Identify the nonpolar molecules in the following group: $SO_2$, $NH_3$, $XeF_2$.

    a) $SO_2$, $NH_3$, and $XeF_2$

    b) $SO_2$ only

    c) $XeF_2$ only

    d) $SO_2$ and $XeF_2$

    e) $SO_2$ and $NH_3$

## 9.3    Valence Bond Theory

H—H

:F̈—F̈:

H—F̈:

Lewis dot structures of $H_2$, $F_2$, and HF

The Lewis theory of chemical bonding provides a relatively simple way for us to visualize the arrangement of electrons in molecules. It is insufficient, however, to explain the differences between the covalent bonds in compounds such as $H_2$, $F_2$, and HF. Although Lewis theory describes the bonds in these three molecules in exactly the same way, they really are quite different from one another, as evidenced by their bond lengths and bond enthalpies listed in Table 9.3. Understanding these differences and why covalent bonds form in the first place requires a bonding model that combines Lewis's notion of atoms sharing electron pairs and the quantum mechanical descriptions of atomic orbitals.

According to *valence bond theory,* atoms share electrons when an atomic orbital on one atom overlaps with an atomic orbital on the other. Each of the overlapping atomic orbitals must

| TABLE 9.3 | Bond Lengths and Bond Enthalpies of $H_2$, $F_2$, and HF | |
|---|---|---|
| | **Bond Length (Å)** | **Bond Enthalpy (kJ/mol)** |
| $H_2$ | 0.74 | 436.4 |
| $F_2$ | 1.42 | 150.6 |
| HF | 0.92 | 568.2 |

$H_2$

$F_2$

HF
Ball-and-stick
models

contain a single, unpaired electron. Furthermore, the two electrons shared by the bonded atoms must have opposite spins [◄◄ Section 6.6]. The nuclei of both atoms are attracted to the shared pair of electrons. It is this mutual attraction for the shared electrons that holds the atoms together.

The H—H bond in $H_2$ forms when the singly occupied 1s orbitals of the two H atoms overlap:

**Student Annotation:** Keep in mind that although there are still just two electrons, each atom "thinks" it owns them both, so when the singly occupied orbitals overlap, both orbitals end up doubly occupied.

Similarly, the F—F bond in $F_2$ forms when the singly occupied 2p orbitals of the two F atoms overlap:

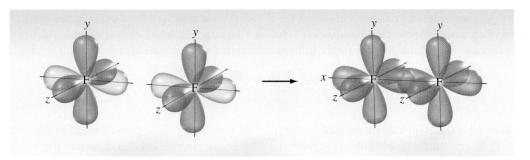

Recall that the ground-state electron configuration of the F atom is [He]$2s^2 2p^5$ [◄◄ Section 6.8]. (The ground-state orbital diagram of F is shown in the margin.)

We can also depict the formation of an H—F bond using the valence bond model. In this case, the singly occupied 1s orbital of the H atom overlaps with the singly occupied 2p orbital of the F atom:

$2s^2$       $2p^5$
Orbital diagram for F

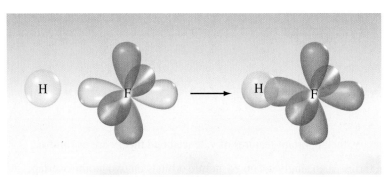

According to the quantum mechanical model, the sizes, shapes, and energies of the 1s orbital of H and the 2p orbital of F are different. Therefore, it is not surprising that the bonds in $H_2$, $F_2$, and HF vary in strength and length.

**Figure 9.3** The change in potential energy of two hydrogen atoms as a function of internuclear distance. The minimum potential energy (−436 kJ/mol) occurs when the distance between the nuclei is 74 pm. The yellow spheres represent the 1s orbitals of hydrogen.

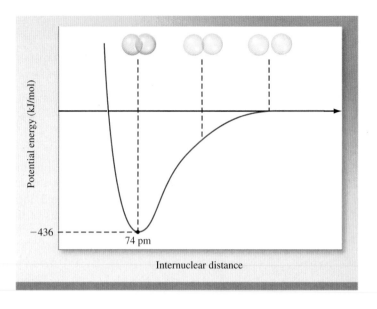

**Student Annotation:** Recall that the enthalpy change for a forward process and that for the reverse process differ only in sign: $\Delta H_{forward} = -\Delta H_{reverse}$ [◄◄ Section 5.3].

*Why* do covalent bonds form? According to valence bond theory, a covalent bond will form between two atoms if the potential energy of the resulting molecule is lower than that of the isolated atoms. Simply put, this means that the formation of covalent bonds is exothermic. While this fact may not seem intuitively obvious, you know that energy must be supplied to a molecule to *break* covalent bonds [◄◄ Section 8.9]. Because the formation of a bond is the reverse process, we should expect energy to be given off. Figure 9.3 illustrates how the potential energy of two hydrogen atoms varies with distance between the nuclei.

Valence bond theory also introduces the concept of *directionality* to chemical bonds. For example, we expect the bond formed by the overlap of a *p* orbital to coincide with the axis along which the *p* orbital lies. Consider the molecule $H_2S$. Unlike the other molecules that we have encountered, $H_2S$ does not have the bond angle that Lewis theory and the VSEPR model would lead us to predict. (With four electron domains on the central atom, we would expect the bond angle to be on the order of 109.5°.) In fact, the H−S−H bond angle is 92°. Looking at this in terms of valence bond theory, the central atom (S) has two unpaired electrons, each of which resides in a 3p orbital. The orbital diagram for the ground-state electron configuration of the S atom is

**Student Annotation:** For you to understand the material in this section and Section 9.4, you must be able to draw orbital diagrams for ground-state electron configurations [◄◄ Section 6.8].

$$S \quad [Ne] \quad \boxed{\uparrow\downarrow} \qquad \boxed{\uparrow\downarrow\ |\ \uparrow\ |\ \uparrow}$$
$$\qquad\qquad\qquad 3s^2 \qquad\qquad 3p^4$$

Remember that *p* orbitals are mutually perpendicular, lying along the *x, y,* and *z* axes [◄◄ Section 6.7]. We can rationalize the observed bond angle by envisioning the overlap of each of the singly occupied 3p orbitals with the 1s orbital of a hydrogen atom:

$H_2S$

In summary, the important features of valence bond theory are as follows:

- A bond forms when singly occupied atomic orbitals on two atoms overlap.
- The two electrons shared in the region of orbital overlap must be of opposite spin.
- Formation of a bond results in a lower potential energy for the system.

Sample Problem 9.4 shows how to use valence bond theory to explain the bonding in a molecule.

---

### Sample Problem 9.4

Hydrogen selenide ($H_2Se$) is a foul-smelling gas that can cause eye and respiratory tract inflammation. The H—Se—H bond angle in $H_2Se$ is approximately 92°. Use valence bond theory to describe the bonding in this molecule.

**Strategy** Consider the central atom's ground-state electron configuration, and determine what orbitals are available for bond formation.

**Setup** The ground-state electron configuration of Se is $[Ar]4s^2 3d^{10} 4p^4$. Its orbital diagram (showing only the 4p orbitals) is

$$\boxed{\uparrow\downarrow}\boxed{\uparrow}\boxed{\uparrow}$$
$$4p^4$$

**Solution** Two of the 4p orbitals are singly occupied and therefore available for bonding. The bonds in $H_2Se$ form as the result of the overlap of a hydrogen 1s orbital with each of these orbitals on the Se atom.

> **Think About It** Because the 4p orbitals on the Se atom are all mutually perpendicular, we should expect the angles between bonds formed by their overlap to be approximately 90°.

**Practice Problem A** Use valence bond theory to describe the bonding in phosphine ($PH_3$), which has H—P—H bond angles of approximately 94°.

**Practice Problem B** For which molecule(s) can we *not* use valence bond theory to explain the bonding: $SO_2$ (O—S—O bond angle ~120°), $CH_4$ (H—C—H bond angles = 109.5°), $AsH_3$ (H—As—H bond angles = 92°)? Explain.

---

### Checkpoint 9.3    Valence Bond Theory

**9.3.1** Which of the following atoms, in its ground state, does not have unpaired electrons? (Select all that apply.)

a) O

b) Be

c) B

d) F

e) Ne

**9.3.2** According to valence bond theory, how many bonds would you expect a nitrogen atom (in its ground state) to form?

a) 2

b) 3

c) 4

d) 5

e) 6

---

## 9.4 Hybridization of Atomic Orbitals

Although valence bond theory is useful and can explain more of our experimental observations than Lewis bond theory, it fails to explain the bonding in many of the molecules that we encounter. According to valence bond theory, for example, an atom must have a singly occupied atomic orbital to form a bond with another atom. How then do we explain the bonding in $BeCl_2$? The central atom, Be, has a ground-state electron configuration of $[He]2s^2$, so it has no unpaired electrons. With no singly occupied atomic orbitals in its ground state, how does Be form two bonds?

Furthermore, in cases where the ground-state electron configuration of the central atom *does* have the required number of unpaired electrons, how do we explain the observed bond angles? Carbon, like sulfur, has two unpaired electrons in its ground state. Using valence bond theory as our guide, we might envision the formation of two covalent bonds with oxygen, as in $CO_2$. If the two unpaired electrons on C (each residing in a 2p orbital) were to form bonds, however, the O—C—O bond angle should be on the order of 90°, like the bond angle in $H_2S$. In fact, the bond angle in $CO_2$ is 180°:

:C̈l—Be—C̈l:

$BeCl_2$

**Student Annotation:** The ground-state orbital diagram for C is:

$$\boxed{\uparrow\downarrow}\boxed{\uparrow\downarrow}\boxed{\uparrow}\boxed{\uparrow}$$
$$1s^2 \quad 2s^2 \qquad 2p^2$$

**Student Annotation:** Remember that Be is
one of the atoms that does not obey the octet
rule [◄◄ Section 8.1]. The Lewis structure of
$BeCl_2$ is

$$:\ddot{\text{C}}\text{l}-\text{Be}-\ddot{\text{C}}\text{l}:$$

$3s^2$    $3p^5$

Orbital diagram for Cl

**Student Annotation:** Hybrid orbitals are
another type of *electron domain*.

180°

Actual bond angle is 180°.

Bond angle should be 90°.

To explain these and other observations, we need to extend our discussion of orbital overlap to
include the concept of **hybridization** or *mixing* of atomic orbitals.

The idea of hybridization of atomic orbitals begins with the molecular geometry and works
backward to explain the bonds and the observed bond angles in a molecule. To extend our discussion of orbital overlap and introduce the concept of hybridization of atomic orbitals, we first
consider beryllium chloride ($BeCl_2$), which has two electron domains on the central atom. Using
its Lewis structure (shown in the margin) and the VSEPR model, we predict that $BeCl_2$ will have a
Cl—Be—Cl bond angle of 180°. If this is true, though, how does Be form two bonds with no
unpaired electrons, and why is the angle between the two bonds 180°?

To answer the first part of the question, we envision the *promotion* of one of the electrons in
the 2s orbital to an empty 2p orbital. Recall that electrons can be promoted from a lower atomic
orbital to a higher one [◄◄ Section 6.3]. The ground-state electron configuration is the one in
which all the electrons occupy orbitals of the lowest possible energy. A configuration in which one
or more electrons occupy a *higher* energy orbital is called an *excited* state. An excited state generally is denoted with an asterisk (e.g., Be* for an excited-state Be atom). Showing only the valence
orbitals, we can represent the promotion of one of the valence electrons of beryllium as

With one of its valence electrons promoted to the 2p subshell, the Be atom now has two unpaired
electrons and therefore can form two bonds. However, the orbitals in which the two unpaired electrons reside are different from each other, so we would expect bonds formed as a result of the overlap of these two orbitals (each with a 3p orbital on a Cl atom) to be different:

Experimentally, though, the bonds in $BeCl_2$ are identical in length and strength.

## Hybridization of *s* and *p* Orbitals

To explain how beryllium forms two identical bonds, we must mix the orbitals in which the
unpaired electrons reside, thus yielding two equivalent orbitals. The mixing of beryllium's 2s
orbital with one of its 2p orbitals, a process known as *hybridization,* yields two *hybrid orbitals*
that are neither *s* nor *p*, but have some character of each. The hybrid orbitals are designated 2sp
or simply *sp*.

Mixing of one *s* orbital and one *p* orbital to yield two *sp* orbitals

(a)

(b)

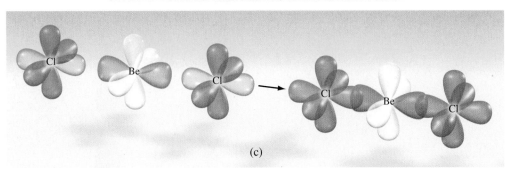

(c)

**Figure 9.4**   (a) An atomic *s* orbital (yellow) and one atomic *p* orbital (blue) combine to form two *sp* hybrid orbitals (green). The realistic hybrid orbital shapes are shown first. The thinner representations are used to keep diagrams clear. (b) The 2*s* orbital and one of the 2*p* orbitals on Be combine to form two *sp* hybrid orbitals. Unoccupied orbitals are shown in white. Like any two electron domains, the hybrid orbitals on Be are 180° apart. (c) The hybrid orbitals on Be each overlap with a singly occupied 3*p* orbital on a Cl atom.

The mathematical combination of the quantum mechanical wave functions for an *s* orbital and a *p* orbital gives rise to two new, equivalent wave functions. As shown in Figure 9.4(a), each *sp* hybrid orbital has one small lobe and one large lobe and, like any two electron domains on an atom, they are oriented in opposite directions with a 180° angle between them. The figure shows the atomic and hybrid orbitals separately for clarity. Note that the hybrid orbitals are shown in two ways: the first is a more realistic shape, whereas the second is a simplified shape that we use to keep the figures clear and make the visualization of orbitals easier. Note also that the representations of hybrid orbitals are green. Figure 9.4(b) shows the locations of the atomic orbitals and the hybrid orbitals, respectively, relative to the beryllium nucleus.

With two *sp* hybrid orbitals, each containing a single unpaired electron, we can see how the Be atom is able to form two identical bonds with two Cl atoms [Figure 9.4(c)]. Each of the singly occupied *sp* hybrid orbitals on the Be atom overlaps with the singly occupied 3*p* atomic orbital on a Cl atom.

We can do a similar analysis of the bonds and the trigonal-planar geometry of boron trifluoride ($BF_3$). The ground-state electron configuration of the B atom is $[He]2s^2 2p^1$, containing just one unpaired electron. Promotion of one of the 2*s* electrons to an empty 2*p* orbital gives the three unpaired electrons needed to explain the formation of *three* bonds. The ground-state and excited-state electron configurations can be represented by

**Student Annotation:** In picturing the shapes of the resulting hybrid orbitals it may help to remember that orbitals, like waves, can combine constructively or destructively. We can think of the large lobe of each *sp* hybrid orbital as the result of a *constructive* combination and the small lobe of each as the result of a *destructive* combination.

**Student Annotation:** The energy required to promote an electron in an atom is more than compensated for by the energy given off when a bond forms.

$$
\text{B} \quad \boxed{\uparrow\downarrow} \quad \boxed{\uparrow\phantom{\downarrow}\phantom{\downarrow}\phantom{\downarrow}} \quad \xrightarrow{\textit{promotion}} \quad \text{B*} \quad \boxed{\uparrow} \quad \boxed{\uparrow\phantom{\downarrow}\uparrow\phantom{\downarrow}}
$$
$$
\phantom{B}\quad 2s^2 \qquad\quad 2p^1 \qquad\qquad\qquad\qquad\qquad 2s^1 \qquad\quad 2p^2
$$

:F̈:
|
B
.F̈.    .F̈.

$BF_3$

**Figure 9.5** (a) An *s* atomic orbital and two *p* atomic orbitals combine to form three *sp*² hybrid orbitals. (b) The three *sp*² hybrid orbitals on B are arranged in a trigonal plane. (Empty atomic orbitals are shown in white.) (c) Hybrid orbitals on B overlap with 2*p* orbitals on F.

(a)

(b)

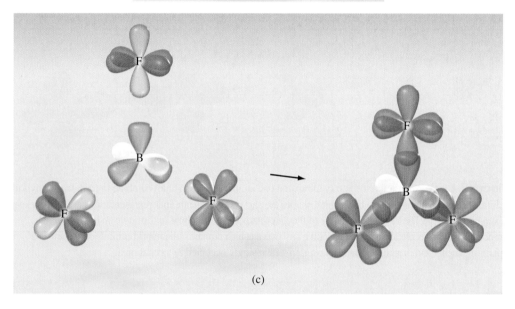
(c)

Because the three bonds in BF₃ are identical, we must hybridize the three singly occupied atomic orbitals (the one *s* and two *p* orbitals) to give three singly occupied hybrid orbitals:

B*  [↑] 2*s*   [↑][↑] 2*p* 2*p* 2*p*   → *hybridization* →   [↑][↑][↑] *sp*² *sp*² *sp*²   [ ] 2*p*

Mixing of one *s* orbital and two *p* orbitals to yield three *sp*² orbitals

Figure 9.5 illustrates the hybridization and bond formation in BF₃.

In both cases (i.e., for BeCl₂ and BF₃), some but not *all* of the *p* orbitals are hybridized. When the remaining unhybridized atomic *p* orbitals do not contain electrons, as in the case of BF₃, they will not be part of the discussion of bonding in this chapter. As we will see in Section 9.5, though, unhybridized atomic orbitals that *do* contain electrons are important in our description of the bonding in a molecule.

We can now apply the same kind of analysis to the methane molecule (CH₄). The Lewis structure of CH₄ has four electron domains around the central carbon atom. This means that we need four hybrid orbitals, which in turn means that four atomic orbitals must be hybridized. The ground-state electron configuration of the C atom contains two unpaired electrons. Promotion of one electron from the 2*s* orbital to the empty 2*p* orbital yields the four unpaired electrons needed for the formation of four bonds:

$$
\begin{array}{c}
\text{H} \\
| \\
\text{H}-\text{C}-\text{H} \\
| \\
\text{H}
\end{array}
$$

CH₄

C   [⇅]   [↑][↑][ ]   $\xrightarrow{promotion}$   C*   [↑]   [↑][↑][↑]

$2s^2$ $2p^2$ $2s^1$ $2p^3$

Hybridization of the *s* orbital and the three *p* orbitals yields four hybrid orbitals designated $sp^3$. We can then place the electrons that were originally in the *s* and *p* atomic orbitals into the $sp^3$ hybrid orbitals.

C*   [↑]   [↑][↑][↑]   $\xrightarrow{hybridization}$   [↑][↑][↑][↑]

2s   2p 2p 2p   $sp^3$ $sp^3$ $sp^3$ $sp^3$

Mixing of one *s* orbital and
three *p* orbitals to yield four $sp^3$ orbitals

The set of four $sp^3$ hybrid orbitals on carbon, like any four electron domains on a central atom, assumes a tetrahedral arrangement. Figure 9.6 illustrates how the hybridization of the C atom results in the formation of the four bonds and the 109.5° bond angles observed in $CH_4$.

## Hybridization of *s*, *p*, and *d* Orbitals

Recall that elements in the third period of the periodic table and beyond do not necessarily obey the octet rule because they have *d* orbitals that can hold additional electrons [◄◄ Section 8.5]. To explain the bonding and geometry of molecules in which there are more than four electron domains on the central atom, we must include *d* orbitals in our hybridization scheme. $PCl_5$, for example, has five electron domains around the P atom. To explain the five bonds in this molecule,

$PCl_5$

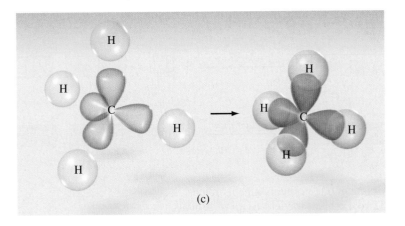

**Figure 9.6**   (a) An *s* atomic orbital and three *p* atomic orbitals combine to form four $sp^3$ hybrid orbitals. (b) The four $sp^3$ hybrid orbitals on C are arranged in a tetrahedron. (c) Hybrid orbitals on C overlap with 1*s* orbitals on H. For clarity, the small lobes of the hybrid orbitals are not shown.

we will need five singly occupied hybrid orbitals. The ground-state electron configuration of the P atom is $[Ne]3s^2 3p^3$, which contains three unpaired electrons. In this case, though, because all three of the $p$ orbitals are occupied, promotion of an electron from the $3s$ orbital to a $3p$ orbital would *not* result in additional unpaired electrons. However, we can promote an electron from the $3s$ orbital to an empty $3d$ orbital, thus forming the five unpaired electrons needed:

**Student Annotation:** Note that the superscript numbers in hybrid orbital notation are used to designate the number of atomic orbitals that have undergone hybridization. When the superscript is 1, it is not shown (analogous to the subscripts in chemical formulas).

Hybridization of the $s$ orbital, the three $p$ orbitals, and one of the $d$ orbitals yields hybrid orbitals that are designated $sp^3 d$. After placing the five electrons in the five hybrid orbitals, we can rationalize the formation of five bonds in the molecule:

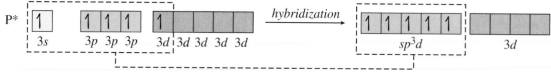

Mixing of one $s$ orbital, three $p$ orbitals,
and one $d$ orbital to yield five $sp^3 d$ orbitals

The $sp^3 d$ orbitals have shapes similar to those we have seen for the $sp$, $sp^2$, and $sp^3$ hybrid orbitals; that is, one large lobe and one small lobe. In addition, the five hybrid orbitals adopt a trigonal bipyramidal arrangement, enabling us to explain the geometry and bond angles in $PCl_5$:

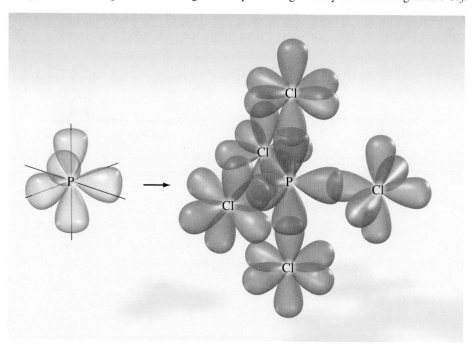

A similar analysis can be done with the $SF_6$ molecule. The ground-state electron configuration of the S atom is $[Ne]3s^2 3p^4$, giving it only two unpaired electrons. To obtain the six unpaired electrons needed to form six S—F bonds, we must promote *two* electrons to empty $d$ orbitals: one from the $3s$ orbital and one from the doubly occupied $3p$ orbital. The resulting hybrid orbitals are designated $sp^3 d^2$.

Mixing of one $s$ orbital, three $p$ orbitals,
and two $d$ orbitals to yield six $sp^3 d^2$ orbitals

The six bonds in $SF_6$ form, therefore, when each $sp^3d^2$ hybrid orbital on the S atom overlaps with a singly occupied $2p$ orbital on an F atom:

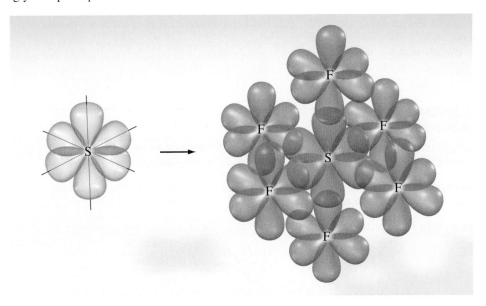

Table 9.4 shows how the number of electron domains on a central atom corresponds to a set of hybrid orbitals. In general, the hybridized bonding in a molecule can be described using the following steps:

1. Draw the Lewis structure.
2. Count the number of electron domains on the central atom. This is the number of hybrid orbitals necessary to account for the molecule's geometry. (This is also the number of *atomic* orbitals that must undergo hybridization.)
3. Draw the ground-state orbital diagram for the central atom.
4. Maximize the number of unpaired valence electrons by promotion.
5. Combine the necessary number of atomic orbitals to generate the required number of hybrid orbitals.
6. Place electrons in the hybrid orbitals, putting one electron in each orbital before pairing any electrons.

**Student Annotation:** The electrons that you place in hybrid orbitals are those that originally resided in the atomic orbitals that have undergone hybridization.

| TABLE 9.4 | Number of Electron Domains and Hybrid Orbitals on Central Atom | | | | |
|---|---|---|---|---|---|
| **Number of Electron Domains on Central Atom** | 2 | 3 | 4 | 5 | 6 |
| **Hybrid Orbitals** | $sp$ | $sp^2$ | $sp^3$ | $sp^3d$ | $sp^3d^2$ |
| **Geometry** | Linear | Trigonal planar | Tetrahedral | Trigonal bipyramidal | Octahedral |

It is important to recognize that we do not use hybrid orbitals to *predict* molecular geometries, but rather to *explain* geometries that are already known. As we saw in Section 9.3, the bonding in many molecules can be explained without the use of hybrid orbitals. Hydrogen sulfide ($H_2S$), for example, has a bond angle of 92°. This bond angle is best explained without the use of hybrid orbitals.

Sample Problem 9.5 shows how to use hybridization to explain the bonding and geometry in a molecule.

### Sample Problem 9.5

Ammonia ($NH_3$) is a trigonal pyramidal molecule with $H-N-H$ bond angles of about 107°. Describe the formation of three equivalent $N-H$ bonds, and explain the angles between them.

**Strategy**  Starting with a Lewis structure, determine the number and type of hybrid orbitals necessary to rationalize the bonding in $NH_3$.

**Setup**  The Lewis structure of $NH_3$ is

$$H-\ddot{N}-H$$
$$|\quad 107°$$
$$H$$

The ground-state electron configuration of the N atom is $[He]2s^2 2p^3$. Its valence orbital diagram is

N [↑↓]   [↑][↑][↑]
$\quad 2s^2 \qquad 2p^3$

**Solution**  Although the N atom has the three unpaired electrons needed to form three $N-H$ bonds, we would expect bond angles of ~90° (not 107°) to form from the overlap of the three mutually perpendicular $2p$ orbitals. Hybridization, therefore, is necessary to explain the bonding in $NH_3$. (Recall from Section 9.3 that valence bond theory alone could be used to explain the bonding in $H_2S$, where the bond angles are ~90°—hybridization was unnecessary.) Although we often need to promote an electron to maximize the number of unpaired electrons, no promotion is necessary for the nitrogen in $NH_3$. We already have the three unpaired electrons necessary, and the promotion of an electron from the $2s$ orbital to one of the $2p$ orbitals would not result in any additional unpaired electrons. Furthermore, there are no empty $d$ orbitals in the second shell. According to the Lewis structure, there are four electron domains on the central atom (three bonds and a lone pair of electrons). Four electron domains on the central atom require four hybrid orbitals, and four hybrid orbitals require the hybridization of four atomic orbitals: one $s$ and three $p$. This corresponds to $sp^3$ hybridization. Because the atomic orbitals involved in the hybridization contain a total of five electrons, we place five electrons in the resulting hybrid orbitals. This means that one of the hybrid orbitals will contain a lone pair of electrons:

Each $N-H$ bond is formed by the overlap between an $sp^3$ hybrid orbital on the N atom and the $1s$ atomic orbital on an H atom. Because there are four electron domains on the central atom, we expect them to be arranged in a tetrahedron. In addition, because one of the electron domains is a lone pair, we expect the $H-N-H$ bond angles to be slightly smaller than the ideal tetrahedral bond angle of 109.5°:

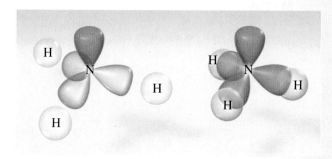

**Think About It** This analysis agrees with the experimentally observed geometry and bond angles of 107° in $NH_3$.

**Practice Problem A** Use hybrid orbital theory to describe the bonding and explain the bond angles in bromine pentafluoride ($BrF_5$).

**Practice Problem B** Use hybrid orbital theory to describe the bonding and explain the bond angles in $BeF_2$:

---

**Checkpoint 9.4** **Hybridization of Atomic Orbitals**

**9.4.1** How many orbitals does a set of $sp^2$ hybrid orbitals contain?

a) 2

b) 3

c) 4

d) 5

e) 6

**9.4.2** How many $p$ atomic orbitals are required to generate a set of $sp^3$ hybrid orbitals?

a) 0

b) 1

c) 2

d) 3

e) 4

---

## 9.5 Hybridization in Molecules Containing Multiple Bonds

The concept of valence bond theory and hybridization can also be used to describe the bonding in molecules containing double and triple bonds, such as ethylene ($C_2H_4$) and acetylene ($C_2H_2$). The Lewis structure of ethylene is

Each carbon atom is surrounded by three electron domains (two single bonds and one double bond). Thus, we expect the hybridization about each C atom to be $sp^2$, just like the B atom in $BF_3$. Applying the procedure described in Section 9.4, we first maximize the number of unpaired electrons by promoting an electron from the $2s$ orbital to the empty $2p$ orbital:

We then hybridize the required number of atomic orbitals, which in this case is three (one for each electron domain on the C atom):

The three equivalent $sp^2$ hybrid orbitals, arranged in a trigonal plane, enable us to explain the three bonds about each C atom. In this case, however, each C atom is left with a singly occupied, *unhybridized* atomic orbital. As we will see, it is the singly occupied *p* orbitals *not* involved in hybridization that give rise to multiple bonds in molecules.

In the bonding schemes that we have described thus far, the overlap of atomic orbitals or hybrid orbitals occurs directly between the two nuclei involved in bonding. Such bonds, in which the shared electron density is concentrated directly along the internuclear axis, are called ***sigma (σ) bonds.*** The ethylene molecule (also known as *ethene*) contains five sigma bonds: one between the two C atoms (the result of the overlap of one of the $sp^2$ hybrid orbitals on each C atom) and four between the C and H atoms (each the result of the overlap of an $sp^2$ hybrid orbital on a C atom and the $1s$ orbital on an H atom). The leftover unhybridized *p* orbital is perpendicular to the plane in which the atoms of the molecule lie. Figure 9.7(a) illustrates the formation and the overlap of $sp^2$ hybrid orbitals and shows the positions of the remaining *p* orbital on each C atom.

Remember that the shapes used to represent the atomic and hybrid orbitals are simplified to make visualization of the molecules easier. The actual shapes of both atomic and hybrid orbitals are such that when the $sp^2$ hybrid orbitals overlap to form a sigma bond between the two C atoms, the remaining unhybridized *p* orbitals also overlap, although to a smaller extent. Figure 9.7(b) shows the overlap of the unhybridized *p* orbitals on the two C atoms in ethylene. The resulting regions of electron density are concentrated above and below the plane of the molecule, in contrast to a sigma bond in which the electron density is concentrated directly along the internuclear axis. Bonds that form from the sideways overlap of *p* orbitals are called ***pi (π) bonds.*** The two regions of overlap shown in Figure 9.7(b) together make up one pi bond. It is the formation of the pi bond that makes the ethylene molecule planar.

A sigma bond and a pi bond together constitute a *double* bond. Because the sideways overlap of *p* orbitals is not as effective as the overlap of hybrid orbitals that point directly toward each other, the contribution of the pi bond to the overall strength of the bond is less than that of the sigma bond. The bond enthalpy of a carbon-carbon double bond (620 kJ/mol) is greater than that of a carbon-carbon single bond (347 kJ/mol), but it is not twice as large [◄◄ Section 8.9].

Sample Problem 9.6 shows how to use a Lewis structure to determine the number of sigma and pi bonds in a molecule.

Multimedia

Chemical bonding—sigma-pi bonding.

Multimedia

Chemical bonding—sigma and pi bonding in benzene.

**Figure 9.7** (a) A sigma bond forms when $sp^2$ hybrid orbitals on the C atoms overlap. Each C atom has one remaining unhybridized *p* orbital. (b) The remaining *p* orbitals overlap to form a pi bond.

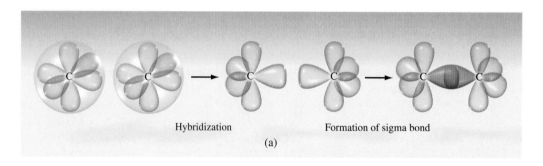

Hybridization          Formation of sigma bond

(a)

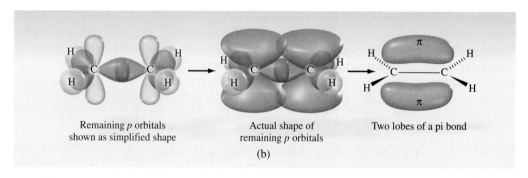

Remaining *p* orbitals          Actual shape of          Two lobes of a pi bond
shown as simplified shape          remaining *p* orbitals

(b)

## Sample Problem 9.6

Thalidomide ($C_{13}H_{10}N_2O_4$) is a sedative and antiemetic that was widely prescribed during the 1950s, although not in the United States, for pregnant women suffering from morning sickness. Its use was largely discontinued when it was determined to be responsible for thousands of devastating birth defects. Determine the number of carbon-carbon sigma bonds and the total number of pi bonds in thalidomide.

**Student Annotation:** For more on the history of thalidomide, see pages 387 and 438.

Thalidomide

**Strategy**  Use the Lewis structure to determine the number of single and double bonds. Then, to convert the number of single and double bonds to the number of sigma and pi bonds, remember that a single bond typically is composed of a sigma bond, whereas a double bond is usually composed of one sigma bond and one pi bond.

**Setup**  There are nine carbon-carbon single bonds and three carbon-carbon double bonds. Overall there are seven double bonds in the molecule (three C=C and four C=O).

**Solution**  Thalidomide contains 12 carbon-carbon sigma bonds and a total of seven pi bonds (three in carbon-carbon double bonds and four in carbon-oxygen double bonds).

**Think About It**  The Lewis structure given for thalidomide is one of two possible resonance structures. Draw the other resonance structure, and count sigma and pi bonds again. Make sure you get the same answer.

**Practice Problem A**  The active ingredient in Tylenol and a host of other over-the-counter pain relievers is acetaminophen ($C_8H_9NO_2$). Determine the total number of sigma and pi bonds in the acetaminophen molecule.

Acetaminophen

**Practice Problem B**  Determine the total number of sigma and pi bonds in a molecule of aspirin ($C_9H_8O_4$).

Aspirin

Because the *p* orbitals that form pi bonds must be parallel to each other, pi bonds restrict the rotation of a molecule in a way that sigma bonds do not. For example, the molecule 1,2-dichloroethane exists as a single isomer. Although we can draw the molecule in several different ways, including the two shown in Figure 9.8(a), all of them are equivalent because the molecule can rotate freely about the sigma bond between the two carbon atoms.

On the other hand, 1,2-dichloroethylene exists as two distinct isomers—*cis* and *trans*—as shown in Figure 9.8(b). The double bond between the carbon atoms consists of one sigma bond and one pi bond. The pi bond restricts rotation about the sigma bond, making the molecules rigid, planar, and not interchangeable. To change one isomer into the other, the pi bond would have to be broken and rotation would have to occur about the sigma bond and the pi bond. This process would require a significant input of energy.

**Student Annotation:** Remember that *isomers* are molecules with the same chemical formula but different structural arrangements of atoms (Section 9.2).

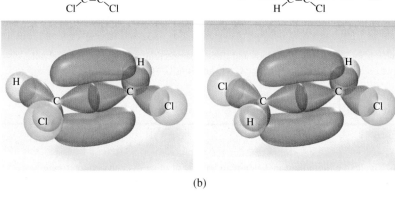

**Figure 9.8** (a) There is free rotation about the C—C single bond. All three Lewis structures represent the same molecule. (b) There is no rotation about the C—C double bond. There are two isomers of CHClCHCl.

H—C≡C—H

C₂H₂

Media Player/MPEG Animation: Figure 9.10, Formation of pi Bonds in Ethylene and Acetylene, pp. 366–67.

The acetylene molecule (C₂H₂) is linear. Because each carbon atom has two electron domains around it in the Lewis structure, the carbon atoms are *sp* hybridized. As before, promotion of an electron first maximizes the number of unpaired electrons:

C $\boxed{1\!\downarrow}$ $\boxed{1\;\;1\;\;\;}$ $\xrightarrow{\text{promotion}}$ C* $\boxed{1}$ $\boxed{1\;\;1\;\;1}$

 $2s^2$ $\quad 2p^2$ $\qquad\qquad\qquad\quad 2s^1$ $\quad\; 2p^3$

The 2s orbital and one of the 2p orbitals then mix to form two *sp* hybrid orbitals:

C* $\boxed{1}$ $\boxed{1\;\;1\;\;1}$ $\xrightarrow{\text{hybridization}}$ $\boxed{1\;\;1}$ $\boxed{1\;\;1}$

 $2s$ $\;\; 2p\;\; 2p\;\; 2p$ $\qquad\qquad\qquad sp\;\; sp$ $\quad 2p$

This leaves two unhybridized *p* orbitals (each containing an electron) on each C atom. Figure 9.9 shows the sigma and pi bonds in the acetylene molecule (also known as *ethyne*). Just as one sigma bond and one pi bond make up a *double* bond, one sigma bond and two pi bonds make up a *triple* bond. Figure 9.10 (pp. 366–67) summarizes the formation of bonds in ethane, ethylene, and acetylene.

Sample Problem 9.7 shows how hybrid orbitals and pi bonds can be used to explain the bonding in formaldehyde, a molecule with a carbon-oxygen double bond.

## Sample Problem 9.7

In addition to its use in aqueous solution as a preservative for laboratory specimens, formaldehyde gas is used as an antibacterial fumigant. Use hybridization to explain the bonding in formaldehyde (CH₂O).

**Strategy** Draw the Lewis structure of formaldehyde, determine the hybridization of the C and O atoms, and describe the formation of the sigma and pi bonds in the molecule.

**Setup** The Lewis structure of formaldehyde is

Ö
‖
H—C—H

The C and O atoms each have three electron domains around them. [Carbon has two single bonds (C—H) and a double bond (C=O); oxygen has a double bond (O=C) and two lone pairs.]

**Solution** Three electron domains correspond to $sp^2$ hybridization. For carbon, promotion of an electron from the $2s$ orbital to the empty $2p$ orbital is necessary to maximize the number of unpaired electrons. For oxygen, no promotion is necessary. Each undergoes hybridization to produce $sp^2$ hybrid orbitals; and each is left with a singly occupied, unhybridized $p$ orbital:

A sigma bond is formed between the C and O atoms by the overlap of one of the $sp^2$ hybrid orbitals from each of them. Two more sigma bonds form between the C atom and the H atoms by the overlap of carbon's remaining $sp^2$ hybrid orbitals with the $1s$ orbital on each H atom. Finally, the remaining $p$ orbitals on C and O overlap to form a pi bond:

**Student Annotation:** The two lone pairs on the O atom are the electrons in the doubly occupied $sp^2$ hybrid orbitals.

**Think About It** Our analysis describes the formation of both a sigma bond and a pi bond between the C and O atoms. This corresponds correctly to the double bond predicted by the Lewis structure.

**Practice Problem A** Use valence bond theory and hybrid orbitals to explain the bonding in hydrogen cyanide (HCN).

**Practice Problem B** Use valence bond theory and hybrid orbitals to explain the bonding in diatomic nitrogen ($N_2$).

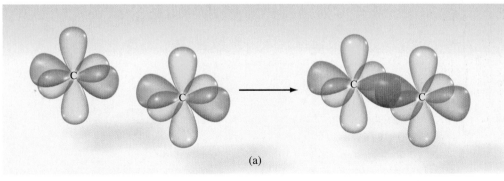

(a)

**Figure 9.9** (a) Formation of the sigma bond in acetylene. (b) Formation of the pi bonds in acetylene.

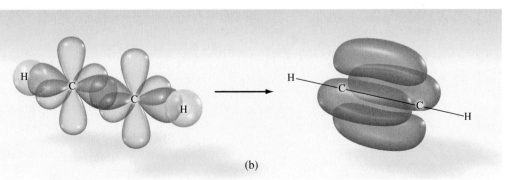

(b)

# Figure 9.10
# Formation of Pi Bonds in Ethylene and Acetylene

2s and 2p atomic orbitals on two C atoms

In ethane ($C_2H_6$) the C atoms are $sp^3$-hybridized.

In ethylene ($C_2H_4$) the C atoms are $sp^2$-hybridized.

Overlap of $sp^2$ hybrid orbitals forms a sigma bond between the two C atoms.

In acetylene ($C_2H_2$) the C atoms are $sp$-hybridized.

Overlap of $sp$ hybrid orbitals forms a sigma bond between the two C atoms.

Overlap of $sp^3$ hybrid orbitals forms a sigma bond between the two C atoms.

Each C atom forms three sigma bonds with H atoms.

Each C atom forms two sigma bonds with H atoms. Each C atom has one leftover unhybridized $p$ orbital.

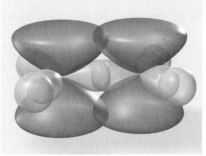

Using the more realistic shape shows how $p$ orbitals overlap.

The parallel, unhybridized $p$ orbitals overlap to form a pi bond with two lobes.

Each C atom forms one sigma bond with an H atom. Each C atom has two leftover unhybridized $p$ orbitals.

Each pair of parallel, unhybridized $p$ orbitals overlaps to form a pi bond with two lobes.

## What's the point?

When carbon is $sp^2$- or $sp$-hybridized, parallel, unhybridized $p$ orbitals interact to form pi bonds. Each pi bond consists of two lobes as a result of the overlap.

| Checkpoint 9.5 | **Hybridization in Molecules Containing Multiple Bonds** |
| --- | --- |

**9.5.1** Which of the following molecules contain one or more pi bonds? (Select all that apply.)

a) $N_2$

b) $Cl_2$

c) $CO_2$

d) $CH_3OH$

e) $CCl_4$

**9.5.2** From left to right, give the hybridization of each carbon atom in the allene molecule ($H_2C=C=CH_2$).

a) $sp^2, sp^2, sp^2$

b) $sp^3, sp^2, sp^3$

c) $sp^2, sp, sp^2$

d) $sp^3, sp, sp^3$

e) $sp^3, sp^3, sp^3$

**9.5.3** Which of the following pairs of atomic orbitals on adjacent nuclei can overlap to form a sigma bond? Consider the $x$ axis to be the internuclear axis.

a) $1s$ and $2s$

b) $1s$ and $2p_x$

c) $2p_y$ and $2p_y$

d) $3p_y$ and $3p_z$

e) $2p_x$ and $3p_x$

**9.5.4** Which of the following pairs of atomic orbitals on adjacent nuclei can overlap to form a pi bond? Consider the $x$ axis to be the internuclear axis.

a) $1s$ and $2s$

b) $2s$ and $2s$

c) $2p_y$ and $2p_y$

d) $3p_y$ and $3p_z$

e) $2p_x$ and $3p_x$

Multimedia

Chemical bonding—paramagnetic liquid oxygen.

**Liquid oxygen is attracted to the poles of a magnet because $O_2$ is paramagnetic.**

**Liquid nitrogen is not attracted to the poles of a magnet because $N_2$ is diamagnetic.**

$$:\ddot{O}=\ddot{O}:$$

$$O_2$$

**Student Annotation:** Recall that a *homonuclear diatomic molecule* is one in which both atoms are the same element [◀◀ Section 2.6].

**Student Annotation:** Remember that the quantum mechanical approach treats atomic orbitals as *wave functions* [◀◀ Section 6.5], and that one of the properties of waves is their capacity for both constructive combination and destructive combination [◀◀ Section 6.1].

| 9.6 | Molecular Orbital Theory |
| --- | --- |

Although the bonding theories we have seen thus far provide simple and effective ways for us to visualize molecules and to predict their shapes and bond angles, Lewis structures and valence bond theory do not enable us to describe or predict some important properties of molecules. Diatomic oxygen, for example, exhibits a property called *paramagnetism.* **Paramagnetic** species are attracted by magnetic fields, whereas *diamagnetic* species are weakly repelled by them. Such magnetic properties are the result of a molecule's electron configuration. Species in which all the electrons are *paired* are diamagnetic, whereas species that contain one or more *unpaired* electrons are paramagnetic. Because $O_2$ exhibits paramagnetism, it must contain unpaired electrons. According to the Lewis structure of $O_2$ (shown in the margin) and the valence bond theory description of $O_2$, however, all the electrons in $O_2$ are paired. Another bonding theory, called molecular orbital theory, is needed to describe the paramagnetism of $O_2$ and other important molecular properties.

According to **molecular orbital theory,** the atomic orbitals involved in bonding actually combine to form new orbitals that are the "property" of the entire molecule, rather than of the atoms forming the bonds. These new orbitals are called **molecular orbitals.** In molecular orbital theory, electrons shared by atoms in a molecule reside in the molecular orbitals.

Molecular orbitals are like atomic orbitals in several ways: they have specific shapes and specific energies, and they can each accommodate a maximum of two electrons. As was the case with atomic orbitals, two electrons residing in the same molecular orbital must have opposite spins, as required by the Pauli exclusion principle. And, like hybrid orbitals, the number of molecular orbitals we get is equal to the number of atomic orbitals we combine.

Our treatment of molecular orbital theory in this book will be limited to descriptions of bonding in diatomic molecules consisting of elements from the first two periods of the periodic table (H through Ne).

### Bonding and Antibonding Molecular Orbitals

To begin our discussion, we consider $H_2$, the simplest homonuclear diatomic molecule. According to *valence* bond theory, an $H_2$ molecule forms when two H atoms are close enough for their $1s$ atomic orbitals to overlap. According to *molecular* orbital theory, two H atoms come together to form $H_2$ when their $1s$ atomic orbitals combine to give molecular orbitals. Figure 9.11 shows the $1s$

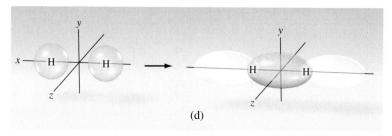

**Figure 9.11** (a) Two *s* atomic orbitals combine to give two sigma molecular orbitals. (b) One of the molecular orbitals is lower in energy than the original atomic orbitals (darker), and (c) one is higher in energy (lighter). The two light yellow lobes make up one molecular orbital. (d) Atomic and molecular orbitals shown relative to the H nuclei.

atomic orbitals of the isolated H atoms and the molecular orbitals that result from their constructive and destructive combinations. The *constructive* combination of the two $1s$ orbitals gives rise to a molecular orbital [Figure 9.11(b)] that lies along the internuclear axis, directly between the two H nuclei. Just as electron density shared between two nuclei in overlapping atomic orbitals drew the nuclei together, electron density in a molecular orbital that lies between two nuclei will draw them together, too. Thus, this molecular orbital is referred to as a ***bonding molecular orbital.***

The *destructive* combination of the $1s$ atomic orbitals also gives rise to a molecular orbital that lies along the internuclear axis, but, as Figure 9.11(c) shows, this molecular orbital, which consists of two lobes, does not lie in between the two nuclei. Electron density in this molecular orbital would actually pull the two nuclei in opposite directions, rather than toward each other. This is referred to as an ***antibonding molecular orbital.***

## $\sigma$ Molecular Orbitals

Molecular orbitals that lie along the internuclear axis (such as the bonding and antibonding molecular orbitals in $H_2$) are referred to as $\sigma$ molecular orbitals. Specifically, the *bonding* molecular orbital formed by the combination of two $1s$ atomic orbitals is designated $\sigma_{1s}$ and the *antibonding* orbital is designated $\sigma_{1s}^*$, where the asterisk distinguishes an antibonding orbital from a bonding orbital. Figure 9.11(d) summarizes the combination of two $1s$ atomic orbitals to yield two molecular orbitals: one bonding and one antibonding.

Like atomic orbitals, molecular orbitals have specific energies. The combination of two atomic orbitals of equal energy, such as two $1s$ orbitals on two H atoms, yields one molecular orbital that is lower in energy (bonding) and one molecular orbital that is higher in energy (antibonding) than the original atomic orbitals. The bonding molecular orbital in $H_2$ is concentrated between the nuclei, along the internuclear axis. Electron density in this molecular orbital both attracts the nuclei and shields them from each other, stabilizing the molecule. Thus, the bonding molecular orbital is lower in energy than the isolated atomic orbitals. In contrast, the antibonding molecular orbital has most of its electron density outside the internuclear region. Electron density in this orbital does not shield one nucleus from the other, which increases the nuclear repulsions and makes the antibonding molecular orbital higher in energy than the isolated atomic orbitals.

Showing all the molecular orbitals in a molecule can make for a very complicated picture. Rather than represent molecules with pictures of their molecular orbitals, we generally use diagrams in which molecular orbitals are represented with boxes placed at the appropriate relative energy levels. Figure 9.12 shows the energies of the $\sigma_{1s}$ and $\sigma_{1s}^*$ molecular orbitals in $H_2$, relative to the energy of the original $1s$ orbitals on two isolated H atoms. Like atomic orbitals, molecular orbitals fill in order of increasing energy. Note that the electrons that originally resided in the atomic orbitals both occupy the lowest-energy molecular orbital, $\sigma_{1s}$, with opposite spins.

We can construct a similar molecular orbital diagram for the hypothetical molecule $He_2$. Like the H atom, the He atom has a $1s$ orbital (unlike H, though, the $1s$ orbital on He has two electrons, not one). The combination of $1s$ orbitals to form molecular orbitals in $He_2$ is essentially the same as what we have described for $H_2$. The placement of electrons is shown in Figure 9.13.

Multimedia

Chemical bonding—formation of molecular orbitals.

**Student Annotation:** The designations $\sigma$ and $\pi$ are used in molecular orbital theory just as they are in valence bond theory: $\sigma$ refers to electron density along the internuclear axis, and $\pi$ refers to electron density that influences both nuclei but that does not lie directly along the internuclear axis.

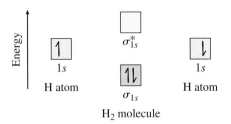

**Figure 9.12**    Relative energies of atomic orbitals in H and molecular orbitals in H$_2$.

**Figure 9.13**    Relative energies of atomic orbitals in He and molecular orbitals in He$_2$.

## Bond Order

With molecular orbital diagrams such as those for H$_2$ and He$_2$, we can begin to see the power of molecular orbital theory. For a diatomic molecule described using molecular orbital theory, we can calculate the **bond order.** The value of the bond order indicates, qualitatively, how *stable* a molecule is. The higher the bond order, the more stable the molecule. Bond order is calculated in the following way:

Equation 9.1    $$\text{bond order} = \frac{\begin{array}{c}\text{number of electrons} \\ \text{in bonding} \\ \text{molecular orbitals}\end{array} - \begin{array}{c}\text{number of electrons} \\ \text{in antibonding} \\ \text{molecular orbitals}\end{array}}{2}$$

In the case of H$_2$, where both electrons reside in the $\sigma_{1s}$ orbital, the bond order is $[(2 - 0)/2] = 1$. In the case of He$_2$, where the two additional electrons reside in the $\sigma_{1s}^{*}$ orbital, the bond order is $[(2 - 2)/2] = 0$. Molecular orbital theory predicts that a molecule with a bond order of zero will not exist and He$_2$, in fact, does *not* exist under ordinary conditions.

We can do similar analyses of the molecules Li$_2$ and Be$_2$. (The Li and Be atoms have ground-state electron configurations of [He]$2s^1$ and [He]$2s^2$, respectively.) The $2s$ atomic orbitals also combine to form the corresponding $\sigma$ and $\sigma^{*}$ molecular orbitals. Figure 9.14 shows the molecular orbital diagrams and bond orders for Li$_2$ and Be$_2$.

As predicted by molecular orbital theory, Li$_2$, with a bond order of 1, is a stable molecule, whereas Be$_2$, with a bond order of 0, does not exist.

## $\pi$ Molecular Orbitals

To consider diatomic molecules beyond Be$_2$, we must also consider the combination of $p$ atomic orbitals. Like $s$ orbitals, $p$ orbitals combine both constructively, to give bonding molecular orbitals that are lower in energy than the original atomic orbitals, and destructively, to give antibonding

**Figure 9.14**    Bond order determination for Li$_2$ and Be$_2$.

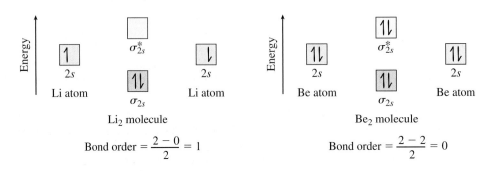

molecular orbitals that are higher in energy than the original atomic orbitals. However, the orientations of $p_x$, $p_y$, and $p_z$ orbitals give rise to two different types of molecular orbitals: $\sigma$ molecular orbitals, in which the regions of electron density in the bonding and antibonding molecular orbitals lie along the internuclear axis, and $\pi$ molecular orbitals, in which the regions of electron density affect both nuclei but do *not* lie along the internuclear axis.

Orbitals that lie along the internuclear axis, as the $2p_x$ orbitals do in Figure 9.15(a), point directly toward each other and combine to form $\sigma$ molecular orbitals. Figure 9.15(b) shows the combination of two $2p_x$ atomic orbitals to give two molecular orbitals designated $\sigma_{2px}$ and $\sigma^*_{2p_x}$. Figure 9.15(c) shows the relative energies of these molecular orbitals.

Orbitals that are aligned parallel to each other, like the $2p_y$ and $2p_z$ orbitals shown in Figure 9.15(a), combine to form $\pi$ molecular orbitals. These bonding molecular orbitals are designated $\pi_{2p_y}$ and $\pi_{2p_z}$; the corresponding antibonding molecular orbitals are designated $\pi^*_{2p_y}$ and $\pi^*_{2p_z}$. Often we refer to the molecular orbitals collectively using the designations $\pi_{2p_{y,z}}$ and $\pi^*_{2p_{y,z}}$. Figure 9.16(a) shows the constructive and destructive combination of parallel $p$ orbitals. Figure 9.16(b) shows the locations of the molecular orbitals resulting from the combination of $p_y$, and $p_z$ orbitals relative to the two atomic nuclei. Again, electron density in the resulting *bonding* molecular orbitals serves to hold the nuclei together, whereas electron density in the *antibonding* molecular orbitals does not.

Just as the $p$ atomic orbitals within a particular shell are higher in energy than the $s$ orbital in the same shell, all the molecular orbitals resulting from the combination of $p$ atomic orbitals are higher in energy than the molecular orbitals resulting from the combination of $s$ atomic orbitals. To understand better the relative energy levels of the molecular orbitals resulting from $p$-orbital combinations, consider the fluorine molecule ($F_2$).

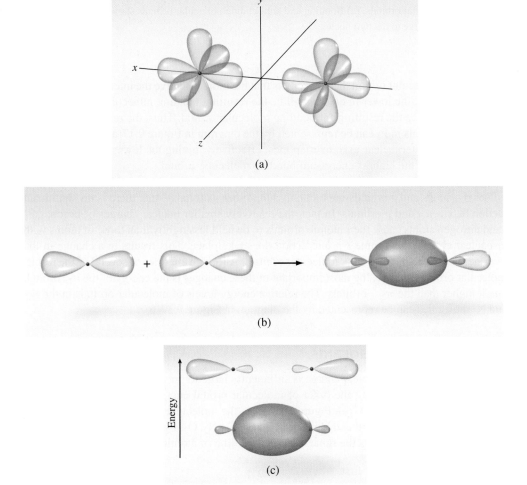

(a)

(b)

(c)

**Figure 9.15** (a) Two sets of $2p$ orbitals. (b) The $p$ atomic orbitals that point toward each other ($p_x$) combine to give bonding and antibonding $\sigma$ molecular orbitals. (c) The antibonding $\sigma$ molecular orbital is higher in energy than the corresponding bonding $\sigma$ molecular orbital.

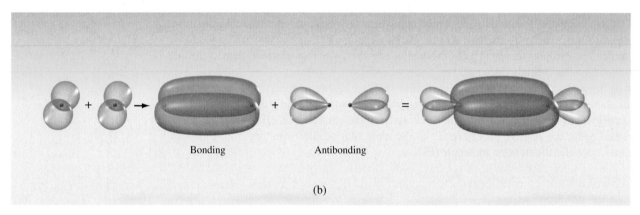

**Figure 9.16**    Parallel *p* atomic orbitals combine to give $\pi$ molecular orbitals. (a) Bonding and antibonding molecular orbitals shown separately. (b) Bonding and antibonding molecular orbitals shown together relative to the two nuclei.

In general, molecular orbital theory predicts that the more effective the interaction or overlap of the atomic orbitals, the lower in energy will be the resulting bonding molecular orbital and the higher in energy will be the resulting antibonding molecular orbital. Thus, the relative energy levels of molecular orbitals in $F_2$ can be represented by the diagram in Figure 9.17(a). The $p_x$ orbitals, which lie along the internuclear axis, overlap most effectively, giving the lowest-energy bonding molecular orbital and the highest-energy antibonding molecular orbital.

The order of orbital energies shown in Figure 9.17(a) assumes that *p* orbitals interact only with other *p* orbitals and *s* orbitals interact only with other *s* orbitals—that there is no significant interaction between *s* and *p* orbitals. In fact, the relatively smaller nuclear charges of boron, carbon, and nitrogen atoms cause their atomic orbitals to be held less tightly than those of atoms with larger nuclear charges, and some *s-p* interaction does take place. This results in a change in the relative energies of the $\sigma_{2p_x}$ and $\pi_{2p_{y,z}}$ molecular orbitals. Although energies of several of the resulting molecular orbitals change, the most important of these changes is the energy of the $\sigma_{2p}$ orbital, making it higher than the $\pi_{2p_{y,z}}$ orbitals. The relative energy levels of molecular orbitals in the $B_2$, $C_2$, and $N_2$ molecules can be represented by the diagram in Figure 9.17(b).

## Molecular Orbital Diagrams

Beginning with oxygen, the nuclear charge is sufficiently large to prevent the interaction of *s* and *p* orbitals. Thus, for $O_2$ and $Ne_2$, the order of molecular orbital energies is the same as that for $F_2$, which is shown in Figure 9.17(a). Figure 9.18 gives the molecular orbital diagrams, magnetic properties, bond orders, and bond enthalpies for $Li_2$, $B_2$, $C_2$, $N_2$, $O_2$, $F_2$, and $Ne_2$. Note that the filling of molecular orbitals follows the same rules as the filling of atomic orbitals [◄◄ Section 6.8]:

- Lower-energy orbitals fill first.
- Each orbital can accommodate a maximum of two electrons with opposite spins.
- Hund's rule is obeyed.

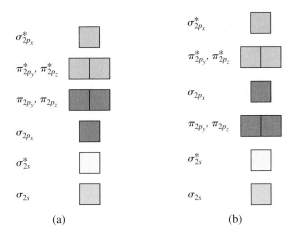

**Figure 9.17**   (a) Ordering of molecular orbital energies for $O_2$ and $F_2$. (b) Ordering of molecular orbital energies for $Li_2$, $B_2$, $C_2$, and $N_2$. Bonding orbitals are darker; antibonding orbitals are lighter.

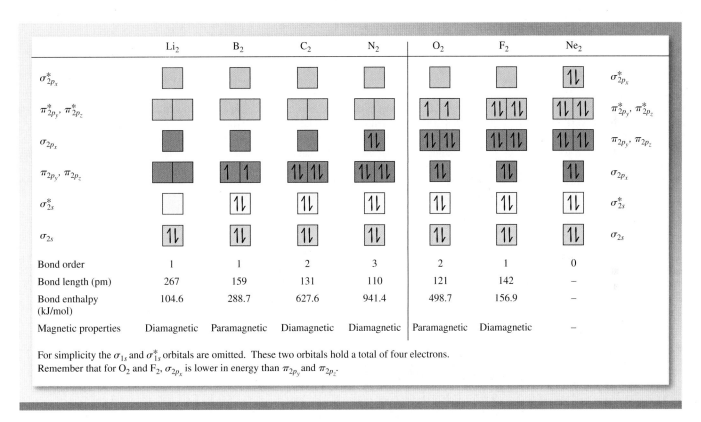

**Figure 9.18**   Molecular orbital diagrams for second-period homonuclear diatomic molecules.

There are several important predictions made by the molecular orbital diagrams in Figure 9.18. First, molecular orbital theory correctly predicts that $Ne_2$, with a bond order of 0, does not exist. Second, it correctly predicts the magnetic properties of the molecules that do exist. Both $B_2$ and $O_2$ are known to be paramagnetic. Third, although bond order is only a qualitative measure of bond strength, the calculated bond orders of the molecules correlate well with the measured bond enthalpies. The $N_2$ molecule, with a bond order of 3, has the largest bond enthalpy of the five molecules. The $B_2$ and $F_2$ molecules, each with a bond order of 1, have the smallest bond enthalpies. Its ability to predict correctly the properties of molecules makes molecular orbital theory a powerful tool in the study of chemical bonding.

Sample Problem 9.8 shows how to use molecular orbital diagrams to determine the magnetic properties and bond order of the superoxide ion.

**Sample Problem 9.8**

The superoxide ion ($O_2^-$) has been implicated in a number of degenerative conditions, including aging and Alzheimer's disease. Using molecular orbital theory, determine whether $O_2^-$ is paramagnetic or diamagnetic, and then calculate its bond order.

**Strategy**  Start with the molecular orbital diagram for $O_2$, add an electron, and then use the resulting diagram to determine the magnetic properties and bond order.

**Setup**  The molecular orbital diagram for $O_2$ is shown in Figure 9.18. The additional electron must be added to the lowest-energy molecular orbital available.

Molecular orbital diagram for $O_2^-$

**Think About It**  Experiments confirm that the superoxide ion is paramagnetic. Also, any time we add one or more electrons to an antibonding molecular orbital, as we did in this problem, we should expect the bond order to decrease. Electrons in antibonding orbitals cause a bond to be less stable.

**Solution**  In this case, either of the two singly occupied $\pi_{2p}^*$ orbitals can accommodate an additional electron. This gives a molecular orbital diagram in which there is one unpaired electron, making $O_2^-$ paramagnetic. The new diagram has six electrons in bonding molecular orbitals and three in antibonding molecular orbitals. We can ignore the electrons in the $\sigma_{2s}$ and $\sigma_{2s}^*$ orbitals because their contributions to the bond order cancel each other. The bond order is $(6 - 3)/2 = 1.5$.

**Practice Problem A**  Use molecular orbital theory to determine whether $N_2^{2-}$ is paramagnetic or diamagnetic, and then calculate its bond order.

**Practice Problem B**  Use molecular orbital theory to determine whether $F_2^{2+}$ is paramagnetic or diamagnetic, and then calculate its bond order.

---

**Checkpoint 9.6**    **Molecular Orbital Theory**

**9.6.1**  Calculate the bond order of $N_2^{2+}$, and determine whether it is paramagnetic or diamagnetic.

a) 2, paramagnetic

b) 2, diamagnetic

c) 3, paramagnetic

d) 3, paramagnetic

e) 1, paramagnetic

**9.6.2**  Which of the following species is paramagnetic? (Select all that apply.)

a) $C_2^{2-}$

b) $O_2^{2+}$

c) $F_2^{2+}$

d) $F_2^{2-}$

e) $C_2^{2+}$

**9.6.3**  Calculate the bond order of $He_2^+$.

a) 0

b) 0.5

c) 1.0

d) 1.5

e) 2

**9.6.4**  Which if any of the following species has a bond order of 0? (Select all that apply.)

a) $B_2^{2+}$

b) $Ne_2^{2+}$

c) $F_2^{2-}$

d) $He_2^{2+}$

e) $H_2^{2-}$

# 9.7 Bonding Theories and Descriptions of Molecules with Delocalized Bonding

The progression of bonding theories in this chapter illustrates the importance of model development. Scientists use models to understand experimental results and to predict future observations. A model is useful as long as it agrees with observation. When it fails to do so, it must be replaced with a new model. What follows is a synopsis of the strengths and weaknesses of the bonding theories presented in Chapters 8 and 9:

### Lewis Theory

*Strength:*  The Lewis theory of bonding enables us to make qualitative predictions about bond strengths and bond lengths. Lewis structures are easy to draw and are widely used by chemists [◄◄ Section 8.9].

*Weakness:*  Lewis structures are two dimensional, whereas molecules are three dimensional. In addition, Lewis theory fails to account for the differences in bonds in compounds such as $H_2$, $F_2$, and HF. It also fails to explain *why* bonds form.

### The Valence-Shell Electron-Pair Repulsion Model

*Strength:*  The VSEPR model enables us to predict the shapes of many molecules and polyatomic ions.

*Weakness:*  Because the VSEPR model is based on the Lewis theory of bonding, it also fails to explain why bonds form.

### Valence Bond Theory

*Strength:*  Valence bond theory describes the formation of covalent bonds as the overlap of atomic orbitals. Bonds form because the resulting molecule has a lower potential energy than the original, isolated atoms.

*Weakness:*  Valence bond theory alone fails to explain the bonding in many molecules such as $BeCl_2$, $BF_3$, and $CH_4$, in which the central atom in its ground state does not have enough unpaired electrons to form the observed number of bonds.

### Hybridization of Atomic Orbitals

*Strength:*  The hybridization of atomic orbitals is not a separate bonding theory; rather, it is an *extension* of valence bond theory. Using hybrid orbitals, we can understand the bonding and geometry of more molecules, including $BeCl_2$, $BF_3$, and $CH_4$.

*Weakness:*  Valence bond theory and hybrid orbitals fail to predict some of the important properties of molecules, such as the paramagnetism of $O_2$.

### Molecular Orbital Theory

*Strength:*  Molecular orbital theory enables us to predict accurately the magnetic and other properties of molecules and ions.

*Weakness:*  Pictures of molecular orbitals can be very complex.

Although molecular orbital theory is in many ways the most powerful of the bonding models, it is also the most complex, so we continue to use the other models when they do an adequate job of explaining or predicting the properties of a molecule. For example, if you need to predict the three-dimensional shape of an $AB_x$ molecule on an exam, you should draw its Lewis structure and apply the VSEPR model. Don't try to draw its molecular orbital diagram. On the other hand, if you need to determine the bond order of a diatomic molecule or ion, you should draw a molecular orbital diagram. In general chemistry, it is best to use the *simplest* theory that can answer a particular question.

Because they remain useful, we don't discard the old models when we develop new ones. In fact, the bonding in some molecules, such as benzene ($C_6H_6$), is best described using a combination of models. Benzene can be represented with two resonance structures [◄◄ Section 8.7].

According to its Lewis structure and valence bond theory, the benzene molecule contains twelve σ bonds (six carbon-carbon, and six carbon-hydrogen) and three π bonds. From experimental evidence, however, we know that benzene does not have three single bonds and three double bonds between carbon atoms. Rather, there are six equivalent carbon-carbon bonds. This is precisely the reason that two different Lewis structures are necessary to represent the molecule. Neither one alone accurately depicts the nature of the carbon-carbon bonds. In fact, the $\pi$ bonds in benzene are **delocalized,** meaning that they are spread out over the entire molecule, rather than confined between two specific atoms. (Bonds that are confined between two specific atoms are called **localized** bonds.) Valence bond theory does a good job of describing the localized $\sigma$ bonds in benzene, but molecular orbital theory does a better job of using delocalized $\pi$ bonds to describe the bonding scheme in benzene.

To describe the $\sigma$ bonds in benzene, begin with a Lewis structure and count the electron domains on the carbon atoms. (Either resonance structure will give the same result.) Each C atom has three electron domains around it (two single bonds and one double bond). Recall from Table 9.4 that an atom that has three electron domains is $sp^2$ hybridized. To obtain the three unpaired electrons necessary on each C atom, one electron from each C atom must be promoted from the doubly occupied $2s$ orbital to an empty $2p$ orbital:

$$\text{C} \;\boxed{\uparrow\downarrow}\; \boxed{\uparrow\;|\;\uparrow\;|\;} \quad \xrightarrow{\textit{promotion}} \quad \text{C*} \;\boxed{\uparrow}\; \boxed{\uparrow\;|\;\uparrow\;|\;\uparrow}$$
$$\qquad\quad 2s^2 \qquad\; 2p^2 \qquad\qquad\qquad\qquad\; 2s^1 \qquad\quad 2p^3$$

This actually creates four unpaired electrons. Next, the orbitals are $sp^2$ hybridized, leaving one singly occupied, unhybridized $2p$ orbital on each C atom:

$$\text{C*} \;\boxed{\uparrow}\;\; \boxed{\uparrow\;|\;\uparrow\;|\;\uparrow} \quad \xrightarrow{\textit{hybridization}} \quad \boxed{\uparrow\;|\;\uparrow\;|\;\uparrow}\;\; \boxed{\uparrow}$$
$$\qquad\;\; 2s \qquad 2p\;\,2p\;2p \qquad\qquad\qquad\qquad sp^2\;sp^2\;sp^2 \qquad 2p$$

The $sp^2$ hybrid orbitals adopt a trigonal planar arrangement and overlap with one another (and with $1s$ orbitals on H atoms) to form the $\sigma$ bonds in the molecule.

The remaining unhybridized $2p$ orbitals (one on each C atom) combine to form molecular orbitals. Because the $p$ orbitals are all *parallel* to one another, only $\pi_{2p}$ and $\pi^*_{2p}$ molecular orbitals form. The combination of these six $2p$ atomic orbitals forms six molecular orbitals: three bonding and three antibonding. These molecular orbitals are delocalized over the entire benzene molecule:

$\sigma$ bonds in benzene

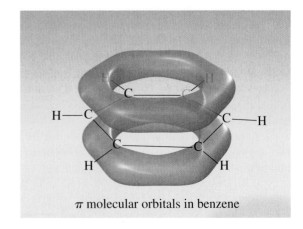

$\pi$ molecular orbitals in benzene

In the ground state, the lower-energy *bonding* molecular orbitals contain all six electrons. The electron density in the delocalized $\pi$ molecular orbitals lies above and below the plane that contains all the atoms and the $\sigma$ bonds in the molecule:

Sample Problem 9.9 shows how to combine valence bond theory and molecular orbital theory to explain the bonding in the carbonate ion.

## Sample Problem 9.9

It takes three resonance structures to represent the carbonate ion ($CO_3^{2-}$):

None of the three, though, is a completely accurate depiction. As with benzene, the bonds that are shown in the Lewis structure as one double and two single are actually three equivalent bonds. Use a combination of valence bond theory and molecular orbital theory to explain the bonding in $CO_3^{2-}$.

**Student Annotation:** Double bonds that appear in different places in different resonance structures represent delocalized $\pi$ bonds.

**Strategy** Starting with the Lewis structure, use valence bond theory and hybrid orbitals to describe the $\sigma$ bonds. Then use molecular orbital theory to describe the delocalized $\pi$ bonding.

**Setup** The Lewis structure of the carbonate ion shows three electron domains around the central C atom, so the carbon must be $sp^2$ hybridized.

**Solution** Each of the $sp^2$ hybrid orbitals on the C atom overlaps with a singly occupied $p$ orbital on an O atom, forming the three $\sigma$ bonds. Each O atom has an additional, singly occupied $p$ orbital, perpendicular to the one involved in $\sigma$ bonding. The unhybridized $p$ orbital on C overlaps with the $p$ orbitals on O to form $\pi$ bonds, which have electron densities above and below the plane of the molecule. Because the species can be represented with resonance structures, we know that the $\pi$ bonds are delocalized.

**Think About It** Although the Lewis structure of $CO_3^{2-}$ shows three electron domains on one of the O atoms, we generally do not treat *terminal* atoms (those with single bonds to only one other atom) as though they are hybridized because it is unnecessary to do so.

**Practice Problem A** Use a combination of valence bond theory and molecular orbital theory to describe the bonding in ozone ($O_3$).

**Practice Problem B** Use a combination of valence bond theory and molecular orbital theory to describe the bonding in the nitrite ion ($NO_2^-$).

---

## Checkpoint 9.7  Bonding Theories and Descriptions of Molecules with Delocalized Bonding

**9.7.1** Which of the following contain one or more delocalized $\pi$ bonds? (Select all that apply.)

a) $O_2$

b) $CO_2$

c) $NO_2^-$

d) $CH_4$

e) $CH_2Cl_2$

**9.7.2** Which of the atoms in $BCl_3$ need hybrid orbitals to describe the bonding in the molecule?

a) all four atoms

b) only the B atom

c) only the three Cl atoms

d) only the B atom and one Cl atom

e) only the B atom and two Cl atoms

**9.7.3** Which of the following can hybrid orbitals be used for? (Select all that apply.)

a) to explain the geometry of a molecule

b) to explain how a central atom can form more bonds than the number of unpaired electrons in its ground-state configuration

c) to predict the geometry of a molecule

d) to explain the magnetic properties of a molecule

e) to predict the magnetic properties of a molecule

**9.7.4** Which of the following enables us to explain the paramagnetism of $O_2$?

a) Lewis theory

b) valence bond theory

c) valence-shell electron-pair repulsion

d) hybridization of atomic orbitals

e) molecular orbital theory

# Applying What You've Learned

Recall from the beginning of the chapter that one of the fundamental scents for which we have olfactory receptors is musky. Musk is a familiar, provocative scent that has been part of the human experience for thousands of years. The primary odorous molecule in musk is muscone, a large, cyclic, organic molecule. The natural source of musk is a gland on the abdomen of the mature male musk deer, a small species of deer native to the Himalayas. The practice of harvesting musk nearly wiped out the musk deer population.

In 1888, Albert Baur, while experimenting with explosives, stumbled onto some synthetic compounds that smelled like musk. Two of these compounds are shown below. (Note the similarity between the structure of musk Baur and that of TNT.) Musk ketone, the molecule that smelled closest to muscone, was far cheaper than natural musk, but it was toxic and hazardous to prepare. In the 1950s, it was replaced with a series of synthetic musk compounds without the $NO_2$ groups. These new compounds are widely used in the fragrance industry.

Muscone

TNT            Musk Baur            Musk Ketone

**Problems:**

a)  How many carbon-carbon sigma bonds are there in musk ketone? How many carbon-carbon or carbon-oxygen pi bonds are there? [◄◄ Sample Problem 9.6]

b)  Determine the hybridization of the carbon atoms circled in red in the musk ketone molecule, and describe their geometries. [◄◄ Sample Problem 9.5]

c)  Of the pi bonds in the musk ketone molecule, which are localized and which are delocalized?

d)  Describe the bonding in the six-membered ring portion of the musk ketone molecule using a combination of valence bond theory and molecular orbital theory. [◄◄ Sample Problem 9.9]

# CHAPTER SUMMARY

## Section 9.1

- According to the *valence-shell electron-pair repulsion (VSEPR)* model, electron pairs in the valence shell of an atom repel one another. An *electron domain* is a lone pair or a bond. Any bond (single, double, or triple) constitutes one electron domain.

- The arrangement of electron domains about a central atom, determined using the VSEPR model, is called the *electron-domain geometry*. The arrangement of *atoms* in a molecule is called the *molecular geometry*. The basic molecular geometries are linear, bent, trigonal planar, tetrahedral, trigonal pyramidal, trigonal bipyramidal, seesaw-shaped, T-shaped, octahedral, square pyramidal, and square planar.

- The *bond angle* is the angle between two adjacent bonds in a molecule or polyatomic ion. A trigonal bipyramid contains two types of bonds: *axial* and *equatorial.*

## Section 9.2

- The polarity of a molecule depends on the polarity of its individual bonds and on its molecular geometry. Even a molecule containing polar bonds may be nonpolar overall if the bonds are distributed symmetrically.

- *Structural isomers* are molecules with the same chemical formula, but different structural arrangements.

## Section 9.3

- According to *valence bond theory,* bonds form between atoms when atomic orbitals overlap, thus allowing the atoms to share valence electrons. A bond forms, furthermore, when the resulting molecule is lower in energy than the original, isolated atoms.

## Section 9.4

- To explain the bonding in some molecules, we need to employ the concept of *hybridization,* in which atomic orbitals mix to form hybrid orbitals.

- To use hybrid-orbital analysis, we must already know the molecular geometry and bond angles in a molecule. Hybrid orbitals are not used to predict molecular geometries.

## Section 9.5

- *Sigma ($\sigma$) bonds* form when the region of orbital overlap lies directly between the two atoms. *Pi ($\pi$) bonds* form when parallel, unhybridized *p* orbitals interact. A *double* bond consists of one sigma bond and one pi bond. A *triple* bond consists of one sigma bond and two pi bonds.

## Section 9.6

- A *paramagnetic* species is one that contains unpaired electrons. A *diamagnetic* species is one in which there are no unpaired electrons. Paramagnetic species are weakly attracted by a magnetic field, whereas diamagnetic species are weakly repelled by a magnetic field.

- According to *molecular orbital theory,* atomic orbitals combine to form new *molecular orbitals* that are associated with the molecule, rather than with individual atoms. Molecular orbitals may be *sigma,* if the orbital lies directly along the internuclear axis, or *pi,* if the orbital does not lie directly along the internuclear axis.

- Molecular orbitals may be *bonding* or *antibonding.* A *bonding molecular orbital* is lower in energy than the isolated atomic orbitals that combined to form it. The corresponding *antibonding molecular orbital* is higher in energy than the isolated atomic orbitals. *Bond order* is a measure of the *strength* of a bond and can be determined using a molecular orbital diagram.

## Section 9.7

- It is generally best to use the bonding theory that most easily describes the bonding in a particular molecule or polyatomic ion. In species that can be represented by two or more resonance structures, the pi bonds are *delocalized,* meaning that they are spread out over the molecule and not constrained to just two atoms. *Localized* bonds are those constrained to two atoms. Many species are best described using a combination of valence bond theory and molecular orbital theory.

# KEY WORDS

# KEY EQUATION

$$9.1 \quad \text{bond order} = \frac{\text{number of electrons in bonding molecular orbitals} - \text{number of electrons in antibonding molecular orbitals}}{2}$$

# QUESTIONS AND PROBLEMS

## Section 9.1: Molecular Geometry

### Review Questions

9.1    How is the geometry of a molecule defined, and why is the study of molecular geometry important?

9.2    Sketch the shape of a linear triatomic molecule, a trigonal planar molecule containing four atoms, a tetrahedral molecule, a trigonal bipyramidal molecule, and an octahedral molecule. Give the bond angles in each case.

9.3    How many atoms are directly bonded to the central atom in a tetrahedral molecule, a trigonal bipyramidal molecule, and an octahedral molecule?

9.4    Discuss the basic features of the VSEPR model. Explain why the magnitude of repulsion decreases in the following order: lone pair–lone pair > lone pair–bonding pair > bonding pair–bonding pair.

9.5    In the trigonal bipyramidal arrangement, why does a lone pair occupy an equatorial position rather than an axial position?

9.6    Explain why the $CH_4$ molecule is not square planar, although its Lewis structure makes it look as though it could be.

### Problems

9.7    Predict the geometries of the following species using the VSEPR method: (a) $PCl_3$, (b) $CHCl_3$, (c) $SiH_4$, (d) $TeCl_4$.

9.8    Predict the geometries of the following species: (a) $AlCl_3$, (b) $ZnCl_2$, (c) $HgBr_2$, (d) $N_2O$ (arrangement of atoms is NNO).

9.9    Predict the geometry of the following molecules and ion using the VSEPR model: (a) $CBr_4$, (b) $BCl_3$, (c) $NF_3$, (d) $H_2Se$, (e) $NO_2^-$.

9.10    Predict the geometry of the following molecules and ion using the VSEPR model: (a) $CH_3I$, (b) $ClF_3$, (c) $H_2S$, (d) $SO_3$, (e) $SO_4^{2-}$.

9.11    Predict the geometry of the following ions using the VSEPR method: (a) $SCN^-$ (arrangement of atoms is SCN), (b) $AlH_4^-$, (c) $SnCl_5^-$, (d) $H_3O^+$, (e) $BeF_4^{2-}$.

9.12    Predict the geometries of the following ions: (a) $NH_4^+$, (b) $NH_2^-$, (c) $CO_3^{2-}$, (d) $ICl_2^-$, (e) $ICl_4^-$.

9.13    Describe the geometry around each of the three central atoms in the $CH_3COOH$ molecule.

9.14    Which of the following species are tetrahedral: $SiCl_4$, $SeF_4$, $XeF_4$, $Cl_4$, $CdCl_4^{2-}$?

## Section 9.2: Molecular Geometry and Polarity

### Review Questions

9.15    Explain why an atom cannot have a permanent dipole moment.

9.16    The bonds in beryllium hydride ($BeH_2$) molecules are polar, and yet the dipole moment of the molecule is zero. Explain.

### Problems

9.17    Determine whether (a) $BrF_5$ and (b) $BCl_3$ are polar.

9.18    Determine whether (a) OCS and (b) $XeF_4$ are polar.

9.19    Which of the molecules shown is polar?

(a)                (b)                (c)                (d)

9.20    Which of the molecules shown is polar?

(a)                (b)                (c)                (d)

## Section 9.3: Valence Bond Theory

### Review Questions

9.21    What is valence bond theory? How does it differ from the Lewis concept of chemical bonding?

9.22    Use valence bond theory to explain the bonding in $Cl_2$ and HCl. Show how the atomic orbitals overlap when a bond is formed.

9.23    According to valence bond theory, how many bonds would you expect each of the following atoms (in the ground state) to form: Be, C?

9.24    According to valence bond theory, how many bonds would you expect each of the following atoms (in the ground state) to form: P, S?

## Section 9.4: Hybridization of Atomic Orbitals

### Review Questions

9.25    What is the hybridization of atomic orbitals? Why is it impossible for an isolated atom to exist in the hybridized state?

9.26    How does a hybrid orbital differ from a pure atomic orbital? Can two $2p$ orbitals of an atom hybridize to give two hybridized orbitals?

9.27    What is the angle between the following two hybrid orbitals on the same atom: (a) $sp$ and $sp$ hybrid orbitals, (b) $sp^2$ and $sp^2$ hybrid orbitals, (c) $sp^3$ and $sp^3$ hybrid orbitals?

### Problems

9.28    Describe the bonding scheme of the $AsH_3$ molecule in terms of hybridization.

9.29    What is the hybridization state of Si (a) in $SiH_4$ and (b) in $H_3Si-SiH_3$?

9.30    Describe the change in hybridization (if any) of the Al atom in the following reaction:

$$AlCl_3 + Cl^- \longrightarrow AlCl_4^-$$

9.31    Consider the reaction

$$BF_3 + NH_3 \longrightarrow F_3B-NH_3$$

Describe the changes in hybridization (if any) of the B and N atoms as a result of this reaction.

9.32    What hybrid orbitals are used by nitrogen atoms in the following species: (a) $NH_3$, (b) $H_2N-NH_2$, (c) $NO_3^-$?

9.33    Describe the hybridization of phosphorus in $PF_5$.

## Section 9.5: Hybridization in Molecules Containing Multiple Bonds

### Visualizing Chemistry
### Figure 9.10

VC 9.1 How is a sigma bond different from a pi bond?

    a) A sigma bond is a bonding molecular orbital; a pi bond is an antibonding molecular orbital.

    b) A sigma bond is a single bond, whereas a pi bond is a double bond.

    c) The electron density in a sigma bond lies along the internuclear axis; that of a pi bond does not.

VC 9.2 Pi bonds form when _____ atomic orbitals on _____ atom(s) overlap.

    a) perpendicular, adjacent

    b) parallel, adjacent

    c) parallel, the same

VC 9.3 Formation of two pi bonds requires the combination of _____ atomic orbitals.

    a) two

    b) four

    c) six

VC 9.4 Why are there no pi bonds in ethane ($C_2H_6$)?

    a) The remaining unhybridized $p$ orbitals do not contain any electrons.

    b) There are no unhybridized $p$ orbitals remaining on either C atom.

    c) The remaining unhybridized $p$ orbitals are not parallel to each other.

### Review Questions

9.34 How would you distinguish between a sigma bond and a pi bond?

9.35 Which of the following pairs of atomic orbitals of adjacent nuclei can overlap to form a sigma bond? Which overlap to form a pi bond? Which cannot overlap (no bond)? Consider the $x$ axis to be the internuclear axis, that is, the line joining the nuclei of the two atoms. (a) $1s$ and $1s$, (b) $1s$ and $2p_x$, (c) $2p_x$ and $2p_y$, (d) $3p_y$ and $3p_y$, (e) $2p_x$ and $2p_x$, (f) $1s$ and $2s$.

### Problems

9.36 What are the hybrid orbitals of the carbon atoms in the following molecules?

    (a) $H_3C-CH_3$

    (b) $H_3C-CH=CH_2$

    (c) $CH_3-C=C-CH_2OH$

    (d) $CH_3CH=O$

    (e) $CH_3COOH$

9.37 Specify which hybrid orbitals are used by carbon atoms in the following species: (a) CO, (b) $CO_2$, (c) $CN^-$.

9.38 The allene molecule ($H_2C=C=CH_2$) is linear (the three C atoms lie on a straight line). What are the hybridization states of the carbon atoms? Draw diagrams to show the formation of sigma bonds and pi bonds in allene.

9.39 What is the hybridization of the central N atom in the azide ion ($N_3^-$)? (The arrangement of atoms is NNN.)

9.40 How many sigma bonds and pi bonds are there in each of the following molecules?

(a)         (b)         (c)

9.41 How many pi bonds and sigma bonds are there in the tetracyanoethylene molecule?

9.42 Tryptophan is one of the 20 amino acids in the human body. Describe the hybridization state of the C and N atoms, and determine the number of sigma and pi bonds in the molecule.

9.43 Benzo($a$)pyrene is a potent carcinogen found in coal and cigarette smoke. Determine the number of sigma and pi bonds in the molecule.

## Section 9.6: Molecular Orbital Theory

### Review Questions

9.44 What is molecular orbital theory? How does it differ from valence bond theory?

9.45 Define the following terms: bonding molecular orbital, antibonding molecular orbital, pi molecular orbital, sigma molecular orbital.

9.46 Sketch the shapes of the following molecular orbitals: $\sigma_{1s}$, $\sigma_{1s}^*$, $\pi_{2p}$, $\pi_{2p}^*$. How do their energies compare?

9.47 Explain the significance of bond order. Can bond order be used for quantitative comparisons of the strengths of chemical bonds?

### Problems

9.48 Explain in molecular orbital terms the changes in H—H internuclear distance that occur as the molecular $H_2$ is ionized first to $H_2^+$ and then to $H_2^{2+}$.

9.49 The formation of $H_2$ from two H atoms is an energetically favorable process. Yet statistically there is less than a 100 percent chance that any two H atoms will undergo the reaction. Apart from energy considerations, how would you account for this observation based on the electron spins in the two H atoms?

9.50 Draw a molecular orbital energy level diagram for each of the following species: $He_2$, HHe, $He_2^+$. Compare their relative stabilities in terms of bond orders. (Treat HHe as a diatomic molecule with three electrons.)

**9.51** Arrange the following species in order of increasing stability: $Li_2$, $Li_2^+$, $Li_2^-$. Justify your choice with a molecular orbital energy level diagram.

9.52 Use molecular orbital theory to explain why the $Be_2$ molecule does not exist.

**9.53** Which of these species has a longer bond, $B_2$ or $B_2^+$? Explain in terms of molecular orbital theory.

**9.54** Acetylene ($C_2H_2$) has a tendency to lose two protons ($H^+$) and form the carbide ion ($C_2^{2-}$), which is present in a number of ionic compounds, such as $CaC_2$ and $MgC_2$. Describe the bonding scheme in the $C_2^{2-}$ ion in terms of molecular orbital theory. Compare the bond order in $C_2^{2-}$ with that in $C_2$.

**9.55** Compare the Lewis and molecular orbital treatments of the oxygen molecule.

9.56 Explain why the bond order of $N_2$ is greater than that of $N_2^+$, but the bond order of $O_2$ is less than that of $O_2^+$.

**9.57** Compare the relative bond orders of the following species and indicate their magnetic properties (that is, diamagnetic or paramagnetic): $O_2$, $O_2^+$, $O_2^-$ (superoxide ion), $O_2^{2-}$ (peroxide ion).

9.58 Use molecular orbital theory to compare the relative stabilities of $F_2$ and $F_2^+$.

**9.59** A single bond is almost always a sigma bond, and a double bond is almost always made up of a sigma bond and a pi bond. There are very few exceptions to this rule. Show that the $B_2$ and $C_2$ molecules are examples of the exceptions.

## Section 9.7: Bonding Theories and Descriptions of Molecules with Delocalized Bonding

### Review Questions

9.60 How does a delocalized molecular orbital differ from a molecular orbital such as that found in $H_2$ or $C_2H_4$? What do you think are the minimum conditions (for example, number of atoms and types of orbitals) for forming a delocalized molecular orbital?

9.61 In Chapter 8 we saw that the resonance concept is useful for dealing with species such as the benzene molecule and the carbonate ion. How does molecular orbital theory deal with these species?

### Problems

9.62 Both ethylene ($C_2H_4$) and benzene ($C_6H_6$) contain the C=C bond. The reactivity of ethylene is greater than that of benzene. For example, ethylene readily reacts with molecular bromine, whereas benzene is normally quite inert toward molecular bromine and many other compounds. Explain this difference in reactivity.

**9.63** Explain why the symbol on the left is a better representation of benzene molecules than that on the right.

9.64 Determine which of these molecules has a more delocalized orbital, and justify your choice. (*Hint:* Both molecules contain two benzene rings. In naphthalene, the two rings are fused together. In biphenyl, the two rings are joined by a single bond, around which the two rings can rotate.)

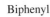
Biphenyl                    Naphthalene

9.65 Nitryl fluoride ($FNO_2$) is very reactive chemically. The fluorine and oxygen atoms are bonded to the nitrogen atom. (a) Write a Lewis structure for $FNO_2$. (b) Indicate the hybridization of the nitrogen atom. (c) Describe the bonding in terms of molecular orbital theory. Where would you expect delocalized molecular orbitals to form?

9.66 Describe the bonding in the nitrate ion $NO_3^-$ in terms of delocalized molecular orbitals.

9.67 What is the state of hybridization of the central O atom in $O_3$? Describe the bonding in $O_3$ in terms of delocalized molecular orbitals.

## Additional Problems

9.68 Which of the following species is not likely to have a tetrahedral shape: (a) $SiBr_4$, (b) $NF_4^+$, (c) $SF_4$, (d) $BeCl_4^{2-}$, (e) $BF_4^-$, (f) $AlCl_4^-$?

**9.69** Draw the Lewis structure of mercury(II) bromide. Is this molecule linear or bent? How would you establish its geometry?

9.70 Although both carbon and silicon are in Group 4A, very few Si=Si bonds are known. Account for the instability of silicon-to-silicon double bonds in general. (*Hint:* Compare the atomic radii of C and Si in Figure 7.6. What effect would the larger size have on pi bond formation?)

**9.71** Predict the geometry of sulfur dichloride ($SCl_2$) and the hybridization of the sulfur atom.

**9.72** Antimony pentafluoride ($SbF_5$) reacts with $XeF_4$ and $XeF_6$ to form ionic compounds, $XeF_3^+SbF_6^-$ and $XeF_5^+SbF_6^-$. Describe the geometries of the cations and anions in these two compounds.

9.73 The molecular model of vitamin C is shown here. (a) Write the molecular formula of the compound. (b) What is the hybridization of each C and O atom? (c) Describe the geometry about each C and O atom.

**9.74** The molecular model of nicotine (a stimulant) is shown here. (a) Write the molecular formula of the compound. (b) What is the hybridization of each C and N atom? (c) Describe the geometry about each C and N atom.

**9.75** Predict the bond angles for the following molecules: (a) $BeCl_2$, (b) $BCl_3$, (c) $CCl_4$, (d) $CH_3Cl$, (e) $Hg_2Cl_2$ (arrangement of atoms: ClHgHgCl), (f) $SnCl_2$, (g) $H_2O_2$, (h) $SnH_4$.

9.76   Briefly compare the VSEPR and hybridization approaches to the study of molecular geometry.

9.77   Draw Lewis structures and give the other information requested for the following molecules: (a) $BF_3$. Shape: planar or nonplanar? (b) $ClO_3^-$. Shape: planar or nonplanar? (c) HCN. Polar or nonpolar? (d) $OF_2$. Polar or nonpolar? (e) $NO_2$. Estimate the ONO bond angle.

9.78   Describe the hybridization state of arsenic in arsenic pentafluoride ($AsF_5$).

9.79   The disulfide bond, $-S-S-$, plays an important role in determining the three-dimensional structure of proteins. Describe the nature of the bond and the hybridization state of the S atoms.

9.80   Draw Lewis structures and give the other information requested for the following: (a) $SO_3$. Polar or nonpolar molecule? (b) $PF_3$. Polar or nonpolar? (c) $F_3SiH$. Polar or nonpolar? (d) $SiH_3^-$. Shape: planar or pyramidal? (e) $Br_2CH_2$. Polar or nonpolar molecule?

9.81   Which of the following molecules are linear: $ICl_2^-$, $IF_2^+$, $OF_2$, $SnI_2$, $CdBr_2$?

9.82   Draw the Lewis structure for the $BeCl_4^{2-}$ ion. Predict its geometry, and describe the hybridization state of the Be atom.

9.83   The $N_2F_2$ molecule can exist in either of the following two forms:

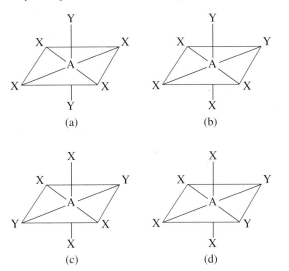

(a) What is the hybridization of N in the molecule? (b) Which structure is polar?

9.84   Cyclopropane ($C_3H_6$) has the shape of a triangle in which a C atom is bonded to two H atoms and two other C atoms at each corner. Cubane ($C_8H_8$) has the shape of a cube in which a C atom is bonded to one H atom and three other C atoms at each corner. (a) Draw Lewis structures of these molecules. (b) Compare the CCC angles in these molecules with those predicted for an $sp^3$-hybridized C atom. (c) Would you expect these molecules to be easy to make?

9.85   The compound 1,2-dichloroethane ($C_2H_4Cl_2$) is nonpolar, while *cis*-dichloroethylene ($C_2H_2Cl_2$) has a dipole moment: The reason for the difference is that groups connected by a single bond can rotate with respect to each other, but no rotation occurs when a double bond connects the groups. On the basis of bonding considerations, explain why rotation occurs in 1,2-dichloroethane but not in *cis*-dichloroethylene.

1,2-dichloroethane          *cis*-dichloroethylene

9.86   Does the following molecule have a dipole moment? Explain.

9.87   So-called greenhouse gases, which contribute to global warming, have a dipole moment or can be bent or distorted during molecular vibration into shapes that have a temporary dipole moment. Which of the following gases are greenhouse gases: $N_2$, $O_2$, $O_3$, CO, $CO_2$, $NO_2$, $N_2O$, $CH_4$, $CFCl_3$?

9.88   (a) From what group must the terminal atoms come in an $AB_x$ molecule where the central atom is from Group 5A, for both the electron-domain geometry and the molecular geometry to be trigonal bipyramidal? (b) From what group must the terminal atoms come in an $AB_x$ molecule where the central atom is from Group 6A, for the electron-domain geometry to be tetrahedral and the molecular geometry to be bent?

9.89   The compound 3'-azido-3'-deoxythymidine, commonly known as AZT, is one of the drugs used to treat AIDS. What are the hybridization states of the C and N atoms in this molecule?

9.90   The following molecules ($AX_4Y_2$) all have an octahedral geometry. Group the molecules that are equivalent to each other.

(a)          (b)

(c)          (d)

9.91   The compounds carbon tetrachloride ($CCl_4$) and silicon tetrachloride ($SiCl_4$) are similar in geometry and hybridization. However, $CCl_4$ does not react with water but $SiCl_4$ does. Explain the difference in their chemical reactivities. (*Hint:* The first step of the reaction is believed to be the addition of a water molecule to the Si atom in $SiCl_4$.)

9.92   Write the ground-state electron configuration for $B_2$. Is the molecule diamagnetic or paramagnetic?

9.93   What is the hybridization of C and of N in this molecule?

9.94   The stable allotropic form of phosphorus is $P_4$, in which each P atom is bonded to three other P atoms. Draw a Lewis structure of this molecule and describe its geometry. At high temperatures, $P_4$ dissociates to form $P_2$ molecules containing a $P=P$ bond. Explain why $P_4$ is more stable than $P_2$.

**9.95**  Use molecular orbital theory to explain the difference between the bond enthalpies of $F_2$ and $F_2^-$. (See Problem 8.121.)

**9.96**  Use molecular orbital theory to explain the bonding in the azide ion ($N_3^-$). (The arrangement of atoms is NNN.)

**9.97**  The ionic character of the bond in a diatomic molecule can be estimated by the formula

$$\frac{\mu}{ed} \times 100\%$$

where $\mu$ is the experimentally measured dipole moment (in C · m), $e$ is the electronic charge, and $d$ is the bond length (in meters). (The quantity $ed$ is the hypothetical dipole moment for the case in which the transfer of an electron from the less electronegative to the more electronegative atom is complete.) Given that the dipole moment and bond length of HF are 1.92 D and 91.7 pm, respectively, calculate the percent ionic character of the molecule. ($1D = 3.336 \times 10^{-30}$ Cm)

**9.98**  Draw three Lewis structures for compounds with the formula $C_2H_2F_2$. Indicate which of the compounds are polar.

**9.99**  Greenhouse gases absorb (and trap) outgoing infrared radiation (heat) from Earth and contribute to global warming. A molecule of a greenhouse gas either possesses a permanent dipole moment or has a changing dipole moment during its vibrational motions. Consider three of the vibrational modes of carbon dioxide

$$\overleftarrow{O}=C=\overrightarrow{O} \qquad \overrightarrow{O}=C=\overrightarrow{O} \qquad \overset{\uparrow}{O}=C=\overset{\uparrow}{O}$$

where the arrows indicate the movement of the atoms. (During a complete cycle of vibration, the atoms move toward one extreme position and then reverse their direction to the other extreme position.) Which of the preceding vibrations are responsible for $CO_2$ behaving as a greenhouse gas?

**9.100**  Aluminum trichloride ($AlCl_3$) is an electron-deficient molecule. It has a tendency to form a dimer (a molecule made up of two $AlCl_3$ units):

$$AlCl_3 + AlCl_3 \longrightarrow Al_2Cl_6$$

(a) Draw a Lewis structure for the dimer. (b) Describe the hybridization state of Al in $AlCl_3$ and $Al_2Cl_6$. (c) Sketch the geometry of the dimer. (d) Do these molecules possess a dipole moment?

**9.101**  Progesterone is a hormone responsible for female sex characteristics. In the usual shorthand structure, each point where lines meet represents a C atom, and most H atoms are not shown. Draw the complete structure of the molecule, showing all C and H atoms. Indicate which C atoms are $sp^2$- and $sp^3$-hybridized.

**9.102**  The molecule benzyne ($C_6H_4$) is a very reactive species. It resembles benzene in that it has a six-membered ring of carbon atoms. Draw a Lewis structure of the molecule and account for the molecule's high reactivity.

**9.103**  Assume that the third-period element phosphorus forms a diatomic molecule, $P_2$, in an analogous way as nitrogen does to form $N_2$. (a) Write the electronic configuration for $P_2$. Use [$Ne_2$] to represent the electron configuration for the first two periods.

(b) Calculate its bond order. (c) What are its magnetic properties (diamagnetic or paramagnetic)?

**9.104**  Consider an $N_2$ molecule in its first excited electronic state, that is, when an electron in the highest occupied molecular orbital is promoted to the lowest empty molecular orbital. (a) Identify the molecular orbitals involved, and sketch a diagram to show the transition. (b) Compare the bond order and bond length of $N_2^*$ with $N_2$, where the asterisk denotes the excited molecule. (c) Is $N_2^*$ diamagnetic or paramagnetic? (d) When $N_2^*$ loses its excess energy and converts to the ground state $N_2$, it emits a photon of wavelength 470 nm, which makes up part of the auroras' lights. Calculate the energy difference between these levels.

**9.105**  The Lewis structure for $O_2$ is

$$\ddot{O}=\ddot{O}$$

Use molecular orbital theory to show that the structure actually corresponds to an excited state of the oxygen molecule.

**9.106**  Draw the Lewis structure of ketene ($C_2H_2O$) and describe the hybridization states of the C atoms. The molecule does not contain O—H bonds. On separate diagrams, sketch the formation of the sigma and pi bonds.

**9.107**  The compound TCDD, or 2,3,7,8-tetrachlorodibenzo-*p*-dioxin, is highly toxic:

It gained considerable notoriety in 2004 when it was implicated in the attempted murder of a Ukranian politician. (a) Describe its geometry, and state whether the molecule has a dipole moment. (b) How many pi bonds and sigma bonds are there in the molecule?

**9.108**  Write the electron configuration of the cyanide ion ($CN^-$). Name a stable molecule that is isoelectronic with the ion.

**9.109**  Carbon monoxide (CO) is a poisonous compound due to its ability to bind strongly to $Fe^{2+}$ in the hemoglobin molecule. The molecular orbitals of CO have the same energy order as those of the $N_2$ molecule. (a) Draw a Lewis structure of CO and assign formal charges. Explain why CO has a rather small dipole moment of 0.12 D. (b) Compare the bond order of CO with that from molecular orbital theory. (c) Which of the atoms (C or O) is more likely to form bonds with the $Fe^{2+}$ ion in hemoglobin?

**9.110**  Which of the following ions possess a dipole moment: (a) $ClF_2^+$, (b) $ClF_2^-$, (c) $IF_4^+$, (d) $IF_4^-$?

**9.111**  Which of the following geometries has a greater stability for tin(IV) hydride ($SnH_4$)?

**9.112**  Carbon dioxide has a linear geometry and is nonpolar. Yet we know that the molecule exhibits bending and stretching motions that create a dipole moment. How would you reconcile these seemingly conflicting descriptions of $CO_2$?

**9.113**  Carbon suboxide ($C_3O_2$) is a colorless pungent-smelling gas. Does this molecule possess a dipole moment? Explain.

# PRE-PROFESSIONAL PRACTICE EXAM QUESTIONS: PHYSICAL SCIENCES

These questions are not based on a descriptive passage.

1. What is the shape of the $ICl_3$ molecule?

   a) Trigonal planar    c) T-shaped
   b) Trigonal pyramidal    d) Tetrahedral

2. Which of the following molecules is nonpolar?

   a) $NCl_3$    c) $PCl_3$
   b) $BCl_3$    d) $BrCl_3$

3. Which of the following has a bond order of 2?

   $$N_2^{2-} \qquad N_2 \qquad N_2^{2+}$$
   $$\text{I} \qquad\quad \text{II} \qquad \text{III}$$

   a) I only    c) III only
   b) II only    d) I and III

4. Which of the following are paramagnetic?

   $$O_2^{2-} \qquad O_2^{-} \qquad O_2^{+} \qquad O_2^{2+}$$
   $$\text{I} \qquad\quad \text{II} \qquad\quad \text{III} \qquad\quad \text{IV}$$

   a) I and II    c) III and IV
   b) II and III    d) I and III

# ANSWERS TO IN-CHAPTER MATERIALS

## Answers to Practice Problems

**9.1A** (a) linear, (b) bent. **9.1B** (a) Group 6A, (b) Group 7A. **9.2A** Bent about O, tetrahedral about each C, trigonal pyramidal about N. All bond angles are ~109.5°. Angles labeled in blue are <109.5°.

$$H-\overset{\cdot\cdot}{\underset{\cdot\cdot}{O}}-\overset{\overset{\displaystyle H}{|}}{\underset{\underset{\displaystyle H}{|}}{C}}-\overset{\overset{\displaystyle H}{|}}{\underset{\underset{\displaystyle H}{|}}{C}}-\overset{\cdot\cdot}{\underset{\underset{\displaystyle H}{|}}{N}}-H$$

**9.2B** $SO_2$ contains double bonds, which are not pushed together by the central atom's lone pair as easily as the single bonds in $NH_3$. **9.3A** (a) polar, (b) nonpolar. **9.3B** (a) Trigonal bipyramidal, square planar, octahedral. (b) Octahedral. **9.4A** Singly occupied $3p$ orbitals from the P atom overlap with $s$ orbitals from H atoms. **9.4B** We cannot use valence bond theory to explain the bonding in $SO_2$ or $CH_4$. In the case of $SO_2$, although the central atom has two unpaired electrons and can form two bonds, the unpaired electrons on S are in $3p$ orbitals. Formation of two bonds by the overlap of two $3p$ orbitals on S would be expected to result in a bond angle of approximately 90°. In the case of $CH_4$, the central atom does not have enough unpaired electrons to form four bonds. **9.5A** Two of the $4p$ electrons in Br are promoted to empty $d$ orbitals. The $s$ orbital, all three $p$ orbitals, and two of the $d$ orbitals hybridize to form six $sp^3d^2$ hybrid orbitals. One of the hybrid orbitals contains the lone pair. Each of the remaining hybrid orbitals contains one electron and overlaps with a singly occupied $2p$ orbital on an F atom. The arrangement of hybrid orbitals is octahedral and the bond angles are ~90°. **9.5B** One of the $2s$ electrons in Be is promoted to an empty $p$ orbital. The $s$ orbital and one $p$ orbital hybridize to form two $sp$ hybrid orbitals. Each hybrid orbital contains one electron and overlaps with a singly occupied $2p$ orbital on an F atom. The arrangement is linear with a bond angle of ~180°. **9.6A** 16 $\sigma$ bonds and 4 $\pi$ bonds. **9.6B** 17 $\sigma$ bonds and 5 $\pi$ bonds. **9.7A** C and N atoms are $sp$-hybridized. The triple bond between C and N is composed of one sigma bond (from overlap of hybrid orbitals) and two pi bonds (from interaction of remaining $p$ orbitals). The single bond between H and C is the result of an $sp$ orbital from C overlapping with an $s$ orbital from H. **9.7B** Each N atom is $sp$-hybridized. One $sp$ orbital on each is singly occupied and one contains a lone pair. The singly occupied $sp$ orbitals overlap to form a sigma bond between the N atoms. The remaining unhybridized $p$ orbitals interact to form two pi bonds. **9.8A** paramagnetic; bond order = 2. **9.8B** paramagnetic; bond order = 2. **9.9A** Two different resonance structures are possible; therefore we consider all three atoms to be $sp^2$-hybridized. One of the hybrid orbitals on the central O atom contains the lone pair, the other two form sigma bonds to the terminal O atoms. Each atom has one remaining unhybridized $p$ orbital. The $p$ orbitals combine to form $\pi$ molecular orbitals. **9.9B** Two different resonance structures are possible; therefore we consider all three atoms to be $sp^2$-hybridized. One of the hybrid orbitals on the central N atom contains the lone pair, the other two form sigma bonds to the terminal O atoms. Each atom has one remaining unhybridized $p$ orbital. The $p$ orbitals combine to form $\pi$ molecular orbitals.

## Answers to Checkpoints

**9.1.1** d. **9.1.2** b. **9.1.3** e. **9.1.4** c. **9.1.5** d. **9.2.1** a. **9.2.2** c. **9.3.1** b, e. **9.3.2** b. **9.4.1** b. **9.4.2** d. **9.5.1** a, c. **9.5.2** c. **9.5.3** a, b, e. **9.5.4** c. **9.6.1** b. **9.6.2** c, e. **9.6.3** b. **9.6.4** a, c, e. **9.7.1** c. **9.7.2** b. **9.7.3** a, b. **9.7.4** e.

## Answers to Applying What You've Learned

**a)** There are 14 carbon-carbon sigma bonds in musk ketone and four pi bonds. **b)** Each of the top two indicated C atoms has three electron domains around it and is therefore $sp^2$-hybridized. The bottom indicated C atom has four electron domains around it and is therefore $sp^3$-hybridized. The geometry about the $sp^2$-hybridized C atoms is trigonal planar, whereas the geometry about the $sp^3$-hybridized C atom is tetrahedral. **c)** Of the pi bonds, only the carbon-carbon and nitrogen-oxygen pi bonds are delocalized. Only the six-membered ring and the nitro groups can be represented with two different resonance structures. **d)** The bonding in the ring can be described as follows: each of the C atoms is $sp^2$-hybridized. The sigma bonds between the C atoms result from the overlap of $sp^2$ hybrid orbitals. Each C atom has one leftover, singly occupied $p$ orbital. As in the case for benzene, the unhybridized $p$ orbitals on all six C atoms combine to form six pi molecular orbitals (three bonding and three antibonding) that lie above and below the plane of the molecule. All six electrons in these pi orbitals reside in the *bonding* molecular orbitals.

# 10
# Organic Chemistry

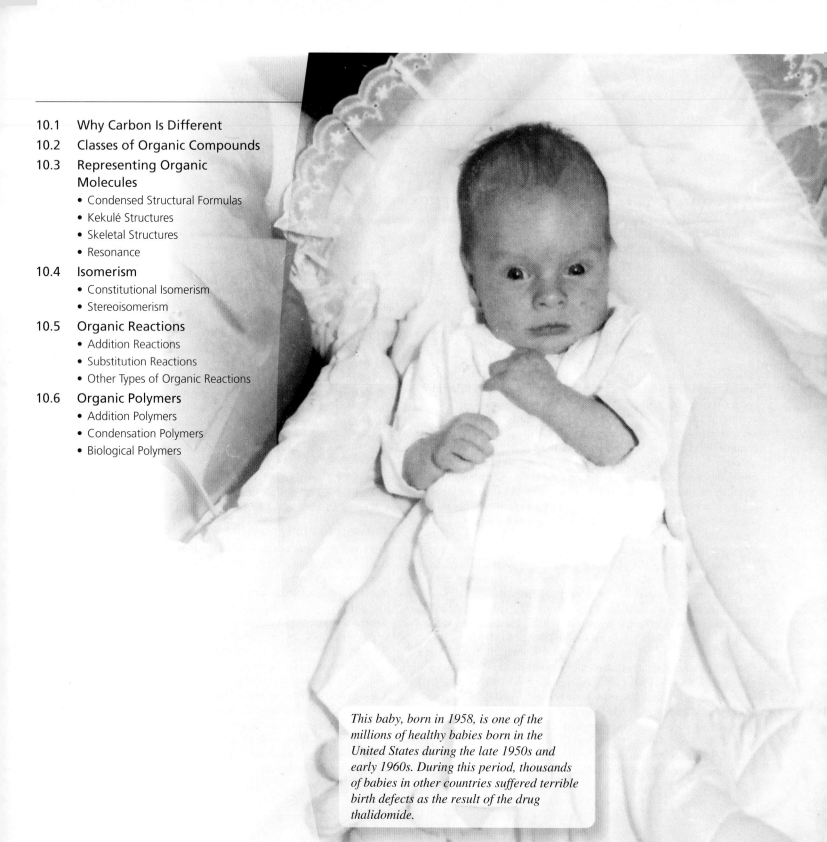

*This baby, born in 1958, is one of the millions of healthy babies born in the United States during the late 1950s and early 1960s. During this period, thousands of babies in other countries suffered terrible birth defects as the result of the drug thalidomide.*

# The Importance of Organic Chemistry to the Development of New Drugs

Beginning in 1957, the drug thalidomide was marketed in 48 countries around the world as a sleeping pill and as an antinausea medicine for pregnant women suffering from morning sickness. By 1962, the drug was shown to have caused horrific birth defects and an untold number of fetal deaths. Thalidomide interferes with spinal cord and limb development, and more than 10,000 babies had been born with severe spinal cord abnormalities and malformed or absent limbs. Many of the victims were born to mothers who reportedly had taken just *one* thalidomide pill early in their pregnancies. At the time, thalidomide was not approved for use in the United States, but the otherwise worldwide tragedy did prompt the U.S. Congress to enact a new law to give the FDA more control over the testing and approval of new drugs.

In August of 1998, the FDA approved thalidomide for the treatment of *erythema nodosum leprosum* (ENL), a painful inflammatory skin condition associated with leprosy. This approval is controversial because of the drug's infamous history, but thalidomide has shown tremendous promise in the treatment of a wide variety of painful and debilitating conditions, including complications from certain cancers, AIDS, and some autoimmune disorders such as lupus and rheumatoid arthritis. Because of the dangers known to be associated with thalidomide, researchers are working on developing *analogues*—drugs that are chemically similar enough to have the same therapeutic benefits but chemically *different* enough *not* to have the undesirable and/or dangerous properties. Two such analogues that are currently being investigated are lenalidomide and CC-4047, shown here:

**Student Annotation:** Thalidomide was not approved for use in the United States thanks in large part to the vigilance of one doctor at the FDA. She was troubled by inadequate research into the safety of the drug and steadfastly refused to approve the drug-maker's application.

Thalidomide                    Lenalidomide                    CC-4047

Scientists who develop new drugs such as lenalidomide and CC-4047 must understand the principles and concepts of *organic chemistry*.

At the end of this chapter, you will be able to answer several questions about the drugs thalidomide, lenalidomide, and CC-4047 [▶▶ Page 427].

## In This Chapter, You Will Learn:

Some of the basic concepts of organic chemistry and how the principles of chemical bonding contribute to the understanding of organic compounds and reactions.

### Before you begin, you should review

- Lewis structures and formal charge [◀◀ Sections 8.5 and 8.6]
- Resonance [◀◀ Section 8.7]
- Molecular geometry and polarity [◀◀ Sections 9.1 and 9.2]

| 10.1 | Why Carbon Is Different |

**Student Annotation:** Carbon-containing compounds that are considered *inorganic* are generally those that are obtained from minerals.

Organic chemistry is usually defined as the study of compounds that contain carbon. This definition is not entirely satisfactory, though, because it would include such things as cyanide and cyanate complexes and metal carbonates, which are considered to be *inorganic.* A somewhat more useful definition of organic chemistry is the study of compounds that contain carbon and hydrogen, although many organic compounds also contain other elements, such as oxygen, sulfur, nitrogen, phosphorus, or the halogens, and many do not contain hydrogen. Examples of organic compounds include the following:

| $CH_4$ | $C_2H_5OH$ | $C_5H_7O_4COOH$ | $CH_3NH_2$ | $CCl_4$ |
|---|---|---|---|---|
| Methane | Ethanol | Ascorbic acid | Methylamine | Carbon tetrachloride |

Early in the study of organic chemistry there was thought to be some fundamental difference between compounds that came from living things, such as plants and animals, and those that came from nonliving things, such as rocks. Compounds obtained from plants or animals were called *organic,* whereas compounds obtained from nonliving sources were called *inorganic.* In fact, until early in the nineteenth century, scientists believed that only nature could produce organic compounds. In 1829, however, Friedrich Wöhler prepared urea, a well-known organic compound, by combining the inorganic substances lead cyanate and aqueous ammonia:

$$Pb(OCN)_2 + 2NH_3 + 2H_2O \xrightarrow{\Delta} \underset{\text{Urea}}{2NH_2CONH_2} + Pb(OH)_2$$

Wöhler's synthesis of urea dispelled the notion that organic compounds were fundamentally different from inorganic compounds—and that they could only be produced by nature. We now know that it is possible to synthesize a wide variety of organic compounds in the laboratory; in fact, many thousands of new organic compounds are produced in research laboratories each year. In this chapter, we will consider several types of organic compounds that are important biologically.

**Student Annotation:** By contrast, there are only about 100,000 known *inorganic* compounds.

Because of its unique nature, carbon is capable of forming millions of different compounds. Carbon's position in the periodic table (Group 4A, Period 2) gives it the following set of unique characteristics:

**Student Annotation:** To form an ion that is isoelectronic with a noble gas, a C atom would have to either gain or lose *four* electrons [◄◄ Section 7.5]—something that is energetically impossible under ordinary conditions. This is not to say that carbon *cannot* form ions. But in the vast majority of its compounds, carbon acquires a complete octet by *sharing* electrons—rather than by gaining or losing them.

- The electron configuration of carbon ($[He]2s^2 2p^2$) effectively prohibits ion formation. This and carbon's electronegativity, which is intermediate between those of metals and nonmetals, cause carbon to complete its octet by sharing electrons. In nearly all its compounds, carbon forms four covalent bonds, which can be oriented in as many as four different directions:

| Methane | Formaldehyde | Carbon dioxide |
|---|---|---|
| (4 $\sigma$ bonds) | (3 $\sigma$ bonds & 1 $\pi$ bond) | (2 $\sigma$ bonds & 2 $\pi$ bonds) |

Boron and nitrogen, carbon's neighbors in Groups 3A and 5A, respectively, usually form covalent compounds, too, but B and N form ions more readily than C.

- Carbon's small atomic radius allows the atoms to approach one another closely, giving rise to short, *strong,* carbon-carbon bonds and *stable* carbon compounds. In addition, carbon atoms that are *sp-* or *sp²*-hybridized approach one another closely enough for their singly occupied, unhybridized *p* orbitals to overlap effectively—giving rise to relatively strong $\pi$ bonds [◄◄ Section 9.5]. Recall that elements in the same group generally exhibit similar chemical behavior [◄◄ Section 2.4]. Silicon atoms, however, are bigger than carbon atoms, so silicon atoms generally cannot approach one another closely enough for their unhybridized *p* orbitals to overlap significantly. As a result, very few compounds exhibit significant $\pi$ bonding between Si atoms:

Ethane

Disilane

- Carbon's valence electrons are in the second shell ($n = 2$), where there are no $d$ orbitals. The valence electrons of silicon, on the other hand, are in the third shell ($n = 3$) where there are $d$ orbitals, which can be occupied or attacked by lone pairs on another substance—resulting in a reaction. This reactivity makes silicon compounds far less stable than the analogous carbon compounds. Ethane ($CH_3-CH_3$), for example, is stable in both water and air, whereas disilane ($SiH_3-SiH_3$) is unstable—breaking down in water and combusting spontaneously in air.

These attributes enable carbon to form chains (straight, branched, and cyclic) containing single, double, and triple carbon-carbon bonds, This, in turn, results in an endless array of organic compounds containing any number and arrangement of carbon atoms. Each carbon atom in a compound can be classified by the number of other carbon atoms to which it is bonded. A carbon atom that is bonded to just one other carbon atom is called a *primary* carbon; one that is bonded to two other carbon atoms is called a *secondary* carbon; one that is bonded to three other carbon atoms is called a *tertiary* carbon; and one that is bonded to four other carbon atoms is called a *quaternary* carbon. These four types of carbon atoms are identified with the labels 1°, 2°, 3°, and 4°, respectively. Each of the four carbon types is labeled in the following structure.

**Student Annotation:** Carbon's formation of chains is called *catenation.*

One of the important organic molecules that we encountered in Chapter 9 is benzene ($C_6H_6$). Organic compounds that are related to benzene, or that contain one or more benzene rings, are called **aromatic** compounds. Organic molecules that do not contain the benzene ring are called **aliphatic** compounds.

Benzene    Phenol    Cinnamic aldehyde

Aromatic compounds

Ethanol    Butyric acid    Acetone

Aliphatic compounds

Organic molecules occur in seemingly limitless variety. In Chapter 2 we encountered *alkanes,* organic compounds consisting of only carbon and hydrogen and containing only single bonds. A variety of different types of organic compounds, each with their own characteristic properties, result from the following:

1. Carbon's ability to form chains by bonding with itself
2. The presence of elements other than carbon and hydrogen
3. Functional groups
4. Multiple bonds

**Student Annotation:** A functional group is a group of atoms that determines many of a molecule's properties [◀◀ Section 2.6].

**Student Annotation:** Recall that *isomers* are different compounds with the same chemical formula [◀◀ Section 9.2].

In this section, we will discuss several types of organic compounds and how we represent them. Consider two isomers of $C_3H_6O$:

$$\underset{\text{Propanal}}{\overset{\displaystyle O}{\underset{\displaystyle \parallel}{CH_3CH_2CH}}} \qquad \underset{\text{Acetone}}{\overset{\displaystyle O}{\underset{\displaystyle \parallel}{CH_3CCH_3}}}$$

Although the two isomers contain exactly the same atoms, their different arrangements of atoms result in two very different compounds. The first is an *aldehyde* called propanal. The second is a *ketone* called acetone. Aldehydes and ketones are two classes of organic compounds. The classes of organic compounds that we will discuss in this chapter are alcohols, carboxylic acids, aldehydes, ketones, esters, amines, and amides.

A class of organic compounds often is represented with a general formula that shows the atoms of the functional group(s) explicitly, and the remainder of the molecule using one or more R's, where R represents an alkyl group. An **alkyl group** is a portion of a molecule that resembles an alkane. In fact, an alkyl group is formed by removing one hydrogen atom from the corresponding alkane. The *methyl group* ($-CH_3$), for example, is formed by removing a hydrogen atom from methane ($CH_4$), the simplest alkane. Methyl groups are found in many organic molecules. Table 10.1 lists some of the simplest alkyl groups. Table 10.2 (p. 394) gives the general formula for each of the classes of organic compounds discussed in this chapter and the Lewis structure of each functional group.

Acetone                              Methylamine                              Toluene

The functional groups in the types of compounds shown in Table 10.2 are the *hydroxy* group (in **alcohols**), the *carboxy* group (in **carboxylic acids**), the $-COOR$ group (in **esters**), the *carbonyl* group (in **aldehydes** and **ketones**), the *amino* group (in **amines**), and the *amide* group (in **amides**). Functional groups determine many of the properties of a compound, including what types of reactions it is likely to undergo. Figure 10.1 shows ball-and-stick models and electrostatic potential maps of the hydroxy, carboxy, carbonyl, amino, and amide functional groups.

A compound consisting of an alkyl group and the functional group $-OH$ is an *alcohol.* The identity of an individual alcohol depends on the identity of R, the alkyl group. For example, when R is the *methyl* group, we have $CH_3OH$. This is methyl alcohol or methanol, also known as wood alcohol. It is highly toxic and can cause blindness or even death in relatively small doses. When R is the *ethyl* group, we have $CH_3CH_2OH$. This is ethyl alcohol or ethanol. Ethanol is the alcohol in alcoholic beverages. When R is the *isopropyl* group, we have (CH$_3$)$_2$CHOH. This is isopropyl alcohol. Isopropyl alcohol, what we commonly call "rubbing alcohol," is widely used as a disinfectant.

Methanol                           Ethanol                          Isopropyl alcohol

| TABLE 10.1 | Alkyl Groups | |
|---|---|---|
| **Name** | **Formula** | **Model** |
| Methyl | −CH$_3$ | |
| Ethyl | −CH$_2$CH$_3$ | |
| Propyl | −CH$_2$CH$_2$CH$_3$ | |
| Isopropyl | −CH(CH$_3$)$_2$ | |
| Butyl | −CH$_2$CH$_2$CH$_2$CH$_3$ | |
| *tert*-butyl | −C(CH$_3$)$_3$ | |
| Pentyl | −CH$_2$CH$_2$CH$_2$CH$_2$CH$_3$ | |
| Isopentyl | −CH$_2$CH$_2$CH(CH$_3$)$_2$ | |
| Hexyl | −CH$_2$CH$_2$CH$_2$CH$_2$CH$_2$CH$_3$ | |
| Heptyl | −CH$_2$CH$_2$CH$_2$CH$_2$CH$_2$CH$_2$CH$_3$ | |
| Octyl | −CH$_2$CH$_2$CH$_2$CH$_2$CH$_2$CH$_2$CH$_2$CH$_3$ | |

*You May Be Wondering...*

# How Are Organic Compounds Named?

Organic compounds are named systematically using International Union of Pure and Applied Chemistry (IUPAC) rules.

## Root Names for Alkanes and Alkyl Groups

| Number of Carbons | Name | Number of Carbons | Name |
|---|---|---|---|
| 1 | Meth- | 6 | Hex- |
| 2 | Eth- | 7 | Hept- |
| 3 | Prop- | 8 | Oct- |
| 4 | But- | 9 | Non- |
| 5 | Pent- | 10 | Dec- |

## Prefixes for Halogen Substituents

| | |
|---|---|
| −F | Fluoro |
| −Cl | Chloro |
| −Br | Bromo |
| −I | Iodo |

## Alkanes

In Chapter 2, we encountered the names of some simple, straight-chain alkanes such as pentane:

$$H-\overset{\underset{|}{H}}{\underset{\underset{H}{|}}{C}}-\overset{\underset{|}{H}}{\underset{\underset{H}{|}}{C}}-\overset{\underset{|}{H}}{\underset{\underset{H}{|}}{C}}-\overset{\underset{|}{H}}{\underset{\underset{H}{|}}{C}}-\overset{\underset{|}{H}}{\underset{\underset{H}{|}}{C}}-H$$

To name substituted alkanes (i.e., those that have **substituents,** which are groups other than −H bonded to the carbons of the chain), we follow a series of steps:

1. Identify the longest continuous carbon chain to get the *parent name.*
2. Number the carbons in the continuous chain, beginning at the end closest to the substituent. Commonly encountered substituents include alkyl groups and halogens.

3. Identify the substituent and use a *number* followed by a dash and a *prefix* to specify its *location* and *identity,* respectively.

Step 1:    The longest continuous carbon chain contains five C atoms.

**2-Methylpentane**

(We could also identify the carbon chain as

but the number of C atoms in the chain is the same either way.) The parent name of a five-carbon chain is *pentane.* (See root names.)

Step 2:    We number the carbon atoms beginning at the end nearest the substituent (shaded in green):

Step 3:    The substituent is a methyl group, −CH₃. It is attached to carbon 2. Therefore, the name is 2-methylpentane.

Sample Problem 10.1 lets you practice naming some simple organic compounds.

---

### Sample Problem 10.1

Give names for the following compounds:

(a)

(b)

(c)

**Strategy** Use the three-step procedure for naming substituted alkanes: (1) name the parent alkane, (2) number the carbons, and (3) name and number the substituent. (Consult Table 2.4 for parent alkane names.)

**Setup** (a) This is a five-carbon chain. We can number the carbons starting at either end because the Cl substituent will be located on carbon 3 either way:

$$
\begin{array}{ccccc}
H & H & Cl & H & H \\
| & | & | & | & | \\
H-C-&C-&C-&C-&C-H \\
| & | & | & | & | \\
H & H & H & H & H \\
\end{array}
$$

(b) This may look like a substituted pentane, too, but the longest carbon chain in this molecule is seven carbons long.

Although Lewis structures appear to be flat and to contain 90° angles, the C atoms in this molecule are all $sp^3$-hybridized (four electron domains around each) and there is free rotation about the C—C bonds [◀◀ Section 9.5]. Thus, the molecule can also be drawn as

The substituent is a methyl group on carbon 4.

(c) This is a substituted *hexane*.

**Solution** (a) 3-chloropentane, (b) 4-methylheptane, (c) 2-methylhexane

> **Think About It** A common error is to misidentify the parent alkane. Double-check to be sure you have identified the *longest* continuous carbon chain in the molecule. Also be sure to number the carbon atoms so as to give the substituent the *lowest* possible number.

**Practice Problem A** Give the systematic IUPAC name for each of the following:

**Practice Problem B** Draw structures for (a) 4-ethyloctane, (b) 2-fluoropentane, and (c) 3-methyldecane.

| TABLE 10.2 | General Formulas for Select Classes of Organic Compounds | | |
|---|---|---|---|
| **Class** | **General Formula** | **Structure** | **Functional Group** |
| Alcohol | ROH | $-\ddot{\text{O}}-\text{H}$ | Hydroxy group |
| Carboxylic acid | RCOOH | $-\overset{\overset{\ddot{\text{O}}}{\parallel}}{\text{C}}-\ddot{\text{O}}-\text{H}$ | Carboxy group |
| Ester | RCOOR′ | $-\overset{\overset{\ddot{\text{O}}}{\parallel}}{\text{C}}-\ddot{\text{O}}-\text{R}'$ | Ester group |
| Aldehyde | RCHO | $-\overset{\overset{\ddot{\text{O}}}{\parallel}}{\text{C}}-\text{H}$ | Carbonyl group |
| Ketone* | RCOR′ | $-\overset{\overset{\ddot{\text{O}}}{\parallel}}{\text{C}}-\text{R}'$ | Carbonyl group |
| Amine† | RNH₂ | $-\overset{\overset{\phantom{.}}{\underset{\underset{\text{H}}{\vert}}{\ddot{\text{N}}}}}{}-\text{H}$ | Amino group (primary, 1°) |
| Amine | RNR′H | $-\overset{\overset{\phantom{.}}{\underset{\underset{\text{H}}{\vert}}{\ddot{\text{N}}}}}{}-\text{R}'$ | Amino group (secondary, 2°) |
| Amine | RNR′R″ | $-\overset{\overset{\phantom{.}}{\underset{\underset{\text{R}''}{\vert}}{\ddot{\text{N}}}}}{}-\text{R}'$ | Amino group (tertiary, 3°) |
| Amide | RCONH₂ | $-\overset{\overset{\ddot{\text{O}}}{\parallel}}{\text{C}}-\underset{\underset{\text{H}}{\vert}}{\ddot{\text{N}}}-\text{H}$ | Amide group (primary, 1°) |
| Amide | RCONR′H | $-\overset{\overset{\ddot{\text{O}}}{\parallel}}{\text{C}}-\underset{\underset{\text{H}}{\vert}}{\ddot{\text{N}}}-\text{R}'$ | Amide group (secondary, 2°) |
| Amide | RCONR′R″ | $-\overset{\overset{\ddot{\text{O}}}{\parallel}}{\text{C}}-\underset{\underset{\text{R}''}{\vert}}{\ddot{\text{N}}}-\text{R}'$ | Amide group (tertiary, 3°) |

*R′ represents a second alkyl group that may or may not be identical to the first alkyl group R. Likewise, R″ represents a third alkyl group that may or may not be identical to R or R′.

†The designations 1°, 2°, and 3° refer to how many R groups are bonded to the N atom.

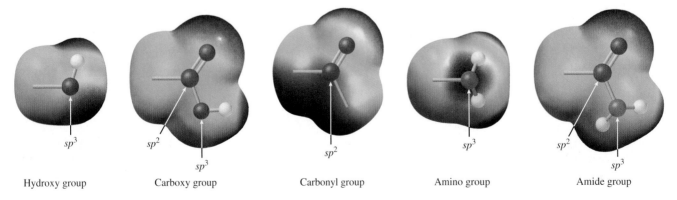

| | | | | |
|---|---|---|---|---|
| Hydroxy group | Carboxy group | Carbonyl group | Amino group | Amide group |

**Figure 10.1**    Models and electrostatic potential maps of the hydroxy, carboxy, carbonyl, amino, and amide functional groups.

Many compounds contain more than one functional group. An *amino acid,* for example, contains both the amine group and the carboxy group.

Alanine

Sample Problem 10.2 lets you practice identifying functional groups in molecules.

## Sample Problem 10.2

Many familiar substances are organic compounds. Some examples include aspartame, the artificial sweetener in the sugar substitute Equal and in many diet sodas; salicylic acid, found in some acne medicines and wart-removal treatments; and amphetamine, a stimulant used to treat narcolepsy, attention-deficit hyperactivity disorder (ADHD), and obesity. Identify the functional group(s) in each molecule:

Aspartame            Salicylic acid            Amphetamine
(a)                      (b)                      (c)

**Strategy**  Look for and identify the combinations of atoms shown in Table 10.2.

**Setup** (a) From right to left, aspartame contains a −COOH group, an −NH$_2$ group, a −CONHR group, and a −COOR group:

(b) Salicylic acid contains an −OH group and a −COOH group:

(c) Amphetamine contains an −NH$_2$ group:

**Solution** (a) From right to left, aspartame contains a *carboxy* group, an *amino* group, an *amide* group, and a −COOR (ester) group.

(b) Salicylic acid contains a *hydroxy* group and a *carboxy* group.

(c) Amphetamine contains an *amino* group.

**Think About It** In part (b), the salicylic acid molecule contains a benzene ring and is therefore *aromatic.* When the hydroxy group is attached to a benzene ring, the resulting aromatic compound is a *phenol,* not an alcohol. Amphetamine has several legitimate medicinal uses, but it is also one of the most commonly misused drugs in the United States. Because it frequently is prescribed to adolescents for ADHD, much of it finds its way into high schools, where its misuse is a serious problem. A closely related compound that frequently makes headlines is *methamphetamine:*

In August of 2005, an issue of *Newsweek* magazine devoted a cover story to methamphetamine and its abuse.

*(Continued)*

**Practice Problem A** Identify the functional groups in each of the following molecules:

(a)                 Ethyl butyrate

(c)                 Tartaric acid

(b)                 Aspirin

**Practice Problem B** Identify the functional groups in each of the following cyclic compounds:

---

## Checkpoint 10.2    Classes of Organic Compounds

**10.2.1** Identify the name of the following compound:

a) 2-Ethylpropane

b) 2-Ethylbutane

c) 2-Methylbutane

d) 2-Methylpentane

e) 3-Methylbutane

**10.2.2** Identify the name of the following compound:

a) 1-Chloroethane

b) 2-Chloropropane

c) 2-Chloromethane

d) 1-Chloropropane

e) 3-Chloroethane

**10.2.3** Identify the name of the following compound:

a) 4-Pentanone

b) 2-Pentanone

c) Ethylpentanal

d) Propylpentanal

e) Propylethanal

**10.2.4** Identify the name of the following compound:

a) 3-Methyl-1-ethanamine

b) 2-Methyl-1-ethanamine

c) 1-Pentanamine

d) 2-Methyl-1-butanamine

e) 3-Methyl-1-butanamine

FAQ *You May Be Wondering...*

# How Do We Name Molecules with More Than One Substituent?

A systematic name must identify a compound unambiguously. Therefore, a special system of rules must be followed to name molecules that contain more than one substituent.

In molecules that contain two or more identical substituents, the prefixes *di, tri, tetra, penta,* and so forth, are used to denote the number of substituents. Numbers are then used to denote their positions. (Note that two substituents may be bonded to the same carbon atom. In this case, the carbon's number is repeated with a comma between the numbers.)

2,3-Dimethylpentane          2,2-Dichlorohexanoic acid

In the case where two or more different substituents are present, the substituent names are alphabetized in the systematic name of the compound. Numbers are used to indicate the positions of the alphabetized substituents. If a prefix is used to denote two or more identical substituents, the prefix is *not* used to determine the alphabetization—only the substituent name is used.

4-Ethyl-2-methylhexane          4-Ethyl-2,2-dimethylhexane

**10.2.5** Identify the functional group(s) in the following molecule:

(Select all that apply.)

a) Hydroxy

b) Carboxy

c) Carbonyl

d) Amino

e) Amide

**10.2.6** Identify the functional group(s) in the following molecule:

(Select all that apply.)

a) Hydroxy

b) Carboxy

c) Carbonyl

d) Amino

e) Amide

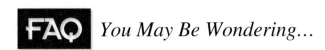

# How Do We Name Compounds with Specific Functional Groups?

## Alcohols

Identify the longest continuous carbon chain that includes the carbon to which the −OH group is attached. This is the parent alkyl group. Name it according to the number of carbons it contains, and change the –*e* ending to –*ol*. Number the C atoms such that the −OH group has the lowest possible number, and, when necessary, use a number to indicate the position of the −OH group:

Ethanol          1-Propanol          2-Butanol

When the chain that bears the −OH group also bears an alkyl substituent, the chain is numbered in the direction that gives the lowest possible number to the carbon attached to −OH.

5-Methyl-3-hexanol

## Carboxylic Acids

Identify the longest continuous carbon chain that includes the carboxy group. Name it according to the number of carbons it contains, and change the –*e* ending to –*oic* acid. Number the C atoms starting with the carbonyl carbon. Use numbers and prefixes to indicate the position and the identity of any substituents.

**Student Annotation:** The carbonyl carbon is the one that is doubly bonded to oxygen.

Many organic compounds have common names in addition to their systematic names. Common names for some of the carboxylic acids shown here are given in parentheses:

Ethanoic acid          Propanoic acid          Butanoic acid          5-Methylhexanoic acid
(acetic acid)          (propionic acid)          (butyric acid)

## Esters

Name esters as derivatives of carboxylic acids by replacing the –*ic* acid ending with –*ate:*

Ethyl acetate                    Methyl propionate
Based on acetic acid          Based on propionic acid

(The first part of an ester's name specifies the substituent that replaces the ionizable hydrogen of the corresponding carboxylic acid.)

## Aldehydes

Identify the longest continuous carbon chain that includes the carbonyl group. Name it according to the number of carbons it contains, and change the –e ending to –al. Number the C atoms starting with the carbonyl carbon. Use numbers and prefixes to indicate the position and the identity of any substituents:

Ethanal (acetaldehyde)   Propanal (propionaldehyde)   Butanal   5-Methylhexanal

## Ketones

Identify the longest continuous carbon chain that includes the carbonyl group. Name it according to the number of carbons it contains, and change the –e ending to –one. If necessary, number the C atoms to give the carbonyl carbon the lowest possible number. Use numbers and prefixes to indicate the position and the identity of any substituents:

Propanone (acetone)   2-Butanone (ethyl methyl ketone)   5-Methyl-3-hexanone

## Primary Amines

Identify the longest continuous carbon chain that includes the carbon to which the $-NH_2$ group is bonded. Name it according to the number of carbons it contains, and change the –e ending to –amine. Number the C atoms starting with the carbon to which the $-NH_2$ group is bonded. Use numbers and prefixes to indicate the position and the identity of any substituents:

Ethanamine (ethylamine)   1-Propanamine (propylamine)   1-Butanamine (butylamine)   5-Methyl-1-hexanamine (5-methylhexylamine)

## Primary Amides

Primary amides are named as derivatives of carboxylic acids, but they can also be named by replacing the –e ending of the corresponding alkane with –amide.

Propanamide (propionamide)   Butanamide (butyramide)

## 10.3 | Representing Organic Molecules

You've learned previously how to represent molecules using *molecular* and *structural* formulas [◄◄ Section 2.6], as well as using Lewis structures [◄◄ Section 8.3]. In this section, we will learn several additional ways to represent molecules—ways that are particularly useful in the study of organic chemistry.

The representation of organic molecules is especially important because the atoms in an organic molecule, unlike those in inorganic compounds, may be arranged in an enormous variety of different ways. For example, there are literally dozens of different ways that a compound containing five carbon atoms, one oxygen atom, and the necessary number of hydrogen atoms can be arranged, with each arrangement representing a unique organic compound. Here are ten possibilities:

### Condensed Structural Formulas

A **condensed structural formula,** or simply a **condensed structure,** shows the same information as a *structural formula,* but in a *condensed* form. For instance, the molecular formula, structural formula, and condensed structural formula of octane are as follows:

$$C_8H_{18}$$
Molecular formula

$$CH_3CH_2CH_2CH_2CH_2CH_2CH_2CH_3$$
Structural formula

$$CH_3(CH_2)_6CH_3$$
Condensed structural formula

Octane

In the condensed structural formula, the identical adjacent groups of atoms (in this case $-CH_2-$ groups) are enclosed in parentheses and subscripted to denote their number.

In molecules where the carbon atoms do not form a single, unbranched chain, branches are indicated using additional parentheses in the condensed structural formula. For example, 2-methylheptane has the same molecular formula ($C_8H_{18}$) as octane. In effect, the chain is one C atom *shorter* than in octane and one of the hydrogen atoms on the second C atom has been replaced by a methyl group. The molecular formula, structural formula, and condensed structural formula of 2-methylheptane are as follows:

$C_8H_{18}$
Molecular formula

$$CH_3$$
$$|$$
$$CH_3CHCH_2CH_2CH_2CH_2CH_3$$
Structural formula

$CH_3CH(CH_3)(CH_2)_4CH_3$
Condensed structural formula

2-Methylheptane

## Kekulé Structures

*Kekulé structures* are similar to Lewis structures except that they do not show lone pairs. For molecules that contain no lone pairs, such as octane, the Lewis and Kekulé structures are identical:

$$H-\overset{\overset{H}{|}}{C}-\overset{\overset{H}{|}}{\underset{\underset{H}{|}}{C}}-\overset{\overset{H}{|}}{\underset{\underset{H}{|}}{C}}-\overset{\overset{H}{|}}{\underset{\underset{H}{|}}{C}}-\overset{\overset{H}{|}}{\underset{\underset{H}{|}}{C}}-\overset{\overset{H}{|}}{\underset{\underset{H}{|}}{C}}-\overset{\overset{H}{|}}{\underset{\underset{H}{|}}{C}}-\overset{\overset{H}{|}}{\underset{\underset{H}{|}}{C}}-H$$

Octane

(Many of the structures already shown in this chapter are Kekulé structures.) The Kekulé structures for several familiar organic molecules are as follows:

Ethanol          Formaldehyde          Acetic acid

## Skeletal Structures

Skeletal structures are especially useful for representing complex organic molecules. A *skeletal structure* consists of straight lines that represent carbon-carbon bonds. The carbon atoms themselves (and the attached hydrogen atoms) are not shown, but you need to know that they are there. The structural formulas and skeletal structures for several hydrocarbons [◄◄ Section 2.6] are as follows:

| Name | Structural Formula | Skeletal Structure |
|------|-------------------|--------------------|
| Pentane | $CH_3CH_2CH_2CH_2CH_3$ | |
| Isopentane | $CH_3CHCH_2CH_3$ with $CH_3$ branch | |
| Neopentane | $CH_3CCH_3$ with $CH_3$ branches | |

The end of each straight line in a skeletal structure corresponds to a carbon atom (unless another atom of a different type is explicitly shown at the end of the line). Additionally, there are as many hydrogen atoms attached to each carbon atom as are necessary to give each carbon atom a total of four bonds.

When a molecule contains an element *other* than carbon or hydrogen, those atoms, called **heteroatoms,** are shown explicitly in the skeletal structure. Furthermore, while the hydrogens attached to carbon atoms typically are not shown, hydrogens attached to heteroatoms *are* shown, as illustrated by the molecule ethylamine in the following:

| Name | Structural Formula | Skeletal Structure |
|---|---|---|
| Propanone (acetone) | $CH_3\overset{\overset{O}{\|\|}}{C}CH_3$ | |
| Ethylamine | $CH_3CH_2NH_2$ | |
| Tetrahydrofuran | $H_2C\overset{\overset{O}{\diagdown \diagup}}{\phantom{.}}CH_2$ $H_2C-CH_2$ | |

Although carbon and hydrogen atoms need not be shown in a skeletal structure, some of the C and H atoms *can* be shown for the purpose of emphasizing a particular part of a molecule. Often when we show a molecule, we choose to emphasize the *functional group(s),* which are largely responsible for the properties and reactivity of the compound [◄◄ Section 2.6].

Sample Problem 10.3 shows how to interpret skeletal structures.

---

### Sample Problem 10.3

Write a molecular formula and a structural formula (or *condensed* structural formula) for the following:

(a)    (b)

**Strategy** Count the C atoms represented and the heteroatoms shown. Determine how many H atoms are present using the octet rule.

**Setup** Each line represents a bond. (Double lines represent double bonds.) Count one C atom at the end of each line unless another atom is shown there. Count the number of H atoms necessary to complete the octet of each C atom.

C atom + 1 H atom    C atom + 3 H atoms        C atom + 3 H atoms

C atom + 3 H atoms    C atom + 1 H atom        C atom (no H atoms)

**Solution** (a) Molecular formula: $C_4H_8$; structural formula: $CH_3(CH)_2CH_3$.

(b) Molecular formula: $C_2H_5NO$; structural formula: $CH_3CONH_2$.

---

**Practice Problem A** Write a structural formula for the compound represented by the following skeletal structure:

**Practice Problem B** Draw the skeletal structure for $(CH_3)_2C=CHNH_2$.

# Resonance

Recall from Chapter 8 that many molecules and ions can be represented by more than one Lewis structure [◄◄ Section 8.7]. Furthermore, two or more equally valid Lewis structures that differ only in the position of their electrons are called *resonance structures*. For example, $SO_3$ can be represented by three different Lewis structures [◄◄ Sample Problem 8.10].

None of these Lewis structures represents the $SO_3$ molecule accurately. The bonds in $SO_3$ are actually *equivalent*—equal in length and strength, which we would not expect if two were single bonds and one were a double bond. Each individual resonance structure implies that electron pairs are *localized* in specific bonds or on specific atoms. The concept of resonance allows us to envision certain electron pairs as *delocalized* over several atoms [◄◄ Section 9.7]. Delocalization of electron pairs imparts additional stability to a molecule (or polyatomic ion), and a species that can be represented by two or more resonance structures is said to be *resonance stabilized*.

Chemists sometimes use curved arrows to specify the differences in positions of electrons in resonance structures. In $SO_3$, for example, the "repositioning" of electrons in the resonance structures can be indicated as follows:

Resonance stabilization is observed in many organic species as well and affects chemical properties such as in the acidic behavior of ethanol ($CH_3CH_2OH$) and ethanoic acid ($CH_3COOH$), also commonly known as *acetic* acid. Each of these molecules has one *ionizable hydrogen atom* [◄◄ Section 2.6], enabling it to behave as a Brønsted acid [◄◄ Section 4.3]. However, the concentration of hydronium ions in a solution of ethanoic acid is hundreds of thousands of times higher than that in a comparable solution of ethanol. The reason for this large discrepancy is that the hydrogen atom on ethanoic acid is far more *easily* ionized than the one on ethanol. *Resonance* helps us explain why.

When a species loses its ionizable hydrogen atom, what remains is an anion. In the case of ethanoic acid and ethanol, the anions are

We can draw a second resonance structure for the anion produced by the ionization of ethanoic acid by repositioning the electron pairs as follows:

The negative charge on the anion resides on an oxygen atom. Which oxygen atom bears the charge depends on which resonance structure we look at. In essence, the greater the number of possible locations for the negative charge, the more stable the anion. And the more stable the anion, the more easily the ionizable hydrogen atom is lost, resulting in the production of more hydronium ions in solution.

It is not possible to draw additional resonance structures for the anion produced by the ionization of ethanol because there is nowhere else to put the lone pairs that reside on the O atom. They cannot be moved in between the O and C atoms because C can only have four electron pairs around it. (We were able to do this with $CH_3COO^-$ because we could also move one of the pairs already around C to the other O atom. See the curved arrows in the preceding resonance structure.)

Sample Problems 10.4 and 10.5 illustrate the use of curved arrow notation and the determination of resonance structures.

## Sample Problem 10.4

Following the curved arrows, draw a resonance structure for the HCONH$^-$ ion.

$$H-\overset{\overset{\ddot{O}}{\|}}{C}-\overset{..}{\underset{..}{N}}-H$$

**Strategy** Move electron pairs as indicated by the curved arrows, and calculate formal charges to reposition the negative charge.

**Setup** To determine the formal charge, we must first calculate the number of valence electrons on the atoms in question. The O and N atoms have *six* and *five* valence electrons, respectively.

**Solution** Repositioning electron pairs as directed by the curved arrows gives

$$H-\overset{\overset{\ddot{O}}{\|}}{C}-\overset{..}{\underset{..}{N}}-H \quad \longleftrightarrow \quad H-\overset{\overset{..}{\underset{}{\ddot{O}:^-}}}{C}=\overset{.}{N}-H$$

The formal charge on N, which had been −1, is now [5 − (3 + 2)] = 0. The formal charge on O, which had been 0, is now [6 − (1 + 6)] = −1. Thus, the negative charge now resides on the oxygen atom.

**Think About It** Count the number of electron pairs around each atom. Remember that the C and N atoms can have no more than four electron pairs. The H atom can have no more than one electron pair.

**Practice Problem A** Follow the curved arrows to draw a second resonance structure for the HCOO$^-$ ion.

$$H-\overset{\overset{\ddot{O}}{\|}}{C}-\ddot{O}:^-$$

**Practice Problem B** Given the following two resonance structures, draw the curved arrows on the first structure that will give rise to the second structure.

$$\left[\begin{matrix} :\ddot{O}: \\ | \\ :\ddot{O}-S-\ddot{O}: \\ | \\ :\ddot{O}: \end{matrix}\right]^{2-} \quad \longleftrightarrow \quad \left[\begin{matrix} :\ddot{O}: \\ \| \\ :\ddot{O}-S=\ddot{O}: \\ | \\ .\ddot{O}. \end{matrix}\right]^{2-}$$

## Sample Problem 10.5

Adenosine triphosphate (ATP) is sometimes called the "universal energy carrier" or "molecular energy currency." It contains two high-energy bonds (shown in red) that, when *hydrolyzed* (broken by the addition of water), release the energy necessary for cell function. Resonance stabilization of the hydrogen phosphate ion is one of the reasons the breakdown of ATP releases energy.

Adenosine triphosphate     +$H_2O$ ⟶     Adenosine diphosphate     +     Inorganic phosphate

Draw all the possible resonance structures for the hydrogen phosphate ion ($HPO_4^{2-}$). Use curved arrows to indicate how electrons are repositioned, and determine the position(s) of the negative charges.

**Strategy** Draw a valid Lewis structure for $HPO_4^{2-}$, and determine whether and where electrons can be repositioned to produce one or more additional structures. Indicate the movement of electrons with curved arrows, and draw all possible resonance structures. Calculate the formal charge on each atom to determine the placement of charges.

**Setup** A valid Lewis structure for the hydrogen phosphate ion is

$$\ddot{\overset{..}{O}} \quad \overset{\overset{\cdot\cdot}{\ddot{O}}}{\underset{\overset{\cdot\cdot}{\ddot{O}}:}{\overset{\parallel}{-\ddot{O}-P-\ddot{O}-H}}}$$

For the purpose of determining formal charges, P and O have five and six valence electrons, respectively.

**Solution** A lone pair can be moved from one of the oxygen atoms to create a double bond to the phosphorus, and a pair of electrons from the *original* double bond can be moved onto *that* oxygen atom. The net result is simply a repositioning of the double bond by moving two electron pairs. This can be done once more, giving a total of three resonance structures for $HPO_4^{2-}$. In each of the resonance structures, the formal charge on phosphorus is $[5 - (5)] = 0$. The formal charge on each singly bonded oxygen is $[6 - (1 + 6)] = -1$, and the formal charge on the doubly bonded oxygen is $[6 - (2 + 4)] = 0$.

$$\ddot{\overset{..}{O}} \longleftrightarrow \ddot{\overset{..}{O}} \longleftrightarrow \ddot{\overset{..}{O}} \longleftrightarrow \ddot{\overset{..}{O}}$$

> **Think About It** ATP can also be hydrolyzed to give AMP (adenosine *mono*phosphate) and *pyrophosphate* ($P_2O_7^{4-}$). Pyrophosphate hydrolyzes, in turn, to give two hydrogen phosphate ions. The oxygen atoms that can help delocalize the negative charges are highlighted:
>
> $$\left[\ddot{O}-\overset{\ddot{O}}{\underset{\ddot{O}}{P}}-\ddot{O}-\overset{\ddot{O}}{\underset{\ddot{O}}{P}}-\ddot{O}\right]^{4-} + H_2O \longrightarrow 2\left[\ddot{O}-\overset{\ddot{O}}{\underset{\ddot{O}}{P}}-\ddot{O}-H\right]^{2-}$$
>
> These structures can also be drawn with one double bond to each phosphorus atom, to minimize formal charges [◀◀ Section 8.8].

**Practice Problem A** Draw a valid Lewis structure for the carbonate ion ($CO_3^{2-}$), and use curved arrows to generate all possible resonance structures.

**Practice Problem B** In which of the following examples do the curved arrows *not* correspond correctly to the repositioning of electrons needed to arrive at the resonance structure shown?

$$\left[\ddot{O}-\ddot{N}=\ddot{O}\right]^- \longleftrightarrow \left[\ddot{O}=\ddot{N}-\ddot{O}\right]^- \qquad \left[\ddot{O}-\ddot{N}=\ddot{O}\right]^- \longleftrightarrow \left[\ddot{O}=\ddot{N}-\ddot{O}\right]^-$$

(a)                                                        (c)

$$\left[\ddot{O}-\ddot{N}=\ddot{O}\right]^- \longleftrightarrow \left[\ddot{O}=\ddot{N}-\ddot{O}\right]^-$$

(b)

## Checkpoint 10.3    Representing Organic Molecules

**10.3.1** Give the molecular formula of the compound represented by

a) $C_5H_{12}O$

b) $C_6H_8O$

c) $C_6H_{12}O$

d) $C_5H_6O$

e) $C_6H_{13}O$

**10.3.2** Give the molecular formula of the compound represented by

a) $C_5H_{12}O$

b) $C_6H_{13}O$

c) $C_6H_{12}O$

d) $C_6H_{14}O$

e) $C_4H_{13}O$

| **Checkpoint 10.3** | **Representing Organic Molecules** *(continued)* |

**10.3.3** Which of the following pairs of species are resonance structures? (Select all that apply.)

a)
$$H-\overset{\overset{\displaystyle H}{|}}{C}=\overset{\overset{\displaystyle :\ddot{O}:}{|}}{C}-\overset{\overset{\displaystyle H}{|}}{\underset{\underset{\displaystyle H}{|}}{C}}-H \longleftrightarrow H-\overset{\overset{\displaystyle H}{|}}{\underset{\underset{\displaystyle H}{|}}{C}}-\overset{\overset{\displaystyle \ddot{O}}{||}}{C}-\overset{\overset{\displaystyle H}{|}}{\underset{\underset{\displaystyle H}{|}}{C}}-H$$

b)
$$H-\overset{\overset{\displaystyle H}{|}}{\underset{\underset{\displaystyle H}{|}}{C}}-\overset{\overset{\displaystyle \ddot{O}}{||}}{C}-\ddot{O}:^{-} \longleftrightarrow H-\overset{\overset{\displaystyle H}{|}}{C}=\overset{\overset{\displaystyle :\ddot{O}:}{|}}{C}-\ddot{O}:^{-}$$

c)
$$H-\overset{\overset{\displaystyle H}{|}}{\underset{\underset{\displaystyle H}{|}}{C}}-\overset{\overset{\displaystyle \ddot{O}}{||}}{C}-\ddot{O}:^{-} \longleftrightarrow H-\overset{\overset{\displaystyle H}{|}}{\underset{\underset{\displaystyle H}{|}}{C}}-\overset{\overset{\displaystyle \ddot{O}}{|}}{C}=\ddot{O}:^{-}$$

d)
$$H-\overset{\overset{\displaystyle H}{|}}{\underset{\underset{\displaystyle H}{|}}{C}}-\overset{\overset{\displaystyle H}{|}}{\underset{\underset{\displaystyle :\ddot{O}:}{|}}{C}}-\overset{\overset{\displaystyle H}{|}}{C}-H \longleftrightarrow H-\overset{\overset{\displaystyle H}{|}}{\underset{\underset{\displaystyle H}{|}}{C}}-\overset{\overset{\displaystyle H}{|}}{\underset{\underset{\displaystyle H}{|}}{C}}-\overset{\overset{\displaystyle H}{|}}{\underset{\underset{\displaystyle H}{|}}{C}}-\ddot{O}-H$$

e)
$$^{-}:\ddot{O}-\overset{\overset{\displaystyle \ddot{O}}{||}}{C}-\overset{\overset{\displaystyle \ddot{O}}{||}}{C}-\ddot{O}:^{-} \longleftrightarrow :\ddot{O}:^{-}\overset{\overset{\displaystyle :\ddot{O}:^{-}}{|}}{}\\ :O=C-C=O:$$

**10.3.4** Which of the following structural formulas represents a species that has two or more resonance structures? (Select all that apply.)

a) HCOOH

b) HCOO⁻

c) $CH_2CH_2$

d) $O_3$

e) $CO_2$

---

| **10.4** | **Isomerism** |

We first encountered the term *isomer* in the context of molecular geometry and molecular polarity [◄◄ Section 9.2]. Isomers are different compounds that have the same chemical formula. In this section, we will discuss the different types of isomerism (namely, constitutional isomerism and stereoisomerism) and their importance to organic chemistry.

### Constitutional Isomerism

Also known as *structural* isomerism, **constitutional isomerism** occurs when the same atoms can be connected in two or more different ways. For example, there are three different ways to arrange the atoms in a compound with the chemical formula $C_5H_{12}$. Constitutional isomers have distinct names and generally have different physical and chemical properties. Table 10.3 lists the three constitutional isomers of $C_5H_{12}$ along with their boiling points for comparison.

### Stereoisomerism

Multimedia

Organic and Biochemistry—structural isomers of hexane.

*Stereoisomers* are those that contain identical bonds but differ in the orientation of those bonds in space. Two types of stereoisomers exist: geometrical isomers and optical isomers. **Geometrical isomers** occur in compounds that have restricted rotation around a bond. For instance, compounds that contain carbon-carbon double bonds can form geometrical isomers. Individual geometrical isomers have the same names but are distinguished by a prefix such as *cis* or *trans*. Dichloroethylene, ethylene in which two of the H atoms (one on each C atom) have been replaced by Cl atoms, exists as two geometrical isomers. The isomer in which the Cl atoms both lie on the same side (above or below, in this example) of the double bond is called the *cis* isomer. The isomer in which the Cl atoms lie on opposite sides of the double bond is called the **trans** isomer. (The compound in which both Cl atoms are attached to the same C atom is a *constitutional* isomer rather than a stereoisomer.) Figure 10.2 depicts the isomers of $C_2H_2Cl_2$.

| TABLE 10.3 | Constitutional Isomers of $C_5H_{12}$ | | | |
|---|---|---|---|---|
| **Name** | **Structural Formula** | **Skeletal Structure** | **Ball-and-Stick Model** | **BP (°C)** |
| Pentane (*n*-pentane) | H H H H H<br>H—C—C—C—C—C—H<br>H H H H H | | | 36.1 |
| Methylbutane (isopentane) | H CH₃ H H<br>H—C—C——C—C—H<br>H H H H | | | 27.8 |
| 2,2-Dimethylpropane (neopentane) | H CH₃ H<br>H—C—C——C—H<br>H CH₃ H | | | 9.5 |

Geometrical isomers usually have different physical and chemical properties. *Trans* isomers tend to be more stable and are generally easier to synthesize than their *cis* counterparts. The existence of *cis* isomers in living systems is a testament to how much better nature is at chemical synthesis than we are. As the information in the box on page 410 demonstrates, geometrical isomerism is sometimes of tremendous biological significance.

Stereoisomers that are mirror images of each other, but are not superimposable, are called **_optical isomers._** Consider the hypothetical organic molecule shown in Figure 10.3. It consists of an $sp^3$-hybridized carbon atom that is bonded to four different groups. Its mirror image appears identical to it, just as your right and left hands appear identical to each other. But, if you have ever

Multimedia

Organic and Biochemistry—chiral molecules.

Cl    Cl       H    Cl       H    Cl
  C=C       C=C       C=C
H    H       Cl    H       H    Cl

*cis*-Dichloroethylene    *trans*-Dichloroethylene    1,1-Dichloroethylene

**Figure 10.2**   Three isomers of dichloroethylene.

(a)

(b)

**Figure 10.3**   (a) Nonsuperimposable mirror images. (b) Despite being mirror images of each other, enantiomers are different compounds.

tried to put a right-handed glove on your left hand, or vice versa, you know that your hands are not identical. Imagine rotating the molecule on the right so that its green and red spheres coincide with those of the molecule on the left. Doing so results in the yellow sphere of one molecule coinciding with the blue sphere on the other. These two molecules are mirror images of each other, but they are not identical.

Molecules with nonsuperimposable mirror images are called ***chiral;*** and a pair of such mirror-image molecules are called ***enantiomers.*** Most of the chemical properties of enantiomers and all their physical properties are identical. Their chemical properties differ only in reactions that involve another chiral species, such as a chiral molecule or a receptor site that is shaped to fit only one enantiomer. Most biochemical processes consist of a series of chemically specific reactions that use chiral receptor sites to facilitate reaction by allowing only the specific reactants to fit (and thus react).

In organic chemistry, it often is necessary to represent tetrahedral molecules (three-dimensional objects) on paper (a two-dimensional surface). By convention, this is done using solid lines to represent bonds that lie in the plane of the page, dashes to represent bonds that point behind the page, and wedges to represent bonds that point in front of the page:

Methane                    1, 2-Dichloroethane

One property of chiral molecules is that the two enantiomers rotate the plane of plane-polarized light in opposite directions; that is, they are *optically active.* Unlike ordinary light, which oscillates in all directions, plane-polarized light oscillates only in a single plane. We use a polarimeter, shown schematically in Figure 10.4, to measure the rotation of polarized light by optical isomers. A beam of unpolarized light first passes through a Polaroid sheet, called the *polarizer,* and then through a sample tube containing a solution of an optically active, chiral species. As the polarized light passes through the sample tube, its plane of polarization is rotated either to the right or to the left. The amount of rotation can be measured by turning the analyzer in the appropriate direction until minimal light transmission is achieved.

**Figure 10.4**    Schematic of a polarimeter.

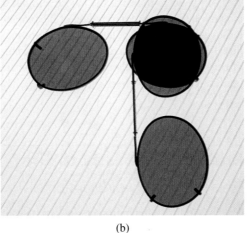

(a)                              (b)

**Figure 10.5** Two pairs of polarized sunglasses. (a) When two polarized lenses overlap with their planes of polarization parallel to each other, light is transmitted. (b) When one pair is rotated so that its plane of polarization is perpendicular to the other, no light is transmitted through the overlapped lenses.

Minimal transmission occurs when the plane of polarization of the light is perpendicular to that of the analyzer through which it is viewed. This effect can be demonstrated using two pairs of polarized sunglasses as shown in Figure 10.5. If the plane of polarization is rotated to the right, the isomer is said to be *dextrorotatory* and is labeled *d*; if it is rotated to the left, the isomer is called *levorotatory* and labeled *l*. Enantiomers always rotate the light by the same amount, but in opposite directions. Thus, in an equimolar mixture of both enantiomers, called a ***racemic mixture,*** the net rotation is zero.

**Student Annotation:** There are several conventions used to designate specific enantiomers. *Dextro-* and *levo-* prefixes refer to the direction of rotation of polarized light. *R* and *S*, the most commonly used designations, are assigned based on the "priority" assigned to each of the four groups attached to the chiral carbon—something that is beyond the scope of this text.

## Bringing Chemistry to Life

### Plane-Polarized Light and 3-D Movies

We see in three dimensions because our eyes view the world from slightly different positions. Our brains synthesize a three-dimensional (3-D) picture based on the two different pictures sent to it by our eyes. Modern 3-D movies make use of this phenomenon to make it seem as though objects on the screen are actually moving toward the viewer.

Three-dimensional movies are filmed using two different cameras at slightly different angles to the action. Thus, there are actually *two* movies that must be shown to us simultaneously. To make sure that our two eyes receive two different perspectives, each movie is projected through a polarizer, which polarizes the two projections in directions perpendicular to each other (Figure 10.6).

Polarized glasses

**Figure 10.6** Two different versions of the same movie are projected onto the screen. Polarized lenses make it so that each eye sees only one version. The result is the perception of three-dimensional action.

If we were to watch the movie without the special glasses provided by the movie house, we would see the blurry combination of the two movies. However, the 3-D glasses consist of polarized lenses, with planes of polarization that are mutually perpendicular. The left lens, polarized in one direction, blocks the image that is polarized perpendicular to it. The right lens, polarized in the other direction, blocks the other image. Our eyes are "tricked" into seeing two different movies, which our brain combines to form one 3-D image. The results of this process can be quite impressive!

## Bringing Chemistry to Life

### Biological Activity of Enantiomers

The importance of enantiomers in living systems cannot be overstated. Numerous processes important to biological function involve one enantiomer of a chiral compound. Many drugs, including thalidomide, are chiral with only one enantiomer having the desired properties. Some such drugs have been manufactured and marketed as racemic mixtures. It has become common, though, for drug companies to invest in *chiral switching,* the preparation of single-isomer versions of drugs originally marketed as racemic mixtures, in an effort to improve on existing therapies and to combat the revenue losses caused by generic drugs. A fairly high-profile example of this is the single-isomer drug Nexium, the so-called purple pill. AstraZeneca, makers of Nexium, held a patent for the drug Prilosec, originally a prescription heartburn medication that is now available over the counter. Prilosec is a racemic mixture of the chiral compound omeprazole. Prior to the 2002 expiration of its patent on Prilosec, AstraZeneca began producing and marketing Nexium, which contains only the therapeutically effective enantiomer (*S*)-omeprazole or *esomeprazole.*

Another example of chiral switching is that of the selective serotonin reuptake inhibitor (SSRI) antidepressant Celexa, which was introduced to the market in 1998 by Forest Laboratories. Celexa is a racemic mixture of (*R*)-citalopram oxalate and (*S*)-citalopram oxalate. While only the (*S*) enantiomer has therapeutic antidepressant properties, both enantiomers contribute to the side effects of the drug and therefore limit effectiveness and patient tolerance. In 2002, the FDA approved Lexapro, a new antidepressant derived from Celexa but from which the therapeutically ineffective (*R*) enantiomer has been removed. The benefits of isolating the active isomer include smaller required dosages, reduced side effects, and a faster and better patient response to the drug.

Although the intellectual property laws regarding single-isomer drug patents are somewhat ambiguous, chiral switching has enabled some pharmaceutical companies to extend the time that they are able to market their popular prescriptions exclusively. Strictly speaking, the FDA does not consider a single enantiomer of an already approved chiral drug to be a "new chemical entity," which is a requirement for obtaining a patent on a compound. Early in the history of chiral switching, however, there was some disagreement among patent examiners regarding what constituted a new chemical entity, and patents were granted on single-isomer drugs that might not be granted today.

## 10.5   Organic Reactions

Coulomb's law, which we first encountered in the box in Section 7.3, measures the force of the attraction between opposite charges. The attraction between regions of opposite charge on neighboring species and the resulting movement of electrons are the basis for our understanding of many organic reactions. We begin by defining what *electrophiles* and *nucleophiles* are, two terms used frequently in organic chemistry.

An ***electrophile*** is a species with a positive or partial positive charge. Literally, an electrophile "loves electrons." Thus, an electrophile is attracted to a region of negative or partial negative charge. An electrophile may be a cation, such as $H^+$, or the positive portion of a polar molecule, such as the H atom in HCl. Electrophiles are *electron-poor.*

A ***nucleophile*** is a species with a negative or partial negative charge. Literally, a nucleophile "loves a nucleus." A nucleophile is attracted to a region of positive or partial positive charge (i.e., an electrophile). A nucleophile may be an anion, such as $Cl^-$, or the negative portion of a polar molecule, such as the Cl atom in HCl. Nucleophiles are *electron-rich.* Electron-rich sites and electron-poor sites are attracted to one another.

**Figure 10.7**   The partial positive charge on H in HCl is attracted to the region of electron density in ethylene's double bond.

## Addition Reactions

The electrostatic potential maps of HCl and $C_2H_4$ shown in Figure 10.7 demonstrate that both molecules have regions of partial positive charge and partial negative charge. For example, HCl is a polar molecule, with H bearing a partial positive charge, due to the large difference in electronegativity between hydrogen and chlorine. Moreover, the carbon-carbon double bond in $C_2H_4$ consists of two pairs of shared electrons, one pair in a sigma bond and one pair in a pi bond [◄◄ Section 9.5], making the double bond a region of partial negative charge. The partial positive charge on the H in the HCl molecule is an electrophile. The double bond in ethylene, a region of relatively high electron density, is a nucleophile.

A reaction takes place when the positive end of the HCl molecule approaches the double bond in ethylene. The pi bond breaks, and the electrons it contained move as indicated by the curved arrows shown in the following equation, forming a sigma bond between the H atom of the HCl molecule and one of the C atoms. As this new bond forms, two things happen:

**Student Annotation:** Although it ionizes completely in aqueous solution [◄◄ Section 4.1], HCl exists as *molecules* in the gas phase.

**Student Annotation:** When an electrophile approaches another species and accepts electrons from it to form a bond, this is called *electrophilic attack*.

**Student Annotation:** Curved arrows are used to illustrate the mechanism by which an organic reaction occurs. Unlike their use in resonance structures, curved arrows in a reaction mechanism correspond to the actual *movement* of electrons.

1. Because there cannot be more than one bond to the H atom, the original bond between H and Cl breaks. Both of the electrons originally shared by H and Cl go with the Cl atom. The resulting intermediate species are shown in square brackets in the following equation (dashed lines represent bonds that are being formed):

The C atom on the right bears a positive charge after the valence electron it originally shared (in the pi bond with the other C atom) is removed from it completely. A species such as this, in which one of the carbons is surrounded by only six electrons, is called a **carbocation**. Although carbon must obey the octet in any stable compound, some reactions involve transient, *intermediate* species in which a carbon atom may be electron deficient—having only three electron pairs around it.

2. The C atom forming the new sigma bond to the H atom changes from $sp^2$-hybridized to $sp^3$-hybridized:

The carbon atom bearing the positive charge is still $sp^2$-hybridized at this point.

The chloride ion produced when both of the electrons originally shared by H and Cl go with Cl is a *nucleophile*. It is attracted to the newly formed positive charge and two of its electrons form a bond to the positively charged C atom as shown by the curved arrows.

The formation of a new sigma bond between the Cl and C atoms causes the hybridization of the second C to change from $sp^2$ to $sp^3$:

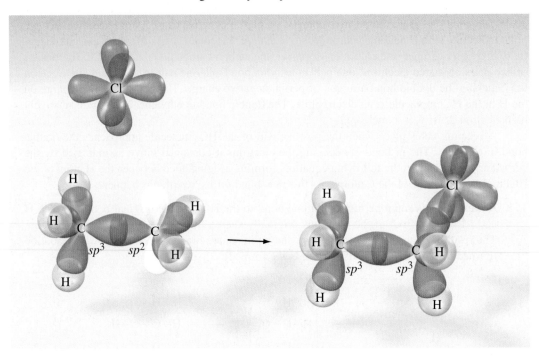

The overall reaction is called an *addition reaction*. Specifically, this is an ***electrophilic addition*** reaction because it begins with the electrophilic attack by HCl on the region of electron density in the double bond.

Addition reactions can also begin with nucleophilic attack, in which case the reaction is called a ***nucleophilic addition.*** In a nucleophilic addition reaction, a bond forms when a nucleophile donates a pair of electrons to an electron-deficient atom. Water, for example, reacts with carbon dioxide to produce carbonic acid. Although the addition reaction happens essentially all at once, it is helpful to think of the movement of electrons as the following stepwise process:

1.

One of the lone pairs on the O atom in water attacks the C atom, which is electron deficient because of the highly electronegative O atoms bonded to it. As the new bond forms, one of the original C–O pi bonds breaks. The electron pair from the broken pi bond is repositioned on the corresponding O atom, leaving the atom with a negative charge.

2.

One of the O–H bonds in water breaks, with both electrons remaining with the O atom. The H atom separates from the water molecule as a proton.

3.

Finally, the negatively charged O atom acquires the proton lost by the water molecule that served as the nucleophile.

A summary of the mechanisms of electrophilic and nucleophilic addition reactions is given in Figure 10.8.

**Student Annotation:** Nucleophilic attack in which only one electron is donated by the attacking species can also occur. These are called *radical* reactions.

(a)

(b)

**Figure 10.8**   (a) Electrophilic addition reaction. (b) Nucleophilic addition reaction. Curved arrows indicate movement of electrons. (Nu represents a nucleophile.)

## Substitution Reactions

Electrophilic and nucleophilic attack can also lead to **substitution reactions** in which one group is replaced by another group. Electrophilic substitution occurs when an electrophile attacks an aromatic molecule and replaces a hydrogen atom. Nucleophilic substitution occurs when a nucleophile replaces another group on a carbon atom. Figure 10.9 shows the general mechanisms for substitution reactions.

**Student Annotation:** Specifically, this type of reaction is called an *electrophilic aromatic substitution* reaction.

The nitration of benzene is an example of an electrophilic substitution reaction:

Nitric acid and sulfuric acid react to produce the nitronium ion ($^+NO_2$), which acts as the electrophile:

The positively charged nitronium ion is attracted to the electron-rich pi bonds of the benzene ring. A bond forms between one of the carbon atoms and the nitronium ion, breaking one of benzene's pi bonds:

The resulting carbocation is stabilized by resonance:

(a)

(b)

**Figure 10.9**   (a) Electrophilic substitution. Benzene is attacked by an electrophile ($E^+$). (b) Nucleophilic substitution reaction.

**Figure 10.10**    Reaction of glucose and ATP to produce glucose-6-phosphate and ADP.

Finally, the electrons in the C—H bond move to the ring, restoring the original pi bond, and the hydrogen atom leaves as a proton, $H^+$:

When necessary, one additional step can convert the —$NO_2$ group into the —$NH_2$ group:

A simple example of nucleophilic substitution is the reaction of an alkyl halide such as methyl bromide with a nucleophile such as the chloride ion. The chloride ion is attracted to the partial posi-tive charge on the C atom in methyl bromide. A lone pair on the chloride ion moves to form a sigma bond between the Cl atom and the C atom. And, because there can be no more than four electron pairs around the C atom, the original C—Br bond breaks, with both of the electrons originally shared by C and Br going to the Br atom. The result is a methyl chloride molecule and a bromide ion:

> **Student Annotation:** The C atom in methyl bromide bears a partial positive charge because it is bonded to the somewhat more electro-negative bromine atom [◄◄ Section 8.4].

Nucleophilic substitutions are particularly important in living systems. The hydrolysis of ATP (adenosine triphosphate), which was described in Section 10.3, is an example of nucleo-philic substitution. In hydrolysis, the oxygen atom in water acts as the nucleophile, attacking the electron-deficient phosphorus atom. A similar reaction happens between glucose and ATP, as shown in Figure 10.10. The P atom that is attacked is electron deficient because of the four highly electronegative O atoms bonded to it. As a bond forms between the attacking O and the P, one of the original P—O bonds breaks. The net result is the replacement of the original —H group on glucose with a —$PO_4^{3-}$ group to give glucose-6-phosphate and ADP (adenosine diphosphate).

> **Student Annotation:** Digestion of proteins also begins with a nucleophilic substitution reaction.

## Bringing Chemistry to Life

### S$_N$1 Reactions

Thalidomide is a chiral drug, but only one of its enantiomers has the desired therapeutic proper-ties. The other enantiomer causes severe birth defects. Thalidomide was originally dispensed as a racemic mixture, giving patients equal amounts of both enantiomers. Unlike some chiral drugs, thalidomide cannot be administered as a single isomer to avoid the undesirable enantiomer. Within hours of administering one enantiomer of thalidomide, both enantiomers are found in roughly equal amounts in the blood. Although the mechanism by which one enantiomer of thalidomide is converted to the other is the subject of some debate, one way that enantiomers can be interconverted is via a nucleophilic substitution reaction.

Nucleophilic substitution reactions fall into two categories, called $S_N1$ reactions and $S_N2$ reactions. (The numbers 1 and 2 refer to a specific aspect of the *kinetics* of the reactions [▶▶ Chapter 14].) The nucleophilic substitution that converts one enantiomer to a mixture of both is an $S_N1$ reaction. An $S_N1$ reaction begins when one of the groups bonded to a carbon "leaves," leaving behind a carbocation [Figure 10.9(b)]. The hybridization of the carbon atom changes from $sp^3$ to $sp^2$ when the carbocation forms.

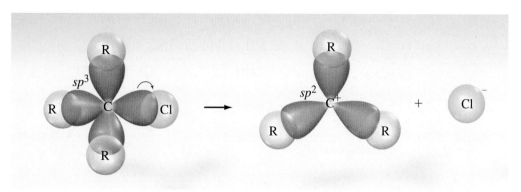

Commonly encountered "leaving groups" include $Cl^-$, $Br^-$, and $I^-$. A carbocation is an unstable species and, being positively charged, is prone to nucleophilic attack. Because the carbocation is *planar* about the carbon that bears the positive charge, it is equally likely that a nucleophile will attack from either side.

**Student Annotation:** Recall that $sp^2$-hybrid orbitals form a trigonal *plane* [◀◀ Table 9.2].

In the reaction of $(CH_3)_3CBr$ with $H_2O$, the product is the same regardless of whether the nucleophile (water) attacks from the front or from the back. When a molecule loses a leaving group from a chiral carbon, however, as would be the case with the thalidomide molecule, attack from one side of the carbocation will yield one enantiomer whereas attack from the other side will yield the other enantiomer. The result is a racemic mixture. The conversion of a single enantiomer to a racemic mixture of both enantiomers is called ***racemization.***

Nucleophilic attack on opposite sides of the carbocation leads
to two different products, which are mirror images of each other.

Sample Problems 10.6 and 10.7 show how to draw mechanisms for addition and substitution reactions.

## Sample Problem 10.6

The electrophilic addition of water to a molecule containing a double bond, known as **hydration,** is a reaction that is important biologically. One example is the enzyme-catalyzed hydration of fumarate to form malate, one of the steps in the citric acid cycle (also known as the Krebs cycle).

**Think About It** This time the product is the same regardless of which C atom forms the C—O bond and which forms the C—H bond, but this is not always the case. If $CH_2CHCH_3$ undergoes hydration, for instance, there are two possible products:

In this case, for reasons beyond the scope of this book, the product that forms is 2-propanol. If you take organic chemistry, you will learn what factors determine which C atom gets the —OH group and which gets the —H.

Using curved arrows to indicate the movement of electrons, draw the mechanism for the electrophilic addition of water to fumarate to yield malate.

**Strategy** For electrophilic addition, draw Lewis structures with the electrophile close to the region of electron density where attack will occur.

**Setup** (a) The electrophile is one of the H atoms in $H_2O$. The site of attack is the pi bond between the C atoms in fumarate:

**Solution**

**Practice Problem A** Using curved arrows to indicate the movement of electrons, draw the mechanism for the electrophilic addition of HI to $C_2H_4$.

**Practice Problem B** In which of the following examples do the curved arrows correspond correctly to the proper movement of electrons in the mechanism of electrophilic addition?

(a)

(b)

(c)

---

## Sample Problem 10.7

Using curved arrows to indicate the movement of electrons, draw the mechanism for each of the following reactions: (a) nucleophilic addition of $CN^-$ to $CH_3CHO$ and (b) electrophilic substitution of benzene with $^+SO_3H$. (Draw all resonance structures for the carbocation intermediate.)

**Strategy** For nucleophilic addition, draw Lewis structures with the nucleophile close to the electron-poor atom where attack will occur (in this case, the carbonyl carbon). For electrophilic substitution, draw Lewis structures with the electrophile close to the site of attack on the benzene ring. Remember that nucleophiles are attracted to and react with electron-poor atoms whereas electrophiles are attracted to and react with electron-rich areas of the molecule. Using this information, and the octet rule, determine which electrons are likely to be involved in the reaction and indicate their repositioning with curved arrows.

**Setup** (a) The nucleophile is $CN^-$. The site of attack is the carbonyl C in $CH_3CHO$, which is electron-poor because it is bonded to the more electronegative O atom.

(b) The electrophile is $^+SO_3H$. The site of attack is the electron-rich, delocalized pi bonds of the benzene ring.

**Solution**

(a)

(b)

*(Continued)*

**Think About It** Additions such as the one in part (a) are carried out in acidic solution. The availability of protons in solution makes the final step the protonation of the negatively charged O atom:

Also, the C atoms in benzene are all equivalent, so the choice of which C will bear the substituent in part (b) is arbitrary. All the following represent the same product:

**Practice Problem A** Draw mechanisms for (a) nucleophilic addition of $H^-$ to $CH_3COCH_3$ and (b) electrophilic substitution of benzene with $Cl^+$.

**Practice Problem B** Draw all the possible products that could result from the following electrophilic substitution reaction:

## Other Types of Organic Reactions

Other important organic reaction types are elimination, oxidation-reduction, and isomerization. An *elimination reaction* is one in which a double bond forms and a molecule such as water is removed. The dehydration of 2-phosphoglycerate to form phosphoenolpyruvate (Figure 10.11), one of the steps in carbohydrate metabolism, is an example of an elimination reaction.

*Oxidation-reduction reactions,* as we learned previously [◄◄ Section 4.4], involve the *loss* and *gain* of electrons, respectively. Determining which species have lost or gained electrons may seem less straightforward with organic reactions than with the inorganic reactions we have encountered, but the following guidelines can help you decide when an organic molecule has been oxidized or reduced:

1. When a molecule gains O or loses H, it is *oxidized.*
2. When a molecule loses O or gains H, it is *reduced.*

An important biological example of oxidation is the enzyme-catalyzed reaction by which ethanol is converted to acetaldehyde in the liver:

$$CH_3CH_2OH \longrightarrow CH_3CHO$$

The ethanol molecule loses two H atoms in this reaction, so it is *oxidized.*

**Figure 10.11** The highlighted atoms are those that constitute the eliminated water molecule.

*Isomerization reactions* are those in which one isomer is converted to another. The interconversion between the sugars aldose and ketose is an example:

Aldose ⇌ Ketose

**Student Annotation:** Many of the organic reactions in living systems require *enzymes*, natural *catalysts*, in order to occur rapidly enough to be useful [▶ Chapter 14].

## Bringing Chemistry to Life

### The Chemistry of Vision

Vision, our ability to perceive light, is the result of an isomerization reaction. Our eyes contain millions of cells called rods that are packed with *rhodopsin,* an 11-*cis*-retinal molecule

11-*cis*-retinal

bonded to a large protein. When visible light strikes rhodopsin, the retinal molecule isomerizes to the all-*trans* isomer:

All-*trans*-retinal

This isomerization causes such a significant change in the structure of retinal that it separates from the protein. These events trigger the electrical impulses that stimulate the optic nerve and result in the brain receiving the signals that we know as vision. The all-*trans*-retinal diffuses away from the protein and is converted back to 11-*cis*-retinal. The regenerated 11-*cis*-retinal can then rebind with the protein. Interestingly, the conversion of the all-*trans* isomer back to the 11-*cis* isomer, which is necessary for us to perceive light, is much slower than the light-induced conversion from *cis* to *trans.* This is why after looking at a bright light, you have a "blind spot" for a period of time.

## Checkpoint 10.5    Organic Reactions

**10.5.1** Identify each species as a nucleophile or an electrophile.

$$
\begin{array}{ccc}
 & & CH_3 \\
 & & | \\
 & & H_3C-\overset{+}{C} \\
 & & | \\
CH_3-\ddot{\underset{..}{O}}{:}^{-} \quad H-\ddot{\underset{..}{S}}{:}^{-} & & CH_3 \\
\text{(i)} \qquad \text{(ii)} & & \text{(iii)}
\end{array}
$$

a) Nucleophile, electrophile, electrophile

b) Nucleophile, nucleophile, nucleophile

c) Electrophile, nucleophile, electrophile

d) Nucleophile, nucleophile, electrophile

e) Electrophile, electrophile, electrophile

**10.5.2** Identify each reaction as addition, substitution, elimination, or isomerization.

(i) $H_2C{=}CHCH_2CH_2CH_3 + Br_2 \longrightarrow BrCH_2-\underset{\underset{Br}{|}}{C}HCH_2CH_2CH_3$

(ii) $CH_3Br + OH^- \longrightarrow CH_3OH + Br^-$

(iii)
$$
\underset{H}{\overset{H_3C}{>}}C{=}C\underset{CH_3}{\overset{H}{<}} \longrightarrow \underset{H}{\overset{H_3C}{>}}C{=}C\underset{H}{\overset{CH_3}{<}}
$$

a) Addition, substitution, isomerization

b) Isomerization, substitution, addition

c) Substitution, addition, isomerization

d) Addition, addition, isomerization

e) Substitution, substitution, addition

# 10.6    Organic Polymers

**Student Annotation:** Polymers typically have very high molar masses—sometimes thousands or even millions of grams.

**Student Annotation:** Some of the other properties that suggested a high-molar-mass solute were low osmotic pressure and negligible freezing-point depression. These are known as *colligative* properties [▶ Chapter 13].

Multimedia

Organic and Biochemistry—natural and synthetic polymers.

***Polymers*** are molecular compounds, either natural or synthetic, that are made up of many *repeating units* called ***monomers.*** The physical properties of these so-called macromolecules differ greatly from those of small, ordinary molecules.

The development of polymer chemistry began in the 1920s with the investigation of the puzzling behavior of some materials including wood, gelatin, cotton, and rubber. For example, when rubber, with the known empirical formula of $C_5H_8$, was dissolved in an organic solvent, the solution displayed several properties, including a higher than expected viscosity, which suggested that the dissolved compound had a very high molar mass. Despite the experimental evidence, though, scientists at the time were not ready to accept the idea that such giant molecules could exist. Instead, they postulated that materials such as rubber consisted of aggregates of small molecular units, like $C_5H_8$ or $C_{10}H_{16}$, held together by intermolecular forces. This misconception persisted for a number of years, until Hermann Staudinger[1] clearly showed that these so-called aggregates were, in fact, enormously large molecules, each of which contained many thousands of atoms held together by covalent bonds.

Once the *structures* of these macromolecules were understood, the way was open for the synthesis of polymers, which now pervade almost every aspect of our daily lives. About 90 percent of today's chemists, including biochemists, work with polymers. In this section, we will discuss the reactions that result in polymer formation and some of the natural polymers that are important to biology.

## Addition Polymers

***Addition polymers*** form when monomers such as ethylene join end to end to make polyethylene. Reactions of this type can be initiated by a *radical*—a species that contains an unpaired electron [◀ Section 8.8]. The mechanism of addition polymerization is as follows:

1. The radical, which is unstable because of its unpaired electron, attacks a carbon atom on an ethylene molecule. This is the *initiation* of the reaction.

2. This attack would result in the carbon atom in question having more than eight electrons around it. To keep the carbon atom from having too many electrons around it, the double bond breaks.

3. One of the electrons in the double bond, together with the electron from the radical, becomes a new bond between the ethylene molecule and the radical.

---

1. Hermann Staudinger (1881–1963). German chemist. One of the pioneers in polymer chemistry. Staudinger was awarded the Nobel Prize in Chemistry in 1953.

4. The other electron remains with the other carbon atom, generating a new radical species.

5. The new radical species, also unstable, attacks another ethylene molecule, causing the same sequence of events to happen again. The generation of a new, highly reactive radical species at each step is known as *propagation* of the reaction.

6. Each step lengthens the chain of carbon atoms and results in the formation of a new radical, continuing the propagation of the reaction. Reactions such as this are known as *chain reactions*. They continue until the system runs out of ethylene molecules or until the radical species encounters another radical species—resulting in *termination* of the reaction.

Initiation step

Propagation steps

Termination step

Some familiar and important addition polymers are listed in Table 10.4.

## Condensation Polymers

Reactions in which two or more molecules become connected with the elimination of a small molecule, often water, are called **condensation reactions** (Figure 10.12). **Condensation polymers** form when molecules with two different functional groups combine, with the elimination of a small molecule, often water. Many condensation polymers are **copolymers,** meaning that they are made up of two or more *different* monomers.

The first synthetic fiber, nylon 66, is a condensation copolymer of two molecules: one with carboxy groups at each end (adipic acid) and one with amine groups at both ends (hexamethylenediamine):

**Student Annotation:** The number 66 refers to the fact that there are six C atoms in each monomer. Other nylons, such as nylon 610, are made from various combinations of molecules similar to adipic acid and hexamethylenediamine.

$$H_2N-(CH_2)_6-NH_2 \quad + \quad HOOC-(CH_2)_4-COOH$$

Hexamethylenediamine          Adipic acid

↓ Condensation

↓ Further condensation reactions

| TABLE 10.4 | Addition Polymers | | |
|---|---|---|---|
| **Name** | **Monomer Unit** | **Structure** | **Uses** |
| Polytetrafluoroethylene (Teflon) | | $\begin{bmatrix} F & F \\ | & | \\ C & C \\ | & | \\ F & F \end{bmatrix}_n$ | Nonstick coatings |
| Polyethylene | | $\begin{bmatrix} H & H \\ | & | \\ C & C \\ | & | \\ H & H \end{bmatrix}_n$ | Plastic bags, bottles, toys |
| Polypropylene | | $\begin{bmatrix} H & H \\ | & | \\ C & C \\ | & | \\ H & CH_3 \end{bmatrix}_n$ | Carpeting, bottles |
| Polyvinylchloride (PVC) | | $\begin{bmatrix} H & H \\ | & | \\ C & C \\ | & | \\ H & Cl \end{bmatrix}_n$ | Water pipes, garden hoses, plastic wrap |
| Polystyrene | | $\begin{bmatrix} H & H \\ | & | \\ C & C \\ | & | \\ H & \end{bmatrix}_n$ | Packing material, insulation, furniture |

**Figure 10.12** (a) Condensation reaction between an alcohol and a carboxylic acid to form an ester. (b) Condensation of two alcohol molecules in the presence of sulfuric acid to form an ether.

$$R-\overset{\overset{\displaystyle O}{\|}}{C}-\ddot{O}H \;+\; H\ddot{O}-R \longrightarrow R-\overset{\overset{\displaystyle O}{\|}}{C}-\ddot{O}-R \;+\; H_2O$$
(a)

$$R-\ddot{O}H \;+\; H\ddot{O}-R \xrightarrow{\;H_2SO_4\;} R-\ddot{O}-R \;+\; H_2O$$
(b)

Nylon was first made by Wallace Carothers[2] at DuPont in 1931. The versatility of nylons is so great that the annual production of nylons and related substances now amounts to several billion pounds.

## Biological Polymers

Naturally occurring polymers include *proteins, polysaccharides,* and *nucleic acids.*

   ***Proteins,*** polymers of amino acids, play an important role in nearly all biological processes. The human body contains an estimated 100,000 different kinds of proteins, each of which has a

---

2. Wallace H. Carothers (1896–1937). American chemist. Besides its enormous commercial success, Carothers's work on nylon is ranked with that of Staudinger in clearly elucidating macromolecular structure and properties. Depressed by the death of his sister and wrongly believing that his life's work had been a failure, Carothers committed suicide at the age of 41.

$$\underset{H}{\overset{H}{\underset{|}{\overset{|}{N}}}}\!\!-\!\!\underset{R_1}{\overset{H}{\underset{|}{\overset{|}{C}}}}\!\!-\!\!\overset{O}{\overset{||}{C}}\!\!-\!\!O^- \;+\; \underset{H}{\overset{H}{\underset{|}{\overset{|}{N}}}}\!\!-\!\!\underset{R_2}{\overset{H}{\underset{|}{\overset{|}{C}}}}\!\!-\!\!\overset{O}{\overset{||}{C}}\!\!-\!\!O^- \;\rightleftharpoons\; \underset{H}{\overset{H}{\underset{|}{\overset{|}{N}}}}\!\!-\!\!\underset{R_1}{\overset{H}{\underset{|}{\overset{|}{C}}}}\!\!-\!\!\overset{O}{\overset{||}{C}}\!\!-\!\!N\!\!-\!\!\underset{R_2}{\overset{H}{\underset{|}{\overset{|}{C}}}}\!\!-\!\!\overset{O}{\overset{||}{C}}\!\!-\!\!O^- \;+\; \overset{H}{\underset{H}{O}}$$

Peptide bond
(amide linkage)

**Figure 10.13** Formation of a peptide bond with elimination of water.

specific physiological function. An amino acid has both the carboxylic acid functional group and the amino functional group. Amino acids are joined together into chains when a condensation reaction occurs between a carboxy group on one molecule and an amino group on another molecule (Figure 10.13).

The bonds that form between amino acids are called ***peptide bonds.*** Very long chains of amino acids assembled in this way are called *proteins,* while shorter chains are called *polypeptides.*

Amino acids consist of a central carbon atom bonded to four different groups: an amino group, a carboxy group, a hydrogen atom, and an additional group (highlighted in Figure 10.14) consisting of carbon, hydrogen, and sometimes other elements such as nitrogen or sulfur. Proteins are made essentially from the 20 different amino acids shown in Figure 10.14. The identity of a protein depends on which of the 20 amino acids it contains and on the order in which the amino acids are assembled.

*Polysaccharides* are polymers of sugars such as glucose and fructose. Starch and cellulose are two polymers of glucose with slightly different linkages—and very different properties. In starch, glucose molecules are connected by what biochemists call α linkages. This enables animals, including humans, to digest the starch in such foods as corn, wheat, potatoes, and rice.

Starch

In cellulose, the glucose molecules are connected by β linkages. Digestion of cellulose requires enzymes that most animals do not have. Species that *do* digest cellulose, such as termites and ruminants (including cattle, sheep, and llamas), do so with the help of enzyme-producing symbiotic bacteria in the gut.

Cellulose

*Nucleic acids,* which are polymers of *nucleotides,* play an important role in protein synthesis. There are two types of nucleic acids: ***deoxyribonucleic acid (DNA)*** and ***ribonucleic acid (RNA).*** Each ***nucleotide*** in a nucleic acid consists of a purine or pyrimidine *base,* a furanose sugar (*deoxyribose* for DNA; *ribose* for RNA), and a phosphate group. Figure 10.15 (p. 426) shows the building blocks of DNA and RNA. The components of a nucleotide are linked together as shown in Figure 10.16 (p. 426). These molecules are among the largest known—they can have molar masses of up to tens of billions of grams. RNA molecules, on the other hand, typically have molar masses on the order of tens of thousands of grams. Despite their sizes, the composition of nucleic acids is relatively simple compared with proteins. Proteins consist of up to 20 different amino acids, whereas DNA and RNA consist of only four different nucleotides each.

Multimedia

Organic and Biochemistry—molecular structure in DNA.

**Figure 10.14** The 20 amino acids essential to living organisms. The shaded area represents the R group.

| Name | Abbreviation | Structure |
|---|---|---|
| Alanine | Ala | $\mathrm{H_3C-\overset{\underset{\displaystyle NH_3^+}{\mid}}{\overset{\displaystyle H}{C}}-COO^-}$ |
| Arginine | Arg | $\mathrm{H_2N-\overset{\underset{\displaystyle NH}{\parallel}}{C}-\overset{\displaystyle H}{N}-CH_2-CH_2-CH_2-\overset{\underset{\displaystyle NH_3^+}{\mid}}{\overset{\displaystyle H}{C}}-COO^-}$ |
| Asparagine | Asn | $\mathrm{H_2N-\overset{\displaystyle O}{\overset{\parallel}{C}}-CH_2-\overset{\underset{\displaystyle NH_3^+}{\mid}}{\overset{\displaystyle H}{C}}-COO^-}$ |
| Aspartic acid | Asp | $\mathrm{HOOC-CH_2-\overset{\underset{\displaystyle NH_3^+}{\mid}}{\overset{\displaystyle H}{C}}-COO^-}$ |
| Cysteine | Cys | $\mathrm{HS-CH_2-\overset{\underset{\displaystyle NH_3^+}{\mid}}{\overset{\displaystyle H}{C}}-COO^-}$ |
| Glutamic acid | Glu | $\mathrm{HOOC-CH_2-CH_2-\overset{\underset{\displaystyle NH_3^+}{\mid}}{\overset{\displaystyle H}{C}}-COO^-}$ |
| Glutamine | Gln | $\mathrm{H_2N-\overset{\displaystyle O}{\overset{\parallel}{C}}-CH_2-CH_2-\overset{\underset{\displaystyle NH_3^+}{\mid}}{\overset{\displaystyle H}{C}}-COO^-}$ |
| Glycine | Gly | $\mathrm{H-\overset{\underset{\displaystyle NH_3^+}{\mid}}{\overset{\displaystyle H}{C}}-COO^-}$ |
| Histidine | His | $\mathrm{HC=C-CH_2-\overset{\underset{\displaystyle NH_3^+}{\mid}}{\overset{\displaystyle H}{C}}-COO^-}$ |
| Isoleucine | Ile | $\mathrm{H_3C-CH_2-\overset{\underset{\displaystyle H}{\mid}}{\overset{\displaystyle CH_3}{C}}-\overset{\underset{\displaystyle NH_3^+}{\mid}}{\overset{\displaystyle H}{C}}-COO^-}$ |

| Name | Abbreviation | Structure |
|---|---|---|
| Leucine | Leu | |
| Lysine | Lys | |
| Methionine | Met | |
| Phenylalanine | Phe | |
| Proline | Pro | |
| Serine | Ser | |
| Threonine | Thr | |
| Tryptophan | Trp | |
| Tyrosine | Tyr | |
| Valine | Val | |

|  | Found only in DNA | Found in both DNA and RNA | Found only in RNA |
|---|---|---|---|
| **Purines** |  | Adenine    Guanine |  |
| **Pyrimidines** | Thymine | Cytosine | Uracil |
| **Sugars** | Deoxyribose |  | Ribose |
| **Phosphate** |  | Phosphate |  |

**Figure 10.15**    The components of the nucleic acids DNA and RNA.

**Figure 10.16**    Structure of a nucleotide, one of the repeating units in DNA.

# Applying What You've Learned

Although it was approved by the FDA in 1998, thalidomide is the most regulated prescription drug in history, because it is known to harm developing fetuses. The drug's manufacturer, Celgene Corporation, has developed the *System for Thalidomide Education and Prescribing Safety* (STEPS) program. In order for physicians to prescribe thalidomide to their patients, they must be registered in the STEPS program. Female patients must have a negative pregnancy test within 24 hours of beginning treatment and must undergo periodic pregnancy testing throughout treatment. Both female and male patients must comply with mandatory contraceptive measures, patient registration, and patient surveys. Moreover, new patients must view an informational video in which a thalidomide victim explains the potential dangers of the drug.

**Problems:**

a)  From the structures given at the beginning of the chapter, identify the functional groups in thalidomide. [◀◀ Sample Problem 10.2]

b)  Write molecular formulas for thalidomide, lenalidomide, and CC-4047. [◀◀ Sample Problem 10.3]

c)  Thalidomide is converted to the drug CC-4047 by substitution of an amino group for one of the H atoms on the aromatic portion of the molecule. Using curved arrows, draw the mechanism for this reaction and all the resonance structures for the carbocation intermediate. [◀◀ Sample Problem 10.7]

# CHAPTER SUMMARY

## Section 10.1

- Organic chemistry is the study of carbon-based substances. Although it was once thought that organic compounds could only be produced by nature, thousands of new organic compounds are now synthesized each year by scientists.

- Carbon's position in the periodic table makes it uniquely able to form long, stable chains—a process called *catenation.*

- *Aromatic* compounds contain one or more benzene rings. *Aliphatic* compounds are organic compounds that do not contain benzene rings.

## Section 10.2

- *Alkyl groups* consist of just carbon and hydrogen. They are derived from the corresponding alkane by removing one hydrogen atom and are represented generically in the formulas of organic compounds with the letter R.

- Functional groups are specific arrangements of atoms that are responsible for the properties and reactivity of organic compounds. Common functional groups and their formulas include

  | | |
  |---|---|
  | *Alcohol* | ROH |
  | *Carboxylic acid* | RCOOH |
  | *Aldehyde* | RCHO |
  | *Ketone* | RCOR′ |
  | *Ester* | RCOOR′ |
  | *Amine* | RNH$_2$, RNHR′, or RNR′R″ |
  | *Amide* | RCONH$_2$, RCONHR′, or RCONR′R″ |

- Many organic compounds contain more than one functional group. A *substituent* in an alkane is a group other than hydrogen that is bonded to the carbon chain.

- An *amino acid* contains both the carboxy group and the amino group.

## Section 10.3

- *Condensed structural formulas* or *condensed structures* abbreviate a series of repeating units, such as $-CH_2CH_2CH_2CH_2-$, into a more compact form, such as $-(CH_2)_4-$. *Kekulé structures* are similar to Lewis structures but do not show the lone pairs on a molecule. *Skeletal structures* use straight lines to represent C—C bonds. They typically do not show the C atoms explicitly except for the purpose of emphasizing a particular functional group. H atoms are not shown in skeletal structures—again, except to emphasize a functional group. The number of H atoms bonded to a C atom must be inferred from the number of C—C bonds in the structure. *Heteroatoms,* atoms other than C and H, are always shown explicitly in a skeletal structure.

## Section 10.4

- *Constitutional isomers* are molecules in which the same atoms are connected differently. *Stereoisomers* are molecules in which the same atoms are connected by the same bonds but the bonds are oriented differently. *Geometrical isomers* arise due to restricted rotation about a carbon-carbon double bond. *Cis* and *trans* isomers are geometrical isomers.

- Molecules that are nonsuperimposable mirror images of each other are *optical isomers.* They are also referred to as *chiral,* and each of the mirror images is called an *enantiomer.* Optical isomers are so called because they rotate the plane of plane-polarized light. The degree of rotation is the same for both enantiomers, but the directions of rotation are opposite each other. An equal mixture of both enantiomers is called a *racemic mixture.* A racemic mixture does not rotate the plane of plane-polarized light.

## Section 10.5

- An *electrophile* generally is a positively charged ion that is attracted to electrons. A *nucleophile* is a negatively charged ion or a partially negatively charged atom in a polar molecule. Nucleophiles and electrophiles are attracted to each other.

- A *carbocation* is an intermediate species in which one of the carbon atoms is surrounded by only six electrons and bears a positive charge.

- *Electrophilic addition* reactions and *nucleophilic addition* reactions involve the addition of a molecule or an ion to another molecule.

- *Substitution reactions* occur when an electrophile replaces a hydrogen on an aromatic ring or when a nucleophile replaces a leaving group on a carbon atom.

- Carbocations are the intermediate species in *racemization,* a nucleophilic substitution reaction in which a single enantiomer is converted into a racemic mixture.

- *Hydration* is the addition of a water molecule across a double bond.

- An *elimination reaction* is one in which a double bond forms and a small molecule, such as water, is eliminated.

- *Isomerization reactions* convert one isomer into another.

## Section 10.6

- *Polymers* are long chains of repeating molecular units called *monomers. Polysaccharides* are polymers of sugars. *Addition polymers* form when a radical species attacks a double bond, forming a new, longer radical species that attacks another double bond, and so on.

- An elimination reaction that joins two molecules is a *condensation reaction. Condensation polymers* form when molecules with two different functional groups undergo a condensation reaction.

- *Copolymers* are polymers that contain more than one type of monomer.

- *Proteins* and *polypeptides* are biological polymers in which the monomers are amino acids. Amino acids are joined by *peptide bonds,* which result from the condensation reaction between the amino group of one amino acid and the carboxyl group of another amino acid.

- *Nucleic acids* are polymers of *nucleotides.* The two types of nucleic acid are *deoxyribonucleic acid (DNA)* and *ribonucleic acid (RNA).* Each nucleotide in a nucleic acid consists of a purine or pyrimidine base, a furanose sugar (deoxyribose for DNA; ribose for RNA), and a phosphate group linked together.

# KEY WORDS

# QUESTIONS AND PROBLEMS

## Section 10.1: Why Carbon Is Different

### Review Questions

10.1 Explain why carbon is able to form so many more compounds than any other element.

10.2 Why was Wöhler's synthesis of urea so important for the development of organic chemistry?

10.3 What are aromatic organic compounds? What are aliphatic organic compounds?

## Section 10.2: Classes of Organic Compounds

### Review Questions

10.4 What are functional groups? Why is it logical and useful to classify organic compounds according to their functional groups?

10.5 Draw the Lewis structure for each of the following functional groups: alcohol, aldehyde, ketone, carboxylic acid, amine.

10.6 Name the classes to which the following compounds belong:

(a) $C_4H_9OH$

(b) $C_2H_5CHO$

(c) $C_6H_5COOH$

(d) $CH_3NH_2$

### Problems

**10.7** Classify each of the following molecules as alcohol, aldehyde, ketone, carboxylic acid, or amine:

(a) $CH_3-CH_2-NH_2$

(b) $CH_3-CH_2-C\overset{\displaystyle O}{\underset{\displaystyle H}{\big\langle}}$

(c) $CH_3-\overset{\displaystyle \underset{\|}{\phantom{.}}}{\underset{\displaystyle O}{C}}-CH_2-CH_3$

(d) $H-\overset{\displaystyle O}{\underset{\|}{C}}-OH$

(e) $CH_3-CH_2CH_2-OH$

10.8 Draw structures for molecules with the following formulas:

(a) $CH_4O$

(b) $C_2H_6O$

(c) $C_3H_6O_2$

(d) $C_3H_8O$

10.9 Name each of the following compounds:

(a)

(b) $CH_3\overset{\displaystyle CH_3}{\underset{\displaystyle CH_3}{C}}CH_2CH_2CH_2\overset{\displaystyle OH}{CH}CH_3$

(c) $ClCHCH_2CH_2\overset{\displaystyle O}{\overset{\|}{C}}H$ with $CH_2CH_3$

10.10 Name each of the following compounds:

(a) (with $NH_2$)

(b) (with O and $NH_2$)

(c)

**10.11** Give the name of the alkane represented by the model shown.

**10.12** The molecular formula corresponding to the model is $C_5H_9ClO_2$. What is the name of the compound?

**10.13** Write structural formulas for each of the following based on their systematic names. The names quoted in parentheses are so-called common names. Common names are not systematic and are more difficult, sometimes impossible, to connect with a unique structure.

(a) 2,2,4-Trimethylpentane ("isooctane")
(b) 3-Methyl-1-butanol ("isoamyl alcohol")
(c) Hexanamide ("caproamide")
(d) 2,2,2-Trichloroethanal ("chloral")

**10.14** Write structural formulas for each of the following:

(a) 3,3-Dimethyl-2-butanone ("pinacolone")
(b) 3-Hydroxybutanal ("acetaldol")
(c) Ethyl pentanoate ("ethyl valerate")
(d) 6-Methyl-2-heptanamine ("isooctylamine")

**10.15** Classify the oxygen-containing groups in the plant hormone abscisic acid.

**10.16** Identify the functional groups in the antipsychotic drug haloperidol.

**10.17** PABA was the active UV-absorbing compound in earlier versions of sunblock creams. What functional groups are present in PABA?

**10.18** Lidocaine ($C_{14}H_{22}N_2O$) is a widely used local anesthetic. Classify its nitrogen-containing functional groups.

## Section 10.3: Representing Organic Molecules

*Problems*

**10.19** Write structural formulas for the following organic compounds: (a) 3-methylhexane, (b) 2,3-dimethylpentane, (c) 2-bromo-4-phenylpentane, (d) 3,4,5-trimethyloctane.

**10.20** Write structural formulas for the following compounds: (a) 1,1,3-trichloro-2-propanol, (b) 3-methyl-3-pentanamine, (c) 3-bromo-1-chloro-2-butanone, (d) propyl-4-bromobutanoate.

**10.21** (a) Convert $CH_3(CH_2)_4C(O)CH_2CO_2H$ to a Kekulé structure and to a skeletal (line) structure.
(b) Convert the following to a condensed structure and to a skeletal (line) structure:

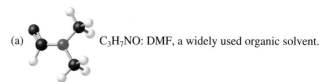

(c) Convert the following to a condensed structure and to a Kekulé structure:

**10.22** (a) Convert $(CH_3)_2C{=}CHCO_2H$ to a Kekulé structure and to a skeletal (line) structure.
(b) Convert the following to a condensed structure and to a skeletal (line) structure:

$$CH_3CH_2NCH_2CH_3$$
$$\underset{\displaystyle CH_3}{|}$$

(c) Convert the skeletal structure of the general anesthetic isoflurane to a condensed structure and to a Kekulé structure:

**10.23** Convert each of the molecular models to a condensed structural formula, a Kekulé structure, and a skeletal (line) structure.

(a) $C_3H_7NO$: DMF, a widely used organic solvent.

(b) $C_6H_8O_7$: Citric acid, which contributes to the tart taste of citrus fruits.

(c) $C_6H_8O_6$: Commonly known as "isoamyl acetate," this ester is largely responsible for the characteristic odor of bananas.

**10.24** Convert each of the molecular models to a condensed structural formula, a Kekulé structure, and a skeletal (line) structure.

(a)  C$_3$H$_6$O$_3$: Dihydroxyacetone, an intermediate in glycolysis, the process by which glucose is converted to energy.

(b)  C$_4$H$_5$Cl: Chloroprene, which gives neoprene on polymerization.

(c)  C$_6$H$_{13}$NO$_2$: Isoleucine, one of the 20 amino acids that occur in proteins.

**10.25** Given a structural formula or skeletal (line) structure, rewrite it in the other style.

(a) CH$_3$CH$_2$CHCH$_2$CH$_2$CH$_2$CH$_3$
　　　　　|
　　　　OH

(b)

(c)

**10.26** Given a structural formula or skeletal (line) structure, rewrite it in the other style.

(a) ClCH$_2$CH$_2$CH$_2$CH=CH$_2$

(b)

(c)

**10.27** Using the curved arrows as a guide to placing the electrons, write a resonance structure for each of the compounds shown. The resonance structure should include formal charges where appropriate.

(a) CH$_3$—C≡N:

(b)

(c)

**10.28** Using the curved arrows as a guide to placing the electrons, write a resonance structure for each of the compounds shown. The resonance structure should include formal charges where appropriate.

(a)

(b)

(c)

**10.29** Use curved arrows to show how the resonance structure on the left can be transformed to the one on the right.

(a)

(b)

(c)

**10.30** Use curved arrows to show how the resonance structure on the left can be transformed to the one on the right.

(a)

(b)

(c)

## Section 10.4: Isomerism

### Review Questions

**10.31**  Alkenes exhibit geometrical isomerism because rotation about the C=C bond is restricted. Explain.

**10.32**  Why is it that alkanes and alkynes, unlike alkenes, have no geometrical isomers?

**10.33**  Define the term *chiral*. What are enantiomers?

**10.34**  What factor determines whether a carbon atom in a compound is chiral?

**10.35**  Fill in the blanks in the given paragraph with the most appropriate term from the following: chiral, *cis,* constitutional isomers, enantiomers, resonance structures, stereoisomers, *trans.*

*Isomers* are different compounds that have the *same molecular formula.* Isomers that have their atoms connected in a different order (branched versus unbranched chain, for example) or a different sequence of bond types (C=CCC versus CC=CC, for example) are termed _____. Isomers with the same order of connections and sequence of bond types, but which differ in the spatial arrangement of the atoms are called _____. This is often seen in compounds where substituents may be on the same or opposite sides of a carbon-carbon double bond. Substituents on the same side are described as _____; those on opposite sides are _____. A different kind of isomerism characterizes a _____ molecule, that is, a molecule with a structure that allows for two nonsuperimposable mirror-image forms. Two nonsuperimposable mirror images are _____ of the other.

### Problems

**10.36**  Write structural formulas for all the $C_5H_{10}$ alkenes, and identify the relationship (constitutional isomer or stereoisomer) of each one to the others. Are any chiral?

**10.37**  Draw all possible structural isomers for the following alkane: $C_7H_{16}$.

**10.38**  Draw all possible isomers for the molecule $C_4H_8$.

**10.39**  Draw all possible isomers for the molecule $C_3H_5Br$.

**10.40**  Which of the following amino acids are chiral:
(a) $CH_3CH(NH_2)COOH$, (b) $CH_2(NH_2)COOH$,
(c) $CH_2(OH)CH(NH_2)COOH$?

**10.41**  Draw all the possible structural isomers for the molecule having the formula $C_7H_7Cl$. All isomers contain one benzene ring.

**10.42**  Draw all the structural isomers of compounds with the formula $C_4H_8Cl_2$. Indicate which isomers are chiral, and give them systematic names.

**10.43**  Indicate the asymmetric carbon atoms in the following compounds:

(a) CH3—CH2—CH—CH—C—NH2 with CH3 above the third carbon, O double-bonded to the C, and NH2 below the fourth carbon.

(b) Cyclopropane ring drawn with H, Br, H on top and H, Br on bottom.

**10.44**  Suppose benzene contained three distinct single bonds and three distinct double bonds. How many different isomers would there be for dichlorobenzene ($C_6H_4Cl_2$)? Draw all your proposed structures.

**10.45**  Write the structural formula of an aldehyde that is a structural isomer of acetone.

**10.46**  How many asymmetric carbon atoms are present in each of the following compounds?

(a) H—C—C—C—Cl with H, H, H on top and H, Cl, H on bottom

(b) $H_3C$—C—C—$CH_2OH$ with OH, $CH_3$ on top and H, H on bottom

(c) ring structure with $CH_2OH$, H, OH, H, HO, H, OH substituents

**10.47**  Draw structures of each of the compounds shown, using wedges and dashes to show stereochemistry.

(a)    ($C_4H_9Br$)

(b)    ($C_3H_7NO_3$)

(c)    ($C_6H_{14}O$)

(d)    ($C_4H_7BrO_2$)

**10.48**  Write the structural formulas of the alcohols with the formula $C_6H_{14}O$ and indicate those that are chiral. Show only the C atoms and the —OH groups.

## Section 10.5: Organic Reactions

### Review Questions

**10.49**  What property distinguishes an electrophile from a nucleophile? Of the two, which would you expect to be more reactive toward a cation? Toward an anion? Give a specific example of an electrophile and a nucleophile.

**10.50** Classify each of the following according to whether you think it reacts as an electrophile or a nucleophile.

(a) $CH_3-\underset{\underset{CH_3}{|}}{\overset{\overset{CH_3}{|}}{C}}-\ddot{\underset{..}{O}}:^-$

(b) $CH_3CH_2\ddot{N}H_2$

(c) $\underset{H_3C}{\overset{}{}}\overset{\overset{CH_3}{|}}{\underset{}{C}}\overset{+}{\underset{CH_3}{}}$

(d) $CH_3-C\equiv\overset{+}{\underset{..}{O}}:$

**10.51** (a) The compound 2-bromopropane [$(CH_3)_2CHBr$] can undergo both substitution and elimination when treated with $CH_3CH_2O^-$, which is a strong base. Predict the organic product in each case, and write a separate chemical equation for each reaction.

(b) The compound 1,2-dibromoethane ($BrCH_2CH_2Br$) was formerly used in large amounts as an agricultural chemical. Write a chemical equation showing how this compound could be prepared from ethylene by an addition reaction.

(c) Certain reactions of aldehydes and ketones begin with isomerization of the aldehyde or ketone to an *enol* isomer. Enols contain an $-OH$ group attached to a carbon-carbon double bond. Write a chemical equation for the isomerization of acetone [$(CH_3)_2C=O$] to its enol isomer.

**10.52** Classify the following reactions according to whether they are addition, substitution, elimination, or isomerization.

(a) [structure] $\xrightarrow{\text{Enzyme}}$ [structure]

(b) $CH_3CH=CH_2 + Cl_2 \xrightarrow{\text{heat}} ClCH_2CH=CH_2 + HCl$

(c) $CH_3CH=CH_2 + Cl_2 \longrightarrow CH_3\underset{\underset{Cl}{|}}{CH}CH_2Cl$

(d) $CH_3\underset{\underset{Cl}{|}}{CH}CH_2Cl + NaSH \longrightarrow CH_3\underset{\underset{Cl}{|}}{CH}CH_2SH + NaCl$

(e) [cyclohexanone] $=O + H-C\equiv N \longrightarrow$ [product with OH and C≡N]

(f) [structure with OH and C≡N] $\xrightarrow[\text{heat}]{H_2SO_4}$ [structure] $-C\equiv N + H_2O$

**10.53** (a) Define *carbocation*.

(b) Which of the following are carbocations?

$CH_3CH_2-\overset{+}{\underset{\underset{H}{|}}{\overset{H}{\diagup}}}\ddot{O}:$    $CH_3-\overset{+}{\ddot{O}}-\overset{+}{CH_2}$    $H_2\overset{+}{C}-CH=CH_2$

(i)           (ii)          (iii)

$CH_3CH_2\overset{+}{CH}_2$     $CH_3-\overset{\overset{CH_3}{|}}{\underset{\underset{CH_3}{|}}{\overset{+}{N}}}-CH_3$

    (iv)         (v)

*Problems*

**10.54** Halogenated hydrocarbons are biodegraded in the natural environment by reactions catalyzed by dehalogenase enzymes. The reaction that takes place with 1,2-dichloroethane begins with nucleophilic substitution involving a carboxylate site of the enzyme.

$$\text{Enzyme}-\overset{\overset{\ddot{O}:}{\|}}{C}\underset{:\ddot{O}:^-}{} + :\ddot{C}l-CH_2CH_2-\ddot{C}l: \longrightarrow$$

$$\text{Enzyme}-\overset{\overset{\ddot{O}:}{\|}}{\underset{:\underset{..}{O}-CH_2CH_2-\ddot{C}l:}{C}} + :\ddot{C}l:^-$$

Expand this equation by adding curved arrows to show the movement of electrons.

**10.55** A common reaction in carbohydrate biochemistry is the conversion of an aldose to a ketose. The glucose to fructose isomerization is a specific example; the equation illustrates the general case:

[structure] $\longrightarrow$ [structure]

The first stage in the reaction is shown in the following equation. Enzymes facilitate the reaction, but for simplicity the overall change can be approximated with water molecules. Use curved arrows to show the flow of electrons in the equation.

[mechanism structures] $\longrightarrow$

[product structures]

**10.56** (a) Benzene reacts with *tert*-butyl cation [$(CH_3)_3C^+$] by a two-step electrophilic aromatic substitution mechanism to yield *tert*-butylbenzene [$C_6H_5C(CH_3)_3$]. Write a chemical equation for each step in the mechanism and use curved arrows to track electron flow.

(b) Electrophilic addition of hydrogen chloride to styrene gives the product shown. Write the mechanism for this reaction including curved arrows.

[benzene]$-CH=CH_2 + HCl \longrightarrow$ [benzene]$-\underset{\underset{Cl}{|}}{CH}CH_3$

**10.57** (a) Acetylide ion undergoes nucleophilic addition to aldehydes and ketones to give the species shown. Subsequent addition of water yields an acetylenic alcohol. Add curved arrows to the equations to show how the reaction occurs.

$$CH_3CH_2\overset{\overset{\ddot{O}}{\|}}{C}H + :C\equiv CH \longrightarrow CH_3CH_2\underset{\underset{H}{|}}{\overset{\overset{:\ddot{O}:^-}{|}}{C}}HC\equiv CH \xrightarrow{\overset{H}{\underset{}{\diagup}}\underset{H-\ddot{O}:}{}}$$

$$CH_3CH_2\underset{\underset{H}{|}}{\overset{\overset{:\ddot{O}H}{|}}{C}}HC\equiv CH + :\ddot{O}-H$$

(b) A nucleophilic site on a molecule can substitute for a halogen elsewhere in the same molecule to form a ring. Use curved arrows to show how the following *cyclization* is related to a conventional nucleophilic substitution.

$$\ddot{S}-CH_2CH_2CH_2CH_2-\ddot{B}r: \longrightarrow \ddot{S} \bigcirc + :\ddot{B}r:^-$$

10.58  Complete the following reaction by including essential unshared electron pairs and formal charges. Use curved arrows to show electron flow.

$$(CH_3)_3B + O\bigcirc \longrightarrow (CH_3)_3B-O\bigcirc$$

**10.59**  The product of the reaction of chlorobenzene with $NaNH_2$ is a very reactive species called *benzyne*. Add necessary electron pairs and formal charges to the net ionic equation, and use curved arrows to show how benzyne is formed. To what general type of reaction does this belong?

$$H_2N^- + \text{(chlorobenzene structure)} \longrightarrow$$

$$H_2N-H + \text{(benzyne structure)} + Cl^-$$

**10.60**  Consider the following reactions of butanal:

(i)  $CH_3CH_2CH_2\overset{O}{\overset{\|}{C}}H \xrightarrow[H_2O]{NaBH_4} CH_3CH_2CH_2CH_2OH$

(ii)  $CH_3CH_2CH_2\overset{O}{\overset{\|}{C}}H \xrightarrow[H_2O]{H_2CrO_4} CH_3CH_2CH_2\overset{O}{\overset{\|}{C}}OH$

In which reaction is butanal oxidized? In which reaction is it reduced?

**10.61**  Esters can be prepared by the acid-catalyzed condensation of a carboxylic acid and an alcohol:

$$CH_3CH_2\overset{O}{\overset{\|}{C}}OH + CH_3OH \longrightarrow CH_3CH_2\overset{O}{\overset{\|}{C}}OCH_3 + H_2O$$

Is this an oxidation-reduction reaction? If so, identify the species being oxidized and the species being reduced.

**10.62**  A compound has the empirical formula $C_5H_{12}O$. Upon controlled oxidation, it is converted into a compound of empirical formula $C_5H_{10}O$, which behaves as a ketone. Draw possible structures for the original compound and the final compound.

**10.63**  Isopropanol is prepared by reacting propylene ($CH_3CHCH_2$) with sulfuric acid, followed by treatment with water. (a) Show the sequence of steps leading to the product. What is the role of sulfuric acid? (b) Draw the structure of an alcohol that is an isomer of isopropanol. (c) Is isopropanol a chiral molecule?

## Section 10.6: Organic Polymers

### Review Questions

**10.64**  Define the following terms: *monomer, polymer, copolymer*.

10.65  Name 10 objects that contain synthetic organic polymers.

**10.66**  Calculate the molar mass of a particular polyethylene sample, $(CH_2-CH_2)_n$, where $n = 4600$.

10.67  Describe the two major mechanisms of organic polymer synthesis.

10.68  What are the steps involved in polymer formation by chain reaction?

10.69  Polysaccharides, proteins, and nucleic acids comprise the three main classes of biopolymers. Compare and contrast them with respect to structure and function. What are the building block units for each? What are the key functional groups involved in linking the units together? In which biopolymer is there the greatest variety of building block structure? In which is there the least?

### Problems

**10.70**  Teflon is formed by a radical addition reaction involving the monomer tetrafluoroethylene. Show the mechanism for this reaction.

10.71  Vinyl chloride ($H_2C=CHCl$), undergoes copolymerization with 1,1-dichloroethylene, ($H_2C=CCl_2$), to form a polymer commercially known as Saran. Draw the structure of the polymer, showing the repeating monomer units.

10.72  Deduce plausible monomers for polymers with the following repeating units:

(a) $(CH_2-CF_2)_n$

(b) $\left( CO-\bigcirc-CONH-\bigcirc-NH \right)_n$

**10.73**  Deduce plausible monomers for polymers with the following repeating units:

(a) $(CH_2-CH=CH-CH_2)_n$
(b) $(CO(CH_2)_6NH)_n$

**10.74**  Draw the structures of the dipeptides that can be formed from the reaction between the amino acids glycine and alanine.

10.75  Draw the structures of the dipeptides that can be formed from the reaction between the amino acids glycine and lysine.

**10.76**  From among the given nucleotides, identify those that occur naturally in RNA and those that occur in DNA. Do any not occur in either RNA or DNA?

(a)

(b)

(c)

(d)

## Additional Problems

**10.77**   Write structural formulas for all the constitutionally isomeric $C_4H_9$ alkyl groups. Check your answers with Table 10.1, and note the names of these groups.

**10.78**   *Ethers* are compounds (excluding esters) that contain the C—O—C functional group. An acceptable way to name them is to list the two groups attached to oxygen in alphabetical order as separate words, followed by the word *ether*. If the two groups are the same, add the prefix *di* to the name of the alkyl group. Thus, $CH_3OCH_2CH_3$ is "ethyl methyl ether" and $CH_3CH_2OCH_2CH_3$ is "diethyl ether."

    Write structural formulas and provide names for all the constitutionally isomeric ethers in which only the $C_4H_9$ alkyl groups from Problem 10.77 are attached to oxygen. Which of these ethers is potentially chiral?

**10.79**   The α-amino acids found in proteins are based on the structural formula:

$$RCHCO_2H$$
$$|$$
$$NH_2$$

From the $C_4H_9$ alkyl groups in Problem 10.77, find the ones that correspond to "R" among the amino acids listed in Figure 10.14. Match the name of the amino acid with the name of the group.

**10.80**   *Alkynes* are hydrocarbons that contain a carbon-carbon triple bond.

(a) Write structural formulas for all the isomeric alkynes of molecular formula $C_5H_8$.
(b) Are any of the alkynes chiral?
(c) Are any of the alkynes stereoisomeric?

**10.81**   Among the many alkenes of molecular formula $C_6H_{12}$, only one is chiral.

(a) Write a structural formula for this alkene.
(b) Place substituents on the tetrahedral carbons so as to represent the two enantiomers of this alkene.

**10.82**   (a) Use VSEPR to predict the geometry at the carbon shown in bold in the carbocation, radical, and anion derived from the *tert*-butyl group. Is the arrangement of bonds to this carbon linear, tetrahedral, trigonal planar, or trigonal pyramidal?

| $^+C(CH_3)_3$ | $\cdot C(CH_3)_3$ | $:\bar{C}(CH_3)_3$ |
|---|---|---|
| *tert*-Butyl   Cation | Radical | Anion |

(b) Which of these species is the most nucleophilic? Which is the most electrophilic?
(c) Predict which species reacts with water to give $HC(CH_3)_3$.
(d) Write a chemical equation for the reaction in part (c), and use curved arrows to show the flow of electrons.

**10.83**   In each of the following pairs, specify whether the two structural formulas represent constitutional isomers, *cis,trans*-stereoisomers, enantiomers, resonance structures, or are simply a different representation of the same structure.

(a)

(b)

(c)

(d)

**10.84**   In each of the following pairs, specify whether the two structural formulas represent constitutional isomers, *cis,trans*-stereoisomers, enantiomers, resonance structures, or are simply a different representation of the same structure.

(a)

(b)

(c)

(d)

**10.85** The electrophile in the following aromatic substitution is $(CH_3)_2CH^+$, and the mechanism involves the intermediate shown. Write two other resonance structures for this intermediate.

**10.86** Two isomeric alkenes are formed in the dehydration of 2-methyl-2-butanol. What are these two alkenes? Are they constitutional isomers or stereoisomers?

**10.87** Electrophilic addition of HCl to *cis*-2-butene gave 2-chlorobutane, which was determined *not* to be optically active when examined with a polarimeter.

Which of the following is the better explanation for the lack of optical activity in the 2-chlorobutane formed in this reaction?

(a) 2-Chlorobutane is not chiral.
(b) Two enantiomers of 2-chlorobutane were formed in equal amounts.

**10.88** Reactions such as the following have been used in carbohydrate synthesis since the nineteenth century.

Classify this procedure according to reaction type.

(a) Electrophilic addition
(b) Electrophilic substitution
(c) Nucleophilic addition
(d) Nucleophilic substitution

**10.89** Excluding compounds that have rings, there are three hydrocarbons that have the molecular formula $C_4H_6$. Write their structural formulas, and specify the hybridization of each carbon in these isomers.

**10.90** Give the structures of the two tertiary amines that are isomers of $CH_3CH_2CH_2CH_2CH_2NH_2$.

**10.91** In which class of compounds, esters or amides, do you think electron donation into the carbonyl group is more pronounced? Explain.

Ester          Amide

**10.92** (a) A plane of symmetry in a molecule is a plane passing through the middle of the molecule that bisects the molecule into two mirror-image halves. If a molecule has a plane of symmetry, it cannot be chiral and cannot be optically active. Which of the following have at least one plane of symmetry? Do any have more than one?

1,1-Dichlorocyclopropane

*cis*-1,2-Dichlorocyclopropane

*trans*-1,2-Dichlorocyclopropane

(b) Specify the relationships (constitutionally isomeric or stereoisomeric) among these compounds.

**10.93** All alkanes give off heat when burned in air. Such *combustion* of alkanes is exothermic, and the sign of $\Delta H°$ is negative. The general equation for the combustion of the alkanes of molecular formula $C_5H_{12}$ is

$$C_5H_{12} + 8O_2 \longrightarrow 5CO_2 + 6H_2O$$

The values of $\Delta H°$ for the combustion of the three pentane isomers are

$CH_3CH_2CH_2CH_2CH_3 = -3536$ kJ/mol
$(CH_3)_2CHCH_2CH_3 = -3529$ kJ/mol
$(CH_3)_4C = -3515$ kJ/mol

(a) What do these data tell you about the effect of chain branching on the relative potential energies and stabilities of these isomers?
(b) Assume this effect is general to predict which one of the 18 $C_8H_{18}$ isomers should be the most stable.

**10.94** Carbon dioxide reacts with sodium hydroxide according to the following equation:

$$CO_2 + 2NaOH \longrightarrow Na_2CO_3 + H_2O$$

The overall reaction is the result of two separate reactions.

Reaction I:   H—Ö:⁻  +  :O=C=O:   ⟶   H\O\C\O:⁻ (with Ö)

Reaction II:

H—Ö:⁻  +  H\O\C(=Ö)\O:⁻   ⟶   :O\C(=Ö)\O:⁻  +  H—Ö\H

(a) Use curved arrows to track the flow of electrons in reaction I.
(b) Use curved arrows to track the flow of electrons in reaction II.
(c) Classify reaction I as electrophilic addition, nucleophilic addition, electrophilic substitution, or acid-base.
(d) Classify reaction II according to the choices in part (c).

**10.95** The combustion of 20.63 mg of compound Y, which contains only C, H, and O, with excess oxygen gave 57.94 mg of $CO_2$ and 11.85 mg of $H_2O$. (a) Calculate how many milligrams of C, H, and O were present in the original sample of Y. (b) Derive the empirical formula of Y. (c) Suggest a plausible structure for Y if the empirical formula is the same as the molecular formula.

**10.96** Match each molecular model with the correct line-wedge-dash structure.

**10.97** Kevlar is a copolymer used in bulletproof vests. It is formed in a condensation reaction between the following two monomers:

H₂N—⟨benzene ring⟩—NH₂      HO—C(=O)—⟨benzene ring⟩—C(=O)—OH

Sketch a portion of the polymer chain showing several monomer units. Write the overall equation for the condensation reaction.

**10.98** Describe the formation of polystyrene.

**10.99** The amino acid glycine can be condensed to form a polymer called polyglycine. Draw the repeating monomer unit.

**10.100** Nylon can be destroyed easily by strong acids. Explain the chemical basis for the destruction. (*Hint:* The products are the starting materials of the polymerization reaction.)

**10.101** How many different tripeptides can be formed by lysine and alanine?

**10.102** Nylon was designed to be a synthetic silk. (a) The average molar mass of a batch of nylon 66 is 12,000 g/mol. How many monomer units are there in this sample? (b) Which part of nylon's structure is similar to a polypeptide's structure? (c) How many different tripeptides (made up of three amino acids) can be formed from the amino acids alanine (Ala), glycine (Gly), and serine (Ser), which account for most of the amino acids in silk?

# PRE-PROFESSIONAL PRACTICE EXAM QUESTIONS: VERBAL REASONING

In 1960, Canadian-born doctor and pharmacologist Frances Kelsey was hired by the FDA to review applications for the approval of new drugs. Her first assignment was an application by the William S. Merrell Company of Cincinnati, Ohio, requesting approval of thalidomide as a sedative and anti-emetic for pregnant women suffering from morning sickness. Kelsey rejected the application, citing the manufacturer's failure to prove the drug's safety, and requested that further studies be done. Although the company initially complied with her request, Kelsey, still unsatisfied with the data, refused the application a second time. Eventually, Merrell appealed to Kelsey's superiors to pressure her to approve the drug. By early 1961, though, a study in England reported that repeated use of the drug might have serious nervous-system side effects. In addition, Kelsey's research in pharmacology early in her career made her question the effects of the drug on a developing fetus. She continued to resist the pressure to approve the drug and was vindicated late in 1961 when a German scientist, Dr. Widukind Lenz, reported that use of thalidomide by pregnant women might be the cause of the epidemic of *phocomelia,* a malformation of limbs in newborn babies, in Germany.

U.S. law at the time allowed drugs that had not yet been approved by the FDA to be distributed to physicians for "experimental use." Never-theless, only 17 thalidomide babies are known to have been born in the United States. If not for the diligence and dedication of Dr. Frances Kelsey, there would almost certainly have been thousands more. In recognition of her incalculable service to the people of the United States, Dr. Kelsey was given the President's Award for Distinguished Federal Civilian Ser-vice, the highest award ever given to a civilian. President John F. Kennedy presented the medal to Kelsey at a ceremony at the White House in Au-gust 1962. As a result of the thalidomide tragedy, new laws were enacted placing a greater burden on drug manufacturers to prove the safety and efficacy of their drugs. In addition, the concept of "informed consent" was introduced, preventing the distribution of unapproved drugs to unsuspect-ing patients.

1.  The main point of the passage is that Frances Kelsey

   a) was born in Canada.
   b) worked for the FDA.
   c) prevented the thalidomide tragedy in the United States.
   d) was awarded the Distinguished Federal Civilian Service medal.

2.  According to the passage, Dr. Kelsey refused to approve thalidomide for distribution in the United States because

   a) her superiors pressured her not to approve it.
   b) she knew that it had caused thousands of birth defects in Europe.
   c) 17 babies had been born with birth defects as a result of the drug.
   d) she was not satisfied with the manufacturer's studies on the safety of the drug.

3.  Based on the passage, what was likely the reason that 17 thalidomide babies were born in the United States?

   a) Physicians in the United States gave the drug to their patients without FDA approval.
   b) Pregnant women arrived from Europe and gave birth in the United States.
   c) U.S. citizens obtained the drug from overseas sources.
   d) Some women in the United States failed to heed the manufacturer's warning about use during pregnancy.

4.  Based on the passage, the author most likely considers Kelsey to be

   a) a celebrity.
   b) an activist.
   c) a hero.
   d) a role model.

# ANSWERS TO IN-CHAPTER MATERIALS

## Answers to Practice Problems

**10.1A** (a) 3-ethylpentane, (b) 2-bromohexane, (c) 2-chlorobutane.
**10.1B** (a) $CH_3CH_2CH_2CHCH_2CH_2CH_2CH_3$, (b) $CH_3CHCH_2CH_2CH_3$,

$$\begin{array}{cc} CH_2 & F \\ | & \\ CH_3 & \end{array}$$

(c) $CH_3CH_2CHCH_2CH_2CH_2CH_2CH_2CH_3$. **10.2A** (a) ester,

$$CH_3$$

(b) carboxy group; ester, (c) 2 hydroxy groups, 2 carboxy groups.
**10.2B** (a) ketone, (b) 3° amide, (c) ester.

**10.3A** $CH_2BrCH_2COOH$. **10.3B**

**10.4A** $H-C=O$

**10.4B**

**10.5A**

**10.5B** c. **10.6A**

**10.6B** b. **10.7A** (a)

**10.7B**

(b)

## Answers to Checkpoints

**10.2.1** c. **10.2.2** b. **10.2.3** b. **10.2.4** e. **10.2.5** b, c. **10.2.6** a, b. **10.3.1** b. **10.3.2** d. **10.3.3** e. **10.3.4** b, d. **10.5.1** d. **10.5.2** a.

## Answers to Applying What You've Learned

**a)** Thalidomide contains several *carbonyl* groups, an *amino* group, and an *amide* group. Note that the amide group on the right could be defined using either of the carbonyl groups adjacent to it. **b)** Molecular formulas: thalidomide, $C_{13}H_{10}N_2O_4$; lenalidomide, $C_{13}H_{13}N_3O_3$; and CC-4047, $C_{13}H_{11}N_3O_4$.

**c)**

# 11

# Gases

*Scuba divers breathe a compressed mixture of gases. For shallow recreational diving, compressed air is generally used. For dives to greater depths, various mixtures of helium, nitrogen, and oxygen are used.*

# How the Properties of Gases Contribute to the Hazards of Scuba Diving

One of the first lessons taught in scuba certification is that divers must never hold their breath during ascent to the surface. Failure to heed this warning can result in serious injury or death. During underwater ascent, the air in a diver's lungs expands. If the air is not expelled, it causes overexpansion and rupture of alveoli—the tiny sacks that normally fill with air on inhalation. This condition, known as "burst lung," can cause air to escape into the chest cavity, where further expansion can collapse the ruptured lung. The potential for these catastrophic injuries is not limited to deep-sea divers, though, as burst lung can occur during a rapid ascent of as little as 3 meters, a depth common in public swimming pools. Furthermore, *spontaneous pneumothorax,* the medical term for this injury, can be caused even without the failure to exhale if there is an air-filled cyst in the diver's lung, or if there is a region of lung tissue blocked by phlegm due to a respiratory tract infection. In addition to the risk of spontaneous pneumothorax, burst lung can result in gas bubbles entering the bloodstream. Expansion of these bubbles during rapid underwater ascent can block circulation, a condition known as *gas embolism,* which can lead to heart attack or stroke.

Because the cause of burst lung in divers is rapid *de*compression, treatment for the resulting conditions, especially gas embolism, usually includes *re*compression in a hyperbaric chamber. Hyperbaric chambers are cylindrical enclosures built to withstand pressures significantly above atmospheric pressure. With the victim of a rapid decompression and, in some cases, medical personnel inside the enclosure, the door is sealed and high-pressure air is pumped in until the interior pressure is high enough to compress the gas bubbles in the victim's system. The victim usually breathes pure oxygen through a mask to help purge the undesirable gases during treatment. The pressure in a typical medical hyperbaric chamber can be increased to as many as six times atmospheric pressure.

Understanding the risks associated with scuba diving and the efforts to prevent and treat related injuries requires knowledge of the behavior of *gases.*

**Pneumothorax**

**Monoplace hyperbaric chamber**

**Student Annotation:** "Monoplace" hyperbaric chambers, which are large enough to accommodate only one person, typically are pressurized with pure oxygen, although monoplace chambers are not used to treat decompression injuries.

At the end of this chapter, you will be able to solve a series of problems involving the properties of gases as they relate to the safety of athletes [▶▶ Page 480].

## In This Chapter, You Will Learn:

About the properties and behavior of gases.

### Before you begin, you should review

- Tracking units [◀◀ Section 1.6]
- Stoichiometry [◀◀ Section 3.3]

**Figure 11.1**    Solid, liquid, and gaseous states of a substance.

# 11.1  Properties of Gases

Recall from Chapter 1 that matter exists in one of three states: solid, liquid, or gas. In fact, most substances that are solid or liquid at room temperature (25°C) *can* exist as gases under appropriate conditions. Water, for instance, *evaporates* under the right conditions. Water vapor is a gas. In this chapter we will explore the nature of gases and how their properties at the molecular level give rise to the macroscopic properties that we observe. Figure 11.1 illustrates the three states of matter at the macroscopic level and at the molecular level.

Relatively few elements exist as gases at room temperature. Those that do are hydrogen, nitrogen, oxygen, fluorine, chlorine, and the noble gases. Of these, the noble gases exist as isolated atoms, whereas the others exist as diatomic molecules [◄◄ Section 2.6]. Figure 11.2 shows where the gaseous elements appear in the periodic table.

Many molecular compounds, most often those with low molar masses, exist as gases at room temperature. Table 11.1 lists some gaseous compounds that may be familiar to you.

**Figure 11.2**    Elements that exist as gases at room temperature.

| TABLE 11.1 | Molecular Compounds That Are Gases at Room Temperature |
|---|---|
| **Molecular Formula** | **Compound Name** |
| HCl | Hydrogen chloride |
| $NH_3$ | Ammonia |
| $CO_2$ | Carbon dioxide |
| $N_2O$ | Dinitrogen monoxide or nitrous oxide |
| $CH_4$ | Methane |
| HCN | Hydrogen cyanide |

(a)

(b)

**Figure 11.3** (a) An empty metal can. (b) When the air is removed by a vacuum pump, atmospheric pressure crushes the can.

## Characteristics of Gases

Gases differ from the condensed phases (solids and liquids) in the following important ways:

1. *A sample of gas assumes both the shape and volume of its container.* Like a liquid, a gas consists of particles (molecules or atoms) that do not have fixed positions in the sample [◄◄ Section 1.2]. As a result, both liquids and gases are able to *flow.* (Recall from Chapter 1 that we refer to liquids and gases collectively as *fluids.*) While a sample of liquid will assume the shape of the part of its container that it occupies, a sample of gas will expand to fill the entire *volume* of its container.

2. *Gases are compressible.* Unlike a solid or a liquid, a gas consists of particles with relatively large distances between them; that is, the distance between any two particles in a gas is much larger than the size of a molecule or atom. Because gas particles are far apart, it is possible to move them closer together by confining them to a smaller volume.

3. *The densities of gases are much smaller than those of liquids and solids and are highly variable depending on temperature and pressure.* The densities of gases are typically expressed in g/L, whereas those of liquids and solids are typically expressed in g/mL or g/cm³.

**Student Annotation:** 1 mL = 1 cm³ [◄◄ Section 1.3].

When we compress a sample of gas, we decrease its volume. Because its mass remains the same, the ratio of mass to volume (density) increases. Conversely, if we increase the volume to which a sample of gas is confined, we decrease its density.

If you have ever seen a hot-air balloon aloft, you have seen a demonstration of how the density of a gas varies with temperature. Hot air is less dense than cold air, so hot air "floats" on cold air, much like oil floats on water.

4. *Gases form homogeneous mixtures (solutions) with one another in any proportion.* Some liquids (e.g., oil and water) do not mix with one another. Gases, on the other hand, because their particles are so far apart, do not interact with one another to any significant degree unless a chemical reaction takes place between them. This allows molecules of different gases to mix uniformly. That is, gases that don't react with each other are mutually *miscible.*

Each of these four characteristics is the result of the properties of gases at the molecular level.

## Gas Pressure: Definition and Units

A sample of gas confined to a container exerts a pressure on the walls of its container. For example, the air in the tires of your car exerts pressure on the inside walls of the tires. In fact, gases exert pressure on everything they touch. Thus, while you may add enough air to increase the pressure inside your tire to 32 pounds per square inch (psi), there is also a pressure of approximately 14.7 psi, called *atmospheric pressure,* acting on the outside of the tire—and on everything else, including your body. The reason you don't feel the pressure of the atmosphere pushing on the outside of your body is that an equal pressure exists inside your body so that there is no net pressure on you.

Atmospheric pressure, the pressure exerted by Earth's atmosphere, can be demonstrated using the empty metal container shown in Figure 11.3(a). Because the container is open to the atmosphere, atmospheric pressure acts on both the internal and external walls of the container. When we attach a vacuum pump to the opening of the container and draw air out of it, however, we reduce the pressure inside the container. When the pressure against the interior walls is reduced, atmospheric pressure crushes the container [Figure 11.3(b)].

**Student Annotation:** Common pencil-type pressure gauges actually measure the difference between internal and external pressure. Thus, if the tire is completely flat, the reading of 0 psi means that the pressure *inside* the tire is the same as that *outside* the tire.

**Pencil-type tire gauge**

| TABLE 11.2 | Units of Pressure Commonly Used in Chemistry | |
|---|---|---|
| **Unit** | **Origin** | **Definition** |
| standard atmosphere (atm) | Pressure at sea level | 1 atm = 101,325 Pa |
| mmHg | Barometer measurement | 1 mmHg = 133.322 Pa |
| torr | Name given to mmHg in honor of Torricelli, the inventor of the barometer | 1 torr = 133.322 Pa |
| bar | Same order of magnitude as atm, but a decimal multiple of Pa | 1 bar = 1 × 10⁵ Pa |

**Student Annotation:** Force is mass × acceleration.

*Pressure* is defined as the force applied per unit area:

$$\text{pressure} = \frac{\text{force}}{\text{area}}$$

The SI unit of force is the **newton (N),** where

$$1 \text{ N} = 1 \text{ kg} \cdot \text{m/s}^2$$

**Student Annotation:** In SI base units, $1 \text{ Pa} = 1 \text{ kg/m} \cdot \text{s}^2$.

The SI unit of pressure is the **pascal (Pa),** defined as 1 newton per square meter.

$$1 \text{ Pa} = 1 \text{ N/m}^2$$

Although the pascal is the SI unit of pressure, there are other units of pressure that are more commonly used. Table 11.2 lists the units in which pressure is most commonly expressed in chemistry and their definitions in terms of pascals. Which of these units you encounter most often will depend on your specific field of study. Use of *atmospheres* (atm) is common in chemistry, although use of the *bar* is becoming increasingly common. Use of *millimeters mercury* (mmHg) is common in medicine and meteorology. We will use all these units in this text.

## Calculation of Pressure

The force experienced by an area exposed to Earth's atmosphere is equal to the weight of the column of air above it. For example, the mass of air above a spot on the ground near sea level, with an area of 1 cm², is approximately 1 kg (Figure 11.4).

The weight of an object that is subject to Earth's gravitational pull is equal to its mass times the gravitational constant, 9.80665 m/s². Thus, the force exerted by this column of air is

$$1 \text{ kg} \times \frac{9.80665 \text{ m}}{\text{s}^2} \approx 10 \text{ kg} \cdot \text{m/s}^2 = 10 \text{ N}$$

Pressure, though, is force per unit area. Specifically, pressure in *pascals* is equal to force in *newtons* per *square meter*. We must first convert area from cm² to m²,

$$1 \text{ cm}^2 \times \left(\frac{1 \text{ m}}{100 \text{ cm}}\right)^2 = 0.0001 \text{ m}^2$$

and then divide force by area,

$$\frac{10 \text{ N}}{0.0001 \text{ m}^2} = 1 \times 10^5 \text{ Pa}$$

This pressure is roughly equal to 1 atm (~1 × 10⁵ Pa), which we would expect at sea level.

We can calculate the pressure exerted by a column of any fluid (gas or liquid) in the same way. In fact, this is how atmospheric pressure is commonly measured—by determining the height of a column of mercury it can support.

## Measurement of Pressure

A simple **barometer,** an instrument used to measure atmospheric pressure, consists of a long glass tube, closed at one end and filled with mercury. The tube is carefully inverted in a container of mercury so that no air enters the tube. When the tube is inverted, and the open end is submerged in

**Figure 11.4** A column of air 1 cm × 1 cm from Earth's surface to the top of the atmosphere weighs approximately 1 kg.

Column of air

**Student Annotation:** Remember that when a unit is raised to a power, any conversion factor you use must also be raised to that power [◄◄ Section 1.6].

**Figure 11.5**   Barometer.

Atmospheric pressure

76 cm

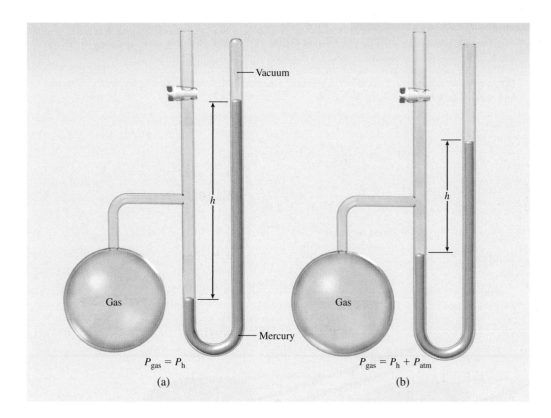

**Figure 11.6**   (a) Closed-tube manometer. The space labeled "vacuum" actually contains a small amount of mercury vapor. (b) Open-tube manometer.

Vacuum

$h$

$h$

Gas

Gas

Mercury

$P_{gas} = P_h$

$P_{gas} = P_h + P_{atm}$

(a)

(b)

the mercury in the container, some of the mercury in the tube will flow out into the container, creating an empty space at the top (closed end) of the tube (Figure 11.5). The weight of the mercury remaining in the tube is supported by atmospheric pressure pushing down on the surface of the mercury in the container. In other words, the pressure exerted by the column of mercury is *equal* to the pressure exerted by the atmosphere. *Standard atmospheric pressure* (1 atm) was originally defined as the pressure that would support a column of mercury exactly 760 mm high at 0°C at sea level. The mmHg unit is also called the torr, after the Italian scientist Evangelista Torricelli,[1] who invented the barometer. Standard atmospheric pressure is given in all the common units of pressure in the margin.

A *manometer* is a device used to measure pressures other than atmospheric pressure. The principle of operation of a manometer is similar to that of a barometer. There are two types of manometers, both of which are shown in Figure 11.6. The closed-tube manometer [Figure 11.6(a)] is normally used to measure pressures below atmospheric pressure, whereas the open-tube manometer [Figure 11.6(b)] is generally used to measure pressures equal to or greater than atmospheric pressure.

**Student Annotation:**
1 atm*
101,325 Pa
760 mmHg*
760 torr*
1.01325 bar
14.7 psi
*These are exact numbers.

---

1. Evangelista Torricelli (1608–1674). Italian mathematician. Torricelli was supposedly the first person to recognize the existence of atmospheric pressure.

The pressure exerted by a column of fluid, such as that in a barometer (Figure 11.5), is given by Equation 11.1.

Equation 11.1                                           $P = hdg$

where $h$ is the height of the column in meters, $d$ is the density of the fluid in kg/m³, and $g$ is the gravitational constant equal to 9.80665 m/s². This equation explains why barometers historically have been constructed using mercury. The height of a column of fluid supported by a given pressure is inversely proportional to the density of the fluid. (At a given $P$, as $d$ goes up, $h$ must go down—and vice versa.) Mercury's high density made it possible to construct barometers and manometers of manageable size. For example, a barometer filled with mercury that stands 1 m tall would have to be over 13 m tall if it were filled with water.

Sample Problem 11.1 shows how to calculate the pressure exerted by a column of fluid.

## Sample Problem 11.1

Calculate the pressure exerted by a column of mercury 70.0 cm high. Express the pressure in pascals and in atmospheres. The density of mercury is 13.5951 g/cm³.

**Strategy** Use Equation 1.1 to calculate pressure. Remember that height must be expressed in meters and density must be expressed in kg/m³.

**Setup**

$$h = 70.0 \text{ cm} \times \frac{1 \text{ m}}{100 \text{ cm}} = 0.700 \text{ m}$$

$$d = \frac{13.5951 \text{ g}}{\text{cm}^3} \times \frac{1 \text{ kg}}{1000 \text{ g}} \times \left(\frac{100 \text{ cm}}{1 \text{ m}}\right)^3 = 1.35951 \times 10^4 \text{ kg/m}^3$$

$$g = 9.80665 \text{ m/s}^2$$

**Solution**

$$\text{pressure} = 0.700 \text{ m} \times \frac{1.35951 \times 10^4 \text{ kg}}{\text{m}^3} \times \frac{9.80665 \text{ m}}{\text{s}^2} = 9.33 \times 10^4 \text{ kg/m} \cdot \text{s}^2 = 9.33 \times 10^4 \text{ Pa}$$

$$9.33 \times 10^4 \text{ Pa} \times \frac{1 \text{ atm}}{101{,}325 \text{ Pa}} = 0.921 \text{ atm}$$

**Think About It** Make sure your units cancel properly in this type of problem. Common errors include forgetting to express height in meters and density in kg/m³. You can avoid these errors by becoming familiar with the value of atmospheric pressure in the various units. A column of mercury slightly less than 760 mm is equivalent to slightly less than 101,325 Pa and slightly less than 1 atm.

**Practice Problem A** What pressure (in atm) is exerted by a column of mercury exactly 1 m high?

**Practice Problem B** What would be the height of a column of water supported by the pressure you calculated in Sample Problem 11.1? Assume that the density of the water is 1.00 g/cm³.

## Checkpoint 11.1    Properties of Gases

11.1.1  Express a pressure of 1.15 atm in units of bar.

    a) 1.17 bar

    b) 1.13 bar

    c) 0.881 bar

    d) 874 bar

    e) $1.51 \times 10^{-3}$ bar

11.1.2  Which of the following is true? (Select all that apply.)

    a) 0.80 atm = 0.80 torr

    b) 4180 mmHg = $5.573 \times 10^5$ Pa

    c) 433 torr = 433 mmHg

    d) 2.300 atm = 1748 torr

    e) 5.5 atm = $1.0 \times 10^5$ Pa

**11.1.3**  Calculate the height of a column of ethanol that would be supported by atmospheric pressure (1 atm). The density of ethanol is 0.789 g/cm³.

a) $1.31 \times 10^4$ m

b) 0.789 m

c) 600 mm

d) 13.1 m

e) 780 mm

**11.1.4**  What pressure is exerted by a column of water 50.0 m high? Assume the density of the water is 1.00 g/cm³.

a) 490 atm

b) 4.84 atm

c) 0.087 atm

d) 50 atm

e) 1.62 atm

**11.1.5**  How do the pressures exerted by the following columns of water compare?

(i)       (ii)       (iii)

a) i = ii < iii

b) i < iii < ii

c) iii < i < ii

d) ii < i = iii

e) iii < i = ii

# 11.2  The Gas Laws

In contrast to the condensed phases, all gases, even those with vastly different chemical compositions, exhibit remarkably similar physical behavior. Numerous experiments carried out in the seventeenth and eighteenth centuries showed that the physical state of a sample of gas can be described completely with just four parameters: temperature ($T$), pressure ($P$), volume ($V$), and number of moles ($n$). Knowing any three of these parameters enables us to calculate the fourth. The relationships between these parameters are known as the **gas laws.**

## Boyle's Law: The Pressure-Volume Relationship

Imagine that you have a plastic syringe filled with air. If you hold your finger tightly against the tip of the syringe and push the plunger with your other hand, decreasing the volume of the air, you will increase the pressure in the syringe. During the seventeenth century Robert Boyle[2] conducted systematic studies of the relationship between gas volume and pressure using a simple apparatus like the one shown in Figure 11.7. The J-shaped tube contains a sample of gas confined by a column of mercury. The apparatus functions as an open-end manometer. When the mercury levels on both sides are equal [Figure 11.7(a)], the pressure of the confined gas is equal to atmospheric pressure. When more mercury is added through the open end, the pressure of the confined gas is increased by an amount proportional to the height of the added mercury—and the volume of the gas decreases. If, for example, as shown in Figure 11.7(b), we *double* the pressure on the confined gas by adding enough mercury to make the difference in mercury levels on the left and right 760 mm (the height of a mercury column that exerts a pressure equal to 1 atm), the volume of the gas is reduced by *half*. If we triple the original pressure on the confined gas by adding more mercury, the volume of the gas is reduced to one-third of its original volume [Figure 11.7(c)].

Table 11.3 gives a set of data typical of Boyle's experiments. Figure 11.8(a) and (b) shows some of the volume data plotted as a function of pressure and as a function of the inverse of pressure, respectively. These data illustrate **Boyle's law,** which states that the pressure of a fixed amount of gas at a constant temperature is inversely proportional to the volume of the gas. This inverse relationship between pressure and volume can be expressed mathematically as follows:

$$V \propto \frac{1}{P}$$

**Multimedia**

Gas Laws—interactive gas laws (interactive).

**Student Annotation:** Because they arise from experiment, these laws are referred to as the *empirical* gas laws.

**Student Annotation:** Remember that the symbol ∝ means "is proportional to."

---

2. Robert Boyle (1627–1691). British chemist and natural philosopher. Although Boyle is commonly associated with the gas law that bears his name, he made many other significant contributions to the fields of chemistry and physics.

**Figure 11.7**   Demonstration of Boyle's law. The volume of a sample of gas is inversely proportional to its pressure. (a) $P$ = 760 mmHg, $V$ = 100 mL. (b) $P$ = 1520 mmHg, $V$ = 50 mL. (c) $P$ = 2280 mmHg, $V$ = 33 mL. Note that the total pressure exerted on the gas is the sum of atmospheric pressure (760 mmHg) and the difference in height of the mercury.

| TABLE 11.3 | | Typical Data from Experiments with the Apparatus of Figure 11.7 | | | | | | | | |
|---|---|---|---|---|---|---|---|---|---|---|
| $P$ (mmHg) | 760 | 855 | 950 | 1045 | 1140 | 1235 | 1330 | 1425 | 1520 | 2280 |
| $V$ (mL) | 100 | 89 | 78 | 72 | 66 | 59 | 55 | 54 | 50 | 33 |
| | Shown in Figure 11.7(a) | | | | | | | | Shown in Figure 11.7(b) | Shown in Figure 11.7(c) |

(a)                                        (b)

**Figure 11.8**   Plots of volume (a) as a function of pressure and (b) as a function of 1/pressure.

or

$$V = k_1 \frac{1}{P} \quad \text{(at constant temperature)} \qquad \text{Equation 11.2(a)}$$

where $k_1$ is a *proportionality constant*. We can rearrange Equation 11.2(a) to get

$$PV = k_1 \quad \text{(at constant temperature)} \qquad \text{Equation 11.2(b)}$$

According to this form of Boyle's law, the *product* of the pressure and the volume of a given sample of gas (at constant temperature) is a *constant*.

Although the individual values of pressure and volume can vary greatly for a given sample of gas, the product of $P$ and $V$ is always equal to the same constant as long as the temperature is held constant and the amount of gas does not change. Therefore, for a given sample of gas under two different sets of conditions at constant temperature, we can write

$$P_1 V_1 = k_1 = P_2 V_2$$

or

$$P_1 V_1 = P_2 V_2 \quad \text{(at constant temperature)} \qquad \text{Equation 11.3}$$

where $V_1$ is the volume at pressure $P_1$ and $V_2$ is the volume at pressure $P_2$.

Sample Problem 11.2 illustrates the use of Boyle's law.

---

### Sample Problem 11.2

If a skin diver takes a breath at the surface, filling his lungs with 5.82 L of air, what volume will the air in his lungs occupy when he dives to a depth where the pressure is 1.92 atm? (Assume constant temperature and that the pressure at the surface is exactly 1 atm.)

**Strategy** Use Equation 11.3 to solve for $V_2$.

**Setup** $P_1 = 1.00$ atm, $V_1 = 5.82$ L, and $P_2 = 1.92$ atm.

**Solution**

$$V_2 = \frac{P_1 \times V_1}{P_2} = \frac{1.00 \text{ atm} \times 5.82 \text{ L}}{1.92 \text{ atm}} = 3.03 \text{ L}$$

**Think About It** At higher pressure, the volume should be smaller. Therefore, the answer makes sense.

**Practice Problem A** Calculate the volume of a sample of gas at 5.75 atm if it occupies 5.14 L at 2.49 atm. (Assume constant temperature.)

**Practice Problem B** At what pressure would a sample of gas occupy 7.86 L if it occupies 3.44 L at 4.11 atm? (Assume constant temperature.)

---

## Charles's and Gay-Lussac's Law: The Temperature-Volume Relationship

Multimedia
Gas Laws—Charles's law.

If you took a helium-filled Mylar balloon outdoors on a cold day, the balloon would shrink somewhat when it came into contact with the cold air. This would occur because the volume of a sample of gas depends on the temperature. A more dramatic illustration is shown in Figure 11.9, where liquid nitrogen is being poured over an air-filled balloon. The large drop in temperature of the air in the balloon (the boiling liquid nitrogen has a temperature of –196°C) results in a significant decrease in its volume, causing the balloon to shrink. Note that the pressure inside the balloon is roughly equal to the external pressure.

The first to study the relationship between gas volume and temperature were French scientists Jacques Charles[3] and Joseph Gay-Lussac.[4] Their studies showed that, at constant pressure,

---

3. Jacques Alexandre Cesar Charles (1746–1823). French physicist. Charles was a gifted lecturer, an inventor of scientific apparatus, and the first person to use hydrogen to inflate balloons.

4. Joseph Louis Gay-Lussac (1778–1850). French chemist and physicist. Like Charles, Gay-Lussac was a balloon enthusiast. He once ascended to an altitude of 20,000 ft to collect air samples for analysis.

**Figure 11.9** (a) Air-filled balloon. (b) Lowering the temperature with liquid nitrogen causes a volume decrease. The pressure inside the balloon, which is roughly equal to the external pressure, remains constant in this process.

(a)                    (b)

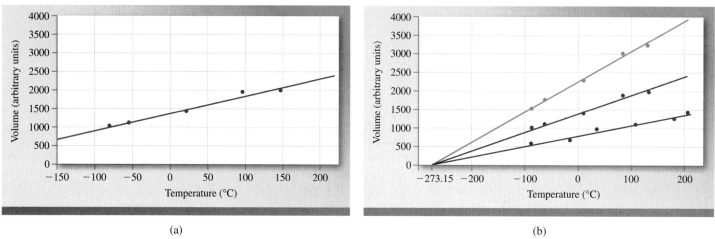

(a)                    (b)

**Figure 11.10** (a) Plot of the volume of a sample of gas as a function of temperature. (b) Plot of the volume of a sample of gas as a function of temperature at three different pressures.

the volume of a gas sample increases when heated and decreases when cooled. Figure 11.10(a) shows a plot of data typical of Charles's and Gay-Lussac's experiments. Note that with pressure held constant, the volume of a sample of gas plotted as a function of temperature yields a straight line. These experiments were carried out at several different pressures [Figure 11.10(b)], each yielding a different straight line. Interestingly, if the lines are extrapolated to zero volume, they all meet at the *x* axis at the temperature –273.15°C. The implication is that a gas sample occupies zero volume at –273.15°C. This is not observed in practice, however, because all gases condense to form liquids or solids before –273.15°C is reached.

In 1848, Lord Kelvin[5] realized the significance of the extrapolated lines all meeting at –273.15°C. He identified –273.15°C as ***absolute zero,*** theoretically the lowest attainable temperature. Then he set up an ***absolute temperature scale,*** now called the ***Kelvin temperature scale,*** with

---

5. William Thomson, Lord Kelvin (1824–1907). Scottish mathematician and physicist. Kelvin did important work in many branches of physics.

absolute zero as the lowest point [◄◄ Section 1.3]. On the Kelvin scale, 1 kelvin (K) is equal in magnitude to 1 degree Celsius. The difference is simply an offset of 273.15. We obtain the absolute temperature by adding 273.15 to the temperature expressed in Celsius, although we often use simply 273 instead of 273.15. Several important points on the two scales match up as follows:

|  | Kelvin Scale (K) | Celsius Scale (°C) |
|---|---|---|
| Absolute zero | 0 K | −273.15°C |
| Freezing point of water | 273.15 K | 0°C |
| Boiling point of water | 373.15 K | 100°C |

> **Student Annotation:** Remember that a kelvin and a degree Celsius have the same magnitude. Thus, while we add 273.15 to the temperature in °C to get the temperature in K, a *change* in temperature in Celsius is *equal* to the change in temperature in K. A temperature of 20°C is the same as 293.15 K. A *change* in temperature of 20°C, however, is the same as a *change* in temperature of 20 K.

The dependence of the volume of a sample of gas on temperature is given by

$$V \propto T$$

or

$$V = k_2 T \quad \text{(at constant pressure)} \qquad \text{Equation 11.4(a)}$$

> **Student Annotation:** Don't forget that volume is proportional to absolute temperature. The volume of a sample of gas at constant pressure doubles if the temperature increases from 100 K to 200 K—but *not* if the temperature increases from 100°C to 200°C!

where $k_2$ is the proportionality constant. We can rearrange Equation 11.4(a) to get

$$\frac{V}{T} = k_2 \quad \text{(at constant pressure)} \qquad \text{Equation 11.4(b)}$$

Equations 11.4(a) and (b) are expressions of **Charles's and Gay-Lussac's law,** often referred to simply as **Charles's law,** which states that the volume of a fixed amount of gas maintained at constant pressure is directly proportional to the absolute temperature of the gas.

Just as we did with the pressure-volume relationship at constant temperature, we can compare two sets of volume-temperature conditions for a given sample of gas at constant pressure. From Equation 11.4 we can write

$$\frac{V_1}{T_1} = k_2 = \frac{V_2}{T_2}$$

or

$$\frac{V_1}{T_1} = \frac{V_2}{T_2} \quad \text{(at constant pressure)} \qquad \text{Equation 11.5}$$

where $V_1$ is the volume of the gas at $T_1$ and $V_2$ is the volume of the gas at $T_2$.
Sample Problem 11.3 shows how to use Charles's law.

---

## Sample Problem 11.3

A sample of argon gas that originally occupied 14.6 L at 25.0°C was heated to 50.0°C at constant pressure. What is its new volume?

**Strategy** Use Equation 11.5 to solve for $V_2$. Remember that temperatures must be expressed in kelvin.

**Setup** $T_1 = 298.15$ K, $V_1 = 14.6$ L, and $T_2 = 323.15$ K.

**Solution**

$$V_2 = \frac{V_1 \times T_2}{T_1} = \frac{14.6 \text{ L} \times 323.15 \text{ K}}{298.15 \text{ K}} = 15.8 \text{ L}$$

> **Think About It** When temperature increases at constant pressure, the volume of a gas sample increases.

---

**Practice Problem A** A sample of gas originally occupies 29.1 L at 0.0°C. What is its new volume when it is heated to 15.0°C? (Assume constant pressure.)

**Practice Problem B** At what temperature (in °C) will a sample of gas occupy 82.3 L if it occupies 50.0 L at 75.0°C? (Assume constant pressure.)

## Avogadro's Law: The Amount-Volume Relationship

In 1811, the Italian scientist Amedeo Avogadro proposed that equal volumes of different gases contain the same number of particles (molecules or atoms) at the same temperature and pressure. This hypothesis gave rise to **Avogadro's law,** which states that the volume of a sample of gas is directly proportional to the number of moles in the sample at constant temperature and pressure:

$$V \propto n$$

or

Equation 11.6(a)    $V = k_3 n$    (at constant temperature and pressure)

Rearranging Equation 11.6(a) gives

Equation 11.6(b)    $\dfrac{V}{n} = k_3$

Equations 11.6(a) and (b) are expressions of Avogadro's law.

As with the other gas laws, we can compare two sets of conditions using Avogadro's law where $n$ and $V$ both change at constant pressure and temperature and write

$$\frac{V_1}{n_1} = k_3 = \frac{V_2}{n_2}$$

or

Equation 11.7    $\dfrac{V_1}{n_1} = \dfrac{V_2}{n_2}$

where $V_1$ is the volume of a sample of gas consisting of $n_1$ moles and $V_2$ is the volume of a sample consisting of $n_2$ moles—under conditions of constant temperature and pressure. Coupled with a balanced chemical equation, Avogadro's law enables us to predict the volumes of gaseous reactants and products. Consider the reaction of $H_2$ and $N_2$ to form $NH_3$:

$$3H_2(g) + N_2(g) \longrightarrow 2NH_3(g)$$

The balanced equation reveals the ratio of combination of reactants in terms of *moles* [◄◄ Section 3.4]. However, because the volume of a gas (at a given temperature and pressure) is directly *proportional* to the number of moles, the balanced equation also reveals the ratio of combination in terms of *volume*. Thus, if we were to combine three volumes (liters, milliliters, etc.) of hydrogen gas with one volume of nitrogen gas, assuming they react completely according to the balanced equation, we would expect two volumes of ammonia gas to be produced (Figure 11.11). The ratio of combination of $H_2$ and $N_2$ (and production of $NH_3$), whether expressed in moles or units of volume, is 3:1:2.

Sample Problem 11.4 shows how to apply Avogadro's law.

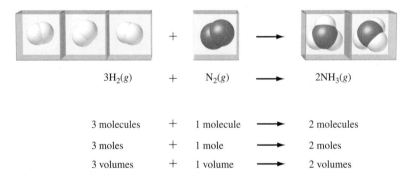

| $3H_2(g)$ | + | $N_2(g)$ | $\longrightarrow$ | $2NH_3(g)$ |
|---|---|---|---|---|
| 3 molecules | + | 1 molecule | $\longrightarrow$ | 2 molecules |
| 3 moles | + | 1 mole | $\longrightarrow$ | 2 moles |
| 3 volumes | + | 1 volume | $\longrightarrow$ | 2 volumes |

**Figure 11.11**    Illustration of Avogadro's law. The volume of a sample of gas is directly proportional to the number of moles.

### Sample Problem 11.4

If we combine 3.0 L of NO and 1.5 L of $O_2$, and they react according to the balanced equation $2NO(g) + O_2(g) \longrightarrow 2NO_2(g)$, what volume of $NO_2$ will be produced? (Assume that the reactants and product are all at the same temperature and pressure.)

**Strategy** Apply Avogadro's law to determine the volume of a gaseous product.

**Setup** Because volume is proportional to the number of moles, the balanced equation determines in what volume ratio the reactants combine and the ratio of product volume to reactant volume. The amounts of reactants given are stoichiometric amounts [◄◄ Section 3.6].

**Solution** According to the balanced equation, the volume of $NO_2$ formed will be equal to the volume of NO that reacts. Therefore, 3.0 L of $NO_2$ will form.

**Think About It** Remember that the coefficients in balanced chemical equations indicate ratios in molecules or moles. Under conditions of constant temperature and pressure, the volume of a gas is proportional to the number of moles. Therefore, the coefficients in balanced equations containing only gases also indicate ratios in liters, provided the reactions occur at constant temperature and pressure.

**Practice Problem A** What volume (in liters) of water vapor will be produced when 34 L of $H_2$ and 17 L of $O_2$ react according to the equation $2H_2(g) + O_2(g) \longrightarrow 2H_2O(g)$?

**Practice Problem B** What volumes (in liters) of carbon monoxide and oxygen gas must react according to the equation $2CO(g) + O_2(g) \longrightarrow 2CO_2(g)$ to form 3.16 L of carbon dioxide?

**Student Annotation:** It is important to recognize that coefficients indicate ratios in liters only in balanced equations in which all the reactants and products are gases. We cannot apply the same approach to reactions in which there are solid, liquid, or aqueous species.

## The Combined Gas Law:
## The Pressure-Temperature-Amount-Volume Relationship

Although the gas laws we have discussed so far are useful, each requires that two of the system's parameters be held constant.

| Problem Type | Relates | Requires Constant |
|---|---|---|
| Boyle's law | $P$ and $V$ | $n$ and $T$ |
| Charles's law | $T$ and $V$ | $n$ and $P$ |
| Avogadro's law | $n$ and $V$ | $P$ and $T$ |

Many of the problems we will encounter involve changes to $P$, $T$, *and* $V$, and, in some cases, also to $n$. To solve such problems, we need a gas law that relates all the variables. By combining Equations 11.3, 11.5, and 11.7, we obtain the ***combined gas law:***

$$\frac{P_1 V_1}{n_1 T_1} = \frac{P_2 V_2}{n_2 T_2} \qquad \text{Equation 11.8(a)}$$

The combined gas law can be used to solve problems where any or all of the variables change. Note that when a problem involves a fixed quantity of gas, Equation 11.8(a) reduces to the more common form of the combined gas law:

$$\frac{P_1 V_1}{T_1} = \frac{P_2 V_2}{T_2} \qquad \text{Equation 11.8(b)}$$

Sample Problem 11.5 illustrates the use of the combined gas law.

### Sample Problem 11.5

If a child releases a 6.25-L helium balloon in the parking lot of an amusement park where the temperature is 28.50°C and the air pressure is 757.2 mmHg, what will the volume of the balloon be when it has risen to an altitude where the temperature is −34.35°C and the air pressure is 366.4 mmHg?

**Strategy** In this case, because there is a fixed amount of gas, we use Equation 11.8(b). The only value we don't know is $V_2$. Temperatures must be expressed in kelvins. We can use any units of pressure, as long as we are consistent.

*(Continued)*

**Think About It** Note that the solution is essentially multiplying the original volume by the ratio of $P_1$ to $P_2$, and by the ratio of $T_2$ to $T_1$. The effect of decreasing external pressure is to increase the balloon volume. The effect of decreasing *temperature* is to *decrease* the volume. In this case, the effect of decreasing pressure predominates and the balloon volume increases significantly.

**Setup** $T_1 = 301.65$ K, $T_2 = 238.80$ K. Solving Equation 11.8(b) for $V_2$ gives

$$V_2 = \frac{P_1 T_2 V_1}{P_2 T_1}$$

**Solution**

$$V_2 = \frac{757.2 \text{ mmHg} \times 238.80 \text{ K} \times 6.25 \text{ L}}{366.4 \text{ mmHg} \times 301.65 \text{ K}} = 10.2 \text{ L}$$

**Practice Problem A** What would be the volume of the balloon in Sample Problem 11.5 if, instead of being released to rise in the atmosphere, it were submerged in a swimming pool to a depth where the pressure is 922.3 mmHg and the temperature is 26.35°C?

**Practice Problem B** The volume of a bubble that starts at the bottom of a lake at 4.55°C increases by a factor of 10 as it rises to the surface where the temperature is 18.45°C and the air pressure is 0.965 atm. Assuming that the density of the lake water is 1.00 g/cm³, determine the depth of the lake. (*Hint:* You will need to use Equation 11.1.)

## Checkpoint 11.2    The Gas Laws

**11.2.1** Given $P_1 = 1.50$ atm, $V_1 = 37.3$ mL, and $P_2 = 1.18$ atm, calculate $V_2$. Assume that $n$ and $T$ are constant.

a) 0.0211 mL

b) 0.0341 mL

c) 29.3 mL

d) 12.7 mL

e) 47.4 mL

**11.2.2** Given $T_1 = 21.5$°C, $V_1 = 50.0$ mL, and $T_2 = 316$°C, calculate $V_2$. Assume that $n$ and $P$ are constant.

a) 100 mL

b) 73.5 mL

c) 25.0 mL

d) 3.40 mL

e) 26.5 mL

**11.2.3** At what temperature will a gas sample occupy 100.0 L if it originally occupies 76.1 L at 89.5°C? Assume constant $P$.

a) 276°C

b) 118°C

c) 203°C

d) 68.1°C

e) 99.6°C

**11.2.4** What volume of $NH_3$ will be produced when 180 mL of $H_2$ reacts with 60.0 mL of $N_2$ according to the following equation:

$$3H_2(g) + N_2(g) \longrightarrow 2NH_3(g)$$

Assume constant $T$ and $P$ for reactants and products.

a) 120 mL

b) 60 mL

c) 180 mL

d) 240 mL

e) 220 mL

**11.2.5** Which diagram could represent the result of increasing the temperature and decreasing the external pressure on a fixed amount of gas in a balloon?

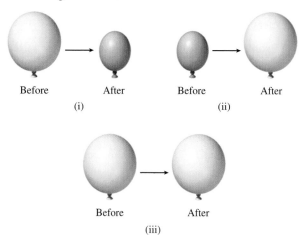

Before     After        Before     After
     (i)                      (ii)

Before     After
    (iii)

a) i only

b) ii only

c) i and ii

d) i, ii, and iii

e) iii only

**11.2.6** Which diagram in question 11.2.5 could represent the result of decreasing both the temperature and the pressure?

a) i only

b) ii only

c) i and ii

d) i, ii, and iii

e) iii only

# 11.3 | The Ideal Gas Equation

Recall that the state of a sample of gas is described completely using the four variables *T, P, V,* and *n*. Each of the gas laws introduced in Section 11.2 relates one variable of a sample of gas to another while the *other* two variables are held constant. In experiments with gases, however, there are usually changes in more than just two of the variables. Therefore, it is useful for us to combine the equations representing the gas laws into a single equation that will enable us to account for changes in any or all of the four variables.

## Deriving the Ideal Gas Equation from the Empirical Gas Laws

Summarizing the gas law equations from Section 11.2:

Multimedia

Gas Laws—ideal gas law (interactive).

$$\text{Boyle's law:} \quad V \propto \frac{1}{P}$$

$$\text{Charles's law:} \quad V \propto T$$

$$\text{Avogadro's law:} \quad V \propto n$$

We can combine these equations into the following general equation that describes the physical behavior of all gases:

$$V \propto \frac{nT}{P}$$

or

$$V = R\frac{nT}{P}$$

where *R* is the proportionality constant. This equation can be rearranged to give:

$$PV = nRT \qquad\qquad \text{Equation 11.9}$$

Equation 11.9 is the most commonly used form of the **ideal gas equation,** which describes the relationship among the four variables *P, V, n,* and *T*. An **ideal gas** is a hypothetical sample of gas whose pressure-volume-temperature behavior is predicted accurately by the ideal gas equation. Although the behavior of *real* gases generally differs slightly from that predicted by Equation 11.9, in most of the cases we will encounter, the differences are usually small enough for us to use the ideal gas equation to make reasonably good predictions about the behavior of gases.

The proportionality constant, *R*, in Equation 11.9 is called the **gas constant.** Its value and units depend on the units in which *P* and *V* are expressed. (The variables *n* and *T* are always expressed in mol and K, respectively.) Recall from Section 11.1 that pressure is commonly expressed in atmospheres, mmHg (torr), pascals, or bar. Volume is typically expressed in liters or milliliters, but can also be expressed in other units, such as m³. Table 11.4 lists several different expressions of the gas constant, *R*.

**Student Annotation:** We will discuss the conditions that result in deviation from ideal behavior in Section 11.7.

**Student Annotation:** When you use *R* in a calculation, use the version that facilitates proper cancellation of units.

| TABLE 11.4 | Various Equivalent Expressions of the Gas Constant, *R* |
|---|---|
| **Numerical Value** | **Unit** |
| 0.08206 | L · atm/K · mol |
| 62.36 | L · torr/K · mol |
| 0.08314 | L · bar/K · mol |
| 8.314 | m³ · Pa/K · mol |
| 8.314 | J/K · mol |
| 1.987 | cal/K · mol |

Note that the product of volume and pressure gives units of *energy* (i.e., joules and calories).

Keep in mind that all these expressions of $R$ are equal to one another, just as 1 yard is equal to 3 ft. They are simply expressed in different units.

One of the simplest uses of the ideal gas equation is the calculation of one of the variables when the other three are already known. For example, we can calculate the volume of 1 mole of an ideal gas at 0°C and 1 atm, conditions known as ***standard temperature and pressure (STP).*** In this case, $n$, $T$, and $P$ are given. $R$ is a constant, leaving $V$ as the only unknown. We can rearrange Equation 11.9 to solve for $V$,

**Student Annotation:** In thermochemistry we often used 25°C as the "standard" temperature—although temperature is *not* actually part of the definition of the *standard state* [◄◄ Section 5.6]. The standard temperature for gases is defined specifically as 0°C.

$$V = \frac{nRT}{P}$$

enter the information that is given, and calculate $V$. Remember that in calculations using the ideal gas equation, temperature must *always* be expressed in kelvins.

**Student Annotation:** In this problem, because they are specified rather than measured, 0°C, 1 mole, and 1 atm are *exact* numbers and do not affect the number of significant figures in the result [◄◄ Section 1.5].

$$V = \frac{(1 \text{ mol})(0.08206 \text{ L} \cdot \text{atm/K} \cdot \text{mol})(273.15 \text{ K})}{1 \text{ atm}} = 22.41 \text{ L}$$

Thus, the volume occupied by 1 mole of an ideal gas at STP is 22.41 L, a volume slightly less than 6 gal.

Sample Problem 11.6 shows how to calculate the molar volume of a gas at a temperature other than 0°C.

---

### Sample Problem 11.6

Calculate the volume of a mole of ideal gas at room temperature (25°C) and 1 atm.

**Strategy** Convert the temperature in °C to temperature in kelvins, and use the ideal gas equation to solve for the unknown volume.

**Think About It** With the pressure held constant, we should expect the volume to increase with increased temperature. Room temperature is higher than the standard temperature for gases (0°C), so the molar volume at room temperature (25°C) *should* be higher than the molar volume at 0°C—and it is.

**Setup** The data given are $n = 1$ mol, $T = 298.15$ K, and $P = 1.00$ atm. Because the pressure is expressed in atmospheres, we use $R = 0.08206$ L · atm/K · mol to solve for volume in liters.

**Solution**

$$V = \frac{(1 \text{ mol})(0.08206 \text{ L} \cdot \text{atm/K} \cdot \text{mol})(298.15 \text{ K})}{1 \text{ atm}} = 24.5 \text{ L}$$

**Practice Problem A** What is the volume of 5.12 moles of an ideal gas at 32°C and 1.00 atm?

**Practice Problem B** At what temperature (in °C) would 1 mole of ideal gas occupy 50.0 L ($P = 1.00$ atm)?

---

## Applications of the Ideal Gas Equation

Using some simple algebraic manipulation, we can solve for variables other than those that appear explicitly in the ideal gas equation. For example, if we know the molar mass of a gas (g/mol), we can determine its density at a given temperature and pressure. Recall from Section 11.1 that the density of a gas is generally expressed in units of g/L. We can rearrange the ideal gas equation to solve for mol/L:

$$\frac{n}{V} = \frac{P}{RT}$$

**Student Annotation:** $n \times \mathcal{M} = m$, where $m$ is mass in grams.

If we then multiply both sides by the molar mass, $\mathcal{M}$, we get

**Student Annotation:** Another way to arrive at Equation 11.10 is to substitute $m/\mathcal{M}$ for $n$ in the ideal gas equation and rearrange to solve for $m/V$ (density):

$$PV = \frac{m}{\mathcal{M}}RT \quad \text{and} \quad \frac{m}{V} = d = \frac{P\mathcal{M}}{RT}$$

$$\mathcal{M} \times \frac{n}{V} = \frac{P}{RT} \times \mathcal{M}$$

where $\mathcal{M} \times n/V$ gives g/L or density, $d$. Therefore,

Equation 11.10                    $$d = \frac{P\mathcal{M}}{RT}$$

**Figure 11.12**   (a) Evacuated flask. (b) Flask filled with gas. The mass of the gas is the difference between the two masses. The density of the gas is determined by dividing mass by volume.

(a)                          (b)

Conversely, if we know the density of a gas, we can determine its molar mass:

$$\mathcal{M} = \frac{dRT}{P}$$                          Equation 11.11

In a typical experiment, in which the molar mass of a gas is determined, a flask of known volume is evacuated and weighed [Figure 11.12(a)]. It is then filled (to a known pressure) with the gas of unknown molar mass and reweighed [Figure 11.12(b)]. The difference in mass is the mass of the gas sample. Dividing by the known volume of the flask gives the density of the gas, and the molar mass can then be determined using Equation 11.11.

   Similarly, the molar mass of a volatile liquid can be determined by placing a small volume of it in the bottom of a flask, the mass and volume of which are known. The flask is then immersed in a hot-water bath, causing the volatile liquid to completely evaporate and its vapor to fill the flask. Because the flask is open, some of the excess vapor escapes. When no more vapor escapes, the flask is capped and removed from the water bath. The flask is then weighed to determine the mass of the vapor. (At this point, some or all of the vapor has condensed but the mass remains the same.) The density of the vapor is determined by dividing the mass of the vapor by the volume of the flask. Equation 11.11 is then used to calculate the molar mass of the volatile liquid.

   Sample Problems 11.7 and 11.8 illustrate the use of Equations 11.10 and 11.11.

**Student Annotation:** Because the flask is open to the atmosphere while the volatile liquid vaporizes, we can use atmospheric pressure as $P$. Also, because the flask is capped at the water bath temperature, we can use the water bath temperature as $T$.

---

### Sample Problem 11.7

Carbon dioxide is effective in fire extinguishers partly because its density is greater than that of air, so $CO_2$ can smother the flames by depriving them of oxygen. (Air has a density of approximately 1.2 g/L at room temperature and 1 atm.) Calculate the density of $CO_2$ at room temperature (25°C) and 1.0 atm.

**Strategy** Use Equation 11.10 to solve for density. Because the pressure is expressed in atm, we should use $R = 0.08206$ L · atm/K · mol. Remember to express temperature in kelvins.

**Setup** The molar mass of $CO_2$ is 44.01 g/mol.

**Solution**

$$d = \frac{P\mathcal{M}}{RT} = \frac{(1 \text{ atm})\left(\dfrac{44.01 \text{ g}}{\text{mol}}\right)}{\left(\dfrac{0.08206 \text{ L} \cdot \text{atm}}{\text{K} \cdot \text{mol}}\right)(298.15 \text{ K})} = 1.8 \text{ g/L}$$

**Think About It** The calculated density of $CO_2$ is greater than that of air under the same conditions (as expected). Although it may seem tedious, it is a good idea to write units for each and every entry in a problem such as this. Unit cancellation is very useful for detecting errors in your reasoning or your solution setup.

**Practice Problem A** Calculate the density of air at 0°C and 1 atm. (Assume that air is 80 percent $N_2$ and 20 percent $O_2$.)

**Practice Problem B** What pressure would be required for helium at 25°C to have the same density as carbon dioxide at 25°C and 1 atm?

## Sample Problem 11.8

A company has just patented a new synthetic alcohol for alcoholic beverages. The new product is said to have all the pleasant properties associated with ethanol but none of the undesirable effects such as hangover, impairment of motor skills, and risk of addiction. The chemical formula is proprietary. You analyze a sample of the new product by placing a small volume of it in a round-bottomed flask with a volume of 511.0 mL and an evacuated mass of 131.918 g. You submerge the flask in a water bath at 100.0°C and allow the volatile liquid to vaporize. You then cap the flask and remove it from the water bath. You weigh it and determine the mass of the vapor in the flask to be 0.768 g. What is the molar mass of the volatile liquid, and what does it mean with regard to the new product? (Assume the pressure in the laboratory is 1 atm.)

**Strategy** Use the measured mass of the vapor and the given volume of the flask to determine the density of the vapor at 1 atm and 100.0°C, and then use Equation 11.8 to determine molar mass.

**Setup** $P = 1$ atm, $V = 0.5110$ L, $R = 0.8206$ L · atm/K · mol, and $T = 373.15$ K.

**Solution**

$$d = \frac{0.768 \text{ g}}{0.5110 \text{ L}} = 1.5029 \text{ g/L}$$

$$\mathcal{M} = \frac{\left(\dfrac{1.5029 \text{ g}}{\cancel{L}}\right)\left(\dfrac{0.08206 \cancel{L} \cdot \cancel{atm}}{\cancel{K} \cdot \text{mol}}\right)(373.15 \cancel{K})}{1 \cancel{atm}} = 46.02 \text{ g/mol}$$

The result is a molar mass suspiciously close to that of ethanol!

**Think About It** Because more than one compound can have a particular molar mass, this method is not definitive for identification. However, in this circumstance, further testing of the proprietary formula certainly would be warranted.

**Practice Problem A** Determine the molar mass of a gas with a density of 1.905 g/L at 80.0°C and 1.00 atm.

**Practice Problem B** A sample of the volatile liquid propyl acetate ($C_5H_{10}O_2$) is analyzed using the procedure and equipment described in Sample Problem 11.8. What will the mass of the 511.0-mL flask be after evaporation of the propyl acetate?

## Checkpoint 11.3    The Ideal Gas Equation

**11.3.1** Calculate the volume occupied by 8.75 moles of an ideal gas at STP.

a) 196 L

b) 268 L

c) 0.718 L

d) 18.0 L

e) 2.56 L

**11.3.2** Calculate the pressure exerted by 10.2 moles of an ideal gas in a 7.5-L vessel at 150°C.

a) 17 atm

b) 31 atm

c) 0.72 atm

d) 1.3 atm

e) 47 atm

**11.3.3** Determine the density of a gas with $\mathcal{M} = 146.07$ g/mol at 1.00 atm and 100.0°C.

a) $6.85 \times 10^{-3}$ g/L

b) 4.77 g/L

c) 146 g/L

d) 30.6 g/L

e) 17.8 g/L

**11.3.4** Determine the molar mass of a gas with $d = 1.963$ g/L at 1.00 atm and 100.0°C.

a) 0.0166 g/mol

b) 60.1 g/mol

c) 16.1 g/mol

d) $6.09 \times 10^3$ g/mol

e) $1.63 \times 10^3$ g/mol

# 11.4  Reactions with Gaseous Reactants and Products

In Chapter 3 we used balanced chemical equations to calculate amounts of reactants and/or products in chemical reactions—expressing those amounts in mass (usually grams). However, in the case of reactants and products that are gases, it is more practical to measure and express amounts in volume (liters or milliliters). This makes the ideal gas equation useful in the stoichiometric analysis of chemical reactions that involve gases.

## Calculating the Required Volume of a Gaseous Reactant

According to Avogadro's law, the volume of a gas at a given temperature and pressure is proportional to the number of moles. Moreover, balanced chemical equations give the ratio of combination of gaseous reactants in both moles and volume (Figure 11.11). Therefore, if we know the volume of one reactant in a gaseous reaction, we can determine the required amount of another reactant (at the same temperature and pressure). For example, consider the reaction of carbon monoxide and oxygen to yield carbon dioxide:

$$2CO(g) + O_2(g) \longrightarrow 2CO_2(g)$$

The ratio of combination of CO and $O_2$ is 2:1, whether we are talking about moles or units of volume. Thus, if we want to determine the stoichiometric amount [◀◀ Section 3.6] of $O_2$ required to combine with a particular volume of CO, we simply use the conversion factor provided by the balanced equation, which can be expressed as any of the following:

$$\frac{1 \text{ mol } O_2}{2 \text{ mol CO}} \quad \text{or} \quad \frac{1 \text{ L } O_2}{2 \text{ L CO}} \quad \text{or} \quad \frac{1 \text{ mL } O_2}{2 \text{ mL CO}}$$

Let's say we want to determine what volume of $O_2$ is required to react completely with 65.8 mL of CO at STP. We could use the ideal gas equation to convert the volume of CO to moles, use the stoichiometric conversion factor to convert to moles $O_2$, and then use the ideal gas equation again to convert moles $O_2$ to volume. But this method involves several unnecessary steps. We get the same result simply by using the conversion factor expressed in milliliters:

$$65.8 \text{ mL CO} \times = \frac{1 \text{ mL } O_2}{2 \text{ mL CO}} = 32.9 \text{ mL } O_2$$

In cases where only one of the reactants is a gas, we *do* need to use the ideal gas equation in our analysis. Recall, for example, the reaction of sodium metal and chlorine gas used to illustrate the Born-Haber cycle [◀◀ Section 8.2]:

$$2Na(s) + Cl_2(g) \longrightarrow 2NaCl(s)$$

Given moles (or more commonly the *mass*) of Na, and information regarding temperature and pressure, we can determine the volume of $Cl_2$ required to react completely:

$$\times \frac{1 \text{ mol Na}}{22.99 \text{ g Na}} = \qquad \times \frac{1 \text{ mol } Cl_2}{2 \text{ mol Na}} = \qquad \times \frac{RT}{P} =$$

| grams Na | → | moles Na | → | moles $Cl_2$ | → | liters $Cl_2$ |

Sample Problem 11.9 shows how to use the ideal gas equation in a stoichiometric analysis.

### Sample Problem 11.9

Sodium peroxide ($Na_2O_2$) is used to remove carbon dioxide from (and add oxygen to) the air supply in spacecrafts. It works by reacting with $CO_2$ in the air to produce sodium carbonate ($Na_2CO_3$) and $O_2$.

$$2Na_2O_2(s) + 2CO_2(g) \longrightarrow 2Na_2CO_3(s) + O_2(g)$$

What volume (in liters) of $CO_2$ (at STP) will react with a kilogram of $Na_2O_2$?

**Strategy** Convert the given mass of $Na_2O_2$ to moles, use the balanced equation to determine the stoichiometric amount of $CO_2$, and then use the ideal gas equation to convert moles of $CO_2$ to liters.

**Setup** The molar mass of $Na_2O_2$ is 77.98 g/mol (1 kg = 1000 g). (Treat the specified mass of $Na_2O_2$ as an exact number.)

*(Continued)*

**Think About It** The answer may seems like an enormous volume of $CO_2$. If you check the cancellation of units carefully in ideal gas equation problems, however, with practice you will develop a sense of whether such a calculated volume is reasonable.

**Solution**

$$1000 \text{ g Na}_2\text{O}_2 \times \frac{1 \text{ mol Na}_2\text{O}_2}{77.98 \text{ g Na}_2\text{O}_2} = 12.82 \text{ mol Na}_2\text{O}_2$$

$$12.82 \text{ mol Na}_2\text{O}_2 \times \frac{2 \text{ mol CO}_2}{2 \text{ mol Na}_2\text{O}_2} = 12.82 \text{ mol CO}_2$$

$$V_{CO_2} = \frac{(12.82 \text{ mol CO}_2)(0.08206 \text{ L} \cdot \text{atm/K} \cdot \text{mol})(273.15 \text{ K})}{1 \text{ atm}} = 287.4 \text{ L CO}_2$$

**Practice Problem A**  What volume (in liters) of $CO_2$ can be consumed at STP by 525 g $Na_2O_2$?

**Practice Problem B**  What mass (in grams) of $Na_2O_2$ is necessary to consume 1.00 L $CO_2$ at STP?

## Determining the Amount of Reactant Consumed Using Change in Pressure

Although none of the empirical gas laws focuses on the relationship between $n$ and $P$ explicitly, we can rearrange the ideal gas equation to show that $n$ is directly proportional to $P$ at constant $V$ and $T$:

Equation 11.12(a)     $n = P \times \left(\dfrac{V}{RT}\right)$   (at constant $V$ and $T$)

Therefore, we can use the change in pressure in a reaction vessel to determine how many moles of a gaseous reactant are consumed in a chemical reaction:

Equation 11.12(b)     $\Delta n = \Delta P \times \left(\dfrac{V}{RT}\right)$   (at constant $V$ and $T$)

where $\Delta n$ is the number of moles of gas consumed and $\Delta P$ is the change in pressure in the reaction vessel. Sample Problem 11.10 shows how to use Equation 11.12(b).

**Student Annotation:** This refers to a reaction in which there is only *one* gaseous reactant and in which none of the products is a gas, such as the reaction described in Sample Problem 11.10. In reactions involving multiple gaseous species, $\Delta n$ refers to the *net* change in number of moles of gas—and the analysis gets somewhat more complicated.

### Sample Problem 11.10

Another air-purification method for enclosed spaces involves the use of "scrubbers" containing aqueous lithium hydroxide, which reacts with carbon dioxide to produce lithium carbonate and water:

$$2\text{LiOH}(aq) + \text{CO}_2(g) \longrightarrow \text{Li}_2\text{CO}_3(s) + \text{H}_2\text{O}(l)$$

Consider the air supply in a submarine with a total volume of $2.5 \times 10^5$ L. The pressure is 0.9970 atm, and the temperature is 25°C. If the pressure in the submarine drops to 0.9891 atm as the result of carbon dioxide being consumed by an aqueous lithium hydroxide scrubber, how many moles of $CO_2$ are consumed?

**Strategy**  Use Equation 11.12(b) to determine $\Delta n$, the number of moles of $CO_2$ consumed.

**Setup**  $\Delta P = 0.9970$ atm $- 0.9891$ atm $= 7.9 \times 10^{-3}$ atm. According to the problem statement, $V = 2.5 \times 10^5$ L and $T = 298.15$ K. For problems in which $P$ is expressed in atmospheres and $V$ in liters, use $R = 0.08206$ L $\cdot$ atm/K $\cdot$ mol.

**Think About It** Careful cancellation of units is *essential*. Note that this amount of $CO_2$ corresponds to 162 moles or 3.9 kg of LiOH. (It's a good idea to verify this yourself.)

**Solution**

$$\Delta n_{CO_2} = 7.9 \times 10^{-3} \text{ atm} \times \frac{2.5 \times 10^5 \text{ L}}{(0.08206 \text{ L} \cdot \text{atm/K} \cdot \text{mol}) \times (298.15 \text{ K})} = 81 \text{ moles CO}_2 \text{ consumed}$$

**Practice Problem A**  Using all the same conditions as those described in Sample Problem 11.10, calculate the number of moles of $CO_2$ consumed if the pressure drops by 0.010 atm.

**Practice Problem B**  By how much would the pressure in the submarine drop if 2.55 kg of LiOH were completely consumed by reaction with $CO_2$? (Assume the same starting $P$, $V$, and $T$ as in Sample Problem 11.10.)

## Predicting the Volume of a Gaseous Product

Using a combination of stoichiometry and the ideal gas equation, we can calculate the volume of gas that we expect to be produced in a chemical reaction. We first use stoichiometry to determine the number of moles produced, and then apply the ideal gas equation to determine what volume will be occupied by that number of moles under the specified conditions.

Sample Problem 11.11 shows how to predict the volume of a gaseous product.

### Sample Problem 11.11

The air bags in cars are inflated when a collision triggers the explosive, highly exothermic decomposition of sodium azide ($NaN_3$):

$$2NaN_3(s) \longrightarrow 2Na(s) + 3N_2(g)$$

A typical driver-side air bag contains about 50 g of $NaN_3$. Determine the volume of $N_2$ gas that would be generated by the decomposition of 50.0 g of sodium azide at 85.0°C and 1.00 atm.

**Strategy** Convert the given mass of $NaN_3$ to moles, use the ratio of the coefficients from the balanced chemical equation to determine the corresponding number of moles of $N_2$ produced, and then use the ideal gas equation to determine the volume of that number of moles at the specified temperature and pressure.

**Setup** The molar mass of $NaN_3$ is 65.02 g/mol.

**Solution**

$$\text{mol } NaN_3 = \frac{50.0 \text{ g } NaN_3}{65.02 \text{ g/mol}} = 0.769 \text{ mol } NaN_3$$

$$0.769 \text{ mol } NaN_3 \times \left(\frac{3 \text{ mol } N_2}{2 \text{ mol } NaN_3}\right) = 1.15 \text{ mol } N_2$$

$$V_{N_2} = \frac{(1.15 \text{ mol } N_2)(0.08206 \text{ L} \cdot \text{atm/K} \cdot \text{mol})(358.15 \text{ K})}{1 \text{ atm}} = 33.9 \text{ L } N_2$$

**Think About It** The calculated volume represents the space between the driver and the steering wheel and dashboard that must be filled by the air bag to prevent injury. Air bags also contain an oxidant that consumes the sodium metal produced in the reaction.

**Practice Problem A** The chemical equation for the metabolic breakdown of glucose ($C_6H_{12}O_6$) is the same as that for the combustion of glucose [◄◄ Section 3.3—Bringing Chemistry to Life box]:

$$C_6H_{12}O_6(aq) + 6O_2(g) \longrightarrow 6CO_2(g) + 6H_2O(l)$$

Calculate the volume of $CO_2$ produced at normal human body temperature (37°C) and 1.00 atm when 10.0 g of glucose is consumed in the reaction.

**Practice Problem B** The passenger-side air bag in a typical car must fill a space approximately four times as large as the driver-side air bag to be effective. Calculate the mass of sodium azide required to fill a 125-L air bag at 85.0°C and 1.00 atm.

### Checkpoint 11.4 Reactions with Gaseous Reactants and Products

11.4.1 Determine the volume of $Cl_2$ gas at STP that will react with 1.00 mole of Na solid to produce NaCl according to the equation,

$$2Na(s) + Cl_2(g) \longrightarrow 2NaCl(s)$$

a) 22.4 L
b) 44.8 L
c) 15.3 L
d) 11.2 L
e) 30.6 L

11.4.2 Determine the mass of $NaN_3$ required for an air bag to produce 100.0 L of $N_2$ gas at 85.0°C and 1.00 atm according to the equation,

$$2NaN_3(s) \longrightarrow 2Na(s) + 3N_2(g)$$

a) 332 g
b) 148 g
c) 221 g
d) 664 g
e) 442 g

## 11.5 | Gas Mixtures

So far our discussion of the physical properties of gases has focused on the behavior of *pure* gaseous substances, even though the gas laws were all developed based on observations of samples of air, which is a *mixture* of gases. In this section, we will consider gas mixtures and their physical behavior. We will restrict our discussion in this section to gases that behave ideally and that do not react with one another.

### Dalton's Law of Partial Pressures

When two or more gaseous substances are placed in a container, each gas behaves as though it occupies the container alone. For example, if we place 1.00 mole of $N_2$ gas in a 5.00-L container at 0°C, it exerts a pressure of

$$P = \frac{(1.00 \text{ mol})(0.08206 \text{ L} \cdot \text{atm/K} \cdot \text{mol})(273.15 \text{ K})}{5.00 \text{ L}} = 4.48 \text{ atm}$$

If we then add a mole of another gas, such as $O_2$, the pressure exerted by $N_2$ does not change. It remains at 4.48 atm. The $O_2$ gas exerts its own pressure, also 4.48 atm. Neither gas is affected by the presence of the other. In a mixture of gases, the pressure exerted by each gas is known as the *partial pressure* ($P_i$) of the gas. We use subscripts to denote partial pressures:

$$P_{N_2} = \frac{(1.00 \text{ mol})(0.08206 \text{ L} \cdot \text{atm/K} \cdot \text{mol})(273.15 \text{ K})}{5.00 \text{ L}} = 4.48 \text{ atm}$$

$$P_{O_2} = \frac{(1.00 \text{ mol})(0.08206 \text{ L} \cdot \text{atm/K} \cdot \text{mol})(273.15 \text{ K})}{5.00 \text{ L}} = 4.48 \text{ atm}$$

and we can solve the ideal gas equation for each component of any gas mixture:

$$P_i = \frac{n_i RT}{V}$$

*Dalton's law of partial pressures* states that the total pressure exerted by a gas mixture is the sum of the partial pressures exerted by each component of the mixture:

$$P_{total} = \Sigma P_i$$

Thus, the total pressure exerted by a mixture of 1.00 mol $N_2$ and 1.00 mol $O_2$ in a 5.00-L vessel at 0°C is

$$P_{total} = P_{N_2} + P_{O_2} = 4.48 \text{ atm} + 4.48 \text{ atm} = 8.96 \text{ atm}$$

Figure 11.13 illustrates Dalton's law of partial pressures.

**Figure 11.13** Schematic illustration of Dalton's law of partial pressures. Total pressure is equal to the sum of partial pressures.

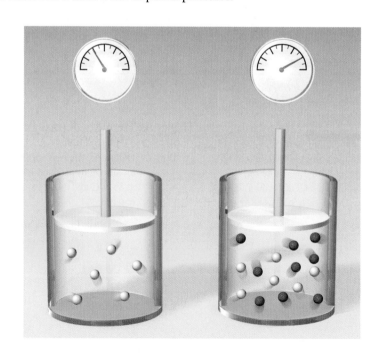

Sample Problem 11.12 shows how to apply Dalton's law of partial pressures.

### Sample Problem 11.12

A 1.00-L vessel contains 0.215 mole of $N_2$ gas and 0.0118 mole of $H_2$ gas at 25.5°C. Determine the partial pressure of each component and the total pressure in the vessel.

**Strategy** Use the ideal gas equation to find the partial pressure of each component of the mixture, and sum the two partial pressures to find the total pressure.

**Setup** $T = 298.65$ K.

**Solution**

$$P_{N_2} = \frac{(0.215 \text{ mol})\left(\frac{0.08206 \text{ L} \cdot \text{atm}}{\text{K} \cdot \text{mol}}\right)(298.65 \text{ K})}{1.00 \text{ L}} = 5.27 \text{ atm}$$

$$P_{H_2} = \frac{(0.0118 \text{ mol})\left(\frac{0.08206 \text{ L} \cdot \text{atm}}{\text{K} \cdot \text{mol}}\right)(298.65 \text{ K})}{1.00 \text{ L}} = 0.289 \text{ atm}$$

$$P_{total} = P_{N_2} + P_{H_2} = 5.27 \text{ atm} + 0.289 \text{ atm} = 5.56 \text{ atm}$$

**Practice Problem A** Determine the partial pressures and the total pressure in a 2.50-L vessel containing the following mixture of gases at 15.8°C: 0.0194 mol He, 0.0411 mol $H_2$, and 0.169 mol Ne.

**Practice Problem B** Determine the number of moles of each gas present in a mixture of $CH_4$ and $C_2H_6$ in a 2.00-L vessel at 25.0°C and 1.50 atm, given that the partial pressure of $CH_4$ is 0.39 atm.

**Think About It** The total pressure in the vessel can also be determined by summing the number of moles of mixture components (0.215 + 0.0118 = 0.227 mol) and solving the ideal gas equation for $P_{total}$:

$$P_{total} = \frac{(0.227 \text{ mol})\left(\frac{0.08206 \text{ L} \cdot \text{atm}}{\text{K} \cdot \text{mol}}\right)(298.65 \text{ K})}{1.00 \text{ L}} = 5.56 \text{ atm}$$

## Mole Fractions

The relative amounts of the components of a gas mixture can be specified using *mole fractions*. The **mole fraction ($\chi_i$)** of a component of a mixture is the number of moles of the component divided by the total number of moles in the mixture:

$$\chi_i = \frac{n_i}{n_{total}}$$

Equation 11.13

**Student Annotation:** Mole fractions do not refer only to gas mixtures. They can be used to specify the concentrations of components of mixtures in any phase. Mole fractions are used extensively in Chapter 12.

There are three things to remember about mole fractions:

1. The mole fraction of a mixture component is always less than 1.
2. The sum of mole fractions for all components of a mixture is always 1.
3. Mole fraction is dimensionless.

In addition, $n$ and $P$ are proportional [Equation 11.12(a)] at a specified $T$ and $V$, so we can determine mole fraction by dividing the partial pressure of a component by the total pressure:

$$\chi_i = \frac{P_i}{P_{total}}$$

Equation 11.14

Rearranging Equations 11.13 and 11.14 gives

$$\chi_i \times n_{total} = n_i$$

Equation 11.15

and

$$\chi_i \times P_{total} = P_i$$

Equation 11.16

Sample Problem 11.13 lets you practice calculations involving mole fractions, partial pressures, and total pressure.

## Sample Problem 11.13

In 1999, the FDA approved the use of nitric oxide (NO) to treat and prevent lung disease, which occurs commonly in premature infants. The nitric oxide used in this therapy is supplied to hospitals in the form of a $N_2$/NO mixture. Calculate the mole fraction of NO in a 10.00-L gas cylinder at room temperature (25°C) that contains 6.022 mol $N_2$ and in which the total pressure is 14.75 atm.

**Strategy** Use the ideal gas equation to calculate the total number of moles in the cylinder. Subtract moles of $N_2$ from the total to determine moles of NO. Divide moles NO by total moles to get mole fraction (Equation 11.14).

**Setup** The temperature is 298.15 K.

**Solution**

**Think About It** To check your work, determine $\chi_{N_2}$ by subtracting $\chi_{NO}$ from 1. Using each mole fraction and the total pressure, calculate the partial pressure of each component using Equation 11.16 and verify that they sum to the total pressure.

$$\text{total moles} = \frac{PV}{RT} = \frac{(14.75 \text{ atm})(10.00 \text{ L})}{\left(\dfrac{0.08206 \text{ L} \cdot \text{atm}}{\text{K} \cdot \text{mol}}\right)(298.15 \text{ K})} = 6.029 \text{ mol}$$

$$\text{mol NO} = \text{total moles} - \text{mol } N_2 = 6.029 - 6.022 = 0.007 \text{ mol NO}$$

$$\chi_{NO} = \frac{n_{NO}}{n_{total}} = \frac{0.007 \text{ mol NO}}{6.029 \text{ mol}} = 0.001$$

**Practice Problem A** Determine the mole fractions and partial pressures of $CO_2$, $CH_4$, and He in a sample of gas that contains 0.250 mole of $CO_2$, 1.29 moles of $CH_4$, and 3.51 moles of He, and in which the total pressure is 5.78 atm.

**Practice Problem B** Determine the partial pressure and number of moles of each gas in a 15.75-L vessel at 30.0°C containing a mixture of xenon and neon gases only. The total pressure in the vessel is 6.50 atm, and the mole fraction of xenon is 0.761.

**Multimedia**

Gas Laws—collecting a gas over water.

## Using Partial Pressures to Solve Problems

The volume of gas produced by a chemical reaction can be measured using an apparatus like the one shown in Figure 11.14. Dalton's law of partial pressures is useful in the analysis of these kinds of experimental results. For example, the decomposition of potassium chlorate ($KClO_3$), the reaction used to generate emergency oxygen supplies on airplanes, produces potassium chloride and oxygen:

$$2KClO_3(s) \longrightarrow 2KCl(s) + 3O_2(g)$$

The oxygen gas is collected over water, as shown in Figure 11.14(a). The volume of water displaced by the gas is equal to the volume of gas produced. (Prior to reading the volume of the gas, the level of the graduated cylinder must be adjusted such that the water levels inside and outside the cylinder are the *same*. This ensures that the *pressure* inside the graduated cylinder is the same as atmospheric pressure [Figure 11.14(b)].) However, because the measured volume contains both the oxygen produced by the reaction *and* water vapor, the pressure exerted inside the graduated cylinder is the sum of the two partial pressures:

$$P_{total} = P_{O_2} + P_{H_2O}$$

By subtracting the partial pressure of water from the total pressure, which is equal to atmospheric pressure, we can determine the partial pressure of oxygen—and thereby determine how many moles are produced by the reaction. We get the partial pressure of water, which depends on temperature, from a table of values. Table 11.5 lists the partial pressure (also known as the *vapor pressure*) of water at various temperatures.

Sample Problem 11.14 shows how to use Dalton's law of partial pressures to determine the amount of gas produced in a chemical reaction and collected over water.

**Figure 11.14** (a) Apparatus for measuring the amount of gas produced in a chemical reaction. (b) When the water levels inside and outside the collection vessel are the same, the pressure inside the vessel is equal to atmospheric pressure.

| TABLE 11.5 | | Vapor Pressure of Water ($P_{H_2O}$) as a Function of Temperature | | | |
|---|---|---|---|---|---|
| T(°C) | P (torr) | T(°C) | P (torr) | T(°C) | P (torr) |
| 0 | 4.6 | 35 | 42.2 | 70 | 233.7 |
| 5 | 6.5 | 40 | 55.3 | 75 | 289.1 |
| 10 | 9.2 | 45 | 71.9 | 80 | 355.1 |
| 15 | 12.8 | 50 | 92.5 | 85 | 433.6 |
| 20 | 17.5 | 55 | 118.0 | 90 | 525.8 |
| 25 | 23.8 | 60 | 149.4 | 95 | 633.9 |
| 30 | 31.8 | 65 | 187.5 | 100 | 760.0 |

## Sample Problem 11.14

Calcium metal reacts with water to produce hydrogen gas [◄◄ Section 7.7]:

$$Ca(s) + 2H_2O(l) \longrightarrow Ca(OH)_2(aq) + H_2(g)$$

Determine the mass of $H_2$ produced at 25°C and 0.967 atm when 525 mL of the gas is collected over water as shown in Figure 11.14.

**Strategy** Use Dalton's law of partial pressures to determine the partial pressure of $H_2$, use the ideal gas equation to determine moles of $H_2$, and then use the molar mass of $H_2$ to convert to mass. (Pay careful attention to units. Atmospheric pressure is given in atmospheres, whereas the vapor pressure of water is tabulated in torr.)

**Setup** $V = 0.525$ L and $T = 298.15$ K. The partial pressure of water at 25°C is 23.8 torr (Table 11.5) or 23.8 torr (1 atm/760 torr) = 0.0313 atm. The molar mass of $H_2$ is 2.016 g/mol.

**Solution**

$$P_{H_2} = P_{total} - P_{H_2O} = 0.967 \text{ atm} - 0.0313 \text{ atm} = 0.936 \text{ atm}$$

$$\text{moles of } H_2 = \frac{(0.9357 \text{ atm})(0.525 \text{ L})}{\left(\frac{0.08206 \text{ L} \cdot \text{atm}}{\text{K} \cdot \text{mol}}\right)(298.15 \text{ K})} = 2.01 \times 10^{-2} \text{ mol}$$

$$\text{mass of } H_2 = (2.008 \times 10^{-2} \text{ mol})(2.016 \text{ g/mol}) = 0.0405 \text{ g } H_2$$

**Think About It** Check unit cancellation carefully, and remember that the densities of gases are relatively low. The mass of approximately half a liter of hydrogen at or near room temperature and 1 atm should be a very small number.

*(Continued)*

Media Player/MPEG Animation: Figure 11.15, Molar Volume of a Gas, pp. 468–469.

**Practice Problem A** Calculate the mass of $O_2$ produced by the decomposition of $KClO_3$ when 821 mL of $O_2$ is collected over water at 30.0°C and 1.015 atm.

**Practice Problem B** Determine the volume of gas collected over water when 0.501 g $O_2$ is produced by the decomposition of $KClO_3$ at 35.0°C and 1.08 atm.

Collection over water can also be used to determine the molar volume of a gas generated in a chemical reaction as shown in Figure 11.15.

## Bringing Chemistry to Life

### Hyperbaric Oxygen Therapy

In 1918, during the Spanish flu epidemic that claimed tens of millions of lives worldwide, physician Orville Cunningham noted that people living at lower elevations appeared to have a greater chance of surviving the flu than those living at higher elevations. Believing this to be the result of increased air pressure, he developed a hyperbaric chamber to treat flu victims. One of Cunningham's earliest and most notable successes was the recovery of a flu-stricken colleague who had been near death. Cunningham subsequently built a hyperbaric chamber large enough to accommodate dozens of patients and treated numerous flu victims, most with success.

In the decades following the Spanish flu epidemic, hyperbaric therapy fell out of favor with the medical community and was largely discontinued. Interest in it was revived when the U.S. military ramped up its underwater activities in the 1940s and hyperbaric chambers were constructed to treat military divers suffering from decompression sickness (DCS), also known as "the bends." Significant advancement in hyperbaric methods began in the 1970s when the Undersea Medical Society (renamed the Undersea and Hyperbaric Medical Society in 1976) became involved in the clinical use of hyperbaric chambers. Today, hyperbaric oxygen therapy (HBOT) is used to treat a wide variety of conditions, including carbon monoxide poisoning, anemia caused by critical blood loss, severe burns, and life-threatening bacterial infections. Once considered an "alternative" therapy and viewed with skepticism, HBOT is now covered by most insurance plans.

Sample Problem 11.15 illustrates the importance of mole fractions and partial pressures in hyperbaric oxygen therapy.

**Student Annotation:** One unfortunate group of patients was undergoing treatment when the power to the chamber was shut off accidentally. All the patients died. At the time, their deaths were attributed to influenza, but they almost certainly died as the result of the unintended rapid decompression.

**Catalina hyperbaric chamber**

### Sample Problem 11.15

The Catalina hyperbaric chamber at the University of Southern California's Wrigley Marine Science Center treats mostly victims of diving accidents. In one treatment protocol, the chamber is pressurized to 6.0 atm with compressed air and the patient breathes a mixture of gases that is 47 percent $O_2$ by volume. In another protocol, the chamber is pressurized with compressed air to 2.8 atm and the patient breathes pure $O_2$. Determine the partial pressure of $O_2$ in each treatment protocol and compare the results.

**Strategy** Because the gas the patient breathes is inside the hyperbaric chamber, its total pressure is the same as the chamber pressure. To obtain the partial pressure of $O_2$, multiply the mole fraction of $O_2$ in the breathing gas in each protocol by the total pressure (Equation 11.16).

**Setup** In the first protocol, the breathing gas is 47 percent $O_2$, so the mole fraction of $O_2$ is 0.47. In the second protocol, where pure $O_2$ is used, the mole fraction of $O_2$ is 1. Use Equation 11.16 to calculate the partial pressure of $O_2$ in each protocol.

$$\chi_i \times P_{total} = P_i$$

**Solution** In the first protocol, the pressure of $O_2$ is

$$0.47 \times 6.0 \text{ atm} = 2.8 \text{ atm}$$

In the second protocol, the pressure of $O_2$ is

$$1 \times 2.8 \text{ atm} = 2.8 \text{ atm}$$

Both protocols produce the same pressure of $O_2$.

**Think About It** Monoplace hyperbaric chambers, which are large enough to accommodate only one person, commonly are pressurized to 2.8 atm with pure $O_2$.

**Practice Problem A** What mole fraction of $O_2$ is necessary for the partial pressure of $O_2$ to be 2.8 atm when the total pressure is 4.6 atm?

**Practice Problem B** What chamber pressure would be required for a patient to receive the therapeutic partial pressure of $O_2$ (2.8 atm) without breathing a special mixture of gases through a mask? Assume that the air used to pressurize the chamber is 21 percent $O_2$ by volume.

---

## Checkpoint 11.5 Gas Mixtures

**11.5.1** What is the partial pressure of He in a 5.00-L vessel at 25°C that contains 0.0410 mole of He, 0.121 mole of Ne, and 0.0922 mole of Ar?

a) 1.24 atm

b) 0.248 atm

c) 0.117 atm

d) 2.87 atm

e) 0.201 atm

**11.5.2** What is the mole fraction of $CO_2$ in a mixture of 0.756 mole of $N_2$, 0.189 mole of $O_2$, and 0.0132 mole of $CO_2$?

a) 0.789

b) 0.0138

c) 0.0140

d) 1.003

e) 0.798

**11.5.3** What is the partial pressure of oxygen in a gas mixture that contains 4.10 moles of oxygen, 2.38 moles of nitrogen, and 0.917 mole of carbon dioxide and that has a total pressure of 2.89 atm?

a) 1.60 atm

b) 3.59 atm

c) 0.391 atm

d) 0.705 atm

e) 0.624 atm

**11.5.4** What mass of acetylene ($C_2H_2$) is produced by the reaction of calcium carbide ($CaC_2$) and water,

$$CaC_2(s) + 2H_2O(l) \longrightarrow C_2H_2(g) + Ca(OH)_2(aq)$$

if 425 mL of the gas is collected over water at 30°C and a pressure of 0.996 atm?

a) 0.016 g

b) 16.3 g

c) 0.019 g

d) 0.424 g

e) 0.443 g

**11.5.5** In the diagram, each color represents a different gas molecule. Calculate the mole fraction of each gas.

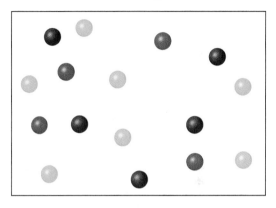

a) $\chi_{red} = 0.5$, $\chi_{blue} = 0.4$, $\chi_{green} = 0.7$

b) $\chi_{red} = 0.05$, $\chi_{blue} = 0.07$, $\chi_{green} = 0.07$

c) $\chi_{red} = 0.3125$, $\chi_{blue} = 0.25$, $\chi_{green} = 0.4375$

d) $\chi_{red} = 0.4167$, $\chi_{blue} = 0.333$, $\chi_{green} = 0.5833$

e) $\chi_{red} = 0.333$, $\chi_{blue} = 0.333$, $\chi_{green} = 0.333$

**11.5.6** Calculate the partial pressure of each gas in the diagram in question 11.5.5 if the total pressure is 8.21 atm.

a) $P_{red} = 0.5$ atm, $P_{blue} = 0.25$ atm, $P_{green} = 4.5$ atm

b) $P_{red} = 2.57$ atm, $P_{blue} = 2.05$ atm, $P_{green} = 3.59$ atm

c) $P_{red} = 3.13$ atm, $P_{blue} = 2.50$ atm, $P_{green} = 2.58$ atm

d) $P_{red} = 2.74$ atm, $P_{blue} = 2.74$ atm, $P_{green} = 2.74$ atm

e) $P_{red} = 3.125$ atm, $P_{blue} = 2.500$ atm, $P_{green} = 4.375$ atm

Figure 11.15

# Molar Volume of a Gas

The reaction between zinc and hydrochloric acid produces hydrogen gas. The net ionic equation for the reaction is

$$Zn(s) + 2H^+(aq) \longrightarrow Zn^{2+}(aq) + H_2(g)$$

The $H_2$ gas evolved in the reaction is collected in the inverted graduated cylinder. Using the balanced equation, we determine how much $H_2$ will be produced.

$$0.072 \text{ g Zn} \times \frac{1 \text{ mol Zn}}{65.41 \text{ g Zn}} = 0.0011 \text{ mol Zn}$$

$$0.0011 \text{ mol Zn} \times \frac{1 \text{ mol } H_2}{1 \text{ mol Zn}} = 0.0011 \text{ mol } H_2$$

The reservoir is inverted and the zinc drops into the 1.0 *M* HCl.

START

A 0.072-g sample of zinc is placed in the reservoir. The vessel contains approximately 5 mL of 1.0 *M* HCl. A graduated cylinder is filled with water and inverted over the tubing to collect the gas produced by the reaction of zinc and acid. The temperature of the water is 25.0°C and the pressure in the room is 748.0 torr.

Zn is the limiting reactant. When all of the zinc has been consumed, the reaction is complete and no more gas is evolved.

We adjust the level of the graduated cylinder to make the water level the same inside and outside of the cylinder. This lets us know that the pressure inside the cylinder is the same as the pressure in the room. When the water levels are the same, we can read the volume of gas collected, which is 26.5 mL.

The pressure inside the cylinder is the sum of two partial pressures: that of the collected $H_2$, and that of water vapor. To determine the partial pressure of $H_2$, which is what we want, we must subtract the vapor pressure of water from the total pressure. Table 11.5 gives the vapor pressure of water at 25.0°C as 23.8 torr. Therefore, the pressure of $H_2$ is $748.0 - 23.8 = 724.2$ torr. Using Equation 11.8(b),

$$\frac{P_1V_1}{T_1} = \frac{P_2V_2}{T_2}$$

we calculate the volume the $H_2$ gas would occupy at STP.

$$\frac{(724.2 \text{ torr})(26.5 \text{ mL})}{298.15 \text{ K}} = \frac{(760 \text{ torr})V_2}{273.15 \text{ K}}$$

$$V_2 = \frac{(724.2 \text{ torr})(26.5 \text{ mL})(273.15 \text{ K})}{(760 \text{ torr})(298.15 \text{ K})} = 23.13 \text{ mL}$$

This is the volume of 0.0011 mol $H_2$. The molar volume is

$$\frac{23.13 \text{ mL}}{0.0011 \text{ mol}} = 2.1 \times 10^4 \text{ mL/mol or 21 L/mol}$$

## What's the point?

The gas collected in the graduated cylinder is a mixture of the gas produced by the reaction and water vapor. We determine the pressure of the gas produced by the reaction by subtracting the tabulated partial pressure of water from the total pressure. Knowing the volume, pressure, and temperature of a sample of gas, in this case $H_2$, we can determine what volume the same sample of gas would occupy at STP. This enables us to determine experimentally the molar volume of $H_2$ at STP, which turns out to be fairly close to the accepted value of 22.4 L. Several sources of error, including uncertainty in the volume of the collected gas, contribute to the result not being exactly 22.4 L.

## 11.6 The Kinetic Molecular Theory of Gases

The gas laws were derived empirically, and they enable us to predict the macroscopic behavior of gases. They do not explain, however, *why* gases behave as they do. The **kinetic molecular theory,** which was put forth in the nineteenth century by a number of physicists, notably Ludwig Boltzmann[6] and James Maxwell,[7] explains how the molecular nature of gases gives rise to their macroscopic properties. The basic assumptions of the kinetic molecular theory are as follows:

1. A gas is composed of particles that are separated by relatively large distances. The volume occupied by individual molecules is negligible.

2. Gas molecules are constantly in random motion, moving in straight paths, colliding with the walls of their container and with one another in perfectly elastic collisions. (Energy is *transferred* but not *lost* in the collisions.)

3. Gas molecules do not exert attractive or repulsive forces on one another.

4. The average kinetic energy, $\overline{E_k}$, of gas molecules in a sample is proportional to the absolute temperature:

$$\overline{E_k} \propto T$$

Recall that kinetic energy is the energy associated with motion [◄◄ Section 5.1]:

$$E_k = \tfrac{1}{2}mu^2$$

Thus, the kinetic energy of an individual gas molecule is proportional to its mass and its velocity squared. When we talk about a group of gas molecules, we determine the average kinetic energy using the *mean square speed, $\overline{u^2}$,* which is the average of the speed squared for all the molecules in the sample:

$$\overline{u^2} = \frac{u_1^2 + u_2^2 + u_3^2 + \cdots + u_N^2}{N}$$

where $N$ is the number of molecules in the sample.

## Application to the Gas Laws

Kinetic molecular theory enables us to understand some of the properties and behavior of gases in the following ways.

### Compressibility

Gases are compressible because molecules in the gas phase are separated by large distances (assumption 1) and can be moved closer together by decreasing the volume occupied by a sample of gas (Figure 11.16).

### Boyle's Law (V ∝ 1/P)

The pressure exerted by a gas is the result of the collisions of gas molecules with the walls of their container (assumption 2). The magnitude of the pressure depends on both the frequency of collision and the speed of molecules when they collide with the walls. Decreasing the volume occupied by a sample of gas increases the frequency of these collisions, thus increasing the pressure (Figure 11.16).

### Charles's Law (V ∝ T)

Heating a sample of gas increases its average kinetic energy (assumption 4). Because the masses of the molecules do not change, an increase in average kinetic energy must be accompanied by an increase in the mean square speed of the molecules. In other words, heating a sample of gas makes the gas molecules move faster. Faster-moving molecules collide more frequently and with greater

---

6. Ludwig Eduard Boltzmann (1844–1906). Austrian physicist. Although Boltzmann was one of the greatest theoretical physicists of all time, his work was not recognized by other scientists in his own lifetime. He suffered from poor health and severe depression and committed suicide in 1906.

7. James Clerk Maxwell (1831–1879). Scottish physicist. Maxwell was one of the great theoretical physicists of the nineteenth century. His work covered many areas in physics, including the kinetic theory of gases, thermodynamics, and electricity and magnetism.

(a)

(b)

**Figure 11.16** Gases can be compressed by decreasing their volume. (a) Before volume decrease. (b) After volume decrease, the increased frequency of collisions between molecules and the walls of their container constitutes higher pressure.

speed at impact, thus increasing the pressure. If the container can expand (as is the case with a balloon or a cylinder with a movable piston), the volume of the gas sample will increase, thereby decreasing the frequency of collisions until the pressure inside the container and the pressure outside are again equal (Figure 11.17).

### Avogadro's Law ($V \propto n$)

Because the magnitude of the pressure exerted by a sample of gas depends on the frequency of the collisions with the container wall, the presence of more molecules would cause an increase in pressure. Again, the container will expand if it can. Expansion of the container will decrease the frequency of collisions until the pressures inside and outside the container are once again equal (Figure 11.18).

### Dalton's Law of Partial Pressures ($P_{total} = \Sigma P_i$)

Gas molecules do not attract or repel one another (assumption 3), so the pressure exerted by one gas is unaffected by the presence of another gas. Consequently, the total pressure exerted by a mixture of gases is simply the sum of the partial pressures of the individual components in the mixture (Figure 11.19).

## Molecular Speed

One of the important outcomes of the kinetic molecular theory is that the total kinetic energy of a mole of gas (any gas) is equal to $\frac{3}{2}RT$. With assumption 4 we saw that the average kinetic energy of one molecule is $\frac{1}{2}m\overline{u^2}$. For 1 mole of the gas we write

$$N_A\left(\frac{1}{2}m\overline{u^2}\right) = \frac{3}{2}RT$$

where $N_A$ is Avogadro's number and $R$ is the gas constant expressed as 8.314 J/K · mol. Because $m \times N_A = \mathcal{M}$, we can rearrange the preceding equation as follows:

$$\overline{u^2} = \sqrt{\frac{3RT}{\mathcal{M}}}$$

Taking the square root of both sides gives

$$\sqrt{\overline{u^2}} = \sqrt{\frac{3RT}{\mathcal{M}}}$$

Heating or cooling a gas at constant pressure

Lower temperature
(Volume decreases)

Higher temperature
(Volume increases)

$$V = \left(\frac{nR}{P}\right)T \qquad \frac{nR}{P} \text{ is constant}$$

(a)

— Charles's Law

Heating or cooling a gas at constant volume

Lower temperature
(Pressure decreases)

Higher temperature
(Pressure increases)

$$P = \left(\frac{nR}{V}\right)T \qquad \frac{nR}{V} \text{ is constant}$$

(b)

**Figure 11.17**    Charles's law. (a) The volume of a sample of gas at constant pressure is proportional to its absolute temperature. (b) The pressure of a sample of gas at constant volume is proportional to its absolute temperature.

Dependence of volume on amount
of gas at constant temperature and pressure

Gas
cylinder

Remove gas
(Volume decreases)

Add
gas molecules
(Volume increases)

Valve

$$V = \left(\frac{RT}{P}\right)n \qquad \frac{RT}{P} \text{ is constant}$$

**Figure 11.18**    Avogadro's law. The volume of a gas at constant temperature and pressure is proportional to the number of moles.

**Figure 11.19** Each component of a gas mixture exerts a pressure independent of the other components. The total pressure is the sum of the individual components' partial pressures.

where $\sqrt{\overline{u^2}}$ is the **root-mean-square (rms) speed ($u_{rms}$)**. The result,

$$u_{rms} = \sqrt{\frac{3RT}{\mathcal{M}}} \qquad \text{Equation 11.17}$$

gives us the root-mean-square speed, which is the speed of a molecule with the average kinetic energy in a gas sample. Equation 11.17 indicates two important things: (1) The root-mean-square speed is directly proportional to the square root of the absolute temperature, and (2) the root-mean-square speed is inversely proportional to the square root of $\mathcal{M}$. Thus, for any two samples of gas at the same temperature, the gas with the larger molar mass will have the lower root-mean-square speed, $u_{rms}$.

Keep in mind that most molecules will have speeds either higher or lower than $u_{rms}$—and that $u_{rms}$ is temperature dependent. James Maxwell studied extensively the behavior of gas molecules at various temperatures. Figure 11.20(a) shows typical Maxwell speed distribution curves for nitrogen gas at three different temperatures. At a given temperature, the distribution curve tells us the number of molecules moving at a certain speed. The maximum of each curve represents the most probable speed—that is, the speed of the largest number of molecules. Note that the most probable speed increases as temperature increases [the maximum shifts toward the right in Figure 11.20(a)]. Furthermore, the curve also begins to flatten out with increasing temperature, indicating that larger numbers of molecules are moving faster.

Figure 11.20(b) shows the speed distributions of three different gases ($Cl_2$, $N_2$, and He) at the same temperature (300 K). The difference in these curves can be explained by noting that lighter molecules, on average, move faster than heavier ones.

Although we can use Equation 11.17 to calculate $u_{rms}$ of a molecule in a particular sample, we will generally find it more useful to compare the $u_{rms}$ values of molecules in different gas samples. For example, we can write Equation 11.17 for two different gases:

$$u_{rms}(1) = \sqrt{\frac{3RT}{\mathcal{M}_1}} \quad \text{and} \quad u_{rms}(2) = \sqrt{\frac{3RT}{\mathcal{M}_2}}$$

We can then determine the $u_{rms}$ of a molecule in one gas relative to that in the other gas:

$$\frac{u_{rms}(1)}{u_{rms}(2)} = \frac{\sqrt{\dfrac{3RT}{\mathcal{M}_1}}}{\sqrt{\dfrac{3RT}{\mathcal{M}_2}}}$$

**Student Annotation:** In Equation 11.17, $R$ must be expressed as 8.314 J/K · mol and $\mathcal{M}$ must be expressed in kg/mol.

**Student Annotation:** Remember that the average kinetic energy of a gas depends on its absolute temperature. Therefore, any two gas samples at the same temperature have the same average kinetic energy.

(a)                                                                (b)

**Figure 11.20**    (a) The distribution of speeds for nitrogen gas at three different temperatures. At higher temperatures more molecules are moving faster. (b) The distribution of speeds for three different gases at the same temperature. On average, lighter molecules move faster than heavier molecules.

Canceling identical terms, when both gases are at the same temperature, we can write

**Equation 11.18**

$$\frac{u_{rms}(1)}{u_{rms}(2)} = \sqrt{\frac{\mathcal{M}_1}{\mathcal{M}_2}}$$

**Student Annotation:** Note that because Equation 11.18 contains the ratio of two molar masses, we can express the molar masses as g/mol or kg/mol. (In Equation 11.17, we had to express molar mass as kg/mol for the units to cancel properly.)

Using Equation 11.18 we can compare $u_{rms}$ values of molecules with different molar masses (at a given temperature). Sample Problem 11.16 shows how this is done.

### Sample Problem 11.16

Determine how much faster a helium atom moves, on average, than a carbon dioxide molecule at the same temperature.

**Strategy**  Use Equation 11.18 and the molar masses of He and $CO_2$ to determine the ratio of their root-mean-square speeds. When solving a problem such as this, it is generally best to label the lighter of the two molecules as molecule 1 and the heavier molecule as molecule 2. This ensures that the result will be greater than 1, which is relatively easy to interpret.

**Setup**  The molar masses of He and $CO_2$ are 4.003 and 44.01 g/mol, respectively.

**Solution**

**Think About It** Remember that the relationship between molar mass and molecular speed (Equation 11.18) is reciprocal. A $CO_2$ molecule has approximately 10 times the mass of an He atom. Therefore, we should expect an He atom, on average, to be moving approximately $\sqrt{10}$ times (~3.2 times) as fast as a $CO_2$ molecule.

$$\frac{u_{rms}(He)}{u_{rms}(CO_2)} = \frac{\sqrt{\dfrac{44.01\ g}{mol}}}{\sqrt{\dfrac{4.003\ g}{mol}}} = 3.316$$

On average, He atoms move 3.316 times faster than $CO_2$ molecules at the same temperature.

**Practice Problem A**  Determine the relative root-mean-square speeds of $O_2$ and $SF_6$ at a given temperature.

**Practice Problem B**  Determine the molar mass and identity of a gas that moves 4.67 times as fast as $CO_2$.

Multimedia

Diffusion of gases.

## Diffusion and Effusion

The random motion of gas molecules gives rise to two readily observable phenomena: diffusion and effusion. **Diffusion** is the mixing of gases as the result of random motion and frequent collisions (Figure 11.21), while **effusion** is the escape of gas molecules from a container to a region of vacuum (Figure 11.22). One of the earliest successes of the kinetic molecular theory was its ability to explain diffusion and effusion.

(a)

(b)

**Figure 11.21** Diffusion is the mixing of gases. (a) Two different gases in separate containers. (b) When the stopcock is opened, the gases mix by diffusion.

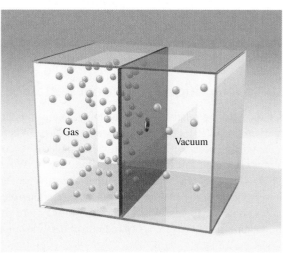

Gas

Vacuum

**Figure 11.22** Effusion is the escape of a gas into a vacuum.

*Graham's law* states that the rate of diffusion or effusion of a gas is inversely proportional to the square root of its molar mass:

$$\text{Rate} \propto \frac{1}{\sqrt{\mathcal{M}}}$$

This is essentially a restatement of Equation 11.17. Thus, lighter gases diffuse and effuse more rapidly than heavier gases.

## Checkpoint 11.6    The Kinetic Molecular Theory of Gases

**11.6.1** Methane ($CH_4$) diffuses approximately 2.4 times as fast as a certain unknown gas. Which of the following could be the unknown gas?

a) $O_2$

b) $C_2H_6$

c) $CH_3Br$

d) $CH_3I$

e) $F_2$

**11.6.2** Which gas effuses faster, He or Ar, and how much faster does it effuse?

a) He effuses 9.98 times as fast as Ar.

b) He effuses 3.16 times as fast as Ar.

c) Ar effuses 9.98 times as fast as He.

d) Ar effuses 3.16 times as fast as He.

e) He and Ar effuse at the same rate.

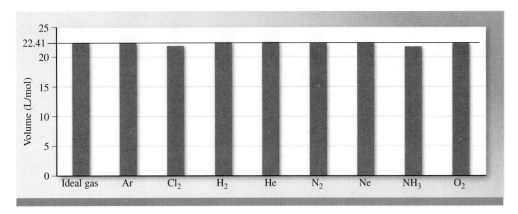

**Figure 11.23**   Molar volumes of some common gases at STP.

# 11.7   Deviation from Ideal Behavior

The gas laws and the kinetic molecular theory assume that molecules in the gas phase occupy negligible volume (assumption 1) and that they do not exert any force on one another, either attractive or repulsive (assumption 3). Gases that behave as though these assumptions were strictly true are said to exhibit *ideal behavior*. Many gases do exhibit ideal or nearly ideal behavior under ordinary conditions. Figure 11.23 shows the molar volumes of some common gases at STP. All are remarkably close to the ideal value of 22.41 L. Although we generally assume that real gases behave ideally, there are conditions, namely, high pressure and low temperature, under which the behavior of a real gas deviates from ideal.

## Factors That Cause Deviation from Ideal Behavior

At high pressures, gas molecules are relatively close together. We can assume that gas molecules occupy no volume only when the distances between molecules are large. When the distances between molecules are reduced, the volume occupied by each individual molecule becomes more significant.

At low temperatures, gas molecules are moving more slowly. We can assume that there are no intermolecular forces between gas molecules, either attractive or repulsive, when the gas molecules are moving very fast and the magnitude of their kinetic energies is much larger than the magnitude of any intermolecular forces. When molecules move more slowly, they have lower kinetic energies and the magnitude of the forces between them becomes more significant.

## The van der Waals Equation

Because there are conditions under which use of the ideal gas equation would result in large errors (i.e., high pressure and/or low temperature), we must use a slightly different approach when gases do not behave ideally. Analyses of real gases that took into account nonzero molecular volumes and intermolecular forces were first carried out by J. D. van der Waals[8] in 1873. Van der Waals's treatment provides us with an interpretation of the behavior of real gases at the molecular level.

Consider the approach of a particular molecule toward the wall of its container (Figure 11.24). The intermolecular attractions exerted by neighboring molecules prevent the molecule from hitting the wall as hard as it otherwise would. This results in the pressure exerted by a real gas being lower than that predicted by the ideal gas equation. Van der Waals suggested that the pressure exerted by an ideal gas, $P_{ideal}$, is related to the experimentally measured pressure, $P_{real}$, by the equation

$$P_{ideal} = P_{real} + \frac{an^2}{V^2}$$

where $a$ is a constant and $n$ and $V$ are the number of moles and volume of the gas, respectively. The correction term for pressure ($an^2/V^2$) can be understood as follows. The intermolecular interaction that gives rise to nonideal behavior depends on how frequently any two molecules encounter each other. The number of such encounters increases with the square of the number of molecules per

**Figure 11.24**   The effect of intermolecular attractions on the pressure exerted by a gas.

---

8. Johannes Diderik van der Waals (1837–1923). Dutch physicist. Van der Waals received the Nobel Prize in Physics in 1910 for his work on the properties of gases and liquids.

| TABLE 11.6 | | Van der Waals Constants of Some Common Gases | | | |
|---|---|---|---|---|---|
| Gas | $a\left(\dfrac{\text{atm} \cdot \text{L}^2}{\text{mol}^2}\right)$ | $b\left(\dfrac{\text{L}}{\text{mol}}\right)$ | Gas | $a\left(\dfrac{\text{atm} \cdot \text{L}^2}{\text{mol}^2}\right)$ | $b\left(\dfrac{\text{L}}{\text{mol}}\right)$ |
| He | 0.034 | 0.0237 | $O_2$ | 1.36 | 0.0318 |
| Ne | 0.211 | 0.0171 | $Cl_2$ | 6.49 | 0.0562 |
| Ar | 1.34 | 0.0322 | $CO_2$ | 3.59 | 0.0427 |
| Kr | 2.32 | 0.0398 | $CH_4$ | 2.25 | 0.0428 |
| Xe | 4.19 | 0.0510 | $CCl_4$ | 20.4 | 0.138 |
| $H_2$ | 0.244 | 0.0266 | $NH_3$ | 4.17 | 0.0371 |
| $N_2$ | 1.39 | 0.0391 | $H_2O$ | 5.46 | 0.0305 |

unit volume ($n/V$), and $a$ is a proportionality constant. The quantity $P_{\text{ideal}}$ is the pressure we would measure if there were no intermolecular attractions.

The other correction concerns the volume occupied by the gas molecules. In the ideal gas equation, $V$ represents the volume of the container. However, each molecule actually occupies a very small but nonzero volume. We can correct for the volume occupied by the gas molecules by subtracting a term, $nb$, from the volume of the container:

$$V_{\text{real}} = V_{\text{ideal}} - nb$$

where $n$ and $b$ are the number of moles and the proportionality constant, respectively.

Incorporating both corrections into the ideal gas equation gives us the **van der Waals equation,** with which we can analyze gases under conditions where ideal behavior is not expected.

experimentally measured pressure      container volume

$$\left(P + \frac{an^2}{V^2}\right)(V - nb) = nRT \qquad\qquad \text{Equation 11.19}$$

corrected       corrected
pressure term   volume term

The van der Waals constants $a$ and $b$ for a number of gases are listed in Table 11.6. The magnitude of $a$ indicates how strongly molecules of a particular type of gas attract one another. The magnitude of $b$ is related to molecular (or atomic) size, although the relationship is not a simple one.

Sample Problem 11.17 shows how to use the van der Waals equation.

## Sample Problem 11.17

A sample of 3.50 moles of $NH_3$ gas occupies 5.20 L at 47°C. Calculate the pressure of the gas (in atm) using (a) the ideal gas equation and (b) the van der Waals equation.

**Strategy** (a) Use the ideal gas equation, $PV = nRT$.

(b) Use Equation 11.19 and $a$ and $b$ values for $NH_3$ from Table 11.6.

**Setup** $T = 320.15$ K, $a = 4.17$ atm · L/mol$^2$, and $b = 0.0371$ L/mol.

**Solution** (a) $P = \dfrac{nRT}{V} = \dfrac{(3.50 \text{ mol})\left(\dfrac{0.08206 \text{ L} \cdot \text{atm}}{\text{K} \cdot \text{mol}}\right)(320.15 \text{ K})}{5.20 \text{ L}} = 17.7 \text{ atm}$

(b) Evaluating the correction terms in the van der Waals equation, we get

$$\frac{an^2}{V^2} = \frac{\left(\dfrac{4.17 \text{ atm} \cdot \text{L}^2}{\text{mol}^2}\right)(3.50 \text{ mol})^2}{(5.20 \text{ L})^2} = 1.89 \text{ atm}$$

$$nb = (3.50 \text{ mol})\left(\frac{0.0371 \text{ L}}{\text{mol}}\right) = 0.130 \text{ L}$$

*(Continued)*

Finally, substituting these results into Equation 11.19, we have

$$(P + 1.89 \text{ atm})(5.20 \text{ L} - 0.130 \text{ L}) = (3.50 \text{ mol})\left(\frac{0.08206 \text{ L} \cdot \text{atm}}{\text{K} \cdot \text{mol}}\right)(320.15 \text{ K})$$

$$P = 16.2 \text{ atm}$$

> **Think About It** As is often the case, the pressure exerted by the real gas sample is lower than predicted by the ideal gas equation.

**Practice Problem A** Using data from Table 11.6, calculate the pressure exerted by 11.9 moles of neon gas in a volume of 5.75 L at 25°C using (a) the ideal gas equation and (b) the van der Waals equation (Equation 11.19). Compare your results.

**Practice Problem B** Calculate the pressure exerted by 0.35 mole of oxygen gas in a volume of 6.50 L at 32°C using (a) the ideal gas equation and (b) the van der Waals equation.

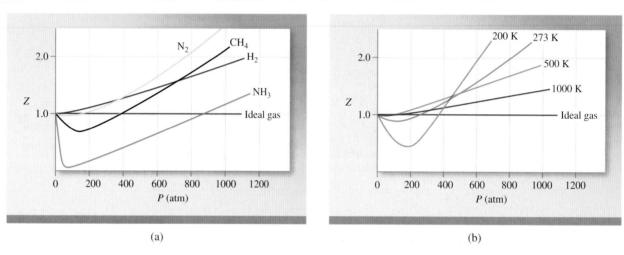

(a)                    (b)

**Figure 11.25** (a) Compressibility factor ($Z$) as a function of pressure for several gases at 0°C. (b) Compressibility factor as a function of pressure for $N_2$ gas at several different temperatures.

One way to measure a gas's deviation from ideal behavior is to determine its compressibility factor, $Z$, where $Z = PV/RT$. For one mole of an ideal gas, $Z$ is equal to 1 at all pressures and temperatures. For real gases, the factors that contribute to nonideal behavior cause the value of $Z$ to deviate from 1. Intermolecular forces help to account for the plots in Figure 11.25(a). Molecules exert both attraction and repulsion on one another. At large separations (low pressures), attraction predominates. In this region, the gas is more compressible than an ideal gas and the curve dips below the horizontal line ($Z < 1$). As molecules are brought closer together under pressure, repulsion begins to play an important role. If the pressure continues to increase, a point is reached when the gas becomes less compressible than an ideal gas because the molecules repel one another and the curve rises above the horizontal line ($Z > 1$). Figure 11.25(b) shows the plots of $Z$ versus pressure at different temperatures. We see that as the temperature increases, the gas behaves more like an ideal gas (the curves become closer to the horizontal line). The increase in the molecules' kinetic energy makes molecular attraction less important.

## Checkpoint 11.7    Deviation from Ideal Behavior

**11.7.1** Which of the following conditions cause deviation from ideal behavior? (Select all that apply.)

a) High pressure

b) Low pressure

c) High temperature

d) Low temperature

e) High volume

**11.7.2** Using the van der Waals equation, calculate the pressure exerted by 1.5 moles of carbon dioxide in a 3.75-L vessel at 10°C.

a) 9.3 atm

b) 8.9 atm

c) 10 atm

d) 8.6 atm

e) –2.4 atm

## FAQ  *You May Be Wondering…*

# What's *Really* the Difference Between Real Gases and Ideal Gases?

As illustrated by the molar volumes in Figure 11.23, the ideal gas equation is remarkably accurate for common gases at or near STP. However, at low temperatures and high pressures, the assumptions that enable us to treat gases as ideal are no longer valid. In these cases, we must consider the effects of attractive forces between the molecules and the nonzero volume that the molecules occupy.

When a gas is cooled and/or compressed, it condenses to a liquid—indicating that there are attractive forces between the molecules [▶◀ Chapter 12]. Even in the gas phase, the attractive forces between molecules can impact the observed behavior of a substance. As illustrated in Figure 11.24, a molecule that is attracted to other molecules in a sample of gas will not strike the wall of the container with as high a velocity as it would if there were no such intermolecular attractions. The pressure term in the van der Waals equation, $P + a(n/V)^2$, is the experimentally determined pressure, $P$, plus a correction for the pressure that we do *not* observe because of attractive forces between the gas molecules. Note that the correction factor depends on the moles-per-unit-volume ($n/V$) squared. The value of the constant $a$ is specific to a particular gas.

When two gas molecules (assumed to be spherical) approach each other, the distance of closest approach is the sum of their radii ($2r$). The volume around each molecule into which the center of another molecule cannot penetrate is called the *excluded volume*. The effect of the excluded volume is to limit the fraction of the container volume actually available for molecules to move about in a gas sample. Thus, the volume term in the van der Waals equation, $V - nb$, is the container volume $V$ minus the correction for the excluded volume, $nb$, where $n$ is the number of moles of the gas and $b$ is the excluded volume per mole of the gas.

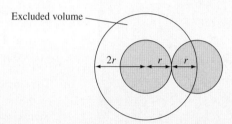

Excluded volume

Sample Problem 11.18 shows how to calculate excluded volume.

---

### Sample Problem 11.18

Consider a gas sample consisting of molecules with radius $r$. (a) Determine the excluded volume defined by two molecules and (b) calculate the excluded volume per mole ($b$) for the gas. Compare the excluded volume per mole with the volume actually occupied by a mole of the molecules.

**Strategy** (a) Use the formula for volume of a sphere to calculate the volume of a sphere of radius $2r$. This is the excluded volume defined by two molecules. (b) Multiply the excluded volume per molecule by Avogadro's number to determine excluded volume per mole.

**Setup** The equation for volume of a sphere is $\frac{4}{3}\pi r^3$; $N_A = 6.022 \times 10^{23}$.

**Solution** (a) Excluded volume defined by two molecules $= \frac{4}{3}\pi(2r)^3 = 8\left(\frac{4}{3}\pi r^3\right)$

(b) Excluded volume per mole $= \dfrac{8\left(\frac{4}{3}\pi r^3\right)}{2 \text{ molecules}} \times \dfrac{N_A \text{ molecules}}{1 \text{ mole}} = \dfrac{4N_A\left(\frac{4}{3}\pi r^3\right)}{1 \text{ mole}}$

The volume actually occupied by a mole of molecules with radius $r$ is $N_A\left(\frac{4}{3}\pi r^3\right)$.

> **Think About It** The excluded volume per mole is four times the volume actually occupied by a mole of molecules.

---

**Practice Problem A** Determine the excluded volume per mole and the total volume of the molecules in a mole for a gas consisting of molecules with radius 165 pm. [Note: To obtain the volume in liters, we must express the radius in decimeters (dm).]

**Practice Problem B** Because the van der Waals constant $b$ is the excluded volume per mole of a gas, we can use the value of $b$ to estimate the radius of a molecule or atom. Consider a gas that consists of molecules, for which the van der Waals constant $b$ is 0.0315 L/mol. Estimate the molecular radius in picometers (pm). Assume that the molecules are spherical.

# Applying What You've Learned

Scuba divers are not the only athletes who can suffer the detrimental effects of sudden changes in pressure. Mountain climbers, too, are susceptible to the dangers of rapid ascent. At high elevation, air pressure is significantly lower than at sea level. A lower total pressure means a lower partial pressure of oxygen, and insufficient oxygen or *hypoxia* can cause altitude sickness. Early symptoms of altitude sickness include headache, dizziness, and nausea. In severe cases, climbers may suffer hallucinations, seizure, coma, and even death.

In 1990, Igor Gamow, a professor of microbiology at the University of Colorado, patented a portable device for high-altitude treatment of altitude sickness. The Gamow Bag is an inflatable cylinder large enough to accommodate an adult mountain climber. The bag is inflated and pressurized with a foot pump, and the afflicted climber remains sealed inside the pressurized bag until symptoms subside enough to begin descent. Although the bag is pressurized to only about 0.14 atm above atmospheric pressure, at very high altitudes this corresponds to a simulated descent on the order of 10,000 ft!

**Student Annotation:** The Gamow Bag earned the praises even of Sir Edmund Hillary, leader of the first expedition to the summit of Mount Everest.

**Gamow Bag**

## Problems:

a)  If a Gamow Bag is pressurized to 0.14 atm above atmospheric pressure at an altitude of 25,000 ft, where atmospheric pressure is 0.37 atm, what height column of mercury would be supported by the pressure inside the pressurized bag?
   [◄◄ Sample Problem 11.1]

b)  The Gamow Bag inflates to a volume of $4.80 \times 10^2$ L. What volume would be occupied by the air in the pressurized bag in part (a) at standard atmospheric pressure? (Assume no change in temperature.) [◄◄ Sample Problem 11.2]

c)  Calculate the density of the air in the pressurized bag in part (a) at 0°C. (Assume that air is 80 percent $N_2$ and 20 percent $O_2$ by volume.) [◄◄ Sample Problem 11.7]

d)  LiOH scrubbers are sometimes used to prevent the buildup of $CO_2$ during use of the bag. What volume of $CO_2$ (at 0°C) can be removed from the pressurized bag in part (a) by 0.50 kg LiOH? [◄◄ Sample Problem 11.9]

e)  Calculate the number of moles of each gas in the pressurized bag in part (a) at 0°C. [◄◄ Sample Problem 11.13]

# CHAPTER SUMMARY

## Section 11.1

- A gas assumes the volume and shape of its container and is compressible. Gases generally have low densities (expressed in g/L) and will mix in any proportions to give homogeneous solutions.

- Gases exert *pressure,* which is the force per unit area. The SI units of force and pressure are the *newton (N)* and the *pascal (Pa),* respectively. Other commonly used units of pressure are atmosphere (atm), mmHg, torr, and bar.

- Pressure can be measured using a *barometer* or a *manometer.*

- *Standard atmospheric pressure* (1 atm) is the pressure exerted by the atmosphere at sea level.

## Section 11.2

- The physical state of a sample of gas can be described using four parameters: temperature ($T$), pressure ($P$), volume ($V$), and number of moles ($n$). Equations relating these parameters are called the *gas laws.*

- *Boyle's law* states that the volume of a sample of gas at constant temperature is inversely proportional to pressure.

- Experiments done by Charles and Gay-Lussac showed that the volume of a gas at constant pressure is directly proportional to temperature. Lord Kelvin used Charles's and Gay-Lussac's data to propose that *absolute zero* is the lowest theoretically attainable temperature. The *absolute temperature scale,* also known as the *Kelvin temperature scale,* is used for all calculations involving gases.

- *Charles's and Gay-Lussac's law,* commonly known as *Charles's law,* states that the volume of a sample of gas at constant pressure is directly proportional to its absolute temperature.

- *Avogadro's law* states that the volume of a sample of gas at constant temperature and pressure is directly proportional to the number of moles.

- The *combined gas law* combines the laws of Boyle, Charles, and Avogadro and relates pressure, volume, temperature, and number of moles without assuming that any of the parameters is constant.

## Section 11.3

- The *ideal gas equation, $PV = nRT$,* makes it possible to predict the behavior of gases. An *ideal gas* is one that behaves in a way predicted by the ideal gas equation. $R$ is the *gas constant,* which may be expressed in a variety of units. The units used to express $R$ depend on the units used to express $P$ and $V$.

- *Standard temperature and pressure (STP)* is defined as 0°C and 1 atm.

- The ideal gas equation can be used to calculate the density of a gas and to interconvert between density and molar mass.

## Section 11.4

- For a reaction occurring at constant temperature and pressure, and involving only gases, the coefficients in the balanced chemical equation apply to units of volume, as well as to numbers of molecules or moles.

- A balanced chemical equation and the ideal gas equation can be used to determine volumes of gaseous reactants and/or products in a reaction.

## Section 11.5

- Each component in a mixture of gases exerts a *partial pressure ($P_i$)* independent of the other mixture components. *Dalton's law of partial pressures* states that the total pressure exerted by a gas *mixture* is the sum of the partial pressures of the components.

- *Mole fraction ($\chi_i$)* is the unitless quotient of the number of moles of a mixture component and the total number of moles in the mixture, $n_i/n_{total}$.

## Section 11.6

- According to the *kinetic molecular theory,* gases are composed of particles with negligible volume that are separated by large distances; the particles are in constant, random motion, and collisions between the particles and between the particles and their container walls are perfectly elastic; there are no attractive or repulsive forces between the particles; and the average kinetic energy of particles in a sample is proportional to the absolute temperature of the sample.

- Kinetic molecular theory can be used to explain the compressibility of gases and the empirical gas laws.

- The *root-mean-square (rms) speed ($u_{rms}$)* of gas molecules in a sample at a given temperature is inversely proportional to the molecular mass.

- According to *Graham's law,* the rates of *diffusion* (mixing of gases) and *effusion* (escape of a gas from a container into a vacuum) are inversely proportional to the square root of the molar mass of the gas.

## Section 11.7

- Deviation from ideal behavior is observed at high pressure and/or low temperature. The *van der Waals equation* makes corrections for the nonzero volume of gas molecules and the attractive forces between molecules.

# KEY WORDS

# KEY EQUATIONS

11.1    $P = hdg$

11.2(a)   $V = k_1 \dfrac{1}{P}$   (at constant temperature)

11.2(b)   $PV = k_1$   (at constant temperature)

11.3    $P_1 V_1 = P_2 V_2$   (at constant temperature)

11.4(a)   $V = k_2 T$   (at constant pressure)

11.4(b)   $\dfrac{V}{T} = k_2$   (at constant pressure)

11.5    $\dfrac{V_1}{T_1} = \dfrac{V_2}{T_2}$   (at constant pressure)

11.6(a)   $V = k_3 n$   (at constant temperature and pressure)

11.6(b)   $\dfrac{V}{n} = k_3$

11.7    $\dfrac{V_1}{n_1} = \dfrac{V_2}{n_2}$

11.8(a)   $\dfrac{P_1 V_1}{n_1 T_1} = \dfrac{P_2 V_2}{n_2 T_2}$

11.8(b)   $\dfrac{P_1 V_1}{T_1} = \dfrac{P_2 V_2}{T_2}$

11.9    $PV = nRT$

11.10   $d = \dfrac{P\mathcal{M}}{RT}$

11.11   $\mathcal{M} = \dfrac{dRT}{P}$

11.12(a)  $n = P \times \left(\dfrac{V}{RT}\right)$   (at constant $V$ and $T$)

11.12(b)  $\Delta n = \Delta P \times \left(\dfrac{V}{RT}\right)$   (at constant $V$ and $T$)

11.13   $\chi_i = \dfrac{n_i}{n_{total}}$

11.14   $\chi_i = \dfrac{P_i}{P_{total}}$

11.15   $\chi_i \times n_{total} = n_i$

11.16   $\chi_i \times P_{total} = P_i$

11.17   $u_{rms} = \sqrt{\dfrac{3RT}{\mathcal{M}}}$

11.18   $\dfrac{u_{rms}(1)}{u_{rms}(2)} = \sqrt{\dfrac{\mathcal{M}_1}{\mathcal{M}_2}}$

11.19   $\left(P + \dfrac{an^2}{V^2}\right)(V - nb) = nRT$

# QUESTIONS AND PROBLEMS

## Section 11.1: Properties of Gases

### Review Questions

11.1    Name five elements and five compounds that exist as gases at room temperature.

11.2    List the physical characteristics of gases.

11.3    Define *pressure* and give the common units for pressure.

11.4    Describe how a barometer and a manometer are used to measure gas pressure.

11.5    Why is mercury a more suitable substance to use in a barometer than water?

11.6 Explain why the height of mercury in a barometer is independent of the cross-sectional area of the tube.

11.7 Would it be easier to drink water with a straw on top of Mt. Everest or at the foot? Explain.

11.8 Is the atmospheric pressure in a mine that is 500 m below sea level greater or less than 1 atm?

11.9 What is the difference between the terms *gas* and *vapor*? At 25°C, which of the following substances in the gas phase should be properly called a gas and which should be called a vapor: molecular chlorine ($Cl_2$), molecular iodine ($I_2$)?

11.10 If the maximum distance that water may be brought up a well by a suction pump is 34 ft (10.3 m), how is it possible to obtain water and oil from hundreds of feet below the surface of Earth?

11.11 Why is it that if the barometer reading falls in one part of the world, it must rise somewhere else?

11.12 Why do astronauts have to wear protective suits when they are on the surface of the moon?

### Problems

**11.13** Convert 375 mmHg to atmospheres, bar, torr, and pascals.

11.14 The atmospheric pressure at the summit of Mt. McKinley is 581 mmHg on a certain day. What is the pressure in atmospheres and in kilopascals?

**11.15** Calculate the height of a column of methanol ($CH_3OH$) that would be supported by atmospheric pressure. The density of methanol is 0.787 g/cm$^3$.

11.16 Calculate the height of a column of ethylene glycol [$CH_2(OH)CH_2(OH)$] that would be supported by atmospheric pressure (1 atm). The density of ethylene glycol is 1.12 g/cm$^3$.

**11.17** What pressure (in atm) is exerted by a column of toluene ($C_7H_8$) 87 m high? The density of toluene is 0.867 g/cm$^3$.

11.18 What pressure (in atm) is exerted by a column of isopropanol ($C_3H_7OH$) 264 m high? The density of isopropanol is 0.785 g/cm$^3$.

## Section 11.2: The Gas Laws

### Review Questions

11.19 State the following gas laws in words and also in the form of an equation: Boyle's law, Charles's law, Avogadro's law. In each case, indicate the conditions under which the law is applicable, and give the units for each quantity in the equation.

11.20 Explain why a helium weather balloon expands as it rises in the air. Assume that the temperature remains constant.

### Problems

**11.21** A gaseous sample of a substance is cooled at constant pressure. Which of the following diagrams best represents the situation if the final temperature is (a) above the boiling point of the substance and (b) below the boiling point but above the freezing point of the substance?

(a)        (b)        (c)        (d)

11.22 Consider the following gaseous sample in a cylinder fitted with a movable piston. Initially there are *n* moles of the gas at temperature *T*, pressure *P*, and volume *V*.

(a)        (b)        (c)        (d)

Choose the cylinder that correctly represents the gas after each of the following changes. (1) The pressure on the piston is tripled at constant *n* and *T*. (2) The absolute temperature is doubled at constant *n* and *P*. (3) *n* more moles of the gas are added at constant *T* and *P*. (4) Absolute temperature is halved at constant *P*.

**11.23** A gas sample occupying a volume of 25.6 mL at a pressure of 0.970 atm is allowed to expand at constant temperature until its pressure reaches 0.541 atm. What is its final volume?

11.24 At 46°C a sample of ammonia gas exerts a pressure of 5.3 atm. What is the pressure when the volume of the gas is reduced to one-fourth of the original value at the same temperature?

**11.25** The volume of a gas is 7.15 L, measured at 1.00 atm. What is the pressure of the gas in mmHg if the volume is changed to 9.25 L? (The temperature remains constant.)

**11.26** A sample of air occupies 3.8 L when the pressure is 1.2 atm. (a) What volume does it occupy at 6.6 atm? (b) What pressure is required to compress it to 0.075 L? (The temperature is kept constant.)

**11.27** A 28.4-L volume of methane gas is heated from 35°C to 72°C at constant pressure. What is the final volume of the gas?

**11.28** Under constant-pressure conditions a sample of hydrogen gas initially at 88°C and 9.6 L is cooled until its final volume is 3.4 L. What is its final temperature?

**11.29** Ammonia burns in oxygen gas to form nitric oxide (NO) and water vapor. How many volumes of NO are obtained from one volume of ammonia at the same temperature and pressure?

**11.30** Molecular chlorine and molecular fluorine combine to form a gaseous product. Under the same conditions of temperature and pressure it is found that one volume of $Cl_2$ reacts with three volumes of $F_2$ to yield two volumes of the product. What is the formula of the product?

## Section 11.3: The Ideal Gas Equation

### Review Questions

11.31 List the characteristics of an ideal gas.

11.32 What are standard temperature and pressure (STP)? What is the significance of STP in relation to the volume of 1 mole of an ideal gas?

11.33 Why is the density of a gas much lower than that of a liquid or solid under atmospheric conditions? What units are normally used to express the density of gases?

## Problems

**11.34**  A sample of nitrogen gas in a 4.5-L container at a temperature of 27°C exerts a pressure of 4.1 atm. Calculate the number of moles of gas in the sample.

**11.35**  Given that 6.9 moles of carbon monoxide gas are present in a container of volume 30.4 L, what is the pressure of the gas (in atm) if the temperature is 82°C?

**11.36**  What volume will 9.8 moles of sulfur hexafluoride ($SF_6$) gas occupy if the temperature and pressure of the gas are 105°C and 9.4 atm, respectively?

**11.37**  A certain amount of gas at 25°C and at a pressure of 0.800 atm is contained in a vessel. Suppose that the vessel can withstand a pressure no higher than 5.00 atm. How high can you raise the temperature of the gas without bursting the vessel?

**11.38**  A gas-filled balloon having a volume of 2.50 L at 1.2 atm and 20°C is allowed to rise to the stratosphere (about 30 km above the surface of Earth), where the temperature and pressure are −23°C and $3.00 \times 10^{-3}$ atm, respectively. Calculate the final volume of the balloon.

**11.39**  The temperature of 2.5 L of a gas initially at STP is raised to 210°C at constant volume. Calculate the final pressure of the gas in atmospheres.

**11.40**  The pressure of 6.0 L of an ideal gas in a flexible container is decreased to one-third of its original pressure, and its absolute temperature is decreased by one-half. What is the final volume of the gas?

**11.41**  A gas evolved during the fermentation of glucose (wine making) has a volume of 0.67 L at 22.5°C and 1.00 atm. What was the volume of this gas at the fermentation temperature of 36.5°C and 1.00 atm pressure?

**11.42**  An ideal gas originally at 0.85 atm and 66°C was allowed to expand until its final volume, pressure, and temperature were 94 mL, 0.60 atm, and 45°C, respectively. What was its initial volume?

**11.43**  Calculate the volume (in liters) of 124.3 g of $CO_2$ at STP.

**11.44**  A gas at 572 mmHg and 35.0°C occupies a volume of 6.15 L. Calculate its volume at STP.

**11.45**  Dry ice is solid carbon dioxide. A 0.050-g sample of dry ice is placed in an evacuated 4.6-L vessel at 30°C. Calculate the pressure inside the vessel after all the dry ice has been converted to $CO_2$ gas.

**11.46**  At STP, 0.280 L of a gas weighs 0.400 g. Calculate the molar mass of the gas.

**11.47**  At 741 torr and 44°C, 7.10 g of a gas occupies a volume of 5.40 L. What is the molar mass of the gas?

**11.48**  Ozone molecules in the stratosphere absorb much of the harmful radiation from the sun. Typically, the temperature and pressure of ozone in the stratosphere are 250 K and $1.0 \times 10^{-3}$ atm, respectively. How many ozone molecules are present in 1.0 L of air under these conditions?

**11.49**  Assuming that air contains 78 percent $N_2$, 21 percent $O_2$, and 1.0 percent Ar, all by volume, how many molecules of each type of gas are present in 1.0 L of air at STP?

**11.50**  A 2.10-L vessel contains 4.65 g of a gas at 1.00 atm and 27.0°C. (a) Calculate the density of the gas in g/L. (b) What is the molar mass of the gas?

**11.51**  Calculate the density of hydrogen bromide (HBr) gas in g/L at 733 mmHg and 46°C.

**11.52**  A certain anesthetic contains 64.9 percent C, 13.5 percent H, and 21.6 percent O by mass. At 120°C and 750 mmHg, 1.00 L of the gaseous compound weighs 2.30 g. What is the molecular formula of the compound?

**11.53**  A compound has the empirical formula $SF_4$. At 20°C, 0.100 g of the gaseous compound occupies a volume of 22.1 mL and exerts a pressure of 1.02 atm. What is the molecular formula of the gas?

## Section 11.4: Reactions with Gaseous Reactants and Products

### Problems

**11.54**  Consider the formation of nitrogen dioxide from nitric oxide and oxygen:

$$2NO(g) + O_2(g) \longrightarrow 2NO_2(g)$$

If 9.0 L of NO is combined with excess $O_2$ at STP, what is the volume in liters of the $NO_2$ produced?

**11.55**  Methane, the principal component of natural gas, is used for heating and cooking. The combustion process is

$$CH_4(g) + 2O_2(g) \longrightarrow CO_2(g) + 2H_2O(l)$$

If 15.0 moles of $CH_4$ react with oxygen, what is the volume of $CO_2$ (in liters) produced at 23.0°C and 0.985 atm?

**11.56**  When coal is burned, the sulfur present in coal is converted to sulfur dioxide ($SO_2$), which is responsible for the acid rain phenomenon:

$$S(s) + O_2(g) \longrightarrow SO_2(g)$$

If 3.15 kg of S reacts with oxygen, calculate the volume of $SO_2$ gas (in mL) formed at 30.5°C and 1.04 atm.

**11.57**  In alcohol fermentation, yeast converts glucose to ethanol and carbon dioxide:

$$C_6H_{12}O_6(s) \longrightarrow 2C_2H_5OH(l) + 2CO_2(g)$$

If 5.97 g of glucose reacts and 1.44 L of $CO_2$ gas is collected at 293 K and 0.984 atm, what is the percent yield of the reaction?

**11.58**  A compound of P and F was analyzed as follows: Heating 0.2324 g of the compound in a 378-$cm^3$ container turned all of it to gas, which had a pressure of 97.3 mmHg at 77°C. Then the gas was mixed with calcium chloride solution, which converted all the F to 0.2631 g of $CaF_2$. Determine the molecular formula of the compound.

**11.59**  A quantity of 0.225 g of a metal M (molar mass = 27.0 g/mol) liberated 0.303 L of molecular hydrogen (measured at 17°C and 741 mmHg) from an excess of hydrochloric acid. Deduce from these data the corresponding equation, and write formulas for the oxide and sulfate of M.

**11.60**  What is the mass of the solid $NH_4Cl$ formed when 73.0 g of $NH_3$ is mixed with an equal mass of HCl? What is the volume of the gas remaining, measured at 14.0°C and 752 mmHg? What gas is it?

**11.61**  Dissolving 3.00 g of an impure sample of calcium carbonate in hydrochloric acid produced 0.656 L of carbon dioxide (measured at 20.0°C and 792 mmHg). Calculate the percent by mass of calcium carbonate in the sample. State any assumptions.

**11.62** Calculate the mass in grams of hydrogen chloride produced when 5.6 L of molecular hydrogen measured at STP react with an excess of molecular chlorine gas.

**11.63** Ethanol ($C_2H_5OH$) burns in air:

$$C_2H_5OH(l) + O_2(g) \longrightarrow CO_2(g) + H_2O(l)$$

Balance the equation and determine the volume of air in liters at 45.0°C and 793 mmHg required to burn 185 g of ethanol. Assume that air is 21.0 percent $O_2$ by volume.

## Section 11.5: Gas Mixtures

### Visualizing Chemistry
### Figure 11.15

**VC 11.1** The molar volume of hydrogen can be determined using the reaction of zinc metal and acid,

$$Zn(s) + 2H^+(aq) \longrightarrow Zn^{2+}(aq) + H_2(g)$$

as shown in Figure 11.15. When the reaction is complete, what does the space above the water in the graduated cylinder contain?

a) $H_2(g)$, $Zn^{2+}(aq)$, and $H_2O(g)$
b) $H_2(g)$ and $H_2O(g)$
a) $H_2(g)$, $H_2O(g)$, and air

**VC 11.2** How would the calculated molar volume be affected if we neglected to subtract the partial pressure of water vapor from the total pressure?

a) It would be greater.
b) It would be smaller.
c) It would not change.

**VC 11.3** How would the calculated molar volume be affected if we neglected to adjust the level of the graduated cylinder prior to reading the volume of gas collected? Assume that the level of water inside the graduated cylinder is higher than the level outside.

a) It would be greater.
b) It would be smaller.
c) It would not change.

**VC 11.4** How would the calculated molar volume be affected if some of the zinc metal failed to drop into the aqueous acid?

a) It would be greater.
b) It would be smaller.
c) It would not change.

### Review Questions

**11.64** State Dalton's law of partial pressures and explain what *mole fraction* is. Does mole fraction have units?

**11.65** What are the approximate partial pressures of $N_2$ and $O_2$ in air at the top of a mountain where atmospheric pressure is 0.8 atm? (See Problem 11.49.)

### Problems

**11.66** A mixture of gases contains 0.31 mol $CH_4$, 0.25 mol $C_2H_6$, and 0.29 mol $C_3H_8$. The total pressure is 1.50 atm. Calculate the partial pressures of the gases.

**11.67** A 2.5-L flask at 15°C contains a mixture of $N_2$, He, and Ne at partial pressures of 0.32 atm for $N_2$, 0.15 atm for He, and 0.42 atm for Ne. (a) Calculate the total pressure of the mixture.

(b) Calculate the volume in liters at STP occupied by He and Ne if the $N_2$ is removed selectively.

**11.68** Dry air near sea level has the following composition by volume: $N_2$, 78.08 percent; $O_2$, 20.94 percent; Ar, 0.93 percent; $CO_2$, 0.05 percent. The atmospheric pressure is 1.00 atm. Calculate (a) the partial pressure of each gas in atmospheres and (b) the concentration of each gas in mol/L at 0°C. (*Hint:* Because volume is proportional to the number of moles present, mole fractions of gases can be expressed as ratios of volumes at the same temperature and pressure.)

**11.69** A mixture of helium and neon gases is collected over water at 28.0°C and 745 mmHg. If the partial pressure of helium is 368 mmHg, what is the partial pressure of neon? (Vapor pressure of water at 28°C = 28.3 mmHg.)

**11.70** A piece of sodium metal reacts completely with water as follows:

$$2Na(s) + 2H_2O(l) \longrightarrow 2NaOH(aq) + H_2(g)$$

The hydrogen gas generated is collected over water at 25.0°C. The volume of the gas is 246 mL measured at 1.00 atm. Calculate the number of grams of sodium used in the reaction. (Vapor pressure of water at 25°C = 0.0313 atm.)

**11.71** A sample of zinc metal reacts completely with an excess of hydrochloric acid:

$$Zn(s) + 2HCl(aq) \longrightarrow ZnCl_2(aq) + H_2(g)$$

The hydrogen gas produced is collected over water at 25.0°C using an arrangement similar to that shown in Figure 11.14(a). The volume of the gas is 7.80 L, and the pressure is 0.980 atm. Calculate the amount of zinc metal in grams consumed in the reaction. (Vapor pressure of water at 25°C = 23.8 mmHg.)

**11.72** Helium is mixed with oxygen gas for deep-sea divers. Calculate the percent by volume of oxygen gas in the mixture if the diver has to submerge to a depth where the total pressure is 5.2 atm. The partial pressure of oxygen is maintained at 0.20 atm at this depth.

**11.73** A sample of ammonia ($NH_3$) gas is completely decomposed to nitrogen and hydrogen gases over heated iron wool. If the total pressure is 866 mmHg after the reaction, calculate the partial pressures of $N_2$ and $H_2$.

**11.74** Consider the three containers shown, all of which have the same volume and are at the same temperature. (a) Which container has the smallest mole fraction of gas A (red)? (b) Which container has the highest partial pressure of gas B (green)? (c) Which container has the highest total pressure?

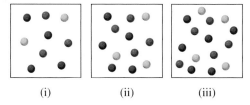

(i)          (ii)          (iii)

**11.75** The volume of the box on the right is twice that of the box on the left. The boxes contain helium atoms (red) and hydrogen molecules (green) at the same temperature. (a) Which box has a higher total pressure? (b) Which box has a higher partial pressure of helium?

## Section 11.6: The Kinetic Molecular Theory of Gases

*Review Questions*

11.76  What are the basic assumptions of the kinetic molecular theory of gases?

11.77  How does the kinetic molecular theory explain Boyle's law, Charles's law, Avogadro's law, and Dalton's law of partial pressures?

11.78  What does the Maxwell speed distribution curve tell us? Does Maxwell's theory work for a sample of 200 molecules? Explain.

11.79  Which of the following statements is correct? (a) Heat is produced by the collision of gas molecules against one another. (b) When a gas is heated at constant volume, the molecules collide with one another more often.

11.80  Uranium hexafluoride ($UF_6$) is a much heavier gas than helium, yet at a given temperature, the average kinetic energies of the samples of the two gases are the same. Explain.

11.81  What is the difference between gas diffusion and effusion?

*Problems*

11.82  Compare the root-mean-square speeds of $O_2$ and $UF_6$ at 65°C.

**11.83**  The temperature in the stratosphere is –23°C. Calculate the root-mean-square speeds of $N_2$, $O_2$, and $O_3$ molecules in this region.

11.84  The average distance traveled by a molecule between successive collisions is called *mean free path*. For a given amount of a gas, how does the mean free path of a gas depend on (a) density, (b) temperature at constant volume, (c) pressure at constant temperature, (d) volume at constant temperature, and (e) size of the atoms?

**11.85**  At a certain temperature the speeds of six gaseous molecules in a container are 2.0, 2.2, 2.6, 2.7, 3.3, and 3.5 m/s. Calculate the root-mean-square speed and the average speed of the molecules. These two average values are close to each other, but the root-mean-square value is always the larger of the two. Why?

11.86  The $^{235}U$ isotope undergoes fission when bombarded with neutrons. However, its natural abundance is only 0.72 percent. To separate it from the more abundant $^{238}U$ isotope, uranium is first converted to $UF_6$, which is easily vaporized above room temperature. The mixture of the $^{235}UF_6$ and $^{238}UF_6$ gases is then subjected to many stages of effusion. Calculate how much faster $^{235}UF_6$ effuses than $^{238}UF_6$.

11.87  An unknown gas evolved from the fermentation of glucose is found to effuse through a porous barrier in 15.0 min. Under the same conditions of temperature and pressure, it takes an equal volume of $N_2$ 12.0 min to effuse through the same barrier. Calculate the molar mass of the unknown gas, and suggest what the gas might be.

11.88  Nickel forms a gaseous compound of the formula $Ni(CO)_x$. What is the value of $x$ given the fact that under the same conditions of temperature and pressure, methane ($CH_4$) effuses 3.3 times faster than the compound?

## Section 11.7: Deviation from Ideal Behavior

*Review Questions*

11.89  Cite two pieces of evidence to show that gases do not behave ideally under all conditions. Under what set of conditions would a gas be expected to behave most ideally: (a) high temperature and low pressure, (b) high temperature and high pressure, (c) low temperature and high pressure, or (d) low temperature and low pressure?

11.90  Figure 11.25(a) shows that at 0°C, with the exception of $H_2$, each of the gases has a pressure at which its compressibility factor is equal to 1—the point at which the curve crosses the ideal gas line. What is the significance of this point? Does each of these gases have a pressure at which the assumptions of ideal behavior (negligible molecular volume and no intermolecular attractions) are valid? Explain.

11.91  Write the van der Waals equation for a real gas. Explain the corrective terms for pressure and volume.

11.92  (a) A real gas is introduced into a flask of volume $V$. Is the corrected volume of the gas greater or less than $V$? (b) Ammonia has a larger $a$ value than neon does (see Table 11.6). What can you conclude about the relative strength of the attractive forces between molecules of ammonia and between atoms of neon?

*Problems*

**11.93**  Using the data shown in Table 11.6, calculate the pressure exerted by 2.50 moles of $CO_2$ confined in a volume of 5.00 L at 450 K. Compare the pressure with that predicted by the ideal gas equation.

11.94  At 27°C, 10.0 moles of a gas in a 1.50-L container exert a pressure of 130 atm. Is this an ideal gas?

## Additional Problems

**11.95**  Discuss the following phenomena in terms of the gas laws: (a) the pressure increase in an automobile tire on a hot day, (b) the "popping" of a paper bag, (c) the expansion of a weather balloon as it rises in the air, (d) the loud noise heard when a lightbulb shatters.

11.96  Under the same conditions of temperature and pressure, which of the following gases would behave most ideally: Ne, $N_2$, or $CH_4$? Explain.

**11.97**  Nitroglycerin, an explosive compound, decomposes according to the equation

$$4C_3H_5(NO_3)_3(s) \longrightarrow 12CO_2(g) + 10H_2O(g) + 6N_2(g) + O_2(g)$$

Calculate the total volume of gases when collected at 1.2 atm and 25°C from $2.6 \times 10^2$ g of nitroglycerin. What are the partial pressures of the gases under these conditions?

11.98  The empirical formula of a compound is CH. At 200°C, 0.145 g of this compound occupies 97.2 mL at a pressure of 0.74 atm. What is the molecular formula of the compound?

11.99  When ammonium nitrite ($NH_4NO_2$) is heated, it decomposes to give nitrogen gas. This property is used to inflate some tennis balls. (a) Write a balanced equation for the reaction. (b) Calculate the quantity (in grams) of $NH_4NO_2$ needed to inflate a tennis ball to a volume of 86.2 mL at 1.20 atm and 22°C.

11.100 The percent by mass of bicarbonate ($HCO_3^-$) in a certain Alka-Seltzer product is 32.5 percent. Calculate the volume of $CO_2$ generated (in mL) at 37°C and 1.00 atm when a person ingests a 3.29-g tablet. (*Hint:* The reaction is between $HCO_3^-$ and HCl acid in the stomach.)

11.101 Three flasks containing gases A (red) and B (blue) are shown here. (a) If the total pressure in (i) is 2.0 atm, what are the pressures in (ii) and (iii)? (b) Calculate the total pressure and the partial pressure of each gas after the valves are opened. The volumes of (i) and (iii) are 2.0 L each, and the volume of (ii) is 1.0 L. The temperature is the same throughout.

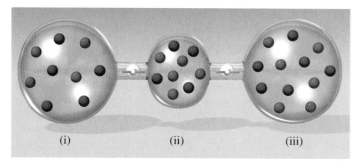

(i)                    (ii)                    (iii)

11.102 The boiling point of liquid nitrogen is −196°C. On the basis of this information alone, do you think nitrogen is an ideal gas at STP?

11.103 In the metallurgical process of refining nickel, the metal is first combined with carbon monoxide to form tetracarbonylnickel, which is a gas at 43°C:

$$Ni(s) + 4CO(g) \longrightarrow Ni(CO)_4(g)$$

This reaction separates nickel from other solid impurities. (a) Starting with 86.4 g of Ni, calculate the pressure of $Ni(CO)_4$ in a container of volume 4.00 L. (Assume the preceding reaction goes to completion.) (b) At temperatures above 43°C, the pressure of the gas is observed to increase much more rapidly than predicted by the ideal gas equation. Explain.

11.104 The partial pressure of carbon dioxide varies with seasons. Would you expect the partial pressure in the Northern Hemisphere to be higher in the summer or winter? Explain.

11.105 A healthy adult exhales about $5.0 \times 10^2$ mL of a gaseous mixture with each breath. Calculate the number of molecules present in this volume at 37°C and 1.1 atm. List the major components of this gaseous mixture.

11.106 Sodium bicarbonate ($NaHCO_3$) is called baking soda because, when heated, it releases carbon dioxide gas, which is responsible for the rising of cookies, some doughnuts, and cakes. (a) Calculate the volume (in liters) of $CO_2$ produced by heating 5.0 g of $NaHCO_3$ at 180°C and 1.3 atm. (b) Ammonium bicarbonate ($NH_4HCO_3$) has also been used for the same purpose. Suggest one advantage and one disadvantage of using $NH_4HCO_3$ instead of $NaHCO_3$ for baking.

11.107 On heating, potassium chlorate ($KClO_3$) decomposes to yield potassium chloride and oxygen gas. In one experiment, a student heated 20.4 g of $KClO_3$ until the decomposition was complete. (a) Write a balanced equation for the reaction. (b) Calculate the volume of oxygen (in liters) if it was collected at 0.962 atm and 18.3°C.

11.108 Some commercial drain cleaners contain a mixture of sodium hydroxide and aluminum powder. When the mixture is poured down a clogged drain, the following reaction occurs:

$$2NaOH(aq) + 2Al(s) + 6H_2O(l) \longrightarrow 2NaAl(OH)_4(aq) + 3H_2(g)$$

The heat generated in this reaction helps melt away obstructions such as grease, and the hydrogen gas released stirs up the solids clogging the drain. Calculate the volume of $H_2$ formed at 23°C and 1.00 atm if 3.12 g of Al are treated with an excess of NaOH.

11.109 The volume of a sample of pure HCl gas was 189 mL at 25°C and 108 mmHg. It was completely dissolved in about 60 mL of water and titrated with an NaOH solution; 15.7 mL of the NaOH solution was required to neutralize the HCl. Calculate the molarity of the NaOH solution.

11.110 Propane ($C_3H_8$) burns in oxygen to produce carbon dioxide gas and water vapor. (a) Write a balanced equation for this reaction. (b) Calculate the number of liters of carbon dioxide measured at STP that could be produced from 7.45 g of propane.

11.111 Consider the following apparatus. Calculate the partial pressures of helium and neon after the stopcock is opened. The temperature remains constant at 16°C.

1.2 L                                    3.4 L
0.63 atm                                 2.8 atm

11.112 Nitric oxide (NO) reacts with molecular oxygen as follows:

$$2NO(g) + O_2(g) \longrightarrow 2NO_2(g)$$

Initially NO and $O_2$ are separated as shown here. When the valve is opened, the reaction quickly goes to completion. Determine what gases remain at the end and calculate their partial pressures. Assume that the temperature remains constant at 25°C.

4.00 L at                                2.00 L at
0.500 atm                                1.00 atm

**11.113** Consider the apparatus shown here. When a small amount of water is introduced into the flask by squeezing the bulb of the medicine dropper, water is squirted upward out of the long glass tubing. Explain this observation. (*Hint:* Hydrogen chloride gas is soluble in water.)

**11.114** Describe how you would measure, by either chemical or physical means, the partial pressures of a mixture of gases of the following composition: (a) $CO_2$ and $H_2$, (b) He and $N_2$.

**11.115** A certain hydrate has the formula $MgSO_4 \cdot xH_2O$. A quantity of 54.2 g of the compound is heated in an oven to drive off the water. If the steam generated exerts a pressure of 24.8 atm in a 2.00-L container at 120°C, calculate $x$.

**11.116** A mixture of $Na_2CO_3$ and $MgCO_3$ of mass 7.63 g is combined with an excess of hydrochloric acid. The $CO_2$ gas generated occupies a volume of 1.67 L at 1.24 atm and 26°C. From these data, calculate the percent composition by mass of $Na_2CO_3$ in the mixture.

**11.117** The apparatus shown here can be used to measure atomic and molecular speeds. Suppose that a beam of metal atoms is directed at a rotating cylinder in a vacuum. A small opening in the cylinder allows the atoms to strike a target area. Because the cylinder is rotating, atoms traveling at different speeds will strike the target at different positions. In time, a layer of the metal will deposit on the target area, and the variation in its thickness is found to correspond to Maxwell's speed distribution. In one experiment it is found that at 850°C some bismuth (Bi) atoms struck the target at a point 2.80 cm from the spot directly opposite the slit. The diameter of the cylinder is 15.0 cm, and it is rotating at 130 revolutions per second. (a) Calculate the speed (in m/s) at which the target is moving. (*Hint:* The circumference of a circle is given by $2\pi r$, where $r$ is the radius.) (b) Calculate the time (in seconds) it takes for the target to travel 2.80 cm. (c) Determine the speed of the Bi atoms. Compare your result in part (c) with the $u_{rms}$ of Bi at 850°C. Comment on the difference.

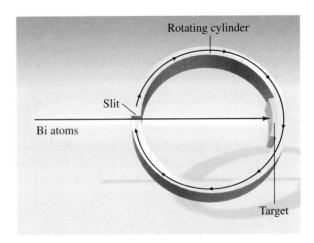

**11.118** If 10.00 g of water is introduced into an evacuated flask of volume 2.500 L at 65°C, calculate the mass of water vaporized. (*Hint:* Assume that the volume of the remaining liquid water is negligible; the vapor pressure of water at 65°C is 187.5 mmHg.)

**11.119** Commercially, compressed oxygen is sold in metal cylinders. If a 120-L cylinder is filled with oxygen to a pressure of 132 atm at 22°C, what is the mass of $O_2$ present? How many liters of $O_2$ gas at 1.00 atm and 22°C could the cylinder produce? (Assume ideal behavior.)

**11.120** The shells of hard-boiled eggs sometimes crack due to the rapid thermal expansion of the shells at high temperatures. Suggest another reason why the shells may crack.

**11.121** Ethylene gas ($C_2H_4$) is emitted by fruits and is known to be responsible for their ripening. Based on this information, explain why a bunch of bananas ripens faster in a closed paper bag than in an open bowl.

**11.122** About $8.0 \times 10^6$ tons of urea [$(NH_2)_2CO$] is used annually as a fertilizer. The urea is prepared at 200°C and under high-pressure conditions from carbon dioxide and ammonia (the products are urea and steam). Calculate the volume of ammonia (in liters) measured at 150 atm needed to prepare 1.0 ton of urea.

**11.123** Some ballpoint pens have a small hole in the main body of the pen. What is the purpose of this hole?

**11.124** The gas laws are vitally important to scuba divers. The pressure exerted by 33 ft of seawater is equivalent to 1 atm pressure. (a) A diver ascends quickly to the surface of the water from a depth of 36 ft without exhaling gas from his lungs. By what factor will the volume of his lungs increase by the time he reaches the surface? Assume that the temperature is constant. (b) The partial pressure of oxygen in air is about 0.20 atm. (Air is 20 percent oxygen by volume.) In deep-sea diving, the composition of air the diver breathes must be changed to maintain this partial pressure. What must the oxygen content (in percent by volume) be when the total pressure exerted on the diver is 4.0 atm? (At constant temperature and pressure, the volume of a gas is directly proportional to the number of moles of gases.)

**11.125** Nitrous oxide ($N_2O$) can be obtained by the thermal decomposition of ammonium nitrate ($NH_4NO_3$). (a) Write a balanced equation for the reaction. (b) In a certain experiment, a student obtains 0.340 L of the gas at 718 mmHg and 24°C. If the gas weighs 0.580 g, calculate the value of the gas constant.

11.126  Two vessels are labeled A and B. Vessel A contains $NH_3$ gas at 70°C, and vessel B contains Ne gas at the same temperature. If the average kinetic energy of $NH_3$ is $7.1 \times 10^{-21}$ J/molecule, calculate the root-mean-square speed of Ne atoms in $m^2/s^2$.

11.127  Which of the following molecules has the largest $a$ value: $CH_4$, $F_2$, $C_6H_6$, Ne?

11.128  The following procedure is a simple though somewhat crude way to measure the molar mass of a gas. A liquid of mass 0.0184 g is introduced into a syringe like the one shown here by injection through the rubber tip using a hypodermic needle. The syringe is then transferred to a temperature bath heated to 45°C, and the liquid vaporizes. The final volume of the vapor (measured by the outward movement of the plunger) is 5.58 mL, and the atmospheric pressure is 760 mmHg. Given that the compound's empirical formula is $CH_2$, determine the molar mass of the compound.

Rubber tip

11.129  In 1995 a man suffocated as he walked by an abandoned mine in England. At that moment there was a sharp drop in atmospheric pressure due to a change in the weather. Suggest what might have caused the man's death.

11.130  Acidic oxides such as carbon dioxide react with basic oxides like calcium oxide (CaO) and barium oxide (BaO) to form salts (metal carbonates). (a) Write equations representing these two reactions. (b) A student placed a mixture of BaO and CaO of combined mass 4.88 g in a 1.46-L flask containing carbon dioxide gas at 35°C and 746 mmHg. After the reactions were complete, she found that the $CO_2$ pressure had dropped to 252 mmHg. Calculate the percent composition by mass of the mixture. Assume that the volumes of the solids are negligible.

11.131  (a) What volume of air at 1.0 atm and 22°C is needed to fill a 0.98-L bicycle tire to a pressure of 5.0 atm at the same temperature? (Note that the 5.0 atm is the gauge pressure, which is the difference between the pressure in the tire and atmospheric pressure. Before filling, the pressure in the tire was 1.0 atm.) (b) What is the total pressure in the tire when the gauge pressure reads 5.0 atm? (c) The tire is pumped by filling the cylinder of a hand pump with air at 1.0 atm and then, by compressing the gas in the cylinder, adding all the air in the pump to the air in the tire. If the volume of the pump is 33 percent of the tire's volume, what is the gauge pressure in the tire after three full strokes of the pump? Assume constant temperature.

11.132  The running engine of an automobile produces carbon monoxide (CO), a toxic gas, at the rate of about 188 g CO per hour. A car is left idling in a poorly ventilated garage that is 6.0 m long, 4.0 m wide, and 2.2 m high at 20°C. (a) Calculate the rate of CO production in mol/min. (b) How long would it take to build up a lethal concentration of CO of 1000 ppmv (parts per million by volume)?

11.133  Interstellar space contains mostly hydrogen atoms at a concentration of about 1 atom/$cm^3$. (a) Calculate the pressure of the H atoms. (b) Calculate the volume (in liters) that contains 1.0 g of H atoms. The temperature is 3 K.

11.134  Atop Mt. Everest, the atmospheric pressure is 210 mmHg and the air density is 0.426 kg/$m^3$. (a) Calculate the air temperature, given that the molar mass of air is 29.0 g/mol. (b) Assuming no change in air composition, calculate the percent decrease in oxygen gas from sea level to the top of Mt. Everest.

11.135  Relative humidity is defined as the ratio (expressed as a percentage) of the partial pressure of water vapor in the air to the equilibrium vapor pressure (see Table 11.5) at a given temperature. On a certain summer day in North Carolina the partial pressure of water vapor in the air is $3.9 \times 10^3$ Pa at 30°C. Calculate the relative humidity.

11.136  Under the same conditions of temperature and pressure, why does 1 L of moist air weigh less than 1 L of dry air? In weather forecasts, an oncoming low-pressure front usually means imminent rainfall. Explain.

11.137  Air entering the lungs ends up in tiny sacs called alveoli. It is from the alveoli that oxygen diffuses into the blood. The average radius of the alveoli is 0.0050 cm, and the air inside contains 14 percent oxygen. Assuming that the pressure in the alveoli is 1.0 atm and the temperature is 37°C, calculate the number of oxygen molecules in one of the alveoli. (Hint: The volume of a sphere of radius $r$ is $\frac{4}{3}\pi r^3$.)

11.138  A student breaks a thermometer and spills most of the mercury (Hg) onto the floor of a laboratory that measures 15.2 m long, 6.6 m wide, and 2.4 m high. (a) Calculate the mass of mercury vapor (in grams) in the room at 20°C. The vapor pressure of mercury at 20°C is $1.7 \times 10^{-6}$ atm. (b) Does the concentration of mercury vapor exceed the air quality regulation of 0.050 mg Hg/$m^3$ of air? (c) One way to deal with small quantities of spilled mercury is to spray sulfur powder over the metal. Suggest a physical and a chemical reason for this action.

11.139  Nitrogen forms several gaseous oxides. One of them has a density of 1.33 g/L measured at 764 mmHg and 150°C. Write the formula of the compound.

11.140  Nitrogen dioxide ($NO_2$) cannot be obtained in a pure form in the gas phase because it exists as a mixture of $NO_2$ and $N_2O_4$. At 25°C and 0.98 atm, the density of this gas mixture is 2.7 g/L. What is the partial pressure of each gas?

11.141  Lithium hydride reacts with water as follows:

$$LiH(s) + H_2O(l) \longrightarrow LiOH(aq) + H_2(g)$$

During World War II, U.S. pilots carried LiH tablets. In the event of a crash landing at sea, the LiH would react with the seawater and fill their life jackets and lifeboats with hydrogen gas. How many grams of LiH are needed to fill a 4.1-L life jacket at 0.97 atm and 12°C?

11.142  The atmosphere on Mars is composed mainly of carbon dioxide. The surface temperature is 220 K, and the atmospheric pressure is about 6.0 mmHg. Taking these values as Martian "STP," calculate the molar volume in liters of an ideal gas on Mars.

11.143  Venus's atmosphere is composed of 96.5 percent $CO_2$, 3.5 percent $N_2$, and 0.015 percent $SO_2$ by volume. Its standard atmospheric pressure is $9.0 \times 10^6$ Pa. Calculate the partial pressures of the gases in pascals.

11.144  A student tries to determine the volume of a bulb like the one shown in Figure 11.12. These are her results: mass of the bulb filled with dry air at 23°C and 744 mmHg = 91.6843 g; mass of evacuated bulb = 91.4715 g. Assume the composition of air is 78 percent $N_2$, 21 percent $O_2$, and 1 percent argon by volume. What is the volume (in mL) of the bulb? (*Hint:* First calculate the average molar mass of air, as shown in Problem 3.153.)

**11.145** Apply your knowledge of the kinetic theory of gases to the following situations. (a) Two flasks of volumes $V_1$ and $V_2$ ($V_2 > V_1$) contain the same number of helium atoms at the same temperature. (i) Compare the root-mean-square (rms) speeds and average kinetic energies of the helium (He) atoms in the flasks. (ii) Compare the frequency and the force with which the He atoms collide with the walls of their containers. (b) Equal numbers of He atoms are placed in two flasks of the same volume at temperatures $T_1$ and $T_2$ ($T_2 > T_1$). (i) Compare the rms speeds of the atoms in the two flasks. (ii) Compare the frequency and the force with which the He atoms collide with the walls of their containers. (c) Equal numbers of He and neon (Ne) atoms are placed in two flasks of the same volume, and the temperature of both gases is 74°C. Comment on the validity of the following statements: (i) The rms speed of He is equal to that of Ne. (ii) The average kinetic energies of the two gases are equal. (iii) The rms speed of each He atom is $1.47 \times 10^3$ m/s.

**11.146** At what temperature will He atoms have the same $u_{rms}$ value as $N_2$ molecules at 25°C?

**11.147** Estimate the distance (in nm) between molecules of water vapor at 100°C and 1.0 atm. Assume ideal behavior. Repeat the calculation for liquid water at 100°C, given that the density of water is 0.96 g/cm³ at that temperature. Comment on your results. (Assume each water molecule to be a sphere with a diameter of 0.3 nm.) (*Hint:* First calculate the number density of water molecules. Next, convert the number density to linear density, that is, the number of molecules in one direction.)

**11.148** Which of the noble gases would not behave ideally under any circumstance? Why?

**11.149** A 5.72-g sample of graphite was heated with 68.4 g of $O_2$ in a 8.00-L flask. The reaction that took place was

$$C(graphite) + O_2(g) \longrightarrow CO_2(g)$$

After the reaction was complete, the temperature in the flask was 182°C. What was the total pressure inside the flask?

**11.150** A 6.11-g sample of a Cu-Zn alloy reacts with HCl acid to produce hydrogen gas. If the hydrogen gas has a volume of 1.26 L at 22°C and 728 mmHg, what is the percent of Zn in the alloy? (*Hint:* Cu does not react with HCl.)

**11.151** A stockroom supervisor measured the contents of a 25.0-gal drum partially filled with acetone on a day when the temperature was 18.0°C and atmospheric pressure was 750 mmHg, and found that 15.4 gal of the solvent remained. After tightly sealing the drum, an assistant dropped the drum while carrying it upstairs to the organic laboratory. The drum was dented, and its internal volume was decreased to 20.4 gal. What is the total pressure inside the drum after the accident? The vapor pressure of acetone at 18.0°C is 400 mmHg. (*Hint:* At the time the drum was sealed, the pressure inside the drum, which is equal to the sum of the pressures of air and acetone, was equal to the atmospheric pressure.)

**11.152** In 2.00 min, 29.7 mL of He effuses through a small hole. Under the same conditions of pressure and temperature, 10.0 mL of a mixture of CO and $CO_2$ effuses through the hole in the same amount of time. Calculate the percent composition by volume of the mixture.

**11.153** A mixture of methane ($CH_4$) and ethane ($C_2H_6$) is stored in a container at 294 mmHg. The gases are burned in air to form $CO_2$ and $H_2O$. If the pressure of $CO_2$ is 356 mmHg, measured at the same temperature and volume as the original mixture, calculate the mole fractions of the gases.

**11.154** Use the kinetic theory of gases to explain why hot air rises.

**11.155** Given that the van der Waals constant $b$ is the excluded volume and that the excluded volume is four times the volume actually occupied by the gas molecules in a sample (see Sample Problem 11.18), determine what percentage of the container volume is actually occupied by $CCl_4(g)$ molecules at (a) STP, (b) 10.0 atm and 273 K, and (c) 50.0 atm and 273 K.

**11.156** A 5.00-mole sample of $NH_3$ gas is kept in a 1.92-L container at 300 K. If the van der Waals equation is assumed to give the correct answer for the pressure of the gas, calculate the percent error made in using the ideal gas equation to calculate the pressure.

**11.157** The root-mean-square speed of a certain gaseous oxide is 493 m/s at 20°C. What is the molecular formula of the compound?

**11.158** Referring to Figure 11.20, we see that the maximum of each speed distribution plot is called the most probable speed ($u_{mp}$) because it is the speed possessed by the largest number of molecules. It is given by $u_{mp} = \sqrt{2RT/M}$. (a) Compare $u_{mp}$ with $u_{rms}$ for nitrogen at 25°C. (b) The following diagram shows the Maxwell speed distribution curves for an ideal gas at two different temperatures $T_1$ and $T_2$. Calculate the value of $T_2$.

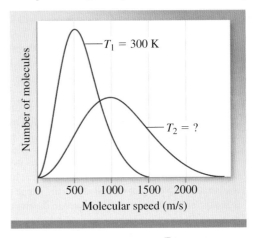

**11.159** A gaseous reaction takes place at constant volume and constant pressure in a cylinder as shown here. Which of the following equations best describes the reaction? The initial temperature ($T_1$) is twice that of the final temperature ($T_2$).

(a) $A + B \longrightarrow C$
(b) $AB \longrightarrow C + D$
(c) $A + B \longrightarrow C + D$
(d) $A + B \longrightarrow 2C + D$

11.160  Assuming ideal behavior, which of the following gases will have the greatest volume at STP? (a) 0.82 mole of He, (b) 24 g of $N_2$, or (c) $5.0 \times 10^{23}$ molecules of $Cl_2$.

🔍 **11.161**  Calculate the density of helium in a helium balloon at 25.0°C. (Assume that the pressure inside the balloon is 1.10 atm.)

11.162  Helium atoms in a closed container at room temperature are constantly colliding with one another and with the walls of their container. Does this "perpetual motion" violate the law of conservation of energy? Explain.

**11.163**  Uranium hexafluoride ($UF_6$) is a much heavier gas than hydrogen, yet at a given temperature, the average kinetic energies of these two gases are the same. Explain.

11.164  Consider the molar volumes shown in Figure 11.23. (a) Explain why $Cl_2$ and $NH_3$ have molar volumes significantly smaller from that of an ideal gas. (b) Explain why $H_2$, He, and Ne have molar

volumes greater than that of an ideal gas. (*Hint:* Look up the boiling points of the gases shown in the figure.)

**11.165**  The plot of $Z$ versus $P$ for a gas at 0°C is shown. Explain the causes of the negative deviation from ideal behavior at lower pressures and the positive deviation from ideal behavior at higher pressures.

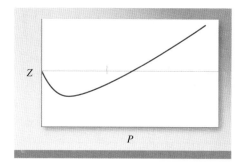

# PRE-PROFESSIONAL PRACTICE EXAM PROBLEMS: PHYSICAL AND BIOLOGICAL SCIENCES

It has been said that every breath we take, on average, contains molecules that were once exhaled by Wolfgang Amadeus Mozart (1756–1791).

1.  Calculate the total number of molecules in the atmosphere. (Assume that the total mass of the atmosphere is $6 \times 10^{18}$ kg and the average molar mass of air is 29.0 g/mol.)

    a) $1 \times 10^{23}$
    b) $1 \times 10^{26}$
    c) $1 \times 10^{29}$
    d) $1 \times 10^{18}$

2.  Assuming the volume of every breath (inhale or exhale) is 0.5 L, calculate the number of molecules exhaled in each breath at body temperature (37°C) and 1 atm.

    a) $1 \times 10^{22}$
    b) $1 \times 10^{21}$
    c) $1 \times 10^{23}$
    d) $6 \times 10^{23}$

3.  Calculate the mass of air exhaled with each breath.

    a) 0.02 g
    b) 0.6 g
    c) 0.2 g
    d) 6 g

4.  If Mozart's life span was exactly 35 years, what is the number of molecules he exhaled in that period (given that an average person breathes 12 times per minute)?

    a) $2 \times 10^8$
    b) $2 \times 10^{29}$
    c) $1 \times 10^{29}$
    d) $3 \times 10^{30}$

# ANSWERS TO IN-CHAPTER PROBLEMS

## Answers to Practice Problems

**11.1A** 1.32 atm. **11.1B** 9.52 m. **11.2A** 2.23 L. **11.2B** 1.80 atm. **11.3A** 30.7 L. **11.3B** 300°C. **11.4A** 34 L. **11.4B** 3.16 L CO, 1.58 L $O_2$. **11.5A** 5.09 L. **11.5B** 85.0 m. A common mistake in this problem is failure to substract the atmospheric pressure (0.965 atm) from the total pressure at the bottom of the lake (9.19 atm). The pressure due to the *water* is only 8.23 atm. **11.6A** 128 L. **11.6B** 336°C. **11.7A** 1.29 g/L. **11.7B** 11 atm. **11.8A** 55.2 g/mol. **11.8B** 133.622 g. **11.9A** 151 L. **11.9B** 3.48 g. **11.10A** $1.0 \times 10^2$ mol. **11.10B** 0.0052 atm. **11.11A** 8.48 L. **11.11B** 184 g. **11.12A** $P_{He} = 0.184$ atm, $P_{H_2} = 0.390$ atm, $P_{Ne} = 1.60$ atm, $P_{total} = 2.18$ atm. **11.12B** 0.032 mol $CH_4$, 0.091 mol $C_2H_6$, 0.123 mol total. **11.13A** $\chi_{CO_2} = 0.0495$, $\chi_{CH_4} = 0.255$, $\chi_{He} = 0.695$, $P_{CO_2} = 0.286$ atm, $P_{CH_4} = 1.47$ atm, $P_{He} = 4.02$ atm. **11.13B** $P_{Xe} = 4.95$ atm, $P_{Ne} = 1.55$ atm,

$n_{Xe} = 3.13$, $n_{Ne} = 0.984$. **11.14A** 1.03 g. **11.14B** 0.386 L. **11.15A** 0.61 atm. **11.15B** 13 atm. **11.16A** 2.137. **11.16B** 2.02 g/mol, $H_2$. **11.17A** 50.6 atm, 51.6 atm. **11.17B** 1.3 atm, 1.3 atm. **11.18A** 0.0453 L/mol, 0.0113 L. **11.18B** 146 pm.

## Answers to Checkpoints

**11.1.1** a. **11.1.2** b, c, d. **11.1.3** d. **11.1.4** b. **11.1.5** e. **11.2.1** e. **11.2.2** a. **11.2.3** c. **11.2.4** a. **11.2.5** b. **11.2.6** d. **11.3.1** a. **11.3.2** e. **11.3.3** b. **11.3.4** b. **11.4.1** d. **11.4.2** b. **11.5.1** e. **11.5.2** b. **11.5.3** a. **11.5.4** d. **11.5.5** c. **11.5.6** b. **11.6.1** c. **11.6.2** b. **11.7.1** a, d. **11.7.2** b.

## Answers to Applying What You've Learned

**a)** 0.39 m. **b)** 245 L. **c)** 0.66 g/L. **d)** $4.6 \times 10^2$ L. **e)** 8.7 mol $N_2$, 2.2 mol $O_2$.

# 12

# Intermolecular Forces and the Physical Properties of Liquids and Solids

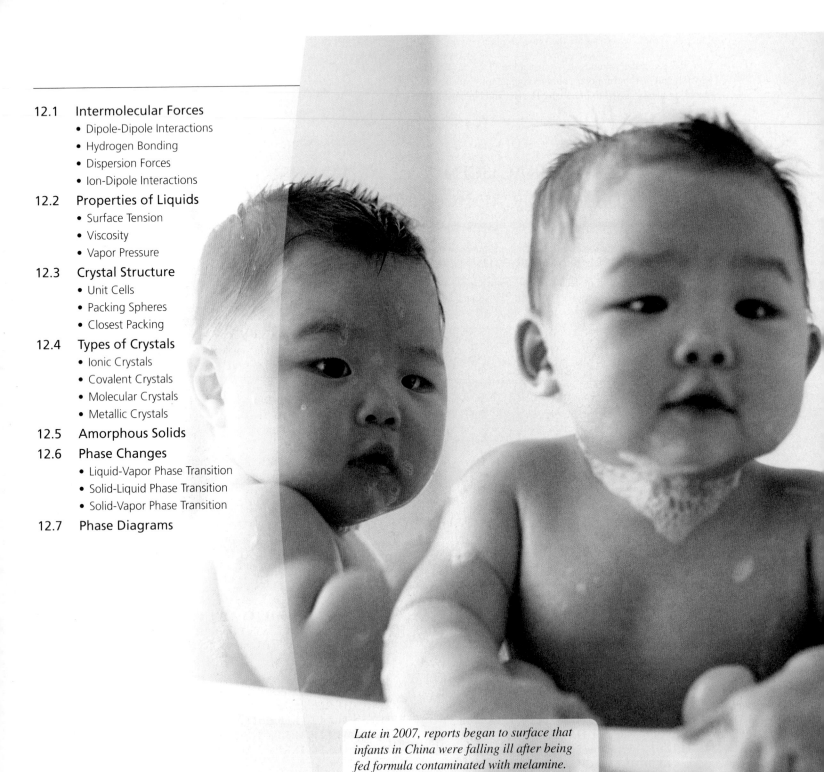

*Late in 2007, reports began to surface that infants in China were falling ill after being fed formula contaminated with melamine. The contamination was the result of criminal activity.*

# How Intermolecular Forces Contributed to the Contaminated Infant Formula Tragedy in China

Life as we know it depends heavily on intermolecular forces—the attractions that keep molecules together. Without them, water—perhaps the most essential ingredient in life—would not be a liquid. On this occasion, though, intermolecular forces caused two contaminants in infant formula, each relatively innocuous by itself, to form a highly toxic complex that caused many thousands of babies in China to develop acute kidney failure.

The contamination turned out to have been the result of foul play. Unscrupulous vendors of dairy products sought to increase their profits by diluting the milk they sold. However, because its protein content would have appeared too low, diluted milk would not have passed the industry quality control tests. To boost the apparent protein content, the vendors added melamine ($C_3H_6N_6$) to the diluted milk. Although melamine is not a protein, its high nitrogen content (66.6 percent by mass) made the milk appear to contain adequate protein. The industry test for protein does not measure protein specifically but rather nitrogen content, which in unadulterated milk would give a measure of the protein content. In addition to being added to the milk illegally, the melamine was itself contaminated with another compound, cyanuric acid ($C_3H_3N_3O_3$). At sufficiently high concentrations, these two contaminants form an insoluble complex that precipitates as an unusual type of stone, causing severe kidney damage. The molecules in the insoluble complex are held together by powerful attractive forces known as *hydrogen bonds* (shown as dashed red lines).

Secrecy surrounding the scandal has made it impossible to know for certain how many babies were poisoned—and how many died. The formation of the toxic melamine-cyanuric acid complex is the result of particularly strong attractive forces between the melamine and cyanuric acid molecules. Attractive forces between molecules (intermolecular forces) and within molecules (intramolecular forces) determine many of the physical properties of liquids and solids.

> At the end of this chapter, you will be able to answer questions about the physical properties of substances used in the treatment and prevention of kidney stones [▶ Page 527].

## In This Chapter, You Will Learn:

About the types of forces that hold the molecules (or atoms or ions) in substances together and about the properties that arise because of these forces.

### Before you begin, you should review

- Molecular geometry and polarity [◀◀ Section 9.2]
- Basic trigonometry [▶▶ Appendix 1]

## 12.1  Intermolecular Forces

In Chapter 11, we learned that gases consist of rapidly moving particles (molecules or atoms), separated by relatively large distances. Liquids and solids, the *condensed* phases [◄◄ margin note on page 6], consist of particles (molecules, atoms, or ions) that are touching one another. The attractive forces that hold particles together in the condensed phases are called *intermolecular forces.* The *magnitude* of intermolecular forces is what determines whether the particles that make up a substance are a gas, liquid, or solid.

Gas

Particles in a gas are separated by large distances and free to move entirely independently of one another.

Liquid

Particles in a liquid are touching one another but free to move about.

Solid

Particles in a solid are essentially locked in place with respect to one another.

**Student Annotation:** The term *intermolecular* is used to refer to attractive forces between molecules, atoms, or ions.

We have already encountered an example of "intermolecular" forces in the form of ionic bonding [◄◄ Section 8.2], where the magnitude of attraction between oppositely charged particles is governed by Coulomb's law [◄◄ Section 7.3]. Because the particles that make up an ionic compound have discrete (*full*) charges, the attractive forces that hold them together are especially powerful. In fact, that's the reason that they're solids at room temperature. The intermolecular forces that we will encounter in this chapter are also the result of Coulombic attractions, but the attractions involve only partial charges [◄◄ Section 8.4] rather than discrete charges and are therefore weaker than the forces involved in ionic bonding.

We will begin our discussion of intermolecular forces with the attractive forces that act between atoms or molecules in a pure substance. These forces are known collectively as *van der Waals forces,* and they include *dipole-dipole interactions,* including *hydrogen bonding* and *dispersion forces.*

**Student Annotation:** Hydrogen bonding constitutes a special type of dipole-dipole interaction and is discussed separately.

### Dipole-Dipole Interactions

*Dipole-dipole interactions* are attractive forces that act between *polar molecules.* Recall that a diatomic molecule containing elements of significantly different electronegativities, such as HCl, has an unequal distribution of electron density and therefore has partial charges, positive ($\delta+$) at one end and negative ($\delta-$) at the other. The partial positive charge on one molecule is attracted to the partial negative charge on a neighboring molecule. Figure 12.1 shows the orientation of polar molecules in a solid and in a liquid. The arrangement is somewhat less orderly in the liquid than it is in the solid.

Because this attractive force between polar molecules is Coulombic, the magnitude of the attractive forces depends on the magnitude of the dipole. In general, the larger the dipole, the larger the attractive force. Certain physical properties such as boiling point *reflect* the magnitude of intermolecular forces. A substance in which the particles are held together by larger intermolecular attractions will require more energy to *separate* the particles and will therefore boil at a higher temperature. Table 12.1 lists several compounds with similar molar masses along with their dipole moments and boiling points.

HCl

Liquid

Solid

**Figure 12.1**  Arrangement of polar molecules in a liquid (left) and in a solid (right).

| TABLE 12.1 | Dipole Moments and Boiling Points of Compounds with Similar Molecular Masses | | |
|---|---|---|---|
| **Compound** | **Structural Formula** | **Dipole Moment (D)** | **Boiling Point (°C)** |
| Propane | $CH_3CH_2CH_3$ | 0.1 | −42 |
| Dimethyl ether | $CH_3OCH_3$ | 1.3 | −25 |
| Methyl chloride | $CH_3Cl$ | 1.9 | −24 |
| Acetaldehyde | $CH_3CHO$ | 2.7 | 21 |
| Acetonitrile | $CH_3CN$ | 2.9 | 82 |

## Hydrogen Bonding

***Hydrogen bonding*** is a special type of dipole-dipole interaction. But, whereas dipole-dipole interactions act between any polar molecules, hydrogen bonding occurs only in molecules that contain H bonded to a small, highly electronegative atom, such as N, O, or F. In a molecule such as HF, which is shown in Figure 12.2, the F atom to which H is bonded draws electron density toward itself. Being small and very electronegative, the F atom draws electron density away from H very effectively. Because H has only one electron, this leaves the hydrogen nucleus practically unshielded—giving H a very large partial positive charge. This large partial positive charge is powerfully attracted to the large partial negative charge (lone pairs) on the small, highly electronegative F atom of a neighboring HF molecule. The result is an especially strong dipole-dipole attraction.

In Sample Problem 8.5 we calculated the partial charges on H and F as +0.41 and −0.41, respectively. The partial charges on H and Cl, by contrast, are only +0.18 and −0.18, respectively. They are smaller in HCl because Cl is larger and less electronegative than F. Because of the very high electronegativity of F, and the high partial positive charges that result, HF exhibits significant hydrogen bonding, whereas HCl does not.

Figure 12.3 (page 496) shows the boiling points of the binary hydrogen compounds of Groups 4A through 7A. Within the series of hydrogen compounds of Group 4A, the boiling point increases with increasing molar mass. For Groups 5A through 7A, the same trend is observed for all but the smallest member of each series, which has what appears to be an unexpectedly high boiling point. This departure from the trend in boiling points illustrates how powerful hydrogen bonding can be. $NH_3$ (Group 5A), HF (Group 7A), and $H_2O$ (Group 6A) have anomalously high boiling points because they exhibit strong hydrogen bonding. $CH_4$ (the smallest member of the Group 4A series) does not exhibit hydrogen bonding because it does not contain an N—H, O—H, or F—H bond; hence, it has the lowest boiling point of that group.

**Student Annotation:** It is a common error to assume that hydrogen bonding occurs in any molecule that contains hydrogen. It *only* occurs to a significant degree in molecules with N—H, O—H, or F—H bonds.

**Student Annotation:** Strictly speaking, HCl does exhibit hydrogen bonding to a very small degree. But only molecules containing an N—H, O—H, or F—H bond exhibit significant hydrogen bonding.

**Student Annotation:** Hydrogen bonding also occurs in mixtures, between solute and solvent molecules that contain N—H, F—H, or O—H bonds.

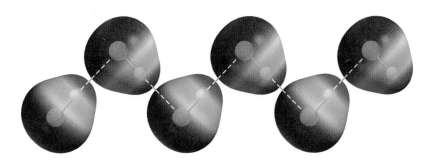

**Figure 12.2**  Hydrogen bonds between HF molecules.

**Figure 12.3**   Boiling points of the hydrogen compounds of elements from Groups 4A through 7A. Although normally the boiling point increases with increasing mass within a group, the *lightest* compound has the *highest* boiling point in Groups 5A through 7A. This departure from the observed trend is due to hydrogen bonding.

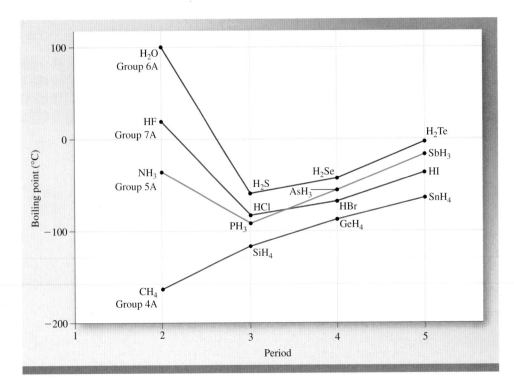

## Bringing Chemistry to Life

### Sickle Cell Disease

Amino acids are the building blocks of proteins. Each amino acid has both amino ($-NH_2$) and carboxy ($-COOH$) functional groups, in addition to a side group that may be polar or nonpolar. Valine (Val), histidine (His), leucine (Leu), threonine (Thr), proline (Pro), and glutamic acid (Glu) are 6 of the 20 amino acids [◄◄ Section 10.6] that make up human proteins. The characteristic side group of each amino acid is shaded to identify it:

Valine (Val)    Histidine (His)    Leucine (Leu)    Threonine (Thr)    Proline (Pro)    Glutamic acid (Glu)

Proteins form when amino acids are joined together with peptide bonds, also known as *amide linkages,* in which the carboxy group on one amino acid connects to the amino group of the next. The carboxy group of the second amino acid connects to the amino group of another, and so on. A water molecule is eliminated with the formation of each peptide bond. What remains of each amino acid after the formation of peptide bonds (and the corresponding loss of water molecules) is referred to as an *amino acid residue,* or simply a *residue.* (Peptide bonds are shown in blue.)

$H_2O$    $H_2O$    $H_2O$    $H_2O$    $H_2O$

The primary structure of a protein refers to the sequence of amino acids that make up the protein chain.

Secondary structure refers to the shape the protein chain adopts as the result of hydrogen bonding between nearby residues. One of the possible secondary structures is a helix.

Tertiary structure refers to the folding of the protein into a characteristic shape, which is stabilized by attractive forces between more distant residues.

Quaternary structure refers to the shape resulting from the assembly of two or more proteins, called subunits, to form a larger protein complex. In the case of hemoglobin, four subunits [two α subunits (purple), each consisting of 141 amino acids; and two β subunits (pink), each consisting of 146 amino acids] make up the larger complex.

Hemoglobin is a *globular* protein, meaning that it assembles into a roughly spherical shape with polar residues covering most of the sphere's surface and nonpolar residues largely confined to the interior of the complex. Distribution of polar residues on the outside of the sphere enhances a globular protein's aqueous solubility.

In normal hemoglobin, the sixth amino acid residue (of 146) in each β subunit is a glutamic acid residue, which is polar. In hemoglobin S, the type associated with sickle cell disease, the glutamic acid residue in the sixth position of each β subunit is replaced by a valine residue, which is nonpolar. This results in a nonpolar group on the surface of the protein complex, where there should be a polar group.

Although most of the surface of the spherical hemoglobin complex is polar, when it is not oxygenated, one of hemoglobin's β subunits has a small nonpolar region on the exterior of the complex. Ordinarily, this is of no consequence.

Nonpolar region

In hemoglobin S, however, the nonpolar valine residue on the surface of one complex is attracted to the nonpolar region on a nearby complex.

This results in hemoglobin forming long chains that precipitate from solution, causing the deformation of red blood cells that is characteristic of sickle cell disease.

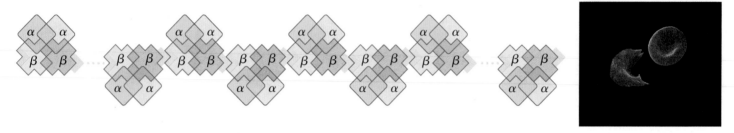

These deformed blood cells clog narrow capillaries, restricting blood flow to the body's organs. The reduced blood flow gives rise to anemia, episodes of debilitating pain, susceptibility to infections, stroke, and eventually premature death.

## Dispersion Forces

Nonpolar gases, such as $N_2$ and $O_2$, can be liquefied under the right conditions of pressure and temperature, so nonpolar molecules must also exhibit attractive intermolecular forces. These intermolecular forces are Coulombic in nature (as are all intermolecular forces), but they differ from other intermolecular forces because they arise from the movement of electrons in nonpolar molecules.

On average, the distribution of electron density in a nonpolar molecule is uniform and symmetrical (which is why a nonpolar molecule has no dipole moment). However, because electrons in a molecule have some freedom to move about, at any given point in time the molecule may have a nonuniform distribution of electron density, giving it a fleeting, temporary dipole—called an ***instantaneous dipole.*** An instantaneous dipole in one molecule can induce dipoles in neighboring molecules. For example, the temporary partial negative charge on a molecule repels the electrons in a molecule next to it. This repulsion polarizes the second molecule, which then acquires a temporary dipole. It in turn polarizes the next molecule and so on, leaving a collection of ordinarily nonpolar molecules with partial positive and negative charges and Coulombic attractions between them as shown in Figure 12.4. The resulting attractive forces are called ***London***[1] ***dispersion forces*** or simply ***dispersion forces.***

**Figure 12.4**   Instantaneous dipoles in ordinarily nonpolar molecules can induce temporary dipoles in neighboring molecules, causing the molecules to be attracted to one another. This type of interaction is responsible for the condensation of nonpolar gases.

---

1. Fritz London (1900–1954). German physicist. London was a theoretical physicist whose major work was on superconductivity in liquid helium.

| TABLE 12.2 | Molar Masses, Boiling Points, and States of the Halogens at Room Temperature | | |
|---|---|---|---|
| Molecule | Molar Mass (g/mol) | Boiling Point (°C) | State (Room Temp.) |
| $F_2$ | 38.0 | −188 | Gas |
| $Cl_2$ | 70.9 | −34 | Gas |
| $Br_2$ | 159.8 | 59 | Liquid |
| $I_2$ | 253.8 | 184 | Solid |

The magnitude of dispersion forces depends on how mobile the electrons in the molecule are. In small molecules, such as $F_2$, the electrons are relatively close to the nuclei and cannot move about very freely; thus, the electron distribution in $F_2$ is not easily *polarized.* In larger molecules, such as $Cl_2$, the electrons are somewhat farther away from the nuclei and therefore move about more freely. The electron density in $Cl_2$ is more easily polarized than that in $F_2$, resulting in larger instantaneous dipoles, larger induced dipoles, and larger intermolecular attractions overall. The trend in magnitude of dispersion forces with increasing molecular size is illustrated by the data in Table 12.2. (It is also illustrated by the correlation between molar mass and boiling point in Figure 12.3.) Dispersion forces act not only between nonpolar molecules, but between *all* molecules.

Sample Problem 12.1 lets you practice determining what kinds of forces exist between particles in liquids and solids.

**Student Annotation:** Molecules that more readily acquire an instantaneous dipole are said to be *polarizable*. In general, larger molecules have greater *polarizability* than small ones.

## Sample Problem 12.1

What kind(s) of intermolecular forces exist in (a) $CCl_4(l)$, (b) $CH_3COOH(l)$, (c) $CH_3COCH_3(l)$, and (d) $H_2S(l)$?

**Strategy** Draw Lewis dot structures and apply VSEPR theory [◄◄ Section 9.2] to determine whether each molecule is polar or nonpolar. Nonpolar molecules exhibit dispersion forces only. Polar molecules exhibit both dipole-dipole interactions and dispersion forces. Polar molecules with N—H, F—H, or O—H bonds exhibit dipole-dipole interactions (including hydrogen bonding) and dispersion forces.

**Setup** The Lewis dot structures for molecules (a) to (d) are:

(a) (b) (c) (d)

**Solution**
(a) $CCl_4$ is nonpolar, so the only intermolecular forces are dispersion forces.

(b) $CH_3COOH$ is polar and contains an O—H bond, so it exhibits dipole-dipole interactions (including hydrogen bonding) and dispersion forces.

(c) $CH_3COCH_3$ is polar but does *not* contain N—H, O—H, or F—H bonds, so it exhibits dipole-dipole interactions and dispersion forces.

(d) $H_2S$ is polar but does not contain N—H, O—H, or F—H bonds, so it exhibits dipole-dipole interactions and dispersion forces.

**Think About It** Being able to draw correct Lewis structures is, once again, vitally important. Review, if you need to, the procedure for drawing them [◄◄ Section 8.5].

**Practice Problem A** What kind(s) of intermolecular forces exist in (a) $CH_3CH_2CH_2CH_2CH_3(l)$, (b) $CH_3CH_2OH(l)$, (c) $H_2CO(l)$, and (d) $O_2(l)$?

**Practice Problem B** What kind(s) of intermolecular forces exist in (a) $CH_2Cl_2(l)$, (b) $CH_3CH_2CH_2OH(l)$, (c) $H_2O_2(l)$, and (d) $N_2(l)$?

## Ion-Dipole Interactions

**Student Annotation:** Hydration [◀◀ Section 4.2] is one example of an ion-dipole interaction.

*Ion-dipole interactions* are Coulombic attractions between ions (either positive or negative) and polar molecules. These interactions occur in mixtures of ionic and polar species such as an aqueous solution of sodium chloride. The magnitude of ion-dipole interactions depends on the charge and the size of the ion, and on the dipole moment and size of the polar molecule. Cations generally interact more strongly with dipoles than anions (of the same magnitude charge) because they tend to be smaller.

Figure 12.5 shows the ion-dipole interaction between the $Na^+$ and $Mg^{2+}$ ions with a water molecule, which has a large dipole moment (1.87 D). Because the $Mg^{2+}$ ion has a *higher charge* and a *smaller size* [◀◀ Section 7.6] than the $Na^+$ ion (the ionic radii of $Mg^{2+}$ and $Na^+$ are 78 and 98 pm, respectively), $Mg^{2+}$ interacts more strongly with water molecules. The properties of solutions are discussed in Chapter 13.

Although the intermolecular forces discussed so far are all *attractive* forces, molecules also exert repulsive forces on one another (when *like* charges approach one another). When two molecules approach each other closely, the repulsions between electrons and between nuclei in the molecules become significant. The magnitude of these repulsive forces rises very steeply as the distance separating the molecules in a condensed phase decreases. This is the reason that liquids and solids are so hard to compress. In these phases, the molecules are already in close contact with one another, and so they greatly resist being compressed further.

**Figure 12.5** Ion-dipole interactions between ions and water molecules. In each case the positive charge on the ion is attracted to the partial negative charge on the oxygen atom in the water molecule. The attraction is stronger when the distance between the two species is smaller.

---

### Checkpoint 12.1    Intermolecular Forces

**12.1.1**   What kind(s) of intermolecular forces exist between benzene molecules ($C_6H_6$)? (Select all that apply.)

    a) Dispersion forces

    b) Dipole-dipole interactions

    c) Hydrogen bonding

    d) Ion-dipole interactions

    e) Ionic bonding

**12.1.2**   Which of the following exhibits significant hydrogen bonding? (Select all that apply.)

    a) HBr

    b) $H_2CF_2$

    c) $H_2$

    d) $H_2O_2$

    e) $CH_3CN$

---

## 12.2    Properties of Liquids

Several of the physical properties of a liquid depend on the magnitude of its intermolecular forces. In this section we consider three such properties: surface tension, viscosity, and vapor pressure.

### Surface Tension

A molecule within a liquid is pulled in all directions by the intermolecular forces between it and the other molecules that surround it. There is no *net* pull in any one direction. A molecule at the surface of the liquid is similarly pulled down and to the sides by neighboring molecules, but Figure 12.6 shows there is no upward pull to balance the downward or inward pull (into the bulk of the liquid). This results in a net pull inward on molecules at the surface, causing the surface of a liquid to tighten like an elastic film, thus minimizing its surface area. The "beading" of water on the surface of a freshly waxed car, shown in Figure 12.7, is an illustration of this phenomenon.

A quantitative measure of the elastic force in the surface of a liquid is the **surface tension,** the amount of energy required to stretch or increase the surface of a liquid by a unit area (for example, by 1 $cm^2$). A liquid with strong intermolecular forces has a high surface tension. Water, for instance, with its strong hydrogen bonds, has a very high surface tension.

Another illustration of surface tension is the *meniscus,* the curved surface of a liquid contained in a narrow tube. Figure 12.8(a) shows the concave surface of water in a graduated cylinder. (You probably know from your laboratory class that you are to read the volume level with the bottom of the meniscus.) This is caused by a thin film of water adhering to the wall of the glass cylinder. The surface tension of water causes this film to contract, and as it does, it pulls the water up the cylinder. This effect, known as **capillary action,** is more pronounced in a cylinder with a very small diameter, such as a capillary tube used to draw a small amount of blood. Two types of forces bring about capil-

**Figure 12.6** Intermolecular forces acting on a molecule in the surface layer of a liquid and in the interior region of the liquid.

(a)

(b)

**Figure 12.7** Surface tension causes water to bead up on a just-washed car.

(a)                                     (b)

**Figure 12.8** (a) When water is measured in a glass graduated cylinder, the adhesive forces are greater than the cohesive forces and the meniscus is concave. (b) When mercury is measured in a glass graduated cylinder, cohesive forces are greater than adhesive forces and the meniscus is convex.

**Blood being drawn from a finger into a capillary tube.**

lary action. One is **cohesion**, the attractions between *like* molecules (in this case, between water molecules). The other is **adhesion**, the attractions between *unlike* molecules (in this case, between water molecules and the molecules that make up the interior surface of the graduated cylinder). If adhesion is stronger than cohesion, as it is in Figure 12.8(a), the contents of the tube will be pulled upward. The upward movement is limited by the weight of the liquid in the tube. In the case of mercury, shown in Figure 12.8(b), the cohesive forces are stronger than the adhesive forces, resulting in a convex meniscus in which the liquid level at the glass wall is lower than that in the middle.

**Student Annotation:** *Cohesion and adhesion are also called cohesive forces and adhesive forces.*

## Viscosity

Another property determined by the magnitude of intermolecular forces in a liquid is viscosity. **Viscosity,** with units of $N \cdot s/m^2$, is a measure of a fluid's resistance to flow. The higher the viscosity, the more slowly a liquid flows. The viscosity of a liquid typically decreases with increasing temperature. The phrase "slow as molasses in winter" refers to the fact that molasses pours more slowly (has a higher viscosity) in cold weather. You may also have noticed that honey and maple syrup seem thinner when they are heated.

Liquids that have strong intermolecular forces have higher viscosities than those that have weaker intermolecular forces. Table 12.3 lists the viscosities of some liquids that may be familiar to you. Water's high viscosity, like its high surface tension, is the result of hydrogen bonding. Note how large the viscosity of glycerol is compared to the other liquids listed in Table 12.3. The structure of glycerol, which is a sweet tasting, syrupy liquid used for a wide variety of things including the manufacture of candy and antibiotics, is

$$
\begin{array}{c}
\text{H} \\
| \\
\text{H}-\text{C}-\text{O}-\text{H} \\
| \\
\text{H}-\text{C}-\text{O}-\text{H} \\
| \\
\text{H}-\text{C}-\text{O}-\text{H} \\
| \\
\text{H}
\end{array}
$$

Glycerol

| TABLE 12.3 | Viscosities of Some Familiar Liquids at 20°C |
| --- | --- |
| **Liquid** | **Viscosity (N·s/m²)** |
| Acetone ($C_3H_6O$) | $3.16 \times 10^{-4}$ |
| Water ($H_2O$) | $1.01 \times 10^{-3}$ |
| Ethanol ($C_2H_5OH$) | $1.20 \times 10^{-3}$ |
| Mercury (Hg) | $1.55 \times 10^{-3}$ |
| Blood | $4 \times 10^{-3}$ |
| Glycerol ($C_3H_8O_3$) | 1.49 |

Like water, glycerol can form hydrogen bonds. Each glycerol molecule has three −OH groups that can participate in hydrogen bonding with other glycerol molecules. Furthermore, because of their shape, the molecules have a great tendency to become entangled rather than to slip past one another as the molecules of less viscous liquids do. These interactions contribute to its high viscosity.

## Vapor Pressure

In Chapter 11 we encountered the term *vapor pressure,* referring to the temperature-dependent partial pressure of water [◄◄ Section 11.5]. In fact, vapor pressure is another property of liquids that depends on the magnitude of intermolecular forces. The molecules in a liquid are in constant motion, and, like the molecules in a gas, they have a distribution of kinetic energies. The most probable kinetic energy for molecules in a sample of liquid increases with increasing temperature, as shown in Figure 12.9. If a molecule at the surface of a liquid has sufficient kinetic energy, it can escape from the liquid phase into the gas phase. This phenomenon is known as **evaporation** or **vaporization.** Consider the apparatus shown in Figure 12.10. As a liquid begins to evaporate, molecules leave the liquid phase and become part of the gas phase in the space above the liquid. Molecules in the gas phase can return to the liquid phase if they strike the liquid surface and again become trapped by intermolecular forces, a process known as **condensation.** Initially, evaporation occurs more rapidly than condensation. As the number of molecules in the gas phase increases, however, so does the rate of condensation. The vapor pressure over the liquid increases until the rate of condensation is equal to the rate of evaporation, which is constant at any given temperature (Figure 12.11). This (or any other) situation, wherein a forward process and reverse process are occurring at the same rate, is called a **dynamic equilibrium.** Although both processes are ongoing (*dynamic*), the number of molecules in the gas phase at any given point in time does not change (*equilibrium*). The pressure exerted by the molecules that have escaped to the gas phase, once the pressure has stopped increasing, is the **equilibrium vapor pressure.**

 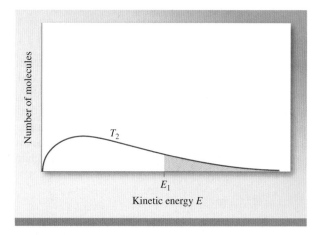

**Figure 12.9**    Kinetic energy distribution curves for molecules in a liquid (a) at temperature $T_1$ and (b) at a higher temperature $T_2$. Note that at the higher temperature, the curve flattens out. The shaded areas represent the number of molecules possessing kinetic energy equal to or greater than a certain kinetic energy $E_1$. The higher the temperature, the greater the number of molecules with high kinetic energy.

**Figure 12.10**    Establishment of equilibrium vapor pressure. (a) Initially, there are no molecules in the gas phase. (b) Molecules enter the gas phase, increasing the total pressure above the liquid. (c) The partial pressure of the liquid continues to increase until the rates of vaporization and condensation are equal. (d) Vaporization and condensation continue to occur at the same rate, and there is no further net change in pressure.

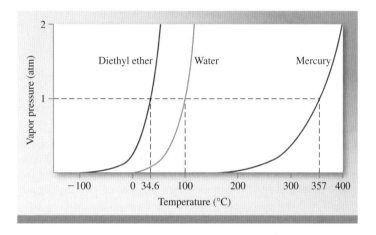

**Figure 12.12**    The increase in vapor pressure with temperature for three liquids. The normal boiling points of the liquids (at 1 atm) are shown on the horizontal axis. The strong metallic bonding in mercury results in its having a very low vapor pressure at room temperature.

The average kinetic energy of molecules in a liquid increases with increasing temperature (see Figure 12.9). At a higher temperature, therefore, a greater percentage of molecules at the liquid surface will possess sufficient kinetic energy to escape into the gas phase. Consequently, as we have already seen with water in Table 11.5, vapor pressure increases with increasing temperature. Figure 12.12 shows the plots of vapor pressure versus temperature for three different liquids.

The plots in Figure 12.12 of vapor pressure as a function of temperature are not linear. However, a linear relationship does exist between the natural log of vapor pressure and the reciprocal of absolute temperature. This relationship is called the ***Clausius*[2]*-Clapeyron*[3] *equation:***

$$\ln P = -\frac{\Delta H_{vap}}{RT} + C \qquad\qquad \text{Equation 12.1}$$

where $\ln P$ is the natural logarithm of the vapor pressure, $\Delta H_{vap}$ is the molar heat of vaporization (kJ/mol), $R$ is the gas constant (8.314 J/K · mol), and $C$ is a constant that must be determined experimentally for each different compound. The Clausius-Clapeyron equation has the form of the general linear equation $y = mx + b$:

$$\ln P = \left(-\frac{\Delta H_{vap}}{R}\right)\left(\frac{1}{T}\right) + C$$
$$\phantom{\ln P = }\;y\;=\qquad mx\qquad + b$$

By measuring the vapor pressure of a liquid at several different temperatures and plotting $\ln P$ versus $1/T$, we can determine the slope of the line, which is equal to $-\Delta H_{vap}/R$. ($\Delta H_{vap}$ is assumed to be independent of temperature.)

**Student Annotation:** The units of $R$ are those that enable us to cancel the units of J/mol (or kJ/mol) associated with $\Delta H_{vap}$ [◄◄ Section 11.3, Table 11.4].

2. Rudolf Julius Emanuel Clausius (1822–1888). German physicist. Clausius's work was mainly in electricity, kinetic theory of gases, and thermodynamics.

3. Benoit Paul Emile Clapeyron (1799–1864). French engineer. Clapeyron made contributions to the thermodynamics of steam engines.

If we know the value of $\Delta H_{vap}$ and the vapor pressure of a liquid at one temperature, we can use the Clausius-Clapeyron equation to calculate the vapor pressure of the liquid at a different temperature. At temperatures $T_1$ and $T_2$, the vapor pressures are $P_1$ and $P_2$. From Equation 12.1 we can write

Equation 12.2
$$\ln P_1 = -\frac{\Delta H_{vap}}{RT_1} + C$$

Equation 12.3
$$\ln P_2 = -\frac{\Delta H_{vap}}{RT_2} + C$$

Subtracting Equation 12.3 from Equation 12.2 we get

$$\ln P_1 - \ln P_2 = -\frac{\Delta H_{vap}}{RT_1} - \left(-\frac{\Delta H_{vap}}{RT_2}\right)$$

$$= \frac{\Delta H_{vap}}{R}\left(\frac{1}{T_2} - \frac{1}{T_1}\right)$$

and finally

Equation 12.4
$$\ln \frac{P_1}{P_2} = \frac{\Delta H_{vap}}{R}\left(\frac{1}{T_2} - \frac{1}{T_1}\right)$$

Sample Problem 12.2 shows how to use Equation 12.4.

## Sample Problem 12.2

Diethyl ether is a volatile, highly flammable organic liquid that today is used mainly as a solvent. (It was used as an anesthetic during the nineteenth century and as a recreational intoxicant early in the twentieth century during prohibition, when ethanol was difficult to obtain.) The vapor pressure of diethyl ether is 401 mmHg at 18°C, and its molar heat of vaporization is 26 kJ/mol. Calculate its vapor pressure at 32°C.

**Strategy** Given the vapor pressure at one temperature, $P_1$, use Equation 12.4 to calculate the vapor pressure at a second temperature, $P_2$.

$$\ln \frac{P_1}{P_2} = \frac{\Delta H_{vap}}{R}\left(\frac{1}{T_2} - \frac{1}{T_1}\right)$$

**Setup** Temperature must be expressed in kelvins, so $T_1 = 291.15$ K and $T_2 = 305.15$ K. Because the molar heat of vaporization is given in kJ/mol, we will have to convert it to J/mol for the units of $R$ to cancel properly: $\Delta H_{vap} = 2.6 \times 10^4$ J/mol. The inverse function of $\ln x$ is $e^x$.

**Solution**

$$\ln \frac{P_1}{P_2} = \frac{2.6 \times 10^4 \text{ J/mol}}{8.314 \text{ J/K} \cdot \text{mol}}\left(\frac{1}{305.15 \text{ K}} - \frac{1}{291.15 \text{ K}}\right)$$

$$= -0.4928$$

$$\frac{P_1}{P_2} = e^{-0.4928} = 0.6109$$

$$\frac{P_1}{0.6109} = P_2$$

$$P_2 = \frac{401 \text{ mmHg}}{0.6109} = 6.6 \times 10^2 \text{ mmHg}$$

**Think About It** It is easy to switch $P_1$ and $P_2$ or $T_1$ and $T_2$ accidentally and get the wrong answer to a problem such as this. One way to help safeguard against this common error is to verify that the vapor pressure is *higher* at the higher temperature.

**Practice Problem A** The vapor pressure of ethanol is $1.00 \times 10^2$ mmHg at 34.9°C. What is its vapor pressure at 55.8°C? ($\Delta H_{vap}$ for ethanol is 39.3 kJ/mol.)

**Practice Problem B** Estimate the molar heat of vaporization of a liquid whose vapor pressure doubles when the temperature is raised from 85°C to 95°C. At what temperature will the vapor pressure be five times the value at 85°?

| **Checkpoint 12.2** | **Properties of Liquids** |

**12.2.1** At what temperature would diethyl ether have a vapor pressure of 250 mmHg? Use the vapor pressure at 18°C and $\Delta H_{vap}$ given in Sample Problem 12.2.

a) 5.6°C

b) 280°C

c) 17°C

d) 6.5°C

e) −270°C

**12.2.2** Given the following information for $C_6F_6$, calculate its $\Delta H_{vap}$. At 300 K, the vapor pressure is 92.47 mmHg, and at 320 K, the vapor pressure is 225.1 mmHg.

a) 208 kJ/mol

b) 411 kJ/mol

c) 16.4 kJ/mol

d) 10.3 kJ/mol

e) 35.5 kJ/mol

**12.2.3** Using the graph, estimate the vapor pressure of the liquid at 100°C.

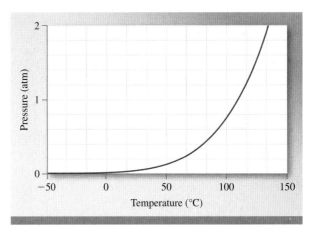

a) 100 atm

b) 0.75 atm

c) 1.00 atm

d) 110 atm

e) 1.50 atm

**12.2.4** Using the result from question 12.2.3 and another point from the graph, estimate $\Delta H_{vap}$ for the liquid.

a) 15 kJ/mol

b) 35 kJ/mol

c) 55 kJ/mol

d) 75 kJ/mol

e) 95 kJ/mol

# 12.3 Crystal Structure

Solids can be categorized as either crystalline or amorphous. A *crystalline solid* possesses rigid and long-range order; its atoms, molecules, or ions occupy specific positions. The arrangement of the particles in a crystalline solid, which we call the *lattice structure,* depends on the nature and the size of the particles involved. The forces responsible for the stability of a crystal can be ionic forces, covalent bonds, van der Waals forces, hydrogen bonds, or a combination of some of these forces. Amorphous solids lack a well-defined arrangement and long-range molecular order. In this section, we will concentrate on the nature of crystalline solids. (Amorphous solids are discussed in Section 12.5.)

**Student Annotation:** Ice is a crystalline solid. Glass is an amorphous solid.

## Unit Cells

A *unit cell* is the basic repeating structural unit of a crystalline solid. Figure 12.13 shows a unit cell and its extension in three dimensions. Each sphere represents an atom, ion, or molecule and is called a *lattice point.* For the purpose of clarity, we will limit our discussion in this section to metal crystals in which each lattice point is occupied by an atom.

Every crystalline solid can be described in terms of one of the seven types of unit cells shown in Figure 12.14. The geometry of the cubic unit cell is particularly simple because all sides and all angles are equal. Any of the unit cells, when repeated in space in all three dimensions, forms the lattice structure characteristic of a crystalline solid.

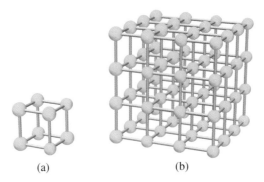

(a)                    (b)

**Figure 12.13**    (a) A single unit cell and (b) a lattice (three-dimensional array) made up of many unit cells. Each sphere represents a lattice point, which may be an atom, a molecule, or an ion.

| Simple cubic | Tetragonal | Orthorhombic | Rhombohedral |
|---|---|---|---|
| $a = b = c$ | $a = b \neq c$ | $a \neq b \neq c$ | $a = b = c$ |
| $\alpha = \beta = \gamma = 90°$ | $\alpha = \beta = \gamma = 90°$ | $\alpha = \beta = \gamma = 90°$ | $\alpha = \beta = \gamma \neq 90°$ |

| Monoclinic | Triclinic | Hexagonal |
|---|---|---|
| $a \neq b \neq c$ | $a \neq b \neq c$ | $a = b \neq c$ |
| $\gamma \neq \alpha = \beta = 90°$ | $\alpha \neq \beta \neq \gamma \neq 90°$ | $\alpha = \beta = 90°, \gamma = 120°$ |

**Figure 12.14**    The seven types of unit cells. Angle $\alpha$ is defined by edges $b$ and $c$, angle $\beta$ by edges $a$ and $c$, and angle $\gamma$ by edges $a$ and $b$.

## Packing Spheres

We can understand the geometric requirements for crystal formation by considering the different ways of packing a number of identical atoms to form an ordered three-dimensional structure. The way the atoms are arranged in layers determines the type of unit cell.

In the simplest case, a layer of atoms can be arranged as shown in Figure 12.15(a). The three-dimensional structure can be generated by placing a layer above and below this layer in such a way that atoms in one layer are directly over the atoms in the layer below. This procedure can be extended to generate many, many layers, as in the case of a crystal. Focusing on the atom labeled with an "*x*," we see that it is in contact with four atoms in its own layer, one atom in the layer above it, and one atom in the layer below it. Each atom in this arrangement is said to have a coordination number of 6 because it has six immediate neighbors. The ***coordination number*** is defined as the number of atoms surrounding an atom in a crystal lattice. The value of the coordination number indicates how tightly the atoms are packed together—the larger the coordination number, the closer the atoms are to one another. The basic repeating unit in the array of atoms is called a ***simple cubic cell*** (scc) [Figure 12.15(b)].

The other types of cubic cells, shown in Figure 12.16, are the ***body-centered cubic cell*** (bcc) and the ***face-centered cubic cell*** (fcc). Unlike the simple cube, the second layer of atoms in the body-centered cubic arrangement fits into the depressions of the first layer and the third layer fits into the depressions of the second layer (Figure 12.17).

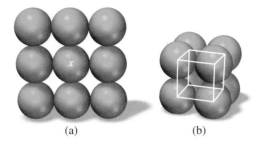

**Figure 12.15** Arrangement of identical spheres in a simple cubic cell. (a) Top view of one layer of spheres. (b) Definition of a simple cubic cell.

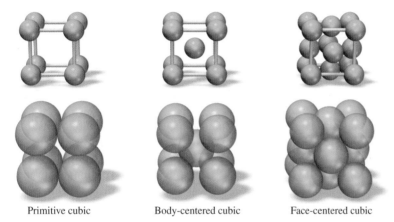

Primitive cubic        Body-centered cubic        Face-centered cubic

**Figure 12.16** Three types of cubic cells. The top view makes it easier to see the locations of the lattice points, but the bottom view is more realistic, with the spheres touching one another.

**Figure 12.17** In the body-centered cubic arrangement, the spheres in each layer rest in the depressions between spheres in the previous layer.

The coordination number of each atom in the bcc structure is 8 (each sphere is in contact with four others in the layer above and four others in the layer below). In the face-centered cubic cell, there are atoms at the center of each of the six faces of the cube, in addition to the eight corner atoms. The coordination number in the face-centered cubic cell is 12 (each sphere is in contact with four others in its own layer, four others in the layer above, and four others in the layer below).

Because every unit cell in a crystalline solid is adjacent to other unit cells, most of a cell's atoms are shared by neighboring cells. (The atom at the center of the body-centered cubic cell is an exception.) In all types of cubic cells, for example, each corner atom belongs to eight unit cells whose corners all touch [Figure 12.18(a)]. An atom that lies on an edge, on the other hand, is shared by four unit cells [Figure 12.18(b)], and a face-centered atom is shared by two unit cells [Figure 12.18(c)]. Because a simple cubic cell has lattice points only at each of the eight corners, and because each corner atom is shared by eight unit cells, there will be the equivalent of only *one* complete atom contained within a simple cubic unit cell (Figure 12.19). A body-centered cubic cell contains the equivalent of two complete atoms, one in the center and eight shared corner atoms. A face-centered cubic cell contains the equivalent of four complete atoms—three from the six face-centered atoms and one from the eight shared corner atoms.

**Figure 12.18** (a) A corner atom in any cell is shared by eight unit cells. (b) An edge atom is shared by four unit cells. (c) A face-centered atom in a cubic cell is shared by two unit cells.

**Figure 12.19** Because each sphere is shared by eight unit cells and there are eight corners in a cube, there is the equivalent of one complete sphere inside a simple cubic unit cell.

## Closest Packing

There is more empty space in the simple cubic and body-centered cubic cells than in the face-centered cubic cell. Closest packing, the most efficient arrangement of atoms, starts with the structure shown in Figure 12.20(a), which we call layer A. Focusing on the only atom that is surrounded completely by other atoms, we see that it has six immediate neighbors in its own layer. In the second layer, which we call layer B, atoms are packed into the depressions between the atoms in the first layer so that all the atoms are as close together as possible [Figure 12.20(b)].

There are two ways that a third layer of atoms can be arranged. They may sit in the depressions between second-layer atoms such that the third-layer atoms lie directly over atoms in the first layer [Figure 12.20(c)]. In this case, the third layer is also labeled A. Alternatively, atoms in the third layer may sit in a *different* set of depressions such that they do not lie directly over atoms in the first layer [Figure 12.20(d)]. In this case, we label the third layer C.

Figure 12.21 shows the exploded views and the structures resulting from these two arrangements. The ABA arrangement [Figure 12.21(a)] is known as the *hexagonal close-packed (hcp) structure,* and the ABC arrangement [Figure 12.21(b)] is the *cubic close-packed (ccp) structure,* which corresponds to the face-centered cube already described. In the hcp structure, the spheres in every other layer occupy the same vertical position (ABABAB . . .), while in the ccp structure, the spheres in every fourth layer occupy the same vertical position (ABCABCA . . .). In both structures, each sphere has a coordination number of 12 (each sphere is in contact with six spheres in

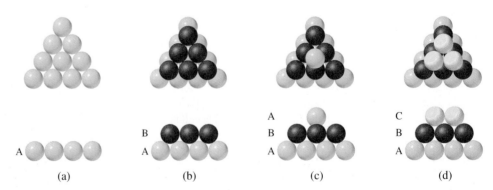

**Figure 12.20** (a) In a close-packed layer, each sphere is in contact with six others. (b) Spheres in the second layer fit into the depressions between the first-layer spheres. (c) In the hexagonal close-packed structure, each third-layer sphere is directly over a first-layer sphere. (d) In the cubic close-packed structure, each third-layer sphere fits into a depression that is directly over a depression in the first layer.

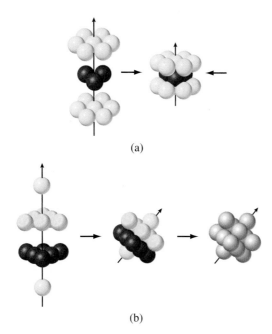

(a)

(b)

**Figure 12.21**   Exploded views of (a) a hexagonal close-packed structure and (b) a cubic close-packed structure. This view is tilted to show the face-centered cubic unit cell more clearly. Note that this arrangement is the same as the face-centered unit cell.

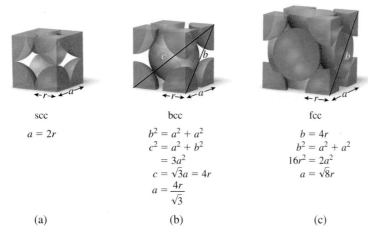

scc

$a = 2r$

bcc

$b^2 = a^2 + a^2$
$c^2 = a^2 + b^2$
$\quad = 3a^2$
$c = \sqrt{3}a = 4r$
$a = \dfrac{4r}{\sqrt{3}}$

fcc

$b = 4r$
$b^2 = a^2 + a^2$
$16r^2 = 2a^2$
$a = \sqrt{8}r$

(a)                    (b)                    (c)

**Figure 12.22**   The relationship between the edge length (*a*) and radius (*r*) of atoms in the (a) simple cubic cell, (b) body-centered cubic cell, and (c) face-centered cubic cell.

its own layer, three spheres in the layer above, and three spheres in the layer below). Both the hcp and ccp structures represent the most efficient way of packing identical spheres in a unit cell, and the coordination number cannot exceed 12.

Many metals form crystals with hcp or ccp structures. For example, magnesium, titanium, and zinc crystallize with their atoms in an hcp array, while aluminum, nickel, and silver crystallize in the ccp arrangement. A substance will crystallize with the arrangement that maximizes the stability of the solid.

Figure 12.22 summarizes the relationship between the atomic radius *r* and the edge length *a* of a simple cubic cell, a body-centered cubic cell, and a face-centered cubic cell. This relationship can be used to determine the atomic radius of a sphere in which the density of the crystal is known.

**Student Annotation:** The noble gases, which are monatomic, crystallize in the ccp structure, with the exception of helium, which crystallizes in the hcp structure.

**FAQ** *You May Be Wondering...*

# How Do We Know the Structures of Crystals?

Virtually all we know about crystal structure has been learned from X-ray diffraction studies. **X-ray diffraction** is the scattering of X rays by the units of a crystalline solid. The scattering, or *diffraction patterns,* produced are used to deduce the arrangement of particles in the solid lattice.

In Section 6.1 we discussed the interference phenomenon associated with waves (see Figure 6.4). Because X rays are a form of electromagnetic radiation (i.e., they are *waves*), they exhibit interference phenomena under suitable conditions. In 1912 Max von Laue[4] correctly suggested that, because the wavelength of X rays is comparable in magnitude to the distances between lattice points in a crystal, the lattice should be able to diffract X rays. An X-ray diffraction pattern is the result of interference in the waves associated with X rays.

Figure 12.23 shows a typical X-ray diffraction setup. A beam of X rays is directed at a mounted crystal. When X-ray photons encounter the electrons in the atoms of a crystalline solid, some of the incoming radiation is reflected, much as visible light is reflected by a mirror; the process is called the scattering of X rays.

To understand how a diffraction pattern arises, consider the scattering of X rays by the atoms in two parallel planes (Figure 12.24). Initially, the two incident rays are in phase with each other (their maxima and minima occur at the same positions). The upper wave is scattered, or reflected, by an atom in the first layer, while the lower wave is scattered by an atom in the second layer. For these two scattered waves to be in phase again, the extra distance traveled by the lower wave (the sum of the distance between points B and C and the distance between points C and D) must be an integral multiple of the wavelength ($\lambda$) of the X ray; that is,

Equation 12.5 $\qquad BC + CD = 2d \sin \theta = n\lambda \qquad n = 1, 2, 3, \ldots$

(a)

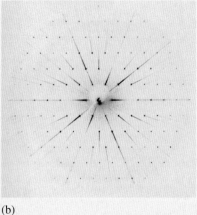

(b)

**Figure 12.23**  (a) Arrangement for obtaining the X-ray diffraction pattern of a crystal. The shield prevents the intense beam of undiffracted X rays from damaging the photographic plate. (b) X-ray diffraction pattern.

---

4. Max Theodor Felix von Laue (1879–1960). German physicist. Von Laue received the Nobel Prize in Physics in 1914 for his discovery of X-ray diffraction.

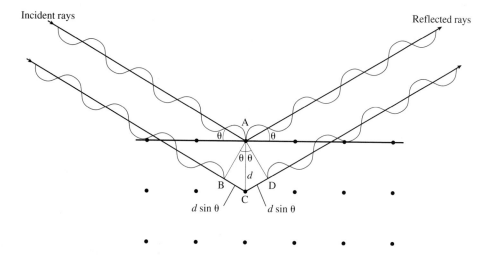

Incident rays                    Reflected rays

**Figure 12.24** Reflection of X rays from two layers of atoms. The lower wave travels a distance $2d \sin \theta$ longer than the upper wave does. For the two waves to be in phase again after reflection, it must be true that $2d \sin \theta = n\lambda$, where $\lambda$ is the wavelength of the X ray and $n = 1$, 2, 3, ... The sharply defined spots in Figure 12.23 are observed only if the crystal is large enough to consist of hundreds of parallel layers.

where $\theta$ is the angle between the X rays and the plane of the crystal and $d$ is the distance between adjacent planes. Equation 12.5 is known as the **Bragg equation,** after William H. Bragg and Sir William L. Bragg.[5] The reinforced waves produce a dark spot on a photographic film for each value of $\theta$ that satisfies the Bragg equation.

## Sample Problem 12.3

X rays of wavelength 0.154 nm strike an aluminum crystal; the rays are reflected at an angle of 19.3°. Assuming that $n = 1$, calculate the spacing between the planes of aluminum atoms (in pm) that is responsible for this angle of reflection.

**Student Annotation:** 1 nm = 1000 pm.

**Strategy** Use Equation 12.5 to solve for $d$.

**Setup** $\theta = 19.3°$, $n = 1$, and $\lambda = 154$ pm.

**Solution**

$$d = \frac{n\lambda}{2 \sin \theta} = \frac{154 \text{ pm}}{2 \sin 19.3°} = 233 \text{ pm}$$

**Think About It** The distance between layers of atoms in a crystal should be similar in magnitude to the wavelength of the X rays diffracted by the crystal (compare 0.154 nm with 0.233 nm).

**Practice Problem A** X rays of wavelength 0.154 nm are diffracted from a crystal at an angle of 14.17°. Assuming that $n = 1$, calculate the distance (in pm) between layers in the crystal.

**Practice Problem B** At what angle will X rays of wavelength 0.154 nm be diffracted from a crystal if the distance (in pm) between layers in the crystal is 188 pm? (Assume $n = 1$.)

The X-ray diffraction technique offers the most accurate method for determining bond lengths and bond angles in molecules in solids. Because X rays are scattered by electrons, chemists can construct an electron-density contour map from the diffraction patterns by using a complex mathematical procedure. Basically, an electron-density contour map tells us the relative electron densities at various locations in a molecule. The densities reach a maximum near the center of each atom. In this manner, we can determine the positions of the nuclei and hence the geometric parameters of the molecule.

5. William Henry Bragg (1862–1942) and Sir William Lawrence Bragg (1890–1972). English physicists, father and son. Both worked on X-ray crystallography. The younger Bragg formulated the fundamental equation for X-ray diffraction. The two shared the Nobel Prize in Physics in 1915.

Sample Problem 12.4 illustrates the relationships between the unit cell type, cell dimensions, and density of a metal.

### Sample Problem 12.4

Gold crystallizes in a cubic close-packed structure (face-centered cubic unit cell) and has a density of 19.3 g/cm$^3$. Calculate the atomic radius of an Au atom in angstroms (Å).

**Strategy** Using the given density and the mass of gold contained within a face-centered cubic unit cell, determine the volume of the unit cell. Then, use the volume to determine the value of $a$, and use the equation supplied in Figure 12.22(c) to find $r$. Be sure to use consistent units for mass, length, and volume.

**Setup** The face-centered cubic unit cell contains a total of four atoms of gold [six faces, each shared by two unit cells, and eight corners, each shared by eight unit cells—Figure 12.24(c)]. $D = m/V$ and $V = a^3$.

**Solution** First, we determine the mass of gold (in grams) contained within a unit cell:

$$m = \frac{4 \text{ atoms}}{\text{unit cell}} \times \frac{1 \text{ mol}}{6.022 \times 10^{23} \text{ atoms}} \times \frac{197.0 \text{ g Au}}{1 \text{ mol Au}} = 1.31 \times 10^{-21} \text{ g/unit cell}$$

Then we calculate the volume of the unit cell in cm$^3$:

$$V = \frac{m}{d} = \frac{1.31 \times 10^{-21} \text{ g}}{19.3 \text{ g/cm}^3} = 6.78 \times 10^{-23} \text{ cm}^3$$

Using the calculated volume and the relationship $V = a^3$ (rearranged to solve for $a$), we determine the length of a side of a unit cell:

$$a = \sqrt[3]{V} = \sqrt[3]{6.78 \times 10^{-23} \text{ cm}^3} = 4.08 \times 10^{-8} \text{ cm}$$

Using the relationship provided in Figure 12.22(c) (rearranged to solve for $r$), we determine the radius of a gold atom in centimeters:

$$r = \frac{a}{\sqrt{8}} = \frac{4.08 \times 10^{-8} \text{ cm}}{\sqrt{8}} = 1.44 \times 10^{-8} \text{ cm}$$

Finally, we convert centimeters to angstroms:

$$1.44 \times 10^{-8} \text{ cm} \times \frac{1 \times 10^{-2} \text{ m}}{1 \text{ cm}} \times \frac{1 \text{ Å}}{1 \times 10^{-10} \text{ m}} = 1.44 \text{ Å}$$

**Think About It** Atomic radii tend to be on the order of 1 Å, so this answer is reasonable.

**Practice Problem A** When silver crystallizes, it forms face-centered cubic cells. The unit cell edge length is 4.087 Å. Calculate the density of silver.

**Practice Problem B** The density of sodium metal is 0.971 g/cm$^3$, and the unit cell edge length is 4.285 Å. Determine the unit cell (simple, body-centered, or face-centered cubic) of sodium metal.

---

### Checkpoint 12.3     Crystal Structure

**12.3.1** Nickel has a face-centered cubic unit cell with an edge length of 352.4 pm. Calculate the density of nickel.

a) 2.227 g/cm$^3$         d) 8.908 g/cm$^3$

b) 4.455 g/cm$^3$         e) 11.14 g/cm$^3$

c) 38.99 g/cm$^3$

**12.3.2** At what angle would you expect X rays of wavelength 0.154 nm to be reflected from a crystal in which the distance between layers is 312 pm? (Assume $n = 1$.)

a) 1.6°         d) 6.8°

b) 29.6°        e) 14.3°

c) 0.25°

## 12.4   Types of Crystals

The structures and properties of crystalline solids, such as melting point, density, and hardness, are determined by the kinds of forces that hold the particles together. We can classify any crystal as one of four types: ionic, covalent, molecular, or metallic.

**Figure 12.25** The unit cell of an ionic compound can be defined either by (a) the positions of anions or (b) the positions of cations.

## Ionic Crystals

Ionic crystals are composed of charged spheres (cations and anions) that are held together by Coulombic attraction. Anions typically are considerably bigger than cations [◄◄ Section 7.6], and the relative sizes and relative numbers of the ions in a compound determine how the ions are arranged in the solid lattice. NaCl adopts a face-centered cubic arrangement as shown in Figure 12.25. Note the positions of ions within the unit cell, and within the lattice overall. Both the $Na^+$ ions and the $Cl^-$ ions adopt face-centered cubic arrangements, and the unit cell defined by the arrangement of cations overlaps with the unit cell defined by the arrangement of anions. Look closely at the unit cell shown in Figure 12.25(a). It is defined as fcc by the positions of the $Cl^-$ ions. Recall that there is the equivalent of four spheres contained in the fcc unit cell (half a sphere at each of six faces and one-eighth of a sphere at each of eight corners). In this case the spheres are $Cl^-$ ions, so the unit cell of NaCl contains four $Cl^-$ ions. Now look at the positions of the $Na^+$ ions. There are $Na^+$ ions centered on each edge of the cube, in addition to one $Na^+$ at the center. Each sphere on the cube's edge is shared by four unit cells, and there are 12 such edges. Thus, the unit cell in Figure 12.25(a) also contains four $Na^+$ ions (one-quarter sphere at each of 12 edges, giving three spheres, and one sphere at the center). The unit cell of an ionic compound always contains the same ratio of cations to anions as the empirical formula of the compound.

Figure 12.26 shows the crystal structures of three ionic compounds: CsCl, ZnS, and $CaF_2$. Cesium chloride [Figure 12.26(a)] has the simple cubic lattice. Despite the apparent similarity of the formulas of CsCl and NaCl, CsCl adopts a different arrangement because the $Cs^+$ ion is much larger than the $Na^+$ ion. Zinc sulfide [Figure 12.26(b)] has the *zincblende* structure, which is based on the face-centered cubic lattice. If the $S^{2-}$ ions occupy the lattice points, the smaller $Zn^{2+}$ ions are arranged tetrahedrally about each $S^{2-}$ ion. Other ionic compounds that have the zincblende structure include CuCl, BeS, CdS, and HgS. Calcium fluoride [Figure 12.26(c)] has the *fluorite* structure. The unit cell in Figure 12.26(c) is defined based on the positions of the cations, rather than the positions of the anions. The $Ca^{2+}$ ions occupy the lattice points, and each $F^-$ ion is surrounded tetrahedrally by four $Ca^{2+}$ ions. The compounds $SrF_2$, $BaF_2$, $BaCl_2$, and $PbF_2$ also have the fluorite structure.

**Student Annotation:** It is important to realize that the lattice points used to define a unit cell must all be identical. We can define the unit cell of NaCl based on the positions of the $Na^+$ ions or the positions of the $Cl^-$ ions.

**Student Annotation:** It is a common mistake to identify the CsCl structure as body-centered cubic. Remember that the lattice points used to define a unit cell must all be identical. In this case, they are all $Cl^-$ ions. CsCl has a simple cubic unit cell.

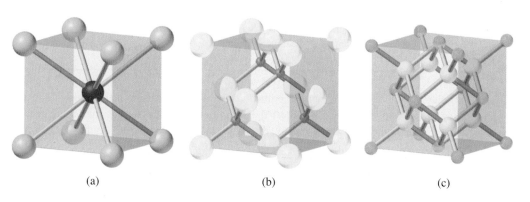

**Figure 12.26** Crystal structures of (a) CsCl, (b) ZnS, and (c) $CaF_2$. In each case, the smaller sphere represents the cation.

Sample Problems 12.5 and 12.6 show how to determine the number of ions in a unit cell and the density of an ionic crystal, respectively.

## Sample Problem 12.5

How many of each ion are contained within a unit cell of ZnS?

**Strategy** Determine the contribution of each ion in the unit cell based on its position.

**Setup** Referring to Figure 12.26, the unit cell has four $Zn^{2+}$ ions completely contained within the unit cell, and $S^{2-}$ ions at each of the eight corners and at each of the six faces. Interior ions (those completely contained within the unit cell) contribute one, those at the corners each contribute one-eighth, and those on the faces each contribute one-half.

**Solution** The ZnS unit cell contains four $Zn^{2+}$ ions (interior) and four $S^{2-}$ ions [$8 \times \frac{1}{8}$(corners) and $6 \times \frac{1}{2}$(faces)].

**Think About It** Make sure that the ratio of cations to anions that you determine for a unit cell matches the ratio expressed in the compound's empirical formula.

**Practice Problem A** Referring to Figure 12.26, determine how many of each ion are contained within a unit cell of $CaF_2$.

**Practice Problem B** Referring to Figure 12.26, determine how many of each ion are contained within a unit cell of CsCl.

## Sample Problem 12.6

The edge length of the NaCl unit cell is 564 pm. Determine the density of NaCl in $g/cm^3$.

**Strategy** Use the number of $Na^+$ and $Cl^-$ ions in a unit cell (four of each) to determine the mass of a unit cell. Calculate volume using the edge length given in the problem statement. Density is mass divided by volume ($d = m/V$). Be careful to use units consistently.

**Student Annotation:** Note that the mass of an atomic *ion* is treated the same as the mass of the parent *atom*. In these cases, the mass of an electron is not significant [◄◄ Section 2.2, Table 2.1].

**Setup** The masses of $Na^+$ and $Cl^-$ ions are 22.99 amu and 35.45 amu, respectively. The conversion factor from amu to grams is

$$\frac{1 \text{ g}}{6.022 \times 10^{23} \text{ amu}}$$

so the masses of the $Na^+$ and $Cl^-$ ions are $3.818 \times 10^{-23}$ g and $5.887 \times 10^{-23}$ g, respectively. The unit cell length is

$$564 \text{ pm} \times \frac{1 \times 10^{-12} \text{ m}}{1 \text{ pm}} \times \frac{1 \text{ cm}}{1 \times 10^{-2} \text{ m}} = 5.64 \times 10^{-8} \text{ cm}$$

**Solution** The mass of a unit cell is $3.882 \times 10^{-22}$ g ($4 \times 3.818 \times 10^{-23}$ g $+ 4 \times 5.887 \times 10^{-23}$ g). The volume of a unit cell is $1.794 \times 10^{-22}$ cm³ [$(5.64 \times 10^{-8}$ cm$)^3$]. Therefore, the density is given by

$$d = \frac{3.882 \times 10^{-22} \text{ g}}{1.794 \times 10^{-22} \text{ cm}^3} = 2.16 \text{ g/cm}^3$$

**Think About It** If you were to hold a cubic centimeter (1 cm³) of salt in your hand, how heavy would you expect it to be? Common errors in this type of problem include errors of unit conversion—especially with regard to length and volume. Such errors can lead to results that are off by many orders of magnitude. Often you can use common sense to gauge whether or not a calculated answer is reasonable. For instance, simply getting the centimeter-meter conversion upside down would result in a calculated density of $2.16 \times 10^{12}$ g/cm³! You *know* that a cubic centimeter of salt doesn't have a mass that large. (That's billions of kilograms!) If the magnitude of a result is not reasonable, go back and check your work.

**Practice Problem A** LiF has the same unit cell as NaCl (fcc). The edge length of the LiF unit cell is 402 pm. Determine the density of LiF in $g/cm^3$.

**Practice Problem B** NiO also adopts the face-centered-cubic arrangement. Given that the density of NiO is 6.67 $g/cm^3$, calculate the length of the edge of its unit cell (in pm).

Most ionic crystals have high melting points, which is an indication of the strong cohesive forces holding the ions together. A measure of the stability of ionic crystals is the lattice energy [◄◄ Section 8.2]; the higher the lattice energy, the more stable the compound. Ionic solids do not conduct electricity because the ions are fixed in position. In the molten (melted) state or when dissolved in water, however, the compound's ions are free to move and the resulting liquid conducts electricity.

## Covalent Crystals

In covalent crystals, atoms are held together in an extensive three-dimensional network entirely by covalent bonds. Well-known examples are two of carbon's allotropes: diamond and graphite. In diamond, each carbon atom is $sp^3$-hybridized and bonded to four other carbon atoms [Figure 12.27(a)]. The strong covalent bonds in three dimensions contribute to diamond's unusual hardness (it is the hardest material known) and very high melting point (3550°C). In graphite, carbon atoms are arranged in six-membered rings [Figure 12.27(b)]. The atoms are all $sp^2$-hybridized, and each atom is bonded to three other atoms. The remaining unhybridized $2p$ orbital on each carbon atom is used in pi bonding. In fact, each layer of graphite has the kind of delocalized molecular orbital that is present in benzene [◄◄ Section 9.7]. Because electrons are free to move around in this extensively delocalized molecular orbital, graphite is a good conductor of electricity in directions along the planes of the carbon atoms. The layers are held together by weak van der Waals forces. The covalent bonds in graphite account for its hardness; however, because the layers can slide past one another, graphite is slippery to the touch and is effective as a lubricant. It is also used as the "lead" in pencils.

Another covalent crystal is quartz ($SiO_2$). The arrangement of silicon atoms in quartz is similar to that of carbon in diamond, but in quartz there is an oxygen atom between each pair of Si atoms. Because Si and O have different electronegativities, the Si—O bond is polar. Nevertheless, $SiO_2$ is similar to diamond in many respects, such as being very hard and having a high melting point (1610°C).

## Molecular Crystals

In a molecular crystal, the lattice points are occupied by molecules, so the attractive forces between them are van der Waals forces and/or hydrogen bonding. An example of a molecular crystal is solid sulfur dioxide ($SO_2$), in which the predominant attractive force is a dipole-dipole interaction. Intermolecular hydrogen bonding is mainly responsible for maintaining the three-dimensional lattice of ice (Figure 12.28). Other examples of molecular crystals are $I_2$, $P_4$, and $S_8$.

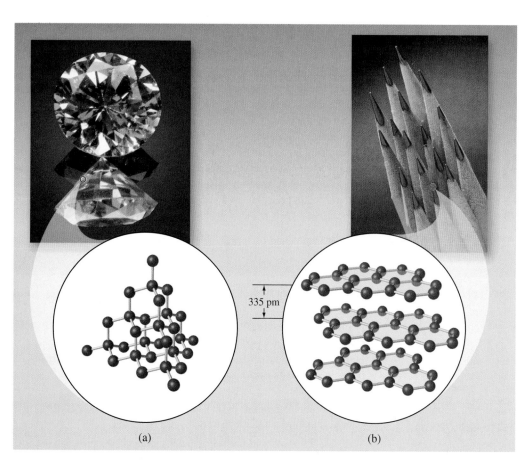

335 pm

(a)        (b)

**Figure 12.27** Structures of (a) diamond and (b) graphite. Note that in diamond, each carbon atom is bonded in a tetrahedral arrangement to four other carbon atoms. In graphite, each carbon atom is bonded in a trigonal planar arrangement to three other carbon atoms. The distance between layers in graphite is 335 pm.

**Figure 12.28**   The three-dimensional structure of ice. The covalent bonds are shown by short solid lines and the weaker hydrogen bonds by long dotted lines between O and H. The empty space in the structure accounts for the low density of ice, relative to liquid water.

**Figure 12.29**   A cross section of a metallic crystal. Each circled positive charge represents the nucleus and inner electrons of a metal atom. The grey area surrounding the positive metal ions indicates the mobile "sea" of electrons.

Except in ice, molecules in molecular crystals are generally packed together as closely as their size and shape allow. Because van der Waals forces and hydrogen bonding are usually quite weak compared with covalent and ionic bonds, molecular crystals are more easily broken apart than ionic and covalent crystals. Indeed, most molecular crystals melt at temperatures below 100°C.

## Metallic Crystals

Every lattice point in a metallic crystal is occupied by an atom of the same metal. Metallic crystals are generally body-centered cubic, face-centered cubic, or hexagonal close-packed. Consequently, metallic elements are usually very dense.

The bonding in metals is quite different from that in other types of crystals. In a metal, the bonding electrons are delocalized over the entire crystal. In fact, metal atoms in a crystal can be imagined as an array of positive ions immersed in a sea of delocalized valence electrons (Figure 12.29). The great cohesive force resulting from delocalization is responsible for a metal's strength, whereas the mobility of the delocalized electrons makes metals good conductors of heat and electricity. Table 12.4 summarizes the properties of the four different types of crystals discussed. Note that the data in Table 12.4 refer to the solid phase of each substance listed.

| TABLE 12.4 | Types of Crystals and Their General Properties | | |
|---|---|---|---|
| **Type of Crystal** | **Cohesive Forces** | **General Properties** | **Examples** |
| Ionic | Coulombic attraction and dispersion forces | Hard, brittle, high melting point, poor conductor of heat and electricity | $NaCl$, $LiF$, $MgO$, $CaCO_3$ |
| Covalent | Covalent bonds | Hard, brittle, high melting point, poor conductor of heat and electricity | C (diamond),* $SiO_2$ (quartz) |
| Molecular† | Dispersion and dipole-dipole forces, hydrogen bonds | Soft, low melting point, poor conductor of heat and electricity | Ar, $CO_2$, $I_2$, $H_2O$, $C_{12}H_{22}O_{11}$ |
| Metallic | Metallic bonds | Variable hardness and melting point, good conductor of heat and electricity | All metallic elements, such as Na, Mg, Fe, Cu |

*Diamond is a good conductor of heat.

†Included in this category are crystals made up of individual atoms.

**Sample Problem 12.7**

The metal iridium (Ir) crystallizes with a face-centered cubic unit cell. Given that the length of the edge of a unit cell is 383 pm, determine the density of iridium in $g/cm^3$.

**Strategy** A face-centered metallic crystal contains four atoms per unit cell [$8 \times \frac{1}{8}$(corners) and $6 \times \frac{1}{2}$(faces)]. Use the number of atoms per cell and the atomic mass to determine the mass of a unit cell. Calculate volume using the edge length given in the problem statement. Density is then mass divided by volume ($d = m/V$). Be sure to make all necessary unit conversions.

**Setup** The mass of an Ir atom is 192.2 amu. The conversion factor from amu to grams is

$$\frac{1\ g}{6.022 \times 10^{23}\ amu}$$

so the mass of an Ir atom is $3.192 \times 10^{-22}$ g. The unit cell length is

$$383\ pm \times \frac{1 \times 10^{-12}\ m}{1\ pm} \times \frac{1\ cm}{1 \times 10^{-2}\ m} = 3.83 \times 10^{-8}\ cm$$

**Solution** The mass of a unit cell is $1.277 \times 10^{-21}$ g ($4 \times 3.192 \times 10^{-22}$ g). The volume of a unit cell is $5.618 \times 10^{-23}\ cm^3$ [$(3.83 \times 10^{-8}\ cm)^3$]. Therefore, the density is given by

$$d = \frac{1.277 \times 10^{-21}\ g}{5.62 \times 10^{-23}\ cm^3} = 22.7\ g/cm^3$$

**Think About It** Metals typically have high densities, so common sense can help you decide whether or not your calculated answer is reasonable.

**Practice Problem A** Aluminum metal crystallizes in a face-centered cubic unit cell. If the length of the cell edge is 404 pm, what is the density of aluminum in $g/cm^3$?

**Practice Problem B** Copper crystallizes in a face-centered cubic lattice. If the density of the metal is 8.96 $g/cm^3$, what is the length of the unit cell edge in picometers?

# 12.5 Amorphous Solids

Solids are most stable in crystalline form. However, if a solid is formed rapidly (e.g., when a liquid is cooled quickly), its atoms or molecules do not have time to align themselves and may become locked in positions other than those of a regular crystal. The resulting solid is said to be amorphous. *Amorphous solids,* such as glass, lack a regular three-dimensional arrangement of atoms. In this section, we will briefly discuss the properties of glass.

Glass is one of civilization's most valuable and versatile materials. It is also one of the oldest—glass articles date back as far as 1000 B.C. *Glass* commonly refers to an optically transparent fusion product of inorganic materials that has cooled to a rigid state without crystallizing. By fusion product we mean that the glass is formed by mixing molten silicon dioxide ($SiO_2$), its chief component, with compounds such as sodium oxide ($Na_2O$), boron oxide ($B_2O_3$), and certain transition metal oxides for color and other properties. In some respects, glass behaves more like a liquid than a solid.

There are about 800 different types of glass in common use today. Figure 12.30 shows two-dimensional schematic representations of crystalline quartz and amorphous quartz glass. Table 12.5 lists the composition and properties of quartz, Pyrex, and soda-lime glass. The color of glass is due largely to the presence of oxides of metal ions—mostly transition metal ions. For example, green glass contains iron(III) oxide ($Fe_2O_3$) or copper(II) oxide (CuO), yellow glass contains uranium(IV) oxide ($UO_2$), blue glass contains cobalt(II) and copper(II) oxides (CoO and CuO), and red glass contains small particles of gold and copper.

# 12.6 Phase Changes

A *phase* is a homogeneous part of a system that is separated from the rest of the system by a well-defined boundary. When an ice cube floats in a glass of water, for example, the liquid water is one

**Figure 12.30** Two-dimensional representation of (a) crystalline quartz and (b) noncrystalline (amorphous) quartz glass. The small spheres represent silicon. In reality, the structure of quartz is three dimensional. Each Si atom is bonded in a tetrahedral arrangement to four O atoms.

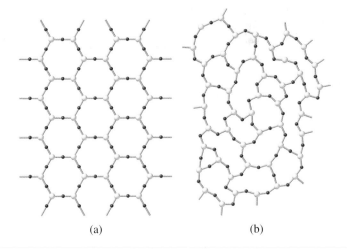

(a)                                    (b)

| TABLE 12.5 | Composition and Properties of Three Types of Glass | |
| --- | --- | --- |
| Pure quartz glass | 100% $SiO_2$ | Low thermal expansion, transparent to a wide range of wavelengths. Used in optical research. |
| Pyrex glass | 60%–80% $SiO_2$, 10%–25% $B_2O_3$, some $Al_2O_3$ | Low thermal expansion; transparent to visible and infrared, but not to ultraviolet light. Used in cookware and laboratory glassware. |
| Soda-lime glass | 75% $SiO_2$, 15% $Na_2O$, 10% CaO | Easily attacked by chemicals and sensitive to thermal shocks. Transmits visible light but absorbs ultraviolet light. Used in windows and bottles. |

phase and the solid water (the ice cube) is another. Although the chemical properties of water are the same in both phases, the physical properties of a solid are different from those of a liquid.

When a substance goes from one phase to another phase, we say that it has undergone a ***phase change.*** Phase changes in a system are generally caused by the addition or removal of energy, usually in the form of heat. Familiar examples of phase changes include the following:

**Student Annotation:** Phase changes are *physical* changes [◀◀ Section 1.4].

| Example | Phase Change |
| --- | --- |
| Freezing of water | $H_2O(l) \longrightarrow H_2O(s)$ |
| Evaporation (or vaporization) of water | $H_2O(l) \longrightarrow H_2O(g)$ |
| Melting (fusion) of ice | $H_2O(s) \longrightarrow H_2O(l)$ |
| Condensation of water vapor | $H_2O(g) \longrightarrow H_2O(l)$ |
| Sublimation of dry ice | $CO_2(s) \longrightarrow CO_2(g)$ |

The establishment of an equilibrium vapor pressure, as described in Section 12.2, involved two of these phase changes: vaporization and condensation. Figure 12.31 summarizes the various types of phase changes.

## Liquid-Vapor Phase Transition

In Section 12.2 we learned that the vapor pressure of a liquid increases with increasing temperature. When the vapor pressure reaches the external pressure, the liquid boils. In fact, the ***boiling point*** of a substance is defined as the temperature at which its vapor pressure equals the external, atmospheric pressure. As a result, the boiling point of a substance varies with the external pressure. At the top of a mountain, for example, where the atmospheric pressure is lower than that at sea level, the vapor pressure of water (or any liquid) reaches the external pressure at a lower temperature. Thus, the boiling point is lower than it would be at sea level.

**Student Annotation:** The temperature at which the vapor pressure of a liquid is equal to 1 atm is called the ***normal boiling point.***

Because the boiling point is defined in terms of the vapor pressure of the liquid, the boiling point is related to the ***molar heat of vaporization ($\Delta H_{vap}$),*** the amount of heat required to vaporize a mole of substance at its boiling point. Indeed, the data in Table 12.6 show that the boiling point generally increases as $\Delta H_{vap}$ increases. Ultimately, both the boiling point and $\Delta H_{vap}$ are determined by the strength of intermolecular forces. For example, argon (Ar) and methane ($CH_4$), which have

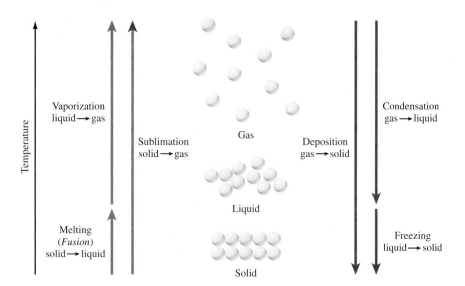

| TABLE 12.6 | Molar Heats of Vaporization for Selected Liquids | |
|---|---|---|
| **Substance** | **Boiling Point (°C)** | **$\Delta H_{vap}$ (kJ/mol)** |
| Argon (Ar) | −186 | 6.3 |
| Benzene ($C_6H_6$) | 80.1 | 31.0 |
| Ethanol ($C_2H_5OH$) | 78.3 | 39.3 |
| Diethyl ether ($C_2H_5OC_2H_5$) | 34.6 | 26.0 |
| Mercury (Hg) | 357 | 59.0 |
| Methane ($CH_4$) | −164 | 9.2 |
| Water ($H_2O$) | 100 | 40.79 |

only relatively weak dispersion forces, have low boiling points and small molar heats of vaporization. Diethyl ether ($C_2H_5OC_2H_5$) has a dipole moment, and the dipole-dipole forces account for its moderately high boiling point and $\Delta H_{vap}$. Both ethanol ($C_2H_5OH$) and water have strong hydrogen bonding, which accounts for their high boiling points and large $\Delta H_{vap}$ values. Strong metallic bonding causes mercury to have the highest boiling point and $\Delta H_{vap}$ of the liquids in Table 12.6. Interestingly, benzene ($C_6H_6$), although nonpolar, has a high *polarizability* due to the distribution of its electrons in delocalized $\pi$ molecular orbitals. The dispersion forces that result can be as strong as (or even stronger than) dipole-dipole forces and/or hydrogen bonds.

The opposite of vaporization is condensation. In principle, a gas can be liquefied (made to condense) either by cooling or by applying pressure. Cooling a sample of gas decreases the kinetic energy of its molecules, so eventually the molecules aggregate to form small drops of liquid. Applying pressure to the gas (compression), on the other hand, reduces the distance between molecules, so they can be pulled together by intermolecular attractions. Many liquefaction processes use a combination of reduced temperature and increased pressure.

Every substance has a ***critical temperature ($T_c$)*** above which its gas phase cannot be liquefied, no matter how great the applied pressure. ***Critical pressure ($P_c$)*** is the minimum pressure that must be applied to liquefy a substance *at* its critical temperature. At temperatures above the critical temperature, there is no fundamental distinction between a liquid and a gas—we simply have a fluid. A fluid at a temperature and pressure that exceed $T_c$ and $P_c$, respectively, is called a ***supercritical fluid.*** Supercritical fluids have some remarkable properties and are used as solvents in a wide variety of industrial applications. The first such large-scale industrial use was the decaffeination of coffee with supercritical $CO_2$.

Table 12.7 lists the critical temperatures and critical pressures of a number of common substances. The critical temperature of a substance reflects the strength of its intermolecular forces. Benzene, ethanol, mercury, and water, which have strong intermolecular forces, also have high critical temperatures compared with the other substances listed in the table.

**Student Annotation:** $T_c$ is the *highest* temperature at which a substance can exist as a liquid.

| TABLE 12.7 | Critical Temperatures and Critical Pressures of Selected Substances | |
|---|---|---|
| Substance | $T_c$ (°C) | $P_c$ (atm) |
| Ammonia ($NH_3$) | 132.4 | 111.5 |
| Argon (Ar) | −122.2 | 6.3 |
| Benzene ($C_6H_6$) | 288.9 | 47.9 |
| Carbon dioxide ($CO_2$) | 31.0 | 73.0 |
| Ethanol ($C_2H_5OH$) | 243 | 63.0 |
| Diethyl ether ($C_2H_5OC_2H_5$) | 192.6 | 35.6 |
| Mercury (Hg) | 1462 | 1036 |
| Methane ($CH_4$) | −83.0 | 45.6 |
| Molecular hydrogen ($H_2$) | −239.9 | 12.8 |
| Molecular nitrogen ($N_2$) | −147.1 | 33.5 |
| Molecular oxygen ($O_2$) | −118.8 | 49.7 |
| Sulfur hexafluoride ($SF_6$) | 45.5 | 37.6 |
| Water ($H_2O$) | 374.4 | 219.5 |

## Solid-Liquid Phase Transition

The transformation of liquid to solid is called *freezing*, and the reverse process is called *melting*, or *fusion*. The *melting point* of a solid or the *freezing point* of a liquid is the temperature at which solid and liquid phases coexist in equilibrium. The normal melting (or freezing) point of a substance is the temperature at which it melts (or freezes) at 1 atm.

The most familiar liquid-solid equilibrium is probably that of water and ice. At 0°C and 1 atm, the dynamic equilibrium is represented by

$$ice \rightleftharpoons water$$

or

$$H_2O(s) \rightleftharpoons H_2O(l)$$

A glass of ice water at 0°C provides a practical illustration of this dynamic equilibrium. As the ice cubes melt to form water, some of the water between ice cubes may freeze, thus joining the cubes together. Remember that in a dynamic equilibrium, forward and reverse processes are occurring at the same rate [◀◀ Section 4.1].

**Student Annotation:** In most cases a glass of ice water would not be a *true* example of a dynamic equilibrium because it would not be kept at 0°C. At room temperature, all the ice eventually melts.

Because molecules are more strongly held in the solid phase than in the liquid phase, heat is required to melt a solid into a liquid. The heating curve in Figure 12.32 shows that when a solid is heated, its temperature increases gradually until point A is reached. At this point, the solid begins to melt. During the melting period (A ⟶ B), the first flat portion of the curve in Figure 12.32, heat is being absorbed by the system, yet its temperature remains constant. The heat helps the molecules overcome the attractive forces in the solid. Once the sample has melted completely (point B), the heat absorbed increases the average kinetic energy of the liquid molecules and the liquid temperature rises (B ⟶ C). The vaporization process (C ⟶ D) can be explained similarly. The temperature remains constant during the period when the increased kinetic energy is used to overcome the cohesive forces in the liquid. When all molecules are in the gas phase, the temperature rises again.

The *molar heat of fusion* ($\Delta H_{fus}$) is the energy, usually expressed in kJ/mol, required to melt 1 mole of a solid. Table 12.8 lists the molar heats of fusion for the substances in Table 12.6. A comparison of the data in the two tables shows that $\Delta H_{fus}$ is smaller than $\Delta H_{vap}$ for each substance. This is consistent with the fact that molecules in a liquid are still fairly closely packed together, so some energy (but not a lot of energy, relatively speaking) is needed to bring about the rearrangement from solid to liquid. When a liquid is vaporized, on the other hand, its molecules become completely separated from one another, so considerably more energy is required to overcome the intermolecular attractive forces.

Cooling a substance has the opposite effect of heating it. If we remove heat from a gas sample at a steady rate, its temperature decreases. As the liquid is being formed, heat is given off

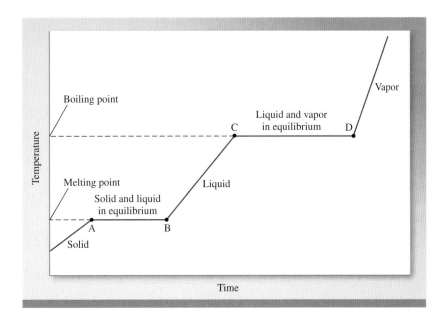

**Figure 12.32** A typical heating curve, from the solid phase through the liquid phase to the gas phase of a substance. Because $\Delta H_{fus}$ is smaller than $\Delta H_{vap}$, a substance melts in less time than it takes to boil. This explains why AB is shorter than CD. The steepness of the solid, liquid, and vapor heating lines is determined by the specific heat of the substance in each state.

| TABLE 12.8 | Molar Heats of Fusion for Selected Substances | |
|---|---|---|
| **Substance** | **Melting Point (°C)** | **$\Delta H_{fus}$ (kJ/mol)** |
| Argon (Ar) | −190 | 1.3 |
| Benzene ($C_6H_6$) | 5.5 | 10.9 |
| Ethanol ($C_2H_5OH$) | −117.3 | 7.61 |
| Diethyl ether ($C_2H_5OC_2H_5$) | −116.2 | 6.90 |
| Mercury (Hg) | −39 | 23.4 |
| Methane ($CH_4$) | −183 | 0.84 |
| Water ($H_2O$) | 0 | 6.01 |

by the system, because its potential energy is decreasing. For this reason, the temperature of the system remains constant over the condensation period (D ⟶ C). After all the vapor has condensed, the temperature of the liquid begins to drop again. Continued cooling of the liquid finally leads to freezing (B ⟶ A).

**Supercooling** is a phenomenon in which a liquid can be temporarily cooled to below its freezing point. Supercooling occurs when heat is removed from a liquid so rapidly that the molecules literally have no time to assume the ordered structure of a solid. A supercooled liquid is unstable. Gentle stirring or the addition to it of a small "seed" crystal of the same substance will cause it to solidify quickly.

## Solid-Vapor Phase Transition

Solids can be vaporized, so solids, too, have a vapor pressure. **Sublimation** is the process by which molecules go directly from the solid phase to the vapor phase. The reverse process, in which molecules go directly from the vapor phase to the solid phase, is called **deposition.** Naphthalene, which is the substance used to make mothballs, has a fairly high vapor pressure for a solid (1 mmHg at 53°C); thus, its pungent vapor quickly permeates an enclosed space. Iodine also sublimes. At room temperature, the violet color of iodine vapor is easily visible in a closed container.

Because molecules are more tightly held in a solid, the vapor pressure of a solid is generally much less than that of the corresponding liquid. The **molar enthalpy of sublimation ($\Delta H_{sub}$)** of a substance is the energy, usually expressed in kilojoules, required to sublime 1 mole of a solid. It is equal to the sum of the molar enthalpies of fusion and vaporization:

$$\Delta H_{sub} = \Delta H_{fus} + \Delta H_{vap} \qquad \text{Equation 12.6}$$

**Solid iodine in equilibrium with its vapor.**

**Figure 12.33**  Heating curve of water.

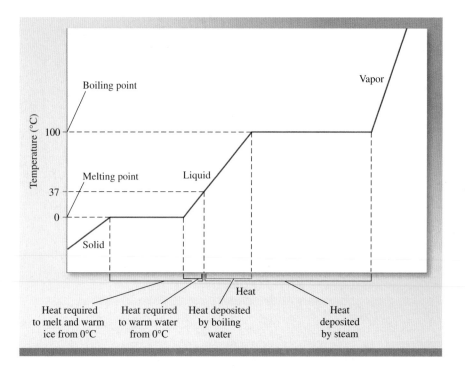

Equation 12.6 is an illustration of Hess's law [◀◀ Section 5.5]. The enthalpy, or heat change, for the overall process is the same whether the substance changes directly from the solid to the vapor phase or if it changes from the solid to the liquid and then to the vapor phase.

## Bringing Chemistry to Life

### The Dangers of Phase Changes

If you have ever suffered a steam burn, you know that it can be far more serious than a burn caused simply by boiling water—even though steam and boiling water are both at the same temperature. A heating curve helps explain why this is so (see Figure 12.33). When boiling water touches your skin, it is cooled to body temperature because it deposits the heat it contains on your skin. The heat deposited on your skin by a sample of boiling water at 100°C can be represented by the orange line under the curve. When an equivalent mass of steam contacts your skin, it first deposits heat as it condenses and *then* cools to body temperature. The heat deposited on your skin by a sample of steam is represented by the red line under the curve. Notice how much more heat is deposited by steam than by liquid water at the same temperature. The steam contains more heat because it has been heated *and* vaporized. The additional heat that was absorbed by the water to vaporize it is what makes a steam burn worse than a burn from boiling water.

A heating curve can also be used to explain why hikers stranded by blizzards are warned not to consume snow in an effort to stay hydrated. When you drink cold water, your body expends energy to warm the water you consume to body temperature. If you consume snow, your body must first expend the energy necessary to melt the snow, and then to warm it. Because a phase change is involved, the amount of energy required to assimilate snow is much greater than the amount necessary to assimilate an equal mass of water—even if the water is ice-cold. This can contribute to *hypothermia*, a potentially dangerous drop in body temperature.

### Sample Problem 12.8

(a) Calculate the amount of heat deposited on the skin of a person burned by 1.00 g of liquid water at 100.0°C and (b) the amount of heat deposited by 1.00 g of steam at 100.0°C. (c) Calculate the amount of energy necessary to warm 100.0 g of water from 0.0°C to body temperature and (d) the amount of heat required to melt 100.0 g of ice at 0.0°C and then warm it to body temperature. (Assume that body temperature is 37.0°C.)

**Strategy** For the purpose of following the sign conventions, we can designate the water as the *system* and the body as the *surroundings*. (a) Heat is transferred from hot water to the skin in a single step: a temperature change. (b) The transfer of heat from steam to the skin takes place in two steps: a phase change and a temperature change. (c) Cold water is warmed to body temperature in a single step: a temperature change. (d) The melting of ice and the subsequent warming of the resulting liquid water takes place in two steps: a phase change and a temperature change. In each case, the heat transferred during a temperature change depends on the mass of the water, the specific heat of water, and the change in temperature. For the phase changes, the heat transferred depends on the amount of water (in moles) and the molar heat of vaporization ($\Delta H_{vap}$) or molar heat of fusion ($\Delta H_{fus}$). In each case, the total energy transferred or required is the sum of the energy changes for the individual steps.

**Setup** The required specific heats ($s$) are 4.184 J/g · °C for water and 1.99 J/g · °C for steam. (Assume that the specific heat values do not change over the range of temperatures in the problem.) From Table 12.6, the molar heat of vaporization ($\Delta H_{vap}$) of water is 40.79 kJ/mol, and from Table 12.8, the molar heat of fusion ($\Delta H_{fus}$) of water is 6.01 kJ/mol. The molar mass of water is 18.02 g/mol. *Note:* The $\Delta H_{vap}$ of water is the amount of heat required to vaporize a mole of water. In this problem, however, we want to know how much heat is deposited when water vapor *condenses,* so we must use the negative, $-40.79$ kJ/mol.

**Solution**
(a) $\Delta T = 37.0°C - 100.0°C = -63.0°C$
From Equation 5.13, we write

$$q = ms\Delta T = 1.00 \text{ g} \times \frac{4.184 \text{ J}}{\text{g} \cdot °C} \times -63.0°C = -2.64 \times 10^2 \text{ J} = -0.264 \text{ kJ}$$

Thus, 1.00 g of water at 100.0°C deposits 0.264 kJ of heat on the skin. (The negative sign indicates that heat is given off by the system and absorbed by the surroundings.)

(b) $\dfrac{1.00 \text{ g}}{18.02 \text{ g/mol}} = 0.0555$ mol water

$$q_1 = n\Delta H_{vap} = 0.0555 \text{ mol} \times \frac{-40.79 \text{ kJ}}{\text{mol}} = -2.26 \text{ kJ}$$

$$q_2 = ms\Delta T = 1.00 \text{ g} \times \frac{4.184 \text{ J}}{\text{g} \cdot °C} \times -63.0°C = -2.64 \times 10^2 \text{ J} = -0.264 \text{ kJ}$$

The overall energy deposited on the skin by 1.00 g of steam is the sum of $q_1$ and $q_2$:

$$-2.26 \text{ kJ} + (-0.264 \text{ kJ}) = -2.53 \text{ kJ}$$

The negative sign indicates that the system (steam) gives off the energy.

(c) $\Delta T = 37.0°C - 0.0°C = 37.0°C$

$$q = ms\Delta T = 100.0 \text{ g} \times \frac{4.184 \text{ J}}{\text{g} \cdot °C} \times 37.0°C = 1.55 \times 10^4 \text{ J} = 15.5 \text{ kJ}$$

The energy required to warm 100.0 g of water from 0.0°C to 37.0°C is 15.5 kJ.

(d) $\dfrac{100.0 \text{ g}}{18.02 \text{ g/mol}} = 5.55$ mol

$$q_1 = n\Delta H_{fus} = 5.55 \text{ mol} \times \frac{6.01 \text{ kJ}}{\text{mol}} = 33.4 \text{ kJ}$$

$$q_2 = ms\Delta T = 100.0 \text{ g} \times \frac{4.184 \text{ J}}{\text{g} \cdot °C} \times 37.0°C = 1.55 \times 10^4 \text{ J} = 15.5 \text{ kJ}$$

The energy required to melt 100.0 g of ice at 0.0°C and warm it to 37.0°C is the sum of $q_1$ and $q_2$:

$$33.4 \text{ kJ} + 15.5 \text{ kJ} = 48.9 \text{ kJ}$$

**Think About It** In problems that include phase changes, the q values corresponding to the phase-change steps will be the largest contributions to the total. If you find that this is not the case in your solution, check to see if you have made the common error of neglecting to convert the q values corresponding to temperature changes from J to kJ.

**Practice Problem A** Calculate the amount of energy (in kJ) necessary to convert 346 g of liquid water from 0°C to water vapor at 182°C.

**Practice Problem B** Determine the final state and temperature of 100 g of water originally at 25.0°C after 50.0 kJ of heat have been added to it.

### Checkpoint 12.6    Phase Changes

**12.6.1**  How much energy (in kJ) is required to convert 25.0 g of liquid water at room temperature (25°C) to steam at 110°C?

   a) 64.9 kJ

   b) 562 kJ

   c) 1339 kJ

   d) 1.34 kJ

   e) 26.9 kJ

**12.6.2**  How much energy (in kJ) is given off when 1.0 g of steam at 100°C cools to room temperature (25°C)?

   a) 0.326 kJ

   b) 316 kJ

   c) 2.58 kJ

   d) 48.9 kJ

   e) 22.1 kJ

## 12.7    Phase Diagrams

The relationships between the phases of a substance can be represented in a single graph known as a phase diagram. A **phase diagram** summarizes the conditions (temperature and pressure) at which a substance exists as a solid, liquid, or gas. Figure 12.34(a) shows the phase diagram of $CO_2$, which is typical of many substances. The graph is divided into three regions, each of which represents a pure phase. The line separating any two regions, called a *phase boundary line,* indicates conditions under which these two phases can exist in equilibrium. The point at which all three phase boundary lines meet is called the **triple point.** The triple point is the only combination of temperature and pressure at which all three phases of a substance can be in equilibrium with one another. The point at which the liquid-vapor phase boundary line abruptly ends is the critical point, corresponding to the critical temperature ($T_c$) and the critical pressure ($P_c$).

In order to understand the information in a phase diagram, consider the dashed lines between numbered points in Figure 12.34(b). If we start with a sample of $CO_2$ at 1 atm and −100°C, the sample is initially a solid (point 1). If we then add heat to the sample at 1 atm, its temperature increases until it reaches −78°C, the sublimation point of $CO_2$ at 1 atm (point 2). When the entire sample has sublimed at −78°C, the temperature of the resulting vapor will begin to increase. We continue adding heat until the temperature of the vapor is −25°C (point 3). At this point, we maintain the temperature at −25°C and begin increasing the pressure until the vapor condenses, which would occur at a pressure of about 70 atm (point 4). With a phase diagram, we can tell in what phase a substance will exist at any given temperature and pressure. Furthermore, we can tell what phase changes will occur as the result of increases or decreases in temperature, pressure, or both.

One of the interesting things about the phase diagram of $CO_2$ is that its triple point occurs above atmospheric pressure −57°C, 5.2 atm. This means that there is no liquid phase at atmospheric pressure, the condition that makes dry ice "dry." At elevated pressure, solid $CO_2$ does melt. In fact, liquid $CO_2$ is used as the solvent in many dry-cleaning operations.

The phase diagram of water (Figure 12.35) is unusual because the solid-liquid phase-boundary line has a negative slope. (Compare this with the solid-liquid phase-boundary line of $CO_2$ in Figure 12.34.) As a result, ice can be liquefied within a narrow temperature range by applying pressure.

> **Student Annotation:** Bismuth, the element pictured on the cover, is another example of a substance whose solid-liquid phase boundary (under ordinary atmospheric conditions) has a negative slope. Like water, bismuth has a liquid density (1.005 g/cm³ at room temperature) that is higher than its solid density (0.9780 g/cm³ at its melting point of 271°C).

**Figure 12.34**    (a) The phase diagram of carbon dioxide. Note that the solid-liquid boundary line has a positive slope. There is no liquid phase below 5.2 atm, so only the solid and vapor phases can exist under ordinary atmospheric conditions. (b) Heating solid $CO_2$ initially at −100°C and 1 atm (point 1) causes it to sublime when it reaches −78°C (point 2). At 25°C, increasing the pressure from 1 atm (point 3) to about 70 (point 4) will cause $CO_2$ to condense to a liquid.

(a)

(b)

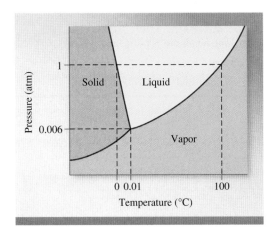

**Figure 12.35**   The phase diagram of water. Each solid line between two phases specifies the conditions of pressure and temperature under which the two phases can exist in equilibrium. The point at which all three phases can exist in equilibrium (0.006 atm and 0.01°C) is called the triple point.

Sample Problem 12.9 lets you practice interpreting the information in a phase diagram.

## Sample Problem 12.9

Using the following phase diagram, (a) determine the normal boiling point and the normal melting point of the substance, (b) determine the physical state of the substance at 2 atm and 110°C, and (c) determine the pressure and temperature that correspond to the triple point of the substance.

**Strategy** Each point on the phase diagram corresponds to a pressure-temperature combination. The normal boiling and melting points are the temperatures at which the substance undergoes phase changes. These points fall on the phase boundary lines. The triple point is where the three phase boundaries meet.

**Setup** By drawing lines corresponding to a given pressure and/or temperature, we can determine the temperature at which a phase change occurs, or the physical state of the substance under specified conditions.

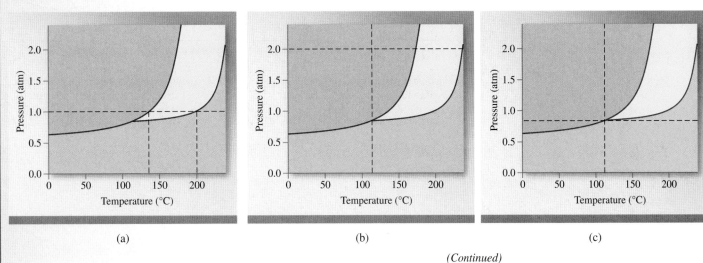

(a)                              (b)                              (c)

*(Continued)*

**Solution**

(a) The normal boiling and melting points are ∼140°C and ∼205°C, respectively.

(b) At 2 atm and 110°C the substance is a solid.

(c) The triple point occurs at ∼0.8 atm and ∼115°C.

> **Think About It** The triple point of this substance occurs at a pressure below atmospheric pressure. Therefore, it will melt rather than sublime when it is heated under ordinary conditions.

**Practice Problem A** Use the following phase diagram to (a) determine the normal boiling point and melting point of the substance, and (b) the physical state of the substance at 1.2 atm and 100°C.

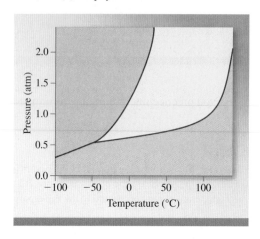

**Practice Problem B** Sketch the phase diagram of a substance using the following data:

| Pressure (atm) | Melting Point (°C) | Boiling Point (°C) | Sublimation Point (°C) |
|---|---|---|---|
| 0.5 | — | — | 0 |
| 1.0 | 60 | 110 | — |
| 1.5 | 75 | 200 | — |
| 2.0 | 105 | 250 | — |
| 2.5 | 125 | 275 | — |

The triple point is at 0.75 atm and 45°C.

---

## Checkpoint 12.7    Phase Diagrams

Refer to the following phase diagrams to answer questions 12.7.1 and 12.7.2.

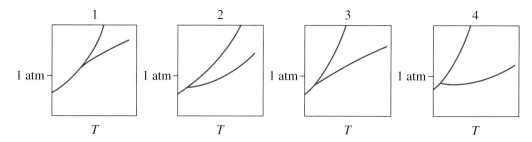

**12.7.1** Which phase diagram corresponds to a substance that will sublime rather than melt as it is heated at 1 atm?

a) 1

b) 1 and 2

c) 1, 2, and 4

d) 3

e) none

**12.7.2** Which phase diagram corresponds to a substance that will liquefy when pressure is increased at a temperature below its freezing point?

a) 1

b) 1 and 2

c) 1, 2, and 4

d) 3

e) none

# Applying What You've Learned

The melamine tragedy in China caused an epidemic of unusual kidney stones in infants, a condition that is ordinarily very rare. In the general population of adults, though, kidney stones are relatively common with roughly 10 percent of adults experiencing symptoms during their lifetimes. The most common type of kidney stone is composed of calcium oxalate ($CaC_2O_4$), and several different approaches can be used for treatment and/or prevention. One method of treatment involves the use of diuretics—drugs that cause elimination of water from the body in the form of urine. Increasing the volume of fluid that passes through the kidneys can help to reduce the amount of calcium available for stone formation. It can also facilitate the passing of a stone that has already formed. Two of the drugs that are prescribed for the treatment and prevention of kidney stones are hydrochlorothiazide and chlorthalidone.

Hydrochlorothiazide                 Chlorthalidone

## Problems:

a) What types of intermolecular forces exist between chlorthalidone molecules and between hydrochlorothiazide molecules? [◄◄ Sample Problem 12.1]

b) Calcium is one of the components of the most common type of kidney stone. Calcium metal crystallizes in a face-centered cubic unit cell with a cell edge length of 558.84 pm. Calculate the radius of a calcium atom in angstroms (Å). [◄◄ Sample Problem 12.3]

c) How many Ca atoms are contained within a unit cell? [◄◄ Sample Problem 12.5]

d) Calculate the density of calcium metal in $g/cm^3$. [◄◄ Sample Problem 12.6]

e) Patients who have suffered from kidney stones often are advised to drink extra water to help prevent the formation of additional stones. An article on WebMD.com recommends drinking at least three quarts (2.84 L) of water every day—nearly 50 percent more than the amount recommended for healthy adults. How much energy must the body expend to warm this amount of water consumed at 10°C to body temperature (37°C)? How much *more* energy would have to be expended if the same quantity of water were consumed as ice at 0°C? $\Delta H_{fus}$ for water is 6.01 kJ/mol. Assume the density and specific heat of water are 1.00 $g/cm^3$ and 4.184 J/g · °C, respectively, and that both quantities are independent of temperature. [◄◄ Sample Problem 12.7].

# CHAPTER SUMMARY

## Section 12.1

- The particles (atoms, molecules, or ions) in the condensed phases (solids and liquids) are held together by **intermolecular forces.** Intermolecular forces are electrostatic attractions between opposite charges or partial charges.

- Intermolecular forces acting between atoms or molecules in a pure substance are called **van der Waals forces** and include **dipole-dipole interactions** (including **hydrogen bonding**) and **London dispersion forces** (also called simply **dispersion forces**).

- Dipole-dipole interactions exist between *polar* molecules, whereas *nonpolar* molecules are held together by *dispersion* forces alone. Dispersion forces are those between **instantaneous dipoles** and *induced* dipoles. When a nonpolar molecule acquires an instantaneous dipole, it is said to be **polarized.** Dispersion forces act between all molecules, nonpolar and polar.

- Hydrogen bonding is an especially strong type of dipole-dipole interaction that occurs in molecules that contain $H-N$, $H-O$, or $H-F$ bonds.

- **Ion-dipole interactions** are those that occur (in solutions) between ions and polar molecules.

## Section 12.2

- Physical properties of liquids that result from intermolecular forces include surface tension, viscosity, and vapor pressure.

- **Surface tension** is the net pull inward on molecules at the surface of a liquid. Surface tension is related to **cohesion,** the attractive forces between molecules within a substance, and **adhesion,** the attractive forces between molecules in a substance and their container. The balance between cohesion and adhesion determines whether a liquid meniscus is concave or convex. It also gives rise to **capillary action,** in which liquid is drawn upward into a narrow tube against gravity.

- **Viscosity** is resistance to flow, reflecting how easily molecules move past one another.

- **Evaporation** (also known as **vaporization**) is the phase change from liquid to vapor. **Condensation** is the phase change from vapor to liquid. In a closed system, when vaporization and condensation are occurring at the same rate, a state of **dynamic equilibrium** exists and the *vapor pressure* is equal to the **equilibrium vapor pressure.** Vapor pressure measures how easily molecules escape to the vapor phase. A **volatile** substance has a high vapor pressure.

- Surface tension, viscosity, and vapor pressure are all temperature dependent. The **Clausius-Clapeyron equation** relates the vapor pressure of a substance to its absolute temperature.

## Section 12.3

- The spheres (molecules, atoms, or ions) in a **crystalline solid** are arranged in a three-dimensional **lattice structure** consisting of a repeating pattern of **unit cells.** The type of unit cell is determined by the positions of the **lattice points.**

- Cubic unit cells may be **simple cubic, body-centered cubic,** or **face-centered cubic,** containing a total of one, two, or four spheres, respectively. The **coordination number** is the number of spheres surrounding each sphere.

- *Closest packing* is the most efficient arrangement of spheres in a solid. It may be *hexagonal* or *cubic.* Each has a coordination number of 12.

- **X-ray diffraction** is used to analyze crystal structure. The wavelength of X rays, the angle of diffraction, and the spacing between atoms in a lattice are related by the **Bragg equation.**

## Section 12.4

- Crystals may be ionic, covalent, molecular, or metallic. Each type of crystalline solid has characteristics determined in part by the types of interactions holding it together.

## Section 12.5

- **Amorphous solids** such as **glass** lack regular three-dimensional structure.

## Section 12.6

- The possible **phase changes** are melting or **fusion** ($s \longrightarrow l$), freezing ($l \longrightarrow s$), *vaporization* ($l \longrightarrow g$), *condensation* ($g \longrightarrow l$), **sublimation** ($s \longrightarrow g$), and **deposition** ($g \longrightarrow s$).

- The **boiling point** of a substance is the temperature at which its vapor pressure equals the external pressure. The **normal boiling point** is the temperature at which its vapor pressure equals 1 atm. Boiling point is pressure dependent. The **molar heat of vaporization** ($\Delta H_{vap}$) is the amount of heat required to vaporize 1 mole of a substance at its boiling point.

- The **critical temperature** ($T_c$) is the temperature above which a gas cannot be liquefied by applying pressure. The **critical pressure** ($P_c$) is the pressure necessary to liquefy a gas at its critical temperature. A substance above its critical temperature and pressure is a **supercritical fluid.**

- The **melting point** or **freezing point** is the temperature at which the solid and liquid phases are in equilibrium. The **molar heat of fusion** ($\Delta H_{fus}$) is the amount of heat required to melt 1 mole of a substance at its melting point. **Supercooling** is the process of rapidly lowering a liquid's temperature below its freezing point.

- The **molar heat of sublimation** ($\Delta H_{sub}$) is equal to the sum of the molar heats of fusion and vaporization: $\Delta H_{sub} = \Delta H_{fus} + \Delta H_{vap}$

## Section 12.7

- A **phase diagram** indicates the phase of a substance under any combination of temperature and pressure. Lines between phases are called *phase boundaries.*

- The **triple point** is where all three phase boundaries meet. This is the temperature and pressure combination at which all three phases are in equilibrium.

# KEY WORDS

| | | | |
|---|---|---|---|
| Adhesion, 501 | Deposition, 521 | Ion-dipole interactions, 500 | Polarized, 499 |
| Amorphous solid, 517 | Dipole-dipole interactions, 494 | Lattice point, 505 | Simple cubic cell, 506 |
| Body-centered cubic cell, 506 | Dispersion forces, 498 | Lattice structure, 505 | Sublimation, 521 |
| Boiling point, 518 | Dynamic equilibrium, 502 | London dispersion forces, 498 | Supercooling, 521 |
| Bragg equation, 511 | Equilibrium vapor pressure, 502 | Melting point, 520 | Supercritical fluid, 519 |
| Capillary action, 500 | Evaporation, 502 | Molar heat of fusion ($\Delta H_{fus}$), 520 | Surface tension, 500 |
| Clausius-Clapeyron equation, 503 | Face-centered cubic cell, 506 | Molar heat of sublimation | Triple point, 524 |
| Cohesion, 501 | Freezing point, 520 | ($\Delta H_{sub}$), 521 | Unit cell, 505 |
| Condensation, 502 | Fusion, 520 | Molar heat of vaporization | van der Waals forces, 494 |
| Coordination number, 506 | Glass, 517 | ($\Delta H_{vap}$), 518 | Vaporization, 502 |
| Critical pressure ($P_c$), 519 | Hydrogen bonding, 495 | Normal boiling point, 518 | Viscosity, 501 |
| Critical temperature ($T_c$), 519 | Instantaneous dipole, 498 | Phase change, 518 | Volatile, 502 |
| Crystalline solid, 505 | Intermolecular forces, 494 | Phase diagram, 524 | X-ray diffraction, 510 |

# KEY EQUATIONS

12.1   $\ln P = -\dfrac{\Delta H_{vap}}{RT} + C$

12.2   $\ln P_1 = -\dfrac{\Delta H_{vap}}{RT_1} + C$

12.3   $\ln P_2 = -\dfrac{\Delta H_{vap}}{RT_2} + C$

12.4   $\ln \dfrac{P_1}{P_2} = \dfrac{\Delta H_{vap}}{R}\left(\dfrac{1}{T_2} - \dfrac{1}{T_1}\right)$

12.5   $BC + CD = 2d \sin \theta = n\lambda \qquad n = 1, 2, 3, \ldots$

12.6   $\Delta H_{sub} = \Delta H_{fus} + \Delta H_{vap}$

# QUESTIONS AND PROBLEMS

## Section 12.1: Intermolecular Forces

### Review Questions

12.1   Give an example for each type of intermolecular force:
(a) dipole-dipole interaction, (b) ion-dipole interaction,
(c) dispersion forces, (d) van der Waals forces.

12.2   Explain the term *polarizability*. What kind of molecules tend
to have high polarizabilities? What is the relationship between
polarizability and intermolecular forces?

12.3   Explain the difference between a temporary dipole moment and a
permanent dipole moment.

12.4   What evidence can you cite that all molecules exert attractive
forces on one another?

12.5   What physical properties are determined by the strength of
intermolecular forces in solids and in liquids?

12.6   Explain why hydrogen bonding is exhibited by some
hydrogen-containing compounds and not by others.

12.7   Describe the types of intermolecular forces that govern the folding
of a protein molecule into its physiologically functioning three-
dimensional state.

### Problems

12.8   The compounds $Br_2$ and $ICl$ are isoelectronic (have the same
number of electrons) and have similar molar masses, yet $Br_2$ melts
at $-7.2°C$ and $ICl$ melts at $27.2°C$. Explain.

**12.9**   If you lived in Alaska, which of the following natural gases could
you keep in an outdoor storage tank in winter: methane ($CH_4$),
propane ($C_3H_8$), or butane ($C_4H_{10}$)? Explain why.

12.10   The binary hydrogen compounds of the Group 4A elements and
their boiling points are: $CH_4$, $-162°C$; $SiH_4$, $-112°C$; $GeH_4$,
$-88°C$; and $SnH_4$, $-52°C$. Explain the increase in boiling points
from $CH_4$ to $SnH_4$.

**12.11**   List the types of intermolecular forces that exist between
molecules (or atoms or ions) in each of the following species:
(a) benzene ($C_6H_6$), (b) $CH_3Cl$, (c) $PF_3$, (d) $NaCl$, (e) $CS_2$.

12.12  Ammonia is both a donor and an acceptor of hydrogen in hydrogen-bond formation. Draw a diagram showing the hydrogen bonding of an ammonia molecule with two other ammonia molecules.

12.13  Which of the following species are capable of hydrogen-bonding among themselves: (a) $C_2H_6$, (b) HI, (c) KF, (d) $BeH_2$, (e) $CH_3COOH$?

12.14  Arrange the following in order of increasing boiling point: RbF, $CO_2$, $CH_3OH$, $CH_3Br$. Explain your reasoning.

12.15  Diethyl ether has a boiling point of 34.5°C, and 1-butanol has a boiling point of 117°C:

$$
\begin{array}{cc}
\underset{\substack{|\\ H}}{\overset{\substack{H \\ |}}{H-C}}-\underset{\substack{|\\ H}}{\overset{\substack{H \\ |}}{C}}-O-\underset{\substack{|\\ H}}{\overset{\substack{H \\ |}}{C}}-\underset{\substack{|\\ H}}{\overset{\substack{H \\ |}}{C}}-H
&
\underset{\substack{|\\ H}}{\overset{\substack{H \\ |}}{H-C}}-\underset{\substack{|\\ H}}{\overset{\substack{H \\ |}}{C}}-\underset{\substack{|\\ H}}{\overset{\substack{H \\ |}}{C}}-\underset{\substack{|\\ H}}{\overset{\substack{H \\ |}}{C}}-OH \\
\text{Diethyl ether} & \text{1-Butanol}
\end{array}
$$

Both of these compounds have the same numbers and types of atoms. Explain the difference in their boiling points.

12.16  Which member of each of the following pairs of substances would you expect to have a higher boiling point: (a) $O_2$ and $Cl_2$, (b) $SO_2$ and $CO_2$, (c) HF and HI?

12.17  Which substance in each of the following pairs would you expect to have the higher boiling point: (a) Ne or Xe, (b) $CO_2$ or $CS_2$, (c) $CH_4$ or $Cl_2$, (d) $F_2$ or LiF, (e) $NH_3$ or $PH_3$? Explain why.

12.18  Explain in terms of intermolecular forces why (a) $NH_3$ has a higher boiling point than $CH_4$ and (b) KCl has a higher melting point than $I_2$.

12.19  What kind of attractive forces must be overcome to (a) melt ice, (b) boil molecular bromine, (c) melt solid iodine, and (d) dissociate $F_2$ into F atoms?

12.20  The following compounds have the same molecular formulas $(C_4H_{10})$. Which one would you expect to have a higher boiling point?

12.21  Explain the difference in the melting points of the following compounds:

m.p. 45°C        m.p. 115°C

(*Hint:* One of the two can form *intra*molecular hydrogen bonds.)

## Section 12.2: Properties of Liquids

### Review Questions

12.22  Explain why liquids, unlike gases, are virtually incompressible.

12.23  What is surface tension? What is the relationship between intermolecular forces and surface tension? How does surface tension change with temperature?

12.24  Despite the fact that stainless steel is much denser than water, a stainless-steel razor blade can be made to float on water. Why?

12.25  Use water and mercury as examples to explain adhesion and cohesion.

12.26  A glass can be filled slightly above the rim with water. Explain why the water does not overflow.

12.27  Draw diagrams showing the capillary action of (a) water and (b) mercury in three tubes of different radii.

12.28  What is viscosity? What is the relationship between intermolecular forces and viscosity?

12.29  Why does the viscosity of a liquid decrease with increasing temperature?

12.30  Why is ice less dense than water?

12.31  Outdoor water pipes have to be drained or insulated in winter in a cold climate. Why?

### Problems

12.32  Predict which of the following liquids has greater surface tension: ethanol $(C_2H_5OH)$ or dimethyl ether $(CH_3OCH_3)$.

12.33  Predict the viscosity of ethylene glycol relative to that of ethanol and glycerol (see Table 12.3).

$$
\begin{array}{c}
CH_2-OH \\
| \\
CH_2-OH \\
\text{Ethylene glycol}
\end{array}
$$

12.34  Vapor pressure measurements at several different temperatures are shown for mercury. Determine graphically the molar heat of vaporization for mercury.

| $T$(°C) | 200 | 250 | 300 | 320 | 340 |
|---|---|---|---|---|---|
| $P$(mmHg) | 17.3 | 74.4 | 246.8 | 376.3 | 557.9 |

12.35  The vapor pressure of benzene $(C_6H_6)$ is 40.1 mmHg at 7.6°C. What is its vapor pressure at 60.6°C? The molar heat of vaporization of benzene is 31.0 kJ/mol.

12.36  The vapor pressure of liquid X is lower than that of liquid Y at 20°C, but higher at 60°C. What can you deduce about the relative magnitude of the molar heats of vaporization of X and Y?

12.37  Estimate the molar heat of vaporization of a liquid whose vapor pressure doubles when the temperature is raised from 75°C to 100°C.

## Section 12.3: Crystal Structure

### Review Questions

12.38  Define the following terms: *crystalline solid, lattice point, unit cell, coordination number, closest packing.*

12.39   Describe the geometries of the following cubic cells: simple cubic, body-centered cubic, face-centered cubic. Which of these structures would give the highest density for the same type of atoms? Which the lowest?

12.40   Classify the solid states in terms of crystal types of the elements in the third period of the periodic table. Predict the trends in their melting points and boiling points.

12.41   The melting points of the oxides of the third-period elements are given in parentheses: $Na_2O$ (1275°C), MgO (2800°C), $Al_2O_3$ (2045°C), $SiO_2$ (1610°C), $P_4O_{10}$ (580°C), $SO_3$ (16.8°C), $Cl_2O_7$ (−91.5°C). Classify these solids in terms of crystal types.

12.42   Define X-ray diffraction. What are the typical wavelengths (in nm) of X rays? (See Figure 6.1.)

12.43   Write the Bragg equation. Define every term and describe how this equation can be used to measure interatomic distances.

### Problems

12.44   What is the coordination number of each sphere in (a) a simple cubic cell, (b) a body-centered cubic cell, and (c) a face-centered cubic cell? Assume the spheres are all the same.

12.45   Calculate the number of spheres that would be found within a simple cubic cell, body-centered cubic cell, and face-centered cubic cell. Assume that the spheres are the same.

12.46   Metallic iron crystallizes in a cubic lattice. The unit cell edge length is 287 pm. The density of iron is 7.87 g/cm³. How many iron atoms are within a unit cell?

12.47   Barium metal crystallizes in a body-centered cubic lattice (the Ba atoms are at the lattice points only). The unit cell edge length is 502 pm, and the density of the metal is 3.50 g/cm³. Using this information, calculate Avogadro's number. [*Hint:* First calculate the volume (in cm³) occupied by 1 mole of Ba atoms in the unit cells. Next calculate the volume (in cm³) occupied by one Ba atom in the unit cell. Assume that 68 percent of the unit cell is occupied by Ba atoms.]

12.48   Vanadium crystallizes in a body-centered cubic lattice (the V atoms occupy only the lattice points). How many V atoms are present in a unit cell?

12.49   Europium crystallizes in a body-centered cubic lattice (the Eu atoms occupy only the lattice points). The density of Eu is 5.26 g/cm³. Calculate the unit cell edge length in picometers.

12.50   Crystalline silicon has a cubic structure. The unit cell edge length is 543 pm. The density of the solid is 2.33 g/cm³. Calculate the number of Si atoms in one unit cell.

12.51   A face-centered cubic cell contains 8 X atoms at the corners of the cell and 6 Y atoms at the faces. What is the empirical formula of the solid?

12.52   When X rays of wavelength 0.090 nm are diffracted by a metallic crystal, the angle of first-order diffraction ($n = 1$) is measured to be 15.2°. What is the distance (in pm) between the layers of atoms responsible for the diffraction?

12.53   The distance between layers in an NaCl crystal is 282 pm. X rays are diffracted from these layers at an angle of 23.0°. Assuming that $n = 1$, calculate the wavelength of the X rays in nanometers.

12.54   Identify the unit cell of molecular iodine ($I_2$) shown here. (*Hint:* Consider the position of iodine molecules, not individual iodine atoms.)

12.55   Shown here is a zinc oxide unit cell. What is the formula of zinc oxide?

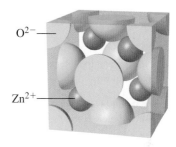

## Section 12.4: Types of Crystals

### Review Questions

12.56   Describe and give examples of the following types of crystals: (a) ionic crystals, (b) covalent crystals, (c) molecular crystals, (d) metallic crystals.

12.57   Why are metals good conductors of heat and electricity? Why does the ability of a metal to conduct electricity decrease with increasing temperature?

### Problems

12.58   A solid is hard, brittle, and electrically nonconducting. Its melt (the liquid form of the substance) and an aqueous solution containing the substance conduct electricity. Classify the solid.

12.59   A solid is soft and has a low melting point (below 100°C). The solid, its melt, and an aqueous solution containing the substance are all nonconductors of electricity. Classify the solid.

12.60   A solid is very hard and has a high melting point. Neither the solid nor its melt conducts electricity. Classify the solid.

12.61   Which of the following are molecular solids and which are covalent solids: $Se_8$, HBr, Si, $CO_2$, C, $P_4O_6$, $SiH_4$?

12.62   Classify the solid state of the following substances as ionic crystals, covalent crystals, molecular crystals, or metallic crystals: (a) $CO_2$, (b) $B_{12}$, (c) $S_8$, (d) KBr, (e) Mg, (f) $SiO_2$, (g) LiCl, (h) Cr.

12.63   Explain why diamond is harder than graphite. Why is graphite an electrical conductor but diamond is not?

## Section 12.5: Amorphous Solids

### Review Questions

12.64   What is an amorphous solid? How does it differ from a crystalline solid?

12.65  Define glass. What is the chief component of glass? Name three types of glass.

## Section 12.6: Phase Changes

### Review Questions

12.66  What is a phase change? Name all possible changes that can occur among the vapor, liquid, and solid phases of a substance.

12.67  What is the equilibrium vapor pressure of a liquid? How is it measured, and how does it change with temperature?

12.68  Use any one of the phase changes to explain what is meant by dynamic equilibrium.

12.69  Define the following terms: (a) molar heat of vaporization, (b) molar heat of fusion, (c) molar heat of sublimation. What are their typical units?

12.70  How is the molar heat of sublimation related to the molar heats of vaporization and fusion? On what law are these relationships based?

12.71  What can we learn about the intermolecular forces in a liquid from the molar heat of vaporization?

12.72  The greater the molar heat of vaporization of a liquid, the greater its vapor pressure. True or false?

12.73  Define boiling point. How does the boiling point of a liquid depend on external pressure? Referring to Table 11.5, what is the boiling point of water when the external pressure is 187.5 mmHg?

12.74  As a liquid is heated at constant pressure, its temperature rises. This trend continues until the boiling point of the liquid is reached. No further rise in temperature of the liquid can be induced by heating. Explain.

12.75  What is critical temperature? What is the significance of critical temperature in condensation of gases?

12.76  What is the relationship between intermolecular forces in a liquid and the liquid's boiling point and critical temperature? Why is the critical temperature of water greater than that of most other substances?

12.77  How do the boiling points and melting points of water and carbon tetrachloride vary with pressure? Explain any difference in behavior of these two substances.

12.78  Why is solid carbon dioxide called dry ice?

12.79  The vapor pressure of a liquid in a closed container depends on which of the following: (a) the volume above the liquid, (b) the amount of liquid present, (c) temperature, (d) intermolecular forces between the molecules in the liquid?

12.80  Wet clothes dry more quickly on a hot, dry day than on a hot, humid day. Explain.

12.81  Which of the following phase transitions gives off more heat: (a) 1 mole of steam to 1 mole of water at 100°C, or (b) 1 mole of water to 1 mole of ice at 0°C?

12.82  A beaker of water is heated to boiling by a Bunsen burner. Would adding another burner raise the temperature of the boiling water? Explain.

12.83  Explain why splashing a small amount of liquid nitrogen (b.p. 77 K) is not as harmful as splashing boiling water on your skin.

### Problems

12.84  Calculate the amount of heat (in kJ) required to convert 150.2 g of water to steam at 100°C.

12.85  How much heat (in kJ) is needed to convert 866 g of ice at $-15$°C to steam at 146°C? (The specific heats of ice and steam are 2.03 and 1.99 J/g · °C, respectively.)

12.86  How is the rate of evaporation of a liquid affected by (a) temperature, (b) the surface area of a liquid exposed to air, (c) intermolecular forces?

**12.87**  The molar heats of fusion and sublimation of molecular iodine are 15.27 and 62.30 kJ/mol, respectively. Estimate the molar heat of vaporization of liquid iodine.

12.88  The following compounds, listed with their boiling points, are liquid at $-10$°C: butane, $-0.5$°C; ethanol, 78.3°C; toluene, 110.6°C. At $-10$°C, which of these liquids would you expect to have the highest vapor pressure? Which the lowest? Explain.

**12.89**  Freeze-dried coffee is prepared by freezing brewed coffee and then removing the ice component with a vacuum pump. Describe the phase changes taking place during these processes.

12.90  A student hangs wet clothes outdoors on a winter day when the temperature is $-15$°C. After a few hours, the clothes are found to be fairly dry. Describe the phase changes in this drying process.

**12.91**  Explain why steam at 100°C causes more serious burns than water at 100°C.

## Section 12.7: Phase Diagrams

### Review Questions

12.92  What is a phase diagram? What useful information can be obtained from studying a phase diagram?

12.93  Explain how water's phase diagram differs from those of most substances. What property of water causes the difference?

### Problems

12.94  The blades of ice skates are quite thin, so the pressure exerted on ice by a skater can be substantial. Explain how this facilitates skating on ice.

**12.95**  A length of wire is placed on top of a block of ice. The ends of the wire extend over the edges of the ice, and a heavy weight is attached to each end. It is found that the ice under the wire gradually melts, so the wire slowly moves through the ice block. At the same time, the water above the wire refreezes. Explain the phase changes that accompany this phenomenon.

12.96  The boiling point and freezing point of sulfur dioxide are $-10$°C and $-72.7$°C (at 1 atm), respectively. The triple point is $-75.5$°C and $1.65 \times 10^{-3}$ atm, and its critical point is at 157°C and 78 atm. On the basis of this information, draw a rough sketch of the phase diagram of $SO_2$.

**12.97**  A phase diagram of water is shown. Label the regions. Predict what would happen as a result of the following changes:
(a) Starting at A, we raise the temperature at constant pressure.
(b) Starting at B, we lower the pressure at constant temperature.
(c) Starting at C, we lower the temperature at constant pressure.

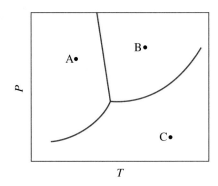

## Additional Problems

**12.98** Name the kinds of attractive forces that must be overcome to (a) boil liquid ammonia, (b) melt solid phosphorus ($P_4$), (c) dissolve CsI in liquid HF, (d) melt potassium metal.

**12.99** Which of the following properties indicates very strong intermolecular forces in a liquid: (a) very low surface tension, (b) very low critical temperature, (c) very low boiling point, (d) very low vapor pressure?

**12.100** At $-35°C$, liquid HI has a higher vapor pressure than liquid HF. Explain.

**12.101** Based on the following properties of elemental boron, classify it as one of the crystalline solids discussed in Section 12.4: high melting point (2300°C), poor conductor of heat and electricity, insoluble in water, very hard substance.

**12.102** Referring to Figure 12.34, determine the stable phase of $CO_2$ at (a) 4 atm and $-60°C$ and (b) 0.5 atm and $-20°C$.

**12.103** Which of the following substances has the highest polarizability: $CH_4$, $H_2$, $CCl_4$, $SF_6$, $H_2S$?

**12.104** Given two complementary strands of DNA containing 100 base pairs each, calculate the ratio of two separate strands to hydrogen-bonded double helix in solution at 300 K. (*Hint:* The formula for calculating this ratio is $e^{-\Delta E/RT}$, where $\Delta E$ is the energy difference between hydrogen-bonded double-strand DNAs and single-strand DNAs and $R$ is the gas constant.) Assume the energy of hydrogen bonds per base pair to be 10 kJ/mol.

**12.105** The average distance between base pairs measured parallel to the axis of a DNA molecule is 3.4 Å. The average molar mass of a pair of nucleotides is 650 g/mol. Estimate the length in centimeters of a DNA molecule of molar mass $5.0 \times 10^9$ g/mol. Roughly how many base pairs are contained in this molecule?

**12.106** A $CO_2$ fire extinguisher is located on the outside of a building in Massachusetts. During the winter months, one can hear a sloshing sound when the extinguisher is gently shaken. In the summertime there is often no sound when it is shaken. Explain. Assume that the extinguisher has no leaks and that it has not been used.

**12.107** What is the vapor pressure of mercury at its normal boiling point (357°C)?

**12.108** A flask of water is connected to a powerful vacuum pump. When the pump is turned on, the water begins to boil. After a few minutes, the same water begins to freeze. Eventually, the ice disappears. Explain what happens at each step.

**12.109** The liquid-vapor boundary line in the phase diagram of any substance always stops abruptly at a certain point. Why?

**12.110** The interionic distances of several alkali halide crystals are as follows:

| Crystal | NaCl | NaBr | NaI | KCl | KBr | KI |
|---|---|---|---|---|---|---|
| Interionic distance (pm) | 282 | 299 | 324 | 315 | 330 | 353 |

Plot lattice energy versus the reciprocal interionic distance. How would you explain the plot in terms of the dependence of lattice energy on the distance of separation between ions? What law governs this interaction? (For lattice energies, see Table 8.1.)

**12.111** Which has a greater density, crystalline $SiO_2$ or amorphous $SiO_2$? Why?

**12.112** A student is given four solid samples labeled W, X, Y, and Z. All have a metallic luster. She is told that the solids could be gold, lead sulfide, mica (which is quartz, or $SiO_2$), and iodine. The results of her investigations are: (a) W is a good electrical conductor; X, Y, and Z are poor electrical conductors. (b) When the solids are hit with a hammer, W flattens out, X shatters into many pieces, Y is smashed into a powder, and Z is not affected. (c) When the solids are heated with a Bunsen burner, Y melts with some sublimation, but X, W, and Z do not melt. (d) In treatment with 6 $M$ $HNO_3$, X dissolves; there is no effect on W, Y, or Z. On the basis of these test results, identify the solids.

**12.113** Which of the following statements are false? (a) Dipole-dipole interactions between molecules are greatest if the molecules possess only temporary dipole moments. (b) All compounds containing hydrogen atoms can participate in hydrogen-bond formation. (c) Dispersion forces exist between all atoms, molecules, and ions.

**12.114** The diagram shows a kettle of boiling water. Identify the phases in regions A and B.

**12.115** The south pole of Mars is covered with solid carbon dioxide, which partly sublimes during the summer. The $CO_2$ vapor recondenses in the winter when the temperature drops to 150 K. Given that the heat of sublimation of $CO_2$ is 25.9 kJ/mol, calculate the atmospheric pressure on the surface of Mars. [*Hint:* Use Figure 12.34 to determine the normal sublimation temperature of dry ice and Equation 12.4, which also applies to sublimations.]

**12.116** The properties of gases, liquids, and solids differ in a number of respects. How would you use the kinetic molecular theory (see Section 11.6) to explain the following observations? (a) Ease of compressibility decreases from gas to liquid to solid. (b) Solids retain a definite shape, but gases and liquids do not. (c) For most substances, the volume of a given amount of material increases as it changes from solid to liquid to gas.

**12.117** Select the substance in each pair that should have the higher boiling point. In each case identify the principal intermolecular forces involved and account briefly for your choice: (a) $K_2S$ or $(CH_3)_3N$, (b) $Br_2$ or $CH_3CH_2CH_2CH_3$.

12.118  A small drop of oil in water assumes a spherical shape. Explain. (*Hint:* Oil is made up of nonpolar molecules, which tend to avoid contact with water.)

**12.119**  Under the same conditions of temperature and density, which of the following gases would you expect to behave less ideally: $CH_4$ or $SO_2$? Explain.

12.120  The fluorides of the second-period elements and their melting points are: LiF, 845°C; $BeF_2$, 800°C; $BF_3$, −126.7°C; $CF_4$, −184°C; $NF_3$, −206.6°C; $OF_2$, −223.8°C; $F_2$, −219.6°C. Classify the type(s) of intermolecular forces present in each compound.

12.121  The standard enthalpy of formation of gaseous molecular iodine is 62.4 kJ/mol. Use this information to calculate the molar heat of sublimation of molecular iodine at 25°C.

12.122  The distance between $Li^+$ and $Cl^-$ is 257 pm in solid LiCl and 203 pm in an LiCl unit in the gas phase. Explain the difference in the bond lengths.

**12.123**  Heat of hydration, that is, the heat change that occurs when ions become hydrated in solution, is largely due to ion-dipole interactions. The heats of hydration for the alkali metal ions are $Li^+$, −520 kJ/mol; $Na^+$, −405 kJ/mol; $K^+$, −321 kJ/mol. Account for the trend in these values.

12.124  If water were a linear molecule, (a) would it still be polar, and (b) would the water molecules still be able to form hydrogen bonds with one another?

**12.125**  Calculate the $\Delta H°$ for the following processes at 25°C: (a) $Br_2(l) \longrightarrow Br_2(g)$ and (b) $Br_2(g) \longrightarrow 2Br(g)$. Comment on the relative magnitudes of these $\Delta H°$ values in terms of the forces involved in each case. (*Hint:* See Table 8.6.)

12.126  Which liquid would you expect to have a greater viscosity, water or diethyl ether? The structure of diethyl ether is shown in Problem 12.15.

**12.127**  A beaker of water is placed in a closed container. Predict the effect on the vapor pressure of the water when (a) its temperature is lowered, (b) the volume of the container is doubled, (c) more water is added to the beaker.

12.128  Ozone ($O_3$) is a strong oxidizing agent that can oxidize all the common metals except gold and platinum. A convenient test for ozone is based on its action on mercury. When exposed to ozone, mercury becomes dull looking and sticks to glass tubing (instead of flowing freely through it). Write a balanced equation for the reaction. What property of mercury is altered by its interaction with ozone?

**12.129**  A sample of limestone ($CaCO_3$) is heated in a closed vessel until it is partially decomposed. Write an equation for the reaction, and state how many phases are present.

12.130  Carbon and silicon belong to Group 4A of the periodic table and have the same valence electron configuration ($ns^2np^2$). Why does silicon dioxide ($SiO_2$) have a much higher melting point than carbon dioxide ($CO_2$)?

**12.131**  Gaseous or highly volatile liquid anesthetics are often preferred in surgical procedures because once inhaled, these vapors can quickly enter the bloodstream through the alveoli and then enter the brain. Several common gaseous anesthetics are shown with their boiling points:

| Halothane | Isoflurane | Enflurane |
|-----------|-----------|-----------|
| 50°C | 48.5°C | 56.5°C |

Based on intermolecular force considerations, explain the advantages of using these anesthetics. (*Hint:* The brain barrier is made of membranes that have a nonpolar interior region.)

12.132  Given the general properties of water and ammonia, comment on the problems that a biological system (as we know it) would have developing in an ammonia medium.

|  | $H_2O$ | $NH_3$ |
|---|---|---|
| Boiling point | 373.15 K | 239.65 K |
| Melting point | 273.15 K | 195.3 K |
| Molar heat capacity | 75.3 J/K · mol | 8.53 J/K · mol |
| Molar heat of vaporization | 40.79 kJ/mol | 23.3 kJ/mol |
| Molar heat of fusion | 6.0 kJ/mol | 5.9 kJ/mol |
| Viscosity | 0.001 N · s/m$^2$ | 0.0254 N · s/m$^2$ (at 240 K) |
| Dipole moment | 1.82 D | 1.46 D |
| Phase at 300 K | Liquid | Gas |

**12.133**  A pressure cooker is a sealed container that allows steam to escape when it exceeds a predetermined pressure. How does this device reduce the time needed for cooking?

12.134  A 1.20-g sample of water is injected into an evacuated 5.00-L flask at 65°C. What percentage of the water will be vapor when the system reaches equilibrium? Assume ideal behavior of water vapor and that the volume of liquid water is negligible. The vapor pressure of water at 65°C is 187.5 mmHg.

**12.135**  What are the advantages of cooking the vegetable broccoli with steam instead of boiling it in water?

12.136  A quantitative measure of how efficiently spheres pack into unit cells is called *packing efficiency,* which is the percentage of the cell space occupied by the spheres. Calculate the packing efficiencies of a simple cubic cell, a body-centered cubic cell, and a face-centered cubic cell. (*Hint:* Refer to Figure 12.24 and use the relationship that the volume of a sphere is $\frac{4}{3}\pi r^3$, where $r$ is the radius of the sphere.)

**12.137**  The phase diagram of helium is shown. Helium is the only known substance that has two different liquid phases: helium-I and helium-II. (a) What is the maximum temperature at which helium-II can exist? (b) What is the minimum pressure at which solid helium can exist? (c) What is the normal boiling point of helium-I? (d) Can solid helium sublime?

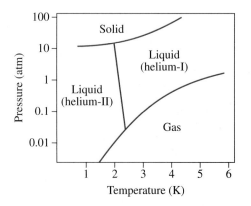

12.138  The phase diagram of sulfur is shown. (a) How many triple points are there? (b) Which is the more stable allotrope under ordinary atmospheric conditions? (c) Describe what happens when sulfur at 1 atm is heated from 80°C to 200°C.

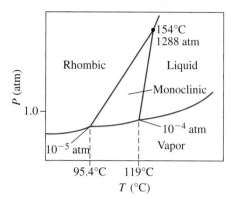

12.139  Provide an explanation for each of the following phenomena: (a) Solid argon (m.p. −189.2°C; b.p. −185.7°C) can be prepared by immersing a flask containing argon gas in liquid nitrogen (b.p. −195.8°C) until it liquefies and then connecting the flask to a vacuum pump. (b) The melting point of cyclohexane ($C_6H_{12}$) increases with increasing pressure exerted on the solid cyclohexane. (c) Certain high-altitude clouds contain water droplets at −10°C. (d) When a piece of dry ice is added to a beaker of water, fog forms above the water.

12.140  Argon crystallizes in the face-centered cubic arrangement at 40 K. Given that the atomic radius of argon is 191 pm, calculate the density of solid argon.

12.141  Given the phase diagram of carbon, answer the following questions: (a) How many triple points are there and what are the phases that can coexist at each triple point? (b) Which has a higher density, graphite or diamond? (c) Synthetic diamond can be made from graphite. Using the phase diagram, how would you go about making diamond?

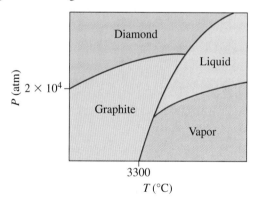

12.142  A chemistry instructor performed the following mystery demonstration. Just before the students arrived in class, she heated some water to boiling in an Erlenmeyer flask. She then removed the flask from the flame and closed the flask with a rubber stopper. After the class commenced, she held the flask in front of the students and announced that she could make the water boil simply by rubbing an ice cube on the outside walls of the flask. To the amazement of everyone, it worked. Give an explanation for this phenomenon.

12.143  Swimming coaches sometimes suggest that a drop of alcohol (ethanol) placed in an ear plugged with water "draws out the water." Explain this action from a molecular point of view.

12.144  Use the concept of intermolecular forces to explain why the far end of a walking cane rises when one raises the handle.

12.145  Why do citrus growers spray their trees with water to protect them from freezing?

12.146  What is the origin of dark spots on the inner glass walls of an old tungsten lightbulb? What is the purpose of filling these lightbulbs with argon gas?

12.147  A student heated a beaker of cold water (on a tripod) with a Bunsen burner. When the gas was ignited, she noticed that there was water condensed on the outside of the beaker. Explain what happened.

12.148  The compound dichlorodifluoromethane ($CCl_2F_2$) has a normal boiling point of −30°C, a critical temperature of 112°C, and a corresponding critical pressure of 40 atm. If the gas is compressed to 18 atm at 20°C, will the gas condense? Your answer should be based on a graphical interpretation.

12.149  Iron crystallizes in a body-centered cubic lattice. The cell length as determined by X-ray diffraction is 286.7 pm. Given that the density of iron is 7.874 g/cm³, calculate Avogadro's number.

12.150  Sketch the cooling curves of water from about 110°C to about −10°C. How would you also show the formation of supercooled liquid below 0°C that then freezes to ice? The pressure is at 1 atm throughout the process. The curves need not be drawn quantitatively.

12.151  The boiling point of methanol is 65.0°C, and the standard enthalpy of formation of methanol vapor is −201.2 kJ/mol. Calculate the vapor pressure of methanol (in mmHg) at 25°C. (Hint: See Appendix 2 for other thermodynamic data of methanol.)

12.152  A sample of water shows the following behavior as it is heated at a constant rate:

If twice the mass of water has the same amount of heat transferred to it, which of the following graphs best describes the temperature variation? Note that the scales for all the graphs are the same.

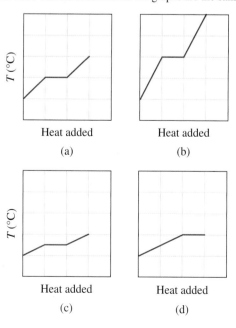

**12.153**    A closed vessel of volume 9.6 L contains 2.0 g of water. Calculate the temperature (in °C) at which only half of the water remains in the liquid phase. (See Table 11.5 for vapor pressures of water at different temperatures.)

**12.154**    The electrical conductance of copper metal decreases with increasing temperature, but that of a CuSO solution increases with increasing temperature. Explain.

**12.155**    Assuming ideal behavior, calculate the density of gaseous HF at its normal boiling point (19.5°C). The experimentally measured density under the same conditions is 3.10 g/L. Account for the difference between your calculated result and the experimental value.

**12.156**    Explain why drivers are advised to use motor oil with lower viscosity in the winter and higher viscosity in the summer.

**12.157**    Which of the following compounds is most likely to exist as a liquid at room temperature: ethane ($C_2H_6$), hydrazine ($N_2H_4$), fluoromethane ($CH_3F$)? Explain.

# PRE-PROFESSIONAL PRACTICE EXAM PROBLEMS:
# PHYSICAL AND BIOLOGICAL SCIENCES

Silicon used in computer chips must have an impurity level below $10^{-9}$ (that is, fewer than one impurity atom for every $10^9$ Si atoms). Silicon is prepared by the reduction of quartz ($SiO_2$) with coke (a form of carbon made by the destructive distillation of coal) at about 2000°C:

$$SiO_2(s) + 2C(s) \longrightarrow Si(l) + 2CO(g)$$

Next, solid silicon is separated from other solid impurities by treatment with hydrogen chloride at 350°C to form gaseous trichlorosilane ($SiCl_3H$):

$$Si(s) + 3HCl(g) \longrightarrow SiCl_3H(g) + H_2(g)$$

Finally, ultrapure Si can be obtained by reversing the preceding reaction at 1000°C:

$$SiCl_3H(g) + H_2(g) \longrightarrow Si(s) + 3HCl(g)$$

1.    The molar heat of vaporization of trichlorosilane is 28.8 kJ/mol. Using this information and the equation

$$\ln \frac{P_1}{P_2} = \frac{\Delta H_{vap}}{R} \left( \frac{1}{T_2} - \frac{1}{T_1} \right)$$

determine the normal boiling point of trichlorosilane.

a) −28.0°C
b) −276°C
c) 30.1°C
d) 275°C

2.    What kind(s) of intermolecular forces exist between trichlorosilane molecules?

a) Dispersion, dipole-dipole, and hydrogen bonding
b) Dispersion and dipole-dipole
c) Ion-ion
d) Dipole-dipole

3.    Each cubic unit cell (edge length $a = 543$ pm) contains eight Si atoms. If there are $1.0 \times 10^{13}$ boron atoms per cubic centimeter in a sample of pure silicon, how many Si atoms are there for every B atom in the sample?

a) $5 \times 10^{22}$
b) $2 \times 10^{-10}$
c) $2 \times 10^{-22}$
d) $5 \times 10^9$

4.    Calculate the density of pure silicon.

a) 2.33 g/cm$^3$
b) 0.292 g/cm$^3$
c) 4.67 g/cm$^3$
d) $3.72 \times 10^{-22}$ g/cm$^3$

# ANSWERS TO IN-CHAPTER MATERIALS

## Answers to Practice Problems

**12.1A** (a) Dispersion forces, (b) dipole-dipole forces, H bonding and dispersion forces, (c) dipole-dipole and dispersion forces, (d) dispersion forces. **12.1B** (a) Dipole-dipole and dispersion forces, (b) dipole-dipole forces, H bonding and dispersion forces, (c) dipole-dipole forces, H bonding and dispersion forces, (d) dispersion forces. **12.2A** 265 mmHg. **12.2B** 75.9 kJ/mol, 109°C. **12.3A** 315 pm. **12.3B** 24.2°. **12.4A** 10.5 g/cm³. **12.4B** Body-centered cubic. **12.5A** 4 Ca, 8 F. **12.5B** 1 Cs, 1 Cl. **12.6A** 2.65 g/cm³. **12.6B** 421 pm. **12.7A** 2.72 g/cm³. **12.7B** 361 pm. **12.8A** 984 kJ. **12.8B** 100°C, liquid and vapor in equilibrium. **12.9A** (a) ~110°C, ~−10°C, (b) liquid.
**12.9B**

## Answers to Checkpoints

**12.1.1** a. **12.1.2** d. **12.2.1** a. **12.2.2** e. **12.2.3** b. **12.2.4** b. **12.3.1** d. **12.3.2** e. **12.6.1** a. **12.6.2** c. **12.7.1** a. **12.7.2** e.

## Answers to Applying What You've Learned

**a)** Both molecules exhibit dipole-dipole forces (including hydrogen bonding) and dispersion forces. **b)** 1.9758 Å. **c)** 4. **d)** 1.526 g/cm³. **e)** $3.2 \times 10^2$ kJ, $1.1 \times 10^3$ kJ.

# 13

# Physical Properties of Solutions

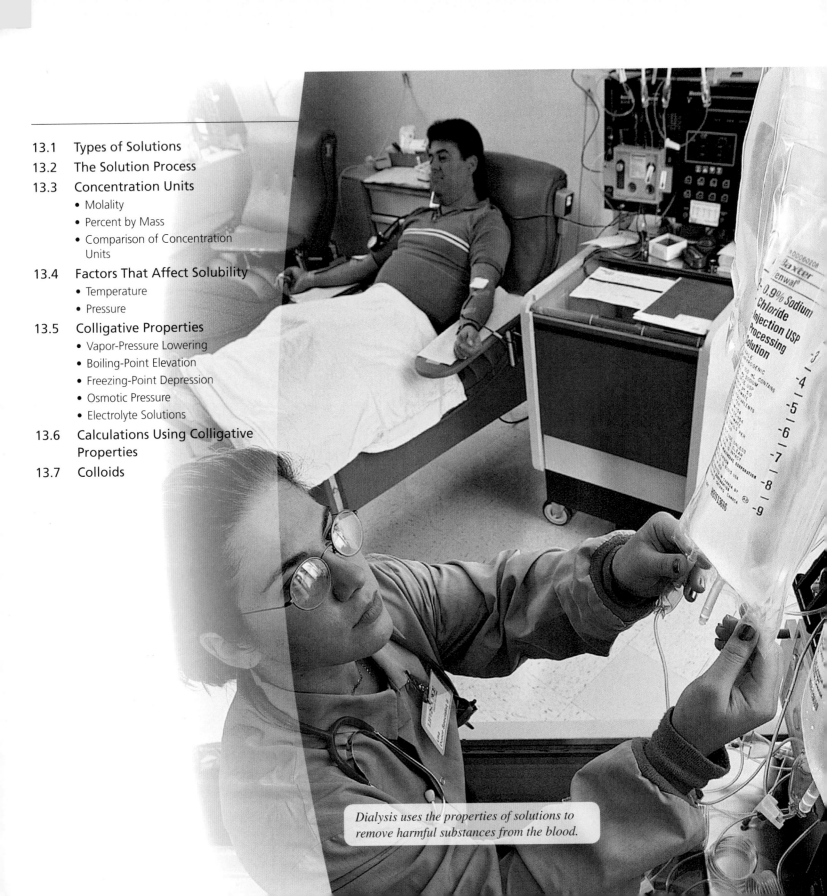

*Dialysis uses the properties of solutions to remove harmful substances from the blood.*

# How an Ordinarily Beneficial Property of Solutions Can Put Dialysis Patients at Risk

In 1993, nine patients who had undergone routine hemodialysis treatment at the University of Chicago Hospitals became seriously ill, and three of them died. The illnesses and deaths were attributed to fluoride poisoning, which occurred when the equipment meant to remove fluoride from the water failed. Although hemodialysis is supposed to remove impurities from the blood, the inadvertent use of fluoridated water in the process actually added a toxin to the patients' blood.

Hemodialysis, often called simply *dialysis,* is the cleansing of toxins from the blood of patients whose kidneys are not functioning properly. It works by routing a patient's blood temporarily through a special filter called a *dialyzer.* Inside the dialyzer, the blood is separated (by an artificial porous membrane) from an aqueous solution called the *dialysate.* The dialysate contains a variety of dissolved substances, typically including sodium chloride, sodium bicarbonate or sodium acetate, calcium chloride, potassium chloride, magnesium chloride, and sometimes glucose. Its composition mimics that of blood plasma. When the two solutions, blood and dialysate, are separated by a porous membrane, the smallest of the dissolved solutes pass through the membrane from the side where the concentration is high to the side where the concentration is low. Because the dialysate contains vital components of blood in concentrations equal to those in blood, no net passage of these substances occurs through the membrane. However, the harmful substances that accumulate in the blood of patients whose kidneys do not function properly pass through the membrane into the dialysate, in which their concentration is initially zero, and are thereby removed from the blood. Properly done, hemodialysis therapy can add years to the life of a patient with kidney failure.

The development and refinement of medical procedures such as hemodialysis require an understanding of the *properties of solutions.*

At the end of this chapter, you will be able to solve a series of problems related to aqueous fluoride solutions [▶▶ Page 569].

## In This Chapter, You Will Learn:

About the energy changes associated with solution formation, how the concentration of a solution is expressed, and how concentration determines certain properties of a solution.

### Before you begin, you should review

- Intermolecular forces [◀◀ Section 12.1]
- Molecular geometry and polarity [◀◀ Section 9.2]
- Solution stoichiometry [◀◀ Section 4.5]

# 13.1    Types of Solutions

As we noted in Section 1.2, a solution is a homogeneous mixture of two or more substances. Recall that a solution consists of a *solvent* and one or more *solutes* [◄◄ Section 4.1]. Although many of the most familiar solutions are those in which a solid is dissolved in a liquid (e.g., salt-water or sugar water), the components of a solution may be solid, liquid, or gas. The possible combinations give rise to seven distinct types of solutions, which we classify by the original states of the solution components. Table 13.1 gives an example of each type.

In this chapter, we will focus on solutions in which the solvent is a liquid; and the liquid solvent we will encounter most often is water. Recall that solutions in which water is the solvent are called *aqueous* solutions [◄◄ Section 4.1].

Solutions can also be classified by the amount of solute dissolved relative to the maximum amount that can be dissolved. A ***saturated solution*** is one that contains the maximum amount of a solute that will dissolve in a solvent at a specific temperature. The amount of solute dissolved in a given volume of a saturated solution is called the ***solubility.*** It is important to realize that solubility refers to a specific solute, a specific solvent, and a specific *temperature.* For example, the solubility of NaCl in water at 20°C is 36 g per 100 mL. The solubility of NaCl at another temperature, or in another solvent, would be different. An ***unsaturated solution*** is one that contains less solute than it has the capacity to dissolve. A ***supersaturated solution,*** on the other hand, contains *more* dissolved solute than is present in a saturated solution (Figure 13.1). It is generally not stable, and eventually the dissolved solute will come out of solution. An example of this phenomenon is shown in Figure 13.2.

**Student Annotation:** The term *solubility* was also defined in Section 4.1.

| **TABLE 13.1** | Types of Solutions | | |
|---|---|---|---|
| **Solute** | **Solvent** | **State of Resulting Solution** | **Example** |
| Gas | Gas | Gas* | Air |
| Gas | Liquid | Liquid | Carbonated water |
| Gas | Solid | Solid | $H_2$ gas in palladium |
| Liquid | Liquid | Liquid | Ethanol in water |
| Liquid | Solid | Solid | Mercury in silver |
| Solid | Liquid | Liquid | Saltwater |
| Solid | Solid | Solid | Brass (Cu/Zn) |

*Gaseous solutions can only contain gaseous solutes.

|     (a)     |     (b)     |     (c)     |     (d)     |     (e)     |

**Figure 13.1**    (a) Many solutions consist of a solid dissolved in water. (b) When all the solid dissolves, the solution is unsaturated. (c) If more solid is added than will dissolve, the solution is saturated. (d) A saturated solution is, by definition, in contact with undissolved solid. (e) Some saturated solutions can be made into supersaturated solutions by heating to dissolve more solid, and cooling carefully to prevent crystallization.

|  (a)  |  (b)  |  (c)  |  (d)  |  (e)  |

**Figure 13.2** In a supersaturated solution, (a) addition of a tiny seed crystal initiates crystallization of excess solute. (b)–(e) Crystallization proceeds rapidly to give a *saturated* solution and the crystallized solid.

# 13.2 The Solution Process

In Chapter 4, we learned guidelines that helped us predict whether or not an ionic solid is soluble in water. We now take a more general look at the factors that determine solubility. This discussion will enable us to understand why so many ionic substances are soluble in water, which is a polar solvent; and it will help us to predict the solubility of ionic and molecular compounds in both polar and nonpolar solvents.

The intermolecular forces that hold molecules together in liquids and solids play a central role in the solution process. When the solute dissolves in the solvent, molecules of the solute disperse throughout the solvent. They are, in effect, *separated* from one another and each solute molecule is *surrounded* by solvent molecules—a process known as ***solvation***. The ease with which solute molecules are separated from one another and surrounded by solvent molecules depends on the relative strengths of three types of interactions:

1. Solute-solute interactions
2. Solvent-solvent interactions
3. Solute-solvent interactions

Unlike the intermolecular forces covered in Chapter 12, all of which were between the molecules, atoms, or ions of a pure substance, solute-solvent interactions are those that exist in a mixture of different substances. Because the components of a mixture can have different properties, there is a greater variety of intermolecular forces to consider. In addition to dispersion forces, which exist between all substances, dipole-dipole forces, which exist between polar molecules, and ion-ion forces, which exist between ions, solutions can exhibit the following intermolecular forces:

| Intermolecular Forces | Example | Model |
|---|---|---|
| *Ion-dipole.* The charge of an ion is attracted to the partial charge on a polar molecule. | NaCl or KI in $H_2O$ | |
| *Dipole-induced dipole.* The partial charge on a polar molecule induces a temporary partial charge on a neighboring nonpolar molecule or atom. | He or $CO_2$ in $H_2O$ | |
| *Ion-induced dipole.* The charge of an ion induces a temporary partial charge on a neighboring nonpolar molecule or atom. | $Fe^{2+}$ and $O_2$ | |

**Student Annotation:** A hemoglobin molecule contains four $Fe^{2+}$ ions. In the early stages of $O_2$ binding, oxygen molecules are attracted to the $Fe^{2+}$ ions by an ion-induced dipole interaction.

**FAQ**    *You May Be Wondering...*

# Why Do Some Things Dissolve and Not Others?

Previously, we learned that the driving force behind some processes is the lowering of the system's potential energy. Recall the minimization of potential energy when two hydrogen atoms are 74 pm apart [◄◄ Section 9.3]. However, because there are substances that dissolve endothermically, meaning that the process increases the system's potential energy, something else must be involved in determining whether or not a substance will dissolve. That something else is *entropy*.

The **entropy** of a system is a measure of how *dispersed* or *spread out* its energy is. Consider two samples of different gases separated by a physical barrier. When we remove the barrier, the gases mix, forming a solution. Under ordinary conditions, we can treat the gases as ideal, meaning we can assume that there are no attractive forces between the molecules in either sample before they mix (no solute-solute or solvent-solvent attractions to break)—and no attractive forces between the molecules in the mixture (no solute-solvent attractions form). The energy of the system does not change—and yet the gases mix spontaneously. The reason such a solution forms is that, although there is no change in the energy of either of the original samples of gas, the energy possessed by each sample of gas spreads out into a larger volume. This increased dispersal of the system's energy is an increase in the *entropy* of the system. There is a natural tendency for entropy to increase; that is, for the energy of a system to become more dispersed—unless there is something preventing that dispersal. Initially, the physical barrier between the two gases prevented their energy from spreading out into the larger volume. It is the increase in entropy that drives the formation of this solution—and many others.

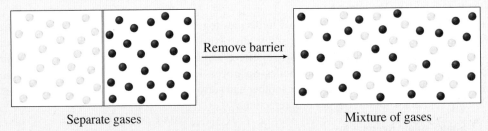

Separate gases                    Mixture of gases

Now consider the case of solid ammonium nitrate (NH₄NO₃), which dissolves in water in an *endothermic* process. In this case, the dissolution *increases* the potential energy of the system. However, the energy possessed by the ammonium nitrate solid spreads out to occupy the volume of the resulting solution causing the entropy of the system to increase. Ammonium nitrate dissolves in water because the favorable increase in the system's entropy outweighs the unfavorable increase in its potential energy. Although the process being endothermic is a barrier to solution formation, it is not enough of a barrier to prevent it.

In some cases, a process is so endothermic that even an increase in entropy is not enough to allow it to happen spontaneously. Sodium chloride (NaCl) is not soluble in a nonpolar solvent such as benzene, for example, because the solvent-solute interactions that would result are too weak to compensate for the energy required to separate the network of positive and negative ions in sodium chloride. The magnitude of the exothermic step in solution formation ($\Delta H_3$) is so small compared to the combined magnitude of the endothermic steps ($\Delta H_1$ and $\Delta H_2$) that the overall process is *very* endothermic and does not happen to any significant degree despite the increase in entropy that would result.

**Student Annotation:** Like the formation of chemical bonds [◄◄ Section 8.9], the formation of intermolecular attractions is exothermic. If that isn't intuitively obvious, think of it this way: It would require energy to *separate* molecules that are attracted to each other. The reverse process, the *combination* of molecules that attract each other, would give off an equal amount of energy [◄◄ Section 5.3].

For simplicity, we can imagine the solution process taking place in the three distinct steps shown in Figure 13.3. Step 1 is the separation of solute molecules from one another, and step 2 is the separation of solvent molecules from one another. Both of these steps require an input of energy to overcome intermolecular attractions, so they are *endothermic*. In step 3 the solvent and solute molecules mix. This process is usually *exothermic*. The enthalpy change for the overall process, $\Delta H_{soln}$, is given by

## Sample Problem 13.1

Determine for each solute whether the solubility will be greater in water, which is polar, or in benzene ($C_6H_6$), which is nonpolar: (a) bromine ($Br_2$), (b) sodium iodide (NaI), (c) carbon tetrachloride ($CCl_4$), and (d) formaldehyde ($CH_2O$).

**Strategy**  Consider the structure of each solute to determine whether or not it is polar. For molecular solutes, start with a Lewis structure and apply the VSEPR theory [◀◀ Section 9.1]. We expect polar solutes, including ionic compounds, to be more soluble in water. Nonpolar solutes will be more soluble in benzene.

**Setup**
(a) Bromine is a homonuclear diatomic molecule and is nonpolar.
(b) Sodium iodide is ionic.
(c) Carbon tetrachloride has the following Lewis structure:

$$:\ddot{\underset{..}{C}}l:$$
$$:\ddot{\underset{..}{C}}l-\overset{|}{\underset{|}{C}}-\ddot{\underset{..}{C}}l:$$
$$:\ddot{\underset{..}{C}}l:$$

With four electron domains around the central atom, we expect a tetrahedral arrangement. A symmetrical arrangement of identical bonds results in a nonpolar molecule.
(d) Formaldehyde has the following Lewis structure:

$$\overset{..}{\underset{..}{O}}$$
$$||$$
$$H\diagup \overset{C}{} \diagdown H$$

Crossed arrows can be used to represent the individual bond dipoles [◀◀ Section 9.2]. This molecule is polar and can form hydrogen bonds with water.

**Solution**
(a) Bromine is more soluble in benzene.
(b) Sodium iodide is more soluble in water.
(c) Carbon tetrachloride is more soluble in benzene.
(d) Formaldehyde is more soluble in water.

**Think About It**  Remember that molecular formula alone is not sufficient to determine the shape or the polarity of a polyatomic molecule. It must be determined by starting with a correct Lewis structure and applying the VSEPR theory [◀◀ Section 9.2].

**Practice Problem A**  Predict whether iodine ($I_2$) is more soluble in liquid ammonia ($NH_3$) or in carbon disulfide ($CS_2$).

**Practice Problem B**  Which of the following should you expect to be more soluble in benzene than in water: $C_3H_8$, HCl, $I_2$, $CS_2$?

$$\Delta H_{soln} = \Delta H_1 + \Delta H_2 + \Delta H_3$$

The overall solution-formation process is exothermic ($\Delta H_{soln} < 0$) when the energy given off in step 3 is greater than the sum of energy required for steps 1 and 2. The overall process is endothermic ($\Delta H_{soln} > 0$) when the energy given off in step 3 is less than the total required for steps 1 and 2. (Figure 13.3 depicts a solution formation that is endothermic overall.)

**Figure 13.3** A molecular view of the solution process portrayed as taking place in three steps: First the solute and solvent molecules are separated (steps 1 and 2, respectively—both endothermic). Then the solvent and solute molecules mix (step 3—exothermic).

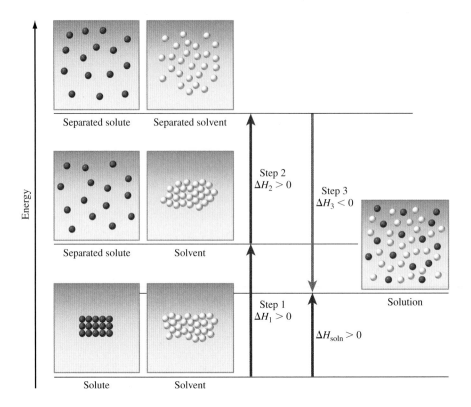

The saying "like dissolves like" is useful in predicting the solubility of a substance in a given solvent. What this expression means is that two substances with intermolecular forces of similar type and magnitude are likely to be soluble in each other. For example, both carbon tetrachloride ($CCl_4$) and benzene ($C_6H_6$) are nonpolar liquids. The only intermolecular forces present in these substances are dispersion forces [◂◂ Section 12.1]. When these two liquids are mixed, they readily dissolve in each other, because the attraction between $CCl_4$ and $C_6H_6$ molecules is comparable in magnitude to the forces between molecules in pure $CCl_4$ and those between molecules in pure $C_6H_6$. Two liquids are said to be **miscible** if they are completely soluble in each other in all proportions. Alcohols such as methanol, ethanol, and 1,2-ethylene glycol are miscible with water because they can form hydrogen bonds with water molecules:

$$
\begin{array}{ccc}
\text{Methanol} & \text{Ethanol} & \text{1,2-Ethylene glycol}
\end{array}
$$

The guidelines listed in Tables 4.2 and 4.3 enable us to predict the solubility of a particular ionic compound in water. When sodium chloride dissolves in water, the ions are stabilized in solution by *hydration,* which involves ion-dipole interactions. In general, ionic compounds are much more soluble in polar solvents, such as water, liquid ammonia, and liquid hydrogen fluoride, than in nonpolar solvents. Because the molecules of nonpolar solvents, such as benzene and carbon tetrachloride, do not have a dipole moment, they cannot effectively solvate the $Na^+$ and $Cl^-$ ions. The predominant intermolecular interaction between ions and nonpolar compounds is an ion-induced dipole interaction, which typically is much weaker than ion-dipole interactions. Consequently, ionic compounds usually have extremely low solubility in nonpolar solvents.

**Student Annotation:** *Solvation* refers in a general way to solute particles being surrounded by solvent molecules. When the solvent is water, we use the more specific term *hydration* [◂◂ Section 4.1].

## Bringing Chemistry to Life

### Vitamin Solubility

Vitamins can be categorized as either water soluble or fat soluble. An excess of a fat-soluble vitamin can be stored in the body, whereas excesses of most water-soluble vitamins are eliminated in the urine.

The difference between water-soluble and fat-soluble vitamins is their molecular structures. Water-soluble vitamins, such as vitamin C ($C_6H_8O_6$), have multiple polar groups that can interact with water to form hydrogen bonds:

Vitamin C

In general, fats are nonpolar molecules containing hydrocarbon chains. Fat-soluble vitamins, such as vitamin E, tend to be more hydrocarbon-like. They may have a polar group or two, but the molecule is predominantly nonpolar:

Vitamin E

**Student Annotation:** In general, fats are nonpolar molecules containing hydrocarbon chains. Molecules that are predominantly nonpolar tend to be more soluble in nonpolar substances such as fats. Polar molecules tend to be more soluble in polar substances such as water. This is an example of what is meant by the expression "like dissolves like."

## Sample Problem 13.2

Predict whether each vitamin will be water soluble or fat soluble.

(a)                              Vitamin D

(b)        Vitamin B$_2$ (riboflavin)          Vitamin B$_6$

**Strategy** Consider the number of polar groups, and determine whether each molecule is predominantly polar or predominantly nonpolar. Predominantly polar molecules tend to be water soluble. Predominantly nonpolar molecules tend to be fat soluble.

*(Continued)*

**Think About It** Elimination of vitamin C causes the urine to be more acidic, which can inhibit the growth of some bacteria responsible for urinary tract infections. Taking larger doses than the official recommended daily amount of 60 mg is recommended by many physicians to lower the incidence of such infections. Fat-soluble vitamins cannot be eliminated in this way, and taking too much can lead to vitamin toxicity.

**Setup** Polar groups include those containing O atoms or N atoms (atoms with lone pairs). Vitamin D contains only one polar group and is therefore predominantly nonpolar. Vitamin B$_2$ contains 10 polar groups (six O atoms and four N atoms) making it predominantly polar.

**Solution**

(a) Fat soluble

(b) Water soluble

**Practice Problem A** Predict whether vitamin B$_6$, also known as pyridoxine, is water soluble or fat soluble.

**Practice Problem B** Predict whether vitamin A is water soluble or fat soluble.

Vitamin A

---

## Checkpoint 13.2    The Solution Process

**13.2.1** Which compounds do you expect to be more soluble in benzene than in water? (Select all that apply.)

    a) SO$_2$      b) CO$_2$      c) Na$_2$SO$_4$      d) C$_2$H$_6$      e) Br$_2$

**13.2.2** Which vitamins (see the given structures) do you expect to be water soluble? (Select all that apply.)

    a) Pantothenic acid only

    b) Nicotinic acid only

    c) Pantothenic acid and nicotinic acid

    d) Vitamin K only

    e) Nicotinic acid and vitamin K

Pantothenic acid                             Nicotinic acid                        Vitamin K

**13.2.3** Which of the following solutions exhibits hydrogen bonding between the solute and solvent? (Select all that apply.)

    a) H$_2$(g) in H$_2$O(l)      b) CH$_3$OH(l) in H$_2$O(l)      c) CO$_2$(g) in H$_2$O(l)      d) NH$_3$(g) in H$_2$O(l)      e) NaCl(s) in H$_2$O(l)

**13.2.4** What best explains why dissolving a gas in water is always exothermic?

    a) There are no solute-solute interactions to break up.

    b) There are no solvent-solvent interactions to break up.

    c) The resulting solute-solvent interactions are especially powerful.

    d) The solubility of gases in water is higher at high temperature.

    e) Water is a polar solvent.

# 13.3 Concentration Units

We learned in Chapter 4 that chemists often express concentration of solutions in units of *molarity*. Recall that molarity, *M*, is defined as the number of moles of solute divided by the number of liters of solution [◀◀ Section 4.5].

$$\text{molarity} = M = \frac{\text{moles of solute}}{\text{liters of solution}}$$

Mole fraction, $\chi$, which is defined as the number of moles of solute divided by the total number of moles, is also an expression of concentration [◀◀ Section 11.5].

$$\text{mole fraction of component A} = \chi_A = \frac{\text{moles of A}}{\text{sum of moles of all components}}$$

In this section, we will learn about *molality* and *percent by mass,* two additional ways to express the concentration of a mixture component. How a chemist expresses concentration depends on the type of problem being solved.

> **Student Annotation:** We have already used percent by mass to describe the composition of a pure substance [◀◀ Section 3.2]. In this chapter, we will use percent by mass to describe *solutions.*

## Molality

*Molality (m)* is the number of moles of solute dissolved in 1 kg (1000 g) of solvent:

$$\text{molality} = m = \frac{\text{moles of solute}}{\text{mass of solvent (in kg)}} \qquad \text{Equation 13.1}$$

For example, to prepare a 1 molal (1 *m*) aqueous sodium sulfate solution, we must dissolve 1 mole (142.0 g) of $Na_2SO_4$ in 1 kg of water.

> **Student Annotation:** It is a common mistake to confuse molarity and molality. Molarity depends on the *volume* of the *solution.* Molality depends on the *mass* of the *solvent.*

## Percent by Mass

The *percent by mass* (also called *percent by weight*) is the ratio of the mass of a solute to the mass of the solution, multiplied by 100 percent. Because the units of mass cancel on the top and bottom of the fraction, any units of mass can be used—provided they are used consistently:

$$\text{percent by mass} = \frac{\text{mass of solute}}{\text{mass of solute} + \text{mass of solvent}} \times 100\% \quad \text{Equation 13.2}$$

For example, we can express the concentration of the 1 *m* aqueous sodium sulfate solution as

$$\text{percent by mass } Na_2SO_4 = \frac{\text{mass of } Na_2SO_4}{\text{mass of } Na_2SO_4 + \text{mass of water}} \times 100\%$$

$$= \frac{142.0 \text{ g}}{1142.0 \text{ g}} \times 100\% = 12.4\%$$

The term *percent* literally means "parts per hundred." If we were to use Equation 13.2 but multiply by 1000 instead of 100, we would get "parts per thousand"; multiplying by 1,000,000 would give "parts per million" or *ppm;* and so on. Parts per million, parts per billion, parts per trillion, and so forth, usually are used to express very low concentrations, such as those of some pollutants in the atmosphere or a body of water. For example, if a 1-kg sample of water is found to contain 3 $\mu$g ($3 \times 10^{-6}$ g) of arsenic, its concentration can be expressed in parts per billion (ppb) as follows:

$$\frac{3 \times 10^{-6} \text{ g}}{1000 \text{ g}} \times 10^9 = 3 \text{ ppb}$$

Sample Problem 13.3 shows how to calculate the concentration in molality and in percent by mass.

**Sample Problem 13.3**

A solution is made by dissolving 170.1 g of glucose ($C_6H_{12}O_6$) in enough water to make a liter of solution. The density of the solution is 1.062 g/mL. Express the concentration of the solution in (a) molality, (b) percent by mass, and (c) parts per million.

**Strategy** Use the molar mass of glucose to determine the number of moles of glucose in a liter of solution. Use the density (in g/L) to calculate the mass of a liter of solution. Subtract the mass of glucose from the mass of solution to determine the mass of water. Use Equation 13.1 to determine the molality. Knowing the mass of glucose and the total mass of solution in a liter, use Equation 13.2 to calculate the percent concentration by mass.

**Setup** The molar mass of glucose is 180.2 g/mol; the density of the solution is 1.062 g/mL.

**Solution**

(a)
$$\frac{170.1 \text{ g}}{180.2 \text{ g/mol}} = 0.9440 \text{ mol glucose per liter of solution}$$

$$1 \text{ liter of solution} \times \frac{1062 \text{ g}}{L} = 1062 \text{ g}$$

$$1062 \text{ g solution} - 170.1 \text{ g glucose} = 892 \text{ g water} = 0.892 \text{ kg water}$$

$$\frac{0.9440 \text{ mol glucose}}{0.892 \text{ kg water}} = 1.06 \ m$$

(b) $\dfrac{170.1 \text{ g glucose}}{1062 \text{ g solution}} \times 100\% = 16.02\%$ glucose by mass

(c) $\dfrac{170.1 \text{ g glucose}}{1062 \text{ g solution}} \times 1{,}000{,}000 = 1.602 \times 10^5$ ppm glucose

> **Think About It** Pay careful attention to units in problems such as this. Most require conversions between grams and kilograms and/or liters and milliliters.

**Practice Problem A** Determine (a) the molality and (b) the percent by mass of urea for a solution prepared by dissolving 5.46 g urea [$(NH_2)_2CO$] in 215 g of water.

**⚡Practice Problem B** Determine the molality of an aqueous solution that is 4.5 percent urea by mass.

## Comparison of Concentration Units

The choice of a concentration unit is based on the purpose of the experiment. For instance, we typically use molarity to express the concentrations of solutions for titrations and gravimetric analyses. *Mole fractions* are used to express the concentrations of gases—and of solutions when we are working with vapor pressures, which we will discuss in Section 13.5.

The advantage of molarity is that it is generally easier to measure the volume of a solution, using precisely calibrated volumetric flasks, than to weigh the solvent. Molality, on the other hand, has the advantage of being temperature independent. The volume of a solution typically increases slightly with increasing temperature, which would change the molarity. The mass of solvent in a solution, however, does *not* change with temperature.

> **Student Annotation:** For very dilute aqueous solutions, molarity and molality have the same value. The mass of a liter of water is 1 kg, and in a very dilute solution, the mass of solute is negligible compared to that of the solvent.

Percent by mass is similar to molality in that it is independent of temperature. Furthermore, because it is defined in terms of the ratio of the mass of solute to the mass of solution, we do not need to know the molar mass of the solute to calculate the percent by mass.

Often it is necessary to convert the concentration of a solution from one unit to another. For example, the same solution may be used for different experiments that require different concentration units for calculations. Suppose we want to express the concentration of a 0.396 *m* aqueous glucose ($C_6H_{12}O_6$) solution (at 25°C) in molarity. We know there is 0.396 mole of glucose in 1000 g of the solvent. We need to determine the *volume* of this solution to calculate molarity. To determine volume, we must first calculate its mass:

$$0.396 \text{ mol } C_6H_{12}O_6 \times \frac{180.2 \text{ g}}{1 \text{ mol } C_6H_{12}O_6} = 71.4 \text{ g } C_6H_{12}O_6$$

$$71.4 \text{ g } C_6H_{12}O_6 + 1000 \text{ g } H_2O = 1071 \text{ g solution}$$

Once we have determined the mass of the solution, we use the *density* of the solution, which is typically determined experimentally, to determine its volume. The density of a 0.396 *m* glucose solution is 1.16 g/mL at 25°C. Therefore, its volume is

$$\text{volume} = \frac{\text{mass}}{\text{density}}$$

$$= \frac{1071 \text{ g}}{1.16 \text{ g/mL}} \times \frac{1 \text{ L}}{1000 \text{ mL}}$$

$$= 0.923 \text{ L}$$

Having determined the volume of the solution, the molarity is given by

$$\text{molarity} = \frac{\text{moles of solute}}{\text{liters of solution}}$$

$$= \frac{0.396 \text{ mol}}{0.923 \text{ L}}$$

$$= 0.429 \text{ mol/L} = 0.429 \text{ } M$$

Sample Problem 13.4 shows how to convert from one unit of concentration to another.

## Sample Problem 13.4

"Rubbing alcohol" is a mixture of isopropyl alcohol ($C_3H_7OH$) and water that is 70 percent isopropyl alcohol by mass (density = 0.79 g/mL at 20°C). Express the concentration of rubbing alcohol in (a) molarity and (b) molality.

**Strategy**
(a) Use density to determine the total mass of a liter of solution, and use percent by mass to determine the mass of isopropyl alcohol in a liter of solution. Convert the mass of isopropyl alcohol to moles, and divide moles by liters of solution to get molarity.

(b) Subtract the mass of $C_3H_7OH$ from the mass of solution to get the mass of water. Divide moles of $C_3H_7OH$ by the mass of water (in kg) to get molality.

**Setup** The mass of a liter of rubbing alcohol is 790 g, and the molar mass of isopropyl alcohol is 60.09 g/mol.

**Solution**

(a)
$$\frac{790 \text{ g solution}}{\text{L solution}} \times \frac{70 \text{ g } C_3H_7OH}{100 \text{ g solution}} = \frac{553 \text{ g } C_3H_7OH}{\text{L solution}}$$

$$\frac{553 \text{ g } C_3H_7OH}{\text{L solution}} \times \frac{1 \text{ mol}}{60.09 \text{ g } C_3H_7OH} = \frac{9.20 \text{ mol } C_3H_7OH}{\text{L solution}} = 9.2 \text{ } M$$

(b) 790 g solution − 553 g $C_3H_7OH$ = 237 g water = 0.237 kg water

$$\frac{9.20 \text{ mol } C_3H_7OH}{0.237 \text{ kg water}} = 39 \text{ } m$$

Rubbing alcohol is 9.2 *M* and 39 *m* in isopropyl alcohol.

**Think About It** Note the large difference between molarity and molality in this case. Molarity and molality are the same (or similar) only for very dilute aqueous solutions.

**Practice Problem A** An aqueous solution that is 16 percent sulfuric acid ($H_2SO_4$) by mass has a density of 1.109 g/mL at 25°C. Determine (a) the molarity and (b) the molality of the solution at 25°C.

**Practice Problem B** Determine the percent sulfuric acid by mass of a 1.49 *m* aqueous solution of $H_2SO_4$.

**Checkpoint 13.3    Concentration Units**

**13.3.1** A solution of a particular concentration is prepared at 20°C. Which of the following does not change when the solution is heated to 75°C? (Select all that apply.)

a) Molarity

b) Molality

c) Percent by mass

d) Mole fraction

e) All of these change.

**13.3.2** What is the molality of a solution prepared by dissolving 6.44 g of naphthalene ($C_{10}H_8$) in 80.1 g benzene?

a) 1.13 $m$

b) 80.4 $m$

c) 0.804 $m$

d) 0.627 $m$

e) 11.7 $m$

**13.3.3** At 20.0°C, a 0.258 $m$ aqueous solution of glucose ($C_6H_{12}O_6$) has a density of 1.0173 g/mL. Calculate the molarity of this solution.

a) 0.258 $M$

b) 0.300 $M$

c) 0.456 $M$

d) 0.251 $M$

e) 0.448 $M$

**13.3.4** At 25.0°C, an aqueous solution that is 25.0 percent $H_2SO_4$ by mass has a density of 1.178 g/mL. Calculate the molarity and the molality of this solution.

a) 3.00 $M$ and 3.40 $m$

b) 3.40 $M$ and 3.40 $m$

c) 3.00 $M$ and 3.00 $m$

d) 3.00 $M$ and 2.98 $m$

e) 3.44 $M$ and 3.14 $m$

# 13.4  Factors That Affect Solubility

Recall that solubility is defined as the maximum amount of solute that will dissolve in a given quantity of solvent at a specific *temperature*. Temperature affects the solubility of most substances. In this section we will consider the effects of temperature on the aqueous solubility of solids and gases, and the effect of *pressure* on the aqueous solubility of gases.

## Temperature

More sugar dissolves in hot tea than in iced tea because the aqueous solubility of sugar, like that of most solid substances, increases as the temperature increases. Figure 13.4 shows the solubility of some common solids in water as a function of temperature. Note how the solubility of a solid and the *change* in solubility over a particular temperature range vary considerably. The relationship between temperature and solubility is complex and often nonlinear.

The relationship between temperature and the aqueous solubility of gases is somewhat simpler than that of solids. Most gaseous solutes become less soluble in water as temperature increases. If you get a glass of water from your faucet and leave it on the kitchen counter for a while, you will see bubbles forming in the water as it warms to room temperature. As the temperature of the water increases, dissolved gases become less soluble and come out of solution—resulting in the formation of bubbles.

One of the more important consequences of the reduced solubility of gases in water at elevated temperature is *thermal pollution*. Hundreds of billions of gallons of water are used every year for industrial cooling, mostly in electric power and nuclear power production. This process heats the water, which is then returned to the rivers and lakes from which it was taken. The increased water temperature has a twofold impact on aquatic life. The rate of metabolism of cold-blooded species such as fish increases with increasing temperature, thereby increasing their need for oxygen. At the same time, the increased water temperature causes a decrease in the solubility of oxygen—making less oxygen available. The result can be disastrous for fish populations.

## Pressure

Although pressure does not influence the solubility of a liquid or a solid significantly, it does greatly affect the solubility of a gas. The quantitative relationship between gas solubility and pres-

**Figure 13.4**   Temperature dependence of the solubility of glucose and several ionic compounds in water.

**Student Annotation:** Henry's law means that if we double the pressure of a gas over a solution (at constant temperature), we double the concentration of gas dissolved in the solution; triple the pressure, triple the concentration; and so on.

sure is given by ***Henry's*[1] *law,*** which states that the solubility of a gas in a liquid is proportional to the pressure of the gas over the solution,

$$c \propto P$$

and is expressed as

$$c = kP \qquad\qquad \text{Equation 13.3}$$

where $c$ is the molar concentration (mol/L) of the dissolved gas, $P$ is the pressure (in atm) of the gas over the solution, and $k$ is a proportionality constant called the ***Henry's law constant.*** Henry's law constants are specific to the gas-solvent combination and vary with temperature. The units of $k$ are mol/L · atm. If there is a mixture of gases over the solution, then $P$ in Equation 13.3 is the partial pressure of the gas in question.

Henry's law can be understood qualitatively in terms of the kinetic molecular theory [◄◄ Section 11.6]. The amount of gas that will dissolve in a solvent depends on how frequently the gas molecules collide with the liquid surface and become trapped by the condensed phase. Suppose we have a gas in dynamic equilibrium [◄◄ Section 12.2] with a solution, as shown in Figure 13.5(a). At any point in time, the number of gas molecules entering the solution is equal to the number of dissolved gas molecules leaving the solution and entering the vapor phase. If the partial pressure of the gas is increased [Figure 13.5(b)], more molecules strike the liquid surface, causing more of them to dissolve. As the concentration of dissolved gas increases, the number of gas molecules leaving the solution also increases. These processes continue until the concentration of dissolved gas in the solution reaches the point again where the number of molecules leaving the solution per second equals the number entering the solution per second.

One interesting application of Henry's law is the production of carbonated beverages. Manufacturers put the "fizz" in soft drinks using pressurized carbon dioxide. The pressure of $CO_2$ applied (typically on the order of 5 atm) is many thousands of times greater than the partial pressure of $CO_2$ in the air. Thus, when a can or bottle of soda is opened, the $CO_2$ dissolved under high-pressure conditions comes out of solution—resulting in the bubbles that make carbonated drinks appealing.

Sample Problem 13.5 illustrates the use of Henry's law.

1. William Henry (1775–1836). English chemist. Henry's major contribution to science was his discovery of the law describing the solubility of gases, which now bears his name.

**Figure 13.5**    A molecular view of Henry's law. When the partial pressure of the gas over the solution increases from (a) to (b), the concentration of the dissolved gas also increases according to Equation 13.3.

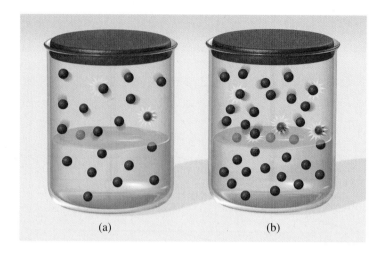

(a)                    (b)

## Sample Problem 13.5

Calculate the concentration of carbon dioxide in a soft drink that was bottled under a partial pressure of 5.0 atm $CO_2$ at 25°C (a) before the bottle is opened and (b) after the soda has gone "flat" at 25°C. The Henry's law constant for $CO_2$ in water at this temperature is $3.1 \times 10^{-2}$ mol/L · atm. Assume that the partial pressure of $CO_2$ in air is 0.0003 atm and that the Henry's law constant for the soft drink is the same as that for water.

**Strategy**  Use Equation 13.3 and the given Henry's law constant to solve for the molar concentration (mol/L) of $CO_2$ at 25°C and the two $CO_2$ pressures given.

**Setup**  At 25°C, the Henry's law constant for $CO_2$ in water is $3.1 \times 10^{-2}$ mol/L · atm.

**Solution**  (a) $c = (3.1 \times 10^{-2}$ mol/L · atm$)(5.0$ atm$) = 1.6 \times 10^{-1}$ mol/L

(b) $c = (3.1 \times 10^{-2}$ mol/L · atm$)(0.0003$ atm$) = 9 \times 10^{-6}$ mol/L

**Think About It**  With a pressure approximately 15,000 times smaller in part (b) than in part (a), we expect the concentration of $CO_2$ to be approximately 15,000 times smaller—and it is.

**Practice Problem A**  Calculate the concentration of $CO_2$ in water at 25°C when the pressure of $CO_2$ over the solution is 4.0 atm.

**Practice Problem B**  Calculate the pressure of $O_2$ necessary to generate an aqueous solution that is $3.4 \times 10^{-2}$ $M$ in $O_2$ at 25°C. The Henry's law constant for $O_2$ in water at 25°C is $1.3 \times 10^{-3}$ mol/L · atm.

## Checkpoint 13.4    Factors That Affect Solubility

13.4.1  The solubility of $N_2$ in water at 25°C and an $N_2$ pressure of 1 atm is $6.8 \times 10^{-4}$ mol/L. Calculate the concentration of dissolved $N_2$ in water under atmospheric conditions where the partial pressure of $N_2$ is 0.78 atm.

a) $6.8 \times 10^{-4}$ $M$

b) $8.7 \times 10^{-4}$ $M$

c) $5.3 \times 10^{-4}$ $M$

d) $1.5 \times 10^{-4}$ $M$

e) $3.1 \times 10^{-3}$ $M$

13.4.2  Calculate the molar concentration of $O_2$ in water at 25°C under atmospheric conditions where the partial pressure of $O_2$ is 0.22 atm. The Henry's law constant for $O_2$ is $1.3 \times 10^{-3}$ mol/L · atm.

a) $2.9 \times 10^{-4}$ $M$

b) $5.9 \times 10^{-3}$ $M$

c) $1.7 \times 10^{-3}$ $M$

d) $1.0 \times 10^{-3}$ $M$

e) $1.3 \times 10^{-3}$ $M$

# 13.5 Colligative Properties

*Colligative properties* are properties that depend on the number of solute particles in solution but do not depend on the nature of the solute particles. That is, colligative properties depend on the concentration of solute particles regardless of whether those particles are atoms, molecules, or ions. The colligative properties are vapor-pressure lowering, boiling-point elevation, freezing-point depression, and osmotic pressure. We begin by considering the colligative properties of relatively dilute solutions ($\leq 0.2\ M$) of nonelectrolytes.

## Vapor-Pressure Lowering

We have seen that a liquid exerts a characteristic vapor pressure [◄◄ Section 12.2]. When a ***non-volatile*** solute (one that does *not* exert a vapor pressure) is dissolved in a liquid, the vapor pressure exerted by the liquid decreases. The difference between the vapor pressure of a pure solvent and that of the corresponding solution depends on the concentration of the solute in the solution. This relationship is expressed by ***Raoult's*** [2] ***law,*** which states that the partial pressure of a solvent over a solution, $P_1$, is given by the vapor pressure of the pure solvent, $P_1^\circ$, times the mole fraction of the solvent in the solution, $\chi_1$:

**Student Annotation:** Table 11.5 gives the vapor pressure of water at various temperatures.

$$P_1 = \chi_1 P_1^\circ \qquad \text{Equation 13.4}$$

In a solution containing only one solute, $\chi_1 = 1 - \chi_2$, where $\chi_2$ is the mole fraction of the solute. Equation 13.4 can therefore be rewritten as

$$P_1 = (1 - \chi_2)P_1^\circ$$

or

$$P_1 = P_1^\circ - \chi_2 P_1^\circ$$

so that

$$P_1^\circ - P_1 = \Delta P = \chi_2 P_1^\circ \qquad \text{Equation 13.5}$$

Thus, the decrease in vapor pressure, $\Delta P$, is directly proportional to the solute concentration expressed as a *mole fraction.*

　　To understand the phenomenon of vapor-pressure lowering, we must understand the degree of *order* associated with the states of matter involved. As we saw in Section 13.2, molecules in the liquid state are fairly highly ordered; that is, they have low entropy. Molecules in the gas phase have significantly less order—they have high entropy. Because there is a natural tendency toward increased entropy, molecules have a certain tendency to leave the region of lower entropy and enter the region of higher entropy. This corresponds to molecules leaving the liquid and entering the gas phase. As we have seen, when a solute is added to a liquid, the liquid's order is disrupted. Thus, the solution has greater entropy than the pure liquid. Because there is a smaller difference in entropy between the solution and the gas phase than there was between the pure liquid and the gas phase, there is a decreased tendency for molecules to leave the solution and enter the gas phase—resulting in a lower vapor pressure exerted by the solvent. This qualitative explanation of vapor-pressure lowering is illustrated in Figure 13.6. The smaller difference in entropy between the solution and gas

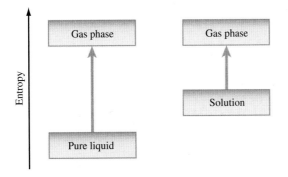

**Figure 13.6** The smaller difference in entropy between the solution and gas phases, relative to that between the pure liquid and gas phases, results in a decreased tendency for solvent molecules to enter the gas phase. This results in a lowering of vapor pressure. The solvent in a solution always exerts a lower vapor pressure than the pure solvent.

2. François Marie Raoult (1839–1901). French chemist. Raoult's work was mainly in solution properties and electrochemistry.

phases, relative to that between the pure liquid and gas phases, results in a decreased tendency for solvent molecules to enter the gas phase. This results in a lowering of vapor pressure. The solvent in a solution will always exert a lower vapor pressure than the pure solvent.

Sample Problem 13.6 shows how to use Raoult's law.

---

### Sample Problem 13.6

Calculate the vapor pressure of water over a solution made by dissolving 225 g of glucose in 575 g of water at 35°C. (At 35°C, $P^{\circ}_{H_2O} = 42.2$ mmHg.)

**Strategy** Convert the masses of glucose and water to moles, determine the mole fraction of water, and use Equation 13.4 to find the vapor pressure over the solution.

**Setup** The molar masses of glucose and water are 180.2 and 18.02 g/mol, respectively.

**Solution**

$$\frac{225 \text{ g glucose}}{180.2 \text{ g/mol}} = (1.25 \text{ mol glucose}) \quad \frac{575 \text{ g water}}{18.02 \text{ g/mol}} = 31.9 \text{ mol water}$$

$$\chi_{water} = \frac{31.9 \text{ mol water}}{1.25 \text{ mol glucose} + 31.9 \text{ mol water}} = 0.962$$

$$P_{H_2O} = \chi_{water} \times P^{\circ}_{H_2O} = 0.962 \times 42.2 \text{ mmHg} = 40.6 \text{ mmHg}$$

The vapor pressure of water over the solution is 40.6 mmHg.

**Think About It** This problem can also be solved using Equation 13.5 to calculate the vapor-pressure lowering, $\Delta P$.

---

**Practice Problem A** Calculate the vapor pressure of a solution made by dissolving 115 g of urea [$(NH_2)_2CO$; molar mass = 60.06 g/mol] in 485 g of water at 25°C. (At 25°C, $P^{\circ}_{H_2O} = 23.8$ mmHg.)

**Practice Problem B** Calculate the mass of urea that should be dissolved in 225 g of water at 35°C to produce a solution with a vapor pressure of 37.1 mmHg. (At 35°C, $P^{\circ}_{H_2O} = 42.2$ mmHg.)

---

If both components of a solution are *volatile* (i.e., have measurable vapor pressure), the vapor pressure of the solution is the sum of the individual partial pressures exerted by the solution components. Raoult's law holds equally well in this case:

$$P_A = \chi_A P^{\circ}_A$$

$$P_B = \chi_B P^{\circ}_B$$

where $P_A$ and $P_B$ are the partial pressures over the solution for components A and B, $P^{\circ}_A$ and $P^{\circ}_B$ are the vapor pressures of the pure substances A and B, and $\chi_A$ and $\chi_B$ are their mole fractions. The total pressure is given by Dalton's law of partial pressures [◄◄ Section 11.5]:

$$P_T = P_A + P_B$$

or

$$P_T = \chi_A P^{\circ}_A + \chi_B P^{\circ}_B$$

For example, benzene and toluene are volatile components that have similar structures and therefore similar intermolecular forces:

Benzene          Toluene

In a solution of benzene and toluene, the vapor pressure of each component obeys Raoult's law. Figure 13.7 shows the dependence of the total vapor pressure ($P_T$) in a benzene-toluene solution on the composition of the solution. Because there are only two components in the solution, we need only express the composition of the solution in terms of the mole fraction of *one* component.

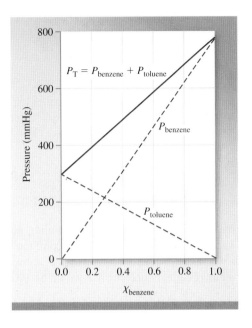

**Figure 13.7** The dependence of partial pressures of benzene and toluene on their mole fractions in a benzene-toluene solution ($\chi_{\text{toluene}} = 1 - \chi_{\text{benzene}}$) at 80°C. This solution is said to be ideal because the vapor pressures obey Raoult's law.

For any value of $\chi_{\text{benzene}}$, the mole fraction of toluene, $\chi_{\text{toluene}}$, is given by the equation $(1 - \chi_{\text{benzene}})$. The benzene-toluene solution is an example of an ***ideal solution,*** which is simply a solution that *obeys* Raoult's law.

Note that for a mixture in which the mole fractions of benzene and toluene are both 0.5, although the liquid mixture is equimolar, the vapor above the solution is not. Because pure benzene has a higher vapor pressure (75 mmHg at 20°C) than pure toluene (22 mmHg at 20°C), the vapor phase over the mixture will contain a higher concentration of the more volatile benzene molecules than it will the less volatile toluene molecules.

## Boiling-Point Elevation

Recall that the *boiling point* of a substance is the temperature at which its vapor pressure equals the external atmospheric pressure [◄◄ Section 12.6]. Because the presence of a nonvolatile solute lowers the vapor pressure of a solution, it also affects the boiling point of the solution—relative to that of the pure liquid. Figure 13.8 shows the phase diagram of water and the changes that occur when a nonvolatile solute is added to it. At any temperature the vapor pressure over a solution is lower than that over the pure liquid, so the liquid-vapor curve for the solution lies below that for the pure solvent. Consequently, the dashed solution curve intersects the horizontal line that marks $P = 1$ atm at a higher temperature than the normal boiling point of the pure solvent. That is, a higher temperature is needed to make the solvent's vapor pressure equal to atmospheric pressure.

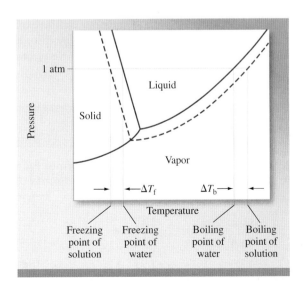

**Figure 13.8** Phase diagram illustrating the boiling-point elevation and freezing-point depression of aqueous solutions. The dashed curves pertain to the solution, and the solid curves to the pure solvent. As this diagram shows, the boiling point of the solution is higher than that of water, and the freezing point of the solution is lower than that of water.

The boiling-point elevation ($\Delta T_b$) is defined as the difference between the boiling point of the solution ($T_b$) and the boiling point of the pure solvent ($T_b^\circ$):

**Student Annotation:** Rearranging this equation, we get the boiling point of the solution by adding $\Delta T_b$ to the boiling point of the pure solvent: $T_b = T_b^\circ + \Delta T_b$.

$$\Delta T_b = T_b - T_b^\circ$$

Because $T_b > T_b^\circ$, $\Delta T_b$ is a positive quantity.

The value of $\Delta T_b$ is proportional to the concentration, expressed in molality, of the solute in the solution.

$$\Delta T_b \propto m$$

or

Equation 13.6    $$\Delta T_b = K_b m$$

where $m$ is the molality of the solution and $K_b$ is the molal boiling-point elevation constant. The units of $K_b$ are °C/$m$. Table 13.2 lists values of $K_b$ for several common solvents. Using the boiling-point elevation constant for water and Equation 13.6, you can show that the boiling point of a 1.00 $m$ aqueous solution of a nonvolatile, nonelectrolyte would be 100.5°C:

$$\Delta T_b = K_b m = (0.52°C/m)(1.00\ m) = 0.52°C$$

**Student Annotation:** The practice of adding salt to water in which food is cooked is for the purpose of increasing the boiling point, thereby cooking the food at a higher temperature.

$$T_b = T_b^\circ + \Delta T_b = 100.0°C + 0.52°C = 100.5°C$$

## Freezing-Point Depression

If you have ever lived in a cold climate, you may have seen roads and sidewalks that were "salted" in the winter. The application of a salt such as NaCl or $CaCl_2$ thaws ice (or prevents its formation) by lowering the freezing point of water.

The phase diagram in Figure 13.8 shows that in addition to shifting the liquid-vapor phase boundary down, the addition of a nonvolatile solute also shifts the solid-liquid phase boundary to the left. Consequently, this dashed line intersects the solid horizontal line at 1 atm at a temperature lower than the freezing point of pure water. The freezing-point depression ($\Delta T_f$) is defined as the difference between the freezing point of the pure solvent and the freezing point of the solution:

**Student Annotation:** We get the freezing point of the solution by *subtracting* $\Delta T_f$ from the freezing point of the pure solvent: $T_f = T_f^\circ - \Delta T_f$.

$$\Delta T_f = T_f^\circ - T_f$$

Because $T_f^\circ > T_f$, $\Delta T_f$ is a positive quantity. Again, the change in temperature is proportional to the molal concentration of the solution:

$$\Delta T_f \propto m$$

or

Equation 13.7    $$\Delta T_f = K_f m$$

where $m$ is the concentration of solute expressed in molality and $K_f$ is the molal freezing-point depression constant (see Table 13.2). Like $K_b$, $K_f$ has units of °C/$m$.

Like boiling-point elevation, freezing-point depression can be explained in terms of differences in entropy. Freezing involves a transition from the more disordered liquid state to the more ordered solid state. For this to happen, energy must be removed from the system. Because a

| **TABLE 13.2** | Molal Boiling-Point Elevation and Freezing-Point Depression Constants of Several Common Solvents | | | |
|---|---|---|---|---|
| **Solvent** | **Normal Boiling Point (°C)** | **$K_b$ (°C/$m$)** | **Normal Freezing Point (°C)** | **$K_f$ (°C/$m$)** |
| Water | 100.0 | 0.52 | 0.0 | 1.86 |
| Benzene | 80.1 | 2.53 | 5.5 | 5.12 |
| Ethanol | 78.4 | 1.22 | −117.3 | 1.99 |
| Acetic acid | 117.9 | 2.93 | 16.6 | 3.90 |
| Cyclohexane | 80.7 | 2.79 | 6.6 | 20.0 |

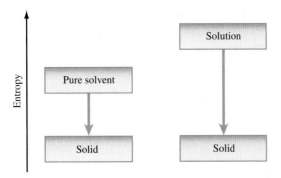

**Figure 13.9** The solution has greater entropy than the pure solvent. The bigger difference in entropy between the solution and the solid means that more energy must be removed from the solution for it to freeze. Thus, the solution freezes at a lower temperature than the pure solvent.

solution has greater disorder than the solvent, there is a bigger difference in entropy between the solution and the solid than there is between the pure solvent and the solid (Figure 13.9). The larger difference in entropy means that more energy must be removed for the liquid-solid transition to happen. Thus, the solution freezes at a lower temperature than does the pure solvent. Boiling-point elevation occurs only when the solute is nonvolatile. Freezing-point depression occurs regardless of the solute's volatility.

Sample Problem 13.7 demonstrates a practical application of freezing-point depression and boiling-point elevation.

**Student Annotation:** When a solution freezes, the solid that separates out is actually pure solvent. The solute remains in the liquid solution.

---

## Sample Problem 13.7

Ethylene glycol [CH$_2$(OH)CH$_2$(OH)] is a common automobile antifreeze. It is water soluble and fairly nonvolatile (b.p. 197°C). Calculate (a) the freezing point and (b) the boiling point of a solution containing 685 g of ethylene glycol in 2075 g of water.

**Strategy** Convert grams of ethylene glycol to moles, and divide by the mass of water in kilograms to get molal concentration. Use molal concentration in Equations 13.7 and 13.6 to determine $\Delta T_f$ and $\Delta T_b$, respectively.

**Setup** The molar mass of ethylene glycol (C$_2$H$_6$O$_2$) is 62.07 g/mol. $K_b$ and $K_f$ for water are 0.52°C/$m$ and 1.86°C/$m$, respectively.

**Solution**

$$\frac{685 \text{ g C}_2\text{H}_6\text{O}_2}{62.07 \text{ g/mol}} = 11.04 \text{ mol C}_2\text{H}_6\text{O}_2 \quad \text{and} \quad \frac{11.04 \text{ mol C}_2\text{H}_6\text{O}_2}{2.075 \text{ kg H}_2\text{O}} = 5.32 \ m \text{ C}_2\text{H}_6\text{O}_2$$

(a) $\Delta T_f = K_f m = (1.86°C/m)(5.32 \ m) = 9.89°C$

The freezing point of the solution is $(0 - 9.89)°C = -9.89°C$.

(b) $\Delta T_b = K_b m = (0.52°C/m)(5.32 \ m) = 2.8°C$

The boiling point of the solution is $(100.0 + 2.8)°C = 102.8°C$.

**Think About It** Because it both lowers the freezing point and raises the boiling point, antifreeze is useful at both temperature extremes.

**Practice Problem A** Calculate the freezing point and boiling point of a solution containing 268 g of ethylene glycol and 1015 g of water.

**Practice Problem B** What mass of ethylene glycol must be added to 1525 g of water to raise the boiling point to 103.9°C?

---

## Osmotic Pressure

Many chemical and biological processes depend on *osmosis,* the selective passage of solvent molecules through a porous membrane from a more dilute solution to a more concentrated one. Figure 13.10 illustrates osmosis. The left compartment of the apparatus contains pure solvent; the right compartment contains a solution made with the same solvent. The two compartments are separated by a *semipermeable membrane,* which allows the passage of solvent molecules but blocks the passage of solute molecules. At the beginning, the liquid levels in the two tubes are equal [Figure 13.10(a)]. As time passes, the level in the right tube rises. It continues to rise until equilibrium

**Figure 13.10**    Osmotic pressure. (a) The levels of the pure solvent (left) and of the solution (right) are equal at the start. (b) During osmosis, the level on the solution side rises as a result of the net flow of solvent from left to right.

(a)                    (b)

Multimedia

Osmosis.

is reached, after which no further net change in levels is observed. The ***osmotic pressure (π)*** of a solution is the pressure required to stop osmosis. As shown in Figure 13.10(b), this pressure can be measured directly from the difference in the final liquid levels.

The osmotic pressure of a solution is directly proportional to the concentration, expressed in *molarity,* of the solute in solution:

$$\pi \propto M$$

and is given by

Equation 13.8                    $\pi = MRT$

where $M$ is the molarity of the solution, $R$ is the gas constant (0.08206 L · atm/K · mol), and $T$ is the absolute temperature. The osmotic pressure ($\pi$) is typically expressed in atmospheres.

Like boiling-point elevation and freezing-point depression, osmotic pressure is directly proportional to the concentration of the solution. This is what we would expect, though, because all colligative properties depend only on the number of solute particles in solution, not on the identity of the solute particles. Two solutions of equal concentration have the same osmotic pressure and are said to be ***isotonic*** to each other.

## Electrolyte Solutions

So far we have discussed the colligative properties of nonelectrolyte solutions. Because electrolytes undergo *dissociation* when dissolved in water [◄◄ Section 4.1], we must consider them separately. Recall, for example, that when NaCl dissolves in water, it dissociates into $Na^+(aq)$ and $Cl^-(aq)$. For every mole of NaCl dissolved, we get two moles of ions in solution. Similarly, when a formula unit of $CaCl_2$ dissolves, we get three ions: one $Ca^{2+}$ ion and two $Cl^-$ ions. Thus, for every mole of $CaCl_2$ dissolved, we get three moles of ions in solution. Colligative properties depend only on the number of dissolved particles—not on the type of particles. This means that a 0.1 *m* solution of NaCl will exhibit a freezing point depression twice that of a 0.1 *m* solution of a nonelectrolyte, such as sucrose. Similarly, we expect a 0.1 *m* solution of $CaCl_2$ to depress the freezing point of water three times as much as a 0.1 *m* sucrose solution. To account for this effect, we introduce and define a quantity called the ***van't Hoff*** [3] ***factor (i),*** which is given by

$$i = \frac{\text{actual number of particles in solution after dissociation}}{\text{number of formula units initially dissolved in solution}}$$

---

3. Jacobus Henricus van't Hoff (1852–1911). Dutch chemist. One of the most prominent chemists of his time, van't Hoff did significant work in thermodynamics, molecular structure and optical activity, and solution chemistry. In 1901 he received the first Nobel Prize in Chemistry.

Thus, $i$ is 1 for all nonelectrolytes. For strong electrolytes such as $NaCl$ and $KNO_3$, $i$ should be 2, and for strong electrolytes such as $Na_2SO_4$ and $CaCl_2$, $i$ should be 3. Consequently, the equations for colligative properties must be modified as follows:

$$\Delta T_f = iK_f m \qquad \text{Equation 13.9}$$

$$\Delta T_b = iK_b m \qquad \text{Equation 13.10}$$

$$\pi = iMRT \qquad \text{Equation 13.11}$$

In reality, the colligative properties of electrolyte solutions are usually smaller than predicted by Equations 13.9 through 13.11, especially at higher concentrations, because of the formation of *ion pairs*. An **ion pair** is made up of one or more cations and one or more anions held together by electrostatic forces (Figure 13.11). The presence of an ion pair reduces the number of particles in solution, thus reducing the observed colligative properties. Tables 13.3 and 13.4 list experimentally measured values of $i$ and those calculated using the van't Hoff factor.

Sample Problem 13.8 demonstrates the experimental determination of a van't Hoff factor.

(a)                                         (b)

**Figure 13.11** (a) Free ions and (b) ion pairs in solution. Ion pairing reduces the number of dissolved particles in a solution, causing a decrease in the observed colligative properties. Furthermore, an ion pair bears no net charge and therefore cannot conduct electricity in solution.

| TABLE 13.3 | Calculated and Measured van't Hoff Factors of 0.0500 $M$ Electrolyte Solutions at 25°C | |
| --- | --- | --- |
| **Electrolyte** | ***i* (Calculated)** | ***i* (Measured)** |
| Sucrose* | 1 | 1.0 |
| HCl | 2 | 1.9 |
| NaCl | 2 | 1.9 |
| MgSO$_4$ | 2 | 1.3 |
| MgCl$_2$ | 3 | 2.7 |
| FeCl$_3$ | 4 | 3.4 |

*Sucrose is a nonelectrolyte. It is listed here for comparison only.

| TABLE 13.4 | Experimentally Measured van't Hoff Factors of Sucrose and NaCl Solutions at 25°C | | |
| --- | --- | --- | --- |
| | **Concentration** | | |
| **Compound** | **0.100 *m*** | **0.00100 *m*** | **0.000100 *m*** |
| Sucrose | 1.00 | 1.00 | 1.00 |
| NaCl | 1.87 | 1.94 | 1.97 |

## Sample Problem 13.8

The osmotic pressure of a 0.0100 $M$ potassium iodide (KI) solution at 25°C is 0.465 atm. Determine the experimental van't Hoff factor for KI at this concentration.

**Strategy** Use osmotic pressure to calculate the molar concentration of KI, and divide by the *nominal* concentration of 0.0100 $M$.

**Setup** $R = 0.08206$ L · atm/K · mol, and $T = 298$ K.

**Solution** Solving Equation 13.8 for $M$,

$$M = \frac{\pi}{RT} = \frac{0.465 \text{ atm}}{(0.08206 \text{ L} \cdot \text{atm/K} \cdot \text{mol})(298 \text{ K})} = 0.0190 \ M$$

$$i = \frac{0.0190 \ M}{0.0100 \ M} = 1.90$$

The experimental van't Hoff factor for KI at this concentration is 1.90.

**Think About It** The calculated van't Hoff factor for KI is 2. The experimentally determined van't Hoff factor must be less than or equal to the calculated value.

**Practice Problem A** The freezing-point depression of a 0.100 $m$ MgSO$_4$ solution is 0.225°C. Determine the experimental van't Hoff factor of MgSO$_4$ at this concentration.

**Practice Problem B** Using the experimental van't Hoff factor from Table 13.4, determine the freezing point of a 0.100 $m$ aqueous solution of NaCl. (Assume that the van't Hoff factors do not change with temperature.)

## Bringing Chemistry to Life

### Intravenous Fluids

Human blood consists of red blood cells (*erythrocytes*), white blood cells (*leukocytes*), and platelets (*thrombocytes*) suspended in plasma, an aqueous solution containing a variety of solutes including salts and proteins. Each red blood cell is surrounded by a protective semipermeable membrane. Inside this membrane, the concentration of dissolved substances is about 0.3 $M$. Likewise, the concentration of dissolved substances in plasma is also about 0.3 $M$. Having the same concentration (and therefore the same osmotic pressure of ~7.6 atm at 37°C) inside and outside the red blood cell prevents a net movement of water into or out of the cell through the protective semipermeable membrane [Figure 13.12(a)]. To maintain this balance of osmotic pressure, fluids that are given

**Figure 13.12** A cell in (a) a hypotonic solution, (b) an isotonic solution, and (c) a hypertonic solution. The cell swells and may eventually burst in (a); it shrinks in (c).

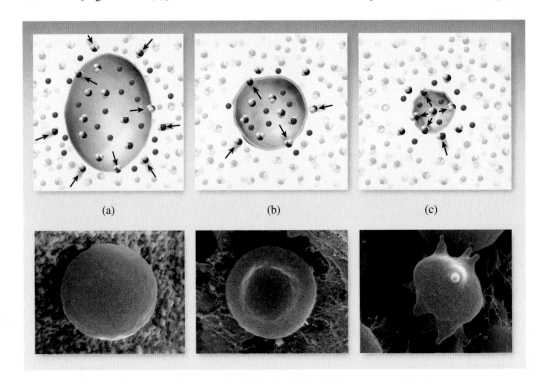

(a)                    (b)                    (c)

intravenously must be *isotonic* to plasma. Five percent dextrose (sugar) and normal saline, which is 0.9% sodium chloride, are two of the most commonly used isotonic intravenous fluids.

A solution that has a lower concentration of dissolved substances than plasma is said to be **hypotonic** to plasma. If a significant volume of pure water were administered intravenously, it would dilute the plasma, lowering its concentration and making it hypotonic to the solution inside the red blood cells. If this were to happen, water would enter the red blood cells via osmosis [Figure 13.12(b)]. The cells would swell and could potentially burst, a process called *hemolysis.* On the other hand, if red blood cells were placed in a solution with a higher concentration of dissolved substances than plasma, a solution said to be **hypertonic** to plasma, water would leave the cells via osmosis [Figure 13.12(c)]. The cells would shrink, a process called *crenation,* which is also potentially dangerous. The osmotic pressure of human plasma must be maintained within a very narrow range to prevent damage to red blood cells. Hypotonic and hypertonic solutions can be administered intravenously to treat specific medical conditions, but the patient must be carefully monitored throughout the treatment.

Interestingly, humans have for centuries exploited the sensitivity of cells to osmotic pressure. The process of "curing" meat with salt or with sugar causes crenation of the bacteria cells that would otherwise cause spoilage.

## Sample Problem 13.9

*Ringer's lactate,* a solution containing several different salts, is often administered intravenously for the initial treatment of trauma patients. One liter of Ringer's lactate contains 0.102 mole of sodium chloride, $4 \times 10^{-3}$ mole of potassium chloride, $1.5 \times 10^{-3}$ mole of calcium chloride, and $2.8 \times 10^{-2}$ mole of sodium lactate. Determine the osmotic pressure of this solution at normal body temperature (37°C). Assume no ion pairing. (The formula of the lactate ion is $CH_3CH_2COO^-$.)

**Strategy** Because the solutes are all ionic, each solute concentration will have to be multiplied by the appropriate van't Hoff factor. (Because we are assuming no ion pairing, we can use calculated van't Hoff factors.) Sum the concentrations for all solutes to determine the total concentration of dissolved particles, and use Equation 13.8 to calculate the osmotic pressure.

**Setup** Because the volume of the solution described is 1 L, the number of moles is also the molarity for each solute. $R = 0.08206$ L · atm/K · mol, $T = 310$ K, and the van't Hoff factors for the solutes in Ringer's lactate are as follows:

$$NaCl(s) \longrightarrow Na^+(aq) + Cl^-(aq) \qquad\qquad i = 2$$
$$KCl(s) \longrightarrow K^+(aq) + Cl^-(aq) \qquad\qquad i = 2$$
$$CaCl_2(s) \longrightarrow Ca^{2+}(aq) + 2Cl^-(aq) \qquad\qquad i = 3$$
$$NaCH_3CH_2COO(s) \longrightarrow Na^+(aq) + CH_3CH_2COO^-(aq) \qquad\qquad i = 2$$

**Solution** The total concentration of ions in solution is the sum of the individual concentrations.

total concentration = $2[NaCl] + 2[KCl] + 3[MgCl_2] + 2[NaCH_3CH_2COO]$

$= 2(0.102 \ M) + 2(4 \times 10^{-3} M) + 3(1.5 \times 10^{-3} \ M) + 2(2.8 \times 10^{-2} \ M)$

$= 2.73 \times 10^{-1} \ M$

$\pi = MRT = (0.273 \ M)(0.08206 \ L \cdot atm/K \cdot mol)(310 \ K) = 6.93$ atm

**Think About It** Ringer's lactate is isotonic with human plasma, which is often the case with fluids administered intravenously.

**Practice Problem A** Determine the osmotic pressure of a solution that is 0.200 *M* in glucose and 0.100 *M* in sodium chloride at 37.0°C. Assume no ion pairing.

**Practice Problem B** Determine the concentration of an aqueous solution that has an osmotic pressure of 5.6 atm at 37°C if (a) the solute is glucose, and (b) the solute is sodium chloride. Assume no ion pairing.

To review:

- Vapor-pressure lowering depends on concentration expressed as *mole fraction,* $\chi$.
- Boiling-point elevation depends on concentration expressed as *molality, m.*
- Freezing-point depression depends on concentration expressed as *molality, m.*
- Osmotic pressure depends on concentration expressed as *molarity, M.*

## Checkpoint 13.5    Colligative Properties

**13.5.1** A solution contains 75.0 g of glucose (molar mass 180.2 g/mol) in 425 g of water. Determine the vapor pressure of water over the solution at 35°C. ($P_{H_2O}^\circ = 42.2$ mmHg at 35°C.)

a) 0.732 mmHg

b) 42.9 mmHg

c) 243 mmHg

d) 41.5 mmHg

e) 42.2 mmHg

**13.5.2** Determine the boiling point and the freezing point of a solution prepared by dissolving 678 g of glucose in 2.0 kg of water. For water, $K_b = 0.52$°C/$m$ and $K_f = 1.86$°C/$m$.

a) 101°C and 3.5°C

b) 99°C and −3.5°C

c) 101°C and −3.5°C

d) 112°C and 6.2°C

e) 88°C and −6.2°C

**13.5.3** Calculate the osmotic pressure of a solution prepared by dissolving 65.0 g of $Na_2SO_4$ in enough water to make 500 mL of solution at 20°C. (Assume no ion pairing.)

a) 0.75 atm

b) 66 atm

c) 44 atm

d) $1 \times 10^{-2}$ atm

e) 22 atm

**13.5.4** A 1.00 $m$ solution of HCl has a freezing point of −3.30°C. Determine the experimental van't Hoff factor for HCl at this concentration.

a) 1.77

b) 2.01

c) 1.90

d) 2

e) 1

---

## Bringing Chemistry to Life

### Hemodialysis

Osmosis refers to the movement of solvent through a membrane from the side where the solute concentration is lower to the side where the solute concentration is higher. Hemodialysis involves a more porous membrane, through which both solvent (water) and small solute particles can pass.

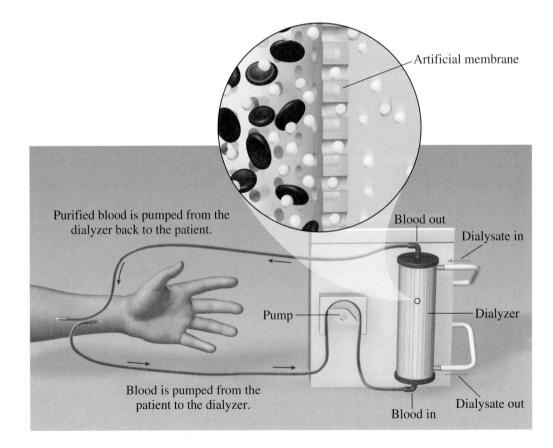

Artificial membrane

Purified blood is pumped from the dialyzer back to the patient.

Blood out

Dialysate in

Pump

Dialyzer

Blood is pumped from the patient to the dialyzer.

Blood in

Dialysate out

The size of the membrane pores is such that only small waste products such as excess potassium ion, creatinine, urea, and extra fluid can pass through. Larger components in blood, such as blood cells and proteins, are too large to pass through the membrane. A solute will pass through the membrane from the side where its concentration is higher to the side where its concentration is lower. The composition of the dialysate ensures that the necessary solutes in the blood (e.g., sodium and calcium ions) are not removed.

Because it is not normally found in blood, fluoride ion, if present in the dialysate, will flow across the membrane into the blood. In fact, this is true of any sufficiently small solute that is not normally found in blood—necessitating requirements for the purity of water used to prepare dialysate solutions that far exceed those for drinking water.

## 13.6 Calculations Using Colligative Properties

The colligative properties of nonelectrolyte solutions provide a means of determining the molar mass of a solute. Although any of the four colligative properties can be used in theory for this purpose, only freezing-point depression and osmotic pressure are used in practice because they show the most pronounced, and therefore the most easily measured, changes. From the experimentally determined freezing-point depression or osmotic pressure, we can calculate the solution's molality or molarity, respectively. Knowing the mass of dissolved solute, we can readily determine its molar mass.

Sample Problems 13.10 and 13.11 illustrate this technique.

**Student Annotation:** These calculations require Equations 13.7 and 13.8, respectively.

---

### Sample Problem 13.10

Quinine was the first drug widely used to treat malaria, and it remains the treatment of choice for severe cases. A solution prepared by dissolving 10.0 g of quinine in 50.0 mL of ethanol has a freezing point 1.55°C below that of pure ethanol. Determine the molar mass of quinine. (The density of ethanol is 0.789 g/mL.) Assume that quinine is a nonelectrolyte.

**Strategy** Use Equation 13.7 to determine the molal concentration of the solution. Use the density of ethanol to determine the mass of solvent. The molal concentration of quinine multiplied by the mass of ethanol (in kg) gives moles of quinine. The mass of quinine (in grams) divided by moles of quinine gives the molar mass.

**Setup**
$$\text{mass of ethanol} = 50.0 \text{ mL} \times 0.789 \text{ g/mL} = 39.5 \text{ g} \quad \text{or} \quad 3.95 \times 10^{-2} \text{ kg}$$

$K_f$ for ethanol (from Table 13.2) is 1.99°C/$m$.

**Solution** Solving Equation 13.7 for molal concentration,

$$m = \frac{\Delta T_f}{K_f} = \frac{1.55°C}{1.99°C/m} = 0.779 \text{ } m$$

The solution is 0.779 $m$ in quinine (i.e., 0.779 mol quinine/kg ethanol solvent).

$$\left(\frac{0.779 \text{ mol quinine}}{\text{kg ethanol}}\right)(3.95 \times 10^{-2} \text{ kg ethanol}) = 0.0308 \text{ mol quinine}$$

$$\text{molar mass of quinine} = \frac{10.0 \text{ g quinine}}{0.0308 \text{ mol quinine}} = 325 \text{ g/mol}$$

**Think About It** Check the result using the molecular formula of quinine: $C_{20}H_{24}N_2O_2$ (324.4 g/mol). Multistep problems such as this one require careful tracking of units at each step.

**Practice Problem A** Calculate the molar mass of naphthalene, the organic compound in "mothballs," if a solution prepared by dissolving 5.00 g of naphthalene in exactly 100 g of benzene has a freezing point 2.00°C below that of pure benzene.

**Practice Problem B** What mass of naphthalene must be dissolved in 2.00 × 10² g of benzene to give a solution with a freezing point 2.50°C below that of pure benzene?

## Sample Problem 13.11

A solution is prepared by dissolving 50.0 g of hemoglobin (Hb) in enough water to make 1.00 L of solution. The osmotic pressure of the solution is measured and found to be 14.3 mmHg at 25°C. Calculate the molar mass of hemoglobin. (Assume that there is no change in volume when the hemoglobin is added to the water.)

**Strategy** Use Equation 13.8 to calculate the molarity of the solution. Because the solution volume is 1 L, the molarity is equal to the number of moles of hemoglobin. Dividing the mass of hemoglobin, which is given in the problem statement, by the number of moles gives the molar mass.

**Setup** $R = 0.08206$ L · atm/K · mol, $T = 298$ K, and $\pi = 14.3$ mmHg/(760 mmHg/atm) $= 1.88 \times 10^{-2}$ atm.

**Solution** Rearranging Equation 13.8 to solve for molarity, we get

$$M = \frac{\pi}{RT} = \frac{1.88 \times 10^{-2} \text{ atm}}{(0.08206 \text{ L} \cdot \text{atm/K} \cdot \text{mol})(298 \text{ K})} = 7.69 \times 10^{-4} \, M$$

Thus, the solution contains $7.69 \times 10^{-4}$ mole of hemoglobin.

$$\text{molar mass of hemoglobin} = \frac{50.0 \text{ g}}{7.69 \times 10^{-4} \text{ mol}} = 6.50 \times 10^{4} \text{ g/mol}$$

**Think About It** Biological molecules can have *very* high molar masses.

**Practice Problem A** A solution made by dissolving 25 mg of insulin in 5.0 mL of water has an osmotic pressure of 15.5 mmHg at 25°C. Calculate the molar mass of insulin. (Assume that there is no change in volume when the insulin is added to the water.)

**Practice Problem B** What mass of insulin must be dissolved in 50.0 mL of water to produce a solution with an osmotic pressure of 16.8 mmHg at 25°C?

The colligative properties of an electrolyte solution can be used to determine percent dissociation. ***Percent dissociation*** is the percentage of dissolved molecules (or formula units, in the case of an ionic compound) that separate into ions in solution. For a strong electrolyte such as NaCl, there should be *complete,* or 100 percent, dissociation. However, the data in Table 13.4 indicate that this is not necessarily the case. An experimentally determined van't Hoff factor smaller than the corresponding calculated value indicates less than 100 percent dissociation. As the experimentally determined van't Hoff factors for NaCl indicate, dissociation of a strong electrolyte is more complete at lower concentration. The ***percent ionization*** of a weak electrolyte, such as a weak acid, also depends on the concentration of the solution.

**Student Annotation:** Recall that the term *dissociation* is used for ionic electrolytes and the term *ionization* is used for molecular electrolytes. In this context, they mean essentially the same thing.

Sample Problem 13.12 shows how to use colligative properties to determine the percent dissociation of a weak electrolyte.

## Sample Problem 13.12

A solution that is 0.100 *M* in hydrofluoric acid (HF) has an osmotic pressure of 2.64 atm at 25°C. Calculate the percent ionization of HF at this concentration.

**Strategy** Use the osmotic pressure and Equation 13.8 to determine the molar concentration of the particles in solution. Compare the concentration of particles to the *nominal* concentration (0.100 *M*) to determine what percentage of the original HF molecules are ionized.

**Setup** $R = 0.08206$ L · atm/K · mol, and $T = 298$ K.

**Solution** Rearranging Equation 13.8 to solve for molarity,

$$M = \frac{\pi}{RT} = \frac{2.64 \text{ atm}}{(0.08206 \text{ L} \cdot \text{atm/K} \cdot \text{mol})(298 \text{ K})} = 0.108 \, M$$

The concentration of dissolved particles is 0.108 *M*. Consider the ionization of HF [◀◀ Section 4.3]:

$$HF(aq) \rightleftharpoons H^+(aq) + F^-(aq)$$

According to this equation, if $x$ HF molecules ionize, we get $x$ H$^+$ ions and $x$ F$^-$ ions. Thus, the total concentration of particles in solution will be the original concentration of HF minus $x$, which gives the concentration of intact HF molecules, plus $2x$, which is the concentration of ions (H$^+$ and F$^-$):

$$(0.100 - x) + 2x = 0.100 + x$$

Therefore, $0.108 = 0.100 + x$ and $x = 0.008$. Because we earlier defined $x$ as the amount of HF ionized, the percent ionization is given by

$$\text{percent ionization} = \frac{0.008\ M}{0.100\ M} \times 100\% = 8\%$$

At this concentration HF is 8 percent ionized.

**Practice Problem A** An aqueous solution that is 0.0100 $M$ in acetic acid (HC$_2$H$_3$O$_2$) has an osmotic pressure of 0.255 atm at 25°C. Calculate the percent ionization of acetic acid at this concentration.

**Practice Problem B** An aqueous solution that is 0.015 $M$ in acetic acid (HC$_2$H$_3$O$_2$) is 3.5 percent ionized at 25°C. Calculate the osmotic pressure of this solution.

> **Think About It** For weak acids, the lower the concentration, the greater the percent ionization. A 0.010 $M$ solution of HF has an osmotic pressure of 0.30 atm, corresponding to 23 percent ionization. A 0.0010 $M$ solution of HF has an osmotic pressure of $3.8 \times 10^{-2}$ atm, corresponding to 56 percent ionization.

---

**Checkpoint 13.6** **Calculations Using Colligative Properties**

13.6.1 A solution made by dissolving 14.2 g of sucrose in 100 g of water exhibits a freezing point depression of 0.77°C. Calculate the molar mass of sucrose.

a) 34 g/mol

b) $3.4 \times 10^2$ g/mol

c) 2.4 g/mol

d) $1.8 \times 10^2$ g/mol

e) 68 g/mol

13.6.2 A 0.010 $M$ solution of the weak electrolyte HA has an osmotic pressure of 0.27 atm at 25°C. What is the percent ionization of the electrolyte at this concentration?

a) 27%

b) 10%

c) 15%

d) 81%

e) 90%

---

# 13.7 Colloids

The solutions discussed so far in this chapter are true homogeneous mixtures. Now consider what happens if we add fine sand to a beaker of water and stir. The sand particles are suspended at first but gradually settle to the bottom of the beaker. This is an example of a heterogeneous mixture. Between the two extremes of homogeneous and heterogeneous mixtures is an intermediate state called a colloidal suspension, or simply, a colloid. A *colloid* is a dispersion of particles of one substance throughout another substance. Colloidal particles are much larger than the normal solute molecules; they range from $1 \times 10^3$ pm to $1 \times 10^6$ pm. Also, a colloidal suspension lacks the homogeneity of a true solution.

Colloids can be further categorized as aerosols (liquid or solid dispersed in gas), foams (gas dispersed in liquid or solid), emulsions (liquid dispersed in another liquid), sols (solid dispersed in liquid or in another solid), and gels (liquid dispersed in a solid). Table 13.5 lists the different types of colloids and gives one or more examples of each.

One way to distinguish a solution from a colloid is by the **Tyndall**[4] **effect.** When a beam of light passes through a colloid, it is scattered by the dispersed phase (Figure 13.13). No such scattering is observed with true solutions because the solute molecules are too small to interact with visible light. Another demonstration of the Tyndall effect is the scattering of light from automobile headlights in fog (Figure 13.14).

> **Student Annotation:** The substance dispersed is called the *dispersed phase;* the substance in which it is dispersed is called the *dispersing medium.*

---

4. John Tyndall (1820–1893). Irish physicist. Tyndall did important work in magnetism and also explained glacier motion.

| TABLE 13.5 | Types of Colloids | | |
|---|---|---|---|
| **Dispersing Medium** | **Dispersed Phase** | **Name** | **Example** |
| Gas | Liquid | Aerosol | Fog, mist |
| Gas | Solid | Aerosol | Smoke |
| Liquid | Gas | Foam | Whipped cream, meringue |
| Liquid | Liquid | Emulsion | Mayonnaise |
| Liquid | Solid | Sol | Milk of magnesia |
| Solid | Gas | Foam | Styrofoam |
| Solid | Liquid | Gel | Jelly, butter |
| Solid | Solid | Solid sol | Alloys such as steel, gemstones (glass with dispersed metal) |

**Student Annotation:** Styrofoam is a registered trademark of the Dow Chemical Company. It refers specifically to extruded polystyrene used for insulation in home construction. "Styrofoam" cups, coolers, and packing peanuts are not really made of Styrofoam.

**Figure 13.13** The Tyndall effect. Light is scattered by colloidal particles (right) but not by dissolved particles (left).

**Figure 13.14** A familiar example of the Tyndall effect: headlights illuminating fog.

Among the most important colloids are those in which the dispersing medium is water. Such colloids can be categorized as **hydrophilic** (*water loving*) or **hydrophobic** (*water fearing*). Hydrophilic colloids contain extremely large molecules such as proteins. In the aqueous phase, a protein like hemoglobin folds in such a way that the hydrophilic parts of the molecule, the parts that can interact favorably with water molecules by ion-dipole forces or hydrogen-bond formation, are on the outside surface (Figure 13.15).

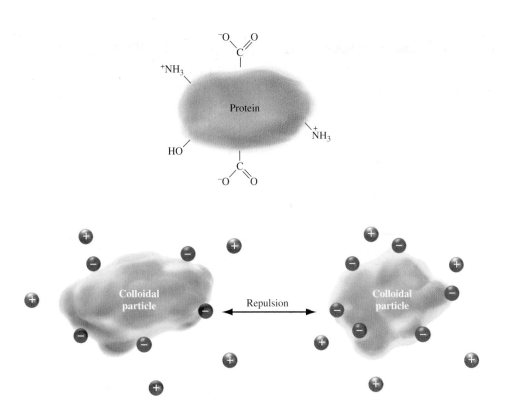

**Figure 13.15** Hydrophilic groups on the surface of a large molecule such as a protein stabilize the molecule in water. Note that all the hydrophilic groups can form hydrogen bonds with water.

**Figure 13.16** Diagram showing the stabilization of hydrophobic colloids. Negative ions are adsorbed onto the surface, and the repulsion between like charges prevents aggregation of the particles.

A hydrophobic colloid normally would not be stable in water, and the particles would clump together, like droplets of oil in water merging to form a film at the water's surface. They can be stabilized, however, by the adsorption of ions on their surface (Figure 13.16). Material that collects on the surface is *adsorbed,* whereas material that passes to the interior is *absorbed.* The adsorbed ions are hydrophilic and can interact with water to stabilize the colloid. In addition, because adsorption of ions leaves the colloid particles *charged,* electrostatic repulsion prevents them from clumping together. Soil particles in rivers and streams are hydrophobic particles that are stabilized in this way. When river water enters the sea, the charges on the dispersed particles are neutralized by the high-salt medium. With the charges on their surfaces neutralized, the particles no longer repel one another and they clump together to form the silt that is seen at the mouth of the river.

Another way hydrophobic colloids can be stabilized is by the presence of other hydrophilic groups on their surfaces. Consider sodium stearate, a soap molecule that has a polar group at one end, often called the "head," and a long hydrocarbon "tail" that is nonpolar (Figure 13.17). The cleansing action of soap is due to the dual nature of the hydrophobic tail and the hydrophilic head. The hydrocarbon tail is readily soluble in oily substances, which are also nonpolar, while the ionic $-COO^-$ group remains outside the oily surface. When enough soap molecules have surrounded an oil droplet, as shown in Figure 13.18, the entire system becomes stabilized in water because the exterior portion is now largely hydrophilic. This is how greasy substances are removed by the action of soap. In general, the process of stabilizing a colloid that would otherwise not stay dispersed is called *emulsification,* and a substance used for such stabilization is called an *emulsifier* or *emulsifying agent.*

**Student Annotation:** A hydrophobic colloid must be *stabilized* to remain suspended in water.

Sodium stearate ($C_{17}H_{35}COO^-Na^+$)

(a)

Hydrophilic head

Hydrophobic tail

(b)

**Figure 13.17** (a) A sodium stearate molecule. (b) The simplified representation of the molecule that shows a hydrophilic head and a hydrophobic tail.

**Figure 13.18**    The mechanism by which soap removes grease. (a) Grease (oily substance) is not soluble in water. (b) When soap is added to water, the nonpolar tails of soap molecules dissolve in grease. (c) The grease can be washed away when the polar heads of the soap molecules stabilize it in water.

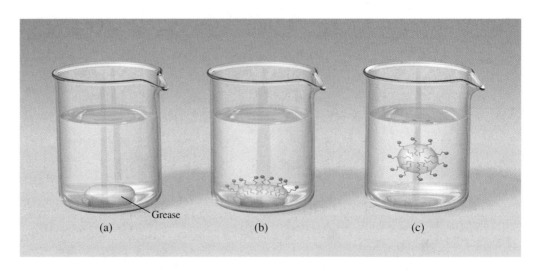

(a)    Grease    (b)    (c)

A mechanism similar to that involving sodium stearate makes it possible for us to digest dietary fat. When we ingest fat, the gallbladder excretes a substance known as bile. Bile contains a variety of substances including bile salts. A *bile salt* is a derivative of cholesterol with an attached amino acid. Like sodium stearate, a bile salt has both a hydrophobic end and a hydrophilic end. (Figure 13.19 shows the bile salt sodium glycocholate.) The bile salts surround fat particles with their hydrophobic ends oriented toward the fat and their hydrophilic ends facing the water, emulsifying the fat in the aqueous medium of the digestive system. This process allows fats to be digested and other nonpolar substances such as fat-soluble vitamins to be absorbed through the wall of the small intestine.

**Student Annotation:** It is being nonpolar that makes some vitamins soluble in fat. Remember the axiom "like dissolves like."

**Figure 13.19**    Structure of sodium glycocholate. The hydrophobic tail of sodium glycocholate dissolves in ingested fats, stabilizing them on the aqueous medium of the digestive system.

# Applying What You've Learned

Despite the use of fluoride in municipal water supplies and many topical dental products, acute fluoride poisoning is relatively rare. Fluoride is now routinely removed from the water used to prepare dialysate solutions. However, water-supply fluoridation became common during the 1960s and 1970s—just when hemodialysis was first being made widely available to patients. This unfortunate coincidence resulted in large numbers of early dialysis patients suffering the effects of fluoride poisoning, before the danger of introducing fluoride via dialysis was recognized. The level of fluoride considered safe for the drinking water supply is based on the presumed ingestion by a healthy person of 14 L of water per week. Many dialysis patients routinely are exposed to as much as *50 times* that volume, putting them at significantly increased risk of absorbing toxic amounts of fluoride.

**Student Annotation:** Elevated levels of fluoride are associated with *osteomalacia*, a condition marked by debilitating bone pain and muscle weakness.

## Problems:

a)  Early fluoridation of municipal water supplies was done by dissolving enough sodium fluoride to achieve a 1-ppm concentration of fluoride ion. Convert 1.0 ppm F$^-$ to *percent by mass* F$^-$ and *molality* of NaF. [◄◄ Sample Problem 13.3]

b)  Calculate the boiling point and freezing point of a solution made by dissolving 4.10 g NaF in 100 mL H$_2$O. (For water, $d = 1$ g/mL.) [◄◄ Sample Problem 13.7]

c)  Many fluoridation facilities now use fluorosilicic acid instead of sodium fluoride. Fluorosilicic acid typically is distributed as a 23% (1.596 *m*) aqueous solution. Calculate the van't Hoff factor of fluorosilicic acid given that a 23% solution has a freezing point of −15.5°C. [◄◄ Sample Problem 13.8]

d)  The density of 23% fluorosilicic acid is 1.19 g/mL. Given that the osmotic pressure of this solution at 25°C is 242 atm, calculate the molar mass of fluorosilicic acid. [◄◄ Sample Problems 13.9 and 13.10]

e)  Hydrofluoric acid (HF) is another compound that can be used in the fluoridation of water. HF is a weak acid that only partially ionizes in solution. If an aqueous solution that is 0.15 *M* in HF has an osmotic pressure of 3.9 atm at 25°C, what is the percent ionization of HF at this concentration? [◄◄ Sample Problem 13.12]

# Chapter Summary

## Section 13.1

- Solutions are homogeneous mixtures of two or more substances, which may be solids, liquids, or gases.

- **Saturated solutions** contain the maximum possible amount of dissolved solute.

- The amount of solute dissolved in a saturated solution is the **solubility** of the solute in the specified solvent at the specified temperature.

- **Unsaturated solutions** contain less than the maximum possible amount of solute.

- **Supersaturated solutions** contain more solute than specified by the solubility.

## Section 13.2

- Substances with similar intermolecular forces tend to be soluble in one another. "Like dissolves like." Two liquids that are soluble in each other are called **miscible.**

- Solution formation may be endothermic or exothermic overall. An increase in **entropy** is the driving force for solution formation. Solute particles are surrounded by solvent molecules in a process called **solvation.**

## Section 13.3

- In addition to molarity (*M*) and mole fraction ($\chi$), **molality (m)** and **percent by mass** are used to express the concentrations of solutions.

- *Molality* is defined as the number of moles of solute per kilogram of solvent. *Percent by mass* is defined as the mass of solute divided by the total mass of the solution, all multiplied by 100 percent.

- Molality and percent by mass have the advantage of being temperature independent. Conversion among molarity, molality, and percent by mass requires solution *density.*

- The units of concentration used depend on the type of problem to be solved.

## Section 13.4

- Increasing the temperature *increases* the solubility of most solids in water and *decreases* the solubility of most gases in water.

- Increasing the pressure increases the solubility of gases in water but does not affect the solubility of solids.

- According to **Henry's law,** the solubility of a gas in a liquid is directly proportional to the partial pressure of the gas over the solution: *c* = *kP.*

- The proportionality constant *k* is the **Henry's law constant.** Henry's law constants are specific to the gas and solvent, and they are temperature dependent.

## Section 13.5

- **Colligative properties** depend on the *number* (but not on the *type*) of dissolved particles. The colligative properties are *vapor-pressure lowering, boiling-point elevation, freezing-point depression,* and *osmotic pressure.*

- A **volatile** substance is one that has a measurable vapor pressure. A **nonvolatile** substance is one that does not have a measurable vapor pressure.

- According to **Raoult's law,** the partial pressure of a substance over a solution is equal to the *mole fraction* ($\chi$) of the substance times its *pure vapor pressure* ($P°$). An **ideal solution** is one that obeys Raoult's law.

- **Osmosis** is the flow of solvent through a **semipermeable membrane,** one that allows solvent molecules but not solute particles to pass, from a more dilute solution to a more concentrated one.

- **Osmotic pressure** ($\pi$) is the pressure required to prevent osmosis from occurring.

- Two solutions with the same osmotic pressure are called **isotonic. Hypotonic** refers to a solution with a *lower* osmotic pressure. **Hypertonic** refers to a solution with a *higher* osmotic pressure. These terms are often used in reference to human plasma, which has an osmotic pressure of 7.6 atm.

- In electrolyte solutions, the number of dissolved particles is increased by dissociation or ionization. The magnitudes of colligative properties are increased by the **van't Hoff factor (i),** which indicates the degree of dissociation or ionization.

- The experimentally determined van't Hoff factor is generally smaller than the calculated value due to the formation of **ion pairs**—especially at high concentrations. Ion pairs are oppositely charged ions that are attracted to each other and effectively become a single "particle" in solution.

## Section 13.6

- Experimentally determined colligative properties can be used to calculate the molar mass of a nonelectrolyte or the **percent dissociation** (or **percent ionization**) of a weak electrolyte.

## Section 13.7

- A **colloid** is a dispersion of particles (about $1 \times 10^3$ pm to $1 \times 10^6$ pm) of one substance in another substance.

- Colloids can be distinguished from true solutions by the **Tyndall effect,** which is the scattering of visible light by colloidal particles.

- Colloids are classified either as **hydrophilic** (water loving) or **hydrophobic** (water fearing).

- Hydrophobic colloids can be stabilized in water by surface interactions with ions or polar molecules.

# Key Words

| | | | |
|---|---|---|---|
| Colligative properties, 553 | Hypotonic, 561 | Osmotic pressure ($\pi$), 558 | Solvation, 541 |
| Colloid, 565 | Ideal solution, 555 | Percent by mass, 547 | Supersaturated solution, 540 |
| Entropy, 542 | Ion pair, 559 | Percent dissociation, 564 | Tyndall effect, 565 |
| Henry's law, 551 | Isotonic, 558 | Percent ionization, 564 | Unsaturated solution, 540 |
| Henry's law constant (*k*), 551 | Miscible, 544 | Raoult's law, 553 | van't Hoff factor (*i*), 558 |
| Hydrophilic, 566 | Molality (*m*), 547 | Saturated solution, 540 | Volatile, 554 |
| Hydrophobic, 566 | Nonvolatile, 553 | Semipermeable membrane, 557 | |
| Hypertonic, 561 | Osmosis, 557 | Solubility, 540 | |

# KEY EQUATIONS

$$13.1 \quad \text{molality} = m = \frac{\text{moles of solute}}{\text{mass of solvent (in kg)}}$$

$$13.2 \quad \text{percent by mass} = \frac{\text{mass of solute}}{\text{mass of solute} + \text{mass of solvent}} \times 100\%$$

$$13.3 \quad c = kP$$

$$13.4 \quad P_1 = \chi_1 P_1^\circ$$

$$13.5 \quad P_1^\circ - P_1 = \Delta P = \chi_2 P_1^\circ$$

$$13.6 \quad \Delta T_b = K_b m$$

$$13.7 \quad \Delta T_f = K_f m$$

$$13.8 \quad \pi = MRT$$

$$13.9 \quad \Delta T_f = iK_f m$$

$$13.10 \quad \Delta T_b = iK_b m$$

$$13.11 \quad \pi = iMRT$$

# QUESTIONS AND PROBLEMS

## Section 13.1: Types of Solutions

### Review Questions

13.1 Distinguish between an unsaturated solution, a saturated solution, and a supersaturated solution.

13.2 Describe the different types of solutions that can be formed by the combination of solids, liquids, and gases. Give examples of each type of solution.

## Section 13.2: The Solution Process

### Review Questions

13.3 Briefly describe the solution process at the molecular level. Use the dissolution of a solid in a liquid as an example.

13.4 Basing your answer on intermolecular force considerations, explain what "like dissolves like" means.

13.5 What is solvation? What factors influence the extent to which solvation occurs? Give two examples of solvation; include one that involves ion-dipole interaction and one in which dispersion forces come into play.

13.6 As you know, some solution processes are endothermic and others are exothermic. Provide a molecular interpretation for the difference.

13.7 Explain why dissolving a solid almost always leads to an increase in disorder.

13.8 Describe the factors that affect the solubility of a solid in a liquid. What does it mean to say that two liquids are miscible?

### Problems

**13.9** Why is naphthalene ($C_{10}H_8$) more soluble than CsF in benzene?

13.10 Explain why ethanol ($C_2H_5OH$) is not soluble in cyclohexane ($C_6H_{12}$).

**13.11** Arrange the following compounds in order of increasing solubility in water: $O_2$, LiCl, $Br_2$, methanol ($CH_3OH$).

13.12 Explain the variations in solubility in water of the listed alcohols:

| Compound | Solubility in Water (g/100 g) at 20°C |
|---|---|
| $CH_3OH$ | ∞ |
| $CH_3CH_2OH$ | ∞ |
| $CH_3CH_2CH_2OH$ | ∞ |
| $CH_3CH_2CH_2CH_2OH$ | 9 |
| $CH_3CH_2CH_2CH_2CH_2OH$ | 2.7 |

*Note:* ∞ means that the alcohol and water are completely miscible in all proportions.

## Section 13.3: Concentration Units

### Review Questions

13.13 Define the following concentration terms and give their units: percent by mass, mole fraction, molarity, molality. Compare their advantages and disadvantages.

13.14 Outline the steps required for conversion between molarity, molality, and percent by mass.

### Problems

**13.15** Calculate the percent by mass of the solute in each of the following aqueous solutions: (a) 5.75 g of NaBr in 67.9 g of solution, (b) 24.6 g of KCl in 114 g of water, (c) 4.8 g of toluene in 39 g of benzene.

**13.16** Calculate the amount of water (in grams) that must be added to (a) 5.00 g of urea ($(NH_2)_2CO$ in the preparation of a 16.2 percent by mass solution, and (b) 26.2 g of $MgCl_2$ in the preparation of a 1.5 percent by mass solution.

**13.17** Calculate the molality of each of the following solutions: (a) 14.3 g of sucrose ($C_{12}H_{22}O_{11}$) in 685 g of water, (b) 7.15 moles of ethylene glycol ($C_2H_6O_2$) in 3505 g of water.

**13.18** Calculate the molality of each of the following aqueous solutions: (a) 2.55 *M* NaCl solution (density of solution = 1.08 g/mL), (b) 45.2 percent by mass KBr solution.

**13.19** Calculate the molalities of the following aqueous solutions: (a) 1.22 $M$ sugar ($C_{12}H_{22}O_{11}$) solution (density of solution = 1.12 g/mL), (b) 0.87 $M$ NaOH solution (density of solution = 1.04 g/mL), (c) 5.24 $M$ NaHCO$_3$ solution (density of solution = 1.19 g/mL).

**13.20** For dilute aqueous solutions in which the density of the solution is roughly equal to that of the pure solvent, the molarity of the solution is equal to its molality. Show that this statement is correct for a 0.010 $M$ aqueous urea (NH$_2$)$_2$CO solution.

**13.21** The alcohol content of hard liquor is normally given in terms of the "proof," which is defined as twice the percentage by volume of ethanol ($C_2H_5OH$) present. Calculate the number of grams of alcohol present in 1.00 L of 75-proof gin. The density of ethanol is 0.798 g/mL.

**13.22** The concentrated sulfuric acid we use in the laboratory is 98.0 percent H$_2$SO$_4$ by mass. Calculate the molality and molarity of the acid solution. The density of the solution is 1.83 g/mL.

**13.23** Calculate the molarity and molality of an NH$_3$ solution made up of 35.0 g of NH$_3$ in 75.0 g of water. The density of the solution is 0.982 g/mL.

**13.24** The density of an aqueous solution containing 15.0 percent of ethanol ($C_2H_5OH$) by mass is 0.984 g/mL. (a) Calculate the molality of this solution. (b) Calculate its molarity. (c) What volume of the solution would contain 0.250 mole of ethanol?

**13.25** Fish breathe the dissolved air in water through their gills. Assuming the partial pressures of oxygen and nitrogen in air to be 0.20 and 0.80 atm, respectively, calculate the mole fractions of oxygen and nitrogen in the air dissolved in water at 298 K. The solubilities of O$_2$ and N$_2$ in water at 298 K are $1.3 \times 10^{-3}$ mol/L · atm and $6.8 \times 10^{-4}$ mol/L · atm, respectively. Comment on your results.

## Section 13.4: Factors That Affect Solubility

### Review Questions

**13.26** How do the solubilities of most ionic compounds in water change with temperature? With pressure?

**13.27** Discuss the factors that influence the solubility of a gas in a liquid.

**13.28** What is thermal pollution? Why is it harmful to aquatic life?

**13.29** What is Henry's law? Define each term in the equation, and give its units. How would you account for the law in terms of the kinetic molecular theory of gases? Give two exceptions to Henry's law.

**13.30** A student is observing two beakers of water. One beaker is heated to 30°C, and the other is heated to 100°C. In each case, bubbles form in the water. Are these bubbles of the same origin? Explain.

**13.31** A man bought a goldfish in a pet shop. Upon returning home, he put the goldfish in a bowl of recently boiled water that had been cooled quickly. A few minutes later the fish was found dead. Explain what happened to the fish.

### Problems

**13.32** A 3.20-g sample of a salt dissolves in 9.10 g of water to give a saturated solution at 25°C. What is the solubility (in g salt/100 g of H$_2$O) of the salt?

**13.33** The solubility of KNO$_3$ is 155 g per 100 g of water at 75°C and 38.0 g at 25°C. What mass (in grams) of KNO$_3$ will crystallize out of solution if exactly 100 g of its saturated solution at 75°C is cooled to 25°C?

**13.34** A 50-g sample of impure KClO$_3$ (solubility = 7.1 g per 100 g H$_2$O at 20°C) is contaminated with 10 percent of KCl (solubility = 25.5 g per 100 g of H$_2$O at 20°C). Calculate the minimum quantity of 20°C water needed to dissolve all the KCl from the sample. How much KClO$_3$ will be left after this treatment? (Assume that the solubilities are unaffected by the presence of the other compound.)

**13.35** A miner working 260 m below sea level opened a carbonated soft drink during a lunch break. To his surprise, the soft drink tasted rather "flat." Shortly afterward, the miner took an elevator to the surface. During the trip up, he could not stop belching. Why?

**13.36** A beaker of water is initially saturated with dissolved air. Explain what happens when He gas at 1 atm is bubbled through the solution for a long time.

**13.37** The solubility of CO$_2$ in water at 25°C and 1 atm is 0.034 mol/L. What is its solubility under atmospheric conditions? (The partial pressure of CO$_2$ in air is 0.0003 atm.) Assume that CO$_2$ obeys Henry's law.

**13.38** The solubility of N$_2$ in blood at 37°C and at a partial pressure of 0.80 atm is $5.6 \times 10^{-4}$ mol/L. A deep-sea diver breathes compressed air with the partial pressure of N$_2$ equal to 4.0 atm. Assume that the total volume of blood in the body is 5.0 L. Calculate the amount of N$_2$ gas released (in liters at 37°C and 1 atm) when the diver returns to the surface of the water, where the partial pressure of N$_2$ is 0.80 atm.

**13.39** A student carried out the following experiment to measure the pressure of carbon dioxide in the space above the carbonated soft drink in a bottle. First, she weighed the bottle (853.5 g). Next, she carefully removed the cap to let the CO$_2$ gas escape. She then reweighed the bottle with the cap (851.3 g). Finally, she measured the volume of the soft drink (452.4 mL). Given that the Henry's law constant for CO$_2$ in water at 25°C is $3.4 \times 10^{-2}$ mol/L · atm, calculate the pressure of CO$_2$ over the soft drink in the bottle before it was opened. Explain why this pressure is only an estimate of the true value.

## Section 13.5: Colligative Properties

### Review Questions

**13.40** What are colligative properties? What is the meaning of the word *colligative* in this context?

**13.41** Give two examples of (a) a volatile liquid and (b) a nonvolatile liquid.

**13.42** Write the equation representing Raoult's law, and express it in words.

**13.43** Use a solution of benzene in toluene to explain what is meant by an ideal solution.

**13.44** Write the equations relating boiling-point elevation and freezing-point depression to the concentration of the solution. Define all the terms, and give their units.

**13.45** How is vapor-pressure lowering related to a rise in the boiling point of a solution?

**13.46** Use a phase diagram to show the difference in freezing points and boiling points between an aqueous urea solution and pure water.

**13.47** What is osmosis? What is a semipermeable membrane?

**13.48** Write the equation relating osmotic pressure to the concentration of a solution. Define all the terms, and specify their units.

13.49 What does it mean when we say that the osmotic pressure of a sample of seawater is 25 atm at a certain temperature?

13.50 Explain why molality is used for boiling-point elevation and freezing-point depression calculations and molarity is used in osmotic pressure calculations.

13.51 Why is the discussion of the colligative properties of electrolyte solutions more involved than that of nonelectrolyte solutions?

13.52 What are ion pairs? What effect does ion-pair formation have on the colligative properties of a solution? How does the ease of ion-pair formation depend on (a) charges on the ions, (b) size of the ions, (c) nature of the solvent (polar versus nonpolar), (d) concentration?

13.53 Indicate which compound in each of the following pairs is more likely to form ion pairs in water: (a) NaCl or $Na_2SO_4$, (b) $MgCl_2$ or $MgSO_4$, (c) LiBr or KBr.

13.54 What is the van't Hoff factor? What information does it provide?

13.55 For most intravenous injections, great care is taken to ensure that the concentration of solutions to be injected is comparable to that of blood plasma. Explain.

### Problems

13.56 A solution is prepared by dissolving 396 g of sucrose ($C_{12}H_{22}O_{11}$) in 624 g of water. What is the vapor pressure of this solution at 30°C? (The vapor pressure of water is 31.8 mmHg at 30°C.)

13.57 How many grams of sucrose ($C_{12}H_{22}O_{11}$) must be added to 552 g of water to give a solution with a vapor pressure 2.0 mmHg less than that of pure water at 20°C? (The vapor pressure of water at 20°C is 17.5 mmHg.)

13.58 The vapor pressure of benzene is 100.0 mmHg at 26.1°C. Calculate the vapor pressure of a solution containing 24.6 g of camphor ($C_{10}H_{16}O$) dissolved in 98.5 g of benzene. (Camphor is a low-volatility solid.)

13.59 The vapor pressures of ethanol ($C_2H_5OH$) and 1-propanol ($C_3H_7OH$) at 35°C are 100 and 37.6 mmHg, respectively. Assume ideal behavior and calculate the partial pressures of ethanol and 1-propanol at 35°C over a solution of ethanol in 1-propanol, in which the mole fraction of ethanol is 0.300.

13.60 The vapor pressure of ethanol ($C_2H_5OH$) at 20°C is 44 mmHg, and the vapor pressure of methanol ($CH_3OH$) at the same temperature is 94 mmHg. A mixture of 30.0 g of methanol and 45.0 g of ethanol is prepared (and can be assumed to behave as an ideal solution). (a) Calculate the vapor pressure of methanol and ethanol above this solution at 20°C. (b) Calculate the mole fraction of methanol and ethanol in the vapor above this solution at 20°C. (c) Suggest a method for separating the two components of the solution.

13.61 How many grams of urea [$(NH_2)_2CO$] must be added to 658 g of water to give a solution with a vapor pressure 2.50 mmHg lower than that of pure water at 30°C? (The vapor pressure of water at 30°C is 31.8 mmHg.)

13.62 What are the boiling point and freezing point of a 3.12 $m$ solution of naphthalene in benzene? (The boiling point and freezing point of benzene are 80.1°C and 5.5°C, respectively.)

13.63 An aqueous solution contains the amino acid glycine ($NH_2CH_2COOH$). Assuming that the acid does not ionize in water, calculate the molality of the solution if it freezes at −1.1°C.

13.64 How many liters of the antifreeze ethylene glycol [$CH_2(OH)CH_2(OH)$] would you add to a car radiator containing 6.50 L of water if the coldest winter temperature in your area is −20°C? Calculate the boiling point of this water-ethylene glycol mixture. (The density of ethylene glycol is 1.11 g/mL.)

13.65 A solution is prepared by condensing 4.00 L of a gas, measured at 27°C and 748 mmHg pressure, into 75.0 g of benzene. Calculate the freezing point of this solution.

13.66 What is the osmotic pressure (in atm) of a 1.57 $M$ aqueous solution of urea [$(NH_2)_2CO$] at 27.0°C?

13.67 Which of the following aqueous solutions has (a) the higher boiling point, (b) the higher freezing point, and (c) the lower vapor pressure: 0.35 $m$ $CaCl_2$ or 0.90 $m$ urea? Explain. Assume complete dissociation.

13.68 Consider two aqueous solutions, one of sucrose ($C_{12}H_{22}O_{11}$) and the other of nitric acid ($HNO_3$). Both solutions freeze at −1.5°C. What other properties do these solutions have in common?

13.69 Arrange the following solutions in order of decreasing freezing point: 0.10 $m$ $Na_3PO_4$, 0.35 $m$ NaCl, 0.20 $m$ $MgCl_2$, 0.15 $m$ $C_6H_{12}O_6$, 0.15 $m$ $CH_3COOH$.

13.70 Arrange the following aqueous solutions in order of decreasing freezing point, and explain your reasoning: 0.50 $m$ HCl, 0.50 $m$ glucose, 0.50 $m$ acetic acid.

13.71 What are the normal freezing points and boiling points of the following solutions: (a) 21.2 g NaCl in 135 mL of water and (b) 15.4 g of urea in 66.7 mL of water?

13.72 At 25°C, the vapor pressure of pure water is 23.76 mmHg and that of seawater is 22.98 mmHg. Assuming that seawater contains only NaCl, estimate its molal concentration.

13.73 Both NaCl and $CaCl_2$ are used to melt ice on roads and sidewalks in winter. What advantages do these substances have over sucrose or urea in lowering the freezing point of water?

13.74 A 0.86 percent by mass solution of NaCl is called "physiological saline" because its osmotic pressure is equal to that of the solution in blood cells. Calculate the osmotic pressure of this solution at normal body temperature (37°C). Note that the density of the saline solution is 1.005 g/mL.

13.75 The osmotic pressure of 0.010 $M$ solutions of $CaCl_2$ and urea at 25°C are 0.605 and 0.245 atm, respectively. Calculate the van't Hoff factor for the $CaCl_2$ solution.

13.76 Calculate the osmotic pressure of a 0.0500 $M$ $MgSO_4$ solution at 25°C. (Hint: See Table 13.3.)

13.77 The tallest trees known are the redwoods in California. Assuming the height of a redwood to be 105 m (about 350 ft), estimate the osmotic pressure required to push water up to the tree top.

13.78 Calculate the difference in osmotic pressure (in atm) at the normal body temperature between the blood plasma of a diabetic patient and that of a healthy adult. Assume that the sole difference between the two people is due to the higher glucose level in the diabetic patient. The glucose levels are 1.75 and 0.84 g/L, respectively. Based on your result, explain why such a patient frequently feels thirsty.

### Section 13.6: Calculations Using Colligative Properties

#### Review Questions

13.79 Describe how you would use freezing-point depression and osmotic pressure measurements to determine the molar mass of a compound. Why are boiling-point elevation and vapor-pressure lowering normally not used for this purpose?

**13.80**  Describe how you would use the osmotic pressure to determine the percent ionization of a weak, monoprotic acid.

## Problems

**13.81**  Pheromones are compounds secreted by the females of many insect species to attract males. One of these compounds contains 80.78 percent C, 13.56 percent H, and 5.66 percent O. A solution of 1.00 g of this pheromone in 8.50 g of benzene freezes at 3.37°C. What are the molecular formula and molar mass of the compound? (The normal freezing point of pure benzene is 5.50°C.)

**13.82**  The elemental analysis of an organic solid extracted from gum arabic (a gummy substance used in adhesives, inks, and pharmaceuticals) showed that it contained 40.0 percent C, 6.7 percent H, and 53.3 percent O. A solution of 0.650 g of the solid in 27.8 g of the solvent diphenyl gave a freezing-point depression of 1.56°C. Calculate the molar mass and molecular formula of the solid. ($K_f$ for diphenyl is 8.00°C/$m$.)

**13.83**  A solution of 2.50 g of a compound having the empirical formula $C_6H_5P$ in 25.0 g of benzene is observed to freeze at 4.3°C. Calculate the molar mass of the solute and its molecular formula.

**13.84**  The molar mass of benzoic acid ($C_6H_5COOH$) determined by measuring the freezing-point depression in benzene is twice what we would expect for the molecular formula, $C_7H_6O_2$. Explain this apparent anomaly.

**13.85**  A solution containing 0.8330 g of a polymer of unknown structure in 170.0 mL of an organic solvent was found to have an osmotic pressure of 5.20 mmHg at 25°C. Determine the molar mass of the polymer.

**13.86**  A quantity of 7.480 g of an organic compound is dissolved in water to make 300.0 mL of solution. The solution has an osmotic pressure of 1.43 atm at 27°C. The analysis of this compound shows that it contains 41.8 percent C, 4.7 percent H, 37.3 percent O, and 16.3 percent N. Calculate the molecular formula of the compound.

**13.87**  A solution of 6.85 g of a carbohydrate in 100.0 g of water has a density of 1.024 g/mL and an osmotic pressure of 4.61 atm at 20.0°C. Calculate the molar mass of the carbohydrate.

**13.88**  A 0.036 $M$ aqueous nitrous acid ($HNO_2$) solution has an osmotic pressure of 0.93 atm at 25°C. Calculate the percent ionization of the acid.

**13.89**  A 0.100 $M$ aqueous solution of the base HB has an osmotic pressure of 2.83 atm at 25°C. Calculate the percent ionization of the base.

## Section 13.7: Colloids

### Review Questions

**13.90**  What are colloids? Referring to Table 13.5, why is there no colloid in which both the dispersed phase and the dispersing medium are gases?

**13.91**  Describe how hydrophilic and hydrophobic colloids are stabilized in water.

**13.92**  Describe and give an everyday example of the Tyndall effect.

## Additional Problems

**13.93**  Lysozyme is an enzyme that cleaves bacterial cell walls. A sample of lysozyme extracted from egg white has a molar mass of 13,930 g. A quantity of 0.100 g of this enzyme is dissolved in

150 g of water at 25°C. Calculate the vapor-pressure lowering, the depression in freezing point, the elevation in boiling point, and the osmotic pressure of this solution. (The vapor pressure of water at 25°C is 23.76 mmHg.)

**13.94**  Solutions A and B have osmotic pressures of 2.4 and 4.6 atm, respectively, at a certain temperature. What is the osmotic pressure of a solution prepared by mixing equal volumes of A and B at the same temperature?

**13.95**  The blood sugar (glucose) level of a diabetic patient is approximately 0.140 g of glucose/100 mL of blood. Every time the patient ingests 40 g of glucose, her blood glucose level rises to approximately 0.240 g/100 mL of blood. Calculate the number of moles of glucose per milliliter of blood and the total number of moles and grams of glucose in the blood before and after consumption of glucose. (Assume that the total volume of blood in her body is 5.0 L.)

**13.96**  Trees in cold climates may be subjected to temperatures as low as −60°C. Estimate the concentration of an aqueous solution in the body of the tree that would remain unfrozen at this temperature. Is this a reasonable concentration? Comment on your result.

**13.97**  A cucumber placed in concentrated brine (saltwater) shrivels into a pickle. Explain.

**13.98**  Two liquids A and B have vapor pressures of 76 and 132 mmHg, respectively, at 25°C. What is the total vapor pressure of the ideal solution made up of (a) 1.00 mole of A and 1.00 mole of B and (b) 2.00 moles of A and 5.00 moles of B?

**13.99**  Determine the van't Hoff factor of $Na_3PO_4$ in a 0.40 $m$ solution whose freezing point is −2.6°C.

**13.100**  A 262-mL sample of a sugar solution containing 1.22 g of the sugar has an osmotic pressure of 30.3 mmHg at 35°C. What is the molar mass of the sugar?

**13.101**  Consider the three mercury manometers shown in the diagram. One of them has 1 mL of water on top of the mercury, another has 1 mL of a 1 $m$ urea solution on top of the mercury, and the third one has 1 mL of a 1 $m$ NaCl solution placed on top of the mercury. Which of these solutions is in the tube labeled X, which is in Y, and which is in Z?

**13.102**  A forensic chemist is given a white powder for analysis. She dissolves 0.50 g of the substance in 8.0 g of benzene. The solution freezes at 3.9°C. Can the chemist conclude that the compound is cocaine ($C_{17}H_{21}NO_4$)? What assumptions are made in the analysis?

**13.103**  "Time-release" drugs have the advantage of releasing the drug to the body at a constant rate so that the drug concentration at any time is not too high as to have harmful side effects or too low as to be ineffective. A schematic diagram of a pill that works on this basis is shown. Explain how it works.

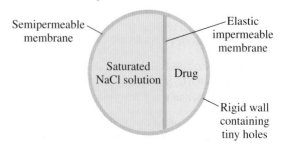

**13.104** A solution of 1.00 g of anhydrous aluminum chloride ($AlCl_3$) in 50.0 g of water freezes at $-1.11°C$. Does the molar mass determined from this freezing point agree with that calculated from the formula? Why?

**13.105** Explain why reverse osmosis is (theoretically) more desirable as a desalination method than distillation or freezing. What minimum pressure must be applied to seawater at 25°C for reverse osmosis to occur? (Treat seawater as a 0.70 $M$ NaCl solution.)

**13.106** What masses of sodium chloride, magnesium chloride, sodium sulfate, calcium chloride, potassium chloride, and sodium bicarbonate are needed to produce 1 L of artificial seawater for an aquarium? The required ionic concentrations are $[Na+]$ = 2.56 $M$, $[K+]$ = 0.0090 $M$, $[Mg^{2+}]$ = 0.054 $M$, $[Ca^{2+}]$ = 0.010 $M$, $[HCO_3^-]$ = 0.0020 $M$, $[Cl^-]$ = 2.60 $M$, $[SO_4^{2-}]$ = 0.051 $M$.

**13.107** The osmotic pressure of blood plasma is approximately 7.5 atm at 37°C. Estimate the total concentration of dissolved species and the freezing point of blood.

**13.108** The antibiotic gramicidin A can transport $Na^+$ ions into a certain cell at the rate of $5.0 \times 10^7$ $Na^+$ ions/channel · s. Calculate the time in seconds to transport enough $Na^+$ ions to increase its concentration by $8.0 \times 10^{-3}$ $M$ in a cell whose intracellular volume is $2.0 \times 10^{-10}$ mL.

**13.109** A protein has been isolated as a salt with the formula $Na_{20}P$ (this notation means that there are 20 $Na^+$ ions associated with a negatively charged protein $P^{20-}$). The osmotic pressure of a 10.0-mL solution containing 0.225 g of the protein is 0.257 atm at 25.0°C. (a) Calculate the molar mass of the protein from these data. (b) Calculate the actual molar mass of the protein.

**13.110** A nonvolatile organic compound Z was used to make up two solutions. Solution A contains 5.00 g of Z dissolved in 100 g of water, and solution B contains 2.31 g of Z dissolved in 100 g of benzene. Solution A has a vapor pressure of 754.5 mmHg at the normal boiling point of water, and solution B has the same vapor pressure at the normal boiling point of benzene. Calculate the molar mass of Z in solutions A and B, and account for the difference.

**13.111** Hydrogen peroxide with a concentration of 3.0% (3.0 g of $H_2O_2$ in 100 mL of solution) is sold in drugstores for use as an antiseptic. For a 10.0-mL 3.0% $H_2O_2$ solution, calculate (a) the oxygen gas produced (in liters) at STP when the compound undergoes complete decomposition and (b) the ratio of the volume of $O_2$ collected to the initial volume of the $H_2O_2$ solution.

**13.112** State which of the alcohols listed in Problem 13.12 you would expect to be the best solvent for each of the following substances, and explain why: (a) $I_2$, (b) KBr, (c) $CH_3CH_2CH_2CH_2CH_3$.

**13.113** Before a carbonated beverage bottle is sealed, it is pressurized with a mixture of air and carbon dioxide. (a) Explain the effervescence that occurs when the cap of the bottle is removed. (b) What causes the fog to form near the mouth of the bottle right after the cap is removed?

**13.114** Iodine ($I_2$) is only sparingly soluble in water (above left photo). Yet upon the addition of iodide ions (for example, from KI), iodine is converted to the triiodide ion, which readily dissolves (above right photo):

$$I_2(s) + I^-(aq) \rightleftharpoons I_3^-(aq)$$

Describe the change in solubility of $I_2$ in terms of the change in intermolecular forces.

**13.115** (a) The root cells of plants contain a solution that is hypertonic in relation to water in the soil. Thus, water can move into the roots by osmosis. Explain why salts such as NaCl and $CaCl_2$ spread on roads to melt ice can be harmful to nearby trees. (b) Just before urine leaves the human body, the collecting ducts in the kidney (which contain the urine) pass through a fluid whose salt concentration is considerably greater than is found in the blood and tissues. Explain how this action helps conserve water in the body.

**13.116** Hemoglobin, the oxygen-transport protein, binds about 1.35 mL of oxygen per gram of the protein. The concentration of hemoglobin in normal blood is 150 g/L blood. Hemoglobin is about 95 percent saturated with $O_2$ in the lungs and only 74 percent saturated with $O_2$ in the capillaries. Calculate the volume of $O_2$ released by hemoglobin when 100 mL of blood flows from the lungs to the capillaries.

**13.117** Two beakers, one containing a 50-mL aqueous 1.0 $M$ glucose solution and the other a 50-mL aqueous 2.0 $M$ glucose solution, are placed under a tightly sealed bell jar at room temperature. What are the volumes in these two beakers at equilibrium?

**13.118** In the apparatus shown, what will happen if the membrane is (a) permeable to both water and the $Na^+$ and $Cl^-$ ions, (b) permeable to water and the $Na^+$ ions but not to the $Cl^-$ ions, (c) permeable to water but not to the $Na^+$ and $Cl^-$ ions?

Membrane

0.01 $M$ NaCl        0.1 $M$ NaCl

**13.119** Concentrated hydrochloric acid is usually available at a concentration of 37.7 percent by mass. What is its molar concentration? (The density of the solution is 1.19 g/mL.)

**13.120** Explain each of the following statements: (a) The boiling point of seawater is higher than that of pure water. (b) Carbon dioxide escapes from the solution when the cap is removed from a carbonated soft drink bottle. (c) Molal and molar concentrations of dilute aqueous solutions are approximately equal. (d) In discussing the colligative properties of a solution (other than osmotic pressure), it is preferable to express the concentration in units of molality rather than in molarity. (e) Methanol (b.p. 65°C) is useful as an antifreeze, but it should be removed from the car radiator during the summer season.

**13.121** A mixture of NaCl and sucrose ($C_{12}H_{22}O_{12}$) of combined mass 10.2 g is dissolved in enough water to make up a 250-mL solution. The osmotic pressure of the solution is 7.32 atm at 23°C. Calculate the mass percent of NaCl in the mixture.

**13.122** A 1.32-g sample of a mixture of cyclohexane ($C_6H_{12}$) and naphthalene ($Cl_{10}H_8$) is dissolved in 18.9 g of benzene ($C_6H_6$). The freezing point of the solution is 2.2°C. Calculate the mass percent of the mixture. (See Table 13.2 for constants.)

**13.123** How does each of the following affect the solubility of an ionic compound: (a) lattice energy, (b) solvent (polar versus nonpolar), (c) enthalpies of hydration of cation and anion?

**13.124** A solution contains two volatile liquids A and B. Complete the following table, in which the symbol ⟷ indicates attractive intermolecular forces.

| Attractive Forces | Deviation from Raoult's Law | $\Delta H_{soln}$ |
|---|---|---|
| A ⟷ A, B ⟷ B > A ⟷ B | | |
| | Negative | |
| | | Zero |

**13.125** The concentration of commercially available concentrated nitric acid is 70.0 percent by mass, or 15.9 *M*. Calculate the density and the molality of the solution.

**13.126** A mixture of ethanol and 1-propanol behaves ideally at 36°C and is in equilibrium with its vapor. If the mole fraction of ethanol in the solution is 0.62, calculate its mole fraction in the vapor phase at this temperature. (The vapor pressures of pure ethanol and 1-propanol at 36°C are 108 and 40.0 mmHg, respectively.)

**13.127** Ammonia ($NH_3$) is very soluble in water, but nitrogen trichloride ($NCl_3$) is not. Explain.

**13.128** For ideal solutions, the volumes are additive. This means that if 5 mL of A and 5 mL of B form an ideal solution, the volume of the solution is 10 mL. Provide a molecular interpretation for this observation. When 500 mL of ethanol ($C_2H_5OH$) is mixed with 500 mL of water, the final volume is less than 1000 mL. Why?

**13.129** Acetic acid is a weak acid that ionizes in solution as follows:

$$CH_3COOH(aq) \rightleftharpoons CH_3COO^-(aq) + H^+(aq)$$

If the freezing point of a 0.106 *m* $CH_3COOH$ solution is −0.203°C, calculate the percent of the acid that has undergone ionization.

**13.130** Aluminum sulfate [$Al_2(SO_4)_3$] is sometimes used in municipal water treatment plants to remove undesirable particles. Explain how this process works.

**13.131** Making mayonnaise involves beating oil into small droplets in water, in the presence of egg yolk. What is the purpose of the egg yolk? (*Hint:* Egg yolk contains lecithins, which are molecules with a polar head and a long nonpolar hydrocarbon tail.)

**13.132** Acetic acid is a polar molecule and can form hydrogen bonds with water molecules. Therefore, it has a high solubility in water. Yet acetic acid is also soluble in benzene ($C_6H_6$), a nonpolar solvent that lacks the ability to form hydrogen bonds. A solution of 3.8 g of $CH_3COOH$ in 80 g $C_6H_6$ has a freezing point of 3.5°C. Calculate the molar mass of the solute, and suggest what its structure might be. (*Hint:* Acetic acid molecules can form hydrogen bonds between themselves.)

**13.133** A 2.6-L sample of water contains 192 μg of lead. Does this concentration of lead exceed the safety limit of 0.050 ppm of lead per liter of drinking water?

**13.134** Fish in the Antarctic Ocean swim in water at about −2°C. (a) To prevent their blood from freezing, what must be the concentration (in molality) of the blood? Is this a reasonable physiological concentration? (b) In recent years, scientists have discovered a special type of protein in the blood of these fish

that, although present in quite low concentrations (≤ 0.001 *m*), has the ability to prevent the blood from freezing. Suggest a mechanism for its action.

**13.135** Why are ice cubes (for example, those you see in the trays in the freezer of a refrigerator) cloudy inside?

**13.136** If a soft drink can is shaken and then opened, the drink escapes violently. However, if after shaking the can, we tap it several times with a metal spoon, no such "explosion" of the drink occurs. Why?

**13.137** Two beakers are placed in a closed container. Beaker A initially contains 0.15 mole of naphthalene ($C_{10}H_8$) in 100 g of benzene ($C_6H_6$), and beaker B initially contains 31 g of an unknown compound dissolved in 100 g of benzene. At equilibrium, beaker A is found to have lost 7.0 g of benzene. Assuming ideal behavior, calculate the molar mass of the unknown compound. State any assumptions made.

**13.138** (a) Derive the equation relating the molality (*m*) of a solution to its molarity (*M*)

$$m = \frac{M}{d - \frac{M\mathcal{M}}{1000}}$$

where *d* is the density of the solution (g/mL) and $\mathcal{M}$ is the molar mass of the solute (g/mol). (*Hint:* Start by expressing the solvent in kilograms in terms of the difference between the mass of the solution and the mass of the solute.) (b) Show that, for dilute aqueous solutions, *m* is approximately equal to *M*.

**13.139** At 27°C, the vapor pressure of pure water is 23.76 mmHg and that of an aqueous solution of urea is 22.98 mmHg. Calculate the molality of urea in the solution.

**13.140** A very long pipe is capped at one end with a semipermeable membrane. How deep (in meters) must the pipe be immersed into the sea for fresh water to begin to pass through the membrane? Assume the water to be at 20°C, and treat it as a 0.70 *M* NaCl solution. The density of seawater is 1.03 g/cm³, and the acceleration due to gravity is 9.81 m/s².

**13.141** A mixture of liquids A and B exhibits ideal behavior. At 84°C, the total vapor pressure of a solution containing 1.2 moles of A and 2.3 moles of B is 331 mmHg. Upon the addition of another mole of B to the solution, the vapor pressure increases to 347 mmHg. Calculate the vapor pressure of pure A and B at 84°C.

**13.142** Use Henry's law and the ideal gas equation to prove the statement that the volume of a gas that dissolves in a given amount of solvent is *independent* of the pressure of the gas. (*Hint:* Henry's law can be modified as *n* = *kP*, where *n* is the number of moles of the gas dissolved in the solvent.)

**13.143** At 298 K, the osmotic pressure of a glucose solution is 10.50 atm. Calculate the freezing point of the solution. The density of the solution is 1.16 g/mL.

**13.144** Valinomycin is an antibiotic. It functions by binding $K^+$ ions and transporting them across the membrane into cells to offset the ionic balance. The molecule is represented here by its skeletal structure in which the end of each straight line corresponds to a carbon atom (unless a different atom is shown at the end of the line). There are as many H atoms attached to each C atom as necessary to give each C atom a total of four bonds. Using the "like dissolves like" principle, explain how the molecule functions. (*Hint:* The $-CH_3$ groups at the two ends of each Y shape are nonpolar.)

**13.145** The diagram here shows vapor pressure curves for pure benzene and a solution of a nonvolatile solute in benzene. Estimate the molality of the solution.

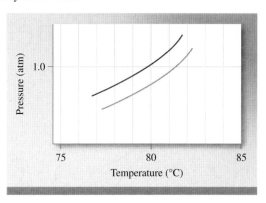

# PRE-PROFESSIONAL PRACTICE EXAM PROBLEMS: PHYSICAL AND BIOLOGICAL SCIENCES

A mixture of two volatile liquids is said to be ideal if each component obeys Raoult's law:

$$P_i = \chi_i P_i^\circ$$

Two volatile liquids A (molar mass 100 g/mol) and B (molar mass 110 g/mol) form an ideal solution. At 55°C, A has a vapor pressure of 98 mmHg and B has a vapor pressure of 42 mmHg. A solution is prepared by mixing equal masses of A and B.

1. Calculate the mole fraction of each component in the solution.

   a) $\chi_A = 0.50$, $\chi_B = 0.50$
   b) $\chi_A = 0.52$, $\chi_B = 0.48$
   c) $\chi_A = 1.0$, $\chi_B = 0.91$
   d) $\chi_A = 0.09$, $\chi_B = 0.91$

2. Calculate the partial pressures of A and B over the solution at 55°C.

   a) $P_A = 98$ mmHg, $P_B = 42$ mmHg
   b) $P_A = 49$ mmHg, $P_B = 21$ mmHg
   c) $P_A = 70$ mmHg, $P_B = 70$ mmHg
   d) $P_A = 51$ mmHg, $P_B = 20$ mmHg

3. Suppose that some of the vapor over the solution at 55°C is condensed to a liquid. Calculate the mole fraction of each component in the condensed liquid.

   a) $\chi_A = 0.50$, $\chi_B = 0.50$
   b) $\chi_A = 0.52$, $\chi_B = 0.48$
   c) $\chi_A = 0.72$, $\chi_B = 0.28$
   d) $\chi_A = 1.0$, $\chi_B = 0.86$

4. Calculate the partial pressures of the components above the condensed liquid at 55°C.

   a) $P_A = 98$ mmHg, $P_B = 42$ mmHg
   b) $P_A = 30$ mmHg, $P_B = 27$ mmHg
   c) $P_A = 71$ mmHg, $P_B = 12$ mmHg
   d) $P_A = 72$ mmHg, $P_B = 28$ mmHg

# ANSWERS TO IN-CHAPTER MATERIALS

## Answers to Practice Problems

**13.1A** $CS_2$. **13.1B** $C_3H_8$, $I_2$, $CS_2$. **13.2A** Water soluble. **13.2B** Fat soluble. **13.3A** (a) 0.423 *m*, (b) 2.48%. **13.3B** 0.78 *m*. **13.4A** (a) 1.8 *M*, (b) 1.9 *m*. **13.4B** 12.8%. **13.5A** 0.12 *M*. **13.5B** 26 atm. **13.6A** 22.2 mmHg. **13.6B** 103 g. **13.7A** f.p. $= -7.91$°C, b.p. $= 102.2$°C. **13.7B** 710 g. **13.8A** $i = 1.21$. **13.8B** f.p. $= -0.35$°C. **13.9A** 10.2 atm. **13.9B** (a) 0.22 M, (b) 0.11 *M*. **13.10A** 128 g/mol. **13.10B** 12.5 g. **13.11A** $6.0 \times 10^3$ g/mol. **13.11B** 0.27 g. **13.12A** 4%. **13.12B** 0.38 atm.

## Answers to Checkpoints

**13.2.1** b, d, e. **13.2.2** c. **13.2.3** b, d. **13.2.4** a. **13.3.1** b, c, d. **13.3.2** d. **13.3.3** d. **13.3.4** a. **13.4.1** c. **13.4.2** a. **13.5.1** d. **13.5.2** c. **13.5.3** b. **13.5.4** a. **13.6.1** b. **13.6.2** b.

## Answers to Applying What You've Learned

**a)** $(1.0 \times 10^{-4})$% $F^-$, $2 \times 10^{-5}$ *m* NaF. **b)** 101.0°C, $-3.6$°C. **c)** 5.22. **d)** 144 g/mol ($H_2SiF_6$). **e)** 6%.

# 14

# Chemical Kinetics

*Methanol is an ingredient in gas-line antifreeze. When ingested, it is converted to formic acid, the substance responsible for methanol's toxic and potentially deadly effects.*

# How Relative Rates of Chemical Reactions Have Been Used to Save Lives

The medical community sometimes uses the relative speeds at which reactions occur to benefit a patient who has ingested a poison, such as methanol. Accidental or intentional ingestion of methanol can cause headache, nausea, blindness, seizures, and even death. In the liver, methanol is metabolized by the enzyme alcohol dehydrogenase (ADH) to yield formaldehyde:

$$\underset{\text{methanol}}{CH_3OH} + NAD^+ \longrightarrow \underset{\text{formaldehyde}}{HCHO} + NADH + H^+$$

**Student Annotation:** NAD$^+$ and NADH are, respectively, the oxidized and reduced forms of nicotinamide adenine dinucleotide, an electron carrier in metabolic oxidation.

The formaldehyde is subsequently converted to formic acid by another enzyme, aldehyde dehydrogenase (ALDH).

$$\underset{\text{formaldehyde}}{HCHO} + NADP^+ + H_2O \longrightarrow \underset{\text{formic acid}}{HCOOH} + NADPH + H^+$$

**Student Annotation:** NADP$^+$ and NADPH are the oxidized and reduced forms of nicotinamide adenine dinucleotide with an added phosphate group.

Formaldehyde and formic acid are the species responsible for the toxic effects of methanol poisoning—including *metabolic acidosis,* in which the blood becomes dangerously acidic.

To treat methanol poisoning, we must prevent ADH from metabolizing any more methanol. Historically, this has been done by administering large quantities of *ethanol,* which has an affinity for ADH of roughly 100 times that of methanol, making the reaction of ADH with *ethanol* significantly faster than the reaction with *methanol*. In this treatment, while the body's supply of ADH is busy converting ethanol to acetaldehyde (a reaction analogous to the conversion of methanol to formaldehyde), the methanol is removed from the body by dialysis. This treatment is not without complications, however, as the quantity of ethanol required typically causes significant inebriation and central nervous system (CNS) depression. The metabolic products of ethanol (acetaldehyde and acetic acid) are also toxic—although generally to a lesser degree than those of methanol. In fact, acetaldehyde is the species responsible for the collection of symptoms known as a hangover.

In 2000, the FDA approved the drug fomepizole, marketed under the name Antizol, for the treatment of methanol poisoning. Fomepizole ($C_4H_6N_2$) has an affinity for ADH approximately 8000 times that of methanol and is used to treat methanol toxicity without the intoxication and CNS depression caused by ethanol. Dialysis is still used to remove methanol from the blood.

Understanding *chemical kinetics* can make it possible to minimize the damage done by undesirable reactions. It also enables us to enhance the speed of desirable reactions.

At the end of this chapter, you will be able to solve a series of problems related to the kinetics of methanol production [▶▶ Page 616].

## In This Chapter, You Will Learn:

How rates of chemical reaction are determined and expressed and about the factors that influence the rates of reactions.

### Before you begin, you should review

- The value of *R* expressed in J/mol [◀◀ Section 11.3, Table 11.4]
- Use of logarithms [▶▶ Appendix 1]

## 14.1   Reaction Rates

Chemical kinetics is the study of how fast reactions take place. Many familiar reactions, such as the initial steps in vision and photosynthesis happen almost instantaneously, whereas others, such as the rusting of iron or the conversion of diamond to graphite, take place on a timescale of days or even millions of years.

Knowledge of kinetics is important to many scientific endeavors, including drug design, pollution control, and food processing. The job of an industrial chemist often is to work on increasing the *rate* of a reaction rather than maximizing its yield or developing a new process.

A chemical reaction can be represented by the general equation

$$\text{reactants} \longrightarrow \text{products}$$

This equation tells us that during the course of a reaction, reactants are consumed while products are formed. We can follow the progress of a reaction by monitoring either the decrease in concentration of the reactants or the increase in the concentrations of the products. The method used to monitor changes in reactant or product concentrations depends on the specific reaction. In a reaction that either consumes or produces a colored species, we can measure the intensity of the color over time with a spectrometer. In a reaction that either consumes or produces a gas, we can measure the change in pressure over time with a manometer. Electrical conductance measurement can be used to monitor the progress if ionic species are consumed or produced.

### Average Reaction Rate

Consider the hypothetical reaction represented by

$$A \longrightarrow B$$

in which A molecules are converted to B molecules. Figure 14.1 shows the progress of this reaction as a function of time.

The decrease in the number of A molecules and the increase in the number of B molecules with time are also shown in Figure 14.2. It is generally convenient to express the rate in terms of the change in concentration with time. Thus, for the reaction A $\longrightarrow$ B, we can express the rate as

$$\text{rate} = -\frac{\Delta[A]}{\Delta t} \qquad \text{or} \qquad \text{rate} = \frac{\Delta[B]}{\Delta t}$$

where $\Delta[A]$ and $\Delta[B]$ are the changes in concentration (molarity) over a time period $\Delta t$. The rate expression containing $\Delta[A]$ has a minus sign because the concentration of A decreases during the time interval—that is, $\Delta[A]$ is a negative quantity. The rate expression containing $\Delta[B]$ does not have a minus sign because the concentration of B increases during the time interval. Rate is always a positive quantity, so when it is expressed in terms of the change in a reactant concentration, a minus sign is needed in the rate expression to make the rate positive. When the rate is expressed in terms of the change in a product concentration, no negative sign is needed to make the rate positive because the product concentration increases with time. Rates calculated in this way are average rates over the time period $\Delta t$.

**Figure 14.1**   The progress of the reaction A $\longrightarrow$ B. Initially, only A molecules (grey spheres) are present. As time progresses, there are more and more B molecules (red spheres).

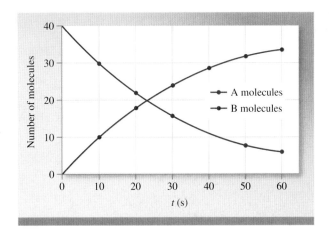

**Figure 14.2** The rate of reaction A ⟶ B represented as the decrease of A molecules with time and as the increase of B molecules with time.

To understand rates of chemical reactions and how they are determined, it is useful to consider some specific reactions. First, we consider the aqueous reaction of molecular bromine ($Br_2$) with formic acid (HCOOH):

$$Br_2(aq) + HCOOH(aq) \longrightarrow 2Br^-(aq) + 2H^+(aq) + CO_2(g)$$

Molecular bromine is reddish-brown, whereas all the other species in the reaction are colorless. As the reaction proceeds, the concentration of bromine decreases and its color fades (Figure 14.3). The decrease in intensity of the color (and, therefore in the concentration of bromine) can be monitored with a spectrometer, which registers the amount of visible light absorbed by bromine (Figure 14.4).

Measuring the bromine concentration at some initial time and then at some final time enables us to determine the average rate of the reaction during that time interval:

$$\text{average rate} = -\frac{\Delta[Br_2]}{\Delta t}$$

$$= -\frac{[Br_2]_{final} - [Br_2]_{initial}}{t_{final} - t_{initial}}$$

**Figure 14.3** From left to right: The decrease in bromine concentration as time elapses is indicated by the loss of color.

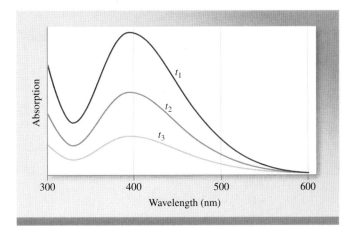

**Figure 14.4** Plot of the absorption of bromine versus wavelength. The maximum absorption of visible light by bromine occurs at 393 nm. As the reaction progresses ($t_1$ to $t_3$), the absorption, which is proportional to [$Br_2$], decreases.

| TABLE 14.1 | Rates of the Reaction of Molecular Bromine and Formic Acid at 25°C | |
|---|---|---|
| Time (s) | [Br$_2$] (M) | Rate (M/s) |
| 0.0 | 0.0120 | $4.20 \times 10^{-5}$ |
| 50.0 | 0.0101 | $3.52 \times 10^{-5}$ |
| 100.0 | 0.00846 | $2.96 \times 10^{-5}$ |
| 150.0 | 0.00710 | $2.49 \times 10^{-5}$ |
| 200.0 | 0.00596 | $2.09 \times 10^{-5}$ |
| 250.0 | 0.00500 | $1.75 \times 10^{-5}$ |
| 300.0 | 0.00420 | $1.48 \times 10^{-5}$ |
| 350.0 | 0.00353 | $1.23 \times 10^{-5}$ |
| 400.0 | 0.00296 | $1.04 \times 10^{-5}$ |

Using data from Table 14.1, we can calculate the average rate over the first 50-s time interval as follows:

$$\text{average rate} = \frac{(0.0101 - 0.0120)\ M}{50.0} = 3.80 \times 10^{-5}\ M/s$$

If we had chosen the first 100 s as our time interval, the average rate would then be given by

$$\text{average rate} = \frac{(0.00846 - 0.0120)\ M}{100.0\ s} = 3.54 \times 10^{-5}\ M/s$$

These calculations demonstrate that the average rate of this reaction depends on the time interval we choose. In other words, the rate changes over time. This is why a plot of the concentration of a reactant or product as a function of time is a curve rather than a straight line [Figure 14.5(a)].

## Instantaneous Rate

If we were to calculate the average rate over shorter and shorter time intervals, we could obtain the *instantaneous rate,* which is the rate for a specific instant in time. Figure 14.6 shows the plot of [Br$_2$] versus time based on the data from Table 14.1. The instantaneous rate is equal to the slope of a tangent to the curve at any particular time. Note that we can pick *any* two points along a tangent to calculate its slope. For a chemist, the instantaneous rate is generally a more useful quantity than the average rate. For the remainder of this chapter, therefore, the term *rate* will be used to mean "instantaneous rate" (unless otherwise stated).

The slope of the tangent, and therefore the reaction *rate,* diminishes with time because the concentration of bromine decreases with time [Figure 14.5(a)]. The data in Table 14.1 show how the rate of this reaction depends on the concentration of bromine. At 50.0 s, for example, when the concentration of bromine is 0.0101 M, the rate is $3.52 \times 10^{-5}$ M/s. When the concentration of

**Figure 14.5**   (a) The plot of [Br$_2$] against time is a curve because the reaction rate changes as [Br$_2$] changes with time. (b) The plot of the reaction rate against [Br$_2$] is a straight line because the rate is proportional to [Br$_2$].

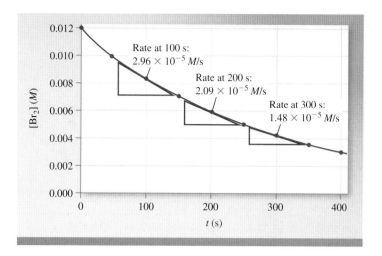

**Figure 14.6** The instantaneous rates of the reaction between molecular bromine and formic acid at $t = 100$, 200, and 300 s are given by the slopes of the tangents at these times.

bromine has been reduced by half (i.e., reduced to 0.00500 $M$), at 250 s, the rate is also reduced by half (i.e., reduced to $1.75 \times 10^{-5}$ $M$/s):

$$\frac{[Br_2]_{50\,s}}{[Br_2]_{250\,s}} \approx 2 \quad \text{and} \quad \frac{\text{rate at 50.0 s}}{\text{rate at 250.0 s}} = \frac{3.52 \times 10^{-5}\,M/s}{1.75 \times 10^{-5}\,M/s} \approx 2$$

Thus, the rate is directly proportional to the concentration of bromine:

$$\text{rate} \propto [Br_2]$$

$$\text{rate} = k[Br_2]$$

where $k$, the proportionality constant, is called the **rate constant.**

Rearranging the preceding equation gives

$$k = \frac{\text{rate}}{[Br_2]}$$

We can use the concentration and rate data from Table 14.1 for any value of $t$ to calculate the value of $k$ for this reaction. For example, using the data for $t = 50.0$ s gives

$$k = \frac{3.52 \times 10^{-5}\,M/s}{0.0101\,M} = 3.49 \times 10^{-3}\,s^{-1} \qquad \text{(at } t = 50.0 \text{ s)}$$

Likewise, using the data for $t = 300.0$ s gives

$$k = \frac{1.48 \times 10^{-5}\,M/s}{0.00420\,M} = 3.52 \times 10^{-3}\,s^{-1} \qquad \text{(at } t = 300 \text{ s)}$$

Slight variations in the calculated values of $k$ are due to experimental deviations in rate measurements. To two significant figures, we get $k = 3.5 \times 10^{-3}\,s^{-1}$ for this reaction, regardless of which line of data we chose from Table 14.1. It is important to note that the value of $k$ does not depend on the concentration of bromine. The rate constant is *constant* at constant temperature. (In Section 14.4 we discuss how $k$ depends on temperature.)

We now consider another specific reaction, the decomposition of hydrogen peroxide:

$$2H_2O_2(aq) \longrightarrow 2H_2O(l) + O_2(g)$$

Because one of the products is a gas, we can monitor the progress of this reaction by measuring the pressure with a manometer. Pressure is converted to concentration using the ideal gas equation:

$$PV = nRT$$

or

$$P_{O_2} = \frac{n}{V}RT = [O_2]\,RT$$

where $n/V$ gives the molarity of the oxygen gas. Rearranging the equation and solving for $[O_2]$, we get

$$[O_2] = \frac{1}{RT}P_{O_2}$$

**Figure 14.7** (a) The rate of hydrogen peroxide decomposition can be measured with a manometer, which (b) shows the increase in the oxygen gas pressure with time.

$$2H_2O_2(aq) \longrightarrow 2H_2O(l) + O_2(g)$$

(a)                (b)

The reaction rate, which is expressed as the rate of oxygen production, can now be written as

**Student Annotation:** Many students prefer simply to convert pressures to concentrations using the ideal gas equation [◄◄ Section 11.3] and then determine the rate using the familiar form of the equation: rate = Δ[O₂]/Δt.

$$\text{rate} = \frac{\Delta[O_2]}{\Delta t} = \frac{1}{RT}\frac{\Delta P_{O_2}}{\Delta t}$$

Determining the rate from pressure data is possible only if the temperature (in kelvins) at which the reaction is carried out is known. Figure 14.7(a) shows the apparatus used to monitor the pressure change in the decomposition of hydrogen peroxide. Figure 14.7(b) shows the plot of oxygen pressure, $P_{O_2}$, versus time. As we did in Figure 14.6, we can draw a tangent to the curve in Figure 14.7(b) to determine the instantaneous rate at any point.

## Stoichiometry and Reaction Rate

For stoichiometrically simple reactions of the type A ⟶ B, the rate can be expressed either in terms of the decrease in reactant concentration with time, $-\Delta[A]/\Delta t$, or in terms of the increase in product concentration with time, $\Delta[B]/\Delta t$—both expressions give the same result. For reactions in which one or more of the stoichiometric coefficients is something other than 1, we must take extra care in expressing the rate. For example, consider again the reaction between molecular bromine and formic acid:

$$Br_2(aq) + HCOOH(aq) \longrightarrow 2Br^-(aq) + 2H^+(aq) + CO_2(g)$$

We have expressed the rate of this reaction in terms of the disappearance of bromine. But what if we chose instead to express the rate in terms of the appearance of bromide ion? According to the balanced equation, 2 moles of $Br^-$ are generated for each mole of $Br_2$ consumed. Thus, $Br^-$ appears at twice the rate that $Br_2$ disappears. To avoid the potential ambiguity of reporting the rate of disappearance or appearance of a specific chemical *species,* we report the ***rate of reaction.*** We determine the rate of reaction such that the result is the same regardless of which species we monitor. For the hypothetical reaction,

$$A \longrightarrow 2B$$

the rate of reaction can be written as either

$$\text{rate} = -\frac{\Delta[A]}{\Delta t} \quad \text{or} \quad \text{rate} = \frac{1}{2}\frac{\Delta[B]}{\Delta t}$$

both of which give the same result. For the bromine and formic acid reaction, we can write the rate of reaction as either

$$\text{rate} = -\frac{\Delta[Br_2]}{\Delta t}$$

as we did earlier, or

$$\text{rate} = \frac{1}{2}\frac{\Delta[Br^-]}{\Delta t}$$

In general, for the reaction

$$aA + bB \longrightarrow cC + dD$$

the rate is given by

$$\text{rate} = -\frac{1}{a}\frac{\Delta[A]}{\Delta t} = -\frac{1}{b}\frac{\Delta[B]}{\Delta t} = \frac{1}{c}\frac{\Delta[C]}{\Delta t} = \frac{1}{d}\frac{\Delta[D]}{\Delta t} \qquad \text{Equation 14.1}$$

Expressing the rate in this fashion ensures that the rate of reaction is the same regardless of which species we measure to monitor the reaction's progress.

Sample Problems 14.1 and 14.2 show how to write expressions for reaction rates and how to take stoichiometry into account in rate expressions.

**Student Annotation:** The rate of change in concentration of each species is divided by the coefficient of that species in the balanced equation.

## Sample Problem 14.1

Write the rate expressions for each of the following reactions:

(a) $I^-(aq) + OCl^-(aq) \longrightarrow Cl^-(aq) + OI^-(aq)$

(b) $2O_3(g) \longrightarrow 3O_2(g)$

(c) $4NH_3(g) + 5O_2(g) \longrightarrow 4NO(g) + 6H_2O(g)$

**Strategy** Use Equation 14.1 to write rate expressions for each of the reactions.

**Setup** For reactions containing gaseous species, progress is generally monitored by measuring pressure. Pressures are converted to molar concentrations using the ideal gas equation, and rate expressions are written in terms of molar concentrations.

**Solution** (a) All the coefficients in this equation are 1. Therefore,

$$\text{rate} = -\frac{\Delta[I^-]}{\Delta t} = -\frac{\Delta[OCl^-]}{\Delta t} = \frac{\Delta[Cl^-]}{\Delta t} = \frac{\Delta[OI^-]}{\Delta t}$$

(b) $\text{rate} = -\dfrac{1}{2}\dfrac{\Delta[O_3]}{\Delta t} = \dfrac{1}{3}\dfrac{\Delta[O_2]}{\Delta t}$

(c) $\text{rate} = -\dfrac{1}{4}\dfrac{\Delta[NH_3]}{\Delta t} = -\dfrac{1}{5}\dfrac{\Delta[O_2]}{\Delta t} = \dfrac{1}{4}\dfrac{\Delta[NO]}{\Delta t} = \dfrac{1}{6}\dfrac{\Delta[H_2O]}{\Delta t}$

**Think About It** Make sure that the change in concentration of each species is divided by the corresponding coefficient in the balanced equation. Also make sure that the rate expressions written in terms of reactant concentrations have a negative sign to make the resulting rate positive.

**Practice Problem A** Write the rate expressions for each of the following reactions:

(a) $CO_2(g) + 2H_2O(g) \longrightarrow CH_4(g) + 2O_2(g)$

(b) $3O_2(g) \longrightarrow 2O_3(g)$

(c) $2NO(g) + O_2(g) \longrightarrow 2NO_2(g)$

**Practice Problem B** Write the balanced equation corresponding to the following rate expressions:

(a) $\text{rate} = -\dfrac{1}{3}\dfrac{\Delta[CH_4]}{\Delta t} = -\dfrac{1}{2}\dfrac{\Delta[H_2O]}{\Delta t} = -\dfrac{\Delta[CO_2]}{\Delta t} = \dfrac{1}{4}\dfrac{\Delta[CH_3OH]}{\Delta t}$

(b) $\text{rate} = -\dfrac{1}{2}\dfrac{\Delta[N_2O_5]}{\Delta t} = \dfrac{1}{2}\dfrac{\Delta[N_2]}{\Delta t} = \dfrac{1}{5}\dfrac{\Delta[O_2]}{\Delta t}$

(c) $\text{rate} = -\dfrac{\Delta[H_2]}{\Delta t} = -\dfrac{\Delta[CO]}{\Delta t} = -\dfrac{\Delta[O_2]}{\Delta t} = \dfrac{\Delta[H_2CO_3]}{\Delta t}$

## Sample Problem 14.2

Consider the reaction

$$4NO_2(g) + O_2(g) \longrightarrow 2N_2O_5(g)$$

At a particular time during the reaction, nitrogen dioxide is being consumed at the rate of 0.00130 $M$/s. (a) At what rate is molecular oxygen being consumed? (b) At what rate is dinitrogen pentoxide being produced?

**Strategy** Determine the rate of reaction using Equation 14.1, and, using the stoichiometry of the reaction, convert to rates of change for the specified individual species.

*(Continued)*

**Setup**

$$\text{rate} = -\frac{1}{4}\frac{\Delta[NO_2]}{\Delta t} = -\frac{\Delta[O_2]}{\Delta t} = \frac{1}{2}\frac{\Delta[N_2O_5]}{\Delta t}$$

We are given

$$\frac{\Delta[NO_2]}{\Delta t} = -0.0013 \text{ } M/s$$

where the minus sign indicates that the concentration of $NO_2$ is decreasing with time. The rate of reaction, therefore, is

$$\text{rate} = -\frac{1}{4}\frac{\Delta[NO_2]}{\Delta t} = -\frac{1}{4}(-0.0013 \text{ } M/s)$$

$$= 3.25 \times 10^{-4} \text{ } M/s$$

**Solution**  (a) $3.25 \times 10^{-4} \text{ } M/s = -\dfrac{\Delta[O_2]}{\Delta t}$

$$\frac{\Delta[O_2]}{\Delta t} = -3.25 \times 10^{-4} \text{ } M/s$$

Molecular oxygen is being consumed at a rate of $3.25 \times 10^{-4} \text{ } M/s$.

(b) $3.25 \times 10^{-4} \text{ } M/s = \dfrac{1}{2}\dfrac{\Delta[N_2O_5]}{\Delta t}$

$$2(3.25 \times 10^{-4} \text{ } M/s) = \frac{\Delta[N_2O_5]}{\Delta t}$$

$$\frac{\Delta[N_2O_5]}{\Delta t} = 6.50 \times 10^{-4} \text{ } M/s$$

Dinitrogen pentoxide is being produced at a rate of $6.50 \times 10^{-4} \text{ } M/s$.

---

**Think About It** Remember that the negative sign in a rate expression indicates that a species is being consumed rather than produced. Rates are always expressed as positive quantities.

---

**Practice Problem A**  Consider the reaction

$$4PH_3(g) \longrightarrow P_4(g) + 6H_2(g)$$

At a particular point during the reaction, molecular hydrogen is being formed at the rate of 0.168 $M/s$. (a) At what rate is $P_4$ being produced? (b) At what rate is $PH_3$ being consumed?

**Practice Problem B**  Consider the following unbalanced equation:

$$A + B \longrightarrow C$$

When C is being formed at the rate of 0.086 $M/s$, A is being consumed at a rate of 0.172 $M/s$ and B is being consumed at a rate of 0.258 $M/s$. Balance the equation based on the relative rates of formation and consumption of products and reactants.

---

## Checkpoint 14.1     Reaction Rates

**14.1.1**  Which expressions are correct for the rate of the following reaction?

$$F_2(g) + 2ClO_2(g) \longrightarrow 2FClO_2(g)$$

(Select all that apply.)

a) rate $= -\dfrac{\Delta[F_2]}{\Delta t}$

b) rate $= -\dfrac{\Delta[ClO_2]}{\Delta t}$

c) rate $= -\dfrac{1}{2}\dfrac{\Delta[F_2]}{\Delta t}$

d) rate $= -\dfrac{1}{2}\dfrac{\Delta[ClO_2]}{\Delta t}$

e) rate $= \dfrac{\Delta[FClO_2]}{\Delta t}$

**14.1.2**  In the same reaction,

$$F_2(g) + 2ClO_2(g) \longrightarrow 2FClO_2(g)$$

if the concentration of $F_2$ is changing at a rate of 0.026 $M/s$, what is the rate of change in the concentration of $ClO_2$?

a) 0.026 $M/s$

b) 0.052 $M/s$

c) 0.013 $M/s$

d) 0.078 $M/s$

e) 0.004 $M/s$

## 14.2 Dependence of Reaction Rate on Reactant Concentration

We saw in Section 14.1 that the rate of reaction between bromine and formic acid is proportional to the concentration of bromine and that the proportionality constant $k$ is the rate constant. We will now explore how the rate, the rate constant, and the reactant concentrations are related.

### The Rate Law

The equation relating the rate of reaction to the concentration of molecular bromine,

$$\text{rate} = k[\text{Br}_2]$$

is an example of a **rate law.** The rate law is an equation that relates the rate of reaction to the concentrations of reactants. For the general reaction,

$$a\text{A} + b\text{B} \longrightarrow c\text{C} + d\text{D}$$

the rate law is

$$\text{rate} = k[\text{A}]^x[\text{B}]^y \qquad \text{Equation 14.2}$$

where $k$ is the rate constant and the exponents $x$ and $y$ are numbers that must be determined experimentally. When we know the values of $k$, $x$, and $y$, we can use Equation 14.2 to calculate the rate of the reaction, given the concentrations of A and B.

In the case of the reaction of molecular bromine and formic acid, the rate law is

$$\text{rate} = k[\text{Br}_2]^x[\text{HCOOH}]^y$$

where $x = 1$ and $y = 0$.

The value of the exponents in the rate law indicate the *order* of the reaction with respect to each reactant. In the reaction of bromine and formic acid, for example, the exponent for the bromine concentration, $x = 1$, means that the reaction is *first order* with respect to bromine. The exponent of 0 for the formic acid concentration indicates that the reaction is *zeroth order* with respect to formic acid. The sum of $x$ and $y$ is called the overall **reaction order.** Thus, the reaction of bromine and formic acid is first order in bromine, zeroth order in formic acid, and first order $(1 + 0 = 1)$ overall.

> **Student Annotation:** It is important to emphasize that the exponents in a rate law must be determined from experimental data and that, in general, they are *not* equal to the coefficients from the chemical equation.

### Experimental Determination of the Rate Law

To see how a rate law is determined from experimental data, consider the following reaction between fluorine and chlorine dioxide:

$$\text{F}_2(g) + 2\text{ClO}_2(g) \longrightarrow 2\text{FClO}_2(g)$$

Rate laws are commonly determined using a table of starting reactant concentrations and initial rates. The **initial rate** is the instantaneous rate at the beginning of the reaction. By varying the starting concentrations of reactants and observing the changes that result in the initial rate, we can determine how the rate depends on each reactant concentration.

Table 14.2 shows the initial rate data for the reaction of fluorine and chlorine dioxide being carried out three different times. Each time, the combination of reactant concentrations is different, and each time the initial rate is different.

To determine the values of the exponents $x$ and $y$ in the rate law,

$$\text{rate} = k[\text{F}_2]^x[\text{ClO}_2]^y$$

| TABLE 14.2 | Initial Rate Data for the Reaction Between $\text{F}_2$ and $\text{ClO}_2$ | | |
|---|---|---|---|
| **Experiment** | **$[\text{F}_2]$ (M)** | **$[\text{ClO}_2]$ (M)** | **Initial Rate (M/s)** |
| 1 | 0.10 | 0.010 | $1.2 \times 10^{-3}$ |
| 2 | 0.10 | 0.040 | $4.8 \times 10^{-3}$ |
| 3 | 0.20 | 0.010 | $2.4 \times 10^{-3}$ |

we must compare two experiments in which one reactant concentration *changes,* and the other remains constant. For example, we first compare the data from experiments 1 and 3:

| Experiment | [F$_2$] (*M*) | [ClO$_2$] (*M*) | Initial Rate (*M*/s) |
|---|---|---|---|
| 1 | 0.10 | 0.010 | 1.2 × 10$^{-3}$ |
| 2 | [F$_2$] doubles  0.10    [ClO$_2$] unchanged 0.040 | | Rate doubles  4.8 × 10$^{-3}$ |
| 3 | 0.20 | 0.010 | 2.4 × 10$^{-3}$ |

When the concentration of fluorine doubles, with the chlorine dioxide concentration held constant, the rate doubles.

$$\frac{[F_2]_3}{[F_2]_1} = \frac{0.20\ M}{0.10\ M} = 2 \qquad \frac{rate_3}{rate_1} = \frac{2.4 \times 10^{-3}\ M/s}{1.2 \times 10^{-3}\ M/s} = 2$$

This indicates that the rate is directly proportional to the concentration of fluorine, and the value of *x* is 1.

**Student Annotation:** Recall that when an exponent is 1, it need not be shown [◀◀ Section 3.3].

$$rate = k\ [F_2][ClO_2]^y$$

Similarly, we can compare experiments 1 and 2:

| Experiment | [F$_2$] (*M*) | [ClO$_2$] (*M*) | Initial Rate (*M*/s) |
|---|---|---|---|
| 1 | [F$_2$] unchanged  0.10    [ClO$_2$] quadruples 0.010 | | Rate quadruples  1.2 × 10$^{-3}$ |
| 2 | 0.10 | 0.040 | 4.8 × 10$^{-3}$ |
| 3 | 0.20 | 0.010 | 2.4 × 10$^{-3}$ |

We find that the rate quadruples when the concentration of chlorine dioxide is quadrupled, but the fluorine concentration is held constant.

$$\frac{[ClO_2]_2}{[ClO_2]_1} = \frac{0.040\ M}{0.010\ M} = 4 \qquad \frac{rate_2}{rate_1} = \frac{4.8 \times 10^{-3} M/s}{1.2 \times 10^{-3} M/s} = 4$$

This indicates that the rate is also directly proportional to the concentration of chlorine dioxide, so the value of *y* is also 1. Thus, we can write the rate law as follows:

$$rate = k[F_2][ClO_2]$$

Because the concentrations of F$_2$ and ClO$_2$ are each raised to the first power, we say that the reaction is first order in F$_2$ and first order in ClO$_2$. The reaction is second order overall.

Knowing the rate law, we can then use the data from any one of the experiments to calculate the rate constant. Using the data for the first experiment in Table 14.2, we can write

$$k = \frac{rate}{[F_2][ClO_2]} = \frac{1.2 \times 10^{-3}\ M/s}{(0.10\ M)(0.010\ M)} = 1.2\ M^{-1} \cdot s^{-1}$$

Table 14.3 contains initial rate data for the hypothetical reaction

$$aA + bB \longrightarrow cC + dD$$

which has the general rate law

$$rate = k[A]^x[B]^y$$

| TABLE 14.3 | Initial Rate Data for the Reaction Between A and B | | |
|---|---|---|---|
| Experiment | [A] (*M*) | [B] (*M*) | Initial Rate (*M*/s) |
| 1 | 0.10 | 0.015 | 2.1 × 10$^{-4}$ |
| 2 | 0.20 | 0.015 | 4.2 × 10$^{-4}$ |
| 3 | 0.10 | 0.030 | 8.4 × 10$^{-4}$ |

Comparing experiments 1 and 2, we see that when [A] doubles, with [B] unchanged, the rate also doubles.

| Experiment | [A] (*M*) | [B] (*M*) | Initial Rate (*M*/s) |
|---|---|---|---|
| 1 | [A] doubles  0.10    [B] unchanged 0.015 | | Rate doubles  2.1 × 10$^{-4}$ |
| 2 | 0.20 | 0.015 | 4.2 × 10$^{-4}$ |
| 3 | 0.10 | 0.030 | 8.4 × 10$^{-4}$ |

Thus, *x* = 1.

| TABLE 14.4 | Units of the Rate Constant $k$ for Reactions of Various Overall Orders | |
|---|---|---|

| Overall Reaction Order | Sample Rate Law | Units of $k$ |
|---|---|---|
| 0 | rate = $k$ | $M \cdot s^{-1}$ |
| 1 | rate = $k[A]$ or rate = $k[B]$ | $s^{-1}$ |
| 2 | rate = $k[A]^2$, rate = $k[B]^2$, or rate = $k[A][B]$ | $M^{-1} \cdot s^{-1}$ |
| 3* | rate = $k[A]^2[B]$ or rate = $k[A][B]^2$ | $M^{-2} \cdot s^{-1}$ |

*Another possibility for a third-order reaction is rate = $k[A][B][C]$, although such reactions are very rare.

Comparing experiments 1 and 3, when [B] doubles with [A] unchanged, we see that the rate quadruples.

| Experiment | [A] $(M)$ | [B] $(M)$ | Initial Rate $(M/s)$ |
|---|---|---|---|
| 1 | 0.10 | 0.015 | $2.1 \times 10^{-4}$ |
| 2 | [A] unchanged 0.20 | [B] doubles 0.015 | Rate quadruples $4.2 \times 10^{-4}$ |
| 3 | 0.10 | 0.030 | $8.4 \times 10^{-4}$ |

Thus, the rate is *not* directly proportional to [B] to the first power, but rather it is directly proportional to [B] to the second power (i.e., $y = 2$):

$$\text{rate} \propto [B]^2$$

The overall rate law is

$$\text{rate} = k[A][B]^2$$

This reaction is therefore first order in A, second order in B, and third order overall.

Once again, knowing the rate law, we can use data from any of the experiments in the table to calculate the rate constant. Using the data from experiment 1, we get

$$k = \frac{\text{rate}}{[A][B]^2} = \frac{2.1 \times 10^{-4} \, M/s}{(0.10 \, M)(0.015 \, M)^2} = 9.3 \, M^{-2} \cdot s^{-1}$$

Note that the units of this rate constant are different from those for the rate constant we calculated for the $F_2$-$ClO_2$ reaction and the bromine reaction. In fact, the units of a rate constant depend on the *overall* order of the reaction. Table 14.4 compares the units of the rate constant for reactions that are zeroth-, first-, second-, and third-order overall.

The following are three important things to remember about the rate law:

1. The exponents in a rate law must be determined from a table of experimental data—in general, they are not related to the stoichiometric coefficients in the balanced chemical equation.

2. Comparing changes in individual reactant concentrations with changes in rate shows how the rate depends on each reactant concentration.

3. Reaction order is always defined in terms of reactant concentrations, never product concentrations.

Sample Problem 14.3 shows how to use initial rate data to determine a rate law.

## Sample Problem 14.3

The gas-phase reaction of nitric oxide with hydrogen at 1280°C is

$$2NO(g) + 2H_2(g) \longrightarrow N_2(g) + 2H_2O(g)$$

From the following data collected at 1280°C, determine (a) the rate law, (b) the rate constant, including units, and (c) the rate of the reaction when [NO] = $4.8 \times 10^{-3} \, M$ and [H$_2$] = $6.2 \times 10^{-3} \, M$.

| Experiment | [NO] $(M)$ | [H$_2$] $(M)$ | Initial Rate $(M/s)$ |
|---|---|---|---|
| 1 | $5.0 \times 10^{-3}$ | $2.0 \times 10^{-3}$ | $1.3 \times 10^{-5}$ |
| 2 | $1.0 \times 10^{-2}$ | $2.0 \times 10^{-3}$ | $5.0 \times 10^{-5}$ |
| 3 | $1.0 \times 10^{-2}$ | $4.0 \times 10^{-3}$ | $1.0 \times 10^{-4}$ |

*(Continued)*

**Strategy** Compare two experiments at a time to determine how the rate depends on the concentration of each reactant.

**Setup** The rate law is rate = $k[NO]^x[H_2]^y$. Comparing experiments 1 and 2, we see that the rate increases by approximately a factor of 4 when [NO] is doubled but [H_2] is held constant. Comparing experiments 2 and 3 shows that the rate doubles when [H_2] doubles but [NO] is held constant.

**Solution** (a) Dividing the rate from experiment 2 by the rate from experiment 1, we get

$$\frac{rate_2}{rate_1} = \frac{5.0 \times 10^{-5}\ M \cdot s^{-1}}{1.3 \times 10^{-5}\ M \cdot s^{-1}} \approx 4 = \frac{k(1.0 \times 10^{-2}\ M)^x(2.0 \times 10^{-3}\ M)^y}{k(5.0 \times 10^{-3}\ M)^x(2.0 \times 10^{-3}\ M)^y}$$

Canceling identical terms in the numerator and denominator gives

$$\frac{(1.0 \times 10^{-2}\ M)^x}{(5.0 \times 10^{-3}\ M)^x} = 2^x = 4$$

Therefore, $x = 2$. The reaction is second order in NO.

Dividing the rate from experiment 3 by the rate from experiment 2, we get

$$\frac{rate_3}{rate_2} = \frac{1.0 \times 10^{-4}\ M \cdot s^{-1}}{5.0 \times 10^{-5}\ M \cdot s^{-1}} = 2 = \frac{k(1.0 \times 10^{-2}\ M)^x(4.0 \times 10^{-3}\ M)^y}{k(1.0 \times 10^{-2}\ M)^x(2.0 \times 10^{-3}\ M)^y}$$

Canceling identical terms in the numerator and denominator gives

$$\frac{(4.0 \times 10^{-3}\ M)^y}{(2.0 \times 10^{-3}\ M)^y} = 2^y = 2$$

Therefore, $y = 1$. The reaction is first order in $H_2$. The overall rate law is

$$rate = k[NO]^2[H_2]$$

(b) We can use data from any of the experiments to calculate the value and units of $k$. Using the data from experiment 1 gives

$$k = \frac{rate}{[NO]^2[H_2]} = \frac{1.3 \times 10^{-5}\ M/s}{(5.0 \times 10^{-3}\ M)^2(2.0 \times 10^{-3}\ M)} = 2.6 \times 10^2\ M^{-2} \cdot s^{-1}$$

(c) Using the rate constant determined in part (b) and the concentrations of NO and $H_2$ given in the problem statement, we can determine the reaction rate as follows:

$$rate = (2.6 \times 10^2\ M^{-2} \cdot s^{-1})(4.8 \times 10^{-3}\ M)^2(6.2 \times 10^{-3}\ M)$$

$$= 3.7 \times 10^{-5}\ M/s$$

**Think About It** The exponent for the concentration of $H_2$ in the rate law is 1, whereas the coefficient for $H_2$ in the balanced equation is 2. It is a common error to try to write a rate law using the stoichiometric coefficients as the exponents. Remember that, in general, the exponents in the rate law are not related to the coefficients in the balanced equation. Rate laws must be determined by examining a table of experimental data.

**Practice Problem A** The reaction of peroxydisulfate ion ($S_2O_8^{2-}$) with iodide ion ($I^-$) is

$$S_2O_8^{2-}(aq) + 3I^-(aq) \longrightarrow 2SO_4^{2-}(aq) + I_3^-(aq)$$

From the following data collected at a certain temperature, determine the rate law and calculate the rate constant, including its units.

| Experiment | $[S_2O_8^{2-}]$ (*M*) | $[I^-]$ (*M*) | Initial Rate (*M/s*) |
|---|---|---|---|
| 1 | 0.080 | 0.034 | $2.2 \times 10^{-4}$ |
| 2 | 0.080 | 0.017 | $1.1 \times 10^{-4}$ |
| 3 | 0.16 | 0.017 | $2.2 \times 10^{-4}$ |

**Practice Problem B** For the following general reaction, rate = $k[A]^2$ and $k = 1.3 \times 10^{-2}\ M^{-1} \cdot s^{-1}$:

$$A + B \longrightarrow 2C$$

Use this information to fill in the missing table entries.

| Experiment | [A] (*M*) | [B] (*M*) | Initial Rate (*M/s*) |
|---|---|---|---|
| 1 | 0.013 | 0.250 | $2.20 \times 10^{-6}$ |
| 2 | 0.026 | 0.250 | |
| 3 | | 0.500 | $2.20 \times 10^{-6}$ |

## Checkpoint 14.2    Dependence of Reaction Rate on Reactant Concentration

Answer Questions 14.2.1 through 14.2.4 using the table of initial rate data for the reaction

$$A + 2B \longrightarrow 2C + D$$

| Experiment | [A] (*M*) | [B] (*M*) | Initial Rate (*M/s*) |
|---|---|---|---|
| 1 | 0.12 | 0.010 | $2.2 \times 10^{-3}$ |
| 2 | 0.36 | 0.010 | $6.6 \times 10^{-3}$ |
| 3 | 0.12 | 0.020 | $2.2 \times 10^{-3}$ |

**14.2.1** What is the rate law for the reaction?

a) rate $= k[A][B]^2$

b) rate $= k[A]^2[B]$

c) rate $= k[A]^3$

d) rate $= k[A]^2$

e) rate $= k[A]$

**14.2.2** Calculate the rate constant.

a) $0.15 \, M^{-1} \cdot s^{-1}$

b) $0.15 \, M \cdot s^{-1}$

c) $0.15 \, s^{-1}$

d) $0.018 \, s^{-1}$

e) $0.018 \, M^{-1} \cdot s^{-1}$

**14.2.3** What is the overall order of the reaction?

a) 0

b) 1

c) 2

d) 3

e) 4

**14.2.4** Determine the rate when [A] = 0.50 *M* and [B] = 0.25 *M*.

a) $9.2 \times 10^{-3}$ *M/s*

b) $2.3 \times 10^{-3}$ *M/s*

c) $4.5 \times 10^{-3}$ *M/s*

d) $5.0 \times 10^{-3}$ *M/s*

e) $1.3 \times 10^{-2}$ *M/s*

**14.2.5** The diagrams represent three experiments in which the reaction A + B $\longrightarrow$ C is carried out with varied initial concentrations of A and B. Determine the rate law for the reaction. (A = green, B = yellow, C = red.)

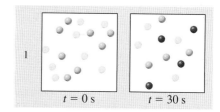

a) rate $= k[A][B]$

b) rate $= k[A]^2[B]$

c) rate $= k[A][B]^2$

d) rate $= k[A]^2$

e) rate $= k[B]$

## 14.3  Dependence of Reactant Concentration on Time

We can use the rate law to determine the rate of a reaction using the rate constant and the reactant concentrations:

$$\underbrace{\text{rate} = k[A]^x[B]^y}_{\text{rate law}}$$

rate → rate constant

A rate law can also be used to determine the remaining concentration of a reactant at a specific time during a reaction. We will illustrate this use of rate laws using reactions that are first order overall and reactions that are second order overall.

## First-Order Reactions

A *first-order reaction* is a reaction whose rate depends on the concentration of one of the reactants raised to the first power. Two examples are the decomposition of ethane ($C_2H_6$) into highly reactive fragments called methyl radicals ($\cdot CH_3$), and the decomposition of dinitrogen pentoxide ($N_2O_5$) into nitrogen dioxide ($NO_2$) and molecular oxygen ($O_2$):

$$C_2H_6 \longrightarrow 2 \cdot CH_3 \qquad\qquad \text{rate} = k[C_2H_6]$$

$$2N_2O_5(g) \longrightarrow 4NO_2(g) + O_2(g) \qquad \text{rate} = k[N_2O_5]$$

In a first-order reaction of the type

$$A \longrightarrow \text{product}$$

the rate can be expressed as the rate of change in reactant concentration,

$$\text{rate} = -\frac{\Delta[A]}{\Delta t}$$

as well as in the form of the rate law:

$$\text{rate} = k[A]$$

Setting these two expressions of the rate equal to each other we get

$$-\frac{\Delta[A]}{\Delta t} = k[A]$$

Applying calculus to the preceding equation, we can show that

Equation 14.3 $\qquad\qquad \ln\dfrac{[A]_t}{[A]_0} = -kt$

where ln is the natural logarithm, and $[A]_0$ and $[A]_t$ are the concentrations of A at times 0 and $t$, respectively. In general, time 0 refers to any specified time during a reaction—not necessarily the beginning of the reaction. Time $t$ refers to any specified time *after* time 0. Equation 14.3 is sometimes called the *integrated rate law*.

In Sample Problem 14.4 we apply Equation 14.3 to a specific reaction.

**Student Annotation:** It is not necessary for you to be able to do the calculus required to arrive at Equation 14.3, but it is very important that you know how to *use* Equation 14.3.

---

### Sample Problem 14.4

The decomposition of hydrogen peroxide is first order in $H_2O_2$.

$$2H_2O_2(aq) \longrightarrow 2H_2O(l) + O_2(g)$$

The rate constant for this reaction at 20°C is $1.8 \times 10^{-5}\ s^{-1}$. If the starting concentration of $H_2O_2$ is 0.75 *M*, determine (a) the concentration of $H_2O_2$ remaining after 3 h and (b) how long it will take for the $H_2O_2$ concentration to drop to 0.10 *M*.

**Strategy** Use Equation 14.3 to find $[H_2O_2]_t$, where $t$ = 3 h, and then solve Equation 14.3 for $t$ to determine how much time must pass for $[H_2O_2]_t$ to equal 0.10 *M*.

**Setup** $[H_2O_2]_0$ = 0.75 *M;* time $t$ for part (a) is (3 h)(60 min/h)(60 s/min) = 10,800 s.

**Solution** (a) $\ln\dfrac{[H_2O_2]_t}{[H_2O_2]_0} = -kt$

$$\ln\frac{[H_2O_2]_t}{0.75\ M} = -(1.8 \times 10^{-5}\ s^{-1})(10,800\ s) = -0.1944$$

**Student Annotation:** The inverse of ln $x$ is $e^x$ [▸▸ Appendix 1].

Take the inverse natural logarithm of both sides of the equation to get

$$\frac{[H_2O_2]_t}{0.75\ M} = e^{-0.1944} = 0.823$$

$$[H_2O_2]_t = (0.823)(0.75\ M) = 0.62\ M$$

The concentration of $H_2O_2$ after 3 h is 0.62 *M*.

(b) $\ln\left(\dfrac{0.10\ M}{0.75\ M}\right) = -2.015 = -(1.8 \times 10^{-5}\ \text{s}^{-1})t$

$\dfrac{2.015}{1.8 \times 10^{-5}\ \text{s}^{-1}} = t = 1.12 \times 10^{5}\ \text{s}$

The time required for the peroxide concentration to drop to 0.10 $M$ is 1.1 × 10$^5$ s or about 31 h.

**Think About It** Don't forget the minus sign in Equation 14.3. If you calculate a reactant concentration at time $t$ that is greater than the concentration at time 0 (or if you get a negative time required for the concentration to drop to a specified level), check your solution for this common error.

**Practice Problem A** The rate constant for the reaction 2A ⟶ B is 7.5 × 10$^{-3}$ s$^{-1}$ at 110°C. The reaction is first order in A. How long (in seconds) will it take for [A] to decrease from 1.25 $M$ to 0.71 $M$?

**Practice Problem B** Refer again to the reaction 2A ⟶ B, for which $k$ = 7.5 × 10$^{-3}$ s$^{-1}$ at 110°C. With a starting concentration of [A] = 2.25 $M$, what will [A] be after 2.0 min?

Equation 14.3 can be rearranged as follows:

$$\ln[A]_t = -kt + \ln[A]_0 \qquad\qquad \text{Equation 14.4}$$

Equation 14.4 has the form of the linear equation $y = mx + b$:

$$\ln[A]_t = (-k)(t) + \ln[A]_0$$
$$y \quad = \quad m \ \ x \ + \quad b$$

Figure 14.8(a) shows the decrease in concentration of reactant A during the course of the reaction. As we saw in Section 14.1, the plot of reactant concentration as a function of time is not a straight line. For a first-order reaction, however, we do get a straight line if we plot the natural log of reactant concentration (ln [A]$_t$) versus time ($y$ versus $x$). The slope of the line is equal to $-k$ [Figure 14.8(b)], so we can determine the rate constant from the slope of this plot.

Sample Problem 14.5 shows how a rate constant can be determined from experimental data.

**Student Annotation:** Because pressure is proportional to concentration, for gaseous reactions [◄◄ Section 11.4] Equations 14.3 and 14.4 can be written as

$$\ln\frac{P_t}{P_0} = -kt$$

and

$$\ln P_t = -kt + \ln P_0$$

respectively, where $P_0$ and $P_t$ are the pressures of reactant A at times 0 and $t$, respectively.

**Student Annotation:** This graphical determination is an alternative to using the method of initial rates to determine the value of $k$.

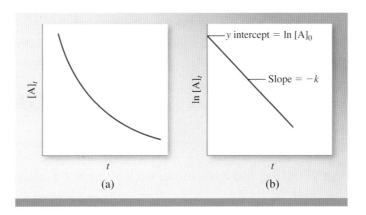

(a)                              (b)

**Figure 14.8**   First-order reaction characteristics: (a) Decrease of reactant concentration with time. (b) A plot of ln [A]$_t$ versus $t$. The slope of the line is equal to $-k$ and the $y$ intercept is equal to ln[A]$_0$.

## Sample Problem 14.5

The rate of decomposition of azomethane is studied by monitoring the partial pressure of the reactant as a function of time:

$$CH_3-N{=}N-CH_3(g) \longrightarrow N_2(g) + C_2H_6(g)$$

The data obtained at 300°C are listed in the following table:

| Time (s) | $P_{\text{azomethane}}$ (mmHg) |
|---|---|
| 0 | 284 |
| 100 | 220 |
| 150 | 193 |
| 200 | 170 |
| 250 | 150 |
| 300 | 132 |

Determine the rate constant of the reaction at this temperature.

*(Continued)*

**Strategy** We can use Equation 14.3 only for first-order reactions, so we must first determine if the decomposition of azomethane is first order. We do this by plotting ln $P$ against time. If the reaction is first order, we can use Equation 14.3 and the data at any two of the times in the table to determine the rate constant.

**Setup** The table expressed as ln $P$ is

| Time (s) | ln $P$ |
|---|---|
| 0 | 5.649 |
| 100 | 5.394 |
| 150 | 5.263 |
| 200 | 5.136 |
| 250 | 5.011 |
| 300 | 4.883 |

Plotting these data gives a straight line, indicating that the reaction is indeed first order. Thus, we can use Equation 14.3 expressed in terms of pressure.

$$\ln \frac{P_t}{P_0} = -kt$$

$P_t$ and $P_0$ can be pressures at any two times during the experiment. $P_0$ need not be the pressure at 0 s—it need only be at the earlier of the two times.

**Solution** Using data from times 100 s and 250 s of the original table ($P_{azomethane}$ versus $t$), we get

$$\ln \frac{150 \text{ mmHg}}{220 \text{ mmHg}} = -k(150 \text{ s})$$

$$\ln 0.682 = -k(150 \text{ s})$$

$$k = 2.55 \times 10^{-3} \text{ s}^{-1}$$

**Think About It** We could equally well have determined the rate constant by calculating the slope of the plot of ln $P$ versus $t$. Using the two points labeled on the plot, we get

$$\text{slope} = \frac{5.011 - 5.394}{250 - 100}$$

$$= -2.55 \times 10^{-3} \text{ s}^{-1}$$

Remember that slope $= -k$, so $k = 2.55 \times 10^{-3} \text{ s}^{-1}$.

**Practice Problem A** Ethyl iodide ($C_2H_5I$) decomposes at a certain temperature in the gas phase as follows:

$$C_2H_5I(g) \longrightarrow C_2H_4(g) + HI(g)$$

From the following data, determine the rate constant of this reaction. Begin by constructing a plot to verify that the reaction is first order.

| Time (min) | [$C_2H_5I$] ($M$) |
|---|---|
| 0 | 0.36 |
| 15 | 0.30 |
| 30 | 0.25 |
| 48 | 0.19 |
| 75 | 0.13 |

**Practice Problem B** Use the calculated $k$ from Practice Problem A to fill in the missing values in the following table:

| Time (min) | [$C_2H_5I$] ($M$) |
|---|---|
| 0 | 0.45 |
| 10 | —— |
| 20 | —— |
| 30 | —— |
| 40 | —— |

We often describe the rate of a reaction using the half-life. The **half-life** ($t_{1/2}$) is the time required for the reactant concentration to drop to *half* its original value. We obtain an expression for $t_{1/2}$ for a first-order reaction as follows:

$$t = \frac{1}{k} \ln \frac{[A]_0}{[A]_t}$$

According to the definition of half-life, $t = t_{1/2}$ when $[A]_t = \frac{1}{2}[A]_0$, so

$$t_{1/2} = \frac{1}{k} \ln \frac{[A]_0}{\frac{1}{2}[A]_0}$$

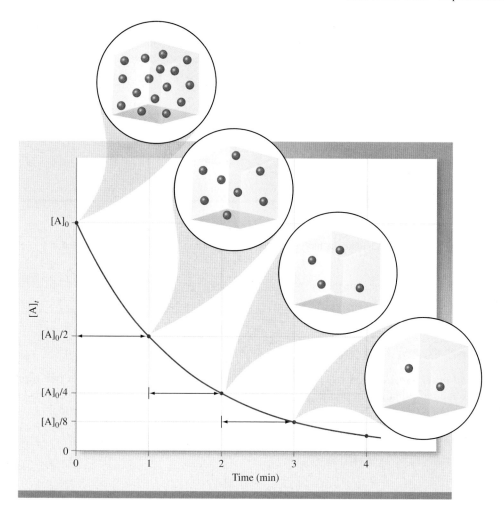

**Figure 14.9**   A plot of [A] versus time for the first-order reaction A $\longrightarrow$ products. The half-life of the reaction is 1 min. The concentration of A is halved every half-life.

Because $\dfrac{[A]_0}{\frac{1}{2}[A]_0} = 2$, and $\ln 2 = 0.693$, the expression for $t_{1/2}$ simplifies to

$$t_{1/2} = \frac{0.693}{k} \qquad \text{Equation 14.5}$$

According to Equation 14.5, the half-life of a first-order reaction is independent of the initial concentration of the reactant. Thus, it takes the same time for the concentration of the reactant to decrease from 1.0 $M$ to 0.50 $M$ as it does for the concentration to decrease from 0.10 $M$ to 0.050 $M$ (Figure 14.9). Measuring the half-life of a reaction is one way to determine the rate constant of a first-order reaction.

The half-life of a first-order reaction is inversely proportional to its rate constant, so a short half-life corresponds to a large rate constant. Consider, for example, two radioactive isotopes used in nuclear medicine: $^{24}$Na ($t_{1/2} = 14.7$ h) and $^{60}$Co ($t_{1/2} = 5.3$ yr). Sodium-24, with the shorter half-life, decays faster. If we started with an equal number of moles of each isotope, most of the sodium-24 would be gone in a week whereas most of the cobalt-60 would remain unchanged.

Sample Problem 14.6 shows how to calculate the half-life of a first-order reaction, given the rate constant.

## Sample Problem 14.6

The decomposition of ethane ($C_2H_6$) to methyl radicals ($CH_3$) is a first-order reaction with a rate constant of $5.36 \times 10^{-4}$ s$^{-1}$ at 700°C:

$$C_2H_6 \longrightarrow 2CH_3$$

Calculate the half-life of the reaction in minutes.

**Strategy** Use Equation 14.5 to calculate $t_{1/2}$ in seconds, and then convert to minutes.

*(Continued)*

**Setup**

$$\text{seconds} \times \frac{1 \text{ minute}}{60 \text{ seconds}} = \text{minutes}$$

**Solution**

$$t_{1/2} = \frac{0.693}{k} = \frac{0.693}{5.36 \times 10^{-4} \text{ s}^{-1}} = 1293 \text{ s}$$

$$1293 \text{ s} \times \frac{1 \text{ min}}{60 \text{ s}} = 21.5 \text{ min}$$

The half-life of ethane decomposition at 700°C is 21.6 min.

> **Think About It** Half-lives and rate constants can be expressed using any units of time and reciprocal time, respectively. Track units carefully when you convert from one unit of time to another.

**Practice Problem A** Calculate the half-life of the decomposition of azomethane, discussed in Sample Problem 14.5.

**Practice Problem B** Calculate the rate constant for the first-order decay of $^{24}$Na ($t_{1/2} = 14.7$ h).

## Second-Order Reactions

A *second-order reaction* is a reaction whose rate depends on the concentration of one reactant raised to the second power or on the product of the concentrations of two different reactants (each raised to the first power). For simplicity, we will consider only the first type of reaction:

$$\text{A} \longrightarrow \text{product}$$

where the rate can be expressed as

$$\text{rate} = -\frac{\Delta[\text{A}]}{\Delta t}$$

or as

$$\text{rate} = k[\text{A}]^2$$

As before, we can combine the two expressions of the rate:

$$-\frac{\Delta[\text{A}]}{\Delta t} = k[\text{A}]^2$$

Again, using calculus, we obtain the following integrated rate law:

**Equation 14.6**
$$\frac{1}{[\text{A}]_t} = kt + \frac{1}{[\text{A}]_0}$$

Thus, for a second-order reaction, we obtain a straight line when we plot the reciprocal of concentration ($1/[\text{A}]_t$) against time (Figure 14.10), and the slope of the line is equal to the rate constant, $k$. As before, we can obtain the expression for the half-life by setting $[\text{A}]_t = \frac{1}{2}[\text{A}]_0$ in Equation 14.6.

$$\frac{1}{\frac{1}{2}[\text{A}]_0} = kt_{1/2} + \frac{1}{[\text{A}]_0}$$

**Figure 14.10** Second-order reaction characteristics: (a) Decrease of reactant concentration with time.

(b) A plot of $\frac{1}{[\text{A}]_t}$ versus $t$. The slope of the line is equal to $k$, and the $y$ intercept is equal to $\frac{1}{[\text{A}]_0}$.

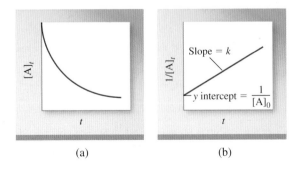

(a)              (b)

Solving for $t_{1/2}$, we obtain

| Equation 14.7 | $t_{1/2} = \dfrac{1}{k[A]_0}$ |
|---|---|

Note that unlike the half-life of a first-order reaction, which is independent of the starting concentration, the half-life of a second-order reaction is inversely proportional to the initial reactant concentration. Determining the half-life at several different initial concentrations is one way to distinguish between first-order and second-order reactions.

Sample Problem 14.7 shows how to use Equations 14.6 and 14.7 to calculate reactant concentrations and the half-life of a second-order reaction.

---

### Sample Problem 14.7

Iodine atoms combine to form molecular iodine in the gas phase:

$$I(g) + I(g) \longrightarrow I_2(g)$$

This reaction is second order and has a rate constant of $7.0 \times 10^9 \, M^{-1} \cdot s^{-1}$ at 23°C. (a) If the initial concentration of I is 0.086 $M$, calculate the concentration after 2.0 min. (b) Calculate the half-life of the reaction when the initial concentration of I is 0.60 $M$ and when the initial concentration of I is 0.42 $M$.

**Strategy** Use Equation 14.6 to determine $[I]_t$ at $t = 2.0$ min; use Equation 14.7 to determine $t_{1/2}$ when $[I]_0 = 0.60 \, M$ and when $[I]_0 = 0.42 \, M$.

**Setup**

$$t = (2.0 \text{ min})(60 \text{ s/min}) = 120 \text{ s}$$

**Solution** (a) $\dfrac{1}{[A]_t} = kt + \dfrac{1}{[A]_0}$

$$= (7.0 \times 10^9 \, M^{-1} \cdot s^{-1})(120 \text{ s}) + \dfrac{1}{0.086 \, M}$$

$$= 8.4 \times 10^{11} \, M^{-1}$$

$$[A]_t = \dfrac{1}{8.4 \times 10^{11} \, M^{-1}} = 1.2 \times 10^{-12} \, M$$

The concentration of atomic iodine after 2.0 min is $1.2 \times 10^{-12} \, M$.

(b) When $[I]_0 = 0.60 \, M$,

$$t_{1/2} = \dfrac{1}{k[A]_0} = \dfrac{1}{(7.0 \times 10^9 \, M^{-1} \cdot s^1)(0.60 \, M)} = 2.4 \times 10^{-10} \text{ s}$$

When $[I]_0 = 0.42 \, M$,

$$t_{1/2} = \dfrac{1}{k[A]_0} = \dfrac{1}{(7.0 \times 10^9 \, M^{-1} \cdot s^{-1})(0.42 \, M)} = 3.4 \times 10^{-10} \text{ s}$$

> **Think About It** (a) Iodine, like the other halogens, exists as diatomic molecules at room temperature. It makes sense, therefore, that atomic iodine would react quickly, and essentially completely, to form $I_2$ at room temperature. The very low remaining concentration of I after 2.0 min makes sense. (b) As expected, the half-life of this second-order reaction is not constant. (A constant half-life is a characteristic of first-order reactions.)

**Practice Problem A** The reaction $2A \longrightarrow B$ is second order in A with a rate constant of 32.0 $M^{-1} \cdot s^{-1}$ at 25.0°C. (a) Starting with $[A]_0 = 0.0075 \, M$, how long will it take for the concentration of A to drop to 0.0018 $M$? (b) Calculate the half-life of the reaction for $[A]_0 = 0.0075 \, M$ and for $[A]_0 = 0.0025 \, M$.

**Practice Problem B** Determine the initial concentration, $[A]_0$, for the reaction in Practice Problem A necessary for the half-life to be (a) 1.50 s, (b) 25.0 s, and (c) 175 s.

---

First- and second-order reactions are the most common reaction types. Reactions of overall order zero exist but are relatively rare. For a zeroth-order reaction

$$A \longrightarrow \text{product}$$

| TABLE 14.5 | Summary of the Kinetics of Zeroth-Order, First-Order, and Second-Order Reactions | | |
|---|---|---|---|
| **Order** | **Rate Law** | **Integrated Rate Law** | **Half-Life** |
| 0 | $rate = k$ | $[A]_t = -kt + [A]_0$ | $\dfrac{[A]_0}{2k}$ |
| 1 | $rate = k[A]$ | $\ln \dfrac{[A]_t}{[A]_0} = -kt$ | $\dfrac{0.693}{k}$ |
| 2 | $rate = k[A]^2$ | $\dfrac{1}{[A]_t} = kt + \dfrac{1}{[A]_0}$ | $\dfrac{1}{k[A]_0}$ |

the rate law is given by

$$rate = k[A]^0 = k$$

Thus, the rate of a ***zeroth-order reaction*** is a constant, independent of reactant concentration. Third-order and higher-order reactions are quite rare and too complex to be covered in this book. Table 14.5 summarizes the kinetics for first-order and second-order reactions of the type A ⟶ product.

---

**Checkpoint 14.3    Dependence of Reactant Concentration on Time**

The first-order decomposition of dinitrogen pentoxide ($N_2O_5$) is represented by

$$2N_2O_5(g) \longrightarrow 4NO_2(g) + O_2(g)$$

Use the table of data to answer questions 14.3.1 and 14.3.2:

| $t(s)$ | $[N_2O_5]$ (M) |
|---|---|
| 0 | 0.91 |
| 300 | 0.75 |
| 600 | 0.64 |
| 1200 | 0.44 |
| 3000 | 0.16 |

**14.3.1**  What is the rate constant for the decomposition of $N_2O_5$?

a) $9 \times 10^{-4}\,s^{-1}$

b) $4 \times 10^{-4}\,s^{-1}$

c) $6 \times 10^{-4}\,s^{-1}$

d) $5 \times 10^{-3}\,s^{-1}$

e) $1 \times 10^{-3}\,s^{-1}$

**14.3.2**  Approximately how long will it take for $[N_2O_5]$ to fall from 0.62 *M* to 0.10 *M*?

a) 1000 s

b) 2000 s

c) 3000 s

d) 4000 s

e) 5000 s

**14.3.3**  Consider the first-order reaction A ⟶ B in which A molecules (blue spheres) are converted to B molecules (yellow spheres). What is the half-life for the reaction?

t = 0 s          t = 10 s

a) 5 s

b) 10 s

c) 15 s

d) 20 s

e) 30 s

**14.3.4**  Which figure represents the numbers of molecules present after 20 s for the reaction in question 14.3.3?

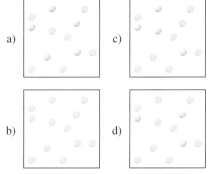

a)          c)

b)          d)

**14.3.5**  Of the plots shown in the left margin, ____ corresponds to a zeroth-order reaction, ____ corresponds to a first-order reaction, and ____ corresponds to a second-order reaction.

a) ii, iii, i

b) i, ii, iii

c) iii, i, ii

d) ii, i, iii

e) i, iii, ii

(i)

(ii)

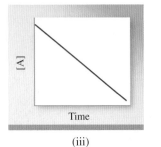

(iii)

# 14.4 Dependence of Reaction Rate on Temperature

Nearly all reactions happen faster at higher temperatures. For example, the time required to cook food depends largely on the boiling point of water. The reaction involved in hard-boiling an egg happens faster at 100°C (about 10 min) than at 80°C (about 30 min). The dependence of reaction rate on temperature is the reason we keep food in a refrigerator—and why food keeps even longer in a freezer. The lower the temperature, the slower the processes that cause food to spoil.

**Student Annotation:** Cookbooks sometimes give alternate directions for cooking at high altitudes, where the lower atmospheric pressure results in water boiling at a lower temperature [◄◄ Section 12.6].

## Collision Theory

Chemical reactions generally occur as a result of collisions between reacting molecules. A greater frequency of collisions usually leads to a higher reaction rate. According to the **collision theory** of chemical kinetics, the reaction rate is directly proportional to the number of molecular collisions per second:

$$\text{rate} \propto \frac{\text{number of collisions}}{\text{s}}$$

Consider the reaction of A molecules with B molecules to form some product. Suppose that each product molecule is formed by the direct combination of an A molecule and a B molecule. If we doubled the concentration of A, then the number of A-B collisions would also double, because there would be twice as many A molecules that could collide with B molecules in any given volume. Consequently, the rate would increase by a factor of 2. Similarly, doubling the concentration of B molecules would increase the rate twofold. Thus, we can express the rate law as

$$\text{rate} = k[\text{A}][\text{B}]$$

The reaction is first order in both A and B and is second order overall.

This view of collision theory is something of a simplification, though, because not every collision between molecules results in a reaction. A collision that *does* result in a reaction is called an **effective collision.** A molecule in motion possesses kinetic energy; the faster it is moving, the greater its kinetic energy. When molecules collide, part of their kinetic energy is converted to *vibrational* energy. If the initial kinetic energies are large, then the colliding molecules will vibrate so strongly as to break some of the chemical bonds. This bond breaking is the first step toward product formation. If the initial kinetic energies are small, the molecules will merely bounce off of each other intact. There is a minimum amount of energy, the **activation energy ($E_a$),** required to initiate a chemical reaction. Without this minimum amount of energy at impact, a collision will be *ineffective;* that is, it will not result in a reaction.

**Student Annotation:** Kinetic energy is the result of motion of the whole molecule, relative to its surroundings. Vibrational energy is the result of motion of the atoms in a molecule, relative to one another.

When molecules react (as opposed to when *atoms* react), having sufficient kinetic energy is not the only requirement for a collision to be effective. Molecules must also be oriented in a way that favors reaction. The reaction between chlorine atoms and nitrosyl chloride (NOCl) illustrates this point:

$$\text{Cl} + \text{NOCl} \longrightarrow \text{Cl}_2 + \text{NO}$$

This reaction is most favorable when a free Cl atom collides directly with the Cl atom in the NOCl molecule [Figure 14.11(a)]. Otherwise, the reactants simply bounce off of each other and no reaction occurs [Figure 14.11(b)].

When molecules collide (in an effective collision), they form an **activated complex** (also called the **transition state**), a temporary species formed by the reactant molecules as a result of the collision. Figure 14.12 shows a potential energy profile for the reaction between Cl and NOCl.

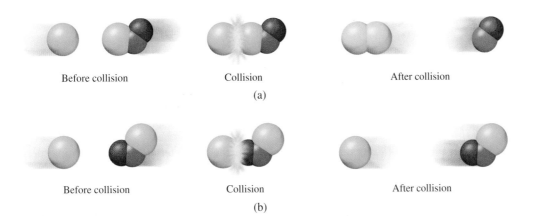

Before collision    Collision    After collision

(a)

Before collision    Collision    After collision

(b)

**Figure 14.11** (a) For an effective collision to take place, the free Cl atom must collide directly with the Cl atom in NOCl. (b) Otherwise, the reactants bounce off of one another and the collision is ineffective—no reaction takes place.

**Figure 14.12**  Energy profile for the reaction of Cl with NOCl. In addition to being oriented properly, reactant molecules must possess sufficient energy to overcome the activation energy.

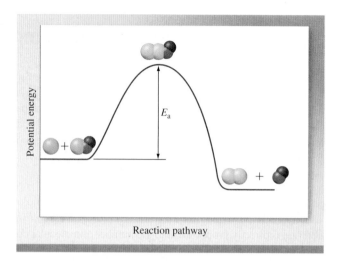

We can think of the activation energy as an energy *barrier* that prevents less energetic molecules from reacting. Because the number of reactant molecules in an ordinary reaction is very large, the speeds, and therefore also the kinetic energies of the molecules, vary greatly. Normally, only a small fraction of the colliding molecules—the fastest-moving ones—have sufficient kinetic energy to exceed the activation energy. These molecules can therefore take part in the reaction. The relationship between rate and temperature should now make sense. According to kinetic molecular theory, the average kinetic energy of a sample of molecules increases as the temperature increases [◄◄ Section 11.6, Figure 11.20]. Thus, a higher percentage of the molecules in the sample have sufficient kinetic energy to exceed the activation energy, and the reaction rate increases.

## The Arrhenius Equation

The dependence of the rate constant of a reaction on temperature can be expressed by the ***Arrhenius equation,***

$$k = Ae^{-E_a/RT}$$    Equation 14.8

**Student Annotation:** Absolute temperature is expressed in *kelvins* [◄◄ Section 1.3].

where $E_a$ is the activation energy of the reaction (in kJ/mol), $R$ is the gas constant (8.314 J/K · mol), $T$ is the absolute temperature, and $e$ is the base of the natural logarithm [▶▶ Appendix 1]. The quantity $A$ represents the collision frequency and is called the frequency factor. It can be treated as a constant for a given reaction over a reasonably wide temperature range. Equation 14.8 shows that the rate constant decreases with increasing activation energy and increases with increasing temperature. This equation can be expressed in a more useful form by taking the natural logarithm of both sides:

$$\ln k = \ln Ae^{-E_a/RT}$$

or

$$\ln k = \ln A - \frac{E_a}{RT}$$    Equation 14.9

which can be rearranged to give the following linear equation:

$$\underset{y}{\ln k} = \underset{m}{\left(-\frac{E_a}{R}\right)}\underset{x}{\left(\frac{1}{T}\right)} + \underset{b}{\ln A}$$    Equation 14.10

Thus, a plot of $\ln k$ versus $1/T$ gives a straight line whose slope is equal to $-E_a/R$ and whose $y$ intercept ($b$) is equal to $\ln A$.

Sample Problem 14.8 demonstrates a graphical method for determining the activation energy of a reaction.

## Sample Problem 14.8

Rate constants for the reaction

$$CO(g) + NO_2(g) \longrightarrow CO_2(g) + NO(g)$$

were measured at four different temperatures. The data are shown in the table. Plot $\ln k$ versus $1/T$, and determine the activation energy (in kJ/mol) for the reaction.

| $k$ ($M^{-1} \cdot s^{-1}$) | $T$ (K) |
|---|---|
| 0.0521 | 288 |
| 0.101 | 298 |
| 0.184 | 308 |
| 0.332 | 318 |

**Strategy** Plot $\ln k$ versus $1/T$, and determine the slope of the resulting line. According to Equation 14.10, slope $= -E_a/R$.

**Setup** $R = 8.314$ J/K $\cdot$ mol. Taking the natural log of each value of $k$ and the inverse of each value of $T$ gives

| $\ln k$ | $1/T$ ($K^{-1}$) |
|---|---|
| $-2.95$ | $3.47 \times 10^{-3}$ |
| $-2.29$ | $3.36 \times 10^{-3}$ |
| $-1.69$ | $3.25 \times 10^{-3}$ |
| $-1.10$ | $3.14 \times 10^{-3}$ |

**Solution** A plot of these data yields the following graph:

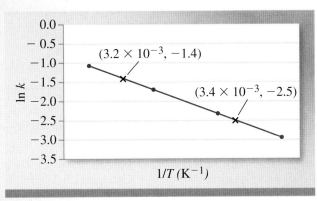

The slope is determined using the $x$ and $y$ coordinates of any two points on the line. Using the points that are labeled on the graph gives

$$\text{slope} = \frac{-1.4 - (-2.5)}{3.2 \times 10^{-3} \text{ K}^{-1} - 3.4 \times 10^{-3} \text{ K}^{-1}} = -5.5 \times 10^3 \text{ K}$$

The value of the slope is $-5.5 \times 10^3$ K. Because slope $= -E_a/R$,

$$E_a = -(\text{slope})(R)$$
$$= -(-5.5 \times 10^3 \text{ K})(8.314 \text{ J/K} \cdot \text{mol})$$
$$= 4.6 \times 10^4 \text{ J/mol} \quad \text{or} \quad 46 \text{ kJ/mol}$$

The activation energy is 46 kJ/mol.

**Think About It** Note that while $k$ has units of $M^{-1} \cdot s^{-1}$, $\ln k$ has no units.

**Practice Problem A** The second-order rate constant for the decomposition of nitrous oxide to nitrogen molecules and oxygen atoms has been determined at various temperatures:

| $k$ ($M^{-1} \cdot s^{-1}$) | $T$ (°C) |
|---|---|
| $1.87 \times 10^{-3}$ | 600 |
| 0.0113 | 650 |
| 0.0569 | 700 |
| 0.244 | 750 |

Determine the activation energy graphically.

**Practice Problem B** Each line in the graph shown at right represents a different reaction. List the reactions in order of increasing activation energy.

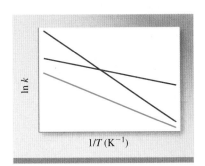

**Student Annotation:** Note that for two different temperatures, the only thing in Equation 14.9 that changes is $k$. The other variables, $A$ and $E_a$ (and of course $R$), are constant.

We can derive an even more useful form of the Arrhenius equation starting with Equation 14.9 written for two different temperatures, $T_1$ and $T_2$:

$$\ln k_1 = \ln A - \frac{E_a}{RT_1}$$

$$\ln k_2 = \ln A - \frac{E_a}{RT_2}$$

Subtracting $\ln k_2$ from $\ln k_1$ gives

$$\ln k_1 - \ln k_2 = \left(\ln A - \frac{E_a}{RT_1}\right) - \left(\ln A - \frac{E_a}{RT_2}\right)$$

$$\ln \frac{k_1}{k_2} = \frac{E_a}{R}\left(-\frac{1}{T_1} + \frac{1}{T_2}\right)$$

Equation 14.11         $$\ln \frac{k_1}{k_2} = \frac{E_a}{R}\left(\frac{1}{T_2} - \frac{1}{T_1}\right)$$

**Student Annotation:** Equations 14.8 through 14.11 are all "Arrhenius equations," but Equation 14.11 is the form you will use most often to solve kinetics problems.

Equation 14.11 enables us to do two things:

1. If we know the rate constant at two different temperatures, we can calculate the activation energy.
2. If we know the activation energy and the rate constant at one temperature, we can determine the value of the rate constant at any other temperature.

Sample Problems 14.9 and 14.10 show how to use Equation 14.11.

---

### Sample Problem 14.9

The rate constant for a particular first-order reaction is given for three different temperatures:

| $T$ (K) | $k$ (s$^{-1}$) |
|---------|----------------|
| 400 | $2.9 \times 10^{-3}$ |
| 450 | $6.1 \times 10^{-2}$ |
| 500 | $7.0 \times 10^{-1}$ |

Using these data, calculate the activation energy of the reaction.

**Strategy**  Use Equation 14.11 to solve for $E_a$.

**Setup**  Solving Equation 14.11 for $E_a$ gives

$$E_a = R\left(\frac{\ln \dfrac{k_1}{k_2}}{\dfrac{1}{T_2} - \dfrac{1}{T_1}}\right)$$

**Solution**  Using the rate constants for 400 K ($T_1$) and 450 K ($T_2$), we get

**Think About It**  A good way to check your work is to use the value of $E_a$ that you calculated (and Equation 14.11) to determine the rate constant at 500 K. Make sure it agrees with the value in the table.

$$E_a = \frac{8.314 \text{ J}}{\text{K} \cdot \text{mol}}\left(\frac{\ln \dfrac{2.9 \times 10^{-3} \text{ s}^{-1}}{6.1 \times 10^{-2} \text{ s}^{-1}}}{\dfrac{1}{450 \text{ K}} - \dfrac{1}{400 \text{ K}}}\right)$$

$$= 91,173 \text{ J/mol} = 91 \text{ kJ/mol}$$

The activation energy of the reaction is 91 kJ/mol.

**Practice Problem A** Use the data in the following table to determine the activation energy of the reaction:

| $T$ (K) | $k$ (s$^{-1}$) |
|---------|----------------|
| 625 | $1.1 \times 10^{-4}$ |
| 635 | $1.5 \times 10^{-4}$ |
| 645 | $2.0 \times 10^{-4}$ |

**⏎Practice Problem B** Based on the data shown in Practice Problem A, what will be the value of $k$ at 655 K?

## Sample Problem 14.10

A certain first-order reaction has an activation energy of 83 kJ/mol. If the rate constant for this reaction is $2.1 \times 10^{-2}$ s$^{-1}$ at 150°C, what is the rate constant at 300°C?

**Strategy** Use Equation 14.11 to solve for $k_2$. Pay particular attention to units in this type of problem.

**Setup** Solving Equation 14.11 for $k_2$ gives

$$k_2 = \frac{k_1}{e^{\left(\frac{E_a}{R}\right)\left(\frac{1}{T_2} - \frac{1}{T_1}\right)}}$$

$E_a = 8.3 \times 10^4$ J, $T_1 = 423$ K, $T_2 = 573$ K, $R = 8.314$ J/K · mol, and $k_1 = 2.1 \times 10^{-2}$ s$^{-1}$.

**Solution**

$$k_2 = \frac{2.1 \times 10^{-2} \text{ s}^{-1}}{e^{\left(\frac{8.3 \times 10^4 \text{ J/mol}}{8.314 \text{ J/K · mol}}\right)\left(\frac{1}{573 \text{ K}} - \frac{1}{423 \text{ K}}\right)}}$$

$$= 1.0 \times 10^1 \text{ s}^{-1}$$

The rate constant at 300°C is 10 s$^{-1}$.

> **Student Annotation:** Note that $E_a$ is converted to joules so that the units will cancel properly. Alternatively, the $R$ could be expressed as 0.008314 kJ/K · mol. Also note again that $T$ must be expressed in kelvins.

> **Think About It** Make sure that the rate constant you calculate at a higher temperature is in fact higher than the original rate constant. According to the Arrhenius equation, the rate constant always increases with increasing temperature. If you get a smaller $k$ at a higher temperature, check your solution for mathematical errors.

**Practice Problem A** Calculate the rate constant at 200°C for a reaction that has a rate constant of $8.1 \times 10^{-4}$ s$^{-1}$ at 90°C and an activation energy of 99 kJ/mol.

**⏎Practice Problem B** Calculate the rate constant at 200°C for a reaction that has a rate constant of $8.1 \times 10^{-4}$ s$^{-1}$ at 90°C and an activation energy of 59 kJ/mol.

## Checkpoint 14.4 Dependence of Reaction Rate on Temperature

Use the table of data collected for a first-order reaction to answer questions 14.4.1 and 14.4.2:

| $T$ (K) | $k$ (s$^{-1}$) |
|---------|----------------|
| 300 | $3.9 \times 10^{-2}$ |
| 310 | $1.1 \times 10^{-1}$ |
| 320 | $2.8 \times 10^{-1}$ |

**14.4.1** What is the activation energy of the reaction?

a) 88 kJ/mol

b) $8.8 \times 10^4$ kJ/mol

c) $8.0 \times 10^4$ kJ/mol

d) $8.8 \times 10^{-3}$ kJ/mol

e) 80 kJ/mol

**14.4.2** What is the rate constant at 80°C?

a) $5.8 \times 10^{-40}$ s$^{-1}$

b) 4.8 s$^{-1}$

c) 2.8 s$^{-1}$

d) $5.2 \times 10^{-3}$ s$^{-1}$

e) $1.7 \times 10^2$ s$^{-1}$

**14.4.3** Which of the following changes when temperature changes?

a) Rate

b) Rate law

c) Rate constant

d) Activation energy

e) Reaction order

A balanced chemical equation does not tell us much about *how* a reaction actually takes place. In many cases, the balanced equation is simply the sum of a series of steps. Consider the following hypothetical example. In the first step of a reaction, a molecule of reactant A combines with a molecule of reactant B to form a molecule of C:

$$A + B \longrightarrow C$$

In the second step, the molecule of C combines with another molecule of B to produce D:

$$C + B \longrightarrow D$$

The overall balanced equation is the sum of these two equations:

$$
\begin{array}{lrcl}
\textit{Step 1:} & A + B & \longrightarrow & \cancel{C} \\
\textit{Step 2:} & \cancel{C} + B & \longrightarrow & D \\
\hline
& A + 2B & \longrightarrow & D
\end{array}
$$

The sequence of steps that sum to give the overall reaction is called the **reaction mechanism.** A reaction mechanism is comparable to the *route* traveled during a trip, whereas the overall balanced chemical equation specifies only the *origin* and the *destination*.

For a specific example of a reaction mechanism, we consider the reaction between nitric oxide and oxygen:

$$2NO(g) + O_2(g) \longrightarrow 2NO_2(g)$$

We know that $NO_2$ does not form as the result of a single collision between two NO molecules and one $O_2$ molecule because $N_2O_2$ is detected during the course of the reaction. We can envision the reaction taking place via the following two steps:

$$2NO(g) \longrightarrow N_2O_2(g)$$

$$N_2O_2(g) + O_2(g) \longrightarrow 2NO_2(g)$$

In the first step, two NO molecules collide to form an $N_2O_2$ molecule. This is followed by the step in which $N_2O_2$ and $O_2$ combine to give two molecules of $NO_2$. The net chemical equation, which represents the overall change, is given by the sum of the first and second steps:

$$
\begin{array}{lrcl}
\textit{Step 1:} & NO + NO & \longrightarrow & N_2O_2 \\
\textit{Step 2:} & N_2O_2 + O_2 & \longrightarrow & 2NO_2 \\
\hline
\textit{Overall reaction:} & 2NO(g) + O_2(g) & \longrightarrow & 2NO_2(g)
\end{array}
$$

Species such as $N_2O_2$ (and C in the hypothetical equation) are called **intermediates** because they appear in the mechanism of the reaction but not in the overall balanced equation. An intermediate is produced in an early step in the reaction and consumed in a later step.

## Elementary Reactions

Each step in a reaction mechanism represents an **elementary reaction,** one that occurs in a single collision of the reactant molecules. The **molecularity** of an elementary reaction is essentially the *number* of reactant molecules involved in the collision. Elementary reactions may be **unimolecular** (one reactant molecule), **bimolecular** (two reactant molecules), or **termolecular** (three reactant molecules). These molecules may be of the same or different types. Each of the elementary steps in the formation of $NO_2$ from NO and $O_2$ is bimolecular, because there are *two* reactant molecules in each step. Likewise, each of the steps in the hypothetical $A + 2B \longrightarrow D$ reaction is bimolecular.

Knowing the steps of a reaction enables us to deduce the rate law. Suppose we have the following elementary reaction:

$$A \longrightarrow \text{products}$$

Because there is only one molecule present, this is a unimolecular reaction. It follows that the larger the number of A molecules present, the faster the rate of product formation. Thus, the rate of a unimolecular reaction is directly proportional to the concentration of A—the reaction is first order in A:

$$\text{rate} = k[A]$$

**Student Annotation:** Termolecular reactions actually are quite rare simply because the simultaneous encounter of three molecules is far less likely than that of two molecules.

**Student Annotation:** The overall equation has three reactant molecules. Because it is not an elementary reaction, though, we do not classify it as a termolecular reaction. Rather, it is the combination of two bimolecular reactions (steps 1 and 2).

For a bimolecular elementary reaction involving A and B molecules,

$$A + B \longrightarrow \text{products}$$

the rate of product formation depends on how frequently A and B collide, which in turn depends on the concentrations of A and B. Thus, we can express the rate as

$$\text{rate} = k[A][B]$$

Therefore, this is a second-order reaction. Similarly, for a bimolecular elementary reaction of the type

$$A + A \longrightarrow \text{products}$$

or

$$2A \longrightarrow \text{products}$$

the rate becomes

$$\text{rate} = k[A]^2$$

which is also a second-order reaction. The preceding examples show that the reaction order for each reactant in an elementary reaction is equal to its stoichiometric coefficient in the chemical equation for that step. In general, we cannot tell just by looking at the balanced equation whether the reaction occurs as shown or in a series of steps. This determination must be made using data obtained experimentally.

## Rate-Determining Step

In a reaction mechanism consisting of more than one elementary step, the rate law for the overall process is given by the **rate-determining step,** which is the *slowest* step in the sequence. An analogy for a process in which there is a rate-determining step is the amount of time required to buy stamps at the post office when there is a long line of customers. The process consists of several steps: waiting in line, requesting the stamps, receiving the stamps, and paying for the stamps. At a time when the line is very long, the amount of time spent waiting in line (step 1) largely determines how much time the overall process takes.

To study reaction mechanisms, we first do a series of experiments to establish initial rates at various reactant concentrations. We then analyze the data to determine the rate constant and overall order of the reaction, and we write the rate law. Finally, we propose a plausible mechanism for the reaction in terms of logical elementary steps. The steps of the proposed mechanism must satisfy two requirements:

1. The sum of the elementary reactions must be the overall balanced equation for the reaction.
2. The rate-determining step must have the same rate law as that determined from the experimental data.

The decomposition of hydrogen peroxide can be facilitated by iodide ions (Figure 14.13). The overall reaction is

$$2H_2O_2(aq) \longrightarrow 2H_2O(l) + O_2(g)$$

By experiment, we find the rate law to be

$$\text{rate} = k[H_2O_2][I^-]$$

Thus, the reaction is first order with respect to both $H_2O_2$ and $I^-$.

The decomposition of $H_2O_2$ is not an elementary reaction, because it does not occur in a single step. If it did, the reaction would be second order in $H_2O_2$ (as a result of the collision of two $H_2O_2$ molecules). What's more, the $I^-$ ion, which is not even part of the overall equation, would not appear in the rate law expression. How can we reconcile these facts? First, we can account for the observed rate law by assuming that the reaction takes place in two separate elementary steps, each of which is bimolecular.

$$\textit{Step 1:} \quad H_2O_2 + I^- \xrightarrow{k_1} H_2O + IO^-$$

$$\textit{Step 2:} \quad H_2O_2 + IO^- \xrightarrow{k_2} H_2O + O_2 + I^-$$

If we further assume that step 1 is the rate-determining step, then the rate of the reaction can be determined from the first step alone:

$$\text{rate} = k_1[H_2O_2][I^-]$$

**Figure 14.13**   The decomposition of hydrogen peroxide is catalyzed by the addition of an iodide salt. Some of the iodide ions are oxidized to molecular iodine, which then reacts with iodide ions to form the brown triiodide ion ($I_3^-$).

**Figure 14.14**   Potential energy profile for a two-step reaction in which the first step is rate determining. R and P represent reactants and products, respectively.

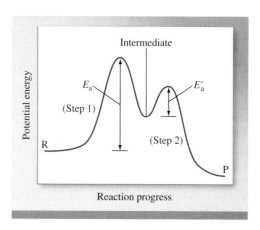

where $k_1 = k$. The $IO^-$ ion is an intermediate because it is produced in the first step and consumed in the second step. It does not appear in the overall balanced equation. The $I^-$ ion also does not appear in the overall equation, but it is consumed in the first step and then produced in the second step. In other words, it is present at the start of the reaction, and it is present at the end. The $I^-$ ion is a *catalyst*, and its function is to *speed up* the reaction. Catalysts are discussed in greater detail in Section 14.6. Figure 14.14 shows the potential energy profile for a reaction like the decomposition of $H_2O_2$. The first step, which is the rate-determining step, has a larger activation energy than the second step. The intermediate, although stable enough to be observed, reacts quickly to form the products. Its existence is only fleeting.

## Experimental Support for Reaction Mechanisms

When you propose a mechanism based on experimental rate data, and your mechanism satisfies the two requirements listed on page 605, you can say that the proposed mechanism is *plausible*—not that it is necessarily *correct*. It is not possible to prove that a mechanism is correct using rate data alone. To determine whether or not a proposed reaction mechanism is actually correct, we must conduct other experiments. In the case of hydrogen peroxide decomposition, we might try to detect the presence of the $IO^-$ ions. If we can detect them, it will support the proposed mechanism. Similarly, for the hydrogen iodide reaction, detection of iodine atoms would lend support to the proposed two-step mechanism. For example, $I_2$ dissociates into atoms when it is irradiated with visible light. Thus, we might predict that the formation of HI from $H_2$ and $I_2$ would speed up as the intensity of light is increased—because that would increase the concentration of I atoms. Indeed, this is just what is observed.

In one case, chemists wanted to know which $C-O$ bond was broken in the reaction between methyl acetate and water to better understand the reaction

$$CH_3-\overset{\overset{O}{\|}}{C}-O-CH_3 \;+\; H_2O \;\longrightarrow\; CH_3-\overset{\overset{O}{\|}}{C}-OH \;+\; CH_3OH$$

The two possibilities of bond breaking are

$$CH_3-\overset{\overset{O}{\|}}{C}\text{-}^{\downarrow}\text{-}O-CH_3 \quad \text{and} \quad CH_3-\overset{\overset{O}{\|}}{C}-O\text{-}^{\downarrow}\text{-}CH_3$$

(a)                    (b)

To distinguish between schemes (a) and (b), chemists used water containing the oxygen-18 isotope instead of ordinary water, which contains the oxygen-16 isotope. When the $^{18}O$ water was used, only the acetic acid formed contained $^{18}O$:

$$CH_3-\overset{\overset{O}{\|}}{C}-^{18}O-H$$

Thus, the reaction must have occurred via bond-breaking scheme (a), because the product formed via scheme (b) would have retained both of its original oxygen atoms and would contain no $^{18}O$.

Another example is photosynthesis, the process by which green plants produce glucose from carbon dioxide and water:

$$6CO_2 + 6H_2O \longrightarrow C_6H_{12}O_6 + 6O_2$$

A question that arose early in the studies of photosynthesis was whether the molecular oxygen produced came from the water, the carbon dioxide, or both. By using water containing only the oxygen-18 isotope, it was concluded that all the oxygen produced by photosynthesis came from the water and none came from the carbon dioxide, because the $O_2$ produced contained only $^{18}O$.

These examples illustrate how creative chemists must be to study reaction mechanisms.

## Checkpoint 14.5   Reaction Mechanisms

Use the following information to answer questions 14.5.1 to 14.5.3:
For the reaction

$$A + B \longrightarrow C + D$$

the experimental rate law is rate $= k[B]^2$.

**14.5.1**   What is the order of the reaction with respect to A and B, respectively?

a) 0 and 1

b) 1 and 0

c) 0 and 2

d) 2 and 1

e) 2 and 0

**14.5.2**   What is the overall order of the reaction?

a) 0

b) 1

c) 2

d) 3

e) 4

**14.5.3**   Which of the following is a plausible mechanism for the reaction?

a) Step 1: $A + B \longrightarrow C + E$ (slow)
   Step 2: $E + A \longrightarrow D$ (fast)

b) Step 1: $B + B \longrightarrow C + E$ (slow)
   Step 2: $E + A \longrightarrow D + B$ (fast)

c) Step 1: $A + A \longrightarrow B + D$ (slow)
   Step 2: $B + B \longrightarrow A + C$ (fast)

d) Step 1: $B + B \longrightarrow C + E$ (slow)
   Step 2: $C + A \longrightarrow D + B$ (fast)

e) Step 1: $A + B \longrightarrow D + E$ (slow)
   Step 2: $E + A \longrightarrow C$ (fast)

**14.5.4**   A plausible mechanism for the reaction

$$H_2 + 2IBr \longrightarrow I_2 + 2HBr$$

is the following:

Step 1: $H_2 + IBr \xrightarrow{k_1} HI + HBr$ (slow)

Step 2: $HI + IBr \xrightarrow{k_2} I_2 + HBr$ (fast)

Select the rate law that will be determined experimentally.

a) rate $= k[H_2]^2$

b) rate $= k[H_2][IBr]^2$

c) rate $= k[IBr]^2$

d) rate $= k[H_2][IBr]$

e) rate $= k[HI][IBr]$

# FAQ *You May Be Wondering...*

# How Can I Tell if a Proposed Reaction Mechanism Is Plausible?

**Student Annotation:** Recall that for an elementary process, the exponents in the rate law are simply the coefficients in the balanced equation.

There are two criteria that must be met for a proposed reaction mechanism to be plausible:

1. The individual steps (elementary processes) must sum to the correct overall reaction.
2. The rate-determining step (the slow step) must have the same rate law as the experimentally determined rate law.

Consider the gas-phase reaction of hydrogen with iodine monochloride to form hydrogen chloride and iodine.

$$H_2 + 2ICl \longrightarrow 2HCl + I_2$$

The experimentally determined rate law for this reaction is rate $= k[H_2][ICl]$. We present four different proposed mechanisms.

## Mechanism 1

Step 1:   $ICl + ICl \longrightarrow I_2 + Cl_2$ (slow)
Step 2:   $Cl_2 + H_2 \longrightarrow 2HCl$

The steps in mechanism 1 sum to the correct overall reaction:

$$ICl + ICl \longrightarrow I_2 + \cancel{Cl_2} \text{ (slow)}$$

$$\cancel{Cl_2} + H_2 \longrightarrow 2HCl$$

$$\overline{\text{Sum: } H_2 + 2ICl \longrightarrow 2HCl + I_2}$$

However, the rate law of the rate-determining step is rate $= k[ICl]^2$. Therefore, mechanism 1 is not plausible. It meets criterion 1, but it does not meet criterion 2.

## Mechanism 2

Step 1:   $H_2 + ICl \longrightarrow HI + HCl$ (slow)
Step 2:   $ICl + HCl \longrightarrow HI + Cl_2$

The rate-determining step in mechanism 2 has the correct rate law:

$$\text{rate} = k[H_2][ICl]$$

However, the steps do not sum to the correct overall reaction:

$$H_2 + ICl \longrightarrow HI + \cancel{HCl} \text{ (slow)}$$

$$ICl + \cancel{HCl} \longrightarrow HI + Cl_2$$

$$\overline{\text{Sum: } H_2 + 2ICl \longrightarrow 2HI + Cl_2}$$

Therefore, mechanism 2 is not plausible. It meets criterion 2, but it does not meet criterion 1.

## Mechanism 3

Step 1:        $H_2 \longrightarrow 2H$ (slow)
Step 2:   $ICl + H \longrightarrow HCl + I$
Step 3:     $H + I \longrightarrow HI$

The rate law for the rate-determining step in mechanism 3 (rate $= k[H_2]$) is not correct.

Furthermore, the steps do not sum to the correct overall reaction:

$$H_2 \longrightarrow \cancel{2}H \text{ (slow)}$$

$$ICl + \cancel{H} \longrightarrow HCl + \cancel{I}$$

$$\cancel{H} + \cancel{I} \longrightarrow HI$$

$$\overline{\text{Sum: } H_2 + ICl \longrightarrow HCl + HI}$$

Therefore, mechanism 3 is not plausible. It meets neither criterion 1 nor criterion 2.

## Mechanism 4

Step 1:   $H_2 + ICl \longrightarrow HCl + HI$ (slow)
Step 2:   $HI + ICl \longrightarrow HCl + I_2$

The rate-determining step in mechanism 4 has the correct rate law:

$$\text{rate} = k[H_2][ICl]$$

Furthermore, the steps in mechanism 4 sum to the correct overall reaction:

$$H_2 + ICl \longrightarrow HCl + \cancel{HI} \text{ (slow)}$$

$$\cancel{HI} + ICl \longrightarrow HCl + I_2$$

$$\overline{\text{Sum: } H_2 + 2ICl \longrightarrow 2HCl + I_2}$$

Therefore, mechanism 4 is plausible. It meets both criteria.

Sample Problem 14.11 lets you practice determining if a proposed reaction mechanism is plausible.

---

### Sample Problem 14.11

The gas-phase decomposition of nitrous oxide ($N_2O$) is believed to occur in two steps:

$$Step\ 1: \quad N_2O \xrightarrow{k_1} N_2 + O$$

$$Step\ 2: \quad N_2O + O \xrightarrow{k_2} N_2 + O_2$$

Experimentally the rate law is found to be rate $= k[N_2O]$. (a) Write the equation for the overall reaction. (b) Identify the intermediate(s). (c) Identify the rate-determining step.

**Strategy** Add the two equations, canceling identical terms on opposite sides of the arrow, to obtain the overall reaction. The canceled terms will be the intermediates if they were first generated and then consumed. Write rate laws for each elementary step; the one that matches the experimental rate law will be the rate-determining step.

**Setup** Intermediates are species that are generated in an earlier step and consumed in a later step. We can write rate laws for elementary reactions simply by using the stoichiometric coefficient for each species as its exponent in the rate law.

$$Step\ 1: \quad N_2O \xrightarrow{k_1} N_2 + O \qquad rate = k[N_2O]$$

$$Step\ 2: \quad N_2O + O \xrightarrow{k_2} N_2 + O_2 \qquad rate = k[N_2O][O]$$

**Solution** (a) $2N_2O \longrightarrow 2N_2 + O_2$

(b) O (atomic oxygen) is the intermediate.

(c) Step 1 is the rate-determining step because its rate law is the same as the experimental rate law: rate $= k[N_2O]$.

> **Think About It** A species that gets canceled when steps are added may be an intermediate or a catalyst. In this case, the canceled species is an intermediate because it was first generated and then consumed. A species that is first consumed and then generated, but doesn't appear in the overall equation, is a catalyst.

**Practice Problem A** The reaction between $NO_2$ and CO to produce NO and $CO_2$ is thought to occur in two steps:

$$Step\ 1: \quad NO_2 + NO_2 \xrightarrow{k_1} NO + NO_3$$

$$Step\ 2: \quad NO_3 + CO \xrightarrow{k_2} NO_2 + CO_2$$

The experimental rate law is rate $= k[NO_2]^2$. (a) Write the equation for the overall reaction. (b) Identify the intermediate(s). (c) Identify the rate-determining step.

**Practice Problem B** Propose a plausible mechanism for the reaction $F_2 + 2ClO_2 \longrightarrow 2FClO_2$, given that the rate law for the reaction is rate $= k[F_2][ClO_2]$.

*You May Be Wondering…*

# What if the First Step in a Mechanism Is Not the Rate-Determining Step?

For first- and second-order reactions, it is reasonably straightforward to propose a plausible mechanism. We simply use the experimentally determined rate law to write the rate-determining step as the first step, and then write one or more additional steps such that the appropriate species will cancel to give the correct overall equation. Sometimes, however, we will encounter a reaction with an experimentally determined rate law that suggests an unlikely scenario. For example, consider the gaseous reaction of nitric oxide with chlorine to produce nitrosyl chloride,

$$2NO(g) + Cl_2(g) \longrightarrow 2NOCl(g)$$

for which the experimentally determined rate law is rate = $k[NO]^2[Cl_2]$. Note that the exponents in the rate law are the same as the coefficients in the balanced equation—making this reaction *third order* overall. Although one possibility is that the reaction is an elementary process, this would require the simultaneous collision of three molecules. As was noted on page 604, such a termolecular reaction is unlikely. A more reasonable mechanism would be one in which a fast first step is followed by the slower, rate-determining step.

*Step 1:*   $NO(g) + Cl_2(g) \xrightarrow{k_1} NOCl_2(g)$   (fast)
*Step 2:*   $NOCl_2(g) + NO(g) \xrightarrow{k_2} 2NOCl(g)$   (slow)

However, when we write the overall rate law using the equation for the rate-determining step, as we have done previously, the resulting rate law

$$\text{rate} = k_2[NOCl_2][NO]$$

includes the concentration of an *intermediate* ($NOCl_2$). Although this rate law is correct for the proposed mechanism, in this form it does not enable us to determine the plausibility of the mechanism by comparison to the experimentally determined rate law. To do this, we must derive a rate law in which only *reactants* from the overall equation appear.

When the intermediate ($NOCl_2$) is not consumed by step 2 as fast as it is produced by step 1, its concentration builds up, causing step 1 to happen in *reverse; that is,* $NOCl_2$ reacts to produce NO and $Cl_2$. This results in the establishment of a dynamic equilibrium [◄◄ Section 4.1], where the forward and reverse of step 1 are occurring at the same rate. We denote this by changing the single reaction arrow in step 1 to equilibrium arrows:

$$NO(g) + Cl_2(g) \underset{k_{-1}}{\overset{k_1}{\rightleftharpoons}} NOCl_2(g)$$

We can write the rate laws for the forward and reverse of step 1, both elementary processes, as

$$\text{rate}_{\text{forward}} = k_1[NO][Cl_2] \quad \text{and} \quad \text{rate}_{\text{reverse}} = k_{-1}[NOCl_2]$$

where $k_1$ and $k_{-1}$ are the individual rate constants for the forward and reverse processes, respectively. Because the two rates are equal, we can write

$$k_1[NO][Cl_2] = k_{-1}[NOCl_2]$$

Rearranging to solve for the concentration of $NOCl_2$ gives

$$\frac{k_1}{k_{-1}}[NO][Cl_2] = [NOCl_2]$$

When we substitute the result into the original rate law for the concentration of $NOCl_2$, we get

$$\text{rate} = k_2[NOCl_2][NO] = k_2\frac{k_1}{k_{-1}}[NO]^2[Cl_2]$$

which agrees with the experimentally determined rate law, rate = $k[NO]^2[Cl_2]$, where $k$ is equal to $\dfrac{k_2k_1}{k_{-1}}$.

Sample Problem 14.12 lets you practice relating the experimentally determined rate law to a reaction mechanism in which a fast first step is followed by a slower, rate-determining step.

---

## 14.6   Catalysis

Multimedia
Catalysis.

Recall from Section 14.5 that the reaction rate for the decomposition of hydrogen peroxide depends on the concentration of iodide ions, even though $I^-$ does not appear in the overall equation. Instead, $I^-$ acts as a catalyst for the reaction. A *catalyst* is a substance that increases the rate of a chemical reaction without itself being consumed. The catalyst may react to form an intermediate, but it is regenerated in a subsequent step of the reaction.

Molecular oxygen is prepared in the laboratory by heating potassium chlorate. The reaction is

$$2KClO_3(s) \longrightarrow 2KCl(s) + 3O_2(g)$$

## Sample Problem 14.12

Consider the gas-phase reaction of nitric oxide and oxygen that was described at the beginning of Section 14.5:

$$2NO(g) + O_2(g) \longrightarrow 2NO_2(g)$$

Show that the following mechanism is plausible. The experimentally determined rate law is rate = $k[NO]^2[O_2]$.

*Step 1:*   $NO(g) + NO(g) \underset{k_{-1}}{\overset{k_1}{\rightleftharpoons}} N_2O_2(g)$   (fast)

*Step 2:*   $N_2O_2(g) + O_2(g) \overset{k_2}{\longrightarrow} 2NO_2(g)$   (slow)

**Strategy** To establish the plausibility of a mechanism, we must compare the rate law of the rate-determining step to the experimentally determined rate law. In this case, the rate-determining step has an intermediate ($N_2O_2$) as one of its reactants, giving us a rate law of rate = $k_2[N_2O_2][O_2]$. Because we cannot compare this directly to the experimental rate law, we must solve for the intermediate concentration in terms of *reactant* concentrations.

**Setup** The first step is a rapidly established equilibrium. Both the forward and reverse of step 1 are elementary processes, which enables us to write their rate laws from the balanced equation.

$$\text{rate}_{\text{forward}} = k_1[NO]^2 \quad \text{and} \quad \text{rate}_{\text{reverse}} = k_{-1}[N_2O_2]$$

**Solution** Because at equilibrium the forward and reverse processes are occurring at the same rate, we can set their rates equal to each other and solve for the intermediate concentration.

$$k_1[NO]^2 = k_{-1}[N_2O_2]$$

$$[N_2O_2] = \frac{k_1[NO]^2}{k_{-1}}$$

Substituting the solution into the original rate law (rate = $k[N_2O_2][O_2]$) gives

$$\text{rate} = k_2\frac{k_1[NO]^2}{k_{-1}}[O_2] = k[NO]^2[O_2] \quad \text{where } k = \frac{k_2k_1}{k_{-1}}$$

> **Think About It** Not all reactions have a single rate-determining step. Analyzing the kinetics of reactions with two or more comparably slow steps is beyond the scope of this book.

**Practice Problem A** Show that the following mechanism is consistent with the experimentally determined rate law of rate = $k[NO]^2[Br_2]$ for the reaction of nitric oxide and bromine: $2NO(g) + Br_2(g) \longrightarrow 2NOBr(g)$.

*Step 1:*   $NO(g) + Br_2(g) \underset{k_{-1}}{\overset{k_1}{\rightleftharpoons}} NOBr_2(g)$   (fast)

*Step 2:*   $NOBr_2(g) + NO(g) \overset{k_2}{\longrightarrow} 2NOBr(g)$   (slow)

**Practice Problem B** The reaction $H_2(g) + I_2(g) \longrightarrow 2HI(g)$ proceeds via a two-step mechanism in which the rate law for the rate-determining step is rate = $k[H_2][I]^2$. Write the mechanism and rewrite the rate law using only reactant concentrations.

However, this thermal decomposition process is very slow in the absence of a catalyst. The rate of decomposition can be increased dramatically by adding a small amount of manganese(IV) dioxide ($MnO_2$), a black powdery substance. All the $MnO_2$ can be recovered at the end of the reaction, just as all the $I^-$ ions remain following the decomposition of $H_2O_2$.

A catalyst speeds up a reaction by providing a set of elementary steps with more favorable kinetics than those that exist in its absence. From Equation 14.8 we know that the rate constant $k$ (and hence the rate) of a reaction depends on the frequency factor ($A$) and the activation energy ($E_a$)—the larger the value of $A$ (or the smaller the value of $E_a$), the greater the rate. In many cases, a catalyst increases the rate by lowering the activation energy for the reaction.

**Figure 14.15** Comparison of the activation energy barriers of (a) an uncatalyzed reaction and (b) the same reaction with a catalyst. A catalyst lowers the energy barrier but does not affect the energies of the reactants or products. Although the reactants and products are the same in both cases, the reaction mechanisms and rate laws are different in (a) and (b).

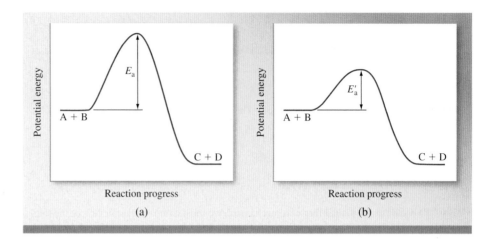

Let's assume that the following reaction has a certain rate constant $k$ and an activation energy $E_a$.

$$A + B \xrightarrow{k} C + D$$

In the presence of a catalyst, however, the rate constant is $k_c$, called the *catalytic rate constant*.

By the definition of a catalyst,

$$\text{rate}_{\text{catalyzed}} > \text{rate}_{\text{uncatalyzed}}$$

Figure 14.15 shows the potential energy profiles for both reactions. The total energies of the reactants (A and B) and those of the products (C and D) are unaffected by the catalyst; the only difference between the two is a lowering of the activation energy from $E_a$ to $E_a'$. Because the activation energy for the reverse reaction is also lowered, a catalyst enhances the rates of the forward and reverse reactions equally.

There are three general types of catalysis, depending on the nature of the rate-increasing substance: heterogeneous catalysis, homogeneous catalysis, and enzyme catalysis.

## Heterogeneous Catalysis

In *heterogeneous catalysis,* the reactants and the catalyst are in different phases. The catalyst is usually a solid, and the reactants are either gases or liquids. Heterogeneous catalysis is by far the most important type of catalysis in industrial chemistry, especially in the synthesis of many important chemicals. Heterogeneous catalysis is also used in the catalytic converters in automobiles.

At high temperatures inside a car's engine, nitrogen and oxygen gases react to form nitric oxide:

$$N_2(g) + O_2(g) \longrightarrow 2NO(g)$$

When released into the atmosphere, NO rapidly combines with $O_2$ to form $NO_2$. Nitrogen dioxide and other gases emitted by automobiles, such as carbon monoxide (CO) and various unburned hydrocarbons, make automobile exhaust a major source of air pollution.

Most new cars are equipped with catalytic converters [Figure 14.16(a)]. An efficient catalytic converter serves two purposes: It oxidizes CO and unburned hydrocarbons to $CO_2$ and $H_2O$, and it converts NO and $NO_2$ to $N_2$ and $O_2$. Hot exhaust gases into which air has been injected are passed through the first chamber of one converter to accelerate the complete burning of hydrocarbons and to decrease CO emissions. Because high temperatures increase NO production, however, a second chamber containing a different catalyst (a transition metal or a transition metal oxide such as CuO or $Cr_2O_3$) and operating at a lower temperature is required to dissociate NO into $N_2$ and $O_2$ before the exhaust is discharged through the tailpipe [Figure 14.16(b)].

## Homogeneous Catalysis

In *homogeneous catalysis* the reactants and the catalyst are dispersed in a single phase, usually liquid. Acid and base catalyses are the most important types of homogeneous catalysis in liquid solution. For example, the reaction of ethyl acetate with water to form acetic acid and ethanol normally occurs too slowly to be measured.

(a)

(b)

$$CH_3COOC_2H_5 + H_2O \longrightarrow CH_3COOH + C_2H_5OH$$

In the absence of the catalyst, the rate law is given by

$$rate = k[CH_3COOC_2H_5]$$

The reaction, however, can be catalyzed by an acid. Often a catalyst is shown above the arrow in a chemical equation:

$$CH_3COOC_2H_5 + H_2O \xrightarrow{H^+} CH_3COOH + C_2H_5OH$$

In the presence of acid, the rate is faster and the rate law is given by

$$rate = k_c[CH_3COOC_2H_5][H^+]$$

Because $k_c > k$ in magnitude, the rate is determined solely by the catalyzed portion of the reaction.

Homogeneous catalysis has several advantages over heterogeneous catalysis. For one thing, the reactions can often be carried out under atmospheric conditions, thus reducing production costs and minimizing the decomposition of products at high temperatures. In addition, homogeneous catalysts can be designed to function selectively for particular types of reactions, and homogeneous catalysts cost less than the precious metals (e.g., platinum and gold) used in heterogeneous catalysis.

## Enzymes: Biological Catalysts

Of all the intricate processes that have evolved in living systems, none is more striking or more essential than enzyme catalysis. **Enzymes** are *biological* catalysts. The amazing fact about enzymes is that not only can they increase the rate of biochemical reactions by factors ranging from $10^6$ to $10^{18}$, but they are also highly specific. An enzyme acts only on certain reactant molecules, called *substrates*, while leaving the rest of the system unaffected. It has been estimated that an average living cell may contain some 3000 different enzymes, each of them catalyzing a specific reaction in which a substrate is converted into the appropriate product(s). Enzyme catalysis is usually homogeneous because the substrate and enzyme are both present in an aqueous solution.

An enzyme is typically a large protein molecule that contains one or more active sites where interactions with substrates take place. These sites are structurally compatible with specific substrate molecules, in much the same way that a key fits a particular lock. In fact, the notion of a rigid enzyme structure that binds only to molecules whose shape exactly matches that of the active site was the basis of an early theory of enzyme catalysis, the so-called lock-and-key theory developed by Emil Fischer[1] in 1894 (Figure 14.17). Fischer's hypothesis accounts for the specificity of

---

1. Emil Fischer (1852–1919). German chemist. Regarded by many as the greatest organic chemist of the nineteenth century. Fischer made many significant contributions in the synthesis of sugars and other important molecules. He was awarded the Nobel Prize in Chemistry in 1902.

**Figure 14.17**   The lock-and-key model of an enzyme's specificity for substrate molecules.

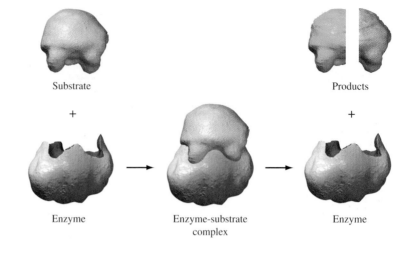

Substrate                                    Products

+                                            +

Enzyme          Enzyme-substrate          Enzyme
                    complex

**Figure 14.18**   Left to right: The binding of glucose molecule (red) to hexokinase (an enzyme in the metabolic pathway). Note how the region at the active site closes around glucose after binding. Often, the geometries of both the substrate and the active site are altered to fit each other.

enzymes, but it contradicts research evidence that a single enzyme binds to substrates of different sizes and shapes. Chemists now know that an enzyme molecule (or at least its active site) has a fair amount of structural flexibility and can modify its shape to accommodate more than one type of substrate. Figure 14.18 shows a molecular model of an enzyme in action.

The mathematical treatment of enzyme kinetics is quite complex, even when we know the basic steps involved in the reaction. A simplified scheme is given by the following elementary steps:

$$E + S \underset{k_{-1}}{\overset{k_1}{\rightleftharpoons}} ES$$

$$ES \xrightarrow{k_2} E + P$$

where E, S, and P represent enzyme, substrate, and product, respectively, and ES is the enzyme-substrate intermediate. It is often assumed that the formation of ES and its decomposition back to enzyme and substrate molecules occur rapidly and that the rate-determining step is the formation of product. Figure 14.19 shows the potential-energy profile for the reaction.

**Figure 14.19**   Comparison of (a) an uncatalyzed reaction and (b) the same reaction catalyzed by an enzyme. The plot in (b) assumes that the catalyzed reaction has a two-step mechanism, in which the second step (ES $\longrightarrow$ E + P) is rate determining.

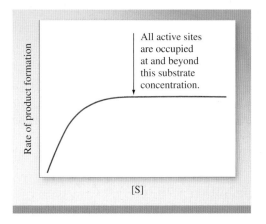

**Figure 14.20** Plot of the rate of product formation versus substrate concentration in an enzyme-catalyzed reaction.

In general, the rate of such a reaction is given by the equation

$$\text{rate} = \frac{\Delta[P]}{\Delta t}$$

$$= k[ES]$$

The concentration of the ES intermediate is itself proportional to the amount of the substrate present, and a plot of the rate versus the concentration of substrate typically yields a curve like that shown in Figure 14.20. Initially the rate rises rapidly with increasing substrate concentration. Above a certain concentration, however, all the active sites are occupied, and the reaction becomes zeroth order in the substrate. In other words, the rate remains the same even though the substrate concentration increases. At and beyond this point, the rate of formation of product depends only on how fast the ES intermediate breaks down, not on the number of substrate molecules present.

## Bringing Chemistry to Life

### Catalysis and Hangovers

We learned at the beginning of the chapter that alcohol dehydrogenase (ADH) and aldehyde dehydrogenase (ALDH) catalyze the metabolism of methanol. In analogous reactions, ADH and ALDH catalyze the metabolism of *ethanol,* the alcohol in alcoholic beverages.

The first reaction converts ethanol to acetaldehyde, which is far more toxic than ethanol:

$$CH_3CH_2OH + NAD^+ \longrightarrow CH_3CHO + NADH + H^+$$

It is acetaldehyde ($CH_3CHO$) that causes the misery associated with a hangover—including headache, nausea, and vomiting.

In the second reaction, acetaldehyde is converted to acetic acid, which is harmless:

$$CH_3CHO + NADP^+ + H_2O \longrightarrow CH_3COOH + NADPH + H^+$$

An effective but painful part of the treatment for alcohol abuse is to administer *disulfiram,* marketed under the name Antabuse. Disulfiram blocks the action of ALDH, preventing the conversion of acetaldehyde to acetic acid. The resulting buildup of acetaldehyde causes the patient to feel very sick almost immediately, making the next cocktail *far* less appealing. The action of disulfiram was discovered accidentally when Danish pharmaceutical researchers who were taking the drug as an experimental treatment for parasitic diseases became very ill every time they consumed alcohol.

# Applying What You've Learned

It takes as little as 5 mL (1 tsp) of methanol to cause permanent blindness or death; and unlike ethanol, methanol can be absorbed in toxic amounts by ingestion, inhalation of vapor, or absorption through the skin. Nevertheless, methanol is present in a number of common household products including antifreeze, windshield-washing fluid, and paint remover.

One method that has been used to synthesize methanol is the combination of carbon monoxide and hydrogen gases at 100°C:

$$CO(g) + 2H_2(g) \longrightarrow CH_3OH(g)$$

This reaction is catalyzed by a nickel compound.

**Problems:**

a)   Write the expression for the rate of this reaction in terms of [CO]. [◄◄ Sample Problem 14.1]

b)   Write the expression for the rate of this reaction in terms of $[H_2]$ and in terms of $[CH_3OH]$. [◄◄ Sample Problem 14.2]

c)   Given the following table of experimental data at 100°C, determine the rate law and the rate constant for the reaction. Then, determine the initial rate of the reaction when the starting concentration of CO is 16.5 $M$. [◄◄ Sample Problem 14.3]

| Experiment | [CO] ($M$) | $[H_2]$ ($M$) | Initial Rate ($M$/s) |
|---|---|---|---|
| 1 | 5.60 | 11.2 | 0.952 |
| 2 | 5.60 | 22.4 | 0.952 |
| 3 | 11.2 | 11.2 | 1.90 |

d)   Calculate the time required for the concentration of CO to be reduced from 16.5 $M$ to 1.91 $M$. [◄◄ Sample Problem 14.4]

e)   Calculate $t_{1/2}$ of the reaction. [◄◄ Sample Problem 14.6]

f)   Given that $k$ is 3.0 s$^{-1}$ at 200°C, calculate $E_a$ of the reaction. [◄◄ Sample Problem 14.9]

g)   Use the calculated value of $E_a$ to determine the value of $k$ at 180°C. [◄◄ Sample Problem 14.10]

# CHAPTER SUMMARY

## Section 14.1

• The rate of a chemical reaction is the change in concentration of reactants or products over time. Rates may be expressed as an *average* rate over a given time interval or as an **instantaneous rate.**

• The **rate constant (k)** is a proportionality constant that relates the **rate of reaction** with the concentration(s) of reactant(s). The rate constant $k$ for a given reaction changes only with temperature.

## Section 14.2

• The **rate law** is an equation that expresses the relationship between rate and reactant concentration(s). In general, the rate law for the reaction of A and B is rate $= k[A]^x[B]^y$.

• The **reaction order** is the power to which the concentration of a given reactant is raised in the rate law equation. The overall reaction order is the sum of the powers to which reactant concentrations are raised in the rate law.

• The **initial rate** is the instantaneous rate of reaction when the reactant concentrations are *starting* concentrations.

• The rate law and reaction order must be determined by comparing changes in the initial rate with changes in starting reactant concentrations. In general, the rate law cannot be determined solely from the balanced equation.

## Section 14.3

• The **integrated rate law** can be used to determine reactant concentrations after a specified period of time. It can also be used to determine how long it will take to reach a specified reactant concentration.

• The rate of a **first-order reaction** is proportional to the concentration of a single reactant. The rate of a **second-order reaction** is proportional to the product of two reactant concentrations ([A][B]), or on the concentration of a single reactant squared ($[A]^2$ or $[B]^2$). The rate of a **zeroth-order reaction** does not depend on reactant concentration.

• The **half-life ($t_{1/2}$)** of a reaction is the time it takes for half of a reactant to be consumed. The half-life is *constant* for first-order reactions, and it can be used to determine the rate constant of the reaction.

## Section 14.4

• **Collision theory** explains why the rate constant, and therefore the reaction rate, increases with increasing temperature. The relationship between temperature and the rate constant is expressed by the **Arrhenius equation.**

• Reactions occur when molecules of sufficient energy (and appropriate orientation) collide. **Effective collisions** are those that result in the formation of an **activated complex,** also called a **transition state.** Only *effective* collisions can result in product formation.

• The **activation energy ($E_a$)** is the minimum energy that colliding molecules must possess in order for the collision to be effective.

## Section 14.5

• A **reaction mechanism** may consist of a series of *steps,* called **elementary reactions.** Unlike rate laws in general, the rate law for an elementary reaction can be written from the balanced equation, using the stoichiometric coefficient for each reactant species as its exponent in the rate law.

• A species that is produced in one step of a reaction mechanism and subsequently consumed in another step is called an **intermediate.** A species that is first consumed and later regenerated is called a *catalyst.* Neither intermediates nor catalysts appear in the overall balanced equation.

• The rate law of each step in a reaction mechanism indicates the **molecularity** or overall *order* of the step. A **unimolecular** step is first order, involving just one molecule; a **bimolecular** step is second order, involving the collision of two molecules; and a **termolecular** step is third order, involving the collision of three molecules. Termolecular processes are relatively rare.

• If one step in a reaction is much slower than all the other steps, it is the **rate-determining step.** The rate-determining step has a rate law identical to the experimental rate law.

## Section 14.6

• A *catalyst* speeds up a reaction, usually by lowering the value of the activation energy. *Catalysis* refers to the process by which a catalyst increases the reaction rate.

• Catalysis may be **heterogeneous,** in which the catalyst and reactants exist in different phases, or **homogeneous,** in which the catalyst and reactants exist in the same phase.

• **Enzymes** are biological catalysts with high specificity for the reactions that they catalyze.

# KEY WORDS

Activated complex, 599

Activation energy ($E_a$), 599

Arrhenius equation, 600

Bimolecular, 604

Catalyst, 610

Collision theory, 599

Effective collision, 599

Elementary reaction, 604

Enzymes, 613

First-order reaction, 592

Half-life ($t_{1/2}$), 594

Heterogeneous catalysis, 612

Homogeneous catalysis, 612

Initial rate, 587

Instantaneous rate, 582

Integrated rate law, 592

Intermediate, 604

Molecularity, 604

Rate constant ($k$), 583

Rate-determining step, 605

Rate law, 587

Rate of reaction, 584

Reaction mechanism, 604

Reaction order, 587

Second-order reaction, 596

Termolecular, 604

Transition state, 599

Unimolecular, 604

Zeroth-order reaction, 598

# KEY EQUATIONS

14.1    $\text{rate} = -\dfrac{1}{a}\dfrac{\Delta[A]}{\Delta t} = -\dfrac{1}{b}\dfrac{\Delta[B]}{\Delta t} = \dfrac{1}{c}\dfrac{\Delta[C]}{\Delta t} = \dfrac{1}{d}\dfrac{\Delta[D]}{\Delta t}$

14.2    $\text{rate} = k[A]^x[B]^y$

14.3    $\ln\dfrac{[A]_t}{[A]_0} = -kt$

14.4    $\ln[A]_t = -kt + \ln[A]_0$

14.5    $t_{1/2} = \dfrac{0.693}{k}$

14.6    $\dfrac{1}{[A]_t} = kt + \dfrac{1}{[A]_0}$

14.7    $t_{1/2} = \dfrac{1}{k[A]_0}$

14.8    $k = Ae^{-E_a/RT}$

14.9    $\ln k = \ln A - \dfrac{E_a}{RT}$

14.10    $\ln k = \left(-\dfrac{E_a}{R}\right)\left(\dfrac{1}{T}\right) + \ln A$

14.11    $\ln\dfrac{k_1}{k_2} = \dfrac{E_a}{R}\left(\dfrac{1}{T_2} - \dfrac{1}{T_1}\right)$

# QUESTIONS AND PROBLEMS

## Section 14.1: Reaction Rates

### Review Questions

14.1    What is meant by the *rate* of a chemical reaction? What are the units of the rate of a reaction?

14.2    Distinguish between average rate and instantaneous rate. Which of the two rates gives us an unambiguous measurement of reaction rate? Why?

14.3    What are the advantages of measuring the initial rate of a reaction?

14.4    Identify two reactions that are very slow (take days or longer to complete) and two reactions that are very fast (reactions that are over in minutes or seconds).

### Problems

14.5    Write the reaction rate expressions for the following reactions in terms of the disappearance of the reactants and the appearance of products:

(a) $H_2(g) + I_2(g) \longrightarrow 2HI(g)$
(b) $5Br^-(aq) + BrO_3^-(aq) + 6H^+(aq) \longrightarrow 3Br_2(aq) + 3H_2O(l)$

14.6    Write the reaction rate expressions for the following reactions in terms of the disappearance of the reactants and the appearance of products:

(a) $2H_2(g) + O_2(g) \longrightarrow 2H_2O(g)$
(b) $4NH_3(g) + 5O_2(g) \longrightarrow 4NO(g) + 6H_2O(g)$

14.7    Consider the reaction

$$2NO(g) + O_2(g) \longrightarrow 2NO_2(g)$$

Suppose that at a particular moment during the reaction nitric oxide (NO) is reacting at the rate of 0.066 *M*/s. (a) At what rate is $NO_2$ being formed? (b) At what rate is molecular oxygen reacting?

14.8    Consider the reaction

$$N_2(g) + 3H_2(g) \longrightarrow 2NH_3(g)$$

Suppose that at a particular moment during the reaction molecular hydrogen is reacting at the rate of 0.082 *M*/s. (a) At what rate is ammonia being formed? (b) At what rate is molecular nitrogen reacting?

## Section 14.2: Dependence of Reaction Rate on Reactant Concentration

### Review Questions

14.9    Explain what is meant by the *rate law* of a reaction.

14.10    Explain what is meant by the *order* of a reaction.

14.11    What are the units for the rate constants of first-order and second-order reactions?

14.12    Consider the zeroth-order reaction: A $\longrightarrow$ product. (a) Write the rate law for the reaction. (b) What are the units for the rate constant? (c) Plot the rate of the reaction versus [A].

14.13    The rate constant of a first-order reaction is 66 $s^{-1}$. What is the rate constant in units of minutes?

14.14 On which of the following properties does the rate constant of a reaction depend: (a) reactant concentrations, (b) nature of reactants, (c) temperature?

## Problems

**14.15** The rate law for the reaction

$$NH_4^+(aq) + NO_2^-(aq) \longrightarrow N_2(g) + 2H_2O(l)$$

is given by rate = $k[NH_4^+][NO_2^-]$. At 25°C, the rate constant is 3.0 $\times 10^{-4}/M \cdot s$. Calculate the rate of the reaction at this temperature if $[NH_4^+] = 0.36 \, M$ and $[NO_2^-] = 0.075 \, M$.

**14.16** Use the data in Table 14.2 to calculate the rate of the reaction at the time when $[F_2] = 0.020 \, M$ and $[ClO_2] = 0.035 \, M$.

**14.17** Consider the reaction

$$A + B \longrightarrow products$$

From the following data obtained at a certain temperature, determine the order of the reaction and calculate the rate constant.

| [A] (*M*) | [B] (*M*) | Rate (*M/s*) |
|---|---|---|
| 1.50 | 1.50 | $3.20 \times 10^{-1}$ |
| 1.50 | 2.50 | $3.20 \times 10^{-1}$ |
| 3.00 | 1.50 | $6.40 \times 10^{-1}$ |

**14.18** Consider the reaction

$$X + Y \longrightarrow Z$$

From the following data, obtained at 360 K, (a) determine the order of the reaction, and (b) determine the initial rate of disappearance of X when the concentration of X is 0.30 *M* and that of Y is 0.40 *M*.

| Initial Rate of Disappearance of X (*M/s*) | [X] (*M*) | [Y] (*M*) |
|---|---|---|
| 0.053 | 0.10 | 0.50 |
| 0.127 | 0.20 | 0.30 |
| 1.02 | 0.40 | 0.60 |
| 0.254 | 0.20 | 0.60 |
| 0.509 | 0.40 | 0.30 |

**14.19** Determine the overall orders of the reactions to which the following rate laws apply: (a) rate = $k[NO_2]^2$, (b) rate = $k$, (c) rate = $k[H_2]^2[Br_2]^{1/2}$, (d) rate = $k[NO]^2[O_2]$.

**14.20** Consider the reaction

$$A \longrightarrow B$$

The rate of the reaction is $1.6 \times 10^{-2} \, M/s$ when the concentration of A is 0.15 *M*. Calculate the rate constant if the reaction is (a) first order in A and (b) second order in A.

**14.21** Cyclobutane decomposes to ethylene according to the equation

$$C_4H_8(g) \longrightarrow 2C_2H_4(g)$$

Determine the order of the reaction and the rate constant based on the following pressures, which were recorded when the reaction was carried out at 430°C in a constant-volume vessel.

| Time (s) | $P_{C_4H_8}$ (mmHg) |
|---|---|
| 0 | 400 |
| 2,000 | 316 |
| 4,000 | 248 |
| 6,000 | 196 |
| 8,000 | 155 |
| 10,000 | 122 |

**14.22** The following gas-phase reaction was studied at 290°C by observing the change in pressure as a function of time in a constant-volume vessel:

$$ClCO_2CCl_3(g) \longrightarrow 2COCl_2(g)$$

Determine the order of the reaction and the rate constant based on the following data.

| Time (s) | P (mmHg) |
|---|---|
| 0 | 15.76 |
| 181 | 18.88 |
| 513 | 22.79 |
| 1164 | 27.08 |

where P is the total pressure.

## Section 14.3: Dependence of Reactant Concentration on Time

### Review Questions

14.23 Write an equation relating the concentration of a reactant A at $t = 0$ to that at $t = t$ for a first-order reaction. Define all the terms, and give their units. Do the same for a second-order reaction.

14.24 Define *half-life*. Write the equation relating the half-life of a first-order reaction to the rate constant.

14.25 Write the equation relating the half-life of a second-order reaction to the rate constant. How does it differ from the equation for a first-order reaction?

14.26 For a first-order reaction, how long will it take for the concentration of reactant to fall to one-eighth its original value? Express your answer in terms of the half-life ($t_{1/2}$) and in terms of the rate constant $k$.

### Problems

**14.27** What is the half-life of a compound if 75 percent of a given sample of the compound decomposes in 60 min? Assume first-order kinetics.

**14.28** The thermal decomposition of phosphine ($PH_3$) into phosphorus and molecular hydrogen is a first-order reaction:

$$4PH_3(g) \longrightarrow P_4(g) + 6H_2(g)$$

The half-life of the reaction is 35.0 s at 680°C. Calculate (a) the first-order rate constant for the reaction and (b) the time required for 95 percent of the phosphine to decompose.

**14.29** The rate constant for the second-order reaction

$$2NOBr(g) \longrightarrow 2NO(g) + Br_2(g)$$

is $0.80/M \cdot s$ at 10°C. (a) Starting with a concentration of 0.086 *M*, calculate the concentration of NOBr after 22 s. (b) Calculate the half-lives when $[NOBr]_0 = 0.072 \, M$ and $[NOBr]_0 = 0.054 \, M$.

**14.30** The rate constant for the second-order reaction

$$2NO_2(g) \longrightarrow 2NO(g) + O_2(g)$$

is $0.54/M \cdot s$ at 300°C. How long (in seconds) would it take for the concentration of $NO_2$ to decrease from 0.65 *M* to 0.18 *M*?

**14.31** The second-order rate constant for the dimerization of a protein (P)

$$P + P \longrightarrow P_2$$

is $6.2 \times 10^{-3}/M \cdot s$ at 25°C. If the concentration of the protein is $2.7 \times 10^{-4} \, M$, calculate the initial rate (*M/s*) of formation of $P_2$. How long (in seconds) will it take to decrease the concentration of P to $2.7 \times 10^{-5} \, M$?

14.32  Consider the first-order reaction X ⟶ Y shown here. (a) What is the half-life of the reaction? (b) Draw pictures showing the number of X (red) and Y (blue) molecules at 20 s and at 30 s.

$t = 0\ \text{s}$          $t = 10\ \text{s}$

**14.33**  The reaction A ⟶ B shown here follows first-order kinetics. Initially different amounts of A molecules are placed in three containers of equal volume at the same temperature. (a) What are the relative rates of the reaction in these three containers? (b) How would the relative rates be affected if the volume of each container were doubled? (c) What are the relative half-lives of the reactions in (i) to (iii)?

(i)              (ii)              (iii)

## Section 14.4: Dependence of Reaction Rate on Temperature

### Review Questions

14.34  Define *activation energy*. What role does activation energy play in chemical kinetics?

14.35  Write the Arrhenius equation, and define all terms.

14.36  Use the Arrhenius equation to show why the rate constant of a reaction (a) decreases with increasing activation energy and (b) increases with increasing temperature.

14.37  The burning of methane in oxygen is a highly exothermic reaction. Yet a mixture of methane and oxygen gas can be kept indefinitely without any apparent change. Explain.

14.38  Sketch a potential energy versus reaction progress plot for the following reactions:
(a) $S(s) + O_2(g) \longrightarrow SO_2(g)$     $\Delta H° = -296\ \text{kJ/mol}$
(b) $Cl_2(g) \longrightarrow Cl(g) + Cl(g)$     $\Delta H° = 243\ \text{kJ/mol}$

14.39  The reaction $H + H_2 \longrightarrow H_2 + H$ has been studied for many years. Sketch a potential energy versus reaction progress diagram for this reaction.

14.40  Over the range of about ±3°C from normal body temperature, the metabolic rate, $M_T$, is given by $M_T = M_{37}(1.1)^{\Delta T}$, where $M_{37}$ is the normal rate (at 37°C) and $\Delta T$ is the change in $T$. Discuss this equation in terms of a possible molecular interpretation.

### Problems

**14.41**  Variation of the rate constant with temperature for the first-order reaction

$$2N_2O_5(g) \longrightarrow 2N_2O_4(g) + O_2(g)$$

is given in the following table. Determine graphically the activation energy for the reaction.

| $T$ (K) | $k$ (s$^{-1}$) |
|---------|----------------|
| 298     | $1.74 \times 10^{-5}$ |
| 308     | $6.61 \times 10^{-5}$ |
| 318     | $2.51 \times 10^{-4}$ |
| 328     | $7.59 \times 10^{-4}$ |
| 338     | $2.40 \times 10^{-3}$ |

**14.42**  Given the same reactant concentrations, the reaction

$$CO(g) + Cl_2(g) \longrightarrow COCl_2(g)$$

at 250°C is $1.50 \times 10^3$ times as fast as the same reaction at 150°C. Calculate the activation energy for this reaction. Assume that the frequency factor is constant.

**14.43**  For the reaction

$$NO(g) + O_3(g) \longrightarrow NO_2(g) + O_2(g)$$

the frequency factor $A$ is $8.7 \times 10^{12}$ s$^{-1}$ and the activation energy is 63 kJ/mol. What is the rate constant for the reaction at 75°C?

**14.44**  The rate constant of a first-order reaction is $4.60 \times 10^{-4}$ s$^{-1}$ at 350°C. If the activation energy is 104 kJ/mol, calculate the temperature at which its rate constant is $8.80 \times 10^{-4}$ s$^{-1}$.

**14.45**  The rate constants of some reactions double with every 10° rise in temperature. Assume that a reaction takes place at 295 K and 305 K. What must the activation energy be for the rate constant to double as described?

**14.46**  The rate at which tree crickets chirp is $2.0 \times 10^2$ per minute at 27°C but only 39.6 per minute at 5°C. From these data, calculate the "activation energy" for the chirping process. (*Hint:* The ratio of rates is equal to the ratio of rate constants.)

14.47  The rate of bacterial hydrolysis of fish muscle is twice as great at 2.2°C as at −1.1°C. Estimate an $E_a$ value for this reaction. Is there any relation to the problem of storing fish for food?

**14.48**  The activation energy for the denaturation of a protein is 396 kJ/mol. At what temperature will the rate of denaturation be 20 percent greater than its rate at 25°C?

**14.49**  Diagram A describes the initial state of reaction

$$H_2 + Cl_2 \longrightarrow 2HCl.$$

        ● H$_2$
        ●● Cl$_2$
        ◑ HCl

Diagram A

Suppose the reaction is carried out at two different temperatures as shown in diagram B. Which picture represents the result at the higher temperature? (The reaction proceeds for the same amount of time at both temperatures.)

(a)              (b)

Diagram B

## Section 14.5: Reaction Mechanisms

### Review Questions

14.50 What do we mean by the *mechanism* of a reaction?

14.51 What is an elementary step? What is the molecularity of a reaction?

14.52 Classify the following elementary reactions as unimolecular, bimolecular, or termolecular:

(a) $2NO + Br_2 \longrightarrow 2NOBr$
(b) $CH_3NC \longrightarrow CH_3CN$
(c) $SO + O_2 \longrightarrow SO_2 + O$

14.53 Reactions can be classified as unimolecular, bimolecular, and so on. Why are there no zero-molecular reactions? Explain why termolecular reactions are rare.

14.54 Determine the molecularity, and write the rate law for each of the following elementary steps:

(a) $X \longrightarrow$ products
(b) $X + Y \longrightarrow$ products
(c) $X + Y + Z \longrightarrow$ products
(d) $X + X \longrightarrow$ products
(e) $X + 2Y \longrightarrow$ products

14.55 What is the rate-determining step of a reaction? Give an everyday analogy to illustrate the meaning of *rate determining*.

14.56 The equation for the combustion of ethane ($C_2H_6$) is

$$2C_2H_6(g) + 7O_2(g) \longrightarrow 4CO_2(g) + 6H_2O(l)$$

Explain why it is unlikely that this equation also represents the elementary step for the reaction.

14.57 Specify which of the following species cannot be isolated in a reaction: activated complex, product, intermediate.

### Problems

14.58 Classify each of the following elementary steps as unimolecular, bimolecular, or termolecular.

(a)

(b)

(c)

14.59 The rate law for the reaction

$$2NO(g) + Cl_2(g) \longrightarrow 2NOCl(g)$$

is given by rate $= k[NO][Cl_2]$. (a) What is the order of the reaction? (b) A mechanism involving the following steps has been proposed for the reaction:

$$NO(g) + Cl_2(g) \longrightarrow NOCl_2(g)$$
$$NOCl_2(g) + NO(g) \longrightarrow 2NOCl(g)$$

If this mechanism is correct, what does it imply about the relative rates of these two steps?

14.60 For the reaction $X_2 + Y + Z \longrightarrow XY + XZ$, it is found that doubling the concentration of $X_2$ doubles the reaction rate,

tripling the concentration of Y triples the rate, and doubling the concentration of Z has no effect. (a) What is the rate law for this reaction? (b) Why is it that the change in the concentration of Z has no effect on the rate? (c) Suggest a mechanism for the reaction that is consistent with the rate law.

14.61 The rate law for the reaction

$$2H_2(g) + 2NO(g) \longrightarrow N_2(g) + 2H_2O(g)$$

is rate $= k[H_2][NO]^2$. Which of the following mechanisms can be ruled out on the basis of the observed rate expression?

*Mechanism I*

$$H_2 + NO \longrightarrow H_2O + N \quad \text{(slow)}$$
$$N + NO \longrightarrow N_2 + O \quad \text{(fast)}$$
$$O + H_2 \longrightarrow H_2O \quad \text{(fast)}$$

*Mechanism II*

$$H_2 + 2NO \longrightarrow N_2O + H_2O \quad \text{(slow)}$$
$$N_2O + H_2 \longrightarrow N_2 + H_2O \quad \text{(fast)}$$

*Mechanism III*

$$2NO \rightleftharpoons N_2O_2 \quad \text{(fast equilibrium)}$$
$$N_2O_2 + H_2 \longrightarrow N_2O + H_2O \quad \text{(slow)}$$
$$N_2O + H_2 \longrightarrow N_2 + H_2O \quad \text{(fast)}$$

14.62 The rate law for the decomposition of ozone to molecular oxygen

$$2O_3(g) \longrightarrow 3O_2(g)$$

is

$$\text{rate} = k\frac{[O_3]^2}{[O_2]}$$

The mechanism proposed for this process is

$$O_3 \underset{k_{-1}}{\overset{k_1}{\rightleftharpoons}} O + O_2$$

$$O + O_3 \overset{k_2}{\longrightarrow} 2O_2$$

Derive the rate law from these elementary steps. Clearly state the assumptions you use in the derivation. Explain why the rate decreases with increasing $O_2$ concentration.

## Section 14.6: Catalysis

### Review Questions

14.63 How does a catalyst increase the rate of a reaction?

14.64 What are the characteristics of a catalyst?

14.65 A certain reaction is known to proceed slowly at room temperature. Is it possible to make the reaction proceed at a faster rate without changing the temperature?

14.66 Most reactions, including enzyme-catalyzed reactions, proceed faster at higher temperatures. However, for a given enzyme, the rate drops off abruptly at a certain temperature. Account for this behavior.

14.67 Are enzyme-catalyzed reactions examples of homogeneous or heterogeneous catalysis? Explain.

14.68 The concentrations of enzymes in cells are usually quite small. What is the biological significance of this fact?

14.69 When fruits such as apples and pears are cut, the exposed areas begin to turn brown. This is the result of an enzyme-catalyzed reaction. Often the browning can be prevented or slowed by adding a few drops of lemon juice. What is the chemical basis of this treatment?

**14.70** The first-order rate constant for the dehydration of carbonic acid

$$H_2CO_3 \longrightarrow CO_2 + H_2O$$

is about $1 \times 10^2 \text{ s}^{-1}$. In view of this rather high rate constant, explain why it is necessary to have the enzyme carbonic anhydrase to enhance the rate of dehydration in the lungs.

### Problems

**14.71** Which two potential-energy profiles represent the same reaction catalyzed and uncatalyzed?

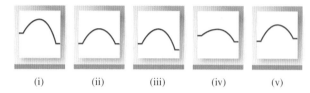

(i)        (ii)        (iii)        (iv)        (v)

**14.72** Consider the following mechanism for the enzyme-catalyzed reaction:

$$E + S \underset{k_{-1}}{\overset{k_1}{\rightleftharpoons}} ES \quad \text{(fast equilibrium)}$$

$$ES \xrightarrow{k_2} E + P \quad \text{(slow)}$$

Derive an expression for the rate law of the reaction in terms of the concentrations of E and S. (*Hint:* To solve for [ES], make use of the fact that, at equilibrium, the rate of the forward reaction is equal to the rate of the reverse reaction.)

### Additional Problems

**14.73** List four factors that influence the rate of a reaction.

**14.74** Suggest experimental means by which the rates of the following reactions could be followed:

(a) $CaCO_3(s) \longrightarrow CaO(s) + CO_2(g)$
(b) $Cl_2(g) + 2Br^-(aq) \longrightarrow Br_2(aq) + 2Cl^-(aq)$
(c) $C_2H_6(g) \longrightarrow C_2H_4(g) + H_2(g)$
(d) $C_2H_5I(g) + H_2O(l) \longrightarrow C_2H_5OH(aq) + H^+(aq) + I^-(aq)$

**14.75** "The rate constant for the reaction

$$NO_2(g) + CO(g) \longrightarrow NO(g) + CO_2(g)$$

is $1.64 \times 10^{-6}/M \cdot s$." What is incomplete about this statement?

**14.76** In a certain industrial process involving a heterogeneous catalyst, the volume of the catalyst (in the shape of a sphere) is $10.0 \text{ cm}^3$. Calculate the surface area of the catalyst. If the sphere is broken down into eight smaller spheres, each having a volume of $1.25 \text{ cm}^3$, what is the total surface area of the spheres? Which of the two geometric configurations of the catalyst is more effective? (The surface area of a sphere is $4\pi r^2$, where $r$ is the radius of the sphere.) Based on your analysis here, explain why it is sometimes dangerous to work in grain elevators.

**14.77** The following pictures represent the progress of the reaction A ⟶ B where the red spheres represent A molecules and the green spheres represent B molecules. Calculate the rate constant of the reaction.

$t = 0 \text{ s}$        $t = 20 \text{ s}$        $t = 40 \text{ s}$

**14.78** The following pictures show the progress of the reaction $2A \longrightarrow A_2$. Determine whether the reaction is first order or second order, and calculate the rate constant.

$t = 0 \text{ min}$        $t = 15 \text{ min}$        $t = 30 \text{ min}$

**14.79** Use the data in Sample Problem 14.5 to determine graphically the half-life of the reaction.

**14.80** The following data were collected for the reaction between hydrogen and nitric oxide at 700°C:

$$2H_2(g) + 2NO(g) \longrightarrow 2H_2O(g) + N_2(g)$$

| Experiment | $[H_2]$ (*M*) | [NO] (*M*) | Initial Rate (*M*/s) |
|---|---|---|---|
| 1 | 0.010 | 0.025 | $2.4 \times 10^{-6}$ |
| 2 | 0.0050 | 0.025 | $1.2 \times 10^{-6}$ |
| 3 | 0.010 | 0.0125 | $0.60 \times 10^{-6}$ |

(a) Determine the order of the reaction. (b) Calculate the rate constant. (c) Suggest a plausible mechanism that is consistent with the rate law. (*Hint:* Assume that the oxygen atom is the intermediate.)

**14.81** When methyl phosphate is heated in acid solution, it reacts with water:

$$CH_3OPO_3H_2 + H_2O \longrightarrow CH_3OH + H_3PO_4$$

If the reaction is carried out in water enriched with $^{18}O$, the oxygen-18 isotope is found in the phosphoric acid product but not in the methanol. What does this tell us about the mechanism of the reaction?

**14.82** The rate of the reaction

$$CH_3COOC_2H_5(aq) + H_2O(l) \longrightarrow$$
$$CH_3COOH(aq) + C_2H_5OH(aq)$$

shows first-order characteristics—that is, rate = $k[CH_3COOC_2H_5]$—even though this is a second-order reaction (first order in $CH_3COOC_2H_5$ and first order in $H_2O$). Explain.

**14.83** Explain why most metals used in catalysis are transition metals.

**14.84** The reaction $2A + 3B \longrightarrow C$ is first order with respect to A and B. When the initial concentrations are [A] = $1.6 \times 10^{-2} M$ and [B] = $2.4 \times 10^{-3} M$, the rate is $4.1 \times 10^{-4} M$/s. Calculate the rate constant of the reaction.

**14.85** The bromination of acetone is acid-catalyzed:

$$CH_3COCH_3 + Br_2 \xrightarrow[\text{catalyst}]{H^+} CH_3COCH_2Br + H^+ + Br^-$$

The rate of disappearance of bromine was measured for several different concentrations of acetone, bromine, and H$^+$ ions at a certain temperature:

| | $[CH_3COCH_3]$ (*M*) | $[Br_2]$ (*M*) | $[H^+]$ (*M*) | Rate of Disappearance of $Br_2$ (*M*/s) |
|---|---|---|---|---|
| (1) | 0.30 | 0.050 | 0.050 | $5.7 \times 10^{-5}$ |
| (2) | 0.30 | 0.10 | 0.050 | $5.7 \times 10^{-5}$ |
| (3) | 0.30 | 0.050 | 0.10 | $1.2 \times 10^{-4}$ |
| (4) | 0.40 | 0.050 | 0.20 | $3.1 \times 10^{-4}$ |
| (5) | 0.40 | 0.050 | 0.050 | $7.6 \times 10^{-5}$ |

(a) What is the rate law for the reaction? (b) Determine the rate constant. (c) The following mechanism has been proposed for the reaction:

$$CH_3-\overset{O}{\overset{||}{C}}-CH_3+H_3O^+ \rightleftharpoons CH_3-\overset{+OH}{\overset{||}{C}}-CH_3+H_2O \quad \text{(fast equilibrium)}$$

$$CH_3-\overset{+OH}{\overset{||}{C}}-CH_3+H_2O \longrightarrow CH_3-\overset{OH}{\overset{|}{C}}=CH_2+H_3O^+ \quad \text{(slow)}$$

$$CH_3-\overset{OH}{\overset{|}{C}}=CH_2+Br_2 \longrightarrow CH_3-\overset{O}{\overset{||}{C}}-CH_2Br+HBr \quad \text{(fast)}$$

Show that the rate law deduced from the mechanism is consistent with that shown in part (a).

**14.86** The decomposition of $N_2O$ to $N_2$ and $O_2$ is a first-order reaction. At 730°C the half-life of the reaction is $3.58 \times 10^3$ min. If the initial pressure of $N_2O$ is 2.10 atm at 730°C, calculate the total gas pressure after one half-life. Assume that the volume remains constant.

**14.87** The reaction $S_2O_8^{2-} + 2I^- \longrightarrow 2SO_4^{2-} + I_2$ proceeds slowly in aqueous solution, but it can be catalyzed by the $Fe^{3+}$ ion. Given that $Fe^{3+}$ can oxidize $I^-$ and $Fe^{2+}$ can reduce $S_2O_8^{2-}$, write a plausible two-step mechanism for this reaction. Explain why the uncatalyzed reaction is slow.

**14.88** What are the units of the rate constant for a third-order reaction?

**14.89** The integrated rate law for the zeroth-order reaction $A \longrightarrow B$ is $[A]_t = [A]_0 - kt$. (a) Sketch the following plots: (i) rate versus $[A]_t$ and (ii) $[A]_t$ versus $t$. (b) Derive an expression for the half-life of the reaction. (c) Calculate the time in half-lives when the integrated rate law is no longer valid, that is, when $[A]_t = 0$.

**14.90** A flask contains a mixture of compounds A and B. Both compounds decompose by first-order kinetics. The half-lives are 50.0 min for A and 18.0 min for B. If the concentrations of A and B are equal initially, how long will it take for the concentration of A to be four times that of B?

**14.91** Referring to Sample Problem 14.5, explain how you would measure the partial pressure of azomethane experimentally as a function of time.

**14.92** The rate law for the reaction $2NO_2(g) \longrightarrow N_2O_4(g)$ is rate = $k[NO_2]^2$. Which of the following changes will change the value of $k$? (a) The pressure of $NO_2$ is doubled. (b) The reaction is run in an organic solvent. (c) The volume of the container is doubled. (d) The temperature is decreased. (e) A catalyst is added to the container.

**14.93** The reaction of $G_2$ with $E_2$ to form 2EG is exothermic, and the reaction of $G_2$ with $X_2$ to form 2XG is endothermic. The activation energy of the exothermic reaction is greater than that of the endothermic reaction. Sketch the potential-energy profile diagrams for these two reactions on the same graph.

**14.94** In the nuclear industry, workers use a rule of thumb that the radioactivity from any sample will be relatively harmless after 10 half-lives. Calculate the fraction of a radioactive sample that remains after this time period. (*Hint:* Radioactive decays obey first-order kinetics.)

**14.95** Briefly comment on the effect of a catalyst on each of the following: (a) activation energy, (b) reaction mechanism,

(c) enthalpy of reaction, (d) rate of forward reaction, (e) rate of reverse reaction.

**14.96** When 6 g of granulated Zn is added to a solution of 2 $M$ HCl in a beaker at room temperature, hydrogen gas is generated. For each of the following changes (at constant volume of the acid) state whether the rate of hydrogen gas evolution will be increased, decreased, or unchanged: (a) 6 g of powdered Zn is used, (b) 4 g of granulated Zn is used, (c) 2 $M$ acetic acid is used instead of 2 $M$ HCl, (d) temperature is raised to 40°C.

**14.97** Strictly speaking, the rate law derived for the reaction in Problem 14.80 applies only to certain concentrations of $H_2$. The general rate law for the reaction takes the form

$$\text{rate} = \frac{k_1[NO]^2[H_2]}{1 + k_2[H_2]}$$

where $k_1$ and $k_2$ are constants. Derive rate law expressions under the conditions of very high and very low hydrogen concentrations. Does the result from Problem 14.80 agree with one of the rate expressions here?

**14.98** A certain first-order reaction is 35.5 percent complete in 4.90 min at 25°C. What is its rate constant?

**14.99** The decomposition of dinitrogen pentoxide has been studied in carbon tetrachloride solvent ($CCl_4$) at a certain temperature:

$$2N_2O_5 \longrightarrow 4NO_2 + O_2$$

| $[N_2O_5]$ ($M$) | Initial Rate ($M/s$) |
|---|---|
| 0.92 | $0.95 \times 10^{-5}$ |
| 1.23 | $1.20 \times 10^{-5}$ |
| 1.79 | $1.93 \times 10^{-5}$ |
| 2.00 | $2.10 \times 10^{-5}$ |
| 2.21 | $2.26 \times 10^{-5}$ |

Determine graphically the rate law for the reaction, and calculate the rate constant.

**14.100** The thermal decomposition of $N_2O_5$ obeys first-order kinetics. At 45°C, a plot of ln $[N_2O_5]$ versus $t$ gives a slope of $-6.18 \times 10^{-4}$ min$^{-1}$. What is the half-life of the reaction?

**14.101** When a mixture of methane and bromine is exposed to light, the following reaction occurs slowly:

$$CH_4(g) + Br_2(g) \longrightarrow CH_3Br(g) + HBr(g)$$

Suggest a reasonable mechanism for this reaction. (*Hint:* Bromine vapor is deep red; methane is colorless.)

**14.102** The rate of the reaction between $H_2$ and $I_2$ to form HI increases with the intensity of visible light. (a) Explain why this fact supports a two-step mechanism. ($I_2$ vapor is purple.) (b) Explain why the visible light has no effect on the formation of H atoms.

**14.103** To prevent brain damage, a standard procedure is to lower the body temperature of someone who has been resuscitated after suffering cardiac arrest. What is the physiochemical basis for this procedure?

**14.104** A gas mixture containing $CH_3$ fragments, $C_2H_6$ molecules, and an inert gas (He) was prepared at 600 K with a total pressure of 5.42 atm. The elementary reaction

$$CH_3 + C_2H_6 \longrightarrow CH_4 + C_2H_5$$

has a second-order rate constant of $3.0 \times 10^4 /M \cdot s$. Given that the mole fractions of $CH_3$ and $C_2H_6$ are 0.00093 and 0.00077, respectively, calculate the initial rate of the reaction at this temperature.

**14.105** Consider the following elementary step:

$$X + 2Y \longrightarrow XY_2$$

(a) Write a rate law for this reaction. (b) If the initial rate of formation of $XY_2$ is $3.8 \times 10^{-3}$ $M$/s and the initial concentrations of X and Y are 0.26 $M$ and 0.88 $M$, respectively, what is the rate constant of the reaction?

**14.106** In recent years, ozone in the stratosphere has been depleted at an alarmingly fast rate by chlorofluorocarbons (CFCs). A CFC molecule such as $CFCl_3$ is first decomposed by UV radiation:

$$CFCl_3 \longrightarrow CFCl_2 + Cl$$

The chlorine radical then reacts with ozone as follows:

$$Cl + O_3 \longrightarrow ClO + O_2$$
$$ClO + O \longrightarrow Cl + O_2$$

(a) Write the overall reaction for the last two steps. (b) What are the roles of Cl and ClO? (c) Why is the fluorine radical not important in this mechanism? (d) One suggestion to reduce the concentration of chlorine radicals is to add hydrocarbons such as ethane ($C_2H_6$) to the stratosphere. How will this work? (e) Draw potential energy versus reaction progress diagrams for the uncatalyzed and catalyzed (by Cl) destruction of ozone: $O_3 + O \longrightarrow 2O_2$. Use the thermodynamic data in Appendix 2 to determine whether the reaction is exothermic or endothermic.

**14.107** Chlorine oxide (ClO), which plays an important role in the depletion of ozone (see Problem 14.106), decays rapidly at room temperature according to the equation

$$2ClO(g) \longrightarrow Cl_2(g) + O_2(g)$$

From the following data, determine the reaction order and calculate the rate constant of the reaction.

| Time (s) | [ClO] ($M$) |
|---|---|
| $0.12 \times 10^{-3}$ | $8.49 \times 10^{-6}$ |
| $0.96 \times 10^{-3}$ | $7.10 \times 10^{-6}$ |
| $2.24 \times 10^{-3}$ | $5.79 \times 10^{-6}$ |
| $3.20 \times 10^{-3}$ | $5.20 \times 10^{-6}$ |
| $4.00 \times 10^{-3}$ | $4.77 \times 10^{-6}$ |

**14.108** A compound X undergoes two *simultaneous* first-order reactions as follows: $X \longrightarrow Y$ with rate constant $k_1$ and $X \longrightarrow Z$ with rate constant $k_2$. The ratio of $k_1/k_2$ at 40°C is 8.0. What is the ratio at 300°C? Assume that the frequency factors of the two reactions are the same.

**14.109** Consider a car fitted with a catalytic converter. The first 5 min or so after it is started are the most polluting. Why?

**14.110** The following scheme in which A is converted to B, which is then converted to C, is known as a consecutive reaction.

$$A \longrightarrow B \longrightarrow C$$

Assuming that both steps are first order, sketch on the same graph the variations of [A], [B], and [C] with time.

**14.111** (a) Consider two reactions, A and B. If the rate constant for reaction B increases by a larger factor than that of reaction A when the temperature is increased from $T_1$ to $T_2$, what can you conclude about the relative values of the activation energies of the two reactions? (b) If a bimolecular reaction occurs every time an A and a B molecule collide, what can you say about the orientation factor and activation energy of the reaction?

**14.112** The rate law for the following reaction

$$CO(g) + NO_2(g) \longrightarrow CO_2(g) + NO(g)$$

is rate $= k[NO_2]^2$. Suggest a plausible mechanism for the reaction, given that the unstable species $NO_3$ is an intermediate.

**14.113** Radioactive plutonium-239 ($t_{1/2} = 2.44 \times 10^5$ yr) is used in nuclear reactors and atomic bombs. If there are $5.0 \times 10^2$ g of the isotope in a small atomic bomb, how long will it take for the substance to decay to $1.0 \times 10^2$ g, too small an amount for an effective bomb?

**14.114** Many reactions involving heterogeneous catalysts are zeroth order; that is, rate $= k$. An example is the decomposition of phosphine ($PH_3$) over tungsten (W):

$$4PH_3(g) \longrightarrow P_4(g) + 6H_2(g)$$

It is found that the reaction is independent of [$PH_3$] as long as phosphine's pressure is sufficiently high ($\geq$ 1 atm). Explain.

**14.115** Thallium(I) is oxidized by cerium(IV) as follows:

$$Tl^+ + 2Ce^{4+} \longrightarrow Tl^{3+} + 2Ce^{3+}$$

The elementary steps, in the presence of Mn(II), are as follows:

$$Ce^{4+} + Mn^{2+} \longrightarrow Ce^{3+} + Mn^{3+}$$
$$Ce^{4+} + Mn^{3+} \longrightarrow Ce^{3+} + Mn^{4+}$$
$$Tl^+ + Mn^{4+} \longrightarrow Tl^{3+} + Mn^{2+}$$

(a) Identify the catalyst, intermediates, and the rate-determining step if the rate law is rate $= k[Ce^{4+}][Mn^{2+}]$. (b) Explain why the reaction is slow without the catalyst. (c) Classify the type of catalysis (homogeneous or heterogeneous).

**14.116** Sucrose ($C_{12}H_{22}O_{11}$), commonly called table sugar, undergoes hydrolysis (reaction with water) to produce fructose ($C_6H_{12}O_6$) and glucose ($C_6H_{12}O_6$):

$$C_{12}H_{22}O_{11} + H_2O \longrightarrow \underset{\text{fructose}}{C_6H_{12}O_6} + \underset{\text{glucose}}{C_6H_{12}O_6}$$

This reaction is of considerable importance in the candy industry. First, fructose is sweeter than sucrose. Second, a mixture of fructose and glucose, called *invert sugar*, does not crystallize, so the candy containing this sugar would be chewy rather than brittle as candy containing sucrose crystals would be. (a) From the following data determine the order of the reaction. (b) How long does it take to hydrolyze 95 percent of sucrose? (c) Explain why the rate law does not include [$H_2O$] even though water is a reactant.

| Time (min) | [$C_{12}H_{22}O_{11}$] ($M$) |
|---|---|
| 0 | 0.500 |
| 60.0 | 0.400 |
| 96.4 | 0.350 |
| 157.5 | 0.280 |

**14.117** The first-order rate constant for the decomposition of dimethyl ether

$$(CH_3)_2O(g) \longrightarrow CH_4(g) + H_2(g) + CO(g)$$

is $3.2 \times 10^{-4}$ s$^{-1}$ at 450°C. The reaction is carried out in a constant-volume flask. Initially only dimethyl ether is present and the pressure is 0.350 atm. What is the pressure of the system after 8.0 min? Assume ideal behavior.

**14.118** At 25°C, the rate constant for the ozone-depleting reaction

$$O(g) + O_3(g) \longrightarrow 2O_2(g)$$

is $7.9 \times 10^{-15}$ cm$^3$/molecule · s. Express the rate constant in units of $1/M \cdot$ s.

**14.119** Consider the following elementary steps for a consecutive reaction:

$$A \xrightarrow{k_1} B \xrightarrow{k_2} C$$

(a) Write an expression for the rate of change of B. (b) Derive an expression for the concentration of B under "steady-state" conditions; that is, when B is decomposing to C at the same rate as it is formed from A.

14.120 Ethanol is a toxic substance that, when consumed in excess, can impair respiratory and cardiac functions by interference with the neurotransmitters of the nervous system. In the human body, ethanol is metabolized by the enzyme alcohol dehydrogenase to acetaldehyde, which causes hangovers. Based on your knowledge of enzyme kinetics, explain why binge drinking (that is, consuming too much alcohol too fast) can prove fatal.

14.121 Strontium-90, a radioactive isotope, is a major product of an atomic bomb explosion. It has a half-life of 28.1 yr. (a) Calculate the first-order rate constant for the nuclear decay. (b) Calculate the fraction of $^{90}$Sr that remains after 10 half-lives. (c) Calculate the number of years required for 99.0 percent of $^{90}$Sr to disappear.

14.122 Consider the potential energy profiles for the following three reactions (from left to right). (1) Rank the rates (slowest to fastest) of the reactions. (2) Calculate $\Delta H$ for each reaction, and determine which reaction(s) are exothermic and which reaction(s) are endothermic. Assume the reactions have roughly the same frequency factors.

Reaction progress    Reaction progress    Reaction progress
        (a)                  (b)                  (c)

14.123 Consider the following potential energy profile for the A $\longrightarrow$ D reaction. (a) How many elementary steps are there? (b) How many intermediates are formed? (c) Which step is rate determining? (d) Is the overall reaction exothermic or endothermic?

Reaction progress

14.124 A factory that specializes in the refinement of transition metals such as titanium was on fire. The firefighters were advised not to douse the fire with water. Why?

14.125 The activation energy for the decomposition of hydrogen peroxide

$$2H_2O_2(aq) \longrightarrow 2H_2O(l) + O_2(g)$$

is 42 kJ/mol, whereas when the reaction is catalyzed by the enzyme catalase, it is 7.0 kJ/mol. Calculate the temperature that would cause the uncatalyzed decomposition to proceed as rapidly as the enzyme-catalyzed decomposition at 20°C. Assume the frequency factor A to be the same in both cases.

14.126 The *activity* of a radioactive sample is the number of nuclear disintegrations per second, which is equal to the first-order rate constant times the number of radioactive nuclei present. The fundamental unit of radioactivity is the *curie* (Ci), where 1 Ci corresponds to exactly $3.70 \times 10^{10}$ disintegrations per second.

This decay rate is equivalent to that of 1 g of radium-226. Calculate the rate constant and half-life for the radium decay. Starting with 1.0 g of the radium sample, what is the activity after 500 yr? The molar mass of Ra-226 is 226.03 g/mol.

14.127 To carry out metabolism, oxygen is taken up by hemoglobin (Hb) to form oxyhemoglobin (HbO$_2$) according to the simplified equation

$$Hb(aq) + O_2(aq) \xrightarrow{\ k\ } HbO_2(aq)$$

where the second-order rate constant is $2.1 \times 10^6/M \cdot s$ at 37°C. For an average adult, the concentrations of Hb and O$_2$ in the blood at the lungs are $8.0 \times 10^{-6}\,M$ and $1.5 \times 10^{-6}\,M$, respectively. (a) Calculate the rate of formation of HbO$_2$. (b) Calculate the rate of consumption of O$_2$. (c) The rate of formation of HbO$_2$ increases to $1.4 \times 10^{-4}\,M/s$ during exercise to meet the demand of the increased metabolism rate. Assuming the Hb concentration to remain the same, what must the oxygen concentration be to sustain this rate of HbO$_2$ formation?

14.128 At a certain elevated temperature, ammonia decomposes on the surface of tungsten metal as follows:

$$2NH_3 \longrightarrow N_2 + 3H_2$$

From the following plot of the rate of the reaction versus the pressure of NH$_3$, describe the mechanism of the reaction.

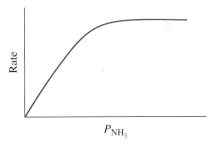

$P_{NH_3}$

14.129 The following expression shows the dependence of the half-life of a reaction ($t_{1/2}$) on the initial reactant concentration [A]$_0$:

$$t_{1/2} \propto \frac{1}{[A]_0^{n-1}}$$

where $n$ is the order of the reaction. Verify this dependence for zeroth-, first-, and second-order reactions.

14.130 Polyethylene is used in many items, including water pipes, bottles, electrical insulation, toys, and mailer envelopes. It is a *polymer*, a molecule with a very high molar mass made by joining many ethylene molecules together. (Ethylene is the basic unit, or monomer, for polyethylene.) The initiation step is

$$R_2 \xrightarrow{\ k_1\ } 2R \cdot \quad \text{(initiation)}$$

The R · species (called a radical) reacts with an ethylene molecule (M) to generate another radical

$$R \cdot + M \longrightarrow M_1 \cdot$$

The reaction of M$_1$ · with another monomer leads to the growth or propagation of the polymer chain:

$$M_1 \cdot + M \xrightarrow{\ k_p\ } M_2 \cdot \quad \text{(propagation)}$$

This step can be repeated with hundreds of monomer units. The propagation terminates when two radicals combine

$$M' \cdot + M'' \cdot \xrightarrow{\ k_t\ } M' - M'' \quad \text{(termination)}$$

The initiator frequently used in the polymerization of ethylene is benzoyl peroxide [(C$_6$H$_5$COO)$_2$]:

$$(C_6H_5COO)_2 \longrightarrow 2C_6H_5COO$$

This is a first-order reaction. The half-life of benzoyl peroxide at 100°C is 19.8 min. (a) Calculate the rate constant (in $min^{-1}$) of the reaction. (b) If the half-life of benzoyl peroxide is 7.30 h, or 438 min, at 70°C, what is the activation energy (in kJ/mol) for the decomposition of benzoyl peroxide? (c) Write the rate laws for the elementary steps in the preceding polymerization process, and identify the reactant, product, and intermediates. (d) What condition would favor the growth of long, high-molar-mass polyethylenes?

**14.131**  The rate constant for the gaseous reaction

$$H_2(g) + I_2(g) \longrightarrow 2HI(g)$$

is $2.42 \times 10^{-2}/M \cdot s$ at 400°C. Initially an equimolar sample of $H_2$ and $I_2$ is placed in a vessel at 400°C, and the total pressure is 1658 mmHg. (a) What is the initial rate ($M$/min) of formation of HI? (b) Determine the rate of formation of HI and the concentration of HI (in molarity) after 10.0 min.

14.132  A protein molecule P of molar mass $\mathcal{M}$ dimerizes when it is allowed to stand in solution at room temperature. A plausible mechanism is that the protein molecule is first denatured (that is, loses its activity due to a change in overall structure) before it dimerizes:

$$P \xrightarrow{\;k\;} P^* \text{ (denatured)} \qquad \text{(slow)}$$
$$2P^* \longrightarrow P_2 \qquad \text{(fast)}$$

where the asterisk denotes a denatured protein molecule. Derive an expression for the average molar mass (of P and $P_2$), $\overline{\mathcal{M}}$, in terms of the initial protein concentration $[P]_0$ and the concentration at time $t$, $[P]_t$, and $\mathcal{M}$. Describe how you would determine $k$ from molar mass measurements.

**14.133**  When the concentration of A in the reaction A $\longrightarrow$ B was changed from 1.20 $M$ to 0.60 $M$, the half-life increased from 2.0 min to 4.0 min at 25°C. Calculate the order of the reaction and the rate constant. (*Hint:* Use the equation in Problem 14.129.)

14.134  At a certain elevated temperature, ammonia decomposes on the surface of tungsten metal as follows:

$$NH_3 \longrightarrow \tfrac{1}{2}N_2 + \tfrac{3}{2}H_2$$

The kinetic data are expressed as the variation of the half-life with the initial pressure of $NH_3$:

| $P$ (mmHg) | $t_{1/2}$ (s) |
|---|---|
| 264 | 456 |
| 130 | 228 |
| 59 | 102 |
| 16 | 60 |

(a) Determine the order of the reaction. (b) How does the order depend on the initial pressure? (c) How does the mechanism of the reaction vary with pressure? (*Hint:* You need to use the equation in Problem 14.129 and plot log $t_{1/2}$ versus log $P$.)

14.135  The activation energy for the reaction

$$N_2O(g) \longrightarrow N_2(g) + O(g)$$

is $2.4 \times 10^2$ kJ/mol at 600 K. Calculate the percentage of the increase in rate from 600 K to 606 K. Comment on your results.

**14.136**  The rate of a reaction was followed by the absorption of light by the reactants and products as a function of wavelengths ($\lambda_1, \lambda_2, \lambda_3$) as time progresses. Which of the following mechanisms is consistent with the experimental data?

(a) A $\longrightarrow$ B, A $\longrightarrow$ C
(b) A $\longrightarrow$ B + C
(c) A $\longrightarrow$ B, B $\longrightarrow$ C + D
(d) A $\longrightarrow$ B, B $\longrightarrow$ C

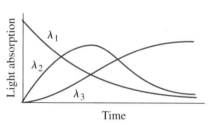

14.137  $\Delta H°$ for the reaction in Problem 14.135 is $-164$ kJ/mol. What is the activation energy ($E_a$) for the reverse reaction?

# PRE-PROFESSIONAL PRACTICE EXAM PROBLEMS: PHYSICAL AND BIOLOGICAL SCIENCES

Metastron, an aqueous solution of $^{89}SrCl_2$, is a drug used to alleviate severe bone pain associated with some metastatic cancers. It contains the radioactive isotope $^{89}Sr$, which emits $\beta$ radiation and has a constant half-life of 50.5 days. Because the strontium ion is chemically similar to the calcium ion, $^{89}Sr$ is incorporated into bone tissue, especially in the bone lesions caused by cancer that has spread. Treatment with Metastron typically involves intravenous injection of 4 mL over the course of about 2 min.

1. Calculate the numerical value of the rate constant ($k$) for the decay of $^{89}Sr$.

   a) 0.0137
   b) 0.0127
   c) 0.693
   d) 0.0198

2. What are the units of the rate constant for the decay of $^{89}Sr$?

   a) Days
   b) Reciprocal days
   c) Reciprocal seconds
   d) The rate constant has no units.

3. What percentage of the original activity of $^{89}Sr$ remains 101 days after an injection of Metastron?

   a) 75 percent
   b) 50 percent
   c) 25 percent
   d) 12 percent

4. How many days must pass following an injection of Metastron for the $\beta$ emission to fall to 10 percent of its original value?

   a) 182
   b) 168
   c) 505
   d) 455

# ANSWERS TO IN-CHAPTER MATERIALS

## Answers to Practice Problems

**14.1A** (a) rate $= -\dfrac{\Delta[CO_2]}{\Delta t} = -\dfrac{1}{2}\dfrac{\Delta[H_2O]}{\Delta t} = \dfrac{\Delta[CH_4]}{\Delta t} = \dfrac{1}{2}\dfrac{\Delta[O_2]}{\Delta t}$,

(b) rate $= -\dfrac{1}{3}\dfrac{\Delta[O_2]}{\Delta t} = \dfrac{1}{2}\dfrac{\Delta[O_3]}{\Delta t}$, (c) rate $= -\dfrac{1}{2}\dfrac{\Delta[NO]}{\Delta t} = -\dfrac{\Delta[O_2]}{\Delta t} =$

$\dfrac{1}{2}\dfrac{\Delta[NO_2]}{\Delta t}$. **14.1B** (a) $3CH_4 + 2H_2O + CO_2 \longrightarrow 4CH_3OH$,

(b) $2N_2O_5 \longrightarrow 2N_2 + 5O_2$, (c) $H_2 + CO + O_2 \longrightarrow H_2CO_3$.
**14.2A** (a) 0.0280 $M$/s. (b) 0.112 $M$/s. **14.2B** $2A + 3B \longrightarrow C$.
**14.3A** rate $= k[S_2O_8^{2-}][I^-]$, $k = 8.1 \times 10^{-2}/M \cdot$ s. **14.3B** 0.013 $M$,
$8.8 \times 10^{-6}$ $M$/s. **14.4A** 75 s. **14.4B** 0.91 $M$.
**14.5A**

$k = 1.4 \times 10^{-2}$/min. **14.5B** 0.39 $M$, 0.34 $M$, 0.30 $M$, 0.26 $M$.
**14.6A** $t_{1/2} = 272$ s. **14.6B** $k = 4.71 \times 10^{-2}$/h. **14.7A** (a) 13.2 s.
(b) $t_{1/2} = 4.2$ s, 13 s. **14.7B** (a) 0.0208 $M$, (b) $1.25 \times 10^{-3}$ $M$,

(c) $1.79 \times 10^{-4}$ $M$. **14.8A** 241 kJ/mol. **14.8B** Red < green < blue.
**14.9A** $1.0 \times 10^2$ kJ/mol. **14.9B** $2.7 \times 10^{-4}$/s. **14.10A** 1.7/s.
**14.10B** $7.6 \times 10^{-2}$/s. **14.11A** (a) $NO_2 + CO \longrightarrow NO + CO_2$,
(b) $NO_3$, (c) step 1 is rate determining. **14.11B** $F_2 + ClO_2 \longrightarrow$
$FClO_2 + F$ (slow), $ClO_2 + F \longrightarrow FClO_2$ (fast). This is one
possibility. Any mechanism in which the slow step has the proper rate
law and the steps add to give the correct overall equation is plausible.
**14.12A** The first step is a rapidly established equilibrium. Setting the
rates of forward and reverse reactions equal to each other gives
$k_1[NO][Br_2] = k_{-1}[NOBr_2]$. Solving for $[NOBr_2]$ gives $k_1[NO][Br_2]/k_{-1}$.
Substituting this into the rate law for the rate-determining step, rate $=$
$k_2[NOBr_2][NO]$, gives rate $= (k_1k_2/k_{-1})[NO]^2[Br_2]$ or $k[NO]^2[Br_2]$.

**14.12B** Step 1: $I_2(g) \underset{k_{-1}}{\overset{k_1}{\rightleftharpoons}} 2I(g)$. Step 2: $H_2(g) + 2I(g) \longrightarrow 2HI(g)$.

Rate $= k[H_2][I_2]$.

## Answers to Checkpoints

**14.1.1** a, d. **14.1.2** b. **14.2.1** e. **14.2.2** d. **14.2.3** b. **14.2.4** a. **14.2.5** b.
**14.3.1** c. **14.3.2** c. **14.3.3** b. **14.3.4** d. **14.3.5** c. **14.4.1** e. **14.4.2** b.
**14.4.3** a, c. **14.5.1** c. **14.5.2** c. **14.5.3** b. **14.5.4** d.

## Answers to Applying What You've Learned

**a)** rate $= -\dfrac{\Delta[CO]}{\Delta t}$. **b)** rate $= -\dfrac{1}{2}\dfrac{\Delta[H_2]}{\Delta t} = \dfrac{\Delta[CH_3OH]}{\Delta t}$.

**c)** rate $= k[CO]$, $k = 0.170$ s$^{-1}$, initial rate when $[CO] = 16.5$ $M = 2.81$ $M$/s.
**d)** 12.7 s. **e)** $t_{1/2} = 4.1$ s. **f)** $E_a = 42$ kJ/mol. **g)** $k = 1.9$ s$^{-1}$ at 180°C.

# 15

# Chemical Equilibrium

*Mountain climbers sometimes fall ill due to the lower oxygen content of air at high altitudes. Long-term exposure to an oxygen-poor environment causes the production of more hemoglobin. The additional red blood cells facilitate the transport of oxygen to the body.*

# How the Principles of Equilibrium Favor Athletes Who "Live High and Train Low"

In chemistry, and in biology, when a necessary ingredient for a process is in short supply, nature sometimes responds in a way that compensates for the shortage. For example, in an isolated population of certain types of fish or frogs consisting of a single sex, some of the individuals will change sexes, enabling the population to reproduce.

An important example of nature responding to a shortage is the production of extra red blood cells in individuals who reside at high elevation—where the concentration of oxygen is lower than at sea level. Extra red blood cells enable the blood to transport adequate oxygen from the lungs to the rest of the body, despite the low oxygen concentration. Enhanced oxygen-carrying capacity of the blood can give an athlete greater aerobic capacity and stamina (at normal oxygen concentrations). This has given rise to attempts by some athletes to increase their red blood count (RBC) artificially—a practice known as "blood doping."

Historically, blood doping has been done with blood infusions or with the drug *erythropoietin*—both of which have been banned by the World Anti-Doping Agency (WADA). In recent years, however, the use of hypoxic sleeping tents has grown in popularity. These tents are kept filled with a mixture of nitrogen and oxygen in which the oxygen concentration is lower than that in natural air. Just as the bodies of people who reside at high elevations naturally produce additional red blood cells to compensate for the lower partial pressure of oxygen in the air, the body of an athlete who spends his or her sleeping hours in a hypoxic tent adjusts in the same way, increasing the athlete's RBC. Those who "live high and train low" are believed to have an advantage over athletes who reside and train at or near sea level. Although this is a natural phenomenon and does not really

constitute blood doping, WADA has issued the opinion that the artificial inducement of this effect is "contrary to the spirit of sport." As of the 2009 edition of the WADA code, the use of hypoxic tents has not been banned, but the issue likely will be revisited in the years to come.

The increase in the RBC in people who live at high altitudes or who sleep in hypoxic tents can be explained using the principles of *chemical equilibrium*.

**Hypoxic sleeping tent**

**Student Annotation:** In Michael Crichton's 1990 novel *Jurassic Park,* frog DNA was used to repair ancient dinosaur DNA to facilitate cloning of the extinct animals. In the story, the scientists believed the cloned animals could not reproduce because the population was designed to be entirely female. However, some of the dinosaurs became males—something known to occur in the frogs from which the DNA for repair had been taken—and the population of dinosaurs grew out of control.

At the end of this chapter, you will be able to solve a series of problems related to the detection of hypoxic RBC enhancement [▶▶ Page 661].

## In This Chapter, You Will Learn:

What constitutes an equilibrium, what factors influence equilibrium, and how knowledge of equilibrium can be used to solve a variety of problems.

### Before you begin, you should review

- Rate laws for elementary reactions [◀◀ Section 14.5]
- The quadratic equation [▶▶ Appendix 1]

## 15.1    The Concept of Equilibrium

Up until now, we have treated chemical equations as though they go to completion; that is, we start with only *reactants* and end up with only *products*. In fact, this is not the case with most chemical reactions. Instead, if we start with only reactants, the reaction will proceed, causing reactant concentrations to decrease (as reactants are consumed) and product concentrations to increase (as products are produced). Eventually, though, the concentrations of reactants and products will stop changing. The reaction will appear to have stopped, and we will be left with a *mixture* of reactants and products.

As an example, consider the decomposition of dinitrogen tetroxide ($N_2O_4$) to yield nitrogen dioxide ($NO_2$):

$$N_2O_4(g) \rightleftharpoons 2NO_2(g)$$

$N_2O_4$ is a colorless gas, whereas $NO_2$ is brown. If we begin by placing a sample of pure $N_2O_4$ in an evacuated flask, the contents of the flask change from colorless to brown as the decomposition produces $NO_2$ (Figure 15.1). At first, the brown color intensifies as the concentration of $NO_2$ increases. After some time has passed, though, the intensity of the brown color stops increasing, indicating that the concentration of $NO_2$ has stopped increasing.

Although the concentration of the product (and of the reactant) has stopped changing, the reaction has not actually stopped. Like most chemical reactions, the decomposition of $N_2O_4$ is a ***reversible process,*** meaning that the products of the reaction can react to form *reactants.* Thus, as the decomposition of $N_2O_4$ continues, the reverse reaction, the combination of $NO_2$ molecules to produce $N_2O_4$, is also occurring. Eventually, the concentrations of both species reach levels where they remain constant because the two processes are occurring at the same rate—and the system is said to have achieved ***equilibrium.***

In this case, both the forward and reverse reactions are *elementary* reactions, so we can write their rate laws from the balanced equation:

$$\text{rate}_{\text{forward}} = k_f[N_2O_4] \quad \text{and} \quad \text{rate}_{\text{reverse}} = k_r[NO_2]^2$$

**Figure 15.1** Reaction of colorless $N_2O_4$ to form brown $NO_2$. Initially, only $N_2O_4$ is present and only the forward reaction (decomposition of $N_2O_4$ to give $NO_2$) is occurring. As $NO_2$ forms, the reverse reaction (recombination of $NO_2$ to give $N_2O_4$) begins to occur. The brown color continues to intensify until the forward and reverse reactions are occurring at the same rate.

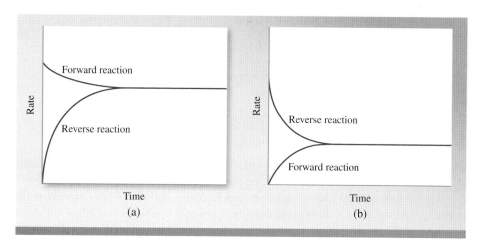

**Figure 15.2**   (a) Starting with just $N_2O_4$, the rate of the reverse reaction (formation of $N_2O_4$) is initially zero. The rate of the reverse reaction rises and the rate of the forward reaction falls until both rates are equal. (b) Starting with $NO_2$, the rate of the forward reaction is initially zero.

where $k_f$ and $k_r$ are the rate constants for the forward and reverse reactions, respectively [◀◀ Section 14.5]. As always, the square brackets denote *molar concentration*. In the experiment shown in Figure 15.1, initially:

- $N_2O_4$ concentration is high, and the rate of the forward reaction is high.
- $NO_2$ concentration is zero, making the rate of the reverse reaction zero.

As the reaction proceeds:

- $N_2O_4$ concentration falls, decreasing the rate of the forward reaction.
- $NO_2$ concentration rises, increasing the rate of the reverse reaction.

Figure 15.2(a) shows how the rates of these forward and reverse reactions change over time, and how they eventually become equal as equilibrium is established.

We could equally well have started the $N_2O_4$-$NO_2$ experiment with pure $NO_2$ in a flask. Figure 15.3 shows how the brown color, initially intense, fades as $NO_2$ combines to form $N_2O_4$.

**Figure 15.3**   $N_2O_4/NO_2$ equilibrium starting with $NO_2$. Initially, only $NO_2$ is present and only the reverse reaction (recombination of $NO_2$ to give $N_2O_4$) is occurring. As $N_2O_4$ forms, the forward reaction (decomposition of $N_2O_4$) begins to occur. The brown color continues to fade until the forward and reverse reactions are occurring at the same rate.

As before, the intensity stops changing after a period of time. When we start the experiment with pure $NO_2$, initially:

- $NO_2$ concentration is high, and the rate of the reverse reaction is high.
- $N_2O_4$ concentration is zero, making the rate of the forward reaction zero.

As the reaction proceeds:

- $NO_2$ concentration falls, decreasing the rate of the reverse reaction.
- $N_2O_4$ concentration rises, increasing the rate of the forward reaction.

Figure 15.2(b) shows how the forward and reverse rates change over time when we start with $NO_2$ instead of $N_2O_4$. We could also conduct this kind of experiment starting with a mixture of $NO_2$ and $N_2O_4$. Again, the forward and reverse reactions would occur, initially at rates determined by the corresponding starting concentrations, and equilibrium would be established when both reactions were occurring at the same rate.

Some important things to remember about equilibrium are:

- Equilibrium is a *dynamic* state—both forward and reverse reactions continue to occur, although there is no net change in reactant and product concentrations over time.
- At equilibrium, the rates of the forward and reverse reactions are equal.
- Equilibrium can be established starting with only *reactants,* with only *products,* or with any *mixture* of reactants and products.

Finally, it is important to understand that the principles of equilibrium apply not only to reversible chemical reactions, but also to reversible physical processes such as phase changes. In this chapter, we will limit our discussion to *chemical* equilibria. We explore reversible *physical* processes in some detail in Chapter 18.

## 15.2 The Equilibrium Constant

We have defined equilibrium as the state in which the opposing forward and reverse reactions are occurring at the same rate:

$$\text{rate}_{\text{forward}} = \text{rate}_{\text{reverse}}$$

Thus, using the $N_2O_4$-$NO_2$ system as an example again,

$$k_f[N_2O_4]_{eq} = k_r[NO_2]^2_{eq}$$

where the subscripted "eq" denotes a concentration at equilibrium. Rearranging this expression gives

$$\frac{k_f}{k_r} = \frac{[NO_2]^2_{eq}}{[N_2O_4]_{eq}}$$

and because the ratio of two constants $(k_f/k_r)$ is also a constant, we have

$$\overbrace{K_c = \frac{[NO_2]^2_{eq}}{[N_2O_4]_{eq}}}^{\text{equilibrium expression}}$$

equilibrium constant

where $K_c$ is the **equilibrium constant** and the equation is known as the **equilibrium expression.** (The subscript "c" stands for *concentration,* referring to the molar concentrations in the expression.) Note the relationship between the equilibrium expression and the balanced chemical equation for the reaction: the numerator contains the product concentration raised to a power equal to its stoichiometric coefficient in the balanced chemical equation:

$$K_c = \frac{[NO_2]^2_{eq}}{[N_2O_4]_{eq}} \qquad N_2O_4(g) \rightleftharpoons 2NO_2(g)$$

| TABLE 15.1 | Initial and Equilibrium Concentrations of $N_2O_4$ and $NO_2$ at 25°C | | | | |
|---|---|---|---|---|---|
| | Initial Concentrations (*M*) | | Equilibrium Concentrations (*M*) | | |
| Experiment | $[N_2O_4]_i$ | $[NO_2]_i$ | $[N_2O_4]$ | $[NO_2]$ | $\dfrac{[NO_2]^2}{[N_2O_4]}$ |
| 1 | 0.670 | 0.00 | 0.643 | 0.0547 | $4.65 \times 10^{-3}$ |
| 2 | 0.446 | 0.0500 | 0.448 | 0.0457 | $4.66 \times 10^{-3}$ |
| 3 | 0.500 | 0.0300 | 0.491 | 0.0475 | $4.60 \times 10^{-3}$ |
| 4 | 0.600 | 0.0400 | 0.594 | 0.0523 | $4.60 \times 10^{-3}$ |
| 5 | 0.000 | 0.200 | 0.0898 | 0.0204 | $4.63 \times 10^{-3}$ |

The denominator, moreover, contains the reactant concentration raised to a power equal to its stoichiometric coefficient in the balanced chemical equation. (In this equation the coefficient of $N_2O_4$ is 1, which generally is not written either as a coefficient or as an exponent.) Table 15.1 lists the starting and equilibrium concentrations of $N_2O_4$ and $NO_2$ in a series of experiments carried out at 25°C. Using the equilibrium concentrations from each of the experiments in the table, the value of the expression $[NO_2]^2/[N_2O_4]$ is indeed constant (the average is $4.63 \times 10^{-3}$) within the limits of experimental error.

## Calculating Equilibrium Constants

In the mid-nineteenth century, Cato Guldberg[1] and Peter Waage[2] studied the equilibrium mixtures of a wide variety of chemical reactions. They observed that at a constant temperature in an equilibrium mixture of reactants and products regardless of the initial concentrations, the *reaction quotient* has a constant value. The **reaction quotient ($Q_c$)** is a fraction with product concentrations in the numerator and reactant concentrations in the denominator—with each concentration raised to a power equal to the corresponding stoichiometric coefficient in the balanced chemical equation. For the general reaction,

$$a\text{A} + b\text{B} \rightleftharpoons c\text{C} + d\text{D}$$

at equilibrium, the reaction quotient $Q_c$ is equal to the equilibrium constant $K_c$.

$$Q_c = \frac{[\text{C}]^c[\text{D}]^d}{[\text{A}]^a[\text{B}]^b} = K_c \quad \text{(at equilibrium)} \qquad \text{Equation 15.1}$$

This expression is known as the **law of mass action.** Like the equilibrium constant $K$, $Q$ is subscripted with a "c" to indicate that the quotient is defined in terms of *concentrations.* For the $N_2O_4$-$NO_2$ system, this approach gives the same equilibrium expression as we got from our kinetics approach: $[NO_2]^2_{eq}/[N_2O_4]_{eq}$. But the law of mass action was developed *empirically* from countless observations of many different reactions—long before the principles of kinetics were developed. Additionally, it applies not only to elementary reactions, but also to more complex reactions that occur via a series of steps. Furthermore, the law of mass action enables us to write the equilibrium expression for any reaction for which we know the balanced equation. Knowing the equilibrium expression for a reaction, we can use equilibrium concentrations to calculate the value of the equilibrium constant.

Sample Problem 15.1 shows how to use equilibrium concentrations and the law of mass action to calculate the value of an equilibrium constant.

**Student Annotation:** In our first example, there was just one reactant and one product. When there are multiple species on either side of the equation, the numerator is the *product* of product concentrations and the denominator is the product of *reactant* concentrations—with each concentration raised to the appropriate power, that is, the stoichiometric coefficient.

---

1. Cato Maximilian Guldberg (1836–1902). Norwegian chemist and mathematician. Guldberg's research was primarily in the field of thermodynamics.

2. Peter Waage (1833–1900). Norwegian chemist. Like his coworker, Guldberg, Waage did research in thermodynamics.

## Sample Problem 15.1

Carbonyl chloride ($COCl_2$), also called phosgene, is a highly poisonous gas that was used on the battlefield in World War I. It is produced by the reaction of carbon monoxide with chlorine gas:

$$CO(g) + Cl_2(g) \rightleftharpoons COCl_2(g)$$

In an experiment conducted at 74°C, the equilibrium concentrations of the species involved in the reaction were as follows: $[CO] = 1.2 \times 10^{-2}\ M$, $[Cl_2] = 0.054\ M$, and $[COCl_2] = 0.14\ M$. (a) Write the equilibrium expression, and (b) determine the value of the equilibrium constant for this reaction at 74°C.

**Strategy**  Use the law of mass action to write the equilibrium expression and plug in the equilibrium concentrations of all three species to evaluate $K_c$.

**Setup**  The equilibrium expression has the form of concentrations of products over concentrations of reactants, each raised to the appropriate power—in the case of this reaction, all the coefficients are 1, so all the powers will be 1.

**Solution**

(a) $K_c = \dfrac{[COCl_2]}{[CO][Cl_2]}$

(b) $K_c = \dfrac{(0.14)}{(1.2 \times 10^{-2})(0.054)} = 216\ \cdot\ \text{or}\ \ 2.2 \times 10^2$

$K_c$ for this reaction at 74°C is $2.2 \times 10^2$.

**Think About It**  When putting the equilibrium concentrations into the equilibrium expression, we leave out the units. It is common practice to express equilibrium constants without units. We will examine the reason why in Section 15.5.

---

**Practice Problem A**  In an analysis of the following reaction at 100°C,

$$Br_2(g) + Cl_2(g) \rightleftharpoons 2BrCl(g)$$

the equilibrium concentrations were found to be $[Br_2] = 2.3 \times 10^{-3}\ M$, $[Cl_2] = 1.2 \times 10^{-2}\ M$, and $[BrCl] = 1.4 \times 10^{-2}\ M$. Write the equilibrium expression, and calculate the equilibrium constant for this reaction at 100°C.

**Practice Problem B**  In another analysis at 100°C involving the same reaction, the equilibrium concentrations of the reactants were found to be $[Br_2] = 4.1 \times 10^{-3}\ M$ and $[Cl_2] = 8.3 \times 10^{-3}\ M$. Determine the value of $[BrCl]$.

---

**Figure 15.4**  The value of the reaction quotient, $Q$, changes as the reaction (experiment 5 from Table 15.1) progresses. When the system reaches equilibrium, $Q$ is equal to the equilibrium constant.

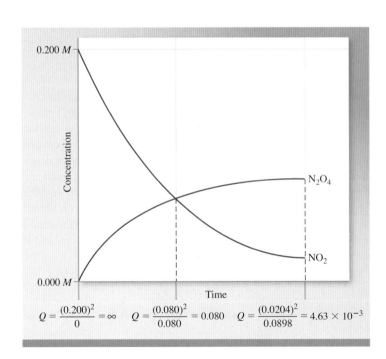

$$Q = \frac{(0.200)^2}{0} = \infty \qquad Q = \frac{(0.080)^2}{0.080} = 0.080 \qquad Q = \frac{(0.0204)^2}{0.0898} = 4.63 \times 10^{-3}$$

Guldberg and Waage's work involved calculating reaction quotients at equilibrium, but the numerical value of the reaction quotient can be determined using starting concentrations or final (equilibrium) concentrations, or concentrations at any time during the reaction (provided that the concentrations are known). The value of $Q_c$ changes as the reaction progresses (Figure 15.4), and $Q_c$ is equal to $K_c$ only when a system is at equilibrium. However, we often find it useful to calculate the value of $Q_c$ for a system that is not at equilibrium. At any point during the progress of a reaction, therefore,

$$Q_c = \frac{[C]^c[D]^d}{[A]^a[B]^b} \qquad \text{Equation 15.2}$$

Sample Problem 15.2 lets you practice writing reaction quotients for a variety of balanced equations.

---

## Sample Problem 15.2

Write reaction quotients for the following reactions:

(a) $N_2(g) + 3H_2(g) \rightleftharpoons 2NH_3(g)$

(b) $H_2(g) + I_2(g) \rightleftharpoons 2HI(g)$

(c) $Ag^+(aq) + 2NH_3(aq) \rightleftharpoons Ag(NH_3)_2^+(aq)$

(d) $2O_3(g) \rightleftharpoons 3O_2(g)$

(e) $Cd^{2+}(aq) + 4Br^-(aq) \rightleftharpoons CdBr_4^{2-}(aq)$

(f) $2NO(g) + O_2(g) \rightleftharpoons 2NO_2(g)$

**Strategy** Use the law of mass action to write reaction quotients.

**Setup** The reaction quotient for each reaction has the form of the concentrations of *products* over the concentrations of *reactants,* each raised to a power equal to its stoichiometric coefficient in the balanced chemical equation.

**Student Annotation:** The molar concentrations in a reaction quotient may be concentrations of *gases* or *aqueous* species.

**Solution**

(a) $Q_c = \dfrac{[NH_3]^2}{[N_2][H_2]^3}$

(d) $Q_c = \dfrac{[O_2]^3}{[O_3]^2}$

(b) $Q_c = \dfrac{[HI]^2}{[H_2][I_2]}$

(e) $Q_c = \dfrac{[CdBr_4^{2-}]}{[Cd^{2+}][Br^-]^4}$

(c) $Q_c = \dfrac{[Ag(NH_3)_2^+]}{[Ag^+][NH_3]^2}$

(f) $Q_c = \dfrac{[NO_2]^2}{[NO]^2[O_2]}$

**Think About It** With practice, writing reaction quotients becomes second nature. Without sufficient practice, it will seem inordinately difficult. It is important that you become proficient at this. It is very often the first step in solving equilibrium problems.

**Practice Problem A** Write the reaction quotient for each of the following reactions:

(a) $2N_2O(g) \rightleftharpoons 2N_2(g) + O_2(g)$

(b) $2NOBr(g) \rightleftharpoons 2NO(g) + Br_2(g)$

(c) $HF(aq) \rightleftharpoons H^+(aq) + F^-(aq)$

(d) $CO(g) + H_2O(g) \rightleftharpoons CO_2(g) + H_2(g)$

(e) $CH_4(g) + 2H_2S(g) \rightleftharpoons CS_2(g) + 4H_2(g)$

(f) $H_2C_2O_4(aq) \rightleftharpoons 2H^+(aq) + C_2O_4^{2-}(aq)$

**Practice Problem B** Write the equation for the equilibrium that corresponds to each of the following reaction quotients:

(a) $Q_c = \dfrac{[HCl]^2}{[H_2][Cl_2]}$

(d) $Q_c = \dfrac{[H^+][ClO^-]}{[HClO]}$

(b) $Q_c = \dfrac{[HF]}{[H^+][F^-]}$

(e) $Q_c = \dfrac{[H^+][HSO_3^-]}{[H_2SO_3]}$

(c) $Q_c = \dfrac{[Cr(OH)_4^-]}{[Cr^{3+}][OH^-]^4}$

(f) $Q_c = \dfrac{[NOBr]^2}{[NO]^2[Br_2]}$

## Magnitude of the Equilibrium Constant

One of the things the equilibrium constant tells us is the extent to which a reaction proceeds at a particular temperature. If we combine stoichiometric amounts of the reactants in a reaction, three outcomes are possible:

1. The reaction will go essentially to completion, and the equilibrium mixture will consist predominantly of products.
2. The reaction will not occur to any significant degree, and the equilibrium mixture will consist predominantly of reactants.
3. The reaction will proceed to a significant degree but will not go to completion, and the equilibrium mixture will contain comparable amounts of both reactants and products.

When the magnitude of $K_c$ is very large, we expect the first outcome. The formation of the $Ag(NH_3)_2^+$ ion is an example of this possibility.

$$Ag^+(aq) + 2NH_3(aq) \rightleftharpoons Ag(NH_3)_2^+(aq) \qquad K_c = 1.5 \times 10^7 \ (25°C)$$

If we were to combine aqueous $Ag^+$ and aqueous $NH_3$ in a mole ratio of 1:2, the resulting equilibrium mixture would contain mostly $Ag(NH_3)_2^+$ with only very small amounts of reactants remaining. A reaction with a very large equilibrium constant is sometimes said to "lie to the right" or to "favor products."

When the magnitude of $K_c$ is very small, we expect the second outcome. The chemical combination of nitrogen and oxygen gases to give nitrogen monoxide is an example of this possibility.

$$N_2(g) + O_2(g) \rightleftharpoons 2NO(g) \qquad K_c = 4.3 \times 10^{-25} \ (\text{at } 25°C)$$

Nitrogen and oxygen gases do not react to any significant extent at room temperature. If we were to place a mixture of $N_2$ and $O_2$ in an evacuated flask and allow the system to reach equilibrium, the resulting mixture would be predominantly $N_2$ and $O_2$ with only a *very* small amount of NO. A reaction with a very small equilibrium constant is said to "lie to the left" or to "favor reactants."

The terms *very large* and *very small* are somewhat arbitrary when applied to equilibrium constants. Generally speaking, an equilibrium constant with a magnitude greater than about $1 \times 10^2$ can be considered large; one with a magnitude smaller than about $1 \times 10^{-2}$ can be considered small.

An equilibrium constant that falls between $1 \times 10^2$ and $1 \times 10^{-2}$ indicates that neither products nor reactants are strongly favored. In this case, a system at equilibrium will contain a *mixture* of reactants and products, the exact composition of which will depend on the stoichiometry of the reaction.

### Checkpoint 15.2    The Equilibrium Constant

**15.2.1**   Select the correct equilibrium expression for the reaction

$$2CO(g) + O_2(g) \rightleftharpoons 2CO_2(g)$$

a) $K_c = \dfrac{2[CO_2]^2}{2[CO]^2[O_2]}$

b) $K_c = \dfrac{(2[CO_2])^2}{(2[CO])^2([O_2])}$

c) $K_c = \dfrac{2[CO_2]}{2[CO][O_2]}$

d) $K_c = \dfrac{2[CO_2]^2}{[2CO]^2[O_2]}$

e) $K_c = \dfrac{[CO_2]^2}{[CO]^2[O_2]}$

**15.2.2**   Determine the value of the equilibrium constant, $K_c$, for the reaction

$$A + 2B \rightleftharpoons C + D$$

if the equilibrium concentrations are $[A] = 0.0115 \ M$, $[B] = 0.0253 \ M$, $[C] = 0.109 \ M$, and $[D] = 0.0110 \ M$.

a) 163

b) 4.12

c) 2.06

d) $6.14 \times 10^{-3}$

e) 0.243

# 15.3 Equilibrium Expressions

The equilibria we have discussed so far have all been *homogeneous;* that is, the reactants and products have all existed in the same phase—either gaseous or aqueous. In these cases, the equilibrium expression consists of writing the product of the product concentrations at equilibrium over the product of the reactant concentrations at equilibrium, with each concentration raised to a power equal to its stoichiometric coefficient in the balanced chemical equation (Equation 15.2). When the species in a reversible chemical reaction are not all in the same phase, the equilibrium is *heterogeneous.*

## Heterogeneous Equilibria

Writing equilibrium expressions for heterogeneous equilibria is also straightforward, but it is slightly different from what we have done so far for homogeneous equilibria. For example, carbon dioxide can combine with elemental carbon to produce carbon monoxide:

$$CO_2(g) + C(s) \rightleftharpoons 2CO(g)$$

The two gases and one solid constitute two separate phases. If we were to write the equilibrium expression for this reaction as we have done previously for homogeneous reactions, including all product concentrations in the numerator and all reactant concentrations in the denominator, we would have

$$K_c^* = \frac{[CO]^2}{[CO_2][C]}$$

($K_c^*$ is superscripted with an asterisk to distinguish it from the equilibrium constant that we are about to derive.) The "concentration" of a solid, however, is a constant. If we were to double the number of moles of elemental carbon (C) in the preceding reaction, we would also double its volume. The ratio of moles to volume, which is how we define the concentration, remains the same. Because the concentration of solid carbon is a constant, it is incorporated into the value of the equilibrium constant and does not appear explicitly in the equilibrium expression.

$$K_c^* \times [C] = \frac{[CO]^2}{[CO_2]}$$

The product of $K_c^*$ and $[C]_{eq}$ gives the *real* equilibrium constant for this reaction. The corresponding equilibrium expression is

$$K_c = \frac{[CO]^2}{[CO_2]}$$

The same argument applies to the concentrations of pure liquids in heterogeneous equilibria. Only gaseous species and aqueous species appear in equilibrium expressions.

Sample Problem 15.3 lets you practice writing equilibrium expressions for heterogeneous equilibria.

---

### Sample Problem 15.3

Write equilibrium expressions for each of the following reactions:

(a) $CaCO_3(s) \rightleftharpoons CaO(s) + CO_2(g)$
(b) $Hg(l) + Hg^{2+}(aq) \rightleftharpoons Hg_2^{2+}(aq)$
(c) $2Fe(s) + 3H_2O(l) \rightleftharpoons Fe_2O_3(s) + 2H_2(g)$
(d) $O_2(g) + 2H_2(g) \rightleftharpoons 2H_2O(l)$.

**Strategy** Use the law of mass action to write the equilibrium expression for each reaction. Only gases and aqueous species appear in the expression.

**Setup**

(a) Only $CO_2$ will appear in the expression.
(b) $Hg^{2+}$ and $Hg_2^{2+}$ will appear in the expression.
(c) Only $H_2$ will appear in the expression.
(d) $O_2$ and $H_2$ will appear in the expression.

*(Continued)*

**Solution**

(a) $K_c = [CO_2]$

(b) $K_c = \dfrac{[Hg_2^{2+}]}{[Hg^{2+}]}$

(c) $K_c = [H_2]^2$

(d) $K_c = \dfrac{1}{[O_2][H_2]^2}$

**Think About It** Like writing equilibrium expressions for homogeneous equilibria, writing equilibrium expressions for heterogeneous equilibria becomes second nature if you practice. The importance of developing this skill now cannot be overstated. Your ability to understand the principles and to solve many of the problems in this and Chapters 16 to 19 depends on your ability to write equilibrium expressions *correctly* and *easily*.

**Practice Problem A** Write equilibrium expressions for each of the following reactions:

(a) $SiCl_4(g) + 2H_2(g) \rightleftharpoons Si(s) + 4HCl(g)$

(b) $Hg^{2+}(aq) + 2Cl^-(aq) \rightleftharpoons HgCl_2(s)$

(c) $Ni(s) + 4CO(g) \rightleftharpoons Ni(CO)_4(g)$

(d) $Zn(s) + Fe^{2+}(aq) \rightleftharpoons Zn^{2+}(aq) + Fe(s)$

**Practice Problem B** Which of the following equilibrium expressions corresponds to a heterogeneous equilibrium? How can you tell?

(a) $K_c = [NH_3][HCl]$

(b) $K_c = \dfrac{[H^+][C_2H_3O_2^-]}{[HC_2H_3O_2]}$

(c) $K_c = \dfrac{[Ag(NH_3)_2^+][Cl^-]}{[NH_3]^2}$

(d) $K_c = [Ba^{2+}][F^-]^2$

## Manipulating Equilibrium Expressions

In our study of enthalpy, we learned that it is possible to manipulate chemical equations to solve thermochemistry problems [◄◄ Section 5.3]. Recall that when we made a change in a thermochemical equation, we had to make the corresponding change in the $\Delta H$ of the reaction. When we reversed a reaction, for example, we had to change the sign of its $\Delta H$. When we change something about how an equilibrium reaction is expressed, we must also make the appropriate changes to the equilibrium expression and the equilibrium constant. Consider the reaction of nitric oxide and oxygen to produce nitrogen dioxide:

$$2NO(g) + O_2(g) \rightleftharpoons 2NO_2(g)$$

The equilibrium expression is

$$K_c = \frac{[NO_2]^2}{[NO]^2[O_2]}$$

and the equilibrium constant at 500 K is $6.9 \times 10^5$. If we were to reverse this equation, writing the equation for the decomposition of $NO_2$ to produce NO and $O_2$,

$$2NO_2(g) \rightleftharpoons 2NO(g) + O_2(g)$$

the new equilibrium expression would be the reciprocal of the original equilibrium expression:

$$K_c' = \frac{[NO]^2[O_2]}{[NO_2]^2}$$

Because the equilibrium expression is the reciprocal of the original, the equilibrium constant is also the reciprocal of the original. At 500 K, therefore, the equilibrium constant for the new equation is $1/(6.9 \times 10^5)$ or $1.5 \times 10^{-6}$.

Alternatively, if instead of reversing the original equation, we were to multiply it by 2,

$$4NO(g) + 2O_2(g) \rightleftharpoons 4NO_2(g)$$

the new equilibrium expression would be

$$K_c'' = \frac{[NO_2]^4}{[NO]^4[O_2]^2}$$

which is the original equilibrium expression squared. Again, because the new equilibrium expression is the square of the original, the new equilibrium constant is the square of the original: $K_c = (6.9 \times 10^5)^2$ or $4.8 \times 10^{11}$.

Finally, if we were to add the reverse of the original equation,

$$2NO_2(g) \rightleftharpoons 2NO(g) + O_2(g)$$

(for which $K_c = 1.6 \times 10^{-6}$ at 500 K) to the equation

$$2H_2(g) + O_2(g) \rightleftharpoons 2H_2O(g)$$

(for which $K_c = 2.4 \times 10^{47}$ at 500 K), the sum of the equations and the equilibrium expression (prior to the cancellation of identical terms) would be

$$2NO_2(g) + 2H_2(g) + O_2(g) \rightleftharpoons 2NO(g) + O_2(g) + 2H_2O(g)$$

and

$$K_c = \frac{[NO]^2[O_2][H_2O]^2}{[NO_2]^2[H_2]^2[O_2]}$$

It may be easier to see before canceling identical terms that for the *sum* of two equations, the equilibrium expression is the *product* of the two corresponding equilibrium expressions.

$$\frac{[NO]^2[O_2][H_2O]^2}{[NO_2]^2[H_2]^2[O_2]} = \frac{[NO]^2[\cancel{O_2}]}{[NO_2]^2} \times \frac{[H_2O]^2}{[H_2]^2[\cancel{O_2}]}$$

Canceling identical terms on the left and right sides of the equation, and on the top and bottom of the equilibrium expression, gives

$$2NO_2(g) + 2H_2(g) \rightleftharpoons 2NO(g) + 2H_2O(g) \quad \text{and} \quad K_c = \frac{[NO]^2[H_2O]^2}{[NO_2]^2[H_2]^2}$$

And, because the new equilibrium expression is the product of the individual expressions, the new equilibrium constant is the product of the individual constants. Therefore, for the reaction

$$2NO_2(g) + 2H_2(g) \rightleftharpoons 2NO(g) + 2H_2O(g)$$

at 500 K, $K_c = (1.6 \times 10^{-6})(2.4 \times 10^{47}) = 3.8 \times 10^{41}$.

Using hypothetical equilibria, Table 15.2 summarizes the various ways that chemical equations can be manipulated and the corresponding changes that must be made to the equilibrium expression and equilibrium constant.

Sample Problem 15.4 shows how to manipulate chemical equations and make the corresponding changes to their equilibrium constants.

**TABLE 15.2**    Manipulation of Equilibrium Constant Expressions

$$A(g) + B(g) \rightleftharpoons 2C(g) \qquad K_{c_1} = 4.39 \times 10^{-3}$$
$$2C(g) \rightleftharpoons D(g) + E(g) \qquad K_{c_2} = 1.15 \times 10^4$$

| Equation | Equilibrium Expression | Relationship to Original $K_c$ | Equilibrium Constant |
|---|---|---|---|
| $2C(g) \rightleftharpoons A(g) + B(g)$<br>Original equation is reversed. | $K_{c_1}' = \dfrac{[A][B]}{[C]^2}$ | $\dfrac{1}{K_{c_1}}$ | $2.28 \times 10^2$<br>New constant is the reciprocal of the original. |
| $2A(g) + 2B(g) \rightleftharpoons 4C(g)$<br>Original equation is multiplied by a number. | $K_{c_1}'' = \dfrac{[C]^4}{[A]^2[B]^2}$ | $(K_{c_1})^2$ | $1.93 \times 10^{-5}$<br>New constant is the original raised to the same number. |
| $\frac{1}{2}A(g) + \frac{1}{2}B(g) \rightleftharpoons C(g)$<br>Original equation is divided by 2. | $K_{c_1}''' = \dfrac{[C]}{[A]^{1/2}[B]^{1/2}}$ | $\sqrt{K_{c_1}}$ | $6.63 \times 10^{-2}$<br>New constant is the square root of the original. |
| $A(g) + B(g) \rightleftharpoons D(g) + E(g)$<br>Two equations are added. | $K_{c_3} = \dfrac{[D][E]}{[A][B]}$ | $K_{c_1} \times K_{c_2}$ | $50.5$<br>New constant is the product of the two original constants. |

*Temperature is the same for both reactions.

## Sample Problem 15.4

The following reactions have the indicated equilibrium constants at 100°C:

(1) $\qquad\qquad$ $2NOBr(g) \rightleftharpoons 2NO(g) + Br_2(g)$ $\qquad$ $K_c = 0.014$

(2) $\qquad\qquad$ $Br_2(g) + Cl_2(g) \rightleftharpoons 2BrCl(g)$ $\qquad$ $K_c = 7.2$

Determine the value of $K_c$ for the following reactions at 100°C:

(a) $2NO(g) + Br_2(g) \rightleftharpoons 2NOBr(g)$ $\qquad$ (d) $2NOBr(g) + Cl_2(g) \rightleftharpoons 2NO(g) + 2BrCl(g)$

(b) $4NOBr(g) \rightleftharpoons 4NO(g) + 2Br_2(g)$ $\qquad$ (e) $NO(g) + BrCl(g) \rightleftharpoons NOBr(g) + \frac{1}{2}Cl_2(g)$

(c) $NOBr(g) \rightleftharpoons NO(g) + \frac{1}{2}Br_2(g)$

**Strategy** Begin by writing the equilibrium expressions for the reactions that are given. Then, determine the relationship of each equation's equilibrium expression to the equilibrium expression of the original equations, and make the corresponding change to the equilibrium constant for each.

**Setup** The equilibrium expressions for the reactions that are given are

$$K_c = \frac{[NO]^2[Br_2]}{[NOBr]^2} \quad \text{and} \quad K_c = \frac{[BrCl]^2}{[Br_2][Cl_2]}$$

(a) This equation is the reverse of original equation 1. Its equilibrium expression is the reciprocal of that for the original equation:

$$2NO(g) + Br_2(g) \rightleftharpoons 2NOBr(g) \qquad K_c = \frac{[NOBr]^2}{[NO]^2[Br_2]}$$

(b) This is original equation 1 multiplied by a factor of 2. Its equilibrium expression is the original expression squared:

$$4NOBr(g) \rightleftharpoons 4NO(g) + 2Br_2(g) \qquad K_c = \left(\frac{[NO]^2[Br_2]}{[NOBr]^2}\right)^2$$

(c) This is original equation 1 multiplied by $\frac{1}{2}$. Its equilibrium expression is the square root of the original:

$$NOBr(g) \rightleftharpoons NO(g) + \frac{1}{2}Br_2(g) \qquad K_c = \sqrt{\frac{[NO]^2[Br_2]}{[NOBr]^2}} \quad \text{or} \quad K_c = \left(\frac{[NO]^2[Br_2]}{[NOBr]^2}\right)^{1/2}$$

(d) This is the sum of original equations 1 and 2. Its equilibrium expression is the product of the two individual expressions:

$$2NOBr(g) + Cl_2(g) \rightleftharpoons 2NO(g) + 2BrCl(g) \qquad K_c = \frac{[NO]^2[BrCl]^2}{[NOBr]^2[Cl_2]}$$

(e) Probably the simplest way to analyze this reaction is to recognize that it is the reverse of the reaction in part (d), multiplied by $\frac{1}{2}$. Its equilibrium expression is the square root of the reciprocal of the expression in part (d):

$$NO(g) + BrCl(g) \rightleftharpoons NOBr(g) + \frac{1}{2}Cl_2(g)$$

$$K_c = \sqrt{\frac{[NOBr]^2[Cl_2]}{[NO]^2[BrCl]^2}} \quad \text{or} \quad K_c = \left(\frac{[NOBr]^2[Cl_2]}{[NO]^2[BrCl]^2}\right)^{1/2}$$

Each equilibrium constant will bear the same relationship to the original as the equilibrium expression bears to the original.

**Solution**

(a) $K_c = 1/0.014 = 71$ $\qquad$ (c) $K_c = (0.014)^{1/2} = 0.12$ $\qquad$ (e) $K_c = (1/0.10)^{1/2} = 3.2$

(b) $K_c = (0.014)^2 = 2.0 \times 10^{-4}$ $\qquad$ (d) $K_c = (0.014)(7.2) = 0.10$

**Think About It** The magnitude of an equilibrium constant reveals whether products or reactants are favored, so the reciprocal relationship between $K_c$ values of forward and reverse reactions should make sense. A very large $K_c$ value means that products are favored. In the reaction of hydrogen ion and hydroxide ion to form water, the value of $K_c$ is very large, indicating that the product, water, is favored.

$$H^+(aq) + OH^-(aq) \rightleftharpoons H_2O(l)$$
$$K_c = 1.0 \times 10^{14} \text{ (at 25°C)}$$

Simply writing the equation backward doesn't change the fact that water is the predominate species. In the reverse reaction, therefore, the favored species is on the reactant side:

$$H_2O(l) \rightleftharpoons H^+(aq) + OH^-(aq)$$
$$K_c = 1.0 \times 10^{-14} \text{ (at 25°C)}$$

As a result, the magnitude of $K_c$ should correspond to reactants being favored; that is, it should be very small.

**Practice Problem A** The following reactions have the indicated equilibrium constants at a particular temperature:

$$N_2(g) + O_2(g) \rightleftharpoons 2NO(g) \qquad K_c = 4.3 \times 10^{-25}$$
$$2NO(g) + O_2(g) \rightleftharpoons 2NO_2(g) \qquad K_c = 6.4 \times 10^9$$

Determine the values of the equilibrium constants for the following equations at the same temperature: (a) $2NO(g) \rightleftharpoons N_2(g) + O_2(g)$, (b) $\frac{1}{2}N_2(g) + \frac{1}{2}O_2(g) \rightleftharpoons NO(g)$, and (c) $N_2(g) + 2O_2(g) \rightleftharpoons 2NO_2(g)$.

**Practice Problem B** Using the data from Practice Problem A, determine the values of the equilibrium constants for the following equations: (a) $4NO(g) \rightleftharpoons N_2(g) + 2NO_2(g)$, (b) $4NO_2(g) \rightleftharpoons 2N_2(g) + 4O_2(g)$, and (c) $2NO(g) + 2NO_2(g) \rightleftharpoons 3O_2(g) + 2N_2(g)$.

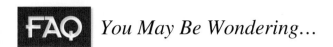 *You May Be Wondering...*

# What if the Equilibrium Expression Contains Only Gases?

When an equilibrium expression contains only gases, we can write an alternate form of the expression in which the concentrations of gases are expressed as partial pressures (atm). Thus, for the equilibrium

$$N_2O_4(g) \rightleftharpoons 2NO_2(g)$$

we can either write the equilibrium expression as

$$K_c = \frac{[NO_2]^2}{[N_2O_4]}$$

or as

$$K_P = \frac{\left(P_{NO_2}\right)^2}{P_{N_2O_4}}$$

where the subscript "$P$" in $K_P$ stands for *pressure*, and $P_{NO_2}$ and $P_{N_2O_4}$ are the equilibrium partial pressures of $NO_2$ and $N_2O_4$, respectively. In general, $K_c$ is not equal to $K_P$ because the partial pressures of reactants and products expressed in atmospheres are not equal to their concentrations expressed in mol/L. However, a simple relationship between $K_P$ and $K_c$ can be derived using the following equilibrium:

$$aA(g) \rightleftharpoons bB(g)$$

where $a$ and $b$ are the stoichiometric coefficients. The equilibrium constant $K_c$ is given by

$$K_c = \frac{[B]^b}{[A]^a}$$

and the expression for $K_P$ is

$$K_P = \frac{(P_B)^b}{(P_A)^a}$$

where $P_A$ and $P_B$ are the partial pressures of A and B. Assuming ideal gas behavior,

$$P_A V = n_A RT$$

$$P_A = \frac{n_A RT}{V} = \left(\frac{n_A}{V}\right)RT$$

and

$$P_A = [A]RT$$

where [A] is the molar concentration of A. Likewise,

$$P_B V = n_B RT$$

$$P_B = \frac{n_B RT}{V} = \left(\frac{n_B}{V}\right)RT$$

and

$$P_B = [B]RT$$

Substituting the expressions for $P_A$ and $P_B$ into the expression for $K_P$ gives

$$K_P = \frac{(P_B)^b}{(P_A)^a}$$

$$= \frac{[B]^b}{[A]^a}(RT)^{b-a}$$

*(Continued)*

which simplifies to

$$K_P = K_c(RT)^{\Delta n}$$

where $\Delta n = b - a$. In general,

Equation 15.3    $\Delta n$ = moles of gaseous products − moles of gaseous reactants

Because pressures are usually expressed in atmospheres, the gas constant $R$ is expressed as $0.08206 \text{ L} \cdot \text{atm/K} \cdot \text{mol}$, and we can write the relationship between $K_P$ and $K_c$ as

Equation 15.4    $K_P = K_c[(0.08206 \text{ L} \cdot \text{atm/K} \cdot \text{mol}) \times T]^{\Delta n}$

$K_P$ is equal to $K_c$ only in the special case where $\Delta n = 0$, as in the following equilibrium reaction:

$$H_2(g) + Br_2(g) \rightleftharpoons 2HBr(g)$$

In this case, Equation 15.4 can be written as

$$K_P = K_c \, [(0.08206 \text{ L} \cdot \text{atm/K} \cdot \text{mol}) \times T]^0$$

$$= K_c$$

Keep in mind that $K_P$ expressions can only be written for reactions in which every species in the equilibrium expression is a gas. (Remember that solids and pure liquids do not appear in the equilibrium expression.)

Sample Problems 15.5 and 15.6 let you practice writing $K_P$ expressions and illustrate the conversion between $K_c$ and $K_P$.

---

## Sample Problem 15.5

Write $K_P$ expressions for (a) $PCl_3(g) + Cl_2(g) \rightleftharpoons PCl_5(g)$, (b) $O_2(g) + 2H_2(g) \rightleftharpoons 2H_2O(l)$, and (c) $F_2(g) + H_2(g) \rightleftharpoons 2HF(g)$.

**Strategy**  Write equilibrium expressions for each equation, expressing the concentrations of the gases in partial pressures.

**Setup**  (a) All the species in this equation are gases, so they will all appear in the $K_P$ expression.
(b) Only the reactants are gases.
(c) All species are gases.

**Think About It**  It isn't necessary for every species in the reaction to be a gas—only those species that appear in the equilibrium expression.

**Solution**  (a) $K_P = \dfrac{(P_{PCl_5})}{(P_{PCl_3})(P_{Cl_2})}$    (b) $K_P = \dfrac{1}{(P_{O_2})(P_{H_2})^2}$    (c) $K_P = \dfrac{(P_{HF})^2}{(P_{F_2})(P_{H_2})}$

---

## Checkpoint 15.3    Equilibrium Expressions

**15.3.1**  Select the correct equilibrium expression for the reaction

$$H^+(aq) + OH^-(aq) \rightleftharpoons H_2O(l)$$

a) $K_c = \dfrac{[H_2O]}{[H^+][OH^-]}$

b) $K_c = \dfrac{[H^+][OH^-]}{[H_2O]}$

c) $K_c = [H^+][OH^-]$

d) $K_c = \dfrac{1}{[H^+][OH^-]}$

e) $K_c = [H^+][OH^-][H_2O]$

**15.3.2**  Select the correct equilibrium expression for the reaction

$$CaO(s) + CO_2(g) \rightleftharpoons CaCO_3(s)$$

a) $K_c = \dfrac{1}{[CO_2]}$

b) $K_c = \dfrac{[CaCO_3]}{[CaO][CO_2]}$

c) $K_c = [CO_2]$

d) $K_c = \dfrac{[CaO][CO_2]}{[CaCO_3]}$

e) $K_c = \dfrac{[CO_2]}{[CaO][CaCO_3]}$

**Practice Problem A** Write $K_P$ expressions for (a) $2CO(g) + O_2(g) \rightleftharpoons 2CO_2(g)$, (b) $CaCO_3(s) \rightleftharpoons CaO(s) + CO_2(g)$, and (c) $N_2(g) + 3H_2(g) \rightleftharpoons 2NH_3(g)$.

**Practice Problem B** Write the equation for the gaseous equilibrium corresponding to each of the following $K_P$ expressions:

(a) $K_P = \dfrac{(P_{NO_3})^2}{(P_{NO_2})^2(P_{O_2})}$ (b) $K_P = \dfrac{(P_{CO_2})(P_{H_2})^4}{(P_{CH_4})(P_{H_2O})^2}$ (c) $K_P = \dfrac{(P_{HI})^2}{(P_{I_2})(P_{H_2})}$

## Sample Problem 15.6

The equilibrium constant, $K_c$, for the reaction

$$N_2O_4(g) \rightleftharpoons 2NO_2(g)$$

is $4.63 \times 10^{-3}$ at 25°C. What is the value of $K_P$ at this temperature?

**Strategy** Use Equation 15.4 to convert from $K_c$ to $K_P$. Be sure to convert temperature in degrees Celsius to kelvins.

**Setup** Using Equation 15.3,

$$\Delta n = 2(NO_2) - 1(N_2O_4) = 1$$

$T = 298$ K.

**Solution** $K_P = \left[ K_c \left( \dfrac{0.08206 \text{ L} \cdot \text{atm}}{\text{K} \cdot \text{mol}} \right) \times T \right]$

$= (4.63 \times 10^{-3})(0.08206 \times 298)$

$= 0.113$

> **Think About It** Note that we have essentially disregarded the units of $R$ and $T$ so that the resulting equilibrium constant, $K_P$, is unitless. Equilibrium constants commonly are treated as unitless quantities.

**Practice Problem A** For the reaction

$$N_2(g) + 3H_2(g) \rightleftharpoons 2NH_3(g)$$

$K_c$ is $2.3 \times 10^{-2}$ at 375°C. Calculate $K_P$ for the reaction at this temperature.

**Practice Problem B** $K_P = 2.79 \times 10^{-5}$ for the reaction in Practice Problem A at 472°C. What is $K_c$ for this reaction at 472°C?

**15.3.3** Given the following information,

$$HF(aq) \rightleftharpoons H^+(aq) + F^-(aq)$$
$$K_c = 6.8 \times 10^{-4} \text{ (at 25°C)}$$

$$H_2C_2O_4(aq) \rightleftharpoons 2H^+(aq) + C_2O_4^{2-}(aq)$$
$$K_c = 3.8 \times 10^{-6} \text{ (at 25°C)}$$

determine the value of $K_c$ for the following reaction at 25°C:

$$C_2O_4^{2-} + 2HF \rightleftharpoons 2F^- + H_2C_2O_4$$

a) $2.6 \times 10^{-9}$

b) $1.8 \times 10^{-12}$

c) $1.2 \times 10^{-1}$

d) $2.6 \times 10^{5}$

e) $6.8 \times 10^{-4}$

**15.3.4** $K_c$ for the reaction

$$Br_2(g) \rightleftharpoons 2Br(g)$$

is $1.1 \times 10^{-3}$ at 1280°C. Calculate the value of $K_P$ for this reaction at this temperature.

a) $1.1 \times 10^{-3}$

b) 18

c) 0.14

d) $9.1 \times 10^{2}$

e) $8.3 \times 10^{-6}$

## 15.4   Using Equilibrium Expressions to Solve Problems

We have already used equilibrium expressions to determine the value of an equilibrium constant using equilibrium concentrations. In this section, we will learn how to use equilibrium expressions to predict the direction of a reaction and to calculate equilibrium concentrations.

### Predicting the Direction of a Reaction

If we start an experiment with only reactants, we know that the reactant concentrations will decrease and the product concentrations will increase; that is, the reaction must proceed in the *forward* direction for equilibrium to be established. Likewise, if we start an experiment with only products, we know that the product concentrations will decrease and the reactant concentrations will increase. In this case, the reaction must proceed in the *reverse* direction to achieve equilibrium. But often we must predict the direction in which a reaction will proceed when we start with a mixture of reactants and products. For this situation, we calculate the value of the reaction quotient, $Q_c$, and compare it to the value of the equilibrium constant, $K_c$.

**Student Annotation:** $Q_c$ is calculated using the initial concentrations of reactants and products. Similarly, $Q_p$ can be calculated using the initial partial pressures of reactants and products.

The equilibrium constant, $K_c$, for the gaseous formation of hydrogen iodide from molecular hydrogen and molecular iodine,

$$H_2(g) + I_2(g) \rightleftharpoons 2HI(g)$$

is 54.3 at 430°C. If we were to conduct an experiment starting with a mixture of 0.243 mole of $H_2$, 0.146 mole of $I_2$, and 1.98 moles of HI in a 1.00-L container at 430°C, would more HI form or would HI be consumed and more $H_2$ and $I_2$ form? Using the starting concentrations, we can calculate the reaction quotient as follows:

**Student Annotation:** Remember that calculating $Q_c$ is just like calculating $K_c$: products over reactants, each raised to the appropriate power—except that the concentrations we use to calculate $Q_c$ are the *starting* concentrations. To calculate $K_c$ we must use *equilibrium* concentrations.

$$Q_c = \frac{[HI]_i^2}{[H_2]_i[I_2]_i} = \frac{(1.98)^2}{(0.243)(0.146)} = 111$$

where the subscript "i" indicates *initial* concentration. Because the reaction quotient does not equal $K_c$ ($Q_c = 111$, $K_c = 54.3$), the reaction is not at equilibrium. To establish equilibrium, the reaction will proceed to the left, consuming HI and producing $H_2$ and $I_2$, decreasing the value of the numerator and increasing the value in the denominator until the value of the reaction quotient equals that of the equilibrium constant. Thus, the reaction proceeds in the reverse direction (from right to left) to reach equilibrium.

There are three possibilities when we compare Q with K:

**Student Annotation:** The comparison of Q with K can refer either to $Q_c$ and $K_c$ or $Q_P$ and $K_P$.

| | |
|---|---|
| $Q < K$ | The ratio of initial concentrations of products to reactants is too small. To reach equilibrium, reactants must be converted to products. The system proceeds in the forward direction (from left to right). |
| $Q = K$ | The initial concentrations are equilibrium concentrations. The system is already at equilibrium, and there will be no net reaction in either direction. |
| $Q > K$ | The ratio of initial concentrations of products to reactants is too large. To reach equilibrium, products must be converted to reactants. The system proceeds in the reverse direction (from right to left). |

Sample Problem 15.7 shows how the value of Q is used to determine the direction of a reaction that is not at equilibrium.

---

### Sample Problem 15.7

At 375°C, the equilibrium constant for the reaction

$$N_2(g) + 3H_2(g) \rightleftharpoons 2NH_3(g)$$

is 1.2. At the start of a reaction, the concentrations of $N_2$, $H_2$, and $NH_3$ are 0.071 $M$, $9.2 \times 10^{-3}$ $M$, and $1.83 \times 10^{-4}$ $M$, respectively. Determine whether this system is at equilibrium, and if not, determine in which direction it must proceed to establish equilibrium.

**Strategy** Use the initial concentrations to calculate $Q_c$, and then compare $Q_c$ with $K_c$.

**Setup**

$$Q_c = \frac{[NH_3]_i^2}{[N_2]_i[H_2]_i^3} = \frac{(1.83 \times 10^{-4})^2}{(0.071)(9.2 \times 10^{-3})^3} = 0.61$$

**Solution**  The calculated value of $Q_c$ is less than $K_c$. Therefore, the reaction is not at equilibrium and must proceed to the right to establish equilibrium.

> **Think About It**  In proceeding to the right, a reaction consumes reactants and produces more products. This increases the numerator in the reaction quotient and decreases the denominator. The result is an increase in $Q_c$ until it is equal to $K_c$, at which point equilibrium will be established.

**Practice Problem A**  The equilibrium constant, $K_c$, for the formation of nitrosyl chloride from nitric oxide and chlorine,

$$2NO(g) + Cl_2(g) \rightleftharpoons 2NOCl(g)$$

is $6.5 \times 10^4$ at 35°C. In which direction will the reaction proceed to reach equilibrium if the starting concentrations of NO, $Cl_2$, and NOCl are $1.1 \times 10^{-3}$ $M$, $3.5 \times 10^{-4}$ $M$, and 1.9 $M$, respectively?

**Practice Problem B**  Calculate $K_P$ for the formation of nitrosyl chloride from nitric oxide and chlorine at 35°C, and determine whether the reaction will proceed to the right or the left to achieve equilibrium when the starting pressures are $P_{NO} = 1.01$ atm, $P_{Cl_2} = 0.42$ atm, and $P_{NOCl} = 1.76$ atm.

## Calculating Equilibrium Concentrations

If we know the equilibrium constant for a reaction, we can calculate the concentrations in the equilibrium mixture from the initial reactant concentrations. Consider the following system involving two organic compounds, *cis-* and *trans-*stilbene:

cis-Stilbene                    trans-Stilbene

The equilibrium constant ($K_c$) for this system is 24.0 at 200°C. If we know that the starting concentration of *cis*-stilbene is 0.850 $M$, we can use the equilibrium expression to determine the equilibrium concentrations of both species. The stoichiometry of the reaction tells us that for every mole of *cis*-stilbene converted, 1 mole of *trans*-stilbene is produced. We will let $x$ be the equilibrium concentration of *trans*-stilbene in mol/L; therefore, the equilibrium concentration of *cis*-stilbene must be $(0.850 - x)$ mol/L. It is useful to summarize these changes in concentrations in an equilibrium table:

|  | *cis*-stilbene $\rightleftharpoons$ | *trans*-stilbene |
|---|---|---|
| *Initial concentration (M):* | 0.850 | 0 |
| *Change in concentration (M):* | $-x$ | $+x$ |
| *Equilibrium concentration (M):* | $0.850 - x$ | $x$ |

We then use the equilibrium concentrations, defined in terms of $x$, in the equilibrium expression:

$$K_c = \frac{[trans\text{-stilbene}]}{[cis\text{-stilbene}]}$$

$$24.0 = \frac{x}{0.850 - x}$$

$$x = 0.816 \, M$$

Having solved for $x$, we calculate the equilibrium concentrations of *cis*- and *trans*-stilbene as follows:

$$[\textit{cis-stilbene}] = (0.850 - x)\,M = 0.034\,M$$

$$[\textit{trans-stilbene}] = x\,M = 0.816\,M$$

A good way to check the answer to a problem such as this is to use the calculated equilibrium concentrations in the equilibrium expression and make sure that we get the correct $K_c$ value.

$$K_c = \frac{0.816}{0.034} = 24$$

Sample Problems 15.8 and 15.9 provide additional examples of this kind of problem.

---

## Sample Problem 15.8

$K_c$ for the reaction of hydrogen and iodine to produce hydrogen iodide,

$$H_2(g) + I_2(g) \rightleftharpoons 2HI(g)$$

is 54.3 at 430°C. What will the concentrations be at equilibrium if we start with 0.240 $M$ concentrations of both $H_2$ and $I_2$?

**Strategy** Construct an equilibrium table to determine the equilibrium concentration of each species in terms of an unknown ($x$); solve for $x$, and use it to calculate the equilibrium molar concentrations.

**Setup** Insert the starting concentrations that we know into the equilibrium table:

$$H_2(g) + I_2(g) \rightleftharpoons 2HI(g)$$

| | | | |
|---|---|---|---|
| *Initial concentration (M):* | 0.240 | 0.240 | 0 |
| *Change in concentration (M):* | | | |
| *Equilibrium concentration (M):* | | | |

**Solution** We define the change in concentration of one of the reactants as $x$. Because there is no product at the start of the reaction, the reactant concentration must decrease; that is, this reaction must proceed in the forward direction to reach equilibrium. According to the stoichiometry of the chemical reaction, the reactant concentrations will both decrease by the same amount ($x$), and the product concentration will increase by twice that amount ($2x$). Combining the initial concentration and the change in concentration for each species, we get expressions (in terms of $x$) for the equilibrium concentrations:

| | $H_2(g)$ + | $I_2(g)$ $\rightleftharpoons$ | $2HI(g)$ |
|---|---|---|---|
| *Initial concentration (M):* | 0.240 | 0.240 | 0 |
| *Change in concentration (M):* | $-x$ | $-x$ | $+2x$ |
| *Equilibrium concentration (M):* | $0.240 - x$ | $0.240 - x$ | $2x$ |

Next, we insert these expressions for the equilibrium concentrations into the equilibrium expression and solve for $x$.

$$K_c = \frac{[HI]^2}{[H_2][I_2]}$$

$$54.3 = \frac{(2x)^2}{(0.240 - x)(0.240 - x)} = \frac{(2x)^2}{(0.240 - x)^2}$$

$$\sqrt{54.3} = \frac{2x}{0.240 - x}$$

$$x = 0.189$$

Using the calculated value of $x$, we can determine the equilibrium concentration of each species as follows:

$$[H_2] = (0.240 - x)\,M = 0.051\,M$$

$$[I_2] = (0.240 - x)\,M = 0.051\,M$$

$$[HI] = 2x = 0.378\,M$$

**Think About It** Always check your answer by inserting the calculated concentrations into the equilibrium expression:

$$\frac{[HI]^2}{[H_2][I_2]} = \frac{(0.378)^2}{(0.051)^2}$$
$$= 54.9 \approx K_c$$

The small difference between the calculated $K_c$ and the one given in the problem statement is due to rounding.

**Practice Problem A** Calculate the equilibrium concentrations of $H_2$, $I_2$, and HI at 430°C if the initial concentrations are $[H_2] = [I_2] = 0$ $M$, and $[HI] = 0.525$ $M$.

**Practice Problem B** Determine the initial concentration of HI if the initial concentrations of $H_2$ and $I_2$ are both 0.10 $M$ and their equilibrium concentrations are both 0.043 $M$ at 430°C.

## Sample Problem 15.9

For the same reaction and temperature as in Sample Problem 15.8, calculate the equilibrium concentrations of all three species if the starting concentrations are as follows: $[H_2] = 0.00623$ $M$, $[I_2] = 0.00414$ $M$, and $[HI] = 0.0424$ $M$.

**Strategy** Using the initial concentrations, calculate the reaction quotient, $Q_c$, and compare it to the value of $K_c$ (given in the problem statement of Sample Problem 15.8) to determine which direction the reaction will proceed to establish equilibrium. Then, construct an equilibrium table to determine the equilibrium concentrations.

**Setup**

$$\frac{[HI]^2}{[H_2][I_2]} = \frac{(0.0424)^2}{(0.00623)(0.00414)} = 69.7$$

Therefore, $Q_c > K_c$, so the system will have to proceed to the *left* (reverse) to reach equilibrium. The equilibrium table is

$$H_2(g) + I_2(g) \rightleftharpoons 2HI(g)$$

| | | | |
|---|---|---|---|
| *Initial concentration (M):* | 0.00623 | 0.00414 | 0.0424 |
| *Change in concentration (M):* | | | |
| *Equilibrium concentration (M):* | | | |

**Solution** Because we know the reaction must proceed from right to left, we know that the concentration of HI will decrease and the concentrations of $H_2$ and $I_2$ will increase. Therefore, the table should be filled in as follows:

| | $H_2(g)$ | + | $I_2(g)$ | $\rightleftharpoons$ | $2HI(g)$ |
|---|---|---|---|---|---|
| *Initial concentration (M):* | 0.00623 | | 0.00414 | | 0.0424 |
| *Change in concentration (M):* | $+x$ | | $+x$ | | $-2x$ |
| *Equilibrium concentration (M):* | $0.00623 + x$ | | $0.00414 + x$ | | $0.0424 - 2x$ |

Next, we insert these expressions for the equilibrium concentrations into the equilibrium expression and solve for $x$.

$$K_c = \frac{[HI]^2}{[H_2][I_2]}$$

$$54.3 = \frac{(0.0424 - 2x)^2}{(0.00623 + x)(0.00414 + x)}$$

It isn't possible to solve this equation the way we did in Sample Problem 15.8 (by taking the square root of both sides) because the concentrations of $H_2$ and $I_2$ are unequal. Instead, we have to carry out the multiplications,

$$54.3(2.58 \times 10^{-5} + 1.04 \times 10^{-2}x + x^2) = 1.80 \times 10^{-3} - 1.70 \times 10^{-1}x + 4x^2$$

Collecting terms we get

$$50.3x^2 + 0.735x - 4.00 \times 10^{-4} = 0$$

This is a quadratic equation of the form $ax^2 + bx + c = 0$. The solution for the quadratic equation (see Appendix 1) is

$$x = \frac{-b \pm \sqrt{b^2 - 4ac}}{2a}$$

*(Continued)*

Here we have $a = 50.3$, $b = 0.735$, and $c = -4.00 \times 10^{-4}$, so

$$x = \frac{-0.735 \pm \sqrt{(0.735)^2 - 4(50.3)(-4.00 \times 10^{-4})}}{2(50.3)}$$

$$x = 5.25 \times 10^{-4} \quad \text{or} \quad x = -0.0151$$

Only the first of these values, $5.25 \times 10^{-4}$, makes sense because concentration cannot be a negative number. Using the calculated value of $x$, we can determine the equilibrium concentration of each species as follows:

$$[H_2] = (0.00623 + x)\,M = 0.00676\,M$$

$$[I_2] = (0.00414 + x)\,M = 0.00467\,M$$

$$[HI] = (0.0424 - 2x)\,M = 0.0414\,M$$

---

**Think About It** Checking this result gives

$$K_c = \frac{[HI]^2}{[H_2][I_2]}$$

$$= \frac{(0.0414)^2}{(0.00676)(0.00467)} = 54.3$$

---

**Practice Problem A** At 1280°C the equilibrium constant $K_c$ for the reaction

$$Br_2(g) \rightleftharpoons 2Br(g)$$

is $1.1 \times 10^{-3}$. If the initial concentrations are $[Br_2] = 6.3 \times 10^{-2}\,M$ and $[Br] = 1.2 \times 10^{-2}\,M$, calculate the concentrations of these two species at equilibrium.

**Practice Problem B** Determine the initial and equilibrium concentrations of Br if the initial and equilibrium concentrations of $Br_2$ are $5.7 \times 10^{-3}\,M$ and $5.3 \times 10^{-3}\,M$, respectively.

---

Here is a summary of the use of initial reactant concentrations to determine equilibrium concentrations:

1. Construct an equilibrium table, and fill in the initial concentrations (including any that are zero).

2. Use initial concentrations to calculate the reaction quotient, $Q$, and compare $Q$ to $K$ to determine the direction in which the reaction will proceed.

**Student Annotation:** If $Q < K$, the reaction will occur as written. If $Q > K$, the reverse reaction will occur.

3. Define $x$ as the amount of a particular species consumed, and use the stoichiometry of the reaction to define (in terms of $x$) the amount of other species consumed or produced.

4. For each species in the equilibrium, add the change in concentration to the initial concentration to get the equilibrium concentration.

5. Use the equilibrium concentrations and the equilibrium expression to solve for $x$.

6. Using the calculated value of $x$, determine the concentrations of all species at equilibrium.

7. Check your work by plugging the calculated equilibrium concentrations into the equilibrium expression. The result should be very close to the $K_c$ stated in the problem.

The same procedure applies to $K_P$.

Sample Problem 15.10 shows how to solve an equilibrium problem using partial pressures.

## Sample Problem 15.10

A mixture of 5.75 atm of $H_2$ and 5.75 atm of $I_2$ is contained in a 1.0-L vessel at 430°C. The equilibrium constant ($K_P$) for the reaction

**Student Annotation:** For this reaction, $\Delta n = 0$; therefore, $K_c$ is equal to $K_p$.

$$H_2(g) + I_2(g) \rightleftharpoons 2HI(g)$$

at this temperature is 54.3. Determine the equilibrium partial pressures of $H_2$, $I_2$, and HI.

**Strategy** Construct an equilibrium table to determine the equilibrium partial pressures.

**Setup** The equilibrium table is

|  | $H_2(g)$ | + | $I_2(g)$ | $\rightleftharpoons$ | $2HI(g)$ |
|---|---|---|---|---|---|
| *Initial partial pressure (atm):* | 5.75 | | 5.75 | | 0 |
| *Change in partial pressure (atm):* | $-x$ | | $-x$ | | $+2x$ |
| *Equilibrium partial pressure (atm):* | $5.75 - x$ | | $5.75 - x$ | | $2x$ |

**Solution** Setting the equilibrium expression equal to $K_P$,

$$54.3 = \frac{(2x)^2}{(5.75 - x)^2}$$

Taking the square root of both sides of the equation gives

$$\sqrt{54.3} = \frac{2x}{5.75 - x}$$

$$7.369 = \frac{2x}{5.75 - x}$$

$$7.369(5.75 - x) = 2x$$

$$42.37 - 7.369x = 2x$$

$$42.37 = 9.369x$$

$$x = 4.52$$

The equilibrium partial pressures are $P_{H_2} = P_{I_2} = 5.75 - 4.52 = 1.23$ atm, and $P_{HI} = 9.04$ atm.

> **Think About It** Plugging the calculated partial pressures into the equilibrium expression gives
>
> $$\frac{(P_{HI})^2}{(P_{H_2})(P_{I_2})} = \frac{(9.04)^2}{(1.23)^2} = 54.0$$
>
> The small difference between this result and the equilibrium constant given in the problem statement is due to rounding.

**Practice Problem A** Determine the equilibrium partial pressures of $H_2$, $I_2$, and HI if we begin the experiment with 1.75 atm each of $H_2$ and $I_2$ at 430°C.

**Practice Problem B** Determine the equilibrium partial pressures of $H_2$, $I_2$, and HI if we begin the experiment with 2.75 atm HI at 430°C.

> **Student Annotation:** When a reaction starts with reactants and products, be sure to calculate Q and compare it to K to determine which direction the reaction will proceed to reach equilibrium.

---

### Checkpoint 15.4 Using Equilibrium Expressions to Solve Problems

Use the following information to answer questions 15.4.1 and 15.4.2:
$K_c$ for the reaction

$$A + B \rightleftharpoons 2C$$

is $1.7 \times 10^{-2}$ at 250°C.

**15.4.1** What will the equilibrium concentrations of A, B, and C be at this temperature if $[A]_i = [B]_i = 0.750\ M$ ($[C]_i = 0$)?

a) $6.1 \times 10^{-3}\ M$, $6.1 \times 10^{-3}\ M$, $0.092\ M$

b) $0.046\ M$, $0.046\ M$, $0.092\ M$

c) $0.70\ M$, $0.70\ M$, $0.046\ M$

d) $0.70\ M$, $0.70\ M$, $0.092\ M$

e) $0.087\ M$, $0.087\ M$, $0.66\ M$

**15.4.2** What will the equilibrium concentrations of A, B, and C be at this temperature if $[C]_i = 0.875\ M$ ($[A]_i = [B]_i = 0$)?

a) $0.41\ M$, $0.41\ M$, $0.82\ M$

b) $0.41\ M$, $0.41\ M$, $0.054\ M$

c) $0.43\ M$, $0.43\ M$, $0.0074\ M$

d) $0.43\ M$, $0.43\ M$, $0.44\ M$

e) $0.43\ M$, $0.43\ M$, $0.43\ M$

**15.4.3** If $K_c = 2$ for the reaction $A_2 + B_2 \rightleftharpoons 2AB$ at a certain temperature, which of the following diagrams represents an equilibrium mixture of $A_2$, $B_2$, and AB? (Select all that apply.)

● = A      ○ = B

(a)      (b)      (c)      (d)

**15.4.4** If $K_c = 3$ for the reaction $X + 2Y \rightleftharpoons Z$ at a certain temperature, then for each of the mixtures of X, Y, and Z shown here, in what direction must the reaction proceed to achieve equilibrium?

X = ●
Y = ○
Z = ●

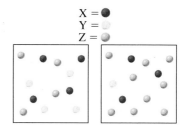

a) Right, right

b) Neither, left

c) Neither, right

d) Left, neither

e) Right, neither

## 15.5  Factors That Affect Chemical Equilibrium

One of the interesting and useful features of chemical equilibria is that they can be manipulated in specific ways to maximize production of a desired substance. Consider, for example, the industrial production of ammonia from its constituent elements by the Haber process:

$$N_2(g) + 3H_2(g) \rightleftharpoons 2NH_3(g)$$

More than 100 million tons of ammonia is produced annually by this reaction, with most of the resulting ammonia being used for fertilizers to enhance crop production. Clearly it would be in the best interest of industry to maximize the yield of $NH_3$. In this section, we will learn about the various ways in which an equilibrium can be manipulated to accomplish this goal.

*Le Châtelier's principle* states that when a *stress* is applied to a system at equilibrium, the system will respond by *shifting* in the direction that minimizes the effect of the stress. In this context, "stress" refers to a disturbance of the system at equilibrium by any of the following means:

- The addition of a reactant or product
- The removal of a reactant or product
- A change in volume of the system, resulting in a change in concentration or partial pressure of the reactants and products
- A change in temperature

"Shifting" refers to the occurrence of either the forward or reverse reaction such that the effect of the stress is partially offset as the system reestablishes equilibrium. An equilibrium that shifts to the right is one in which more products are produced by the forward reaction. An equilibrium that shifts to the left is one in which more reactants are produced by the reverse reaction. Using Le Châtelier's principle, we can predict the direction in which an equilibrium will shift, given the specific stress that is applied.

### Addition or Removal of a Substance

Again using the Haber process as an example,

$$N_2(g) + 3H_2(g) \rightleftharpoons 2NH_3(g)$$

consider a system at 700 K, in which the equilibrium concentrations are as follows:

$$[N_2] = 2.05\ M \qquad [H_2] = 1.56\ M \qquad [NH_3] = 1.52\ M$$

**Student Annotation:** Remember that at equilibrium, the reaction quotient, $Q_c$, is equal to the equilibrium constant, $K_c$.

Using these concentrations in the reaction quotient expression, we can calculate the value of $K_c$ for the reaction at this temperature as follows:

$$Q_c = \frac{[NH_3]^2}{[N_2][H_2]^3} = \frac{(1.52)^2}{(2.05)(1.56)^3} = 0.297 = K_c$$

If we were to apply stress to this system by adding more $N_2$, increasing its concentration from 2.05 $M$ to 3.51 $M$, the system would no longer be at equilibrium. To see that this is true, use the new concentration of nitrogen in the reaction quotient expression. The new calculated value of $Q_c$ (0.173) is no longer equal to the value of $K_c$ (0.297).

$$Q_c = \frac{[NH_3]^2}{[N_2][H_2]^3} = \frac{(1.52)^2}{(3.51)(1.56)^3} = 0.173 \neq K_c$$

**Student Annotation:** While reestablishing equilibrium causes a decrease in the $N_2$ concentration, the final concentration will still be higher than that in the original equilibrium mixture. The system responds to the stress of the added reactant by consuming part of it.

For this system to reestablish equilibrium, the net reaction will have to shift in such a way that $Q_c$ is again equal to $K_c$, which is constant at a given temperature. Recall from Section 15.4 that when $Q$ is less than $K$, the reaction proceeds to the right to achieve equilibrium. Likewise, an equilibrium that is stressed in such a way that $Q$ *becomes* less than $K$ will *shift* to the right to reestablish equilibrium. This means that the forward reaction, the consumption of $N_2$ and $H_2$ to produce $NH_3$, will occur. The result will be a net decrease in the concentrations of $N_2$ and $H_2$ (thus making the denominator of the reaction quotient smaller), and a net increase in the concentration of $NH_3$ (thus making the numerator larger). When the concentrations of all species are such that $Q_c$ is again equal to $K_c$, the system will have established a new *equilibrium position*, meaning that it will have shifted in one direction or the other, resulting in a new equilibrium concentration for each species. Figure 15.5 shows how the concentrations of $N_2$, $H_2$, and $NH_3$ change when $N_2$ is added to the original equilibrium mixture.

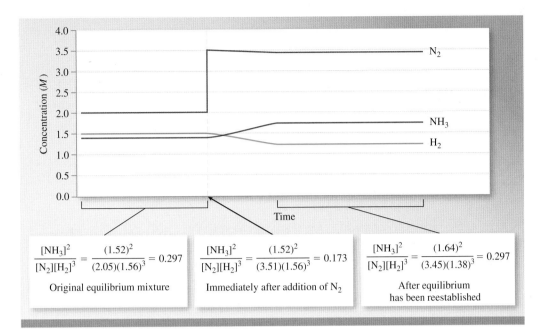

**Figure 15.5** Adding more of a reactant to a system at equilibrium causes the equilibrium position to shift toward product. The system responds to the addition of $N_2$ by consuming some of the added $N_2$ (and some of the other reactant, $H_2$) to produce more $NH_3$.

$$\frac{[NH_3]^2}{[N_2][H_2]^3} = \frac{(1.52)^2}{(2.05)(1.56)^3} = 0.297$$

Original equilibrium mixture

$$\frac{[NH_3]^2}{[N_2][H_2]^3} = \frac{(1.52)^2}{(3.51)(1.56)^3} = 0.173$$

Immediately after addition of $N_2$

$$\frac{[NH_3]^2}{[N_2][H_2]^3} = \frac{(1.64)^2}{(3.45)(1.38)^3} = 0.297$$

After equilibrium
has been reestablished

addition   addition

$$N_2(g) + 3H_2(g) \rightleftharpoons 2NH_3(g)$$

removal

(a)

addition

$$N_2(g) + 3H_2(g) \rightleftharpoons 2NH_3(g)$$

removal   removal

(b)

**Figure 15.6** (a) Addition of a reactant or removal of a product will cause an equilibrium to shift to the right. (b) Addition of a product or removal of a reactant will cause an equilibrium to shift to the left.

Conversely, if we were to remove $N_2$ from the original equilibrium mixture, the lower concentration in the denominator of the reaction quotient would result in $Q_c$ being greater than $K_c$. In this case the reaction will shift to the left. That is, the reverse reaction will take place, thereby increasing the concentrations of $N_2$ and $H_2$ and decreasing the concentration of $NH_3$ until $Q_c$ is once again equal to $K_c$.

The addition or removal of $NH_3$ will cause a shift in the equilibrium, too. The addition of $NH_3$ will cause a shift to the left; the removal of $NH_3$ will cause a shift to the right. Figure 15.6(a) shows the additions and removals that cause this equilibrium to shift to the right. Figure 15.6(b) shows those that cause it to shift to the left.

In essence, a system at equilibrium will respond to the addition of a species by consuming some of that species, and it will respond to the removal of a species by producing more of that species. It is important to remember that the addition or removal of a species from an equilibrium mixture does not change the value of the equilibrium constant, $K$. Rather, it changes temporarily the value of the reaction quotient, $Q$. Furthermore, to cause a shift in the equilibrium, the species added or removed must be one that appears in the reaction quotient expression. In the case of a heterogeneous equilibrium, altering the amount of a solid or liquid species does not change the position of the equilibrium because doing so does not change the value of $Q$.

Sample Problem 15.11 shows the effects of stress on a system at equilibrium.

### Sample Problem 15.11

Hydrogen sulfide ($H_2S$) is a contaminant commonly found in natural gas. It is removed by reaction with oxygen to produce elemental sulfur.

$$2H_2S(g) + O_2(g) \rightleftharpoons 2S(s) + 2H_2O(g)$$

For each of the following scenarios, determine whether the equilibrium will shift to the right, shift to the left, or neither: (a) addition of $O_2(g)$, (b) removal of $H_2S(g)$, (c) removal of $H_2O(g)$, and (d) addition of $S(s)$.

**Strategy** Use Le Châtelier's principle to predict the direction of shift for each case. Remember that the position of the equilibrium is only changed by the addition or removal of a species that appears in the reaction quotient expression.

*(Continued)*

**Setup**  Begin by writing the reaction quotient expression:

$$Q_c = \frac{[H_2O]^2}{[H_2S]^2[O_2]}$$

Because sulfur is a solid, it does not appear in the expression. Changes in the concentration of any of the other species will cause a change in the equilibrium position. Addition of a reactant or removal of a product that appears in the expression for $Q_c$ will shift the equilibrium to the right:

addition     addition

$$2H_2S(g) + O_2(g) \rightleftharpoons 2S(s) + 2H_2O(g)$$

removal

Removal of a reactant or addition of a product that appears in the expression for $Q_c$ will shift the equilibrium to the left:

addition

$$N_2(g) + 3H_2(g) \rightleftharpoons 2NH_3(g)$$

removal     removal

**Think About It** In each case, analyze the effect the change will have on the value of $Q_c$. In part (a), for example, $O_2$ is added, so its concentration increases. Looking at the reaction quotient expression, we can see that a larger concentration of oxygen corresponds to a larger overall denominator—giving the overall fraction a smaller value. Thus, $Q$ will temporarily be smaller than $K$ and the reaction will have to shift to the right, consuming some of the added $O_2$ (along with some of the $H_2S$ in the mixture) to reestablish equilibrium.

**Solution**

(a) Shift to the right

(b) Shift to the left

(c) Shift to the right

(d) No change

**Practice Problem A**  For each change indicated, determine whether the equilibrium

$$PCl_3(g) + Cl_2(g) \rightleftharpoons PCl_5(g)$$

will shift to the right, shift to the left, or neither: (a) addition of $PCl_3(g)$, (b) removal of $PCl_3(g)$, (c) removal of $PCl_5(g)$, and (d) removal of $Cl_2(g)$.

**Practice Problem B**  What can be added to the equilibrium that will (a) shift it to the left, (b) shift it to the right, (c) not shift it in either direction?

$$AgCl(s) + 2NH_3(aq) \rightleftharpoons Ag(NH_3)_2^+(aq) + Cl^-(aq)$$

## Changes in Volume and Pressure

If we were to start with a gaseous system at equilibrium in a cylinder with a movable piston, we could change the volume of the system, thereby changing the concentrations of the reactants and products.

Consider again the equilibrium between $N_2O_4$ and $NO_2$:

$$N_2O_4(g) \rightleftharpoons 2NO_2(g)$$

At 25°C the equilibrium constant for this reaction is $4.63 \times 10^{-3}$. Suppose we have an equilibrium mixture of 0.643 $M$ $N_2O_4$ and 0.0547 $M$ $NO_2$ in a cylinder fitted with a movable piston. If we push down on the piston, the equilibrium will be disturbed and will shift in the direction that minimizes the effect of this disturbance. Consider what happens to the concentrations of both species if we decrease the volume of the cylinder by half. Both concentrations are initially *doubled*: $[N_2O_4]$ = 1.286 $M$ and $[NO_2]$ = 0.1094 $M$. If we plug the new concentrations into the reaction quotient expression, we get

$$Q_c = \frac{[NO_2]_{eq}^2}{[N_2O_4]_{eq}} = \frac{(0.1094)^2}{1.286} = 9.31 \times 10^{-3}$$

which is not equal to $K_c$, so the system is no longer at equilibrium. Because $Q_c$ is greater than $K_c$, the equilibrium will have to shift to the left for equilibrium to be reestablished (Figure 15.7).

**Figure 15.7** The effect of a volume decrease (pressure increase) on the $N_2O_4(g) \rightleftharpoons 2NO_2(g)$ equilibrium. When volume is decreased, the equilibrium is driven toward the side with the smallest number of moles of gas.

In general, a decrease in volume of a reaction vessel will cause a shift in the equilibrium in the direction that minimizes the total number of moles of gas. Conversely, an increase in volume will cause a shift in the direction that maximizes the total number of moles of gas.

Sample Problem 15.12 shows how to predict the equilibrium shift that will be caused by a volume change.

---

## Sample Problem 15.12

For each reaction, predict in what direction the equilibrium will shift when the volume of the reaction vessel is decreased.

(a) $PCl_5(g) \rightleftharpoons PCl_3(g) + Cl_2(g)$

(b) $2PbS(s) + 3O_2(g) \rightleftharpoons 2PbO(s) + 2SO_2(g)$

(c) $H_2(g) + I_2(g) \rightleftharpoons 2HI(g)$

**Strategy** Determine which direction minimized the number of moles of gas in the reaction. Count only moles of *gas*.

**Setup** We have (a) 1 mole of gas on the reactant side and 2 moles of gas on the product side, (b) 3 moles of gas on the reactant side and 2 moles of gas on the product side, and (c) 2 moles of gas on each side.

**Solution**

(a) Shift to the left

(b) Shift to the right

(c) No shift

> **Student Annotation:** It is a common error to count all the species in a reaction to determine which side has fewer moles. To determine what direction shift a volume change will cause, it is only the number of moles of *gas* that matters.

> **Think About It** When there is no difference in the number of moles of gas, changing the volume of the reaction vessel will change the concentrations of reactant(s) and product(s)—but the system will remain at equilibrium. ($Q$ will remain equal to $K$.)

**Practice Problem A** For each reaction, predict the direction of shift caused by increasing the volume of the reaction vessel.

(a) $2NOCl(g) \rightleftharpoons 2NO(g) + Cl_2(g)$

(b) $CaCO_3(s) \rightleftharpoons CaO(s) + CO_2(g)$

(c) $Zn(s) + 2H^+(aq) \rightleftharpoons Zn^{2+}(aq) + H_2(g)$

**Practice Problem B** For the following equilibrium, give an example of a stress that will cause a shift to the right, a stress that will cause a shift to the left, and one that will cause no shift:

$$H_2(g) + F_2(g) \rightleftharpoons 2HF(g)$$

---

It is possible to change the total pressure of a system without changing its volume—by adding an inert gas such as helium to the reaction vessel. Because the total volume remains the same, the concentrations of reactant and product gases do not change. Therefore, the equilibrium is not disturbed and no shift will occur.

## Changes in Temperature

A change in concentration or volume may alter the position of an equilibrium (i.e., the relative amounts of reactants and products), but it does not change the value of the equilibrium constant. Only a change in temperature can alter the value of the equilibrium constant. To understand why, consider the following reaction:

$$N_2O_4(g) \rightleftharpoons 2NO_2(g)$$

The forward reaction is endothermic (absorbs heat, $\Delta H > 0$):

$$\text{Heat} + N_2O_4(g) \rightleftharpoons 2NO_2(g) \qquad \Delta H° = 58.0 \text{ kJ/mol}$$

If we treat heat as though it were a reactant, we can use Le Châtelier's principle to predict what will happen if we add or remove heat. Increasing the temperature (adding heat) will shift the reaction in the forward direction because heat appears on the reactant side. Lowering the temperature (removing heat) will shift the reaction in the reverse direction. Consequently, the equilibrium constant, given by

$$K_c = \frac{[NO_2]^2}{[N_2O_4]}$$

(a)

increases when the system is heated and decreases when the system is cooled (Figure 15.8). A similar argument can be made in the case of an exothermic reaction, where heat can be considered to be a product. Increasing the temperature of an exothermic reaction causes the equilibrium constant to decrease, shifting the equilibrium toward reactants.

Another example of this phenomenon is the equilibrium between the following ions:

$$\underset{\text{blue}}{CoCl_4^{2-}} + 6H_2O \rightleftharpoons \underset{\text{pink}}{Co(H_2O)_6^{2+}} + 4Cl^- + \text{heat}$$

The reaction as written, the formation of $Co(H_2O)_6^{2+}$, is exothermic. Thus, the reverse reaction, the formation of $CoCl_4^{2-}$, is endothermic. On heating, the equilibrium shifts to the left and the solution turns blue. Cooling favors the exothermic reaction [the formation of $Co(H_2O)_6^{2+}$] and the solution turns pink (Figure 15.9).

In summary, a temperature increase favors an endothermic reaction, and a temperature decrease favors an exothermic process. Temperature affects the position of an equilibrium by changing the value of the equilibrium constant. Figures 15.10 and 15.11 illustrate the effects of various stresses on systems at equilibrium.

## Catalysis

A catalyst speeds up a reaction by lowering the reaction's activation energy [◄◄ Section 14.6]. However, a catalyst lowers the activation energy of the forward and reverse reactions to the same extent (see Figure 14.15). The presence of a catalyst, therefore, does not alter the equilibrium constant, nor does it shift the position of an equilibrium system. Adding a catalyst to a reaction mixture that is not at equilibrium will simply cause the mixture to reach equilibrium sooner. The same equilibrium mixture could be obtained without the catalyst, but it might take a much longer time.

(b)

**Figure 15.8** (a) $N_2O_4$-$NO_2$ equilibrium. (b) Because the reaction is endothermic, at higher temperature, the $N_2O_4(g) \rightleftharpoons 2NO_2(g)$ equilibrium shifts toward product, making the reaction mixture darker.

**Figure 15.9** (a) An equilibrium mixture of $CoCl_4^{2-}$ ions and $Co(H_2O)_6^{2+}$ ions appears violet. (b) Heating favors the formation of $CoCl_4^{2-}$, making the solution look more blue. (c) Cooling favors the formation of $Co(H_2O)_6^{2+}$, making the solution look more pink.

(a)              (b)              (c)

## Bringing Chemistry to Life

### Hemoglobin Production at High Altitude

As we learned at the end of Chapter 11, rapid ascent to a high altitude can cause altitude sickness. The symptoms of altitude sickness, including dizziness, headache, and nausea, are caused by hypoxia, an insufficient oxygen supply to body tissues. In severe cases, without prompt treatment, a victim may slip into a coma and die. And yet a person staying at a high altitude for weeks or months can recover gradually from altitude sickness, adjust to the low oxygen content in the atmosphere, and live and function normally.

Media Player/MPEG Animation: Figures 15.10 and 15.11, Le Châtelier's Principle, pp. 656–659.

The combination of oxygen with the hemoglobin (Hb) molecule, which carries oxygen through the blood, is a complex reaction, but for our purposes it can be represented by the following simplified equation:*

$$Hb(aq) + O_2(aq) \rightleftharpoons HbO_2(aq)$$

where $HbO_2$ is oxyhemoglobin, the hemoglobin-oxygen complex that actually transports oxygen to tissues. The equilibrium expression for this process is

$$K_c = \frac{[HbO_2]}{[Hb][O_2]}$$

At an altitude of 3 km, the partial pressure of oxygen is only about 0.14 atm, compared with 0.20 atm at sea level. According to Le Châtelier's principle, a decrease in oxygen concentration will shift the hemoglobin-oxyhemoglobin equilibrium from right to left. This change depletes the supply of oxyhemoglobin, causing hypoxia. Over time, the body copes with this problem by producing more hemoglobin molecules. As the concentration of Hb increases, the equilibrium gradually shifts back toward the right (toward the formation of oxyhemoglobin). It can take several weeks for the increase in hemoglobin production to meet the body's oxygen needs adequately. A return to full capacity may require several years to occur. Studies show that long-time residents of high-altitude areas have high hemoglobin levels in their blood—sometimes as much as 50 percent more than individuals living at sea level. The production of more hemoglobin and the resulting increased capacity of the blood to deliver oxygen to the body have made high-altitude training and hypoxic tents popular among some athletes.

---

*Biological processes such as this are not true equilibria, but rather they are *steady-state* situations. In a steady state, the constant concentrations of reactants and products are not the result of forward and reverse reactions occurring at the same rate. Instead, reactant concentration is replenished by a previous reaction and product concentration is maintained by a subsequent reaction. Nevertheless, many of the principles of equilibrium, including Le Châtelier's principle, still apply.

---

**Checkpoint 15.5**   **Factors That Affect Chemical Equilibrium**

**15.5.1** Which of the following equilibria will shift to the right when $H_2$ is added? (Select all that apply.)

a) $2H_2 + O_2 \rightleftharpoons 2H_2O$

b) $2HI \rightleftharpoons H_2 + I_2$

c) $H_2 + CO_2 \rightleftharpoons H_2O + CO$

d) $2NaHCO_3 \rightleftharpoons Na_2CO_3 + H_2O + CO_2$

e) $2CO + O_2 \rightleftharpoons 2CO_2$

**15.5.2** Which of the following will cause the equilibrium

$$C(graphite) + CO_2(g) \rightleftharpoons 2CO(g)$$

to shift to the right? (Select all that apply.)

a) Decreasing the volume

b) Increasing the volume

c) Adding more C(graphite)

d) Adding more $CO_2(g)$

e) Removing CO(g) as it forms

**15.5.3** Which of the following equilibria will shift to the left when the temperature is increased? [$\Delta H$ (kJ/mol) values are given in parentheses.] (Select all that apply.)

a) $S + H_2 \rightleftharpoons H_2S$      $\Delta H$ (−20)

b) $C + H_2O \rightleftharpoons CO + H_2$      $\Delta H$ (131)

c) $H_2 + CO_2 \rightleftharpoons H_2O + CO$      $\Delta H$ (41)

d) $MgO + CO_2 \rightleftharpoons MgCO_3$      $\Delta H$ (−117)

e) $2CO + O_2 \rightleftharpoons 2CO_2$      $\Delta H$ (−566)

**15.5.4** For which of the following reactions will a change in volume *not* affect the position of the equilibrium? (Select all that apply.)

a) $MgO(s) + CO_2(g) \rightleftharpoons MgCO_3(s)$

b) $H_2(g) + Cl_2(g) \rightleftharpoons 2HCl(g)$

c) $BaCO_3(s) \rightleftharpoons BaO(s) + CO_2(g)$

d) $Br_2(l) + H_2(g) \rightleftharpoons 2HBr(g)$

e) $C(graphite) + CO_2(g) \rightleftharpoons 2CO(g)$

*(Continued on page 660)*

Figure 15.10    Le Châtelier's principle.

# Effect of Addition to an Equilibrium Mixture

$$H_2(g) + I_2(g) \rightleftharpoons 2HI(g)$$

[H$_2$] = 0.112 M
[I$_2$] = 0.112 M
[HI] = 0.825 M

[H$_2$] = 0.112 M
[I$_2$] = 0.112 M
[HI] = 0.825 M

[H$_2$] = 0.112 M
[I$_2$] = 0.112 M
[HI] = 0.825 M

$$Q_c = \frac{[HI]^2}{[H_2][I_2]} = \frac{(0.825)^2}{(0.112)(0.112)} = 54.3$$

$$Q_c = K_c$$

$$Q_c = \frac{[HI]^2}{[H_2][I_2]} = \frac{(0.825)^2}{(0.112)(0.112)} = 54.3$$

$$Q_c = K_c$$

$$Q_c = \frac{[HI]^2}{[H_2][I_2]} = \frac{(0.825)^2}{(0.112)(0.112)} = 54.3$$

$$Q_c = K_c$$

Add a reactant—I$_2$(g).

Add a product—HI(g).

Add a species not involved
in the equilibrium—He(g).

[H$_2$] = 0.112 M
[I$_2$] = 0.499 M
[HI] = 0.825 M

[H$_2$] = 0.112 M
[I$_2$] = 0.112 M
[HI] = 3.17 M

[H$_2$] = 0.112 M
[I$_2$] = 0.112 M
[HI] = 0.825 M
[He] = 0.565 M

$$Q_c = \frac{(0.825)^2}{(0.112)(0.499)} = 12$$

$$Q_c \neq K_c$$

$$Q_c = \frac{(3.17)^2}{(0.112)(0.112)} = 801$$

$$Q_c \neq K_c$$

$$Q_c = \frac{(0.825)^2}{(0.112)(0.112)} = 54.3$$

$$Q_c = K_c$$

Equilibrium shifts toward product.

Equilibrium shifts toward reactant.

Equilibrium does not shift in either direction.

[H$_2$] = 0.0404 M
[I$_2$] = 0.427 M
[HI] = 0.968 M

[H$_2$] = 0.362 M
[I$_2$] = 0.362 M
[HI] = 2.67 M

[H$_2$] = 0.112 M
[I$_2$] = 0.112 M
[HI] = 0.825 M
[He] = 0.565 M

$$Q_c = \frac{(0.968)^2}{(0.0404)(0.427)} = 54.3$$

$$Q_c = K_c$$

$$Q_c = \frac{(2.67)^2}{(0.362)(0.362)} = 54.4$$

$$Q_c = K_c$$

## What's the point?

Adding a reactant to an equilibrium mixture shifts the
equilibrium toward the product side of the equation.
Adding a product shifts the equilibrium toward the
reactant side. Adding a species that is neither a reactant
nor a product does not cause a shift in the equilibrium.

# Effect of Temperature Change

$$N_2O_4(g) \rightleftharpoons 2NO_2(g) \quad \Delta H° = 58.04 \text{ kJ/mol}$$

Temperature decrease drives an endothermic equilibrium toward reactants.

Temperature increase drives an endothermic equilibrium toward products.

$$H_2(g) + I_2(g) \rightleftharpoons 2HI(g) \quad \Delta H° = -9.4 \text{ kJ/mol}$$

Temperature increase drives an exothermic equilibrium toward reactants.

Temperature decrease drives an exothermic equilibrium toward products.

## What's the point?

Increasing the temperature of an equilibrium mixture causes a shift toward the product side for an endothermic reaction, and a shift toward the reactant side for an exothermic reaction.

# Figure 15.11 Le Châtelier's principle.

# Effect of Volume Change

$$[N_2O_4] = 1.08\ M$$
$$[NO_2] = 0.0707\ M$$

$$Q_c = \frac{[NO_2]^2}{[N_2O_4]} = \frac{(0.0707)^2}{1.08} = 4.63 \times 10^{-3}$$

$$Q_c = K_c$$

$$[N_2O_4] = 0.540\ M$$
$$[NO_2] = 0.0354\ M$$

$$Q_c = \frac{(0.0354)^2}{0.540} = 2.3 \times 10^{-3}$$

$$Q_c \neq K_c$$

$$[N_2O_4] = 2.16\ M$$
$$[NO_2] = 0.141\ M$$

$$Q_c = \frac{(0.141)^2}{2.16} = 9.2 \times 10^{-3}$$

$$Q_c \neq K_c$$

## What's the point?

Increasing the volume causes a shift toward the side with the *larger* number of moles of gas. Decreasing the volume of an equilibrium mixture causes a shift toward the side of the equation with the *smaller* number of moles of gas.

$$[N_2O_4] = 0.533\ M$$
$$[NO_2] = 0.0497\ M$$

$$Q_c = \frac{(0.0497)^2}{0.533} = 4.6 \times 10^{-3}$$

$$Q_c = K_c$$

$$[N_2O_4] = 2.18\ M$$
$$[NO_2] = 0.100\ M$$

$$Q_c = \frac{(0.100)^2}{2.18} = 4.6 \times 10^{-3}$$

$$Q_c = K_c$$

$$[H_2] = 0.112\ M$$
$$[I_2] = 0.112\ M$$
$$[HI] = 0.825\ M$$

$$Q_c = \frac{[HI]^2}{[H_2][I_2]} = \frac{(0.825)^2}{(0.112)(0.112)} = 54.3$$

$$Q_c = K_c$$

$$[H_2] = 0.056\ M$$
$$[I_2] = 0.056\ M$$
$$[HI] = 0.413\ M$$

$$Q_c = \frac{(0.413)^2}{(0.056)(0.056)} = 54.3$$

$$Q_c = K_c$$

$$[H_2] = 0.224\ M$$
$$[I_2] = 0.224\ M$$
$$[HI] = 1.65\ M$$

$$Q_c = \frac{(1.65)^2}{(0.224)(0.224)} = 54.3$$

$$Q_c = K_c$$

## What's the point?

For an equilibrium with equal numbers of gaseous moles on both sides, a change in volume does not cause the equilibrium to shift in either direction.

$$[H_2] = 0.056\ M$$
$$[I_2] = 0.056\ M$$
$$[HI] = 0.413\ M$$

$$Q_c = \frac{(0.413)^2}{(0.056)(0.056)} = 54.3$$

$$Q_c = K_c$$

$$[H_2] = 0.224\ M$$
$$[I_2] = 0.224\ M$$
$$[HI] = 1.65\ M$$

$$Q_c = \frac{(1.65)^2}{(0.224)(0.224)} = 54.3$$

$$Q_c = K_c$$

*You May Be Wondering...*

# What Happens to the Units in Equilibrium Constants?

We have said that it is common to disregard units and write equilibrium constants as dimensionless quantities. The reason is as follows: Prior to being used in an equilibrium expression, each molar concentration is divided by a reference concentration of 1 $M$, and each partial pressure is divided by a reference pressure of 1 atm. The reference concentration (1 $M$) and reference pressure (1 atm) are known as the *standard states* for aqueous and gaseous species, respectively. This cancels the units without changing the value of the concentration or pressure in question. The dimensionless result of dividing a concentration by a reference value is called the *activity*. Activities commonly are used by chemists in place of concentrations, but a detailed discussion of activities is beyond the scope of this book. Using as an example the equilibrium between $N_2O_4$ and $NO_2$,

$$N_2O_4(g) \rightleftharpoons 2NO_2(g)$$

and data from experiment 3 in Table 15.1, the equilibrium concentrations of $N_2O_4$ and $NO_2$ (at 25°C) are 0.491 $M$ and 0.0475 $M$, respectively. We can calculate the value of $K_c$ as follows:

$$K_c = \frac{\left(\dfrac{0.0475\ M}{1\ M}\right)^2}{\dfrac{0.491\ M}{1\ M}} = 4.60 \times 10^{-3}$$

We can then convert the molar concentrations to partial pressures using the ideal gas equation:

$$P_{N_2O_4} = \frac{n_{N_2O_4}}{V}RT = [N_2O_4](0.08206)(298) = 12.0\ \text{atm}$$

$$P_{NO_2} = \frac{n_{NO_2}}{V}RT = [NO_2](0.08206)(298) = 1.16\ \text{atm}$$

Finally, we can calculate the value of $K_P$ as

$$K_P = \frac{\left(\dfrac{1.16\ \text{atm}}{1\ \text{atm}}\right)^2}{\dfrac{12.0\ \text{atm}}{1\ \text{atm}}} = 0.112$$

In each case, dividing by the reference value eliminates the units without changing the numerical value of the result. We will not show the division by reference values explicitly, but you should realize that this is the reason why there are no units associated with equilibrium constants.

Division of each entry in an equilibrium expression by a reference value actually enables us to use molar concentrations and partial pressures in the same equilibrium expression. For example, we can write the equilibrium expression for the reaction

$$Zn(s) + 2H^+(aq) \rightleftharpoons Zn^{2+}(aq) + H_2(g)$$

as

$$K = \frac{\left(\dfrac{[Zn^{2+}]}{1\ M}\right)\left(\dfrac{P_{H_2}}{1\ \text{atm}}\right)}{\left(\dfrac{[H^+]}{1\ M}\right)^2}$$

Each of the entries in the expression is divided by the appropriate reference value, making the result dimensionless. Note that this equilibrium constant is neither a $K_c$ nor a $K_P$. It is referred to simply as $K$ or as $K_{eq}$, where the subscripted *eq* denotes *equilibrium*. We will encounter the use of "mixed" equilibrium constants such as this in Chapters 19 and 20.

---

**Checkpoint 15.5**    **Factors That Affect Chemical Equilibrium** *(continued)*

**15.5.5**  The diagram shows the gaseous reaction $2A \rightleftharpoons A_2$ at equilibrium. How will the numbers of A and $A_2$ change if the volume of the container is increased at constant temperature?

   a) $A_2$ will increase and A will decrease.

   b) A will increase and $A_2$ will decrease.

   c) $A_2$ and A will both decrease.

   d) $A_2$ and A will both increase.

   e) Neither $A_2$ nor A will change.

**15.5.6**  The diagrams show equilibrium mixtures of $A_2$, $B_2$, and AB at two different temperatures $(T_2 > T_1)$. Is the reaction $A_2 + B_2 \rightleftharpoons 2AB$ endothermic or exothermic?

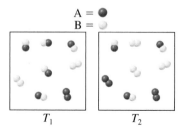

   a) Endothermic

   b) Exothermic

   c) Neither

   d) There is not enough information to determine.

# Applying What You've Learned

At present, the World Anti-Doping Agency has no way to detect the use of hypoxic sleeping tents, other than inspection of an athlete's home. Imagine, however, that a biochemical analysis company is developing a way to determine whether an elevated red blood cell count in an athlete's blood is the result of such a practice. A key substance in the detection process is the proprietary compound OD17X, which is produced by the combination of two other proprietary compounds, OD1A and OF2A. The aqueous reaction is represented by the equation

$$OD1A(aq) + OF2A(aq) \rightleftharpoons OD17X(aq) \qquad \Delta H = 29 \text{ kJ/mol}$$

**Problems:**

a)  Write the equilibrium expression for the given reaction. [◄◄ Sample Problem 15.3]

b)  Given the equilibrium concentrations at room temperature of [OD1A] = 2.12 $M$, [OF2A] = 1.56 $M$, and [OD17X] = $1.01 \times 10^{-4}$ $M$, calculate the value of the equilibrium constant ($K_c$) at room temperature. [◄◄ Sample Problem 15.2]

c)  The OD17X produced is precipitated with another proprietary substance according to the equation

$$OD17X(aq) + A771A(aq) \rightleftharpoons OD17X\text{-}A77(s)$$

   $K_c$ for the precipitation equilibrium is $1.00 \times 10^6$ at room temperature. Write the equilibrium expression for the sum of the two reactions, and determine the value of the overall equilibrium constant. [◄◄ Sample Problem 15.4]

d)  Determine whether a mixture with [OD1A] = 3.00 $M$, [OF2A] = 2.50 $M$, and [OD17X] = $2.7 \times 10^{-4}$ $M$ is at equilibrium and, if not, in which direction it will have to proceed to achieve equilibrium. [◄◄ Sample Problem 15.7]

e)  Because the production of OD17X is endothermic, it can be enhanced by increasing the temperature. At 250°C, the equilibrium constant for the reaction

$$OD1A(aq) + OF2A(aq) \rightleftharpoons OD17X(aq)$$

   is $3.8 \times 10^2$. If a synthesis at 250°C begins with 1.00 $M$ of each reactant, what will be the equilibrium concentrations of reactants and products? [◄◄ Sample Problem 15.8]

# CHAPTER SUMMARY

## Section 15.1

- By our definition of a chemical reaction, reactants are consumed and products are produced. This is known as the *forward* reaction. A *reversible process* is one in which the products can also be consumed to produce reactants, a process known as the *reverse* reaction.

- *Equilibrium* is the condition where the forward and reverse reactions are occurring at the same rate and there is no net change in the reactant and product concentrations over time.

- In theory, equilibrium can be established starting with just reactants, with just products, or with a combination of reactants and products.

## Section 15.2

- The *reaction quotient* $(Q_c)$ is the product of the *product concentrations* over the product of the *reactant concentrations,* with each concentration raised to a power equal to the corresponding stoichiometric coefficient in the balanced chemical equation. This is known as the **law of mass action.**

- At equilibrium, the reaction quotient is equal to a constant value, the **equilibrium constant $(K_c)$.**

- At equilibrium, concentrations in the reaction quotient are *equilibrium* concentrations and the quotient is called the **equilibrium expression.**

- $K_c$ is constant at constant temperature. The value of $K_c$ can be calculated by plugging equilibrium concentrations into the equilibrium expression.

- A large equilibrium constant $(K_c > 10^3)$ indicates that products are favored at equilibrium. A small equilibrium constant $(K_c < 10^{-3})$ indicates that reactants are favored at equilibrium.

## Section 15.3

- Solids and liquids do not appear in the equilibrium expression for a heterogeneous reaction.

- When chemical equations that represent equilibria are reversed, multiplied, combined with other equations, or any combination of these processes, the corresponding changes must be made to the equilibrium constants.

- Equilibrium expressions that contain only gases can be written either as $K_c$ expressions or as $K_P$ expressions. $K_P$ expressions have the same form as $K_c$ expressions but contain partial pressures rather than molar concentrations. The reaction quotient, $Q$, can also be expressed in terms of the pressures of products and reactants. In this case it is labeled $Q_P$.

- $K_c$ and $K_P$ are not usually equal. The values are the same *only* when the reaction results in no net change in the number of moles of gas.

## Section 15.4

- We can predict in which direction a reaction must proceed to achieve equilibrium by comparing the values of $Q_c$ and $K_c$ (or of $Q_P$ and $K_P$).

- Starting concentrations can be used, along with the equilibrium expression and equilibrium constant, to determine equilibrium concentrations.

## Section 15.5

- According to **Le Châtelier's principle,** a system at equilibrium will react to stress by shifting in the direction that will partially offset the effect of the stress.

- The stresses that can be applied to a system at equilibrium include the addition or removal of a substance, changes in the volume of the reaction vessel, and changes in the temperature.

- The addition of a catalyst does not shift the equilibrium in either direction.

# KEY WORDS

# KEY EQUATIONS

$$15.1 \quad Q_c = \frac{[C]^c[D]^d}{[A]^a[B]^b} = K_c \text{ (at equilibrium)}$$

$$15.2 \quad Q_c = \frac{[C]^c[D]^d}{[A]^a[B]^b}$$

15.3    $\Delta n$ = moles of gaseous products − moles of gaseous reactants

$$15.4 \quad K_P = K_c \, [(0.08206 \text{ L} \cdot \text{atm/K} \cdot \text{mol}) \times T]^{\Delta n}$$

# QUESTIONS AND PROBLEMS

## Section 15.1: The Concept of Equilibrium

### Review Questions

15.1 Define *equilibrium*. Give two examples of a dynamic equilibrium.

15.2 Which of the following statements is correct about a reacting system at equilibrium: (a) the concentrations of reactants are equal to the concentrations of products, (b) the rate of the forward reaction is equal to the rate of the reverse reaction.

15.3 Consider the reversible reaction A $\rightleftharpoons$ B. Explain how equilibrium can be reached by starting with only A, only B, or a mixture of A and B.

## Section 15.2: The Equilibrium Constant

### Review Questions

15.4 What is the law of mass action?

15.5 Briefly describe the importance of equilibrium in the study of chemical reactions.

15.6 Define *reaction quotient*. How does it differ from the equilibrium constant?

15.7 Write reaction quotients for the following reactions: (a) $2NO(g) + O_2(g) \rightleftharpoons N_2O_4(g)$, (b) $S(s) + 3F_2(g) \rightleftharpoons SF_6(g)$, (c) $Co^{3+}(aq) + 6NH_3(aq) \rightleftharpoons Co(NH_3)_6^{3+}(aq)$, (d) $HCOOH(aq) \rightleftharpoons HCOO^-(aq) + H^+(aq)$.

15.8 Write the equation for the reaction that corresponds to each of the following reaction quotients:

(a) $Q_c = \dfrac{[H_2]^2[S_2]}{[H_2S]^2}$

(b) $Q_c = \dfrac{[NO_2]^2[Cl_2]}{[NClO_2]^2}$

(c) $Q_c = \dfrac{[HgI_4^{2-}]}{[Hg^{2+}][I^-]^4}$

(d) $Q_c = \dfrac{[NO]^2[Br_2]}{[NOBr]^2}$

### Problems

15.9 Consider the reaction

$$2NO(g) + 2H_2(g) \rightleftharpoons N_2(g) + 2H_2O(g)$$

At a certain temperature, the equilibrium concentrations are [NO] = 0.31 $M$, [H$_2$] = 0.16 $M$, [N$_2$] = 0.082 $M$, and [H$_2$O] = 4.64 $M$. (a) Write the equilibrium expression for the reaction. (b) Determine the value of the equilibrium constant.

15.10 The equilibrium constant for the reaction

$$2SO_2(g) + O_2(g) \rightleftharpoons 2SO_3(g)$$

is $2.8 \times 10^2$ at a certain temperature. If [SO$_2$] = 0.0124 $M$ and [O$_2$] = 0.031 $M$, what is [SO$_3$]?

15.11 Consider the following equilibrium process at 700°C:

$$2H_2(g) + S_2(g) \rightleftharpoons 2H_2S(g)$$

Analysis shows that there are 2.50 moles of H$_2$, $1.35 \times 10^{-5}$ mole of S$_2$, and 8.70 moles of H$_2$S present in a 12.0-L flask. Calculate the equilibrium constant $K_c$ for the reaction.

15.12 The equilibrium constant for the reaction

$$2H_2(g) + CO(g) \rightleftharpoons CH_3OH(g)$$

is $1.6 \times 10^{-2}$ at a certain temperature. If there are $1.17 \times 10^{-2}$ mole of H$_2$ and $3.46 \times 10^{-3}$ mole of CH$_3$OH at equilibrium in a 5.60-L flask, what is the concentration of CO?

## Section 15.3: Equilibrium Expressions

### Review Questions

15.13 Define *homogeneous* equilibrium and *heterogeneous* equilibrium. Give two examples of each.

15.14 What do the symbols $K_c$ and $K_P$ represent?

15.15 Write the expressions for the equilibrium constants $K_P$ of the following thermal decomposition reactions:

(a) $2NaHCO_3(s) \rightleftharpoons Na_2CO_3(s) + CO_2(g) + H_2O(g)$
(b) $2CaSO_4(s) \rightleftharpoons 2CaO(s) + 2SO_2(g) + O_2(g)$

15.16 Write equilibrium constant expressions for $K_c$, and for $K_P$, if applicable, for the following processes:

(a) $2CO_2(g) \rightleftharpoons 2CO(g) + O_2(g)$
(b) $3O_2(g) \rightleftharpoons 2O_3(g)$
(c) $CO(g) + Cl_2(g) \rightleftharpoons COCl_2(g)$
(d) $H_2O(g) + C(s) \rightleftharpoons CO(g) + H_2(g)$
(e) $HCOOH(aq) \rightleftharpoons H^+(aq) + HCOO^-(aq)$
(f) $2HgO(s) \rightleftharpoons 2Hg(l) + O_2(g)$

15.17 Write the equilibrium constant expressions for $K_c$ and for $K_P$, if applicable, for the following reactions:

(a) $2NO_2(g) + 7H_2(g) \rightleftharpoons 2NH_3(g) + 4H_2O(l)$
(b) $2ZnS(s) + 3O_2(g) \rightleftharpoons 2ZnO(s) + 2SO_2(g)$
(c) $C(s) + CO_2(g) \rightleftharpoons 2CO(g)$
(d) $C_6H_5COOH(aq) \rightleftharpoons C_6H_5COO^-(aq) + H^+(aq)$

15.18 Write the equation relating $K_c$ to $K_P$, and define all the terms.

15.19 What is the rule for writing the equilibrium constant for the overall reaction that is the sum of two or more reactions?

15.20 Give an example of a multiple equilibria reaction.

### Problems

15.21 The equilibrium constant for the reaction A $\rightleftharpoons$ B is $K_c = 10$ at a certain temperature. (1) Starting with only reactant A, which of the diagrams shown here best represents the system at equilibrium? (2) Which of the diagrams best represents the system at equilibrium if $K_c = 0.10$? Explain why you can calculate $K_c$ in each case without knowing the volume of the container. The grey spheres represent the A molecules, and the green spheres represent the B molecules.

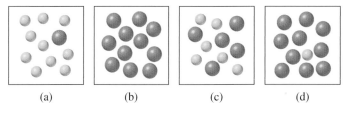

(a)　　　　　(b)　　　　　(c)　　　　　(d)

15.22 The following diagrams represent the equilibrium state for three different reactions of the type

$$A + X \rightleftharpoons AX \ (X = B, C, \text{ or } D)$$

$A + B \rightleftarrows AB$    $A + C \rightleftarrows AC$    $A + D \rightleftarrows AD$

(a) Which reaction has the largest equilibrium constant?
(b) Which reaction has the smallest equilibrium constant?

**15.23**  The equilibrium constant ($K_c$) for the reaction

$$2HCl(g) \rightleftharpoons H_2(g) + Cl_2(g)$$

is $4.17 \times 10^{-34}$ at 25°C. What is the equilibrium constant for the reaction

$$H_2(g) + Cl_2(g) \rightleftharpoons 2HCl(g)$$

at the same temperature?

**15.24**  What is $K_P$ at 1273°C for the reaction

$$2CO(g) + O_2(g) \rightleftharpoons 2CO_2(g)$$

if $K_c$ is $2.24 \times 10^{22}$ at the same temperature?

**15.25**  The equilibrium constant $K_P$ for the reaction

$$2SO_3(g) \rightleftharpoons 2SO_2(g) + O_2(g)$$

is $1.8 \times 10^{-5}$ at 350°C. What is $K_c$ for this reaction?

**15.26**  Consider the reaction

$$N_2(g) + O_2(g) \rightleftharpoons 2NO(g)$$

If the equilibrium partial pressures of $N_2$, $O_2$, and NO are 0.15, 0.33, and 0.050 atm, respectively, at 2200°C, what is $K_P$?

**15.27**  A reaction vessel contains $NH_3$, $N_2$, and $H_2$ at equilibrium at a certain temperature. The equilibrium concentrations are $[NH_3] = 0.25\ M$, $[N_2] = 0.11\ M$, and $[H_2] = 1.91\ M$. Calculate the equilibrium constant $K_c$ for the synthesis of ammonia if the reaction is represented as

(a) $N_2(g) + 3H_2(g) \rightleftharpoons 2NH_3(g)$
(b) $\frac{1}{2}N_2(g) + \frac{3}{2}H_2(g) \rightleftharpoons NH_3(g)$

**15.28**  The equilibrium constant $K_c$ for the reaction

$$I_2(g) \rightleftharpoons 2I(g)$$

is $3.8 \times 10^{-5}$ at 727°C. Calculate $K_c$ and $K_P$ for the equilibrium

$$2I(g) \rightleftharpoons I_2(g)$$

at the same temperature.

**15.29**  At equilibrium, the pressure of the reacting mixture

$$CaCO_3(s) \rightleftharpoons CaO(s) + CO_2(g)$$

is 0.105 atm at 350°C. Calculate $K_P$ and $K_c$ for this reaction.

**15.30**  The equilibrium constant $K_P$ for the reaction

$$PCl_5(g) \rightleftharpoons PCl_3(g) + Cl_2(g)$$

is 1.05 at 250°C. The reaction starts with a mixture of $PCl_5$, $PCl_3$, and $Cl_2$ at pressures of 0.177, 0.223, and 0.111 atm, respectively, at 250°C. When the mixture comes to equilibrium at that temperature, which pressures will have decreased and which will have increased? Explain why.

**15.31**  Ammonium carbamate ($NH_4CO_2NH_2$) decomposes as follows:

$$NH_4CO_2NH_2(s) \rightleftharpoons 2NH_3(g) + CO_2(g)$$

Starting with only the solid, it is found that when the system reaches equilibrium at 40°C, the total gas pressure ($NH_3$ and $CO_2$) is 0.363 atm. Calculate the equilibrium constant $K_P$.

**15.32**  Pure phosgene gas ($COCl_2$), $3.00 \times 10^{-2}$ mol, was placed in a 1.50-L container. It was heated to 800 K, and at equilibrium the pressure of CO was found to be 0.497 atm. Calculate the equilibrium constant $K_P$ for the reaction

$$CO(g) + Cl_2(g) \rightleftharpoons COCl_2(g)$$

**15.33**  Consider the equilibrium

$$2NOBr(g) \rightleftharpoons 2NO(g) + Br_2(g)$$

If nitrosyl bromide (NOBr) is 34 percent dissociated at 25°C and the total pressure is 0.25 atm, calculate $K_P$ and $K_c$ for the dissociation at this temperature.

**15.34**  The following equilibrium constants have been determined for hydrosulfuric acid at 25°C:

$$H_2S(aq) \rightleftharpoons H^+(aq) + HS^-(aq) \qquad K_c' = 9.5 \times 10^{-8}$$
$$HS^-(aq) \rightleftharpoons H^+(aq) + S^{2-}(aq) \qquad K_c'' = 1.0 \times 10^{-19}$$

Calculate the equilibrium constant for the following reaction at the same temperature:

$$H_2S(aq) \rightleftharpoons 2H^+(aq) + S^{2-}(aq)$$

**15.35**  The following equilibrium constants have been determined for oxalic acid at 25°C:

$$H_2C_2O_4(aq) \rightleftharpoons H^+(aq) + HC_2O_4^-(aq) \qquad K_c' = 6.5 \times 10^{-2}$$
$$HC_2O_4^-(aq) \rightleftharpoons H^+(aq) + C_2O_4^{2-}(aq) \qquad K_c'' = 6.1 \times 10^{-5}$$

Calculate the equilibrium constant for the following reaction at the same temperature:

$$H_2C_2O_4(aq) \rightleftharpoons 2H^+(aq) + C_2O_4^{2-}(aq)$$

**15.36**  The following equilibrium constants were determined at 1123 K:

$$C(s) + CO_2(g) \rightleftharpoons 2CO(g) \qquad K_P' = 1.3 \times 10^{14}$$
$$CO(g) + Cl_2(g) \rightleftharpoons COCl_2(g) \qquad K_P'' = 6.0 \times 10^{-3}$$

Write the equilibrium constant expression $K_P$, and calculate the equilibrium constant at 1123 K for

$$C(s) + CO_2(g) + 2Cl_2(g) \rightleftharpoons 2COCl_2(g)$$

**15.37**  At a certain temperature, the following reactions have the constants shown:

$$S(s) + O_2(g) \rightleftharpoons SO_2(g) \qquad K_c' = 4.2 \times 10^{52}$$
$$2S(s) + 3O_2(g) \rightleftharpoons 2SO_3(g) \qquad K_c'' = 9.8 \times 10^{128}$$

Calculate the equilibrium constant $K_c$ for the following reaction at that temperature:

$$2SO_2(g) + O_2(g) \rightleftharpoons 2SO_3(g)$$

## Section 15.4: Using Equilibrium Expressions to Solve Problems

### Review Questions

**15.38**  Outline the steps for calculating the concentrations of reacting species in an equilibrium reaction.

### Problems

**15.39**  The equilibrium constant $K_P$ for the reaction

$$2SO_2(g) + O_2(g) \rightleftharpoons 2SO_3(g)$$

is $5.60 \times 10^4$ at 350°C. The initial pressures of $SO_2$, $O_2$, and $SO_3$ in a mixture are 0.350, 0.762, and 0 atm, respectively, at 350°C. When the mixture reaches equilibrium, is the total pressure less than or greater than the sum of the initial pressures?

**15.40**  For the synthesis of ammonia

$$N_2(g) + 3H_2(g) \rightleftharpoons 2NH_3(g)$$

the equilibrium constant $K_c$ at 375°C is 1.2. Starting with $[H_2]_0 = 0.76\ M$, $[N_2]_0 = 0.60\ M$, and $[NH_3]_0 = 0.48\ M$, which gases will have increased in concentration and which will have decreased in concentration when the mixture comes to equilibrium?

**15.41** For the reaction

$$H_2(g) + CO_2(g) \rightleftharpoons H_2O(g) + CO(g)$$

at 700°C, $K_c = 0.534$. Calculate the number of moles of $H_2$ that are present at equilibrium if a mixture of 0.300 mole of CO and 0.300 mole of $H_2O$ is heated to 700°C in a 10.0-L container.

**15.42** At 1000 K, a sample of pure $NO_2$ gas decomposes:

$$2NO_2(g) \rightleftharpoons 2NO(g) + O_2(g)$$

The equilibrium constant $K_P$ is 158. Analysis shows that the partial pressure of $O_2$ is 0.25 atm at equilibrium. Calculate the pressure of NO and $NO_2$ in the mixture.

**15.43** The equilibrium constant $K_c$ for the reaction

$$H_2(g) + Br_2(g) \rightleftharpoons 2HBr(g)$$

is $2.18 \times 10^6$ at 730°C. Starting with 3.20 moles of HBr in a 12.0-L reaction vessel, calculate the concentrations of $H_2$, $Br_2$, and HBr at equilibrium.

**15.44** The dissociation of molecular iodine into iodine atoms is represented as

$$I_2(g) \rightleftharpoons 2I(g)$$

At 1000 K, the equilibrium constant $K_c$ for the reaction is $3.80 \times 10^{-5}$. Suppose you start with 0.0456 mole of $I_2$ in a 2.30-L flask at 1000 K. What are the concentrations of the gases at equilibrium?

**15.45** The equilibrium constant $K_c$ for the decomposition of phosgene ($COCl_2$) is $4.63 \times 10^{-3}$ at 527°C:

$$COCl_2(g) \rightleftharpoons CO(g) + Cl_2(g)$$

Calculate the equilibrium partial pressures of all the components, starting with pure phosgene at 0.760 atm.

**15.46** Consider the following equilibrium process at 686°C:

$$CO_2(g) + H_2(g) \rightleftharpoons CO(g) + H_2O(g)$$

The equilibrium concentrations of the reacting species are [CO] = 0.050 $M$, [$H_2$] = 0.045 $M$, [$CO_2$] = 0.086 $M$, and [$H_2O$] = 0.040 $M$. (a) Calculate $K_c$ for the reaction at 686°C. (b) If we add $CO_2$ to increase its concentration to 0.50 mol/L, what will the concentrations of all the gases be when equilibrium is reestablished?

**15.47** Consider the heterogeneous equilibrium process:

$$C(s) + CO_2(g) \rightleftharpoons 2CO(g)$$

At 700°C, the total pressure of the system is found to be 4.50 atm. If the equilibrium constant $K_P$ is 1.52, calculate the equilibrium partial pressures of $CO_2$ and CO.

**15.48** The equilibrium constant $K_c$ for the reaction

$$H_2(g) + CO_2(g) \rightleftharpoons H_2O(g) + CO(g)$$

is 4.2 at 1650°C. Initially 0.80 mol $H_2$ and 0.80 mol $CO_2$ are injected into a 5.0-L flask. Calculate the concentration of each species at equilibrium.

**15.49** The aqueous reaction

$$\text{L-glutamate} + \text{pyruvate} \rightleftharpoons \alpha\text{-ketoglutarate} + \text{L-alanine}$$

is catalyzed by the enzyme L-glutamate–pyruvate aminotransferase. At 300 K, the equilibrium constant for the reaction is 1.11. Predict whether the forward reaction will occur if the concentrations of the reactants and products are [L-glutamate] = $3.0 \times 10^{-5}$ $M$, [pyruvate] = $3.3 \times 10^{-4}$ $M$, [$\alpha$-ketoglutarate] = $1.6 \times 10^{-2}$ $M$, and [L-alanine] = $6.25 \times 10^{-3}$ $M$.

## Section 15.5: Factors That Affect Chemical Equilibrium

**Visualizing Chemistry**

**Figure 15.10 and Figure 15.11**

Solid calcium carbonate decomposes to form solid calcium oxide and carbon dioxide gas.

$$CaCO_3(s) \rightleftharpoons CaO(s) + CO_2(g) \qquad \Delta H° = 177.8 \text{ kJ/mol}$$

Consider an equilibrium mixture of calcium carbonate and its decomposition products in a container.

**VC 15.1** Which of the following changes will result in the formation of more CaO(s)?

   a) Addition of more $CaCO_3(s)$
   b) Addition of more $CO_2(g)$
   c) Removal of some of the $CO_2(g)$

**VC 15.2** Which of the following changes will result in the formation of more $CaCO_3(s)$?

   a) Addition of more CaO(s)
   b) Addition of more $CO_2(g)$
   c) Removal of some of the $CaCO_3(s)$

**VC 15.3** Which of the following changes will not cause the equilibrium to shift in either direction? (Select all that apply.)

   a) Removal of some of the $CaCO_3(s)$
   b) Removal of some of the $CO_2(g)$
   c) Addition of more CaO(s)

**VC 15.4** Which of the following will be affected by a change in the system's temperature? (Select all that apply.)

   a) The amount of $CaCO_3(s)$
   b) The amount of CaO(s)
   c) The amount of $CO_2(g)$

A sample of calcium carbonate is in equilibrium with its decomposition products in a closed system: $CaCO_3(s) \rightleftharpoons CaO(s) + CO_2(g)$.

**VC 15.5** If the volume of the system is increased at constant temperature, relative to the original equilibrium mixture, which of the following will be true when equilibrium is reestablished? (Select all that apply.)

   a) There will be a higher concentration of $CO_2(g)$.
   b) There will be a lower concentration of $CO_2(g)$.
   c) There will be more $CO_2(g)$.

**VC 15.6** If the volume of the system is decreased at constant temperature, relative to the original equilibrium mixture, which of the following will be true when equilibrium is reestablished? (Select all that apply.)

   a) There will be more $CaCO_3(s)$.
   b) There will be more CaO(s).
   c) There will be more $CO_2(g)$.

A sample of gaseous hydrogen iodide is in equilibrium with its decomposition products in a closed system: $2HI(g) \rightleftharpoons H_2(g) + I_2(g)$.

**VC 15.7** If the volume of the system is increased at constant temperature, relative to the original equilibrium mixture, which of the following will be true when equilibrium is reestablished?

   a) The concentration of HI will increase and the concentrations of $H_2$ and $I_2$ will decrease.
   b) The concentration of all three species will be lower.
   c) The concentration of all three species will be unchanged.

VC 15.8  If the volume of the system is decreased at constant temperature, relative to the original equilibrium mixture, which of the following will be true when equilibrium is reestablished?

a) There will be more HI.
b) There will be less HI.
c) The amount of HI will be unchanged.

## Review Questions

15.50  Explain Le Châtelier's principle. How does this principle enable us to maximize the yields of desirable reactions and minimize the effect of undesirable ones?

15.51  Use Le Châtelier's principle to explain why the equilibrium vapor pressure of a liquid increases with increasing temperature.

15.52  List four factors that can shift the position of an equilibrium. Only one of these factors can alter the value of the equilibrium constant. Which one is it?

15.53  Does the addition of a catalyst have any effects on the position of an equilibrium?

## Problems

15.54  Consider the following equilibrium system involving $SO_2$, $Cl_2$, and $SO_2Cl_2$ (sulfuryl dichloride):

$$SO_2(g) + Cl_2(g) \rightleftharpoons SO_2Cl_2(g)$$

Predict how the equilibrium position would change if (a) $Cl_2$ gas were added to the system, (b) $SO_2Cl_2$ were removed from the system, (c) $SO_2$ were removed from the system. The temperature remains constant in each case.

15.55  Heating solid sodium bicarbonate in a closed vessel establishes the following equilibrium:

$$2NaHCO_3(s) \rightleftharpoons Na_2CO_3(s) + H_2O(g) + CO_2(g)$$

What would happen to the equilibrium position if (a) some of the $CO_2$ were removed from the system, (b) some solid $Na_2CO_3$ were added to the system, (c) some of the solid $NaHCO_3$ were removed from the system? The temperature remains constant.

15.56  Consider the following equilibrium systems:

(a) $A \rightleftharpoons 2B$      $\Delta H° = 20.0$ kJ/mol
(b) $A + B \rightleftharpoons C$      $\Delta H° = -5.4$ kJ/mol
(c) $A \rightleftharpoons B$      $\Delta H° = 0.0$ kJ/mol

Predict the change in the equilibrium constant $K_c$ that would occur in each case if the temperature of the reacting system were raised.

15.57  What effect does an increase in pressure have on each of the following systems at equilibrium? The temperature is kept constant, and, in each case, the reactants are in a cylinder fitted with a movable piston.

(a) $A(s) \rightleftharpoons 2B(s)$
(b) $2A(l) \rightleftharpoons B(l)$
(c) $A(s) \rightleftharpoons B(g)$
(d) $A(g) \rightleftharpoons B(g)$
(e) $A(g) \rightleftharpoons 2B(g)$

15.58  Consider the equilibrium

$$2I(g) \rightleftharpoons I_2(g)$$

What would be the effect on the position of equilibrium of (a) increasing the total pressure on the system by decreasing its volume, (b) adding $I_2$ to the reaction mixture, and (c) decreasing the temperature?

15.59  Consider the following equilibrium process:

$$PCl_5(g) \rightleftharpoons PCl_3(g) + Cl_2(g) \qquad \Delta H° = 92.5 \text{ kJ/mol}$$

Predict the direction of the shift in equilibrium when (a) the temperature is raised, (b) more chlorine gas is added to the reaction mixture, (c) some $PCl_3$ is removed from the mixture, (d) the pressure on the gases is increased, (e) a catalyst is added to the reaction mixture.

15.60  Consider the reaction

$$2SO_2(g) + O_2(g) \rightleftharpoons 2SO_3(g) \qquad \Delta H° = -198.2 \text{ kJ/mol}$$

Comment on the changes in the concentrations of $SO_2$, $O_2$, and $SO_3$ at equilibrium if we were to (a) increase the temperature, (b) increase the pressure, (c) increase $SO_2$, (d) add a catalyst, (e) add helium at constant volume.

15.61  In the uncatalyzed reaction

$$N_2O_4(g) \rightleftharpoons 2NO_2(g)$$

the pressure of the gases at equilibrium are $P_{N_2O_4} = 0.377$ atm and $P_{NO_2} = 1.56$ atm at 100°C. What would happen to these pressures if a catalyst were added to the mixture?

15.62  Consider the gas-phase reaction

$$2CO(g) + O_2(g) \rightleftharpoons 2CO_2(g)$$

Predict the shift in the equilibrium position when helium gas is added to the equilibrium mixture (a) at constant pressure and (b) at constant volume.

15.63  Consider the following equilibrium reaction in a closed container:

$$CaCO_3(s) \rightleftharpoons CaO(s) + CO_2(g)$$

What will happen if (a) the volume is increased, (b) some CaO is added to the mixture, (c) some $CaCO_3$ is removed, (d) some $CO_2$ is added to the mixture, (e) a few drops of an NaOH solution are added to the mixture, (f) a few drops of an HCl solution are added to the mixture (ignore the reaction between $CO_2$ and water), (g) temperature is increased?

15.64  The following diagrams show the reaction $A + B \rightleftharpoons AB$ at two different temperatures. Is the forward reaction endothermic or exothermic?

600 K        800 K

15.65  The following diagrams show an equilibrium mixture of $O_2$ and $O_3$ at temperatures $T_1$ and $T_2$ ($T_2 > T_1$). (a) Write an equilibrium equation showing the forward reaction to be exothermic. (b) Predict how the number of $O_2$ and $O_3$ molecules would change if the volume were decreased at constant temperature.

$T_1$        $T_2$

## Additional Problems

**15.66** Consider the statement: "The equilibrium constant of a reacting mixture of solid $NH_4Cl$ and gaseous $NH_3$ and $HCl$ is 0.316." List three important pieces of information that are missing from this statement.

**15.67** Pure nitrosyl chloride ($NOCl$) gas was heated to 240°C in a 1.00-L container. At equilibrium, the total pressure was 1.00 atm and the $NOCl$ pressure was 0.64 atm.

$$2NOCl(g) \rightleftharpoons 2NO(g) + Cl_2(g)$$

(a) Calculate the partial pressures of $NO$ and $Cl_2$ in the system.
(b) Calculate the equilibrium constant $K_P$.

**15.68** The equilibrium constant ($K_P$) for the formation of the air pollutant nitric oxide ($NO$) in an automobile engine at 530°C is $2.9 \times 10^{-11}$.

$$N_2(g) + O_2(g) \rightleftharpoons 2NO(g)$$

(a) Calculate the partial pressure of $NO$ under these conditions if the partial pressures of nitrogen and oxygen are 3.0 and 0.012 atm, respectively. (b) Repeat the calculation for atmospheric conditions where the partial pressures of nitrogen and oxygen are 0.78 and 0.21 atm and the temperature is 25°C. (The $K_P$ for the reaction is $4.0 \times 10^{-31}$ at this temperature.) (c) Is the formation of $NO$ endothermic or exothermic? (d) What natural phenomenon promotes the formation of $NO$? Why?

**15.69** Baking soda (sodium bicarbonate) undergoes thermal decomposition as follows:

$$2NaHCO_3(s) \rightleftharpoons Na_2CO_3(s) + CO_2(g) + H_2O(g)$$

Would we obtain more $CO_2$ and $H_2O$ by adding extra baking soda to the reaction mixture in (a) a closed vessel or (b) an open vessel?

**15.70** Consider the following reaction at equilibrium:

$$A(g) \rightleftharpoons 2B(g)$$

From the data shown here, calculate the equilibrium constant (both $K_P$ and $K_c$) at each temperature. Is the reaction endothermic or exothermic?

| Temperature (°C) | [A] (M) | [B] (M) |
|---|---|---|
| 200 | 0.0125 | 0.843 |
| 300 | 0.171 | 0.764 |
| 400 | 0.250 | 0.724 |

**15.71** The equilibrium constant $K_P$ for the reaction

$$2H_2O(g) \rightleftharpoons 2H_2(g) + O_2(g)$$

is $2 \times 10^{-42}$ at 25°C. (a) What is $K_c$ for the reaction at the same temperature? (b) The very small value of $K_P$ (and $K_c$) indicates that the reaction overwhelmingly favors the formation of water molecules. Explain why, despite this fact, a mixture of hydrogen and oxygen gases can be kept at room temperature without any change.

**15.72** Consider the following reacting system:

$$2NO(g) + Cl_2(g) \rightleftharpoons 2NOCl(g)$$

What combination of temperature and pressure would maximize the yield of nitrosyl chloride ($NOCl$)? [*Hint:* $\Delta H_f^{\circ}(NOCl) = 51.7$ kJ/mol.]

**15.73** At a certain temperature and a total pressure of 1.2 atm, the partial pressures of an equilibrium mixture

$$2A(g) \rightleftharpoons B(g)$$

are $P_A = 0.60$ atm and $P_B = 0.60$ atm. (a) Calculate the $K_P$ for the reaction at this temperature. (b) If the total pressure were increased to 1.5 atm, what would be the partial pressures of A and B at equilibrium?

**15.74** The decomposition of ammonium hydrogen sulfide

$$NH_4HS(s) \rightleftharpoons NH_3(g) + H_2S(g)$$

is an endothermic process. A 6.1589-g sample of the solid is placed in an evacuated 4.000-L vessel at exactly 24°C. After equilibrium has been established, the total pressure inside is 0.709 atm. Some solid $NH_4HS$ remains in the vessel. (a) What is the $K_P$ for the reaction? (b) What percentage of the solid has decomposed? (c) If the volume of the vessel were doubled at constant temperature, what would happen to the amount of solid in the vessel?

**15.75** Consider the reaction

$$2NO(g) + O_2(g) \rightleftharpoons 2NO_2(g)$$

At 430°C, an equilibrium mixture consists of 0.020 mole of $O_2$, 0.040 mole of $NO$, and 0.96 mole of $NO_2$. Calculate $K_P$ for the reaction, given that the total pressure is 0.20 atm.

**15.76** When heated, ammonium carbamate decomposes as follows:

$$NH_4CO_2NH_2(s) \rightleftharpoons 2NH_3(g) + CO_2(g)$$

At a certain temperature, the equilibrium pressure of the system is 0.318 atm. Calculate $K_P$ for the reaction.

**15.77** A mixture of 0.47 mole of $H_2$ and 3.59 moles of $HCl$ is heated to 2800°C. Calculate the equilibrium partial pressures of $H_2$, $Cl_2$, and $HCl$ if the total pressure is 2.00 atm. For the reaction

$$H_2(g) + Cl_2(g) \rightleftharpoons 2HCl(g)$$

$K_P$ is 193 at 2800°C.

**15.78** When heated at high temperatures, iodine vapor dissociates as follows:

$$I_2(g) \rightleftharpoons 2I(g)$$

In one experiment, a chemist finds that when 0.054 mole of $I_2$ was placed in a flask of volume 0.48 L at 587 K, the degree of dissociation (that is, the fraction of $I_2$ dissociated) was 0.0252. Calculate $K_c$ and $K_P$ for the reaction at this temperature.

**15.79** One mole of $N_2$ and three moles of $H_2$ are placed in a flask at 375°C. Calculate the total pressure of the system at equilibrium if the mole fraction of $NH_3$ is 0.21. The $K_P$ for the reaction is $4.31 \times 10^{-4}$.

**15.80** At 1130°C, the equilibrium constant ($K_c$) for the reaction

$$2H_2S(g) \rightleftharpoons 2H_2(g) + S_2(g)$$

is $2.25 \times 10^{-4}$. If $[H_2S] = 4.84 \times 10^{-3}$ M and $[H_2] = 1.50 \times 10^{-3}$ M, calculate $[S_2]$.

**15.81** A quantity of 6.75 g of $SO_2Cl_2$ was placed in a 2.00-L flask. At 648 K, there is 0.0345 mole of $SO_2$ present. Calculate $K_c$ for the reaction

$$SO_2Cl_2(g) \rightleftharpoons SO_2(g) + Cl_2(g)$$

**15.82** The following diagram represents a gas-phase equilibrium mixture for the reaction $AB \rightleftharpoons A + B$ at a certain temperature. Describe what would happen to the system after each of the following changes: (a) the temperature is decreased, (b) the volume is increased, (c) He atoms are added to the mixture at constant volume, (d) a catalyst is added to the mixture.

**15.83**  The formation of $SO_3$ from $SO_2$ and $O_2$ is an intermediate step in the manufacture of sulfuric acid, and it is also responsible for the acid rain phenomenon. The equilibrium constant $K_P$ for the reaction

$$2SO_2(g) + O_2(g) \rightleftharpoons 2SO_3(g)$$

is 0.13 at 830°C. In one experiment, 2.00 mol $SO_2$ and 2.00 mol $O_2$ were initially present in a flask. What must the total pressure at equilibrium be to have an 80.0 percent yield of $SO_3$?

**15.84**  Consider the dissociation of iodine:

$$I_2(g) \rightleftharpoons 2I(g)$$

A 1.00-g sample of $I_2$ is heated to 1200°C in a 500-mL flask. At equilibrium, the total pressure is 1.51 atm. Calculate $K_P$ for the reaction at this temperature.

**15.85**  Consider the following reaction at 1600°C:

$$Br_2(g) \rightleftharpoons 2Br(g)$$

When 1.05 moles of $Br_2$ are put in a 0.980-L flask, 1.20 percent of the $Br_2$ undergoes dissociation. Calculate the equilibrium constant $K_c$ for the reaction.

**15.86**  Eggshells are composed mostly of calcium carbonate ($CaCO_3$) formed by the reaction

$$Ca^{2+}(aq) + CO_3^{2-}(aq) \rightleftharpoons CaCO_3(s)$$

The carbonate ions are supplied by carbon dioxide produced as a result of metabolism. Explain why eggshells are thinner in the summer when the rate of panting by chickens is greater. Suggest a remedy for this situation.

**15.87**  The equilibrium constant $K_P$ for the following reaction is $4.31 \times 10^{-4}$ at 375°C:

$$N_2(g) + 3H_2(g) \rightleftharpoons 2NH_3(g)$$

In a certain experiment a student starts with 0.862 atm of $N_2$ and 0.373 atm of $H_2$ in a constant-volume vessel at 375°C. Calculate the partial pressures of all species when equilibrium is reached.

**15.88**  A quantity of 0.20 mole of carbon dioxide was heated to a certain temperature with an excess of graphite in a closed container until the following equilibrium was reached:

$$C(s) + CO_2(g) \rightleftharpoons 2CO(g)$$

Under these conditions, the average molar mass of the gases was 35 g/mol. (a) Calculate the mole fractions of CO and $CO_2$. (b) What is $K_P$ if the total pressure is 11 atm? (*Hint:* The average molar mass is the sum of the products of the mole fraction of each gas and its molar mass.)

**15.89**  When dissolved in water, glucose (corn sugar) and fructose (fruit sugar) exist in equilibrium as follows:

$$\text{fructose} \rightleftharpoons \text{glucose}$$

A chemist prepared a 0.244 $M$ fructose solution at 25°C. At equilibrium, it was found that its concentration had decreased to 0.113 $M$. (a) Calculate the equilibrium constant for the reaction. (b) At equilibrium, what percentage of fructose was converted to glucose?

**15.90**  At room temperature, solid iodine is in equilibrium with its vapor through sublimation and deposition. Describe how you would use radioactive iodine, in either solid or vapor form, to show that there is a dynamic equilibrium between these two phases.

**15.91**  At 1024°C, the pressure of oxygen gas from the decomposition of copper(II) oxide (CuO) is 0.49 atm:

$$4CuO(s) \rightleftharpoons 2Cu_2O(s) + O_2(g)$$

(a) What is $K_P$ for the reaction? (b) Calculate the fraction of CuO that will decompose if 0.16 mole of it is placed in a 2.0-L flask at 1024°C. (c) What would the fraction be if a 1.0-mole sample of CuO were used? (d) What is the smallest amount of CuO (in moles) that would establish the equilibrium?

**15.92**  A mixture containing 3.9 moles of NO and 0.88 mole of $CO_2$ was allowed to react in a flask at a certain temperature according to the equation

$$NO(g) + CO_2(g) \rightleftharpoons NO_2(g) + CO(g)$$

At equilibrium, 0.11 mole of $CO_2$ was present. Calculate the equilibrium constant $K_c$ of this reaction.

**15.93**  The equilibrium constant $K_c$ for the reaction

$$H_2(g) + I_2(g) \rightleftharpoons 2HI(g)$$

is 54.3 at 430°C. At the start of the reaction, there are 0.714 mole of $H_2$, 0.984 mole of $I_2$, and 0.886 mole of HI in a 2.40-L reaction chamber. Calculate the concentrations of the gases at equilibrium.

**15.94**  When heated, a gaseous compound A dissociates as follows:

$$A(g) \rightleftharpoons B(g) + C(g)$$

In an experiment, A was heated at a certain temperature until its equilibrium pressure reached $0.14P$, where $P$ is the total pressure. Calculate the equilibrium constant $K_P$ of this reaction.

**15.95**  When a gas was heated under atmospheric conditions, its color deepened. Heating above 150°C caused the color to fade, and at 550°C the color was barely detectable. However, at 550°C, the color was partially restored by increasing the pressure of the system. Which of the following best fits the preceding description: (a) a mixture of hydrogen and bromine, (b) pure bromine, (c) a mixture of nitrogen dioxide and dinitrogen tetroxide. (*Hint:* Bromine has a reddish color, and nitrogen dioxide is a brown gas. The other gases are colorless.) Justify your choice.

**15.96**  Both $Mg^{2+}$ and $Ca^{2+}$ are important biological ions. One of their functions is to bind to the phosphate group of ATP molecules or amino acids of proteins. For Group 2A metals in general, the equilibrium constant for binding to the anions increases in the order $Ba^{2+} < Sr^{2+} < Ca^{2+} < Mg^{2+}$. What property of the Group 2A metal cations might account for this trend?

**15.97**  The equilibrium constant $K_c$ for the following reaction is 1.2 at 375°C.

$$N_2(g) + 3H_2(g) \rightleftharpoons 2NH_3(g)$$

(a) What is the value of $K_P$ for this reaction? (b) What is the value of the equilibrium constant $K_c$ for $2NH_3(g) \rightleftharpoons N_2(g) + 3H_2(g)$? (c) What is $K_c$ for $\frac{1}{2}N_2(g) + \frac{3}{2}H_2(g) \rightleftharpoons NH_3(g)$? (d) What are the values of $K_P$ for the reactions described in parts (b) and (c)?

**15.98**  In this chapter we learned that a catalyst has no effect on the position of an equilibrium because it speeds up both the forward and reverse rates to the same extent. To test this statement, consider a situation in which an equilibrium of the type

$$2A(g) \rightleftharpoons B(g)$$

is established inside a cylinder fitted with a weightless piston. The piston is attached by a string to the cover of a box containing a catalyst. When the piston moves upward (expanding against atmospheric pressure), the cover is lifted and the catalyst is exposed to the gases. When the piston moves downward, the box is closed. Assume that the catalyst speeds up the forward reaction ($2A \longrightarrow B$) but does not affect the reverse process ($B \longrightarrow 2A$). Suppose the catalyst is suddenly exposed to the equilibrium system as shown on the next page. Describe what would happen subsequently. How does this "thought experiment" convince you that no such catalyst can exist?

**15.99** A sealed glass bulb contains a mixture of $NO_2$ and $N_2O_4$ gases. Describe what happens to the following properties of the gases when the bulb is heated from 20°C to 40°C: (a) color, (b) pressure, (c) average molar mass, (d) degree of dissociation (from $N_2O_4$ to $NO_2$), (e) density. Assume that volume remains constant. (*Hint:* $NO_2$ is a brown gas; $N_2O_4$ is colorless.)

**15.100** At 20°C, the vapor pressure of water is 0.0231 atm. Calculate $K_P$ and $K_c$ for the process

$$H_2O(l) \rightleftharpoons H_2O(g)$$

**15.101** Industrially, sodium metal is obtained by electrolyzing molten sodium chloride. The reaction at the cathode is $Na^+ + e^- \longrightarrow Na$. We might expect that potassium metal would also be prepared by electrolyzing molten potassium chloride. However, potassium metal is soluble in molten potassium chloride and therefore is hard to recover. Furthermore, potassium vaporizes readily at the operating temperature, creating hazardous conditions. Instead, potassium is prepared by the distillation of molten potassium chloride in the presence of sodium vapor at 892°C:

$$Na(g) + KCl(l) \rightleftharpoons NaCl(l) + K(g)$$

In view of the fact that potassium is a stronger reducing agent than sodium, explain why this approach works. (The boiling points of sodium and potassium are 892°C and 770°C, respectively.)

**15.102** In the gas phase, nitrogen dioxide is actually a mixture of nitrogen dioxide ($NO_2$) and dinitrogen tetroxide ($N_2O_4$). If the density of such a mixture is 2.3 g/L at 74°C and 1.3 atm, calculate the partial pressures of the gases and $K_P$ for the dissociation of $N_2O_4$.

**15.103** A 2.50-mole sample of NOCl was initially in a 1.50-L reaction chamber at 400°C. After equilibrium was established, it was found that 28.0 percent of the NOCl had dissociated:

$$2NOCl(g) \rightleftharpoons 2NO(g) + Cl_2(g)$$

Calculate the equilibrium constant $K_c$ for the reaction.

**15.104** About 75 percent of hydrogen for industrial use is produced by the *steam-reforming* process. This process is carried out in two stages called primary and secondary reforming. In the primary stage, a mixture of steam and methane at about 30 atm is heated over a nickel catalyst at 800°C to give hydrogen and carbon monoxide:

$$CH_4(g) + H_2O(g) \rightleftharpoons CO(g) + 3H_2(g) \quad \Delta H° = 206 \text{ kJ/mol}$$

The secondary stage is carried out at about 1000°C, in the presence of air, to convert the remaining methane to hydrogen:

$$CH_4(g) + \tfrac{1}{2}O_2(g) \rightleftharpoons CO(g) + 2H_2(g) \quad \Delta H° = 35.7 \text{ kJ/mol}$$

(a) What conditions of temperature and pressure would favor the formation of products in both the primary and secondary stages? (b) The equilibrium constant $K_c$ for the primary stage is 18 at 800°C. (i) Calculate $K_P$ for the reaction. (ii) If the partial pressures of methane and steam were both 15 atm at the start, what are the pressures of all the gases at equilibrium?

**15.105** Photosynthesis can be represented by

$$6CO_2(g) + 6H_2O(l) \rightleftharpoons C_6H_{12}O_6(s) + 6O_2(g)$$
$$\Delta H° = 2801 \text{ kJ/mol}$$

Explain how the equilibrium would be affected by the following changes: (a) partial pressure of $CO_2$ is increased, (b) $O_2$ is removed from the mixture, (c) $C_6H_{12}O_6$ (glucose) is removed from the mixture, (d) more water is added, (e) a catalyst is added, (f) temperature is decreased.

**15.106** Consider the decomposition of ammonium chloride at a certain temperature:

$$NH_4Cl(s) \rightleftharpoons NH_3(g) + HCl(g)$$

Calculate the equilibrium constant $K_P$ if the total pressure is 2.2 atm at that temperature.

**15.107** Water is a very weak electrolyte that undergoes the following ionization (called *autoionization*):

$$H_2O(l) \underset{k_{-1}}{\overset{k_1}{\rightleftharpoons}} H^+(aq) + OH^-(aq)$$

(a) If $k_1 = 2.4 \times 10^{-5} \text{ s}^{-1}$ and $k_{-1} = 1.3 \times 10^{11}/M \cdot \text{s}$, calculate the equilibrium constant $K$ where $K = [H^+][OH^-]/[H_2O]$. (b) Calculate the product $[H^+][OH^-]$, $[H^+]$, and $[OH^-]$. (*Hint:* Calculate the concentration of liquid water using its density, 1.0 g/mL.)

**15.108** Consider the following reaction, which takes place in a single elementary step:

$$2A + B \underset{k_{-1}}{\overset{k_1}{\rightleftharpoons}} A_2B$$

If the equilibrium constant $K_c$ is 12.6 at a certain temperature and if $k_1 = 5.1 \times 10^{-2} \text{ s}^{-1}$, calculate the value of $k_{-1}$.

**15.109** At 25°C, the equilibrium partial pressures of $NO_2$ and $N_2O_4$ are 0.15 atm and 0.20 atm, respectively. If the volume is doubled at constant temperature, calculate the partial pressures of the gases when a new equilibrium is established.

**15.110** In 1899 the German chemist Ludwig Mond developed a process for purifying nickel by converting it to the volatile nickel tetracarbonyl $[Ni(CO)_4]$ (b.p. = 42.2°C):

$$Ni(s) + 4CO(g) \rightleftharpoons Ni(CO)_4(g)$$

(a) Describe how you can separate nickel and its solid impurities. (b) How would you recover nickel? $[\Delta H°_f$ for $Ni(CO)_4$ is $-602.9$ kJ/mol.]

**15.111** Consider the equilibrium reaction described in Problem 15.30. A quantity of 2.50 g of $PCl_5$ is placed in an evacuated 0.500-L flask and heated to 250°C. (a) Calculate the pressure of $PCl_5$, assuming it does not dissociate. (b) Calculate the partial pressure of $PCl_5$ at equilibrium. (c) What is the total pressure at equilibrium? (d) What is the degree of dissociation of $PCl_5$? (The degree of dissociation is given by the fraction of $PCl_5$ that has undergone dissociation.)

**15.112** Consider the equilibrium system $3A \rightleftharpoons B$. Sketch the changes in the concentrations of A and B over time for the following situations: (a) initially only A is present, (b) initially only B is present, (c) initially both A and B are present (with A in higher concentration). In each case, assume that the concentration of B is higher than that of A at equilibrium.

**15.113** The vapor pressure of mercury is 0.0020 mmHg at 26°C. (a) Calculate $K_c$ and $K_P$ for the process $Hg(l) \rightleftharpoons Hg(g)$. (b) A chemist breaks a thermometer and spills mercury onto the floor of a laboratory measuring 6.1 m long, 5.3 m wide, and 3.1 m high. Calculate the mass of mercury (in grams) vaporized at equilibrium and the concentration of mercury vapor (in mg/m³). Does this concentration exceed the safety limit of 0.05 mg/m³? (Ignore the volume of furniture and other objects in the laboratory.)

**15.114**  At 25°C, a mixture of $NO_2$ and $N_2O_4$ gases are in equilibrium in a cylinder fitted with a movable piston. The concentrations are $[NO_2] = 0.0475\ M$ and $[N_2O_4] = 0.487\ M$. The volume of the gas mixture is halved by pushing down on the piston at constant temperature. Calculate the concentrations of the gases when equilibrium is reestablished. Will the color become darker or lighter after the change? [*Hint:* $K_c$ for the dissociation of $N_2O_4$ is $4.63 \times 10^{-3}$. $N_2O_4(g)$ is colorless, and $NO_2(g)$ has a brown color.]

**15.115**  A student placed a few ice cubes in a drinking glass with water. A few minutes later she noticed that some of the ice cubes were fused together. Explain what happened.

**15.116**  Consider the potential-energy diagrams for two types of reactions $A \rightleftharpoons B$. In each case, answer the following questions for the system at equilibrium. (a) How would a catalyst affect the forward and reverse rates of the reaction? (b) How would a catalyst affect the energies of the reactant and product? (c) How would an increase in temperature affect the equilibrium constant? (d) If the only effect of a catalyst is to lower the activation energies for the forward and reverse reactions, show that the equilibrium constant remains unchanged if a catalyst is added to the reacting mixture.

**15.117**  The equilibrium constant $K_c$ for the reaction

$$2NH_3(g) \rightleftharpoons N_2(g) + 3H_2(g)$$

is 0.83 at 375°C. A 14.6-g sample of ammonia is placed in a 4.00-L flask and heated to 375°C. Calculate the concentrations of all the gases when equilibrium is reached.

**15.118**  The dependence of the equilibrium constant of a reaction on temperature is given by the van't Hoff equation:

$$\ln K = \frac{-\Delta H°}{RT} + C$$

where $C$ is a constant. The following table gives the equilibrium constant ($K_P$) for the reaction at various temperatures.

$$2NO(g) + O_2(g) \rightleftharpoons 2NO_2(g)$$

| $K_P$ | 138 | 5.12 | 0.436 | 0.0626 | 0.0130 |
|---|---|---|---|---|---|
| $T(K)$ | 600 | 700 | 800 | 900 | 1000 |

Determine graphically the $\Delta H°$ for the reaction.

**15.119**  Consider the reaction between $NO_2$ and $N_2O_4$ in a closed container:

$$N_2O_4(g) \rightleftharpoons 2NO_2(g)$$

Initially, 1 mole of $N_2O_4$ is present. At equilibrium, $x$ mole of $N_2O_4$ has dissociated to form $NO_2$. (a) Derive an expression for $K_P$ in terms of $x$ and $P$, the total pressure. (b) How does the expression in part (a) help you predict the shift in equilibrium due to an increase in $P$? Does your prediction agree with Le Châtelier's principle?

**15.120**  (a) Use the van't Hoff equation in Problem 15.118 to derive the following expression, which relates the equilibrium constants at two different temperatures:

$$\ln \frac{K_1}{K_2} = \frac{\Delta H°}{R}\left(\frac{1}{T_2} - \frac{1}{T_1}\right)$$

How does this equation support the prediction based on Le Châtelier's principle about the shift in equilibrium with temperature? (b) The vapor pressures of water are 31.82 mmHg at 30°C and 92.51 mmHg at 50°C. Calculate the molar heat of vaporization of water.

**15.121**  The $K_P$ for the reaction

$$SO_2Cl_2(g) \rightleftharpoons SO_2(g) + Cl_2(g)$$

is 2.05 at 648 K. A sample of $SO_2Cl_2$ is placed in a container and heated to 648 K, while the total pressure is kept constant at 9.00 atm. Calculate the partial pressures of the gases at equilibrium.

**15.122**  The "boat" form and the "chair" form of cyclohexane ($C_6H_{12}$) interconvert as shown here:

Boat                Chair

In this representation, the H atoms are omitted and a C atom is assumed to be at each intersection of two lines (bonds). The conversion is first order in each direction. The activation energy for the chair ⟶ boat conversion is 41 kJ/mol. If the frequency factor is $1.0 \times 10^{12}\ s^{-1}$, what is $k_1$ at 298 K? The equilibrium constant $K_c$ for the reaction is $9.83 \times 10^3$ at 298 K.

**15.123**  Consider the following reaction at a certain temperature

$$A_2 + B_2 \rightleftharpoons 2AB$$

The mixing of 1 mole of $A_2$ with 3 moles of $B_2$ gives rise to $x$ mole of AB at equilibrium. The addition of 2 more moles of $A_2$ produces another $x$ mole of AB. What is the equilibrium constant for the reaction?

**15.124**  Iodine is sparingly soluble in water but much more so in carbon tetrachloride ($CCl_4$). The equilibrium constant, also called the partition coefficient, for the distribution of $I_2$ between these two phases

$$I_2(aq) \rightleftharpoons I_2(CCl_4)$$

is 83 at 20°C. (a) A student adds 0.030 L of $CCl_4$ to 0.200 L of an aqueous solution containing 0.032 g of $I_2$. The mixture at 20°C is shaken, and the two phases are then allowed to separate. Calculate the fraction of $I_2$ remaining in the aqueous phase. (b) The student now repeats the extraction of $I_2$ with another 0.030 L of $CCl_4$. Calculate the fraction of the $I_2$ from the original solution that remains in the aqueous phase. (c) Compare the result in part (b) with a single extraction using 0.060 L of $CCl_4$. Comment on the difference.

**15.125**  For the purpose of determining $K_P$ using Equation 15.4, what is $\Delta n$ for the following equation?

$$C_3H_8(g) + 5O_2(g) \rightleftharpoons 3CO_2(g) + 4H_2O(l)$$

**15.126**  Industrial production of ammonia from hydrogen and nitrogen gases is done using the Haber process.

$$N_2(g) + 3H_2(g) \rightleftharpoons 2NH_3(g) \qquad \Delta H° = -92.6\ kJ/mol$$

Based on your knowledge of the principles of equilibrium, what would the optimal temperature and pressure conditions be for production of ammonia on a large scale? Are the same conditions also optimal from the standpoint of kinetics? Explain.

**15.127**  For which of the following reactions is $K_c$ equal to $K_P$? For which can we not write a $K_P$ expression?

(a) $4NH_3(g) + 5O_2(g) \rightleftharpoons 4NO(g) + 6H_2O(g)$

(b) $CaCO_3(s) \rightleftharpoons CaO(s) + CO_2(g)$

(c) $Zn(s) + 2H^+(aq) \rightleftharpoons Zn^{2+}(aq) + H_2(g)$

(d) $PCl_3(g) + 3NH_3(g) \rightleftharpoons 3HCl(g) + P(NH_2)_3(g)$

(e) $NH_3(g) + HCl(g) \rightleftharpoons NH_4Cl(s)$

(f) $NaHCO_3(s) + H^+(aq) \rightleftharpoons H_2O(l) + CO_2(g) + Na^+(aq)$

(g) $H_2(g) + F_2(g) \rightleftharpoons 2HF(g)$

(h) $C(graphite) + CO_2(g) \rightleftharpoons 2CO(g)$

## PRE-PROFESSIONAL PRACTICE EXAM PROBLEMS: PHYSICAL AND BIOLOGICAL SCIENCES

Lime (CaO) is used to prevent $SO_2$ from escaping from the smokestacks of coal-burning power plants via the formation of solid $CaSO_4 \cdot 2H_2O$ (gypsum). One of the important reactions in the overall process is the decomposition of $CaCO_3$:

$$CaCO_3(s) \rightleftharpoons CaO(s) + CO_2(g)$$

which has an equilibrium constant ($K_c$) of $3.0 \times 10^{-6}$ at 725°C.

1.  Calculate the value of $K_P$ for the decomposition of $CaCO_3$ at 725°C.

    a) $3.0 \times 10^{-6}$      c) $2.0 \times 10^{-2}$
    b) $2.5 \times 10^{-4}$      d) $3.7 \times 10^{-8}$

2.  If a 12.0-g sample of solid $CaCO_3$ is placed in an evacuated vessel at 725°C, what will the pressure of $CO_2$ be when the system reaches equilibrium?

    a) $3.0 \times 10^{-6}$ atm      c) $1.7 \times 10^{-3}$ atm
    b) $3.3 \times 10^{5}$ atm      d) 1.0 atm

3.  Which of the following actions will cause an increase in the pressure of $CO_2$ in the vessel described in Question 2?

    a) Addition of He gas
    b) Addition of $SO_2$ gas
    c) Addition of more $CaCO_3$ solid
    d) Increasing the volume of the vessel

4.  If a 12.0-g sample of solid $CaCO_3$ is placed in a vessel at 725°C in which the pressure of $CO_2$ is 2.5 atm, what mass of CaO will form?

    a) 6.72 g      c) 6.00 g
    b) 12.0 g      d) None

## ANSWERS TO IN-CHAPTER MATERIALS

### Answers to Practice Problems

**15.1A** $K_c = \dfrac{[BrCl]^2}{[Br_2][Cl_2]}$, $K_c = 7.1$. **15.1B** 0.016 $M$.

**15.2A** (a) $Q_c = \dfrac{[N_2]^2[O_2]}{[N_2O]^2}$, (b) $Q_c = \dfrac{[NO]^2[Br_2]}{[NOBr]^2}$, (c) $Q_c = \dfrac{[H^+][F^-]}{[HF]}$,

(d) $Q_c = \dfrac{[CO_2][H_2]}{[CO][H_2O]}$, (e) $Q_c = \dfrac{[CS_2][H_2]^4}{[CH_4][H_2S]^2}$, (f) $Q_c = \dfrac{[H^+]^2[C_2O_4^{2-}]}{[H_2C_2O_4]}$.

**15.2B** (a) $H_2 + Cl_2 \rightleftharpoons 2HCl$, (b) $H^+ + F^- \rightleftharpoons HF$, (c) $Cr^{3+} + 4OH^- \rightleftharpoons Cr(OH)_4^-$, (d) $HClO \rightleftharpoons H^+ + ClO^-$, (e) $H_2SO_3 \rightleftharpoons H^+ + HSO_3^-$, (f) $2NO + Br_2 \rightleftharpoons 2NOBr$.

**15.3A** (a) $K_c = \dfrac{[HCl]^4}{[SiCl_4][H_2]^2}$, (b) $K_c = \dfrac{1}{[Hg^{2+}][Cl^-]^2}$, (c) $K_c = \dfrac{[Ni(CO)_4]}{[CO]^4}$,

(d) $K_c = \dfrac{[Zn^{2+}]}{[Fe^{2+}]}$. **15.3B** The expressions in (a), (c), and (d) correspond

to heterogeneous equilibria. It would be impossible to write a balanced chemical equation using only the species in each of these expressions, indicating that there are species in each reaction (solids or liquids) that do not appear in the equilibrium expression. **15.4A** (a) $2.3 \times 10^{24}$, (b) $6.6 \times 10^{-13}$, (c) $2.8 \times 10^{-15}$. **15.4B** (a) $1.5 \times 10^{34}$, (b) $1.3 \times 10^{29}$,

(c) $8.5 \times 10^{38}$. **15.5A** (a) $K_p = \dfrac{(P_{CO_2})^2}{(P_{CO})^2(P_{O_2})}$, (b) $K_p = P_{CO_2}$,

(c) $K_p = \dfrac{(P_{NH_3})^2}{(P_{N_2})(P_{H_2})^3}$. **15.5B** (a) $2NO_2 + O_2 \rightleftharpoons 2NO_3$,

(b) $CH_4 + 2H_2O \rightleftharpoons CO_2 + 4H_2$, (c) $I_2 + H_2 \rightleftharpoons 2HI$.
**15.6A** $8.1 \times 10^{-6}$. **15.6B** 0.104. **15.7A** Left. **15.7B** Right. **15.8A** $[H_2] = [I_2] = 0.056$ $M$; $[HI] = 0.413$ $M$. **15.8B** 0.20 $M$. **15.9A** $[Br] = 8.4 \times 10^{-3}$ $M$, $[Br_2] = 6.5 \times 10^{-2}$ $M$. **15.9B** $[Br] = 1.6 \times 10^{-3}$ $M$ initially and $2.4 \times 10^{-3}$ $M$ at equilibrium. **15.10A** $[P_{H_2}] = [P_{I_2}] = 0.37$ atm, $P_{HI} = 2.75$ atm. **15.10B** $P_{H_2} = P_{I_2} = 0.294$ atm, $P_{HI} = 2.16$ atm. **15.11A** (a) Right, (b) left, (c) right, (d) left. **15.11B** (a) $Ag(NH_3)_2^+$ $(aq)$ or $Cl^-(aq)$, (b) $NH_3(aq)$, (c) $AgCl(s)$. **15.12A** (a) Right, (b) right, (c) right. **15.12B** Right shift: remove HF; left shift: remove $H_2$; no shift: reduce volume of container.

### Answers to Checkpoints

**15.2.1** e. **15.2.2** a. **15.3.1** d. **15.3.2** a. **15.3.3** c. **15.3.4** c. **15.4.1** d. **15.4.2** b. **15.4.3** a, b, c. **15.4.4** e. **15.5.1** a, c. **15.5.2** b, d, e. **15.5.3** a, d, e. **15.5.4** b. **15.5.5** b. **15.5.6** b.

### Answers to Applying What You've Learned

a) $\dfrac{[OD17X]}{[OD1A][OF2A]}$. b) $3.05 \times 10^{-5}$. c) $\dfrac{1}{[OD1A][OF2A][A771A]}$, 30.5.

d) $Q_c = 3.6 \times 10^{-5}$; therefore, the mixture is not at equilibrium. It will have to proceed to the left (reverse reaction) to achieve equilibrium. e) At equilibrium, $[OD1A] = 0.05$ $M$, $[OF2A] = 0.05$ $M$, $[OD17X] = 0.95$ $M$.

# 16

# Acids and Bases

*A forest devastated by acid rain.*

# How Human Activity Has Impacted the Acidity of Rain

Rainwater in unpolluted areas is slightly acidic. The cause is carbon dioxide in the air, which dissolves in raindrops and reacts to form the weak acid, *carbonic* acid ($H_2CO_3$). In areas where the air is polluted by the burning of fossil fuels, rain can be very acidic. This phenomenon, known as *acid rain*, was first discovered in 1852 by Scottish chemist Robert Angus Smith (1817–1884) in Manchester, England. At that time, the industrial revolution was well under way and the British economy relied heavily on the use of coal to generate steam. The two main atmospheric contributors to the acidity of acid rain are sulfur dioxide ($SO_2$) and oxides of nitrogen ($NO_x$), which react to produce, respectively, sulfuric acid ($H_2SO_4$) and nitric acid ($HNO_3$)—both *strong* acids. $SO_2$ and $NO_x$ are both produced by the burning of sulfur-bearing coal.

Although acid rain was first discovered in 1852, scientific and societal efforts to understand and remedy its causes did not emerge for another century. U.S. scientists only began to study the causes and effects of acid rain extensively in the late 1960s. Public awareness of the phenomenon was heightened in the 1970s when results of studies at the Hubbard Brook Experimental Forest in New Hampshire were published, detailing the devastating impact of acid rain on the ecology of the region.

Despite the dire circumstance described by atmospheric scientists and ecologists in the last half of the twentieth century, there appears to be reason for optimism with regard to acid rain. The Acid Rain Program, part of the 1990 amendment to the Clean Air Act (originally enacted in 1963), has contributed to reducing the acidity of rainwater in industrial areas of the United States to levels below those seen in the 1960s. Reducing sulfur emissions from coal-burning power plants has involved using coal containing less sulfur, fitting smokestacks with chemical scrubbers to remove $SO_2$, and developing alternative sources of energy. Significant reductions in $NO_x$ emissions have been achieved through the use of catalytic converters, which have been standard equipment on automobiles in the United States since the mid-1970s.

Understanding the causes of acid rain and working on prevention and remediation requires an understanding of the properties of *acids* and *bases*.

**Student Annotation:** a) "$NO_x$" is a collective term used to mean NO and $NO_2$.

At the end of this chapter, you will be able to solve a series of problems related to the acidity of rainwater [▶▶ Page 715].

## In This Chapter, You Will Learn:

More about the properties of acids and bases and how those properties are related to molecular structure.

### Before you begin, you should review

- The list of strong acids and the list of strong bases [◀◀ Section 4.3, Table 4.4]
- How to solve equilibrium problems [◀◀ Section 15.4]

## 16.1  Brønsted Acids and Bases

In Chapter 4 we learned that a Brønsted acid is a substance that can donate a proton and a Brønsted base is a substance that can accept a proton [◄◄ Section 4.3]. In this chapter we extend our discussion of Brønsted acid-base theory to include conjugate acids and conjugate bases.

When a Brønsted acid donates a proton, what remains of the acid is known as a *conjugate base.* For example, in the ionization of HCl in water,

$$HCl(aq) + H_2O(l) \rightleftharpoons H_3O^+(aq) + Cl^-(aq)$$

$$\text{acid} \qquad\qquad\qquad\qquad\qquad\qquad \text{conjugate base}$$

HCl donates a proton to water, producing the hydronium ion ($H_3O^+$) and the chloride ion ($Cl^-$), which is the conjugate base of HCl. The two species, HCl and $Cl^-$, are known as a *conjugate acid-base pair* or simply a *conjugate pair.* Table 16.1 lists the conjugate bases of several familiar species.

Conversely, when a Brønsted base *accepts* a proton, the newly formed *protonated* species is known as a *conjugate acid.* When ammonia ($NH_3$) ionizes in water,

$$NH_3(aq) + H_2O(l) \rightleftharpoons NH_4^+(aq) + OH^-(aq)$$

$$\text{base} \qquad\qquad\qquad\qquad\qquad \text{conjugate acid}$$

$NH_3$ accepts a proton from water to become the ammonium ion $(NH)_4^+$. The ammonium ion is the conjugate acid of ammonia. Table 16.2 lists the conjugate acids of several common species.

Any reaction that we describe using Brønsted acid-base theory involves an acid and a base. The acid donates the proton, and the base accepts it. Furthermore, the products of such a reaction are always a conjugate base and a conjugate acid. It is useful to identify and label each species in a Brønsted acid-base reaction. For the ionization of HCl in water, the species are labeled as follows:

And for the ionization of $NH_3$ in water,

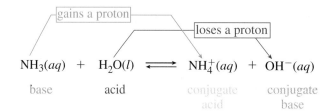

Sample Problems 16.1 and 16.2 let you practice identifying conjugate pairs and the species in a Brønsted acid-base reaction.

| TABLE 16.1 | Conjugate Bases of Some Common Species |
|---|---|

| Species | Conjugate Base |
|---|---|
| $CH_3COOH$ | $CH_3COO^-$ |
| $H_2O$ | $OH^-$ |
| $NH_3$ | $NH_2^-$ |
| $H_2SO_4$ | $HSO_4^-$ |

| TABLE 16.2 | Conjugate Acids of Some Common Species |
|---|---|

| Species | Conjugate Acid |
|---|---|
| $NH_3$ | $NH_4^+$ |
| $H_2O$ | $H_3O^+$ |
| $OH^-$ | $H_2O$ |
| $H_2NCONH_2$ (urea) | $H_2NCONH_3^+$ |

**Think About It** A species does not need to be what we think of as an acid for it to have a conjugate base. For example, we would not refer to the hydroxide ion ($OH^-$) as an acid—but it does have a conjugate base, the oxide ion ($O^{2-}$). Furthermore, a species that can either lose or gain a proton, such as $HCO_3^-$, has both a conjugate base ($CO_3^{2-}$) and a conjugate acid ($H_2CO_3$).

### Sample Problem 16.1

What is (a) the conjugate base of $HNO_3$, (b) the conjugate acid of $O^{2-}$, (c) the conjugate base of $HSO_4^-$, and (d) the conjugate acid of $HCO_3^-$?

**Strategy** To find the conjugate base of a species, *remove* a proton from the formula. To find the conjugate acid of a species, *add* a proton to the formula.

**Setup** The word *proton,* in this context, refers to $H^+$. Thus, the formula and the charge will both be affected by the addition or removal of $H^+$.

**Solution**

(a) $NO_3^-$       (c) $SO_4^{2-}$

(b) $OH^-$       (d) $H_2CO_3$

**Practice Problem A** What is (a) the conjugate acid of $ClO_4^-$, (b) the conjugate acid of $S^{2-}$, (c) the conjugate base of $H_2S$, and (d) the conjugate base of $H_2C_2O_4$?

**Practice Problem B** $HSO_3^-$ is the conjugate acid of what species? $HSO_3^-$ is the conjugate base of what species?

---

## Sample Problem 16.2

Label each of the species in the following equations as an acid, base, conjugate base, or conjugate acid:

(a) $HF(aq) + NH_3(aq) \rightleftharpoons F^-(aq) + NH_4^+(aq)$

(b) $CH_3COO^-(aq) + H_2O(l) \rightleftharpoons CH_3COOH(aq) + OH^-(aq)$

**Strategy** In each equation, the reactant that loses a proton is the acid and the reactant that gains a proton is the base. Each product is the conjugate of one of the reactants. Two species that differ only by a proton constitute a conjugate pair.

**Setup** (a) HF loses a proton and becomes $F^-$; $NH_3$ gains a proton and becomes $NH_4^+$.

(b) $CH_3COO^-$ gains a proton to become $CH_3COOH$; $H_2O$ loses a proton to become $OH^-$.

**Solution**

(a) $\underset{\text{acid}}{HF(aq)} + \underset{\text{base}}{NH_3(aq)} \rightleftharpoons \underset{\substack{\text{conjugate} \\ \text{base}}}{F^-(aq)} + \underset{\substack{\text{conjugate} \\ \text{acid}}}{NH_4^+(aq)}$

(b) $\underset{\text{base}}{CH_3COO^-(aq)} + \underset{\text{acid}}{H_2O(l)} \rightleftharpoons \underset{\text{conjugate acid}}{CH_3COOH(aq)} + \underset{\text{conjugate base}}{OH^-(aq)}$

> **Think About It** In a Brønsted acid-base reaction, there is always an acid and a base, and whether a substance behaves as an acid or a base depends on what it is combined with. Water, for example, behaves as a base when combined with HCl but behaves as an acid when combined with $NH_3$.

**Practice Problem A** Identify and label the species in each reaction.

(a) $NH_4^+(aq) + H_2O(l) \rightleftharpoons NH_3(aq) + H_3O^+(aq)$

(b) $CN^-(aq) + H_2O(l) \rightleftharpoons HCN(aq) + OH^-(aq)$

**Practice Problem B** (a) Write an equation in which $HSO_4^-$ reacts (with water) to form its conjugate base. (b) Write an equation in which $HSO_4^-$ reacts (with water) to form its conjugate acid.

---

**Checkpoint 16.1**   **Brønsted Acids and Bases**

16.1.1  Which of the following pairs of species are conjugate pairs? (Select all that apply.)

a) $H_2S$ and $S^{2-}$

b) $NH_2^-$ and $NH_3$

c) $O_2$ and $H_2O_2$

d) HBr and $Br^-$

e) HCl and $OH^-$

16.1.2  Which of the following species does *not* have a conjugate base? (Select all that apply.)

a) $HC_2O_4^-$

b) $OH^-$

c) $O^{2-}$

d) $CO_3^{2-}$

e) HClO

---

# 16.2 The Acid-Base Properties of Water

Water is often referred to as the "universal solvent," because it is so common and so important to life on Earth. In addition, most of the acid-base chemistry that you will encounter takes place in aqueous solution. In this section, we take a closer look at water's ability to act as either a Brønsted acid (as in the ionization of $NH_3$) or a Brønsted base (as in the ionization of HCl). A species that can behave either as a Brønsted acid or a Brønsted base is called **amphoteric.**

Water is a very weak electrolyte, but it does undergo ionization to a small extent:

$$H_2O(l) \rightleftharpoons H^+(aq) + OH^-(aq)$$

This reaction is known as the *autoionization of water.* Because we can represent the aqueous proton as either $H^+$ or $H_3O^+$ [◀◀ Section 4.3], we can also write the autoionization of water as

$$2H_2O(l) \rightleftharpoons H_3O^+(aq) + OH^-(aq)$$

acid　　　　base　　　　conjugate　　conjugate
　　　　　　　　　　　　　acid　　　　base

where one water molecule acts as an acid and the other acts as a base.

As indicated by the double arrow in the equation, the reaction is an equilibrium. The equilibrium expression for the autoionization of water is

$$K_w = [H_3O^+][OH^-] \quad \text{or} \quad K_w = [H^+][OH^-]$$

Because the autoionization of water is an important equilibrium that you will encounter frequently in the study of acids and bases, we use the subscript "w" to indicate that the equilibrium constant is that specifically for the autoionization of *water.* It is important to realize, though, that $K_w$ is simply a $K_c$ for a *specific reaction.* We will frequently replace the c in $K_c$ expressions with a letter or a series of letters to indicate the specific type of reaction to which the $K_c$ refers. For example, $K_c$ for the ionization of a weak acid is called $K_a$, and $K_c$ for the ionization of a weak base is called $K_b$. In Chapter 17 we will encounter $K_{sp}$, where "sp" stands for "solubility product." Each specially subscripted $K$ is simply a $K_c$ for a specific type of reaction.

In pure water, autoionization is the only source of $H_3O^+$ and $OH^-$, and the stoichiometry of the reaction tells us that their concentrations are equal. At 25°C, the concentrations of hydronium and hydroxide ions in pure water are $[H_3O^+] = [OH^-] = 1.0 \times 10^{-7}$ *M*. Using the equilibrium expression, we can calculate the value of $K_w$ at 25°C as follows:

$$K_w = [H_3O^+][OH^-] = (1.0 \times 10^{-7})(1.0 \times 10^{-7}) = 1.0 \times 10^{-14}$$

Furthermore, in any aqueous solution at 25°C, the product of $H_3O^+$ and $OH^-$ concentrations is equal to $1.0 \times 10^{-14}$.

Equation 16.1　　　$K_w = [H_3O^+][OH^-] = 1.0 \times 10^{-14}$ (at 25°C)

Although their product is a constant, the individual concentrations of hydronium and hydroxide can be influenced by the addition of an acid or a base. The relative amounts of $H_3O^+$ and $OH^-$ determine whether a solution is neutral, acidic, or basic.

- When $[H_3O^+] = [OH^-]$, the solution is neutral.
- When $[H_3O^+] > [OH^-]$, the solution is acidic.
- When $[H_3O^+] < [OH^-]$, the solution is basic.

Sample Problem 16.3 shows how to use Equation 16.1.

---

### Sample Problem 16.3

The concentration of hydronium ions in stomach acid is 0.10 *M*. Calculate the concentration of hydroxide ions in stomach acid at 25°C.

**Strategy** Use the value of $K_w$ to determine $[OH^-]$ when $[H_3O^+] = 0.10$ *M*.

**Setup** $K_w = [H_3O^+][OH^-] = 1.0 \times 10^{-14}$ at 25°C. Rearranging Equation 16.1 to solve for $[OH^-]$,

$$[OH^-] = \frac{1.0 \times 10^{-14}}{[H_3O^+]}$$

**Solution**

$$[OH^-] = \frac{1.0 \times 10^{-14}}{0.10} = 1.0 \times 10^{-13} \text{ } M$$

**Think About It** Remember that equilibrium constants are temperature dependent. The value of $K_w$ is $1.0 \times 10^{-14}$ only at 25°C.

**Practice Problem A** The concentration of hydroxide ions in the antacid milk of magnesia is $5.0 \times 10^{-4}$ *M*. Calculate the concentration of hydronium ions at 25°C.

**Practice Problem B** The value of $K_w$ at normal body temperature (37°C) is $2.8 \times 10^{-14}$. Calculate the concentration of hydroxide ions in stomach acid at body temperature. ($[H_3O^+] = 0.10$ *M*.)

---

**Multimedia**
Chemical Equilibrium—equilibrium (interactive).

**Student Annotation:** Recall that in a heterogeneous equilibrium such as this, liquids and solids do not appear in the equilibrium expression [◀◀ Section 15.3].

**Student Annotation:** The constant $K_w$ is sometimes referred to as the *ion-product constant.*

**Student Annotation:** Recall that we disregard the units when we substitute concentrations into an equilibrium expression [◀◀ Section 15.5].

**Student Annotation:** Because the product of $H_3O^+$ and $OH^-$ concentrations is a constant, we cannot alter the concentrations independently. Any change in one also affects the other.

## Checkpoint 16.2   The Acid-Base Properties of Water

**16.2.1** Calculate $[OH^-]$ in a solution in which $[H_3O^+] = 0.0012\ M$ at 25°C.

a) $1.2 \times 10^{-3}\ M$

b) $8.3 \times 10^{-17}\ M$

c) $1.0 \times 10^{-14}\ M$

d) $8.3 \times 10^{-12}\ M$

e) $1.2 \times 10^{11}\ M$

**16.2.2** Calculate $[H_3O^+]$ in a solution in which $[OH^-] = 0.25\ M$ at 25°C.

a) $4.0 \times 10^{-14}\ M$

b) $1.0 \times 10^{-14}\ M$

c) $2.5 \times 10^{13}\ M$

d) $1.0 \times 10^{-7}\ M$

e) $4.0 \times 10^{-7}\ M$

# 16.3   The pH Scale

The acidity of an aqueous solution depends on the concentration of hydronium ions, $[H_3O^+]$. This concentration can range over many orders of magnitude, which can make reporting the numbers cumbersome. To describe the acidity of a solution, rather than report the molar concentration of hydronium ions, we typically use the more convenient pH scale. The **pH** of a solution is defined as the negative base-10 logarithm of the hydronium ion concentration (in mol/L).

$$pH = -\log[H_3O^+] \quad \text{or} \quad pH = -\log[H^+] \qquad \text{Equation 16.2}$$

> **Student Annotation:** Equation 16.2 converts numbers that can span an enormous range (~$10^1$ to $10^{-14}$) to numbers generally ranging from ~1 to 14.

The pH of a solution is a dimensionless quantity, so the units of concentration must be removed from $[H_3O^+]$ before taking the logarithm. Because $[H_3O^+] = [OH^-] = 1.0 \times 10^{-7}\ M$ in pure water at 25°C, the pH of pure water at 25°C is

$$-\log(1.0 \times 10^{-7}) = 7.00$$

> **Student Annotation:** A word about significant figures: When we take the log of a number with two significant figures, we report the result to two places past the decimal point. Thus, pH 7.00 has two significant figures, not three.

Remember, too, that a solution in which $[H_3O^+] = [OH^-]$ is neutral. At 25°C, therefore, a neutral solution has pH 7.00. An acidic solution, one in which $[H_3O^+] > [OH^-]$, has pH $<$ 7.00, whereas a basic solution, in which $[H_3O^+] < [OH^-]$, has pH $>$ 7.00. Table 16.3 shows the calculation of pH for solutions ranging from 0.10 $M$ to $1.0 \times 10^{-14}\ M$.

In the laboratory, pH is measured with a pH meter (Figure 16.1). Table 16.4 lists the pH values of a number of common fluids. Note that the pH of body fluids varies greatly, depending on the location and function of the fluid. The low pH (high acidity) of gastric juices is vital for digestion of food, whereas the higher pH of blood is required to facilitate the transport of oxygen.

| TABLE 16.3 | Benchmark pH Values for a Range of Hydronium Ion Concentrations at 25°C | | |
|---|---|---|---|
| $[H_3O^+](M)$ | $-\log[H_3O^+]$ | pH | |
| 0.10 | $-\log(1.0 \times 10^{-1})$ | 1.00 | |
| 0.010 | $-\log(1.0 \times 10^{-2})$ | 2.00 | |
| $1.0 \times 10^{-3}$ | $-\log(1.0 \times 10^{-3})$ | 3.00 | |
| $1.0 \times 10^{-4}$ | $-\log(1.0 \times 10^{-4})$ | 4.00 | |
| $1.0 \times 10^{-5}$ | $-\log(1.0 \times 10^{-5})$ | 5.00 | |
| $1.0 \times 10^{-6}$ | $-\log(1.0 \times 10^{-6})$ | 6.00 | Acidic |
| $1.0 \times 10^{-7}$ | $-\log(1.0 \times 10^{-7})$ | 7.00 | Neutral |
| $1.0 \times 10^{-8}$ | $-\log(1.0 \times 10^{-8})$ | 8.00 | Basic |
| $1.0 \times 10^{-9}$ | $-\log(1.0 \times 10^{-9})$ | 9.00 | |
| $1.0 \times 10^{-10}$ | $-\log(1.0 \times 10^{-10})$ | 10.00 | |
| $1.0 \times 10^{-11}$ | $-\log(1.0 \times 10^{-11})$ | 11.00 | |
| $1.0 \times 10^{-12}$ | $-\log(1.0 \times 10^{-12})$ | 12.00 | |
| $1.0 \times 10^{-13}$ | $-\log(1.0 \times 10^{-13})$ | 13.00 | |
| $1.0 \times 10^{-14}$ | $-\log(1.0 \times 10^{-14})$ | 14.00 | |

**Figure 16.1** A pH meter is commonly used in the laboratory to determine the pH of a solution. Although many pH meters have a range of 1 to 14, pH values can actually be less than 1 and greater than 14.

**TABLE 16.4    pH Values of Some Common Fluids**

| Fluid | pH | Fluid | pH |
|---|---|---|---|
| Stomach acid | 1.5 | Saliva | 6.4−6.9 |
| Lemon juice | 2.0 | Milk | 6.5 |
| Vinegar | 3.0 | Pure water | 7.0 |
| Grapefruit juice | 3.2 | Blood | 7.35−7.45 |
| Orange juice | 3.5 | Tears | 7.4 |
| Urine | 4.8−7.5 | Milk of magnesia | 10.6 |
| Rainwater (in clean air) | 5.5 | Household ammonia | 11.5 |

A measured pH can be used to determine experimentally the concentration of hydronium ion in solution. Solving Equation 16.2 for $[H_3O^+]$ gives

**Student Annotation:** $10^x$ is the inverse function of log. (It is usually the second function on the same key.) You must be comfortable performing these operations on your calculator.

Equation 16.3                      $$[H_3O^+] = 10^{-pH}$$

Sample Problems 16.4 and 16.5 illustrate calculations involving pH.

---

**Sample Problem 16.4**

**Think About It** When a hydronium ion concentration falls between two "benchmark" concentrations in Table 16.3, the pH falls between the two corresponding pH values. In part (c), for example, the hydronium ion concentration ($8.8 \times 10^{-11}$ $M$) is greater than $1.0 \times 10^{-11}$ $M$ but less than $1.0 \times 10^{-10}$ $M$. Therefore, we expect the pH to be between 11.00 and 10.00.

| $[H_3O^+]$ $(M)$ | $-\log [H_3O^+]$ | pH |
|---|---|---|
| $1.0 \times 10^{-10}$ | $-\log (1.0 \times 10^{-10})$ | 10.00 |
| $8.8 \times 10^{-11}$* | $-\log (8.8 \times 10^{-11})$ | 10.06† |
| $1.0 \times 10^{-11}$ | $-\log (1.0 \times 10^{-11})$ | 11.00 |

*$[H_3O^+]$ between two benchmark values
†pH between two benchmark values

Recognizing the benchmark concentrations and corresponding pH values is a good way to determine whether or not your calculated result is reasonable.

Determine the pH of a solution at 25°C in which the hydronium ion concentration is (a) $3.5 \times 10^{-4}$ $M$, (b) $1.7 \times 10^{-7}$ $M$, and (c) $8.8 \times 10^{-11}$ $M$.

**Strategy** Given $[H_3O^+]$, use Equation 16.2 to solve for pH.

**Setup**

(a) pH $= -\log (3.5 \times 10^{-4})$

(b) pH $= -\log (1.7 \times 10^{-7})$

(c) pH $= -\log (8.8 \times 10^{-11})$

**Solution**

(a) pH $= 3.46$

(b) pH $= 6.77$

(c) pH $= 10.06$

**Practice Problem A** Determine the pH of a solution at 25°C in which the hydronium ion concentration is (a) $3.2 \times 10^{-9}$ $M$, (b) $4.0 \times 10^{-8}$ $M$, and (c) $5.6 \times 10^{-2}$ $M$.

**Practice Problem B** Determine the pH of a solution at 25°C in which the hydroxide ion concentration is (a) $8.3 \times 10^{-15}$ $M$, (b) $3.3 \times 10^{-4}$ $M$, and (c) $1.2 \times 10^{-3}$ $M$.

---

**Sample Problem 16.5**

Calculate the hydronium ion concentration in a solution at 25°C in which the pH is (a) 4.76, (b) 11.95, and (c) 8.01.

**Strategy** Given pH, use Equation 16.3 to calculate $[H_3O^+]$.

**Setup**

(a) $[H_3O^+] = 10^{-4.76}$

(b) $[H_3O^+] = 10^{-11.95}$

(c) $[H_3O^+] = 10^{-8.01}$

**Solution**

(a) $[H_3O^+] = 1.7 \times 10^{-5} M$

(b) $[H_3O^+] = 1.1 \times 10^{-12} M$

(c) $[H_3O^+] = 9.8 \times 10^{-9} M$

**Practice Problem A** Calculate the hydronium ion concentration in a solution at 25°C in which the pH is (a) 9.90, (b) 1.45, and (c) 7.01.

**Practice Problem B** Calculate the hydroxide ion concentration in a solution at 25°C in which the pH is (a) 11.89, (b) 2.41, and (c) 7.13.

**Think About It** If you use the calculated hydronium ion concentrations to recalculate pH, you will get numbers slightly different from those given in the problem. In part (a), for example, $-\log (1.7 \times 10^{-5}) = 4.77$. The small difference between this and 4.76 (the pH given in the problem) is due to a rounding error. Remember that a concentration derived from a pH with two digits to the right of the decimal point can have only two significant figures. Note also that the benchmarks can be used equally well in this circumstance. A pH between 4 and 5 corresponds to a hydronium ion concentration between $1 \times 10^{-4} M$ and $1 \times 10^{-5} M$.

A **pOH** scale analogous to the pH scale can be defined using the negative base-10 logarithm of the *hydroxide* ion concentration of a solution, $[OH^-]$.

$$pOH = -\log [OH^-] \qquad \text{Equation 16.4}$$

Rearranging Equation 16.4 to solve for hydroxide ion concentration gives

$$[OH^-] = 10^{-pOH} \qquad \text{Equation 16.5}$$

Now consider again the $K_w$ equilibrium expression for water at 25°C:

$$[H_3O^+][OH^-] = 1.0 \times 10^{-14}$$

Taking the negative logarithm of both sides, we obtain

$$-\log ([H_3O^+][OH^-]) = -\log (1.0 \times 10^{-14})$$

$$-(\log [H_3O^+] + \log [OH^-]) = 14.00$$

$$-\log [H_3O^+] - \log [OH^-] = 14.00$$

$$(-\log [H_3O^+]) + (-\log [OH^-]) = 14.00$$

And from the definitions of pH and pOH, we see that at 25°C

$$pH + pOH = 14.00 \qquad \text{Equation 16.6}$$

Equation 16.6 provides another way to express the relationship between the hydronium ion concentration and the hydroxide ion concentration. On the pOH scale, 7.00 is neutral, numbers greater than 7.00 indicate that a solution is acidic, and numbers less than 7.00 indicate that a solution is basic. Table 16.5 lists pOH values for a range of hydroxide ion concentrations at 25°C.

| TABLE 16.5 | Benchmark pOH Values for a Range of Hydroxide Ion Concentrations at 25°C | |
|---|---|---|
| **$[OH^-]$ (M)** | **pOH** | |
| 0.10 | 1.00 | |
| $1.0 \times 10^{-3}$ | 3.00 | |
| $1.0 \times 10^{-5}$ | 5.00 | Basic |
| $1.0 \times 10^{-7}$ | 7.00 | Neutral |
| $1.0 \times 10^{-9}$ | 9.00 | Acidic |
| $1.0 \times 10^{-11}$ | 11.00 | |
| $1.0 \times 10^{-13}$ | 13.00 | |

Sample Problems 16.6 and 16.7 illustrate calculations involving pOH.

---

**Think About It** Remember that the pOH scale is, in essence, the *reverse* of the pH scale. On the pOH scale, numbers below 7 indicate a basic solution, whereas numbers above 7 indicate an acidic solution. The pOH benchmarks (abbreviated in Table 16.5) work the same way the pH benchmarks do. In part (a), for example, a hydroxide ion concentration between $1 \times 10^{-4}\ M$ and $1 \times 10^{-5}\ M$ corresponds to a pOH between 4 and 5:

| $[OH^-]\ (M)$ | pOH |
|---|---|
| $1.0 \times 10^{-4}$ | 4.00 |
| $3.7 \times 10^{-5}*$ | 4.43† |
| $1.0 \times 10^{-5}$ | 5.00 |

*$[OH^-]$ between two benchmark values
†pOH between two benchmark values

---

## Sample Problem 16.6

Determine the pOH of a solution at 25°C in which the hydroxide ion concentration is (a) $3.7 \times 10^{-5}\ M$, (b) $4.1 \times 10^{-7}\ M$, and (c) $8.3 \times 10^{-2}\ M$.

**Strategy** Given $[OH^-]$, use Equation 16.4 to calculate pOH.

**Setup**
(a) $pOH = -\log(3.7 \times 10^{-5})$
(b) $pOH = -\log(4.1 \times 10^{-7})$
(c) $pOH = -\log(8.3 \times 10^{-2})$

**Solution**
(a) $pOH = 4.43$
(b) $pOH = 6.39$
(c) $pOH = 1.08$

---

**Practice Problem A** Determine the pOH of a solution at 25°C in which the hydroxide ion concentration is (a) $5.7 \times 10^{-12}\ M$, (b) $7.3 \times 10^{-3}\ M$, and (c) $8.5 \times 10^{-6}\ M$.

**Practice Problem B** Determine the pH of a solution at 25°C in which the hydroxide ion concentration is (a) $2.8 \times 10^{-8}\ M$, (b) $9.9 \times 10^{-9}\ M$, and (c) $1.0 \times 10^{-11}\ M$.

---

## Sample Problem 16.7

Calculate the hydroxide ion concentration in a solution at 25°C in which the pOH is (a) 4.91, (b) 9.03, and (c) 10.55.

**Strategy** Given pOH, use Equation 16.5 to calculate $[OH^-]$.

**Setup**
(a) $[OH^-] = 10^{-4.91}$
(b) $[OH^-] = 10^{-9.03}$
(c) $[OH^-] = 10^{-10.55}$

**Solution**
(a) $[OH^-] = 1.2 \times 10^{-5}\ M$
(b) $[OH^-] = 9.3 \times 10^{-10}\ M$
(c) $[OH^-] = 2.8 \times 10^{-11}\ M$

---

**Think About It** Use the benchmark pOH values to determine whether these solutions are reasonable. In part (a), for example, the pOH between 4 and 5 corresponds to $[OH^-]$ between $1 \times 10^{-4}\ M$ and $1 \times 10^{-5}\ M$.

---

**Practice Problem A** Calculate the hydroxide ion concentration in a solution at 25°C in which the pOH is (a) 13.02, (b) 5.14, and (c) 6.98.

**Practice Problem B** Calculate the hydronium ion concentration in a solution at 25°C in which the pOH is (a) 2.74, (b) 10.31, and (c) 12.40.

---

## Bringing Chemistry to Life

### Antacids and the pH Balance in Your Stomach

An average adult produces between 2 and 3 L of gastric juice daily. Gastric juice is an acidic digestive fluid secreted by glands in the mucous membrane that lines the stomach. It contains hydrochloric acid (HCl), among other substances. The pH of gastric juice is about 1.5, which corresponds to a hydrochloric acid concentration of 0.03 *M*—a concentration strong enough to dissolve zinc metal!

The inside lining of the stomach is made up of *parietal* cells, which are fused together to form tight junctions. The interiors of the cells are protected from the surroundings by cell membranes. These membranes allow water and neutral molecules to pass in and out of the stomach, but they usually block the movement of ions such as $H^+$, $Na^+$, $K^+$, and $Cl^-$. The $H^+$ ions come from carbonic acid ($H_2CO_3$) formed as a result of the hydration of $CO_2$, an end product of metabolism:

$$CO_2(g) + H_2O(l) \rightleftharpoons H_2CO_3(aq)$$

$$H_2CO_3(aq) \rightleftharpoons H^+(aq) + HCO_3^-(aq)$$

These reactions take place in the blood plasma bathing the cells in the mucosa. By a process known as *active transport,* $H^+$ ions move across the membrane into the stomach interior. (Active transport processes are aided by enzymes.) To maintain electrical balance, an equal number of $Cl^-$ ions also move from the blood plasma into the stomach. Once in the stomach, most of these ions are prevented by cell membranes from diffusing back into the blood plasma.

The purpose of the highly acidic medium within the stomach is to digest food and to activate certain digestive enzymes. Eating stimulates $H^+$ ion secretion. A small fraction of these ions normally are reabsorbed by the mucosa, causing a number of tiny hemorrhages. About half a million cells are shed by the lining every minute, and a healthy stomach is completely relined every few days. However, if the acid content is excessively high, the constant influx of $H^+$ ions through the membrane back to the blood plasma can cause muscle contraction, pain, swelling, inflammation, and bleeding.

One way to temporarily reduce the $H^+$ ions concentration in the stomach is to take an antacid. The major function of antacids is to neutralize excess HCl in gastric juice. The following table lists the active ingredients of some popular antacids. The reactions by which these antacids neutralize stomach acid are as follows:

$$NaHCO_3(aq) + HCl(aq) \longrightarrow NaCl(aq) + H_2O(l) + CO_2(g)$$

$$CaCO_3(aq) + 2HCl(aq) \longrightarrow CaCl_2(aq) + H_2O(l) + CO_2(g)$$

$$MgCO_3(aq) + 2HCl(aq) \longrightarrow MgCl_2(aq) + H_2O(l) + CO_2(g)$$

$$Mg(OH)_2(s) + 2HCl(aq) \longrightarrow MgCl_2(aq) + 2H_2O(l)$$

$$Al(OH)_2NaCO_3(s) + 4HCl(aq) \longrightarrow AlCl_3(aq) + NaCl(aq) + 3H_2O(l) + CO_2(g)$$

### Active Ingredients in Some Common Antacids

| Commercial Name | Active Ingredients |
| --- | --- |
| Alka-Seltzer | Aspirin, sodium bicarbonate, citric acid |
| Milk of magnesia | Magnesium hydroxide |
| Rolaids | Dihydroxy aluminum sodium carbonate |
| TUMS | Calcium carbonate |
| Maalox | Sodium bicarbonate, magnesium carbonate |

The $CO_2$ released by most of these reactions increases gas pressure in the stomach, causing the person to belch. The fizzing that takes place when Alka-Seltzer dissolves in water is caused by carbon dioxide, which is released by the reaction between citric acid and sodium bicarbonate:

$$3NaHCO_3(aq) + H_3C_6H_5O_7(aq) \longrightarrow 3CO_2(g) + 3H_2O(l) + Na_3C_6H_5O_7(aq)$$

This effervescence helps to disperse the ingredients and enhances the palatability of the solution.

## Checkpoint 16.3    The pH Scale

**16.3.1** Determine the pH of a solution at 25°C in which $[H^+] = 6.35 \times 10^{-8}$ *M*.

a) 7.65

b) 6.80

c) 7.20

d) 6.35

e) 8.00

**16.3.2** Determine $[H^+]$ in a solution at 25°C if pH = 5.75.

a) $1.8 \times 10^{-6}$ *M*

b) $5.6 \times 10^{-9}$ *M*

c) $5.8 \times 10^{-6}$ *M*

d) $2.4 \times 10^{-9}$ *M*

e) $1.0 \times 10^{-6}$ *M*

*(Continued)*

**16.3.3** Determine the pOH of a solution at 25°C in which $[OH^-] = 4.65 \times 10^{-3} M$.

a) 11.67

b) 13.68

c) 0.32

d) 4.65

e) 2.33

**16.3.4** Determine $[OH^-]$ in a solution at 25°C if pH = 10.50.

a) $3.2 \times 10^{-11} M$

b) $3.2 \times 10^{-4} M$

c) $1.1 \times 10^{-2} M$

d) $7.1 \times 10^{-8} M$

e) $8.5 \times 10^{-7} M$

## 16.4    Strong Acids and Bases

Most of this chapter and Chapter 17 deal with equilibrium and the application of the principles of equilibrium to a variety of reaction types. In the context of our discussion of acids and bases, however, it is necessary to review the ionization of strong acids and the dissociation of strong bases. These reactions generally are not treated as *equilibria* but rather as processes that go to completion. This makes the determination of pH for a solution of strong acid or strong base relatively simple.

### Strong Acids

There are many different acids, but as we learned in Chapter 4, relatively few qualify as *strong*.

| Strong Acid | Ionization Reaction |
|---|---|
| Hydrochloric acid | $HCl(aq) + H_2O(l) \longrightarrow H_3O^+(aq) + Cl^-(aq)$ |
| Hydrobromic acid | $HBr(aq) + H_2O(l) \longrightarrow H_3O^+(aq) + Br^-(aq)$ |
| Hydroiodic acid | $HI(aq) + H_2O(l) \longrightarrow H_3O^+(aq) + I^-(aq)$ |
| Nitric acid | $HNO_3(aq) + H_2O(l) \longrightarrow H_3O^+(aq) + NO_3^-(aq)$ |
| Chloric acid | $HClO_3(aq) + H_2O(l) \longrightarrow H_3O^+(aq) + ClO_3^-(aq)$ |
| Perchloric acid | $HClO_4(aq) + H_2O(l) \longrightarrow H_3O^+(aq) + ClO_4^-(aq)$ |
| Sulfuric acid | $H_2SO_4(aq) + H_2O(l) \longrightarrow H_3O^+(aq) + HSO_4^-(aq)$ |

It is a good idea to commit this short list of strong acids to memory.

Because the ionization of a strong acid is complete, the concentration of hydronium ion at equilibrium is equal to the starting concentration of the strong acid. For instance, if we prepare a 0.10 *M* solution of HCl, the concentration of hydronium ion in the solution is 0.10 *M*. All the HCl ionizes, and no HCl molecules remain. Thus, at equilibrium (when the ionization is complete), $[HCl] = 0 M$ and $[H_3O^+] = [Cl^-] = 0.10 M$. Therefore, the pH of the solution (at 25°C) is

$$pH = -\log(0.10) = 1.00$$

This is a very low pH, which is consistent with a relatively concentrated solution of a strong acid.

Sample Problems 16.8 and 16.9 let you practice relating the concentration of a strong acid to the pH of an aqueous solution.

**Student Annotation:** We indicate that ionization of a strong acid is *complete* by using a single arrow ($\longrightarrow$) instead of the double, equilibrium arrow ($\rightleftharpoons$) in the equation.

**Multimedia**

Acids and Bases—the dissociation of strong and weak acids (interactive).

**Student Annotation:** Remember that although sulfuric acid has two ionizable protons, only the first ionization is complete.

**Student Annotation:** As we will see in Section 16.5, a solution of equal concentration but containing a weak acid has a higher pH.

| Sample Problem 16.8 |
|---|

Calculate the pH of an aqueous solution at 25°C that is (a) 0.035 *M* in HI, (b) $1.2 \times 10^{-4} M$ in $HNO_3$, and (c) $6.7 \times 10^{-5} M$ in $HClO_4$.

**Strategy**  HI, $HNO_3$, and $HClO_4$ are all strong acids, so the concentration of hydronium ion in each solution is the same as the stated concentration of the acid. Use Equation 16.2 to calculate pH.

**Setup**  (a) $[H_3O^+] = 0.035 M$

(b) $[H_3O^+] = 1.2 \times 10^{-4} M$

(c) $[H_3O^+] = 6.7 \times 10^{-5} M$

**Solution** (a) pH = −log (0.035) = 1.46

(b) pH = −log (1.2 × 10⁻⁴) = 3.92

(c) pH = −log (6.7 × 10⁻⁵) = 4.17

**Practice Problem A** Calculate the pH of an aqueous solution at 25°C that is
(a) 0.081 $M$ in HI, (b) 8.2 × 10⁻⁶ $M$ in $HNO_3$, and (c) 5.4 × 10⁻³ $M$ in $HClO_4$.

**Practice Problem B** Calculate the pOH of an aqueous solution at 25°C that is
(a) 0.011 $M$ in $HNO_3$, (b) 3.5 × 10⁻³ $M$ in HBr, and (c) 9.3 × 10⁻¹⁰ $M$ in HCl.

---

**Think About It** Again, note that when a
hydronium ion concentration falls between two
of the benchmark concentrations in Table 16.3,
the pH falls between the two corresponding pH
values. In part (b), for example, the hydronium ion
concentration of 1.2 × 10⁻⁴ $M$ is greater than 1.0 ×
10⁻⁴ $M$ and less than 1.0 × 10⁻³ $M$. Therefore, we
expect the pH to be between 4.00 and 3.00.

| $[H_3O^+]$ $(M)$ | $-\log [H_3O^+]$ | pH |
|---|---|---|
| 1.0 × 10⁻³ | −log (1.0 × 10⁻³) | 3.00 |
| 1.2 × 10⁻⁴* | −log (1.2 × 10⁻³) | 3.92† |
| 1.0 × 10⁻⁴ | −log (1.0 × 10⁻⁴) | 4.00 |

*$[H_3O^+]$ between two benchmark values
†pH between two benchmark values

Being comfortable with the benchmark
hydronium ion concentrations and the
corresponding pH values will help you avoid
some of the common errors in pH calculations.

---

## Sample Problem 16.9

Calculate the concentration of HCl in a solution at 25°C that has pH (a) 4.95, (b) 3.45,
and (c) 2.78.

**Strategy** Use Equation 16.3 to convert from pH to the molar concentration of
hydronium ion. In a strong acid solution, the molar concentration of hydronium ion is
equal to the acid concentration.

**Setup** (a) $[HCl] = [H_3O^+] = 10^{-4.95}$

(b) $[HCl] = [H_3O^+] = 10^{-3.45}$

(c) $[HCl] = [H_3O^+] = 10^{-2.78}$

**Solution** (a) 1.1 × 10⁻⁵ $M$

(b) 3.5 × 10⁻⁴ $M$

(c) 1.7 × 10⁻³ $M$

**Student Annotation:** When we take the
inverse log of a number with two digits to the
right of the decimal point, the result has two
significant figures.

**Think About It** As pH decreases,
acid concentration increases.

**Practice Problem A** Calculate the concentration of $HNO_3$ in a solution at 25°C that has pH
(a) 2.06, (b) 1.77, and (c) 6.01.

**Practice Problem B** Calculate the concentration of HBr in a solution at 25°C that has pOH
(a) 9.19, (b) 12.18, and (c) 10.96.

---

## Strong Bases

The list of strong bases is also fairly short. It consists of the hydroxides of alkali metals (Group
1A) and the hydroxides of the heaviest alkaline earth metals (Group 2A). The dissociation of a
strong base is, for practical purposes, complete. Equations representing dissociations of the strong
bases are as follows:

Multimedia

Acids and Bases—ionization of a strong
base and weak base (interactive).

**Group 1A hydroxides**

LiOH($aq$) $\longrightarrow$ Li⁺($aq$) + OH⁻($aq$)

NaOH($aq$) $\longrightarrow$ Na⁺($aq$) + OH⁻($aq$)

KOH($aq$) $\longrightarrow$ K⁺($aq$) + OH⁻($aq$)

RbOH($aq$) $\longrightarrow$ Rb⁺($aq$) + OH⁻($aq$)

CsOH($aq$) $\longrightarrow$ Cs⁺($aq$) + OH⁻($aq$)

**Group 2A hydroxides**

Ca(OH)₂($aq$) $\longrightarrow$ Ca²⁺($aq$) + 2OH⁻($aq$)

Sr(OH)₂($aq$) $\longrightarrow$ Sr²⁺($aq$) + 2OH⁻($aq$)

Ba(OH)₂($aq$) $\longrightarrow$ Ba²⁺($aq$) + 2OH⁻($aq$)

**Student Annotation:** Recall that Ca(OH)₂
and Sr(OH)₂ are not very soluble, but what
*does* dissolve dissociates completely
[◀◀ Section 4.3, Table 4.4].

Again, because the reaction goes to completion, the pH of such a solution is relatively easy to
calculate. In the case of a Group 1A hydroxide, the hydroxide ion concentration is simply the
starting concentration of the strong base. In a solution that is 0.018 $M$ in NaOH, for example,

$[OH^-] = 0.018\ M$. Its pH can be calculated in two ways. We can either use Equation 16.1 to determine hydronium ion concentration,

$$[H_3O^+][OH^-] = 1.0 \times 10^{-14}$$

$$[H_3O^+] = \frac{1.0 \times 10^{-14}}{[OH^-]} = \frac{1.0 \times 10^{-14}}{0.018} = 5.56 \times 10^{-13}\ M$$

and then Equation 16.2 to determine pH:

$$pH = -\log(5.56 \times 10^{-13}\ M) = 12.25$$

or we can calculate the pOH with Equation 16.3,

$$pOH = -\log(0.018) = 1.75$$

and use Equation 16.6 to convert to pH:

$$pH + pOH = 14.00$$

$$pH = 14.00 - 1.75 = 12.25$$

Both methods give the same result.

In the case of a Group 2A metal hydroxide, we must be careful to account for the reaction stoichiometry. For instance, if we prepare a solution that is $1.9 \times 10^{-4}\ M$ in barium hydroxide, the concentration of hydroxide ion at equilibrium (after complete dissociation) is $2(1.9 \times 10^{-4}\ M)$ or $3.8 \times 10^{-4}\ M$—twice the original concentration of $Ba(OH)_2$. Once we have determined the hydroxide ion concentration, we can determine pH as before:

$$[H_3O^+] = \frac{1.0 \times 10^{-14}}{[OH^-]} = \frac{1.0 \times 10^{-14}}{3.8 \times 10^{-4}} = 2.63 \times 10^{-11}\ M$$

and

$$pH = -\log(2.63 \times 10^{-11}\ M) = 10.58$$

or

$$pOH = -\log(3.8 \times 10^{-4}) = 3.42$$

$$pH + pOH = 14.00$$

$$pH = 14.00 - 3.42 = 10.58$$

Sample Problems 16.10 and 16.11 illustrate calculations involving hydroxide ion concentration, pOH, and pH.

---

### Sample Problem 16.10

Calculate the pOH of the following aqueous solutions at 25°C: (a) 0.013 $M$ LiOH, (b) 0.013 $M$ Ba(OH)$_2$, (c) $9.2 \times 10^{-5}$ $M$ KOH.

**Strategy** LiOH, Ba(OH)$_2$, and KOH are all strong bases. Use reaction stoichiometry to determine hydroxide ion concentration and Equation 16.4 to determine pOH.

**Setup** (a) The hydroxide ion concentration is simply equal to the concentration of the base. Therefore, $[OH^-] = [LiOH] = 0.013\ M$.

(b) The hydroxide ion concentration is twice that of the base:

$$Ba(OH)_2(aq) \longrightarrow Ba^{2+}(aq) + 2OH^-(aq)$$

Therefore, $[OH^-] = 2 \times [Ba(OH)_2] = 2(0.013\ M) = 0.026\ M$.

(c) The hydroxide ion concentration is equal to the concentration of the base. Therefore, $[OH^-] = [KOH] = 9.2 \times 10^{-5}\ M$.

**Solution** (a) $pOH = -\log(0.013) = 1.89$

(b) $pOH = -\log(0.026) = 1.59$

(c) $pOH = -\log(9.2 \times 10^{-5}) = 4.04$

**Think About It** These are basic pOH values, which is what we should expect for the solutions described in the problem. Note that while the solutions in parts (a) and (b) have the same base concentration, they do not have the same hydroxide concentration and therefore do not have the same pOH.

**Practice Problem A** Calculate the pOH of the following aqueous solutions at 25°C:
(a) 0.15 *M* NaOH, (b) 8.4 × 10⁻³ *M* RbOH, (c) 1.7 × 10⁻⁵ *M* CsOH.

**Practice Problem B** Calculate the pH of the following aqueous solutions at 25°C:
(a) 9.5 × 10⁻⁸ *M* NaOH, (b) 6.1 × 10⁻² *M* LiOH, (c) 6.1 × 10⁻² *M* Ba(OH)₂.

---

## Sample Problem 16.11

An aqueous solution of a strong base has pH 8.15 at 25°C. Calculate the original concentration of base in the solution (a) if the base is NaOH and (b) if the base is Ba(OH)₂.

**Strategy** Use Equation 16.6 to convert from pH to pOH and Equation 16.5 to determine the hydroxide ion concentration. Consider the stoichiometry of dissociation in each case to determine the concentration of the base itself.

**Setup**

$$pOH = 14.00 - 8.15 = 5.85$$

(a) The dissociation of 1 mole of NaOH produces 1 mole of OH⁻. Therefore, the concentration of the base is *equal* to the concentration of hydroxide ion.

(b) The dissociation of 1 mole of Ba(OH)₂ produces 2 moles of OH⁻. Therefore, the concentration of the base is only one-half the concentration of hydroxide ion.

**Solution**

$$[OH^-] = 10^{-5.85} = 1.41 \times 10^{-6}\ M$$

(a) $[NaOH] = [OH^-] = 1.4 \times 10^{-6}\ M$

(b) $[Ba(OH)_2] = \frac{1}{2}[OH^-] = 7.1 \times 10^{-7}\ M$

> **Student Annotation:** Remember to keep an additional significant figure or two until the end of the problem—to avoid *rounding error* [◀ Section 1.5].

> **Think About It** Alternatively, we could determine the hydroxide ion concentration using Equation 16.3,
>
> $$[H_3O^+] = 10^{-8.15} = 7.1 \times 10^{-9}\ M$$
>
> and Equation 16.1,
>
> $$[OH^-] = \frac{1.0 \times 10^{-14}}{7.1 \times 10^{-9}\ M}$$
> $$= 1.4 \times 10^{-6}\ M$$
>
> Once [OH⁻] is known, the solution is the same as shown previously.

**Practice Problem A** An aqueous solution of a strong base has pH 8.98 at 25°C. Calculate the concentration of base in the solution (a) if the base is LiOH and (b) if the base is Ba(OH)₂.

**Practice Problem B** An aqueous solution of a strong base has pOH 1.76 at 25°C. Calculate the concentration of base in the solution (a) if the base is NaOH and (b) if the base is Ba(OH)₂.

---

## Checkpoint 16.4    Strong Acids and Bases

**16.4.1** Calculate the pH of a 0.075 *M* solution of perchloric acid (HClO₄) at 25°C.

    a) 12.88

    b) 7.75

    c) 6.25

    d) 1.12

    e) 7.00

**16.4.2** What is the concentration of HBr in a solution with pH 5.89 at 25°C?

    a) 7.8 × 10⁻⁹ *M*

    b) 1.3 × 10⁻⁶ *M*

    c) 5.9 × 10⁻¹⁴ *M*

    d) 8.1 × 10⁻⁷ *M*

    e) 1.0 × 10⁻⁷ *M*

**16.4.3** What is the pOH of a solution at 25°C that is 1.3 × 10⁻³ *M* in Ba(OH)₂?

    a) 2.89

    b) 2.59

    c) 3.19

    d) 11.11

    e) 11.41

**16.4.4** What is the concentration of KOH in a solution at 25°C that has pOH 3.31?

    a) 2.0 × 10⁻¹¹ *M*

    b) 3.3 × 10⁻¹ *M*

    c) 3.3 × 10⁻⁷ *M*

    d) 4.5 × 10⁻⁴ *M*

    e) 4.9 × 10⁻⁴ *M*

*(Continued)*

**Checkpoint 16.4**   **Strong Acids and Bases** (*continued*)

**16.4.5**   What is the pH of a solution at 25°C that is 0.0095 *M* in LiOH?

    a) 11.68

    b) 2.02

    c) 11.98

    d) 1.72

    e) 12.28

**16.4.6**   What is the concentration of $Ca(OH)_2$ in a solution at 25°C if the pH is 9.01?

    a) $1.0 \times 10^{-5}$ *M*

    b) $5.1 \times 10^{-6}$ *M*

    c) $2.0 \times 10^{-5}$ *M*

    d) $9.8 \times 10^{-9}$ *M*

    e) $4.9 \times 10^{-9}$ *M*

**16.4.7**   Which diagram best represents a solution of sulfuric acid $(H_2SO_4)$?

(a)

(b)

(c)

(d)

---

# 16.5   Weak Acids and Acid Ionization Constants

Most acids are **weak acids,** which ionize only to a limited extent in water. At equilibrium, an aqueous solution of a weak acid contains a mixture of aqueous acid molecules, hydronium ions, and the corresponding conjugate base. The degree to which a weak acid ionizes depends on the *concentration* of the acid and the *equilibrium constant* for the ionization.

## The Ionization Constant, $K_a$

Consider a weak monoprotic acid HA. Its ionization in water is represented by

$$HA(aq) + H_2O(l) \rightleftharpoons H_3O^+(aq) + A^-(aq)$$

or by

$$HA(aq) \rightleftharpoons H^+(aq) + A^-(aq)$$

The equilibrium expression for this reaction is

$$K_a = \frac{[H_3O^+][A^-]}{[HA]} \quad \text{or} \quad K_a = \frac{[H^+][A^-]}{[HA]}$$

where $K_a$ is the equilibrium constant for the reaction. More specifically, $K_a$ is called the *acid ionization constant.* Although all weak acids ionize less than 100 percent, they vary in strength. The magnitude of $K_a$ indicates how strong a weak acid is. A large $K_a$ value indicates a stronger acid, whereas a small $K_a$ value indicates a weaker acid. For example, acetic acid $(CH_3COOH)$ and hydrofluoric acid (HF) are both weak acids, but HF is the stronger acid of the two, as evidenced by its larger $K_a$ value. Solutions of equal concentration of the two acids do *not* have the same pH. The pH of the HF solution is lower.

| Solution (at 25°C) | $K_a$ | pH |
|---|---|---|
| 0.10 *M* HF | $7.1 \times 10^{-4}$ | 2.09 |
| 0.10 *M* $CH_3COOH$ | $1.8 \times 10^{-5}$ | 2.87 |

Table 16.6 lists a number of weak acids and their $K_a$ values at 25°C in order of decreasing acid strength.

## Calculating pH from $K_a$

Calculating the pH of a weak acid solution is an equilibrium problem, which we solve using the methods introduced in Chapter 15. Figure 16.2 on pages 688 and 689 shows in detail how an

| TABLE 16.6 | Ionization Constants of Some Weak Acids at 25°C | | |
|---|---|---|---|
| **Name of Acid** | **Formula** | **Structure** | $K_a$ |
| Hydrofluoric acid | HF | H—F | $7.1 \times 10^{-4}$ |
| Nitrous acid | $HNO_2$ | O=N—O—H | $4.5 \times 10^{-4}$ |
| Formic acid | HCOOH | H—C(=O)—O—H | $1.7 \times 10^{-4}$ |
| Benzoic acid | $C_6H_5COOH$ | ⬡—C(=O)—O—H | $6.5 \times 10^{-5}$ |
| Acetic acid | $CH_3COOH$ | $CH_3$—C(=O)—O—H | $1.8 \times 10^{-5}$ |
| Hydrocyanic acid | HCN | H—C≡N | $4.9 \times 10^{-10}$ |
| Phenol | $C_6H_5OH$ | ⬡—O—H | $1.3 \times 10^{-10}$ |

equilibrium table is used to determine the pH of a weak acid. Suppose we want to determine the pH of a 0.50 $M$ HF solution at 25°C. The ionization of HF is represented by

$$HF(aq) + H_2O(l) \rightleftharpoons H_3O^+(aq) + F^-(aq)$$

The equilibrium expression for this reaction is

$$K_a = \frac{[H_3O^+][F^-]}{[HF]} = 7.1 \times 10^{-4}$$

We construct an equilibrium table and enter the starting concentrations of all species in the equilibrium expression.

**Student Annotation:** Remember that solids and pure liquids do not appear in the equilibrium expression [◀◀ Section 15.3].

$$HF(aq) + H_2O(l) \rightleftharpoons H_3O^+(aq) + F^-(aq)$$

| | HF | | $H_3O^+$ | $F^-$ |
|---|---|---|---|---|
| *Initial concentration (M):* | 0.50 | | 0 | 0 |
| *Change in concentration (M):* | | | | |
| *Equilibrium concentration (M):* | | | | |

Using the reaction stoichiometry, we determine the changes in all species:

$$HF(aq) + H_2O(l) \rightleftharpoons H_3O^+(aq) + F^-(aq)$$

| | HF | | $H_3O^+$ | $F^-$ |
|---|---|---|---|---|
| *Initial concentration (M):* | 0.50 | | 0 | 0 |
| *Change in concentration (M):* | $-x$ | | $+x$ | $+x$ |
| *Equilibrium concentration (M):* | | | | |

Finally, we express the equilibrium concentration of each species in terms of $x$.

$$HF(aq) + H_2O(l) \rightleftharpoons H_3O^+(aq) + F^-(aq)$$

| | HF | | $H_3O^+$ | $F^-$ |
|---|---|---|---|---|
| *Initial concentration (M):* | 0.50 | | 0 | 0 |
| *Change in concentration (M):* | $-x$ | | $+x$ | $+x$ |
| *Equilibrium concentration (M):* | $0.50 - x$ | | $x$ | $x$ |

These equilibrium concentrations are then entered into the equilibrium expression to give

$$K_a = \frac{(x)(x)}{0.50 - x} = 7.1 \times 10^{-4}$$

# Figure 16.2

# Using Equilibrium Tables to Solve Problems

## Using $K_a$ and Concentration to Determine pH of a Weak Acid

**START**

Determine the pH of a weak acid with a concentration of 0.10 *M* and a $K_a$ of $2.5 \times 10^{-5}$.

The concentrations will change by an unknown amount, *x*.
• [HA] will decrease by *x*
• [H$^+$] and [A$^-$] will increase by *x*
We enter the anticipated changes in concentrations in the middle row of the table.

Many equilibrium problems can be solved using an equilibrium table. The table is constructed with a column under each species in the equilibrium equation, and three rows labeled i (initial), c (change), and e (equilibrium). (This is why such a table is sometimes referred to as an "ice" table.)

HA

We use pH to determine the equilibrium concentration of H$^+$:

$$[H^+] = 10^{-pH} = 10^{-3.82} = 1.5 \times 10^{-4} M$$

Because the ionization of HA produces equal amounts of H$^+$ and A$^-$, the equilibrium concentration of A$^-$ is also $1.5 \times 10^{-4}$ *M*. We enter these concentrations in the bottom row of the table.

## Using pH and Concentration to Determine $K_a$ of a Weak Acid

Determine the $K_a$ of a weak acid if a 0.12 *M* solution has a pH of 3.82.

**START**

The equilibrium concentrations, which we enter in the last row of the table, are expressed in terms of the unknown $x$. We write the equilibrium expression for the reaction,

$$K_a = \frac{[H^+][A^-]}{[HA]}$$

and enter the $K_a$ value and the equilibrium concentrations from the last row in the table:

$$2.5 \times 10^{-5} = \frac{(x)(x)}{(0.10 - x)} = \frac{x^2}{(0.10 - x)}$$

Because the magnitude of $K_a$ is small, we expect the amount of HA ionized ($x$) to be small compared to 0.10. Therefore, we can assume that $0.10 - x \approx 0.10$. This simplifies the solution to

$$2.5 \times 10^{-5} = \frac{x^2}{0.10}$$

Solving for $x$ gives

$$(2.5 \times 10^{-5})(0.10) = x^2$$

$$2.5 \times 10^{-6} = x^2$$

$$\sqrt{2.5 \times 10^{-6}} = x$$

$$x = 1.6 \times 10^{-3} \, M$$

We used $x$ to represent $[H^+]$ at equilibrium. ($x$ is also equal to $[A^-]$ at equilibrium and to the change in [HA].) Therefore, we determine pH by taking the negative log of $x$.

$$pH = -\log[H^+] = -\log(1.6 \times 10^{-3}) = 2.80$$

 **CHECK** Verify that $\dfrac{1.6 \times 10^{-3} \, M}{0.10 \, M} \times 100\% < 5\%$

and that $\dfrac{(1.6 \times 10^{-3})^2}{(0.10 - 1.6 \times 10^{-3})} \approx K_a$

$$HA \rightleftharpoons H^+ + A^-$$

To complete the last row of the table, we also need the equilibrium value of [HA]. The amount of weak acid that ionized is equal to the amount of $H^+$ produced by the ionization. Therefore, at equilibrium,

$$[HA] = 0.12 \, M - 1.5 \times 10^{-4} \, M \approx 0.12 \, M$$

With the bottom row of the table complete, we can use these equilibrium concentrations to calculate the value of $K_a$.

$$K_a = \frac{[H^+][A^-]}{[HA]} = \frac{(1.5 \times 10^{-4})(1.5 \times 10^{-4})}{(0.12)} = 1.9 \times 10^{-7}$$

 **CHECK** Verify that when $[HA] = 0.12 \, M$ and $K_a = 1.9 \times 10^{-7}$, $[H^+] = \sqrt{(1.9 \times 10^{-7})(0.12)} = 1.5 \times 10^{-4} \, M$ and pH = 3.82.

## What's the point?

- Starting with the molar concentration and $K_a$ of a weak monoprotic acid, we can use an equilibrium table to determine pH.
- Starting with pH and molar concentration, we can use an equilibrium table to determine the $K_a$ of a weak monoprotic acid.

Rearranging this expression, we get

$$x^2 + 7.1 \times 10^{-4}x - 3.55 \times 10^{-4} = 0$$

This is a quadratic equation, which we can solve using the quadratic formula given in Appendix 1. In the case of a weak acid, however, often we can use a shortcut to simplify the calculation. Because HF is a weak acid, and weak acids ionize only to a slight extent, $x$ must be small compared to 0.50. Therefore, we can make the following approximation:

$$0.50 - x \approx 0.50$$

Now the equilibrium expression becomes

$$\frac{x^2}{0.50 - x} \approx \frac{x^2}{0.50} = 7.1 \times 10^{-4}$$

Rearranging, we get

$$x^2 = (0.50)(7.1 \times 10^{-4}) = 3.55 \times 10^{-4}$$
$$x = \sqrt{3.55 \times 10^{-4}} = 1.9 \times 10^{-2}\, M$$

Thus, we have solved for $x$ without having to use the quadratic equation. At equilibrium we have

$$[HF] = (0.50 - 0.019)\, M = 0.48\, M$$

$$[H_3O^+] = 0.019\, M$$

$$[F^-] = 0.019\, M$$

and the pH of the solution is

$$pH = -\log(0.019) = 1.72$$

This shortcut gives a good approximation as long as the magnitude of $x$ is significantly smaller than the initial acid concentration. As a rule, it is acceptable to use this shortcut if the calculated value of $x$ is less than 5 percent of the initial acid concentration. In this case, the approximation is acceptable because

$$\frac{0.019\, M}{0.50\, M} \times 100\% = 3.8\%$$

This is the formula for the *percent ionization* of the acid [◄◄ Section 13.6]. Recall that the percent ionization of a weak electrolyte, such as a weak acid, depends on concentration. Consider a more dilute solution of HF, one that is 0.050 $M$. Using the preceding procedure to solve for $x$, we would get $6.0 \times 10^{-3}\, M$. The following test shows, however, that this answer is *not* a valid approximation because it is greater than 5 percent of 0.050 $M$:

$$\frac{6.0 \times 10^{-3}\, M}{0.050\, M} \times 100\% = 12\%$$

**Student Annotation:** In many cases, use of the quadratic equation can be avoided with a method called *successive approximation*, which is presented in Appendix 1.

In this case, we must solve for $x$ using the quadratic equation [◄◄ Sample Problem 15.9].
Sample Problem 16.12 shows how to use $K_a$ to determine the pH of a weak acid solution.

### Sample Problem 16.12

The $K_a$ of hypochlorous acid (HClO) is $3.5 \times 10^{-8}$. Calculate the pH of a solution at 25°C that is 0.0075 $M$ in HClO.

**Strategy** Construct an equilibrium table, and express the equilibrium concentration of each species in terms of $x$. Solve for $x$ using the approximation shortcut, and evaluate whether or not the approximation is valid. Use Equation 16.2 to determine pH.

**Setup**

| | $HClO(aq) +$ | $H_2O(l) \rightleftharpoons$ | $H_3O^+(aq) +$ | $ClO^-(aq)$ |
|---|---|---|---|---|
| *Initial concentration (M):* | 0.0075 | | 0 | 0 |
| *Change in concentration (M):* | $-x$ | | $+x$ | $+x$ |
| *Equilibrium concentration (M):* | $0.0075 - x$ | | $x$ | $x$ |

**Solution** These equilibrium concentrations are then substituted into the equilibrium expression to give

**Student Annotation:** Applying the 5 percent test indicates that the approximation shortcut is valid in this case: $(1.62 \times 10^{-5}/0.0075) \times 100\% < 5\%$.

$$K_a = \frac{(x)(x)}{0.0075 - x} = 3.5 \times 10^{-8}$$

Assuming that $0.0075 - x \approx 0.0075$,

$$\frac{x^2}{0.0075} = 3.5 \times 10^{-8}$$

$$x^2 = (3.5 \times 10^{-8})(0.0075)$$

Solving for $x$, we get

$$x = \sqrt{2.625 \times 10^{-10}} = 1.62 \times 10^{-5}\ M$$

According to the equilibrium table, $x = [H_3O^+]$. Therefore,

$$pH = -\log(1.62 \times 10^{-5}) = 4.79$$

**Think About It** We learned in Section 16.2 that the concentration of hydronium ion in pure water at 25°C is $1.0 \times 10^{-7}\ M$, yet we use $0\ M$ as the starting concentration to solve for the pH of a solution of weak acid:

| | $HA(aq) \rightleftharpoons$ | $H^+(aq)$ | $+ A^-(aq)$ |
|---|---|---|---|
| *Initial concentration (M):* | | 0 | 0 |
| *Change in concentration (M):* | | | |
| *Equilibrium concentration (M):* | | | |

The reason for this is that the *actual* concentration of hydronium ion in pure water is insignificant compared to the amount produced by the ionization of the weak acid. We could use the actual concentration of hydronium as the initial concentration, but doing so would not change the result because $(x + 1.0 \times 10^{-7})\ M \approx x\ M$. In solving problems of this type, we neglect the small concentration of $H^+$ due to the autoionization of water.

**Practice Problem A** Calculate the pH at 25°C of a $0.18\ M$ solution of a weak acid that has $K_a = 9.2 \times 10^{-6}$.

**Practice Problem B** Calculate the pH at 25°C of a $0.065\ M$ solution of a weak acid that has $K_a = 1.2 \times 10^{-5}$.

## Percent Ionization

When we first encountered acids [◄◄ Section 4.1], we learned that a strong acid is one that ionizes completely. Weak acids ionize only partially. The degree to which a weak acid ionizes is a measure of its strength, just as the magnitude of its ionization constant ($K_a$) is a measure of its strength. A quantitative measure of the degree of ionization is *percent ionization*, which, for a weak, monoprotic acid (HA) is calculated as follows:

$$\text{percent ionization} = \frac{[H^+]_{eq}}{[HA]_0} \times 100\% \qquad \text{Equation 16.7}$$

in which the hydrogen ion concentration, $[H^+]_{eq}$, is the concentration at equilibrium and the weak acid concentration, $[HA]_0$, is the *original* concentration of the acid, which is not necessarily the same as the concentration at equilibrium.

We have learned previously how to use colligative properties to determine the percent ionization of a weak electrolyte, such as a weak acid [◄◄ Section 13.6]. Now that we have an understanding of equilibrium and of ionization constants, we can predict the percent ionization of a weak acid. Moreover, we will discover that the percent ionization is not constant for a given weak acid, but in fact depends on the acid's concentration.

**Student Annotation:** In fact, a weak acid's original concentration is never *exactly* equal to its equilibrium concentration. However, sometimes the amount that ionizes is so small compared to the original concentration that, to an appropriate number of significant figures, the weak acid concentration does not change and $[HA]_0 \approx [HA]_{eq}$.

Consider a $0.10\ M$ solution of benzoic acid, for which the ionization constant ($K_a$) is $6.5 \times 10^{-5}$. Using the procedure described in Sample Problem 16.12, we can determine the equilibrium concentrations of benzoic acid, $H^+$, and the benzoate ion.

| | $C_6H_5COOH(aq) + H_2O(l) \rightleftharpoons$ | | $H_3O^+(aq)$ | $+ C_6H_5COO^-(aq)$ |
|---|---|---|---|---|
| *Initial concentration (M):* | 0.100 | | 0 | 0 |
| *Change in concentration (M):* | $-x$ | | $+x$ | $+x$ |
| *Equilibrium concentration (M):* | $0.100 - x$ | | $x$ | $x$ |

Solving for $x$ gives $0.0025\ M$. Therefore, at equilibrium, $[C_6H_5COOH] = 0.097\ M$ and $[H^+] = [C_6H_5COO^-] = 0.0025\ M$. The percent ionization of benzoic acid at this concentration is

$$\frac{0.0025\ M}{0.100\ M} \times 100\% = 2.5\%$$

Now consider what happens when we dilute this equilibrium mixture by adding enough water to double the volume. The concentrations of all three species are cut in half: $[C_6H_5COOH] = 0.049\ M$, $[H^+] = 0.0013\ M$, and $[C_6H_5COO^-] = 0.0013\ M$. If we plug these new concentrations

**Figure 16.3**   Percent ionization of a weak acid depends on the original acid concentration. As concentration approaches zero, ionization approaches 100 percent. The dashed grey line represents 100 percent ionization, which is characteristic of a strong acid.

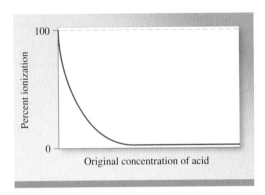

into the equilibrium expression and calculate the reaction quotient ($Q$), we get a number different from $K_a$. In fact, we get $(0.0013^2/0.049) = 3.4 \times 10^{-5}$, which is *smaller* than $K_a$. When $Q$ is smaller than $K$, the reaction must proceed to the right to reestablish equilibrium [◄◄ Section 15.4]. Proceeding to the right, in the case of weak acid ionization, corresponds to more of the acid ionizing—meaning that its percent ionization increases.

We can solve again for percent ionization starting with half the original concentration of benzoic acid.

$$C_6H_5COOH(aq) + H_2O(l) \rightleftharpoons H_3O^+(aq) + C_6H_5COO^-(aq)$$

| | | | |
|---|---|---|---|
| *Initial concentration (M):* | 0.050 | 0 | 0 |
| *Change in concentration (M):* | $-x$ | $+x$ | $+x$ |
| *Equilibrium concentration (M):* | $0.050 - x$ | $x$ | $x$ |

This time, solving for $x$ gives 0.0018 $M$. Therefore, at equilibrium, $[C_6H_5COOH] = 0.048\ M$, and $[H^+] = [C_6H_5COO^-] = 0.0018\ M$. The percent ionization of benzoic acid at this concentration is

$$\frac{0.0018\ M}{0.0050\ M} \times 100\% = 3.6\%$$

Figure 16.3 shows the dependence of percent ionization of a weak acid on concentration. Note that as concentration approaches zero, percent ionization approaches 100.

Sample Problem 16.13 lets you practice calculating percent ionization of weak acid solutions.

---

**Sample Problem 16.13**

Determine pH and percent ionization for acetic acid solutions at 25°C with concentrations (a) 0.15 $M$, (b) 0.015 $M$, and (c) 0.0015 $M$.

**Strategy**   Using the procedure described in Sample Problem 16.12, we construct an equilibrium table and for each concentration of acetic acid, we solve for the equilibrium concentration of $H^+$. We use Equation 16.2 to find pH, and Equation 16.7 to find percent ionization.

**Setup**   From Table 16.6, the ionization constant, $K_a$, for acetic acid is $1.8 \times 10^{-5}$.

**Setup**

(a)
$$CH_3COOH(aq) + H_2O(l) \rightleftharpoons H_3O^+(aq) + CH_3COO^-(aq)$$

| | | | |
|---|---|---|---|
| *Initial concentration (M):* | 0.15 | 0 | 0 |
| *Change in concentration (M):* | $-x$ | $+x$ | $+x$ |
| *Equilibrium concentration (M):* | $0.15 - x$ | $x$ | $x$ |

Solving for $x$ gives $[H^+] = 0.0016\ M$ and pH $= -\log(0.0016) = 2.78$.

$$\text{percent ionization} = \frac{0.0016\ M}{0.15\ M} \times 100\% = 1.1\%$$

(b) Solving in the same way as part (a) gives: $[H^+] = 5.2 \times 10^{-4}\ M$ and pH $= 3.28$.

$$\text{percent ionization} = \frac{5.2 \times 10^{-4}\ M}{0.015\ M} \times 100\% = 3.5\%$$

(c) Solving the quadratic equation, or using successive approximation [▶▶ Appendix 1] gives $[H^+] = 1.6 \times 10^{-4}\ M$ and pH = 3.78.

$$\text{percent ionization} = \frac{1.6 \times 10^{-4}\ M}{0.0015\ M} \times 100\% = 11\%$$

> **Think About It** Percent ionization also increases as concentration decreases for weak bases [▶▶ Section 16.6].

**Practice Problem A** Determine the pH and percent ionization for hydrocyanic acid (HCN) solutions of concentration (a) 0.25 $M$, (b) 0.0075 $M$, and (c) $8.3 \times 10^{-5}\ M$.

**Practice Problem B** At what concentration does hydrocyanic acid exhibit (a) 0.05 percent ionization, (b) 0.10 percent ionization, and (c) 0.15 percent ionization?

## Using pH to Determine $K_a$

In Chapter 15 we learned that we can determine the value of an equilibrium constant using equilibrium concentrations [◀◀ Section 15.2]. Using a similar approach, we can use the pH of a weak acid solution to determine the value of $K_a$. Suppose we want to determine the $K_a$ of a weak acid (HA) and we know that a 0.25 $M$ solution of the acid has a pH of 3.47 at 25°C. The first step is to use pH to determine the equilibrium hydronium ion concentration. Using Equation 16.3, we get

$$[H_3O^+] = 10^{-3.47} = 3.39 \times 10^{-4}\ M$$

> **Student Annotation:** Remember that we keep extra significant figures until the end of a multistep problem to minimize rounding error [◀◀ Section 1.5].

We use the starting concentration of the weak acid and the equilibrium concentration of the hydronium ion to construct an equilibrium table and determine the equilibrium concentrations of all three species.

|  | HA(aq) + | H₂O(l) ⇌ | H₃O⁺(aq) + | A⁻(aq) |
|---|---|---|---|---|
| Initial concentration (M): | 0.25 | | 0 | 0 |
| Change in concentration (M): | $-3.39 \times 10^{-4}$ | | $+3.39 \times 10^{-4}$ | $+3.39 \times 10^{-4}$ |
| Equilibrium concentration (M): | 0.2497 | | $3.39 \times 10^{-4}$ | $3.39 \times 10^{-4}$ |

These equilibrium concentrations are substituted into the equilibrium expression to give

$$K_a = \frac{(3.39 \times 10^{-4})^2}{0.2497} = 4.6 \times 10^{-7}$$

Therefore, the $K_a$ of this weak acid is $4.6 \times 10^{-7}$.

Sample Problem 16.14 shows how to determine $K_a$ using pH.

### Sample Problem 16.14

Aspirin (acetylsalicylic acid, $HC_9H_7O_4$) is a weak acid. It ionizes in water according to the equation

$$HC_9H_7O_4(aq) + H_2O(aq) \rightleftharpoons H_3O^+(aq) + C_9H_7O_4^-(aq)$$

A 0.10 $M$ aqueous solution of the aspirin has a pH of 2.27 at 25°C. Determine the $K_a$ of aspirin.

**Strategy** Determine the hydronium ion concentration from the pH. Use the hydronium ion concentration to determine the equilibrium concentrations of the other species, and plug the equilibrium concentrations into the equilibrium expression to evaluate $K_a$.

**Setup** Using Equation 16.3, we have

$$[H_3O^+] = 10^{-2.27} = 5.37 \times 10^{-3}\ M$$

To calculate $K_a$, though, we also need the equilibrium concentrations of $C_9H_7O_4^-$ and $HC_9H_7O_4$. The stoichiometry of the reaction tells us that $[C_9H_7O_4^-] = [H_3O^+]$. Furthermore, the amount of aspirin that has *ionized* is equal to the amount of hydronium ion in solution. Therefore, the equilibrium concentration of aspirin is $(0.10 - 5.37 \times 10^{-3})\ M = 0.095\ M$.

Aspirin

*(Continued)*

$$HC_9H_7O_4(aq) + H_2O(l) \rightleftharpoons H_3O^+(aq) + C_9H_7O_4^-(aq)$$

| | | | |
|---|---|---|---|
| Initial concentration (M): | 0.10 | 0 | 0 |
| Change in concentration (M): | −0.005 | +5.37 × 10⁻³ | +5.37 × 10⁻³ |
| Equilibrium concentration (M): | 0.095 | 5.37 × 10⁻³ | 5.37 × 10⁻³ |

**Solution** Substitute the equilibrium concentrations into the equilibrium expression as follows:

$$K_a = \frac{[H_3O^+][C_9H_7O_4^-]}{[HC_9H_7O_4]} = \frac{(5.37 \times 10^{-3})^2}{0.095} = 3.0 \times 10^{-4}$$

The $K_a$ of aspirin is $3.0 \times 10^{-4}$.

> **Think About It** Check your work by using the calculated value of $K_a$ to solve for the pH of a 0.10 M solution of aspirin.

**Practice Problem A** Calculate the $K_a$ of a weak acid if a 0.065 M solution of the acid has a pH of 2.96 at 25°C.

**Practice Problem B** Calculate the $K_a$ of a weak acid if a 0.015 M solution of the acid has a pH of 5.03 at 25°C.

---

## Checkpoint 16.5    Weak Acids and Acid Ionization Constants

**16.5.1** The $K_a$ of a weak acid is $5.5 \times 10^{-4}$. What is the pH of a 0.63 M solution of this acid at 25°C?

a) 3.26

b) $1.83 \times 10^{-2}$

c) 1.73

d) 1.63

e) 0.201

**16.5.2** A 0.042 M solution of a weak acid has pH 4.01 at 25°C. What is the $K_a$ of this acid?

a) $9.5 \times 10^{-9}$

b) $9.8 \times 10^{-5}$

c) 0.91

d) $4.2 \times 10^{-2}$

e) $2.3 \times 10^{-7}$

**16.5.3** The diagrams show solutions of three different weak acids with the general formula HA. List the acids in order of increasing $K_a$ value.

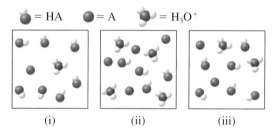

a) i < ii < iii

b) i < iii < ii

c) ii < i < iii

d) ii < iii < i

e) iii < ii < i

## 16.6    Weak Bases and Base Ionization Constants

Just as most acids are weak, most bases are also weak. The ionization of a *weak base* is incomplete and is treated in the same way as the ionization of a weak acid. In this section, we will see how the ionization constant for a weak base, $K_b$, is related to the pH of an aqueous solution.

### The Ionization Constant, $K_b$

The ionization of a weak base can be represented by the equation

$$B(aq) + H_2O(l) \rightleftharpoons HB^+(aq) + OH^-(aq)$$

where B is the weak base and $HB^+$ is its conjugate acid. The equilibrium expression for the ionization is

$$K_b = \frac{[HB^+][OH^-]}{[B]}$$

| **TABLE 16.7** | Ionization Constants of Some Weak Bases at 25°C | | |
|---|---|---|---|
| **Name of Base** | **Formula** | **Structure** | $K_b$ |
| Ethylamine | $C_2H_5NH_2$ | $CH_3-CH_2-\overset{..}{N}-H$ <br> $\quad\quad\quad\quad\quad |$ <br> $\quad\quad\quad\quad\quad H$ | $5.6 \times 10^{-4}$ |
| Methylamine | $CH_3NH_2$ | $CH_3-\overset{..}{N}-H$ <br> $\quad\quad\quad |$ <br> $\quad\quad\quad H$ | $4.4 \times 10^{-4}$ |
| Ammonia | $NH_3$ | $H-\overset{..}{N}-H$ <br> $\quad\quad |$ <br> $\quad\quad H$ | $1.8 \times 10^{-5}$ |
| Pyridine | $C_5H_5N$ | ⬡N: | $1.7 \times 10^{-9}$ |
| Aniline | $C_6H_5NH_2$ | ⬡$-\overset{..}{N}-H$ <br> $\quad\quad\quad |$ <br> $\quad\quad\quad H$ | $3.8 \times 10^{-10}$ |
| Urea | $H_2NCONH_2$ | $\quad\quad\quad\quad O$ <br> $\quad\quad\quad\quad ||$ <br> $H-\overset{..}{N}-C-\overset{..}{N}-H$ <br> $\quad\quad |\quad\quad\quad |$ <br> $\quad\quad H\quad\quad\quad H$ | $1.5 \times 10^{-14}$ |

where $K_b$ is the equilibrium constant—specifically known as the **base ionization constant.** Like weak acids, weak bases vary in strength. Table 16.7 lists a number of common weak bases and their ionization constants. The ability of any one of these substances to act as a base is the result of the lone pair of electrons on the nitrogen atom. The presence of this lone pair is what enables a compound to accept a proton, which is what makes a compound a Brønsted base.

## Calculating pH from $K_b$

Solving problems involving weak bases requires the same approach we used for weak acids. It is important to remember, though, that solving for $x$ in a typical weak base problem gives us the hydroxide ion concentration rather than the hydronium ion concentration.

Sample Problem 16.15 shows how to use $K_b$ to calculate the pH of a weak base solution.

### Sample Problem 16.15

What is the pH of a 0.040 $M$ ammonia solution at 25°C?

**Strategy** Construct an equilibrium table, and express equilibrium concentrations in terms of the unknown $x$. Plug these equilibrium concentrations into the equilibrium expression, and solve for $x$. From the value of $x$, determine the pH.

**Setup**

$$NH_3(aq) + H_2O(l) \rightleftharpoons NH_4^+(aq) + OH^-(aq)$$

| | $NH_3$ | $H_2O$ | $NH_4^+$ | $OH^-$ |
|---|---|---|---|---|
| *Initial concentration (M):* | 0.040 | | 0 | 0 |
| *Change in concentration (M):* | $-x$ | | $+x$ | $+x$ |
| *Equilibrium concentration (M):* | $0.040 - x$ | | $x$ | $x$ |

**Solution** The equilibrium concentrations are substituted into the equilibrium expression to give

$$K_b = \frac{[NH_4^+][OH^-]}{[NH_3]} = \frac{(x)(x)}{0.040 - x} = 1.8 \times 10^{-5}$$

*(Continued)*

Assuming that $0.040 - x \approx 0.040$ and solving for $x$ gives

$$\frac{(x)(x)}{0.040 - x} \approx \frac{(x)(x)}{0.040} = 1.8 \times 10^{-5}$$

$$x^2 = (1.8 \times 10^{-5})(0.040) = 7.2 \times 10^{-7}$$

$$x = \sqrt{7.2 \times 10^{-7}} = 8.5 \times 10^{-4} \, M$$

**Student Annotation:** Applying the 5 percent test indicates that the approximation shortcut is valid in this case: $(8.49 \times 10^{-4}/0.040) \times 100\% = 2\%$.

According to the equilibrium table, $x = [OH^-]$. Therefore, $pOH = -\log(x)$:

$$-\log(8.5 \times 10^{-4}) = 3.07$$

and $pH = 14.00 - pOH = 14.00 - 3.07 = 10.93$. The pH of a 0.040 $M$ solution of $NH_3$ at 25°C is 10.93.

**Think About It** It is a common error in $K_b$ problems to forget that $x$ is the *hydroxide* ion concentration rather than the *hydronium* ion concentration. Always make sure that the pH you calculate for a solution of base is a basic pH, that is, a pH greater than 7.

**Practice Problem A** Calculate the pH at 25°C of a 0.0028 $M$ solution of a weak base with a $K_b$ of $6.8 \times 10^{-8}$.

**Practice Problem B** Calculate the pH at 25°C of a 0.16 $M$ solution of a weak base with a $K_b$ of $2.9 \times 10^{-11}$.

## Using pH to Determine $K_b$

Just as we can use pH to determine the $K_a$ of a weak acid, we can also use it to determine the $K_b$ of a weak base. Sample Problem 16.16 demonstrates this procedure.

### Sample Problem 16.16

Caffeine, the stimulant in coffee and tea, is a weak base that ionizes in water according to the equation

$$C_8H_{10}N_4O_2(aq) + H_2O(l) \rightleftharpoons HC_8H_{10}N_4O_2^+(aq) + OH^-(aq)$$

A 0.15 $M$ solution of caffeine at 25°C has a pH of 8.45. Determine the $K_b$ of caffeine.

**Strategy** Use pH to determine pOH, and pOH to determine the hydroxide ion concentration. From the hydroxide ion concentration, use reaction stoichiometry to determine the other equilibrium concentrations and plug those concentrations into the equilibrium expression to evaluate $K_b$.

**Setup**

$$pOH = 14.00 - 8.45 = 5.55$$

$$[OH^-] = 10^{-5.55} = 2.82 \times 10^{-6} \, M$$

Based on the reaction stoichiometry, $[HC_8H_{10}N_4O_2^+] = [OH^-]$, and the amount of hydroxide ion in solution at equilibrium is equal to the amount of caffeine that has ionized. At equilibrium, therefore,

$$[C_8H_{10}N_4O_2] = (0.15 - 2.82 \times 10^{-6}) \, M \approx 0.15 \, M$$

|  | $C_8H_{10}N_4O_2(aq)$ | $+ \ H_2O(l)$ | $\rightleftharpoons$ | $HC_8H_{10}N_4O_2^+(aq)$ | $+ \ OH^-(aq)$ |
|---|---|---|---|---|---|
| *Initial concentration (M):* | 0.15 | | | 0 | 0 |
| *Change in concentration (M):* | $-2.82 \times 10^{-6}$ | | | $+2.82 \times 10^{-6}$ | $+2.82 \times 10^{-6}$ |
| *Equilibrium concentration (M):* | 0.15 | | | $2.82 \times 10^{-6}$ | $2.82 \times 10^{-6}$ |

**Think About It** Check your answer by using the calculated $K_b$ to determine the pH of a 0.15 $M$ solution.

**Solution** Plugging the equilibrium concentrations into the equilibrium expression gives

$$K_b = \frac{[HC_8H_{10}N_4O_2^+][OH^-]}{[C_8H_{10}N_4O_2]} = \frac{(2.82 \times 10^{-6})^2}{0.15} = 5.3 \times 10^{-11}$$

**Practice Problem A** Determine the $K_b$ of a weak base if a 0.50 $M$ solution of the base has a pH of 9.59 at 25°C.

**Practice Problem B** Determine the $K_b$ of a weak base if a 0.35 $M$ solution of the base has a pH of 11.84 at 25°C.

| **Checkpoint 16.6** | **Weak Bases and Base Ionization Constants** |
|---|---|

**16.6.1**  What is the pH of a 0.63 $M$ solution of weak base at 25°C if $K_b = 9.5 \times 10^{-7}$?

   a) 3.11

   b) 10.89

   c) 7.00

   d) 1.12

   e) 12.88

**16.6.2**  A 0.12 $M$ solution of a weak base has a pH of 10.76 at 25°C. Determine $K_b$.

   a) $2.5 \times 10^{-21}$

   b) $8.3 \times 10^{8}$

   c) $4.0 \times 10^{-8}$

   d) $2.8 \times 10^{-6}$

   e) $1.0 \times 10^{-14}$

**16.6.3**  The diagrams show solutions of three different weak bases. List the bases in order of increasing $K_b$ value.

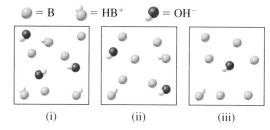

   a) i < ii < iii

   b) i < iii < ii

   c) ii < i < iii

   d) ii < iii < i

   e) iii < ii < i

# 16.7  Conjugate Acid-Base Pairs

At the beginning of this chapter, we introduced the concept of conjugate acids and conjugate bases. In this section, we examine the properties of conjugate acids and bases, independent of their parent compounds.

## The Strength of a Conjugate Acid or Base

When a strong acid such as HCl dissolves in water, it ionizes completely because its conjugate base (Cl⁻) has essentially *no* affinity for the H⁺ ion in solution.

$$HCl(aq) \longrightarrow H^+(aq) + Cl^-(aq)$$

Because the chloride ion has no affinity for the H⁺ ion, it does not act as a Brønsted base in water. If we dissolve a chloride salt such as NaCl in water, for example, the Cl⁻ ions in solution would not accept protons from the water:

$$Cl^-(aq) + H_2O(l) \xrightarrow{\quad\times\quad} HCl(aq) + OH^-(aq)$$

**Student Annotation:** The products HCl and OH⁻ do not actually form when Cl⁻ and $H_2O$ are combined.

The chloride ion, which is the conjugate base of a strong acid, is an example of a *weak conjugate base*.

Now consider the case of a weak acid. When HF dissolves in water, the ionization happens only to a *limited* degree because the conjugate base, F⁻, has a strong affinity for the H⁺ ion:

$$HF(aq) \rightleftharpoons H^+(aq) + F^-(aq)$$

This equilibrium lies far to the left ($K_a = 7.1 \times 10^{-4}$). Because the fluoride ion has a strong affinity for the H⁺ ion, it acts as a Brønsted base in water. If we were to dissolve a fluoride salt, such as NaF in water, the F⁻ ions in solution would, to some extent, accept protons from water:

$$F^-(aq) + H_2O(l) \rightleftharpoons HF(aq) + OH^-(aq)$$

**Student Annotation:** Note that one of the products of the reaction of a conjugate base with water is always the corresponding weak acid.

The fluoride ion, which is the conjugate base of a weak acid, is an example of a *strong conjugate base*.

Conversely, a strong base has a *weak conjugate acid* and a weak base has a *strong conjugate acid*. For example, $H_2O$ is the weak conjugate acid of the strong base OH⁻, whereas the ammonium ion ($NH_4^+$) is the strong conjugate acid of the weak base ammonia ($NH_3$). When an ammonium salt is dissolved in water, the ammonium ions donate protons to the water molecules:

$$NH_4^+(aq) + H_2O(l) \rightleftharpoons NH_3(aq) + H_3O^+(aq)$$

In general, there is a reciprocal relationship between the strength of an acid or base and the strength of its conjugate.

| Acid | Example | Conjugate base | Formula | | Base | Example | Conjugate acid | Formula |
|---|---|---|---|---|---|---|---|---|
| strong | $HNO_3$ | weak conjugate | $NO_3^-$ | | strong | $OH^-$ | weak conjugate | $H_2O$ |
| weak | HCN | strong conjugate | $CN^-$ | | weak | $NH_3$ | strong conjugate | $NH_4^+$ |

It is important to recognize that the words *strong* and *weak* do not mean the same thing in the context of conjugate acids and conjugate bases as they do in the context of acids and bases in general. A strong conjugate reacts with water—either accepting a proton from it or donating a proton to it—to a small but measurable extent. A strong conjugate acid acts as a weak Brønsted acid in water; and a strong conjugate base acts as a weak Brønsted base in water. A weak conjugate does not react with water to any measurable extent.

## The Relationship Between $K_a$ and $K_b$ of a Conjugate Acid-Base Pair

Because it accepts a proton from water to a small extent, what we refer to as a "strong conjugate base" is actually a weak Brønsted base. Therefore, every strong conjugate base has an ionization constant, $K_b$. Likewise, every strong conjugate acid, because it acts as a weak Brønsted base, has an ionization constant, $K_a$.

A simple relationship between the ionization constant of a weak acid ($K_a$) and the ionization constant of its conjugate base ($K_b$) can be derived as follows, using acetic acid as an example:

$$CH_3COOH(aq) \rightleftharpoons H^+(aq) + CH_3COO^-(aq)$$
$$\text{acid} \qquad\qquad\qquad \text{conjugate base}$$

$$K_a = \frac{[H^+][CH_3COO^-]}{[CH_3COOH]}$$

The conjugate base, $CH_3COO^-$, reacts with water according to the equation

$$CH_3COO^-(aq) + H_2O(l) \rightleftharpoons CH_3COOH(aq) + OH^-(aq)$$

and the base ionization equilibrium expression is written as

$$K_b = \frac{[CH_3COOH][OH^-]}{[CH_3COO^-]}$$

As for any chemical equations, we can add these two equilibria and cancel identical terms:

$$CH_3COOH(aq) \rightleftharpoons H^+(aq) + CH_3COO^-(aq)$$
$$+ \ CH_3COO^-(aq) + H_2O(l) \rightleftharpoons CH_3COOH(aq) + OH^-(aq)$$
$$\overline{H_2O(l) \rightleftharpoons H^+(aq) + OH^-(aq)}$$

The sum is the autoionization of water. In fact, this is the case for any weak acid and its conjugate base:

$$HA \rightleftharpoons H^+ + A^-$$
$$+ \ A^- + H_2O \rightleftharpoons HA + OH^-$$
$$\overline{H_2O \rightleftharpoons H^+ + OH^-}$$

or for any weak base and its conjugate acid.

$$B + H_2O \rightleftharpoons HB^+ + OH^-$$
$$+ \ HB^+ + H_2O \rightleftharpoons B + H_3O^+$$
$$\overline{2H_2O \rightleftharpoons H_3O^+ + OH^-}$$

Recall that when we add two equilibria, the equilibrium constant for the *net* reaction is the product of the equilibrium constants for the individual equations [◄◄ Section 15.3]. Thus, for any conjugate acid-base pair,

$$K_a \times K_b = K_w \qquad\qquad \text{Equation 16.8}$$

Equation 16.8 gives the quantitative basis for the reciprocal relationship between the strength of an acid and that of its conjugate base (or between the strength of a base and that of its conjugate acid). Because $K_w$ is a constant, $K_b$ must decrease if $K_a$ increases, and vice versa.

Sample Problem 16.17 shows how to determine ionization constants for conjugates.

---

## Sample Problem 16.17

Determine (a) $K_b$ of the acetate ion ($CH_3COO^-$), (b) $K_a$ of the methylammonium ion ($CH_3NH_3^+$), (c) $K_b$ of the fluoride ion ($F^-$), and (d) $K_a$ of the ammonium ion ($NH_4^+$).

**Strategy** Each species listed is either a conjugate base or a conjugate acid. Determine the identity of the acid corresponding to each conjugate base and the identity of the base corresponding to each conjugate acid; then, consult Tables 16.6 and 16.7 for their ionization constants. Use the tabulated ionization constants and Equation 16.8 to calculate each indicated $K$ value.

**Setup** (a) A $K_b$ value is requested, indicating that the acetate ion is a conjugate base. To identify the corresponding Brønsted acid, add a proton to the formula to get $CH_3COOH$ (acetic acid). The $K_a$ of acetic acid (from Table 16.6) is $1.8 \times 10^{-5}$.

(b) A $K_a$ value is requested, indicating that the methylammonium ion is a conjugate acid. Determine the identity of the corresponding Brønsted base by removing a proton from the formula to get $CH_3NH_2$ (methylamine). The $K_b$ of methylamine (from Table 16.7) is $4.4 \times 10^{-4}$.

(c) $F^-$ is the conjugate base of HF; $K_a = 7.1 \times 10^{-4}$.

(d) $NH_4^+$ is the conjugate acid of $NH_3$; $K_b = 1.8 \times 10^{-5}$. Solving Equation 16.8 separately for $K_a$ and $K_b$ gives, respectively,

$$K_a = \frac{K_w}{K_b} \qquad \text{and} \qquad K_b = \frac{K_w}{K_a}$$

**Solution** (a) Conjugate base $CH_3COO^-$: $K_b = \dfrac{1.0 \times 10^{-14}}{1.8 \times 10^{-5}} = 5.6 \times 10^{-10}$

(b) Conjugate acid $CH_3NH_3^+$: $K_a = \dfrac{1.0 \times 10^{-14}}{4.4 \times 10^{-4}} = 2.3 \times 10^{-11}$

(c) Conjugate base $F^-$: $K_b = \dfrac{1.0 \times 10^{-14}}{7.1 \times 10^{-4}} = 1.4 \times 10^{-11}$

(d) Conjugate acid $NH_4^+$: $K_a = \dfrac{1.0 \times 10^{-14}}{1.8 \times 10^{-5}} = 5.6 \times 10^{-10}$

**Think About It** Because the conjugates of weak acids and bases have ionization constants, salts containing these ions have an effect on the pH of a solution. In Section 16.10 we will use the ionization constants of conjugate acids and conjugate bases to calculate pH for solutions containing dissolved salts.

**Practice Problem A** Determine (a) $K_b$ of the benzoate ion ($C_6H_5COO^-$), (b) $K_b$ of the ascorbate ion ($HC_6H_6O_6^-$), and (c) $K_a$ of the ethylammonium ion ($C_2H_5NH_3^+$).

**Practice Problem B** Determine (a) $K_b$ of the weak base B whose conjugate acid $HB^+$ has $K_a = 8.9 \times 10^{-4}$ and (b) $K_a$ of the weak acid HA whose conjugate base has $K_b = 2.1 \times 10^{-8}$.

| **Checkpoint 16.7** | **Conjugate Acid-Base Pairs** |

**16.7.1**  Calculate the $K_b$ of the cyanide ion ($CN^-$).

a) $4.9 \times 10^{-10}$

b) $2.0 \times 10^{-5}$

c) $4.9 \times 10^{-24}$

d) $1.0 \times 10^{-7}$

e) $2.2 \times 10^{-5}$

**16.7.2**  Which of the anions listed is the strongest base? (See Table 16.6.)

a) Ascorbate ion ($HC_6H_6O_6^-$)

b) Benzoate ion ($C_6H_5COO^-$)

c) Nitrite ion ($NO_2^-$)

d) Phenolate ion ($C_6H_5O^-$)

e) Formate ion ($HCOO^-$)

**16.7.3**  The diagrams show solutions of three different weak acids with formulas HX, HY, and HZ. List the anions $X^-$, $Y^-$, and $Z^-$ in order of increasing $K_b$ value.

a) $Z^- < Y^- < X^-$

b) $Y^- < Z^- < X^-$

c) $Z^- < X^- < Y^-$

d) $X^- < Y^- < Z^-$

e) $X^- < Z^- < Y^-$

## 16.8  Diprotic and Polyprotic Acids

**Student Annotation:** *A triprotic acid has $K_{a_1}$, $K_{a_2}$, and $K_{a_3}$.*

Diprotic and polyprotic acids undergo successive ionizations, losing one proton at a time [◄◄ Section 4.3], and each ionization has a $K_a$ associated with it. Ionization constants for a diprotic acid are designated $K_{a_1}$ and $K_{a_2}$. We write a separate equilibrium expression for each ionization, and we may need two or more equilibrium expressions to calculate the concentrations of species in solution at equilibrium. For carbonic acid ($H_2CO_3$), for example, we write

$$H_2CO_3(aq) \rightleftharpoons H^+(aq) + HCO_3^-(aq) \qquad K_{a_1} = \frac{[H^+][HCO_3^-]}{[H_2CO_3]}$$

$$HCO_3^-(aq) \rightleftharpoons H^+(aq) + CO_3^{2-}(aq) \qquad K_{a_2} = \frac{[H^+][CO_3^{2-}]}{[HCO_3^-]}$$

Note that the conjugate base in the first ionization is the acid in the second ionization. Table 16.8 shows the ionization constants of several diprotic acids and one polyprotic acid. For a given acid, the first ionization constant is much larger than the second ionization constant, and so on. This trend makes sense because it is easier to remove a proton from a neutral species than from one that is negatively charged, and it is easier to remove a proton from a species with a single negative charge than from one with a double negative charge.

Sample Problem 16.18 shows how to calculate equilibrium concentrations of all species in solution for an aqueous solution of a diprotic acid.

| **Sample Problem 16.18** |

Oxalic acid ($H_2C_2O_4$) is a poisonous substance used mainly as a bleaching agent. Calculate the concentrations of all species present at equilibrium in a 0.10 $M$ solution at 25°C.

**Strategy**  Follow the same procedure for each ionization as for the determination of equilibrium concentrations for a monoprotic acid. The conjugate base resulting from the first ionization is the acid for the second ionization, and its starting concentration is the equilibrium concentration from the first ionization.

**Setup**  The ionizations of oxalic acid and the corresponding ionization constants are

$$H_2C_2O_4(aq) \rightleftharpoons H^+(aq) + HC_2O_4^-(aq) \qquad K_{a_1} = 6.5 \times 10^{-2}$$

$$HC_2O_4^-(aq) \rightleftharpoons H^+(aq) + C_2O_4^{2-}(aq) \qquad K_{a_2} = 6.1 \times 10^{-5}$$

Construct an equilibrium table for each ionization, using $x$ as the unknown in the first ionization and $y$ as the unknown in the second ionization.

$$H_2C_2O_4(aq) \rightleftharpoons H^+(aq) + HC_2O_4^-(aq)$$

| | | | |
|---|---|---|---|
| *Initial concentration (M):* | 0.10 | 0 | 0 |
| *Change in concentration (M):* | $-x$ | $+x$ | $+x$ |
| *Equilibrium concentration (M):* | $0.10 - x$ | $x$ | $x$ |

The equilibrium concentration of the hydrogen oxalate ion ($HC_2O_4^-$) after the first ionization becomes the starting concentration for the second ionization. Additionally, the equilibrium concentration of $H^+$ is the starting concentration for the second ionization.

$$HC_2O_4^-(aq) \rightleftharpoons H^+(aq) + C_2O_4^{2-}(aq)$$

| | | | |
|---|---|---|---|
| *Initial concentration (M):* | $x$ | $x$ | 0 |
| *Change in concentration (M):* | $-y$ | $+y$ | $+y$ |
| *Equilibrium concentration (M):* | $x - y$ | $x + y$ | $y$ |

**Solution**

$$K_{a_1} = \frac{[H^+][HC_2O_4^-]}{[H_2C_2O_4]}$$

$$6.5 \times 10^{-2} = \frac{x^2}{0.10 - x}$$

Applying the approximation and neglecting $x$ in the denominator of the expression gives

$$6.5 \times 10^{-2} \approx \frac{x^2}{0.10}$$

$$x^2 = 6.5 \times 10^{-3}$$

$$x = 8.1 \times 10^{-2} \, M$$

*(Continued)*

**TABLE 16.8** Ionization Constants of Some Diprotic and Polyprotic Acids at 25°C

| Name of Acid | Formula | Structure | $K_{a_1}$ | $K_{a_2}$ | $K_{a_3}$ |
|---|---|---|---|---|---|
| Sulfuric acid | $H_2SO_4$ | $H-O-\overset{\overset{O}{\|\|}}{\underset{\underset{O}{\|\|}}{S}}-O-H$ | Very large | $1.3 \times 10^{-2}$ | |
| Oxalic acid | $H_2C_2O_4$ | $H-O-\overset{\overset{O}{\|\|}}{C}-\overset{\overset{O}{\|\|}}{C}-O-H$ | $6.5 \times 10^{-2}$ | $6.1 \times 10^{-5}$ | |
| Sulfurous acid | $H_2SO_3$ | $H-O-\overset{\overset{O}{\|\|}}{S}-O-H$ | $1.3 \times 10^{-2}$ | $6.3 \times 10^{-8}$ | |
| Ascorbic acid (vitamin C) | $H_2C_6H_6O_6$ | (structure) | $8.0 \times 10^{-5}$ | $1.6 \times 10^{-12}$ | |
| Carbonic acid | $H_2CO_3$ | $H-O-\overset{\overset{O}{\|\|}}{C}-O-H$ | $4.2 \times 10^{-7}$ | $4.8 \times 10^{-11}$ | |
| Hydrosulfuric acid* | $H_2S$ | $H-S-H$ | $9.5 \times 10^{-8}$ | $1 \times 10^{-19}$ | |
| Phosphoric acid | $H_3PO_4$ | $H-O-\overset{\overset{O}{\|\|}}{\underset{\underset{H}{\|}}{\underset{O}{\|}}{P}}-O-H$ | $7.5 \times 10^{-3}$ | $6.2 \times 10^{-8}$ | $4.2 \times 10^{-13}$ |

*The second ionization constant of $H_2S$ is very low and difficult to measure. The value in this table is an estimate.

Testing the approximation,

$$\frac{8.1 \times 10^{-2} \, M}{0.10 \, M} \times 100\% = 81\%$$

Clearly the approximation is not valid, so we must solve the following quadratic equation:

$$x^2 + 6.5 \times 10^{-2} x - 6.5 \times 10^{-3} = 0$$

The result is $x = 0.054 \, M$. Thus, after the first ionization, the concentrations of species in solution are

$$[\text{H}^+] = 0.054 \, M$$

$$[\text{HC}_2\text{O}_4^-] = 0.054 \, M$$

$$[\text{H}_2\text{C}_2\text{O}_4] = (0.10 - 0.054) \, M = 0.046 \, M$$

Rewriting the equilibrium table for the second ionization, using the calculated value of $x$, gives the following:

$$\text{HC}_2\text{O}_4^-(aq) \rightleftharpoons \text{H}^+(aq) + \text{C}_2\text{O}_4^{2-}(aq)$$

| | HC$_2$O$_4^-$ | H$^+$ | C$_2$O$_4^{2-}$ |
|---|---|---|---|
| *Initial concentration (M):* | 0.054 | 0.054 | 0 |
| *Change in concentration (M):* | $-y$ | $+y$ | $+y$ |
| *Equilibrium concentration (M):* | $0.054 - y$ | $0.054 + y$ | $y$ |

$$K_{a_2} = \frac{[\text{H}^+][\text{C}_2\text{O}_4^{2-}]}{[\text{HC}_2\text{O}_4^-]}$$

$$6.1 \times 10^{-5} = \frac{(0.054 + y)(y)}{0.054 - y}$$

Assuming that $y$ is very small and applying the approximations $0.054 + y \approx 0.054$ and $0.054 - y \approx 0.054$ gives

$$\frac{(0.054)(y)}{0.054} = y = 6.1 \times 10^{-5} \, M$$

We must test the approximation as follows to see if it is valid:

$$\frac{6.1 \times 10^{-5} \, M}{0.054 \, M} \times 100\% = 0.11\%$$

This time, because the ionization constant is much smaller, the approximation is valid. At equilibrium, the concentrations of all species are

$$[\text{H}_2\text{C}_2\text{O}_4] = 0.046 \, M$$

$$[\text{HC}_2\text{O}_4^-] = (0.054 - 6.1 \times 10^{-5}) \, M = 0.054 \, M$$

$$[\text{H}^+] = (0.054 + 6.1 \times 10^{-5}) \, M = 0.054 \, M$$

$$[\text{C}_2\text{O}_4^{2-}] = 6.1 \times 10^{-5} \, M$$

**Think About It** Note that the second ionization did not contribute significantly to the H$^+$ concentration. Therefore, we could determine the pH of this solution by considering only the first ionization. This is true in general for polyprotic acids where $K_{a_1}$ is at least $1000 \times K_{a_2}$. [It is necessary to consider the second ionization to determine the concentration of oxalate ion (C$_2$O$_4^{2-}$).]

**Practice Problem A** Calculate the concentrations of H$_2$C$_2$O$_4$, HC$_2$O$_4^-$, C$_2$O$_4^{2-}$, and H$^+$ ions in a 0.20 $M$ oxalic acid solution at 25°C.

**Practice Problem B** Calculate the concentrations of H$_2$SO$_4$, HSO$_4^-$, SO$_4^{2-}$, and H$^+$ ions in a 0.14 $M$ sulfuric acid solution at 25°C.

## Checkpoint 16.8    Diprotic and Polyprotic Acids

**16.8.1** Calculate the equilibrium concentration of CO$_3^{2-}$ in a 0.050 $M$ solution of carbonic acid at 25°C.

a) $4.2 \times 10^{-7} \, M$

b) $4.8 \times 10^{-11} \, M$

c) $1.5 \times 10^{-4} \, M$

d) 0.050 $M$

e) 0.049 $M$

**16.8.2** What is the pH of a 0.40 $M$ solution of phosphoric acid at 25°C?

a) 5.48

b) 1.26

c) 1.29

d) 12.74

e) 0.80

**16.8.3** List the molecular and ionic species in order of increasing concentration in a solution of ascorbic acid ($H_2C_6H_6O_6$).

a) $[H_2C_6H_6O_6] < [HC_6H_6O_6^-] < [C_6H_6O_6^{2-}] = [H^+]$

b) $[H_2C_6H_6O_6] < [HC_6H_6O_6^-] < [C_6H_6O_6^{2-}] < [H^+]$

c) $[C_6H_6O_6^{2-}] < [HC_6H_6O_6^-] = [H^+] < [H_2C_6H_6O_6]$

d) $[H^+] = [H_2C_6H_6O_6] < [HC_6H_6O_6^-] < [C_6H_6O_6^{2-}]$

e) $[C_6H_6O_6^{2-}] = [H^+] < [HC_6H_6O_6^-] < [H_2C_6H_6O_6]$

**16.8.4** Which is true for any polyprotic acid?

a) $K_{a_2} > K_{a_1}$

b) $K_{a_2} < K_{a_1}$

c) $K_{a_2} = K_{a_1}$

d) $K_{a_2} = \dfrac{K_{a_1}}{K_w}$

e) $K_{a_2} > \dfrac{K_w}{K_{a_1}}$

## 16.9 Molecular Structure and Acid Strength

The strength of an acid is measured by its tendency to ionize:

$$HX \longrightarrow H^+ + X^-$$

Two factors influence the extent to which the acid undergoes ionization. One is the strength of the H—X bond. The stronger the bond, the more difficult it is for the HX molecule to break up and hence the weaker the acid. The other factor is the polarity of the H—X bond. The difference in the electronegativities between H and X results in a polar bond like

$$\overset{\delta+ \quad \delta-}{H-X}$$

If the bond is highly polarized (i.e., if there is a large accumulation of positive and negative charges on the H and X atoms, respectively), HX will tend to break up into $H^+$ and $X^-$ ions. A high degree of polarity, therefore, gives rise to a stronger acid. In this section, we consider the roles of bond strength and bond polarity in determining the strength of an acid.

### Hydrohalic Acids

The halogens form a series of binary acids called the hydrohalic acids (HF, HCl, HBr, and HI). Table 16.9 shows that of this series only HF is a weak acid ($K_a = 7.1 \times 10^{-4}$). The data in the table indicate that the predominant factor in determining the strength of the hydrohalic acids is bond strength. HF has the largest bond enthalpy, making its bond the most difficult to break. In this series of binary acids, acid strength increases as bond strength decreases. The strength of the acids increases as follows:

$$HF \ll HCl < HBr < HI$$

**Student Annotation:** The polarity of the H—X bond actually decreases from H—F to H—I, largely because F is the most electronegative element. This would suggest that HF would be the strongest of the hydrohalic acids. Based on the data in Table 16.9, however, bond enthalpy is the more important factor in determining the strengths of these acids.

### Oxoacids

An oxoacid, as we learned in Chapter 2, contains hydrogen, oxygen, and a central, nonmetal atom [◄◄ Section 2.7]. As the Lewis structures in Figure 16.4 show, oxoacids contain one or more O—H bonds. If the central atom is an electronegative element, or is in a high oxidation state, it will attract electrons, causing the O—H bond to be more polar. This makes it easier for the hydrogen to be lost as $H^+$, making the acid stronger.

| TABLE 16.9 | Bond Enthalpies for Hydrogen Halides and Acid Strengths for Hydrohalic Acids | | |
|---|---|---|---|
| **Bond** | **Bond Enthalpy (kJ/mol)** | **Acid Strength** | |
| H—F | 562.8 | Weak | |
| H—Cl | 431.9 | Strong | |
| H—Br | 366.1 | Strong | |
| H—I | 298.3 | Strong | |

**Figure 16.4** Lewis structures of some common oxoacids. Recall that there is more than one possible Lewis structure for oxoacids in which the central atom is from period 3 or below on the periodic table [◄◄ Section 8.8].

$$H-\ddot{O}-\overset{\overset{\ddot{O}}{\|}}{C}-\ddot{O}-H \qquad H-\ddot{O}-N=\ddot{O}: \qquad H-\ddot{O}-\overset{\overset{\ddot{O}}{\|}}{N}-\ddot{O}:$$

Carbonic acid             Nitrous acid             Nitric acid

$$H-\ddot{O}-\overset{\overset{\ddot{O}}{\|}}{\underset{\underset{H}{|}}{P}}-\ddot{O}-H \qquad H-\ddot{O}-\overset{\overset{\ddot{O}}{\|}}{\underset{\underset{H}{|}}{P}}-\ddot{O}-H \qquad H-\ddot{O}-\overset{\overset{\ddot{O}}{\|}}{\underset{\underset{.\ddot{O}.}{}}{S}}-\ddot{O}-H$$

Phosphorous acid         Phosphoric acid         Sulfuric acid

**Figure 16.5** Lewis structures of the oxoacids of chlorine. The oxidation number of the Cl atom is shown in parentheses. Note that although hypochlorous acid is written as HClO, the H atom is bonded to the O atom.

$$H-\ddot{O}-\ddot{C}l: \qquad\qquad H-\ddot{O}-\overset{\overset{\ddot{O}}{\|}}{\ddot{C}}l-\ddot{O}:$$

Hypochlorous acid (+1)        Chlorous acid (+3)

$$H-\ddot{O}-\overset{\overset{:\ddot{O}:}{|}}{Cl}-\ddot{O}: \qquad\qquad H-\ddot{O}-\overset{\overset{:\ddot{O}:}{|}}{\underset{\underset{:\ddot{O}:}{|}}{Cl}}-\ddot{O}:$$

Chloric acid (+5)        Perchloric acid (+7)

To compare their strengths, it is convenient to divide the oxoacids into two groups:

1. *Oxoacids having different central atoms that are from the same group of the periodic table and that have the same oxidation number.* Two examples are

$$H-\ddot{O}-\overset{\overset{:\ddot{O}:}{|}}{Cl}-\ddot{O}: \qquad\qquad H-\ddot{O}-\overset{\overset{:\ddot{O}:}{|}}{Br}-\ddot{O}:$$

$$HClO_3 \qquad\qquad HBrO_3$$

Within this group, acid strength increases with increasing electronegativity of the central atom. Cl and Br have the same oxidation number in these acids, +5. However, because Cl is more electronegative than Br, it attracts the electron pair it shares with oxygen (in the Cl—O—H group) to a greater extent than Br does (in the corresponding Br—O—H group). Consequently, the O—H bond is more polar in chloric acid than in bromic acid and ionizes more readily. The relative acid strengths are

$$HClO_3 > HBrO_3$$

2. *Oxoacids having the same central atom but different numbers of oxygen atoms.* Within this group, acid strength increases with increasing oxidation number of the central atom. Consider the oxoacids of chlorine shown in Figure 16.5. In this series the ability of chlorine to draw electrons away from the OH group (thus making the O—H bond more polar) increases with the number of electronegative O atoms attached to Cl. Thus, $HClO_4$ is the strongest acid because it has the largest number of oxygen atoms attached to Cl. The acid strength decreases as follows:

**Student Annotation:** As the number of attached oxygen atoms increases, the oxidation number of the central atom also increases [◄◄ Section 4.4].

$$HClO_4 > HClO_3 > HClO_2 > HClO$$

Sample Problem 16.19 compares acid strengths based on molecular structure.

---

### Sample Problem 16.19

Predict the relative strengths of the oxoacids in each of the following groups: (a) HClO, HBrO, and HIO; (b) $HNO_3$ and $HNO_2$.

**Strategy** In each group, compare the electronegativities or oxidation numbers of the central atoms to determine which O—H bonds are the most polar. The more polar the O—H bond, the more readily it is broken and the stronger the acid.

**Setup**

(a) In a group with different central atoms, we must compare electronegativities. The electronegativities of the central atoms in this group decrease as follows: Cl > Br > I.

(b) These two acids have the same central atom but differ in the number of attached oxygen atoms. In a group such as this, the greater the number of attached oxygen atoms, the higher the oxidation number of the central atom and the stronger the acid.

**Student Annotation:** Another way to compare the strengths of these two is to remember that $HNO_3$ is one of the seven strong acids. $HNO_2$ is not.

**Solution**

(a) Acid strength decreases as follows: HClO > HBrO > HIO.

(b) $HNO_3$ is a stronger acid than $HNO_2$.

> **Think About It** Four of the strong acids are oxoacids: $HNO_3$, $HClO_4$, $HClO_3$, and $H_2SO_4$.

**Practice Problem A** Indicate which is the stronger acid: (a) $HBrO_3$ or $HBrO_4$; (b) $H_2SeO_4$ or $H_2SO_4$.

**Practice Problem B** Based on the information in this section, which is the predominant factor in determining the strength of an oxoacid: electronegativity of the central atom or oxidation state of the central atom?

## Carboxylic Acids

So far our discussion has focused on inorganic acids. A particularly important group of organic acids is the *carboxylic acids,* whose Lewis structures can be represented by

$$R-\overset{\overset{\displaystyle \ddot{O}}{\|}}{C}-\ddot{O}-H$$

where R is part of the acid molecule and the shaded portion represents the carboxyl group, −COOH.

The conjugate base of a carboxylic acid, called a carboxylate anion, $RCOO^-$, can be represented by more than one resonance structure:

$$R-\overset{\overset{\displaystyle \ddot{O}}{\|}}{C}-\ddot{O}\text{:}^- \longleftrightarrow R-\overset{\overset{\displaystyle :\ddot{O}\text{:}^-}{\mid}}{C}=\ddot{O}\text{:}$$

**Student Annotation:** You learned in Chapter 4 [◄◄ Section 4.1] that carboxylic acid formulas are often written with the ionizable H atom last, in order to keep the functional group together. You should recognize the formulas for organic acids written either way. For example, acetic acid may be written as $HC_2H_3O_2$ or as $CH_3COOH$.

In terms of molecular orbital theory [◄◄ Section 9.6], we attribute the stability of the anion to its ability to spread out or delocalize the electron density over several atoms. The greater the extent of electron delocalization, the more stable the anion and the greater the tendency for the acid to undergo ionization—that is, the stronger the acid.

The strength of carboxylic acids depends on the nature of the R group. Consider, for example, acetic acid and chloroacetic acid:

$$H-\overset{\overset{\displaystyle H}{\mid}}{\underset{\underset{\displaystyle H}{\mid}}{C}}-\overset{\overset{\displaystyle \ddot{O}}{\|}}{C}-\ddot{O}-H \qquad\qquad H-\overset{\overset{\displaystyle Cl}{\mid}}{\underset{\underset{\displaystyle H}{\mid}}{C}}-\overset{\overset{\displaystyle \ddot{O}}{\|}}{C}-\ddot{O}-H$$

$$\text{Acetic acid } (K_a = 1.8 \times 10^{-5}) \qquad \text{Chloroacetic acid } (K_a = 1.4 \times 10^{-3})$$

The presence of the electronegative Cl atom in chloroacetic acid shifts the electron density toward the R group, thereby making the O−H bond more polar. Consequently, there is a greater tendency for chloroacetic acid to ionize:

$$H-\overset{\overset{\displaystyle :\ddot{Cl}\text{:}}{\mid}}{\underset{\underset{\displaystyle H}{\mid}}{C}}-\overset{\overset{\displaystyle \ddot{O}}{\|}}{C}-\ddot{O}-H \rightleftharpoons \left[ H-\overset{\overset{\displaystyle :\ddot{Cl}\text{:}}{\mid}}{\underset{\underset{\displaystyle H}{\mid}}{C}}-\overset{\overset{\displaystyle \ddot{O}}{\|}}{C}-\ddot{O}\text{:} \right]^- + H^+$$

Chloroacetic acid is the stronger of the two acids.

## 16.10    Acid-Base Properties of Salt Solutions

In Section 16.7, we saw that the conjugate base of a weak acid acts as a weak Brønsted base in water. Consider a solution of the salt sodium fluoride (NaF). Because NaF is a strong electrolyte, it dissociates completely in water to give a solution of sodium cations ($Na^+$) and fluoride anions ($F^-$). The fluoride ion, which is the conjugate base of hydrofluoric acid, reacts with water to produce hydrofluoric acid and hydroxide ion:

$$F^-(aq) + H_2O(l) \rightleftharpoons HF(aq) + OH^-(aq)$$

This is a specific example of **salt hydrolysis,** in which ions produced by the dissociation of a salt react with water to produce either hydroxide ions or hydronium ions—thus impacting pH. Using our knowledge of how ions from a dissolved salt interact with water, we can determine (based on the identity of the dissolved salt) whether a solution will be neutral, basic, or acidic. Note in the preceding example that sodium ions ($Na^+$) do not hydrolyze and thus have no impact on the pH of the solution.

### Basic Salt Solutions

Sodium fluoride is a salt that dissolves to give a basic solution. In general, an anion that is the conjugate base of a weak acid reacts with water to produce hydroxide ion. Other examples include the acetate ion ($CH_3COO^-$), the nitrite ion ($NO_2^-$), the sulfite ion ($SO_3^{2-}$), and the hydrogen carbonate ion ($HCO_3^-$). Each of these anions undergoes hydrolysis to produce the corresponding weak acid and hydroxide ion:

$$A^-(aq) + H_2O(l) \rightleftharpoons HA(aq) + OH^-(aq)$$

We can therefore make the qualitative prediction that a solution of a salt in which the anion is the conjugate base of a weak acid will be basic. We calculate the pH of a basic salt solution the same way we calculate the pH of any weak base solution, using the $K_b$ value for the anion. The necessary $K_b$ value is calculated using the tabulated $K_a$ value of the corresponding weak acid (see Table 16.6).

Sample Problem 16.20 shows how to calculate the pH of a basic salt solution.

### Sample Problem 16.20

Calculate the pH of a 0.10 *M* solution of sodium fluoride (NaF) at 25°C.

**Strategy** A solution of NaF contains $Na^+$ ions and $F^-$ ions. The $F^-$ ion is the conjugate base of the weak acid, HF. Use the $K_a$ value for HF ($7.1 \times 10^{-4}$, from Table 16.6) and Equation 16.8 to determine $K_b$ for $F^-$:

$$K_b = \frac{K_w}{K_a} = \frac{1.0 \times 10^{-14}}{7.1 \times 10^{-4}} = 1.4 \times 10^{-11}$$

Then, solve this pH problem like any equilibrium problem, using an equilibrium table.

**Setup** It's always a good idea to write the equation corresponding to the reaction that takes place along with the equilibrium expression:

$$F^-(aq) + H_2O(l) \rightleftharpoons HF(aq) + OH^-(aq) \qquad K_b = \frac{[HF][OH^-]}{[F^-]}$$

Construct an equilibrium table, and determine, in terms of the unknown *x*, the equilibrium concentrations of the species in the equilibrium expression:

|  | $F^-(aq)$ | $+ H_2O(l)$ | $\rightleftharpoons$ | $HF(aq)$ | $+ OH^-(aq)$ |
|---|---|---|---|---|---|
| *Initial concentration (M):* | 0.10 | | | 0 | 0 |
| *Change in concentration (M):* | $-x$ | | | $+x$ | $+x$ |
| *Equilibrium concentration (M):* | $0.10 - x$ | | | $x$ | $x$ |

**Solution** Substituting the equilibrium concentrations into the equilibrium expression and using the shortcut to solve for $x$, we get

$$1.4 \times 10^{-11} = \frac{x^2}{0.10 - x} \approx \frac{x^2}{0.10}$$

$$x = \sqrt{(1.4 \times 10^{-11})(0.10)} = 1.2 \times 10^{-6} \, M$$

According to our equilibrium table, $x = [OH^-]$. In this case, the autoionization of water makes a significant contribution to the hydroxide ion concentration so the total concentration will be the sum of $1.2 \times 10^{-6} \, M$ (from the ionization of $F^-$) and $1.0 \times 10^{-7} \, M$ (from the autoionization of water). Therefore, we calculate the pOH first as

$$pOH = -\log (1.2 \times 10^{-6} + 1.0 \times 10^{-7}) = 5.95$$

and then the pH,

$$pH = 14.00 - pOH = 14.00 - 5.95 = 8.05$$

The pH of a 0.10 $M$ solution of NaF at 25°C is 8.05.

> **Think About It** It's easy to mix up pH and pOH in this type of problem. Always make a qualitative prediction regarding the pH of a salt solution first, and then check to make sure that your calculated pH agrees with your prediction. In this case, we would predict a basic pH because the anion in the salt ($F^-$) is the conjugate base of a weak acid (HF). The calculated pH, 8.05, is indeed basic.

**Practice Problem A** Determine the pH of a 0.15 $M$ solution of sodium acetate ($CH_3COONa$) at 25°C.

**Practice Problem B** Determine the concentration of a solution of sodium fluoride (NaF) that has pH 8.51 at 25°C.

## Acidic Salt Solutions

When the cation of a salt is the conjugate acid of a weak base, a solution of the salt will be acidic. For example, when ammonium chloride dissolves in water, it dissociates to give a solution of ammonium ions and chloride ions:

$$NH_4Cl(s) \xrightarrow{\text{H}_2\text{O}} NH_4^+(aq) + Cl^-(aq)$$

The ammonium ion is the conjugate acid of the weak base ammonia ($NH_3$). It acts as a weak Brønsted acid, reacting with water to produce hydronium ion:

$$NH_4^+(aq) + H_2O(l) \rightleftharpoons NH_3(aq) + H_3O^+(aq)$$

We would therefore predict that a solution containing the ammonium ion is acidic. To calculate the pH, we must determine the $K_a$ for $NH_4^+$ using the tabulated $K_b$ value for $NH_3$ and Equation 16.8. Because $Cl^-$ is the weak conjugate base of the strong acid HCl, $Cl^-$ does not hydrolyze and therefore has no impact on the pH of the solution.

Sample Problem 16.21 shows how to calculate the pH of an acidic salt solution.

### Sample Problem 16.21

Calculate the pH of a 0.10 $M$ solution of ammonium chloride ($NH_4Cl$) at 25°C.

**Strategy** A solution of $NH_4Cl$ contains $NH_4^+$ cations and $Cl^-$ anions. The $NH_4^+$ ion is the conjugate acid of the weak base $NH_3$. Use the $K_b$ value for $NH_3$ ($1.8 \times 10^{-5}$ from Table 16.7) and Equation 16.8 to determine $K_a$ for $NH_4^+$.

$$K_a = \frac{K_w}{K_b} = \frac{1.0 \times 10^{-14}}{1.8 \times 10^{-5}} = 5.6 \times 10^{-10}$$

**Setup** Again, we write the balanced chemical equation and the equilibrium expression:

$$NH_4^+(aq) + H_2O(l) \rightleftharpoons NH_3(aq) + H_3O^+(aq) \qquad K_a = \frac{[NH_3][H_3O^+]}{[NH_4^+]}$$

*(Continued)*

Next, construct a table to determine the equilibrium concentrations of the species in the equilibrium expression:

$$NH_4^+(aq) + H_2O(l) \rightleftharpoons NH_3(aq) + H_3O^+(aq)$$

|  | $NH_4^+(aq)$ | $+ H_2O(l)$ | $\rightleftharpoons NH_3(aq)$ | $+ H_3O^+(aq)$ |
|---|---|---|---|---|
| *Initial concentration (M):* | 0.10 | | 0 | 0 |
| *Change in concentration (M):* | $-x$ | | $+x$ | $+x$ |
| *Equilibrium concentration (M):* | $0.10 - x$ | | $x$ | $x$ |

**Solution**  Substituting the equilibrium concentrations into the equilibrium expression and using the shortcut to solve for $x$, we get

$$5.6 \times 10^{-10} = \frac{x^2}{0.10 - x} \approx \frac{x^2}{0.10}$$

$$x = \sqrt{(5.6 \times 10^{-10})(0.10)} = 7.5 \times 10^{-6}\,M$$

According to the equilibrium table, $x = [H_3O^+]$. The pH can be calculated as follows:

$$pH = -\log(7.5 \times 10^{-6}) = 5.12$$

The pH of a 0.10 *M* solution of ammonium chloride (at 25°C) is 5.12.

**Think About It**  In this case, we would predict an acidic pH because the cation in the salt ($NH_4^+$) is the conjugate acid of a weak base ($NH_3$). The calculated pH is acidic.

**Practice Problem A**  Determine the pH of a 0.25 *M* solution of pyridinium nitrate ($C_5H_6NNO_3$) at 25°C. [Pyridinium nitrate dissociates in water to give pyridinium ions ($C_5H_6N^+$), the conjugate acid of *pyridinium* (see Table 16.7), and nitrate ions ($NO_3^-$).]

**Practice Problem B**  Determine the concentration of a solution of ammonium chloride ($NH_4Cl$) that has pH 5.37 at 25°C.

The metal ion in a dissolved salt can also react with water to produce an acidic solution. The extent of hydrolysis is greatest for the small and highly charged metal cations such as $Al^{3+}$, $Cr^{3+}$, $Fe^{3+}$, $Bi^{3+}$, and $Be^{2+}$. For example, when aluminum chloride dissolves in water, each $Al^{3+}$ ion becomes associated with six water molecules (Figure 16.6).

Consider one of the bonds that forms between the metal ion and an oxygen atom from one of the six water molecules in $Al(H_2O)_6^{3+}$:

$$Al \rightleftharpoons O \begin{array}{c} H \\ \\ H \end{array}$$

The positively charged $Al^{3+}$ ion draws electron density toward itself, increasing the polarity of the O—H bonds. Consequently, the H atoms have a greater tendency to ionize than those in water molecules not associated with the $Al^{3+}$ ion. The resulting ionization process can be written as

$$Al(H_2O)_6^{3+}(aq) + H_2O(l) \rightleftharpoons Al(OH)(H_2O)_5^{2+}(aq) + H_3O^+(aq)$$

or as

$$Al(H_2O)_6^{3+}(aq) \rightleftharpoons Al(OH)(H_2O)_5^{2+}(aq) + H^+(aq)$$

**Figure 16.6**  The six $H_2O$ molecules surround the $Al^{3+}$ ion in an octahedral arrangement. The attraction of the small $Al^{3+}$ ion for the lone pairs on the oxygen atoms is so great that the O—H bonds in an $H_2O$ molecule attached to the metal cation are weakened, allowing the loss of a proton ($H^+$) to an incoming $H_2O$ molecule. This hydrolysis of the metal cation makes the solution acidic.

The equilibrium constant for the metal cation hydrolysis is given by

$$K_a = \frac{[Al(OH)(H_2O)_5^{2+}][H^+]}{[Al(H_2O)_6^{3+}]} = 1.3 \times 10^{-5}$$

$Al(OH)(H_2O)_5^{2+}$ can undergo further ionization:

$$Al(OH)(H_2O)_5^{2+}(aq) \rightleftharpoons Al(OH)_2(H_2O)_4^+(aq) + H^+(aq)$$

and so on. It is generally sufficient, however, to take into account only the first stage of hydrolysis when determining the pH of a solution that contains metal ions.

## Neutral Salt Solutions

The extent of hydrolysis is greatest for the smallest and most highly charged metal ions because a compact, highly charged ion is more effective in polarizing the O−H bond and facilitating ionization. This is why relatively large ions of low charge, including the metal cations of Groups 1A and 2A (the cations of the strong bases), do not undergo significant hydrolysis ($Be^{2+}$ is an exception). Thus, most metal cations of Groups 1A and 2A do not impact the pH of a solution.

Similarly, anions that are conjugate bases of strong acids do not hydrolyze to any significant degree. Consequently, a salt composed of the cation of a strong base and the anion of a strong acid, such as NaCl, produces a neutral solution.

To summarize, the pH of a salt solution can be predicted qualitatively by identifying the ions in solution and determining which of them, if any, undergoes significant hydrolysis.

**Student Annotation:** The metal cations of the strong bases are those of the alkali metals: ($Li^+$, $Na^+$, $K^+$, $Rb^+$, and $Cs^+$) and those of the heavy alkaline earth metals ($Sr^{2+}$ and $Ba^{2+}$).

| | **Examples** |
|---|---|
| A cation that will make a solution acidic is<br>• the conjugate acid of a weak base.<br>• a small, highly charged metal ion (other than from Group 1A or 2A). | $NH_4^+$, $CH_3NH_3^+$, $C_2H_5NH_3^+$<br>$Al^{3+}$, $Cr^{3+}$, $Fe^{3+}$, $Bi^{3+}$ |
| An anion that will make a solution basic is<br>• the conjugate base of a weak acid. | $CN^-$, $NO_2^-$, $CH_3COO^-$ |
| A cation that will not affect the pH of a solution is<br>• a Group 1A or heavy Group 2A cation (except $Be^{2+}$). | $Li^+$, $Na^+$, $Ba^{2+}$ |
| An anion that will not affect the pH of a solution is<br>• the conjugate base of a strong acid. | $Cl^-$, $NO_3^-$, $ClO_4^-$ |

Sample Problem 16.22 lets you practice predicting the pH of salt solutions.

---

### Sample Problem 16.22

Predict whether a 0.10 $M$ solution of each of the following salts will be basic, acidic, or neutral:
(a) LiI, (b) $NH_4NO_3$, (c) $Sr(NO_3)_2$, (d) $KNO_2$, (e) NaCN.

**Strategy** Identify the ions present in each solution, and determine which, if any, will impact the pH of the solution.

**Setup** (a) Ions in solution: $Li^+$ and $I^-$. $Li^+$ is a Group 1A cation; $I^-$ is the conjugate base of the strong acid HI. Therefore, neither ion hydrolyzes to any significant degree.

(b) Ions in solution: $NH_4^+$ and $NO_3^-$. $NH_4^+$ is the conjugate acid of the weak base $NH_3$; $NO_3^-$ is the conjugate base of the strong acid $HNO_3$. In this case, the cation will hydrolyze, making the pH acidic:

$$NH_4^+(aq) + H_2O(l) \rightleftharpoons NH_3(aq) + H_3O^+(aq)$$

(c) Ions in solution: $Sr^{2+}$ and $NO_3^-$. $Sr^{2+}$ is a heavy Group 2A cation; $NO_3^-$ is the conjugate base of the strong acid $HNO_3$. Neither ion hydrolyzes to any significant degree.

(d) Ions in solution: $K^+$ and $NO_2^-$. $K^+$ is a Group 1A cation; $NO_2^-$ is the conjugate base of the weak acid $HNO_2$. In this case, the anion hydrolyzes, thus making the pH basic:

$$NO_2^-(aq) + H_2O(l) \rightleftharpoons HNO_2(aq) + OH^-(aq)$$

(e) Ions in solution: $Na^+$ and $CN^-$. $Na^+$ is a Group 1A cation; $CN^-$ is the conjugate base of the weak acid HCN. In this case, too, the anion hydrolyzes, thus making the pH basic:

$$CN^-(aq) + H_2O(l) \rightleftharpoons HCN(aq) + OH^-(aq)$$

*(Continued)*

**Solution**

(a) Neutral

(b) Acidic

(c) Neutral

(d) Basic

(e) Basic

**Think About It** It's very important that you are able to identify the ions in solution correctly. If necessary, review the formulas and charges of the common polyatomic ions [◄◄ Section 2.7, Table 2.8].

**Practice Problem A** Predict whether a 0.10 $M$ solution of each of the following salts will be basic, acidic, or neutral: (a) $CH_3COOLi$, (b) $C_5H_5NHCl$, (c) KF, (d) $KNO_3$, (e) $KClO_4$.

**Practice Problem B** In addition to those given in Sample Problem 16.22 and Practice Problem A, identify two salts that will dissolve to give (a) an acidic solution, (b) a basic solution, and (c) a neutral solution.

## Salts in Which Both the Cation and the Anion Hydrolyze

So far we have considered salts in which only one ion undergoes hydrolysis. In some salts, both the cation and the anion hydrolyze. Whether a solution of such a salt is basic, acidic, or neutral depends on the relative strengths of the weak acid and the weak base. Although the process of calculating the pH in these cases is more complex than in cases where only one ion hydrolyzes, we can make qualitative predictions regarding pH using the values of $K_b$ (of the salt's anion) and $K_a$ (of the salt's cation):

- When $K_b > K_a$, the solution is basic.
- When $K_b < K_a$, the solution is acidic.
- When $K_b \approx K_a$, the solution is neutral or nearly neutral.

The salt $NH_4NO_2$, for example, dissociates in solution to give $NH_4^+$ ($K_a = 5.6 \times 10^{-10}$) and $NO_2^-$ ($K_b = 2.2 \times 10^{-11}$). Because $K_a$ for the ammonium ion is larger than $K_b$ for the nitrite ion, we would expect the pH of an ammonium nitrite solution to be slightly acidic.

## Checkpoint 16.10    Acid-Base Properties of Salt Solutions

**16.10.1** Calculate the pH of a 0.075 $M$ solution of $NH_4NO_3$ at 25°C.

a) 5.19

b) 8.81

c) 7.00

d) 2.93

e) 11.07

**16.10.2** Calculate the pH of a 0.082 $M$ solution of NaCN at 25°C.

a) 5.20

b) 8.80

c) 7.00

d) 2.89

e) 11.11

**16.10.3** Which of the following salts will produce a basic solution when dissolved in water? (Select all that apply.)

a) Sodium hypochlorite (NaClO)

b) Potassium fluoride (KF)

c) Lithium carbonate ($Li_2CO_3$)

d) Barium chloride ($BaCl_2$)

e) Ammonium iodide ($NH_4I$)

**16.10.4** Which of the following salts will produce a neutral solution when dissolved in water? (Select all that apply.)

a) Calcium chlorite [$Ca(ClO_2)_2$]

b) Potassium iodide (KI)

c) Lithium nitrate ($LiNO_3$)

d) Barium cyanide [$Ba(CN)_2$]

e) Ammonium iodide ($NH_4I$)

**16.10.5** The diagrams represent solutions of three salts NaX (X = A, B, or C). Arrange the three X⁻ anions in order of increasing base strength. (The Na⁺ ions are not shown.)

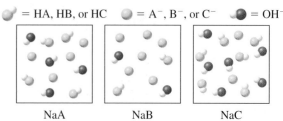

= HA, HB, or HC    = A⁻, B⁻, or C⁻    = OH⁻

NaA        NaB        NaC

a) A⁻ < B⁻ < C⁻

b) A⁻ < C⁻ < B⁻

c) B⁻ < A⁻ < C⁻

d) C⁻ < B⁻ < A⁻

e) C⁻ < A⁻ < B⁻

# 16.11 Acid-Base Properties of Oxides and Hydroxides

As we saw in Chapter 8, oxides can be classified as acidic, basic, or amphoteric. Thus, our discussion of acid-base reactions would be incomplete if we did not examine the properties of these compounds.

## Oxides of Metals and Nonmetals

Figure 16.7 shows the formulas of a number of oxides of the main group elements in their highest oxidation states. All alkali metal oxides and all alkaline earth metal oxides except BeO are basic. Beryllium oxide and several metallic oxides in Groups 3A and 4A are amphoteric. Nonmetallic oxides in which the oxidation number of the main group element is high are acidic (e.g., $N_2O_5$, $SO_3$, and $Cl_2O_7$), but those in which the oxidation number of the main group element is low (e.g., CO and NO) show no measurable acidic properties. No nonmetallic oxides are known to have basic properties.

The basic metallic oxides react with water to form metal hydroxides:

$$Na_2O(s) + H_2O(l) \longrightarrow 2NaOH(aq)$$

$$BaO(s) + H_2O(l) \longrightarrow Ba(OH)_2(aq)$$

The reactions between acidic oxides and water are as follows:

$$CO_2(g) + H_2O(l) \rightleftharpoons H_2CO_3(aq)$$

$$SO_3(g) + H_2O(l) \rightleftharpoons H_2SO_4(aq)$$

$$N_2O_5(g) + H_2O(l) \rightleftharpoons 2HNO_3(aq)$$

$$P_4O_{10}(g) + 6H_2O(l) \rightleftharpoons 4H_3PO_4(aq)$$

$$Cl_2O_7(g) + H_2O(l) \rightleftharpoons 2HClO_4(aq)$$

The reaction between $CO_2$ and $H_2O$ explains why pure water gradually becomes acidic when it is exposed to air, which contains $CO_2$. The pH of rainwater exposed only to unpolluted air is slightly acidic. The reaction between $SO_3$ and $H_2O$ is largely responsible for acid rain.

Reactions between acidic oxides and bases and those between basic oxides and acids resemble normal acid-base reactions in that the products are a salt and water:

$$CO_2(g) + 2NaOH(aq) \longrightarrow Na_2CO_3(aq) + H_2O(l)$$

$$BaO(s) + 2HNO_3(aq) \longrightarrow Ba(NO_3)_2(aq) + H_2O(l)$$

**Figure 16.7** Oxides of the main group elements in their highest oxidation states.

Aluminum oxide ($Al_2O_3$) is amphoteric. Depending on the reaction conditions, it can behave either as an acidic oxide or as a basic oxide. For example, $Al_2O_3$ acts as a base with hydrochloric acid to produce a salt ($AlCl_3$) and water:

$$Al_2O_3(s) + 6HCl(aq) \longrightarrow 2AlCl_3(aq) + 3H_2O(l)$$

and acts as an acid with sodium hydroxide:

$$Al_2O_3(s) + 2NaOH(aq) + 3H_2O(l) \longrightarrow 2NaAl(OH)_4(aq)$$

Only a salt, sodium aluminum hydroxide [$NaAl(OH)_4$, which contains the $Na^+$ and $Al(OH)_4^-$ ions] is formed in the reaction with sodium hydroxide—no water is produced. Nevertheless, the reaction is still classified as an acid-base reaction because $Al_2O_3$ neutralizes NaOH.

Some transition metal oxides in which the metal has a high oxidation number act as acidic oxides. Two examples are manganese(VII) oxide ($Mn_2O_7$) and chromium(VI) oxide ($CrO_3$), both of which react with water to produce acids:

$$Mn_2O_7(l) + H_2O(l) \longrightarrow 2HMnO_4(aq)$$
permanganic acid

$$CrO_3(s) + H_2O(l) \longrightarrow H_2CrO_4(aq)$$
chromic acid

## Basic and Amphoteric Hydroxides

All the alkali and alkaline earth metal hydroxides, except $Be(OH)_2$, are basic. $Be(OH)_2$, $Al(OH)_3$, $Sn(OH)_2$, $Pb(OH)_2$, $Cr(OH)_3$, $Cu(OH)_2$, $Zn(OH)_2$, and $Cd(OH)_2$ are amphoteric. All amphoteric hydroxides are insoluble, but beryllium hydroxide reacts with both acids and bases as follows:

$$Be(OH)_2(s) + 2H^+(aq) \longrightarrow Be^{2+}(aq) + 2H_2O(l)$$

$$Be(OH)_2(s) + 2OH^-(aq) \longrightarrow Be(OH)_4^{2-}(aq)$$

Aluminum hydroxide reacts with both acids and bases in a similar fashion:

$$Al(OH)_3(s) + 3H^+(aq) \longrightarrow Al^{3+}(aq) + 3H_2O(l)$$

$$Al(OH)_3(s) + OH^-(aq) \longrightarrow Al(OH)_4^-(aq)$$

## 16.12  Lewis Acids and Bases

So far we have discussed acid-base properties in terms of the Brønsted theory. For example, a Brønsted base is a substance that must be able to accept protons. By this definition, both the hydroxide ion and ammonia are bases:

In each case, the atom to which the proton becomes attached possesses at least one unshared pair of electrons. This characteristic property of $OH^-$, $NH_3$, and other Brønsted bases suggests a more general definition of acids and bases.

In 1932 G. N. Lewis defined what we now call a ***Lewis base*** as a substance that can donate a pair of electrons. A ***Lewis acid*** is a substance that can accept a pair of electrons. In the protonation of ammonia, for example, $NH_3$ acts as a Lewis base because it donates a pair of electrons to the proton $H^+$, which acts as a Lewis acid by accepting the pair of electrons. A Lewis acid-base reaction, therefore, is one that involves the donation of a pair of electrons from one species to another.

The significance of the Lewis concept is that it is more general than other definitions. Lewis acid-base reactions include many reactions that do not involve Brønsted acids. Consider, for example, the reaction between boron trifluoride ($BF_3$) and ammonia to form an adduct compound:

The B atom in $BF_3$ is $sp^2$-hybridized [◄◄ Section 9.4]. The vacant, unhybridized $2p_z$ orbital accepts the pair of electrons from $NH_3$. Thus, $BF_3$ functions as an acid according to the Lewis definition, even though it does not contain an ionizable proton. A *coordinate covalent* bond [◄◄ Section 8.8] is formed between the B and N atoms. In fact, every Lewis acid-base reaction results in the formation of a coordinate covalent bond.

Boric acid is another Lewis acid containing boron. Boric acid (a weak acid used in eyewash) is an oxoacid with the following structure:

Boric acid does not ionize in water to produce $H^+$. Instead, it produces $H^+$ in solution by taking a hydroxide ion away from water.

$$B(OH)_3(aq) + H_2O(l) \rightleftharpoons B(OH)_4^-(aq) + H^+(aq)$$

In this Lewis acid-base reaction, boric acid accepts a pair of electrons from the hydroxide ion that is derived from the water molecule, leaving behind the hydrogen ion.

The hydration of carbon dioxide to produce carbonic acid,

$$CO_2(g) + H_2O(l) \rightleftharpoons H_2CO_3(aq)$$

can be explained in terms of Lewis acid-base theory as well. The first step involves the donation of a lone pair on the O atom in $H_2O$ to the C atom in $CO_2$. An orbital is vacated on the C atom to accommodate the lone pair by relocation of the electron pair in one of the C–O pi bonds, changing the hybridization of the oxygen atom from $sp^2$ to $sp^3$:

As a result, $H_2O$ is a Lewis base and $CO_2$ is a Lewis acid. Finally, a proton is transferred onto the O atom bearing the negative charge to form $H_2CO_3$:

Other examples of Lewis acid-base reactions are

$$Ag^+(aq) + 2NH_3(aq) \rightleftharpoons Ag(NH_3)_2^+(aq)$$

$$Cd^{2+}(aq) + 4I^-(aq) \rightleftharpoons CdI_4^{2-}(aq)$$

$$Ni(s) + 4CO(g) \rightleftharpoons Ni(CO)_4(g)$$

The hydration of metal ions is in itself a Lewis acid-base reaction. When copper(II) sulfate ($CuSO_4$) dissolves in water, each $Cu^{2+}$ ion becomes associated with six water molecules as $Cu(H_2O)_6^{2+}$. In this case, $Cu^{2+}$ acts as the acid, accepting electrons, whereas $H_2O$ acts as the base, donating electrons.

Sample Problem 16.23 shows how to classify Lewis acids and bases.

## Sample Problem 16.23

Identify the Lewis acid and Lewis base in each of the following reactions:

(a) $C_2H_5OC_2H_5 + AlCl_3 \rightleftharpoons (C_2H_5)_2OAlCl_3$

(b) $Hg^{2+}(aq) + 4CN^-(aq) \rightleftharpoons Hg(CN)_4^{2-}(aq)$

**Strategy** Determine which species in each reaction accepts a pair of electrons (Lewis acid) and which species donates a pair of electrons (Lewis base).

**Setup** (a) It can be helpful to draw Lewis structures of the species involved:

$$
\begin{array}{cc}
CH_3CH_2 & Cl \\
\diagdown & | \\
\ddot{O} & Al \\
\diagup & \diagup \quad \diagdown \\
CH_3CH_2 & Cl \qquad Cl
\end{array}
$$

(b) Metal ions act as Lewis acids, accepting electron pairs from anions or molecules with lone pairs.

**Student Annotation:** An electron-deficient molecule is one with less than a complete octet around the central atom.

**Solution** (a) The Al is $sp^2$-hybridized in $AlCl_3$ with an empty $2p_z$ orbital. It is electron deficient, sharing only six electrons. Therefore, the Al atom has the capacity to gain two electrons to complete its octet. This property makes $AlCl_3$ a Lewis acid. On the other hand, the lone pairs on the oxygen atom in $C_2H_5OC_2H_5$ make the compound a Lewis base:

$$
\begin{array}{cc}
CH_3CH_2 & Cl \\
\diagdown & | \\
\ddot{O} \longrightarrow & Al \\
\diagup & \diagup \quad \diagdown \\
CH_3CH_2 & Cl \qquad Cl
\end{array}
$$

**Think About It** In Lewis acid-base reactions, the acid is usually a cation or an electron-deficient molecule, whereas the base is an anion or a molecule containing an atom with lone pairs.

(b) $Hg^{2+}$ accepts four pairs of electrons from the $CN^-$ ions. Therefore, $Hg^{2+}$ is the Lewis acid and $CN^-$ is the Lewis base.

**Practice Problem A** Identify the Lewis acid and Lewis base in the following reaction:

$$Co^{3+}(aq) + 6NH_3(aq) \rightleftharpoons Co(NH_3)_6^{3+}(aq)$$

**Practice Problem B** Write formulas for the Lewis acid and Lewis base that react to form $H_2NAlBr_3$.

## Checkpoint 16.12    Lewis Acids and Bases

**16.12.1** Which of the following cannot act as a Lewis base? (Select all that apply.)

a) $NH_3$

b) $OH^-$

c) $CH_4$

d) $Fe^{2+}$

e) $Al^{3+}$

**16.12.2** Which of the following is a Lewis acid but not a Brønsted acid? (Select all that apply.)

a) $H_2O$

b) $BCl_3$

c) $OH^-$

d) $Al^{3+}$

e) $NH_3$

# Applying What You've Learned

As was noted at the beginning of the chapter, rain in unpolluted areas is slightly acidic because it contains carbonic acid ($H_2CO_3$). Carbonic acid is a weak, diprotic acid that ionizes to give $H^+$ and the hydrogen carbonate ion:

$$H_2CO_3(aq) \rightleftharpoons H^+(aq) + HCO_3^-(aq)$$

a)  Give the formulas for the conjugate acid and the conjugate base of the hydrogen carbonate ion (also known as the bicarbonate ion). [◀◀ Sample Problem 16.1]

b)  Determine $[H^+]$ and pH for a raindrop in which the carbonic acid concentration is $1.8 \times 10^{-5}$ M. Assume that carbonic acid is the only acid in the raindrop and that the second ionization is negligible. [◀◀ Sample Problem 16.12]

c)  Calculate $[OH^-]$ and pOH of the raindrop in part (b). [◀◀ Sample Problems 16.3 and 16.6]

d)  What concentration of HCl would have the same pH as the raindrop in part (b)? [◀◀ Sample Problem 16.9]

e)  Determine $K_b$ for the hydrogen carbonate ion ($HCO_3^-$) and the pH of a 0.10 M solution of $NaHCO_3$. [◀◀ Sample Problems 16.17 and 16.21]

f)  Carbonic acid undergoes a second ionization to produce additional $H^+$ and the carbonate ion:

$$HCO_3^-(aq) \rightleftharpoons H^+(aq) + CO_3^{2-}(aq)$$

Calculate the concentrations of all species ($H_2CO_3$, $H^+$, $HCO_3^-$, and $CO_3^{2-}$) in the raindrop. [◀◀ Sample Problem 16.18]

g)  Sulfuric acid ($H_2SO_4$) accounts for as much as 80 percent of the acid in acid rain. Its first ionization is complete, producing $H^+$ and the hydrogen sulfate ion:

$$H_2SO_4(aq) \rightleftharpoons H^+(aq) + HSO_4^-(aq)$$

Its second ionization, which produces additional $H^+$ and the sulfate ion, has an ionization constant ($K_{a_2}$) of $1.3 \times 10^{-2}$.

$$HSO_4^-(aq) \rightleftharpoons H^+(aq) + SO_4^{2-}(aq)$$

Calculate the concentrations of all species in a raindrop in which the sulfuric acid concentration is $4.00 \times 10^{-5}$ M. Assume that sulfuric acid is the only acid present. [◀◀ Sample Problem 16.18]

h)  At low concentrations, the percent ionization of a weak acid such as the hydrogen sulfate ion can be quite high. What is the percent ionization of the hydrogen sulfate ion at the concentration calculated in part (g)? [◀◀ Sample Problem 16.13]

# CHAPTER SUMMARY

## Section 16.1

- A Brønsted acid donates a proton; a Brønsted base accepts a proton.

- When a Brønsted acid donates a proton, the anion that remains is a *conjugate base.*

- When a Brønsted base accepts a proton, the resulting cation is a *conjugate acid.*

- The combination of a Brønsted acid and its conjugate base (or the combination of a Brønsted base and its conjugate acid) is called a *conjugate pair.*

## Section 16.2

- Water is *amphoteric,* meaning it can act both as a Brønsted acid and a Brønsted base.

- Pure water undergoes *autoionization* (to a very small extent), resulting in concentrations of $H^+$ and $OH^-$ of $1.0 \times 10^{-7}$ $M$ at 25°C.

- $K_w$ is the equilibrium constant for the autoionization of water, also called the *ion-product constant.* $K_w = [H^+][OH^-] = 1.0 \times 10^{-14}$ at 25°C.

## Section 16.3

- The *pH* scale measures acidity: $pH = -\log [H^+]$.

- pH = 7.00 is neutral, pH < 7.00 is acidic, and pH > 7.00 is basic.

- The *pOH* scale is analogous to the pH scale, but it measures *basicity:* $pOH = -\log [OH^-]$.

- pOH = 7.00 is neutral, pOH < 7.00 is basic, and pOH > 7.00 is acidic.

- pH + pOH = 14.00 (at 25°C).

## Section 16.4

- There are seven strong acids: $HCl$, $HBr$, $HI$, $HNO_3$, $HClO_3$, $HClO_4$, and $H_2SO_4$.

- Strong acids ionize completely in aqueous solution. (Only the first ionization of the diprotic acid $H_2SO_4$ is complete.)

- The strong bases are the Group 1A and the heaviest Group 2A hydroxides: $LiOH$, $NaOH$, $KOH$, $RbOH$, $CsOH$, $Ca(OH)_2$, $Sr(OH)_2$, and $Ba(OH)_2$.

## Section 16.5

- A *weak acid* ionizes only partially. The *acid ionization constant, $K_a$,* is the equilibrium constant that indicates to what extent a weak acid ionizes.

- We solve for the pH of a solution of weak acid using the concentration of the acid, the $K_a$ value, and an equilibrium table.

- We can also determine the $K_a$ of a weak acid if we know the initial acid concentration and the pH at equilibrium.

## Section 16.6

- A *weak base* ionizes only partially. The *base ionization constant, $K_b$,* is the equilibrium constant that indicates to what extent a weak base ionizes.

- We solve for the pH of a solution of weak base using the concentration of the base, the $K_b$ value, and an equilibrium table.

- We can also determine the $K_b$ of a weak base if we know the concentration and the pH at equilibrium.

## Section 16.7

- The conjugate base of a strong acid is a *weak conjugate base,* meaning that it does not react with water.

- The conjugate base of a weak acid is a *strong conjugate base,* meaning that it acts as a weak Brønsted base in water.

- The conjugate base of a strong base is a *weak conjugate acid,* meaning that it does not react with water.

- The conjugate acid of a weak base is a *strong conjugate acid,* meaning that it acts as a weak Brønsted acid in water.

- For any conjugate acid-base pair, $K_a \times K_b = K_w$.

## Section 16.8

- Diprotic and polyprotic acids have more than one proton to donate. They undergo stepwise ionizations. Each ionization has a $K_a$ value associated with it.

- The $K_a$ values for stepwise ionizations become progressively smaller.

- In most cases, it is only necessary to consider the first ionization of an acid to determine pH. To determine the concentrations of other species at equilibrium, it may be necessary to consider subsequent ionizations.

## Section 16.9

- The strength of an acid is affected by molecular structure.

- Polar and weak bonds to the ionizable hydrogen lead to a stronger acid.

- Resonance stabilization of the conjugate base favors the ionization process, resulting in a stronger acid.

## Section 16.10

- Salts dissolve in water to give neutral, acidic, or basic solutions depending on their constituent ions. *Salt hydrolysis* is the reaction of an ion with water to produce hydronium or hydroxide ions.

- Cations that are strong conjugate acids such as $NH_4^+$ make a solution more acidic.

- Anions that are strong conjugate bases such as $F^-$ make a solution more basic. Anions that are conjugate bases of strong acids have no effect on pH.

- Small, highly charged metal ions hydrolyze to give acidic solutions.

## Section 16.11

- Oxides of metals generally are basic; oxides of nonmetals generally are acidic.

- Metal hydroxides may be basic or amphoteric.

## Section 16.12

- Lewis theory provides more general definitions of acids and bases.

- A *Lewis acid* accepts a pair of electrons; a *Lewis base* donates a pair of electrons.

- A Lewis acid is generally electron-poor and need not have a hydrogen atom.

- A Lewis base is an anion or a molecule with one or more lone pairs of electrons.

# KEY WORDS

Acid ionization constant ($K_a$), 686

Amphoteric, 675

Autoionization of water, 676

Base ionization constant ($K_b$), 695

Conjugate acid, 674

Conjugate base, 674

Conjugate pair, 674

Ion-product constant, 676

Lewis acid, 712

Lewis base, 712

pH, 677

pOH, 679

Salt hydrolysis, 706

Strong conjugate acid, 697

Strong conjugate base, 697

Weak acid, 686

Weak base, 694

Weak conjugate acid, 697

Weak conjugate base, 697

# KEY EQUATIONS

16.1   $K_w = [H_3O^+][OH^-] = 1.0 \times 10^{-14}$   (at 25°C)

16.2   $pH = -\log [H_3O^+]$   or   $pH = -\log [H^+]$

16.3   $[H_3O^+] = 10^{-pH}$   or   $[H^+] = 10^{-pH}$

16.4   $pOH = -\log [OH^-]$

16.5   $[OH^-] = 10^{-pOH}$

16.6   $pH + pOH = 14.00$

16.7   Percent ionization $= \dfrac{[H^+]_{eq}}{[HA]_0} \times 100\%$

16.8   $K_a \times K_b = K_w$

# QUESTIONS AND PROBLEMS

## Section 16.1: Brønsted Acids and Bases

### Review Questions

16.1   Define Brønsted acids and bases. Give an example of a conjugate pair in an acid-base reaction.

16.2   For a species to act as a Brønsted base, an atom in the species must possess a lone pair of electrons. Explain why this is so.

### Problems

16.3   Classify each of the following species as a Brønsted acid or base, or both: (a) $H_2O$, (b) $OH^-$, (c) $H_3O^+$, (d) $NH_3$, (e) $NH_4^+$, (f) $NH_2^-$, (g) $NO_3^-$, (h) $CO_3^{2-}$, (i) HBr, (j) HCN.

16.4   Identify the acid-base conjugate pairs in each of the following reactions:

(a) $CH_3COO^- + HCN \rightleftharpoons CH_3COOH + CN^-$
(b) $HCO_3^- + HCO_3^- \rightleftharpoons H_2CO_3 + CO_3^{2-}$
(c) $H_2PO_4^- + NH_3 \rightleftharpoons HPO_4^{2-} + NH_4^+$
(d) $HClO + CH_3NH_2 \rightleftharpoons CH_3NH_3^+ + ClO^-$
(e) $CO_3^{2-} + H_2O \rightleftharpoons HCO_3^- + OH^-$

16.5   Write the formulas of the conjugate bases of the following acids: (a) $HNO_2$, (b) $H_2SO_4$, (c) $H_2S$, (d) HCN, (e) HCOOH (formic acid).

16.6   Write the formula for the conjugate acid of each of the following bases: (a) $HS^-$, (b) $HCO_3^-$, (c) $CO_3^{2-}$, (d) $H_2PO_4^-$, (e) $HPO_4^{2-}$, (f) $PO_4^{3-}$, (g) $HSO_4^-$, (h) $SO_4^{2-}$, (i) $SO_3^{2-}$.

16.7   Write the formula for the conjugate base of each of the following acids: (a) $CH_2ClCOOH$, (b) $HIO_4$, (c) $H_3PO_4$, (d) $H_2PO_4^-$, (e) $HPO_4^{2-}$, (f) $H_2SO_4$, (g) $HSO_4^-$, (h) $HIO_3$, (i) $HSO_3^-$, (j) $NH_4^+$, (k) $H_2S$, (l) $HS^-$, (m) HClO.

16.8   Oxalic acid ($H_2C_2O_4$) has the following structure:

$$O=C-OH$$
$$|$$
$$O=C-OH$$

An oxalic acid solution contains the following species in varying concentrations: $H_2C_2O_4$, $HC_2O_4^-$, $C_2O_4^{2-}$, and $H^+$. (a) Draw Lewis structures of $HC_2O_4^-$ and $C_2O_4^{2-}$. (b) Which of the four species can act only as acids, which can act only as bases, and which can act as both acids and bases?

## Section 16.2: The Acid-Base Properties of Water

### Review Questions

16.9   Write the equilibrium expression for the autoionization of water and an equation relating $[H^+]$ and $[OH^-]$ in solution at 25°C.

16.10   In Section 15.3 we learned that when we multiply a chemical equation by 2, we must square its equilibrium constant. Explain why $K_w$ is the same ($1.0 \times 10^{-14}$ at 25°C) whether we start with one water molecule or two.

16.11   The equilibrium constant for the autoionization of water

$$H_2O(l) \rightleftharpoons H^+(aq) + OH^-(aq)$$

is $1.0 \times 10^{-14}$ at 25°C and $3.8 \times 10^{-14}$ at 40°C. Is the forward process endothermic or exothermic?

16.12   Define the term *amphoteric*.

16.13   Compare the magnitudes of $[H^+]$ and $[OH^-]$ in aqueous solutions that are acidic, basic, and neutral.

## Problems

16.14   Calculate the $OH^-$ concentration in an aqueous solution at 25°C with each of the following $H_3O^+$ concentrations: (a) $1.13 \times 10^{-4}$ *M*, (b) $4.55 \times 10^{-8}$ *M*, (c) $7.05 \times 10^{-11}$ *M*, (d) $3.13 \times 10^{-2}$ *M*.

**16.15**   Calculate the $H_3O^+$ concentration in an aqueous solution at 25°C with each of the following $OH^-$ concentrations: (a) $2.50 \times 10^{-2}$ *M*, (b) $1.67 \times 10^{-5}$ *M*, (c) $8.62 \times 10^{-3}$ *M*, (d) $1.75 \times 10^{-12}$ *M*.

16.16   The value of $K_w$ at 50°C is $5.48 \times 10^{-14}$. Calculate the $OH^-$ concentration in each of the aqueous solutions from Problem 16.14 at 50°C.

**16.17**   The value of $K_w$ at 100°C is $5.13 \times 10^{-13}$. Calculate the $H_3O^+$ concentration in each of the aqueous solutions from Problem 16.15 at 100°C.

## Section 16.3: The pH Scale

### Review Questions

16.18   Define pH. Why do chemists normally choose to discuss the acidity of a solution in terms of pH rather than hydrogen ion concentration $[H^+]$?

16.19   The pH of a solution is 6.7. From this statement alone, can you conclude that the solution is acidic? If not, what additional information would you need? Can the pH of a solution be zero or negative? If so, give examples to illustrate these values.

16.20   Define pOH. Write the equation relating pH and pOH.

### Problems

**16.21**   Calculate the concentration of $OH^-$ ions in a $1.4 \times 10^{-3}$ *M* HCl solution.

16.22   Calculate the concentration of $H^+$ ions in a 0.62 *M* NaOH solution.

**16.23**   Calculate the pH of each of the following solutions: (a) 0.0010 *M* HCl, (b) 0.76 *M* KOH.

**16.24**   Calculate the pH of each of the following solutions: (a) $2.8 \times 10^{-4}$ *M* $Ba(OH)_2$, (b) $5.2 \times 10^{-4}$ *M* $HNO_3$.

**16.25**   Calculate the hydrogen ion concentration in mol/L for solutions with the following pH values: (a) 2.42, (b) 11.21, (c) 6.96, (d) 15.00.

16.26   Calculate the hydrogen ion concentration in mol/L for each of the following solutions: (a) a solution whose pH is 5.20, (b) a solution whose pH is 16.00, (c) a solution whose hydroxide concentration is $3.7 \times 10^{-9}$ *M*.

16.27   Complete the following table for a solution:

| pH | $[H^+]$ | Solution is |
|---|---|---|
| <7 | | |
| | $<1.0 \times 10^{-7}$ *M* | |
| | | Neutral |

16.28   Fill in the word *acidic, basic,* or *neutral* for the following solutions:
(a) pOH > 7; solution is _____.
(b) pOH = 7; solution is _____.
(c) pOH < 7; solution is _____.

**16.29**   The pOH of a solution is 9.40 at 25°C. Calculate the hydrogen ion concentration of the solution.

**16.30**   Calculate the number of moles of KOH in 5.50 mL of a 0.360 *M* KOH solution. What is the pOH of the solution at 25°C?

**16.31**   How much NaOH (in grams) is needed to prepare 546 mL of solution with a pH of 10.00 at 25°C?

**16.32**   A solution is made by dissolving 18.4 g of HCl in enough water to make 662 mL of solution. Calculate the pH of the solution at 25°C.

## Section 16.4: Strong Acids and Bases

### Review Questions

16.33   Without referring to the text, write the formulas of four strong acids and four strong bases.

16.34   Which of the following statements are true regarding a 1.0 *M* solution of a strong acid HA at 25°C? (Choose all that apply.)
(a) $[A^-] > [H^+]$
(b) The pH is 0.00.
(c) $[H^+] = 1.0$ *M*
(d) $[HA] = 1.0$ *M*

16.35   Why are ionizations of strong acids and strong bases generally not treated as equilibria?

### Problems

**16.36**   Calculate the pH of an aqueous solution at 25°C that is (a) 0.12 *M* in HCl, (b) 2.4 *M* in $HNO_3$, and (c) $3.2 \times 10^{-4}$ *M* in $HClO_4$.

**16.37**   Calculate the pH of an aqueous solution at 25°C that is (a) 1.02 *M* in HI, (b) 0.035 *M* in $HClO_4$, and (c) $1.5 \times 10^{-6}$ *M* in HCl.

16.38   Calculate the concentration of HBr in a solution at 25°C that has a pH of (a) 0.12, (b) 2.46, and (c) 6.27.

**16.39**   Calculate the concentration of $HNO_3$ in a solution at 25°C that has a pH of (a) 4.21, (b) 3.55, and (c) 0.98.

**16.40**   Calculate the pOH and pH of the following aqueous solutions at 25°C: (a) 0.066 *M* KOH, (b) 5.43 *M* NaOH, (c) 0.74 *M* $Ba(OH)_2$.

**16.41**   Calculate the pOH and pH of the following aqueous solutions at 25°C: (a) 1.24 *M* LiOH, (b) 0.22 *M* $Ba(OH)_2$, (c) 0.085 *M* NaOH.

**16.42**   An aqueous solution of a strong base has a pH of 9.78 at 25°C. Calculate the concentration of the base if the base is (a) LiOH and (b) $Ba(OH)_2$.

**16.43**   An aqueous solution of a strong base has a pH of 11.04 at 25°C. Calculate the concentration of the base if the base is (a) KOH and (b) $Ba(OH)_2$.

## Section 16.5: Weak Acids and Acid Ionization Constants

### Visualizing Chemistry
### Figure 16.2

Three weak acid solutions are shown here.

(i)            (ii)           (iii)

VC 16.1  Which weak acid has the largest $K_a$ value?

    a) i
    b) ii
    c) iii

VC 16.2  Which weak acid has the highest pH?

    a) i
    b) ii
    c) iii

VC 16.3  For which weak acid solution can we most likely neglect $x$ in the denominator of the equilibrium expression in the determination of pH?

    a) i
    b) ii
    c) iii

VC 16.4  In the event that we cannot neglect $x$ in the denominator of the equilibrium expression, _____ to solve for pH.

    a) we cannot use an equilibrium table
    b) we must use the quadratic equation
    c) it is unnecessary

### Review Questions

16.44  Explain what is meant by the strength of an acid.

16.45  What does the ionization constant tell us about the strength of an acid?

16.46  List the factors on which the $K_a$ of a weak acid depends.

16.47  Why do we normally not quote $K_a$ values for strong acids such as HCl and HNO₃? Why is it necessary to specify temperature when giving $K_a$ values?

16.48  Which of the following solutions has the highest pH: (a) 0.40 $M$ HCOOH, (b) 0.40 $M$ HClO₄, (c) 0.40 $M$ CH₃COOH?

16.49  Without referring to the text, write the formulas of four weak acids.

### Problems

16.50  Classify each of the following species as a weak or strong acid: (a) HNO₃, (b) HF, (c) H₂SO₄, (d) HSO₄⁻, (e) H₂CO₃, (f) HCO₃⁻, (g) HCl, (h) HCN, (i) HNO₂.

16.51  Classify each of the following species as a weak or strong base: (a) LiOH, (b) CN⁻, (c) H₂O, (d) ClO₄⁻, (e) NH₂⁻.

16.52  Which of the following statements are true for a 0.10 $M$ solution of a weak acid HA? (Choose all that apply.)

    (a) The pH is 1.00.
    (b) $[H^+] \gg [A^-]$.
    (c) $[H^+] = [A^-]$.
    (d) The pH is less than 1.

16.53  The $K_a$ for benzoic acid is $6.5 \times 10^{-5}$. Calculate the pH of a 0.10 $M$ aqueous solution of benzoic acid at 25°C.

16.54  The $K_a$ for hydrofluoric acid is $7.1 \times 10^{-4}$. Calculate the pH of a 0.15 $M$ aqueous solution of hydrofluoric acid at 25°C.

16.55  Calculate the pH of an aqueous solution at 25°C that is 0.095 $M$ in hydrocyanic acid (HCN). ($K_a$ for hydrocyanic acid = $4.9 \times 10^{-10}$.)

16.56  Calculate the pH of an aqueous solution at 25°C that is 0.34 $M$ in phenol (C₆H₅OH). ($K_a$ for phenol = $1.3 \times 10^{-10}$.)

16.57  Determine the percent ionization of the following solutions of formic acid at 25°C: (a) 0.016 $M$, (b) $5.7 \times 10^{-4}$ $M$, (c) 1.75 $M$.

16.58  Determine the percent ionization of the following solutions of phenol: (a) 0.56 $M$, (b) 0.25 $M$, (c) $1.8 \times 10^{-6}$ $M$.

16.59  A 0.015 $M$ solution of a monoprotic acid is 0.92 percent ionized. Calculate the ionization constant for the acid.

16.60  Calculate the concentration at which a monoprotic acid with $K_a = 4.5 \times 10^{-5}$ will be 2.5 percent ionized.

16.61  Calculate the $K_a$ of a weak acid if a 0.19 $M$ aqueous solution of the acid has a pH of 4.52 at 25°C.

16.62  The pH of an aqueous acid solution is 6.20 at 25°C. Calculate the $K_a$ for the acid. The initial acid concentration is 0.010 $M$.

16.63  What is the original molarity of a solution of formic acid (HCOOH) whose pH is 3.26 at 25°C? ($K_a$ for formic acid = $1.7 \times 10^{-4}$.)

16.64  What is the original molarity of a solution of a weak acid whose $K_a$ is $3.5 \times 10^{-5}$ and whose pH is 5.26 at 25°C?

16.65  In biological and medical applications, it is often necessary to study the autoionization of water at 37°C instead of 25°C. Given that $K_w$ for water is $2.5 \times 10^{-14}$ at 37°C, calculate the pH of pure water at this temperature.

## Section 16.6: Weak Bases and Base Ionization Constants

### Review Questions

16.66  Compare the pH values for 0.10 $M$ solutions of NaOH and of NH₃ to illustrate the difference between a strong base and a weak base.

16.67  Which of the following has a higher pH: (a) 1.0 $M$ NH₃, (b) 0.20 $M$ NaOH ($K_b$ for NH₃ = $1.8 \times 10^{-5}$)?

### Problems

16.68  Calculate the pH for each of the following solutions at 25°C: (a) 0.10 $M$ NH₃, (b) 0.050 $M$ C₅H₅N (pyridine). ($K_b$ for pyridine = $1.7 \times 10^{-9}$.)

16.69  The pH of a 0.30 $M$ solution of a weak base is 10.66 at 25°C. What is the $K_b$ of the base?

**16.70** What is the original molarity of an aqueous solution of ammonia ($NH_3$) whose pH is 11.22 at 25°C? (Use the $K_b$ for ammonia provided in Problem 16.67.)

**16.71** Calculate the pH at 25°C of a 0.61 M aqueous solution of a weak base B with a $K_b$ of $1.5 \times 10^{-4}$.

**16.72** Determine the $K_b$ of a weak base if a 0.19 M aqueous solution of the base at 25°C has a pH of 10.88.

**16.73** The following diagrams represent aqueous solutions of three different monoprotic acids: HA, HB, and HC. (a) Which conjugate base ($A^-$, $B^-$, or $C^-$) has the smallest $K_b$ value? (b) Which anion is the strongest base? The water molecules have been omitted for clarity.

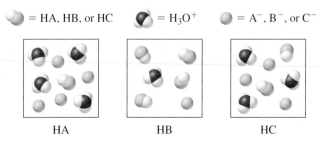

HA          HB          HC

## Section 16.7: Conjugate Acid-Base Pairs

### Review Questions

**16.74** Write the equation relating $K_a$ for a weak acid and $K_b$ for its conjugate base. Use $NH_3$ and its conjugate acid $NH_4^+$ to derive the relationship between $K_a$ and $K_b$.

**16.75** From the relationship $K_aK_b = K_w$, what can you deduce about the relative strengths of a weak acid and its conjugate base?

### Problems

**16.76** The following diagrams represent solutions of three salts NaX (X = A, B, or C). (a) Which $X^-$ has the weakest conjugate acid? (b) Arrange the three $X^-$ anions in order of decreasing base strength. The $Na^+$ ion and water molecules have been omitted for clarity.

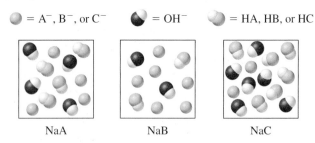

NaA          NaB          NaC

**16.77** Calculate $K_b$ for each of the following ions: $CN^-$, $F^-$, $CH_3COO^-$, $HCO_3^-$.

## Section 16.8: Diprotic and Polyprotic Acids

### Review Questions

**16.78** Write all the species (except water) that are present in a phosphoric acid solution. Indicate which species can act as a Brønsted acid, which as a Brønsted base, and which as both a Brønsted acid and a Brønsted base.

### Problems

**16.79** (1) Which of the following diagrams represents a solution of a weak diprotic acid? (2) Which diagrams represent chemically implausible situations? (The hydrated proton is shown as a hydronium ion. Water molecules are omitted for clarity.)

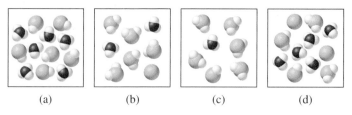

(a)          (b)          (c)          (d)

**16.80** The first and second ionization constants of a diprotic acid $H_2A$ are $K_{a_1}$ and $K_{a_2}$ at a certain temperature. Under what conditions will $[A^{2-}] = K_{a_1}$?

**16.81** Compare the pH of a 0.040 M HCl solution with that of a 0.040 M $H_2SO_4$ solution. (*Hint:* $H_2SO_4$ is a strong acid; $K_a$ for $HSO_4^- = 1.3 \times 10^{-2}$.)

**16.82** What are the concentrations of $HSO_4^-$, $SO_4^{2-}$, and $H^+$ in a 0.20 M $KHSO_4$ solution? (*Hint:* $H_2SO_4$ is a strong acid; $K_a$ for $HSO_4^- = 1.3 \times 10^{-2}$.)

**16.83** Calculate the concentrations of $H^+$, $HCO_3^-$, and $CO_3^{2-}$ in a 0.025 M $H_2CO_3$ solution.

**16.84** Calculate the pH at 25°C of a 0.25 M aqueous solution of phosphoric acid ($H_3PO_4$). ($K_{a_1}$, $K_{a_2}$, and $K_{a_3}$ for phosphoric acid are $7.5 \times 10^{-3}$, $6.25 \times 10^{-8}$, and $4.8 \times 10^{-13}$, respectively.)

**16.85** Calculate the pH at 25°C of a 0.25 M aqueous solution of oxalic acid ($H_2C_2O_4$). ($K_{a_1}$ and $K_{a_2}$ for oxalic acid are $6.5 \times 10^{-2}$ and $6.1 \times 10^{-5}$, respectively.)

## Section 16.9: Molecular Structure and Acid Strength

### Review Questions

**16.86** List four factors that affect the strength of an acid.

**16.87** How does the strength of an oxoacid depend on the electronegativity and oxidation number of the central atom?

### Problems

**16.88** Predict the relative acid strengths of the following compounds: $H_2O$, $H_2S$, and $H_2Se$.

**16.89** Compare the strengths of the following pairs of acids: (a) $H_2SO_4$ and $H_2SeO_4$, (b) $H_3PO_4$ and $H_3AsO_4$.

**16.90** Which of the following is the stronger acid: $CH_2ClCOOH$ or $CHCl_2COOH$? Explain your choice.

**16.91** Consider the following compounds:

$$\text{Phenol} \qquad CH_3—O—H$$

Phenol          Methanol

Experimentally, phenol is found to be a stronger acid than methanol. Explain this difference in terms of the structures of the conjugate bases. (*Hint:* A more stable conjugate base favors ionization. Only one of the conjugate bases can be stabilized by resonance.)

## Section 16.10: Acid-Base Properties of Salt Solutions

### Review Questions

16.92    Define salt hydrolysis. Categorize salts according to how they affect the pH of a solution.

16.93    Explain why small, highly charged metal ions are able to undergo hydrolysis.

16.94    $Al^{3+}$ is not a Brønsted acid, but $Al(H_2O)_6^{3+}$ is. Explain.

16.95    Specify which of the following salts will undergo hydrolysis: KF, $NaNO_3$, $NH_4NO_2$, $MgSO_4$, KCN, $C_6H_5COONa$, RbI, $Na_2CO_3$, $CaCl_2$, HCOOK.

### Problems

16.96    Predict the pH ($>7$, $<7$, or $\approx 7$) of aqueous solutions containing the following salts: (a) KBr, (b) $Al(NO_3)_3$, (c) $BaCl_2$, (d) $Bi(NO_3)_3$.

16.97    Predict whether the following solutions are acidic, basic, or nearly neutral: (a) NaBr, (b) $K_2SO_3$, (c) $NH_4NO_2$, (d) $Cr(NO_3)_3$.

16.98    A certain salt, MX (containing the $M^+$ and $X^-$ ions), is dissolved in water, and the pH of the resulting solution is 7.0. What can you say about the strengths of the acid and the base from which the salt is derived?

16.99    In a certain experiment, a student finds that the pHs of 0.10 $M$ solutions of three potassium salts KX, KY, and KZ are 7.0, 9.0, and 11.0, respectively. Arrange the acids HX, HY, and HZ in order of increasing acid strength.

16.100    Predict whether a solution containing the salt $K_2HPO_4$ will be acidic, neutral, or basic.

16.101    Predict the pH ($>7$, $<7$, or $\approx 7$) of a $NaHCO_3$ solution.

16.102    Calculate the pH of a 0.36 $M$ $CH_3COONa$ solution. ($K_a$ for acetic acid = $1.8 \times 10^{-5}$.)

16.103    Calculate the pH of a 0.42 $M$ $NH_4Cl$ solution. ($K_b$ for ammonia = $1.8 \times 10^{-5}$.)

16.104    Calculate the pH of a 0.082 $M$ NaF solution. ($K_a$ for HF = $7.1 \times 10^{-4}$.)

16.105    Calculate the pH of a 0.91 $M$ $C_2H_5NH_3I$ solution. ($K_b$ for $C_2H_5NH_2$ = $5.6 \times 10^{-4}$.)

## Section 16.11: Acid-Base Properties of Oxides and Hydroxides

### Review Questions

16.106    Classify the following oxides as acidic, basic, amphoteric, or neutral: (a) $CO_2$, (b) $K_2O$, (c) CaO, (d) $N_2O_5$, (e) CO, (f) NO, (g) $SnO_2$, (h) $SO_3$, (i) $Al_2O_3$, (j) BaO.

16.107    Write equations for the reactions between (a) $CO_2$ and NaOH($aq$), (b) $Na_2O$ and $HNO_3$($aq$).

### Problems

16.108    Explain why metal oxides tend to be basic if the oxidation number of the metal is low and tend to be acidic if the oxidation number of the metal is high. (*Hint:* Metallic compounds in which the oxidation numbers of the metals are low are more ionic than those in which the oxidation numbers of the metals are high.)

16.109    Arrange the oxides in each of the following groups in order of increasing basicity: (a) $K_2O$, $Al_2O_3$, BaO, (b) $CrO_3$, CrO, $Cr_2O_3$.

16.110    $Zn(OH)_2$ is an amphoteric hydroxide. Write balanced ionic equations to show its reaction with (a) HCl, (b) NaOH [the product is $Zn(OH)_4^{2-}$].

16.111    $Al(OH)_3$ is insoluble in water. It dissolves in concentrated NaOH solution. Write a balanced ionic equation for this reaction. What type of reaction is this?

## Section 16.12: Lewis Acids and Bases

### Review Questions

16.112    What are the Lewis definitions of an acid and a base? In what way are they more general than the Brønsted definitions?

16.113    In terms of orbitals and electron arrangements, what must be present for a molecule or an ion to act as a Lewis acid (use $H^+$ and $BF_3$ as examples)? What must be present for a molecule or ion to act as a Lewis base (use $OH^-$ and $NH_3$ as examples)?

### Problems

16.114    Classify each of the following species as a Lewis acid or a Lewis base: (a) $CO_2$, (b) $H_2O$, (c) $I^-$, (d) $SO_2$, (e) $NH_3$, (f) $OH^-$, (g) $H^+$, (h) $BCl_3$.

16.115    Describe the following reaction in terms of the Lewis theory of acids and bases:

$$AlCl_3(s) + Cl^-(aq) \longrightarrow AlCl_4^-(aq)$$

16.116    Which would be considered a stronger Lewis acid: (a) $BF_3$ or $BCl_3$, (b) $Fe^{2+}$ or $Fe^{3+}$? Explain.

16.117    All Brønsted acids are Lewis acids, but the reverse is not true. Give two examples of Lewis acids that are not Brønsted acids.

16.118    Identify the Lewis acid and the Lewis base in the following reactions:

(a) $5CO(g) + Fe(s) \longrightarrow Fe(CO)_5(l)$
(b) $NH_3(g) + BCl_3(g) \longrightarrow Cl_3BNH_3(s)$
(c) $Hg^{2+}(aq) + 4I^-(aq) \longrightarrow HgI_4^{2-}(aq)$

16.119    Identify the Lewis acid and the Lewis base in the following reactions:

(a) $AlBr_3(s) + Br^-(aq) \longrightarrow AlBr_4^-(aq)$
(b) $6CO(g) + Cr(s) \longrightarrow Cr(CO)_6(s)$
(c) $Cu^{2+}(aq) + 4CN^-(aq) \longrightarrow Cu(CN)_4^{2-}(aq)$

## Additional Problems

16.120    $H_2SO_4$ is a strong acid, but $HSO_4^-$ is a weak acid. Account for the difference in strength of these two related species.

16.121    Which of the following diagrams best represents a strong acid, such as HCl, dissolved in water? Which represents a weak acid? Which represents a very weak acid? (The hydrated proton is shown as a hydronium ion. Water molecules are omitted for clarity.)

(a)          (b)          (c)          (d)

**16.122**  Predict the direction that predominates in this reaction:

$$F^-(aq) + H_2O(l) \rightleftharpoons HF(aq) + OH^-(aq)$$

**16.123**  Predict the products and tell whether the following reaction will occur to any measurable extent:

$$CH_3COOH(aq) + Cl^-(aq) \longrightarrow$$

**16.124**  In a 0.080 $M$ $NH_3$ solution, what percent of the $NH_3$ is present as $NH_4^+$?

**16.125**  Calculate the pH and percent ionization of a 0.88 $M$ $HNO_2$ solution at 25°C.

**16.126**  A typical reaction between an antacid and the hydrochloric acid in gastric juice is

$$NaHCO_3(s) + HCl(aq) \rightleftharpoons NaCl(aq) + H_2O(l) + CO_2(g)$$

Calculate the volume (in liters) of $CO_2$ generated from 0.350 g of $NaHCO_3$ and excess gastric juice at 1.00 atm and 37.0°C.

**16.127**  To which of the following would the addition of an equal volume of 0.60 $M$ NaOH lead to a solution having a lower pH: (a) water, (b) 0.30 $M$ HCl, (c) 0.70 $M$ KOH, (d) 0.40 $M$ $NaNO_3$?

**16.128**  The pH of a 0.0642 $M$ solution of a monoprotic acid is 3.86. Is this a strong acid?

**16.129**  Like water, liquid ammonia undergoes autoionization:

$$NH_3 + NH_3 \rightleftharpoons NH_4^+ + NH_2^-$$

(a) Identify the Brønsted acids and Brønsted bases in this reaction. (b) What species correspond to $H^+$ and $OH^-$, and what is the condition for a neutral solution?

**16.130**  HA and HB are both weak acids although HB is the stronger of the two. Will it take a larger volume of a 0.10 $M$ NaOH solution to neutralize 50.0 mL of 0.10 $M$ HB than would be needed to neutralize 50.0 mL of 0.10 $M$ HA?

**16.131**  A solution contains a weak monoprotic acid HA and its sodium salt NaA both at 0.1 $M$ concentration. Show that $[OH^-] = K_w/K_a$.

**16.132**  The three common chromium oxides are CrO, $Cr_2O_3$, and $CrO_3$. If $Cr_2O_3$ is amphoteric, what can you say about the acid-base properties of CrO and $CrO_3$?

**16.133**  Use the data in Table 16.6 to calculate the equilibrium constant for the following reaction:

$$HCOOH(aq) + OH^-(aq) \rightleftharpoons HCOO^-(aq) + H_2O(l)$$

**16.134**  Use the data in Table 16.6 to calculate the equilibrium constant for the following reaction:

$$CH_3COOH(aq) + NO_2^-\,aq) \rightleftharpoons CH_3COO^-(aq) + HNO_2(aq)$$

**16.135**  Most of the hydrides of Group 1A and Group 2A metals are ionic (the exceptions are $BeH_2$ and $MgH_2$, which are covalent compounds). (a) Describe the reaction between the hydride ion ($H^-$) and water in terms of a Brønsted acid-base reaction. (b) The same reaction can also be classified as a redox reaction. Identify the oxidizing and reducing agents.

**16.136**  Calculate the pH of a 0.20 $M$ ammonium acetate ($CH_3COONH_4$) solution.

**16.137**  Novocaine, used as a local anesthetic by dentists, is a weak base ($K_b = 8.91 \times 10^{-6}$). What is the ratio of the concentration of the base to that of its acid in the blood plasma (pH = 7.40) of a patient? (As an approximation, use the $K_a$ values at 25°C.)

**16.138**  Which of the following is the stronger base: $NF_3$ or $NH_3$? (*Hint:* F is more electronegative than H.)

**16.139**  Which of the following is a stronger base: $NH_3$ or $PH_3$? (*Hint:* The N—H bond is stronger than the P—H bond.)

**16.140**  The ion product of $D_2O$ is $1.35 \times 10^{-15}$ at 25°C. (a) Calculate pD where pD = $-\log[D^+]$. (b) For what values of pD will a solution be acidic in $D_2O$? (c) Derive a relation between pD and pOD.

**16.141**  Give an example of (a) a weak acid that contains oxygen atoms, (b) a weak acid that does not contain oxygen atoms, (c) a neutral molecule that acts as a Lewis acid, (d) a neutral molecule that acts as a Lewis base, (e) a weak acid that contains two ionizable H atoms, (f) a conjugate acid-base pair, both of which react with HCl to give carbon dioxide gas.

**16.142**  What is the pH of 250.0 mL of an aqueous solution containing 0.616 g of a strong acid?

**16.143**  (a) Use VSEPR to predict the geometry of the hydronium ion ($H_3O^+$). (b) The O atom in $H_2O$ has two lone pairs and in principle can accept two $H^+$ ions. Explain why the species $H_4O^{2+}$ does not exist. What would be its geometry if it did exist?

**16.144**  HF is a weak acid, but its strength increases with concentration. Explain. (*Hint:* $F^-$ reacts with HF to form $HF_2^-$. The equilibrium constant for this reaction is 5.2 at 25°C.)

**16.145**  When chlorine reacts with water, the resulting solution is weakly acidic and reacts with $AgNO_3$ to give a white precipitate. Write balanced equations to represent these reactions. Explain why manufacturers of household bleaches add bases such as NaOH to their products to increase their effectiveness.

**16.146**  When the concentration of a strong acid is not substantially higher than $1.0 \times 10^{-7}$ $M$, the ionization of water must be taken into account in the calculation of the solution's pH. (a) Derive an expression for the pH of a strong acid solution, including the contribution to $[H^+]$ from $H_2O$. (b) Calculate the pH of a $1.0 \times 10^{-7}$ $M$ HCl solution.

**16.147**  Calculate the pH of a 2.00 $M$ $NH_4CN$ solution.

**16.148**  Calculate the concentrations of all species in a 0.100 $M$ $H_3PO_4$ solution.

**16.149**  In the vapor phase, acetic acid molecules associate to a certain extent to form dimers:

$$2CH_3COOH(g) \rightleftharpoons (CH_3COOH)_2(g)$$

At 51°C, the pressure of a certain acetic acid vapor system is 0.0342 atm in a 360-mL flask. The vapor is condensed and neutralized with 13.8 mL of 0.0568 $M$ NaOH. (a) Calculate the degree of dissociation ($\alpha$) of the dimer under these conditions:

$$(CH_3COOH)_2 \rightleftharpoons 2CH_3COOH$$

(b) Calculate the equilibrium constant $K_P$ for the reaction in part (a).

**16.150**  Calculate the concentrations of all the species in a 0.100 $M$ $Na_2CO_3$ solution.

**16.151** Henry's law constant for $CO_2$ at 38°C is $2.28 \times 10^{-3}$ mol/ L · atm. Calculate the pH of a solution of $CO_2$ at 38°C in equilibrium with the gas at a partial pressure of 3.20 atm.

16.152 Hydrocyanic acid (HCN) is a weak acid and a deadly poisonous compound—in the gaseous form (hydrogen cyanide), it is used in gas chambers. Why is it dangerous to treat sodium cyanide with acids (such as HCl) without proper ventilation?

**16.153** How many grams of NaCN would you need to dissolve in enough water to make exactly 250 mL of solution with a pH of 10.00?

**16.154** A solution of formic acid (HCOOH) has a pH of 2.53. How many grams of formic acid are there in 100.0 mL of the solution?

**16.155** Calculate the pH of a 1-L solution containing 0.150 mole of $CH_3COOH$ and 0.100 mole of HCl.

16.156 A 1.87-g sample of Mg reacts with 80.0 mL of a HCl solution whose pH is −0.544. What is the pH of the solution after all the Mg has reacted? Assume constant volume.

**16.157** You are given two beakers, one containing an aqueous solution of strong acid (HA) and the other an aqueous solution of weak acid (HB) of the same concentration. Describe how you would compare the strengths of these two acids by (a) measuring the pH, (b) measuring electrical conductance, and (c) studying the rate of hydrogen gas evolution when these solutions are combined with an active metal such as Mg or Zn.

16.158 Use Le Châtelier's principle to predict the effect of the following changes on the extent of hydrolysis of sodium nitrite ($NaNO_2$) solution: (a) HCl is added, (b) NaOH is added, (c) NaCl is added, (d) the solution is diluted.

**16.159** A 0.400 $M$ formic acid (HCOOH) solution freezes at −0.758°C. Calculate the $K_a$ of the acid at that temperature. (*Hint:* Assume that molarity is equal to molality. Carry out your calculations to three significant figures and round off to two for $K_a$.)

16.160 The disagreeable odor of fish is mainly due to organic compounds ($RNH_2$) containing an amino group, $-NH_2$, where R is the rest of the molecule. Amines are bases just like ammonia. Explain why putting some lemon juice on fish can greatly reduce the odor.

**16.161** A solution of methylamine ($CH_3NH_2$) has a pH of 10.64. How many grams of methylamine are there in 100.0 mL of the solution?

16.162 Describe the hydration of $SO_2$ as a Lewis acid-base reaction.

**16.163** Both the amide ion ($NH_2^-$) and the nitride ion ($N^{3-}$) are stronger bases than the hydroxide ion and hence do not exist in aqueous solutions. (a) Write equations showing the reactions of these ions with water, and identify the Brønsted acid and base in each case. (b) Which of the two is the stronger base?

16.164 The atmospheric sulfur dioxide ($SO_2$) concentration over a certain region is 0.12 ppm by volume. Calculate the pH of the rainwater due to this pollutant. Assume that the dissolution of $SO_2$ does not affect its pressure. ($K_a$ for $H_2SO_3 = 1.3 \times 10^{-2}$.)

**16.165** Explain the action of smelling salt, which is ammonium carbonate [$(NH_4)_2CO_3$]. (*Hint:* The thin film of aqueous solution that lines the nasal passages is slightly basic.)

16.166 About half of the hydrochloric acid produced annually in the United States (3.0 billion pounds) is used in metal pickling. This process involves the removal of metal oxide layers from metal surfaces to prepare them for coating. (a) Write the overall and net ionic equations for the reaction between iron(III) oxide, which represents the rust layer over iron, and HCl. Identify the Brønsted acid and base. (b) Hydrochloric acid is also used to remove scale (which is mostly $CaCO_3$) from water pipes. Hydrochloric acid reacts with calcium carbonate in two stages; the first stage forms the bicarbonate ion, which then reacts further to form carbon dioxide. Write equations for these two stages and for the overall reaction. (c) Hydrochloric acid is used to recover oil from the ground. It dissolves rocks (often $CaCO_3$) so that the oil can flow more easily. In one process, a 15 percent (by mass) HCl solution is injected into an oil well to dissolve the rocks. If the density of the acid solution is 1.073 g/mL, what is the pH of the solution?

16.167 Which of the following does not represent a Lewis acid-base reaction?

(a) $H_2O + H^+ \longrightarrow H_3O^+$
(b) $NH_3 + BF_3 \longrightarrow H_3NBF_3$
(c) $PF_3 + F_2 \longrightarrow PF_5$
(d) $Al(OH)_3 + OH^- \longrightarrow Al(OH)_4^-$

16.168 Determine whether each of the following statements is true or false. If false, explain why the statement is wrong. (a) All Lewis acids are Brønsted acids. (b) The conjugate base of an acid always carries a negative charge. (c) The percent ionization of a base increases with its concentration in solution. (d) A solution of barium fluoride is acidic.

**16.169** How many milliliters of a strong monoprotic acid solution at pH = 4.12 must be added to 528 mL of the same acid solution at pH = 5.76 to change its pH to 5.34? Assume that the volumes are additive.

16.170 Hemoglobin (Hb) is a blood protein that is responsible for transporting oxygen. It can exist in the protonated form as $HbH^+$. The binding of oxygen can be represented by the simplified equation

$$HbH^+ + O_2 \rightleftharpoons HbO_2 + H^+$$

(a) What form of hemoglobin is favored in the lungs where oxygen concentration is highest? (b) In body tissues, where the cells release carbon dioxide produced by metabolism, the blood is more acidic due to the formation of carbonic acid. What form of hemoglobin is favored under this condition? (c) When a person hyperventilates, the concentration of $CO_2$ in his or her blood decreases. How does this action affect the given equilibrium? Frequently a person who is hyperventilating is advised to breathe into a paper bag. Why does this action help the individual?

**16.171** A 20.27-g sample of a metal carbonate ($MCO_3$) is combined with 500 mL of a 1.00 $M$ HCl solution. The excess HCl acid is then neutralized by 32.80 mL of 0.588 $M$ NaOH. Identify M.

16.172 Calculate the pH of a solution that is 1.00 $M$ HCN and 1.00 $M$ HF. Compare the concentration (in molarity) of the $CN^-$ ion in this solution with that in a 1.00 $M$ HCN solution. Comment on the difference.

**16.173**    Tooth enamel is largely hydroxyapatite [$Ca_3(PO_4)_3OH$]. When it dissolves in water (a process called *demineralization*), it dissociates as follows:

$$Ca_5(PO_4)_3OH \longrightarrow 5Ca^{2+} + 3PO_4^{3-} + OH^-$$

The reverse process, called *remineralization,* is the body's natural defense against tooth decay. Acids produced from food remove the $OH^-$ ions and thereby weaken the enamel layer. Most toothpastes contain a fluoride compound such as NaF or $SnF_2$. What is the function of these compounds in preventing tooth decay?

**16.174**    Use the van't Hoff equation (see Problem 15.118) and the data in Appendix 2 to calculate the pH of water at its normal boiling point.

**16.175**    At 28° and 0.982 atm, gaseous compound HA has a density of 1.16 g/L. A quantity of 2.03 g of this compound is dissolved in water and diluted to exactly 1 L. If the pH of the solution at 25°C is 5.22 (due to ionization of HA), calculate the $K_a$ of the acid.

**16.176**    A 10.0-g sample of white phosphorus was burned in an excess of oxygen. The product was dissolved in enough water to make 500.0 mL of solution. Calculate the pH of the solution at 25°C.

# PRE-PROFESSIONAL PRACTICE EXAM PROBLEMS: PHYSICAL AND BIOLOGICAL SCIENCES

The following questions are not based on a descriptive passage.

1.    Which of the following best describes the titration of a weak acid with a strong base?

   a) Low starting pH, and pH ≈ 7 at equivalence point
   b) High starting pH, and pH ≈ 7 at equivalence point
   c) High starting pH, and pH < 7 at equivalence point
   d) Low starting pH, and pH > 7 at equivalence point

2.    What is the pH of a solution made by combining 150 mL of 0.125 $M$ HCl and 150 mL of 0.120 $M$ $Ba(OH)_2$ at 25°C?

   a) 1.240
   b) 12.760
   c) 2.602
   d) 11.398

3.    What volume of 0.085 $M$ NaOH is necessary to titrate to the equivalence point 25.0 mL of 0.15 $M$ HCN at 25°C?

   a) 44 mL
   b) 88 mL
   c) 22 mL
   d) 25 mL

4.    Determine the pH at the equivalence point for the titration in Question 3.

   a) 2.98
   b) 11.02
   c) 8.71
   d) 5.29

# Answers to In-Chapter Materials

## Answers to Practice Problems

**16.1A** (a) $HClO_4$, (b) $HS^-$, (c) $HS^-$, (d) $HC_2O_4^-$. **16.1B** $SO_3^{2-}$, $H_2SO_3$.
**16.2A** (a) $NH_4^+$ acid, $H_2O$ base, $NH_3$ conjugate base, $H_3O^+$ conjugate
acid; (b) $CN^-$ base, $H_2O$ acid, $HCN$ conjugate acid, $OH^-$ conjugate base.
**16.2B** (a) $HSO_4^- + H_2O \longrightarrow H_3O^+ + SO_4^{2-}$, (b) $HSO_4^- + H_2O \longrightarrow$
$H_2SO_4 + OH^-$. **16.3A** $2.0 \times 10^{-11}$ M. **16.3B** $2.8 \times 10^{-13}$ M. **16.4A**
(a) 8.49, (b) 7.40, (c) 1.25. **16.4B** (a) $-0.08$, (b) 10.52, (c) 11.08. **16.5A**
(a) $1.3 \times 10^{-10}$ M, (b) $3.5 \times 10^{-2}$ M, (c) $9.8 \times 10^{-8}$ M. **16.5B** (a) $7.8 \times$
$10^{-3}$ M, (b) $2.6 \times 10^{-12}$ M, (c) $1.3 \times 10^{-7}$ M. **16.6A** (a) 11.24, (b) 2.14,
(c) 5.07. **16.6B** (a) 6.45, (b) 6.00, (c) 3.00. **16.7A** (a) $9.5 \times 10^{-14}$ M,
(b) $7.2 \times 10^{-6}$ M, (c) $1.0 \times 10^{-7}$ M. **16.7B** (a) $5.5 \times 10^{-12}$ M, (b) $2.0 \times$
$10^{-4}$ M, (c) $2.5 \times 10^{-2}$ M. **16.8A** (a) 1.09, (b) 5.09, (c) 2.27. **16.8B**
(a) 12.04, (b) 11.54, (c) 7.00. **16.9A** (a) $8.7 \times 10^{-3}$ M, (b) $1.7 \times 10^{-2}$ M,
(c) $9.8 \times 10^{-7}$ M. **16.9B** (a) $1.5 \times 10^{-5}$ M, (b) $1.5 \times 10^{-2}$ M, (c) $9.1 \times$
$10^{-4}$ M. **16.10A** (a) 0.82, (b) 2.08, (c) 4.77. **16.10B** (a) 7.29, (b) 12.79,
(c) 13.09. **16.11A** (a) $9.5 \times 10^{-6}$ M, (b) $4.8 \times 10^{-6}$ M. **16.11B** (a) $1.7 \times$
$10^{-2}$ M, (b) $8.7 \times 10^{-3}$ M. **16.12A** 2.89. **16.12B** 3.05. **16.13A** (a) 4.96,
0.0044%, (b) 5.72, 0.026%, (c) 6.70, 0.24%. **16.13B** (a) $2.0 \times 10^{-3}$ M,
(b) $4.9 \times 10^{-4}$ M, (c) $2.2 \times 10^{-4}$ M. **16.14A** $1.9 \times 10^{-5}$. **16.14B** $5.8 \times$
$10^{-9}$. **16.15A** 9.14. **16.15B** 8.33. **16.16A** $3.0 \times 10^{-9}$. **16.16B** $1.4 \times 10^{-4}$.
**16.17A** (a) $1.5 \times 10^{-10}$, (b) $1.2 \times 10^{-10}$, (c) $1.8 \times 10^{-11}$. **16.17B** (a) $1.1 \times$
$10^{-11}$, (b) $4.8 \times 10^{-7}$. **16.18A** $[H_2C_2O_4] = 0.11$ M, $[HC_2O_4^-] = 0.086$ M,
$[C_2O_4^{2-}] = 6.1 \times 10^{-5}$ M, $[H^+] = 0.086$ M. **16.18B** $[H_2SO_4] = 0$ M,
$[HSO_4^-] = 0.13$ M, $[SO_4^{2-}] = 0.011$ M, $[H^+] = 0.15$ M. **16.19A** (a) $HBrO_4$,
(b) $H_2SO_4$. **16.19B** Electronegativity. **16.20A** 8.96. **16.20B** 0.75 M. **16.21A**
2.92. **16.21B** 0.032 M. **16.22A** (a) Basic, (b) acidic, (c) basic, (d) neutral,
(e) neutral. **16.22B** (a) $NH_4Br$ and $CH_3NH_3I$, (b) $NaHPO_4$ and $KOBr$,
(c) $NaI$ and $KBr$. **16.23A** Lewis acid: $Co^{3+}$, Lewis base: $NH_3$. **16.23B** $NH_3$
and $AlBr_3$.

## Answers to Checkpoints

**16.1.1** b, d. **16.1.2** c, d. **16.2.1** d. **16.2.2** a. **16.3.1** c. **16.3.2** a. **16.3.3** e.
**16.3.4** b. **16.4.1** d. **16.4.2** b. **16.4.3** b. **16.4.4** e. **16.4.5** c. **16.4.6** b.
**16.4.7** b. **16.5.1** c. **16.5.2** e. **16.5.3** b. **16.6.1** b. **16.6.2** d. **16.6.3** e.
**16.7.1** b. **16.7.2** d. **16.7.3** a. **16.8.1** b. **16.8.2** c. **16.8.3** c. **16.8.4** b.
**16.10.1** a. **16.10.2** e. **16.10.3** a, b, c. **16.10.4** b, c. **16.10.5** c.
**16.12.1** c, d, e. **16.12.2** b, d.

## Answers to Applying What You've Learned

**a)** $H_2CO_3$, $CO_3^{2-}$. **b)** $[H^+] = 2.6 \times 10^{-6}$ M, pH = 5.59. **c)** $[OH^-] =$
$3.9 \times 10^{-9}$ M, pOH = 8.41. **d)** $2.6 \times 10^{-6}$ M. **e)** $K_b = 2.4 \times 10^{-8}$, pH =
9.69. **f)** $[H_2CO_3] = 1.5 \times 10^{-5}$ M, $[H^+] = [HCO_3^-] = 2.6 \times 10^{-6}$ M,
$[CO_3^{2-}] = 4.8 \times 10^{-11}$ M. **g)** $[H_2SO_4] = [HSO_4^-] = 0$ M, $[H^+] =$
$8.0 \times 10^{-5}$ M, $[SO_4^{2-}] = 4.0 \times 10^{-5}$ M. **h)** To an appropriate number
of significant figures, the hydrogen sulfate ion is 100 percent ionized.

# 17

# Acid-Base Equilibria and Solubility Equilibria

*Saliva becomes more acidic when high-sugar foods are eaten. Brushing regularly with fluoride toothpaste helps prevent cavities by making tooth enamel less soluble in acid.*

# Equilibrium and Tooth Decay

Teeth are protected by a hard enamel layer about 2 mm thick that is composed of a mineral called *hydroxyapatite* [$Ca_5(PO_4)_3OH$]. *Demineralization* is the process by which hydroxyapatite dissolves in the saliva. Because phosphates of alkaline earth metals such as calcium are insoluble [◄◄ Section 4.2, Table 4.3], this process happens only to a very small extent:

$$Ca_5(PO_4)_3OH(s) \rightleftharpoons 5Ca^{2+}(aq) + 3PO_4^{3-}(aq) + OH^-(aq)$$

The reverse process, called *remineralization*, is the body's natural defense against tooth decay:

$$5Ca^{2+}(aq) + 3PO_4^{3-}(aq) + OH^-(aq) \rightleftharpoons Ca_5(PO_4)_3OH(s)$$

Bacteria in the mouth break down some of the food we eat to produce organic acids such as acetic acid and lactic acid. Acid production is greatest from foods that are high in sugar. Thus, after a sugary snack, the $H^+$ concentration in the mouth increases, causing $OH^-$ ions in the saliva to be consumed.

$$H^+(aq) + OH^-(aq) \longrightarrow H_2O(l)$$

Removal of $OH^-$ from the saliva draws the $Ca_5(PO_4)_3OH$ dissolution equilibrium to the right, promoting demineralization. Once the protective enamel layer is weakened, tooth decay begins. The best way to prevent tooth decay is to eat a diet low in sugar and brush immediately after every meal.

**Student Annotation:** In children, the growth of the enamel layer (mineralization) occurs faster than demineralization; in adults, demineralization and remineralization occur at roughly equal rates.

At the end of this chapter, you will be able to solve a series of problems involving the solubility of hydroxyapatite, the principal component of tooth enamel[►► Page 766].

## In This Chapter, You Will Learn:

More about the behavior of acids and bases and about the factors that affect the aqueous solubility of ionic compounds.

### Before you begin, you should review

- Determination of pH [◄◄ Section 16.3]
- Manipulating equilibrium expressions [◄◄ Section 15.3]

Up until now, we have discussed the properties of solutions containing a single solute. In this section, we will examine how the properties of a solution change when a second solute is introduced.

Recall that a system at equilibrium will shift in response to being stressed and that stress can be applied in a variety of ways, including the addition of a reactant or a product [◄◄ Section 15.5]. Consider a liter of solution containing 0.10 mole of acetic acid. Using the $K_a$ for acetic acid ($1.8 \times 10^{-5}$) and an equilibrium table [◄◄ Section 16.5], the pH of this solution at 25°C can be determined:

$$CH_3COOH(aq) \rightleftharpoons H^+(aq) + CH_3COO^-(aq)$$

| | | | |
|---|---|---|---|
| *Initial concentration (M):* | 0.10 | 0 | 0 |
| *Change in concentration (M):* | $-x$ | $+x$ | $+x$ |
| *Equilibrium concentration (M):* | $0.10 - x$ | $x$ | $x$ |

Assuming that $(0.10 - x)\,M \approx 0.10\,M$ and solving for $x$, we get $1.34 \times 10^{-3}$. Therefore, $[CH_3COOH] = 0.09866\,M$, $[H^+] = [CH_3COO^-] = 1.34 \times 10^{-3}\,M$ and pH = 2.87.

Now consider what happens when we add 0.050 mole of sodium acetate ($CH_3COONa$) to the solution. Sodium acetate dissociates completely in aqueous solution to give sodium ions and acetate ions:

$$CH_3COONa(aq) \xrightarrow{H_2O} Na^+(aq) + CH_3COO^-(aq)$$

Thus, by adding sodium acetate, we have increased the concentration of acetate ion. Because acetate ion is a product in the ionization of acetic acid, the addition of acetate ion causes the equilibrium to shift to the left. The net result is a reduction in the percent ionization of acetic acid.

addition

$$CH_3COOH(aq) \xleftarrow{\phantom{xxx}} H^+(aq) + CH_3COO^-(aq)$$

Equilibrium is driven toward reactant.

Shifting the equilibrium to the left consumes not only some of the added acetate ion, but also some of the hydrogen ion. This causes the pH to change (in this case the pH increases).

Sample Problem 17.1 shows how an equilibrium table can be used to calculate the pH of a solution of acetic acid after the addition of sodium acetate.

> **Student Annotation:** The percent ionization of acetic acid is
> $$\frac{1.34 \times 10^{-3}\,M}{0.10\,M} \times 100\% = 1.3\%$$

> **Student Annotation:** By adding sodium acetate, we also add sodium ions to the solution. However, sodium ions do not interact with water or with any of the other species present [◄◄ Section 16.10].

### Sample Problem 17.1

Determine the pH at 25°C of a solution prepared by adding 0.050 mole of sodium acetate to 1.0 L of 0.10 $M$ acetic acid. (Assume that the addition of sodium acetate does not change the volume of the solution.)

**Strategy** Construct a new equilibrium table to solve for the hydrogen ion concentration.

**Setup** We use the stated concentration of acetic acid, 0.10 $M$, and $[H^+] \approx 0\,M$ as the initial concentrations in the table:

$$CH_3COOH(aq) \rightleftharpoons H^+(aq) + CH_3COO^-(aq)$$

| | | | |
|---|---|---|---|
| *Initial concentration (M):* | 0.10 | 0 | 0.050 |
| *Change in concentration (M):* | $-x$ | $+x$ | $+x$ |
| *Equilibrium concentration (M):* | $0.10 - x$ | $x$ | $0.050 + x$ |

> **Student Annotation:** Remember that prior to the ionization of a weak acid, the concentration of hydrogen ion in water at 25°C is $1.0 \times 10^{-7}\,M$. However, because this concentration is insignificant compared to the concentration resulting from the ionization, we can neglect it in our equilibrium table.

**Solution** Substituting the equilibrium concentrations, in terms of the unknown $x$, into the equilibrium expression gives

$$1.8 \times 10^{-5} = \frac{(x)(0.050 + x)}{0.10 - x}$$

Because we expect $x$ to be very small (even smaller than $1.34 \times 10^{-3}\,M$—see above), because the ionization of $CH_3COOH$ is suppressed by the presence of $CH_3COO^-$, we assume

$$(0.10 - x)\,M \approx 0.10\,M \quad \text{and} \quad (0.050 - x)\,M \approx 0.050\,M$$

Therefore, the equilibrium expression simplifies to

$$1.8 \times 10^{-5} = \frac{(x)(0.050)}{0.10}$$

and $x = 3.6 \times 10^{-5} M$. According to the equilibrium table, $[H^+] = x$, so pH $= -\log(3.6 \times 10^{-5}) = 4.44$.

**Student Annotation:** In this case, the percent ionization of acetic acid is

$$\frac{3.6 \times 10^{-5} M}{0.10 M} \times 100\% = 0.036\%$$

This is considerably smaller than the percent ionization prior to the addition of sodium acetate.

**Practice Problem A** Determine the pH at 25°C of a solution prepared by dissolving 0.075 mole of sodium acetate in 1.0 L of 0.25 $M$ acetic acid. (Assume that the addition of sodium acetate does not change the volume of the solution.)

**Practice Problem B** Determine the pH at 25°C of a solution prepared by dissolving 0.35 mole of ammonium chloride in 1.0 L of 0.25 $M$ aqueous ammonia.

**Think About It** The equilibrium concentrations of $CH_3COOH$, $CH_3COO^-$, and $H^+$ are the same regardless of whether we add sodium acetate to a solution of acetic acid, add acetic acid to a solution of sodium acetate, or dissolve both species at the same time. We could have constructed an equilibrium table starting with the equilibrium concentrations in the 0.10 $M$ acetic acid solution:

|  | $CH_3COOH(aq)$ | $\rightleftharpoons$ $H^+(aq)$ + | $CH_3COO^-(aq)$ |
|---|---|---|---|
| *Initial concentration (M):* | 0.09866 | $1.34 \times 10^{-3}$ | $5.134 \times 10^{-2}$ |
| *Change in concentration (M):* | $+y$ | $-y$ | $-y$ |
| *Equilibrium concentration (M):* | $0.09866 + y$ | $1.34 \times 10^{-3} - y$ | $5.134 \times 10^{-2} - y$ |

In this case, the reaction proceeds to the left. (The acetic acid concentration increases, and the concentrations of hydrogen and acetate ions decrease.) Solving for $y$ gives $1.304 \times 10^{-3} M$. $[H^+] = 1.34 \times 10^{-3} - y = 3.6 \times 10^{-5} M$ and pH $= 4.44$. We get the same pH either way.

An aqueous solution of a weak electrolyte contains both the weak electrolyte and its ionization products, which are ions. If a soluble salt that contains one of those ions is added, the equilibrium shifts to the left, thereby *suppressing* the ionization of the weak electrolyte. In general, when a compound containing an ion in common with a dissolved substance is added to a solution at equilibrium, the equilibrium shifts to the left. This phenomenon is known as the ***common ion effect.***

**Student Annotation:** This is the sum of the equilibrium concentration of acetate ion in a 0.10 $M$ solution of acetic acid ($1.34 \times 10^{-3}$ $M$) and the added acetate ion (0.050 $M$).

**Student Annotation:** Treating the problem as though both $CH_3COOH$ and $CH_3COO^-$ are added at the same time and the reaction proceeds to the right simplifies the solution.

## Checkpoint 17.1   The Common Ion Effect

**17.1.1** Which of the following would cause a decrease in the percent ionization of nitrous acid ($HNO_2$) when added to a solution of nitrous acid at equilibrium? (Select all that apply.)

a) $NaNO_2$

b) $H_2O$

c) $Ca(NO_2)_2$

d) $HNO_3$

e) $NaNO_3$

**17.1.2** What is the pH of a solution prepared by adding 0.05 mole of NaF to 1.0 L of 0.1 $M$ HF at 25°C? (Assume that the addition of NaF does not change the volume of the solution.) ($K_a$ for HF = $7.1 \times 10^{-4}$.)

a) 2.1

b) 2.8

c) 1.4

d) 4.6

e) 7.3

**Student Annotation:** The *common ion* can also be $H^+$ or $OH^-$. For example, addition of a strong acid to a solution of a weak acid suppresses ionization of the weak acid. Similarly, addition of a strong base to a solution of weak base suppresses ionization of the weak base.

# 17.2   Buffer Solutions

A solution that contains a weak acid and its conjugate base (or a weak base and its conjugate acid) is a *buffer solution* or simply a *buffer.* Buffer solutions, by virtue of their composition, *resist* changes in pH upon addition of small amounts of either an acid or a base. The ability to resist pH change is very important to chemical and biological systems, including the human body. The pH of blood is about 7.4, whereas that of gastric juices is about 1.5. Each of these pH values is crucial for proper enzyme function and the balance of osmotic pressure, and each is maintained within a very narrow pH range by a buffer.

**Student Annotation:** Any solution of a weak acid contains some conjugate base. In a buffer solution, though, the amounts of weak acid and conjugate base must be *comparable,* meaning that the conjugate base must be supplied by a dissolved salt.

## Calculating the pH of a Buffer

Consider a solution that is 1.0 $M$ in acetic acid and 1.0 $M$ sodium acetate. If a small amount of acid is added to this solution, it is consumed completely by the acetate ion,

$$H^+(aq) + CH_3COO^-(aq) \longrightarrow CH_3COOH(aq)$$

**Student Annotation:** Remember that sodium acetate is a strong electrolyte [◄◄ Section 4.1], so it dissociates completely in water to give sodium ions and acetate ions:

$$CH_3COONa \longrightarrow Na^+(aq) + CH_3COO^-(aq)$$

thus converting a strong acid ($H^+$) to a weak acid ($CH_3COOH$). Addition of a strong acid lowers the pH of a solution. However, a buffer's ability to convert a strong acid to a weak acid minimizes the effect of the addition on the pH.

Similarly, if a small amount of a base is added, it is consumed completely by the acetic acid,

$$CH_3COOH(aq) + OH^-(aq) \longrightarrow CH_3COO^-(aq) + H_2O(l)$$

thus converting a strong base ($OH^-$) to a weak base ($CH_3COO^-$). Addition of a strong base increases the pH of a solution. Again, however, a buffer's ability to convert a strong base to a weak base minimizes the effect of the addition on pH.

To illustrate the function of a buffer, suppose that we have 1 L of the acetic acid–sodium acetate solution described previously. We can calculate the pH of the buffer using the procedure in Section 17.1:

**Multimedia**

Acids and Bases—effect of addition of a strong acid and a strong base on a buffer.

$$CH_3COOH(aq) \rightleftharpoons H^+(aq) + CH_3COO^-(aq)$$

| | $CH_3COOH$ | $H^+$ | $CH_3COO^-$ |
|---|---|---|---|
| Initial concentration (M): | 1.0 | 0 | 1.0 |
| Change in concentration (M): | $-x$ | $+x$ | $+x$ |
| Equilibrium concentration (M): | $1.0 - x$ | $x$ | $1.0 + x$ |

The equilibrium expression is

$$K_a = \frac{(x)(1.0 + x)}{1.0 - x}$$

**Student Annotation:** The forward reaction is suppressed by the presence of the common ion, $CH_3COO^-$, and the reverse process is suppressed by the presence of $CH_3COOH$.

Because it is reasonable to assume that $x$ will be very small,

$$(1.0 - x)\,M \approx 1.0\,M \qquad \text{and} \qquad (1.0 + x)\,M \approx 1.0\,M$$

Thus, the equilibrium expression simplifies to

$$1.8 \times 10^{-5} = \frac{(x)(1.0)}{1.0} = x$$

At equilibrium, therefore, $[H^+] = 1.8 \times 10^{-5}\,M$ and pH = 4.74.

Now consider what happens when we add 0.10 mole of HCl to the buffer. (We assume that the addition of HCl causes no change in the volume of the solution.) The reaction that takes place when we add a strong acid is the conversion of $H^+$ to $CH_3COOH$. The added acid is all consumed, along with an equal amount of acetate ion. We keep track of the amounts of acetic acid and acetate ion when a strong acid (or base) is added by writing the starting amounts above the equation and the final amounts (after the added substance has been consumed) below the equation:

**Student Annotation:** As long as the amount of strong acid added to the buffer does not exceed the amount of conjugate base originally present, *all* the added acid will be consumed and converted to weak acid.

| Upon addition of $H^+$: | 1.0 mol | 0.1 mol | 1.0 mol |
|---|---|---|---|
| | $CH_3COO^-(aq)$ + | $H^+(aq) \longrightarrow$ | $CH_3COOH(aq)$ |
| After $H^+$ has been consumed: | 0.9 mol | 0 mol | 1.1 mol |

We can use the resulting amounts of acetic acid and acetate ion to construct a new equilibrium table:

$$CH_3COOH(aq) \rightleftharpoons H^+(aq) + CH_3COO^-(aq)$$

| | $CH_3COOH$ | $H^+$ | $CH_3COO^-$ |
|---|---|---|---|
| Initial concentration (M): | 1.1 | 0 | 0.9 |
| Change in concentration (M): | $-x$ | $+x$ | $+x$ |
| Equilibrium concentration (M): | $1.1 - x$ | $x$ | $0.9 + x$ |

We can solve for pH as we have done before, assuming that $x$ is small enough to be neglected,

$$1.8 \times 10^{-5} = \frac{(x)(0.9 + x)}{1.1 - x} \approx \frac{(x)(0.9)}{1.1}$$

$$x = 2.2 \times 10^{-5}\,M$$

**Student Annotation:** Had we added 0.10 mole of HCl to 1 L of pure water, the pH would have gone from 7.00 to 1.00!

Thus, when equilibrium is reestablished, $[H^+] = 2.2 \times 10^{-5}\,M$ and pH = 4.66—a change of only 0.08 pH unit.

In the determination of the pH of a buffer such as the one just described, we always neglect the small amount of weak acid that ionizes $(x)$ because ionization is suppressed by the presence of a common ion. Similarly, we ignore the hydrolysis of the acetate ion because of the presence of acetic acid. This enables us to derive an expression for determining the pH of a buffer. We begin with the equilibrium expression

$$K_a = \frac{[H^+][A^-]}{[HA]}$$

Rearranging to solve for $[H^+]$ gives

$$[H^+] = \frac{K_a[HA]}{[A^-]}$$

Taking the negative logarithm of both sides, we obtain

$$-\log [H^+] = -\log K_a - \log \frac{[HA]}{[A^-]}$$

or

$$-\log [H^+] = -\log K_a + \log \frac{[A^-]}{[HA]}$$

Thus,

$$pH = pK_a + \log \frac{[A^-]}{[HA]} \qquad \text{Equation 17.1}$$

where

$$pK_a = -\log K_a \qquad \text{Equation 17.2}$$

Equation 17.1 is known as the **Henderson-Hasselbalch equation.** Its more general form is

$$pH = pK_a + \log \frac{[\text{conjugate base}]}{[\text{weak acid}]} \qquad \text{Equation 17.3}$$

In the case of our acetic acid $(1.0\ M)$ and sodium acetate $(1.0\ M)$ buffer, the concentrations of weak acid and conjugate base are *equal.* When this is true, the log term in the Henderson-Hasselbalch equation is zero and the pH is numerically equal to the $pK_a$. In the case of an acetic acid–acetate ion buffer, $pK_a = -\log 1.8 \times 10^{-5} = 4.74$.

After the addition of 0.10 mole of HCl, we determined that the concentrations of acetic acid and acetate ion were $1.1\ M$ and $0.9\ M$, respectively. Using these concentrations in the Henderson-Hasselbalch equation gives

$$pH = 4.74 + \log \frac{[CH_3COO^-]}{[CH_3COOH]}$$

$$= 4.74 + \log \frac{0.9\ M}{1.1\ M}$$

$$= 4.74 + (-0.087) = 4.65$$

The small difference between this pH and the 4.66 calculated using an equilibrium table is due to differences in rounding. Figure 17.1 illustrates how a buffer solution resists drastic changes in pH.

Sample Problem 17.2 shows how the Henderson-Hasselbalch equation is used to determine the pH of a buffer after the addition of a strong base.

Media Player/MPEG Animation:
Figure 17.1, Buffer Solutions, pp. 732–733.

Figure 17.1

# Buffer Solutions

When we add 0.001 mol of strong acid, it is completely consumed by the acetate ion in the buffer.

Before reaction:  0.001 mol        0.010 mol            0.010 mol

$$H^+(aq) + CH_3COO^-(aq) \longrightarrow CH_3COOH(aq)$$

After reaction:      0 mol            0.009 mol            0.011 mol

0.100 *M* CH$_3$COOH
0.100 *M* CH$_3$COO$^-$

$$pH = 4.74 + \log \frac{[CH_3COO^-]}{[CH_3COOH]} = 4.74$$

Water
pH = 7.00

**START**

**START**

The buffer solution is 0.100 *M* in acetic acid and 0.100 *M* in sodium acetate. 100 mL of this buffer contains (0.100 mol/L)(0.10 L) = 0.010 mol each acetic acid and acetate ion.

When we add 0.001 mol of strong base, it is completely consumed by the acetic acid in the buffer.

Before reaction:  0.001 mol        0.010 mol            0.010 mol

$$OH^-(aq) + CH_3COOH(aq) \longrightarrow H_2O(l) + CH_3COO^-(aq)$$

After reaction:    0 mol            0.009 mol            0.011 mol

We can calculate the new pH using the Henderson-Hasselbalch equation:

$$pH = 4.74 + \log \frac{0.009}{0.011} = 4.65$$

There is nothing in pure water to consume strong acid. Therefore, its pH drops drastically.

$$pH = -\log \frac{0.001 \text{ mol}}{0.10 \text{ L}} = 2.00$$

There is nothing in pure water to consume strong base. Therefore, its pH rises drastically.

$$pOH = -\log \frac{0.001 \text{ mol}}{0.10 \text{ L}} = 2.00, \ pH = 12.00$$

We can calculate the new pH using the Henderson-Hasselbalch equation:

$$pH = 4.74 + \log \frac{0.011}{0.009} = 4.83$$

## What's the point?

A buffer contains both a weak acid and its conjugate base.*
Small amounts of strong acid or strong base are consumed by
the buffer components, thereby preventing drastic pH changes.
Pure water does not contain species that can consume acid
or base. Even a very small addition of either acid or base causes
a large change in pH.

*A buffer could also be prepared using a weak base and its
conjugate acid.

## Sample Problem 17.2

Starting with 1.00 L of a buffer that is 1.00 $M$ in acetic acid and 1.00 $M$ in sodium acetate, calculate the pH after the addition of 0.100 mole of NaOH. (Assume that the addition does not change the volume of the solution.)

**Strategy** Added base will react with the acetic acid component of the buffer, converting $OH^-$ to $CH_3COO^-$:

$$CH_3COOH(aq) + OH^-(aq) \longrightarrow H_2O(l) + CH_3COO^-(aq)$$

Write the starting amount of each species above the equation and the final amount of each species below the equation. Use the final amounts as concentrations in Equation 17.1.

**Setup**

| | | | |
|---|---|---|---|
| Upon addition of $OH^-$: | 1.00 mol | 0.10 mol | 1.00 mol |

$$CH_3COOH(aq) + OH^-(aq) \longrightarrow H_2O(l) + CH_3COO^-(aq)$$

| | | | |
|---|---|---|---|
| After $OH^-$ has been consumed: | 0.90 mol | 0 mol | 1.10 mol |

**Solution**

$$pH = 4.74 + \log \frac{1.10\ M}{0.90\ M}$$

$$= 4.74 + \log \frac{1.10\ M}{0.90\ M} = 4.83$$

Thus, the pH of the buffer after addition of 0.10 mole of NaOH is 4.83.

**Think About It** Always do a "reality check" on a calculated pH. Although a buffer does minimize the effect of added base, the pH does increase. If you find that you've calculated a lower pH after the addition of a base, check for errors like mixing up the weak acid and conjugate base concentrations or losing track of a minus sign.

**Practice Problem A** Calculate the pH of 1 L of a buffer that is 1.0 $M$ in acetic acid and 1.0 $M$ in sodium acetate after the addition of 0.25 mole of NaOH.

**Practice Problem B** How much HCl must be added to a liter of buffer that is 1.5 $M$ in acetic acid and 0.75 $M$ in sodium acetate to result in a buffer pH of 4.10?

## Preparing a Buffer Solution with a Specific pH

A solution is only a buffer if it has the capacity to resist pH change when either an acid or a base is added. If the concentrations of a weak acid and conjugate base differ by more than a factor of 10, the solution does not have this capacity. Therefore, we consider a solution a buffer, and can use Equation 17.1 to calculate its pH, only if the following condition is met:

$$10 \geq \frac{[\text{conjugate base}]}{[\text{weak acid}]} \geq 0.1$$

Consequently, the log term in Equation 17.1 can only have values from $-1$ to 1, and the pH of a buffer cannot be more than one pH unit different from the $pK_a$ of the weak acid it contains. This is known as the *range* of the buffer, where pH = $pK_a \pm 1$. This enables us to select the appropriate conjugate pair to prepare a buffer with a specific, desired pH.

First, we choose a weak acid whose $pK_a$ is close to the desired pH. Next, we substitute the pH and $pK_a$ values into Equation 17.1 to obtain the necessary ratio of [conjugate base]/[weak acid]. This ratio can then be converted to molar quantities for the preparation of the buffer.

Sample Problem 17.3 demonstrates this procedure.

| Weak Acid | $K_a$ | $pK_a$ |
|---|---|---|
| HF | $7.1 \times 10^{-4}$ | 3.15 |
| $HNO_2$ | $4.5 \times 10^{-4}$ | 3.35 |
| HCOOH | $1.7 \times 10^{-4}$ | 3.77 |
| $C_6H_5COOH$ | $6.5 \times 10^{-5}$ | 4.19 |
| $CH_3COOH$ | $1.8 \times 10^{-5}$ | 4.74 |
| HCN | $4.9 \times 10^{-10}$ | 9.31 |
| $C_6H_5OH$ | $1.3 \times 10^{-10}$ | 9.89 |

## Sample Problem 17.3

Select an appropriate weak acid from the table in the margin, and describe how you would prepare a buffer with a pH of 9.50.

**Strategy** Select an acid with a $pK_a$ within one pH unit of 9.50. Use the $pK_a$ of the acid and Equation 17.1 to calculate the necessary ratio of [conjugate base]/[weak acid]. Select concentrations of the buffer components that yield the calculated ratio.

**Setup** Two of the acids listed in Table 16.6 have p$K_a$ values in the desired range: hydrocyanic acid (HCN, p$K_a$ = 9.31) and phenol ($C_6H_5OH$, p$K_a$ = 9.89).

**Solution** Plugging the values for phenol into Equation 17.1 gives

$$9.50 = 9.89 + \log \frac{[C_6H_5O^-]}{[C_6H_5OH]}$$

$$9.50 - 9.89 = \log \frac{[C_6H_5O^-]}{[C_6H_5OH]} = -0.39$$

$$\frac{[C_6H_5O^-]}{[C_6H_5OH]} = 10^{-0.39} = 0.41$$

Therefore, the ratio of $[C_6H_5O^-]$ to $[C_6H_5OH]$ must be 0.41 to 1. One way to achieve this would be to dissolve 0.41 mole of $C_6H_5ONa$ and 1.00 mole of $C_6H_5OH$ in 1 L of water.

**Think About It** There is an infinite number of combinations of [conjugate base] and [weak acid] that will give the necessary ratio. Note that this pH could also be achieved using HCN and a cyanide salt. For most purposes, it is best to use the least toxic compounds available.

**Practice Problem A** Select an appropriate acid from Table 16.6, and describe how you would prepare a buffer with pH = 4.5.

**Practice Problem B** What range of pH values could be achieved with a buffer consisting of nitrous acid ($HNO_2$) and sodium nitrite ($NO_2^-$)?

---

## Bringing Chemistry to Life

### Maintaining the pH of Blood

There are about 5 L of blood in the average adult. Circulating blood keeps cells alive by providing them with oxygen and nutrients and by removing carbon dioxide and other waste materials. The efficiency of this enormously complex system relies on buffers.

The two primary components of blood are *blood plasma* and red blood cells, or *erythrocytes*. Blood plasma contains many compounds, including proteins, metal ions, and inorganic phosphates. The erythrocytes contain hemoglobin molecules, as well as the enzyme carbonic anhydrase, which catalyzes both the formation and the decomposition of carbonic acid ($H_2CO_3$):

$$CO_2(aq) + H_2O(l) \rightleftharpoons H_2CO_3(aq)$$

The substances inside the erythrocytes are protected from extracellular fluid (blood plasma) by a semipermeable cell membrane that allows only certain molecules to diffuse through it.

The pH of blood plasma is maintained at about 7.40 by several buffer systems, the most important of which is the $HCO_3^-/H_2CO_3$ system. In the erythrocyte, where the pH is 7.25, the principal buffer systems are $HCO_3^-/H_2CO_3$ and hemoglobin. The hemoglobin molecule is a complex protein molecule (molar mass of 65,000 g) that contains a number of ionizable protons. As a very rough approximation, we can treat it as a monoprotic acid in the form HHb:

$$HHb(aq) \rightleftharpoons H^+(aq) + Hb^-(aq)$$

where HHb represents the hemoglobin molecule and $Hb^-$ is the conjugate base of HHb. Oxyhemoglobin ($HHbO_2$), formed by the combination of oxygen with hemoglobin, is a stronger acid than HHb:

$$HHbO_2(aq) \rightleftharpoons H^+(aq) + HbO_2^-(aq)$$

Carbon dioxide produced by metabolic processes diffuses into the erythrocyte, where it is rapidly converted to $H_2CO_3$ by carbonic anhydrase:

$$CO_2(aq) + H_2O(l) \rightleftharpoons H_2CO_3(aq)$$

The ionization of the carbonic acid,

$$H_2CO_3(aq) \rightleftharpoons H^+(aq) + HCO_3^-(aq)$$

has two important consequences. First, the bicarbonate ion diffuses out of the erythrocyte and is carried by the blood plasma to the lungs. This is the major mechanism for removing carbon dioxide. Second, the $H^+$ ions shift the equilibrium in favor of the un-ionized oxyhemoglobin molecules:

$$H^+(aq) + HbO_2^-(aq) \rightleftharpoons HHbO_2(aq)$$

Because $HHbO_2$ releases oxygen more readily than does its conjugate base ($HbO_2^-$), the formation of the acid promotes the following reaction from left to right:

$$HHbO_2(aq) \rightleftharpoons HHb(aq) + O_2(aq)$$

The $O_2$ molecules diffuse out of the erythrocyte and are taken up by other cells to carry out metabolism.

When the venous blood returns to the lungs, the preceding processes are reversed. The bicarbonate ions now diffuse into the erythrocyte, where they react with hemoglobin to form carbonic acid:

$$HHb(aq) + HCO_3^-(aq) \rightleftharpoons Hb^-(aq) + H_2CO_3(aq)$$

Most of the acid is then converted to $CO_2$ by carbonic anhydrase:

$$H_2CO_3(aq) \rightleftharpoons H_2O(l) + CO_2(aq)$$

The carbon dioxide diffuses to the lungs and is eventually exhaled. The formation of the $Hb^-$ ions (due to the reaction between HHb and $HCO_3^-$) also favors the uptake of oxygen at the lungs,

$$Hb^-(aq) + O_2(aq) \rightleftharpoons HbO_2^-(aq)$$

because $Hb^-$ has a greater affinity for oxygen than does HHb.

When the arterial blood flows back to the body tissues, the entire cycle is repeated.

## Checkpoint 17.2    Buffer Solutions

**17.2.1** Which of the following combinations can be used to prepare a buffer?

a) HCl/NaCl

b) HF/KF

c) $NH_3/NH_4Cl$

d) $HNO_3/HNO_2$

e) $NaNO_2/HNO_3$

**17.2.2** What is the pH of a buffer that is 0.76 $M$ in HF and 0.98 $M$ in NaF?

a) 3.26

b) 3.04

c) 3.15

d) 10.85

e) 10.74

**17.2.3** Consider 1 L of a buffer that is 0.85 $M$ in formic acid (HCOOH) and 1.4 $M$ in sodium formate (HCOONa). Calculate the pH after the addition of 0.15 mol HCl. (Assume the addition causes no volume change.)

a) 4.11

b) 3.99

c) 3.87

d) 10.13

e) 10.01

**17.2.4** Consider 1 L of a buffer that is 1.5 $M$ in hydrocyanic acid (HCN) and 1.2 $M$ in sodium cyanide (NaCN). Calculate the pH after the addition of 0.25 mol NaOH. (Assume the addition causes no volume change.)

a) 9.21

b) 9.37

c) 9.04

d) 4.63

e) 4.96

**17.2.5** The solutions shown contain one or more of the following compounds: $H_2A$ (a weak diprotic acid), NaHA, and $Na_2A$ (the sodium salts of $HA^-$ and $A^{2-}$). (For clarity, the sodium ions and water molecules are not shown.) Which solutions are buffers? (Select all that apply.)

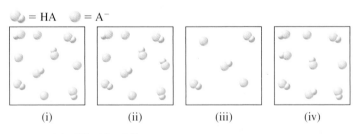

(i)    (ii)    (iii)    (iv)

a) i < iii < iv < ii

b) iv < ii < iii < i

c) i < ii < iv < iii

d) iv < ii < i = iii

e) i = iii < iv < ii

**17.2.6** The solutions shown contain a combination of the weak acid HA and its sodium salt NaA. (For clarity, the sodium ions and water molecules are not shown.) Arrange the solutions in order of increasing pH.

# 17.3 | Acid-Base Titrations

In Section 4.6 we introduced acid-base titrations as a form of chemical analysis. Having discussed buffer solutions, we can now look in more detail at the quantitative aspects of acid-base titrations. We will consider three types of reactions: (1) titrations involving a strong acid and a strong base, (2) titrations involving a weak acid and a strong base, and (3) titrations involving a strong acid and a weak base. Titrations involving a weak acid and a weak base are complicated by the hydrolysis of both the cation and the anion of the salt formed. These titrations will not be discussed here. Figure 17.2 shows the experimental setup for monitoring the pH over the course of an acid-base titration.

## Strong Acid–Strong Base Titrations

The reaction between the strong acid HCl and the strong base NaOH can be represented by

$$NaOH(aq) + HCl(aq) \longrightarrow NaCl(aq) + H_2O(l)$$

or by the net ionic equation,

$$OH^-(aq) + H^+(aq) \longrightarrow H_2O(l)$$

Consider the addition of a 0.100 $M$ NaOH solution (from a burette) to a vessel containing 25.0 mL of 0.100 $M$ HCl. For convenience, we will use only three significant figures for volume and concentration and two significant figures for pH. Figure 17.3 shows the titration curve—the plot of pH as a function of titrant volume added.

Multimedia

Acids and Bases—titration of HCl with NaOH.

**Student Annotation:** Recall that only the digits to the right of the decimal point are significant in a pH value.

**Student Annotation:** The *titrant* is the solution that is added from the burette.

**Figure 17.2**   A pH meter is used to monitor an acid-base titration.

**Figure 17.3**   Titration curve (pH as a function of volume titrant added) of a strong acid–strong base titration. A 0.100 $M$ NaOH solution, the titrant, is added from a burette to 25.0 mL of a 0.100 $M$ HCl solution in an Erlenmeyer flask.

Before the addition of NaOH begins, the pH of the acid is given by $-\log (0.100)$, or 1.00. When NaOH is added, the pH of the solution increases slowly at first. Near the equivalence point, the pH begins to rise steeply, and at the equivalence point, when equimolar amounts of acid and base have reacted, the curve rises almost vertically. In a strong acid–strong base titration, both the hydrogen ion and hydroxide ion concentrations are very small at the equivalence point (roughly $1 \times 10^{-7}$ $M$); consequently, the addition of a single drop of the base causes a large increase in $[OH^-]$ and a steep rise in the pH of the solution. Beyond the equivalence point, the pH again increases slowly with the continued addition of NaOH.

It is possible to calculate the pH of the solution at every stage of titration. Here are three sample calculations.

1. Consider the addition of 10.0 mL of 0.100 $M$ NaOH to 25.0 mL of 0.100 $M$ HCl: The total volume of the solution is 35.0 mL. The number of millimoles of NaOH in 10.0 mL is

$$10.0 \text{ mL} \times \frac{0.100 \text{ mmol NaOH}}{1 \text{ mL}} = 1.00 \text{ mmol}$$

The number of millimoles of HCl originally present in 25.0 mL of solution is

$$25.0 \text{ mL} \times \frac{0.100 \text{ mmol HCl}}{1 \text{ mL}} = 2.50 \text{ mmol}$$

Thus, the amount of HCl left after partial neutralization is $2.50 - 1.00$, or 1.50 mmol. Next, we determine the resulting concentration of $H^+$. We have 1.50 mmol in 35.0 mL:

$$\frac{1.50 \text{ mmol HCl}}{35.0 \text{ mL}} = 0.0429 \text{ } M$$

Thus $[H^+] = 0.0429$ $M$, and the pH of the solution is

$$pH = -\log (0.04289) = 1.37$$

2. Consider the addition of 25.0 mL of 0.100 $M$ NaOH to 25.0 mL of 0.100 $M$ HCl: This is a straightforward calculation, because it involves a complete neutralization reaction and neither ion in the salt (NaCl) undergoes hydrolysis [◄◄ Section 16.10]. At the equivalence point, $[H^+] = [OH^-] = 1.00 \times 10^{-7}$ $M$ and the pH of the solution is 7.000.

3. Consider the addition of 35.0 mL of 0.100 $M$ NaOH to 25.0 mL of 0.100 $M$ HCl: The total volume of the solution is now 60.0 mL. The number of millimoles of NaOH added is

$$35.0 \text{ mL} \times \frac{0.100 \text{ mmol NaOH}}{1 \text{ mL}} = 3.50 \text{ mmol}$$

There are 2.50 mmol of HCl in 25.0 mL of solution. After complete neutralization of HCl, 2.50 mmol of NaOH have been consumed, and the number of millimoles of NaOH remaining is $3.5 - 2.5$ or 1.00 mmol. The concentration of NaOH in 60.0 mL of solution is

$$\frac{1.00 \text{ mmol NaOH}}{60.0 \text{ mL}} = 0.0167 \text{ } M$$

Thus $[OH^-] = 0.0167$ $M$ and $pOH = -\log (0.0167) = 1.78$. The pH of the solution is $14.00 - 1.78$ or 12.22.

Table 17.1 lists the data at nine different points during a strong acid–strong base titration along with the calculated pH at each point.

## Weak Acid–Strong Base Titrations

Consider the neutralization reaction between acetic acid (a weak acid) and sodium hydroxide (a strong base):

$$CH_3COOH(aq) + NaOH(aq) \longrightarrow CH_3COONa(aq) + H_2O(l)$$

This equation can be simplified to

$$CH_3COOH(aq) + OH^-(aq) \longrightarrow CH_3COO^-(aq) + H_2O(l)$$

The acetate ion that results from this neutralization undergoes hydrolysis [◄◄ Section 16.10] as follows:

$$CH_3COO^-(aq) + H_2O(l) \rightleftharpoons CH_3COOH(aq) + OH^-(aq)$$

| TABLE 17.1 | Determination of pH at Several Different Points in a Strong Acid–Strong Base Titration | | | | |
|---|---|---|---|---|---|
| Volume OH⁻ added (mL) | OH⁻ added (mmol) | H⁺ remaining (mmol) | Total volume (mL) | [H⁺] (mol/L) | pH |
| 0 | 0 | 2.5 | 25.0 | 0.100 | 1.000 |
| 5.0 | 0.50 | 2.0 | 30.0 | 0.0667 | 1.176 |
| 10.0 | 1.0 | 1.5 | 35.0 | 0.0429 | 1.364 |
| 15.0 | 1.5 | 1.0 | 40.0 | 0.0250 | 1.602 |
| 20.0 | 2.0 | 0.5 | 45.0 | 0.0111 | 1.955 |
| 25.0 | 2.5 | 0 | 50.0 | $1.00 \times 10^{-7}$ | 7.000 |

| Volume OH⁻ added (mL) | OH⁻ added (mmol) | Excess OH⁻ (mmol) | Total volume (mL) | [OH⁻] (mol/L) | pOH | pH |
|---|---|---|---|---|---|---|
| 30.0 | 3.0 | 0.5 | 55.0 | 0.0091 | 2.04 | 11.96 |
| 35.0 | 3.5 | 1.0 | 60.0 | 0.0167 | 1.78 | 12.223 |

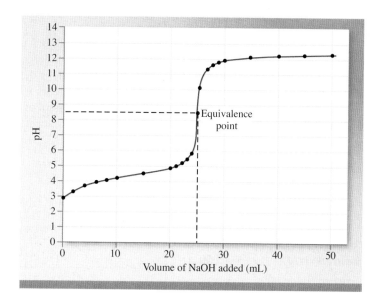

**Figure 17.4**  Titration curve of a weak acid–strong base titration. A 0.100 M NaOH solution is added from a burette to 25.0 mL of a 0.100 M $CH_3COOH$ solution in an Erlenmeyer flask. Because of the hydrolysis of the salt formed, the pH at the equivalence point is greater than 7.

At the equivalence point, therefore, when we only have sodium acetate in solution, the pH will be greater than 7 as a result of the OH⁻ formed by hydrolysis of the acetate ion.

The curve for titration of 25.0 mL of 0.1 M acetic acid with 0.10 M sodium hydroxide is shown in Figure 17.4. Note how the shape of the curve differs from the one in Figure 17.3. Compared to the curve for titration of a strong acid with a strong base, the curve for titration of a weak acid with a strong base has a higher initial pH, a more gradual change in pH as base is added, and a shorter vertical region near the equivalence point.

Again, it is possible to calculate the pH at every stage of the titration. Here are four sample calculations.

1. Prior to the addition of any base, the pH is determined by the ionization of acetic acid. We use its concentration (0.10 M) and its $K_a$ ($1.8 \times 10^{-5}$) to calculate the H⁺ concentration using an equilibrium table:

$$CH_3COOH(aq) \rightleftharpoons H^+(aq) + CH_3COO^-(aq)$$

| | CH₃COOH(aq) | H⁺(aq) | CH₃COO⁻(aq) |
|---|---|---|---|
| *Initial concentration (M):* | 0.10 | 0 | 0 |
| *Change in concentration (M):* | $-x$ | $+x$ | $+x$ |
| *Equilibrium concentration (M):* | $0.10 - x$ | $x$ | $x$ |

$$K_a = \frac{[H^+][CH_3COO^-]}{[CH_3COOH]} = \frac{x^2}{0.10 - x} = 1.8 \times 10^{-5}$$

We can neglect $x$ on the bottom of the equation [◀◀ Section 16.5]. Solving for $x$,

$$\frac{x^2}{0.10} = 1.8 \times 10^{-5}$$

$$x^2 = (1.8 \times 10^{-5})(0.10) = 1.8 \times 10^{-6}$$

$$x = \sqrt{1.8 \times 10^{-6}} = 1.34 \times 10^{-3}\, M$$

gives $[H^+] = 1.34 \times 10^{-3}\, M$ and pH = 2.87.

2. After the first addition of base, some of the acetic acid has been converted to acetate ion via the reaction

$$CH_3COOH(aq) + OH^-(aq) \longrightarrow CH_3COO^-(aq) + H_2O(l)$$

With significant amounts of both acetic acid and acetate ion in solution, we now treat the solution as a buffer and use the Henderson-Hasselbalch equation to calculate the pH.

After the addition of 10.0 mL of base, the solution contains 1.5 mmol of acetic acid and 1.0 mmol of acetate ion (see Table 17.2).

**Student Annotation:** Remember that we can use mol or mmol amounts in place of concentrations in the Henderson-Hasselbalch equation.

$$pH = 4.74 + \log\frac{1.0\ mmol}{1.5\ mmol} = 4.56$$

Each of the points between the beginning of the titration and the equivalence point can be calculated in this way.

3. At the equivalence point, all the acetic acid has been neutralized and we are left with acetate ion in solution. (There is also sodium ion, which does not undergo hydrolysis and therefore does not impact the pH of the solution.) At this point, pH is determined by the concentration and the $K_b$ of acetate ion. The equivalence point occurs when 25.0 mL of base has been added, making the total volume 50.0 mL. The 2.5 mmol of acetic acid (see Table 17.2) has all been converted to acetate ion. Therefore, the concentration of acetate ion is

$$[CH_3COO^-] = \frac{2.5\ mmol}{50.0\ mL} = 0.050\, M$$

As we did at the beginning of the titration, we construct an equilibrium table:

$$CH_3COO^-(aq) + H_2O(l) \rightleftharpoons OH^-(aq) + CH_3COOH(aq)$$

| | $CH_3COO^-$ | | $OH^-$ | $CH_3COOH$ |
|---|---|---|---|---|
| *Initial concentration (M):* | 0.050 | | 0 | 0 |
| *Change in concentration (M):* | $-x$ | | $+x$ | $+x$ |
| *Equilibrium concentration (M):* | $0.050 - x$ | | $x$ | $x$ |

**Student Annotation:** $K_a \times K_b = K_w$ [◀◀ Section 16.7].

The $K_b$ for acetate ion is $5.6 \times 10^{-10}$.

$$K_b = \frac{[OH^-][CH_3COOH]}{[CH_3COO^-]} = \frac{x^2}{0.050 - x} = 5.6 \times 10^{-10}$$

As before, we can neglect $x$ in the denominator of the equation. Solving for $x$,

$$\frac{x^2}{0.050} = 5.6 \times 10^{-10}$$

$$x^2 = (5.6 \times 10^{-10})(0.050) = 2.8 \times 10^{-11}$$

$$x = \sqrt{2.78 \times 10^{-11}} = 5.3 \times 10^{-6}\, M$$

gives $[OH^-] = 5.3 \times 10^{-6}\, M$, pOH = 5.28, and pH = 8.72.

4. After the equivalence point, the curve for titration of a weak acid with a strong base is identical to the curve for titration of a strong acid with a strong base. Because all the acetic acid has been consumed, there is nothing in solution to consume the additional added $OH^-$, and the pH levels off between 12 and 13.

Table 17.2 lists the data for the titration of 25.0 mL of 0.10 $M$ acetic acid with 0.10 $M$ NaOH.

Sample Problem 17.4 shows how to calculate the pH for the titration of a weak acid with a strong base.

| TABLE 17.2 | Determination of pH at Several Different Points in a CH₃COOH-NaOH Titration |

| Volume OH⁻ added (mL) | OH⁻ added (mmol) | CH₃COOH remaining | CH₃COO⁻ produced | pH |
|---|---|---|---|---|
| 0 | 0 | 2.5 | 0.0 | 2.87* |
| 5.0 | 0.50 | 2.0 | 0.50 | 4.14 |
| 10.0 | 1.0 | 1.5 | 1.0 | 4.56 |
| 15.0 | 1.5 | 1.0 | 1.5 | 4.92 |
| 20.0 | 2.0 | 0.5 | 2.0 | 5.34 |
| 25.0 | 2.5 | 0.0 | 2.5 | 8.72† |

| Volume OH⁻ added (mL) | OH⁻ added (mmol) | Excess OH⁻ (mmol) | Total volume (mL) | [OH⁻] (mol/L) | pOH | pH |
|---|---|---|---|---|---|---|
| 30.0 | 3.0 | 0.5 | 55.0 | 0.0091 | 2.04 | 11.96 |
| 35.0 | 3.5 | 1.0 | 60.0 | 0.017 | 1.78 | 12.22 |

*$[CH_3COOH] = 0.10\ M$, $K_a = 1.8 \times 10^{-5}$.

†$[CH_3COO^-] = 0.050\ M$, $K_b = 5.6 \times 10^{-10}$.

**Student Annotation:** Prior to the addition of any base, *and* at the equivalence point, this is an *equilibrium* problem that is solved using a *concentration*, an *ionization constant*, and an *equilibrium table*.

**Student Annotation:** *Prior* to the equivalence point, pH is determined using the Henderson-Hasselbalch equation:

$$pH = pK_a + \log \frac{[\text{conjugate base}]}{[\text{weak acid}]}$$

**Student Annotation:** After the equivalence point, the titration curve of a weak acid is identical to that of a strong acid.

---

## Sample Problem 17.4

Calculate the pH in the titration of 50.0 mL of 0.120 *M* acetic acid by 0.240 *M* sodium hydroxide after the addition of (a) 10.0 mL of base, (b) 25.0 mL of base, and (c) 35.0 mL of base.

**Strategy** The reaction between acetic acid and sodium hydroxide is

$$CH_3COOH(aq) + OH^-(aq) \longrightarrow H_2O(l) + CH_3COO^-(aq)$$

Prior to the equivalence point [part (a)], the solution contains both acetic acid and acetate ion, making the solution a buffer. We can solve part (a) using Equation 17.1, the Henderson-Hasselbalch equation. At the equivalence point [part (b)], all the acetic acid has been neutralized and we have only acetate ion in solution. We must determine the concentration of acetate ion and solve part (b) as an equilibrium problem, using the $K_b$ for acetate ion. After the equivalence point [part (c)], all the acetic acid has been neutralized and there is nothing to consume the additional added base. We must determine the concentration of excess hydroxide ion in the solution and solve for pH using Equations 16.4 and 16.6.

**Setup** Remember that *M* can be defined as either mol/L or mmol/mL [◀◀ Section 4.5]. For this type of problem, it simplifies the calculations to use millimoles rather than moles. $K_a$ for acetic acid is $1.8 \times 10^{-5}$, so $pK_a = 4.74$. $K_b$ for acetate ion is $5.6 \times 10^{-10}$.

(a) The solution originally contains (0.120 mmol/mL)(50.0 mL) = 6.00 mmol of acetic acid. A 10.0-mL amount of base contains (0.240 mmol/mL)(10.0 mL) = 2.40 mmol of base. After the addition of 10.0 mL of base, 2.40 mmol of OH⁻ has neutralized 2.40 mmol of acetic acid, leaving 3.60 mmol of acetic acid and 2.40 mmol acetate ion in solution:

Upon addition of OH⁻:        6.00 mmol      2.40 mmol          0 mmol

$$CH_3COOH(aq) + OH^-(aq) \rightleftharpoons H_2O(l) + CH_3COO^-(aq)$$

After OH⁻ has been consumed:    3.60 mmol     0 mmol           2.40 mmol

(b) After the addition of 25.0 mL of base, the titration is at the equivalence point. We calculate the pH using the concentration and the $K_b$ of acetate ion.

(c) After the addition of 35.0 mL of base, the titration is past the equivalence point and we solve for pH by determining the concentration of excess hydroxide ion.

**Solution** (a) $pH = pK_a + \log \dfrac{2.40}{3.60} = 4.74 - 0.18 = 4.56$

*(Continued)*

(b) At the equivalence point, we have 6.0 mmol of acetate ion in the total volume. We determine the total volume by calculating what volume of 0.24 $M$ base contains 6.0 mmol:

$$(\text{volume})(0.240 \text{ mmol/mL}) = 6.00 \text{ mmol}$$

$$\text{volume} = \frac{6.00 \text{ mmol}}{0.240 \text{ mmol/mL}} = 25.0 \text{ mL}$$

Therefore, the equivalence point occurs when 25.0 mL of base has been added, making the total volume 50.0 mL + 25.0 mL = 75.0 mL. The concentration of acetate ion at the equivalence point is therefore

$$\frac{6.00 \text{ mmol CH}_3\text{COO}^-}{75.0 \text{ mL}} = 0.0800 \ M$$

We can construct an equilibrium table using this concentration and solve for pH using the ionization constant for $CH_3COO^-$ ($K_b = 5.6 \times 10^{-10}$):

$$CH_3COO^-(aq) + H_2O(l) \rightleftharpoons OH^-(aq) + CH_3COOH(aq)$$

| | $CH_3COO^-(aq) + H_2O(l) \rightleftharpoons$ | $OH^-(aq) +$ | $CH_3COOH(aq)$ |
|---|---|---|---|
| *Initial concentration (M):* | 0.0800 | 0 | 0 |
| *Change in concentration (M):* | $-x$ | $+x$ | $+x$ |
| *Equilibrium concentration (M):* | $0.0800 - x$ | $x$ | $x$ |

Using the equilibrium expression and assuming that $x$ is small enough to be neglected,

$$K_b = \frac{[CH_3COOH][OH^-]}{[CH_3COO^-]} = \frac{(x)(x)}{0.0800 - x} \approx \frac{x^2}{0.0800} = 5.6 \times 10^{-10}$$

$$x = \sqrt{4.48 \times 10^{-11}} = 6.7 \times 10^{-6} \ M$$

According to the equilibrium table, $x = [OH^-]$, so $[OH^-] = 6.7 \times 10^{-6} \ M$. At equilibrium, therefore, pOH = $-\log (6.7 \times 10^{-6}) = 5.17$ and pH = $14.00 - 5.17 = 8.83$.

(c) After the equivalence point, we must determine the concentration of excess base and calculate pOH and pH using Equations 16.4 and 16.6. A 35.0-mL amount of the base contains (0.240 mmol/mL)(35.0 mL) = 8.40 mmol of $OH^-$. After neutralizing the 6.00 mmol of acetic acid originally present in the solution, this leaves 8.40 − 6.00 = 2.40 mmol of excess $OH^-$. The total volume is 50.0 + 35.0 = 85.0 mL. Therefore, $[OH^-]$ = 2.40 mmol/85.0 mL = 0.0280 $M$, pOH = $-\log (0.0280) = 1.553$, and pH = $14.000 - 1.553 = 12.447$.
   In summary, (a) pH = 4.56, (b) pH = 8.83, and (c) pH = 12.447.

> **Think About It** For each point in a titration, decide first what species are in solution and what *type* of problem it is. If the solution contains only a weak acid (or weak base), as is the case before any titrant is added, or if it contains only a conjugate base (or conjugate acid), as is the case at the equivalence point, when pH is determined by salt hydrolysis, it is an *equilibrium* problem that requires a concentration, an ionization constant, and an equilibrium table. If the solution contains comparable concentrations of both members of a conjugate pair, which is the case at points prior to the equivalence point, it is a *buffer* problem and is solved using the Henderson-Hasselbalch equation. If the solution contains excess titrant, either a strong base or strong acid, it is simply a pH problem requiring only a concentration.

**Practice Problem A** For the titration of 10.0 mL of 0.15 $M$ acetic acid with 0.10 $M$ sodium hydroxide, determine the pH when (a) 10.0 mL of base has been added, (b) 15.0 mL of base has been added, and (c) 20.0 mL of base has been added.

**Practice Problem B** For the titration of 25.0 mL of 0.20 $M$ hydrofluoric acid with 0.20 $M$ sodium hydroxide, determine the volume of base added when pH is (a) 2.85, (b) 3.15, and (c) 11.89. [To solve part c, you may want to review the approach in Sample Problem 4.10 (page 145).]

## Strong Acid–Weak Base Titrations

Consider the titration of HCl, a strong acid, with $NH_3$, a weak base:

$$HCl(aq) + NH_3(aq) \longrightarrow NH_4Cl(aq)$$

or simply

$$H^+(aq) + NH_3(aq) \longrightarrow NH_4^+(aq)$$

The pH at the equivalence point is less than 7 because the ammonium ion acts as a weak Brønsted acid:

$$NH_4^+(aq) + H_2O(l) \rightleftharpoons NH_3(aq) + H_3O^+(aq)$$

or simply

$$NH_4^+(aq) \rightleftharpoons NH_3(aq) + H^+(aq)$$

**Figure 17.5** Titration curve of a strong acid–weak base titration. A 0.100 $M$ HCl solution is added from a burette to 25.0 mL of a 0.100 $M$ NH$_3$ solution in an Erlenmeyer flask. As a result of salt hydrolysis, the pH at the equivalence point is lower than 7.

| Volume HCl added (mL) | pH |
|---|---|
| 0.0 | 11.13 |
| 5.0 | 9.86 |
| 10.0 | 9.44 |
| 15.0 | 9.08 |
| 20.0 | 8.66 |
| 22.0 | 8.39 |
| 24.0 | 7.88 |
| 25.0 | 5.28 |
| 26.0 | 2.70 |
| 28.0 | 2.22 |
| 30.0 | 2.00 |
| 35.0 | 1.70 |
| 40.0 | 1.52 |
| 45.0 | 1.40 |
| 50.0 | 1.30 |

Because of the volatility of an aqueous ammonia solution, it is more convenient to use hydrochloric acid as the titrant (i.e., to add HCl solution from the burette). Figure 17.5 shows the titration curve for this experiment.

Analogous to the titration of a weak acid with a strong base, the initial pH is determined by the concentration and the $K_b$ of ammonia:

$$NH_3(aq) + H_2O(l) \rightleftharpoons NH_4^+(aq) + OH^-(aq)$$

Consider the titration of 25.0 mL of 0.10 $M$ NH$_3$ with 0.10 $M$ HCl. We calculate the initial pH by constructing an equilibrium table and solving for $x$:

$$NH_3(aq) + H_2O(l) \rightleftharpoons NH_4^+(aq) + OH^-(aq)$$

| | $NH_3(aq) + H_2O(l) \rightleftharpoons$ | | $NH_4^+(aq) +$ | $OH^-(aq)$ |
|---|---|---|---|---|
| *Initial concentration (M):* | 0.10 | | 0 | 0 |
| *Change in concentration (M):* | $-x$ | | $+x$ | $+x$ |
| *Equilibrium concentration (M):* | $0.10 - x$ | | $x$ | $x$ |

Using the equilibrium expression and assuming that $x$ is small enough to be neglected,

$$K_b = \frac{[NH_4^+][OH^-]}{[NH_3]} = \frac{(x)(x)}{0.10 - x} \approx \frac{x^2}{0.10} = 1.8 \times 10^{-5}$$

$$x^2 = 1.8 \times 10^{-6}$$

$$x = \sqrt{1.8 \times 10^{-6}} = 1.3 \times 10^{-3}$$

$$pH = 11.11$$

The pH at the equivalence point is calculated using the concentration and $K_a$ of the conjugate base of NH$_3$, the NH$_4^+$ ion, and an equilibrium table.

Sample Problem 17.5 shows how this is done.

## Sample Problem 17.5

Calculate the pH at the equivalence point when 25.0 mL of 0.100 $M$ NH$_3$ is titrated with 0.100 $M$ HCl.

**Strategy** The reaction between NH$_3$ and HCl is

$$NH_3(aq) + H^+(aq) \longrightarrow NH_4^+(aq)$$

At the equivalence point, all the NH$_3$ has been converted to NH$_4^+$. Therefore, we must determine the concentration of NH$_4^+$ at the equivalence point and use the $K_a$ for NH$_4^+$ to solve for pH using an equilibrium table.

*(Continued)*

**Setup** The solution originally contains (0.100 mmol/mL)(25.0 mL) = 2.50 mmol $NH_4^+$. At the equivalence point, 2.50 mmol of HCl has been added. The volume of 0.100 $M$ HCl that contains 2.50 mmol is

$$(\text{volume})(0.100 \text{ mmol/mL}) = 2.50 \text{ mmol}$$

$$\text{volume} = \frac{2.50 \text{ mmol}}{0.100 \text{ mmol/mL}} = 25.0 \text{ mL}$$

It takes 25.0 mL of titrant to reach the equivalence point, so the total solution volume is 25.0 + 25.0 = 50.0 mL. At the equivalence point, all the $NH_3$ originally present has been converted to $NH_4^+$. The concentration of $NH_4^+$ is (2.50 mmol)/(50.0 mL) = 0.0500 $M$. We must use this concentration as the starting concentration of ammonium ion in our equilibrium table.

**Solution**

$$NH_4^+(aq) + H_2O(l) \rightleftharpoons NH_3(aq) + H_3O(aq)$$

| | $NH_4^+$ | | $NH_3$ | $H_3O$ |
|---|---|---|---|---|
| Initial concentration (*M*): | 0.0500 | | 0 | 0 |
| Change in concentration (*M*): | $-x$ | | $+x$ | $+x$ |
| Equilibrium concentration (*M*): | $0.0500 - x$ | | $x$ | $x$ |

The equilibrium expression is

$$K_a = \frac{[NH_3][H^+]}{[NH_4^+]} = \frac{(x)(x)}{0.0500} - x \approx \frac{x^2}{0.0500} = 5.6 \times 10^{-10}$$

$$x^2 = 2.8 \times 10^{-11}$$

$$x = \sqrt{2.8 \times 10^{-11}} = 5.3 \times 10^{-6} M$$

$[H^+] = x = 5.3 \times 10^{-6} M$. At equilibrium, therefore, pH $= -\log (5.3 \times 10^{-6}) = 5.28$.

**Think About It** In the titration of a weak base with a strong acid, the species in solution at the equivalence point is the conjugate acid. Therefore, we should expect an *acidic* pH. Once all the $NH_3$ has been converted to $NH_4^+$, there is no longer anything in the solution to consume added acid. Thus, the pH after the equivalence point depends on the number of millimoles of $H^+$ added and not consumed divided by the new total volume.

**Practice Problem A** Calculate the pH at the equivalence point in the titration of 50.0 mL of 0.10 $M$ methylamine (see Table 16.7) with 0.20 $M$ HCl.

**Practice Problem B** A 50.0-mL quantity of a 0.20 $M$ solution of one of the weak bases in Table 16.7 is titrated with 0.050 $M$ HCl. At the equivalence point, the pH is 2.99. Identify the weak base.

## Acid-Base Indicators

The equivalence point is the point at which the acid has been neutralized completely by the added base. The equivalence point in a titration can be determined by monitoring the pH over the course of the titration, or it can be determined using an *acid-base indicator*. An acid-base indicator is usually a weak organic acid or base for which the ionized and un-ionized forms are different colors.

Consider a weak organic acid that we will refer to as HIn. To be an effective acid-base indicator, HIn and its conjugate base, In⁻, must have distinctly different colors. In solution, the acid ionizes to a small extent:

$$HIn(aq) \rightleftharpoons H^+(aq) + In^-(aq)$$

In a sufficiently acidic medium, the ionization of HIn is suppressed according to Le Châtelier's principle, and the preceding equilibrium shifts to the left. In this case, the color of the solution will be that of HIn. In a basic medium, on the other hand, the equilibrium shifts to the right and the color of the solution will be that of the conjugate base, In⁻.

**Student Annotation:** The *endpoint* is where the color changes. The *equivalence point* is where neutralization is complete. Experimentally, we use the endpoint to estimate the equivalence point.

The ***endpoint*** of a titration is the point at which the color of the indicator changes. Not all indicators change color at the same pH, however, so the choice of indicator for a particular titration depends on the strength of the acid (and the base) used in the titration. To use the endpoint to determine the equivalence point of a titration, we must select an appropriate indicator.

The endpoint of an indicator does not occur at a specific pH; rather, there is a range of pH over which the color change occurs. In practice, we select an indicator whose color change occurs

**Figure 17.6**   Titration curve of a strong acid with a strong base (blue) and titration curve of a weak acid with a strong base (red). The indicator phenolphthalein can be used to determine the equivalence point of either titration. Methyl red can be used for the strong acid–strong base titration but cannot be used for the weak acid–strong base titration because its color change does not coincide with the steepest part of the curve.

**Figure 17.7**   Solutions containing extracts of red cabbage (obtained by boiling the cabbage in water) produce different colors when treated with an acid and a base. The pH of the solutions increases from left to right.

over a pH range that coincides with the steepest part of the titration curve. Consider the information in Figure 17.6, which shows the titration curves for hydrochloric acid and acetic acid—each being titrated with sodium hydroxide. Either of the indicators shown can be used for the titration of a strong acid with a strong base because both endpoints coincide with the steepest part of the HCl-NaOH titration curve. However, methyl red changes from red to yellow over the pH range of 4.2 to 6.3. This endpoint occurs significantly *before* the equivalence point in the titration of acetic acid, which occurs at about pH 8.7. Therefore, methyl red is *not* a suitable indicator for use in the titration of acetic acid with sodium hydroxide. Phenolphthalein, on the other hand, *is* a suitable indicator for the $CH_3COOH$-NaOH titration.

Many acid-base indicators are plant pigments. For example, boiling red cabbage in water extracts pigments that exhibit a variety of colors at different pH values (Figure 17.7).

Table 17.3 lists a number of indicators commonly used in acid-base titrations. The choice of indicator for a particular titration depends on the strength of the acid and base to be titrated.

Sample Problem 17.6 illustrates this point.

| TABLE 17.3 | Some Common Acid-Base Indicators | | |
|---|---|---|---|
| | **Color** | | |
| **Indicator** | **In Acid** | **In Base** | **pH Range** |
| Thymol blue | Red | Yellow | 1.2–2.8 |
| Bromophenol blue | Yellow | Bluish purple | 3.0–4.6 |
| Methyl orange | Orange | Yellow | 3.1–4.4 |
| Methyl red | Red | Yellow | 4.2–6.3 |
| Chlorophenol blue | Yellow | Red | 4.8–6.4 |
| Bromothymol blue | Yellow | Blue | 6.0–7.6 |
| Cresol red | Yellow | Red | 7.2–8.8 |
| Phenolphthalein | Colorless | Reddish pink | 8.3–10.0 |

## Sample Problem 17.6

Which indicator (or indicators) listed in Table 17.3 would you use for the acid-base titrations shown in (a) Figure 17.3, (b) Figure 17.4, and (c) Figure 17.5?

**Strategy** Determine the pH range that corresponds to the steepest part of each titration curve and select an indicator (or indicators) that changes color within that range.

**Setup** (a) The titration curve in Figure 17.3 is for the titration of a strong acid with a strong base. The steep part of the curve spans a pH range of about 4 to 10.

(b) Figure 17.4 shows the curve for the titration of a weak acid with a strong base. The steep part of the curve spans a pH range of about 7 to 10.

(c) Figure 17.5 shows the titration of a weak base with a strong acid. The steep part of the curve spans a pH range of about 7 to 3.

**Solution** (a) Most of the indicators listed in Table 17.3, with the exceptions of thymol blue, bromophenol blue, and methyl orange, would work for the titration of a strong acid with a strong base.

(b) Cresol red and phenolphthalein are suitable indicators.

(c) Bromophenol blue, methyl orange, methyl red, and chlorophenol blue are all suitable indicators.

**Think About It** If we don't select an appropriate indicator, the endpoint (color change) will not coincide with the equivalence point.

**Practice Problem A** Referring to Table 17.3, specify at least one indicator that would be suitable for the following titrations: (a) $CH_3NH_2$ with HBr, (b) $HNO_3$ with NaOH, (c) $HNO_2$ with KOH.

**Practice Problem B** For which of the bases in Table 16.7 could you titrate a 0.1 $M$ solution of base with 0.1 $M$ nitric acid using the indicator thymol blue?

## Checkpoint 17.3    Acid-Base Titrations

**17.3.1** For which of the following titrations will the pH at the equivalence point *not* be neutral? (Select all that apply.)

a) HCN with NaOH

b) HBr with NaOH

c) $NH_3$ with $HNO_3$

d) $HNO_2$ with KOH

e) HF with KOH

**17.3.2** Calculate the pH at the equivalence point in the titration of 30 mL of 0.25 $M$ $CH_3COOH$ with 0.25 $M$ KOH.

a) 7.00

b) 5.08

c) 8.92

d) 2.82

e) 11.18

**17.3.3** Calculate the pH after the addition of 25 mL of 0.10 $M$ NaOH to 50 mL of 0.10 $M$ HF.

a) 3.15

b) 9.31

c) 12.52

d) 1.48

e) 11.87

**17.3.4** Calculate the pH after the addition of 35 mL of 0.10 $M$ NaOH to 30 mL of 0.10 $M$ HCN.

a) 11.89

b) 2.11

c) 12.22

d) 1.78

e) 13.00

For Questions 17.3.5 to 17.3.8, refer to the following diagrams. The solutions shown represent various points in the titration of the weak acid HA with NaOH. (For clarity, the sodium ions and water molecules are not shown.)

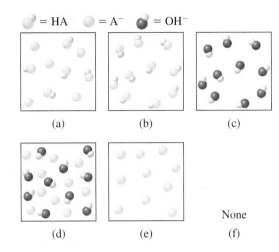

= HA  = A⁻  = OH⁻

(a) (b) (c)

(d) (e) (f) None

**17.3.5** Which diagram corresponds to the beginning of the titration, prior to the addition of any NaOH?

**17.3.6** Which diagram corresponds to the equivalence point of the titration? (Ignore salt hydrolysis.)

**17.3.7** Which diagram illustrates the solution *after* the equivalence point, when excess NaOH has been added?

**17.3.8** Which diagram corresponds to the point at which the pH is equal to the p$K_a$ of the weak acid?

# 17.4 Solubility Equilibria

The solubility of ionic compounds is important in industry, medicine, and everyday life. For example, barium sulfate ($BaSO_4$), an insoluble compound that is opaque to X rays, is used to diagnose ailments of the digestive tract. Tooth decay begins when tooth enamel, which is mainly made of hydroxyapatite [$Ca_5(PO_4)_3OH$], is made more soluble in saliva by the presence of acid.

The general rules for predicting the solubility of ionic compounds in water were introduced in Section 4.2. While these rules are useful, they do not allow us to make quantitative predictions about how much of a given ionic compound will dissolve in water. To develop a quantitative approach, we must start with the principles of chemical equilibrium. Unless otherwise stated, the solvent is water in the following discussion and the temperature is 25°C.

> **Student Annotation:** The compounds described as "insoluble" in Chapter 4 [◄◄ Section 4.2, Table 4.3] are actually very *slightly* soluble—each to a different degree.

## Solubility Product Expression and $K_{sp}$

Consider a saturated solution of silver chloride that is in contact with undissolved solid silver chloride. The equilibrium can be represented as

$$AgCl(s) \rightleftharpoons Ag^+(aq) + Cl^-(aq)$$

Although AgCl is not very soluble, *all* the AgCl that does dissolve in water dissociates completely into $Ag^+$ and $Cl^-$ ions. We can write the equilibrium expression for the dissociation of AgCl as

$$K_{sp} = [Ag^+][Cl^-]$$

where $K_{sp}$ is called the **solubility product constant.** ($K_{sp}$ is just another specially subscripted $K_c$, where "sp" stands for "solubility product.")

Because each AgCl unit contains only one $Ag^+$ and one $Cl^-$ ion, its solubility product expression is particularly simple to write. Many ionic compounds dissociate into more than two ions. Table 17.4 lists a number of slightly soluble ionic compounds along with equations representing their dissolution equilibria and their solubility product constants. (Compounds deemed "soluble" by the solubility rules in Chapter 4 are not listed for the same reason we did not list $K_a$ values for the strong acids in Table 16.6.) In general, the magnitude of $K_{sp}$ indicates the solubility of an ionic compound—the smaller the $K_{sp}$ value, the less soluble the compound. To make a direct comparison of $K_{sp}$ values, however, we must compare salts with similar formulas, such as AgCl with ZnS (one cation, one anion) or $CaF_2$ with $Fe(OH)_2$ (one cation, two anions).

> **Student Annotation:** A solubility product equilibrium expression is like any other equilibrium expression: $K$ is equal to the concentrations of products over the concentrations of reactants, each raised to its coefficient from the balanced chemical equation.
>
> Thus, for the process
>
> $$MX_n(s) \rightleftharpoons M^{n+}(aq) + nX^-(aq)$$
>
> the $K_{sp}$ expression is
>
> $$K_{sp} = [M^{n+}][X^-]^n$$
>
> $MX_n$ does not appear in the expression because, as for any heterogeneous equilibrium, the equilibrium expression does not include pure liquids or solids [◄◄ Section 15.3].

| TABLE 17.4 | Solubility Products of Some Slightly Soluble Ionic Compounds at 25°C | |
|---|---|---|

| Compound | Dissolution Equilibrium | $K_{sp}$ |
|---|---|---|
| Aluminum hydroxide | $Al(OH)_3(s) \rightleftharpoons Al^{3+}(aq) + 3OH^-(aq)$ | $1.8 \times 10^{-33}$ |
| Barium carbonate | $BaCO_3(s) \rightleftharpoons Ba^{2+}(aq) + CO_3^{2-}(aq)$ | $8.1 \times 10^{-9}$ |
| Barium fluoride | $BaF_2(s) \rightleftharpoons Ba^{2+}(aq) + 2F^-(aq)$ | $1.7 \times 10^{-6}$ |
| Barium sulfate | $BaSO_4(s) \rightleftharpoons Ba^{2+}(aq) + SO_4^{2-}(aq)$ | $1.1 \times 10^{-10}$ |
| Bismuth sulfide | $Bi_2S_3(s) \rightleftharpoons 2Bi^{3+}(aq) + 3S^{2-}(aq)$ | $1.6 \times 10^{-72}$ |
| Cadmium sulfide | $CdS(s) \rightleftharpoons Cd^{2+}(aq) + S^{2-}(aq)$ | $8.0 \times 10^{-28}$ |
| Calcium carbonate | $CaCO_3(s) \rightleftharpoons Ca^{2+}(aq) + CO_3^{2-}(aq)$ | $8.7 \times 10^{-9}$ |
| Calcium fluoride | $CaF_2(s) \rightleftharpoons Ca^{2+}(aq) + 2F^-(aq)$ | $4.0 \times 10^{-11}$ |
| Calcium hydroxide | $Ca(OH)_2(s) \rightleftharpoons Ca^{2+}(aq) + 2OH^-(aq)$ | $8.0 \times 10^{-6}$ |
| Calcium phosphate | $Ca_3(PO_4)_2(s) \rightleftharpoons 3Ca^{2+}(aq) + 2PO_4^{3-}(aq)$ | $1.2 \times 10^{-26}$ |
| Calcium sulfate | $CaSO_4(s) \rightleftharpoons Ca^{2+}(aq) + SO_4^{2-}(aq)$ | $2.4 \times 10^{-5}$ |
| Chromium(III) hydroxide | $Cr(OH)_3(s) \rightleftharpoons Cr^{3+}(aq) + 3OH^-(aq)$ | $3.0 \times 10^{-29}$ |
| Cobalt(II) sulfide | $CoS(s) \rightleftharpoons Co^{2+}(aq) + S^{2-}(aq)$ | $4.0 \times 10^{-21}$ |
| Copper(I) bromide | $CuBr(s) \rightleftharpoons Cu^+(aq) + Br^-(aq)$ | $4.2 \times 10^{-8}$ |
| Copper(I) iodide | $CuI(s) \rightleftharpoons Cu^+(aq) + I^-(aq)$ | $5.1 \times 10^{-12}$ |
| Copper(II) hydroxide | $Cu(OH)_2(s) \rightleftharpoons Cu^{2+}(aq) + 2OH^-(aq)$ | $2.2 \times 10^{-20}$ |
| Copper(II) sulfide | $CuS(s) \rightleftharpoons Cu^{2+}(aq) + S^{2-}(aq)$ | $6.0 \times 10^{-37}$ |
| Iron(II) hydroxide | $Fe(OH)_2(s) \rightleftharpoons Fe^{2+}(aq) + 2OH^-(aq)$ | $1.6 \times 10^{-14}$ |
| Iron(III) hydroxide | $Fe(OH)_3(s) \rightleftharpoons Fe^{3+}(aq) + 3OH^-(aq)$ | $1.1 \times 10^{-36}$ |
| Iron(III) phosphate | $FePO_4(s) \rightleftharpoons Fe^{3+}(aq) + PO_4^{3-}(aq)$ | $1.3 \times 10^{-22}$ |
| Iron(II) sulfide | $FeS(s) \rightleftharpoons Fe^{2+}(aq) + S^{2-}(aq)$ | $6.0 \times 10^{-19}$ |
| Lead(II) bromide | $PbBr_2(s) \rightleftharpoons Pb^{2+}(aq) + 2Br^-(aq)$ | $6.6 \times 10^{-6}$ |
| Lead(II) carbonate | $PbCO_3(s) \rightleftharpoons Pb^{2+}(aq) + CO_3^{2-}(aq)$ | $3.3 \times 10^{-14}$ |
| Lead(II) chloride | $PbCl_2(s) \rightleftharpoons Pb^{2+}(aq) + 2Cl^-(aq)$ | $2.4 \times 10^{-4}$ |
| Lead(II) chromate | $PbCrO_4(s) \rightleftharpoons Pb^{2+}(aq) + CrO_4^{2-}(aq)$ | $2.0 \times 10^{-14}$ |
| Lead(II) fluoride | $PbF_2(s) \rightleftharpoons Pb^{2+}(aq) + 2F^-(aq)$ | $4.0 \times 10^{-8}$ |
| Lead(II) iodide | $PbI_2(s) \rightleftharpoons Pb^{2+}(aq) + 2I^-(aq)$ | $1.4 \times 10^{-8}$ |
| Lead(II) sulfate | $PbSO_4(s) \rightleftharpoons Pb^{2+}(aq) + SO_4^-(aq)$ | $1.8 \times 10^{-8}$ |
| Lead(II) sulfide | $PbS(s) \rightleftharpoons Pb^{2+}(aq) + S^{2-}(aq)$ | $3.4 \times 10^{-28}$ |
| Magnesium carbonate | $MgCO_3(s) \rightleftharpoons Mg^{2+}(aq) + CO_3^{2-}(aq)$ | $4.0 \times 10^{-5}$ |
| Magnesium hydroxide | $Mg(OH)_2(s) \rightleftharpoons Mg^{2+}(aq) + 2OH^-(aq)$ | $1.2 \times 10^{-11}$ |
| Manganese(II) sulfide | $MnS(s) \rightleftharpoons Mn^{2+}(aq) + S^{2-}(aq)$ | $3.0 \times 10^{-14}$ |
| Mercury(I) bromide | $Hg_2Br_2(s) \rightleftharpoons Hg_2^{2+}(aq) + 2Br^-(aq)$ | $6.4 \times 10^{-23}$ |
| Mercury(I) chloride | $Hg_2Cl_2(s) \rightleftharpoons Hg_2^{2+}(aq) + 2Cl^-(aq)$ | $3.5 \times 10^{-18}$ |
| Mercury(I) sulfate | $Hg_2SO_4(s) \rightleftharpoons Hg_2^{2+}(aq) + SO_4^{2-}(aq)$ | $6.5 \times 10^{-7}$ |
| Mercury(II) sulfide | $HgS(s) \rightleftharpoons Hg^{2+}(aq) + S^{2-}(aq)$ | $4.0 \times 10^{-54}$ |
| Nickel(II) sulfide | $NiS(s) \rightleftharpoons Ni^{2+}(aq) + S^{2-}(aq)$ | $1.4 \times 10^{-24}$ |
| Silver bromide | $AgBr(s) \rightleftharpoons Ag^+(aq) + Br^-(aq)$ | $7.7 \times 10^{-13}$ |
| Silver carbonate | $Ag_2CO_3(s) \rightleftharpoons 2Ag^+(aq) + CO_3^{2-}(aq)$ | $8.1 \times 10^{-12}$ |

| TABLE 17.4 | Solubility Products of Some Slightly Soluble Ionic Compounds at 25°C *(continued)* | |
|---|---|---|

| Compound | Dissolution Equilibrium | $K_{sp}$ |
|---|---|---|
| Silver chloride | $AgCl(s) \rightleftharpoons Ag^+(aq) + Cl^-(aq)$ | $1.6 \times 10^{-10}$ |
| Silver chromate | $Ag_2CrO_4(s) \rightleftharpoons 2Ag^+(aq) + CrO_4^{2-}(aq)$ | $1.2 \times 10^{-12}$ |
| Silver iodide | $AgI(s) \rightleftharpoons Ag^+(aq) + I^-(aq)$ | $8.3 \times 10^{-17}$ |
| Silver sulfate | $Ag_2SO_4(s) \rightleftharpoons 2Ag^+(aq) + SO_4^{2-}(aq)$ | $1.5 \times 10^{-5}$ |
| Silver sulfide | $Ag_2S(s) \rightleftharpoons 2Ag^+(aq) + S^{2-}(aq)$ | $6.0 \times 10^{-51}$ |
| Strontium carbonate | $SrCO_3(s) \rightleftharpoons Sr^{2+}(aq) + CO_3^{2-}(aq)$ | $1.6 \times 10^{-9}$ |
| Strontium(II) hydroxide | $Sr(OH)_2(s) \rightleftharpoons Sr^{2+}(aq) + 2OH^-(aq)$ | $3.2 \times 10^{-4}$ |
| Strontium sulfate | $SrSO_4(s) \rightleftharpoons Sr^{2+}(aq) + SO_4^{2-}(aq)$ | $3.8 \times 10^{-7}$ |
| Tin(II) sulfide | $SnS(s) \rightleftharpoons Sn^{2+}(aq) + S^{2-}(aq)$ | $1.0 \times 10^{-26}$ |
| Zinc hydroxide | $Zn(OH)_2(s) \rightleftharpoons Zn^{2+}(aq) + 2OH^-(aq)$ | $1.8 \times 10^{-14}$ |
| Zinc sulfide | $ZnS(s) \rightleftharpoons Zn^{2+}(aq) + S^{2-}(aq)$ | $3.0 \times 10^{-23}$ |

## Calculations Involving $K_{sp}$ and Solubility

There are two ways to express the solubility of a substance: *molar solubility,* which is the number of moles of solute in 1 L of a saturated solution (mol/L), and *solubility,* which is the number of grams of solute in 1 L of a saturated solution (g/L). Both of these expressions refer to concentrations of saturated solutions at a particular temperature (usually 25°C).

Often we know the value of $K_{sp}$ for a compound and are asked to calculate the compound's molar solubility. The procedure for solving such a problem is essentially identical to the procedure for solving weak acid or weak base equilibrium problems:

1. Construct an equilibrium table.
2. Fill in what we know.
3. Figure out what we don't know.

For example, the $K_{sp}$ of silver bromide (AgBr) is $7.7 \times 10^{-13}$. We can construct an equilibrium table and fill in the starting concentrations of $Ag^+$ and $Br^-$ ions:

| $AgBr(s) \rightleftharpoons$ | $Ag^+(aq) +$ | $Br^-(aq)$ |
|---|---|---|
| *Initial concentration (M):* | 0 | 0 |
| *Change in concentration (M):* | | |
| *Equilibrium concentration (M):* | | |

Let $s$ be the molar solubility (in mol/L) of AgBr. Because one unit of AgBr yields one $Ag^+$ cation and one $Br^-$ anion, both $[Ag^+]$ and $[Br^-]$ are equal to $s$ at equilibrium:

| $AgBr(s) \rightleftharpoons$ | $Ag^+(aq) +$ | $Br^-(aq)$ |
|---|---|---|
| *Initial concentration (M):* | 0 | 0 |
| *Change in concentration (M):* | $+s$ | $+s$ |
| *Equilibrium concentration (M):* | $s$ | $s$ |

The equilibrium expression is

$$K_{sp} = [Ag^+][Br^-]$$

Therefore,

$$7.7 \times 10^{-13} = (s)(s)$$

and

$$s = \sqrt{7.7 \times 10^{-13}} = 8.8 \times 10^{-7} \, M$$

**Student Annotation:** Be careful not to confuse the terms *solubility* and $K_{sp}$. Solubility is the concentration of a saturated solution. $K_{sp}$ is an equilibrium constant.

Thus, the molar solubility of AgBr is $8.8 \times 10^{-7}$ M. Furthermore, we can express this solubility in g/L by multiplying the molar solubility by the molar mass of AgBr:

$$\frac{8.8 \times 10^{-7} \text{ mol AgBr}}{1 \text{ L}} \times \frac{187.8 \text{ g}}{1 \text{ mol AgBr}} = 1.7 \times 10^{-4} \text{ g/L}$$

Sample Problem 17.7 demonstrates this approach.

---

## Sample Problem 17.7

Calculate the solubility of copper(II) hydroxide [$Cu(OH)_2$] in g/L.

**Strategy**  Write the dissociation equation for $Cu(OH)_2$, and look up its $K_{sp}$ value in Table 17.4. Solve for molar solubility using the equilibrium expression. Convert molar solubility to solubility in g/L using the molar mass of $Cu(OH)_2$.

**Setup**  The equation for the dissociation of $Cu(OH)_2$ is

$$Cu(OH)_2(s) \rightleftharpoons Cu^{2+}(aq) + 2OH^-(aq)$$

and the equilibrium expression is $K_{sp} = [Cu^{2+}][OH^-]^2$. According to Table 17.4, $K_{sp}$ for $Cu(OH)_2$ is $2.2 \times 10^{-20}$. The molar mass of $Cu(OH)_2$ is 97.57 g/mol.

**Solution**

|  | $Cu(OH)_2(s) \rightleftharpoons$ | $Cu^{2+}(aq) +$ | $2OH^-(aq)$ |
|---|---|---|---|
| *Initial concentration (M):* |  | 0 | 0 |
| *Change in concentration (M):* |  | $+s$ | $+2s$ |
| *Equilibrium concentration (M):* |  | $s$ | $2s$ |

Therefore,

$$2.2 \times 10^{-20} = (s)(2s)^2 = 4s^3$$

$$s = \sqrt[3]{\frac{2.2 \times 10^{-20}}{4}} = 1.8 \times 10^{-7} \text{ M}$$

The molar solubility of $Cu(OH)_2$ is $1.8 \times 10^{-7}$ M. Multiplying by its molar mass gives

$$\text{solubility of } Cu(OH)_2 = \frac{1.8 \times 10^{-7} \text{ mol } Cu(OH)_2}{1 \text{ L}} \times \frac{97.57 \text{ g } Cu(OH)_2}{1 \text{ mol } Cu(OH)_2}$$

$$= 1.7 \times 10^{-5} \text{ g/L}$$

**Think About It**  Common errors arise in this type of problem when students neglect to raise an entire term to the appropriate power. For example, $(2s)^2$ is equal to $4s^2$ (not $2s^2$).

**Practice Problem A**  Calculate the molar solubility and the solubility in g/L of each salt at 25°C: (a) AgCl, (b) SnS, (c) SrCO_3.

**Practice Problem B**  Calculate the molar solubility and the solubility in g/L of each salt at 25°C: (a) PbF_2, (b) Ag_2CO_3, (c) Bi_2S_3.

---

We can also use molar solubility to determine the value of $K_{sp}$. Sample Problem 17.8 illustrates this procedure.

---

## Sample Problem 17.8

The solubility of calcium sulfate ($CaSO_4$) is measured experimentally and found to be 0.67 g/L. Calculate the value of $K_{sp}$ for calcium sulfate.

**Strategy**  Convert solubility to molar solubility using the molar mass of $CaSO_4$, and substitute the molar solubility into the equilibrium expression to determine $K_{sp}$.

**Setup** The molar mass of $CaSO_4$ is 136.2 g/mol. The molar solubility of $CaSO_4$ is

$$\text{molar solubility of } CaSO_4 = \frac{0.67 \text{ g } CaSO_4}{1 \text{ L}} \times \frac{1 \text{ mol } CaSO_4}{136.2 \text{ g } CaSO_4}$$

$$s = 4.9 \times 10^{-3} \text{ mol/L}$$

The equation and the equilibrium expression for the dissociation of $CaSO_4$ are

$$CaSO_4(s) \rightleftharpoons Ca^{2+}(aq) + SO_4^{2-}(aq) \quad \text{and} \quad K_{sp} = [Ca^{2+}][SO_4^{2-}]$$

**Solution** Substituting the molar solubility into the equilibrium expression gives

$$K_{sp} = (s)(s) = (4.9 \times 10^{-3})^2 = 2.4 \times 10^{-5}$$

> **Think About It** The $K_{sp}$ for $CaSO_4$ is relatively large (compared to many of the $K_{sp}$ values in Table 17.4). In fact, sulfates are listed as soluble compounds in Table 4.2, but calcium sulfate is listed as an insoluble exception. Remember that the term *insoluble* really refers to compounds that are *slightly* soluble, and that different sources may differ with regard to how soluble a compound must be to be considered "soluble."

**Practice Problem A** Given the solubility, calculate the solubility product constant ($K_{sp}$) of each salt at 25°C: (a) $PbCrO_4$, $s = 4.0 \times 10^{-5}$ g/L; (b) $BaC_2O_4$, $s = 0.29$ g/L; (c) $MnCO_3$, $s = 4.2 \times 10^{-6}$ g/L.

**Practice Problem B** Given the solubility, calculate the solubility product constant ($K_{sp}$) of each salt at 25°C: (a) $Ag_2SO_3$, $s = 4.6 \times 10^{-3}$ g/L; (b) $Hg_2I_2$, $s = 1.5 \times 10^{-7}$ g/L; (c) $Zn_3(PO_4)_2$, $s = 5.9 \times 10^{-5}$ g/L.

## Predicting Precipitation Reactions

For the dissociation of an ionic solid in water, any one of the following conditions may exist: (1) the solution is unsaturated, (2) the solution is saturated, or (3) the solution is supersaturated. For concentrations of ions that do not correspond to equilibrium conditions, we use the reaction quotient ($Q$) [◄◄ Section 15.2] to predict when a precipitate will form. Note that $Q$ has the same form as $K_{sp}$ except that the concentrations of ions are not equilibrium concentrations. For example, if we mix a solution containing $Ag^+$ ions with one containing $Cl^-$ ions, we write

$$Q = [Ag^+]_i[Cl^-]_i$$

The subscript "i" denotes that these are *initial* concentrations and do not necessarily correspond to those at equilibrium. If $Q$ is less than or equal to $K_{sp}$, no precipitate will form. If $Q$ is greater than $K_{sp}$, AgCl will precipitate out. (Precipitation will continue until the product of the ion concentrations is equal to $K_{sp}$.)

The ability to predict whether precipitation will occur often has practical value. In industrial and laboratory preparations, we can adjust the concentrations of ions until the ion product exceeds $K_{sp}$ to obtain a desired ionic compound (in the form of a precipitate). The ability to predict precipitation reactions is also useful in medicine. Kidney stones, which can be extremely painful, consist largely of calcium oxalate ($CaC_2O_4$, $K_{sp} = 2.3 \times 10^{-9}$). The normal physiological concentration of calcium ions in blood plasma is about $5 \times 10^{-3}$ $M$. Oxalate ions ($C_2O_4^{2-}$), derived from oxalic acid present in vegetables such as rhubarb and spinach, react with the calcium ions to form insoluble calcium oxalate, which can gradually build up in the kidneys. Proper adjustment of a patient's diet can help to reduce precipitate formation.

Sample Problem 17.9 demonstrates the steps involved in predicting precipitation reactions.

### Sample Problem 17.9

Predict whether a precipitate will form when each of the following is added to 650 mL of 0.0080 $M$ $K_2SO_4$: (a) 250 mL of 0.0040 $M$ $BaCl_2$; (b) 175 mL of 0.15 $M$ $AgNO_3$; (c) 325 mL of 0.25 $M$ $Sr(NO_3)_2$. (Assume volumes are additive.)

**Strategy** For each part, identify the compound that might precipitate and look up its $K_{sp}$ value in Table 17.4 or Appendix 3. Determine the concentrations of each compound's constituent ions, and use them to determine the value of the reaction quotient, $Q_{sp}$; then compare each reaction quotient with the value of the corresponding $K_{sp}$. If the reaction quotient is greater than $K_{sp}$, a precipitate will form.

*(Continued)*

**Setup** The compounds that might precipitate and their $K_{sp}$ values are (a) $BaSO_4$, $K_{sp} = 1.1 \times 10^{-10}$; (b) $Ag_2SO_4$, $K_{sp} = 1.5 \times 10^{-5}$; (c) $SrSO_4$, $K_{sp} = 3.8 \times 10^{-7}$.

**Solution** (a) Concentrations of the constituent ions of $BaSO_4$ are:

$$[Ba^{2+}] = \frac{250 \text{ mL} \times 0.0040 \, M}{650 \text{ mL} + 250 \text{ mL}} = 0.0011 \, M \quad \text{and} \quad [SO_4^{2-}] = \frac{650 \text{ mL} \times 0.0080 \, M}{650 \text{ mL} + 250 \text{ mL}} = 0.0058 \, M$$

Using these concentrations in the equilibrium expression, $[Ba^{2+}][SO_4^{2-}]$, gives a reaction quotient of $(0.0011)(0.0058) = 6.4 \times 10^{-6}$, which is greater than the $K_{sp}$ of $BaSO_4$ ($1.1 \times 10^{-10}$). Therefore, $BaSO_4$ will precipitate.

(b) Concentrations of the constituent ions of $Ag_2SO_4$ are:

$$[Ag^{+}] = \frac{175 \text{ mL} \times 0.15 \, M}{650 \text{ mL} + 175 \text{ mL}} = 0.0032 \, M \quad \text{and} \quad [SO_4^{2-}] = \frac{650 \text{ mL} \times 0.0080 \, M}{650 \text{ mL} + 175 \text{ mL}} = 0.0063 \, M$$

Using these concentrations in the equilibrium expression, $[Ag^{+}]^2[SO_4^{2-}]$, gives a reaction quotient of $(0.032)^2(0.0063) = 6.5 \times 10^{-6}$, which is less than the $K_{sp}$ of $Ag_2SO_4$ ($1.5 \times 10^{-5}$). Therefore, $Ag_2SO_4$ will not precipitate.

(c) Concentrations of the constituent ions of $SrSO4$ are:

$$[Sr^{2+}] = \frac{325 \text{ mL} \times 0.25 \, M}{650 \text{ mL} + 325 \text{ mL}} = 0.083 \, M \quad \text{and} \quad [SO_4^{2-}] = \frac{650 \text{ mL} \times 0.0080 \, M}{650 \text{ mL} + 325 \text{ mL}} = 0.0053 \, M$$

Using these concentrations in the equilibrium expression, $[Sr^{2+}][SO_4^{2-}]$, gives a reaction quotient of $(0.083)(0.0053) = 4.4 \times 10^{-4}$, which is greater than the $K_{sp}$ of $SrSO_4$ ($3.8 \times 10^{-7}$). Therefore, $SrSO_4$ will precipitate.

**Think About It** Students sometimes have difficulty deciding what compound might precipitate. Begin by writing down the constituent ions in the two solutions before they are combined. Consider the two possible combinations: the cation from the first solution and the anion from the second, or vice versa. You can consult the information in Tables 4.2 and 4.3 to determine whether one of the combinations is insoluble. Also keep in mind that only an *insoluble* salt will have a tabulated $K_{sp}$ value.

**Practice Problem A** Predict whether a precipitate will form from each of the following combinations: (a) 25 mL of $1 \times 10^{-5} \, M$ $Co(NO_3)_2$ and 75 mL of $5 \times 10^{-4} \, M$ $Na_2S$; (b) 500 mL of $7.5 \times 10^{-4} \, M$ $AlCl_3$ and 100 mL of $1.7 \times 10^{-5} \, M$ $Hg_2(NO_3)_2$; (c) 1.5 L of 0.025 $M$ $BaCl_2$ and 1.25 L of 0.014 $M$ $Pb(NO_3)_2$.

**Practice Problem B** What is the maximum mass (in grams) of each of the following soluble salts that can be added to 150 mL of 0.050 $M$ $BaCl_2$ without causing a precipitate to form: (a) $(NH_4)_2SO_4$, (b) $Pb(NO_3)_2$, (c) NaF? (Assume that the addition of solid causes no change in volume.)

---

## Checkpoint 17.4    Solubility Equilibria

**17.4.1** Using the $K_{sp}$ for aluminum hydroxide [$Al(OH)_3$], calculate its molar solubility.

a) $2.9 \times 10^{-9} \, M$

b) $7.7 \times 10^{-12} \, M$

c) $4.2 \times 10^{-17} \, M$

d) $8.4 \times 10^{-12} \, M$

e) $3.8 \times 10^{-9} \, M$

**17.4.2** What precipitate will form if 0.10 $M$ solutions of $Pb(NO_3)_2$ and NaI are mixed?

a) $Pb(NO_3)_2$

b) NaI

c) $PbI_2$

d) $NaNO_3$

e) None

**17.4.3** The diagrams represent saturated solutions of MX, MY, and MZ, each in equilibrium with its solid. Each solution also contains the soluble nitrate salt of $M^{n+}$. Arrange the solutions in order of increasing $K_{sp}$. (For clarity, the water molecules and nitrate ions are not shown.)

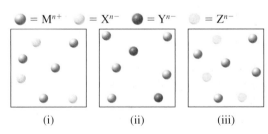

(i)        (ii)        (iii)

a) i < iii < ii

b) ii < iii < i

c) i = ii < iii

d) iii < ii = i

e) ii < i < iii

**17.4.4** The diagram represents a saturated solution of the salt $MA_2$. Which soluble salt will cause the precipitation of the greatest quantity of $MA_2$ when 0.1 mole is dissolved in the saturated solution? (For clarity, the water molecules are not shown.)

$\bullet = M^{2+}$   $\bullet = A^-$

a) NaA

b) $BaA_2$

c) $AlA_3$

d) $M(NO_3)_2$

e) $M_3(PO_4)_2$

**17.4.5** Diagrams 1 and 2 represent a saturated solution of the slightly soluble salt $M_2B$ and a solution of the soluble salt $MNO_3$, respectively. Which of the solutions of the soluble salt $Na_2B$ can be added to the solution of $MNO_3$ without causing a precipitate to form? Select all that apply. Assume that the solution volumes are equal, and that they are additive when combined. (For clarity, the water molecules, nitrate ions, and sodium ions are not shown.)

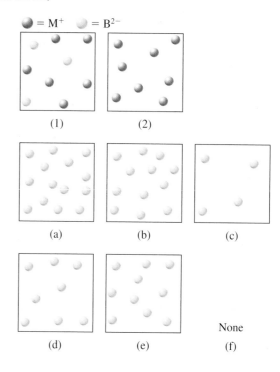

$\bullet = M^+$    $\bullet = B^{2-}$

(1)          (2)

(a)          (b)          (c)

(d)          (e)          None (f)

17.5 # Factors Affecting Solubility

In this section we will examine the effect of several factors on solubility including the common ion effect, pH, and the formation of complex ions.

## The Common Ion Effect

The solubility product is an equilibrium constant, and precipitation of an ionic compound from solution occurs whenever the ion product exceeds the $K_{sp}$ for that substance. In a saturated solution of AgCl, for example, the ion product $[Ag^+][Cl^-]$ is equal to $K_{sp}$. The solubility of AgCl in water can be calculated as follows, using the procedure introduced in Section 17.4:

$$1.6 \times 10^{-10} = [Ag^+][Cl^-]$$

In a solution in which AgCl is the only solute, $[Ag^+] = [Cl^-]$. Therefore,

$$1.6 \times 10^{-10} = s^2$$

and $s = 1.3 \times 10^{-5}\ M$. Thus, the solubility of AgCl in water at 25°C is $1.3 \times 10^{-5}\ M$.

Now suppose we want to determine the solubility of AgCl in a solution already containing a solute that has an ion in common with AgCl. For example, consider dissolving AgCl in a 0.10 $M$ solution of $AgNO_3$. In this case, the $Ag^+$ and $Cl^-$ concentrations at equilibrium will not be equal. In fact, the $Ag^+$ ion concentration will be equal to 0.10 $M$ *plus* the concentration contributed by AgCl. The equilibrium expression is

$$1.6 \times 10^{-10} = [Ag^+][Cl^-] = (0.10 + s)(s)$$

Because we expect $s$ to be very small, we can simplify this calculation as follows:

$$(0.10 + s)\ M \approx 0.10\ M$$

**Student Annotation:** $s$ still represents the concentration of $Cl^-$ at equilibrium and the *solubility* of AgCl.

Therefore,

$$1.6 \times 10^{-10} = 0.10s$$

**Student Annotation:** The common ion effect is an example of Le Châtelier's principle [◄◄ Section 15.5].

and $s = 1.6 \times 10^{-9}$ M. Thus, AgCl is *significantly* less soluble in 0.10 M AgNO$_3$ than in pure water—due to the common ion effect. Figure 17.8 (pp. 756–57) illustrates the common ion effect. Sample Problem 17.10 shows how the common ion effect affects solubility.

Media Player/MPEG Animation: Figure 17.8, Common Ion Effect, pp. 756–757.

---

### Sample Problem 17.10

Calculate the molar solubility of silver chloride in a solution that is $6.5 \times 10^{-3}$ M in silver nitrate.

**Strategy**  Silver nitrate is a strong electrolyte that dissociates completely in water. Therefore, the concentration of Ag$^+$ before any AgCl dissolves is $6.5 \times 10^{-3}$ M. Use the equilibrium expression, the $K_{sp}$ for AgCl, and an equilibrium table to determine how much AgCl will dissolve.

**Setup**  The dissolution equilibrium and the equilibrium expression are

$$\text{AgCl}(s) \rightleftharpoons \text{Ag}^+(aq) + \text{Cl}^-(aq) \qquad 1.6 \times 10^{-10} = [\text{Ag}^+][\text{Cl}^-]$$

**Solution**

| | AgCl(s) $\rightleftharpoons$ | Ag$^+$(aq) | + Cl$^-$(aq) |
|---|---|---|---|
| *Initial concentration (M):* | | $6.5 \times 10^{-3}$ | 0 |
| *Change in concentration (M):* | | $+s$ | $+s$ |
| *Equilibrium concentration (M):* | | $6.5 \times 10^{-3} + s$ | $s$ |

Substituting these concentrations into the equilibrium expression gives

$$1.6 \times 10^{-10} = (6.5 \times 10^{-3} + s)(s)$$

We expect $s$ to be very small, so

$$6.5 \times 10^{-3} + s \approx 6.5 \times 10^{-3}$$

and

$$1.6 \times 10^{-10} = (6.5 \times 10^{-3})(s)$$

Thus

$$s = \frac{1.6 \times 10^{-10}}{6.5 \times 10^{-3}} = 2.5 \times 10^{-8} \text{ M}$$

**Think About It**  The molar solubility of AgCl in water is $\sqrt{1.6 \times 10^{-10}} = 1.3 \times 10^{-5}$ M. The presence of $6.5 \times 10^{-3}$ M AgNO$_3$ reduces the solubility of AgCl by a factor of ~500.

Therefore, the molar solubility of AgCl in $6.5 \times 10^{-3}$ M AgNO$_3$ is $2.5 \times 10^{-8}$ M.

**Practice Problem A**  Calculate the molar solubility of AgI in (a) pure water and (b) 0.0010 M NaI.

**Practice Problem B**  Arrange the following salts in order of increasing molar solubility in 0.0010 M AgNO$_3$: AgBr, Ag$_2$CO$_3$, AgCl, AgI, Ag$_2$S.

---

## pH

The solubility of a substance can also depend on the pH of the solution. Consider the solubility equilibrium of magnesium hydroxide:

$$\text{Mg(OH)}_2(s) \rightleftharpoons \text{Mg}^{2+}(aq) + 2\text{OH}^-(aq)$$

According to Le Châtelier's principle, adding OH$^-$ ions (increasing the pH) shifts the equilibrium from right to left, thereby decreasing the solubility of Mg(OH)$_2$. (This is actually another example of the common ion effect.) On the other hand, adding H$^+$ ions (decreasing the pH) shifts the equilibrium from left to right, and the solubility of Mg(OH)$_2$ *increases*. Thus, insoluble bases tend to dissolve in acidic solutions. Similarly, insoluble acids tend to dissolve in basic solutions.

To examine the effect of pH on the solubility of $Mg(OH)_2$, we first calculate the pH of a saturated $Mg(OH)_2$ solution:

$$K_{sp} = (s)(2s)^2 = 4s^3$$

$$4s^3 = 1.2 \times 10^{-11}$$

$$s^3 = 3.0 \times 10^{-12}$$

$$s = 1.4 \times 10^{-4} \, M$$

At equilibrium, therefore,

$$[OH^-] = 2(1.4 \times 10^{-4} \, M) = 2.8 \times 10^{-4} \, M$$

$$pOH = -\log (2.8 \times 10^{-4}) = 3.55$$

$$pH = 14.00 - 3.55 = 10.45$$

In a solution with a pH of *less* than 10.45, the solubility of $Mg(OH)_2$ would increase. The dissolution process and the effect of additional $H^+$ ions are summarized as follows:

$$Mg(OH)_2(s) \rightleftharpoons Mg^{2+}(aq) + 2OH^-(aq)$$

$$2H^+(aq) + 2OH^-(aq) \longrightarrow 2H_2O(l)$$

*Overall:* $\quad Mg(OH)_2(s) + 2H^+(aq) \rightleftharpoons Mg^{2+}(aq) + 2H_2O(l)$

If the pH of the medium were higher than 10.45, $[OH^-]$ would be higher and the solubility of $Mg(OH)_2$ would decrease because of the common ion $(OH^-)$ effect.

The pH also influences the solubility of salts that contain a basic anion. For example, the solubility equilibrium for $BaF_2$ is

$$BaF_2(s) \rightleftharpoons Ba^{2+}(aq) + 2F^-(aq)$$

and

$$K_{sp} = [Ba^{2+}][F^-]^2$$

In an acidic medium, the high $[H^+]$ will shift the following equilibrium to the left, consuming $F^-$:

$$HF(aq) \rightleftharpoons H^+(aq) + F^-(aq)$$

As the concentration of $F^-$ decreases, the concentration of $Ba^{2+}$ must increase to satisfy the equality $K_{sp} = [Ba^{2+}][F^-]^2$ and maintain the state of equilibrium. Thus, more $BaF_2$ dissolves. The process and the effect of pH on the solubility of $BaF_2$ can be summarized as follows:

$$BaF_2(s) \rightleftharpoons Ba^{2+}(aq) + 2F^-(aq)$$

$$2H^+(aq) + 2F^-(aq) \longrightarrow 2HF(aq)$$

*Overall:* $\quad BaF_2(s) + 2H^+(aq) \rightleftharpoons Ba^{2+}(aq) + 2HF(aq)$

The solubilities of salts containing anions that do *not* hydrolyze, such as $Cl^-$, $Br^-$, and $NO_3^-$, are unaffected by pH.

Sample Problem 17.11 demonstrates the effect of pH on solubility.

## Sample Problem 17.11

Which of the following compounds will be more soluble in acidic solution than in water: (a) CuS, (b) AgCl, (c) $PbSO_4$?

**Strategy** For each salt, write the dissociation equilibrium equation and determine whether it produces an anion that will react with $H^+$. Only an anion that is the conjugate base of a weak acid will react with $H^+$.

**Setup**
(a) $CuS(s) \rightleftharpoons Cu^{2+}(aq) + S^{2-}(aq)$
$S^{2-}$ is the conjugate base of the weak acid $HS^-$. $S^{2-}$ reacts with $H^+$ as follows:

$$S^{2-}(aq) + H^+(aq) \longrightarrow HS^-(aq)$$

*(Continued on p. 758)*

Figure 17.8

# Common Ion Effect

We prepare a saturated solution by adding AgCl to water and stirring.

In the resulting saturated solution, the concentrations of $Ag^+$ and $Cl^-$ are equal, and the product of their concentrations is equal to $K_{sp}$.

$$[Ag^+][Cl^-] = 1.6 \times 10^{-10}$$

Therefore, the concentrations are
$$[Ag^+] = 1.3 \times 10^{-5}\,M$$
$$\text{and } [Cl^-] = 1.3 \times 10^{-5}\,M$$

Neither AgCl nor a saturated solution of AgCl is purple. The color has been used in this illustration to clarify the process.

START

After filtering off the solid AgCl, we dissolve enough NaCl to make the concentration of $Cl^-$ = 1.0 $M$.

Because the concentration of $Cl^-$ is now larger, the product of $Ag^+$ and $Cl^-$ concentrations is no longer equal to $K_{sp}$.

$$[Ag^+][Cl^-] = (1.3 \times 10^{-5}\,M)(1.0\,M) > 1.6 \times 10^{-10}$$

In any solution saturated with AgCl at 25°C, the product of $[Ag^+]$ and $[Cl^-]$ must equal the $K_{sp}$ of AgCl. Therefore, AgCl will precipitate until the product of ion concentrations is again $1.6 \times 10^{-10}$.

Note that this causes nearly all the dissolved AgCl to precipitate. With a $Cl^-$ concentration of 1.0 $M$, the highest possible concentration of $Ag^+$ is $1.6 \times 10^{-10}\,M$.

$$[Ag^+](1.0\,M) = 1.6 \times 10^{-10}$$
$$\text{therefore, } [Ag^+] = 1.6 \times 10^{-10}\,M$$

The amount of AgCl precipitated is exaggerated for emphasis. The actual amount of AgCl would be extremely small.

## What's the point?

When two salts contain the same ion, the ion they both contain is called the "common ion." The solubility of a slightly soluble salt such as AgCl can be decreased by the addition of a *soluble* salt with a common ion. In this example, AgCl is precipitated by adding NaCl. AgCl could also be precipitated by adding a soluble salt containing the $Ag^+$ ion, such as $AgNO_3$.

(b) $AgCl(s) \rightleftharpoons Ag^+(aq) + Cl^-(aq)$

   $Cl^-$ is the conjugate base of the strong acid HCl. $Cl^-$ does not react with $H^+$.

(c) $PbSO_4(s) \rightleftharpoons Pb^{2+}(aq) + SO_4^{2-}(aq)$

   $SO_4^{2-}$ is the conjugate base of the weak acid $HSO_4^-$. It reacts with $H^+$ as follows:

$$SO_4^{2-}(aq) + H^+(aq) \longrightarrow HSO_4^-(aq)$$

A salt that produces an anion that reacts with $H^+$ will be more soluble in acid than in water.

**Solution** CuS and $PbSO_4$ are more soluble in acid than in water. (AgCl is no more or less soluble in acid than in water.)

> **Think About It** When a salt dissociates to give the conjugate base of a weak acid, $H^+$ ions in an acidic solution *consume* a product (base) of the dissolution. This drives the equilibrium to the right (more solid dissolves) according to Le Châtelier's principle.

**Practice Problem A** Determine if the following compounds are more soluble in acidic solution than in pure water: (a) $Ca(OH)_2$, (b) $Mg_3(PO_4)_2$, (c) $PbBr_2$.

**Practice Problem B** Other than those in Sample Problem 17.11 and those in Practice Problem A, list three salts that are more soluble in acidic solution than in pure water.

## Complex Ion Formation

A **complex ion** is an ion containing a central metal cation bonded to one or more molecules or ions. Complex ions are crucial to many chemical and biological processes. Here we will consider the effect of complex ion formation on solubility. In Chapter 22 we will discuss the chemistry of complex ions in more detail.

> **Student Annotation:** Lewis acid-base reactions in which a metal cation combines with a Lewis base result in the formation of complex ions.

   *Transition* metals have a particular tendency to form complex ions. For example, a solution of cobalt(II) chloride ($CoCl_2$) is pink because of the presence of the $Co(H_2O)_6^{2+}$ ions (Figure 17.9). When HCl is added, the solution turns blue because the complex ion $CoCl_4^{2-}$ forms:

$$Co^{2+}(aq) + 4Cl^-(aq) \rightleftharpoons CoCl_4^{2-}(aq)$$

Copper(II) sulfate ($CuSO_4$) dissolves in water to produce a blue solution. The hydrated copper(II) ions are responsible for this color; many other sulfates (e.g., $Na_2SO_4$) are colorless. Adding a few drops of concentrated ammonia solution to a $CuSO_4$ solution causes the formation of a light-blue precipitate, copper(II) hydroxide:

$$Cu^{2+}(aq) + 2OH^-(aq) \longrightarrow Cu(OH)_2(s)$$

The $OH^-$ ions are supplied by the ammonia solution. If more $NH_3$ is added, the blue precipitate redissolves to produce a beautiful dark-blue solution, this time due to the formation of the complex ion $Cu(NH_3)_4^{2+}$ (Figure 17.10):

$$Cu(OH)_2(s) + 4NH_3(aq) \rightleftharpoons Cu(NH_3)_4^{2+}(aq) + 2OH^-(aq)$$

Thus, the formation of the complex ion $Cu(NH_3)_4^{2+}$ increases the solubility of $Cu(OH)_2$.

   A measure of the tendency of a metal ion to form a particular complex ion is given by the **formation constant ($K_f$)** (also called the stability constant), which is the equilibrium constant for the complex ion formation. The larger $K_f$ is, the more stable the complex ion is. Table 17.5 (page 762) lists the formation constants of a number of complex ions.

   The formation of the $Cu(NH_3)_4^{2+}$ ion can be expressed as

$$Cu^{2+}(aq) + 4NH_3(aq) \rightleftharpoons Cu(NH_3)_4^{2+}(aq)$$

The corresponding formation constant is

$$K_f = \frac{[Cu(NH_3)_4^{2+}]}{[Cu^{2+}][NH_3]^4} = 5.0 \times 10^{13}$$

The large value of $K_f$ in this case indicates that the complex ion is very stable in solution and accounts for the very low concentration of copper(II) ions at equilibrium.

   Recall that $K$ for the sum of two reactions is the product of the individual $K$ values [◄◄ Section 15.3]. The dissolution of silver chloride is represented by the equation

$$AgCl(s) \rightleftharpoons Ag^+(aq) + Cl^-(aq)$$

**Figure 17.9** (Left) An aqueous cobalt(II) chloride solution. The pink color is due to the presence of $Co(H_2O)_6^{2+}$ ions. (Right) After the addition of HCl solution, the solution turns blue because of the formation of the complex $CoCl_4^{2-}$ ions.

**Figure 17.10** (Left) An aqueous solution of copper(II) sulfate. (Center) After the addition of a few drops of concentrated aqueous ammonia solution, a light-blue precipitate of $Cu(OH)_2$ is formed. (Right) When more concentrated aqueous ammonia solution is added, the $Cu(OH)_2$ precipitate dissolves to form the dark-blue complex ion $Cu(NH_3)_4^{2+}$.

The sum of this equation and the one representing the formation of $Ag(NH_3)_2^+$ is

$$AgCl(s) \rightleftharpoons \cancel{Ag^+(aq)} + Cl^-(aq) \qquad K_{sp} = 1.6 \times 10^{-10}$$

$$\underline{\cancel{Ag^+(aq)} + 2NH_3(aq) \rightleftharpoons Ag(NH_3)_2^+(aq)} \qquad K_f = 1.5 \times 10^7$$

$$AgCl(s) + 2NH_3(aq) \rightleftharpoons Ag(NH_3)_2^+(aq) + Cl^-(aq)$$

and the corresponding equilibrium constant is $(1.6 \times 10^{-10})(1.5 \times 10^7) = 2.4 \times 10^{-3}$. This is significantly larger than the $K_{sp}$ value, indicating that much more AgCl will dissolve in the presence of aqueous ammonia than in pure water. In general, the effect of complex ion formation generally is to *increase* the solubility of a substance.

**Student Annotation:** Formation of a complex ion *consumes* the metal ion produced by the dissociation of a salt, increasing the salt's solubility simply due to Le Châtelier's principle [◀◀ Section 15.5].

# How Do I Solve Equilibrium Problems Involving Complex Ion Formation?

Many of the equilibrium problems we encounter involve equilibrium constants, such as $K_a$, $K_b$, and $K_{sp}$, that are very small. Often, this enables us to neglect the unknown $x$ in the denominator of the equilibrium expression, which simplifies the math necessary to solve the problem [◀◀ Section 16.5]. The solution of an equilibrium problem involving complex ion formation is complicated both by the magnitude of $K_f$ and by the stoichiometry of the reaction. Consider the combination of aqueous copper(II) ions and ammonia to form the complex ion $Cu(NH_3)_4^{2+}$:

$$Cu^{2+}(aq) + 4NH_3(aq) \rightleftharpoons Cu(NH_3)_4^{2+}(aq) \qquad K_f = 5.0 \times 10^{14}$$

**Student Annotation:** The term *free* is used to refer to a metal ion that is *not* part of a complex ion.

Let's say we wish to determine the molar concentration of free copper(II) ion in solution when 0.10 mole of $Cu(NO_3)_2$ is dissolved in a liter of 3.0 $M$ $NH_3$. We cannot solve this with the same approach we used to determine the pH of a weak acid solution. Not only can we not neglect $x$, the amount of copper(II) ion consumed in the reaction, but having to raise the ammonia concentration to the fourth power in the equilibrium expression results in an equation that is not easily solved. Another approach is needed.

Because the magnitude of $K_f$ is so large, we begin by assuming that *all* the copper(II) ion is consumed to form the complex ion. Then we consider the equilibrium in terms of the *reverse* reaction; that is, the dissociation of $Cu(NH_3)_4^{2+}$, for which the equilibrium constant is the reciprocal of $K_f$.

$$Cu(NH_3)_4^{2+}(aq) \rightleftharpoons Cu^{2+}(aq) + 4NH_3(aq) \qquad K = 2.0 \times 10^{-15}$$

Now we construct an equilibrium table and, because this $K$ is so small, we can expect $x$ (the amount of complex ion that dissociates) to be insignificant compared to the concentration of the complex ion and the concentration of ammonia. [Note that the concentration of ammonia, which had been 3.0 $M$, has been diminished by $4 \times 0.10$ $M$ due to the amount required to complex 0.10 mole of copper(II) ion.]

$$Cu(NH_3)_4^{2+}(aq) \rightleftharpoons Cu^{2+}(aq) + 4NH_3(aq)$$

|  | $Cu(NH_3)_4^{2+}$ | $Cu^{2+}$ | $4NH_3$ |
|---|---|---|---|
| *Initial concentration (M):* | 0.10 | 0 | 2.6 |
| *Change in concentration (M):* | $-x$ | $+x$ | $+4x$ |
| *Equilibrium concentration (M):* | $0.10 - x$ | $x$ | $2.6 + 4x$ |

We can neglect $x$ with respect to the concentrations of $Cu(NH_3)_4^{2+}$ and $NH_3$ ($0.10 - x \approx 0.10$ and $2.6 + 4x \approx 2.6$), and the solution becomes

$$\frac{[Cu^{2+}][NH_3]^4}{[Cu(NH_3)_4^{2+}]} = \frac{x(2.6)^4}{0.10} = 2.0 \times 10^{-15}$$

and $x = 4.4 \times 10^{-18}$ $M$. Note that because the formation constant is so large, the amount of copper that remains uncomplexed is extremely small. As always, it is a good idea to check the answer by plugging it into the equilibrium expression:

$$\frac{(4.4 \times 10^{-18})[2.6 + 4(4.4 \times 10^{-18})]^4}{0.10 - 4.4 \times 10^{-18}} = 2.0 \times 10^{-15}$$

Sample Problem 17.12 lets you practice applying this approach to a complex-ion formation equilibrium problem.

## Sample Problem 17.12

In the presence of aqueous cyanide, cadmium(II) forms the complex ion $Cd(CN)_4^{2-}$. Determine the molar concentration of free (uncomplexed) cadmium(II) ion in solution when 0.20 mole of $Cd(NO_3)_2$ is dissolved in a liter of 2.0 $M$ sodium cyanide (NaCN).

**Strategy** Because formation constants are typically very large, we begin by assuming that all the $Cd^{2+}$ ion is consumed and converted to complex ion. We then determine how much $Cd^{2+}$ is produced by the subsequent dissociation of the complex ion, a process for which the equilibrium constant is the reciprocal of $K_f$.

**Setup** From Table 17.5, the formation constant ($K_f$) for the complex ion $Cd(CN)_4^{2-}$ is $7.1 \times 10^{16}$. The reverse process,

$$Cd(CN)_4^{2-}(aq) \rightleftharpoons Cd^{2+}(aq) + 4CN^-(aq)$$

has an equilibrium constant of $1/K_f = 1.4 \times 10^{-17}$. The equilibrium expression for the dissociation is

$$1.4 \times 10^{-17} = \frac{[Cd^{2+}][CN^-]^4}{[Cd(CN)_4^{2-}]}$$

The formation of complex ion will consume some of the cyanide originally present. Stoichiometry indicates that four $CN^-$ ions are required to react with one $Cd^{2+}$ ion. Therefore, the concentration of $CN^-$ that we enter in the top row of the equilibrium table will be $[2.0\ M - 4(0.20\ M)] = 1.2\ M$.

**Solution** We construct an equilibrium table:

$$Cd(CN)_4^{2-}(aq) \rightleftharpoons Cd^{2+}(aq) + 4CN^-(aq)$$

| | | | |
|---|---|---|---|
| Initial concentration (M): | 0.20 | 0 | 1.2 |
| Change in concentration (M): | $-x$ | $+x$ | $+4x$ |
| Equilibrium concentration (M): | $0.20 - x$ | $x$ | $1.2 + 4x$ |

and, because the magnitude of $K$ is so *small*, we can neglect $x$ with respect to the initial concentrations of $Cd(CN)_4^{2-}$ and $CN^-$ ($0.20 - x \approx 0.20$ and $1.2 + 4x \approx 1.2$), so the solution becomes

$$\frac{[Cd^{2+}][CN^-]^4}{[Cd(CN)_4^{2-}]} = \frac{x(1.2)^4}{0.20} = 1.4 \times 10^{-17}$$

and $x = 1.4 \times 10^{-18}\ M$.

**Think About It** When you assume that all the metal ion is consumed and converted to complex ion, it's important to remember that some of the complexing agent (in this case, $CN^-$ ion) is consumed in the process. Don't forget to adjust its concentration accordingly before entering it in the top row of the equilibrium table.

**Practice Problem A** In the presence of aqueous ammonia, cobalt(III) forms the complex ion $Co(NH_3)_6^{3+}$. Determine the molar concentration of free cobalt(III) ion in solution when 0.15 mole of $Co(NO_3)_3$ is dissolved in a liter of 2.5 $M$ aqueous ammonia.

**Practice Problem B** Use information from Tables 17.4 and 17.5 to determine the molar solubility of chromium(III) hydroxide in a buffered solution with pH = 11.45.

| TABLE 17.5 | Formation Constants of Selected Complex Ions in Water at 25°C | |
|---|---|---|
| **Complex Ion** | **Equilibrium Expression** | **Formation Constant ($K_f$)** |
| $Ag(NH_3)_2^+$ | $Ag^+ + 2NH_3 \rightleftharpoons Ag(NH_3)_2^+$ | $1.5 \times 10^7$ |
| $Ag(CN)_2^-$ | $Ag^+ + 2CN^- \rightleftharpoons Ag(CN)_2^-$ | $1.0 \times 10^{21}$ |
| $Cu(CN)_4^{2-}$ | $Cu^{2+} + 4CN^- \rightleftharpoons Cu(CN)_4^{2-}$ | $1.0 \times 10^{25}$ |
| $Cu(NH_3)_4^{2+}$ | $Cu^{2+} + 4NH_3 \rightleftharpoons Cu(NH_3)_4^{2+}$ | $5.0 \times 10^{13}$ |
| $Cd(CN)_4^{2-}$ | $Cd^{2+} + 4CN^- \rightleftharpoons Cd(CN)_4^{2-}$ | $7.1 \times 10^{16}$ |
| $CdI_4^{2-}$ | $Cd^{2+} + 4I^- \rightleftharpoons CdI_4^{2-}$ | $2.0 \times 10^6$ |
| $HgCl_4^{2-}$ | $Hg^{2+} + 4Cl^- \rightleftharpoons HgCl_4^{2-}$ | $1.7 \times 10^{16}$ |
| $HgI_4^{2-}$ | $Hg^{2+} + 4I^- \rightleftharpoons HgI_4^{2-}$ | $2.0 \times 10^{30}$ |
| $Hg(CN)_4^{2-}$ | $Hg^{2+} + 4CN^- \rightleftharpoons Hg(CN)_4^{2-}$ | $2.5 \times 10^{41}$ |
| $Co(NH_3)_6^{3+}$ | $Co^{3+} + 6NH_3 \rightleftharpoons Co(NH_3)_6^{3+}$ | $5.0 \times 10^{31}$ |
| $Zn(NH_3)_4^{2+}$ | $Zn^{2+} + 4NH_3 \rightleftharpoons Zn(NH_3)_4^{2+}$ | $2.9 \times 10^9$ |
| $Cr(OH)_4^-$ | $Cr^{3+} + 4OH^- \rightleftharpoons Cr(OH)_4^-$ | $8 \times 10^{29}$ |

Finally, there is a class of hydroxides, called amphoteric hydroxides, which can react with both acids and bases. Examples are $Al(OH)_3$, $Pb(OH)_2$, $Cr(OH)_3$, $Zn(OH)_2$, and $Cd(OH)_2$. $Al(OH)_3$ reacts with acids and bases as follows:

$$Al(OH)_3(s) + 3H^+(aq) \longrightarrow Al^{3+}(aq) + 3H_2O(l)$$

$$Al(OH)_3(s) + OH^-(aq) \rightleftharpoons Al(OH)_4^-(aq)$$

The increase in solubility of $Al(OH)_3$ in a basic medium is the result of the formation of the complex ion $Al(OH)_4^-$ in which $Al(OH)_3$ acts as the Lewis acid and $OH^-$ acts as the Lewis base. Other amphoteric hydroxides react similarly with acids and bases.

## Checkpoint 17.5    Factors Affecting Solubility

**17.5.1** Calculate the molar solubility of AgCl in 0.10 $M$ $CaCl_2$.

a) $1.6 \times 10^{-10} M$

b) $1.6 \times 10^{-9} M$

c) $8.0 \times 10^{-10} M$

d) $1.3 \times 10^{-5} M$

e) $2.6 \times 10^{-8} M$

**17.5.2** Which of the following substances will be more soluble in *acidic* solution than in pure water? (Select all that apply.)

a) $PbCO_3$

b) $AgS$

c) $AgI$

d) $Fe(OH)_3$

e) $CaF_2$

**17.5.3** In which of the solutions would the slightly soluble salt $MC_2$ be *most* soluble, and in which would it be *least* soluble? (For clarity, water molecules and counter ions in soluble salts are not shown.)

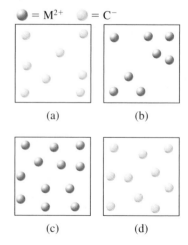

⬤ = $M^{2+}$    ⦾ = $C^-$

(a)          (b)

(c)          (d)

# 17.6 Separation of Ions Using Differences in Solubility

In chemical analysis, it sometimes is necessary to remove one type of ion from solution by precipitation while leaving other ions in solution. For instance, the addition of sulfate ions to a solution containing both potassium and barium ions causes $BaSO_4$ to precipitate out, thereby removing most of the $Ba^{2+}$ ions from the solution. The other "product," $K_2SO_4$, is soluble and will remain in solution. The $BaSO_4$ precipitate can be separated from the solution by filtration.

## Fractional Precipitation

Even when both products are insoluble, we can still achieve some degree of separation by choosing the proper reagent to bring about precipitation. Consider a solution that contains $Cl^-$, $Br^-$, and $I^-$ ions. One way to separate these ions is to convert them to insoluble silver halides. As the $K_{sp}$ values in the margin show, the solubility of the silver halides decreases from AgCl to AgI. Thus, when a soluble compound such as silver nitrate is slowly added to this solution, AgI begins to precipitate first, followed by AgBr, and then AgCl. This practice is known as *fractional precipitation.*

Sample Problem 17.13 describes the separation of only two ions ($Cl^-$ and $Br^-$), but the procedure can be applied to a solution containing more than two different types of ions.

Multimedia

Solutions Precipitation—fractional crystallization.

**Student Annotation:**

| Compound | $K_{sp}$ |
|----------|----------|
| AgCl | $1.6 \times 10^{-10}$ |
| AgBr | $7.7 \times 10^{-13}$ |
| AgI | $8.3 \times 10^{-17}$ |

---

## Sample Problem 17.13

Silver nitrate is added slowly to a solution that is 0.020 $M$ in $Cl^-$ ions and 0.020 $M$ in $Br^-$ ions. Calculate the concentration of $Ag^+$ ions (in mol/L) required to initiate the precipitation of AgBr without precipitating AgCl.

**Strategy** Silver nitrate dissociates in solution to give $Ag^+$ and $NO_3^-$ ions. Adding $Ag^+$ ions in sufficient amount will cause the slightly soluble ionic compounds AgCl and AgBr to precipitate from solution. Knowing the $K_{sp}$ values for AgCl and AgBr (and the concentrations of $Cl^-$ and $Br^-$ already in solution), we can use the equilibrium expressions to calculate the maximum concentration of $Ag^+$ that can exist in solution without exceeding $K_{sp}$ for each compound.

**Setup** The solubility equilibria, $K_{sp}$ values, and equilibrium expressions for AgCl and AgBr are

$$AgCl(s) \rightleftharpoons Ag^+(aq) + Cl^-(aq) \qquad K_{sp} = 1.6 \times 10^{-10} = [Ag^+][Cl^-]$$

$$AgBr(s) \rightleftharpoons Ag^+(aq) + Br^-(aq) \qquad K_{sp} = 7.7 \times 10^{-13} = [Ag^+][Br^-]$$

Because the $K_{sp}$ for AgBr is smaller (by a factor of more than 200), AgBr should precipitate first; that is, it will require a lower concentration of added $Ag^+$ to begin precipitation. Therefore, we first solve for $[Ag^+]$ using the equilibrium expression for AgBr to determine the minimum $Ag^+$ concentration necessary to initiate precipitation of AgBr. We then solve for $[Ag^+]$ again, using the equilibrium expression for AgCl to determine the *maximum* $Ag^+$ concentration that can exist in the solution without initiating the precipitation of AgCl.

**Solution** Solving the AgBr equilibrium expression for $Ag^+$ concentration, we have

$$[Ag^+] = \frac{K_{sp}}{[Br^-]}$$

and

$$[Ag^+] = \frac{7.7 \times 10^{-13}}{0.020} = 3.9 \times 10^{-11} \, M$$

For AgBr to precipitate from solution, the silver ion concentration must exceed $3.9 \times 10^{-11} \, M$. Solving the AgCl equilibrium expression for the $Ag^+$ concentration, we have

$$[Ag^+] = \frac{K_{sp}}{[Cl^-]}$$

*(Continued)*

<table>
<tr><td>

**Think About It** If we continue adding $AgNO_3$ until the $Ag^+$ concentration is high enough to begin the precipitation of AgCl, the concentration of $Br^-$ remaining in solution can also be determined using the $K_{sp}$ expression:

$$[Br^-] = \frac{K_{sp}}{[Ag^+]} = \frac{7.7 \times 10^{-13}}{8.0 \times 10^{-9}}$$
$$= 9.6 \times 10^{-5}\ M$$

Thus, by the time AgCl begins to precipitate, $(9.6 \times 10^{-5}\ M) \div (0.020\ M) = 0.0048$, so less than 0.5 percent of the original bromide ion remains in the solution.

</td></tr>
</table>

and

$$[Ag^+] = \frac{1.6 \times 10^{-10}}{0.020} = 8.0 \times 10^{-9}\ M$$

For AgCl *not* to precipitate from solution, the silver ion concentration must stay below $8.0 \times 10^{-9}\ M$. Therefore, to precipitate the $Br^-$ ions without precipitating the $Cl^-$ from this solution, the $Ag^+$ concentration must be greater than $3.9 \times 10^{-11}\ M$ and less than $8.0 \times 10^{-9}\ M$.

**Practice Problem A** Lead(II) nitrate is added slowly to a solution that is 0.020 $M$ in $Cl^-$ ions. Calculate the concentration of $Pb^{2+}$ ions (in mol/L) required to initiate the precipitation of $PbCl_2$. ($K_{sp}$ for $PbCl_2$ is $2.4 \times 10^{-4}$.)

**Practice Problem B** Calculate the concentration of $Ag^+$ (in mol/L) necessary to initiate the precipitation of (a) AgCl and (b) $Ag_3PO_4$ from a solution in which $[Cl^-]$ and $[PO_4^{3-}]$ are each 0.10 $M$. ($K_{sp}$ for $Ag_3PO_4$ is $1.8 \times 10^{-18}$.)

## Qualitative Analysis of Metal Ions in Solution

The principle of selective precipitation can be used to identify the types of ions present in a solution. This practice is called *qualitative analysis.* There are about 20 common cations that can be analyzed readily in aqueous solution. These cations can be divided into five groups according to the solubility products of their insoluble salts (Table 17.4). Because an unknown solution may contain from 1 to all 20 ions, any analysis must be carried out systematically from group 1 through group 5. The general procedure for separating these 20 ions is as follows:

- *Group 1 cations.* When dilute HCl is added to the unknown solution, only the $Ag^+$, $Hg_2^{2+}$, and $Pb^{2+}$ ions precipitate as insoluble chlorides. The other ions, whose chlorides are soluble, remain in solution.
- *Group 2 cations.* After the chloride precipitates have been removed by filtration, hydrogen sulfide is added to the unknown solution, which is acidic due to the addition of HCl. Metal ions from group 2 react to produce metal sulfides:

$$M^{2+}(aq) + H_2S(aq) \rightleftharpoons MS(s) + 2H^+(aq)$$

In the presence of $H^+$, this equilibrium shifts to the *left.* Therefore, only the metal sulfides with the *smallest* $K_{sp}$ values precipitate under acidic conditions. These are $Bi_2S_3$, CdS, CuS, and SnS (see Table 17.4). The solution is then filtered to remove the insoluble sulfides.
- *Group 3 cations.* At this stage, sodium hydroxide is added to the solution to make it basic. In a basic solution, the metal sulfide equilibrium shifts to the right and the more soluble sulfides (CoS, FeS, MnS, NiS, ZnS) now precipitate out of solution. The $Al^{3+}$ and $Cr^{3+}$ ions actually precipitate as the hydroxides $Al(OH)_3$ and $Cr(OH)_3$, rather than as the sulfides, because the hydroxides are less soluble. The solution is filtered again to remove the insoluble sulfides and hydroxides.
- *Group 4 cations.* After all the group 1, 2, and 3 cations have been removed from solution, sodium carbonate is added to the basic solution to precipitate $Ba^{2+}$, $Ca^{2+}$, and $Sr^{2+}$ ions as $BaCO_3$, $CaCO_3$, and $SrCO_3$. These precipitates, too, are removed from solution by filtration.
- *Group 5 cations.* At this stage, the only cations possibly remaining in solution are $Na^+$, $K^+$, and $NH_4^+$. The presence of $NH_4^+$ ions can be determined by adding sodium hydroxide:

$$NaOH(aq) + NH_4^+(aq) \longrightarrow Na^+(aq) + H_2O(l) + NH_3(g)$$

The ammonia gas is detected either by its characteristic odor or by observing a wet piece of red litmus paper turning blue when placed above (not in contact with) the solution. To confirm the presence of $Na^+$ and $K^+$ ions, a flame test is usually used in which a piece of platinum wire (chosen because platinum is inert) is dipped into the original solution and then held over a Bunsen burner flame. $Na^+$ ions emit a yellow flame when heated in this manner, whereas $K^+$ ions emit a violet flame (Figure 17.11). Figure 17.12 summarizes this scheme for separating metal ions.

Multimedia

Chemical Analysis—flame tests of metals.

**Figure 17.11** Flame tests for sodium (yellow flame) and potassium (violet flame).

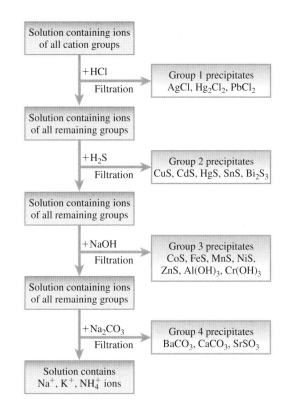

**Figure 17.12** A flowchart for the separation of cations in qualitative analysis.

---

| **Checkpoint 17.6** | **Separation of Ions Using Differences in Solubility** |
|---|---|

**17.6.1** A solution is 0.10 $M$ in $Br^-$, $CO_3^{2-}$, $Cl^-$, $I^-$, and $SO_4^{2-}$ ions. Which compound will precipitate first as silver nitrate is added to the solution?

a) AgBr

b) $Ag_2CO_3$

c) AgCl

d) AgI

e) $Ag_2SO_4$

**17.6.2** Barium nitrate is added slowly to a solution that is 0.10 $M$ in $SO_4^{2-}$ ions and 0.10 $M$ in $F^-$ ions. Calculate the concentration of $Ba^{2+}$ ions (in mol/L) required to initiate the precipitation of $BaSO_4$ without precipitating $BaF_2$.

a) $1.7 \times 10^{-6}\ M$

b) $1.1 \times 10^{-9}\ M$

c) $1.7 \times 10^{-4}\ M$

d) $1.7 \times 10^{-5}\ M$

e) $1.1 \times 10^{-8}\ M$

# Applying What You've Learned

Most toothpastes contain fluoride, which helps to reduce tooth decay. The $F^-$ ions in toothpaste replace some of the $OH^-$ ions during the remineralization process:

$$5Ca^{2+}(aq) + 3PO_4^{3-}(aq) + F^-(aq) \rightleftharpoons Ca_5(PO_4)_3F(s)$$

Because $F^-$ is a weaker base than $OH^-$, the modified enamel, called *fluoroapatite*, is more resistant to the acid produced by bacteria.

**Problems:**

a)  Calculate the molar solubility of hydroxyapatite given that its $K_{sp}$ is $2 \times 10^{-59}$.
    [◀◀ Sample Problem 17.7]

b)  Calculate the $K_{sp}$ of fluoroapatite given that its molar solubility is $7 \times 10^{-8}$ $M$.
    [◀◀ Sample Problem 17.8]

c)  Calculate the molar solubility of fluoroapatite in an aqueous solution in which the concentration of fluoride ion is 0.10 $M$. [◀◀ Sample Problem 17.10]

d)  Calculate the molar solubility of hydroxyapatite in a buffered aqueous solution with pH = 4.0. [◀◀ Sample Problem 17.10]

# CHAPTER SUMMARY

## Section 17.1

- The presence of a common ion suppresses the ionization of a weak acid or weak base. This is known as the ***common ion effect.***

- A common ion is added to a solution in the form of a *salt.*

## Section 17.2

- A solution that contains significant concentrations of both members of a conjugate acid-base pair (weak acid−conjugate base or weak base−conjugate acid) is a *buffer solution* or simply a ***buffer.***

- Buffer solutions resist pH change upon addition of small amounts of strong acid or strong base. Buffers are important to biological systems.

- The pH of a buffer can be calculated using an equilibrium table or with the ***Henderson-Hasselbalch equation.***

- The $pK_a$ of a weak acid is $-\log K_a$. When the weak acid and conjugate base concentrations in a buffer solution are equal, $pH = pK_a$.

- We can prepare a buffer with a specific pH by choosing a weak acid with a $pK_a$ close to the desired pH.

## Section 17.3

- The titration curve of a strong acid−strong base titration has a long, steep region near the equivalence point.

- Titration curves for weak acid−strong base or weak base−strong acid titrations have a significantly shorter steep region.

- The pH at the equivalence point of a strong acid−strong base titration is 7.00.

- The pH at the equivalence point of a weak acid−strong base titration is above 7.00.

- The pH at the equivalence point of a weak base−strong acid titration is below 7.00.

- Acid-base indicators are usually weak organic acids that exhibit two different colors depending on the pH of the solution. The ***endpoint*** of a titration is the point at which the color of the indicator changes. It is used to estimate the *equivalence point* of a titration.

- The indicator used for a particular titration should exhibit a color change in the pH range corresponding to the steep region of the titration curve.

## Section 17.4

- The ***solubility product constant ($K_{sp}$)*** is the equilibrium constant that indicates to what extent a slightly soluble ionic compound dissociates in water.

- $K_{sp}$ can be used to determine ***molar solubility*** or ***solubility*** in g/L, and vice versa.

- $K_{sp}$ can also be used to predict whether or not a precipitate will form when two solutions are mixed.

## Section 17.5

- Solubility is affected by common ions, pH, and complex ion formation. The ***formation constant ($K_f$)*** indicates to what extent complex ions form.

- A salt that dissociates to give a strong conjugate base such as fluoride ion will be more soluble in acidic solution than in pure water.

- A salt that dissociates to give hydroxide ion will be more soluble at lower pH and less soluble at higher pH.

- The solubility of an ionic compound increases when the formation of a ***complex ion*** consumes one of the products of dissociation.

## Section 17.6

- Ions can be separated using ***fractional precipitation.***

- Fractional precipitation schemes can be designed based on $K_{sp}$ values.

- Groups of cations can be identified through the use of selective precipitation. This is the basis of ***qualitative analysis.***

# KEY WORDS

Buffer, 729

Common ion effect, 729

Complex ion, 758

Endpoint, 744

Formation constant ($K_f$), 758

Fractional precipitation, 763

Henderson-Hasselbalch equation, 731

Molar solubility, 749

Qualitative analysis, 764

Solubility, 749

Solubility product constant ($K_{sp}$), 747

# KEY EQUATIONS

17.1   $\displaystyle pH = pK_a + \log \frac{[A^-]}{[HA]}$

17.2   $pK_a = -\log K_a$

17.3   $\displaystyle pH = pK_a + \log \frac{[\text{conjugate base}]}{[\text{weak acid}]}$

# QUESTIONS AND PROBLEMS

## Section 17.1: The Common Ion Effect

### Review Questions

17.1 Use Le Châtelier's principle to explain how the common ion effect affects the pH of a weak acid solution.

17.2 Describe the effect on pH (increase, decrease, or no change) that results from each of the following additions: (a) potassium acetate to an acetic acid solution, (b) ammonium nitrate to an ammonia solution, (c) sodium formate (HCOONa) to a formic acid (HCOOH) solution, (d) potassium chloride to a hydrochloric acid solution, (e) barium iodide to a hydroiodic acid solution.

17.3 Define $pK_a$ for a weak acid. What is the relationship between the value of the $pK_a$ and the strength of the acid?

17.4 The $pK_a$'s of two monoprotic acids HA and HB are 5.9 and 8.1, respectively. Which of the two is the stronger acid?

### Problems

17.5 Determine the pH of (a) a 0.40 $M$ CH$_3$COOH solution and (b) a solution that is 0.40 $M$ CH$_3$COOH and 0.20 $M$ CH$_3$COONa.

17.6 Determine the pH of (a) a 0.20 $M$ NH$_3$ solution, and (b) a solution that is 0.20 $M$ NH$_3$ and 0.30 $M$ NH$_4$Cl.

## Section 17.2: Buffer Solutions

### Visualizing Chemistry
### Figure 17.1

VC 17.1 Which pair of substances can be dissolved together to prepare a buffer solution?

| CH$_3$COOH | CH$_3$COONa | NaOH | HCl | NaCl |
|:---:|:---:|:---:|:---:|:---:|
| (i) | (ii) | (iii) | (iv) | (v) |

a) i/ii
b) i/ii, i/iii, ii/iv
c) i/ii, i/iii, ii/iii, iv/v

VC 17.2 Consider the buffer shown in Figure 17.1, in which [CH$_3$COOH] = 0.10 $M$, [CH$_3$COO$^-$] = 0.10 $M$, and pH = 4.74. How would the pH of the buffer change if we doubled the volume by adding 100 mL of water?

a) The pH would increase.
b) The pH would decrease.
c) The pH would not change.

VC 17.3 Consider a buffer similar to the one shown in Figure 17.1, in which [CH$_3$COOH] = 0.50 $M$ and [CH$_3$COO$^-$] = 0.75 $M$. Which substance can we add more of without causing a drastic change in pH?

a) HCl
b) NaOH
c) Neither

VC 17.4 According to Figure 17.1, the reaction that occurs when strong acid is added to the buffer is

$$H^+(aq) + CH_3COO^-(aq) \longrightarrow CH_3COOH(aq)$$

Estimate the value of $K$ for this reaction.

a) $1 \times 10^{14}$
b) $6 \times 10^{-10}$
c) $6 \times 10^4$

### Review Questions

17.7 What is a buffer solution? What must a solution contain to be a buffer?

17.8 Using only a pH meter, water, and a graduated cylinder, how would you distinguish between an acid solution and a buffer solution at the same pH?

### Problems

17.9 Which of the following solutions can act as a buffer: (a) KCl/HCl, (b) KHSO$_4$/H$_2$SO$_4$, (c) Na$_2$HPO$_4$/NaH$_2$PO$_4$, (d) KNO$_2$/HNO$_2$?

17.10 Which of the following solutions can act as a buffer: (a) KCN/HCN, (b) Na$_2$SO$_4$/NaHSO$_4$, (c) NH$_3$/NH$_4$NO$_3$, (d) NaI/HI?

17.11 Calculate the pH of the buffer system made up of 0.15 $M$ NH$_3$/0.35 $M$ NH$_4$Cl.

17.12 Calculate the pH of the following two buffer solutions: (a) 2.0 $M$ CH$_3$COONa/2.0 $M$ CH$_3$COOH, (b) 0.20 $M$ CH$_3$COONa/0.20 $M$ CH$_3$COOH. Which is the more effective buffer? Why?

17.13 The pH of a bicarbonate-carbonic acid buffer is 8.00. Calculate the ratio of the concentration of carbonic acid (H$_2$CO$_3$) to that of the bicarbonate ion (HCO$_3^-$).

17.14 What is the pH of the buffer 0.10 $M$ Na$_2$HPO$_4$/0.15 $M$ KH$_2$PO$_4$?

17.15 The pH of a sodium acetate–acetic acid buffer is 4.50. Calculate the ratio [CH$_3$COO$^-$]/[CH$_3$COOH].

17.16 The pH of blood plasma is 7.40. Assuming the principal buffer system is HCO$_3^-$/H$_2$CO$_3$, calculate the ratio [HCO$_3^-$]/[H$_2$CO$_3$]. Is this buffer more effective against an added acid or an added base?

17.17 Calculate the pH of the 0.20 $M$ NH$_3$/0.20 $M$ NH$_4$Cl buffer. What is the pH of the buffer after the addition of 10.0 mL of 0.10 $M$ HCl to 65.0 mL of the buffer?

17.18 Calculate the pH of 1.00 L of the buffer 1.00 $M$ CH$_3$COONa/1.00 $M$ CH$_3$COOH before and after the addition of (a) 0.080 mol NaOH and (b) 0.12 mol HCl. (Assume that there is no change in volume.)

17.19 A diprotic acid, H$_2$A, has the following ionization constants: $K_{a_1} = 1.1 \times 10^{-3}$ and $K_{a_2} = 2.5 \times 10^{-6}$. To make up a buffer solution of pH 5.80, which combination would you choose: NaHA/H$_2$A or Na$_2$A/NaHA?

17.20 A student is asked to prepare a buffer solution at pH 8.60, using one of the following weak acids: HA ($K_a = 2.7 \times 10^{-3}$), HB ($K_a = 4.4 \times 10^{-6}$), HC ($K_a = 2.6 \times 10^{-9}$). Which acid should the student choose? Why?

17.21 The following diagrams contain one or more of the compounds: H$_2$A, NaHA, and Na$_2$A, where H$_2$A is a weak diprotic acid. (1) Which of the solutions can act as buffer solutions? (2) Which solution is the most effective buffer solution? Water molecules and Na$^+$ ions have been omitted for clarity.

= H$_2$A    = HA$^-$    = A$^{2-}$

(a)          (b)          (c)          (d)

17.22   The following diagrams represent solutions containing a weak
        acid HA (p$K_a$ = 5.0) and its sodium salt NaA. (1) Which solution
        has the lowest pH? Which has the highest pH? (2) How many
        different species are present after the addition of two $H^+$ ions to
        solution (a)? (3) How many different species are present after the
        addition of two $OH^-$ ions to solution (b)?

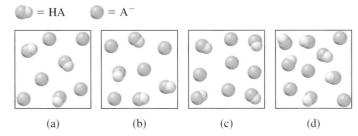

    ● = HA      ● = $A^-$

        (a)           (b)           (c)           (d)

## Section 17.3: Acid-Base Titrations

### Review Questions

17.23   Briefly describe what happens in an acid-base titration.

17.24   Sketch titration curves for the following acid-base titrations:
        (a) HCl versus NaOH, (b) HCl versus $CH_3NH_2$, (c) $CH_3COOH$
        versus NaOH. In each case, the base is added to the acid in an
        Erlenmeyer flask. Your graphs should show the pH on the y axis
        and the volume of base added on the x axis.

17.25   Explain how an acid-base indicator works in a titration. What are the
        criteria for choosing an indicator for a particular acid-base titration?

17.26   The amount of indicator used in an acid-base titration must be
        small. Why?

### Problems

17.27   A 0.2688-g sample of a monoprotic acid neutralizes 16.4 mL of
        0.08133 M KOH solution. Calculate the molar mass of the acid.

17.28   A 5.00-g quantity of a diprotic acid was dissolved in water and
        made up to exactly 250 mL. Calculate the molar mass of the acid
        if 25.0 mL of this solution required 11.1 mL of 1.00 M KOH for
        neutralization. Assume that both protons of the acid were titrated.

17.29   In a titration experiment, 12.5 mL of 0.500 M $H_2SO_4$ neutralizes
        50.0 mL of NaOH. What is the concentration of the NaOH
        solution?

17.30   In a titration experiment, 20.4 mL of 0.883 M HCOOH neutralizes
        19.3 mL of $Ba(OH)_2$. What is the concentration of the $Ba(OH)_2$
        solution?

17.31   A 0.1276-g sample of an unknown monoprotic acid was
        dissolved in 25.0 mL of water and titrated with a 0.0633 M NaOH
        solution. The volume of base required to bring the solution to the
        equivalence point was 18.4 mL. (a) Calculate the molar mass of
        the acid. (b) After 10.0 mL of base had been added during the
        titration, the pH was determined to be 5.87. What is the $K_a$ of the
        unknown acid?

17.32   A solution is made by mixing exactly 500 mL of 0.167 M NaOH
        with exactly 500 mL of 0.100 M $CH_3COOH$. Calculate the
        equilibrium concentrations of $H^+$, $CH_3COOH$, $CH_3COO^-$, $OH^-$,
        and $Na^+$.

17.33   Calculate the pH at the equivalence point for the following
        titration: 0.20 M HCl versus 0.20 M methylamine ($CH_3NH_2$).

17.34   Calculate the pH at the equivalence point for the following
        titration: 0.10 M HCOOH versus 0.10 M NaOH.

17.35   A 25.0-mL solution of 0.100 M $CH_3COOH$ is titrated with a
        0.200 M KOH solution. Calculate the pH after the following
        additions of the KOH solution: (a) 0.0 mL, (b) 5.0 mL,
        (c) 10.0 mL, (d) 12.5 mL, (e) 15.0 mL.

17.36   A 10.0-mL solution of 0.300 M $NH_3$ is titrated with a 0.100 M
        HCl solution. Calculate the pH after the following additions
        of the HCl solution: (a) 0.0 mL, (b) 10.0 mL, (c) 20.0 mL,
        (d) 30.0 mL, (e) 40.0 mL.

17.37   Referring to Table 17.3, specify which indicator or indicators you
        would use for the following titrations: (a) HCOOH versus NaOH,
        (b) HCl versus KOH, (c) $HNO_3$ versus $CH_3NH_2$.

17.38   A student carried out an acid-base titration by adding NaOH
        solution from a burette to an Erlenmeyer flask containing an
        HCl solution and using phenolphthalein as the indicator. At
        the equivalence point, she observed a faint reddish-pink color.
        However, after a few minutes, the solution gradually turned
        colorless. What do you suppose happened?

17.39   The ionization constant $K_a$ of an indicator HIn is $1.0 \times 10^{-6}$. The
        color of the nonionized form is red and that of the ionized form is
        yellow. What is the color of this indicator in a solution whose pH
        is 4.00?

17.40   The $K_a$ of a certain indicator is $2.0 \times 10^{-6}$. The color of HIn
        is green and that of $In^-$ is red. A few drops of the indicator are
        added to an HCl solution, which is then titrated against an NaOH
        solution. At what pH will the indicator change color?

17.41   The following diagrams represent solutions at various stages in
        the titration of a weak base B (such as $NH_3$) with HCl. Identify the
        solution that corresponds to (1) the initial stage before the addition
        of HCl, (2) halfway to the equivalence point, (3) the equivalence
        point, (4) beyond the equivalence point. Is the pH greater than,
        less than, or equal to 7 at the equivalence point? Water and $Cl^-$
        ions have been omitted for clarity.

    ● = B      ● = $BH^+$      ● = $H_3O^+$

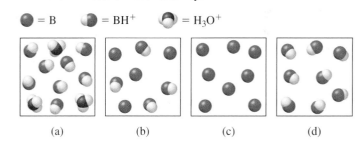

        (a)           (b)           (c)           (d)

17.42   The following diagrams represent solutions at various stages in
        the titration of a weak acid HA with NaOH. Identify the solution
        that corresponds to (1) the initial stage before the addition of
        NaOH, (2) halfway to the equivalence point, (3) the equivalence
        point, (4) beyond the equivalence point. Is the pH greater than,
        less than, or equal to 7 at the equivalence point? Water and $Na^+$
        ions have been omitted for clarity.

    ● = HA      ● = $A^-$      ● = $OH^-$

        (a)           (b)           (c)           (d)

## Section 17.4: Solubility Equilibria

*Review Questions*

17.43   Use $BaSO_4$ to distinguish between the terms *solubility* and *solubility product.*

17.44   Why do we usually not quote the $K_{sp}$ values for soluble ionic compounds?

17.45   Write balanced equations and solubility product expressions for the solubility equilibria of the following compounds: (a) CuBr, (b) $ZnC_2O_4$, (c) $Ag_2CrO_4$, (d) $Hg_2Cl_2$, (e) $AuCl_3$, (f) $Mn_3(PO_4)_2$.

17.46   Write the solubility product expression for the ionic compound $A_xB_y$.

17.47   How can we predict whether a precipitate will form when two solutions are mixed?

17.48   Silver chloride has a larger $K_{sp}$ than silver carbonate (see Table 17.4). Does this mean that AgCl also has a larger molar solubility than $Ag_2CO_3$?

*Problems*

**17.49**   Calculate the concentration of ions in the following saturated solutions: (a) [I⁻] in AgI solution with $[Ag^+] = 9.1 \times 10^{-9}\ M$, (b) $[Al^{3+}]$ in $Al(OH)_3$ solution with $[OH^-] = 2.9 \times 10^{-9}\ M$.

**17.50**   From the solubility data given, calculate the solubility products for the following compounds: (a) $SrF_2$, $7.3 \times 10^{-2}$ g/L, (b) $Ag_3PO_4$, $6.7 \times 10^{-3}$ g/L.

**17.51**   The molar solubility of $MnCO_3$ is $4.2 \times 10^{-6}\ M$. What is $K_{sp}$ for this compound?

**17.52**   The solubility of an ionic compound MX (molar mass = 346 g) is $4.63 \times 10^{-3}$ g/L. What is $K_{sp}$ for this compound?

**17.53**   The solubility of an ionic compound $M_2X_3$ (molar mass = 288 g) is $3.6 \times 10^{-17}$ g/L. What is $K_{sp}$ for this compound?

17.54   Using data from Table 17.4, calculate the molar solubility of $CaF_2$.

**17.55**   What is the pH of a saturated zinc hydroxide solution?

**17.56**   The pH of a saturated solution of a metal hydroxide MOH is 9.68. Calculate the $K_{sp}$ for this compound.

**17.57**   If 20.0 mL of 0.10 *M* $Ba(NO_3)_2$ is added to 50.0 mL of 0.10 *M* $Na_2CO_3$, will $BaCO_3$ precipitate?

**17.58**   A volume of 75 mL of 0.060 *M* NaF is mixed with 25 mL of 0.15 *M* $Sr(NO_3)_2$. Calculate the concentrations in the final solution of $NO_3^-$, $Na^+$, $Sr^{2+}$, and $F^-$. ($K_{sp}$ for $SrF_2 = 2.0 \times 10^{-10}$.)

## Section 17.5: Factors Affecting Solubility

   **Visualizing Chemistry**
**Figure 17.8**

VC 17.5   Which of the following would cause precipitation of the largest amount of AgCl when 0.1 mole is dissolved in the saturated solution shown in Figure 17.8?

   a) NaCl
   b) CsCl
   c) Both would cause precipitation of the same amount.

VC 17.6   Which of the following would cause precipitation of the largest amount of AgCl when 0.1 g is dissolved in the saturated solution shown in Figure 17.8?

   a) NaCl
   b) CsCl
   c) Both would cause precipitation of the same amount.

VC 17.7   How would the concentration of silver ion in the last view of Figure 17.8 be different if we dissolved the sodium chloride *before* dissolving the silver chloride?

   a) It would be higher.
   b) It would be lower.
   c) It would not change.

VC 17.8   How would the concentration of silver ion in the last view of Figure 17.8 be different if we added silver nitrate to the saturated solution of AgCl instead of sodium chloride?

   a) It would be higher.
   b) It would be lower.
   c) It would not change.

*Review Questions*

17.59   How does the common ion effect influence solubility equilibria? Use Le Châtelier's principle to explain the decrease in solubility of $CaCO_3$ in an $Na_2CO_3$ solution.

17.60   The molar solubility of AgCl in $6.5 \times 10^{-3}\ M$ $AgNO_3$ is $2.5 \times 10^{-8}\ M$. In deriving $K_{sp}$ from these data, which of the following assumptions are reasonable? (a) $K_{sp}$ is the same as solubility. (b) $K_{sp}$ of AgCl is the same in $6.5 \times 10^{-3}\ M$ $AgNO_3$ as in pure water. (c) Solubility of AgCl is independent of the concentration of $AgNO_3$. (d) $[Ag^+]$ in solution does not change significantly upon the addition of AgCl to $6.5 \times 10^{-3}\ M$ $AgNO_3$. (e) $[Ag^+]$ in solution after the addition of AgCl to $6.5 \times 10^{-3}\ M$ $AgNO_3$ is the same as it would be in pure water.

17.61   Give an example to illustrate the general effect of complex ion formation on solubility.

*Problems*

**17.62**   How many grams of $CaCO_3$ will dissolve in $3.0 \times 10^2$ mL of 0.050 *M* $Ca(NO_3)_2$?

**17.63**   The solubility product of $PbBr_2$ is $8.9 \times 10^{-6}$. Determine the molar solubility in (a) pure water, (b) 0.20 *M* KBr solution, and (c) 0.20 *M* $Pb(NO_3)_2$ solution.

**17.64**   Calculate the molar solubility of AgCl in a 1.00-L solution containing 10.0 g of dissolved $CaCl_2$.

**17.65**   Calculate the molar solubility of $BaSO_4$ in (a) water and (b) a solution containing 1.0 *M* $SO_4^{2-}$ ions.

**17.66**   Which of the following ionic compounds will be more soluble in acid solution than in water: (a) $BaSO_4$, (b) $PbCl_2$, (c) $Fe(OH)_3$, (d) $CaCO_3$?

**17.67**   Which of the following will be more soluble in acid solution than in pure water: (a) CuI, (b) $Ag_2SO_4$, (c) $Zn(OH)_2$, (d) $BaC_2O_4$, (e) $Ca_3(PO_4)_2$?

**17.68**   Compare the molar solubility of $Mg(OH)_2$ in water and in a solution buffered at a pH of 9.0.

**17.69** Calculate the molar solubility of $Fe(OH)_2$ in a solution buffered at (a) a pH of 8.00 and (b) a pH of 10.00.

**17.70** The solubility product of $Mg(OH)_2$ is $1.2 \times 10^{-11}$. What minimum $OH^-$ concentration must be attained (for example, by adding NaOH) to decrease the Mg concentration in a solution of $Mg(NO_3)_2$ to less than $1.0 \times 10^{-10}$ M?

**17.71** Calculate whether or not a precipitate will form if 2.00 mL of 0.60 M $NH_3$ is added to 1.0 L of $1.0 \times 10^{-3}$ M $FeSO_4$.

**17.72** If 2.50 g of $CuSO_4$ is dissolved in $9.0 \times 10^2$ mL of 0.30 M $NH_3$, what are the concentrations of $Cu^{2+}$, $Cu(NH_3)_4^{2+}$, and $NH_3$ at equilibrium?

**17.73** Calculate the concentrations of $Cd^{2+}$, $Cd(CN)_4^{2-}$, and $CN^-$ at equilibrium when 0.50 g of $Cd(NO_3)_2$ dissolves in $5.0 \times 10^2$ mL of 0.50 M NaCN.

**17.74** If NaOH is added to 0.010 M $Al^{3+}$, which will be the predominant species at equilibrium: $Al(OH)_3$ or $Al(OH)_4^-$? The pH of the solution is 14.00. [$K_f$ for $Al(OH)_4^- = 2.0 \times 10^{33}$.]

**17.75** Calculate the molar solubility of AgI in a 1.0 M $NH_3$ solution.

**17.76** Both $Ag^+$ and $Zn^{2+}$ form complex ions with $NH_3$. Write balanced equations for the reactions. However, $Zn(OH)_2$ is soluble in 6 M NaOH, and AgOH is not. Explain.

**17.77** Explain, with balanced ionic equations, why (a) $CuI_2$ dissolves in ammonia solution, (b) AgBr dissolves in NaCN solution, and (c) $HgCl_2$ dissolves in KCl solution.

## Section 17.6: Separation of Ions Using Differences in Solubility

### Review Questions

**17.78** Outline the general procedure of qualitative analysis.

**17.79** Give two examples of metal ions in each group (1 through 5) in the qualitative analysis scheme.

### Problems

**17.80** Solid NaI is slowly added to a solution that is 0.010 M in $Cu^+$ and 0.010 M in $Ag^+$. (a) Which compound will begin to precipitate first? (b) Calculate $[Ag^+]$ when CuI just begins to precipitate. (c) What percent of $Ag^+$ remains in solution at this point?

**17.81** Find the approximate pH range suitable for the separation of $Fe^{3+}$ and $Zn^{2+}$ ions by precipitation of $Fe(OH)_3$ from a solution that is initially 0.010 M in both $Fe^{3+}$ and $Zn^{2+}$.

**17.82** In a group 1 analysis, a student obtained a precipitate containing both AgCl and $PbCl_2$. Suggest one reagent that would enable the student to separate AgCl(s) from $PbCl_2(s)$.

**17.83** In a group 1 analysis, a student adds HCl acid to the unknown solution to make $[Cl^-] = 0.15$ M. Some $PbCl_2$ precipitates. Calculate the concentration of $Pb^{2+}$ remaining in solution.

**17.84** Both KCl and $NH_4Cl$ are white solids. Suggest one reagent that would enable you to distinguish between these two compounds.

**17.85** Describe a simple test that would allow you to distinguish between $AgNO_3(s)$ and $Cu(NO_3)_2(s)$.

## Additional Problems

**17.86** The buffer range is defined by the equation pH = $pK_a \pm 1$. Calculate the range of the ratio [conjugate base]/[acid] that corresponds to this equation.

**17.87** The $pK_a$ of the indicator methyl orange is 3.46. Over what pH range does this indicator change from 90 percent HIn to 90 percent $In^-$?

**17.88** Sketch the titration curve of a weak acid with a strong base like the one shown in Figure 17.4. On your graph, indicate the volume of base used at the equivalence point and also at the half-equivalence point, that is, the point at which half of the acid has been neutralized. Explain how the measured pH at the half-equivalence point can be used to determine $K_a$ of the acid.

**17.89** A 200-mL volume of NaOH solution was added to 400 mL of a 2.00 M $HNO_2$ solution. The pH of the mixed solution was 1.50 units greater than that of the original acid solution. Calculate the molarity of the NaOH solution.

**17.90** The $pK_a$ of butyric acid (HBut) is 4.7. Calculate $K_b$ for the butyrate ion ($But^-$).

**17.91** A solution is made by mixing exactly 500 mL of 0.167 M NaOH with exactly 500 mL 0.100 M HCOOH. Calculate the equilibrium concentrations of $H^+$, HCOOH, $HCOO^-$, $OH^-$, and $Na^+$.

**17.92** Tris [tris(hydroxymethyl)aminomethane] is a common buffer for studying biological systems: (a) Calculate the pH of the tris buffer after mixing 15.0 mL of 0.10 M HCl solution with 25.0 mL of 0.10 M tris. (b) This buffer was used to study an enzyme-catalyzed reaction. As a result of the reaction, 0.00015 mole of $H^+$ was consumed. What is the pH of the buffer at the end of the reaction? (c) What would be the final pH if no buffer were present?

$$HOCH_2-\underset{\underset{HOCH_2}{|}}{\overset{\overset{HOCH_2}{|}}{C}}-NH_3^+ \xrightleftharpoons{pK_a = 8.1} HOCH_2-\underset{\underset{HOCH_2}{|}}{\overset{\overset{HOCH_2}{|}}{C}}-NH_2 + H^+$$

**17.93** $Cd(OH)_2$ is an insoluble compound. It dissolves in excess NaOH in solution. Write a balanced ionic equation for this reaction. What type of reaction is this?

**17.94** A student mixes 50.0 mL of 1.00 M $Ba(OH)_2$ with 86.4 mL of 0.494 M $H_2SO_4$. Calculate the mass of $BaSO_4$ formed and the pH of the mixed solution.

**17.95** For which of the following reactions is the equilibrium constant called a solubility product?
(a) $Zn(OH)_2(s) + 2OH^-(aq) \rightleftharpoons Zn(OH)_4^{2-}(aq)$
(b) $3Ca^{2+}(aq) + 2PO_4^{3-}(aq) \rightleftharpoons Ca_3(PO_4)_2(s)$
(c) $CaCO_3(s) + 2H^+(aq) \rightleftharpoons Ca^{2+}(aq) + H_2O(l) + CO_2(g)$
(d) $PbI_2(s) \rightleftharpoons Pb^{2+}(aq) + 2I^-(aq)$

**17.96** A 2.0-L kettle contains 116 g of boiler scale ($CaCO_3$). How many times would the kettle have to be completely filled with distilled water to remove all the deposit at 25°C?

**17.97** Equal volumes of 0.12 M $AgNO_3$ and 0.14 M $ZnCl_2$ solution are mixed. Calculate the equilibrium concentrations of $Ag^+$, $Cl^-$, $Zn^{2+}$, and $NO_3^-$.

**17.98** Find the approximate pH range suitable for separating $Mg^{2+}$ and $Zn^{2+}$ by the precipitation of $Zn(OH)_2$ from a solution that is initially 0.010 M in $Mg^{2+}$ and $Zn^{2+}$.

**17.99**  Calculate the solubility (in g/L) of $Ag_2CO_3$.

**17.100**  A volume of 25.0 mL of 0.100 $M$ HCl is titrated against a 0.100 $M$ $CH_3NH_2$ solution added to it from a burette. Calculate the pH values of the solution after (a) 10.0 mL of $CH_3NH_2$ solution has been added, (b) 25.0 mL of $CH_3NH_2$ solution has been added, (c) 35.0 mL of $CH_3NH_2$ solution has been added.

**17.101**  The molar solubility of $Pb(IO_3)_2$ in a 0.10 $M$ $NaIO_3$ solution is $2.4 \times 10^{-11}$ mol/L. What is $K_{sp}$ for $Pb(IO_3)_2$?

**17.102**  When a KI solution was added to a solution of mercury(II) chloride, a precipitate [mercury(II) iodide] formed. A student plotted the mass of the precipitate versus the volume of the KI solution added and obtained the following graph. Explain the shape of the graph.

**17.103**  Barium is a toxic substance that can seriously impair heart function. For an X ray of the gastrointestinal tract, a patient drinks an aqueous suspension of 20 g $BaSO_4$. If this substance were to equilibrate with the 5.0 L of the blood in the patient's body, what would be $[Ba^{2+}]$? For a good estimate, we may assume that the $K_{sp}$ of $BaSO_4$ at body temperature is the same as at 25°C. Why is $Ba(NO_3)_2$ not chosen for this procedure?

**17.104**  The $pK_a$ of phenolphthalein is 9.10. Over what pH range does this indicator change from 95 percent HIn to 95 percent $In^-$?

**17.105**  Solid NaBr is slowly added to a solution that is 0.010 $M$ in $Cu^+$ and 0.010 $M$ in $Ag^+$. (a) Which compound will begin to precipitate first? (b) Calculate $[Ag^+]$ when CuBr just begins to precipitate. (c) What percent of $Ag^+$ remains in solution at this point?

**17.106**  Cacodylic acid is $(CH_3)_2AsO_2H$. Its ionization constant is $6.4 \times 10^{-7}$. (a) Calculate the pH of 50.0 mL of a 0.10 $M$ solution of the acid. (b) Calculate the pH of 25.0 mL of 0.15 $M$ $(CH_3)_2AsO_2Na$. (c) Mix the solutions in parts (a) and (b). Calculate the pH of the resulting solution.

**17.107**  Radiochemical techniques are useful in estimating the solubility product of many compounds. In one experiment, 50.0 mL of a 0.010 $M$ $AgNO_3$ solution containing a silver isotope with a radioactivity of 74,025 counts per min per mL was mixed with 100 mL of a 0.030 $M$ $NaIO_3$ solution. The mixed solution was diluted to 500 mL and filtered to remove all the $AgIO_3$ precipitate. The remaining solution was found to have a radioactivity of 44.4 counts per min per mL. What is the $K_{sp}$ of $AgIO_3$?

**17.108**  The molar mass of a certain metal carbonate, $MCO_3$, can be determined by adding an excess of HCl acid to react with all the carbonate and then "back-titrating" the remaining acid with NaOH. (a) Write an equation for these reactions. (b) In a certain experiment, 20.00 mL of 0.0800 $M$ HCl was added to a 0.1022-g sample of $MCO_3$. The excess HCl required 5.64 mL of 0.1000 $M$ NaOH for neutralization. Calculate the molar mass of the carbonate and identify M.

**17.109**  Acid-base reactions usually go to completion. Confirm this statement by calculating the equilibrium constant for each of the following cases: (a) A strong acid reacting with a strong base. (b) A strong acid reacting with a weak base ($NH_3$). (c) A weak acid ($CH_3COOH$) reacting with a strong base. (d) A weak acid ($CH_3COOH$) reacting with a weak base ($NH_3$). (*Hint:* Strong acids exist as $H^+$ ions and strong bases exist as $OH^-$ ions in solution. You need to look up $K_a$, $K_b$, and $K_w$.)

**17.110**  Calculate $x$, the number of molecules of water in oxalic acid hydrate ($H_2C_2O_4 \cdot xH_2O$), from the following data: 5.00 g of the compound is made up to exactly 250 mL solution, and 25.0 mL of this solution requires 15.9 mL of 0.500 $M$ NaOH solution for neutralization.

**17.111**  Describe how you would prepare a 1-L 0.20 $M$ $CH_3COONa$/0.20 $M$ $CH_3COOH$ buffer system by (a) mixing a solution of $CH_3COOH$ with a solution of $CH_3COONa$, (b) mixing a solution of $CH_3COOH$ with a solution of NaOH, and (c) mixing a solution of $CH_3COONa$ with a solution of HCl.

**17.112**  Phenolphthalein is the common indicator for the titration of a strong acid with a strong base. (a) If the $pK_a$ of phenolphthalein is 9.10, what is the ratio of the nonionized form of the indicator (colorless) to the ionized form (reddish pink) at pH 8.00? (b) If 2 drops of 0.060 $M$ phenolphthalein are used in a titration involving a 50.0-mL volume, what is the concentration of the ionized form at pH 8.00? (Assume that 1 drop = 0.050 mL.)

**17.113**  Of the reactions depicted, which best represents (a) what occurs when strong acid is added to a buffer solution, and (b) what occurs when strong base is added to a buffer solution?

> ● = HA   ● = $A^-$   ○ = $H_3O^+$   ● = $OH^-$   ○ = $H_2O$

(a) ● + ● ⟶ ○ + ●

(b) ● + ○ ⟶ ● + ○

(c) ● + ○ ⇌ ● + ○

(d) ● + ○ ⇌ ● + ●

(e) ○ + ● ⟶ ○ + ○

**17.114**  Oil paintings containing lead(II) compounds as constituents of their pigments darken over the years. Suggest a chemical reason for the color change.

**17.115**  What reagents would you employ to separate the following pairs of ions in solution: (a) $Na^+$ and $Ba^{2+}$, (b) $K^+$ and $Pb^{2+}$, (c) $Zn^{2+}$ and $Hg^{2+}$?

**17.116**  Look up the $K_{sp}$ values for $BaSO_4$ and $SrSO_4$ in Table 17.4. Calculate the concentrations of $Ba^{2+}$, $Sr^{2+}$, and $SO_4^{2-}$ in a solution that is saturated with both compounds.

**17.117**  In principle, amphoteric oxides, such as $Al_2O_3$ and BeO can be used to prepare buffer solutions because they possess both acidic and basic properties (see Section 16.11). Explain why these compounds are of little practical use as buffer components.

**17.118**  $CaSO_4$ ($K_{sp} = 2.4 \times 10^{-5}$) has a larger $K_{sp}$ value than that of $Ag_2SO_4$ ($K_{sp} = 1.4 \times 10^{-5}$). Does it necessarily follow that $CaSO_4$ also has greater solubility (g/L)? Explain.

**17.119** When lemon juice is added to tea, the color becomes lighter. In part, the color change is due to dilution, but the main reason for the change is an acid-base reaction. What is the reaction? (*Hint:* Tea contains "polyphenols," which are weak acids, and lemon juice contains citric acid.)

**17.120** How many milliliters of 1.0 $M$ NaOH must be added to 200 mL of 0.10 $M$ NaH$_2$PO$_4$ to make a buffer solution with a pH of 7.50?

**17.121** The maximum allowable concentration of Pb$^{2+}$ ions in drinking water is 0.05 ppm (that is, 0.05 g of Pb$^{2+}$ in 1 million grams of water). Is this guideline exceeded if an underground water supply is at equilibrium with the mineral anglesite (PbSO$_4$) ($K_{sp} = 1.6 \times 10^{-8}$)?

**17.122** One of the most common antibiotics is penicillin G (benzylpenicillinic acid), which has the following structure:

It is a weak monoprotic acid:

$$\text{HP} \rightleftharpoons \text{H}^+ + \text{P}^- \qquad K_a = 1.64 \times 10^{-3}$$

where HP denotes the parent acid and P$^-$ the conjugate base. Penicillin G is produced by growing molds in fermentation tanks at 25°C and a pH range of 4.5 to 5.0. The crude form of this antibiotic is obtained by extracting the fermentation broth with an organic solvent in which the acid is soluble. (a) Identify the acidic hydrogen atom. (b) In one stage of purification, the organic extract of the crude penicillin G is treated with a buffer solution at pH = 6.50. What is the ratio of the conjugate base of penicillin G to the acid at this pH? Would you expect the conjugate base to be more soluble in water than the acid? (c) Penicillin G is not suitable for oral administration, but the sodium salt (NaP) is because it is soluble. Calculate the pH of a 0.12 $M$ NaP solution formed when a tablet containing the salt is dissolved in a glass of water.

**17.123** Which of the following solutions has the highest [H$^+$]: (a) 0.10 $M$ HF, (b) 0.10 $M$ HF in 0.10 $M$ NaF, (c) 0.10 $M$ HF in 0.10 $M$ SbF$_5$? (*Hint:* SbF$_5$ reacts with F$^-$ to form the complex ion SbF$_6^-$.)

**17.124** Distribution curves show how the fractions of a nonionized acid and its conjugate base vary as a function of the pH of the medium. Plot distribution curves for CH$_3$COOH and its conjugate base CH$_3$COO$^-$ in solution. Your graph should show fraction as the $y$ axis and pH as the $x$ axis. What are the fractions and pH at the point where these two curves intersect?

**17.125** Calcium oxalate is a major component of kidney stones. Predict whether the formation of kidney stones can be minimized by increasing or decreasing the pH of the fluid present in the kidney. The pH of normal kidney fluid is about 8.2. [The first and second acid ionization constants of oxalic acid (H$_2$C$_2$O$_4$) are $6.5 \times 10^{-2}$ and $6.1 \times 10^{-5}$, respectively. The solubility product of calcium oxalate is $3.0 \times 10^{-9}$.]

**17.126** Water containing Ca$^{2+}$ and Mg$^{2+}$ ions is called *hard water* and is unsuitable for some household and industrial use because these ions react with soap to form insoluble salts, or curds. One way to remove the Ca$^{2+}$ ions from hard water is by adding washing soda (Na$_2$CO$_3 \cdot$ 10H$_2$O). (a) The molar solubility of CaCO$_3$ is

$9.3 \times 10^{-5}$ $M$. What is its molar solubility in a 0.050 $M$ Na$_2$CO$_3$ solution? (b) Why are Mg$^{2+}$ ions not removed by this procedure? (c) The Mg$^{2+}$ ions are removed as Mg(OH)$_2$ by adding slaked lime [Ca(OH)$_2$] to the water to produce a saturated solution. Calculate the pH of a saturated Ca(OH)$_2$ solution. (d) What is the concentration of Mg$^{2+}$ ions at this pH? (e) In general, which ion (Ca$^{2+}$ or Mg$^{2+}$) would you remove first? Why?

**17.127** Amino acids are building blocks of proteins. These compounds contain at least one amino group (—NH$_2$) and one carboxyl group (—COOH). Consider glycine (NH$_2$CH$_2$COOH). Depending on the pH of the solution, glycine can exist in one of three possible forms:

Fully protonated: $^+$NH$_3$—CH$_2$—COOH
Dipolar ion: $^+$NH$_3$—CH$_2$—COO$^-$
Fully ionized: NH$_2$—CH$_2$—COO$^-$

Predict the predominant form of glycine at pH 1.0, 7.0, and 12.0. The p$K_a$ of the carboxyl group is 2.3 and that of the ammonium group (—NH$_3^+$) is 9.6.

**17.128** Consider the ionization of the following acid-base indicator:

$$\text{HIn}(aq) \rightleftharpoons \text{H}^+(aq) + \text{In}^-(aq)$$

The indicator changes color according to the ratios of the concentrations of the acid to its conjugate base. When [HIn]/[In$^-$] $\geq$ 10, color of acid (HIn) predominates. When [HIn]/[In$^-$] $\leq$ 0.1, color of conjugate base (In$^-$) predominates. Show that the pH range over which the indicator changes from the acid color to the base color is pH = p$K_a \pm 1$, where $K_a$ is the ionization constant of the acid HIn.

**17.129** One way to distinguish a buffer solution with an acid solution is by dilution. (a) Consider a buffer solution made of 0.500 $M$ CH$_3$COOH and 0.500 $M$ CH$_3$COONa. Calculate its pH and the pH after it has been diluted 10-fold. (b) Compare the result in part (a) with the pHs of a 0.500 $M$ CH$_3$COOH solution before and after it has been diluted 10-fold.

**17.130** (a) Referring to Figure 17.4, describe how you would determine the p$K_b$ of the base. (b) Derive an analogous Henderson-Hasselbalch equation relating pOH to p$K_b$ of a weak base B and its conjugate acid HB$^+$. Sketch a titration curve showing the variation of the pOH of the base solution versus the volume of a strong acid added from a burette. Describe how you would determine the p$K_b$ from this curve.

**17.131** AgNO$_3$ is added slowly to a solution that contains 0.1 $M$ each of Br$^-$, CO$_3^{2-}$, and SO$_4^{2-}$ ions. What compound will precipitate first and what compound will precipitate last?

**17.132** The diagrams represent solutions of MX, which may also contain one or both of the soluble salts, MNO$_3$ and NaX. (Na$^+$ and NO$_3^-$ ions are not shown.) If (a) represents a saturated solution of MX, classify each of the other solutions as unsaturated, saturated, or supersaturated.

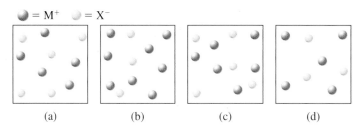

(a)       (b)       (c)       (d)

**17.133** Which of the following compounds, when added to water, will increase the solubility of CdS: (a) LiNO$_3$, (b) Na$_2$SO$_4$, (c) KCN, (d) NaClO$_3$?

**17.134** The titration curve shown here represents the titration of a weak diprotic acid ($H_2A$) versus NaOH. (a) Label the major species present at the marked points. (b) Estimate the $pK_{a_1}$ and $pK_{a_2}$ values of the acid. Assume that any salt hydrolysis is negligible.

Volume of NaOH added

**17.135** A 1.0-L saturated silver carbonate solution at 5°C is filtered to remove undissolved solid and treated with enough hydrochloric acid to decompose the dissolved compound. The carbon dioxide generated is collected in a 19-mL vial and exerts a pressure of 114 mmHg at 25°C. What is the $K_{sp}$ of silver carbonate at 5°C?

**17.136** Draw distribution curves for an aqueous carbonic acid solution. Your graph should show fraction of species present as the $y$ axis and pH as the $x$ axis. Note that at any pH, only two of the three species ($H_2CO_3$, $HCO_3^-$, and $CO_3^{2-}$) are present in appreciable concentrations. Use the $K_a$ values in Table 16.8.

**17.137** A sample of 0.96 L of HCl gas at 372 mmHg and 22°C is bubbled into 0.034 L of 0.57 $M$ $NH_3$. What is the pH of the resulting solution? Assume the volume of solution remains constant and that the HCl is totally dissolved in the solution.

**17.138** Histidine is one of the 20 amino acids found in proteins. Shown here is a fully protonated histidine molecule, where the numbers denote the $pK_a$ values of the acidic groups:

$$9.17 \quad\quad O \quad\quad 1.82$$
$$H_3N-CH-C-OH$$
$$| \quad\quad CH_2$$
$$6.00$$
$$HN$$
$$\quad\quad NH$$

(a) Show stepwise ionization of histidine in solution. (*Hint:* The $H^+$ ion will first come off from the strongest acid group followed by the next strongest acid group and so on.) (b) A dipolar ion is one in which the species has an equal number of positive and negative charges. Identify the dipolar ion in part (a). (c) The pH at which the dipolar ion predominates is called the isoelectric point, denoted by p$I$. The isoelectric point is the average of the $pK_a$ values leading to and following the formation of the dipolar ion. Calculate the p$I$ of histidine. (d) The histidine group plays an important role in buffering blood (the pH of blood is about 7.4). Which conjugate acid-base pair shown in part (a) is responsible for maintaining the pH of blood?

# PRE-PROFESSIONAL PRACTICE EXAM PROBLEMS:
# PHYSICAL AND BIOLOGICAL SCIENCES

Aqueous acid reacts with carbonate ions to produce carbonic acid, which produces carbon dioxide. A 1.0-L saturated silver carbonate solution at 5°C is treated with enough hydrochloric acid to consume all the carbonate in solution. The carbon dioxide generated is collected in a 19-mL vial and exerts a pressure of 114 mmHg at 25°C.

1. Which of the following reactions represents the overall process of acid and carbonate reacting to give carbon dioxide?

   a) $CO_3^{2-} + 2H^+ \longrightarrow H_2CO_3$
   b) $H_2CO_3 \rightleftharpoons H_2O + CO_2$
   c) $CO_3^{2-} + 2H^+ \longrightarrow H_2O + CO_2$
   d) $Ag_2CO_3 + 2H^+ \longrightarrow 2Ag^+ + CO_2 + H_2O$

2. What is the $K_{sp}$ of silver carbonate at 5°C?

   a) $2.5 \times 10^{-11}$
   b) $5.4 \times 10^{-8}$
   c) $6.3 \times 10^{-12}$
   d) $2.7 \times 10^{-8}$

3. At 25°C, the $K_{sp}$ of silver carbonate is $8.1 \times 10^{-12}$. Based on this and the answer to Question 1, what can be said about the dissolution of silver carbonate?

   a) It is endothermic.
   b) It is exothermic.
   c) It is neither exothermic nor endothermic.
   d) It produces hydrogen gas.

4. Which of the following, if added to a saturated solution of $Ag_2CO_3$, would increase the solubility of $Ag_2CO_3$?

   a) $Na_2CO_3$
   b) $NaHCO_3$
   c) $AgNO_3$
   d) $HNO_3$

# ANSWERS TO IN-CHAPTER MATERIALS

## Answers to Practice Problems

**17.1A** 4.22. **17.1B** 9.11. **17.2A** 5.0. **17.2B** 0.33 mole. **17.3A** Dissolve 0.6 mol $CH_3COONa$ and 1 mol $CH_3COOH$ in enough water to make 1 L of solution. **17.3B** 2.35–4.35. **17.4A** (a) 5.04, (b) 8.76, (c) 12.2. **17.4B** (a) 8.3 mL, (b) 12.5 mL, (c) 27.0 mL. **17.5A** 5.91. **17.5B** Aniline. **17.6A** (a) Bromophenol blue, methyl orange, methyl red, or chlorophenol, (b) any but thymol blue, bromophenol blue, or methyl orange, (c) cresol red or phenolphthalein. **17.6B** Pyridine or aniline. **17.7A** (a) $1.3 \times 10^{-5}$ $M$, $1.8 \times 10^{-3}$ g/L; (b) $1.0 \times 10^{-3}$ $M$, $1.5 \times 10^{-11}$ g/L; (c) $4.0 \times 10^{-5}$ $M$, $5.9 \times 10^{-3}$ g/L. **17.7B** (a) $2.2 \times 10^{-3}$ $M$, 0.53 g/L; (b) $1.3 \times 10^{-4}$ $M$, $3.5 \times 10^{-2}$ g/L; (c) $1.7 \times 10^{-15}$ $M$, $8.8 \times 10^{-13}$ g/L. **17.8A** (a) $1.5 \times 10^{-14}$, (b) $1.7 \times 10^{-6}$, (c) $1.3 \times 10^{-15}$. **17.8B** (a) $1.5 \times 10^{-14}$, (b) $4.8 \times 10^{-29}$, (c) $9.0 \times 10^{-33}$. **17.9A** (a) Yes, (b) yes, (c) no. **17.9B** (a) $4.4 \times 10^{-8}$ g, (b) 1.2 g, (c) 0.037 g. **17.10A** (a) $9.1 \times 10^{-9}$ $M$, (b) $8.3 \times 10^{-14}$ $M$. **17.10B** $Ag_2S < AgI < AgBr < AgCl < Ag_2CO_3$. **17.11A** (a) Yes, (b) yes, (c) no. **17.11B** Any salts containing the $CO_3^{2-}$ ion, the $OH^-$ ion, the $S^{2-}$ ion, or any anion of a weak acid such as the $SO_3^{2-}$ ion or the $F^-$ ion. **17.12A** $1.8 \times 10^{-34}$ $M$. **17.12B** $6.8 \times 10^{-2}$ $M$. **17.13A** 0.60 $M$. **17.13B** (a) $1.6 \times 10^{-9}$ $M$, (b) $2.6 \times 10^{-6}$ $M$.

## Answers to Checkpoints

**17.1.1** a, c, d. **17.1.2** b. **17.2.1** b, c, e. **17.2.2** a. **17.2.3** c. **17.2.4** b. **17.2.5** a, b. **17.2.6** d. **17.3.1** a, c, d, e. **17.3.2** c. **17.3.3** a. **17.3.4** a. **17.3.5** b. **17.3.6** e. **17.3.7** d. **17.3.8** a. **17.4.1** a. **17.4.2** c. **17.4.3** e. **17.4.4** c. **17.4.5** b, c, d, e. **17.5.1** c. **17.5.2** a, b, d, e. **17.5.3** b, d. **17.6.1** d. **17.6.2** b.

## Answers to Applying What You've Learned

**a)** $9 \times 10^{-8}$ $M$. **b)** $3 \times 10^{-60}$ $M$. **c)** $1 \times 10^{-8}$ $M$. **d)** $2 \times 10^{-7}$ $M$.

# 18

# Entropy, Free Energy, and Equilibrium

*"Plotkin's Entropy" by artist Laurel Latto.*

# How Living Systems Must Obey the Laws of Thermodynamics

The painting "Plotkin's Entropy," by the artist Laurel Latto, is an abstract depiction of an extended polypeptide chain as it is synthesized in a biological cell. Under normal physiological conditions, the polypeptide spontaneously folds into a unique three-dimensional structure called the native protein, which performs various functions. Because the original chain can assume many possible configurations while the native protein can have only one specific arrangement, the folding process is accompanied by a decrease in entropy of the system. (Note that solvent molecules, water in this case, can also play a role in affecting the entropy change.) In accord with the second law of thermodynamics, any spontaneous process must result in an increase in the entropy of the universe. It follows, therefore, that there must be an increase in the entropy of the surroundings that outweighs the decrease in the entropy of the system. The intramolecular attractions between amino acid residues cause the folding of the polypeptide chain to be exothermic. The energy produced by the process spreads out, increasing molecular motion in the surroundings—thereby increasing the entropy of the surroundings.

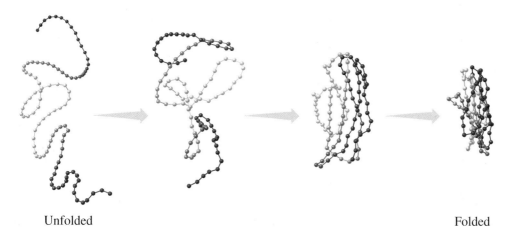

Unfolded

Folded

At the end of this chapter, you will be able to answer several questions regarding entropy changes [▶▶ Page 806].

## In This Chapter, You Will Learn:

About the three laws of thermodynamics and how to use thermodynamic quantities to determine whether or not a process is expected to be spontaneous.

### Before you begin, you should review

- System and surroundings [◀◀ Section 5.1]
- Hess's law [◀◀ Section 5.5]
- Equilibrium expressions [◀◀ Section 15.3]

# 18.1  Spontaneous Processes

**Student Annotation:** The conditions that most often are specified are *temperature*, *pressure*, and in the case of a solution, *concentration*.

An understanding of thermodynamics enables us to predict whether or not a reaction will occur when reactants are combined. This is important in the synthesis of new compounds in the laboratory, the manufacturing of chemicals on an industrial scale, and the understanding of natural processes such as cell function. A process that *does* occur under a specific set of conditions is called a ***spontaneous process.*** One that does *not* occur under a specific set of conditions is called ***nonspontaneous.*** Table 18.1 lists examples of familiar spontaneous processes and their nonspontaneous counterparts. These examples illustrate what we know intuitively: Under a given set of conditions, a process that occurs spontaneously in one direction does not also occur spontaneously in the opposite direction.

Processes that result in a decrease in the energy of a system often are spontaneous. For example, the combustion of methane is exothermic:

$$CH_4(g) + 2O_2(g) \longrightarrow CO_2(g) + 2H_2O(l) \qquad \Delta H° = -890.4 \text{ kJ/mol}$$

Thus, the energy of the system is lowered because heat is given off during the course of the reaction. Likewise, in the acid-base neutralization reaction,

$$H^+(aq) + OH^-(aq) \longrightarrow H_2O(l) \qquad \Delta H° = -56.2 \text{ kJ/mol}$$

heat is given off, lowering the energy of the system. Each of these processes is spontaneous, and each results in a lowering of the system's energy.

Now consider the melting of ice:

$$H_2O(s) \longrightarrow H_2O(l) \qquad \Delta H° = 6.01 \text{ kJ/mol}$$

In this case, the process is endothermic and yet it is also spontaneous at temperatures above 0°C. Conversely, the freezing of water is an *exothermic* process:

$$H_2O(l) \longrightarrow H_2O(s) \qquad \Delta H° = -6.01 \text{ kJ/mol}$$

Yet it is *not* spontaneous at temperatures above 0°C.

Based on the first two examples, and many others like them, we might conclude that exothermic processes tend to be spontaneous and, indeed, a negative $\Delta H$ does *favor* spontaneity. The last two examples, however, make it clear that the sign of $\Delta H$ *alone* is insufficient to predict spontaneity in every circumstance. For the remainder of this chapter, we will examine the *two* factors that determine whether or not a process is spontaneous under a given set of conditions.

# 18.2  Entropy

To predict the spontaneity of a chemical or physical process, we need to know both the change in *enthalpy* [◂◂ Section 5.3] and the change in *entropy* associated with the process. We first encountered the concept of entropy in our discussion of solution formation [◂◂ Section 13.2]. We will now look in more detail at what entropy is, and why it matters.

| **TABLE 18.1** | Familiar Spontaneous and Nonspontaneous Processes | |
|---|---|
| **Spontaneous** | **Nonspontaneous** |
| Ice melting at room temperature | Water freezing at room temperature |
| Sodium metal reacting violently with water to produce sodium hydroxide and hydrogen gas [◂◂ Section 7.7] | Sodium hydroxide reacting with hydrogen gas to produce sodium metal and water |
| A ball rolling downhill | A ball rolling uphill |
| The rusting of iron at room temperature | The conversion of rust back to iron metal at room temperature |
| Water freezing at −10°C | Ice melting at −10°C |

# A Qualitative Description of Entropy

Qualitatively, the **entropy (S)** of a system is a measure of how *spread out* or how *dispersed* the system's energy is. The simplest interpretation of this is how spread out a system's energy is in *space*. In other words, for a given system, the greater the volume it occupies, the greater its entropy. This interpretation explains how the process in Figure 18.1 occurs spontaneously despite there being no enthalpy change. Because they are moving, the gas molecules that were originally confined to one side of the container possess *motional energy*. In the absence of a barrier preventing it, the motional energy of molecules will spread out to occupy a larger volume. The dispersal of a system's motional energy to occupy a larger volume when the barrier is removed constitutes an *increase* in the system's entropy. Just as spontaneity is favored by a process being exothermic, spontaneity is also favored by an increase in the system's entropy. Whether it is the enthalpy change, the entropy change, or both, for a process to be spontaneous, *something* must favor spontaneity.

**Student Annotation:** Motional energy includes *translational* energy, in which the entire molecule moves through space [◄◄ Section 5.1]; *rotational* energy, in which the molecule spins about an axis running through its center of mass; and *vibrational* energy, in which atoms of a molecule move relative to one another.

# A Quantitative Definition of Entropy

At this point, it is useful to introduce the mathematical definition of entropy proposed by Ludwig Boltzmann:

$$S = k \ln W \qquad \text{Equation 18.1}$$

where $k$ is the Boltzmann constant ($1.38 \times 10^{-23}$ J/K) and $W$ is the number of energetically equivalent different ways the molecules in a system can be arranged. To illustrate what this means, let's consider a simplified version of the process shown in Figure 18.1. Prior to the removal of the barrier between the left and right sides of the container, at any given instant, each molecule has a particular location, somewhere in the left side of the container. To narrow down the possible locations of the molecules, we imagine that each side of the container is divided into a number of equal smaller volumes called *cells*. In the simplest scenario, with just one molecule in the system, the number of possible locations of the molecule is equal to the number of cells. If the system contains *two* molecules, the number of possible arrangements is equal to the number of cells *squared*. (Note that a cell may contain more than one molecule.) Each time we increase the number of molecules by one, the number of possible arrangements increases by a factor equal to the number of cells. In general, for a volume consisting of $X$ cells, and containing $N$ molecules, the number of possible arrangements, $W$, is given by the equation

**Student Annotation:** The Boltzmann constant is equal to the gas constant, $R$ (in J/K · mol), divided by Avogadro's constant, $N_A$.

**Student Annotation:** The number of possible arrangements is sometimes called the number of *microstates*.

$$W = X^N \qquad \text{Equation 18.2}$$

Figure 18.2 illustrates this for a simple case involving just two molecules. We imagine the container is divided into four cells each with volume $v$. Initially, both molecules are confined to the left side, which consists of two cells. With two molecules in two cells, there are $2^2 = 4$ possible arrangements of the molecules [Figure 18.2(a)]. When the barrier is removed, doubling the volume available to the molecules, the number of cells also doubles. With four cells available, there are $4^2 = 16$ possible arrangements of the molecules. Eight of the sixteen arrangements have the molecules on opposite sides of the container [Figure 18.2(b)]. Of the other eight arrangements, four have both molecules on the left side [as shown in Figure 18.2(a)], and four have both molecules on the right side (not shown). There are three different states possible for this system.

**Figure 18.1** A spontaneous process. The rapidly moving gas molecules originally confined to one side of a container spread out to fill the whole container when the barrier is removed.

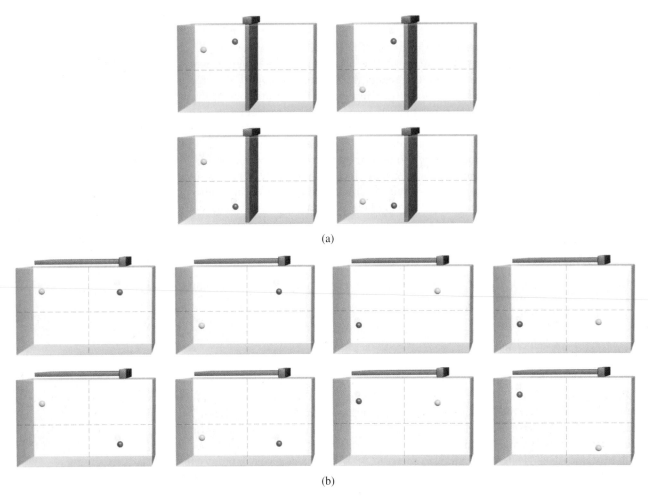

**Figure 18.2**   (a) Before the barrier is removed, the molecules are both in the left side of the container, which we imagine is divided into two cells of equal volume. There are four possible arrangements of two molecules in two cells. (b) When the barrier between the two sides of the container is removed, the volume (and the number of cells) available to the molecules doubles. The new number of possible arrangements is $4^2 = 16$, eight of which have the molecules in opposite sides of the container—the most probable outcome.

1. One molecule on each side (eight possible arrangements)
2. Both molecules on the left (four possible arrangements)
3. Both molecules on the right (four possible arrangements)

**Student Annotation:** The state with the largest number of possible arrangements has the greatest entropy.

The most *probable* state is the one with the *largest* number of possible arrangements. In this case, the most probable state is the one with one molecule on each side of the container. The same principle applies to systems with larger numbers of molecules. Increasing the number of molecules increases the number of possible arrangements, but the most probable state will be the one in which the gas molecules are divided evenly between the two sides of the container.

## 18.3   Entropy Changes in a System

### Calculating $\Delta S_{sys}$

The change in entropy of a system is the difference between the entropy of the final state and the entropy of the initial state.

Equation 18.3                    $\Delta S_{sys} = S_{final} - S_{initial}$

Using Equation 18.1, we can write an expression for the entropy of each state,

$$\Delta S_{sys} = k \ln W_{final} - k \ln W_{initial} = k \ln \frac{W_{final}}{W_{initial}}$$

Combining this result with Equation 18.2 gives

$$\Delta S_{sys} = k \ln \frac{(X_{final})^N}{(X_{initial})^N} = k \ln \left(\frac{X_{final}}{X_{initial}}\right)^N = k N \ln \left(\frac{X_{final}}{X_{initial}}\right)$$

> **Student Annotation:** Remember that the difference between two logs is equal to the log of the corresponding quotient:
>
> $$\ln A - \ln B = \ln \frac{A}{B}$$
>
> and that $\ln A^x = x \ln A$ [▶▶ Appendix 1].

Because $X$ is the number of cells, and the volume of each cell is $v$, the total volume is related to the number of cells by

$$V = Xv \qquad \text{or} \qquad X = \frac{V}{v}$$

We substitute $V_{final}/v$ for $X_{final}$ and $V_{initial}/v$ for $X_{initial}$ to get

$$\Delta S_{sys} = k N \ln \frac{V_{final}/v}{V_{initial}/v} = k N \ln \frac{V_{final}}{V_{initial}}$$

Finally, because the Boltzmann constant, $k$, is the gas constant, $R$, divided by Avogadro's constant,

$$k = \frac{R}{N_A}$$

and because the number of molecules, $N$, is the product of the number of moles, $n$, and Avogadro's constant, $N_A$;

$$N = n \times N_A$$

$$kN = \left(\frac{R}{N_A} \times n \times N_A\right) = nR$$

the equation becomes

$$\Delta S_{sys} = nR \ln \frac{V_{final}}{V_{initial}} \qquad \text{Equation 18.4}$$

Sample Problem 18.1 shows how to use Equation 18.4 to calculate the entropy change for a process like the one shown in Figure 18.1, the expansion of an ideal gas at constant temperature.

## Sample Problem 18.1

Determine the change in entropy for 1.0 mole of an ideal gas originally confined to one-half of a 5.0-L container when the gas is allowed to expand to fill the entire container at constant temperature.

**Strategy** This is the isothermal expansion of an ideal gas. Because the molecules spread out to occupy a greater volume, we expect there to be an increase in the entropy of the system. Use Equation 18.3 to solve for $\Delta S_{sys}$.

**Setup** $R = 8.314$ J/K · mol, $n = 1.0$ mole, $V_{final} = 5.0$ L, and $V_{initial} = 2.5$ L.

**Solution**

$$\Delta S_{sys} = nR \ln \frac{V_{final}}{V_{initial}} = 1.0 \text{ mol} \times \frac{8.314 \text{ J}}{\text{K} \cdot \text{mol}} \times \ln \frac{5.0 \text{ L}}{2.5 \text{ L}} = 5.8 \text{ J/K}$$

> **Think About It** Remember that for a process to be spontaneous, *something* must favor spontaneity. If the process is spontaneous but not exothermic (in this case, there is no enthalpy change), then we should expect $\Delta S_{sys}$ to be positive.

**Practice Problem A** Determine the change in entropy ($\Delta S_{sys}$), for the expansion of 0.10 mole of an ideal gas from 2.0 L to 3.0 L at constant temperature.

**Practice Problem B** To what fraction of its original volume must a 0.50-mole sample of ideal gas be compressed at constant temperature for $\Delta S_{sys}$ to be $-6.7$ J/K?

# Standard Entropy, $S°$

Although Equation 18.1 provides a quantitative definition of entropy, we seldom use it or Equation 18.3 to calculate the entropy change for a real process because of the difficulty involved in determining $W$, the number of different possible arrangements (Equation 18.2) in a macroscopic system. Instead, for processes other than isothermal expansion or compression of an ideal gas (for which we can use Equation 18.4), we routinely determine entropy changes using tabulated values.

Using calorimetry [◄◄ Section 5.4], it is possible to determine the *absolute* value of the entropy of a substance, $S$; something we cannot do with either energy or enthalpy. (Recall that while we can determine $\Delta U$ and $\Delta H$ for a process that a system undergoes, we cannot determine the absolute values of either $U$ or $H$ for a system [◄◄ Sections 5.2 and 5.3].) **Standard entropy** is the absolute entropy of a substance at 1 atm. (Tables of standard entropy values typically are the values at 25°C because so many processes are carried out at room temperature—although temperature is *not* part of the standard state definition and therefore must be specified.) Table 18.2 lists standard entropies of a few elements and compounds. Appendix 2 provides a more extensive listing. The units of entropy are J/K · mol. We use joules rather than kilojoules because entropy values typically are quite small. The entropies of substances (elements and compounds) are always positive (i.e., $S > 0$), even for elements in their standard states. (Recall that the standard *enthalpy* of formation, $\Delta H_f°$, for elements in their standard states is arbitrarily defined as zero, and for compounds it may be either positive or negative [◄◄ Section 5.6].)

Referring to Table 18.2, we can identify several important trends:

- For a given substance, the standard entropy is greater in the liquid phase than in the solid phase. [Compare the standard entropies of Na(s) and Na(l).] This results from there being greater molecular motion in a liquid, resulting in many possible arrangements of atoms in the liquid phase; whereas the positions of atoms in the solid are fixed.
- For a given substance, the standard entropy is greater in the gas phase than in the liquid phase. [Compare the standard entropies of Na(l) and Na(g) and those of $H_2O(l)$ and $H_2O(g)$.] This results from there being much greater molecular motion in a gas, resulting in many more possible arrangements of atoms in the gas phase than in the liquid phase—in part because the gas phase occupies a much greater volume than either of the condensed phases.
- For two monatomic species, the one with the larger molar mass has the greater standard entropy. [Compare the standard entropies of He(g) and Ne(g).]
- For two substances in the same phase, and with similar molar masses, the substance with the more complex molecular structure has the greater standard entropy. [Compare the standard entropies of $O_3(g)$ and $F_2(g)$.] The more complex a molecular structure, the more different types of motion the molecule can exhibit. A diatomic molecule such as $F_2$, for example, exhibits only one type of vibration, whereas a bent triatomic molecule such as $O_3$ exhibits three different types of vibrations. Each mode of motion contributes to the total number of available energy levels within which a system's energy can be dispersed. Figure 18.3 illustrates the ways in which the $F_2$ and $O_3$ molecules can rotate and vibrate.
- In cases where an element exists in two or more allotropic forms, the form in which the atoms are more mobile has the greater entropy. [Compare the standard entropies of C(diamond) and C(graphite). In diamond, the carbon atoms occupy fixed positions in a three-dimensional array. In graphite, although the carbon atoms occupy fixed positions within the two-dimensional sheets (See Figure 12.27), the sheets are free to move with respect to one another, which increases the mobility and, therefore, total number of possible arrangements of atoms within the solid.]

| TABLE 18.2 | Standard Entropy Values ($S°$) for Some Substances at 25°C | | |
|---|---|---|---|
| **Substance** | **$S°$ (J/K · mol)** | **Substance** | **$S°$ (J/K · mol)** |
| $H_2O(l)$ | 69.9 | C(diamond) | 2.4 |
| $H_2O(g)$ | 188.7 | C(graphite) | 5.69 |
| Na(s) | 51.05 | $O_2(g)$ | 205.0 |
| Na(l) | 57.56 | $O_3(g)$ | 237.6 |
| Na(g) | 153.7 | $F_2(g)$ | 203.34 |
| He(g) | 126.1 | Au(s) | 47.7 |
| Ne(g) | 146.2 | Hg(l) | 77.4 |

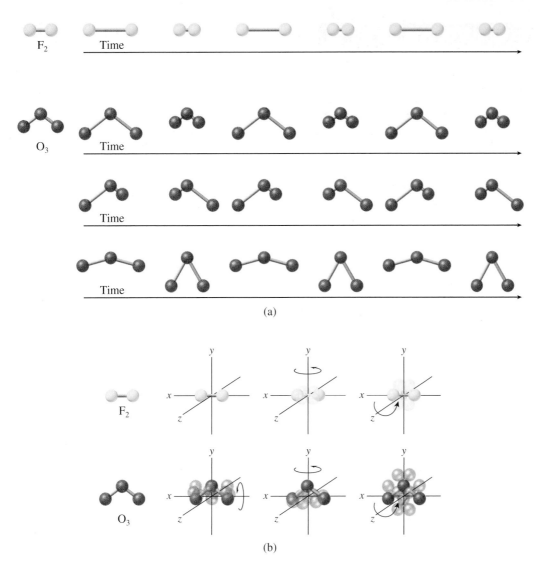

(a)

(b)

**Figure 18.3** In addition to translational motion, molecules exhibit both *vibrations,* in which the atoms' positions relative to one another change, and *rotations,* in which the molecule rotates about its center of mass. (a) A diatomic molecule such as fluorine only exhibits one type of vibration. A bent, triatomic molecule such as ozone exhibits three types of vibration. (b) A diatomic molecule exhibits two different rotations, whereas a bent, triatomic molecule exhibits three different rotations. (Note that rotation of $F_2$ about the $x$ axis would cause no change in the positions of either atom in the molecule.)

Now let's consider a process represented by the following chemical equation:

$$aA + bB \longrightarrow cC + dD$$

Just as the enthalpy change of a reaction is the difference between the enthalpies of the products and reactants (Equation 5.12), the entropy change is the difference between the entropies of the products and reactants:

$$\Delta S_{rxn}^{\circ} = [cS^{\circ}(C) + dS^{\circ}(D)] - [aS^{\circ}(A) + bS^{\circ}(B)] \qquad \text{Equation 18.5}$$

Or, using $\Sigma$ to represent summation and $m$ and $n$ to represent the stoichiometric coefficients of the reactants and products, respectively, Equation 18.5 can be generalized as follows:

$$\Delta S_{rxn}^{\circ} = \Sigma nS^{\circ}(\text{products}) - \Sigma mS^{\circ}(\text{reactants}) \qquad \text{Equation 18.6}$$

The standard entropy values of a large number of substances have been measured in J/K · mol. To calculate the standard entropy change for a reaction ($\Delta S_{rxn}^{\circ}$), we look up the standard entropies of the products and reactants and use Equation 18.5.

Sample Problem 18.2 demonstrates this approach.

**Student Annotation:** The *reaction* generally is the *system.* Therefore, $\Delta S_{rxn}^{\circ}$ is $\Delta S_{sys}^{\circ}$.

## Sample Problem 18.2

From the standard entropy values in Appendix 2, calculate the standard entropy changes for the following reactions at 25°C:

(a) $CaCO_3(s) \longrightarrow CaO(s) + CO_2(g)$

(b) $N_2(g) + 3H_2(g) \longrightarrow 2NH_3(g)$

(c) $H_2(g) + Cl_2(g) \longrightarrow 2HCl(g)$

**Strategy** Look up standard entropy values and use Equation 18.5 to calculate $\Delta S^{\circ}_{rxn}$. Just as we did when we calculated standard enthalpies of reaction, we consider stoichiometric coefficients to be dimensionless—giving $\Delta S^{\circ}_{rxn}$ units of J/K · mol.

**Student Annotation:** Recall that here *per mole* means per mole of reaction as written [◄◄ Section 5.6].

**Setup** From Appendix 2, $S^{\circ}[CaCO_3(s)] = 92.9$ J/K · mol, $S^{\circ}[CaO(s)] = 39.8$ J/K · mol, $S^{\circ}[CO_2(g)] = 213.6$ J/K · mol, $S^{\circ}[N_2(g)] = 191.5$ J/K · mol, $S^{\circ}[H_2(g)] = 131.0$ J/K · mol, $S^{\circ}[NH_3(g)] = 193.0$ J/K · mol, $S^{\circ}[Cl_2(g)] = 223.0$ J/K · mol, and $S^{\circ}[HCl(g)] = 187.0$ J/K · mol.

**Solution**

(a) $\Delta S^{\circ}_{rxn} = [S^{\circ}(CaO) + S^{\circ}(CO_2)] - [S^{\circ}(CaCO_3)]$

$\qquad = [(39.8 \text{ J/K} \cdot \text{mol}) + (213.6 \text{ J/K} \cdot \text{mol})] - (92.9 \text{ J/K} \cdot \text{mol})$

$\qquad = 160.5 \text{ J/K} \cdot \text{mol}$

(b) $\Delta S^{\circ}_{rxn} = [2S^{\circ}(NH_3)] - [S^{\circ}(N_2) + 3S^{\circ}(H_2)]$

$\qquad = (2)(193.0 \text{ J/K} \cdot \text{mol}) - [(191.5 \text{ J/K} \cdot \text{mol}) + (3)(131.0 \text{ J/K} \cdot \text{mol})]$

$\qquad = -198.5 \text{ J/K} \cdot \text{mol}$

(c) $\Delta S^{\circ}_{rxn} = [2S^{\circ}(HCl)] - [S^{\circ}(H_2) + S^{\circ}(Cl_2)]$

$\qquad = (2)(187.0 \text{ J/K} \cdot \text{mol}) - [(131.0 \text{ J/K} \cdot \text{mol}) + (223.0 \text{ J/K} \cdot \text{mol})]$

$\qquad = 20.0 \text{ J/K} \cdot \text{mol}$

**Think About It** Remember to multiply each standard entropy value by the correct stoichiometric coefficient. Like Equation 5.12, Equation 18.5 can only be used with a *balanced* chemical equation.

**Practice Problem A** Calculate the standard entropy change for the following reactions at 25°C. Predict first whether each one will be positive, negative, or too close to call.

(a) $2CO_2(g) \longrightarrow 2CO(g) + O_2(g)$

(b) $3O_2(g) \longrightarrow 2O_3(g)$

(c) $2NaHCO_3(s) \longrightarrow Na_2CO_3(s) + H_2O(l) + CO_2(g)$

**Practice Problem B** In each of the following reactions, there is one species for which the standard entropy is not listed in Appendix 2. In each case, using the values that *are* in Appendix 2 and the $\Delta S^{\circ}_{rxn}$ that is given, determine the value of the missing standard entropy at 25°C: (a) $K(s) \longrightarrow K(l)$, $\Delta S^{\circ}_{rxn} = 7.9$ J/K · mol, (b) $2S(\text{rhombic}) + Cl_2(g) \longrightarrow S_2Cl_2(g)$, $\Delta S^{\circ}_{rxn} = 44.74$ J/K · mol, (c) $O_2(g) + 2MgF_2(s) \longrightarrow 2MgO(s) + 2F_2(g)$, $\Delta S^{\circ}_{rxn} = 140.76$ J/K · mol.

## Qualitatively Predicting the Sign of $\Delta S^{\circ}_{sys}$

Equation 18.5 enables us to calculate $\Delta S^{\circ}_{rxn}$ for a process when the standard entropies of the products and reactants are known. However, sometimes it's useful just to know the *sign* of $\Delta S^{\circ}_{rxn}$. Although multiple factors can influence the sign of $\Delta S^{\circ}_{rxn}$, the outcome is often dominated by a single factor, which can be used to make a qualitative prediction. Several processes that lead to an increase in entropy are

- Melting
- Vaporization or sublimation
- Temperature increase
- Reaction resulting in a greater number of gas molecules

When a solid is melted, the molecules have greater energy and are more mobile. They go from being in fixed positions in the solid, to being free to move about in the liquid. As we saw in the discussion of standard entropy, this leads to many more possible arrangements of the molecules and, therefore, greater entropy. The same rationale holds for the vaporization or sublimation of a substance. There is a dramatic increase in energy/mobility, and in the number of possible arrangements of a system's molecules when the molecules go from a condensed phase to the gas phase. Therefore, there is a much larger increase in the system's entropy, relative to the solid-to-liquid transition.

When the temperature of a system is increased, the energy of the system's molecules increases. To visualize this, recall from the discussion of kinetic molecular theory that increasing the temperature of a gas increases its average kinetic energy. This corresponds to an increase in the average speed of the gas molecules and a spreading out of the range of molecular speeds. [See Figure 11.20(a).] If we think of each of the possible molecular speeds within the range as a discrete energy level, we can see that at higher temperatures, there is a greater number of possible molecular speeds and, therefore, a greater number of energy states available to the molecules in the system. With a greater number of available energy states, there is a greater number of possible arrangements of molecules *within* those states and, therefore, a greater entropy.

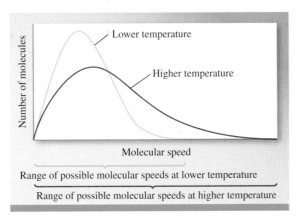

Because the entropy of a substance in the gas phase is always significantly greater than its entropy in either the liquid or solid phase, a reaction that results in an increase in the number of gas molecules causes an increase in the system's entropy. For reactions that do not involve gases, an increase in the number of solid, liquid, or aqueous molecules also usually causes an entropy increase.

By considering these factors, we can usually make a reasonably good prediction of the sign of $\Delta S^\circ_{rxn}$ for a physical or chemical process, without having to look up the absolute entropy values for the species involved. Figure 18.4 summarizes the factors that can be used to compare entropies and illustrates several comparisons.

In addition to melting, vaporization/sublimation, temperature increase, and reactions that increase the number of gas molecules, which can always be counted upon to result in an entropy increase, the process of *dissolving* a substance often leads to an increase in entropy. In the case of a molecular solute, such as sucrose (sugar), dissolving causes dispersal of the molecules (and consequently, of the system's *energy*) into a larger volume—resulting in an increase in entropy. In the case of an ionic solute, the analysis is slightly more complicated. We saw in our discussion of solution formation [◄◄ Section 13.2] that the dissolution of ammonium nitrate ($NH_4NO_3$) is spontaneous, even though it is endothermic, because the system's entropy increases when the ionic solid dissociates and is dispersed throughout the solution. In general, this is the case for ionic solutes in which the charges on ions are small. (In the case of $NH_4NO_3$, they are $+1$ and $-1$.) However, when ions are dispersed in water, they become hydrated (surrounded by water molecules in a specific arrangement [◄◄ Figure 4.4 page 124]). This leads to a *decrease* in the entropy of the *water*, as hydration reduces the mobility of some of the water molecules by fixing them in positions around the dissolved ions. When the charges on ions are low, the increase in entropy of the solute typically outweighs the decrease in entropy of the water—resulting in an overall *increase* in the entropy of the system—as is the case with $NH_4NO_3$. By contrast, when highly charged ions such as $Al^{3+}$ and $Fe^{3+}$ are hydrated, the decrease in entropy of the water can actually outweigh the increase in entropy of the solute, leading to an overall *decrease* in entropy of the system. Table 18.3 lists the changes in entropy associated with the spontaneous dissolution of several ionic solids.

Media Player/MPEG Animation: Figure 18.4, Factors That Influence the Entropy of a System, pp. 786–787.

| TABLE 18.3 | Entropy Changes for the Dissolution ($\Delta S^\circ_{soln}$) of Some Ionic Solids at 25°C | |
| --- | --- | --- |
| **Dissolution Equation** | | $\Delta S^\circ_{soln}$ (J/K · mol) |
| $NH_4NO_3(s) \longrightarrow NH_4^+(aq) + NO_3^-(aq)$ | | 108.1 |
| $AlCl_3(s) \longrightarrow Al^{3+}(aq) + 3Cl^-(aq)$ | | $-253.2$ |
| $FeCl_3(s) \longrightarrow Fe^{3+}(aq) + 3Cl^-(aq)$ | | $-266.1$ |

**Student Annotation:** Note that the subscript "rxn" in $\Delta S^\circ_{rxn}$ changes to "soln" to refer specifically to the *solution* process.

## Figure 18.4

# Factors That Influence the Entropy of a System

## Volume Change

Quantum mechanical analysis shows that the spacing between translational energy levels is inversely proportional to the volume of the container. Thus, when the volume is increased, more energy levels become available within which the system's energy can be dispersed.

## Temperature Change

At higher temperatures, molecules have greater kinetic energy—making more energy levels accessible. This increases the number of energy levels within which the system's energy can be dispersed, causing entropy to increase.

## Molecular Complexity

Unlike atoms, which exhibit only translational motion, molecules can also exhibit rotational and vibrational motions. The greater a molecule's complexity, the greater the number of possible ways it can rotate and vibrate. The ozone molecule ($O_3$), for example, is more complex than the fluorine molecule ($F_2$) and exhibits more different kinds of vibrations and rotations. (See Figure 18.3.) This results in more energy levels within which the system's energy can be dispersed. The number and spacing of additional energy levels have been simplified to keep the illustration clear.

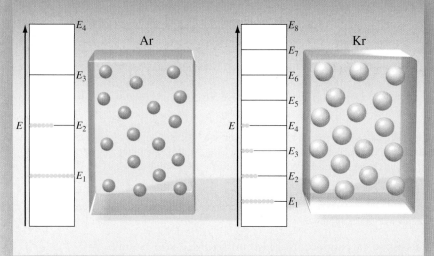

## Molar Mass

The energy levels for a substance with a larger molar mass are more closely spaced. Kr, for example, has roughly twice the molar mass of Ar. Thus, Kr has roughly twice as many energy levels within which the system's energy can be dispersed.

## Phase Change

Because of greater mobility, there are many more different possible arrangements ($W$) of molecules in the liquid phase than there are in the solid phase; and there are many, *many* more different possible arrangements of molecules in the gas phase than there are in the liquid phase. Entropy of a substance increases when it is melted ($s \rightarrow l$), vaporized ($l \rightarrow g$), or sublimed ($s \rightarrow g$).

## Chemical Reaction

When a chemical reaction produces more gas molecules than it consumes, the number of different possible arrangements of molecules ($W$) increases and entropy increases.

# What's the point?

Although several factors can influence the entropy of a system or the entropy change associated with a process, often one factor dominates the outcome. Each of these comparisons shows a qualitative illustration of one of the important factors.

Sample Problem 18.3 lets you practice making qualitative predictions of the sign of $\Delta S^\circ_{rxn}$.

---

### Sample Problem 18.3

For each process, determine the sign of $\Delta S$ for the system: (a) decomposition of $CaCO_3(s)$ to give $CaO(s)$ and $CO_2(g)$, (b) heating bromine vapor from 45°C to 80°C, (c) condensation of water vapor on a cold surface, (d) reaction of $NH_3(g)$ and $HCl(g)$ to give $NH_4Cl(s)$, and (e) dissolution of sugar in water.

**Strategy** Consider the change in energy/mobility of atoms and the resulting change in number of possible positions that each particle can occupy in each case. An increase in the number of arrangements corresponds to an increase in entropy and therefore a positive $\Delta S$.

**Setup** Increases in entropy generally accompany solid-to-liquid, liquid-to-gas, and solid-to-gas transitions; the dissolving of one substance in another; a temperature increase; and reactions that increase the net number of moles of gas.

**Solution** $\Delta S$ is (a) positive, (b) positive, (c) negative, (d) negative, and (e) positive.

> **Think About It** For reactions involving only liquids and solids, predicting the sign of $\Delta S^\circ$ can be more difficult, but in many such cases an increase in the total number of molecules and/or ions is accompanied by an increase in entropy.

**Practice Problem A** For each of the following processes, determine the sign of $\Delta S$: (a) crystallization of sucrose from a supersaturated solution, (b) cooling water vapor from 150°C to 110°C, (c) sublimation of dry ice.

**Practice Problem B** Make a qualitative prediction of the sign of $\Delta H^\circ_{soln}$ for $AlCl_3(s)$ and the dissolution of $FeCl_3(s)$. See Table 18.3 on page 785. Explain your reasoning.

---

### Checkpoint 18.3    Entropy Changes in a System

**18.3.1** For which of the following physical processes is $\Delta S$ negative? (Select all that apply.)

a) Freezing ethanol

b) Evaporating water

c) Mixing carbon tetrachloride and benzene

d) Heating water

e) Condensing bromine vapor

**18.3.2** For which of the following chemical reactions is $\Delta S$ negative? (Select all that apply.)

a) $2O_3(g) \longrightarrow 3O_2(g)$

b) $4Fe(s) + 3O_2(g) \longrightarrow 2Fe_2O_3(s)$

c) $2H_2O_2(aq) \longrightarrow 2H_2O(l) + O_2(g)$

d) $2Li(s) + 2H_2O(l) \longrightarrow 2LiOH(aq) + H_2(g)$

e) $2NH_3(g) \longrightarrow N_2(g) + 3H_2(g)$

**18.3.3** Consider the gas-phase reaction of $A_2$ (blue) and $B_2$ (orange) to form $AB_3$. What are the correct balanced equation and the sign of $\Delta S$ for the reaction?

a) $A_2 + B_2 \longrightarrow AB_3$, negative

b) $2A_2 + 3B_2 \longrightarrow 4AB_3$, positive

c) $2A_2 + 3B_2 \longrightarrow 4AB_3$, negative

d) $A_2 + 3B_2 \longrightarrow 2AB_3$, negative

e) $A_2 + 3B_2 \longrightarrow 2AB_3$, positive

---

## 18.4    Entropy Changes in the Universe

Recall that the *system* typically is the part of the universe we are investigating (e.g., the reactants and products in a chemical reaction). The *surroundings* are everything else [◄◄ Section 5.1]. Together, the system and surroundings make up the *universe*. We have seen that the *dispersal* or *spreading out* of a system's energy corresponds to an increase in the system's entropy. Moreover, an increase in the system's entropy is one of the factors that determines whether or not a process is spontaneous. However, correctly predicting the spontaneity of a process requires us to consider entropy changes in both the system and the surroundings.

Consider the following processes:

- An ice cube spontaneously melts in a room where the temperature is 25°C. In this case, the motional energy of the air molecules at 25°C is transferred to the ice cube (at 0°C), causing the ice to melt. There is no temperature change during a phase change. However, because the molecules are more mobile and there are many more different possible arrangements in *liquid* water than there are in ice, there is an increase in the entropy of the system. In this case, because the process of melting is endothermic, heat is transferred *from* the surroundings *to* the system and the temperature of the surroundings decreases. The slight decrease in temperature causes a small decrease in molecular motion and a decrease in the entropy of the surroundings.

  $\Delta S_{sys}$ is positive.
  $\Delta S_{surr}$ is negative.

- A cup of hot water spontaneously cools to room temperature as the motional energy of the water molecules spreads out to the cooler surrounding air. Although the loss of energy from the system and corresponding temperature decrease cause a *decrease* in the entropy of the *system*, the increased temperature of the surrounding air causes an *increase* in the entropy of the *surroundings*.

  $\Delta S_{sys}$ is negative.
  $\Delta S_{surr}$ is positive.

Thus, it is not just the entropy of the *system* that determines if a process is spontaneous, the entropy of the *surroundings* is also important. There are also examples of spontaneous processes in which $\Delta S_{sys}$ and $\Delta S_{surr}$ are *both* positive. The decomposition of hydrogen peroxide produces water and oxygen gas, $2H_2O_2(l) \longrightarrow 2H_2O(l) + O_2(g)$. Because the reaction results in an increase in the number of gas molecules, we know that there is an increase in the entropy of the system. However, this is an exothermic reaction, meaning that it also gives off heat to the surroundings. An increase in temperature of the surroundings causes an increase in the entropy of the surroundings as well. (Note that there are no spontaneous processes in which $\Delta S_{sys}$ and $\Delta S_{surr}$ are both negative, which will become clear shortly.)

## Calculating $\Delta S_{surr}$

When an exothermic process takes place, the heat transferred from the system to the surroundings increases the temperature of the molecules in the surroundings. Consequently, there is an increase in the number of energy states accessible to the molecules in the surroundings and the entropy of the surroundings increases. Conversely, in an endothermic process, heat is transferred from the surroundings to the system, decreasing the entropy of the surroundings. Remember that for constant-pressure processes, the heat released or absorbed, $q$, is equal to the enthalpy change of the system, $\Delta H_{sys}$ [◄◄ Section 5.3]. The change in entropy for the surroundings, $\Delta S_{surr}$, is directly proportional to $\Delta H_{sys}$:

$$\Delta S_{surr} \propto -\Delta H_{sys}$$

The minus sign indicates that a negative enthalpy change in the system (an *exothermic* process) corresponds to a positive entropy change in the surroundings. For an *endothermic* process, the enthalpy change in the system is a positive number and corresponds to a negative entropy change in the surroundings.

In addition to being directly proportional to $\Delta H_{sys}$, $\Delta S_{surr}$ is inversely proportional to temperature:

$$\Delta S_{surr} \propto \frac{1}{T}$$

Combining the two expressions gives

$$\Delta S_{surr} = \frac{-\Delta H_{sys}}{T} \qquad \text{Equation 18.7}$$

# The Second Law of Thermodynamics

We have seen that both the system and surroundings can undergo changes in entropy during a process. The sum of the entropy changes for the system and the surroundings is the entropy change for the universe overall:

$$\Delta S_{univ} = \Delta S_{sys} + \Delta S_{surr} \qquad \text{Equation 18.8}$$

The *second law of thermodynamics* says that for a process to be spontaneous as written (in the forward direction), $\Delta S_{univ}$ must be positive. Therefore, the system may undergo a *decrease* in entropy, as long as the surroundings undergoes a larger *increase* in entropy, and vice versa. A process for which $\Delta S_{univ}$ is *negative* is not spontaneous as written.

In some cases, $\Delta S_{univ}$ is neither positive nor negative but is equal to zero. This happens when the entropy changes of the system and surroundings are equal in magnitude and opposite in sign and describes a specific type of process known as an *equilibrium* process. An *equilibrium process* is one that does not occur spontaneously in either the net forward or net reverse direction but can be made to occur by the addition or removal of energy to a system at equilibrium. An example of

an equilibrium process is the melting of ice at 0°C. (Remember that at 0°C, ice and liquid water are in equilibrium with each other [◄◄ Section 12.6].)

With Equations 18.6 and 18.7, we can calculate the entropy changes for both the system and surroundings in a process. We can then use the second law of thermodynamics (Equation 18.8) to determine if the process is spontaneous or nonspontaneous as written or if it is an equilibrium process.

Consider the synthesis of ammonia at 25°C:

$$N_2(g) + 3H_2(g) \longrightarrow 2NH_3(g) \qquad \Delta H^\circ_{rxn} = -92.6 \text{ kJ/mol}$$

From Sample Problem 18.2(b), we have $\Delta S^\circ_{sys} = -199 \text{ J/K} \cdot \text{mol}$, and substituting $\Delta H^\circ_{sys}$ ($-92.6$ kJ/mol) into Equation 18.7, we get

$$\Delta S_{surr} = \frac{-(-92.6 \times 1000) \text{ J/mol}}{298 \text{ K}} = 311 \text{ J/K} \cdot \text{mol}$$

The entropy change for the universe is

$$\Delta S^\circ_{univ} = \Delta S^\circ_{sys} + \Delta S^\circ_{surr}$$

$$= -199 \text{ J/K} \cdot \text{mol} + 311 \text{ J/K} \cdot \text{mol}$$

$$= 112 \text{ J/K} \cdot \text{mol}$$

Because $\Delta S^\circ_{univ}$ is positive, the reaction will be spontaneous at 25°C. Keep in mind, though, that just because a reaction is spontaneous does not mean that it will occur at an observable rate. The synthesis of ammonia is, in fact, extremely slow at room temperature. Thermodynamics can tell us whether or not a reaction will occur spontaneously under specific conditions, but it does not tell us how fast it will occur.

Sample Problem 18.4 lets you practice identifying spontaneous, nonspontaneous, and equilibrium processes.

---

## Sample Problem 18.4

Determine if each of the following is a spontaneous process, a nonspontaneous process, or an equilibrium process at the specified temperature: (a) $H_2(g) + I_2(g) \longrightarrow 2HI(g)$ at 0°C, (b) $CaCO_3(s) \longrightarrow CaO(s) + CO_2(g)$ at 200°C, (c) $CaCO_3(s) \longrightarrow CaO(s) + CO_2(g)$ at 1000°C, (d) $Na(s) \longrightarrow Na(l)$ at 98°C. (Assume that the thermodynamic data in Appendix 2 do not vary with temperature.)

**Strategy** For each process, use Equation 18.6 to determine $\Delta S^\circ_{sys}$ and Equations 5.12 and 18.7 to determine $\Delta H^\circ_{sys}$ and $\Delta S^\circ_{surr}$. At the specified temperature, the process is *spontaneous* if $\Delta S_{sys}$ and $\Delta S_{surr}$ sum to a positive number, *nonspontaneous* if they sum to a negative number, and an *equilibrium process* if they sum to zero. Note that because the *reaction* is the *system*, $\Delta S_{rxn}$ and $\Delta S_{sys}$ are used interchangeably.

**Setup** From Appendix 2,

(a) $S°[H_2(g)] = 131.0$ J/K · mol, $S°[I_2(g)] = 260.57$ J/K · mol, $S°[HI(g)] = 206.3$ J/K · mol; $\Delta H_f°[H_2(g)] = 0$ kJ/mol, $\Delta H_f°[I_2(g)] = 62.25$ kJ/mol, $\Delta H_f°[HI(g)] = 25.9$ kJ/mol.

(b), (c) In Sample Problem 18.2(a), we determined that for this reaction, $\Delta S_{rxn}° = 160.5$ J/K · mol, $\Delta H_f°[CaCO_3(s)] = -1206.9$ kJ/mol, $\Delta H_f°[CaO(s)] = -635.6$ kJ/mol, $\Delta H_f°[CO_2(g)] = -393.5$ kJ/mol.

(d) $S°[Na(s)] = 51.05$ J/K · mol, $S°[Na(l)] = 57.56$ J/K · mol; $\Delta H_f°[Na(s)] = 0$ kJ/mol, $\Delta H_f°[Na(l)] = 2.41$ kJ/mol.

**Solution**

(a) $\Delta S_{rxn}° = [2S°(HI)] - [S°(H_2) + S°(I_2)]$

$$= (2)(206.3 \text{ J/K · mol}) - [131.0 \text{ J/K · mol} + 260.57 \text{ J/K · mol}] = 21.03 \text{ J/K · mol}$$

$\Delta H_{rxn}° = [2\Delta H_f°(HI)] - [\Delta H_f°(H_2) + \Delta H_f°(I_2)]$

$$= (2)(25.9 \text{ kJ/mol}) - [0 \text{ kJ/mol} + 62.25 \text{ kJ/mol}] = -10.5 \text{ kJ/mol}$$

$$\Delta S_{surr} = \frac{-\Delta H_{rxn}}{T} = \frac{-(-10.5 \text{ kJ/mol})}{273 \text{ K}} = 0.0385 \text{ kJ/K · mol} = 38.5 \text{ J/K · mol}$$

$$\Delta S_{univ} = \Delta S_{sys} + \Delta S_{surr} = 21.03 \text{ J/K · mol} + 38.5 \text{ J/K · mol} = 59.5 \text{ J/K · mol}$$

$\Delta S_{univ}$ is positive; therefore the reaction is spontaneous at 0°C.

(b), (c) $\Delta S_{rxn}° = 160.5$ J/K · mol

$\Delta H_{rxn}° = [\Delta H_f°(CaO) + \Delta H_f°(CO_2)] - [\Delta H_f°(CaCO_3)]$

$$= [-635.6 \text{ kJ/mol} + (-393.5 \text{ kJ/mol})] - (-1206.9 \text{ kJ/mol}) = 177.8 \text{ kJ/mol}$$

(b)    $T = 200°C$ and

$$\Delta S_{surr} = \frac{-\Delta H_{sys}}{T} = \frac{-(177.8 \text{ kJ/mol})}{473 \text{ K}} = -0.376 \text{ kJ/K · mol} = -376 \text{ J/K · mol}$$

$$\Delta S_{univ} = \Delta S_{sys} + \Delta S_{surr} = 160.5 \text{ J/K · mol} + (-376 \text{ J/K · mol}) = -216 \text{ J/K · mol}$$

$\Delta S_{univ}$ is negative, therefore the reaction is nonspontaneous at 200°C.

(c)    $T = 1000°C$ and

$$\Delta S_{surr} = \frac{-\Delta H_{sys}}{T} = \frac{-(177.8 \text{ kJ/mol})}{1273 \text{ K}} = -0.1397 \text{ kJ/K · mol} = -139.7 \text{ J/K · mol}$$

$$\Delta S_{univ} = \Delta S_{sys} + \Delta S_{surr} = 160.5 \text{ J/K · mol} + (-139.7 \text{ J/K · mol}) = 20.8 \text{ J/K · mol}$$

In this case, $\Delta S_{univ}$ is positive; therefore, the reaction is spontaneous at 1000°C.

(d) $\Delta S_{rxn}° = S°[Na(l)] - S°[Na(s)] = 57.56$ J/K · mol $- 51.05$ J/K · mol $= 6.51$ J/K · mol

$\Delta H_{rxn}° = \Delta H_f°[Na(l)] - \Delta H_f°[Na(s)] = 2.41$ kJ/mol $- 0$ kJ/mol $= 2.41$ kJ/mol

$$\Delta S_{surr} = \frac{-\Delta H_{rxn}}{T} = \frac{-(2.41 \text{ kJ/mol})}{371 \text{ K}} = -0.0650 \text{ kJ/K · mol} = -6.50 \text{ J/K · mol}$$

$$\Delta S_{univ} = \Delta S_{sys} + \Delta S_{surr} = 6.51 \text{ J/K · mol} + (-6.50 \text{ J/K · mol}) = 0.01 \text{ J/K · mol} \approx 0$$

**Student Annotation:** The small difference between the magnitudes of $\Delta S_{sys}$ and $\Delta S_{surr}$ results from thermodynamic values not being entirely independent of temperature. The tabulated values of $S°$ and $\Delta H_f°$ are for 25°C.

$\Delta S_{univ}$ is zero; therefore, the reaction is an equilibrium process at 98°C. In fact, this is the melting point of sodium.

**Practice Problem A** For each of the following, calculate $\Delta S_{univ}$ and identify the process as a spontaneous process, a nonspontaneous process, or an equilibrium process at the specified temperature: (a) $CO_2(g) \longrightarrow CO_2(aq)$ at 25°C, (b) $N_2O_4(g) \longrightarrow 2NO_2(g)$ at 10.4°C, (c) $PCl_3(l) \longrightarrow PCl_3(g)$ at 61.2°C. (Assume that the thermodynamic data in Appendix 2 do not vary with temperature.)

**Practice Problem B** (a) Calculate $\Delta S_{univ}$ and determine if the reaction $H_2O_2(l) \longrightarrow H_2O_2(g)$ is spontaneous, nonspontaneous, or an equilibrium process at 163°C. (b) The reaction $NH_3(g)$ + $HCl(g) \longrightarrow NH_4Cl(s)$ is spontaneous in the forward direction at room temperature but, because it is exothermic, becomes less spontaneous with increasing temperature. Determine the temperature at which it is no longer spontaneous in the forward direction. (c) Determine the boiling point of $Br_2$. (Assume that the thermodynamic data in Appendix 2 do not vary with temperature.)

**Think About It** Remember that standard enthalpies of formation have units of kJ/mol, whereas standard absolute entropies have units of J/K · mol. Make sure that you convert kilojoules to joules, or vice versa, before combining the terms.

## The Third Law of Thermodynamics

Finally, we consider the third law of thermodynamics briefly in connection with the determination of standard entropy. We have related the entropy of a system to the number of possible arrangements of the system's molecules. The larger the number of possible arrangements, the larger the entropy. Imagine a pure, perfect crystalline substance at absolute zero (0 K). Under these conditions, there is essentially no molecular motion and, because the molecules occupy fixed positions in the solid, there is only one way to arrange the molecules. From Equation 18.1, we write

$$S = k \ln W = k \ln 1 = 0$$

According to the ***third law of thermodynamics,*** the entropy of a perfect crystalline substance is *zero* at absolute zero. As temperature increases, molecular motion increases, causing an increase in the number of possible arrangements of the molecules and in the number of accessible energy states, among which the system's energy can be dispersed. (See Figure 18.4.) This results in an increase in the system's entropy. Thus, the entropy of any substance at any temperature above 0 K is greater than zero. If the crystalline substance is impure or imperfect in any way, then its entropy is greater than zero even at 0 K because without perfect crystalline order, there is more than one possible arrangement of molecules.

The significance of the third law of thermodynamics is that it enables us to determine experimentally the *absolute* entropies of substances. Starting with the knowledge that the entropy of a pure crystalline substance is zero at 0 K, we can measure the increase in entropy of the substance as it is heated. The change in entropy of a substance, $\Delta S$, is the difference between the final and initial entropy values:

$$\Delta S = S_{\text{final}} - S_{\text{initial}}$$

where $S_{\text{initial}}$ is zero if the substance starts at 0 K. Therefore, the measured *change* in entropy is equal to the *absolute* entropy of the substance at the final temperature.

$$\Delta S = S_{\text{final}}$$

The entropy values arrived at in this way are called *absolute* entropies because they are *true* values—unlike standard enthalpies of formation, which are derived using an arbitrary reference. Because the tabulated values are determined at 1 atm, we usually refer to absolute entropies as *standard* entropies, $S°$. Figure 18.5 shows the increase in entropy of a substance as temperature increases from absolute zero. At 0 K, it has a zero entropy value (assuming that it is a perfect crystalline substance). As it is heated, its entropy increases gradually at first because of greater

**Figure 18.5**   Entropy increases in a substance as temperature increases from absolute zero.

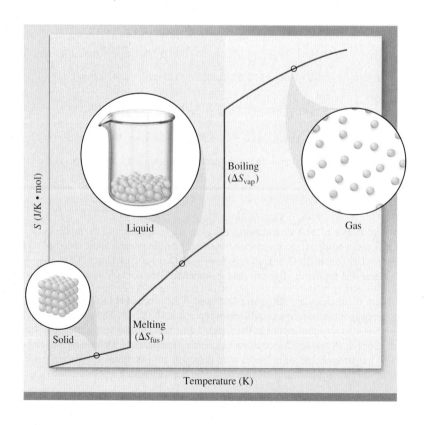

molecular motion within the crystal. At the melting point, there is a large increase in entropy as the solid is transformed into the liquid. Further heating increases the entropy of the liquid again due to increased molecular motion. At the boiling point, there is a large increase in entropy as a result of the liquid-to-vapor transition. Beyond that temperature, the entropy of the gas continues to increase with increasing temperature.

**Checkpoint 18.4**    **Entropy Changes in the Universe**

**18.4.1** Using data from Appendix 2, calculate $\Delta S°$ (in J/K · mol) for the following reaction:

$$2NO(g) + O_2(g) \longrightarrow 2NO_2(g)$$

a) 145.3 J/K · mol

b) −145.3 J/K · mol

c) 59.7 J/K · mol

d) −59.7 J/K · mol

e) −421.2 J/K · mol

**18.4.2** Using data from Appendix 2, calculate $\Delta S°$ (in J/K · mol) for the following reaction:

$$CH_4(g) + 2O_2(g) \longrightarrow CO_2(g) + 2H_2O(l)$$

a) 107.7 J/K · mol

b) −107.7 J/K · mol

c) 2.6 J/K · mol

d) 242.8 J/K · mol

e) −242.8 J/K · mol

**18.4.3** The diagrams show a spontaneous chemical reaction. What can we deduce about $\Delta S_{surr}$ for this process?

a) $\Delta S_{surr}$ is positive.

b) $\Delta S_{surr}$ is negative.

c) $\Delta S_{surr}$ is zero.

d) There is not enough information to deduce the sign of $\Delta S_{surr}$.

# 18.5 Predicting Spontaneity

## Gibbs Free-Energy Change, ΔG

According to the second law of thermodynamics, $\Delta S_{univ} > 0$ for a spontaneous process. What we are usually concerned with and usually *measure*, however, are the properties of the system rather than those of the surroundings or those of the universe overall. Therefore, it is convenient to have a thermodynamic function that enables us to determine whether or not a process is spontaneous by considering the system alone.

We begin with Equation 18.8. For a spontaneous process,

$$\Delta S_{univ} = \Delta S_{sys} + \Delta S_{surr} > 0$$

Substituting $-\Delta H_{sys}/T$ for $\Delta S_{surr}$, we write

$$\Delta S_{univ} = \Delta S_{sys} + \left(-\frac{\Delta H_{sys}}{T}\right) > 0$$

Multiplying both sides of the equation by $T$ gives

$$T\Delta S_{univ} = T\Delta S_{sys} - \Delta H_{sys} > 0$$

Now we have an equation that expresses the second law of thermodynamics (and predicts whether or not a process is spontaneous) in terms of only the *system*. We no longer need to consider the surroundings. For convenience, we can rearrange the preceding equation, multiply through by −1, and replace the > sign with a < sign:

$$-T\Delta S_{univ} = \Delta H_{sys} - T\Delta S_{sys} < 0$$

According to this equation, a process carried out at constant pressure and temperature is spontaneous if the changes in enthalpy and entropy of the system are such that $\Delta H_{sys} - T\Delta S_{sys}$ is less than zero.

To express the spontaneity of a process more directly, we introduce another thermodynamic function called the **Gibbs**[1] **free energy (G),** or simply **free energy.**

Equation 18.9

$$G = H - TS$$

Each of the terms in Equation 18.9 pertains to the system. $G$ has units of energy just as $H$ and $TS$ do. Furthermore, like enthalpy and entropy, free energy is a state function. The change in free energy, $\Delta G$, of a system for a process that occurs at constant temperature is

Equation 18.10

$$\Delta G = \Delta H - T\Delta S$$

Equation 18.10 enables us to predict the spontaneity of a process using the change in enthalpy, the change in entropy, and the absolute temperature. At constant temperature and pressure, for processes that are spontaneous as written (in the forward direction), $\Delta G$ is negative. For processes that are not spontaneous as written but that are spontaneous in the reverse direction, $\Delta G$ is positive. For systems at equilibrium, $\Delta G$ is zero.

**Student Annotation:** In this context, *free energy* is the energy available to do work. Thus, if a particular process is accompanied by a release of usable energy (i.e., if $\Delta G$ is negative), this fact alone guarantees that it is spontaneous, and there is no need to consider what happens to the rest of the universe.

- $\Delta G < 0$     The reaction is spontaneous in the forward direction (and nonspontaneous in the reverse direction).

- $\Delta G > 0$     The reaction is nonspontaneous in the forward direction (and spontaneous in the reverse direction).

- $\Delta G = 0$     The system is at equilibrium.

Often we can predict the sign of $\Delta G$ for a process if we know the signs of $\Delta H$ and $\Delta S$. Table 18.4 shows how we can use Equation 18.10 to make such predictions.

Based on the information in Table 18.4, you may wonder what constitutes a "low" or a "high" temperature. For the example given in the table, 0°C is the temperature that divides high from low. Water freezes spontaneously at temperatures below 0°C, and ice melts spontaneously at temperatures above 0°C. At 0°C, a system of ice and water is at equilibrium. The temperature that divides "high" from "low" depends, though, on the individual reaction. To determine that temperature, we must set $\Delta G$ equal to 0 in Equation 18.10 (i.e., the equilibrium condition):

$$0 = \Delta H - T\Delta S$$

Rearranging to solve for $T$ yields

$$T = \frac{\Delta H}{\Delta S}$$

The temperature that divides high from low for a particular reaction can now be calculated if the values of $\Delta H$ and $\Delta S$ are known.

Sample Problem 18.5 demonstrates the use of this approach.

| TABLE 18.4 | Predicting the Sign of $\Delta G$ Using Equation 18.10 and the Signs of $\Delta H$ and $\Delta S$ | | | |
|---|---|---|---|---|
| **When $\Delta H$ Is** | **And $\Delta S$ Is** | **$\Delta G$ Will Be** | **And the Process Is** | **Example** |
| Negative | Positive | Negative | Always spontaneous | $2H_2O_2(aq) \longrightarrow 2H_2O(l) + O_2(g)$ |
| Positive | Negative | Positive | Always nonspontaneous | $3O_2(g) \longrightarrow 2O_3(g)$ |
| Negative | Negative | Negative when $T\Delta S < \Delta H$ <br> Positive when $T\Delta S > \Delta H$ | Spontaneous at low $T$ <br> Nonspontaneous at high $T$ | $H_2O(l) \longrightarrow H_2O(s)$ (freezing of water) |
| Positive | Positive | Negative when $T\Delta S > \Delta H$ <br> Positive when $T\Delta S < \Delta H$ | Spontaneous at high $T$ <br> Nonspontaneous at low $T$ | $2HgO(s) \longrightarrow 2Hg(l) + O_2(g)$ |

---

1. Josiah Willard Gibbs (1839–1903). American physicist. One of the founders of thermodynamics. Gibbs was a modest and private individual who spent almost all his professional life at Yale University. Because he published most of his work in obscure journals, Gibbs never gained the eminence that his contemporary and admirer James Maxwell did. Even today, very few people outside of chemistry and physics have ever heard of Gibbs.

## Sample Problem 18.5

According to Table 18.4, a reaction will be spontaneous only at high temperatures if both $\Delta H$ and $\Delta S$ are positive. For a reaction in which $\Delta H = 199.5$ kJ/mol and $\Delta S = 476$ J/K · mol, determine the temperature (in °C) above which the reaction is spontaneous.

**Strategy** The temperature that divides high from low is the temperature at which $\Delta H = T\Delta S$ ($\Delta G = 0$). Therefore, we use Equation 18.10, substituting 0 for $\Delta G$ and solving for $T$ to determine temperature in kelvins; we then convert to degrees Celsius.

**Setup**

$$\Delta S = \left(\frac{476 \text{ J}}{\text{K} \cdot \text{mol}}\right)\left(\frac{1 \text{ kJ}}{1000 \text{ J}}\right) = 0.476 \text{ kJ/K} \cdot \text{mol}$$

**Solution**

$$T = \frac{\Delta H}{\Delta S} = \frac{199.5 \text{ kJ/mol}}{0.476 \text{ kJ/K} \cdot \text{mol}} = 419 \text{ K}$$

$$= (419 - 273) = 146°C$$

> **Think About It** Spontaneity is favored by a release of energy ($\Delta H$ being negative) and by an increase in entropy ($\Delta S$ being positive). When both quantities are positive, as in this case, only the entropy change favors spontaneity. For an endothermic process such as this, which requires the input of heat, it should make sense that adding more heat by increasing the temperature will shift the equilibrium to the right, thus making it "more spontaneous."

**Practice Problem A** A reaction will be spontaneous only at low temperatures if both $\Delta H$ and $\Delta S$ are negative. For a reaction in which $\Delta H = -380.1$ kJ/mol and $\Delta S = -95.00$ J/K · mol, determine the temperature (in °C) below which the reaction is spontaneous.

**Practice Problem B** Given that the reaction $4Fe(s) + 3O_2(g) + 6H_2O(l) \longrightarrow 4Fe(OH)_3(s)$ is spontaneous at temperatures below 1950°C, estimate the standard entropy of $Fe(OH)_3(s)$.

## Standard Free-Energy Changes, $\Delta G°$

The **standard free energy of reaction** ($\Delta G°_{rxn}$) is the free-energy change for a reaction when it occurs under standard-state conditions—that is, when reactants in their standard states are converted to products in *their* standard states. The conventions used by chemists to define the standard states of pure substances and solutions are

**Student Annotation:** The introduction of the term $\Delta G°$ enables us to write Equation 18.10 as
$$\Delta G° = \Delta H° - T\Delta S°$$

- Gases        1 atm pressure
- Liquids      Pure liquid
- Solids       Pure solid
- Elements     The most stable allotropic form at 1 atm and 25°C
- Solutions    1 molar concentration

To calculate $\Delta G°_{rxn}$, we start with the general equation

$$aA + bB \longrightarrow cC + dD$$

The standard free-energy change for this reaction is given by

$$\Delta G°_{rxn} = [c\Delta G°_f(C) + d\Delta G°_f(D)] - [a\Delta G°_f(A) + b\Delta G°_f(B)] \quad \text{Equation 18.11}$$

Equation 18.11 can be generalized as follows:

$$\Delta G°_{rxn} = \Sigma n\Delta G°_f \text{ (products)} - \Sigma m\Delta G°_f \text{ (reactants)} \qquad \text{Equation 18.12}$$

where $m$ and $n$ are stoichiometric coefficients. The term $\Delta G°_f$ is the **standard free energy of formation** of a compound—that is, the free-energy change that occurs when 1 mole of the compound is synthesized from its constituent elements, each in its standard state. For the combustion of graphite,

$$C(\text{graphite}) + O_2(g) \longrightarrow CO_2(g)$$

the standard free-energy change (from Equation 18.12) is

$$\Delta G°_{rxn} = [\Delta G°_f(CO_2)] - [\Delta G°_f(C, \text{graphite}) + \Delta G°_f(O_2)]$$

As with standard enthalpy of formation, the standard free energy of formation of any element (in its most stable allotropic form at 1 atm) is defined as zero. Thus,

$$\Delta G_f^\circ(\text{C, graphite}) = 0 \quad \text{and} \quad \Delta G_f^\circ(\text{O}_2) = 0$$

Therefore, the standard free-energy change for the reaction in this case is equal to the standard free energy of formation of $CO_2$:

$$\Delta G_{rxn}^\circ = \Delta G_f^\circ(\text{CO}_2)$$

Appendix 2 lists the values of $\Delta G_f^\circ$ at 25°C for a number of compounds. Sample Problem 18.6 demonstrates the calculation of standard free-energy changes.

---

### Sample Problem 18.6

Calculate the standard free-energy changes for the following reactions at 25°C:

(a) $CH_4(g) + 2O_2(g) \longrightarrow CO_2(g) + 2H_2O(l)$

(b) $2MgO(s) \longrightarrow 2Mg(s) + O_2(g)$

**Strategy** Look up the $\Delta G_f^\circ$ values for the reactants and products in each equation, and use Equation 18.12 to solve for $\Delta G_{rxn}^\circ$.

**Setup** From Appendix 2, we have the following values: $\Delta G_f^\circ[\text{CH}_4(g)] = -50.8$ kJ/mol, $\Delta G_f^\circ[\text{CO}_2(g)] = -394.4$ kJ/mol, $\Delta G_f^\circ[\text{H}_2\text{O}(l)] = -237.2$ kJ/mol, and $\Delta G_f^\circ[\text{MgO}(s)] = -569.6$ kJ/mol. All the other substances are elements in their standard states and have, by definition, $\Delta G_f^\circ = 0$.

**Solution**

(a) $\Delta G_{rxn}^\circ = (\Delta G_f^\circ[\text{CO}_2(g)] + 2\Delta G_f^\circ[\text{H}_2\text{O}(l)]) - (\Delta G_f^\circ[\text{CH}_4(g)] + 2\Delta G_f^\circ[\text{O}_2(g)])$

$\qquad = [(-394.4 \text{ kJ/mol}) + (2)(-237.2 \text{ kJ/mol})] - [(-50.8 \text{ kJ/mol}) + (2)(0 \text{ kJ/mol})]$

$\qquad = -818.0 \text{ kJ/mol}$

(b) $\Delta G_{rxn}^\circ = (2\Delta G_f^\circ[\text{Mg}(s)] + \Delta G_f^\circ[\text{O}_2(g)]) - (2\Delta G_f^\circ[\text{MgO}(s)])$

$\qquad = [(2)(0 \text{ kJ/mol}) + (0 \text{ kJ/mol})] - [(2)(-569.6 \text{ kJ/mol})]$

$\qquad = 1139 \text{ kJ/mol}$

**Practice Problem A** Calculate the standard free-energy changes for the following reactions at 25°C:

(a) $H_2(g) + Br_2(l) \longrightarrow 2HBr(g)$

(b) $2C_2H_6(g) + 7O_2(g) \longrightarrow 4CO_2(g) + 6H_2O(l)$

**Practice Problem B** For each reaction, determine the value of $\Delta G_f^\circ$ that is not listed in Appendix 2:

(a) $Li_2O(s) + 2HCl(g) \longrightarrow 2LiCl(s) + H_2O(g) \qquad \Delta G_{rxn}^\circ = -244.88 \text{ kJ/mol}$

(b) $Na_2O(s) + 2HI(g) \longrightarrow 2NaI(s) + H_2O(l) \qquad \Delta G_{rxn}^\circ = -435.44 \text{ kJ/mol}$

> **Think About It** Note that, like standard enthalpies of formation ($\Delta H_f^\circ$), standard free energies of formation ($\Delta G_f^\circ$) depend on the *state* of matter. Using water as an example, $\Delta G_f^\circ[\text{H}_2\text{O}(l)] = -237.2$ kJ/mol and $\Delta G_f^\circ[\text{H}_2\text{O}(g)] = -228.6$ kJ/mol. Always double-check to make sure you have selected the right value from the table.

---

## Using $\Delta G$ and $\Delta G°$ to Solve Problems

> **Student Annotation:** The sign of $\Delta G°$ does indicate whether or not a process is spontaneous when *all* reactants and products are in their standard states, but this is very seldom the case.

It is the sign of $\Delta G$, the free-energy change, *not* the sign of $\Delta G°$, the standard free-energy change, that indicates whether or not a process will occur spontaneously under a given set of conditions. What the *sign* of $\Delta G°$ tells us is the same thing that the magnitude of the equilibrium constant ($K$) tells us [◄◄ Section 15.2]. A negative $\Delta G°$ value corresponds to a large $K$ value (products favored at equilibrium), whereas a positive $\Delta G°$ value corresponds to a small $K$ value (reactants favored at equilibrium).

Like equilibrium constants, $\Delta G°$ values change with temperature. One of the uses of Equation 18.10 is to determine the temperature at which a particular equilibrium will begin to favor a desired product. For example, calcium oxide (CaO), also called quicklime, is an extremely valuable inorganic substance with a variety of industrial uses, including water treatment and pollution control. It is prepared by heating limestone ($CaCO_3$), which decomposes at a high temperature:

$$CaCO_3(s) \rightleftharpoons CaO(s) + CO_2(g)$$

The reaction is reversible, and under the right conditions, CaO and $CO_2$ readily recombine to form $CaCO_3$ again. To prevent this from happening in the industrial preparation, the system is never maintained at equilibrium; rather, $CO_2$ is constantly removed as it forms, shifting the equilibrium from left to right, thus promoting the formation of calcium oxide.

An important piece of information for the chemist responsible for maximizing CaO production is the temperature at which the decomposition equilibrium of $CaCO_3$ begins to favor products. We can make a reliable estimate of that temperature as follows. First we calculate $\Delta H°$ and $\Delta S°$ for the reaction at 25°C, using the data in Appendix 2. To determine $\Delta H°$, we apply Equation 5.19:

$$\Delta H° = [\Delta H_f°(CaO) + \Delta H_f°(CO_2)] - [\Delta H_f°(CaCO_3)]$$

$$= [(-635.6 \text{ kJ/mol}) + (-393.5 \text{ kJ/mol})] - (-1206.9 \text{ kJ/mol})$$

$$= 177.8 \text{ kJ/mol}$$

Next we apply Equation 18.6 to find $\Delta S°$:

$$\Delta S° = [S°(CaO) + S°CO_2)] - S°(CaCO_3)$$

$$= [(39.8 \text{ J/K} \cdot \text{mol}) + (213.6 \text{ J/K} \cdot \text{mol})] - (92.9 \text{ J/K} \cdot \text{mol})$$

$$= 160.5 \text{ J/K} \cdot \text{mol}$$

From Equation 18.10, we can write

$$\Delta G° = \Delta H° - T\Delta S°$$

and we obtain

$$\Delta G° = (177.8 \text{ kJ/mol}) - (298 \text{ K})(0.1605 \text{ kJ/K} \cdot \text{mol})$$

$$= 130.0 \text{ kJ/mol}$$

Because $\Delta G°$ is a large positive number, the reaction does *not* favor product formation at 25°C (298 K). And, because $\Delta H°$ and $\Delta S°$ are both positive, we know that $\Delta G°$ will be negative (product formation will be favored) at high temperatures. We can determine what constitutes a high temperature for this reaction by calculating the temperature at which $\Delta G°$ is zero.

$$0 = \Delta H° - T\Delta S°$$

or

$$T = \frac{\Delta H°}{\Delta S°}$$

$$= \frac{(177.8 \text{ kJ/mol})(1000 \text{ J/kJ})}{0.1605 \text{ kJ/K} \cdot \text{mol}}$$

$$= 1108 \text{ K } (835°C)$$

At temperatures higher than 835°C, $\Delta G°$ becomes negative, indicating that the reaction would then favor the formation of CaO and $CO_2$. At 840°C (1113 K), for example,

$$\Delta G° = \Delta H° - T\Delta S°$$

$$= 177.8 \text{ kJ/mol} - (1113 \text{ K})(0.1605 \text{ kJ/K} \cdot \text{mol})\left(\frac{1 \text{ kJ}}{1000 \text{ J}}\right)$$

$$= -0.8 \text{ kJ/mol}$$

At still higher temperatures, $\Delta G°$ becomes increasingly negative, thus favoring product formation even more. Note that in this example we used the $\Delta H°$ and $\Delta S°$ values at 25°C to calculate changes to $\Delta G°$ at much higher temperatures. Because both $\Delta H°$ and $\Delta S°$ actually change with temperature, this approach does not give us a truly accurate value for $\Delta G°$, but it does give us a reasonably good estimate.

Equation 18.10 can also be used to calculate the change in entropy that accompanies a phase change. At the temperature at which a phase change occurs (i.e., the melting point or boiling point of a substance), the system is at equilibrium ($\Delta G = 0$). Therefore, Equation 18.10 becomes

$$0 = \Delta H - T\Delta S$$

or

$$\Delta S = \frac{\Delta H}{T}$$

**Student Annotation:** Be careful with units in problems of this type. $S°$ values are tabulated using joules, whereas $\Delta H_f°$ values are tabulated using kilojoules.

Consider the ice-water equilibrium. For the ice-to-water transition, $\Delta H$ is the molar heat of fusion (see Table 12.8) and $T$ is the melting point. The entropy change is therefore

$$\Delta S_{\text{ice}\longrightarrow\text{water}} = \frac{6010\ \text{J/mol}}{273\ \text{K}} = 22.0\ \text{J/K}\cdot\text{mol}$$

Thus, when 1 mole of ice melts at 0°C, there is an increase in entropy of 22.0 J/K · mol. The increase in entropy is consistent with the increase in microstates from solid to liquid. Conversely, for the water-to-ice transition, the decrease in entropy is given by

$$\Delta S_{\text{water}\longrightarrow\text{ice}} = \frac{-6010\ \text{J/mol}}{273\ \text{K}} = -22.0\ \text{J/K}\cdot\text{mol}$$

The same approach can be applied to the water-to-steam transition. In this case, $\Delta H$ is the heat of vaporization and $T$ is the boiling point of water.

Sample Problem 18.7 examines the phase transitions in benzene.

### Sample Problem 18.7

The molar heats of fusion and vaporization of benzene are 10.9 and 31.0 kJ/mol, respectively. Calculate the entropy changes for the solid-to-liquid and liquid-to-vapor transitions for benzene. At 1 atm pressure, benzene melts at 5.5°C and boils at 80.1°C.

**Strategy**  The solid-liquid transition at the melting point and the liquid-vapor transition at the boiling point are *equilibrium* processes. Therefore, because $\Delta G$ is zero at equilibrium, in each case we can use Equation 18.10, substituting 0 for $\Delta G$ and solving for $\Delta S$, to determine the entropy change associated with the process.

**Setup**  The melting point of benzene is 5.5 + 273.15 = 278.7 K and the boiling point is 80.1 + 273.15 = 353.3 K.

**Solution**

**Think About It**  For the same substance, $\Delta S_{\text{vap}}$ is always significantly larger than $\Delta S_{\text{fus}}$. The change in number of microstates is always bigger in a liquid-to-gas transition than in a solid-to-liquid transition.

$$\Delta S_{\text{fus}} = \frac{\Delta H_{\text{fus}}}{T_{\text{melting}}} = \frac{10.9\ \text{kJ/mol}}{278.7\ \text{K}}$$

$$= 0.0391\ \text{kJ/K}\cdot\text{mol}\quad\text{or}\quad 39.1\ \text{J/K}\cdot\text{mol}$$

$$\Delta S_{\text{vap}} = \frac{\Delta H_{\text{vap}}}{T_{\text{boiling}}} = \frac{31.0\ \text{kJ/mol}}{353.3\ \text{K}}$$

$$= 0.0877\ \text{kJ/K}\cdot\text{mol}\quad\text{or}\quad 87.7\ \text{J/K}\cdot\text{mol}$$

**Practice Problem A**  The molar heats of fusion and vaporization of argon are 1.3 and 6.3 kJ/mol, respectively, and argon's melting point and boiling point are −190°C and −186°C, respectively. Calculate the entropy changes for the fusion and vaporization of argon.

**Practice Problem B**  Using data from Appendix 2 and assuming that the tabulated values do not change with temperature, (a) calculate $\Delta H^{\circ}_{\text{fus}}$ and $\Delta S^{\circ}_{\text{fus}}$ for sodium metal and determine the melting temperature of sodium, and (b) calculate $\Delta H^{\circ}_{\text{vap}}$ and $\Delta S^{\circ}_{\text{vap}}$ for sodium metal and determine the boiling temperature of sodium.

### Checkpoint 18.5    Predicting Spontaneity

**18.5.1**  A reaction for which $\Delta H$ and $\Delta S$ are both negative is

a) nonspontaneous at all temperatures

b) spontaneous at all temperatures

c) spontaneous at high temperatures

d) spontaneous at low temperatures

e) at equilibrium

**18.5.2**  At what temperature (in °C) does a reaction go from being nonspontaneous to spontaneous if it has $\Delta H = 171$ kJ/mol and $\Delta S = 161$ J/K · mol?

a) 270°C

b) 670°C

c) 1100°C

d) 790°C

e) 28°C

**18.5.3** Using data from Appendix 2, calculate $\Delta G°$ (in kJ/mol) at 25°C for the reaction:

$$CH_4(g) + 2O_2(g) \longrightarrow CO_2(g) + 2H_2O(g)$$

a) $-580.8$ kJ/mol

b) $580.8$ kJ/mol

c) $-572.0$ kJ/mol

d) $-800.8$ kJ/mol

e) $-818.0$ kJ/mol

**18.5.4** Calculate $\Delta S_{sub}$ (in J/K · mol) for the sublimation of iodine in a closed flask at 45°C.

$$I_2(s) \longrightarrow I_2(g)$$

$\Delta H_{sub} = 62.4$ kJ/mol.

a) $1.4$ J/K · mol

b) $196$ J/K · mol

c) $1387$ J/K · mol

d) $0.196$ J/K · mol

e) $721$ J/K · mol

# 18.6 Free Energy and Chemical Equilibrium

Reactants and products in a chemical reaction are almost always in something *other* than their standard states—that is, solutions usually have concentrations other than 1 *M* and gases usually have pressures other than 1 atm. To determine whether or not a reaction is spontaneous, therefore, we must take into account the actual concentrations and/or pressures of the species involved. And although we can determine $\Delta G°$ from tabulated values, we need to know $\Delta G$ to determine spontaneity.

**Student Annotation:** Even for a reaction that *starts* with all reactants and products in their standard states, as soon as the reaction begins, the concentrations of all species change and standard-state conditions no longer exist.

## Relationship Between $\Delta G$ and $\Delta G°$

The relationship between $\Delta G$ and $\Delta G°$, which is derived from thermodynamics, is

$$\Delta G = \Delta G° + RT \ln Q \qquad \text{Equation 18.13}$$

where $R$ is the gas constant (8.314 J/K · mol or $8.314 \times 10^{-3}$ kJ/K · mol), $T$ is the absolute temperature at which the reaction takes place, and $Q$ is the reaction quotient [◀◀ Section 15.2]. Thus, $\Delta G$ depends on two terms: $\Delta G°$ and $RT \ln Q$. For a given reaction at temperature $T$, the value of $\Delta G°$ is fixed but that of $RT \ln Q$ can vary because $Q$ varies according to the composition of the reaction mixture.

**Student Annotation:** The $Q$ used in Equation 18.13 can be either $Q_c$ (for reactions that take place in solution) or $Q_P$ (for reactions that take place in the gas phase).

Consider the following equilibrium:

$$H_2(g) + I_2(g) \rightleftharpoons 2HI(g)$$

Using Equation 18.12 and information from Appendix 2, we find that $\Delta G°$ for this reaction at 25°C is 2.60 kJ/mol. The value of $\Delta G$, however, depends on the pressures of all three gaseous species. If we start with a mixture of gases in which $P_{H_2} = 2.0$ atm, $P_{I_2} = 2.0$ atm, and $P_{HI} = 3.0$ atm, the reaction quotient, $Q_P$, is

$$Q_P = \frac{(P_{HI})^2}{(P_{H_2})(P_{I_2})} = \frac{(3.0)^2}{(2.0)(2.0)} = \frac{9.0}{4.0}$$
$$= 2.25$$

Using this value in Equation 18.13 gives

$$\Delta G = \frac{2.60 \text{ kJ}}{\text{mol}} + \left( \frac{8.314 \times 10^{-3} \text{ kJ}}{\text{K} \cdot \text{mol}} \right)(298 \text{ K})(\ln 2.25)$$
$$= 4.3 \text{ kJ/mol}$$

Because $\Delta G$ is positive, we conclude that, starting with these concentrations, the forward reaction will not occur spontaneously as written. Instead, the *reverse* reaction will occur spontaneously and the system will reach equilibrium by consuming part of the HI initially present and producing more $H_2$ and $I_2$.

If, on the other hand, we start with a mixture of gases in which $P_{H_2} = 2.0$ atm, $P_{I_2} = 2.0$ atm, and $P_{HI} = 1.0$ atm, the reaction quotient, $Q_P$, is

$$Q_P = \frac{(P_{HI})^2}{(P_{H_2})(P_{I_2})} = \frac{(1.0)^2}{(2.0)(2.0)} = \frac{1}{4}$$
$$= 0.25$$

Using this value in Equation 18.13 gives

$$\Delta G = \frac{2.6 \text{ kJ}}{\text{mol}} + \left(\frac{8.314 \times 10^{-3} \text{ kJ}}{\text{K} \cdot \text{mol}}\right)(298 \text{ K})(\ln 0.25)$$

$$= -0.8 \text{ kJ/mol}$$

With a negative value for $\Delta G$, the reaction will be spontaneous as written—in the forward direction. In this case, the system will achieve equilibrium by consuming some of the $H_2$ and $I_2$ to produce more HI.

Sample Problem 18.8 uses $\Delta G°$ and the reaction quotient to determine in which direction a reaction is spontaneous.

---

### Sample Problem 18.8

The equilibrium constant, $K_P$, for the reaction

$$N_2O_4(g) \rightleftharpoons 2NO_2(g)$$

is 0.113 at 298 K, which corresponds to a standard free-energy change of 5.4 kJ/mol. In a certain experiment, the initial pressures are $P_{N_2O_4} = 0.453$ atm and $P_{NO_2} = 0.122$ atm. Calculate $\Delta G$ for the reaction at these pressures, and predict the direction in which the reaction will proceed spontaneously to establish equilibrium.

**Strategy**  Use the partial pressures of $N_2O_4$ and $NO_2$ to calculate the reaction quotient $Q_P$, and then use Equation 18.13 to calculate $\Delta G$.

**Setup**  The reaction quotient expression is

$$Q_P = \frac{(P_{NO_2})^2}{P_{N_2O_4}} = \frac{(0.122)^2}{0.453} = 0.0329$$

**Solution**

$$\Delta G = \Delta G° + RT \ln Q_P$$

$$= \frac{5.4 \text{ kJ}}{\text{mol}} + \left(\frac{8.314 \times 10^{-3} \text{ kJ}}{\text{K} \cdot \text{mol}}\right)(298 \text{ K})(\ln 0.0329)$$

$$= 5.4 \text{ kJ/mol} - 8.46 \text{ kJ/mol}$$

$$= -3.1 \text{ kJ/mol}$$

**Think About It**  Remember, a reaction with a positive $\Delta G°$ value can be spontaneous if the starting concentrations of reactants and products are such that $Q < K$.

Because $\Delta G$ is negative, the reaction proceeds spontaneously from left to right to reach equilibrium.

---

**Practice Problem A**  $\Delta G°$ for the reaction

$$H_2(g) + I_2(g) \rightleftharpoons 2HI(g)$$

is 2.60 kJ/mol at 25°C. Calculate $\Delta G$, and predict the direction in which the reaction is spontaneous if the starting concentrations are $P_{H_2} = 3.5$ atm, $P_{I_2} = 1.5$ atm, and $P_{HI} = 1.75$ atm.

**Practice Problem B**  What is the minimum partial pressure of $I_2$ required for the preceding reaction to be spontaneous in the forward direction at 25°C if the partial pressures of $H_2$ and HI are 3.5 and 1.75 atm, respectively?

---

## Relationship Between $\Delta G°$ and $K$

By definition, $\Delta G = 0$ and $Q = K$ at equilibrium, where $K$ is the equilibrium constant. Thus, $\Delta G = \Delta G° + RT \ln Q$ (Equation 18.13) becomes

$$0 = \Delta G° + RT \ln K$$

**Student Annotation:** In this equation and in Equation 18.14, $K$ is $K_c$ for reactions that take place in solution and $K_P$ for reactions that take place in the gas phase.

or

Equation 18.14                $$\Delta G° = -RT \ln K$$

According to Equation 18.14, then, the larger $K$ is, the more negative $\Delta G°$ is. For chemists, Equation 18.14 is one of the most important equations in thermodynamics because it enables us to find the equilibrium constant of a reaction if we know the change in standard free energy, and vice versa.

It is significant that Equation 18.14 relates the equilibrium constant to the standard free-energy change, $\Delta G°$, rather than to the actual free-energy change, $\Delta G$. The actual free-energy change of the system varies as the reaction progresses and becomes zero at equilibrium. On the other hand, $\Delta G°$, like $K$, is a constant for a particular reaction at a given temperature. Figure 18.6 shows plots of the free energy of a reacting system versus the extent of the reaction for two reactions. Table 18.5 summarizes the relationship between the *magnitude* of an equilibrium constant and the *sign* of the corresponding $\Delta G°$. Remember this important distinction: It is the sign of $\Delta G$ and not that of $\Delta G°$ that determines the direction of reaction spontaneity. The sign of $\Delta G°$ only tells us the relative amounts of products and reactants when equilibrium is reached, *not* the direction the reaction must go to reach equilibrium.

**Multimedia**
Chemical Equilibrium—equilibrium.

**Multimedia**
Chemical Kinetics—interactive reaction coordinate diagram.

**Student Annotation:** The sign of $\Delta G°$ tells us the same thing that the magnitude of $K$ tells us. The sign of $\Delta G$ tells us the same thing as the comparison of $Q$ and $K$ values [◀◀ Section 15.4].

(a)

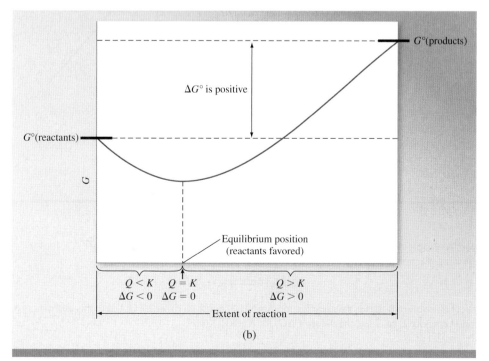

(b)

**Figure 18.6** (a) $\Delta G° < 0$. At equilibrium, there is a significant conversion of reactants to products. (b) $\Delta G° > 0$. At equilibrium, reactants are favored over products. In both cases, the net reaction toward equilibrium is from left to right (reactants to products) if $Q < K$ and right to left (products to reactants) if $Q > K$. At equilibrium, $Q = K$.

| TABLE 18.5 | | Relationship Between $K$ and $\Delta G°$ as Predicted by Equation 18.14 |
|---|---|---|

| $K$ | ln $K$ | $\Delta G°$ | **Result at Equilibrium** |
|---|---|---|---|
| > 1 | Positive | Negative | Products are favored. |
| = 1 | 0 | 0 | Neither products nor reactants are favored. |
| < 1 | Negative | Positive | Reactants are favored. |

For reactions with very large or very small equilibrium constants, it can be very difficult—sometimes impossible—to determine $K$ values by measuring the concentrations of the reactants and products. Consider, for example, the formation of nitric oxide from molecular nitrogen and molecular oxygen:

$$N_2(g) + O_2(g) \rightleftharpoons 2NO(g)$$

At 25°C, the equilibrium constant, $K_P$, is

$$K_P = \frac{(P_{NO})^2}{(P_{N_2})(P_{O_2})} = 4.0 \times 10^{-31}$$

The very small value of $K_P$ means that the concentration of NO at equilibrium will be exceedingly low and, for all intents and purposes, impossible to measure directly. In such a case, the equilibrium constant is more conveniently determined using $\Delta G°$, which can be calculated either from tabulated $\Delta G_f°$ values or from $\Delta H°$ and $\Delta S°$.

Sample Problems 18.9 and 18.10 show how to use $\Delta G°$ to calculate $K$ and how to use $K$ to calculate $\Delta G°$, respectively.

---

**Sample Problem 18.9**

Using data from Appendix 2, calculate the equilibrium constant, $K_P$, for the following reaction at 25°C:

$$2H_2O(l) \rightleftharpoons 2H_2(g) + O_2(g)$$

**Strategy** Use data from Appendix 2 and Equation 18.12 to calculate $\Delta G°$ for the reaction. Then use Equation 18.14 to solve for $K_P$.

**Setup**

$$\Delta G° = (2\Delta G_f°[H_2(g)] + \Delta G_f°[O_2(g)]) - (2\Delta G_f°[H_2O(l)])$$

$$= [(2)(0 \text{ kJ/mol}) + (2)(0 \text{ kJ/mol})] - [(2)(-237.2 \text{ kJ/mol})]$$

$$= 474.4 \text{ kJ/mol}$$

**Solution**

$$\Delta G° = -RT \ln K_P$$

$$\frac{474.4 \text{ kJ}}{\text{mol}} = -\left(\frac{8.314 \times 10^{-3} \text{ kJ}}{\text{K} \cdot \text{mol}}\right)(298 \text{ K}) \ln K_P$$

$$-191.5 = \ln K_P$$

$$K_P = e^{-191.5}$$

$$= 7 \times 10^{-84}$$

**Think About It** This is an extremely small equilibrium constant, which is consistent with the large, positive value of $\Delta G°$. We know from everyday experience that water does not decompose spontaneously into its constituent elements at 25°C.

**Practice Problem A** Using data from Appendix 2, calculate the equilibrium constant, $K_P$, for the following reaction at 25°C:

$$2O_3(g) \rightleftharpoons 3O_2(g)$$

**Practice Problem B** $K_f$ for the complex ion $Ag(NH_3)_2^+$ is $1.5 \times 10^7$ at 25°C. Using this and data from Appendix 2, calculate the value of $\Delta G_f°$ for $Ag(NH_3)_2^+(aq)$.

## Sample Problem 18.10

The equilibrium constant, $K_{sp}$, for the dissolution of silver chloride in water at 25°C,

$$AgCl(s) \rightleftharpoons Ag^+(aq) + Cl^-(aq)$$

is $1.6 \times 10^{-10}$. Calculate $\Delta G°$ for the process.

**Strategy** Use Equation 18.14 to calculate $\Delta G°$.

**Setup** $R = 8.314 \times 10^{-3}$ kJ/K · mol and $T = (25 + 273) = 298$ K.

**Solution**

$$\Delta G° = -RT \ln K_{sp}$$

$$= -\left(\frac{8.314 \times 10^{-3} \text{ kJ}}{\text{K} \cdot \text{mol}}\right)(298 \text{ K}) \ln (1.6 \times 10^{-10})$$

$$= 55.9 \text{ kJ/mol}$$

**Think About It** The relatively large, positive $\Delta G°$, like the very small $K$ value, corresponds to a process that lies very far to the left. Note that the $K$ in Equation 18.14 can be any type of $K_c$ ($K_a$, $K_b$, $K_{sp}$, etc.) or $K_P$.

**Practice Problem A** Calculate $\Delta G°$ for the process:

$$BaF_2(s) \rightleftharpoons Ba^{2+}(aq) + 2F^-(aq)$$

The $K_{sp}$ of $BaF_2$ at 25°C is $1.7 \times 10^{-6}$.

**Practice Problem B** $K_{sp}$ for $Co(OH)_2$ at 25°C is $3.3 \times 10^{-16}$. Using this and data from Appendix 2, calculate the value of $\Delta G_f°$ for $Co(OH)_2(s)$.

## Checkpoint 18.6  Free Energy and Chemical Equilibrium

**18.6.1** For the reaction

$$A(aq) + B(aq) \rightleftharpoons C(aq)$$

$\Delta G° = -1.95$ kJ/mol at 25°C. What is $\Delta G$ (in kJ/mol) at 25°C when the concentrations are [A] = [B] = 0.315 $M$ and [C] = 0.405 $M$?

a) −1.95 kJ/mol

b) 1.53 kJ/mol

c) −5.43 kJ/mol

d) 8.16 kJ/mol

e) −12.1 kJ/mol

**18.6.2** Consider the reaction

$$X(g) + Y(g) \rightleftharpoons Z(g)$$

for which $\Delta G° = -1.2$ kJ/mol at 25°C. Without knowing the pressures of the reactants and the product, we know that this reaction

a) is spontaneous as written.

b) is nonspontaneous as written.

c) favors reactants at equilibrium.

d) favors products at equilibrium.

e) is at equilibrium.

**18.6.3** The $\Delta G°$ for the reaction

$$N_2(g) + 3H_2(g) \rightleftharpoons 2NH_3(g)$$

is −33.3 kJ/mol at 25°C. What is the value of $K_P$?

a) $2 \times 10^{-6}$

b) $7 \times 10^5$

c) $3 \times 10^{-70}$

d) 1

e) $1 \times 10^{-1}$

**18.6.4** The $K_{sp}$ for iron(III) hydroxide [Fe(OH)$_3$] is $1.1 \times 10^{-36}$ at 25°C. For the process

$$Fe(OH)_3(s) \rightleftharpoons Fe^{3+}(aq) + 3OH^-(aq)$$

determine $\Delta G°$ (in kJ/mol) at 25°C.

a) $-2.73 \times 10^{-36}$ kJ/mol

b) −17.2 kJ/mol

c) 17.2 kJ/mol

d) −205 kJ/mol

e) 205 kJ/mol

## 18.7   Thermodynamics in Living Systems

Many biochemical reactions have a positive $\Delta G°$ value, yet they are essential to the maintenance of life. In living systems, these reactions are coupled to an energetically favorable process, one that has a negative $\Delta G°$ value. The principle of coupled reactions is based on a simple concept: we can use a thermodynamically favorable reaction to drive an unfavorable one. Suppose, for example, that we want to extract zinc from a zinc sulfide (ZnS). The following reaction will not work because it has a large positive $\Delta G°$ value:

$$ZnS(s) \longrightarrow Zn(s) + S(s) \qquad \Delta G° = 198.3 \text{ kJ/mol}$$

On the other hand, the combustion of sulfur to form sulfur dioxide is favored because of its large negative $\Delta G°$ value:

$$S(s) + O_2(g) \longrightarrow SO_2(g) \qquad \Delta G° = -300.1 \text{ kJ/mol}$$

By coupling the two processes, we can bring about the separation of zinc from zinc sulfide. In practice, this means heating ZnS in air so that the tendency of S to form $SO_2$ will promote the decomposition of ZnS:

$$
\begin{aligned}
ZnS(s) &\longrightarrow Zn(s) + S(s) & \Delta G° &= 198.3 \text{ kJ/mol} \\
S(s) + O_2(g) &\longrightarrow SO_2(g) & \Delta G° &= -300.1 \text{ kJ/mol} \\
\hline
ZnS(s) + O_2(g) &\longrightarrow Zn(s) + SO_2(g) & \Delta G° &= -101.8 \text{ kJ/mol}
\end{aligned}
$$

Coupled reactions play a crucial role in our survival. In biological systems, enzymes facilitate a wide variety of nonspontaneous reactions. In the human body, for example, food molecules, represented by glucose ($C_6H_{12}O_6$), are converted to carbon dioxide and water during metabolism, resulting in a substantial release of free energy:

$$C_6H_{12}O_6(s) + 6O_2(g) \longrightarrow 6CO_2(g) + 6H_2O(l) \qquad \Delta G° = -2880 \text{ kJ/mol}$$

In a living cell, this reaction does not take place in a single step; rather, the glucose molecule is broken down with the aid of enzymes in a series of steps. Much of the free energy released along the way is used to synthesize adenosine triphosphate (ATP) from adenosine diphosphate (ADP) and phosphoric acid (Figure 18.7):

$$ADP + H_3PO_4 \longrightarrow ATP + H_2O \qquad \Delta G° = 31 \text{ kJ/mol}$$

The function of ATP is to store free energy until it is needed by cells. Under appropriate conditions, ATP undergoes hydrolysis to give ADP and phosphoric acid, with a release of 31 kJ/mol of free energy, which can be used to drive energetically unfavorable reactions, such as protein synthesis.

Proteins are polymers made of amino acids. The stepwise synthesis of a protein molecule involves the joining of individual amino acids. Consider the formation of the dipeptide (a unit composed of two amino acids) alanylglycine from alanine and glycine. This reaction represents the first step in the synthesis of a protein molecule:

$$\text{alanine} + \text{glycine} \longrightarrow \text{alanylglycine} \qquad \Delta G° = 29 \text{ kJ/mol}$$

**Figure 18.7**   Structures of ATP and ADP.

Adenosine triphosphate
(ATP)

Adenosine diphosphate
(ADP)

**Figure 18.8** Schematic representation of ATP synthesis and coupled reactions in living systems. The conversion of glucose to carbon dioxide and water during metabolism releases free energy. The released free energy is used to convert ADP to ATP. The ATP molecules are then used as an energy source to drive unfavorable reactions such as protein synthesis from amino acids.

The positive $\Delta G°$ value means this reaction does not favor the formation of product, so only a little of the dipeptide would be formed at equilibrium. With the aid of an enzyme, however, the reaction is coupled to the hydrolysis of ATP as follows:

$$ATP + H_2O + alanine + glycine \longrightarrow ADP + H_3PO_4 + alanylglycine$$

The overall free-energy change is given by $\Delta G° = -31$ kJ/mol $+ 29$ kJ/mol $= -2$ kJ/mol, which means that the coupled reaction now favors the formation of product and an appreciable amount of alanylglycine will be formed under these conditions. Figure 18.8 shows the ATP-ADP interconversions that act as energy storage (from metabolism) and free-energy release (from ATP hydrolysis) to drive essential reactions.

# Applying What You've Learned

When a folded protein in solution is heated to a high enough temperature, its polypeptide chain will unfold to become the denatured protein—a process known as "denaturation." The temperature at which most of the protein unfolds is called the "melting" temperature. The melting temperature of a certain protein is found to be 63°C, and the enthalpy of denaturation is 510 kJ/mol.

**Problems:**

a) Estimate the entropy of denaturation, assuming that the denaturation is a single-step equilibrium process; that is, folded protein $\rightleftharpoons$ denatured protein. [◄◄ Sample Problem 18.7] The single polypeptide protein chain has 98 amino acids. Calculate the entropy of denaturation per amino acid.

b) Assuming that $\Delta H$ and $\Delta S$ do not change with temperature, determine $\Delta G$ for the denaturation at 20°C. [◄◄ Sample Problem 18.5]

c) Assuming that the $\Delta G$ value from part (b) is $\Delta G°$ for the denaturation, determine the value of the equilibrium constant for the process at 20°C. [◄◄ Sample Problem 18.9]

# Chapter Summary

## Section 18.1

- A *spontaneous process* is one that occurs under a specified set of conditions.

- A *nonspontaneous process* is one that does *not* occur under a specified set of conditions.

- Spontaneous processes do not necessarily happen quickly.

## Section 18.2

- *Entropy* is a thermodynamic state function that measures how dispersed or spread out a system's energy is.

## Section 18.3

- Entropy change for a process can be calculated using standard entropy values or can be predicted qualitatively based on factors such as temperature, phase, and number of molecules.

- Whether or not a process is spontaneous depends on the change in *enthalpy* and the change in *entropy* of the system.

- Tabulated *standard entropy* values are absolute values.

## Section 18.4

- According to the *second law of thermodynamics,* the entropy change for the universe is positive for a *spontaneous* process and zero for an *equilibrium process.*

- According to the *third law of thermodynamics,* the entropy of a perfectly crystalline substance at 0 K is zero.

## Section 18.5

- The *Gibbs free energy (G)* or simply the *free energy* of a system is the energy available to do work.

- The *standard free energy of reaction ($\Delta G^\circ_{rxn}$)* for a reaction tells us whether the equilibrium lies to the right (negative $\Delta G^\circ_{rxn}$) or to the left (positive $\Delta G^\circ_{rxn}$).

- *Standard free energies of formation ($\Delta G^\circ_f$)* can be used to calculate standard free energies of reaction.

## Section 18.6

- The free-energy change ($\Delta G$) is determined using the standard free-energy change ($\Delta G^\circ$) and the reaction quotient ($Q$).

- The sign of $\Delta G$ tells us whether the reaction is spontaneous under the conditions described.

- $\Delta G^\circ$ is related to the equilibrium constant, $K$. A negative $\Delta G^\circ$ corresponds to a large $K$; a positive $\Delta G^\circ$ corresponds to a small $K$.

## Section 18.7

- In living systems, thermodynamically favorable reactions provide the free energy needed to drive necessary but thermodynamically unfavorable reactions.

# Key Words

# Key Equations

18.1 $\quad S = k \ln W$

18.2 $\quad W = X^N$

18.3 $\quad \Delta S_{sys} = S_{final} - S_{initial}$

18.4 $\quad \Delta S_{sys} = nR \ln \dfrac{V_{final}}{V_{initial}}$

18.5 $\quad \Delta S^\circ_{rxn} = [cS^\circ(C) + dS^\circ(D)] - [aS^\circ(A) + bS^\circ(B)]$

18.6 $\quad \Delta S^\circ_{rxn} = \Sigma nS^\circ(\text{products}) - \Sigma mS^\circ(\text{reactants})$

18.7 $\quad \Delta S_{surr} = \dfrac{-\Delta H_{sys}}{T}$

18.8 $\quad \Delta S_{univ} = \Delta S_{sys} + \Delta S_{surr}$

18.9 $\quad G = H - TS$

18.10 $\quad \Delta G = \Delta H - T\Delta S$

18.11 $\quad \Delta G^\circ_{rxn} = [c\Delta G^\circ_f(C) + d\Delta G^\circ_f(D)] - [a\Delta G^\circ_f(A) + b\Delta G^\circ_f(B)]$

18.12 $\quad \Delta G^\circ_{rxn} = \Sigma n\Delta G^\circ_f(\text{products}) - \Sigma m\Delta G^\circ_f(\text{reactants})$

18.13 $\quad \Delta G = \Delta G^\circ + RT \ln Q$

18.14 $\quad \Delta G^\circ = -RT \ln K$

# QUESTIONS AND PROBLEMS

## Section 18.1: Spontaneous Processes

### Review Questions

18.1    Explain what is meant by a *spontaneous process.* Give two examples each of spontaneous and nonspontaneous processes.

18.2    Which of the following processes are spontaneous and which are nonspontaneous: (a) dissolving table salt (NaCl) in hot soup, (b) climbing Mt. Everest, (c) spreading fragrance in a room by removing the cap from a perfume bottle, (d) separating helium and neon from a mixture of the gases?

18.3    Which of the following processes are spontaneous and which are nonspontaneous at a given temperature?

(a) $NaNO_3(s) \xrightarrow{H_2O} NaNO_3(aq)$    saturated soln
(b) $NaNO_3(s) \xrightarrow{H_2O} NaNO_3(aq)$    unsaturated soln
(c) $NaNO_3(s) \xrightarrow{H_2O} NaNO_3(aq)$    supersaturated soln

## Section 18.2: Entropy

### Review Questions

18.4    Describe what is meant by the term *entropy.* What are the units of entropy?

18.5    What is the relationship between entropy and the number of possible arrangements of molecules in a system?

### Problems

18.6    Referring to the setup in Figure 18.2, determine the number of possible arrangements, $W$, and calculate the entropy before and after removal of the barrier if the number of molecules is (a) 10, (b) 50, (c) 100.

**18.7**    In the setup shown, a container is divided into eight cells and contains two molecules. Initially, both molecules are confined to the left side of the container. (a) Determine the number of possible arrangements before and after removal of the central barrier. (b) After the removal of the barrier, how many of the arrangements correspond to the state in which both molecules are in the left side of the container? How many correspond to the state in which both molecules are in the right side of the container? How many correspond to the state in which the molecules are in opposite sides of the container? Calculate the entropy for each state and comment on the most probable state of the system after removal of the barrier.

## Section 18.3: Entropy Changes in a System

**Visualizing Chemistry**
**Figure 18.4**

VC 18.1    Consider two gas samples at STP: one consisting of a mole of $F_2$ gas ($S° = 203.34$ J/K · mol) and one consisting of a mole of F gas ($S° = 158.7$ J/K · mol). What factors account for the difference in standard entropies of these two species?

| Volume | Molar mass | Temperature | Phase | Molecular complexity |
|--------|-----------|-------------|-------|----------------------|
| (i) | (ii) | (iii) | (iv) | (v) |

a) i, ii, iii, and iv
b) ii and v
c) ii, iv, and v

VC 18.2    Now consider the reaction $F_2(g) \longrightarrow 2F(g)$ at constant temperature and pressure. What factors contribute to the entropy increase associated with the reaction?

| Volume increase | Molar mass increase | Increased number of molecules | Phase change | Increase in molecular complexity |
|-----------------|---------------------|-------------------------------|--------------|----------------------------------|
| (i) | (ii) | (iii) | (iv) | (v) |

a) i and iii
b) i, ii, and iii
c) i, iv, and v

VC 18.3    Which of the following best describes why entropy always increases with temperature?

a)  As temperature increases, the number of molecules increases.
b)  As temperature increases, energy states become more closely spaced.
c)  As temperature increases, the molecules become more energetic and can access more energy states.

VC 18.4    Which of the following best explains why entropy typically increases with molar mass?

a)  As molar mass increases, the number of molecules increases.
b)  As molar mass increases, energy states become more closely spaced.
c)  As molar mass increases, the molecules become more energetic and can access more energy states.

### Review Questions

18.8    How does the entropy of a system change for each of the following processes?

(a) A solid melts.
(b) A liquid freezes.
(c) A liquid boils.
(d) A vapor is converted to a solid.
(e) A vapor condenses to a liquid.
(f) A solid sublimes.
(g) A solid dissolves in water.

18.9  How does the entropy of a system change for each of the following processes?

 (a) Bromine liquid vaporizes.
 (b) Water freezes to form ice.
 (c) Naphthalene, the key component of mothballs, sublimes.
 (d) Sugar crystals form from a supersaturated solution.
 (e) A block of lead melts.
 (f) Iodine vapor condenses to form solid iodine.
 (g) Carbon tetrachloride dissolves in liquid benzene.

18.10  Predict whether the entropy change is positive or negative for each of the following reactions. Give reasons for your predictions.

 (a) $2KClO_4(s) \longrightarrow 2KClO_3(s) + O_2(g)$
 (b) $H_2O(g) \longrightarrow H_2O(l)$
 (c) $2Na(s) + 2H_2O(l) \longrightarrow 2NaOH(aq) + H_2(g)$
 (d) $N_2(g) \longrightarrow 2N(g)$

18.11  State whether the sign of the entropy change expected for each of the following processes will be positive or negative, and explain your predictions.

 (a) $PCl_3(l) + Cl_2(g) \longrightarrow PCl_5(g)$
 (b) $2HgO(s) \longrightarrow 2Hg(l) + O_2(g)$
 (c) $H_2(g) \longrightarrow 2H(g)$
 (d) $U(s) + 3F_2(g) \longrightarrow UF_6(s)$

### Problems

18.12  Calculate $\Delta S_{sys}$ for (a) the isothermal expansion of 2.0 moles of an ideal gas from 10.0 L to 15.0 L, (b) the isothermal expansion of 1.5 moles of an ideal gas from 20.0 L to 22.5 L, and (c) the isothermal compression of 5.0 moles of an ideal gas from 100.0 L to 75.0 L.

**18.13**  Calculate $\Delta S_{sys}$ for (a) the isothermal compression of 0.0050 mole of an ideal gas from 112 mL to 52.5 mL, (b) the isothermal compression of 0.015 mole of an ideal gas from 225 mL to 22.5 mL, and (c) the isothermal expansion of 22.1 moles of an ideal gas from 122 L to 275 L.

18.14  For each pair of substances listed here, choose the one having the larger standard entropy value at 25°C. The same molar amount is used in the comparison. Explain the basis for your choice. (a) Li(s) or Li(l), (b) $C_2H_5OH(l)$ or $CH_3OCH_3(l)$ (*Hint:* Which molecule can hydrogen-bond?), (c) Ar(g) or Xe(g), (d) CO(g) or $CO_2(g)$, (e) $O_2(g)$ or $O_3(g)$, (f) $NO_2(g)$ or $N_2O_4(g)$.

**18.15**  Arrange the following substances (1 mole each) in order of increasing entropy at 25°C: (a) Ne(g), (b) $SO_2(g)$, (c) Na(s), (d) NaCl(s), (e) $H_2(g)$. Give the reasons for your arrangement.

18.16  Using the data in Appendix 2, calculate the standard entropy changes for the following reactions at 25°C:

 (a) S(rhombic) + $O_2(g) \longrightarrow SO_2(g)$
 (b) $MgCO_3(s) \longrightarrow MgO(s) + CO_2(g)$
 (c) $2C_2H_6(g) + 7O_2(g) \longrightarrow 4CO_2(g) + 6H_2O(l)$

18.17  Using the data in Appendix 2, calculate the standard entropy changes for the following reactions at 25°C:

 (a) $H_2(g) + CuO(s) \longrightarrow Cu(s) + H_2O(g)$
 (b) $2Al(s) + 3ZnO(s) \longrightarrow Al_2O_3(s) + 3Zn(s)$
 (c) $CH_4(g) + 2O_2(g) \longrightarrow CO_2(g) + 2H_2O(l)$

### Section 18.4: Entropy Changes in the Universe

### Review Questions

18.18  State the second law of thermodynamics in words, and express it mathematically.

18.19  State the third law of thermodynamics in words, and explain its usefulness in calculating entropy values.

### Problems

18.20  Calculate $\Delta S_{surr}$ for each of the reactions in Problem 18.16 and determine if each reaction is spontaneous at 25°C.

**18.21**  Calculate $\Delta S_{surr}$ for each of the reactions in Problem 18.17 and determine if each reaction is spontaneous at 25°C.

18.22  Using data from Appendix 2, calculate $\Delta S_{rxn}^{\circ}$ and $\Delta S_{surr}$ for each of the reactions in Problem 18.10 and determine if each reaction is spontaneous at 25°C.

**18.23**  Using data from Appendix 2, calculate $\Delta S_{rxn}^{\circ}$ and $\Delta S_{surr}$ for each of the reactions in Problem 18.11 and determine if each reaction is spontaneous at 25°C.

### Section 18.5: Predicting Spontaneity

### Review Questions

18.24  Define *free energy*. What are its units?

18.25  Why is it more convenient to predict the direction of a reaction in terms of $\Delta G_{sys}$ instead of $\Delta S_{univ}$? Under what conditions can $\Delta G_{sys}$ be used to predict the spontaneity of a reaction?

18.26  What is the significance of the sign of $\Delta G_{sys}$?

18.27  From the following combinations of $\Delta H$ and $\Delta S$, predict if a process will be spontaneous at a high or low temperature: (a) both $\Delta H$ and $\Delta S$ are negative, (b) $\Delta H$ is negative and $\Delta S$ is positive, (c) both $\Delta H$ and $\Delta S$ are positive, (d) $\Delta H$ is positive and $\Delta S$ is negative.

### Problems

18.28  Calculate $\Delta G^{\circ}$ for the following reactions at 25°C:

 (a) $N_2(g) + O_2(g) \longrightarrow 2NO(g)$
 (b) $H_2O(l) \longrightarrow H_2O(g)$
 (c) $2C_2H_2(g) + 5O_2(g) \longrightarrow 4CO_2(g) + 2H_2O(l)$

 (*Hint:* Look up the standard free energies of formation of the reactants and products in Appendix 2.)

18.29  Calculate $\Delta G^{\circ}$ for the following reactions at 25°C:

 (a) $2Mg(s) + O_2(g) \longrightarrow 2MgO(s)$
 (b) $2SO_2(g) + O_2(g) \longrightarrow 2SO_3(g)$
 (c) $2C_2H_6(g) + 7O_2(g) \longrightarrow 4CO_2(g) + 6H_2O(l)$

 (See Appendix 2 for thermodynamic data.)

18.30  From the values of $\Delta H$ and $\Delta S$, predict which of the following reactions would be spontaneous at 25°C: reaction A: $\Delta H = 10.5$ kJ/mol, $\Delta S = 30$ J/K · mol; reaction B: $\Delta H = 1.8$ kJ/mol, $\Delta S = -113$ J/K · mol. If either of the reactions is nonspontaneous at 25°C, at what temperature might it become spontaneous?

**18.31** Find the temperatures at which reactions with the following $\Delta H$ and $\Delta S$ values would become spontaneous: (a) $\Delta H = -126$ kJ/mol, $\Delta S = 84$ J/K · mol; (b) $\Delta H = -11.7$ kJ/mol, $\Delta S = -105$ J/K · mol.

**18.32** The molar heats of fusion and vaporization of ethanol are 7.61 and 26.0 kJ/mol, respectively. Calculate the molar entropy changes for the solid-liquid and liquid-vapor transitions for ethanol. At 1 atm pressure, ethanol melts at $-117.3°C$ and boils at $78.3°C$.

**18.33** The molar heats of fusion and vaporization of mercury are 23.4 and 59.0 kJ/mol, respectively. Calculate the molar entropy changes for the solid-liquid and liquid-vapor transitions for mercury. At 1 atm pressure, mercury melts at $-38.9°C$ and boils at $357°C$.

**18.34** Consider the formation of a dimeric protein:

$$2P \longrightarrow P_2$$

At 25°C, we have $\Delta H° = 17$ kJ/mol and $\Delta S° = 65$ J/K · mol. Is the dimerization favored at this temperature? Comment on the effect of lowering the temperature. Does your result explain why some enzymes lose their activities under cold conditions?

**18.35** Use the values listed in Appendix 2 to calculate $\Delta G°$ for the following alcohol fermentation:

$$C_6H_{12}O_6(s) \longrightarrow 2C_2H_5OH(l) + 2CO_2(g)$$

**18.36** As an approximation, we can assume that proteins exist either in the native (physiologically functioning) state or the denatured state. The standard molar enthalpy and entropy of the denaturation of a certain protein are 512 kJ/mol and 1.60 kJ/K · mol, respectively. Comment on the signs and magnitudes of these quantities, and calculate the temperature at which the denaturation becomes spontaneous.

**18.37** Certain bacteria in the soil obtain the necessary energy for growth by oxidizing nitrites to nitrates:

$$2NO_2^- + O_2 \longrightarrow 2NO_3^-$$

Given that the standard Gibbs free energies of formation of $NO_2^-$ and $NO_3^-$ are $-34.6$ and $-110.5$ kJ/mol, respectively, calculate the amount of Gibbs free energy released when 1 mole of $NO_2^-$ is oxidized to 1 mole of $NO_3^-$.

## Section 18.6: Free Energy and Chemical Equilibrium

### Review Questions

**18.38** Explain the difference between $\Delta G$ and $\Delta G°$.

**18.39** Explain why Equation 18.14 is of great importance in chemistry.

**18.40** Fill in the missing entries in the following table:

| $K$ | ln $K$ | $\Delta G°$ | Result at equilibrium |
|-----|--------|-------------|----------------------|
| $<1$ | | | |
| | | 0 | |
| | | | Products are favored |

### Problems

**18.41** Calculate $K_P$ for the following reaction at 25°C:

$$H_2(g) + I_2(g) \rightleftharpoons 2HI(g) \qquad \Delta G° = 2.60 \text{ kJ/mol}$$

**18.42** For the autoionization of water at 25°C,

$$H_2O(l) \rightleftharpoons H^+(aq) + OH^-(aq)$$

$K_w$ is $1.0 \times 10^{-14}$. What is $\Delta G°$ for the process?

**18.43** Consider the following reaction at 25°C:

$$Fe(OH)_2(s) \rightleftharpoons Fe^{2+}(aq) + 2OH^-(aq)$$

Calculate $\Delta G°$ for the reaction. $K_{sp}$ for $Fe(OH)_2$ is $1.6 \times 10^{-14}$.

**18.44** Calculate $\Delta G°$ and $K_P$ for the following equilibrium reaction at 25°C:

$$2H_2O(g) \rightleftharpoons 2H_2(g) + O_2(g)$$

**18.45** (a) Calculate $\Delta G°$ and $K_P$ for the following equilibrium reaction at 25°C.

$$PCl_5(g) \rightleftharpoons PCl_3(g) + Cl_2(g)$$

(b) Calculate $\Delta G$ for the reaction if the partial pressures of the initial mixture are $P_{PCl_5} = 0.0029$ atm, $P_{PCl_3} = 0.27$ atm, and $P_{Cl_2} = 0.40$ atm.

**18.46** The equilibrium constant ($K_P$) for the reaction

$$H_2(g) + CO_2(g) \rightleftharpoons H_2O(g) + CO(g)$$

is 4.40 at 2000 K. (a) Calculate $\Delta G°$ for the reaction. (b) Calculate $\Delta G$ for the reaction when the partial pressures are $P_{H_2} = 0.25$ atm, $P_{CO_2} = 0.78$ atm, $P_{H_2O} = 0.66$ atm, and $P_{CO} = 1.20$ atm.

**18.47** Consider the decomposition of calcium carbonate:

$$CaCO_3(s) \rightleftharpoons CaO(s) + CO_2(g)$$

Calculate the pressure in atm of $CO_2$ in an equilibrium process (a) at 25°C and (b) at 800°C. Assume that $\Delta H° = 177.8$ kJ/mol and $\Delta S° = 160.5$ J/K · mol for the temperature range.

**18.48** The equilibrium constant $K_P$ for the reaction

$$CO(g) + Cl_2(g) \rightleftharpoons COCl_2(g)$$

is $5.62 \times 10^{35}$ at 25°C. Calculate $\Delta G_f°$ for $COCl_2$ at 25°C.

**18.49** At 25°C, $\Delta G°$ for the process

$$H_2O(l) \rightleftharpoons H_2O(g)$$

is 8.6 kJ/mol. Calculate the vapor pressure of water at this temperature.

**18.50** Calculate $\Delta G°$ for the process

$$C(\text{diamond}) \rightleftharpoons C(\text{graphite})$$

Is the formation of graphite from diamond favored at 25°C? If so, why is it that diamonds do not become graphite on standing?

## Section 18.7: Thermodynamics in Living Systems

### Review Questions

**18.51** What is a coupled reaction? What is its importance in biological reactions?

**18.52** What is the role of ATP in biological reactions?

### Problems

**18.53** Referring to the metabolic process involving glucose on page 804, calculate the maximum number of moles of ATP that can be synthesized from ADP from the breakdown of 1 mole of glucose.

**18.54** In the metabolism of glucose, the first step is the conversion of glucose to glucose 6-phosphate:

$$\text{glucose} + H_3PO_4 \longrightarrow \text{glucose 6-phosphate} + H_2O$$
$$\Delta G° = 13.4 \text{ kJ/mol}$$

Because $\Delta G°$ is positive, this reaction does not favor the formation of products. Show how this reaction can be made to proceed by coupling it with the hydrolysis of ATP. Write an equation for the coupled reaction, and estimate the equilibrium constant for the coupled process.

## Additional Problems

**18.55** Predict the signs of $\Delta H$, $\Delta S$, and $\Delta G$ of the system for the following processes at 1 atm: (a) ammonia melts at $-60°C$, (b) ammonia melts at $-77.7°C$, (c) ammonia melts at $-100°C$. (The normal melting point of ammonia is $-77.7°C$.)

**18.56** Calculate $\Delta G$ for the reaction

$$H_2O(l) \rightleftharpoons H^+(aq) + OH^-(aq)$$

at 25°C for the following initial concentrations:

(a) $[H^+] = 1.0 \times 10^{-7}\ M$, $[OH^-] = 1.0 \times 10^{-7}\ M$
(b) $[H^+] = 1.0 \times 10^{-3}\ M$, $[OH^-] = 1.0 \times 10^{-4}\ M$
(c) $[H^+] = 1.0 \times 10^{-12}\ M$, $[OH^-] = 2.0 \times 10^{-8}\ M$
(d) $[H^+] = 3.5\ M$, $[OH^-] = 4.8 \times 10^{-4}\ M$

**18.57** Which of the following thermodynamic functions are associated only with the first law of thermodynamics: $S$, $U$, $G$, and $H$?

**18.58** A student placed 1 g of each of three compounds A, B, and C in a container and found that after 1 week no change had occurred. Offer some possible explanations for the fact that no reactions took place. Assume that A, B, and C are totally miscible liquids.

**18.59** The enthalpy change in the denaturation of a certain protein is 125 kJ/mol. If the entropy change is 397 J/K · mol, calculate the minimum temperature at which the protein would denature spontaneously.

**18.60** Consider the following facts: Water freezes spontaneously at $-5°C$ and 1 atm, and ice has a lower entropy than liquid water. Explain how a spontaneous process can lead to a decrease in entropy.

**18.61** Ammonium nitrate ($NH_4NO_3$) dissolves spontaneously and endothermically in water. What can you deduce about the sign of $\Delta S$ for the solution process?

**18.62** Calculate the equilibrium pressure of $CO_2$ due to the decomposition of barium carbonate ($BaCO_3$) at 25°C.

**18.63** (a) Trouton's rule states that the ratio of the molar heat of vaporization of a liquid ($\Delta H_{vap}$) to its boiling point in kelvins is approximately 90 J/K · mol. Use the following data to show that this is the case and explain why Trouton's rule holds true:

| | $T_{bp}$ (°C) | $\Delta H_{vap}$ (kJ/mol) |
|---|---|---|
| Benzene | 80.1 | 31.0 |
| Hexane | 68.7 | 30.8 |
| Mercury | 357 | 59.0 |
| Toluene | 110.6 | 35.2 |

(b) Use the values in Table 12.6 to calculate the same ratio for ethanol and water. Explain why Trouton's rule does not apply to these two substances as well as it does to other liquids.

**18.64** Referring to Problem 18.63, explain why the ratio is considerably smaller than 90 J/K · mol for liquid HF.

**18.65** Carbon monoxide (CO) and nitric oxide (NO) are polluting gases contained in automobile exhaust. Under suitable conditions, these gases can be made to react to form nitrogen ($N_2$) and the less harmful carbon dioxide ($CO_2$). (a) Write an equation for this reaction. (b) Identify the oxidizing and reducing agents.

(c) Calculate the $K_P$ for the reaction at 25°C. (d) Under normal atmospheric conditions, the partial pressures are $P_{N_2} = 0.80$ atm, $P_{CO_2} = 3.0 \times 10^{-4}$ atm, $P_{CO} = 5.0 \times 10^{-5}$ atm, and $P_{NO} = 5.0 \times 10^{-7}$ atm. Calculate $Q_P$, and predict the direction toward which the reaction will proceed. (e) Will raising the temperature favor the formation of $N_2$ and $CO_2$?

**18.66** For reactions carried out under standard-state conditions, Equation 18.10 takes the form $\Delta G° = \Delta H° - T\Delta S°$. (a) Assuming $\Delta H°$ and $\Delta S°$ are independent of temperature, derive the equation

$$\ln \frac{K_2}{K_1} = \frac{\Delta H°}{R}\left(\frac{T_2 - T_1}{T_1 T_2}\right)$$

where $K_1$ and $K_2$ are the equilibrium constants at $T_1$ and $T_2$, respectively. (b) Given that at 25°C $K_c$ is $4.63 \times 10^{-3}$ for the reaction

$$N_2O_4(g) \rightleftharpoons 2NO_2(g) \qquad \Delta H° = 58.0\ kJ/mol$$

calculate the equilibrium constant at 65°C.

**18.67** The $K_{sp}$ of AgCl is given in Table 17.4. What is its value at 60°C? [*Hint:* You need the result of Problem 18.66(a) and the data in Appendix 2 to calculate $\Delta H°$.]

**18.68** Under what conditions does a substance have a standard entropy of zero? Can an element or a compound ever have a negative standard entropy?

**18.69** Water gas, a mixture of $H_2$ and CO, is a fuel made by combining steam with red-hot coke (a by-product of coal distillation):

$$H_2O(g) + C(s) \rightleftharpoons CO(g) + H_2(g)$$

From the data in Appendix 2, estimate the temperature at which the reaction begins to favor the formation of products.

**18.70** Consider the following Brønsted acid-base reaction at 25°C:

$$HF(aq) + Cl^-(aq) \rightleftharpoons HCl(aq) + F^-(aq)$$

(a) Predict whether $K$ will be greater or smaller than 1. (b) Does $\Delta S°$ or $\Delta H°$ make a greater contribution to $\Delta G°$? (c) Is $\Delta H°$ likely to be positive or negative?

**18.71** The pH of gastric juice is about 1.00 and that of blood plasma is 7.40. Calculate the Gibbs free energy required to secrete a mole of $H^+$ ions from blood plasma to the stomach at 37°C.

**18.72** Crystallization of sodium acetate from a supersaturated solution occurs spontaneously (see page 541). Based on this, what can you deduce about the signs of $\Delta S$ and $\Delta H$?

**18.73** Consider the thermal decomposition of $CaCO_3$:

$$CaCO_3(s) \rightleftharpoons CaO(s) + CO_2(g)$$

The equilibrium vapor pressures of $CO_2$ are 22.6 mmHg at 700°C and 1829 mmHg at 950°C. Calculate the standard enthalpy of the reaction. [*Hint:* See Problem 18.66(a).]

**18.74** A certain reaction is spontaneous at 72°C. If the enthalpy change for the reaction is 19 kJ/mol, what is the *minimum* value of $\Delta S$ (in J/K · mol) for the reaction?

**18.75** Predict whether the entropy change is positive or negative for each of these reactions:

(a) $Zn(s) + 2HCl(aq) \rightleftharpoons ZnCl_2(aq) + H_2(g)$
(b) $O(g) + O(g) \rightleftharpoons O_2(g)$
(c) $NH_4NO_3(s) \rightleftharpoons N_2O(g) + 2H_2O(g)$
(d) $2H_2O_2(l) \rightleftharpoons 2H_2O(l) + O_2(g)$

**18.76** The reaction $NH_3(g) + HCl(g) \longrightarrow NH_4Cl(s)$ proceeds spontaneously at 25°C even though there is a decrease in entropy in the system (gases are converted to a solid). Explain.

**18.77** Use the following data to determine the normal boiling point, in kelvins, of mercury. What assumptions must you make to do the calculation?

$$Hg(l): \Delta H_f^\circ = 0 \text{ (by definition)}$$
$$S^\circ = 77.4 \text{ J/K} \cdot \text{mol}$$

$$Hg(g): \Delta H_f^\circ = 60.78 \text{ kJ/mol}$$
$$S^\circ = 174.7 \text{ J/K} \cdot \text{mol}$$

**18.78** The molar heat of vaporization of ethanol is 39.3 kJ/mol, and the boiling point of ethanol is 78.3°C. Calculate $\Delta S$ for the vaporization of 0.50 mole of ethanol.

**18.79** A certain reaction is known to have a $\Delta G^\circ$ value of $-122$ kJ/mol. Will the reaction necessarily occur if the reactants are mixed together?

**18.80** In the Mond process for the purification of nickel, carbon monoxide is combined with heated nickel to produce $Ni(CO)_4$, which is a gas and can therefore be separated from solid impurities:

$$Ni(s) + 4CO(g) \rightleftharpoons Ni(CO)_4(g)$$

Given that the standard free energies of formation of $CO(g)$ and $Ni(CO)_4(g)$ are $-137.3$ and $-587.4$ kJ/mol, respectively, calculate the equilibrium constant of the reaction at 80°C. Assume that $\Delta G_f^\circ$ is temperature independent.

**18.81** Calculate $\Delta G^\circ$ and $K_P$ for the following processes at 25°C:

(a) $H_2(g) + Br_2(l) \rightleftharpoons 2HBr(g)$
(b) $\frac{1}{2}H_2(g) + \frac{1}{2}Br_2(l) \rightleftharpoons HBr(g)$

Account for the differences in $\Delta G^\circ$ and $K_P$ obtained for parts (a) and (b).

**18.82** Calculate the pressure of $O_2$ (in atm) over a sample of NiO at 25°C if $\Delta G^\circ = 212$ kJ/mol for the reaction

$$NiO(s) \rightleftharpoons Ni(s) + \frac{1}{2}O_2(g)$$

**18.83** Comment on the statement: "Just talking about entropy increases its value in the universe."

**18.84** For a reaction with a negative $\Delta G^\circ$ value, which of the following statements is false? (a) The equilibrium constant $K$ is greater than one. (b) The reaction is spontaneous when all the reactants and products are in their standard states. (c) The reaction is always exothermic.

**18.85** Consider the reaction

$$N_2(g) + O_2(g) \rightleftharpoons 2NO(g)$$

Given that $\Delta G^\circ$ for the reaction at 25°C is 173.4 kJ/mol, (a) calculate the standard free energy of formation of NO and (b) calculate $K_P$ of the reaction. (c) One of the starting substances in smog formation is NO. Assuming that the temperature in a running automobile engine is 1100°C, estimate $K_P$ for the given reaction. (d) As farmers know, lightning helps to produce a better crop. Why?

**18.86** Heating copper(II) oxide at 400°C does not produce any appreciable amount of Cu:

$$CuO(s) \rightleftharpoons Cu(s) + \frac{1}{2}O_2(g) \qquad \Delta G^\circ = 127.2 \text{ kJ/mol}$$

However, if this reaction is coupled to the conversion of graphite to carbon monoxide, it becomes spontaneous. Write an equation for the coupled process, and calculate the equilibrium constant for the coupled reaction.

**18.87** Consider the decomposition of magnesium carbonate:

$$MgCO_3(s) \rightleftharpoons MgO(s) + CO_2(g)$$

Calculate the temperature at which the decomposition begins to favor products. Assume that both $\Delta H^\circ$ and $\Delta S^\circ$ are independent of temperature.

**18.88** (a) Over the years, there have been numerous claims about "perpetual motion machines," machines that will produce useful work with no input of energy. Explain why the first law of thermodynamics prohibits the possibility of such a machine existing. (b) Another kind of machine, sometimes called a "perpetual motion of the second kind," operates as follows. Suppose an ocean liner sails by scooping up water from the ocean and then extracting heat from the water, converting the heat to electric power to run the ship, and dumping the water back into the ocean. This process does not violate the first law of thermodynamics, for no energy is created—energy from the ocean is just converted to electric energy. Show that the second law of thermodynamics prohibits the existence of such a machine.

**18.89** The activity series in Table 4.5 shows that reaction (a) is spontaneous whereas reaction (b) is nonspontaneous at 25°C:

(a) $Fe(s) + 2H^+ \longrightarrow Fe^{2+}(aq) + H_2(g)$
(b) $Cu(s) + 2H^+ \longrightarrow Cu^{2+}(aq) + H_2(g)$

Use the data in Appendix 2 to calculate the equilibrium constant for these reactions and hence confirm that the activity series is correct.

**18.90** The rate constant for the elementary reaction

$$2NO(g) + O_2(g) \longrightarrow 2NO_2(g)$$

is $7.1 \times 10^9/M^2 \cdot s$ at 25°C. What is the rate constant for the reverse reaction at the same temperature?

**18.91** The following reaction was described as the cause of sulfur deposits formed at volcanic sites.

$$2H_2S(g) + SO_2(g) \rightleftharpoons 3S(s) + 2H_2O(g)$$

It may also be used to remove $SO_2$ from powerplant stack gases. (a) Identify the type of redox reaction it is. (b) Calculate the equilibrium constant ($K_P$) at 25°C, and comment on whether this method is feasible for removing $SO_2$. (c) Would this procedure become more effective or less effective at a higher temperature?

**18.92** Describe two ways that you could determine $\Delta G^\circ$ of a reaction.

**18.93** The following reaction represents the removal of ozone in the stratosphere:

$$2O_3(g) \rightleftharpoons 3O_2(g)$$

Calculate the equilibrium constant ($K_P$) for this reaction. In view of the magnitude of the equilibrium constant, explain why this reaction is not considered a major cause of ozone depletion in the absence of human-made pollutants such as the nitrogen oxides and CFCs. Assume the temperature of the stratosphere is $-30$°C and $\Delta G_f^\circ$ is temperature independent.

**18.94** When a native protein in solution is heated to a high enough temperature, its polypeptide chain will unfold to become the denatured protein. The temperature at which a large portion of the protein unfolds is called the melting temperature. The melting temperature of a certain protein is found to be 46°C, and the enthalpy of denaturation is 382 kJ/mol. Estimate the entropy of denaturation, assuming that the denaturation is a two-state process; that is, native protein ⟶ denatured protein. The single polypeptide protein chain has 122 amino acids. Calculate the entropy of denaturation per amino acid. Comment on your result.

**18.95** A 74.6-g ice cube floats in the Arctic Sea. The pressure and temperature of the system and surroundings are at 1 atm and 0°C, respectively. Calculate $\Delta S_{sys}$, $\Delta S_{surr}$, and $\Delta S_{univ}$ for the melting of the ice cube. What can you conclude about the nature of the process from the value of $\Delta S_{univ}$? (The molar heat of fusion of water is 6.01 kJ/mol.)

18.96 Comment on the feasibility of extracting copper from its ore chalcocite ($Cu_2S$) by heating:

$$Cu_2S(s) \longrightarrow 2Cu(s) + S(s)$$

Calculate the $\Delta G°$ for the overall reaction if this process is coupled to the conversion of sulfur to sulfur dioxide and $\Delta G_f°(Cu_2S) = -86.1$ kJ/mol.

**18.97** Active transport is the process in which a substance is transferred from a region of lower concentration to one of higher concentration. This is a nonspontaneous process and must be coupled to a spontaneous process, such as the hydrolysis of ATP. The concentrations of $K^+$ ions in the blood plasma and in nerve cells are 15 m$M$ and 400 m$M$, respectively (1 m$M = 1 \times 10^{-3} M$). Use Equation 18.13 to calculate $\Delta G$ for the process at the physiological temperature of 37°C:

$$K^+(15 \text{ m}M) \longrightarrow K^+(400 \text{ m}M)$$

In this calculation, the $\Delta G°$ term can be set to zero. What is the justification for this step?

**18.98** Large quantities of hydrogen are needed for the synthesis of ammonia. One preparation of hydrogen involves the reaction between carbon monoxide and steam at 300°C in the presence of a copper-zinc catalyst:

$$CO(g) + H_2O(g) \rightleftharpoons CO_2(g) + H_2(g)$$

Calculate the equilibrium constant ($K_P$) for the reaction and the temperature at which the reaction favors the formation of CO and $H_2O$. Will a larger $K_P$ be attained at the same temperature if a more efficient catalyst is used?

**18.99** Consider two carboxylic acids (acids that contain the —COOH group): $CH_3COOH$ (acetic acid, $K_a = 1.8 \times 10^{-5}$) and $CH_2ClCOOH$ (chloroacetic acid, $K_a = 1.4 \times 10^{-3}$). (a) Calculate $\Delta G°$ for the ionization of these acids at 25°C. (b) From the equation $\Delta G° = \Delta H° - T\Delta S°$, we see that the contributions to the $\Delta G°$ term are an enthalpy term ($\Delta H°$) and a temperature times entropy term ($T\Delta S°$). These contributions are listed here for the two acids:

| | $\Delta H°$ (kJ/mol) | $T\Delta S°$ (kJ/mol) |
|---|---|---|
| $CH_3COOH$ | −0.57 | −27.6 |
| $CH_2ClCOOH$ | −4.7 | −21.1 |

Which is the dominant term in determining the value of $\Delta G°$ (and hence $K_a$ of the acid)? (c) What processes contribute to $\Delta H°$? (Consider the ionization of the acids as a Brønsted acid-base reaction.) (d) Explain why the $T\Delta S°$ term is more negative for $CH_3COOH$.

18.100 Many hydrocarbons exist as structural isomers, which are compounds that have the same molecular formula but different structures. For example, both butane and isobutane have the same molecular formula of $C_4H_{10}$ (see Problem 12.20 on page 530). Calculate the mole percent of these molecules in an equilibrium mixture at 25°C, given that the standard free energy of formation of butane is −15.9 kJ/mol and that of isobutane

is −18.0 kJ/mol. Does your result support the notion that straight-chain hydrocarbons (that is, hydrocarbons in which the C atoms are joined along a line) are less stable than branch-chain hydrocarbons?

**18.101** One of the steps in the extraction of iron from its ore (FeO) is the reduction of iron(II) oxide by carbon monoxide at 900°C:

$$FeO(s) + CO(g) \rightleftharpoons Fe(s) + CO_2(g)$$

If CO is allowed to react with an excess of FeO, calculate the mole fractions of CO and $CO_2$ at equilibrium. State any assumptions.

18.102 Derive the equation

$$\Delta G = RT \ln \frac{Q}{K}$$

where $Q$ is the reaction quotient, and describe how you would use it to predict the spontaneity of a reaction.

**18.103** The sublimation of carbon dioxide at −78°C is given by

$$CO_2(s) \longrightarrow CO_2(g) \qquad \Delta H_{sub} = 25.2 \text{ kJ/mol}$$

Calculate $\Delta S_{sub}$ when 84.8 g of $CO_2$ sublimes at this temperature.

18.104 A student looked up the $\Delta G_f°$, $\Delta H_f°$, and $\Delta S°$ values for $CO_2$ in Appendix 2. Plugging these values into Equation 18.10, the student found that $\Delta G_f° \neq \Delta H_f° - T\Delta S°$ at 298 K. What is wrong with this approach?

18.105 Consider the following reaction at 298 K:

$$2H_2(g) + O_2(g) \longrightarrow 2H_2O(l) \qquad \Delta H° = -571.6 \text{ kJ/mol}$$

Calculate $\Delta S_{sys}$, $\Delta S_{surr}$, and $\Delta S_{univ}$ for the reaction.

**18.106** The concentration of glucose inside a cell is 0.12 m$M$ and that outside a cell is 12.3 m$M$. Calculate the Gibbs free-energy change for the transport of 3 moles of glucose into the cell at 37°C.

18.107 Which of the following are not state functions: S, H, q, w, T?

18.108 Which of the following is not accompanied by an increase in the entropy of the system: (a) mixing of two gases at the same temperature and pressure, (b) mixing of ethanol and water, (c) discharging a battery, (d) expansion of a gas followed by compression to its original temperature, pressure, and volume?

18.109 Hydrogenation reactions (for example, the process of converting C=C bonds to C—C bonds in the food industry) are facilitated by the use of a transition metal catalyst, such as Ni or Pt. The initial step is the adsorption, or binding, of hydrogen gas onto the metal surface. Predict the signs of $\Delta H$, $\Delta S$, and $\Delta G$ when hydrogen gas is adsorbed onto the surface of Ni metal.

18.110 Give a detailed example of each of the following, with an explanation: (a) a thermodynamically spontaneous process, (b) a process that would violate the first law of thermodynamics, (c) a process that would violate the second law of thermodynamics, (d) an irreversible process, (e) an equilibrium process.

**18.111** At 0 K, the entropy of carbon monoxide crystal is not zero but has a value of 4.2 J/K · mol, called the residual entropy. According to the third law of thermodynamics, this means that the crystal does not have a perfect arrangement of the CO molecules. (a) What would be the residual entropy if the arrangement were totally random? (b) Comment on the difference between the result in part (a) and 4.2 J/K · mol. (Hint: Assume that each CO molecule has two choices for orientation, and use Equation 18.1 to calculate the residual entropy.)

18.112 The following diagram shows the variation of the equilibrium constant with temperature for the reaction

$$I_2(g) \rightleftharpoons 2I(g)$$

Calculate $\Delta G°$, $\Delta H°$, and $\Delta S°$ for the reaction at 872 K. (*Hint:* See Problem 18.66.)

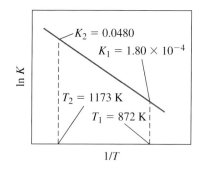

18.113 The standard enthalpy of formation and the standard entropy of gaseous benzene are 82.93 kJ/mol and 269.2 J/K · mol, respectively. Calculate $\Delta H°$, $\Delta S°$, and $\Delta G°$ for the given process at 25°C. Comment on your answers.

$$C_6H_6\,(l) \longrightarrow C_6H_6(g)$$

18.114 Consider the gas-phase reaction between $A_2$ (green) and $B_2$ (red) to form AB at 298 K:

$$A_2(g) + B_2(g) \rightleftharpoons 2AB(g) \qquad \Delta G° = -3.4 \text{ kJ/mol}$$

(a) Which of the following reaction mixtures is at equilibrium?
(b) Which of the following reaction mixtures has a negative $\Delta G$ value?
(c) Which of the following reaction mixtures has a positive $\Delta G$ value?

The partial pressures of the gases in each frame are equal to the number of $A_2$, $B_2$, and AB molecules times 0.10 atm. Round your results to two significant figures.

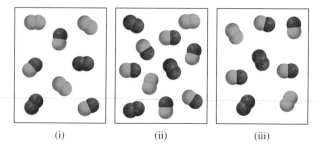

(i)        (ii)        (iii)

# PRE-PROFESSIONAL PRACTICE EXAM PROBLEMS: PHYSICAL AND BIOLOGICAL SCIENCES

In chemistry, the standard state for a solution is 1 *M*. This means that each solute concentration expressed in molarity is divided by 1 *M*. In biological systems, however, we define the standard state for the $H^+$ ions to be $1 \times 10^{-7}$ *M*. Consequently, the change in the standard Gibbs free energy according to these two conventions will be different for reactions involving the uptake or release of $H^+$ ions, depending on which convention is used. We will therefore replace $\Delta G°$ with $\Delta G°'$, where the prime denotes that it is the standard Gibbs free-energy change for a biological process.

1. What is most likely the reason for the standard state for $H^+$ being defined as $1 \times 10^{-7}$ *M* in biological systems?

   a) It is an arbitrary convention.
   b) Physiological conditions reduce all concentrations by a factor of $10^7$.
   c) Physiological conditions increase all concentrations by a factor of $10^7$.
   d) Biological systems have a pH of about 7.

2. Consider the reaction

$$A + B \longrightarrow C + xH^+$$

   where $x$ is a stoichiometric coefficient. Using the equation

$$\Delta G = \Delta G° + RT \ln Q$$

   determine the relationship between $\Delta G°'$ and $\Delta G°$, keeping in mind that $\Delta G$ is the same for a process regardless of which convention is used.

   a) $\Delta G°' = \Delta G° + RT \ln (1 \times 10^7)$
   b) $\Delta G°' = \Delta G° + RT \ln (1 \times 10^{-7})^x$
   c) $\Delta G°' = \Delta G° + e^{-7/RT}$
   d) $\Delta G°' = \Delta G° + RT \times e^{-7}$

3. Determine the relationship between $\Delta G°'$ and $\Delta G°$ for the reverse process:

$$C + xH^+ \longrightarrow A + B$$

   a) $\Delta G°' = \Delta G° + RT \ln (1 \times 10^7)^x$
   b) $\Delta G°' = \Delta G° + RT \ln (1 \times 10^{-7})$
   c) $\Delta G°' = \Delta G° + e^{7/RT}$
   d) $\Delta G°' = \Delta G° + RT \times e^7$

4. $NAD^+$ and NADH are the oxidized and reduced forms of nicotinamide adenine dinucleotide, two key compounds in the metabolic pathways. For the oxidation of NADH

$$NADH + H^+ \longrightarrow NAD^+ + H_2$$

   $\Delta G°$ is $-21.8$ kJ/mol at 298 K. Calculate $\Delta G°'$ for this process, and calculate $\Delta G$ using either the chemical or the biological convention when [NADH] $= 1.5 \times 10^{-2}$ *M*, [$H^+$] $= 3.0 \times 10^{-5}$ *M*, [NAD] $= 4.6 \times 10^{-3}$ *M*, and $P_{H_2} = 0.010$ atm.

   a) 18.1 kJ/mol, 50.3 kJ/mol
   b) 21.8 kJ/mol, 29.6 kJ/mol
   c) 18.1 kJ/mol, 10.3 kJ/mol
   d) 21.8 kJ/mol, 10.3 kJ/mol

# ANSWERS TO IN-CHAPTER MATERIALS

## Answers to Practice Problems

**18.1A** 0.34 J/K. **18.1B** $\frac{1}{5}$. **18.2A** (a) 173.6 J/K · mol, (b) −139.8 J/K · mol, (c) 215.3 J/K · mol. **18.2B** (a) $S°[K(l)]$ = 71.5 J/K · mol, (b) $S°[S_2Cl_2(g)]$ = 331.5 J/K · mol, (c) $S°[MgF_2(s)]$ = 57.24 J/K · mol. **18.3A** (a) Negative, (b) negative, (c) positive. **18.3B** The sign of $\Delta H°$ for both dissolution processes is negative. Something must favor spontaneity; if not entropy change, then enthalpy change. Because these processes both involve decreases in the system's entropy, they must be exothermic, or they could not be spontaneous. **18.4A** (a) $\Delta S_{univ}$ = −27.2 J/K · mol, nonspontaneous, (b) $\Delta S_{univ}$ = −28.1 J/K · mol, nonspontaneous, (c) $\Delta S_{univ}$ = 0 J/K · mol, equilibrium. **18.4B** (a) $\Delta S_{univ}$ = 5.2 J/K · mol, spontaneous, (b) 346°C, (c) 58°C. **18.5A** 3728°C. **18.5B** 108 J/K · mol. **18.6A** (a) −106 kJ/mol, (b) −2935 kJ/mol. **18.6B** (a) $\Delta G_f°[Li_2O(s)]$ = −561.2 kJ/mol, (b) $\Delta G_f°[NaI(s)]$ = −286.1 kJ/mol. **18.7A** $\Delta S_{fus}$ = 16 J/K · mol, $\Delta S_{vap}$ = 72 J/K · mol. **18.7B** (a) $\Delta H_{fus}°$ = 2.41 kJ/mol, $\Delta S_{fus}°$ = 6.51 J/K · mol, $T_{melting}$ = 97°C. (b) $\Delta H_{vap}°$ = 105.3 kJ/mol, $\Delta S_{vap}°$ = 96.1 J/K · mol, $T_{boiling}$ = 823°C. **18.8A** 1.3 kJ/mol, reverse reaction is spontaneous. **18.8B** 2.5 atm. **18.9A** $2 \times 10^{57}$. **18.9B** −16.8 kJ/mol. **18.10A** 32.9 kJ/mol. **18.10B** −454.4 kJ/mol.

## Answers to Checkpoints

**18.3.1** a, e. **18.3.2** b. **18.3.3** d. **18.4.1** b. **18.4.2** e. **18.4.3** b. **18.5.1** d. **18.5.2** d. **18.5.3** d. **18.5.4** b. **18.6.1** b. **18.6.2** d. **18.6.3** b. **18.6.4** e.

## Answers to Applying What You've Learned

**a)** Positive. **b)** 1061.4 J/K · mol. **c)** 1114.9 kJ/mol. **d)** 1075.4°C.

# 19

# Electrochemistry

*The pain associated with biting on a piece of aluminum foil is the result of an electrochemical reaction.*

# How Electrochemistry Can Cause Dental Pain

Anyone who accidentally bites on a piece of aluminum foil in such a way that the foil touches an amalgam dental filling will experience a momentary sharp pain. The pain is actually the result of electrical stimulation of the nerve of a tooth caused by a current flowing between the aluminum foil and the metal in the filling. Historically, the material most commonly used to fill cavities is known as *dental amalgam*. (An *amalgam* is a substance made by combining mercury with one or more other metals.) Dental amalgam consists of liquid mercury mixed in roughly equal parts with an alloy powder containing silver, tin, copper, and sometimes smaller amounts of other metals such as zinc.

What happens is this: the aluminum and the dental filling act as *electrodes* in a galvanic cell. Aluminum acts as the anode, and the amalgam acts as the cathode. In effect, when the aluminum comes into contact with the amalgam, an electrochemical cell is established in the mouth, causing a current to flow. This current stimulates the nerve of the tooth, causing a *very* unpleasant sensation.

The pain caused by having aluminum foil contact an amalgam filling is the result of *electrochemistry*.

Used with permission of Atlantic Feature Syndicate for Mark Parisi.

At the end of this chapter, you will be able to solve a series of problems related to the electrochemistry associated with dental pain [▶▶ Page 848].

## In This Chapter, You Will Learn:

How chemical reactions can produce electric energy and how reaction conditions affect the energy production. You will also learn how electric energy can be used to drive chemical reactions that otherwise would not occur.

### Before you begin, you should review

- Oxidation numbers [◀◀ Section 4.4]
- The reaction quotient, $Q$ [◀◀ Section 15.2]
- Gibbs free energy [◀◀ Section 18.5]

# 19.1　Balancing Redox Reactions

In Chapter 4 we briefly discussed oxidation-reduction or "redox" reactions, those in which electrons are transferred from one species to another. In this section we review how to identify a reaction as a redox reaction and look more closely at how such reactions are balanced.

A redox reaction is one in which there are *changes* in oxidation states, which we identify using the rules introduced in Chapter 4. The following are examples of redox reactions:

$$2KClO_3(s) \longrightarrow 2KCl(s) + 3O_2(g)$$

$$\overset{+1\;+5\;-2}{\phantom{2KClO_3}} \qquad \overset{+1\;-1}{\phantom{2KCl}} \qquad \overset{0}{\phantom{3O_2}}$$

$$CH_4(g) + 2O_2(g) \longrightarrow CO_2(g) + 2H_2O(l)$$

$$\overset{-4\;+1}{\phantom{CH_4}} \qquad \overset{0}{\phantom{2O_2}} \qquad \overset{+4\;-2}{\phantom{CO_2}} \qquad \overset{+1\;-2}{\phantom{2H_2O}}$$

$$Sn(s) + Cu^{2+}(aq) \longrightarrow Cu(s) + Sn^{2+}(aq)$$

$$\overset{0}{\phantom{Sn}} \qquad \overset{+2}{\phantom{Cu}} \qquad \overset{0}{\phantom{Cu}} \qquad \overset{+2}{\phantom{Sn}}$$

Equations for redox reactions, such as those shown here, can be balanced by inspection, the method of balancing introduced in Chapter 3 [◄◄ Section 3.3], but remember that redox equations must be balanced for mass (number of atoms) *and* for charge (number of electrons) [◄◄ Section 4.4]. In this section we introduce the *half-reaction method* to balance equations that cannot be balanced simply by inspection.

Consider the aqueous reaction of the iron(II) ion ($Fe^{2+}$) with the dichromate ion ($Cr_2O_7^{2-}$):

$$Fe^{2+} + Cr_2O_7^{2-} \longrightarrow Fe^{3+} + Cr^{3+}$$

Because there is no species containing oxygen on the product side of the equation, it would not be possible to balance this equation simply by adjusting the coefficients of reactants and products. However, there are two things about the reaction that make it possible to *add* species to the equation to balance it—without changing the chemical reaction it represents:

- The reaction takes place in aqueous solution, so we can add $H_2O$ as needed to balance the equation.
- This particular reaction takes place in acidic solution, so we can add $H^+$ as needed to balance the equation. (Some reactions take place in basic solution, enabling us to add $OH^-$ as needed for balancing. We will learn more about this shortly.)

After writing the unbalanced equation, we balance it stepwise as follows:

1. Separate the unbalanced reaction into **half-reactions.** A half-reaction is an oxidation or a reduction that occurs as part of the overall redox reaction.

   *Oxidation:*　　　　$Fe^{2+} \longrightarrow Fe^{3+}$

   *Reduction:*　　　　$Cr_2O_7^{2-} \longrightarrow Cr^{3+}$

2. Balance each of the half-reactions with regard to atoms other than O and H. In this case, no change is required for the oxidation half-reaction. We adjust the coefficient of the chromium(III) ion to balance the reduction half-reaction.

   *Oxidation:*　　　　$Fe^{2+} \longrightarrow Fe^{3+}$

   *Reduction:*　　　　$Cr_2O_7^{2-} \longrightarrow 2Cr^{3+}$

3. Balance both half-reactions for O by adding $H_2O$. Again, the oxidation in this case requires no change, but we must add seven water molecules to the product side of the reduction.

   *Oxidation:*　　　　$Fe^{2+} \longrightarrow Fe^{3+}$

   *Reduction:*　　　　$Cr_2O_7^{2-} \longrightarrow 2Cr^{3+} + 7H_2O$

4. Balance both half-reactions for H by adding $H^+$. Once again, the oxidation in this case requires no change, but we must add 14 hydrogen ions to the reactant side of the reduction.

   *Oxidation:*　　　　$Fe^{2+} \longrightarrow Fe^{3+}$

   *Reduction:*　$14H^+ + Cr_2O_7^{2-} \longrightarrow 2Cr^{3+} + 7H_2O$

5. Balance both half-reactions for charge by adding electrons. To do this, we determine the total charge on each side and add electrons to make the total charges equal. In the case of the oxidation, there is a charge of $+2$ on the reactant side and a charge of $+3$ on the product side. Adding one electron to the product side makes the charges equal.

$$\text{Oxidation:} \quad \underbrace{Fe^{2+}}_{} \longrightarrow \underbrace{Fe^{3+} + e^-}_{}$$
$$\text{Total charge:} \quad +2 \qquad\qquad +2$$

In the case of the reduction, there is a total charge of $[(14)(+1) + (1)(-2)] = +12$ on the reactant side and a total charge of $[(2)(+3)] = +6$ on the product side. Adding six electrons to the reactant side makes the charges equal.

$$\text{Reduction:} \quad \underbrace{6e^- + 14H^+ + Cr_2O_7^{2-}}_{} \longrightarrow \underbrace{2Cr^{3+} + 7H_2O}_{}$$
$$\text{Total charge:} \qquad\qquad +6 \qquad\qquad\qquad +6$$

6. If the number of electrons in the balanced oxidation half-reaction is not the same as the number of electrons in the balanced reduction half-reaction, multiply one or both of the half-reactions by the number(s) required to make the number of electrons the same in both. In this case, with one electron in the oxidation and six in the reduction, multiplying the oxidation by 6 accomplishes this.

$$\text{Oxidation:} \qquad\qquad 6(Fe^{2+} \longrightarrow Fe^{3+} + e^-)$$
$$6Fe^{2+} \longrightarrow 6Fe^{3+} + 6e^-$$
$$\text{Reduction:} \quad 6e^- + 14H^+ + Cr_2O_7^{2-} \longrightarrow 2Cr^{3+} + 7H_2O$$

7. Finally, add the balanced half-reactions back together and cancel the electrons, in addition to any other identical terms that appear on both sides.

$$6Fe^{2+} \longrightarrow 6Fe^{3+} + 6e^-$$
$$\underline{6e^- + 14H^+ + Cr_2O_7^{2-} \longrightarrow 2Cr^{3+} + 7H_2O}$$
$$6Fe^{2+} + 14H^+ + Cr_2O_7^{2-} \longrightarrow 6Fe^{3+} + 2Cr^{3+} + 7H_2O$$

A final check shows that the resulting equation is balanced both for mass and for charge.

Some redox reactions occur in basic solution. When this is the case, balancing by the half-reaction method is done exactly as described for reactions in acidic solution, but it requires two additional steps:

8. For each $H^+$ ion in the final equation, add one $OH^-$ ion to each side of the equation, combining the $H^+$ and $OH^-$ ions to produce $H_2O$.

9. Make any additional cancellations made necessary by the new $H_2O$ molecules.

Sample Problem 19.1 shows how to use the half-reaction method to balance a reaction that takes place in basic solution.

## Sample Problem 19.1

Permanganate ion and iodide ion react in basic solution to produce manganese(IV) oxide and molecular iodine. Use the half-reaction method to balance the equation:

$$MnO_4^- + I^- \longrightarrow MnO_2 + I_2$$

**Strategy** The reaction takes place in basic solution, so apply steps 1 through 9 to balance for mass and for charge.

**Setup** Identify the oxidation and reduction half-reactions by assigning oxidation numbers.

$$MnO_4^- + I^- \longrightarrow MnO_2 + I_2$$
$$\overset{(+7)(-2)}{\phantom{x}} \quad \overset{(-1)}{\phantom{x}} \qquad \overset{(+4)(-2)}{\phantom{x}} \quad \overset{(0)}{\phantom{x}}$$

**Solution**

Step 1. Separate the unbalanced reaction into half-reactions.

$$\text{Oxidation:} \quad I^- \longrightarrow I_2$$
$$\text{Reduction:} \quad MnO_4^- \longrightarrow MnO_2$$

*(Continued)*

Step 2.    Balance each half-reaction for mass, excluding O and H.

$$2I^- \longrightarrow I_2$$

$$MnO_4^- \longrightarrow MnO_2$$

Step 3.    Balance both half-reactions for O by adding $H_2O$.

$$2I^- \longrightarrow I_2$$

$$MnO_4^- \longrightarrow MnO_2 + 2H_2O$$

Step 4.    Balance both half-reactions for H by adding $H^+$.

$$2I^- \longrightarrow I_2$$

$$4H^+ + MnO_4^- \longrightarrow MnO_2 + 2H_2O$$

Step 5.    Balance the total charge of both half-reactions by adding electrons.

$$2I^- \longrightarrow I_2 + 2e^-$$

$$3e^- + 4H^+ + MnO_4^- \longrightarrow MnO_2 + 2H_2O$$

Step 6.    Multiply the half-reactions to make the numbers of electrons the same in both.

$$3(2I^- \longrightarrow I_2 + 2e^-)$$

$$2(3e^- + 4H^+ + MnO_4^- \longrightarrow MnO_2 + 2H_2O)$$

Step 7.    Add the half-reactions back together, cancelling electrons.

$$6I^- \longrightarrow 3I_2 + \cancel{6e^-}$$

$$\cancel{6e^-} + 8H^+ + 2MnO_4^- \longrightarrow 2MnO_2 + 4H_2O$$

$$\overline{8H^+ + 2MnO_4^- + 6I^- \longrightarrow 2MnO_2 + 3I_2 + 4H_2O}$$

Step 8.    For each $H^+$ ion in the final equation, add one $OH^-$ ion to each side of the equation, combining the $H^+$ and $OH^-$ ions to produce $H_2O$.

$$8H^+ + 2MnO_4^- + 6I^- \longrightarrow 2MnO_2 + 3I_2 + 4H_2O$$

$$\underline{+\ 8OH^- \hspace{6cm} +\ 8OH^-}$$

$$\boxed{8H_2O} + 2MnO_4^- + 6I^- \longrightarrow 2MnO_2 + 3I_2 + \boxed{4H_2O} + 8OH^-$$

Step 9.    Carry out any cancellations made necessary by the additional $H_2O$ molecules.

$$4H_2O + 2MnO_4^- + 6I^- \longrightarrow 2MnO_2 + 3I_2 + 8OH^-$$

> **Think About It**  Verify that the final equation is balanced for mass and for charge. Remember that electrons cannot appear in the overall balanced equation.

**Practice Problem A**  Use the half-reaction method to balance the following equation in basic solution:

$$CN^- + MnO_4^- \longrightarrow CNO^- + MnO_2$$

**✋ Practice Problem B**  Use the half-reaction method to balance the following equation in acidic solution:

$$Fe^{2+} + MnO_4^- \longrightarrow Fe^{3+} + Mn^{2+}$$

### Checkpoint 19.1    Balancing Redox Reactions

19.1.1    Which of the following equations does *not* represent a redox reaction? (Select all that apply.)

a) $NH_3 + HCl \longrightarrow NH_4Cl$

b) $2H_2O_2 \longrightarrow 2H_2O + O_2$

c) $2O_3 \longrightarrow 3O_2$

d) $3NO_2 + H_2O \longrightarrow NO + 2HNO_3$

e) $LiCl \longrightarrow Li^+ + Cl^-$

19.1.2    $MnO_4^-$ and $C_2O_4^{2-}$ react in basic solution to form $MnO_2$ and $CO_3^{2-}$. What are the coefficients of $MnO_4^-$ and $C_2O_4^{2-}$ in the balanced equation?

a) 1 and 1

b) 2 and 1

c) 2 and 3

d) 2 and 6

e) 2 and 2

# 19.2 Galvanic Cells

**Media Player/ MPEG Content**

Electrochemistry—operation of voltaic cell.

When zinc metal is placed in a solution containing copper(II) ions, Zn is oxidized to $Zn^{2+}$ ions whereas $Cu^{2+}$ ions are reduced to Cu [◄◄ Section 4.4]:

$$Zn(s) + Cu^{2+}(aq) \longrightarrow Zn^{2+}(aq) + Cu(s)$$

The electrons are transferred directly from the reducing agent, Zn, to the oxidizing agent, $Cu^{2+}$, in solution. However, if we physically separate two half-reactions from each other, we can arrange it such that the electrons must travel through a wire to pass from the Zn atoms to the $Cu^{2+}$ ions. As the reaction progresses, it generates a flow of electrons through the wire and thereby generates electricity.

> **Student Annotation:** This reaction is shown in Figure 4.5.

The experimental apparatus for generating electricity through the use of a spontaneous reaction is called a ***galvanic cell.*** Figure 19.1 shows the essential components of a galvanic cell. A zinc bar is immersed in an aqueous $ZnSO_4$ solution in one container, and a copper bar is immersed in an aqueous $CuSO_4$ solution in another container. The cell operates on the principle that the oxidation of Zn to $Zn^{2+}$ and the reduction of $Cu^{2+}$ to Cu can be made to take place simultaneously in separate locations with the transfer of electrons between them occurring through an external wire. The zinc and copper bars are called ***electrodes.*** By definition, the ***anode*** in a galvanic cell is the electrode at which *oxidation* occurs and the ***cathode*** is the electrode at which *reduction* occurs. (Each combination of container, electrode, and solution is called a ***half-cell.***) This particular arrangement of electrodes and electrolytes is called a Daniell cell.

> **Student Annotation:** A *galvanic* cell can also be called a *voltaic* cell. Both terms refer to a cell in which a spontaneous chemical reaction generates a flow of electrons.

The half-reactions for the galvanic cell shown in Figure 19.1 are

*Oxidation:*    $Zn(s) \longrightarrow Zn^{2+}(aq) + 2e^-$

*Reduction:*  $Cu^{2+}(aq) + 2e^- \longrightarrow Cu(s)$

Media Player/MPEG Animation: Figure 19.1, Construction of a Galvanic Cell, pp. 822–823.

To complete the electric circuit, and allow electrons to flow through the external wire, the solutions must be connected by a conducting medium through which the cations and anions can move from one half-cell to the other. This requirement is satisfied by a ***salt bridge,*** which, in its simplest form, is an inverted U tube containing an inert electrolyte solution, such as KCl or $NH_4NO_3$. The ions in the salt bridge must not react with the other ions in solution or with the electrodes (see Figure 19.1). During the course of the redox reaction, electrons flow through the external wire from the anode (Zn electrode) to the cathode (Cu electrode). In the solution, the cations ($Zn^{2+}$, $Cu^{2+}$, and $K^+$) move toward the cathode, while the anions ($SO_4^{2-}$ and $Cl^-$) move toward the anode. Without the salt bridge connecting the two solutions, the buildup of positive charge in the anode compartment (due to the departure of electrons and the resulting formation of $Zn^{2+}$) and the buildup of negative charge in the cathode compartment (created by the arrival of electrons and the reduction of $Cu^{2+}$ ions to Cu) would quickly prevent the cell from operating.

> **Student Annotation:** This is the origin of the terms *cathode* and *anode:*
> • Cations move toward the cathode.
> • Anions move toward the anode.

An electric current flows from the anode to the cathode because there is a difference in electrical potential energy between the electrodes. This flow of electric current is analogous to the flow of water down a waterfall, which occurs because there is a difference in the gravitational potential energy, or the flow of gas from a high-pressure region to a low-pressure region. Experimentally the difference in electrical potential between the anode and the cathode is measured by a voltmeter (Figure 19.2) and the reading (in volts) is called the ***cell potential*** ($E_{cell}$). The potential of a cell depends not only on the nature of the electrodes and the ions in solution, but also on the concentrations of the ions and the temperature at which the cell is operated.

> **Student Annotation:** The volt is a derived SI unit: 1 V = 1 J/1 C.

The conventional notation for representing galvanic cells is the cell diagram. For the cell shown in Figure 19.1, if we assume that the concentrations of $Zn^{2+}$ and $Cu^{2+}$ ions are 1 *M*, the cell diagram is

$$Zn(s)\,|\,Zn^{2+}(1\ M)\,\|\,Cu^{2+}(1\ M)\,|\,Cu(s)$$

> **Student Annotation:** The terms *cell potential, cell voltage, cell electromotive force,* and *cell emf* are used interchangeably and are all symbolized the same way with $E_{cell}$.

The single vertical line represents a phase boundary. For example, the zinc electrode is a solid and the $Zn^{2+}$ ions are in solution. Thus, we draw a line between Zn and $Zn^{2+}$ to show the phase boundary. The double vertical lines denote the salt bridge. By convention, the anode is written first, to the left of the double lines, and the other components appear in the order in which we would encounter them in moving from the anode to the cathode (from left to right in the cell diagram).

# Figure 19.1
# Construction of a Galvanic Cell

We add a salt bridge, a tube containing a solution of a strong electrolyte—in this case $Na_2SO_4$. Having this solution in electrical contact with the two solutions in the beakers allows ions to migrate toward the electrodes, ensuring that the two compartments remain electrically neutral.

The two metal pieces are the electrodes in the galvanic cell. We connect the electrodes with a length of wire routed through a voltmeter and a switch so that we can complete the circuit when we have completed construction of the cell.

To make the reaction between zinc and copper more useful, we can construct a galvanic cell. In one beaker, we place a piece of zinc metal in a 1.00 $M$ solution of $Zn^{2+}$ ions. In the other, we place a piece of copper metal in a 1.00 $M$ solution of $Cu^{2+}$ ions.

As shown in Figure 4.5, when zinc metal (Zn) is immersed in a solution containing copper ions ($Cu^{2+}$), the zinc is oxidized to zinc ions ($Zn^{2+}$), and copper ions are reduced to copper metal (Cu).

$$Zn(s) + Cu^{2+}(aq) \longrightarrow Zn^{2+}(aq) + Cu(s)$$

This is an oxidation-reduction reaction in which electrons flow spontaneously from zinc metal to the copper ions in solution. The lightening of the blue color indicates that the concentration of $Cu^{2+}$ has decreased. Copper metal is deposited on the solid zinc surface. Some of the zinc metal has gone into solution as $Zn^{2+}$ ions, which do not impart any color in the solution.

before          after

When we close the switch, we complete the circuit; the voltmeter indicates the initial potential of the cell: 1.10 V.

When we replace the voltmeter with a lightbulb, electrons flow from the zinc electrode (the anode) to the copper electrode (the cathode). The flow of electrons lights the bulb. Anions in the salt bridge migrate toward the anode; cations migrate toward the cathode.

As the reaction proceeds, zinc metal from the anode is oxidized, increasing the $Zn^{2+}$ concentration in the beaker on the left; and additional copper metal is deposited on the cathode, decreasing the $Cu^{2+}$ concentration in the beaker on the right. As the concentrations of both ions change, the potential of the cell decreases. After allowing the reaction to proceed for a time, we can reinsert the voltmeter to measure the decreased voltage.

## What's the point?

Zinc is a stronger reducing agent than copper, so there is a natural tendency for electrons to flow from zinc metal to copper ions. We can harness this flow of electrons by forcing the half-reactions to occur in separate compartments. Electrons still flow from Zn to $Cu^{2+}$, but they must flow through the wire connecting the two electrodes. The potential of the cell decreases as the reaction proceeds, as the reading on the voltmeter shows.

**Multimedia**

Electrochemistry—an interactive voltaic cell.

## 19.3    Standard Reduction Potentials

When the concentrations of the $Cu^{2+}$ and $Zn^{2+}$ ions are both 1.0 $M$, the cell potential of the cell described in Section 19.2 is 1.10 V at 25°C (see Figure 19.2). This measured potential is related to the half-reactions that take place in the anode and cathode compartments. The overall cell potential is the difference between the electric potentials at the Zn and Cu electrodes—known as the *half-cell potentials*. Just as it is impossible for only one of the half-reactions to occur independently, it is impossible to measure the potential of just a single half-cell. However, if we arbitrarily define the potential value of a particular half-cell as zero, we can use it to determine the *relative* potentials of other half-cells. Then we can use these relative potentials to determine overall *cell* potentials. The hydrogen electrode, shown in Figure 19.3, serves as the reference for this purpose. Hydrogen gas is bubbled into a hydrochloric acid solution at 25°C. The platinum electrode has two functions. First, it provides a surface on which the dissociation (and oxidation) of hydrogen molecules can take place:

$$H_2 \longrightarrow 2H^+ + 2e^-$$

Second, it serves as an electrical conductor to the external circuit.

**Figure 19.2**    The galvanic cell described in Figure 19.1. Note the U tube (salt bridge) connecting the two beakers. When the concentrations of $Zn^{2+}$ and $Cu^{2+}$ are 1 molar (1 $M$) at 25°C, the cell voltage is 1.10 V.

**Figure 19.3**    A hydrogen electrode operating under standard-state conditions. Hydrogen gas at 1 atm is bubbled through a 1 $M$ HCl solution. The electrode itself is made of platinum.

**Figure 19.4** (a) A cell consisting of a zinc electrode and a hydrogen electrode. (b) A cell consisting of a copper electrode and a hydrogen electrode. Both cells are operating under standard-state conditions. Note that in (a) the SHE is the cathode, but in (b) it is the anode.

Under standard-state conditions [◄◄ Section 18.5], when the pressure of $H_2$ is 1 atm and the concentration of HCl solution is 1 $M$, the potential for the reduction of $H^+$ at 25°C is defined as *exactly* zero:

$$2H^+(1\ M) + 2e^- \longrightarrow H_2\,(1\ \text{atm}) \qquad E° = 0\ \text{V}$$

As before, the ° superscript denotes standard-state conditions, and $E°$ is the **standard reduction potential;** that is, $E°$ is the potential associated with a reduction half-reaction at an electrode when the ion concentration is 1 $M$ and the gas pressure is 1 atm. Because the hydrogen electrode is used to determine all other electrode potentials, it is called the **standard hydrogen electrode (SHE).**

A half-cell potential is measured by constructing a galvanic cell in which one of the electrodes is the SHE. The measured voltage is then used to determine the potential for the other half-cell. Figure 19.4(a) shows a galvanic cell with a zinc electrode and a SHE. When the circuit is completed in this cell, electrons flow from the zinc electrode to the SHE, thereby oxidizing Zn to $Zn^{2+}$ and reducing $H^+$ ions to $H_2$. In this case, therefore, the zinc electrode is the anode (where oxidation takes place) and the SHE is the cathode (where reduction takes place). The cell diagram is

$$Zn(s)\,|\,Zn^{2+}(1\ M)\,\|\,H^+(1\ M)\,|\,H_2(1\ \text{atm})\,|\,Pt(s)$$

The measured potential for this cell is 0.76 V at 25°C. We can write the half-cell reactions as follows:

*Anode (oxidation):* $\qquad\qquad Zn(s) \longrightarrow Zn^{2+}(1\ M) + 2e^-$

*Cathode (reduction):* $\quad 2H^+(1\ M) + 2e^- \longrightarrow H_2(1\ \text{atm})$

*Overall:* $\qquad\qquad\quad Zn(s) + 2H^+(1\ M) \longrightarrow Zn^{2+}(1\ M) + H_2(1\ \text{atm})$

By convention, the standard cell potential, $E°_{cell}$, which is composed of a contribution from the anode and a contribution from the cathode, is given by

$$E°_{cell} = E°_{cathode} - E°_{anode} \qquad\qquad \text{Equation 19.1}$$

where $E°_{cathode}$ and $E°_{anode}$ are the standard reduction potentials of the cathode and anode, respectively. For the Zn-SHE cell, we write

$$E°_{cell} = E°_{H^+/H_2} - E°_{Zn^{2+}/Zn}$$

$$0.76\ \text{V} = 0 - E°_{Zn^{2+}/Zn}$$

where the subscript "$H^+/H_2$" means "$2H^+ + 2e^- \longrightarrow H_2$" and the subscript "$Zn^{2+}/Zn$" means "$Zn^{2+} + 2e^- \longrightarrow Zn$". Thus, the standard reduction potential of zinc, $E_{Zn^{2+}/Zn}$ is $-0.76$ V.

The standard reduction potential of copper can be determined in a similar fashion, by using a cell with a copper electrode and a SHE [Figure 19.4(b)]. In this case, electrons flow from the SHE to the copper electrode when the circuit is completed; that is, the *copper* electrode is the cathode and the SHE is the *anode.*

> **Student Annotation:** One indication of the direction of electron flow is the fact that as the reaction proceeds, the mass of the zinc electrode decreases as a result of the oxidation half-reaction:
>
> $$Zn(s) \longrightarrow Zn^{2+}(aq) + 2e^-$$

> **Student Annotation:** Reduction of $Cu^{2+}$ ions causes the mass of the Cu electrode to increase.

The cell diagram is

$$Pt(s) \mid H_2(1 \text{ atm}) \mid H^+(1 \text{ } M) \parallel Cu^{2+}(1 \text{ } M) \mid Cu(s)$$

and the half-cell reactions are

Anode (oxidation):                    $H_2(1 \text{ atm}) \longrightarrow 2H^+(1 \text{ } M) + 2e^-$

Cathode (reduction):          $\underline{Cu^{2+}(1 \text{ } M) + 2e^- \longrightarrow Cu(s)}$

Overall:                $H_2(1 \text{ atm}) + Cu^{2+}(1 \text{ } M) \longrightarrow 2H^+(1 \text{ } M) + Cu(s)$

Under standard-state conditions and at 25°C, the measured potential for the cell is 0.34 V, so we write

$$E^\circ_{cell} = E^\circ_{cathode} - E^\circ_{anode}$$

$$= E^\circ_{Cu^{2+}/Cu} - E^\circ_{H^+/H_2}$$

$$0.34 \text{ V} = E^\circ_{Cu^{2+}/Cu} - 0$$

Thus, the standard reduction potential of copper, $E^\circ_{Cu^{2+}/Cu}$, is 0.34 V, where the subscript "$Cu^{2+}/Cu$" means "$Cu^{2+} + 2e^- \longrightarrow Cu$."

Having determined the standard reduction potentials of Zn and Cu, we can use Equation 19.1 to calculate the cell potential for the Daniell cell described in Section 19.2:

$$E^\circ_{cell} = E^\circ_{cathode} - E^\circ_{anode}$$

$$= E^\circ_{Cu^{2+}/Cu} - E^\circ_{Zn^{2+}/Zn}$$

$$= 0.34 \text{ V} - (-0.76 \text{ V})$$

$$= 1.10 \text{ V}$$

As in the case of $\Delta G^\circ$ [◀◀ Section 18.5], we can use the sign of $E^\circ$ to predict whether a reaction lies to the right or to the left. A positive $E^\circ$ means that the redox reaction will favor the formation of products at equilibrium. Conversely, a negative $E^\circ$ indicates that reactants will be favored at equilibrium. We will examine how $E^\circ_{cell}$, $\Delta G^\circ$, and $K$ are related in Section 19.4.

Table 19.1 lists standard reduction potentials, in order of decreasing reduction potential, for a number of half-cell reactions. To avoid ambiguity, all half-cell reactions are shown as reductions. A galvanic cell is composed of two half-cells, and therefore two half-cell reactions. Whether or not a particular half-cell reaction occurs as a reduction in a galvanic cell depends on how its reduction potential compares to that of the other half-cell reaction. If it has the greater (or more *positive*) reduc-

**Student Annotation:** The half-reaction with the greater reduction potential has the greater potential to occur as a reduction.

tion potential of the two, it will occur as a reduction. If it has the smaller (or more *negative*) reduction potential of the two, it will occur in the reverse direction, as an oxidation.

Consider the example of the cell described in Section 19.2. The two half-cell reactions and their standard reduction potentials are

$$Cu^{2+} + 2e^- \longrightarrow Cu \qquad E^\circ = 0.34 \text{ V}$$

$$Zn^{2+} + 2e^- \longrightarrow Zn \qquad E^\circ = -0.76 \text{ V}$$

The Cu half-reaction, having a greater (more positive) reduction potential, is the half-reaction that will occur as a reduction:

**Student Annotation:** The electrode where reduction occurs is the cathode. Therefore, $E^\circ_{cathode} = 0.34 \text{ V}$.

$$Cu^{2+} + 2e^- \longrightarrow Cu$$

The Zn half-reaction has a smaller (less positive) reduction potential and will occur, instead, as an oxidation:

**Student Annotation:** The electrode where oxidation occurs is the anode. Therefore, $E^\circ_{anode} = -0.76 \text{ V}$.

$$Zn \longrightarrow Zn^{2+} + 2e^-$$

Adding the two half-reactions gives the overall cell reaction:

$$Zn + Cu^{2+} \longrightarrow Zn^{2+} + Cu$$

Consider, however, what would happen if we were to construct a galvanic cell combining the Zn half-cell with an Mn half-cell. The reduction potential of Mn is −1.18 V.

$$Zn^{2+} + 2e^- \longrightarrow Zn \qquad E^\circ = -0.76 \text{ V}$$

$$Mn^{2+} + 2e^- \longrightarrow Mn \qquad E^\circ = -1.18 \text{ V}$$

## TABLE 19.1    Standard Reduction Potentials at 25°C*

| Half-Reaction | $E°$(V) |
|---|---|
| $F_2(g) + 2e^- \longrightarrow 2F^-(aq)$ | +2.87 |
| $O_3(g) + 2H^+(aq) + 2e^- \longrightarrow O_2(g) + H_2O(l)$ | +2.07 |
| $Co^{3+}(aq) + e^- \longrightarrow Co^{2+}(aq)$ | +1.82 |
| $H_2O_2(aq) + 2H^+(aq) + 2e^- \longrightarrow 2H_2O(l)$ | +1.77 |
| $PbO_2(s) + 4H^+(aq) + SO_4^{2-}(aq) + 2e^- \longrightarrow PbSO_4(s) + 2H_2O(l)$ | +1.70 |
| $Ce^{4+}(aq) + e^- \longrightarrow Ce^{3+}(aq)$ | +1.61 |
| $MnO_4^-(aq) + 8H^+(aq) + 5e^- \longrightarrow Mn^{2+}(aq) + 4H_2O(l)$ | +1.51 |
| $Au^{3+}(aq) + 3e^- \longrightarrow Au(s)$ | +1.50 |
| $Cl_2(g) + 2e^- \longrightarrow 2Cl^-(aq)$ | +1.36 |
| $Cr_2O_7^{2-}(aq) + 14H^+(aq) + 6e^- \longrightarrow 2Cr^{3+}(aq) + 7H_2O(l)$ | +1.33 |
| $MnO_2(s) + 4H^+(aq) + 2e^- \longrightarrow Mn^{2+}(aq) + 2H_2O(l)$ | +1.23 |
| $O_2(g) + 4H^+(aq) + 4e^- \longrightarrow 2H_2O(l)$ | +1.23 |
| $Br_2(l) + 2e^- \longrightarrow 2Br^-(aq)$ | +1.07 |
| $NO_3^-(aq) + 4H^+(aq) + 3e^- \longrightarrow NO(g) + 2H_2O(l)$ | +0.96 |
| $2Hg^{2+}(aq) + 2e^- \longrightarrow Hg_2^{2+}(aq)$ | +0.92 |
| $Hg_2^{2+}(aq) + 2e^- \longrightarrow 2Hg(l)$ | +0.85 |
| $Ag^+(aq) + e^- \longrightarrow Ag(s)$ | +0.80 |
| $Fe^{3+}(aq) + e^- \longrightarrow Fe^{2+}(aq)$ | +0.77 |
| $O_2(g) + 2H^+(aq) + 2e^- \longrightarrow H_2O_2(aq)$ | +0.68 |
| $MnO_4^-(aq) + 2H_2O(l) + 3e^- \longrightarrow MnO_2(s) + 4OH^-(aq)$ | +0.59 |
| $I_2(s) + 2e^- \longrightarrow 2I^-(aq)$ | +0.53 |
| $O_2(g) + 2H_2O(l) + 4e^- \longrightarrow 4OH^-(aq)$ | +0.40 |
| $Cu^{2+}(aq) + 2e^- \longrightarrow Cu(s)$ | +0.34 |
| $AgCl(s) + e^- \longrightarrow Ag(s) + Cl^-(aq)$ | +0.22 |
| $SO_4^{2-}(aq) + 4H^+(aq) + 2e^- \longrightarrow SO_2(g) + 2H_2O(l)$ | +0.20 |
| $Cu^{2+}(aq) + e^- \longrightarrow Cu^+(aq)$ | +0.15 |
| $Sn^{4+}(aq) + 2e^- \longrightarrow Sn^{2+}(aq)$ | +0.13 |
| $2H^+(aq) + 2e^- \longrightarrow H_2(g)$ | 0.00 |
| $Pb^{2+}(aq) + 2e^- \longrightarrow Pb(s)$ | −0.13 |
| $Sn^{2+}(aq) + 2e^- \longrightarrow Sn(s)$ | −0.14 |
| $Ni^{2+}(aq) + 2e^- \longrightarrow Ni(s)$ | −0.25 |
| $Co^{2+}(aq) + 2e^- \longrightarrow Co(s)$ | −0.28 |
| $PbSO_4(s) + 2e^- \longrightarrow Pb(s) + SO_4^{2-}(aq)$ | −0.31 |
| $Cd^{2+}(aq) + 2e^- \longrightarrow Cd(s)$ | −0.40 |
| $Fe^{2+}(aq) + 2e^- \longrightarrow Fe(s)$ | −0.44 |
| $Cr^{3+}(aq) + 3e^- \longrightarrow Cr(s)$ | −0.74 |
| $Zn^{2+}(aq) + 2e^- \longrightarrow Zn(s)$ | −0.76 |
| $2H_2O(l) + 2e^- \longrightarrow H_2(g) + 2OH^-(aq)$ | −0.83 |
| $Mn^{2+}(aq) + 2e^- \longrightarrow Mn(s)$ | −1.18 |
| $Al^{3+}(aq) + 3e^- \longrightarrow Al(s)$ | −1.66 |
| $Be^{2+}(aq) + 2e^- \longrightarrow Be(s)$ | −1.85 |
| $Mg^{2+}(aq) + 2e^- \longrightarrow Mg(s)$ | −2.37 |
| $Na^+(aq) + e^- \longrightarrow Na(s)$ | −2.71 |
| $Ca^{2+}(aq) + 2e^- \longrightarrow Ca(s)$ | −2.87 |
| $Sr^{2+}(aq) + 2e^- \longrightarrow Sr(s)$ | −2.89 |
| $Ba^{2+}(aq) + 2e^- \longrightarrow Ba(s)$ | −2.90 |
| $K^+(aq) + e^- \longrightarrow K(s)$ | −2.93 |
| $Li^+(aq) + e^- \longrightarrow Li(s)$ | −3.05 |

*Increasing strength as oxidizing agent* (left margin, upward arrow)

*Increasing strength as reducing agent* (right margin, downward arrow)

*For all half-reactions the concentration is 1 M for dissolved species and the pressure is 1 atm for gases. These are the standard-state values.

In this case, the greater (less negative) reduction potential is that of Zn, so the Zn half-reaction will occur as a reduction and the Zn electrode will be the cathode. The Mn electrode will be the anode. The cell potential is therefore

$$E^\circ_{cell} = E^\circ_{cathode} - E^\circ_{anode}$$

$$= E^\circ_{Zn^{2+}/Zn} - E^\circ_{Mn^{2+}/Mn}$$

$$= (-0.76 \text{ V}) - (-1.18 \text{ V})$$

$$= 0.42 \text{ V}$$

The overall cell reaction is

$$Mn + Zn^{2+} \longrightarrow Mn^{2+} + Zn$$

By using Equation 19.1, we can predict the direction of an overall cell reaction.

It is important to understand that the standard reduction potential is an *intensive* property (like temperature and density), not an *extensive* property (like mass and volume) [◄◄ Section 1.4]. This means that the value of the standard reduction potential does *not* depend on the amount of a substance involved. Therefore, when it is necessary to multiply one of the half-reactions by a coefficient to balance the overall equation, the value of $E^\circ$ for the half-reaction remains the same. Consider a galvanic cell made up of a Zn half-cell and an Ag half-cell:

$$Zn(s) \mid Zn^{2+}(1 \text{ } M) \parallel Ag^+(1 \text{ } M) \mid Ag(s)$$

The half-cell reactions are

$$Ag^+ + e^- \longrightarrow Ag \qquad E^\circ = 0.80 \text{ V}$$

$$Zn^{2+} + 2e^- \longrightarrow Zn \qquad E^\circ = -0.76 \text{ V}$$

The Ag half-reaction, with the more positive standard reduction potential, will occur as a reduction, and the Zn half-reaction will occur as an oxidation. Balancing the equation for the overall cell reaction requires multiplying the reduction (the Ag half-reaction) by 2:

$$2(Ag^+ + e^- \longrightarrow Ag)$$

We can then add the two half-reactions and cancel the electrons to get the overall, balanced equation:

$$2Ag^+ + 2e^- \longrightarrow 2Ag$$
$$\underline{+ \ Zn \longrightarrow Zn^{2+} + 2e^-}$$
$$2Ag^+ + Zn \longrightarrow 2Ag + Zn^{2+}$$

The standard cell potential can be calculated using Equation 19.1:

$$E^\circ_{cell} = E^\circ_{cathode} - E^\circ_{anode}$$

$$= E^\circ_{Ag^+/Ag} - E^\circ_{Zn^{2+}/Zn}$$

$$= 0.80 \text{ V} - (-0.76 \text{ V})$$

$$= 1.56 \text{ V}$$

**Student Annotation:** It is a common error to multiply $E^\circ$ values by the same number as the half-cell equation. Think of the reduction potential as the height of a waterfall. Just as water falls from a higher elevation to a lower elevation, electrons move from an electrode with a higher potential toward one with a lower potential. The amount of water falling does not affect the net change in its elevation. Likewise, the number of electrons moving from higher potential to lower potential does not affect the size of the change in potential.

Although we multiplied the Ag half-reaction by 2, we did not multiply its standard reduction potential by 2.

Table 19.1 is essentially an extended version of the activity series [◄◄ Section 4.4]. Sample Problem 19.2 illustrates how to use standard reduction potentials to predict the direction of the overall reaction in a galvanic cell.

### Sample Problem 19.2

A galvanic cell consists of an Mg electrode in a 1.0 *M* Mg(NO$_3$)$_2$ solution and a Cd electrode in a 1.0 *M* Cd(NO$_3$)$_2$ solution. Determine the overall cell reaction, and calculate the standard cell potential at 25°C.

**Strategy**  Use the tabulated values of $E^\circ$ to determine which electrode is the cathode and which is the anode, combine cathode and anode half-cell reactions to get the overall cell reaction, and use Equation 19.1 to calculate $E^\circ_{cell}$.

**Setup** The half-cell reactions and their standard reduction potentials are

$$Mg^{2+} + 2e^- \longrightarrow Mg \qquad E° = -2.37 \text{ V}$$

$$Cd^{2+} + 2e^- \longrightarrow Cd \qquad E° = -0.40 \text{ V}$$

Because the Cd half-cell reaction has the greater (less negative) standard reduction potential, it will occur as the reduction. The Mg half-cell reaction will occur as the oxidation. Therefore, $E°_{cathode} = -0.40$ V and $E°_{anode} = -2.37$ V.

**Solution** Adding the two half-cell reactions together gives the overall cell reaction:

$$Mg \longrightarrow Mg^{2+} + 2e^-$$

$$Cd^{2+} + 2e^- \longrightarrow Cd$$

$$\textit{Overall:} \quad Mg + Cd^{2+} \longrightarrow Mg^{2+} + Cd$$

The standard cell potential is

$$E°_{cell} = E°_{cathode} - E°_{anode}$$

$$= E°_{Cd^{2+}/Cd} - E°_{Mg^{2+}/Mg}$$

$$= (-0.40 \text{ V}) - (-2.37 \text{ V})$$

$$= 1.97 \text{ V}$$

**Think About It** If you ever calculate a negative voltage for a galvanic cell potential, you have done something wrong—check your work. Under standard-state conditions, the overall cell reaction will proceed in the direction that gives a positive $E°_{cell}$.

**Practice Problem A** Determine the overall cell reaction and $E°_{cell}$ (at 25°C) of a galvanic cell made of a Cd electrode in a 1.0 $M$ Cd(NO$_3$)$_2$ solution and a Pb electrode in a 1.0 $M$ Pb(NO$_3$)$_2$ solution.

**Practice Problem B** A galvanic cell with $E°_{cell} = 0.30$ V can be constructed using an iron electrode in a 1.0 $M$ Fe(NO$_3$)$_2$ solution, and either a tin electrode in a 1.0 $M$ Sn(NO$_3$)$_2$ solution, or a chromium electrode in a 1.0 $M$ Cr(NO$_3$)$_3$ solution—even though Sn$^{2+}$/Sn and Cr$^{3+}$/Cr have different standard reduction potentials. Explain and give the overall balanced reaction for each cell.

Standard reduction potentials can also be used to determine what, if any, redox reaction will take place when reactants are combined in the same beaker—rather than divided into half-cells. Unlike a galvanic cell, in which we use reduction potentials to determine if the reaction proceeds in the *forward* or *reverse* direction, when the reaction is not divided into half-cells, we use reduction potentials to determine *if* the reaction will occur as written—or *not*.

Sample Problem 19.3 illustrates this technique.

## Sample Problem 19.3

Predict what redox reaction will take place, if any, when molecular bromine (Br$_2$) is added to (a) a 1 $M$ solution of NaI and (b) a 1 $M$ solution of NaCl. (Assume a temperature of 25°C.)

**Strategy** In each case, write the equation for the redox reaction that *might* take place and use $E°$ values to determine whether or not the proposed reaction will actually occur.

**Setup** From Table 19.1:

$$Br_2(l) + 2e^- \longrightarrow 2Br^-(aq) \qquad E° = 1.07 \text{ V}$$

$$I_2(s) + 2e^- \longrightarrow 2I^-(aq) \qquad E° = 0.53 \text{ V}$$

$$Cl_2(g) + 2e^- \longrightarrow 2Cl^-(aq) \qquad E° = 1.36 \text{ V}$$

**Solution** (a) If a redox reaction is to occur, it will be the oxidation of I$^-$ ions by Br$_2$:

$$Br_2(l) + 2I^-(aq) \longrightarrow 2Br^-(aq) + I_2(s)$$

Because the reduction potential of Br$_2$ is greater than that of I$_2$, Br$_2$ will be reduced to Br$^-$ and I$^-$ will be oxidized to I$_2$. Thus, the preceding reaction *will* occur.

*(Continued)*

(b) In this case, the proposed reaction is the reduction of $Br_2$ by $Cl^-$ ions:

$$Br_2(l) + 2Cl^-(aq) \longrightarrow 2Br^-(aq) + Cl_2(g)$$

However, because the reduction potential of $Br_2$ is smaller than that of $Cl_2$, this reaction will *not* occur. $Cl_2$ is more readily reduced than $Br_2$, so $Br_2$ is not reduced by $Cl^-$.

**Practice Problem A** Determine what redox reaction, if any, occurs (at 25°C) when lead metal (Pb) is added to (a) a 1.0 $M$ solution of $NiCl_2$ and (b) a 1.0 $M$ solution of HCl.

**⊞Practice Problem B** Would it be safer to store a cobalt(II) chloride solution in a tin container or an iron container? Explain.

---

## Checkpoint 19.3    Standard Reduction Potentials

**19.3.1** Calculate $E°_{cell}$ at 25°C for a galvanic cell made of a Cr electrode in a solution that is 1.0 $M$ in $Cr^{3+}$ and an Au electrode in a solution that is 1.0 $M$ in $Au^{3+}$.

a) −2.03 V

b) 1.11 V

c) −1.11 V

d) 2.24 V

e) 0.76 V

**19.3.2** Calculate $E°_{cell}$ at 25°C for a galvanic cell made of a Cr electrode in a solution that is 1.0 $M$ in $Cr^{3+}$ and an Ag electrode in a solution that is 1.0 $M$ in $Ag^+$.

a) 0.06 V

b) −0.06 V

c) 1.54 V

d) 3.14 V

e) 3.02 V

**19.3.3** What redox reaction, if any, will occur at 25°C when Al metal is placed in a solution that is 1.0 $M$ in $Cr^{3+}$?

a) $Al^{3+} + Cr \longrightarrow Al + Cr^{3+}$

b) $Cr^{3+} + Al \longrightarrow Cr + Al^{3+}$

c) $Al^{3+} + Cr^{3+} \longrightarrow Al + Cr$

d) $Al + Cr \longrightarrow Al^{3+} + Cr^{3+}$

e) No redox reaction will occur.

**19.3.4** What redox reaction, if any, will occur at 25°C when Zn metal is placed in a solution that is 1.0 $M$ in $Sr^{2+}$?

a) $Zn^{2+} + Sr \longrightarrow Zn + Sr^{2+}$

b) $Sr^{2+} + Zn \longrightarrow Sr + Zn^{2+}$

c) $Zn^{2+} + Sr^{2+} \longrightarrow Zn + Sr$

d) $Zn + Sr \longrightarrow Zn^{2+} + Sr^{2+}$

e) No redox reaction will occur.

---

| 19.4 | Spontaneity of Redox Reactions Under Standard-State Conditions |
|---|---|

We now look at how $E°_{cell}$ is related to thermodynamic quantities such as $\Delta G°$ and $K$. In a galvanic cell, chemical energy is converted to electric energy. The electric energy produced in a galvanic cell is the product of the cell potential and the total electric charge (in coulombs) that passes through the cell:

$$\text{electric energy} = \text{volts} \times \text{coulombs}$$

$$= \text{joules}$$

The total charge is determined by the number of moles of electrons ($n$) that pass through the circuit. By definition,

$$\text{total charge} = nF$$

where $F$, the Faraday[1] constant, is the electric charge contained in 1 mole of electrons. One faraday is equivalent to 96,485.3 C, although we usually round the number to three significant figures. Thus,

$$1 F = 96,500 \text{ C/mol } e^-$$

---

1. Michael Faraday (1791–1867). English chemist and physicist. Faraday is regarded by many as the greatest experimental scientist of the nineteenth century. He started as an apprentice to a bookbinder at the age of 13, but became interested in science after reading a book on chemistry. Faraday invented the electric motor and was the first person to demonstrate the principle governing electric generators. Besides making notable contributions to the fields of electricity and magnetism, Faraday also worked on optical activity and discovered and named benzene.

Because

$$1 \text{ J} = 1 \text{ C} \times 1 \text{ V}$$

we can also express the units of faraday as

$$1 \ F = 96,500 \text{ J/V} \cdot \text{mol } e^-$$

The measured cell potential is the maximum voltage that the cell can produce. This value is used to calculate the maximum amount of electric energy that can be obtained from the chemical reaction. This energy is used to do electrical work ($w_{\text{electrical}}$), so

$$w_{\text{max}} = w_{\text{electrical}}$$
$$= -nFE_{\text{cell}}$$

where $w_{\text{max}}$ is the maximum amount of work that can be done. The negative sign on the right-hand side indicates that the electrical work is done *by* the system *on* the surroundings. In Chapter 18 we defined *free energy* as the energy available to do *work*. Specifically, the change in free energy, $\Delta G$, represents the maximum amount of useful work that can be obtained from a reaction:

$$\Delta G = w_{\text{max}}$$

Therefore, we can write

$$\Delta G = -nFE_{\text{cell}} \qquad\qquad \text{Equation 19.2}$$

Both $n$ and $F$ are positive quantities and $\Delta G$ is negative for a spontaneous process, so $E_{\text{cell}}$ must be positive for a spontaneous process. For reactions in which reactants and products are in their standard states, Equation 19.2 becomes

$$\Delta G^\circ = -nFE^\circ_{\text{cell}} \qquad\qquad \text{Equation 19.3}$$

Equation 19.3 makes it possible to relate $E^\circ_{\text{cell}}$ to the equilibrium constant, $K$, of a redox reaction. In Section 18.6 we saw that the standard free-energy change, $\Delta G^\circ$, for a reaction is related to its equilibrium constant as follows [◄◄ Section 18.6, Equation 18.14]:

$$\Delta G^\circ = -RT \ln K$$

Therefore, if we combine Equations 18.14 and 19.3, we get

$$-nFE^\circ_{\text{cell}} = -RT \ln K$$

Solving for $E^\circ_{\text{cell}}$ gives

$$E^\circ_{\text{cell}} = \frac{RT}{nF} \ln K \qquad\qquad \text{Equation 19.4}$$

When $T = 298$ K and $n$ moles of electrons are transferred per mole of reaction, Equation 19.4 can be simplified by inserting the values for $R$ and $F$:

$$E^\circ_{\text{cell}} = \frac{\left(8.314 \, \dfrac{\text{J}}{\text{K} \cdot \text{mol}}\right)(298 \text{ K})}{(n)\left(96,500 \, \dfrac{\text{J}}{\text{V} \cdot \text{mol}}\right)}$$
$$= \frac{0.0257 \text{ V}}{n} \ln K$$

And, by converting to the base-10 logarithm of $K$, we get

$$E^\circ_{\text{cell}} = \frac{0.0592 \text{ V}}{n} \log K \qquad \text{(at 25°C)} \qquad\qquad \text{Equation 19.5}$$

Thus, if we know any one of the three quantities $\Delta G^\circ$, $K$, or $E^\circ_{\text{cell}}$, we can convert to the others by using Equations 18.14, 19.3, and 19.5.

Sample Problems 19.4 and 19.5 demonstrate the interconversions among $\Delta G°$, $K$, and $E°_{cell}$. For simplicity, the subscript "cell" is not shown.

## Sample Problem 19.4

Calculate the standard free-energy change for the following reaction at 25°C:

$$2Au(s) + 3Ca^{2+}(aq)(1.0\ M) \rightleftharpoons 2Au^{3+}(aq)(1.0\ M) + 3Ca(s)$$

**Strategy** Use $E°$ values from Table 19.1 to calculate $E°$ for the reaction, and then use Equation 19.3 to calculate the standard free-energy change.

**Setup** The half-cell reactions are

$$Cathode\ (reduction):\quad 3Ca^{2+}(aq) + 6e^- \longrightarrow 3Ca(s)$$

$$Anode\ (oxidation):\qquad\qquad 2Au(s) \longrightarrow 2Au^{3+}(aq) + 6e^-$$

From Table 19.1, $E°_{Ca^{2+}/Ca} = -2.87$ V and $E°_{Au^{3+}/Au} = 1.50$ V.

**Solution**

$$E°_{cell} = E°_{cathode} - E°_{anode}$$

$$= E°_{Ca^{2+}/Ca} - E°_{Au^{3+}/Au}$$

$$= -2.87\ V - 1.50\ V$$

$$= -4.37\ V$$

Next, substitute this value of $E°$ into Equation 19.3 to obtain $\Delta G°$:

$$\Delta G° = -nFE°$$

The overall reaction shows that $n = 6$, so

$$\Delta G° = -(6\ e^-)(96{,}500\ \text{J/V} \cdot \text{mol}\ e^-)(-4.37\ V)$$

$$= 2.53 \times 10^6\ \text{J/mol}$$

$$= 2.53 \times 10^3\ \text{kJ/mol}$$

**Think About It** The large positive value of $\Delta G°$ indicates that reactants are favored at equilibrium, which is consistent with the fact that $E°$ for the reaction is negative.

**Practice Problem A** Calculate $\Delta G°$ for the following reaction at 25°C:

$$3Mg(s) + 2Al^{3+}(aq) \rightleftharpoons 3Mg^{2+}(aq) + 2Al(s)$$

**Practice Problem B** Calculate $\Delta G°$ for the following reaction at 25°C:

$$Pb(s) + Ni^{2+}(aq) \rightleftharpoons Pb^{2+}(aq) + Ni(s)$$

## Sample Problem 19.5

Calculate the equilibrium constant for the following reaction at 25°C:

$$Sn(s) + 2Cu^{2+}(aq) \rightleftharpoons Sn^{2+}(aq) + 2Cu^+(aq)$$

**Strategy** Use $E°$ values from Table 19.1 to calculate $E°$ for the reaction, and then calculate the equilibrium constant using Equation 19.5 (rearranged to solve for $K$).

**Setup** The half-cell reactions are

$$Cathode\ (reduction):\quad 2Cu^{2+}(aq) + 2e^- \longrightarrow 2Cu^+(aq)$$

$$Anode\ (oxidation):\qquad\qquad Sn(s) \longrightarrow Sn^{2+}(aq) + 2e^-$$

From Table 19.1, $E°_{Cu^{2+}/Cu^+} = 0.15$ V and $E°_{Sn^{2+}/Sn} = -0.14$ V.

**Solution**

$$E^\circ_{cell} = E^\circ_{cathode} - E^\circ_{anode}$$

$$= E^\circ_{Cu^{2+}/Cu^+} - E^\circ_{Sn^{2+}/Sn}$$

$$= 0.15 \text{ V} - (-0.14 \text{ V})$$

$$= 0.29 \text{ V}$$

Solving Equation 19.5 for $K$ gives

$$K = 10^{nE^\circ/0.0592 \text{ V}}$$

$$= 10^{(2)(0.29 \text{ V})/0.0592 \text{ V}}$$

$$= 6 \times 10^9$$

> **Think About It** A positive standard cell potential corresponds to a large equilibrium constant.

**Practice Problem A** Calculate the equilibrium constant for the following reaction at 25°C:

$$2Ag(s) + Fe^{2+}(aq) \rightleftharpoons 2Ag^+(aq) + Fe(s)$$

**Practice Problem B** Like equilibrium constants, $E^\circ_{cell}$ values are temperature dependent. At 80°C, $E^\circ_{cell}$ for the cell diagram shown is 0.18 V.

$$Pt \mid H_2(g) \mid HCl(aq) \parallel AgCl(s) \mid Ag(s)$$

The corresponding cell reaction is

$$H_2(g) + 2AgCl(s) \rightleftharpoons 2Ag(s) + 2H^+(aq) + 2Cl^-(aq)$$

Calculate the equilibrium constant for this reaction at 80°C.

---

**Checkpoint 19.4**    **Spontaneity of Redox Reactions Under Standard-State Conditions**

**19.4.1** Calculate $K$ at 25°C for the following reaction:

$$Fe^{2+}(aq) + Ni(s) \rightleftharpoons Fe(s) + Ni^{2+}(aq)$$

a) $6 \times 10^{-4}$

b) $4 \times 10^{-7}$

c) $3 \times 10^6$

d) $1 \times 10^{-13}$

e) $2 \times 10^3$

**19.4.2** Calculate $\Delta G^\circ$ at 25°C for the following reaction:

$$3Cu^{2+}(aq) + Cr(s) \rightleftharpoons 3Cu^+(aq) + Cr^{3+}(aq)$$

a) $2.6 \times 10^2$ kJ/mol

b) $-2.6 \times 10^2$ kJ/mol

c) $1 \times 10^{45}$ kJ/mol

d) $-86$ kJ/mol

e) $86$ kJ/mol

---

## 19.5  Spontaneity of Redox Reactions Under Conditions Other Than Standard State

So far we have focused on redox reactions in which the reactants and products are in their standard states. Standard-state conditions, though, are difficult to come by and usually impossible to maintain. Just as there is an equation that relates $\Delta G$ to $\Delta G^\circ$ [◄◄ Section 18.6, Equation 18.13], there is an equation that relates $E$ to $E^\circ$. We will now derive this equation.

### The Nernst Equation

Consider a redox reaction of the type

$$aA + bB \longrightarrow cC + dD$$

From Equation 18.13,

$$\Delta G = \Delta G^\circ + RT \ln Q$$

Because $\Delta G = -nFE$ and $\Delta G° = -nFE°$, the equation can be expressed as

$$-nFE = -nFE° + RT \ln Q$$

Dividing the equation through by $-nF$, we get

Equation 19.6 $$E = E° - \frac{RT}{nF} \ln Q$$

where $Q$ is the reaction quotient [◄◄ Section 15.2]. Equation 19.6 is known as the **Nernst**[2] **equation.** At 298 K, Equation 19.6 can be rewritten as

$$E = E° - \frac{0.0257 \text{ V}}{n} \ln Q$$

or using the base-10 logarithm of $Q$ as

Equation 19.7 $$E = E° - \frac{0.0592 \text{ V}}{n} \log Q$$

During the operation of a galvanic cell, electrons flow from the anode to the cathode, resulting in product formation and a decrease in reactant concentration. Thus, $Q$ increases, which means that $E$ decreases. Eventually, the cell reaches equilibrium. At equilibrium, there is no net transfer of electrons, so $E = 0$ and $Q = K$, where $K$ is the equilibrium constant.

The Nernst equation enables us to calculate $E$ as a function of reactant and product concentrations in a redox reaction. For example, for the cell pictured in Figure 19.1,

$$Zn(s) + Cu^{2+}(aq) \longrightarrow Zn^{2+}(aq) + Cu(s)$$

The Nernst equation for this cell at 25°C can be written as

$$E = 1.10 \text{ V} - \frac{0.0592 \text{ V}}{2} \log \frac{[Zn^{2+}]}{[Cu^{2+}]}$$

If the ratio $[Zn^{2+}]/[Cu^{2+}]$ is less than 1, $\log ([Zn^{2+}]/[Cu^{2+}])$ is a negative number, making the second term on the right-hand side of the preceding equation a positive quantity. Under this condition, $E$ is *greater* than the standard potential $E°$. If the ratio is greater than 1, $E$ is *smaller* than $E°$.

Sample Problem 19.6 shows how to use the Nernst equation.

## Sample Problem 19.6

Predict whether the following reaction will occur spontaneously as written at 298 K:

$$Co(s) + Fe^{2+}(aq) \longrightarrow Co^{2+}(aq) + Fe(s)$$

assuming $[Co^{2+}] = 0.15 \ M$ and $[Fe^{2+}] = 0.68 \ M$.

**Strategy** Use $E°$ values from Table 19.1 to determine $E°$ for the reaction, and use Equation 19.7 to calculate $E$. If $E$ is positive, the reaction will occur spontaneously.

**Setup** From Table 19.1,

Cathode (reduction):  $Fe^{2+}(aq) + 2e^- \longrightarrow Fe(s)$

Anode (oxidation):        $Co(s) \longrightarrow Co^{2+}(aq) + 2e^-$

$$E°_{cell} = E°_{cathode} - E°_{anode}$$
$$= E°_{Fe^{2+}/Fe} - E°_{Co^{2+}/Co}$$
$$= -0.44 \text{ V} - (-0.28 \text{ V})$$
$$= -0.16 \text{ V}$$

The reaction quotient, $Q$, for the reaction is $[Co^{2+}]/[Fe^{2+}]$. Therefore, $Q = (0.15/0.68) = 0.22$.

---

2. Walter Hermann Nernst (1864–1941). German chemist and physicist. Nernst's work was mainly on electrolyte solutions and thermodynamics. He also invented an electric piano. Nernst was awarded the Nobel Prize in Chemistry in 1920 for his contribution to thermodynamics.

**Solution** From Equation 19.7,

$$E = E° - \frac{0.0592 \text{ V}}{n} \log Q$$

$$= -0.16 \text{ V} - \frac{0.0592 \text{ V}}{2} \log 0.22$$

$$= -0.14 \text{ V}$$

The negative $E$ value indicates that the reaction is *not* spontaneous as written under the conditions described.

**Think About It** For this reaction to be spontaneous as written, the ratio of $[Fe^{2+}]$ to $[Co^{2+}]$ would have to be enormous. We can determine the required ratio by first setting $E$ equal to zero:

$$0 \text{ V} = -0.16 \text{ V} - \frac{0.0592 \text{ V}}{2} \log Q$$

$$-\frac{(0.16 \text{ V})(2)}{0.0592 \text{ V}} = \log Q$$

$$\log Q = -5.4$$

$$Q = 10^{-5.4} = \frac{[Co^{2+}]}{[Fe^{2+}]} = 4 \times 10^{-6}$$

For $E$ to be positive, therefore, the ratio of $[Fe^{2+}]$ to $[Co^{2+}]$, the reciprocal of $Q$, would have to be greater than $3 \times 10^5$ to 1.

**Practice Problem A** Will the following reaction occur spontaneously at 298 K if $[Fe^{2+}] = 0.60$ $M$ and $[Cd^{2+}] = 0.010$ $M$?

$$Cd(s) + Fe^{2+}(aq) \longrightarrow Cd^{2+}(aq) + Fe(s)$$

**Practice Problem B** Consider the electrochemical cell in Practice Problem 19.5B (page 833), for which $E°_{cell} = 0.18$ V at 80°C.

$$H_2(g) + 2AgCl(s) \rightleftharpoons 2Ag(s) + 2H^+(aq) + 2Cl^-(aq)$$

If pH = 1.05 in the anode compartment, and $[Cl^-] = 2.5$ $M$ in the cathode compartment, determine the partial pressure of $H_2$ necessary in the anode compartment for the cell potential to be 0.27 V at 80°C.

## Concentration Cells

Because electrode potential depends on ion concentrations, it is possible to construct a galvanic cell from two half-cells composed of the same material but differing in ion concentrations. Such a cell is called a *concentration cell.*

Consider a galvanic cell consisting of a zinc electrode in 0.10 $M$ zinc sulfate in one compartment and a zinc electrode in 1.0 $M$ zinc sulfate in the other compartment (Figure 19.5). According to Le Châtelier's principle, the tendency for the reduction

$$Zn^{2+}(aq) + 2e^- \longrightarrow Zn(s)$$

to occur increases with increasing concentration of $Zn^{2+}$ ions. Therefore, reduction should occur in the more concentrated compartment and oxidation should take place on the more dilute side. The cell diagram is

$$Zn(s) \,|\, Zn^{2+}(0.10 \text{ M}) \,\|\, Zn^{2+}(1.0 \text{ M}) \,|\, Zn(s)$$

and the half-cell reactions are

*Oxidation:*             $Zn(s) \longrightarrow Zn^{2+}(0.10 \text{ M}) + 2e^-$

*Reduction:*   $Zn^{2+}(1.0 \text{ M}) + 2e^- \longrightarrow Zn(s)$

*Overall:*            $Zn^{2+}(1.0 \text{ M}) \longrightarrow Zn^{2+}(0.10 \text{ M})$

The cell potential is

$$E = E° - \frac{0.0592 \text{ V}}{2} \log \frac{[Zn^{2+}]_{dilute}}{[Zn^{2+}]_{concentrated}}$$

where the subscripts "dilute" and "concentrated" refer to the 0.10 $M$ and 1.0 $M$ concentrations, respectively. The $E°$ for this cell is zero (because the same electrode-ion combination is used in both half-cells; i.e., $E°_{cathode} = E°_{anode}$), so

$$E = 0 - \frac{0.0592 \text{ V}}{2} \log \frac{0.10}{1.0}$$

$$= 0.030 \text{ V}$$

The cell potential for a concentration cell is typically *small* and decreases continually during the operation of the cell as the concentrations in the two compartments approach each other. When the concentrations of the ions in the two compartments are equal, $E$ becomes zero and no further change occurs.

**Figure 19.5** A concentration cell. Oxidation will occur in the container with the lower $Zn^{2+}$ concentration. Reduction will occur in the container with the higher $Zn^{2+}$ concentration.

$Zn(s) \longrightarrow Zn^{2+}(aq) + 2e^-$

$2e^-$ lost per Zn atom oxidized

Zn

$Zn^{2+}$

$Zn^{2+}(aq) + 2e^- \longrightarrow Zn(s)$

$2e^-$ gained per $Zn^{2+}$ ion reduced

$Zn^{2+}$

Zn

Anode (−)
Zn

Salt bridge

Cathode (+)
Zn

0.0296

$Zn^{2+}$
(0.10 *M*)

$Zn^{2+}$
(1.0 *M*)

## Bringing Chemistry to Life

### Biological Concentration Cells

**ECG strip**

**Patient with ECG electrodes**

A biological cell can be compared to a concentration cell for the purpose of calculating its membrane potential. Membrane potential is the electric potential that exists across the membrane of various kinds of cells, including muscle cells and nerve cells. This potential is responsible for the propagation of nerve impulses and heartbeat. A membrane potential is established whenever there are unequal concentrations of the same type of ion in the interior and exterior of a cell—and the membrane is permeable to the ion. For example, the concentrations of $Na^+$ ions in the interior and exterior of a nerve cell are $1.5 \times 10^{-2}$ *M* and $1.5 \times 10^{-1}$ *M*, respectively. Treating the situation as a concentration cell and applying the Nernst equation for a single ion, we can write

$$E_{Na^+} = E^\circ_{Na^+} - \frac{0.0592 \text{ V}}{1} \log \frac{[Na^+]_{ex}}{[Na^+]_{in}}$$

$$= -(0.0592 \text{ V}) \log \frac{1.5 \times 10^{-1}}{1.5 \times 10^{-2}}$$

$$= 0.059 \text{ V} \quad \text{or} \quad 59 \text{ mV}$$

where the subscripts "ex" and "in" denote "exterior" and "interior," respectively. We have set $E^\circ_{Na} = 0$ because the same type of ion is involved. Thus, an electric potential of 59 mV exists across the membrane due to the unequal concentrations of $Na^+$ ions.

When a nerve cell is stimulated, there is a large change in the membrane permeability, decreasing the membrane potential temporarily to about 34 mV. This sudden change in the potential is called the *action potential*. Once created, the action potential propagates along the nerve fiber until it reaches either a synaptic junction (the connection between nerve cells) or a neuromuscular junction (the connection between a nerve cell and a muscle cell). In the muscle cell of the heart, an action potential is generated during the heartbeat. This potential produces enough current to be detected by electrodes placed on the chest. The amplified signals can be recorded either on a moving chart or displayed on an oscilloscope. The record, called an *electrocardiogram* (ECG, also known as an EKG, where K is from the German word *kardio* for heart), is a valuable tool in the diagnosis of heart disease.

*You May be Wondering…*

# How Are Very Small Solution Concentrations Measured?

In Chapter 17, we studied solubility equilibria and learned how to use tabulated $K_{sp}$ values to determine solubilities of ionic compounds that are only very slightly soluble. The tabulated $K_{sp}$ values are determined by measuring the concentration of one of a compound's constituent ions in a saturated solution of the compound. For example, to determine the $K_{sp}$ of silver bromide (AgBr), we must measure the concentration of $Ag^+$, or the concentration of $Br^-$. Because the $K_{sp}$ of silver bromide is $7.7 \times 10^{-13}$, the concentration of either ion in a saturated solution is $8.8 \times 10^{-7}$ M. You may have wondered how such a small concentration is measured. It cannot be done using visible spectrophotometry, the method described in Chapter 4; not only because the concentration is so low, but also because a solution AgBr is colorless—meaning that it does not absorb visible light. In fact, concentrations of ions in these cases are measured using concentration cells.

Sample Problem 19.7 illustrates this process.

---

### Sample Problem 19.7

An electrochemical cell is constructed for the purpose of determining the $K_{sp}$ of silver cyanide (AgCN) at 25°C. One half-cell consists of a silver electrode in a 1.00 M solution of silver nitrate. The other half-cell consists of a silver electrode in a saturated solution of silver cyanide. The cell potential is measured and found to be 0.470 V. Determine the concentration of silver ion in the saturated silver cyanide solution and the value of $K_{sp}$ for AgCN.

**Strategy** Use Equation 19.7 to solve for the unknown concentration of silver ion. The half-cell with the higher $Ag^+$ (1.0 M $AgNO_3$) concentration will be the cathode; the half-cell with the lower, unknown $Ag^+$ concentration (saturated AgCN solution) will be the anode. The overall reaction is $Ag^+(1.0\ M) \longrightarrow Ag^+(x\ M)$.

**Setup** Because this is a concentration cell, $E°_{cell} = 0$ V. The reaction quotient, $Q$, is $(x\ M)/(1.00\ M)$; and the value of $n$ is 1.

**Solution**

$$E_{cell} = E°_{cell} - \frac{0.0592\ V}{n} \log Q$$

$$0.470\ V = 0 - \frac{0.0592\ V}{1} \log \frac{x}{1.00}$$

$$-7.939 = \log \frac{x}{1.00}$$

$$10^{-7.939} = 1.15 \times 10^{-8} = \log \frac{x}{1.00}$$

$$x = 1.15 \times 10^{-8}$$

Therefore, $[Ag^+] = 1.15 \times 10^{-8}$ M and $K_{sp}$ for AgCN $= x^2 = 1.3 \times 10^{-16}$.

> **Think About It** Remember that in a saturated solution of a salt that dissociates into two ions, the ion concentrations are equal to each other and each ion concentration is equal to the square root of $K_{sp}$ [◄◄ Section 17.4].

---

**Practice Problem A** An electrochemical cell is constructed for the purpose of determining the $K_{sp}$ of copper(I) chloride, CuCl, at 25°C. One half-cell consists of a copper electrode in a 1.00 M solution of copper(I) nitrate. The other half-cell consists of a copper electrode in a saturated solution of copper(I) chloride. The measured cell potential is 0.175 V. Determine the concentration of copper(I) ion in the saturated copper(I) chloride solution and the value of $K_{sp}$ for CuCl.

**Practice Problem B** The $K_{sp}$ of copper(II) ferrocyanide ($Cu_2[Fe(CN)_6]$) is $1.3 \times 10^{-16}$ at 25°C. Determine the potential of a concentration cell in which one half-cell consists of a copper electrode in 1.00 M copper(II) nitrate, and the other consists of a copper electrode in a saturated solution of $Cu_2[Fe(CN)_6]$. Ferrocyanide, ($[Fe(CN)_6]^{4-}$), is a *complex ion* [◄◄ Section 17.5].

## Checkpoint 19.5    Spontaneity of Redox Reactions Under Conditions Other Than Standard State

**19.5.1**  Calculate $E$ at 25°C for a galvanic cell based on the reaction,

$$Zn(s) + Cu^{2+}(aq) \longrightarrow Zn^{2+}(aq) + Cu(s),$$

in which $[Zn^{2+}] = 0.55\ M$ and $[Cu^{2+}] = 1.02\ M$.

a) 1.10 V

b) 1.11 V

c) 1.09 V

d) 0.0118 V

e) 1.03 V

**19.5.2**  Calculate the cell potential at 25°C of a galvanic cell consisting of an Ag electrode in 0.15 $M$ AgNO$_3$ and an Ag electrode in 1.0 $M$ AgNO$_3$.

a) 0.0 V

b) 0.049 V

c) −0.049 V

d) 0.024 V

e) −0.024 V

**19.5.3**  Calculate $E^\circ_{cell}$ at 80°C for a galvanic cell based on the reaction

$$H_2(g) + 2AgCl(s) \rightleftharpoons 2Ag(s) + 2H^+(aq) + 2Cl^-(aq)$$

in which $[H^+] = 0.10\ M$, $[Cl^-] = 1.5\ M$, and $P = 1.25$ atm. ($E^\circ_{cell} = 0.18$ V at 80°C.)

a) 0.12 V

b) 0.16 V

c) 0.18 V

d) 0.22 V

e) 0.24 V

**19.5.4**  Which of these would cause an increase in the cell potential of the electrochemical cell in Question 19.5.2?

(i) Adding AgNO$_3$ to the cathode compartment

(ii) Adding saturated AgCl solution to the anode compartment

(iii) Adding NaCl to the anode compartment

a) i and iii

b) ii and iii

c) iii only

d) i, ii, and iii

e) None of these

To answer Questions 19.5.5 and 19.5.6, consider the following cell diagram:

$$Pb(s) \mid Pb(NO_3)_2(0.60\ M) \parallel AgNO_3(0.40\ M) \mid Ag(s)$$

**19.5.5**  Determine the initial value of $E_{cell}$ under the conditions shown in the cell diagram; and determine the initial value of $E_{cell}$ if $[Ag^+]$ were increased by a factor of 4 at 25°C.

a) 0.91 V, 0.88 V

b) 0.91 V, 0.95 V

c) 0.93 V, 0.95 V

d) 0.93 V, 0.96 V

e) 0.92 V, 0.88 V

**19.5.6**  Which of the following would cause a decrease in the cell potential?

(i) Adding NH$_3$ to the cathode compartment

(ii) Adding NH$_3$ to the anode compartment

(iii) Adding NaNO$_3$ to the anode compartment

a) i only

b) i and iii

c) iii only

d) i, ii, and iii

e) None of these

## 19.6    Batteries

A **battery** is a galvanic cell, or a series of connected galvanic cells, that can be used as a portable, self-contained source of direct electric current. In this section we examine several types of batteries.

### Dry Cells and Alkaline Batteries

The most common batteries, *dry cells* and *alkaline batteries,* are those used in flashlights, toys, and certain portable electronics such as CD players. The two are similar in appearance, but differ in the spontaneous chemical reaction responsible for producing a voltage. Although the reactions that take place in these batteries are somewhat complex, the reactions shown here approximate the overall processes.

A dry cell, so named because it has no fluid component, consists of a zinc container (the anode) in contact with manganese dioxide and an electrolyte (Figure 19.6). The electrolyte consists of ammonium chloride and zinc chloride in water, to which starch is added to thicken the solution to a paste so that it is less likely to leak. A carbon rod, immersed in the electrolyte in the center of the cell, serves as the cathode. The cell reactions are:

*Anode:*  $$Zn(s) \longrightarrow Zn^{2+}(aq) + 2e^-$$

*Cathode:*  $$2NH_4^+(aq) + 2MnO_2(s) + 2e^- \longrightarrow Mn_2O_3(s) + 2NH_3(aq) + H_2O(l)$$

*Overall:*  $$Zn(s) + 2NH_4^+(aq) + 2MnO_2(s) \longrightarrow Zn^{2+}(aq) + Mn_2O_3(s) + 2NH_3(aq) + H_2O(l)$$

The voltage produced by a dry cell is about 1.5 V.

An alkaline battery is also based on the reduction of manganese dioxide and the oxidation of zinc. However, the reactions take place in a *basic* medium, hence the name *alkaline* battery. The anode consists of powdered zinc suspended in a gel, which is in contact with a concentrated solution of KOH. The cathode is a mixture of manganese dioxide and graphite. The anode and cathode are separated by a porous barrier (Figure 19.7).

*Cathode:*  $$2MnO_2(s) + 2H_2O(l) + 2e^- \longrightarrow 2MnO(OH)(s) + 2OH^-(aq)$$

*Anode:*  $$Zn(s) + 2OH^-(aq) \longrightarrow Zn(OH)_2(s) + 2e^-$$

*Overall:*  $$Zn(s) + 2MnO_2(s) + 2H_2O(l) \longrightarrow Zn(OH)_2(s) + 2MnO(OH)(s) + 2OH^-(aq)$$

Alkaline batteries are more expensive than dry cells and offer superior performance and shelf life.

**Figure 19.6** Interior view of the type of dry cell used in flashlights and other small devices.

## Lead Storage Batteries

The lead storage battery commonly used in automobiles consists of six identical cells joined together in series. Each cell has a lead anode and a cathode made of lead dioxide ($PbO_2$) packed on a metal plate (Figure 19.8). Both the cathode and the anode are immersed in an aqueous solution of sulfuric acid, which acts as the electrolyte. The cell reactions are

*Anode:*  $$Pb(s) + SO_4^{2-}(aq) \longrightarrow PbSO_4(s) + 2e^-$$

*Cathode:*  $$PbO_2(s) + 4H^+(aq) + SO_4^{2-}(aq) + 2e^- \longrightarrow PbSO_4(s) + 2H_2O(l)$$

*Overall:*  $$Pb(s) + PbO_2(s) + 4H^+(aq) + 2SO_4^{2-}(aq) \longrightarrow 2PbSO_4(s) + 2H_2O(l)$$

Under normal operating conditions, each cell produces 2 V. A total of 12 V from the six cells is used to power the ignition circuit of the automobile and its other electric systems. The lead storage battery can deliver large amounts of current for a short time, such as the time it takes to start the engine.

Unlike dry cells and alkaline batteries, the lead storage battery is rechargeable. Recharging the battery means reversing the normal electrochemical reaction by applying an external voltage at the cathode and the anode. (This kind of process is called *electrolysis,* which we discuss in Section 19.7.)

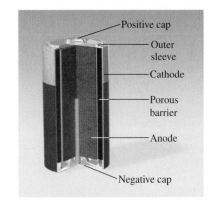

**Figure 19.7** Interior view of an alkaline battery.

**Figure 19.8** Interior view of a lead storage battery. Under normal operating conditions, the concentration of the sulfuric acid solution is about 38 percent by mass.

## Lithium-Ion Batteries

Sometimes called "the battery of the future," lithium-ion batteries have several advantages over other battery types. The overall reaction that takes place in the lithium-ion battery is

$$
\begin{array}{ll}
\textit{Anode:} & \text{Li}(s) \longrightarrow \text{Li}^+ + e^- \\
\textit{Cathode:} & \underline{\text{Li}^+ + \text{CoO}_2 + e^- \longrightarrow \text{LiCoO}_2(s)} \\
\textit{Overall:} & \text{Li}(s) + \text{CoO}_2 \longrightarrow \text{LiCoO}_2(s)
\end{array}
$$

> **Student Annotation:** Lithium has the largest negative reduction potential of all the metals, making it a powerful reducing agent.

The overall cell potential is 3.4 V, which is relatively large. Lithium is also the lightest metal—only 6.941 g of Li (its molar mass) are needed to produce 1 mole of electrons. Furthermore, a lithium-ion battery can be recharged hundreds of times. These qualities make lithium batteries suitable for use in portable devices such as cell phones, digital cameras, and laptop computers.

## Fuel Cells

Fossil fuels are a major source of energy, but the conversion of fossil fuel into electric energy is a highly inefficient process. Consider the combustion of methane:

$$\text{CH}_4(g) + 2\text{O}_2(g) \longrightarrow \text{CO}_2(g) + 2\text{H}_2\text{O}(l) + \text{energy}$$

To generate electricity, heat produced by the reaction is first used to convert water to steam, which then drives a turbine, which then drives a generator. A significant fraction of the energy released in the form of heat is lost to the surroundings at each step (even the most efficient power plant converts only about 40 percent of the original chemical energy into electricity). Because combustion reactions are redox reactions, it is more desirable to carry them out directly by electrochemical means, thereby greatly increasing the efficiency of power production. This objective can be accomplished by a device known as a ***fuel cell,*** a galvanic cell that requires a continuous supply of reactants to keep functioning.

> **Student Annotation:** Strictly speaking, a fuel cell is not a battery because it is not self-contained.

In its simplest form, a hydrogen-oxygen fuel cell consists of an electrolyte solution, such as a potassium hydroxide solution, and two inert electrodes. Hydrogen and oxygen gases are bubbled through the anode and cathode compartments (Figure 19.9), where the following reactions take place:

$$
\begin{array}{ll}
\textit{Anode:} & 2\text{H}_2(g) + 4\text{OH}^-(aq) \longrightarrow 4\text{H}_2\text{O}(l) + 4e^- \\
\textit{Cathode:} & \underline{\text{O}_2(g) + 2\text{H}_2\text{O}(l) + 4e^- \longrightarrow 4\text{OH}^-(aq)} \\
\textit{Overall:} & 2\text{H}_2(g) + \text{O}_2(g) \longrightarrow 2\text{H}_2\text{O}(l)
\end{array}
$$

**Figure 19.9** A hydrogen-oxygen fuel cell. The Ni and NiO embedded in the porous carbon electrodes are catalysts.

Using $E°$ values from Table 19.1, the standard cell potential is calculated as follows:

$$E°_{cell} = E°_{cathode} - E°_{anode}$$

$$= 0.40 \text{ V} - (-0.83 \text{ V})$$

$$= 1.23 \text{ V}$$

Thus, the cell reaction is spontaneous under standard-state conditions. Note that the reaction is the same as the hydrogen combustion reaction, but the oxidation and reduction are carried out separately at the anode and the cathode. Like platinum in the standard hydrogen electrode, the electrodes serve two purposes. They serve as electrical conductors, and they provide the necessary surfaces for the initial decomposition of the molecules into atomic species, which must take place before electrons can be transferred. Electrodes that serve this particular purpose are called "electrocatalysts." Metals such as platinum, nickel, and rhodium also make good electrocatalysts.

In addition to the $H_2$-$O_2$ system, a number of other fuel cells have been developed. Among these is the propane-oxygen fuel cell. The corresponding half-cell reactions are

*Anode:* $\quad\quad\quad\quad\quad C_3H_8(g) + 6H_2O(l) \longrightarrow 3CO_2(g) + 20H^+(aq) + 20e^-$

*Cathode:* $\quad 5O_2(g) + 20H^+(aq) + 20e^- \longrightarrow 10H_2O(l)$

*Overall:* $\quad\quad\quad\quad\quad C_3H_8(g) + 5O_2(g) \longrightarrow 3CO_2(g) + 4H_2O(l)$

The overall reaction is identical to the burning of propane in oxygen.

Unlike batteries, fuel cells do not store chemical energy. Reactants must be constantly resupplied as they are consumed, and products must be removed as they form. However, properly designed fuel cells may be as much as 70 percent efficient, which is about twice as efficient as an internal combustion engine. In addition, fuel-cell generators are free of the noise, vibration, heat transfer, thermal pollution, and other problems normally associated with conventional power plants. Nevertheless, fuel cells are not yet in widespread use. One major problem is the expense of electrocatalysts that can function efficiently for long periods of time without contamination. One notable application of fuel cells is their use in space vehicles. Hydrogen-oxygen fuel cells provide electric power (and drinking water!) for space flight.

# 19.7 Electrolysis

**Multimedia**

Electrochemistry—the electrolysis of potassium iodide.

In Section 19.6, we mentioned that lead storage batteries are rechargeable and that recharging means reversing the electrochemical processes by which the battery ordinarily operates through the application of an external voltage. This process, the use of electric energy to drive a non-spontaneous chemical reaction, is called *electrolysis.* An *electrolytic cell* is one used to carry out electrolysis. The same principles apply to the processes in both galvanic and electrolytic cells. In this section, we discuss three examples of electrolysis based on those principles. We then examine some of the quantitative aspects of electrolysis.

**Student Annotation:**

| Cell Type | Chemical Reaction | Electric Energy |
|---|---|---|
| Galvanic | Spontaneous | Produced |
| Electrolytic | Nonspontaneous | Consumed |

## Electrolysis of Molten Sodium Chloride

In its molten (melted) state, sodium chloride, an ionic compound, can be electrolyzed to separate it into its constituent elements, sodium and chlorine. Figure 19.10(a) is a diagram of a Downs cell, which is used for the large-scale electrolysis of NaCl. In molten NaCl, the cations and anions are the $Na^+$ and $Cl^-$ ions, respectively. Figure 19.10(b) is a simplified diagram showing the reactions that occur at the electrodes. The electrolytic cell contains a pair of electrodes connected to the battery. The battery serves to push electrons in the direction they would not flow spontaneously. The electrode toward which the electrons are pushed is the cathode, where reduction takes place. The electrode away from which electrons are drawn is the anode, where oxidation takes place. The reactions at the electrodes are

*Anode (oxidation):* $\quad\quad\quad\quad 2Cl^-(l) \longrightarrow Cl_2(g) + 2e^-$

*Cathode (reduction):* $\quad 2Na^+(l) + 2e^- \longrightarrow 2Na(l)$

*Overall:* $\quad\quad\quad\quad 2Na^+(l) + 2Cl^-(l) \longrightarrow 2Na(l) + Cl_2(g)$

This process is a major industrial source of pure sodium metal and chlorine gas.

**Figure 19.10**   (a) A practical arrangement called a Downs cell for the electrolysis of molten NaCl (m.p. = 801°C). The sodium metal formed at the cathodes is in the liquid state. Because liquid sodium metal is lighter than molten NaCl, the sodium floats to the surface, as shown, and is collected. Chlorine gas forms at the anode and is collected at the top. (b) A simplified diagram showing the electrode reactions during the electrolysis of molten NaCl. The battery is needed to drive the nonspontaneous reaction.

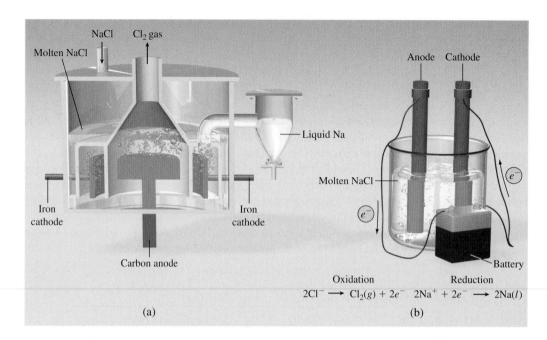

Oxidation
$$2Cl^- \longrightarrow Cl_2(g) + 2e^-$$

Reduction
$$2Na^+ + 2e^- \longrightarrow 2Na(l)$$

(a)                                    (b)

Using data from Table 19.1, we estimate $E^\circ_{cell}$ to be $-4$ V for this process. The negative standard reduction potential indicates that for the process to occur as written, a minimum of approximately 4 V must be supplied by the battery to drive the reaction in the desired direction. In practice, an even higher voltage is required because of inefficiencies in the electrolytic process and because of overvoltage, a phenomenon we will discuss later in this section.

## Electrolysis of Water

Under ordinary atmospheric conditions (1 atm and 25°C), water will not spontaneously decompose to form hydrogen and oxygen gas because the standard free-energy change for the reaction is a large positive quantity:

$$2H_2O(l) \longrightarrow 2H_2(g) + O_2(g) \qquad \Delta G^\circ = 474.4 \text{ kJ/mol}$$

However, this reaction can be made to occur in an electrolytic cell like the one shown in Figure 19.11. This cell consists of a pair of electrodes made of a nonreactive metal, such as platinum, immersed in water. When the electrodes are connected to the battery, nothing happens because there are not enough ions in pure water to carry much of an electric current. The reaction occurs readily in a 0.1 $M$ $H_2SO_4$ solution, however, because there is a sufficient ion concentration to conduct electricity. Immediately, gas bubbles begin to appear at both electrodes.

The processes at the electrodes are

*Anode:*     $2H_2O(l) \longrightarrow O_2(g) + 4H^+(aq) + 4e^-$

*Cathode:*   $\underline{4H^+(aq) + 4e^- \longrightarrow 2H_2(g)}$

*Overall:*   $2H_2O(l) \longrightarrow O_2(g) + 2H_2(g)$

Note that there is no net consumption of the acid.

## Electrolysis of an Aqueous Sodium Chloride Solution

An aqueous sodium chloride solution is the most complicated of the three examples of electrolysis considered here because NaCl($aq$) contains several species that could be oxidized and reduced. The reductions that might occur at the cathode are

$$2H^+(aq) + 2e^- \longrightarrow H_2(g) \qquad\qquad E^\circ = 0.00 \text{ V}$$

$$2H_2O(l) + 2e^- \longrightarrow H_2(g) + 2OH^-(aq) \qquad E^\circ = -0.83 \text{ V}$$

or

$$Na^+(aq) + e^- \longrightarrow Na(s) \qquad\qquad E^\circ = -2.71 \text{ V}$$

**Figure 19.11** Apparatus for small-scale electrolysis of water. The volume of hydrogen gas generated at the cathode is twice that of oxygen gas generated at the anode.

We can rule out the reduction of $Na^+$ ion because of the large negative $E°$ value. Under standard-state conditions, the reduction of $H^+$ is more apt to occur than the reduction of $H_2O$. However, in a solution of NaCl, the $H^+$ concentration is very low, making the reduction of $H_2O$ the more probable reaction at the cathode.

The oxidation reactions that might occur at the anode are

$$2Cl^-(aq) \longrightarrow Cl_2(g) + 2e^-$$

or

$$2H_2O(l) \longrightarrow O_2(g) + 4H^+(aq) + 4e^-$$

Referring to Table 19.1, we find

$$Cl_2(g) + 2e^- \longrightarrow 2Cl^-(aq) \qquad E° = 1.36 \text{ V}$$

$$O_2(g) + 4H^+(aq) + 4e^- \longrightarrow 2H_2O(l) \qquad E° = 1.23 \text{ V}$$

The standard reduction potentials of the two reactions are not very different, but the values do suggest that the oxidation of $H_2O$ *should* occur more readily. However, by experiment we find that the gas produced at the anode is $Cl_2$, not $O_2$. In the study of electrolytic processes, we sometimes find that the voltage required for a reaction is considerably higher than the electrode potentials would indicate. The ***overvoltage*** is the difference between the calculated voltage and the actual voltage required to cause electrolysis. The overvoltage for $O_2$ formation is quite high. Under normal operating conditions, therefore, $Cl_2$ gas forms at the anode instead of $O_2$.

Thus, the half-cell reactions in the electrolysis of aqueous sodium chloride are

| | |
|---|---|
| *Anode (oxidation):* | $2Cl^-(aq) \longrightarrow Cl_2(g) + 2e^-$ |
| *Cathode (reduction):* | $2H_2O(l) + 2e^- \longrightarrow H_2(g) + 2OH^-(aq)$ |
| *Overall:* | $2H_2O(l) + 2Cl^-(aq) \longrightarrow H_2(g) + Cl_2(g) + 2OH^-(aq)$ |

As the overall reaction shows, the concentration of the $Cl^-$ ions decreases during electrolysis and that of the $OH^-$ ions increases. Therefore, in addition to $H_2$ and $Cl_2$, the useful by-product NaOH can be obtained by evaporating the aqueous solution at the end of the electrolysis.

Electrolysis has many important applications in industry, mainly in the extraction and purification of metals. We discuss some of these applications in Chapter 23.

**Student Annotation:** Keep in mind that in the electrolysis of aqueous solutions, the water itself may be oxidized and/or reduced.

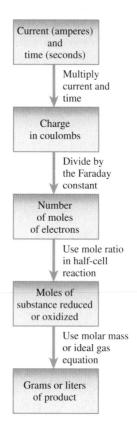

**Figure 19.12** Steps involved in calculating amounts of substances reduced or oxidized in electrolysis.

## Quantitative Applications of Electrolysis

The quantitative treatment of electrolysis was developed primarily by Faraday. He observed that the mass of product formed (or reactant consumed) at an electrode was proportional to both the amount of electricity transferred at the electrode and the molar mass of the substance being produced (or consumed). In the electrolysis of molten NaCl, for example, the cathode reaction tells us that one Na atom is produced when one $Na^+$ ion accepts an electron from the electrode. To reduce 1 mole of $Na^+$ ions, we must supply an Avogadro's number ($6.02 \times 10^{23}$) of electrons to the cathode. On the other hand, stoichiometry tells us that it takes 2 moles of electrons to reduce 1 mole of $Mg^{2+}$ ions and 3 moles of electrons to reduce 1 mole of $Al^{3+}$ ions:

$$Na^+ + e^- \longrightarrow Na$$

$$Mg^{2+} + 2e^- \longrightarrow Mg$$

$$Al^{3+} + 3e^- \longrightarrow Al$$

In an electrolysis experiment, we generally measure the current (in amperes) that passes through an electrolytic cell in a given period of time. The relationship between charge (in coulombs) and the current is

$$1 \text{ C} = 1 \text{ A} \times 1 \text{ s}$$

That is, a coulomb is the quantity of electric charge passing any point in the circuit in 1 s when the current is 1 A. Therefore, if we know the current (in amperes) and how long it is applied (in seconds), we can calculate the charge (in coulombs). Knowing the charge enables us to determine the number of moles of electrons. And knowing the number of moles of electrons allows us to use stoichiometry to determine the number of moles of product. Figure 19.12 shows the steps involved in calculating the quantities of substances produced in electrolysis.

To illustrate this approach, consider an electrolytic cell in which molten $CaCl_2$ is separated into its constituent elements, Ca and $Cl_2$. Suppose a current of 0.452 A is passed through the cell for 1.50 h. How much product will be formed at each electrode? The first step is to determine which species will be oxidized at the anode and which species will be reduced at the cathode. Here the choice is straightforward because we have only $Ca^{2+}$ and $Cl^-$ ions. The cell reactions are

*Anode (oxidation):*     $2Cl^-(l) \longrightarrow Cl_2(g) + 2e^-$

*Cathode (reduction):*     $\underline{Ca^{2+}(l) + 2e^- \longrightarrow Ca(l)}$

*Overall:*     $Ca^{2+}(l) + 2Cl^-(l) \longrightarrow Ca(l) + Cl_2(g)$

The quantities of calcium metal and chlorine gas formed depend on the number of electrons that pass through the electrolytic cell, which in turn depends on *charge*, or current × time:

$$\text{coulombs} = 0.452 \text{ A} \times 1.50 \text{ h} \times \frac{3600 \text{ s}}{1 \text{ h}} \times \frac{1 \text{ C}}{1 \text{ A} \cdot \text{s}} = 2.441 \times 10^3 \text{ C}$$

Because 1 mol $e^- = 96{,}500$ C and 2 mol $e^-$ are required to reduce 1 mole of $Ca^{2+}$ ions, the mass of Ca metal formed at the cathode is calculated as follows:

$$\text{grams Ca} = (2.441 \times 10^3 \text{ C}) \left( \frac{1 \text{ mol } e^-}{96{,}500 \text{ C}} \right) \left( \frac{1 \text{ mol Ca}}{2 \text{ mol } e^-} \right) \left( \frac{40.08 \text{ g Ca}}{1 \text{ mol Ca}} \right) = 0.507 \text{ g Ca}$$

The anode reaction indicates that 1 mole of chlorine is produced per 2 mol $e^-$ of electricity. Hence, the mass of chlorine gas formed is

$$\text{grams Cl}_2 = (2.441 \times 10^3 \text{ C}) \left( \frac{1 \text{ mol } e^-}{96{,}500 \text{ C}} \right) \left( \frac{1 \text{ mol Cl}_2}{2 \text{ mol } e^-} \right) \left( \frac{70.90 \text{ g Cl}_2}{1 \text{ mol Cl}_2} \right) = 0.897 \text{ g Cl}_2$$

Sample Problem 19.8 applies this approach to electrolysis in an aqueous solution.

### Sample Problem 19.8

A current of 1.26 A is passed through an electrolytic cell containing a dilute sulfuric acid solution for 7.44 h. Write the half-cell reactions, and calculate the volume of gases generated at STP.

**Strategy** As shown in Figure 19.12, we can use current and time to determine charge. We can then convert charge to moles of electrons, and use the balanced half-reactions to determine how many moles of product form at each electrode. Finally, we can convert moles to volume.

**Setup** The half-cell reactions for the electrolysis of water are

$$Anode: \quad 2H_2O(l) \longrightarrow O_2(g) + 4H^+(aq) + 4e^-$$
$$Cathode: \quad 4H^+(aq) + 4e^- \longrightarrow 2H_2(g)$$
$$\overline{Overall: \quad \quad 2H_2O(l) \longrightarrow O_2(g) + 2H_2(g)}$$

Remember that STP for gases means 273 K and 1 atm.

**Solution**

$$coulombs = (1.26\ A)(7.44\ h)\left(\frac{3600\ s}{1\ h}\right)\left(\frac{1\ C}{1\ A \cdot s}\right) = 3.375 \times 10^4\ C$$

At the anode:

$$moles\ O_2 = (3.375 \times 10^4\ C)\left(\frac{1\ mol\ e^-}{96,500\ C}\right)\left(\frac{1\ mol\ O_2}{4\ mol\ e^-}\right) = 0.0874\ mol\ O_2$$

The volume of 0.0874 mol $O_2$ at STP is given by

$$V = \frac{nRT}{P}$$
$$= \frac{(0.0874\ mol)(0.08206\ L \cdot atm/K \cdot mol)(273.15)}{1\ atm} = 1.96\ L\ O_2$$

Similarly, for hydrogen we write

$$moles\ H_2\ (3.375 \times 10^4\ C)\left(\frac{1\ mol\ e^-}{96,500\ C}\right)\left(\frac{1\ mol\ H_2}{2\ mol\ e^-}\right) = 0.175\ mol\ H_2$$

The volume of 0.175 mol $H_2$ at STP is given by

$$V = \frac{nRT}{P}$$
$$= \frac{(0.175\ mol)(0.08206\ L \cdot atm/K \cdot mol)(273.15)}{1\ atm} = 3.92\ L\ H_2$$

> **Think About It** The volume of $H_2$ is twice that of $O_2$ (see Figure 19.11), which is what we would expect based on Avogadro's law (at the same temperature and pressure, volume is directly proportional to the number of moles of gas: $V \propto n$) [◀◀ Section 11.2].

**Practice Problem A** A constant current of 0.912 A is passed through an electrolytic cell containing molten $MgCl_2$ for 18 h. What mass of Mg is produced?

**Practice Problem B** A constant current is passed through an electrolytic cell containing molten $MgCl_2$ for 12 h. If 4.83 L of $Cl_2$ (at STP) is produced at the anode, what is the current in amperes?

---

## Checkpoint 19.7 Electrolysis

**19.7.1** In the electrolysis of molten $CaCl_2$, a current of 1.12 A is passed through the cell for 3.0 h. What is the mass of Ca produced at the cathode?

    a) 2.51 g

    b) 1.26 g

    c) 5.02 g

    d) 10.0 g

    e) $2.42 \times 10^5$ g

**19.7.2** How long will a current of 0.995 A need to be passed through water (containing $H_2SO_4$) for 5.00 L of $O_2$ to be produced at STP?

    a) 6.0 h

    b) 8.2 h

    c) 1.5 h

    d) 3.0 h

    e) 24.0 h

**19.7.3** The diagram shows an electrolytic cell being powered by a galvanic cell. Identify each of the electrodes from left to right as an anode or a cathode.

    a) Cathode, anode, anode, cathode

    b) Cathode, anode, cathode, anode

    c) Anode, cathode, anode, cathode

    d) Anode, anode, cathode, cathode

    e) Anode, cathode, cathode, anode

## 19.8 Corrosion

The term **corrosion** generally refers to the deterioration of a metal by an electrochemical process. There are many examples of corrosion, including rust on iron, tarnish on silver, and the green layer that forms on copper and brass. In this section we discuss the processes involved in corrosion and some of the measures taken to prevent it.

The formation of rust on iron requires oxygen and water. Although the reactions involved are quite complex and not completely understood, the main steps are believed to be as follows. A region of the metal's surface serves as the anode, where the following oxidation occurs:

$$Fe(s) \longrightarrow Fe^{2+}(aq) + 2e^-$$

The electrons given up by iron reduce atmospheric oxygen to water at the cathode, which is another region of the same metal's surface:

$$O_2(g) + 4H^+(aq) + 4e^- \longrightarrow 2H_2O(l)$$

The overall redox reaction is

$$2Fe(s) + O_2(g) + 4H^+(aq) \longrightarrow 2Fe^{2+}(g) + 2H_2O(l)$$

With data from Table 19.1, the standard potential for this process can be calculated as follows:

$$E^\circ_{cell} = E^\circ_{cathode} - E^\circ_{anode}$$
$$= 1.23\ V - (-0.44\ V)$$
$$= 1.67\ V$$

Note that this reaction occurs in an *acidic* medium; the $H^+$ ions are supplied in part by the reaction of atmospheric carbon dioxide with water to form the weak acid, carbonic acid ($H_2CO_3$).

The $Fe^{2+}$ ions formed at the anode are further oxidized by oxygen as follows:

$$4Fe^{2+}(aq) + O_2(g) + (4 + 2x)H_2O(l) \longrightarrow 2Fe_2O_3 \cdot xH_2O(s) + 8H^+(aq)$$

This hydrated form of iron(III) oxide is known as rust. The amount of water associated with the iron(III) oxide varies, so we represent the formula as $Fe_2O_3 \cdot xH_2O$.

Figure 19.13 shows the mechanism of rust formation. The electric circuit is completed by the migration of electrons and ions; this is why rusting occurs so rapidly in saltwater. In cold climates, salts (NaCl or $CaCl_2$) spread on roadways to melt ice and snow are a major cause of rust formation on automobiles.

Other metals also undergo oxidation. Aluminum, for example, which is used to make airplanes, beverage cans, and aluminum foil, has a much greater tendency to oxidize than does iron. Unlike the corrosion of iron, though, corrosion of aluminum produces an insoluble layer of protective coating ($Al_2O_3$) that prevents the underlying metal from additional corrosion.

**Media Player/ MPEG Content**

Electrochemistry—the production of aluminum.

**Student Annotation:** From Table 19.1:

$Fe^{2+}(aq) + 2e^- \longrightarrow Fe(s)\ E^\circ = -0.44\ V$
$Al^{3+}(aq) + 3e^- \longrightarrow Al(s)\ E^\circ = -1.66\ V$

Al, with the more negative reduction potential, is less likely to be reduced (more likely to be oxidized) than Fe.

**Figure 19.13** The electrochemical process involved in rust formation. The $H^+$ ions are supplied by $H_2CO_3$, which forms when $CO_2$ from air dissolves in water.

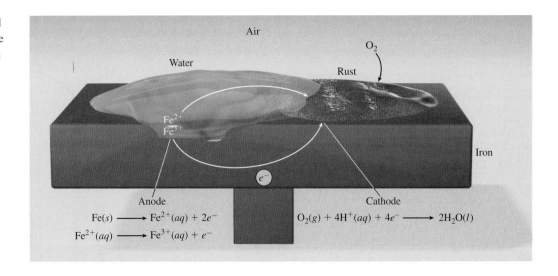

Coinage metals such as copper and silver also corrode, but much more slowly than either iron or aluminum:

$$Cu(s) \longrightarrow Cu^{2+}(aq) + 2e^-$$

$$Ag(s) \longrightarrow Ag^+(aq) + e^-$$

In ordinary atmospheric exposure, copper forms a layer of copper carbonate ($CuCO_3$), a green substance referred to as *patina*, that protects the metal underneath from further corrosion. Likewise, silverware that comes into contact with foodstuffs develops a layer of silver sulfide ($Ag_2S$).

Multimedia

Electrochemistry—chrome plating.

A number of methods have been devised to protect metals from corrosion. Most of these methods are aimed at preventing rust formation. The most obvious approach is to coat the metal surface with paint to prevent exposure to the substances necessary for corrosion. If the paint is scratched or otherwise damaged, however, thus exposing even the smallest area of bare metal, rust will form under the paint layer. The surface of iron metal can be made inactive by a process called *passivation*. A thin oxide layer is formed when the metal is treated with a strong oxidizing agent such as concentrated nitric acid. A solution of sodium chromate is often added to cooling systems and radiators to prevent rust formation.

The tendency for iron to oxidize is greatly reduced when it is alloyed with certain other metals. For example, in stainless steel, an alloy of iron and chromium, a layer of chromium oxide forms that protects the iron from corrosion.

An iron container can be covered with a layer of another metal such as tin or zinc. A "tin" can is made by applying a thin layer of tin over iron. Rust formation is prevented as long as the tin layer remains intact. However, once the surface has been breached by a scratch or a dent, rusting occurs rapidly. If we look up the standard reduction potentials in Table 19.1, we find that when tin and iron are in contact with each other, tin, with its greater reduction potential, acts as a cathode. Iron acts as an anode and is therefore oxidized.

$$Sn^{2+}(aq) + 2e^- \longrightarrow Sn(s) \qquad E^\circ = -0.14 \text{ V}$$

$$Fe^{2+}(aq) + 2e^- \longrightarrow Fe(s) \qquad E^\circ = -0.44 \text{ V}$$

Zinc-plating, or **galvanization,** protects iron from corrosion by a different mechanism. According to Table 19.1,

$$Zn^{2+}(aq) + 2e^- \longrightarrow Zn(s) \qquad E^\circ = -0.76 \text{ V}$$

so zinc is *more* easily oxidized than iron. Like aluminum, zinc oxidizes to form a protective coating. Even when the zinc layer is compromised, though, and the underlying iron is exposed, zinc is still the more easily oxidized of the two metals and will act as the anode. Iron will be the cathode, thereby remaining reduced.

Galvanization is one example of *cathodic protection,* the process by which a metal is protected by being made the cathode in what amounts to a galvanic cell. Another example is the use of zinc or magnesium bars to protect underground storage tanks and ships. When a steel tank or hull is connected to a more easily oxidized metal, corrosion of the steel is prevented.

# Applying What You've Learned

The pain caused by having aluminum foil contact an amalgam dental filling results from the filling being made the cathode in an electrochemical cell. Another type of discomfort can result from the filling being made the anode in an electrochemical cell. This occurs when the filling touches a metal with a greater reduction potential than the components of the amalgam, such as gold. When an amalgam filling comes into contact with a gold inlay, the tin in the filling (the most easily oxidized of the major amalgam components) is *oxidized*—creating an unpleasant metallic taste in the mouth. A simplified, unbalanced equation for the redox reaction that takes place is

$$Sn(s) + O_2(g) \longrightarrow Sn^{2+}(aq) + H_2O(l)$$

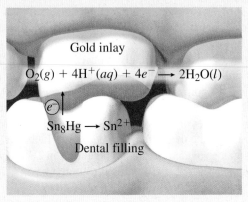

Gold inlay touching amalgam dental filling.

**Problems:**

a)  Balance (in acidic medium) the equation for the oxidation of tin from an amalgam filling.
    [◄◄ Sample Problem 19.1]

b)  Calculate the standard cell potential for the reaction in part (a).
    [◄◄ Sample Problem 19.2]

c)  Which of the components of dental amalgam (mercury, silver, tin, copper, or zinc) would be oxidized when a filling is brought into contact with lead?
    [◄◄ Sample Problem 19.3]

d)  Calculate the standard free-energy change for the reaction in part (a).
    [◄◄ Sample Problem 19.4]

e)  Calculate the equilibrium constant at 25°C for the reaction in part (a).
    [◄◄ Sample Problem 19.5]

# CHAPTER SUMMARY

## Section 19.1

- Redox reactions are those in which oxidation numbers change. **Half-reactions** are the separated oxidation and reduction reactions that make up the overall redox reaction.

- Redox equations can be balanced via the half-reaction method, which allows for the addition of $H_2O$ to balance O, $H^+$ to balance H, and $OH^-$ for reactions taking place in basic solution.

## Section 19.2

- An electrochemical cell in which a spontaneous chemical reaction generates a flow of electrons through a wire is called a **galvanic cell.**

- Half-reactions in a galvanic cell take place in separate compartments called **half-cells.** Half-cells contain **electrodes** in solutions and are connected via an external wire and by a **salt bridge.**

- The electrode at which reduction occurs is called the **cathode;** the electrode at which oxidation occurs is called the **anode.**

- The difference in electric potential between the cathode and the anode is the **cell potential ($E_{cell}$).**

## Section 19.3

- We use **standard reduction potentials ($E°$)** to calculate the standard cell voltage or *standard cell potential ($E°_{cell}$).*

- Half-cell potentials are measured relative to the **standard hydrogen electrode (SHE),** the half-reaction for which has an arbitrarily defined standard reduction potential of zero.

## Section 19.4

- $E°_{cell}$ is related to the standard free-energy change ($\Delta G°$) and to the equilibrium constant, $K$. A positive $E°_{cell}$ corresponds to a negative ($\Delta G°$) value and a large $K$ value.

## Section 19.5

- $E_{cell}$ under other than standard-state conditions is determined from $E°_{cell}$ and the reaction quotient, $Q$, using the **Nernst equation.**

- A **concentration cell** has the same type of electrode and the same ion in solution (at different concentrations) in the anode and cathode compartments.

## Section 19.6

- **Batteries** are portable, self-contained sources of electric energy consisting of galvanic cells—or a series of galvanic cells.

- **Fuel cells** are not really batteries but also supply electric energy via a spontaneous redox reaction. Reactants must be supplied constantly for a fuel cell to operate.

## Section 19.7

- **Electrolysis** is the use of electric energy to drive a nonspontaneous redox reaction. An electrochemical cell used for this purpose is called an **electrolytic cell.**

- The voltage that must actually be supplied to drive a nonspontaneous redox reaction is greater than the calculated amount because of **overvoltage.**

- Electrolysis is used to recharge lead storage batteries, separate compounds into their constituent elements, and separate and purify metals.

- We can calculate the amount of a substance produced in electrolysis if we know the current applied to the cell and the length of time for which it is applied.

## Section 19.8

- **Corrosion** is the undesirable oxidation of metals.

- Corrosion can be prevented by coating the metal surface with paint, a less easily oxidized metal, or a more easily oxidized metal such as zinc.

- The use of a more easily oxidized metal is known as *cathodic protection,* wherein the metal being protected is made the cathode in a galvanic cell. **Galvanization** is the cathodic protection of iron or steel using zinc.

# KEY WORDS

# KEY EQUATIONS

19.1   $E^\circ_{cell} = E^\circ_{cathode} - E^\circ_{anode}$

19.2   $\Delta G = -nFE_{cell}$

19.3   $\Delta G^\circ = -nFE^\circ_{cell}$

19.4   $E^\circ_{cell} = \dfrac{RT}{nF} \ln K$

19.5   $E^\circ_{cell} = \dfrac{0.0592\text{ V}}{n} \log K$   (at 25°C)

19.6   $E = E^\circ - \dfrac{RT}{nF} \ln Q$

19.7   $E = E^\circ - \dfrac{0.0592\text{ V}}{n} \log Q$

# QUESTIONS AND PROBLEMS

## Section 19.1: Balancing Redox Equations

### Problems

**19.1**   Balance the following redox equations by the half-reaction method:

(a) $H_2O_2 + Fe^{2+} \longrightarrow Fe^{3+} + H_2O$ (in acidic solution)
(b) $Cu + HNO_3 \longrightarrow Cu^{2+} + NO + H_2O$ (in acidic solution)
(c) $CN^- + MnO_4^- \longrightarrow CNO^- + MnO_2$ (in basic solution)
(d) $Br_2 \longrightarrow BrO_3^- + Br^-$ (in basic solution)
(e) $S_2O_3^{2-} + I_2 \longrightarrow I^- + S_4O_6^{2-}$ (in acidic solution)

**19.2**   Balance the following redox equations by the half-reaction method:

(a) $Mn^{2+} + H_2O_2 \longrightarrow MnO_2 + H_2O$ (in basic solution)
(b) $Bi(OH)_3 + SnO_2^{2-} \longrightarrow SnO_3^{2-} + Bi$ (in basic solution)
(c) $Cr_2O_7^{2-} + C_2O_4^{2-} \longrightarrow Cr^{3+} + CO_2$ (in acidic solution)
(d) $ClO_3^- + Cl^- \longrightarrow Cl_2 + ClO_2$ (in acidic solution)
(e) $Mn^{2+} + BiO_3^- \longrightarrow Bi^{3+} + MnO_4^-$ (in acidic solution)

## Section 19.2: Galvanic Cells

### Visualizing Chemistry
### Figure 19.1

**VC 19.1**  In the first scene of the animation, when a zinc bar is immersed in an aqueous copper sulfate solution, solid copper deposits on the bar. What reaction would take place if a *copper* bar were immersed in an aqueous *zinc* sulfate solution?

a) No reaction would take place.
b) Solid copper would still deposit on the bar.
c) Solid zinc would deposit on the bar.

**VC 19.2**  What causes the change in the potential of the galvanic cell in Figure 19.1 as the cell operates?

a) Changes in the sizes of the zinc and copper electrodes.
b) Changes in the concentrations of zinc and copper ions.
c) Changes in the volumes of solutions in the half-cells.

**VC 19.3**  Why does the color of the blue solution in the galvanic cell (Figure 19.1) fade as the cell operates?

a) Blue $Cu^{2+}$ ions are replaced by colorless $Zn^{2+}$ ions solution.
b) Blue $Cu^{2+}$ ions are removed from solution by reduction.
c) Blue $Cu^{2+}$ ions are removed from solution by oxidation.

**VC 19.4**  What happens to the mass of the copper electrode in the galvanic cell in Figure 19.1 as the cell operates?

a) It increases.
b) It decreases.
c) It does not change.

### Review Questions

**19.3**   Define the following terms: *anode, cathode, cell voltage, electromotive force, standard reduction potential.*

**19.4**   Describe the basic features of a galvanic cell. Why are the two components of the cell separated from each other?

**19.5**   What is the function of a salt bridge? What kind of electrolyte should be used in a salt bridge?

**19.6**   What is a cell diagram? Write the cell diagram for a galvanic cell consisting of an Al electrode placed in a 1 $M$ $Al(NO_3)_3$ solution and an Ag electrode placed in a 1 $M$ $AgNO_3$ solution.

**19.7**   What is the difference between the half-reactions discussed in redox processes in Chapter 4 and the half-cell reactions discussed in Section 19.2?

## Section 19.3: Standard Reduction Potentials

### Review Questions

**19.8**   Discuss the spontaneity of an electrochemical reaction in terms of its standard emf ($E^\circ_{cell}$).

**19.9**   After operating a Daniell cell (see Figure 19.1) for a few minutes, the cell emf begins to drop. Explain.

## Problems

**19.10** Calculate the standard emf of a cell that uses the $Mg/Mg^{2+}$ and $Cu/Cu^{2+}$ half-cell reactions at 25°C. Write the equation for the cell reaction that occurs under standard-state conditions.

**19.11** Calculate the standard emf of a cell that uses $Ag/Ag^+$ and $Al/Al^{3+}$ half-cell reactions. Write the cell reaction that occurs under standard-state conditions.

**19.12** Predict whether $Fe^{3+}$ can oxidize $I^-$ to $I_2$ under standard-state conditions.

**19.13** Which of the following reagents can oxidize $H_2O$ to $O_2(g)$ under standard-state conditions: $H^+(aq)$, $Cl^-(aq)$, $Cl_2(g)$, $Cu^{2+}(aq)$, $Pb^{2+}(aq)$, $MnO_4^-(aq)$ (in acid)?

**19.14** Consider the following half-reactions:

$$MnO_4^-(aq) + 8H^+(aq) + 5e^- \longrightarrow Mn^{2+}(aq) + 4H_2O(l)$$

$$NO_3^-(aq) + 4H^+(aq) + 3e^- \longrightarrow NO(g) + 2H_2O(l)$$

Predict whether $NO_3^-$ ions will oxidize $Mn^{2+}$ to $MnO_4^-$ under standard-state conditions.

**19.15** Predict whether the following reactions would occur spontaneously in aqueous solution at 25°C. Assume that the initial concentrations of dissolved species are all 1.0 $M$.

(a) $Ca(s) + Cd^{2+}(aq) \longrightarrow Ca^{2+}(aq) + Cd(s)$
(b) $2Br^-(aq) + Sn^{2+}(aq) \longrightarrow Br_2(l) + Sn(s)$
(c) $2Ag(s) + Ni^{2+}(aq) \longrightarrow 2Ag^+(aq) + Ni(s)$
(d) $Cu^+(aq) + Fe^{3+}(aq) \longrightarrow Cu^{2+}(aq) + Fe^{2+}(aq)$

**19.16** Which species in each pair is a better oxidizing agent under standard-state conditions: (a) $Br_2$ or $Au^{3+}$, (b) $H_2$ or $Ag^+$, (c) $Cd^{2+}$ or $Cr^{3+}$, (d) $O_2$ in acidic media or $O_2$ in basic media?

**19.17** Which species in each pair is a better reducing agent under standard-state conditions: (a) Na or Li, (b) $H_2$ or $I_2$, (c) $Fe^{2+}$ or Ag, (d) $Br^-$ or $Co^{2+}$?

## Section 19.4: Spontaneity of Redox Reactions Under Standard-State Conditions

### Review Questions

19.18 Use the information in Table 2.1, and calculate the Faraday constant.

19.19 Write the equations relating $\Delta G°$ and $K$ to the standard emf of a cell. Define all the terms.

19.20 Compare the ease of measuring the equilibrium constant electrochemically with that by chemical means (see Equation 18.14).

### Problems

**19.21** What is the equilibrium constant for the following reaction at 25°C?

$$Mg(s) + Zn^{2+}(aq) \rightleftharpoons Mg^{2+}(aq) + Zn(s)$$

**19.22** The equilibrium constant for the reaction

$$Sr(s) + Mg^{2+}(aq) \rightleftharpoons Sr^{2+}(aq) + Mg(s)$$

is $2.69 \times 10^{12}$ at 25°C. Calculate $E°$ for a cell made up of $Sr/Sr^{2+}$ and $Mg/Mg^{2+}$ half-cells.

**19.23** Use the standard reduction potentials to find the equilibrium constant for each of the following reactions at 25°C:

(a) $Br_2(l) + 2I^-(aq) \rightleftharpoons 2Br^-(aq) + I_2(s)$
(b) $2Ce^{4+}(aq) + 2Cl^-(aq) \rightleftharpoons Cl_2(g) + 2Ce^{3+}(aq)$
(c) $5Fe^{2+}(aq) + MnO_4^-(aq) + 8H^+(aq) \rightleftharpoons Mn^{2+}(aq) + 4H_2O + 5Fe^{3+}(aq)$

**19.24** Calculate $\Delta G°$ and $K_c$ for the following reactions at 25°C:

(a) $Mg(s) + Pb^{2+}(aq) \rightleftharpoons Mg^{2+}(aq) + Pb(s)$
(b) $O_2(g) + 4H^+(aq) + 4Fe^{2+}(aq) \rightleftharpoons 2H_2O(l) + 4Fe^{3+}(aq)$
(c) $2Al(s) + 3I_2(s) \rightleftharpoons 2Al^{3+}(aq) + 6I^-(aq)$

**19.25** Under standard-state conditions, what spontaneous reaction will occur in aqueous solution among the ions $Ce^{4+}$, $Ce^{3+}$, $Fe^{3+}$, and $Fe^{2+}$? Calculate $\Delta G°$ and $K_c$ for the reaction.

**19.26** Given that $E° = 0.52$ V for the reduction $Cu^+(aq) + e^- \longrightarrow Cu(s)$, calculate $E°$, $\Delta G°$, and $K$ for the following reaction at 25°C:

$$2Cu^+(aq) \rightleftharpoons Cu^{2+}(aq) + Cu(s)$$

## Section 19.5: Spontaneity of Redox Reactions Under Conditions Other Than Standard State

### Review Questions

19.27 Write the Nernst equation, and explain all the terms.

19.28 Write the Nernst equation for the following processes at some temperature $T$:

(a) $Mg(s) + Sn^{2+}(aq) \rightleftharpoons Mg^{2+}(aq) + Sn(s)$
(b) $2Cr(s) + 3Pb^{2+}(aq) \rightleftharpoons 2Cr^{3+}(aq) + 3Pb(s)$

### Problems

**19.29** What is the potential of a cell made up of $Zn/Zn^{2+}$ and $Cu/Cu^{2+}$ half-cells at 25°C if $[Zn^{2+}] = 0.25$ $M$ and $[Cu^{2+}] = 0.15$ $M$?

**19.30** Calculate $E°$, $E$, and $\Delta G$ for the following cell reactions.

(a) $Mg(s) + Sn^{2+}(aq) \rightleftharpoons Mg^{2+}(aq) + Sn(s)$
$[Mg^{2+}] = 0.045$ $M$, $[Sn^{2+}] = 0.035$ $M$
(b) $3Zn(s) + 2Cr^{3+}(aq) \rightleftharpoons 3Zn^{2+}(aq) + 2Cr(s)$
$[Cr^{3+}] = 0.010$ $M$, $[Zn^{2+}] = 0.0085$ $M$

**19.31** Calculate the standard potential of the cell consisting of the $Zn/Zn^{2+}$ half-cell and the SHE. What will the emf of the cell be if $[Zn^{2+}] = 0.45$ $M$, $P_{H_2} = 2.0$ atm, and $[H^+] = 1.8$ $M$?

**19.32** What is the emf of a cell consisting of a $Pb^{2+}/Pb$ half-cell and a $Pt/H^+/H_2$ half-cell if $[Pb^{2+}] = 0.10$ $M$, $[H] = 0.050$ $M$, and $P_{H_2} = 1.0$ atm?

**19.33** Referring to the arrangement in Figure 19.1, calculate the $[Cu^{2+}]/[Zn^{2+}]$ ratio at which the following reaction is spontaneous at 25°C:

$$Cu(s) + Zn^{2+}(aq) \longrightarrow Cu^{2+}(aq) + Zn(s)$$

**19.34** Calculate the emf of the following concentration cell:

$$Mg(s) \,|\, Mg^{2+}(0.24\ M) \,\|\, Mg^{2+}(0.53\ M) \,|\, Mg(s)$$

## Section 19.6: Batteries

### Review Questions

19.35 What is a battery? Describe several types of batteries.

**19.36** Explain the differences between a primary galvanic cell—one that is not rechargeable—and a storage cell (for example, the lead storage battery), which is rechargeable.

**19.37** Discuss the advantages and disadvantages of fuel cells over conventional power plants in producing electricity.

### Problems

**19.38** The hydrogen-oxygen fuel cell is described in Section 19.6.
(a) What volume of $H_2(g)$, stored at 25°C at a pressure of 155 atm, would be needed to run an electric motor drawing a current of 8.5 A for 3.0 h? (b) What volume (in liters) of air at 25°C and 1.00 atm will have to pass into the cell per minute to run the motor? Assume that air is 20 percent $O_2$ by volume and that all the $O_2$ is consumed in the cell. The other components of air do not affect the fuel-cell reactions. Assume ideal gas behavior.

**19.39** Calculate the standard emf of the propane fuel cell discussed on page 841 at 25°C, given that $\Delta G_f^\circ$ for propane is −23.5 kJ/mol.

## Section 19.7: Electrolysis

### Review Questions

**19.40** What is the difference between a galvanic cell (such as a Daniell cell) and an electrolytic cell?

**19.41** What is Faraday's contribution to quantitative electrolysis?

**19.42** Define the term *overvoltage*. How does overvoltage affect electrolytic processes?

### Problems

**19.43** The half-reaction at an electrode is

$$Mg^{2+}(molten) + 2e^- \longrightarrow Mg(s)$$

Calculate the number of grams of magnesium that can be produced by supplying 1.00 F to the electrode.

**19.44** Consider the electrolysis of molten barium chloride ($BaCl_2$).
(a) Write the half-reactions. (b) How many grams of barium metal can be produced by supplying 0.50 A for 30 min?

**19.45** Considering only the cost of electricity, would it be cheaper to produce a ton of sodium or a ton of aluminum by electrolysis?

**19.46** If the cost of electricity to produce magnesium by the electrolysis of molten magnesium chloride is $155 per ton of metal, what is the cost (in dollars) of the electricity necessary to produce (a) 10.0 tons of aluminum, (b) 30.0 tons of sodium, and (c) 50.0 tons of calcium?

**19.47** One of the half-reactions for the electrolysis of water is

$$2H_2O(l) \longrightarrow O_2(g) + 4H^+(aq) + 4e^-$$

If 0.076 L of $O_2$ is collected at 25°C and 755 mmHg, how many faradays of electricity had to pass through the solution?

**19.48** How many faradays of electricity are required to produce (a) 0.84 L of $O_2$ at exactly 1 atm and 25°C from aqueous $H_2SO_4$ solution, (b) 1.50 L of $Cl_2$ at 750 mmHg and 20°C from molten NaCl, and (c) 6.0 g of Sn from molten $SnCl_2$?

**19.49** Calculate the amounts of Cu and $Br_2$ produced in 1.0 h at inert electrodes in a solution of $CuBr_2$ by a current of 4.50 A.

**19.50** In the electrolysis of an aqueous $AgNO_3$ solution, 0.67 g of Ag is deposited after a certain period of time. (a) Write the half-reaction for the reduction of $Ag^+$. (b) What is the probable oxidation half-reaction? (c) Calculate the quantity of electricity used (in coulombs).

**19.51** A steady current was passed through molten $CoSO_4$ until 2.35 g of metallic cobalt was produced. Calculate the number of coulombs of electricity used.

**19.52** A constant electric current flows for 3.75 h through two electrolytic cells connected in series. One contains a solution of $AgNO_3$ and the second a solution of $CuCl_2$. During this time, 2.00 g of silver is deposited in the first cell. (a) How many grams of copper are deposited in the second cell? (b) What is the current flowing (in amperes)?

**19.53** What is the hourly production rate of chlorine gas (in kg) from an electrolytic cell using aqueous NaCl electrolyte and carrying a current of $1.500 \times 10^3$ A? The anode efficiency for the oxidation of $Cl^-$ is 93.0 percent.

**19.54** Chromium plating is applied by electrolysis to objects suspended in a dichromate solution, according to the following (unbalanced) half-reaction:

$$Cr_2O_7^{2-}(aq) + e^- + H^+(aq) \longrightarrow Cr(s) + H_2O(l)$$

How long (in hours) would it take to apply a chromium plating $1.0 \times 10^{-2}$ mm thick to a car bumper with a surface area of $0.25$ m$^2$ in an electrolytic cell carrying a current of 25.0 A? (The density of chromium is 7.19 g/cm$^3$.)

**19.55** The passage of a current of 0.750 A for 25.0 min deposited 0.369 g of copper from a $CuSO_4$ solution. From this information, calculate the molar mass of copper.

**19.56** A quantity of 0.300 g of copper was deposited from a $CuSO_4$ solution by passing a current of 3.00 A through the solution for 304 s. Calculate the value of the Faraday constant.

**19.57** In a certain electrolysis experiment, 1.44 g of Ag were deposited in one cell (containing an aqueous $AgNO_3$ solution), while 0.120 g of an unknown metal X was deposited in another cell (containing an aqueous $XCl_3$ solution) in series with the $AgNO_3$ cell. Calculate the molar mass of X.

**19.58** One of the half-reactions for the electrolysis of water is

$$2H^+(aq) + 2e^- \longrightarrow H_2(g)$$

If 0.845 L of $H_2$ is collected at 25°C and 782 mmHg, how many faradays of electricity had to pass through the solution?

## Section 19.8: Corrosion

### Review Questions

**19.59** Steel hardware, including nuts and bolts, is often coated with a thin plating of cadmium. Explain the function of the cadmium layer.

**19.60** "Galvanized iron" is steel sheet that has been coated with zinc; "tin" cans are made of steel sheet coated with tin. Discuss the functions of these coatings and the electrochemistry of the corrosion reactions that occur if an electrolyte contacts the scratched surface of a galvanized iron sheet or a tin can.

**19.61** Tarnished silver contains $Ag_2S$. The tarnish can be removed by placing silverware in an aluminum pan containing an inert electrolyte solution, such as NaCl. Explain the electrochemical principle for this procedure. [The standard reduction potential for the half-cell reaction $Ag_2S(s) + 2e^- \longrightarrow 2Ag(s) + S^{2-}(aq)$ is −0.71 V.]

**19.62** How does the tendency of iron to rust depend on the pH of the solution?

## Additional Problems

**19.63** For each of the following redox reactions, (i) write the half-reactions, (ii) write a balanced equation for the whole reaction, (iii) determine in which direction the reaction will proceed spontaneously under standard-state conditions:

(a) $H_2(g) + Ni^{2+}(aq) \longrightarrow H^+(aq) + Ni(s)$
(b) $MnO_4^-(aq) + Cl^-(aq) \longrightarrow Mn^{2+}(aq) + Cl_2(g)$ (in acid solution)
(c) $Cr(s) + Zn^{2+}(aq) \longrightarrow Cr^{3+}(aq) + Zn(s)$

**19.64** The oxidation of 25.0 mL of a solution containing $Fe^{2+}$ requires 26.0 mL of 0.0250 $M$ $K_2Cr_2O_7$ in acidic solution. Balance the following equation, and calculate the molar concentration of $Fe^{2+}$:

$$Cr_2O_7^{2-} + Fe^{2+} + H^+ \longrightarrow Cr^{3+} + Fe^{3+}$$

**19.65** The $SO_2$ present in air is mainly responsible for the phenomenon of acid rain. The concentration of $SO_2$ can be determined by titrating against a standard permanganate solution as follows:

$$5SO_2 + 2MnO_4^- + 2H_2O \longrightarrow 5SO_4^{2-} + 2Mn^{2+} + 4H^+$$

Calculate the number of grams of $SO_2$ in a sample of air if 7.37 mL of 0.00800 $M$ $KMnO_4$ solution is required for the titration.

**19.66** A sample of iron ore weighing 0.2792 g was dissolved in an excess of a dilute acid solution. All the iron was first converted to Fe(II) ions. The solution then required 23.30 mL of 0.0194 $M$ $KMnO_4$ for oxidation to Fe(III) ions. Calculate the percent by mass of iron in the ore.

**19.67** The concentration of a hydrogen peroxide solution can be conveniently determined by titration against a standardized potassium permanganate solution in an acidic medium according to the following unbalanced equation:

$$MnO_4^- + H_2O_2 \longrightarrow O_2 + Mn^{2+}$$

(a) Balance this equation. (b) If 36.44 mL of a 0.01652 $M$ $KMnO_4$ solution is required to completely oxidize 25.00 mL of an $H_2O_2$ solution, calculate the molarity of the $H_2O_2$ solution.

**19.68** Oxalic acid ($H_2C_2O_4$) is present in many plants and vegetables. (a) Balance the following equation in acid solution:

$$MnO_4^- + C_2O_4^{2-} \longrightarrow Mn^{2+} + CO_2$$

(b) If a 1.00-g sample of plant matter requires 24.0 mL of 0.0100 $M$ $KMnO_4$ solution to reach the equivalence point, what is the percent by mass of $H_2C_2O_4$ in the sample?

**19.69** Calcium oxalate ($CaC_2O_4$) is insoluble in water. This property has been used to determine the amount of $Ca^{2+}$ ions in blood. The calcium oxalate isolated from blood is dissolved in acid and titrated against a standardized $KMnO_4$ solution as described in Problem 19.68. In one test it is found that the calcium oxalate isolated from a 10.0-mL sample of blood requires 24.2 mL of $9.56 \times 10^{-4}$ $M$ $KMnO_4$ for titration. Calculate the number of milligrams of calcium per milliliter of blood.

**19.70** Complete the following table. State whether the cell reaction is spontaneous, nonspontaneous, or at equilibrium.

| $E$ | $\Delta G$ | Cell Reaction |
|---|---|---|
| $> 0$ | | |
| | $> 0$ | |
| $= 0$ | | |

**19.71** From the following information, calculate the solubility product of AgBr:

$Ag^+(aq) + e^- \longrightarrow Ag(s)$     $E° = 0.80$ V
$AgBr(s) + e^- \longrightarrow Ag(s) + Br^-(aq)$     $E° = 0.07$ V

**19.72** Consider a galvanic cell composed of the SHE and a half-cell using the reaction $Ag^+(aq) + e^- \longrightarrow Ag(s)$. (a) Calculate the standard cell potential. (b) What is the spontaneous cell reaction under standard-state conditions? (c) Calculate the cell potential when $[H^+]$ in the hydrogen electrode is changed to (i) $1.0 \times 10^{-2}$ $M$ and (ii) $1.0 \times 10^{-5}$ $M$, all other reagents being held at standard-state conditions. (d) Based on this cell arrangement, suggest a design for a pH meter.

**19.73** A galvanic cell consists of a silver electrode in contact with 346 mL of 0.100 $M$ $AgNO_3$ solution and a magnesium electrode in contact with 288 mL of 0.100 $M$ $Mg(NO_3)_2$ solution. (a) Calculate $E$ for the cell at 25°C. (b) A current is drawn from the cell until 1.20 g of silver has been deposited at the silver electrode. Calculate $E$ for the cell at this stage of operation.

**19.74** Explain why chlorine gas can be prepared by electrolyzing an aqueous solution of NaCl but fluorine gas cannot be prepared by electrolyzing an aqueous solution of NaF.

**19.75** Calculate the emf of the following concentration cell at 25°C:

$$Cu(s) \mid Cu^{2+}(0.080\ M) \parallel Cu^{2+}(1.2\ M) \mid Cu(s)$$

**19.76** The cathode reaction in the Leclanché cell is given by

$$2MnO_2(s) + Zn^{2+}(aq) + 2e^- \longrightarrow ZnMn_2O_4(s)$$

If a Leclanché cell produces a current of 0.0050 A, calculate how many hours this current supply will last if there is initially 4.0 g of $MnO_2$ present in the cell. Assume that there is an excess of $Zn^{2+}$ ions.

**19.77** For a number of years, it was not clear whether mercury(I) ions existed in solution as $Hg^+$ or as $Hg_2^{2+}$. To distinguish between these two possibilities, we could set up the following system:

$$Hg(l) \mid soln\ A \parallel soln\ B \mid Hg(l)$$

where soln A contained 0.263 g mercury(I) nitrate per liter and soln B contained 2.63 g mercury(I) nitrate per liter. If the measured emf of such a cell is 0.0289 V at 18°C, what can you deduce about the nature of the mercury(I) ions?

**19.78** An aqueous KI solution to which a few drops of phenolphthalein have been added is electrolyzed using an apparatus like the one shown here:

Describe what you would observe at the anode and the cathode. (*Hint:* Molecular iodine is only slightly soluble in water, but in the presence of $I^-$ ions, it forms the brown color of $I_3^-$ ions. See Problem 13.114.)

**19.79** A piece of magnesium metal weighing 1.56 g is placed in 100.0 mL of 0.100 $M$ $AgNO_3$ at 25°C. Calculate $[Mg^{2+}]$ and $[Ag^+]$ in solution at equilibrium. What is the mass of the magnesium left? The volume remains constant.

**19.80** Describe an experiment that would enable you to determine which is the cathode and which is the anode in a galvanic cell using copper and zinc electrodes.

**19.81** An acidified solution was electrolyzed using copper electrodes. A constant current of 1.18 A caused the anode to lose 0.584 g after $1.52 \times 10^3$ s. (a) What is the gas produced at the cathode, and what is its volume at STP? (b) Given that the charge of an electron is $1.6022 \times 10^{-19}$ C, calculate Avogadro's number. Assume that copper is oxidized to $Cu^{2+}$ ions.

**19.82** In a certain electrolysis experiment involving $Al^{3+}$ ions, 60.2 g of Al is recovered when a current of 0.352 A is used. How many minutes did the electrolysis last?

**19.83** Consider the oxidation of ammonia:

$$4NH_3(g) + 3O_2(g) \longrightarrow 2N_2(g) + 6H_2O(l)$$

(a) Calculate the $\Delta G°$ for the reaction. (b) If this reaction were used in a fuel cell, what would the standard cell potential be?

**19.84** When an aqueous solution containing gold(III) salt is electrolyzed, metallic gold is deposited at the cathode and oxygen gas is generated at the anode. (a) If 9.26 g of Au is deposited at the cathode, calculate the volume (in liters) of $O_2$ generated at 23°C and 747 mmHg. (b) What is the current used if the electrolytic process took 2.00 h?

**19.85** In an electrolysis experiment, a student passes the same quantity of electricity through two electrolytic cells, one containing a silver salt and the other a gold salt. Over a certain period of time, the student finds that 2.64 g of Ag and 1.61 g of Au are deposited at the cathodes. What is the oxidation state of gold in the gold salt?

**19.86** People living in cold-climate countries where there is plenty of snow are advised not to heat their garages in the winter. What is the electrochemical basis for this recommendation?

**19.87** Given that

$$2Hg^{2+}(aq) + 2e^- \longrightarrow Hg_2^{2+}(aq) \qquad E° = 0.92 \text{ V}$$

$$Hg_2^{2+}(aq) + 2e^- \longrightarrow 2Hg(l) \qquad E° = 0.85 \text{ V}$$

calculate $\Delta G°$ and $K$ for the following process at 25°C:

$$Hg_2^{2+}(aq) \longrightarrow Hg^{2+}(aq) + Hg(l)$$

(The preceding reaction is an example of a *disproportionation reaction* in which an element in one oxidation state is both oxidized and reduced.)

**19.88** Fluorine ($F_2$) is obtained by the electrolysis of liquid hydrogen fluoride (HF) containing potassium fluoride (KF). (a) Write the half-cell reactions and the overall reaction for the process. (b) What is the purpose of KF? (c) Calculate the volume of $F_2$ (in liters) collected at 24.0°C and 1.2 atm after electrolyzing the solution for 15 h at a current of 502 A.

**19.89** A 300-mL solution of NaCl was electrolyzed for 6.00 min. If the pH of the final solution was 12.24, calculate the average current used.

**19.90** Industrially, copper is purified by electrolysis. The impure copper acts as the anode, and the cathode is made of pure copper. The electrodes are immersed in a $CuSO_4$ solution. During electrolysis, copper at the anode enters the solution as $Cu^{2+}$ while $Cu^{2+}$ ions are reduced at the cathode. (a) Write half-cell reactions and the overall reaction for the electrolytic process. (b) Suppose the anode was contaminated with Zn and Ag. Explain what happens to these impurities during electrolysis. (c) How many hours will it take to obtain 1.00 kg of Cu at a current of 18.9 A?

**19.91** An aqueous solution of a platinum salt is electrolyzed at a current of 2.50 A for 2.00 h. As a result, 9.09 g of metallic Pt is formed at the cathode. Calculate the charge on the Pt ions in this solution.

**19.92** Consider a galvanic cell consisting of a magnesium electrode in contact with 1.0 M $Mg(NO_3)_2$ and a cadmium electrode in contact with 1.0 M $Cd(NO_3)_2$. Calculate $E°$ for the cell, and draw a diagram showing the cathode, anode, and direction of electron flow.

**19.93** A current of 6.00 A passes through an electrolytic cell containing dilute sulfuric acid for 3.40 h. If the volume of $O_2$ gas generated at the anode is 4.26 L (at STP), calculate the charge (in coulombs) on an electron.

**19.94** Gold will not dissolve in either concentrated nitric acid or concentrated hydrochloric acid. However, the metal does dissolve in a mixture of the acids (one part $HNO_3$ and three parts HCl by volume), called *aqua regia*. (a) Write a balanced equation for this reaction. (*Hint:* Among the products are $HAuCl_4$ and $NO_2$.) (b) What is the function of HCl?

**19.95** Explain why most useful galvanic cells give voltages of no more than 1.5 to 2.5 V. What are the prospects for developing practical galvanic cells with voltages of 5 V or more?

**19.96** A silver rod and a SHE are dipped into a saturated aqueous solution of silver oxalate ($Ag_2C_2O_4$), at 25°C. The measured potential difference between the rod and the SHE is 0.589 V, the rod being positive. Calculate the solubility product constant for silver oxalate.

**19.97** Zinc is an amphoteric metal; that is, it reacts with both acids and bases. The standard reduction potential is $-1.36$ V for the reaction

$$Zn(OH)_4^{2-}(aq) + 2e^- \longrightarrow Zn(s) + 4OH^-(aq)$$

Calculate the formation constant ($K_f$) for the reaction

$$Zn^{2+}(aq) + 4OH^-(aq) \rightleftharpoons Zn(OH)_4^{2-}(aq)$$

**19.98** Use the data in Table 19.1 to determine whether or not hydrogen peroxide will undergo disproportionation in an acid medium: $2H_2O_2 \longrightarrow 2H_2O + O_2$.

**19.99** The magnitudes (but *not* the signs) of the standard reduction potentials of two metals X and Y are

$$Y^{2+} + 2e^- \longrightarrow Y \qquad |E°| = 0.34 \text{ V}$$

$$X^{2+} + 2e^- \longrightarrow X \qquad |E°| = 0.25 \text{ V}$$

where the $\parallel$ notation denotes that only the magnitude (but not the sign) of the $E°$ value is shown. When the half-cells of X and Y are connected, electrons flow from X to Y. When X is connected to a SHE, electrons flow from X to SHE. (a) Are the $E°$ values of the half-reactions positive or negative? (b) What is the standard emf of a cell made up of X and Y?

**19.100** A galvanic cell is constructed as follows. One half-cell consists of a platinum wire immersed in a solution containing 1.0 M $Sn^{2+}$ and 1.0 M $Sn^{4+}$; the other half-cell has a thallium rod immersed in a solution of 1.0 M $Tl^+$. (a) Write the half-cell reactions and the overall reaction. (b) What is the equilibrium constant at 25°C? (c) What is the cell voltage if the $Tl^+$ concentration is increased 10-fold? ($E°_{Tl^+/Tl} = -0.34$ V.)

**19.101** Given the standard reduction potential for $Au^{3+}$ in Table 19.1 and

$$Au^+(aq) + e^- \longrightarrow Au(s) \qquad E° = 1.69 \text{ V}$$

answer the following questions. (a) Why does gold not tarnish in air? (b) Will the following disproportionation occur spontaneously?

$$3Au^+(aq) \longrightarrow Au^{3+}(aq) + 2Au(s)$$

(c) Predict the reaction between gold and fluorine gas.

19.102 The ingestion of a very small quantity of mercury is not considered too harmful. Would this statement still hold if the gastric juice in your stomach were mostly nitric acid instead of hydrochloric acid?

19.103 When 25.0 mL of a solution containing both $Fe^{2+}$ and $Fe^{3+}$ ions is titrated with 23.0 mL of 0.0200 $M$ $KMnO_4$ (in dilute sulfuric acid), all the $Fe^{2+}$ ions are oxidized to $Fe^{3+}$ ions. Next, the solution is treated with Zn metal to convert all the $Fe^{3+}$ ions to $Fe^{2+}$ ions. Finally, 40.0 mL of the same $KMnO_4$ solution is added to the solution to oxidize the $Fe^{2+}$ ions to $Fe^{3+}$. Calculate the molar concentrations of $Fe^{2+}$ and $Fe^{3+}$ in the original solution.

19.104 Consider the Daniell cell in Figure 19.1. When viewed externally, the anode appears negative and the cathode positive (electrons are flowing from the anode to the cathode). Yet in solution anions are moving toward the anode, which means that it must appear positive to the anions. Because the anode cannot simultaneously be negative and positive, give an explanation for this apparently contradictory situation.

19.105 Use the data in Table 19.1 to show that the decomposition of $H_2O_2$ (a disproportionation reaction) is spontaneous at 25°C:

$$2H_2O_2(aq) \longrightarrow 2H_2O(l) + O_2(g)$$

19.106 The concentration of sulfuric acid in the lead-storage battery of an automobile over a period of time has decreased from 38.0 percent by mass (density = 1.29 g/mL) to 26.0 percent by mass (1.19 g/mL). Assume the volume of the acid remains constant at 724 mL. (a) Calculate the total charge in coulombs supplied by the battery. (b) How long (in hours) will it take to recharge the battery back to the original sulfuric acid concentration using a current of 22.4 A?

19.107 Consider a Daniell cell operating under non-standard-state conditions. Suppose that the cell's reaction is multiplied by 2. What effect does this have on each of the following quantities in the Nernst equation: (a) $E$, (b) $E°$, (c) $Q$, (d) $\ln Q$, (e) $n$?

19.108 A spoon was silver-plated electrolytically in an $AgNO_3$ solution. (a) Sketch a diagram for the process. (b) If 0.884 g of Ag was deposited on the spoon at a constant current of 18.5 mA, how long (in min) did the electrolysis take?

19.109 Comment on whether $F_2$ will become a stronger oxidizing agent with increasing $H^+$ concentration.

19.110 In recent years there has been much interest in electric cars. List some advantages and disadvantages of electric cars compared to automobiles with internal combustion engines.

19.111 Calculate the pressure of $H_2$ (in atm) required to maintain equilibrium with respect to the following reaction at 25°C,

$$Pb(s) + 2H^+(aq) \rightleftharpoons Pb^{2+}(aq) + H_2(g)$$

given that $[Pb^{2+}]$ = 0.035 $M$ and the solution is buffered at pH 1.60.

19.112 A piece of magnesium ribbon and a copper wire are partially immersed in a 0.1 $M$ HCl solution in a beaker. The metals are joined externally by another piece of metal wire. Bubbles are seen to evolve at both the Mg and Cu surfaces. (a) Write equations representing the reactions occurring at the metals. (b) What visual evidence would you seek to show that Cu is not oxidized to $Cu^{2+}$? (c) At some stage, NaOH solution is added to the beaker to neutralize the HCl acid. Upon further addition of NaOH, a white precipitate forms. What is it?

19.113 The zinc-air battery shows much promise for electric cars because it is lightweight and rechargeable:

The net transformation is $Zn(s) + \frac{1}{2}O_2(g) \longrightarrow ZnO(s)$. (a) Write the half-reactions at the zinc-air electrodes, and calculate the standard emf of the battery at 25°C. (b) Calculate the emf under actual operating conditions when the partial pressure of oxygen is 0.21 atm. (c) What is the energy density (measured as the energy in kilojoules that can be obtained from 1 kg of the metal) of the zinc electrode? (d) If a current of $2.1 \times 10^5$ A is to be drawn from a zinc-air battery system, what volume of air (in liters) would need to be supplied to the battery every second? Assume that the temperature is 25°C and the partial pressure of oxygen is 0.21 atm.

19.114 Calculate $E°$ for the reactions of mercury with (a) 1 $M$ HCl and (b) 1 $M$ $HNO_3$. Which acid will oxidize Hg to $Hg_2^{2+}$ under standard-state conditions? Can you identify which pictured test tube contains $HNO_3$ and Hg and which contains HCl and Hg?

19.115 Because all alkali metals react with water, it is not possible to measure the standard reduction potentials of these metals directly as in the case of, say, zinc. An indirect method is to consider the following hypothetical reaction:

$$Li^+(aq) + \frac{1}{2}H_2(g) \longrightarrow Li(s) + H^+(aq)$$

Using the appropriate equation presented in this chapter and the thermodynamic data in Appendix 2, calculate $E°$ for $Li^+(aq) + e^- \longrightarrow Li(s)$ at 298 K. Use 96,485.338 C/mol $e^-$ for the Faraday constant. Compare your result with that listed in Table 19.1.

19.116 A galvanic cell using $Mg/Mg^{2+}$ and $Cu/Cu^{2+}$ half-cells operates under standard-state conditions at 25°C, and each compartment has a volume of 218 mL. The cell delivers 0.22 A for 31.6 h. (a) How many grams of Cu are deposited? (b) What is the $[Cu^{2+}]$ remaining?

19.117 Given the following standard reduction potentials, calculate the ion-product, $K_w$, for water at 25°C:

$$2H^+(aq) + 2e^- \longrightarrow H_2(g) \qquad E° = 0.00 \text{ V}$$
$$2H_2O(l) + 2e^- \longrightarrow H_2(g) + 2OH^-(aq) \qquad E° = -0.83 \text{ V}$$

19.118 Compare the pros and cons of a fuel cell, such as the hydrogen-oxygen fuel cell, and a coal-fired power station for generating electricity.

**19.119** Lead storage batteries are rated by ampere-hours, that is, the number of amperes they can deliver in an hour. (a) Show that 1 Ah = 3600 C. (b) The lead anodes of a certain lead-storage battery have a total mass of 406 g. Calculate the maximum theoretical capacity of the battery in ampere-hours. Explain why in practice we can never extract this much energy from the battery. (*Hint:* Assume all the lead will be used up in the electrochemical reaction, and refer to the electrode reactions on page 839.) (c) Calculate $E^\circ_{cell}$ and $\Delta G^\circ$ for the battery.

**19.120** Use Equations 18.10 and 19.3 to calculate the emf values of the Daniell cell at 25°C and 80°C. Comment on your results. What assumptions are used in the derivation? (*Hint:* You need the thermodynamic data in Appendix 2.)

**19.121** A construction company is installing an iron culvert (a long cylindrical tube) that is 40.0 m long with a radius of 0.900 m. To prevent corrosion, the culvert must be galvanized. This process is carried out by first passing an iron sheet of appropriate dimensions through an electrolytic cell containing $Zn^{2+}$ ions, using graphite as the anode and the iron sheet as the cathode. If the voltage is 3.26 V, what is the cost of electricity for depositing a layer 0.200 mm thick if the efficiency of the process is 95 percent? The electricity rate is $0.12 per kilowatt hour (kWh), where 1 W = 1 J/s and the density of Zn is 7.14 g/cm$^3$.

**19.122** A $9.00 \times 10^2$ mL amount of 0.200 $M$ $MgI_2$ solution was electrolyzed. As a result, hydrogen gas was generated at the cathode and iodine was formed at the anode. The volume of hydrogen collected at 26°C and 779 mmHg was $1.22 \times 10^3$ mL. (a) Calculate the charge in coulombs consumed in the process. (b) How long (in min) did the electrolysis last if a current of 7.55 A was used? (c) A white precipitate was formed in the process. What was it, and what was its mass in grams? Assume the volume of the solution was constant.

**19.123** Based on the following standard reduction potentials,

$$Fe^{2+}(aq) + 2e^- \longrightarrow Fe(s) \qquad E^\circ_1 = -0.44 \text{ V}$$
$$Fe^{3+}(aq) + e^- \longrightarrow Fe^{2+}(aq) \qquad E^\circ_2 = 0.77 \text{ V}$$

calculate the standard reduction potential for the half-reaction

$$Fe^{3+}(aq) + 3e^- \longrightarrow Fe(s) \qquad E^\circ_3 = ?$$

**19.124** To remove the tarnish ($Ag_2S$) on a silver spoon, a student carried out the following steps. First, she placed the spoon in a large pan filled with water so the spoon was totally immersed. Next, she added a few tablespoonfuls of baking soda (sodium bicarbonate), which readily dissolved. Finally, she placed some aluminum foil at the bottom of the pan in contact with the spoon and then heated the solution to about 80°C. After a few minutes, the spoon was removed and rinsed with cold water. The tarnish was gone, and the spoon regained its original shiny appearance. (a) Describe with equations the electrochemical basis for the procedure. (b) Adding NaCl instead of $NaHCO_3$ would also work because both compounds are strong electrolytes. What is the added advantage of using $NaHCO_3$? (*Hint:* Consider the pH of the solution.) (c) What is the purpose of heating the solution? (d) Some commercial tarnish removers contain a fluid (or paste) that is a dilute HCl solution. Rubbing the spoon with the fluid will also remove the tarnish. Name two disadvantages of using this procedure compared to the one described previously.

**19.125** Calculate the equilibrium constant for the following reaction at 298 K:

$$Zn(s) + Cu^{2+}(aq) \rightleftharpoons Zn^{2+}(aq) + Cu(s)$$

**19.126** Cytochrome-c is a protein involved in biological electron transfer processes. The redox half-reaction is shown by the reduction of the $Fe^{3+}$ ion to the $Fe^{2+}$ ion:

$$\text{cyt } c(Fe^{3+}) + e^- \longrightarrow \text{cyt } c(Fe^{2+}) \qquad E^\circ = 0.254 \text{ V}$$

Calculate the number of moles of cyt $c(Fe^{3+})$ formed from cyt $c(Fe^{2+})$ with the Gibbs free energy derived from the oxidation of 1 mole of glucose.

**19.127** The nitrite ion ($NO_2^-$) in soil is oxidized to the nitrate ion ($NO_3^-$) by the bacterium *Nitrobacter agilis* in the presence of oxygen. The half-reactions are

$$NO_3^- + 2H^+ + 2e^- \longrightarrow NO_2^- + H_2O \qquad E^\circ = 0.42 \text{ V}$$
$$O_2 + 4H^+ + 4e^- \longrightarrow 2H_2O \qquad E^\circ = 1.23 \text{ V}$$

Calculate the yield of ATP synthesis per mole of nitrite oxidized. (*Hint:* Refer to Section 18.7.)

**19.128** Fluorine is a highly reactive gas that attacks water to form HF and other products. Follow the procedure in Problem 19.115 to show how you can determine indirectly the standard reduction potential for fluorine. Compare your result with the value in Table 19.1.

**19.129** As discussed in Section 19.5, the potential of a concentration cell diminishes as the cell operates and the concentrations in the two compartments approach each other. When the concentrations in both compartments are the same, the cell ceases to operate. At this stage, is it possible to generate a cell potential by adjusting a parameter other than concentration? Explain.

# PRE-PROFESSIONAL PRACTICE EXAM PROBLEMS: PHYSICAL AND BIOLOGICAL SCIENCES

A galvanic cell is constructed by immersing a piece of copper wire in 25.0 mL of a 0.20 $M$ $CuSO_4$ solution and a zinc strip in 25.0 mL of a 0.20 $M$ $ZnSO_4$ solution. $Cu^{2+}$ ions react with aqueous $NH_3$ to form the complex ion $Cu(NH_3)_4^{2+}$:

$$Cu^{2+}(aq) + 4NH_3(aq) \longrightarrow Cu(NH_3)_4^{2+}(aq)$$

1. Using the equation

$$E = E° - \frac{0.0592 \text{ V}}{n} \log Q$$

calculate the emf of the cell at 25°C.

a) 0.0 V
b) 1.10 V
c) 0.90 V
d) 1.30 V

2. What would happen if a small amount of concentrated $NH_3$ solution were added to the $CuSO_4$ solution?

a) Nothing.
b) Emf would increase.
c) Emf would decrease.
d) Not enough information to determine.

3. What would happen if a small amount of concentrated $NH_3$ solution were added to the $ZnSO_4$ solution?

a) Nothing.
b) Emf would increase.
c) Emf would decrease.
d) Not enough information to determine.

4. In a separate experiment, 25.0 mL of 3.00 $M$ $NH_3$ is added to the $CuSO_4$ solution. If the emf of the cell is 0.68 V at equilibrium, calculate the formation constant ($K_f$) of $Cu(NH_3)_4^{2+}$.

a) $9.4 \times 10^{22}$
b) $1.1 \times 10^{-23}$
c) $1.5 \times 10^{-14}$
d) $1.5 \times 10^{14}$

# ANSWERS TO IN-CHAPTER MATERIALS

## Answers to Practice Problems

**19.1A** $2MnO_4^- + H_2O + 3CN^- \longrightarrow 2MnO_2 + 2OH^- + 3CNO^-$.
**19.1B** $MnO_4^- + 5Fe^{2+} + 8H^+ \longrightarrow Mn^{2+} + 5Fe^{3+} + 4H_2O$.
**19.2A** $Cd + Pb^{2+} \longrightarrow Cd^{2+} + Pb$, $E°_{cell} = 0.27$ V. **19.2B** In one cell, the iron electrode is the anode and the overall reaction is $Fe(s) + Sn^{2+}(aq) \longrightarrow Fe^{2+}(aq) + Sn(s)$. $E°_{cell} = E°_{Sn^{2+}/Sn} - E°_{Fe^{2+}/Fe} = (-0.14$ V$) - (-0.44$ V$) = 0.30$ V. In the other cell, the iron electrode is the cathode and the overall reaction is $3Fe^{2+}(aq) + 2Cr(s) \longrightarrow 3Fe(s) + 2Cr^{3+}(aq)$. $E°_{cell} = E°_{Fe^{2+}/Fe} - E°_{Cr^{3+}/Cr} = (-0.44$ V$) - (-0.74$ V$) = 0.30$ V. **19.3A** (a) No reaction, (b) $2H^+ + Pb \longrightarrow H_2 + Pb^{2+}$.
**19.3B** Cobalt has a more positive reduction potential than iron. Cobalt ion would, therefore, be *reduced* in the presence of iron metal; and the iron in an iron container would oxidize to $Fe^{2+}$. Metal that is oxidized goes from the solid phase to the aqueous phase, meaning that the container would effectively *dissolve*. Cobalt has a less positive reduction potential than tin. Tin metal would not be oxidized by $Co^{2+}$ ion, and the container would remain intact. A cobalt(II) chloride solution would be more safely stored in a tin container than in an iron container. **19.4A** $-411$ kJ/mol.
**19.4B** 23.2 kJ/mol. **19.5A** $1 \times 10^{-42}$. **19.5B** $1.4 \times 10^5$. **19.6A** Yes, the reaction is spontaneous. **19.6B** 18 atm. **19.7A** $[Cu^+] = 1.11 \times 10^{-3}$ $M$, $K_{sp} = 1.2 \times 10^{-6}$. **19.7B** 0.16 V. **19.8A** 7.44 g Mg. **19.8B** 0.96 A.

## Answers to Checkpoints

**19.1.1** a, c, e. **19.1.2** c. **19.3.1** d. **19.3.2** c. **19.3.3** b. **19.3.4** e. **19.4.1** b.
**19.4.2** b. **19.5.1** b. **19.5.2** b. **19.5.3** e. **19.5.4** d. **19.5.5** b. **19.5.6** a.
**19.7.1** a. **19.7.2** e. **19.7.3** c.

## Answers to Applying What You've Learned

**a)** $2Sn + O_2 + 4H^+ \longrightarrow 2Sn^{2+} + 2H_2O$. **b)** 1.37 V. **c)** Tin and zinc.
**d)** $-5.3 \times 10^2$ kJ/mol. **e)** $5 \times 10^{92}$.

# 20

# Nuclear Chemistry

The CT scan shows a brain tumor that might be difficult or impossible to treat by conventional surgical methods. Nuclear medicine, including boron neutron capture therapy (BNCT), enables doctors to treat cancers of this type.

# How Nuclear Chemistry Is Used to Treat Cancer

Brain tumors are some of the most difficult cancers to treat because the site of the malignant growth makes surgical excision difficult or impossible. Likewise, conventional radiation therapy using X rays or $\gamma$ rays from outside the skull is usually not effective. An ingenious approach to this problem is *boron neutron capture therapy* (BNCT). This technique involves first administering a boron-10 compound that is selectively taken up by tumor cells and then applying a beam of low-energy neutrons to the tumor site. $^{10}$B captures a neutron to produce $^{11}$B, which disintegrates via the following nuclear reaction:

$$^{10}_{5}\text{B} + {}^{1}_{0}\text{n} \longrightarrow {}^{7}_{3}\text{Li} + {}^{4}_{2}\alpha$$

The highly energetic particles produced by this reaction destroy the tumor cells in which the $^{10}$B is concentrated. Because the particles are confined to just a few micrometers, they preferentially destroy tumor cells without damaging neighboring normal cells.

BNCT is a highly promising treatment and is an active area of research. One of the major goals of the research is to develop suitable compounds to deliver $^{10}$B to the desired site. For such a compound to be effective, it must meet several criteria. It must have a high affinity for tumor cells, be able to pass through membrane barriers to reach the tumor site, and have minimal toxic effects on the human body.

This is one example of how *nuclear chemistry* is important in the treatment of cancer.

At the end of this chapter, you will be able to answer questions about isotopes used in nuclear medicine [▶▶ Page 885].

## In This Chapter, You Will Learn:

Some of the fundamentals of nuclear chemistry and how nuclear reactions are important to living systems and to society.

### Before you begin, you should review

- Radioactivity [◀◀ Section 2.2]

- Atomic number, mass number, and isotopes [◀◀ Section 2.3]

- First-order kinetics [◀◀ Section 14.3]

## 20.1 Nuclei and Nuclear Reactions

With the exception of hydrogen ($_1^1$H), all nuclei contain protons and neutrons. Some nuclei are unstable and undergo *radioactive decay,* emitting particles and/or electromagnetic radiation [◂◂ Section 2.2]. Spontaneous emission of particles or electromagnetic radiation is known as **radioactivity.** All elements having an atomic number greater than 83 are unstable and are therefore radioactive. Polonium-210 ($_{84}^{210}$Po), for example, decays spontaneously to Pb by emitting an $\alpha$ particle.

Another type of nuclear process, known as **nuclear transmutation,** results from the bombardment of nuclei by neutrons, protons, or other nuclei. An example of a nuclear transmutation is the conversion of atmospheric $_7^{14}$N to $_6^{14}$C and $_1^1$H, which results when the nitrogen isotope is bombarded by neutrons (from the sun). In some cases, heavier elements are synthesized from lighter elements. This type of transmutation occurs naturally in outer space, but it can also be achieved artificially, as we will see in Section 20.4.

Radioactive decay and nuclear transmutation are *nuclear reactions,* which differ significantly from ordinary chemical reactions. Table 20.1 summarizes the differences.

To discuss nuclear reactions in any depth, we must understand how to write and balance nuclear equations. Writing a nuclear equation differs somewhat from writing equations for chemical reactions. In addition to writing the symbols for the various chemical elements, we must also explicitly indicate the number of subatomic particles in *every* species involved in the reaction.

The symbols for subatomic particles are as follows:

$$\underset{\text{proton}}{_1^1\text{H} \ \ \text{or} \ \ _1^1\text{p}} \qquad \underset{\text{neutron}}{_0^1\text{n}} \qquad \underset{\text{electron}}{_{-1}^0 e \ \ \text{or} \ \ _{-1}^0\beta} \qquad \underset{\text{positron}}{_{+1}^0 e \ \ \text{or} \ \ _{+1}^0\beta} \qquad \underset{\alpha \text{ particle}}{_2^4\alpha \ \ \text{or} \ \ _2^4\text{He}}$$

In accordance with the notation introduced in Section 2.3, the superscript in each case denotes the mass number (the total number of neutrons and protons present) and the subscript is the atomic number (the number of protons). Thus, the "atomic number" of a proton is 1, because there is one proton present, and the "mass number" is also 1, because there is one proton but no neutrons present. On the other hand, the mass number of a neutron is 1, but its atomic number is zero, because there are no protons present. For the electron, the mass number is zero (there are neither protons nor neutrons present), but the atomic number is −1, because the electron possesses a unit negative charge.

The symbol $_{-1}^0 e$ represents an electron in or from an atomic orbital. The symbol $_{-1}^0\beta$, on the other hand, represents an electron that, although physically identical to any other electron, comes from a nucleus (in a decay process in which a neutron is converted to a proton and an electron) and not from an atomic orbital. The *positron* has the same mass as the electron, but bears a charge of +1. The $\alpha$ particle has two protons and two neutrons, so its atomic number is 2 and its mass number is 4.

**Student Annotation:** An $\alpha$ particle is identical to a helium-4 nucleus and can be represented either as $_2^4\alpha$ or $_2^4$He.

In balancing any nuclear equation, we must balance the total of all atomic numbers and the total of all mass numbers for the products and reactants. If we know the atomic numbers and mass numbers of all but one of the species in a nuclear equation, we can identify the *unknown* species by applying these rules, as shown in Sample Problem 20.1.

| TABLE 20.1 | Comparison of Chemical Reactions and Nuclear Reactions |
|---|---|

| Chemical Reactions | Nuclear Reactions |
|---|---|
| 1. Atoms are rearranged by the breaking and forming of chemical bonds. | 1. Elements are converted to other elements (or isotopes). |
| 2. Only electrons in atomic or molecular orbitals are involved in the reaction. | 2. Protons, neutrons, electrons, and other subatomic particles such as $\alpha$ particles may be involved. |
| 3. Reactions are accompanied by the absorption or release of relatively small amounts of energy. | 3. Reactions are accompanied by the absorption or release of tremendous amounts of energy. |
| 4. Rates of reaction are influenced by temperature, pressure, concentration, and catalysts. | 4. Rates of reaction normally are not affected by temperature, pressure, or catalysts. |

### Sample Problem 20.1

Identify the missing species X in each of the following nuclear equations:

(a) $^{212}_{84}Po \longrightarrow ^{208}_{82}Pb + X$

(b) $^{90}_{38}Sr \longrightarrow X + ^{0}_{-1}\beta$

(c) $X \longrightarrow ^{18}_{8}O + ^{0}_{+1}\beta$

**Strategy** Determine the mass number for the unknown species, X, by summing the mass numbers on both sides of the equation:

$$\Sigma \text{ reactant mass numbers} = \Sigma \text{ product mass numbers}$$

Similarly, determine the atomic number for the unknown species:

$$\Sigma \text{ reactant atomic numbers} = \Sigma \text{ product atomic numbers}$$

Use the mass number and atomic number to determine the identity of the unknown species.

**Setup** (a) 212 = (208 + mass number of X); mass number of X = 4. 84 = (82 + atomic number of X); atomic number of X = 2.

(b) 90 = (mass number of X + 0); mass number of X = 90. 38 = [atomic number of X + (−1)]; atomic number of X = 39.

(c) Mass number of X = (18 + 0); mass number of X = 18. Atomic number of X = (8 + 1); atomic number of X = 9.

**Solution** (a) $X = ^{4}_{2}\alpha$: $^{208}_{82}Po \longrightarrow ^{208}_{82}Pb + ^{4}_{2}\alpha$

(b) $X = ^{90}_{39}Y$: $^{90}_{38}Sr \longrightarrow ^{90}_{39}Y + ^{0}_{-1}\beta$

(c) $X = ^{18}_{9}F$: $^{18}_{9}F \longrightarrow ^{18}_{8}O + ^{0}_{+1}\beta$

**Think About It** The rules of summation that we apply to balance nuclear equations can be thought of as the *conservation of mass number* and the *conservation of atomic number*.

**Practice Problem A** Identify X in each of the following nuclear equations:

(a) $^{78}_{33}As \longrightarrow X + ^{0}_{-1}\beta$

(b) $^{1}_{1}H + ^{4}_{2}He \longrightarrow X$

(c) $^{258}_{100}Fm \longrightarrow ^{257}_{100}Fm + X$

**Practice Problem B** Identify X in each of the following nuclear equations:

(a) $X + ^{0}_{-1}\beta \longrightarrow ^{244}_{94}Pu$

(b) $^{238}_{92}U \longrightarrow X + ^{4}_{2}He$

(c) $X \longrightarrow ^{14}_{7}N + ^{0}_{-1}e$

### Checkpoint 20.1 Nuclei and Nuclear Reactions

**20.1.1** Identify the species X in the following nuclear equation:

$$^{222}_{86}Rn \longrightarrow X + ^{4}_{2}\alpha$$

a) $^{226}_{84}Po$

b) $^{226}_{88}Ra$

c) $^{212}_{84}Po$

d) $^{218}_{84}Po$

e) $^{218}_{88}Ra$

**20.1.2** Identify the species X in the following nuclear equation:

$$^{15}_{8}O \longrightarrow X + ^{0}_{-1}\beta$$

a) $^{15}_{9}F$

b) $^{14}_{8}O$

c) $^{16}_{9}F$

d) $^{15}_{7}N$

e) $^{14}_{7}N$

## 20.2 Nuclear Stability

The nucleus occupies a very small portion of the total volume of an atom, but it contains most of the atom's mass because both the protons and the neutrons reside there. In studying the stability of the atomic nucleus, it is helpful to know something about its density, because it tells us how tightly the particles are packed together. As a sample calculation, let us assume that a nucleus has a radius of $5 \times 10^{-3}$ pm and a mass of $1 \times 10^{-22}$ g. These figures correspond roughly to a nucleus containing 30 protons and 30 neutrons. Density is mass/volume, and we can calculate the volume from the known radius (the volume of a sphere is $\frac{4}{3}\pi r^3$, where $r$ is the radius of the sphere). First we convert the picometer units to centimeters. Then, we calculate the density in g/cm$^3$:

$$r = (5 \times 10^{-3} \text{ pm}) \left( \frac{1 \times 10^{-12} \text{ m}}{1 \text{ pm}} \right) \left( \frac{100 \text{ cm}}{1 \text{ m}} \right) = 5 \times 10^{-13} \text{ cm}$$

$$\text{density} = \frac{\text{mass}}{\text{volume}} = \frac{1 \times 10^{-22} \text{ g}}{\frac{4}{3}\pi r^3} = \frac{1 \times 10^{-22} \text{ g}}{\frac{4}{3}\pi (5 \times 10^{-13} \text{ cm})^3}$$

$$= 2 \times 10^{14} \text{ g/cm}^3$$

This is an exceedingly high density. The highest density known for an element is 22.6 g/cm$^3$, for osmium (Os). Thus, the average atomic nucleus is roughly $9 \times 10^{12}$ (or 9 *trillion*) times as dense as the densest element known!

The enormously high density of the nucleus means that some very strong force is needed to hold the particles together so tightly. From *Coulomb's law* we know that like charges repel and unlike charges attract one another. We would thus expect the protons to repel one another strongly, particularly when we consider how close they must be to each other. This indeed is so. However, in addition to the repulsion, there are also short-range attractions between proton and proton, proton and neutron, and neutron and neutron. The stability of any nucleus is determined by the difference between coulombic repulsion and the short-range attraction. If repulsion outweighs attraction, the nucleus disintegrates, emitting particles and/or radiation. If attractive forces prevail, the nucleus is stable.

## Patterns of Nuclear Stability

The principal factor that determines whether a nucleus is stable is the *neutron-to-proton ratio (n/p)*. For stable atoms of elements having low atomic number ($\leq$20), the n/p value is close to 1. As the atomic number increases, the neutron-to-proton ratios of the stable nuclei also increase. This deviation at higher atomic numbers arises because more neutrons are needed to counteract the strong repulsion among the protons and stabilize the nucleus. The following rules are useful in gauging whether or not a particular nucleus is expected to be stable:

1. There are more stable nuclei containing 2, 8, 20, 50, 82, or 126 protons or neutrons than there are containing other numbers of protons or neutrons. For example, there are 10 stable isotopes of tin (Sn) with the atomic number 50 and only 2 stable isotopes of antimony (Sb) with the atomic number 51. The numbers 2, 8, 20, 50, 82, and 126 are called *magic numbers.*

2. There are many more stable nuclei with even numbers of both protons and neutrons than with odd numbers of these particles (Table 20.2).

3. All isotopes of the elements with atomic numbers higher than 83 are radioactive.

4. All isotopes of technetium (Tc, Z = 43) and promethium (Pm, Z = 61) are radioactive.

**Student Annotation:** The significance of these numbers for nuclear stability is similar to the numbers of electrons associated with the very stable noble gases (i.e., 2, 10, 18, 36, 54, and 86 electrons).

**Student Annotation:** Of the two stable isotopes of antimony mentioned in rule 1, both have even numbers of neutrons: $^{121}_{51}\text{Sb}$ and $^{123}_{51}\text{Sb}$.

Figure 20.1 shows a plot of the number of neutrons versus the number of protons in various isotopes. The stable nuclei are located in an area of the graph known as the *belt of stability*. Most radioactive nuclei lie outside this belt. Above the belt of stability, the nuclei have higher neutron-to-proton ratios than those within the belt (for the same number of protons). To lower this ratio (and hence move down toward the belt of stability), these nuclei undergo the following process, called *β-particle emission*:

$$^1_0\text{n} \longrightarrow ^1_1\text{p} + ^{\ \ 0}_{-1}\beta$$

| TABLE 20.2 | Number of Stable Isotopes with Even and Odd Numbers of Protons and Neutrons | |
|---|---|---|
| **Protons** | **Neutrons** | **Number of Stable Isotopes** |
| Odd | Odd | 4 |
| Odd | Even | 50 |
| Even | Odd | 53 |
| Even | Even | 164 |

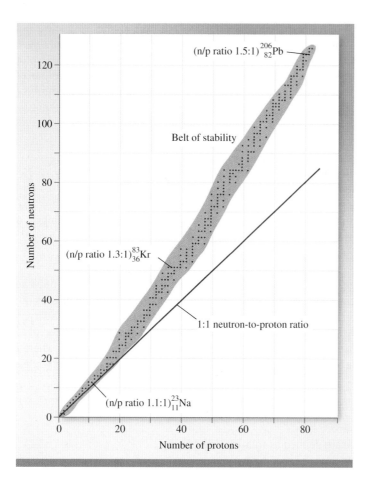

**Figure 20.1**   Plot of neutrons versus protons for various stable isotopes, represented by dots. The straight line represents the points at which the neutron-to-proton ratio is 1. The shaded area represents the belt of stability.

Beta-particle emission leads to an increase in the number of protons in the nucleus and a simultaneous decrease in the number of neutrons. Some examples are

$$\ce{^{14}_{6}C} \longrightarrow \ce{^{14}_{7}N} + \ce{^{0}_{-1}\beta}$$

$$\ce{^{40}_{19}K} \longrightarrow \ce{^{40}_{20}Ca} + \ce{^{0}_{-1}\beta}$$

$$\ce{^{97}_{40}Zr} \longrightarrow \ce{^{97}_{41}Nb} + \ce{^{0}_{-1}\beta}$$

Below the belt of stability, the nuclei have lower neutron-to-proton ratios than those in the belt (for the same number of protons). To increase this ratio (and hence move up toward the belt of stability), these nuclei may emit a positron,

$$\ce{^{1}_{1}p} \longrightarrow \ce{^{1}_{0}n} + \ce{^{0}_{+1}\beta}$$

An example of positron emission is

$$\ce{^{38}_{19}K} \longrightarrow \ce{^{38}_{18}Ar} + \ce{^{0}_{+1}\beta}$$

Alternatively, a nucleus may undergo electron capture,

$$\ce{^{1}_{1}p} + \ce{^{0}_{-1}e} \longrightarrow \ce{^{1}_{0}n}$$

*Electron capture* is the capture of an electron—usually a 1*s* electron—by the nucleus. The captured electron combines with a proton in the nucleus to form a *neutron* so that the atomic number decreases by 1 while the mass number remains the same. Electron capture has the same net effect on the nucleus as positron emission. Examples of electron capture are

$$\ce{^{37}_{18}Ar} + \ce{^{0}_{-1}e} \longrightarrow \ce{^{37}_{17}Cl}$$

$$\ce{^{55}_{26}Fe} + \ce{^{0}_{-1}e} \longrightarrow \ce{^{55}_{25}Mn}$$

## Nuclear Binding Energy

A quantitative measure of nuclear stability is the ***nuclear binding energy,*** which is the energy required to break up a nucleus into its component protons and neutrons. This quantity represents the conversion of mass to energy that occurs during an exothermic nuclear reaction.

The concept of nuclear binding energy evolved from studies of nuclear properties showing that the masses of nuclei are always less than the sum of the masses of the *nucleons,* which is a general term for the protons and neutrons in a nucleus. For example, the $^{19}_9F$ isotope has an atomic mass of 18.99840 amu. The nucleus has 9 protons and 10 neutrons and therefore a total of 19 nucleons. Using the known masses of the $^1_1H$ atom (1.007825 amu) and the neutron (1.008665 amu), we can carry out the following analysis. The mass of 9 $^1_1H$ atoms (i.e., the mass of 9 protons and 9 electrons) is

$$9 \times 1.007825 \text{ amu} = 9.070425 \text{ amu}$$

and the mass of 10 neutrons is

$$10 \times 1.008665 \text{ amu} = 10.08665 \text{ amu}$$

Therefore, the atomic mass of an $^{19}_9F$ atom calculated from the known numbers of electrons, protons, and neutrons is

$$9.070425 \text{ amu} + 10.08665 \text{ amu} = 19.15708 \text{ amu}$$

This value is larger than 18.99840 amu (the measured mass of $^{19}_9F$) by 0.15868 amu.

The difference between the mass of an atom and the sum of the masses of its protons, neutrons, and electrons is called the **mass defect.** According to relativity theory, the loss in mass shows up as energy (heat) given off to the surroundings. Thus, the formation of $^{19}_9F$ is *exothermic.* According to *Einstein's mass-energy equivalence relationship* ($E = mc^2$, where $E$ is energy, $m$ is mass, and $c$ is the velocity of light), we can calculate the amount of energy released. We start by writing

Equation 20.1 $$\Delta E = (\Delta m)c^2$$

where $\Delta E$ and $\Delta m$ are defined as follows:

$$\Delta E = \text{energy of product} - \text{energy of reactants}$$

$$\Delta m = \text{mass of product} - \text{mass of reactants}$$

Thus, the change in mass is

$$\Delta m = 18.99840 \text{ amu} - 19.15708 \text{ amu}$$

$$= -0.15868 \text{ amu}$$

or

$$\Delta m = (-0.15868 \text{ amu})\left(\frac{1 \text{ kg}}{6.0221418 \times 10^{26} \text{ amu}}\right)$$

$$= -2.6349 \times 10^{-28} \text{ kg}$$

Because $^{19}_9F$ has a mass that is less than the mass calculated from the number of electrons and nucleons present, $\Delta m$ is a negative quantity. Consequently, $\Delta E$ is also a negative quantity; that is, energy is released to the surroundings as a result of the formation of the fluorine-19 nucleus. We calculate $\Delta E$ as follows:

$$\Delta E = (-2.6349 \times 10^{-28} \text{ kg})(2.99792458 \times 10^8 \text{ m/s})^2$$

$$= -2.3681 \times 10^{-11} \text{ kg} \cdot \text{m}^2/\text{s}^2$$

$$= -2.3681 \times 10^{-11} \text{ J}$$

This is the amount of energy released when one fluorine-19 nucleus is formed from 9 protons and 10 neutrons. The nuclear binding energy of the nucleus is $2.3682 \times 10^{-11}$ J, which is the amount of energy needed to decompose the nucleus into separate protons and neutrons. In the formation of 1 mole of fluorine nuclei, for instance, the energy released is

$$\Delta E = (-2.3681 \times 10^{-11} \text{ J})(6.0221418 \times 10^{23}/\text{mol})$$

$$= -1.4261 \times 10^{13} \text{ J/mol}$$

$$= -1.4261 \times 10^{10} \text{ kJ/mol}$$

**Figure 20.2** Plot of nuclear binding energy per nucleon versus mass number.

The nuclear binding energy, therefore, is $1.4261 \times 10^{10}$ kJ for 1 mole of fluorine-19 nuclei, which is a tremendously large quantity when we consider that the enthalpies of ordinary chemical reactions are on the order of only 200 kJ. The procedure we have followed can be used to calculate the nuclear binding energy of any nucleus.

As we have noted, nuclear binding energy is an indication of the stability of a nucleus. When comparing the stability of any two nuclei, however, we must account for the fact that they have different numbers of nucleons. It makes more sense, therefore, to compare nuclei using the *nuclear binding energy per nucleon:*

$$\text{nuclear binding energy per nucleon} = \frac{\text{nuclear binding energy}}{\text{number of nucleons}}$$

For the fluorine-19 nucleus,

$$\text{nuclear binding energy per nucleon} = \frac{2.3681 \times 10^{-11} \text{ J}}{19 \text{ nucleons}}$$

$$= 1.2464 \times 10^{-12} \text{ J/nucleon}$$

The nuclear binding energy per nucleon makes it possible to compare the stability of all nuclei on a common basis. Figure 20.2 shows the variation of nuclear binding energy per nucleon plotted against mass number. As you can see, the curve rises rather steeply. The highest binding energies per nucleon belong to elements with intermediate mass numbers—between 40 and 100—and are greatest for elements in the iron, cobalt, and nickel region (the Group 8B elements) of the periodic table. This means that the net attractive forces among the particles (protons and neutrons) are greatest for the nuclei of these elements.

Nuclear binding energy and nuclear binding energy per nucleon are calculated for an iodine nucleus in Sample Problem 20.2.

**Student Annotation:** Note that when we report a nuclear binding energy per mole, we give just the magnitude without the negative sign.

**Student Annotation:** In general, the greater the nuclear binding energy per nucleon, the more stable the nucleus.

## Sample Problem 20.2

The atomic mass of $^{127}_{53}$I is 126.904473 amu. Calculate the nuclear binding energy of this nucleus and the corresponding nuclear binding energy per nucleon.

**Strategy** To calculate the nuclear binding energy, we first determine the difference between the mass of the nucleus and the mass of all the protons and neutrons, which yields the mass defect. Next, we must apply Einstein's mass-energy relationship [$\Delta E = (\Delta m)c^2$].

**Solution** There are 53 protons and 74 neutrons in the iodine nucleus. The mass of 53 $^1_1$H atoms is

$$53 \times 1.007825 \text{ amu} = 53.41473 \text{ amu}$$

and the mass of 74 neutrons is

$$74 \times 1.008665 \text{ amu} = 74.64121 \text{ amu}$$

*(Continued)*

Therefore, the predicted mass for $_{53}^{127}I$ is $53.41473 + 74.64121 = 128.05594$ amu, and the mass defect is

$$\Delta m = 126.904473 \text{ amu} - 128.05594 \text{ amu}$$

$$= -1.15147 \text{ amu}$$

$$= (-1.15147 \text{ amu})\left(\frac{1 \text{ kg}}{6.0221418 \times 10^{26} \text{ amu}}\right)$$

$$= -1.91206 \times 10^{-27} \text{ kg}$$

The energy released is

$$\Delta E = (\Delta m)c^2$$

$$= (-1.91206 \times 10^{-27} \text{ kg})(2.99792458 \times 10^8 \text{ m/s})^2$$

$$= 1.71847 \times 10^{-10} \text{ kg} \cdot \text{m}^2/\text{s}^2$$

$$= 1.71847 \times 10^{-10} \text{ J}$$

Thus the nuclear binding energy is $1.71847 \times 10^{-10}$ J. The nuclear binding energy per nucleon is obtained as follows:

$$\frac{1.71847 \times 10^{-10} \text{ J}}{127 \text{ nucleons}} = 1.35313 \times 10^{-12} \text{ J/nucleon}$$

**Practice Problem A**  Calculate the nuclear binding energy (in joules) and the nuclear binding energy per nucleon of $_{83}^{209}Bi$ (208.980374 amu).

**Practice Problem B**  The nuclear binding energy for $_{79}^{197}Au$ is $1.2683 \times 10^{-12}$ J/nucleon. Determine the mass of a $_{79}^{197}Au$ atom.

---

## Checkpoint 20.2    Nuclear Stability

**20.2.1**  Determine the binding energy per nucleon in a $^{238}U$ nucleus (238.0507847 amu).

a) $2.891 \times 10^{-10}$ J/nucleon

b) $1.215 \times 10^{-12}$ J/nucleon

c) $1.186 \times 10^{-12}$ J/nucleon

d) $2.823 \times 10^{-10}$ J/nucleon

e) $3.212 \times 10^{-24}$ J/nucleon

**20.2.2**  What is the energy associated with a mass defect of 2.000 amu?

a) $3.321 \times 10^{-27}$ J

b) $2.989 \times 10^{-7}$ J

c) $1.800 \times 10^{17}$ J

d) $2.989 \times 10^{-10}$ J

e) $3.309 \times 10^{-45}$ J

**20.2.3**  What is the change in mass (in kg) for the following reaction? $\Delta H°$ for the reaction is $-890.4$ kJ/mol.

a) $1.479 \times 10^{-21}$ kg

b) $1.479 \times 10^{-24}$ kg

c) $1.645 \times 10^{-35}$ kg

d) $4.932 \times 10^{-33}$ kg

e) $9.907 \times 10^{-15}$ kg

**20.2.4**  What type of radioactive decay will the isotopes $^{13}B$ and $^{188}Au$ most likely undergo?

a) Beta emission, positron emission

b) Beta emission, beta emission

c) Positron emission, beta emission

d) Positron emission, positron emission

---

## 20.3  Natural Radioactivity

Nuclei that do not lie within the belt of stability, as well as nuclei with more than 83 protons, tend to be unstable. The spontaneous emission by unstable nuclei of particles or electromagnetic radiation, or both, is known as radioactivity. The main types of radioactivity are the emission of $\alpha$

**Figure 20.3**   Decay series for uranium-238. (Times shown are half-lives.)

particles (doubly charged helium nuclei, $He^{2+}$); the emission of $\beta$ particles (electrons of nuclear origin); the emission of $\gamma$ rays, which are very-short-wavelength (0.1 nm to 104 nm) electromagnetic waves; the emission of positrons; and electron capture.

The disintegration of a radioactive nucleus often is the beginning of a ***radioactive decay series,*** which is a *sequence* of nuclear reactions that ultimately result in the formation of a stable isotope. Figure 20.3 shows the decay series of naturally occurring uranium-238, which involves 14 steps. This decay scheme, known as the *uranium decay series,* also shows the half-lives of all the nuclei involved.

It is important to be able to balance the nuclear reaction for each of the steps in a radioactive decay series. For example, the first step in the uranium decay series is the decay of uranium-238 to thorium-234, with the emission of an $\alpha$ particle. Hence, the reaction is represented by

$$^{238}_{92}U \longrightarrow \, ^{234}_{90}Th + \, ^{4}_{2}\alpha$$

The next step is represented by

$$^{234}_{90}Th \longrightarrow \, ^{234}_{91}Pa + \, ^{0}_{-1}\beta$$

and so on. In a discussion of radioactive decay steps, the beginning radioactive isotope is called the *parent* and the product isotope is called the *daughter.* Thus, $^{238}_{92}U$ is the parent in the first step of the uranium decay series, and $^{234}_{90}Th$ is the daughter.

## Kinetics of Radioactive Decay

All radioactive decays obey first-order kinetics. Therefore, the rate of radioactive decay at any time $t$ is given by

$$\text{rate of decay at time } t = kN$$

where $k$ is the first-order rate constant and $N$ is the number of radioactive nuclei present at time $t$. According to Equation 14.3, the number of radioactive nuclei at time zero ($N_0$) and time $t$ ($N_t$) is

$$\ln \frac{N_t}{N_0} = -kt$$

and the corresponding half-life of the reaction is given by Equation 14.5:

$$t_{1/2} = \frac{0.693}{k}$$

The half-lives and, therefore, the rate constants of radioactive isotopes vary greatly from nucleus to nucleus. Looking at Figure 20.3, for example, we find that $^{238}_{92}U$ and $^{214}_{84}Po$ are two extreme cases:

**Multimedia**

Nuclear Chemistry—radioactive half-life.

$$^{238}_{92}U \longrightarrow {}^{234}_{90}Th + {}^{4}_{2}\alpha \qquad t_{1/2} = 4.51 \times 10^9 \text{ yr}$$

$$^{214}_{84}Po \longrightarrow {}^{210}_{82}Pb + {}^{4}_{2}\alpha \qquad t_{1/2} = 1.6 \times 10^{-4} \text{ s}$$

These two rate constants, after conversion to the same time unit, differ by many orders of magnitude. Furthermore, the rate constants are unaffected by changes in environmental conditions such as temperature and pressure. These highly unusual features are not seen in ordinary chemical reactions (see Table 20.1).

## Dating Based on Radioactive Decay

The half-lives of radioactive isotopes have been used as "atomic clocks" to determine the ages of certain objects. Some examples of dating by radioactive decay measurements will be described here.

The carbon-14 isotope is produced when atmospheric nitrogen is bombarded by cosmic rays:

$$^{14}_{7}N + {}^{1}_{0}n \longrightarrow {}^{14}_{6}C + {}^{1}_{1}H$$

The radioactive carbon-14 isotope decays according to the equation

$$^{14}_{6}C \longrightarrow {}^{14}_{7}N + {}^{0}_{-1}\beta$$

This reaction is the basis of radiocarbon or "carbon-14" dating. To determine the age of an object, we measure the *activity* (disintegrations per second) of $^{14}C$ and compare it to the activity of $^{14}C$ in living matter.

Sample Problem 20.3 shows how to use radiocarbon dating to determine the age of an artifact.

---

### Sample Problem 20.3

A wooden artifact is found to have a $^{14}C$ activity of 9.1 disintegrations per second. Given that the $^{14}C$ activity of an equal mass of fresh-cut wood has a constant value of 15.2 disintegrations per second, determine the age of the artifact. The half-life of carbon-14 is 5715 years.

**Strategy** The activity of a radioactive sample is proportional to the number of radioactive nuclei. Thus, we can use Equation 14.3 with activity in place of concentration:

$$\ln \frac{{}^{14}C \text{ activity in artifact}}{{}^{14}C \text{ activity in fresh-cut wood}} = -kt$$

To determine $k$, though, we must solve Equation 14.5, using the value of $t_{1/2}$ for carbon-14 (5715 years) given in the problem statement.

**Setup** Solving Equation 14.5 for $k$ gives

$$k = \frac{0.693}{5715 \text{ yr}} = 1.21 \times 10^{-4} \text{ yr}^{-1}$$

**Solution**

$$\ln \frac{9.1 \text{ disintegrations per second}}{15.2 \text{ disintegrations per second}} = -1.21 \times 10^{-4} \text{ yr}^{-1} (t)$$

$$t = \frac{-0.513}{-1.21 \times 10^{-4} \text{ yr}^{-1}} = 4240 \text{ yr}$$

Therefore, the age of the artifact is $4.2 \times 10^3$ years.

**Think About It** Carbon dating cannot be used for objects older than about 60,000 years (about 10 half-lives). After that much time has passed, the activity of carbon-14 has fallen to a level too low to be measured reliably.

**Practice Problem A** A piece of linen cloth found at an ancient burial site is found to have a $^{14}C$ activity of 4.8 disintegrations per minute. Determine the age of the cloth. Assume that the carbon-14 activity of an equal mass of living flax (the plant from which linen is made) is 14.8 disintegrations per minute.

**Practice Problem B** What would be the $^{14}C$ activity in a 2500-year-old wooden object? Assume that the $^{14}C$ activity of an equal mass of fresh-cut wood is 13.9 disintegrations per second.

Because some of the intermediate products in the uranium decay series have very long half-lives (see Figure 20.3), this series is particularly suitable for estimating the age of rocks found on Earth and of extraterrestrial objects. The half-life for the first step ($^{238}_{92}U$ to $^{234}_{90}Th$) is $4.51 \times 10^9$ years. This is about 20,000 times the second largest value (i.e., $2.47 \times 10^5$ years), which is the half-life for $^{234}_{92}U$ to $^{230}_{90}Th$. As a good approximation, therefore, we can assume that the half-life for the *overall* process (i.e., from $^{238}_{92}U$ to $^{206}_{82}Pb$) is equal to the half-life of the first step:

$$^{238}_{92}U \longrightarrow {}^{206}_{82}Pb + 8\,{}^4_2\alpha + 6\,{}^0_{-1}\beta \qquad t_{1/2} = 4.51 \times 10^9 \text{ yr}$$

In naturally occurring uranium minerals, we should and do find some lead-206 formed by radioactive decay. Assuming that no lead was present when the mineral was formed and that the mineral has not undergone chemical changes that would allow the lead-206 isotope to be separated from the parent uranium-238, it is possible to estimate the age of the rocks from the mass ratio of $^{206}_{82}Pb$ to $^{238}_{92}U$. According to the preceding nuclear equation, 1 mol (206 g) of lead is formed for every 1 mol (238 g) of uranium that undergoes complete decay. If only half a mole of uranium-238 has undergone decay, the mass ratio $^{206}Pb/^{238}U$ becomes

$$\frac{206 \text{ g}/2}{238 \text{ g}/2} = 0.866$$

and the process would have taken a half-life of $4.51 \times 10^9$ years to complete (Figure 20.4). Ratios lower than 0.866 mean that the rocks are less than $4.51 \times 10^9$ years old, and higher ratios suggest a greater age. Interestingly, studies based on the uranium series, as well as other decay series, put the age of the oldest rocks and, therefore, probably the age of Earth itself, at $4.5 \times 10^9$, or 4.5 billion, years.

One of the most important dating techniques in geochemistry is based on the radioactive decay of potassium-40. Radioactive potassium-40 decays by several different modes, but the one relevant for dating is that of electron capture:

$$^{40}_{19}K + {}^0_{-1}e \longrightarrow {}^{40}_{18}Ar \qquad t_{1/2} = 1.2 \times 10^9 \text{ yr}$$

The accumulation of gaseous argon-40 is used to gauge the age of a specimen. When a potassium-40 atom in a mineral decays, argon-40 is trapped in the lattice of the mineral and can escape only if the material is melted. Melting, therefore, is the procedure for analyzing a mineral sample in the laboratory. The amount of argon-40 present can be conveniently measured with a mass spectrometer (see page 48). Knowing the ratio of argon-40 to potassium-40 in the mineral and the half-life of decay makes it possible to establish the ages of rocks ranging from millions to billions of years old.

Sample Problem 20.4 shows how to use radioisotopes to determine the age of a specimen.

**Figure 20.4** After one half-life, half of the original uranium-238 has been converted to lead-206.

**Sample Problem 20.4**

A rock is found to contain 5.51 mg of $^{238}U$ and 1.63 mg of $^{206}Pb$. Determine the age of the rock ($t_{1/2}$ of $^{238}U = 4.51 \times 10^9$ yr).

*(Continued)*

**Strategy** We must first determine what mass of $^{238}$U decayed to produce the measured amount of $^{206}$Pb and then use it to determine the original mass of $^{238}$U. Knowing the initial and final masses of $^{238}$U, we can use Equation 14.3 to solve for $t$.

**Setup**

$$1.63 \text{ mg } ^{206}\text{Pb} \times \frac{238 \text{ mg } ^{238}\text{U}}{206 \text{ mg } ^{206}\text{Pb}} = 1.88 \text{ mg } ^{238}\text{U}$$

Thus, the original mass of $^{238}$U was 5.51 mg + 1.88 mg = 7.39 mg. The rate constant, $k$, is determined using Equation 14.5 and $t_{1/2}$ for $^{238}$U:

$$k = \frac{0.693}{4.51 \times 10^9} = 1.54 \times 10^{-10} \text{ yr}^{-1}$$

**Solution**

$$\ln \frac{5.51 \text{ mg}}{7.39 \text{ mg}} = -1.54 \times 10^{-10} \text{ yr}^{-1} \, (t)$$

$$t = \frac{-0.294}{-1.54 \times 10^{-10} \text{ yr}^{-1}} = 1.91 \times 10^9 \text{ yr}$$

The rock is 1.9 billion years old.

> **Think About It** This is slightly more complicated than the radiocarbon problem. We cannot use the measured masses of the two isotopes in Equation 14.3 because they are masses of different elements.

**Practice Problem A** Determine the age of a rock that contains 12.75 mg of $^{238}$U and 1.19 mg of $^{206}$Pb.

**Practice Problem B** How much $^{206}$Pb will be in a rock sample that is $1.3 \times 10^8$ years old and that contains 3.25 mg of $^{238}$U?

---

## Checkpoint 20.3　　**Natural Radioactivity**

**20.3.1** Determine the age of a rock found to contain 4.31 mg $^{238}$U and 2.47 mg of $^{206}$Pb.

a) $7.19 \times 10^9$ years

b) $3.62 \times 10^9$ years

c) $1.18 \times 10^8$ years

d) $3.30 \times 10^9$ years

e) $1.78 \times 10^8$ years

**20.3.2** Determine the $^{14}$C activity in disintegrations per second (dps) of a wooden artifact that is 23,000 years old. (Assume that fresh-cut wood of equal mass has an activity of 15.2 dps.)

a) 16.3 dps

b) 15.2 dps

c) 0.935 dps

d) 11.5 dps

e) 4.31 dps

**20.3.3** Iron-59 decays to cobalt via beta emission with a half-life of 45.1 days. From the diagram, determine how many half-lives have elapsed since the sample was pure iron-59.

a) 2

b) 3

c) 0.25

d) 8

e) 14

● = Cobalt　　● = Iron-59

**20.3.4** Isotope X decays to isotope Y with a half-life of 45 days. Which diagram most closely represents the sample of X after 105 days?

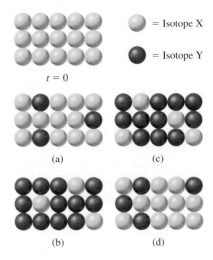

○ = Isotope X

● = Isotope Y

$t = 0$

(a)　　(c)

(b)　　(d)

## 20.4 Nuclear Transmutation

The scope of nuclear chemistry would be rather narrow if study were limited to natural radioactive elements. An experiment performed by Rutherford in 1919, however, suggested the possibility of producing radioactivity artificially. When he bombarded a sample of nitrogen with $\alpha$ particles, the following reaction took place:

$$^{14}_{7}\text{N} + ^{4}_{2}\alpha \longrightarrow ^{17}_{8}\text{O} + ^{1}_{1}\text{p}$$

An oxygen-17 isotope was produced with the emission of a proton. This reaction demonstrated for the first time the feasibility of converting one element into another, by the process of nuclear transmutation. Nuclear transmutation differs from radioactive decay in that transmutation is brought about by the *collision* of two particles.

The preceding reaction can be abbreviated as $^{14}_{7}\text{N}(\alpha,\text{p})\,^{17}_{8}\text{O}$. In the parentheses the bombarding particle is written first, followed by the emitted particle.

Sample Problem 20.5 shows how to use this notation to represent nuclear transmutations.

---

### Sample Problem 20.5

Write the balanced nuclear equation for the reaction represented by $^{56}_{26}\text{Fe}(\text{d},\alpha)\,^{54}_{25}\text{Mn}$, where d represents a deuterium nucleus.

**Strategy** The species written first is a reactant. The species written last is a product. Within the parentheses, the bombarding particle (a reactant) is written first, followed by the emitted particle (a product).

**Setup** The bombarding and emitted particles are represented by $^{2}_{1}\text{H}$ and $^{4}_{2}\alpha$, respectively.

**Solution**

$$^{56}_{26}\text{Fe} + ^{2}_{1}\text{H} \longrightarrow ^{54}_{25}\text{Mn} + ^{4}_{2}\alpha$$

**Think About It** Check your work by summing the mass numbers and the atomic numbers on both sides of the equation.

**Practice Problem A** Write an equation for the process represented by $^{106}_{46}\text{Pd}(\alpha,\text{p})\,^{109}_{47}\text{Ag}$.

**Practice Problem B** Write the abbreviated form of the following process:

$$^{33}_{17}\text{Cl} + ^{1}_{0}\text{n} \longrightarrow ^{31}_{15}\text{P} + ^{3}_{2}\text{He}$$

---

Particle accelerators made it possible to synthesize the so-called *transuranium elements,* elements with atomic numbers greater than 92. Neptunium ($Z = 93$) was first prepared in 1940. Since then, 24 other transuranium elements have been synthesized. All isotopes of these elements are radioactive. Table 20.3 lists the transuranium elements that have been reported and some of the reactions by which they have been produced.

Although light elements are generally not radioactive, they can be made so by bombarding their nuclei with appropriate particles. As we saw in the previous section, the radioactive carbon-14 isotope can be prepared by bombarding nitrogen-14 with neutrons. Tritium ($^{3}_{1}\text{H}$) is prepared according to the following bombardment:

$$^{6}_{3}\text{Li} + ^{1}_{0}\text{n} \longrightarrow ^{3}_{1}\text{H} + ^{4}_{2}\alpha$$

Tritium decays with the emission of $\beta$ particles:

$$^{3}_{1}\text{H} \longrightarrow ^{3}_{2}\text{He} + ^{0}_{-1}\beta \qquad t_{1/2} = 12.5 \text{ yr}$$

Many synthetic isotopes are prepared by using neutrons as projectiles. This approach is particularly convenient because neutrons carry no charges and therefore are not repelled by the targets—the nuclei. In contrast, when the projectiles are positively charged particles (e.g., protons or $\alpha$ particles), they must have considerable kinetic energy to overcome the electrostatic repulsion between themselves and the target nuclei. The synthesis of phosphorus from aluminum is one example:

$$^{27}_{13}\text{Al} + ^{4}_{2}\alpha \longrightarrow ^{30}_{15}\text{P} + ^{1}_{0}\text{n}$$

| TABLE 20.3 | Preparation of the Transuranium Elements | | |
|---|---|---|---|
| **Atomic Number** | **Name** | **Symbol** | **Preparation*** |
| 93 | Neptunium | Np | $^{238}_{92}U + ^{1}_{0}n \longrightarrow ^{239}_{93}Np + ^{0}_{-1}\beta$ |
| 94 | Plutonium | Pu | $^{239}_{93}Np \longrightarrow ^{239}_{94}Pu + ^{0}_{-1}\beta$ |
| 95 | Americium | Am | $^{239}_{94}Pu + ^{1}_{0}n \longrightarrow ^{240}_{95}Am + ^{0}_{-1}\beta$ |
| 96 | Curium | Cm | $^{239}_{94}Pu + ^{4}_{2}\alpha \longrightarrow ^{242}_{96}Cm + ^{1}_{0}n$ |
| 97 | Berkelium | Bk | $^{241}_{95}Am + ^{4}_{2}\alpha \longrightarrow ^{243}_{97}Bk + 2^{1}_{0}n$ |
| 98 | Californium | Cf | $^{242}_{96}Cm + ^{4}_{2}\alpha \longrightarrow ^{245}_{98}Cf + ^{1}_{0}n$ |
| 99 | Einsteinium | Es | $^{238}_{92}U + 15^{1}_{0}n \longrightarrow ^{253}_{99}Es + 7^{0}_{-1}\beta$ |
| 100 | Fermium | Fm | $^{238}_{92}U + 17^{1}_{0}n \longrightarrow ^{255}_{100}Fm + 8^{0}_{-1}\beta$ |
| 101 | Mendelevium | Md | $^{253}_{99}Es + ^{4}_{2}\alpha \longrightarrow ^{256}_{101}Md + ^{1}_{0}n$ |
| 102 | Nobelium | No | $^{246}_{96}Cm + ^{12}_{6}C \longrightarrow ^{254}_{102}No + 4^{1}_{0}n$ |
| 103 | Lawrencium | Lr | $^{252}_{98}Cf + ^{10}_{5}B \longrightarrow ^{257}_{103}Lr + 5^{1}_{0}n$ |
| 104 | Rutherfordium | Rf | $^{249}_{98}Cf + ^{12}_{6}C \longrightarrow ^{257}_{104}Rf + 4^{1}_{0}n$ |
| 105 | Dubnium | Db | $^{249}_{98}Cf + ^{15}_{7}N \longrightarrow ^{260}_{105}Db + 4^{1}_{0}n$ |
| 106 | Seaborgium | Sg | $^{249}_{98}Cf + ^{18}_{8}O \longrightarrow ^{263}_{106}Sg + 4^{1}_{0}n$ |
| 107 | Bohrium | Bh | $^{209}_{83}Bi + ^{54}_{24}Cr \longrightarrow ^{262}_{107}Bh + ^{1}_{0}n$ |
| 108 | Hassium | Hs | $^{208}_{82}Pb + ^{58}_{26}Fe \longrightarrow ^{265}_{108}Hs + ^{1}_{0}n$ |
| 109 | Meitnerium | Mt | $^{209}_{83}Bi + ^{58}_{26}Fe \longrightarrow ^{266}_{109}Mt + ^{1}_{0}n$ |
| 110 | Darmstadtium | Ds | $^{208}_{82}Pb + ^{62}_{28}Ni \longrightarrow ^{269}_{110}Ds + ^{1}_{0}n$ |
| 111 | Roentgenium | Rg | $^{209}_{83}Bi + ^{64}_{28}Ni \longrightarrow ^{272}_{111}Rg + ^{1}_{0}n$ |
| 112 | Ununbium | Uub | $^{208}_{82}Pb + ^{70}_{30}Zn \longrightarrow ^{277}_{112}Uub + ^{1}_{0}n$ |
| 113 | Ununtrium | Uut | $^{288}_{115}Uup \longrightarrow ^{284}_{113}Uut + ^{4}_{2}\alpha$ |
| 114 | Ununquadium | Uuq | $^{244}_{94}Pu + ^{48}_{20}Ca \longrightarrow ^{289}_{114}Uuq + 3^{1}_{0}n$ |
| 115 | Ununpentium | Uup | $^{243}_{95}Am + ^{48}_{20}Ca \longrightarrow ^{288}_{115}Uup + 3^{1}_{0}n$ |
| 116 | Ununhexium | Uuh | $^{248}_{96}Cm + ^{48}_{20}Ca \longrightarrow ^{292}_{116}Uuh + 4^{1}_{0}n$ |
| 118 | Ununoctium | Uuo | $^{249}_{98}Cf + ^{48}_{20}Ca \longrightarrow ^{294}_{118}Uuo + 3^{1}_{0}n$ |

*Some of the transuranium elements have been prepared by more than one method.

**Figure 20.5**    Schematic diagram of a cyclotron particle accelerator. The particle (an ion) to be accelerated starts at the center and is forced to move in a spiral path through the influence of electric and magnetic fields until it emerges at a high velocity. The magnetic fields are perpendicular to the plane of the dees (so-called because of their shape), which are hollow and serve as electrodes.

**Figure 20.6** Section of a particle accelerator.

A *particle accelerator* uses electric and magnetic fields to increase the kinetic energy of charged species so that a reaction will occur (Figure 20.5). Alternating the polarity (i.e., + and −) on specially constructed plates causes the particles to accelerate along a spiral path. When they have the energy necessary to initiate the desired nuclear reaction, they are guided out of the accelerator into a collision with a target substance.

Various designs have been developed for particle accelerators, one of which accelerates particles along a linear path of about 3 km (Figure 20.6). It is now possible to accelerate particles to a speed well above 90 percent of the speed of light. (According to Einstein's theory of relativity, it is impossible for a particle to move *at* the speed of light. The only exception is the photon, which has a zero rest mass.) The extremely energetic particles produced in accelerators are employed by physicists to smash atomic nuclei to fragments. Studying the debris from such disintegrations provides valuable information about nuclear structure and binding forces.

---

**Checkpoint 20.4** **Nuclear Transmutation**

**20.4.1** Identify the balanced nuclear equation for the reaction represented by $^{98}_{42}\text{Mo}(\text{d,n})^{99}_{43}\text{Tc}$.

a) $^{98}_{42}\text{Mo} + ^1_1\text{H} \longrightarrow ^{99}_{43}\text{Tc}$

b) $^{98}_{42}\text{Mo} \longrightarrow ^{99}_{43}\text{Tc} + ^2_1\text{H} + ^1_0\text{n}$

c) $^{98}_{42}\text{Mo} + ^1_0\text{n} \longrightarrow ^{99}_{43}\text{Tc} + ^2_1\text{H}$

d) $^{98}_{42}\text{Mo} + ^2_1\text{H} + ^1_0\text{n} \longrightarrow ^{99}_{43}\text{Tc}$

e) $^{98}_{42}\text{Mo} + ^2_1\text{H} \longrightarrow ^{99}_{43}\text{Tc} + ^1_0\text{n}$

**20.4.2** Identify the correct abbreviated form of the equation

$$^{125}_{53}\text{I} + ^{\ 0}_{-1}\beta \longrightarrow ^{125}_{52}\text{Te} + \gamma$$

a) $^{125}_{53}\text{I}(\gamma,\beta)^{125}_{52}\text{Te}$

b) $^{125}_{53}\text{I}(\beta,\gamma)^{125}_{52}\text{Te}$

c) $^{125}_{52}\text{Te}(\beta,\gamma)^{125}_{53}\text{I}$

d) $^{125}_{52}\text{I}(\beta,\gamma)^{125}_{53}\text{Te}$

e) $^{125}_{52}\text{I}(\gamma,\beta)^{125}_{53}\text{Te}$

---

# 20.5 Nuclear Fission

*Nuclear fission* is the process in which a heavy nucleus (mass number > 200) divides to form smaller nuclei of intermediate mass and one or more neutrons. Because the heavy nucleus is less stable than its products (see Figure 20.2), this process releases a large amount of energy.

The first nuclear fission reaction to be studied was that of uranium-235 bombarded with slow neutrons, whose speed is comparable to that of air molecules at room temperature. Under these conditions, uranium-235 undergoes fission, as shown in Figure 20.7. Actually, this reaction is very complex: more than 30 different elements have been found among the fission products (Figure 20.8). A representative reaction is

$$^{235}_{92}\text{U} + ^1_0\text{n} \longrightarrow ^{90}_{38}\text{Sr} + ^{143}_{54}\text{Xe} + 3^1_0\text{n}$$

Media Player/MPEG Animation: Figure 20.7, Nuclear Fission and Fusion, pp. 874–875.

Figure 20.7

# Nuclear Fission and Fusion

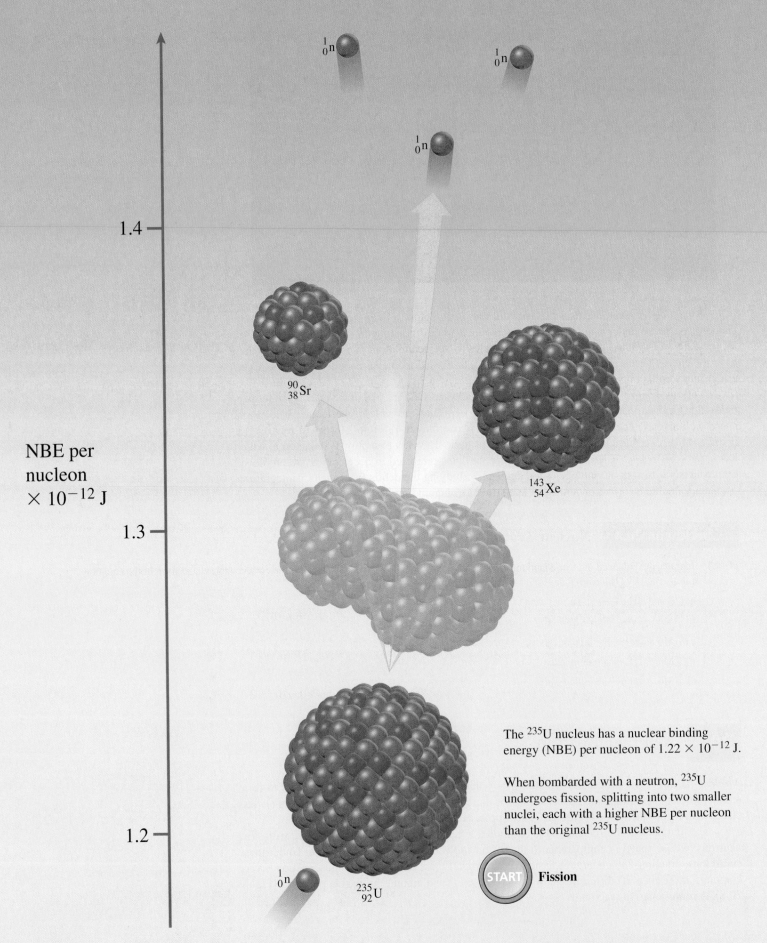

NBE per nucleon $\times 10^{-12}$ J

${}^{1}_{0}n$

${}^{1}_{0}n$

${}^{1}_{0}n$

1.4 —

${}^{90}_{38}Sr$

${}^{143}_{54}Xe$

1.3 —

The ${}^{235}U$ nucleus has a nuclear binding energy (NBE) per nucleon of $1.22 \times 10^{-12}$ J.

When bombarded with a neutron, ${}^{235}U$ undergoes fission, splitting into two smaller nuclei, each with a higher NBE per nucleon than the original ${}^{235}U$ nucleus.

1.2 —

${}^{1}_{0}n$

${}^{235}_{92}U$

START  **Fission**

START Fusion

The $^2$H and $^3$H nuclei have the following nuclear binding energies per nucleon:

$$^2\text{H}: 0.185 \times 10^{-12} \text{ J}$$
$$^3\text{H}: 0.451 \times 10^{-12} \text{ J}$$

At very high temperatures, the $^2$H and $^3$H nuclei undergo fusion to produce a $^4$He nucleus and a neutron. The $^4$He nucleus has a significantly higher NBE per nucleon: $1.13 \times 10^{-12}$ J

## What's the point?

Large nuclei, such as $^{235}$U, can achieve greater nuclear stability by splitting into smaller nuclei with greater NBE per nucleon. Small nuclei achieve stability by undergoing fusion to produce a larger nucleus with a greater NBE per nucleon. Note that different scales are used to show the change in NBE per nucleon for the two processes. There is a much greater change in NBE per nucleon in the fusion process than in the fission process. As with chemical reactions, nuclear reactions are favored when the products are more stable than the reactants.

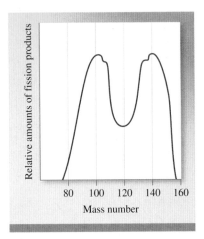

**Figure 20.8**    Relative yields of the products resulting from the fission of $^{235}U$ as a function of mass number.

Multimedia

Nuclear chemistry—nuclear chain reaction.

Multimedia

Nuclear chemistry—nuclear fission.

| TABLE 20.4 | Nuclear Binding Energies of $^{235}U$ and Its Fission Products |
|---|---|

| | **Nuclear Binding Energy** |
|---|---|
| $^{235}U$ | $2.82 \times 10^{-10}$ J |
| $^{90}Sr$ | $1.23 \times 10^{-10}$ J |
| $^{143}Xe$ | $1.92 \times 10^{-10}$ J |

Although many heavy nuclei can be made to undergo fission, only the fission of naturally occurring uranium-235 and of the artificial isotope plutonium-239 has any practical importance. Table 20.4 lists the nuclear binding energies of uranium-235 and its fission products. As the table shows, the binding energy per nucleon for uranium-235 is less than the sum of the binding energies for strontium-90 and xenon-143. Therefore, when a uranium-235 nucleus is split into two smaller nuclei, a certain amount of energy is released.

Let's estimate the magnitude of this energy. The difference between the binding energies of the reactants and products is $(1.23 \times 10^{-10} + 1.92 \times 10^{-10})$ J $-$ $(2.82 \times 10^{-10})$ J, or $3.3 \times 10^{-11}$ J per uranium-235 nucleus. For 1 mole of uranium-235, the energy released would be $(3.3 \times 10^{-11})(6.02 \times 10^{23})$, or $2.0 \times 10^{13}$ J. This is an *extremely* exothermic reaction, considering that the heat of combustion of 1 ton of coal is only about $5 \times 10^7$ J.

The significant feature of uranium-235 fission is not just the enormous amount of energy released, but the fact that more neutrons are produced than are originally captured in the process. This property makes possible a **nuclear chain reaction,** which is a self-sustaining sequence of nuclear fission reactions. The neutrons generated during the initial stages of fission can induce fission in other uranium-235 nuclei, which in turn produce more neutrons, and so on. In less than a second, the reaction can become uncontrollable, liberating a tremendous amount of heat to the surroundings. Figure 20.9 shows two types of fission reactions. For a chain reaction to occur, enough uranium-235 must be present in the sample to capture the neutrons. Otherwise, many of the neutrons will escape from the sample and the chain reaction will not occur. In this situation the mass of the sample is said to be *subcritical*. Figure 20.9 shows what happens when the amount of the fissionable material is equal to or greater than the **critical mass,** the minimum mass of fissionable material required to generate a self-sustaining nuclear chain reaction. In this case, most of the neutrons will be captured by uranium-235 nuclei, and a chain reaction will occur.

The first application of nuclear fission was in the development of the atomic bomb. How is such a bomb made and detonated? The crucial factor in the bomb's design is the determination of the critical mass for the bomb. A small atomic bomb is equivalent to 20,000 tons of TNT (trinitrotoluene). Because 1 ton of TNT releases about $4 \times 10^9$ J of energy, 20,000 tons would produce $8 \times 10^{13}$ J. Recall that 1 mole, or 235 g, of uranium-235 liberates $2.0 \times 10^{13}$ J of energy when it undergoes fission. Thus, the mass of the isotope present in a small bomb must be at least

$$(235 \text{ g}) \left( \frac{8 \times 10^{13} \text{ J}}{2.0 \times 10^{13} \text{ J}} \right) \approx 1 \text{ kg}$$

An atomic bomb is never assembled with the critical mass already present. Instead, the critical mass is formed by using a conventional explosive, such as TNT, to force the fissionable sections together, as shown in Figure 20.10. Neutrons from a source at the center of the device trigger the nuclear chain reaction. Uranium-235 was the fissionable material in the bomb dropped

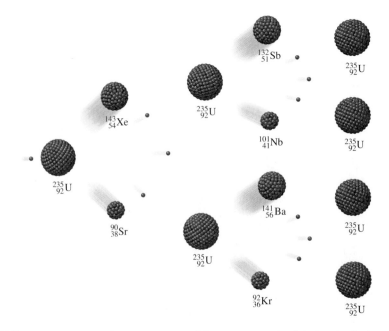

**Figure 20.9**    If a critical mass is present, many of the neutrons emitted during the fission process will be captured by other $^{235}U$ nuclei and a chain reaction will occur.

on Hiroshima, Japan, on August 6, 1945. Plutonium-239 was used in the bomb exploded over Nagasaki 3 days later. The fission reactions generated were similar in these two cases, as was the extent of the destruction.

A peaceful but controversial application of nuclear fission is the generation of electricity using heat from a controlled chain reaction in a nuclear reactor. Currently, nuclear reactors provide about 20 percent of the electric energy in the United States. This is a small but by no means negligible contribution to the nation's energy production. Several different types of nuclear reactors are in operation; we will briefly discuss the main features of three of them, along with their advantages and disadvantages.

Most of the nuclear reactors in the United States are *light water reactors.* Figure 20.11 is a schematic diagram of such a reactor, and Figure 20.12 shows the refueling process in the core of a nuclear reactor.

**Figure 20.10** Schematic diagram of an atomic bomb. The TNT explosives are set off first. The explosion forces the sections of fissionable material together to form an amount considerably larger than the critical mass.

**Figure 20.11** Schematic diagram of a nuclear fission reactor. The fission process is controlled by cadmium or boron rods. The heat generated by the process is used to produce steam for the generation of electricity via a heat exchange system.

**Figure 20.12** Refueling the core of a nuclear reactor.

Multimedia

Nuclear chemistry—operation of a nuclear power plant.

**Figure 20.13**    Uranium oxide ($U_3O_8$).

An important aspect of the fission process is the speed of the neutrons. Slow neutrons split uranium-235 nuclei more efficiently than do fast ones. Because fission reactions are highly exothermic, the neutrons produced usually move at high velocities. For greater efficiency, they must be slowed down before they can be used to induce nuclear disintegration. To accomplish this goal, scientists use **moderators,** which are substances that can reduce the kinetic energy of neutrons. A good moderator must satisfy several requirements: It should be nontoxic and inexpensive (as very large quantities of it are necessary), and it should resist conversion into a radioactive substance by neutron bombardment. Furthermore, it is advantageous for the moderator to be a fluid so that it can also be used as a coolant. No substance fulfills all these requirements, although water comes closer than many others that have been considered. Nuclear reactors that use light water ($H_2O$) as a moderator are called *light* water reactors because $_1^1H$ is the lightest isotope of the element hydrogen.

The nuclear fuel consists of uranium, usually in the form of its oxide, $U_3O_8$ (Figure 20.13). Naturally occurring uranium contains about 0.7 percent of the uranium-235 isotope, which is too low a concentration to sustain a small-scale chain reaction. For effective operation of a light water reactor, uranium-235 must be enriched to a concentration of 3 or 4 percent. In principle, the main difference between an atomic bomb and a nuclear reactor is that the chain reaction that takes place in a nuclear reactor is kept under control at all times. The factor limiting the rate of the reaction is the number of neutrons present. This can be controlled by lowering cadmium or boron *control rods* between the fuel elements. These rods capture neutrons according to the equations

$$^{113}_{48}\text{Cd} + ^1_0\text{n} \longrightarrow ^{114}_{48}\text{Cd} + \gamma$$

$$^{10}_{5}\text{B} + ^1_0\text{n} \longrightarrow ^7_3\text{Li} + ^4_2\alpha$$

where $\gamma$ denotes gamma rays. Without the control rods, the reactor core would melt from the heat generated and release radioactive materials into the environment. Nuclear reactors have rather elaborate cooling systems that absorb the heat given off by the nuclear reaction and transfer it outside the reactor core, where it is used to produce enough steam to drive an electric generator. In this respect, a nuclear power plant is similar to a conventional power plant that burns fossil fuel. In both cases, large quantities of cooling water are needed to condense steam for reuse. Thus, most nuclear power plants are built near a river or a lake. Unfortunately, this method of cooling causes thermal pollution.

Another type of nuclear reactor uses $D_2O$, or *heavy* water, as the moderator, rather than $H_2O$. Deuterium absorbs neutrons much less efficiently than does ordinary hydrogen. Because fewer neutrons are absorbed, the reactor is more efficient and does not require enriched uranium. More neutrons leak out of the reactor, too, though this is not a serious disadvantage.

The main advantage of a heavy water reactor is that it eliminates the need for building expensive uranium enrichment facilities. However, $D_2O$ must be prepared by either fractional distillation or electrolysis of ordinary water, which can be very expensive considering the amount of water used in a nuclear reactor. In countries where hydroelectric power is abundant, the cost of producing $D_2O$ by electrolysis can be reasonably low. At present, Canada is the only nation successfully using heavy water nuclear reactors. The fact that no enriched uranium is required in a heavy water reactor allows a country to enjoy the benefits of nuclear power without undertaking work that is closely associated with weapons technology.

A **breeder reactor** uses uranium fuel, but unlike a conventional nuclear reactor, it produces more fissionable materials than it uses.

When uranium-238 is bombarded with fast neutrons, the following reactions take place:

$$^{238}_{92}\text{U} + ^1_0\text{n} \longrightarrow ^{239}_{92}\text{U}$$

$$^{239}_{92}\text{U} \longrightarrow ^{239}_{93}\text{Np} + ^{\ 0}_{-1}\beta \qquad t_{1/2} = 23.4 \text{ min}$$

$$^{239}_{93}\text{Np} \longrightarrow ^{239}_{94}\text{Pu} + ^{\ 0}_{-1}\beta \qquad t_{1/2} = 2.35 \text{ days}$$

In this manner, the nonfissionable uranium-238 is transmuted into the fissionable isotope plutonium-239 (Figure 20.14).

In a typical breeder reactor, nuclear fuel containing uranium-235 or plutonium-239 is mixed with uranium-238 so that breeding takes place within the core. For every uranium-235 (or plutonium-239) nucleus undergoing fission, more than one neutron is captured by uranium-238 to generate plutonium-239. Thus, the stockpile of fissionable material can be steadily increased as the starting nuclear fuels are consumed. It takes about 7 to 10 years to regenerate the sizable amount of material needed to refuel the original reactor and to fuel another reactor of comparable size. This interval is called the *doubling time.*

**Figure 20.14**    Radioactive plutonium oxide ($PuO_2$) has a red glow.

Another *fertile* isotope is $^{232}_{90}\text{Th}$. Upon capturing slow neutrons, thorium is transmuted to uranium-233, which, like uranium-235, is a fissionable isotope:

$$^{232}_{90}\text{Th} + ^{1}_{0}\text{n} \longrightarrow ^{233}_{90}\text{Th}$$

$$^{233}_{90}\text{Th} \longrightarrow ^{233}_{91}\text{Pa} + ^{0}_{-1}\beta \qquad t_{1/2} = 22 \text{ min}$$

$$^{233}_{91}\text{Pa} \longrightarrow ^{233}_{92}\text{U} + ^{0}_{-1}\beta \qquad t_{1/2} = 27.4 \text{ days}$$

Uranium-233 ($t_{1/2} = 1.6 \times 10^5$ years) is stable enough for long-term storage.

Although the amounts of uranium-238 and thorium-232 in Earth's crust are relatively plentiful (4 ppm and 12 ppm by mass, respectively), the development of breeder reactors has been very slow. To date, the United States does not have a single operating breeder reactor, and only a few have been built in other countries, such as France and Russia. One problem is economics; breeder reactors are more expensive to build than conventional reactors. There are also more technical difficulties associated with the construction of such reactors. As a result, the future of breeder reactors, in the United States at least, is rather uncertain.

Many people, including environmentalists, regard nuclear fission as a highly undesirable method of energy production. Many fission products such as strontium-90 are dangerous radioactive isotopes with long half-lives. Plutonium-239, used as a nuclear fuel and produced in breeder reactors, is one of the most toxic substances known. It is an $\alpha$-emitter with a half-life of 24,400 years.

Accidents, too, present many dangers. An accident at the Three Mile Island reactor in Pennsylvania in 1979 first brought the potential hazards of nuclear plants to public attention. In this instance, very little radiation escaped the reactor, but the plant remained closed for more than a decade while repairs were made and safety issues addressed. Only a few years later, on April 26, 1986, a reactor at the Chernobyl nuclear plant in Ukraine surged out of control. The fire and explosion that followed released much radioactive material into the environment. People working near the plant died within weeks as a result of the exposure to the intense radiation. The long-term effect of the radioactive fallout from this incident has not yet been clearly assessed, although agriculture and dairy farming were affected by the fallout. The number of potential cancer deaths attributable to the radiation contamination is estimated to be between a few thousand and more than 100,000.

In addition to the risk of accidents, the problem of radioactive waste disposal has not been satisfactorily resolved even for safely operated nuclear plants. Many suggestions have been made as to where to store or dispose of nuclear waste, including burial underground, burial beneath the ocean floor, and storage in deep geologic formations. But none of these sites has proved absolutely safe in the long run. Leakage of radioactive wastes into underground water, for example, can endanger nearby communities. The ideal disposal site would seem to be the sun, where a bit more radiation would make little difference, but this kind of operation requires space technology that is 100 percent reliable.

Because of the hazards, the future of nuclear reactors is clouded. What was once hailed as the ultimate solution to our energy needs in the twenty-first century is now being debated and questioned by both the scientific community and the general public. It seems likely that the controversy will continue for some time.

## 20.6 Nuclear Fusion

In contrast to the nuclear fission process, ***nuclear fusion,*** the combining of small nuclei into larger ones, is largely exempt from the waste disposal problem.

Figure 20.2 showed that for the lightest elements, nuclear stability increases with increasing mass number. This behavior suggests that if two light nuclei combine or fuse together to form a larger, more stable nucleus, an appreciable amount of energy will be released in the process. This is the basis for ongoing research into the harnessing of nuclear fusion for the production of energy.

Nuclear fusion occurs constantly in the sun. The sun is made up mostly of hydrogen and helium. In its interior, where temperatures reach about 15 million degrees Celsius, the following fusion reactions are believed to take place:

$$^{1}_{1}\text{H} + ^{2}_{1}\text{H} \longrightarrow ^{3}_{2}\text{He}$$

$$^{3}_{2}\text{He} + ^{3}_{2}\text{He} \longrightarrow ^{4}_{2}\text{He} + 2^{1}_{1}\text{H}$$

$$^{1}_{1}\text{H} + ^{1}_{1}\text{H} \longrightarrow ^{2}_{1}\text{H} + ^{0}_{-1}\beta$$

Because fusion reactions take place only at very high temperatures, they are often called ***thermonuclear reactions.***

A major concern in choosing the proper nuclear fusion process for energy production is the temperature necessary to carry out the process. Some promising reactions are listed here:

| Reaction | Energy Released |
|---|---|
| $^2_1H + ^2_1H \longrightarrow ^3_1H + ^1_1H$ | $6.3 \times 10^{-13}$ J |
| $^2_1H + ^3_1H \longrightarrow ^4_2He + 2^1_0n$ | $2.8 \times 10^{-12}$ J |
| $^6_3Li + ^2_1H \longrightarrow 2^4_2He$ | $3.6 \times 10^{-12}$ J |

These reactions must take place at extremely high temperatures, on the order of 100 million degrees Celsius, to overcome the repulsive forces between the nuclei. The first reaction is particularly attractive because the world's supply of deuterium is virtually inexhaustible. The total volume of water on Earth is about $1.5 \times 10^{21}$ L. Because the natural abundance of deuterium is 0.015 percent, the total amount of deuterium present is roughly $4.5 \times 10^{21}$ g, or $5.0 \times 10^{15}$ tons. Although it is expensive to prepare deuterium, the cost is minimal compared to the value of the energy released by the reaction.

In contrast to the fission process, nuclear fusion looks like a very promising energy source, at least on paper. Although thermal pollution would be a problem, fusion has the following advantages: (1) the fuels are cheap and almost inexhaustible and (2) the process produces little radioactive waste. If a fusion machine were turned off, it would shut down completely and instantly, without any danger of a meltdown.

If nuclear fusion is so great, why isn't there even one fusion reactor producing energy? Although we possess the scientific knowledge to design such a reactor, the technical difficulties have not yet been solved. The basic problem is finding a way to hold the nuclei together long enough, and at the appropriate temperature, for fusion to occur. At temperatures of about 100 million degrees Celsius, molecules cannot exist, and most or all of the atoms are stripped of their electrons. This state of matter, a gaseous mixture of positive ions and electrons, is called *plasma*. The problem of containing this plasma is a formidable one. No solid container can exist at such temperatures, unless the amount of plasma is small, but then the solid surface would immediately cool the sample and quench the fusion reaction. One approach to solving this problem is to use *magnetic confinement*. Because plasma consists of charged particles moving at high speeds, a magnetic field will exert a force on it. As Figure 20.15 shows, the plasma moves through a doughnut-shaped tunnel, confined by a complex magnetic field. Thus, the plasma never comes in contact with the walls of the container.

Another promising design employs high-power lasers to initiate the fusion reaction. In test runs, a number of laser beams transfer energy to a small fuel pellet, heating it and causing it to *implode*—that is, to collapse inward from all sides and compress into a small volume (Figure 20.16). Consequently, fusion occurs. Like the magnetic confinement approach, laser fusion presents a number of technical difficulties that still need to be overcome before it can be put to practical use on a large scale.

The technical problems inherent in the design of a nuclear fusion reactor do not affect the production of a *hydrogen bomb*, also called a *thermonuclear* bomb. In this case, the objective is all power and no control. Hydrogen bombs do not contain gaseous hydrogen or gaseous deuterium; they contain solid lithium deuteride (LiD), which can be packed very tightly. The detonation of a hydrogen bomb occurs in two stages—first a fission reaction and then a fusion reaction. The required temperature for fusion is achieved with an atomic bomb. Immediately after the

**Figure 20.15** A magnetic plasma confinement design called a tokamak.

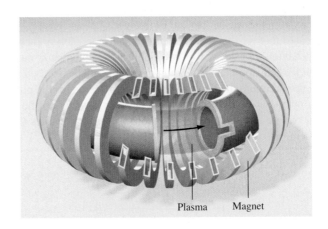

Plasma    Magnet

atomic bomb explodes, the following fusion reactions occur, releasing vast amounts of energy (Figure 20.17):

$$^6_3\text{Li} + ^2_1\text{H} \longrightarrow 2\,^4_2\alpha$$

$$^2_1\text{H} + ^2_1\text{H} \longrightarrow ^3_1\text{H} + ^1_1\text{H}$$

There is no critical mass in a fusion bomb, and the force of the explosion is limited only by the quantity of reactants present. Thermonuclear bombs are described as being "cleaner" than atomic bombs because the only radioactive isotopes they produce are tritium, which is a weak $\beta$-particle emitter ($t_{1/2} = 12.5$ years), and the products of the fission starter. Their damaging effects on the environment can be aggravated, however, by incorporating in the construction some nonfissionable material such as cobalt. Upon bombardment by neutrons, cobalt-59 is converted to cobalt-60, which is a very strong $\gamma$-ray emitter with a half-life of 5.2 years. The presence of radioactive cobalt isotopes in the debris or fallout from a thermonuclear explosion would be fatal to those who survived the initial blast.

**Figure 20.16** This small-scale fusion reaction was carried out at the Lawrence Livermore National Laboratory using one of the world's most powerful lasers, Nova.

## 20.7 Uses of Isotopes

Radioactive and stable isotopes alike have many applications in science and medicine. We have previously described the use of isotopes in the study of reaction mechanisms [◄◄ Section 14.5] and in dating artifacts (page 868). In this section we will discuss a few more examples.

### Chemical Analysis

The formula of the thiosulfate ion is $S_2O_3^{2-}$. For some years, chemists were uncertain as to whether the two sulfur atoms occupied equivalent positions in the ion. The thiosulfate ion is prepared by treating the sulfite ion with elemental sulfur:

$$SO_3^{2-}(aq) + S(s) \longrightarrow S_2O_3^{2-}(aq)$$

When thiosulfate is treated with dilute acid, the reaction is reversed. The sulfite ion is re-formed, and elemental sulfur precipitates:

$$S_2O_3^{2-}(aq) \xrightarrow{\text{H}^+} SO_3^{2-}(aq) + S(s)$$

If this sequence is started with elemental sulfur enriched with the radioactive sulfur-35 isotope, the isotope acts as a "label" for S atoms. All the labels are found in the sulfur precipitate; none of them appears in the final sulfite ions. As a result, the two atoms of sulfur in $S_2O_3^{2-}$ are not structurally equivalent, as would be the case if the structure were

$$\left[\ddot{\text{O}}-\ddot{\text{S}}-\ddot{\text{O}}-\ddot{\text{S}}-\ddot{\text{O}}\!:\right]^{2-}$$

If the sulfur atoms were equivalent, the radioactive isotope would be present in both the elemental sulfur precipitate and the sulfite ion. Based on spectroscopic studies, we now know that the structure of the thiosulfate ion is

$$\left[\begin{array}{c}\ddot{\text{S}}\\\|\\:\ddot{\text{O}}-\text{S}-\ddot{\text{O}}:\\\|\\.\ddot{\text{O}}.\end{array}\right]^{2-}$$

**Figure 20.17** Explosion of a thermonuclear bomb.

**Multimedia**

Nuclear chemistry—nuclear medical techniques.

The study of photosynthesis is also rich with isotope applications. The overall photosynthesis reaction can be represented as

$$6CO_2 + 6H_2O \longrightarrow C_6H_{12}O_6 + 6O_2$$

In Section 14.5 we learned that the $^{18}O$ isotope was used to determine the source of $O_2$. The radioactive $^{14}C$ isotope helped to determine the path of carbon in photosynthesis. Starting with $^{14}CO_2$, it was possible to isolate the intermediate products during photosynthesis and measure the amount of radioactivity of each carbon-containing compound. In this manner the path from $CO_2$ through various intermediate compounds to carbohydrate could be clearly charted. Isotopes, especially radioactive isotopes that are used to trace the path of the atoms of an element in a chemical or biological process, are called *tracers*.

**Figure 20.18**    $^{123}$I image of a normal brain (top) and the brain of an Alzheimer's victim (bottom).

## Isotopes in Medicine

Tracers are also used for diagnosis in medicine. Sodium-24 (a $\beta$-emitter with a half-life of 14.8 h) injected into the bloodstream as a salt solution can be monitored to trace the flow of blood and detect possible constrictions or obstructions in the circulatory system. Iodine-131 (a $\beta$-emitter with a half-life of 8 days) has been used to test the activity of the thyroid gland. A malfunctioning thyroid can be detected by giving the patient a drink of a solution containing a known amount of Na$^{131}$I and measuring the radioactivity just above the thyroid to see if the iodine is absorbed at the normal rate. Another radioactive isotope of iodine, iodine-123 (a $\gamma$-ray emitter), is used to image the brain (Figure 20.18). In each of these cases, though, the amount of radioisotope used must be kept small to prevent the patient from suffering permanent damage from the high-energy radiation.

Technetium, the first artificially prepared element, is one of the most useful elements in nuclear medicine. Although technetium is a transition metal, all its isotopes are radioactive. In the laboratory it is prepared by the nuclear reactions

$$^{98}_{42}\text{Mo} + ^{1}_{0}\text{n} \longrightarrow ^{99}_{42}\text{Mo}$$

$$^{99}_{42}\text{Mo} \longrightarrow ^{99m}_{43}\text{Tc} + ^{0}_{-1}\beta$$

where the superscript "m" denotes that the technetium-99 isotope is produced in its *excited* nuclear state. This isotope has a half-life of about 6 hours, decaying by $\gamma$ radiation to technetium-99 in its nuclear *ground* state. Thus, it is a valuable diagnostic tool. The patient either drinks or is injected with a solution containing $^{99m}$Tc. By detecting the $\gamma$ rays emitted by $^{99m}$Tc, doctors can obtain images of organs such as the heart, liver, and lungs.

A major advantage of using radioactive isotopes as tracers is that they are easy to detect. Their presence even in very small amounts can be detected by photographic techniques or by devices known as counters. Figure 20.19 is a diagram of a Geiger counter, an instrument widely used in scientific work and medical laboratories to detect radiation.

## 20.8    Biological Effects of Radiation

In this section we will examine briefly the effects of radiation on biological systems. But first we must define the quantitative measures of radiation. The fundamental unit of radioactivity is the *curie* (Ci); 1 Ci corresponds to exactly $3.70 \times 10^{10}$ nuclear disintegrations per second. This decay rate is equivalent to that of 1 g of radium. A *millicurie* (mCi) is one-thousandth of a curie. Thus, 10 mCi of a carbon-14 sample is the quantity that undergoes $(10 \times 10^{-3})(3.70 \times 10^{10}) = 3.70 \times 10^{8}$ disintegrations per second.

The intensity of radiation depends on the number of disintegrations as well as on the energy and type of radiation emitted. One common unit for the absorbed dose of radiation is the *rad* (radiation *a*bsorbed *d*ose), which is the amount of radiation that results in the absorption of $1 \times 10^{-5}$ J per gram of irradiated material. The biological effect of radiation depends on the part of the body

**Figure 20.19**    Schematic diagram of a Geiger counter. Radiation ($\alpha$, $\beta$, or $\gamma$ rays) entering through the window ionizes the argon gas to generate a small current flow between the electrodes. This current is amplified and is used to flash a light or operate a counter with a clicking sound.

irradiated and the type of radiation. For this reason, the rad is often multiplied by a factor called the *RBE* (*r*elative *b*iological *e*ffectiveness). The product is called a *rem* (*r*oentgen *e*quivalent for *m*an):

$$\text{number of rems} = \text{number of rads} \times 1 \text{ RBE}$$

**Multimedia**

Nuclear Chemistry—alpha, beta, and gamma emission (interactive).

Of the three types of nuclear radiation, $\alpha$ particles usually have the least-penetrating power. Beta particles are more penetrating than $\alpha$ particles, but less so than $\gamma$ rays.

Gamma rays have very short wavelengths and high energies. Furthermore, because they carry no charge, they cannot be stopped by shielding materials as easily as $\alpha$ and $\beta$ particles. If $\alpha$- or $\beta$-emitters are ingested or inhaled, however, their damaging effects are greatly aggravated because the organs will be constantly subject to damaging radiation at close range. For example, strontium-90, a $\beta$-emitter, can replace calcium in bones, where it does the greatest damage.

Table 20.5 lists the average amounts of radiation an American receives every year. For short-term exposures to radiation, a dosage of 50 to 200 rems will cause a decrease in white blood cell counts and other complications, while a dosage of 500 rems or greater may result in death within weeks. Current safety standards permit nuclear workers to be exposed to no more than 5 rems per year and specify a maximum of 0.5 rem of human-made radiation per year for the general public.

The chemical basis of radiation damage is that of ionizing radiation. Radiation (of either particles or $\gamma$ rays) can remove electrons from atoms and molecules in its path, leading to the formation of ions and radicals. **Radicals** (also called *free radicals*) are molecular fragments having one or more unpaired electrons; they are usually short lived and highly reactive. When water is irradiated with $\gamma$ rays, for example, the following reactions take place:

$$H_2O \xrightarrow{\text{radiation}} H_2O^+ + e^-$$

$$H_2O^+ + H_2O \longrightarrow H_3O^+ + \cdot OH$$
$$\text{hydroxyl radical}$$

The electron (in the hydrated form) can subsequently react with water or with a hydrogen ion to form atomic hydrogen, and with oxygen to produce the superoxide ion ($O_2^-$) (a radical):

$$e^- + O_2 \longrightarrow \cdot O_2^-$$

In the tissues the superoxide ions and other free radicals attack cell membranes and a host of organic compounds, such as enzymes and DNA molecules. Organic compounds can themselves be directly ionized and destroyed by high-energy radiation.

It has long been known that exposure to high-energy radiation can induce cancer in humans and other animals. Cancer is characterized by uncontrolled cellular growth. On the other hand, it is also well established that cancer cells can be destroyed by proper radiation treatment. In radiation therapy, a compromise is sought. The radiation to which the patient is exposed must be sufficient to destroy cancer cells without killing too many normal cells and, it is hoped, without inducing another form of cancer.

Radiation damage to living systems is generally classified as *somatic* or *genetic*. Somatic injuries are those that affect the organism during its own lifetime. Sunburn, skin rash, cancer, and cataracts are examples of somatic damage. Genetic damage means inheritable changes or gene mutations. For example, a person whose chromosomes have been damaged or altered by radiation may have deformed offspring.

| **TABLE 20.5** | Average Yearly Radiation Doses for Americans | |
|---|---|
| **Source** | **Dose (mrem/yr)\*** |
| Cosmic rays | 20–50 |
| Ground and surroundings | 25 |
| Human body[†] | 26 |
| Medical and dental X rays | 50–75 |
| Air travel | 5 |
| Fallout from weapons tests | 5 |
| Nuclear waste | 2 |
| Total | 133–188 |

\*1 mrem = millirem = $1 \times 10^{-3}$ rem.

[†]The radioactivity in the body comes from food and air.

## Bringing Chemistry to Life

### Radioactivity in Tobacco

"SURGEON GENERAL'S WARNING: Smoking Is Hazardous to Your Health." Warning labels such as this appear on every package of cigarettes sold in the United States. The link between cigarette smoke and cancer has long been established. There is, however, another cancer-causing mechanism in smokers. The culprit in this case is a radioactive environmental pollutant present in the tobacco leaves from which cigarettes are made.

The soil in which tobacco is grown is heavily treated with phosphate fertilizers, which are rich in uranium and its decay products. Consider a particularly important step in the uranium-238 decay series:

$$^{226}_{88}\text{Ra} \longrightarrow \, ^{222}_{86}\text{Rn} + \, ^{4}_{2}\alpha$$

The product formed, radon-222, is an unreactive gas. (Radon is the only gaseous species in the uranium-238 decay series.) Radon-222 emanates from radium-226 and is present at high concentrations in soil gas and in the surface air layer under the vegetation canopy provided by the field of growing tobacco. In this layer some of the daughters of radon-222, such as polonium-218 and lead-214, become firmly attached to the surface and interior of tobacco leaves. As Figure 20.3 shows, the next few decay reactions leading to the formation of lead-210 proceed rapidly. Gradually, the concentration of radioactive lead-210 can build to quite a high level.

During the combustion of a cigarette, tiny insoluble smoke particles are inhaled and deposited in the respiratory tract of the smoker and are eventually transported and stored at sites in the liver, spleen, and bone marrow. Measurements indicate a high lead-210 content in these particles. The lead-210 content is not high enough to be hazardous *chemically* (it is insufficient to cause *lead poisoning*), but it is hazardous because it is radioactive. Because of its long half-life (20.4 years), lead-210 and its radioactive daughters—bismuth-210 and polonium-210—continue to build up in the body of a smoker over the years. Constant exposure of the organs and bone marrow to $\alpha$- and $\beta$-particle radiation increases the probability the smoker will develop cancer. The overall impact on health is similar to that caused by indoor radon gas.

# Applying What You've Learned

In addition to BNCT, another promising treatment for brain tumors is brachytherapy using iodine-125. In brachytherapy, "seeds" containing $^{125}I$ are implanted directly into the tumor. As the radioisotope decays, $\gamma$ rays destroy the tumor cells. Careful implantation prevents the radiation from harming nearby healthy cells.

**Problems:**

a) $^{125}I$ is produced by a two-step process in which $^{124}Xe$ nuclei are bombarded with neutrons to produce $^{125}Xe$—a process called *neutron activation*. $^{125}Xe$ then decays by electron capture to produce $^{125}I$, which also decays by electron capture. Write nuclear equations for the two steps that produce $^{125}I$ from $^{124}Xe$, and identify the product of the electron-capture decay of $^{125}I$. [◄◄ Sample Problem 20.1]

b) The mass of an $^{125}I$ nucleus is 124.904624 amu. Calculate the nuclear binding energy and the nuclear binding energy per nucleon. [◄◄ Sample Problem 20.2]

c) The half-life of $^{125}I$ is 59.4 days. How long will it take for the activity of implanted $^{125}I$ seeds to fall to 5.00 percent of its original value? [◄◄ Sample Problem 20.3]

d) Iridium-192 is another isotope used in brachytherapy. It is produced by a nuclear transmutation. Identify the target nucleus X, and write the balanced nuclear equation for the reaction represented by $^{191}X(n,\gamma)^{192}Ir$. [◄◄ Sample Problem 20.5]

Brachytherapy seeds (shown with a penny to illustrate their size).

# CHAPTER SUMMARY

## Section 20.1

- Spontaneous emission of particles or radiation from unstable nuclei is known as *radioactivity.* Unstable nuclei emit $\alpha$ particles, $\beta$ particles, *positrons,* or $\gamma$ rays.

- *Nuclear transmutation* is the conversion of one nucleus to another. Nuclear reactions are balanced by summing the mass numbers and the atomic numbers.

## Section 20.2

- Stable nuclei with low atomic numbers have neutron-to-proton ratios close to 1. Heavier stable nuclei have higher ratios. Nuclear stability is favored by certain numbers of nucleons including even numbers and "magic" numbers.

- The difference between the actual mass of a nucleus and the mass calculated by summing the masses of the individual nucleons is the *mass defect.*

- *Nuclear binding energy,* determined by using Einstein's equation $E = mc^2$, is a measure of nuclear stability.

## Section 20.3

- Uranium-238 is the parent of a natural *radioactive decay series* that can be used to determine the ages of rocks. Radiocarbon dating is done using carbon-14.

## Section 20.4

- *Transuranium elements* are created by bombarding other elements with accelerated neutrons, protons, $\alpha$ particles, or other nuclei.

## Section 20.5

- *Nuclear fission* is the splitting of a large nucleus into two smaller nuclei and one or more neutrons. When the free neutrons are captured efficiently by other nuclei, a *nuclear chain reaction* can occur in which the fission process is sustained. The minimum amount of fissionable material required to sustain the reaction is known as the *critical mass.*

- Nuclear reactors use the heat from a controlled nuclear fission reaction to produce power. Fission is controlled, in part, by *moderators*— materials that limit the speed of liberated neutrons but that do not themselves undergo fission when bombarded with neutrons. The three important types of reactors are light water reactors, heavy water reactors, and *breeder reactors.* Breeder reactors produce more fissionable material than they consume.

## Section 20.6

- *Nuclear fusion,* the type of reaction that occurs in the sun, is the combination of two light nuclei to form one heavier nucleus. Fusion reactions are sometimes referred to as *thermonuclear reactions* because they take place only at very high temperatures.

## Section 20.7

- Radioactive isotopes are easy to detect and thus make excellent *tracers* in chemical reactions and in medical procedures.

## Section 20.8

- High-energy radiation damages living systems by causing ionization and the formation of *radicals,* or *free radicals,* which are chemical species with unpaired electrons.

# KEY WORDS

| | | | |
|---|---|---|---|
| Breeder reactor, 878 | Nuclear binding energy, 863 | Nuclear transmutation, 860 | Radioactivity, 860 |
| Critical mass, 876 | Nuclear chain reaction, 876 | Positron, 860 | Thermonuclear reaction, 880 |
| Mass defect, 864 | Nuclear fission, 873 | Radical, 883 | Tracers, 881 |
| Moderator, 878 | Nuclear fusion, 879 | Radioactive decay series, 867 | Transuranium elements, 871 |

# KEY EQUATION

$$20.1 \quad \Delta E = (\Delta m)c^2$$

# QUESTIONS AND PROBLEMS

## Section 20.1: Nuclei and Nuclear Reactions

### Review Questions

20.1 How do nuclear reactions differ from ordinary chemical reactions?

20.2 What are the steps in balancing nuclear equations?

20.3 What is the difference between $_{-1}^{0}e$ and $_{-1}^{0}\beta$?

20.4 What is the difference between an electron and a positron?

### Problems

20.5 Complete the following nuclear equations, and identify X in each case:

(a) $_{12}^{26}Mg + _{1}^{1}p \longrightarrow \alpha + X$
(b) $_{27}^{59}Co + _{1}^{2}H \longrightarrow _{27}^{60}Co + X$
(c) $_{92}^{235}U + _{0}^{1}n \longrightarrow _{36}^{94}Kr + _{56}^{139}Ba + 3X$
(d) $_{24}^{53}Cr + _{2}^{4}\alpha \longrightarrow _{0}^{1}n + X$
(e) $_{8}^{20}O \longrightarrow _{9}^{20}F + X$

20.6 Complete the following nuclear equations, and identify X in each case:

(a) $_{53}^{135}I \longrightarrow _{54}^{135}Xe + X$
(b) $_{19}^{40}K \longrightarrow _{-1}^{0}\beta + X$
(c) $_{27}^{59}Co + _{0}^{1}n \longrightarrow _{25}^{56}Mn + X$
(d) $_{92}^{235}U + _{0}^{1}n \longrightarrow _{40}^{99}Zr + _{52}^{135}Te + 2X$

## Section 20.2: Nuclear Stability

### Review Questions

20.7 State the general rules for predicting nuclear stability.

20.8 What is the belt of stability?

20.9 Why is it impossible for the isotope $_{2}^{2}He$ to exist?

20.10 Define *nuclear binding energy, mass defect,* and *nucleon.*

20.11 How does Einstein's equation, $E = mc^2$, enable us to calculate nuclear binding energy?

20.12 Why is it preferable to use nuclear binding energy per nucleon for a comparison of the stabilities of different nuclei?

### Problems

20.13 The radius of a uranium-235 nucleus is about $7.0 \times 10^{-3}$ pm. Calculate the density of the nucleus in $g/cm^3$. (Assume the atomic mass is 235 amu.)

20.14 For each pair of isotopes listed, predict which one is less stable: (a) $_{3}^{6}Li$ or $_{3}^{9}Li$, (b) $_{11}^{23}Na$ or $_{11}^{25}Na$, (c) $_{20}^{48}Ca$ or $_{21}^{48}Sc$.

20.15 For each pair of elements listed, predict which one has more stable isotopes: (a) Co or Ni, (b) F or Se, (c) Ag or Cd.

20.16 In each pair of isotopes shown, indicate which one you would expect to be radioactive: (a) $_{10}^{20}Ne$ or $_{10}^{17}Ne$, (b) $_{20}^{40}Ca$ or $_{20}^{45}Ca$, (c) $_{42}^{95}Mo$ or $_{43}^{92}Tc$, (d) $_{80}^{195}Hg$ or $_{80}^{196}Hg$, (e) $_{83}^{209}Bi$ or $_{96}^{242}Cm$.

20.17 Given that

$$H(g) + H(g) \longrightarrow H_2(g) \qquad \Delta H° = -436.4 \text{ kJ/mol}$$

calculate the change in mass (in kg) per mole of $H_2$ formed.

20.18 Estimates show that the total energy output of the sun is $5 \times 10^{26}$ J/s. What is the corresponding mass loss in kg/s of the sun?

20.19 Calculate the nuclear binding energy (in joules) and the binding energy per nucleon of the following isotopes: (a) $_{3}^{7}Li$ (7.01600 amu) and (b) $_{17}^{35}Cl$ (34.96885 amu).

20.20 Calculate the nuclear binding energy (in joules) and the binding energy per nucleon of the following isotopes: (a) $_{2}^{4}He$ (4.002603 amu) and (b) $_{74}^{184}W$ (183.950928 amu).

20.21 Given that the nuclear binding energy of $^{48}Cr$ is $1.37340 \times 10^{-12}$ J/nucleon, calculate the mass of a single $^{48}Cr$ atom.

20.22 Given that the nuclear binding energy of $^{192}Ir$ is $1.27198 \times 10^{-12}$ J/nucleon, calculate the mass of a single $^{192}Ir$ atom.

## Section 20.3: Natural Radioactivity

### Review Questions

20.23 Discuss factors that lead to nuclear decay.

20.24 Outline the principle for dating materials using radioactive isotopes.

### Problems

20.25 Fill in the blanks in the following radioactive decay series:

(a) $^{232}Th \xrightarrow{\alpha} \underline{\quad} \xrightarrow{\beta} \underline{\quad} \xrightarrow{\beta} {}^{228}Th$
(b) $^{235}U \xrightarrow{\alpha} \underline{\quad} \xrightarrow{\beta} \underline{\quad} \xrightarrow{\alpha} {}^{227}Ac$
(c) $\underline{\quad} \xrightarrow{\alpha} {}^{233}Pa \xrightarrow{\beta} \underline{\quad} \xrightarrow{\alpha} \underline{\quad}$

20.26 A radioactive substance undergoes decay as follows:

| Time (days) | Mass (g) |
|---|---|
| 0 | 500 |
| 1 | 389 |
| 2 | 303 |
| 3 | 236 |
| 4 | 184 |
| 5 | 143 |
| 6 | 112 |

Calculate the first-order decay constant and the half-life of the reaction.

20.27 The radioactive decay of Tl-206 to Pb-206 has a half-life of 4.20 min. Starting with $5.00 \times 10^{22}$ atoms of Tl-206, calculate the number of such atoms left after 42.0 min.

20.28 A freshly isolated sample of $^{90}Y$ was found to have an activity of $9.8 \times 10^5$ disintegrations per minute at 1:00 P.M. on December 3, 2006. At 2:15 P.M. on December 17, 2006, its activity was measured again and found to be $2.6 \times 10^4$ disintegrations per minute. Calculate the half-life of $^{90}Y$.

**20.29** A wooden artifact has a $^{14}C$ activity of 18.9 disintegrations per minute, compared to 27.5 disintegrations per minute for live wood. Given that the half-life of $^{14}C$ is 5715 years, determine the age of the artifact.

**20.30** In the thorium decay series, thorium-232 loses a total of six $\alpha$ particles and four $\beta$ particles in a 10-stage process. What is the final isotope produced?

**20.31** Consider the decay series A $\longrightarrow$ B $\longrightarrow$ C $\longrightarrow$ D where A, B, and C are radioactive isotopes with half-lives of 4.50 s, 15.0 days, and 1.00 s, respectively, and D is nonradioactive. Starting with 1.00 mole of A, and none of B, C, or D, calculate the number of moles of A, B, C, and D left after 30 days.

**20.32** The activity of radioactive carbon-14 decay of a piece of charcoal found at a volcanic site is 11.2 disintegrations per second. If the activity of carbon-14 decay in an equal mass of living matter is 18.3 disintegrations per second, what is the age of the charcoal? (See Problem 20.29 for the half-life of carbon-14.)

**20.33** The age of some animal bones was determined by carbon-14 dating to be $8.4 \times 10^3$ years old. Calculate the activity of the carbon-14 in the bones in disintegrations per minute per gram, given that the original activity was 15.3 disintegrations per minute per gram. (See Problem 20.29 for the half-life of carbon-14.)

**20.34** Given that the half-life of $^{238}U$ is $4.51 \times 10^9$ years, determine the age of a rock found to contain 1.09 mg $^{238}U$ and 0.08 mg $^{206}Pb$.

**20.35** Determine the ratio of $^{238}U$ to $^{206}Pb$ in a rock that is $1.7 \times 10^8$ years old. (See Problem 20.34 for the half-life of $^{238}U$.)

## Section 20.4: Nuclear Transmutation

### Review Questions

**20.36** What is the difference between radioactive decay and nuclear transmutation?

**20.37** How is nuclear transmutation achieved in practice?

### Problems

**20.38** Write balanced nuclear equations for the following reactions, and identify X: (a) $X(p,\alpha)^{12}_{6}C$, (b) $^{27}_{13}Al(d,\alpha)X$, (c) $^{55}_{25}Mn(n,\gamma)X$.

**20.39** Write the abbreviated forms for the following reactions:
(a) $^{14}_{7}N + ^{4}_{2}\alpha \longrightarrow ^{17}_{8}O + ^{1}_{1}p$
(b) $^{9}_{4}Be + ^{4}_{2}\alpha \longrightarrow ^{12}_{6}C + ^{1}_{0}n$
(c) $^{238}_{92}U + ^{2}_{1}H \longrightarrow ^{238}_{93}Np + 2^{1}_{0}n$

**20.40** Write balanced nuclear equations for the following reactions, and identify X: (a) $^{80}_{34}Se(d,p)X$, (b) $X(d,2p)^{9}_{3}Li$, (c) $^{10}_{5}B(n,\alpha)X$.

**20.41** Write the abbreviated forms for the following reactions:
(a) $^{40}_{20}Ca + ^{2}_{1}H \longrightarrow ^{41}_{20}Ca + ^{1}_{1}p$
(b) $^{32}_{16}S + ^{1}_{0}n \longrightarrow ^{32}_{15}P + ^{1}_{1}p$
(c) $^{239}_{94}Pu + ^{4}_{2}\alpha \longrightarrow ^{242}_{96}Cm + ^{1}_{0}n$

**20.42** Describe how you would prepare astatine-211, starting with bismuth-209.

**20.43** A long-cherished dream of alchemists was to produce gold from cheaper and more abundant elements. This dream was finally realized when $^{198}_{80}Hg$ was converted into gold by neutron bombardment. Write a balanced equation for this reaction.

## Section 20.5: Nuclear Fission

### Visualizing Chemistry
#### Figure 20.7

**VC 20.1** The fission of $^{235}U$ can result in a variety of products, including those shown in Figure 20.7. Which of the following equations does *not* represent another possible fission process?
a) $^{1}_{0}n + ^{235}_{92}U \longrightarrow ^{137}_{52}Te + ^{97}_{40}Zr + 2^{1}_{0}n$
b) $^{1}_{0}n + ^{235}_{92}U \longrightarrow ^{142}_{56}Ba + ^{91}_{36}Kr + 3^{1}_{0}n$
c) $^{1}_{0}n + ^{235}_{92}U \longrightarrow ^{137}_{55}Cs + ^{90}_{37}Rb + 3^{1}_{0}n$

**VC 20.2** How many neutrons are produced in the fission reaction shown?
$$^{1}_{0}n + ^{239}_{94}Pu \longrightarrow ^{109}_{44}Ru + ^{129}_{50}Sn + \underline{\quad}^{1}_{0}n$$
a) 1
b) 2
c) 3

**VC 20.3** The fission of $^{235}U$ shown in Figure 20.7 is represented by the equation
$$^{1}_{0}n + ^{235}_{92}U \longrightarrow ^{143}_{54}Xe + ^{90}_{38}Sr + 3^{1}_{0}n$$
How does the combined mass of products compare to the combined mass of reactants for this process?
a) The combined mass of products is smaller than the combined mass of reactants.
b) The combined mass of products is larger than the combined mass of reactants.
c) The combined mass of products is equal to the combined mass of reactants.

**VC 20.4** The fusion of $^{2}_{1}H$ and $^{3}_{1}H$ shown in Figure 20.7 is represented by the equation
$$^{2}_{1}H + ^{3}_{1}H \longrightarrow ^{4}_{2}He + ^{1}_{0}n$$
How does the combined mass of products compare to the combined mass of reactants for this process?
a) The combined mass of products is smaller than the combined mass of reactants.
b) The combined mass of products is larger than the combined mass of reactants.
c) The combined mass of products is equal to the combined mass of reactants.

### Review Questions

**20.44** Define *nuclear fission, nuclear chain reaction,* and *critical mass.*

**20.45** Which isotopes can undergo nuclear fission?

**20.46** Explain how an atomic bomb works.

**20.47** Explain the functions of a moderator and a control rod in a nuclear reactor.

**20.48** Discuss the differences between a light water and a heavy water nuclear fission reactor. What are the advantages of a breeder reactor over a conventional nuclear fission reactor?

**20.49** No form of energy production is without risk. Make a list of the risks to society involved in fueling and operating a conventional coal-fired electric power plant, and compare them with the risks of fueling and operating a nuclear fission-powered electric plant.

## Section 20.6: Nuclear Fusion

*Review Questions*

20.50  Define *nuclear fusion, thermonuclear reaction,* and *plasma.*

20.51  Why do heavy elements such as uranium undergo fission whereas light elements such as hydrogen and lithium undergo fusion?

20.52  How does a hydrogen bomb work?

20.53  What are the advantages of a fusion reactor over a fission reactor? What are the practical difficulties in operating a large-scale fusion reactor?

## Section 20.7: Uses of Isotopes

*Problems*

20.54  Describe how you would use a radioactive iodine isotope to demonstrate that the following process is in dynamic equilibrium:

$$PbI_2(s) \rightleftharpoons Pb^{2+}(aq) + 2I^-(aq)$$

20.55  Consider the following redox reaction:

$$IO_4^-(aq) + 2I^-(aq) + H_2O(l) \longrightarrow$$
$$I_2(s) + IO_3^-(aq) + 2OH^-(aq)$$

When $KIO_4$ is added to a solution containing iodide ions labeled with radioactive iodine-128, all the radioactivity appears in $I_2$ and none in the $IO_3^-$ ion. What can you deduce about the mechanism for the redox process?

20.56  Explain how you might use a radioactive tracer to show that ions are not completely motionless in crystals.

20.57  Each molecule of hemoglobin, the oxygen carrier in blood, contains four Fe atoms. Explain how you would use the radioactive $^{59}_{26}Fe$ ($t_{1/2}$ = 46 days) to show that the iron in a certain food is converted into hemoglobin.

## Section 20.8: Biological Effects of Radiation

*Review Questions*

20.58  List the factors that affect the intensity of radiation from a radioactive element.

20.59  What are *rad* and *rem,* and how are they related?

20.60  Explain, with examples, the difference between somatic and genetic radiation damage.

20.61  Compare the extent of radiation damage done by $\alpha$, $\beta$, and $\gamma$ sources.

## Additional Problems

20.62  How does a Geiger counter work?

20.63  Strontium-90 is one of the products of the fission of uranium-235. This strontium isotope is radioactive, with a half-life of 28.1 years. Calculate how long (in years) it will take for 1.00 g of the isotope to be reduced to 0.200 g by decay.

20.64  Nuclei with an even number of protons and an even number of neutrons are more stable than those with an odd number of protons and/or an odd number of neutrons. What is the significance of the even numbers of protons and neutrons in this case?

20.65  Tritium ($^3H$) is radioactive and decays by electron emission. Its half-life is 12.5 years. In ordinary water the ratio of $^1H$ to $^3H$ atoms is $1.0 \times 10^{17}$ to 1. (a) Write a balanced nuclear equation for tritium decay. (b) How many disintegrations will be observed per minute in a 1.00-kg sample of water?

20.66  (a) What is the activity, in millicuries, of a 0.500-g sample of $^{237}_{93}Np$? (This isotope decays by $\alpha$-particle emission and has a half-life of $2.20 \times 10^6$ years.) (b) Write a balanced nuclear equation for the decay of $^{237}_{93}Np$.

20.67  The following equations are for nuclear reactions that are known to occur in the explosion of an atomic bomb. Identify X.

(a) $^{235}_{92}U + ^1_0n \longrightarrow ^{140}_{56}Ba + 3^1_0n + X$
(b) $^{235}_{92}U + ^1_0n \longrightarrow ^{144}_{55}Cs + ^{90}_{37}Rb + 2X$
(c) $^{235}_{92}U + ^1_0n \longrightarrow ^{87}_{35}Br + 3^1_0n + X$
(d) $^{235}_{92}U + ^1_0n \longrightarrow ^{160}_{62}Sm + ^{72}_{30}Zn + 4X$

20.68  Calculate the nuclear binding energies (in J/nucleon) for the following species: (a) $^{10}B$ (10.0129 amu), (b) $^{11}B$ (11.009305 amu), (c) $^{14}N$ (14.003074 amu), (d) $^{56}Fe$ (55.93494 amu).

20.69  Write complete nuclear equations for the following processes: (a) tritium ($^3H$) undergoes $\beta$ decay, (b) $^{242}Pu$ undergoes $\alpha$-particle emission, (c) $^{131}I$ undergoes $\beta$ decay, (d) $^{251}Cf$ emits an $\alpha$ particle.

20.70  The nucleus of nitrogen-18 lies above the stability belt. Write the equation for a nuclear reaction by which nitrogen-18 can achieve stability.

20.71  Why is strontium-90 a particularly dangerous isotope for humans? The half-life of strontium-90 is 29.1 years. Calculate the radioactivity in millicuries of 15.6 mg of $^{90}Sr$.

20.72  How are scientists able to tell the age of a fossil?

20.73  After the Chernobyl accident, people living close to the nuclear reactor site were urged to take large amounts of potassium iodide as a safety precaution. What is the chemical basis for this action?

20.74  To detect bombs that may be smuggled onto airplanes, the Federal Aviation Administration (FAA) will soon require all major airports in the United States to install thermal neutron analyzers. The thermal neutron analyzer will bombard baggage with low-energy neutrons, converting some of the nitrogen-14 nuclei to nitrogen-15, with simultaneous emission of $\gamma$ rays. Because nitrogen content is usually high in explosives, detection of a high dosage of $\gamma$ rays will suggest that a bomb may be present. (a) Write an equation for the nuclear process. (b) Compare this technique with the conventional X-ray detection method.

20.75  Astatine, the last member of Group 7A, can be prepared by bombarding bismuth-209 with $\alpha$ particles. (a) Write an equation for the reaction. (b) Represent the equation in the abbreviated form as discussed in Section 20.4.

20.76  Explain why achievement of nuclear fusion in the laboratory requires a temperature of about 100 million degrees Celsius, which is much higher than that in the interior of the sun (15 million degrees Celsius).

20.77  The carbon-14 decay rate of a sample obtained from a young tree is 0.260 disintegration per second per gram of the sample. Another wood sample prepared from an object recovered at an archaeological excavation gives a decay rate of 0.186 disintegration per second per gram of the sample. What is the age of the object?

20.78  Tritium contains one proton and two neutrons. There is no significant proton-proton repulsion present in the nucleus. Why, then, is tritium radioactive?

**20.79** The usefulness of radiocarbon dating is limited to objects no older than 60,000 years. What percent of the carbon-14, originally present in the sample, remains after this period of time?

**20.80** The radioactive potassium-40 isotope decays to argon-40 with a half-life of $1.2 \times 10^9$ years. (a) Write a balanced equation for the reaction. (b) A sample of moon rock is found to contain 18 percent potassium-40 and 82 percent argon by mass. Calculate the age of the rock in years. (Assume that all the argon in the sample is the result of potassium decay.)

**20.81** Nuclear waste disposal is one of the major concerns of the nuclear industry. In choosing a safe and stable environment to store nuclear wastes, consideration must be given to the heat released during nuclear decay. As an example, consider the $\beta$ decay of $^{90}$Sr (89.907738 amu):

$$^{90}_{38}\text{Sr} \longrightarrow {}^{90}_{39}\text{Y} + {}^{0}_{-1}\beta \qquad t_{1/2} = 28.1 \text{ yr}$$

The $^{90}$Y (89.907152 amu) further decays as follows:

$$^{90}_{39}\text{Y} \longrightarrow {}^{90}_{40}\text{Zr} + {}^{0}_{-1}\beta \qquad t_{1/2} = 64 \text{ h}$$

Zirconium-90 (89.904703 amu) is a stable isotope. (a) Use the mass defect to calculate the energy released (in joules) in each of the preceding two decays. (The mass of the electron is $5.4857 \times 10^{-4}$ amu.) (b) Starting with 1 mole of $^{90}$Sr, calculate the number of moles of $^{90}$Sr that will decay in a year. (c) Calculate the amount of heat released (in kJ) corresponding to the number of moles of $^{90}$Sr decayed to $^{90}$Zr in part (b).

**20.82** Both barium (Ba) and radium (Ra) are members of Group 2A and are expected to exhibit similar chemical properties. However, Ra is not found in barium ores. Instead, it is found in uranium ores. Explain.

**20.83** As a result of being exposed to the radiation released during the Chernobyl nuclear accident, the dose of iodine-131 in a person's body is 7.4 mC ($1 \text{ mC} = 1 \times 10^{-3}$ Ci). Use the relationship rate $= \lambda N$ to calculate the number of atoms of iodine-131 to which this radioactivity corresponds. (The half-life of $^{131}$I is 8.1 days.)

**20.84** Which of the following poses a greater health hazard: a radioactive isotope with a short half-life or a radioactive isotope with a long half-life? Explain. [Assume the same type of radiation ($\alpha$ or $\beta$) and comparable energetics per particle emitted.]

**20.85** From the definition of curie, calculate Avogadro's number, given that the molar mass of $^{226}$Ra is 226.03 g/mol and that it decays with a half-life of $1.6 \times 10^3$ years.

**20.86** Since 1994, elements 110, 111, 112, and 114 have been synthesized. Element 110 was created by bombarding $^{208}$Pb with $^{62}$Ni, element 111 was created by bombarding $^{209}$Bi with $^{64}$Ni, element 112 was created by bombarding $^{208}$Pb with $^{66}$Zn, and element 114 was created by bombarding $^{244}$Pu with $^{48}$Ca. Write an equation for each synthesis. Predict the chemical properties of these elements. Use W for element 110, X for element 111, Y for element 112, and Z for element 114.

**20.87** (a) Calculate the energy released when a U-238 isotope decays to Th-234. The atomic masses are as follows: U-238: 238.05078 amu; Th-234: 234.03596 amu; and He-4: 4.002603 amu. (b) The energy released in part (a) is transformed into the kinetic energy of the recoiling Th-234 nucleus and the $\alpha$ particle. Which of the two will move away faster? Explain.

**20.88** A person received an anonymous gift of a decorative cube, which he placed on his desk. A few months later he became ill and died shortly afterward. After investigation, the cause of his death was linked to the box. The box was air-tight and had no toxic chemicals on it. What might have killed the man?

**20.89** Identify two of the most abundant radioactive elements that exist on Earth. Explain why they are still present. (You may wish to consult a website such as that of the University of Sheffield and WebElements Ltd, UK, webelements.com.)

**20.90** Sources of energy on Earth include fossil fuels, geothermal power, gravity, hydroelectric power, nuclear fission, nuclear fusion, the sun, and wind. Which of these have a "nuclear origin," either directly or indirectly?

**20.91** Cobalt-60 is an isotope used in diagnostic medicine and cancer treatment. It decays with $\gamma$-ray emission. Calculate the wavelength of the radiation in nanometers if the energy of the $\gamma$ ray is $2.4 \times 10^{-13}$ J/photon.

**20.92** Americium-241 is used in smoke detectors because it has a long half-life (458 years) and its emitted $\alpha$ particles are energetic enough to ionize air molecules. Using the given schematic diagram of a smoke detector, explain how it works.

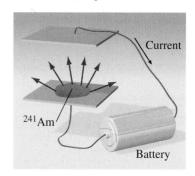

**20.93** The constituents of wine contain, among others, carbon, hydrogen, and oxygen atoms. A bottle of wine was sealed about 6 years ago. To confirm its age, which of the isotopes would you choose in a radioactive dating study? The half-lives of the isotopes are: $^{13}$C: 5715 years; $^{15}$O: 124 s; $^{3}$H: 12.5 years. Assume that the activities of the isotopes were known at the time the bottle was sealed.

**20.94** Name two advantages of a nuclear-powered submarine over a conventional submarine.

**20.95** In 1997 a scientist at a nuclear research center in Russia placed a thin shell of copper on a sphere of highly enriched uranium-235. Suddenly, there was a huge burst of radiation, which turned the air blue. Three days later, the scientist died of radiation exposure. Explain what caused the accident. (*Hint:* Copper is an effective metal for reflecting neutrons.)

**20.96** A radioactive isotope of copper decays as follows:

$$^{64}\text{Cu} \longrightarrow {}^{64}\text{Zn} + {}^{0}_{-1}\beta \qquad t_{1/2} = 12.8 \text{ h}$$

Starting with 84.0 g of $^{64}$Cu, calculate the quantity of $^{64}$Zn produced after 18.4 h.

**20.97** A 0.0100-g sample of a radioactive isotope with a half-life of $1.3 \times 10^9$ years decays at the rate of $2.9 \times 10^4$ disintegrations per minute. Calculate the molar mass of the isotope.

**20.98** The half-life of $^{27}$Mg is 9.50 min. (a) Initially there were $4.20 \times 10^{12}$ $^{27}$Mg nuclei present. How many $^{27}$Mg nuclei are left 30.0 min later? (b) Calculate the $^{27}$Mg activities (in Ci) at $t = 0$ and $t = 30.0$ min. (c) What is the probability that any one $^{27}$Mg nucleus decays during a 1-s interval? What assumption is made in this calculation?

**20.99** (a) Assuming nuclei are spherical in shape, show that the radius ($r$) of a nucleus is proportional to the cube root of mass number ($A$). (b) In general, the radius of a nucleus is given by $r = r_0 A^{1/3}$, where $r_0$, the proportionality constant, is given by $1.2 \times 10^{-15}$ m. Calculate the volume of the $^{238}$U nucleus.

20.100 Modern designs of atomic bombs contain, in addition to uranium or plutonium, small amounts of tritium and deuterium to boost the power of explosion. What is the role of tritium and deuterium in these bombs?

**20.101** The quantity of a radioactive material is often measured by its activity (measured in curies or millicuries) rather than by its mass. In a brain scan procedure, a 70-kg patient is injected with 20.0 mCi of $^{99m}$Tc, which decays by emitting $\gamma$-ray photons with a half-life of 6.0 h. Given that the RBE of these photons is 0.98 and only two-thirds of the photons are absorbed by the body, calculate the rem dose received by the patient. Assume all the $^{99m}$Tc nuclei decay while in the body. The energy of a $\gamma$-ray photon is $2.29 \times 10^{-14}$ J.

**20.102** In 2006, an ex-KGB agent was murdered in London. The investigation following the agent's death revealed that he was poisoned with the radioactive isotope $^{210}$Po, which had apparently been added to his food. (a) $^{210}$Po is prepared by bombarding $^{209}$Bi with neutrons. Write an equation for the reaction. (b) The half-life of $^{210}$Po is 138 days. It decays by $\alpha$ particle emission. Write the equation for the decay process. (c) Calculate the energy of an emitted $\alpha$ particle. Assume both the parent and daughter nuclei have zero kinetic energy. The atomic masses of $^{210}$Po, $^{206}$Pb, and $^4_2\alpha$ are 209.98286, 205.97444, and 4.00150 amu, respectively. (d) Ingestion of 1 $\mu$g of $^{210}$Po could prove fatal. What is the total energy released by this quantity of $^{210}$Po over the course of 138 days?

**20.103** Alpha particles produced by radioactive decay eventually pick up electrons from their surroundings to form helium atoms. Calculate the volume (in mL) of He collected at STP when 1.00 g of pure $^{226}$Ra is stored in a closed container for 125 years. (Assume that there are five $\alpha$ particles generated per $^{226}$Ra as it decays to $^{206}$Pb.)

20.104 An electron and a positron are accelerated to nearly the speed of light before colliding in a particle accelerator. The resulting collision produces an exotic particle having a mass many times that of a proton. Does this result violate the law of conservation of mass? Explain.

# PRE-PROFESSIONAL PRACTICE EXAM PROBLEMS: PHYSICAL AND BIOLOGICAL SCIENCES

The radioactive isotope $^{238}$Pu, used in pacemakers, decays by emitting an $\alpha$ particle with a half-life of 86 years. The energy of the emitted $\alpha$ particle is $9.0 \times 10^{-13}$ J, which is the energy per decay. After 10 years, the activity of the isotope decreases by 8.0 percent. (Power is measured in watts or J/s.)

1. What is the rate constant for the decay?

   a) $2.6 \times 10^{-10}$ s$^{-1}$
   b) 86 year$^{-1}$
   c) $2.7 \times 10^{-9}$ s$^{-1}$
   d) 0.012 year$^{-1}$

2. Identify the equation for the decay process.

   a) $^{238}$Pu $\longrightarrow$ $^{236}$Pu $+ \alpha$
   b) $^{238}$Pu $+ \alpha \longrightarrow$ $^{234}$U
   c) $^{238}$Pu $\longrightarrow$ $^{234}$U $+ \alpha$
   d) $^{238}$Pu $\longrightarrow$ $^{234}$Pu $+ \alpha$

3. Assuming that all the $\alpha$-particle energy is used to run the pacemaker, calculate the power output at $t = 0$ and $t = 10$ years. Initially, 1.0 mg of $^{238}$Pu was present in the pacemaker.

   a) 0.081 mW, 0.075 mW
   b) $2.6 \times 10^{-10}$ mW, $2.4 \times 10^{-10}$ mW
   c) 1.0 mW, 0.92 mW
   d) 0.58 mW, 0.53 mW

4. How long will it take for the power output to fall to 80 percent of its original value?

   a) 28 years
   b) 69 years
   c) 86 years
   d) 43 years

# ANSWERS TO IN-CHAPTER MATERIALS

## Answers to Practice Problems

**20.1A** (a) $^{78}_{34}$Se, (b) $^5_3$Li, (c) $^1_0$n. **20.1B** (a) $^{244}_{95}$Am, (b) $^{234}_{90}$Th, (c) $^{14}_6$C.
**20.2A** $2.6280 \times 10^{-10}$ J; $1.2574 \times 10^{-12}$ J/nucleon. **20.2B** 196.9665 amu.
**20.3A** $9.3 \times 10^3$ yr. **20.3B** $1.0 \times 10^1$ dps. **20.4A** $6.6 \times 10^8$ yr. **20.4B** $5.7 \times 10^{-2}$ mg. **20.5A** $^{106}_{46}$Pd $+ ^4_2$He $\longrightarrow$ $^1_1$H $+ ^{109}_{47}$Ag. **20.5B** $^{33}_{17}$Cl(n,$^3_2$He)$^{31}_{15}$P.

## Answers to Checkpoints

**20.1.1** d. **20.1.2** a. **20.2.1** b. **20.2.2** d. **20.2.3** c. **20.2.4** a. **20.3.1** d. **20.3.2** c.
**20.3.3** b. **20.3.4** b. **20.4.1** e. **20.4.2** b.

## Answers to Applying What You've Learned

a) $^{124}_{54}$Xe $+ ^1_0$n $\longrightarrow$ $^{125}_{54}$Xe, $^{125}_{54}$Xe $+ ^0_{-1}e \longrightarrow$ $^{125}_{53}$I; the product of electron-capture decay of $^{125}_{53}$I is $^{125}_{52}$Te. b) $1.36 \times 10^{-12}$ J/nucleon.
c) 257 days. d) X is $^{191}_{77}$Ir; $^{191}_{77}$Ir $+ ^1_0$n $\longrightarrow$ $^{192}_{77}$Ir $+ \gamma$.

# 21

# Environmental Chemistry

*Earth's atmosphere contains a region of increased ozone concentration, commonly known as the "ozone layer." In recent decades, pollutants have contributed to the destruction of ozone, potentially reducing the ozone layer's ability to protect life on Earth from harmful radiation.*

# How Human-made Molecules Have Impacted the Environment

The presence of ozone ($O_3$) in the atmosphere is responsible for the absorption of light that otherwise would significantly alter life on Earth. The wavelengths of light that are absorbed by stratospheric ozone are known to cause cancer, genetic mutations, and the destruction of plant life. Ozone is destroyed naturally by the absorption of this short-wavelength light—and is regenerated by another solar process. The balance of ozone destruction and regeneration can be disrupted, however, by the presence of substances not found naturally in the atmosphere.

In 1973 F. Sherwood "Sherry" Rowland and Mario Molina, chemistry professors at the University of California–Irvine, began studying the chemical behavior of chlorofluoro-carbons (CFCs) in the atmosphere. They discovered that although CFC molecules were extraordinarily stable in the troposphere, the very stability that made them attractive as coolants and propellants also allowed them to survive the gradual diffusion into the stratosphere where they would ultimately be broken down by high-energy ultraviolet radiation. Rowland and Molina proposed that chlorine atoms liberated in the breakdown of CFCs could potentially catalyze the destruction of large amounts of ozone in the stratosphere. The work of Rowland and Molina, along with other atmospheric scientists, provoked a debate among the scientific and international communities regarding the fate of the ozone layer—and the planet. The Montreal Protocol, once referred to by United Nations Secretary-General Kofi Annan as "perhaps the single most successful international agreement to date," is an international treaty designed to protect the ozone layer. Originally signed in 1987, it calls for a halt to the production of substances believed to contribute to the destruction of stratospheric ozone.

In 1995 Rowland and Molina, along with Dutch atmospheric chemist Paul Crutzen, were awarded the Nobel Prize in Chemistry for their elucidation of the role of human-made chemicals in the catalytic destruction of stratospheric ozone.

At the end of this chapter, you will be able to answer questions about an environmentally important molecule [▶▶ Page 913].

## In This Chapter, You Will Learn:

About some of the facets of environmental chemistry.

### Before you begin, you should review

- Bond enthalpy [◀◀ Section 8.9]
- Catalysis [◀◀ Section 14.6]
- Molecular polarity [◀◀ Section 9.2]

# 21.1  Earth's Atmosphere

Earth is unique among the planets of our solar system in having an atmosphere that is chemically active and rich in oxygen. Mars, for example, has a much thinner atmosphere that is about 90 percent carbon dioxide. Jupiter has no solid surface; it is made up, instead, of 90 percent hydrogen, 9 percent helium, and 1 percent other substances.

It is generally believed that three billion or four billion years ago, Earth's atmosphere consisted mainly of ammonia, methane, and water. There was little, if any, free oxygen present. Ultraviolet (UV) radiation from the sun probably penetrated the atmosphere, rendering the surface of Earth sterile. However, the same UV radiation may have triggered the chemical reactions (perhaps beneath the surface) that eventually led to life on Earth. Primitive organisms used energy from the sun to break down carbon dioxide (produced by volcanic activity) to obtain carbon, which they incorporated in their own cells. The major by-product of this process, called *photosynthesis,* is oxygen. Another important source of oxygen is the *photodecomposition* of water vapor by UV light. Over time, the more reactive gases such as ammonia and methane have largely disappeared, and today our atmosphere consists

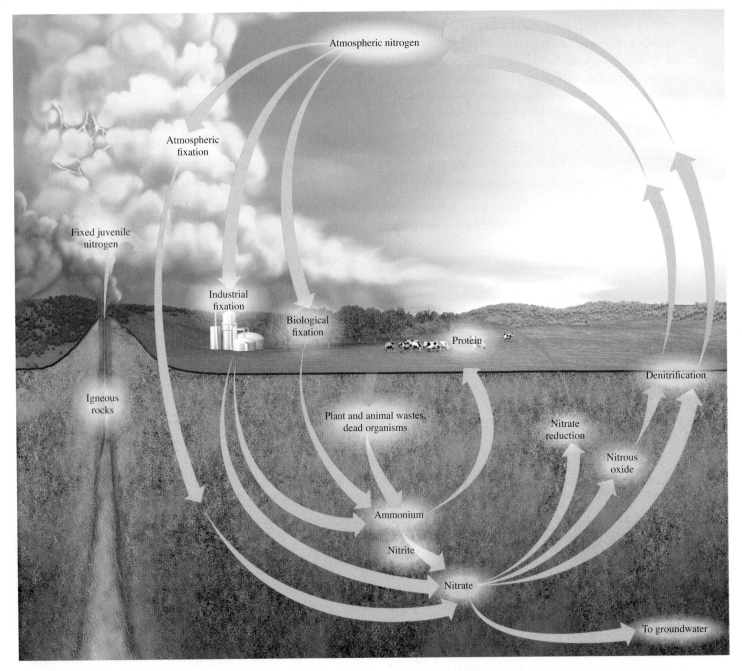

**Figure 21.1**    The nitrogen cycle. Although the supply of nitrogen in the Atmosphere is virtually inexhaustible, it must be combined with hydrogen or oxygen before it can be assimilated by higher plants, which in turn are consumed by animals. Juvenile nitrogen is nitrogen that has not previously participated in the nitrogen cycle.

mainly of oxygen and nitrogen gases. Biological processes determine to a great extent the atmospheric concentrations of these gases, one of which is reactive (oxygen) and the other unreactive (nitrogen).

Table 21.1 shows the composition of dry air at sea level. The total mass of the atmosphere is about $5.3 \times 10^{18}$ kg. Water is excluded from this table because its concentration in air can vary drastically from location to location.

Figure 21.1 shows the major processes involved in the cycle of nitrogen in nature. Molecular nitrogen, with its triple bond, is a very stable molecule. However, through biological and industrial *nitrogen fixation* (the conversion of molecular nitrogen into nitrogen compounds), atmospheric nitrogen gas is converted into nitrates and other compounds suitable for assimilation by algae and plants. Another important mechanism for producing nitrates from nitrogen gas is lightning. The steps are

$$N_2(g) + O_2(g) \xrightarrow{\text{electric energy}} 2NO(g)$$

$$2NO(g) + O_2(g) \longrightarrow 2NO_2(g)$$

$$2NO_2(g) + H_2O(l) \longrightarrow HNO_2(aq) + HNO_3(aq)$$

About 30 million tons of $HNO_3$ are produced this way annually. Nitric acid is converted to nitrate salts in the soil. These nutrients are taken up by plants, which in turn are ingested by animals. Animals use the nutrients from plants to make proteins and other essential biomolecules. Denitrification reverses nitrogen fixation to complete the cycle. For example, certain anaerobic organisms decompose animal wastes as well as dead plants and animals to produce free molecular nitrogen from nitrates.

The main processes of the global oxygen cycle are shown in Figure 21.2. This cycle is complicated by the fact that oxygen takes so many different chemical forms. Atmospheric oxygen is

| TABLE 21.1 | Composition of Dry Air at Sea Level |
| --- | --- |
| **Gas** | **Composition (% by Volume)** |
| $N_2$ | 78.03 |
| $O_2$ | 20.99 |
| Ar | 0.94 |
| $CO_2$ | 0.033 |
| Ne | 0.0015 |
| He | 0.000524 |
| Kr | 0.00014 |
| Xe | 0.000006 |

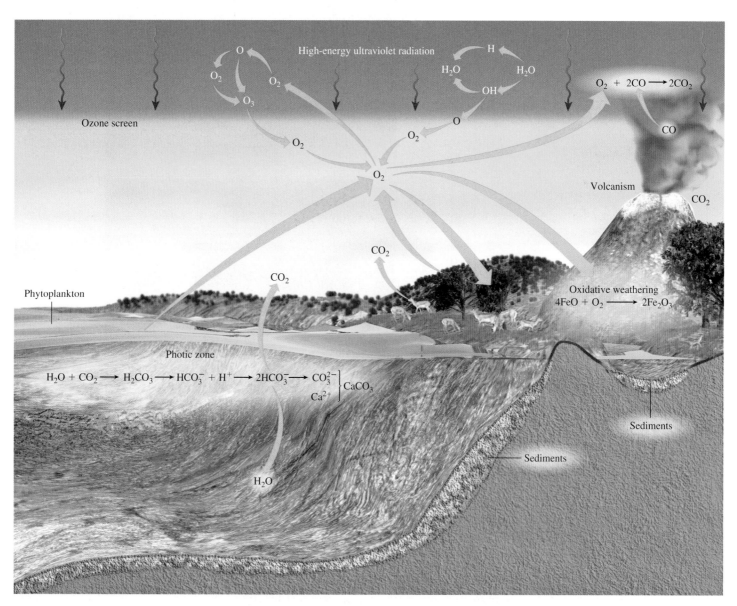

**Figure 21.2** The oxygen cycle. The cycle is complicated because oxygen appears in so many chemical forms and combinations, primarily as molecular oxygen, in water, and in organic and inorganic compounds.

**Figure 21.3** Regions of Earth's atmosphere. Notice the variation of temperature with altitude. Most of the phenomena shown here are discussed in the chapter.

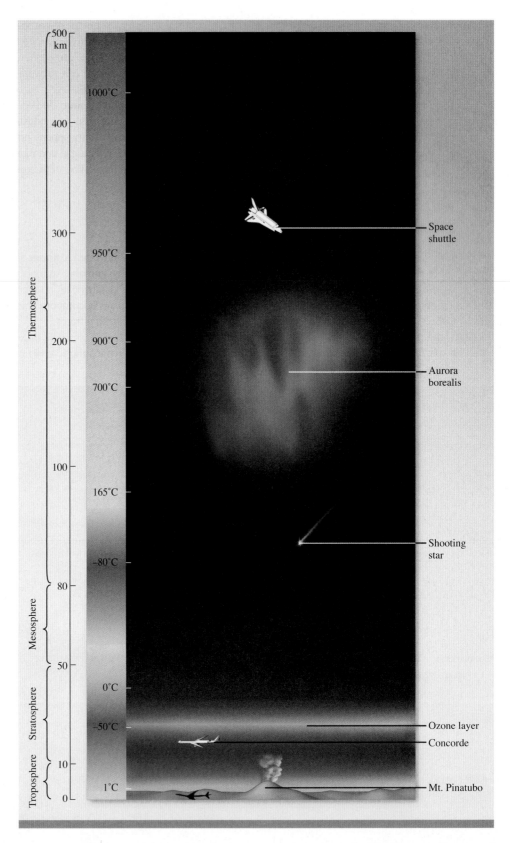

removed through respiration and various industrial processes (mostly combustion), which produce carbon dioxide. Photosynthesis is the major mechanism by which molecular oxygen is regenerated from carbon dioxide and water.

Scientists divide the atmosphere into several different layers according to temperature variation and composition (Figure 21.3). As far as visible events are concerned, the most active region is the ***troposphere,*** the layer of the atmosphere closest to Earth's surface. The troposphere contains about 80 percent of the total mass of air and practically all the atmosphere's water vapor. The troposphere is the thinnest layer of the atmosphere (10 km), but it is where all the dramatic events of weather—rain, lightning, hurricanes—occur. Temperature decreases almost linearly with increasing altitude in this region.

Above the troposphere is the *stratosphere,* which consists of nitrogen, oxygen, and ozone. In the stratosphere, the air temperature *increases* with altitude. This warming effect is the result of exothermic reactions triggered by UV radiation from the sun (see Section 21.3). One of the products of this reaction sequence is ozone ($O_3$), which serves to prevent harmful UV rays from reaching Earth's surface.

In the *mesosphere,* which is above the stratosphere, the concentration of ozone and other gases is low, and the temperature decreases again with increasing altitude. The *thermosphere,* or *ionosphere,* is the uppermost layer of the atmosphere. The increase in temperature in this region is the result of the bombardment of molecular oxygen and nitrogen and atomic species by energetic particles, such as electrons and protons, from the sun. Typical reactions are

$$N_2 \longrightarrow 2N \qquad \Delta H° = 941.4 \text{ kJ/mol}$$

$$N \longrightarrow N^+ + e^- \qquad \Delta H° = 1400 \text{ kJ/mol}$$

$$O_2 \longrightarrow O_2^+ + e^- \qquad \Delta H° = 1176 \text{ kJ/mol}$$

In reverse, these processes liberate the equivalent amount of energy, mostly as heat. Ionized particles are responsible for the reflection of radio waves back toward Earth.

**Figure 21.4** Aurora borealis, commonly referred to as the northern lights.

## 21.2 Phenomena in the Outer Layers of the Atmosphere

In this section we will discuss two dazzling phenomena that occur in the outer regions of the atmosphere. One is a natural event. The other is a curious by-product of human space travel.

### Aurora Borealis and Aurora Australis

Violent eruptions on the surface of the sun, called *solar flares,* result in the ejection of myriad electrons and protons into space, where they disrupt radio transmission and provide us with spectacular celestial light shows known as *auroras* (Figure 21.4). These electrons and protons collide with the molecules and atoms in Earth's upper atmosphere, causing them to become ionized and electronically excited. Eventually, the excited molecules and ions return to the ground state with the emission of light. For example, an excited oxygen atom emits photons at wavelengths of 558 nm (green) and between 630 and 636 nm (red):

$$O^* \longrightarrow O + h\nu$$

where the asterisk denotes an electronically excited species and $h\nu$ denotes the emitted photon [◀◀ Section 6.2]. Similarly, the blue and violet colors often observed in auroras result from the transition in the ionized nitrogen molecule:

$$N_2^+{}^* \longrightarrow N_2^+ + h\nu$$

The wavelengths for this transition fall between 391 and 470 nm.

The incoming streams of solar protons and electrons are oriented by Earth's magnetic field so that most auroral displays occur in doughnut-shaped zones about 2000 km in diameter centered on the North and South Poles. *Aurora borealis* is the name given to this phenomenon in the Northern Hemisphere. In the Southern Hemisphere, it is called *aurora australis.* Sometimes, the number of solar particles is so immense that auroras are also visible from locations as far south as Olympia, Washington.

Sample Problem 21.1 shows how to determine the maximum wavelength capable of breaking a chemical bond.

---

### Sample Problem 21.1

The bond enthalpy of $O_2$ is 498.7 kJ/mol. Calculate the maximum wavelength (in nm) of a photon that can cause the dissociation of an $O_2$ molecule.

**Strategy** We want to calculate the wavelength of a photon that will break an $O{=}O$ bond. Therefore, we need the amount of energy in one bond. The bond energy of $O_2$ is given in units of kJ/mol. The units needed for the energy of one bond are J/molecule. Once we know the energy in one bond, we can calculate the minimum frequency and maximum wavelength needed to dissociate one $O_2$ molecule.

*(Continued)*

**Setup** The conversion steps are

kJ/mol ⟶ J/molecule ⟶ frequency of photon ⟶ wavelength of photon

**Solution** First we calculate the energy required to dissociate one $O_2$ molecule:

$$\text{energy per molecule} = \frac{498.7 \times 10^3 \text{ J}}{1 \text{ mol}} \times \frac{1 \text{ mol}}{6.022 \times 10^{23} \text{ molecules}} = 8.281 \times 10^{-19} \frac{\text{J}}{\text{molecule}}$$

The energy of the photon is given by $E = h\nu$ (Equation 6.2). Therefore,

$$\nu = \frac{E}{h} = \frac{8.281 \times 10^{-19} \text{ J}}{6.63 \times 10^{-34} \text{ J} \cdot \text{s}} = 1.25 \times 10^{15} \text{ s}^{-1}$$

Finally, we calculate the wavelength of the photon, given by $\lambda = c/\nu$ (Equation 6.1), as follows:

$$\lambda = \frac{3.00 \times 10^8 \text{ m/s}}{1.25 \times 10^{15} \text{ s}^{-1}} = 2.40 \times 10^{-7} \text{ m} = 240 \text{ nm}$$

**Think About It** In principle, any photon with a wavelength of 240 nm or *shorter* can dissociate an $O_2$ molecule.

**Practice Problem A** Calculate the wavelength (in nm) of a photon needed to dissociate an $O_3$ molecule:

$$O_3 \longrightarrow O + O_2 \qquad \Delta H° = 107.2 \text{ kJ/mol}$$

**Practice Problem B** Which of the following gaseous species is dissociated by visible light: $CS_2$, $F_2$, HI, ClF, HCN? (See Table 8.6 on p. 326.)

## The Mystery Glow of Space Shuttles

A human-made light show that baffled scientists for several years is produced by space shuttles orbiting Earth. In 1983, astronauts first noticed an eerie orange glow on the outside surface of their spacecraft at an altitude about 300 km above Earth (Figure 21.5).

The light, which usually extends about 10 cm away from the protective silica heat tiles and other surface materials, is most pronounced on the parts of the shuttle facing its direction of travel. This fact led scientists to postulate that collision between oxygen atoms in the atmosphere and the fast-moving shuttle somehow produced the orange light. Spectroscopic measurements of the glow, as well as laboratory tests, strongly suggested that nitric oxide (NO) and nitrogen dioxide ($NO_2$) also played a part. It is believed that oxygen atoms interact with nitric oxide adsorbed on (i.e., bound to) the shuttle's surface to form electronically excited nitrogen dioxide:

$$O + NO \longrightarrow NO_2^*$$

As the $NO_2^*$ leaves the shell of the spacecraft, it emits photons at a wavelength of 680 nm (orange):

$$NO_2^* \longrightarrow NO_2 + h\nu$$

Support for this explanation came inadvertently in 1991, when astronauts aboard *Discovery* released various gases, including carbon dioxide, neon, xenon, and nitric oxide, from the cargo bay in the course of an unrelated experiment. Expelled one at a time, these gases scattered onto the surface of the shuttle's tail. The nitric oxide caused the normal shuttle glow to intensify markedly, but the other gases had no effect on it.

What is the source of the nitric oxide on the outside of the spacecraft? Scientists believe that some of it may come from the exhaust gases emitted by the shuttle's rockets and that some of it is present in the surrounding atmosphere. The shuttle glow does not harm the vehicle, but it does interfere with spectroscopic measurements on distant objects made from the spacecraft.

**Figure 21.5** The glowing tail section of the space shuttle viewed from inside the vehicle.

---

**Checkpoint 21.2**    **Phenomena in the Outer Layers of the Atmosphere**

**21.2.1** What maximum wavelength (in nm) of light is necessary to break the bond in a nitrogen molecule ($N_2$)? (The bond energy of $N_2$ is 941.4 kJ/mol.)

a) 127 nm

b) 211 nm

c) 236 nm

d) 942 nm

e) 941 nm

**21.2.2** What process gives rise to the aurora borealis and the aurora australis?

a) Formation of oxygen and nitrogen molecules

b) Collisions between nitrogen atoms

c) Collisions between oxygen atoms

d) Emission of photons of visible light

e) Absorption of photons of visible light

# 21.3 Depletion of Ozone in the Stratosphere

As mentioned earlier, ozone in the stratosphere prevents UV radiation emitted by the sun from reaching Earth's surface. The formation of ozone in this region begins with the *photodissociation* of oxygen molecules by solar radiation at wavelengths below 240 nm:

$$O_2 \xrightarrow{\text{UV} <240 \text{ nm}} O + O \qquad \text{Equation 21.1}$$

The highly reactive O atoms combine with oxygen molecules to form ozone as follows:

$$O + O_2 + M \longrightarrow O_3 + M \qquad \text{Equation 21.2}$$

where M is some inert substance such as $N_2$. The role of M in this exothermic reaction is to absorb some of the excess energy released and prevent the spontaneous decomposition of the $O_3$ molecule. The energy that is not absorbed by M is given off as heat. (As the M molecules themselves become de-excited, they release more heat to the surroundings.) In addition, ozone itself absorbs UV light between 200 and 300 nm:

$$O_3 \xrightarrow{\text{UV}} O + O_2 \qquad \text{Equation 21.3}$$

The process continues when O and $O_2$ recombine to form $O_3$ as shown in Equation 21.2, further warming the stratosphere.

If all the stratospheric ozone were compressed into a single layer at STP on Earth, that layer would be only about 3 mm thick! Although the concentration of ozone in the stratosphere is very low, it is sufficient to filter out (i.e., absorb) solar radiation in the 200- to 300-nm range (see Equation 21.3). In the stratosphere, it acts as our protective shield against UV radiation, which can induce skin cancer, cause genetic mutations, and destroy crops and other forms of vegetation.

The formation and destruction of ozone by natural processes is a dynamic equilibrium that maintains a constant concentration of ozone in the stratosphere. Since the mid-1970s scientists have been concerned about the harmful effects of CFCs on the ozone layer. Generally known by the trade name Freons, CFCs were first synthesized in the 1930s. Some of the common ones are $CFCl_3$ (Freon 11), $CF_2Cl_2$ (Freon 12), $C_2F_3Cl_3$ (Freon 113), and $C_2F_4Cl_2$ (Freon 114). Because these compounds are readily liquefied, relatively inert, nontoxic, noncombustible, and volatile, they have been used as coolants in refrigerators and air conditioners, in place of highly toxic liquid sulfur dioxide ($SO_2$) and ammonia ($NH_3$). Large quantities of CFCs are also used in the manufacture of disposable foam products such as cups and plates, as aerosol propellants in spray cans, and as solvents to clean newly soldered electronic circuit boards (Figure 21.6). In 1977, the peak year of production, nearly $1.5 \times 10^6$ tons of CFCs were produced in the United States. Most of the CFCs produced for commercial and industrial use are eventually discharged into the atmosphere.

Because of their relative inertness, the CFCs slowly diffuse unchanged up to the stratosphere, where UV radiation of wavelengths between 175 and 220 nm causes them to decompose:

$$CFCl_3 \longrightarrow CFCl_2 + Cl$$

$$CF_2Cl_2 \longrightarrow CF_2Cl + Cl$$

The reactive chlorine atoms then undergo the following reactions:

$$Cl + O_3 \longrightarrow ClO + O_2 \qquad \text{Equation 21.4}$$

$$ClO + O \longrightarrow Cl + O_2 \qquad \text{Equation 21.5}$$

The overall result (the sum of Equations 21.4 and 21.5) is the net removal of an $O_3$ molecule from the stratosphere:

$$O_3 + O \longrightarrow 2O_2 \qquad \text{Equation 21.6}$$

The oxygen atoms in Equation 21.5 are supplied by the photochemical decomposition of molecular oxygen and ozone described earlier. The Cl atom plays the role of a catalyst in the reaction mechanism scheme represented by Equations 21.4 and 21.5 because it is not used up and therefore can take part in many such reactions. In fact, one Cl atom can destroy up to 100,000 $O_3$ molecules

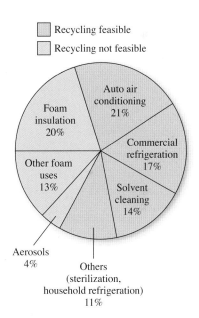

☐ Recycling feasible
☐ Recycling not feasible

**Figure 21.6** Uses of CFCs. Since 1978, the use of aerosol propellants has been banned in the United States.

**Figure 21.7** The variations in concentrations of ClO and $O_3$ with latitude.

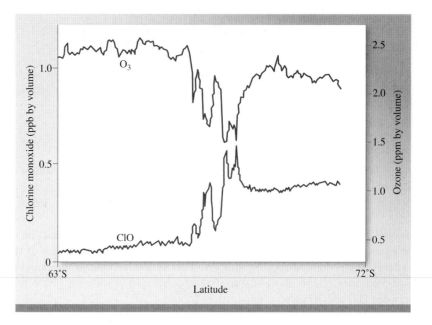

before it is removed by some other reaction. The ClO (chlorine monoxide) species is an intermediate because it is produced in the first elementary step (Equation 21.4) and consumed in the second step (Equation 21.5). The preceding mechanism for the destruction of ozone has been supported by the detection of ClO in the stratosphere in recent years (Figure 21.7). As can be seen, the concentration of $O_3$ decreases in regions that have high amounts of ClO.

Another group of compounds that can destroy stratospheric ozone are the nitrogen oxides, generally denoted as $NO_x$. (Examples of $NO_x$ are NO and $NO_2$.) These compounds come from the exhausts of high-altitude supersonic aircraft and from human and natural activities on Earth. Solar radiation decomposes a substantial amount of the other nitrogen oxides to nitric oxide (NO), which participates in the destruction of ozone as follows:

$$O_3 \longrightarrow O_2 + O$$

$$NO + O_3 \longrightarrow NO_2 + O_2$$

$$\underline{NO_2 + O \longrightarrow NO + O_2}$$

$$\textit{Overall:} \qquad 2O_3 \longrightarrow 3O_2$$

In this case, NO is the catalyst and $NO_2$ is the intermediate. Nitrogen dioxide also reacts with chlorine monoxide to form chlorine nitrate:

$$ClO + NO_2 \longrightarrow ClONO_2$$

Chlorine nitrate is relatively stable and serves as a "chlorine reservoir," which plays a role in the depletion of the stratospheric ozone over the North and South Poles.

## Polar Ozone Holes

In the mid-1980s, evidence began to accumulate that an "Antarctic ozone hole" developed in late winter, depleting the stratospheric ozone over Antarctica by as much as 50 percent (Figure 21.8). In the stratosphere, a stream of air known as the "polar vortex" circles Antarctica in winter. Air trapped within this vortex becomes extremely cold during the polar night. This condition leads to the formation of ice particles known as polar stratospheric clouds (PSCs) (Figure 21.9). Acting as a heterogeneous catalyst, these PSCs provide a surface for reactions converting HCl (emitted from Earth) and chlorine nitrate to more reactive chlorine molecules:

$$HCl + ClONO_2 \longrightarrow Cl_2 + HNO_3$$

By early spring, the sunlight splits molecular chlorine into chlorine atoms,

$$Cl_2 + h\nu \longrightarrow 2Cl$$

which then attack ozone as shown earlier.

**Figure 21.8**   In recent years, scientists have found that the ozone layer in the stratosphere over the South Pole has become thinner. This map, based on data collected over a number of years, shows the depletion of ozone in blue.

The situation is not as severe in the warmer Arctic region, where the vortex does not persist quite as long. Studies have shown that ozone levels in this region have declined between 4 and 8 percent in the past decade. Volcanic eruptions, such as that of Mount Pinatubo in the Philippines in 1991, inject large quantities of dust-sized particles and sulfuric acid aerosols into the atmosphere. These particles can perform the same catalytic function as the ice crystals at the South Pole. As a result, the Arctic hole is expected to grow larger for several years following an eruption.

Recognizing the serious implications of the loss of ozone in the stratosphere, nations throughout the world have acknowledged the need to drastically curtail or totally stop the production of CFCs. In 1978 the United States was one of the few countries to ban the use of CFCs in hair sprays and other aerosols. The Montreal protocol was signed by most industrialized nations in 1987, setting targets for cutbacks in CFC production and the complete elimination of these substances by the year 2000. Although progress has been made in this respect, many nations have not been able to abide by the treaty because of the importance of CFCs to their economies. Recycling could play a significant supplementary role in preventing CFCs already in appliances from escaping into the atmosphere. As Figure 21.6 shows, more than half of the CFCs in use are recoverable.

An intense effort is under way to find CFC substitutes that are not harmful to the ozone layer. One of the promising candidates is called hydrochlorofluorocarbon-123, or HCFC-123 ($CF_3CHCl_2$). The presence of the hydrogen atom makes the compound more susceptible to oxidation in the lower atmosphere, so it never reaches the stratosphere. Specifically, it is attacked by the hydroxyl radical in the troposphere:

$$CF_3CHCl_2 + OH \longrightarrow CF_3CCl_2 + H_2O$$

The $CF_3CCl_2$ fragment reacts with oxygen, eventually decomposing to $CO_2$, water, and hydrogen halides that are removed by rainwater. Unfortunately, the same hydrogen atom also makes the compound more active biologically than the CFCs. Laboratory tests have shown that HCFC-123 can cause tumors in rats, although its toxic effect on humans is not known. Another promising group of compounds that can substitute for CFCs are the hydrofluorocarbons (HFCs). Because they do not contain chlorine, HFCs will not promote the destruction of ozone even if they diffuse to the stratosphere. Examples of these compounds are $CF_3CFH_2$, $CF_3CF_2H$, $CF_3CH_3$, and $CF_2HCH_3$. In particular, $CF_3CFH_2$ is already widely used in place of CFCs in air conditioning and refrigeration applications.

**Figure 21.9**   Polar stratospheric clouds containing ice particles can catalyze the formation of Cl atoms and lead to the destruction of ozone.

Although it is unclear whether the CFCs already released to the atmosphere will eventually result in catastrophic damage to life on Earth, it is conceivable that the depletion of ozone can be slowed by reducing the availability of Cl atoms. Indeed, some chemists have suggested sending a fleet of planes to spray 50,000 tons of ethane ($C_2H_6$) or propane ($C_3H_8$) high over the South Pole in an attempt to heal the hole in the ozone layer. Being a reactive species, the chlorine atom would react with the hydrocarbons as follows:

$$Cl + C_2H_6 \longrightarrow HCl + C_2H_5$$

$$Cl + C_3H_8 \longrightarrow HCl + C_3H_7$$

The products of these reactions would not affect the ozone concentration. A less realistic plan is to rejuvenate the ozone layer by producing large quantities of ozone and releasing it into the stratosphere from airplanes. Technically this solution is feasible, but it would be enormously costly and it would require the collaboration of many nations.

Having discussed the chemistry in the outer regions of Earth's atmosphere, we will focus in Sections 21.4 through 21.8 on events closer to us—that is, in the troposphere.

## 21.4  Volcanoes

Volcanic eruptions, Earth's most spectacular natural displays of energy, are instrumental in forming large parts of Earth's crust. The upper mantle, immediately under the crust, is nearly molten. A slight increase in heat, such as that generated by the movement of one crustal plate under another, melts the rock. The molten rock, called *magma,* rises to the surface and generates some types of volcanic eruptions (Figure 21.10).

An active volcano emits gases, liquids, and solids. The gases spewed into the atmosphere include primarily $N_2$, $CO_2$, HCl, HF, $H_2S$, and water vapor. It is estimated that volcanoes are the source of about two-thirds of the sulfur in the air. On the slopes of Mount St. Helens, which last erupted in 1980, deposits of elemental sulfur are visible near the eruption site. At high temperatures, the hydrogen sulfide gas given off by a volcano is oxidized by air:

$$2H_2S(g) + 3O_2(g) \longrightarrow 2SO_2(g) + 2H_2O(g)$$

Some of the $SO_2$ is reduced by more $H_2S$ from the volcano to elemental sulfur and water:

$$2H_2S(g) + SO_2(g) \longrightarrow 3S(s) + 2H_2O(g)$$

**Figure 21.10**    A volcanic eruption.

The rest of the $SO_2$ is released into the atmosphere, where it reacts with water to form acid rain (see Section 21.6).

The tremendous force of a volcanic eruption carries a sizable amount of gas into the stratosphere. There $SO_2$ is oxidized to $SO_3$, which is eventually converted to sulfuric acid aerosols in a series of complex reactions. In addition to destroying ozone in the stratosphere (see page 901), these aerosols can also affect climate. Because the stratosphere is above the atmospheric weather patterns, the aerosol clouds often persist for more than a year. They absorb solar radiation and thereby cause a drop in temperature at Earth's surface. However, this cooling effect is local rather than global, because it depends on the site and frequency of volcanic eruptions.

## 21.5  The Greenhouse Effect

Although carbon dioxide is only a trace gas in Earth's atmosphere, with a concentration of about 0.033 percent by volume (see Table 21.1), it plays a critical role in controlling our climate. The so-called *greenhouse effect* describes the trapping of heat near Earth's surface by gases in the atmosphere, particularly carbon dioxide. The glass roof of a greenhouse transmits visible sunlight and absorbs some of the outgoing infrared (IR) radiation, thereby trapping the heat. Carbon dioxide acts somewhat like a glass roof, except that the temperature rise in the greenhouse is due mainly to the restricted air circulation inside. Calculations show that if the atmosphere did not contain carbon dioxide, Earth would be 30°C cooler!

Figure 21.11 shows the carbon cycle in our global ecosystem. The transfer of carbon dioxide to and from the atmosphere is an essential part of the carbon cycle. Carbon dioxide is produced

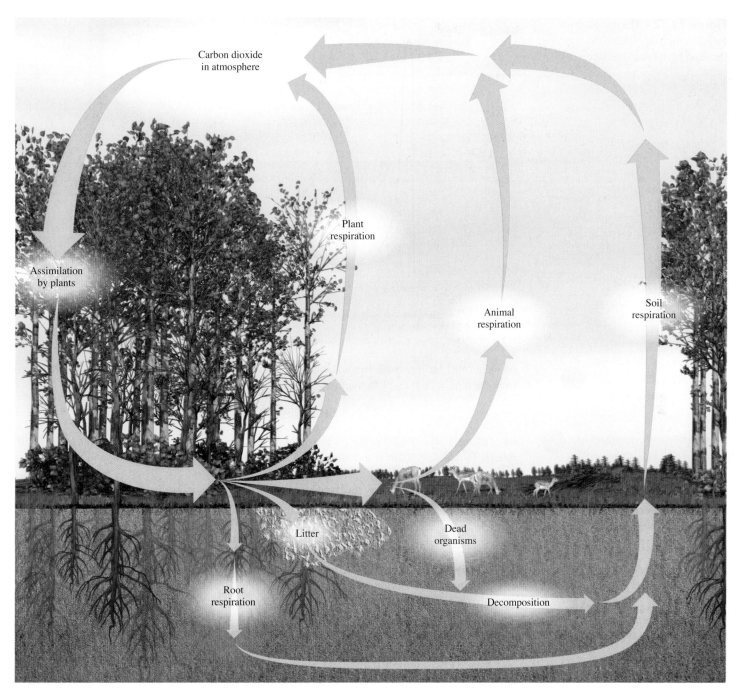

**Figure 21.11** The carbon cycle.

when any form of carbon or a carbon-containing compound is burned in an excess of oxygen. Many carbonates give off $CO_2$ when heated, and all give off $CO_2$ when treated with acid:

$$CaCO_3(s) \longrightarrow CaO(s) + CO_2(g)$$

$$CaCO_3(s) + 2HCl(aq) \longrightarrow CaCl_2(aq) + H_2O(l) + CO_2(g)$$

Carbon dioxide is also a by-product of the fermentation of sugar:

$$\underset{\text{glucose}}{C_6H_{12}O_6(aq)} \xrightarrow{\text{yeast}} \underset{\text{ethanol}}{2C_2H_5OH(aq)} + 2CO_2(g)$$

Carbohydrates and other complex carbon-containing molecules are consumed by animals, which respire and release $CO_2$ as an end product of metabolism:

$$C_6H_{12}O_6(aq) + 6O_2(g) \longrightarrow 6CO_2(g) + 6H_2O(l)$$

**Figure 21.12**   The incoming radiation from the sun and the outgoing radiation from Earth's surface.

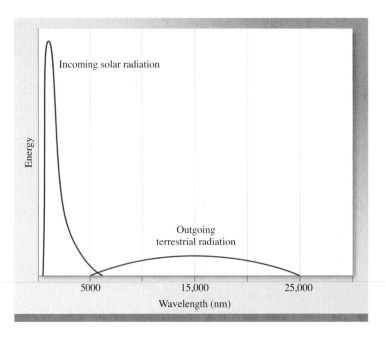

**Figure 21.13**   Vibration motion of a diatomic molecule. Chemical bonds can be stretched and compressed like a spring.

Stable form

Stretched

Compressed

As mentioned earlier, another major source of $CO_2$ is volcanic activity. Carbon dioxide is removed from the atmosphere by photosynthetic plants and certain microorganisms:

$$6CO_2(g) + 6H_2O(l) \longrightarrow C_6H_{12}O_6(aq) + 6O_2(g)$$

After plants and animals die, the carbon in their tissues is oxidized to $CO_2$ and returns to the atmosphere. In addition, there is a dynamic equilibrium between atmospheric $CO_2$ and carbonates in the oceans and lakes.

The solar radiant energy received by Earth is distributed over a band of wavelengths between 100 and 5000 nm, but much of it is concentrated in the 400- to 700-nm range, which is the visible region of the spectrum (Figure 21.12). By contrast, the thermal radiation emitted by Earth's surface is characterized by wavelengths longer than 4000 nm (the IR region) because of the much lower average surface temperature compared to that of the sun. The outgoing IR radiation can be absorbed by water and carbon dioxide, but not by nitrogen and oxygen.

All molecules vibrate, even at the lowest temperatures. The energy associated with molecular vibration is quantized, much like the electronic energies of atoms and molecules. To vibrate more energetically, a molecule must absorb a photon of a specific wavelength in the IR region. First, however, its dipole moment *must* change during the course of a vibration. [Recall that the dipole moment of a molecule is the product of the charge and the distance between charges (see page 311).] Figure 21.13 shows how a diatomic molecule can vibrate. If the molecule is homonuclear like $N_2$ and $O_2$, there can be no change in the dipole moment; the molecule has a zero dipole moment no matter how far apart or close together the two atoms are. We call such molecules IR-inactive because they *cannot* absorb IR radiation. On the other hand, all heteronuclear diatomic molecules are IR-active; that is, they all can absorb IR radiation because their dipole moments constantly change as the bond lengths change.

A *polyatomic* molecule can vibrate in more than one way. Water, for example, can vibrate in three different ways as shown in Figure 21.14. Because water is a polar molecule, any of these vibrations results in a change in dipole moment because there is a change in bond length. Therefore, an $H_2O$ molecule is IR-active. Carbon dioxide has a linear geometry and is nonpolar. Figure 21.15 shows two of the four ways a $CO_2$ molecule can vibrate. One of them [Figure 21.15(a)] symmetrically displaces atoms from the center of gravity and will not create a dipole moment, but the other vibration [Figure 21.15(b)] is IR-active because the dipole moment changes from zero to a maximum value in one direction and then reaches the same maximum value when it changes to the other extreme position.

Upon receiving a photon in the IR region, a molecule of $H_2O$ or $CO_2$ is promoted to a higher vibrational energy level:

$$H_2O + h\nu \longrightarrow H_2O^*$$

$$CO_2 + h\nu \longrightarrow CO_2^*$$

**Figure 21.14**   The three different modes of vibration of a water molecule. Each mode of vibration can be imagined by moving the atoms along the arrows and then reversing the direction of motion.

(a)          (b)

**Figure 21.15**   Two of the four ways a carbon dioxide molecule can vibrate. The vibration in (a) does not result in a change in dipole moment, but the vibration in (b) renders the molecule IR-active.

(the asterisk denotes a vibrationally excited molecule). These energetically excited molecules soon lose their excess energy either by collision with other molecules or by spontaneous emission of radiation. Part of this radiation is emitted to outer space and part returns to Earth's surface.

Although the total amount of water vapor in our atmosphere has not altered noticeably over the years, the concentration of $CO_2$ has been rising steadily since the turn of the century as a result of the burning of fossil fuels (petroleum, natural gas, and coal). Figure 21.16 shows the percentages of $CO_2$ emitted due to human activities in the United States in 1998, and Figure 21.17 shows the variation of carbon dioxide concentration over a period of years, as measured in Hawaii. In the Northern Hemisphere, the seasonal oscillations are caused by the removal of carbon dioxide by photosynthesis during the growing season and its buildup during the fall and winter months. The trend is toward an increase in $CO_2$. The current rate of increase is about 1 ppm (1 part $CO_2$ per million parts air) by volume per year, which is equivalent to $9 \times 10^9$ tons of $CO_2$! Scientists have estimated that by the year 2010 the $CO_2$ concentration will exceed preindustrial levels by about 25 percent.

In addition to $CO_2$ and $H_2O$, other greenhouse gases, such as the CFCs, $CH_4$, $NO_x$, and $N_2O$ also contribute appreciably to the warming of the atmosphere. Figure 21.18 shows the gradual increase in temperature over the years, and Figure 21.19 shows the relative contributions of the greenhouse gases to global warming.

It is predicted by some meteorologists that should the buildup of greenhouse gases continue at its current rate, Earth's average temperature will increase by about 1° to 3°C in the twenty-first century. Although a temperature increase of a few degrees may seem insignificant, it is actually large enough to disrupt the delicate thermal balance on Earth and could cause glaciers and ice caps to melt. Consequently, the sea level would rise and coastal areas would be flooded. Predicting weather trends is extremely difficult, though, and there are other potentially moderating factors

**Figure 21.16**   Sources of carbon dioxide emission in the United States. Note that not all the emitted $CO_2$ enters the atmosphere. Some of it is taken up by carbon dioxide "sinks," such as the ocean.

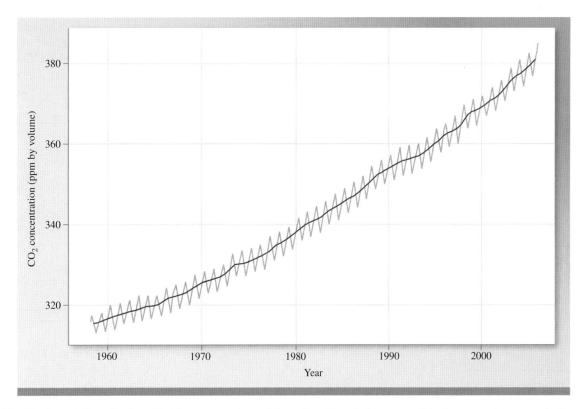

**Figure 21.17**   Yearly variation of carbon dioxide concentration at Mauna Loa, Hawaii, the source of the longest running record of atmospheric $CO_2$. The general trend clearly points to an increase of carbon dioxide in the atmosphere.

**Figure 21.18** Temperature rise on Earth's surface from 1880 to 1992. Zero represents the average temperature for the years 1951–1980.

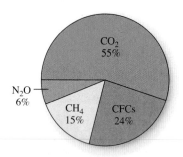

**Figure 21.19** Contribution to global warming by various greenhouse gases. The concentrations of CFCs and methane are much lower than that of carbon dioxide. However, because they can absorb IR radiation much more effectively than $CO_2$, they make a significant contribution to the overall warming effect.

to take into account before concluding that global warming is inevitable and irreversible. For example, the ash from volcanic eruptions diffuses upward and can stay in the atmosphere for years. By reflecting incoming sunlight, volcanic ash can cause a cooling effect. Furthermore, the warming effect of CFCs in the troposphere is offset by its action in the stratosphere. Because ozone is a polar polyatomic molecule, it is also an effective greenhouse gas. A decrease in ozone brought about by CFCs actually produces a noticeable drop in temperature.

To combat the greenhouse effect, we must lower carbon dioxide emissions. This can be done by improving energy efficiency in automobiles and in household heating and lighting, and by developing nonfossil fuel energy sources, such as photovoltaic cells. Nuclear energy is a viable alternative, but its use is highly controversial due to the difficulty of disposing of radioactive waste and the fact that nuclear power stations are more prone to accidents than conventional power stations (see Chapter 20). The proposed phasing out of CFCs, the most potent greenhouse gas, will help to slow down the warming trend. The recovery of methane gas generated at landfills and the reduction of natural gas leakages are other steps we could take to control $CO_2$ emissions. Finally, the preservation of the Amazon jungle, tropical forests in Southeast Asia, and other large forests is vital to maintaining the steady-state concentration of $CO_2$ in the atmosphere. Converting forests to farmland for crops and grassland for cattle may do irreparable damage to the delicate ecosystem and permanently alter the climate pattern on Earth.

**Student Annotation:** Global climate change is the subject of the Academy Award–winning 2006 documentary, *An Inconvenient Truth*, presented by former vice president Al Gore.

### Sample Problem 21.2

Which of the following qualify as greenhouse gases: CO, NO, $NO_2$, $Cl_2$, $H_2$, Ne?

**Strategy** To behave as a greenhouse gas, either the molecule must possess a dipole moment or some of its vibrational motions must generate a temporary dipole moment.

**Setup** The necessary conditions immediately rule out homonuclear diatomic molecules and atomic species.

**Think About It** $CO_2$, the best-known greenhouse gas, is *nonpolar*. It is only necessary for at least one of a molecule's vibrational modes to induce a *temporary* dipole for it to act as a greenhouse gas.

**Solution** Only CO, NO, and $NO_2$, which are all polar molecules, qualify as greenhouse gases. Both $Cl_2$ and $H_2$ are homonuclear diatomic molecules, and Ne is atomic. These three species are all IR-inactive.

**Practice Problem A** Which of the following is a more effective greenhouse gas: CO or $H_2O$? Explain.

**Practice Problem B** Both $O_2$ and $O_3$ exhibit molecular vibration. Despite this, only $O_3$ acts as a greenhouse gas. Explain.

**Checkpoint 21.5** **The Greenhouse Effect**

**21.5.1** Which of the following can act as a greenhouse gas? (Select all that apply.)

a) $CH_4$

b) $N_2$

c) Rn

d) $O_3$

e) Xe

**21.5.2** The greenhouse effect is

a) caused by depletion of stratospheric ozone.

b) entirely the result of human activity.

c) a natural phenomenon that has been enhanced by human activity.

d) the absorption of the sun's energy by molecules and atoms in the upper atmosphere.

e) responsible for the aurora borealis and the aurora australis.

# 21.6 Acid Rain

Every year acid rain causes hundreds of millions of dollars' worth of damage to stone buildings and statues throughout the world. The term *stone leprosy* is used by some environmental chemists to describe the corrosion of stone by acid rain (Figure 21.20). Acid rain is also toxic to vegetation and aquatic life. Many well-documented cases show dramatically how acid rain has destroyed agricultural and forest lands and killed aquatic organisms.

Precipitation in the northeastern United States has an average pH of about 4.3 (Figure 21.21). Because atmospheric $CO_2$ in equilibrium with rainwater would not be expected to result in a pH less than 5.5, sulfur dioxide ($SO_2$) and, to a lesser extent, nitrogen oxides from auto emissions are believed to be responsible for the high acidity of rainwater. Acidic oxides, such as $SO_2$, react with water to give the corresponding acids. There are several sources of atmospheric $SO_2$. Nature itself contributes much $SO_2$ in the form of volcanic eruptions. Also, many metals exist combined with sulfur in nature. Extracting the metals often entails *smelting,* or *roasting,* the ores—that is, heating the metal sulfide in air to form the metal oxide and $SO_2$. For example,

$$2ZnS(s) + 3O_2(g) \longrightarrow 2ZnO(s) + 2SO_2(g)$$

The metal oxide can be reduced more easily than the sulfide (by a more reactive metal or in some cases by carbon) to the free metal.

Although smelting is a major source of $SO_2$, the burning of fossil fuels in industry, in power plants, and in homes accounts for most of the $SO_2$ emitted to the atmosphere (Figure 21.22). The

**Figure 21.20** The effect of acid rain on a marble statue. The photos were taken just a few decades apart.

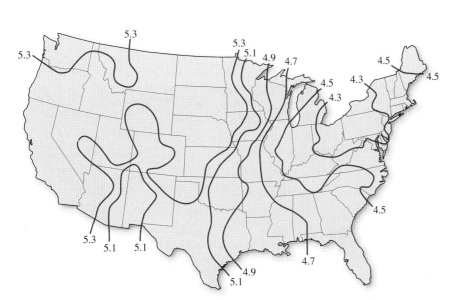

**Figure 21.21** Mean precipitation pH in the United States in 1994. Most $SO_2$ comes from the midwestern states. Prevailing winds carry the acid droplets formed over the Northeast. Nitrogen oxides also contribute to acid rain formation.

**Figure 21.22** Sulfur dioxide and other air pollutants being released into the atmosphere from a coal-burning power plant.

sulfur content of coal ranges from 0.5 to 5 percent by mass, depending on the source of the coal. The sulfur content of other fossil fuels is similarly variable. Oil from the Middle East, for instance, is low in sulfur, whereas that from Venezuela has a high sulfur content. To a lesser extent, the nitrogen-containing compounds in oil and coal are converted to nitrogen oxides, which can also acidify rainwater.

All in all, some 50 million to 60 million tons of $SO_2$ are released into the atmosphere each year! In the troposphere, $SO_2$ is almost all oxidized to $H_2SO_4$ in the form of aerosol, which ends up in wet precipitation or acid rain. The mechanism for the conversion of $SO_2$ to $H_2SO_4$ is quite complex and not fully understood. The reaction is believed to be initiated by the hydroxyl radical (OH):

$$OH + SO_2 \longrightarrow HOSO_2$$

The $HOSO_2$ radical is further oxidized to $SO_3$:

$$HOSO_2 + O_2 \longrightarrow HO_2 + SO_3$$

The sulfur trioxide formed would then rapidly react with water to form sulfuric acid:

$$SO_3 + H_2O \longrightarrow H_2SO_4$$

$SO_2$ can also be oxidized to $SO_3$ and then converted to $H_2SO_4$ on particles by heterogeneous catalysis. Eventually, the acid rain can corrode limestone and marble ($CaCO_3$). A typical reaction is

$$CaCO_3(s) + H_2SO_4(aq) \longrightarrow CaSO_4(s) + H_2O(l) + CO_2(g)$$

Sulfur dioxide can also attack calcium carbonate directly:

$$2CaCO_3(s) + 2SO_2(g) + O_2(g) \longrightarrow 2CaSO_4(s) + 2CO_2(g)$$

There are two ways to minimize the effects of $SO_2$ pollution. The most direct approach is to remove sulfur from fossil fuels before combustion, but this is technologically difficult to accomplish. A cheaper but less efficient way is to remove $SO_2$ as it is formed. For example, in one process powdered limestone is injected into the power plant boiler or furnace along with the coal (Figure 21.23). At high temperatures, the following decomposition occurs:

$$\underset{\text{limestone}}{CaCO_3(s)} \longrightarrow \underset{\text{quicklime}}{CaO(s)} + CO_2(g)$$

The quicklime reacts with $SO_2$ to form calcium sulfite and some calcium sulfate:

$$CaO(s) + SO_2(g) \longrightarrow CaSO_3(s)$$

$$2CaO(s) + 2SO_2(g) + O_2(g) \longrightarrow 2CaSO_4(s)$$

**Figure 21.23** Common procedure for removing $SO_2$ from burning fossil fuel. Powdered limestone decomposes into CaO, which reacts with $SO_2$ to form $CaSO_3$. The remaining $SO_2$ is combined with an aqueous suspension of CaO to form $CaSO_3$.

To remove any remaining $SO_2$, an aqueous suspension of quicklime is injected into a purification chamber prior to the gases' escape through the smokestack. Quicklime is also added to lakes and soils in a process called *liming* to reduce their acidity (Figure 21.24). Installing a sulfuric acid plant near a metal ore refining site is also an effective way to cut $SO_2$ emission because the $SO_2$ produced by roasting metal sulfides can be captured for use in the synthesis of sulfuric acid. This is a very sensible way to turn what is a pollutant in one process into a starting material for another process!

**Figure 21.24** Spreading calcium oxide (CaO) over acidified soil. This process is called liming.

# 21.7 Photochemical Smog

The word *smog* was coined to describe the combination of smoke and fog that shrouded London during the 1950s. The primary cause of this noxious cloud was sulfur dioxide. Today, however, *photochemical smog,* which is formed by the reactions of automobile exhaust in the presence of sunlight, is much more common.

Automobile exhaust consists mainly of NO, CO, and various unburned hydrocarbons. These gases are called *primary pollutants* because they set in motion a series of photochemical reactions that produce *secondary pollutants.* It is the secondary pollutants—chiefly $NO_2$ and $O_3$—that are responsible for the buildup of smog.

Nitric oxide is the product of the reaction between atmospheric nitrogen and oxygen at high temperatures inside an automobile engine:

$$N_2(g) + O_2(g) \longrightarrow 2NO(g)$$

Once released into the atmosphere, nitric oxide is oxidized to nitrogen dioxide:

$$2NO(g) + O_2(g) \longrightarrow 2NO_2(g)$$

Sunlight causes the photochemical decomposition of $NO_2$ (at a wavelength shorter than 400 nm) into NO and O:

$$NO_2(g) + h\nu \longrightarrow NO(g) + O(g)$$

Atomic oxygen is a highly reactive species that can initiate a number of important reactions, one of which is the formation of ozone:

$$O(g) + O_2(g) + M \longrightarrow O_3(g) + M$$

where M is some inert substance such as $N_2$. Ozone attacks the C=C linkage in rubber:

$$\underset{R}{\overset{R}{\diagdown}}C=C\underset{R}{\overset{R}{\diagup}} + O_3 \longrightarrow \begin{matrix} R & O & R \\ \diagdown & \diagup \diagdown & \diagup \\ C & & C \\ \diagup \diagdown & & \diagdown \diagup \\ R & O-O & R \end{matrix} \xrightarrow{H_2O} \underset{R}{\overset{R}{\diagdown}}C=O + O=C\underset{R}{\overset{R}{\diagup}} + H_2O_2$$

where R represents groups of C and H atoms. In smog-ridden areas, this reaction can cause automobile tires to crack. Similar reactions are also damaging to lung tissues and other biological substances.

Ozone can be formed also by a series of very complex reactions involving unburned hydrocarbons, nitrogen oxides, and oxygen. One of the products of these reactions is peroxyacetyl nitrate (PAN):

$$CH_3-\underset{\underset{O}{\|}}{C}-O-O-NO_2$$

PAN is a powerful lachrymator, or tear producer, and causes breathing difficulties.

Figure 21.25 shows typical variations with time of primary and secondary pollutants. Initially, the concentration of $NO_2$ is quite low. As soon as solar radiation penetrates the atmosphere, though, more $NO_2$ is formed from NO and $O_2$. The concentration of ozone remains fairly constant at a low level in the early morning hours. As the concentration of unburned hydrocarbons and aldehydes increases in the air, the concentrations of $NO_2$ and $O_3$ also rise rapidly. The actual amounts depend on the location, traffic, and weather conditions, but their presence is always accompanied by haze (Figure 21.26). The oxidation of hydrocarbons produces various organic intermediates, such as alcohols and carboxylic acids, which are all less volatile than the hydrocarbons themselves. These substances eventually condense into small droplets of liquid. The dispersion of these droplets in air, called an *aerosol,* scatters sunlight and reduces visibility. This interaction also makes the air look hazy.

**Figure 21.25**    Typical variations with time in concentration of air pollutants on a smoggy day.

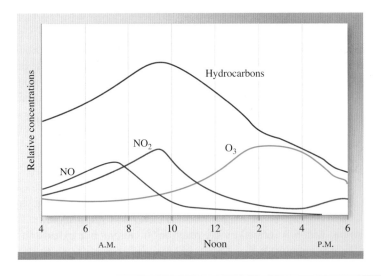

As the mechanism of photochemical smog formation has become better understood, major efforts have been made to reduce the buildup of primary pollutants. Most automobiles now are equipped with catalytic converters designed to oxidize CO and unburned hydrocarbons to $CO_2$ and $H_2O$ and to reduce NO and $NO_2$ to $N_2$ and $O_2$ [◀◀ Section 14.6]. More efficient automobile engines and better public transportation systems would also help to decrease air pollution in urban areas. A recent technological innovation to combat photochemical smog is to coat automobile radiators and air conditioner compressors with a platinum catalyst. So equipped, a running car can purify the air that flows under the hood by converting ozone and carbon monoxide to oxygen and carbon dioxide:

$$O_3(g) + CO(g) \xrightarrow{\text{Pt}} O_2(g) + CO_2(g)$$

In a city like Los Angeles, where the number of miles driven in one day equals nearly 300 million, this approach would significantly improve the air quality and reduce the "high-ozone level" warnings frequently issued to its residents.

**Figure 21.26**    A smoggy day in a big city.

## 21.8    Indoor Pollution

Difficult as it is to avoid air pollution outdoors, it is no easier to avoid pollution indoors. The air quality in homes and in the workplace is affected by human activities, by construction materials, and by other factors in our immediate environment. The common indoor pollutants are radon, carbon monoxide, carbon dioxide, and formaldehyde.

### The Risk from Radon

In a highly publicized case about 35 years ago, an employee reporting for work at a nuclear power plant in Pennsylvania set off the plant's radiation monitor. Astonishingly, the source of his contamination turned out not to be the plant, but radon in his home!

A lot has been said and written about the potential dangers of radon as an air pollutant. Just what is radon? Where does it come from? And how does it affect our health?

Radon is a member of Group 8A (the noble gases). It is an intermediate product of the radioactive decay of uranium-238 (see Figure 20.3). All isotopes of radon are radioactive, but radon-222 is the most hazardous because it has the longest half-life—3.8 days. Radon, which accounts for slightly over half the background radioactivity on Earth, is generated mostly from the phosphate minerals of uranium (Figure 21.27).

Since the 1970s, high levels of radon have been detected in homes built on reclaimed land above uranium mill tailing deposits. The colorless, odorless, and tasteless radon gas enters a building through tiny cracks in the basement floor. It is slightly soluble in water, so it can be spread in different media. Radon-222 is an $\alpha$-emitter. When it decays, it produces radioactive polonium-214 and polonium-218, which can build up to high levels in an enclosed space. These solid radioactive particles can adhere to airborne dust and smoke, which are inhaled into the lungs and deposited in the respiratory tract. Over a long period of time, the $\alpha$ particles emitted by polonium and its decay products, which are also radioactive, can cause lung cancer.

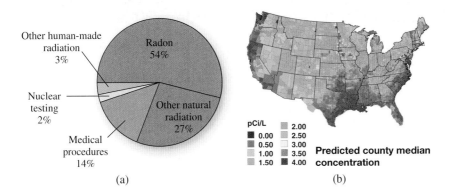

(a)

(b)

What can be done to combat radon pollution indoors? The first step is to measure the radon level in the basement with a reliable test kit. Short-term and long-term kits are available (Figure 21.28). The short-term tests use activated charcoal to collect the decay products of radon over a period of several days. The container is sent to a laboratory where a technician measures the radioactivity ($\gamma$ rays) from radon-decay products lead-214 and bismuth-214. Knowing the length of exposure, the lab technician back-calculates to determine radon concentration. The long-term test kits use a piece of special polymer film on which an $\alpha$ particle will leave a "track." After several months' exposure, the film is etched with a sodium hydroxide solution and the number of tracks counted. Knowing the length of exposure enables the technician to calculate the radon concentration. If the radon level is unacceptably high, then the house must be regularly ventilated. This precaution is particularly important in recently built houses, which are well insulated. A more effective way to prevent radon pollution is to reroute the gas before it gets into the house (e.g., by installing a ventilation duct to draw air from beneath the basement floor to the outside).

Currently there is considerable controversy regarding the health effects of radon. The first detailed studies of the effects of radon on human health were carried out in the 1950s when it was recognized that uranium miners suffered from an abnormally high incidence of lung cancer. Some scientists have challenged the validity of these studies because the miners were also smokers. It seems quite likely that there is a synergistic effect between radon and smoking on the development of lung cancer. Radon decay products will adhere not only to tobacco tar deposits in the lungs, but also to the solid particles in cigarette smoke, which can be inhaled by smokers and nonsmokers. More systematic studies are needed to evaluate the environmental impact of radon. In the meantime, the Environmental Protection Agency (EPA) has recommended remedial action where the radioactivity level due to radon exceeds 4 picocuries (pCi) per liter of air. [A curie corresponds to $3.70 \times 10^{10}$ disintegrations of radioactive nuclei per second; a picocurie is a trillionth of a curie, or $3.70 \times 10^{-2}$ disintegrations per second (dps).]

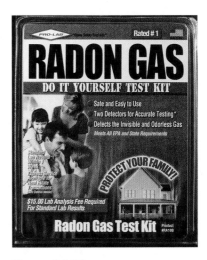

**Figure 21.28** Home radon detector.

---

### Sample Problem 21.3

The half-life of Rn-222 is 3.8 days. Starting with 1.0 g of Rn-222, how much will be left after 10 half-lives?

**Strategy** All radioactive decays obey first-order kinetics, making the half-life independent of the initial concentration.

**Setup** Because the question involves an integral number of half-lives, we can deduce the amount of Rn-222 remaining without using Equation 14.3.

**Solution** After one half-life, the amount of Rn left is $0.5 \times 1.0$ g, or 0.5 g. After two half-lives, only 0.25 g of Rn remains. Generalizing the fraction of the isotope left after $n$ half-lives as $(1/2)^n$, where $n = 10$, we write

$$\text{quantity of Rn-222 left} = 1.0 \text{ g} \times (1/2)^{10}$$
$$= 9.8 \times 10^{-4} \text{ g}$$

**Think About It** An alternative solution is to calculate the first-order rate constant from the half-life and use Equation 14.3:

$$\ln \frac{N_t}{N_0} = -kt$$

where $N$ is the mass of Rn-222. Try this and verify that your answers are the same. (Since most of the kinetics problems we encounter do not involve an integral number of half-lives, we generally use Equation 14.3 as the *first* approach to solving them.)

**Practice Problem A** The concentration of Rn-222 in the basement of a house is $1.8 \times 10^{-6}$ mol/L. Assume the air remains static, and calculate the concentration of the radon after 2.4 days.

**Practice Problem B** How long will it take for the radioactivity due to radon to fall to a level considered acceptable by the EPA if the starting activity is $2.25 \times 10^3$ dps/L?

## Carbon Dioxide and Carbon Monoxide

Both carbon dioxide ($CO_2$) and carbon monoxide (CO) are products of combustion.

In the presence of an abundant supply of oxygen, $CO_2$ is formed; in a limited supply of oxygen, both CO and $CO_2$ are formed. The indoor sources of these gases are gas cooking ranges, woodstoves, space heaters, tobacco smoke, human respiration, and exhaust fumes from cars (in garages). Carbon dioxide is not a toxic gas, but it does have an asphyxiating effect. In airtight buildings, the concentration of $CO_2$ can reach as high as 2000 ppm by volume (compared with 3 ppm outdoors). Workers exposed to high concentrations of $CO_2$ in skyscrapers and other sealed environments become fatigued more easily and have difficulty concentrating. Adequate ventilation is the solution to $CO_2$ pollution.

Like $CO_2$, CO is a colorless and odorless gas, but it differs from $CO_2$ in that it is highly poisonous. The toxicity of CO lies in its unusual ability to bind very strongly to hemoglobin, the oxygen carrier in blood. Both $O_2$ and CO bind to the Fe(II) ion in hemoglobin, but the affinity of hemoglobin for CO is about 200 times greater than it is for $O_2$. Hemoglobin molecules with tightly bound CO (called carboxyhemoglobin) cannot carry the oxygen needed for metabolic processes. At a concentration of 70 ppm, CO can cause drowsiness and headache; at higher concentrations, death may result when about half the hemoglobin molecules become complexed with CO. The best first-aid response to CO poisoning is to remove the victim immediately to an area with a plentiful oxygen supply or to give mouth-to-mouth resuscitation.

## Formaldehyde

Formaldehyde ($CH_2O$) is a rather disagreeable-smelling liquid used as a preservative for laboratory specimens. Industrially, formaldehyde resins are used as bonding agents in building and furniture construction materials such as plywood and particle board. In addition, urea-formaldehyde insulation foams are used to fill wall cavities. The resins and foams slowly break down to release free formaldehyde, especially under acid and humid conditions. Low concentrations of formaldehyde in the air can cause drowsiness, nausea, headaches, and other respiratory ailments. Laboratory tests show that breathing high concentrations of formaldehyde can induce cancers in animals, but whether it has a similar effect in humans is unclear. The safe standard of formaldehyde in indoor air has been set at 0.1 ppm by volume.

Because formaldehyde is a reducing agent, devices have been constructed to remove it by means of a redox reaction. Indoor air is circulated through an air purifier containing an oxidant such as $Al_2O_3/KMnO_4$, which converts formaldehyde to the less harmful and less volatile formic acid (HCOOH). Proper ventilation is the best way to remove formaldehyde. However, care should be taken not to remove the air from a room too quickly without replenishment, because a reduced pressure would cause the formaldehyde resins to decompose faster, resulting in the release of *more* formaldehyde.

---

**Checkpoint 21.8**    **Indoor Pollution**

21.8.1  The risk posed by radon gas in homes is

a) respiratory distress

b) nausea

c) dizziness

d) coma

e) lung cancer

21.8.2  What mass of a 1.0-g sample of radon remains after 30 days? (The half-life of radon is 3.8 days.)

a) 0.030 g

b) 0.0040 g

c) 0.99 g

d) 0.97 g

e) 0.0010 g

# Applying What You've Learned

The research of Paul Crutzen, the third recipient of the Nobel Prize for Chemistry in 1995, involved the effect of nitric oxide (NO) on the destruction of stratospheric ozone. Unlike CFCs, which may take 50 to 100 years to diffuse into the upper atmosphere, nitric oxide is introduced directly to the stratosphere in the exhaust of high-altitude aircraft. Early in the 1970s, the United States considered construction of a large fleet of supersonic transport airplanes (SSTs), similar to the Concorde. Environmentalists argued, based in part on the work of Paul Crutzen, that to do so would significantly endanger the ozone layer.

**Problems:**

a) The bond enthalpy of NO is 630.6 kJ/mol. Determine the maximum wavelength of light required to break the bond in an NO molecule. [◄◄ Sample Problem 21.1]

b) Can nitric oxide act as a greenhouse gas? Explain. [◄◄ Sample Problem 21.2]

# CHAPTER SUMMARY

## Section 21.1

- Earth's atmosphere is made up mainly of nitrogen and oxygen, plus a number of other trace gases. Molecular nitrogen in the atmosphere is incorporated into other compounds via **nitrogen fixation.**

- The regions of the atmosphere, from Earth's surface outward, are the **troposphere,** the **stratosphere,** the **mesosphere,** the **thermosphere,** and the **ionosphere.**

- The chemical processes that go on in the atmosphere are influenced by solar radiation, volcanic eruption, and human activities.

## Section 21.2

- In the outer regions of the atmosphere, the bombardment of molecules and atoms by solar particles gives rise to the *aurora borealis* in the Northern Hemisphere and the *aurora australis* in the Southern Hemisphere. The glow on a space shuttle is caused by excitation of molecules adsorbed on the shuttle's surface.

## Section 21.3

- Ozone in the stratosphere absorbs harmful UV radiation in the 200- to 300-nm range and protects life underneath. For many years, chlorofluorocarbons have been destroying the ozone layer.

## Section 21.4

- Volcanic eruptions can lead to air pollution, deplete ozone in the stratosphere, and affect climate.

## Section 21.5

- Carbon dioxide's ability to absorb infrared radiation enables it to trap some of the outgoing heat from Earth, warming its surface—a phenomenon known as the **greenhouse effect.** Other gases such as the CFCs and methane also contribute to the greenhouse effect. Global warming refers to the result of the *enhanced* greenhouse effect caused by human activities.

## Section 21.6

- Sulfur dioxide, and to a lesser extent nitrogen oxides, generated mainly from the burning of fossil fuels and from the roasting of metal sulfides, causes acid rain.

## Section 21.7

- **Photochemical smog** is formed by the photochemical reaction of automobile exhaust in the presence of sunlight. It is a complex reaction involving nitrogen oxides, ozone, and hydrocarbons.

## Section 21.8

- Indoor air pollution is caused by radon, a radioactive gas formed during uranium decay; carbon monoxide and carbon dioxide, products of combustion; and formaldehyde, a volatile organic substance released from resins used in construction materials.

# KEY WORDS

| | | | |
|---|---|---|---|
| Greenhouse effect, 902 | Mesosphere, 897 | Photochemical smog, 909 | Thermosphere, 897 |
| Ionosphere, 897 | Nitrogen fixation, 895 | Stratosphere, 897 | Troposphere, 896 |

# QUESTIONS AND PROBLEMS

## Section 21.1: Earth's Atmosphere

### Review Questions

21.1   Describe the regions of Earth's atmosphere.

21.2   Briefly outline the main processes of the nitrogen and oxygen cycles.

21.3   Explain why, for maximum performance, supersonic airplanes need to fly at a high altitude (in the stratosphere).

21.4   Jupiter's atmosphere consists mainly of hydrogen (90 percent) and helium (9 percent). How does this mixture of gases contrast with the composition of Earth's atmosphere? Why does the composition differ?

### Problems

**21.5**   Referring to Table 21.1, calculate the mole fraction of $CO_2$ and its concentration in parts per million by volume.

21.6   Calculate the partial pressure of $CO_2$ (in atm) in dry air when the atmospheric pressure is 754 mmHg.

**21.7**   Describe the processes that result in the warming of the stratosphere.

21.8   Calculate the mass (in kg) of nitrogen, oxygen, and carbon dioxide gases in the atmosphere. Assume that the total mass of air in the atmosphere is $5.25 \times 10^{21}$ g.

## Section 21.2: Phenomena in the Outer Layers of the Atmosphere

### Review Questions

21.9  What process gives rise to the aurora borealis and aurora australis?

21.10  Why can astronauts not release oxygen atoms to test the mechanism of shuttle glow?

### Problems

**21.11**  The highly reactive OH radical (a species with an unpaired electron) is believed to be involved in some atmospheric processes. Table 8.6 lists the bond enthalpy for the oxygen-to-hydrogen bond in OH as 460 kJ/mol. What is the longest wavelength (in nm) of radiation that can bring about the following reaction?

$$OH(g) \longrightarrow O(g) + H(g)$$

21.12  The green color observed in the aurora borealis is produced by the emission of a photon by an electronically excited oxygen atom at 558 nm. Calculate the energy difference between the two levels involved in the emission process.

## Section 21.3: Depletion of Ozone in the Stratosphere

### Review Questions

21.13  Briefly describe the absorption of solar radiation in the stratosphere by $O_2$ and $O_3$ molecules.

21.14  Explain the processes that have a warming effect on the stratosphere.

21.15  List the properties of CFCs, and name four major uses of these compounds.

21.16  How do CFCs and nitrogen oxides destroy ozone in the stratosphere?

21.17  What causes the polar ozone holes?

21.18  How do volcanic eruptions contribute to ozone destruction?

21.19  Describe ways to curb the destruction of ozone in the stratosphere.

21.20  Discuss the effectiveness of some of the CFC substitutes.

### Problems

**21.21**  Given that the quantity of ozone in the stratosphere is equivalent to a 3.0-mm-thick layer of ozone on Earth at STP, calculate the number of ozone molecules in the stratosphere and their mass in kilograms. (*Hint:* The radius of Earth is 6371 km and the surface area of a sphere is $4\pi r^2$, where $r$ is the radius.)

21.22  Referring to the answer in Problem 21.21, and assuming that the level of ozone in the stratosphere has already fallen 6.0 percent, calculate the number of kilograms of ozone that would have to be manufactured on a daily basis so that we could restore the ozone to the original level in 100 years. If ozone is made according to the process $3O_2(g) \longrightarrow 2O_3(g)$, how many kilojoules of energy would be required?

21.23  Both Freon-11 and Freon-12 are made by the reaction of carbon tetrachloride ($CCl_4$) with hydrogen fluoride. Write equations for these reactions.

21.24  Why are CFCs not decomposed by UV radiation in the troposphere?

**21.25**  The average bond enthalpies of the C—Cl and C—F bonds are 340 and 485 kJ/mol, respectively. Based on this information, explain why the C—Cl bond in a CFC molecule is preferentially broken by solar radiation at 250 nm.

21.26  Like CFCs, certain bromine-containing compounds such as $CF_3Br$ can also participate in the destruction of ozone by a similar mechanism starting with the Br atom:

$$CF_3Br \longrightarrow CF_3 + Br$$

Given that the average C—Br bond energy is 276 kJ/mol, estimate the longest wavelength required to break this bond. Will this compound be decomposed in the troposphere only or in both the troposphere and stratosphere?

**21.27**  Draw Lewis structures for chlorine nitrate ($ClONO_2$) and chlorine monoxide (ClO).

21.28  Draw Lewis structures for HCFC-123 ($CF_3CHCl_2$) and $CF_3CFH_2$.

## Section 21.4: Volcanoes

### Review Questions

21.29  What are the effects of volcanic eruptions on climate?

21.30  Classify the reaction between $H_2S$ and $SO_2$ that leads to the formation of sulfur at the site of a volcanic eruption.

## Section 21.5: The Greenhouse Effect

### Review Questions

21.31  What is the greenhouse effect? What is the criterion for classifying a gas as a greenhouse gas?

21.32  Why is more emphasis placed on the role of carbon dioxide in the greenhouse effect than on that of water?

21.33  Describe three human activities that generate carbon dioxide. List two major mechanisms for the uptake of carbon dioxide.

21.34  Deforestation contributes to the greenhouse effect in two ways. What are they?

21.35  How does an increase in world population enhance the greenhouse effect?

21.36  Is ozone a greenhouse gas? If so, sketch three ways an ozone molecule can vibrate.

21.37  What effects do CFCs and their substitutes have on Earth's temperature?

21.38  Why are CFCs more effective greenhouse gases than methane and carbon dioxide?

## Problems

**21.39**  The annual production of zinc sulfide (ZnS) is $4.0 \times 10^4$ tons. Estimate the number of tons of $SO_2$ produced by roasting it to extract zinc metal.

**21.40**  Calcium oxide or quicklime (CaO) is used in steelmaking, cement manufacture, and pollution control. It is prepared by the thermal decomposition of calcium carbonate:

$$CaCO_3(s) \longrightarrow CaO(s) + CO_2(g)$$

Calculate the yearly release of $CO_2$ (in kg) to the atmosphere if the annual production of CaO in the United States is $1.7 \times 10^{10}$ kg.

**21.41**  The molar heat capacity of a diatomic molecule is 29.1 J/K · mol. Assuming the atmosphere contains only nitrogen gas and there is no heat loss, calculate the total heat intake (in kJ) if the atmosphere warms up by 3°C during the next 50 years. Given that there are $1.8 \times 10^{20}$ moles of diatomic molecules present, how many kilograms of ice (at the North and South Poles) will this quantity of heat melt at 0°C? (The molar heat of fusion of ice is 6.01 kJ/mol.)

**21.42**  As mentioned in the chapter, spraying the stratosphere with hydrocarbons such as ethane and propane should eliminate Cl atoms. What is the drawback of this procedure if used on a large scale for an extended period of time?

## Section 21.6: Acid Rain

### Review Questions

**21.43**  Name the gas that is largely responsible for the acid rain phenomenon.

**21.44**  List three detrimental effects of acid rain.

**21.45**  Briefly discuss two industrial processes that lead to acid rain.

**21.46**  Discuss ways to curb acid rain.

**21.47**  Water and sulfur dioxide are both polar molecules, and their geometry is similar. Why is $SO_2$ not considered a major greenhouse gas?

**21.48**  Describe the removal of $SO_2$ by CaO (to form $CaSO_3$) in terms of a Lewis acid-base reaction.

### Problems

**21.49**  An electric power station annually burns $3.1 \times 10^7$ kg of coal containing 2.4 percent sulfur by mass. Calculate the volume of $SO_2$ emitted at STP.

**21.50**  The concentration of $SO_2$ in the troposphere over a certain region is 0.16 ppm by volume. The gas dissolves in rainwater as follows:

$$SO_2(g) + H_2O(l) \rightleftharpoons H^+(aq) + HSO_3^-(aq)$$

Given that the equilibrium constant for the preceding reaction is $1.3 \times 10^{-2}$, calculate the pH of the rainwater. Assume that the reaction does not affect the partial pressure of $SO_2$.

## Section 21.7: Photochemical Smog

### Review Questions

**21.51**  What is photochemical smog? List the factors that favor the formation of photochemical smog.

**21.52**  What are primary and secondary pollutants?

**21.53**  Identify the gas that is responsible for the brown color of photochemical smog.

**21.54**  The safety limits of ozone and carbon monoxide are 120 ppb by volume and 9 ppm by volume, respectively. Why does ozone have a lower limit?

**21.55**  Suggest ways to minimize the formation of photochemical smog.

**21.56**  In which region of the atmosphere is ozone beneficial? In which region is it detrimental?

### Problems

**21.57**  Assume that the formation of nitrogen dioxide:

$$2NO(g) + O_2(g) \longrightarrow 2NO_2(g)$$

is an elementary reaction. (a) Write the rate law for this reaction. (b) A sample of air at a certain temperature is contaminated with 2.0 ppm of NO by volume. Under these conditions, can the rate law be simplified? If so, write the simplified rate law. (c) Under the conditions described in part (b), the half-life of the reaction has been estimated to be $6.4 \times 10^3$ min. What would the half-life be if the initial concentration of NO were 10 ppm?

**21.58**  The gas-phase decomposition of peroxyacetyl nitrate (PAN) obeys first-order kinetics:

$$CH_3COOONO_2 \longrightarrow CH_3COOO + NO_2$$

with a rate constant of $4.9 \times 10^{-4}\ s^{-1}$. Calculate the rate of decomposition (in $M$/s) if the concentration of PAN is 0.55 ppm by volume. Assume STP conditions.

**21.59**  On a smoggy day in a certain city, the ozone concentration was 0.42 ppm by volume. Calculate the partial pressure of ozone (in atm) and the number of ozone molecules per liter of air if the temperature and pressure were 20.0°C and 748 mmHg, respectively.

**21.60**  Which of the following settings is the most suitable for photochemical smog formation: (a) Gobi desert at noon in June, (b) New York City at 1 P.M. in July, (c) Boston at noon in January? Explain your choice.

## Section 21.8: Indoor Pollution

### Review Questions

**21.61**  List the major indoor pollutants and their sources.

**21.62**  What is the best way to deal with indoor pollution?

**21.63**  Why is it dangerous to idle a car's engine in a poorly ventilated place, such as the garage?

**21.64**  Describe the properties that make radon an indoor pollutant. Would radon be more hazardous if $^{222}Rn$ had a longer half-life?

### Problems

**21.65**  A concentration of $8.00 \times 10^2$ ppm by volume of CO is considered lethal to humans. Calculate the minimum mass of CO (in grams) that would become a lethal concentration in a closed room 17.6 m long, 8.80 m wide, and 2.64 m high. The temperature and pressure are 20.0°C and 756 mmHg, respectively.

**21.66**  A volume of 5.0 L of polluted air at 18.0°C and 747 mmHg is passed through lime water [an aqueous suspension of $Ca(OH)_2$], so that all the carbon dioxide present is precipitated as $CaCO_3$. If the mass of the $CaCO_3$ precipitate is 0.026 g, calculate the percentage by volume of $CO_2$ in the air sample.

## Additional Problems

**21.67** Briefly describe the harmful effects of the following substances: $O_3$, $SO_2$, $NO_2$, CO, $CH_3COOONO_2$ (PAN), Rn.

**21.68** The equilibrium constant ($K_P$) for the reaction

$$N_2(g) + O_2(g) \rightleftharpoons 2NO(g)$$

is $4.0 \times 10^{-31}$ at 25°C and $2.6 \times 10^{-6}$ at 1100°C, the temperature of a running car's engine. Is this an endothermic or exothermic reaction?

**21.69** As stated in the chapter, carbon monoxide has a much higher affinity for hemoglobin than oxygen does. (a) Write the equilibrium constant expression ($K_c$) for the following process:

$$CO(g) + HbO_2(aq) \rightleftharpoons O_2(g) + HbCO(aq)$$

where $HbO_2$ and HbCO are oxygenated hemoglobin and carboxyhemoglobin, respectively. (b) The composition of a breath of air inhaled by a person smoking a cigarette is $1.9 \times 10^{-6}$ mol/L CO and $8.6 \times 10^{-3}$ mol/L $O_2$. Calculate the ratio of [HbCO] to [HbO$_2$], given that $K_c$ is 212 at 37°C.

**21.70** Instead of monitoring carbon dioxide, suggest another gas that scientists could study to substantiate the fact that $CO_2$ concentration is steadily increasing in the atmosphere.

**21.71** In 1991 it was discovered that nitrous oxide ($N_2O$) is produced in the synthesis of nylon. This compound, which is released into the atmosphere, contributes *both* to the depletion of ozone in the stratosphere and to the greenhouse effect. (a) Write equations representing the reactions between $N_2O$ and oxygen atoms in the stratosphere to produce nitric oxide (NO), which is then oxidized by ozone to form nitrogen dioxide. (b) Is $N_2O$ a more effective greenhouse gas than carbon dioxide? Explain. (c) One of the intermediates in nylon manufacture is adipic acid [$HOOC(CH_2)_4COOH$]. About $2.2 \times 10^9$ kg of adipic acid is consumed every year. It is estimated that for every mole of adipic acid produced, 1 mole of $N_2O$ is generated. What is the maximum number of moles of $O_3$ that can be destroyed as a result of this process per year?

**21.72** A glass of water initially at pH 7.0 is exposed to dry air at sea level at 20°C. Calculate the pH of the water when equilibrium is reached between atmospheric $CO_2$ and $CO_2$ dissolved in the water, given that Henry's law constant for $CO_2$ at 20°C is 0.032 mol/L · atm. (*Hint:* Assume no loss of water due to evaporation, and use Table 21.1 to calculate the partial pressure of $CO_2$. Your answer should correspond roughly to the pH of rainwater.)

**21.73** A 14-m by 10-m by 3.0-m basement had a high radon content. On the day the basement was sealed off from its surroundings so that no exchange of air could take place, the partial pressure of $^{222}$Rn was $1.2 \times 10^{-6}$ mmHg. Calculate the number of $^{222}$Rn isotopes ($t_{1/2} = 3.8$ days) at the beginning and end of 31 days. Assume STP conditions.

**21.74** Ozone in the troposphere is formed by the following steps:

$$NO_2 \longrightarrow NO + O \qquad (1)$$

$$O + O_2 \longrightarrow O_3 \qquad (2)$$

The first step is initiated by the absorption of visible light ($NO_2$ is a brown gas). Calculate the longest wavelength required for step 1 at 25°C. (*Hint:* You need to first calculate $\Delta H$ and hence $\Delta E$ for step 1. Next, determine the wavelength for decomposing $NO_2$ from $\Delta E$.)

**21.75** Although the hydroxyl radical (OH) is present only in a trace amount in the troposphere, it plays a central role in its chemistry because it is a strong oxidizing agent and can react with many pollutants as well as some CFC substitutes. The hydroxyl radical is formed by the following reactions:

$$O_3 \xrightarrow{\lambda\,=\,320\ nm} O^* + O_2$$

$$O + H_2O \longrightarrow 2OH$$

where O* denotes an electronically excited atom. (a) Explain why the concentration of OH is so small even though the concentrations of $O_3$ and $H_2O$ are quite large in the troposphere. (b) What property makes OH a strong oxidizing agent? (c) The reaction between OH and $NO_2$ contributes to acid rain. Write an equation for this process. (d) The hydroxyl radical can oxidize $SO_2$ to $H_2SO_4$. The first step is the formation of a neutral $HSO_3$ species, followed by its reaction with $O_2$ and $H_2O$ to form $H_2SO_4$ and the hydroperoxyl radical ($HO_2$). Write equations for these processes.

**21.76** The equilibrium constant ($K_P$) for the reaction $2CO(g) + O_2(g) \rightleftharpoons 2CO_2(g)$ is $1.4 \times 10^{90}$ at 25°C. Given this enormous value, why doesn't CO convert totally to $CO_2$ in the troposphere?

**21.77** A person was found dead of carbon monoxide poisoning in a well-insulated cabin. Investigation showed that he had used a blackened bucket to heat water on a butane burner. The burner was found to function properly with no leakage. Explain, with an appropriate equation, the cause of his death.

**21.78** The carbon dioxide level in the atmosphere today is often compared with that in preindustrial days. Explain how scientists use tree rings and air trapped in polar ice to arrive at the comparison.

**21.79** What is ironic about the following cartoon?

**21.80** Calculate the standard enthalpy of formation ($\Delta H_f^\circ$) of ClO from the following bond energies: $Cl_2$: 242.7 kJ/mol; $O_2$: 498.7 kJ/mol; ClO: 206 kJ/mol.

**21.81** Methyl bromide ($CH_3Br$, b.p. = 3.6°C) is used as a soil fumigant to control insects and weeds. It is also a marine by-product. Photodissociation of the C—Br bond produces Br atoms that can react with ozone similar to Cl, except more effectively. Do you expect $CH_3Br$ to be photolyzed in the troposphere? The bond enthalpy of the C—Br bond is about 293 kJ/mol.

**21.82** The effective incoming solar radiation per unit area on Earth is 342 W/m$^2$. Of this radiation, 6.7 W/m$^2$ is absorbed by $CO_2$ at 14,993 nm in the atmosphere. How many photons at this wavelength are absorbed per second in 1 m$^2$ by $CO_2$? (1 W = 1 J/s.)

**21.83** As stated in the chapter, about 50 million tons of sulfur dioxide is released into the atmosphere every year. (a) If 20 percent of the $SO_2$ is eventually converted to $H_2SO_4$, calculate the number of 1000-lb marble statues the resulting acid rain can damage. As an estimate, assume that the acid rain only destroys the surface layer of each statue, which is made up of 5 percent of its total mass. (b) What is the other undesirable result of the acid rain damage?

**21.84** Peroxyacetyl nitrate (PAN) undergoes thermal decomposition as follows:

$$CH_3(CO)OONO_2 \longrightarrow CH_3(CO)OO + NO_2$$

The rate constant is $3.0 \times 10^{-4}\ s^{-1}$ at 25°C. At the boundary between the troposphere and stratosphere, where the temperature is about −40°C, the rate constant is reduced to $2.6 \times 10^{-7}\ s^{-1}$. (a) Calculate the activation energy for the decomposition of PAN. (b) What is the half-life of the reaction (in min) at 25°C?

**21.85** How are past temperatures determined from ice cores obtained from the Arctic or Antarctica? (*Hint:* Look up the stable isotopes of hydrogen and oxygen. How does energy required for vaporization depend on the masses of $H_2O$ molecules containing different isotopes? How would you determine the age of an ice core?)

**21.86** The balance between $SO_2$ and $SO_3$ is important in understanding acid rain formation in the troposphere. From the following information at 25°C

$$S(s) + O_2(g) \rightleftharpoons SO_2(g) \qquad K_1 = 4.2 \times 10^{52}$$
$$2S(s) + 3O_2(g) \rightleftharpoons 2SO_3(g) \qquad K_2 = 9.8 \times 10^{128}$$

calculate the equilibrium constant for the reaction

$$2SO_2(g) + O_2(g) \rightleftharpoons 2SO_3(g)$$

# PRE-PROFESSIONAL PRACTICE EXAM PROBLEMS: PHYSICAL AND BIOLOGICAL SCIENCES

Indoor air pollutants, including radon gas, can pose significant health risks. Radon is a radioactive gas found in the soils and rocks of Earth's crust. It finds its way into homes through cracks in the foundation or basement, and sometimes through the water supply. Radon is invisible and odorless, and becomes a health hazard when it is allowed to build up inside the home. The isotope of radon commonly found in homes is radon-222, a product of the uranium-238 decay chain, formed by the $\alpha$ decay of radium-226. Radon-222, in turn, decays by $\alpha$ emission.

1. Select the equation that correctly represents the $\alpha$ decay of radium-226.

   a) $^{226}Ra \longrightarrow {}^{226}Rn + \alpha$
   b) $^{226}Ra \longrightarrow {}^{222}Rn + \alpha$
   c) $^{226}Ra + \alpha \longrightarrow {}^{226}Rn$
   d) $^{226}Ra + \alpha \longrightarrow {}^{222}Rn$

2. How many neutrons does a radium-226 nucleus contain?

   a) 226
   b) 88
   c) 113
   d) 138

3. What is the product when radon-222 decays by $\alpha$ emission?

   a) Radon-218
   b) Radium-218
   c) Polonium-218
   d) Polonium-226

4. What would the product be if radon-222 decayed by $\beta$ emission?

   a) Francium-222
   b) Radon-223
   c) Radon-221
   d) Astatine-222

# ANSWERS TO IN-CHAPTER MATERIALS

## Answers to Practice Problems

**21.1A** 1120 nm. **21.1B** HI, $F_2$, ClF. **21.2A** $H_2O$. **21.2B** To act as a greenhouse gas, a molecule must be IR-active. To be IR-active, a molecule must undergo a change in dipole moment as the result of one or more of its vibrations. **21.3A** $1.2 \times 10^{-6}$ mol/L. **21.3B** 52.8 days.

## Answers to Checkpoints

**21.2.1** a. **21.2.2** d. **21.5.1** a, d. **21.5.2** c. **21.8.1** e. **21.8.2** b.

## Answers to Applying What You've Learned

**a)** 190 nm. **b)** Yes, because it is a polar molecule and therefore is IR-active.

# 22

# Coordination Chemistry

Lead paint, known for its brightness and durability, was commonly used to paint homes, fences, and interior walls. Its use in homes was banned in 1978 because of the health risks associated with exposure to lead.

# How Coordination Chemistry Is Used to Treat Lead Poisoning

Although the Consumer Product Safety Commission (CPSC) banned the residential use of lead-based paint in 1978, millions of children remain at risk for exposure to lead from deteriorating paint in older homes. Lead poisoning is especially harmful to children under the age of 5 years because it interferes with growth and development and it has been shown to lower IQ. Symptoms of chronic exposure to lead include diminished appetite, nausea, malaise, and convulsions. *Blood lead level* (BLL), expressed as micrograms per deciliter ($\mu$g/dL), is used to monitor the effect of chronic exposure. A BLL $\leq$ 10 $\mu$g/dL is considered normal; a BLL > 45 $\mu$g/dL requires medical and environmental intervention. At high levels (>70 $\mu$g/dL), lead can cause seizures, coma, and death.

Treatment for lead poisoning involves *chelation therapy,* in which a chelating agent is administered orally, intravenously, or intramuscularly. Chelating agents form strong coordinate-covalent bonds to metal ions, forming stable, water-soluble complex ions that are easily removed from the body via the urine. One of the drugs commonly used for this purpose is dimercaptosuccinic acid (DMSA), marketed under the name Chemet. Chelation therapy relies on *coordination chemistry.*

Dimercaptosuccinic acid (DMSA)

At the end of this chapter, you will be able to answer questions about some important coordination compounds [▶▶ Page 939].

## In This Chapter You, Will Learn:

About the properties of coordination compounds and how, through the use of chelates, coordination chemistry is used to solve a variety of medical and other societal problems.

### Before you begin, you should review

- Lewis acids and bases [◀◀ Section 16.12]
- The shapes of *d* orbitals [◀◀ Section 6.7]

# 22.1    Coordination Compounds

*Coordination compounds* contain *coordinate covalent* bonds [◄◄ Section 8.8] formed by the reactions of metal ions with groups of *anions* or *polar molecules*. The metal ion in these kinds of reactions acts as a Lewis acid, accepting electrons, whereas the anions or polar molecules act as Lewis bases, donating pairs of electrons to form bonds to the metal ion. Thus, a coordinate covalent bond is a covalent bond in which one of the atoms donates *both* of the electrons that constitute the bond. Often a coordination compound consists of a *complex* ion and one or more *counter* ions. In writing formulas for such coordination compounds, we use square brackets to separate the complex ion from the counter ion.

$$K_2[PtCl_6]$$

This compound consists of the complex ion $PtCl_6^{2-}$ and two $K^+$ counter ions.

> **Student Annotation:** A complex ion is one in which a metal cation is covalently bound to one or more molecules or ions [◄◄ Section 17.5].

Some coordination compounds, such as $Fe(CO)_5$, do not contain complex ions. Most but not all of the metals in coordination compounds are transition metals. Our understanding of the nature of coordination compounds stems from the classic work of Alfred Werner,[1] who prepared and characterized many coordination compounds. In 1893, at the age of 26, Werner proposed what is now commonly referred to as Werner's coordination theory.

> **Student Annotation:** We can use the term *coordination complex* to refer to a compound, such as $Fe(CO)_5$, or to a complex ion.

Nineteenth-century chemists were puzzled by a certain class of reactions that seemed to violate valence theory. For example, the valences of the elements in cobalt(III) chloride and those in ammonia seem to be completely satisfied, and yet these two substances react to form a stable compound having the formula $CoCl_3 \cdot 6NH_3$. To explain this behavior, Werner postulated that most elements exhibit two types of valence: *primary valence* and *secondary valence*. In modern terminology, primary valence corresponds to the oxidation number and secondary valence to the coordination number of the element. In $CoCl_3 \cdot 6NH_3$, according to Werner, cobalt has a primary valence of 3 and a secondary valence of 6.

Today we use the formula $[Co(NH_3)_6]Cl_3$ to indicate that the ammonia molecules and the cobalt atom form a complex ion; the chloride ions are not part of the complex but are counter ions, held to the complex ion by Coulombic attraction.

## Properties of Transition Metals

Transition metals are those that either have incompletely filled *d* subshells or form ions with incompletely filled *d* subshells (Figure 22.1). Incompletely filled *d* subshells give rise to several notable properties, including distinctive colors, the formation of paramagnetic compounds, catalytic activity, and the tendency to form complex ions. The most common transition metals are scandium through copper, which occupy the fourth row of the periodic table. Table 22.1 lists the electron configurations and some of the properties of these metals.

> **Student Annotation:** The Group 2B metals—Zn, Cd, and Hg—do not fit either of these criteria. As a result, they are *d*-block metals, but they are not actually transition metals.

Multimedia

Chemical Reactions—oxidation states of vanadium.

Most of the transition metals exhibit a close-packed structure in which each atom has a coordination number of 12. Furthermore, these elements have relatively small atomic radii. The combined effect of closest packing and small atomic size results in strong metallic bonds. Therefore,

**Figure 22.1**    The transition metals (shown in green). Note that although the Group 2B elements (Zn, Cd, Hg) are described as transition metals by some chemists, neither the metals nor their ions possess incompletely filled *d* subshells.

---

1. Alfred Werner (1866–1919). Swiss chemist. Werner started as an organic chemist but did his most notable work in coordination chemistry. For his theory of coordination compounds, Werner was awarded the Nobel Prize in Chemistry in 1913.

| **TABLE 22.1** | Electron Configurations and Other Properties of the Fourth-Period Transition Metals | | | | | | | | |
|---|---|---|---|---|---|---|---|---|---|
| | Sc | Ti | V | Cr | Mn | Fe | Co | Ni | Cu |
| **Electron configuration** | | | | | | | | | |
| M | $4s^23d^1$ | $4s^23d^2$ | $4s^23d^3$ | $4s^13d^5$ | $4s^23d^5$ | $4s^23d^6$ | $4s^23d^7$ | $4s^23d^8$ | $4s^13d^{10}$ |
| $M^{2+}$ | — | $3d^2$ | $3d^3$ | $3d^4$ | $3d^5$ | $3d^6$ | $3d^7$ | $3d^8$ | $3d^9$ |
| $M^{3+}$ | [Ar] | $3d^1$ | $3d^2$ | $3d^3$ | $3d^4$ | $3d^5$ | $3d^6$ | $3d^7$ | $3d^8$ |
| **Electronegativity** | 1.3 | 1.5 | 1.6 | 1.6 | 1.5 | 1.8 | 1.9 | 1.9 | 1.9 |
| **Ionization energy (kJ/mol)** | | | | | | | | | |
| First | 631 | 658 | 650 | 652 | 717 | 759 | 760 | 736 | 745 |
| Second | 1235 | 1309 | 1413 | 1591 | 1509 | 1561 | 1645 | 1751 | 1958 |
| Third | 2389 | 2650 | 2828 | 2986 | 3250 | 2956 | 3231 | 3393 | 3578 |
| **Radius (pm)** | | | | | | | | | |
| M | 162 | 147 | 134 | 130 | 135 | 126 | 125 | 124 | 128 |
| $M^{2+}$ | — | 90 | 88 | 85 | 80 | 77 | 75 | 69 | 72 |
| $M^{3+}$ | 81 | 77 | 74 | 64 | 66 | 60 | 64 | | |

**Student Annotation:** The trends in radii, electronegativity, and ionization energy for the first row of transition metals are somewhat different from those for main group elements. This is largely the effect of *shielding* of the 4s electrons by the 3d electrons.

**Student Annotation:** The electron configurations of the first-row transition metals and their ions were discussed in Section 6.9 and Section 7.5, respectively.

| **TABLE 22.2** | Physical Properties of Elements K to Zn | | | | | | | | | | | |
|---|---|---|---|---|---|---|---|---|---|---|---|---|
| | 1A | 2A | | Transition Metals | | | | | | | | 2B |
| | K | Ca | Sc | Ti | V | Cr | Mn | Fe | Co | Ni | Cu | Zn |
| Atomic radius (pm) | 235 | 197 | 162 | 147 | 134 | 130 | 135 | 126 | 125 | 124 | 128 | 138 |
| Melting point (°C) | 63.7 | 838 | 1539 | 1668 | 1900 | 1875 | 1245 | 1536 | 1495 | 1453 | 1083 | 419.5 |
| Boiling point (°C) | 760 | 1440 | 2730 | 3260 | 3450 | 2665 | 2150 | 3000 | 2900 | 2730 | 2595 | 906 |
| Density (g/cm³) | 0.86 | 4.51 | 3.0 | 4.51 | 6.1 | 7.19 | 7.43 | 7.86 | 8.9 | 8.9 | 8.96 | 7.14 |

transition metals have higher densities, higher melting points and boiling points, and higher heats of fusion and vaporization than the main group and Group 2B metals (Table 22.2).

Transition metals exhibit variable oxidation states in their compounds. Figure 22.2 shows the oxidation states of the first row of transition metals. Note that all these metals can exhibit the oxidation state +3 and *nearly* all can exhibit the oxidation state +2. Of these two, the +2 oxidation state is somewhat more common for the heavier elements. The highest oxidation state for a transition metal is +7, exhibited by manganese ($4s^23d^5$). Transition metals exhibit their highest oxidation states in compounds that contain highly electronegative elements such as oxygen and fluorine—for example, $V_2O_5$, $CrO_3$, and $Mn_2O_7$.

**Student Annotation:** The oxidation state of O in each of these compounds is −2, making those of V, Cr, and Mn +5, +6, and +7, respectively [◄◄ Section 4.4].

| | | | +7 | | | | | |
|---|---|---|---|---|---|---|---|---|
| | | +6 | +6 | +6 | | | | |
| | +5 | +5 | +5 | +5 | | | | |
| | +4 | +4 | +4 | +4 | +4 | +4 | | |
| +3 | +3 | +3 | +3 | +3 | +3 | +3 | +3 | +3 |
| | +2 | +2 | +2 | +2 | +2 | +2 | +2 | +2 |
| | | | | | | | | +1 |
| Sc | Ti | V | Cr | Mn | Fe | Co | Ni | Cu |

**Figure 22.2** Oxidation states of the first-row transition metals. The most stable oxidation numbers are shown in red. The zero oxidation state is encountered in some compounds, such as $Ni(CO)_4$ and $Fe(CO)_5$.

| TABLE 22.3 | Common Ligands |
|---|---|

| Name | Structure |
|---|---|
| Monodentate | |
| Ammonia | $H-\overset{..}{N}-H$ with $H$ below |
| Carbon monoxide | $:C\equiv O:$ |
| Chloride ion | $:\overset{..}{\underset{..}{Cl}}:^-$ |
| Cyanide ion | $[:C\equiv N:]^-$ |
| Thiocyanate ion | $[:\overset{..}{\underset{..}{S}}-C\equiv N:]^-$ |
| Water | $H-\overset{..}{\underset{..}{O}}-H$ |
| Bidentate | |
| Ethylenediamine | $H_2\overset{..}{N}-CH_2-CH_2-\overset{..}{N}H_2$ |
| Oxalate ion | (structure of oxalate ion) |
| Polydentate | |
| Ethylenediaminetetraacetate ion (EDTA) | (structure of EDTA) |

## Ligands

The molecules or ions that surround the metal in a complex ion are called **ligands** (Table 22.3). The formation of covalent bonds between ligands and a metal can be thought of as a Lewis acid-base reaction. (Recall that a Lewis base is a species that donates a pair of electrons [◄◄ Section 16.12].) To be a ligand, a molecule or ion must have at least one unshared pair of valence electrons, as these examples illustrate:

$$H-\overset{\overset{..}{O}}{\underset{}{}}-H \qquad H-\overset{..}{N}-H \text{ (with H)} \qquad :\overset{..}{\underset{..}{Cl}}:^- \qquad :C\equiv O:$$

Therefore, ligands play the role of Lewis bases. The transition metal, on the other hand, acts as a Lewis acid, accepting (and sharing) pairs of electrons from the Lewis bases.

The atom in a ligand that is bound directly to the metal atom is known as the **donor atom.** For example, *nitrogen* is the donor atom in the $[Cu(NH_3)_4]^{2+}$ complex ion:

(structure of the complex ion $[Cu(NH_3)_4]^{2+}$)

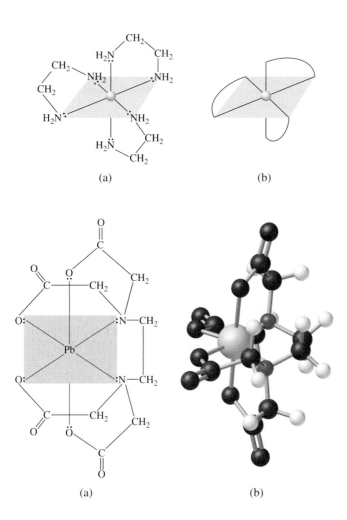

**Figure 22.3** (a) Structure of a metal-ethylenediamine complex cation, such as $[Co(en)_3]^{2+}$. Each ethylenediamine molecule provides two N donor atoms and is therefore a bidentate ligand. (b) Simplified structure of the same complex cation.

**Figure 22.4** (a) EDTA complex of lead. The complex bears a net charge of 2− because each O donor atom has one negative charge and the lead ion carries two positive charges. Only the lone pairs that participate in bonding are shown. Note the octahedral geometry about the $Pb^{2+}$ ion. (b) Molecular model of the $Pb^{2+}$-EDTA complex. The light green sphere is the $Pb^{2+}$ ion.

The **coordination number** in a coordination compound refers to the number of donor atoms surrounding the central metal atom in a complex ion. The coordination number of $Cu^{2+}$ in $[Cu(NH_3)_4]^{2+}$ is 4. The most common coordination numbers are 4 and 6, although coordination numbers of 2 and 5 are also known.

Depending on the number of donor atoms a ligand possesses, it is classified as monodentate (1 donor atom), bidentate (2 donor atoms), or polydentate (> 2 donor atoms). Table 22.3 lists some common ligands. Figure 22.3 shows how ethylenediamine, sometimes abbreviated "en," forms two bonds to a metal atom.

Bidentate and polydentate ligands are also called **chelating agents** because of their ability to hold the metal atom like a claw (from the Greek *chele,* meaning "claw"). One example is EDTA (Figure 22.4), a polydentate ligand used to treat metal poisoning. Six donor atoms enable EDTA to form a very stable complex ion with lead. This stable complex enables the body to remove lead from the blood.

The oxidation state of a transition metal in a complex ion is determined using the known charges of the *ligands* and the known *overall* charge of the complex ion. In the complex ion $[PtCl_6]^{2-}$, for example, each chloride ion ligand has an oxidation number of −1. For the overall charge of the ion to be −2, the Pt must have an oxidation number of +4.

Sample Problem 22.1 shows how to determine transition metal oxidation states in coordination compounds.

**Student Annotation:** Remember that the oxidation number of a monoatomic ion is equal to the charge [◂◂ Section 4.4].

---

## Sample Problem 22.1

Determine the oxidation state of the central metal atom in each of the following compounds:
(a) $[Ru(NH_3)_5(H_2O)]Cl_2$, (b) $[Cr(NH_3)_6](NO_3)_3$, and (c) $Fe(CO)_5$.

**Strategy** Identify the components of each compound, and use known oxidation states and charges to determine the oxidation state of the metal.

*(Continued)*

**Setup** (a) $[Ru(NH_3)_5(H_2O)]Cl_2$ consists of a complex ion (the part of the formula enclosed in square brackets) and two $Cl^-$ counter ions. Because the overall charge on the compound is zero, the complex ion is $[Ru(NH_3)_5(H_2O)]^{2+}$. There are six ligands: five ammonia molecules and one water molecule. Each molecule has a zero charge (i.e., each ligand is neutral), so the charge on the metal is equal to the overall charge on the complex ion.

(b) $[Cr(NH_3)_6](NO_3)_3$ consists of a complex ion and three $NO_3^-$ ions, making the complex ion $[Cr(NH_3)_6]^{3+}$. Each of the six ammonia molecule ligands is neutral (i.e., each has a zero charge), making the charge on the metal equal to the overall charge on the complex ion.

(c) $Fe(CO)_5$ does not contain a complex ion. The ligands are CO molecules, which have a zero charge, so the central metal also has a zero charge.

**Solution** (a) +2

(b) +3

(c) 0

> **Think About It** To solve a problem like this, you must be able to recognize the common polyatomic ions and you must know their charges.

**Practice Problem A** Give oxidation numbers for the metals in (a) $K[Au(OH)_4]$ and (b) $K_4[Fe(CN)_6]$.

**Practice Problem B** The oxidation state of cobalt in each of the following species is +3. Determine the overall charge on each complex: $[Co(NH_3)_4Cl_2]^?$, $[Co(CN)_6]^?$, $[Co(NH_3)_3Br_3]^?$, $[Co(en)_2Cl_2]^?$, $[Co(en)_3]^?$, $[Co(NH_3)_6]^?$, $[Co(NH_3)_5Cl]^?$.

## Nomenclature of Coordination Compounds

Now that we have discussed the various types of ligands and the oxidation numbers of metals, our next step is to learn how to name coordination compounds. The rules for naming ionic coordination compounds are as follows:

1.  The cation is named before the anion, as in other ionic compounds. The rule holds regardless of whether the complex ion bears a net positive or a net negative charge. In the compounds $K_2[Fe(CN)_6]$ and $[Co(NH_3)_4]Cl$, for example, we name the $K^+$ and $[Co(NH_3)_4]^+$ cations first, respectively.

2.  Within a complex ion, the ligands are named first, in alphabetical order, and the metal ion is named last.

3.  The names of anionic ligands end with the letter *o*, whereas neutral ligands are usually called by the names of the molecules. The exceptions are $H_2O$ (aqua), CO (carbonyl), and $NH_3$ (ammine). Table 22.4 lists some common ligands and their nomenclature.

| TABLE 22.4 | Names of Common Ligands in Coordination Compounds |
|---|---|
| **Ligand** | **Name of Ligand in Coordination Compound** |
| Bromide ($Br^-$) | Bromo |
| Chloride ($Cl^-$) | Chloro |
| Cyanide ($CN^-$) | Cyano |
| Hydroxide ($OH^-$) | Hydroxo |
| Oxide ($O^{2-}$) | Oxo |
| Carbonate ($CO_3^{2-}$) | Carbonato |
| Nitrite ($NO_2^-$) | Nitro |
| Oxalate ($C_2O_4^{2-}$) | Oxalato |
| Ammonia ($NH_3$) | Ammine |
| Carbon monoxide (CO) | Carbonyl |
| Water ($H_2O$) | Aqua |
| Ethylenediamine | Ethylenediamine |
| Ethylenediaminetetraacetate | Ethylenediaminetetraacetate |

| TABLE 22.5 | Names of Anions Containing Metal Atoms |
|---|---|
| **Metal** | **Name of Metal in Anionic Complex** |
| Aluminum | Aluminate |
| Chromium | Chromate |
| Cobalt | Cobaltate |
| Copper | Cuprate |
| Gold | Aurate |
| Iron | Ferrate |
| Lead | Plumbate |
| Manganese | Manganate |
| Molybdenum | Molybdate |
| Nickel | Nickelate |
| Silver | Argentate |
| Tin | Stannate |
| Tungsten | Tungstate |
| Zinc | Zincate |

4. When two or more of the same ligand are present, use Greek prefixes *di, tri, tetra, penta,* and *hexa,* to specify their number. Thus, the ligands in the cation $[Co(NH_3)_4Cl_2]^+$ are "tetraamminedichloro." (Note that prefixes are *not* used for the purpose of *alphabetizing* the ligands.)

5. The oxidation number of the metal is indicated in Roman numerals immediately following the name of the metal. For example, the Roman numeral III is used to indicate the $+3$ oxidation state of chromium in $[Cr(NH_3)_4Cl_2]^+$, which is called tetraamminedichlorochromium(III) ion.

6. If the complex is an anion, its name ends in *–ate.* In $K_4[Fe(CN)_6]$, for example, the anion $[Fe(CN)_6]^{4-}$ is called hexacyanoferrate(II) ion. Note that the Roman numeral indicating the oxidation state of the metal *follows* the suffix *–ate.* Table 22.5 lists the names of anions containing metal atoms.

Sample Problems 22.2 and 22.3 apply these rules to the nomenclature of coordination compounds.

**Student Annotation:** When the name of a ligand already contains a Greek prefix, a different set of prefixes is used to denote the number of the ligand:

| 2 | *bis* |
| 3 | *tris* |
| 4 | *tetrakis* |

Two ethylenediamine ligands, for example, would be specified by *bis*(ethylenediamine).

**Student Annotation:** Note that there is no space between the compound name and the parenthetical Roman numeral.

---

## Sample Problem 22.2

Write the names of the following coordination compounds: (a) $[Co(NH_3)_4Cl_2]Cl$ and (b) $K_3[Fe(CN)_6]$.

**Strategy** For each compound, name the cation first and the anion second. Refer to Tables 22.4 and 22.5 for the names of ligands and anions containing metal atoms.

**Setup** (a) The cation is a complex ion containing four ammonia molecules and two chloride ions. The counter ion is chloride ($Cl^-$), so the charge on the complex cation is $+1$, making the oxidation state of cobalt $+3$.

(b) The cation is $K^+$, and the anion is a complex ion containing six cyanide ions. The charge on the complex ion is $-3$, making the oxidation state of iron $+3$.

**Solution** (a) Tetraamminedichlorocobalt(III) chloride

(b) Potassium hexacyanoferrate(III)

**Think About It** When the anion is a *complex* ion, its name must end in *–ate,* followed by the metal's oxidation state in Roman numerals. Also, do not use prefixes to denote numbers of counter ions.

---

**Practice Problem A** Give the correct name for (a) $[Co(NH_3)_4Br_2]Cl$, (b) $[Cr(H_2O)_4Cl_2]Cl$, and (c) $K_2[CuCl_4]$.

**Practice Problem B** Give the correct name for (a) $Na_3[Fe(CN)_6]$, (b) $[Cr(en)_2Cl_2]Br$, and (c) $[Co(en)_3]Cl_3$. [*Hint:* For parts (b) and (c), read the Student Annotation above regarding the use of prefixes.]

## Sample Problem 22.3

Write formulas for the following compounds: (a) pentaamminechlorocobalt(III) chloride and (b) dichloro*bis*(ethylenediamine)platinum(IV) nitrate.

**Strategy** If you can't remember them yet, refer to Tables 22.4 and 22.5 for the names of ligands and anions containing metal atoms.

**Setup** (a) There are six ligands: five $NH_3$ molecules and one $Cl^-$ ion. The oxidation state of cobalt is +3, making the overall charge on the complex ion +2. Therefore, there are two chloride ions as counter ions.

(b) There are four ligands: two bidentate ethylenediamines and two $Cl^-$ ions. The oxidation state of platinum is +4, making the overall charge on the complex ion +2. Therefore, there are two nitrate ions as counter ions.

**Solution** (a) $[Co(NH_3)_5Cl]Cl_2$

(b) $[Pt(en)_2Cl_2](NO_3)_2$

**Think About It** Although ligands are alphabetized in a compound's name, they do not necessarily appear in alphabetical order in the compound's formula.

**Practice Problem A** Write the formulas for (a) pentaaquabromoruthenium(II) nitrate, (b) potassium tetrabromodichloroplatinate(IV), and (c) sodium hexanitrocobaltate(III).

**Practice Problem B** Write the formulas for (a) *bis*(ethylenediamine)oxalatovanadium(IV) chloride, (b) dibromo*bis*(ethylenediamine)chromium(III) nitrate, and (c) *tris*(ethylenediamine)platinum(IV) sulfate. (*Hint:* Read the Student Annotation on page 927.)

## Checkpoint 22.1    Coordination Compounds

**22.1.1** Select the correct name for the compound $[Cu(NH_3)_4]Cl_2$.

a) Coppertetraammine dichloride

b) Tetraamminecopper(II) chloride

c) Tetraaminedichlorocuprate(II)

d) Dichlorotetraaminecopper(II)

e) Tetraaminedichlorocopper(II)

**22.1.2** Select the correct name for the compound $K_3[FeF_6]$.

a) Tripotassiumironhexafluoride

b) Hexafluorotripotassiumferrate(III)

c) Hexafluoroiron(III) potassium

d) Potassium hexafluoroferrate(III)

e) Potassium ironhexafluorate

**22.1.3** Select the correct formula for pentaamine nitrocobalt(III).

a) $[Co(NH_3)_5NO_2]^{3+}$

b) $[Co(NH_3)_5NO_2]^{2+}$

c) $Co(NH_3)_5NO_2$

d) $[Co(NH_3)_5](NO_2)$

e) $[Co(NH_3)_5](NO_2)_2$

**22.1.4** Select the correct formula for tetraaquadichlorochromium(III) chloride.

a) $[Cr(H_2O)_4Cl_2]Cl_3$

b) $[Cr(H_2O)_4Cl_2]Cl_2$

c) $[Cr(H_2O)_4Cl_2]Cl$

d) $[Cr(H_2O)_4]Cl_3$

e) $[Cr(H_2O)_4]Cl_2$

## 22.2    Structure of Coordination Compounds

The geometry of a coordination compound often plays a significant role in determining its properties. Figure 22.5 shows four different geometric arrangements for metal atoms with monodentate ligands. In these diagrams we see that structure and the coordination number of the metal relate to each other as follows:

| Coordination Number | Structure |
|---|---|
| 2 | Linear |
| 4 | Tetrahedral or square planar |
| 6 | Octahedral |

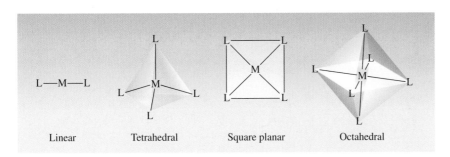

**Figure 22.5** Common geometries of complex ions. In each case M is a metal and L is a monodentate ligand.

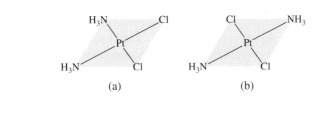

**Figure 22.6** The (a) *cis* and (b) *trans* isomers of diamminedichloroplatinum(II). Note that the two Cl atoms are adjacent to each other in the *cis* isomer and diagonally across from each other in the *trans* isomer.

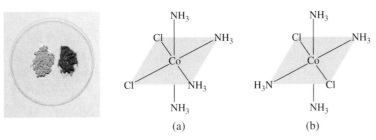

**Figure 22.7** The (a) *cis* and (b) *trans* isomers of tetraamminedichlorocobalt(III) ion, $[Co(NH_3)_4Cl_2]^+$. The ion has only two geometric isomers.

In studying the geometry of coordination compounds, we sometimes find that there is more than one way to arrange the ligands around the central atom. Such compounds in which ligands are arranged differently, known as **stereoisomers**, have distinctly different physical and chemical properties. Coordination compounds may exhibit two types of stereoisomerism: *geometric* and *optical.*

   **Geometrical isomers** are stereoisomers that cannot be interconverted without breaking chemical bonds. Geometric isomers come in pairs. We use the terms *cis* and *trans* to distinguish one geometric isomer of a compound from the other. *Cis* means that two particular atoms (or groups of atoms) are adjacent to each other, and *trans* means that the atoms (or groups of atoms) are on opposite sides in the structural formula. The *cis* and *trans* isomers of coordination compounds generally have quite different colors, melting points, dipole moments, and chemical reactivities. Figure 22.6 shows the *cis* and *trans* isomers of diamminedichloroplatinum(II). Note that although the types of bonds are the same in both isomers (two Pt—N and two Pt—Cl bonds), the spatial arrangements are different. Another example is the tetraamminedichlorocobalt(III) ion, shown in Figure 22.7.

   **Optical isomers** are nonsuperimposable mirror images. (*Superimposable* means that if one structure is laid over the other, the positions of all the atoms will match.) Like geometric isomers, optical isomers come in pairs. However, the optical isomers of a compound have *identical* physical and chemical properties, such as melting point, boiling point, dipole moment, and chemical reactivity toward molecules that are not *themselves* optical isomers. Optical isomers differ from each other, though, in their interactions with plane-polarized light, as we will see.

   The structural relationship between two optical isomers is analogous to the relationship between your left and right hands. If you place your left hand in front of a mirror, the image you see will look like your right hand (Figure 22.8). Your left hand and right hand are mirror images of each other. They are nonsuperimposable, however, because when you place your left hand over your right hand (with both palms facing down), they do not match. This is why a right-handed glove will not fit comfortably on your left hand.

   Figure 22.9 shows the *cis* and *trans* isomers of dichlorobis(ethylenediamine)cobalt(III) ion and the mirror image of each. Careful examination reveals that the *trans* isomer and its mirror image are superimposable, but the *cis* isomer and its mirror image are not. Thus, the *cis* isomer and its mirror image are *optical isomers.*

   Optical isomers are described as *chiral* (from the Greek word for "hand") because, like your left and right hands, chiral molecules are nonsuperimposable. Isomers that are superimposable

**Student Annotation:** In general, stereoisomers are compounds that are made up of the same types and numbers of atoms, bonded together in the same sequence, but with different spatial arrangements.

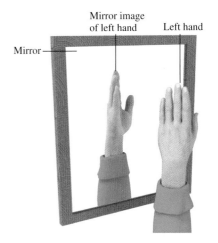

**Figure 22.8** A left hand and its mirror image.

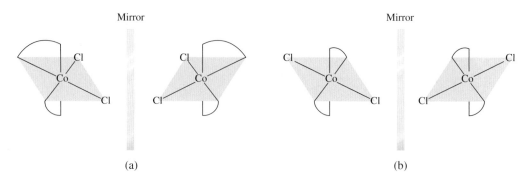

(a)                                                    (b)

**Figure 22.9**    The (a) *cis* and (b) *trans* isomers of dichlorobis(ethylenediamine)cobalt(III) ion and their mirror images. If you could rotate the mirror image in (b) 90° clockwise about the vertical position and place the ion over the *trans* isomer, you would find that the two are superimposable. No matter how you rotate the *cis* isomer and its mirror image in (a), however, you cannot superimpose one on the other.

with their mirror images are said to be *achiral.* Chiral molecules play a vital role in enzyme reactions in biological systems. Many drug molecules are chiral, although only one of a pair of chiral isomers is biologically effective [◄◄ Section 10.4].

Chiral molecules are said to be optically active because of their ability to rotate the plane of polarization of polarized light as it passes through them. Unlike ordinary light, which vibrates in all directions, plane-polarized light vibrates only in a single plane. We use a ***polarimeter*** to measure the rotation of polarized light by optical isomers (Figure 22.10). A beam of unpolarized light first passes through a Polaroid sheet, called the polarizer, and then through a sample tube containing a solution of an optically active, chiral compound. As the polarized light passes through the sample tube, its plane of polarization is rotated either to the right (clockwise) or to the left (counterclockwise). This rotation can be measured directly by turning the analyzer in the appropriate direction until minimal light transmission is achieved (Figure 22.11). If the plane of polarization is rotated to the right, the isomer is said to be ***dextrorotatory*** and the isomer is labeled *d;* if the rotation is to the left, the isomer is ***levorotatory*** and the isomer is labeled *l.* The *d* and *l* isomers of a chiral substance, called ***enantiomers,*** always rotate the plane of polarization by the same amount, but in opposite directions. Thus, in an equimolar mixture of two enantiomers, called a ***racemic mixture,*** the net rotation is zero.

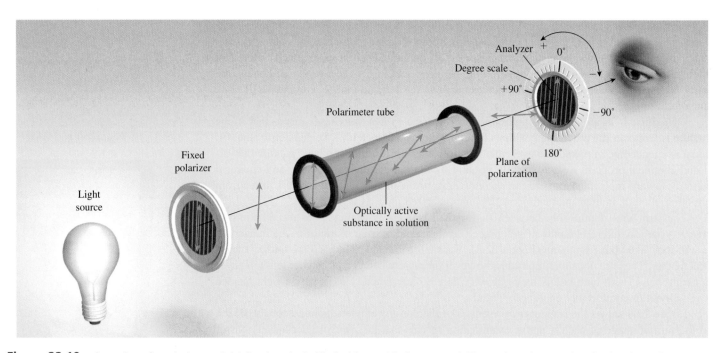

**Figure 22.10**    Operation of a polarimeter. Initially, the tube is filled with an achiral compound. The analyzer is rotated so that its plane of polarization is perpendicular to that of the polarizer. Under this condition, no light reaches the observer. Next, a chiral compound is placed in the tube as shown. The plane of polarization of the polarized light is rotated as it travels through the tube so that some light reaches the observer. Rotating the analyzer (either to the left or to the right) until no light reaches the observer again allows the angle of optical rotation to be measured.

**Figure 22.11** Polarized lenses. No light passes through the lenses when they are rotated so that their planes of polarization are perpendicular.

## 22.3 Bonding in Coordination Compounds: Crystal Field Theory

A satisfactory theory of bonding in coordination compounds must account for properties such as color and magnetism, as well as stereochemistry and bond strength. No single theory as yet does all this for us. Rather, several different approaches have been applied to transition metal complexes. We will consider only one of them here—crystal field theory—because it accounts for both the color and magnetic properties of many coordination compounds.

We will begin our discussion of crystal field theory with the most straightforward case—namely, complex ions with octahedral geometry. Then we will see how it is applied to tetrahedral and square-planar complexes.

### Crystal Field Splitting in Octahedral Complexes

Crystal field theory explains the bonding in complex ions purely in terms of electrostatic forces. In a complex ion, two types of electrostatic interaction come into play. One is the attraction between the positive metal ion and the negatively charged ligand or the negatively charged end of a polar ligand. This is the force that binds the ligands to the metal. The second type of interaction is the electrostatic repulsion between the lone pairs on the ligands and the electrons in the $d$ orbitals of the metals.

The $d$ orbitals have different orientations [◀◀ Section 6.7], but in the absence of an external disturbance, they all have the same energy. In an octahedral complex, a central metal atom is surrounded by six lone pairs of electrons (on the six ligands), so all five $d$ orbitals experience electrostatic repulsion. The magnitude of this repulsion depends on the orientation of the $d$ orbital that is involved. Take the $d_{x^2-y^2}$ orbital as an example. In Figure 22.12, we see that the lobes of this orbital point toward the corners of the octahedron along the $x$ and $y$ axes, where the lone-pair electrons are positioned. Thus, an electron residing in this orbital would experience a greater repulsion from the ligands than an electron would in the $d_{xy}$, $d_{yz}$, or $d_{xz}$ orbitals. For this reason, the energy of the $d_{x^2-y^2}$ orbital is increased relative to the $d_{xy}$, $d_{yz}$, and $d_{xz}$ orbitals. The $d_{z^2}$ orbital's energy is also greater, because its lobes are pointed at the ligands along the $z$ axis. As a result of these metal-ligand interactions, the five $d$ orbitals in an octahedral complex are split between two sets of energy levels: a higher level with two orbitals ($d_{x^2-y^2}$ and $d_{z^2}$) having the same energy, and a lower level with three equal-energy orbitals ($d_{xy}$, $d_{yz}$, and $d_{xz}$), as shown in Figure 22.13. The **crystal field splitting ($\Delta$)** is the energy difference between two sets of $d$ orbitals in a metal atom when ligands are present. The magnitude of $\Delta$ depends on the metal and the nature of the ligands; it has a direct effect on the color and magnetic properties of complex ions.

**Figure 22.12** The five *d* orbitals in an octahedral environment. The metal atom (or ion) is at the center of the octahedron, and the six lone pairs on the donor atoms of the ligands are at the corners.

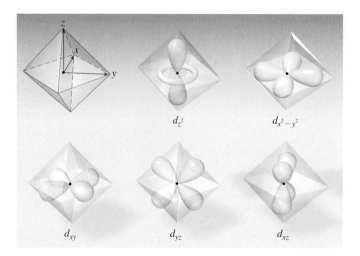

**Figure 22.13** Crystal field splitting between *d* orbitals in an octahedral complex.

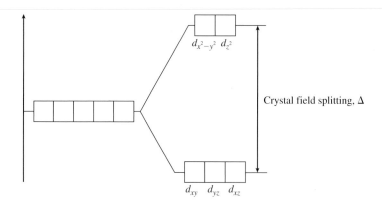

## Color

In Chapter 6 we learned that white light, such as sunlight, is a combination of all colors. A substance appears black if it absorbs all the visible light that strikes it. If it absorbs no visible light, it is white or colorless. An object appears green if it absorbs all light but reflects the green component. An object also looks green if it reflects all colors except red, the *complementary* color of green (Figure 22.14).

What has been said of reflected light also applies to *transmitted* light (i.e., the light that passes *through* the medium, such as a solution). Consider the hydrated cupric ion ($[Cu(H_2O)_6]^{2+}$); it absorbs light in the orange region of the spectrum, so a solution of $CuSO_4$ appears blue to us. Recall from Chapter 6 that when the energy of a photon is equal to the difference between the ground state and an excited state, absorption occurs as the photon strikes the atom (or ion or compound), and an electron is promoted to a higher level. Using these concepts, we can calculate the energy change involved in the electron transition. The energy of a photon is given by

$$E = h\nu$$

where $h$ represents Planck's constant ($6.63 \times 10^{-34}$ J · s) and $\nu$ is the frequency of the radiation, which is $5.00 \times 10^{14}$ s$^{-1}$ for a wavelength of 600 nm. Here $E = \Delta$, so we have

$$\Delta = h\nu$$

$$= (6.63 \times 10^{-34} \text{ J} \cdot \text{s})(5.00 \times 10^{14} \text{ s})$$

$$= 3.32 \times 10^{-19} \text{ J}$$

This value is very small, but it is the energy absorbed by only *one* ion. If the wavelength of the photon absorbed by an ion lies outside the visible region, then the transmitted light looks the same (to us) as the incident light—white—and the ion appears colorless.

The best way to measure crystal field splitting is to use spectroscopy to determine the wavelength at which light is absorbed. The $[Ti(H_2O)_6]^{3+}$ ion provides a straightforward example, because $Ti^{3+}$ has only one $3d$ electron (Figure 22.15). The $[Ti(H_2O)_6]^{3+}$ ion absorbs light in the visible region of the spectrum (Figure 22.16). The wavelength corresponding to maximum

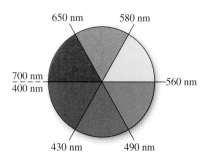

**Figure 22.14** A color wheel with appropriate wavelengths. Complementary colors, such as red and green, are on opposite sides of the wheel.

(a)

(b)

**Figure 22.15** (a) The process of photon absorption, and (b) a graph of the absorption spectrum of $[Ti(H_2O)_6]^{3+}$. The energy of the incoming photon is equal to the crystal field splitting. The maximum absorption peak in the visible region occurs at 498 nm.

absorption is 498 nm [Figure 22.15(b)]. To calculate the crystal field splitting energy, we start by writing

$$\Delta = h\nu$$

Next, recall that

$$\nu = \frac{c}{\lambda}$$

where $c$ is the speed of light and $\lambda$ is the wavelength. Therefore,

$$\Delta = \frac{hc}{\lambda} = \frac{(6.63 \times 10^{-34} \text{ J} \cdot \text{s})(3.00 \times 10^8 \text{ m/s})}{(498 \text{ nm})(1 \times 10^{-9} \text{ m/nm})} = 3.99 \times 10^{-19} \text{ J}$$

This is the energy required to excite *one* $[Ti(H_2O)_6]^{3+}$ ion. To express this energy difference in the more convenient units of kJ/mol, we write

$$\Delta = (3.99 \times 10^{-19} \text{ J/ion})(6.02 \times 10^{23} \text{ ions/mol})$$

$$= 240,000 \text{ J/mol}$$

$$= 240 \text{ kJ/mol}$$

Aided by spectroscopic data for a number of complexes, all having the same metal ion but different ligands, chemists calculated the crystal field splitting for each ligand and established the following **spectrochemical series,** which is a list of ligands arranged in increasing order of their abilities to split the *d* orbital energy levels:

$$I^- < Br^- < Cl^- < OH^- < F^- < H_2O < NH_3 < en < CN^- < CO$$

These ligands are arranged in the order of increasing value of $\Delta$. CO and $CN^-$ are called *strong-field ligands,* because they cause a large splitting of the *d* orbital energy levels. The halide ions and hydroxide ion are *weak-field ligands,* because they split the *d* orbitals to a lesser extent.

## Magnetic Properties

The magnitude of the crystal field splitting also determines the magnetic properties of a complex ion. The $[Ti(H_2O)_6]^{3+}$ ion, having only one *d* electron, is always paramagnetic. However, for an ion with several *d* electrons, the situation is less immediately clear. Consider, for example, the octahedral complexes $[FeF_6]^{3-}$ and $[Fe(CN)_6]^{3-}$ (Figure 22.17). The electron configuration of $Fe^{3+}$ is $[Ar]3d^5$, and there are two possible ways to distribute the five *d* electrons among the *d* orbitals. According to Hund's rule [◄◄ Section 6.8], maximum stability is reached when the

**Figure 22.16** Colors of some of the first-row transition metal ions in solution. From left to right: $Ti^{3+}$, $Cr^{3+}$, $Mn^{2+}$, $Fe^{3+}$, $Co^{2+}$, $Ni^{2+}$, $Cu^{2+}$. The $Sc^{3+}$ and $V^{5+}$ ions are colorless.

**Figure 22.17** Energy-level diagrams for the $Fe^{3+}$ ion and for the $[FeF_6]^{3-}$ and $[Fe(CN)_6]^{3-}$ complex ions.

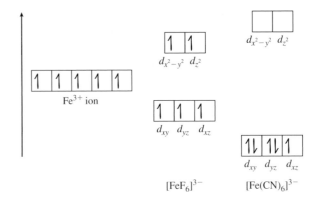

electrons are placed in five separate orbitals with parallel spins. This arrangement can be achieved only at a cost, however, because two of the five electrons must be promoted to the higher-energy $d_{x^2-y^2}$ and $d_{z^2}$ orbitals. No such energy investment is needed if all five electrons enter the $d_{xy}$, $d_{yz}$, and $d_{xz}$ orbitals. According to Pauli's exclusion principle [◀◀ Section 6.8], there will be only one unpaired electron present in this case.

Figure 22.18 shows the distribution of electrons among $d$ orbitals that results in low- and high-spin complexes. The actual arrangement of the electrons is determined by the amount of

**Figure 22.18** Orbital diagrams for the high-spin and low-spin octahedral complexes corresponding to the electron configurations of (a) $d^4$, (b) $d^5$, (c) $d^6$, and (d) $d^7$.

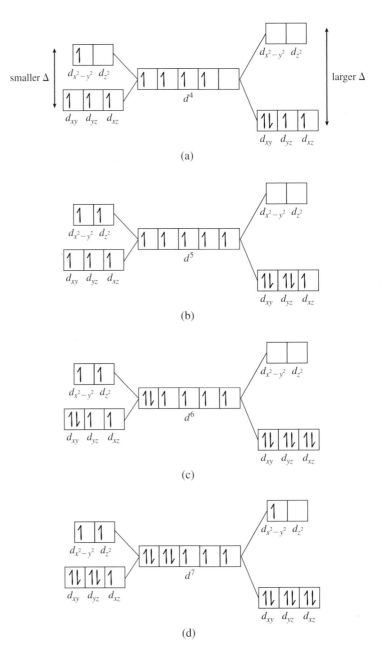

stability gained by having maximum parallel spins versus the investment in energy required to promote electrons to higher $d$ orbitals. Because $F^-$ is a weak-field ligand, the five $d$ electrons enter five separate $d$ orbitals with parallel spins to create a high-spin complex. The cyanide ion is a strong-field ligand, though, so it is energetically preferable for all five electrons to be in the lower orbitals, thus forming a low-spin complex. High-spin complexes are more paramagnetic than low-spin complexes.

The actual number of unpaired electrons (or spins) in a complex ion can be found by magnetic measurements, and in general, experimental findings support predictions based on crystal field splitting. However, a distinction between low- and high-spin complexes can be made only if the metal ion contains more than three and fewer than eight $d$ electrons, as shown in Figure 22.18.

Sample Problem 22.4 shows how to determine the number of spins in an octahedral complex.

---

## Sample Problem 22.4

Predict the number of unpaired spins in the $[Cr(en)_3]^{2+}$ ion.

**Strategy** The magnetic properties of a complex ion depend on the strength of the ligands. Strong-field ligands, which cause a high degree of splitting among the $d$ orbital energy levels, result in low-spin complexes. Weak-field ligands, which cause only a small degree of splitting among the $d$ orbital energy levels, result in high-spin complexes.

**Setup** The electron configuration of $Cr^{2+}$ is $[Ar]3d^4$; and en is a strong-field ligand.

**Solution** Because en is a *strong*-field ligand, we expect $[Cr(en)_3]^{2+}$ to be a low-spin complex. According to Figure 22.18, all four electrons will be placed in the lower-energy $d$ orbitals ($d_{xy}$, $d_{yz}$, and $d_{xz}$) and there will be a total of two unpaired spins.

> **Think About It** It is easy to draw the wrong conclusion regarding high- and low-spin complexes. Remember that the term *high spin* refers to the number of spins (*unpaired electrons*), not to the energy levels of the $d$ orbitals. The greater the energy gap between the lower-energy and higher-energy $d$ orbitals, the greater the chance that the complex will be *low* spin.

**Practice Problem A** How many unpaired spins are in $[Mn(H_2O)_6]^{2+}$? (*Hint:* $H_2O$ is a weak-field ligand.)

**✔ Practice Problem B** Transition metal ions can have as few as one $d$ electron or as many as nine $d$ electrons. Which numbers of $d$ electrons (1–9) result in the same number of unpaired spins in both the high-spin and low-spin states?

---

## Tetrahedral and Square-Planar Complexes

So far we have concentrated on octahedral complexes. The splitting of the $d$ orbital energy levels in tetrahedral and square-planar complexes, though, can also be accounted for satisfactorily by the crystal field theory. In fact, the splitting pattern for a tetrahedral ion is just the reverse of that for octahedral complexes. In this case, the $d_{xy}$, $d_{yz}$, and $d_{xz}$ orbitals are more closely directed at the ligands and therefore have more energy than the $d_{x^2-y^2}$ and $d_{z^2}$ orbitals (Figure 22.19). Most tetrahedral complexes are high-spin complexes. Presumably, the tetrahedral arrangement reduces the magnitude of the metal-ligand interactions, resulting in a smaller $\Delta$ value. This is a reasonable assumption because the number of ligands is smaller in a tetrahedral complex.

As Figure 22.20 shows, the splitting pattern for square-planar complexes is the most complicated. The $d_{x^2-y^2}$ orbital possesses the highest energy (as in the octahedral case), and the $d_{xy}$ orbital is the next highest. However, the relative placement of the $d_{z^2}$ and the $d_{xz}$ and $d_{yz}$ orbitals cannot be determined simply by inspection and must be calculated.

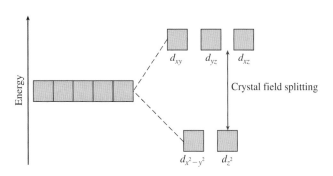

**Figure 22.19** Crystal field splitting between $d$ orbitals in a tetrahedral complex.

**Figure 22.20** Energy-level diagram for a square-planar complex. Because there are more than two energy levels, we cannot define crystal field splitting as we can for octahedral and tetrahedral complexes.

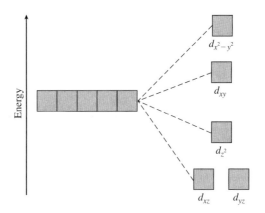

---

**Checkpoint 22.3** | **Bonding in Coordination Compounds: Crystal Field Theory**

22.3.1 How many unpaired spins would you expect the $[Mn(CO)_6]^{2+}$ ion to have?

a) 0

b) 1

c) 2

d) 3

e) 5

22.3.2 Which of the following metal ions can potentially form both low-spin and high-spin complexes? (Select all that apply.)

a) $Ti^{2+}$

b) $Cu^+$

c) $Fe^{2+}$

d) $Ni^{2+}$

e) $Cr^{3+}$

---

# 22.4 Reactions of Coordination Compounds

Complex ions undergo *ligand exchange* (or *substitution*) reactions in solution. The rates of these reactions vary widely, depending on the nature of the metal ion and the ligands.

In studying ligand exchange reactions, it is often useful to distinguish between the stability of a complex ion and its tendency to react, which we call *kinetic lability*. Stability in this context is a thermodynamic property, which is measured in terms of the species' formation constant $K_f$ [◂◂ Section 17.5]. For example, we say that the complex ion tetracyanonickelate(II) is *stable* because it has a large formation constant ($K_f = 1 \times 10^{30}$):

$$Ni^{2+} + 4CN^- \rightleftharpoons [Ni(CN)_4]^{2-}$$

By using cyanide ions labeled with the radioactive isotope carbon-14, chemists have shown that $[Ni(CN)_4]^{2-}$ undergoes ligand exchange very rapidly in solution. The following equilibrium is established almost as soon as the species are mixed:

$$[Ni(CN)_4]^{2-} + 4{*}CN^- \rightleftharpoons [Ni({*}CN)_4]^{2-} + 4CN^-$$

where the asterisk denotes a $^{14}C$ atom. Complexes like the tetracyanonickelate(II) ion are termed *labile complexes* because they undergo *rapid* ligand exchange reactions. Thus, a thermodynamically *stable* species (i.e., one that has a *large* formation constant) is not necessarily unreactive.

A complex that is thermodynamically *unstable* in acidic solution is $[Co(NH_3)_6]^{3+}$. The equilibrium constant for the following reaction is about $1 \times 10^{20}$:

$$[Co(NH_3)_6]^{3+} + 6H^+ + 6H_2O \rightleftharpoons [Co(H_2O)_6]^{3+} + 6NH_4^+$$

When equilibrium is reached, the concentration of the $[Co(NH_3)_6]^{3+}$ ion is very low. This reaction requires several days to complete, however, because the $[Co(NH_3)_6]^{3+}$ ion is so inert. This is an example of an *inert complex*—a complex ion that undergoes very slow exchange reactions (on the order of hours or even days). It shows that a thermodynamically unstable species is not necessarily chemically reactive. The rate of reaction is determined by the energy of activation, which is high in this case.

Most complex ions containing $Co^{3+}$, $Cr^{3+}$, and $Pt^{2+}$ are kinetically inert. Because they exchange ligands very slowly, they are easy to study in solution. As a result, our knowledge of the bonding, structure, and isomerism of coordination compounds has come largely from studies of these compounds.

# 22.5 Applications of Coordination Compounds

Coordination compounds are found in living systems and have many uses in the home, in industry, and in medicine. We briefly describe a few examples in this section.

## Metallurgy

The extraction of silver and gold by the formation of cyanide complexes and the purification of nickel by converting the metal to the gaseous compound $Ni(CO)_4$ are typical examples of the use of coordination compounds in metallurgical processes.

## Chelation Therapy

Earlier we mentioned that chelation therapy is used in the treatment of lead poisoning. Other metals, such as arsenic and mercury, can also be removed using chelating agents.

## Chemotherapy

Several platinum-containing coordination compounds, including cisplatin [$Pt(NH_3)_2Cl_2$] and carboplatin [$Pt(NH_3)_2(OCO)_2C_4H_6$], can effectively inhibit the growth of cancerous cells. The mechanism for the action of cisplatin is the *chelation* of DNA, the molecule that contains the genetic code. During cell division, the double-stranded DNA unwinds into two single strands, which must be accurately copied for the new cells to be identical to their parent cell. X-ray studies show that cisplatin binds to DNA by forming cross-links in which the two chlorides on cisplatin are replaced by nitrogen atoms in the adjacent guanine bases on the same strand of the DNA. (Guanine is one of the four bases in DNA [◀◀ Section 10.6, Figure 10.15].) This causes a bend in the double-stranded structure at the binding site. It is believed that this structural distortion is a key factor in inhibiting replication. The damaged cell is then destroyed by the body's immune system. Because the binding of cisplatin to DNA requires both Cl atoms to be on the same side of the complex, the *trans* isomer of the compound is totally ineffective as an anticancer drug.

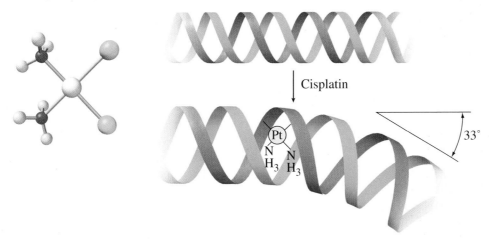

## Chemical Analysis

Although EDTA has a great affinity for a large number of metal ions (especially 2+ and 3+ ions), other chelates are more selective in binding. Dimethylglyoxime, for example, forms an insoluble brick-red solid with $Ni^{2+}$ and an insoluble bright-yellow solid with $Pd^{2+}$. These characteristic colors are used in qualitative analysis to identify nickel and palladium. Furthermore, the quantities of ions present can be determined by gravimetric analysis [◀◀ Section 4.6] as follows: To a solution containing $Ni^{2+}$ ions, say, we add an excess of dimethylglyoxime reagent, and a brick-red precipitate forms. The precipitate is then filtered, dried, and weighed. Knowing the formula of the complex (Figure 22.21), we can readily calculate the amount of nickel in the original solution.

**Figure 22.21** Structure of nickel dimethylglyoxime. Note that the overall structure is stabilized by hydrogen bonds.

## Detergents

The cleansing action of soap in hard water is hampered by the reaction of the $Ca^{2+}$ ions in the water with the soap molecules to form insoluble salts or curds. In the late 1940s the detergent industry introduced a "builder," usually sodium tripolyphosphate, to circumvent this problem. The tripolyphosphate ion is an effective chelating agent that forms stable, soluble complexes with $Ca^{2+}$ ions. Sodium tripolyphosphate revolutionized the detergent industry. Because phosphates are plant nutrients, however, wastewater containing phosphates discharged into rivers and lakes causes algae to grow, resulting in oxygen depletion. Under these conditions, most or all aquatic life eventually succumbs. This process is called *eutrophication*. Consequently, many states have banned phosphate detergents since the 1970s, and manufacturers have reformulated their products to eliminate phosphates.

INGREDIENTS: Water, Butylene Glycol, Mineral Oil Alcohol,Propylene Glycol Dicaprylate/Dicaprate, Isoparaffin, Glyceryl Stearate, Tri-PPG-3 Myristyl Ether Citrate, Cetearyl Alcohol, Dimethicone, Methylparaben, Polysorbate 60, DMDM Hydantoin, Carbomer, Ethylparaben, Propylparaben, Disodium EDTA, Sodium Hydroxide, Xanthan Gum, Butylparaben, Titanium Dioxide

QUESTIONS? Call **1-877-543-8477,**

## Sequestrants

In addition to its use in medicine and chemical analysis, EDTA is used as a food additive to sequester metal ions. EDTA sequesters copper, iron, and nickel ions that would otherwise catalyze the oxidation reactions that cause food to spoil. EDTA is a common preservative in a wide variety of consumer products.

**Multimedia**

Organic and Biochemistry—oxygen binding in hemoglobin.

# Bringing Chemistry to Life

## The Coordination Chemistry of Oxygen Transport

Because of its central function as an oxygen carrier for metabolic processes, hemoglobin is probably the most studied of all the proteins. The molecule contains four folded long chains called subunits. Hemoglobin carries oxygen in the blood from the lungs to the tissues, where it delivers the oxygen molecules to myoglobin. Myoglobin, which is made up of only one subunit, stores oxygen for metabolic processes in the muscle.

The porphine molecule forms an important part of the hemoglobin structure. Upon coordination to a metal, the $H^+$ ions that are bonded to two of the four nitrogen atoms in porphine are displaced. Complexes derived from porphine are called porphyrins, and the iron-porphyrin combination is called the heme group. The iron in the heme group has an oxidation state of +2; it is coordinated to the four nitrogen atoms in the porphine group and also to a nitrogen donor atom in a ligand that is attached to the protein. The sixth ligand is a water molecule, which binds to the $Fe^{2+}$ ion on the other side of the ring to complete the octahedral complex. This hemoglobin molecule is called deoxyhemoglobin and imparts a bluish tinge to venous blood. The water ligand can be replaced readily by molecular oxygen (in the lungs) to form red *oxy*hemoglobin, which is found in arterial blood. Each subunit contains a heme group, so each hemoglobin molecule can bind up to four $O_2$ molecules.

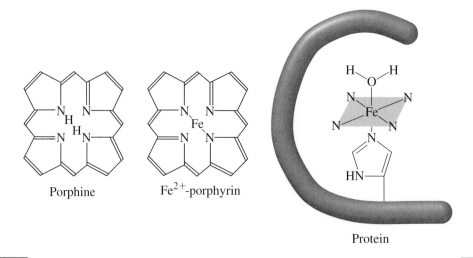

Porphine          $Fe^{2+}$-porphyrin                    Protein

# Applying What You've Learned

Elevated BLL and other heavy metal poisoning can be treated with one of several chelating agents, including DMSA and EDTA. EDTA is administered intravenously as either the sodium salt (Endrate) or as the calcium disodium salt (Versenate). Endrate is not approved for the treatment of lead poisoning because of its high affinity for calcium. It is approved, however, for treating *hyper-calcemia,* a condition in which there is excess calcium in the blood—usually as a result of bone cancer. The accidental use of Endrate during treatment for lead poisoning resulted in the death of a 2-year-old girl in February of 2005. The girl's death was attributed to sudden cardiac arrest caused by the removal of too much calcium from her blood.

Chelation therapy works by the administration of a ligand, which binds to metal ions already in the body. Many drugs, including cisplatin [◀◀ Chapter 3, Applying What You've Learned], are themselves coordination compounds in which the central metal ion binds to electron-rich sites (such as oxygen or sulfur atoms) in biological molecules.

## Problems:

a) Determine the oxidation state of platinum in cisplatin, $[Pt(NH_3)_2Cl_2]$.
   [◀◀ Sample Problem 22.1]

b) Give the systematic name for cisplatin. [◀◀ Sample Problem 22.2]

c) Write the formula for the compound potassium hexachloroplatinate(IV).
   [◀◀ Sample Problem 22.3]

# CHAPTER SUMMARY

## Section 22.1

- *Coordination compounds* contain coordinate covalent bonds between a metal ion (often a transition metal ion) and two or more polar molecules or ions.

- The molecules or anions that surround a metal in a coordination complex are called **ligands.**

- Many coordination compounds consist of a *complex ion* and a *counter* ion.

- Transition metals are those that have incompletely filled *d* subshells—or that *give rise* to ions with incompletely filled *d* subshells.

- Transition metals exhibit variable oxidation states ranging from +1 to +7.

- To act as a ligand, a molecule or ion must have at least one unshared pair of electrons. The atom that bears the unshared pair of electrons is the **donor atom.**

- Ligands are classified as *monodentate, bidentate,* or *polydentate,* based on the number of donor atoms they contain. Bidentate and polydentate ions are also known as **chelating agents.**

- The **coordination number** is the number of donor atoms surrounding a metal in a complex.

- Ionic coordination compounds are named by first naming the cation and then the anion. Complex ions are named by listing the ligands in alphabetical order, followed by the metal and its oxidation state (as a Roman numeral). When the complex ion is the anion, the anion's name ends in *–ate.*

## Section 22.2

- The coordination number largely determines the geometry of a coordination complex.

- Coordination compounds containing different arrangements of the same ligands are **stereoisomers.** The two types of stereoisomerism are geometric and optical.

- *Geometrical isomers* contain the same atoms and bonds arranged differently in space.

- *Optical isomers* are nonsuperimposable mirror images. We call a *pair* of optical isomers **enantiomers.** The rotation of polarized light is measured with a **polarimeter.**

- Enantiomers rotate the plane of plane-polarized light in opposite directions. The enantiomer that rotates it to the right is called **dextrorotatory** and is labeled *d.* The enantiomer that rotates it to the left is called **levorotatory** and is labeled *l.* An equal mixture of a pair of enantiomers, called a **racemic mixture,** does not cause any net rotation of plane-polarized light.

## Section 22.3

- Ligands in a coordination complex cause the energy levels of the *d* orbitals on a metal to split. The difference in energy between the lower and higher *d* orbital energy levels is called the **crystal field splitting ($\Delta$).**

- The magnitude of $\Delta$ depends on the nature of the ligands in the complex. The **spectrochemical series** orders some common ligands in order of increasing *field strength.*

- *Strong*-field ligands give rise to a larger $\Delta$ value; *weak*-field ligands yield a smaller $\Delta$ value.

- Crystal field splitting sometimes changes the number of unpaired electrons, and therefore the magnetic properties, of a metal.

- Complexes containing transition metals with $d^4$, $d^5$, $d^6$, or $d^7$ configurations may be *high spin* or *low spin.* In high-spin complexes, the number of unpaired electrons is maximized because $\Delta$ is small; in low-spin complexes, the number of unpaired electrons is minimized because $\Delta$ is large.

## Section 22.4

- Complex ions undergo ligand exchange in solution. The rate at which ligand exchange occurs is a measure of a complex's *kinetic lability* and does not necessarily correspond directly to the complex's *thermodynamic stability.*

## Section 22.5

- Coordination chemistry is important in many biological, medical, and industrial processes.

# KEY WORDS

| | | | |
|---|---|---|---|
| Chelating agent, 925 | Dextrorotatory, 930 | Levorotatory, 930 | Racemic mixture, 930 |
| Coordination compound, 922 | Donor atom, 924 | Ligand, 924 | Spectrochemical series, 933 |
| Coordination number, 925 | Enantiomers, 930 | Optical isomers, 929 | Stereoisomers, 929 |
| Crystal field splitting ($\Delta$), 931 | Geometrical isomers, 929 | Polarimeter, 930 | |

# QUESTIONS AND PROBLEMS

## Section 22.1: Coordination Compounds

### Review Questions

22.1    What distinguishes a transition metal from a main group metal?

22.2    Why is zinc not considered a transition metal?

22.3    Explain why atomic radii decrease very gradually from scandium to copper.

22.4    Without referring to the text, write the ground-state electron configurations of the first-row transition metals. Explain any irregularities.

22.5    Write the electron configurations of the following ions: $V^{5+}$, $Cr^{3+}$, $Mn^{2+}$, $Fe^{3+}$, $Cu^{2+}$, $Sc^{3+}$, $Ti^{4+}$.

22.6    Why do transition metals have more oxidation states than other elements?

22.7    Give the highest oxidation states for scandium to copper.

22.8    Define the following terms: *coordination compound, ligand, donor atom, coordination number, chelating agent.*

22.9    Describe the interaction between a donor atom and a metal atom in terms of a Lewis acid-base reaction.

## Problems

**22.10** Complete the following statements for the complex ion $[Co(en)_2(H_2O)CN]^{2+}$. (a) en is the abbreviation for _____. (b) The oxidation number of Co is _____. (c) The coordination number of Co is _____. (d) _____ is a bidentate ligand.

**22.11** Complete the following statements for the complex ion $[Cr(C_2O_4)_2(H_2O)_2]^-$. (a) The oxidation number of Cr is _____. (b) The coordination number of Cr is _____. (c) _____ is a bidentate ligand.

**22.12** Give the oxidation numbers of the metals in the following species: (a) $K_3[Fe(CN)_6]$, (b) $K_3[Cr(C_2O_4)_3]$, (c) $[Ni(CN)_4]^{2-}$.

**22.13** Give the oxidation numbers of the metals in the following species: (a) $Na_2MoO_4$, (b) $MgWO_4$, (c) $Fe(CO)_5$.

**22.14** What are the systematic names for the following ions and compounds?
(a) $[Co(NH_3)_4Cl_2]^+$      (c) $[Co(en)_2Br_2]^+$
(b) $[Cr(NH_3)_3Cl_3]$      (d) $[Co(NH_3)_6]Cl_3$

**22.15** What are the systematic names for the following ions and compounds?
(a) $cis$-$Co(en)_2Cl_2]^+$      (c) $[Co(NH_3)_5Cl]Cl_2$
(b) $[Pt(NH_3)_5Cl]Cl_3$

**22.16** Write the formulas for each of the following ions and compounds: (a) tetrahydroxozincate(II), (b) pentaaquachlorochromium(III) chloride, (c) tetrabromocuprate(II), (d) ethylenediaminetetraacetatoferrate(II).

**22.17** Write the formulas for each of the following ions and compounds: (a) dichlorobis(ethylenediamine)chromium(III), (b) pentacarbonyliron(0), (c) potassium tetracyanocuprate(II), (d) tetraammineaquachlorocobalt(III) chloride.

## Section 22.2: Structure of Coordination Compounds

### Review Questions

**22.18** Define the following terms: *stereoisomers, geometric isomers, optical isomers, plane-polarized light.*

**22.19** Specify which of the following structures can exhibit geometric isomerism: (a) linear, (b) square planar, (c) tetrahedral, (d) octahedral.

**22.20** What determines whether a molecule is chiral? How does a polarimeter measure the chirality of a molecule?

**22.21** Explain the following terms: *enantiomers, racemic mixtures.*

### Problems

**22.22** The complex ion $[Ni(CN)_2Br_2]^{2-}$ has a square-planar geometry. Draw the structures of the geometric isomers of this complex.

**22.23** How many geometric isomers are in the following species: (a) $[Co(NH_3)_2Cl_4]^-$, (b) $[Co(NH_3)_3Cl_3]$?

**22.24** Draw structures of all the geometric and optical isomers of each of the following cobalt complexes:
(a) $[Co(NH_3)_6]^{3+}$      (b) $[Co(NH_3)_5Cl]^{2+}$      (c) $[Co(C_2O_4)_3]^{3-}$

**22.25** Draw structures of all the geometric and optical isomers of each of the following cobalt complexes: (a) $[Co(NH_3)_4Cl_2]^+$, (b) $[Co(en)_3]^{3+}$.

## Section 22.3: Bonding in Coordination Compounds: Crystal Field Theory

### Review Questions

**22.26** Briefly describe crystal field theory.

**22.27** Define the following terms: *crystal field splitting, high-spin complex, low-spin complex, spectrochemical series.*

**22.28** What is the origin of color in a coordination compound?

**22.29** Compounds containing the $Sc^{3+}$ ion are colorless, whereas those containing the $Ti^{3+}$ ion are colored. Explain.

**22.30** For the same type of ligands, explain why the crystal field splitting for an octahedral complex is always greater than that for a tetrahedral complex.

## Problems

**22.31** Transition metal complexes containing $CN^-$ ligands are often yellow in color, whereas those containing $H_2O$ ligands are often green or blue. Explain.

**22.32** Predict the number of unpaired electrons in the following complex ions: (a) $[Cr(CN)_6]^{4-}$, (b) $[Cr(H_2O)_6]^{2+}$.

**22.33** The absorption maximum for the complex ion $[Co(NH_3)_6]^{3+}$ occurs at 470 nm. (a) Predict the color of the complex, and (b) calculate the crystal field splitting in kJ/mol.

**22.34** From each of the following pairs, choose the complex that absorbs light at a longer wavelength: (a) $[Co(NH_3)_6]^{2+}$, $[Co(H_2O)_6]^{2+}$; (b) $[FeF_6]^{3-}$, $[Fe(CN)_6]^{3-}$; (c) $[Cu(NH_3)_4]^{2+}$, $[CuCl_4]^{2-}$.

**22.35** A solution made by dissolving 0.875 g of $Co(NH_3)_4Cl_3$ in 25.0 g of water freezes at $-0.56°C$. Calculate the number of moles of ions produced when 1 mole of $Co(NH_3)_4Cl_3$ is dissolved in water, and suggest a structure for the complex ion present in this compound.

**22.36** In biological systems, the $Cu^{2+}$ ions coordinated with S atoms tend to form tetrahedral complexes, whereas those coordinated with N atoms tend to form octahedral complexes. Explain.

**22.37** Plastocyanin, a copper-containing protein found in photosynthetic systems, is involved in electron transport, with the copper ion switching between the +1 and +2 oxidation states. The copper ion is coordinated with two histidine residues, a cysteine residue, and a methionine residue in a tetrahedral configuration. How does the crystal field splitting ($\Delta$) change between these two oxidation states?

## Section 22.4: Reactions of Coordination Compounds

### Review Questions

**22.38** Define the terms *labile complex* and *inert complex*.

**22.39** Explain why a thermodynamically stable species may be chemically reactive and a thermodynamically unstable species may be unreactive.

### Problems

**22.40** Oxalic acid ($H_2C_2O_4$) is sometimes used to clean rust stains from sinks and bathtubs. Explain the chemistry underlying this cleaning action.

**22.41** The $[Fe(CN)_6]^{3-}$ complex is more labile than the $[Fe(CN)_6]^{4-}$ complex. Suggest an experiment that would prove that $[Fe(CN)_6]^{3-}$ is a labile complex.

**22.42** Aqueous copper(II) sulfate solution is blue in color. When aqueous potassium fluoride is added, a green precipitate is formed. When aqueous potassium chloride is added instead, a bright-green solution is formed. Explain what is happening in these two cases.

**22.43** When aqueous potassium cyanide is added to a solution of copper(II) sulfate, a white precipitate, soluble in an excess of potassium cyanide, is formed. No precipitate is formed when hydrogen sulfide is bubbled through the solution at this point. Explain.

**22.44** A concentrated aqueous copper(II) chloride solution is bright green in color. When diluted with water, the solution becomes light blue. Explain.

**22.45** In a dilute nitric acid solution, $Fe^{3+}$ reacts with thiocyanate ion ($SCN^-$) to form a dark-red complex:

$$[Fe(H_2O)_6]^{3+} + SCN^- \rightleftharpoons H_2O + [Fe(H_2O)_5NCS]^{2+}$$

The equilibrium concentration of $[Fe(H_2O)_5NCS]^{2+}$ may be determined by how darkly colored the solution is (measured by a spectrometer). In one such experiment, 1.0 mL of 0.20 $M$ $Fe(NO_3)_3$ was mixed with 1.0 mL of $1.0 \times 10^{-3}$ $M$ KSCN and 8.0 mL of dilute $HNO_3$. The color of the solution quantitatively indicated that the $[Fe(H_2O)_5NCS]^{2+}$ concentration was $7.3 \times 10^{-5}$ $M$. Calculate the formation constant for $[Fe(H_2O)_5NCS]^{2+}$.

## Section 22.5: Applications of Coordination Compounds

### Review Question

**22.46** Describe and give examples of the applications of coordination compounds.

## Additional Problems

**22.47** Suffocation victims usually look purple, but a person poisoned by carbon monoxide often has rosy cheeks. Explain.

**22.48** As we read across the first-row transition metals from left to right, the +2 oxidation state becomes more stable in comparison with the +3 state. Why is this so?

**22.49** Which is a stronger oxidizing agent in aqueous solution, $Mn^{3+}$ or $Cr^{3+}$? Explain your choice.

**22.50** Carbon monoxide binds to Fe in hemoglobin some 200 times more strongly than oxygen. This is the reason why CO is a toxic substance. The metal-to-ligand sigma bond is formed by donating a lone pair from the donor atom to an empty $sp^3d^2$ orbital on Fe. (a) On the basis of electronegativities, would you expect the C or O atom to form the bond to Fe? (b) Draw a diagram illustrating the overlap of the orbitals involved in the bonding.

**22.51** What are the oxidation states of Fe and Ti in the ore ilmenite ($FeTiO_3$)? (*Hint:* Look up the ionization energies of Fe and Ti in Table 22.1; the fourth ionization energy of Ti is 4180 kJ/mol.)

**22.52** A student has prepared a cobalt complex that has one of the following three structures: $[Co(NH_3)_6]Cl_3$, $[Co(NH_3)_5Cl]Cl_2$, or $[Co(NH_3)_4Cl_2]Cl$. Explain how the student would distinguish between these possibilities by an electrical conductance experiment. At the student's disposal are three strong electrolytes—NaCl, $MgCl_2$, and $FeCl_3$—which may be used for comparison purposes.

**22.53** Chemical analysis shows that hemoglobin contains 0.34 percent of Fe by mass. What is the minimum possible molar mass of hemoglobin? The actual molar mass of hemoglobin is about 65,000 g. How do you account for the discrepancy between your minimum value and the actual value?

**22.54** Explain the following facts: (a) Copper and iron have several oxidation states, whereas zinc has only one. (b) Copper and iron form colored ions, whereas zinc does not.

**22.55** A student in 1895 prepared three coordination compounds containing chromium, with the following properties:

| Formula | Color | Cl⁻ Ions in Solution per Formula Unit |
|---|---|---|
| (a) $CrCl_3 \cdot 6H_2O$ | Violet | 3 |
| (b) $CrCl_3 \cdot 6H_2O$ | Light green | 2 |
| (c) $CrCl_3 \cdot 6H_2O$ | Dark green | 1 |

Write modern formulas for these compounds, and suggest a method for confirming the number of $Cl^-$ ions present in solution in each case. (*Hint:* Some of the compounds may exist as hydrates.)

**22.56** The formation constant for the reaction $Ag^+ + 2NH_3 \rightleftharpoons$ $[Ag(NH_3)_2]^+$ is $1.5 \times 10^7$, and that for the reaction $Ag^+ +$

$2CN^- \rightleftharpoons [Ag(CN)_2]^-$ is $1.0 \times 10^{21}$ at 25°C (see Table 17.5). Calculate the equilibrium constant and $\Delta G°$ at 25°C for the reaction

$$[Ag(NH_3)_2]^+ + 2CN^- \rightleftharpoons [Ag(CN)_2]^- + 2NH_3$$

**22.57** From the standard reduction potentials listed in Table 19.1 for $Zn/Zn^{2+}$ and $Cu/Cu^{2+}$, calculate $\Delta G°$ and the equilibrium constant for the reaction

$$Zn(s) + 2Cu^{2+}(aq) \longrightarrow Zn^{2+}(aq) + 2Cu^+(aq)$$

**22.58** Using the standard reduction potentials listed in Table 19.1 and the *Handbook of Chemistry and Physics,* show that the following reaction is favorable under standard-state conditions:

$$2Ag(s) + Pt^{2+}(aq) \longrightarrow 2Ag^+(aq) + Pt(s)$$

What is the equilibrium constant of this reaction at 25°C?

**22.59** The $Co^{2+}$-porphyrin complex is more stable than the $Fe^{2+}$-porphyrin complex. Why, then, is iron the metal ion in hemoglobin (and other heme-containing proteins)?

**22.60** What are the differences between geometric isomers and optical isomers?

**22.61** Oxyhemoglobin is bright red, whereas deoxyhemoglobin is purple. Show that the difference in color can be accounted for qualitatively on the basis of high-spin and low-spin complexes. (*Hint:* $O_2$ is a strong-field ligand. See the Bringing Chemistry to Life box on page 938.)

**22.62** Hydrated $Mn^{2+}$ ions are practically colorless (see Figure 22.16) even though they possess five $3d$ electrons. Explain. (*Hint:* Electronic transitions in which there is a change in the number of unpaired electrons do not occur readily.)

**22.63** Which of the following hydrated cations are colorless: $Fe^{2+}(aq)$, $Zn^{2+}(aq)$, $Cu^+(aq)$, $Cu^{2+}(aq)$, $V^{5+}(aq)$, $Ca^{2+}(aq)$, $Co^{2+}(aq)$, $Sc^{3+}(aq)$, $Pb^{2+}(aq)$? Explain your choice.

**22.64** Aqueous solutions of $CoCl_2$ are generally either light pink or blue. Low concentrations and low temperatures favor the pink form, whereas high concentrations and high temperatures favor the blue form. Adding hydrochloric acid to a pink solution of $CoCl_2$ causes the solution to turn blue; the pink color is restored by the addition of $HgCl_2$. Account for these observations.

**22.65** Suggest a method that would allow you to distinguish between *cis*-$Pt(NH_3)_2Cl_2$ and *trans*-$Pt(NH_3)_2Cl_2$.

**22.66** You are given two solutions containing $FeCl_2$ and $FeCl_3$ at the same concentration. One solution is light yellow, and the other one is brown. Identify these solutions based only on color.

**22.67** The label of a certain brand of mayonnaise lists EDTA as a food preservative. How does EDTA prevent the spoilage of mayonnaise?

**22.68** The compound 1,1,1-trifluoroacetylacetone (tfa) is a bidentate ligand:

$$\overset{O}{\overset{\|}{CF_3C}}CH_2\overset{O}{\overset{\|}{C}}CH_3$$

It forms a tetrahedral complex with $Be^{2+}$ and a square-planar complex with $Cu^{2+}$. Draw structures of these complex ions, and identify the type of isomerism exhibited by these ions.

**22.69** How many geometric isomers can the following square-planar complex have?

$$\begin{array}{c} a \quad\quad b \\ \diagdown\;/ \\ Pt \\ \diagup\;\diagdown \\ d \quad\quad c \end{array}$$

**22.70** $[Pt(NH_3)_2Cl_2]$ is found to exist in two geometric isomers designated I and II, which react with oxalic acid as follows:

$$I + H_2C_2O_4 \longrightarrow [Pt(NH_3)_2C_2O_4]$$
$$II + H_2C_2O_4 \longrightarrow [Pt(NH_3)_2(HC_2O_4)_2]$$

Comment on the structures of I and II.

€22.71 Commercial silver-plating operations frequently use a solution containing the complex $Ag(CN)_2^-$ ion. Because the formation constant ($K_f$) is quite large, this procedure ensures that the free $Ag^+$ concentration in solution is low for uniform electrodeposition. In one process, a chemist added 9.0 L of 5.0 M NaCN to 90.0 L of 0.20 M $AgNO_3$. Calculate the concentration of free $Ag^+$ ions at equilibrium. See Table 17.5 for $K_f$ value.

22.72 Draw qualitative diagrams for the crystal field splittings in (a) a linear complex ion $ML_2$, (b) a trigonal-planar complex ion $ML_3$, and (c) a trigonal-bipyramidal complex ion $ML_5$.

€22.73 (a) The free Cu(I) ion is unstable in solution and has a tendency to disproportionate:

$$2Cu^+(aq) \rightleftharpoons Cu^{2+}(aq) + Cu(s)$$

Use the information in Table 19.1 (page 827) to calculate the equilibrium constant for the reaction. (b) Based on your results in part (a), explain why most Cu(I) compounds are insoluble.

€22.74 Consider the following two ligand exchange reactions:

$$[Co(H_2O)_6]^{3+} + 6NH_3 \rightleftharpoons [Co(NH_3)_6]^{3+} + 6H_2O$$

$$[Co(H_2O)_6]^{3+} + 3en \rightleftharpoons [Co(en)_3]^{3+} + 6H_2O$$

(a) Which of the reactions should have a larger $\Delta S°$? (b) Given that the Co—N bond strength is approximately the same in both complexes, which reaction will have a larger equilibrium constant? Explain your choices.

22.75 The $Cr^{3+}$ ion forms octahedral complexes with two neutral ligands X and Y. The color of $CrX_6^{3+}$ is blue while that of $CrY_6^{3+}$ is yellow. Which is a stronger field ligand, X or Y?

22.76 Explain the difference between these two compounds: $CrCl_3 \cdot 6H_2O$ and $[Cr(H_2O)_6]Cl_3$.

€22.77 The $K_f$ for the formation of the complex ion between $Pb^{2+}$ and $EDTA^{4-}$,

$$Pb^{2+} + EDTA^{4-} \rightleftharpoons [Pb(EDTA)]^{2-}$$

is $1.0 \times 10^{18}$ at 25°C. Calculate $[Pb^{2+}]$ at equilibrium in a solution containing $1.0 \times 10^{-3}$ M $Pb^{2+}$ and $2.0 \times 10^{-3}$ M $EDTA^{4-}$.

22.78 Locate the transition metal atoms and ions in the periodic table shown here. Atoms: (a) $[Kr]5s^24d^5$. (b) $[Xe]6s^24f^{14}5d^4$. Ions: (c) $[Ar]3d^3$ (ion with charge 4+). (d) $[Xe]4f^{14}5d^8$ (ion with charge 3+).

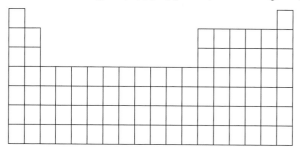

€22.79 Manganese forms three low-spin complex ions with the cyanide ion with the formulas $[Mn(CN)_6]^{5-}$, $[Mn(CN)_6]^{4-}$, and $[Mn(CN)_6]^{3-}$. For each complex ion, determine the oxidation number of Mn and the number of unpaired $d$ electrons.

22.80 Copper is known to exist in the +3 oxidation state, which is believed to be involved in some biological electron-transfer reactions. (a) Would you expect this oxidation state of copper to be stable? Explain. (b) Name the compound $K_3CuF_6$ and predict the geometry and magnetic properties of the complex ion. (c) Most of the known Cu(III) compounds have square planar geometry. Are these compounds diamagnetic or paramagnetic?

# PRE-PROFESSIONAL PRACTICE EXAM PROBLEMS: PHYSICAL AND BIOLOGICAL SCIENCES

The following questions are not based on a passage.

1. What are the correct names for the complex ions $[Ni(H_2O)_6]^{2+}$ and $[Ni(NH_3)_6]^{2+}$?

   a) Hexaaquanickel(VI) ion and hexaamminenickel(VI) ion
   b) Nickel(II) hexaaqua ion and nickel(II) hexaammine ion
   c) Hexaaquanickel(II) ion and hexaamminenickel(II) ion
   d) Hexaaquanickelate and hexaamminenickelate

2. One of the complex ions in Question 1 is green, and the other is purple. Based on the spectrochemical series, which is which?

   $I^- < Br^- < Cl^- < OH^- < F^- < H_2O < NH_3 < en < CN^- < CO$

   a) $[Ni(H_2O)_6]^{2+}$ is green and $[Ni(NH_3)_6]^{2+}$ is purple.
   b) $[Ni(H_2O)_6]^{2+}$ is purple and $[Ni(NH_3)_6]^{2+}$ is green.
   c) Both ions should be the same color because they contain the same metal.
   d) There is not enough information to determine which is which.

3. How many geometric isomers are possible for the complex ion $[Ni(H_2O)_4(NH_3)_2]^{2+}$?

   a) 1    b) 2    c) 4    d) 6

4. How many ions in solution would result from dissolving $[Ni(H_2O)_4(NH_3)_2]Cl_2$?

   a) 2    b) 3    c) 8    d) 11

# ANSWERS TO IN-CHAPTER MATERIALS

## Answers to Practice Problems

**22.1A** (a) K: +1, Au: +3, (b) K: +1, Fe: +2. **22.1B** (a) +1, −3, 0, +1, +3, +3, +2. **22.2A** (a) tetraamminedibromocobalt(III) chloride, (b) tetra-aquadichlorochromium(III) chloride, (c) potassium tetrachlorocuprate(II). **22.2B** (a) sodium hexacyanoferrate(III), (b) dichlorobis(ethylenediamine)-chromium(III) chloride, (c) tris(ethylenediamine)cobalt(III) chloride. **22.3A** (a) $[Ru(H_2O)_5Br]NO_3$, (b) $K_2[PtBr_4Cl_2]$, (c) $Na_3[Co(NO_2)_6]$.

**22.3B** (a) $[V(en)_2C_2O_4]Cl_2$, (b) $[Cr(en)_2Br_2]NO_3$, (c) $[Pt(en)_3](SO_4)_2$. **22.4A** 5. **22.4B** 1, 2, 3, 8, and 9.

## Answers to Checkpoints

**22.1.1** b. **22.1.2** d. **22.1.3** b. **22.1.4** c. **22.3.1** b. **22.3.2** c.

## Answers to Applying What You've Learned

**a)** +2. **b)** *cis*-Diamminedichloroplatinum(II). **c)** $K_2[PtCl_6]$.

# 23

# Metallurgy and the Chemistry of Metals

*In 1943, because copper was needed for the war effort, pennies were made from steel. A very small number of copper pennies were made in 1943, forty of which are known to still exist. The copper penny in this photograph is a counterfeit. However, according to the United States Mint, genuine copper 1943 pennies have sold for as much as $82,500.*

# The Importance of Metals in Human Biology

Calcium, potassium, and sodium are the three most abundant metals in the human body. Present in a much smaller amount, copper is nonetheless vitally important to human health. Copper-dependent enzymes known as *cuproenzymes* play a critical role in a host of biochemical processes, including cellular energy production, connective tissue formation, and reactions essential to normal functioning of the brain and central nervous system. Two hereditary diseases that involve errors of copper metabolism are Menkes disease and Wilson's disease.

In Menkes disease, copper is not absorbed in the intestine and is not available for distribution throughout the body. Consequently, cuproenzymes such as *cytochrome c oxidase, lysyl oxidase,* and *dopamine beta hydroxylase* cannot function normally. Sufferers exhibit severe developmental delay; subnormal body temperature; kinky, steel-colored hair; seizures; and ultimately the degeneration of muscle, bone, and organs. Baby boys born with the most severe form of Menkes disease typically do not live beyond their third year of life.

**Student Annotation:** Menkes disease is an X-linked recessive condition and nearly all victims are male.

Wilson's disease causes the body to *retain* copper. Although people with Wilson's disease are able to absorb copper in the intestine, the liver of a person with Wilson's disease does not release copper into the bile as it should. Eventually, damaged by the buildup of copper, the liver releases copper directly into the blood, where it is carried throughout the body. Excess copper causes damage to the kidneys, brain, and eyes. Unlike Menkes disease, Wilson's disease can be managed and treated. Sufferers must follow a strict low-copper diet, avoiding such foods as mushrooms, nuts, and chocolate. In addition, they must undergo lifelong periodic chelation therapy.

Menkes and Wilson's diseases illustrate the importance of *metals* to living systems.

At the end of this chapter, you will have an appreciation for the importance of metals and the many roles they play in human health [▶▶ Page 962].

## In This Chapter, You Will Learn:

How some metals are produced and about their chemistry.

### Before you begin, you should review

- Periodic trends and metallic character [◀◀ Section 7.4]
- Electrolysis [◀◀ Section 19.7]

| TABLE 23.1 | Natural Sources of Common Metals |
|---|---|
| **Type** | **Minerals** |
| Uncombined metals | Ag, Au, Bi, Cu, Pd, Pt |
| Carbonates | $BaCO_3$ (witherite), $CaCO_3$ (calcite, limestone), $MgCO_3$ (magnesite), $CaCO_3 \cdot MgCO_3$ (dolomite), $PbCO_3$ (cerussite), $ZnCO_3$ (smithsonite) |
| Halides | $CaF_2$ (fluorite), NaCl (halite), KCl (sylvite), $Na_3AlF_6$ (cryolite) |
| Oxides | $Al_2O_3 \cdot 2H_2O$ (bauxite), $Al_2O_3$ (corundum), $Fe_2O_3$ (hematite), $Fe_3O_4$ (magnetite), $Cu_2O$ (cuprite), $MnO_2$ (pyrolusite), $SnO_2$ (cassiterite), $TiO_2$ (rutile), ZnO (zincite) |
| Phosphates | $Ca_3(PO_4)_2$ (phosphate rock), $Ca_5(PO_4)_3OH$ (hydroxyapatite) |
| Silicates | $Be_3Al_2Si_6O_{18}$ (beryl), $ZrSiO_4$ (zircon), $NaAlSi_3O_8$ (albite), $Mg_3(Si_4O_{10})(OH)_2$ (talc) |
| Sulfides | $Ag_2S$ (argentite), CdS (greenockite), $Cu_2S$ (chalcocite), $FeS_2$ (pyrite), HgS (cinnabar), PbS (galena), ZnS (sphalerite) |
| Sulfates | $BaSO_4$ (barite), $CaSO_4$ (anhydrite), $PbSO_4$ (anglesite), $SrSO_4$ (celestite), $MgSO_4 \cdot 7H_2O$ (epsomite) |

**Figure 23.1**  Metals and their best-known minerals. Lithium is found in spodumene ($LeAlSi_2O_6$), and beryllium in beryl (see Table 23.1). The rest of the alkaline earth metals are found in minerals that are carbonates and sulfates. The minerals for Sc, Y, and La are the phosphates. Some metals have more than one type of important mineral. For example, in addition to the sulfide, iron is found as the oxides hematite ($Fe_2O_3$) and magnetite ($Fe_3O_4$); and aluminum, in addition to the oxide, is found in beryl ($Be_3Al_2Si_6O_{18}$). Technetium (Tc) is a synthetic element.

## 23.1 Occurrence of Metals

Most metals come from minerals. A *mineral* is a naturally occurring substance with a range of chemical composition. A mineral deposit concentrated enough to allow economical recovery of a desired metal is known as *ore*. Table 23.1 lists the principal types of minerals, and Figure 23.1 shows a classification of metals according to their minerals.

The most abundant metals, which exist as minerals in Earth's crust, are aluminum, iron, calcium, magnesium, sodium, potassium, titanium, and manganese (see page 47). Seawater is a rich source of some metal ions, including $Na^+$, $Mg^{2+}$, and $Ca^{2+}$. Furthermore, vast areas of the ocean floor are covered with *manganese nodules,* which are made up mostly of manganese, along with iron, nickel, copper, and cobalt in a chemically combined state (Figure 23.2).

**Figure 23.2**  Magnesium nodules on the ocean floor.

## Bringing Chemistry to Life

### The Importance of Molybdenum

Molybdenum is one of the less abundant metals in Earth's crust. Its importance to human health was recognized relatively recently. Like copper, it is essential in trace amounts for proper enzyme function, and it is toxic in large amounts. Molybdenum deficiency in humans has been described and is known to cause visual problems, rapid heart rate, and coma. However, the number of people affected by a lack of dietary molybdenum is very small. A deficiency that *has* affected a large group of people is molybdenum deficiency in the *soil*. In the north China province of Honan, there is a small region (Lin Xian) where for generations the incidence of esophageal cancer has been one of the highest in the world. It turns out that the soil in this region is naturally low in molybdenum. Without molybdenum, the enzyme *nitrate reductase,* found in nitrogen-fixing bacteria, was inactive. Thus, instead of being converted to amines, *nitrates* in the soil were converted to *nitrosamines,* compounds that are known to cause cancer. The addition of molybdenum to the soil may result in lower levels of nitrosamines in the diets of Lin Xian residents and, it is hoped, a lower incidence of esophageal cancer.

**Lin Xian in the Honan province in China.**

## 23.2  Metallurgical Processes

*Metallurgy* is the science and technology of separating metals from their ores and of compounding alloys. An *alloy* is a solid solution either of two or more metals, or of a metal or metals with one or more *nonmetals.* The three principal steps in the recovery of a metal from its ore are (1) preparation of the ore, (2) production of the metal, and (3) purification of the metal.

### Preparation of the Ore

In the preliminary treatment of an ore, the desired mineral is separated from waste materials—usually clay and silicate minerals—which are collectively called the *gangue.* One very useful method for carrying out such a separation is called *flotation.*

In this process, the ore is finely ground and added to water containing oil and detergent. The liquid mixture is then beaten or blown to form a froth. The oil preferentially wets the mineral particles, which are then carried to the top in the froth, while the gangue settles to the bottom. The froth is skimmed off, allowed to collapse, and dried to recover the mineral particles.

Another physical separation process makes use of the magnetic properties of certain minerals. *Ferromagnetic* metals are strongly attracted to magnets. The mineral magnetite ($Fe_3O_4$), in particular, can be separated from the gangue by using a strong electromagnet. Cobalt is another ferromagnetic metal.

Mercury forms amalgams with a number of metals. An *amalgam* is an alloy of mercury with another metal or metals [◀◀ Chapter 19]. Mercury can therefore be used to extract metal from ore. Mercury dissolves the silver and gold in an ore to form a liquid amalgam, which is easily separated from the remaining ore. The gold or silver is recovered by distilling off the mercury.

### Production of Metals

Because metals in their combined forms always have positive oxidation numbers, the production of a free metal is a reduction process. Preliminary operations may be necessary to convert the ore to a chemical state more suitable for reduction. For example, an ore may be *roasted* to drive off volatile impurities and at the same time to convert the carbonates and sulfides to the corresponding oxides, which can be reduced more conveniently to yield the pure metals:

$$CaCO_3(s) \longrightarrow CaO(s) + CO_2(g)$$

$$2PbS(s) + 3O_2(g) \longrightarrow 2PbO(s) + 2SO_2(g)$$

This last equation demonstrates that the conversion of sulfides to oxides is a major source of sulfur dioxide, a notorious air pollutant.

How a pure metal is obtained by reduction from its combined form depends on the standard reduction potential of the metal. Table 23.2 outlines the reduction processes for several metals.

| TABLE 23.2 | Reduction Processes for Some Common Metals |
| --- | --- |

<table>
<tr><td></td><td>**Metal**</td><td>**Reduction Process**</td></tr>
<tr><td rowspan="4">Decreasing activity of metals →</td><td>Lithium, sodium, magnesium, calcium</td><td>Electrolytic reduction of the molten chloride</td></tr>
<tr><td>Aluminum</td><td>Electrolytic reduction of anhydrous oxide (in molten cryolite)</td></tr>
<tr><td>Chromium, manganese, titanium, vanadium, iron, zinc</td><td>Reduction of the metal oxide with a more electropositive metal, or reduction with coke and carbon monoxide</td></tr>
<tr><td>Mercury, silver, platinum, copper, gold</td><td>These metals occur in the free (uncombined) state, or they can be obtained by roasting their sulfides</td></tr>
</table>

Most major metallurgical processes now in use involve *pyrometallurgy,* procedures carried out at high temperatures. The reduction in these procedures may be accomplished either chemically or electrolytically.

## Chemical Reduction

A more electropositive metal can be used as a reducing agent to separate a less electropositive metal from its compound at high temperatures:

$$V_2O_5(s) + 5Ca(l) \longrightarrow 2V(l) + 5CaO(s)$$

$$TiCl_4(g) + 2Mg(l) \longrightarrow Ti(s) + 2MgCl_2(l)$$

$$Cr_2O_3(s) + 2Al(s) \longrightarrow 2Cr(l) + Al_2O_3(s)$$

$$3Mn_3O_4(s) + 8Al(s) \longrightarrow 9Mn(l) + 4Al_2O_3(s)$$

In some cases, even molecular hydrogen can be used as a reducing agent, as in the preparation of tungsten (used as filaments in lightbulbs) from tungsten(VI) oxide:

$$WO_3(s) + 3H_2(g) \longrightarrow W(s) + 3H_2O(g)$$

## Electrolytic Reduction

Electrolytic reduction is suitable for very electropositive metals, such as sodium, magnesium, and aluminum. The process is usually carried out on the anhydrous molten oxide or halide of the metal:

$$2MO(l) \longrightarrow 2M \text{ (at cathode)} + O_2 \text{ (at anode)}$$

$$2MCl(l) \longrightarrow 2M \text{ (at cathode)} + Cl_2 \text{ (at anode)}$$

We will describe the specific procedures later in this section.

## The Metallurgy of Iron

Multimedia

Chemical Reactions—smelting of iron.

Iron exists in Earth's crust in many different minerals, such as iron pyrite ($FeS_2$), siderite ($FeCO_3$), hematite ($Fe_2O_3$), and magnetite ($Fe_3O_4$, often represented as $FeO \cdot Fe_2O_3$). Of these, hematite and magnetite are particularly suitable for the extraction of iron. The metallurgical processing of iron involves the chemical reduction of the minerals by carbon (in the form of coke) in a blast furnace (Figure 23.3). The concentrated iron ore, limestone ($CaCO_3$), and coke are introduced into the furnace from the top. A blast of hot air is forced up the furnace from the bottom—hence, the name *blast furnace.* The oxygen gas reacts with the carbon in the coke to form mostly carbon monoxide and some carbon dioxide. These reactions are highly exothermic, and as the hot CO and $CO_2$ gases rise, they react with the iron oxides in different temperature zones as shown in Figure 23.3. The key steps in the extraction of iron are

$$3Fe_2O_3(s) + CO(g) \longrightarrow 2Fe_3O_4(s) + CO_2(g)$$

$$Fe_3O_4(s) + CO(g) \longrightarrow 3FeO(s) + CO_2(g)$$

$$FeO(s) + CO(g) \longrightarrow Fe(l) + CO_2(g)$$

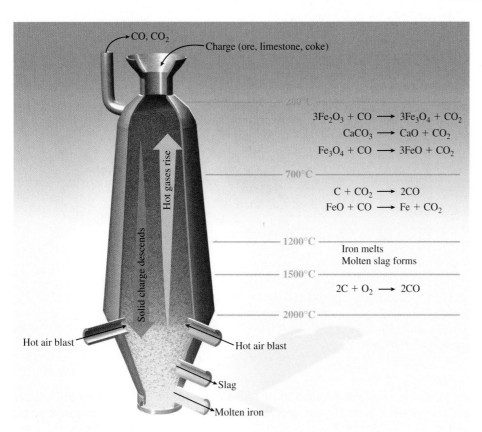

**Figure 23.3** A blast furnace. Iron ore, limestone, and coke are introduced at the top of the furnace. Iron is obtained from the ore by reduction with carbon.

The limestone decomposes in the furnace as follows:

$$CaCO_3(s) \longrightarrow CaO(s) + CO_2(g)$$

The calcium oxide then reacts with the impurities in the iron, which are mostly sand ($SiO_2$) and aluminum oxide ($Al_2O_3$):

$$CaO(s) + SiO_2(s) \longrightarrow CaSiO_3(l)$$

$$CaO(s) + Al_2O_3(s) \longrightarrow Ca(AlO_2)_2(l)$$

The mixture of calcium silicate and calcium aluminate that remains molten at the furnace temperature is known as *slag*.

By the time the ore works its way down to the bottom of the furnace, most of it has already been reduced to iron. The temperature of the lower part of the furnace is above the melting point of impure iron, and so the molten iron at the lower level can be run off to a receiver. The slag, because it is less dense, forms the top layer above the molten iron and can be run off at that level, as shown in Figure 23.3.

Iron extracted in this way contains many impurities and is called *pig iron;* it may contain up to 5 percent carbon and some silicon, phosphorus, manganese, and sulfur. Some of the impurities stem from the silicate and phosphate minerals, while carbon and sulfur come from coke. Pig iron is granular and brittle. It has a relatively low melting point (about 1180°C), so it can be cast in various forms; for this reason it is also called *cast iron.*

## Steelmaking

Steel manufacturing is one of the most important metal industries. In the United States, the annual consumption of steel is well above 100 million tons. Steel is an iron alloy that contains from 0.03 to 1.4 percent carbon plus various amounts of other elements. The wide range of useful mechanical properties associated with steel is primarily a function of the chemical composition and heat treatment of a particular type of steel.

Whereas the production of iron is basically a reduction process (converting iron oxides to metallic iron), the conversion of iron to steel is essentially an oxidation process in which the unwanted impurities are removed from the iron by reaction with oxygen gas. One of several methods used in steelmaking is the *basic oxygen process*. Because of its ease of operation

**Figure 23.4**   The basic oxygen process of steelmaking. The capacity of a typical vessel is 100 tons of cast iron.

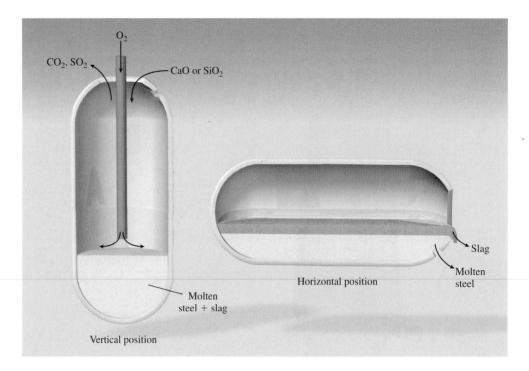

and the relatively short time (about 20 minutes) required for each large-scale (hundreds of tons) conversion, the basic oxygen process is by far the most common means of producing steel today.

Figure 23.4 shows the basic oxygen process. Molten iron from the blast furnace is poured into an upright cylindrical vessel. Pressurized oxygen gas is introduced via a water-cooled tube above the molten metal. Under these conditions, manganese, phosphorus, and silicon, as well as excess carbon, react with oxygen to form oxides. These oxides are then reacted with the appropriate fluxes (e.g., CaO or SiO$_2$) to form slag. The type of flux chosen depends on the composition of the iron. If the main impurities are silicon and phosphorus, a basic flux such as CaO is added to the iron:

$$SiO_2(s) + CaO(s) \longrightarrow CaSiO_3(l)$$

$$P_4O_{10}(l) + 6CaO(s) \longrightarrow 2Ca_3(PO_4)_2(l)$$

On the other hand, if manganese is the main impurity, then an acidic flux such as SiO$_2$ is needed to form the slag:

$$MnO(s) + SiO_2(s) \longrightarrow MnSiO_3(l)$$

The molten steel is sampled at intervals. When the desired blend of carbon and other impurities has been reached, the vessel is rotated to a horizontal position so that the molten steel can be tapped off (Figure 23.5).

The properties of steel depend not only on its chemical composition but also on the heat treatment. At high temperatures, iron and carbon in steel combine to form iron carbide (Fe$_3$C), called *cementite:*

$$3Fe(s) + C(s) \longrightarrow Fe_3C(s)$$

The forward reaction is endothermic, so the formation of cementite is favored at high temperatures. When steel containing cementite is cooled slowly, the preceding equilibrium shifts to the left and the carbon separates as small particles of graphite, which give the steel a grey color. (Very slow decomposition of cementite also takes place at room temperature.) If the steel is cooled rapidly, equilibrium is not attained and the carbon remains largely in the form of cementite (Fe$_3$C). Steel containing cementite is light in color, and it is harder and more brittle than that containing graphite.

Heating the steel to some appropriate temperature for a short time and then cooling it rapidly to give it the desired mechanical properties is known as "tempering." In this way, the ratio of carbon present as graphite and as cementite can be varied within rather wide limits. Table 23.3 lists the composition, properties, and uses of various types of steel.

**Figure 23.5**   Steelmaking.

| TABLE 23.3 | Types of Steel | | | | | | | | |
|---|---|---|---|---|---|---|---|---|---|

| | Composition (Percent by Mass)* | | | | | | | | |
|---|---|---|---|---|---|---|---|---|---|
| **Type** | **C** | **Mn** | **P** | **S** | **Si** | **Ni** | **Cr** | **Others** | **Uses** |
| Plain | 1.35 | 1.65 | 0.04 | 0.05 | 0.06 | — | — | Cu (0.2–0.6) | Sheet products, tools |
| High-strength | 0.25 | 1.65 | 0.04 | 0.05 | 0.15–0.9 | 0.4–1.0 | 0.3–1.3 | Cu (0.01–0.08) | Construction, steam turbines |
| Stainless | 0.03–1.2 | 1.0–10 | 0.04–0.06 | 0.03 | 1–3 | 1–22 | 4.0–27 | — | Kitchen utensils, razor blades |

*A single number indicates the maximum amount of the substance present.

## Purification of Metals

Metals prepared by reduction usually need further treatment to remove impurities. The extent of purification depends on how the metal will be used. Three common purification procedures are distillation, electrolysis, and zone refining.

### Distillation

Metals that have low boiling points, such as mercury, magnesium, and zinc, can be separated from other metals by fractional distillation. One well-known method of fractional distillation is the *Mond*[1] *process* for the purification of nickel. Carbon monoxide gas is passed over the impure nickel metal at about 70°C to form the volatile tetracarbonylnickel (b.p. 43°C), a highly toxic substance, which is separated from the less volatile impurities by distillation:

$$Ni(s) + 4CO(g) \longrightarrow Ni(CO)_4(g)$$

Pure metallic nickel is recovered from $Ni(CO)_4$ by heating the gas at 200°C:

$$Ni(CO)_4(g) \longrightarrow Ni(s) + 4CO(g)$$

The carbon monoxide that is released is recycled back into the process.

### Electrolysis

Electrolysis is another important purification technique. The copper metal obtained by roasting copper sulfide usually contains impurities such as zinc, iron, silver, and gold. The more electropositive metals are removed by an electrolysis process in which the impure copper acts as the anode and *pure* copper acts as the cathode in a sulfuric acid solution containing $Cu^{2+}$ ions (Figure 23.6). The reactions are

*Anode (oxidation):* $\qquad\qquad\qquad Cu(s) \longrightarrow Cu^{2+}(aq) + 2e^-$

*Cathode (reduction):* $\quad Cu^{2+}(aq) + 2e^- \longrightarrow Cu(s)$

**Figure 23.6** Electrolytic purification of copper.

---

1. Ludwig Mond (1839–1909). British chemist of German origin. Mond made many important contributions to industrial chemistry. His method for purifying nickel by converting it to the volatile $Ni(CO)_4$ compound has been described as having given "wings" to the metal.

**Figure 23.7**   Copper cathodes used in the electrorefining process.

Reactive metals in the copper anode, such as iron and zinc, are also oxidized at the anode and enter the solution as $Fe^{2+}$ and $Zn^{2+}$ ions. They are not reduced at the cathode, however. The less electropositive metals, such as gold and silver, are not oxidized at the anode. Eventually, as the copper anode dissolves, these metals fall to the bottom of the cell. Thus, the net result of this electrolysis process is the transfer of copper from the anode to the cathode. Copper prepared this way has a purity greater than 99.5 percent (Figure 23.7).

### Zone Refining

Another often-used method of obtaining extremely pure metals is zone refining. In this process, a metal rod containing a few impurities is drawn through an electric heating coil that melts the metal (Figure 23.8). Most impurities dissolve in the molten metal. As the metal rod emerges from the heating coil, it cools and the pure metal crystallizes, leaving the impurities in the molten metal portion that is still in the heating coil. (This is analogous to the freezing of seawater, in which the solid that separates is mostly pure solvent—water. In zone refining, the liquid metal acts as the solvent and the impurities act as the solutes.) When the molten zone carrying the impurities, now at increased concentration, reaches the end of the rod, it is allowed to cool and is then cut off. Repeating this procedure a number of times results in metal with a purity greater than 99.99 percent.

## 23.3   Band Theory of Conductivity

To gain a better understanding of the conductivity properties of metals, we must also apply our knowledge of quantum mechanics. The model we will use to study metallic bonding is *band theory,* so called because it states that delocalized electrons move freely through "bands" formed by overlapping molecular orbitals. We will also apply band theory to certain elements that are semiconductors.

### Conductors

Metals are characterized by high electrical conductivity. Consider magnesium, for example. The electron configuration of Mg is $[Ne]3s^2$, so each atom has two valence electrons in the $3s$ orbital. In a metallic crystal, the atoms are packed closely together, so the energy levels of each magnesium atom are affected by the immediate neighbors of the atom as a result of orbital overlaps. According to molecular orbital theory [◄◄ Section 9.6], the interaction between two atomic orbitals leads to the formation of a bonding and an antibonding molecular orbital. Because the number of atoms in even a small piece of magnesium is enormously large (on the order of $10^{20}$ atoms), the number of molecular orbitals they form is also very large. These molecular orbitals are so closely spaced on the energy scale that they are more appropriately described as a "band" (Figure 23.9). The closely spaced *filled* energy levels make up the *valence band.* The upper half of the energy

**Figure 23.8**   Zone-refining technique for purifying metals. Top to bottom: An impure metal rod is moved slowly through a heating coil. As the metal rod moves forward, the impurities dissolve in the molten portion of the metal while pure metal crystallizes out in front of the molten zone.

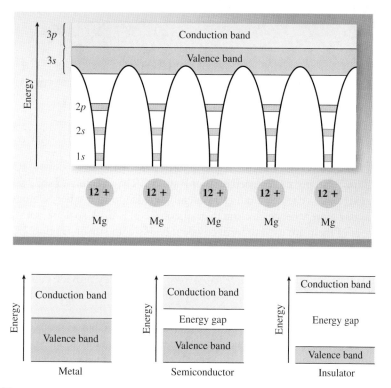

**Figure 23.9** Formation of conduction bands in magnesium. The electrons in the 1s, 2s, and 2p orbitals are localized on each Mg atom. However, the 3s and 3p orbitals overlap to form delocalized molecular orbitals. Electrons in these orbitals can travel throughout the metal, and this accounts for the electrical conductivity of the metal.

**Figure 23.10** Comparison of the energy gaps between the valence band and the conduction band in a metal, a semiconductor, and an insulator. In a metal the energy gap is virtually nonexistent; in a semiconductor the energy gap is small; and in an insulator the energy gap is very large, thus making the promotion of an electron from the valence band to the conduction band difficult.

levels corresponds to the empty, delocalized molecular orbitals formed by the overlap of the 3p orbitals. This set of closely spaced *empty* levels is called the *conduction band.*

We can imagine a metallic crystal as an array of positive ions immersed in a sea of delocalized valence electrons. The great cohesive force resulting from the delocalization is partly responsible for the strength noted in most metals. Because the valence band and the conduction band are adjacent to each other, the amount of energy needed to promote a valence electron to the conduction band is negligible. There the electron can travel freely through the metal, because the conduction band is void of electrons. This freedom of movement explains why metals are good *conductors*—that is, they are capable of conducting an electric current.

Why don't substances like wood and glass conduct electricity as metals do? Figure 23.10 provides an answer to this question. Basically, the electrical conductivity of a solid depends on the spacing of the energy bands and the extent to which they are occupied. In magnesium and other metals, the valence bands are adjacent to the conduction bands, so these metals readily act as conductors. In wood and glass, on the other hand, the gap between the valence band and the conduction band is considerably greater than it is in a metal. Consequently, much more energy is needed to excite an electron into the conduction band. Lacking this energy, electrons cannot move freely. Therefore, glass and wood are *insulators,* ineffective conductors of electricity.

## Semiconductors

*Semiconductors* are elements that normally are *not* conductors, but will conduct electricity at elevated temperatures or when combined with a small amount of certain other elements. The Group 4A elements silicon and germanium are especially suited for this purpose. The use of semiconductors in transistors and solar cells, to name two applications, has revolutionized the electronic industry in recent decades, leading to the increased miniaturization of electronic equipment.

The energy gap between the filled and empty bands of these solids is much smaller than it is for insulators (see Figure 23.10). If the energy needed to excite electrons from the valence band into the conduction band is provided, the solid becomes a conductor. This behavior is opposite that of the metals. A metal's ability to conduct electricity decreases with increasing temperature, because the enhanced vibration of atoms at higher temperatures tends to disrupt the flow of electrons.

The ability of a semiconductor to conduct electricity can also be enhanced by adding small amounts of certain impurities to the element, a process called *doping.* Consider what happens, for

**Figure 23.11**    (a) Silicon crystal doped with phosphorus. (b) Silicon crystal doped with boron. Note the formation of a negative center in (a) and a positive center in (b).

example, when a trace amount of boron or phosphorus is added to solid silicon. (Only about five out of every million Si atoms are replaced by B or P atoms.) The structure of solid silicon is similar to that of diamond; that is, each Si atom is covalently bonded to four other Si atoms. Phosphorus ($[Ne]3s^23p^3$) has one more valence electron than silicon ($[Ne]3s^23p^2$), so there is a valence electron left over after four of them are used to form covalent bonds with silicon (Figure 23.11).

This extra electron can be removed from the phosphorus atom by applying a voltage across the solid. The free electron can move through the structure and function as a conduction electron. Impurities of this type are known as *donor impurities,* because they provide conduction electrons. Solids containing donor impurities are called **n-type semiconductors,** where *n* stands for negative (the charge of the "extra" electron).

The opposite effect occurs if boron is added to silicon. A boron atom has three valence electrons ($1s^22s^22p^1$), one less than silicon. Thus, for every boron atom in the silicon crystal, there is a single *vacancy* in a bonding orbital. It is possible, though, to excite a valence electron from a nearby Si into this vacant orbital. A vacancy created at that Si atom can then be filled by an electron from a neighboring Si atom, and so on. In this manner, electrons can move through the crystal in one direction while the vacancies, or "positive holes," move in the opposite direction, and the solid becomes an electrical conductor. Impurities that are electron deficient are called *acceptor impurities.* Semiconductors that contain acceptor impurities are called **p-type semiconductors,** where *p* stands for positive.

In both the *p*-type and *n*-type semiconductors, the energy gap between the valence band and the conduction band is effectively reduced, so only a small amount of energy is needed to excite the electrons. Typically, the conductivity of a semiconductor is increased by a factor of 100,000 or so by the presence of impurity atoms.

The growth of the semiconductor industry since the early 1960s has been truly remarkable. Today semiconductors are essential components of nearly all electronic equipment, ranging from radios and television sets to pocket calculators and computers. One of the main advantages of solid-state devices over vacuum-tube electronics is that the former can be made on a single "chip" of silicon no larger than the cross section of a pencil eraser. Consequently, much more equipment can be packed into a small volume—a point of particular importance in space travel, as well as in handheld calculators and microprocessors (computers-on-a-chip).

## 23.4  Periodic Trends in Metallic Properties

Metals are lustrous in appearance, solid at room temperature (with the exception of mercury), good conductors of heat and electricity, malleable (can be hammered flat), and ductile (can be drawn into wires). Figure 23.12 shows the positions of the representative metals and the Group

**Figure 23.12**    Main group metals (green) and Group 2B metals (blue) according to their positions in the periodic table.

| 1A 1 | | | | | | | | | | | | | | | | | 8A 18 |
|---|---|---|---|---|---|---|---|---|---|---|---|---|---|---|---|---|---|
| H | 2A 2 | | | | | | | | | | | 3A 13 | 4A 14 | 5A 15 | 6A 16 | 7A 17 | He |
| Li | Be | | | | | | | | | | | B | C | N | O | F | Ne |
| Na | Mg | 3B 3 | 4B 4 | 5B 5 | 6B 6 | 7B 7 | 8B 8 | 8B 9 | 8B 10 | 1B 11 | 2B 12 | Al | Si | P | S | Cl | Ar |
| K | Ca | Sc | Ti | V | Cr | Mn | Fe | Co | Ni | Cu | Zn | Ga | Ge | As | Se | Br | Kr |
| Rb | Sr | Y | Zr | Nb | Mo | Tc | Ru | Rh | Pd | Ag | Cd | In | Sn | Sb | Te | I | Xe |
| Cs | Ba | La | Hf | Ta | W | Re | Os | Ir | Pt | Au | Hg | Tl | Pb | Bi | Po | At | Rn |
| Fr | Ra | Ac | Rf | Db | Sg | Bh | Hs | Mt | Ds | Rg | | | | | | | |

2B metals in the periodic table. (The transition metals are discussed in Chapter 22.) As we saw in Chapter 8, the electronegativity of elements increases from left to right across a period and from bottom to top in a group [◄◄ Section 8.4, Figure 8.6]. The metallic character of metals increases in just the opposite directions—that is, from right to left across a period and from top to bottom in a group. Because metals generally have low electronegativities, they tend to form cations and almost always have positive oxidation numbers in their compounds. However, beryllium and magnesium in Group 2A and the metals in Group 3A and beyond also form covalent compounds.

In Sections 23.5 through 23.7 we will study the chemistry of selected metals from Group 1A (the alkali metals), Group 2A (the alkaline earth metals), and Group 3A (aluminum).

# 23.5 The Alkali Metals

As a group, the alkali metals (the Group 1A elements) are the most electropositive (or the least electronegative) elements known. They exhibit many similar properties, some of which are listed in Table 23.4. Based on their electron configurations, we expect the oxidation number of these elements in their compounds to be $+1$ because the cations would be isoelectronic with the preceding noble gases. This is indeed the case.

The alkali metals have low melting points and are soft enough to be sliced with a knife. These metals all possess a body-centered crystal structure with low packing efficiency. This accounts for their low densities among metals. In fact, lithium is the lightest metal known. Because of their great chemical reactivity, the alkali metals never occur naturally in elemental form; instead, they are found combined with halide, sulfate, carbonate, and silicate ions. In this section we will describe the chemistry of two members of Group 1A—sodium and potassium. The chemistry of lithium, rubidium, and cesium is less important; all isotopes of francium, the last member of the group, are radioactive.

Sodium and potassium are about equally abundant in nature. They occur in silicate minerals such as albite ($NaAlSi_3O_8$) and orthoclase ($KAlSi_3O_8$). Over long periods of time (on a geologic scale), silicate minerals are slowly decomposed by wind and rain, and their sodium and potassium ions are converted to more soluble compounds. Eventually rain leaches these compounds out of the soil and carries them to the sea. Yet when we look at the composition of seawater, we find that the concentration ratio of sodium to potassium is about 28 to 1. The reason for this uneven distribution is that potassium is essential to plant growth, while sodium is not. Thus, plants take up many of the potassium ions along the way, while sodium ions are free to move on to the sea. Other minerals that contain sodium or potassium are halite ($NaCl$), shown in Figure 23.13, Chile saltpeter ($NaNO_3$), and sylvite ($KCl$). Sodium chloride is also obtained from rock salt.

Metallic sodium is most conveniently obtained from *molten* sodium chloride by electrolysis in the Downs cell (review Figure 19.10). The melting point of sodium chloride is rather high (801°C), and much energy is needed to keep large amounts of the substance molten. Adding a suitable substance, such as $CaCl_2$, lowers the melting point to about 600°C—a more convenient temperature for the electrolysis process.

**Figure 23.13**    Halite (NaCl).

| TABLE 23.4 | Properties of Alkali Metals | | | | |
|---|---|---|---|---|---|
| | **Li** | **Na** | **K** | **Rb** | **Cs** |
| Valence electron configuration | $2s^1$ | $3s^1$ | $4s^1$ | $5s^1$ | $6s^1$ |
| Density (g/cm³) | 0.534 | 0.97 | 0.86 | 1.53 | 1.87 |
| Melting point (°C) | 179 | 97.6 | 63 | 39 | 28 |
| Boiling point (°C) | 1317 | 892 | 770 | 688 | 678 |
| Atomic radius (pm) | 155 | 190 | 235 | 248 | 267 |
| Ionic radius (pm)* | 60 | 95 | 133 | 148 | 169 |
| Ionization energy (kJ/mol) | 520 | 496 | 419 | 403 | 375 |
| Electronegativity | 1.0 | 0.9 | 0.8 | 0.8 | 0.7 |
| Standard reduction potential (V)† | −3.05 | −2.71 | −2.93 | −2.93 | −2.92 |

\* Refers to the cation $M^+$, where M denotes an alkali metal atom.

† The half-reaction is $M^+(aq) + e^- \longrightarrow M(s)$.

Metallic potassium cannot be easily prepared by the electrolysis of molten KCl because it is too soluble in the molten KCl to float to the top of the cell for collection. Moreover, it vaporizes readily at the operating temperatures, creating hazardous conditions. Instead, it is usually obtained by the distillation of molten KCl in the presence of sodium vapor at 892°C. The reaction that takes place at this temperature is

$$Na(g) + KCl(l) \longrightarrow NaCl(l) + K(g)$$

This reaction may seem strange given that potassium is a stronger reducing agent than sodium (see Table 23.4). Potassium has a lower boiling point (770°C) than sodium (892°C), however, so it is more volatile at 892°C and distills off more easily. According to Le Châtelier's principle, constantly removing the potassium vapor drives the reaction to the right, ensuring metallic potassium is recovered.

**Multimedia**

Periodic Table—properties of the alkali and alkaline earth metals.

Sodium and potassium are both extremely reactive, but potassium is the more reactive of the two. Both react with water to form the corresponding hydroxides. In a limited supply of oxygen, sodium burns to form sodium oxide ($Na_2O$). In the presence of excess oxygen, however, sodium forms the pale-yellow peroxide:

$$2Na(s) + O_2(g) \longrightarrow Na_2O_2(s)$$

Sodium peroxide reacts with water to give an alkaline solution and hydrogen peroxide:

$$Na_2O_2(s) + 2H_2O(l) \longrightarrow 2NaOH(aq) + H_2O_2(aq)$$

Like sodium, potassium forms the peroxide. In addition, potassium also forms the superoxide when it burns in air:

$$K(s) + O_2(g) \longrightarrow KO_2(s)$$

When potassium superoxide reacts with water, oxygen gas is evolved:

$$2KO_2(s) + 2H_2O(l) \longrightarrow 2KOH(aq) + O_2(g) + H_2O_2(aq)$$

This reaction is utilized in breathing equipment (Figure 23.14). Exhaled air contains both moisture and carbon dioxide. The moisture reacts with $KO_2$ in the apparatus to generate oxygen gas as shown in the preceding reaction. Furthermore, $KO_2$ also reacts with exhaled $CO_2$, which produces more oxygen gas:

$$4KO_2(s) + 2CO_2(g) \longrightarrow 2K_2CO_3(s) + 3O_2(g)$$

Thus, a person using the apparatus can continue to breathe oxygen without being exposed to toxic fumes outside.

Sodium and potassium metals dissolve in liquid ammonia to produce beautiful blue solutions:

$$Na \xrightarrow{NH_3} Na^+ + e^-$$

$$K \xrightarrow{NH_3} K^+ + e^-$$

**Figure 23.14**  Self-contained breathing apparatus.

Both the cation and the electron exist in the solvated form, and the solvated electrons are responsible for the characteristic blue color of such solutions. Metal-ammonia solutions are powerful reducing agents (because they contain free electrons); they are useful in synthesizing both organic and inorganic compounds. It was discovered that the hitherto unknown alkali metal *anions,* $M^-$, are also formed in such solutions. This means that an ammonia solution of an alkali metal contains ion pairs such as $Na^+Na^-$ and $K^+K^-$! (In each case, the metal cation exists as a complex ion with *crown ether,* an organic compound with a high affinity for cations.) In fact, these "salts" are so stable that they can be isolated in crystalline form. This finding is of considerable theoretical interest, because it shows clearly that the alkali metals can have an oxidation number of $-1$, although $-1$ is not found in ordinary compounds.

Sodium and potassium are essential elements of living matter. Sodium ions and potassium ions are present in intracellular and extracellular fluids, and they are essential for osmotic balance and enzyme functions. We now describe the preparations and uses of several of the important compounds of sodium and potassium.

### Sodium Chloride

The source, properties, and uses of sodium chloride were discussed in Chapter 7.

## Sodium Carbonate

Sodium carbonate (called soda ash) is used in all kinds of industrial processes, including water treatment and the manufacture of soaps, detergents, medicines, and food additives. Today about half of all $Na_2CO_3$ produced is used in the glass industry. Sodium carbonate ranks eleventh among the chemicals produced in the United States. For many years, $Na_2CO_3$ was produced by the *Solvay*[2] *process*, in which ammonia is first dissolved in a saturated solution of sodium chloride. Bubbling carbon dioxide into the solution precipitates sodium bicarbonate as follows:

$$NH_3(aq) + NaCl(aq) + H_2CO_3(aq) \longrightarrow NaHCO_3(s) + NH_4Cl(aq)$$

Sodium bicarbonate is then separated from the solution and heated to give sodium carbonate:

$$2NaHCO_3(s) \longrightarrow Na_2CO_3(s) + CO_2(g) + H_2O(g)$$

However, the rising cost of ammonia and the pollution problem resulting from the by-products have prompted chemists to look for other sources of sodium carbonate. One is the mineral *trona* $[Na_5(CO_3)_2(HCO_3) \cdot 2H_2O]$, large deposits of which have been found in Wyoming. When trona is crushed and heated, it decomposes as follows:

$$2Na_5(CO_3)_2(HCO_3) + 2H_2O(s) \longrightarrow 5Na_2CO_3(s) + CO_2(g) + 3H_2O(g)$$

The sodium carbonate obtained this way is dissolved in water, the solution is filtered to remove the insoluble impurities, and the sodium carbonate is crystallized as $Na_2CO_3 \cdot 10H_2O$. Finally, the hydrate is heated to give pure, anhydrous sodium carbonate.

## Sodium Hydroxide and Potassium Hydroxide

The properties of sodium hydroxide and potassium hydroxide are very similar. These hydroxides are prepared by the electrolysis of aqueous NaCl and KCl solutions; both hydroxides are strong bases and very soluble in water. Sodium hydroxide is used in the manufacture of soap and many organic and inorganic compounds. Potassium hydroxide is used as an electrolyte in some storage batteries, and aqueous potassium hydroxide is used to remove carbon dioxide and sulfur dioxide from air.

## Sodium Nitrate and Potassium Nitrate

Large deposits of sodium nitrate (*Chile saltpeter*) are found in Chile. It decomposes with the evolution of oxygen at about 500°C:

$$2NaNO_3(s) \longrightarrow 2NaNO_2(s) + O_2(g)$$

Potassium nitrate (*saltpeter*) is prepared beginning with the "reaction"

$$KCl(aq) + NaNO_3(aq) \longrightarrow KNO_3(aq) + NaCl(aq)$$

This process is carried out just below 100°C. Because $KNO_3$ is the least soluble salt at room temperature, it is separated from the solution by fractional crystallization. Like $NaNO_3$, $KNO_3$ decomposes when heated.

Gunpowder consists of potassium nitrate, wood charcoal, and sulfur in the approximate proportions of 6:1:1 by mass. When gunpowder is heated, the reaction is

$$2KNO_3(s) + S(l) + 3C(s) \longrightarrow K_2S(s) + N_2(g) + 3CO_2(g)$$

The sudden formation of hot nitrogen and carbon dioxide gases causes an explosion.

## 23.6 The Alkaline Earth Metals

The alkaline earth metals are somewhat less electropositive and less reactive than the alkali metals. Except for the first member of the family, beryllium, which resembles aluminum (a Group 3A metal) in some respects, the alkaline earth metals have similar chemical properties. Because their $M^{2+}$ ions attain the stable electron configuration of the preceding noble gas, the oxidation number

---

2. Ernest Solvay (1838–1922). Belgian chemist. Solvay's main contribution to industrial chemistry was the development of the process for the production of sodium carbonate that now bears his name.

| TABLE 23.5 | Properties of Alkaline Earth Metals | | | | |
|---|---|---|---|---|---|
| | **Be** | **Mg** | **Ca** | **Sr** | **Ba** |
| Valence electron configuration | $2s^2$ | $3s^2$ | $4s^2$ | $5s^2$ | $6s^2$ |
| Density (g/cm$^3$) | 1.86 | 1.74 | 1.55 | 2.6 | 3.5 |
| Melting point (°C) | 1280 | 650 | 838 | 770 | 714 |
| Boiling point (°C) | 2770 | 1107 | 1484 | 1380 | 1640 |
| Atomic radius (pm) | 112 | 160 | 197 | 215 | 222 |
| Ionic radius (pm)* | 31 | 65 | 99 | 113 | 135 |
| First ionization energy (kJ/mol) | 899 | 738 | 590 | 548 | 502 |
| Second ionization energy (kJ/mol) | 1757 | 1450 | 1145 | 1058 | 958 |
| Electronegativity | 1.5 | 1.2 | 1.0 | 1.0 | 0.9 |
| Standard reduction potential (V)† | −1.85 | −2.37 | −2.87 | −2.89 | −2.90 |

\* Refers to the cation $M^{2+}$, where M denotes an alkali earth metal atom.

† The half-reaction is $M^{2+}(aq) + 2e^- \longrightarrow M(s)$.

of alkaline earth metals in the combined form is almost always +2. Table 23.5 lists some common properties of these metals. Radium is not included in the table because all radium isotopes are radioactive and it is difficult and expensive to study the chemistry of this Group 2A element.

## Magnesium

Magnesium is the sixth most plentiful element in Earth's crust (about 2.5 percent by mass). Among the principal magnesium ores are brucite [$Mg(OH)_2$], dolomite ($CaCO_3 \cdot MgCO_3$) (Figure 23.15), and epsomite ($MgSO_4 \cdot 7H_2O$). Seawater is a good source of magnesium—there are about 1.3 g of magnesium in each kilogram of seawater. As is the case with most alkali and alkaline earth metals, metallic magnesium is obtained by electrolysis, in this case from its molten chloride, $MgCl_2$ (obtained from seawater).

The chemistry of magnesium is intermediate between that of beryllium and the heavier Group 2A elements. Magnesium does not react with cold water but does react slowly with steam:

$$Mg(s) + H_2O(g) \longrightarrow MgO(s) + H_2(g)$$

It burns brilliantly in air to produce magnesium oxide and magnesium nitride:

$$2Mg(s) + O_2(g) \longrightarrow 2MgO(s)$$

$$3Mg(s) + N_2(g) \longrightarrow Mg_3N_2(s)$$

This property makes magnesium (in the form of thin ribbons or fibers) useful in flash photography and flares.

Magnesium oxide reacts very slowly with water to form magnesium hydroxide, a white solid suspension called *milk of magnesia,* which is used to treat acid indigestion:

$$MgO(s) + H_2O(l) \longrightarrow Mg(OH)_2(s)$$

Magnesium is a typical alkaline earth metal in that its hydroxide is a strong base. [The only alkaline earth hydroxide that is not a strong base is $Be(OH)_2$, which is amphoteric.]

The major uses of magnesium are in lightweight structural alloys, for cathodic protection; in organic synthesis; and in batteries. Magnesium is essential to plant and animal life, and $Mg^{2+}$ ions are not toxic. It is estimated that the average adult ingests about 0.3 g of magnesium ions daily. Magnesium plays several important biological roles. It is present, for instance, in intracellular and extracellular fluids, and magnesium ions are essential for the proper functioning of a number of enzymes. Magnesium is also present in the green plant pigment chlorophyll, which plays an important part in photosynthesis.

## Calcium

Earth's crust contains about 3.4 percent calcium by mass. Calcium occurs in limestone, calcite, chalk, and marble as $CaCO_3$; in dolomite as $CaCO_3 \cdot MgCO_3$ (see Figure 23.15); in gypsum as

**Figure 23.15**   Dolomite ($CaCO_3 \cdot MgCO_3$).

$CaSO_4 \cdot 2H_2O$; and in fluorite as $CaF_2$ (Figure 23.16). Metallic calcium is best prepared by the electrolysis of molten calcium chloride ($CaCl_2$).

As we read down Group 2A from beryllium to barium, metallic properties increase. Unlike beryllium and magnesium, calcium (like strontium and barium) reacts with cold water to yield the corresponding hydroxide, although the rate of reaction is much slower than those involving the alkali metals:

$$Ca(s) + 2H_2O(l) \longrightarrow Ca(OH)_2(aq) + H_2(g)$$

Calcium hydroxide [$Ca(OH)_2$] is commonly known as slaked lime or hydrated lime. Lime (CaO), which is also referred to as quicklime, is one of the oldest materials known to humankind. Quicklime is produced by the thermal decomposition of calcium carbonate:

$$CaCO_3(s) \longrightarrow CaO(s) + CO_2(g)$$

whereas slaked lime is produced by the reaction between quicklime and water:

$$CaO(s) + H_2O(l) \longrightarrow Ca(OH)_2(aq)$$

**Figure 23.16** Fluorite ($CaF_2$).

Quicklime is used in metallurgy (see Section 23.2) and in the removal of $SO_2$ when fossil fuel is burned. Slaked lime is used in water treatment. For many years, farmers have used lime to lower the acidity of the soil for their crops (a process called *liming*). Nowadays lime is also applied to lakes affected by acid rain.

Metallic calcium has rather limited uses. It serves mainly as an alloying agent for metals like aluminum and copper and in the preparation of beryllium metal from its compounds. It is also used as a dehydrating agent for organic solvents.

Calcium is an essential element in living matter. It is the major component of bones and teeth; the calcium ion is present in a complex phosphate salt called hydroxyapatite [$Ca_5(PO_4)_3OH$]. A characteristic function of $Ca^{2+}$ ions in living systems is the activation of a variety of metabolic processes, including a vital role in heart action, blood clotting, muscle contraction, and nerve impulse transmission.

## 23.7 Aluminum

Aluminum is the most abundant metal and the third most plentiful element in Earth's crust (7.5 percent by mass). The elemental form does not occur in nature; instead, its principal ore is bauxite ($Al_2O_3 \cdot 2H_2O$). Other minerals containing aluminum are orthoclase ($KAlSi_3O_8$), beryl ($Be_3Al_2Si_6O_{18}$), cryolite ($Na_3AlF_6$), and corundum ($Al_2O_3$) (Figure 23.17).

Aluminum is usually prepared from bauxite, which is frequently contaminated with silica ($SiO_2$), iron oxides, and titanium(IV) oxide. The ore is first heated in sodium hydroxide solution to convert the silica into soluble silicates:

$$SiO_2(s) + 2OH^-(aq) \longrightarrow SiO_3^{2-}(aq) + H_2O(l)$$

At the same time, aluminum oxide is converted to the aluminate ion ($AlO_2^-$):

$$Al_2O_3(s) + 2OH^-(aq) \longrightarrow 2AlO_2^-(aq) + H_2O(l)$$

Iron oxide and titanium oxide are unaffected by this treatment and are filtered off. Next, the solution is treated with acid to precipitate the insoluble aluminum hydroxide:

$$AlO_2^-(aq) + H_3O^+(aq) \longrightarrow Al(OH)_3(s)$$

After filtration, the aluminum hydroxide is heated to obtain aluminum oxide:

$$2Al(OH)_3(s) \longrightarrow Al_2O_3(s) + 3H_2O(g)$$

**Figure 23.17** Corundum ($Al_2O_3$).

Anhydrous aluminum oxide, or *corundum*, is reduced to aluminum by the **Hall**[3] **process.** Figure 23.18 shows a Hall electrolytic cell, which contains a series of carbon anodes. The cathode is also made of carbon and constitutes the lining inside the cell. The key to the Hall process is the use of

---

3. Charles Martin Hall (1863–1914). American inventor. While Hall was an undergraduate at Oberlin College, he became interested in finding an inexpensive way to extract aluminum. Shortly after graduation, when he was only 22 years old, Hall succeeded in obtaining aluminum from aluminum oxide in a backyard woodshed. Amazingly, the same discovery was made at almost the same moment in France by Paul Héroult, another 22-year-old inventor working in a similar makeshift laboratory.

**Figure 23.18**   Electrolytic production of aluminum based on the Hall process.

cryolite ($Na_3AlF_6$; m.p. 1000°C) as the solvent for aluminum oxide (m.p. 2045°C). The mixture is electrolyzed to produce aluminum and oxygen gas:

*Anode (oxidation):*         $3[2O^{2-} \longrightarrow O_2(g) + 4e^-]$

*Cathode (reduction):*    $4[Al^{3+} + 3e^- \longrightarrow Al(l)]$

*Overall:*                          $2Al_2O_3 \longrightarrow 4Al(l) + 3O_2(g)$

Oxygen gas reacts with the carbon anodes (at elevated temperatures) to form carbon monoxide, which escapes as a gas. The liquid aluminum metal (m.p. 660.2°C) sinks to the bottom of the vessel, from which it can be drained from time to time during the procedure.

Aluminum is one of the most versatile metals known. It has a low density (2.7 g/cm³) and high tensile strength (i.e., it can be stretched or drawn out). Aluminum is malleable, it can be rolled into thin foils, and it is an excellent electrical conductor. Its conductivity is about 65 percent that of copper. However, because aluminum is cheaper and lighter than copper, it is widely used in high-voltage transmission lines. Although aluminum's chief use is in aircraft construction, the pure metal itself is too soft and weak to withstand much strain. Its mechanical properties are greatly improved by alloying it with small amounts of metals such as copper, magnesium, and manganese, as well as silicon. Aluminum is not used by living systems and is generally considered to be nontoxic.

As we read across the periodic table from left to right in a given period, metallic properties gradually decrease. Thus, although aluminum is considered an active metal, it does not react with water as do sodium and calcium. Aluminum reacts with hydrochloric acid and with strong bases as follows:

$$2Al(s) + 6HCl(aq) \longrightarrow 2AlCl_3(aq) + 3H_2(g)$$

$$2Al(s) + 2NaOH(aq) + 2H_2O(l) \longrightarrow 2NaAlO_2(aq) + 3H_2(g)$$

Aluminum readily forms the oxide $Al_2O_3$ when exposed to air:

$$4Al(s) + 3O_2(g) \longrightarrow 2Al_2O_3(s)$$

A tenacious film of this oxide protects metallic aluminum from further corrosion and accounts for some of the unexpected inertness of aluminum. Aluminum oxide has a very large exothermic enthalpy of formation ($\Delta H_f^\circ = -1670$ kJ/mol). This property makes aluminum suitable for use in solid propellants for rockets such as those used for some space shuttles. When a mixture of aluminum and ammonium perchlorate ($NH_4ClO_4$) is ignited, aluminum is oxidized to $Al_2O_3$, and the heat liberated in the reaction causes the gases that are formed to expand with great force. This action lifts the rocket.

The great affinity of aluminum for oxygen is illustrated nicely by the reaction of aluminum powder with a variety of metal oxides, particularly the transition metal oxides, to produce the corresponding metals. A typical reaction is

$$2Al(s) + Fe_2O_3(s) \longrightarrow Al_2O_3(l) + 2Fe(l) \qquad \Delta H^\circ = -852 \text{ kJ/mol}$$

**Figure 23.19**   The temperature of a thermite reaction can reach 3000°C.

which can result in temperatures approaching 3000°C. This transformation, which is used in the welding of steel and iron, is called the *thermite reaction* (Figure 23.19).

Aluminum chloride exists as a dimer:

$$\begin{array}{ccccc} Cl & & Cl & & Cl \\ & \diagdown & \diagup\diagdown & \diagup & \\ & Al & \longrightarrow & Al & \\ & \diagup & \diagup\diagdown & \diagdown & \\ Cl & & Cl & & Cl \end{array}$$

Each of the bridging chlorine atoms forms a normal covalent bond and a *coordinate* covalent bond (each indicated by an arrow) with two aluminum atoms. Each aluminum atom is assumed to be $sp^3$-hybridized, so the vacant $sp^3$ hybrid orbital can accept a lone pair from the chlorine atom (Figure 23.20). Aluminum chloride undergoes hydrolysis as follows:

$$AlCl_3(s) + 3H_2O(l) \longrightarrow Al(OH)_3(s) + 3HCl(aq)$$

Aluminum hydroxide, like $Be(OH)_2$, is amphoteric:

$$Al(OH)_3(s) + 3H^+(aq) \longrightarrow Al^{3+}(aq) + 3H_2O(l)$$

$$Al(OH)_3(s) + OH^-(aq) \longrightarrow Al(OH)_4^-(aq)$$

In contrast to the boron hydrides, which are a well-defined series of compounds, aluminum hydride is a polymer in which each aluminum atom is surrounded octahedrally by bridging hydrogen atoms (Figure 23.21).

When an aqueous mixture of aluminum sulfate and potassium sulfate is evaporated slowly, crystals of $KAl(SO_4)_2 \cdot 12H_2O$ are formed. Similar crystals can be formed by substituting $Na^+$ or $NH_4^+$ for $K^+$, and $Cr^{3+}$ or $Fe^{3+}$ for $Al^{3+}$. These compounds are called *alums*, and they have the general formula

$$M^+M^{3+}(SO_4)_2 \cdot 12H_2O \qquad M^+: K^+, Na^+, NH_4^+$$
$$M^{3+}: Al^{3+}, Cr^{3+}, Fe^{3+}$$

Alums are examples of double salts—that is, salts that contain two different cations.

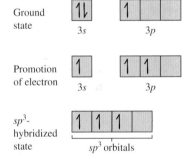

**Figure 23.20**   The $sp^3$ hybridization of an Al atom in $Al_2Cl_6$. Each Al atom has one vacant $sp^3$ hybrid orbital that can accept a lone pair from the bridging Cl atom.

**Figure 23.21**   Structure of aluminum hydride. Note that this compound is a polymer. Each Al atom is surrounded in an octahedral arrangement by six bridging H atoms.

# Applying What You've Learned

Most health problems related to copper are the result of errors in copper metabolism. However, although it is rare, copper deficiency can result from a diet that is poor in copper. Symptoms of dietary copper deficiency include anemia (a deficiency of red blood cells) and neutropenia (a deficiency of a particular type of white blood cell).

The fact that copper is essential to human health was first demonstrated with a group of children in Peru. One patient's ordeal was detailed by Cordano and Graham in the journal *Pediatrics* in 1966. During her first few years of life, the patient was hospitalized several times with anemia, neutropenia, osteoporosis, and multiple fractures. At age 6, over a period of 3 months, she received 20 blood transfusions for her severe anemia, which had not responded to treatment. When Dr. Cordano became aware of the patient's history, he initiated treatment with copper supplementation. The patient never required another transfusion and after 6 months on copper supplements, at age 7, she walked for the first time in her life.

**Writing Prompt:**
Research the subject on the Web, and write a 500-word essay on the causes, diagnosis, and treatment of dietary copper deficiency. Include a specific case study.

# CHAPTER SUMMARY

## Section 23.1

- Depending on their reactivities, metals exist in nature in either the free or combined state. (More reactive metals are found combined with other elements.) Most metals are found in **minerals.** Minerals with high metal content are called **ores.**

## Section 23.2

- **Metallurgy** involves recovering metal from ores. The three stages of metal recovery are *preparation, separation,* and *purification.* An **alloy** is a solid mixture of one or more metals, sometimes also containing one or more *nonmetals.* An **amalgam** is a mixture of mercury and one or more other metals.

- The methods commonly used for purifying metals are distillation, electrolysis, and zone refining. **Pyrometallurgy** refers to metallurgical processes carried out at high temperatures.

## Section 23.3

- Metallic bonds can be thought of as the force between positive ions immersed in a sea of electrons. In terms of **band theory,** the atomic orbitals merge to form energy bands.

- A substance is a **conductor** when electrons can be readily promoted to the conduction band, where they are free to move through the substance. In an **insulator,** the energy gap between the valence band and the conduction band is so large that electrons cannot be promoted into the conduction band.

## Section 23.4

- Metals typically are good conductors and are malleable and ductile. Metallic character increases from top to bottom in a group and decreases from left to right across a period.

## Section 23.5

- The alkali metals are the most reactive of all the metallic elements. They have an oxidation state of $+1$ in their compounds. Under special conditions, some of them can form anions with an oxidation state of $-1$.

## Section 23.6

- The alkaline earth metals are somewhat less reactive than the alkali metals. They almost always have an oxidation number of $+2$ in their compounds. The properties of the alkaline earth elements become increasingly metallic from top to bottom in their group.

## Section 23.7

- Aluminum ordinarily does not react with water due to a protective coating of aluminum oxide; its hydroxide is amphoteric. The **Hall process** is used to reduce aluminum oxide to aluminum.

- **Semiconductors** are substances that normally are not conductors but will conduct electricity at elevated temperatures or when combined with a small amount of certain other elements. Semiconductors in which an electron-rich impurity is added to enhance conduction are known as **n-type semiconductors.** Semiconductors in which an electron-poor impurity is added to enhance conduction are known as **p-type semiconductors.**

# KEY WORDS

Alloy, 947

Amalgam, 947

Band theory, 952

Conductor, 953

Hall process, 959

Insulator, 953

Metallurgy, 947

Mineral, 946

*n*-type semiconductor, 954

Ore, 946

*p*-type semiconductor, 954

Pyrometallurgy, 948

Semiconductor, 953

# QUESTIONS AND PROBLEMS

## Section 23.1: Occurrence of Metals

### Review Questions

23.1  Define the terms *mineral* and *ore.*

23.2  List three metals that are usually found in an uncombined state in nature and three metals that are always found in a combined state in nature.

### Problems

23.3  Write chemical formulas for the following minerals: (a) calcite, (b) dolomite, (c) fluorite, (d) halite, (e) corundum, (f) magnetite, (g) beryl, (h) galena, (i) epsomite, (j) anhydrite.

23.4  Name the following minerals: (a) $MgCO_3$, (b) $Na_3AlF_6$, (c) $Al_2O_3$, (d) $Ag_2S$, (e) $HgS$, (f) $ZnS$, (g) $SrSO_4$, (h) $PbCO_3$, (i) $MnO_2$, (j) $TiO_2$.

## Section 23.2: Metallurgical Processes

### Review Questions

23.5  Define the terms *metallurgy, alloy,* and *amalgam.*

23.6  Describe the main steps involved in the preparation of an ore.

23.7  What does *roasting* mean in metallurgy? Why is roasting a major source of air pollution and acid rain?

23.8    Describe with examples the chemical and electrolytic reduction processes used in the production of metals.

23.9    Describe the main steps used to purify metals.

23.10    Describe the extraction of iron in a blast furnace.

23.11    Briefly discuss the steelmaking process.

23.12    Briefly describe the zone refining process.

## Problems

**23.13**    In the Mond process for the purification of nickel, CO is passed over metallic nickel to give $Ni(CO)_4$:

$$Ni(s) + 4CO(g) \rightleftharpoons Ni(CO)_4(g)$$

Given that the standard free energies of formation of $CO(g)$ and $Ni(CO)_4(g)$ are $-137.3$ and $-587.4$ kJ/mol, respectively, calculate the equilibrium constant of the reaction at 80°C. (Assume $\Delta G_f^\circ$ to be independent of temperature.)

**23.14**    Copper is purified by electrolysis (see Figure 23.6). A 5.00-kg anode is used in a cell where the current is 37.8 A. How long (in hours) must the current run to dissolve this anode and electroplate it onto the cathode?

**23.15**    Consider the electrolytic procedure for purifying copper described in Figure 23.6. Suppose that a sample of copper contains the following impurities: Fe, Ag, Zn, Au, Co, Pt, and Pb. Which of the metals will be oxidized and dissolved in solution and which will be unaffected and simply form the sludge that accumulates at the bottom of the cell?

23.16    How would you obtain zinc from sphalerite (ZnS)?

**23.17**    A certain mine produces $2.0 \times 10^8$ kg of copper from chalcopyrite ($CuFeS_2$) each year. The ore contains only 0.80 percent Cu by mass. (a) If the density of the ore is 2.8 g/cm³, calculate the volume (in cm³) of ore removed each year. (b) Calculate the mass (in kg) of $SO_2$ produced by roasting (assume chalcopyrite to be the only source of sulfur).

**23.18**    Starting with rutile ($TiO_2$), explain how you would obtain pure titanium metal. (*Hint:* First convert $TiO_2$ to $TiCl_4$. Next, reduce $TiCl_4$ with Mg. Look up physical properties of $TiCl_4$, Mg, and $MgCl_2$ in a chemistry handbook.)

**23.19**    Which of the following compounds would require electrolysis to yield the free metals: $Ag_2S$, $CaCl_2$, NaCl, $Fe_2O_3$, $Al_2O_3$, $TiCl_4$?

23.20    Although iron is only about two-thirds as abundant as aluminum in Earth's crust, mass for mass it costs only about one-quarter as much to produce. Why?

## Section 23.3: Band Theory of Conductivity

### Review Questions

23.21    Define the following terms: *conductor, insulator, semiconducting elements, donor impurities, acceptor impurities, n-type semiconductors, p-type semiconductors.*

23.22    Briefly discuss the nature of bonding in metals, insulators, and semiconducting elements.

23.23    Describe the general characteristics of *n*-type and *p*-type semiconductors.

### Problems

**23.24**    State whether silicon would form *n*-type or *p*-type semiconductors with the following elements: Ga, Sb, Al, As.

## Section 23.4: Periodic Trends in Metallic Properties

### Review Questions

23.25    Discuss the general properties of metals.

23.26    Use periodic trends in ionization energy and electronegativity to show how the metallic character changes within a group.

23.27    Use periodic trends in ionization energy and electronegativity to show how the metallic character changes across a period.

## Section 23.5: The Alkali Metals

### Review Questions

23.28    How is sodium prepared commercially?

23.29    Why is potassium usually not prepared electrolytically from one of its salts?

23.30    Describe the uses of the following compounds: NaCl, $Na_2CO_3$, NaOH, KOH, $KO_2$.

23.31    Under what conditions do sodium and potassium form $Na^-$ and $K^-$ ions?

### Problems

**23.32**    Complete and balance the following equations:
    (a) $K(s) + H_2O(l) \longrightarrow$
    (b) $NaH(s) + H_2O(l) \longrightarrow$
    (c) $Na(s) + O_2(g) \longrightarrow$
    (d) $K(s) + O_2(g) \longrightarrow$

**23.33**    Write a balanced equation for each of the following reactions: (a) sodium reacts with water, (b) an aqueous solution of NaOH reacts with $CO_2$, (c) solid $Na_2CO_3$ reacts with an HCl solution, (d) solid $NaHCO_3$ reacts with an HCl solution, (e) solid $NaHCO_3$ is heated, (f) solid $Na_2CO_3$ is heated.

23.34    Sodium hydride (NaH) can be used as a drying agent for many organic solvents. Explain how it works.

**23.35**    Calculate the volume of $CO_2$ at 10.0°C and 746 mmHg pressure obtained by treating 25.0 g of $Na_2CO_3$ with an excess of hydrochloric acid.

## Section 23.6: The Alkaline Earth Metals

### Review Questions

23.36    List the common ores of magnesium and calcium.

23.37    How are magnesium and calcium obtained commercially?

### Problems

**23.38**    From the thermodynamic data in Appendix 2, calculate the $\Delta H^\circ$ values for the following decompositions:
    (a) $MgCO_3(s) \longrightarrow MgO(s) + CO_2(g)$
    (b) $CaCO_3(s) \longrightarrow CaO(s) + CO_2(g)$

    Which of the two compounds is more easily decomposed by heat?

**23.39** Starting with magnesium and concentrated nitric acid, describe how you would prepare magnesium oxide. [*Hint:* First convert Mg to $Mg(NO_3)_2$. Next, MgO can be obtained by heating $Mg(NO_3)_2$.]

23.40 Describe two ways of preparing magnesium chloride.

**23.41** The second ionization energy of magnesium is only about twice as great as the first, but the third ionization energy is 10 times as great. Why does it take so much more energy to remove the third electron?

**23.42** List the sulfates of the Group 2A metals in order of increasing solubility in water. Explain the trend. (*Hint:* You need to consult a chemistry handbook.)

**23.43** Helium contains the same number of electrons in its outer shell as do the alkaline earth metals. Explain why helium is inert whereas the Group 2A metals are not.

**23.44** When exposed to air, calcium first forms calcium oxide, which is then converted to calcium hydroxide, and finally to calcium carbonate. Write a balanced equation for each step.

**23.45** Write chemical formulas for (a) quicklime and (b) slaked lime.

## Section 23.7: Aluminum

### Review Questions

23.46 Describe the Hall process for preparing aluminum.

23.47 What action renders aluminum inert?

### Problems

23.48 Before Hall invented his electrolytic process, aluminum was produced by the reduction of its chloride with an active metal. Which metals would you use for the production of aluminum in that way?

**23.49** With the Hall process, how many hours will it take to deposit 664 g of Al at a current of 32.6 A?

**23.50** Aluminum forms the complex ions $AlCl_4^-$ and $AlF_6^{3-}$. Describe the shapes of these ions. $AlCl_6^{3-}$ does not form. Why? (*Hint:* Consider the relative sizes of $Al^{3+}$, $F^-$, and $Cl^-$ ions.)

**23.51** The overall reaction for the electrolytic production of aluminum by means of the Hall process may be represented as

$$Al_2O_3(s) + 3C(s) \longrightarrow 2Al(l) + 3CO(g)$$

At 1000°C, the standard free-energy change for this process is 594 kJ/mol. (a) Calculate the minimum voltage required to produce 1 mole of aluminum at this temperature. (b) If the actual voltage applied is exactly three times the ideal value, calculate the energy required to produce 1.00 kg of the metal.

**23.52** In basic solution, aluminum metal is a strong reducing agent and is oxidized to $AlO_2^-$. Give balanced equations for the reaction of Al in basic solution with the following: (a) $NaNO_3$, to give ammonia; (b) water, to give hydrogen; (c) $Na_2SnO_3$, to give metallic tin.

**23.53** Write a balanced equation for the thermal decomposition of aluminum nitrate to form aluminum oxide, nitrogen dioxide, and oxygen gas.

23.54 Describe some of the properties of aluminum that make it one of the most versatile metals known.

**23.55** The pressure of gaseous $Al_2Cl_6$ increases more rapidly with temperature than predicted by the ideal gas equation even though $Al_2Cl_6$ behaves like an ideal gas. Explain.

**23.56** Starting with aluminum, describe with balanced equations how you would prepare (a) $Al_2Cl_6$, (b) $Al_2O_3$, (c) $Al_2(SO_4)_3$, (d) $NH_4Al(SO_4)_2 \cdot 12H_2O$.

**23.57** Explain the change in bonding when $Al_2Cl_6$ dissociates to form $AlCl_3$ in the gas phase.

## Additional Problems

23.58 In steelmaking, nonmetallic impurities such as P, S, and Si are removed as the corresponding oxides. The inside of the furnace is usually lined with $CaCO_3$ and $MgCO_3$, which decompose at high temperatures to yield CaO and MgO. How do CaO and MgO help in the removal of the nonmetallic oxides?

**23.59** When 1.164 g of a certain metal sulfide was roasted in air, 0.972 g of the metal oxide was formed. If the oxidation number of the metal is +2, calculate the molar mass of the metal.

23.60 An early view of metallic bonding assumed that bonding in metals consisted of localized, shared electron-pair bonds between metal atoms. What evidence would help you to argue against this viewpoint?

**23.61** Referring to Figure 23.6, would you expect $H_2O$ and H to be reduced at the cathode and $H_2O$ oxidized at the anode?

**23.62** A 0.450-g sample of steel contains manganese as an impurity. The sample is dissolved in acidic solution and the manganese is oxidized to the permanganate ion $MnO_4^-$. The $MnO_4^-$ ion is reduced to $Mn^{2+}$ by reacting with 50.0 mL of 0.0800 $M$ $FeSO_4$ solution. The excess $Fe^{2+}$ ions are then oxidized to $Fe^{3+}$ by 22.4 mL of 0.0100 $M$ $K_2Cr_2O_7$. Calculate the percent by mass of manganese in the sample.

**23.63** Given that $\Delta G_f^\circ(Fe_2O_3) = -741.0$ kJ/mol and that $\Delta G_f^\circ(Al_2O_3) = -1576.4$ kJ/mol, calculate $\Delta G^\circ$ for the following reactions at 25°C:

(a) $2Fe_2O_3(s) \longrightarrow 4Fe(s) + 3O_2(g)$
(b) $2Al_2O_3(s) \longrightarrow 4Al(s) + 3O_2(g)$

23.64 Use compounds of aluminum as examples to explain what is meant by amphoterism.

**23.65** When an inert atmosphere is needed for a metallurgical process, nitrogen is frequently used. However, in the reduction of $TiCl_4$ by magnesium, helium is used. Explain why nitrogen is not suitable for this process.

23.66 It has been shown that $Na_2$ species form in the vapor phase. Describe the formation of the "disodium molecule" in terms of a molecular orbital energy level diagram. Would you expect the alkaline earth metals to exhibit a similar property?

**23.67** Explain each of the following statements: (a) An aqueous solution of $AlCl_3$ is acidic. (b) $Al(OH)_3$ is soluble in NaOH solution but not in $NH_3$ solution.

23.68 Write balanced equations for the following reactions: (a) the heating of aluminum carbonate, (b) the reaction between $AlCl_3$ and K, (c) the reaction between solutions of $Na_2CO_3$ and $Ca(OH)_2$.

**23.69** Write a balanced equation for the reaction between calcium oxide and dilute HCl solution.

**23.70** What is wrong with the following procedure for obtaining magnesium?

$$MgCO_3(s) \longrightarrow MgO(s) + CO_2(g)$$

$$MgO(s) + CO(g) \longrightarrow Mg(s) + CO_2(g)$$

**23.71** Explain why most metals have a flickering appearance.

**23.72** Predict the chemical properties of francium, the last member of Group 1A.

**23.73** Describe a medicinal or health-related application for each of the following compounds: NaF, $Li_2CO_3$, $Mg(OH)_2$, $CaCO_3$, $BaSO_4$.

**23.74** The following are two reaction schemes involving magnesium. *Scheme I:* When magnesium burns in oxygen, a white solid (A) is formed. A dissolves in 1 *M* HCl to give a colorless solution (B). Upon addition of $Na_2CO_3$ to B, a white precipitate is formed (C). On heating, C decomposes to D and a colorless gas is generated (E). When E is passed through limewater [an aqueous suspension of $Ca(OH)_2$], a white precipitate appears (F). *Scheme II:* Magnesium reacts with 1 *M* $H_2SO_4$ to produce a colorless solution (G). Treating G with an excess of NaOH produces a white precipitate (H). H dissolves in 1 *M* $HNO_3$ to form a colorless solution. When the solution is slowly evaporated, a white solid (I) appears. On heating I, a brown gas is given off. Identify A−I, and write equations representing the reactions involved.

**23.75** Lithium and magnesium exhibit a diagonal relationship in some chemical properties. How does lithium resemble magnesium in its reaction with oxygen and nitrogen? Consult a handbook of chemistry and compare the solubilities of carbonates, fluorides, and phosphates of these metals.

**23.76** To prevent the formation of oxides, peroxides, and superoxides, alkali metals are sometimes stored in an inert atmosphere. Which of the following gases should not be used for lithium: Ne, Ar, $N_2$, Kr? Why?

**23.77** Which of the following metals is not found in the free state in nature: Ag, Cu, Zn, Au, Pt?

**23.78** After heating, a metal surface (such as that of a cooking pan or skillet) develops a color pattern like an oil slick on water. Explain.

**23.79** A sample of 10.00 g of sodium reacts with oxygen to form 13.83 g of sodium oxide ($Na_2O$) and sodium peroxide ($Na_2O_2$). Calculate the percent composition of the mixture.

**23.80** The electrical conductance of copper metal decreases with temperature, but that of a $CuSO_4$ solution increases with temperature. Explain.

**23.81** As stated in the chapter, potassium superoxide ($KO_2$) is a useful source of oxygen employed in breathing equipment. Calculate the pressure at which oxygen gas stored at 20°C would have the same density as the oxygen gas provided by $KO_2$. The density of $KO_2$ at 20°C is 2.15 $g/cm^3$.

**23.82** Chemical tests of four metals A, B, C, and D show the following results: (a) Only B and C react with 0.5 *M* HCl to give $H_2$ gas. (b) When B is added to a solution containing the ions of the other metals, metallic A, C, and D are formed. (c) A reacts with 6 *M* $HNO_3$, but D does not. Arrange the metals in the increasing order as reducing agents. Suggest four metals that fit these descriptions.

# PRE-PROFESSIONAL PRACTICE EXAM PROBLEMS: VERBAL REASONING

Copper is essential for a wide range of biochemical processes, which are vital for human health. However, copper is also potentially toxic. Under physiological conditions, copper exists in two different oxidation states: $Cu^+$ (cuprous) and $Cu^{2+}$ (cupric). It is the one-electron shift back and forth between these two oxidation states that makes copper essential to the function of cuproenzymes. However, this electron transfer also contributes to the potential toxicity of copper in the body. Cycling between $Cu^+$ and $Cu^{2+}$ can generate hydroxyl radicals ($\cdot OH$), highly reactive oxygen species that can damage biological molecules such as DNA and proteins. Recently, intracellular proteins have been discovered that appear to prevent the toxic effects of copper ions. These so-called copper chaperones escort copper ions directly to the cuproenzymes that need them, keeping the cells free of "unchaperoned" copper ions, which would otherwise be involved in the formation of hydroxyl radicals.

Menkes disease results from an inability to absorb copper in the intestine. Lack of availability of copper results in reduced activity of cuproenzymes. Symptoms of Menkes disease include mental retardation and abnormalities in connective tissue. Victims of Menkes disease are almost always male, and they typically do not survive beyond the age of 3. In recent years, an experimental treatment for Menkes disease has been reported. The treatment involves injection of a copper compound (copper histadine) to supply copper to the body's cells. One report details the effects of this treatment on four patients who exceeded the life expectancy of Menkes disease sufferers by as much as 20 years with regular injections of copper histadine. Although the mental retardation was largely mitigated by the treatment, abnormalities in connective tissue were not and so although the patients survived, they were severely disabled by their symptoms.

1. According to the passage, the potential toxicity of copper in the human body results from

   a) copper ions damaging DNA and proteins.
   b) increased activity of cuproenzymes.
   c) the formation of hydroxyl radicals.
   d) excess copper accumulating in the brain.

2. What is the function of a copper chaperone?

   a) To convert copper atoms into copper ions.
   b) To convert $Cu^+$ ions to $Cu^{2+}$ ions.
   c) To convert $Cu^{2+}$ ions to $Cu^+$ ions.
   d) To escort copper ions to the cuproenzymes that require them.

3. According to the passage, what is the cause of Menkes disease?

   a) Abnormalities in connective tissue
   b) Failure to absorb copper in the intestine
   c) Excess copper in the cells
   d) Faulty cuproenzymes

4. According to the passage, victims of Menkes disease

   a) are typically male.
   b) typically live up to 20 years.
   c) can be cured with copper histidine injections.
   d) all of the above.

# 24

# Nonmetallic Elements and Their Compounds

*Arsenic, the poison of many murder mysteries, can cause severe abdominal pain, intense thirst, vomiting, convulsions, and ultimately death. Despite its toxicity, it has been used historically in medicine, cosmetics, and pigments.*

# Toxicity of Arsenic(III) Compounds

Arsenic has always been the poison of choice for murder mysteries. In the 1960s, samples of Napoleon's hair were analyzed and found to contain a high level of arsenic. It was naturally concluded that he was intentionally poisoned. Studies in recent years, however, have shown that Napoleon's exposure to arsenic may have been environmental. The wallpaper in his drawing room was found to contain the green pigment copper arsenite ($CuHAsO_4$). The mold growing on the papers could have converted the compound to the volatile, toxic trimethyl arsine [$(CH_3)_3As$], which Napoleon then ingested.

Unlike the first two elements in Group 5A, nitrogen and phosphorus, arsenic is not an essential element in the human body. Moreover, elemental arsenic itself is not all that harmful. The commonly used poison is actually arsenic(III) oxide ($As_2O_3$), a white compound that dissolves in water, has no taste, and if administered over a period of time, is hard to detect. The toxicity of As(III) inorganic compounds lies in their ability to bind to the sulfhydryl group ($-SH$) of proteins and enzymes, thus impairing their normal functions. Arsenite also inhibits enzyme activities in the mitochondria and uncouples oxidative phosphorylation. This results in a decrease in ATP production and an increase in harmful reactive oxygen species (ROS) such as hydrogen peroxide and superoxide ion. The accumulation of ROS can lead to DNA damage and initiate carcinogenic processes.

Arsenic is a *metalloid*. The metalloids and the nonmetallic elements and their compounds exhibit chemistry that varies considerably.

At the end of this chapter, you will have an appreciation for the widely varied properties of nonmetals and the important chemistry they exhibit [▶▶ Page 992].

## In This Chapter, You Will Learn:

Some of the properties of metalloids and nonmetals and which compounds they form.

### Before you begin, you should review

- General trends in chemical properties [◀◀ Section 7.7]

## 24.1    General Properties of Nonmetals

Properties of nonmetals are more varied than those of metals. Hydrogen, oxygen, nitrogen, fluorine, chlorine, and the noble gases are all gases in the elemental state, whereas only bromine is a liquid. All the remaining nonmetals are solids at room temperature. Unlike metals, nonmetals are poor conductors of heat and electricity, and when they form compounds, nonmetals can exhibit either positive or negative oxidation numbers.

A small group of elements, called *metalloids*, have properties characteristic of both metals and nonmetals. The metalloids boron, silicon, germanium, and arsenic are semiconducting elements (see Section 20.3).

Nonmetals are more electronegative than metals [◄◄ Section 8.4]. The electronegativity of elements increases from left to right across any period and from bottom to top in any group in the periodic table (see Figure 8.6). With the exception of hydrogen, the nonmetals are concentrated in the upper right-hand corner of the periodic table (Figure 24.1). Compounds formed by a combination of metals with nonmetals tend to be *ionic*, having a metallic cation and a nonmetallic anion.

In this chapter we will discuss the chemistry of a number of common and important nonmetallic elements—namely, hydrogen; carbon (Group 4A); nitrogen and phosphorus (Group 5A); oxygen and sulfur (Group 6A); and the halogens: fluorine, chlorine, bromine, and iodine (Group 7A).

## 24.2    Hydrogen

Hydrogen is the simplest element known—its most common atomic form contains only one proton and one electron. The atomic form of hydrogen exists only at very high temperatures, however. Normally, elemental hydrogen is a diatomic molecule, the product of an exothermic reaction between H atoms:

$$H(g) + H(g) \longrightarrow H_2(g) \qquad \Delta H^\circ = -436.4 \text{ kJ/mol}$$

Molecular hydrogen is a colorless, odorless, and nonpoisonous gas. At 1 atm, liquid hydrogen has a boiling point of $-252.9°C$ (20.3 K).

Hydrogen is the most abundant element in the universe, accounting for about 70 percent of the universe's total mass. It is the tenth most abundant element in Earth's crust, where it is found in combination with other elements. Unlike Jupiter and Saturn, Earth does not have a strong enough gravitational pull to retain the lightweight $H_2$ molecules, so hydrogen is not found in our atmosphere.

The ground-state electron configuration of H is $1s^1$. It resembles the alkali metals (Group 1A) in that it can be oxidized to the $H^+$ ion, which exists in aqueous solutions in the hydrated form. On the other hand, hydrogen resembles the *halogens* (Group 7A) in that it forms the hydride ion ($H^-$), which is isoelectronic with *helium* ($1s^2$). Hydrogen is found in a large number of covalent compounds. It also has the unique capacity, when bonded to small, electronegative atoms, for hydrogen-bond formation [◄◄ Section 12.1].

**Student Annotation:** Remember that the halogens also form anions that are isoelectronic with the noble gases [◄◄ Section 7.5].

**Figure 24.1**    Main group nonmetallic elements (in blue) and metalloids (in orange).

| 1A 1 | | | | | | | | | | | | | | | | | 8A 18 |
|---|---|---|---|---|---|---|---|---|---|---|---|---|---|---|---|---|---|
| 1 H | 2A 2 | | | | | | | | | | | 3A 13 | 4A 14 | 5A 15 | 6A 16 | 7A 17 | 2 He |
| 3 Li | 4 Be | | | | | | | | | | | 5 B | 6 C | 7 N | 8 O | 9 F | 10 Ne |
| 11 Na | 12 Mg | 3B 3 | 4B 4 | 5B 5 | 6B 6 | 7B 7 | 8 | ─8B─ 9 | 10 | 1B 11 | 2B 12 | 13 Al | 14 Si | 15 P | 16 S | 17 Cl | 18 Ar |
| 19 K | 20 Ca | 21 Sc | 22 Ti | 23 V | 24 Cr | 25 Mn | 26 Fe | 27 Co | 28 Ni | 29 Cu | 30 Zn | 31 Ga | 32 Ge | 33 As | 34 Se | 35 Br | 36 Kr |
| 37 Rb | 38 Sr | 39 Y | 40 Zr | 41 Nb | 42 Mo | 43 Tc | 44 Ru | 45 Rh | 46 Pd | 47 Ag | 48 Cd | 49 In | 50 Sn | 51 Sb | 52 Te | 53 I | 54 Xe |
| 55 Cs | 56 Ba | 57 La | 72 Hf | 73 Ta | 74 W | 75 Re | 76 Os | 77 Ir | 78 Pt | 79 Au | 80 Hg | 81 Tl | 82 Pb | 83 Bi | 84 Po | 85 At | 86 Rn |
| 87 Fr | 88 Ra | 89 Ac | 104 Rf | 105 Db | 106 Sg | 107 Bh | 108 Hs | 109 Mt | 110 Ds | 111 Rg | 112 – | 113 – | 114 – | 115 – | 116 – | 117 – | 118 – |

Hydrogen gas plays an important role in industrial processes. About 95 percent of the hydrogen produced is produced at or near the plant where it is used for industrial processes, such as the synthesis of ammonia. The large-scale industrial preparation is the reaction between propane (from natural gas and also as a product of oil refineries) and steam in the presence of a catalyst at 900°C:

$$C_3H_8(g) + 3H_2O(g) \longrightarrow 3CO(g) + 7H_2(g)$$

In another process, steam is passed over a bed of red-hot coke:

$$C(s) + H_2O(g) \longrightarrow CO(g) + H_2(g)$$

The mixture of carbon monoxide and hydrogen gas produced in this reaction is commonly known as *water gas*. Because both CO and $H_2$ burn in air, water gas was used as a fuel for many years. But because CO is poisonous, water gas has been replaced by natural gases, such as methane and propane.

Small quantities of hydrogen gas can be prepared conveniently in the laboratory by combining zinc with dilute hydrochloric acid (Figure 24.2):

$$Zn(s) + 2HCl(aq) \longrightarrow ZnCl_2(aq) + H_2(g)$$

Hydrogen gas can also be produced by the reaction between an alkali metal or an alkaline earth metal (Ca or Ba) and water (see Section 7.7), but these reactions are too violent to be suitable for the laboratory preparation of hydrogen gas. Very pure hydrogen gas can be obtained by the electrolysis of water, but this method consumes too much energy to be practical on a large scale.

## Binary Hydrides

*Binary hydrides* are compounds containing hydrogen and another element, either a metal or a nonmetal. Depending on the structure and properties, these hydrides are broadly divided into three types: (1) ionic hydrides, (2) covalent hydrides, and (3) interstitial hydrides.

*Ionic hydrides* are formed when molecular hydrogen combines directly with any alkali metal or with the alkaline earth metals Ca, Sr, or Ba:

$$2Li(s) + H_2(g) \longrightarrow 2LiH(s)$$

$$Ca(s) + H_2(g) \longrightarrow CaH_2(s)$$

All ionic hydrides are solids that have the high melting points characteristic of ionic compounds. The anion in these compounds is the hydride ion ($H^-$), which is a very strong Brønsted base. It readily accepts a proton from a proton donor such as water:

$$H^-(aq) + H_2O(l) \longrightarrow OH^-(aq) + H_2(g)$$

**Figure 24.3** Binary hydrides of the representative elements. In cases in which hydrogen forms more than one compound with the same element, only the formula of the simplest hydride is shown. The properties of many of the transition metal hydrides are not well characterized.

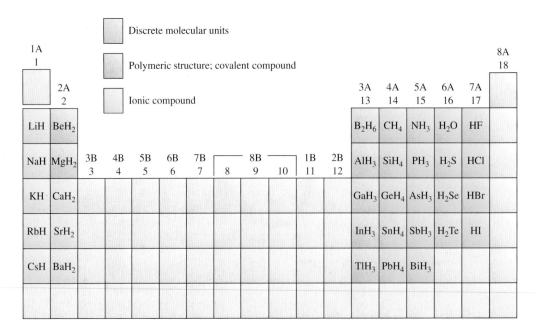

Because of their high reactivity with water, ionic hydrides are frequently used to remove traces of water from organic solvents.

In *covalent hydrides,* the hydrogen atom is covalently bonded to the atom of another element. There are two types of covalent hydrides—those containing discrete molecular units, such as $CH_4$ and $NH_3$, and those having complex polymeric structures, such as $(BeH_2)_x$ and $(AlH_3)_x$, where $x$ is a very large number.

Figure 24.3 shows the binary ionic and covalent hydrides of the main group elements. The physical and chemical properties of these compounds change from *ionic* to *covalent* across a given period. Consider, for example, the hydrides of the second-period elements: LiH, $BeH_2$, $B_2H_6$, $CH_4$, $NH_3$, $H_2O$, and HF. LiH is an ionic compound with a high melting point (680°C). The structure of $BeH_2$ (in the solid state) is polymeric; it is a covalent compound. The molecules $B_2H_6$ and $CH_4$ are nonpolar. In contrast, $NH_3$, $H_2O$, and HF are all polar molecules in which the hydrogen atom is the *positive* end of the polar bond. Of this group of hydrides ($NH_3$, $H_2O$, and HF), only HF is acidic in water.

As we move down any group in Figure 24.3, the compounds change from *covalent* to *ionic.* In Group 2A, for example, $BeH_2$ and $MgH_2$ are covalent, but $CaH_2$, $SrH_2$, and $BaH_2$ are ionic.

Molecular hydrogen forms a number of hydrides with transition metals. In some of these compounds, the ratio of hydrogen atoms to metal atoms is *not* a constant. Such compounds are called *interstitial hydrides.* Depending on conditions, for example, the formula for titanium hydride can vary between $TiH_{1.8}$ and $TiH_2$.

Many of the interstitial hydrides have metallic properties such as electrical conductivity. It is known, however, that hydrogen is definitely bonded to the metal in these compounds, although the exact nature of the bonding is often unclear.

Molecular hydrogen interacts in a unique way with palladium (Pd). Hydrogen gas is readily adsorbed onto the surface of the palladium metal, where it dissociates into atomic hydrogen. The H atoms then "dissolve" into the metal. On heating and under the pressure of $H_2$ gas on one side of the metal, these atoms diffuse through the metal and recombine to form molecular hydrogen, which emerges as the gas from the other side. Because no other gas behaves in this way with palladium, this process has been used to separate hydrogen gas from other gases on a small scale.

## Isotopes of Hydrogen

Hydrogen has three isotopes: $^1_1H$ (hydrogen), $^2_1H$ (deuterium, symbol D), and $^3_1H$ (tritium, symbol T). The natural abundances of the stable hydrogen isotopes are hydrogen, 99.985 percent; and deuterium, 0.015 percent. Tritium is a radioactive isotope with a half-life of about 12.5 years.

Table 24.1 compares some of the common properties of $H_2O$ with those of $D_2O$. Deuterium oxide, or *heavy water* as it is commonly called, is used in some nuclear reactors as a coolant and a moderator of nuclear reactions [◄◄ Section 20.5]. $D_2O$ can be separated from $H_2O$ by fractional distillation because $H_2O$ boils at a lower temperature, as Table 24.1 shows. Another technique

| TABLE 24.1 | Properties of $H_2O$ and $D_2O$ | | |
|---|---|---|---|
| **Property** | | **$H_2O$** | **$D_2O$** |
| Molar mass (g/mol) | | 18.02 | 20.03 |
| Melting point (°C) | | 0 | 3.8 |
| Boiling point (°C) | | 100 | 101.4 |
| Density at 4°C (g/cm³) | | 1.000 | 1.108 |

for separating $D_2O$ is the electrolysis of water. Because $H_2$ gas is formed about eight times as fast as $D_2$ during electrolysis, the water remaining in the electrolytic cell becomes progressively enriched with $D_2O$. Interestingly, the Dead Sea, which for thousands of years has entrapped water that has no outlet other than through evaporation, has a higher $[D_2O]/[H_2O]$ ratio than water found elsewhere.

Although $D_2O$ resembles $H_2O$ chemically in most respects, it is still a toxic substance. The reason is that deuterium is heavier than hydrogen, so its compounds often react more slowly than those of the lighter isotope. Drinking $D_2O$ instead of $H_2O$ on a regular basis could prove fatal because of the slower rate of transfer of D compared with that of H in the acid-base reactions involved in enzyme catalysis. This *kinetic isotope effect* is also manifest in acid ionization constants. For example, the ionization constant of acetic acid

$$CH_3COOH(aq) \rightleftharpoons CH_3COO(aq) + H^+(aq) \qquad K_a = 1.8 \times 10^{-5}$$

is about three times as large as that of *deuterated* acetic acid, in which the ionizable hydrogen atom is replaced by a deuterium atom:

$$CH_3COOD(aq) \rightleftharpoons CH_3COO(aq) + D^+(aq) \qquad K_a = 6 \times 10^{-6}$$

## Hydrogenation

*Hydrogenation* is the addition of hydrogen to compounds containing multiple bonds, usually $C=C$ and $C\equiv C$ bonds. A simple example of hydrogenation is the conversion of ethylene to ethane:

Ethylene          Ethane

This reaction is quite slow under normal conditions, but the rate can be greatly increased by the presence of a catalyst such as nickel or platinum.

Hydrogenation is an extremely important process in the food industry. Vegetable oils have considerable nutritional value, but some oils must be hydrogenated before we can use them because of their unsavory flavor and their inappropriate molecular structures (i.e., there are too many $C=C$ bonds present). Upon exposure to air, these *polyunsaturated* molecules (i.e., molecules with many $C=C$ bonds) undergo oxidation to yield unpleasant-tasting products (oil that has oxidized is said to be *rancid*). In the hydrogenation process, a small amount of nickel (about 0.1 percent by mass) is added to the oil and the mixture is exposed to hydrogen gas at high temperature and pressure. Afterward, the nickel is removed by filtration. Hydrogenation reduces the number of double bonds in the molecule but does not completely eliminate them. If all the double bonds are eliminated, the oil becomes hard and brittle. Under controlled conditions, suitable cooking oils and margarine may be prepared by the hydrogenation of vegetable oils extracted from cottonseed, corn, and soybeans.

## The Hydrogen Economy

The world's fossil fuel reserves are being depleted at an alarmingly fast rate. Faced with this dilemma, scientists have made intensive efforts in recent years to develop a method of obtaining hydrogen gas as an alternative energy source. Hydrogen gas could replace gasoline to power automobiles (after considerable modification of the engine) or be used with oxygen gas in fuel cells to generate electricity (see page 840). One major advantage of using hydrogen gas in these ways is that the reactions are essentially free of pollutants; the end product formed in a hydrogen-powered engine or in a fuel cell would be water, just as in the burning of hydrogen gas in air:

$$2H_2(g) + O_2(g) \longrightarrow 2H_2O(l)$$

Despite these attractive features, though, the success of a hydrogen economy would depend on how cheaply we could produce hydrogen gas and how easily we could store it.

Although electrolysis of water consumes too much energy for large-scale application, if scientists can devise a more practical method of "splitting" water molecules, we could obtain vast amounts of hydrogen from seawater. One approach that is currently in the early stages of development would use solar energy. In this scheme, a catalyst (a complex molecule containing one or more transition metal atoms, such as ruthenium) absorbs a photon from solar radiation and becomes energetically excited. In its excited state, the catalyst is capable of reducing water to molecular hydrogen.

Some of the interstitial hydrides we have discussed would make suitable storage compounds for hydrogen. The reactions that form these hydrides are usually reversible, so hydrogen gas can be obtained simply by reducing the pressure of the hydrogen gas above the metal. The advantages of using interstitial hydrides are as follows: (1) many metals have a high capacity to take up hydrogen gas—sometimes up to three times as many hydrogen atoms as there are metal atoms; and (2) because these hydrides are solids, they can be stored and transported more easily than gases or liquids.

## 24.3 Carbon

Although it constitutes only about 0.09 percent by mass of Earth's crust, carbon is an essential element of living matter. It is found free in the form of diamond and graphite, and it is also a component of natural gas, petroleum, and coal. (Coal is a natural dark-brown to black solid used as a fuel; it is formed from fossilized plants and consists of amorphous carbon with various organic and some inorganic compounds.) Carbon combines with oxygen to form carbon dioxide in the atmosphere and occurs as carbonate in limestone and chalk.

Diamond and graphite are *allotropes* of carbon. Figure 24.4 shows the phase diagram of carbon. Although graphite is the stable form of carbon at 1 atm and 25°C, owners of diamond jewelry need not be alarmed, because the rate of the spontaneous process

$$C(\text{diamond}) \longrightarrow C(\text{graphite}) \qquad \Delta G° = -2.87 \text{ kJ/mol}$$

is extremely slow. In fact, millions of years may pass before a diamond turns to graphite.

Synthetic diamond can be prepared from graphite by applying very high pressures and temperatures. Figure 24.5 shows a synthetic diamond and its starting material, graphite. Synthetic diamonds generally lack the optical properties of natural diamonds. They are useful, however, as abrasives and in cutting concrete and many other hard substances, including metals and alloys. Graphite is used as a lubricant and as the "lead" in pencils.

Carbon has the unique ability to form long chains (consisting of more than 50 C atoms) and stable rings with five or six members. This phenomenon is called *catenation,* the linking of like atoms. Carbon's versatility is responsible for the millions of organic compounds (made up of carbon and hydrogen and other elements such as oxygen, nitrogen, and the halogens) found on Earth [◀◀ Chapter 10].

Carbon combines with metals to form ionic compounds called *carbides,* such as $CaC_2$ and $Be_2C$, in which carbon is in the form of $C_2^{2-}$ or $C^{4-}$ ions. These ions are strong Brønsted bases and react with water as follows:

$$C_2^{2-}(aq) + 2H_2O(l) \longrightarrow 2OH^-(aq) + C_2H_2(g)$$
$$C^{4-}(aq) + 4H_2O(l) \longrightarrow 4OH^-(aq) + CH_4(g)$$

Carbon also forms a covalent compound with silicon. Silicon carbide (SiC) is called *carborundum* and is prepared as follows:

$$SiO_2(s) + 3C(s) \longrightarrow SiC(s) + 2CO(g)$$

Carborundum is also formed by heating silicon with carbon at 1500°C. Carborundum is almost as hard as diamond, and it has the diamond structure; that is, each carbon atom is bonded tetrahedrally to four Si atoms, and vice versa. It is used mainly for cutting, grinding, and polishing metals and glasses.

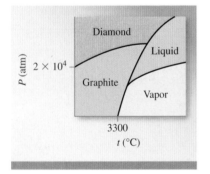

**Figure 24.4**   Phase diagram of carbon. Note that under atmospheric conditions, graphite is the stable form of carbon.

**Figure 24.5**   A synthetic diamond and the starting material—graphite.

Another important class of carbon compounds, the *cyanides,* contain the anion group :C≡N:⁻. Cyanide ions are *extremely* toxic because they bind almost irreversibly to the Fe(III) ion in cytochrome oxidase, a key enzyme in metabolic processes. Hydrogen cyanide, which has the aroma of bitter almonds, is even more dangerous because of its volatility (b.p. 26°C). A few tenths of 1 percent by volume of HCN in air can cause death within minutes. Hydrogen cyanide can be prepared by treating sodium cyanide or potassium cyanide with acid:

$$NaCN(s) + HCl(aq) \longrightarrow NaCl(aq) + HCN(aq)$$

Because HCN (in solution it is called *hydrocyanic acid*) is a very weak acid ($K_a = 4.9 \times 10^{-10}$), most of the HCN produced in this reaction is in the nonionized form and leaves the solution as hydrogen cyanide gas. For this reason, acids should never be mixed with metal cyanides in the laboratory without proper ventilation.

**Student Annotation:** Hydrogen cyanide (HCN) is the gas used in gas-chamber executions.

Cyanide ions are used to extract gold and silver. Although these metals are usually found in the uncombined state in nature, in other metal ores they may be present in relatively small concentrations and are more difficult to extract. In a typical process, the crushed ore is treated with an aqueous cyanide solution in the presence of air to dissolve the gold by forming the soluble complex ion $[Au(CN)_2]^-$:

$$4Au(s) + 8CN^-(aq) + O_2(g) + 2H_2O(l) \longrightarrow 4[Au(CN)_2]^-(aq) + 4OH^-(aq)$$

The complex ion $[Au(CN)_2]^-$ (along with some cation, such as $Na^+$) is separated from other insoluble materials by filtration and treated with an electropositive metal such as zinc to recover the gold:

$$Zn(s) + 2[Au(CN)_2]^-(aq) \longrightarrow [Zn(CN)_4]^{2-}(aq) + 2Au(s)$$

Figure 24.6 shows an aerial view of a "cyanide pond" used for the extraction of gold.

Of the several oxides of carbon, the most important are carbon monoxide (CO) and carbon dioxide ($CO_2$). Carbon monoxide is a colorless, odorless gas formed by the incomplete combustion of carbon or carbon-containing compounds:

$$2C(s) + O_2(g) \longrightarrow 2CO(g)$$

Carbon monoxide is used in metallurgical processes for extracting nickel in organic synthesis and in the production of hydrocarbon fuels with hydrogen. Industrially, it is prepared by passing steam over heated coke. Carbon monoxide burns readily in oxygen to form carbon dioxide:

$$2CO(g) + O_2(g) \longrightarrow 2CO_2(g) \qquad \Delta H° = -566 \text{ kJ/mol}$$

Carbon monoxide is *not* an acidic oxide (it differs from carbon dioxide in that regard), and it is only slightly soluble in water.

Carbon dioxide is a colorless and odorless gas. Unlike carbon monoxide, $CO_2$ is nontoxic—although it is a simple asphyxiant. It is an acidic oxide. Carbon dioxide is used in beverages, in fire extinguishers, and in the manufacture of baking soda ($NaHCO_3$) and soda ash ($Na_2CO_3$). Solid carbon dioxide, called *dry ice,* is used as a refrigerant.

**Figure 24.6** A cyanide pond for extracting gold from metal ore.

# 24.4 Nitrogen and Phosphorus

## Nitrogen

About 78 percent of air by volume is nitrogen. The most important mineral sources of nitrogen are saltpeter ($KNO_3$) and Chile saltpeter ($NaNO_3$). Nitrogen is an essential element of life because it is a component of proteins and nucleic acids.

Molecular nitrogen is obtained by the fractional distillation of air (the boiling points of liquid nitrogen and liquid oxygen are $-196°C$ and $-183°C$, respectively). In the laboratory, very pure nitrogen gas can be prepared by the thermal decomposition of ammonium nitrite:

$$NH_4NO_2(s) \longrightarrow 2H_2O(g) + N_2(g)$$

The $N_2$ molecule contains a triple bond and is very stable with respect to dissociation into atomic species. However, nitrogen forms a large number of compounds with hydrogen and oxygen in which the oxidation number of nitrogen varies from $-3$ to $+5$ (Table 24.2). Most nitrogen compounds are covalent; when heated with certain metals, however, nitrogen forms ionic nitrides containing the $N^{3-}$ ion:

$$6Li(s) + N_2(g) \longrightarrow 2Li_3N(s)$$

| TABLE 24.2 | Common Compounds of Nitrogen | | |
|---|---|---|---|
| **Oxidation Number** | **Compound** | **Formula** | **Structure** |
| −3 | Ammonia | $NH_3$ | $H-\ddot{N}-H$ with $H$ below |
| −2 | Hydrazine | $N_2H_4$ | $H-\ddot{N}-\ddot{N}-H$ with $H$ $H$ below |
| −1 | Hydroxylamine | $NH_2OH$ | $H-\ddot{N}-O-H$ with $H$ below |
| 0 | Nitrogen (for reference) | $N_2$ | $:N\equiv N:$ |
| +1 | Nitrous oxide | $N_2O$ | $:N\equiv N-\ddot{O}:$ |
| +2 | Nitric oxide | $NO$ | $:\dot{N}=\ddot{O}:$ |
| +3 | Nitrous acid | $HNO_2$ | $:\ddot{O}=\ddot{N}-\ddot{O}-H$ |
| +4 | Nitrogen dioxide | $NO_2$ | $:\ddot{O}-\dot{N}=\ddot{O}:$ |
| +5 | Nitric acid | $HNO_3$ | $:\ddot{O}=N-\ddot{O}-H$ with $:\ddot{O}:$ below |

The nitride ion is a very strong Brønsted base and reacts with water to produce ammonia and hydroxide ions:

$$N^{3-}(aq) + 3H_2O(l) \longrightarrow NH_3(g) + 3OH^-(aq)$$

Ammonia is one of the best-known nitrogen compounds. It is prepared industrially from nitrogen and hydrogen by the Haber process. It can be prepared in the laboratory by treating ammonium chloride with sodium hydroxide:

$$NH_4Cl(aq) + NaOH(aq) \longrightarrow NaCl(aq) + H_2O(l) + NH_3(g)$$

Ammonia is a colorless gas (b.p. $-33.4°C$) with an irritating odor. Most of the ammonia produced annually in the United States is used in fertilizers.

Liquid ammonia, like water, undergoes autoionization:

$$2NH_3(l) \rightleftharpoons NH_4^+ + NH_2^-$$

or simply

$$NH_3(l) \rightleftharpoons H^+ + NH_2^-$$

where $NH_2^-$ is called the *amide ion*. Both the $H^+$ and $NH_2^-$ ions are solvated with $NH_3$ molecules—an example of an ion-dipole interaction [◄◄ Section 12.1]. At 50°C, the ion product $[H^+][NH_2^-]$ is about $1 \times 10^{-33}$, considerably smaller than the $1 \times 10^{-14}$ for water at 25°C. Nevertheless, liquid ammonia is a suitable solvent for many electrolytes, especially when a more basic medium is required or if the solutes react with water.

Another important hydride of nitrogen is hydrazine:

$$\begin{matrix} H & & H \\ \diagdown\ddot{\phantom{.}} & & \ddot{\phantom{.}}\diagup \\ \ddot{N}-\ddot{N} & & \\ \diagup & & \diagdown \\ H & & H \end{matrix}$$

Each N atom is $sp^3$-hybridized. Hydrazine is a colorless liquid that smells like ammonia. It melts at 2°C and boils at 114°C.

Hydrazine is a base that can be protonated to give the $N_2H_5^+$ and $N_2H_6^{2+}$ ions. A reducing agent, it can reduce $Fe^{3+}$ to $Fe^{2+}$, $MnO_4^-$ to $Mn^{2+}$, and $I_2$ to $I^-$. Its reaction with oxygen is highly exothermic:

$$N_2H_4(l) + O_2(g) \longrightarrow N_2(g) + 2H_2O(l) \qquad \Delta H° = -666.6 \text{ kJ/mol}$$

Hydrazine and its derivative methylhydrazine [$N_2H_3(CH_3)$], together with the oxidizer dinitrogen tetroxide ($N_2O_4$), are used as rocket fuels. Hydrazine also plays a role in polymer synthesis and in the manufacture of pesticides.

There are many nitrogen oxides, but the three particularly important ones are nitrous oxide, nitric oxide, and nitrogen dioxide. Nitrous oxide ($N_2O$) is a colorless gas with a pleasing odor and sweet taste. It is prepared by heating ammonium nitrate to about 270°C:

$$NH_4NO_3(s) \longrightarrow N_2O(g) + 2H_2O(g)$$

Nitrous oxide resembles molecular oxygen in that it supports combustion. It does so because it decomposes when heated to form molecular nitrogen and molecular oxygen:

$$2N_2O(g) \longrightarrow 2N_2(g) + O_2(g)$$

It is chiefly used as an anesthetic in dental procedures and other minor surgery. Nitrous oxide is also called "laughing gas" because a person inhaling the gas becomes somewhat giddy. No satisfactory explanation has yet been proposed for this unusual physiological response.

Nitric oxide (NO) is a colorless gas. The reaction of $N_2$ and $O_2$ in the atmosphere,

$$N_2(g) + O_2(g) \rightleftharpoons 2NO(g) \qquad \Delta G° = 173.4 \text{ kJ/mol}$$

is a form of *nitrogen fixation*. The equilibrium constant for the preceding reaction is very small at room temperature: $K_P$ is only $4.0 \times 10^{-31}$ at 25°C, so very little NO will form at that temperature. However, the equilibrium constant increases rapidly with temperature, such as in a running auto engine. An appreciable amount of nitric oxide is formed in the atmosphere by the action of lightning. In the laboratory, the gas can be prepared by the reduction of dilute nitric acid with copper:

$$3Cu(s) + 8HNO_3(aq) \longrightarrow 3Cu(NO_3)_2(aq) + 4H_2O(l) + 2NO(g)$$

The nitric oxide molecule is paramagnetic, containing one unpaired electron. It can be represented by the following resonance structures:

**Student Annotation:** The NO molecule does not obey the octet rule [◀ Section 8.8].

$$\ddot{N}{=}\ddot{O}: \longleftrightarrow {}^{-}\ddot{N}{=}\overset{+}{\ddot{O}}:$$

Unlike nitrous oxide and nitric oxide, nitrogen dioxide is a highly toxic, yellow-brown gas with a choking odor. In the laboratory, nitrogen dioxide is prepared by the action of concentrated nitric acid on copper (Figure 24.7):

$$Cu(s) + 4HNO_3(aq) \longrightarrow Cu(NO_3)_2(aq) + 2H_2O(l) + 2NO_2(g)$$

Nitrogen dioxide is paramagnetic, and it has a strong tendency to dimerize to dinitrogen tetroxide, which is diamagnetic:

$$2NO_2 \rightleftharpoons N_2O_4$$

This reaction occurs in both the gas phase and the liquid phase.

Nitrogen dioxide is an *acidic* oxide; it reacts rapidly with cold water to form both nitrous acid ($HNO_2$) and nitric acid ($HNO_3$):

$$2NO_2(g) + H_2O(l) \longrightarrow HNO_2(aq) + HNO_3(aq)$$

This is a disproportionation reaction in which the oxidation number of nitrogen changes from +4 (in $NO_2$) to +3 (in $HNO_2$) and +5 (in $HNO_3$). This reaction is quite different from that between $CO_2$ and $H_2O$, in which only one acid (carbonic acid) is formed.

Nitric acid is one of the most important inorganic acids. It is a liquid (b.p. 82.6°C), but it does not exist as a pure liquid because it decomposes spontaneously to some extent as follows:

$$4HNO_3(l) \longrightarrow 4NO_2(g) + 2H_2O(l) + O_2(g)$$

The concentrated nitric acid used in the laboratory is 68 percent $HNO_3$ by mass (density 1.42 g/cm$^3$), which corresponds to 15.7 $M$.

Nitric acid is a powerful oxidizing agent. The oxidation number of N in $HNO_3$ is +5. The most common reduction products of nitric acid are $NO_2$ (oxidation number of N = +4), NO (oxidation number of N = +2), and $NH_4^+$ (oxidation number of N = −3). Nitric acid can oxidize metals both above and below hydrogen in the activity series. Copper, for example, is oxidized by concentrated nitric acid.

In the presence of a strong reducing agent, such as zinc metal, nitric acid can be reduced all the way to the ammonium ion:

$$4Zn(s) + 10H^+(aq) + NO_3^-(aq) \longrightarrow 4Zn^{2+}(aq) + NH_4^+(aq) + 3H_2O(l)$$

**Figure 24.7** The production of $NO_2$ gas when copper reacts with concentrated nitric acid.

Concentrated nitric acid does not oxidize gold. However, when the acid is added to concentrated hydrochloric acid in a 1:3 ratio by volume (one part $HNO_3$ to three parts HCl), the resulting solution, called *aqua regia,* can oxidize gold as follows:

$$Au(s) + 3HNO_3(aq) + 4HCl(aq) \longrightarrow HAuCl_4(aq) + 3H_2O(l) + 3NO_2(g)$$

The oxidation of Au is promoted by the complexing ability of the $Cl^-$ ion (to form the $AuCl_4^-$ ion).

Concentrated nitric acid also oxidizes a number of nonmetals to their corresponding oxoacids:

$$P_4(s) + 20HNO_3(aq) \longrightarrow 4H_3PO_4(aq) + 20NO_2(g) + 4H_2O(l)$$

$$S(s) + 6HNO_3(aq) \longrightarrow H_2SO_4(aq) + 6NO_2(g) + 2H_2O(l)$$

Nitric acid is used in the manufacture of fertilizers, dyes, drugs, and explosives.

## Phosphorus

Like nitrogen, phosphorus is a member of the Group 5A family, and in some respects the chemistry of phosphorus resembles that of nitrogen. Phosphorus occurs most commonly in nature as *phosphate rocks,* which are mostly calcium phosphate [$Ca_3(PO_4)_2$] and fluoroapatite [$Ca_5(PO_4)_3F$] (Figure 24.8). Elemental phosphorus can be obtained by heating calcium phosphate with coke and silica sand:

$$2Ca_3(PO_4)_2(s) + 10C(s) + 6SiO_2(s) \longrightarrow 6CaSiO_3(s) + 10CO(g) + P_4(s)$$

There are several allotropic forms of phosphorus, but only white phosphorus and red phosphorus are of importance. White phosphorus consists of discrete tetrahedral $P_4$ molecules (Figure 24.9). A solid (m.p. 44.2°C), white phosphorus is insoluble in water but quite soluble in carbon disulfide ($CS_2$) and organic solvents such as chloroform ($CHCl_3$). White phosphorus is a highly toxic substance. It bursts into flames spontaneously when exposed to air; hence, it is used in incendiary bombs and grenades:

$$P_4(s) + 5O_2(g) \longrightarrow P_4O_{10}(s)$$

The high reactivity of white phosphorus is attributed to structural strain: the P—P bonds are compressed in the tetrahedral $P_4$ molecule. White phosphorus was once used in matches, but because of its toxicity it has been replaced by tetraphosphorus trisulfide ($P_4S_3$).

When heated in the absence of air, white phosphorus is slowly converted to red phosphorus at about 300°C:

$$nP_4(\text{white phosphorus}) \longrightarrow (P_4)_n(\text{red phosphorus})$$

Red phosphorus has a polymeric structure (see Figure 24.9) and is more stable and less volatile than white phosphorus.

The most important hydride of phosphorus is phosphine ($PH_3$), a colorless, very poisonous gas formed by heating white phosphorus in concentrated sodium hydroxide:

$$P_4(s) + 3NaOH(aq) + 3H_2O(l) \longrightarrow 3NaH_2PO_2(aq) + PH_3(g)$$

**Figure 24.8** Fluoroapatite [$Ca_5(PO_4)_3F$].

**Figure 24.9** The structures of white and red phosphorus. Red phosphorus is believed to have a chain structure, as shown.

White phosphorus

Red phosphorus

Phosphine is moderately soluble in water and more soluble in carbon disulfide and organic solvents. Its aqueous solution is neutral, unlike that of ammonia. In liquid ammonia, phosphine dissolves to give $NH_4^+PH_2^-$. Phosphine is a strong reducing agent; it reduces many metal salts to the corresponding metals. The gas burns in air:

$$PH_3(g) + 2O_2(g) \longrightarrow H_3PO_4(s)$$

Phosphorus forms binary compounds with halogens—namely, the trihalides ($PX_3$) and the pentahalides ($PX_5$), where X denotes a halogen atom. In contrast, nitrogen can form only trihalides ($NX_3$). Unlike nitrogen, phosphorus has a $3d$ subshell, which can be used for valence-shell expansion. We can explain the bonding in $PCl_5$ by assuming that phosphorus undergoes $sp^3d$ hybridization of its $3s$, $3p$, and $3d$ orbitals [◄◄ Section 9.4]. The five $sp^3d$ hybrid orbitals also account for the trigonal bipyramidal geometry of the $PCl_5$ molecule (see Table 9.4).

Phosphorus trichloride is prepared by heating white phosphorus in chlorine:

$$P_4(l) + 6Cl_2(g) \longrightarrow 4PCl_3(g)$$

A colorless liquid (b.p. 76°C), $PCl_3$ is hydrolyzed according to the following equation:

$$PCl_3(l) + 3H_2O(l) \longrightarrow H_3PO_3(aq) + 3HCl(g)$$

In the presence of an excess of chlorine gas, $PCl_3$ is converted to phosphorus pentachloride, which is a light-yellow solid:

$$PCl_3(l) + Cl_2(g) \longrightarrow PCl_5(s)$$

X-ray studies indicate that solid phosphorus pentachloride exists as $[PCl_4^+][PCl_6^-]$, in which the $PCl_4^+$ ion has a *tetrahedral* geometry and the $PCl_6^-$ ion has an *octahedral* geometry. In the gas phase, $PCl_5$ (which has trigonal bipyramidal geometry) is in equilibrium with $PCl_3$ and $Cl_2$:

$$PCl_5(g) \rightleftharpoons PCl_3(g) + Cl_2(g)$$

Phosphorus pentachloride reacts with water as follows:

$$PCl_5(s) + 4H_2O(l) \longrightarrow H_3PO_4(aq) + 5HCl(aq)$$

The two important oxides of phosphorus are tetraphosphorus hexaoxide ($P_4O_6$) and tetraphosphorus decaoxide ($P_4O_{10}$), shown in Figure 24.10. The oxides are obtained by burning white phosphorus in limited and excess amounts of oxygen gas, respectively:

$$P_4(s) + 3O_2(g) \longrightarrow P_4O_6(s)$$
$$P_4(s) + 5O_2(g) \longrightarrow P_4O_{10}(s)$$

Both oxides are acidic; that is, they are converted to acids in water. The compound $P_4O_{10}$ is a white flocculent powder (m.p. 420°C) that has a great affinity for water:

$$P_4O_{10}(s) + 6H_2O(l) \longrightarrow 4H_3PO_4(aq)$$

For this reason, it is often used for drying gases and for removing water from solvents.

There are many oxoacids containing phosphorus. Some examples are phosphorous acid ($H_3PO_3$), phosphoric acid ($H_3PO_4$), hypophosphorous acid ($H_3PO_2$), and triphosphoric acid ($H_5P_3O_{10}$) (the structures of which are shown in Figure 24.11). Phosphoric acid, also called orthophosphoric acid, is a weak triprotic acid. It is prepared industrially by the reaction of calcium phosphate with sulfuric acid:

$$Ca_3(PO_4)_2(s) + 3H_2SO_4(aq) \longrightarrow 2H_3PO_4(aq) + 3CaSO_4(s)$$

Phosphorus

Oxygen

$P_4O_6$

$P_4O_{10}$

**Figure 24.10** The structures of $P_4O_6$ and $P_4O_{10}$. Note the tetrahedral arrangement of the P atoms in $P_4O_{10}$.

**Figure 24.11**  Structures of some common phosphorus-containing oxoacids.

$$
\begin{array}{cc}
\text{H–O–P–O–H} & \text{H–O–P–H} \\
\text{Phosphorous acid (H}_3\text{PO}_3) & \text{Hypophosphorous acid (H}_3\text{PO}_2) \\[2ex]
\text{H–O–P–O–H} & \text{H–O–P–O–P–O–P–O–H} \\
\text{Phosphoric acid (H}_3\text{PO}_4) & \text{Triphosphoric acid (H}_5\text{P}_3\text{O}_{10})
\end{array}
$$

In the pure form, phosphoric acid is a colorless solid (m.p. 42.2°C). The phosphoric acid we use in the laboratory is usually an 82% $H_3PO_4$ solution (by mass). Phosphoric acid and phosphates have many commercial applications in detergents, fertilizers, flame retardants, and toothpastes, and as buffers in carbonated beverages.

Like nitrogen, phosphorus is an element that is essential to life. It constitutes only about 1 percent by mass of the human body, but it is a very important 1 percent. About 23 percent of the human skeleton is mineral matter. The phosphorus content of this mineral matter, calcium phosphate $[Ca_3(PO_4)_2]$, is 20 percent. Our teeth are basically $Ca_3(PO_4)_2$ and $Ca_5(PO_4)_3OH$. Phosphates are also important components of the genetic materials deoxyribonucleic acid (DNA) and ribonucleic acid (RNA).

## 24.5  Oxygen and Sulfur

### Oxygen

Oxygen is by far the most abundant element in Earth's crust, constituting about 46 percent of its mass. In addition, the atmosphere contains about 21 percent molecular oxygen by volume (23 percent by mass). Like nitrogen, oxygen in the free state is a diatomic molecule ($O_2$). In the laboratory, oxygen gas can be obtained by heating potassium chlorate:

$$2KClO_3(s) \longrightarrow 2KCl(s) + 3O_2(g)$$

The reaction is usually catalyzed by manganese(IV) dioxide ($MnO_2$). Pure oxygen gas can be prepared by electrolyzing water (page 843). Industrially, oxygen gas is prepared by the fractional distillation of liquefied air. Oxygen gas is colorless and odorless.

Oxygen is a building block of practically all biomolecules, accounting for about a fourth of the atoms in living matter. Molecular oxygen is the essential oxidant in the metabolic breakdown of food molecules. Without it, a human being cannot survive for more than a few minutes.

Although oxygen has two allotropes, $O_2$ and $O_3$, when we speak of molecular oxygen, we normally mean $O_2$. Ozone ($O_3$) is less stable than $O_2$. The $O_2$ molecule is paramagnetic because it contains two unpaired electrons (see Section 10.7).

A strong oxidizing agent, molecular oxygen is one of the most widely used industrial chemicals. Its main uses are in the steel industry and in sewage treatment. Oxygen is also used as a bleaching agent for pulp and paper, in medicine to ease breathing difficulties, in oxyacetylene torches, and as an oxidizing agent in many inorganic and organic reactions.

Oxygen forms three types of oxides: the normal oxide (or simply the *oxide*), which contains the $O^{2-}$ ion; the peroxide, which contains the $O_2^{2-}$ ion; and the superoxide, which contains the $O_2^-$ ion:

$$:\!\ddot{O}\!:^{2-} \qquad :\!\ddot{O}\!:\!\ddot{O}\!:^{2-} \qquad :\!\ddot{O}\!:\!\ddot{O}\!:^-$$

$$\text{Oxide} \quad \text{Peroxide} \quad \text{Superoxide}$$

The ions are all strong Brønsted bases and react with water as follows:

*Oxide:*  $O^{2-}(aq) + H_2O(l) \longrightarrow 2OH^-(aq)$

*Peroxide:*  $2O_2^{2-}(aq) + 2H_2O(l) \longrightarrow O_2(g) + 4OH^-(aq)$

*Superoxide:*  $4O_2^-(aq) + 2H_2O(l) \longrightarrow 3O_2(g) + 4OH^-(aq)$

The reaction of $O^{2-}$ with water is a *hydrolysis* reaction, but those involving $O_2^{2-}$ and $O_2^-$ are *redox* processes.

The nature of bonding in oxides changes across any period in the periodic table. Oxides of elements on the left side of the periodic table, such as those of the alkali metals and alkaline earth metals, are generally ionic solids with high melting points. Oxides of the metalloids and of the metallic elements toward the middle of the periodic table are also solids, but they have much less ionic character. Oxides of nonmetals are covalent compounds that generally exist as liquids or gases at room temperature. The acidic character of the oxides increases from left to right. Consider the oxides of the third-period elements:

$$\underbrace{Na_2O \qquad MgO}_{Basic} \qquad \underbrace{Al_2O_3}_{Amphoteric} \qquad \underbrace{SiO_2 \qquad P_4O_{10} \qquad SO_3 \qquad Cl_2O_7}_{Acidic}$$

The basicity of the oxides increases as we move down a particular group. MgO does not react with water, for example, but reacts with acid as follows:

$$MgO(s) + 2H^+(aq) \longrightarrow Mg^{2+}(aq) + H_2O(l)$$

On the other hand, BaO, which is more basic, undergoes hydrolysis to yield the corresponding hydroxide:

$$BaO(s) + H_2O(l) \longrightarrow Ba(OH)_2(aq)$$

The best-known peroxide is hydrogen peroxide ($H_2O_2$). It is a colorless, syrupy liquid (m.p. $-0.9°C$), prepared in the laboratory by the action of cold dilute sulfuric acid on barium peroxide octahydrate:

$$BaO_2(s) + 8H_2O(s) + H_2SO_4(aq) \longrightarrow BaSO_4(s) + H_2O_2(aq) + 8H_2O(l)$$

The structure of hydrogen peroxide is shown in Figure 24.12. Using the VSEPR method, we see that the H—O and O—O bonds are bent about each oxygen atom in a configuration similar to the structure of water. The lone-pair–bonding-pair repulsion is greater in $H_2O_2$ than in $H_2O$, so the H—O—O angle is only 97° (compared with 104.5° for H—O—H in $H_2O$). Hydrogen peroxide is a polar molecule ($\mu = 2.16$ D). Hydrogen peroxide readily decomposes when heated or exposed to sunlight or even in the presence of dust particles or certain metals, including iron and copper:

$$2H_2O_2(l) \longrightarrow 2H_2O(l) + O_2(g) \qquad \Delta H° = -196.4 \text{ kJ/mol}$$

This is a *disproportionation* reaction. The oxidation number of oxygen changes from $-1$ to $-2$ and 0.

Hydrogen peroxide is miscible with water in all proportions due to its ability to hydrogen-bond with water. Dilute hydrogen peroxide solutions (3 percent by mass), available in drugstores, are used as mild antiseptics; more concentrated $H_2O_2$ solutions are employed as bleaching agents for textiles, fur, and hair. The high heat of decomposition of hydrogen peroxide also makes it a suitable component in rocket fuel.

Hydrogen peroxide is a strong oxidizing agent; it can oxidize $Fe^{2+}$ ions to $Fe^{3+}$ ions in an acidic solution:

$$H_2O_2(aq) + 2Fe^{2+}(aq) + 2H^+(aq) \longrightarrow 2Fe^{3+}(aq) + 2H_2O(l)$$

It also oxidizes $SO_3^{2-}$ ions to $SO_4^{2-}$ ions:

$$H_2O_2(aq) + SO_3^{2-}(aq) \longrightarrow SO_4^{2-}(aq) + H_2O(l)$$

In addition, hydrogen peroxide can act as a reducing agent toward substances that are stronger oxidizing agents than itself. For example, hydrogen peroxide reduces silver oxide to metallic silver,

$$H_2O_2(aq) + Ag_2O(s) \longrightarrow 2Ag(s) + H_2O(l) + O_2(g)$$

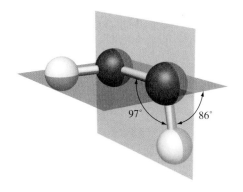

**Figure 24.12** The structure of $H_2O_2$.

**Figure 24.13**   The preparation of $O_3$ from $O_2$ by electrical discharge. The outside of the outer tube and the inside of the inner tube are coated with metal foils that are connected to a high-voltage source. (The metal foil on the inside of the inner tube is not shown.) During the electrical discharge, $O_2$ gas is passed through the tube. The $O_3$ gas formed exits from the upper right-hand tube, along with some unreacted $O_2$ gas.

and permanganate ($MnO_4^-$) to manganese(II) in an acidic solution,

$$5H_2O_2(aq) + 2MnO_4^-(aq) + 6H^+(aq) \longrightarrow 2Mn^{2+}(aq) + 5O_2(g) + 8H_2O(l)$$

If we want to determine hydrogen peroxide concentration, this reaction can be carried out as a redox titration, using a standard permanganate solution.

There are relatively few known superoxides (i.e., compounds containing the $O_2^-$ ion). In general, only the most reactive alkali metals (K, Rb, and Cs) form superoxides.

Both the peroxide ion and the superoxide ion are by-products of metabolism. Because these ions are highly reactive, they can inflict great damage on living cells. Fortunately, our bodies are equipped with the enzymes catalase, peroxidase, and superoxide dismutase, which convert these toxic substances to water and molecular oxygen.

Ozone is a rather toxic, light-blue gas (b.p. $-111.3°C$). Its pungent odor is noticeable around sources of significant electrical discharges (such as a subway train). Ozone can be prepared from molecular oxygen, either photochemically or by subjecting $O_2$ to an electrical discharge (Figure 24.13):

$$3O_2(g) \longrightarrow 2O_3(g) \qquad \Delta G° = 326.8 \text{ kJ/mol}$$

Because the standard free energy of formation of ozone is a large positive quantity ($\Delta G_f° = 163.4$ kJ/mol), ozone is less stable than molecular oxygen. The ozone molecule has a bent structure in which the bond angle is $116.5°$:

Ozone is mainly used to purify drinking water; to deodorize air and sewage gases; and to bleach waxes, oils, and textiles.

Ozone is a very powerful oxidizing agent—its oxidizing power is exceeded only by that of molecular fluorine (see Table 19.1). For example, ozone can oxidize sulfides of many metals to the corresponding sulfates:

$$4O_3(g) + PbS(s) \longrightarrow PbSO_4(s) + 4O_2(g)$$

Ozone oxidizes all the common metals except gold and platinum. In fact, a convenient test for ozone is based on its action on mercury. When exposed to ozone, mercury loses its metallic luster and sticks to glass tubing (instead of flowing freely through it). This behavior is attributed to the change in surface tension caused by the formation of mercury(II) oxide:

$$O_3(g) + 3Hg(l) \longrightarrow 3HgO(s)$$

The beneficial effect of ozone in the stratosphere and its undesirable action in smog formation were discussed in Chapter 21.

## Sulfur

Although sulfur is not a very abundant element (it constitutes only about 0.06 percent of Earth's crust by mass), it is readily available because it occurs commonly in nature in the elemental form.

The largest known reserves of sulfur are found in sedimentary deposits. In addition, sulfur occurs widely in gypsum ($CaSO_4 \cdot 2H_2O$) and various sulfide minerals such as pyrite ($FeS_2$) (Figure 24.14). Sulfur is also present in natural gas as $H_2S$, $SO_2$, and other sulfur-containing compounds.

Sulfur is extracted from underground deposits by the *Frasch*[1] *process,* shown in Figure 24.15. In this process, superheated water (liquid water heated to about 160°C under high pressure to prevent it from boiling) is pumped down the outermost pipe to melt the sulfur. Next, compressed air is forced down the innermost pipe. Liquid sulfur mixed with air forms an emulsion that is less dense than water and therefore rises to the surface as it is forced up the middle pipe. Sulfur produced in this manner, which amounts to about 10 million tons per year, has a purity of about 99.5 percent.

There are several allotropic forms of sulfur, the most important being the rhombic and monoclinic forms. Rhombic sulfur is thermodynamically the most stable form; it has a puckered $S_8$ ring structure:

**Figure 24.14** Pyrite ($FeS_2$).

$S_8$

Multimedia

Nonmetallic elements—the extraction of sulfur.

It is a yellow, tasteless, and odorless solid (m.p. 112°C) that is insoluble in water but soluble in carbon disulfide. When heated, it is slowly converted to monoclinic sulfur (m.p. 119°C), which also consists of the $S_8$ units. When liquid sulfur is heated above 150°C, the rings begin to break up, and the entangling of the sulfur chains results in a sharp increase in the liquid's viscosity. Further heating tends to rupture the chains, so the viscosity decreases.

Like nitrogen, sulfur shows a wide variety of oxidation numbers in its compounds (Table 24.3). The best-known hydrogen compound of sulfur is hydrogen sulfide, which is prepared by the action of an acid on a sulfide; for example,

$$FeS(s) + H_2SO_4(aq) \longrightarrow FeSO_4(aq) + H_2S(g)$$

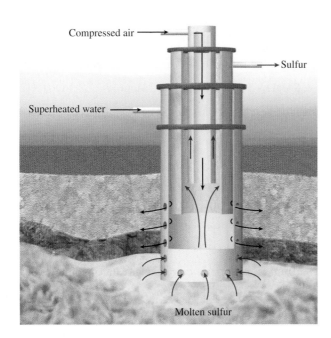

Compressed air

Sulfur

Superheated water

Molten sulfur

**Figure 24.15** The Frasch process. Three concentric pipes are inserted into a hole drilled down to the sulfur deposit. Superheated water is forced down the outer pipe into the sulfur, causing it to melt. Molten sulfur is then forced up the middle pipe by compressed air.

---

1. Herman Frasch (1851–1914). German chemical engineer. Besides inventing the process for obtaining pure sulfur, Frasch developed methods for refining petroleum.

| TABLE 24.3 | Common Compounds of Sulfur | | |
|---|---|---|---|
| **Oxidation Number** | **Compound** | **Formula** | **Structure** |
| $-2$ | Hydrogen sulfide | $H_2S$ | |
| $0$ | Sulfur (for reference) | $S_8$ | |
| $+1$ | Disulfur dichloride | $S_2Cl_2$ | |
| $+2$ | Sulfur dichloride | $SCl_2$ | |
| $+4$ | Sulfur dioxide | $SO_2$ | |
| $+6$ | Sulfur trioxide | $SO_3$ | |

Nowadays, hydrogen sulfide used in qualitative analysis is prepared by the hydrolysis of thioacetamide:

$$CH_3CSNH_2 + 2H_2O + H^+ \longrightarrow CH_3COOH + H_2S + NH_4^+$$

Thioacetamide          Acetic acid

Hydrogen sulfide is a colorless gas (b.p. $-60.2°C$) that smells like rotten eggs. (The odor of rotten eggs actually does come from hydrogen sulfide, which is formed by the bacterial decomposition of sulfur-containing proteins.) Hydrogen sulfide is a highly toxic substance that, like hydrogen cyanide, attacks respiratory enzymes. It is a very weak diprotic acid (see Table 16.8). In basic solution, $H_2S$ is a reducing agent. For example, it is oxidized by permanganate to elemental sulfur:

$$3H_2S(aq) + 2MnO_4^-(aq) \longrightarrow 3S(s) + 2MnO_2(s) + 2H_2O(l) + 2OH^-(aq)$$

Sulfur has two important oxides: sulfur dioxide ($SO_2$) and sulfur trioxide ($SO_3$). Sulfur dioxide is formed when sulfur burns in air:

$$S(s) + O_2(g) \longrightarrow SO_2(g)$$

In the laboratory, it can be prepared by the action of an acid on a sulfite; for example,

$$2HCl(aq) + Na_2SO_3(aq) \longrightarrow 2NaCl(aq) + H_2O(l) + SO_2(g)$$

or by the action of concentrated sulfuric acid on copper:

$$Cu(s) + 2H_2SO_4(aq) \longrightarrow CuSO_4(aq) + 2H_2O(l) + SO_2(g)$$

Sulfur dioxide (b.p. $-10°C$) is a pungent, colorless gas that is quite toxic. An acidic oxide, it reacts with water as follows:

$$SO_2(g) + H_2O(l) \rightleftharpoons H^+(aq) + HSO_3^-(aq)$$

Sulfur dioxide is slowly oxidized to sulfur trioxide, but the reaction rate can be greatly enhanced by a platinum or vanadium oxide catalyst:

$$2SO_2(g) + O_2(g) \longrightarrow 2SO_3(g)$$

Sulfur trioxide dissolves in water to form sulfuric acid:

$$SO_3(g) + H_2O(l) \longrightarrow H_2SO_4(aq)$$

The contributing role of sulfur dioxide to acid rain is discussed on page 908.

Sulfuric acid is the world's most important industrial chemical. It is prepared industrially by first burning sulfur in air:

$$S(s) + O_2(g) \longrightarrow SO_2(g)$$

Next is the key step of converting sulfur dioxide to sulfur trioxide:

$$2SO_2(g) + O_2(g) \longrightarrow 2SO_3(g)$$

Vanadium(V) oxide ($V_2O_5$) is the catalyst used for the second step. Because the sulfur dioxide and oxygen molecules react in contact with the surface of solid $V_2O_5$, the process is referred to as the *contact process.*

Sulfuric acid is a diprotic acid. It is a colorless, viscous liquid (m.p. 10.4°C). The concentrated sulfuric acid we use in the laboratory is 98 percent $H_2SO_4$ by mass (density = 1.84 g/cm$^3$), which corresponds to a concentration of 18 *M*. The oxidizing strength of sulfuric acid depends on its temperature and concentration. A cold, dilute sulfuric acid solution reacts with metals above hydrogen in the activity series, thereby liberating molecular hydrogen in a displacement reaction:

$$Mg(s) + H_2SO_4(aq) \longrightarrow MgSO_4(aq) + H_2(g)$$

This is a typical reaction of an active metal with an acid. The strength of sulfuric acid as an oxidizing agent is greatly enhanced when it is both hot and concentrated. In such a solution, the oxidizing agent is actually the sulfate ion rather than the hydrated proton, $H^+(aq)$. Thus, copper reacts with concentrated sulfuric acid as follows:

$$Cu(s) + 2H_2SO_4(aq) \longrightarrow CuSO_4(aq) + SO_2(g) + 2H_2O(l)$$

Depending on the nature of the reducing agents, the sulfate ion may be further reduced to elemental sulfur or the sulfide ion. For example, the reduction of $H_2SO_4$ by HI yields $H_2S$ and $I_2$:

$$8HI(aq) + H_2SO_4(aq) \longrightarrow H_2S(aq) + 4I_2(s) + 4H_2O(l)$$

Concentrated sulfuric acid oxidizes nonmetals. For example, it oxidizes carbon to carbon dioxide and sulfur to sulfur dioxide:

$$C(s) + 2H_2SO_4(aq) \longrightarrow CO_2(g) + 2SO_2(g) + 2H_2O(l)$$

$$S(s) + 2H_2SO_4(aq) \longrightarrow 3SO_2(g) + 2H_2O(l)$$

Carbon disulfide, a colorless, flammable liquid (b.p. 46°C), is formed by heating carbon and sulfur to a high temperature:

$$C(s) + 2S(l) \longrightarrow CS_2(l)$$

It is only slightly soluble in water. Carbon disulfide is a good solvent for sulfur, phosphorus, iodine, and nonpolar substances such as waxes and rubber.

Another interesting compound of sulfur is sulfur hexafluoride ($SF_6$), which is prepared by heating sulfur in an atmosphere of fluorine:

$$S(l) + 3F_2(g) \longrightarrow SF_6(g)$$

Sulfur hexafluoride is a nontoxic, colorless gas (b.p. 63.8°C). It is the most inert of all sulfur compounds; it resists attack even by molten KOH. The structure and bonding of $SF_6$ were discussed in Chapters 8 and 9.

# 24.6 The Halogens

The halogens—fluorine, chlorine, bromine, and iodine—are reactive nonmetals. Table 24.4 lists some of the properties of these elements. Although all halogens are highly reactive and toxic, the magnitude of reactivity and toxicity generally decreases from fluorine to iodine. The chemistry of fluorine differs from that of the rest of the halogens in the following ways:

1. Fluorine is the most reactive of all the halogens. The difference in reactivity between fluorine and chlorine is greater than that between chlorine and bromine. Table 24.4 shows that

| TABLE 24.4 | Properties of the Halogens | | | |
|---|---|---|---|---|
| Property | F | Cl | Br | I |
| Valence electron configuration | $2s^2 2p^5$ | $3s^2 3p^5$ | $4s^2 4p^5$ | $5s^2 5p^5$ |
| Melting point (°C) | $-223$ | $-102$ | $-7$ | 114 |
| Boiling point (°C) | $-187$ | $-35$ | 59 | 183 |
| Appearance* | Pale-yellow gas | Yellow-green gas | Red-brown liquid | Dark-violet vapor Dark metallic-looking solid |
| Atomic radius (pm) | 72 | 99 | 114 | 133 |
| Ionic radius (pm)† | 136 | 181 | 195 | 216 |
| Ionization energy (kJ/mol) | 1680 | 1251 | 1139 | 1003 |
| Electronegativity | 4.0 | 3.0 | 2.8 | 2.5 |
| Standard reduction potential (V)* | 2.87 | 1.36 | 1.07 | 0.53 |
| Bond enthalpy (kJ/mol)* | 150.6 | 242.7 | 192.5 | 151.0 |

\* These values and descriptions apply to the diatomic species $X_2$, where X represents a halogen atom. The half-reaction is
$X_2(g) + 2e^- \longrightarrow 2X^-(aq)$.

† Refers to the anion $X^-$.

the F—F bond is considerably weaker than the Cl—Cl bond. The weak bond in $F_2$ can be explained in terms of the lone pairs on the F atoms:

$$:\ddot{F}-\ddot{F}:$$

The small size of the F atoms (see Table 24.4) allows a close approach of the three lone pairs on each of the F atoms, resulting in a greater repulsion than that found in $Cl_2$, which consists of larger atoms.

2. Hydrogen fluoride (HF) has a relatively high boiling point (19.5°C) as a result of strong intermolecular hydrogen bonding, whereas all other hydrogen halides have much lower boiling points.

3. Hydrofluoric acid is a weak acid, whereas all other hydrohalic acids (HCl, HBr, and HI) are strong acids.

4. Fluorine reacts with cold sodium hydroxide solution to produce oxygen difluoride as follows:

$$2F_2(g) + 2NaOH(aq) \longrightarrow 2NaF(aq) + H_2O(l) + OF_2(g)$$

The same reaction with chlorine or bromine, on the other hand, produces a halide and a hypohalite:

$$X_2(g) + 2NaOH(aq) \longrightarrow NaX(aq) + NaXO(aq) + H_2O(l)$$

where X stands for Cl or Br. Iodine does not react under the same conditions.

5. Silver fluoride (AgF) is soluble. All other silver halides (AgCl, AgBr, and AgI) are insoluble.

The element astatine also belongs to the Group 7A family. However, all isotopes of astatine are radioactive; its longest-lived isotope is astatine-210, which has a half-life of 8.3 h. As a result, it is both difficult and expensive to study astatine in the laboratory.

The halogens form a very large number of compounds. In the elemental state, they form diatomic molecules ($X_2$). In nature, however, because of their high reactivity, halogens are always found combined with other elements. Chlorine, bromine, and iodine occur as halides in seawater, and fluorine occurs in the minerals fluorite ($CaF_2$) and cryolite ($Na_3AlF_6$).

## Preparation and General Properties of the Halogens

Because fluorine and chlorine are strong oxidizing agents, they must be prepared by electrolysis rather than by chemical oxidation of the fluoride and chloride ions. Electrolysis does not work for aqueous solutions of fluorides, however, because fluorine is a stronger oxidizing agent than oxygen. From Table 19.1 (page 827), we find that

$$F_2(g) + 2e^- \longrightarrow 2F^-(aq) \qquad E° = 2.87 \text{ V}$$

$$O_2(g) + 4H^+(aq) + 4e^- \longrightarrow 2H_2O(l) \qquad E° = 1.23 \text{ V}$$

If $F_2$ were formed by the electrolysis of an aqueous fluoride solution, it would immediately oxidize water to oxygen. For this reason, fluorine is prepared by electrolyzing liquid hydrogen fluoride containing potassium fluoride to increase its conductivity, at about 70°C (Figure 24.16):

$$\textit{Anode (oxidation):} \qquad 2F^- \longrightarrow F_2(g) + 2e^-$$

$$\textit{Cathode (reduction):} \quad 2H^+ + 2e^- \longrightarrow H_2(g)$$

$$\textit{Overall reaction:} \qquad 2HF(l) \longrightarrow H_2(g) + F_2(g)$$

Chlorine gas ($Cl_2$) is prepared industrially by the electrolysis of molten NaCl or by the *chlor-alkali process,* the electrolysis of a concentrated aqueous NaCl solution (called brine). (*Chlor* denotes chlorine, and *alkali* denotes an alkali metal, such as sodium.) Two of the common cells used in the chlor-alkali process are the mercury cell and the diaphragm cell. In both cells, the overall reaction is

$$2NaCl(aq) + 2H_2O(l) \xrightarrow{\text{electrolysis}} 2NaOH(aq) + H_2(g) + Cl_2(g)$$

As you can see, this reaction yields two useful by-products, NaOH and $H_2$. The cells are designed to separate the molecular chlorine from the sodium hydroxide solution and the molecular hydrogen to prevent side reactions such as

$$2NaOH(aq) + Cl_2(g) \longrightarrow NaOCl(aq) + NaCl(aq) + H_2O(l)$$

and

$$H_2(g) + Cl_2(g) \longrightarrow 2HCl(g)$$

These reactions must be prevented because they consume the desired products and can be dangerous because a mixture of $H_2$ and $Cl_2$ is explosive.

Figure 24.17 shows the mercury cell used in the chlor-alkali process. The cathode is a liquid mercury pool at the bottom of the cell, and the anode is made of either graphite or titanium coated

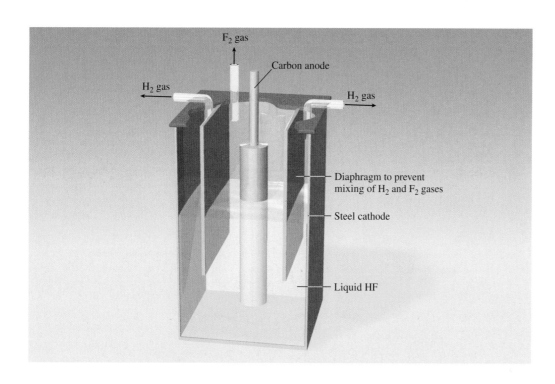

**Figure 24.16** Electrolytic cell for the preparation of fluorine gas. Note that because $H_2$ and $F_2$ form an explosive mixture, these gases must be separated from each other.

**Figure 24.17**    Mercury cell used in the chlor-alkali process. The cathode contains mercury. The sodium-mercury amalgam is treated with water outside the cell to produce sodium hydroxide and hydrogen gas.

**Figure 24.18**    The industrial manufacture of chlorine gas.

with platinum. Brine is continuously passed through the cell as shown in the diagram. The electrode reactions are

$$\textit{Anode (oxidation):} \qquad 2Cl^-(aq) \longrightarrow Cl_2(g) + 2e^-$$

$$\underline{\textit{Cathode (reduction):} \quad 2Na(aq) + 2e^- \xrightarrow{\text{Hg}(l)} 2NaHg}$$

$$\textit{Overall reaction:} \qquad 2NaCl(aq) \longrightarrow 2Na/Hg + Cl_2(g)$$

where Na/Hg denotes the formation of sodium amalgam. The chlorine gas generated this way is very pure. The sodium amalgam does not react with the brine solution but decomposes as follows when treated with pure water outside the cell:

$$2Na/Hg + 2H_2O(l) \longrightarrow 2NaOH(aq) + H_2(g) + 2Hg(l)$$

The by-products are sodium hydroxide and hydrogen gas. Although the mercury is cycled back into the cell for reuse, some of it is always discharged with waste solutions into the environment, resulting in mercury pollution. This is a major drawback of the mercury cell. Figure 24.18 shows the industrial manufacture of chlorine gas.

The half-cell reactions in a diaphragm cell are shown in Figure 24.19. The asbestos diaphragm is permeable to the ions but not to the hydrogen and chlorine gases and so prevents the gases from mixing. During electrolysis, a positive pressure is applied on the anode side of the compartment to prevent the migration of the OH$^-$ ions from the cathode compartment. Periodically, fresh brine solution is added to the cell and the sodium hydroxide solution is run off as shown. The diaphragm cell presents no pollution problems. Its main disadvantage is that the sodium hydroxide solution is contaminated with unreacted sodium chloride.

**Figure 24.19**    Diaphragm cell used in the chlor-alkali process.

In the laboratory, chlorine, bromine, and iodine can be prepared by heating the alkali halides (NaCl, KBr, or KI) in concentrated sulfuric acid in the presence of manganese(IV) oxide. A representative reaction is

$$MnO_2(s) + 2H_2SO_4(aq) + 2NaCl(aq) \longrightarrow MnSO_4(aq) + Na_2SO_4(aq) + 2H_2O(l) + Cl_2(g)$$

## Compounds of the Halogens

Most of the halides can be categorized as either ionic or covalent. The fluorides and chlorides of many metallic elements, especially those belonging to the alkali metal and alkaline earth metal (except beryllium) families, are *ionic* compounds. Most of the halides of *nonmetals* such as sulfur and phosphorus are *covalent* compounds. The oxidation numbers of the halogens can vary from $-1$ to $+7$. The only exception is fluorine. Because it is the most electronegative element, fluorine can have only two oxidation numbers, 0 (as in $F_2$) and $-1$, in all its compounds.

The hydrogen halides, an important class of halogen compounds, can be formed by the direct combination of the elements:

$$H_2(g) + X_2(g) \rightleftharpoons 2HX(g)$$

where X denotes a halogen atom. These reactions (especially the ones involving $F_2$ and $Cl_2$) can occur with explosive violence. Industrially, hydrogen chloride is produced as a by-product in the manufacture of chlorinated hydrocarbons:

$$C_2H_6(g) + Cl_2(g) \longrightarrow C_2H_5Cl(g) + HCl(g)$$

In the laboratory, hydrogen fluoride and hydrogen chloride can be prepared by combining the metal halides with concentrated sulfuric acid:

$$CaF_2(s) + H_2SO_4(aq) \longrightarrow 2HF(g) + CaSO_4(s)$$

$$2NaCl(s) + H_2SO_4(aq) \longrightarrow 2HCl(g) + Na_2SO_4(aq)$$

Hydrogen bromide and hydrogen iodide cannot be prepared this way because they are oxidized to elemental bromine and iodine. For example, the reaction between NaBr and $H_2SO_4$ is

$$2NaBr(s) + 2H_2SO_4(aq) \longrightarrow Br_2(l) + SO_2(g) + Na_2SO_4(aq) + 2H_2O(l)$$

Instead, hydrogen bromide is prepared by first reacting bromine with phosphorus to form phosphorus tribromide:

$$P_4(s) + 6Br_2(l) \longrightarrow 4PBr_3(l)$$

Next, $PBr_3$ is treated with water to yield HBr:

$$PBr_3(l) + 3H_2O(l) \longrightarrow 3HBr(g) + H_3PO_3(aq)$$

Hydrogen iodide can be prepared in a similar manner.

HF is so highly reactive that it attacks silica and silicates:

$$6HF(aq) + SiO_2(g) \longrightarrow H_2SiF_6(aq) + 2H_2O(l)$$

This property makes HF suitable for etching glass and is the reason that hydrogen fluoride must be kept in plastic or inert metal (e.g., Pt) containers. Hydrogen fluoride is used in the manufacture of Freons (see Chapter 17); for example,

$$CCl_4(l) + HF(g) \longrightarrow CFCl_3(g) + HCl(g)$$

$$CFCl_3(g) + HF(g) \longrightarrow CF_2Cl_2(g) + HCl(g)$$

It is also important in the production of aluminum. Hydrogen chloride is used in the preparation of hydrochloric acid, inorganic chlorides, and in various metallurgical processes. Hydrogen bromide and hydrogen iodide do not have any major industrial uses.

Aqueous solutions of hydrogen halides are acidic. The strength of the acids increases as follows:

$$HF \ll HCl < HBr < HI$$

The halogens also form a series of oxoacids with the following general formulas:

| HXO | HXO$_2$ | HXO$_3$ | HXO$_4$ |
|-----|---------|---------|---------|
| Hypohalous acid | Halous acid | Halic acid | Perhalic acid |

(for example, HClO, hypochlorous acid; HClO$_2$, chlorous acid; HClO$_3$, chloric acid; and HClO$_4$, perchloric acid).

Chlorous acid (HClO$_2$) is the only known *halous* acid. All the halogens except fluorine form halic and perhalic acids. The Lewis structures of the chlorine oxoacids are

$$H\!:\!\ddot{O}\!:\!\ddot{C}l\!: \qquad H\!:\!\ddot{O}\!:\!\ddot{C}l\!:\!\ddot{O}\!: \qquad H\!:\!\ddot{O}\!:\!\ddot{C}l\!:\!\ddot{O}\!: \qquad H\!:\!\ddot{O}\!:\!\ddot{C}l\!:\!\ddot{O}\!:$$

| Hypochlorous acid | Chlorous acid | Chloric acid | Perchloric acid |

For a given halogen, the acid strength decreases from perhalic acid to hypohalous acid [◄◄ Section 16.9].

Table 24.5 lists some of the halogen compounds. Periodic acid (HIO$_4$) does not appear because this compound cannot be isolated in the pure form. Instead the formula H$_5$IO$_6$ is often used to represent periodic acid.

## Uses of the Halogens

Applications of the halogens and their compounds are widespread in industry, health care, and other areas. One such application is fluoridation, the practice of adding small quantities of fluorides (about 1 ppm by mass) such as NaF to drinking water to reduce dental caries.

One of the most important inorganic fluorides is uranium hexafluoride (UF$_6$), which is essential to the gaseous diffusion process for separating isotopes of uranium (U-235 and U-238). Industrially, fluorine is used to produce polytetrafluoroethylene, a polymer better known as Teflon:

$$+(CF_2-CF_2)_n$$

where $n$ is a large number. Teflon is used in electrical insulators, high-temperature plastics, cooking utensils, and so on.

Chlorine plays an important biological role in the human body, where the chloride ion is the principal anion in intracellular and extracellular fluids. Chlorine is widely used as an industrial bleaching agent for paper and textiles. Ordinary household laundry bleach contains the active ingredient sodium hypochlorite (about 5 percent by mass), which is prepared by combining chlorine gas with a cold solution of sodium hydroxide:

$$Cl_2(g) + 2NaOH(aq) \longrightarrow NaCl(aq) + NaClO(aq) + H_2O(l)$$

Chlorine is also used to purify water and disinfect swimming pools. When chlorine dissolves in water, it undergoes the following reaction:

$$Cl_2(g) + H_2O(l) \longrightarrow HCl(aq) + HClO(aq)$$

It is thought that the ClO$^-$ ions destroy bacteria by oxidizing life-sustaining compounds within them.

Chlorinated methanes, such as carbon tetrachloride and chloroform, are useful organic solvents. Large quantities of chlorine are used to produce insecticides, such as DDT. However, in

| TABLE 24.5 | Common Compounds of Halogens* | | | |
|---|---|---|---|---|
| **Compound** | **F** | **Cl** | **Br** | **I** |
| Hydrogen halide | HF (−1) | HCl (−1) | HBr (−1) | HI (−1) |
| Oxides | OF$_2$ (−1) | Cl$_2$O (+1) | Br$_2$O (+1) | I$_2$O$_5$ (+5) |
| | | ClO$_2$ (+4) | BrO$_2$ (+4) | |
| | | Cl$_2$O$_7$ (+7) | | |
| Oxoacids | HFO (−1) | HClO (+1) | HBrO (+1) | HIO (+1) |
| | | HClO$_2$ (+3) | | |
| | | HClO$_3$ (+5) | HBrO$_3$ (+5) | HIO$_3$ (+5) |
| | | HClO$_4$ (+7) | HBrO$_4$ (+7) | H$_5$IO$_6$ (+7) |

* The number in parentheses indicates the oxidation number of the halogen.

view of the damage they inflict on the environment, the use of many of these compounds is either totally banned or greatly restricted in the United States. Chlorine is also used to produce polymers such as poly(vinyl chloride).

So far as we know, bromine compounds occur naturally only in some marine organisms. Seawater is about $1 \times 10^{-3}\, M\, Br^-$, so it is the main source of bromine. Bromine is used to prepare ethylene dibromide ($BrCH_2CH_2Br$), which is used as an insecticide and as a scavenger for lead (i.e., to combine with lead) in gasoline to keep lead deposits from clogging engines. Studies have shown that ethylene dibromide is a very potent carcinogen.

Bromine combines directly with silver to form silver bromide (AgBr), which is used in photographic films.

Iodine is not used as widely as the other halogens. A 50 percent (by mass) alcohol solution of iodine, known as *tincture of iodine,* is used medicinally as an antiseptic. Iodine is an essential constituent of the thyroid hormone thyroxine:

**Figure 24.20**   Cloud seeding using AgI particles.

Iodine deficiency in the diet may result in enlargement of the thyroid gland (known as goiter). Iodized table salt sold in the United States usually contains 0.01 percent KI or NaI, which is more than sufficient to satisfy the 1 mg of iodine per week required for the formation of thyroxine in the human body.

Silver iodide (AgI) is a pale-yellow solid that darkens when exposed to light. In this respect, it is similar to silver bromide. Silver iodide is sometimes used in cloud seeding, a process for inducing rainfall on a small scale (Figure 24.20). The advantage of using silver iodide is that enormous numbers of nuclei (i.e., small particles of silver iodide on which ice crystals can form) become available. About $10^{15}$ nuclei are produced from 1 g of AgI by vaporizing an acetone solution of silver iodide in a hot flame. The nuclei are then dispersed into the clouds from an airplane.

# Applying What You've Learned

King George III of Britain (1738–1820) suffered from periodic physical and mental illness throughout his adult life. Several of the episodes were severe enough to render the king temporarily unable to rule. Although most of the king's symptoms are now attributed to *porphyria*, a hereditary metabolic disorder, a 2004 analysis of samples of the king's hair revealed high levels of arsenic. One possibility that has been raised regarding the source of the arsenic is the *antimony* used in the king's treatment for chronic illness. Antimony is an element known since ancient times and used in cosmetics (stibnite, the most common antimony ore was used as eye liner during biblical times) and in medicine.

**Writing Prompt:**
Research arsenic and antimony, and write a 500-word essay describing their history and explaining how medicinal antimony might have been the source of the arsenic found in the hair samples from Britain's King George III.

# CHAPTER SUMMARY

## Section 24.1

- Properties of nonmetals vary more than properties of metals.

- Nonmetal elements may be solid, liquid, or gaseous and may exhibit both positive and negative oxidation numbers in their compounds.

## Section 24.2

- Hydrogen atoms contain one proton and one electron. They are the simplest atoms.

- Hydrogen combines with many metals and nonmetals to form hydrides; some hydrides are ionic and some are covalent.

- There are three isotopes of hydrogen: $^1_1H$, $^2_1H$ (deuterium), and $^3_1H$ (tritium).

- "Heavy water" ($D_2O$) contains deuterium.

## Section 24.3

- The important inorganic compounds of carbon are the *carbides;* the *cyanides,* most of which are extremely toxic; *carbon monoxide,* also toxic and a major air pollutant; the *carbonates* and *bicarbonates;* and *carbon dioxide,* an end product of metabolism and a component of the global carbon cycle.

## Section 24.4

- Elemental nitrogen ($N_2$) contains a triple bond and is very stable.

- Compounds in which nitrogen has oxidation numbers from $-3$ to $+5$ are formed between nitrogen and hydrogen and/or oxygen atoms.

- Ammonia ($NH_3$) is widely used in fertilizers.

- White phosphorus ($P_4$) is highly toxic, very reactive, and flammable; the polymeric red phosphorus $[(P_4)_n]$ is more stable.

- Phosphorus forms oxides and halides with oxidation numbers of $+3$ and $+5$ and several oxoacids. The phosphates are the most important phosphorus compounds.

## Section 24.5

- Elemental oxygen ($O_2$) is paramagnetic and contains two unpaired electrons.

- Oxygen forms ozone ($O_3$), oxides ($O^{2-}$), peroxides ($O_2^{2-}$), and superoxides ($O_2^-$).

- The most abundant element in Earth's crust, oxygen is essential for life on Earth.

- Sulfur is obtained from Earth's crust as a molten liquid via the Frasch process.

- Sulfur exists in a number of allotropic forms and has a variety of oxidation numbers in its compounds.

- Sulfuric acid is the cornerstone of the chemical industry. It is produced from sulfur via sulfur dioxide and sulfur trioxide by means of the contact process.

## Section 24.6

- The halogens are toxic and reactive elements that are found only in compounds with other elements.

- Fluorine and chlorine are strong oxidizing agents and are prepared by electrolysis.

- With the exception of fluorine, the halogens may have both negative and positive oxidation states.

- Fluorine's oxidation state in its compounds is always $-1$.

# QUESTIONS AND PROBLEMS

## Section 24.1: General Properties of Nonmetals

*Review Questions*

24.1  Without referring to Figure 24.1, state whether each of the following elements are metals, metalloids, or nonmetals: (a) Cs, (b) Ge, (c) I, (d) Kr, (e) W, (f) Ga, (g) Te, (h) Bi.

24.2  List two chemical and two physical properties that distinguish a metal from a nonmetal.

24.3  Make a list of physical and chemical properties of chlorine ($Cl_2$) and magnesium. Comment on their differences with reference to the fact that one is a metal and the other is a nonmetal.

24.4  Carbon is usually classified as a nonmetal. However, the graphite used in "lead" pencils conducts electricity. Look at a pencil, and list two nonmetallic properties of graphite.

## Section 24.2: Hydrogen

*Review Questions*

24.5  Explain why hydrogen has a unique position in the periodic table.

24.6  Describe two laboratory and two industrial preparations for hydrogen.

24.7  Hydrogen exhibits three types of bonding in its compounds. Describe each type of bonding with an example.

24.8  What are interstitial hydrides?

24.9  Give the name of (a) an ionic hydride and (b) a covalent hydride. In each case describe the preparation and give the structure of the compound.

24.10  Describe what is meant by the "hydrogen economy."

## Problems

**24.11** Elements number 17 and 20 form compounds with hydrogen. Write the formulas for these two compounds, and compare their chemical behavior in water.

**24.12** Give an example of hydrogen as (a) an oxidizing agent and (b) a reducing agent.

**24.13** Compare the physical and chemical properties of the hydrides of each of the following elements: Na, Ca, C, N, O, Cl.

**24.14** Suggest a physical method that would enable you to separate hydrogen gas from neon gas.

**24.15** Write a balanced equation to show the reaction between $CaH_2$ and $H_2O$. How many grams of $CaH_2$ are needed to produce 26.4 L of $H_2$ gas at 20°C and 746 mmHg?

**24.16** How many kilograms of water must be processed to obtain 2.0 L of $D_2$ at 25°C and 0.90 atm pressure? Assume that deuterium abundance is 0.015 percent and that recovery is 80 percent.

**24.17** Predict the outcome of the following reactions:

(a) $CuO(s) + H_2(g) \longrightarrow$
(b) $Na_2O(s) + H_2(g) \longrightarrow$

**24.18** Starting with $H_2$, describe how you would prepare (a) HCl, (b) $NH_3$, and (c) LiOH.

## Section 24.3: Carbon

### Review Questions

**24.19** Give an example of a carbide and a cyanide.

**24.20** How are cyanide ions used in metallurgy?

**24.21** Briefly discuss the preparation and properties of carbon monoxide and carbon dioxide.

**24.22** What is coal?

**24.23** Describe two chemical differences between CO and $CO_2$.

**24.24** Describe the reaction between $CO_2$ and $OH^-$ in terms of a Lewis acid-base reaction.

### Problems

**24.25** Draw a Lewis structure for the $C_2^{2-}$ ion.

**24.26** Balance the following equations:

(a) $Be_2C(s) + H_2O(l) \longrightarrow$
(b) $CaC_2(s) + H_2O(l) \longrightarrow$

**24.27** Unlike $CaCO_3$, $Na_2CO_3$ does not readily yield $CO_2$ when heated. On the other hand, $NaHCO_3$ undergoes thermal decomposition to produce $CO_2$ and $Na_2CO_3$. (a) Write a balanced equation for the reaction. (b) How would you test for the $CO_2$ evolved? [*Hint:* Treat the gas with limewater, an aqueous solution of $Ca(OH)_2$.]

**24.28** Two solutions are labeled A and B. Solution A contains $Na_2CO_3$, and solution B contains $NaHCO_3$. Describe how you would distinguish between the two solutions if you were provided with an $MgCl_2$ solution. (*Hint:* You need to know the solubilities of $MgCO_3$ and $MgHCO_3$.)

**24.29** Magnesium chloride is dissolved in a solution containing sodium bicarbonate. On heating, a white precipitate is formed. Explain what causes the precipitation.

**24.30** A few drops of concentrated ammonia solution added to a calcium bicarbonate solution cause a white precipitate to form. Write a balanced equation for the reaction.

**24.31** Sodium hydroxide is hygroscopic—that is, it absorbs moisture when exposed to the atmosphere. A student placed a pellet of NaOH on a watch glass. A few days later, she noticed that the pellet was covered with a white solid. What is the identity of this solid? (*Hint:* Air contains $CO_2$.)

**24.32** A piece of red-hot magnesium ribbon will continue to burn in an atmosphere of $CO_2$ even though $CO_2$ does not support combustion. Explain.

**24.33** Is carbon monoxide isoelectronic with nitrogen ($N_2$)?

## Section 24.4: Nitrogen and Phosphorus

### Review Questions

**24.34** Describe a laboratory and an industrial preparation of nitrogen gas.

**24.35** What is meant by *nitrogen fixation*? Describe a process for fixation of nitrogen on an industrial scale.

**24.36** Describe an industrial preparation of phosphorus.

**24.37** Why is the $P_4$ molecule unstable?

### Problems

**24.38** Nitrogen can be obtained by (a) passing ammonia over red-hot copper(II) oxide and (b) heating ammonium dichromate [one of the products is Cr(III) oxide]. Write a balanced equation for each preparation.

**24.39** Write balanced equations for the preparation of sodium nitrite by (a) heating sodium nitrate and (b) heating sodium nitrate with carbon.

**24.40** Sodium amide ($NaNH_2$) reacts with water to produce sodium hydroxide and ammonia. Describe this reaction as a Brønsted acid-base reaction.

**24.41** Write a balanced equation for the formation of urea, $[(NH_2)_2CO]$, from carbon dioxide and ammonia. Should the reaction be run at a high or low pressure to maximize the yield?

**24.42** Some farmers feel that lightning helps produce a better crop. What is the scientific basis for this belief?

**24.43** Explain why nitric acid can be reduced but not oxidized.

**24.44** At 620 K, the vapor density of ammonium chloride relative to hydrogen ($H_2$) under the same conditions of temperature and pressure is 14.5, although, according to its formula mass, it should have a vapor density of 26.8. How would you account for this discrepancy?

**24.45** Write a balanced equation for each of the following processes: (a) On heating, ammonium nitrate produces nitrous oxide. (b) On heating, potassium nitrate produces potassium nitrite and oxygen gas. (c) On heating, lead nitrate produces lead(II) oxide, nitrogen dioxide ($NO_2$), and oxygen gas.

**24.46**  Explain why, under normal conditions, the reaction of zinc with nitric acid does not produce hydrogen.

**24.47**  Potassium nitrite can be produced by heating a mixture of potassium nitrate and carbon. Write a balanced equation for this reaction. Calculate the theoretical yield of $KNO_2$ produced by heating 57.0 g of $KNO_3$ with an excess of carbon.

**24.48**  Predict the geometry of nitrous oxide ($N_2O$), by the VSEPR method, and draw resonance structures for the molecule. (*Hint:* The atoms are arranged as NNO.)

**24.49**  Consider the reaction

$$N_2(g) + O_2(g) \rightleftharpoons 2NO(g)$$

Given that the $\Delta G°$ for the reaction at 298 K is 173.4 kJ/mol, calculate (a) the standard free energy of formation of NO, (b) $K_P$ for the reaction, and (c) $K_c$ for the reaction.

**24.50**  From the data in Appendix 2, calculate $\Delta H°$ for the synthesis of NO (which is the first step in the manufacture of nitric acid) at 25°C:

$$4NH_3(g) + 5O_2(g) \longrightarrow 4NO(g) + 6H_2O(l)$$

**24.51**  When 1.645 g of white phosphorus is dissolved in 75.5 g of $CS_2$, the solution boils at 46.709°C, whereas pure $CS_2$ boils at 46.300°C. The molal boiling-point elevation constant for $CS_2$ is 2.34°C/$m$. Calculate the molar mass of white phosphorus, and give the molecular formula.

**24.52**  Explain why two N atoms can form a double bond or a triple bond, whereas two P atoms normally can form only a single bond.

**24.53**  Dinitrogen pentoxide is a product of the reaction between $P_4O_{10}$ and $HNO_3$. Write a balanced equation for this reaction. Calculate the theoretical yield of $N_2O_5$ if 79.4 g of $P_4O_{10}$ is combined with an excess of $HNO_3$. (*Hint:* One of the products is $HPO_3$.)

**24.54**  Starting with elemental phosphorus ($P_4$), show how you would prepare phosphoric acid.

**24.55**  What is the hybridization of phosphorus in the phosphonium ion ($PH_4^+$)?

**24.56**  Explain why (a) $NH_3$ is more basic than $PH_3$, (b) $NH_3$ has a higher boiling point than $PH_3$, (c) $PCl_5$ exists but $NCl_5$ does not, and (d) $N_2$ is more inert than $P_4$.

## Section 24.5: Oxygen and Sulfur

### Review Questions

**24.57**  Describe one industrial and one laboratory preparation of $O_2$.

**24.58**  Give an account of the various kinds of oxides that exist, and illustrate each type by two examples.

**24.59**  Hydrogen peroxide can be prepared by treating barium peroxide with sulfuric acid. Write a balanced equation for this reaction.

**24.60**  Describe the Frasch process for obtaining sulfur.

**24.61**  Describe the contact process for the production of sulfuric acid.

**24.62**  How is hydrogen sulfide generated in the laboratory?

### Problems

**24.63**  One of the steps involved in the depletion of ozone in the stratosphere by nitric oxide may be represented as

$$NO(g) + O_3(g) \longrightarrow NO_2(g) + O_2(g)$$

From the data in Appendix 2, calculate $\Delta G°$, $K_P$, and $K_c$ for the reaction at 25°C.

**24.64**  Draw molecular orbital energy level diagrams for $O_2$, $O_2^-$, and $O_2^{2-}$.

**24.65**  Hydrogen peroxide is unstable and decomposes readily:

$$2H_2O_2(aq) \longrightarrow 2H_2O(l) + O_2(g)$$

This reaction is accelerated by light, heat, or a catalyst. (a) Explain why hydrogen peroxide sold in drugstores comes in dark bottles. (b) The concentrations of aqueous hydrogen peroxide solutions are normally expressed as percent by mass. In the decomposition of hydrogen peroxide, how many liters of oxygen gas can be produced at STP from 15.0 g of a 7.50% hydrogen peroxide solution?

**24.66**  Oxygen forms double bonds in $O_2$, but sulfur forms single bonds in $S_8$. Explain.

**24.67**  What are the oxidation numbers of O and F in HFO?

**24.68**  In 2004 about 48 million tons of sulfuric acid was produced in the United States. Calculate the amount of sulfur (in grams and moles) used to produce that amount of sulfuric acid.

**24.69**  Sulfuric acid is a dehydrating agent. Write balanced equations for the reactions between sulfuric acid and the following substances: (a) HCOOH, (b) $H_3PO_4$, (c) $HNO_3$, (d) $HClO_3$. (*Hint:* Sulfuric acid is not decomposed by the dehydrating action.)

**24.70**  Calculate the amount of $CaCO_3$ (in grams) that would be required to react with 50.6 g of $SO_2$ emitted by a power plant.

**24.71**  $SF_6$ exists, but $OF_6$ does not. Explain.

**24.72**  Explain why $SCl_6$, $SBr_6$, and $SI_6$ cannot be prepared.

**24.73**  The bad smell of water containing hydrogen sulfide can be removed by the action of chlorine. The reaction is

$$H_2S(aq) + Cl_2(aq) \longrightarrow 2HCl(aq) + S(s)$$

If the hydrogen sulfide content of contaminated water is 22 ppm by mass, calculate the amount of $Cl_2$ (in grams) required to remove all the $H_2S$ from $2.0 \times 10^2$ gal of water. (1 gallon = 3.785 L.)

**24.74**  Compare the physical and chemical properties of $H_2O$ and $H_2S$.

**24.75**  Concentrated sulfuric acid reacts with sodium iodide to produce molecular iodine, hydrogen sulfide, and sodium hydrogen sulfate. Write a balanced equation for the reaction.

**24.76**  Describe two reactions in which sulfuric acid acts as an oxidizing agent.

## Section 24.6: The Halogens

### Review Questions

**24.77**  Describe an industrial method for preparing each of the halogens.

**24.78**  Name the major uses of the halogens.

## Problems

**24.79** Draw structures for (a) $(HF)_2$ and (b) $HF_2^-$.

**24.80** Sulfuric acid is a weaker acid than hydrochloric acid. Yet hydrogen chloride is evolved when concentrated sulfuric acid is added to sodium chloride. Explain.

**24.81** Use the VSEPR method to predict the geometries of the following species: (a) $I_3^-$, (b) $SiCl_4$, (c) $PF_5$, (d) $SF_4$.

**24.82** Metal chlorides can be prepared in a number of ways:
(a) direct combination of metal and molecular chlorine,
(b) reaction between metal and hydrochloric acid, (c) acid-base neutralization, (d) metal carbonate treated with hydrochloric acid, (e) precipitation reaction. Give an example for each type of preparation.

**24.83** A 375-gal tank is filled with water containing 167 g of bromine in the form of $Br^-$ ions. How many liters of $Cl_2$ gas at 1.00 atm and 20°C will be required to oxidize all the bromide to molecular bromine?

**24.84** Hydrogen fluoride can be prepared by the action of sulfuric acid on sodium fluoride. Explain why hydrogen bromide cannot be prepared by the action of the same acid on sodium bromide.

**24.85** What volume of bromine ($Br_2$) vapor measured at 100°C and 700 mmHg pressure would be obtained if 2.00 L of dry chlorine ($Cl_2$), measured at 15°C and 760 mmHg, was absorbed by a potassium bromide solution?

**24.86** Aqueous copper(II) sulfate solution is blue. When aqueous potassium fluoride is added to the $CuSO_4$ solution, a green precipitate is formed. If aqueous potassium chloride is added instead, a bright-green solution is formed. Explain what happens in each case.

**24.87** Iodine pentoxide ($I_2O_5$) is sometimes used to remove carbon monoxide from the air by forming carbon dioxide and iodine. Write a balanced equation for this reaction, and identify species that are oxidized and reduced.

**24.88** Show that chlorine, bromine, and iodine are very much alike by giving an account of their behavior (a) with hydrogen, (b) in producing silver salts, (c) as oxidizing agents, and (d) with sodium hydroxide. (e) In what respects is fluorine not a typical halogen element?

## Additional Problems

**24.89** Write a balanced equation for each of the following reactions:
(a) Heating phosphorous acid yields phosphoric acid and phosphine ($PH_3$). (b) Lithium carbide reacts with hydrochloric acid to give lithium chloride and methane. (c) Bubbling HI gas through an aqueous solution of $HNO_2$ yields molecular iodine and nitric oxide. (d) Hydrogen sulfide is oxidized by chlorine to give HCl and $SCl_2$.

**24.90** Both $N_2O$ and $O_2$ support combustion. Suggest one physical and one chemical test to distinguish between the two gases.

**24.91** (a) Which of the following compounds has the greatest ionic character: $PCl_5$, $SiCl_4$, $CCl_4$, $BCl_3$? (b) Which of the following ions has the smallest ionic radius: $F^-$, $C^{4-}$, $N^{3-}$, $O^{2-}$? (c) Which of the following atoms has the highest ionization energy: F, Cl, Br, I? (d) Which of the following oxides is most acidic: $H_2O$, $SiO_2$, $CO_2$?

**24.92** Describe the bonding in the $C_2^{2-}$ ion in terms of the molecular orbital theory.

**24.93** What is the change in oxidation number for the following reaction?

$$3O_2 \longrightarrow 2O_3$$

**24.94** Starting with deuterium oxide ($D_2O$), describe how you would prepare (a) NaOD, (b) DCl, (c) $ND_3$, (d) $C_2D_2$, (e) $CD_4$, and (f) $D_2SO_4$.

**24.95** Solid $PCl_5$ exists as $[PCl_4^+][PCl_6^-]$. Draw Lewis structures for these ions. Describe the hybridization state of the P atoms.

**24.96** Consider the Frasch process. (a) How is it possible to heat water well above 100°C without turning it into steam? (b) Why is water sent down the outermost pipe? (c) Why would excavating a mine and digging for sulfur be a dangerous procedure for obtaining the element?

**24.97** As we saw in Section 23.2, the reduction of iron oxides is accomplished by using carbon monoxide as a reducing agent. Starting with coke in a blast furnace, the following equilibrium plays a key role in the extraction of iron:

$$C(s) + CO_2(g) \longrightarrow 2CO(g)$$

Use the data in Appendix 2 to calculate the equilibrium constant at 25°C and 100°C. Assume $\Delta H°$ and $\Delta S°$ to be independent of temperature.

**24.98** Lubricants used in watches usually consist of long-chain hydrocarbons. Oxidation by air forms solid polymers that eventually destroy the effectiveness of the lubricants. It is believed that one of the initial steps in the oxidation is removal of a hydrogen atom (hydrogen abstraction). By replacing the hydrogen atoms at reactive sites with deuterium atoms, it is possible to substantially slow down the overall oxidation rate. Why? (*Hint:* Consider the kinetic isotope effect.)

**24.99** How are lightbulbs frosted? (*Hint:* Consider the action of hydrofluoric acid on glass, which is made of silicon dioxide.)

**24.100** Life evolves to adapt to its environment. In this respect, explain why life most frequently needs oxygen for survival, rather than the more abundant nitrogen.

**24.101** A 10.0-g sample of white phosphorus was burned in an excess of oxygen. The product was dissolved in enough water to make 500 mL of solution. Calculate the pH of the solution at 25°C.

**24.102** Predict the physical and chemical properties of astatine, a radioactive element and the last member of Group 7A.

**24.103** Assuming ideal behavior, calculate the density of gaseous HF at its normal boiling point (19.5°C). The experimentally measured density under the same conditions is 3.10 g/L. Account for the discrepancy between your calculated value and the experimental result.

**24.104** Ammonium nitrate is the most important nitrogen-containing fertilizer in the world. Given only air and water as starting materials and any equipment and catalyst at your disposal, describe how you would prepare ammonium nitrate. State conditions under which you can increase the yield in each step.

# PRE-PROFESSIONAL PRACTICE EXAM PROBLEMS: VERBAL REASONING

Iodine deficiency results in a condition known as goiter, which is characterized by an enlarged thyroid, often appearing as a large, bulbous protrusion on the neck. Although it is relatively rare today, goiter was once common in regions where the soil and food supply are iodine poor, including the northern half of the contiguous United States. In 1918, 30 percent of the men registering for the World War I draft in Michigan were found to have significant goiter symptoms. Many of the young men were so ill as to be disqualified from enlisting in the Army.

In 1922, concerned about the high incidence of goiter in his home state, David Murray Cowie, a professor of pediatrics at the University of Michigan, began a campaign for the addition of a small amount of sodium iodide or potassium iodide to the table salt used by all Americans. Dr. Cowie was actually modeling a successful public health program that had all but eliminated iodine deficiency disorders (IDD) in Switzerland. In 1924 the Morton Salt Company began distributing iodized salt nationwide.

1. The main point of the passage is that

   a) goiter was once common but is now rare because of iodized salt.
   b) goiter is caused by iodine deficiency.
   c) David Murray Cowie was a pediatrician in Michigan.
   d) men suffering from goiter were disqualified from serving in the Army.

2. The reason dietary iodine is added to table salt is most likely that

   a) sodium iodide closely resembles sodium chloride.
   b) citizens of all socioeconomic strata use table salt.
   c) sodium iodide is not water soluble.
   d) elemental iodine is toxic.

3. The most likely reason it took 2 years to have iodized salt made available nationwide is that

   a) shipping took longer without interstate highways.
   b) the Morton Salt Company didn't exist before 1924.
   c) public health officials had to be convinced that iodization of salt was safe and effective.
   d) Michigan was not yet part of the United States.

4. According to the passage, goiter

   a) can be caused by a low-salt diet.
   b) can be cured by the addition of iodide to the diet.
   c) is potentially debilitating.
   d) was a more serious public health problem in Switzerland than in the United States.

# 25

# Modern Materials

*Left to right: Richard R. Schrock (USA), Yves Chauvin (France), and Robert H. Grubbs (USA) shared the 2005 Nobel Prize in Chemistry for their work on the metathesis method of organic synthesis.*

# Modern Materials Chemistry and the 2005 Nobel Prize

The 2005 Nobel Prize in Chemistry was awarded to Yves Chauvin, Robert H. Grubbs, and Richard R. Schrock for the development of the metathesis method in organic synthesis. In the metathesis method, double bonds in organic molecules are broken, and then new double bonds are formed. In *olefin metathesis,* a carbon-carbon double bond is broken in each of two olefins. The subsequent formation of two new carbon-carbon double bonds results in products in which the compounds have exchanged substituent groups with each other:

**Student Annotation:** *Metathesis* is a word meaning "an exchange of position," and *olefin* is an old-fashioned word for an alkene, a molecule with a carbon-carbon double bond.

Note that each reactant molecule has one R group and one R′ group, whereas each product molecule has two *identical* groups (i.e., two R groups or two R′ groups). This process requires a metal catalyst and is believed to occur by the formation of a four-membered ring:

The metathesis method can be applied to many industrial chemical processes, including those that potentially can produce pharmaceuticals for the treatment of such conditions as bacterial and viral infections, cancer, and Alzheimer's disease.

At the end of this chapter, you will be able to answer questions about modern polymeric materials [▶▶ Page 1022].

## In This Chapter, You Will Learn:

About some of the chemistry involved in the development of modern materials.

### Before you begin, you should review

- Functional groups [◀◀ Section 10.2]
- Organic polymers [◀◀ Section 10.6]

Bottles that can be recycled are made of thermoplastic polymers.

# 25.1 | Polymers

Many biological molecules, such as DNA, starch, and proteins, have very large molecular masses. These molecules are **polymers** because they are made up of many smaller parts linked together [◄◄ Section 10.6]. The prefix *poly* comes from a Greek root meaning "many," and the small molecules that make up the individual building blocks of polymers are called **monomers.** Many natural polymers, such as silk and rubber, now have synthetic analogues that are manufactured by the chemical industry.

Polymers are used in many medical applications, such as flexible wound dressings, surgical implants, and prosthetics. Some of these polymers are **thermoplastic,** which means that they can be melted and reshaped, or heated and bent. Others are **thermosetting,** which means that their shape is determined as part of the chemical process that formed the polymer. Thermosetting polymers cannot be reshaped easily and are not easily recycled, whereas thermoplastic polymers can be melted down and cast into new shapes in different products.

## Addition Polymers

The simplest type of polymerization reaction involves the bonding of monomer molecules by movement of electrons from a multiple bond into new single bonds between molecules. This type of polymerization is called **addition polymerization** [◄◄ Section 10.6]. A molecule with a carbon-carbon double bond is called an *alkene,* and a molecule with a carbon-carbon triple bond is called an *alkyne.* The synthesis of polyethylene from ethylene ($C_2H_4$) illustrates an addition polymerization. The arrows with half-pointed heads show movement of a single electron. The process of "splitting open" the double bond is initiated by a molecule with a single O—O bond that can be broken in half by heating. The free radical initiator molecule attaches to one of the ethylene carbon atoms and breaks the pi bond between the two carbon atoms in ethylene. Figure 25.1 shows the scheme of addition polymerization with respect to the double bonds opening up to form single bonds between monomer units to form polyethylene from ethylene.

The structure of polyethylene can be represented as $-(CH_2-CH_2)_n$, where $n$ is the number of ethylene monomer molecules that reacted (which could be hundreds, thousands, or more). The end of the chain contains a carbon atom with an unpaired electron (i.e., $-CH_2CH_2\cdot$), though, so it must be closed by bonding with another atom. This might be a hydrogen atom (e.g., $-CH_2CH_2-H$), but it might also be part of the initiator molecule (e.g., $-CH_2CH_2-OR$). Although the residue of the initiator molecule has very different chemical properties than the main polymer chain, it will not significantly affect the polymer's properties because the polyethylene chain is so long.

Although the structure of a polyethylene chain is drawn in Figure 25.1 as a straight line, the bond angles around a central carbon atom with four bonds are approximately 109.5° each [◄◄ Section 9.1]. Thus, the "straight" chain really zigzags, instead. The polymer chain bends, too, because there is free rotation about carbon-carbon single bonds. These molecular-level features manifest themselves in such macroscopic properties as flexibility.

It is possible, depending on the reaction conditions, for branches to form off the main polyethylene chain during polymerization. Polymer chains with many branches cannot pack together as efficiently as "straight" or unbranched chains, so branched-chain polyethylene is more flexible and lower in density than straight-chain polyethylene. The difference between branched and unbranched polyethylene is shown schematically in Figure 25.2.

Polyethylene consisting primarily of unbranched chains is known as high-density polyethylene (HDPE), whereas polyethylene that consists primarily of branched chains is known as low-density polyethylene (LDPE). HDPE is used in plastic bottles and containers where it is important to maintain shape, whereas LDPE is used in plastic food bags and wraps.

Natural rubber formed from latex is an addition polymer. The monomer unit is isoprene, a molecule with two C=C double bonds. In the reaction that forms *polyisoprene,* a pair of

Multimedia

Organic and Biochemistry—natural and synthetic polymers.

**Figure 25.1** Addition polymerization to form polyethylene from ethylene. Each curved, single-headed arrow represents the movement of one electron.

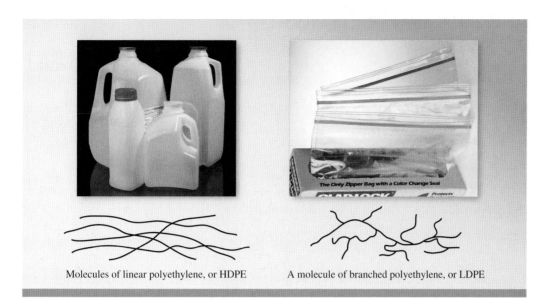

Molecules of linear polyethylene, or HDPE      A molecule of branched polyethylene, or LDPE

**Figure 25.2** Linear polyethylene is used to make plastic milk bottles. Plastic bags and wrap like Glad Cling Wrap are made from branched polyethylene.

electrons moves from one double bond to form a new double bond between the middle two carbon atoms, and the other double bond opens up to form a carbon-carbon single bond to the next monomer unit.

Isoprene

cis-Polyisoprene

Natural rubber is not very durable—it only retains its useful properties in a relatively narrow temperature range. To address this problem, Charles Goodyear[1] developed industrial *vulcanization* in 1839, a process in which sulfur is used to create linkages between individual polymer chains in the rubber. By heating the rubber with sulfur ($S_8$) under carefully controlled conditions, some of the carbon-hydrogen bonds are broken and replaced with carbon-sulfur bonds. The sulfur atoms are bonded to one or more additional sulfur atoms and then to another polymer chain through another sulfur-carbon bond (Figure 25.3). These linkages, called ***cross-links,*** make the rubber much stronger and more rigid than natural polyisoprene. Rubber is an ***elastomer,*** a material that can stretch or bend and then return to its original shape as long as the limits of its elasticity are not exceeded. The cross-linking of the polyisoprene chains plays an important role in maintaining rubber's elastomer properties.

Goodyear, the current supplier of tires to NASCAR teams, varies the composition of the rubber in its tires for each race depending on the expected tire wear and banking of the track. The left- and right-side tires can even be of different compositions depending on the track layout and

**Student Annotation:** Each tire used in the NASCAR race series costs approximately $400. A single race car can go through as many as 32 tires in one 500-mile race, depending on track conditions. There are dozens of races throughout the year, and even more tires are used up during practices and prerace qualifying. It takes a lot of tires to compete in NASCAR!

1. Charles Goodyear (1800–1860). American chemist. Goodyear was the first person to realize the potential of natural rubber. His vulcanization process made the use of rubber practical, contributing greatly to the development of the automobile industry.

**Figure 25.3**    Vulcanization.

cis-Polyisoprene        Sulfur        Cross-linked polyisoprene

conditions. Softer tire compounds offer more grip but wear out faster, whereas harder compounds last longer but grip less.

Many other molecules containing carbon-carbon double bonds can form addition polymers. Styrene [Figure 25.4(a)] is an analogue of ethylene in which a hydrogen atom has been replaced with a $C_6H_5$ (a *phenyl*) group. The polymerization of styrene forms *polystyrene* (Figure 25.5). As a foam (with air or some other gas blown into the solid), polystyrene is used in disposable cups and plates as well as Styrofoam insulation and craft materials. Vinyl chloride [Figure 25.4(b)], an analogue of ethylene in which one of the hydrogen atoms has been replaced by a chlorine atom, is another monomer that can be used to form a polymer with many common uses. Poly(vinyl chloride) (PVC) is used in pipes and fittings, some plastic wraps, and the classic "vinyl" records that were the predecessors to musical recordings on cassettes and CDs.

(a)                    (b)                    (c)

**Figure 25.4**    (a) Styrene, (b) vinyl chloride, and (c) tetrafluoroethylene (the monomer of Teflon).

**Figure 25.5**    Structure of polystyrene.

| TABLE 25.1 | Some Common Addition Polymers | | |
|---|---|---|---|
| **Polymer** | **Structure** | **Monomer** | **Uses** |
| Polyethylene | $+CH_2-CH_2+_n$ | $H_2C{=}CH_2$ | Plastic bags and wraps, bottles, toys, Tyvek wrap |
| Polystyrene | | | Disposable cups and plates, insulation, packing materials |
| Poly(vinyl chloride) | | | Pipes and fittings, plastic wraps, records |
| Polyacrylonitrile | | | Fibers for carpeting and clothing (Orlon) |
| Polybutadiene | | | Synthetic rubber |

Polytetrafluoroethylene, more commonly known as the trademarked brand name Teflon, is formed from the addition polymerization of tetrafluoroethylene. Tetrafluoroethylene, shown in Figure 25.4(c), is an analogue of ethylene in which fluorine atoms have replaced all four of the hydrogen atoms. Teflon is a very good electrical insulator, so it is commonly used to coat wires. It is probably best known, though, as a nonstick substance used to coat bakeware, frying pans, and pots. It is also used in films that can be inserted into threaded joints between metal pipes to make it easier to unscrew the connection when necessary. Because Teflon is chemically inert, it is not possible to cross-link the chains like an elastomer. The structures of some addition polymers, including the structures of their respective monomers and what they are typically used for, are summarized in Table 25.1.

The diversity of polymers increases greatly when different monomers are used in the same polymer chain. *Copolymers* are polymers made from two or more different monomers [◄◄ Section 10.6]. Different combinations of the two or more monomers give rise to different types of copolymers. These different copolymers are summarized in Table 25.2.

| TABLE 25.2 | Types of Copolymers |
|---|---|
| **Copolymer Type** | **Representation*** |
| Random | ... −A−A−B−A−B−B−B−A−A−B−A−B−A−B−A−B−B−A− ... |
| Alternating | ... −A−B−A−B−A−B−A−B−A−B−A−B−A−B−A−B− ... |
| Block | ... −A−A−A−A−A−A−B−B−B−B−B−B−A−A−A−A−A−A− ... |
| Graft | ... −A−A−A−A−A−A−A−A−A−A−A−A−A−A−A−A−A−A− ... <br>     \| <br>     B−B−B−B−B−B−B−B−B ... |

*A and B represent different monomers (e.g., ethylene and styrene).

Sample Problem 25.1 lets you practice deducing the structure of an addition polymer based on the structure of its monomer.

### Sample Problem 25.1

Polypropylene is used in applications that range from indoor-outdoor carpeting to soda bottles to bank notes. It is produced by the addition polymerization of propylene, $H_2C=CH-CH_3$. Draw the structure of polypropylene, showing at least two repeating monomer units, and write its general formula.

**Strategy** Addition polymers form via a free-radical reaction in which the pi electrons in the carbon-carbon double bond of a monomer molecule are used to form carbon-carbon single bonds to other monomer molecules. Draw the structural formula of propylene such that the double bond can be "opened up" to form single bonds between consecutive monomer units.

**Setup** The structural formula of propylene is

$$\begin{array}{ccc} H & & H \\ \diagdown & & \diagup \\ & C=C & \\ \diagup & & \diagdown \\ H & & CH_3 \end{array}$$

By drawing three adjacent propylene molecules, we can rearrange the bonds to show three repeating units.

$$\begin{array}{ccc} H & \quad H \quad & H \quad H \quad H \quad H \\ C=C & C=C & C=C \\ H \quad CH_3 & H \quad CH_3 & H \quad CH_3 \end{array}$$

**Solution** The structure of polypropylene showing three monomer units is

$$\begin{array}{cccccc} H & H & H & H & H & H \\ | & | & | & | & | & | \\ -C-C-C-C-C-C- \\ | & | & | & | & | & | \\ H & CH_3 & H & CH_3 & H & CH_3 \end{array}$$

Thus, the general formula of polypropylene is

$$\left( \begin{array}{cc} H & H \\ | & | \\ C-C \\ | & | \\ H & CH_3 \end{array} \right)_n$$

**Think About It** The general formula looks like the structure of the monomer, except that the carbon-carbon double bond has been opened up to form bonds to monomer units on each side. This means the answer is reasonable. In fact, the opposite process (i.e., restoring the carbon-carbon double bond) can be used to determine the structure of the monomer used to create an addition polymer.

**Practice Problem A** Saran Wrap, the original plastic film, is polyvinylidene chloride, which is formed by the addition polymerization of vinylidene chloride, $H_2C=CCl_2$. Draw the structure of polyvinylidene chloride, showing at least two repeating units, and write its general structure.

**Practice Problem B** The general structure of polymethyl methacrylate, an addition polymer used in Plexiglas and Lucite, is as follows:

$$\left( \begin{array}{cc} H & CH_3 \\ | & | \\ C-C \\ | & | \\ H & C-OCH_3 \\ & \| \\ & O \end{array} \right)_n$$

Draw the structure of the monomer used to make this polymer.

Butadiene ($CH_2=CH-CH=CH_2$) and styrene, two monomers listed in Table 25.1, can form a block copolymer called poly(styrene-butadiene-styrene) or SBS rubber [Figure 25.6(a)], in which there are alternating blocks of polybutadiene and polystyrene. They also form a graft polymer known as high-impact polystyrene (HIPS) [Figure 25.6(b)], in which chains of polybutadiene branch off of the main polystyrene chain.

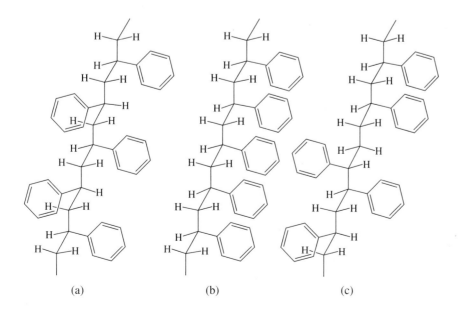

**Figure 25.6**   Structures of (a) SBS rubber and (b) HIPS. SBS consists of alternating blocks of polybutadiene and polystyrene, whereas HIPS consists of a polystyrene backbone with polybutadiene branches.

Polymers containing a vinyl group ($-CH_2-CHR-$) can exist as optical isomers because the carbon atom bonded to the R group is tetrahedral and bonded to four *different* groups [◄◄ Section 10.4]. *Tacticity* is the term that describes the relative arrangements of such chiral carbon atoms within a polymer. Polymers in which all the substituents (i.e., the R groups) are in the same relative orientation (i.e., on the same side of the polymer chain) are said to be *isotactic.* Polymers in which the substituents alternate positions along the polymer chain are said to be *syndiotactic.* Finally, polymers in which the substituents are oriented randomly along the polymer chain are said to be *atactic.* The tacticity of polystyrene is shown in Figure 25.7. Polymers that are formed by free-radical polymerization processes, such as polystyrene and PVC, are usually atactic. Special catalysts have been developed, though, that can be used to synthesize isotactic and syndiotactic polymers.

## Condensation Polymers

*Condensation polymers* form when monomers with two different functional groups combine, resulting in the elimination of a small molecule (usually water) [◄◄ Section 10.6]. Proteins are biological condensation polymers that can have molecular weights in the thousands or millions of grams per mole. Proteins form between amino acid monomers. Each amino acid contains an amino group ($-NH_2$) and a carboxy group ($-COOH$), and the condensation reaction occurs between the amino group of one amino acid and the carboxy group of another. (See Figure 10.13.) One water molecule is produced as each new C−N bond is formed.

**Figure 25.7**   Structures of (a) isotactic, (b) syndiotactic, and (c) atactic polymers.

(a)              (b)              (c)

The C$-$N bond formed between an amino group and a carboxy group is called an amide bond because RCONR′R″ is the general form of an amide [◄◄ Section 10.2, Table 10.2]. Polymers in which the monomers are connected by amide linkages are called *polyamides.* The amide bond between amino acids is more specifically called a peptide bond, and proteins are routinely referred to as polypeptides. Although polypeptides can form from a single amino acid (e.g., polyvaline), most proteins in organisms consist of chains of different amino acids. As a result, most proteins are random copolymers (Table 25.2).

Nylon 6,6 is a synthetic condensation polymer formed from hexamethylenediamine, a molecule with two amino groups, and adipic acid, which has two carboxy groups. Water is the small molecule eliminated in this condensation reaction.

$$n H_2N\!-\!(CH_2)_6\!-\!NH_2 + n OH\!-\!\overset{O}{\overset{\|}{C}}\!-\!(CH_2)_4\!-\!\overset{O}{\overset{\|}{C}}\!-\!OH \longrightarrow \left[\!-\!NH\!-\!(CH_2)_6\!-\!NH\!-\!\overset{O}{\overset{\|}{C}}\!-\!(CH_2)_4\!-\!\overset{O}{\overset{\|}{C}}\!-\!\right]_n$$
$$+ 2n H_2O$$

Both hexamethylenediamine and adipic acid contain six carbon atoms, which is why the resulting polymer is called nylon 6,6. Like proteins, nylon 6,6 contains amide linkages between monomers, so nylon 6,6 is a polyamide, too. Also like proteins, nylon 6,6 is a copolymer, although nylon 6,6 is an *alternating* copolymer (Table 25.2), not a *random* copolymer.

Dacron is the trade name for the condensation copolymer formed from ethylene glycol (a diol—a molecule with two alcohol groups) and *para*-terephthalic acid (a diacid—a molecule with two carboxy groups). The condensation reaction removes the hydrogen atom from the alcohol on one monomer and the OH group from the carboxy group on a second monomer, producing water.

$$n HOCH_2CH_2OH + n \;\overset{O}{\overset{\|}{C}}\!\!\!-\!\!\!\bigcirc\!\!\!-\!\!\!\overset{O}{\overset{\|}{C}} \longrightarrow \left(\!-\!OCH_2CH_2O\!-\!\overset{O}{\overset{\|}{C}}\!-\!\bigcirc\!-\!\overset{O}{\overset{\|}{C}}\!-\!\right)_n$$
$$+ 2n H_2O$$

The bond that forms between each ethylene glycol monomer and each *para*-terephthalic acid monomer is called an ester linkage because RCOOR′ is the general form of an ester [◄◄ Section 10.2, Table 10.2]. Polymers in which the monomers are connected by ester linkages are called *polyesters.*

Deoxyribonucleic acid (DNA; see Figure 25.8), which stores the genetic information for every known organism and has a molecular weight as large as billions of grams per mole, is a biological condensation copolymer. If your body contained anything but a tiny fraction of a mole of DNA, you would be extremely heavy! DNA is a copolymer of a five-carbon sugar called deoxyribose, four different molecular bases (adenine, thymine, guanine, and cytosine), and phosphoric acid ($H_3PO_4$). Ribonucleic acid (RNA) is a biological condensation copolymer analogous to DNA, except that RNA is made from a five-carbon sugar called ribose and the four different molecular bases are adenine, uracil, guanine, and cytosine. RNA molecules vary greatly in size. They are large molecules, but still much smaller than DNA molecules.

Sample Problem 25.2 shows how to determine the monomers in a copolymer.

**Figure 25.8**    Structure of DNA.

---

**Sample Problem 25.2**

Kevlar, a condensation polymer used in bulletproof vests, has the following general structure:

$$\left(\!-\!NH\!-\!\bigcirc\!-\!NH\!-\!\overset{O}{\overset{\|}{C}}\!-\!\bigcirc\!-\!\overset{O}{\overset{\|}{C}}\!-\!\right)_n$$

Write the structure of the two monomers that form Kevlar.

**Strategy** Identify the condensation linkages, split them apart, and then add H to one of the resulting exposed bonds and OH to the other.

**Setup** Kevlar contains an *amide* (linkage between $\overset{O}{\overset{\|}{C}}$ and N). We split the C$-$N bond, adding an H to the exposed N and an OH to the exposed C to produce an amine and a carboxylic acid, respectively.

**Solution**  To determine what the monomers are, first remove the C—N bond:

Next, add $H_2O$ as H to the N and OH to the C:

Finally, add the components of another water molecule to the open ends of the molecules—H to N and OH to C, as before:

> **Think About It**  To determine whether the monomers in the final answer are reasonable, recombine them in a condensation reaction to see if they re-form Kevlar, the polymer in the problem.

**Practice Problem A**  Nylon 6, like nylon 6,6, is a condensation polymer (a polyamide). Unlike nylon 6,6, however, nylon 6 is *not* a copolymer because it is synthesized from a single monomer called 6-aminohexanoic acid:

$$H_2NCH_2(CH_2)_3CH_2\overset{\overset{\textstyle O}{\|}}{C}OH$$

Draw the structure of at least three repeating units of nylon 6.

**Practice Problem B**  Kodel is a polymer used to make stain-resistant carpets.

Draw the structure of the monomer. Is this an additional polymer or a condensation polymer?

## Bringing Chemistry to Life

### Electrically Conducting Polymers

You have seen so far that it is possible to polymerize organic compounds with carbon-carbon double bonds (alkenes). The resulting addition polymers (e.g., polyethylene) have carbon-carbon single bonds. It is also possible, though, to polymerize organic compounds with carbon-carbon triple bonds. The simplest of these compounds (known collectively as *alkynes*) is acetylene ($C_2H_2$), which is commonly used in welding torches. The polymerization of acetylene is very similar to the addition polymerization of ethylene. That is, a free-radical initiator molecule attaches to one of the carbon atoms and breaks one of the pi bonds between the two carbon atoms in acetylene. The resulting polymer, polyacetylene, thus retains a carbon-carbon double bond:

> **Student Annotation:** The systematic name of acetylene is ethyne, but the common name *acetylene* is far more commonly used, which is somewhat unfortunate since the *-ene* ending makes it sound like it is an alkene with a double bond instead of an alkyne with a triple bond!

$$H-C\equiv C-H \quad H-C\equiv C-H \quad H-C\equiv C-H \longrightarrow \underset{\underset{\textstyle H}{|}}{-}\overset{\overset{\textstyle H}{|}}{C}=\underset{\underset{\textstyle H}{|}}{C}-\overset{\overset{\textstyle H}{|}}{C}=\underset{\underset{\textstyle H}{|}}{C}-\overset{\overset{\textstyle H}{|}}{C}=\overset{\overset{\textstyle H}{|}}{C}-$$

The generalized form of the reaction can be represented as follows:

$$n\, H-C\equiv C-H \longrightarrow \left(\!\!\overset{\overset{\textstyle H}{|}}{C}=\underset{\underset{\textstyle H}{|}}{C}\!\!\right)_{\!\!n}$$

The structure of polyacetylene contains alternating carbon-carbon single and carbon-carbon double bonds. Recall from Chapter 9 that the carbon atoms in a $C=C$ double bond are $sp^2$ hybridized and that it is the leftover $p$ orbital on each carbon atom that overlaps to form the pi bond. Each carbon atom in polyacetylene is identical (structurally and electronically), so there is extended overlap of the $p$ orbitals throughout the polymer chain. The $p$ electrons are *delocalized*—that is, they can move throughout the network of overlapping orbitals that extend the length of the polymer chain. By being able to move electrons from one end of the polymer chain to the other, polyacetylene can conduct electricity much like a wire. That is, polyacetylene is a plastic that conducts electricity!

For comparison, consider polyethylene and Teflon. Both are polyalkenes in which the polymer chain contains only carbon-carbon single bonds. Since all the electrons are localized in sigma bonds between the carbon atoms, there are no delocalized electrons available to carry charge throughout the polymer chain. The presence of electrons in *delocalized* orbitals, such as those in polyacetylene, is the key to electrical conductivity.

Sample Problem 25.3 shows how to determine the structure of an electrically conducting polymer.

## Sample Problem 25.3

Propyne ($HC\equiv CCH_3$) can be used to form an electrically conducting polymer. Draw the structure of polypropyne, showing at least three repeating units, and write the general formula for the polymer.

**Strategy** Addition polymers, whether they are synthesized from alkenes or alkynes, form via a free-radical reaction in which one pair of pi electrons in the carbon-carbon multiple bond of a monomer molecule is used to form carbon-carbon single bonds to other monomer molecules. Draw the structural formula of propyne such that the triple bond can be "opened up" to form single bonds between consecutive monomer units.

**Setup** The structural formula of propyne can be determined from the formula given and by analogy to acetylene. The only difference between propyne and acetylene is the $CH_3$ group in place of one of the hydrogen atoms:

$$H-C\equiv C-CH_3$$

By drawing three adjacent propyne molecules, we can rearrange the bonds to show three repeating units:

$$\text{---}H-C\equiv C-CH_3\text{---}H-C\equiv C-CH_3\text{---}H-C\equiv C-CH_3\text{---}$$

**Solution** The structure of polypropylene showing three monomer units is

$$
\begin{array}{ccc}
CH_3 & CH_3 & CH_3 \\
| & | & | \\
-C=C-C=C-C=C- \\
| & | & | \\
H & H & H
\end{array}
$$

The repeating unit is $-CH=C(CH_3)-$, so the general structure is

$$
\left(\!\!\begin{array}{c}
CH_3 \\
| \\
C=C \\
| \\
H
\end{array}\!\!\right)_n
$$

**Think About It** The repeating polymer unit in the general structure resembles the original monomer except that there is a carbon-carbon double bond in place of a carbon-carbon triple bond. If each of the carbon-carbon single bonds at the ends of the repeating unit in parentheses were split open, then one electron from each bond could be combined to re-form the second pi bond in the carbon-carbon triple bond of the monomer.

**Practice Problem** Draw the structures of poly(1-butyne) and poly(1-butene), showing three or more repeating units. How are they different? How are they similar? Which of them, if any, do you think will conduct electricity?

$$H-C\equiv C-CH_2CH_3 \qquad H_2C=CH-CH_2CH_3$$
$$\text{1-butyne} \qquad\qquad \text{1-butene}$$

**Checkpoint 25.1** **Polymers**

**25.1.1** Classify the following copolymer:

−B−B−B−B−B−A−A−A−A−A−B−B−B−B−B−A−A−A−A−A−

(Select all that apply.)

a) Block

b) Random

c) Graft

d) Alternating

e) All the above

**25.1.2** What feature is common to molecules that can undergo polymerization?

a) Fluorine

b) Hydrogen bonds

c) Sulfur

d) Multiple bonds

e) Lone pairs

# 25.2 Ceramics and Composite Materials

## Ceramics

The use of ceramics in the form of pottery dates back to antiquity. Along with common ceramic substances like brick and cement, modern advanced ceramics are found in electronic devices and on the exterior of spaceships. All these **ceramics** are polymeric inorganic compounds that share the properties of hardness, strength, and high melting points. Ceramics are usually formed by melting and then solidifying inorganic substances (including clays).

Most ceramic materials are electrical insulators, but some conduct electricity very well at very cold temperatures. Ceramics are also good heat insulators, which is why the outer layer of the space shuttle (shown in Figure 25.9) is made of ceramic tiles. These ceramic tiles can withstand the very hot temperatures of re-entry into the atmosphere (where temperatures can reach 1650°C), while keeping the underlying metal structural body of the orbiter (and the astronauts within) relatively cool.

Ceramics can be prepared by heating a slurry of a powder of the inorganic substance in water to a very high temperature under high pressure. This process, called **sintering,** bonds the

**Student Annotation:** The larger the lattice energy of an ionic compound, the higher the melting point [◄◄ Section 8.2, Table 8.1]. Ceramics all have large lattice energies.

Multimedia

Chemical Reactions—thermite reaction.

**Figure 25.9** Captain Wendy Lawrence inspects the ceramic tiles on the exterior of the space shuttle.

particles to each other, thus producing the finished ceramic. Sintering is a relatively easy process to perform, but the resulting particle size is irregular, so the solid may contain cracks, spaces, and imperfections. Imperfections in the interior of a solid ceramic can be impossible to detect, but they make weak spots in the substance that can cause the ceramic piece to fail. To avoid these problems, the *sol-gel process* is frequently employed for modern ceramics that are used in structural applications. The sol-gel process produces particles of nearly uniform size that are much more likely to produce a solid ceramic without gaps or cracks.

The first step in the sol-gel process is the preparation of an alkoxide of the metal or metalloid that is going to be made into the ceramic. This can be illustrated with the yttrium(III) ion, which is used to prepare a yttrium-oxygen ceramic:

$$Y + 3CH_3CH_2OH \longrightarrow Y(OCH_3CH_2)_3 + 3H^+$$
$$\text{yttrium alkoxide}$$

The resulting metal alkoxide is dissolved in alcohol. Water is then added to generate the metal hydroxide and regenerate the alcohol:

$$Y(OCH_3CH_2)_3 + 3H_2O \longrightarrow Y(OH)_3 + 3CH_3CH_2OH$$
$$\text{yttrium alkoxide} \qquad\qquad \text{yttrium hydroxide} \qquad \text{ethanol}$$

The metal hydroxide, once formed, undergoes condensation polymerization to form a chain with bridging oxygen atoms between the metal atoms:

Just as in condensation polymerization between organic molecules, a small molecule (water, in this case) is produced by the combination of atoms from the two monomers. Repeated condensation produces an insoluble chain of metal atoms connected by oxygen atoms. The polymer chain may contain branches (i.e., it may be three dimensional) because all three of the hydroxide groups attached to the central metal ion are identical (and are thus equally likely to react). The suspension of the metal oxide-hydroxide polymer is called a *gel*. The gel is carefully heated to remove the liquid, and what remains is a collection of tiny, remarkably uniformly sized particles. Sintering of material produced by the sol-gel process produces a ceramic with relatively few imperfections.

## Composite Materials

A **composite material** is made from two or more substances with different properties that remain separate in the bulk material. Each contributes properties to the overall material, though, such that the composite exhibits the best properties of each of its components. One of the oldest examples of a human-made composite material is the formation of bricks from mud and straw, a process that dates back to biblical times.

Modern composite materials commonly include reinforcing fibers, analogous to the straw in mud bricks, in a polymer matrix called a resin. Fiberglass, Kevlar, or carbon fibers can be used to give strength to the composite, whereas polyester, polyamide, or epoxy forms the matrix that holds the fibers together. These types of composites are called *polymer matrix composites*. Composites made from a metal and a ceramic, organic polymer, or another metal are called *metal matrix composites*. Carbide drill tips are made with a combination of softer cobalt and tougher tungsten carbide. Toyota has used a metal matrix composite in the engine block of some of its cars, and some bicycles are made with aluminum metal matrix composites. Composites made of a ceramic as the primary reinforcing material accompanied by organic polymers are called *ceramic matrix composites*. A common biological ceramic matrix composite is bone. Bones consist of reinforcing fibers made of collagen (a protein) surrounded by a matrix of hydroxyapatite [$Ca_5(PO_4)_3OH$].

Another type of composite material that does not fit neatly into one of these categories is a reinforced carbon-carbon composite (RCC), which consists of carbon fibers in a graphite matrix. To prevent oxidation, a silicon carbide coating is applied. RCC is used on the nose cones of missiles, but it is most commonly known as the coating on the nose of the space shuttle. It is believed that the Space Shuttle *Columbia* disaster in 2003 occurred because the ceramic tiles and RCC panel material were both damaged on liftoff by the impact of falling foam insulation. The extreme heat of reentry into the atmosphere damaged the orbiter's structure through the breaches in the ceramic tiles and the RCC panel, and the orbiter and all seven astronauts on board were lost.

Carbon fibers can be woven into fabrics and threads to be used as the structural components of vehicles and for sporting equipment such as bicycles, tennis rackets, and skateboards.

# 25.3 Liquid Crystals

*Liquid crystals* are substances that exhibit properties of both liquids, such as the ability to flow and to take on the shape of a container, and those of crystals, such as a regular arrangement of particles in a lattice. Some substances exhibit liquid crystal behavior when they are melted from a solid. As the temperature increases, the solid liquefies but retains some order in one or two dimensions. At a higher temperature, the ordered liquid becomes a more conventional liquid in which there is no consistent orientation of the molecules.

Liquids are *isotropic,* because their properties are independent of the direction of testing. Because particles in a liquid are free to rotate, there are molecules present in every possible orientation, so the bulk sample gives the same measurements regardless of the direction in which the measurements are taken or observations are made. Liquid crystals are *anisotropic* because the properties they display depend on the direction (orientation) of the measurement. A box full of pencils, all neatly aligned, is an analogy for anisotropy, because the pencils are long and narrow. Looking at the pencils end on is different than looking at them from the side. What you see depends on the direction from which you view the pencils. Similarly, what you see when you look at anisotropic molecules depends on the direction from which you view them.

Frederick Reinitzer discovered the first compound to exhibit liquid crystal behavior in 1888. Cholesteryl benzoate was observed to form an ordered liquid crystalline phase when melted, and this liquid crystalline phase became an ordinary liquid at higher temperatures:

Cholesteryl benzoate

The structure of cholesteryl benzoate is fairly rigid due to the presence of the fused rings and $sp^2$-hybridized carbon atoms. The molecule is relatively long compared to its width, too. Rigidity and this particular shape makes it possible for the cholesteryl benzoate molecules to arrange themselves in an orderly manner, in much the way that pencils, chopsticks, or tongue depressors can be arranged.

There are three different types of ordering (called nematic, smectic, and cholesteric) that a liquid crystal can adopt (see Figure 25.10). A *nematic* liquid crystal contains molecules ordered in one dimension only. Nematic molecules are all aligned parallel to one another, but there is no organization into layers or rows. A *smectic* liquid crystal is ordered in two dimensions. Smectic molecules are aligned parallel to one other and are further arranged in layers that are parallel to one another. *Cholesteric* liquid crystal molecules are parallel to each other within each layer, but each layer is rotated with respect to the layers above and below it. The layers are rotated because there are repulsive intermolecular forces between layers.

Liquid crystalline substances were curiosities for many years but have found applications in many fields. Liquid crystal displays (LCDs) in calculators, watches, and laptop computers are possible, for example, because polarized light can be transmitted through liquid crystals in one phase but not transmitted through another (Figure 25.11). Polarized light is produced by passing ordinary

**Student Annotation:** Recall from Chapter 9 that there is free rotation about carbon-carbon single bonds (i.e., bonds between $sp^3$-hybridized carbon atoms), but not about carbon-carbon double bonds (i.e., bonds between $sp^2$-hybridized carbon atoms).

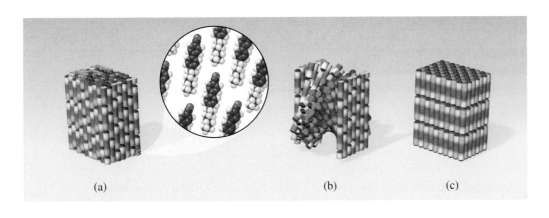

**Figure 25.10** (a) Nematic, (b) cholesteric, and (c) smectic liquid crystals.

(a)       (b)       (c)

**Figure 25.11**   Liquid crystal display. When the current is on, the liquid crystal molecules are aligned and polarized light cannot pass through. When the current is off, the molecules are not aligned and light can pass through.

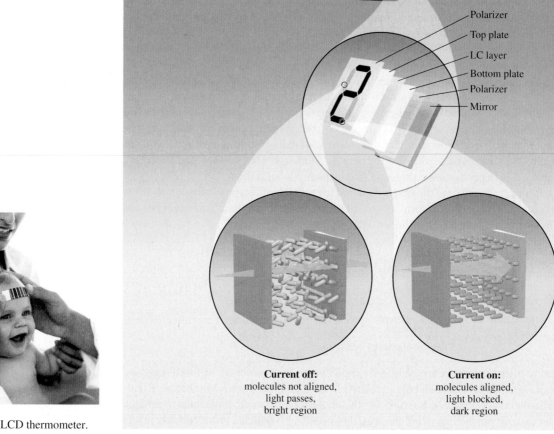

Polarizer
Top plate
LC layer
Bottom plate
Polarizer
Mirror

**Current off:**
molecules not aligned,
light passes,
bright region

**Current on:**
molecules aligned,
light blocked,
dark region

**Figure 25.12**   LCD thermometer.

light (which oscillates in all directions perpendicular to the beam) through a filter that allows only the light waves oscillating in one particular plane to pass through. This plane-polarized light can be rotated by a twisted liquid crystal such that it is allowed to pass through another filter when a voltage is applied to the liquid. If the light passes through the liquid when the voltage is off, though, the light will *not* pass through. This principle can be used in calculator and watch displays because each digit is made up of no more than the seven segments shown in Figure 25.11, and the specific digit depends on which segments are bright and which are dark. LCDs in laptop computers and televisions are more sophisticated, but still operate on the same basic principle.

Liquid crystals can also be used in thermometers (Figure 25.12). The spacing between crystal layers depends on temperature, and the wavelength of light reflected by the crystal depends on this spacing. Thus, the color reflected by the liquid crystal will change with temperature, and this color can be used to indicate the temperature to which the liquid crystal is exposed.

Sample Problem 25.4 lets you practice determining whether or not a molecule might exhibit liquid crystal properties.

**Student Annotation:** A material that changes color as the temperature changes is called *thermochromic.*

## Sample Problem 25.4

Would the following molecule make a good liquid crystal? Why or why not?

$$CH_3-\overset{\overset{\text{O}}{\|}}{C}-\!\!\!\langle\rangle\!\!\!-\overset{\overset{\text{O}}{\|}}{C}-\!\!\!\langle\rangle\!\!\!-CH_2-CH_2-CH_2-CH_2-CH_2-CH_3$$

**Strategy**  Cholesteryl benzoate exhibits liquid crystal properties because it has structurally rigid regions (fused rings and $sp^2$-hybridized carbon atoms) and because it is relatively long compared to its width. Examine the structure in question to see if it has rigid regions and to see if it is longer in one dimension than in another.

**Setup** Carbon atoms that are $sp^2$-hybridized contribute to the rigidity of a molecule's shape.

**Solution** The left-hand portion of the molecule contains an $sp^2$- hybridized carbon atom, a benzene ring, another $sp^2$-hybridized carbon atom, and another benzene ring. These features are relatively rigid and should allow the molecule to maintain its shape when heated. The chain of $CH_2$ groups ending in a $CH_3$ group is not rigid due to free rotation about the carbon-carbon single bonds (i.e., $sp^3$-hybridized carbon atoms). The overall shape of the molecule is longer than it is wide, though, which combined with a large portion that is rigid should make the molecule a good candidate for liquid crystal behavior.

> **Think About It** Double bonds in a molecular structure indicate the presence of pi bonds. Recall that it is the pi bonds that restrict rotation about bonds in a molecule [◄◄ Section 9.5] and lend rigidity to the structure.

**Practice Problem A** Which compound would you expect to exhibit liquid crystal behavior, and why?

**Practice Problem B** This compound does not possess liquid crystal properties. What changes are needed (for example, addition of a group) to convert it to a liquid crystal?

---

**Checkpoint 25.3** **Liquid Crystals**

**25.3.1** Which of the following is a good analogy for anisotropy? (Select all that apply.)

a) Glass of water

b) Bucket of ice cubes

c) Box of spaghetti

d) Box of rice

e) Box of macaroni

**25.3.2** What characteristics make a molecule likely to exhibit liquid crystal properties? (Select all that apply.)

a) Long narrow shape

b) Flexibility

c) Rigidity

d) Low molar mass

e) High molar mass

---

# 25.4 Biomedical Materials

Many modern materials are finding uses in medical applications. Replacement joints, dental implants, and artificial organs all contain modern polymers, composites, and ceramics. To function successfully in a biomedical application, though, a material must first be compatible with the living system. The human body very easily recognizes foreign substances and attacks them in an effort to rid them from the body. Thus, a biomaterial must be designed with enough similarity to the body's own systems that the body will accept the material as its own. Additionally, if the polymer, ceramic, or composite contains leftover chemicals from its manufacture, these contaminants may lead to adverse reactions with the body over the lifetime of the medical implant. It is important, therefore, that the substance be as pure as possible.

The physical properties of a biomedical material are important, too, because the longer the material lasts, the less often the medical device has to be replaced during the lifetime of the patient. Most biomedical materials must possess great strength and flexibility to perform in the body. For example, the materials in artificial joints and heart valves must be able to flex many times without breaking. Materials used in dental implants must show great hardness, moreover, so as not to crack during biting and chewing. It takes years of research to develop successful biomedical materials that meet these needs.

## Dental Implants

Dental implant materials have been used for many years. The oldest dental fillings were made of various materials including lead (which fell out of favor due to its softness—the danger of lead poisoning was as yet unknown), tin, platinum, gold, and aluminum. The remains of Confederate soldiers have been shown to include fillings made of a lead-tungsten mixture, probably from shotgun pellets; tin-iron; a mercury amalgam; gold; and even radioactive thorium. (The dentist probably thought he was using tin, which is similar in appearance.) Many modern fillings are made of dental *amalgam* [◄◄ Chapter 19], a solution of several metals in mercury. Modern dental amalgam consists of 50 percent mercury and 50 percent of an alloy powder that usually contains (in order of abundance) silver, tin, copper, and zinc. These metals tend to expand slightly with age, causing fissures and cracks in the tooth that may require further intervention (e.g., crowns, root canals, or tooth replacement). Some people consider amalgam fillings to be unsightly, too.

Dental composite materials are now used that have several advantages over traditional amalgams. The composites can be made in a wide range of colors, for example, to match the color of the other teeth. The existing healthy portion of the tooth can be etched with acid, moreover, to create pores into which the composite material can bond. With traditional dental amalgam, the dentist must create indentations in the healthy tooth to hold the amalgam in place. Destroying a portion of the healthy tooth is undesirable and can be avoided with the use of the composite.

The dental composite consists of a matrix (made from a methacrylate resin) and a silica filler. The composite material is applied to the cavity in a putty-like consistency and then is dried and polymerized by a photochemical reaction initiated by light of a particular wavelength. Because the light does not penetrate very far into the composite, the thickness of the applied composite is critical. If the layer of composite is too thick, some of the composite will remain soft. A properly constructed dental composite filling will last more than 10 years. Over time, though, composite fillings tend to shrink, leaving breaches that can cause leakage, a situation that must be addressed to prevent further tooth decay.

Porcelain ceramic fillings and crowns are very common, but the ceramic has two disadvantages: it is very hard and can wear on the opposing teeth, and it is brittle and may crack if subjected to great force. Most dentists, therefore, do not recommend porcelain ceramic crowns and fillings for the molar teeth, which do the bulk of the grinding work during chewing.

A material used in dentistry must have properties that maximize both patient comfort and the lifetime of the implant. Dental fillings and tooth replacements must be resistant to acids, for example, because many foods (such as citrus fruits and soft drinks) contain acids directly. Any food containing carbohydrates can produce acids, though, if traces are left in the mouth because the bacteria that reside there consume carbohydrates, producing acids in the process. These acids eat away at the natural hydroxyapatite in teeth, creating caries (cavities). Materials like dental amalgam, gold, and dental composite resist attack by acids in the mouth.

A dental material should also have low thermal conductivity; that is, it should conduct heat poorly. This is especially important in applications where the implant material is in contact with the nerve inside the tooth. If the implant material conducts heat well, then whenever hot or cold foods come in contact with the implant, the hot or cold will be transmitted through the material to the nerve, causing pain. Metals are good thermal conductors, so dental amalgams must not touch nerves.

Finally, dental materials should resist wear (so they last a long time) and they should resist expansion and contraction as the temperature fluctuates. The close spacing and precision fits involved in dental fillings and implants would result in discomfort and possibly failure of the implant if the material expanded or contracted appreciably with changes in temperature.

## Soft Tissue Materials

Burn victims who have lost large amounts of skin do not have enough cells to grow new skin, so a synthetic substitute must be used. The most promising artificial skin material is based on a polymer of lactic acid and glycolic acid. Both of these compounds contain an alcohol group

(−OH) and a carboxy group (−COOH), so they can form a polyester copolymer in a condensation reaction that mirrors the formation of Dacron polyester (Section 25.1):

$$\underset{\text{glycolic acid}}{\text{HOCH}_2\overset{\overset{\displaystyle O}{\|}}{\text{C}}}\boxed{\text{OH  H}}\underset{\text{lactic acid}}{-\text{OCH}\overset{\overset{\displaystyle O}{\|}}{\text{C}}-\text{OH}}\underset{\underset{\text{CH}_3}{|}}{\phantom{-}}\longrightarrow\left(-\text{OCH}_2\overset{\overset{\displaystyle O}{\|}}{\text{C}}-\text{OCH}\overset{\overset{\displaystyle O}{\|}}{\text{C}}-\underset{\underset{\text{CH}_3}{|}}{\phantom{}}\right)_n + \text{H}_2\text{O}$$

This copolymer forms the structural mesh that supports the growth of skin tissue cells from a source other than the patient. Once the skin cells grow on the structural mesh, the artificial skin is applied to the patient, and the copolymer mesh eventually disappears as the ester linkages are hydrolyzed. During this treatment, the patients must take drugs that will suppress their immune systems, which ordinarily would attack these substances as foreign to prevent their bodies from rejecting the new skin.

**Student Annotation:** Hydrolysis is essentially the opposite of a condensation reaction.

*Sutures,* commonly known as stitches, are now made of the same lactic acid–glycolic acid copolymer as the structural mesh of artificial skin. Before, sutures were made of materials that did not degrade so they eventually had to be removed again. This type of suture is still used in cases where the degrading suture is deemed a risk or there might be a need to reopen the incision at a later time. Biodegradable sutures are routinely used during operations on animals, too, which decreases the number of trips to the veterinarian, thereby reducing the stress on the pet and its owner.

When portions of blood vessels need to be replaced or circumvented in the body, it may or may not be possible to obtain a suitable vascular graft from another part of the patient's body. Dacron polyester (Section 25.1) can be woven into a tubular shape for this purpose, and pores can be worked into the material so that the graft can integrate itself into the tissue around it, as well as to allow capillaries to connect. Unfortunately, the body's immune system recognizes this as a foreign substance and blood platelets stick to the walls of the Dacron graft, causing clotting.

Artificial hearts and heart valves are potential solutions to the shortage of donor organs. Artificial hearts have not yet been perfected to the point that they can work on a permanent basis, but valve replacement is a common procedure that can successfully prolong a patient's life. The most common replacement heart valve is fixed in position by surrounding the valve with lactic acid–glycolic acid copolymer, the same polyester that is used in skin grafts and biodegradable sutures. The polymer is woven into a mesh ring that allows the body tissue to grow into the ring and hold it in place.

Researchers today are using *electrospinning* to place nanofibers directly onto wounds. (Nanofibers are products of nanotechnology; see Section 25.5.) Electrospinning uses a high-voltage electric field to create a jet of polymer solution that transforms into a spiraling dry fiber only nanometers in diameter. The wound dressings applied by direct electrospinning have high rates of moisture transmission and bacterial resistance. Enzymes can be added to the solution containing the nanofiber polymer, and these enzymes can be released into the skin after the dressing is applied. Proteins can also be cross-linked onto the nanofiber, making possible many applications of the nanofiber wound dressing.

## Artificial Joints

Joint replacements, such as artificial knees, elbows, and hips, require major surgery but have become relatively commonplace because suitable materials have become available for not only the bone replacement but also the contact surfaces. In knee replacement, for example, the goal is usually to replace worn-out cartilage surfaces (which may have resulted from arthritis or injury). It is possible to replace shattered bones that cannot heal back to their original state as well.

The materials originally used in joint replacements were ivory, metals, and glass, but they are fast being replaced by polymeric plastics and ceramics. Polymethyl methacrylate (PMMA), which was the subject of Practice Problem 25.1B in Section 25.1, and polyethylene have been used in many total joint replacements. However, these joints tend to fail over time when they are implanted in younger, more active patients, so efforts are under way to improve the wear quality of the PMMA and polyethylene parts.

Although metallurgical research provided new substances that were much stronger and less likely to break, including titanium-based alloys, total joint replacements were predicted to last no more than 20 years in a patient before needing to be replaced again. Ultrahigh-molecular-weight polyethylene, treated with radiation to initiate cross-linking of the polymer chains and stored under inert nitrogen gas to stop free-radical damage to the chain from the radiation, is now commonly used in many replacement joints, with the expectation of improved wear. It remains to be seen whether ultrahigh-molecular-weight polyethylene will prolong the life of these devices, since many of the joints have been in service in patients for less than a decade.

Other surfaces have also been researched to reduce wear. Metal-on-metal bearing surfaces suffered early setbacks due to the manufacturing tolerances of the cobalt-based alloy used. As the bearing surfaces wore, they produced metal sludge, and high friction between components caused loosening of the joint. Modern metal-on-metal bearing surfaces have improved as the manufacturing has become more consistent and produced components lower in friction.

Ceramic-on-ceramic bearing surfaces have also been developed. Some ceramics have been prone to cracking, but many have shown great promise for long-lasting use. When a ceramic component in a replacement joint does crack, however, it can be disastrous because the entire joint may need to be replaced and further healthy bone may need to be sacrificed to install the new replacement joint.

Despite the progress made so far on developing safe, durable, and useful artificial joints, much work remains to be done. One long-term issue that has yet to be researched extensively, for example, is the effect of the release of metal cations from the joint into the body over the lifetime of the joint replacement. This may become a more urgent health issue as the lifetimes of the artificial joints increases and the ages at which patients get the replacements decreases.

## 25.5 Nanotechnology

Atoms and molecules have sizes on the order of a tenth of a nanometer (where 1 nm $= 10^{-9}$ m) [◄◄ Section 2.2]. *Nanotechnology* is the development and study of such small-scale materials and objects. An atomic force microscope [Figure 25.13(a)] is one way to image the surface of a material at the atomic or molecular level.

A sharp-pointed stylus is moved over the surface of the substance under study, and the intermolecular forces between the material and the stylus force the stylus up and down. The stylus is mounted on a probe that reflects a laser beam into a detector. From the differences in deflection of the laser beam, the electronics attached to the probe can create an image of the surface of the object [Figure 25.13(b)].

The scanning tunneling microscope (STM) works on a similar principle but only for samples that conduct electricity. The STM measures the peak and valley heights of the sample from the flow of electric current. The tip of the STM must move up and down to keep the current constant, and the up-and-down movements are translated into an image of the atomic and molecular structure of the substance [Figure 25.13(c)].

### Graphite, Buckyballs, and Nanotubes

**Student Annotation:** Recall that *allotropes* are different forms of the same element [◄◄ Section 2.6].

Carbon exists as several allotropes, one of which is graphite. Graphite consists of sheets of carbon atoms that are all $sp^2$-hybridized [◄◄ Section 9.4]:

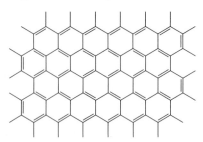

Intermolecular forces [◄◄ Chapter 11] hold the sheets together. Because of the delocalization of the electrons in the extended network of $p$ orbitals perpendicular to the graphite sheet, graphite is an electrical conductor within its sheets, but not between them. Electrical conductivity is much more common for metals than it is for nonmetals like graphite.

Graphite and diamond were long thought to be the only two allotropes of carbon. In 1985, though, a compound consisting of 60 carbon atoms was isolated from graphite rods that had been vaporized with an intense pulse of laser light. The structure of the compound ($C_{60}$) was identical to

(a)                              (b)                              (c)

**Figure 25.13**  (a) Atomic force microscope (AFM). (b) AFM image of a yeast cell. (c) STM image of iron atoms arranged to display the Chinese characters for *atom* on a copper surface.

that of a soccer ball, with 12 pentagonal and 20 hexagonal faces and carbon atoms at the corners of each face:

This pattern also resembles an architectural structure called a geodesic dome, so the $C_{60}$ molecule was named *buckminsterfullerene* in honor of R. Buckminster Fuller, the designer of the geodesic dome. The nickname for $C_{60}$ is the *buckyball.*

Since the discovery of $C_{60}$, elongated and elliptical cages of 70 and 80 carbon atoms have also been discovered. These molecules, called *fullerenes* as a group, have been proposed to exist in such exotic places as stars and interstellar media and have been observed in such mundane places as deposits of chimney soot.

A single sheet of graphite is known as *graphene.* A graphene is essentially a two-dimensional carbon crystal consisting of an array of six-membered rings. Graphenes are extraordinarily strong (hundreds of times stronger than steel), they conduct electricity, and their shapes can be customized by the introduction of five-membered and seven-membered rings at specific locations. Because of their unique combination of properties, graphenes currently are the subject of a great deal of research. Proposed uses of graphenes include replacement of copper interconnects in integrated circuits, production of transparent conduction electrodes—useful for touchscreens and liquid crystal displays—and ultrahigh-sensitivity detection devices.

Unless otherwise constrained, graphene sheets will wrap around on themselves forming cylinders with diameters on the scale of nanometers. Such cylinders are called a *carbon nanotubes.* Two types of nanotube structures are known, as shown in Figure 25.14. The differences in arrangement of hexagons give the two kinds of nanotubes different electrical conductivities.

Carbon nanotubes of different diameters can nest inside each other, producing multiple-walled nanotubes. Most nanotubes are single-walled, though, and they can be as narrow as 1 nm in diameter. These nanotubes are remarkably strong—stronger, in fact, than a comparably sized piece of *steel*! Because nanotubes consist only of carbon atoms, they can come into close contact, thereby allowing intermolecular forces to hold large numbers of tubes together. This, in turn, makes it possible to align groups of tubes, potentially forming long strands and fibers. Lightweight fibers and strands with tremendous strength have many applications in such things as auto parts, medical implants, and sports equipment.

Besides carbon, other elements have been fashioned into nanotubes, too. Molybdenum sulfide nanotubes have been synthesized, for example, with openings between atoms that are just the right size to permit some gas molecules to pass through, but not others. This may have potential for the nanoscale storage of gases for fuel cells.

**Student Annotation:** An analogy is to take a piece of paper and roll it so that you have a tube.

**Figure 25.14**    Carbon nanotubes.

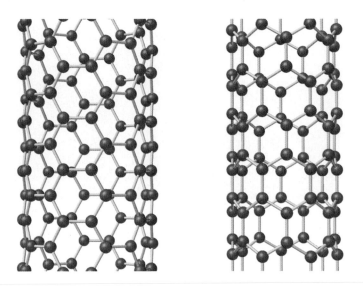

Many universities have developed large-scale facilities for nanotechnology research in collaboration with industrial partners. Since the lower limits of size have been reached in the production of silicon chip–based electronic circuits, nanotechnology may provide the pathway to circuits that are much smaller than have ever been imagined.

## 25.6 Semiconductors

Recall from Section 9.6 that bonds between atoms can be described as resulting from the combination of *atomic* orbitals to form *molecular* orbitals. In a bulk sample of metal containing many, many atoms, there are many, many molecular orbitals. These molecular orbitals are so close in energy that instead of forming individual bonding and antibonding orbitals for each pair of atomic orbitals that combine, they form a band of bonding levels and a band of antibonding levels. The energies of the bonding and antibonding bands depend on the energies of the atomic orbitals that combined to form them in the first place. As a result, the band structure of a bulk sample depends on the original atoms' energy levels. The band energies and the gaps (or lack of gaps) between the bonding band (called the ***valence band***) and the antibonding band (called the ***conduction band***) make it possible to classify a substance as an electrical conductor, a semiconductor, or an electrical insulator based on the behavior of the electrons in the bands (Figure 25.15).

The valence band is filled with the valence (bonding) electrons, whereas the conduction band is empty or only partially filled with electrons. It is the conduction band that allows electrons to move between atoms. The gap between the valence band and the conduction band can vary from nothing (for an electrical conductor) to a small amount (for a semiconductor) to a very large amount (for an electrical insulator). Thus, the size of the band gap determines the conduction behavior of the material.

**Figure 25.15**    Energy bands in metals (conductors), semiconductors, and insulators.

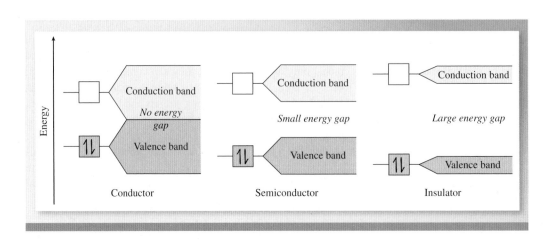

Metals have no band gap, so they are good conductors of electricity. The valence band and conduction band actually overlap or are one and the same in metallic conductors, so there is no energy barrier to the movement of electrons from one atom to another in a metal. ***Semiconductors*** have a band gap, but it is relatively small, so there is limited movement of electrons from the valence band to the conduction band. ***Nonconductors*** (electrical insulators) have large band gaps, so it is nearly impossible to promote electrons from the valence band to the conduction band.

Silicon, germanium, and carbon in the form of graphite are the only elemental semiconductors at room temperature. All three of these elements are in Group 4A and have four valence electrons. (Tin and lead are also in Group 4A, but they are metals, not semiconductors.) Other semiconductors consist of combinations of elements whose valence electron count totals 8. For example, gallium (Group 3A) and phosphorus (Group 5A) form a semiconductor because gallium contributes three valence electrons and phosphorus contributes five, giving a total of eight valence electrons. Semiconductors have also been formed between Group 2B elements (particularly Zn and Cd) and Group 6A elements.

Sample Problem 25.5 lets you practice identifying combinations of elements that can exhibit semiconductor properties.

**Student Annotation:** The group number for a main group element corresponds to the number of valence electrons that the atom has [◄◄ Section 8.5].

## Sample Problem 25.5

State whether each of the following combinations of elements could form a semiconductor:
(a) Ga-Se, (b) In-P, (c) Cd-Te.

**Strategy** Count the valence electrons in each type of atom. If they total eight for the two elements, then the combination will probably form a semiconductor.

**Setup** Locate each element in the periodic table. All of them are main group elements, so we can use their group numbers to directly determine the number of valence electrons each element contributes: (a) Ga is in Group 3A, and Se is in Group 6A. (b) In is in Group 3A, and P is in Group 5A. (c) Cd is in Group 2B, and Te is in Group 6A.

**Solution** (a) Ga has three valence electrons, and Se has six. This gives a total of nine, which is too many to form a semiconductor.

(b) In (three valence electrons) and P (five valence electrons) combine for a total of eight, so In-P should be a semiconductor.

(c) Cd (two valence electrons) and Te (six valence electrons) also combine for a total of eight, so Cd-Te should be a semiconductor, too.

**Think About It** Semiconductors consisting of Group 3 and Group 5 elements are used in the contact layers of light-emitting diodes (LEDs).

**Practice Problem A** Name an element in each case that could be combined with each of the following elements to form a semiconductor: (a) O, (b) Sb, (c) Zn.

**Practice Problem B** For each of the elements C, As, and Ga, indicate which can be combined with silicon to produce a semiconductor, which can be added to silicon as a dopant to produce a *p*-type semiconductor, and which can be added to silicon as a dopant to produce an *n*-type semiconductor.

The conductivity of a semiconductor can be enhanced greatly by ***doping,*** the addition of very small quantities of an element with one more or one fewer valence electron than the natural semiconductor. For the purpose of simplicity, we will consider a pure silicon semiconductor. Silicon is a Group 4A element, so it has four valence electrons per atom. A small (parts per million) amount of phosphorus (Group 5A, five valence electrons) can be added, thus doping the silicon with phosphorus. Since each phosphorus atom has an "extra" electron relative to the pure semiconductor, these extra electrons must reside in the conduction band, where they increase its conductivity. This type of doped semiconductor is called an ***n-type semiconductor*** because the semiconductivity has been enhanced by the addition of negative particles, the extra electrons.

It is also possible to dope a semiconductor with an element that has fewer valence electrons than the semiconductor itself. For example, a silicon semiconductor could be doped with small amounts (again, parts per million) of gallium (Group 3A, three valence electrons). Now, instead of an "extra" electron, there is a "hole" (one less electron) in every place that a gallium atom has replaced a silicon atom. These holes are effectively positive charges (because each is the absence of an electron), so this type of material is called a ***p-type semiconductor.*** A *p*-type semiconductor

**Figure 25.16**    The effects of doping on the conduction of semiconductors.

**Figure 25.17**    LED arrays in traffic signals. Many communities have replaced the red and green signals with LED arrays that are much more energy efficient than the incandescent bulbs used previously. The yellow lights are being replaced more slowly, however, because LED arrays are expensive and the short duration of yellow lights would not necessarily allow an overall cost saving when the energy efficiency is factored in.

has increased conductivity because the holes (which are in the valence band) move through the silicon rather than electrons. The energy required to move an electron from the valence band into a hole (also in the valence band) is considerably less than the energy needed to promote an electron from the valence band to the conduction band of a semiconductor (Figure 25.16). Thus, the movement of the holes results from the movement of charge, and conductivity is the movement of charge.

A combination of *p*-type and *n*-type semiconductors can be used to create a solar cell in which the holes and extra electrons in the semiconductors are at equilibrium in the dark. At equilibrium, the positive holes in the *n*-type layer have offset the movement of electrons into the *p*-type layer. Light falling on the solar cell causes the electrons that have entered the *p*-type layer to go back to the *n*-type layer, and other electrons to pass through a wire from the *n*-type layer to the *p*-type layer. This movement of electrons constitutes an electric current.

***Diodes,*** which are electronic devices that restrict the flow of electrons in a circuit to one direction, work in essentially the opposite direction from solar cells. A light-emitting diode (LED) consists of *n*-type and *p*-type semiconductor layers placed in contact. When a small voltage is applied, the "extra" electrons from the *n*-type side combine with the holes in the *p*-type side, thus emitting energy (light) whose wavelength depends on the band gap. The band gap is different for different semiconductors (e.g., Ga-As versus Ga-P), so the specific combination of semiconductor materials can be chosen to make the band gap correspond to a desired color of light. LEDs are finding increasing use in applications previously dominated by traditional incandescent lightbulbs such as emergency exit signs, car tail lights, and traffic signals (Figure 25.17).

---

**Checkpoint 25.6**    **Semiconductors**

**25.6.1**    Semiconductors are substances that

a) have a large band gap.

b) have a small band gap.

c) have no band gap.

d) do not conduct electricity.

e) conduct electricity only when combined with other elements.

**25.6.2**    Elements that are semiconductors at room temperature are

a) members of Group 2A.

b) members of Group 3A.

c) members of Group 4A.

d) combinations of elements from Group 3A and Group 5A.

e) combinations of elements from Group 2A and Group 6A.

---

## 25.7    Superconductors

No ordinary electrical conductor is perfect. Even metals have some resistance to the flow of electrons, which wastes energy in the form of heat. ***Superconductors*** have no resistance to the flow of electrons and thus could be very useful for the transmission of electricity over the long distances between power plants and cities and towns. The first superconductor was discovered in 1911 by the Dutch physicist H. Kamerlingh-Onnes. He received the 1913 Nobel Prize in Physics for showing that mercury was a superconductor at 4 K, the temperature of liquid helium. Liquid helium is very expensive, however, so there were few feasible applications of a mercury superconductor.

Since that time, more superconductors have been discovered including a lanthanum-, barium-, copper-, and oxygen-containing *ceramic* compound, and a series of copper oxides ("cuprates"), all of which become superconducting below 77 K, the temperature of liquid nitrogen (a much more common and less expensive refrigerant than liquid helium). The temperature below which an element, compound, or material becomes superconducting is called the *superconducting transition temperature*, $T_c$. The higher the $T_c$, the more useful the superconductor.

By 1987, superconductivity was observed in another ceramic, yttrium barium copper oxide (YBCO) ($YBa_2Cu_3O_7$). Commonly called a "1-2-3 compound" because of the 1:2:3 ratio of yttrium to barium to copper, YBCO has a $T_c$ of 93 K (above the temperature of liquid nitrogen). Although 93 K is still cold, it qualifies as quite warm in the world of superconductors, so YBCO is considered to be the first *high-temperature superconductor*. YBCO was first prepared by combining the three metal carbonates at high temperature (1000 to 1300 K). The copper ions are present in a mixture of +2 and +3 oxidation states:

$$4BaCO_3 + Y_2(CO_3)_3 + 4CuCO_3 + Cu_2(CO_3)_3 \longrightarrow 2YBa_2Cu_3O_7 + 14 CO_2$$

YBCO must be sintered carefully to obtain the optimal superconductivity.

Superconducting wires were first made from $Bi_2Sr_2CaCu_2O_8$, known as BSCCO-2212, a high-temperature superconductor with $T_c = 95$ K. It can be prepared in wire forms because it has layers of bismuth and oxygen atoms that YBCO does not have. BSCCO-2212 is so named because its formula consists of two bismuth, two strontium, one calcium, and two copper atoms. Other BSCCO compounds exist, too, and they all have a general formula of $Bi_2Sr_2Ca_nCu_{n+1}O_{2n+6}$. BSCCO-2223, for example, which has the formula $Bi_2Sr_2Ca_2Cu_3O_{10}$ (where $n = 2$), is a superconductor with $T_c = 107$ K.

Superconductors at or below $T_c$ exhibit the *Meissner effect*, the exclusion of magnetic fields. In Figure 25.18, for example, a magnet is shown levitating above a YBCO pellet, a superconductor at or below −180°C, its $T_c$.

Magnetic levitation is being researched for use in trains, because the amount of friction between the train and the tracks would be drastically reduced in a magnetically levitated ("maglev") train versus one running on ordinary tracks and wheels. The engineering challenge to a maglev train is that the superconductor must be kept at or below its $T_c$ for the magnetic levitation to occur, and the highest temperatures obtained for superconducting substances are approximately 140 K—warm on the superconductivity scale but still very cold by ordinary standards.

Superconductivity in metals can be explained satisfactorily by BCS theory, first proposed by John Bardeen, Leon Neil Cooper, and John Robert Schrieffer in 1957. They received the Nobel Prize in Physics in 1972 for their work. BCS theory treats superconductivity using quantum mechanical effects, proposing that electrons with opposite spin can pair due to fundamental attractive forces between the electrons. At temperatures below $T_c$, the paired electrons resist energetic interference from other atoms and experience no resistance to flow. Superconductivity in ceramics has yet to be satisfactorily explained.

**Student Annotation:** Allex Müller and Georg Bednorz were awarded the 1987 Nobel Prize in Physics "for their important breakthrough in the discovery of superconductivity in ceramic materials."

**Student Annotation:** The decomposition of a metal carbonate to the metal oxide and carbon dioxide gas is a very common reaction.

**Student Annotation:** Many a Nobel Prize has been won by scientists who research and develop modern materials!

**Student Annotation:** You might be skeptical that two negatively charged electrons would attract each other, but there is evidence that all subatomic particles do exert some attractive forces on each other.

**Figure 25.18** The Meissner effect. A magnet levitates above a YBCO pellet at the temperature of liquid nitrogen (−196°C).

# Applying What You've Learned

Now that you know more about polymers, we can apply some of the concepts you've learned to ring-opening metathesis polymerization (ROMP), a polymerization reaction that utilizes the metathesis reaction introduced at the beginning of the chapter. The general mechanism for ROMP is as follows:

The pi bond from the carbon-carbon double bond in the ring of the monomer reacts with the metal-carbon double bond in the transition metal catalyst (M=CHR) to first form a metallacyclobutane complex. The metallacyclobutane complex then breaks up to form a new carbon-carbon double bond and a new metal-carbon double bond. The new carbon-carbon double bond is the first in the polymer chain, and the new metal-carbon double bond is available to react with additional monomers and thus grow the chain. As long as the metal from the catalyst is attached to the polymer via this metal-carbon double bond, the polymer is said to be "living" because it will continue to grow as more monomer is added. The process we will consider is the polymerization of two cyclic alkenes (cyclobutene and 3-chlorocyclobutene) using ROMP with an appropriate transition metal catalyst (abbreviated as M=CHR):

Cyclobutene    3-Chlorocyclobutene

**Problems:**

a)  What is the formula of the polymer formed by polymerizing first fifty cyclobutene molecules and then seventy 3-chlorocyclobutene molecules? [◄◄ Sample Problem 25.1]

b)  What type of polymer (see Table 25.2) is this? [◄◄ Sample Problem 25.2]

c)  Would this polymer conduct electricity? [◄◄ Sample Problem 25.3]

# CHAPTER SUMMARY

## Section 25.1

- *Polymers* are made up of many repeating units, called *monomers,* that are linked together by covalent bonds.

- *Thermoplastic* polymers can be melted and reshaped, whereas *thermosetting* polymers assume their final shape as part of the chemical reaction that forms them in the first place.

- *Addition polymers* are formed via a radical reaction in which a pi bond in an alkene or alkyne is opened up to form new bonds to adjacent monomers.

- Polymer chains can be *cross-linked* to provide added durability and flexibility.

- *Elastomers,* such as natural rubber, can stretch and bend and then return to their original shape.

- *Copolymers* contain more than one type of monomer. Copolymers can be random, alternating, block, or graft, depending on the sequence of the different monomers and the way that the monomers are attached in the polymer.

- *Tacticity* describes the relative geometric arrangements of the groups attached to the chiral carbon atoms in vinyl polymers. In *isotactic* polymers, all the groups attached to the chiral carbon atoms have identical arrangements. In *syndiotactic* polymers, they alternate positions along the polymer chain. In *atactic* polymers, they have random arrangements.

- *Condensation polymers* form when two different groups on monomers react, thus producing a covalent bond between monomers and a small molecule (usually water) that is expelled. The two reacting groups may be on the same monomer (as in the amino acids that form polypeptides and proteins), or they may be on two different monomers (as in hexamethylenediamine and adipic acid, which form nylon 6,6). Most condensation polymers are really copolymers.

- Amines and organic acids form condensation polymers known as *polyamides.* Alcohols and organic acids form condensation polymers known as *polyesters.* Amino acids form polyamide polymers known as polypeptides. Long, straight-chain polypeptides are known as proteins.

- DNA and RNA are condensation polymers consisting of sugars, organic bases, and phosphate ions.

## Section 25.2

- *Ceramics* are polymeric inorganic compounds that are hard and strong and have high melting points.

- Ceramics are *sintered* to bond the particles to each other and to close any gaps that may exist in the material.

- The sol-gel process is used to prepare ceramics for structural applications because it produces particles of nearly uniform size that are much more likely to produce a solid ceramic without gaps or cracks.

- *Composite materials* contain several components each of which contributes properties to the overall material. Modern composites include reinforcing fibers and a surrounding matrix. The reinforcing fibers may be made of polymers, metals, or ceramics.

## Section 25.3

- *Liquid crystals* are substances that maintain an ordered arrangement in the liquid phase. Molecules containing structurally rigid regions (e.g., rings and multiple bonds) and molecules that are longer in one dimension than in another seem to form the best liquid crystals. Although ordinary liquids are *isotropic,* liquid crystals are *anisotropic,* meaning that their apparent properties depend on the direction from which we view them.

- Liquid crystals may be *nematic, smectic,* or *cholesteric,* depending on the type of ordering the molecules can adopt. Nematic liquid crystals contain order in only one dimension (parallel molecules), smectic liquid crystals contain order in two dimensions (parallel molecules in parallel layers), and cholesteric liquid crystals contain layers that are rotated with respect to one another. Within each cholesteric layer, the molecules are parallel.

## Section 25.4

- Dental amalgam consists of several metals dissolved in an alloy with mercury. Dental composite materials consist of a methacrylate polymer in a silica matrix. The composite can be varied in color to match existing teeth, making the repair more aesthetically pleasing than metallic amalgam.

- Biodegradable *sutures* and artificial skin are made of synthetic polyesters that support the growth of tissue around the polymer and then hydrolyze into the body.

- Artificial joint replacements can be made of metal, ceramics, and organic polymers.

## Section 25.5

- *Nanotechnology* is the study and development of objects with sizes on the order of a nanometer (one billionth of a meter).

- Pure carbon exists as diamond, graphite, *fullerenes,* and *carbon nanotubes.*

- Two-dimensional sheets of $sp^2$-hybridized carbon atoms are called *graphenes.*

## Section 25.6

- Electrical conductors (e.g., metals) have no gap between their *valence band* and *conduction band,* so electrons are easily promoted from the valence to the conduction band. Semiconductors, which conduct electricity more than nonconductors (insulators) but less than metallic conductors, have a small gap. The gap in *nonconductors* is too large for electrons to be promoted from the valence to the conduction band.

- *Semiconductors* generally have a total of eight valence electrons per formula unit. Thus, they typically form between two Group 4A elements, one Group 3A element and one Group 5A element, or one Group 2B element and one Group 6A element.

- Semiconductors can be *doped* with an element with fewer valence electrons to produce a *p-type semiconductor,* or with an element with more valence electrons to produce an *n-type semiconductor.*

- *p*-Type and *n*-type semiconductors can be used together in solar cells and *diodes.*

## Section 25.7

- *Superconductors* offer no resistance to the flow of electrons. The earliest known superconductors only exhibited the behavior at liquid helium temperature (4 K), but modern *high-temperature superconductors* conduct at temperatures as high as 140 K.

- The highest temperature at which a superconductor exhibits superconductivity is called the *superconducting transition temperature* ($T_c$).

- Superconducting substances exhibit the *Meissner effect,* the exclusion of magnetic fields in the substance's volume. This can be demonstrated by levitating a magnet above a superconductor.

# KEY WORDS

Addition polymerization, 1000

Anisotropic, 1011

Atactic, 1005

Carbon nanotube, 1017

Ceramic, 1009

Cholesteric, 1011

Composite material, 1010

Condensation polymer, 1005

Conduction band, 1018

Copolymer, 1003

Cross-link, 1001

Diode, 1020

Doping, 1019

Elastomer, 1001

Fullerene, 1017

Graphene, 1017

High-temperature superconductor, 1021

Isotactic, 1005

Isotropic, 1011

Liquid crystal, 1011

Meissner effect, 1021

Monomer, 1000

Nanotechnology, 1016

Nematic, 1011

Nonconductor, 1019

$n$-Type semiconductor, 1019

Polyamide, 1006

Polyester, 1006

Polymer, 1000

$p$-Type semiconductor, 1019

Semiconductor, 1019

Sintering, 1009

Smectic, 1011

Superconducting transition temperature ($T_c$), 1021

Superconductor, 1020

Suture, 1015

Syndiotactic, 1005

Tacticity, 1005

Thermoplastic, 1000

Thermosetting, 1000

Valence band, 1018

# QUESTIONS AND PROBLEMS

## Section 25.1: Polymers

### Review Questions

25.1   Bakelite, the first commercially produced polymer, contains monomer units of phenol and formaldehyde. If an item made of Bakelite were broken, it could not be melted down and re-formed. Is Bakelite a thermoplastic or thermosetting polymer? Can Bakelite be recycled?

25.2   How are SBS rubber [Figure 25.6(a)] and nylon 6,6 similar? How are they different?

### Problems

**25.3**   What type of monomer unit is needed to make an electrically conducting organic polymer?

25.4   Draw the structure of the addition polymer formed by 1-butene:

$$H_2C = \overset{\overset{\displaystyle H}{|}}{C} - \underset{\underset{\displaystyle H_2}{|}}{C} - CH_3$$

Show at least two repeating units.

**25.5**   Draw the structure of the alternating copolymer of these two compounds, showing at least two repeating units.

$$\underset{H_3C}{\overset{H_3C}{\diagdown}} C = C \underset{CH_3}{\overset{CH_2}{\diagup}} \qquad H - C \equiv C - H$$

25.6   Draw the structure of an alternating copolymer of these two amino acids, showing at least two repeating units:

$$H_2N - \underset{\underset{\displaystyle SH}{\underset{\displaystyle |}{CH_2}}}{\overset{\displaystyle O}{\underset{\displaystyle |}{CH}}} - \overset{\displaystyle O}{\overset{\displaystyle \|}{C}} - OH \qquad H_2N - \underset{\underset{\displaystyle CH_3}{\underset{\displaystyle |}{CH - OH}}}{\overset{\displaystyle O}{\underset{\displaystyle |}{CH}}} - \overset{\displaystyle O}{\overset{\displaystyle \|}{C}} - OH$$

## Section 25.2: Ceramics and Composite Materials

### Review Questions

25.7   What types of atoms and bonds are present in the polymer "backbone" in a ceramic containing a transition metal?

25.8   Describe two natural types of composite materials and the substances that they contain.

### Problems

**25.9**   What steps would be used to form a ceramic material from scandium and ethanol ($CH_3CH_2OH$) using the sol-gel process?

25.10   Amorphous silica ($SiO_2$) can be formed in uniform spheres from a sol-gel process using methanol ($CH_3OH$) solvent. List the chemical reactions involved in this process.

**25.11**   Bakelite, described in Review Question 25.1, is manufactured using materials like cellulose paper or glass fibers that are compressed and heated among the monomer units. Is Bakelite a simple polymer? If not, what classification best fits Bakelite?

## Section 25.3: Liquid Crystals

### Review Questions

25.12 Is a normal liquid isotropic or anisotropic? How is an anisotropic material different from an isotropic material?

25.13 Describe how nematic, smectic, and cholesteric liquid crystals are similar and different.

### Problems

25.14 Would each of these molecules be likely to form a liquid crystal? If not, why not?

(a)

(b)

(c)

25.15 Would long-chain hydrocarbons or addition polymers of alkene molecules form liquid crystals? Why or why not?

25.16 Would an ionic compound form a liquid crystal? Why or why not?

## Section 25.4: Biomedical Materials

### Review Questions

25.17 Name three major concerns that must be addressed when proposing any material to be used in medical applications inside the body.

25.18 Name a necessary chemical property and a necessary physical property of dental materials.

### Problems

25.19 Is the polymer commonly used in artificial skin material (shown here) an addition polymer or a condensation polymer? Is it a block, alternating, graft, or random copolymer?

25.20 What are some advantages and disadvantages of the use of ceramics in replacement joint bearing surfaces relative to metals?

25.21 How do amalgam and composite tooth fillings differ in the ways they fail over time?

## Section 25.5: Nanotechnology

### Review Questions

25.22 How does an STM measure the peak and valley heights along the surface of a sample?

25.23 Name four allotropic forms of carbon.

### Problems

25.24 How does an AFM differ from an STM? Why can graphite possibly be investigated with either instrument whereas diamond can only be investigated with an AFM?

25.25 What is the hybridization of the carbon atoms in a single-walled nanotube?

25.26 What structural or bonding aspect of graphite and carbon nanotubes allows their use as semiconductors, whereas diamond is an electrical insulator?

25.27 What type of intermolecular forces holds the sheets of carbon atoms together in graphite? What type of intermolecular forces hold carbon nanotubes together in strands and fibers?

## Section 25.6: Semiconductors

### Review Questions

25.28 How do the band gaps differ among conductors, semiconductors, and insulators?

25.29 The only elemental semiconductors (Si, Ge, and graphite) are members of Group 4A. Why are Sn and Pb, which are also in Group 4A, not classified as semiconductors?

### Problems

25.30 State whether each combination of elements would be expected to produce a semiconductor: (a) Ga and Sb, (b) As and N, (c) B and P, (d) Zn and Sb, (e) Cd and S.

25.31 What type of semiconductor would be formed (a) if Si were doped with Sb and (b) if Ge were doped with B?

25.32 Describe the movement of electrons in a solar battery that uses $p$-type and $n$-type semiconductors when light shines on the solar cell.

## Section 25.7: Superconductors

### Review Questions

25.33 Why is YBCO called a "high-temperature" semiconductor when it exhibits superconductivity only below $-180°C$, which is very cold?

25.34 How might the Meissner effect be used in a real-world application? What are the engineering challenges to this application with the current superconductors?

### Problems

25.35 What is the molecular formula of BSCCO-2201?

25.36 Superconductivity at room temperature is sometimes a subject of science fiction or fantasy. For example, in the movie *Terminator 2: Judgment Day,* a cyborg has a CPU that contains a room-temperature superconductor. How might a computer with such a CPU differ in design from current computers?

**25.37**  What are the electron configurations of the copper ions in YBCO? Are both ions expected oxidation states for copper?

## Additional Problems

**25.38**  Draw representations of block copolymers and graft copolymers using the symbols A and B for two types of monomer units. Are the categories exclusive? Using a third monomer type, see if you can draw a representation of a polymer that is both a block and a graft copolymer.

**25.39**  Draw the structures of the two monomer units in this polymer:

$$\left[ NH\!-\!\left(CH_2\right)_8\!\!-\!NH\!-\!\overset{\overset{\displaystyle O}{\parallel}}{C}\!-\!\left(CH_2\right)_4\!\!-\!\overset{\overset{\displaystyle O}{\parallel}}{C} \right]_n$$

**25.40**  Is the polymer in Problem 25.39 a condensation polymer or an addition polymer? Is it a polyester, a polyamide, or something else?

**25.41**  What types of bonding (covalent, ionic, network, metallic) are present in a plastic polymer? In a ceramic material?

**25.42**  Draw representations of isotactic, syndiotactic, and atactic polyacrylonitrile (for the monomer unit, see Table 25.1).

**25.43**  Fluoride ion is commonly used in drinking water supplies and in toothpaste (in the form of sodium fluoride) to prevent tooth decay. The fluoride ions replace hydroxide ions in the enamel mineral hydroxyapatite [$Ca_5(PO_4)_3OH$, the same mineral as in bone], forming fluoroapatite [$Ca_5(PO_4)_3F$]. Why does replacing the hydroxide ion with fluoride ion prevent tooth decay? How does this relate to the properties of materials used in artificial fillings?

**25.44**  Would the metal alkoxide in the sol-gel process be acidic or basic?

**25.45**  Would the compound shown form a liquid crystal? Why or why not?

**25.46**  The semiconductor band gap energy determines the color of an LED. Knowing what you learned in Chapter 6 about the energies of photons of different colors of light, which semiconductor would have the larger band gap: the one in a green exit sign's LEDs, or the one in a red traffic signal?

# PRE-PROFESSIONAL PRACTICE EXAM PROBLEMS: PHYSICAL AND BIOLOGICAL SCIENCES

These questions are not based on a passage.

1.  Which of the following would form an *n*-type semiconductor: (i) doping Ge with As, (ii) doping Si with B, (iii) doping graphite with P?

    a) i, ii, and iii
    b) i and iii
    c) i and ii
    d) iii only

2.  What is unusual about the copper in YBCO?

    a) It exists as a mixture of +1 and +2 oxidation states.
    b) It exists as a mixture of +1 and 0 oxidation states.
    c) It exists as a mixture of +1 and +3 oxidation states.
    d) It exists as a mixture of +2 and +3 oxidation states.

3.  A semiconductor can be produced by the combination of two elements with a total of how many valence electrons?

    a) 6
    b) 8
    c) 12
    d) 16

4.  Metals are good conductors of electricity because they

    a) have no band gap.
    b) have a large band gap.
    c) have a small band gap.
    d) combine with nonmetals to form semiconductors.

# ANSWERS TO IN-CHAPTER MATERIALS

## Answers to Practice Problems

**25.1A**

$$-\overset{\underset{|}{H}}{\underset{\overset{|}{H}}{C}}-\overset{\underset{|}{Cl}}{\underset{\overset{|}{Cl}}{C}}-\overset{\underset{|}{H}}{\underset{\overset{|}{H}}{C}}-\overset{\underset{|}{Cl}}{\underset{\overset{|}{Cl}}{C}}-$$

**25.1B**

$$\underset{H}{\overset{H}{C}}=\underset{COCH_3}{\overset{CH_3}{C}}$$

**25.2A** $-NHCH_2(CH_2)_3CH_2\overset{O}{\overset{\|}{C}}NHCH_2(CH_2)_3CH_2\overset{O}{\overset{\|}{C}}NHCH_2(CH_2)_3CH_2\overset{O}{\overset{\|}{C}}-$

$-(CH_2CCl_2)-$

**25.2B** Condensation.

$$H-O-CH_2-\hexagon-CH_2-O-\overset{O}{\overset{\|}{C}}-\phenyl-\overset{O}{\overset{\|}{C}}-OH$$

**25.3** $-CH=C(CH_2CH_3)-CH=C(CH_2CH_3)-CH=C(CH_2CH_3)-$

$-CH_2-CH(CH_2CH_3)-CH_2-CH(CH_2CH_3)-CH_2-CH(CH_2CH_3)-$

The two structures are similar in that they both have ethyl groups on alternating carbons. They are different in that poly(1-butyne) is a conjugated system and poly(1-butene) is not. The first structure is more likely to be an electrically conducting polymer. **25.4A** The top molecule would be more likely to exhibit liquid crystal behavior because it is long, thin, and has conjugated rings that limit free rotation. **25.4B** Addition of an alkyl chain makes the molecule longer, contributing to its ability to exhibit liquid crystal behavior.

**25.5A** (a) Elements with two valence electrons, e.g., Be, Sr, Cd, (b) elements with three valence electrons, e.g., B, Ga, In, (c) elements with six valence electrons, e.g., O, Se, Te. **25.5B** Carbon can be combined with silicon to produce a semiconductor. Arsenic can be used to dope silicon to produce an *n*-type semiconductor; and gallium can be used to dope silicon to produce a *p*-type semiconductor.

## Answers to Checkpoints

**25.1.1** a. **25.1.2** d. **25.3.1** c. **25.3.2** a, c. **25.6.1** b. **25.6.2** c.

## Answers to Applying What You've Learned

**a)** $(CHCH_2CH_2CH)_{50}(CHCHClCH_2CH)_{70}$. **b)** Block. **c)** No.

# Appendix 1

## Scientific Notation

Chemists often deal with numbers that are either extremely large or extremely small. For example, in 1 g of the element hydrogen there are roughly

$$602,200,000,000,000,000,000,000$$

hydrogen atoms. Each hydrogen atom has a mass of only

$$0.00000000000000000000000166 \text{ g}$$

These numbers are cumbersome to handle, and it is easy to make mistakes when using them in arithmetic computations. Consider the following multiplication:

$$0.0000000056 \times 0.00000000048 = 0.000000000000000002688$$

It would be easy for us to miss one zero or add one more zero after the decimal point. Consequently, when working with very large and very small numbers, we use a system called *scientific notation*. Regardless of their magnitude, all numbers can be expressed in the form

$$N \times 10^n$$

where $N$ is a number between 1 and 10 and $n$, the exponent, is a positive or negative integer (whole number). Any number expressed in this way is said to be written in scientific notation.

Suppose that we are given a certain number and asked to express it in scientific notation. Basically, this assignment calls for us to find $n$. We count the number of places that the decimal point must be moved to give the number $N$ (which is between 1 and 10). If the decimal point has to be moved to the left, then $n$ is a positive integer; if it has to be moved to the right, $n$ is a negative integer. The following examples illustrate the use of scientific notation:

1. Express 568.762 in scientific notation:

$$568.762 = 5.68762 \times 10^2$$

Note that the decimal point is moved to the left by two places and $n = 2$.

2. Express 0.00000772 in scientific notation:

$$0.00000772 = 7.72 \times 10^{-6}$$

Here the decimal point is moved to the right by six places and $n = -6$.

Keep in mind the following two points. First, $n = 0$ is used for numbers that are not expressed in scientific notation. For example, $74.6 \times 10^0$ ($n = 0$) is equivalent to 74.6. Second, the usual practice is to omit the superscript when $n = 1$. Thus the scientific notation for 74.6 is $7.46 \times 10$ and not $7.46 \times 10^1$

Next, we consider how scientific notation is handled in arithmetic operations.

## Addition and Subtraction

To add or subtract using scientific notation, we first write each quantity—say $N_1$ and $N_2$—with the same exponent $n$. Then we combine $N_1$ and $N_2$; the exponents remain the same. Consider the following examples:

$$(7.4 \times 10^3) + (2.1 \times 10^3) = 9.5 \times 10^3$$

$$(4.31 \times 10^4) + (3.9 \times 10^3) = (4.31 \times 10^4) + (0.39 \times 10^4)$$

$$= 4.70 \times 10^4$$

$$(2.22 \times 10^{-2}) - (4.10 \times 10^{-3}) = (2.22 \times 10^{-2}) - (0.41 \times 10^{-2})$$

$$= 1.81 \times 10^{-2}$$

## Multiplication and Division

To multiply numbers expressed in scientific notation, we multiply $N_1$ and $N_2$ in the usual way, but *add* the exponents together. To divide using scientific notation, we divide $N_1$ and $N_2$ as usual and subtract the exponents. The following examples show how these operations are performed:

$$(8.0 \times 10^4) \times (5.0 \times 10^2) = (8.0 \times 5.0)(10^{4+2})$$

$$= 40 \times 10^6$$

$$= 4.0 \times 10^7$$

$$(4.0 \times 10^{-5}) \times (7.0 \times 10^3) = (4.0 \times 7.0)(10^{-5+3})$$

$$= 28 \times 10^{-2}$$

$$= 2.8 \times 10^{-1}$$

$$\frac{6.9 \times 10^7}{3.0 \times 10^{-5}} = \frac{6.9}{3.0} \times 10^{7-(-5)}$$

$$= 2.3 \times 10^{12}$$

$$\frac{8.5 \times 10^4}{5.0 \times 10^9} = \frac{8.5}{5.0} \times 10^{4-9}$$

$$= 1.7 \times 10^{-5}$$

# Basic Trigonometry

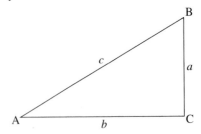

In the triangle shown, A, B, and C are angles (C = 90°) and $a$, $b$, and $c$ are side lengths. To calculate unknown angles or sides, we use the following relationships:

$$a^2 + b^2 = c^2$$

$$\sin A = \frac{a}{c}$$

$$\cos A = \frac{b}{c}$$

$$\tan A = \frac{a}{b}$$

# Logarithms

## Common Logarithms

The concept of logarithms is an extension of the concept of exponents, which is discussed on page A-1. The common, or base-10, logarithm of any number is the power to which 10 must be raised to equal the number. The following examples illustrate this relationship:

| Logarithm | Exponent |
|---|---|
| $\log 1 = 0$ | $10^0 = 1$ |
| $\log 10 = 1$ | $10^1 = 10$ |
| $\log 100 = 2$ | $10^2 = 100$ |
| $\log 10^{-1} = -1$ | $10^{-1} = 0.1$ |
| $\log 10^{-2} = -2$ | $10^{-2} = 0.01$ |

In each case the logarithm of the number can be obtained by inspection.

Because the logarithms of numbers are exponents, they have the same properties as exponents. Thus, we have

| Logarithm | Exponent |
|---|---|
| $\log AB = \log A + \log B$ | $10^A \times 10^B = 10^{A+B}$ |
| $\log \dfrac{A}{B} = \log A - \log B$ | $\dfrac{10^A}{10^B} = 10^{A-B}$ |

Furthermore, $\log A^n = n \log A$.

Now suppose we want to find the common logarithm of $6.7 \times 10^{-4}$. On most electronic calculators, the number is entered first and then the log key is pressed. This operation gives us

$$\log 6.7 \times 10^{-4} = -3.17$$

Note that there are as many digits *after* the decimal point as there are significant figures in the original number. The original number has two significant figures, and the "17" in $-3.17$ tells us that the log has two significant figures. The "3" in $-3.17$ serves only to locate the decimal point in the number $6.7 \times 10^{-4}$. Other examples are

| Number | Common Logarithm |
|---|---|
| 62 | 1.79 |
| 0.872 | $-0.0595$ |
| $1.0 \times 10^{-7}$ | $-7.00$ |

Sometimes (as in the case of pH calculations) it is necessary to obtain the number whose logarithm is known. This procedure is known as taking the antilogarithm; it is simply the reverse of taking the logarithm of a number. Suppose in a certain calculation we have pH = 1.46 and are asked to calculate $[H^+]$. From the definition of pH (pH = $-\log [H^+]$) we can write

$$[H^+] = 10^{-1.46}$$

Many calculators have a key labeled $\log^{-1}$ or INV log to obtain antilogs. Other calculators have a $10^x$ or $y^x$ key (where $x$ corresponds to $-1.46$ in our example and $y$ is 10 for base-10 logarithm). Therefore, we find that $[H^+] = 0.035 \, M$.

## Natural Logarithms

Logarithms taken to the base $e$ instead of 10 are known as natural logarithms (denoted by ln or $\log_e$); $e$ is equal to 2.7183. The relationship between common logarithms and natural logarithms is as follows:

$$\log 10 = 1 \qquad 10^1 = 10$$

$$\ln 10 = 2.303 \qquad e^{2.303} = 10$$

Thus,

$$\ln x = 2.303 \log x$$

To find the natural logarithm of 2.27, say, we first enter the number on the electronic calculator and then press the ln key to get

$$\ln 2.27 = 0.820$$

If no ln key is provided, we can proceed as follows:

$$2.303 \log 2.27 = 2.303 \times 0.356$$

$$= 0.820$$

Sometimes we may be given the natural logarithm and asked to find the number it represents. For example,

$$\ln x = 59.7$$

On many calculators, we simply enter the number and press the $e$ key:

$$e^{59.7} = 8 \times 10^{25}$$

## The Quadratic Equation

A quadratic equation takes the form

$$ax^2 + bx + c = 0$$

If coefficients $a$, $b$, and $c$ are known, then $x$ is given by

$$x = \frac{-b \pm \sqrt{b^2 - 4ac}}{2a}$$

Suppose we have the following quadratic equation:

$$2x^2 + 5x - 12 = 0$$

Solving for $x$, we write

$$x = \frac{-5 \pm \sqrt{(5)^2 - 4(2)(-12)}}{2(2)}$$

$$= \frac{-5 \pm \sqrt{25 + 96}}{4}$$

Therefore,

$$x = \frac{-5 + 11}{4} = \frac{3}{2}.$$

and

$$x = \frac{-5 - 11}{4} = -4$$

## Successive Approximation

In the determination of hydrogen ion concentration in a weak acid solution, use of the quadratic equation can sometimes be avoided using a method known as *successive approximation*. Consider the example of a 0.0150 $M$ solution of hydrofluoric acid (HF). The $K_a$ for HF is $7.10 \times 10^{-4}$. To determine the hydrogen ion concentration in this solution, we construct an equilibrium table and enter the initial concentrations, the expected change in concentrations, and the equilibrium concentrations of all species:

$$\text{HF}(aq) \rightleftharpoons \text{H}^+(aq) + \text{F}^-(aq)$$

|  | HF | H⁺ | F⁻ |
|---|---|---|---|
| *Initial concentration (M):* | 0.0150 | 0 | 0 |
| *Change in concentration (M):* | $-x$ | $+x$ | $+x$ |
| *Equilibrium concentration (M):* | $0.0150 - x$ | $x$ | $x$ |

Using the rule that $x$ can be neglected if the initial acid concentration divided by the $K_a$ is greater than 100, we find that in this case $x$ cannot be neglected ($0.0150/7.1 \times 10^{-4} \approx 21$). Successive approximation involves first neglecting $x$ with respect to initial acid concentration

$$\frac{x^2}{0.0150 - x} \approx \frac{x^2}{0.0150} = 7.10 \times 10^{-4}$$

and solving for $x$:

$$x^2 = (0.0150)(7.10 \times 10^{-4}) = 1.07 \times 10^{-5}$$

$$x = \sqrt{1.07 \times 10^{-5}} = 0.00326 \ M$$

We then solve for $x$ again, this time using the calculated value of $x$ on the bottom of the fraction:

$$\frac{x^2}{0.0150 - x} = \frac{x^2}{0.0150 - 0.00326} = 7.10 \times 10^{-4}$$

$$x^2 = (0.0150 - 0.00326)(7.10 \times 10^{-4}) = 8.33 \times 10^{-6}$$

$$x = \sqrt{8.33 \times 10^{-6}} = 0.00289 \ M$$

Note that the calculated value of $x$ decreased from 0.00326 to 0.00289. We now use the new calculated value of $x$ on the bottom of the fraction and solve for $x$ again:

$$\frac{x^2}{0.0150 - x} = \frac{x^2}{0.0150 - 0.00289} = 7.10 \times 10^{-4}$$

$$x^2 = (0.0150 - 0.00289)(7.10 \times 10^{-4}) = 8.60 \times 10^{-6}$$

$$x = \sqrt{8.60 \times 10^{-6}} = 0.00293 \ M$$

This time the value of $x$ increased slightly. We use the new calculated value and solve for $x$ again.

$$\frac{x^2}{0.0150 - x} = \frac{x^2}{0.0150 - 0.00293} = 7.10 \times 10^{-4}$$

$$x^2 = (0.0150 - 0.00293)(7.10 \times 10^{-4}) = 8.57 \times 10^{-6}$$

$$x = \sqrt{8.57 \times 10^{-6}} = 0.00293 \ M$$

This time we find that the answer is still 0.00293 $M$, so there is no need to repeat the process. In general, we apply the method of successive approximation until the value of $x$ obtained does not differ from the value obtained in the previous step. The value of $x$ determined using successive approximation is the same value we would get if we were to use the quadratic equation.

# Appendix 2

## THERMODYNAMIC DATA AT 1 ATM AND 25°C*

| INORGANIC SUBSTANCES | | | |
| --- | --- | --- | --- |
| Substance | $\Delta H^\circ_f$ (kJ/mol) | $\Delta G^\circ_f$ (kJ/mol) | $S^\circ$ (J/K · mol) |
| Ag(s) | 0 | 0 | 42.7 |
| $Ag^+(aq)$ | 105.9 | 77.1 | 73.9 |
| AgCl(s) | −127.0 | −109.7 | 96.1 |
| AgBr(s) | −99.5 | −95.9 | 107.1 |
| AgI(s) | −62.4 | −66.3 | 114.2 |
| $AgNO_3(s)$ | −123.1 | −32.2 | 140.9 |
| Al(s) | 0 | 0 | 28.3 |
| $Al^{3+}(aq)$ | −524.7 | −481.2 | −313.38 |
| $Al_2O_3(s)$ | −1669.8 | −1576.4 | 50.99 |
| As(s) | 0 | 0 | 35.15 |
| $AsO_4^{3-}(aq)$ | −870.3 | −635.97 | −144.77 |
| $AsH_3(g)$ | 171.5 | | |
| $H_3AsO_4(s)$ | −900.4 | | |
| Au(s) | 0 | 0 | 47.7 |
| $Au_2O_3(s)$ | 80.8 | 163.2 | 125.5 |
| AuCl(s) | −35.2 | | |
| $AuCl_3(s)$ | −118.4 | | |
| B(s) | 0 | 0 | 6.5 |
| $B_2O_3(s)$ | −1263.6 | −1184.1 | 54.0 |
| $H_3BO_3(s)$ | −1087.9 | −963.16 | 89.58 |
| $H_3BO_3(aq)$ | −1067.8 | −963.3 | 159.8 |
| Ba(s) | 0 | 0 | 66.9 |
| $Ba^{2+}(aq)$ | −538.4 | −560.66 | 12.55 |
| BaO(s) | −558.2 | −528.4 | 70.3 |
| $BaCl_2(s)$ | −860.1 | −810.66 | 125.5 |
| $BaSO_4(s)$ | −1464.4 | −1353.1 | 132.2 |
| $BaCO_3(s)$ | −1218.8 | −1138.9 | 112.1 |
| Be(s) | 0 | 0 | 9.5 |
| BeO(s) | −610.9 | −581.58 | 14.1 |
| $Br_2(l)$ | 0 | 0 | 152.3 |

*The thermodynamic quantities of ions are based on the reference states that $\Delta H^\circ_f[H^+(aq)] = 0$, $\Delta G^\circ_f[H^+(aq)] = 0$, and $S^\circ[H^+(aq)] = 0$.

| Substance | $\Delta H_f^\circ$ (kJ/mol) | $\Delta G_f^\circ$ (kJ/mol) | $S^\circ$ (J/K · mol) |
|---|---|---|---|
| $Br_2(g)$ | 30.7 | 3.14 | 245.13 |
| $Br^-(aq)$ | −120.9 | −102.8 | 80.7 |
| $HBr(g)$ | −36.2 | −53.2 | 198.48 |
| C(graphite) | 0 | 0 | 5.69 |
| C(diamond) | 1.90 | 2.87 | 2.4 |
| $CO(g)$ | −110.5 | −137.3 | 197.9 |
| $CO_2(g)$ | −393.5 | −394.4 | 213.6 |
| $CO_2(aq)$ | −412.9 | −386.2 | 121.3 |
| $CO_3^{2-}(aq)$ | −676.3 | −528.1 | −53.1 |
| $HCO_3^-(aq)$ | −691.1 | −587.1 | 94.98 |
| $H_2CO_3(aq)$ | −699.7 | −623.2 | 187.4 |
| $CS_2(g)$ | 115.3 | 65.1 | 237.8 |
| $CS_2(l)$ | 87.3 | 63.6 | 151.0 |
| $HCN(aq)$ | 105.4 | 112.1 | 128.9 |
| $CN^-(aq)$ | 151.0 | 165.69 | 117.99 |
| $(NH_2)_2CO(s)$ | −333.19 | −197.15 | 104.6 |
| $(NH_2)_2CO(aq)$ | −319.2 | −203.84 | 173.85 |
| $Ca(s)$ | 0 | 0 | 41.6 |
| $Ca(g)$ | 179.3 | 145.5 | 154.8 |
| $Ca^{2+}(aq)$ | −542.96 | −553.0 | −55.2 |
| $CaO(s)$ | −635.6 | −604.2 | 39.8 |
| $Ca(OH)_2(s)$ | −986.6 | −896.8 | 83.4 |
| $CaF_2(s)$ | −1214.6 | −1161.9 | 68.87 |
| $CaCl_2(s)$ | −794.96 | −750.19 | 113.8 |
| $CaSO_4(s)$ | −1432.69 | −1320.3 | 106.69 |
| $CaCO_3(s)$ | −1206.9 | −1128.8 | 92.9 |
| $Cd(s)$ | 0 | 0 | 51.46 |
| $Cd^{2+}(aq)$ | −72.38 | −77.7 | −61.09 |
| $CdO(s)$ | −254.6 | −225.06 | 54.8 |
| $CdCl_2(s)$ | −389.1 | −342.59 | 118.4 |
| $CdSO_4(s)$ | −926.17 | −820.2 | 137.2 |
| $Cl_2(g)$ | 0 | 0 | 223.0 |
| $Cl(g)$ | 121.7 | 105.7 | 165.2 |
| $Cl^-(aq)$ | −167.2 | −131.2 | 56.5 |
| $HCl(g)$ | −92.3 | −95.27 | 187.0 |
| $Co(s)$ | 0 | 0 | 28.45 |
| $Co^{2+}(aq)$ | −67.36 | −51.46 | 155.2 |
| $CoO(s)$ | −239.3 | −213.38 | 43.9 |
| $Cr(s)$ | 0 | 0 | 23.77 |
| $Cr^{2+}(aq)$ | −138.9 | | |
| $Cr_2O_3(s)$ | −1128.4 | −1046.8 | 81.17 |
| $CrO_4^{2-}(aq)$ | −863.16 | −706.26 | 38.49 |
| $Cr_2O_7^{2-}(aq)$ | −1460.6 | −1257.29 | 213.8 |

*(Continued)*

| Substance | $\Delta H_f^\circ$ (kJ/mol) | $\Delta G_f^\circ$ (kJ/mol) | $S^\circ$ (J/K · mol) |
|---|---|---|---|
| Cs(s) | 0 | 0 | 82.8 |
| Cs(g) | 76.50 | 49.53 | 175.6 |
| $Cs^+$(aq) | −247.69 | −282.0 | 133.05 |
| CsCl(s) | −442.8 | −414.4 | 101.2 |
| Cu(s) | 0 | 0 | 33.3 |
| $Cu^+$(aq) | 51.88 | 50.2 | −26.4 |
| $Cu^{2+}$(aq) | 64.39 | 64.98 | −99.6 |
| CuO(s) | −155.2 | −127.2 | 43.5 |
| $Cu_2O$(s) | −166.69 | −146.36 | 100.8 |
| CuCl(s) | −134.7 | −118.8 | 91.6 |
| $CuCl_2$(s) | −205.85 | ? | ? |
| CuS(s) | −48.5 | −49.0 | 66.5 |
| $CuSO_4$(s) | −769.86 | −661.9 | 113.39 |
| $F_2$(g) | 0 | 0 | 203.34 |
| F(g) | 80.0 | 61.9 | 158.7 |
| $F^-$(aq) | −329.1 | −276.48 | −9.6 |
| HF(g) | −271.6 | −270.7 | 173.5 |
| Fe(s) | 0 | 0 | 27.2 |
| $Fe^{2+}$(aq) | −87.86 | −84.9 | −113.39 |
| $Fe^{3+}$(aq) | −47.7 | −10.5 | −293.3 |
| FeO(s) | −272.0 | −255.2 | 60.8 |
| $Fe_2O_3$(s) | −822.2 | −741.0 | 90.0 |
| $Fe(OH)_2$(s) | −568.19 | −483.55 | 79.5 |
| $Fe(OH)_3$(s) | −824.25 | ? | ? |
| H(g) | 218.2 | 203.2 | 114.6 |
| $H_2$(g) | 0 | 0 | 131.0 |
| $H^+$(aq) | 0 | 0 | 0 |
| $OH^-$(aq) | −229.94 | −157.30 | −10.5 |
| $H_2O$(g) | −241.8 | −228.6 | 188.7 |
| $H_2O$(l) | −285.8 | −237.2 | 69.9 |
| $H_2O_2$(g) | −136.1 | −105.5 | 232.9 |
| $H_2O_2$(l) | −187.6 | −118.1 | 109.6 |
| Hg(l) | 0 | 0 | 77.4 |
| $Hg^{2+}$(aq) | | −164.38 | |
| HgO(s) | −90.7 | −58.5 | 72.0 |
| $HgCl_2$(s) | −230.1 | | |
| $Hg_2Cl_2$(s) | −264.9 | −210.66 | 196.2 |
| HgS(s) | −58.16 | −48.8 | 77.8 |
| $HgSO_4$(s) | −704.17 | | |
| $Hg_2SO_4$(s) | −741.99 | −623.92 | 200.75 |
| $I_2$(g) | 62.25 | 19.37 | 260.57 |
| $I_2$(s) | 0 | 0 | 116.7 |
| I(g) | 106.8 | 70.21 | 180.67 |
| $I^-$(aq) | −55.9 | −51.67 | 109.37 |
| HI(g) | 25.9 | 1.30 | 206.3 |

| Substance | $\Delta H_f^\circ$ (kJ/mol) | $\Delta G_f^\circ$ (kJ/mol) | $S^\circ$ (J/K · mol) |
|---|---|---|---|
| K(s) | 0 | 0 | 63.6 |
| $K^+$(aq) | −251.2 | −282.28 | 102.5 |
| KOH(s) | −425.85 | | |
| KCl(s) | −435.87 | −408.3 | 82.68 |
| $KClO_3$(s) | −391.20 | −289.9 | 142.97 |
| $KClO_4$(s) | −433.46 | −304.18 | 151.0 |
| KBr(s) | −392.17 | −379.2 | 96.4 |
| KI(s) | −327.65 | −322.29 | 104.35 |
| $KNO_3$(s) | −492.7 | −393.1 | 132.9 |
| Li(s) | 0 | 0 | 28.0 |
| Li(g) | 159.3 | 126.6 | 138.8 |
| $Li^+$(aq) | −278.46 | −293.8 | 14.2 |
| LiCl(s) | −408.3 | −384.0 | 59.30 |
| $Li_2O$(s) | −595.8 | ? | ? |
| LiOH(s) | −487.2 | −443.9 | 50.2 |
| Mg(s) | 0 | 0 | 32.5 |
| Mg(g) | 150 | 115 | 148.55 |
| $Mg^{2+}$(aq) | −461.96 | −456.0 | −117.99 |
| MgO(s) | −601.8 | −569.6 | 26.78 |
| $Mg(OH)_2$(s) | −924.66 | −833.75 | 63.1 |
| $MgCl_2$(s) | −641.8 | −592.3 | 89.5 |
| $MgSO_4$(s) | −1278.2 | −1173.6 | 91.6 |
| $MgCO_3$(s) | −1112.9 | −1029.3 | 65.69 |
| Mn(s) | 0 | 0 | 31.76 |
| $Mn^{2+}$(aq) | −218.8 | −223.4 | −83.68 |
| $MnO_2$(s) | −520.9 | −466.1 | 53.1 |
| N(g) | 470.4 | 455.5 | 153.3 |
| $N_2$(g) | 0 | 0 | 191.5 |
| $N_3^-$(aq) | 245.18 | ? | ? |
| $NH_3$(g) | −46.3 | −16.6 | 193.0 |
| $NH_4^+$(aq) | −132.80 | −79.5 | 112.8 |
| $NH_4Cl$(s) | −315.39 | −203.89 | 94.56 |
| $NH_3$(aq) | −80.3 | −26.5 | 111.3 |
| $N_2H_4$(l) | 50.4 | | |
| NO(g) | 90.4 | 86.7 | 210.6 |
| $NO_2$(g) | 33.85 | 51.8 | 240.46 |
| $N_2O_4$(g) | 9.66 | 98.29 | 304.3 |
| $N_2O$(g) | 81.56 | 103.6 | 219.99 |
| $HNO_2$(aq) | −118.8 | −53.6 | |
| $HNO_3$(l) | −173.2 | −79.9 | 155.6 |
| $NO_3^-$(aq) | −206.57 | −110.5 | 146.4 |
| Na(s) | 0 | 0 | 51.05 |
| Na(l) | 2.41 | 0.50 | 57.56 |
| Na(g) | 107.7 | 77.3 | 153.7 |

*(Continued)*

| Substance | $\Delta H_f^\circ$ (kJ/mol) | $\Delta G_f^\circ$ (kJ/mol) | $S^\circ$ (J/K · mol) |
|---|---|---|---|
| $Na^+(aq)$ | −239.66 | −261.87 | 60.25 |
| $NaOH(aq)$ | −469.6 | −419.2 | 49.8 |
| $Na_2O(s)$ | −415.9 | −376.56 | 72.8 |
| $NaCl(s)$ | −410.9 | −384.0 | 72.38 |
| $NaI(s)$ | −288.0 | | |
| $Na_2SO_4(s)$ | −1384.49 | −1266.8 | 149.49 |
| $NaNO_3(s)$ | −466.68 | −365.89 | 116.3 |
| $Na_2CO_3(s)$ | −1130.9 | −1047.67 | 135.98 |
| $NaHCO_3(s)$ | −947.68 | −851.86 | 102.09 |
| $Ni(s)$ | 0 | 0 | 30.1 |
| $Ni^{2+}(aq)$ | −64.0 | −46.4 | −159.4 |
| $NiO(s)$ | −244.35 | −216.3 | 38.58 |
| $Ni(OH)_2(s)$ | −538.06 | −453.1 | 79.5 |
| $O(g)$ | 249.4 | 230.1 | 160.95 |
| $O_2(g)$ | 0 | 0 | 205.0 |
| $O_3(aq)$ | −12.09 | 16.3 | 110.88 |
| $O_3(g)$ | 142.2 | 163.4 | 237.6 |
| $P(white)$ | 0 | 0 | 44.0 |
| $P(red)$ | −18.4 | 13.8 | 29.3 |
| $PCl_3(l)$ | −319.7 | −272.3 | 217.1 |
| $PCl_3(g)$ | −288.07 | −269.6 | 311.7 |
| $PCl_5(g)$ | −374.9 | −305.0 | 364.5 |
| $PO_4^{3-}(aq)$ | −1284.07 | −1025.59 | −217.57 |
| $P_4O_{10}(s)$ | −3012.48 | | |
| $PH_3(g)$ | 9.25 | 18.2 | 210.0 |
| $HPO_4^{2-}(aq)$ | −1298.7 | −1094.1 | −35.98 |
| $H_2PO_4^-(aq)$ | −1302.48 | −1135.1 | 89.1 |
| $Pb(s)$ | 0 | 0 | 64.89 |
| $Pb^{2+}(aq)$ | 1.6 | −24.3 | 21.3 |
| $PbO(s)$ | −217.86 | −188.49 | 69.45 |
| $PbO_2(s)$ | −276.65 | −218.99 | 76.57 |
| $PbCl_2(s)$ | −359.2 | −313.97 | 136.4 |
| $PbS(s)$ | −94.3 | −92.68 | 91.2 |
| $PbSO_4(s)$ | −918.4 | −811.2 | 147.28 |
| $Pt(s)$ | 0 | 0 | 41.84 |
| $PtCl_4^{2-}(aq)$ | −516.3 | −384.5 | 175.7 |
| $Rb(s)$ | 0 | 0 | 69.45 |
| $Rb(g)$ | 85.8 | 55.8 | 170.0 |
| $Rb^+(aq)$ | −246.4 | −282.2 | 124.27 |
| $RbBr(s)$ | −389.2 | −378.1 | 108.3 |
| $RbCl(s)$ | −435.35 | −407.8 | 95.90 |
| $RbI(s)$ | −328 | −326 | 118.0 |
| $S(rhombic)$ | 0 | 0 | 31.88 |
| $S(monoclinic)$ | 0.30 | 0.10 | 32.55 |

| Substance | $\Delta H_f^\circ$ (kJ/mol) | $\Delta G_f^\circ$ (kJ/mol) | $S^\circ$ (J/K · mol) |
|---|---|---|---|
| $SO(g)$ | 5.01 | −19.9 | 221.8 |
| $SO_2(g)$ | −296.4 | −300.4 | 248.5 |
| $SO_3(g)$ | −395.2 | −370.4 | 256.2 |
| $SO_3^{2-}(aq)$ | −624.25 | −497.06 | 43.5 |
| $SO_4^{2-}(aq)$ | −907.5 | −741.99 | 17.15 |
| $H_2S(g)$ | −20.15 | −33.0 | 205.64 |
| $HSO_3^-(aq)$ | −627.98 | −527.3 | 132.38 |
| $HSO_4^-(aq)$ | −885.75 | −752.87 | 126.86 |
| $H_2SO_4(l)$ | −811.3 | ? | ? |
| $SF_6(g)$ | −1096.2 | ? | ? |
| $Si(s)$ | 0 | 0 | 18.70 |
| $SiO_2(s)$ | −859.3 | −805.0 | 41.84 |
| $Sr(s)$ | 0 | 0 | 54.39 |
| $Sr^{2+}(aq)$ | −545.5 | −557.3 | −39.33 |
| $SrCl_2(s)$ | −828.4 | −781.15 | 117.15 |
| $SrSO_4(s)$ | −1444.74 | −1334.28 | 121.75 |
| $SrCO_3(s)$ | −1218.38 | −1137.6 | 97.07 |
| $U(s)$ | 0 | 0 | 50.21 |
| $UF_6(g)$ | −2147 | −2064 | 378 |
| $Zn(s)$ | 0 | 0 | 41.6 |
| $Zn^{2+}(aq)$ | −152.4 | −147.2 | −106.48 |
| $ZnO(s)$ | −348.0 | −318.2 | 43.9 |
| $ZnCl_2(s)$ | −415.89 | −369.26 | 108.37 |
| $ZnS(s)$ | −202.9 | −198.3 | 57.7 |
| $ZnSO_4(s)$ | −978.6 | −871.6 | 124.7 |

## ORGANIC SUBSTANCES

| Substance | Formula | $\Delta H_f^\circ$ (kJ/mol) | $\Delta G_f^\circ$ (kJ/mol) | $S^\circ$ (J/K · mol) |
|---|---|---|---|---|
| Acetic acid(l) | $CH_3COOH$ | −484.2 | −389.45 | 159.8 |
| Acetaldehyde(g) | $CH_3CHO$ | −166.35 | −139.08 | 264.2 |
| Acetone(l) | $CH_3COCH_3$ | −246.8 | −153.55 | 198.7 |
| Acetylene(g) | $C_2H_2$ | 226.6 | 209.2 | 200.8 |
| Benzene(l) | $C_6H_6$ | 49.04 | 124.5 | 172.8 |
| Butane(g) | $C_4H_{10}$ | −124.7 | −15.7 | 310.0 |
| Ethanol(l) | $C_2H_5OH$ | −276.98 | −174.18 | 161.0 |
| Ethane(g) | $C_2H_6$ | −84.7 | −32.89 | 229.5 |
| Ethylene(g) | $C_2H_4$ | 52.3 | 68.1 | 219.5 |
| Formic acid(l) | $HCOOH$ | −409.2 | −346.0 | 129.0 |
| Glucose(s) | $C_6H_{12}O_6$ | −1274.5 | −910.56 | 212.1 |
| Methane(g) | $CH_4$ | −74.85 | −50.8 | 186.2 |
| Methanol(l) | $CH_3OH$ | −238.7 | −166.3 | 126.8 |
| Propane(g) | $C_3H_8$ | −103.9 | −23.5 | 269.9 |
| Sucrose(s) | $C_{12}H_{22}O_{11}$ | −2221.7 | −1544.3 | 360.2 |

# Appendix 3

## SOLUBILITY PRODUCT CONSTANTS AT 25°C

| Compound | Dissolution Equilibrium | $K_{sp}$ |
|---|---|---|
| **Bromides** | | |
| Copper(I) bromide | $CuBr(s) \rightleftharpoons Cu^+(aq) + Br^-(aq)$ | $4.2 \times 10^{-8}$ |
| Lead(II) bromide | $PbBr_2(s) \rightleftharpoons Pb^{2+}(aq) + 2Br^-(aq)$ | $6.6 \times 10^{-6}$ |
| Mercury(I) bromide | $Hg_2Br_2(s) \rightleftharpoons Hg_2^{2+}(aq) + 2Br^-(aq)$ | $6.4 \times 10^{-23}$ |
| Silver bromide | $AgBr(s) \rightleftharpoons Ag^+(aq) + Br^-(aq)$ | $7.7 \times 10^{-13}$ |
| **Carbonates** | | |
| Barium carbonate | $BaCO_3(s) \rightleftharpoons Ba^{2+}(aq) + CO_3^{2-}(aq)$ | $8.1 \times 10^{-9}$ |
| Calcium carbonate | $CaCO_3(s) \rightleftharpoons Ca^{2+}(aq) + CO_3^{2-}(aq)$ | $8.7 \times 10^{-9}$ |
| Lead(II) carbonate | $PbCO_3(s) \rightleftharpoons Pb^{2+}(aq) + CO_3^{2-}(aq)$ | $3.3 \times 10^{-14}$ |
| Magnesium carbonate | $MgCO_3(s) \rightleftharpoons Mg^{2+}(aq) + CO_3^{2-}(aq)$ | $4.0 \times 10^{-5}$ |
| Silver carbonate | $Ag_2CO_3(s) \rightleftharpoons 2Ag^+(aq) + CO_3^{2-}(aq)$ | $8.1 \times 10^{-12}$ |
| Strontium carbonate | $SrCO_3(s) \rightleftharpoons Sr^{2+}(aq) + CO_3^{2-}(aq)$ | $1.6 \times 10^{-9}$ |
| **Chorides** | | |
| Lead(II) chloride | $PbCl_2(s) \rightleftharpoons Pb^{2+}(aq) + 2Cl^-(aq)$ | $2.4 \times 10^{-4}$ |
| Mercury(I) chloride | $Hg_2Cl_2(s) \rightleftharpoons Hg_2^{2+}(aq) + 2Cl^-(aq)$ | $3.5 \times 10^{-18}$ |
| Silver chloride | $AgCl(s) \rightleftharpoons Ag^+(aq) + Cl^-(aq)$ | $1.6 \times 10^{-10}$ |
| **Chromates** | | |
| Lead(II) chromate | $PbCrO_4(s) \rightleftharpoons Pb^{2+}(aq) + CrO_4^{2-}(aq)$ | $2.0 \times 10^{-14}$ |
| Silver(I) chromate | $Ag_2CrO_4(s) \rightleftharpoons 2Ag^+(aq) + CrO_4^{2-}(aq)$ | $1.2 \times 10^{-12}$ |
| **Fluorides** | | |
| Barium fluoride | $BaF_2(s) \rightleftharpoons Ba^{2+}(aq) + 2F^-(aq)$ | $1.7 \times 10^{-6}$ |
| Calcium fluoride | $CaF_2(s) \rightleftharpoons Ca^{2+}(aq) + 2F^-(aq)$ | $4.0 \times 10^{-11}$ |
| Lead(II) fluoride | $PbF_2(s) \rightleftharpoons Pb^{2+}(aq) + 2F^-(aq)$ | $4.0 \times 10^{-8}$ |

| Compound | Dissolution Equilibrium | $K_{sp}$ |
|---|---|---|
| **Hydroxides** | | |
| Aluminum hydroxide | $Al(OH)_3(s) \rightleftharpoons Al^{3+}(aq) + 3OH^-(aq)$ | $1.8 \times 10^{-33}$ |
| Calcium hydroxide | $Ca(OH)_2(s) \rightleftharpoons Ca^{2+}(aq) + 2OH^-(aq)$ | $8.0 \times 10^{-6}$ |
| Chromium(III) hydroxide | $Cr(OH)_3(s) \rightleftharpoons Cr^{3+}(aq) + 3OH^-(aq)$ | $3.0 \times 10^{-29}$ |
| Copper(II) hydroxide | $Cu(OH)_2(s) \rightleftharpoons Cu^{2+}(aq) + 2OH^-(aq)$ | $2.2 \times 10^{-20}$ |
| Iron(II) hydroxide | $Fe(OH)_2(s) \rightleftharpoons Fe^{2+}(aq) + 2OH^-(aq)$ | $1.6 \times 10^{-14}$ |
| Iron(III) hydroxide | $Fe(OH)_3(s) \rightleftharpoons Fe^{3+}(aq) + 3OH^-(aq)$ | $1.1 \times 10^{-36}$ |
| Magnesium hydroxide | $Mg(OH)_2(s) \rightleftharpoons Mg^{2+}(aq) + 2OH^-(aq)$ | $1.2 \times 10^{-11}$ |
| Strontium hydroxide | $Sr(OH)_2(s) \rightleftharpoons Sr^{2+}(aq) + 2OH^-(aq)$ | $3.2 \times 10^{-4}$ |
| Zinc hydroxide | $Zn(OH)_2(s) \rightleftharpoons Zn^{2+}(aq) + 2OH^-(aq)$ | $1.8 \times 10^{-14}$ |
| **Iodides** | | |
| Copper(I) iodide | $CuI(s) \rightleftharpoons Cu^+(aq) + I^-(aq)$ | $5.1 \times 10^{-12}$ |
| Lead(II) iodide | $PbI_2(s) \rightleftharpoons Cu^{2+}(aq) + 2I^-(aq)$ | $1.4 \times 10^{-8}$ |
| Silver iodide | $AgI(s) \rightleftharpoons Ag^+(aq) + I^-(aq)$ | $8.3 \times 10^{-17}$ |
| **Phosphates** | | |
| Calcium phosphate | $Ca_3(PO_4)_2(s) \rightleftharpoons 3Ca^{2+}(aq) + 2PO_4^{3-}(aq)$ | $1.2 \times 10^{-26}$ |
| Iron(III) phosphate | $FePO_4(s) \rightleftharpoons Fe^{3+}(aq) + PO_4^{3-}(aq)$ | $1.3 \times 10^{-22}$ |
| **Sulfates** | | |
| Barium sulfate | $BaSO_4(s) \rightleftharpoons Ba^{2+}(aq) + SO_4^{2-}(aq)$ | $1.1 \times 10^{-10}$ |
| Calcium sulfate | $CaSO_4(s) \rightleftharpoons Ca^{2+}(aq) + SO_4^{2-}(aq)$ | $2.4 \times 10^{-5}$ |
| Lead(II) sulfate | $PbSO_4(s) \rightleftharpoons Pb^{2+}(aq) + SO_4^{2-}(aq)$ | $1.8 \times 10^{-8}$ |
| Mercury(I) sulfate | $Hg_2SO_4(s) \rightleftharpoons Hg_2^{2+}(aq) + SO_4^{2-}(aq)$ | $6.5 \times 10^{-7}$ |
| Silver sulfate | $Ag_2SO_4(s) \rightleftharpoons 2Ag^+(aq) + SO_4^{2-}(aq)$ | $1.5 \times 10^{-5}$ |
| Strontium sulfate | $SrSO_4(s) \rightleftharpoons Sr^{2+}(aq) + SO_4^{2-}(aq)$ | $3.8 \times 10^{-7}$ |
| **Sulfides** | | |
| Bismuth sulfide | $Bi_2S_3(s) \rightleftharpoons 2Bi^{3+}(aq) + 3S^{2-}(aq)$ | $1.6 \times 10^{-72}$ |
| Cadmium sulfide | $CdS(s) \rightleftharpoons Cd^{2+}(aq) + S^{2-}(aq)$ | $8.0 \times 10^{-28}$ |
| Cobalt(II) sulfide | $CoS(s) \rightleftharpoons Co^{2+}(aq) + S^{2-}(aq)$ | $4.0 \times 10^{-21}$ |
| Copper(II) sulfide | $CuS(s) \rightleftharpoons Cu^{2+}(aq) + S^{2-}(aq)$ | $6.0 \times 10^{-37}$ |
| Iron(II) sulfide | $FeS(s) \rightleftharpoons Fe^{2+}(aq) + S^{2-}(aq)$ | $6.0 \times 10^{-19}$ |
| Lead(II) sulfide | $PbS(s) \rightleftharpoons Pb^{2+}(aq) + S^{2-}(aq)$ | $3.4 \times 10^{-28}$ |
| Manganese(II) sulfide | $MnS(s) \rightleftharpoons Mn^{2+}(aq) + S^{2-}(aq)$ | $3.0 \times 10^{-14}$ |
| Mercury(II) sulfide | $HgS(s) \rightleftharpoons Hg^{2+}(aq) + S^{2-}(aq)$ | $4.0 \times 10^{-54}$ |
| Nickel(II) sulfide | $NiS(s) \rightleftharpoons Ni^{2+}(aq) + S^{2-}(aq)$ | $1.4 \times 10^{-24}$ |
| Silver sulfide | $Ag_2S(s) \rightleftharpoons 2Ag^+(aq) + S^{2-}(aq)$ | $6.0 \times 10^{-51}$ |
| Tin(II) sulfide | $SnS(s) \rightleftharpoons Sn^{2+}(aq) + S^{2-}(aq)$ | $1.0 \times 10^{-26}$ |
| Zinc sulfide | $ZnS(s) \rightleftharpoons Zn^{2+}(aq) + S^{2-}(aq)$ | $3.0 \times 10^{-23}$ |

# Glossary

## A

**absolute temperature scale.** A scale based on $-273.15°C$ (absolute zero) being the lowest point. (11.2)

**absolute zero.** Theoretically the lowest obtainable temperature: $-273.15°C$ or 0 K. (11.2)

**absorbance.** Negative base ten logarithm of transmittance. (4.5)

**absorption spectrum.** Plot of absorbance as a function of wavelength of incident light. (4.5)

**accuracy.** The closeness of a measurement to the true or accepted value. (1.5)

**acid.** See *Arrhenius acid, Brønsted acid,* and *Lewis acid.*

**acid ionization constant ($K_a$).** The equilibrium constant that indicates to what extent a weak acid ionizes. (16.5)

**actinide series.** Series of elements that has partially filled $5f$ and/or $6d$ subshells. (6.9)

**activated complex.** A transient species that forms when molecules collide in an effective collision. (14.4)

**activation energy ($E_a$).** The minimum amount of energy to begin a chemical reaction. (14.4)

**activity series.** A list of metals arranged from top to bottom in order of decreasing ease of oxidation. (4.4)

**actual yield.** The amount of product actually obtained from a reaction. (3.7)

**addition polymer.** A large molecule that forms when small molecules known as monomers join together. (10.6)

**addition polymerization.** Process by which monomers combine to form polymers without the elimination of small molecules, such as water. (25.1)

**adhesion.** The attractions between unlike molecules. (12.2)

**alcohol.** A compound consisting of an alkyl group and the functional group $-OH$. (10.2)

**aldehyde.** A compound containing a hydrogen atom bonded to a carbonyl group. (10.2)

**aliphatic.** Describes organic molecules that do not contain the benzene ring. (10.1)

**alkali metal.** An element from Group 1A, with the exception of H (i.e., Li, Na, K, Rb, Cs, and Fr). (2.4)

**alkaline earth metal.** An element from Group 2A (Be, Mg, Ca, Sr, Ba, and Ra). (2.4)

**alkane.** Hydrocarbons having the general formula $C_nH_{2n+2}$, where $n = 1, 2, \ldots$ (2.6)

**alkyl group.** A portion of a molecule that resembles an alkane. (10.2)

**allotrope.** One of two or more distinct forms of an element. (2.6)

**alloy.** Homogeneous mixture of two or more metals. (23.2)

**alpha particle.** A helium ion with a positive charge of $+2$. (2.2)

**alpha ray.** See *alpha particle.*

**amalgam.** A substance made by combining mercury with one or more other metals. (Chapter 19, Opening Essay; 23.2)

**amide.** An organic molecule that contains an amide group. (10.2)

**amine.** An organic molecule that contains an amino group. (10.2)

**amino acid.** A compound that contains both an amino group and a carboxy group. (10.2)

**amorphous solid.** A solid that lacks a regular three-dimensional arrangement of atoms. (12.5)

**amphoteric.** Describes an oxide that displays both acidic and basic properties. (7.7, 16.2)

**amplitude.** The vertical distance from the midline of a wave to the top of the peak or the bottom of the trough. (6.1)

**angular momentum quantum number ($\ell$).** Describes the shape of the atomic orbital. (6.6)

**anion.** An ion with a negative charge. (2.7)

**anisotropic.** Dependent upon the axis of measurement. (25.3)

**anode.** The electrode at which oxidation occurs. (19.2)

**antibonding molecular orbital.** A molecular orbital that is higher in energy than the atomic orbitals that combined to produce it. (9.6)

**aqueous.** Dissolved in water. (3.3)

**aromatic.** Describes organic compounds that are related to benzene or that contain one or more benzene rings. (10.1)

**Arrhenius acid.** Substance that increases $H^+$ concentration when added to water. (2.6, 4.3)

**Arrhenius base.** Substance that increases $OH^-$ concentration when added to water. (4.1, 4.3)

**Arrhenius equation.** An equation that gives the dependence of the rate constant of a reaction on temperature: $k = Ae^{-E_a/RT}$. (14.4)

**atactic.** Describes polymers in which the substituents are oriented randomly along the polymer chain. (25.2)

**atom.** The basic unit of an element that can enter into chemical combination. (2.2)

**atomic ion.** Atom that has lost or gained one or more electrons, giving it a positive or negative charge. (2.7)

**atomic mass.** The mass of the atom given in atomic mass units (amu). (2.5)

**atomic mass unit (amu).** A mass exactly equal to one-twelfth the mass of one carbon-12 atom. (2.5)

**atomic number ($Z$).** The number of protons in the nucleus of each atom of an element. (2.3)

**atomic orbital.** The wave function of an electron in an atom. (6.5)

**atomic radius.** Metallic: One-half the distance between the nuclei in the two adjacent atoms of the same element in a metal. Covalent: One-half the distance between the nuclei of the two identical atoms in a diatomic molecule. (7.4)

**atomic weight.** The average atomic mass. (2.5)

**Aufbau principle.** The process by which the periodic table can be built up by successively adding one proton to the nucleus and one electron to the appropriate atomic orbital. (6.8)

**autoionization of water.** Ionization of water molecules to give $H^+$ and $OH^-$ ions. (16.2)

**Avogadro's law.** The volume of a sample of gas ($V$) is directly proportional to the number of moles ($n$) in the sample at constant temperature and pressure: $V \propto n$. (11.2)

**Avogadro's number ($N_A$).** The number of atoms in exactly 12 g of carbon-12: $6.022 \times 10^{23}$. (3.4)

**axial.** Describes the two bonds that form an axis perpendicular to the trigonal plane. (9.1)

## B

**band theory.** A theory wherein atomic orbitals merge to form energy bands. (23.3)

**barometer.** An instrument used to measure atmospheric pressure. (11.1)

**base.** See *Arrhenius base, Brønsted base,* and *Lewis base.*

**base ionization constant ($K_b$).** The equilibrium constant that indicates to what extent a weak base ionizes. (16.6)

**battery.** A portable, self-contained source of electric energy consisting of galvanic cells or a series of galvanic cells. (19.6)

**Beer-Lambert law.** Equation relating absorbance to molar absorptivity, solution concentration, and distance that light travels through a solution. (4.5)

**beta particle.** An electron. (2.2)

**beta ray.** See *beta particle*.

**bimolecular.** Describes a reaction in which two reactant molecules collide. (14.5)

**binary compound.** A substance that consists of just two different elements. (2.6)

**blackbody radiation.** The electromagnetic radiation emitted from a heated solid. (6.2)

**body-centered cubic cell.** A unit cell with one atom at the center of the cube and one atom at each of the eight corners. (12.3)

**boiling point.** The temperature at which vapor pressure equals atmospheric pressure. (12.6)

**bond angle.** The angle between two adjacent A$-$B bonds. (9.1)

**bond enthalpy.** The enthalpy change associated with breaking a particular bond in 1 mole of gaseous molecules. (8.9)

**bonding molecular orbital.** A molecular orbital that is lower in energy than the atomic orbitals that combined to produce it. (9.6)

**bond order.** A number based on the number of electrons in bonding and antibonding molecular orbitals that indicates, qualitatively, how stable a bond is. (9.6)

**Born-Haber cycle.** The cycle that relates the lattice energy of an ionic compound to quantities that can be measured. (8.2)

**Boyle's law.** The pressure of a fixed amount of gas at a constant temperature is inversely proportional to the volume of the gas: $P \propto 1/V$. (11.2)

**Bragg equation.** An equation relating the wavelength of X rays, the angle of diffraction, and the spacing between atoms in a lattice. (12.3)

**breeder reactor.** A nuclear reactor that produces more fissionable material than it consumes. (20.5)

**Brønsted acid.** A substance that donates a proton ($H^+$). (4.3, 16.1)

**Brønsted base.** A substance that accepts a proton ($H^+$). (4.3, 16.1)

**buffer.** A solution that contains significant concentrations of both members of a conjugate pair (weak acid/conjugate base or weak base/conjugate acid). (17.2)

# C

**calorimetry.** The measurement of heat changes. (5.4)

**capillary action.** The movement of liquid up a narrow tube as the result of adhesive forces. (12.2)

**carbocation.** A species in which one of the carbons is surrounded by only six electrons. (10.5)

**carbon nanotube.** A tube made of carbon atoms with dimensions on the order of nanometers. (25.5)

**carboxylic acid.** An organic acid that contains a carboxy group. (10.2)

**catalyst.** A substance that increases the rate of a chemical reaction without itself being consumed. (14.6)

**catenation.** The formation of long carbon chains. (10.1)

**cathode.** The electrode at which reduction occurs. (19.2)

**cation.** An ion with a positive charge. (2.7)

**cell potential ($E_{cell}$).** The difference in electric potential between the cathode and the anode. (19.2)

**ceramics.** Polymeric inorganic compounds that share the properties of hardness, strength, and high melting points. (25.2)

**chalcogens.** Elements in Group 6A (O, S, Se, Te, and Po). (2.4)

**Charles and Gay-Lussac's law.** See *Charles's law*.

**Charles's law.** The volume of a fixed amount of gas ($V$) maintained at constant pressure is directly proportional to absolute temperature ($T$): $V \propto T$. (11.2)

**chelating agent.** A polydentate ligand that forms complex ions with metal ions in solution. (22.1)

**chemical change.** A process in which one or more substances are changed into one or more new substances. (1.4)

**chemical energy.** Energy stored within the structural units (molecules or polyatomic ions) of chemical substances. (5.1)

**chemical equation.** Chemical symbols used to represent a chemical reaction. (3.3)

**chemical formula.** Chemical symbols and numerical subscripts used to denote the composition of the substance. (2.6)

**chemical property.** Any property of a substance that cannot be studied without converting the substance into some other substance. (1.4)

**chemistry.** The study of matter and the changes it undergoes. (1.1)

**chiral.** Describes molecules with nonsuperimposable mirror images. (10.4)

**cholesteric.** Describes molecules that are parallel to each other within each layer, but where each layer is rotated with respect to the layers above and below it. (25.3)

**cis.** Describes the isomer in which two substituents both lie on the same side of a double bond. (10.4)

**Clausius-Clapeyron equation.** A linear relationship that exists between the natural log of vapor pressure and the reciprocal of absolute temperature. (12.2)

**closed system.** A system that can exchange energy (but not mass) with the surroundings. (5.2)

**cohesion.** The attraction between like molecules. (12.2)

**colligative properties.** Properties that depend on the number of solute particles in solution but do not depend on the nature of the solute particles. (13.5)

**collision theory.** The reaction rate is directly proportional to the number of molecular collisions per second. (14.4)

**colloid.** A dispersion of particles of one substance throughout another substance. (13.7)

**combination reaction.** A reaction in which two or more reactants combine to form a single product. (3.7)

**combustion.** Burning in air. (3.5)

**combustion analysis.** An experimental determination of an empirical formula by a reaction with oxygen to produce carbon dioxide and water. (3.5)

**combustion reaction.** A reaction in which a substance burns in the presence of oxygen. (3.7)

**common ion effect.** The presence of a common ion suppresses the ionization of a weak acid or weak base. (17.1)

**complex ion.** Charged species consisting of a central metal cation bonded to two or more anions or polar molecules. (17.5)

**composite material.** A material made from two or more substances with different properties that remain separate in the bulk material. (25.2)

**compound.** A substance composed of atoms of two or more elements chemically united in fixed proportions. (1.2)

**concentration.** Amount of solute relative to the volume of a solution or to the amount of solvent in a solution. (4.1)

**concentration cell.** A cell that has the same type of electrode and the same ion in solution (at different concentrations) in the anode and cathode compartments. (19.5)

**condensation.** The phase transition from gas to liquid. (12.2)

**condensation polymer.** A large molecule that forms when small molecules undergo condensation reactions. (10.6, 25.1)

**condensation reaction.** An elimination reaction in which two or more molecules become connected with the elimination of a small molecule, often water. (10.6)

**condensed structural formula.** Shows the same information as a structural formula, but in a condensed form. (10.3)

**condensed structure.** Chemical structure simplified by using abbreviations for repeating structural units. (10.3)

**conduction band.** The antibonding band. (25.6)

**conductor.** A substance through which electrons move freely. (23.3)

**conjugate acid.** The cation that remains when a Brønsted base accepts a proton. (16.1)

**conjugate base.** The anion that remains when a Brønsted acid donates a proton. (16.1)

**conjugate pair.** The combination of a Brønsted acid and its conjugate base (or the combination of a Brønsted base and its conjugate acid). (16.1)

**constitutional isomers.** Compounds with the same chemical formula but different structures. (10.4)

**conversion factor.** A fraction in which the same quantity is expressed one way in the numerator and another way in the denominator. (1.6)

**coordinate covalent bond.** A covalent bond in which one of the atoms donates both electrons. (8.8)

**coordination compound.** A compound that contains coordinate covalent bonds between a metal ion (often a transition metal ion) and two or more polar molecules or ions. (22.1)

**coordination number.** The number of atoms surrounding an atom in a crystal lattice. (12.3) The number of donor atoms surrounding a metal in a complex. (22.1)

**copolymer.** A polymer made of two or more different monomers. (10.6, 25.1)

**corrosion.** The undesirable oxidation of metals. (19.8)

**Coulomb's law.** The force ($F$) between two charged objects ($Q_1$ and $Q_2$) is directly proportional to the product of the two charges and inversely proportional to the distance ($d$) between the objects squared. (7.3)

**covalent bond.** A shared pair of electrons. (8.3)

**covalent bonding.** Two atoms sharing a pair of electrons. (8.3)

**covalent radius.** Half the distance between adjacent, identical nuclei in a molecule. (7.4)

**critical mass.** The minimum amount of fissionable material required to sustain a reaction. (20.5)

**critical pressure ($P_c$).** The minimum pressure that must be applied to liquefy a substance at its critical temperature. (12.6)

**critical temperature ($T_c$).** The temperature at which the gas phase cannot be liquefied, no matter how great the applied pressure. (12.6)

**cross-link.** A bond that forms between a functional group off the backbone of the polymer chain that interacts with another functional group of a second polymer strand creating a new covalent bond and causing the polymer to be stronger and more rigid. (25.1)

**crystal field splitting ($\Delta$).** The difference in energy between the lower and higher $d$-orbital energy levels. (22.3)

**crystalline solid.** A solid that possesses rigid and long-range order; its atoms, molecules, or ions occupy specific positions. (12.3)

# D

**Dalton's law of partial pressures.** The total pressure exerted by a gas mixture is the sum of the partial pressures exerted by each component of the mixture. (11.5)

**dative bond.** A covalent bond in which one of the atoms donates both electrons. (8.8)

**de Broglie wavelength.** A wavelength calculated using the following equation: $\lambda = h/mu$. (6.4)

**decomposition reaction.** A reaction in which one reactant forms two or more products. (3.7)

**degenerate.** Having an equal energy. (6.8)

**delocalized.** Spread out over the molecule or part of the molecule, rather than confined between two specific atoms. (9.7)

**density.** The ratio of mass to volume. (1.3)

**deoxyribonucleic acid (DNA).** Biological polymer arranged in a double-helix shape, consisting of two long chains of nucleotides in which the sugar is deoxyribose. (10.6)

**deposition.** The phase change from gas to solid. (12.6)

**dextrorotatory.** The term used to describe the enantiomer that rotates the plane-polarized light to the right. (22.2)

**diagonal relationships.** Similarities in chemical properties of elements that are in different groups but that are positioned diagonally to one another in the periodic table. (7.7)

**diamagnetic.** A species without unpaired electrons that is weakly repelled by magnetic fields. (9.6)

**diatomic molecule.** A molecule that contains two atoms. (2.6)

**diffusion.** The mixing of gases. (11.6)

**dilution.** The process of preparing a less concentrated solution from a more concentrated one. (4.5)

**dimensional analysis.** The use of conversion factors in problem solving. (1.6)

**diode.** An electronic device that restricts the flow of electrons in a circuit to one direction. (25.6)

**dipole moment ($\mu$).** A quantitative measure of the polarity of a bond. (8.4)

**dipole-dipole interactions.** Attractive forces that act between polar molecules. (12.1)

**diprotic acid.** An acid with two ionizable protons. (4.3)

**dispersion forces.** See *London dispersion forces.*

**displacement.** An atom or ion in a compound is replaced by an atom of another element. (4.4)

**displacement reaction.** Reaction in which two reactants trade components (double displacement) or where a component of a reactant is removed (single displacement). (4.4)

**disproportionation.** When an element undergoes both oxidation and reduction in the same reaction. (4.4)

**dissociation.** The process by which an ionic compound, upon dissolution, breaks apart into its constituent ions. (4.1)

**donor atom.** The atom that bears the unshared pair of electrons. (22.1)

**doping.** The addition of very small quantities of an element with one more or one fewer valence electron than the natural semiconductor. (25.6)

**$d$ orbital.** Atomic orbital in which the angular momentum quantum number ($\ell$) is 2. (6.6)

**double bond.** A multiple bond in which the atoms share two pairs of electrons. (8.3)

**dynamic equilibrium.** Occurs when a forward process and reverse process are occurring at the same rate. (12.2)

# E

**effective collision.** A collision that results in a reaction. (14.4)

**effective nuclear charge ($Z_{eff}$).** The actual magnitude of positive charge that is "experienced" by an electron in the atom. (7.3)

**effusion.** The escape of a gas from a container into a vacuum. (11.6)

**elastomer.** A material that can stretch or bend and then return to its original shape as long as the limits of its elasticity are not exceeded. (25.1)

**electricity.** The movement of electrons. (4.1)

**electrode.** A piece of conducting metal in an electrochemical cell at which either oxidation or reduction takes place. (19.2)

**electrolysis.** The use of electric energy to drive a nonspontaneous redox reaction. (19.7)

**electrolyte.** A substance that dissolves in water to yield a solution that conducts electricity. (4.1)

**electrolytic cell.** An electrochemical cell used for electrolysis. (19.7)

**electromagnetic spectrum.** Consists of radio waves, microwave radiation, infrared radiation, visible light, ultraviolet radiation, X rays, and gamma rays. (6.1)

**electromagnetic wave.** A wave that has an electric field component and a magnetic field component. (6.1)

**electron.** A negatively charged subatomic particle found outside the nucleus of all atoms. (2.2)

**electron affinity ($EA$).** The energy released (the negative of the enthalpy change, $\Delta H$) when an atom in the gas phase accepts an electron. (7.4)

**electron configuration.** The distribution of electrons in the atomic orbitals of an atom. (6.8)

**electron density.** The probability that an electron will be found in a particular region of an atom. (6.5)

**electron domain.** A lone pair or a bond, regardless of whether the bond is single, double, or triple. (9.1)

**electron-domain geometry.** The arrangement of electron domains (bonds and lone pairs) around a central atom. (9.1)

**electron spin quantum number ($m_s$).** The fourth quantum number that differentiates two electrons in the same orbital. (6.6)

**electronegativity.** The ability of an atom in a compound to draw electrons to itself. (8.4)

**electrophile.** A region of positive or partial positive charge. (10.5)

**electrophilic addition.** An addition reaction that begins when an electrophile approaches a region of electron density. (10.5)

**electrostatic energy.** The potential energy that results from the interaction of charged particles. (5.1)

**element.** A substance that cannot be separated into simpler substances by chemical means. (1.2)

**elementary reaction.** A reaction that occurs in a single collision of the reactant molecules. (14.5)

**elimination reaction.** A reaction in which a double bond forms and a molecule such as water is removed. (10.5)

**emission spectrum.** The light emitted, either as a continuum or in discrete lines, by a substance in an excited electronic state. (6.3)

**empirical formula.** The chemical formula that conveys with the smallest possible whole numbers the ratio of combination of elements in a compound. (2.6)

**enantiomers.** Molecules that are mirror images of each other but cannot be superimposed. (10.4, 22.2)

**endothermic process.** A process that absorbs heat. (5.1)

**endpoint.** The point at which the color of the indicator changes. (4.6, 17.3)

**energy.** The capacity to do work or transfer heat. (5.1)

**enthalpy ($H$).** A thermodynamic quantity defined by the equation $H = U + PV$. The change in enthalpy, $\Delta H$, is equal to the heat exchanged between the system and surroundings at constant pressure $q_P$. (5.3)

**enthalpy of reaction ($\Delta H_{rxn}$).** The difference between the enthalpy of the products and the enthalpy of the reactants. (5.3)

**entropy ($S$).** A thermodynamic state function that describes how dispersed a system's energy is. (18.2)

**enzymes.** Biological catalysts. (14.6)

**equatorial.** The three bonds that are arranged in a trigonal plane in a trigonal bipyramidal geometry. (9.1)

**equilibrium.** A state in which forward and reverse processes are occurring at the same rate. (15.1)

**equilibrium constant ($K$).** A number equal to the ratio of the equilibrium concentrations of products to the equilibrium concentrations of reactants, with each concentration raised to the power of its stoichiometric coefficient. (15.2)

**equilibrium expression.** The quotient of product concentrations and reactant concentrations, each raised to the power of its stoichiometric coefficient. (15.2)

**equilibrium process.** A process that can be made to occur by the addition or removal of energy but does not happen on its own. (18.3)

**equilibrium vapor pressure.** The pressure exerted by the molecules that have escaped to the gas phase, once the pressure has stopped increasing. (12.2)

**equivalence point.** The point in a titration where the reaction (e.g., neutralization) is complete. (4.6)

**ester.** An organic molecule containing a $-COOR$ group. (10.2)

**evaporation.** The phase change from liquid to solid at a temperature below the boiling point. (12.2)

**excess reactant.** The reactant present in a greater amount than necessary to react with all the limiting reactant. (3.7)

**excited state.** A state that is higher in energy than the ground state. (6.3)

**exothermic process.** A process that gives off heat. (5.1)

**extensive property.** A property that depends on the amount of matter involved. (1.4)

## F

**face-centered cubic cell.** A cubic unit cell with one atom on each of the six faces and one atom at each of the eight corners. (12.3)

**family.** The elements in a vertical column of the periodic table. (2.4)

**first law of thermodynamics.** Energy can be converted from one form to another, but cannot be created or destroyed. (5.2)

**first-order reaction.** A reaction whose rate depends on the reactant concentration raised to the first power. (14.3)

**$f$ orbital.** Atomic orbital in which the angular momentum quantum number ($\ell$) is 3. (6.6)

**formal charge.** Method of electron "bookkeeping" in which shared electrons are divided equally between the atoms that share them. (8.6)

**formation constant ($K_f$).** The equilibrium constant that indicates to what extent complex-ion formation reactions occur. (17.5)

**formula mass.** The mass of a formula unit. (3.1)

**formula weight.** See *formula mass.*

**fractional precipitation.** The separation of a mixture based upon the components' solubilities. (17.6)

**free energy.** The energy available to do work. (18.4)

**free radical.** A molecule with an odd number of electrons. (8.8)

**freezing point.** The temperature at which a liquid undergoes the phase transition to solid. (12.6)

**frequency ($\nu$).** The number of waves that pass through a particular point in 1 s. (6.1)

**fuel cell.** A voltaic cell in which reactants must be continually supplied. (19.6)

**fullerenes.** Molecules with elongated and elliptical cages of 70 and 80 carbon atoms. (25.5)

**functional group.** The part of a molecule characterized by a special arrangement of atoms that is largely responsible for the chemical behavior of the parent molecule. (2.6)

**fusion.** The phase transition from solid to liquid (melting) (12.6) or the combination of light nuclei to form heavier nuclei. (20.5)

## G

**galvanic cell.** An electrochemical cell in which a spontaneous chemical reaction generates a flow of electrons. (19.2)

**galvanization.** The cathodic protection of iron or steel using zinc. (19.8)

**gamma rays.** High-energy radiation. (2.2)

**gas constant ($R$).** The proportionality constant that appears in the ideal gas equation. (11.3)

**gas laws.** Equations that relate the volume of a gas sample to its other parameters: temperature ($T$), pressure ($P$), and number of moles ($n$). (11.2)

**geometrical isomers.** Molecules that contain the same atoms and bonds arranged differently in space. (10.4, 22.2)

**Gibbs free energy.** The energy available to do work. (18.4)

**glass.** Commonly refers to an optically transparent fusion product of inorganic materials that has cooled to a rigid state without crystallizing. (12.5)

**Graham's law.** The rates of diffusion and effusion are inversely proportional to the square root of the molar mass of the gas. (11.6)

**graphene.** Two-dimensional sheet of $sp^2$-hybridized carbon atoms. (25.5)

**gravimetric analysis.** An analytical technique based on the measurement of mass. (4.6)

**greenhouse effect.** Describes the trapping of heat near Earth's surface by gases in the atmosphere, particularly carbon dioxide. (21.5)

**ground state.** The lowest energy state of an atom. (6.3)

**group.** The elements in a vertical column of the periodic table. (2.4)

## H

**half-cell.** One compartment of an electrochemical cell containing an electrode immersed in a solution. (19.2)

**half-life ($t_{1/2}$).** The time required for the reactant concentration to drop to half its original value. (14.3)

**half-reaction.** The separated oxidation and reduction reactions that make up the overall redox reaction. (19.1)

**half-reaction method.** Balancing an oxidation-reduction equation by separating the oxidation and the reduction, balancing them separately, and adding them back together. (4.4)

**Hall process.** Electrolytic reduction of aluminum from anhydrous aluminum oxide (corundum). (23.7)

**halogens.** The elements in Group 7A (F, Cl, Br, I, and At). (2.4)

**heat.** The transfer of thermal energy between two bodies that are at different temperatures. (5.1)

**heat capacity (*C*).** The amount of heat required to raise the temperature of an object by 1°C. (5.4)

**Heisenberg uncertainty principle.** It is impossible to know simultaneously both the momentum (*p*) (defined as mass times velocity, $m \times u$) and the position (*x*) of a particle with certainty. (6.5)

**Henderson-Hasselbalch equation.** pH = p$K_a$ + log ([conjugate base]/[weak acid]). (17.2)

**Henry's law.** The solubility of a gas in a liquid is proportional to the pressure of the gas over the solution. (13.4)

**Henry's law constant (*k*).** The proportionality constant that is specific to the gas-solvent combination and varies with temperature. (13.4)

**Hess's law.** The change in enthalpy that occurs when reactants are converted to products in a reaction is the same whether the reaction takes place in one step or in a series of steps. (5.5)

**heteroatom.** Any atom in an organic molecule other than carbon or hydrogen. (10.3)

**heterogeneous catalysis.** A catalysis in which the reactants and the catalyst are in different phases. (14.6)

**heterogeneous mixture.** A mixture in which the composition varies. (1.2)

**heteronuclear.** Containing two or more different elements. (2.6)

**high-temperature superconductor.** A material that shows no resistance to the flow of electrons at an unusually high temperature. (25.7)

**homogeneous catalysis.** A catalysis in which the reactants and the catalyst are in the same phase. (14.6)

**homogeneous mixture.** A mixture in which the composition is uniform. Also called a solution. (1.2)

**homonuclear.** Containing atoms of only one element. (2.6)

**Hund's rule.** The most stable arrangement of electrons in orbitals of equal energy is the one in which the number of electrons with the same spin is maximized. (6.8)

**hybridization.** The mixing of atomic orbitals. (9.4)

**hydrate.** A compound with a specific number of water molecules within its solid structure. (2.7)

**hydration.** The process by which water molecules surround solute particles in an aqueous solution. (4.2)

**hydrocarbon.** A compound containing only carbon and hydrogen. (2.6)

**hydrogen bonding.** A special type of dipole-dipole interaction that occurs only in molecules that contain H bonded to a small, highly electronegative atom, such as N, O, or F. (12.1)

**hydronium ion.** A hydrated proton ($H_3O^+$). (4.3)

**hydrophilic.** Water-loving. (13.7)

**hydrophobic.** Water-fearing. (13.7)

**hypertonic.** Describes a solution with a higher concentration of dissolved substances than plasma. (13.5)

**hypothesis.** A tentative explanation for a set of observations. (1.1)

**hypotonic.** Describes a solution that has a lower concentration of dissolved substances than plasma. (13.5)

# I

**ideal gas.** A hypothetical sample of gas whose pressure-volume-temperature behavior is predicted accurately by the ideal gas equation. (11.3)

**ideal gas equation.** An equation that describes the relationship among the four variables *P*, *V*, *n*, and *T*. (11.3)

**ideal solution.** A solution that obeys Raoult's law. (13.5)

**indicator.** A substance that has a distinctly different color in acidic and basic media. (4.6)

**initial rate.** The instantaneous rate at the beginning of a reaction. (14.2)

**inorganic compounds.** Compounds that do not contain carbon or that are derived from nonliving sources. (2.6)

**instantaneous dipole.** A fleeting nonuniform distribution of electron density in a molecule without a permanent dipole. (12.1)

**instantaneous rate.** The reaction rate at a specific time. (14.1)

**insulator.** A substance that does not conduct electricity. (23.3)

**integrated rate law.** ln ($[A]_t/[A]_0$) = $-kt$. (14.3)

**intensive property.** A property that does not depend on the amount of matter involved. (1.4)

**intermediate.** A chemical species that is produced in one step of a reaction mechanism and consumed in a subsequent step. (14.5)

**intermolecular forces.** The attractive forces that hold particles together in the condensed phases. (12.1)

**International System of Units.** See *SI units.* (1.3)

**ion.** Atom or molecule that has lost or gained one or more electrons giving it a positive or negative charge. (2.7)

**ion-dipole interactions.** Coulombic attractions between ions and polar molecules. (12.1)

**ionic bonding.** An electrostatic attraction that holds oppositely charged ions together in an ionic compound. (8.2)

**ionic compound.** Substance consisting of ions held together by electrostatic attraction. (2.7)

**ionic equation.** Chemical equation in which all strong electrolytes are shown as ions. (4.2)

**ionic radius.** The radius of a cation or an anion. (7.6)

**ionizable hydrogen atom.** A hydrogen atom that can be lost as a hydrogen ion, ($H^+$). (2.6)

**ionization.** The process by which a molecular compound forms ions when it dissolves. (4.1)

**ionization energy (*IE*).** The minimum energy required to remove an electron from an atom in the gas phase. (7.4)

**ionosphere.** See *thermosphere.*

**ion pair.** Ions in solution that are held together by electrostatic forces. (13.5)

**isoelectronic.** Describes two or more species with identical electron configurations. (7.5)

**isoelectronic series.** A series of two or more species that have identical electron configurations but different nuclear charges. (7.6)

**isolated system.** A system that can exchange neither energy nor mass with the surroundings. (5.2)

**isomerization reaction.** A reaction in which one isomer is converted to another. (10.5)

**isotactic.** Polymers in which all the substituents (i.e., the R groups) are in the same relative orientation (i.e., on the same side of the polymer chain). (25.1)

**isotonic.** Equal in concentration and osmotic pressure. (13.5)

**isotope.** Atoms that have the same atomic number (*Z*) but different mass numbers (*A*). (2.3)

**isotropic.** Independent of the axis of measurement. (25.3)

# J

**joule.** SI unit of energy, 1 kg · m²/s². (5.1)

# K

**Kekulé structure.** A structure similar to a Lewis structure in which the lone pairs may not be shown. (10.3)

**kelvin.** The SI base unit of temperature. (1.3)

**Kelvin temperature scale.** A temperature scale offset from the Celsius scale by 273.15. One kelvin (1 K) is equal in magnitude to one degree Celsius (1°C). (11.2)

**ketone.** An organic compound consisting of two R groups bonded to a carbonyl. (10.2)

**kinetic energy.** The energy that results from motion. (5.1)

**kinetic molecular theory.** A theory that explains how the molecular nature of gases gives rise to their macroscopic properties. (11.6)

# L

**Lanthanide (rare earth) series.** A series of 14 elements that have incompletely filled 4*f* subshells or that readily give rise to cations that have incompletely filled 4*f* subshells. (6.9)

**lattice.** A three-dimensional array of cations and anions. (2.6)

**lattice energy.** The amount of energy required to convert a mole of ionic solid to its constituent ions in the gas phase. (8.2)

**lattice points.** Positions occupied by atoms, ions, or molecules in a unit cell. (12.3)

**lattice structure.** The arrangement of the particles in a crystalline solid. (12.3)

**law.** A concise verbal or mathematical statement of a reliable relationship between phenomena. (1.1)

**law of conservation of energy.** First law of thermodynamics stating that energy can be neither created nor destroyed. (5.1)

**law of conservation of mass.** An alternate statement of the first law of thermodynamics stating that matter can be neither created nor destroyed. (2.1)

**law of definite proportions.** Different samples of a given compound always contain the same elements in the same mass ratio. (2.1)

**law of mass action.** For a reversible reaction at equilibrium and a constant temperature, the reaction quotient, $Q$, has a constant value, $K$ (the equilibrium constant). (15.2)

**law of multiple proportions.** Different compounds made up of the same elements differ in the number of atoms of each kind that combine. (2.1)

**Le Châtelier's principle.** When a stress is applied to a system at equilibrium, the system will respond by shifting in the direction that minimizes the effect of the stress. (15.5)

**levorotatory.** The term used to describe the enantiomer that rotates the plane-polarized light to the left. (22.2)

**Lewis acid.** A species that can accept a pair of electrons. (16.12)

**Lewis base.** A species that can donate a pair of electrons. (16.12)

**Lewis dot symbol.** An elemental symbol surrounded by dots, where each dot represents a valence electron. (8.1)

**Lewis structure.** A representation of covalent bonding in which shared electron pairs are shown either as dashes or as pairs of dots between two atoms, and lone pairs are shown as pairs of dots on individual atoms. (8.4)

**Lewis theory of bonding.** A chemical bond involves atoms sharing electrons. (8.3)

**ligand.** A molecule or anion that can form coordinate bonds to a metal to form a coordination complex. (22.1)

**limiting reactant.** The reactant that is completely consumed and determines the amount of product formed. (3.7)

**line spectra.** The emission or absorption of light only at discrete wavelengths. (6.3)

**liquid crystal.** A substance that exhibits properties of both a liquid, such as the ability to flow and to take on the shape of a container, and those of a crystal, such as a regular arrangement of particles in a lattice. (25.3)

**localized.** Describes electrons that are shared between two specific atoms and that cannot be repositioned to generate additional resonance structures. (9.7)

**London dispersion forces.** Attractive forces that act between all molecules, including nonpolar molecules, resulting from the formation of instantaneous dipoles and induced dipoles. (12.1)

**lone pair.** A pair of valence electrons that are not involved in covalent bond formation. (8.3)

# M

**magnetic quantum number ($m_\ell$).** Describes the orientation of an orbital in space. (6.6)

**main group elements.** Elements in the $s$ and $p$ blocks of the periodic table. (7.2)

**manometer.** A device used to measure the pressure of gases relative to atmospheric pressure. (11.1)

**mass.** A measure of the amount of matter in an object or sample. (1.3)

**mass defect.** The difference between the actual mass of a nucleus and the mass calculated by summing the masses of the individual nucleons. (20.2)

**mass number ($A$).** The number of neutrons and protons present in the nucleus of an atom of an element. (2.3)

**matter.** Anything that occupies space and has mass. (1.1)

**Meissner effect.** The exclusion of magnetic fields. (25.7)

**melting point.** The temperature at which solid and liquid phases are in equilibrium. (12.6)

**mesosphere.** The region located above the stratosphere in which the concentration of ozone and other gases is low and temperature decreases again with increasing altitude. (21.1)

**metal.** Element with a tendency to lose electrons, located left of the zigzag line on the periodic table. (2.4)

**metallic radius.** Half the distance between the nuclei of two adjacent, identical metal atoms. (7.4)

**metalloid.** Elements with properties intermediate between metals and nonmetals. (7.4)

**metallurgy.** The preparation, separation, and purification of metals. (23.2)

**microstate.** A specific microscopic configuration of a system. (18.2)

**mineral.** A naturally occurring substance with a characteristic chemical composition and specific physical properties. (23.1)

**miscible.** Mutually soluble in any proportions. (13.1)

**mixture.** A combination of two or more substances in which the substances retain their distinct identities. (1.2)

**moderator.** A material that limits the speed of liberated neutrons but does not itself undergo fission when bombarded with neutrons. (20.5)

**molality ($m$).** The number of moles of solute dissolved in 1 kg (1000 g) of solvent. (13.3)

**molar absorptivity.** Proportionality constant relating absorbance to solution concentration and distance that light travels through a solution. (4.5)

**molar concentration.** See *molarity (M)*.

**molar heat of fusion ($\Delta H_{fus}$).** The energy, usually expressed in kJ/mol, required to melt 1 mole of a solid. (12.6)

**molar heat of sublimation ($\Delta H_{sub}$).** The energy, usually expressed in kilojoules, required to sublime 1 mole of a solid. (12.6)

**molar heat of vaporization ($\Delta H_{vap}$).** The amount of heat required to vaporize a mole of substance at its boiling point. (12.6)

**molarity ($M$).** The number of moles of solute per liter of solution. (4.5)

**molar mass.** The mass in grams of 1 mole of the substance. (3.4)

**molar solubility.** The number of moles of solute in one liter of saturated solution (mol/L). (17.4)

**mole (mol).** The amount of a substance that contains as many elementary entities (atoms, molecules, formula units, etc.) as there are atoms in exactly 0.012 kg (12 g) of carbon-12. (3.4)

**molecular equation.** A chemical equation written with all compounds represented by their chemical formulas. (4.2)

**molecular formula.** A chemical formula that gives the number of atoms of each element in a molecule. (2.6)

**molecular geometry.** The arrangement of bonded atoms. (9.1)

**molecularity.** The number of molecules involved in a specific step in a reaction mechanism. (14.5)

**molecular mass.** The sum of the atomic masses (in amu) of the atoms that make up a molecule. (3.1)

**molecular orbital.** An orbital that results from the interaction of the atomic orbitals of the bonding atoms. (9.6)

**molecular orbital theory.** A theory that describes the orbitals in a molecule as bonding and antibonding combinations of atomic orbitals. (9.6)

**molecular weight.** Average molecular mass. (3.1)

**molecule.** A combination of two or more atoms in a specific arrangement held together by chemical bonds. (2.6)

**mole fraction ($\chi_i$).** The number of moles of a component divided by the total number of moles in a mixture. (11.5)

**monatomic ion.** An ion that contains only one atom. (2.7)

**monomer.** A small molecule that can be linked in large numbers to form a large molecule (polymer). (10.6, 25.1)

**monoprotic acid.** An acid with one ionizable proton. (4.3)

**multiple bond.** A chemical bond in which two atoms share two or more pairs of electrons. (8.3)

# N

**nanotechnology.** The development and study of extremely small-scale materials and objects. (25.5)

**nematic.** Describes an arrangement in which molecules are all aligned parallel to one another but with no organization into layers or rows. (25.3)

**Nernst equation.** An equation relating the emf of a galvanic cell with the standard emf and the concentrations of reactants and products. (19.5)

**net ionic equation.** Chemical equation from which spectator ions have been removed. (4.2)

**neutralization reaction.** A reaction between an acid and a base. (4.3)

**neutron.** An electrically neutral subatomic particle with a mass slightly greater than that of a proton. (2.2)

**newton (N).** The SI unit of force. (11.1)

**nitrogen fixation.** The conversion of molecular nitrogen into nitrogen compounds. (21.1)

**noble gas core.** A representation in an electron configuration that shows in brackets the most recently completed noble gas. (6.9)

**noble gases.** Elements in Group 8A (He, Ne, Ar, Kr, Xe, and Rn). (2.4)

**node.** A collection of points at which electron density in an atom is zero. (6.4)

**nonconductor.** A substance that does not conduct electricity. (25.6)

**nonelectrolyte.** A substance that dissolves in water to yield a solution that does not conduct electricity. (4.1)

**nonmetal.** Element with a tendency to gain electrons, located in the upper right portion of the periodic table. (2.4)

**nonpolar.** Having a uniform distribution of electron density. (8.4)

**nonspontaneous process.** A process that does not occur under a specified set of conditions. (18.1)

**nonvolatile.** Having no measurable vapor pressure. (13.5)

**normal boiling point.** Temperature at which a substance boils at 1 atm pressure. (12.6)

***n*-type semiconductor.** Semiconductors in which an electron-rich impurity is added to enhance conduction. (23.3, 25.6)

**nuclear binding energy.** The energy required to separate the nucleons in a nucleus. (20.2)

**nuclear chain reaction.** A self-sustaining reaction sequence of fission reactions. (20.5)

**nuclear fission.** The splitting of a large nucleus into smaller nuclei and one or more neutrons. (20.5)

**nuclear fusion.** The combination of two light nuclei to form one heavier nucleus. (20.6)

**nuclear transmutation.** The conversion of one nucleus to another. (20.1)

**nucleic acid.** Macromolecule formed by polymerization of nucleotides. (10.6)

**nucleons.** Protons, neutrons, and electrons. (2.2)

**nucleophile.** A region of negative or partial negative charge. (10.5)

**nucleophilic addition.** An addition reaction that begins when a nucleophile donates a pair of electrons to an electron-deficient atom. (10.5)

**nucleotide.** A structural unit consisting of a sugar (ribose or deoxyribose) bonded to both a cyclic-amine base and a phosphate group. (10.6)

**nucleus.** The central core of the atom that contains the protons and neutrons. (2.2)

# O

**octet rule.** Atoms will lose, gain, or share electrons to achieve a noble gas electron configuration. (8.3)

**open system.** A system that can exchange mass and energy with its surroundings. (5.2)

**optical isomers.** Nonsuperimposable mirror images. (10.4, 22.2)

**ore.** A mineral deposit concentrated enough to allow economical recovery of a desired metal. (23.1)

**organic compounds.** Compounds containing carbon and hydrogen, sometimes in combination with other elements such as oxygen, nitrogen, sulfur, and the halogens. (2.6)

**osmosis.** The selective passage of solvent molecules through a porous membrane from a more dilute solution to a more concentrated one. (13.5)

**osmotic pressure ($\pi$).** The pressure required to stop osmosis. (13.5)

**overvoltage.** The difference between the electrode potential and the actual voltage required to cause electrolysis. (19.7)

**oxidation.** Loss of electrons. (4.4)

**oxidation number.** See *oxidation state.*

**oxidation-reduction reaction.** See *redox reaction.*

**oxidation state.** The charge an atom would have if electrons were transferred completely. (4.4)

**oxidizing agent.** A species that accepts electrons. (4.4)

**oxoacid.** An acid consisting of one or more ionizable protons and an oxoanion. (2.7)

**oxoanion.** A polyatomic anion that contains one or more oxygen atoms bonded to a central atom. (2.7)

# P

**paramagnetic.** A species with unpaired electrons that are attracted by magnetic fields. (9.6)

**partial pressure ($P_i$).** The pressure exerted by a component in a gas mixture. (11.5)

**pascal (Pa).** The SI unit of pressure. (11.1)

**Pauli exclusion principle.** No two electrons in an atom can have the same four quantum numbers. (6.8)

**peptide bond.** The bond that forms between amino acids. (10.6)

**percent by mass.** The ratio of the mass of an individual component to the total mass, multiplied by 100 percent. (13.3)

**percent composition by mass.** The percent of the total mass contributed by each element in a compound. (3.2)

**percent dissociation.** The percentage of dissolved molecules (or formula units, in the case of an ionic compound) that separate into ions in solution. (13.6)

**percent ionization.** Quantitative description of the degree to which an electrolyte exists as ions in solution. (13.6)

**percent yield.** The ratio of actual yield to theoretical yield, multiplied by 100 percent. (3.7)

**period.** A horizontal row of the periodic table. (2.4)

**periodic table.** A chart in which elements having similar chemical and physical properties are grouped together. (2.4)

**pH.** A scale used to measure acidity. $pH = -\log [H^+]$. (16.3)

**phase change.** When a substance goes from one phase to another phase. (12.6)

**phase diagram.** Summarizes the conditions (temperature and pressure) at which a substance exists as a solid, a liquid, or a gas. (12.7)

**photochemical smog.** Air pollution resulting from the interaction of sunlight with nitrogen monoxide and carbon monoxide from automobile exhaust. (21.7)

**photoelectric effect.** A phenomenon in which electrons are ejected from the surface of a metal exposed to light of at least a certain minimum frequency. (6.2)

**photon.** A particle of light. (6.2)

**physical change.** A process in which the state of matter changes but the identity of the matter does not change. (1.4)

**physical property.** A property that can be observed and measured without changing the identity of a substance. (1.4)

**pi ($\pi$) bond.** Bonds that form from the interaction of parallel *p* orbitals. (9.5)

**pOH.** A scale used to measure basicity. $pOH = -\log [OH^-]$. (16.3)

**polar.** Having a nonuniform electron density. (8.4)

**polar covalent bond.** Bonds in which electrons are unequally shared. (8.4)

**polarimeter.** Device used to measure the angle of rotation of plane-polarized light caused by an optically active compound. (22.3)

**polarized.** A molecule in which a dipole moment has been induced. (12.1)

**polyamide.** Polymers in which the monomers are connected by amide linkages. (25.1)

**polyatomic.** Molecules containing more than two atoms. (2.6)

**polyatomic ion.** Molecule that has lost or gained one or more electrons giving it a positive or negative charge. (2.7)

**polyester.** Polymers in which the monomers are connected by ester linkages. (25.1)

**polymer.** Molecular compounds, either natural or synthetic, that are made up of many repeating units called monomers. (10.6, 25.1)

**polypeptide.** Short chains of amino acids. (10.6)

**polyprotic acid.** An acid with more than two ionizable protons. (2.7, 4.3)

**polysaccharides.** Biological polymer consisting of sugars. (10.6)

*p* **orbital.** Atomic orbital in which the angular momentum quantum number ($\ell$) is 1. (6.6)

**positron.** A subatomic particle with the same mass as an electron, but with a positive charge. (20.1)

**potential energy.** The energy possessed by an object by virtue of its position. (5.1)

**precipitate.** An insoluble solid product that separates from a solution. (4.2)

**precipitation reaction.** A chemical reaction in which a precipitate forms. (4.2)

**precision.** The closeness of agreement of two or more measurements of the same quantity. (1.5)

**pressure.** The force applied per unit area. (11.1)

**principal quantum number (*n*).** Designates the size of the orbital. (6.6)

**product.** A substance that forms in a chemical reaction. (3.3)

**protein.** A polymer of amino acids. (10.6)

**proton.** A positively charged particle in the nucleus of an atom. (2.2)

*p*-**type semiconductor.** A semiconductor in which an electron-poor impurity is added to enhance conduction. (23.3, 25.6)

**pyrometallurgy.** Metallurgic processes carried out at high temperatures. (23.2)

# Q

**qualitative analysis.** The determination of the types of ions present in a solution. (17.6)

**qualitative property.** A property of a system that can be determined by general observation. (1.4)

**quantitative property.** A property of a system that can be measured and expressed with a number. (1.4)

**quantum.** The smallest quantity of energy that can be emitted (or absorbed) in the form of electromagnetic radiation. (6.2)

**quantum numbers.** Numbers required to describe the arrangement of electrons in an atom. (6.6)

# R

**racemic mixture.** An equimolar mixture of enantiomers that does not rotate the plane of plane-polarized light. (10.4, 22.2)

**racemization.** The conversion of a single enantiomer to a racemic mixture of both enantiomers. (10.5)

**radiation.** The emission and transmission of energy through space in the form of waves. (2.2)

**radical.** A chemical species with an odd number of electrons. (8.8, 20.8)

**radioactive decay series.** A sequence of nuclear reactions that ultimately result in the formation of a stable isotope. (20.3)

**radioactivity.** The spontaneous emission of particles or radiation from unstable nuclei. (2.2, 20.1)

**Raoult's law.** The partial pressure of a solvent over a solution, $P_i$, is given by the vapor pressure of the pure solvent, $P°$, times the mole fraction of the solvent in the solution, $\chi_i$. (13.5)

**rate constant (*k*).** The proportionality constant in a rate law. (14.1)

**rate-determining step.** The slowest step in a reaction mechanism. (14.5)

**rate law.** An equation relating the rate of reaction to the concentrations of reactants. (14.2)

**rate of reaction.** The change in concentration of reactants or products per unit time. (14.1)

**reactant.** A substance that is consumed in a chemical reaction. (3.3)

**reaction mechanism.** Series of steps by which a chemical reaction occurs. (14.5)

**reaction order.** The sum of the powers to which all reactant concentrations appearing in the rate law are raised. (14.2)

**reaction quotient (*Q*).** A fraction with product concentrations in the numerator and reactant concentrations in the denominator, each raised to its stoichiometric coefficient. (15.2)

**redox reaction.** A chemical reaction in which electrons are transferred from one reactant to another. (4.4)

**reducing agent.** A species that can donate electrons. (4.4)

**reduction.** A gain of electrons. (4.4)

**resonance structures.** Two or more equally valid Lewis structures for a single molecule that differ only in the positions of electrons. (8.7)

**reversible process.** A process in which the products can react to form reactants. (15.1)

**ribonucleic acid (RNA).** Biological polymer consisting of nucleotides in which the sugar is ribose. (10.6)

**root-mean-square speed ($\mu_{rms}$).** The molecular speed that is inversely proportional to the molecular mass. (11.6)

# S

**salt.** An ionic compound made up of the cation from a base and the anion from an acid. (4.3)

**salt bridge.** An inverted U tube containing an inert electrolyte solution, such as KCl or $NH_4NO_3$, that maintains electrical neutrality in an electrochemical cell. (19.2)

**salt hydrolysis.** The reaction of a salt's constituent ions with water to produce either hydroxide ions or hydronium ions. (16.10)

**saturated solution.** A solution that contains the maximum amount of a solute that will dissolve in a solvent at a specific temperature. (13.1)

**scientific method.** A systematic approach to experimentation. (1.1)

**second law of thermodynamics.** The entropy of the universe increases in a spontaneous process and remains unchanged in an equilibrium process. (18.2)

**second-order reaction.** A reaction whose rate depends on the concentration of one reactant raised to the second power or on the product of the concentrations of two different reactants each raised to the first power. (14.3)

**semiconductor.** A substance that normally does not conduct electricity but that will conduct at elevated temperatures or when combined with a small amount of certain other elements. (23.3, 25.6)

**semipermeable membrane.** A membrane that allows the passage of solvent molecules but blocks the passage of solute molecules. (13.5)

**shielding.** The partial obstruction of nuclear charge by core electrons. (7.3)

**sigma ($\sigma$) bond.** A bond in which the shared electron density is concentrated directly along the internuclear axis. (9.5)

**significant figures.** Meaningful digits in a measured or calculated value. (1.5)

**simple cubic cell.** The basic repeating unit in which there is one atom at each of the eight corners in a cube. (12.3)

**single bond.** A pair of electrons shared by two atoms. (8.3)

**sintering.** A method used to form objects by heating a finely divided substance. (25.2)

**SI units.** International System of Units. A system of units based on metric units. (1.3)

**skeletal structure.** A structure in which straight lines represent carbon-carbon bonds. (10.3)

**smectic.** Containing molecules ordered in two dimensions. The molecules are aligned parallel to one other and are further arranged in layers that are parallel to one another. (25.3)

**solubility.** The maximum amount of solute that will dissolve in a given quantity of solvent at a specific temperature. (4.2, 13.1, 17.4)

**solubility product constant ($K_{sp}$).** The equilibrium constant that indicates to what extent a slightly soluble ionic compound dissolves in water. (17.4)

**solute.** The dissolved substance in a solution. (4.1)

**solution.** A homogeneous mixture consisting of a solvent and one or more solutes. (4.1)

**solvent.** A substance in a solution that is present in the largest amount. (4.1)

*s* **orbital.** Atomic orbital in which the angular momentum quantum number ($\ell$) is 0. (6.6)

**specific heat (s).** The amount of heat required to raise the temperature of 1 g of a substance by 1°C. (5.4)

**spectator ion.** An ion that does not participate in the reaction and appears on both the reactant and product side in the ionic equation. (4.2)

**spectrochemical series.** A list of ligands arranged in increasing order of their abilities to split the d orbital energy levels. (22.3)

**spontaneous process.** A process that occurs under a specified set of conditions. (18.1)

**standard atmospheric pressure.** The pressure that would support a column of mercury exactly 760 mm high at 0°C. (11.1)

**standard enthalpy of formation ($\Delta H_f^\circ$).** The heat change that results when 1 mole of a compound is formed from its constituent elements in their standard states. (5.6)

**standard enthalpy of reaction ($\Delta H_{rxn}^\circ$).** The enthalpy of a reaction carried out under standard conditions. (5.6)

**standard entropy ($S^\circ$).** The absolute entropy of a substance at 1 atm. (18.2)

**standard free energy of formation ($\Delta G_f^\circ$).** The free-energy change that occurs when 1 mole of the compound forms from its constituent elements in their standard states. (18.4)

**standard free energy of reaction ($\Delta G_{rxn}^\circ$).** The free-energy change for a reaction when it occurs under standard-state conditions. (18.4)

**standard hydrogen electrode (SHE).** A half-cell based on the half-reaction $2H^+(1\ M) + 2e^- \longrightarrow H_2(1\ atm)$, which has an arbitrarily defined standard reduction potential of zero. (19.3)

**standard reduction potential ($E_{red}^\circ$).** The potential associated with a reduction half-reaction at an electrode when the ion concentration is 1 M and the gas pressure is 1 atm. (19.3)

**standard solution.** A solution of precisely known concentration. (4.6)

**standard temperature and pressure (STP).** 0°C and 1 atm. (11.3)

**state function.** Properties that are determined by the state of the system, independent of how the state was achieved. (5.2)

**state of a system.** The values of all relevant macroscopic properties, such as composition, energy, temperature, pressure, and volume. (5.2)

**stereoisomers.** Molecules that contain identical bonds but differ in the orientation of those bonds in space. (10.4, 22.2)

**stoichiometric amounts.** Quantities of reactants in the same relative amounts as those represented in the balanced chemical equation. (3.6)

**stoichiometric coefficients.** The numeric values written to the left of each species in a chemical equation to balance the equation. (3.3)

**stratosphere.** The region of the atmosphere located above the troposphere and consisting of nitrogen, oxygen, and ozone. (21.1)

**strong conjugate acid.** The conjugate acid of a weak base. Acts as a weak Brønsted acid in water. (16.7)

**strong conjugate base.** The conjugate base of a weak acid. Acts as a weak Brønsted base in water. (16.7)

**strong electrolyte.** An electrolyte that ionizes or dissociates completely. (4.1)

**structural formula.** A chemical formula that shows the general arrangement of atoms within the molecule. (2.6)

**structural isomer.** Molecules that have the same chemical formula but different arrangements of atoms. (9.2)

**sublimation.** The phase change from solid to gas. (12.6)

**substance.** Matter with a definite (constant) composition and distinct properties. (1.2)

**substituent.** A group other than −H bonded to the carbons of an organic molecule. (10.2)

**substitution reaction.** One group is replaced by another group by electrophilic or nucleophilic attack. (10.5)

**superconducting transition temperature ($T_c$).** The temperature below which an element, compound, or material becomes superconducting. (25.7)

**superconductor.** A substance with no resistance to the flow of electrons. (25.7)

**supercooling.** A phenomenon in which a liquid can be temporarily cooled to below its freezing point. (12.6)

**supercritical fluid.** A fluid at a temperature and pressure that exceed $T_c$ and $P_c$. (12.6)

**supersaturated solution.** A solution that contains more dissolved solute than is present in a saturated solution. (13.1)

**surface tension.** The amount of energy required to stretch or increase the surface of a liquid by a unit area. (12.2)

**surroundings.** The part of the universe not included in the system. (5.1)

**suture.** Thread used by physicians to close wounds and/or incisions. (25.4)

**syndiotactic.** Polymers in which the substituents alternate positions along the polymer chain. (25.1)

**system.** The specific part of the universe that is of interest to us. (5.1)

# T

**tacticity.** Describes the relative arrangements of chiral carbon atoms within a polymer. (25.1)

**termolecular.** Involving three reactant molecules. (14.5)

**theoretical yield.** The maximum amount of product that can be obtained from a reaction. (3.7)

**theory.** A unifying principle that explains a body of experimental observations and the laws that are based on them. (1.1)

**thermal energy.** The energy associated with the random motion of atoms and molecules. (5.1)

**thermochemical equation.** A chemical equation that includes the enthalpy change. (5.3)

**thermochemistry.** The study of the heat associated with chemical reactions and physical processes. (5.1)

**thermodynamics.** The scientific study of the interconversion of heat and other kinds of energy. (5.1)

**thermonuclear reaction.** Generally refers to a fusion reaction. (20.6)

**thermoplastic.** Polymers that can be melted and reshaped. (25.1)

**thermosetting.** Polymers that assume their final shape as part of the chemical reaction that forms them. (25.1)

**thermosphere.** The outermost layer of the atmosphere. (21.1)

**third law of thermodynamics.** The entropy of a pure crystalline solid is zero at absolute zero (0 K). (18.3)

**titration.** The gradual addition of a solution of known concentration to another solution of unknown concentration until the chemical reaction between the two solutions is complete. (4.6)

**tracers.** Radioactive isotopes that are used to trace the path of the atoms of an element in a chemical or biological process. (20.7)

*trans.* The isomer in which two substituents lie on opposite sides of a double bond. (10.4)

**transition element.** An element that has—or readily forms one or more ions that have—an incompletely filled d subshell (Groups 3B to 8B or Group 1B). (2.4)

**transition metals.** The elements in Group 1B and Groups 3B to 8B. (2.4)

**transition state.** See *activated complex.*

**transmittance.** Ratio of the intensity of light transmitted through a sample (l) to the intensity of the incident light ($l_0$). (4.5)

**transuranium elements.** Elements with atomic numbers greater than 92, created by bombarding other elements with accelerated neutrons, protons, alpha particles, or other nuclei. (20.4)

**triple bond.** A multiple bond in which the atoms share three pairs of electrons. (8.3)

**triple point.** The point at which all three phase boundary lines meet. (12.7)

**triprotic acid.** An acid molecule with three ionizable protons. (4.3)

**troposphere.** The layer of the atmosphere closest to Earth's surface. (21.1)

**Tyndall effect.** The scattering of visible light by colloidal particles. (13.7)

# U

**unimolecular.** Describes a reaction involving one reactant molecule. (14.5)

**unit cell.** The basic repeating structural unit of a crystalline solid. (12.3)

**unsaturated solution.** A solution that contains less solute than it has the capacity to dissolve. (13.1)

# V

**valence band.** The bonding band. (25.6)

**valence bond theory.** Atoms share electrons when an atomic orbital on one atom overlaps with an atomic orbital on the other. (9.3)

**valence electrons.** The outermost electrons of an atom. (7.2)

**valence-shell electron-pair repulsion (VSEPR).** A model that accounts for electron pairs in the valence shell of an atom repelling one another. (9.1)

**van der Waal's equation.** An equation relating the volume of a real gas to the other parameters, $P$, $T$, and $n$. (11.7)

**van der Waal's forces.** The attractive forces that hold particles together in the condensed phases that include dipole-dipole interactions (including hydrogen bonding) and dispersion forces. (12.1)

**van't Hoff factor ($i$).** The ratio of the actual number of particles in solution after dissociation to the number of formula units initially dissolved in solution. (13.5)

**vaporization.** The phase change from liquid to gas at the boiling point. (12.6)

**viscosity.** A measure of a fluid's resistance to flow. (12.2)

**visible spectrophotometry.** Measurement of absorption of visible light for qualitative and quantitative analysis. (4.5)

**volatile.** Describes a substance that has a high vapor pressure. (12.2)

# W

**wavelength ($\lambda$).** The distance between identical points on successive waves. (6.1)

**weak acid.** An acid that ionizes only partially. (16.5)

**weak base.** A base that ionizes only partially. (16.6)

**weak conjugate acid.** A conjugate acid of a strong base. Does not react with water. (16.7)

**weak conjugate base.** A conjugate base of a strong acid. Does not react with water. (16.7)

**weak electrolyte.** A compound that produces ions upon dissolving but exists in solution predominantly as molecules that are not ionized. (4.1)

# X

**X-ray diffraction.** A method of using X rays to bombard a crystalline sample to determine the structure of the crystal. (12.3)

# Z

**zeroth-order reaction.** A constant rate, independent of reactant concentration. (14.3)

# Answers

## Chapter 1

**1.5** (a) Law. (b) Theory. (c) Hypothesis. **1.7** (a) C and O. (b) F and H.
(c) N and H. (d) O. **1.13** (a) K. (b) Sn. (c) Cr. (d) B. (e) Ba. (f) Pu.
(g) S. (h) Ar. (i) Hg. **1.15** (a) Homogeneous mixture. (b) Element.
(c) Compound. (d) Homogeneous mixture. (e) Heterogeneous mixture.
(f) Homogeneous mixture. (g) Heterogeneous mixture. **1.17** (a) Element.
(b) Compound. (c) Compound. (d) Element. **1.23** 3.12 g/mL.
**1.25** (a) 35°C. (b) −11°C. (c) 39°C. (d) 1011°C. (e) −459.67°F.
**1.27** 2.52 mL. **1.29** (a) 388.36 K. (b) $3.10 \times 10^2$ K. (c) $6.30 \times 10^2$ K.
**1.35** (a) Quantitative. (b) Qualitative. (c) Qualitative. (d) Qualitative.
(e) Qualitative. **1.37** (a) Physical change. (b) Chemical change.
(c) Physical change. (d) Chemical change. (e) Physical change.
**1.39** 99.9 g; 20°C; 11.35 g/cm$^3$. **1.45** (a) 0.0152. (b) 0.0000000778. (c)
0.000001. (d) 1600.1. **1.47** (a) $1.8 \times 10^{-2}$. (b) $1.14 \times 10^{10}$. (c) $-5 \times 10^4$.
(d) $1.3 \times 10^3$. **1.49** (a) One. (b) Three. (c) Three. (d) Four. (e) Three.
(f) One. (g) One or two. **1.51** (a) 1.28. (b) $3.18 \times 10^{-3}$ mg. (c) $8.14 \times$
$10^7$ dm. **1.53** Tailor Z's measurements are the most accurate. Tailor Y's
measurements are the least accurate. Tailor X's measurements are the
most precise. Tailor Y's measurements are the least precise. **1.55** (a)
$1.10 \times 10^8$ mg. (b) $6.83 \times 10^{-5}$ m$^3$. (c) $7.2 \times 10^3$ L. (d) $6.24 \times 10^{-8}$ lb.
**1.57** $3.1557 \times 10^7$ s. **1.59** (a) 81 in/s. (b) $1.2 \times 10^2$ m/min. (c) 7.4 km/h.
**1.61** 88 km/h. **1.63** $3.7 \times 10^{-3}$ g Pb. **1.65** (a) $1.85 \times 10^{-7}$ m. (b) $1.4 \times$
$10^{17}$ s. (c) $7.12 \times 10^{-5}$ m$^3$. (d) $8.86 \times 10^4$ L. **1.67** $6.25 \times 10^{-4}$ g/cm$^3$.
**1.69** 0.88 s. **1.71** (a) 2.5 cm. (b) 2.55 cm. **1.73** (a) Chemical. (b) Chemical.
(c) Physical. (d) Physical. (e) Chemical. **1.75** (a) $8.08 \times 10^4$ g. (b) $1.4 \times$
$10^{-6}$ g. (c) 39.9 g. **1.77** 31.35 cm$^3$. **1.79** 10.50 g/cm$^3$. **1.81** 11.4 g/cm$^3$.
**1.83** %Error (°F) = 0.1%; %Error (°C) = 0.3%. **1.85** −40°F = −40°C.
**1.87** $5 \times 10^2$ mL/breath. **1.89** $4.8 \times 10^{19}$ kg NaCl = $5.3 \times 10^{16}$ tons NaCl.
**1.91** The density of the crucible is equal to the density of pure platinum.
**1.93** (a) 75.0 g Au. (b) A troy ounce is heavier than an ounce. **1.95** (a) 0.5%.
(b) 3.1%. **1.97** 208.0 s = 3 min 28.0 s. **1.99** $1.77 \times 10^6$ g Cu. **1.101** $6.0 \times$
$10^{12}$ g Au; $\$1.3 \times 10^{14}$. **1.103** $7.3 \times 10^{21}$ kg Si. **1.105** $9.5 \times 10^{10}$ kg CO$_2$.
**1.107** (a) homogeneous. (b) heterogeneous. **1.109** $1.1 \times 10^2$ yr. **1.111** $2.3 \times$
$10^4$ kg NaF/yr; 99% NaF wasted. **1.113** $4.0 \times 10^{-19}$ g/L. **1.115** Density =
7.20 g/cm$^3$; $r$ = 0.853 cm. **1.117** It would be more difficult to prove that the
unknown substance is an element. Most compounds would decompose on
heating, making them easy to identify. **1.119** Gently heat the liquid to see if
any solid remains after the liquid evaporates. Also, collect the vapor and
then compare the densities of the condensed liquid with the original liquid.
**1.121** The volume occupied by the ice is larger than the volume of the glass
bottle. The glass bottle would break.

## Chapter 2

**2.15** 0.12 mi. **2.21** 145. **2.23** $^{15}_{7}$N: protons = 7, electrons = 7, neutrons = 8;
$^{33}_{16}$S: protons = 16, electrons = 16, neutrons = 17; $^{63}_{29}$Cu: protons = 29,
electrons = 29, neutrons = 34; $^{84}_{38}$Sr: protons = 38, electrons = 38,
neutrons = 46; $^{130}_{56}$Ba: protons = 56, electrons = 56, neutrons = 74;

$^{186}_{74}$W: protons = 74, electrons = 74, neutrons = 112; $^{202}_{80}$Hg: protons = 80,
electrons = 80, neutrons = 122. **2.25** (a) $^{186}_{74}$W. (b) $^{201}_{80}$Hg. (c) $^{76}_{34}$Se.
(d) $^{239}_{94}$Pu. **2.27** (a) 19. (b) 34. (c) 75. (d) 192. **2.35** Metallic character
(a) increases as you progress down a group of the periodic table and
(b) decreases from the left to right across the periodic table. **2.37** Na and K;
N and P; F and Cl. **2.39** Iron: Fe, period 4, upper-left square of Group 8B;
Iodine: I, period 5, Group 7A; Sodium: Na, period 3, Group 1A;
Phosphorus: P, period 3, Group 5A; Sulfur: S, period 3, Group 6A;
Magnesium: Mg, period 3, Group 2A. **2.45** 207.2 amu. **2.47** $^6$Li = 7.5%,
$^7$Li = 92.5%. **2.49** $5.1 \times 10^{24}$ amu. **2.59** (a) Polyatomic, elemental form, not
a compound. (b) Polyatomic, compound. (c) Diatomic, compound. **2.61**
Elements: N$_2$, S$_8$, H$_2$; Compounds: NH$_3$, NO, CO, CO$_2$, SO$_2$. **2.63** (a) CN.
(b) CH. (c) C$_9$H$_{20}$. (d) P$_2$O$_5$. (e) BH$_3$. **2.65** C$_3$H$_7$NO$_2$. **2.67** (a) Nitrogen
trichloride. (b) Iodine heptafluoride. (c) Tetraphosphorus hexoxide.
(d) Disulfur dichloride. **2.69** (a) NF$_3$: nitrogen trifluoride. (b) PBr$_5$:
phosphorus pentabromide. (c) SCl$_2$: sulfur dichloride. **2.75** Na$^+$: 11 protons,
10 electrons; Ca$^{2+}$: 20 protons, 18 electrons; Al$^{3+}$: 13 protons, 10 electrons;
Fe$^{2+}$: 26 protons, 24 electrons; I$^-$: 53 protons, 54 electrons; F$^-$: 9 protons,
10 electrons; S$^{2-}$: 16 protons, 18 electrons; O$^{2-}$: 8 protons, 10 electrons;
N$^{3-}$: 7 protons, 10 electrons. **2.77** (a) Na$_2$O. (b) FeS. (c) Co$_2$(SO$_4$)$_3$.
(d) BaF$_2$. **2.79** Ionic: LiF, BaCl$_2$, KCl; Molecular: SiCl$_4$, B$_2$H$_6$, C$_2$H$_4$. **2.81**
(a) Potassium dihydrogen phosphate. (b) Potassium hydrogen phosphate.
(c) Hydrogen bromide. (d) Hydrobromic acid. (e) Lithium carbonate.
(f) Potassium dichromate. (g) Ammonium nitrite. (h) Hydrogen iodate
(in water, iodic acid). (i) Phosphorus pentafluoride. (j) Tetraphosphorus
hexoxide. (k) Cadmium iodide. (l) Strontium sulfate. (m) Aluminum
hydroxide. **2.83** (a) RbNO$_2$. (b) K$_2$S. (c) NaHS. (d) Mg$_3$(PO$_4$)$_2$. (e) CaHPO$_4$.
(f) PbCO$_3$. (g) SnF$_2$. (h) (NH$_4$)$_2$SO$_4$. (i) AgClO$_4$. (j) BCl$_3$. **2.85** (a)
Mg(NO$_3$)$_2$. (b) Al$_2$O$_3$. (c) LiH. (d) Na$_2$S. **2.87** Acid: compound that produces
H$^+$; Base: compound that produces OH$^-$; Oxoacids: acids that contain
oxygen; Oxoanions: the anions that remain when oxoacids lose H$^+$ ions;
Hydrates: ionic solids that have water molecules in their formulas. **2.89** (c)
Changing the electrical charge of an atom usually has a major effect on its
chemical properties. The two electrically neutral carbon isotopes should
have nearly identical chemical properties. **2.91** I$^-$. **2.93** NaCl is an ionic
compound; it doesn't consist of molecules. **2.95** (a) Molecule and
compound. (b) Element and molecule. (c) Element. (d) Molecule and
compound. (e) Element. (f) Element and molecule. (g) Element and
molecule. (h) Molecule and compound. (i) Compound, not molecule.
(j) Element. (k) Element and molecule. (l) Compound, not molecule. **2.97** It
establishes a standard mass unit that permits the measurement of masses of
all other isotopes relative to carbon-12. **2.99** $^{11}_{5}$B, protons = 5, neutrons = 6,
electrons = 5, net charge = 0; $^{54}_{26}$Fe$^{2+}$, protons = 26, neutrons = 28,
electrons = 24, net charge = +2; $^{31}_{15}$P$^{3-}$, protons = 15, neutrons = 16,
electrons = 18, net charge = −3; $^{196}_{79}$Au, protons = 79, neutrons = 117,
electrons = 79, net charge = 0; $^{222}_{86}$Rn, protons = 86, neutrons = 136,
electrons = 86, net charge = 0. **2.101** (a) Li$^+$. (b) S$^{2-}$. (c) I$^-$. (d) N$^{3-}$.
(e) Al$^{3+}$. (f) Cs$^+$. (g) Mg$^{2+}$. **2.103** Group 7A, binary: HF, hydrofluoric
acid; HCl, hydrochloric acid; HBr, hydrobromic acid; HI, hydroiodic acid.

Group 7A, oxoacids: $HClO_4$, perchloric acid; $HClO_3$, chloric acid; $HClO_2$, chlorous acid; $HClO$, hypochlorous acid; $HBrO_3$, bromic acid; $HBrO_2$, bromous acid; $HBrO$, hypobromous acid; $HIO_4$, periodic acid; $HIO_3$, iodic acid; $HIO$, hypoiodous acid. Examples of oxoacids containing other Group A-block elements are: $H_3BO_3$, boric acid; $H_2CO_3$, carbonic acid; $HNO_3$, nitric acid; $HNO_2$, nitrous acid; $H_3PO_4$, phosphoric acid; $H_3PO_3$, phosphorous acid; $H_3PO_2$, hypophosphorous acid; $H_2SO_4$, sulfuric acid; $H_2SO_3$, sulfurous acid. Binary acids formed from other Group A-block elements other than Group 7A: $H_2S$, hydrosulfuric acid. **2.105** $^4_2He$: protons = 2, neutrons = 2, neutrons/protons = 1.00; $^{20}_{10}Ne$: protons = 10, neutrons = 10, neutrons/protons = 1.00; $^{40}_{18}Ar$: protons = 18, neutrons = 22, neutrons/protons = 1.22; $^{84}_{36}Kr$: protons = 36, neutrons = 48, neutrons/protons = 1.33; $^{132}_{54}Xe$: protons = 54, neutrons = 78, neutrons/protons = 1.44. The neutron/proton ratio increases with increasing atomic number. **2.107** Cu, Ag, and Au are fairly chemically unreactive. This makes them especially suitable for making coins and jewelry that you want to last a very long time. **2.109** MgO and SrO. **2.111** (a) Berkelium (Berkeley, CA); Europium (Europe); Francium (France); Scandium (Scandinavia); Ytterbium (Ytterby, Sweden); Yttrium (Ytterby, Sweden). (b) Einsteinium (Albert Einstein); Fermium (Enrico Fermi); Curium (Marie and Pierre Curie); Mendelevium (Dmitri Mendeleev); Lawrencium (Ernest Lawrence); Meitnerium (Lise Meitner). (c) Arsenic, Cesium, Chlorine, Chromium, Iodine. **2.113** The mass of fluorine reacting with hydrogen and deuterium would be the same. The ratio of F atoms to hydrogen (or deuterium) atoms is 1:1 in both compounds. This does not violate the law of definite proportions. When the law of definite proportions was formulated, scientists did not know of the existence of isotopes. **2.115** (a) Br. (b) Rn. (c) Se. (d) Rb. (e) Pb. **2.117** $Mg^{2+}$, $HCO_3^-$, $Mg(HCO_3)_2$, Magnesium bicarbonate; $Sr^{2+}$, $Cl^-$, $SrCl_2$, Strontium chloride; $Fe^{3+}$, $NO_2^-$, $Fe(NO_2)_3$, Iron(III) nitrite; $Mn^{2+}$, $ClO_3^-$, $Mn(ClO_3)_2$, Manganese(II) chlorate; $Sn^{4+}$, $Br^-$, $SnBr_4$, Tin(IV) bromide; $Co^{2+}$, $PO_4^{3-}$, $Co_3(PO_4)_2$, Cobalt(II) phosphate; $Hg_2^{2+}$, $I^-$, $Hg_2I_2$, Mercury(I) iodide; $Cu^+$, $CO_3^{2-}$, $Cu_2CO_3$, Copper(I) carbonate; $Li^+$, $N^{3-}$, $Li_3N$, Lithium nitride; $Al^{3+}$, $S^{2-}$, $Al_2S_3$, Aluminum sulfide. **2.119** $1.908 \times 10^{-8}$ g. The predicted change (loss) in mass is too small a quantity to measure. Therefore, for all practical purposes, the law of conservation of mass is assumed to hold for ordinary chemical processes.

**2.121** $CH_4$: H—C—H. $C_2H_6$: H—C—C—H. $C_3H_8$: H—C—C—C—H .

$C_4H_{10}$: H—C—C—C—C—H, H—C—C—C—H .

$C_5H_{12}$: H—C—C—C—C—C—H, H—C—C—C—C—H,

H—C——C——C—H.

**2.123** (a) Yes. (b) Acetylene: any formula with C:H = 1:1 (CH, $C_2H_2$, etc.); Ethane: any formula with C:H = 1:3 ($CH_3$, $C_2H_6$, etc.). **2.125** Manganese, Mn. **2.127** Chloric acid, nitrous acid, hydrocyanic acid, and sulfuric acid.

# Chapter 3

**3.3** (a) 50.48 amu. (b) 92.02 amu. (c) 64.07 amu. (d) 84.16 amu. (e) 34.02 amu. (f) 342.3 amu. (g) 17.03 amu. **3.5** (a) 16.04 amu. (b) 46.01 amu. (c) 80.07 amu. (d) 78.11 amu. (e) 149.9 amu. (f) 174.27 amu. (g) 310.2 amu. **3.9** 78.77% Sn; 21.23% O. **3.11** (d) Ammonia, $NH_3$. **3.13** 39.89% Ca, 18.50% P, 41.41% O, 0.20% H. **3.19** (a) KOH + $H_3PO_4$ $\longrightarrow$ $K_3PO_4$ + $H_2O$. (b) Zn + AgCl $\longrightarrow$ $ZnCl_2$ + Ag. (c) $NaHCO_3$ $\longrightarrow$ $Na_2CO_3$ + $H_2O$ + $CO_2$. (d) $NH_4NO_2$ $\longrightarrow$ $N_2$ + $H_2O$. (e) $CO_2$ + KOH $\longrightarrow$ $K_2CO_3$ + $H_2O$.
**3.21** (a) Potassium and water react to form potassium hydroxide and hydrogen. (b) Barium hydroxide and hydrochloric acid react to form barium chloride and water. (c) Copper and nitric acid react to form copper nitrate, nitrogen monoxide and water. (d) Aluminum and sulfuric acid react to form aluminum sulfate and hydrogen. (e) Hydrogen iodide reacts to form hydrogen and iodine. **3.23** (a) $2N_2O_5$ $\longrightarrow$ $2N_2O_4$ + $O_2$. (b) $2KNO_3$ $\longrightarrow$ $2KNO_2$ + $O_2$. (c) $NH_4NO_3$ $\longrightarrow$ $N_2O$ + $2H_2O$. (d) $NH_4NO_2$ $\longrightarrow$ $N_2$ + $2H_2O$. (e) $2NaHCO_3$ $\longrightarrow$ $Na_2CO_3$ + $H_2O$ + $CO_2$. (f) $P_4O_{10}$ + $6H_2O$ $\longrightarrow$ $4H_3PO_4$. (g) $2HCl$ + $CaCO_3$ $\longrightarrow$ $CaCl_2$ + $H_2O$ + $CO_2$. (h) $2Al$ + $3H_2SO_4$ $\longrightarrow$ $Al_2(SO_4)_3$ + $3H_2$. (i) $CO_2$ + $2KOH$ $\longrightarrow$ $K_2CO_3$ + $H_2O$. (j) $CH_4$ + $2O_2$ $\longrightarrow$ $CO_2$ + $2H_2O$. (k) $Be_2C$ + $4H_2O$ $\longrightarrow$ $2Be(OH)_2$ + $CH_4$. (l) $3Cu$ + $8HNO_3$ $\longrightarrow$ $3Cu(NO_3)_2$ + $2NO$ + $4H_2O$. (m) $S$ + $6HNO_3$ $\longrightarrow$ $H_2SO_4$ + $6NO_2$ + $2H_2O$. (n) $2NH_3$ + $3CuO$ $\longrightarrow$ $3Cu$ + $N_2$ + $3H_2O$. **3.25** (d) $3A$ + $2B$ $\longrightarrow$ $2C$ + D. **3.31** $5.8 \times 10^3$ light-yr. **3.33** $9.96 \times 10^{-15}$ mol Co. **3.35** $3.01 \times 10^3$ g Au. **3.37** (a) $1.244 \times 10^{-22}$ g/As atom. (b) $9.746 \times 10^{-23}$ g/Ni atom. **3.39** $2.98 \times 10^{22}$ Cu. **3.41** Two atoms of lead. **3.43** 409 g/mol. **3.45** $3.01 \times 10^{22}$ C atoms, $6.02 \times 10^{22}$ H atoms, $3.01 \times 10^{22}$ O atoms. **3.47** $2.1 \times 10^9$ molecules. **3.49** (a) 80.56% C, 7.51% H, 11.93% O. (b) $2.11 \times 10^{21}$ molecules $C_9H_{10}O$. **3.51** $C_2H_3NO_5$. **3.53** 39.3 g S. **3.55** 5.97 g F. **3.57** (a) $CH_2O$. (b) KCN. **3.59** $C_8H_{10}N_4O_2$. **3.61** $6.12 \times 10^{21}$ molecules. **3.65** $C_{10}H_{20}O$. **3.67** $CH_2O$. **3.69** Empirical formula: $C_3H_2Cl$; Molecular formula: $C_6H_4Cl_2$. **3.71** $C_3H_7O_2NS$. **3.75** 1.01 mol $Cl_2$. **3.77** $2.0 \times 10^1$ mol $CO_2$. **3.79** (a) $2NaHCO_3$ $\longrightarrow$ $Na_2CO_3$ + $CO_2$ + $H_2O$. (b) 78.3 g $NaHCO_3$. **3.81** 255.9 g $C_2H_5OH$; 0.324 L. **3.83** 0.294 mol KCN. **3.85** $NH_4NO_3(s)$ $\longrightarrow$ $N_2O(g)$ + $2H_2O(g)$. (b) $2.0 \times 10^1$ g $N_2O$. **3.87** 18.0 g $O_2$. **3.93** 6 mol $NH_3$ produced; 1 mol $H_2$ left. **3.95** $4NO_2$ + $O_2$ $\longrightarrow$ $2N_2O_5$. The limiting reactant is $NO_2$. **3.97** HCl is the limiting reactant; 23.4 g $Cl_2$ are produced. **3.99** (a) 7.05 g $O_2$. (b) 92.9%. **3.101** $3.48 \times 10^3$ g $C_6H_{14}$. **3.103** 8.55 g $S_2Cl_2$; 76.6%. **3.105** (a) Combustion. (b) Combination. (c) Decomposition. **3.107** Diagram (b). **3.109** $C_3H_2ClF_5O$; 184.50 g/mol. **3.111** $Cl_2O_7$. **3.113** (a) 0.212 mol O. (b) 0.424 mol O. **3.115** 39.6% NO. **3.117** 30.20% C, 5.069% H, 44.57% Cl, 20.16% S. **3.119** 700 g. **3.121** (a) $Zn(s)$ + $H_2SO_4(aq)$ $\longrightarrow$ $ZnSO_4(aq)$ + $H_2(g)$. (b) 64.2%. (c) We assume that the impurities are inert and do not react with the sulfuric acid to produce hydrogen. **3.123** $1.7 \times 10^{15}$ g $CO_2$/yr. **3.125** 0.0011 mol chlorophyll. **3.127** (a) $6.532 \times 10^4$ g. (b) $7.6 \times 10^2$ g HG. **3.129** (a) $4.24 \times 10^{22}$ $K^+$ ions, $4.24 \times 10^{22}$ $Br^-$ ions. (b) $4.58 \times 10^{22}$ $Na^+$ ions, $2.29 \times 10^{22}$ $SO_4^{2-}$ ions. (c) $4.34 \times 10^{22}$ $Ca^{2+}$ ions, $2.89 \times 10^{22}$ $PO_4^{3-}$ ions. **3.131** (a) 0.307 g salicylic acid. (b) 0.410 g salicylic acid. (c) 12.08 g aspirin; 90.2%. **3.133** (a) $C_3H_7NO$. (b) $C_6H_{14}N_2O_2$. **3.135** $6.022 \times 10^{23}$ amu = 1 g. **3.137** 16.00 amu. **3.139** (e) 0.50 mol $Cl_2$. **3.141** $PtCl_2$ and $PtCl_4$. **3.143** (a) Compound X: $MnO_2$; Compound Y: $Mn_3O_4$. (b) $3MnO_2$ $\longrightarrow$ $Mn_3O_4$ + $O_2$. **3.145** $6.1 \times 10^5$ tons $H_2SO_4$. **3.147** $Mg_3N_2$, magnesium nitride. **3.149** $PbC_8H_{20}$. **3.151** (a) $4.3 \times 10^{22}$ Mg atoms. (b) $1.6 \times 10^2$ pm. **3.153** 28.97 g/mol. **3.155** $3.1 \times 10^{23}$ molecules/mol. **3.157** (a) $C_3H_8(g)$ + $3H_2O(g)$ $\longrightarrow$ $3CO(g)$ + $7H_2(g)$. (b) 909 kg $H_2$. **3.159** (a) $0.47/kg. (b) 0.631 kg $K_2O$. **3.161** $BaBr_2$. **3.163** 32.17% NaCl, 20.09% $Na_2SO_4$, 47.75% $NaNO_3$. **3.165** 0.709 g $NO_2$ are produced. $O_3$ is the limiting reactant. 0.0069 mol NO are left. **3.167** (a) $NH_4NO_2$ is the *only* reactant. When a reaction involves only a single reactant, we usually do not describe that reactant as the "limiting reactant" because it is understood that this is the case. (b) In principle, two or more reactants can be limiting reactants if they are exhausted simultaneously. In practice, it is virtually impossible to measure the reactants so precisely that two or more run out at the same time.

# Chapter 4

**4.7** Diagram (c). **4.9** (a) Strong electrolyte. (b) Nonelectrolyte. (c) Weak electrolyte. (d) Strong electrolyte. **4.11** (a) Nonconducting. (b) Conducting. (c) Conducting. **4.13** Since HCl dissolved in water conducts electricity, HCl(*aq*) must actually exist as $H^+(aq)$ cations and $Cl^-(aq)$ anions. Since HCl dissolved in benzene solvent does not conduct electricity, then we must assume that the HCl molecules in benzene solvent do not ionize, but rather exist as un-ionized molecules. **4.17** Diagram (c). **4.19** (a) Insoluble. (b) Insoluble. (c) Soluble. (d) Soluble. **4.21** (a) $2Ag^+(aq) + 2NO_3^-(aq) + 2Na^+(aq) + SO_4^{2-}(aq)$ $\longrightarrow Ag_2SO_4(s) + 2Na^+(aq) + 2NO_3^-(aq); 2Ag^+(aq) + SO_4^{2-}(aq)$ $\longrightarrow Ag_2SO_4(s)$. (b) $Ba^{2+}(aq) + 2Cl^-(aq) + Zn^{2+}(aq) + SO_4^{2-}(aq)$ $\longrightarrow BaSO_4(s) + Zn^{2+}(aq) + 2Cl^-(aq); Ba^{2+}(aq) + SO_4^{2-}(aq) \longrightarrow$ $BaSO_4(s)$. (c) $2NH_4^+(aq) + CO_3^{2-}(aq) + Ca^{2+}(aq) + 2Cl^-(aq) \longrightarrow$ $CaCO_3(s) + 2NH_4^+(aq) + 2Cl^-(aq); Ca^{2+}(aq) + CO_3^{2-}(aq) \longrightarrow$ $CaCO_3(s)$. **4.23** (a) No precipitate forms. (b) $Ba^{2+}(aq) + SO_4^{2-}(aq)$ $\longrightarrow BaSO_4(s)$. **4.31** (a) Brønsted base. (b) Brønsted base. (c) Brønsted acid. (d) Brønsted acid and Brønsted base. **4.33** (a) $HC_2H_3O_2(aq) +$ $KOH(aq) \longrightarrow HC_2H_3O_2(aq) + H_2O(l); Ionic: HC_2H_3O_2(aq) +$ $K^+(aq) + OH^-(aq) \longrightarrow C_2H_3O_2^-(aq) + K^+(aq) + H_2O(l);$ *Net ionic:* $HC_2H_3O_2(aq) + OH^-(aq) \longrightarrow C_2H_3O_2^-(aq) + H_2O(l)$. (b) $H_2CO_3(aq) + 2NaOH(aq) \longrightarrow Na_2CO_3(aq) + 2H_2O(l), Ionic:$ $H_2CO_3(aq) + 2Na^+(aq) + 2OH^-(aq) \longrightarrow 2Na^+(aq) + CO_3^{2-}(aq) +$ $2H_2O(l); Net ionic: H_2CO_3(aq) + 2OH^-(aq) \longrightarrow CO_3^{2-}(aq) +$ $2H_2O(l)$. (c) $2HNO_3(aq) + Ba(OH)_2(aq) \longrightarrow Ba(NO_3)_2(aq) +$ $2H_2O(l), Ionic: 2H^+(aq) + 2NO_3^-(aq) + Ba^{2+}(aq) + 2OH^-(aq) \longrightarrow$ $Ba^{2+}(aq) + 2NO_3^-(aq) + 2H_2O(l); Net ionic: 2H^+(aq) + 2OH^-(aq)$ $\longrightarrow 2H_2O(l) \text{ or } H^+(aq) + OH^-(aq) \longrightarrow H_2O(l)$. **4.41** (a) $2Sr \longrightarrow$ $2Sr^{2+} + 4e^-$, Sr is the reducing agent; $O_2 + 4e^- \longrightarrow 2O^{2-}$, $O_2$ is the oxidizing agent. (b) $2Li \longrightarrow 2Li^+ + 2e^-$, Li is the reducing agent; $H_2$ $+ 2e^- \longrightarrow 2H^-$, $H_2$ is the oxidizing agent. (c) $2Cs \longrightarrow 2Cs^+ +$ $2e^-$, Cs is the reducing agent; $Br_2 + 2e^- \longrightarrow 2Br^-$, $Br_2$ is the oxidizing agent. (d) $3Mg \longrightarrow 3Mg^{2+} + 6e^-$, Mg is the reducing agent; $N_2 + 6e^- \longrightarrow 2N^{3-}$, $N_2$ is the oxidizing agent. **4.43** $H_2S$ ($-2$), $S^{2-}$ ($-2$), $HS^-$ ($-2$) $< S_8$ ($0$) $< SO_2$ ($+4$) $< SO_3$ ($+6$), $H_2SO_4$ ($+6$). **4.45** (a) $+1$. (b) $+7$. (c) $-4$. (d) $-1$. (e) $-2$. (f) $+6$. (g) $+6$. (h) $+7$. (i) $+4$. (j) $0$. (k) $+5$. (l) $-1/2$. (m) $+5$. (n) $+3$. **4.47** (a) $+1$. (b) $-1$. (c) $+3$. (d) $+3$. (e) $+4$. (f) $+6$. (g) $+2$. (h) $+4$. (i) $+2$. (j) $+3$. (k) $+5$. **4.49** If nitric acid is a strong oxidizing agent and zinc is a strong reducing agent, then zinc metal will probably reduce nitric acid when the two react; that is, N will gain electrons and the oxidation number of N must decrease. Since the oxidation number of nitrogen in nitric acid is $+5$, then the nitrogen-containing product must have a smaller oxidation number for nitrogen. The only compound in the list that doesn't have a nitrogen oxidation number less than $+5$ is $N_2O_5$. This is never a product of the reduction of nitric acid. **4.51** Molecular oxygen is a powerful oxidizing agent. In $SO_3$, the oxidation number of the element bound to oxygen (S) is at its maximum value ($+6$); the sulfur cannot be oxidized further. The other elements bound to oxygen in this problem have less than their maximum oxidation number and can undergo further oxidation. Only $SO_3$ does not react with molecular oxygen. **4.53** (a) Decomposition. (b) Displacement. (c) Decomposition. (d) Combination. **4.59** 232 g KI. **4.61** $6.00 \times 10^{-3}$ mol $MgCl_2$. **4.63** (a) 1.16 *M*. (b) 0.608 *M*. (c) 1.78 *M*. **4.65** (a) 136 mL. (b) 62.2 mL. (c) 47 mL. **4.67** Dilute 323 mL of the 2.00 *M* HCl solution to a final volume of 1.00 L. **4.69** Dilute 3.00 mL of the 4.00 *M* $HNO_3$ solution to a final volume of 60.0 mL. **4.71** (a) $BaCl_2$: 0.300 *M* $Cl^-$; NaCl: 0.566 *M* $Cl^-$; $AlCl_3$: 3.606 *M* $Cl^-$. (b) 1.28 *M* $Sr(NO_3)_2$. **4.73** 2.325 *M*. **4.81** 0.215 g AgCl. **4.83** 0.165 g NaCl; $Ag^+(aq) + Cl^-(aq)$ $\longrightarrow$ AgCl(*s*). **4.85** NaOH: Diagram (d); $Ba(OH)_2$: Diagram (c). **4.87** (a) 42.78 mL. (b) 158.5 mL. (c) 79.23 mL. **4.89** (a) Redox. (b) Precipitation. (c) Acid-base. (d) Combination. (e) Redox. (f) Redox. (g) Precipitation. (h) Redox. (i) Redox. (j) Redox. **4.91** (d) 0.20 *M* $Mg(NO_3)_2$ (greatest concentration of ions). **4.93** 773 mL. **4.95** (a) Weak electrolyte. (b) Strong

electrolyte. (c) Strong electrolyte. (d) Nonelectrolyte. **4.97** (a) $C_2H_5ONH_2$ molecules. (b) $K^+$ and $F^-$ ions. (c) $NH_4^+$ and $NO_3^-$ ions. (d) $C_3H_7OH$ molecules. **4.99** 1146 g/mol. **4.101** 1.28 *M*. **4.103** 43.4 g $BaSO_4$. **4.105** (a) Mix a solution containing $Mg^{2+}$ ions such as $MgCl_2(aq)$ or $Mg(NO_3)_2(aq)$ with a solution containing hydroxide ions such as NaOH(*aq*). $Mg(OH)_2$ will precipitate and can be collected by filtration. (b) 0.78 L. **4.107** 1.72 *M*. **4.109** (1) Electrolysis to ascertain if hydrogen and oxygen were produced, (2) The reaction with an alkali metal to see if a base and hydrogen gas were produced, and (3) The dissolution of a metal oxide to see if a base was produced (or a nonmetal oxide to see if an acid was produced). **4.111** Diagram (a) showing $Ag^+$ and $NO_3^-$. The reaction is $AgOH(aq) + HNO_3(aq) \longrightarrow H_2O(l) + AgNO_3(aq)$. **4.113** (a) Check with litmus paper, combine with carbonate or bicarbonate to see if $CO_2$ gas is produced, combine with a base and check for neutralization with an indicator. (b) Titrate a known quantity of acid with a standard NaOH solution. (c) Visually compare the conductivity of the acid with a standard NaCl solution of the same molar concentration. **4.115** (a) $Pb(NO_3)_2(aq) + Na_2SO_4(aq) \longrightarrow PbSO_4(s) + 2NaNO_3(aq)$; net ionic: $Pb^{2+}(aq) + SO_4^{2-}(aq) \longrightarrow PbSO_4(s)$. (b) $6.34 \times 10^{-5}$ *M*. **4.117** (a) $HI(aq) + KOH(aq) \longrightarrow KI(aq) + H_2O(l)$, evaporate to dryness. (b) $2HI(aq) + K_2CO_3(aq) \longrightarrow 2KI(aq) + CO_2(g) + H_2O(l)$, evaporate to dryness. **4.119** (a) Combine any soluble magnesium salt with a soluble hydroxide, filter the precipitate. (b) Combine any soluble silver salt with any soluble iodide salt, filter the precipitate. (c) Combine any soluble barium salt with any soluble phosphate salt, filter the precipitate. **4.121** (a) Add $Na_2SO_4$. (b) Add KOH. (c) Add $AgNO_3$. (d) Add $Ca(NO_3)_2$. (e) Add $Mg(NO_3)_2$. **4.123** Reaction 1: $SO_3^{2-}(aq) + H_2O_2(aq)$ $\longrightarrow SO_4^{2-}(aq) + H_2O(l)$; Reaction 2: $SO_4^{2-}(aq) + Ba^{2+}(aq) \longrightarrow$ $BaSO_4(s)$. **4.125** $Cl_2O$ ($+1$), $Cl_2O_3$ ($+3$), $ClO_2$ ($+4$), $Cl_2O_6$ ($+6$), $Cl_2O_7$ ($+7$). **4.127** 4.99 grains. **4.129** (a) $CaF_2(s) + H_2SO_4(aq) \longrightarrow CaSO_4(s)$ $+ 2HF(g); 2NaCl(s) + H_2SO_4(aq) \longrightarrow Na_2SO_4(aq) + 2HCl(g)$. (b) The sulfuric acid would oxidize the $Br^-$ and $I^-$ ions to $Br_2$ and $I_2$. (c) $PBr_3(l) + 3H_2O(l) \longrightarrow 3HBr(g) + H_3PO_3(aq)$. **4.131** Electric furnace method: $P_4(s) + 5O_2(g) \longrightarrow P_4O_{10}(s)$ (redox), $P_4O_{10}(s) + 6H_2O(l)$ $\longrightarrow 4H_3PO_4(aq)$ (acid-base); Wet process: $Ca_5(PO_4)_3F(s) +$ $5H_2SO_4(aq) \longrightarrow HF(aq) + 3H_3PO_4(aq) + 5CaSO_4(s)$ (acid-base and precipitation). **4.133** (a) $4KO_2(s) + 2CO_2(g) \longrightarrow 2K_2CO_3(s) + 3O_2(g)$. (b) $-1/2$. (c) 34.4 L air. **4.135** No. The oxidation number of all oxygen atoms is zero. **4.137** (a) Acid: $H_3O^+$, base: $OH^-$.

(b) Acid $NH_4^+$; base $NH_2^-$.

**4.139** When a solid dissolves in solution, the volume of the solution usually changes. **4.141** (a) The precipitate $CaSO_4$ formed over Ca preventing the Ca from reacting with the sulfuric acid. (b) Aluminum is protected by a tenacious oxide layer with the composition $Al_2O_3$. (c) These metals react more readily with water: $2Na(s) + 2H_2O(l) \longrightarrow$ $2NaOH(aq) + H_2(g)$. (d) The metal should be placed below Fe and above H. (e) Any metal above Al in the activity series will react with $Al^{3+}$. Metals from Mg to Li will work. **4.143** (a) (1) $Cu(s) + 4HNO_3(aq)$ $\longrightarrow Cu(NO_3)_2(aq) + 2NO_2(g) + 2H_2O(l)$ (redox); (2) $Cu(NO_3)_2(aq)$ $+ 2NaOH(aq) \longrightarrow Cu(OH)_2(s) + 2NaNO_3(aq)$ (precipitation); (3) $Cu(OH)_2(s) \longrightarrow CuO(s) + H_2O(g)$ (decomposition); (4) $CuO(s) +$ $H_2SO_4(aq) \longrightarrow CuSO_4(aq) + H_2O(l)$ (acid-base); (5) $CuSO_4(aq) +$ $Zn(s) \longrightarrow Cu(s) + ZnSO_4(aq)$ (redox); (6) $Zn(s) + 2HCl(aq) \longrightarrow$ $ZnCl_2(aq) + H_2(g)$ (redox). (b) (1) 194 g $Cu(NO_3)_2$; (2) 101 g $Cu(OH)_2$; (3) 82.1 g CuO; (4) 165 g $CuSO_4$; (5) 65.6 g Cu. (c) All of the reaction steps are clean and almost quantitative; therefore, the recovery yield should be high. **4.145** $[Na^+] = 0.5295$ *M*, $[NO_3^-] = 0.4298$ *M*, $[OH^-] =$ 0.09968 *M*, $[Mg^{2+}] \approx 0$ *M*. **4.147** 1.41 *M* $KMnO_4$. **4.149** 0.00666 *M* $C_6H_{12}O_6$. **4.151** $Cu^{2+}(aq) + S^{2-}(aq) \longrightarrow CuS(s)$, $2.31 \times 10^{-4}$ *M* $Cu^{2+}$.

## Chapter 5

**5.13** 48 J. **5.15** 577 J (work done *on* the system). **5.23** (a) 0. (b) −9.5 J. (c) −18 J. **5.25** 4.51 kJ/g. **5.27** $4.80 \times 10^2$ kJ. **5.29** For processes occurring in solution (cellular fluids), there is usually almost no volume change, so $P\Delta V \approx 0$ and $\Delta U + P\Delta V \approx \Delta U$ (see Equation 5.10). **5.35** 728 kJ. **5.37** 50.7°C. **5.39** 26.3°C. **5.41** 47.8°C/day; 4.1 kg water/day. **5.45** 0.30 kJ/mol. **5.47** −238.7 kJ/mol. **5.53** $CH_4(g)$ and $H(g)$. **5.55** $\Delta H_f^\circ[H_2O(l)]$. **5.57** $\Delta H_f^\circ[H_2O(l)]$ is more negative than $\Delta H_f^\circ[H_2O_2(l)]$. **5.59** 177.8 kJ/mol. **5.61** (a) −571.6 kJ/mol. (b) −2599 kJ/mol. **5.63** (a) −724 kJ/mol. (b) $-1.37 \times 10^3$ kJ/mol. (c) $-2.01 \times 10^3$ kJ/mol. **5.65** −3924 kJ/mol. **5.67** $-3.43 \times 10^4$ kJ. **5.69** In a chemical reaction the same elements and the same numbers of atoms are always on both sides of the equation. This provides a consistent reference that allows the energy change in the reaction to be interpreted in terms of the chemical or physical changes that have occurred. In a nuclear reaction the same elements are not always on both sides of the equation and no common reference point exists. **5.71** −44.35 kJ/mol. **5.73** −175.3 kJ. **5.75** (a) Although we cannot measure $\Delta H_{rxn}^\circ$ for this reaction, the reverse process is the combustion of glucose. We could easily measure $\Delta H_{rxn}^\circ$ for this combustion by burning a mole of glucose in a bomb calorimeter. (b) $1.1 \times 10^{19}$ kJ. **5.77** −350.7 kJ/mol. **5.79** 0.492 J/g · °C. **5.81** The first reaction, which is exothermic, can be used to promote the second reaction, which is endothermic. Thus, the two gases are produced alternately. **5.83** 446 mol $H_2$. **5.85** 0.167 mol ethane. **5.87** 5.60 kJ/mol. **5.89** (a). **5.91** (a) 0 J. (b) −9.1 J. **5.93** (a) The more closely packed, the greater the mass of food. Heat capacity depends on both the mass and specific heat: $C = ms$. The heat capacity of the food is greater than the heat capacity of air; hence, the cold in the freezer will be retained longer. (b) Tea and coffee are mostly water; whereas, soup might contain vegetables and meat. Water has a higher heat capacity than the other ingredients in soup; therefore, coffee and tea retain heat longer than soup. **5.95** $-1.84 \times 10^3$ kJ. **5.97** $3.0 \times 10^9$ atomic bombs. **5.99** (a) $2LiOH(aq) + CO_2(g) \longrightarrow Li_2CO_3(aq) + H_2O(l)$. (b) 1.1 kg $CO_2$, 1.2 kg LiOH. **5.101** $-5.2 \times 10^6$ kJ. **5.103** $-3.60 \times 10^2$ kJ/mol Zn or $-3.60 \times 10^2$ kJ/2 mol $Ag^+$. **5.105** We assume that when the car is stopped, its kinetic energy is completely converted into heat (friction of the brakes and friction between the tires and the road). Thus, $q = (1/2)mu^2$, and the amount of heat generated must be proportional to the braking distance, $d$: $d \propto q$ or $d \propto u^2$. Therefore, as $u$ increases to $2u$, $d$ increases to $(2u)^2 = 4u^2$, which is proportional to $4d$. **5.107** Water has a larger specific heat than air. Thus cold, damp air can extract more heat from the body than cold, dry air. By the same token, hot, humid air can deliver more heat to the body. **5.109** $5.8 \times 10^2$ m. **5.111** This law does not hold for Hg because it is a liquid. **5.113** $\Delta H_f^\circ(C_2H_2) = 226.6$ kJ/mol, $\Delta H_f^\circ(C_6H_6) = 49.0$ kJ/mol, $\Delta H_{rxn}^\circ = -630.8$ kJ/mol. **5.115** 4.1 cents. **5.117** (a) $CaC_2(s) + 2H_2O(l) \longrightarrow Ca(OH)_2(s) + C_2H_2(g)$. (b) $1.51 \times 10^6$ J. **5.119** (a) glucose: 31 kJ, sucrose: 33 kJ. (b) glucose: 15 m, sucrose: 16 m. **5.121** $w = 0$, $\Delta U = -5153$ kJ/mol. **5.123** −564.2 kJ/mol. **5.125** 96.21%. **5.127** 58.1°C. **5.129** −9.78 kJ/mol.

**5.131**

|     | $q$ | $w$ | $\Delta U$ | $\Delta H$ |
|-----|-----|-----|------------|------------|
| (a) | −   | 0   | −          | −          |
| (b) | −   | −   | −          | −          |
| (c) | +   | −   | ?          | +          |
| (d) | +   | 0   | +          | +          |
| (e) | +   | −   | 0          | 0          |

**5.133** 43.5°C/day; 4.1 kg water/day. **5.135** 23.6°C.

## Chapter 6

**6.5** (a) $3.5 \times 10^3$ nm. (b) $5.30 \times 10^{14}$ Hz. **6.7** $7.0 \times 10^2$ s. **6.9** $3.26 \times 10^7$ nm, microwave. **6.15** $2.82 \times 10^{-19}$ J. **6.17** (a) $4.6 \times 10^7$ nm, not in the visible region. (b) $4.3 \times 10^{-24}$ J/photon. (c) 2.6 J/mol. **6.19** $1.29 \times 10^{-15}$ J. **6.21** Infrared photons have insufficient energy to cause the

chemical changes. **6.23** $v = 3.23 \times 10^{20}$ s$^{-1}$, $\lambda = 9.29 \times 10^{-13}$ m. **6.29** Analyze the emitted light by passing it through a prism. **6.31** Excited atoms of the chemical elements emit the same characteristic frequencies or lines in a terrestrial laboratory, in the sun, or in a star many light-years distant from Earth. **6.33** $3.027 \times 10^{-19}$ J. **6.35** $v = 1.60 \times 10^{14}$ Hz, $\lambda = 1.88 \times 10^3$ nm. **6.37** 5. **6.41** 0.565 nm. **6.43** $9.96 \times 10^{-32}$ cm. **6.51** $\Delta u \geq 4.38 \times 10^{-26}$ m/s. This uncertainty is far smaller than can be measured. **6.55** $\ell = 1$, $m_\ell = -1$, 0, and 1; $\ell = 0$, $m_\ell = 0$. **6.57** 4s, 4p, 4d, and 4f subshells; 1, 3, 5, and 7 orbitals, respectively. **6.63** (a) $n = 2$, $\ell = 1$, $m_\ell = 1$, 0, or −1. (b) $n = 3$, $\ell = 0$, $m_\ell = 0$. (c) $n = 5$, $\ell = 2$, $m_\ell = 2$, 1, 0, −1, or −2. **6.65** A 2s orbital is larger than a 1s orbital and exhibits a node. Both have the same spherical shape. The 1s orbital is lower in energy than the 2s. **6.67** In H, energy depends only on $n$, but for all other atoms, energy depends on $n$ and $\ell$. **6.69** (a) 2s. (b) 3. (c) equal. (d) equal. (e) 5s. **6.75** In the $n$th shell, there are $n^2$ orbitals and $2n^2$ electrons. **6.77** (a) 3. (b) 6. (c) 0. **6.79** Al: two 2p electrons missing, $1s^22s^22p^63s^23p^1$; B: too many 2p electrons, $1s^22s^22p^1$; F: too many 2p electrons, $1s^22s^22p^5$. **6.81** 1s: (1, 0, 0, +1/2) and (1, 0, 0, −1/2), 2s: (2, 0, 0, +1/2) and (2, 0, 0, −1/2), 2p (2, 1, −1, +1/2), (2, 1, −1, −1/2), (2, 1, 0, +1/2), (2, 1, 0, −1/2), (2, 1, +1, +1/2), and (2, 1, +1, −1/2), 3s (3, 0, 0, +1/2) and (3, 0, 0, −1/2). The element is Mg. **6.91** [Kr]$5s^24d^5$. **6.93** Ge: [Ar]$4s^23d^{10}4p^2$; Fe: [Ar]$4s^23d^6$; Zn: [Ar]$4s^23d^{10}$; Ni: [Ar]$4s^23d^8$; W: [Xe]$6s^24f^{14}5d^4$; Tl: [Xe]$6s^24f^{14}5d^{10}6p^1$. **6.95** Part (b) is correct in the view of contemporary quantum theory. Bohr's explanation of emission and absorption line spectra appears to have universal validity. Parts (a) and (c) are artifacts of Bohr's early planetary model of the hydrogen atom and are *not* considered to be valid today. **6.97** (a) 4. (b) 6. (c) 10. (d) 1. (e) 2. **6.99** In the photoelectric effect, light of sufficient energy shining on a metal surface causes electrons to be ejected (photoelectrons). Since the electrons are charged particles, the metal surface becomes positively charged as more electrons are lost. After a long enough period of time, the positive surface charge becomes large enough to start attracting the ejected electrons back toward the metal with the result that the kinetic energy of the departing electrons becomes smaller. **6.101** (a) $1.20 \times 10^{18}$ photons. (b) $3.76 \times 10^8$ W. **6.103** $\lambda = 419$ nm. This wavelength is in the visible region of the electromagnetic spectrum. Since water is continuously being struck by visible radiation *without* decomposition, it seems unlikely that photodissociation of water by this method is feasible. **6.105** $3.0 \times 10^{19}$ photons/s. **6.107** He$^+$: $n = 3 \longrightarrow 2$: $\lambda = 164$ nm; $n = 4 \longrightarrow 2$: $\lambda = 121$ nm; $n = 5 \longrightarrow 2$: $\lambda = 108$ nm; $n = 6 \longrightarrow 2$: $\lambda = 103$ nm. H: $n = 3 \longrightarrow 2$: $\lambda = 656$ nm; $n = 4 \longrightarrow 2$: $\lambda = 486$ nm; $n = 5 \longrightarrow 2$: $\lambda = 434$ nm; $n = 6 \longrightarrow 2$: $\lambda = 410$ nm. All the Balmer transitions for He$^+$ are in the ultraviolet region; whereas, the transitions for H are all in the visible region. **6.109** $\frac{1}{\lambda_1} = \frac{1}{\lambda_2} + \frac{1}{\lambda_3}$. **6.111** (a) He, $1s^2$. (b) N, $1s^22s^22p^3$. (c) Na, $1s^22s^22p^63s^1$. (d) As, [Ar]$4s^23d^{10}4p^3$. (e) Cl, [Ne]$3s^23p^5$.

**6.113** (a) $\underset{1s^2}{\uparrow\downarrow}$ $\underset{2s^2}{\uparrow\downarrow}$ $\underset{2p^5}{\uparrow\downarrow\ \uparrow\downarrow\ \uparrow}$. (b) [Ne] $\underset{3s^2}{\uparrow\downarrow}$ $\underset{3p^3}{\uparrow\ \uparrow\ \uparrow}$.

(c) [Ar] $\underset{4s^2}{\uparrow\downarrow}$ $\underset{3d^7}{\uparrow\downarrow\ \uparrow\downarrow\ \uparrow\ \uparrow\ \uparrow}$.

**6.115** $\lambda = 0.596$ m; Microwave/radio region. **6.117** (a) False. (b) False. (c) True. (d) False. (e) True. **6.119** $4.9 \times 10^{23}$ photons; 34 $H_2O$ molecules/photon. **6.121** $2.2 \times 10^5$ J. **6.123** (b) and (d) are allowed transitions. Any of the transitions in Figure 6.11 is possible as long as $\ell$ for the final state differs from $\ell$ of the initial state by 1. **6.125** 17.4 pm. **6.127** $\lambda = 0.382$ pm, $v = 7.86 \times 10^{20}$ s$^{-1}$. **6.129** $n_i = 1$: $\Delta E = 1.96 \times 10^{-17}$ J; $n_i = 5$: $\Delta E = 7.85 \times 10^{-19}$ J; transition from $n = 5$ to $n = 1$: $\Delta E = -1.88 \times 10^{-17}$ J; $\lambda = 10.6$ nm. **6.131** $3.9 \times 10^5$ m/s. **6.133** (a) We note that the maximum solar radiation centers around 500 nm. Thus, over billions of years, organisms have adjusted their development to capture energy at or near this wavelength. The two most notable cases are photosynthesis and vision. (b) Astronomers record blackbody radiation curves from stars and compare them with those obtained from objects at different temperatures in the laboratory. Because the shape of the curve and the wavelength corresponding to the maximum depend on the temperature of an object, astronomers can reliably determine the temperature at the surface of a star from the closest matching curve and wavelength. **6.135** 1.06 nm.

## Chapter 7

**7.17** Sulfur, $1s^2 2s^2 2p^6 3s^2 3p^4$. **7.19** (a) and (d); (b) and (e); (c) and (f). **7.21** (a) Group 1A or 1. (b) Group 5A or 15. (c) Group 8A or 18. (d) Group 8B or 10. **7.25** (a) $\sigma = 2$ and $Z_{eff} = +4$. (b) $2s$, $Z_{eff} = +3.22$; $2p$, $Z_{eff} = +3.14$. The values are lower than those in part (a) because the $2s$ and $2p$ electrons actually do shield each other somewhat. **7.33** Na > Mg > Al > P > Cl. **7.35** Fluorine. **7.37** Left to right: S, Se, Ca, K. **7.39** The atomic radius is largely determined by how strongly the outer-shell electrons are held by the nucleus. The larger the effective nuclear charge, the more strongly the electrons are held and the smaller the atomic radius. For the second period, the atomic radius of Li is largest because the $2s$ electron is well shielded by the filled $1s$ shell. The effective nuclear charge that the outermost electrons feel increases across the period as a result of incomplete shielding by electrons in the same shell. Consequently, the orbital containing the electrons is compressed and the atomic radius decreases. **7.41** K < Ca < P < F < Ne. **7.43** The Group 3A elements (such as Al) all have a single electron in the outermost $p$ subshell, which is well shielded from the nuclear charge by the inner electrons and the $ns^2$ electrons. Therefore, less energy is needed to remove a single $p$ electron than to remove a paired $s$ electron from the same principal energy level (such as for Mg). **7.45** 496 kJ/mol is paired with $1s^2 2s^2 2p^6 3s^1$. 2080 kJ/mol is paired with $1s^2 2s^2 2p^6$, a very stable noble gas configuration. **7.47** $8.40 \times 10^6$ kJ/mol. **7.49** Cl. **7.51** Alkali metals have a valence electron configuration of $ns^1$ so they can accept another electron in the $ns$ orbital. On the other hand, alkaline earth metals have a valence electron configuration of $ns^2$. Alkaline earth metals have little tendency to accept another electron, as it would have to go into a higher energy $p$ orbital. **7.57** Fe. **7.59** (a) [Ne]. (b) [Ne]. (c) [Ar]. (d) [Ar]. (e) [Ar]. (f) $[Ar]3d^6$. (g) $[Ar]3d^9$. (h) $[Ar]3d^{10}$. **7.61** (a) $Cr^{3+}$. (b) $Sc^{3+}$. (c) $Rh^{3+}$. (d) $Ir^{3+}$. **7.63** $Be^{2+}$ and He; $N^{3-}$ and $F^-$; $Fe^{2+}$ and $Co^{3+}$; $S^{2-}$ and Ar. **7.67** (a) Cl. (b) $Na^+$. (c) $O^{2-}$. (d) $Al^{3+}$. (e) $Au^{3+}$. **7.69** The $Cu^+$ ion is larger than $Cu^{2+}$ because it has one more electron. **7.71** The binding of a cation to an anion results from electrostatic attraction. As the $+2$ cation gets smaller (from $Ba^{2+}$ to $Mg^{2+}$), the distance between the opposite charges decreases and the electrostatic attraction increases. **7.75** $-199.7°C$. **7.77** Since ionization energies decrease going down a column in the periodic table, francium should have the lowest first ionization energy of all the alkali metals. As a result, Fr should be the most reactive of all the Group 1A elements toward water and oxygen. The reaction with oxygen would probably be similar to that of K, Rb, or Cs. **7.79** The Group 1B elements are much less reactive than the Group 1A elements. The 1B elements are more stable because they have much higher ionization energies resulting from incomplete shielding of the nuclear charge by the inner $d$ electrons. The $ns^1$ electron of a Group 1A element is shielded from the nucleus more effectively by the completely filled noble gas core. Consequently, the outer $s$ electrons of 1B elements are more strongly attracted by the nucleus. **7.81** (a) $Li_2O(s) + H_2O(l) \longrightarrow 2LiOH(aq)$. (b) $CaO(s) + H_2O(l) \longrightarrow Ca(OH)_2(aq)$. (c) $SO_3(g) + H_2O(l) \longrightarrow H_2SO_4(aq)$. **7.83** BaO. As we move down a column, the metallic character of the elements increases. **7.85** (a) Br. (b) N. (c) Rb. (d) Mg. **7.87** $O^{2-} < F^- < Na^+ < Mg^{2+}$. **7.89** M = K, X = Br. **7.91** $O^+$ and N; $S^{2-}$ and Ar; $N^{3-}$ and Ne; $As^{3+}$ and Zn; $Cs^+$ and Xe. **7.93** (a) and (d). **7.95** Fluorine is a yellow-green gas that attacks glass; chlorine is a pale yellow gas; bromine is a fuming red liquid; iodine is a dark, metallic-looking solid. **7.97** F. **7.99** $H^-$. Since $H^-$ has only one proton compared to two protons for He, the nucleus of $H^-$ will attract the two electrons less strongly compared to He. **7.101** $Li_2O$, lithium oxide, basic; BeO, beryllium oxide, amphoteric; $B_2O_3$, diboron trioxide, acidic; $CO_2$, carbon dioxide, acidic; $N_2O_5$, dinitrogen pentoxide, acidic. **7.103** Hydrogen has just one valence electron but it is also just one electron short of a noble gas configuration.

Thus, it can behave like a member of Group 1A or like a member of Group 7A. **7.105** 0.66. **7.107** 77.5%. **7.109** $6.94 \times 10^{-19}$ J/electron. If there are no other electrons with lower kinetic energy, then this is the electron from the valence shell. UV light of the *longest* wavelength (lowest energy) that can still eject electrons should be used. **7.111** X must belong to Group 4A; it is probably Sn or Pb because it is not a very reactive metal (it is certainly not reactive like an alkali metal). Y is a nonmetal since it does *not* conduct electricity. Since it is a light yellow solid, it is probably phosphorus (Group 5A). Z is an alkali metal since it reacts with air to form a basic oxide or peroxide. **7.113** (a) $IE_1 = 3s^1$ electron, $IE_2 = 2p^6$ electron, $IE_3 = 2p^5$ electron, $IE_4 = 2p^4$ electron, $IE_5 = 2p^3$ electron, $IE_6 = 2p^2$ electron, $IE_7 = 2p^1$ electron, $IE_8 = 2s^2$ electron, $IE_9 = 2s^1$ electron, $IE_{10} = 1s^2$ electron, $IE_{11} = 1s^1$ electron. (b) Each break ($IE_1 \longrightarrow IE_2$ and $IE_9 \longrightarrow IE_{10}$) represents the transition to another shell ($n = 3 \longrightarrow 2$ and $n = 2 \longrightarrow 1$).

**7.115** Considering electron configurations, $Fe^{2+}[Ar]3d^6 \longrightarrow Fe^{3+}[Ar]3d^5$, $Mn^{2+}[Ar]3d^5 \longrightarrow Mn^{3+}[Ar]3d^4$, a half-filled shell has extra stability. In oxidizing $Fe^{2+}$, the product is a $d^5$-half-filled shell. In oxidizing $Mn^{2+}$, a $d^5$-half-filled shell electron is being lost, which requires more energy. **7.117** The electron configuration of titanium is $[Ar]4s^2 3d^2$. $K_2TiO_4$ is unlikely to exist because of the oxidation state of Ti of $+6$. Ti in an oxidation state greater than $+4$ is unlikely because of the very high ionization energies needed to remove the fifth and sixth electrons. **7.119** (a) $2KClO_3(s) \longrightarrow 2KCl(s) + 3O_2(g)$. (b) $N_2(g) + 3H_2(g) \longrightarrow 2NH_3(g)$ (industrial); $NH_4Cl(s) + NaOH(aq) \longrightarrow NH_3(g) + NaCl(aq) + H_2O(l)$. (c) $CaCO_3(s) \longrightarrow CaO(s) + CO_2(g)$ (industrial); $CaCO_3(s) + 2HCl(aq) \longrightarrow CaCl_2(aq) + H_2O(l) + CO_2(g)$. (d) $Zn(s) + H_2SO_4(aq) \longrightarrow ZnSO_4(aq) + H_2(g)$. (e) Same as (c), (first equation). **7.121** Examine a solution of $Na_2SO_4$, which is colorless. This shows that the $SO_4^{2-}$ ion is colorless. Thus the blue color is due to $Cu^{2+}(aq)$. **7.123** $Z_{eff}$ increases from left to right across the table, so electrons are held more tightly. (This explains the electron affinity values of C and O.) Nitrogen has a zero value of electron affinity because of the stability of the half-filled $2p$ subshell (that is, N has little tendency to accept another electron). **7.125** Once an atom gains an electron forming a negative ion, adding additional electrons is typically an unfavorable process due to electron-electron repulsions. Second and third electron affinities do not occur spontaneously and are therefore difficult to measure. **7.127** (a) It was determined that the periodic table was based on atomic number, not atomic mass. (b) Argon: 39.95 amu; Potassium: 39.10 amu. **7.129** Element 119: $[Rn]7s^2 5f^{14} 6d^{10} 7p^6 8s^1$. **7.131** 2A. There is a large jump from the second to the third ionization energy, indicating a change in the principal quantum number $n$. **7.133** (a) $SiH_4$, $GeH_4$, $SnH_4$, $PbH_4$. (b) RbH should be more ionic than NaH. (c) $Ra(s) + 2H_2O(l) \longrightarrow Ra(OH)_2(aq) + H_2(g)$. (d) Be (diagonal relationship). **7.135** See sections 7.1 and 7.2. **7.137** (c) Carbon. **7.139** Li: $Z_{eff} = 1.26$, $Z_{eff}/n = 0.630$; Na: $Z_{eff} = 1.84$, $Z_{eff}/n = 0.613$; K: $Z_{eff} = 2.26$, $Z_{eff}/n = 0.565$. As we move down a group, $Z_{eff}$ increases. This is what we would expect because shells with larger $n$ values are less effective at shielding the outer electrons from the nuclear charge. The $Z_{eff}/n$ values are fairly constant, meaning that the screening per shell is about the same. **7.141** Nitrogen. Lithium forms a stable nitride ($Li_3N$).

## Chapter 8

**8.19** (a) Decreases the ionic bond energy. (b) Triples the ionic bond energy. (c) Increases the bond energy by a factor of 4. (d) Increases the bond energy by a factor of 2.

**8.21**

(a) Na· + :F̈·  ⟶  Na⁺:F̈:⁻   (b) 2K· + ·S̈·  ⟶  2K⁺:S̈:²⁻

(c) Ḃa + ·Ö·  ⟶  Ba²⁺:Ö:²⁻   (d) Ȧl· + ·N̈·  ⟶  Al³⁺:N̈:³⁻

**8.23** 860 kJ/mol. **8.33** (a) $BF_3$, boron trifluoride, covalent. (b) KBr, potassium bromide, ionic. **8.37** C—H (ΔEN = 0.4) < Br—H (ΔEN = 0.7) < F—H (ΔEN = 1.9) < Li—Cl (ΔEN = 2.0) < Na—Cl (ΔEN = 2.1) < K—F (ΔEN = 3.2). **8.39** Cl—Cl < Br—Cl < Si—C < Cs—F.
**8.41** (a) Covalent. (b) Polar covalent. (c) Ionic. (d) Polar covalent.

**8.43**

(a) :F̈—Ö—F̈:   (b) :F̈—N̈=N̈—F̈:   (c) H—Si—Si—H (with H above and below each Si)

(d) ⁻:Ö—H   (e) H—C(—Cl:)—C(=O above)(—Ö:⁻)—H (glycine-type structure)   (f) H—C—N⁺(—H)—H (with H atoms)

**8.45**

(a) :F̈—Cl̈(—F̈:)—F̈:   (b) H—S̈e—H   (c) H—N̈(—Ö—H)—H

(d) :Cl̈—P(—Cl̈:)(:O: above)—Cl̈:   (e) H—C(H,H)—C(H,H)—Br̈:   (f) :Cl̈—N̈(—Cl̈: below)—Cl̈:

(g) H—C(—H above)(—H below)—H with H—N̈—H

**8.47** (a) Ö=N⁺=Ö   (b) S̈=C=N̈:⁻  or  ⁻:S̈—C≡N:

(c) ⁻:S̈—S̈:⁻   (d) :F̈—Cl̈⁺—F̈:

**8.49** (a) Neither oxygen atom has a complete octet, and the left-most hydrogen atom shows two bonds (4 electrons). Hydrogen can hold only two electrons in its valence shell.

(b) H—C(—H below)(H :O: above)—C(=O)—Ö—H

**8.53** (a) H—C(:Ö:⁻)=Ö  ⟷  H—C(:O:)—Ö:⁻

(b) H₂C=N⁺(:Ö:⁻)—Ö:⁻  ⟷  H₂C—N⁺(:Ö:)—Ö:  ⟷  H₂C—N(:O:)=N... (three resonance forms)

**8.55** H—N̈=N⁺=N̈:⁻  ⟷  H—N̈—N⁺≡N:  ⟷  H—N̈=N⁺=N̈:²⁻

**8.57** Ö=C=N̈:⁻  ⟷  ⁻:Ö—C≡N:  ⟷  :O≡C—N̈:²⁻

**8.59** (adenine-type resonance structures with H—N̈—H, ring of N and C atoms, C—H, N—H)

**8.67** No. :Cl̈—Be—Cl̈: To make an octet on Be is not plausible,

⁺Cl̈=Be²⁻=Cl̈⁺ **8.69** No. Cl—Sb(—Cl)(Cl below)(Cl above)—Cl **8.71** Coordinate covalent bond,

:Cl̈—Al(—Cl̈: above and below)—Cl̈:  +  :Cl̈:⁻  ⟶  :Cl̈—Al̈(:Cl̈: above and below)—Cl̈: **8.73** Completed octet on

S: [:Ö—S̈—Ö: with :Ö: below]²⁻ ; zero formal charge on S: [:Ö—S̈—Ö: with :O: below]²⁻

**8.77** 303.0 kJ/mol. **8.79** (a) −2759 kJ/mol. (b) −2855.4 kJ/mol.
**8.81** −651 kJ/mol. **8.83** Ionic: RbCl, $KO_2$. Covalent: $PF_5$, $BrF_3$, $CI_4$.
**8.85** Ionic: NaF, $MgF_2$, $AlF_3$. Covalent: $SiF_4$, $PF_5$, $SF_6$, $ClF_3$. **8.87** KF is an ionic compound. It is a solid at room temperature made up of K⁺ and F⁻ ions. It has a high melting point, and it is a strong electrolyte. Benzene, $C_6H_6$, is a covalent compound that exists as discrete molecules. It is a liquid at room temperature. It has a low melting point, is insoluble in water, and is a nonelectrolyte.

**8.89** ⁻N̈=N⁺=N̈⁻  ⟷  :N≡N⁺—N̈:²⁻  ⟷  ²⁻:N̈—N⁺≡N:

**8.91** (a) $AlCl_4^-$. (b) $AlF_6^{3-}$. (c) $AlCl_3$. **8.93** $CF_2$ would be very unstable because carbon does not have an octet. $LiO_2$ would not be stable because the lattice energy between Li⁺ and superoxide $O_2^-$ would be too low to stabilize the solid. $CsCl_2$ requires a $Cs^{2+}$ cation. The second ionization energy is too large to be compensated by the increase in lattice energy. $PI_5$ appears to be a reasonable species. However, the iodine atoms are too large to have five of them "fit" around a single P atom. **8.95** (a) False. (b) True. (c) False. (d) False. **8.97** −67 kJ/mol. **8.99** $N_2$, since it has a triple bond. **8.101** $CH_4$ and $NH_4^+$; $N_2$ and CO; $C_6H_6$ and $B_3N_3H_6$.
**8.103**

(structure: H—N(—H₂C—H)—C(=O above)—C(=O)—Ö—H ring fused indole-type structure with C—H, N—H around five- and six-membered rings)

**8.105**   H—N̈:⁻(—H) + H—Ö:(—H)  ⟶  H—N̈(—H)—H + ⁻:Ö—H

**8.107** The central iodine atom in $I_3^-$ has *ten* electrons surrounding it: two bonding pairs and three lone pairs. The central iodine has an expanded octet. Elements in the second period such as fluorine cannot have an expanded octet as would be required for $F_3^-$.

**8.109**   H—C(—H below)(—H above)—N̈=C=Ö  ⟷  H—C(—H below)(—H above)—N⁺≡C—Ö:⁻

**8.111** Form (a) is the most important structure with no formal charges and all satisfied octets. (b) is likely not as important as (a) because of the positive formal charge on O. Forms (c) and (d) do not satisfy the octet rule for all atoms and are likely not important.

**8.113** (a) :F̈—C(:Cl̈: above)(:Cl̈: below)—Cl̈:   (b) :F̈—C(:F̈: above)(:Cl̈: below)—Cl̈:   (c) H—C(:F̈: above)(:F: below)—Cl̈:   (d) :F̈—C(:F̈:)(:F̈:)—C(:F̈:)(:F̈:)—H

**8.115** (a) −9.2 kJ/mol. (b) −9.2 kJ/mol.

**8.117** (a) $:C≡O:^+$  (b) $:N≡O:^+$  (c) $:C≡N:$  (d) $:N≡N:$
**8.119** True. Each noble gas atom already has completely filled $ns$ and $np$ subshells. **8.121** (a) 114 kJ/mol. (b) The bond in $F_2^-$ is weaker.
**8.123** (a) $:N=Ö ⟷ :N=Ö^+$ The first structure is the most important. (b) No. **8.125** The arrows indicate coordinate covalent bonds.

This dimer does not possess a dipole moment.
**8.127** (a) $·Ö—H$ (b) To complete its octet, the OH radical has a strong tendency to form a bond with a H atom. (c) −46 kJ/mol. (d) 260 nm.
**8.129** 347 kJ/mol. **8.131** EN(O) = 3.2 (Pauling 3.5); EN(F) = 4.4 (Pauling 4.0); EN(Cl) = 3.5 (Pauling 3.0). **8.133** (1) You could estimate the lattice energy of the solid by trying to measure its melting point. $Mg^+O^-$ would have a lattice energy (and, therefore, a melting point) similar to that of $Na^+Cl^-$. This lattice energy and melting point are much lower than those of $Mg^{2+}O^{2-}$. (2) You could determine the magnetic properties of the solid. An $Mg^+O^-$ solid would be paramagnetic while $Mg^{2+}O^{2-}$ solid is diamagnetic. See Chapter 9 of the text.
**8.135**

(a) $H—^H \quad H^+ ⟷ H^+ \quad ^H_H ⟷ H——H$ (b) −413 kJ/mol.

**8.137**

$:N≡N^+—N^-=N=N^- ⟷ ^-N=N^+=N—N^+≡N: ⟷ :N≡N^+—N^-—N^+≡N:$

**8.139** C—C: 347 kJ/mol; N—N: 193 kJ/mol; O—O: 142 kJ/mol. Lone pairs appear to weaken the bond.

## Chapter 9

**9.7** (a) Trigonal pyramidal. (b) Tetrahedral. (c) Tetrahedral. (d) See-saw.
**9.9** (a) Tetrahedral. (b) Trigonal planar. (c) Trigonal pyramidal. (d) Bent.
(e) Bent. **9.11** (a) Linear. (b) Tetrahedral. (c) Trigonal bipyramidal.
(d) Trigonal pyramidal. (e) Tetrahedral. **9.13** Carbon at the center of $H_3C—$: electron domain geometry = tetrahedral, molecular geometry = tetrahedral; Carbon at center of $—CO—OH$: electron-domain geometry = trigonal planar, molecular geometry = trigonal planar; Oxygen in $—O—H$: electron domain geometry = tetrahedral, molecular geometry = bent.
**9.17** (a) Polar. (b) Nonpolar. **9.19** Only (c) is polar. **9.29** (a) $sp^3$. (b) $sp^3$.
**9.31** In $BF_3$, B is $sp^2$ hybridized. In $NH_3$, N is $sp^3$ hybridized. In $F_3B—NH_3$, B and N are both $sp^3$ hybridized. **9.33** $sp^3d$. **9.37** (a) $sp$. (b) $sp$.
(c) $sp$. **9.39** $sp$. **9.41** Nine $\sigma$ bonds and nine $\pi$ bonds. **9.43** 36 $\sigma$ bonds and 10 $\pi$ bonds. **9.49** In order for the two hydrogen atoms to combine to form a $H_2$ molecule, the electrons must have opposite spins. If two H atoms collide and their electron spins are parallel, no bond will form. **9.51** $Li_2^-$ = $Li_2^+ < Li_2$. (Both $Li_2^+$ and $Li_2^-$ have bond order of ½. $Li_2$ has a bond order of 1.) **9.53** $B_2^+$, with a bond order of ½. ($B_2$ has a bond order of 1.)
**9.55** The Lewis diagram has all electrons paired (incorrect) and a double bond (correct). The MO diagram has two unpaired electrons (correct), and a bond order of 2 (correct). **9.57** $O_2$: bond order = 2, paramagnetic; $O_2^+$: bond order = 2.5, paramagnetic; $O_2^-$: bond order = 1.5, paramagnetic; $O_2^{2-}$: bond order = 1, diamagnetic. **9.59** The two shared electrons that make up the single bond in $B_2$ both lie in pi molecular orbitals and constitute a pi bond. The four shared electrons that make up the double bond in $C_2$ all lie in pi molecular orbitals and constitute two pi bonds. **9.63** The left symbol shows three delocalized double bonds (correct). The right symbol shows three localized double bonds and three single bonds (incorrect). **9.65** (a) $Ö=N—Ö^- ⟷ ^-Ö=N—Ö$

(b) $sp^2$. (c) Sigma bonds join the nitrogen atom to the fluorine and oxygen atoms. There is a pi molecular orbital delocalized over the N and O

atoms. **9.67** The central oxygen atom is $sp^2$ hybridized. The unhybridized $2p_z$ orbital on the central oxygen overlaps with the $2p_z$ orbitals on the two terminal atoms. **9.69** $:Br—Hg—Br:$ Linear. You could establish the geometry of $HgBr_2$ by measuring its dipole moment. **9.71** Bent; $sp^3$.
**9.73** (a) $C_6H_8O_6$. (b) Five central O atoms $sp^3$; hybridization of the sixth (double bonded) peripheral O atom is $sp^2$. Three C atoms $sp^2$; three C atoms $sp^3$. (c) The five central O atoms bent. The three $sp^2$ C atoms are trigonal planar. The three $sp^3$ C atoms are tetrahedral. **9.75** (a) 180°.
(b) 120°. (c) 109.5°. (d) 109.5°. (e) 180°. (f) 120°. (g) 109.5°. (h) 109.5°.

**9.77** (a) , planar. (b) $\left[:Ö—Cl—Ö:\atop :Ö:\right]^-$, nonplanar.

(c) $H—C≡N:$, polar. (d) $:F^{Ö}F:$, polar. (e) $O^N O$, greater than 120°. Experimental value is around 135°. **9.79** The S—S bond is a normal 2-electron shared pair covalent bond. Each S is $sp^3$ hybridized, so the X—S—S angle is about 109°. **9.81** Only $ICl_2^-$ and $CdBr_2$ are linear. **9.83** (a) $sp^2$. (b) The molecule on the right is polar. **9.85** Rotation about the sigma bond in 1,2-dichloroethane does not destroy the bond, so the bond is free to rotate. Thus, the molecule is nonpolar because the C—Cl bond moments cancel each other because of the averaging effect brought about by rotation. The $\pi$ bond between the C atoms in cis-dichloroethylene prevents rotation (in order to rotate, the $\pi$ bond must be broken). Therefore, the molecule is polar. **9.87** All except homonuclear diatomic molecules $N_2$ and $O_2$.
**9.89**

The carbon atoms and nitrogen atoms marked with an asterisk (C* and N*) are $sp^2$ hybridized; unmarked carbon atoms and nitrogen atoms are $sp^3$ hybridized; and the nitrogen atom marked with (#) is $sp$ hybridized.
**9.91** C has no d orbitals but Si does (3d). Thus, $H_2O$ molecules can add to Si in hydrolysis (valence-shell expansion). **9.93** The carbons are all $sp^2$ hybridized. The nitrogen double bonded to carbon in the ring is $sp^2$ hybridized. The other nitrogens are $sp^3$ hybridized. **9.95** $F_2$ has 8 electrons in bonding orbitals and 6 electrons in antibonding orbitals, giving it a bond order of 1. $F_2^-$ has 8 electrons in bonding orbitals and 7 electrons in antibonding orbitals, giving it a bond order of 1/2. **9.97** 43.6%. **9.99** The second ("asymmetric stretching mode") and third ("bending mode").
**9.101**

The four carbons marked with an asterisk are $sp^2$ hybridized. The remaining carbons are $sp^3$ hybridized. **9.103** (a) $[Ne_2](\sigma_{3s})^2(\sigma_{3s}^*)^2(\pi_{3p_y})^2(\pi_{3p_z})^2(\sigma_{3p_x})^2$. (b) 3. (c) Diamagnetic. **9.105** In the Lewis structure, all the electrons are paired. From molecular orbital theory, the electrons would be arranged as follows: $[He_2](\sigma_{2s})^2(\sigma_{2s}^*)^2(\sigma_{2p_x})^2(\pi_{2p_y})^2(\pi_{2p_z})^2(\pi_{2p_y}^*)^1(\pi_{2p_z}^*)^1$. This shows two unpaired electrons. To pair these electrons, energy is required to flip the spin of one, thus making the Lewis structure an excited state. **9.107** (a) Although the O atoms are $sp^3$ hybridized, they are locked in a planar structure by the benzene rings. The molecule is symmetrical and therefore is not polar. (b) 20 $\sigma$ bonds and 6 $\pi$ bonds. **9.109** (a) $:\overset{-}{C}\equiv\overset{+}{O}:$ The electronegativity difference between O and C suggests that electron density should concentrate on the O atom, but assigning formal charges places a negative charge on the C atom. Therefore, we expect CO to have a small dipole moment. (b) CO is isoelectronic with $N_2$, bond order 3. This agrees with the triple bond in the Lewis structure. (c) Since C has a negative formal charge, it is more likely to form bonds with $Fe^{2+}$. ($OC-Fe^{2+}$ rather than $CO-Fe^{2+}$). **9.111** Tetrahedral. **9.113** No.

# Chapter 10

**10.7** (a) Amine. (b) Aldehyde. (c) Ketone. (d) Carboxylic acid. (e) Alcohol. **10.9** (a) 3-ethyl-2,4,4-trimethylhexane. (b) 6,6-dimethyl-2-heptanol. (c) 4-chlorohexanal. **10.11** 3,5-dimethyloctane. **10.13** (a) $(CH_3)_3CCH_2CH(CH_3)_2$. (b) $HO(CH_2)_2CH(CH_3)_2$. (c) $CH_3(CH_2)_4C(O)NH_2$. (d) $Cl_3CCHO$.

**10.15**

A = Carbonyl (Ketone)
B = Carboxy (Caroboxylic acid)
C = Hydroxy (Alcohol)

**10.17**

**10.19** (a) $CH_3CH_2CHCH_2CH_2CH_3$ with $CH_3$    (b) $CH_3CHCHCH_2CH_3$ with $CH_3$

(c) $CH_3CHCH_2CHCH_3$ with Br and $C_6H_5$    (d) $CH_3CH_2CHCHCHCH_2CH_2CH_3$ with $CH_3$ $CH_3$ $CH_3$

**10.21** (a) Kekule:

Skeletal (line):
(b) Condensed: $(C_2H_5)_2CHCH_2CO_2C(CH_3)_3$,

Skeletal line:
(c) Condensed: $(CH_3)_2CHCH_2NHCH(CH_3)_2$,

Kekule:

**10.23** (a) Condensed structural: $(CH_3)_2NCHO$,

Kekule:    Line:

(b) Condensed structural: $(CH_2COOH)_2C(OH)COOH$,

Kekule:    Line:

(c) Condensed structural: $(CH_3)_2CH(CH_2)_2OC(O)CH_3$,

Kekule:

Line:

**10.25** (a)    OH    (b) $(CH_3)_3CCH_2CHCH_2CH$ with O and Br

(c)

**10.27** (a) $CH_3-C\equiv N: \longleftrightarrow CH_3-\overset{+}{C}=\overset{-}{\ddot{N}}:$

(b)

(c)

**10.29** (a)

(b)

(c)

**10.37**

**10.39** (structures: Br–CH=CH–CH₃; Br cis alkene; allyl bromide CH₂=CH–CH₂–Br; 2-bromopropene with branch; cyclopropyl–Br)

**10.41** (structures: benzyl chloride CH₂Cl; 4-chlorotoluene; 3-chlorotoluene; 2-chlorotoluene)

**10.43** (a) $CH_3-CH_2-\overset{*}{CH}-\overset{*}{CH}-\overset{O}{\overset{\|}{C}}-NH_2$ with $CH_3$ and $NH_2$ substituents (b) cyclopropane ring with Br, H substituents, starred carbons

**10.45** $CH_3CH_2\overset{O}{\overset{\|}{C}}-H$

**10.47** (a) $H_3C$, Br, H, CH₃ tetrahedral structure (b) HO, H, OH, H₂N, O structure (c) H, OH, CH₃, H₃C structure (d) Br, H, H₃C, HO, H, O structure

**10.55** (mechanism with water, oxygen, proton transfer arrows)

**10.57** (a) $CH_3CH_2\overset{\ddot{O}:}{\overset{\|}{C}}H + :\bar{C}\equiv CH \longrightarrow$
$CH_3CH_2\overset{:\ddot{O}:^-}{CH}C\equiv CH \xrightarrow{H-\ddot{O}:} CH_3CH_2\overset{:\ddot{O}H}{CH}C\equiv CH + :\ddot{O}-H$

(b) $:\ddot{S}-CH_2CH_2CH_2CH_2-\ddot{B}r: \longrightarrow :S: \text{(tetrahydrothiophene ring)} + :\ddot{B}r:^-$

**10.59** $H_2\ddot{N}:^- +$ (chlorobenzene ring with H's) $\longrightarrow H_2\ddot{N}-H +$ (benzyne ring) $+ :\ddot{C}l:^-$ ;
elimination reaction.

**10.61** This is not an oxidation-reduction reaction. There are no changes in oxidation states for any of the atoms.

**10.63** (a) Sulfuric acid is a catalyst.

$H^+$ + (propene) $\longrightarrow$ (carbocation with $:\ddot{O}-H$, H) $\longrightarrow$ (isopropanol) $+ H^+$

(b) n-propanol: $HO$ (propyl chain) (c) No.

**10.71** $\begin{bmatrix} \overset{H}{\underset{H}{C}} - \overset{H}{\underset{Cl}{C}} - \overset{H}{\underset{H}{C}} - \overset{Cl}{\underset{Cl}{C}} \end{bmatrix}_n$

**10.73** (a) $CH_2=CHCH=CH_2$

(b) $\overset{O}{\overset{\|}{HO-C}}-CH_2-CH_2-CH_2-CH_2-CH_2-CH_2-NH_2$

**10.75**

$H_2N-\overset{H}{\underset{H}{CH}}-\overset{O}{\overset{\|}{C}}-\overset{H}{\underset{}{N}}-\overset{(CH_2)_4}{\underset{NH_2}{CH}}-\overset{O}{\overset{\|}{C}}-OH$ GlyLys

$H_2N-\overset{(CH_2)_4}{\underset{NH_2}{CH}}-\overset{O}{\overset{\|}{C}}-\overset{H}{\underset{}{N}}-\overset{H}{\underset{H}{CH}}-\overset{O}{\overset{\|}{C}}-OH$ LysGly

**10.77**

$CH_3CH_2CH_2CH_2-$ Butyl  $CH_3\overset{}{\underset{CH_3}{CH}}CH_2CH_3$ sec-Butyl  $CH_3\overset{}{\underset{CH_3}{CH}}CH_2-$ Isobutyl  $CH_3\overset{CH_3}{\underset{CH_3}{C}}-$ tert-Butyl

**10.79** $CH_3\overset{CH_3}{\underset{}{CH}}CH_2\overset{}{\underset{NH_2}{CH}}CO_2H$ Leucine  $CH_3CH_2\overset{CH_3}{\underset{}{CH}}\overset{}{\underset{NH_2}{CH}}CO_2H$ Isoleucine

**10.81** (a) $CH_3CH_2\overset{}{\underset{CH_3}{CH}}CH=CH_2$

(b) (two tetrahedral structures: H₃C, H, C₂H₅, C₂H₃ and H, CH₃, C₂H₅, C₂H₃)

**10.83** (a) *Cis/trans* stereoisomers. (b) Constitutional isomers. (c) Resonance structures. (d) Different representations of the same structure.

**10.85** (two resonance structures of isopropyl-dimethyl cyclohexadienyl cation)

**10.87** (b).

**10.89** $\underset{sp^3\ sp^3}{CH_3CH_2C\equiv CH}$ $\underset{sp\ \ sp}{}$  $\underset{sp^3\ \ \ \ \ sp^3}{CH_3C\equiv CCH_3}$ $\underset{sp\ sp}{}$  $\underset{\text{all } sp^2}{H_2C=CHCH=CH_2}$

**10.91** Since N is less electronegative than O, electron donation in the amide would be more pronounced. **10.93** (a) The more negative $\Delta H$ implies stronger alkane bonds; branching decreases the total bond enthalpy (and overall stability) of the alkane. (b) The least highly branched isomer (*n*-octane). **10.95** (a) 15.81 mg C, 1.32 mg H, 3.49 mg O.

(b) $C_6H_6O$. (c) $HO-$(phenyl ring) ; (furan ring with vinyl group)

**10.97** [polymer structure: condensation of terephthalic acid and a diamine]

$$\left[ HO-\overset{O}{\overset{\|}{C}}-\bigcirc-\overset{O}{\overset{\|}{C}}-OH \right]_n + \left[ NH_2-\bigcirc-NH_2 \right]_n \longrightarrow \left[ \overset{O}{\overset{\|}{C}}-\bigcirc-\overset{O}{\overset{\|}{C}}-NH-\bigcirc-NH \right]_n + nH_2O$$

**10.99** [polymer structure]

$$\left[ \overset{O}{\overset{\|}{C}}-\overset{H}{\overset{|}{N}}-\overset{H}{\overset{|}{C}} \right]$$

**10.101** 8 different tripeptides: Lys—Lys—Lys, Lys—Lys—Ala, Lys—Ala—Lys, Ala—Lys—Lys, Lys—Ala—Ala, Ala—Lys—Ala, Ala—Ala—Lys, Ala—Ala—Ala.

## Chapter 11

**11.13** 0.493 atm, 0.500 bar, 375 torr, $5.00 \times 10^4$ Pa. **11.15** 13.1 m. **11.17** 7.3 atm. **11.21** (a) Diagram (d). (b) Diagram (b). **11.23** 45.9 mL. **11.25** 587 mmHg. **11.27** 31.8 L. **11.29** 1 volume of NO. **11.35** 6.6 atm. **11.37** 1590°C. **11.39** 1.8 atm. **11.41** 0.70 L. **11.43** 63.31 L. **11.45** $6.1 \times 10^{-3}$ atm. **11.47** 35.0 g/mol. **11.49** $2.1 \times 10^{22}$ $N_2$ molecules, $5.6 \times 10^{22}$ $O_2$ molecules, $2.7 \times 10^{20}$ $O_2$ molecules. **11.51** 2.98 g/L. **11.53** $SF_4$. **11.55** $3.70 \times 10^2$ L. **11.57** 88.9%. **11.59** $M(s) + 3HCl(aq) \longrightarrow 1.5H_2(g) + MCl_3(aq)$, $M_2O_3$; $M_2(SO_4)_3$. **11.61** 94.7%. Assuming the impurity (or impurities) do not produce $CO_2$. **11.63** $C_2H_5OH(l) + 3O_2(g) \longrightarrow 2CO_2(g) + 3H_2O(l)$; $1.44 \times 10^3$ L air. **11.67** (a) 0.89 atm. (b) 1.4 L. **11.69** 349 mmHg. **11.71** 19.8 g Zn. **11.73** $P_{N_2} = 217$ mmHg, $P_{H_2} = 650$ mmHg. **11.75** (a) Box 2. (b) Box 2. **11.83** $u_{rms}(N_2) = 472$ m/s, $u_{rms}(O_2) = 441$ m/s, $u_{rms}(O_3) = 360$ m/s. **11.85** RMS = 2.8 m/s, Average speed = 2.7 m/s. The root-mean-square value is always greater than the average value, because squaring favors the larger values compared to just taking the average value. **11.87** 43.8 g/mol, $CO_2$. **11.93** 18.0 atm; Ideal: 18.5 atm. **11.95** (a) Neither the amount of gas in the tire nor its volume changes appreciably. The pressure is proportional to the temperature. Therefore, as the temperature rises, the pressure increases. (b) As the paper bag is hit, its volume decreases so that its pressure increases. The popping sound occurs when the bag is broken. (c) As the balloon rises, the pressure outside decreases steadily, and the balloon expands. (d) The pressure inside the bulb is less than 1 atm. **11.97** $1.7 \times 10^2$ L. $P_{CO_2} = 0.49$ atm, $P_{H_2O} = 0.41$ atm, $P_{N_2} = 0.25$ atm, and $P_{O_2} = 0.041$ atm. **11.99** (a) $NH_4NO_2(s) \longrightarrow N_2(g) + 2H_2O(l)$. (b) 0.273 g. **11.101** (a) $P_{ii} = 4.0$ atm, $P_{iii} = 2.67$ atm. (b) 2.67 atm, $P_A = 1.33$ atm, $P_B = 1.33$ atm. **11.103** (a) 9.54 atm. (b) If the tetracarbonylnickel gas starts to decompose significantly above 43°C, then for every mole of tetracarbonylnickel gas decomposed, four moles of carbon monoxide would be produced. **11.105** $1.3 \times 10^{22}$ molecules. $CO_2$, $O_2$, $N_2$ and $H_2O$. **11.107** (a) $2KClO_3(s) \longrightarrow 2KCl(s) + 3O_2(g)$. (b) 6.21 L. **11.109** 0.0701 $M$. **11.111** $P_{He} = 0.16$ atm, $P_{Ne} = 2.0$ atm. **11.113** When the water enters the flask from the dropper, some hydrogen chloride dissolves, creating a partial vacuum. Pressure from the atmosphere forces more water up the vertical tube. **11.115** 7. **11.117** (a) 61.2 m/s. (b) $4.58 \times 10^{-4}$ s. (c) 328 m/s. $u_{rms} = 366$ m/s. The magnitudes of the speeds are comparable, but not identical. The rms speed is derived for 3-dimensional motion and is not applicable to the speed distributions in directed molecular beams. Also, the experiment is measuring a most-probable speed, not an rms speed. **11.119** 20.9 kg $O_2$; $1.58 \times 10^4$ L $O_2$. **11.121** The fruit ripens more rapidly because the quantity (partial pressure) of ethylene gas inside the bag increases. **11.123** As the pen is used the amount of ink decreases, increasing the volume inside the pen. As the volume increases, the pressure inside the pen decreases. The hole is needed to equalize the pressure as the volume inside the pen increases. **11.125** (a) $NH_4NO_3(s) \longrightarrow N_2O(g) + 2H_2O(l)$. (b) 0.0821 L · atm/K · mol. **11.127** $C_6H_6$. **11.129** The gases inside the mine were a mixture of carbon dioxide, carbon monoxide, methane, and other harmful compounds. The low atmospheric pressure caused the gases to flow out of the mine (the gases in the mine were at a higher pressure), and the man suffocated. **11.131** (a) 4.90 L. (b) 6.0 atm. (c) 1 atm. **11.133** (a) $4 \times 10^{-22}$ atm. (b) $6 \times 10^{20}$ L/g of H. **11.135** 91%. **11.137** $1.7 \times 10^{12}$. **11.139** $NO_2$. **11.141** 1.4 g. **11.143** $P_{CO_2} = 8.7 \times 10^6$ Pa, $P_{N_2} = 3.2 \times 10^5$ Pa, $P_{SO_2} = 1.4 \times 10^3$ Pa. **11.145** (a) (i) He atoms in both flasks have the same rms speed. (ii) The atoms in the larger volume strike the walls of their container with the same average force as those in the smaller volume, but pressure in the larger volume will be less. (b) (i) $u_2$ is greater than $u_1$ by a factor of $\sqrt{T_2/T_1}$. (ii) Both the frequency of collision and the average force of collision of the atoms at $T_2$ will be greater than those at $T_1$. (c) (i) False. The rms speed of He is greater than that of Ne because He is less massive. (ii) True. The average kinetic energies of the two gases are equal. (iii) The rms speed of the whole collection of atoms is $1.47 \times 10^3$ m/s, but the speeds of the individual atoms are random and constantly changing. **11.147** Vapor: 3.7 nm; Liquid: 0.31 nm. Water molecules are packed very closely together in the liquid, but much farther apart in steam. **11.149** 9.98 atm. **11.151** 1.41 atm. **11.153** $\chi_{methane} = 0.789$, $\chi_{ethane} = 0.211$. **11.155** (a) 0.154%. (b) 1.5%. (c) 7.7%. **11.157** NO. **11.159** Reaction (b). **11.161** $1.80 \times 10^{-4}$ g/mL. **11.163** The kinetic energy of a particle depends on its mass and its speed (KE = $mu^2/2$). At a given temperature, the more massive $UF_6$ molecules move more slowly than do the $H_2$ molecules. **11.165** Intermolecular attractions tend to lower Z at low pressures and excluded volume tends to increase Z at high pressures.

## Chapter 12

**12.9** Butane would be a liquid in winter (boiling point $-44.5$°C), and on the coldest days even propane would become a liquid (boiling point $-0.5$°C). Only methane would remain gaseous (boiling point $-161.6$°C). **12.11** (a) Dispersion. (b) Dispersion and dipole-dipole. (c) Dispersion and dipole-dipole. (d) Dispersion and ionic. (e) Dispersion. **12.13** (e) $CH_3COOH$. **12.15** 1-butanol has greater intermolecular forces because it can form hydrogen bonds. **12.17** (a) Xe, it is larger and therefore has stronger dispersion forces. (b) $CS_2$, it is larger and therefore has stronger dispersion forces. (c) $Cl_2$, it is larger and therefore has stronger dispersion forces. (d) LiF, it is an ionic compound, and the ion-ion attractions are much stronger than the dispersion forces between $F_2$ molecules. (e) $NH_3$, it can form hydrogen bonds and $PH_3$ cannot. **12.19** (a) Dispersion and dipole-dipole, including hydrogen bonding. (b) Dispersion only. (c) Dispersion only. (d) Covalent bonds. **12.21** The compound with $-NO_2$ and $-OH$ groups on adjacent carbons can form hydrogen bonds with itself (*intra*molecular hydrogen bonds). Such bonds do not contribute to *inter*molecular attraction and do not help raise the melting point of the compound. The other compound, with the $-NO_2$ and $-OH$ groups on opposite sides of the ring, can form only *inter*molecular hydrogen bonds; therefore, it will take a higher temperature to escape into the gas phase. **12.33** Ethylene glycol has two $-OH$ groups, allowing it to exert strong intermolecular forces through hydrogen bonding. Its viscosity should fall between ethanol (1 OH group) and glycerol (3 OH groups). **12.35** 331 mm Hg. **12.37** 29.9 kJ/mol. **12.45** Simple cubic: one sphere; body-centered cubic: two spheres; face-centered cubic: four spheres. **12.47** $6.20 \times 10^{23}$ atoms/mol. **12.49** 458 pm. **12.51** $XY_3$. **12.53** 0.220 nm. **12.55** ZnO. **12.59** Molecular solid. **12.61** Molecular: $Se_8$, HBr, $CO_2$, $P_4O_6$, and $SiH_4$; covalent: Si and C. **12.63** Diamond: each carbon atom is covalently bonded to four other carbon atoms. Because these bonds are strong and uniform, diamond is a very hard substance. Graphite: the carbon atoms in each layer are linked by strong bonds, but the layers are bound by weak dispersion forces. As a result, graphite may be cleaved easily between layers and is not hard. In graphite, all atoms are $sp^2$ hybridized; each atom is covalently bonded to three other atoms. The remaining unhybridized $2p$ orbital is used in pi bonding, forming a delocalized molecular orbital. The electrons are free to move around in this extensively delocalized molecular orbital, making graphite a good conductor of electricity in directions along the planes of carbon atoms. **12.85** $2.72 \times 10^3$ kJ. **12.87** 47.03 kJ/mol. **12.89** Two phase changes occur in this process. First, the liquid is turned to solid (freezing), then the solid ice is turned to gas (sublimation). **12.91** When steam condenses to liquid water at 100°C, it releases a large amount of heat equal to the enthalpy of vaporization. Thus, steam at 100°C exposes one to more heat than an equal amount of water at 100°C. **12.95** Initially, the ice melts because of the increase in pressure. As the wire sinks into the ice, the water above the wire refreezes. Eventually the wire actually moves completely through the ice block without cutting it in half. **12.97** (a) Ice would melt. (If heating continues, the liquid water would eventually boil and become a vapor.) (b) Liquid water would vaporize. (c) Water vapor would solidify without becoming a liquid. **12.99** (d). **12.101** Covalent. **12.103** $CCl_4$. **12.105** 0.26 cm; $7.69 \times 10^6$ base pairs. **12.107** 760 mmHg. **12.109** It has reached the critical point; the point of critical temperature ($T_c$) and critical pressure ($P_c$). **12.111** Crystalline $SiO_2$. Its regular structure results in a more efficient packing. **12.113** (a) and (b). **12.115** $8.3 \times 10^{-3}$ atm. **12.117** (a) $K_2S$. Ionic forces are much stronger than the dipole-dipole forces in $(CH_3)_3N$. (b) $Br_2$. Both molecules are nonpolar; but $Br_2$ has a larger mass. **12.119** $SO_2$ will behave less ideally because it is polar and has greater intermolecular forces. **12.121** 62.4 kJ/mol. **12.123** Smaller ions can approach polar water molecules more closely, resulting in larger ion-dipole interactions. The greater the ion-dipole interaction, the larger is the heat of hydration. **12.125** (a) 30.7 kJ/mol. (b) 192.5 kJ/mol. It requires more energy to break the bond than to vaporize the molecule. **12.127** (a) Decreases. (b) No change. (c) No change. **12.129** $CaCO_3(s) \longrightarrow CaO(s) + CO_2(g)$. Initial state: one solid phase, final state: two solid phase components and one gas phase component. **12.131** The molecules are all polar. The F atoms can form H-bonds with water and other $-OH$ and $-NH$ groups in the membrane, so water solubility plus easy attachment to the membrane would allow these molecules to pass the blood-brain barrier. **12.133** The time required to cook food depends on the boiling point of the water in which it is cooked. The boiling point of water increases when the pressure inside the cooker increases. **12.135** (a) Extra heat produced when steam condenses at 100°C. (b) Avoids extraction of ingredients by boiling in water. **12.137** (a) ~2.3 K. (b) ~10 atm. (c) ~5 K. (d) No. **12.139** (a) Pumping allows Ar atoms to escape, thus removing heat from the liquid phase. Eventually the liquid freezes. (b) The slope of the solid-liquid line of cyclohexane is positive. Therefore, its melting point increases with pressure. (c) These droplets are super-cooled liquids. (d) When the dry ice is added to water, it sublimes. The cold $CO_2$ gas generated causes nearby water vapor to condense, hence the appearance of fog. **12.141** (a) Two triple points: Diamond/graphite/liquid and graphite/liquid/vapor. (b) Diamond. (c) Apply high pressure at high temperature. **12.143** Ethanol mixes well

with water. The mixture has a lower surface tension and readily flows out of the ear channel. **12.145** When water freezes it releases heat, helping keep the fruit warm enough not to freeze. Also, a layer of ice is a thermal insulator. **12.147** The fuel source for the Bunsen burner is most likely methane gas. When methane burns in air, carbon dioxide and water are produced. The water vapor produced during the combustion condenses to liquid water when it comes in contact with the outside of the cold beaker. **12.149** $6.019 \times 10^{23}$ **12.151** 127 mmHg. **12.153** 55°C. **12.155** 0.833 g/L. The hydrogen-bonding interactions in HF are relatively strong, and since the ideal gas equation ignores intermolecular forces, it underestimates significantly the density of HF gas near its boiling point. **12.157** Fluoromethane. Of the three compounds, only fluoromethane has a permanent dipole moment.

## Chapter 13

**13.9** "Like dissolves like." Naphthalene and benzene are nonpolar, whereas CsF is ionic. **13.11** $O_2 < Br_2 < LiCl < CH_3OH$. **13.15** (a) 8.47%. (b) 17.7%. (c) 11%. **13.17** (a) 0.0610 *m*. (b) 2.04 *m*. **13.19** (a) 1.7 *m*. (b) 0.87 *m*. (c) 7.0 *m*. **13.21** $3.0 \times 10^2$ g. **13.23** 18.3 *M*; 27.4 *m*. **13.25** $\chi(N_2) = 0.677$, $\chi(O_2) = 0.323$. Due to the greater solubility of oxygen, it has a larger mole fraction in solution than it does in the air. **13.33** 45.9 g. **13.35** According to Henry's law, the solubility of a gas in a liquid increases as the pressure increases ($c = kP$). The soft drink tastes flat at the bottom of the mine because the carbon dioxide pressure is greater and the dissolved gas is not released from the solution. As the miner goes up in the elevator, the atmospheric carbon dioxide pressure decreases and dissolved gas is released from his stomach. **13.37** $1.0 \times 10^{-5}$ mol/L. **13.39** 3.3 atm. This pressure is only an estimate since we ignored the amount of $CO_2$ that was present in the unopened container in the gas phase. **13.57** $1.3 \times 10^3$ g. **13.59** $P_{ethanol} = 30$ mmHg; $P_{1\text{-propanol}} = 26.3$ mmHg. **13.61** 187 g. **13.63** 0.59 *m*. **13.65** −5.4°C. **13.67** (a) $CaCl_2$. (b) Urea. (c) $CaCl_2$. $CaCl_2$ is an ionic compound and is therefore an electrolyte in water. Assuming that $CaCl_2$ completely dissociates, the total ion concentration will be $3 \times 0.35 = 1.05$ *m*, which is larger than the urea (nonelectrolyte) concentration of 0.90 *m*. **13.69** 0.15 *m* $C_6H_{12}O_6 > 0.15$ *m* $CH_3COOH > 0.10$ *m* $Na_3PO_4 > 0.20$ *m* $MgCl_2 > 0.35$ *m* NaCl. **13.71** Boiling point: 102.8°C, Freezing point: −10.0°C. (b) Boiling point: 102.0°C, Freezing point: −7.14°C. **13.73** Both NaCl and $CaCl_2$ are strong electrolytes. Urea and sucrose are nonelectrolytes. The NaCl or $CaCl_2$ will yield more particles per mole of the solid dissolved, resulting in greater freezing point depression. Also, sucrose and urea would make a mess when the ice melts. **13.75** 2.47. **13.77** 9.16 atm. **13.81** 282.5 g/mol; $C_{19}H_{38}O$. **13.83** $4.3 \times 10^2$ g/mol; $C_{24}H_{20}P_4$. **13.85** $1.75 \times 10^4$ g/mol. **13.87** 342 g/mol. **13.89** 15.7%. **13.93** $\Delta P = 2.05 \times 10^{-5}$ mmHg; $\Delta T_f = 8.9 \times 10^{-5}$ °C; $\Delta T_b = 2.5 \times 10^{-5}$ °C; $\pi = 0.889$ mmHg. **13.95** Before: $7.77 \times 10^{-6}$ mol glucose/mL blood; $3.89 \times 10^{-3}$ mol glucose; 7.00 g glucose. After: $1.33 \times 10^{-5}$ mol glucose/mL blood; $6.66 \times 10^{-2}$ mol glucose; 12.0 g glucose. **13.97** Water migrates through the semipermeable cell walls of the cucumber into the concentrated salt solution. **13.99** 3.5. **13.101** X = water, Y = NaCl, Z = urea. **13.103** The pill is in a hypotonic solution. Consequently, by osmosis, water moves across the semipermeable membrane into the pill. The increase in pressure pushes the elastic membrane to the right, causing the drug to exit through the small holes at a constant rate. **13.105** Reverse osmosis involves no phase changes and is usually cheaper than distillation or freezing. 34 atm. **13.107** 0.295 *M*; −0.55°C. **13.109** (a) $2.14 \times 10^3$ g/mol. (b) $4.50 \times 10^4$ g/mol. **13.111** (a) 0.099 L. (b) 9.9. **13.113** (a) At reduced pressure, the solution is supersaturated with $CO_2$. (b) As the escaping $CO_2$ expands it cools, condensing water vapor in the air to form fog. **13.115** (a) Runoff of the salt solution into the soil increases the salinity of the soil. If the soil becomes hypertonic relative to the tree cells, osmosis would reverse, and the tree would lose water to the soil and eventually die of dehydration. (b) Assuming the collecting duct acts as a semipermeable membrane, water would flow from the urine into the hypertonic fluid, thus returning water to the body. **13.117** 33 mL, 67 mL. **13.119** 12.3 *M*. **13.121** 14.2%. **13.123** (a) Solubility decreases with increasing lattice

energy. (b) Ionic compounds are more soluble in a polar solvent.
(c) Solubility increases with enthalpy of hydration of the cation and
anion. **13.125** 1.43 g/mL; 37.0 *m*. **13.127** $NH_3$ can form hydrogen bonds
with water; $NCl_3$ cannot. **13.129** 3%. **13.131** Egg yolk contains lecithins
that solubilize oil in water (See Figure 13.18 of the text). The nonpolar oil
becomes soluble in water because the nonpolar tails of lecithin dissolve in
the oil, and the polar heads of the lecithin molecules dissolve in polar
water (like dissolves like). **13.133** Yes. **13.135** As the water freezes,
dissolved minerals in the water precipitate from solution. The minerals
refract light and create an opaque appearance. **13.137** $1.8 \times 10^2$ g/mol.
**13.139** 1.9 *m*. **13.141** $P_B^\circ = 4.0 \times 10^2$ mmHg; $P_A^\circ = 1.9 \times 10^2$ mmHg.
**13.143** −0.737°C. **13.145** About 0.4 molal.

## Chapter 14

**14.5** (a) Rate $= -\dfrac{\Delta[H_2]}{\Delta t} = -\dfrac{\Delta[I_2]}{\Delta t} = \dfrac{1}{2}\dfrac{\Delta[HI]}{\Delta t}$ .

(b) Rate $= -\dfrac{1}{5}\dfrac{\Delta[Br^-]}{\Delta t} = -\dfrac{\Delta[BrO_3^-]}{\Delta t} = -\dfrac{1}{6}\dfrac{\Delta[H^+]}{\Delta t} = \dfrac{1}{3}\dfrac{\Delta[Br_2]}{\Delta t}$ .

**14.7** (a) 0.066 *M*/s. (b) 0.033 *M*/s. **14.15** $8.1 \times 10^{-6}$ *M*/s. **14.17** The
reaction is first order in A and first order overall; $k = 0.213$ s$^{-1}$. **14.19**
(a) 2. (b) 0. (c) 2.5. (d) 3. **14.21** First order; $k = 1.19 \times 10^{-4}$ s$^{-1}$. **14.27**
30 min. **14.29** (a) 0.034 *M*. (b) 17 s; 23 s. **14.31** $4.5 \times 10^{-10}$ *M*/s; $5.4 \times 10^6$ s. **14.33** (a) 4:3:6. (b) The relative rates would be unaffected, each
absolute rate would decrease by 50%. (c) 1:1:1. **14.41** 103 kJ/mol. **14.43**
$3.0 \times 10^3$ s$^{-1}$. **14.45** 51.8 kJ/mol. **14.47** $1.3 \times 10^2$ kJ/mol. For maximum
freshness, fish should be frozen immediately after capture and kept frozen
until cooked. **14.49** Diagram (a). **14.59** (a) Second-order. (b) The first
step is the slower (rate-determining) step. **14.61** Mechanism I can be
discarded. Mechanisms II and III are possible. **14.71** (i) and (iv). **14.73**
Temperature, energy of activation, concentration of reactants, and a
catalyst. **14.75** Temperature must be specified. **14.77** 0.035 s$^{-1}$.
**14.79** 272 s. **14.81** Since the methanol contains no oxygen-18, the oxygen
atom must come from the phosphate group and not the water. The
mechanism must involve a bond-breaking process like:

$$CH_3 - O \overset{2}{-}\overset{\underset{\displaystyle O}{\|}}{P} - O - H$$
$$\underset{\displaystyle O-H}{|}$$

**14.83** Most transition metals have several stable oxidation states. This
allows the metal atoms to act as either a source or a receptor of electrons
in a broad range of reactions. **14.85** (a) rate $= k[CH_3COCH_3][H^+]$.
(b) $3.8 \times 10^{-3}$ $M^{-1} \cdot s^{-1}$. (c) $k = k_1 k_2/k_{-1}$. **14.87** (I) $Fe^{3+}$ oxidizes $I^-$:
$2Fe^{3+} + 2I^- \longrightarrow 2Fe^{2+} + I_2$; (II) $Fe^{2+}$ reduces $S_2O_8^{2-}$: $2Fe^{2+} + S_2O_8^{2-} \longrightarrow 2Fe^{3+} + 2SO_4^{2-}$; Overall Reaction: $2I^- + S_2O_8^{2-} \longrightarrow I_2 + 2SO_4^{2-}$. ($Fe^{3+}$ undergoes a redox cycle: $Fe^{3+} \longrightarrow Fe^{2+} \longrightarrow Fe^{3+}$).
The uncatalyzed reaction is slow because both $I^-$ and $S_2O_8^{2-}$ are
negatively charged, which makes their mutual approach unfavorable.
**14.89** (a) (i) rate $= k[A]^0 = k$,

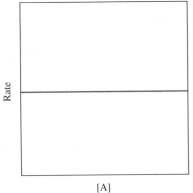

[A]

(ii) $[A] = -kt + [A]_0$,

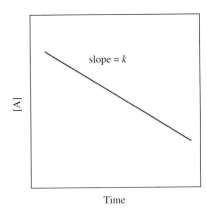

slope = k

[A]

Time

(b) $t_{1/2} = \dfrac{[A]_0}{2k}$ . (c) $t = 2t_{1/2}$. **14.91** There are three gases present and we
can measure only the total pressure of the gases. To measure the partial
pressure of azomethane at a particular time, we must withdraw a sample
of the mixture, analyze and determine the mole fractions. Then,
$P_{azomethane} = P_T \chi_{azomethane}$.

**14.93**

PE

$E_2 + G_2$

$G_2 + X_2$

2XG

2EG

Reaction progress

**14.95** (a) A catalyst works by changing the reaction mechanism, thus
lowering the activation energy. (b) A catalyst changes the reaction
mechanism. (c) A catalyst does not change the enthalpy of reaction.
(d) A catalyst increases the forward rate of reaction. (e) A catalyst
increases the reverse rate of reaction.

**14.97** At very high $[H_2]$: $k_2[H_2] \gg 1$, rate $= \dfrac{k_1[NO]^2[H_2]}{k_2[H_2]} = \dfrac{k_1}{k_2}[NO]^2$ .

At very low $[H_2]$: $k_2[H_2] \ll 1$, rate $= \dfrac{k_1[NO]^2[H_2]}{1} = k_1[NO]^2[H_2]$ .

The result from Problem 14.80 agrees with the rate law determined for low
$[H_2]$. **14.99** Rate $= k[N_2O_5]$; $k = 1.0 \times 10^{-5}$ s$^{-1}$. **14.101** The red bromine
vapor absorbs photons of blue light and dissociates to form bromine atoms:
$Br_2 \longrightarrow 2Br\cdot$. The bromine atoms collide with methane molecules and
abstract hydrogen atoms: $Br\cdot + CH_4 \longrightarrow HBr + \cdot CH_3$. The methyl
radical then reacts with $Br_2$, giving the observed product and regenerating a
bromine atom to start the process over again: $\cdot CH_3 + Br_2 \longrightarrow CH_3Br + Br\cdot$, $Br\cdot + CH_4 \longrightarrow HBr + \cdot CH_3$, and so on. **14.103** Lowering the
temperature would slow all chemical reactions, which would be especially
important for those that might damage the brain. **14.105** (a) rate $= k[X][Y]^2$. (b) 0.019 $M^{-2}s^{-1}$. **14.107** Second-order; $k = 2.4 \times 10^7$ $M^{-1}s^{-1}$.
**14.109** During the first five minutes or so the engine is relatively cold, so
the exhaust gases will not fully react with the components of the catalytic
converter. Remember, for almost all reactions, the rate of reaction increases
with temperature. **14.111** (a) The activation energy of reaction B is larger
than that of reaction A. (b) $E_a \approx 0$. Orientation factor is not important.
**14.113** $5.7 \times 10^5$ yr. **14.115** (a) Catalyst: $Mn^{2+}$; intermediates: $Mn^{3+}$,
$Mn^{4+}$; first step is rate-determining. (b) Without the catalyst, the reaction
would be a termolecular one involving 3 cations! ($Tl^+$ and two $Ce^{4+}$). The
reaction would be slow. (c) The catalyst is a homogeneous catalyst because
it has the same phase (aqueous) as the reactants. **14.117** 0.45 atm. **14.119**
(a) $\Delta[B]/\Delta t = k_1[A] - k_2[B]$. (b) $[B] = (k_1/k_2)[A]$. **14.121** (a) $k = 0.0247$
yr$^{-1}$. (b) $9.8 \times 10^{-4}$. (c) 187 yr. **14.123** (a) Three. (b) Two. (c) The third
step. (d) Exothermic. **14.125** $1.8 \times 10^3$ K. **14.127** (a) $2.5 \times 10^{-5}$ *M*/s.
(b) $2.5 \times 10^{-5}$ *M*/s. (c) $8.3 \times 10^{-6}$ *M*.

**14.129** $n = 0$, $t_{1/2} = C\dfrac{1}{[A]_0^{-1}} = C[A]_0$;

$n = 1$, $t_{1/2} = C\dfrac{1}{[A]_0^0} = C$;

$n = 2$, $t_{1/2} = C\dfrac{1}{[A]_0}$.

**14.131** (a) $1.13 \times 10^{-3}$ $M$/min. (b) $6.83 \times 10^{-4}$ $M$/min; $8.8 \times 10^{-3}$ $M$.
**14.133** Second-order, $k = 0.42/M \cdot$ min. **14.135** 60% increase. The result shows the profound effect of an exponential dependence. **14.137** $4.1 \times 10^2$ kJ/mol.

## Chapter 15

**15.9** (a) $\dfrac{[N_2][H_2O]^2}{[NO]^2[H_2]^2}$. (b) $7.2 \times 10^2$. **15.11** $1.08 \times 10^7$.

**15.21** (1) Diagram (a). (2) Diagram (d). Volume will cancel from the $K_c$ expression. **15.23** $2.40 \times 10^{33}$. **15.25** $3.5 \times 10^{-7}$. **15.27** (a) $8.2 \times 10^{-2}$.
(b) 0.29. **15.29** $K_P = 0.105$, $K_c = 2.05 \times 10^{-3}$. **15.31** $K_P = 7.09 \times 10^{-3}$.
**15.33** $K_P = 9.6 \times 10^{-3}$, $K_c = 3.9 \times 10^{-4}$. **15.35** $4.0 \times 10^{-6}$. **15.37** $5.6 \times 10^{23}$. **15.39** The equilibrium pressure is less than the original pressure.
**15.41** 0.173 mol $H_2$. **15.43** $[H_2] = [Br_2] = 1.80 \times 10^{-4}$ $M$; [HBr] = 0.267 $M$. **15.45** $P_{COCl_2} = 0.408$ atm; $P_{CO} = P_{Cl_2} = 0.352$ atm. **15.47** $P_{CO}$ = 1.96 atm; $P_{CO_2}$ = 2.54 atm. **15.49** The forward reaction will not occur.
**15.55** (a) The equilibrium would shift to the right. (b) The equilibrium would be unaffected. (c) The equilibrium would be unaffected. **15.57**
(a) No effect. (b) No effect. (c) Shift to the left. (d) No effect. (e) Shift to the left. **15.59** (a) Shift to the right. (b) Shift to the left. (c) Shift to the right. (d) Shift to the left. (e) A catalyst has no effect on equilibrium position. **15.61** No change. **15.63** (a) Shift to the right. (b) No effect.
(c) No effect. (d) Shift to the left. (e) Shift to the right. (f) Shift to the left.
(g) Shift to the right. **15.65** (a) $2O_3(g) \rightleftharpoons 3O_2(g)$, $\Delta H^\circ_{rxn} = -284.4$ kJ/mol. (b) Equilibrium would shift to the left. The number of $O_3$ molecules would increase and the number of $O_2$ molecules would decrease. **15.67**
(a) $P_{NO} = 0.24$ atm; $P_{Cl_2} = 0.12$ atm. (b) $K_P = 0.017$. **15.69** (a) No effect.
(b) More $CO_2$ and $H_2O$. **15.71** (a) $K_c = 8 \times 10^{-44}$. (b) A mixture of $H_2$ and $O_2$ can be kept at room temperature because of a very large activation energy. **15.73** (a) $K_P = 1.7$. (b) $P_A = 0.69$ atm, $P_B = 0.81$ atm. **15.75** $K_P = 1.5 \times 10^5$. **15.77** $P_{H_2} = 0.28$ atm; $P_{Cl_2} = 0.051$ atm; $P_{HCl} = 1.67$ atm. **15.79** $5.0 \times 10^1$ atm. **15.81** 0.038. **15.83** $3.3 \times 10^2$ atm. **15.85** $K_c = 6.28 \times 10^{-4}$. **15.87** $P_{N_2} = 0.860$ atm, $P_{H_2} = 0.366$ atm, $P_{NH_3} = 4.40 \times 10^{-3}$ atm. **15.89** (a) $K_c = 1.16$. (b) 53.7%. **15.91** (a) $K_P = 0.49$.
(b) 0.23 (23%). (c) 0.037 (3.7%). (d) Greater than 0.037 mol. **15.93** $[H_2]$ = 0.07 $M$, $[I_2] = 0.18$ $M$, [HI] = 0.83 $M$. **15.95** (c). $N_2O_4$(colorless) $\longrightarrow$ $2NO_2$(brown) is consistent with the observations. The reaction is endothermic so heating darkens the color. Above 150°C, the $NO_2$ breaks up into colorless NO and $O_2$: $2NO_2(g) \longrightarrow 2NO(g) + O_2(g)$. An increase in pressure shifts the equilibrium back to the left, restoring the color by producing $NO_2$. **15.97** (a) $K_P = 4.2 \times 10^{-4}$. (b) 0.83. (c) 1.1.
(d) $2.3 \times 10^3$; $2.1 \times 10^{-2}$. **15.99** (a) Color deepens. (b) Increases.
(c) Decreases. (d) Increases. (e) Unchanged. **15.101** Potassium is more volatile than sodium. Therefore, its removal shifts the equilibrium from left to right. **15.103** $3.5 \times 10^{-2}$. **15.105** (a) Shifts to the right. (b) Shifts to the right. (c) No change. (d) No change. (e) No change. (f) Shifts to the left. **15.107** (a) $1.8 \times 10^{-16}$. (b) $[H^+][OH^-] = 1.0 \times 10^{-14}$; $[H^+]$ = $[OH^-] = 1.0 \times 10^{-7}$ $M$. **15.109** $P_{N_2O_4} = 0.09$ atm; $P_{NO_2} = 0.10$ atm.
**15.111** (a) 1.03 atm. (b) 0.39 atm. (c) 1.67 atm. (d) 0.620 (62.0%).
**15.113** (a) $K_P = 2.6 \times 10^{-6}$; $K_c = 1.1 \times 10^{-7}$. (b) 2.2 g; 22 mg/m³;
Yes. **15.115** There is a temporary dynamic equilibrium between the melting of ice cubes and the freezing of water between the ice cubes.

**15.117** $[NH_3] = 0.042$ $M$; $[N_2] = 0.086$ $M$;

$[H_2] = 0.26$ $M$. **15.119** (a) $K_P = \dfrac{\left(\dfrac{4x^2}{1+x}\right)P}{1-x} = \dfrac{4x^2}{1-x^2}P$. (b) If $P$ increases,

the fraction $\dfrac{4x^2}{1-x^2}$ (and therefore $x$) must decrease. Equilibrium shifts to

the left to produce less $NO_2$ and more $N_2O_4$ as predicted. **15.121** $P_{SO_2Cl_2}$ = 3.58 atm, $P_{SO_2} = P_{Cl_2} = 2.71$ atm. **15.123** 4.0. **15.125** −3. **15.127** $K_P = K_c$: (d), (g); cannot write a $K_P$: (c), (f).

## Chapter 16

**16.3** (a) Both. (b) Base. (c) Acid. (d) Base. (e) Acid. (f) Base. (g) Base.
(h) Base. (i) Acid. (j) Acid. **16.5** (a) $NO_2^-$. (b) $HSO_4^-$. (c) $HS^-$. (d) $CN^-$.
(e) $HCOO^-$. **16.7** (a) $CH_2ClCOO^-$. (b) $IO_4^-$. (c) $H_2PO_4^-$. (d) $HPO_4^{2-}$.
(e) $PO_4^{3-}$. (f) $HSO_4^-$. (g) $SO_4^{2-}$. (h) $IO_3^-$. (i) $SO_3^{2-}$. (j) $NH_3$. (k) $HS^-$.
(l) $S^{2-}$. (m) $OCl^-$. **16.15** (a) $4.0 \times 10^{-13}$ $M$. (b) $6.0 \times 10^{-10}$ $M$. (c) $1.2 \times 10^{-12}$ $M$. (d) $5.7 \times 10^{-3}$ $M$. **16.17** (a) $2.05 \times 10^{-11}$ $M$. (b) $3.07 \times 10^{-8}$ $M$. (c) $5.95 \times 10^{-11}$ $M$. (d) $2.93 \times 10^{-1}$ $M$. **16.21** $7.1 \times 10^{-12}$ $M$.
**16.23** (a) 3.00. (b) 13.89. **16.25** (a) $3.8 \times 10^{-3}$ $M$. (b) $6.2 \times 10^{-12}$ $M$.
(c) $1.1 \times 10^{-7}$ $M$. (d) $1.0 \times 10^{-15}$ $M$. **16.27** pH < 7: $[H^+] > 1.0 \times 10^{-7}$ $M$, solution is acid; pH > 7: $[H^+] < 1.0 \times 10^{-7}$ $M$, solution is basic; pH = 7: $[H^+] = 1.0 \times 10^{-7}$ $M$, solution is neutral. **16.29** $2.5 \times 10^{-5}$ $M$. **16.31** $2.2 \times 10^{-3}$ g. **16.37** (a) −0.009. (b) 1.46. (c) 5.82. **16.39**
(a) $6.2 \times 10^{-5}$ $M$. (b) $2.8 \times 10^{-4}$ $M$. (c) 0.10 $M$. **16.41** (a) pOH = −0.093, pH = 14.09. (b) pOH = 0.36, pH = 13.64. (c) pOH = 1.07, pH = 12.93. **16.43** (a) $1.1 \times 10^{-3}$ $M$. (b) $5.5 \times 10^{-4}$ $M$. **16.51** (a) Strong base. (b) Weak base. (c) Weak base. (d) Weak base. (e) Strong base.
**16.53** 2.59. **16.55** 5.17. **16.57** (a) 10%. (b) 84%. (c) 0.97%. **16.59**
$1.3 \times 10^{-6}$. **16.61** $4.8 \times 10^{-9}$. **16.63** $2.3 \times 10^{-3}$ $M$. **16.65** 6.80. **16.69**
$6.97 \times 10^{-7}$. **16.71** 11.98. **16.73** (a) $A^-$. (b) $B^-$. **16.77** $K_b(CN^-)$ = $2.0 \times 10^{-5}$; $K_b(F^-) = 1.4 \times 10^{-11}$; $K_b(CH_3COO^-) = 5.6 \times 10^{-10}$;
$K_b(HCO_3^-) = 2.4 \times 10^{-8}$. **16.79** (1) Diagram (c). (2) Diagrams (b) and (d). **16.81** pH (0.040 $M$ HCl) = 1.40; pH (0.040 $M$ $H_2SO_4$) = 1.31.
**16.83** $[H^+] = [HCO_3^-] = 1.0 \times 10^{-4}$ $M$, $[CO_3^{2-}] = 4.8 \times 10^{-11}$ $M$.
**16.85** pH = 1.00. **16.89** (a) $H_2SO_4 > H_2SeO_4$. (b) $H_3PO_4 > H_3AsO_4$.
**16.91** The conjugate bases are $C_6H_5O^-$ from phenol and $CH_3O^-$ from methanol. The $C_6H_5O^-$ is stabilized by resonance:

The $CH_3O^-$ ion has no such resonance stabilization. A more stable conjugate base means an increase in the strength of the acid. **16.97**
(a) Neutral. (b) Basic. (c) Acidic. (d) Acidic. **16.99** HZ < HY < HX.
**16.101** pH > 7. **16.103** 4.82. **16.105** 5.39. **16.109** (a) $Al_2O_3 < BaO <$ $K_2O$. (b) $CrO_3 < Cr_2O_3 < CrO$. **16.111** $Al(OH)_3(s) + OH^-(aq) \longrightarrow$ $Al(OH)_4^-(aq)$. This is a Lewis acid-base reaction. **16.115** $AlCl_3$ is a Lewis acid with an incomplete octet of electrons and $Cl^-$ is the Lewis base donating a pair of electrons. **16.117** $CO_2$, $SO_2$, and $BCl_3$. **16.119** (a) $AlBr_3$ is the Lewis acid; $Br^-$ is the Lewis base. (b) Cr is the Lewis acid; CO is the Lewis base. (c) $Cu^{2+}$ is the Lewis acid; $CN^-$ is the Lewis base.
**16.121** A strong acid: Diagram (b); a weak acid: Diagram (c); a very weak acid: Diagram (d). **16.123** $CH_3COO^-(aq)$ and HCl$(aq)$. This reaction will not occur to any measurable extent. **16.125** pH = 1.70;
% ionization = 2.3%. **16.127** (c). **16.129** (a) For the forward reaction $NH_4^+$ and $NH_3$ are the acid and base, respectively. For the reverse reaction $NH_3$ and $NH_2^-$ are the acid and base, respectively. (b) $H^+$ corresponds to $NH_4^+$; $OH^-$ corresponds to $NH_2^-$. For the neutral

solution, $[NH_4^+] = [NH_2^-]$. **16.131** $K_a = \dfrac{[H^+][A^-]}{[HA]}$, $[HA] \approx 0.1\ M$,

$[A^-] \approx 0.1\ M$. Therefore, $K_a = [H^+] = \dfrac{K_w}{[OH^-]}$ and $[OH^-] = \dfrac{K_w}{K_a}$.

**16.133** $1.7 \times 10^{10}$. **16.135** (a) $H^-$ (base$_1$) + $H_2O$ (acid$_2$) $\longrightarrow$ $OH^-$ (base$_2$) + $H_2$ (acid$_1$). (b) $H^-$ is the reducing agent and $H_2O$ is the oxidizing agent. **16.137** $2.8 \times 10^{-2}$. **16.139** $PH_3$ is a weaker base than $NH_3$. **16.141** (a) $HNO_2$. (b) HF. (c) $BF_3$. (d) $NH_3$. (e) $H_2SO_3$. (f) $HCO_3^-$ and $CO_3^{2-}$. **16.143** (a) Trigonal pyramidal. (b) $H_4O^{2+}$ does not exist because the positively charged $H_3O^+$ has no affinity to accept the positive $H^+$ ion. If $H_4O^{2+}$ existed, it would have a tetrahedral geometry. **16.145** $Cl_2(g) + H_2O(l) \rightleftharpoons HCl(aq) + HClO(aq)$; $HCl(aq) + AgNO_3(aq) \rightleftharpoons AgCl(s) + HNO_3(aq)$. In the presence of $OH^-$ ions, the first equation is shifted to the right: $H^+$ (from HCl) + $OH^- \longrightarrow H_2O$. Therefore, the concentration of HClO increases. (The 'bleaching action' is due to $ClO^-$ ions.) **16.147** 11.80. **16.149** (a) 0.181 or 18.1%. (b) $4.63 \times 10^{-3}$. **16.151** 4.26. **16.153** $7.2 \times 10^{-3}$ g. **16.155** 1.000. **16.157** (a) The pH of the solution of HA would be lower. (b) The electrical conductance of the HA solution would be greater. (c) The rate of hydrogen evolution from the HA solution would be greater. **16.159** $1.4 \times 10^{-4}$. **16.161** $2.7 \times 10^{-3}$ g. **16.163** (a) $NH_2^-$ (base) + $H_2O$ (acid) $\longrightarrow$ $NH_3 + OH^-$; $N^{3-}$ (base) + $3H_2O$ (acid) $\longrightarrow$ $NH_3 + 3OH^-$. (b) $N^{3-}$. **16.165** When the smelling salt is inhaled, some of the powder dissolves in the basic solution. The ammonium ions react with the base as follows: $NH_4^+(aq) + OH^-(aq) \longrightarrow NH_3(aq) + H_2O$. It is the pungent odor of ammonia that prevents a person from fainting. **16.167** (c). **16.169** 21 mL. **16.171** Magnesium. **16.173** Both NaF and $SnF_2$ provide $F^-$ ions in solution. The $F^-$ ions replace $OH^-$ ions during the remineralization process $5Ca^{2+} + 3PO_4^{3-} + F^- \longrightarrow Ca_5(PO_4)_3F$ (fluorapatite). Because $F^-$ is a weaker base than $OH^-$, fluorapatite is more resistant to attacks by acids compared to hydroxyapatite. **16.175** $K_a = 5.2 \times 10^{-10}$.

## Chapter 17

**17.5** (a) 2.57. (b) 4.44. **17.9** (c) and (d). **17.11** 8.89. **17.13** 0.024. **17.15** 0.58. **17.17** 9.25; 9.18. **17.19** $Na_2A/NaHA$. **17.21** (1) Solutions (a), (b), and (c); (2) Solution (a). **17.27** 202 g/mol. **17.29** 0.25 M. **17.31** (a) $1.10 \times 10^2$ g/mol. (b) $1.6 \times 10^{-6}$. **17.33** 5.82. **17.35** (a) 2.87. (b) 4.56. (c) 5.34. (d) 8.78. (e) 12.10. **17.37** (a) Cresol red or phenolphthalein. (b) Most of the indicators in Table 17.3, except thymol blue and, to a lesser extent, bromophenol blue and methyl orange. (c) Bromophenol blue, methyl orange, methyl red, or chlorophenol blue. **17.39** Red. **17.41** (1) Diagram (c), (2) Diagram (b), (3) Diagram (d), (4) Diagram (a). The pH at the equivalence point is below 7 (acidic). **17.49** (a) $[I^-] = 9.1 \times 10^{-9}\ M$. (b) $[Al^{3+}] = 7.4 \times 10^{-8}\ M$. **17.51** $1.8 \times 10^{-11}$. **17.53** $3.3 \times 10^{-93}$. **17.55** 9.52. **17.57** Yes. **17.63** (a) 0.013 M or $1.3 \times 10^{-2}\ M$. (b) $2.2 \times 10^{-4}\ M$. (c) $3.3 \times 10^{-3}\ M$. **17.65** (a) $1.0 \times 10^{-5}\ M$. (b) $1.1 \times 10^{-10}\ M$. **17.67** (b), (c), (d), and (e) **17.69** (a) 0.016 M or $1.6 \times 10^{-2}\ M$. (b) $1.6 \times 10^{-6}\ M$. **17.71** A precipitate of $Fe(OH)_2$ will form. **17.73** $[Cd^{2+}] = 1.1 \times 10^{-18}\ M$, $[Cd(CN)_4^{2-}] = 4.2 \times 10^{-3}\ M$, $[CN^-] = 0.48\ M$. **17.75** $3.5 \times 10^{-5}\ M$. **17.77** $Cu^{2+}(aq) + 4NH_3(aq) \rightleftharpoons [Cu(NH_3)_4]^{2+}(aq)$. (b) $Ag^+(aq) + 2CN^-(aq) \rightleftharpoons [Ag(CN)_2]^-(aq)$. (c) $Hg^{2+}(aq) + 4Cl^-(aq) \rightleftharpoons [HgCl_4]^{2-}(aq)$. **17.81** Greater than 2.68 but less than 8.11. **17.83** 0.011 M. **17.85** Chloride ion will precipitate $Ag^+$ but not $Cu^{2+}$. So, dissolve some solid in $H_2O$ and add HCl. If a precipitate forms, the salt was $AgNO_3$. A flame test will also work. $Cu^{2+}$ gives a green flame test. **17.87** 2.51 to 4.41. **17.89** 1.3 M. **17.91** $[Na^+] = 0.0835\ M$, $[HCOO^-] = 0.0500\ M$; $[OH^-] = 0.0335\ M$; $[H^+] = 3.0 \times 10^{-13}\ M$; $[HCOOH] = 8.8 \times 10^{-11}\ M$. **17.93** $Cd(OH)_2(s) + 2OH^-(aq) \rightleftharpoons Cd(OH)_4^{2-}(aq)$. This is a Lewis acid-base reaction. **17.95** (d). **17.97** $[Ag^+] = 2.0 \times 10^{-9}\ M$; $[Cl^-] = 0.080\ M$; $[Zn^{2+}] = 0.070\ M$; $[NO_3^-] = 0.060\ M$. **17.99** 0.035 g/L. **17.101** $2.4 \times 10^{-13}$. **17.103** $[Ba^{2+}] = 1.0 \times 10^{-5}\ M$. $Ba(NO_3)_2$ is too soluble to be used for this purpose. **17.105** (a) AgBr. (b) $1.8 \times 10^{-7}\ M$. (c) 0.0018%.

**17.107** $3.0 \times 10^{-8}$. **17.109** (a) $1.0 \times 10^{14}$. (b) $1.8 \times 10^9$. (c) $1.8 \times 10^9$. (d) $3.2 \times 10^4$. **17.111** (a) Mix 500 mL of 0.40 M $CH_3COOH$ with 500 mL of 0.40 M $CH_3COONa$. (b) Mix 500 mL of 0.80 M $CH_3COOH$ with 500 mL of 0.40 M NaOH. (c) Mix 500 mL of 0.80 M $CH_3COONa$ with 500 mL of 0.40 M HCl. **17.113** (a) Figure (b). (b) Figure (a). **17.115** (a) Add sulfate. (b) Add sulfide. (c) Add iodide. **17.117** They are insoluble in water. **17.119** The ionized polyphenols have a dark color. In the presence of citric acid from lemon juice, the anions are converted to the lighter-colored acids. **17.121** Yes. **17.123** (c). **17.125** Decreasing the pH would increase the solubility of calcium oxalate and should help minimize the formation of calcium oxalate kidney stones. **17.127** At pH = 1.0: $^+NH_3-CH_2-COOH$. At pH = 7.0: $^+NH_3-CH_2-COO^-$. At pH = 12.0: $NH_2-CH_2-COO^-$. **17.129** (a) Before dilution: pH = 4.74. After a 10-fold dilution: pH = 4.74. (b) Before dilution: pH = 2.52. After dilution, pH = 3.02. **17.131** $AgBr(s)$ will precipitate first. $Ag_2SO_4(s)$ will precipitate last. **17.133** (c). **17.135** $8.8 \times 10^{-12}$. **17.137** 4.75.

## Chapter 18

**18.7** (a) With barrier: 16; Without barrier: 64. (b) 16; 16; 32. Both particles on one side: $S = 3.83 \times 10^{-23}$ J/K; particles on opposite sides: $S = 4.78 \times 10^{-23}$. The most probable state is the one with the larger entropy; that is, the state in which the particles are on opposite sides. **18.13** (a) $\Delta S_{sys} = -0.031$ J/K. (b) $\Delta S_{sys} = -0.29$ J/K. (c) $\Delta S_{sys} = 1.5 \times 10^2$ J/K. **18.15** (c) < (d) < (e) < (a) < (b). **18.17** (a) 47.5 J/K · mol. (b) $-12.5$ J/K · mol. (c) $-242.8$ J/K · mol. **18.21** (a) 291 J/K · mol; spontaneous. (b) $2.10 \times 10^3$ J/K · mol; spontaneous. (c) $\Delta S_{surr} = 2.99 \times 10^3$ J/K · mol; spontaneous. **18.23** (a) $\Delta S_{sys} = -75.6$ J/K · mol; $\Delta S_{surr} = 185$ J/K · mol; spontaneous. (b) $\Delta S_{sys} = 215.8$ J/K · mol; $\Delta S_{surr} = -304$ J/K · mol; not spontaneous. (c) $\Delta S_{sys} = 98.2$ J/K · mol; $\Delta S_{surr} = -1.46 \times 10^3$ J/K · mol; not spontaneous. (d) $\Delta S_{sys} = -282$ J/K · mol; $\Delta S_{surr} = 7.20 \times 10^3$ J/K · mol; spontaneous. **18.29** (a) $-1139$ kJ/mol. (b) $-140.0$ kJ/mol. (c) $-2935$ kJ/mol. **18.31** (a) All temperatures. (b) Below 111 K. **18.33** $\Delta S_{fus} = 99.9$ J/K · mol; $\Delta S_{vap} = 93.6$ J/K · mol. **18.35** $\Delta G_{rxn}^\circ = -226.6$ kJ/mol. **18.37** 75.9 kJ. **18.41** 0.35. **18.43** 79 kJ/mol. **18.45** (a) $\Delta G_{rxn}^\circ = 35.4$ kJ/mol; $K_P = 6.2 \times 10^{-7}$. (b) 44.6 kJ/mol. **18.47** (a) $1.6 \times 10^{-23}$ atm. (b) 0.535 atm. **18.49** $3.1 \times 10^{-2}$ atm or 23.6 mmHg. **18.53** 93 ATP molecules. **18.55** $\Delta H_{fus} > 0$. $\Delta S_{fus} > 0$. (a) $\Delta G_{fus} < 0$. (b) $\Delta G_{fus} = 0$. (c) $\Delta G_{fus} > 0$. **18.57** U and H. **18.59** 42°C. **18.61** $\Delta S$ must be positive ($\Delta S > 0$). **18.63** (a) Benzene: $\Delta S_{vap} = 87.8$ J/K · mol; hexane: $\Delta S_{vap} = 90.1$ J/K · mol; mercury: $\Delta S_{vap} = 93.7$ J/K · mol; toluene: $\Delta S_{vap} = 91.8$ J/K · mol. Trouton's rule is a statement about $\Delta S_{vap}^\circ$. In most substances, the molecules are in constant and random motion in both the liquid and gas phases, so $\Delta S_{vap}^\circ \approx 90$ J/K · mol. (b) Ethanol: $\Delta S_{vap} = 111.9$ J/K · mol; water: $\Delta S_{vap} = 109.4$ J/K · mol. In ethanol and water, there are fewer possible arrangements of the molecules due to the network of H-bonds, so $\Delta S_{vap}^\circ$ is greater. **18.65** (a) $2CO + 2NO \longrightarrow 2CO_2 + N_2$. (b) The oxidizing agent is NO; the reducing agent is CO. (c) $K_P = 3 \times 10^{120}$. (d) $Q_P = 1.2 \times 10^{14}$. To the right. (e) No. **18.67** $2.6 \times 10^{-9}$. **18.69** 976 K = 703°C. **18.71** 38 kJ. **18.73** 174 kJ/mol. **18.75** (a) Positive. (b) Negative. (c) Positive. (d) Positive. **18.77** 625 K. We assume that $\Delta H^\circ$ and $\Delta S^\circ$ do not depend on temperature. **18.79** No. A negative $\Delta G^\circ$ tells us that a reaction has the potential to happen, but gives no indication of the rate. **18.81** (a) $\Delta G^\circ = -106.4$ kJ/mol; $K_P = 4 \times 10^{18}$. (b) $\Delta G^\circ = -53.2$ kJ/mol; $K_P = 2 \times 10^9$. The $K_P$ in (a) is the square of the $K_P$ in (b). Both $\Delta G^\circ$ and $K_P$ depend on the number of moles of reactants and products specified in the balanced equation. **18.83** Talking involves various biological processes (to provide the necessary energy) that lead to an increase in the entropy of the universe. Since the overall process (talking) is spontaneous, the entropy of the universe must increase. **18.85** (a) 86.7 kJ/mol. (b) $4 \times 10^{-31}$. (c) $3 \times 10^{-6}$. (d) Lightning supplies the energy necessary to drive this reaction, converting the two most abundant gases in the atmosphere into NO(g). The NO gas dissolves in the rain, which carries it into the soil where it is converted into nitrate and nitrite by bacterial action. This "fixed" nitrogen is a necessary nutrient for plants. **18.87** $T > 673.2$ K. **18.89** (a) $7.6 \times 10^{14}$. (b) $4.1 \times 10^{-12}$. The activity series is correct. The very large value of $K$ for reaction (a) indicates that *products* are highly favored; whereas, the very small value of $K$ for reaction (b) indicates that *reactants* are highly favored. **18.91** (a) Disproportionation redox reaction. (b) $8.2 \times 10^{15}$. This method is feasible for removing $SO_2$. (c) Less effective. **18.93** $1.8 \times 10^{70}$. Due to the large magnitude of $K$, you would expect this reaction to be spontaneous in the forward direction. However, this reaction has a large activation energy, so the rate of reaction is extremely slow. **18.95** $\Delta S_{sys} = 91.1$ J/K; $\Delta S_{surr} = -91.1$ J/K; $\Delta S_{univ} = 0$. The system is at equilibrium. **18.97** 8.5 kJ/mol. **18.99** (a) $CH_3COOH$: 27 kJ/mol; $CH_2ClCOOH$: 16 kJ/mol. (b) The *system's* entropy change dominates. (c) The breaking and making of specific $O-H$ bonds. Other contributions include solvent separation and ion solvation. (d) The $CH_3COO^-$ ion, which is smaller than $CH_2ClCOO^-$, can participate in hydration to a greater extent, leading to solutions with fewer possible arrangements. **18.101** $\chi_{CO} = 0.45$; $\chi_{CO_2} = 0.55$. We assumed that $\Delta G°$ calculated from $\Delta G_f°$ values was temperature independent. **18.103** 249 J/K. **18.105** $\Delta S_{sys} = -327$ J/K · mol; $\Delta S_{surr} = 1918$ J/K · mol; $\Delta S_{univ} = 1591$ J/K · mol. **18.107** $q$, and $w$. **18.109** $\Delta G$ must be negative. $\Delta S$ must be negative. $\Delta H$ must be negative. **18.111** (a) $S = 5.76$ J/K · mol. (b) The fact that the actual residual entropy is 4.2 J/K · mol means that the orientation is not totally random. **18.113** $\Delta H° = 33.89$ kJ/mol; $\Delta S° = 96.4$ J/K · mol; $\Delta G° = 5.2$ kJ/mol. $\Delta H°$ is positive because this is an endothermic process. We also expect $\Delta S°$ to be positive because this is a liquid $\longrightarrow$ vapor phase change. $\Delta G°$ is positive because we are at a temperature that is below the boiling point of benzene (80.1°C).

## Chapter 19

**19.1** (a) $2H^+ + H_2O_2 + 2Fe^{2+} \longrightarrow 2Fe^{3+} + 2H_2O$. (b) $6H^+ + 2HNO_3 + 3Cu \longrightarrow 3Cu^{2+} + 2NO + 4H_2O$. (c) $3CN^- + 2MnO_4^- + H_2O \longrightarrow 3CNO^- + 2MnO_2 + 2OH^-$. (d) $6OH^- + 3Br_2 \longrightarrow BrO_3^- + 3H_2O + 5Br^-$. (e) $2S_2O_3^{2-} + I_2 \longrightarrow S_4O_6^{2-} + 2I^-$. **19.11** $Al(s) + 3Ag^+(1.0\ M) \longrightarrow Al^{3+}(1.0\ M) + 3Ag(s)$; $E°_{cell} = 2.46$ V. **19.13** $Cl_2(g)$ and $MnO_4^-(aq)$. **19.15** (a) Spontaneous. (b) Not spontaneous. (c) Not spontaneous. (d) Spontaneous. **19.17** (a) Li. (b) $H_2$. (c) $Fe^{2+}$. (d) $Br^-$. **19.21** $3 \times 10^{54}$. **19.23** (a) $2 \times 10^{18}$. (b) $3 \times 10^8$. (c) $3 \times 10^{62}$. **19.25** $Ce^{4+}(aq) + Fe^{2+}(aq) \longrightarrow Ce^{3+}(aq) + Fe^{3+}(aq)$; $\Delta G° = -81$ kJ/mol; $K = 2 \times 10^{14}$. **19.29** 1.09 V. **19.31** $E°_{cell} = 0.76$ V; $E_{cell} = 0.78$ V. **19.33** $6.0 \times 10^{-38}$. **19.39** 1.09 V. **19.43** 12.2 g. **19.45** Sodium. **19.47** 0.012 F. **19.49** 5.33 g Cu; 13.4 g $Br_2$. **19.51** $7.70 \times 10^3$ C. **19.53** 1.84 kg $Cl_2$/h. **19.55** 63.3 g/mol. **19.57** 27.0 g/mol. **19.63** (a) Half-reactions: $H_2(g) \longrightarrow 2H^+(aq) + 2e^-$, $Ni^{2+}(aq) + 2e^- \longrightarrow Ni(s)$; Balanced equation: $H_2(g) + Ni^{2+}(aq) \longrightarrow 2H^+(aq) + Ni(s)$. The reaction will proceed to the left. (b) Half-reactions: $5e^- + 8H^+(aq) + MnO_4^-(aq) \longrightarrow Mn^{2+}(aq) + 4H_2O$; $2Cl^-(aq) \longrightarrow Cl_2(g) + 2e^-$. Balanced equation: $16H^+(aq) + 2MnO_4^-(aq) + 10Cl^-(aq) \longrightarrow 2Mn^{2+}(aq) + 8H_2O(l) + 5Cl_2(g)$. The reaction will proceed to the right. (c) Half-reactions: $Cr(s) \longrightarrow Cr^{3+}(aq) + 3e^-$, $Zn^{2+}(aq) + 2e^- \longrightarrow Zn(s)$. Balanced equation: $2Cr(s) + 3Zn^{2+}(aq) \longrightarrow 2Cr^{3+}(aq) + 3Zn(s)$. The reaction will proceed to the left. **19.65** 0.00944 g $SO_2$. **19.67** (a) $2MnO_4^- + 6H^+ + 5H_2O_2 \longrightarrow 2Mn^{2+} + 8H_2O + 5O_2$. (b) 0.0602 M. **19.69** 0.232 mg $Ca^{2+}$/mL blood. **19.71** $5 \times 10^{-13}$. **19.73** (a) 3.14 V. (b) 3.13 V. **19.75** 0.035 V. **19.77** Mercury(I) is $Hg_2^{2+}$. **19.79** 1.44 g Mg; $[Ag^+] = 7 \times 10^{-55}$ M; $[Mg^{2+}] = 0.0500$ M. **19.81** (a) $H_2$, 0.206 L. (b) $6.09 \times 10^{23}$ $e^-$/mol $e^-$. **19.83** (a) $-1356.8$ kJ/mol. (b) 1.17 V. **19.85** +3. **19.87** $\Delta G° = 6.8$ kJ/mol; $K = 0.064$. **19.89** 1.4 A. **19.91** +4. **19.93** $1.60 \times 10^{-19}$ C/$e^-$. **19.95** Cells of higher voltage require very reactive oxidizing and reducing agents, which are difficult to handle. (From Table 19.1 of the text, we see that 5.92 V is the theoretical limit of a cell made up of $Li^+$/Li and $F_2$/$F^-$ electrodes under standard-state conditions.) Batteries made up of several cells in series are easier to use. **19.97** $K_f = 2 \times 10^{20}$. **19.99** (a) $E°_{red}$ for X is negative. $E°_{red}$ for Y is positive. (b) 0.59 V. **19.101** (a) Gold does not tarnish in air because the

reduction potential for oxygen is not sufficiently positive to result in the oxidation of gold. (b) Yes. (c) $2Au + 3F_2 \longrightarrow 2AuF_3$. **19.103** $[Fe^{2+}] = 0.0920$ M; $[Fe^{3+}] = 0.0680$ M. **19.105** $H_2O_2(aq) + 2H^+(aq) + 2e^- \longrightarrow 2H_2O(l)$, $E°_{cathode} = 1.77$ V. $H_2O_2(aq) \longrightarrow O_2(g) + 2H^+(aq) + 2e^-$, $E°_{anode} = 0.68$ V. $E°_{cell} = E°_{cathode} - E°_{anode} = 1.09$ V. The decomposition is spontaneous. **19.107** (a) Unchanged. (b) Unchanged. (c) Squared. (d) Doubled. (e) Doubled. **19.109** As $[H^+]$ increases, $F_2$ does become a stronger oxidizing agent. **19.111** $4.4 \times 10^2$ atm. **19.113** (a) Anode: $Zn \longrightarrow Zn^{2+} + 2e^-$; Cathode: $\frac{1}{2}O_2 + 2e^- \longrightarrow O^{2-}$; 1.65 V. (b) 1.63 V. (c) $4.86 \times 10^3$ kJ/kg Zn. (d) 64 L of air. **19.115** $-3.05$ V. **19.117** $1 \times 10^{-14}$. **19.119** (a) 1A·h = 1A $\times$ 3600s = 3600 C. (b) 105 A·h. This ampere·hour cannot be fully realized because the concentration of $H_2SO_4$ keeps decreasing. (c) $E°_{cell} = 2.01$ V; $\Delta G° = -388$ kJ/mol. **19.121** \$217. **19.123** $-0.037$ V. **19.125** $2 \times 10^{37}$. **19.127** 5 mol ATP/mol $NO_2^-$. **19.129** A small non-zero emf will appear if the temperatures of the two half-cells are different.

## Chapter 20

**20.5** (a) $^{23}_{11}Na$. (b) $^1_1p$ or $^1_1H$. (c) $^1_0n$. (d) $^{56}_{26}Fe$. (e) $^0_{-1}\beta$. **20.13** $2.72 \times 10^{14}$ g/cm$^3$. **20.15** (a) Ni. (b) Se. (c) Cd. **20.17** $-4.85 \times 10^{-12}$ kg/mol $H_2$. **20.19** (a) $6.30 \times 10^{-12}$ J; $9.00 \times 10^{-13}$ J/nucleon. (b) $4.78 \times 10^{-11}$ J; $1.37 \times 10^{-12}$ J/nucleon. **20.21** $7.963 \times 10^{-26}$ kg. **20.25** (a) $^{232}_{90}Th \xrightarrow{\alpha} {}^{228}_{88}Ra \xrightarrow{\beta} {}^{228}_{89}Ac \xrightarrow{\beta} {}^{228}_{90}Th$.

(b) $^{235}_{92}U \xrightarrow{\alpha} {}^{231}_{90}Th \xrightarrow{\beta} {}^{231}_{91}Pa \xrightarrow{\alpha} {}^{227}_{89}Ac$.

(c) $^{237}_{93}Np \xrightarrow{\alpha} {}^{233}_{91}Pa \xrightarrow{\beta} {}^{233}_{92}U \xrightarrow{\alpha} {}^{229}_{90}Th$.

**20.27** $4.88 \times 10^{19}$ atoms. **20.29** $3.09 \times 10^3$ yr. **20.31** No A remains, 0.25 mole of B, no C is left, 0.75 mole of D. **20.33** 5.5 dpm. **20.35** 43:1. **20.39** (a) $^{14}N(\alpha,p)^{17}O$. (b) $^9Be(\alpha,n)^{12}C$. (c) $^{238}U(d,2n)^{238}Np$. **20.41** (a) $^{40}Ca(d,p)^{41}Ca$. (b) $^{32}S(n,p)^{32}P$. (c) $^{239}Pu(\alpha,n)^{242}Cm$. **20.43** $^{198}_{80}Hg + {}^1_0n \longrightarrow {}^{199}_{80}Hg \longrightarrow {}^{198}_{79}Au + {}^1_1p$. **20.55** The fact that the radioisotope appears only in the $I_2$ shows that the $IO_3^-$ is formed only from the $IO_4^-$. **20.57** Add iron-59 to the person's diet, and allow a few days for the iron-59 isotope to be incorporated into the person's body. Isolate red blood cells from a blood sample and monitor radioactivity from the hemoglobin molecules present in the red blood cells. **20.63** 65.3 yr. **20.65** (a) $^3_1H \longrightarrow {}^3_2He + {}^0_{-1}\beta$ (b) 70.5 dpm. **20.67** (a) $^{93}_{36}Kr$. (b) $^1_0n$. (c) $^{146}_{57}La$. (d) $^1_0n$. **20.69** (a) $^3_1H \longrightarrow {}^3_2He + {}^0_{-1}\beta$. (b) $^{242}_{94}Pu \longrightarrow {}^4_2\alpha + {}^{238}_{92}U$. (c) $^{131}_{53}I \longrightarrow {}^{131}_{54}Xe + {}^0_{-1}\beta$. (d) $^{251}_{98}Cf \longrightarrow {}^{247}_{96}Cm + {}^4_2\alpha$. **20.71** Because both Ca and Sr belong to Group 2A, radioactive strontium that has been ingested into the human body becomes concentrated in bones (replacing Ca) and can damage blood cell production. $2.12 \times 10^3$ mCi. **20.73** Normally the human body concentrates iodine in the thyroid gland. The purpose of the large doses of KI is to displace radioactive iodine from the thyroid and allow its excretion from the body. **20.75** (a) $^{209}_{83}Bi + {}^4_2\alpha \longrightarrow {}^{211}_{85}At + 2{}^1_0n$. (b) $^{209}_{83}Bi(\alpha, 2n)^{211}_{85}At$. **20.77** $2.77 \times 10^3$ yr. **20.79** 0.070%. **20.81** (a) $5.59 \times 10^{-15}$ J; $2.84 \times 10^{-13}$ J. (b) 0.024 mol. (c) $4.26 \times 10^6$ kJ. **20.83** $2.8 \times 10^{14}$ iodine-131 atoms. **20.85** $6.1 \times 10^{23}$ atoms/mol. **20.87** (a) $1.83 \times 10^{-12}$ J. (b) The $\alpha$ particle will move away faster because it is smaller. **20.89** U-238, $t_{1/2} = 4.5 \times 10^9$ yr and Th-232, $t_{1/2} = 1.4 \times 10^{10}$ yr. They are still present because of their long half-lives. **20.91** $8.3 \times 10^{-4}$ nm. **20.93** Only $^3H$ has a suitable half-life. The other half-lives are either too long or too short to determine the time span of 6 years accurately. **20.95** A small-scale chain reaction (fission of $^{235}U$) took place. Copper played the crucial role of reflecting neutrons from the splitting uranium-235 atoms back into the uranium sphere to trigger the chain reaction. Note that a sphere has the most appropriate geometry for such a chain reaction. In fact, during the implosion process prior to an atomic explosion, fragments of uranium-235 are pressed roughly into a sphere for the chain reaction to occur (see Section 20.5 of the text). **20.97** $2.1 \times 10^2$ g/mol. **20.99** (a) $r = r_0 A^{1/3}$, where $r_0$ is a proportionality constant. (b) $1.7 \times 10^{-42}$ m$^3$. **20.101** 0.49 rem. **20.103** 3.4 mL.

## Chapter 21

**21.5** $3.30 \times 10^{-4}$; 330 ppm. **21.7** In the stratosphere, the air temperature rises with altitude. This warming effect is the result of exothermic reactions triggered by UV radiation from the sun. **21.11** 260 nm. **21.21** $4.0 \times 10^{37}$ molecules; $3.2 \times 10^{12}$ kg $O_3$. **21.23** $CCl_4 + HF \longrightarrow$ $HCl + CFCl_3$ (Freon-11); $CFCl_3 + HF \longrightarrow HCl + CF_2Cl_2$ (Freon-12). **21.25** $E = 479$ kJ/mol. Solar radiation preferentially breaks the $C-Cl$ bond. There is not enough energy to break the $C-F$ bond.

**21.27** $\overset{..}{\underset{..}{Cl}}-\overset{..}{\underset{..}{O}}-\overset{+}{N}-\overset{..}{\underset{..}{O}}{}^{-}$ ; $\overset{..}{\underset{..}{Cl}}-\overset{..}{\underset{..}{O}}\cdot$ . **21.39** $2.6 \times 10^4$ tons. **21.41** $1.6 \times 10^{19}$ kJ; $4.8 \times 10^{16}$ kg. **21.49** $5.2 \times 10^8$ L. **21.57** (a) rate = $k[NO]^2[O_2]$. (b) rate = $k'[NO]^2$, where $k' = k[O_2]$. (c) $1.3 \times 10^3$ min. **21.59** $4.1 \times 10^{-7}$ atm; $1 \times 10^{16}$ molecules/L. **21.65** 378 g. **21.67** $O_3$: greenhouse gas, toxic to humans, attacks rubber; $SO_2$: toxic to humans, forms acid rain; $NO_2$: forms acid rain, destroys ozone; $CO$: toxic to humans; PAN: a powerful lachrymator, causes breathing difficulties; Rn: causes lung cancer. **21.69** (a) $K_c = \dfrac{[O_2][HbCO]}{[CO][HbO_2]}$ (b) $4.7 \times 10^{-2}$. **21.71** (a) $N_2O + O \longrightarrow 2NO$; $2NO + 2O_3 \longrightarrow 2NO_2 + 2O_2$; overall: $N_2O + O + 2O_3 \longrightarrow 2NO_2 + 2O_2$. (b) $N_2O$ is a more effective greenhouse gas than $CO_2$ because it has a permanent dipole. (c) $3.0 \times 10^{10}$ mol. **21.73** $1.8 \times 10^{19}$ Rn atoms; $6.4 \times 10^{16}$ Rn atoms. **21.75** (a) Its small concentration is the result of the high reactivity of the OH radical. (b) OH has an unpaired electron; free radicals are always good oxidizing agents. (c) $OH + NO_2 \longrightarrow HNO_3$. (d) $OH + SO_2 \longrightarrow HSO_3$; $HSO_3 + O_2 + H_2O \longrightarrow H_2SO_4 + HO_2$. **21.77** The blackened bucket has a large deposit of elemental carbon. When heated over the burner, it forms poisonous carbon monoxide. $C + CO_2 \longrightarrow 2CO$. A smaller amount of CO is also formed as follows: $2C + O_2 \longrightarrow 2CO$. **21.79** The use of the aerosol liberates CFCs that destroy the ozone layer. **21.81** Yes. Light of wavelength 409 nm (visible) or shorter will break the $C-Br$ bond. **21.83** (a) $6.2 \times 10^8$. (b) The $CO_2$ liberated from limestone contributes to global warming. **21.85** Most water molecules contain oxygen-16, but a small percentage of water molecules contain oxygen-18. The ratio of the two isotopes in the ocean is essentially constant, but the ratio in the water vapor evaporated from the oceans is temperature-dependent, with the vapor becoming slightly enriched with oxygen-18 as temperature increases. The water locked up in ice cores provides a historical record of this oxygen-18 enrichment, and thus ice cores contain information about past global temperatures.

## Chapter 22

**22.11** (a) +3. (b) 6. (c) Oxalate ion ($C_2O_4^{2-}$). **22.13** (a) Na: +1; Mo: +6. (b) Mg: +2; W: +6. (c) Fe: 0. **22.15** (a) cis-dichlorobis(ethylenediamine) cobalt(III). (b) pentamminechloroplatinum(IV) chloride. (c) pentamminechlorocobalt(III) chloride. **22.17** (a) $[Cr(en)_2Cl_2]^+$. (b) $Fe(CO)_5$. (c) $K_2[Cu(CN)_4]$. (d) $[Co(NH_3)_4(H_2O)Cl]Cl_2$. **22.23** (a) Two. (b) Two.

**22.25** (a)  (b)

trans        cis

**22.31** Cyanide ion is a very strong-field ligand. It causes a larger crystal field splitting than water, resulting in the absorption of higher energy (shorter wavelength) radiation when a d electron is excited to a higher energy d orbital. **22.33** (a) Orange. (b) 255 kJ/mol. **22.35** Two moles. $[Co(NH_3)_4Cl_2]Cl$. Refer to Problem 22.25 (a) for a diagram of the structure of the complex ion. **22.37** $\Delta$ would be greater for the higher oxidation state. **22.41** Use a radioactive label such as $^{14}CN^-$ (in NaCN). Add NaCN to a solution of $K_3Fe(CN)_6$. Isolate some of the $K_3Fe(CN)_6$ and check its radioactivity. If the complex shows radioactivity, then it must mean that the $CN^-$ ion has participated in the exchange reaction.

**22.43** $Cu(CN)_2$ is the white precipitate. It is soluble in KCN(aq), due to formation of $[Cu(CN)_4]^{2-}$ , so the concentration of $Cu^{2+}$ is too small for $Cu^{2+}$ ions to precipitate with sulfide. **22.45** $1.4 \times 10^2$. **22.47** The purple color is caused by the build-up of deoxyhemoglobin. When either oxyhemoglobin or deoxyhemoglobin takes up CO, the carbonylhemoglobin takes on a red color, the same as oxyhemoglobin. **22.49** $Mn^{3+}$ is $3d^4$ and $Cr^{3+}$ is $3d^5$. Therefore, $Mn^{3+}$ has a greater tendency to accept an electron and is a stronger oxidizing agent. The $3d^5$ electron configuration of $Cr^{3+}$ is a stable configuration. **22.51** $Ti^{3+}$; $Fe^{3+}$. **22.53** $1.6 \times 10^4$ g hemoglobin/mol Fe. The discrepancy between our minimum value and the actual value can be explained by realizing that there are four iron atoms per mole of hemoglobin. **22.55** (a) $[Cr(H_2O)_6]Cl_3$, number of ions: 4. (b) $[Cr(H_2O)_5Cl]Cl_2 \cdot H_2O$, number of ions: 3. (c) $[Cr(H_2O)_4Cl_2]Cl \cdot 2H_2O$, number of ions: 2. Compare the compounds with equal molar amounts of NaCl, $MgCl_2$, and $FeCl_3$ in an electrical conductance experiment. The solution that has similar conductance to the NaCl solution contains (c); the solution with the conductance similar to $MgCl_2$ contains (b); and the solution with conductance similar to $FeCl_3$ contains (a). **22.57** $\Delta G° = -1.8 \times 10^2$ kJ/mol; $K = 6 \times 10^{30}$. **22.59** Iron is much more abundant than cobalt. **22.61** Oxyhemoglobin absorbs higher energy light than deoxyhemoglobin. Oxyhemoglobin is diamagnetic (low spin), while deoxyhemoglobin is paramagnetic (high spin). These differences occur because oxygen ($O_2$) is a strong-field ligand. **22.63** Complexes are expected to be colored when the highest occupied orbitals have between one and nine d electrons. $Zn^{2+}$, $Cu^+$, and $Pb^{2+}$ are $d^{10}$ ions. $V^{5+}$, $Ca^{2+}$, and $Sc^{3+}$ are $d^0$ ions. **22.65** Dipole moment measurement. Only the cis isomer has a dipole moment. **22.67** EDTA sequesters metal ions (like $Ca^{2+}$ and $Mg^{2+}$), which are essential for the growth and function of bacteria. **22.69** 3. **22.71** $2.2 \times 10^{-20}$ M. **22.73** (a) $2.7 \times 10^6$. (b) Free $Cu^+$ ions are unstable in solution. Therefore, the only stable compounds containing $Cu^+$ ions are insoluble. **22.75** Y. **22.77** 0.0 M. **22.79** $[Mn(CN)_6]^{5-}$: Mn is +1, one unpaired d electron. $[Mn(CN)_6]^{4-}$: Mn is +2, one unpaired d electron. $[Mn(CN)_6]^{3-}$: Mn is +3, two unpaired d electrons.

## Chapter 23

**23.3** (a) $CaCO_3$. (b) $CaCO_3 \cdot MgCO_3$. (c) $CaF_2$. (d) NaCl. (e) $Al_2O_3$. (f) $Fe_3O_4$. (g) $Be_3Al_2Si_6O_{18}$. (h) PbS. (i) $MgSO_4 \cdot 7H_2O$. (j) $CaSO_4$. **23.13** $K_P = 4.5 \times 10^5$. **23.15** Ag, Pt, and Au will not be oxidized, but the other metals will. **23.17** (a) $8.9 \times 10^{12}$ cm$^3$. (b) $4.0 \times 10^8$ kg. **23.19** Al, Na, and Ca. **23.33** (a) $2Na(s) + 2H_2O(l) \longrightarrow 2NaOH(aq) + H_2(g)$. (b) $2NaOH(aq) + CO_2(g) \longrightarrow Na_2CO_3(aq) + H_2O(l)$. (c) $Na_2CO_3(s) + 2HCl(aq) \longrightarrow 2NaCl(aq) + CO_2(g) + H_2O(l)$. (d) $NaHCO_3(aq) + HCl(aq) \longrightarrow NaCl(aq) + CO_2(g) + H_2O(l)$. (e) $2NaHCO_3(s) \longrightarrow Na_2CO_3(s) + CO_2(g) + H_2O(g)$. (f) No reaction. **23.35** 5.59 L. **23.39** $3Mg(s) + 8HNO_3(aq) \longrightarrow 3Mg(NO_3)_2(aq) + 4H_2O(l) + 2NO(g)$. The magnesium nitrate is recovered from solution by evaporation, dried, and heated in air to obtain magnesium oxide: $2Mg(NO_3)_2(s) \longrightarrow 2MgO(s) + 4NO_2(g) + O_2(g)$. **23.41** The electron configuration of magnesium is $[Ne]3s^2$. The 3s electrons are outside the neon core (shielded), so they have relatively low ionization energies. Removing the third electron means separating an electron from the neon (closed shell) core, which requires a great deal more energy. **23.43** Even though helium and the Group 2A metals have $ns^2$ outer electron configurations, helium has a closed shell noble gas configuration and the Group 2A metals do not. The electrons in He are much closer to and more strongly attracted by the nucleus. Hence, the electrons in He are not easily removed. Helium is inert. **23.45** (a) CaO(s). (b) $Ca(OH)_2(s)$. **23.49** 60.7 h. **23.51** (a) 1.03 V. (b) $3.32 \times 10^4$ kJ/mol. **23.53** $4Al(NO_3)_3(s) \longrightarrow 2Al_2O_3(s) + 12NO_2(g) + 3O_2(g)$. **23.55** The "bridge" bonds in $Al_2Cl_6$ break at high temperature: $Al_2Cl_6(g) \rightleftharpoons 2AlCl_3(g)$. This increases the number of molecules in the gas phase and causes the pressure to be higher than expected for pure $Al_2Cl_6$. **23.57** In $Al_2Cl_6$, each aluminum atom is surrounded by 4 bonding pairs of electrons ($AB_4$-type molecule), and therefore each aluminum atom is $sp^3$ hybridized. VSEPR analysis shows

$AlCl_3$ to be an $AB_3$-type molecule (no lone pairs on the central atom). The geometry should be trigonal planar, and the aluminum atom should therefore be $sp^2$ hybridized. **23.59** 65.4 g/mol. **23.61** Copper(II) ion is more easily reduced than either water or hydrogen ion. Copper metal is more easily oxidized than water. Water should not be affected by the copper purification process under standard conditions. **23.63** (a) 1482 kJ/mol. (b) 3152.8 kJ/mol. **23.65** $Mg(s)$ reacts with $N_2(g)$ at high temperatures to produce $Mg_3N_2(s)$. $Ti(s)$ also reacts with $N_2(g)$ at high temperatures to produce $TiN(s)$. **23.67** (a) In water the aluminum(III) ion causes an increase in the concentration of hydrogen ion (lower pH). This results from the effect of the small diameter and high charge (3+) of the aluminum ion on surrounding water molecules. The aluminum ion draws electrons in the $O-H$ bonds to itself, thus allowing easy formation of $H^+$ ions. (b) $Al(OH)_3$ is an amphoteric hydroxide. It will dissolve in strong base with the formation of a complex ion. $Al(OH)_3(s) + OH^-(aq) \longrightarrow$ $Al(OH)_4^-(aq)$. The concentration of $OH^-$ in aqueous ammonia is too low for this reaction to occur. **23.69** $CaO(s) + 2HCl(aq) \longrightarrow CaCl_2(aq) +$ $H_2O(l)$. **23.71** Metals have closely spaced energy levels and a very small energy gap between filled and empty levels. Consequently, many electronic transitions can take place with absorption and subsequent emission of light continually occurring. Some of these transitions fall in the visible region of the spectrum and give rise to the flickering appearance. **23.73** NaF: cavity prevention. $Li_2CO_3$: antidepressant. $Mg(OH)_2$: laxative. $CaCO_3$: calcium supplement; antacid. $BaSO_4$: radiocontrast agent. **23.75** Both Li and Mg form oxides ($Li_2O$ and MgO). Other Group 1A metals (Na, K, etc.) also form peroxides and superoxides. In Group 1A, only Li forms nitride ($Li_3N$), like Mg ($Mg_3N_2$). Li resembles Mg in that its carbonate, fluoride, and phosphate have low solubilities. **23.77** Zn. **23.79** 87.66%; 12.34%. **23.81** 727 atm.

## Chapter 24

**24.11** HCl; $CaH_2$. A water solution of HCl is called hydrochloric acid. Calcium hydride will react according to the equation $CaH_2(s) +$ $2H_2O(l) \longrightarrow Ca(OH)_2(aq) + 2H_2(g)$. **24.13** NaH: Ionic compound, reacts with water as follows: $NaH(s) + H_2O(l) \longrightarrow NaOH(aq) +$ $H_2(g)$; $CaH_2$: Ionic compound, reacts with water as follows: $CaH_2(s) +$ $2H_2O(l) \longrightarrow Ca(OH)_2(aq) + 2H_2(g)$; $CH_4$: Covalent compound, unreactive, burns in air or oxygen: $CH_4(g) + 2O_2(g) \longrightarrow CO_2(g) +$ $2H_2O(l)$; $NH_3$: Covalent compound, weak base in water: $NH_3(aq) +$ $H_2O(l) \rightleftharpoons NH_4^+(aq) + OH^-(aq)$; $H_2O$: Covalent compound, forms strong intermolecular hydrogen bonds, good solvent for both ionic compounds and substances capable of forming hydrogen bonds; HCl: Covalent compound (polar), acts as a strong acid in water: $HCl(g) +$ $H_2O(l) \longrightarrow H_3O^+(aq) + Cl^-(aq)$. **24.15** $CaH_2(s) + 2H_2O(l) \longrightarrow$ $Ca(OH)_2(aq) + 2H_2(g)$; 22.7 g. **24.17** (a) $CuO(s) + H_2(g) \longrightarrow Cu(s) +$ $H_2O(l)$. (b) No reaction. **24.25** $[:C \equiv C:]^{2-}$. **24.27** (a) $2NaHCO_3(s) \longrightarrow$ $Na_2CO_3(s) + H_2O(g) + CO_2(g)$. (b) $Ca(OH)_2(aq) + CO_2(g) \longrightarrow$ $CaCO_3(s) + H_2O(l)$. The visual proof is the formation of a white precipitate of $CaCO_3$. **24.29** Heat causes bicarbonates to decompose according to the reaction: $2HCO_3^- \longrightarrow CO_3^{2-} + H_2O + CO_2$. Generation of carbonate ion causes precipitation of the insoluble $MgCO_3$. **24.31** $NaHCO_3$ plus some $Na_2CO_3$. **24.33** Yes. **24.39** (a) $2NaNO_3(s) \longrightarrow$ $2NaNO_2(s) + O_2(g)$. (b) $NaNO_3(s) + C(s) \longrightarrow NaNO_2(s) + CO(g)$. **24.41** $NH_3(g) + CO_2(g) \longrightarrow (NH_2)_2CO(s) + H_2O(l)$. The reaction should be run at high pressure. **24.43** The oxidation state of N in nitric acid is +5, the highest oxidation state for N. N can be reduced to an oxidation state −3. **24.45** (a) $NH_4NO_3(s) \longrightarrow N_2O(g) + 2H_2O(l)$. (b) $2KNO_3(s) \longrightarrow 2KNO_2(s) + O_2(g)$. (c) $Pb(NO_3)_2(s) \longrightarrow$

$PbO(s) + 2NO_2(g) + O_2(g)$. **24.47** $KNO_3(s) + C(s) \longrightarrow KNO_2(s) +$ $CO(g)$; 48.0 g. **24.49** (a) 86.7 kJ/mol. (b) $4 \times 10^{-31}$. (c) $4 \times 10^{-31}$. **24.51** 125 g/mol; $P_4$. **24.53** $4HNO_3(aq) + P_4O_{10}(s) \longrightarrow 2N_2O_5(g) +$ $4HPO_3(l)$; 60.4 g. **24.55** $sp^3$. **24.63** $G° = -198.3$ kJ/mol; $K_P = 6 \times 10^{34}$; $K_c = K_P$. **24.65** (a) To exclude light. (b) 0.371 L. **24.67** F: −1; O: 0. **24.69** (a) $HCOOH(l) \rightleftharpoons CO(g) + H_2O(l)$. (b) $4H_3PO_4(l) \rightleftharpoons$ $P_4O_{10}(s) + 6H_2O(l)$. (c) $2HNO_3(l) \rightleftharpoons N_2O_5(g) + H_2O(l)$. (d) $2HClO_3(l) \rightleftharpoons Cl_2O_5(l) + H_2O(l)$. **24.71** To form $OF_6$ there would have to be six bonds (twelve electrons) around the oxygen atom. This would violate the octet rule. Since oxygen does not have $d$ orbitals, it cannot have an expanded octet. **24.73** 35 g. **24.75** $9H_2SO_4(aq) +$ $8NaI(aq) \longrightarrow 4I_2(s) + H_2S(g) + 4H_2O(l) + 8NaHSO_4(aq)$. **24.79** (a) $H-\ddot{F}:\cdots H-\ddot{F}:$ (b) $[:\ddot{F}:\cdots H \cdots \ddot{F}:]^-$ **24.81** (a) Linear. (b) Tetrahedral. (c) Trigonal bipyramidal. (d) See-saw. **24.83** 25.3 L. **24.85** 2.81 L. **24.87** $I_2O_5(s) + 5CO(g) \longrightarrow 5CO_2(g) + I_2(s)$. Iodine is reduced; carbon is oxidized. **24.89** (a) $2H_3PO_3(aq) \longrightarrow H_3PO_4(aq) + PH_3(g) + O_2(g)$. (b) $Li_4C(s) + 4HCl(aq) \longrightarrow 4LiCl(aq) + CH_4(g)$. (c) $2HI(g) +$ $2HNO_2(aq) \longrightarrow I_2(s) + 2NO(g) + 2H_2O(l)$. (d) $H_2S(g) + 2Cl_2(g)$ $\longrightarrow 2HCl(g) + SCl_2(l)$. **24.91** (a) $SiCl_4$. (b) $F^-$. (c) F. (d) $CO_2$. **24.93** There is no change in oxidation number; it is zero for both compounds.

**24.95** $PCl_4^+:$ $\begin{bmatrix} & Cl & \\ & | & \\ Cl - & P & - Cl \\ & | & \\ & Cl & \end{bmatrix}^+$ , $sp^3$. $PCl_6^-:$ $\begin{bmatrix} Cl & Cl & Cl \\ & \backslash | / & \\ & P & \\ & / | \backslash & \\ Cl & Cl & Cl \end{bmatrix}^-$ , $sp^3d^2$.

**24.97** 25°C: $K = 9.61 \times 10^{-22}$; 100°C: $K = 1.2 \times 10^{-15}$. **24.99** The glass is etched by the reaction: $6HF(aq) + SiO_2(s) \longrightarrow H_2SiF_6(aq) +$ $2H_2O(l)$. This process gives the glass a frosted appearance. **24.101** 1.18. **24.103** 0.833 g/L. The molar mass derived from the observed density is 74.41, which suggests that the molecules are associated to some extent in the gas phase. This makes sense due to strong hydrogen bonding in HF.

## Chapter 25

**25.3** The monomer must have a triple bond.

**25.5** $\begin{bmatrix} H & C_2H_5 & H & \\ | & | & | & \\ -C - & C - & C = & C- \\ | & | & | & \\ CH_3 & CH_3 & & H \end{bmatrix}_n$ or $\begin{bmatrix} H & C_2H_5 & & \\ | & | & | & | \\ -C - & C - & C = & C- \\ | & | & | & | \\ CH_3 & CH_3 & H & H \end{bmatrix}_n$

**25.9** (1) Produce the alkoxide: $Sc(s) + 2C_2H_5OH(l) \longrightarrow$ $Sc(OC_2H_5)(alc) + 2H^+(alc)$ ("alc" indicates a solution in alcohol); (2) Hydrolyze to produce hydroxide pellets: $Sc(OC_2H_5)(alc) +$ $2H_2O(l) \longrightarrow Sc(OH)_2(s) + 2C_2H_5OH(alc)$; (3) Sinter pellets to produce ceramic: $Sc(OH)_2(s) \longrightarrow ScO(s) + 2H_2O(g)$. **25.11** Bakelite is best described as a thermosetting composite polymer. **25.15** No. These polymers are too flexible, and liquid crystals require long, relatively rigid molecules. **25.19** As shown, it is an alternating condensation copolymer of the polyester class. **25.21** Metal amalgams expand with age; composite fillings tend to shrink. **25.25** $sp^2$. **25.27** Dispersion forces; dispersion forces. **25.31** (a) n-type. (b) p-type. **25.35** $Bi_2Sr_2CuO_6$. **25.37** Two are +2 ($[Ar]3d^9$), one is +3 ($[Ar]3d^8$). The +3 oxidation state is unusual for copper. **25.39** $H_2N-(CH_2)_8-NH_2$   $HO-\overset{\overset{O}{\|}}{C}-(CH_2)_4-\overset{\overset{O}{\|}}{C}-OH$

**25.41** Plastic polymer: covalent bonds, disulfide (covalent) bonds, H-bonds and dispersion forces. Ceramics: ionic and network covalent bonds. **25.43** Fluoroapatite is less soluble than hydroxyapatite, particularly in acidic solutions. Dental fillings must also be insoluble. **25.45** The molecule is long, flat and rigid, so it should form a liquid crystal.

# Answers

## To Pre-Professional Practice Exam Problems

**Chapter 1**

1. b
2. d
3. c
4. a

**Chapter 2**

1. c
2. d
3. a
4. b

**Chapter 3**

1. d
2. b
3. b
4. c

**Chapter 4**

1. a
2. b
3. a
4. c

**Chapter 5**

1. d
2. a
3. c
4. b

**Chapter 6**

1. b
2. a
3. c
4. c

**Chapter 7**

1. b
2. a
3. b
4. d

**Chapter 8**

1. c
2. a
3. a
4. b

**Chapter 9**

1. c
2. b
3. b
4. d

**Chapter 10**

1. c
2. d
3. a
4. c

**Chapter 11**

1. b
2. a
3. b
4. d

**Chapter 12**

1. c
2. b
3. d
4. a

**Chapter 13**

1. b
2. d
3. c
4. c

**Chapter 14**

1. a
2. b
3. c
4. b

**Chapter 15**

1. b
2. a
3. b
4. d

**Chapter 16**

1. d
2. b
3. a
4. b

**Chapter 17**

1. c
2. c
3. a
4. d

**Chapter 18**

1. d
2. b
3. a
4. c

**Chapter 19**

1. b
2. c
3. d
4. d

**Chapter 20**

1. a
2. c
3. d
4. a

**Chapter 21**

1. b
2. d
3. c
4. a

**Chapter 22**

1. c
2. a
3. b
4. b

**Chapter 23**

1. c
2. d
3. b
4. a

**Chapter 24**

1. a
2. b
3. c
4. b

**Chapter 25**

1. b
2. d
3. b
4. d

# Credits

## Chapter 1

**Opener:** Washington State History Research Center; **p. 3:** © Chris Livingston/Getty; **1.1a:** © Steve Allen/Getty; **1.1b:** © The McGraw-Hill Companies, Inc./Charles D. Winters/Timeframe Photography Inc.; **1.1c:** © Stockbyte/PunchStock; **1.1d:** © Corbis RF; **1.5, 1.7a-b:** © The McGraw-Hill Companies, Inc./Charles D. Winters/Timeframe Photography Inc.; **p. 12:** © David Tietz/Editorial Image, LLC; **p. 14:** © Joe Skipper/Reuters/Corbis.

## Chapter 2

**Opener, p. 35:** © David Tietz/Editorial Image, LLC; **2.1a:** © SSPL/The Image Works; **2.1b:** © Photo Researchers; **2.1c:** © Barry Bland/Alamy; **2.5a-c, 2.17:** © The McGraw-Hill Companies, Inc./Charles D. Winters/Timeframe Photography Inc.; **p. 67(both):** © David Tietz/Editorial Image, LLC.

## Chapter 3

**Opener:** © Ezra Shaw/Getty; **p. 77:** Copyrighted images courtesy of Bedford Laboratories, a Division of Ben Venue Laboratories, Inc. (www.bedfordlabs.com); **3.1:** © David Tietz/Editorial Image, LLC; **p. 85:** © The McGraw-Hill Companies, Inc./John Flournoy, photographer; **3.2:** © Richard Megna/Fundamental Photographs; **p. 100:** © The McGraw-Hill Companies, Inc./Charles D. Winters/Timeframe Photography Inc.

## Chapter 4

**Opener:** © Photodisc/Getty RF; **p. 117:** © 67photo/Alamy; **4.1a-c:** © The McGraw-Hill Companies, Inc./Stephen Frisch, photographer; **p. 120(bottom):** © David Tietz/Editorial Image, LLC; **4.3(all):** © The McGraw-Hill Companies, Inc./Charles D. Winters/Timeframe Photography Inc.; **4.5:** © David Tietz/Editorial Image, LLC; **Sample Problem 4.5a-c, 4.6a-b, p. 135, 4.9(both):** © The McGraw-Hill Companies, Inc./Charles D. Winters/Timeframe Photography Inc.; **4.11(both):** © David Tietz/Editorial Image, LLC; **4.12a-b:** © The McGraw-Hill Companies, Inc./Charles D. Winters /Timeframe Photography

Inc.; **4.15(both):** © Richard Megna/Fundamental Photographs; **p. 159:** © Jack Sullivan/Alamy.

## Chapter 5

**Opener:** © Keith Leighton/Alamy; **p. 171:** © David Tietz/Editorial Image, LLC; **5.1:** © Bettmann/Corbis; **p. 194, p. 200(top left, top right):** © Brian Moeskau; **p. 200(bottom):** © David Tietz/Editorial Image, LLC.

## Chapter 6

**Opener:** © Olivier Voisin/Photo Researchers; **6.1(hand):** © Jim Wehtje/Getty; **p. 216:** © Brian Moeskau; **p. 218:** © SPL/Photo Researchers; **6.6a:** © Doug Menuez/Getty; **6.6b:** © The McGraw-Hill Companies, Inc./Charles D. Winters/Timeframe Photography Inc.; **6.8:** Courtesy of Sargent-Welch; **6.13a-b:** © Dr. Stanley Flegler/Visuals Unlimited; **p. 247:** © Phanie/Photo Researchers.

## Chapter 7

**Opener:** © GARO/PHANIE/Photo Researchers; **p. 257(both):** © David Tietz/Editorial Image, LLC; **7.14(lithium, sodium, potassium):** © The McGraw-Hill Companies, Inc./Charles D. Winter/Timeframe Photography Inc.; **7.14(rubidium, cesium, beryllium, magnesium):** © The McGraw-Hill Companies, Inc./Stephen Frisch, photographer; **7.15(calcium):** © The McGraw-Hill Companies, Inc./Charles D. Winters /Timeframe Photography Inc.; **7.15(strontium):** © Ed R. Degginger/Color-Pic, Inc.; **7.15(barium, boron):** © The McGraw-Hill Companies, Inc./Stephen Frisch, photographer; **7.16(aluminum):** © The McGraw-Hill Companies, Inc./Charles D. Winters/Timeframe Photography Inc.; **7.16(gallium, indium):** © The McGraw-Hill Companies, Inc./Stephen Frisch, photographer; **7.17(graphite, diamond, silicon, germanium, tin, lead), 7.18( nitrogen, white phosphorus, red phosphorus):** © The McGraw-Hill Companies, Inc./Charles D. Winters/Timeframe Photography Inc.; **7.18(arsenic):** © The McGraw-Hill Companies, Inc./Stephen Frisch, photographer; **7.18(antimony):** © The McGraw-

Hill Companies, Inc./Charles D. Winters/Timeframe Photography Inc.; **7.18(bismuth):** © The McGraw-Hill Companies, Inc./Stephen Frisch, photographer; **p. 283(bottom):** © Heinrich Pniok http://www.pse-mendelejew.de; **7.19(oxygen):** © David Tietz/Editorial Image, LLC; **7.19(sulfur):** © Richard Treptow/Photo Researchers; **7.19(selenium):** © The McGraw-Hill Companies, Inc./Charles D. Winters/Timeframe Photography Inc.; **7.19(tellurium), 7.20(fluorine):** © The McGraw-Hill Companies, Inc./Stephen Frisch, photographer; **7.20(chlorine):** © Charles D. Winters/Photo Researchers; **7.20(bromine):** © The McGraw-Hill Companies, Inc./Stephen Frisch, photographer; **7.20(iodine):** © The McGraw-Hill Companies, Inc./Charles D. Winters /Timeframe Photography Inc.; **7.21(helium, neon, argon, krypton, xenon):** © The McGraw-Hill Companies, Inc./Stephen Frisch, photographer; **p. 286:** © 3D4Medical.com/Getty; **p. 288(both):** © The McGraw-Hill Companies, Inc./Charles D. Winters /Timeframe Photography Inc.

## Chapter 8

**Opener:** © Photodisc/Getty RF; **p. 297:** © David Tietz/Editorial Image, LLC; **8.2(all):** © Richard Megna/Fundamental Photographs; **p. 321(top):** © Eliot J. Schechter/Stringer/AFP/Getty; **p. 321(bottom):** © Wesley Bocxe/Photo Researchers.

## Chapter 9

**Opener:** © Radius Images/Corbis; **9.1a-e:** © The McGraw-Hill Companies, Inc./Stephen Frisch, photographer; **p. 368(both):** © The McGraw-Hill Companies, Inc./Charles D. Winters/Timeframe Photography Inc.

## Chapter 10

**Opener:** Courtesy of Julia Burdge; **p. 391 (methanol):** © David Tietz/Editorial Image, LLC; **p. 391(ethanol):** © John A. Rizzo/Getty; **p. 391(isopropyl):** © David Tietz/Editorial Image, LLC; **10.5a-b:** © The McGraw-Hill Companies, Inc./Charles D. Winters /Timeframe Photography Inc.

# CREDITS

## Chapter 11

Opener: © Martin Strmiska/Alamy; **p. 441:** © Kike Calvo/VW/The Image Works; **11.3a-b:** © The McGraw-Hill Companies, Inc./Charles D. Winters/Timeframe Photography Inc.; **p. 443(bottom):** © David Tietz/Editorial Image, LLC; **11.9a-b:** © The McGraw-Hill Companies, Inc./Charles D. Winters/Timeframe Photography Inc.; **p. 446:** Courtesy of USC Catalina Hyperbaric Chamber; **p. 480:** © David Samuel Robbins/Corbis.

## Chapter 12

Opener: © Blend Images/Alamy; **p. 498:** © Dr. Stanley Flegler/Visuals Unlimited; **12.7a:** © Purestock/PunchStock; **12.7b:** © Ingram Publishing/SuperStock; **12.8:** © Yoav Levy/PhotoTake; **12.23b:** © Dr. M.B. Hursthouse/Photo Researchers; **12.27a-b, 12.28:** © The McGraw-Hill Companies, Inc./Charles D. Winters/Timeframe Photography Inc.; **p. 521:** © The McGraw-Hill Companies, Inc./Ken Karp, photographer.

## Chapter 13

Opener: © Digital Vision Ltd.; **13.1a-e, 13.2a-e:** © The McGraw-Hill Companies, Inc./Charles D. Winters/Timeframe Photography Inc.; **13.12a-c:** © David M. Phillips/Photo Researchers; **13.13:** © Kip Peticolas/Fundamental Photographs; **13.14:** © Photodisc/Getty RF; **PR 13.114(both):** © David Tietz/Editorial Image, LLC; **PR 13.134:** © Bill Curtsinger/National Geographic/Getty.

## Chapter 14

Opener: © David Tietz/Editorial Image, LLC; **14.3, 14.7:** © The McGraw-Hill Companies, Inc./Ken Karp, photographer; **14.13(both):** © The McGraw-Hill Companies, Inc./Charles D. Winters /Timeframe Photography Inc.; **p. 615:** © Corbis RF.

## Chapter 15

Opener: © Brand X/JupiterImages; **p. 629:** Courtesy of Hypoxico Inc.; **15.1(all), 15.3(all):** © Richard Megna/Fundamental Photographs; **15.8a-b, 15.9a-c:** © The McGraw-Hill Companies, Inc./Charles D. Winters/Timeframe Photography Inc.

## Chapter 16

Opener: © Tom Stoddart Archive/Hulton/Getty; **16.1:** © The McGraw-Hill Companies, Inc./Charles D. Winters/Timeframe Photography Inc.

## Chapter 17

Opener: © White Rock/Getty; **p. 727:** © The McGraw-Hill Companies, Inc.; **17.2, 17.7, 17.9, 17.10:** © The McGraw-Hill Companies, Inc./Ken Karp, photographer; **17.11:** © The McGraw-Hill Companies, Inc./Stephen Frisch, photographer.

## Chapter 18

Opener: © Laurel Latto.

## Chapter 19

Opener: © Spyros Bourboulis/First Light/Getty; **19.2:** © The McGraw-Hill Companies, Inc./Ken Karp, photographer; **p. 836(top):** © The McGraw-Hill Companies, Inc./Charles D. Winters/Timeframe Photography Inc.; **p. 836(bottom):** © Ragnar Schmuck/Zefa/Corbis; **19.11:** © The McGraw-Hill Companies, Inc./Stephen Frisch, photographer; **Questions and Problems 19.78:** © David Tietz/Editorial Image, LLC; **Questions and Problems 19.114:** © The McGraw-Hill Companies, Inc./Ken Karp, photographer.

## Chapter 20

Opener: © ISM/PhotoTake; **20.6:** © Nick Wall/Photo Researchers; **20.12:** © Vanderlei Almeida/AFP/Getty; **20.13:** © E.R. Degginger/Color-pic Inc.; **20.14:** © Peter Essick/Aurora/Getty; **20.16:** © Lawrence Livermore National Labs; **20.17:** © Corbis; **20.18:** © Mediscan/Corbis; **p. 885:** © David Tietz/Editorial Image, LLC.

## Chapter 21

Opener: © Digital Vision/PunchStock; **21.4:** © Dennis Fast/VisualWritten/The Image Works; **21.5:** NASA; **21.8:** © German Remote Sensing Data Center; **21.9:** © Time Life Pictures/NASA/Time Life Pictures/Getty; **21.10:** © Jim Sugar/Corbis; **21.20(top):** © NYC Parks Photo Archive/Fundamental Photographs; **21.20(bottom):** © Kristen Brochmann/Fundamental Photographs; **21.22:** © Corbis RF; **21.24:** © Owen Franken; **21.26:** © Kent Knudson/PhotoLink/Getty; **21.28:** © David Tietz/Editorial Image, LLC.

## Chapter 22

Opener: © Robert George Young/Photographer's Choice/Getty; **p. 921(top):** The Advertising Archives; **p. 921(bottom), 22.11:** © David Tietz/Editorial Image, LLC; **22.16:** © The McGraw-Hill Companies, Inc./Charles D. Winters/Timeframe Photography Inc.; **p. 938:** © David Tietz/Editorial Image, LLC.

## Chapter 23

Opener: © David Tietz/Editorial Image, LLC; **23.2:** © Peter Ryan/Photo Researchers; **p. 947:** © Lin Yueseng/Panorama/The Image Works; **23.5:** © Yang Liu/Corbis; **23.7:** © Maximilian Stock Ltd/Photo Researchers; **23.13:** © David Tietz/Editorial Image, LLC; **23.14:** © Radius Images/Corbis; **23.15, 23.16, 23.17:** © The McGraw-Hill Companies, Inc./Charles D. Winters/Timeframe Photography Inc.; **23.19:** © Phil Degginger/Color-Pic, Inc.

## Chapter 24

Opener: Smithsonian Institution, National Museum of American History; **24.5:** © David Tietz/Editorial Image, LLC; **24.6:** © Noeline Kelly/Corbis; **24.7:** © The McGraw-Hill Companies, Inc./Charles D. Winters /Timeframe Photography Inc.; **24.8, 24.14:** © David Tietz/Editorial Image, LLC; **24.18:** © Dynamic Graphics Group/Creatas/Alamy; **24.20:** © Inga Spence/Visuals Unlimited.

## Chapter 25

Opener: © Jonas Ekstromer/AFP/Getty; **p. 1000:** © Tracy Montana/PhotoLink/Getty; **25.2(left):** © David Tietz/Editorial Image, LLC; **25.2(right):** © Richard Megna/Fundamental Photographs; **p. 1001(bottom):** AP Photo/Robert E. Klein; **p. 1002:** © Comstock Images/Alamy; **25.9:** NASA; **25.12:** © Stockbyte/Getty; **p. 1016:** © Jim Wehtje/Photodisc/Getty; **25.13a:** © Tek Image/Photo Researchers; **25.13b:** © Dr. Ben Oostra/Visuals Unlimited; **25.13c:** Reprint Courtesy of International Business Machines Corporation copyright 1993 © International Business Machines Corporation; **p. 1017 (bottom left):** © Phil Degginger/Color-Pic, Inc.; **p. 1017(bottom right):** © Getty Royalty Free; **25.17:** © David Tietz/Editorial Image, LLC; **25.18:** © IBM Research/Peter Arnold, Inc.

# Index

## Fundamental Constants

| | |
|---|---|
| Avogadro's number ($N_A$) | $6.0221418 \times 10^{23}$ |
| Electron charge ($e$) | $1.6022 \times 10^{-19}$ C |
| Electron mass | $9.109387 \times 10^{-28}$ g |
| Faraday constant ($F$) | $96{,}485.3$ C/mol $e^-$ |
| Gas constant ($R$) | $0.08206$ L $\cdot$ atm/K $\cdot$ mol |
| | $8.314$ J/K $\cdot$ mol |
| | $62.36$ L $\cdot$ torr/K $\cdot$ mol |
| | $1.987$ cal/K $\cdot$ mol |
| Planck's constant ($h$) | $6.6256 \times 10^{-34}$ J $\cdot$ s |
| Proton mass | $1.672623 \times 10^{-24}$ g |
| Neutron mass | $1.674928 \times 10^{-24}$ g |
| Speed of light in a vacuum | $2.99792458 \times 10^8$ m/s |

## Some Prefixes Used with SI Units

| | | | | |
|---|---|---|---|---|
| tera (T) | $10^{12}$ | | centi (c) | $10^{-2}$ |
| giga (G) | $10^{9}$ | | milli (m) | $10^{-3}$ |
| mega (M) | $10^{6}$ | | micro ($\mu$) | $10^{-6}$ |
| kilo (k) | $10^{3}$ | | nano (n) | $10^{-9}$ |
| deci (d) | $10^{-1}$ | | pico (p) | $10^{-12}$ |

## Useful Conversion Factors and Relationships

1 lb = 453.6 g

1 in = 2.54 cm (exactly)

1 mi = 1.609 km

1 km = 0.6215 mi

1 pm = $1 \times 10^{-12}$ m = $1 \times 10^{-10}$ cm

1 atm = 760 mmHg = 760 torr = 101,325 N/m$^2$ = 101,325 Pa

1 cal = 4.184 J (exactly)

1 L $\cdot$ atm = 101.325 J

1 J = 1 C $\times$ 1 V

$$?°C = (°F - 32°F) \times \frac{5°C}{9°F}$$

$$?°F = \frac{9°F}{5°C} \times (°C) + 32°F$$

$$?K = (°C + 273.15°C)\left(\frac{1K}{1°C}\right)$$